Mastering Endovascular Techniques

George Geroulakos • Efthymios Avgerinos
Jean Pierre Becquemin • Gregory C. Makris
Alberto Froio
Editors

Mastering Endovascular Techniques

Tips and Tricks in Endovascular Surgery

Springer

Editors
George Geroulakos
Department of Vascular Surgery
Attikon University Hospital, National and Kapodistrian University of Athens
Athens, Greece

Department of Surgery
Imperial College of Science Technology and Medicine
London, UK

Jean Pierre Becquemin
Henri Mondor Hôspital, APHP Créteil
Hôpital Privé Paris East, Ramsay Santé
Créteil, France

Alberto Froio
Department of Vascular Surgery Fondazione
IRCCS San Gerardo Monza,
School of Medicine and Surgery
University of Milano-Bicocca
Monza, Italy

Efthymios Avgerinos
Department of Vascular Surgery
Attikon Hospital, National and Kapodistrian University of Athens
Athens, Greece

Gregory C. Makris
Vascular and Interventional Radiology
Guy's and St Thomas' Hospital, NHS Foundation Trust
London, UK

ISBN 978-3-031-42734-3 ISBN 978-3-031-42735-0 (eBook)
https://doi.org/10.1007/978-3-031-42735-0

This Springer imprint is published by the registered company Springer Nature Switzerland AG
The registered company address is: Gewerbestrasse 11, 6330 Cham, Switzerland

To

Professor Brian Hopkinson
pioneer in Endovascular Surgery, humanitarian, charismatic academic, teacher, mentor, and friend.
Thank you Brian for supporting me on my first trainee post at the QMC in Nottingham, UK, and beyond.

George Geroulakos

Foreword

Porto, January 2024

According to the official UEMS definition, Vascular Surgery is the clinical and scientific discipline concerned with the diagnosis, treatment and prevention of diseases affecting arteries, veins and lymphatics excluding vessels of the brain and heart.

Nowadays Vascular Surgery is not just a mono-speciality but is also a key component of multi-specialist teams working in the treatment of various pathologies.

Proper diagnosis and treatment are, of course, dependent on adequate training of dedicated surgeons, and since vascular surgery is a speciality highly dependent on maximum aptitude, it is becoming abundantly clear that the practice of vascular surgery strictly requires a well-structured training programme, considering the inherent risks associated with vascular medicine malpractice.

The treatment of vascular diseases by minimally invasive techniques have undergone explosive advances in the last decade. This progress in understanding and available treatments has come with the cost of increasing complexity.

The Editors of this book are to be congratulated for bringing together many world leading experts in the endovascular techniques to produce a text which provides a cutting-edge view of the current state of the different tips and tricks in endovascular surgery. "Mastering Endovascular Techniques" will be an invaluable up-to-date source of information and a must-have volume for all specialists interested in this field and in particular for trainees. This book will help candidates to better prepare for the FEBVS examination and I gladly endorse this initiative.

Armando Mansilha
Secretary General
UEMS Section and Board of Vascular Surgery

Foreword

It is a distinct pleasure for me to write a foreword for this book that provides a description of and a tribute to the mastery of endovascular techniques in all aspects of vascular disease. It has been close to 30 years since I recognized the potential impact of endovascular treatments on the specialty of vascular surgery. At that time, I estimated that in 10–15 years 40–70% of the open operations we were then performing would be replaced or improved by endovascular procedures. Most others at the time thought it was a ridiculous prediction. Surprisingly however, I proved to be wrong on the low side, as more than 70% of today's vascular procedures are exclusively or partially performed via endovascular approaches. In addition, technology, research, evidence, and techniques are rapidly evolving and constantly improving. As a result, a dedicated Master's program in Endovascular Techniques and a dedicated "Master of Endovascular Techniques" volume are the natural consequence of the "endovascular revolution" or this evolution toward a much-needed focus on endovascular education.

This book represents a nearly all-inclusive treatise for the initiation and development of endovascular skills and techniques. It also addresses cutting-edge innovation in vascular territories previously considered outside the realm of endovascular treatment (e.g., the aortic arch, pedal arteries, deep veins, and pulmonary arteries). Importantly, it also addresses the medical judgment issues necessary in the application or restriction of the techniques described and illustrated. The target readership includes all vascular surgeons, angiologists, interventional radiologists, and cardiologists and all their vascular trainees who need to master vascular interventions. Other specialists with an interest in endovascular treatments will also find this book useful.

The chapter authors have been carefully selected to represent a roster of multidisciplinary innovative thought leaders in the endovascular field. Acknowledging the role of multidisciplinary care and that not one specialty is best in endovascular skills, the editors have a reasonably balanced representation of an international faculty of vascular surgeons, interventional radiologists, interventional cardiologists, and neuro-interventionalists.

The future of endovascular treatments is bright, and they will certainly evolve even more in the years and decades to come. Evolution never stops. Darwin wrote, "I have called this principle, by which each slight variation, if useful, is preserved, by the term Natural Selection." Natural selection in techniques, devices, and methodologies will continue to apply ensuring that our patients receive the best possible treatment for their vascular conditions. This is progress and this is what this book is all about. It is a testament to the ingenuity and constant pursuit of excellence on the part of care givers and their engineering and industry colleagues everywhere.

I congratulate the editors of this volume for their effort. I am sure that this book will inspire and guide current practitioners as well as the next generation of endovascular specialists.

Cleveland Clinic
Cleveland, OH, USA

New York University Medical Center
New York, NY, USA

Frank J. Veith

Preface

This book is a unique collection of real-life case scenarios that highlight the diversity of problems and solutions that may be encountered in endovascular surgery.

The ambition of this book is to provide the reader with a clinical, scientific, and pragmatic approach to endovascular techniques. It is specifically aimed at vascular surgery and interventional radiology residents who wish to take the European Vascular Surgery Certification Examination, the American Boards in Vascular Surgery, the European Board of Interventional Radiology examination, and the national training examinations. It will also be of great interest to established surgeons and interventional radiologists who will be able to find in it an update of scientific data and help in their daily practices.

The different chapters have been organized according to the pathology and locations of arterial and venous cardiovascular diseases and the approaches for fistulas.

Each chapter follows a uniform organizational chart, including: (1) the presentation of a clinical case on which the reader can reflect on the diagnosis and the course of action to be taken, (2) a review of the literature oriented towards the clinical case and describing the different explorations, therapeutic choices, and their results, (3) the authors' choices are then reported and accompanied by justification of their choice and the result, and (4) finally, the "toolbox" describes the materials used by the authors to carry out the acts described. Extensive references are provided to reassure the reader that the management follows the principles of evidence-based medicine.

Currently the evolution of technology and medical equipment, the publication of innovative research in medical journals and its rapid dissemination by the Internet, the participation in numerous international congresses bringing together recognized experts, and the webinars organized by vascular societies or by public and private institutions fulfill the need for continuous medical education. We believe that in this background, traditional textbook teaching remains relevant. Our book accomplishes our goals by summarizing in a single volume the entire vascular pathology that can be managed by endovascular techniques. No doubt it will feature prominently in your office and facilitate your practice.

We are grateful to our international faculty of authors who have contributed generously their knowledge and time to this project and our publisher Springer for advice and support.

These recommendations from the military strategist SUN TZU (around 500 years BC) should be kept in mind for the successful strategy of an endovascular intervention.

"The General who wins his battles spends a lot of time preparing for them. Whoever loses them is in too much of a hurry. Long deliberations lead to victory, haste to defeat, leave without thinking and you will be crushed." These recommendations from the military strategist Sun Tzu (around 500 years BC) should be kept in mind for the successful strategy of an endovascular intervention.

Athens, Greece — George Geroulakos
Pittsburgh, USA — Efthymios Avgerinos
Paris, France — Jean Pierre Becquemin
London, UK — Grigoris C. Makris
Milan, Italy — Alberto Froio

Contents

List of Contributors

Iannis Ben Abdallah Centre constitutif des malformations artério veineuses de l'enfant et de l'adulte de l'hôpital Lariboisière, Filière FAVA-multi, Paris, France

Assistance Publique - Hôpitaux de Paris, Paris, France

Université Paris Cité, Paris, France

Othman Abdul-Malak University of Pittsburgh Medical Center, Pittsburgh, PA, USA

Said Abisi Guy's and St Thomas' NHS Foundation Trust, London, UK

Irfan Ahmed London, UK

Alberto Froio School of Medicine and Surgery, University of Milano-Bicocca, Milan, Italy

Vascular Surgery Department, Fondazione IRCCS San Gerardo, Monza, Italy

Kirill Alektoroff Department of Diagnostic and Interventional Neuroradiology, Klinikum Bremen-Mitte/Bremen-Ost, Bremen, Germany

Fanny S. Alie-Cusson Division of Vascular Surgery, Department of Surgery, University of Pittsburgh Medical Center, Pittsburgh, PA, USA

Tariq Ali Norfolk and Norwich University Hospitals, Norwich, UK

Abdel Kader Allouni Department of Interventional Radiology, University Hospitals of North Midlands NHS Trust, Stoke-on-Trent, UK

Ahmed Al-Nowfal London, UK

Pierre Alric Department of Thoracic and Vascular Surgery, Arnaud de Villeneuve Hospital, Montpellier, France

Seyed Ameli-Renani St George's University Hospitals NHS Foundation Trust, London, UK

Elizabeth A. Andraska Division of Vascular Surgery, Department of Surgery, University of Pittsburgh Medical Center, Pittsburgh, PA, USA

Constantine N. Antonopoulos Department of Vascular Surgery, National and Kapodistrian University of Athens, "Attikon" University Hospital, Athens, Greece

Anastasios Apostolos First Department of Cardiology, Medical School, National and Kapodistrian University of Athens, Hippokration Hospital, Athens, Greece

Nikolaos Achilleas Arkoudis 2nd Department of Radiology, National and Kapodistrian University, "ATTIKON" University General Hospital, Athens, Greece

Miltiadis Krokidis Athens National and Kapodistrian University of Athens, Athens, Greece

Efthymios Avgerinos Clinic for Vascular and Endovascular Surgery, Athens Medical Center, Athens, Greece

Attikon Hospital, Department of Vascular Surgery, University of Athens & Athens Medical Center Clinic of Vascular and Endovascular Surgery, Athens, Greece

Nizar Abu Bakr Rhein Main Vascular Center, Department of Vascular and Endovascular Surgery, Asklepios Clinics Langen, Wiesbaden and Seligenstadt, Langen, Germany

P. Banugo Guy's and St Thomas' NHS Foundation Trust, London, UK

Mohammed Barbati Universitätsklinikum Aachen, Klinik für Gefäßchirurgie, Aachen, Deutschland

Thomas Barge Department of Radiology, John Radcliffe Hospital, Oxford, UK

Department of Interventional Radiology, Oxford University Hospitals NHS Foundation Trust, Oxford, UK

David Beckett The Whiteley Clinic, Guildford, UK

Jean Pierre Becquemin Cardio Vascular Department, Henri Mondor Hospital University Paris Val de Marne, Creteil, France

Institut Vasculaire Paris Est, Hôpital Privé Paul d'Egine, Ramsay Group Champigny, Champigny sur Marne, France

Ailin Belloni School of Medicine and Surgery, University of Milano-Bicocca, Milan, Italy

Vascular Surgery Department, Fondazione IRCCS San Gerardo, Monza, Italy

Katrine Jessica Beltramolli School of Medicine and Surgery, University of Milano-Bicocca, Milan, Italy

Vascular Surgery Department, Fondazione IRCCS San Gerardo, Monza, Italy

Kyla M. Bennett Division of Vascular Surgery, University of Wisconsin-Madison School of Medicine and Public Health, Madison, WI, USA

C. Bicknell Imperial Vascular Unit, London, UK

H. Bidd Guy's and St Thomas' NHS Foundation Trust, London, UK

Zaeem Billah Division of Vascular and Interventional Radiology, Kaiser Permanente, Los Angeles, CA, USA

Theodosios Bisdas Clinic for Vascular Surgery, Athens Medical Center, Athens, Greece

Daniele Bissacco Section of Vascular Surgery, Cardio Thoracic Vascular Department, Fondazione IRCCS Cà Granda Ospedale Maggiore Policlinico, Milan, Italy

Stephen Black St. Thomas' Hospital, Vascular Surgery, London, UK

Raphaël Blanc Departement of Interventional Neuroradiology, Hopital Fondation Rothschild, Paris, France

William Boisseau Departement of Interventional Neuroradiology, Hopital Fondation Rothschild, Paris, France

Marine Bordet Department of Vascular and Endovascular Surgery, University Hospital of Lyon, Lyon, France

Michel Bosiers Bern University Hospital, University of Bern, Bern, Switzerland

Julien Boucherit Departement of Interventional Neuroradiology, Hopital Fondation Rothschild, Paris, France

Vassilios Bouris Clinic for Vascular and Endovascular Surgery, Athens Medical Center, Athens, Greece

Annouk Bisdorff Bresson Centre constitutif des malformations artério veineuses de l'enfant et de l'adulte de l'hôpital Lariboisière, Filière FAVA-multi, Paris, France

Assistance Publique - Hôpitaux de Paris, Paris, France

Elias Brountzos 2nd Department of Radiology, National and Kapodistrian University, "ATTIKON" University General Hospital, Athens, Greece

Ludovic Canaud Department of Thoracic and Vascular Surgery, Arnaud de Villeneuve Hospital, Montpellier, France

Stefania Caputo Vascular Surgery Unit, University Hospital, IRCCS Policlinico S. Orsola, University of Bologna—DIMEC, Bologna, Italy

Pierre Cerceau Centre constitutif des malformations artério veineuses de l'enfant et de l'adulte de l'hôpital Lariboisière, Filière FAVA-multi, Paris, France

Assistance Publique - Hôpitaux de Paris, Paris, France

Rabih Chaer Division of Vascular Surgery, University of Pittsburgh Medical Center, Pittsburgh, PA, USA

Rabih A. Chaer Division of Vascular Surgery, Department of Surgery, University of Pittsburgh Medical Center, Pittsburgh, PA, USA

Laurent Chiche Department of Vascular and Endovascular Surgery, Pitié Salpêtrière University Hospital, Paris, France

Frédéric Cochennec Department of Vascular and Endovascular Surgery, Pitié Salpêtrière University Hospital, Paris, France

Raphaël Coscas Department of Vascular Surgery, Ambroise Paré University Hospital, Versailles Saint-Quentin and Paris-Saclay Universities, Boulogne-Billancourt, France

Lucinda Cruddas Imperial Vascular Unit, Imperial College Healthcare NHS Trust, St Mary's Hospital, London, UK

Ilenia D'Alessio Vascular Surgery Department, Fondazione "De Gasperis", Grande Ospedale Metropolitano Niguarda, Milan, Italy

Firouza Dahi Rhein Main Vascular Center, Department of Vascular and Endovascular Surgery, Asklepios Clinics Langen, Wiesbaden and Seligenstadt, Langen, Germany

Deep Das Department of Neurology, Bangur Institute of Neurosciences, Institute of Post Graduate Medical Education and Research, Kolkata, India

Woodlands Multi-speciality Hospital and C K Birla Hospitals, Kolkata, India

Gianmarco de Donato Vascular Surgery Unit, Department of Medicine, Surgery and Neuroscience, University of Siena, Siena, Italy

Athanasios Diamantopoulos Guy's and St Thomas' NHS Foundation Trust, London, UK

Maurizio Domanin Section of Vascular Surgery, Cardio Thoracic Vascular Department, Fondazione IRCCS Cà Granda Ospedale Maggiore Policlinico, Milan, Italy

Department of Clinical Sciences and Community Health, University of Milan, Milan, Italy

Konstantinos P. Donas Rhein Main Vascular Center, Department of Vascular and Endovascular Surgery, Asklepios Clinics Langen, Wiesbaden and Seligenstadt, Langen, Germany

Maria Drakopoulou First Department of Cardiology, Medical School, National and Kapodistrian University of Athens, Hippokration Hospital, Athens, Greece

Arpan Dutta Neurology Bangur Institute of Neurosciences, Institute of Post Graduate Medical Education and Research, Kolkata, India

Sotirios Economides Greek Atomic Energy Commission (EEAE), Athens, Greece

Mohamed. Ragab Elnemr Guy's and St Thomas' NHS Foundation Trust, London, UK

Cairo National Institute of Diabetes and Endocrinology, Cairo, Egypt

Division Of Vascular and Endovascular Surgery, Charleston Area Medical Center, West Virginia, USA

Mohammad Eslami University of Pittsburgh Medical Center, Pittsburgh, PA, USA

Division Of Vascular and Endovascular Surgery, Charleston Area Medical Center, West Virginia, USA

Gianluca Faggioli Vascular Surgery Unit, University Hospital, IRCCS Policlinico S. Orsola, University of Bologna—DIMEC, Bologna, Italy

Francesca Maria Feroldi Vascular Surgery Unit, University Hospital, IRCCS Policlinico S. Orsola, University of Bologna—DIMEC, Bologna, Italy

Pauline Fillet Vascular and Endovascular Surgery Unit, Clinique de Genolier, Swiss Medical Network, Genolier, Switzerland

Department of Vascular and Endovascular Surgery, La Cote Healthcare Group, Morges, Lausanne, Switzerland

Alberto Froio School of Medicine and Surgery, University of Milano-Bicocca, Milan, Italy

Department of Vascular Surgery, Fondazione IRCCS San Gerardo, Monza, Italy

Romain Gallet Cardiology Department, Hôpitaux Universitaires Henri Mondor, Créteil, France

Enrico Gallitto Vascular Surgery Unit, University Hospital, IRCCS Policlinico S. Orsola, University of Bologna—DIMEC, Bologna, Italy

Mauro Gargiulo Vascular Surgery Unit, University Hospital, IRCCS Policlinico S. Orsola, University of Bologna—DIMEC, Bologna, Italy

Efstratios Georgakarakos Department of Vascular Surgery, "Democritus" University of Thrace, University Hospital of Alexandroupolis, Alexandroupolis, Greece

George Geroulakos Department of Vascular Surgery, Attikon University Hospital, National and Kapodistrian University of Athens, Athens, Greece

Department of Surgery, Imperial College of Science Technology and Medicine, London, UK

Athanasios Giannoukas Vascular Surgery Department, Faculty of Medicine, Larissa University Hospital, School of Health Sciences, University of Thessaly, Larissa, Greece

Antoine Girault Department of Vascular and Thoracic Surgery, Hôpital Marie Lannelongue, Groupe Hospitalier Paris-Saint Joseph, Université Paris-Saclay, Paris, France

Panos Gkoutzios Interventional Radiology, Guys and St Thomas' NHS Foundation Trust, London, UK

Guy's & St Thomas' NHS Foundation Trust- Interventional Radiology Department, London, UK

G. Gorbunov Cardiovascular Surgery Department, North Western I.I Mechnikov State Medical University, St. Petersburg, Russia

Yann Gouëffic Department of Vascular and Endovascular Surgery, Groupe Hospitalier Paris St Joseph, Paris, France

Mark Gregory Guy's and St. Thomas Hospital, NHS Foundation Trust, London, UK

Subhadeep Gupta Neurology Bangur Institute of Neurosciences, Institute of Post Graduate Medical Education and Research, Kolkata, India

V. Gurevich North Western Scientific Clinical Center of Federal Medical Biological Agency of Russia, Atherosclerosis Treatment, St. Petersburg, Russia

Salim G. Habib Division of Vascular Surgery, University of Pittsburgh Medical Center (UPMC), Heart and Vascular Institute, Pittsburgh, PA, USA

T. Hamwi Department of Vascular Surgery, Ludwig-Maximillian University Hospital, Munich, Germany

Natasha Hasemaki Second Department of Vascular Surgery, Laiko Hospital, Athens National and Kapodistrian University of Athens, Athens, Greece

Department of Vascular Surgery, Laiko Hospital, Athens, Greece

A. Hertault Hopital Privé de Villeneuve d'Ascq, Lille, France

Quentin Holay Departement of Interventional Neuroradiology, Hopital Fondation Rothschild, Paris, France

Houman Jalaie Universitätsklinikum Aachen, Klinik für Gefäßchirurgie, Aachen, Deutschland

Peter Jamieson Department of Interventional Radiology, University Hospitals of North Midlands NHS Trust, Stoke-on-Trent, UK

Silvan Jungi Bern University Hospital, University of Bern, Bern, Switzerland

Ioannis Kachrimanidis First Department of Cardiology, Medical School, National and Kapodistrian University of Athens, Hippokration Hospital, Athens, Greece

Ioannis D. Kakisis National and Kapodistrian University of Athens, Department of Vascular Surgery, Attikon University Hospital, Athens, Greece

Konstantinos Kalogeras Third Department of Cardiology, Sotiria Hospital, Medical School, Athens, Greece

Thomas Kalogirou Vascular Unit, 5th Department of Surgery, Hippocratio Hospital, Medical School, Aristotle University of Thessaloniki, Thessaloniki, Greece

Kartik Kansagra Division of Vascular and Interventional Radiology, Kaiser Permanente, Los Angeles, CA, USA

Christos Karkos Vascular Unit, 5th Department of Surgery, Hippocratio Hospital, Medical School, Aristotle University of Thessaloniki, Thessaloniki, Greece

Narayan Karunanithy School of Biomedical Engineering and Imaging Sciences, King's College London, London, UK

Konstantinos Katsanos School of Medicine, Patras University Hospital, Rion, Greece

Athanasios Katsargyris Second Department of Vascular Surgery, Laiko Hospital, Athens National and Kapodistrian University of Athens, Athens, Greece

Athens University Medical School, Athens, Greece

Majid Kazemtash Rhein Main Vascular Center, Department of Vascular and Endovascular Surgery, Asklepios Clinics Langen, Wiesbaden and Seligenstadt, Langen, Germany

Basil Zia Khan Department of Interventional Radiology, Oxford University Hospitals NHS Foundation Trust, Oxford, UK

N. Khan Guy's and St Thomas' NHS Foundation Trust,London, UK

G. Khubulava Cardiovascular Surgery Department, The First I.P. Pavlov St. Petersburg Medical University, St. Petersburg, Russia

Angela A. Kokkosis Vascular Surgery, Stony Brook University Medical Center, Stony Brook, NY, USA

Leonidas Koliastasis First Department of Cardiology, Medical School, National and Kapodistrian University of Athens, Hippokration Hospital, Athens, Greece

Dimitrios Kontothanassis Department of Phlebology, Istituto Flebologico Italiano, Padova, Italy

General and Vascular Surgery, Phlebology Department, Istituto Flebologico Italiano, Padova, Italy

General and Vascular Surgeon CEO Istituto Flebologico Italiano, Padova, Italy

Drosos Kotelis Bern University Hospital, University of Bern, Bern, Switzerland

George S. Koutsias Aristotle University of Thessaloniki, Thessaloniki, Greece

Rahul Kumar G S Neuroscience Clinic and Research Center, Interventional Neurology, Patna, India

Nicos Labropoulos Surgery and Radiology, Vascular Laboratory, Stony Brook Medicine, Stony Brook, NY, USA

Claude Laurian Centre constitutif des malformations artério veineuses de l'enfant et de l'adulte de l'hôpital Lariboisière, Filière FAVA-multi, Paris, France

Andreas M. Lazaris Vascular Surgery Department, Attikon University Hospital, National and Kapodistrian University of Athens, Athens, Greece

Giulia Lerva Vascular Surgery Department, Fondazione "De Gasperis", Grande Ospedale Metropolitano Niguarda, Milan, Italy

Nathan L. Liang Division of Vascular Surgery, Department of Surgery, University of Pittsburgh Medical Center, Pittsburgh, PA, USA

Christos D. Liapis Vascular and Endovascular Surgery Clinic, Athens Medical Center, Athens, Greece

Antonino Logiacco Vascular Surgery Unit, University Hospital, IRCCS Policlinico S. Orsola, University of Bologna—DIMEC, Bologna, Italy

Elena Lucertini Ospedali Riuniti di Livorno, Livorno, Italy

Michael Madigan University of Pittsburgh Medical Center, Pittsburgh, PA, USA

Agnese Magni School of Medicine and Surgery, University of Milano-Bicocca, Milan, Italy

Vascular Surgery Department, Fondazione IRCCS San Gerardo, Monza, Italy

Vladimir Makaloski Bern University Hospital, University of Bern, Bern, Switzerland

Gregory Makris Guys' and St Thomas' NHS Foundation Trust, London, UK

Alexandros Mallios Department of Vascular Surgery, Groupe Hospitalier Paris Saint Joseph, Paris, France

Department of Surgery, Centre Hospitalier de Chartres, Chartres, France

I. Mandal Department of Radiology, John Radcliffe Hospital, Oxford, UK

Tim J. Mandigers Department of Vascular Surgery, Fondazione IRCCS Cà Granda Ospedale Maggiore Policlinico, Milan, Italy

Department of Vascular Surgery, University Medical Center Utrecht, Utrecht, The Netherlands

Dinesh Manoharan Department of Vascular Radiology, Sheffield Vascular Institute, Sheffield Teaching Hospitals NHS Foundation Trust, Sheffield, UK

Martina Mariani School of Medicine and Surgery, University of Milano-Bicocca, Milan, Italy

Vascular Surgery Department, Fondazione IRCCS San Gerardo, Monza, Italy

Guy Martin Imperial Vascular Unit, Imperial College Healthcare NHS Trust, St Mary's Hospital, London, UK

Claudine Massoni Centre constitutif des malformations artério veineuses de l'enfant et de l'adulte de l'hôpital Lariboisière, Filière FAVA-multi, Paris, France

Ryan McEnaney University of Pittsburgh Medical Center, Pittsburgh, PA, USA

Antoine Millon Department of Vascular and Endovascular Surgery, University Hospital of Lyon, Lyon, France

Robert Morgan Department of Interventional Radiology, St Georges Hospital, London, UK

Ornella Moschovaki-Zeiger 2nd Department of Radiology, National and Kapodistrian University, "ATTIKON" University General Hospital, Athens, Greece

Konstantinos G. Moulakakis Department of Vascular Surgery, University of Patras Medical School, Patras, Greece

Spyridon N. Mylonas Department of Vascular and Endovascular Surgery, University of Cologne, Cologne, Germany

Y. Nakatis Vascular Surgery Department, North Western Scientific Clinical Center of Federal Medical Biological Agency of Russia, St. Petersburg, Russia

Bahaa Nasr Department of Vascular and Endovascular Surgery, Brest University Hospital, Brest, France

A. S. Niekamp Miami Cardiac and Vascular Institute, Baptist Health South Florida, Miami, FL, USA

Georgios Oikonomou First Department of Cardiology, Medical School, National and Kapodistrian University of Athens, Hippokration Hospital, Athens, Greece

Konstantinos Palialexis 2nd Department of Radiology, National and Kapodistrian University, "ATTIKON" University General Hospital, Athens, Greece

Panagiotis Papanagiotou Department of Diagnostic and Interventional Neuroradiology, Klinikum Bremen-Mitte/Bremen-Ost, Bremen, Germany

Department of Radiology, Aretaieion University Hospital, National and Kapodistrian University of Athens, Athens, Greece

Nikos Paraskevas Centre constitutif des malformations artério veineuses de l'enfant et de l'adulte de l'hôpital Lariboisière, Filière FAVA-multi, Paris, France

Assistance Publique - Hôpitaux de Paris, Paris, France

Michael Parker Norfolk and Norwich University Hospitals, Norwich, UK

Savino Pasquadibisceglie School of Medicine and Surgery, University of Milano-Bicocca, Milan, Italy

Vascular Surgery Department, Fondazione IRCCS San Gerardo, Monza, Italy

Edoardo Pasqui Vascular Surgery Unit, Department of Medicine, Surgery and Neuroscience, University of Siena, Siena, Italy

Shyamal Patel St George's University Hospitals NHS Foundation Trust, London, UK

Quentin Pellenc Vascular and Endovascular Surgery Unit, Clinique de Genolier, Swiss Medical Network, Genolier, Switzerland

Department of Vascular and Endovascular Surgery, La Cote Healthcare Group, Morges, Lausanne, Switzerland

M. Pichlmaier Department of Cardiac Surgery, Ludwig-Maximillian University Hospital, Munich, Germany

A. L. Pouncey Imperial Vascular Unit, London, UK

Calin-Gheorghe Precup Department of Vascular and Endovascular Surgery, University Hospital of Lyon, Lyon, France

Anastasios Psyllas Rhein Main Vascular Center, Department of Vascular and Endovascular Surgery, Asklepios Clinics Langen, Wiesbaden and Seligenstadt, Langen, Germany

Celia Riga Imperial Vascular Unit, Imperial College Healthcare NHS Trust, St Mary's Hospital, London, UK

Marie Josee van Rijn Erasmus Medical Center, Rotterdam, The Netherlands

Tarun Sabharwal Guy's and St. Thomas Hospital, NHS Foundation Trust, London, UK

Kalyan Sajja Life Hospital, Guntur, India

S. A. Salazar Florida International University Herbert Wertheim College of Medicine, Miami, FL, USA

Karim M. Salem Division of Vascular Surgery, Department of Surgery, University of Pittsburgh Medical Center, Pittsburgh, PA, USA

Division of Vascular Surgery, Department of Surgery, Rutgers New Jersey Medical School, Newark, NJ, USA

Athanasios Saratzis Department of Cardiovascular Sciences, University of Leicester, Leicester, UK

Mohamed Hosny Sayed Guy's and St Thomas' NHS Foundation Trust, London, UK

Cairo University Hospitals, Cairo, Egypt

Vittorio Maria Segramora Vascular Surgery Department, Fondazione IRCCS San Gerardo, Monza, Italy

Dana B. Semaan Division of Vascular Surgery, University of Pittsburgh Medical Center (UPMC), Heart and Vascular Institute, Pittsburgh, PA, USA

George S. Sfyroeras Vascular Surgery Department, Attikon University Hospital, Athens, Greece

J. Shea Imperial Vascular Unit, London, UK

Austin Shinagawa Division of Vascular and Interventional Radiology, Kaiser Permanente, Los Angeles, CA, USA

Caleb Solivio California University of Science and Medicine, Colton, CA, USA

Konstantinos Spanos Vascular Surgery Department, Faculty of Medicine, Larissa University Hospital, School of Health Sciences, University of Thessaly, Larissa, Greece

P. Spath Department of Vascular Surgery, Ludwig-Maximillian University Hospital, Munich, Germany

Stavros Spiliopoulos 2nd Department of Radiology, National and Kapodistrian University, "ATTIKON" University General Hospital, Athens, Greece

Natalie Sridharan University of Pittsburgh Medical Center, Pittsburgh, PA, USA

Trilochan Srivastava SMS Medical College and Hospital, Jaipur, India

J. Stana Department of Vascular Surgery, Ludwig-Maximillian University Hospital, Munich, Germany

Gemmi Sufali Vascular Surgery Unit, University Hospital, IRCCS Policlinico S. Orsola, University of Bologna—DIMEC, Bologna, Italy

A. Svetlikov Division of Faculty Surgery, Cardiovascular Surgery Department, North Western Scientific Clinical Center of Federal Medical Biological Agency of Russia, St. Petersburg State University, St. Petersburg, Russia

Gergana T. Taneva Rhein Main Vascular Center, Department of Vascular and Endovascular Surgery, Asklepios Clinics Langen, Wiesbaden and Seligenstadt, Langen, Germany

Apostolos K. Tassiopoulos Division of Vascular and Endovascular Surgery, Stony Brook Medicine, New York, USA

Emmanuel Teiger Cardiology Department, Hôpitaux Universitaires Henri Mondor, Créteil, France

Iakovos Theodoulou Guy's and St Thomas' NHS Foundation Trust, London, UK

Barts NHS Health Trust, The Royal Hospital, London, UK

Narayanan Thulasidasan Interventional Radiology, Guys and St Thomas' NHS Foundation Trust, London, UK

Guy's & St Thomas' NHS Foundation Trust- Interventional Radiology Department, London, UK

Department of Vascular Surgery, Ramsay Group, Champigny sur Marne, France

Valerio Tolva Vascular Surgery Department, Fondazione "De Gasperis", Grande Ospedale Metropolitano Niguarda, Milan, Italy

Konstantinos Toutouzas First Department of Cardiology, Medical School, National and Kapodistrian University of Athens, Hippokration Hospital, Athens, Greece

Georgios Trantalis First Department of Cardiology, Medical School, National and Kapodistrian University of Athens, Hippokration Hospital, Athens, Greece

Santi Trimarchi Section of Vascular Surgery, Cardio Thoracic Vascular Department, Fondazione IRCCS Cà Granda Ospedale Maggiore Policlinico, Milan, Italy

Department of Clinical Sciences and Community Health, University of Milan, Milan, Italy

N. Tsilimparis Department of Vascular Surgery, Ludwig-Maximillian University Hospital, Munich, Germany

Costas Tsioufis First Department of Cardiology, Medical School, National and Kapodistrian University of Athens, Hippokration Hospital, Athens, Greece

T. Turlejski Department of Radiology, John Radcliffe Hospital, Oxford, UK

Vasileios Tzilalis 401 General Military Hospital, Athens, Greece

Raman Uberoi Department of Interventional Radiology, Oxford University Hospitals NHS Foundation Trust, Oxford, UK

Geogy Vatakencherry Division of Vascular and Interventional Radiology, Kaiser Permanente, Los Angeles, CA, USA

Manolis Vavuranakis Third Department of Cardiology, National and Kapodistrian University of Athens, Sotiria Hospital, Medical School, Athens, Greece

Paolo Verlato School of Medicine and Surgery, University of Milano-Bicocca, Milan, Italy

Vascular Surgery Department, Fondazione IRCCS San Gerardo, Monza, Italy

Demetrios V. Vlahakos National and Kapodistrian University of Athens, School of Medicine, Athens, Greece

Lena L. Vodovotz University of Pittsburgh Medical School, Pittsburgh, PA, USA

Panagiotis Volteas Division of Vascular and Endovascular Surgery, Stony Brook Medicine, Stony Brook, NY, USA

Vascular and Endovascular Surgery, Stony Brook Medicine, New York, USA

Gregory Walker Division of Neurology, Department of Medicine, Royal Columbian Hospital, New Westminster, BC, Canada

Angie White The Whiteley Clinic, Guildford, UK

Mark S. Whiteley The Whiteley Clinic, Guildford, UK

Iosif Xenogiannis Second Department of Cardiology, Attikon University Hospital, National and Kapodistrian University of Athens Medical School, Athens, Greece

Bowen Xie University of Pittsburgh Medical Center, Pittsburgh, PA, USA

Theodore H. Yuo Division of Vascular Surgery, University of Pittsburgh Medical Center (UPMC), Heart and Vascular Institute, Pittsburgh, PA, USA

Part I

Introduction in Endovascular Techniques

Biological Effects of Ionizing Radiations

1

Sotirios Economides

Introduction

The first biological effects of ionizing radiation were reported soon after their discovery. A lot of data on the health effects that might be caused by the exposure to ionizing radiation have been collected from occupationally exposed workers (e.g., radiologists, technologists, miners, etc.) as well as from the survivors of the atomic bombs in Hiroshima and Nagasaki.

The process associated with the appearance of the biological effects is triggered by the energy deposited by the ionizing radiations during their interactions with matter. Part of this energy may appear as heat which is biologically harmless. However, there is a part of this energy which will be absorbed by the cell, leading to direct and indirect biological effects.

Direct effects are associated with the killing of significant number of cells by ionizing radiation that may lead to biological effects which appear weeks or months effects after irradiation (deterministic effects). On the other hand, indirect effects of ionizing radiation may be caused by modifications in the genetic material of cells and appear several years after their irradiation.

To assess the biological effects due to the exposure to ionizing radiation, the doses received by an individual are considered cumulatively (i.e., each dose received is added to the previous ones). Moreover, this should be taken into account in the design of systems of radiation protection and the establishment of dose limits and constraints for the exposed workers and the public.

Direct and Indirect Effects of Ionizing Radiation

The appearance of biological effects to living organisms due to the exposure to ionizing radiations is the result of the damage that may be induced to cells and more specifically to their DNA. This damage is associated to either direct or indirect effects.

The direct effects concern mainly the damage caused to atoms or molecules (i.e., proteins, RNA, DNA) when the cells are directly hit by ionizing radiation. This damage may be associated to single or multiple breaks to the chromosomal strands of the DNA [1]. While most of the damage to single-stranded DNA may be repaired, lesions to double-stranded DNA may lead to cell death due to deletions, mutations, and chromosomal alterations.

The indirect effects of ionizing radiation are caused by free radicals which are produced through the radiolysis of cell water and can cause chemical modifications to their DNA, [1, 2]. These DNA lesions are capable to generate oxidized DNA forms which activate an inflammatory cascade [3].

Moreover, cells in temporal or spatial distance from a cell damaged by ionizing radiation may present delayed effects such as the bystander effect and the genomic instability.

The bystander effect concerns the spread of the effect caused by the ionizing radiation through the communication of the damaged cells with their neighboring cells [1]. This effect can be attributed either to gap junctional intercellular communication or release of soluble factors from the affected cells [4, 5].

The genomic instability is associated to delayed effects (e.g., reproductive death) that may appear to the descendants of cells which have been affected by ionizing radiation. Irradiated cells damaged but not killed by the ionizing radiation may exhibit an increased mutation rate in their descendants, several generations after the initial exposure to ionizing radiation [1, 6, 7].

S. Economides (✉)
Greek Atomic Energy Commission (EEAE), Athens, Greece
e-mail: sotiris.economides@eeae.gr

G. Geroulakos et al. (eds.), *Mastering Endovascular Techniques*, https://doi.org/10.1007/978-3-031-42735-0_1

Repair Mechanisms

The repair characteristics concerning the damages caused to the DNA (i.e., simple, and multiple strand breaks) are strongly associated to the biological effects which may appear due to exposure to ionizing radiation [8, 9]. In addition, the response of the DNA strongly depends on the radiation dose induced the damage and the associated dose rate, as well as on the quality (energy) of the radiation. In this respect, simple DNA strand breaks can be corrected by the cell's repair mechanisms while double breaks are difficult to correct and may lead to mutations or the death of the cell.

Apoptosis is one additional cell "repair" mechanism which reduces the occurrence of surviving cells which may carry mutations. It can be considered as a form of programmed cell death that occurs in multicellular organisms after certain cell changes (e.g., blebbing, cell shrinkage, nuclear fragmentation, chromatin condensation, DNA fragmentation, etc.) [8, 10].

Assessment of Radiation Risks

The assessment of the risks associated with the appearance of biological effects induced by ionizing radiations should be based on certain quantities such as:

- **Absorbed dose (*D*)**: Is the mean energy deposited in a tissue or organ volume. It is well defined at any point in matter and measurable. The radiation damage induced to tissues or organs depends on the absorbed dose which is expressed in Grays (Gy).
- **Equivalent dose (H_T)**: Is defined as the product of the corresponding radiation weighting factor (w_R) for a radiation *R* with the absorbed dose. Radiation weighting factor (w_R) values (Table 1.1) are associated with the Relative Biological Effectiveness (RBE) of the different types of ionizing radiations as well as on biophysical considerations and judgements [8].
- **Effective dose (*E*)**: Is equal to the product of the corresponding tissue weighting factor value w_T (Table 1.2) and the equivalent dose H_T for a certain type of radiation. *E* is measured in Sieverts (Sv) and used to assess the potential of an exposure to cause biological effects considering the type of radiation and the radiosensitivity of different tissues and organs.

To assess the likelihood for tissue reactions due to exposure to ionizing radiation, the absorbed dose (*D*) to an organ or tissue, weighted by an appropriate RBE value for this organ or tissue should be used [11–13]. The RBE values depend on

- the type and energy of ionizing radiation,
- the dose and dose rate of the exposure, and,
- the tissue or organ irradiated.

Table 1.1 Radiation weighting factors (w_R) values for the calculation of equivalent dose H_T [8]

Radiation type	Radiation weighting factor (w_R)
Photons	1
Electrons	1
Protons	2
Alpha particles	20
Neutrons	Continuous function of neutron energy

Table 1.2 Tissue weighting factors (w_T) values for the calculation of effective dose *E* [8]

Tissue	Tissue weighting factor (w_T)
Bone marrow, colon, lungs, stomach, breast, remainder tissues	0.12
Gonads	0.08
Bladder, esophagus, liver, thyroid	0.04
Bone surface, brain, salivary glands, skin	0.01

For the assessment of external exposures at workplaces usually the effective dose (*E*) values are used. It is indirectly estimated by measuring personal dose equivalent at 10 mm below a specified point of the body (H_p (10)), assuming uniform whole-body irradiation [8].

The use of effective dose to assess patients' exposure is not suggested, as often only parts of an organ or the human body are irradiated during medical exposures, and the age distribution of patients differs from that of the general public.

To optimize radiological protection (mainly occupational exposure) collective dose quantities could be used. These quantities consider the exposure of a group of individuals for a certain period or during an operation in designated radiation areas. Collective effective dose, *S* (Sv) is the sum of all individual effective doses over a certain period or during an operation under consideration. Its estimation assumes a linear dose effect relationship for stochastic effects without a threshold (i.e., LNT model). On this basis it is possible to regard effective doses as additive [8].

Deterministic Effects

As deterministic are considered those biological effects of ionizing radiation the severity of which increases proportionally with the dose received above a threshold value (Fig. 1.1). Examples of deterministic effects are skin injuries, hair loss, cataract, etc. For interventional radiology procedures, skin is the tissue of main concern as it receives the highest radiation doses during fluoroscopically guided procedures.

Deterministic effects can result from the impairment of the integrity and function of organs and tissues due to the exposure to high ionizing radiation doses. These doses may cause the death of a significant number of cells, sufficient to result in tissue reactions.

The response of tissues and organs to ionizing radiation strongly depends on their structure. In a few cases, the loss of cells after irradiation is mediated by apoptosis, while in others, it is caused by reproductive failure of regenerative stem cells.

Most non-proliferating mature cells do not die from irradiation, but from natural senescence. This is one of the main reasons for the existence of a threshold dose for deterministic effects to appear. This threshold dose differs among different individuals, organs or tissues and is subject to biological variation (i.e., cellular radiosensitivity, function of differentiated cells, cellular composition, cell renewal capacity, etc.) [8].

Early tissue reactions may appear hours or a few weeks after the exposure to ionizing radiation. They can be of inflammatory type such as erythema, and subsequent reactions due to cell loss (e.g., mucositis, and desquamatory reactions in epithelial tissues). More prone to early reactions are tissues with rapidly proliferating cell systems (e.g., hematopoietic tissue, cells lining the gastrointestinal tract, the basal cell layer in the skin, male germ cells, etc.) [8].

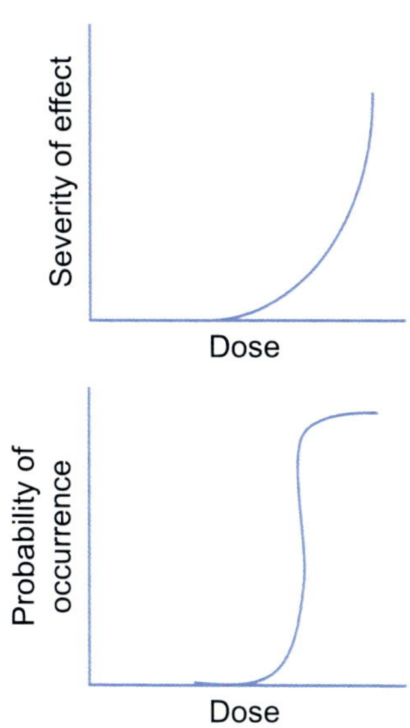

Fig. 1.1 Probability of occurrence and severity for deterministic effects as a function of dose

Table 1.3 Range of doses (exposure to acute low LET uniform whole-body radiation) associated with the appearance of specific deterministic effects and death [8]

Absorbed dose (Gy)	Syndrome	Time of death after exposure (days)
3–5 ($LD_{50/60}$)[a]	Damage to bone marrow	30–60
5–15	Damage to the gastrointestinal tract	7–20
5–15	Damage to the lungs and kidney	60–150
>15	Damage to nervous system	<5, dose-dependent

[a]The LD50/60 is the dose necessary to kill 50% of the exposed population in 60 days

Late tissue reactions may appear several months or even years after the exposure to radiation. They may be associated to damages to blood vessels or connective tissue elements that are essential for the function of organs and tissues. Late reactions can be either "generic" if they occur due to the damage caused to the target tissue (e.g., vascular occlusions) [8] or "consequential" if they occur because of early reactions (e.g., dermal necrosis, intestinal strictures, etc.) [13].

For exposures at dose rates lower than around 0.1 Gy/h there is repair of cellular damage while for high LET irradiations the injury is less repairable. Furthermore, some cells seem to be hypersensitive to doses less than 0.5 Gy, but not at higher doses [14].

At doses more than about 5 Gy, additional effects may occur, including severe gastrointestinal (stem cell and endothelial capillary cell) damage which, when combined with hemopoietic damage, causes death in 1–2 weeks (Table 1.3). At even higher doses toward 50 Gy and above, there is acute damage in the nervous and cardiovascular systems resulting in death after a few days [15].

Stochastic Effects

Exposing cells, even to very low doses of ionizing radiation, may lead to the damage of their genetic material. Killing of a small number of cells due to this exposure, in most cases, have no further impact. However, the induction of genetic modifications or transformations to a single cell leading to malignancy, may have serious consequences (i.e., development of radiation induced cancer many years later, heritable disease in future generations and developmental effects under certain conditions) [8].

The above effects are called stochastic and may appear even at very low radiation doses. As they are associated with a likelihood of occurrence, there is no threshold dose for

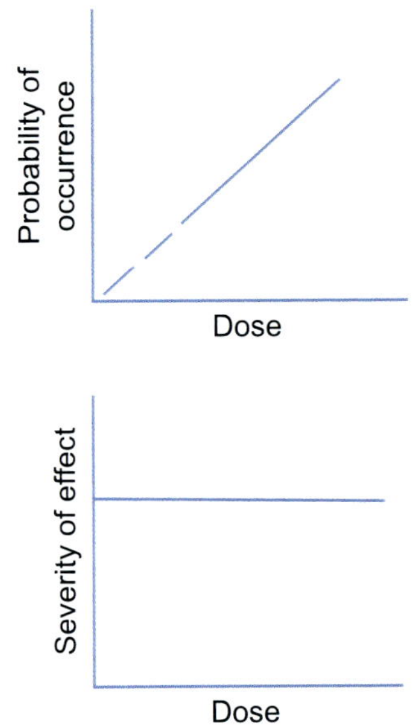

Fig. 1.2 Probability of occurrence and severity of stochastic effects as a function of dose

their appearance. The risk of stochastic effects increases in the range of low doses linearly with dose while their severity is not affected [8, 15]. In addition, their severity does not increase with an increase in the received dose (Fig. 1.2).

For low radiation doses and/or dose rates, the risk for stochastic effects is substantially lower, as there is a greater likelihood for the induced damage to be repaired. However, there is still a risk of long-term effects (e.g., cancer) that may appear years or even decades later. This risk is higher for children and adolescents, due to their increased radiosensitivity compared to adults.

Risk of Cancer

Ionizing radiation can induce most, but not all cancer types. The latency period for the appearance of cancer after irradiation is approximately a decade, but shorter (2–5 years) for leukemia and thyroid cancer. The phases of tumors' development due to cell mutations after their exposure to ionizing radiation are as follows [8, 15–17]:

(a) Tumor initiation: Entry of a normal cell into pre-neoplastic state that can lead to cancer.
(b) Tumor promotion: Growth and development of a pre-neoplastic clone of initiated cells.
(c) Malignant conversion: Change from a pre-neoplastic state to one where cancer development is likely.
(d) Tumor progression: Cells gain properties that allow more rapid development and acquire of invasive characteristics.

Epidemiological studies on populations exposed to ionizing radiation, such as atomic bomb survivors or radiotherapy patients, indicate that a significant increase in the risk of cancer at doses above 100 mSv. Additionally, recent studies on individuals who underwent medical exposures during childhood have indicated that the risk of cancer may increase even at doses in the range of 50–100 mSv.

For doses lower than about 100 mSv an increase in dose will be proportional to the likelihood of incurring cancer. The associated detriment adjusted nominal risk coefficients for cancer are 5.5×10^{-2} Sv^{-1} for the whole population and 4.1×10^{-2} Sv^{-1} for adult workers [8].

Hereditary Effects

When ionizing radiation hits germ cells, it may damage their genetic material and result in malformations, metabolic disorders, immune deficiencies etc. in the descendants of exposed individuals that may appear after many generations.

The detriment adjusted probability coefficients for heritable effects up to the second generation are 0.2×10^{-2} Sv^{-1} for the whole population and 0.1×10^{-2} Sv^{-1} for adult workers [8]. So far, no association between radiation exposure and the occurrence of hereditary damage has been observed in humans.

The International Commission on Radiological Protection (ICRP) assumes that an acute dose of 1 gray (Gy) to parental gonads leads to one additional severe disease caused by radiation-induced mutations in 500 births. This genetic risk may last for up to two generations [8].

Other Diseases Caused by Ionizing Radiation

The results of the analysis of data from the Japanese atomic bomb survivors and cancer patients who underwent radiotherapy indicate that there is statistical evidence that non-cancer effects (i.e., heart disease, stroke, digestive disorders, and respiratory disease) can be induced by exposures at effective doses of the order of 1 Sv. These effects seem to be directly associated with the level of the absorbed dose.

Radiotherapy to the heart at high doses (>10 Gy) increases the risk of radiation-related heart diseases (e.g., pericarditis, valvular disease, cardiomyopathy, etc.) after a minimal latency period of 1–2 years (although acute pericarditis may develop as soon as some weeks after). Furthermore, there is a dose–response relationship in late cardiovascular disease mortality, including both heart disease and stroke after at least a decade. Such an effect could neither been confirmed nor excluded at doses lower than 0.5 Gy [17–19].

Effects of Ionizing Radiation to the Embryo/Fetus

The embryo is sensitive to ionizing radiation during the preimplantation period of embryonic developments. However, for doses of a few tens of mGy the probability of lethal effects of ionizing radiation are very low as well as of health risks after birth [20, 21].

Regarding the appearance of malformations due to the exposure of the embryo/fetus to ionizing radiation they are gestational age-dependent, with the period of maximum radiosensitivity to be that of major organogenesis and for doses higher than about 100 mGy.

More specifically:

1st–2nd week post conception: During this phase, the number of cells of the fertilized egg is small. Therefore, exposure of the fetus to high doses of ionizing radiation may cause its failed implantation in the uterine lining or death. In this case, the pregnancy will not be perceived.

If the pregnancy continues, it is considered that the child to be born will not be harmed due to its irradiation, without, however, the radiation consequences being completely ruled out. In any case, this is considered as a low-risk phase.

3rd–8th week post conception: During this phase of organogenesis, there is a likelihood of malformations for doses to the fetus higher than 100 mSv. This likelihood increases during the proliferation of fetal cells and the differentiation of the developing organs. The probability of end results is 0.015% per 1 mSv.

8th–15th week post conception: It seems that there is a dose threshold of at least 300 mGy for the induction of severe mental retardation due to exposure of fetus to ionizing radiation in this most radiosensitive period. The impact on IQ of low doses (few tens of mGy) to the fetus is insignificant in most cases.

According to epidemiological studies, the risk for cancer after the fetal/embryo exposure to ionizing radiation is similar to the risk after an exposure in early childhood [8]. However, recent evidence on in utero exposure indicates that the lifetime risks of solid cancers (but not leukemia) are considerably lower than those for exposure in childhood, i.e., at most, about three times that of the population as a whole [22].

References

1. Elgazzar AH, Kazem N. Biological effects of ionizing radiation. In: Elgazzar A, editor. The pathophysiologic basis of nuclear medicine. Cham: Springer; 2015. https://doi.org/10.1007/978-3-319-06112-2_21.
2. Lomax M, Folkes L, O'Neill P. Biological consequences of radiation-induced DNA damage: relevance to radiotherapy. Clin Oncol. 2013;25(10):578–85. https://doi.org/10.1016/j.clon.2013.06.007.
3. Ermakov A, Konkova M, Kostyuk S, et al. Oxidized extracellular DNA as a stress signal in human cells. Oxidative Med Cell Longev. 2013;2013:649747. https://doi.org/10.1155/2013/649747.
4. Azzam EI, de Toledo SM, Little JB. Direct evidence for the participation of gap junction-mediated intercellular communication in the transmission of damage signals from alpha-particle irradiated to nonirradiated cells. Proc Natl Acad Sci U S A. 2001;98:473–8. https://doi.org/10.1073/pnas.98.2.473.
5. Ramesh R, Marrogi AJ, Munshi A, Abboud CN, Freeman SM. In vivo analysis of the 'bystander effect': a cytokine cascade. Exp Hematol. 1996;24:829–38.
6. Morgan WF. Non-targeted and delayed effects of exposure to ionizing radiation: II. Radiation-induced genomic instability and bystander effects in vivo, clastogenic factors and transgenerational effects. Radiat Res. 2003;59:581–96. https://doi.org/10.1667/0033-7587(2003)159[0567:nadeoe]2.0.co;2.
7. Suzuki K, Ojima M, Kodama S, Watanabe M. Radiation-induced DNA damage and delayed induced genomic instability. Oncogene. 2003;22(45):6988–93. https://doi.org/10.1038/sj.onc.1206881.
8. ICRP. Low-dose extrapolation of radiation-related cancer risk. ICRP Publication 99. Ann ICRP. 2005;35(4):1–140.
9. Green D. Means to an end: apoptosis and other cell death mechanisms. Cold Spring Harbor: Cold Spring Harbor Laboratory Press; 2011.
10. ICRP. RBE for deterministic effects. ICRP Publication 58. Ann ICRP. 1990;20(4)
11. National Council on Radiation Protection and Measurements (NCRP). The relative biological effectiveness of radiations of different quality. Report No. 104. 1990.
12. ICRP. Relative biological effectiveness (RBE), quality factor (Q), and radiation weighting factor (w_R). ICRP Publication 92. Ann ICRP. 2003;33(4):1.
13. Dörr W, Hendry JH. Consequential late effects in normal tissues. Radiother Oncol. 2001;61(3):223–31. https://doi.org/10.1016/S0167-8140(01)00429-7.
14. ICRP. ICRP statement on tissue reactions/early and late effects of radiation in normal tissues and organs—threshold doses for tissue reactions in a radiation protection context. ICRP Publication 118. Ann ICRP. 2012;41(1/2):1.
15. United Nations Scientific Committee on the Effects of Atomic Radiation (UNSCEAR). Sources and effects of ionizing radiation. 2000 Report UNSCEAR Reports. 2000. https://doi.org/10.18356/49c437f9-en.
16. National Council on Radiation Protection and Measurements (NCRP). Evaluation of the linear-non threshold dose-response model for ionizing radiation. Report No. 136. 2001.
17. Preston DL, Shimizu Y, Pierce DA, Suyama A, Mabuchi K. Studies of mortality of atomic bomb survivors. Report 13: solid cancer and noncancer disease mortality: 1950–1997. Radiat Res. 2003;160(4):381–407. https://doi.org/10.1667/rr3049.
18. UNSCEAR. Effects of ionizing radiation: 2006 report to the general assembly. Annex B. 2006.
19. Shimizu Y, Kodama K, Nishi N, Kasagi F, Suyama A, Soda M, et al. Radiation exposure and circulatory disease risk: Hiroshima and Nagasaki atomic bomb survivor data, 1950–2003. BMJ. 2010;340:b5349. https://doi.org/10.1136/bmj.b5349.
20. ICRP. Biological effects after prenatal irradiation (embryo and fetus). ICRP Publication 90. Ann ICRP. 2003;33:1–2.
21. European Commission. Guidance for protection of unborn children and infants irradiated due to parental medical exposures. Radiation Protection 100. Directorate-General Environment, Nuclear Safety and Civil Protection. 1998.
22. Preston DL, Cullings H, Suyama A, Funamoto S, Nishi N, Soda M, Mabuchi K, Kodama K, Kasagi F, Shore RE. Solid cancer incidence in atomic bomb survivors exposed in utero or as young children. J Natl Cancer Inst. 2008;100(6):428–36. https://doi.org/10.1093/jnci/djn045.

Principles and Guidelines for Radiation Protection

2

A. Hertault

Why Is Radiation Protection Essential?

Physics Principles and Image Formation

X-rays are electromagnetic waves consisting of photons propagating in a straight line in vacuum. They are charged in energy, expressed in electron volts (eV). On the spectrum of electromagnetic waves, X-rays have a wavelength between 0.03 and 10 nm, meaning they are between the ultraviolet and gamma rays. X-rays used in the medical field carry an energy between 10 and 150 keV. For comparison, the energy transmitted by visible light is only around 2 eV [1].

When an X-ray beam crosses matter, several phenomena are possible [2] (Fig. 2.1):

- No interaction: the X-ray beam passes through the matter without losing energy.
- Scattering: the X-ray beam is redirected in another direction. This can be done without loss of energy, or by transferring a part of its energy, resulting in an ionization phenomenon at the atomic level. This last mechanism is called the "Compton effect".
- Absorption: the whole energy of the beam is absorbed, with ionization of the matter. This mechanism is called the "photoelectric effect".

Within an X-ray generator, electrons are accelerated and crushed against an anode. This produces 99% of heat (explaining the overheating phenomena encountered with imaging devices, and the need for cooling systems) and the remaining 1% are photons that will form the X-ray beam [3]. At the exit of the generator, the X-ray beam is homogeneous. As it passes through the patient, the beam will lose part of its energy by the mechanisms of diffusion and absorption explained above [4]. When it reaches the detector, the beam has become inhomogeneous, and will form the radiological image as we know it (Fig. 2.2).

In practice, obtaining an X-ray image is inseparable from an ionization of the tissues, which will be at the origin of the biological toxicity of X-rays. In addition, the patient will become a secondary source of radiation due to the scattering phenomenon, which will expose nearby health care workers.

Biologic Risks Associated to X-Ray Exposure

Exposure to X-rays will cause damage to biological tissues by a direct mechanism of alteration of molecules or DNA, but also indirectly by ionization of water molecules into free radicals such as hydrogen peroxide. Biological effects related to X-ray exposure can be divided into two main categories. On the one hand, the deterministic effects, and on the other hand, the stochastic effects.

Deterministic Effects

Deterministic effects are related to cell death following X-ray exposure. They are threshold-dependent and appear only after being exposed to a certain dose. Their occurrence can therefore be predicted. Their severity is correlated to the intensity of the exposure, which means that their severity can also be predicted. Their appearance is delayed from a few weeks to a few months after exposure. The main manifestation of deterministic effects found in medicine is radiation dermatitis. This is a skin burn that can be found at the point of contact between the X-ray beam and the patient's skin (most often on the patient's back). This complication must be systematically sought in the weeks following an exposure if the dose delivered to the skin was greater than 2 Gy. In more severe exposures (>5 Gy), a chronic wound with necrosis of the dermis may be found and may require reconstructive surgery [5].

A. Hertault (✉)
Hopital Privé de Villeneuve d'Ascq, Lille, France

G. Geroulakos et al. (eds.), *Mastering Endovascular Techniques*, https://doi.org/10.1007/978-3-031-42735-0_2

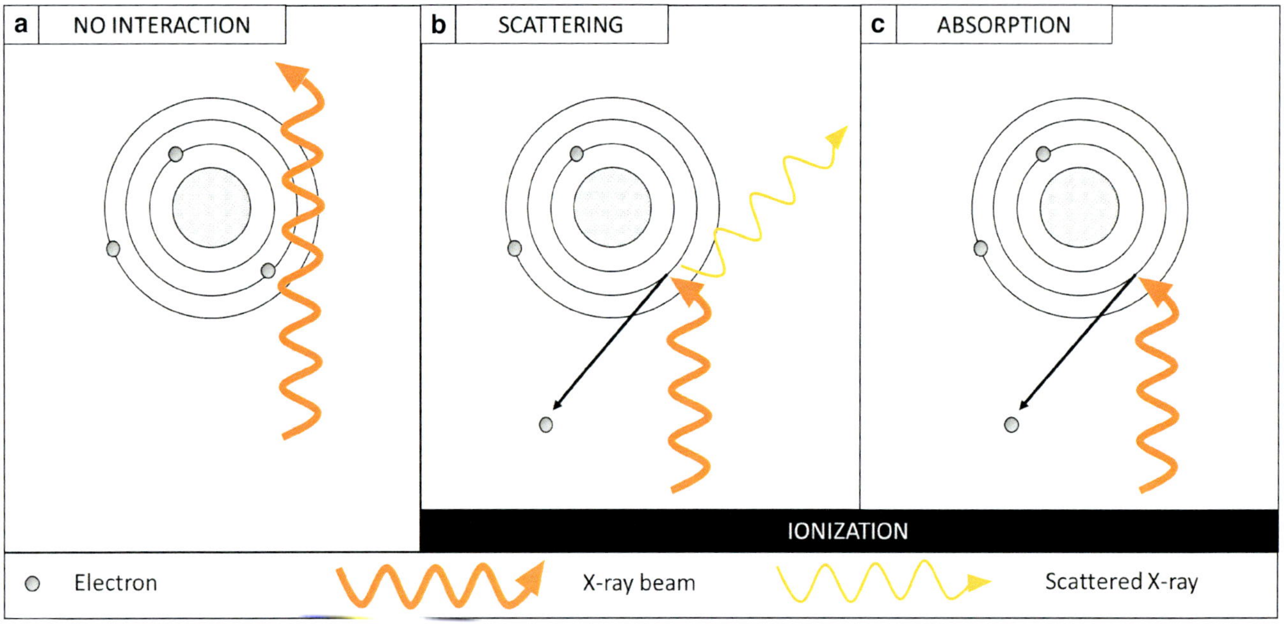

Fig. 2.1 When the X-ray beam passes through the tissue, three types of situations are possible. The beam can pass through the material without interaction, and therefore without energy loss (**a**). The beam can interact with electrons at the atomic level and be reflected in another direction, with or without energy transfer (**b**). In case of energy transfer (Compton effect), an electron is ejected with an ionisation phenomenon. Finally, the entire energy of the beam can be absorbed at the atomic level, again with ionisation (**c**). This is the photoelectric effect

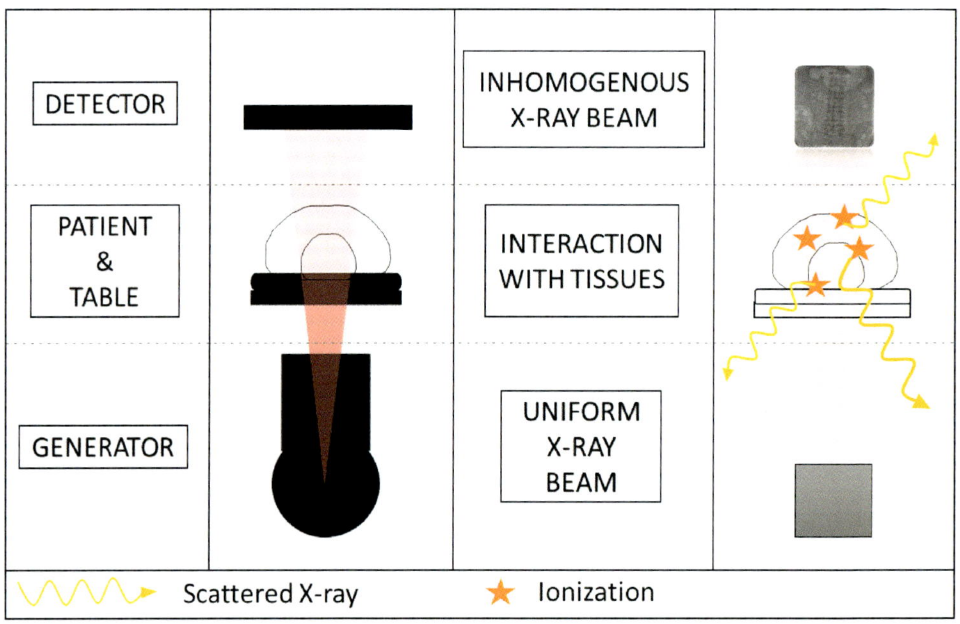

Fig. 2.2 At the output of the X-ray generator, the beam is uniform. As it passes through the tissue, scattering and absorption mechanisms come into effect. When it reaches the detector, the beam is inhomogeneous and allows the formation of the X-ray image

In caregivers, the main deterministic effect found is radiation-induced cataract, responsible for the appearance of typical posterior subcapsular opacities. Previously, this complication was considered to occur after intense exposures (>5 Gy), but in recent years it has been considered that it could occur at much lower thresholds (around 0.5 Gy) [6]. Observational studies suggest that between interventional and non-interventional cohorts of cardiologists, the risk of observing posterior subcapsular opacities would be up to 6 times greater [7, 8].

The incidence of deterministic effects following a medical procedure is estimated to be between 1/10,000 and 1/100,000 cases, but the true incidence is unknown due to underdiagnosis and underreporting of this complication [9].

Stochastic Effects

Stochastic effects are related to DNA alteration. Unlike deterministic effects, they are not triggered above a certain threshold, which means that they can be induced even by small exposures. On the other hand, it is known that their probability of occurrence increases with life expectancy and that a young subject will be at greater risk of developing a complication related to a stochastic effect than an elderly subject. Similarly, the severity is not correlated with the intensity of the exposure. They can occur years after an exposure. The main complications related to stochastic effects are solid cancers, leukemia and transmission of genetic abnormalities to offspring.

The possibility of stochastic effects is of particular concern for caregivers, who are repeatedly exposed to small doses of radiation. Available data suggest an increased incidence of thyroid, breast, brain, and skin melanomas in occupationally exposed personnel [10–12]. With some studies even reporting a risk of cancer that could be 20% higher in exposed professionals compared to their unexposed colleagues [13, 14]. Unfortunately, the lack of large-scale cohort follow-up of exposed workers over a long period makes it impossible to assess the real risk of developing complications related to stochastic effects. The precautionary principle must therefore be applied.

What are the Key Principles of Radiation Protection?

The "As Low As Reasonable Achievable" or ALARA rule governs the principles of radiation protection [15]. It can be broken down into three main items:

- Justification: any use of X-rays must have an expected benefit greater than the theoretical risk incurred, which in medicine overlaps with the traditional benefit/risk balance. Thus, since exposure of a young subject is associated with a greater stochastic risk than for an elderly subject, examinations or procedures requiring X-rays should be avoided as much as possible. Similarly, whenever possible, an examination or technique that does not require the use of X-rays should be preferred when similar results can be expected (e.g., ultrasound should be preferred to CT in the diagnosis of acute appendicitis). It is interesting to note here that the benefit/risk balance is less clear for caregivers who are exposed to the risk but for a collective rather than individual benefit.
- Optimization: This principle brings together all the elements that will make it possible to reduce exposure when it is necessary. It is often broken down into three parts: (1) time: limiting as much as possible the exposure time, (2) distance: increasing as much as possible the distance with the source, and (3) shielding: protecting oneself. In daily practice, it is based on the combination of good practices and the optimization of the performance of imaging equipment.
- Dose limits: This principle aims to set annual doses that are considered "acceptable" for exposed workers.

Focus on Optimization: Tips and Tricks for Daily Practice

X-rays are odorless, colorless and soundless. Understanding that one is exposed and at risk therefore requires increasing one's awareness to radiation. In this respect, passive and active dosimeters are of considerable interest. Retrieving monthly or annual dosimetry reports and comparing them to the data available in the literature allows for a better understanding of the potential for progress in radiation protection. It is also interesting to compare the radiation reports provided by the imaging equipment and to compare them to the data in the literature for a given type of procedure. In recent years, active dosimeter systems have been available for fitting over lead shielding. These dosimeters provide real time information on the exposure of the operators during the procedures, thanks to a simple color code, and thus allow to adjust one's behavior in order to limit one's exposure [15]. Dose archiving softwares (DACS) are also commercialized and allow a more detailed analysis of the use of imaging systems in order to improve their settings and usage [16].

Methods to reduce the use of X-rays during interventional procedures can be grouped into two main categories: On the one hand, tricks that will allow to limit the number of images produced, and on the other hand, those that will allow to reduce the amount of X-rays necessary to produce an image.

Limit the Number of Produced Images

Limit Fluoroscopy Time

The operator must be trained to acquire images only during times of interest and know how to release the pedal as soon as it is possible. It is also necessary to remain vigilant not to acquire images when not looking at the monitors. Modern imaging technologies, such as image fusion, allow the positioning of the C-arm, the choice of working angulations or the adjustment of collimation without having to resort to image acquisition, which makes it possible to significantly reduce the number of images acquired, and thus the irradiation during a procedure [16–19].

Use Pulse Mode

The use of a continuous scopy mode should be avoided, in favor of a pulsed scopy mode, i.e. based on the production of X-rays over short periods, interspersed with free intervals. Any procedure should be started in pulsed mode, with the lowest possible image rate to allow the procedure to be performed safely. If necessary, the image rate can be temporarily increased during the procedure to increase the image resolution but be sure to return to the lowest image rate as soon as possible. Reducing pulse rates from 30 to 15 or 7.5 pulses/s decreases fluoroscopy dose by 47% and 72% respectively [20, 21].

Have the Main Operator Control the Imaging Equipment

Errors in communication or understanding can lead to unnecessary acquisitions or non-optimised use of X-rays. It is advised that the primary operator keep control of the image production to avoid this [22].

Limit the Dose Required to Produce Images

Set Up the Imaging Equipment

Imaging systems are equipped with an image quality adjustment algorithm, called Automatic Exposure (or Brightness) Control (AEC or ABC), allowing it to maintain a constant image quality, if necessary by increasing the amount of X-rays delivered [9, 23, 24]. When the W-ray beam crosses more flesh (in obese patients or in steep angulations for example) or when radiodense materials, such as surgical tools, are in the field, the algorithm will automatically increase the beam energy, without the operator control, to maintain image quality. To make the best use of an imaging device, it is necessary to define with the supplier the use that will be made of the imaging equipment, and to set up the device so that the ABC algorithm with the best compromise between image quality and delivered dose can be chosen for each type of procedure.

Use Low Dose or Half Dose Settings

Imaging systems are equipped with options to modulate the delivered dose, usually called low-dose (or half-dose) modes. It is recommended to always start a procedure in low-dose mode, and temporary revert to normal mode for the most complex procedure times if necessary [25, 26].

Use Collimation

Collimation limits the size of the field of view (FOV) to the area of interest by using metal apertures to adjust the beam size, thus limiting the exposure of surrounding tissue to X-rays. It is generally considered that the reduction in radiation generated by collimation is equivalent to the area hidden, which means that a 30% reduction in FOV allows a 30% reduction in delivered dose. In addition, it improves image quality by limiting the background noise related to scattered radiation [20, 27–29].

Limit C-Arm Angulation

As the C-arm is moved from anteroposterior to lateral incidence, the thickness of tissue to be penetrated generally becomes greater, resulting in increased attenuation of beam power and decreased image quality. However, imaging systems are configured to maintain constant image quality, so the system will increase the beam power and delivered dose in compensation. It is therefore essential to return to an anteroposterior position as soon as possible [28, 30] and to define the least extreme working angles during the procedure. In this respect, 3D workstations are particularly useful for planning these angulations before the procedure [31].

Optimize the Image Chain Geometry

To make the best use of the X-ray beam characteristics, it is necessary to raise the operating table as high as possible and lower the detector as close to the patient as possible. In this configuration, the beam can widen enough to have a larger FOV and a delivered dose spread over a broader area of skin [32] (knowing that the FOV can then be adjusted by collimation). In addition, the closer the detector is to the patient, the less chance the scattered radiation has of reaching it and generating background at that level [24, 32].

Limit Use of Digital Subtraction Angiography (DSA) (and Other Angiographic Modes)

The imaging system can produce images in fluoroscopy and angiography modes. Fluoroscopy is an image quality mode adapted to allow guidance during interventional procedures, with a reasonable delivered dose. Angiography modes, including subtraction, are diagnostic modes requiring a more powerful X-ray beam. It is therefore recommended that the use of angiography modes be limited as much as possible and that fluoroscopy modes be preferred [9, 27, 33]. Modern imaging systems allow subtraction in fluoroscopy, as well as digital storage of fluoroscopy loops.

Avoid Magnification

Magnification is the action of focusing the X-ray beam on a smaller area, in order to deliver more energy and thus obtain a better image quality. This will increase the dose delivered to the patient's skin, as well as the scattered radiation for the operators. It is generally recommended to avoid magnification if possible, or to use it only at specific times during the procedure. Modern imaging systems are equipped with a digital zoom option, which allows the image size to be increased without increasing the radiation dose, but at the cost of a decrease in image quality.

Impact of Equipment

Modern imaging systems are equipped with technology that allows for better image quality at a lower radiation cost. Compared to traditional analogue image intensifier, digital flat panel detectors offer a much higher sensitivity to X-rays with a better signal-to-noise ratio and limited geometric distortion [34]. Digital image processing also makes it possible to reduce irradiation through the use of software, some of which have been mentioned above, such as image fusion, digital zoom, subtraction and registration of fluoroscopy sequences, etc.

Individual Protection Measures and Shielding

The measures outlined above can reduce the intensity of the primary X-ray beam and scattered radiation. However, it is necessary to complement this with individual protective measures to minimize exposure.

The Inverse Square Law

The energy carried by the scattered radiation decreases in inverse proportion to the distance, the so-called inverse square law where $X = 1/d^2$ (X = exposure, d = distance). The first measure to reduce exposure is therefore to move away from the source of the radiation (i.e., the patient who acts as the source of the scattered radiation for operators) [27]. Thus, it is recommended to take a few steps back when possible, to perform angiography sequences from the control room using a power injector or to work with long sheaths when appropriate.

Positioning Around the Table

The scattered radiation is most intense at the point of beam entry, i.e. in the normal working position under the operating table. According to the same principle, when working with lateral angles, the radiation received is more important on the side of the generator, for example at the left oblique angle for an operator located at the right groin of the patient [32]. When possible, another oblique angle should be used or the operator should move.

Shielding

Personal or collective protective equipment uses X-ray filtering materials, such as lead, to attenuate the scattered radiation. As lead is relatively heavy, most modern equipment is made of a composite alloy of other lighter attenuating metals (barium, aluminum, tungsten, etc.). Their protection efficiency is expressed in lead thickness equivalent.

Wearable Aprons

Leaded aprons are the most common form of personal protective equipment and can reduce by up to 90% the operator's exposure [35]. For most interventional procedures, an apron offering at least 35 mm lead equivalent protection on the front and 25 mm on the back is sufficient. It is imperative that the aprons are properly fitted to the operator's body to avoid large gaps in the neck, back or underarms that would leave entire areas of the body unprotected. In order to achieve a more acceptable weight, most aprons use the overlap area to provide adequate protection. For example, an apron with a lead equivalent of 0.25 mm provides protection of 0.25 + 0.25 = 0.50 mm lead equivalent at the overlap. It is therefore imperative that the apron closes properly, otherwise the lateral parts of the abdomen are less well protected.

Wearing a two-piece apron (skirt and top) allows for better weight distribution, improved comfort and tolerance, and reduced osteoarticular complications.

Lead and its equivalents can crack or split if folded, considerably altering the effectiveness of the protection. It is therefore essential to ensure that aprons are stored carefully and that their condition must be checked regularly.

Thyroïd Collar

Although the radiosensitivity of the thyroid and the risk of cancer decreases after the age of 20, the thyroid collar also protects other potentially radiosensitive organs in the neck such as the trachea or carotid arteries. It should be worn systematically [36] and quality control should be carried out annually.

Leg Shields

DNA alterations can be found in circulating cells of operators after an endovascular procedure without a table-mounted shield or additional tibial protection [37]. Therefore, the use of table-mounted shields (on both sides of the table), leaded trolley or additional leg protection is necessary. This type of protection can reduce the scattered radiation reaching the lower part of the body by >90% [38].

Eye and Upper Body Protections

Given the specific risk of radiation-induced cataract, dedicated eye protection is essential. Eye protection includes glasses with or without a corrective lens, fit-over glasses that can be positioned over the prescription glasses and visors. Regardless of the type of protection chosen, it is important to understand that scattered radiation generally arrives from the side, so lead eyewear must cover the side of the eye toward the temple to be effective [39].

The operator's skull and body can also act as a secondary source of radiation by reflecting some of the scattered radiation back to the eye, making it useful to place a larger screen between the operator and the patient. Ceiling-mounted lead shields can reduce up to 80% of the scattered radiation to the upper part of the operator's body during femoral access procedures [38]. Their use should therefore be systematic.

Controversial Protections

Leaded protective caps with a lead thickness of 0.5 mm are heavy (>1 kg) and their discomfort limits their use. The new generation of composite caps have shown inconsistent results in terms of radiation protection effectiveness in the literature [40, 41] and their use remains infrequent.

The protective drapes to be placed on the patient are sterile and contain a bismuth or tungsten alloy. They are placed directly on the sterile field, in order to block the scattered radiation emanating from the patient. They make it possible to partially limit the diffusion to the operator, in particular in unprotected areas such as the hands [42]. Their main limitation remains that they must be positioned at a distance from the acquisition field, so as not to be in the FOV (which would lead the auto-exposure algorithm to increase the dose in compensation).

Leaded gloves are not commonly used because of their stiffness and discomfort. They theoretically reduce radiation to the hands, with an attenuation efficiency of 15–30% [43, 44]. Their use is debated because the radiation passing through them may end trapped between the layers, which would result in an increased delivered dose to the hand. Furthermore, it is essential to be careful never to position them in the primary beam, as their attenuation would cause the auto-exposure algorithm to increase the power of the beam.

What You Should Definitely Remember

X-rays are used in medicine for diagnostic or therapeutic purposes. However, the contact between X-rays and tissues is responsible for ionization phenomena that cause biological complications. Patients, located in the primary beam, are at risk of radiation dermatitis in case of excessive exposure. Health care workers, regularly exposed to scattered radiation, are at risk of cancer and radiation-induced cataract. Because of these risks, it is essential to apply the benefit/risk balance to any use of X-rays.

During X-ray guided procedures, every effort should be made to limit the number of images acquired, to adjust the power of the primary beam, to reduce the exposure of the health care workers and to protect themselves in an optimal way. Together, these measures significantly reduce patient and staff exposure and create a safer working environment.

References

1. Hall EJ, Giaccia AJ. Radiobiology for the radiologist. Philadelphia: Lippincott Williams & Wilkins; 2006. https://books.google.fr/books?id=6HhjwRyqBzgC
2. Bushberg JT. The essential physics of medical imaging. Philadelphia: Lippincott Williams & Wilkins; 2002. p. 965.
3. Russo P. Handbook of X-ray imaging: physics and technology. Boca Raton: CRC Press; 2017. p. 2926.
4. Hendee WR, Ritenour ER. Medical imaging physics. New York: Wiley; 2003. p. 535.
5. Koenig TR, Wolff D, Mettler FA, Wagner LK. Skin injuries from fluoroscopically guided procedures: part 1, characteristics of radiation injury. AJR Am J Roentgenol. 2001;177(1):3–11.
6. Chodick G, Bekiroglu N, Hauptmann M, Alexander BH, Freedman DM, Doody MM, et al. Risk of cataract after exposure to low doses of ionizing radiation: a 20-year prospective cohort study among US radiologic technologists. Am J Epidemiol. 2008;168(6):620–31.
7. Elmaraezy A, Ebraheem Morra M, Tarek Mohammed A, Al-Habaa A, Elgebaly A, Abdelmotaleb Ghazy A, et al. Risk of cataract among interventional cardiologists and catheterization lab staff: a systematic review and meta-analysis. Catheter Cardiovasc Interv. 2017;90(1):1–9.
8. Karatasakis A, Brilakis HS, Danek BA, Karacsonyi J, Martinez-Parachini JR, Nguyen-Trong PKJ, et al. Radiation-associated lens changes in the cardiac catheterization laboratory: results from the IC-CATARACT (CATaracts attributed to RAdiation in the CaTh lab) study. Catheter Cardiovasc Interv. 2018;91(4):647–54.
9. Hertault A, Maurel B, Midulla M, Bordier C, Desponds L, Kilani MS, et al. Editor's choice—Minimizing radiation exposure during endovascular procedures: basic knowledge, literature review, and reporting standards. Eur J Vasc Endovasc Surg. 2015;50(1):21–36.
10. Zielinski JM, Garner MJ, Band PR, Krewski D, Shilnikova NS, Jiang H, et al. Health outcomes of low-dose ionizing radiation exposure among medical workers: a cohort study of the Canadian national dose registry of radiation workers. Int J Occup Med Environ Health. 2009;22(2):149–56.
11. Preston DL, Kitahara CM, Freedman DM, Sigurdson AJ, Simon SL, Little MP, et al. Breast cancer risk and protracted low-to-moderate dose occupational radiation exposure in the US Radiologic Technologists Cohort, 1983–2008. Br J Cancer. 2016;115(9):1105–12.
12. Rajaraman P, Doody MM, Yu CL, Preston DL, Miller JS, Sigurdson AJ, et al. Cancer risks in U.S. Radiologic Technologists working with fluoroscopically guided interventional procedures, 1994–2008. AJR Am J Roentgenol. 2016;206(5):1101–8. quiz 1109
13. Andreassi MG, Piccaluga E, Guagliumi G, Del Greco M, Gaita F, Picano E. Occupational health risks in cardiac catheterization laboratory workers. Circ Cardiovasc Interv. 2016;9(4):e003273.
14. Wang JX, Zhang LA, Li BX, Zhao YC, Wang ZQ, Zhang JY, et al. Cancer incidence and risk estimation among medical X-ray workers in China, 1950–1995. Health Phys. 2002;82(4):455–66.
15. Murat D, Wilken-Tergau C, Gottwald U, Nemitz O, Uher T, Schulz E. Effects of real-time dosimetry on staff radiation exposure in the cardiac catheterization laboratory. J Invasive Cardiol. 2021;33(5):E337–41.
16. Hertault A, Rhee R, Antoniou GA, Adam D, Tonda H, Rousseau H, et al. Radiation dose reduction during EVAR: results from a prospective multicentre study (the REVAR study). Eur J Vasc Endovasc Surg. 2018;56(3):426–33.
17. Maurel B, Martin-Gonzalez T, Chong D, Irwin A, Guimbretière G, Davis M, et al. A prospective observational trial of fusion imaging in infrarenal aneurysms. J Vasc Surg. 2018;68(6):1706–1713.e1.
18. McNally MM, Scali ST, Feezor RJ, Neal D, Huber TS, Beck AW. Three-dimensional fusion computed tomography decreases

radiation exposure, procedure time, and contrast use during fenestrated endovascular aortic repair. J Vasc Surg. 2015;61(2):309–16.
19. Ahmad W, Obeidi Y, Majd P, Brunkwall JS. The 2D–3D registration method in image fusion is accurate and helps to reduce the used contrast medium, radiation, and procedural time in standard EVAR procedures. Ann Vasc Surg. 2018;51:177–86.
20. Ketteler ER, Brown KR. Radiation exposure in endovascular procedures. J Vasc Surg. 2011;53(1 Suppl):35S–8S.
21. Miller DL, Balter S, Noonan PT, Georgia JD. Minimizing radiation-induced skin injury in interventional radiology procedures. Radiology. 2002;225(2):329.
22. Peach G, Sinha S, Black SA, Morgan RA, Loftus IM, Thompson MM, et al. Operator-controlled imaging significantly reduces radiation exposure during EVAR. Eur J Vasc Endovasc Surg. 2012;44(4):395–8.
23. Kim KP, Miller DL, Balter S, Kleinerman RA, Linet MS, Kwon D, et al. Occupational radiation doses to operators performing cardiac catheterization procedures. Health Phys. 2008;94(3):211–27.
24. Lipsitz EC, Veith FJ, Ohki T, Heller S, Wain RA, Suggs WD, et al. Does the endovascular repair of aortoiliac aneurysms pose a radiation safety hazard to vascular surgeons? J Vasc Surg. 2000;32(4):704–10.
25. Dias NV, Billberg H, Sonesson B, Törnqvist P, Resch T, Kristmundsson T. The effects of combining fusion imaging, low-frequency pulsed fluoroscopy, and low-concentration contrast agent during endovascular aneurysm repair. J Vasc Surg. 2016;63(5):1147–55.
26. Bicknell CD. Occupational radiation exposure and the vascular interventionalist. Eur J Vasc Endovasc Surg. 2013;46(4):431.
27. Killewich LA, Falls G, Mastracci TM, Brown KR. Factors affecting radiation injury. J Vasc Surg. 2011;53(1 Suppl):9S–14S.
28. Panuccio G, Greenberg RK, Wunderle K, Mastracci TM, Eagleton MG, Davros W. Comparison of indirect radiation dose estimates with directly measured radiation dose for patients and operators during complex endovascular procedures. J Vasc Surg. 2011;53(4):885–894.e1. discussion 894
29. Tonnessen BH, Pounds L. Radiation physics. J Vasc Surg. 2011;53(1 Suppl):6S–8S.
30. Kirkwood ML, Arbique GM, Guild JB, Timaran C, Chung J, Anderson JA, et al. Surgeon education decreases radiation dose in complex endovascular procedures and improves patient safety. J Vasc Surg. 2013;58(3):715–21.
31. Stahlberg E, Planert M, Panagiotopoulos N, Horn M, Wiedner M, Kleemann M, et al. Pre-operative simulation of the appropriate C-arm position using computed tomography post-processing software reduces radiation and contrast medium exposure during EVAR procedures. Eur J Vasc Endovasc Surg. 2017;53(2):269–74.
32. Haqqani OP, Agarwal PK, Halin NM, Iafrati MD. Minimizing radiation exposure to the vascular surgeon. J Vasc Surg. 2012;55(3):799–805.
33. Durán A, Hian SK, Miller DL, Le Heron J, Padovani R, Vano E. Recommendations for occupational radiation protection in interventional cardiology. Catheter Cardiovasc Interv. 2013;82(1):29–42.
34. Livingstone RS, Chase D, Varghese A, George PV, George OK. Transition from image intensifier to flat panel detector in interventional cardiology: impact of radiation dose. J Med Phys. 2015;40(1):24.
35. Badawy MK, Deb P, Chan R, Farouque O. A review of radiation protection solutions for the staff in the cardiac catheterisation laboratory. Heart Lung Circ. 2016;25(10):961–7.
36. National Council on Radiation Protection and Measurements. Radiation dose management for fluoroscopically-guided interventional medical procedures. NCRP Report No. 168. Bethesda, Maryland: National Council on Radiation Protection and Measurements.
37. El-Sayed T, Patel AS, Cho JS, Kelly JA, Ludwinski FE, Saha P, et al. Radiation-induced DNA damage in operators performing endovascular aortic repair. Circulation. 2017;136(25):2406–16.
38. Fetterly KA, Magnuson DJ, Tannahill GM, Hindal MD, Mathew V. Effective use of radiation shields to minimize operator dose during invasive cardiology procedures. JACC Cardiovasc Interv. 2011;4(10):1133–9.
39. Kirkwood ML, Klein A, Guild J, Arbique G, Xi Y, Tsai S, et al. Novel modification to leaded eyewear results in significant operator eye radiation dose reduction. J Vasc Surg. 2020;72(6):2139–44.
40. Reeves RR, Ang L, Bahadorani J, Naghi J, Dominguez A, Palakodeti V, et al. Invasive cardiologists are exposed to greater left sided cranial radiation: the BRAIN study (BRAIN radiation exposure and attenuation during invasive cardiology procedures). JACC Cardiovasc Interv. 2015;8(9):1197–206.
41. Kirkwood ML, Arbique GM, Guild JB, Zeng K, Xi Y, Rectenwald J, et al. Radiation brain dose to vascular surgeons during fluoroscopically guided interventions is not effectively reduced by wearing lead equivalent surgical caps. J Vasc Surg. 2018;68(2):567–71.
42. Vlastra W, Delewi R, Sjauw KD, Beijk MA, Claessen BE, Streekstra GJ, et al. Efficacy of the RADPAD protection drape in reducing operators' radiation exposure in the catheterization laboratory. Circ Cardiovasc Interv. 2017;10(11):e006058.
43. Wagner LK, Mulhern OR. Radiation-attenuating surgical gloves: effects of scatter and secondary electron production. Radiology. 1996;200(1):45–8.
44. Kamusella P, Scheer F, Lüdtke CW, Wiggermann P, Wissgott C, Andresen R. Interventional angiography: radiation protection for the examiner by using lead-free gloves. J Clin Diagn Res JCDR. 2017;11(7):TC26–9.

3 Principles of Anaesthesia in Endovascular Procedures

N. Khan, P. Banugo, and H. Bidd

Case Study

A 73-year-old male patient with a history of hypertension, type 2 diabetes, and peripheral vascular disease presented for elective repair of a 6.5 cm juxtarenal aneurysm. The plan was to perform a fenestrated endovascular aneurysm repair (FEVAR).

After placement of the stent graft, and following removal of the femoral sheaths, the patient suddenly became hypotensive, with a fall in systolic blood pressure from 110 to 85 mmHg. Immediate management included intravenous administration of a 250 mL fluid bolus and boluses of metaraminol to restore his blood pressure. An arterial blood gas sample revealed a stable haemoglobin level, a potassium of 5.6 mmol/L (previously 4.2 mmol/L), a lactate of 2.7 mmol/L, and an elevated $PaCO_2$ level. This picture was assumed to be consistent with an ischaemic-reperfusion event following lower limb reperfusion. The ventilation rate on the anaesthetic machine was increased, from 12 to 18 breaths/min, to facilitate carbon dioxide removal, and fluid resuscitation was continued.

Within 20 min, all parameters had normalised. The patient was subsequently woken up and extubated without event.

Introduction

The ever-increasing scope of interventional radiology has facilitated more complex patients being treated and managed with an endovascular technique. Where previously vascular disease, in particular arterial, was treated with an open technique, endovascular approaches can now be used to stabilise and treat major pathology, often being reserved for patients deemed high-risk and not suitable for an open approach. Naturally, the anaesthetic approach has also evolved to match the surgical insult. In this chapter, we will discuss the anaesthetic considerations for patients undergoing endovascular procedures, both arterial and venous, and peripheral compared to central pathology.

N. Khan (✉) · P. Banugo · H. Bidd
Guy's and St Thomas' NHS Foundation Trust, London, UK
e-mail: nazia.khan@gstt.nhs.uk; pele.banugo@gstt.nhs.uk; heena.bidd@gstt.nhs.uk

Pre-operative Assessment

In the pre-operative setting, it is important to evaluate the burden of disease and work towards risk factor modification to reduce perioperative morbidity and mortality.

Peripheral vascular disease represents significant systemic illness and is likely associated with other co-morbidities, such as coronary artery disease, diabetes, renal impairment, and frailty. From the data we can see common co-morbidities can increase mortality by 3% up to 9% (Table 3.1).

Table 3.1 Outcome by comorbidities [1]

Comorbidity	30-day mortality (%)	Number of patients with comorbidity
Respiratory disease	3.7	1810
Ischaemic heart disease	3.8	1457
Cancer	3.8	1417
Arrhythmia	5.7	1029
Diabetes (non-insulin)	2.9	1005
Cerebrovascular accident	4.4	591
Diabetes (insulin)	4.1	386
Congestive cardiac failure	8.2	243
Documented cirrhosis	8.9	123

Data from a UK prospective audit of 13,513 patients having inpatient surgery, March 1–7, 2010 (several patients had more than one comorbidity)

G. Geroulakos et al. (eds.), *Mastering Endovascular Techniques*, https://doi.org/10.1007/978-3-031-42735-0_3

Cardiac Morbidity

Coronary artery disease has been seen in up to 50% of patients having vascular surgery, with cardiac complication rates as high as 13.5% [2]. Comprehensive guidelines from the American College of Cardiology/American Heart Association (ACC/AHA) recommend a stepwise approach through a targeted history, assessing stability of disease, routine investigations such as pro-BNP levels and electrocardiogram (ECG) and risk stratification using the Duke Activity Status Index [3]. Invasive investigations such as coronary angiography, though useful in accurately quantifying coronary flow, are not readily available and unlikely to be warranted in asymptomatic patients. Multiple studies have shown no benefit in performing prophylactic angioplasty.

Unstable angina, significant heart failure, poorly controlled atrial fibrillation or hypertension can be managed with pharmacological intervention in both the acute and elective setting. However, any optimisation programme is likely to take 6–12 weeks and therefore needs to be weighed up against the urgency of surgery [4–6].

Endocrine Morbidity

Prevalence of peripheral arterial disease (PAD) in diabetic patients is estimated at 20% in the over 40-year-olds, increasing to 29% in the over 50s [7]. Though there have been no specific studies in diabetic patients having endovascular procedures, when comparing all surgical patients, the diabetic group continues to have a higher incidence of post-operative complications and longer length of stay than the non-diabetic group. Adequate diabetic control, measured through HbA1c <8.5% is imperative in the elective scenario. For those having time critical surgery, diabetes control remains a priority and can be achieved through a variable rate infusion if the patient is likely to miss more than one meal. Often, these patients present with critical limb ischaemia needing limb revascularisation and each admission should be used as an opportunity to improve diabetic control and reduce the risk of further complications [8].

Renal Morbidity

Pre-morbid renal failure, pre-operative elevated creatinine and post-operative acute kidney injury are all independent risk factors for post-operative complications. In endovascular practise, measured volumes of contrast and avoidance of nephrotoxic medication should be used in patients identified as at risk of renal impairment [9]. Patients with known end-stage renal disease should have their dialysis appropriately timed with their endovascular repair. It is best performed the day before their surgery to minimise haemodynamic instability caused by anaesthesia and dialysis. It is essential that fluid and electrolyte balance is as near to normal as possible. Careful consideration must be given to patients having fistuloplasty or thrombectomy of their fistula as there is an associated rise in potassium on improving blood flow thus it is vital to aim for a low to normal potassium level pre-operatively.

Frailty

Frailty refers to a "distinctive health state related to the ageing process in which multiple body systems gradually lose their in-built reserves". The incidence of frailty in patients aged 65 years or more is about 10%, rising to between 25 and 50% after the age of 85 years. The clinical impact of frailty has less favourable outcomes in almost all studies compared to the non-frail group, with studies showing longer hospital stays, higher readmission rates, and decreased survival rates [4, 10].

Understanding Risk

Risk stratification tools can vary from a simple graded tool, such as the American Society of Anaesthesiologists' Physical Status Score (ASA-PSS), to the Physiological and Operative Severity Score for the enUmeration of Mortality and Morbidity or POSSUM score. Some are disease specific, i.e., STOPBANG score for obstructive sleep apnoea, whilst others are predictors of 30-day mortality, like the SORT scoring system [1, 11, 12].

Risk refers to the likelihood of a complication or adverse outcome occurring during or after the procedure. Vascular patients are likely to have a plethora of medical problems thus the benefits of any intervention needs to be balanced against this. Though there is no single tool that can accurately predict the risk of endovascular procedures, it can be estimated using scoring systems whilst bearing in mind the level of surgical insult and burden of disease. Any discussion should be centred on these three elements and balanced against the urgency of the case.

Peripheral Limb Revascularisation

Peripheral limb revascularisation can refer to angioplasty, venoplasty, thrombectomy, or fistuloplasty in the elective, urgent or acute setting. When planning the anaesthetic technique for patients requiring limb revascularisation, it is important to consider: the site and duration of the procedure, position of the patient, pre-morbid state of the patient, degree of symptoms the patient is suffering with, and the urgency of the case.

Site and Duration

For both upper and lower limb work, the more distal the site of stenosis, the more amenable it is to a regional technique. The use of single shot nerve blocks is becoming common practise. Insertion of a catheter (centrally or peripherally) through which local anaesthesia can be continuously infused, offers intra-operative anaesthesia and better post-operative pain relief and side effect profile, then the analgesic options available on the ward.

The use of regional anaesthesia must take into consideration what anticoagulant medications the patient is currently taking, the patient's coagulation profile, and the use of intra-operative heparin. In extremely high-risk patients, the balance of risk may favour a regional approach over a general anaesthetic.

The sole use of regional anaesthesia lends itself well to procedures less than 2 h in duration. If it is anticipated that the procedure will last several hours, the addition of sedation or general anaesthesia should be considered as the limiting factor for patient cooperation becomes discomfort in non-operated sites, e.g., neck and leg stiffness and pain.

Position

Most endovascular repairs are done with the patient in the supine position. However, if the patient is required to be prone, general anaesthesia may be required to tolerate this position for prolonged periods of time.

Symptomatology

In patients with peripheral arterial disease, rest pain is a common symptom. Patients often manage these symptoms by hanging their foot off the end of the bed, which can make performing any endovascular procedure quite challenging. Revascularisation can also give rise to reperfusion pain. Options for management of these symptoms are primarily regional nerve block. Alternatively, the patient may require a general anaesthetic to facilitate the procedure.

Endovascular Aneurysm Repair

The first successful endovascular aneurysm repair (EVAR) was performed in Buenos Aires in 1990, by Dr. Juan C. Parodi and Dr. Julio Palmaz, under local anaesthesia [13]. As device design has improved, and expertise has developed, more complex and technically challenging aneurysms are being treated endovascularly, accounting for almost 80% of elective abdominal aortic aneurysms treated.

Surgical Access

Irrespective of the level of aortic disease (infrarenal, juxta-renal, supra-renal, thoracic, or aortic arch), the surgical approach is via either percutaneous or standard surgical incisions in one or both groins, although brachial access, and less commonly subclavian access, is occasionally necessary.

Anaesthetic Technique

Since exposure of the diseased aorta and aortic-cross clamping is not needed, the endovascular approach is much less invasive than open repair and widens the options for anaesthesia to include local anaesthesia, central neuraxial blockage (spinal or epidural anaesthesia), and sedation, in addition to general anaesthesia.

Local Anaesthesia

An awake approach is appropriate for only a select cohort of patients. Suitability should be assessed during the pre-operative assessment and is influenced by patient-related factors, such as the ability to lie flat for a prolonged period and the ability to follow breathing commands intra-operatively. Combining a short-acting local anaesthetic (e.g. lidocaine) with a long-acting drug (e.g. bupivacaine) allows for rapid-onset loss of sensation at the surgical site followed by prolonged action lasting several hours. Haemodynamic instability is negligible with this approach. In an emergency situation, e.g. 'ruptured AAA', local anaesthetic can be considered in especially unstable patients as a starting point, with conversion to a general anaesthetic, if needed, once the patient is more stable. Patients should be warned about the possibility of ischaemic pain in the buttocks and legs. This is due to prolonged occlusion of femoral and iliac arteries.

Central Neuraxial Blockade

Endovascular aneurysm repair can be performed under either spinal anaesthesia or epidural anaesthesia, although a combined spinal epidural approach may be the preferred. The onset of the spinal component takes 10–15 min and provides effective anaesthesia for approximately 90–120 min. For longer surgery, a combined spinal epidural approach can be used with the epidural component being 'topped up' with bupivacaine to extend the period of anaesthesia. The epidural component can further be used to provide analgesia in the post-operative period, if necessary. It is important to follow the latest guidance on central neuraxial procedures in patients on anticoagulants. Adherence to this reduces the risk of epidural haematoma formation. At least 60 min should elapse between the performance of neuraxial blockade and administration of unfractionated heparin intraoperatively [14].

Sedation

Sedation is a technique that uses pharmacological agents to make a patient comfortable enough to tolerate minor or minimally invasive procedures. Sedation can be delivered on a spectrum from light to deep, titrating to the desired effect. There are many contributing factors to the effectiveness of sedation; the choice of pharmacological agent used, the theatre environment, and the willingness of the patient, are all vital elements that need to be carefully balanced to achieve optimal sedation. A single agent, or multiple agents, can be administered as a bolus dose or via an infusion, influenced by the type of procedure and severity of pain likely to be experienced. Sedation can also be used in conjunction with regional anaesthesia to enhance the patient experience.

General Anaesthesia

This approach should be reserved for procedures that are likely to take in excess of 3–4 h, or if a patient is unable to tolerate the procedure under alternative options. Total intravenous anaesthesia (TIVA) with tracheal intubation works well, allowing for stable intraoperative conditions with smooth, rapid emergence from anaesthesia.

Simple Versus Complex EVAR

Procedure complexity and duration, patient comorbidities and patient preference all influence the choice of anaesthetic technique. Complex EVARs are likely to take 3–4 h and are typically performed under general anaesthesia unless the patient is considered to be frail and high-risk, when local anaesthetic with sedation may be considered more appropriate.

The risk of contrast-induced nephropathy increases with the dose of contrast used. Patients at risk include those with preoperative renal impairment, cardiac failure, patients having para-renal stents sited, and patients undergoing prolonged, complex procedures [15]. Every effort should be made to limit the amount of contrast used. The patient must continue to be hydrated well into the post-operative period and nephrotoxic drugs should be avoided. No other treatments have been proven to reduce the risk of acute kidney injury.

Large bore venous access is mandatory for all EVARs. Central venous access is rarely necessary but should be considered with the more complex procedures. Invasive blood pressure monitoring, patient warming, and urinary catheter insertion should always be instituted.

Intra-operative blood loss is minimal for simple infrarenal EVARS but the risk increases with increased complexity of the procedure, thus it is vital to have cross matched blood available and access to a rapid infusion device. The need for brachial access should be discussed at team brief and peripheral lines placed on the contralateral side. Aim for a rapid wake up post procedure to allow early assessment of neurological function, especially if spinal cord ischaemia is a risk.

Aortic Arch Branched Repairs and Thoracic Endovascular Aortic Repair (TEVAR)

In patients not suitable for an open arch repair and cardiopulmonary bypass, or deep circulatory arrest, an endovascular method can be used. To assist this approach, rapid ventricular pacing may be necessary [16]. Rapid pacing is performed through insertion of a temporary pacing wire into the internal jugular or subclavian vein to the right ventricle. A short period of rapid pacing will cause profound hypotension, limiting blood flow through the aortic arch and facilitating deployment of the endograft into the ascending aorta.

Similarly, some thoracic endovascular aneurysm repairs (TEVAR) may also need graft deployment into the arch of aorta and rapid ventricular pacing, in certain circumstances.

Spinal Cord Ischaemia

Spinal cord ischaemia refers to the loss of perfusion to the spinal cord during thoraco-abdominal aneurysm repair. This rare, but devastating, risk is more commonly associated with an open surgical approach, however, the incidence during TEVAR remains at 0–10.3%, average of 4.5% [17, 18]. Patients typically present with symptoms of incontinence and impaired motor function with some preserved sensory function in the immediate post-operative period. It is possible to mitigate some of this risk through staging surgery, maintaining physiological parameters, and insertion of a spinal drain.

In patients having a complex thoracic repair, a staged approach to surgery over several months allows collateral blood supply to form and provides extra protection to the spinal cord if it should suffer from significant aortic blood loss or ischaemia intra-operatively.

$$\text{Spinal Cord Perfusion} = \text{Mean Arterial Pressure}\left(\text{MAP}\right) - \text{Spinal Cord Pressure}.$$

Furthermore, maintaining spinal cord perfusion in the perioperative period, particularly intra-operatively, and in the immediate post-operative phase can reduce the risk of spinal cord ischaemia. This can be readily achieved through manipulation of the mean arterial pressure to a therapeutic level of 90 or 100 mmHg (with vasopressor support). In addition, other physiological parameters should also be optimised to assist in supporting blood pressure. This can include aiming for a haemoglobin >100 g/L [19].

Finally, insertion of a spinal drain can help control the volume of cerebrospinal fluid (CSF) and in turn spinal cord pressure [20]. Drain insertion can either be done prophylactically (at the time of induction of anaesthesia, prior to surgery commencing), or therapeutically, in the post-operative period. As with all central neuraxial catheter insertions, attention must be given to the timing of anticoagulation and dual antiplatelet therapy.

Prophylactic Versus Therapeutic (Reactive)

Prophylactically inserting a spinal drain enables rapid management of deteriorating neurology, by allowing immediate and controlled drainage of CSF to relieve pressure off the spinal cord.

Therapeutic insertion avoids the risks of insertion of the drain unless absolutely necessary, though risks insertion in an anticoagulated patient (patients are heparinised intra-operatively), and possibly in difficult settings e.g. in critical care with an agitated patient.

Spinal drains carry significant risks to the patient, including post-dural puncture headaches, vertebral canal haematoma, infection, intracranial haematoma, and death. Appropriate counselling must be offered for both the risk of spinal cord ischaemia and spinal drain insertion prior to surgery [21, 22].

Ruptured Aneurysm

For this emergency procedure, anaesthetic considerations are similar to elective procedures. Decision of anaesthetic technique will be guided by assessment of the patient's clinical parameters and pre-morbid state, likelihood to co-operate, and anatomical concerns governing surgical approach. Preferentially, a local anaesthetic technique with sedation provides the most cardio-stable option but requires co-operation by the patient.

Anaesthetic considerations include massive blood loss and associated transfusion; coagulopathy; electrolyte imbalance; hypothermia; blood pressure control through glycerine trinitrate infusion or vasopressor support.

The ideal location for this procedure is a hybrid theatre, which can accommodate immediate conversion to an open repair. There are some drawbacks, however. Such an environment may be remote to main theatres and unfamiliar to non-vascular anaesthetic practitioners. Induction of anaesthesia must be done on the fluoroscopy table, which cannot tilt or be adjusted to a more desirable position for induction of anaesthesia.

Post-operative Care

Analgesia

Due to the minimally invasive nature of endovascular treatment, post-operative pain is usually minimal. Simple analgesics may be sufficient. However, revascularisation of a limb may lead to additional pain, such as reperfusion pain, that requires stronger opiate-based analgesics. Severe post-operative pain may signify a vascular complication and necessitates a review from the operating team.

Position

Initial patient position in recovery is determined by the site of surgical access. If access was achieved via the groin, the patient will need to be supine for at least 2 h, or a duration in keeping with local policy.

Level of Care

Peripheral arterial and venous cases that require post-operative heparin infusions may require additional monitoring but for most patients, ward-based care is appropriate. All endovascular aneurysm repairs will need critical care admission post-operatively for organ perfusion monitoring and support. Complex EVARs, carry a greater perioperative risk than infra-renal EVARS [23], including an increased risk of renal impairment, post-operative blood transfusion, ventilator dependence, and increased length of critical care stay. Patients with spinal drains need to be nursed in a level 2 monitored area with careful attention paid to the drainage system and regular monitoring for neurological deficits. It is vital that practise is supported by management protocols that are readily available and visible to all staff caring for these patients.

References

1. Minto G, Biccard B. Assessment of the high-risk perioperative patient. Continuing Education in Anaesthesia, Critical Care and Pain. 2014;14(1):12–7. https://doi.org/10.1093/bjaceaccp/mkt020.
2. O'Neil-Callahan K, Katsimaglis G, Tepper MR, Ryan J, Mosby C, Ioannidis JPA, et al. Statins decrease perioperative cardiac complications in patients undergoing noncardiac vascular surgery: the statins for risk reduction in surgery (StaRRS) study. J Am Coll Cardiol. 2005;45(3):336–42.
3. Fleisher LA, Beckman JA, Brown KA, Calkins H, Chaikof E, Fleischmann KE, et al. ACC/AHA 2007 guidelines on perioperative cardiovascular evaluation and care for Noncardiac Surgery: executive summary: a report of the American College of Cardiology/American Heart Association task force on practice guidelines (writing committee to revise the 2002 guidelines on perioperative cardiovascular evaluation for noncardiac surgery): developed in collaboration with the American Society of Echocardiography, American Society of Nuclear Cardiology, Heart Rhythm Society, Society of Cardiovascular Anesthesiologists, Society for Cardiovascular Angiography and Interventions, Society for Vascular Medicine and Biology, and Society for Vascular Surgery. Circulation. 2007;116(17):1971–96.
4. Sluvot DP, Lipsitz EC. Surgical technique and peripheral arterial disease. Circulation. 2012;126(9):1127–38. https://doi.org/10.1161/CIRCULATIONAHA.111.059048.
5. Alleway R, Butt A, Ellis D, Freeth H, Kelly K, Jarman D, Joy S, Majid W, Mayet S, Nwosu E, Shotton H. Knowing the risk a review of the peri-operative care of surgical patients a report by the National Confidential Enquiry into patient outcome and death. 2011.
6. Dhaliwal G, Mukherjee D. Peripheral arterial disease: epidemiology, natural history, diagnosis and treatment. Int J Angiol. 2007;16(02):36–44. https://doi.org/10.1055/s-0031-1278244.
7. Thiruvoipati T. Peripheral artery disease in patients with diabetes: epidemiology, mechanisms, and outcomes. World J Diabetes. 2015;6(7):961. https://doi.org/10.4239/wjd.v6.i7.961.
8. Biscetti F, Cecchini AL, Rando MM, Nardella E, Gasbarrini A, Massetti M, Flex A. Principal predictors of major adverse limb events in diabetic peripheral artery disease: a narrative review. Atherosclerosis Plus. 2021;46:1–14. https://doi.org/10.1016/j.athplu.2021.10.003.
9. Novis BK, Roizen MF, Aronson S, Thisted RA. Association of preoperative risk factors with postoperative acute renal failure. Anesth Analg. 1994;78(1):143–9. https://doi.org/10.1213/00000539-199401000-00023.
10. Fit for frailty part 2 of the guide entitled, Designing, managing and commissioning services for older people with the development of this guide. What is frailty? 2017. http://www.bgs.org.uk/campaigns/fff/fff_full.pdf.
11. Chung F, Abdullah HR, Liao P. STOP-Bang Questionnaire. Chest. 2016;149(3):631–8. https://doi.org/10.1378/chest.15-0903.
12. www.sortsurgery.com.
13. Parodi JC. The challenges of medical innovation. J Vasc Surg. 2020;71(1):212–5. https://doi.org/10.1016/j.jvs.2019.07.104.
14. Kietaibl S, Ferrandis R, Godier A, Llau J, Lobo C, Macfarlane AJ, Schlimp CJ, Vandermeulen E, Volk T, von Heymann C, Wolmarans M, Afshari A. Regional anaesthesia in patients on antithrombotic drugs. Eur J Anaesthesiol. 2022;39(2):100–32. https://doi.org/10.1097/EJA.0000000000001600.
15. Wong GTC, Lee EYP, Irwin MG. Contrast induced nephropathy in vascular surgery. Br J Anaesth. 2016;117:ii63–73. https://doi.org/10.1093/bja/aew213.
16. Josesph G, Thompson VS, Kota A. Eur J Vasc Endovasc Surg. 2021;61:338–9.
17. Uchida N. How to prevent spinal cord injury during endovascular repair of thoracic aortic disease. Gen Thorac Cardiovasc Surg. 2014;62(7):391–7. https://doi.org/10.1007/s11748-014-0395-9.
18. Etz DC, Luehr M, Aspern KV, Misfeld M, Gudehus S, Ender J, Koelbel T, Debus ES, Mohr F-W. Spinal cord ischemia in open and endovascular thoracoabdominal aortic aneurysm repair: new concepts. J Cardiovasc Surg. 2014;55(2 Suppl 1):159 68.
19. Scott DA, Denton MJ. Spinal cord protection in aortic endovascular surgery. Br J Anaesth. 2016;117:ii26–31. https://doi.org/10.1093/bja/aew217.
20. LeMaire SA, Price MD, Green SY, Zarda S, Coselli JS. Results of open thoracoabdominal aortic aneurysm repair. Ann Cardiothorac Surg. 2012;1(3):286–92. https://doi.org/10.3978/j.issn.2225-319X.2012.08.16.
21. Kotelis D, Bianchini C, Kovacs B, Müller T, Bischoff M, Böckler D. Early experience with automatic pressure-controlled cerebrospinal fluid drainage during thoracic endovascular aortic repair. J Endovasc Ther. 2015;22(3):368–72. https://doi.org/10.1177/1526602815579904.
22. Wynn MM, Mell MW, Tefera G, Hoch JR, Acher CW. Complications of spinal fluid drainage in thoracoabdominal aortic aneurysm repair: a report of 486 patients treated from 1987 to 2008. J Vasc Surg. 2009;49(1):29–34.; discussion 34–5. https://doi.org/10.1016/j.jvs.2008.07.076.
23. Ultee KHJ, Zettervall SL, Soden PA, Darling J, Verhagen HJM, Schermerhorn ML. Perioperative outcome of endovascular repair for complex abdominal aortic aneurysms. J Vasc Surg. 2017;65(6):1567–75. https://doi.org/10.1016/j.jvs.2016.10.123.

4 The Angiographic Suite: Setup and Ergonomics

Bowen Xie and Michael Madigan

Case Presentation
A 70-year-old gentleman with a history of coronary artery disease status post coronary artery bypass grafts ×3 presented as a transfer to our institution for a ruptured infrarenal abdominal aortic aneurysm with hypotension and back pain. The patient arrived intubated and was brought emergently to the hybrid operating room for endovascular repair. Both the anesthesia and operative team were present in the operating room to receive the patient on transfer.

Continued at page 28

Introduction

The use of minimally invasive interventions to treat vascular disease has increased over the last few decades [1] due to decreased length of stay and post-operative morbidity with comparable results to open surgical procedures [2]. The angiography suite now serves as the primary location for vascular surgeons and other interventionalists to provide appropriate and expedited care as a part of their clinical practice. Though endovascular procedures have shown benefits for patients, interventionalists must recognize and address potential hazards associated with angiography and take the necessary precautions to reduce radiation exposure and workplace-associated injuries. This chapter serves to describe the process of setting up an efficient and safe angiography suite, as well as coordinating care to maximize utility of the suite in a hospital system.

B. Xie · M. Madigan (✉)
University of Pittsburgh Medical Center, Pittsburgh, PA, USA
e-mail: xieb@upmc.edu; madiganmc@upmc.edu

Layout

There are a variety of factors to consider when planning for the installation of an angiography suite or hybrid operating room (OR) including, but not limited to, location, room set-up, adjunctive equipment, supplies, and imaging system [3, 4].

Location

Ideally, the angiography suite or hybrid room should be placed either near other interventional suites or the ORs in order to facility transportation of equipment, tools, and personnel. If the interventional suites are in a separate location from the ORs, it is preferred that a hybrid room be located next to the other ORs to allow for the immediate availability of surgical instruments and equipment, as well as anesthesia and operating room staff [5]. It is noted that this may not be possible due to structural or engineering limitations of the facility.

Room Set-Up

Booms

One or more moveable booms should be installed in order to maximize working space, organize equipment, and improve efficiency. The moveable aspect of the booms allows for flexibility of the angiography suite in order to accommodate various types interventions. Furthermore, the availability of electrical outlets, connectors, and gas/air outlets also facilitate the workflow. A separate boom may be necessary for the anesthesia team in order to further improve efficiency and organization.

Lighting

It is important to have both appropriate surgical lighting as well as ambient lighting during vascular interventional pro-

G. Geroulakos et al. (eds.), *Mastering Endovascular Techniques*, https://doi.org/10.1007/978-3-031-42735-0_4

cedures. The lights will typically be ceiling mounted to improve adjustability and provide adequate lighting during open surgical approaches [6]. However, care must be taken in order to avoid frequent collisions with the display monitors and the imaging system. The ambient room lighting must provide the ability to dim as this is often helpful during fluoroscopy for improved visualization.

Monitors

Adequate imaging visualization is crucial in the setting of endovascular procedures. As such, high-quality monitors are essential for appropriate visualization by the operator and the assistant personnel. Ceiling mounted monitors provide the ability for greater range of motion in order to adjust and improve operator ergonomics. Multiple monitors are preferred to visualize the intervention as well as vitals and anesthesia metrics during the procedure.

Operating Table

There are a variety of factors to consider with operating tables. First, radiolucency of the table is necessary during fluoroscopic procedures with carbon fiber being the preferred material due to its lightweight and radiolucent properties. Floating tabletops will provide the ability for fast and precise movements during interventional procedures [7]. The addition of lead skirts surrounding the table will also help to reduce radiation scatter to the operator and assistants. Additional positioning aids, such as cushioned padding and radiolucent arm boards, are also necessary for specific vascular interventions such as dialysis access interventions or upper extremity fluoroscopy. Finally, the addition of side rails will allow for the mounting of surgical retractors that are sometimes necessary for more extensive open surgical interventions.

Storage

Adequate storage capacity with high-level organization is imperative in the angiography suite in order to ensure quick access to the necessary equipment and tools for a variety of interventions. Built-in storage cabinets will not only reduce workspace clutter, but also improve organizational design. Commonly used sheaths, catheters, wires, balloons, stents, closure devices and surgical equipment should be stored within these cabinets for ease of access during both fluoroscopic and hybrid procedures.

Control Room

The presence of a control room separate from the sterile operating field is ideal for additional visualization of images, operative case planning, supervision, and observation for trainees.

Adjunctive Equipment

Ultrasound

The use of ultrasound guidance for venous and arterial percutaneous access has been shown to reduce the incidence of complications at the entry site [8, 9]. Furthermore, duplex ultrasound is required for performance of certain interventions, such a venous ablation procedure [10].

Intravascular Ultrasound

The use of intravascular ultrasound serves as a important supplement in the setting of venous as well as aortic dissection interventions due to its ability to provide intraluminal visualization both pre and post-stent placement [4]. It is also a necessary adjunct to performing interventions for iliac venous occlusive disease [11]. Others have also suggested benefit for some peripheral interventions to more accurately identify stenoses and to assist with sizing for angioplasty or stent deployment [12, 13].

Power Injector

Utilization of a power injector allows for the operator to be positioned further from the radiation source and thereby reducing the exposure to the operative staff without sacrificing image quality or intervention time [14]. In the case of aortic imaging, the increased injection rate may also offer improved timing and visualization of critical structures while at the same time minimizing contrast load. The power injector can be mounted onto the operating table for added efficiency.

Carbon Dioxide Injector

With the prevalence of chronic kidney disease and contrast allergies within the population requiring vascular interventions, carbon dioxide (CO_2) has established a role as a substitute for standard iodonated contrast. Therefore, the use of a CO_2 injector proves to be a necessary adjunct in modern day vascular interventions [15].

Suction Thrombectomy Systems

Endovascular intervention for acute arterial and venous thrombus are increasingly utilized in modern vascular interventions [16, 17]. With the development of catheter-directed suction thrombectomy, there has been increased utilization of this technology in the setting of acute limb ischemia, iliofemoral deep venous thrombosis, pulmonary embolism, and even mesenteric or visceral vessel thrombosis. The purchase of the suction thrombectomy system may potentially reduce the long-term costs of repeat catheter-directed interventions in exchange for an increased overhead cost of the initial purchase.

Commonly Used Supplies

In comparison to open surgical procedures, which utilize a standard set of reusable instruments and clamps for most cases, there is a significant added cost to having adequate disposable supplies for endovascular interventions. Physicians within the system and practice should therefore collaborate in order to standardize and simplify the necessary equipment and supplies. This can allow for in-bulk purchasing or contract agreements with supply companies to reduce the cost of supplies. Table 4.1 summarizes the broad categories of necessary equipment and supplies needed for a range of endovascular procedures [4].

Imaging System

There are a wide variety of imaging systems available for use. Rather than focusing on the advantages and disadvantages of specific systems, this section will highlight various features that may be important for optimal imaging during endovascular interventions.

Mounting

A ceiling-mounted systems provides the benefit of reducing the amount of space occupied, however, this comes in exchange for potential interference with lights, booms, and monitors as well as the potential decreased sterility due to dust or particle trapping within the system itself. Floor-mounted systems may occupy additional floor space, but can often times be moved during non-fluoroscopic portions of the operation and are suggested to be more hygienic by some authors [3].

Table 4.1 Commonly used equipment for various procedure types

Arterial and venous interventions	Dialysis interventions	Miscellaneous
Access needle, wire, and micro sheath	High-pressure balloons	Vena cava filters
Sheaths	Cutting balloons	Filter retrieval kits
Flush/marker catheters	Tunneled dialysis catheters	
Selective/crossing catheters and wires		
Stiff working wires		
Re-entry devices		
Angioplasty balloons		
Over-the-wire Fogarty catheters		
Snares		
Drug-coated balloons		
Bare metal stents		
Covered stents		
Drug-coated stents		
Inflation device		
Coils/embolization devices		
Closure devices		

Image Detection

Most, if not all, fixed imaging systems now come equipped with flat panel detectors, which produce higher image quality and resolution are compared to image intensifiers, with some systems providing resolution at four times the conventional systems (over 2000 pixels resolution) [18]. Flat panel detectors also produce a larger field of view, which may prove to be essential in the setting of advanced endovascular interventions, such as fenestrated endovascular aortic aneurysm repair. The dimensions of the flat panel detector among commonly used systems are 30 × 38 cm for maximum field of view with certain broad detectors reaching up to 41 × 41 cm [18–20]. Additionally, these imaging systems may be equipped with 3D fusion imaging and reconstruction (i.e., EVAR Assist 2® from GE, SmartCT Angio® from Philips, and EVAR-3D Guidance® from Siemens) [21–23] to provide the ability to improve real-time intervention planning, to obtain real-time 3D image acquisition, as well as to decrease radiation exposure, contrast volume, and operative time by outlining vessels and optimizing angulation for precise device deployment.

The Benefits of a Hybrid Operating Room

The advancement in endovascular technologies has helped change the treatment paradigm of vascular therapy toward novel approaches that require both open surgical and endovascular interventions to address complex vascular disease. This approach has led to the development of a new working environment [24]. An updated, conventional angiography suite can aide in delivery of modern vascular care. Furthermore, a number of studies have demonstrated that a hybrid room with an endovascular suite integrated within a traditional OR offers significant benefits over both a conventional angiography suite and the use of a portable C-arm imaging system [25–27]. The benefits of a hybrid OR and limitations of a portable C-arm in a traditional operating room are listed below in Tables 4.2 and 4.3. A hybrid OR with an integrated imaging system provides the ability to seamlessly perform open surgical procedures and endovascular procedures during the same intervention. Often, complex endovascular interventions are assisted by an open surgical intervention, such as femoral endarterectomy, or vessel exposure, such as a brachial artery cutdown, to allow for endovascular interventions that would otherwise be difficult via a standard percutaneous approach. Furthermore, the ability to convert to an open surgical intervention in setting of unanticipated access or other complications during

Table 4.2 The benefits of a hybrid operating room

	Benefits of a hybrid operating room
Sterility	• Improved air circulation, positive air pressure control to reduce air-borne contamination • Optimal temperature and humidity control • Easily washable ceilings, walls, and floors to minimize potential harboring environment for bacteria and blood-borne pathogens
Ability to convert	• Ability/equipment for immediate open surgical conversion and treatment of suboptimal results/complications • Ability/equipment for direct surgical exposure of artery/graft in anatomy unsuitable for percutaneous access
Anesthesia	• Adequate space for anesthetic equipment in cases extensive cardiopulmonary monitoring and control

Table 4.3 Limitations of OR adaptations via portable C-arm system

	Limitations of a portable C-arm
Efficiency	• Reliance on additional personnel to operate portable equipment • Difficulty maneuvering portable equipment to obtain optimal imaging angles
Imaging capacity	• Decreased image intensifier size leading to smaller field of view • Decreased monitor resolution and size of displays in portable systems • Decreased tissue penetration/image quality in high complexity interventions • Decreased projection angles of X-ray source/image intensifier leading to suboptimal visualization of planes during complex interventions
Radiation exposure	• Increased radiation scatter to operator due to fix image intensifier distance in portable C-arm systems • Lack of available ceiling mounted shields and table lead aprons to reduce scatter radiation

endovascular intervention provides an efficient manner to address these potential issues. The well-controlled conditions within the OR also serve to improve sterility and potentially reduce the risk of infection of implantable devices [28]. A mounted system provides the benefit of improved ergonomics and decreased radiation exposure to both the patient and the operator due to the improved imaging capabilities and safety measures when compared to the portable system used as an adjunct in the OR.

Examples of Hybrid Room Set-Up

Below we provide an example of a set up during an endovascular aortic repair of a ruptured aneurysm in the hybrid OR at our institution (Figs. 4.1 and 4.2). The layout of the hybrid room has been set up with the floor-mounted imaging system at the head of the table adjacent to the anesthesia staff in order to maximize operative space and field of view. Multiple ultrasound machines are available for both the anesthesia staff as well as the interventionalists to establish intravascular access for resuscitation, intraoperative monitoring, and procedural access. A Foley kit and bovie pads are also available for pre-procedural placement. The angiography table is placed in a neutral position with the monitor at, or below the eye level of the interventionalist.

A micro-puncture vascular access kit has been opened and prepped in anticipation of percutaneous access to the common femoral arteries. In addition to basic wires (hydrophobic and hydrophilic), both selective and flush catheters have been opened and prepped in order to obtain an aortogram. Sheaths of various sizes (5–8 Fr) are available with additional larger bore sheaths and stiff wires on standby in case an aortic occlusion balloon is necessary for proximal control. The remainder of the stent grafts, closure devices and adjunctive supplies remain unopened, but within the room for quick and easy access. Furthermore, a tray of vascular surgical instruments along with bovie and suction equipment have been laid out in anticipation of potential open surgical exposure of access vessels in the setting of poor anatomy for percutaneous access or closure.

As highlighted by the clinical scenario, it is imperative to have a trained team with experience in both endovascular and open interventions assisting in a hybrid procedure, including circulating nurses, scrub techs, scrub nurses, and interventional assistants, such as physician assistants, residents, fellows, or surgical assistants in order to expedite the operative intervention and optimize patient outcomes.

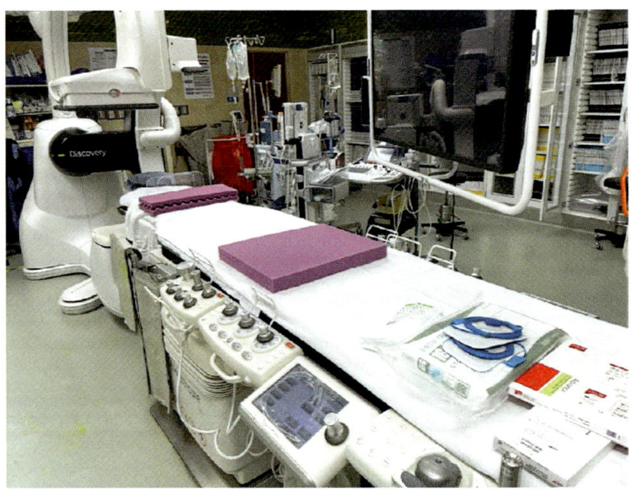

Fig. 4.1 Hybrid operating room design and set-up prior to endovascular abdominal aneurysm repair (EVAR)

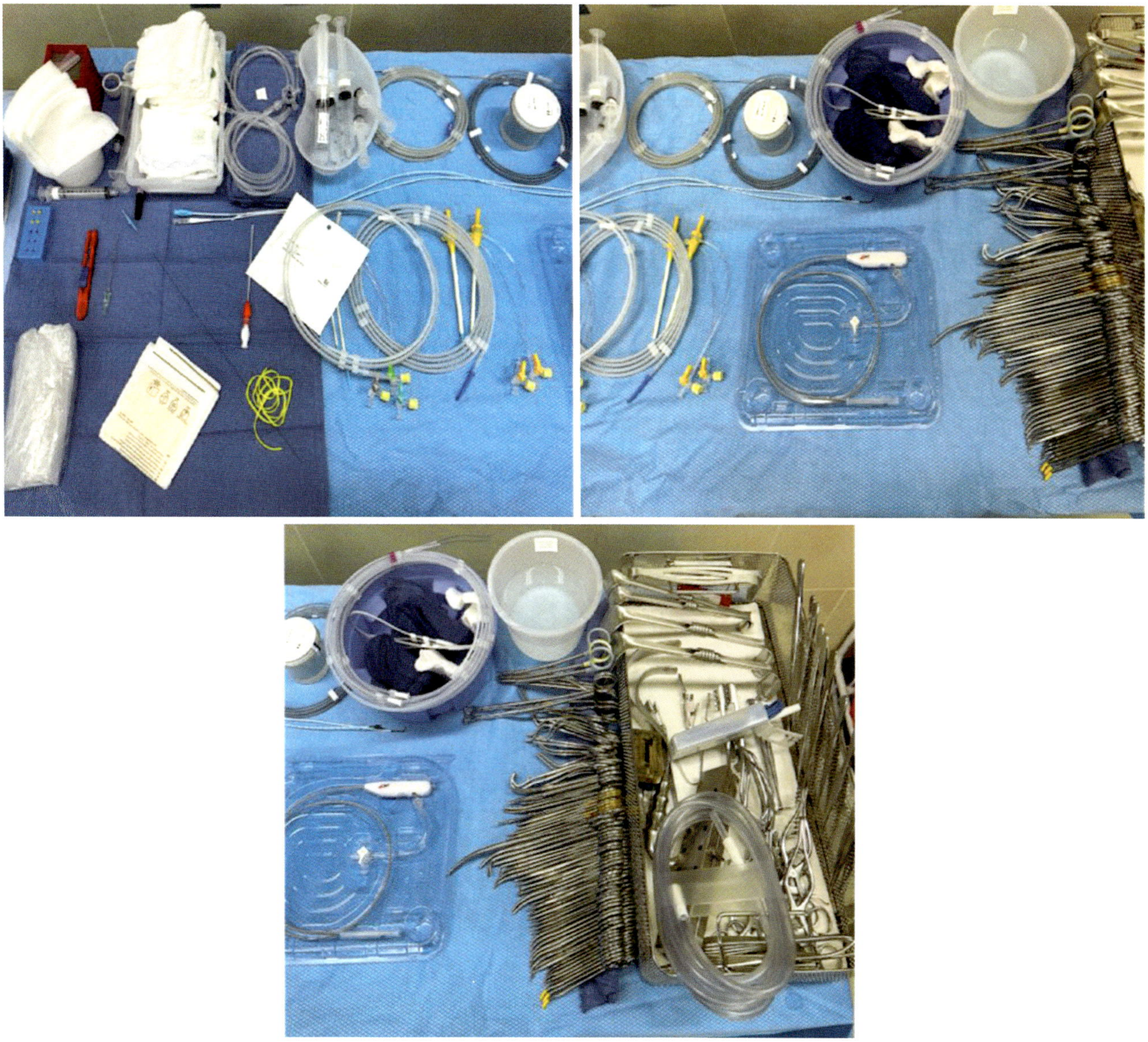

Fig. 4.2 Vascular intervention supplies and surgical instruments for EVAR

Coordination with Other Specialties

Given the high overhead costs of constructing an angiography suite or hybrid operating room and the potential loss of revenue associated with only part-time use, it may be essential to coordinate with other interventional services (i.e., interventional radiology, cardiology, and cardiac surgery) in order to maximize the utility of the room with a variety of other interventional procedures such as cardiac catheterization and transcatheter aortic valve replacement. Therefore, an appropriate professional relationships should be established between various specialties in order to establish a comprehensive and interdisciplinary environment for cardiovascular intervention and patient care.

Ergonomics

While significant advances have been made in terms of radiation safety in the domain of fluoroscopic interventions, much less has been done to improve ergonomics. Interventionalists are at a high risk of orthopedic injuries often due to poor body mechanics and posture secondary to the burden of protective equipment and the poor ergonomic design of interventional suites [29–32]. Prior data from interventional radiologist and cardiologists have shown a high incidence of spinal complaints (cervical, thoracic, and lumbar) as well as axial skeleton complaints with an association to decreased longevity of their interventional careers [33]. This can contribute to work absences, poor overall health, and even affect procedural outcomes.

Effect of Ergonomic Challenges

A recently published study of interventional cardiologist demonstrated that approximately 9–33% of all participants reported missing work days due to their musculoskeletal complaints [34]. The increased need for leave of absences interrupts the workflow and may also lead to early retirement thus decreasing the long-term workforce. This trend is similarly reflected among vascular surgeons, with the majority of those surveyed by the Society of Vascular Surgery reporting pain after a day of open or endovascular procedures and a significant portion seeking medical care or even surgical intervention [35]. Furthermore, there is growing evidence suggesting that female surgeons and interventionalists are at an increased risk of musculoskeletal injuries compared to their male counterparts, particularly in the upper extremities [36, 37]. Finally, the pain associated with musculoskeletal injuries has been associated with self-reported burnout amongst vascular surgeons in a study by Davila et al. [38], which can have long-term detrimental physical and psychological effects on providers. Norasi et al. demonstrated both the neck and the trunk were at high risk of developing musculoskeletal disorders due to awkward posturing during surgical procedures. There was a more prominent risk in open operative procedures, especially when using visual aids such as loupes [39].

Radiation Shielding

The term "interventional disc disease" has been developed over the years to emphasize the trend of orthopedic work-related complaints. These complaints were more prominent amongst those using fluoroscopy as compared to other operative surgical specialties not using fluoroscopy suggesting a correlation with chronic lead apron use [33]. Appropriate fit and weight distribution of lead garments are essential to reduce the likelihood of musculoskeletal injury and fatigue. The introduction of two-piece garments as well as wrap-around lead has reduced axial strain and improved weight distribution. Additionally, the transition to composite or lead-free alternative protective garments further reduces overall weight, however, there have been some reported differences in terms of the amount of shielding provided compared to their lead equivalents [40, 41]. The development of a fully suspended radiation protection system equipped with radiation shielding from head to calf that eliminates the weight bearing component on the operator while maintaining mobility and maneuverability holds some promise for the future [42].

Room Design

Adjustments within the operating and procedure rooms must be tailored to maximize proceduralist ergonomics. Display monitors during fluoroscopic procedures as well as other imaging adjuncts should be placed in front of the interventionalist and in line with their shoulders allowing for neutral neck position with a slight downward (15°) angle thus reducing cervical spine strain [34]. Table height should also be adjusted in order to minimize prolonged muscle activation. Studies have shown that table and hand positioning at 70–80% of elbow height appears to reduce back discomfort [43]. Furthermore, additional adjuncts such as cushioned floor mats and the use of shoe insoles may also have a preventative effect on axial skeletal strain. Figure 4.3 demonstrates the example of ergonomics within the angiography suite as highlighted by Knuttinen et al. [34] Table height is adjusted according to the operator at near elbow height, duplex ultrasound and display monitors are positioned in the line of vision of the operator at a slight downward angle to avoid excessive neck strain during the intervention. Finally, while mobile lead shielding is essential in reducing radiation dose, it is important to account for postural changes involved in the use of the shielding in order to reduce the incidence of ergonomically unfavorable positions. Ergonomic strategies are summarized in Table 4.4 [34].

Case Presentation

Continued from page 23

Once the patient was appropriately positioned with adequate access, his abdomen and groins were prepped in the standard sterile fashion. Bilateral femoral artery access was obtained simultaneously via micro-puncture access kits under duplex ultrasound guidance. After appropriate sheath placement, all operators moved to the right side of the patient and the monitor and table were adjusted for optimal visualization. The room lights were dimmed appropriately to improve visualization of the display without impeding the scrubs techs' abilities to prepare the instruments and devices on the back table. The monitors were positioned at just below operator eye level to reduce cervical strain during the intervention. Each operator and assistant also wore fitted two-piece lead aprons to reduce thoracic and lumbar spine strain during the procedure. The floating tabletop in the hybrid suit allowed for rapid patient movement during the intervention to adequately visualize both proximally into the aorta and distally into the iliac arteries. A 32 cm Medtronic Endurant® endograft was deployed with iliac extensions bilaterally (Fig. 4.4). The storage capacity of the hybrid suite allowed

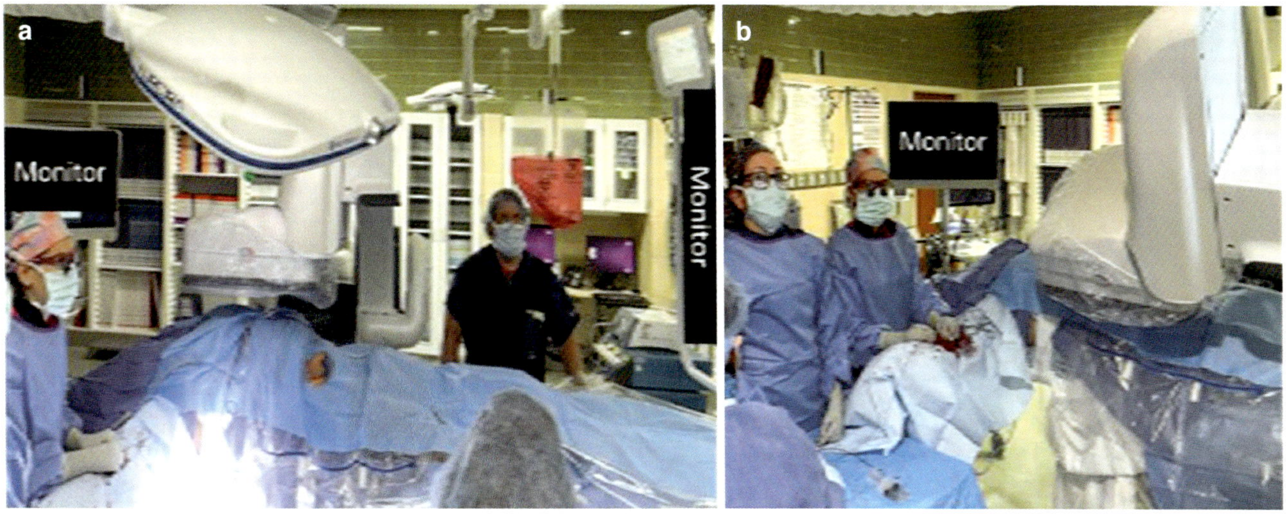

Fig. 4.3 Proper ergonomic positioning during lower extremity angiography via right brachial approach, (**a**) side view of intra-procedural positioning with monitor in line of vision, table at elbow height and slight downward tilt of neck, (**b**) frontal view of intra-procedural posturing

Table 4.4 Ergonomic strategies to reduce the risk of musculoskeletal injuries

Ergonomic strategies
Position monitor directly in front of the operator and in line with shoulders
Position monitor height at a 15° downward viewing angle
Adjust table height to at or just below elbow level
Avoid repetitive truncal motions or prolonged truncal rotation

for rapid access to the necessary catheters and balloons to ensure appropriate cannulation of endograft and deployment of the iliac limbs. The use of a power injector along with the improved magnification and resolution of the integrated imaging system allowed for precise deployment of the endograft while minimizing contrast use and radiation. Both femoral accesses were closed with the pre-deployed ProGlides® without complications, though surgical instruments were opened and prepared in the hybrid suite for expedited femoral cutdowns and arterial repair in case of closure device failure. The patient was brought ICU for resuscitation and recovery in stable condition. He was extubated on post-operative day one and only had a transient acute kidney injury that resolved during his hospital course with appropriate urine output throughout. He was discharged home on post-operative day 6.

Acknowledgments Research reported in this publication was supported by the National Heart, Lung, And Blood Institute of the National Institutes of Health under Award Number T32HL098036. The content is solely the responsibility of the authors and does not necessarily represent the official views of the National Institutes of Health.

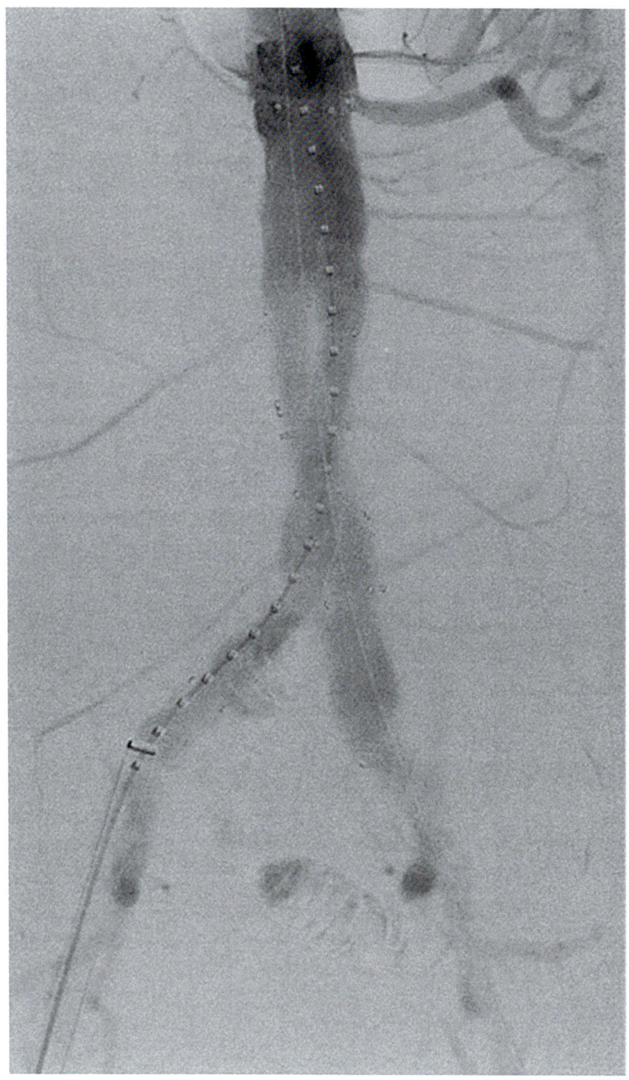

Fig. 4.4 Completion angiography after successful EVAR of ruptured abdominal aortic aneurysm

References

1. Jacob-Brassard J, Al-Omran M, Stukel T, Mamdani M, Lee D, Papia G, et al. Trends in lower extremity revascularization and amputation for peripheral artery disease over the last two decades: a population-based time series analysis. J Vasc Surg. 2021;74(5):e437. https://doi.org/10.1016/j.jvs.2021.08.024.
2. Buck DB, van Herwaarden JA, Schermerhorn ML, Moll FL. Endovascular treatment of abdominal aortic aneurysms. Nat Rev Cardiol. 2014;11(2):112–23. https://doi.org/10.1038/nrcardio.2013.196.
3. Nollert G, Wich S. Planning a cardiovascular hybrid operating room: the technical point of view. Heart Surg Forum. 2009;12(3):E125–30. https://doi.org/10.1532/HSF98.20091033.
4. Moore WSLPFOGS. Moore's vascular and endovascular surgery: a comprehensive review. 2019.
5. Bonatti J, Schachner T, Bonaros N, Jonetzko P, Öhlinger A, Ruetzler E, et al. Simultaneous hybrid coronary revascularization using totally endoscopic left internal mammary artery bypass grafting and placement of rapamycin eluting stents in the same interventional session. Cardiology. 2008;110(2):92–5. https://doi.org/10.1159/000110486.
6. Curlin J, Herman CK. Current state of surgical lighting. Surg J. 2020;6(02):e87–97.
7. Hamm C, Bösenberg H, Brennecke R, Daschner F, Dziekan G, Erbel R, et al. Guidelines for equipping and managing heart catheter rooms (1st revision). Issued by the governing body of the German Society of Cardiology-Heart and Cardiovascular Research. Revised by order of the Committee of Clinical Cardiology. Zeitschrift fur Kardiologie. 2001;90(5):367–76.
8. Kumar A, Chuan A. Ultrasound guided vascular access: efficacy and safety. Best Pract Res Clin Anaesthesiol. 2009;23(3):299–311. https://doi.org/10.1016/j.bpa.2009.02.006.
9. Balceniuk MD, Cardelli F, Ayers BC, Zhao P, Raman K, Ellis JL, et al. Regional variation in usage of ultrasound-guided access in the vascular quality initiative. J Vasc Surg. 2019;70(2):e23–e4. https://doi.org/10.1016/j.jvs.2019.06.021.
10. Lo RC, Fokkema MTM, Curran T, Darling J, Hamdan AD, Wyers M, et al. Routine use of ultrasound-guided access reduces access site-related complications after lower extremity percutaneous revascularization. J Vasc Surg. 2015;61(2):405–12. https://doi.org/10.1016/j.jvs.2014.07.099.
11. Montminy ML, Thomasson JD, Tanaka GJ, Lamanilao LM, Crim W, Raju S. A comparison between intravascular ultrasound and venography in identifying key parameters essential for iliac vein stenting. J Vasc Surg Venous Lymphat Disord. 2019;7(6):801–7. https://doi.org/10.1016/j.jvsv.2019.03.015.
12. Scoccianti M, Verbin CS, Kopchok GE, Back MR, Donayre CE, Sinow RM, et al. Intravascular ultrasound guidance for peripheral vascular interventions. J Endovasc Ther. 1994;1(1):71–80. https://doi.org/10.1177/152660289500100110.
13. Hitchner E, Zayed M, Varu V, Lee G, Aalami O, Zhou W. A prospective evaluation of using IVUS during percutaneous superficial femoral artery interventions. Ann Vasc Surg. 2015;29(1):28–33. https://doi.org/10.1016/j.avsg.2014.07.026.
14. Larsen AS, Osteras BH. Step back from the patient: reduction of radiation dose to the operator by the systematic use of an automatic power injector for contrast media in an interventional angiography suite. Acta Radiol. 2012;53(3):330–4. https://doi.org/10.1258/ar.2011.110560.
15. Shaw DR, Kessel DO. The current status of the use of carbon dioxide in diagnostic and interventional angiographic procedures. Cardiovasc Intervent Radiol. 2006;29(3):323–31. https://doi.org/10.1007/s00270-005-0092-2.
16. Sista Akhilesh K, Horowitz James M, Tapson Victor F, Rosenberg M, Elder Mahir D, Schiro Brian J, et al. Indigo aspiration system for treatment of pulmonary embolism. J Am Coll Cardiol Interv. 2021;14(3):319–29. https://doi.org/10.1016/j.jcin.2020.09.053.
17. Kwok CHR, Fleming S, Chan KKC, Tibballs J, Samuelson S, Ferguson J, et al. Aspiration thrombectomy versus conventional catheter-directed thrombolysis as first-line treatment for noniatrogenic acute lower limb ischemia. J Vasc Interv Radiol. 2018;29(5):607–13. https://doi.org/10.1016/j.jvir.2017.11.030.
18. Solutions AH. Philips Allura Xper FD20 Cath/Angio System. 2022. https://avantehs.com/p/philips-allura-xper-fd20-cathangio-system/13790. Accessed 6 Apr 2022.
19. GE: Discovery™ IGS 7 with AutoRight™. 2022. https://www.gehealthcare.com/products/interventional-image-guided-systems/igs-for-hybrid-or/discovery-igs-7#imaging. Accessed 6 Apr 2022.
20. ECRI: Evaluation: Siemens Artis pheno Floor-Mounted Angiography System for Hybrid OR Applications. 2022. https://www.ecri.org/search-results/member-preview/hdjournal/pages/evaluation-siemens-artis-pheno-for-hybrid-or#:~:text=Detector%3A%20The%20Artis%20pheno%20zen40HDR,cm%20down%20to%2011%20cm. Accessed 6 Apr 2022.
21. GE: EVAR ASSIST 2. 2022. https://www.gehealthcare.com/products/interventional-image-guided-systems/evar-assist. Accessed 6 Apr 2022.
22. Philips: SmartCT Angio. 2022. https://www.usa.philips.com/healthcare/product/HCOPT18/smartct-angio-imaging-technology#features. Accessed 6 Apr 2022.
23. Siemens: EVAR-3D Guidance. 2022. https://www.siemens-healthineers.com/en-us/clinical-specialities/surgery/surgical-disciplines/evar3d. Accessed 6 Apr 2022.
24. Zhou W, Lin PH, Bush RL, Lumsden AB. Endovascular training of vascular surgeons: have we made progress? Semin Vasc Surg. 2006;19(2):122–6. https://doi.org/10.1053/j.semvascsurg.2006.03.010.
25. Calligaro KD, Dougherty MJ, Patterson DE, Raviola CA, DeLaurentis DA. Value of an endovascular suite in the operating room. Ann Vasc Surg. 1998;12(3):296–8. https://doi.org/10.1007/s100169900157.
26. Sikkink CJ, Reijnen MM, Zeebregts CJ. The creation of the optimal dedicated endovascular suite. Eur J Vasc Endovasc Surg. 2008;35(2):198–204. https://doi.org/10.1016/j.ejvs.2007.08.009.
27. Kaneko T, Davidson MJ. Use of the hybrid operating room in cardiovascular medicine. Circulation. 2014;130(11):910–7. https://doi.org/10.1161/CIRCULATIONAHA.114.006510.
28. Eliason JL, Guzman RJ, Passman MA, Naslund TC. Infected endovascular graft secondary to coil embolization of endoleak: a demonstration of the importance of operative sterility. Ann Vasc Surg. 2002;16(5):562–5. https://doi.org/10.1007/s10016-001-0275-5.
29. Moore B, vanSonnenberg E, Casola G, Novelline RA. The relationship between back pain and lead apron use in radiologists. Am J Roentgenol. 1992;158(1):191–3. https://doi.org/10.2214/ajr.158.1.1530763.
30. Ross AM, Segal J, Borenstein D, Jenkins E, Cho S. Prevalence of spinal disc disease among interventional cardiologists. Am J Cardiol. 1997;79(1):68–70. https://doi.org/10.1016/S0002-9149(96)00678-9.
31. Goldstein JA, Balter S, Cowley M, Hodgson J, Klein LW. Occupational hazards of interventional cardiologists: prevalence of orthopedic health problems in contemporary practice. Catheter Cardiovasc Interv. 2004;63(4):407–11. https://doi.org/10.1002/ccd.20201.
32. Machan L. Risks of back injury. J Vasc Interv Radiol. 2002;2(13):P130–1.
33. Benjamin JL, Meisinger QC. Ergonomics in the development and prevention of musculoskeletal injury in interventional radi-

ologists. Tech Vasc Interv Radiol. 2018;21(1):16–20. https://doi.org/10.1053/j.tvir.2017.12.004.
34. Knuttinen MG, Zurcher KS, Wallace A, Doe C, Naidu SG, Money SR, et al. Ergonomics in IR. J Vasc Interv Radiol. 2021;32(2):235–41. https://doi.org/10.1016/j.jvir.2020.11.001.
35. Wohlauer M, Coleman DM, Sheahan MG, Meltzer AJ, Halloran B, Hallbeck S, et al. Physical pain and musculoskeletal discomfort in vascular surgeons. J Vasc Surg. 2021;73(4):1414–21. https://doi.org/10.1016/j.jvs.2020.07.097.
36. Adams SR, Hacker MR, McKinney JL, Elkadry EA, Rosenblatt PL. Musculoskeletal pain in gynecologic surgeons. J Minim Invasive Gynecol. 2013;20(5):656–60. https://doi.org/10.1016/j.jmig.2013.04.013.
37. Sutton E, Irvin M, Zeigler C, Lee G, Park A. The ergonomics of women in surgery. Surg Endosc. 2014;28(4):1051–5. https://doi.org/10.1007/s00464-013-3281-0.
38. Davila VJ, Meltzer AJ, Hallbeck MS, Stone WM, Money SR. Physical discomfort, professional satisfaction, and burnout in vascular surgeons. J Vasc Surg. 2019;70(3):913–20.e2. https://doi.org/10.1016/j.jvs.2018.11.026.
39. Norasi H, Tetteh E, Money SR, Davila VJ, Meltzer AJ, Morrow MM, et al. Intraoperative posture and workload assessment in vascular surgery. Appl Ergon. 2021;92:103344. https://doi.org/10.1016/j.apergo.2020.103344.
40. Lu H, Boyd C, Dawson J. Lightweight lead aprons: the emperor's new clothes in the angiography suite? Eur J Vasc Endovasc Surg. 2019;57(5):730–9. https://doi.org/10.1016/j.ejvs.2019.01.031.
41. Fakhoury E, Provencher J-A, Subramaniam R, Finlay DJ. Not all lightweight lead aprons and thyroid shields are alike. J Vasc Surg. 2019;70(1):246–50. https://doi.org/10.1016/j.jvs.2018.07.055.
42. Marichal DA, Anwar T, Kirsch D, Clements J, Carlson L, Savage C, et al. Comparison of a suspended radiation protection system versus standard lead apron for radiation exposure of a simulated interventionalist. J Vasc Interv Radiol. 2011;22(4):437–42. https://doi.org/10.1016/j.jvir.2010.12.016.
43. van Veelen MA, Kazemier G, Koopman J, Goossens RHM, Meijer DW. Assessment of the ergonomically optimal operating surface height for laparoscopic surgery. J Laparoendosc Adv Surg Tech. 2002;12(1):47–52. https://doi.org/10.1089/109264202753486920.

Basic Tools and Terminology for Endovascular Interventions

5

Peter Jamieson and Abdel Kader Allouni

Basic Tools

There is a wealth of equipment that can be used for endovascular procedures, a list that only grows with new innovations and new medical devices companies. As an endovascular operator, it is important to be familiar with a wide range of equipment, though becoming skilled with every device on the market is unachievable. What follows is an overview of the groups of devices available, their characteristics and some examples of each category, though one should note that other devices are available from alternative manufacturers, with similar features and applications.

Needles

The vast majority of endovascular procedures begin with a needle inserted into a vessel under ultrasound guidance. Needles vary in length and diameter, with a general rule that shorter and wider needles are stiffer, whereas longer and thinner needles are more flexible and can bend in the tissues, making it more difficult to accurately hit a deep target.

Most vascular access needles are one-piece, hollow, bevelled needles, permitting passage of a guidewire into the vessel immediately following vessel single wall puncture, though the operator should be careful to ensure good, pulsatile flow from the needle hub to ensure the bevel is not across the vessel wall, risking dissection when the wire is introduced. A small notch or marker on the hub usually corresponds to the side the bevel faces.

Some needles are filled with an inner trocar that provides extra rigidity and prevents the entrainment of air. In the past these were used for double-wall vessel puncture (front and back wall), though they are now rarely used for vessel access and are more often used for non-vascular interventional procedures.

Vascular access is gained through use of the Seldinger technique, whereby the vessel is punctured with a hollow needle, a guidewire is inserted through the needle into the vessel, the needle is removed and a sheath with dilator is inserted over the guidewire. Finally, removal of the dilator (and often the guidewire) leaves just the sheath in situ. More information on guidewires and sheaths can be found below.

Needle outer diameter is measured in gauge (G)—the lower the gauge, the larger the needle. 19 G approximates to 0.038 of an inch, 1 mm or 3 French.

- 25 G (orange) needle for superficial local anaesthetic injection
- 23 G (blue) or 21 G (green) needle for deeper local anaesthetic injection
- 22G (micro-puncture) needle permits a 0.018 wire. This is more frequently used for access to small arteries such as the radial or infra-popliteal arteries, but can be used for any vascular access. A dilator/sheath can be placed over the 0.018 wire, or it can be exchanged through a 4 Fr dilator for a 0.035 wire.
- 18–19 G needle permits a 0.035 wire (as for standard sheath J-wires)

Dilators

Dilators are stiff plastic tubes with tapered tips that can be advanced over a guidewire. They function to widen the vessel wall and the overlying soft tissues. Serial dilatation with increasingly wide dilators can upsize even tiny micro-puncture holes to a width large enough to accommodate a large-bore access sheath for aortic stent-graft placement.

P. Jamieson · A. K. Allouni (✉)
Department of Interventional Radiology, University Hospitals of North Midlands NHS Trust, Stoke-on-Trent, UK
e-mail: peter.jamieson@uhnm.nhs.uk

G. Geroulakos et al. (eds.), *Mastering Endovascular Techniques*, https://doi.org/10.1007/978-3-031-42735-0_5

Sheaths

Sheaths are haemostatic conduits that are generally stationary within the access site of the vessel and prevent trauma caused by repeated manipulation of endovascular catheters or other devices and maintain haemostasis during catheter exchanges. Sheaths are composed of a thin layer of stiff hollow plastic. They are often placed at the start of endovascular procedures once access to the vessel has been gained by a needle and secured with a wire. They are inserted over a dilator, over a wire, and the dilator is removed once the sheath is in position.

At the external end of a sheath is a one-way valve, which prevents egress of blood but allows the passage of wires and catheters. From the hub extends a side-arm, on which there is a 3-way tap to allow flushing.

Sheaths are typically labelled in French (Fr) size, which represents their inner diameter (i.e. the size of device they accommodate). For example, a 4 Fr sheath accommodates a 4 Fr catheter. The external diameter of a sheath is normally 1.5–2 Fr larger than its internal diameter. 3 Fr = 1 mm.

Sheaths are available in diameters from 4–30 Fr. The sheath should be selected in advance based on the intentions of the procedure, considering the target position of the sheath tip and the size of the devices likely required during the procedure. Standard sheaths are 10–12 cm in length, but longer sheaths, up to 110 cm in length, are available, maintaining secure access closer to the site of therapy, particularly useful in contralateral femoral or arm access, visceral or carotid interventions or infra-popliteal interventions.

Braided sheaths (e.g. Destination Guiding Sheath, Terumo) incorporate stiff rings or coils, often stainless steel, that provide increased resistance to kinking or "ovalisation" when compared with conventional sheaths, and should be considered when the sheath is to cross tight corners, such as the aortic bifurcation. Sheaths are also available with shaped rather than conventional straight tips, such as the renal double curve (RDC) sheath, that intends to support access to the visceral vessels.

Guidewires

Guidewires are used to navigate vessels to reach a target lesion or vessel segment and to support the passage of a catheter or device to the target to allow treatment. They are constructed of a stiff inner "mandrel" core wrapped in a fine wire or coated with a hydrophilic coating that makes them more manoeuvrable when wet.

The physical properties of wires are interrelated and are generally characterised by their pushability, trackability, steerability, torque, supportability and opacity.

- Pushability is the amount of force needed to advance the wire, which is related to both the axial stiffness of the wire and the degree of surface friction.
- Trackability is the ability of the wire to follow the tip along a vessel, related to flexibility and surface coating.
- Steerability is the capacity of the wire tip to be navigate vessels, intimately related to pushability and torque. Torque is translation of a turn of the wire by the operator into rotational movement at the tip of the wire.
- Supportability is the capacity of the wire to support the passage of a device and its successful deployment. Stiffer wires provide greater support for devices, but have a greater propensity to cause damage to the vessel and are less useful for vessel navigation.
- Opacity is the level of visibility of a wire under fluoroscopic imaging—this can be enhanced with the inclusion of platinum or gold, often most critical at the wire tip.

Most wires are composed of a stiff inner core which is wrapped by a fine wire or coated with a polymer. Steel core wires are stiff, whereas the inclusion of Nitinol in a wire increases flexibility and allows the wire to spring back into shape after navigating a tortuous vessel segment.

Guide wires usually have a floppy tip, created by the tapering of the stiff inner core, that enables navigation and prevents vessel damage, with a stiff body to enable pushability and "torquability".

Some wires are coated with a polymer, such as silicone or polytetrafluoroethylene (PTFE), to reduce friction. In a fluid environment, hydrophilic coatings reduce friction for easier movement in small, stenosed or tortuous vessels. Hydrophilic wires are slippery when wet and sticky when dry, so are more difficult to handle. Manipulation of these wires is helped by using a torque device (see below).

Radiopaque markers are commonly used to improve the visibility of wires on fluoroscopic X-ray imaging. These are often precious metals, such as gold marker bands or the inclusion of a platinum wire.

Wires can be spun between the thumb and index finger (or with a torque device) to direct the angled tip in the desired direction so that branches can be engaged, or avoided. Continuously spinning the wire while advancing it within a vessel prevents the wire from catching on to small branches or atherosclerotic plaques. When advancing or removing catheters over a wire, the back-end of the wire should be fixed to prevent inadvertent wire advancement or loss of position.

Wire Shapes and Sizes

The tips of the wires come in various configurations to help navigate, such as straight or differing degrees of curves, including a "J" curve. Some tips can also be shaped by the operator using a hard edge.

Wire diameters are measured in thousandths of an inch, usually between 0.014 and 0.038 inches, referred to colloquially as '014', '018' and '035'. Thinner wires, such as 014 and 018, are generally used in smaller vessels and they allow the use of smaller profile catheters and balloons.

Lengths are measured in centimetres, ranging from 80 to 450 cm.

Length of wires varies, with a standard wire being around 145 cm long, but exchange wire length should allow for catheter exchange, so must be at least twice the length of the catheter or device.

Wire Types

Access Wires—simple short wires used for access and then exchanged.

Navigation/Manoeuvring Wires—flexible wires, often hydrophilic, with shaped tips that can be steered to sub-select vessels.

Support Wires—stiff, retain position and provide stability for catheter exchanges and for inflating balloons or deploying devices such as stents, whilst maintaining access. Exchange wires are usually longer, often 260–300 cm.

Graduated Wires—increase in stiffness from proximal to distal, e.g. TAD/Jindo which transitions from a 0.018 diameter floppy tip to an 0.035 rigid shaft, useful in renal or mesenteric stenting.

Weighted tip wires***

Wire Problems

Shearing—Coated wires should not be passed through access needles because of the risk of shearing off the coating within the vessel.

Kinking—If a dilator or sheath is advanced at too acute an angle or over a wire with insufficient tension then a bend or kink can develop in a non-hydrophilic wire. This makes passing a catheter over the wire difficult. Applying sufficient retraction on the wire as the dilator or sheath is advanced, and matching the puncture angle when advancing sheaths are essential to prevent this difficulty. When this occurs, the wire should be exchanged.

Common Wire Examples

Wire Type	Examples	Indications	Notes
Standard/access wires	J-wire (e.g. Rosen wire, Cook Medical) Bentson wire (Cook Medical)	Seldinger access, device delivery or exchange. Bentson wires can cross moderate stenoses or push coils	J wires have 3, 5 or 15 mm J-shaped tips. Benston wires have a straight, floppy atraumatic tip
Moderate stiffness	Amplatz Stiff (Boston)	Exchanges	Available in up to 3 m length. Floppy tip
Very stiff	Amplatz Superstiff or Ultrastiff (Boston)	Difficult access, exchanges, device delivery	
Extra stiff	Lunderquist (Cook), Meier (Boston)	Thoracic or abdominal aorta stent graft delivery	Expensive. Great stability and very visible tip
Steerable—hydrophilic	Glidewire (Terumo), Roadrunner (Merit), Zipwire (Boston)	Navigation, traversing disease	Stiff Glidewire (Terumo) combines support and steerability. Glidewires (Terumo) are available in 0.018 and 0.035 sizes
Shapeable tip, microcatheter support/ exchange	V14/V18 (Boston Scientific), Fathom (***)	Working/traversing wire in 0.014 or 0.018 systems	For when smaller calibre devices are required
Weighted tip	Approach CTO Microwire (Cook Medical)	Traversing chronic total occlusions	Provided in different weights of tip, measured in grams

Standard wires: These are soft wires that can be used for initial access, exchanges and device delivery. 'J' wires—3 mm, 5 mm or 15 mm J-shaped tip. The tip shape reduces the risk of cannulating small collaterals or causing intimal damage when advanced. Bentson wire is a straight, floppy-tipped standard wire that is useful for the above indications, but can also be used for crossing moderate stenoses or pushing coils.

Moderate stiffness: e.g. Amplatz (Boston) Stiff, commonly used for difficult access and exchanges.

Very stiff: e.g. Amplatz (Boston) Superstiff or Ultrastiff, commonly used for device delivery.

Extra stiff: e.g. Lunderquist (Cook) and Meier (Boston) Wires, commonly used for thoracic or abdominal aortic endograft delivery.

Steerable wires: Glidewire (Terumo)/Roadrunner (Merit)/Zip (Boston) for crossing tight stenosis or tortuous vessels. Stiff Glidewire (Terumo) combines support and steerability.

- V18 control wire (Boston)—has a radio-opaque, shapeable, hydrophilic tip to increase manoeuvrability and permit vessel selection, but a stiff main shaft to improve pushability and facilitate catheter exchange. Typically used in smaller vessels such as the tibial arteries.

Catheters

Guide Catheters

Catheters are flexible tubes that are typically inserted over a guidewire through a sheath. Their shape facilitates cannulation of vessels and they can be used to inject contrast media to allow for the acquisition of digital subtraction angiograms (DSA), or to inject medications, such as thrombolytic agents. Catheters come in a variety of shapes and sizes to suit different anatomical structures.

The main practical distinctions between catheters are whether they are end- or side-hole, hydrophilic or non-hydrophilic and the shape of their tip. Their propensity to retain their shape and their "torquability" are characteristics of the material they are made of and their thickness. For example, Teflon catheters have a low co-efficient of friction but are at higher risk of kinking and polyethylene catheters are softer and more flexible, yet retain torsional rigidity and therefore torque. Catheters are visible under fluoroscopy as they are impregnated with barium sulphate.

Fr sizes in catheters refer to their outer diameter, as opposed to sheaths, where the size refers to inner diameter. Most modern catheters are 5 Fr or smaller. 5 Fr catheters tend to have the same size of lumen as 4 Fr catheters but with thicker walls, giving them greater stiffness, torquability and stability, but necessitating a large access hole. Higher flow rates can be achieved through shorter, wider catheters, and thicker catheters with narrower lumens are less likely to burst. Catheters typically burst at the hub, but a kink increases the risk of the catheter bursting inside the patient.

In general, catheters should be advanced over a wire; otherwise, their tips can scrape along the vessel wall and cause trauma.

Microcatheter

More slender than traditional catheters, 1–8–3 Fr in diameter. They are normally used co-axially through larger 4 or 5 Fr catheters, typically over 0.010–0.018 guidewires, though some microcatheters can be advanced without a wire in certain scenarios. Microcatheters are more flexible and thus can gain access to smaller and more tortuous vessels and are commonly used within visceral and cerebral vessels for procedures such as embolisation. Support catheters are of similar calibre, but are stiffer, with utility in the below-knee arteries where they enhance the operator's ability to cross lesions with small guidewires.

Common Catheter Examples

Catheter type	Examples	Indications	Notes
Forward-facing, angled tip	MPA, Berenstein (I or II), Van Shee, vertebral	General vessel navigation/cannulation, including arch vessels	Vary in length and severity of angulation
Non-selective	Pigtail	Angiography/venography, including by injection pump	Pigtail-shaped tip with multiples side-holes
Wall-bracing	Cobra	Crossing aortic bifurcation, visceral vessel cannulation	The shape helps anchor the catheter by increasing apposition to the vessel wall
Reverse-angled	Simmons (I or II), SOS Omni, sidewinder, rim, USC	Visceral vessel cannulation, aortic arch vessel cannulation, crossing aortic bifurcation	Need forming once inside the patient, provide stability access to visceral vessels
Hydrophilic	Glidecath or Glidecobra (Terumo)	When struggling to track a non-hydrophilic catheter	Some stability is sacrificed in return for increased trackability
Microcatheter	Progreat Microcatheter (Terumo), Direxion (Boston Scientific)	Selective vessel cannulation, typically for embolisation (coils, particles, microvascular plugs)	Accommodate an 0.014 or 0.018 wire. Fit co-axially through a 4 Fr or 5 Fr catheter. 1.8–3 Fr in diameter

Catheter type	Examples	Indications	Notes
Support microcatheter	CXI Support Catheter (Cook Medical)	Traversing occlusions or stenoses in small vessels	Often fit co-axially through a 4 Fr catheter over an 0.014 or 0.018 wire. Provide support for lesion traversing in small arteries
Thrombolysis			

Torque Devices

Small barrel shaped devices that provide a secure grip point for wires that are hard to grip and rotate (hydrophilic or monofilament). The wire is passed through these devices, and they grasp the wire either using a spring mechanism or by tightening a micro chuck.

Taps/Hubs/Connectors

Flow switch—a short, linear device with male and female Luer lock ends that can be attached to syringes and catheters, through which a wire can be passed. There is a central switch that can be closed, preventing the passage of wires and the egress of blood.

Tuohy–Bohrst—a Y-shaped connected with a luer lock connector on one end and a haemostatic valve and tap for flushing on the other end. The haemostatic valve can be tightened to differing degrees depending on the width of the wire or catheter within it. A flow switch is often attached to the tap of the Tuohy–Bohrst. This device helps when using a smaller wire or catheter within a larger catheter in which the entire lumen is not occupied and would normally allow blood flow around the inner wire, for example, a 0.018 wire or a microcatheter within a 4 Fr catheter. This maintains haemostasis at the back of the catheter and allows flushing, preventing the clot formation that may otherwise occur.

Removable haemostatic valve (e.g. Cook Medical). These enable the conversion of a guide catheter to a sheath and are also useful for clot aspiration from a catheter. Destination Guide Sheaths (Terumo) include a removable valve.

Snares

Snares are devices with loops at their distal end that can be pulled tight to capture a wire, catheter or other device. The most common snares are gooseneck snares, a single loop snare, but other shapes such as the "polyhedral" snares exist and they can improve the operator's chance of capturing small objects. Snares are commonly used to capture guidewires, allowing the reversal of access (e.g. from retrograde to antegrade), facilitating alternative approaches to the navigation of tight bends or crossing of resistant occlusions. Snares can also be used to retrieve migrated foreign bodies, such as coils or guidewires, or to strip fibrin sheaths from longstanding central venous catheters.

Balloons

Angioplasty Balloons

Dilatation balloons, most commonly used for angioplasty, but also with utility in the gastrointestinal and genitourinary systems, are a form of catheter with a non-compliant or semi-compliant plastic balloon at its distal end. Non-compliant balloons do not conform readily to the vessel contour and therefore exert more pressure on a lesion than on normal vessel wall. A radio-opaque marker indicates each end of the "working length" of the balloon, though the operator should be aware that a short tapering section of balloon will project beyond the markers at both ends.

Balloons are labelled by their diameter at nominal pressure and by their length when fully inflated. They should be selected based on the target vessel diameter and lesion length, with consideration of the pressure likely to be required for successful vessel treatment (for example, venous stenoses may require treatment with a high pressure balloon).

Balloons are available on long or short delivery systems, which describe the length of the catheter delivering the balloon. Short delivery systems, typically 75–80 cm in length, are most commonly used, but longer systems, such as those 135 cm in length may be required for contralateral angioplasty, or when access is obtained from the arm.

Balloons are more frequently "over the wire", similar to a conventional catheter, where the lumen of the catheter runs along its entire length, but alternatively can be "monorail" or "rapid exchange", whereby the lumen (and therefore the wire) exit the balloon part way along its shaft, permitting these rapid exchange balloons to be used over shorted wires than for a conventional balloon.

Further information provided includes the thickness of the wire accommodated by the balloon and the size of sheath required to insert them. Smaller profile balloons that can be inserted through a 4 Fr sheath often require an "014" or "018" exchange wire; balloons that can track over an "035" wire tend to require a 5 Fr sheath or larger.

Balloon packaging will specify a nominal pressure in atmospheres (ATM), which represents the pressure at which the balloon will form its labelled diameter, but balloons can be further inflated up to a maximum "burst" pressure to treat resistant stenoses, at which point the balloon will be of a larger diameter, typically 5–10% greater than its nominal diameter. The burst pressure is the maximum pressure the balloon can withstand without rupturing.

Balloons should be inflated with dilute contrast to permit visualisation under fluoroscopy, with care to keep air present within the balloon minimal in case of balloon rupture and distal embolisation. Balloons inflate from outside-in.

A few special case balloons exist for use in certain scenarios:

- High pressure balloons have higher burst pressures and can be used to treat resistant stenoses, such as venous stenoses occurring in high flow arterio-venous fistulae. They tend to require a larger access sheath for delivery.
- Cutting balloons have multiple thin longitudinal blades along the length of the balloon that score the intima when the balloon is inflated. They should be inflated multiple times within the lesion, rotated slightly between each inflation to score the intima in multiple places, before further angioplasty with a conventional balloon. Care must be taken when handling these balloons to avoid injury to the operator.
- Drug-eluting balloons are coated with a medication that prevents smooth muscle hyperplasia at the site of angioplasty, an example of which is the Ranger Drug-Coated Balloon (Boston Scientific), which is coated in Paclitaxel. Drug-coated balloons have been shown to reduce the risk of re-stenosis at the site of angioplasty, though, at the time of writing, their use is limited in many centres, due to previous data showing an increase in mortality following their use.

Compliant Balloons

Compliant balloons are used less frequently, but have particular utility when vessel occlusion is required, for example, a Le Maitre balloon, which can be used to occlude the aorta to prevent bleeding in pelvic surgery or traumatic scenarios. Compliant balloons are also used to mould stent grafts to the surrounding vessel wall, most commonly during endovascular aneurysm repairs.

Inflation Devices

Inflation devices are filled with dilute contrast and used to inflate angioplasty balloons to a pressure displayed on a gauge. Through use of a corkscrew method of inflation and a quick-release mechanism, they allow the generation of far greater inflation pressures than could otherwise be managed with human strength and a conventional syringe.

Stents

Stents function by exerting outward radial force and, unlike balloons, remain in place once the procedure has concluded, maintaining this radial force. Stents used in vascular procedures are metallic (plastic stents are sometimes used in biliary or urological procedures). There are two forms of bare metal stent, described below.

Self-expanding stents are made of nitinol, a nickel–titanium alloy that has memory of their intended shape and will attempt to form this when no longer constrained. In their packaging, they are contained by a plastic delivery sheath that allows the device to remain low profile for insertion through access sheaths and positioning in the correct location. A release mechanism exposes the stent and allows it to expand to its former diameter, continuing to exert outward radial force wherever this nominal diameter cannot be reached.

Self-expanding stents are more flexible and conform better to tortuous anatomy than balloon-expandable stents. They are simple to deploy but tend to shorten as they are deployed.

Balloon-expandable stents are loaded onto a balloon catheter—the stent is deployed by dilation of the balloon, taking the stent to its nominal diameter, where it exerts higher radial force than a self-expanding stent does. They can be deployed more accurately than self-expanding and are therefore useful for iliac origin disease. Balloon-expandable stents are non-conformable and will straighten the anatomy they are deployed into. They can be crushed or fracture if placed in an area where they are forced to flex, such as close to the groin, and should not be used in these scenarios. It should also be noted that balloon-expandable stents are not covered prior to deployment like self-expanding stents are, so there is risk of the stent becoming dislodged if being advanced through a tight stenosis. This can be avoided by "unsheathing" the stent within the stenosis.

Covered stents (stent grafts)—Stents can also be covered with material, typically a synthetic polymer or fabric. Once again, they can be self-expanding or balloon-mounted. Covered stents can be used in scenarios where vessel re-lining is required, such as treatment of a focal bleeding point or aneurysm exclusion, including EVAR. Due to the presence of fabric, as well as metallic stents, they are larger profile than bare stents and require larger delivery sheaths.

As with balloons, stents are also available in *drug-eluting* forms, such as the Eluvia drug-eluting stent that utilises paclitaxel, with some studies showing increased long-term patency when used in atherosclerotic disease of the femoropopliteal segment. As for their balloon counterparts, historical controversies regarding potential long term mortality risks cast a shadow over their use, dissuading many practitioners from employing them in vivo, and therefore they are mostly limited to coronary use.

Type	Examples	Indications	Notes
Bare metal, self-expanding	Smart (Cordis), Luminexx (Bard)	Standard stent placement, e.g. iliac	More suitable for tortuous vessels than balloon-expandable
Bare metal, self-expanding, flexible	Supera (Abbott), Biomimic (***)	Stenting in	Length dependent on vessel preparation/deployment
Bare metal, balloon-expandable	Palmaz	Iliac vessels, visceral arteries	Better suited to straight anatomy. Can be positioned more accurately than self-expanding (e.g. close to bifurcation)
Covered, self-expanding	Viabahn (***)	Exclusion of bleeding, aneurysms, fistulae	
Covered, balloon-expandable	VBX (***)	Exclusion of bleeding, aneurysms, fistulae	
Thoracic aorta covered stent graft	TAG Conformable (Gore), Valiant (Medtronic)	Thoracic aortic aneurysm or dissection	
Abdominal aorta covered stent graft	Endurant (Medtronic), Excluder (Gore), *** (Cordis)	Thoracic aortic, common iliac aneurysms	Typically a bifurcated system deployed from both groins
Drug-eluting stent	Eluvia (Boston Scientific)	Resistant femoropopliteal lesions	Still not widely used due to concerns regarding potentially increased all-cause mortality

Re-entry Devices

Re-entry devices are catheters that can aid the return into the lumen of an artery from the sub-intimal space. Wires and catheters can sometimes breach the intima of a vessel when attempts are made to cross occlusions or tight stenoses. Though this sometimes takes place inadvertently, while the operator is attempting an intraluminal approach, it is a legitimate way of crossing occlusions and the operator may perform this intentionally.

However, in order to angioplasty or stent the lesion, the operator must regain access to the lumen of the artery beyond the occlusion. Often this is straightforward with a standard hydrophilic wire (e.g. Terumo glide wire) and catheter (e.g. Berenstein), but on occasion this can prove tricky. Re-entry devices are typically comprised of a sharp needle towards the end of a catheter that can be extended near-perpendicularly from the catheter when in position. A marker alerts the operator to the side of the catheter the needle will extend from, and this can be orientated towards the lumen. The intention is that this needle punctures the lumen and then can guide a wire into it. The re-entry device is best used early in the attempt to break back into the vessel before the sub-intimal space becomes "baggy", to increase the probability of the needle engaging and puncturing the intima.

Examples of re-entry devices include the Outback (Cordis) and Pioneer (Medtronic) catheters.

Embolisation Materials

Embolisation is the use of a material that induces intentional blockage or reduction of blood flow. Embolic agents vary in their permanency, size of material, size of vessel appropriate for treatment, radiopacity and mechanism of action.

Coils are metallic devices made of titanium, platinum, stainless steel or nitinol, that attempt to "coil" into a pre-determined shape at the point of deployment. They are often covered with Dacron fibres that help to cause flow obstruction, but ultimately rely on the patient's own coagulation to cause thrombosis. Coils should be tightly packed in order to be most effective and should be oversized by approximately 15% to improve stability within vessels.

Coils are described by their shape, diameter and length:

- Shape refers to the shape the coil forms when released and allowed to form the shape it has memory of. Helical are most common, but a variety of shapes exist for different purposes.
- The deployed diameter refers to the width the coil when formed. Coils are generally oversized by ~25% in relation to the vessel diameter. The coil wire diameter determines which catheter is required for deployment—coils are available in standard 0.035 diameters or micro-coil sizes such as 0.018 or even 0.010, as with guidewires.
- Length describes the undeployed length of the coil, which gives the operator an idea of the amount of coil that will

need to be pushed from the catheter to complete coil placement. When deployed, they are much shorter due to the memory shape they take on.

Various types of coil detachment exist. The simplest (and cheapest) are those that are pushed through a catheter and are immediately released on exit from the catheter, known as "pushable" coils. This type of coil can also be flushed from the catheter using saline. "Detachable" coils feature some sort of detachment device, which include interlocking mechanisms or electrolytic or mechanical breakage of an internal wire. Detachable coils are considered when precise placement is required and there is greater risk of coil migration or non-target embolisation of a vital arterial branch, as stability of the coil can be tested prior to detachment. Detachable coils are more expensive.

Vascular plugs (e.g. Amplatzer vascular plugs, Abbott) are devices that expand to a pre-determined 3D geometrical shape when unsheathed, much like a self-expanding stent, but instead are made of a fine, partially occlusive mesh that promotes thrombosis. They are quick to deploy and can be recaptured until the point of detachment, performed by unscrewing the lock in an anticlockwise direction. Plugs typically require larger delivery sheaths than coils and, like coils, they rely on the patient's own clotting to induce thrombosis, the timing of which is unpredictable. Amplatzer I and II plugs exist for treatment of vessels from 2–17 mm diameter, requiring delivery catheters of 5–9 Fr respectively. Amplatzer IV plugs are smaller in their delivery calibre and can treat vessels of 2.5–6 mm through a 0.038 guidewire-compatible catheter.

Microvascular plugs (Medtronic) are similar devices, but instead of mesh have occlusive fabric and therefore do not rely on thrombus formation. They are lower profile than Amplatzer plugs, which allows for placement through a microcatheter in more distal or tortuous vessels, up to 5–7 mm in diameter.

Particles are small non-biodegradable materials, most commonly formed of polyvinyl alcohol (PVA), which can be uniform (microspheres) or non-uniform. Uniform particles (e.g. Embospheres, Merit) are more expensive but are less likely to cause catheter clogging. Particles mechanically occlude vessels, activate thrombin and cause ingrowth of fibroblasts, leading to fairly permanent occlusion.

Particles are mixed with a solution containing 40% contrast, which suspends them and prevents aggregation, and then are injected following selective vessel cannulation. The smaller the particles used, the more distal the vasculature in the vascular bed is blocked. Particles are commonly used when ischaemia of hyperplastic tissue or tumour is desired, such as the prostate, uterine fibroids, or vascular bone metastases prior to surgery. Care must be taken to avoid "non-target embolisation", which occurs when the particles travel to and block an unintended vessel.

Particles can be impregnated with chemotherapy agents and used to perform chemoembolisation. For example, PVA particles sized to between 100–500 μm can be soaked in doxorubicin and used to embolise hepatocellular carcinoma. A similar principle is utilised for selective internal radiation therapy (SIRT), where particles are instead radioactive and deliver local radiotherapy.

Gelfoam is a sponge-like material made of porcine gelatin. It expands when in contact with blood, forming temporary occlusion that later breaks down allowing vessel re-canalisation within a few weeks. This temporary embolisation has utility in scenarios such as in gastrointestinal bleeding, such as those caused by an ulcer, or in traumatic bleeding, where cessation of bleeding until the underlying abnormality has had time to heal is all that is required. Gelfoam can be injected as a slurry, formed from aggressive mixing of Gelfoam strips with contrast in a syringe external to the patient, or as rolled-up strips known as "torpedoes".

Liquid embolics solidify or polymerise upon contact with blood. Glue (cyanoacrylate), indistinct from the agent used as skin glue, forms a cast within vessels shortly following injection. A very effective and fast-acting embolic agent, it requires a greater level of skill and care to use to avoid complication. It is emulsified with lipiodol (Guerbet), a poppy-seed oil that acts as both a contrast agent and for extending the glue polymerisation time, so that the embolic can be visualised under fluoroscopy. A 1:1 mixture creates a very viscous embolic that will form quickly with little transit, whereas greater concentrations of lipiodol allow for a more liquid form that will travel further from the catheter tip. The delivery catheter must be flushed with dextrose prior to injection of glue, as normal saline will induce premature polymerisation.

Onyx (Medtronic) is a newer embolic liquid. It is formed from a combination of ethylene vinyl-alcohol copolymer (EVOH), dimethyl-sulfoxide (DMSO) and tantalum powder, creating a radio-opaque, injectable embolic that forms a spongy coherent substance when the DMSO diffuses away in blood. It comes in different formulations, Onyx 18, which is less viscous and therefore penetrates further, and Onyx 20 or Onyx 34, which are in turn more viscous and more useful in higher flow scenarios. Prior to Onyx injection, the delivery catheter "dead-space" should be filled with DMSO to prevent premature polymer solidification.

Finally, *sclerosants* act to induce endothelial damage, leading to inflammation, thrombosis and eventually fibrosis. Sodium tetradecyl sulfate (STS) is a synthetic detergent commonly used to treat varicose veins, and also used by some operators in testicular vein embolisation for the treatment of a varicocoele. It is mixed with air to form a foam-like substance. It is not radio-opaque, but can be indirectly visualised through the formation of a column of contrast that the foam displaces. Alcohol (ethanol) can be used in a similar fashion, also inducing endothelial damage and thrombosis; however it must be used with great care to avoid toxicity and non-target embolisation.

Pressure measurement devices allow for the measurement of vessel pressure. They are fixed to the end of a catheter and provide live luminal pressure readings reflecting the position of the catheter tip. Possible use cases include the assessment of portal or hepatic vein pressures in the context of trans-jugular portosystemic shunting, or the evaluation of angiographically dubious stenoses. If a stenosis is significant, there should be a pressure rise across the stenosis and a pressure drop beyond it, when compared with proximal pressures.

Closure Devices

Conventional vessel closure for percutaneous access is with digital compression. However, closure devices can quicken vessel closure and reduce the time of, or obviate the need for, digital compression. Closure devices can shorten the time required before ambulation and shorten post-operative stays, increasing efficiency within the IR department. Closure devices can also permit the closure of large holes than would be possible with digital compression alone. A variety of closure device types exist:

Suture-mediated—Perclose (Abbott), Prostar

- Can be pre-deployed for large bore access as "pre-closure", where sutures are then tightened at the conclusion of the procedure.

Plug closure—Angioseal (Terumo), Mynx Grip/Control (Cordis), Manta (Teleflex)

- These devices place a collagen plug on the external aspect of the vessel defect, pushed against a luminal object that is held against the internal aspect of the vessel defect with traction.
- The Angioseal (Terumo) device uses a luminal footplate that dissolves 3 months following placement. The Mynx (Cordis) device uses a luminal balloon, that is deflated and removed following plug placement.

Clip closure—Starclose (Abbott)

- A nitinol clip is fixed to the external aspect of the arterial wall and seals the vessel defect.

Each device will have its own individual contraindications and complications and unique method of deployment. Heavy calcific disease at the puncture site and a small access vessel are the most common contraindications to closure device usage.

Thrombectomy, Embolectomy and Atherectomy Devices

- Catheters
  - Side-hole
  - Ultrasonic
- Mechanical
  - Stent retrieval devices
  - Angiojet

Embolic Protection Devices

Embolic protection devices protect the intracerebral vessels from emboli during carotid interventions. Filters are the most common devices but can be difficult to place in tortuous vessels and may induce spasm. Balloon protection is also available but risks causing cerebral ischaemia while inflated.

Filters:

- Fixed basket:
  - Symmetrical = Accunet, Emboshield Pro
  - Asymmetrical = Filterwire EZ
- Bare wire:
  - Symmetrical = Emboshield Pro
  - Asymmetrical = Spider
- Balloon protection:
  - Proximal = Mo.Ma, NPS
  - Distal = Percusurge

Inferior Vena Cava Filters

Venous Ablation Devices

Atherectomy Devices

Intravascular Ultrasound

Central Venous Catheters and Ports

Tunnelled central venous catheters (CVCs) and implantable central venous access ports provide reliable central venous access for the delivery of prolonged courses of cytotoxic therapy or intravenous antibiotics and for blood sampling. Tunnelled CVCs are also used for the administration of parenteral nutrition or to facilitate haemodialysis.

Ports require a more involved procedure to install, but are entirely internal once implanted, reducing the risks of infection and migration, and requiring fewer limitations in terms of bathing or sports.

Tunnelled CVCs are composed of silicone or polyurethane and feature a cuff that sits close to the exit of the tunnel, inducing a fibrotic reaction that seals the subcutaneous tissues to the cuff, therefore preventing line migration. Haemodialysis lines are most commonly two separate lines, with tips placed approximately 3 cm from each other, whereas Hickman lines, used for venous sampling and administration of medication, can be single or dual lumen within a single line.

Ports feature the same silicone or polyurethane line that sits within the chosen vein, but instead is attached to a titanium or plastic chamber with a self-sealing septum that sits immediately below the skin.

Other Components of the Interventional Radiology Trolley

- Sterile drapes for the procedural trolley.
- Antiseptic scrub for the procedural site, such as chlorhexidine soaked gauze, or ChloraPrep sticks.
- Sterile drapes for the patient. Typical these contain a fenestration for arterial access, with specific shapes for femoral, brachial or radial access. Standard rectangular drapes can be used to extend where necessary.
- Sterile coverings for protective lead screens, the image intensifier and any other controls.
- Sterile ultrasound probe covers.
- A closed system with a 3-way tap, connected to heparinised saline, contrast and a waste drainage bag for egress.
- Basin containing heparinised saline for wires/catheters.
- Gauze.
- Syringes.
  - At least 2 for heparinised saline flushes, 2 for contrast and 1 for lidocaine. These syringes should be visually distinct, at a minimum through labelling, but now more commonly by syringe colour or shape.
  - Smaller syringes will also be required if microcatheters are in use.
- Scalpel—for skin incision.
- Mosquito/forceps—for use separating superficial soft tissues. Can also be used to fix lines to drapes.
- Towel clips, for fixation of ultrasound probes, lines, drapes.
- Flow-switch—device that can be attached to catheters and opened to allow passage of wires and fluids or closed to prevent this occurring.
- Sharps pad, for safe storage of sharps.

Personal Protective Equipment

- Lead gowns.
  - Two piece gowns with a separate top and skirt spread the load onto the hips and reduce the weight transferred through the back.
  - 0.25–0.35 mm lead-equivalent thickness, often with a wraparound design that doubles the protection on the front of the body
- Lead thyroid shield.
- Lead glasses.
- Lead caps and gloves are also available, but their use is more heterogeneous.
- Appropriate radiation monitoring.
- Sterile gloves and gown.
- Face mask or shield to reduce the risk of splash injury.
- Hanging, lead-lined Perspex screen that can be easily re-positioned.
- Floor-standing lead screen.
- Table-mounted lead apron.

Common Medications Used During Endovascular Procedures

Local anaesthetic is adequate for the majority of endovascular procedures, with different agents providing a range of action onset and duration. The same agents can also be administered as regional anaesthesia, such as a sciatic nerve block that can enable a patient with considerable rest pain to remain still during lower limb revascularisation. Co-administration with adrenaline/epinephrine serves to delay absorption (reducing the risk of systemic toxicity) and prolong the duration of action.

- Lidocaine has a rapid onset of action (<5 min) and a duration of action of 1–2 h. It is most commonly used as a 1% preparation but exists up to 4% strength. Maximum dose is 4.5 mg/kg (300 mg in a 67 kg individual, or 30 mL of 1% lidocaine), or 7 mg/kg when delivered with epinephrine.
- Bupivacaine has a slightly longer onset of action (8 min) and is more painful on administration but provides anaesthesia for 4–8 h.

Antiplatelets

There are no clear recommendations regarding the optimal regimen of antiplatelets with regards to peripheral arterial disease, but the Transatlantic Inter-Society Consensus (TASC) guidelines do recommend antiplatelet therapy is started prior to and continued following endovascular or surgical vascular procedures. Many centres will commence 3–6 months of dual antiplatelet therapy following infra-inguinal stent placement. Unless the appropriate antiplatelet is commenced 1 week prior to the procedure, a loading dose of 300 mg of aspirin or clopidogrel should be given on the day of the procedure, followed by a maintenance dose of 75 mg daily.

Heparin

- For prevention of thromboembolism.
- Often administered during angioplasty or stent deployment procedures following secured arterial access, due to the risks conferred by temporary occlusion of blood flow during angioplasty and the risks of distal embolisation. Dosing is typically a bolus of 50 IU/kg up to a maximum of 5000 IU, with 1000 IU maintenance dosing hourly, but more accurate maintenance of anticoagulation can be assisted through measurement of activated clotting times (ACT) and maintenance in a range of 300–400 s.
- Short-acting heparin is also used as bridging therapy for patients on long-term anticoagulation, or for treatment for those with acute ischaemic events, and cessation of these takes place in the hours prior to the procedure, depending on the mode of administration.
- A rare but significant complication of heparin administration is immune-mediated heparin-induced thrombocytopaenia, which can cause life-threatening thromboembolic complications. Bleeding is a risk, as with any anticoagulant.

Heparinised saline is used for irrigation of angiographic equipment and the flushing of angiographic catheters and sheaths. It is formed by mixing 1 IU of heparin per mL of normal saline. It can be set up with a pressure bag for continuous flushing, with particular utility when long sheaths are used.

Tissue plasminogen activator (TPA, Alteplase) is a thrombolytic medication used in the treatment of acute limb ischaemia, as well as for acute myocardial infarction, pulmonary embolism and ischaemic stroke. It can be infused directly into thrombus through use of a side-holed catheter that ideally traverses the thrombus. Thrombolysis is contraindicated in patients with known active bleeding, recent surgery or a known intracranial neoplasm. *Streptokinase* and urokinase are older fibrinolytics, but its use is more historical due to the significant risk of allergic reactions.

Antibiotics may be required as prophylaxis, most commonly in the context of interventions involving a prosthetic graft, or in some centres, for tunnelled line insertions. Local policy for antibiotic choice should be consulted.

Vasodilators can be used therapeutically to treat arterial or venous spasm, or prophylactically prior to intervention within vessels high-risk for spasm, typically smaller vessels such as the infrapopliteal or visceral arteries.

- Glyceryl Dinitrate (Isoket)
- Glyceryl Trinitrate (GTN)

Antihypertensives

Hypertension is common in patients with peripheral arterial disease (PAD) and is more likely to be poorly treatment is PAD patients than the general population. It presents a considerable cardiovascular risk factor and contributes to progression of PAD.

Conscious Sedation

- Midazolam—benzodiazepine—induces sedation, anxiolysis and causes amnesia. Can be administered IV, typically as a bolus of 0.5–2 mg and maintained with further boluses 25% of the original dose every 30–60 min. It can also be given intramuscularly pre-procedure.
- Fentanyl citrate—a potent analgesic and sometimes co-administered with benzodiazepines for sedation—administered IV, typically in 25–50 µg aliquots every 15–70 min. Use with caution in patients pre-disposed to respiratory depression.
- Safety—conscious sedation, judicious use, especially when benzodiazepines and opiates are used in conjunction, as there is considerable risk of respiratory depression.
  - CO_2 capnography, or end-tidal carbon dioxide, provides earlier warning signs of respiratory depression than oxygen saturation and should be utilised when sedatives are administered.
  - Oxygen should be available along with masks and delivery conduits such as masks or nasal prongs.
  - Airway adjuncts and/or supraglottic airway devices and ventilation equipment may be required in the event of over-sedation.
  - Reversal agents—naloxone 0.4–2 mg IV, repeated at 2–3 min intervals (opiates), flumazenil (benzodiazepines).

Contrast Media

The most common contrast media used during endovascular procedures are low-osmolality iodine-based, such as Iohexol (Omnipaque, GE Healthcare), often used at strengths of 240–350 mg of iodine per mL. These agents provide the highest quality invasive angiographic imaging. Consideration the patient's allergy history and have access to medication and equipment required in the treatment of anaphylaxis.

Consider the patient's renal function and the need for pre- and post-procedural hydration.

Carbon dioxide-based angiography is an alternative to iodinated contrast in patients with iodinated contrast allergy or renal dysfunction, but is more complex to use, incurs higher radiation dose, is limited by overlapping bowel gas and is often more uncomfortable for patients.

Gadolinium-based contrasts agents, such as those used in MR imaging (e.g. Dotarem, Guerbet), can be used in a similar manner to iodinated contrast in patients that are iodine-allergic, but is of lower density and only safe in much smaller volumes (10–20 mL). It is not safe in patients with significant renal dysfunction due to the risk of nephrogenic systemic fibrosis.

Anaphylaxis

- Indicated by soft tissue angio-oedema, bronchospasm, rash and cardio-respiratory compromise.
- Call for help and the local resuscitation team.
- 0.5 mg Adrenaline IM (0.5 mL of a 1:1000 preparation). Repeat after 5 min if no response. Lower doses in children <12 years.
- Give oxygen and maintain the airway with adjuncts if required.
- IV fluid bolus of 500–1000 mL.
- Nebulised adrenaline or nebulised salbutamol/ipratropium may be required for persistent upper airway obstruction or bronchospasm respectively.

Imaging Equipment

Do we want to include this? E.g. generators, X-ray tubes, tables, c-arms.

While ultrasound is used for vascular access, and occasionally for other percutaneous procedures such as the injection of thrombin, most endovascular procedures are performed under the guidance of fluoroscopy. **Pulsed fluoroscopy**, typically between 2–7 frames per second, provides "real-time" X-rays images. **Digital subtraction angiography** is a technique by which an initial mask image is subtracted from the live image, therefore showing only what is new or has changed since the mask image, most importantly contrast agents.

Generators provide electrical energy for X-ray generation and control X-rays. The quality of images is determined by various factors such as tube voltage, tube current, pulse width, pulse rate, focal spot size, beam filtration and dose rate. Dose rate (measured in μGy per image) is determined by kVp, mA and pulse width, changes in which reduce image noise but increase dose to the patient (and the operator). Generators adjust radiation by adjusting tube current, pulse width and tube voltage, in that order, optimising radiation output for different types of angiographic studies and patient sizes.

X-ray tubes are formed of a tungsten filament cathode and a tungsten anode disc, converting kinetic energy of electrons into X-rays (and heat). Collimators are attached to the X-ray tube and comprise beam-blocking facets that shape the beam at exit from the tube, and wedge filters that can alter the strength of emitted radiation to optimise image quality and reduce patient dose. The **image chain** is more often a flat-panel detector than an image intensifier on more modern machines. It converts the received pattern of transmitted X-rays into a live image that is displayed in real time.

The **gantry stand** holds the X-ray tube and the image chain. The C-arm is the most common design used for new equipment. Single-plane gantry stands can be ceiling or floor-mounted, while in biplane configurations the lateral plane is mounted on ceiling rails and the frontal plane assembly is mounted on the floor. Rotational angiography and cone-beam CT are possible with gantries that have a true isocenter.

Patient tables are often composed of carbon fibre to permit maximal strength with minimal beam attenuation. Tabletops move axially relative to the patient and can be raised or lowered, which alters magnification, field of view and dose. Many tables also permit a degree of tilt.

Displays should show live fluoroscopy with a second screen or partition that displays a reference image or roadmap, with information on patient dose and screening times. A biplane system will require live and reference imaging for both the frontal and lateral projections. There should also be provision to display patient observations, and some will offer display or alternative imaging such as previous cross-sectional imaging or radiographs.

A **contrast pump injector** can control contrast injection for angiography. On modern machines, it is linked to the X-ray machine to allow synchronisation for digital subtraction angiography, with the ability to alter pump or X-ray delay to match DSA timing with the expected progression of contrast. The machine can control injection volume and injection rate.

Basic Terminology

Acute Limb Ischaemia Plaque rupture or an embolus leads to abrupt lack of blood flow to the arms or legs, presenting with symptoms such as pain, coldness and pallor.

Amputation Surgical removal of a limb or other body part.

Aneurysm A bulge or swelling in a blood vessel that can rupture and cause life-threatening bleeding. A true aneurysm is contained by all three layers of wall; a pseudoaneurysm is not.

Ankle Brachial Pressure Index (ABPI) A test used to diagnose peripheral artery disease by comparing blood pressure measurements in the arm and ankle.

Angiodysplasia A condition in which abnormal blood vessels in the gastrointestinal tract can cause bleeding.

Angiography A medical imaging technique used to visualise blood vessels using X-rays, CT or MRI. X-ray and CT angiography utilises a contrast agent, MR angiography can be performed with or without a contrast agent.

Angioplasty A procedure used to widen the lumen of narrowed or blocked blood vessels using a balloon catheter that is inflated at the site of the stenosis or blockage. Short for percutaneous transluminal angioplasty (PTA).

Antiplatelets Medications such as aspirin, clopidogrel and ticagrelor that prevent platelets from sticking together and forming clots.

Arteriovenous Fistula An connection between an artery and a vein, which can be intentional, typically used for dialysis access, or can be abnormal and cause problems such as bleeding or cardiac failure.

Artery A blood vessel that carries oxygenated blood away from the heart to the body's tissues and organs.

Atherectomy A procedure used to remove plaque from the walls of arteries, typically using a catheter-based device.

Balloon Catheter A catheter with a balloon at the end that can be inflated to widen the lumen of narrowed or blocked blood vessels during angioplasty, to occlude vessels or to mould stent grafts.

Bypass A surgical procedure used to reroute blood flow around a blocked or narrowed blood vessel. Autologous veins or prosthetic grafts of plastics such as PTFE are used.

Catheter A thin, flexible tube normally advanced over a guidewire, used to deliver contrast agents, fluid, or medications and to provide support for guidewires.

Chemoembolisation A treatment for cancer in which chemotherapy drugs are delivered directly to a tumour through arteries, often through use of embolisation particles impregnated with the chemotherapy agent, commonly used for hepatocellular carcinoma. Also known as transarterial chemoembolisation, or TACE.

Chronic Limb Threatening Ischaemia (CLTI) Also known as critical limb ischaemia. Longstanding peripheral artery stenoses or occlusions lead to ischaemic symptoms in the limbs such as pain at rest, non-healing wounds or gangrene.

Claudication Pain or cramping in the legs caused by inadequate blood flow to the muscles during exercise. It ceases when resting.

Coil Embolisation A procedure used to treat bleeding, aneurysms, or other vascular abnormalities, in which small metal coils are deposited, causing mechanical obstruction and promoting clotting in order to block blood flow.

Computed Tomography (CT) A medical imaging technique that uses X-rays and computer processing to create detailed cross-sectional images of the body. In the context of endovascular imaging, iodinated contrast is almost always given.

Contrast Agent A substance injected into the body to make certain tissues or blood vessels more visible during medical imaging. Iodinated contrast is used for CT, whereas gadolinium-based contrast is used for MRI. Catheter angiography is normally performed with iodinated contrast, though gadolinium can also be used, a carbon dioxide can be used as a negative contrast agent.

Catheter-Directed Thrombolysis A procedure to dissolve blood clots using a catheter and a fibrinolytic agent such as alteplase. This is often through a slow infusion that lasts for 24 h or more.

Central Venous Catheter A catheter placed in a large vein in the neck, chest, or groin to administer medications, fluids, or nutrition, the facilitate dialysis, or to permit venous sampling. Central venous catheters can be tunnelled under the skin for a short distance to reduce infection risk and increase longevity.

Deep Vein Thrombosis A blood clot that forms in a deep vein, most commonly in the legs, presenting with pain, limb swelling and/or erythema.

Digital Subtraction Angiography A type of imaging that involves the computerised subtraction of a mask image from future images that contain a contrast agent, demonstrating the contrast flow in more clarity.

Dissection A tear in the intima, the inner layer of an artery or vein. This can cause separation of the intima from the other layers of the vessel wall, potentially narrowing or obstructing the lumen.

Embolisation A procedure used to block or reduce blood flow to a selected part of the body, using substances such as particles, coils or plugs.

Embolectomy A surgical procedure used to remove a blood clot from an artery, often using a balloon catheter.

Embolus A blood clot, air bubble, fat locule or foreign material that travels through the bloodstream and blocks a blood vessel.

Endoleak A complication that can occur after endovascular repair of an aneurysm, most commonly abdominal aortic aneurysms, in which perfusion of the aneurysm sac persists following treatment and can present a continued risk of rupture.

Endovascular Aneurysm Repair A minimally invasive procedure used to repair an aneurysm from within the blood vessel, typically using a stent graft (covered stent).

Fibroid A non-malignant tumour that grows in the uterus and can present with menorrhagia or pressure symptoms.

In some circumstances, they can be treated with uterine artery embolisation.

Fluoroscopy A medical imaging technique that uses X-rays to create real-time images of the body.

Gastrointestinal Bleeding Bleeding that occurs in the digestive tract, typically due to ulcers, angiodysplasia or cancer. It can present with haematemesis, rectal bleeding or haemodynamic instability.

Graft A synthetic or natural material used to replace or repair damaged tissue or blood vessels. This can refer to surgical bypass grafts or to covered stents that are often referred to as "stent grafts".

Haematoma A collection of blood outside of a blood vessel, typically caused by vessel injury.

Haemorrhage Excessive bleeding, either internally or externally.

Heparin A medication that prevents blood clots from forming by inhibiting the activity of clotting factors.

Hypertension High blood pressure, which can cause damage to blood vessels and increase the risk of cardiovascular disease.

Inferior Vena Cava (IVC) Filter A device placed in the IVC to prevent blood clots from reaching the lungs and causing a pulmonary embolus.

Interventional Radiology A medical specialty that uses imaging techniques to guide minimally invasive procedures to diagnose and treat a variety of conditions.

Intravascular Ultrasound (IVUS) An imaging technique that uses an ultrasound probe on the tip of a catheter to create images of the inside of blood vessels.

Ischaemia A lack of blood flow to a part of the body, leading to hypoxia, typically due to narrowing or occlusion of blood vessels or due to hypovolaemia.

Magnetic Resonance Imaging (MRI) A medical imaging technique that uses a strong magnetic field to create signal emission from the body, which is captured and processed to form cross sectional images. It provides better contrast resolution than CT without the limitation of ionising radiation, but images are of lower spatial resolution and take longer to acquire.

Mycotic Aneurysm An aneurysm caused by a bacterial infection.

Necrosis Death of tissue due to lack of blood flow and tissue hypoxia.

Percutaneous A medical procedure that is performed through the skin rather than through an open incision.

Peripheral Arterial Disease (PAD) A condition in which the blood vessels in the legs or arms become narrowed or blocked, causing pain, numbness or tissue loss.

Plaque A build-up of cholesterol, calcium, and other substances in the walls of arteries, leading to luminal narrowing and reduced blood flow.

Portal Hypertension A condition in which high blood pressure in the portal vein which carries blood from the digestive tract to the liver can cause complications such as variceal bleeding. This is most commonly secondary to liver cirrhosis, but can also occur due to portal vein thrombosis. Transjugular intrahepatic portosystemic shunt (TIPSS) (see below) is a potential treatment.

Pulmonary Embolism (PE) A blood clot that travels to the lungs, causing shortness of breath and chest pain and can be life-threatening.

Renal Artery Stenosis Narrowing of the renal arteries, which can lead to reduced renal function and/or hypertension.

Sclerotherapy A treatment that uses an injection of a sclerosing agent to obliterate vascular abnormalities such as varicose veins, varicocoeles or vascular malformations.

Selective Internal Radiation Therapy (SIRT) A procedure used to treat hepatocellular carcinoma by delivering radioactive particles directly to the tumour through the hepatic arteries.

Sheath A thin, hollow, rigid plastic tube that sits within a vessel to provide secure vessel access, reduced vessel trauma during wire and catheter exchange, maintain haemostasis and permit injection of saline, contrast or other medications.

Stent A small metal or plastic mesh tube that can be left within a narrowed or blocked blood vessel to hold it open. Vascular stents are metal and can be covered (with a PTFE covering) or uncovered.

Thrombectomy A procedure to remove a blood clot from a blood vessel, often using an aspiration catheter or stent-based system. Mechanical thrombectomy devices exist for use in the peripheral vasculature.

Transjugular Intrahepatic Portosystemic Shunt (TIPSS) A procedure to re-direct blood flow from the portal venous system into the systemic veins, with the intention of reducing portal pressure and mitigating the potential complications of portal hypertension such as variceal bleeding. A stent between a portal and hepatic vein is used to achieve this.

Tumour Embolisation A procedure used to block the blood supply to a tumour, typically using small beads or other particles that are injected into the arteries feeding the tumour. Chemoembolisation is a form of this, but bland embolisation is also useful, such as pre-operative embolisation of hypervascular tumours or metastases to reduce bleeding during surgery.

Ultrasound A medical imaging technique that uses sound waves to create images of the body, displayed in real time. It can be used to guide procedures, such as vascular access, and for assessments, such as lower limb vascular assessments or monitoring of aneurysms.

Ultrasound Duplex A form of imaging that combines ultrasound images with Doppler shift, providing information regarding the direction and velocity of flow.

Uterine Artery Embolisation (UAE) A procedure used to treat uterine fibroids by blocking the blood supply to them using small particles injected into the uterine arteries.

Varicose Veins Enlarged, tortuous veins that are commonly found in the legs.

Vascular Malformation An abnormal formation of blood vessels, typically congenital, but may progress or present later in life, causing symptoms such as pain, swelling, or bleeding.

Vein A blood vessel that carries blood back to the heart.

Venous Insufficiency A condition in which the valves in the veins do not function properly, leading to the pooling of blood and the development of leg swelling, discolouration and/or ulceration.

References

1. Kandarpa Handbook. https://www.ncbi.nlm.nih.gov/pmc/articles/PMC6735447/. https://www.resus.org.uk/library/additional-guidance/guidance-anaphylaxis/emergency-treatment.
2. TASC. Complex Cases Schillinger. Learning Vascular and Interventional Radiology, Mun~oz and Ribes. Pocketbook Clinical IR Warhadpande. https://www.ahajournals.org/doi/10.1161/CIRCULATIONAHA.122.059606.

Principles, Sites and Techniques of Vascular Access

6

Dinesh Manoharan and Narayan Karunanithy

Case Presentation

A 55-year-old woman presented to the emergency department with severe abdominal pain, cachexia and diarrhoea. Previously, she had been diagnosed with chronic mesenteric ischemia as a result of occlusion of the superior and inferior mesenteric arteries, and had undergone stenting of the inferior mesenteric artery. Following a repeat CT scan and endoscopy, it was confirmed that there was an occlusion of the superior mesenteric artery and evidence of ischemic colitis as a result of poor perfusion. Furthermore, she had bilateral iliac and common femoral artery occlusions making endovascular approach challenging (Fig. 6.1). The treatment plan for her was discussed in the multidisciplinary team meeting. In view of the fact that she was an unsuitable candidate for surgery, we decided to perform endovascular stenting of the superior mesenteric artery through the radial approach.

Continued at page 56

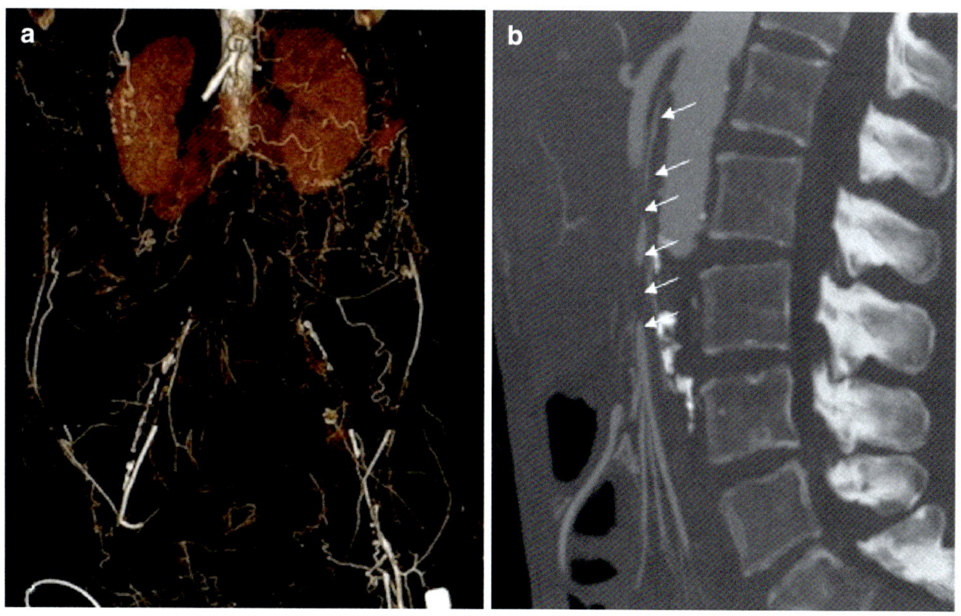

Fig. 6.1 (**a**) Volume rendered image from the CT scan shows aortoiliac and femoral occlusion. (**b**) Maximum intensity projection images showed long segment stenosis and occlusion of the superior mesenteric artery

D. Manoharan (✉)
Department of Vascular Radiology, Sheffield Vascular Institute, Sheffield Teaching Hospitals NHS Foundation Trust, Sheffield, UK
e-mail: dinesh.manoharan@nhs.net

N. Karunanithy
School of Biomedical Engineering and Imaging Sciences, King's College London, London, UK
e-mail: Narayan.Karunanithy@gstt.nhs.uk

G. Geroulakos et al. (eds.), *Mastering Endovascular Techniques*, https://doi.org/10.1007/978-3-031-42735-0_6

Introduction

In 1847, German physiologist Carl Ludwig succeeded in inserting the first catheter into an artery to study its waveform [1]. After nearly a century, another German physician, Werner Forssmann, performed the first human cardiac catheterization by inserting a urinary catheter into his antecubital vein [2]. By the 1950s, angiography by needle puncture had become an established procedure, although it was carried out using large bore needles. As a result, its use was restricted to large arteries, vessel puncture remained difficult, and there were many complications. It was in 1953 when a Swedish interventional radiologist named Sven-Ivar Seldinger developed the technique that bears his name that resulted in the development of a safe method for accessing arteries [3].

The "Seldinger Technique" involves five steps and is the basis for any vascular access procedure in modern medicine (Fig. 6.2).

Step 1: Puncture the vessel of interest using a Trocar needle

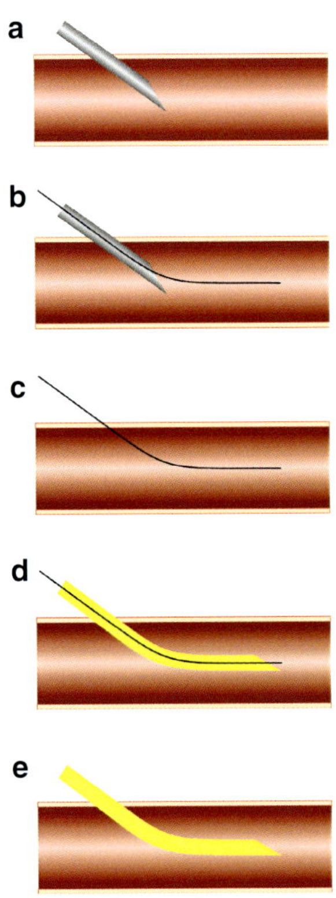

Fig. 6.2 The steps of Seldinger technique. (**a**) Puncture the vessel of interest using a Trocar needle. (**b**) Insertion of a guidewire. (**c**) Removal of the trocar needle. (**d**) Insertion of a sheath/catheter/cannula over the guidewire. (**e**) Removal of the guidewire. Reprinted by permission from Cupr78up, CC BY-SA 3.0 <https://creativecommons.org/licenses/by-sa/3.0>, via Wikimedia Commons. https://commons.wikimedia.org/wiki/File:Seldinger.png

Table 6.1 Suggested toolkit for vascular access

	Manufacturer	Size/length
Arterial access (CFA/PA/SCA)		
Access Needle	Any	18G/7 cm
Micropuncture access set	Cook	Needle 21G/7 cm Wire 0.18 in/40 cm Outer catheter 4 F/10 cm
Standard sheaths	Any	Depending on the size of the vessel and needs
Fixed Core wire (J tip)	Cook	0.021 in/180 cm
Arterial access RA/BA		
Transradial introducer Kit	Terumo	Needle: 22G/21G/20G/38 mm Sheath: 4 to 7 Fr/10 cm Guidewire: 0.018/0.021/0.25/45 cm
Arterial access (Tibials/Pedal)		
Micropuncture Pedal Introducer access set	Cook	Needle 21G/4 cm Wire 0.18 in/40 cm Outer catheter 4 F/7 cm
Venous access (larger veins)		
Access Needle	Any	18G/7 cm
Fixed Core wire (J tip)	Cook	0.021 in/180 cm
Standard sheaths	Any	Depending on the size of the vessel and needs
Venous access (smaller veins)		
Micropuncture access set	Cook	Needle 21G/7 cm Wire 0.18 in /40 cm Outer catheter 4 F/10 cm

Step 2: Insert a guidewire and advance it into the lumen
Step 3: Remove the trocar needle over the guidewire
Step 4: Insert a Sheath/catheter/cannula over the guidewire into the vessel lumen
Step 5: Remove the guidewire, and use the sheath/cannula/catheter remaining in-situ to perform the intended procedure.

Nowadays, it is possible to access arteries and veins at various sites throughout the body for a range of procedures. This chapter describes the techniques for safe and reliable vascular access, key tips & tricks and complications to be aware of. A list of suggested hardware required is summarized in Table 6.1.

Arterial Access

Common Femoral Artery

The common femoral artery (CFA) is the continuation of the external iliac artery under the inguinal ligament after the take-off of the deep and superficial inferior epigastric arteries (also known as the hypogastric artery) [4]. The CFA passes above the medial aspect of the femoral head before

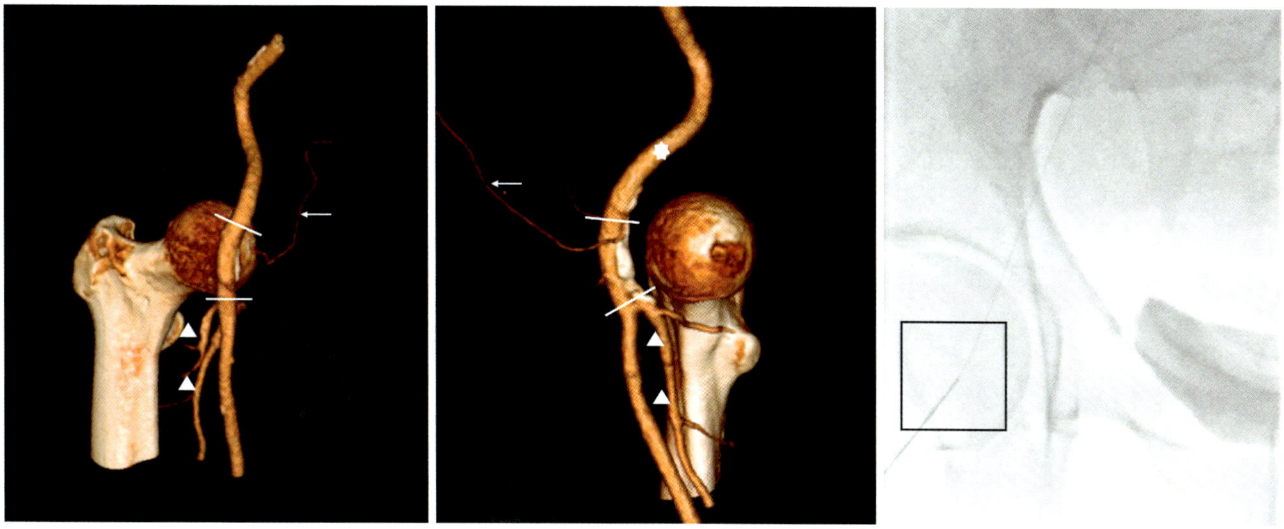

Fig. 6.3 (**a** & **b**): Volume rendered coronal (**a**) and sagittal (**b**) images of the right common femoral artery (CFA, between the white lines) over the femoral head. The inferior epigastric artery (white arrow) and profunda femoral arteries (arrowheads) are shown. The external iliac artery (star in **b**) can be seen diving deep into the pelvis superior to the femoral head. (**c**) Fluoroscopic image showing the ideal access site of the CFA over the femoral head (Black box)

dividing into the superficial and profunda femoral arteries. The ideal access site for the CFA is over the femoral head, which can be used to compress against for haemostasis after the procedure (Fig. 6.3). Additionally, the presence of a fibrous femoral sheath overlying this segment contributes to the prevention of pseudoaneurysm formation. It is important to note that access above this site has no posterior support (artery dives into the pelvis) and thus the risk of haemodynamically significant bleeding after a suprainguinal puncture is much higher (Fig. 6.3b). Typically the CFA lies between the femoral nerve laterally and the common femoral vein (CFV) medially. However, the CFA can overlap the CFV and hence posing a potential risk of arteriovenous fistulas [5].

Technique

Access can be obtained by palpation only, using ultrasound or fluoroscopy [6]. Blind access is done by puncturing the CFA at the point of maximum pulsation. This site corresponds to the midpoint of CFA in 92.7% of cases [7]. Use of ultrasound (US) or fluoroscopy is more accurate to locate the femoral head with US-guided access proven to have fewer complications [8, 9]. Local anaesthesia (lidocaine only for short procedures and lidocaine mixed with bupivacaine for prolonged procedures) is administered to the skin and soft tissues up to the adventitia of the anterior wall of the artery. Typically, 18 G needles are used for access, and the skin entry site is 2 cm below the middle third of the femoral head. From this site of skin entry, the access needle should penetrate the CFA over the middle third of the femoral head at about a 45-degree angle for typical 'retrograde access'(Fig. 6.3c). A double wall penetration technique that penetrates the posterior wall of the artery and hits the femoral head is useful when performing access without US. Following this, the needle is withdrawn until blood flows back and the guidewire is inserted. A single wall puncture can now be easily performed with US guidance, which reduces the possibility of bleeding during a puncture. Furthermore, US can help avoid diseased segments of vessels to prevent complications such as dissection, plaque shift and closure device-related events [9]. US guidance also has a greater first-attempt success rate, lower risk of venous puncture, shorter time to access and lower number of attempts [10].

Tips and Problem Solving

There is an increased risk of bleeding complications following the procedure in patients with a large body habitus [11, 12]. It may be possible to reduce complications in such patients by taping the belly so that it does not overhang the groin (or using dedicated panniculus retractor kit), using contralateral access rather than antegrade access, and using closure devices.

When accessing arteries, it is best to avoid heavily calcified wall segments. This is because achieving haemostasis after a procedure is more difficult as the arteries are more rigid, the needle makes irregular tears in the vessel wall and closure devices do not work or will be suboptimal. Additionally, displacement of plaques could result in arterial occlusion.

For antegrade access, in addition to the above challenges, the possibility of the wire entering the profunda femoral rather than the superficial femoral artery can be encountered. Use of US guidance is highly recommended. Under US guidance, the wire can be manipulated into the SFA. Alternatively, an angiogram through the dilator to show the bifurcation point, slowly withdrawing the dilator to just above the bifurcation and then manipulating an angled hydrophilic wire into the SFA will allow suitable access to be achieved.

Radial Artery

Since Campeau first described radial artery access in 1989, it has become increasingly popular for cardiac and neurological interventions. [13]. In recent years, it has gained wide usage for body interventions as well as more studies have shown it to be safe and effective. In RIVAL, ACCESS, RIFLE-STEACS, and MATRIX randomized trials involving more than 16,000 patients, radial artery access was associated with fewer access-related complications without any difference in stroke incidence [14–18]. As it is very easy to achieve haemostasis, radial access is particularly beneficial for patients with large body habitus, thrombocytopenia, or taking anticoagulants [19]. Further, placing the arm abducted 45–90° on an arm board has been shown to reduce the radiation exposure to the operator threefold, and radial access has also been associated with better patient satisfaction, likely as a result of early mobility after the procedure [20].

Radial access does have some limitations. As a result of the smaller calibre of the artery resulting in a greater sheath/vessel ratio, there is a slightly higher risk of vessel thrombosis and spasm. The risk is negated using the techniques described below. Another limitation is the relatively lower pushability and torqueability owing to the longer device lengths. However, newer dedicated devices are in development to address these challenges. Additionally, performing intra-procedural dynaCT may not be possible with the arm abducted and hence it may be more appropriate to keep the hand across the body during this segment of the procedure. (Fig. 6.4).

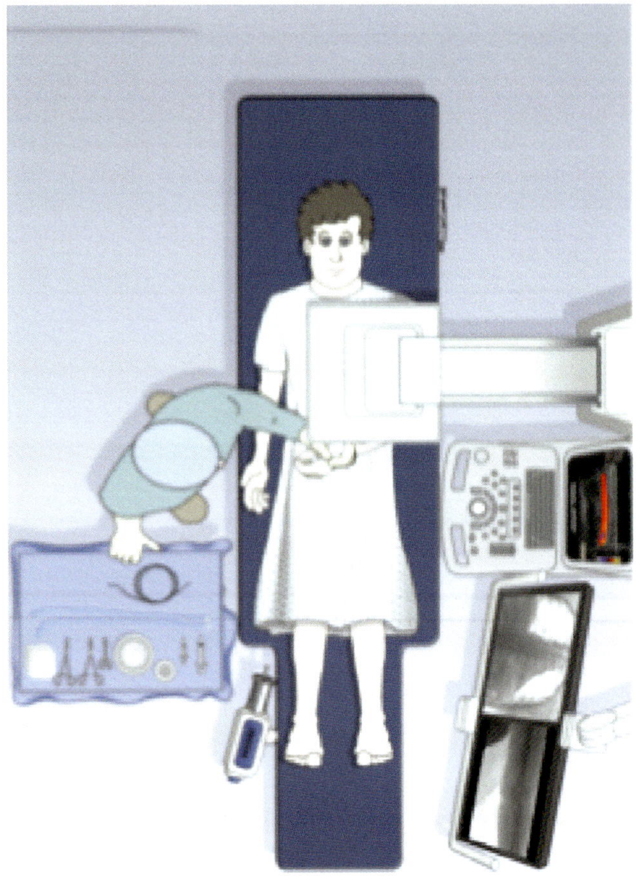

Fig. 6.4 Illustration showing alternate room set-up for left distal radial artery access. Reprinted by permission from Springer Nature: Springer Nature, CardioVascular and Interventional Radiology, Left Distal Transradial Access (ldTRA): A Comparative Assessment of Conventional and Distal Radial Artery Size, Anastasia Hadjivassiliou et al, License number: 5391931202441(2020)

Pre-Procedure Assessment

While Allen Test and Barbeau Test were historically used to assess the palmar arch, recent studies have shown that there is no significant difference in thumb capillary lactate levels, grip strength, or incidence of ischemia between individuals with normal and abnormal results [17, 21]. Hence, an abnormal Allen or Barbeau test should not preclude radial access. Nonetheless, before access can be gained, a US evaluation must be performed to determine the calibre and rule out radial loops and blocked arteries.

Technique

Choosing between right and left radial access depends on the operator's preferences and the procedure involved. Usually, left radial access is used since it follows a straighter path into the thoracic aorta and avoids the arch, reducing the risk of stroke. It is possible to access the radial artery proximally (at the wrist) or distally (at the snuff box) (Fig. 6.5). As a result of its larger calibre, it is more common to use the proximal artery, but some cases of radial occlusion have been associated with it since it causes complete occlusion of the outflow throughout the procedure [22]. Since the superficial palmar arch arises before the access site, radial occlusion is rare in the distal radial artery access.

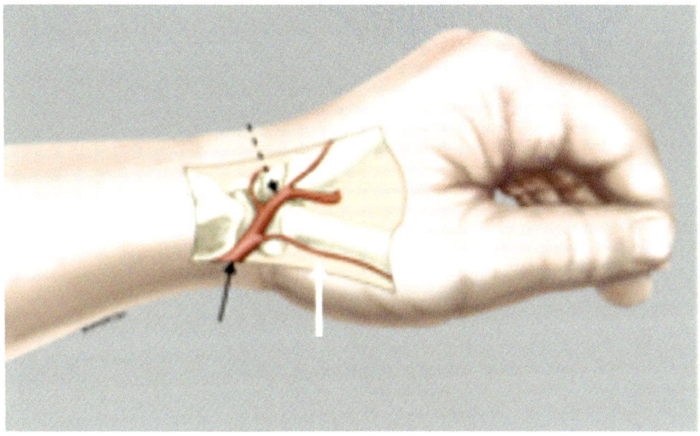

Fig. 6.5 Radial artery anatomy at the wrist demonstrating the access sites at the proximal (solid black arrow) and distal (dotted arrow) site at the anatomical snuff box distal to the superficial radial artery take-off (white arrow). Adapted by permission from Springer Nature: Springer Nature, CardioVascular and Interventional Radiology, Left Distal Transradial Access (ldTRA): A Comparative Assessment of Conventional and Distal Radial Artery Size, Anastasia Hadjivassiliou et al, License number: 5391931202441(2020)

Local anaesthetic (lignocaine) mixed with some isosorbide dinitrate (or Glyceyl Trinitrate) is typically administered to reduce vasospasm during access. A short bevelled 21G needle usually found in dedicated radial access sets should be used. The needles found in routine micropuncture access typically have longer bevels, making them unsuitable for radial arteries with small diameters. Single-wall access is recommended under direct US guidance. The guidewire is advanced slowly and no resistance to advancing it should be encountered. If resistance is encountered, the guidewire should be removed and the needle bevel adjusted to confirm satisfactory pulsatile flow. Dedicated radial access sheaths do not require a skin incision for insertion over the guidewire. Once the sheath is in situ a cocktail of drugs is given to prevent radial artery spasm and thrombosis. At our institute, the "radial cocktail" consists of 3000 IU of heparin, 400 ug of Isosorbide Dinitrate and 2.5 mg of Verapamil. The cocktail is drawn up in a 20 mL luer lock syringe. Blood is first aspirated into the syringe to allow haemodilution of up to 20 mL total volume and then very slowly injected through the sheath.

On completion, a radial access closure device is placed to achieve haemostasis. A protocol for the gradual deflation of the balloon is determined.

Tips and Problem Solving

Approximately 15% of patients experience radial artery spasms, which are almost always transient [23]. It manifests as forearm discomfort and difficulty in manipulating wires and catheters. Predictors of such an event include radial artery anomaly, small radial artery diameter, female sex, multiple prior interventions, multiple access puncture attempts, and pain during radial artery cannulation [24–26]. The condition is often resolved with the administration of nitrates. Second alternative access should be used if the spasm does not resolve optimally despite the drugs. Further manipulation against resistance may result in rare and serious complications such as perforation, sheath entrapment, or an avulsion endarterectomy [27].

Brachial Artery Access

The brachial artery offers all the advantages of radial access, as well as the possibility of inserting a larger sheath. However, it has a high risk of complications. This is because the axillary and brachial arteries are in an enclosed space called the medial brachial fascial compartment [28]. Within this compartment are the median and ulnar nerves which often get compressed when a hematoma forms in the compartment. Access of the brachial artery in the antecubital fossa may be safer as the nerves are further apart from the brachial artery at this level [28]. Reported rates of brachial artery access major complications are approximately 5% [29]

Technique

Radial and ulnar pulses should be checked before the procedure. Again, the left brachial is preferred due to the straighter course for body interventions and the avoidance of the arch. Nevertheless, the right side may be preferred if the patient is left-handed, has a subclavian occlusion, or because it is easier to access for the operator. A 21G micropuncture needle is used to puncture the artery overlying the humerus a few centimetres above the antecubital crease. Single wall puncture should be used, and multiple attempts avoided to prevent bleeding risk that may lead to compartment syndrome. After the procedure haemostasis is usually achieved by manual compression. Closure devices can be used if the artery is of large calibre but it increases the risk of occlusion [30]. We apply a compression bandage after haemostasis is obtained for four hours, checking the pulses in the brachial and radial arteries at regular intervals. It is important to educate the patients to report any swelling, pulsatile mass, pain at the puncture site or arm, tingling or weakness they may experience. Any neurological symptoms should be taken seriously and acted on immediately to prevent long-term sequelae.

Popliteal Access

The popliteal artery is usually used as an alternate method to cross a chronic total occlusion when antegrade approach fails. Its disadvantages include the need to often place the patient in a prone position, difficulties in achieving haemostasis, and inability to use most closure devices. As newer techniques such as tibial access have emerged that do not present these issues, popliteal access should be reserved for exceptional situations. It can also be accessed when there is a single vessel run-off to the foot. US is mostly needed as it is difficult to palpate and to avoid injury to the sciatic and tibial nerves which run in close relation to the vessel. Haemostasis can be usually obtained by manual compression and use of closure devices is off-label currently.

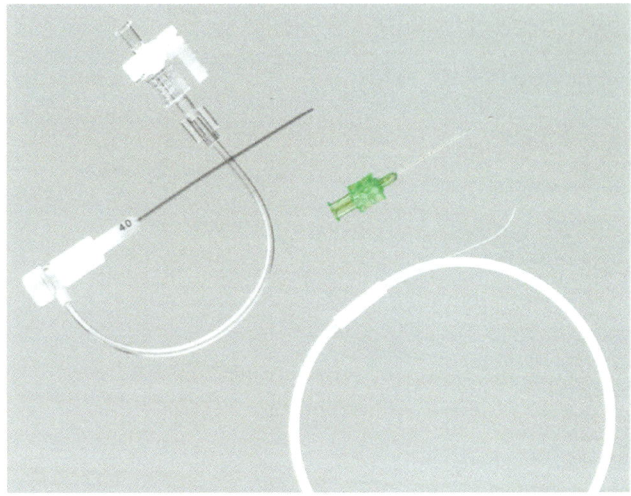

Fig. 6.6 Pedal access set from Cook Medical Inc. (Limerick, IR)

Tibial/Pedal Access

In 1990, Iyer et al. described the use of retrograde access of the tibial vessels using surgical cut down when conventional antegrade access to cross the occlusion failed [31]. Surgical cut down has been superseded by percutaneous approaches. The rate of major complications is reported to be 2–3% and is higher when larger sheaths are used [32].

Technique

Tibial and pedal vessels are small and often calcified in the cohort of patients where use is necessitated. Hence the vessel tends to slip away from the puncture needle. Additionally, overuse of local anaesthetic should be avoided to prevent vessel compression. US is vital although fluoroscopy may be helpful in most cases in heavily calcified vessels. Posterior tibial artery near the ankle may be straightened with dorsiflexion, while dorsalis pedis and anterior tibial arteries may be straightened by plantar flexion. As far as possible, access should be gained to the tibial and pedal vessels in a non-diseased segment.

The use of a short-bevelled needle with an echogenic tip facilitates good needle visualization. They are usually available as part of a pedal access kit (Cook Medical Inc. (Limerick, Ireland) (Fig. 6.6). A short bevelled echogenic tip 21G needle with a 4 cm length is used. After obtaining access, a 0.018 wire is advanced under fluoroscopic guidance followed by the insertion of a 4F sheath or dilator. As the access is only required to cross the occlusion, larger sheaths are avoided as intervention is then performed from the antegrade access [33]. After the occlusion is crossed and antegrade access obtained, the dilator can be removed, and haemostasis is typically achieved with gentle manual compression only. Occasionally additional prolonged inflation of a balloon across the access site may be required.

Tips and Tricks

Heparinization of the patient is essential during pedal/tibial access to prevent vessel occlusion. Target Activated clotting time should be more than 200s. Although the benefit of giving a spasmodic agent is unproven, we believe that such an agent (Nitroglycerin and/or verapamil) may be administered when it is necessary to insert a sheath [34]. Upon removal of the sheath, some amount of spasm is common, so an anti-spasmodic medication may be administered just before removal.

Venous Access

Venous access can be obtained in the periphery or central site. Central access is needed when peripheral access is inadequate or when there is a need for long-term administration of drugs, parenteral nutrition, dialysis etc. Venous access is also used to perform various interventions such as venoplasties, transjugular intrahepatic portosystemic shunt creations, etc. Some common sites of venous access are described in this chapter.

Internal Jugular Vein

The internal jugular vein (IJV) arises at the base of the skull and terminates at the level of the clavicle where it joins the subclavian vein. It lies within the carotid sheath in close proximity to the common carotid artery. The relation of the

IJV to the carotid artery is important, as inadvertent arterial injury is a relatively common complication. Right IJV has a straighter course to the heart and is usually preferred for venous access compared to the left.

Technique

Blind access to the IJV without image guidance is possible but fraught with complications and not recommended [35, 36]. The neck is gently extended, and the head is turned to the opposite side. It is important not to flex the neck by placing large pillows underneath since this will shorten the neck and make access more difficult. Generally, the puncture site is at the triangle formed by the two heads of the sternocleidomastoid and the clavicle. Since the IJV is superficial and often collapsible with minimal pressure, it is often necessary to place the US probe gently and to use less local anaesthesia. Intraluminal access is confirmed by gentle aspiration of blood. Standard Seldinger's technique is then used to insert a sheath or a catheter. It is important to confirm the passage of the guidewire into the vein under US as inadvertent carotid or vertebral access may lead to stroke. Advancing the guidewire into the IVC is recommended before placing the sheath or catheter to prevent arrhythmias and inadvertent cardiac injury.

Tips and Tricks

When a patient is dehydrated or malnourished, the vein may be collapsed and difficult to access. In such cases placing the patient completely supine or even in a slight Trendelenburg position or Valsalva might help distend the vein. In the event of an accidental puncture of the carotid artery, the needle should be withdrawn immediately, and manual pressure should be applied. If a sheath or catheter has been inserted, it should be left in place and only removed after careful consideration of the optimum endovascular or surgical method for doing so [37].

Femoral Vein

The common femoral vein is accessed over the femoral head. This can be done by inserting the needle just medial to the maximum femoral artery pulsation in emergencies. However, in most instances, US-guided access is the standard. Asking the patient to perform a Valsalva manoeuvre or placing them in a reverse Trendelenburg position may help in distending the vein. Standard Seldinger's technique is then used. It was previously thought that insertion of non-tunnelled lines in the femoral vein was associated with increased risk of Central Line-Associated Bloodstream Infection (CLABSI); however, recent studies have questioned this notion [38]. Multiple studies have now shown that femoral and internal Juglar lines have similar higher infection rates compared to subclavian lines [39, 40].

Subclavian Vein

The use of US has significantly improved the chance of success and reduced complication rates with subclavian vein access [41]. Subclavian vein lines do have lower infection rates as the exit site is away from the oral mucosa [38–40]. However, the concern with subclavian vein lines is the mechanical sheer-stress the line will be subject to at the thoracic inlet and the potential for accelerating subclavian vein stenosis making future arm venous access (e.g. haemodialysis) complex. In addition, there is a small but significant risk of pneumothorax and inadvertent arterial injury.

Technique

There are two main approaches for subclavian access via either supra or infraclavicular approaches. The infraclavicular approach is commonly used and the subclavian vein is accessed just below the middle one-third of the clavicle. Technically this segment of the subclavian vein overlies the 1st rib or just lateral to it (axillary vein). Access here has less chance of pneumothorax because of the protection from the underlying rib. The supraclavicular approach is underlying the medial one-third of the clavicle.

A recent systematic review by Nazir et al showed that supraclavicular approach has a higher success rate [42]. However, most of the studies in this review did not use ultrasound for access. A recent randomized trial by Kim et al. concluded that supraclavicular approach took less time to attain a puncture and the complication rate was similar as well [43]. However, we feel that accessing the supraclavicular portion of the subclavian vein should be done with due caution. Unlike infraclavicular approaches, the subclavian vessels are within the thoracic cavity at this site and in direct relation to the adjacent low-pressure pleural cavity [44]. Therefore, if the underlying vein is torn or the artery punctured there is no tamponade effect from the underlying tissues. Moreover, this site cannot be manually compressed or easily accessed by a surgical cut-down. There is also an added risk of pericardial tamponade as the pericardium often extends over the proximal branches of aorta [45].

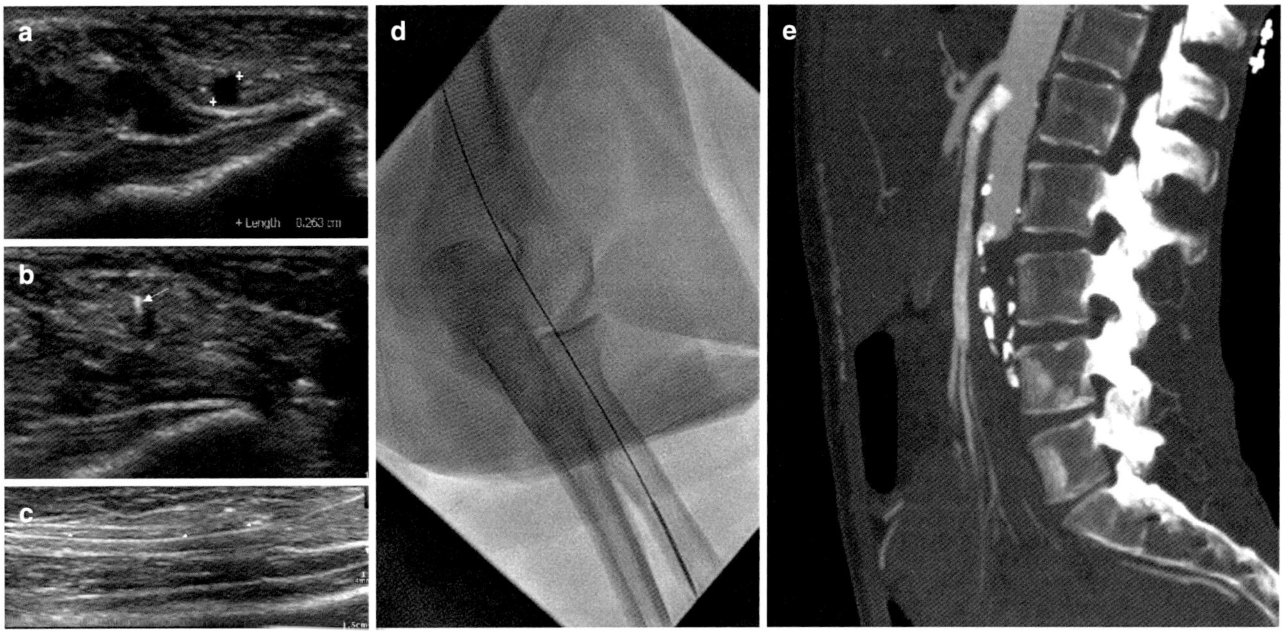

Fig. 6.7 (**a**) US image of the left radial artery at the wrist showing suitable calibre for safe access. Ultrasound guided access of was performed using a 21G needed in the centre of the artery (**b**) and guidewire was inserted (**c**). After insertion of a 6 F sheath, the radial access was used to successfully stent the superior mesenteric artery (**d** and **e**)

Upper Arm Veins

The basilic and cephalic veins are commonly accessed for insertion of peripherally inserted central catheter placements. They may also be used for performing interventions such as central venoplasties and testicular vein embolization. Access to these vessels is usually straightforward using US and Seldinger's technique.

Conclusion

Performing vascular access is a core skill every interventionalist should be familiar with. To achieve safe and successful vascular access, it is crucial to have a thorough understanding of anatomy, be familiar with the use of imaging, and be able to manage complications.

Case Presentation

Continued from page 49

Morphine patient controlled analgesia infusion was started and the entire procedure was performed using local anaesthesia alone. An ultrasound assessment of the left proximal radial artery at the wrist revealed a diameter of 2.63 mm which was suitable for the hardware to be used (Fig. 6.7a). A type-A curve was observed in the Barbeau test. A mixture containing 2 mL of 1% lignocaine and Glyceryl trinitrate was administered around the radial artery and skin as local anaesthetic. A short bevelled 21G needle found in the Terumo RadifocusTM Introducer II Transradial Kit was used to puncture the artery in the centre using ultrasound guidance (Fig. 6.7b). Following this, a 6 Fr sheath was placed and access was secured. We administered the radial cocktail as described above. Through the radial access, the superior mesenteric artery was successfully stented (Fig. 6.7c and d). The sheath was then removed and the haemostasis was secured with a radial band. Upon discharge, she was no longer in pain and her radial access site was alright with good radial pulse.

References

1. Pierre L, Pasrija D, Keenaghan M. Arterial lines. In: StatPearls [Internet]. Treasure Island (FL): StatPearls Publishing; 2022. [cited 2022 Jul 17]. Available from: http://www.ncbi.nlm.nih.gov/books/NBK499989/.
2. Heiss HW, Hurst JW. Werner forssmann: A German problem with the Nobel prize. Clin Cardiol. 1992;15(7):547–9.
3. Seldinger SI. Catheter replacement of the needle in percutaneous arteriography: a new technique. Acta Radiol. 1953;39(5):368–76.
4. Schnyder G, Sawhney N, Whisenant B, Tsimikas S, Turi ZG. Common femoral artery anatomy is influenced by demographics and comorbidity: implications for cardiac and peripheral invasive studies. Catheter Cardiovasc Interv Off J Soc Card Angiogr Interv. 2001;53(3):289–95.
5. Baum PA, Matsumoto AH, Teitelbaum GP, Zuurbier RA, Barth KH. Anatomic relationship between the common femoral artery and vein: CT evaluation and clinical significance. Radiology. 1989;173(3):775–7.
6. Barbetta I, van den Berg JC. Access and hemostasis: femoral and popliteal approaches and closure devices—why, what, when, and how? Semin Interv Radiol. 2014;31(4):353–60.

7. Grier D, Hartnell G. Percutaneous femoral artery puncture: practice and anatomy. Br J Radiol. 1990;63(752):602–4.
8. Fitts J, Ver Lee P, Hofmaster P, Malenka D, Northern New England Cardiovascular Study Group. Fluoroscopy-guided femoral artery puncture reduces the risk of PCI-related vascular complications. J Interv Cardiol. 2008;21(3):273–8.
9. Sorrentino S, Nguyen P, Salerno N, Polimeni A, Sabatino J, Makris A, et al. Standard versus ultrasound-guided cannulation of the femoral artery in patients undergoing invasive procedures: a meta-analysis of randomized controlled trials. J Clin Med. 2020;9(3):677.
10. Seto AH, Abu-Fadel MS, Sparling JM, Zacharias SJ, Daly TS, Harrison AT, et al. Real-time ultrasound guidance facilitates femoral arterial access and reduces vascular complications: FAUST (Femoral Arterial Access With Ultrasound Trial). JACC Cardiovasc Interv. 2010;3(7):751–8.
11. Nguyen P, Makris A, Hennessy A, Jayanti S, Wang A, Park K, et al. Outcomes in femoral access patients with large abdominal circumference. Heart Lung Circ. 2019;1(28):S415–6.
12. Kim M, Kim MA, Kim HL, Lee WJ, Lim WH, Seo JB, et al. Body mass index and the risk of low femoral artery puncture in coronary angiography under fluoroscopy guidance. Medicine (Baltimore). 2018;97(9):e0070.
13. Campeau L. Percutaneous radial artery approach for coronary angiography. Catheter Cardiovasc Diagn. 1989;16(1):3–7.
14. Kiemeneij F, Laarman GJ, Odekerken D, Slagboom T, van der Wieken R. A randomized comparison of percutaneous transluminal coronary angioplasty by the radial, brachial and femoral approaches: the access study. J Am Coll Cardiol. 1997;29(6):1269–75.
15. Mehta SR, Jolly SS, Cairns J, Niemela K, Rao SV, Cheema AN, et al. Effects of radial versus femoral artery access in patients with acute coronary syndromes with or without ST-segment elevation. J Am Coll Cardiol. 2012;60(24):2490–9.
16. Romagnoli E, Biondi-Zoccai G, Sciahbasi A, Politi L, Rigattieri S, Pendenza G, et al. Radial versus femoral randomized investigation in ST-segment elevation acute coronary syndrome: the RIFLE-STEACS (Radial Versus Femoral Randomized Investigation in ST-Elevation Acute Coronary Syndrome) study. J Am Coll Cardiol. 2012;60(24):2481–9.
17. Valgimigli M, Gagnor A, Calabró P, Frigoli E, Leonardi S, Zaro T, et al. Radial versus femoral access in patients with acute coronary syndromes undergoing invasive management: a randomised multicentre trial. Lancet. 2015;385(9986):2465–76.
18. Mason PJ, Shah B, Tamis-Holland JE, Bittl JA, Cohen MG, Safirstein J, et al. An update on radial artery access and best practices for transradial coronary angiography and intervention in acute coronary syndrome: a scientific statement from the American Heart Association. Circ Cardiovasc Interv. 2018;11(9):e000035.
19. Titano JJ, Biederman DM, Marinelli BS, Patel RS, Kim E, Tabori NE, et al. Safety and feasibility of transradial access for visceral interventions in patients with thrombocytopenia. Cardiovasc Intervent Radiol. 2016;39(5):676–82.
20. Yamada R, Bracewell S, Bassaco B, Camacho J, Anderson MB, Conrad A, et al. Transradial versus transfemoral arterial access in liver cancer embolization: randomized trial to assess patient satisfaction. J Vasc Interv Radiol. 2018;29(1):38–43.
21. Bertrand OF, Carey PC, Gilchrist IC. Allen or no Allen: that is the question! J Am Coll Cardiol. 2014;63(18):1842–4.
22. Nairoukh Z, Jahangir S, Adjepong D, Malik BH. Distal radial artery access: the future of cardiovascular intervention. Cureus. 12(3):e7201.
23. Kristić I, Lukenda J. Radial artery spasm during transradial coronary procedures. J Invasive Cardiol. 2011;23(12):527–31.
24. Ruiz-Salmerón RJ, Mora R, Vélez-Gimón M, Ortiz J, Fernández C, Vidal B, et al. Radial artery spasm in transradial cardiac catheterization. Assessment of factors related to its occurrence, and of its consequences during follow-up. Rev Esp Cardiol. 2005;58(5):504–11.
25. Deftereos S, Giannopoulos G, Kossyvakis C, Driva M, Kaoukis A, Raisakis K, et al. Radial artery flow-mediated dilation predicts arterial spasm during transradial coronary interventions. Catheter Cardiovasc Interv. 2011;77(5):649–54.
26. Rathore S, Stables RH, Pauriah M, Hakeem A, Mills JD, Palmer ND, et al. Impact of length and hydrophilic coating of the introducer sheath on radial artery spasm during transradial coronary intervention: a randomized study. JACC Cardiovasc Interv. 2010;3(5):475–83.
27. Dieter RS, Akef A, Wolff M. Eversion endarterectomy complicating radial artery access for left heart catheterization. Catheter Cardiovasc Interv. 2003;58(4):478–80.
28. Lee KS, Sos TA. Brachial artery access. Tech Vasc Interv Radiol. 2015;18(2):87–92.
29. Mantripragada K, Abadi K, Echeverry N, Shah S, Snelling B. Transbrachial access site complications in endovascular interventions: a systematic review of the literature. Cureus. 14(6):e25894.
30. Lupattelli T, Clerissi J, Clerici G, Minnella DP, Casini A, Losa S, et al. The efficacy and safety of closure of brachial access using the AngioSeal closure device: experience with 161 interventions in diabetic patients with critical limb ischemia. J Vasc Surg. 2008;47(4):782–8.
31. Iyer SS, Dorros G, Zaitoun R, Lewin RF. Retrograde recanalization of an occluded posterior tibial artery by using a posterior tibial cutdown: two case reports. Catheter Cardiovasc Diagn. 1990;20(4):251–3.
32. Grözinger G, Hallecker J, Grosse U, Syha R, Ketelsen D, Brechtel K, et al. Tibiopedal and distal femoral retrograde vascular access for challenging chronic total occlusions: predictors for technical success, and complication rates in a large single-center cohort. Eur Radiol. 2021;31(1):535–42.
33. Giannopoulos S, Palena LM, Armstrong EJ. Technical success and complication rates of retrograde arterial access for endovascular therapy for critical limb ischaemia: a systematic review and meta-analysis. Eur J Vasc Endovasc Surg. 2021;61(2):270–9.
34. Sanghvi KA, Kusick J, Krathen C. Retrograde tibio-pedal access for revascularization of lower-extremity peripheral artery disease using a 6 Fr slender sheath: the "Pedal-First" pilot project. Cath Lab Dig. 2018;26(10). Available from https://www.hmpgloballearningnetwork.com/site/cathlab/article/retrograde-tibio-pedal-access-revascularization-lower-extremity-peripheral-artery-disease. [cited 2022 Aug 28].
35. Saugel B, Scheeren TWL, Teboul JL. Ultrasound-guided central venous catheter placement: a structured review and recommendations for clinical practice. Crit Care. 2017;28(21):225.
36. Brass P, Hellmich M, Kolodziej L, Schick G, Smith AF. Ultrasound guidance versus anatomical landmarks for internal jugular vein catheterization. Cochrane Database Syst Rev. 2015; [cited 2022 Sep 4];(1). Available from: https://www.cochranelibrary.com/cdsr/doi/10.1002/14651858.CD006962.pub2/full
37. Ripal T, Gandhi MD. Management of inadvertent placement of a central line in the carotid artery with a closure device and embolic protection device. Cath Lab Dig. 2016; [cited 2022 Sep 3];24(11). Available from: https://www.hmpgloballearningnetwork.com/site/cathlab/article/management-inadvertent-placement-central-line-carotid-artery-closure-device-embolic
38. Marik PE, Flemmer M, Harrison W. The risk of catheter-related bloodstream infection with femoral venous catheters as compared to subclavian and internal jugular venous catheters: a systematic review of the literature and meta-analysis. Crit Care Med. 2012;40(8):2479–85.
39. Parienti JJ, Mongardon N, Mégarbane B, Mira JP, Kalfon P, Gros A, et al. Intravascular complications of central venous catheterization by insertion site. N Engl J Med. 2015;373(13):1220–9.
40. Youn SH, Lee JCJ, Kim Y, Moon J, Choi Y, Jung K. Central venous catheter-related infection in severe trauma patients. World J Surg. 2015;39(10):2400–6.

41. Brass P, Hellmich M, Kolodziej L, Schick G, Smith AF. Ultrasound guidance versus anatomical landmarks for subclavian or femoral vein catheterization. Cochrane Database Syst Rev. 2015; [cited 2022 Sep 4];(1). Available from: https://www.cochranelibrary.com/cdsr/doi/10.1002/14651858.CD011447/full
42. Nazir A, Niazi K, Zaidi SMJ, Ali M, Maqsood S, Malik J, et al. Success rate and complications of the supraclavicular approach for central venous access: a systematic review. Cureus. 2022;14(4):e23781.
43. Kim YJ, Ma S, Yoon HK, Lee HC, Park HP, Oh H. Supraclavicular versus infraclavicular approach for ultrasound-guided right subclavian venous catheterisation: a randomised controlled non-inferiority trial. Anaesthesia. 2022;77(1):59–65.
44. Bannon MP, Heller SF, Rivera M. Anatomic considerations for central venous cannulation. Risk Manag Healthc Policy. 2011;13(4):27–39.
45. Bodenham A. Reducing major procedural complications from central venous catheterisation. Anaesthesia. 2011;66(1):6–9.

Techniques and Devices for Vascular Closure

7

S. A. Salazar and A. S. Niekamp

Case Presentation
A 73-year-old male presents for elective repair of a 6.5 cm infrarenal abdominal aortic aneurysm (AAA) diagnosed on a surveillance CT-scan from an outpatient interventional radiology clinic. The patient weighs 82 kilograms and after review of the CT-scan, it was noted that the common femoral arteries were widely patent bilaterally without significant atherosclerotic disease. The arteries are also located 7 mm from the skin surface. The patient also has a history of chronic obstructive pulmonary disease, coronary artery disease, diabetes, hypertension, and hyperlipidemia. Vascular surgery was consulted for evaluation for potential open surgical repair. The patient was deemed not a surgical candidate and the decision was made to proceed with endovascular repair of the AAA. The decision for percutaneous access was made as opposed to a surgical cutdown due to the patient's favorable anatomy.

Background

With the widespread adoption and evolution of minimally invasive techniques, hemostasis following percutaneous vascular access is an important consideration for interventionists, such as interventional radiologists, interventional cardiologists, and vascular surgeons. For any vascular port-of-entry, appropriate measures are necessary to ensure the proper closure of the vessel and mitigate the incidence of complications. Historically, the gold standard for achieving hemostasis after vessel puncture has always been manual compression of the access site. However, there is a variety of factors that limit the utility of manual compression, including time demands for healthcare staff, the need for extended bedrest, and in some cases large patient body habitus. Successful hemostasis with manual compression requires labor-intensive, consistent pressure applied at the access site often for upwards of 15–20 min and patient bedrest in the range of 4–8 h, with time dependent on the use of anticoagulation or size of the vascular sheath deployed. However, though manual compression is considered a safe option and remains the standard of care [1], alternatives exist in the modern era to help avoid these drawbacks.

Vascular closure devices (VCDs) were first introduced in the 1990s, with the first of these devices VasoSeal (St. Jude Medical, St. Paul, MN) receiving FDA approval in 1995 [2]. Since then, they have become increasingly popular among interventionists. This class of devices allow for the closure of arteriotomy sites following percutaneous vascular access using several different mechanisms of action. Although various studies have shown that VCDs do not significantly decrease the risk of vascular complications, their main benefits lie in the reduction of time needed for hemostasis and patient bedrest [3]. This is thought to result in decreased healthcare costs and time demands on healthcare staff, as well as improvement in patient comfort and satisfaction in the postoperative period [4].

Today, a large variety of VCDs are available on the market. As such, it is important for any interventionist to be aware of the indications of available devices and to know which patients would benefit most from their use.

Indications for Intervention

Application of VCDs is indicated in any patient requiring percutaneous vascular access closure who would benefit from reduction in time to hemostasis and ambulation. In patients undergoing procedures requiring large bore access,

The original version of the chapter has been revised. A correction to this chapter can be found at https://doi.org/10.1007/978-3-031-42735-0_83

S. A. Salazar (✉)
Florida International University Herbert Wertheim College of Medicine, Miami, FL, USA
e-mail: ssala060@med.fiu.edu

A. S. Niekamp (✉)
Miami Cardiac and Vascular Institute, Baptist Health South Florida, Miami, FL, USA
e-mail: AndrewSN@baptisthealth.net

G. Geroulakos et al. (eds.), *Mastering Endovascular Techniques*, https://doi.org/10.1007/978-3-031-42735-0_7

such as endovascular aortic aneurysm repair (EVAR), thoracic endovascular aortic aneurysm repair (TEVAR), and transcatheter aortic valve replacement (TAVR), these devices are routinely used and are particularly beneficial due to the larger diameter of the vascular sheath complicating closure [5–7]. Patients treated with anticoagulation are also excellent candidates for VCD use due to medication effects prolonging time to hemostasis with manual compression [1]. Individuals in which extended bed-rest is difficult, such as patients with musculoskeletal pain, pressure ulcers, orthopnea, or dementia, may benefit from the decreased time to ambulation as well [3].

Preoperative Preparation

Timing of Intervention

Needless to say, vessel closure in the setting of endovascular procedures is performed once all necessary aspects of the case are complete and the interventionist deems it is appropriate to close. However, it is important to note that some commonly used VCDs, such as the suture-mediated Perclose ProGlide and Prostar XL (Abbott Vascular, IL, USA), must be deployed prior to the procedure ("preclose" technique), with the subsequent suture knot tightened after completion of the case [8, 9]. In contrast, other devices such as Angio-Seal (Terumo Medical, BE, EU) are able to be applied post-procedurally [10]. Thus, it is necessary for the interventionist to have a suitable plan for vessel closure prior to the procedure, become familiar with the devices at their disposal, and understand how they are meant to be deployed.

Imaging

The most commonly selected site to obtain peripheral vascular access remains the common femoral artery, although alternative access sites are becoming increasingly popular. For patients undergoing diagnostic or interventional procedures, evaluation of the access site with angiography is typically recommended prior to VCD deployment [11]. Information obtained via angiography, such as the presence of vessel calcification, small luminal diameter at the access site, and vessel tortuosity, are all important considerations to assess if VCD usage is safe. In addition to understanding the target anatomy, patients undergoing large bore intervention such as EVAR or TAVR may also have a pre-operative CT to assist in vascular mapping and determining approach to vascular access [12].

Ultrasound capability should be available in the procedure room to assist with proper vascular access (i.e. to avoid multiple sticks, confirm location) and to guide appropriate VCD placement post-procedurally. With certain VCDs, such as the Manta Vascular Closure Device (Teleflex, PA, USA), radio-opaque markers are present to help demarcate the access site [13].

Risk Assessment

Among patients in whom a VCD is being considered, there are many factors affecting successful device deployment and rate of vascular complications. Proper patient selection is essential and the interventionist should exercise their own clinical judgement based on the instructions for use (IFU) and profile of the device being implemented, their own experience with the particular device, and finally individual case characteristics. Overall though, given the number of VCDs currently available, there are no absolute contraindications for most devices and they are each better tailored for certain cases. Relative contraindications for the usage of VCDs include access site above the inguinal ligament, vessel wall calcification, severe peripheral arterial disease, or small luminal diameter at the access site (i.e. distal to the common femoral artery bifurcation). In most existing evidence, patients with these access site characteristics are excluded due to their presumed high-risk [3]. Use of VCDs should also be avoided in patients who develop an overlying hematoma during the procedure.

The following are characteristics that may predispose to complications or adverse outcomes associated with VCDs:

Patient

- Obesity.
- Female gender.
- Peripheral vascular disease (PVD).
- Tortuous vessels.
- Presence of vessel calcification at the access site.
- Development of overlying hematoma.
- Emergency procedure.
- Undergoing multiple procedures (i.e. long procedure time).

Operator/Center

- Operator inexperience.
- Forceful manipulation of access site.
- Multiple attempts at vascular access.
- Access obtained through the lateral vessel wall.
- Inappropriate access in the iliac arteries (i.e. above the inguinal ligament).
- Small luminal diameter at the access site.
- Lack of presence of a multidisciplinary team (i.e. surgery, anesthesia).
- Lack of availability of blood products.

Other

- Large bore vascular access for EVAR, TAVR, etc. (sheath diameter > 8 F).
- Inaccurate vessel measurement (i.e. pre-operative CT).
- Safety profile of VCD.
- Device failure.

Endovascular Strategy and Technique

Basic Endovascular Toolkit

- Specific vascular closure device kit.
- Vascular sheath.
- Guidewire.
- Hollow needle.

A list of suggested materials is included in Table 7.1.

Currently, there are sixteen VCD products available on the market. The two broadest categories into which these devices can be sorted include those with passive or active mechanisms. Passive VCDs use methods to mechanically assist with manual compression or to apply pro-thrombotic factors to the access site and adjacent tissue tract. In contrast, active VCDs involve the placement of implanted materials such as sutures or vessel clips in order to obtain hemostasis. With the exception of a few, all of these devices are indicated for closure of femoral artery access sites – although their use has been described in other locations with varying degrees of success.

Table 7.1 Suggested endovascular toolkit for management of vascular closure following EVAR

	Manufacturer	Size/length
Vascular closure device		
Perclose Proglide or Prostyle	Abbott vascular	5–21F (max OD 26F)
Prostar XL	Abbott vascular	8.5–24F
Manta vascular closure device 18F	Teleflex	15–18F (max OD 25F)
Wires		
Standard access wire	Any	≤0.035″
Sheaths		
Standard large-bore access sheath	Any	18–24F
Stent-graft	Any	Sized to access vessel
Thrombolytic catheters	Any	
Thrombectomy devices	Any	

This list is based on the author's experience and includes items needed during device deployment and treatment of possible major complications

The majority of VCDs are indicated for the closure of 5–8 F puncture sites. In cases of large bore access, there are currently five devices indicated to close puncture sites greater than 8 F. These include the Perclose ProGlide, Perclose Prostyle, Prostar XL, Vascade MVP Venous Vascular Closure System (VVCS), and Manta Vascular Closure Device. A review of available devices and decision-making regarding technique will be discussed in this section.

Suture-Based Closure Device: Perclose ProGlide and Perclose ProStyle

The Perclose ProGlide (Abbott Vascular, IL, USA; Fig. 7.1) device delivers a single pre-tied polypropylene monofilament suture at the arteriotomy site [3, 10]. Puncture sizes in the range of 5–8 F can be closed using a single device; however, multiple devices can be employed to close sites greater than 8 and up to 21 F by arterial sheaths and up to 24 F by venous sheaths – allowing it to be applied in cases necessitating large bore access. The device uses a "pre-close" system, where the sutures are deployed prior to the procedure and subsequently tightened upon completion, and allows for maintenance of guidewire access.

Perclose ProStyle (Abbott Vascular, IL, USA) represents a newer iteration of the popular Perclose ProGlide device. Changes include higher tensile strength needles and differences in design to improve ease of use; however, it employs the same suture deployment mechanism as described in its predecessor. The maximum guidewire compatibility for both of these devices is 0.038 inches.

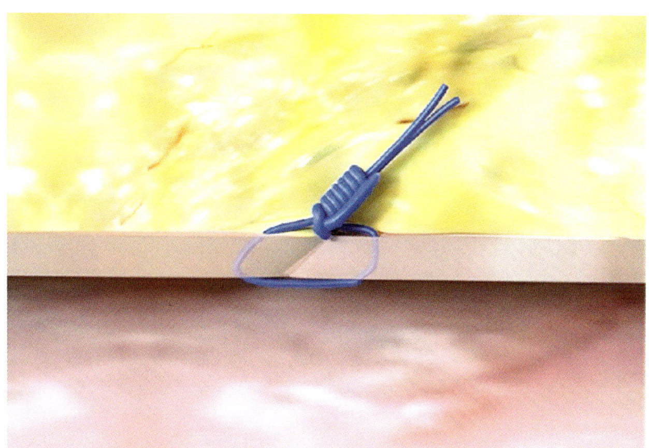

Fig. 7.1 Arteriotomy site with a single pre-tied suture knot deployed using Perclose ProGlide device

Suture-Based Closure Device: Prostar XL

The Prostar XL device (Abbott Vascular, IL, USA) deploys two braided polyester sutures at the access site and is indicated for closure of common femoral access sites between 8.5–10 F with a single use [3, 14, 15]. Unlike Perclose ProGlide, the suture deployment mechanism can be repeated to place more sutures at the access site without the need for additional devices. ProStar XL can be applied for up to 24 F vascular sheaths and employs a similar "pre-close" technique to the Perclose devices. The maximum guidewire compatibility is 0.035 inches.

Clip-Based Closure Device: StarClose SE

StarClose SE (Abbott Vascular, IL, USA; Fig. 7.2) is a clip-based VCD that closes the arteriotomy site using an implantable 4 mm nitinol clip [3, 14, 15]. The disc shaped metal clip left behind remains extravascular. The device results in immediate mechanical closure of the site without inflammatory response, allowing for hemostasis to be confirmed on the table. Considered simple to use relative to other devices, it can be used to close 5–6 F access sites – although it has been used in up to 8 F access sites - and has a maximum wire compatibility of 0.035 inches.

Plug-Based Closure Device: Exoseal

Exoseal (Cordis Corp., NJ, USA) is a plug-based device that deploys a synthetic, bioabsorbable polyglycolic acid (PGA) plug [3, 15]. This device is delivered through a preexisting procedural sheath (with a maximum length up to 12 cm) and has a visual marker to ensure the plug is placed in the intended location extravascular to the femoral artery access site. There is no intravascular component to this device and the plug is typically resorbed within 60–90 days. This device is indicated for closure of 5–7 F puncture sites.

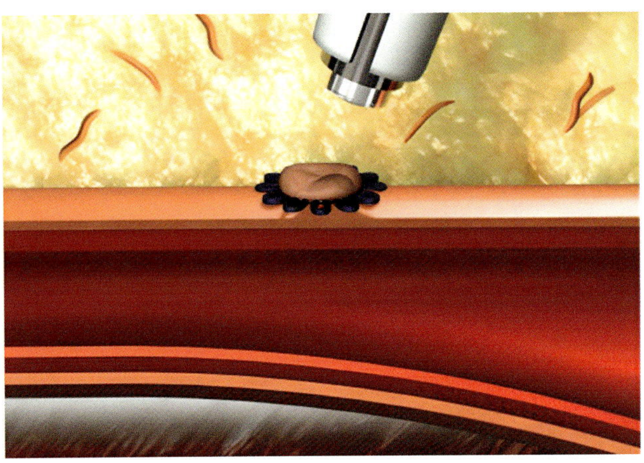

Fig. 7.2 Extravascular clip deployed at femoral arteriotomy site using StarClose SE device

Plug-Based Closure Devices: MynxGrip and Mynx Control Vascular Closure Devices

The MynxGrip and Mynx Control Vascular Closure Devices (Cordis Corp, NJ, USA; Fig. 7.3) both function by delivering a polyethylene glycol (PEG) based sealant to the extravascular space above the access site [16, 17]. Simultaneously, a semi-compliant balloon is inflated in the vessel lumen, promoting hemostasis and allowing the sealant to expand across the puncture surface and tissue tract. The sealant is completely reabsorbed within 30 days. Both devices can be utilized with 5–7 F puncture sites and procedural sheaths up to a length of 12 cm. The Mynx Control includes a streamlined deployment system, whereas the MynxGrip can be used for both arteriotomy and venotomy sites.

Manual Assistive Closure Devices: Cardiva Catalyst II and III

Cardiva Catalyst II (Haemonetics, MA, USA; Fig. 7.4) is a manual assistive device available for closure of 5–7 F puncture sites [3, 15]. After introduction through the pre-existing vascular sheath, the device houses a disc technology that is retracted and opposed to the arteriotomy site to provide temporary hemostasis. Afterwards, a hemostatic coating on the wire is exposed to the site and tissue tract, activating the coagulation cascade and promoting platelet aggregation.

A version of Catalyst II, the Catalyst III (Haemonetics, MA, USA) is intended for use in patients receiving intraprocedural heparin. With this device, the same technology is employed with the addition of protamine sulfate, allowing for heparin deactivation [3, 15]. Both of these devices,

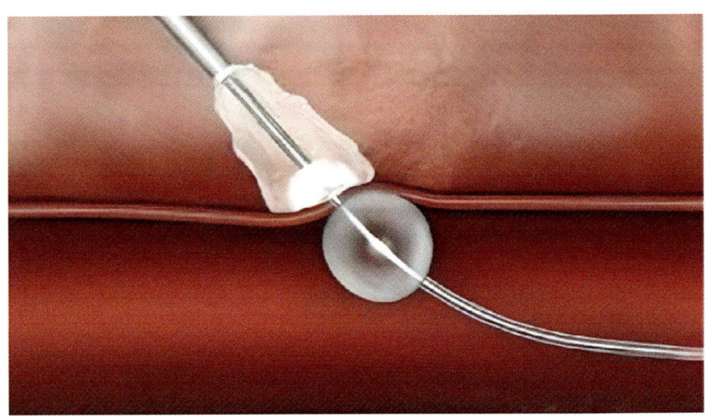

Fig. 7.3 MynxGrip device with semi-compliant balloon inflated in vessel lumen and delivery of PEG-based sealant above vessel

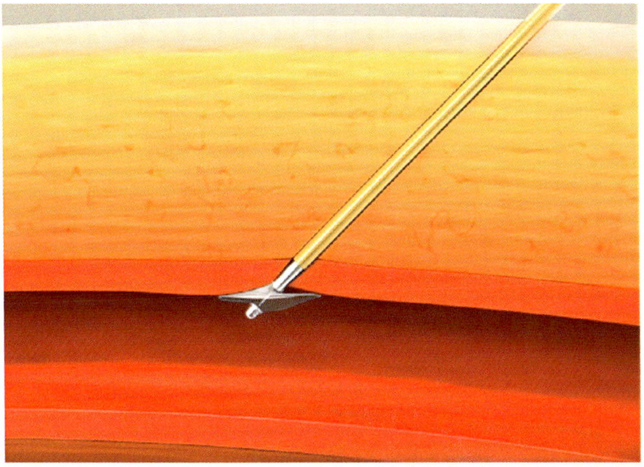

Fig. 7.4 Disc apposition against vessel wall with Cardiva Catalyst II device. Hemostatic coating is later exposed within tissue tract

although having been shown to reduce time to hemostasis, require continued need for manual compression.

Collagen Plug-Based Closure Devices: Vascade Vascular Closure System (VCS) and Vascade MVP Venous Vascular Closure System (VVCS)

The Vascade Vascular Closure System (Haemonetics, MA, USA) is a device which uses a similar deployment mechanism to the Cardiva Catalyst devices. Following introduction through the vascular sheath and disc apposition, a collagen plug is deployed into the tissue tract. This plug rapidly expands causing mechanical hemostasis and promoting coagulation. Over time, the collagen plug is resorbed with nothing left behind [14]. This device is indicated for closure of 5–7 F access sites.

With the Vascade MVP Venous Vascular Closure System (Haemonetics, MA, USA), this technology is extended to femoral venotomy sites with large bore access, applicable to sheaths with 6–12 F inner diameters and up to 15 F outer diameters [18].

Plug-Based Closure Devices: AngioBlock and AngioTen

AngioBlock and AngioTen (InvaMed, AN, TR) are two VCDs that facilitate hemostasis by utilizing a delivery sheath to deploy a polymer compound over the arterial adventitia. AngioBlock is suitable for closure of 5–8 F arterial access sites and has a maximum guidewire compatibility of 0.035 (5, 6 F) or 0.038 (7, 8 F). AngioTen can be used for closure of 5–9 F or 10–14 F arterial access sites dependent on the model.

Plug-Based Closure Device: Manta Vascular Closure Device

The Manta Vascular Closure Device (Teleflex, PA, USA; Fig. 7.5) is designed specifically for closure following large bore femoral arterial access [13]. Prior to the procedure, access site depth is measured to determine appropriate position of the device at time of closure. Hemostasis is achieved through the placement of a non-thrombogenic, bioresorbable intraluminal anchor and collagen plug. Two models are available which can be utilized for closure of access sites of 10–14 F (maximum outer diameter of 18 F) or 15–18 F (maximum outer diameter of 25 F). This device requires swapping of the pre-existing vascular sheath over the guidewire.

Collagen Plug-Based Closure Devices: Angio-Seal Evolution and Angio-Seal VIP Vascular Closure Devices

Angio-Seal Evolution and Angio-Seal VIP Vascular Closure Devices (Terumo Medical, BE, EU; Fig. 7.6) are the latest iterations of the popular Angio-Seal devices. These devices created a mechanical seal at the arteriotomy site between an intraluminal anchor and extravascular collagen plug held together by an absorbable suture [10]. The device components are fully resorbed in 60–90 days. Angio-Seal Evolution uses an automated collagen compaction system which increases ease of use, while the Angio-Seal VIP creates a larger collagen footprint on the vessel by twisting down on the arterial wall. Both of these devices can be used in less than 6 F or less than 8 F puncture sites dependent on the

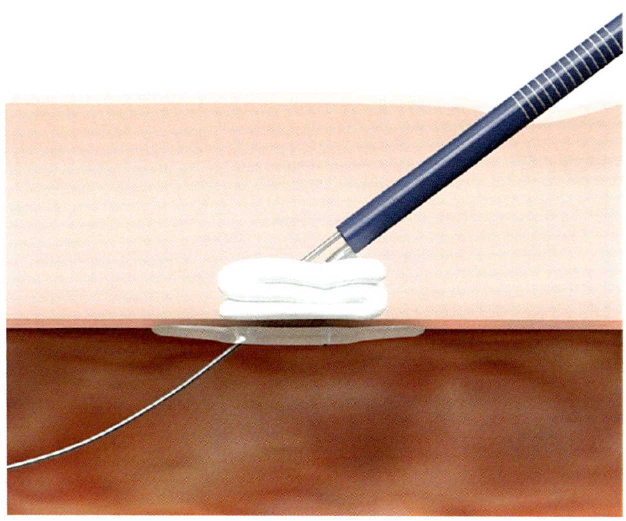

Fig. 7.5 Deployment of Manta Vascular Closure Device with intraluminal anchor and overlying collagen plug, sandwiching the arteriotomy site

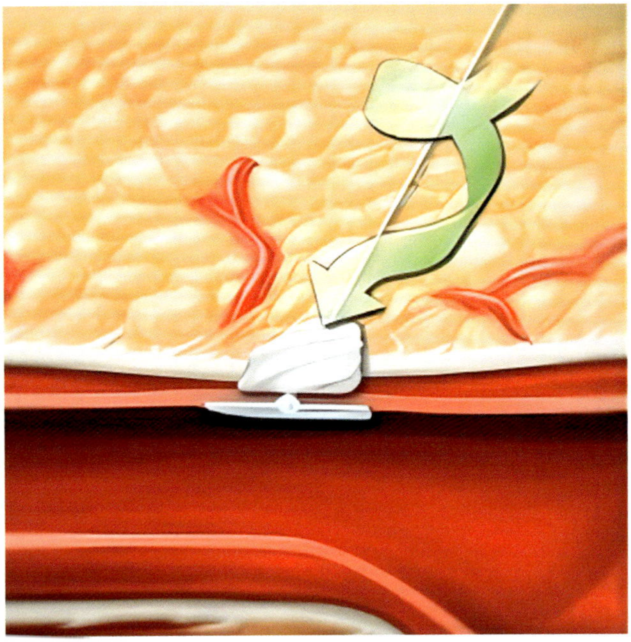

Fig. 7.6 Components of deployed Angio-Seal Evolution device including bioabsorbable anchor, collagen plug, and suture at arteriotomy site

model and are compatible with a maximum guidewire thickness of 0.035 inches or 0.038 inches, respectively.

Plug-Based Closure Device: FemoSeal Vascular Closure Device

The FemoSeal Vascular Closure Device (Terumo Medical, BE, EU) is a device that deploys a bioresorbable, mechanical seal which sandwiches the arteriotomy site between two polymer discs [14]. These polymer discs are held together by an absorbable suture, and all components are resorbed within 90 days. This device can be used in access sites less than 7 F in size and has a maximum guidewire compatibility of up to 0.038 inches.

Decision-Making Regarding Technique Selection

Currently, the potential techniques available to interventionists for vascular closure include MC, a plethora of VCDs, and surgical closure – the latter of which will not be reviewed in this chapter. Among these, VCDs represent the most popular option due their safety profile and ease of use, avoiding the often unnecessary time constraints of MC and the invasiveness of surgical cutdown.

Clinical evidence suggests that all VCDs – whether suture-based, clip-based, or plug-based – are associated with high rates of successful closure and decreased time to hemostasis, ambulation, and hospital discharge compared to MC. However, while there is an abundance of clinical trials assessing the effectiveness of specific VCDs relative to MC, there is a lack of data comparing specific VCDs head-to-head. The same is true for studies assessing for the incidence of complications [19]. As such, the decision-making around choosing a specific device is in large part based on operator preference, patient case, and individual practice availability. Simply put, no particular device is best for all situations and clinical judgement is recommended prior to implementation.

Initial considerations should include technical aspects of the device, such as whether the VCD is compatible with the guidewire and pre-existing vascular sheath, as well as the size of sheath used during the procedure. As discussed previously, selection may be limited by the number of devices available for closure of large bore access sites. As an additional rule of thumb, the smaller the profile of the VCD, the less likely it is to result in complications.

Operator experience with particular devices is also important however, as this affects personal comfort and outcomes. As an example, suture-mediated devices like the Prostar XL usually require a greater number of steps for deployment and are documented to have a longer learning curve relative to other devices [20].

Lastly, VCDs are thought to decrease healthcare costs by increasing departmental throughput and decreasing demands on hospital personnel - although data supporting this idea is still being accumulated [1].

Anticipated Complications and Management

Overall, the incidence of complications with VCDs is rare and similar to that of MC. Care should be taken to ensure that device instructions are carefully followed. However, the complication profile of VCDs varies by device and is different from those associated with MC.

The commonly recognized complications of MC include hematoma, hemorrhage, pseudoaneurysm, thrombus, and arteriovenous (AV) fistula. These complications can occur with VCDs, in addition to retroperitoneal bleeding, ischemia (as a result of thrombus formation, luminal narrowing, or device embolization), infection, and foreign body reactions [3, 15, 21]. Knowledge of these complications and having the appropriate tools to address them is essential in reducing adverse outcomes for patients, particularly in those considered to have higher pre-operative risk due to obesity, large bore access, use of anticoagulation, PVD, tortuous vessels, etc.

Notably, in a Cochrane Review published in 2016, collagen plug-based VCDs were noted to have a lower rates of groin hematoma and pseudoaneurysm when compared to MC. Rates of overall complications, such as retroperitoneal hemorrhage, AV fistula, femoral artery thrombosis, limb ischemia, and deep vein thrombosis, were noted to be no different across all VCD types in comparison to MC [19]. In a separate review article, use of VCDs appeared to be associated with a small risk of groin infection [14]. In terms of large bore closure devices, Perclose ProGlide and Manta have been shown to have similar outcomes in terms of complications [5, 22, 23]. Devices which contain an intraluminal component, such as the Angio-Seal, also pose a theoretical risk for component embolization – although these events have been rarely reported [24].

Recommendations for the management of complications include maintenance of access in the contralateral femoral artery to angiographically check vessel patency post-VCD placement. Ultrasound can also be used to this effect, and to assist in proper deployment. In the event of a major complication such as retroperitoneal bleeding, balloon inflation in the proximal ipsilateral iliac artery or distal aorta can be performed through the contralateral access to control bleeding. A suitably sized stent graft should be readily available to facilitate repair, as well as blood transfusion products and surgical back-up if necessary. Access above the inguinal ligament is a factor that can predispose to the development of retroperitoneal hemorrhage [25]. In the case of thrombus or distal embolization, devices to perform either catheter directed-thrombolysis or aspiration thrombectomy are recommended to be at the operator's disposal. Manual compression can also be used to manage challenging patients and in the case of device failure.

Perioperative Care and Surveillance

In patients receiving manual compression, the duration of pressure to obtain hemostasis is typically 15–20 min but can be greater based on the previously mentioned factors of intraprocedural heparin and size of the vascular sheath. Following this, the patient should undergo a period of bedrest ranging from 4–24 h to prevent complications such as re-bleeding.

In patients treated with VCDs, both of these times are decreased to varying extent depending on the closure device utilized. As an example, patients receiving Cardiva Catalyst devices may still require a degree of manual compression to achieve hemostasis due to the nature of the device, while those treated with StarClose SE mostly achieve immediate hemostasis [3, 14, 15]. Bedrest requirements also vary. Otherwise, additional perioperative management is driven by the specific intervention performed. In the absence of complications, no follow-up imaging is required for these devices.

Case Presentation

The patient was brought to the angiography suite and placed under monitored anesthesia care. The patient was prepped and draped in the standard sterile surgical fashion with both groin sites prepped for access. Ultrasound was used to assess the patency of the common femoral arteries and under ultrasound guidance; access was obtained into the right common femoral artery. An 0.035 guidewire was advanced into the aorta and an 8F dilator was used to dilate the arteriotomy site. After dilatation, a ProGlide device was deployed into the artery at the 10 o'clock position. Once the sutures of the device were harvested and secured, a second ProGlide device was also deployed, this time at the 2 o'clock position. The sutures of the second device were secured and an 8-French sheath was then advanced into the arteriotomy site. Eventually, this arteriotomy site would be dilated to accept a 24 French sheath. The process was also repeated on the left common femoral arteriotomy to accept a 16-French sheath. Following the successful completion of the endovascular repair of the AAA, the sutures from the ProGlide devices successfully achieved hemostasis in both groin access sites, thus avoiding the need for a surgical cutdown in this patient.

References

1. Krishnasamy VP, Hagar MJ, Scher DJ, Sanogo ML, Gabriel GE, Sarin SN. Vascular closure devices: technical tips, complications, and management. Tech Vasc Interv Radiol. 2015;18(2):100–12.
2. Barbetta I, van den Berg JC. Access and hemostasis: femoral and popliteal approaches and closure devices-why, what, when, and how? Semin Intervent Radiol. 2014;31(4):353–60.
3. Patel R, Muller-Hulsbeck S, Morgan R, Uberoi R. Vascular closure devices in interventional radiology practice. Cardiovasc Intervent Radiol. 2015;38(4):781–93.
4. Pang N, Gao J, Zhang B, Guo M, Zhang N, Sun M, et al. Vascular closure devices versus manual compression in cardiac interventional procedures: systematic review and meta-analysis. Cardiovasc Ther. 2022;2022:8569188.
5. Mahalwar G, Shariff M, Datla S, Agrawal A, Rathore SS, Arif TB, et al. Meta-analysis of ProGlide versus MANTA vascular closure devices for large-bore access site management. Indian Heart J. 2022;74(3):251–5.
6. Vierhout BP, Pol RA, El Moumni M, Zeebregts CJ. Editor's choice – Arteriotomy closure devices in EVAR, TEVAR, and TAVR: a systematic review and meta-analysis of randomised clinical trials and cohort studies. Eur J Vasc Endovasc Surg. 2017;54(1):104–15.
7. Montalto C, Munafò AR, Arzuffi L, Soriano F, Mangieri A, Nava S, et al. Large-bore arterial access closure after transcatheter aortic valve replacement: a systematic review and network meta-analysis. Eur Heart J Open. 2022;2(4):oeac043.
8. Del Prete A, Della Rocca DG, Calcagno S, Di Pietro R, Del Prete G, Biondi-Zoccai G, et al. Perclose Proglide™ for vascular closure. Futur Cardiol. 2021;17(2):269–82.
9. Maniotis C, Andreou C, Karalis I, Koutouzi G, Agelaki M, Koutouzis M. A systematic review on the safety of Prostar XL versus ProGlide after TAVR and EVAR. Cardiovasc Revasc Med. 2017;18(2):145–50.

10. Bhogal S, Waksman R. Vascular closure: the ABC's. Curr Cardiol Rep. 2022;24(4):355–64.
11. Sheth RA, Walker TG, Saad WE, Dariushnia SR, Ganguli S, Hogan MJ, et al. Society of Interventional Radiology Standards of Practice Committee. Quality improvement guidelines for vascular access and closure device use. J Vasc Interv Radiol. 2014;25(1):73–84.
12. Rajiah P, Schoenhagen P. The role of computed tomography in pre-procedural planning of cardiovascular surgery and intervention. Insights Imaging. 2013;4(5):671–89.
13. Moccetti F, Brinkert M, Seelos R, Ockert S, Bossard M, Cuculi F, Kobza R, et al. Insights from a multidisciplinary introduction of the MANTA vascular closure device. JACC Cardiovasc Interv. 2019;12(17):1730–6.
14. Noori VJ, Eldrup-Jørgensen J. A systematic review of vascular closure devices for femoral artery puncture sites. J Vasc Surg. 2018;68(3):887–99.
15. Schwartz BG, Burstein S, Economides C, Kloner RA, Shavelle DM, Mayeda GS. Review of vascular closure devices. J Invasive Cardiol. 2010;22(12):599–607.
16. Jakobsen L, Holm NR, Maeng M, Thim T, Kristensen SD, Mogensen LH, Christiansen EH, et al. Comparison of MynxGrip vascular closure device and manual compression for closure after femoral access angiography: a randomized controlled trial: the closure devices used in every day practice study, CLOSE-UP III trial. BMC Cardiovasc Disord. 2022;22(1):68.
17. Diamantopoulos A, Nourzaie R, Mulholland D, Dolan SG, Ahmed I, Gkoutzios P, et al. Safety and efficacy of the Mynx control vascular closure device in peripheral arterial procedures: a prospective study. Vascular. 2023;31(2):369–74.
18. Natale A, Mohanty S, Liu PY, Mittal S, Al-Ahmad A, De Lurgio DB, et al. Venous vascular closure system versus manual compression following multiple access electrophysiology procedures: the AMBULATE trial. JACC Clin Electrophysiol. 2020;6(1):111–24.
19. Robertson L, Andras A, Colgan F, Jackson R. Vascular closure devices for femoral arterial puncture site haemostasis. Cochrane Database Syst Rev. 2016;3:CD009541.
20. Balzer JO, Scheinert D, Diebold T, Haufe M, Vogl TJ, Biamino G. Postinterventional transcutaneous suture of femoral artery access sites in patients with peripheral arterial occlusive disease: a study of 930 patients. Catheter Cardiovasc Interv. 2001;53(2):174–81.
21. Tavris DR, Dey S, Albrecht-Gallauresi B, Brindis RG, Shaw R, Weintraub W, et al. Risk of local adverse events following cardiac catheterization by hemostasis device use - phase II. J Invasive Cardiol. 2005;17(12):644–50.
22. Sedhom R, Dang AT, Elwagdy A, Megaly M, Elgendy IY, Zahr F, et al. Outcomes with plug-based versus suture-based vascular closure device after transfemoral transcatheter aortic valve replacement: a systematic review and meta-analysis. Catheter Cardiovasc Interv. 2023;101(4):817–27.
23. Sakata T, Kuno T, Fujisaki T, Yokoyama Y, Misumida N, Sugiura T, et al. Selection of vascular closure devices in transcatheter aortic valve replacement: systematic review and network meta-analysis. Cardiovasc Revasc Med. 2023;46:78–84.
24 Lasam G, Oaks JB, Brensilver J. Angio-Seal™ embolization: a rare etiology of an acute distal limb ischemia. Case Rep Vasc Med. 2017;2017:9613863.
25. Sajnani N, Bogart DB. Retroperitoneal hemorrhage as a complication of percutaneous intervention: report of 2 cases and review of the literature. Open Cardiovasc Med J. 2013;7:16–22.

Contrast Media, Nephropathy and Prophylaxis

8

Demetrios V. Vlahakos

Case Report

A 76-year old obese male patient was admitted to our hospital with symptoms and signs of pulmonary edema, nausea, vomiting, weakness, fatigue and oliguria.

He had a long history of hypertension, well-controlled with ramipril, amlodipine and bisoprolol and diabetes mellitus for 20 years with adequate control by metformin and insulin (average glycated hemoglobin A1c of 7%). He has been a smoker for the last 40 years and his hypercholesterolemia was treated with atorvastatin.

Seven days ago, he was hospitalized in another facility, where he received contrast media for angiography and peripheral vascular intervention due to intermittent claudication. At that time, he had a baseline serum creatinine of 2 mg/dL, an estimated glomerular filtration rate (MDRD equation) of 34.7 mL/min/1.73 m^2 and proteinuria of 1.8 g/24 h. Two days after the procedure he was discharged home on his usual medical regimen without reevaluation of renal function after administration of the contrast medium.

On admission to our unit his blood pressure was 170/80 mmHg and heart rate 96 beats per min. He had oliguria and symptoms and signs of pulmonary edema (dyspnea, tachypnea, use of auxiliary respiratory muscles, rales in the lower fields of both lungs). The laboratory investigation showed hematocrit 38.5%, hemoglobin 13.1 g/dL, white blood cell count 10,200/μL, blood urea nitrogen 111 mg/dL and serum creatinine 7.8 mg/dL. Arterial blood gas examination was consistent with metabolic acidosis and profound hypoxia. Chest X-ray was diagnostic of pulmonary edema. Kidney size was normal (11 cm) on ultrasound examination.

He was started on hemodialysis due to pulmonary edema, metabolic acidosis, and uremia and continued regular hemodialysis sessions for approximately 3 weeks, followed by renal recovery and cessation of dialysis treatments. Five weeks after the procedure, his renal function returned to baseline.

Introduction

Contrast media are water-soluble agents containing iodine, which enhances the visibility of vascular structures and organs during radiographic procedures. They are usually classified as ionic or nonionic. Ionic agents were developed first, are still in widespread use, their iodine can be dissociated in water leading to higher concentration of ions, hyperosmolality and additional complications. In non-ionic compounds, iodine is covalently bound to the molecule, does not dissociate, and have fewer side effects. Contrast media can be administered either intravenously (IV) or intraarterially (IA) to help making valuable diagnoses and guide treatment in a variety of patients.

Although well-tolerated, in the vast majority of cases, contrast media can cause discomfort, allergic reactions, kidney damage, and thyroid problems. Immediate mild hypersensitivity reactions (<3%) have included skin rashes, flushing, urticaria, pruritus, rhinorrhea, nausea, vomiting, diaphoresis, coughing, and dizziness, while more severe reactions (<0.04%) have included facial and laryngeal edema, bronchospasm, angioedema, hypotension and cardiac arrhythmias, pulmonary edema, seizures, and death (<1 death per 100,000 patients). Delayed adverse reactions have included skin rash and redness, sometimes associated with nausea, vomiting, dizziness and either hyperthyroidism or hypothyroidism due to the exposure to iodine material [1, 2].

Iodinated contrast media are excreted via the kidney and can be seen in renal calyces within 2 min after intravenous or even earlier after intraarterial administration. Both animal experiments and clinical observations have documented that the contrast material can cause kidney damage leading to decreased kidney function and acute kidney injury (AKI). They exert their cytotoxic effect directly on the proximal tubules of the kidney, and indirectly by enhancing the gen-

D. V. Vlahakos (✉)
National and Kapodistrian University of Athens, School of Medicine, Athens, Greece
e-mail: vlahakos@otenet.gr

G. Geroulakos et al. (eds.), *Mastering Endovascular Techniques*, https://doi.org/10.1007/978-3-031-42735-0_8

eration of reactive oxygen species and by jeopardizing renal microcirculation via increasing endothelin levels and decreasing nitric oxide generation [3, 4].

Terminology

Acute kidney injury (AKI) after intravascular administration of contrast media for diagnostic procedures or therapeutic angiographic interventions is the third leading cause of iatrogenic AKI and has been well recognized over the past 50 years. This entity was initially called contrast-induced nephropathy (CIN) and was seen more often with the older high-osmolar contrast media. However, both the tolerance and safety of contrast media have improved dramatically after the development of newer low- or iso-osmolar, ionic or non-ionic molecules. The terminology has eventually changed to contrast-induced AKI. However, because along with the contrast material other comorbidities and predisposing factors can contribute to the development of AKI and because it is impossible to determine the degree of the involvement of contrast media in AKI on clinical grounds, more general terms, such as contrast-associated AKI or post-contrast AKI have been adopted.

The usual course of contrast associated AKI consists of a transient asymptomatic elevation in serum creatinine shortly after iodinated contrast medium administration. There is not a standard definition of cutoff values used to define contrast associated AKI. Thus, contrast-associated AKI is defined, as an absolute increase in serum creatinine levels by more than 0.3 or 0.5 mg/dL from baseline, or a relative increase in serum creatinine by >25 or >50% from baseline value within 48–72 h after administration of contrast media. Although no definition of AKI has been designed specifically for contrast-associated AKI, both the European Society of Urogenital Radiology and the American College of Radiology have adopted the definition of AKI by the Acute Kidney Injury Network (AKIN) and suggest to define contrast-associated AKI, as an absolute increase in serum creatinine levels by ≥0.3 mg/dL from baseline or a relative increase in serum creatinine by ≥50% from baseline within 48–72 h from contrast infusion. Contrast associated AKI is usually non-oliguric, the creatinine peaks within 3–4 days, and often returns to baseline within 7–10 days. It is unusual for patients to develop permanent renal dysfunction.

Incidence and Risk Factors

The incidence of contrast associated AKI varies among the different studies based on the comorbidities, the type of procedure performed, and the threshold increase in serum creatinine used to define AKI. Observational studies with propensity score matching control groups found a similar risk of AKI with contrast-enhanced procedures compared with those without contrast administration [5, 6].

However, this kind of studies could not detect higher rates of contrast-associated AKI in populations with low risk for AKI. Despite the efforts to use propensity score to match for known confounders, the possibility still exists that patients who did not receive contrast media had a higher baseline risk for kidney injury. At any rate, the incidence of AKI following administration of contrast media seems to be smaller than previously thought. From a total of 985,737 consecutive patients participating in the National Cardiovascular Data Registry Cath-PCI registry from June 2009 through June 2011, 69,658 (7.1%) experienced AKI and 3005 (0.3%) required dialysis. As CKD progresses, there is a sharp increase in the rates of contrast-associated AKI (8% in mild, 12.9% in moderate and 26.6% in severe CKD) [7].

In the PRESERVE trial, which included 5177 patients at high risk due to chronic kidney disease, all hydrated prior to contrast administration, contrast-associated nephropathy was reported in less than 10% of patients [8].

In the AMACING trial including 660 patients with mild to moderate chronic kidney disease, (none with eGFR < 30 mL/min/1.73 m^2 and 65% with eGFR 46–59 mL/min/1.73 m^2) randomized to prophylactic hydration or no hydration prior to intravenous contrast material administration, contrast-associated AKI developed in less than 3% of patients in both groups [9].

The risk of AKI has been shown to be significantly higher with high-osmolar CM compared to low-osmolar contrast media and depends on the volume of contrast media administered during the procedure. For instance, for each 100 mL increase in volume of contrast media administered at the time of primary coronary intervention the risk of AKI increased by 12%. Similarly, in a study with 10,065 consecutive patients undergoing percutaneous coronary intervention from 2000 to 2008, a gradual increase of the risk for contrast-associated AKI was observed depending on the volume of contrast media above the maximum allowable contrast dose (5 mL × body weight [kg])/baseline serum creatinine [mg/dL]) [10, 11].

By contrast, ultra-low contrast volumes of 27 mL reduced rates of contrast-associated AKI in patients with CKD and median serum creatinine of 2.1 mg/dL and mean eGFR of 31 ± 10 mL/min/1.73 m^2 [12].

Contrast media agents in our days are universally low- or iso-osmolar solutions. Twenty years ago, NEPHRIC Study showed lower risk for contrast associated AKI with iodixanol (iso-osmolar) vs ioexol (low-osmolar) in diabetic patients with CKD. Similarly, a recent analysis of randomized controlled trials comparing the nephrotoxic effects between iso- and low-osmolar contrast media in diabetic patients with or without CKD suggested the use of iso-osmolar contrast media in diabetic patients with CKD [13, 14].

Other studies and a subsequent meta-analysis of 25 studies fail to reveal a clear advantage of iso-osmolar contrast compared with modern low-osmolar contrast agents for the prevention of post-contrast AKI after intravenous application. However, in patients, with renal insufficiency, intra-arterial application of the low-osmolar iohexol was associated with a greater risk of contrast medium-induced nephropathy compared with iodixanol. No significant differences were found between iodixanol and other low-osmolar contrast media [15].

Besides the type (low- vs iso-osmolar agents), volume, frequency of administration in a short time interval (48–72 h) and route of administration (intravenous vs intraarterial), many patient-specific factors have been considered as risk factors for the development of contrast induced AKI, most of them associated with the severity of underlying kidney damage (number of remaining functioning nephrons) and the adequacy of blood supply and oxygenation of renal parenchyma (low cardiac output due to either dehydration or heart failure). Thus, the most accepted patient-related risk factors have included advanced age, pre-existing renal insufficiency (especially if eGFR <30 mL/min/1.73 m^2), a prolonged history of diabetes mellitus and/or hypertension, multiple myeloma, anemia, proteinuria and hypoalbuminemia, low cardiac output and congestive heart failure with left ventricular ejection fraction less than 40%, the use of diuretics or dehydration from any cause, hypotension, administration of other nephrotoxic medications (non-steroidal anti-inflammatory agents, antibiotics etc.) and renal transplantation.

Prophylaxis

Screening of patients at risk for contrast-associated AKI is necessary to identify patients that would need preventing measures, as well as careful post-procedural management. Patients less than 60 years of age without diabetes, hypertension or known problem with the kidneys are at an extremely low risk of AKI and can be given contrast material without further intervention or change in medical regimen, other than avoidance of volume depletion.

The decision to administer contrast in high-risk patients should always be a matter of clinical judgment and physician needs to clarify if contrast is diagnostically imperative and its administration outweighs the risks of post-contrast AKI and other complications. This is particularly true in patients requiring closely spaced contrast-enhanced studies.

Among the various strategies studied to reduce the incidence of post-contrast acute kidney injury in at-risk patients, the best is considered to be the peri-exposure volume expansion. However, the ideal route, rate, and volume for volume expansion therapy have not yet been clearly determined, since conflicting results have been published. Some studies showed that hydrating patients before contrast media administration significantly reduces the risk of contrast-associated AKI and others claimed that hydration has minimal impact on preventing contrast-associated AKI [16].

Theoretically, the intravenous route of crystalloid solutions should be preferred to the oral route since most of the free water administered per os will move into the cells or remain in the intracellular space and only less than 1/10 of the administered volume will stay in the intravascular space to support the circulation. This notion was proven correct in some, but not all the studies [17, 18]. Isotonic intravenous fluids (0.9% normal saline, lactated Ringer's solution) are preferred and are better than 0.45% half-normal saline [19]. There is no clear superiority of 0.9% normal saline or sodium bicarbonate solutions for prophylaxis in this setting [20].

Recent randomized studies showed low risk of contrast associated AKI in patients with eGFR ≥30 and particularly >45 mL/min/1.73 m^2. However, these studies were performed in heterogeneous populations and most patients received low-volume iodinated contrast by intravenous administration [9]. In patients with multiple risk-factors and those at higher-risk, especially if eGFR is <30 mL/min/1.73 m^2 pre-procedural prophylaxis with intravenous volume expansion therapy should be utilized.

The optimal IV volume expansion protocol is unknown and ideally should be tailored to the patient's volume status and medical conditions. For inpatients an infusion of 0.9% normal saline at 1 mL/kg/h IV in patients with normal cardiac output and at 0.5 mL/kg/h in patients with congestive heart failure beginning 8–12 h prior to contrast administration and continuing 8–12 h afterwards is warranted. For outpatients and emergency situations an IV bolus of 0.9% normal saline at 3 mL/kg/h and a total volume of 500 mL should be administered prior to contrast administration followed by 1 mL/kg/h for 8–12 h afterwards. Of note, forced diuresis and matched hydration seems to reduce contrast-associated AKI more than intravenous hydration alone in patients undergoing coronary interventions [21].

Although the risk of giving IV contrast to dialysis patients may be relatively small, these risks should be weighed against the likely diagnostic benefit of contrast administration. The common belief that dialysis patients require early post-procedural dialysis was based in earlier observation with high-osmolar contrast media that could provoke cardiac decompensation and arrhythmias. Presently with the use of iso- or low-osmolar contrast media this belief is unsupported and in fact, an episode of hypotension during dialysis session immediately after coronary angioplasty and stenting could be catastrophic. On the other hand, a pre-procedural dialysis session could be appropriate, if a large dose of contrast is anticipated in patients with heart failure. At any rate, the nephrology service should be readily available for consultation in all hospitalized dialysis patients.

Outcome

The usual course of post-contrast acute kidney injury consists of a transient asymptomatic elevation in serum creatinine. Serum creatinine usually begins to rise within 24 h of intravascular iodinated contrast medium administration, peaks within 3–4 days, and often returns to baseline within 7-10 days. It is unusual for patients to develop permanent renal dysfunction.

In general, the incidence of post contrast AKI in patients with mild-to-moderate CKD is low (<3%) and no specific prophylaxis is necessary. In a single-center, randomized trial (AMACING study) similar rates of contrast-associated AKI were found among 660 patients with eGFR >30 mL/min/1.73 m^2) randomized to receive prophylactic intravenous isotonic saline (incidence 2.7%) or no intravenous fluids (incidence 2.6%). Adverse events including heart failure, hyponatremia, and arrhythmia were more common in the group receiving intravenous fluid (5.5% versus 0%). However, this trial has some potentially important limitations, since no patients had an eGFR <30 mL/min/1.73 m^2, two thirds of patients had eGFR >45 mL/min/1.73 m^2, the average volume of contrast material was 90 mL, and was given intravenously in the majority of patients [9].

In patients with underlying renal disease, diabetes and/or hypertension for many years, a subgroup about 20–30% will be expected to have substantial impairment of the renal function after contrast administration. Dialysis may be required in less than 1–3% of non-diabetic low-risk patients, but in diabetic patients with renal impairment, dialysis may be required in anywhere from 10 to 15% of cases. Of those who do required dialysis, at least 20% may end up on permanent dialysis [22, 23].

Several studies have shown that patients who receive intravascular iodinated contrast and develop contrast-associated AKI tend to have longer hospital stays, higher mortality, and higher incidence of cardiac and neurologic long-term events than patients who do not develop AKI [24–28]. However, these results must be interpreted with caution because no control group with patients not receiving radiocontrast material was included. It is possible that the increased morbidity and mortality seen in patients with contrast-associated AKI is because the intravascular iodinated contrast exposure unmasked some other factors and underlying comorbid diseases present in these patients. At any rate, the possibility of development various damages after contrast administration highlights the importance of avoiding contrast when other imaging modalities are available.

In conclusion, contrast-associated AKI is the third most common iatrogenic cause of AKI. The usual course of contrast-associated AKI is an asymptomatic elevation in serum creatinine that returns to baseline within 7–10 days. Contrast-associated AKI rarely progress to permanent renal dysfunction. Currently, that contrast media are universally low- or iso-osmolar, contrast-associated AKI rarely affects stable patients with eGFR >30 mL/min/1.73 m^2. The risk of contrast-associated AKI increases substantially in elderly individuals and in patients with diabetes and pre-existing severe CKD. Among a variety of measures that have been attempted to prevent contrast-associated AKI, pre-hydration with crystalloids and the reduction of the volume of contrast media are generally recommended.

References

1. American College of Radiology. Committee on Drugs and Contrast Media. ACR manual on contrast media, 2022. https://www.acr.org/-/media/ACR/Files/Clinical-Resources/Contrast_Media.pdf
2. European Society of Urogenital Radiology. ESUR guidelines on contrast media v10.0. http://www.esur.org/guidelines/. Accessed 05 July 2020.
3. Kusirisin P, Chattipakorn SC, Chattipakorn N. Contrast-induced nephropathy and oxidative stress: mechanistic insights for better interventional approaches. J Transl Med. 2020;18:400. https://doi.org/10.1186/s12967-020-02574-8.
4. Mamoulakis C, Fragkiadoulaki I, Karkala P, Georgiadis G, et al. Contrast-induced nephropathy in an animal model: evaluation of novel biomarkers in blood and tissue samples. Toxicol Rep. 2019;6:395–400.
5. McDonald JS, McDonald RJ, Comin J, Williamson EE, Katzberg RW, Murad MH, Kallmes DF. Frequency of acute kidney injury following intravenous contrast medium administration: a systematic review and meta-analysis. Radiology. 2013;267:119–28.
6. Williams LS, Walker GR, Loewenherz JW, Gidel LT. Association of contrast and acute kidney injury in the critically ill: a propensity-matched study. Chest. 2020;157:866–76.
7. Tsai TT, Patel UD, Chang TI, Kennedy KF, et al. Contemporary incidence, predictors, and outcomes of acute kidney injury in patients undergoing percutaneous coronary interventions: insights from the NCDR Cath-PCI registry. JACC. 2014;7(1):1–9.
8. Weisbord SD, Gallagher M, Jneid H, Garcia S, Cass A, Thwin SS, Conner TA, Chertow GM, Bhatt DL, Shunk K, Parikh CR, McFalls EO, Brophy M, Ferguson R, Wu H, Androsenko M, Myles J, Kaufman J, Palevsky PM, PRESERVE Trial Group. Outcomes after angiography with sodium bicarbonate and acetylcysteine. N Engl J Med. 2018;378(7):603–14.
9. Nijssen EC, Rennenberg RJ, Nelemans PJ, Essers BA, Janssen MM, Vermeeren MA, Ommen VV, Wildberger JE. Prophylactic hydration to protect renal function from intravascular iodinated contrast material in patients at high risk of contrast-induced nephropathy (AMACING): a prospective, randomized, phase 3, controlled, open-label, non-inferiority trial. Lancet. 2017;389(10076):1312–22.
10. Rihal CS, Textor SC, Grill DE, Berger PB, Ting HH, Best PJ, Singh M, Bell MR, Barsness GW, Mathew V, Garratt KN, Holmes DR Jr. Incidence and prognostic importance of acute renal

failure after percutaneous coronary intervention. Circulation. 2002;105(19):2259–64.

11. Brown JR, Robb JF, Lock CA, et al. Does safe dosing of iodinated contrast prevent contrast-induced acute kidney injury. Circulation. 2010;3:346–50. By contrast, ultra-low contrast volumes of 27 ml reduced rates of contrast-associated AKI in patients with CKD (median serum creatinine of 2.1 mg/dl and mean eGFR of 31+/–10 ml/min/1.73 m^2) (12).
12. Kane GC, Doyle BJ, Lerman A, et al. Ultra-low contrast volumes reduce rates of contrast-induced nephropathy in patients with chronic kidney disease undergoing coronary angiography. JACC. 2008;1:89–90.
13. Aspelin P, Aubry P, Fransson S, et al. Nephrotoxic effects in high-risk patients undergoing angiography. N Engl J Med. 2003;348:491–9.
14. Zhao F, Lei R, Yang SK, Luo M, et al. Comparative effect of iso-osmolar versus low-osmolar contrast media on the incidence of contrast-induced acute kidney injury in diabetic patients: a systematic review and meta-analysis. Cancer Imaging. 2019;19 https://doi.org/10.1186/s40644-019-0224-6.
15. Heinrich MC, Häberle L, Müller V, Bautz W, Uder M. Nephrotoxicity of iso-osmolar iodixanol compared with nonionic low-osmolar contrast media: meta-analysis of randomized controlled trials. Radiology. 2009;250(1):68–86.
16. Zaki HA, Bashir K, Iftikhar H, Alhatemi M, Elmoheen A. Evaluating the effectiveness of pretreatment with intravenous fluid in reducing the risk of developing contrast-induced nephropathy: a systematic review and meta-analysis. Cureus. 2022;14(5):e24825.
17. Trivedi HS, Moore H, Nasr S, et al. A randomized prospective trial to assess the role of saline hydration on the development of contrast nephrotoxicity. Nephron Clin Pract. 2003;93(1):C29–34.
18. Kong DG, Hou YF, Ma LL, Yao DK, Wang LX. Comparison of oral and intravenous hydration strategies for the prevention of contrast-induced nephropathy in patients undergoing coronary angiography or angioplasty: a randomized clinical trial. Acta Cardiol. 2012;67(5):565–9. Isotonic intravenous fluids (0.9% normal saline, lactated Ringers solution) are preferred and are better than 0.45% half-normal saline (19).
19. Mueller C, Buerkle G, Buettner HJ, et al. Prevention of contrast media-associated nephropathy: randomized comparison of 2 hydration regimens in 1620 patients undergoing coronary angioplasty. Arch Intern Med. 2002;162:329–36.
20. Mueller C, Seidensticker P, Buettner HJ, Perruchoud AP, Staub D, Christ A, Buerkle G. Incidence of contrast nephropathy in patients receiving comprehensive intravenous and oral hydration. Swiss Med Wkly. 2005;135(19-20):286.
21. Mirza AJ, Ali K, Huwez F, Taha AY, et al. Contrast induced nephropathy: efficacy of matched hydration and forced diuresis for prevention in patients with impaired renal function undergoing coronary procedures–CINEMA trial. Int J Cardiol Heart Vasc. 2022;39:100959.
22. Hiremath S, Akbari A, Wells GA, Chow BJW. Are iso-osmolar, as compared to low-osmolar, contrast media cost-effective in patients undergoing cardiac catheterization? An economic analysis. Int Urol Nephrol. 2018;50(8):1477–82.
23. Ellis JH, Khalatbari S, Yosef M, et al. Influence of clinical factors on risk of contrast-induced nephrotoxicity from IV iodinated low-osmolality contrast material in patients with a low estimated glomerular filtration rate. AJR Am J Roentgenol. 2019;213:W188.
24. McCullough PA, Wolyn R, Rocher LL, Levin RN, O'Neill WW. Acute renal failure after coronary intervention: incidence, risk factors, and relationship to mortality. Am J Med. 1997;103:368–75.
25. Wi J, Ko YG, Kim JS, Kim BK, et al. Impact of contrast-induced acute kidney injury with transient or persistent renal dysfunction on long-term outcomes of patients with acute myocardial infarction undergoing percutaneous coronary intervention. Heart. 2011;97(21):1753–7.
26. Brown JR, Malenka DJ, DeVries JT, Robb JF, Jayne JE, Friedman BJ, et al. Transient and persistent renal dysfunction are predictors of survival after percutaneous coronary intervention: insights from the Dartmouth Dynamic Registry. Catheter Cardiovasc Interv. 2008;72:347–54.
27. Gruberg L, Mintz GS, Mehran R, et al. The prognostic implications of further renal function deterioration within 48 h of interventional coronary procedures in patients with pre-existent chronic renal insufficiency. J Am Coll Cardiol. 2000;36:1542–8.
28. Cheng W, Wu X, Liu Q, Wang HS, et al. Post-contrast acute kidney injury in a hospitalized population: short-, mid-, and long-term outcome and risk factors for adverse events. Eur Radiol. 2020;30(6):3516–27.

Balloons and Percutaneous Transluminal Balloon Angioplasty

9

Ioannis D. Kakisis

Case Presentation

A 73-year-old male presented to the outpatient clinic complaining of severe pain in his right foot, which was exacerbated when he was lying down and was relieved when he stood up or sat in a chair. His medical history included dyslipidemia, hypertension and 60 pack-years of smoking. On physical examination, the foot was pale and cold. Pulses were present in the femoral artery but absent in the popliteal, the posterior tibial and the dorsalis pedal artery. The ankle-brachial pressure index was 0.3. A duplex ultrasound scan revealed occlusion of the superficial femoral artery at the adductor canal, occlusion of the anterior tibial and segmental, short occlusions of the posterior tibial artery. The diagnosis was confirmed by computed tomography angiography and the patient was scheduled for percutaneous transluminal angioplasty of the right SFA.

Continued at page 79

Background

Percutaneous transluminal balloon angioplasty represents the basic endovascular technique for the treatment of arterial or venous stenoses serving either as a standalone procedure or supplemented by additional techniques such as stenting or atherectomy. The first endovascular balloon catheter was invented by Fogarty et al. in 1963 but it was designed as "A method for extraction of arterial emboli and thrombi" and not for angioplasty [1]. The idea of "Transluminal Treatment of Arteriosclerotic Obstruction" was conceived by Dotter and Judkins who described, in 1964, a transluminal technique for recanalization using tapered, radiopaque, Teflon dilating catheters of progressively larger diameter which were slipped over a guidewire [2]. Ten years later, their conclusion-prophecy that "the interest and ingenuity of others will lead to refinements of technique" was fulfilled by Grüntzig and Hopff, who described a balloon catheter specifically designed for the percutaneous recanalization of chronic arterial occlusions [3].

Mechanism of Action

The original view that balloon angioplasty exerts its action by compression and lateral displacement of the atheromatous material was not verified by histological studies or by intravascular ultrasound. It is now clear that the atherosclerotic plaque is incompressible. The main mechanism of luminal enlargement appears to be overstretching of the arterial wall, beyond the point of elastic return [4, 5]. Stretching of the arterial wall leads to rupture of the intima and the media, which contributes to the plastic deformity of the wall, and detachment of the plaque edges from the media (dissection) [4, 5]. Plaque rupture, carrying the risk of distal embolization, may also occur and is inevitable in cases of circumferential lesions. Healing of the disrupted arterial layers follows by the formation of a neointima and scar tissue [5].

Balloon Technology

A balloon catheter consists of two basic components, the shaft and the balloon, both of which have been the field of continuous research and development [6, 7].

The Shaft

The shaft configuration can be either over-the-wire (OTW) or rapid exchange (RX), also called single operator exchange (SOE) or monorail. The OTW balloon catheters have a coaxial design, where the guidewire tracks along the whole length of the catheter, entering at the distal end of the balloon and

I. D. Kakisis (✉)
National and Kapodistrian University of Athens, Department of Vascular Surgery, Attikon University Hospital, Athens, Greece

G. Geroulakos et al. (eds.), *Mastering Endovascular Techniques*, https://doi.org/10.1007/978-3-031-42735-0_9

exiting at the proximal end of the catheter. Therefore, the proximal hub of the balloon catheter has two ports, one for the wire and one for the inflation of the balloon. The OTW configuration offers good support, translated into good pushability and trackability, but requires a long guidewire for the exchange of the balloon catheter, calculated as at least the distance between the vascular access point and the lesion plus the length of the balloon catheter. OTW balloon catheters are the most commonly used catheters in peripheral endovascular interventions.

In the monorail, RX or SOE balloon catheters, the guidewire enters at the distal end of the balloon and has a short coaxial route within the catheter exiting about 20-25 cm proximally to the balloon instead of the end of the catheter. The remaining part of the balloon catheter has only one lumen and the proximal hub only one port, for the inflation of the balloon. This design requires shorter guidewires, offers easier catheter exchange by a single operator and has a lower profile. On the other hand, it offers less pushability and trackability. RX balloon cathes are most commonly used for small diameter arteries such as the coronary and the tibial arteries, or when the crossing profile is of utmost importance, such as in the carotid arteries.

The Balloon

Profile

The crossing profile of a balloon catheter is defined as the maximum diameter of the balloon region of the catheter. At this point, the thickness of the folded balloon is added to the diameter of the inner tube of the shaft which defines the guidewire lumen. Consequently, balloon catheters supported by 0.014-inch guidewires have a lower crossing profile than balloon catheters with 0.035-inch guidewire design. Lower crossing profiles allow easier passage though tight stenoses and occlusions and require sheaths or guiding catheters of lower diameter as well.

Inflation and Deflation Time

The inflation and deflation time of the balloon is directly proportional to the resistance (R) of the inflation tube of the catheter shaft, which, according to Poiseuille's law, is directly proportional to the length (L) of the shaft and the viscosity (η) of the mixture of saline and contract medium, and inversely proportional to the radius (r) of the inflation tube to the fourth power:

$$R = 8\eta L / \pi r^4$$

Therefore, the radius of the inflation tube represents the major determinant of the inflation and deflation time. An inflation tube of small diameter secures a low crossing profile of the balloon catheter, but this comes at the cost of longer inflation and deflation times. The viscosity of the liquid, on the other hand, is defined by the ratio of saline and contrast medium. More contrast medium leads to better visibility of the inflated balloon but, again, this comes at a cost of longer inflation and deflation times.

Compliance

Compliance of a balloon is defined as the change in balloon diameter (Δd) divided by the change in balloon pressure (Δp) that caused the diameter change:

$$C = \Delta d / \Delta p$$

Non-compliant balloons are typically made of PET (polyethylene terephthalate) or nylon. Inflating these balloons above their nominal pressure (pressure at which the balloon reaches its nominal diameter) will cause a sharp increase in the inflation pressure with minimal change in the diameter of the balloon. Thus, the balloon will expand uniformly over its longitudinal axis exerting high radial force in the stenotic area and minimal force in the normal artery, proximally and distally to the lesion. Non-compliant balloons are therefore preferable for angioplasties of stenotic lesions.

Compliant balloons are typically made of polyolefin copolymer (POC), polyethylene, polyurethane or silicone. These balloons are inflated by volume instead of pressure, meaning that continuing inflation will cause expansion of the balloon with a relatively small increase in the inflation pressure. The balloon will expand in the direction of least resistance which, in the case of a stenotic lesion, means that the balloon will acquire a dog-bone shape causing deformation and stress injury to the adjacent normal artery. Their conformability, on the other hand, makes them ideal for temporary occlusion of large vessels or to expand vascular prostheses, securing apposition of the graft fabric and the arterial wall.

Trackability, Pushability, and Crossability

Trackability is the ability of the balloon catheter to be advanced to the target lesion through the tortuous vessel anatomy. It depends on several factors, including the flexibility and the profile of the balloon catheter, as well as the friction between the balloon catheter and the guiding catheter on the outside and the guide-wire on the inside.

Pushability is the ability of the balloon catheter to transmit to its distal tip the force applied to the proximal part of the catheter. High force transmission means that the operator can feel the resistance at the catheter tip which allows him/her to adjust the pushing force he/she applies so that crossing of the lesion is achieved without the risk of vessel injury. A low balloon catheter profile with a high axial stiffness and hydrophilic coating contribute to a high pushability.

Crossability is the ability of the balloon catheter to pass through the lesion. Factors affecting crossability include the design of the catheter tip, the crossing profile, the way the balloon is folded, the hydrophilic coating and the kink resistance of the catheter.

Technique

Vascular Access

Prerequisites for using an artery as an access point to the arterial system include the relatively superficial position of the artery, so that it will be easily palpated and punctured, the presence of a bone behind the artery, so that it will be easily pressed at the end of the procedure, and the relatively large calibre of the artery, so that the required sheath will be safely inserted. The most commonly used artery, fulfilling all these conditions, is the common femoral artery (ipsilateral or contralateral to the lesion), whereas the brachial, axillary, radial, popliteal and even the anterior and posterior tibial arteries can also be used. The selection of the appropriate access site for each procedure among the aforementioned options is based on its proximity to the lesion, the calibre of the sheath to be used and the availability of the access artery, which depends on the patency of the artery and the condition of the overlying skin.

Retrograde Femoral Access

The retrograde femoral access is the most common approach for angioplasty of the ipsilateral common and external iliac artery, the carotids and the renal arteries, whereas it can also be used for angioplasties of the contralateral limb. Advantages of the common femoral artery (CFA), as an access point, include the superficial position, the large calibre, the presence of the femoral head behind it, the proximity to most vascular lesions and the abundant working space over the legs of the patient.

The puncture site should be above the femoral bifurcation, which is located approximately at the level of the lower border of the femoral head, and below the inguinal ligament, which is located approximately at the level of the upper border of the femoral head. Puncture above the inguinal ligament carries the risk of retroperitoneal haemorrhage, since the artery goes deep in the pelvis and without an osseous structure behind it to allow manual compression. Puncture of the superficial femoral artery, on the other hand, carries the risk of vascular complications such as acute occlusion, arteriovenous fistula and pseudoaneurysm, since the origin of the superficial femoral artery is frequently calcified, has a smaller diameter, the femoral vein lies behind it, so that both the artery and the vein can be simultaneously punctured, and there is no bone behind them to allow manual compression.

Previous open surgery on the CFA may make advancement of the sheath difficult through the scar tissue. Useful tricks to overcome this difficulty include the use of a stiff wire, instead of the standard one, as well as the pass of sheath dilators of progressively larger diameter to create a track through the scar tissue. The scar tissue, on the other hand will prevent the formation of post-procedural hematomas.

A frequent problem encountered in retrograde punctures of the CFA is that, in the case of severe stenoses or occlusions of the ipsilateral common or external iliac arteries, the femoral pulse will be weak or absent. Anatomic landmarks (puncture at the midpoint of the inguinal ligament, which extends between the anterior superior iliac spine and the pubic tubercle) as well as radiographic landmarks (puncture 1 cm lateral to the most medial cortex of the femoral head or puncture of a radiographically visible calcified artery) could be used, although the most accurate method is undoubtedly the catheterization of the CFA under ultrasonographic guidance.

Antegrade Femoral Access

Antegrade femoral access offers the shortest way to the ipsilateral superficial femoral, the popliteal and the infragenicular arteries. On the other hand, antegrade catheterization of the CFA is more difficult especially in obese patients where an assistant should retract the abdomen out of the way. Failure to puncture the CFA proximal enough to its bifurcation may lead to passage of the entry guidewire into the deep femoral artery. In such a situation, a catheter with a short, angulated tip could be advanced into the deep femoral artery and then slowly withdrawn, with continuous injection of contrast media under oblique fluoroscopy, until its tip faces the superficial femoral artery (SFA). A guidewire could then be advanced into the SFA.

Retrograde Popliteal Access

The patient is placed in a prone position and the popliteal artery is punctured under ultrasonographic guidance. Care should be taken to avoid simultaneous puncture of the overlying popliteal vein. Retrograde popliteal access allows simultaneous angioplasty of tandem lesions in the iliac arteries and the SFA, angioplasty of lesions in the CFA and flush occlusions of the SFA. It is also indicated in cases of difficult or failed attempts to catheterize the CFA. The main disadvantage of the popliteal approach is that the prone position may not be well tolerated by obese patients.

Retrograde Tibial Access

The anterior and posterior tibial arteries can be cannulated using a micropuncture kit under ultrasonographic or fluoroscopic guidance. The most common indication is the recanalization of chronic total occlusive lesions with simultaneous antegrade and retrograde subintimal angioplasty when re-entry cannot be achieved by the antegrade approach.

Upper extremity Artery Access

Angioplasty through the upper extremity arteries offers the advantage of immediate mobilization of the patient as well as easier advancement of balloons and stents in arteries originating from the aorta at an acute angle, such as the superior mesenteric artery. Disadvantages include the smaller diameter of these arteries compared to the CFA, as well as their close proximity with nerves, such as the median nerve in the case of the brachial artery or the brachial plexus in the case of the axillary artery. Direct nerve injury or, more frequently, pressure by a hematoma may result in peripheral paresthesia.

Lesion Crossing

Crossing of the lesion with a wire is the most critical part of the procedure. A 0.035-inch, straight or angled, floppy-tipped, hydrophilic coated guidewire supported by an angled diagnostic catheter is most commonly used. Smaller diameter guidewires such as 0.014″ or 0.018″ can also be used, especially in tight stenoses or occlusions or in smaller diameter arteries. Apart from offering support, the catheter adds steerability, whereas specific catheter shapes can be used to steer the guidewire into specific arteries.

The lesion is crossed carefully under fluoroscopic guidance so that the wire will not be advanced into a dissection plane inadvertently. If this happens, there are two options. The guidewire can be withdrawn and another attempt to cross the lesion transluminally can be made or the guidewire can be advanced past the lesion in the subintimal plane and re-entry into the lumen be attempted at the other end of the lesion. Once the lesion is crossed, either transluminally or subintimally, the hydrophilic guidewire may be exchanged with a stiffer one to provide better support. It is of utmost importance that, once the guidewire is in place, extreme care is taken to maintain the guidewire position during subsequent exchanges of catheters, balloons and stents.

Balloon Angioplasty

The balloon catheter is selected based on the following factors:

- the balloon diameter should be slightly larger than the diameter of the target artery, except for severely calcified lesions where a more conservative approach may be selected to avoid artery rupture,
- the balloon length should be slightly longer that the lesion length, so that it does not cause injury to the adjacent normal artery, proximally and distally to the lesion,
- the shaft length is selected based on the distance between the access point and the lesion.

Under fluoroscopic guidance, the balloon is advanced across the lesion. Correct positioning is achieved by fluoroscopic reference to anatomic structures, by road mapping, or by markers placed on the screen by the operator at the time of the diagnostic angiography. The balloon is inflated to the rated nominal pressure using an inflation device loaded with dilute contrast (30–50% contrast in saline). Air should be removed from both the inflation device and the balloon and the inflation should be performed with the inflation device facing downwards to prevent air from entering the balloon. Apart from reducing the visibility of the inflated balloon, this would also carry the risk of infusion of air within the arteries in case of balloon rupture. As the balloon is inflated under fluoroscopic guidance, a waist is seen at the point of maximum stenosis, which usually resolves with further inflation of the balloon. This is an indication for the adequacy of the angioplasty. The optimal inflation time has not been determined and depends on the interventionalist's preference and the type of the lesion. An inflation of up to 60 s may be adequate for most atherosclerotic lesions, whereas several inflations of a few minutes each may be necessary for intimal hyperplasia and recurrent lesions. After deflation and withdrawal of the balloon, an angiogram is performed to verify the adequacy of the result. In cases of residual stenosis or dissection, reinflation of the balloon for a longer time may be performed and if this fails to address the problem, stenting may be considered.

Pain during balloon inflation, attributed to stretching of the adventitia, is common and could be considered as a sign of sufficient arterial dilation. The pain should resolve upon deflation of the balloon. Continued pain could be a sign of arterial rupture and should prompt immediate angiography. The possibility of rupture or dissection underscores the importance of the maintenance of the guidewire position across the lesion until completion angiography has been performed.

Complications

Access Site Complications

Hematoma

Hematoma at the vascular access site represents the most common complication of percutaneous angioplasty procedures, occurring in up to 9% of the cases [8]. Risk factors for this complication can be divided into patient-related and operator-dependent. Patient-related risk factors include increasing age, female gender, extremes in weight, use of antiplatelets or anticoagulants, renal failure and other comorbidities such as diabetes, peripheral artery disease, hyperlipidemia, hypertension and New York Heart Association class III or IV heart failure [9]. Operator-dependent risk factors include the access technique, multiple puncture attempts, posterior vessel wall puncture, the size of the sheath, the use of thrombolysis, the use of vascular closure devices, and post-procedure manual compression [8]. Ultrasound-guided access appears to be the safest technique, allowing the operator to directly locate the CFA and identify the healthier spot for needle puncture, avoiding calcified plaques, back wall injury and inadvertent puncture of adjacent veins.

Hematomas are usually small and cause limited patient discomfort. Large hematomas, however, can cause nerve compression, especially in cases of upper extremity access, deep vein thrombosis, due to compression of the adjacent vein, and even compartment syndrome. In cases of high puncture at the groin, above the inguinal ligament, bleeding control by manual compression may be impossible, leading to retroperitoneal haemorrhage. Retroperitoneal haemorrhage is a life-threatening complication as it does not present with overt bleeding and may not be diagnosed until the patient develops abdominal or back pain, hypotension and tachycardia. Diagnosis is confirmed by computed tomography angiography which, apart from the presence of the retroperitoneal hematoma, will disclose whether there is active bleeding from the distal external iliac artery. In such cases bleeding control can be achieved by either open surgical repair or by the endovascular placement of a covered stent.

Pseudoaneurysm

Pseudoaneurysms at the vascular access site represent the second most common complication, developing in 0.9–4.9% of percutaneous angioplasty procedures [10]. The risk factors are similar to those for hematoma, with a large sheath size, inadequate compression, puncture of the deep femoral artery and the use of antiplatelets or anticoagulants being the most common.

The clinical presentation of pseudoaneurysms is also similar to that of a hematoma, with small pseudoaneurysms causing minimal discomfort, whereas large pseudoaneurysms may cause compression symptoms to the adjacent structures, including veins, nerves and the skin. Differential diagnosis between a hematoma and a pseudoaneurysm is made by palpation, revealing a pulsatile mass in the case of a pseudoaneurysm, and by auscultation, revealing a systolic bruit. The diagnosis is confirmed by duplex ultrasound, revealing a characteristic turbulent flow within the aneurysm (the yin yang sign) and an equally characteristic "to and fro" flow at the pseudoaneurysm neck (Fig. 9.1).

The treatment of choice is ultrasound-guided thrombin injection, with open surgical repair reserved for pseudoaneurysms in which thrombin injection is contraindicated, namely pseudoaneurysms with a neck diameter of >3 mm, which increases the risk of distal embolization in case of thrombin injection, or with concomitant arteriovenous fistula, which poses the risk of pulmonary embolism. Ultrasound-guided manual compression can also be attempted but is time consuming, painful for the patient and labour-intensive for the

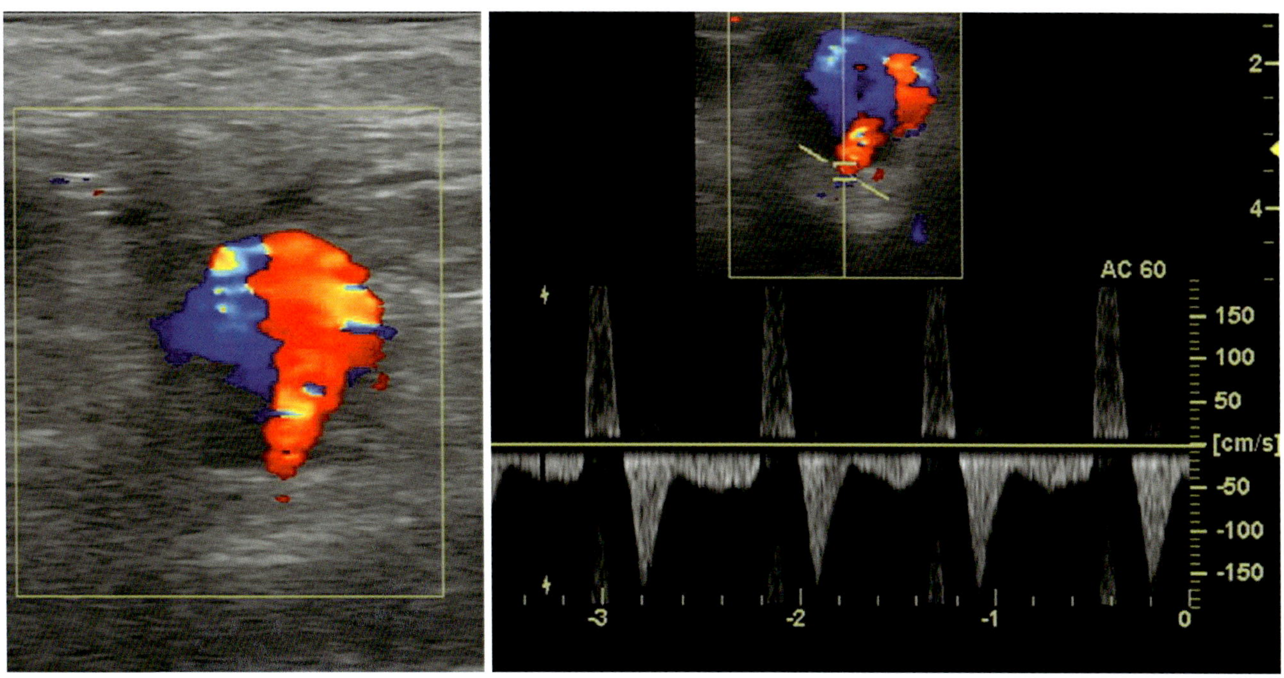

Fig. 9.1 Duplex ultrasound of a pseudoaneurysm of the common femoral artery, revealing a characteristic turbulent flow within the aneurysm (the yin yang sign) and a "to and fro" flow at the pseudoaneurysm neck

operator. The need for the ultrasound during manual compression is to help in the regulation of the amount of downward force applied on the pseudoaneurysm, which should be high enough to interrupt flow within the pseudoaneurysm but not that high to stop flow in the underlying artery. The success rate of ultrasound-guided manual compression is reported to range between 60 and 95% [11], whereas the success rate of ultrasound-guided thrombin injection exceeds 95% [10].

Arteriovenous Fistula

Arteriovenous fistulas at the vascular access site occur in 0.25–2.8% of percutaneous angioplasty procedures [12]. They occur due to a simultaneous puncture of an artery and a vein by the access needle. Since the common femoral artery and the common femoral vein lie side by side, it is impossible to puncture both vessels simultaneously. Arteriovenous fistulas usually occur between the superficial femoral or deep femoral artery and the corresponding veins or the common femoral vein. However, an arteriovenous fistula between the common femoral artery and the superficial circumflex iliac vein may also occur [12].

Patient-related risk factors include female gender and hypertension, whereas procedural risk factors include emergency procedures, the use of anticoagulation and a left-sided puncture [12, 13]. Diagnosis is suspected by clinical examination revealing the presence of a thrill or a bruit and is verified by a colour duplex ultrasound.

Since these fistulas occur due to the simultaneous puncture of an artery and a vein by an 18G needle, the outer diameter of which is 1.27 mm, they are usually small in size and rarely symptomatic. Moreover, the natural history of arteriovenous fistulas is usually benign, resulting in spontaneous resolution in 38–81% of the cases by 2–12 months [13, 14]. Symptoms may be arterial (limb ischemia), venous (limb oedema) or cardiac (high output heart failure). In such cases, closure of the arteriovenous fistula will be needed, usually by open surgery or by a covered stent in appropriately selected cases [15].

Access Site Thrombosis

Access site thrombosis is rare, occurring in less than 0.3% of peripheral angioplasty procedures [16, 17]. It is most commonly due to local dissection, caused by the puncture of a heavily calcified common femoral artery. An increased risk of thrombotic access site complications when closure devices are used has also been reported [16], but has not been confirmed in a large meta-analysis by the Cochrane Database of Systematic Reviews [18].

Access site thrombosis will present with acute limb ischemia or with worsening of a preexisting limb ischemia. The treatment of choice is open surgery (thromboendarterectomy), although an endovascular intervention through a remote access site is also an option.

Angioplasty Site Complications

Dissection

Dissection may occur during crossing of the lesion with the wire or during the inflation of the balloon. Some degree of dissection during the inflation of the balloon is inevitable since the arterial wall is stretched whereas the atherosclerotic plaque cannot. The classification system of the National Heart, Lung and Blood Institute (NHLBI) for coronary dissections after balloon angioplasty can be adopted for peripheral artery dissections [19]. Dissections in this scheme are graded based on their angiographic appearances as types A through F. Type A dissections represent minor radiolucent areas within the arterial lumen during contrast injection. Type B dissections are parallel tracts or a double lumen separated by a radiolucent area. In both type A and B dissections, there is minimal or no persistence of contrast after the dye has cleared from the lumen. Type C dissections appear as contrast outside the arterial lumen ("extraluminal cap") with persistence of contrast after dye clearance. Type D dissections represent spiral ("barber shop pole") filling defects, frequently with excessive contrast staining of the false lumen. Type E dissections appear as persistent filling defects within the arterial lumen. Type F dissections represent those that lead to total occlusion of the artery.

Minor dissections that are non-obstructive (types A and B) may be left untreated, whereas flow-limiting dissections (types C-F) need to be treated, usually by bailout stenting. Risk factors for severe dissection after balloon angioplasty of the superficial femoral artery include a vessel diameter < 5 mm, lesion length > 15 cm, and occlusion or degree of stenosis ≥70% [20, 21], whereas a prolonged balloon inflation time (>3 min) and the use of long balloons appear to be protective [22, 23].

The reported incidence of flow-limiting dissections after femoropopliteal balloon angioplasty ranges widely between 7.4 and 53% [24, 25]. Apart from being an obvious cause of procedural failure, flow-limiting dissections are associated with a significantly higher restenosis and target lesion revascularization rate [20, 26].

Distal Embolization

Distal embolization is a limb threatening complication of balloon angioplasty of the peripheral arteries. The incidence is reported to be less than 5%, when clinical and angiographic criteria are used for the diagnosis, although embolic signals by Doppler ultrasound are detected in all patients during both wire crossing and angioplasty [27, 28]. Risk factors include the presence of thrombus, TASC-D lesions and prior history of amputation [28]. Distal protection with a filter basket has been proposed as a preventive measure, but its cost-effectiveness remains to be proven. Therapeutic options include aspiration thrombectomy, mechanical thrombectomy and catheter-directed thrombolysis, whereas surgical thrombectomy may be needed when the aforementioned endovascular means have failed to restore patency.

Arterial Rupture

Arterial rupture during balloon angioplasty is a life-threatening complication, occurring in less than 1% of cases. Acute pain, that does not resolve after balloon deflation, and hypotension raise the suspicion of arterial rupture, the diagnosis of which is verified by immediate angiography showing contrast medium extravasation. Risk factors include the use of an oversized or a cutting balloon and recanalization of occlusions and high-grade stenoses in heavily calcified arteries [29, 30]. The external iliac artery appears to be the most susceptible to rupture [30]. Treatment consists of reinflating the balloon to about 1 atm across the rupture to achieve temporary hemostasis. This will give us enough time to select the appropriate stent-graft, unpack, flush and place it so that definitive control of bleeding is achieved. Surgical repair is reserved for cases where the implantation of a covered-stent is impossible or has failed to seal the bleeding.

Case Presentation

Continued from page 73

The patient was taken to the angiographic suite and placed in a supine position on the interventional table. Under local anaesthesia and ultrasonographic guidance an antegrade femoral access was obtained and a 5F sheath was inserted. Diagnostic angiography demonstrated the presence of atherosclerotic disease, causing moderate stenosis in the proximal two thirds of the SFA and a 6 cm in length occlusion of the SFA at the adductor canal (Fig. 9.2). The lesion was crossed with a 0.035-inch angled, floppy-tipped, hydrophilic

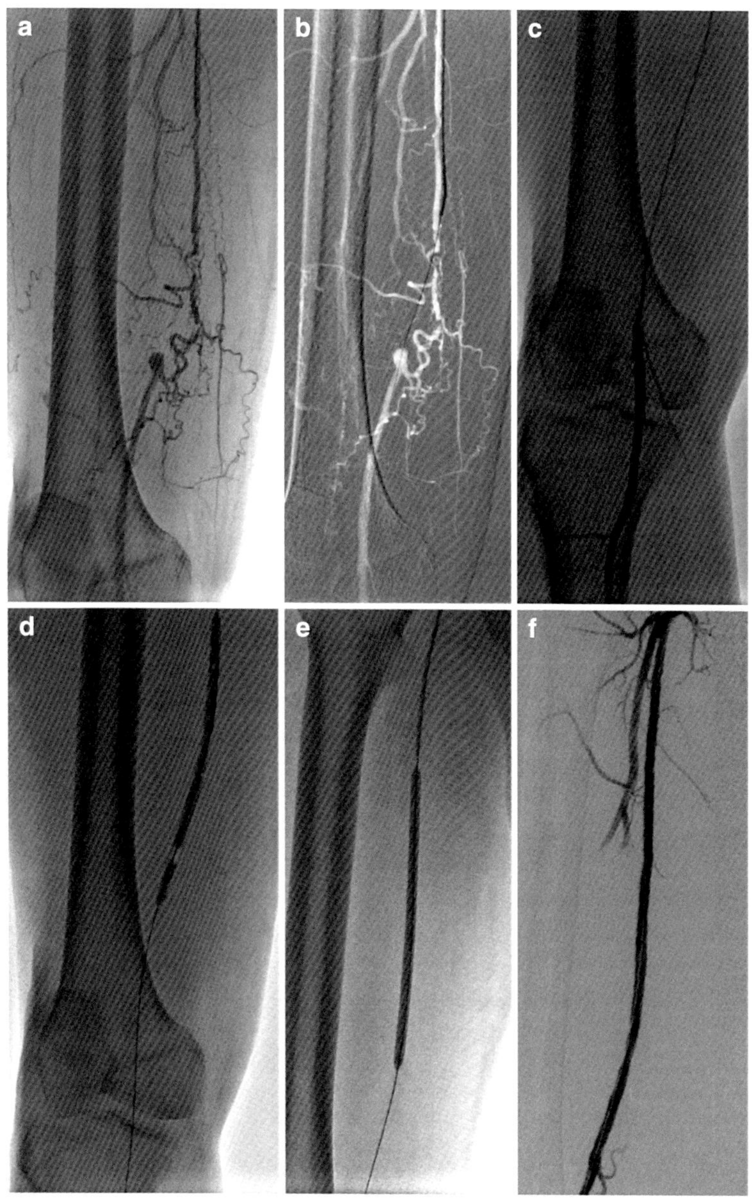

Fig. 9.2 (**a**) Diagnostic angiography confirming the presence of a 6 cm in length occlusion of the SFA at the adductor canal, (**b**) crossing of the lesion with a 0.035-inch angled, floppy-tipped, hydrophilic coated guidewire supported by an angled diagnostic catheter, (**c**) injection of contrast medium verifies the intraluminal position of the wire and catheter after crossing of the lesion, (**d**) and (**e**) balloon angioplasty with a 6 mm in diameter and 15 cm in length balloon, (**f**) completion angiography confirming the re-established SFA patency

coated guidewire supported by an angled diagnostic catheter. After crossing of the lesion, the catheter was advanced to the popliteal artery and the wire was removed. Injection of contrast medium verified the intraluminal position of the catheter. The wire was advanced again through the catheter, the catheter was removed and balloon angioplasty of the whole length of the SFA followed with a 6 mm in diameter and 15 cm in length balloon. Completion angiography confirmed the unrestricted patency of the SFA and the absence of distal embolization.

The patient reported immediate pain relief. On physical examination, the pink colour of the foot was restored, the capillary refill time was less than a second and there were palpable pulses in the popliteal artery.

The patient was placed on best medical therapy, including clopidogrel, blood pressure and lipid control. He was referred to the smoking cessation clinic and was advised to walk for at least 30 min every day. He was discharged on the following day.

References

1. Fogarty TJ, Cranley JJ, Krause RJ, Strasser ES, Hafner CD. A method for extraction of arterial emboli and thrombi. Surg Gynecol Obstet. 1963;116:241–4.
2. Dotter CT, Judkins MP. Transluminal treatment of arteriosclerotic obstruction. Description of a new technic and a preliminary report of its application. Circulation. 1964;30:654–70. https://doi.org/10.1161/01.cir.30.5.654.
3. Grüntzig A, Hopff H. Perkutane Rekanalisation chronischer arterieller Verschlüsse mit einem neuen Dilatationskatheter. Modifikation der Dotter-Technik. Dtsch Med Wochenschr. 1974;6(99):2502–11. https://doi.org/10.1055/s-0028-1108161.
4. The SH, Gussenhoven EJ, Zhong Y, Li W, van Egmond F, Pieterman H, van Urk H, Gerritsen GP, Borst C, Wilson RA, et al. Effect of balloon angioplasty on femoral artery evaluated with intravascular ultrasound imaging. Circulation. 1992;86:483–93. https://doi.org/10.1161/01.cir.86.2.483. PMID: 1386293.
5. Castaneda-Zuniga WR, Formanek A, Tadavarthy M, Vlodaver Z, Edwards JE, Zollikofer C, Amplatz K. The mechanism of balloon angioplasty. Radiology. 1980;135:565–71. https://doi.org/10.1148/radiology.135.3.7384437. PMID: 7384437
6. Schmidt W, Schmitz KP. Devices. In: Lanzer P, editor. Mastering endovascular techniques: a guide to excellence. Philadelphia: Lippincott Williams & Wilkins; 2007. p. 114–35.
7. Sullivan TM, Rizvi AZ. Technique: endovascular therapeutic. In: Cronenwett JL, Johnston KW, editors. Rutherford's vascular surgery. 16th ed. Philadelphia, PA: Elsevier; 2014. p. 1322–37.
8. Sambol E, McKinsey. Local complications: endovascular. In: Cronenwett JL, Johnston KW, editors. Rutherford's vascular surgery. 16th ed. Philadelphia, PA: Elsevier; 2014. p. 704–22.
9. Sobolev M, Slovut DP, Lee Chang A, Shiloh AL, Eisen LA. Ultrasound-guided catheterization of the femoral artery: a systematic review and meta-analysis of randomized controlled trials. J Invasive Cardiol. 2015;27:318–23. PMID: 26136279.
10. Delf J, Ramachandran S, Mustafa S, Saeed A, Kandiyil N. Factors associated with pseudoaneurysm development and necessity for reintervention: a single centre study. Br J Radiol. 2019;92:20180893. https://doi.org/10.1259/bjr.20180893. PMID: 30982331; PMCID: PMC6592087.
11. Lange P, Houe T, Helgstrand UJ. The efficacy of ultrasound-guided compression of iatrogenic femoral pseudo-aneurysms. Eur J Vasc Endovasc Surg. 2001;21:248–50. https://doi.org/10.1053/ejvs.2001.1325. PMID: 11352684.
12. Middleton WD, Robinson KA. Analysis and classification of postcatheterization femoral arteriovenous fistulas based on color Doppler examinations. J Ultrasound Med. 2022;41:207–16. https://doi.org/10.1002/jum.15696. PMID: 33720458.
13. Kelm M, Perings SM, Jax T, Lauer T, Schoebel FC, Heintzen MP, Perings C, Strauer BE. Incidence and clinical outcome of iatrogenic femoral arteriovenous fistulas: implications for risk stratification and treatment. J Am Coll Cardiol. 2002;40:291–7. https://doi.org/10.1016/s0735-1097(02)01966-6. PMID: 12106934.
14. Toursarkissian B, Allen BT, Petrinec D, Thompson RW, Rubin BG, Reilly JM, Anderson CB, Flye MW, Sicard GA. Spontaneous closure of selected iatrogenic pseudoaneurysms and arteriovenous fistulae. J Vasc Surg. 1997;25:803–9. https://doi.org/10.1016/s0741-5214(97)70209-x. PMID: 9152307.
15. Zilinyi RS, Sethi SS, Parikh MA, Parikh SA. Iatrogenic arteriovenous fistula following femoral access precipitating high-output heart failure. JACC Case Rep. 2021;3:421–4. https://doi.org/10.1016/j.jaccas.2020.12.031. PMID: 34317549; PMCID: PMC8311044.
16. Noori VJ, Eldrup-Jørgensen J. A systematic review of vascular closure devices for femoral artery puncture sites. J Vasc Surg. 2018;68:887–99. https://doi.org/10.1016/j.jvs.2018.05.019.
17. Ramirez JL, Zarkowsky DS, Sorrentino TA, Hicks CW, Vartanian SM, Gasper WJ, Conte MS, Iannuzzi JC. Antegrade common femoral artery closure device use is associated with decreased complications. J Vasc Surg. 2020;72:1610–1617.e1. https://doi.org/10.1016/j.jvs.2020.01.052.
18. Robertson L, Andras A, Colgan F, Jackson R. Vascular closure devices for femoral arterial puncture site haemostasis. Cochrane Database Syst Rev. 2016;3:CD009541. https://doi.org/10.1002/14651858.CD009541.pub2. PMID: 26948236
19. Rogers JH, Lasala JM. Coronary artery dissection and perforation complicating percutaneous coronary intervention. J Invasive Cardiol. 2004;16:493–9. PMID: 15353832.
20. Fujihara M, Takahara M, Sasaki S, Nanto K, Utsunomiya M, Iida O, Yokoi Y. Angiographic dissection patterns and patency outcomes after balloon angioplasty for superficial femoral artery disease. J Endovasc Ther. 2017;24:367–75. https://doi.org/10.1177/1526602817698634.
21. Hong H, Park UJ, Roh YN, Kim HT. Predictive factors of severe dissection after balloon angioplasty for femoropopliteal artery disease. Ann Vasc Surg. 2021;77:109–15. https://doi.org/10.1016/j.avsg.2021.05.048. PMID: 34428439.
22. Horie K, Tanaka A, Taguri M, Kato S, Inoue N. Impact of prolonged inflation times during plain balloon angioplasty on angiographic dissection in femoropopliteal lesions. J Endovasc Ther. 2018;25:683–91. https://doi.org/10.1177/1526602818799733. PMID: 30203701.
23. Tan M, Urasawa K, Koshida R, Haraguchi T, Kitani S, Igarashi Y, Sato K. Comparison of angiographic dissection patterns caused by long vs short balloons during balloon angioplasty of chronic Femoropopliteal occlusions. J Endovasc Ther. 2018;25:192–200. https://doi.org/10.1177/1526602818756610. PMID: 29429382.
24. Fanelli F, Cannavale A, Boatta E, Corona M, Lucatelli P, Wlderk A, Cirelli C, Salvatori FM. Lower limb multilevel treatment with drug-eluting balloons: 6-month results from the DEBELLUM randomized trial. J Endovasc Ther. 2012;19:571–80. https://doi.org/10.1583/JEVT-12-3926MR.1. PMID: 23046320.

25. Werk M, Albrecht T, Meyer DR, Ahmed MN, Behne A, Dietz U, Eschenbach G, Hartmann H, Lange C, Schnorr B, Stiepani H, Zoccai GB, Hänninen EL. Paclitaxel-coated balloons reduce restenosis after femoro-popliteal angioplasty: evidence from the randomized PACIFIER trial. Circ Cardiovasc Interv. 2012;5:831–40. https://doi.org/10.1161/CIRCINTERVENTIONS.112.971630. PMID: 23192918.
26. Kobayashi N, Hirano K, Yamawaki M, Araki M, Sakai T, Sakamoto Y, Mori S, Tsutsumi M, Honda Y, Ito Y. Simple classification and clinical outcomes of angiographic dissection after balloon angioplasty for femoropopliteal disease. J Vasc Surg. 2018;67(4):1151–8. https://doi.org/10.1016/j.jvs.2017.08.092. PMID: 29242063.
27. Lam RC, Shah S, Faries PL, McKinsey JF, Kent KC, Morrissey NJ. Incidence and clinical significance of distal embolization during percutaneous interventions involving the superficial femoral artery. J Vasc Surg. 2007;46(6):1155–9. https://doi.org/10.1016/j.jvs.2007.07.058. PMID: 18154991.
28. Shammas NW, Shammas GA, Dippel EJ, Jerin M, Shammas WJ. Predictors of distal embolization in peripheral percutaneous interventions: a report from a large peripheral vascular registry. J Invasive Cardiol. 2009;21:628–31. PMID: 19966364.
29. Sobrinho G, Albino JP. Delayed rupture of the external iliac artery after balloon angioplasty and stent placement. J Vasc Interv Radiol. 2008;19:460–2. https://doi.org/10.1016/j.jvir.2007.11.013. PMID: 18295711.
30. Allaire E, Melliere D, Poussier B, Kobeiter H, Desgranges P, Becquemin JP. Iliac artery rupture during balloon dilatation: what treatment? Ann Vasc Surg. 2003;17:306–14. https://doi.org/10.1007/s10016-001-0404-1. PMID: 12712371.

The Specificities of the Common Femoral Artery Anatomy, Calcification and Treatment

10

Bahaa Nasr and Yann Gouëffic

Introduction

The latest advances in percutaneous endovascular management of peripheral artery disease and the recent introduction of drug-eluting devices have led to a dramatic increase in the number of patients treated percutaneously [1, 2]. However, common femoral artery (CFA) bifurcation is a notable exception because of the strategic location of the vessel, the clinical concern for possible stent fracture, and the heavily eccentric calcified character of the CFA plaques.

Repeated studies have shown good medium-term outcome of endovascular treatment of the CFA bifurcation. Stricker et al. showed a primary patency rate of 83% at 24 months follow up. [3] In a prospective, randomized trial (TECCO trial), Gouëffic et al. reported a 30-days perioperative morbidity and mortality rate significantly higher in the open surgery group [4]. At 2 years, there were similar rates of clinical improvement, primary patency and freedom from target lesion revascularization in both arms [4].

The aim of this paper is to provide a comprehensive review of the literature on CFA anatomy and calcifications in relation to endovascular management of the CFA bifurcation.

CFA Anatomy

The common femoral artery follows the external iliac artery passing under the inguinal ligament. It then divides into the deep and superficial femoral arteries. Some physicians consider hip motion a contraindication to stent placement across the inguinal ligament for fear of kinking, fracture or development of intimal hyperplasia leading to device occlusion [5–7]. The CFA is relatively fixed between the inguinal ligament above and the Sartorius muscle below [8]. Lopez et al. found that the deformation of the CFA is the result of a movement involving bending and torsion of the axis of the artery at two different points along its course. These deformations occur below the inguinal ligament and above the Sartorius muscle [8]. Tijani et al. analyzed the deformations of the iliofemoral axis during the movements of flexion and of extension of the hip and evaluated the impact of the implantation of a CFA stent on the deformations observed [9]. In this study, the authors performed bilateral angiography to assess the deformities of the external iliac artery, common femoral artery and femoral bifurcation before and after stent implantation. They demonstrated that the CFA is a fixed segment between two points of mobility at the junction between the external iliac artery and the CFA proximally and the junction between the CFA and its bifurcation distally. The flexion of the hip joint did not reveal new points of deformation in the CFA segment from the front or the side. As far as, the implantation of a nitinol self-expandable stent in the CFA did not modify these observations. These results are consistent with the stent fractures seen in the various clinical trials. Indeed, stent fractures are often found in endovascular treatment of extensive lesions between the external iliac and common femoral arteries [4, 10].

The interwoven design of the Supera helical interwoven nitinol stent (Abbott Vascular, Santa Clara, California) appears to be an ideal candidate for this location. It has significantly greater radial strength and crush resistance than other self-expanding stents, making it theoretically attractive as a stent of choice for the frequent bending and extrinsic compression encountered in this vessel segment. This stent has demonstrated good patency rates associated with reduced stent fractures in femoropopliteal lesions [11–14]. Given the significant differences in stent design and fracture rates between the newer generations of stents, it is likely that newer generation stents may be preferred for the treatment of

B. Nasr
Department of Vascular and Endovascular Surgery, Brest University Hospital, Brest, France

Y. Gouëffic (✉)
Department of Vascular and Endovascular Surgery, Groupe Hospitalier Paris St Joseph, Paris, France
e-mail: ygoueffic@ghpsj.fr

G. Geroulakos et al. (eds.), *Mastering Endovascular Techniques*, https://doi.org/10.1007/978-3-031-42735-0_10

CFA stenosis. Unfortunately, vascular injury at the distal ends of the stent caused by hip motion could lead to the development of neointimal hyperplasia. Drug-eluting self-expanding stents may offer an advantage through a potential reduction in restenosis.

Bifurcation lesions are one of the more complex lesion subgroups that we are confronted with. Treating CFA by endovascular repair remains challenging, especially when the lesions are extended to the bifurcation and require complex endovascular technics. CFA bifurcation stenting can contribute to hemodynamic disturbances and therefore potential restenosis [15].

Azema et al. classified CFA occlusive lesions into four groups [10]. Type I represents lesions that are located at the iliac external artery and are extended to the CFA. Type II lesions are limited to the CFA. Type III represents lesions located at the CFA and its bifurcation. Type IV lesions represent proximal and distal stenosis bypass anastomosis. Rabellino et al. proposed a new classification by subclassifying type III CFA lesions and adding a new group including lesions that started at the external iliac artery extending into CFA and affecting its bifurcation [16]. The Bifurcation lesions of the CFA could also be classified according to the Medina classification, a classification developed for coronary arteries that was applied for the femoral bifurcation [17].

The cornerstone of a successful procedure is to optimally restore flow in the femoral bifurcation. In a morphological analysis of the CFA bifurcation, we reported a mean angle between the superficial femoral artery (SFA) and the deep femoral artery (DFA) of 28.47° ± 5.51°, a mean angle between the CFA and the DFA of 149.85° ± 9.85°, and a mean angle between the CFA and the SFA of 167.23° ± 9.80° [18]. We have shown a positive correlation between the bifurcation angle and the diameter of the mother vessel, meaning that the bifurcation angle is less acute when the CFA diameter is larger [18]. In these cases, T-stenting is the appropriate technique because the second stent, implanted through the struts of the first stent, completely covers the ostium of the second branch without covering the ostium of the main branch. However, the kissing stent technique is ideally performed when the bifurcation angle is acute and when the proximal CFA is free of disease and there is no need to deploy a stent proximally. We also found a 25% diameter reduction between the mother vessel and the larger daughter vessel. This reduction results in incomplete apposition of the stent struts, which prevents optimal flow in the CFA bifurcation and may increase in-stent restenosis and occlusion. To avoid this risk, "Tour Eiffel" technic could be used. Dedicated stents should be designed with diameter reduction and flow law in mind to avoid apposition failure between the mother and daughter vessels and to ensure optimal flow in the daughter vessels.

CFA Calcifications

The main problem underlying atherosclerotic disease of the CFA is the typically high calcium concentration, and the very challenging endovascular procedures when CFA is highly calcified. However, little is known about the calcification process at the femoral level. Previous studies have reported large differences in histomorphological composition with respect to lipids, intraplaque haemorrhage, inflammatory cells and calcifications. Plaque characteristics depend on its location in the arterial bed [19, 20]. Vascular calcification begins as microcalcifications, which can progress over time to macrocalcifications that can occur as a calcified plate called sheet calcification, or as osteoid metaplasia [21, 22]. Davaine et al. performed histological analysis to characterize the cellular populations present in the CFA atherosclerotic lesions [23]. Forty-three femoral plaques from femoral endarterectomy specimens were analyzed. 65% showed osteoid metaplasia, which included osteoblasts, osteoclast-like cells and bone marrow (Fig. 10.1). Azeez et al. reported a high relative amount of nodular calcification in femoral plaques [24]. The showed that a > 90% stenosed vessels contained 22.4 ± 12.3% of nodular and 14.5 ± 11.8% of sheet calcification [24]. Nodular calcification has been suggested to emerge by fracture from sheet calcification [25]. This possibly explains the highest prevalence of nodular calcification in the CFA, which is subjected to most dynamic movement, in comparison to more stagnant coronary and carotid arteries.

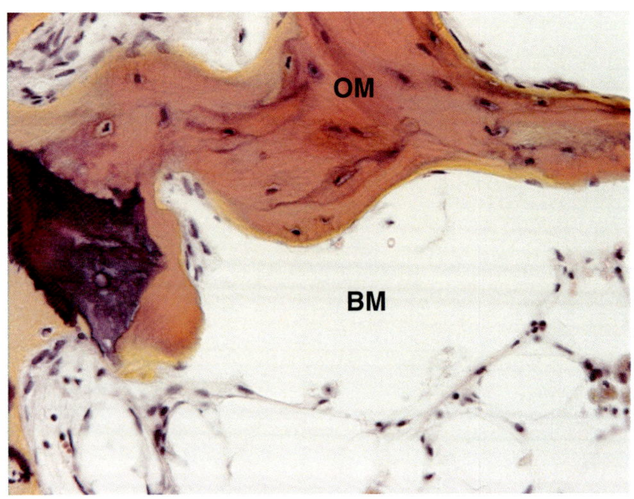

Fig. 10.1 Magnification of the area of osteoid metaplasia (OM) with osteocytes (*), and bone marrow (BM) in a common femoral plaque

CFA calcifications will have an impact in the vessel preparation, which is a key step in ensuring good subsequent treatment. Angioplasty is the most common technique. The interest of arterial preparation by balloon angioplasty has been shown by several studies [26, 27]. The randomized Zilver PTX trial showed a trend towards better patency at 5 years in the drug-eluting-stent (DES) and preparation group versus the DES without preparation group (72.4 vs 64.9%) [26]. However, balloon angioplasty works by applying constant pressure to superficial soft tissue, which can lead to excessive force, causing elastic recoil, frequent dissection and loss of blood flow. Other solutions could be considered to combat calcification. High-pressure balloons can be used in addition to simple angioplasty to remove the most refractory stenoses. Atherectomy devices allow lumen gain and it is mandatory when using drug-coated devices. As in open surgery, atherectomy aims to destroy and/or remove atheromatous plaque. Atherectomy can be directional (SilverHawk®, TurboHawk®, Hawkone®, Pantheris®), orbital (Diamondback360 R® Peripheral Orbital Atherectomy), rotative (Jetstream®, Phoenix®) or laser (Turbo-Elite® laser atherectomy catheter). Plaque modification could lead to better drug penetration into the arterial wall and improved drug uptake [28]. For eccentric calcifications, the directional atherectomy system is a good indication. However, for circumferential calcified lesions, a rotational atherectomy system or endovascular lithotripsy technique is preferred. Debulking may be particularly useful in the treatment of CFA lesions, given the high rates of osteoid metaplasia observed in the CFA atherosclerotic plaques. Cioppa et al. reported the outcomes of 30 patients with severely calcified obstructions of the CFA treated by directional atherectomy and drug-coated balloon [29]. At 12 months, they observed a primary patency rate of 93.4% and a freedom from target lesion revascularization (TLR) of 96.7%. Bailout stenting was needed in 10% of treated cases. In a retrospective study comparing the directional atherectomy with antirestenotic therapy (DAART) in the form of DCB vs DCB angioplasty alone, Stavroulakis et al. showed that CFA preparation with directional atherectomy was not associated with statistically significant higher primary patency or freedom from TLR compared to DCB angioplasty alone at 12 months [30].

Intravascular lithotripsy (IVL®; Shockwave Medical, Santa Clara, USA) was recently introduced as an endovascular option to treat the typically high calcium burden lesions. The use of IVL showed promising results in femoropopliteal and below-the-knee occlusive lesions [31, 32]. This technic is based on the use of pulsatile sonic pressure waves that interact with calcium, creating significant shear stress that has the ability to fracture the calcium. Brodmann et al. reported good acute results using the peripheral intravascular lithotripsy system use in calcified CFA lesions [33]. The mean diameter stenosis after IVL treatment was 21.3%, corresponding to an acute mean lumen gain of 3.1 ± 1.3 mm. The IVL advantage is that it allows angioplasty to be performed with less pressure, thus avoiding dissection. It also makes the artery wall more compliant, making it easier to treat with a DCB or a self-expanding stent. Stavroulakis et al. analyzed 33 cases of severely calcified CFA atherosclerotic disease treated with intravascular lithotripsy [34]. They showed the combination of IVL and DCB angioplasty for calcified CFA disease was associated with low risk for periprocedural complications, acceptable 12 months clinical outcomes, and low rates of reinterventions [34].

The CFA Treatment

Percutaneous transluminal angioplasty. After vessel preparation, the treatment is chosen according to the preparation technique and the operator's preferences. Percutaneous transluminal angioplasty (PTA) is the most commonly used technique in the treatment of femoropopliteal lesions. When used alone to treat CFA lesions, PTA was associated with a high rate of provisional stenting for suboptimal angioplasty results or for bail-out reasons. Bonvini et al. had a 38.1% of provisional stent placement [6]. At 12 months, they described a restenosis rate of 19.5% and a TLR rate of 14.1%. In a review of the literature, Wong et al. reported that CFA lesions treated with PTA alone showed primary patency rates of 59% to 88% at 1 year and 83% at 2 years [35]. Poorer outcomes of PTA are related to the eccentric calcifications of the CFA lesions, the intimal tears and dissections related to the high shear forces applied by the high balloon pressures [36].

Drug-coated balloon angioplasty. Drug-coated therapies have been developed as an attractive alternative for femoropopliteal lesions with the aim of reducing reintervention rates by preventing restenosis. It has the advantage of vascular intima hyperplasia inhibition without leaving anything behind. However, as with PTA, DCB do not prevent elastic recoil and eventually lead to flow-limiting dissections. Stavroulakis et al. compared the directional atherectomy with DCB (DAART) vs DCB angioplasty alone in the treatment of CFA occlusive lesions [30]. They treated 47 lesions (DCB 26 vs DAART 21), with a mean lesion length of 39 ± 14 mm for DCB group and 34 ± 16 mm for DAART group. Vessel calcification scores were comparable between groups. They reported a provisional stenting rate of 3.8% in the DCB group and 4.7% in the DAART group. The 12-months PP (88% vs 68%) and freedom from TLR (89% vs 75%) were higher in the DAART group without statistically significant difference.

Scaffolding strategy. Despite improvements in endovascular techniques and materials, the CFA was widely regarded as a "no stent zone". Stenting was reserved for salvaging suboptimal angioplasty results in patients who were not candidates for more definitive surgical treatment. Literature review of CFA lesions treated with stenting showed excellent mid-term outcomes [3, 4]. In a retrospective study, Stricker et al. reported the mid-term outcomes of CFA stenting in 27 occlusive lesions [3]. The PP was 87% and 83% at 12 and 24 months. In a single centre retrospective study, Thiney et al. evaluated the PTA of the CFA lesions [37]. Primary stenting was performed in 95% of cases. At 2-years, they reported a PP rate of 92.5%. The prospective randomized French TECCO trial, which compared CFA stenting with CFA endarterectomy, showed significantly lower morbidity and mortality in the endovascular group (6.4% vs 26%; p = 0.005) [4]. It also showed no difference in PP and reintervention rates at 2 years in both arms [4].

The "simple" stenting technique, using a single stent in the CFA or a single stent between the CFA and one of the daughter's arteries, must be distinguished from the "complex" stenting techniques. The choice of technique depends on the location of the lesion and its classification. Simple stenting can be used for Azema I and II lesions and in case of occlusion of one of the daughter arteries (SFA or DFA). We will describe the different complex stenting techniques used for Azema III type bifurcation lesions.

The "V" and the "simultaneous kissing stents" technique (Fig. 10.2). The "V stenting" technique consists of the implantation of 2 stents together. One stent is advanced into the DFA, the other into the SFA, and the 2 stents touch each other forming a proximal carina. When this carina extends 5 mm or more into the CFA, this technique is called "kissing stent". The most suitable lesions for this technique are ostial lesions. The stents used in this technique are often balloon expandable stents. In order to facilitate implantation, it is preferable to work with two 0.014 guides or to implant the 2 stents simultaneously by performing a retrograde puncture of the homolateral SFA. The main advantage of this technic is to never lose the vessel access. As far as, there is no need to re-cross a stent when a final kissing inflation is performed. Disadvantages include the difficulty of placing another stent proximally if needed and the gap that may be left between the kissing stent and the proximal stent.

The "T stenting" technique (Fig. 10.3). This technique consists of positioning a stent between the CFA and the DFA. If a balloon stent is used, the POT (Proximal optimization technique) can be performed at the proximal part of the stent in the CFA. This allows the stent to be applied to the CFA walls and the stent mesh to be opened at this point. The guidewire is left in place and a second guidewire is used to catheterize the SFA through the previous stent struts. The stent struts are opened by a 4 mm balloon allowing the passage and implantation of a second stent at the SFA. A final kissing balloon angioplasty allows for proper stent application and remodeling.

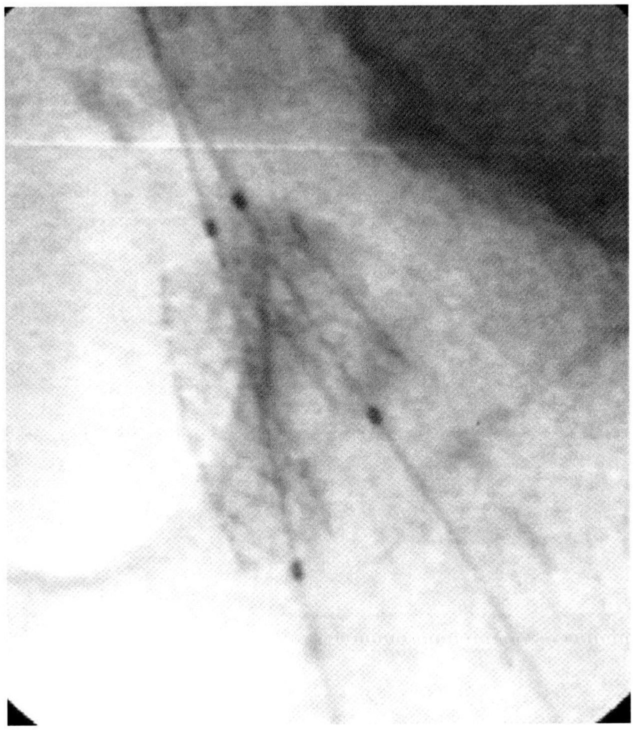

Fig. 10.2 The "V" and the "simultaneous kissing stents" technique

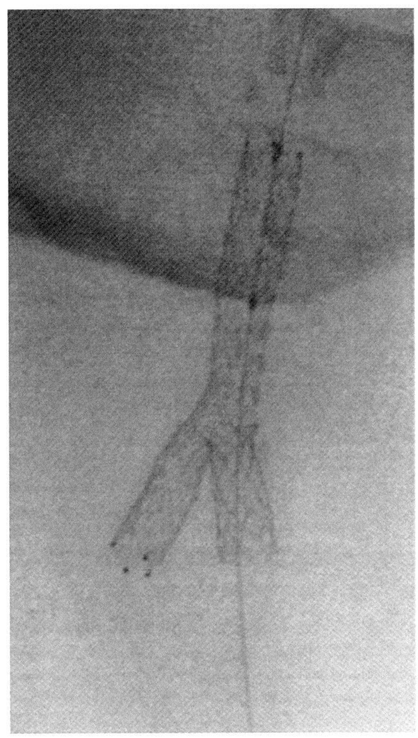

Fig. 10.3 The "T stenting" technique

The "Tour Eiffel" technique (Fig. 10.4). This technique consists of placing a self-expanding stent in the CFA. A kissing angioplasty with two 0.014 balloons is performed at the

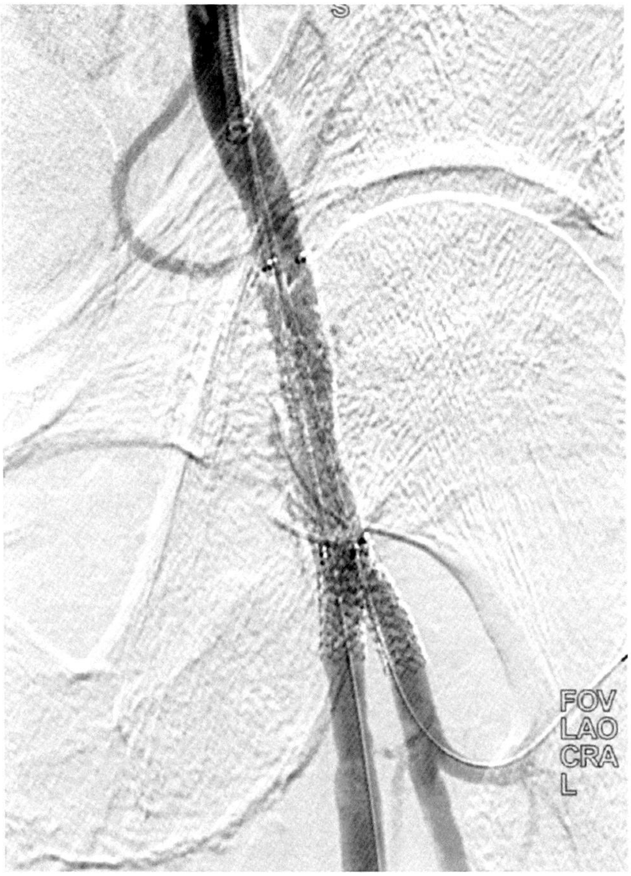

Fig. 10.4 The "Tour Eiffel" technique

ostium of both daughter arteries, extending into the existing self-expanding stent. Finally, a kissing stent is made at the level of the daughter arteries. The kissing stent can be made with two balloon stents, which protrude very slightly into the self-expanding stent.

Despite the anatomical location of the CFA and concerns about mechanical stress during hip flexion, several studies have shown a low rate of stent fracture of less than 3% [3, 37, 38]. However, the main limitation is that stenting of the CFA may preclude an important access site for percutaneous procedures. Despite these limitations, stenting with new generation devices with increased flexibility and radial force may be a treatment option in highly selected patients and in severely calcified lesions. The Supera™ stent (Abott Vascular Inc., Santa Clara, CA, USA) has shown very good results in CFA treatment [39, 40]. The interwoven design of this self-expanding stent enables it to resist the risk of calcium-related crushing, to be flexible enough to be positioned at the iliofemoral junction lesions, and to maintain any subsequent access at this level. The prospective, multicentric VMI-CFA study evaluating the Supera™ stent in the treatment of occlusive lesions of the CFA showed 100% primary patency at 6 months [41].

Discussion

CFA endarterectomy was historically recommended as the standard of care of the CFA occlusive lesions, given its excellent, durable patency and long-term freedom from reintervention. However, open surgery is associated with high rates of morbidity and mortality, particularly in obese patients and those with other cardiopulmonary morbidities. Studies have reported increased complication rates. In a large registry, Nguyen et al. reported a morbidity and mortality rate of 15%, including a death rate of 3.4% [42]. In a randomized trial comparing surgery with bioresorbable stents, Linni et al. observed a surgical site infection rate of 18% [43]. Endovascular treatment of CFA has emerged as an alternative to open surgery with a low rate of local and overall complications as a major advantage [4]. The specificities of the CFA are its anatomy and the bulky, eccentric calcifications. As far as, femoral bifurcation involvement is common, with disease extending into the proximal DFA and SFA. In a retrospective observational study describing the morphology and the composition of CFA lesions, Kaneta et al. showed that 65% of lesions were Azema III lesions [44]. In terms of calcification, a total 33% CFAs had a calcium burden greater than 1.1 cm^3, which was identified as a potential threshold for the need to stent the lesion [44].

An analysis based on the TECCO trial comparing clinical outcomes of simple and complex stenting techniques, showed no significant difference between the two groups, regarding the primary patency rates at 24 months (simple lesions 86.3% vs complex lesions 79%; $p = 0.66$) and freedom from TLR (simple lesions 93.3% vs complex lesions 82%; $p = 0.34$) [45]. To our knowledge, this is the only study from a randomized trial that compares all endovascular treatment strategies for the CFA bifurcation. However, as reported in the coronary literature, bifurcation lesions are one of the more complex lesion subgroups that we are increasingly confronted with. In this field, many techniques have been used and described in randomized studies and meta-analysis; [46, 47] despite this great interest for these lesions, the optimal stenting strategy remains a subject of debate. Atherectomy devices combined with drug coating balloon could be a promising alternative to CFA bifurcation stenting. PESTO trial is a prospective, multicentre randomized study aiming to compare the performance of directional atherectomy and DCB against open surgical reconstruction in CFA lesions (ClinicalTrials.gov Identifier: NCT02517827). However, distal embolization remains a major disadvantage with atherectomy and hence use of embolic protection devices is recommended [48].

The use of self-expandable stent in the CFA decrease the extrinsic compression and make possible a redo CFA puncture through the stent mesh. Balloon expandable stents should be preserved for DFA and SFA ostial stenosis to allow precise ostial positioning of the stent. Paris et al. showed that CFA stenting did not preclude future CFA vascular access [49]. By scaffolding the vessel wall with high radial outward force, stents improve procedural outcomes, by avoiding the problems of elastic recoil, residual stenosis, and dissection after balloon angioplasty. However, beyond the acute treatment phase, scaffolding the vessel can cause inflammation and lead to intimal hyperplasia formation and in-stent restenosis [50]. Therefore, the "combination strategy" of DCB and BMS seems to be a promising therapy. Supera helical interwoven nitinol stent (Abott Vascular, Santa Clara, California) in combination with a DCB and drug-eluting self-expanding stents could offer an advantage in term of crush resistance and potential reduction of restenosis. For the femoropopliteal lesions, the RAPID trial compared patency rates of the LEGFLOW DCB (Cardionovum GmbH, Bonn, Germany) in combination with a SUPERA stent to the SUPERA stent alone [51]. The primary patency at 1 year was 68.3% in the DCB + Supera stent group vs 62.0% in the Supera stent group ($p = 0.9$) in the intention-to-treat analysis and 74.7% in the DCB + Supera stent group vs 62% in the control group ($p = 0.273$) in the per-protocol analysis [51].

Technological innovations in the treatment of calcified plaque, the use of stents with specific characteristics for this area, and the promising short- and medium-term results allow us to consider endovascular treatment of occlusive lesions of the femoral bifurcation as a first-line treatment, especially in patients at high risk of complications of scarpa. However, we need further studies with large numbers of patients to define the optimal stenting technique and stent type to be used in Azema type III lesions. We also need additional long-term data to ensure that CFA stenting does not limit future surgical or endovascular options in cases of stent failure. Finally, we need randomized trial comparing efficacy of CFA stenting and "leave nothing behind" strategy for the treatment of CFA disease.

Conclusion

A decade ago, CFA endarterectomy was considered the only way to treat occlusive lesions of the CFA, as it was widely considered a "no stent zone". However, the excellent short and medium term results of the randomized trial, the development of new techniques for vessel preparation and the improvement of the stents used have allowed a move towards endovascular first-line treatment.

References

1. DeRubertis BG, Faries PL, McKinsey JF, Chaer RA, Pierce M, Karwowski J, et al. Shifting paradigms in the treatment of lower extremity vascular disease: a report of 1000 percutaneous interventions. Ann Surg. 2007;246(3):415–22; discussion 422-424. https://doi.org/10.1097/SLA.0b013e31814699a2.
2. Gallagher KA, Meltzer AJ, Ravin RA, Graham A, Shrikhande G, Connolly PH, et al. Endovascular management as first therapy for chronic total occlusion of the lower extremity arteries: comparison of balloon angioplasty, stenting, and directional atherectomy. J Endovasc Ther. 2011;18(5):624–37. https://doi.org/10.1583/11-3539.1.
3. Stricker H, Jacomella V. Stent-assisted angioplasty at the level of the common femoral artery bifurcation: midterm outcomes. J Endovasc Ther. 2004;11(3):281–6. https://doi.org/10.1583/03-1169.1.
4. Gouëffic Y, Della Schiava N, Thaveau F, Rosset E, Favre J-P, Salomon du Mont L, et al. Stenting or surgery for De Novo common femoral artery stenosis. JACC Cardiovasc Interv. 2017;10(13):1344–54. https://doi.org/10.1016/j.jcin.2017.03.046.
5. Park SI, Won JH, Kim BM, Kim JK, Lee DY. The arterial folding point during flexion of the hip joint. Cardiovasc Intervent Radiol. 2005;28(2):173–7. https://doi.org/10.1007/s00270-004-0190-6.
6. Bonvini RF, Rastan A, Sixt S, Beschorner U, Noory E, Schwarz T, et al. Angioplasty and provisional stent treatment of common femoral artery lesions. J Vasc Interv Radiol. 2013;24(2):175–83. https://doi.org/10.1016/j.jvir.2012.10.020.
7. Soga Y, Tomoi Y, Sato K, Iida O, Yokoi H. Clinical outcome after endovascular treatment for isolated common femoral and popliteal artery disease. Cardiovasc Interv Ther. 2013;28(3):250–7. https://doi.org/10.1007/s12928-013-0164-1.
8. Lopez JF, Magne JL, Champetier J. The femoral artery and flexion of the hip joint. Surg Radiol Anat. 1989;11(4):275–81. https://doi.org/10.1007/BF02098696.
9. Tijani Y, Burgaud M, Hamel A, Raux M, Nasr B, Gouëffic Y. The common femoral artery is a fixed arterial segment. Ann Vasc Surg. 2021;73:51–4. https://doi.org/10.1016/j.avsg.2020.10.049.
10. Azéma L, Davaine JM, Guyomarch B, Chaillou P, Costargent A, Patra P, et al. Endovascular repair of common femoral artery and concomitant arterial lesions. Eur J Vasc Endovasc Surg. 2011;41(6):787–93. https://doi.org/10.1016/j.ejvs.2011.02.025.
11. Bishu K, Armstrong EJ. Supera self-expanding stents for endovascular treatment of femoropopliteal disease: a review of the clinical evidence. Vasc Health Risk Manag. 2015;11:387–95. https://doi.org/10.2147/VHRM.S70229.
12. Dumantepe M, Seren M, Fazlıogullari O, Ayoglu U, Teymen B. Treatment of complex atherosclerotic femoropopliteal artery disease with a self-expanding interwoven nitinol stent: midterm results. Vascular. 2021;29(5):711–9. https://doi.org/10.1177/1708538114568884.
13. Garcia L, Jaff MR, Metzger C, Sedillo G, Pershad A, Zidar F, et al. Wire-interwoven nitinol stent outcome in the superficial femoral and proximal popliteal arteries: twelve-month results of the SUPERB trial. Circ Cardiovasc Interv. 2015;8(5):e000937. https://doi.org/10.1161/CIRCINTERVENTIONS.113.000937.
14. Nasr B, Gouailler F, Marret O, Guillou M, Chaillou P, Guyomarc'h B, et al. Treatment of long femoropopliteal lesions with self-expanding interwoven nitinol stent: 24 month outcomes of the STELLA-SUPERA trial. J Endovasc Ther. 2023;30(1):98–105. https://doi.org/10.1177/15266028221075227.
15. Finet G, Gilard M, Perrenot B, Rioufol G, Motreff P, Gavit L, et al. Fractal geometry of arterial coronary bifurcations: a quantitative coronary angiography and intravascular ultrasound analysis. EuroIntervention. 2008;3(4):490–8. https://doi.org/10.4244/eijv3i4a87.

16. Rabellino M, Valle Raleigh J, Chiabrando JG, Di Caro V, Chas J, Garagoli F, et al. Novel common femoral artery lesion classification in patients undergoing endovascular revascularization. Cardiovasc Intervent Radiol. 2022;45(4):438–47. https://doi.org/10.1007/s00270-021-03011-z.
17. Bonvini RF, Rastan A, Sixt S, Noory E, Schwarz T, Frank U, et al. Endovascular treatment of common femoral artery disease: medium-term outcomes of 360 consecutive procedures. J Am Coll Cardiol. 2011;58(8):792–8. https://doi.org/10.1016/j.jacc.2011.01.070.
18. Nasr B, Didier R, Carret M, Gilard M, Gouny P, Gouëffic Y. Stent size optimization in the femoral bifurcation using a fractal model: a morphological analysis. Ann Vasc Surg. 2021;72:57–65. https://doi.org/10.1016/j.avsg.2020.12.002.
19. Herisson F, Heymann M-F, Chétiveaux M, Charrier C, Battaglia S, Pilet P, et al. Carotid and femoral atherosclerotic plaques show different morphology. Atherosclerosis. 2011;216(2):348–54. https://doi.org/10.1016/j.atherosclerosis.2011.02.004.
20. Aboyans V, Ricco J-B, Bartelink M-LEL, Björck M, Brodmann M, Cohnert T, et al. Editor's choice - 2017 ESC guidelines on the diagnosis and treatment of peripheral arterial diseases, in collaboration with the European Society for Vascular Surgery (ESVS). Eur J Vasc Endovasc Surg. 2018;55(3):305–68. https://doi.org/10.1016/j.ejvs.2017.07.018.
21. Laivuori M, Tolva J, Lokki AI, Linder N, Lundin J, Paakkanen R, et al. Osteoid metaplasia in femoral artery plaques is associated with the clinical severity of lower extremity artery disease in men. Front Cardiovasc Med. 2020;7:594192. https://doi.org/10.3389/fcvm.2020.594192.
22. Cardoso L, Weinbaum S. Microcalcifications, their genesis, growth, and biomechanical stability in fibrous cap rupture. Adv Exp Med Biol. 2018;1097:129–55. https://doi.org/10.1007/978-3-319-96445-4_7.
23. Davaine J-M, Quillard T, Chatelais M, Guilbaud F, Brion R, Guyomarch B, et al. Bone like arterial calcification in femoral atherosclerotic lesions: prevalence and role of osteoprotegerin and pericytes. Eur J Vasc Endovasc Surg. 2016;51(2):259–67. https://doi.org/10.1016/j.ejvs.2015.10.004.
24. Azeez M, Laivuori M, Tolva J, Linder N, Lundin J, Albäck A, et al. High relative amount of nodular calcification in femoral plaques is associated with milder lower extremity arterial disease. BMC Cardiovasc Disord. 2022;22(1):563. https://doi.org/10.1186/s12872-022-02945-7.
25. Sakamoto A, Virmani R, Finn AV, Gupta A. Calcified nodule as the cause of acute coronary syndrome: connecting bench observations to the bedside. Cardiology. 2018;139(2):101–4. https://doi.org/10.1159/000485119.
26. Dake MD, Ansel GM, Jaff MR, Ohki T, Saxon RR, Smouse HB, et al. Durable clinical effectiveness with paclitaxel-eluting stents in the femoropopliteal artery: 5-year results of the Zilver PTX randomized trial. Circulation. 2016;133(15):1472–83.; discussion 1483. https://doi.org/10.1161/CIRCULATIONAHA.115.016900.
27. Zorger N, Manke C, Lenhart M, Finkenzeller T, Djavidani B, Feuerbach S, et al. Peripheral arterial balloon angioplasty: effect of short versus long balloon inflation times on the morphologic results. J Vasc Interv Radiol. 2002;13(4):355–9. https://doi.org/10.1016/s1051-0443(07)61736-9.
28. Tellez A, Dattilo R, Mustapha JA, Gongora CA, Hyon CM, Palmieri T, et al. Biological effect of orbital atherectomy and adjunctive paclitaxel-coated balloon therapy on vascular healing and drug retention: early experimental insights into the familial hypercholesterolaemic swine model of femoral artery stenosis. EuroIntervention. 2014;10(8):1002–8. https://doi.org/10.4244/EIJY14M10_03.
29. Cioppa A, Stabile E, Salemme L, Popusoi G, Pucciarelli A, Iacovelli F, et al. Combined use of directional atherectomy and drug-coated balloon for the endovascular treatment of common femoral artery disease: immediate and one-year outcomes. EuroIntervention. 2017;12(14):1789–94. https://doi.org/10.4244/EIJ-D-15-00187.
30. Stavroulakis K, Schwindt A, Torsello G, Beropoulis E, Stachmann A, Hericks C, et al. Directional atherectomy with antirestenotic therapy vs drug-coated balloon angioplasty alone for common femoral artery atherosclerotic disease. J Endovasc Ther. 2018;25(1):92–9. https://doi.org/10.1177/1526602817748319.
31. Brodmann M, Werner M, Brinton TJ, Illindala U, Lansky A, Jaff MR, et al. Safety and performance of lithoplasty for treatment of calcified peripheral artery lesions. J Am Coll Cardiol. 2017;70(7):908–10. https://doi.org/10.1016/j.jacc.2017.06.022.
32. Brodmann M, Holden A, Zeller T. Safety and feasibility of intravascular lithotripsy for treatment of below-the-knee arterial STENOSES. J Endovasc Ther. 2018;25(4):499–503. https://doi.org/10.1177/1526602818783989.
33. Brodmann M, Schwindt A, Argyriou A, Gammon R. Safety and feasibility of intravascular lithotripsy for treatment of common femoral artery stenoses. J Endovasc Ther. 2019;26(3):283–7. https://doi.org/10.1177/1526602819844998.
34. Stavroulakis K, Torsello G, Chlouverakis G, Bisdas T, Damerau S, Tsilimparis N, et al. Intravascular lithotripsy and drug-coated balloon angioplasty for severely calcified common femoral artery atherosclerotic disease. J Endovasc Ther. 2023:15266028231158312. https://doi.org/10.1177/15266028231158313.
35. Wong G, Lahsaei S, Aoun J, Garcia LA. Management of common femoral artery occlusive disease: a review of endovascular treatment strategies and outcomes. Catheter Cardiovasc Interv. 2019;93(3):514–21. https://doi.org/10.1002/ccd.27983.
36. Fitzgerald PJ, Ports TA, Yock PG. Contribution of localized calcium deposits to dissection after angioplasty. An observational study using intravascular ultrasound. Circulation. 1992;86(1):64–70. https://doi.org/10.1161/01.cir.86.1.64.
37. Thiney P-O, Millon A, Boudjelit T, Della Schiava N, Feugier P, Lermusiaux P. Angioplasty of the common femoral artery and its bifurcation. Ann Vasc Surg. 2015;29(5):960–7. https://doi.org/10.1016/j.avsg.2015.02.001.
38. Nasr B, Kaladji A, Vent P-A, Chaillou P, Costargent A, Quillard T, et al. Long-term outcomes of common femoral artery stenting. Ann Vasc Surg. 2017;40:10–8. https://doi.org/10.1016/j.avsg.2016.07.088.
39. Scheinert D, Scheinert S, Sax J, Piorkowski C, Bräunlich S, Ulrich M, et al. Prevalence and clinical impact of stent fractures after femoropopliteal stenting. J Am Coll Cardiol. 2005;45(2):312–5. https://doi.org/10.1016/j.jacc.2004.11.026.
40. Werner M, Paetzold A, Banning-Eichenseer U, Scheinert S, Piorkowski M, Ulrich M, et al. Treatment of complex atherosclerotic femoropopliteal artery disease with a self-expanding interwoven nitinol stent: midterm results from the Leipzig SUPERA 500 registry. EuroIntervention. 2014;10(7):861–8. https://doi.org/10.4244/EIJV10I7A147.
41. Deloose K, Martins I, Neves C, Callaert J. Endovascular treatment for the common femoral artery: is there a challenger to open surgery? J Cardiovasc Surg. 2019;60(1):8–13. https://doi.org/10.23736/S0021-9509.18.10787-7.
42. Nguyen B-N, Amdur RL, Abugideiri M, Rahbar R, Neville RF, Sidawy AN. Postoperative complications after common femoral endarterectomy. J Vasc Surg. 2015;61(6):1489–1494.e1. https://doi.org/10.1016/j.jvs.2015.01.024.
43. Linni K, Ugurluoglu A, Hitzl W, Aspalter M, Hölzenbein T. Bioabsorbable stent implantation vs. common femoral artery endarterectomy: early results of a randomized trial. J Endovasc Ther. 2014;21(4):493–502. https://doi.org/10.1583/14-4699R.1.
44. Kaneta G, Husain S, Musto L, Hamakarim T, Elsharkawi A, Littlejohn S, et al. Editor's choice - eligibility of common femoral artery atherosclerotic disease for endovascular treatment - the

CONFESS study. Eur J Vasc Endovasc Surg. 2022;64(6):684–91. https://doi.org/10.1016/j.ejvs.2022.08.034.
45. Nasr B, Della Schiava N, Thaveau F, Rosset E, Favre J-P, Salomon du Mont L, et al. The common femoral artery bifurcation lesions: clinical outcome of simple versus complex stenting techniques - an analysis based on the TECCO trial. Ann Vasc Surg. 2020;64:2–10. https://doi.org/10.1016/j.avsg.2019.10.031.
46. Sawaya FJ, Lefèvre T, Chevalier B, Garot P, Hovasse T, Morice M-C, et al. Contemporary approach to coronary bifurcation lesion treatment. JACC Cardiovasc Interv. 2016;9(18):1861–78. https://doi.org/10.1016/j.jcin.2016.06.056.
47. Latib A, Colombo A. Bifurcation disease: what do we know, what should we do? JACC Cardiovasc Interv. 2008;1(3):218–26. https://doi.org/10.1016/j.jcin.2007.12.008.
48. Zeller T, Langhoff R, Rocha-Singh KJ, Jaff MR, Blessing E, Amann-Vesti B, et al. Directional Atherectomy followed by a paclitaxel-coated balloon to inhibit restenosis and maintain vessel patency: twelve-month results of the DEFINITIVE AR study. Circ Cardiovasc Interv. 2017;10(9):e004848. https://doi.org/10.1161/CIRCINTERVENTIONS.116.004848.
49. Paris CL, White CJ, Collins TJ, Jenkins JS, Reilly JP, Grise MA, et al. Catheter-based therapy of common femoral artery atherosclerotic disease. Vasc Med. 2011;16(2):109–12. https://doi.org/10.1177/1358863X11404280.
50. Laird JR, Katzen BT, Scheinert D, Lammer J, Carpenter J, Buchbinder M, et al. Nitinol stent implantation versus balloon angioplasty for lesions in the superficial femoral artery and proximal popliteal artery: twelve-month results from the RESILIENT randomized trial. Circ Cardiovasc Interv. 2010;3(3):267–76. https://doi.org/10.1161/CIRCINTERVENTIONS.109.903468.
51. de Boer SW, van den Heuvel DAF, de Vries-Werson DAB, Vos JA, Fioole B, Vroegindeweij D, et al. Short-term results of the RAPID randomized trial of the legflow paclitaxel-eluting balloon with supera stenting vs supera stenting alone for the treatment of intermediate and long superficial femoral artery lesions. J Endovasc Ther. 2017;24(6):783–92. https://doi.org/10.1177/1526602817725062.

Re-Entry and Recanalisation Techniques in the Endovascular Management of Peripheral Arterial Disease

11

Gregory Makris, Michael Parker, and Tariq Ali

Case Presentation

An 84-year-old female presented to the vascular surgeons with dry gangrene of the right hallux. CT angiography demonstrated diffuse calcific stenoses of the superficial femoral artery (SFA) with an occluded popliteal artery and single-vessel run-off to the foot via the anterior tibial (AT) artery.

Under local anaesthesia, an antegrade right common femoral artery (CFA) puncture was performed with a standard access needle before a 4Fr catheter was advanced to the level of the lesser trochanter to perform digital subtraction angiography (DSA). This demonstrated a long-segment occlusion of the SFA-popliteal artery with reconstitution of flow at the level of P3. Multifocal moderate SFA stenoses were also seen with severe focal stenosis at the origin of the AT. Variant anatomy was noted with the peroneal artery arising directly from the popliteal artery above the TPT. The AT was the sole runoff vessel (Fig. 11.1).

An 0.035″ wire was advanced to P1 through the catheter before exchanging the catheter for a 6Fr destination sheath. 5000 IU heparin IA was administered at the time of insertion. Attempting to cross the SFA/popliteal occlusion with a selection of 0.018″ and 0.014″ wires proved unsuccessful, and therefore the decision was made to use a retrograde puncture of the DP with a long micropuncture kit.

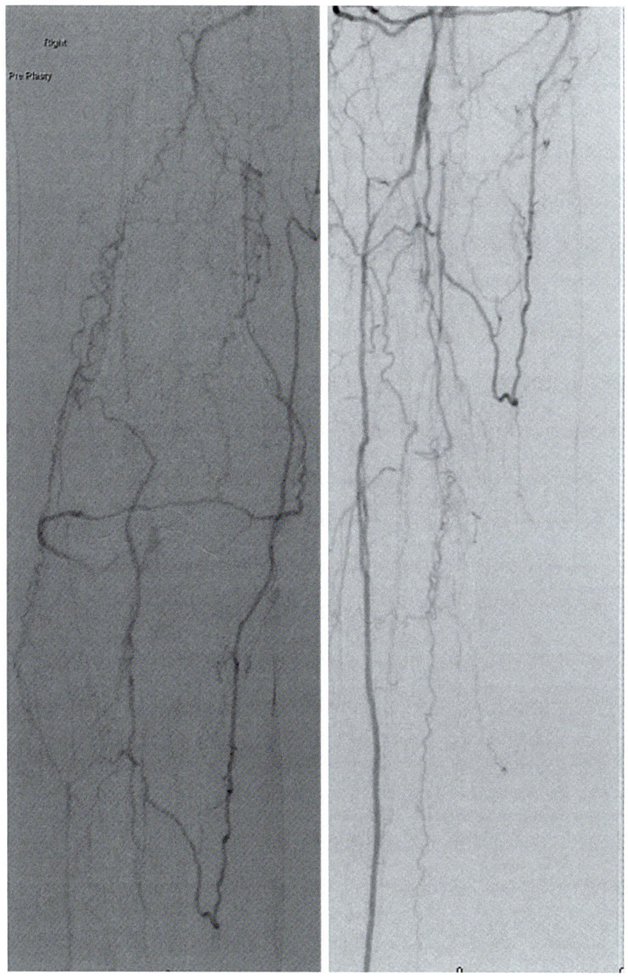

Fig. 11.1 LEFT: Demonstration of the SFA-popliteal occlusion. RIGHT: Short focal proximal AT severe stenosis

G. Makris (✉)
Guys' and St Thomas' NHS Foundation Trust, London, UK
e-mail: g.makris@aibs.gr

M. Parker · T. Ali
Norfolk and Norwich University Hospitals, Norwich, UK
e-mail: MICHAEL.PARKER@nnuh.nhs.uk; TARIQ.ALI@nnuh.nhs.uk

G. Geroulakos et al. (eds.), *Mastering Endovascular Techniques*, https://doi.org/10.1007/978-3-031-42735-0_11

Background

Peripheral arterial disease (PAD) of the lower limb affects approximately 230 million people worldwide and is associated with multiple negative health outcomes both locally (ulcers, amputation) and distally (stroke, ischaemic heart disease). PAD is a progressive disease characterised by stenosis of arteries outside the coronary, abdominal, and cerebral circulation. The lower limbs are affected more frequently than the upper limbs with symptoms ranging from intermittent claudication to critical limb ischaemia; both of which confer increased morbidity and mortality upon patients. The estimated overall prevalence of PAD in Western societies is 15%, increasing to 30% in older individuals. The estimated associated mortality rate of PAD is increasing, with 40,000 associated deaths in 2013 (an increase of 155% from 1990), according to the Global Burden of Disease Study [5–7].

The major risk factors for PAD include smoking, hyper/dyslipidaemia, diabetes mellitus (DM), obesity, and a family history of PAD. Research from the National Health and Nutrition Examination Survey demonstrated that 95% of patients with PAD had at least one of the aforementioned major risk factors and 70% had two or more. The American Heart Association (AHA) and American College of Cardiology (ACC) have identified four patient groups who are at increased risk of PAD [8, 12]:

1. Patients aged 65 or over.
2. Patients aged 50 to 64 with major risk factors for atherosclerosis (smoking history, DM, hyperlipidaemia, hypertension, or a family history of PAD).
3. Patients under 50 years of age with DM and one or more additional risk factors for atherosclerosis.
4. Patients with known atherosclerotic disease in another vascular bed (e.g. coronary artery disease, abdominal aortic aneurysm, mesenteric artery stenosis).

PAD of the lower limb can be broken down into *four clinical categories*: Asymptomatic, intermittent claudication, acute, and chronic limb ischaemia. Up to 50% of patients with PAD are asymptomatic, often with symptoms masked by concomitant medical comorbidities, such as decreased mobility and peripheral neuropathy secondary to DM. Intermittent claudication is the presence of pain or cramping in the limb upon exertion which is often predictable and reproducible and relieved by rest. Symptoms are most felt in the calves but, depending on the level of the stenosis, may also occur in the thighs or buttocks. Intermittent claudication is experienced by approximately one-third of patients with lower-limb PAD [9–11].

Critical limb ischaemia (CLI) is characterised by ischaemic pain at rest, limb ulceration, and gangrene over a period of over 2 weeks' duration. Rest pain is usually relieved by the patient placing their foot in a dependent position, such as hanging it over the edge of a bed. The diagnosis of CLI is dependent on both clinical signs and symptoms with objective evidence of PAD such as the ankle-brachial pressure index (ABPI) <0.9, transcutaneous oxygen tension ($TcPO_2$) <55 mmHg, or skin perfusion pressure < 50 mmHg. It is worth noting that in the presence of diabetes mellitus a normal or raised ABPI should not be used to exclude PAD due to confounding effects of calcified vessels giving a falsely raised score [13, 14].

The Rutherford and Fontaine Classification systems are used to quantify the severity of symptoms of PAD [12]. However, the Fontaine system is usually used for research purposes as opposed to in routine clinical practice. The Rutherford test aligns the severity of clinical symptoms based on the effects of walking for 5 min on a treadmill at 2 miles per hour at a 12% incline with severity based solely on the patient's symptoms (Tables 11.1 and 11.2):

Finally, acute limb ischaemia (ALI) is defined by a sudden-onset loss of limb perfusion within 2 weeks of a vascular insult. It is a medical emergency which can result in permanent disability, tissue loss, amputation, or death. Speed is of the essence in managing ALI, as tissue loss can be seen within 4–6 h from the initial event. The commonest symptoms of ALI are pain, pallor, paraesthesia, absent or decreased pulses, paralysis, and poikilothermia. Common causes of ALI are in-situ thrombus or embolus obstructing an already stenosed vessel, trauma, dissection, or occlusion of a stent or bypass graft [15–18].

It is worth noting that anatomical assessment with angiography (direct or cross-sectional) or with a duplex ultrasound is only indicated in symptomatic patients who would be considered for revascularisation. Patients who would not be treated with revascularisation are not advised to have imaging, as it would not change their management.

Table 11.1 The Rutherford classification system

Grade	Category	Presentation
0	0	Asymptomatic
1	1	Mild claudication
1	2	Moderate claudication
1	3	Severe claudication
2	4	Ischaemic rest pain
2	5	Minor tissue loss
2	6	Major tissue loss

Table 11.2 The Fontaine classification system

Stage	Clinical findings
I	Asymptomatic
IIa	Mild claudication
IIb	Moderate or severe claudication
III	Ischaemic rest pain
IV	Ulceration or gangrene

Conservative Management

Management of PAD should also encompass the patient's increased risk of ischaemic heart disease. Patients should be offered smoking cessation advice, antihypertensives, statins, glycaemic control (in the presence of DM), and antiplatelet therapy. Evidence suggests that ACE inhibitors and angiotensin-II receptor antagonists significantly reduce cardiovascular events in patients with PAD [8, 19, 23].

The use of single-antiplatelet therapy in patients with asymptomatic PAD and an ABPI <0.9 is controversial, with no clear evidence to suggest that it confers improved outcomes. In saying that, it is hypothesised that the use of antiplatelet agents may benefit these patients due to their increased risk of cardiovascular events. Single antiplatelet therapy with either aspirin or clopidogrel has been demonstrated to significantly decrease cardiovascular events and death secondary to vascular events in this patient population. The Clopidogrel Versus Aspirin in Patients at Risk of Ischaemic Events (CAPRIE) trial demonstrated that clopidogrel was superior to aspirin in reducing both cardiovascular risk and bleeding events in patients with symptomatic PAD. However, the VOYAGER study demonstrated that major adverse limb and cardiovascular risks were significantly reduced when aspirin was combined with Rivaroxaban, although with a slightly increased risk of bleeding when compared with Aspin alone.

Indications for Intervention

Revascularisation is indicated in patients who have been advised on the benefits of lifestyle modification, a supervised exercise programme has not significantly improved symptoms, and imaging has demonstrated the patient is suitable for revascularisation [1]. In the majority of cases, a duplex ultrasound is sufficient in patients with lower-limb PAD to assess for suitability for revascularisation; however, should there be any ambiguity then contrast-enhanced MRI or CT is advised in this patient group before proceeding to intervention. It is strongly advised that patients undergo an MDT assessment prior to proceeding to treatment for chronic PAD.

Treatment Considerations

Imaging findings for lower-limb PAD are classified as per the Trans-Atlantic Inter-Society Consensus on Management of Peripheral Arterial Disease (TASC II) for both above- and below the knee disease (Tables 11.3 and 11.4):

According to the TASC II recommendations, endovascular management is preferred for Type A and B lesions, and surgery for Type D. Ideally Type C lesions should be surgically managed, however treatment strategies for both type B and C lesions should be designed with the patient's preferences, comorbidities, and local practice and experience in consideration.

Table 11.3 The TASC classification of radiological findings of above-knee PAD [2]

Classification	Findings
Type A	• Unilateral or bilateral common iliac artery (CIA) stenosis, or • Unilateral or bilateral single stenosis of the external iliac artery (EIA) <3 cm, or • Single stenosis <10 cm or single occlusion <5 cm within the common femoral (CFA) or popliteal artery
Type B	• <3 cm stenosis of the infrarenal aorta, or • Single or multiple stenosis totalling 3–10 cm involving the EIA but not extending into the CFA, or • Unilateral EIA occlusion not involving the origins of the EIA or CFA
Type C	• Bilateral CIA occlusions, or • Bilateral EIA stenosis 3–10 cm in length and not involving the CFA, or • Unilateral EIA stenosis extending into the CFA, or • Unilateral EIA occlusion involving the origins of the IIA and/or CFA, or • Heavily calcifie unilateral EIA occlusion with or without involvement of the origins of the IIA and/or CFA, or • Multiple CFA/popliteal stenoses or occlusions >15 cm in total length, or • Recurrent stenoses or occlusions which require >2 interventions
Type D	• Infrarenal aortic occlusion, or • Diffuse disease involving the aorta and both iliac arteries requiring treatment, or • Multiple stenoses/occlusions involving the unilateral CIA, EIA, and/or CFA, or • Bilateral EIA occlusions • Chronic total CFA or superficial femoral artery (SFA) >20 cm in length and involving the popliteal artery and proximal crural vessels

Table 11.4 The TASC classification of radiological findings of below-knee PAD [3]

Classification	Findings
Type A	• Single focal stenosis <5 cm in length in the target tibial artery with occlusion or stenosis of equal or worse severity in the other tibial arteries
Type B	• Multiple stenoses each <5 cm with a total length < 10 cm, or single occlusion <3 cm in length in the target tibial artery with occlusion or stenosis of equal or worse severity in the other tibial arteries
Type C	• Multiple stenoses in the target tibial artery and/or single occlusion with a total length > 10 cm with occlusion or stenosis of equal or worse severity in the other tibial arteries
Type D	• multiple stenoses in the target tibial artery >10 cm in total length with dense calcification and non-visible collaterals with occlusions or dense calcification in the other tibial arteries

Pre-Procedural Assessment

Prior to each procedure, a full medical assessment is required to ensure that the patient is fit for treatment. A full blood count, coagulation screen, and renal function are advised to ensure there are no contraindications to treatment. It is advisable for the patient to have a Hb >80, Platelets >80, and INR of <1.5, however in emergency situations these can be overridden by the supervising clinician should clinical need preclude waiting for intervention. The main risks from treating a patient with thrombocytopaenia or a raised INR are from bleeding, either intra-procedurally or when trying to achieve haemostasis at the end. The importance of renal function comes when considering the potentially nephrotoxic effects of iodinated contrast medium, especially in long cases in which multiple angiograms may be required.

It is also inadvisable to perform an arterial puncture on a patient with significant hypertension (systolic blood pressure > 200) due to difficulties in achieving haemostasis post-procedurally, and in elective patients, it is also prudent to not proceed should the patient be systemically unwell due to the risk for potential complications intra- or post-procedurally.

Regarding anticoagulation such as warfarin or DOACS, the drugs should be stopped pre-procedurally to decrease the chance of haemorrhagic complications. Warfarin is commonly stopped 7 days' pre-procedurally whereas DOACS such as Rivaroxaban or Apixaban should be stopped for 2 days' pre-procedurally. These patients can be given bridging low molecular weight heparin (LMWH) should bridging be required. There is no need to stop antiplatelet agents such as aspirin or clopidogrel.

Endovascular Strategy and Technique

The BASIL trial compared outcomes in patients with lower limb ischaemia treated with either a surgery-first or angioplasty-first strategy with a primary endpoint of amputation-free survival. It suggested that outcomes were similar for both treatments (in suitable patients) with the consideration that percutaneous angioplasty was a more cost-effective option.

The success of percutaneous transluminal angioplasty (PTA) is multifactorial, including the type and severity of PAD, lesion length, vessel quality, associated diseases (DM, IHD), and persistent risk factors such as smoking and hypertension [4]. The probability of technical success for PTA is greater in the iliac arteries and decreases in more distal vessels. Evidence suggests that the technical success of PTA and primary stenting in aortoiliac occlusive disease is 96% with primary and secondary patency rates of 63% and 86% respectively, which is comparable with surgical bypass. Furthermore, this patient subgroup had a 26% increase in walking distance in the 2 years' post-intervention with associated increased physical function and decreased limb pain. In the femoral arteries, stenting is reserved for patients who have a technical failure of PTA or who have recurrent stenosis in fewer than 3 months' post-procedure.

It is often advisable that patients suffering from intermittent claudication are treated with less aggressive management strategies than those with CLI. Lesions are defined as significant should the stenosis comprise >50% of the vessel lumen. Lesions which cause less than 50% luminal narrowing are usually not deemed clinically significant and therefore are unlikely to benefit from PTA or surgical intervention.

Basic Endovascular Toolkit

The vast majority of procedures are performed under local anaesthesia. However, general anaesthesia is employed if the procedure is done in combination with surgery or if the patient is non-compliant.

For a standard lower limb PTA, the commonly used kit includes:

- Vascular access needle.
- Standard access wire (usually "J" tip).
- Selective catheter(s).
- Vascular access sheath(s) – range in size but typically 4Fr-7Fr sheaths are used for endovascular procedures, depending on adjuncts required during the procedure such as stents, re-entry catheters, and atherectomy devices.
- Angioplasty balloons sized to the lesion on pre-procedural imaging.
- Inflation device.
- Bare-metal stents sized to 10% larger than the vessel diameter of varying lengths. Should iliac work be undertaken, then it is sensible to have access to appropriately sized covered stents, especially in heavily calcified vessels.
- Closure devices such as Prostyle™ or Angioseal™; however, with the latter precludes further intervention at the same access vessel in the next 90 days.

General Access

Access is primarily via the common femoral artery at the level of the mid-femoral head within the centre of the anterior aspect of the vessel, or "12 o'clock". In the presence of complex aorto-iliac occlusions then upper limb access can be considered, for example via the brachial artery. The puncture should be performed under ultrasound guidance in all circumstances; however, whether this is done via a transverse or longitudinal plane is down to operator preference. Should the ipsilateral groin be deemed unsuitable, such as in the presence of a heavily stenosed CFA, surgical scarring, obesity, or active infection, then a contralateral retrograde puncture should be considered.

Once access has been achieved, direct catheter angiography is performed with either iodinated contrast medium or CO_2. In the case of iliac disease, this is commonly performed with a pigtail catheter and a pump to deliver contrast. In infra-inguinal disease, angiography is usually performed through contrast administration via a selective catheter placed at the level of the lesser trochanter. In instances of severe disease then it is possible the tibial vessels are poorly opacified and therefore in such patients the catheter can be advanced distally (e.g. to the level of the popliteal artery) before angiography is performed to assess the tibial vessels.

Once the sheath has been advanced over the guidewire, it should be flushed with heparinised saline and intra-arterial heparin administered. 3000 IU heparin is usually sufficient for supra-popliteal lesions. Some practitioners advise 5000 IU heparin to be administered for a potentially complex case or if the tibial vessels are to be treated.

Chronic Total Limb Occlusion (CTO)

CTOs can be crossed with either a luminal or subintimal approach, with little data to suggest one strategy is significantly beneficial over the other. Sometimes in CTO an intraluminal crossing is impossible, and therefore a subintimal plane must be created. Regrettably, this confers the risk of the guidewire ending in an extraluminal space and consequent technical failure. Should a subintimal plane be utilised then primary stenting is often necessitated due to persistent stenosis, elastic recoil, or dissection. It is worth noting that in most cases primary patency rates in CTO do not often match the high technical success rates, with many patients requiring reintervention. Despite this, long-term limb salvage rates remain high in this patient cohort, with rates of >80% in many series.

To successfully treat a CTO one must cross the proximal portion of the lesion, the lesion itself, and then ensure that they re-enter the lumen distally. Crossing the proximal "cap" of the occlusion can be difficult if it is calcified, fibrotic, or if there are adjacent collateral vessels which the Glidewire may cannulate preferentially. Initial attempts should be made to cross the occlusion luminally, usually utilising a "drilling" approach with the Glidewire with the operator spinning it between their thumb and forefinger to try to help the wire find a niche and tract through which it can pass. Should this be successful, then the wire is accompanied with a support catheter (such as a Navicross™) to add stability and help advance it further. It is worth considering the use of a smaller calibre wire (0.018 or 0.014) should a 0.035 wire be unsuccessful. If there are difficulties passing a catheter once purchase has been achieved within the lesion, then segmental balloon angioplasty can be used to try to open up the channel to allow distal progression.

Should luminal attempts be unsuccessful, then a subintimal approach can be utilised, thus creating a neo-lumen between the intima and adventitia to allow blood flow distally. Evidence suggests that despite subintimal angioplasty not affording as favourable primary and secondary patency at 12- and 24 months (59.1% vs. 78.4% and 56.8% vs. 76.5% respectively), it still affords high limb salvage rates of 95%. It is worth noting, however, that the evidence used to support these claims are of relatively low quality and further randomised control trials are required to support subintimal angioplasty over other techniques.

This technique is dependent on the operator entering the subintimal space with a Glidewire and support catheter. It is a recognised technique to then form a "loop" with the Glidewire and use this to dissect the subintimal tract, followed by the support catheter (Fig. 11.2). The use of a loop is arguably safer as it decreases the chance of the tip of the wire becoming extraluminal. Ideally, the wire will spontaneously become luminal once the CTO has been crossed, however in some cases this will not occur, and a re-entry device, such as an Outback® (Cordis, FL USA) catheter, is required to break back into the luminal space. Once it has been proven that one is intra-luminal through the delivery of contrast through the support catheter, then the support catheter can be exchanged for an appropriately sized angioplasty balloon to dilate the channel and allow for increased blood flow distally. Should there be a technical failure due to persistent stenosis or elastic recoil then primary stenting should be considered with a bare-metal stent (unless the treated vessel ruptures). The stent should be long enough to cover the entire length of the CTO from "normal vessel to normal vessel". Regarding sizing, it is advised that the diameter of the stent is 10% larger than the measured vessel diameter with post-dilation to 1 mm below the stent size; for example, a 5 mm vessel would require a 6 mm stent which is post-dilated with a 5 mm angioplasty balloon. Unlike primary angioplasty, when post-dilating a stent the balloon is simply required to be inflated to the nominal pressure and then immediately deflated.

Subintimal Tracking and Re-entry (STAR) Technique

A modification of the STAR technique is Direct Tip Injection in Occlusive Lesions (DIOL). This method utilises the

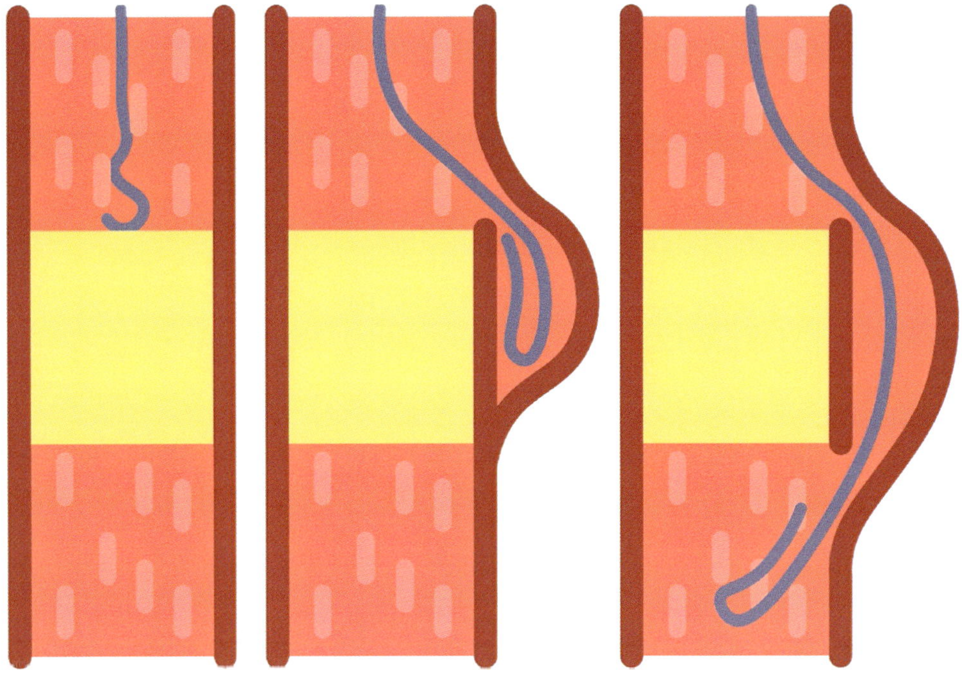

Fig. 11.2 Subintimal wire passage

hydro-dissection of the subintimal plane via the injection of contrast media through a micro- or balloon catheter. This technique allows the operator to visualise the created micro-channel as well as creating space to advance the necessary kit. Should this be utilised in an antegrade fashion, then it may result in iatrogenic dissections which extend beyond the target re-entry point, to compress the true lumen, and compromise flow distally in the crural vessels. In these instances, a smaller, gentle injection of 1-2 mL of contrast media should be used. Conversely, when using an antegrade approach, forceful injection of 5 mL of contrast media may be required to both visualise the tracts and increase the space within the subintimal plane. Once this has been achieved then attempts to re-enter the lumen and complete the procedure may be attempted.

Sharp Recanalisation

Chronic total occlusions frequently have a calcified proximal "cap" which prevents the interventionalist from traversing the lesion with conventional guidewires and catheters. Should it prove impossible to cross the lesion then sharp recanalisation can be considered. In these circumstances, the guidewire is removed and turned around, with the practitioner advancing the stiff, back end of the wire into the occlusion. The rationale of this is to pierce the stiff cap before the wire is removed once more and the soft end is used to cross the lesion and continue with the procedure. One should be satisfied that the cap has been penetrated by either feeling a "give" in the wire once it has progressed or noting its movement through the occlusion on fluoroscopic imaging.

Once you are sure you have traversed the proximal cap, advance your crossing catheter, (e.g. Navicross®), just into the occlusion before turning the wire back around to its correct direction. Once you have done so rapidly rotate the wire within your fingers to try to find a channel through which you can advance your equipment. If this is unsuccessful, then a loop can be formed within the wire to dissect the subintimal plane. Loops are not only useful in giving your wire stability to progress the wire, but also, they have a decreased chance of vessel perforation and the wire ending up in the perivascular soft tissues. This is not necessarily a serious consequence; however, guidewires prefer to follow a path of least resistance and, should this be through a perforated vessel wall, then you may be unable to cross the lesion and therefore have to either perform a new puncture or abandon the procedure entirely.

There are undoubtedly risks in performing such a manoeuvre, primarily that using the stiff end of the wire confers a risk of both vessel perforation and dissection; potentially conferring increased morbidity and mortality upon the patient. Usually, vessel perforations at the level of the SFA and below are of little clinical concern due to the immediate tamponade effect of adjacent musculature and fascial planes. Perforations at the level of the CFA and above can be catastrophic due to a lack of adjacent compressive tissues, resulting in pseudoaneurysm formation and potential death from exsanguination.

CTO Crossing Devices

Re-entry Devices

It is often difficult to re-enter the true lumen when utilising a subintimal approach, with reported failure rates of up to 26% and, when successful, it may only be possible to become luminal distal to the vessel reconstitution point, with consequent sacrifice of significant collateral vessels. These challenges have inspired the development of re-entry devices such as the Outback ® (Cordis), BeBack (Bentley), Pioneer (Philips), Offroad™ (Boston Sci) and Enteer™ (Medtronic) catheters, to name a few. Evidence suggests that the use of re-entry devices have similar outcomes in subintimal angioplasty when compared to standard catheters (88.3% when using re-entry devices vs. 92.5% without) with similar rates of complications.

Furthermore, using these devices can reduce the procedural time (and thus radiation dose), the number of guidewires required, and the volume of contrast media used when compared to other techniques such as subintimal arterial flossing with antegrade and retrograde intervention (SAFARI). This confers not only a benefit to patients but also to healthcare providers given the decreased cost of treating CTO.

Outback ® (Cordis)

This catheter is designed to work over a 0.014 wire with the Outback requiring a 6Fr sheath to accommodate the device. The catheter itself has a hydrophilic coating to afford easier passage through the subintimal plane and a 22G nitinol re-entry cannula. The Outback catheter has "L" and "T" radiopaque markers to allow for the orientation of the catheter under fluoroscopic guidance, thereby mitigating the need for intravascular ultrasound (IVUS).

Once the CTO has been crossed subintimally with an 0.014 wire, advance the tip of the device to the desired re-entry point. Once you are satisfactorily positioned, rotate the catheter so that the "L" marker is facing the true lumen. At this point, the image intensifier must be positioned so that you have a 90-degree orthogonal view of the vessel before rotating the device so the "T" marker *en masse*. At this point, the wire should be withdrawn by approximately 5 centimetres into the device before the nitinol cannula is advanced into the true lumen. When you are satisfied you are luminal once more, retract the cannula and remove the Outback® device before proceeding to treat the occlusion.

Good technical results with the Outback ® system have been reported in the treatment of lower limb CTO; however, difficulties persist with becoming luminal once more in the presence of significantly calcified distal re-entry sites.

BeBack (Bentley)™ Catheter

The BeBack™ Crossing Catheter (Bentley) is similar in design and function to the above-described OutBack re-entry device. It comes in two calibres, a 2.9Fr catheter requiring an 0.014 wire and 5Fr sheath, or a 4Fr catheter which needs at least a 0.018 wire and 6Fr sheath. As with the OutBack system, the GoBack Crossing Catheter can be used entirely under fluoroscopic guidance without the need for IVUS. Once a lesion has been crossed subintimally, the nitinol needle at the tip can be deployed to up to 11 mm in the 4Fr system, and 7 mm in the 2.9Fr system. The needle initially exits the catheter tip in a straight line, however, is designed to curve to allow for directional deployment as it is advanced further. To facilitate precise deployment, there is a "C" shaped radiopaque marker within the body of the catheter, with the open portion of the "C" corresponding to the needle angle. As with the other re-entry devices, once the needle has penetrated the lumen, the guidewire can be advanced, and the re-entry catheter exchanged for the necessary kit to complete the patient's treatment.

Pioneer Plus (Philips)

The Pioneer Plus Re-entry Device (Philips US) is a re-entry catheter which has a 20 MHz intravascular ultrasound transducer at its tip. It is a monorail catheter using a 0.014 system and requiring a 6Fr sheath. The rationale for having an adjunctive IVUS attached to the catheter tip is to allow for accurate assessment of the intima at the level of re-entry and to demonstrate the true lumen before the advancement of the nitinol cannula through the subintimal layer before becoming luminal once more. Reported outcomes with the Pioneer device are good, with evidence suggesting that this re-entry system has an average re-entry time of 6–10 min with technical success rates of 95–100%.

The mechanism of action is similar to the Outback ® catheter. Firstly, one must traverse the CTO via the subintimal plane before determining a suitable re-entry site. Once this has been achieved, IVUS is performed to allow to direct the cannula towards the true lumen prior to the advancement of a non-hydrophilic 0.014 wire into the vessel lumen. This can be ascertained as the true lumen is at the 12 o'clock position on IVUS, and the lumen diameter can be measured with the reticule displayed on the IVUS monitor. Once the operator is happy with both the position and diameter of the true lumen, the red handle is twisted to the correct needle deployment length before advancing the blue needle deployment ring quickly to advance the nitinol cannula into the true lumen. When the cannula is within the vessel lumen, the needle is retracted and the catheter removed before the CTO is treated with angioplasty or stenting, depending on technical success.

OffRoad™ Re-entry Catheter System

The OffRoad™ catheter (Boston Scientific, USA) uniquely has a 5.4 mm conical angioplasty balloon just proximal to its tip. When inflated, this allows the operator to dilate the subintimal tract and create more space to manoeuvre the piercing needle into the vessel lumen. The Re-ROUTE trial demonstrated that this catheter was effective in navigating approximately 85% of CTOs. The catheter itself operates via a dual-wire system: the main catheter requires a 5Fr sheath and is designed for use with a 0.035 guidewire. The microcatheter lancet slides through the main catheter body with a diameter of 2.4Fr and facilitates an 0.018 wire (or smaller). As with the other described re-entry devices, the catheter is to be advanced past the occlusion in the subintimal plane before re-entry and passage of necessary kit. Bear in mind with this dual-wire system that it will be the 0.18 wire which ultimately ends up in the true lumen when considering which equipment, you will use for further intervention.

Enteer Re-entry System

The Enteer™ system (Medtronic) is similar in function to that of the aforementioned OffRoad re-entry device in that it has an angioplasty balloon at its distal end. This device, however, has a straight angioplasty balloon which is more similar to a standard PTA balloon in shape when compared to the conical OffRoad balloon. The balloon is inflated to dilate the subintimal tract. As with many of the other systems, it is an over-the-wire system which utilises either a 0.014 or a 0.018 guidewire. The supplied wire with this system has a 28° curvature at the working end to allow for the operator to direct it towards and puncture the true lumen.

SAFARI and CART Techniques

Subintimal arterial flossing with antegrade and retrograde intervention (SAFARI) is a treatment consideration when other re-entry techniques fail. Firstly, imaging must be obtained to ascertain the feasibility of retrograde arterial access, be that ultrasound or through direct angiography. The foot vessels are most frequently used, the dorsalis pedis is commonly accessed within the level of the mid-navicular or slightly above the ankle mortice, and the posterior tibial artery posterior to the medial malleolus or immediately superior to the ankle joint. Fluoroscopic access is usually achieved at the lower third of the interosseous membrane. It should remain a consideration that haemostasis is more difficult to achieve after accessing the peroneal artery due to the lack of a deep bony prominence to manually compress against with a subsequent risk of compartment syndrome secondary to haemorrhage.

A micropuncture needle is used to puncture the chosen vessel before the advancement of the supplied dilator (usually 3Fr) over the guidewire. In some instances, a longer micropuncture dilator can be used in instances of more heavily calcified vessels to disrupt calcific plaques and facilitate the passage of your kit. Once access has been secured, an exchange-length (300 cm) 0.018 or 0.014 wire can be advanced retrogradely towards the occlusion. In some instances, the occlusion can be crossed luminally from below; however, in many cases, subintimal recanalisation is required from this retrograde approach. Should the vessel be amenable to it, then a 4Fr sheath or catheter can be used to add stability to a crossing catheter. It is often a sensible idea to administer GTN in 100mcg-200mcg aliquots (should the patient's blood pressure permit) and to maintain an ACT over 250 s to decrease the likelihood of thrombosis formation in smaller vessels, particularly those with a lumen diameter less than 1.5 mm.

Once the retrograde wire has crossed the occlusion and re-entered the true lumen then it may be snared from the antegrade access before being pulled through the sheath, giving continuous wire access throughout the length of the vessel. Should the antegrade catheter already be within the same subintimal plane as your retrograde wire, then this too can be cannulated to allow wire passage. Woefully, in many instances both accesses, (despite being in the same subintimal plane), do not communicate with one another. Should this occur, there are a few techniques to help navigate out of this predicament. *Firstly*, a new antegrade access can be attempted to have communicating subintimal tracts or a re-entry catheter can be used to attempt to enter the true lumen. If these are unsuccessful, then a reverse controlled antegrade and retrograde subintimal tracking (CART) technique can be trialled. This procedure involves the passage of a 4 mm angioplasty balloon over the antegrade wire and inflating it to disrupt the obstructing intimal layers to allow through-and-through access. One final consideration is to attempt sharp recanalisation using the back end of a guidewire, however, this is not a risk-free manoeuvre as previously discussed in this chapter (Fig. 11.3).

Percutaneous Deep Venous Arterialisation

Percutaneous deep venous arterialisation is a novel treatment for patients with CLI who are not suitable for bypass surgery or standard PTA due to a lack of suitable distal runoff vessels. The crux of the treatment is to supply oxygenated blood to the capillary bed using a vein as the delivery method via an iatrogenic arteriovenous fistula.

The procedure entails the use of a proximal, antegrade arterial puncture and a distal, retrograde venous puncture. Both angiography and venography are performed to prove the patient's anatomy is suitable and to demonstrate the shortest point between the target artery and vein to facilitate the formation of the fistula. Usually, a 7Fr sheath is sufficient in the arterial segment, and a 5Fr sheath for the vein.

Once the shortest distance has been identified, catheters are advanced within the artery and vein to sit adjacent to one

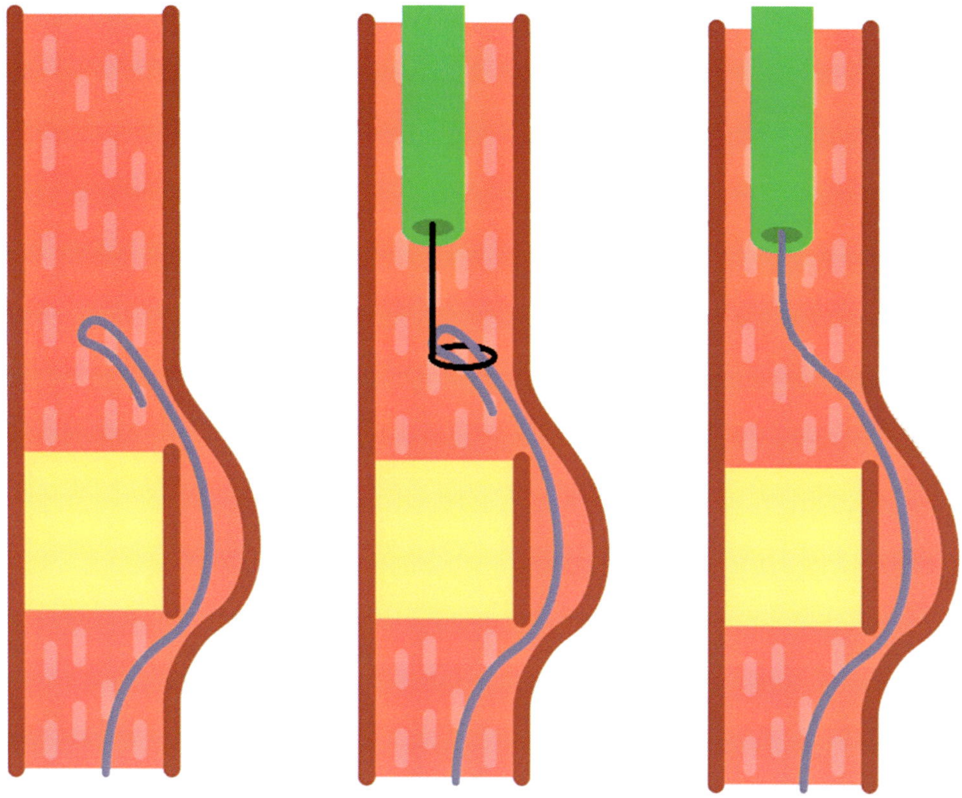

Fig. 11.3 The SAFARI technique demonstrating snaring a subintimal wire to allow for a through-and-through wire

another at this point. A re-entry catheter is then used under fluoroscopic of IVUS guidance to exit the artery and enter the vein. Once this through-and-through access has been obtained, a guidewire is advanced from the arterial to the venous segment. Valvulotomy is performed on the vein, often with a suitably sized angioplasty balloon before the covered stents are deployed from the arterial anastomosis through the venous segment to occlude any collateral vessels. Should persistent, large collaterals be present then these can be selectively embolised.

The evidence for the success of percutaneous deep venous arterialisation is somewhat limited, primarily due to it being a rarely performed procedure. The available data suggests primary patency rates of approximately 30% at 6 months and 40% at 10-month follow-up, with limb salvage rates of up to 70%. This preliminary data therefore may make this an attractive option in patients with no other option.

Possible Complications

Complications from PTA are most commonly from achieving haemostasis at the end of the procedure or intra-operative complications secondary to technical aspects during the procedure itself. Evidence suggests that major complications such as vessel thrombosis or rupture occur in approximately 5% of cases, with 2.5% of patients undergoing PTA requiring surgery and an overall limb loss and mortality rate of 0.2%.

Improper haemostasis can be secondary to the puncture being oblique within the vessel wall or it being too proximal or distal and therefore not facilitating adequate compression once the sheath has been removed. In these instances, complications can include a haematoma at the access site or pseudoaneurysm formation. Manual haemostasis is the most common method of closure of the arteriotomy; however, ultrasound-guided compression can be of use in decreasing such complications, especially should a brachial artery puncture be performed. Should a pseudoaneurysm form then manual compression can be utilised in the immediate post-procedural period, however, should it be large (>3 cm) or persistent then ultrasound-guided thrombin injection or surgical repair may be required.

Complications may also arise from vascular injury during the procedure, namely dissection or rupture. Dissection is in essence the stripping of the intima or media of the arterial wall and is easily identified fluoroscopically as a flap within the lumen. The commonest causes of dissection are either angioplasty or improper manoeuvring of the kit, such as advancing a catheter without the use of a guidewire. In most instances a dissection can be managed with prolonged angioplasty, however, should this not resolve the situation then an uncovered stent can be deployed over the dissection flap. Care must be taken to occlude the entrance of the false lumen to prevent distal extension of the dissection. Should a dissection be flow-limiting then it can occlude the vessel with consequent acute limb ischaemia if it is not appropriately managed.

Vessel rupture is often secondary to either oversizing the angioplasty balloon or performing PTA on a heavily diseased (calcified) vessel. The management of a vessel rupture is dependent on the location, as an iliac rupture can result in life-threatening retroperitoneal haemorrhage; however, a rupture within the SFA or crural vessels is unlikely to confer such serious consequences due to tamponade from the adjacent fascial planes. Clinically a rupture can present as signs of haemodynamic compromise with persistent pain post-angioplasty. Radiologically a persistent extravascular contrast "blush" will be seen adjacent to the rupture demonstrating contrast extravasation. Should a rupture occur within the iliac arteries then covered stenting is advised as a matter of haste. One should anticipate this potential complication and have suitable stents (such as a Viabahn™) present in the room as well as a sheath large enough to facilitate its deployment. Upon discovering a rupture, the angioplasty balloon should be deployed to cover it immediately and tamponade the bleed whilst the stent is prepared.

Thrombosis and embolisation are potentially limb-threatening consequences of endovascular management of PAD, hence why these patients receive intra-arterial heparin once the sheath is inserted. Thrombus formation can be difficult to discern from vasospasm at the time of the procedure, with both affording an irregular appearance to the vessel lumen. Should a complete occlusive thrombus occur then there will be an abrupt termination of contrast flow distal to it, with contrast reflux seen proximally. In a non-occlusive thrombus, the initial management attempts can be to aspiration with a catheter or, prolonged balloon angioplasty to try to compress it against the vessel wall. Chemical thrombolysis involves intra-arterial Alteplase® at a dose of 5 mg at the time of the thrombosis before consideration of performing catheter-directed thrombolysis with a Cragg-Macnamara catheter over 24 h and up to 72 h. Patients who require catheter-directed thrombolysis will need to be admitted to the intensive care unit for monitoring due to the risk of life-threatening haemorrhage. Contraindications to thrombolysis include recent intracerebral haemorrhage, intracranial neoplasm, active bleeding, and uncontrolled hypertension. Should the patient not be suitable for catheter-directed thrombolysis, then Rheolytic thrombectomy (e.g., with an Angiojet™ system) can be performed which confers decreased morbidity and mortality to catheter-directed thrombolysis, however, this treatment is dependent on the operator being able to advance a guidewire past the thrombus itself. In instances where catheter-directed thrombolysis and mechanical thrombectomy are not practical then surgical thrombectomy or bypass is advised should the patient be fit.

Perioperative Care and Surveillance

The perioperative care of patients who have undergone PTA or stenting for PAD is often dependent on the complexity of the procedure, the size of the sheath used, and whether a closure device was used. In instances of manual haemostasis, then prolonged bed rest is advised (up to 4 h) to decrease the chances of significant haemorrhage from the arteriotomy site. Recovery staff are advised to check the patient's observations and observe the puncture site frequently during this period to assess for haematoma formation or active bleeding which may require urgent intervention. The use of a closure device (such as Angioseal™ or Prostyle™) speeds up the recovery process allowing for expedited mobilisation as the arteriotomy site has been mechanically closed.

Other considerations regarding recovery are if the patient has received any sedation such as fentanyl or midazolam or if the case was performed under general anaesthesia. In such instances, further monitoring is advised in the post-procedural period to ensure there are no complications from the use of such medications, such as aspiration or respiratory suppression.

In the absence of complications, and if the patient is deemed functionally fit to do so, then most patients can be discharged home as a day case once their recovery period is over.

Case Presentation

An 0.018 guidewire was navigated into the DP micropuncture catheter, through the AT and into the distal SFA via a subintimal approach. The distal sheath was cannulated, and the wire passed proximally to allow for a through-and-through wire. Angioplasty was performed of the AT with a 2 mm × 220 mm and subsequently 3 mm × 120 mm angioplasty balloon to good effect. The SFA-popliteal occlusion was addressed with a 5 mm × 120 mm angioplasty balloon before deployment of a 5 mm × 120 mm stent distally and a 6 mm × 80 mm stent proximally and post-dilated with a 5 mm angioplasty balloon. Satisfactory post-stenting angiographic appearances. Manual haemostasis was achieved with no immediate complications (Fig. 11.4).

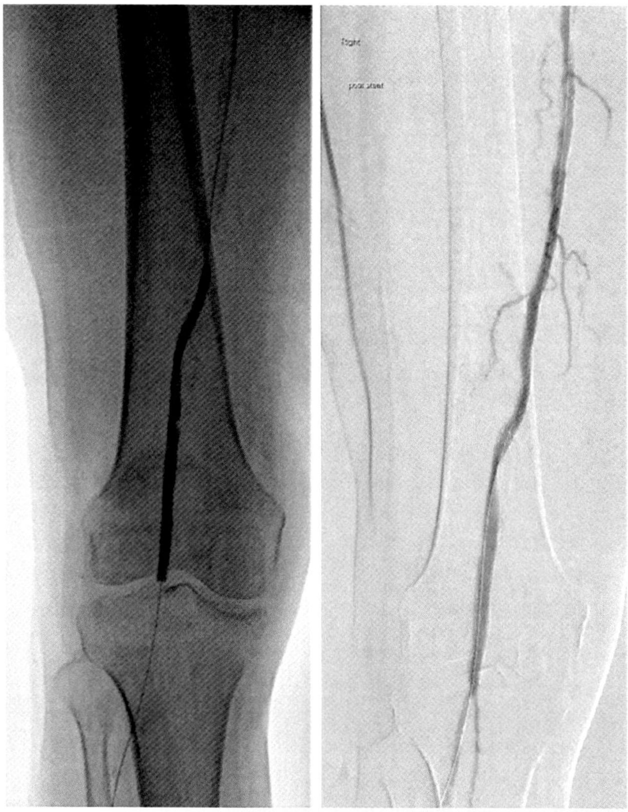

Fig. 11.4 LEFT: The through-and-through wire from the DP puncture through to the CFA sheath facilitating angioplasty of the SFA-popliteal occlusion. RIGHT: Post-angioplasty angiographic run demonstrating patency throughout

References

1. White C, Gray W. Endovascular therapies for peripheral arterial disease: An evidence-based review. Circulation. 2007;116(19):2203–15.
2. Lazaris A, Salas C, Tsiamis A, et al. Factors affecting patency of subintimal infrainguinal angioplasty in patients with critical lower limb ischemia. Eur J Vasc Endovasc Surg. 2006;32:668–74.
3. Kokkinidis D, Katsaros I, Jonnalagadda A, et al. Use, safety, and effectiveness of subintimal angioplasty and re-entry devices for the treatment of Femoropopliteal chronic Total limb occlusions: a systematic review of 87 studies and 4,665 patients. Cardiovasc Revasc Med. 2020;21(1):34–45.
4. Del Giudice C, Van Den Heuvel D, Wille J, et al. Percutaneous deep venous arterialization for severe critical limb ischemia in patients with no option of revascularization: early experience from two European centers. Cardiovasc Interv Radiol. 2018;41:1474–80.
5. Criqui M, Matshushita K, Aboyans V, et al. Lower extremity peripheral artery disease: contemporary epidemiology, management gaps, and future directions: a scientific statement from the American Heart Association circulation. Circulation. 2021;144:e171–91.
6. Shu J, Gaetano S. Update on peripheral artery disease: epidemiology and evidence-based facts. Atherosclerosis. 2018;2018(05):033.
7. Conte S, Vale P. Peripheral arterial disease. Heart Lung Circ. 2018;27(4):427–32.
8. Gerhard-Herman MD, Gornik HL, Barrett C, et al. AHA/ACC guideline on the management of patients with lower extremity peripheral artery disease: a report of the American College of Cardiology/American Heart Association task force on clinical practice guidelines. J Am Coll Cardiol. 2016;69(11):e71–e126.
9. Tummala S, Scherbel D. Clinical assessment of peripheral arterial disease in the office: what do the guidelines say? Semin Interv Radiol. 2018;35(5):365–77.
10. Berger J, Davies M, Clement D, Waltham M. Overview of lower extremity peripheral artery disease. UpToDate; 2018.
11. Bailey M, Griffin K, Scott J. Clinical assessment of patients with peripheral arterial disease. Semin Interv Radiol. 2014;31(04):292–9.
12. Gerhard-Herman M, Gornik L, Barrett C, et al. AHA/ACC guideline on the management of patients with lower extremity peripheral artery disease: executive summary: a report of the American College of Cardiology/American Heart Association Task Force on Clinical Practice Guidelines. Circulation. 2016;135(12):e686–725.
13. Harissis H, Koliousi E, Matsagas M, et al. Measurement of transcutaneous oxygen tension in limbs with arteriovenous hemodialysis access. Dial Transplant. 2008;37(2):67–70.
14. Pitts J. Skin perfusion pressure: a case study demonstrating microcirculatory blood flow. J Diagn Med Sonogr. 2014;30(4):213.
15. NICE Clinical Guidance (2012) Peripheral arterial disease: diagnosis and management.
16. Yusuf S, Teo K, Pogue J, et al. Telmisartan, ramipril, or both in patients at high risk for vascular events. N Engl J Med. 2008;358(15):1547–59.
17. Antithrombotic Trialists' Collaboration. Collaborative meta-analysis of randomised trials of antiplatelet therapy for prevention of death, myocardial infarction, and stroke in high-risk patients. Br Med J. 2002;324(7329):71–86.
18. CAPRIE Steering Committee. A randomised, blinded, trial of clopidogrel versus aspirin in patients at risk of ischaemic events (CAPRIE). Lancet. 1996;348(9038):1329–39.
19. Norgren L, Hiatt W, Dormandy J, et al. Inter-society consensus for the Management of Peripheral Arterial Disease (TASC II). J Vasc Surg. 2007;45 Suppl S:S5–67.
20. TASC Steering Committee, Jaff MR, White CJ, Hiatt WR, Fowkes GR, Dormandy J, Razavi M, Reekers J, Norgren L. An update on methods for revascularization and expansion of the TASC lesion classification to include below-the-knee arteries: A supplement to the inter-society consensus for the management of peripheral arterial disease (TASC II): The TASC steering committee. Catheter Cardiovasc Interv. 2015;86(4):611–25.
21. Bonaca MP, Bauersachs RM, Anand SS, Debus ES, Nehler MR, Patel MR, et al. N Engl J Med. 2020;382(21):1994–2004.
22. Papavassilou V, Walker S, Boilia A, et al. Techniques for the endovascular management of complications following lower limb percutaneous transluminal angioplasty. Eur J Vasc Endovasc Surg. 2003;25(2):125–30.
23. Franzone A, Ferrone M, Carotenuto G, et al. The role of atherectomy in the treatment of lower extremity peripheral artery disease. BMC Surg. 2012;12:1–9.

Deep Venous Arterialisation

12

Basil Zia Khan, Thomas Barge, and Raman Uberoi

Deep Venous Arterialisation Case at Oxford

A 53-year-old male patient with a past medical history of type II diabetes mellitus, asthma, severe peripheral vascular disease, and right middle toe amputation had repeated hospital admissions for infected diabetic foot ulcers.

The patient underwent a total of three right-sided infra popliteal angioplasty intervention. The initial angioplasty demonstrated a patent superficial femoral (SFA) and popliteal artery with a three-vessel run off; however, there were multifocal stenoses limiting flow within the posterior tibial artery (PTA). The anterior tibial artery (ATA) and peroneal arteries also demonstrated multifocal stenoses without limiting flow. The proximal PTA underwent a 2 mm angioplasty. Despite treatment, there was no improvement in ulcer healing, subsequent infra popliteal angioplasties demonstrated progression of disease with new occlusion of the distal ATA and PTA. These were not treatable with conventional endovascular techniques and a decision was made to perform a deep venous arterialisation (DVA).

Anterograde arterial access was made at the SFA and retrograde venous access was made into the distal PTV. The proximal PTA was dilated. This was followed by creation of a fistula between the proximal PTA and PTV using outback. Two 5 × 10 cm Viabahn stents were placed extending across the fistula into the distal PTV. Subsequent venoplasty up to 5 mm was done to the level of the deep venous arch to destroy the valves. During the procedure the fistula occluded which was treated successfully with thromboaspiration and further plasty. Final run demonstrated adequate flow through the DVA circuit (Fig. 12.1).

B. Z. Khan · T. Barge · R. Uberoi (✉)
Department of Interventional Radiology, Oxford University Hospitals NHS Foundation Trust, Oxford, UK
e-mail: basil.Kahn@ouh.nhs.uk; Tom.barge@ouh.nhs.uk; raman.uberoi@ouh.nhs.uk

G. Geroulakos et al. (eds.), *Mastering Endovascular Techniques*, https://doi.org/10.1007/978-3-031-42735-0_12

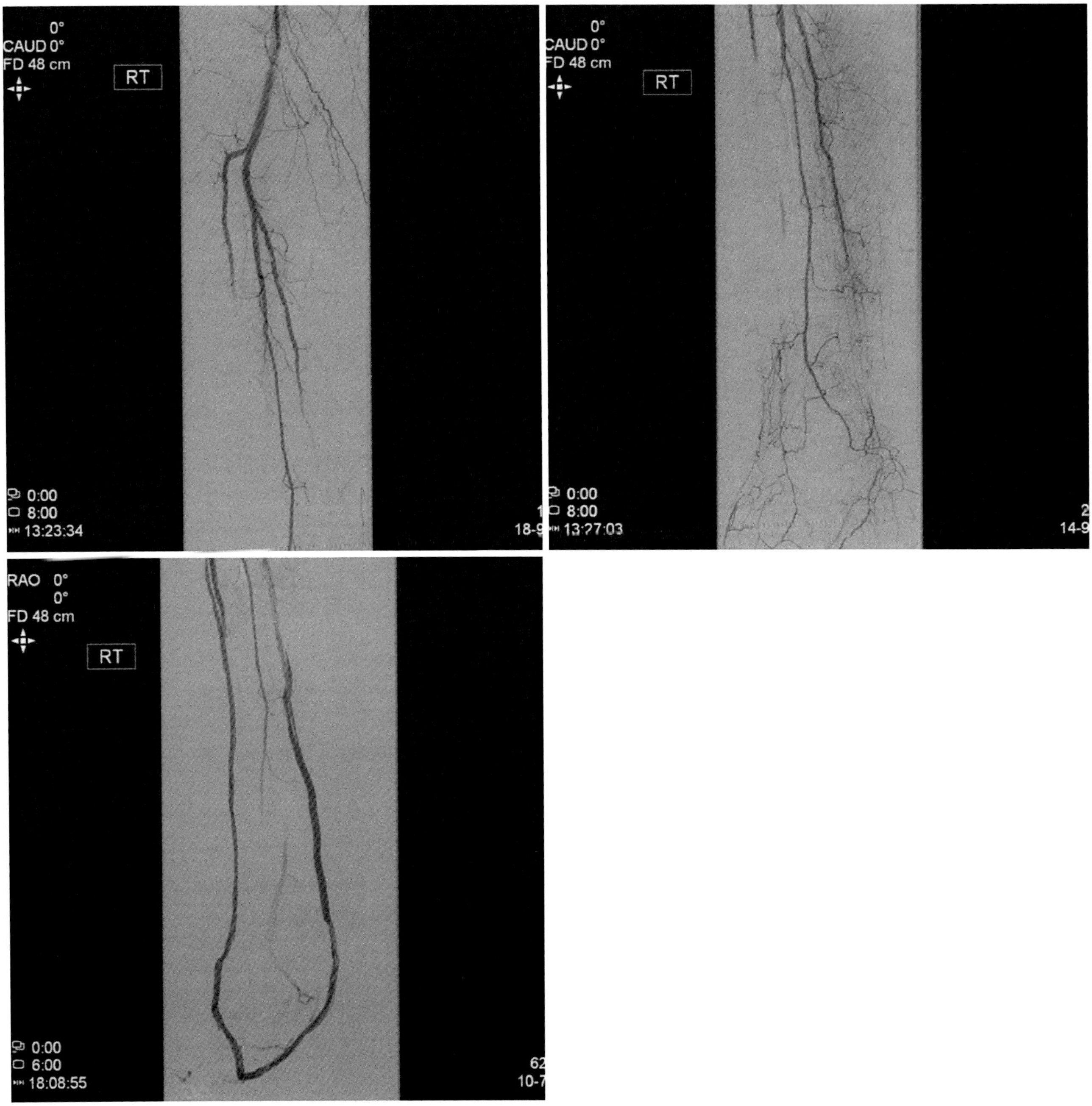

Fig. 12.1 (**a** and **b**) demonstrate vascular appearances prior to intervention. (**c**) demonstrates a satisfactory appearance of a patent DVA circuit

Introduction

Chronic limb-threatening ischaemia (CLTI) is a common cause for patients to present to an interventional radiology service. Management revolves around improving arterial inflow, ultimately to improve tissue perfusion. In the lower limb iliofemoral disease is commonly treated either endovascularly or via surgical bypass, whilst endovascular therapy dominates in the infra-popliteal segments. Angioplasty, including re-canalisation of chronic total occlusion of crural vessels, is the mainstay of treatment options. However, in some patients, there may be no suitable targets for re-canalisation. This is usually due to a combination of total occlusion of all crural arteries, and no suitable distal run-off into the foot, usually due to end stage plantar atherosclerotic disease. Blood flow to the foot is via small, unnamed collaterals, with little macroscopic flow demonstrated on angiography, giving the appearances of a so called 'desert foot'. It is this patient cohort that deep venous arterialisation (DVA) may be suitable.

DVA involves the creation of a fistulous connection between a crural artery and vein. The aim is to cause retrograde flow down a crural vein into the foot, fundamentally to increase the oxygenation of the blood pool in the capillary bed. The extensive venous collateral drainage pathways from the lower limb facilitate outflow, with reversal of flow in the arterialised vein not significantly impacting on venous drainage. The origins of DVA dates back to 1881, when Francois Frank, a French physiologist performed the first arteriovenous anastomosis in dogs [1], soon followed the first in human procedure in 1894 [2]. The first pedal vein arterialisation was recorded in 1912 [3]. Since its introduction over a century ago the procedure has been refined, with Sheil publishing the first case series of DVA procedures in 1977 [3]. A total of 6 patients with CLTI underwent a DVA where a connection was made between the long saphenous vein and dorsal venous arch [4], reporting improvement in pain and tissue healing in 5 out of the 6 patients. Whilst early DVAs were done using an open surgical technique, they are now mostly done using an endovascular approach.

Table 12.1 Wound grading from WIfI classification adapted from Mills JL Sr et al. [5]

Wound	
0	Pain at rest, no ulcer or gangrene
1	Small shallow ulcer on leg/foot salvageable with single–/double-digit amputation or skin coverage. No gangrene
2	Deep ulcer exposing bone/tendon, joint in foot or leg. Salvageable with > three-digit amputation or trans metatarsal (TMA) amputation with skin coverage. Shallow heel ulcer with no calcaneal involvement. Gangrene limited to digits
3	Deep ulcer involving forefoot and mid foot. Full thickness heel ulcer with/without calcaneal involvement. Extensive gangrene. Treatable with reconstruction of foot or non-traditional TMA

Table 12.2 Ischaemia grading from WIfI classification adapted from Mills JL Sr et al. [5]

Ischaemia			
Grade	Ankle systolic pressure (mmHg)	Toe pressure (mmHg)	Ankle-brachial index
0	100 and above	≥60	≥ 0.80
1	70–100	40–59	0.6–0.79
2	50–70	30–39	0.4–0.59
3	<50	<30	≤0.39

Table 12.3 Foot infection grading from WIfI classification adapted from Mills JL Sr et al. [5]

Foot Infection	
0	No signs or symptoms
1	Local features of infection confined to skin
2	Localised infection extending to deeper structures
3	Localised infection with features of systemic inflammatory response syndrome

Indications, Contraindications and Work Up for DVA

DVA has mainly been performed on patients with CLTI where standard revascularisation options have failed or are not technically suitable. Candidates for DVA are often placed in classes 5 and 6 of the Rutherford criteria or stages 3 and 4 of the Fontaine criteria.

In 2014, a new classification system known as Wound, Ischaemia, Foot Infection (WIfI) was introduced by the Society for Vascular Surgery (SVS). The aim of this classification is to provide staging of a patient with CLTI in terms of amputation risk and benefit of revascularisation. The higher the overall WIfl score the greater the amputation risk and benefit of revascularisation (Tables 12.1, 12.2, and 12.3).

Candidates for DVA must demonstrate good arterial inflow up to the point of crossover/occlusion hence the importance of pre interventional angiographic assessment of the foot and leg. Inflow may be optimised during the DVA procedure, but significant proximal arterial disease may be a contraindication to DVA. Furthermore, structures such as the deep venous arch must be patent and good venous outflow must be established. This can be achieved via duplex ultrasound assessment, magnetic resonance imaging or conventional venography and a lack of acceptable venous outflow would be a contraindication. Extensive tissue loss, or the combination of tissue loss and significant rest pain may also be relative contraindications.

DVA may be performed under local anaesthesia and conscious sedation, or under general anaesthesia. Patients typically require an overnight stay, although as experience in the procedure increases, it may be possible to perform these as a day case. Given the cohort of patients DVA is performed on, it may be challenging for them to remain still for a prolonged period during the procedure, which can be long. Consequently, general or regional anaesthesia may be required. This may be a complicating factor due to the frequently encountered medical co-morbidities, which may deem some patient's unacceptably high risk. An alternative consideration is the use of regional limb nerve blocks, which may provide excellent pain relief for the patient as well as ensuring the operative limb remains still, which is crucial for certain technical steps for the procedure.

Patients require standard workup for angioplasty and/or anaesthesia review pre-operative.

Open Surgical Approach to DVA

DVAs were initially performed with an open surgical technique and may at times still be performed. Common arterial targets include the tibial, popliteal, superficial or common femoral arteries, and venous targets include the posterior tibial vein or dorsal venous arch. The arterial and venous access points may then be connected by means of a venous conduit, harvested from other areas of the body, mostly the great saphenous or cephalic veins. The idea conduit should be at least 3 mm in diameter [6]. Following harvesting of the conduit, the vein is reversed and tunnelled before anastomosing with the inflow artery and target vein. The site of anastomosis between the conduit and target vein is usually done at the level of the ankle. Polytetrafluorethylene (PTFE) conduits are an alternative to harvested veins and less commonly used to achieve a bypass. After a satisfactory anastomosis has been achieved, valvulotomy is performed on the distal target veins.

Endovascular Approach to DVA

Unlike in open techniques, the mainstay of endovascular DVA is to create a fistulous connection between an adjacent artery and vein. Most common sites for this are the posterior tibial artery and vein, and less commonly the anterior tibial vessels. Venous access is performed. This can either be directly into the PT or AT vein, or alternatively into a pedal vein. If accessing the PT vein, pedal access has the advantage that wire access across the distal PT at the ankle is achieved, which facilitates venoplasty of the valve at the level of the ankle, which can be a challenging site to otherwise deal with. A wire is then navigated to the cross-over point at which point the anastomosis will be created. Antegrade arterial access is obtained, typically via the ipsilateral common femoral artery, and a wire navigated to the cross-over point. The fistulous connection is then created by means of a re-entry device. Both fluoroscopic-guided catheters (e.g. Outback, Cordis) and Intravascular ultrasound (IVUS)-guided (e.g. Pioneer Plus, Phillips) catheters have been used. These systems are inserted down the arterial access and utilise a side-angled needle to puncture trough the arterial wall into the vein (Figs. 12.2, 12.3, and 12.4).

An angioplasty balloon can be inflated in the vein to act as a target for needle puncture. Once needle access into the vein has been achieved, a through-and-through wire between the arterial and venous access can be achieved. This may require the use of an endovascular snare from the venous access to capture and retrieve the wire.

The arteriotomy site is then serially pre-dilated, typically from 2 mm to 3–4 mm to facilitate passage of further devices. In some situations, if the crural artery is significantly diseased, atherectomy may be performed to maximise inflow.

Valvotomy must then be performed to ensure retrograde flow down the arterialised vein must next be performed. This is typically performed by venoplasty across the veins, and may require use of high pressure, sculpting or cutting balloons to achieve sufficient valvotomy (Fig. 12.5).

A covered stent graft is then deployed across the arteriotomy to maintain the fistulous connection. At our institution, we use a covered self-expanding stent, usually a 5 mm Viabahn (Gore). The stent may extend distally in the vein to just above the level of the ankle. The stent is then post dilated

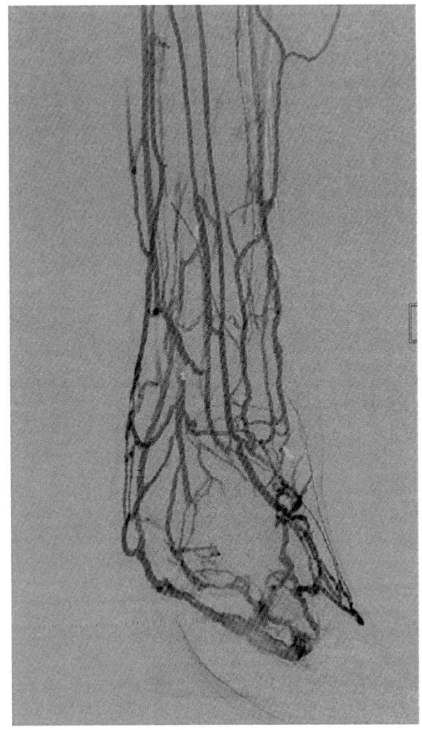

Fig. 12.2 Venogram of the foot outflow obtained following pedal venous access during an endovascular DVA

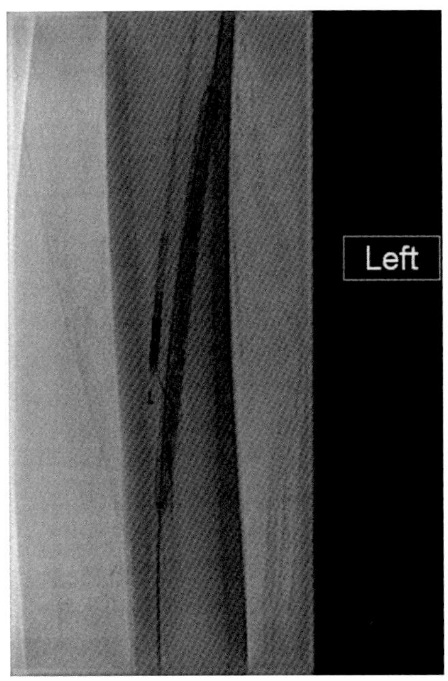

Fig. 12.3 An Outback catheter has been positioned in the posterior tibial artery and an angioplasty balloon inflated in the adjacent posterior tibial vein. The Outback needle has then been advanced under fluoroscopy to puncture the balloon and obtain fistulous access

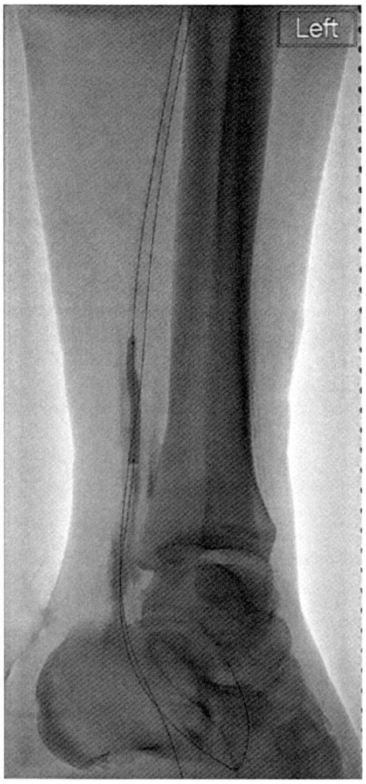

Fig. 12.4 Angioplasty of the fistulous tract created by the re-entry catheter

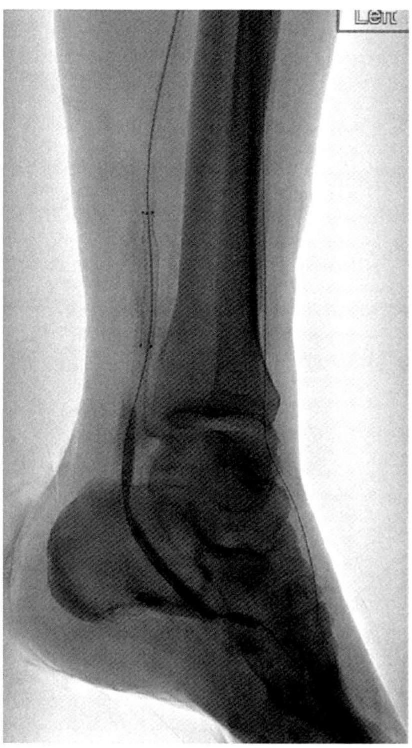

Fig. 12.5 A wire has been navigate from the arterial access through the fistulous connection between the posterior tibial artery and vein (note the presence of a stent) and around the pedal venous arch into the anterior tibial vein. Venoplasty is then performed to disrupt the valves in the arterialised vein. Note the waisting of the balloon at the site of the posterior tibial valve at the level of the ankle—often a relatively resistant site

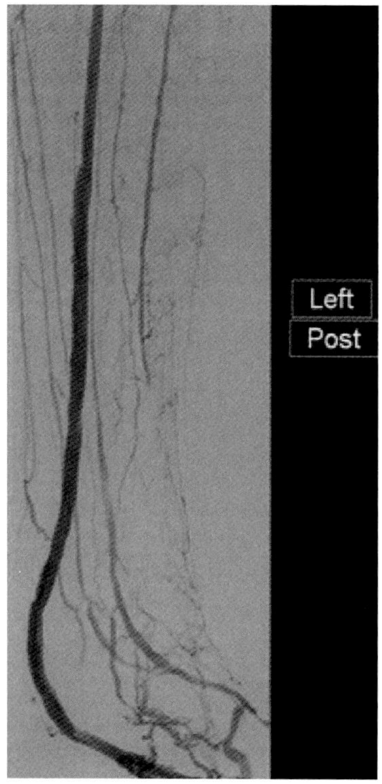

Fig. 12.6 Completion angiography via injection from the posterior tibial artery showing the complete DVA circuit with a stented posterior tibial vein. Contrast can be seen passing through the pedal venous loop into the outflow anterior tibial vein

across the arteriotomy and into the vein to 5 mm, and proximally at the arterial side to 4 mm.

Venous outflow must then be optimised. Now the fistulous connection is secure, the venous access can be removed. Low pressure ballooning across the venous access site may be required for haemostasis. A wire can then be navigated around a pedal venous loop, ideally into an outflow vein, such as the long saphenous vein. Further valvotomy may be required to optimise the vascular loop. Arterial access can then be secured and the patient observed as per local protocols (Fig. 12.6).

The LimFlow system was introduced in 2012 and has been designed as a complete kit to help perform percutaneous DVA. It was first used in a patient in 2013. Since then the LimFlow system has undergone three various studies. These are the PROMISE I, ALPS Post Market and the ongoing PROMISE II studies. Centres across the world have become more accustomed to this system since its introduction a decade ago. The LimFlow kit consist of the following:

1. Limflow arterial catheter
2. Limflow venous catheter (consists of a radio opaque mesh which once expanded enables better visualisation of the target vein)

3. Limflow Valvulotome (forward facing hook with a cutting basket allows safer removal of venous valves as compared to traditional high pressure venoplasty which can damage the venous walls)
4. Limflow Stents (Crossing and Extension)

The LimFlow system is currently in its second iteration since its introduction. The second generation system no longer utilises US guidance to ensure alignment of the arterial and venous catheters, instead it relies on fluoroscopic guidance only. It is currently undergoing further study in the PROMISE II trial which is looking into wound healing rates and amputation free survival in patients undergoing DVA for CLTI.

Post Procedural Care Aspects for Endovascular DVA

It is generally accepted that an endovascular DVA takes 4–6 weeks to mature. Surveillance and medical optimisation is key during this period. Patients require close follow-up in the immediate aftermath of the procedure to check for any reversible complications. This consists of clinical examination of the foot to assess the state of wounds and sonographic assessment of the lower limb vessels and stent to assess patency and flow. Volume flow (VF) and peak systolic velocity (PSV) have been identified as measurements to take when performing an US follow-up. This practice has been adopted from data gathered on follow-up of infra-inguinal bypasses and arteriovenous fistulas due to the similar anatomical and physiological changes secondary to this intervention. Schreve et al. recommends taking VF and PSV measurements at five points, these include the inflow artery, proximal third, middle third, distal third parts of the stent and the distal out flow vein greater than 3 cm from the distal end of the stent [7]. They also suggest that the VF and PSV values are most reliable at the middle third of the venous stent (Table 12.4).

As with any endovascular procedure involving placement of PTFE stents for prolonged periods, it is recommended to place patients on a combination of dual platelet therapy. The evidence for this is applied from studies investigating long-term effects of placing stents in lower limb vessels and not on a specific randomised controlled trial investigating best post procedural pharmacological practice on patient with a DVA.

Table 12.4 shows VF and PSF values used in follow-up of DVA to assess circuit [7]

Prediction	Volume flow (mL/min)	Peak systolic flow (cm/s)
Patency	>364	>99
Occlusion/stenosis	<195	<55

Outcomes and Complications of Endovascular DVA

Common parameters used to assess outcomes in patients undergoing DVA include patency of the circuit, limb salvage, amputation-free survival (AFS), wound healing (WH) and resolution of rest pain. Current literature indicates patency at 6 months ranges from 29% to 40% [8, 9]. The PROMISE I early feasibility trial was a single arm, prospective multicentre study investigating outcomes of DVAs performed with the use of the LimFlow system. Table 12.5 below shows results from the study.

At 24 months the PROMISE I trial also demonstrated a limb salvage rate of 77% [10]. Another study suggested limb salvage rates as high as 86% [11] with one study suggesting 100% resolution in rest pain [8]. The ALPS multicentre study investigated the midterm results of percutaneous DVA with the LimFlow system in 32 patients. It reported a median time of 4.9 months to achieve complete wound healing [12]. Table 12.6 highlights the results from the ALPS study at 6, 12 and 24 months. Another smaller study consisting of five patients suggested a median wound healing time of 39 weeks [8].

Table 12.7 shows results from the DEPARTURE study which also investigated 12-month outcomes of percutaneous DVA in 18 patients.

The DEPARTURE study also reported that median time to complete wound healing was 234 days [13]. The PROMISE II study is currently on going and is further investigating AFS and WH in patients undergoing DVA with the LimFlow system for no options CLTI.

Reported complications in current literature include in-stent thrombosis/stenoses and failure to revascularise the

Table 12.5 demonstrates AFS and WH results from the PROMISE I study [10]

Follow-up (months)	AFS (%)	Wound healing (%)
6	74	67
12	70	75
24	59	92

Table 12.6 shows AFS, LS and WH results from the ALPS study [12]

Follow-up (Months)	AFS (%)	Limb Salvage (%)	Wound Healing (%)
6	83.9	86.8	36.6
12	71	79.8	68.2
24	67.2	79.8	72.7

Table 12.7 shows AFS, LS and WH from the DEPARTURE study [13]

Follow-up (months)	AFS (%)	Limb Salvage (%)	Wound Healing (%)
6	55.6	72.2	23.0
12	49.4	72.2	53.2

foot. Schmidt et al. reported a median time of 2.6 months to occlude the DVA circuit [12]. Reintervention may be required to try and address this. Stent stenosis or poor outflow due to persistent venous valves have also been reported as reasons for reintervention aside from thrombotic occlusion. Current literature suggests a reintervention rate between 53% and 71% [9, 12]. It is also possible that despite adequate arterialisation of the more proximal aspect of the lower limb vein, there is insufficient flow to the more distal venous beds due to competing venous outflow. In these circumstances, coiling or ablation of the competing veins has been performed, to optimise arterialised flow to the foot.

Conclusion and Future Directions

DVA is an evolving technique, which shows promise for patients with CLTI where conventional revascularisation techniques have failed. It is a challenging field and should only be performed by operators experienced in endovascular CLTI therapies and in carefully selected patients. There in ongoing innovation in this field, including in the development of a specifically developed system for formation of the arteriovenous conduit and valvotomy (LimFlow), although this is not yet commercially available. Ongoing research, including assessment of the heath economic impact of this technique will be valuable in helping to assess which patients will most benefit.

Equipment Required to Perform a DVA (Table 12.8)

Table 12.8 Example of equipment used to perform DVA

Item	Function
Introducer set • Lidocaine 1% • Scalpel • Introducer needle • Dilator • Guidewire • Femoral access sheath	Establish and secure arterial and venous access
Guidewires • Terumo—Glide wire advantage hydrophilic coated (180, 300 cm) • Abbott hi-torque command (190/300 cm) • Boston Scientific—V18 control wire (110, 150, 200, 300 cm)	Manoeuvre to site of arteriovenous fistula
Flush and selective catheters • Outback Cordis re-entry catheter • Pioneer plus IVUS guided re-entry catheter	Cannulating and manoeuvre to site of arteriovenous fistula Create fistula between artery and vein
Balloon catheters	Plasty of arteriovenous fistula, destroy venous valves
Covered stents • GORE VIABAHN	Ensure patency of the DVA circuit

References

1. Halstead AE, Vaughan RT. Arteriovenous anastomosis in the treatment of gangrene in the extremities. Surg Gynecol Obstet. 1912;24:1–19.
2. Francois-Franck M. Note à propos de la communication de M Raimond Petit sur la susture artério-veneuse. Compt Rend Hebd Soc Biol. 1896;48:150.
3. Lechareas S, Sritharan K, Williams RGM. Early and eighteen month clinical outcomes of first UK case of percutaneous deep vein arterialisation (pDVA) to treat "no option" chronic limb-threatening ischemia using the LimFlow system. CVIR Endovasc. 2021;4(1):62. https://doi.org/10.1186/s42155-021-00252-4.
4. Sheil GR. Treatment of critical ischaemia of the lower limb by venous arterialization: an interim report. Br J Surg. 1977;64(3):197–9. https://doi.org/10.1002/bjs.1800640314.
5. Mills JL Sr, Conte MS, Armstrong DG, Pomposelli FB, Schanzer A, Sidawy AN, Andros G. Society for Vascular Surgery Lower Extremity Guidelines Committee. The Society for Vascular Surgery Lower Extremity Threatened Limb Classification System: risk stratification based on wound, ischemia, and foot infection (WIfI). J Vasc Surg. 2014;59(1):220–34. https://doi.org/10.1016/j.jvs.2013.08.003. Epub 2013 Oct 12
6. Ho VT, Gologorsky R, Kibrik P, Chandra V, Prent A, Lee J, Dua A. Open, percutaneous, and hybrid deep venous arterialization technique for no-option foot salvage. J Vasc Surg. 2020;71(6):2152–60. https://doi.org/10.1016/j.jvs.2019.10.085. Epub 2019 Dec 31
7. Schreve MA, Huizing E, Kum S, de Vries JPM, de Borst GJ, Ünlü Ç. Volume flow and peak systolic velocity of the arteriovenous circuit in patients after percutaneous deep venous arterialization. Diagnostics (Basel). 2020;10(10):760. https://doi.org/10.3390/diagnostics10100760.
8. Del Giudice C, Van Den Heuvel D, Wille J, Mirault T, Messas E, Ferraresi R, Kum S, Sapoval M. Percutaneous deep venous arterialization for severe critical limb ischemia in patients with no option of revascularization: early experience from two European centers. Cardiovasc Intervent Radiol. 2018;41(10):1474–80. https://doi.org/10.1007/s00270-018-2020-2. Epub 2018 Jul 17. Erratum in: Cardiovasc Intervent Radiol. 2018 Jul 30.
9. Kum S, Huizing E, Schreve MA, Ünlü Ç, Ferraresi R, Samarakoon LB, van den Heuvel DA. Percutaneous deep venous arterialization in patients with critical limb ischemia. J Cardiovasc Surg. 2018;59(5):665–9. https://doi.org/10.23736/S0021-9509.18.10569-6. Epub 2018 May 22
10. Clair DG, Mustapha JA, Shishehbor MH, Schneider PA, Henao S, Bernardo NN, Deaton DH. PROMISE I: early feasibility study of the LimFlow system for percutaneous deep vein arterialization in no-option chronic limb-threatening ischemia: 12-month results. J Vasc Surg. 2021;74(5):1626–35. https://doi.org/10.1016/j.jvs.2021.04.057. Epub 2021 May 18
11. Mustapha JA, Saab FA, Clair D, Schneider P. Interim results of the PROMISE I trial to investigate the LimFlow system of percutaneous deep vein arterialization for the treatment of critical limb ischemia. J Invasive Cardiol. 2019;31(3):57–63.
12. Schmidt A, Schreve MA, Huizing E, Del Giudice C, Branzan D, Ünlü Ç, Varcoe RL, Ferraresi R, Kum S. Midterm outcomes of percutaneous deep venous arterialization with a dedicated system for patients with no-option chronic limb-threatening ischemia: the ALPS multicenter study. J Endovasc Ther. 2020;27(4):658–65. https://doi.org/10.1177/1526602820922179. Epub 2020 May 18
13. Nakama T, Ichihashi S, Ogata K, Kojima S, Muraishi M, Obunai K, Watanabe H. Twelve-month clinical outcomes of percutaneous deep venous arterialization with alternative techniques and ordinary endovascular therapy devices for patients with chronic limb-threatening ischemia: results of the DEPARTURE Japan study. Cardiovasc Intervent Radiol. 2022;45(5):622–32. https://doi.org/10.1007/s00270-022-03095-1. Epub 2022 Mar 11

Robotics and Endovascular Surgery: Current Status

13

Lucinda Cruddas, Guy Martin, and Celia Riga

Introduction

Cardiovascular disease is a major cause of morbidity and mortality in the Western world, and was responsible for 17.9 million deaths in 2016 [1]. The disease primarily affects coronary and peripheral blood vessels, and may result in a wide spectrum of complications including myocardial infarction, stroke, limb loss and ultimately death. Minimally invasive interventional techniques have revolutionised the management of cardiovascular disease, and have a wide and varied use across all aspects of disease treatment including ablative techniques for the treatment of arrhythmias, stroke intervention with mechanical thrombectomy, aneurysms with stenting, and peripheral arterial disease with a range of revascularisation technologies.

In minimally invasive endovascular surgery, percutaneous access is established to guide catheters and guidewires through vessels in the cardiovascular system. Fluoroscopy enables real-time visualisation of catheters and guidewires for safe manipulation and therapeutic action. Stenosis of cardiac and peripheral arteries can be treated with an inflatable balloon and expandable stents. Aneurysms are treated with stent grafts which relieve pressure from the fragile aneurysmal vessel wall. In cardiac physiology, ablation catheters apply ultrasound or radiofrequency energy to ablate tissue responsible for arrhythmias. Endovascular surgery requires precision and stability. The operator is guided by 2D fluoroscopic images, and axial force and torque delivered at the proximal end of catheters and guidewires. Force and haptic feedback to the operator is limited. Vessel anatomy and catheter manoeuvrability also present challenges in navigating anatomy and stability and reaching target sites. These challenges may be overcome by the advent of robotic technologies which aim to provide stability and precision and reduce tremor and human error.

Robotic endovascular surgery and associated technologies are the logical next stage of endovascular development and have been at the forefront of surgical innovation over the past few decades. The first robotic system was the Da Vinci, approved in 2000 by the Food and Drug Administration (FDA). The Da Vinci is licensed for use in general laparoscopic, urological, thoracoscopic and thoracoscopically assisted procedures. Laparoscopic robotic surgery has upscaled exponentially, with multiple systems available, which are employed internationally. Robot-assisted laparoscopic prostatectomy has achieved parity with standard laparoscopic techniques [2], and in 2020 approximately 1,243,000 laparoscopic robotic procedures were performed [3].

Robotic systems in endovascular surgery remain in development, but have the potential to continue to revolutionise treatment of aneurysmal and occlusive disease. Robotic endovascular technologies use steerable catheters which are currently in use with endovascular intervention, but with remotely controlled navigation systems. Additionally, 3D projection of endovascular technologies has been a focus of further development. Endovascular robotics remains in early stages compared with laparoscopic robotics due to the younger endovascular technologies which continue to be refined, in parallel with robotic innovations. The advent of steerable catheters was an early foundation for endovascular robotics. They were designed to overcome the challenges in navigating vascular anatomy, which continues to drive endovascular and robotic innovations. Robotic platforms enable rotation in 30-degree increments and axial movement in single-millimetre increments to enable greater stability, precision and navigation [3].

This chapter will review existing endovascular technologies including steerable catheters. Robotic technologies will be examined, including approved technologies and those still in development. The challenges of current designs and further developments will also be explored in order to provide

L. Cruddas · G. Martin · C. Riga (✉)
Imperial Vascular Unit, Imperial College Healthcare NHS Trust, St Mary's Hospital, London, UK
e-mail: lucinda.cruddas@nhs.net; guy.martin@nhs.net; c.riga@imperial.ac.uk

G. Geroulakos et al. (eds.), *Mastering Endovascular Techniques*, https://doi.org/10.1007/978-3-031-42735-0_13

research, development and clinical priorities to further translate and scale robotic technologies into routine surgical practice.

Endovascular Surgery

Endovascular techniques are used in the treatment of aneurysmal disease; cardiac arrhythmias; coronary and peripheral vascular stenosis due to atheromatous disease or dissection; and thrombectomy in stroke. Minimally invasive endovascular surgery has many potential benefits compared to open techniques, including fewer wound complications, less post-operative pain, faster recovery and reduced hospital stay length [4, 5]. Cardiac and respiratory complications may be reduced, and mortality rates may be lower [6]. Minimally invasive technology may also enable the treatment of patients who are not fit for a more substantial open procedure, or in whom there is no technical option for open repair. Endovascular techniques still face a number of challenges, however. In order to manipulate catheters and guidewires to distant targets, operators are dependent on visual cues from 2D fluoroscopy and haptic feedback. Navigation of anatomy can be challenging especially in tortuous and angulated vessels. Large doses of contrast agent may be required, which can result in nephrotoxicity. Contact between guidewires and catheters with the vessel wall can cause complications including dissection, embolisation, perforation and thrombosis. Additionally, even with the use of lead aprons, the operator is exposed to significant radiation, which may be associated with an increased incidence of occupational cancers and cataracts [7]. Lead aprons and prolonged standing can also cause orthopaedic complications for the operator. These challenges have driven the need for further innovation in order to improve the safety, outcomes and applicability of minimally invasive techniques.

Innovation in endovascular surgery has been achieved by advancement in catheter design. Safe catheterisation requires stability and manoeuvrability. Key trends in the advancement of catheter technology include the development of controlled steerable tips and the incorporation of haptic feedback. These innovations have also been supported by significant developments in supporting wire technology, with a wide range of disease- or technique-specific wires now widely available. Intravascular devices are also an area of development. Complex aneurysmal disease and dissection can be treated by complex stent grafts which include branched, fenestrated and custom-made designs. Additionally, arterial calcified plaques can be treated with drug-coated balloons to reduce the risk of restenosis, or with atherectomy to remove calcified, fibrotic tissue. Intra-arterial lithotripsy can be employed in calcified coronary arteries prior to stent deployment during percutaneous coronary intervention.

Surgeon skills must develop in parallel to new technologies. Endovascular training requires standardised assessment to ensure safe employment of new technologies and proficiency in complex endovascular procedures. Advances in technology have been accompanied by developments in virtual reality which can provide objective performance feedback.

Steerable Catheters

Steerable catheters are the foundation for endovascular robotic technologies. They were developed to overcome the challenges of conventional catheters and are widely used. Steerable technologies enable the operator to vary the shape of the catheter distally. This provides a greater degree of control over direction of motion, and subsequent improved navigation of anatomy [8]. Steerable catheters are categorised according to their actuation mechanism and can be either magnetically or actively driven. Active devices include: hydraulically driven, pull wire and smart-material actuated [8]. Magnetic catheters and guidewires have a magnetic tip which can be deflected by a surrounding magnetic field, generated by a large permanent magnet within the operating room. Magnetic catheters have a soft tip which is safe within vessel walls. They are, however, expensive to install and maintain [9]. Pull wire catheters exist in multiple commercial forms and are commonly used in cardiac ablation and mapping. They use actuator tendon control handles, attached to an elastic nitinol catheter [10]. Commercially available steerable catheters which have been integrated with robotic technology include: Polaris XTM Steerable Diagnostic Catheters (Boston Scientific Inc., Marlborough, USA) and Artisan Extend Control Catheters (Hansen Medical Inc., Mountain View, USA) [11].

Hydraulic and smart-material technologies have not become commercial technologies. Hydraulic pressure-driven catheter technologies work by injecting a solution into hydraulic bellow segments, to create variations in pressure. Pressure changes modify the length of the bellows, bending them in a single plane and steering the distal catheter tip [11]. Challenges in controlling the continuous bending of individual segments have meant that hydraulically driven catheters are not a mainstream technology [9]. Smart-material actuated catheters use variations in temperature to deflect a shape memory alloy (SMA) to guide the catheter tip [12]. However, this technology carries the risk of overheating.

Force and pressure on catheter tips must be monitored and fed back to the operator to prevent procedure complications and ensure technical success. This is particularly relevant in cardiac ablation and for inflation pressure of balloon catheters [13]. Fibre optic-based sensors are commercially available for cardiac procedures but not for alternative interventions [14, 15]. The TactiCath catheter (Endosense SA, Geneva, Switzerland) is an example of one of the main commercially available ablation catheters with optical fibres in the distal end to measure magnitude and force applied at the catheter tip [16].

Manual steerable catheters are a safe, established technology and used across a range of endovascular procedures [17, 18]. They are considered superior to conventional fixed catheters as their flexibility enables navigation of challenging anatomy, greater precision and reduced catheter changes. This results in greater technical success. Patient and operator safety is also improved by the reduced need for contrast and radiation. There is a reduced dependence on operator skill and a flatter learning curve [8]. Limitations of steerable catheters include: unpredictable friction; short shafts; large diameters; low torsion; coupled tip and shaft movements; costs; and limited compatibility with adjunct technologies [19]. Robotically remote-controlled catheters have been the next stage in steerable catheter development, in order to further improve steerability, precision and stability.

Remote-Controlled Catheter Navigation Systems

Remote-controlled endovascular navigation systems are available commercially. These technologies include electromechanically actuated platforms: Sensei robotic navigation system (previously Auris Surgical Robotics, San Carlos CA, now acquired by Johnson & Johnson, New Brunswick, New Jersey), Magellan robotic catheter system (previously Auris Surgical Robotics, San Carlos, CA, now acquired by Johnson & Johnson, New Brunswick, New Jersey), CorPath vascular robotic system (Corindus Vascular Robotics, MA, USA) and Amigo remote catheter system (Catheter Precision Inc., NJ, USA). A magnetically controlled robotic system is also commercially available: Niobe magnetic navigation system (MNS; stereotaxis USA). Sensei and Magellan (Johnson & Johnson, New Brunswick, New Jersey) are currently the main commercial platforms in clinical practice and have been acquired by Auris Health Inc. (Redwood City, CA, USA).

Sensei/Magellan Robotic System

The Sensei robotic catheter was approved by the FDA in 2007 for cardiac mapping and ablation, and is now also in use for both standard and complex endovascular aortic aneurysm repair (EVAR) [20–23]. A steerable guide catheter is controlled by instruction from a "master-slave" remote workstation. Steerable catheters are inserted into a custom-designed sheath, the Artisan Extend Control Catheter, which is manipulated via the remote control of a 3-dimesion of freedom (3-DoF) joystick [24–26]. Any cardiac or mapping catheter can be accommodated in the custom-steerable sheath. The CoHesion visualisation module enables a visual display of forces and tactile feedback by vibration of the motion controller using 3D electroanatomic mapping Ensite (St Jude Medical, MN, USA) and a distal force measurement system, IntelliSense [9]. The sheath rigidity may, however, increase the risk of mechanical complications [27]. A rigid sheath can cause increased vessel wall trauma and may be associated with bleeding, haematoma, dissection, pseudo aneurysm, vessel rupture and thromboembolism.

The Magellan robotic system succeeded Sensei and received FDA approval in 2012 (Fig. 13.1a, b, c) [20]. Similar to the Sensei robotic catheter, the Magellan has a robotic arm which controls a steerable catheter. The catheter has an inner leader and an outer sheath, controlled by pull wires via a remote operator, with a 7-DoF joystick or navigation buttons. Catheter insertion, withdrawal, movement, rotation and angulation are controlled remotely [24]. The Magellan robotic system is associated with greater stability, fewer catheter tip movements, greater steerability and greater manoeuvrability than with conventional techniques [24, 26]. Robotic cannulation of peripheral vessels is faster compared to manual cannulation, although it lacks force sensing and feedback. Disadvantages also include the dependence on custom single-use 14F sheaths and long set-up times [24]. Robotic set-up times are up to 15 min, whilst conventional set-up times are negligible. The aim is for faster target vessel cannulation to offset longer set-up times. Faster procedure times and increased accuracy could also offset extra costs. Production of the Magellan robotic system was stopped in 2016 [9].

Amigo System (Catheter Robotics Inc. NJ, USA)

The Amigo system (Catheter Robotics Inc. NJ, UDS) is an FDA-approved remotely controlled electromechanical robotic system (Figs. 13.2 and 13.3). Its intended use is for cardiac mapping and radiofrequency ablation. A robotic arm is attached to a standard electrophysiology table and used to control standard commercially available steerable catheters with 3-DoF (rotation, tip deflection, insertion/withdrawal). Linear motion, deflection and rotation can be produced [27]. The Amigo system has been shown to be associated with less operator radiation exposure than conventional techniques. Patient outcomes are comparable to manual techniques. Contact force is integrated, with a manual override system [30]. The Amigo is, however, large and unwieldly [27, 30–32].

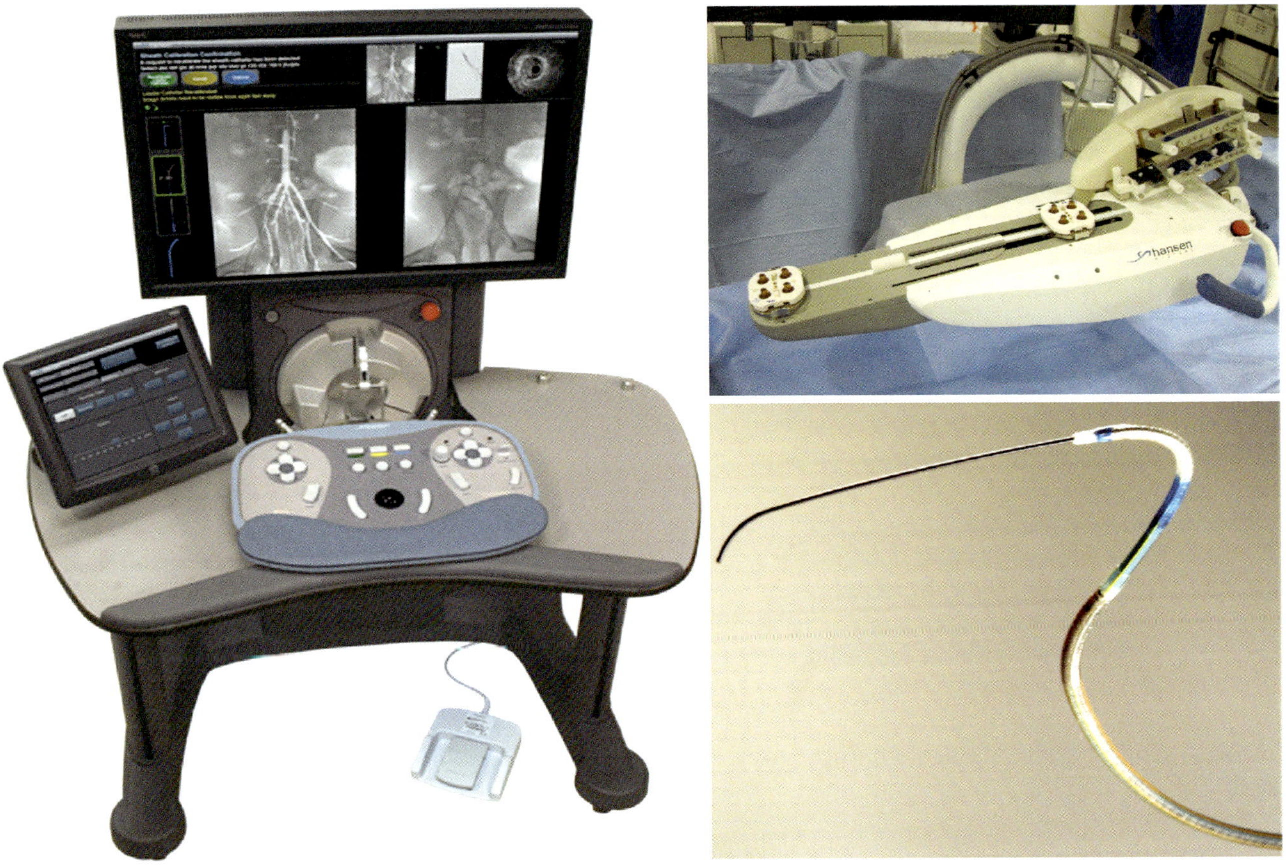

Fig. 13.1 The Magellan robot. (**a**) Magellan robot user interface. (**b**) Magellan robotic arm. (**c**) Magellan steerable catheter

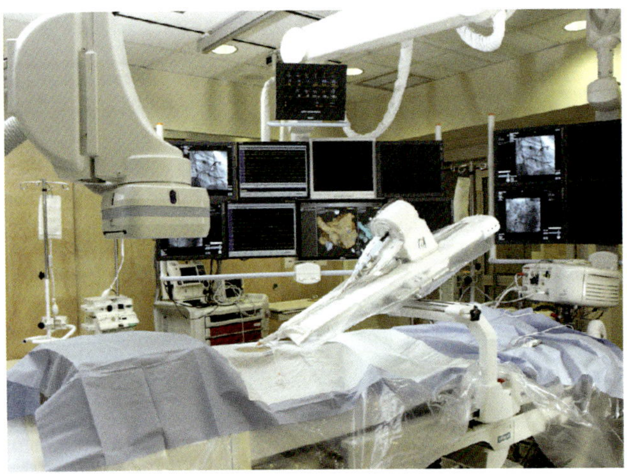

Fig. 13.2 Amigo [28]

CorPath (Corindus Vascular Robotics, MA, USA)

The CorPath 200 system received FDA approval in 2012 for remote-controlled stenting and ballooning in coronary interventions (Fig. 13.3). It was succeeded by CorPath GRX, which received FDA approval for PCI and peripheral arterial intervention in 2016 and 2018 respectively [9]. It uses electromechanical technology for remote control of standard catheters, guidewires and balloon/stent catheters, and is compatible with angioplasty balloons and stent delivery systems [33]. The CorPath 200 has a bedside robotic unit with a single-use cassette on a robotic drive and an articulating arm. Standard guidewires and catheters can be moved and exchanged with proximal force sensing, and are controlled by touch screen and joystick controls in a remote workstation. The catheter can be advanced in 1 mm increments, and rotated in 30-degree increments. The physician must, however, manually perform initial catheter placement, and guide and deploy devices [32]. CorPath 200 is associated with greater technical success and subsequent faster procedure times and less fluoroscopy use [34]. There are currently no large studies to assess the use of CorPath GRX, however [31].

Magnetic Robotic Navigation Systems: Niobe (Stereotaxis, St. Louis, MO)

Magnetic technology can be used to remotely control catheters. In the Niobe system, two external magnets are placed on either side of the patient to create a magnetic field

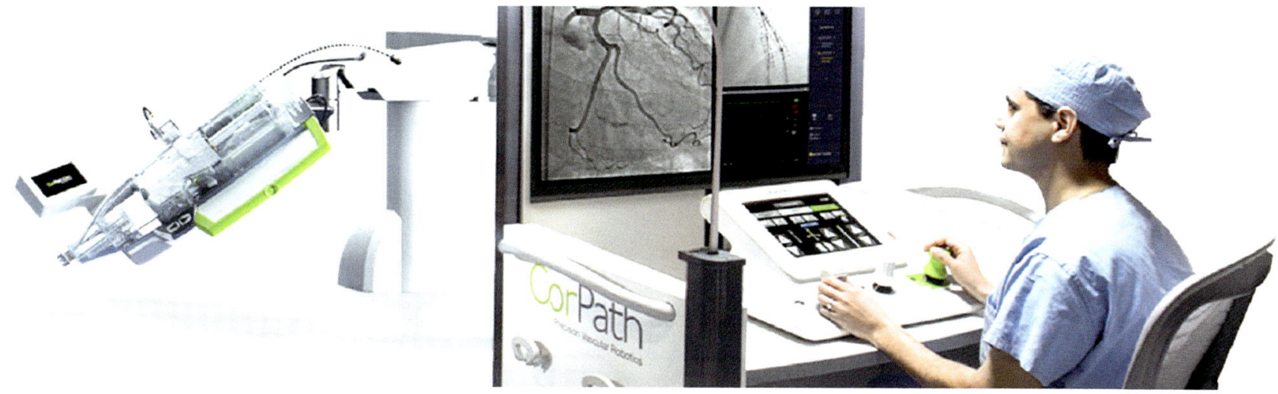

Fig. 13.3 CorPath [29]

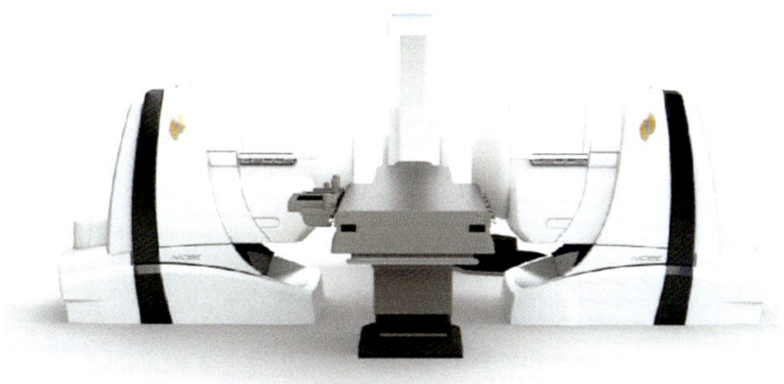

Fig. 13.4 Niobe [35]

(Fig. 13.4). Bespoke catheters and guidewires with magnetic tips are controlled by the external magnets which vary their intensity and position to deflect the catheter tip with 3-DoF [36]. External magnet orientation is controlled remotely via a mouse and joystick, and catheters are advanced by a mechanical motor drive. The Niobe system has been used in cardiac mapping and ablation, and peripheral arterial intervention [9, 32]. Magnetic catheters are less stiff than electromechanical catheters and therefore may be associated with a lower risk of vessel wall injury and perforation [9, 27]. However, the system requires a dedicated room and magnetically compatible equipment. It is also incompatible with real-time MRI [31].

The Catheter Guidance Control and Imaging (CGCI) system (Magnetecs Inc., CA, USA) is currently in development for therapeutic interventions in cardiology, neurology and obstetrics and gynaecology [37]. It uses eight electromagnets in a semi-circle configuration to manipulate a magnetic field and steer a magnetised catheter (Maxwell mapping catheter) [9]. The magnetic field can manipulate instantaneous real-time catheter movements, which are precise, rapid and stable. Control is either by manual or automatic control mode. The CGCI system integrates fluoroscopy, intracardiac echocardiac images, a multichannel ECG recording system and real-time mapping into a single 3D display for visualisation of anatomy and equipment location. The catheter tip possesses tissue-contact sensing filters to guide catheter movement. The operator benefits from real-time 3D visual feedback, and both operator and patient benefit from reduced fluoroscopic exposure, faster procedure times, reduced training times and higher success rates [37].

Initial data from robotic endovascular technologies is promising. Compared with conventional methods, robotic procedures are associated with greater stability and catheter control. Procedure times are shorter and fluoroscopy exposure is subsequently less. Costs are higher, but there is the potential to offset these costs with faster procedure times and fewer complications. Limitations include: lack of haptic feedback; lack of compatibility with all conventional materials; and need for manual deployment of devices. These limitations have provided the foundations for further developments.

Current Experiences in Robotic Endovascular Surgery

Aortic Arch Cannulation

Robotic cannulation of the aortic arch in thoracic endovascular aortic repair (TEVAR) has been demonstrated to be safe and effective in patients [38]. Eleven patients underwent robotic and manual cannulation of the aortic arch, and robotic cannulation was associated with significantly fewer high-intensity transient signals, detected by intraoperative transcranial Doppler (TCS), compared to manual techniques. This demonstrated that improved manoeuvrability, stability and control associated with the robotic system could result in less cerebral embolisation during TEVAR.

Visceral Artery Cannulation

The Magellan robotic system was used to perform a fenestrated endovascular aneurysm repair (FEVAR) for an asymptomatic 73 mm juxtarenal abdominal aortic aneurysm using the Anaconda device (Vascutek, Inchinnan, United Kingdom) with three customised fenestrations [39]. The left renal artery was targeted robotically, whilst the two right renal vessels were cannulated manually. The heavily calcified left renal artery was successfully cannulated robotically, taking only 3 min, with an additional 2 min for guidewire exchange. This was faster than the manual cannulation of the right renal arteries (48 min for both arteries). Magellan set-up time was 5 min. This case study demonstrates the effectiveness and feasibility of the Magellan robotic system with accuracy and reduced radiation as notable benefits. Figure 13.5 demonstrates intraoperative robotic sheath and catheter positioning, whilst Fig. 13.6 demonstrates real-time master–robotic interface.

Robotic cannulation of renal and visceral target vessels has been demonstrated in a number of patients undergoing complex endovascular aortic procedures including fenestrated, branched and chimney stents [40] (Fig. 13.7). Robotic

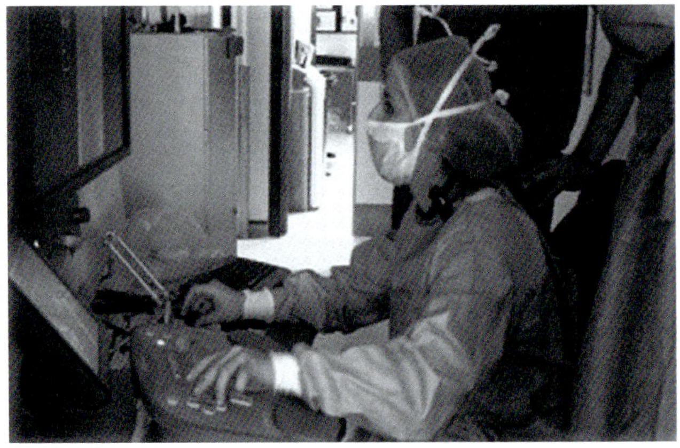

Fig. 13.6 Magellan user interface performing robotic cannulation of the left renal artery in fenestrated endovascular aneurysm repair

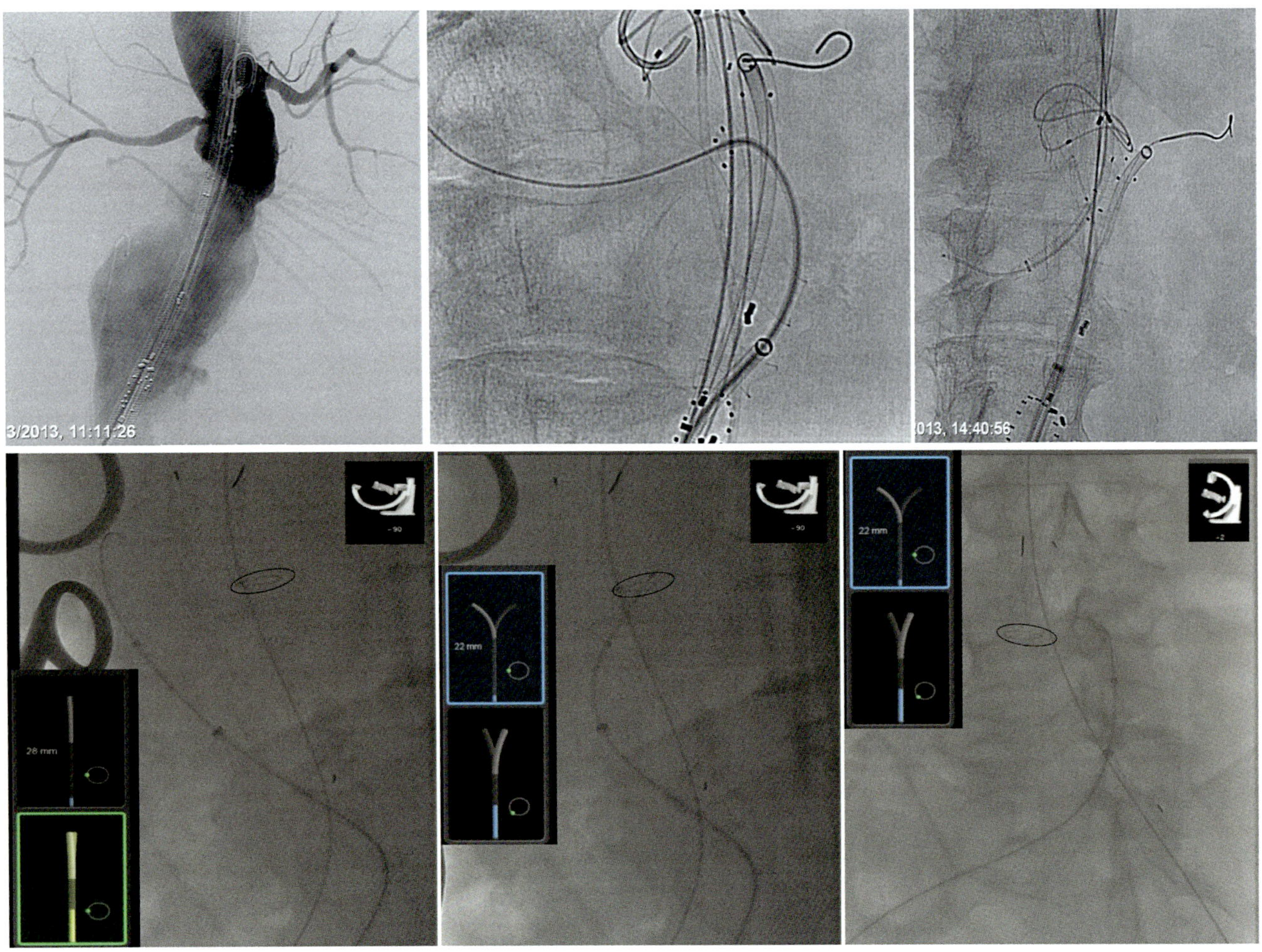

Fig. 13.5 (**a**) Aneurysm with anaconda main body under angiography. (**b**) Robotic cannulation of the left renal artery. (**c**) Magellan sheath positioning

navigation was employed in thirty-seven target vessels in fifteen patients, with a successful cannulation rate of 81%. All branched and chimney stent target vessels were accessed successfully by the robotic system. Mean cannulation time was 4 min and 23 s and there were no robotic-related intraoperative or post-operative adverse events. This further confirms the feasibility for robotic-controlled systems to perform safe cannulation of target vessels in complex endovascular aortic procedures.

Contralateral Gate Cannulation

A multicentre retrospective cohort study has demonstrated the effectiveness of robotic contralateral gate cannulation in endovascular aortic aneurysm repair (EVAR) [41]. Fourteen robotic cases were compared to eleven manual cases. Endovascular catheter path-length was significantly shorter for robotic techniques, compared to manual techniques. Median cannulation time was significantly shorter for manual cannulation compared to robotic. Efficiency ratio (path length/aorto-iliac centrelines) was significantly lower in robotic cases.

Embolisation

Magellan robot-assisted bilateral uterine embolisation has been performed in five female patients [42]. Robotic bilateral internal iliac and uterine artery cannulation was successful in all patients, and procedure success rates were 100%. No intraoperative or post-operative complications were recorded and all patients reported a significant improvement in symptoms at 6kl-month follow-up.

Embolisation of a saccular renal artery aneurysm has also been performed using the Magellan robot [43]. This case demonstrated the utility of the Magellan in high-risk procedures with complex vascular anatomy, due to its precision and stability. Figure 13.8 demonstrates intraoperative robotic sheath and catheter cannulation.

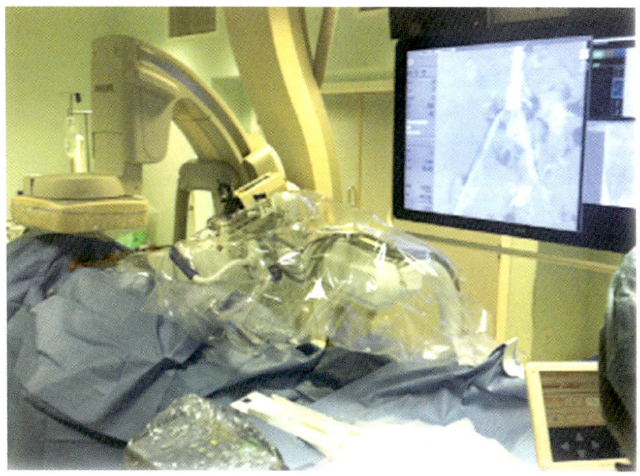

Fig. 13.7 Magellan robot performing cannulation of left renal artery in fenestrated endovascular aneurysm repair

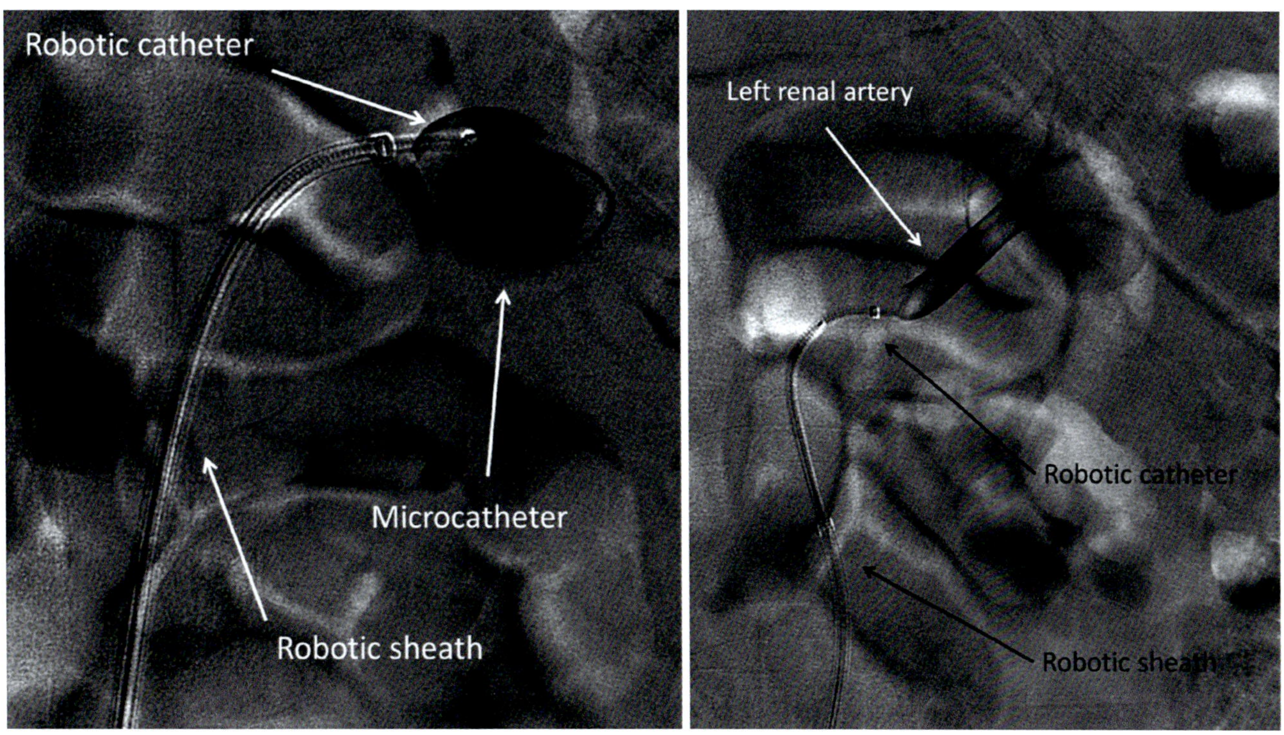

Fig. 13.8 Magellan sheath and catheter cannulation in renal artery aneurysm [43]

Developments in Existing Robotic Platforms

In addition to the currently commercially available robotic systems, a number of systems have been proposed to overcome the deficits and shortcomings of existing technologies. These include: compatibility with conventional materials; haptic feedback to the operator; autonomous catheter function; and an authentic master–slave interface. These technologies will be reviewed.

Use of Standard Catheters

Existing robotic systems are mostly incompatible with conventional catheters, guidewires, balloons and stents, and the requirement for system-specific equipment leads to substantial cost increases. A number of systems have been explored to incorporate conventional catheters into endovascular robotic technologies. Robotic systems have traditionally been compatible with specialised, specific catheters which have been designed alongside the robotic system to enable maximum degrees of freedom and stability. It has been recognised that this is costly and impractical, and the use of standard catheters and wires should be incorporated into new robotic models. In magnetic robotic endovascular navigation systems, the need for specialised equipment may be unavoidable.

System for Endovascular Telerobotic Access (SETA) is a prototype robot-assisted system compatible with conventional catheters [44, 45]. It has a slave manipulator comprising two translational stages and one steering stage, which enable simultaneous manipulation of conventional guidewires and catheters. The slave mechanism can both steer and insert or withdraw both catheters and guidewires. A second mechanism provides linear drive to the catheters. The actuators were constructed using friction wheels whilst the steering stage was constructed with a gripper-pulley system. The slave has torque sensors coupled to actuators to provide force feedback to the surgeon, whilst the controller uses a haptic device to command the position and velocity of the slave [44, 45]. SETA also possesses technology to remove tremor. SETA has currently only been evaluated in phantom vasculature. Time taken to complete procedures and amount of contrast fluid used during the procedure were comparable to manual intervention. SETA was rated as easy to use and performance was equivalent to manual intervention. It is low cost and easy to construct. Drawbacks include lack of natural twisting action and steering accuracy [44, 45].

Cercenelli and colleagues developed a telerobotic system (TS) to remotely navigate and reposition standard steerable catheters with deflectable tips, within the heart [46]. The study was performed in animal subjects. The system has a driving unit, control unit and remotely controlled user interface. A translation stage acts as an actuator of the distal catheter tip. The controller is via a 2-DoF joystick. Forces at the proximal tip are displayed graphically at the user interface. The system does not require a dedicated guide catheter and does not necessitate bulky machinery. In vivo study demonstrated reduced time for catheter positioning and navigation compared to manual procedure [46].

Further research in EM technology has included the development of the CathROB prototype robotic system [27]. It has been effective in cardiac mapping and ablation. CathROB uses electromagnetic technology to remotely control steerable catheters with 3-DoF. It has force sensing technology which supports both catheter navigation function and operator haptic feedback. It is compact, has a fast set-up time and does not require a specialised room. The system is compatible with standard sheaths, catheters and deployment devices. Further studies are required to demonstrate its safety and cost effectiveness prior to widespread use.

Force Sensing and Haptic Feedback

Haptic feedback utilises touch to provide information to users; the lack of haptic feedback with endovascular robotic technologies potentially poses a safety risk to the patient. Haptic feedback is not currently embedded within any commercial robotic systems. Commercial haptic interaction devices are, however, available, including the Geomagic Touch device (3D Systems Corp, Rock Hill, SC, USA) [47]. These devices, however, are limited by providing haptic feedback in a single linear direction, delivering unknown consistency of feedback and having high costs. Haptic feedback requires accurate localisation of catheters, accurate estimation of catheter tip deformation, and a computerised system capable of force estimation by integrating catheter deformation and live x-ray images. The estimated force must then be translated into a recognisable tactile force to the operator in order to support accurate and atraumatic navigation and therapy.

Haptic feedback can be either sensor-based (direct) or model-based (sensorless), and feedback can be either from the proximal or distal end of the catheter. A number of sensor-based feedback models have been demonstrated in endovascular models and virtual reality, using both distal and proximal feedback [48–56]. Model-based haptic estimation has also been demonstrated in endovascular models and virtual reality, using both distal and proximal feedback [57–59]. In sensor-based feedback, the distal catheter tip has sensors and feedback is replicated in a master controller. In model-based feedback, forces are estimated using a computer model. Force sensor-based devices include miniaturised piezoelectric force sensors and strain gage sensors, or piezo resistive force sensors [59–61]. These are limited by electromagnetic interference from other devices and electrolytic environment, and so led to the development of optical fibre-based sensors, and light-intensity modulators (LIM) which can measure contact forces along the catheter body and tip [62–68]. Sensor-based catheters provide important, instant haptic feedback but are limited by size, material flexibility and electromagnetic compatibility. Information is limited to a specific point on the catheter tip, which may not be in contact with tissue at a given time. All such technologies continue to be limited by high costs, however, and therefore have not been adopted in routine clinical practice.

Model-based haptic feedback incorporates sensors, a mechanical model of the catheter/guidewire, live imaging and the patient's anatomy. Feedback is constructed using force/torque sensors and orientation/position coders. Various models of catheters exist, with no clear model having an advantage [69]. Additionally, model-based haptic feedback has largely been investigated in virtual environments. Whilst sensor-based haptic feedback has been shown to have shorter response time and less computational cost, model-based haptic feedback is less susceptible to noise, is easy to use and has lower costs [69]. Model-based haptic feedback necessitates a compromise between accuracy and computational efficiency.

Autonomous Catheter Positioning

A natural progression from catheter tip sensors is autonomous catheter positioning based on feedback. This technology is still in its infancy. Ganji and colleagues [70] developed a robotic steerable cardiac catheter, with 3-DoF and autonomous steering properties to navigate the cardiac chambers in animal models. The distal shaft of the deflectable catheter possessed magnetic sensors which, in conjunction with the kinematic catheter, could perform semi-autonomous catheter positioning. Force control-based robotic systems are a further development to improve catheter navigation, independent of the operator. In vitro cardiac studies have demonstrated the potential for catheters to use information from 3-dimenional ultrasound (3DUS) and tissue interactions, to either maintain a constant force or follow a predetermined specific force instruction [71]. These technologies require further research and in vivo studies.

Autonomous positioning is the ultimate goal of robotic endovascular technologies to remove human error. Autonomous positioning requires: validated anatomy of the target area; real-time visualisation of catheter/wire location; tissue feedback; feedback/alert system to the operators; ability to autonomously plan a new route when obstacles are encountered; and the ability to switch back to manual control. Autonomous positioning would be the ultimate amalgamation of current robotic technologies. Refinement and development of steerable catheters with haptic feedback and 3D live imaging are vital to realise this goal. The anatomical environment of endovascular surgeries remains complex, however, and rules cannot cover the range of surgical complexities encountered. Robotic reinforcement learning may overcome this, but more research is required.

Master–Slave Interface

Robotic systems have traditionally been based on a remote master workstation, controlled by a joystick. Remote control skills are independent of the traditional manual skills required for endovascular procedures. This results in operators having to learn two independent skills: manipulation of conventional catheters as well as manipulation via a joystick. This could result in increased training times. The use of a joystick, however, is simple, intuitive and achieves the same outcome whilst removing some of the complexities of conventional catheter manipulation.

Systems have been developed whereby a master catheter is manipulated by the surgeon to mimic conventional catheter movements, and these actions are translated to a slave catheter. Thakur and colleagues [72] designed a model where the master catheter was manipulated by conventional axial and radial motion. Although the system demonstrated evidence of motion replication in vitro, it lacked force feedback to the surgeon. Subsequent systems have been developed with force feedback [49, 60, 73]. The Hapcath system [59] used this technology, but also measured forces at the guidewire tip using a piezo resistive force sensor and was successfully demonstrated in vitro. Force feedback has been shown to reduce the magnitude and duration of force on vessel wells in vitro [49]. Further research by Yang and colleagues [47] demonstrated a novel surgical robot controller which mimics the traditional handling of catheters and guidewires, with force feedback. This model demonstrated reduced procedure time and reduced force feedback denoting reduced risk of vessel wall injury and greater safety in vitro.

Remote Control over Significant Geographical Distance

Remote control over large geographical distances is an additional area of research. This could overcome the unequal distribution of endovascular surgeons and resources. Guo and colleagues [74] demonstrated a cloud server which could effectively be used for remote control endovascular robotic procedures. Additionally, Madder and colleagues demonstrated that the CorPath GRX could safely perform robotic endovascular interventions in in vivo animal models when manipulated from a separate geographical location [75]. The system was also shown to be effective in performing PCI on a model between Boston and San Francisco (over 3085 miles) via wired networks and 5G wireless networks [76].

New Technologies in Robotic Surgeries

In addition to improving existing robotic technologies, research has focused on novel techniques for remote catheter control and adjunctive technologies, the most important of which is imaging. These are explored in more detail below.

Electromagnetic Endovascular Navigation Tracking Systems (EM ENS)

EM ENS use electromagnetic coils in device tips to provide digital data on coil position within an electromagnetic surgical field. Voltage data is converted to digital data and visualised on a preoperative 3D CT scan. Preoperative imaging is combined with intraoperative MRI, CT or fluoroscopy. Two systems are currently available: Aurora (Northern Digital, Waterloo, Ontario, Canada); and StealthStation TREON Plus (Medtronic Inc., Minneapolis, MN) [26]. Aurora ENS has been applied in a range of models, using a range of devices [77–83], and demonstrated good precision. It was not, however, associated with reduced procedure time or radiation exposure [56]. Evidence for StealthStation ENS is limited to two studies using a rigid model, and nonclinical wire [81, 82]. It demonstrated good accuracy, improved performance scores and reduced fluoroscopy exposure. More research is required into these systems.

Ultrasound-Based Localisation Tracking Systems (ULS)

Ultrasound has been postulated as an alternative to magnetic catheter tips for localisation and tracking [26]. ULS uses a piezoelectric ultrasound transmitter at the tip of a catheter. Signals are sent to triangulated sensors on the body's surface. Real-time catheter tip position can then be demonstrated on a preoperative 3D CT image, without real-time imaging. There are currently two studies demonstrating the use of ULS in vitro, with potential for use in endovascular aortic procedures [84, 85]. These studies demonstrated feasibility and also a method for aperture error compensation [85]. This technology requires further refinement, miniaturisation and extension to other devices beyond a 9F ULS catheter. ULS could potentially negate the need for intraoperative contrast use and x-ray. Due to the complex nature of procedures, fluoroscopy may, however, be required in addition to ULS to ensure procedure safety. Combining these techniques will minimise fluoroscopy use.

3D Rotational Angiography (3DRA)

Traditional fluoroscopy provides the operator with a 2D image. 3DRA constructs a 3D image prior to intervention, with the aim of improving diagnostic accuracy and therapeutic intervention [7]. 3DRA reconstruction negates the need for intraoperative contrast use, and reduces radiation exposure [86]. A prototype navigator has been developed which calibrates pre-operative 3DRA images with catheters and guidewires with 5-DoF electromagnetic sensors [87]. Real-time tracking can be performed during the procedure, combining live fluoroscopy with 3DRA. Limitations include the static nature of the 3DRA, which does not account for natural movements of breathing and vessel deformation when in contact with devices and wires, and continued dependence on fluoroscopy and radiation. Combining robotic technologies with 3DRA, however, may provide a solution to real-time catheter tracking, without the need for fluoroscopy.

MR Endovascular Robotic Platform

Robotic platforms depend on fluoroscopic-guided intervention. This continues to expose patients and surgeons to ionising radiation, and patients to the risk of contrast-induced nephropathy. MR has potential advantages. Firstly, MR endovascular robotic platforms may use real-time data from either MR or fluoroscopy, potentially minimising and eliminating the need for contrast and radiation. It is also superior in imaging soft tissue, such as vasculature, and providing 3D imaging. Early studies [88, 89] have demonstrated that time to complete complex procedures was comparable to manual techniques. Simple procedures and overall total procedure time, however, were significantly longer with the MR endovascular robotic platform, compared with manual techniques. MR scanners are also ergonomically cumbersome and patients cannot easily be accessed within MR. MR robotic endovascular intervention is currently only compatible with steerable catheters for cardiac procedures.

An MRI-guided robotic platform was designed for manipulation of standard catheters and guidewires using multimodal imaging (fluoroscopy and MRI) [88]. The system used a master controller which mimicked manual movements, and was guided by haptic and visual feedback. The platform required conventional manual access of the arterial system. Following access, instruments were connected to the robotic slave, and controlled remotely. Preliminary preclinical trials revealed a reduction in forces applied to the phantom compared with conventional manual techniques. Maximal forces were also reduced. Complex task navigation was completed faster.

An early-stage MR robotic prototype has been proposed [89]. It included an interventional room with an MR safe slave robot, and a control workstation with a navigation system and master device. Motor control was generated by pneumatic pressure patterns. Real-time imaging from MR or fluoroscopy guided controller intervention. Master controllers were adapted to mimic conventional manual instrument handling, and had a 2-DoF motion catheter and guidewire, which performed linear and rotational movement, and torque and force feedback. The slave robot mimicked the master control with 6-DoF. Guidewire and catheter carriers performed linear force transmission, driven by pneumatic clamp and linear actuator. Rotary platforms performed catheter and guidewire rotation; however, a significant limitation was that catheters and guidewires could not be manipulated simultaneously. The slave robot set-up was fast, intuitive and cost effective compared to manual techniques. The system was tested by seven senior surgical registrars or consultants on a thoracic phantom, using either robotic or manual execution. Robotic cannulation was successful in 90–100% of cases, and forces were similar to other experimental data. Cannulation time was slower but still within an average of 2 min. Technical limitations were related to the catheter buckling and limited linear operational workspace in the interventional room. Further research is required to integrate devices into MRI suites, incorporate MR safe steerable catheters and guidewires, and automate procedural sub-tasks.

Microrobots

The development of microrobots has seen a paradigm shift in robotic endovascular surgery, with a greater focus on the potential independent functional capabilities of robots. Microrobots are mobile robots measuring a few hundred micrometres [90]. Their autonomous function and small size may further the scope of endovascular practice and facilitate access of distant, difficult-access vessels. Their independent function could help to overcome human error and limitations in controller dexterity. There are two types of microrobots: tethered ferromagnetic soft robots which use magnetic steering, and soft polymer screen type which propel by rotation around a helical axis. Soft robots can deform and therefore have infinite DoFs. Design of microrobots has been challenging. Microrobots are an existing area of research in the field of oncology, specifically to deliver targeted therapy to cancerous cells [91].

A new design of microrobot has combined magnetic actuation of ferromagnetic soft robots [92–94]. These robots can deform and change shape within seconds by magnetic manipulation. This would enable navigation of difficult anatomy in vivo, and in vitro experiments have demonstrated successful navigation of tortuous and aneurysmal anatomy in a cerebrovascular phantom. These microrobots also released therapeutic agents at specified locations [94]. A self-lubricating hydrogel skin reduced the risk of vessel wall damage. Other therapeutic indications of microrobots could be to break down thrombosis/emboli by mechanically rubbing against them. This has been demonstrated in vitro, using magnetic actuation of helical robots under ultrasound guidance [95].

Microrobot technology remains in its infancy and there are obstacles which need to be addressed in its development. Delivery and retrieval of microrobots needs to be defined, as well as visualisation in vivo. The microrobot will likely enter the body via injection and migrate to its target site. Device removal would also ideally be via injection, although limited device lifespan may also be an option, with microrobots degrading either naturally or via stimulation after achieving their therapeutic goal [96]. Microrobots must also be integrated with a feedback and controller system. Control may be extracorporeal or intrinsic, with intrinsic control affording greater autonomy and the potential for machine learning. Currently, the focus of microrobots has been to deliver targeted cell therapies and local drug delivery. Endovascular therapeutics including stenting or angioplasty are yet to be explored.

The design of microrobots also has a number of limitations. Devices may be lost, or controllers may lose navigational control of devices in vivo. This could result in vessel wall injury. Microrobots could also heat fragile tissues. Manipulation and navigation of atherosclerotic plaques may result in distal embolisation. Measurement of success of intervention also needs to be recordable and defined.

Advantages and Challenges of Robotic Technologies

Benefits to Patients

Robotic technology has many potential advantages over standard endovascular techniques. Stability, steerability and accuracy in both catheter positioning and navigation have been shown to be superior in robotic techniques compared to manual techniques [27, 31, 36]. These advantages mean fewer catheter movements, fewer catheter exchanges and shorter navigation times [9]. These improvements may confer fewer complications and better outcomes for patients, and have been demonstrated in both simple and complex procedures. Robotics techniques have been shown to be superior to manual techniques across a range of contexts including in complex anatomy, cannulation and simple therapeutic procedures [30]. Benefits have been demonstrated across a range of endovascular areas, including transcatheter aortic valve implantation (TAVI) [97], coronary intervention [33, 98], peripheral vascular intervention [34] and aortic stent grafting [24, 99, 100]. Patients also benefit from reduced radiation exposure and reduced contrast burden [33, 36].

Benefits to Surgeon

Surgeons, as well as the whole theatre team, benefit from reduced radiation exposure [34, 36, 38]. Fluoroscopic-guided aortic procedures were shown to be associated with an acute DNA damage response in operators [101]. This may result in an increased risk of brain and neck tumours, basal cell carcinomas and cataracts amongst operators [102–104]. A reduction in radiation exposure could justify higher costs and complexities associated with endovascular robotics. Remote control also creates the potential for a surgeon to perform procedures from a remote location, which may widen accessibility to treatments for patients in different geographical locations. Surgeon comfort may also be improved as they may remain seated throughout the procedure, and do not require lead aprons, resulting in fewer orthopaedic complications, improved comfort and potentially enhanced performance [24].

Robotic technology has been shown to be intuitive and easy to learn. This confers a shorter learning curve compared to manual techniques, and robotic technology has been shown to improve accuracy and economy of movement [36, 105]. Surgeons benefit from the development of new technical skills. Further advances in robotic technology may also be associated with greater catheter automation, thereby reducing the surgeon's technical workload. Robotic technologies remove human tremor for greater precision and reduce the risks associated with human error, thus giving rise to improved surgical outcomes.

Technical Challenges

Robotic technology must be easily integrated into existing practice and augment existing techniques. It should be compatible with commercial standard catheters in order to minimise costs. Systems must be able to simultaneously manipulate catheters and guidewires, and be remotely controlled. Catheters must have dexterity to reach their target. Mechanisms should provide catheter stability, manoeuvrability, safety and practicality. The systems must be safe, and surgeons should experience the same operative forces and haptic feedback as in manual procedures. Developments should also improve the occupational safety of surgeons [38]. These criteria are not currently satisfied by existing technologies. This chapter has highlighted the need for further developments in haptic feedback technology, initial catheter placement and conventional catheter use. Specialised robotic catheters are costly, and also incompatible with conventional balloon and stent systems. This means that robotic target cannulation may be followed by exchange for standard devices for therapeutic purposes. This negates much of the advantages conferred by robotic systems.

In spite of reduced radiation exposure, even with robotic technologies, individual support staff are still required in the operating room to manage equipment. Alternative approaches for reducing radiation and contrast may therefore also need to be considered.

Sustainability

Robotic technologies are expensive, bulky and impractical in the current clinical setting and thus are not currently widespread. They require a specialised room, and are time consuming to set up. These costs may be offset by faster procedure time and reduced complications. In order to be cost effective, further developments should focus on making robotic technologies compatible with conventional catheters, guidewires, stents and balloons. In addition, hospitals need to complement the implementation of robotic technologies with adequate training facilities. The introduction of training facilities must be accompanied by clear, transparent and robust technical assessment. This requires technology capable of assessing catheter performance, and the creation of clear, unified training standards.

Current cost analysis of endovascular robotic systems is difficult due to limited application of endovascular robotic technologies. Cost-benefit analysis of widespread surgical laparoscopic robots may be of value. One study [106] demonstrated that, in the US, once the initial cost of a da Vinci robot is covered, the profits from high-volume laparoscopic radical prostatectomies are maintained if switched to robotic-assisted surgeries. However, initial costs are high, with one da Vinci robot costing $1.5 million, with an additional $112,000 annual service contract and $200 of disposables per case. The da Vinci has demonstrated that despite initial high costs, robotic technologies do not significantly increase costs once the robot is purchased and in high-volume use. This data is promising, although dependent on increased efficiency and high caseloads. Potential cost benefits should also be considered alongside the potential reduced risk of cancers for operators.

Training

Advantages have been demonstrated in the acquisition of robotic technical skills compared to manual skills. It is important, however, that alongside competency in robotic technologies, surgeons are also equipped with skills to manage complications of new technologies. Robotic skills should be developed alongside conventional manual skills. In order to monitor development of robotic skills, training requirements and objective assessment are required. Currently, certification is industry based. The following parameters need to be assessed in the training setting: technical success; procedural time; fluoroscopy exposure; number of vessel wall contacts; force of vessel wall contacts; and motion-based metrics. Assessment is completed by a grading system such as Objective Structured Assessment of Technical Skills (OSATS) [9]. These assessments can, however, be too generalised and oversimplified. Language models have also been employed, such as the Hidden Markov Models (HMMs). Learning algorithms and learning models are required for training and clear assessment of skills. Learning algorithms may also contribute to the development of automation of catheter function.

Conclusions

Endovascular intervention has revolutionised the management of cardiovascular disease. The field continues to develop and evolve with the advent of robotic endovascular technologies. Robotic- assisted intervention for endovascular surgery may still be in development but is an area of great potential growth and further innovation. Robotic technologies have been shown to improve accuracy, stability and steerability compared with manual procedure. These benefits have been demonstrated across a range of procedures. Existing commercial robotic technologies are limited by lack of haptic feedback, incompatibility with conventional catheters and a master–slave interface which is based on a different skill set to manual practice. Surgeon skills are not being fully utilised and costs remain high. Endovascular robotics has also witnessed a re-evaluation of real-time imaging, moving away from 2D fluoroscopy imaging, to alternatives including 3DRA and MRI. Alternative technologies have been explored including electromagnetic coils, ultrasound and microrobots. Robotic technologies hold great promise for future development. Technical superiority must, however, be matched by cost effectiveness, accessibility and thorough training assessment.

References

1. World Health Statistics. Monitoring health for the SDGs, sustainable development goals. Geneva, Switzerland: World Health Organisation; 2020; www.who.int/publications/i/item/9789240005105. Accessed 18 May 2022.
2. Dehn T. Robotic surgery. Ann R Coll Surg Engl. 2010;92:5–8.
3. Puschel A, Schafmayer C, Grob J. Robot-assisted techniques in vascular and endovascular surgery. Langenbeck's. Arch Surg. 2022;407:1789.
4. Cooper MA, Hutfless DL, Segev A, et al. Hospital level underutilization of minimally invasive surgery in the United States: retrospective review. Br Med J. 2014:349.
5. Fagogenis G, Mencattello M, Machaidze Z, et al. Autonomous robotic intracardiac catheter navigation using haptic vison. Science. Robotics. 2019;4(29):eaaw1977.
6. Lovegrove RE, Javid M, Magee TR, Galland RB. A meta-analysis of 21 178 patients undergoing open or endovascular repair of abdominal aortic aneurysm. Br J Surg. 2008;95:677–84.
7. Klein LW, Miller DL, et al. Occupational health hazards in the interventional laboratory: time for a safer environment. Catheter Cardiovasc Interv. 2009;73:432–8.
8. Fu Y, Liu H, Huang W, et al. Steerable catheters in minimally invasive surgery. Int J Med Robot. 2009;5:381–91.
9. Raffii-Tari H, Payne CJ, Yang G-Z. Current and emerging robot-assisted endovascular catheterisation technologies: a review. Ann Biomed Eng. 2014;42(4):697–715.
10. Canarillo DB, Milne CF, Carlson CR, Zinn MR, Salisbury JK. Mechanics modelling of tendon-driven continuum manipulators. IEEE Trans Robot. 2008;24:1262–73.
11. Hu X, Chen A, Luo Y, Zhang C, Zhang E. Steerable catheters for minimally invasive surgery: a review and future directions. Computer Assist Surg. 2018;23(1):21–41.
12. Lim G, Park K, Sugihara K, Nubanu K, Esashi M. Future of active catheters. Sensors Actuators A Phys. 1996;56:113–21.
13. Strandman C, Smith L, Tenerz L, Hok B. A production process of silicon sensor elements for a fibre optic pressure sensor. Sensors Actuators A Phys. 1997;63:69–74.
14. Polygerinos P, Seneviratne LD, Razavi R, Schaeffer T, Althoefer K. Triaxial catheter-tip force sensor for MRI-guided cardiac procedures. IEEE ASME Trans Mechatron. 2013;18:386–96.
15. Polygerinos P, Zbyszewski PD, Schaeffter T, Razavi R. Seneviratne, Althoefer K. MRI-compatible fiber-optic force sensors for catheterization. In: Medical Image Computing and Computer-Assisted Intervention-MICCAI; 2013. p. 369–77.
16. Yokoyama K, et al. Novel force contact sensor incorporated in irrigates radiofrequency ablation catheter predicts lesion size and incidence of steam pop and thrombus. Circ Arrhythm Electrophysiol. 2008;1:254–362.
17. Oberhuber A, Duran M, Ertas N, et al. Implantation of an iliac branch device after EVAR via a femoral approach using a steerable sheath. J Endovasc Ther. 2015;22:610–2.
18. Priokowski C, Eitel C, Rolf S, et al. Steerable versus nonsteerable sheath technology in atrial fibrillation a prospective, randomised study. Circ Arrhythm Electrophysiol. 2011;4:157–65.
19. Ali A, et al. Catheter steering in interventional cardiology: mechanical analysis and novel solution. Proc Inst Mech Eng H. 2019;233(12):1207–18.
20. Duran C, Estrada S, O'Malley M, et al. Kinematics effectively delineate accomplished users o92f endovascular robotics with a physical training model. J Vasc Surg. 2015;61(2):535–41.
21. Dello Russo A, Fassini G, Conti S, et al. Analysis of catheter contact force during atrial fibrillation ablation using robotic navigation system: results from a randomised study. J Interv Card Electrophysiol. 2016;46:97–103.
22. Carrell T, Dastur N, Salter R, Taylor P. Use of a remotely steerable "robotic" catheter in a branched endovascular aortic graft. J Vasc Surg. 2012;55:223–5.
23. Riga CV, Bicknell CD, Cheshire N, Hamady M. Initially clinical application of a robotically steerable catheter system in endovascular aneurysm repair. J Endovasc Ther. 2009;16:149–53.
24. Riga CV, Cheshire NJW, Hamady MD, Bicknell CD. The role of robotic endovascular catheters in fenestrated stent grafting. J Vasc Surg. 2010;514(4):810–20.
25. Guiudice CD, Pellerin O, Nouri Neville M, et al. Comparison of two endovascular steerable robotic catheters for percutaneous

robot-assisted fibroid embolisation. Cardiovasc Intervent Radiol. 2018;41:483–8.
26. De Ruiter QMD, Moll FL, van Herwaarden JA. Current state in tracking and robotic navigation systems for application in endovascular aortic aneurysm repair. J Vasc Surg. 2015;61(1):256–64.
27. Cercenelli L, Bortolani B, Marcelli E. CathROB: a highly compact and remote catheter navigation system; 2017. p. 2712453.
28. Catheter Robotics Inc. Amigo remote catheter system. 2021. http://catheterrobotics.com/images/AmigoBrochure.pdf.
29. Corindus Inc. CorPath GRX How it works. 2021. https://www.corindus.com/corpath-grx/how-it-works.
30. Shaikh Z, Eilenberg MF, Cohen TJ. The amigo remote catheter system: from concept to bedside. J Innov Card Rhythm Manag. 2017;8(8):2795–802.
31. Rueda MA, Theodoreli Riga C, Hamady MS. Robotics in interventional radiology: past; present and future. Arab J Interv Radiol. 2018;2(2):56–63.
32. Gunuz S, Albadawi H, Oklu R. Robotic devices for minimally invasive endovascular interventions: a new Dawn for interventional radiology. Adv Intell Syst. 2021;3:2000181.
33. Weisz G, Metzger C, Caputo R. Safety and feasibility of robotic percutaneous coronary intervention. PRECISE (percutaneous robotically enhanced coronary intervention) study. J Am Coll Cardiol. 2013;61(15):1596–600.
34. Mahmud E, Schmid F, Kalmar P, et al. Feasibility and safety of robotic peripheral vascular interventions. Results of the RAPID trial. J Am Coll Cardiol Intv. 2016;9(19):2058–64.
35. Stereotaxis Inc. Niobe robotic navigation system. 2021. https://www.stereotaxis.com/products/#!/niobe.
36. Antoniou G, Riga CV, Mayer EK. Clinical applications of robotic technology in vascular and endovascular surgery. J Vasc Surg. 2011;493-499:493.
37. CGCI Overview. Our technology. Magnetecs: guiding medical technology. 2022. https://www.magnetecs.com/overview.php.
38. Perera AH, et al. Robotic arch catheter placement reduces cerebral embolization during thoracic endovascular aortic repair. Eur J Vasc Endovasc Surg. 2017;53(3):362–9.
39. Riga C, Bicknell C, Rolls A, Cheshire N, Hamady M. Robot-assisted fenestrated endovascular aneurysm repair (FEVAR) using the Magellan system. J Vasc Interv Radiol. 2013;24(2):191–6.
40. Cochennec F, et al. Feasibility and safety of renal and visceral target vessel cannulation using robotically steerable catheters during complex endovascular aortic procedures. J Endovasc Ther. 2015;22(2):187–93.
41. Cheung S, et al. Comparison of manual versus robot-assisted contralateral gate cannulation in patients undergoing endovascular aneurysm repair. Int J Comput Assist Radiol Surg. 2020;15(12):2071–8.
42. Rolls AE, et al. Robot-assisted uterine artery embolization: a first in-woman safety evaluation of the Magellan system. J Vasc Interv Radiol. 2014;25(12):1841–8.
43. Aziz A, Ham S. Coil embolization of a renal aneurysm using a minimally invasive endovascular robotic system. Vasc Endovasc Surg. 2018;52(3):207–11.
44. Srimathveeravalli G, Kesavadas T, Li X. Design and fabrication of a robotic mechanism for remote steering and positional of interventional devices. Int J Med Robot. 2010;6:160–70.
45. IEEE/ICME International Conference on Complex Medical Engineering. 2007. 82–86.
46. Cercenelli L, Marcelli E, Plicchi g. Initial experience with a Telerobotic system to remotely navigate and automatically reposition standard steerable EP catheters. Biomed Eng. 2007;53(5):523–9.
47. Yang C, Guo S, Bao X. An isomorphic interactive device for the interventional surgical robot after in vivo study. Micromachines. 2022;13(1):111.
48. Guo S, Xiao N, Gao B. A novel robot-assisted catheter surgery system with force feedback. In: Selected topics in micro/nano-robotics for biomedical applications. Berlin, Germany: Springer; 2013. p. 175–90.
49. Payne CJ, Rafii-Tari H, Yang G-Z. A force feedback system for endovascular catheterisation. In: Proceedings of the IEEE/RSJ international conference on intelligent robots and systems; 2012. p. 1298–304.
50. Yin X, Guo S, Hirata H, Ishihara H. Design and experimental evaluation of a teleoperated haptic robot–assisted catheter operating system. J Intell Mater Syst Struct. 2016;27(1):3–16.
51. Tavallaei MA, et al. Design, development and evaluation of a compact telerobotic catheter navigation system. Int J Med Robot Comput Assist Surg. 2016;12(3):442–52.
52. Zhang L, Guo S, Yu H, Song Y. Performance evaluation of a strain-gauge force sensor for a haptic robot-assisted catheter operating system. Microsyst Technol. 2017;23:5041–50.
53. Guo S, Wang Y, Xiao N, Li Y, Jiang Y. Study on real-time force feedback for a master–slave interventional surgical robotic system. Biomed Microdevices. 2018;20(2):1–10.
54. Wang Y, Guo S, Xiao N, Li Y, Jiang Y. Online measuring and evaluation of guidewire inserting resistance for robotic interventional surgery systems. Microsyst Technol. 2018;24:3467–77.
55. Su H, et al. Fiber-optic force sensors for MRI-guided interventions and rehabilitation: a review. IEEE Sensors J. 2017;17(7):1952–63.
56. Gannot I. Optical fibers and sensors for medical applications, vol. 4253. Bellingham, WA: SPIE; 2001.
57. Sharma T, Naik S, Langevine J, Gill B, Zhang JX. Aligned PVDF-TrFE nanofibers with high-density PVDF nanofibers and PVDF core–shell structures for endovascular pressure sensing. IEEE Trans Biomed Eng. 2015;62(1):188–95.
58. Guo S, Qin M, Xiao N, Wang Y, Peng W, Bao X. High precise haptic device for the robotic catheter navigation system. In: Proceedings of International Conference on mechatronics, robotics and tics and manufacturing; 2016. p. 2524–9.
59. Hooshiar A, Razban M, Bandari NM, Dargahi J. Sensing principle for real-time characterization of viscoelasticity in the beating myocardial tissue. In: IEEE international conference on computational intelligence and virtual environments for measurement systems and applications; 2017. p. 72–7.
60. Mei T, Budelmann C, Kern TA, Sindlinger S, Minamisava C, Werthschutzky R. Intravascular palpation and haptic feedback during angioplasty. In: Proceedings - 3rd joint eurohaptics conference and symposium on haptic interfaces for virtual environment and teleoperator systems; 2009. p. 380–1.
61. Park JW, et al. Development of a force-reflecting robotic platform for cardiac catheter navigation. Artif Organs. 2010;34(11):1034–9.
62. Ahmadi R, Dargahi J, Packirisamy M, Cecere R. A new hybrid catheter-tip tactile sensor with relative hardness measuring capability for use in catheter-based heart surgery. In: Proceedings of IEEE Sensors; 2010. p. 1592–5.
63. Polygerinos P, Puangmali P, Schaeffter T, Razavi R, Seneviratne LD, Althoefer K. Novel miniature MRI-compatible fiber-optic force sensor for cardiac catheterization procedures. In: Proceedings of International Conference on Robotics and Automation; 2010. p. 2598–603.
64. Polygerinos P, Ataollahi A, Schaeffter T, Razavi R, Seneviratne LD, Althoefer K. MRI-compatible intensity-modulated force sensor for cardiac catheterization procedures. IEEE Trans Biomed Eng. 2011;58(3):721–6.
65. Puangmali P, Liu H, Seneviratne LD, Dasgupta P, Althoefer K. Miniature 3-axis distal force sensor for minimally invasive surgical palpation. IEEE/ASME Trans Mechatronics. 2012;17(4):646–56.

66. Noh Y, et al. A three-axial body force sensor for flexible manipulators. In: Proceedings of International Conference on Robotics and Automation; 2014. p. 6388–93.
67. Thuruthel TG, Shih B, Laschi C, Tolley MT. Soft robot perception using embedded soft sensors and recurrent neural networks. Sci. Robot. 2019;4(26):eaav1488.
68. Issatayeva A, Beisenova A, Molardi C, Tosi D. Distributed strain sensing medical catheter for epidural anesthesia. In: Optics in health care and biomedical optics VIII, vol. 10820; 2018. p. 539–40.
69. Hooshiar A, Najarian S, Dargahi J. Haptic Telerobotic cardiovascular intervention: a review of approaches, methods and future perspectives. IEEE Rev Biomed Eng. 2020;13:32–50.
70. Ganji Y, Janabi-Sharifi F, Cheema AN. Robot-assisted catheter manipulation for intracardiac navigation. Int J Comput Assist Radiol Surg. 2009;4:307–15.
71. Kesner SB, Howe RD. Force control of flexible catheter robots for beating heart surgery. In: 2011 IEEE international conference on robotics and automation; 2011. p. 1589–94.
72. Thakur Y, Bax S, Holdsworth W, Drangova M. Design and performance evaluation of a remote catheter navigation system. IEEE Trans Biomed Eng. 2009;56:1901–8.
73. Tanimoto M, Arai F, Fukuda T, et al. Telesurgery system for intravascular neurosurgery. In: Medical Image and Computing and Computer-Assisted Intervention-MICCAI; 2000. p. 29–39.
74. Guo Y, Guo S, Yang C, editors. Feasibility study on cloud communication operation for an interventional surgery robot. Beijing, China: 2020 IEEE International Conference on Mechatronics and Automation (ICMA); 2O2O.
75. Madder RD, VanOosterhout S, Mulder A, et al. Feasibility of robotic telestenting over long geographic distances: a pre-clinical ex vivo and in vivo study. EuroIntervention. 2019;15:e510–2.
76. Madder RD, VanOosterhout S, Parker J. Robotic telestenting performance in transcontinental and regional pre-clinical models. Catheter Cardiovasc Interv. 2021;97:E327–32.
77. Abi-Jaoudeh N, Glossop N, Dake M, et al. Electromagnetic navigation for thoracic aortic stent-graft deployment: a pilot study in swine. J Vasc Interv Radiol. 2010;21:888–95.
78. De Lambert A, Esneault S, Lucas A. Electromagnetic tracking for registration and navigation in endovascular aneurysm repair: a phantom study. J Endovasc Ther. 2012;43:684–9.
79. Manstad-Hulaas F, Ommendal S, Tangen GA. Side-branched AAA stent graft insertion using navigation technology: a phantom study. Eur Surg Res. 2007;39:364–71.
80. Manstad-Hulaas F, Tangen GA, Gruionu LG, Aadahl P, Hernes TA. Three-dimensional endovascular navigation with electromagnetic tracking: ex vivo and in vivo accuracy. J Endovasc Ther. 2011;18:230–40.
81. Manstad-Hulaas F, Tangen GA, Dahl T, Hernes TA, Aadahl P. Three-dimensional electromagnetic navigation vs. fluoroscopy for endovascular aneurysm repair: a prospective feasibility study in patients. J Endovasc Ther. 2021;19:70–8.
82. Cochennec F, Riga C, Hamady M, Cheshire N, Bicknell C. Improved catheter navigation with 3D electromagnetic guidance. J Endovasc Ther. 2013;20:39–47.
83. Sidhu R, Weir-McCall J, Cochennec F, et al. Evaluation of an electromagnetic 3D navigation system to facilitate endovascular tasks: a feasibility study. J Endovasc Ther. 2012;43:22–9.
84. Mung JC, Huang SG, Moos JM, Yen JT, Weaver fa. Stereotactic endovascular aortic navigation with a novel ultrasonic-based three-dimensional localisation system. J Vasc Surg. 2013;2013(57):1637.
85. Mung J, Han S, Yen JT. Design and in vitro evaluation of a real-time catheter localisation system using time of flight measurements from seven 3.5 MHz single element ultrasound transducers towards abdominal aortic aneurysm procedures. Ultrasonics. 2011;51:758–75.
86. Schueler BA, Kallmes DF, Cloft HJ. 3D cerebral angiography: radiation dose comparison with digital subtraction angiography. Ann J Neuroradiol. 2005;26(8):1898–901.
87. Gunduz S, Albadawi H, Oklu R. Robotic devices for minimally invasive endovascular interventions: a new Dawn FOR interventional radiology. Adv Intell Syst. 2021;3:2000181.
88. Abdelaziz M, et al. Toward a versatile robotic platform for fluoroscopy and MRI guided endovascular interventions: a preclinical study. In: IEEE/RSJ International Conference on Intelligent Robots and Systems (IROS); 2019. p. 5411–8.
89. Kundrat D, et al. An MR-safe endovascular robotic platform: design, control, and ex-vivo evaluation. IEEE Trans Biomed Eng. 2021;68(10):3110–21.
90. Lum GZ, Ye Z. Dong Xiaoguang. Shape-programmable magnetic soft matter. Proc Natl Acad Sci. 2016;113(41):E6007–15.
91. Schmidt CK, Medina-Sanchez M, Edmondson RJ, Schmidt OG. Engineering microrobots for targeted cancer therapies from a medical perspective. Nat Commun. 2020;11:1–18.
92. Hu W, Lum GZ, Mastrangeli M, Sitti M. Small-scale soft-bodied robot with multimodal locomotion. Nature. 2018;554:81–5.
93. Kim Y, Yuk H, Zhao R, Chester SA, Zhao X. Printing ferromagnetic domains for untethered fast-transforming soft materials. Nature. 2018;558(7709):274–9.
94. Kim Y, Parada GA, Liu S, Zhao X. Ferromagnetic soft continuum robots. Sci Robot. 2019;4:1–15.
95. Khalil SM, Mahdy D, Sharkawy AE, et al. Mechanical rubbing of blood clots using helical robots under ultrasound guidance. IEEE Robot Autom Lett. 2018;3(2):1112–9.
96. Field RD, Anandakumaran PN, Sia SK. Soft medical microrobots: design components and system integration. Appl Phys Rev. 2019;6:1–20.
97. Rippel RA, Rolls AR, Riga CV, et al. The use of robotic endovascular catheters in the facilitation of transcatheter aortic valve implantation. Eur J Cardiothorac Surg. 2014;45:836–41.
98. Beyar R, Gruberg L, Deleanu D, et al. Remote-control percutaneous coronary interventions: concept, validation, and first in humans pilot clinical trial. J Am Coll Cardiol. 2006;47(2):296–300.
99. Riga CV, Bicknell CD, Rolls A, Cheshire NJ, Hamady MS. Robotic-assisted fenestrated endovascular aneurysm repair (FEVAR) using the Magellan system. J Vasc Interv Radiol. 2013;24(2):191–6.
100. Riga CV, Bicknell CD, Wallace D, Hamady M, Cheshire N. Robot-assisted antegrade in-situ fenestrated stent grafting. Cardiovasc Intervent Radiol. 2009;32:522–4.
101. El-Sayed T, Patel A, Cho J, et al. Radiation-induced DNA damage in operators performing endovascular aortic repair. Circulation. 2017;136:2406–16.
102. Roguin A, Goldstein J, Bar O, et al. Brain and neck tumors among physicians performing interventional procedures. Am J Cardiol. 2013;111:1368–72.
103. Yoshinaga S, Hauptmann M, Sigurdson A, et al. Nonmelanoma skin cancer in relation to ionizing radiation exposure among U.S. radiologic technologists. Int J Cancer. 2005;115:828–34.
104. Bitarafan Rajabi A, Noohi F, Hashemi H, et al. Ionizing radiation-induced cataract in interventional cardiology staff. Res Cardiovasc Med. 2015;4(1):1–6.
105. Riga C, et al. Advanced catheter technology: is this the answer to overcoming the long learning curve in complex endovascular procedures. Eur J Vasc Surg. 2011;42:531–8.
106. Steinberg PL, Merguerian MD, Bihrle W III, et al. A da Vinci Robot system can make sense for a mature laparoscopic prostatectomy program. JSLS. 2008;12(1):9–12.

Part II

Peripheral Occlusive Disease

14 Embolic Materials and Embolization Techniques

Elena Lucertini and Miltiadis Krokidis Athens

Case Presentation

Case Background

A chest CT scan of a 49-year-old female performed for other reasons revealed as incidental finding the presence of an arteriovenous communication between the right pulmonary artery and the right pulmonary vein. When the patient was asked, she recalled two previous episodes of transient cerebral ischemic events in the last 10 years with negative cerebral CT scans. She also referred to suffering from regular nosebleeds since her adolescence and that her mother and her younger sister also suffer from regular nosebleeds. The diagnosis of hereditary haemorrhagic telangiectasia (otherwise known as Osler-Weber-Rendu disease) was made and a decision to proceed with embolization of the pulmonary arteriovenous malformation (AVM) followed.

Background

Transcatheter arterial embolization is one of the major applications of endovascular embolization and it is defined as the deliberate occlusion of arterial vessels for the treatment of diseased or injured vasculature.

Arterial embolization has been used since the beginning of the twentieth century, initially with the aim of reducing tumour blood flow before surgery in hypervascular tumours and afterwards also to treat arteriovenous malformation (AVM) and arteriovenous fistulas (AVF) [1, 2].

At present, endovascular embolization has been extended to multiple clinical areas in both emergency and non-emergency settings. In an emergency, embolization is performed to stop active bleeding from an arterial source, due to trauma, ruptured aneurysm, or even due to diffuse anti-coagulation [3–7]. In an elective setting, transcatheter embolization is used to seal non-ruptured aneurysms, and to treat AVMs and endoleaks after endovascular repair of abdominal aortic aneurysms [5–11]. Furthermore, embolization has offered minimally invasive oncologic solutions for the treatment of hepatocellular carcinoma (HCC) or liver metastatic disease but also revolutionized the approach to benign pathologies like fibroids and prostate hyperplasia that can now be treated without surgical interventions [12–15].

Indications for Intervention

The main indication for the treatment of embolization is the necessity of selective occlusion of a very specific vascular area or the delivery of drugs via an intra-arterial route in a designated anatomical area or organ. Patients may have a very good general status for an elective procedure or equally may be severely unwell when embolization procedures are performed with a life-saving purpose, e.g., bleeding after pelvic trauma or postpartum. In an elective setting it is preferred to avoid performing embolization in patients with a bleeding risk and to postpone the procedure until their status is optimized. Nevertheless, embolization in emergency may be performed even in patients with deranged clotting with the use of appropriate tools, e.g., closure devices for the arterial closure and embolic materials that do not depend on clotting function, like glue.

Preoperative Preparation

Timing of Intervention

Embolization may be performed as an elective or as an emergency procedure. A 24-h embolization service is required to be available at all tertiary care surgical and trauma centres,

E. Lucertini
Ospedali Riuniti di Livorno, Livorno, Italy

M. K. Athens (✉)
National and Kapodistrian University of Athens, Athens, Greece

G. Geroulakos et al. (eds.), *Mastering Endovascular Techniques*, https://doi.org/10.1007/978-3-031-42735-0_14

and in case bleeding from an arterial source is documented, the procedure is expected to take place within 30–60 min while the patient is resuscitated.

Anticoagulation

The coagulation status of the patient needs to be assessed prior to an embolization procedure to select appropriate materials. This is usually performed by asking about the intake of any anticoagulation drugs (e.g. heparin, clopidogrel) and if or when they have been stopped and by performing the required blood tests (e.g. INR, platelet count) prior to the procedure.

Imaging

The procedure is performed under digital subtraction angiography (DSA). However, adequate imaging prior to the procedure is required for planning. Computed tomography (CT) angiography is usually used for that purpose. Pictures need to be obtained in arterial phase after bolus tracking of intravenous contrast material. Use of Maximum Intensity Projection (MIP) reconstruction may be helpful particularly when embolization of small vessels is required (e.g. gastrointestinal bleeding). In some cases, magnetic resonance (MR) angiography may be used instead of CT in combination with the imaging of the organ that required embolization (e.g. MR imaging of uterine fibroids prior to treatment). Intraoperative imaging with cone bean CT may also be required to confirm embolization target (e.g. prostate embolization).

Materials and Techniques

Endovascular access is obtained, and a standard endovascular kit is used in most cases that consists of the following:

- Standard access wires and sheaths (ranging from 4 to 12Fr).
- Selective 4 or 5Fr catheters.
- 0.035″ hydrophilic guidewire
- 0.027″– 0.021″ microcatheters
- 0.021″– 0.010″ microwires
- Embolization material.

The first embolization materials used historically were blood clots, small muscle fragments and stainless-steel pellets [16]. Recently, with the advancement of medical technology, a wide range of embolic agents have been developed, and adequate knowledge of the different types and features is mandatory for embolization specialists.

Mechanical Embolic Agents

Coils

Coils are one of the first embolic agents used; they aim to mechanically occlude vessels or aneurysms. The first embolic coils were made from small pieces of Teflon guidewires that were cut after removal of the inner core; they were introduced in 1976 by Gianturco and named after him [17]. The first generation of coils were developed as separate devices, with thrombogenic coating, but none of them could be retrieved after deployment and were therefore named as "pushable" coils [18]. Coils that could be retracted after deployment within an electrolytic mechanism – therefore "retrievable" – were developed in 1989 by Guglielmi, allowing repositioning in case of suboptimal positioning or size. Controlled detaching of coils was a real revolution in embolization techniques. After Guglielmi, many other detachable coils were developed, using electrolytic or mechanical detaching systems [5].

Apart from the type of deployment system other features also distinguish coils and in particular size, shape, flexibility and bioactivity. All coils are developed for long-term implantation and are therefore made of bio-inert materials, to avoid local or systemic adverse responses. The most common material for coils is platinum, but nitinol has also been recently introduced. Platinum is a radiopaque, highly thrombogenic, very flexible and compliant material, allowing a good packing density with low trauma risk on vessel or aneurysm wall and permitting an optimal conformation into the shapes of irregular aneurysms [19]. Nitinol on the other hand is more of a shape-memory material, mainly used in stent production; it is very flexible and offers higher radial force compared to platinum [20].

In addition to the traditional bio-inert bare metallic coils, hydro coils, bioactive coils and polymer- or protein-coated coils have been produced. Bioactive and coated coils were developed with the aim of increasing their thrombogenic effects, consequently obtaining faster occlusion of aneurysms or vessels, while hydrocoils are coated by a hydrogel layer in order to improve packing density and obtain larger filling volumes [21].

Plugs

Plugs are mechanical occlusion devices produced with shape-memory polymers specifically designed to occlude vessels quickly with a single re-sheathable device. Plugs determine mechanical obliteration of vessels and consequent deceleration of blood flow, finally causing thrombosis [16, 22].

Stents and Flow Diverters

Stents play a very important role in embolization procedures as they are not only able to reconstitute the wall of a damaged or aneurismatic vessel, but they can also be of help in other widespread embolization techniques such as coiling using a "stent-assisted coiling" technique. Stents find their greatest use in the treatment of bleeding caused by damage to a main arterial or venous vessel, which cannot be completely occluded without causing potential distal organ ischemia, and in the embolization of both visceral and cerebral aneurysms.

In the first case, covered-mesh stents are commonly used to reconstitute the wall of the damaged vessel, stopping the bleeding and ensuring the maintenance of the blood flow through the treated artery or vein. Necessary conditions for this type of procedure are to have adequate distal and proximal stent landing zones, without excessive size discrepancy of the proximal and distal landing areas, not to have significant tortuosity of the parent vessel, and to guarantee that the placement of the covered stent does not lead to occlusion of some essential vascular branch, for example when the damage is located close to an arterial bifurcation [23–25].

Another device similar to stents are the flow-diverter devices (FDDs), which can be placed in the artery covering the aneurysm neck. The wall of FDDs blocks the blood flow where it is turbulence, i.e. into the aneurismatic sac, but maintains flow in the main vessel lumen. FDDs can be used alone or in addition to coiling, depending on the case [26, 27]. They were initially introduced to treat cerebral aneurysms, but their use has been extended to the treatment of visceral aneurysms. However, there is no long-term information about their effectiveness in dealing with such large visceral aneurysms, especially in terms of flow-diverting power [28, 29].

The "stent-assisted coiling" technique is particularly useful for wide-necked aneurysms with potential high risk of coils displacement into the parent vessel. In this case, non-covered stents or FDDs may be used, depending on the case, particularly if side branches that need to be preserved are present. Coils can be placed after stent deployment by passing through the stent mesh with a microcatheter or with a second microcatheter parallel to the one used to deliver the stent, previously "jailed" between stent and artery wall into the aneurysmatic sac, with the so-called jailing technique [23, 25].

Detachable Balloons

Detachable balloons were initially used for endovascular aneurysm treatment, but they soon were phased out. Nevertheless, they are still used for the embolization of traumatic or congenital single-hole fistulas.

Although their use alone is rather limited, balloons still represent a good support for other embolization techniques. In aneurysms embolization, the use of a temporary balloon protection across the aneurysm neck considerably facilitates its filling with spirals or plugs, avoiding at the same time their escape from the sac during the procedure. Another application of balloons in embolization techniques is in neurovascular procedures, where balloons may be used to occlude the flow distally to the aneurysm, to prevent clot migration during the procedure. In some cases, however, even temporary arterial occlusion obtained with a balloon may lead to the exclusion and thrombosis of a pseudoaneurysm sac [16, 30].

Particulate Embolic Agents

Particulate embolic agents are very commonly used in clinical practice. They include irregularly shaped or calibrated particles, which can be natural or synthetic and permanent or biodegradable. Particulates can be spherical such as microspheres or non-spherical such as polyvinyl alcohol (PVA) and they are available in a number of sizes from 40 to 1200 μm [16, 22].

Polyvynil Alcohol (PVA) Particulates

PVA particulates are commonly used for embolization, especially for uterine fibroids, hyperplastic prostatic gland and hypervascular tumours. PVA are prepared by mechanical fragmentation by rasping and blending or punching PVA plugs out of the compressed foam, followed by sieving to separate particles with different size ranges, between 100 μm and 1100 μm. Despite sieving, PVA are considered non-calibrated because of their irregular shape, which can promote aggregation, and because of the possible passage of elongated particles through the sieves along their long axis during the sieving process, resulting in some variability in particles size. At the end of the process, PVA particulates are a dry powder which needs to be hydrated with saline solution and contrast medium before use.

After being delivered through a catheter, PVA particulates determine immediate mechanical occlusion of blood vessels, initially by attaching to the vessel wall and then, within the first few hours or days, by thrombus formation and finally by inflammation-induced vessel remodelling, wall necrosis and fibrosis, determining a permanent vessel occlusion. However, in some cases recanalization may occur. In such cases, repeated embolization with PVA of the same vascular territory is still possible.

The biggest disadvantage of PVA is their high aggregation, due to the wide range of sizes within the particle group and due to their irregular and hydrophobic surface, which can cause blockage within the catheter or in the more proximal vessels (it is important to consider that the dimension of particles does not necessarily correlate with the diameter of occluded arteries); hence sometimes it is difficult to predict the embolization behaviour [14, 31].

Calibrated Microspheres

Microspheres, unlike PVA particles, are symmetrical and very precisely sized (ranging from 40 to 1200 μm) with a smooth surface that reduces the risk of particle aggregation and allows a more precise, size-dependent and predictable embolization. Nevertheless, special attention needs to be paid to potential particle reflux and risk of non-target embolization during injection. The mechanism of embolization with microspheres is similar to PVA particles, blocking the flow into the vessel and inducing inflammation and fibrosis, though the induced inflammatory reaction is lower with microspheres compared to PVA. Embolization performed with microspheres is considered permanent; however, re-embolization may be possible if required.

In addition to size, further features to consider are rigidity and elastic recovery, which respectively indicate the resistance to compression and the time required for the microsphere to regain its original shape.

Microspheres are available in prefilled syringes, which need to be diluted with fluid (usually saline solution) and contrast medium before injection. Depending on their use, sizing of microspheres varies; for uterine fibroids embolization with larger microspheres is often performed (usually up to 500–700 μm), while in tumour embolization, smaller (100–150 μm) microparticles are usually preferred to obtain a deeper and more distal microvascular embolization. Moreover, for oncologic treatment, various chemotherapeutic drugs may be encapsulated into designated microspheres to enhance their therapeutic effect [13].

There are several types of microparticles available produced by different medical device companies. The first microspheres were introduced by Biosphere Medical and they were the trisacryl gelatin (TAGM) microparticles named Embosphere, available in a range of sizes from 40 to 1200 μm. After them, many other microspheres were introduced by Biosphere Medical itself (EmboGold, Quadrasphere, Hepasphere), Boston Scientific (Contour SE, Bead Block, DC Bead) and CeloNova BioSciences (Embozene). The different types of Embosphere differ in rigidity, viscoelasticity, elastic recovery, level of occlusion and generated inflammation. Contour SE are the most occlusive particles, with lower rigidity, viscoelasticity and elastic recovery, while Embosphere have the highest rigidity and elastic recovery and the lowest level of occlusion. No significant differences are reported in inflammation level for particles with size >500 μm, which is lower compared to smaller spheres. In microspheres <500 μm, however, Embosphere are those with the higher inflammatory power [14, 31, 32].

Autologous Clots and Tissues

Blood clots are the most commonly used temporary autologous embolic agents, including natural clots, which are temporary embolic agents, or modified clots. Clots can be heated to retard the lysis or acidified in order to resist fibrinolysis. Oxidized cellulose-modified blood or thrombin-modified blood can also be used, as well as aged blood, possibly with tantalum powder to make it radiopaque. Autologous tissues include muscle and subcutaneous tissues, directly taken from patients before the procedure and then cut into small pieces and suspended into saline solution. Unlike blood clots, autologous tissues are considered permanent embolic agents. Autologous embolic agents, especially blood clots, are cheap and highly biocompatible; however, they are rarely used in clinical practice, reserved for those rare cases in which a rapid lysis of the clot to restore circulation within hours is desired [16, 22].

Gelatin Foam

Gelatin foams are biodegradable embolic agents, very useful for temporary embolization. They may be used as haemostatic embolic agents, for example for pre-operative embolization or for younger patients where permanent embolics may not be the most preferable option. They mechanically produce vessel obstruction and thrombus formation with a similar mechanism to that of fibrin, causing haemostasis. Gelatin embolics are available in different configurations such as sponge or sheets, which can be manually cut into smaller pieces up to 1–2 mm and then diluted with saline solution and iodinated contrast to make them radiopaque. They can be used in both endovascular and extravascular procedures. When placed in soft tissues, gelatin foam usually biodegrades completely within 4–6 weeks [16, 22].

Liquid Embolic Agents

Liquid embolic materials allow homogeneous and complete filling of the target vessels, reducing the risk of recanalization of the embolized area compared to mechanical and particulate embolics. On the other hand, liquid embolic agents are considered slightly more challenging to handle and the learning curve is somewhat steeper. They are preferably used for the treatment of AVMs, AVFs and type II endoleaks and they are increasingly used for bleeding control. Potential risks like retrograde embolization, entrapment of the microcatheter tip and venous embolization need to be considered [9, 30, 33].

Cyanoacrylates

After the replacement of isobutyl cyanoacrylate because of suspected carcinogenesis, n-butyl cyanoacrylate (NBCA), commonly known as "Glue", became one of the most used embolic agents in neurovascular interventions [34]. NBCA is in fluid state at the temperature of the room, and it polymerizes almost immediately upon contact with ionic fluids such as blood or saline solution, but also heparin. This feature makes

NBCA an excellent embolic agent; however, attention should be paid to avoid early polymerization. Recommended injection rate is slow but continuous, to permit the glue to reach with the flow the most distal part of the vessel or the nidus of an AVM before it polymerizes; if NBCA is deposited too proximally to the arterial feeder, the risk of recanalization is high.

NBCA can also be mixed with other agents such as Lipiodol in various concentrations but most commonly 1:3 or 1:4. This mixture modifies the polymerization time. Mixture of glue with metals like tantalum or tungsten powder offers higher radiopacity [9, 30].

The most commonly used cyanoacrylates are Histoacryl (widely used for years for the embolization of cerebral AVMs and dural AVFs with good results, although it has not yet been approved for intravascular use), Glubran 2 (GEM SRL, Viareggio, Italy), composed by a combination of NBCA and metacryloxysulpholane to increase the polymerization time and at the same time reduce inflammatory and cytotoxic effects, and Purefill (Balt, Montmorency, France), composed by n-hexyl cyanoacrylate with a very low adhesive power, allowing longer injection time [35, 36].

Copolymers

Copolymers are non-adhesive liquid embolic agents widely used in neuro-interventional procedures such as AVMs or dural fistulas embolization, but recently increasingly also in "peripheral" embolization. Copolymers are always combined with dimethyl sulfoxide (DMSO), which prevents their premature precipitation. Most commonly used copolymers are ethylene vinyl alcohol (EVOH) based, and these produce occlusion of vessels by precipitation, a mechanism described as similar to hardening of lava flow, and then solidify though the dissipation of DMSO [9].

Copolymers should be slowly injected over several minutes to create a proximal plug to facilitate anterograde embolization and minimize backflow. Sometimes, considerably long waiting periods may elapse between injections, disrupting the progression of the embolic agent and forcing the operator to prematurely abort the procedure. The innovative "pressure cooker technique" was introduced by Chapot et al. to enable a comprehensive, forceful and continuous injection of Onyx and avoid reflux, by creating a plug of coils and glue between the tip and the detachment zone of a previously placed microcatheter. The application of this technique was then extended to all the other copolymers [9, 33, 37].

Compared to adhesive embolics like cyanoacylates, non-adhesive embolic agents have a significantly lower risk of trapping the microcatheter's tip in the glue with consequent inability to retrieve it. This phenomenon may still occur with copolymers, but only when the reflux of embolic material has exceeded a certain distance from the tip of the microcatheter. To overcome this problem, some microcatheters with detachable tips have been produced. Moreover, copolymers precipitate within a longer period compared to the polymerizing time of cyanoacrylates, allowing for longer injection times, more controlled embolization, and subsequent better filling of the target vessel.

Before the injection of copolymers, the dead space of the microcatheter must always be filled with DMSO, in order to prevent precipitation of the agent within the microcatheter lumen. As DMSO can damage the material of the catheter, DMSO-resistant catheters and syringes have been developed and are recommended as the first choice in such cases.

There are four currently available copolymers: Onyx (Medtronic, Irvine, CA, USA), Menox (Meril Life Sciences, India Pvt. Ltd), Squid (Balt, Montmorency, France); and PHIL (MicroVention, Aliso Viejo, CA, USA. Onyx, Menox and Squid are all EVOH copolymers made radiopaque with tantalum suspension, while PHIL is a different type of copolymer. The embolic material choice is based on the operator's preference, availability, and level of expertise with each agent, as well as the characteristics of the target lesion [9, 38].

Anticipated Complications

Post-embolization syndrome is the most common complication after particle embolization of a large area of the body (e.g. fibroids, HCC). This consists of a flu-like syndrome with fatigue and low temperature fever that is usually self-limiting and lasts approximately a week. In case fever persists for longer, abscess formation of the organ of the embolized area should be considered. In such cases, cross-sectional imaging is required to confirm abscess formation and treatment with intravenous antibiotics or percutaneous abscess drainage may even be required. Another possible complication is non-target embolization. This might lead to a series of undesirable outcomes (e.g. bowel ischemia, skin ulceration) and should be avoided by all means. With the use of liquid embolics there is always the risk of catheter entrapment in the embolized area. In such cases, the catheter needs to be left in situ in some vascular territories, whereas in some cases a surgical approach may be required.

Perioperative Care and Surveillance

After embolization the patient needs to follow the rules of the vascular access that was used. The sheath diameter and the use or not of closure devices will determine the hours that the patient would stay in bed. However, if radial or brachial access is used the patient may be immediately mobilized. The patient may also be mobilized after 2 h in case of procedures with venous access. In some cases, a very strict antalgic regimen may be required post procedure (e.g. in the case of fibroid embolization a patient-controlled anaesthesia pump is required). The patient is usually discharged the next day for most elective embolization procedures; however, this is subject to the severity of his condition. Follow-up visits are usually arranged differently for each specific case.

Case Presentation Continuation

Procedure

The procedure was performed under general anaesthesia. Access from the right common femoral vein was obtained. A 5 Fr sheath was inserted, and selective catheterization of the right pulmonary artery was performed, after navigation of the catheter via the cardiac chambers. Angiographic control confirmed the presence of a pulmonary arteriovenous communication in the right lower pulmonary lobe (Fig. 14.1a). A 65 cm sheath was inserted in the right lower lobe pulmonary artery. Selective catheterization of the fistula followed with a 2.7 Fr microcatheter system and embolization with detachable microcoils followed (Fig. 14.1b). The initial coils were used as scaffold to prevent migration in the arterial circulation via the pulmonary veins (Fig. 14.1c). Complete exclusion of the malformation was achieved with several microcoils that were densely packed within the lesion (Fig. 14.2). The patient was very well after the procedure and was discharged the next morning.

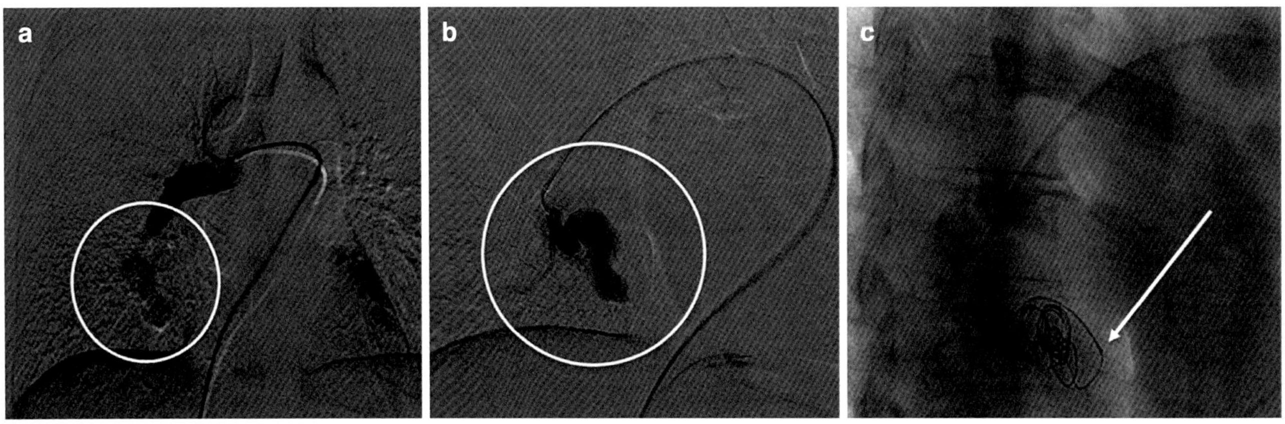

Fig. 14.1 (**a**) Digital subtraction angiography after catheterization of the right pulmonary artery that confirms the presence of the malformation at the lower right lobe (white circle). (**b**) Selective catheterization of the malformation with a microcatheter. (**c**) Deployment of the first detachable coil that would act as scaffold for the rest. Appropriate sizing is required to avoid coil migration at that stage given the very high flow of the malformation

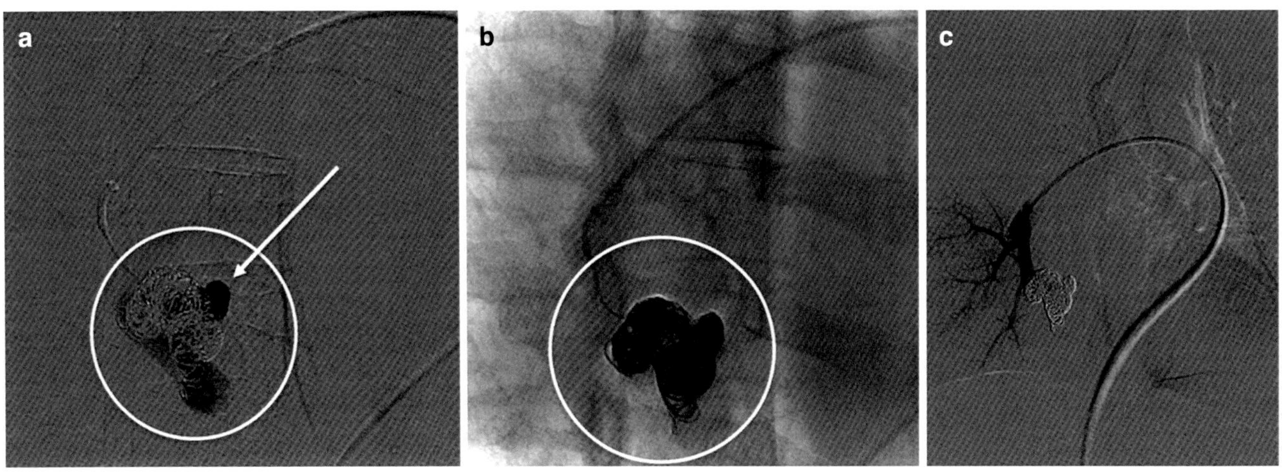

Fig. 14.2 (**a**) Angiographic picture after deployment of several coils reveals further component of the lesion that needs to be embolized (arrow). (**b**) Further embolization of the lesion (white circle) and coil packing followed. (**c**) Angiographic picture confirms the satisfactory outcome with no flow in the malformation and preservation of the flow in the rest of the parenchyma

Toolkit

Suggested Toolkit for Embolization with Micro Coils

Vascular access kit: Micropuncture, then 4–6 Fr sheath
Access catheter: 4 or 5Fr angled catheter (vertebral, cobra 1 or 2, headhunter)

Access wire: 0.035″ inch hydrophilic wire, 180 or 260 cm

Microcatheter: 2.2-27Fr with or without 0.016–0.021″ inch microwire

Microwire: Hydrophilic 0.016–0.021″ angled (20-45°)

Microcolis: 0.016–0.018″ pushable or detachable

References

1. Dawbarn RHM. The starvation operation for malignancy in the external carotid area. J Am Med Assoc. 1904;13(792):5.
2. Brooks B. The treatment of traumatic arteriovenous fistula. South Med J. 1930;23(1000):6.
3. Touma L, Cohen S, Cassinotto C, Reinhold C, Barkun A, Tran VT, et al. Transcatheter arterial embolization of spontaneous soft tissue hematomas: a systematic review. Cardiovasc Intervent Radiol. 2019;42(3):335–43.
4. Kolber MK, Shukla PA, Kumar A, Silberzweig JE. Ethylene vinyl alcohol copolymer (onyx) embolization for acute hemorrhage: a systematic review of peripheral applications. J Vasc Interv Radiol. 2015;26(6):809–15.
5. Guglielmi G. History of endovascular endosaccular occlusion of brain aneurysms: 1965-1990. Interv Neuroradiol. 2007;13(3):217–24.
6. Barrionuevo P, Malas MB, Nejim B, Haddad A, Morrow A, Ponce O, et al. A systematic review and meta-analysis of the management of visceral artery aneurysms. J Vasc Surg. 2019;70(5):1694–9.
7. Chaer RA, Abularrage CJ, Coleman DM, Eslami MH, Kashyap VS, Rockman C, et al. The Society for Vascular Surgery clinical practice guidelines on the management of visceral aneurysms. J Vasc Surg. 2020;72(1):3S–39S.
8. Park HS, Do YS, Park KB, Kim DI, Kim YW, Kim MJ, et al. Ethanol embolotherapy of hand arteriovenous malformations. J Vasc Surg. 2011;53(3):725–31.
9. Vollherbst DF, Chapot R, Bendszus M, Möhlenbruch MA. Glue, onyx, squid or PHIL? Liquid embolic agents for the embolization of cerebral arteriovenous malformations and Dural arteriovenous fistulas. Clin Neuroradiol. 2022;32(1):25–38.
10. Scallan O, Kribs S, Power AH, DeRose G, Duncan A, Dubois L. Onyx versus coil embolization for the treatment of type II endoleaks. J Vasc Surg. 2021;73(6):1966–72.
11. Ameli-Renani S, Pavlidis V, Morgan RA. Secondary Endoleak management following TEVAR and EVAR. Cardiovasc Intervent Radiol. 2020;43(12):1839–54. https://doi.org/10.1007/s00270-020-02572-9.
12. Kulik L, El-Serag HB. Epidemiology and Management of Hepatocellular Carcinoma. Gastroenterology. 2019;156(2):477–91.
13. Raoul JL, Forner A, Bolondi L, Cheung TT, Kloeckner R, de Baere T. Updated use of TACE for hepatocellular carcinoma treatment: how and when to use it based on clinical evidence. Cancer Treat Rev. 2019;72:28–36.
14. Siskin GP, Beck A, Schuster M, Mandato K, Englander M, Herr A. Leiomyoma infarction after uterine artery embolization: a prospective randomized study comparing Tris-acryl gelatin microspheres versus polyvinyl alcohol microspheres. J Vasc Interv Radiol. 2008;19(1):58–65.
15. Naidu SG, Narayanan H, Saini G, Segaran N, Alzubaidi SJ, Patel IJ, et al. Prostate artery embolization—review of indications, patient selection, techniques and results. J Clin Med. 2021;10(21):5139.
16. Hu J, Albadawi H, Chong BW, Deipolyi AR, Sheth RA, Khademhosseini A, et al. Advances in biomaterials and Technologies for Vascular Embolization. Adv Mater. 2019;31:e1901071.
17. Wallace S, Gianturco C, Anderson JH, GoldsteiN HM, Davis LJ, Bree RL. Therapeutic vascular occlusion utilizing steel coil technique: clinical applications. Am J Roentgenol. 1976;127(3):381–7.
18. Hilal SK, Khandji A, Solomon RW. Obliteration of intracranial aneurysms with pre-shaped highly thrombogenic coils. Radiology. 1989;173:250–7.
19. Guglielmi G, Vinuela F, Sepetka I, Macellari V. Electrothrombosis of saccular aneurysms via endovascular approach. Part 1: electrochemical basis, technique, and experimental results. J Neurosurg. 1991;75(1):1–7.
20. Kónya A, Maxin M, Wright KC. New embolization coil containing a nitinol wire core: preliminary in vitro and in vivo experiences. J Vasc Interv Radiol. 2001;12(7):869–77.
21. Reinges MHT, Krings T, Drexler AY, Ludolph A, Sellhaus B, Bovi M, et al. Bare, bio-active and hydrogel-coated coils for endovascular treatment of experimentally induced aneurysms: long-term histological and scanning electron microscopy results. Interv Neuroradiol. 2010;16(2):139–50.
22. Leyon JJ, Littlehales T, Rangarajan B, Hoey ET, Ganeshan A. Endovascular embolization: review of currently available embolization agents. Curr Probl Diagn Radiol. 2014;43(1):35–53.
23. Bechan RS, Sprengers ME, Majoie CB, Peluso JP, Sluzewski M, Van Rooij WJ. Stent-assisted coil embolization of intracranial aneurysms : complications in acutely ruptured versus unruptured aneurysms. AJNR. 2016;37:502.
24. Ching KC, Santos E, Mccluskey KM, Orons PD, Bandi R, Friend CJ, et al. Covered stents and coil embolization for treatment of Postpancreatectomy arterial hemorrhage. J Vasc Interv Radiol. 2016;27(1):73–9. https://doi.org/10.1016/j.jvir.2015.09.024.
25. Irie K, Negoro M, Hayakawa M, Eayashi J, Kanno T. Stent assisted coil embolization: the treatment of wide-necked, dissecting, and fusiform aneurysms. Interv Neuroradiol. 2003;9:255–61.
26. Dandapat S, Mendez-Ruiz A, Martínez-Galdámez M, Macho J, Derakhshani S, Foa Torres G, et al. Review of current intracranial aneurysm flow diversion technology and clinical use. J Neuro Interv Surg. 2021;13(1):54–62.
27. Briganti F, Leone G, Marseglia M, Mariniello G, Caranci F, Brunetti A, et al. Endovascular treatment of cerebral aneurysms using flow-diverter devices: a systematic review, vol. 28; 2015. p. 365.
28. Rabuffi P, Bruni A, Antonuccio EGM, Ambrogi C, Vagnarelli S. Treatment of visceral artery aneurysms and pseudoaneurysms with the use of cerebral flow diverting stents: initial experience. CVIR Endovasc. 2020;3(1):48.
29. Tipaldi MA, Krokidis M, Orgera G, Pignatelli M, Ronconi E, Laurino F, et al. Endovascular management of giant visceral artery aneurysms. Sci Rep. 2021;11:1–6.
30. Brassel F, Meila D. Evolution of embolic agents in interventional neuroradiology. Clin Neuroradiol. 2015;25:333–9.
31. Laurent A. Microspheres and nonspherical particles for embolization. Tech Vasc Interv Radiol. 2007;10(4):248–56.
32. Stampfl S, Bellemann N, Stampfl U, Sommer CM, Thierjung H, Lopez-Benitez R, et al. Arterial distribution characteristics of Embozene particles and comparison with other spherical embolic agents in the porcine acute embolization model. J Vasc Interv Radiol. 2009;20(12):1597–607.

33. Lanzino G, Burrows AM, Cloft HJ. Onyx embolization. J Neurosurg. 2014;120:375–6.
34. Brothers MF, Kaufmann JCE, Fox AJ, Deveikis JP. N-butyl 2-cyanoacrylate - substitute for IBCA in interventional neuroradiology: histopathologic and polymerization time studies. Am J Neuroradiol. 1989;10(4):777–86.
35. Leonardi M, Barbara C, Simonetti L, Giardino R, Nicoli Aldini N, Fini M, et al. Glubran 2: a new acrylic glue for neuroradiological endovascular use: experimental study on animals. Interv Neuroradiol. 2002;8(3):245–50.
36. Izaaryene J, Saeed Kilani M, Rolland PH, Gaubert JY, Jacquier A, Bartoli JM, et al. Preclinical study on an animal model of a new non-adhesive cyanoacrylate (Purefill®) for arterial embolization. Diagn Interv Imaging. 2016;97(11):1109–16.
37. Chapot R, Stracke P, Velasco A, Nordmeyer H, Heddier M, Stauder M, et al. The pressure cooker technique for the treatment of brain AVMs. J Neuroradiol. 2014;41(1):87–91.
38. Sirakov S, Sirakov A, Minkin K, Penkov M, Ninov K, Hristov H, et al. Initial experience with the new ethylene vinyl alcohol copolymer based liquid embolic agent Menox in the endovascular treatment of cerebral arteriovenous malformations. J Neuro Interv Surg. 2019;11(10):1040–4.

15 Interventions for Aorto-Iliac Disease

Athanasios Saratzis

Case Presentation A 72-year-old lady presented with a left-sided lower-limb arterial ulceration (three toes were affected – dry gangrene) and bilateral lower-limb rest pain (chronic limb-threatening ischaemia). She had a history of type 2 diabetes mellitus, obesity, ex-smoking habit, precious bilateral superficial femoral artery subintimal angioplasties (patent – performed 18 months ago), and had previously undergone an anterior resection for an early-stage colonic (sigmoid) malignancy with a curative intent. She was already on Clopidogrel 75 mg once daily and high-dose statin therapy (known to have intermittent claudication). Her baseline computed tomographic (CT) angiography disclosed a long calcified occlusion of the left common and external iliac artery, an occlusion of the proximal right common iliac artery, and an occluded distal aorta below the renal arteries; the inferior mesenteric artery was also occluded (note previous colectomy).

Background and Introduction

Lower-limb peripheral arterial disease (PAD) affects 20% of people over 55 years of age [1] and is the main cause of lower-limb amputations in developed countries [2]. Often, patients with PAD develop severe leg pain at rest or necrosis (e.g. ulceration or gangrene), referred to as chronic limb-threatening ischaemia (CLTI) [3]; CLTI is limb and life threatening and requires urgent revascularisation or patients will either lose their leg and/or die [3]. The aorta and iliac arteries are two of the most common sites affected in those with CLTI [4]. Patients with CLTI and aorto-iliac disease typically have severe occlusions of their aorta or iliac axis over long arterial segments, which makes their revascularisation technically complex. Some 46,891 patients underwent revascularisation in the NHS for CLTI in 2019 – 20,828 of them for aorto-iliac disease [5], based on the National Vascular Registry and Hospital Episode Statistics (HES). Patients with CLTI and aorto-iliac occlusive disease can have revascularisation either with open surgery or more recently developed minimally invasive techniques in the form of endovascular treatment. Endovascular treatment in this instance usually involves a combination of angioplasty, deployment of stents, or other techniques [3].

An international multidisciplinary expert group in PAD and CLTI, the TransAtlantic Inter-Society Consensus (TASC) force, classified aorto-iliac PAD based on severity of arterial stenoses or occlusions from grade A (less severe) to grade D (most severe) [6]. For TASC A/B disease [2], endovascular treatment is the universally established and accepted strategy both in the NHS and globally [2]. For TASC C/D disease, there is uncertainty regarding the effectiveness of open surgery vs. endovascular treatment, reflected in all guidelines [2]. The fact that endovascular treatments (stents, angioplasties, other endovascular procedures) have seen great advances in the last 20 years has added even more uncertainty regarding the management of patients with TASC C/D aorto-iliac PAD (Fig. 15.1).

Both open surgery and endovascular treatment in these patients with occlusive aorto-iliac disease can be associated with significant complications. This is due to the complex technical nature of any revascularisation procedure when someone has severe occlusive arterial disease and the patients' co-morbidities, who are often frail when presenting for revascularisation. Open surgery is associated with more blood loss, longer duration of in-hospital stay after the intervention, a 3% early mortality rate, and

A. Saratzis (✉)
Department of Cardiovascular Sciences, University of Leicester, Leicester, UK
e-mail: as875@le.ac.uk

G. Geroulakos et al. (eds.), *Mastering Endovascular Techniques*, https://doi.org/10.1007/978-3-031-42735-0_15

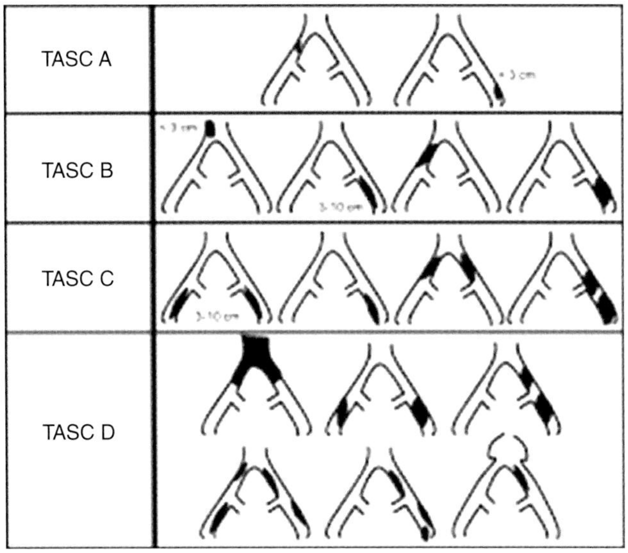

Fig. 15.1 TASC classification for patients with aorto-iliac arterial disease. TASC C and D patients have severe occlusive disease and are more likely to present with CLTI, i.e., severe symptoms; these patients will be eligible to take part in EVOCC as there is treatment uncertainty in this group regarding superiority of surgery over newer endovascular (minimally invasive) treatment

many peri-operative complications, mostly in the form of major cardiovascular events [7]. Endovascular treatment is less invasive, with a 1% early mortality rate and typically no need for post-operative admission to intensive care [4, 7]. Endovascular treatment of severe AIOD has evolved considerably with the evolution of iliac stenting techniques [8]. Kissing iliac stents are associated with acceptable patency rates, comparable in some cases with open surgery; however, no high-quality randomised controlled trials comparing open vs. endovascular AIOD reconstructions are available [9, 10]. Furthermore, a variety of novel endovascular therapeutic approaches have been reported for AIOD, involving the use of both covered and bare metal stents in the aorta or iliacs. The Covered versus Balloon Expandable Stent Trial (COBEST) is the only high quality randomised trial in this clinical context; COBEST showed a benefit when a covered stent was used occlusive iliac lesions compared with bare metal stents, in terms of stent patency [11, 12]. However, the study only evaluated the performance of iliac scaffolds and did not assess clinical or cost-effectiveness.

Various parameters may influence patency when treating severe AIOD with endovascular means, including stent-positioning, stent-design and the discrepancy between the stented lumen and aortic lumen or "radial mismatch" [11, 13, 14]. The latter might cause flow perturbations and thrombus formation. Choosing a configuration with the lowest radial mismatch is the hypothesis that led to the advent of the Covered Endovascular Repair of the Aortic Bifurcation (CERAB) technique, which aims to achieve a more "anatomical" distal aortic reconstruction with three covered stents [14–16], compared to kissing iliac stents or reconstruction without a covered aortic stent. To date, results from CERAB series up to 3 years have demonstrated acceptable results [14–16]. In an analysis comparing reconstruction using a covered aortic stent alongside covered iliac stents for advanced aorto-iliac disease, we showed that the CERAB configuration (covered stents) is associated with less re-interventions over a medium term follow-up [17]. The currently available CERAB studies and those relating to other similar techniques published to date do have a number of limitations, however. They are mostly single centre cohort studies from high-volume experienced centres with relatively limited follow-up. Results have not been replicated in non-tertiary centres, and longer-term outcomes are lacking. Additionally, the currently available endovascular literature in AIOD has not quantified the degree of calcium in atherosclerotic plaques, and anatomical data are not reported uniformly. A direct comparison between bare metal and covered aortic stents in complex AIOD has also not been attempted.

A recent multicentre registry led by our group recently provided real-world practice from 12 European centres regarding the use of iliac and aortic stents in patients with severe occlusive aorto-iliac disease. Overall, more than 200 patients underwent endovascular revascularisation for AIOD, of which 32% had aortic chronic total occlusion (CTO), 46% had iliac CTO, and 48% presented with CLTI. In addition, most patients had severe aortic and/or iliac calcification. Analysis of mid-term outcomes (17 months of follow-up), revascularisation, or major amputation between patients with bare metal vs. covered aortic stents. The only difference was that covered aortic stents were associated with improved freedom from TLR if combined with covered iliac stents in a multivariable conditional regression. This is in line with the COBEST trial, which did show better patency when using covered stents for occlusive iliac disease. At the moment, a large scale pragmatic randomised trial assessing clinical and cost-effectiveness of endovascular treatments for severe occlusive AIOD is urgently needed. Until then, patients having endovascular procedures for TASC C and D lesions should be followed-up carefully and receive best medical therapy as per current guidance. Further, we have recently published a meta-analysis of 66 observational studies (9319 patients), where we found that primary and secondary patency of CERAB and kissing stenting at 1 and 3 years for TASC C and D aorto-iliac disease is similar to that of open surgery (aorto-bi-femoral bypass) [7] – Table 15.1.

Table 15.1 Results from a meta-analysis of 66 observational studies detailing outcomes for TASC C and TASC D aorto-iliac occlusive disease following endovascular treatment or open (bypass) surgery - please note this is non-randomised data. The table summarises weighted proportion for each treatment strategy. Standard endovascular treatment refers to kissing iliac stenting [7]

Patency (95% confidence interval)	Standard endovascular treatment (*n* = 3204)	CERAB (*n* = 240)	Open bypass surgery (*n* = 5875)
Patency			
1 year primary	90% (88–93%)	88% (83–92%)	96% (95–97%)
1 year secondary	96% (94–98%)	97% (94–99%)	97% (95–97%)
3 years primary	78% (73–84%)	82% (75–88%)	93% (91–94%)
3 years secondary	93% (89–96%)	97% (92–99%)	97% (95–98%)
5 years primary	71% (60–82%)	N/A	88% (86–90%)
5 years secondary	89% (80–95%)	N/A	95% (93–97%)

Pre-Operative Preparation

All patients should undergo a structured anaesthetic assessment, given that endovascular aorto-iliac procedures for TASC D and D lesions can be challenging technically and lengthy. Further, angioplasty and post-dilatation of the iliac lesions can be painful. Depending on local expertise and the patient's co-morbidities, the procedures can be performed under general or loco-regional anaesthetic. There is no high-quality evidence focusing on outcomes following these endovascular procedures based on type of anaesthetic. In our recent cohort studies in the UK and Europe, the type of anaesthetic did not impact on peri-operative outcomes or major adverse cardiac events for patients undergoing CERAB or kissing stenting [4, 7]. In our meta-analysis of 9319 patients with TASC C and D aorto-iliac disease undergoing open or endovascular reconstruction, type of anaesthetic also did not influence peri-operative outcomes [7]. These findings might, however, reflect the fact that none of the available evidence in this context is of randomised nature and selection or reporting bias is highly prevalent in the available literature [7]. Before proceeding with reconstruction of their aorto-iliac axis, patients should be commenced on Clopidogrel, high-dose statin therapy, and blood pressure medication (where necessary), based on local guidance. Best medical therapy is a paramount constituent of treatment as antithrombotics, statin therapy, optimum glycemic and blood pressure control have been shown not only to improve long-term outcomes but also reduce the chance of peri-operative events, such as acute kidney injury [18]. Cardiopulmonary exercise testing has been shown to have an important role in planning aortic procedures; however its role in PAD remains unclear [19]. Further, we strongly advocate the initiation of supervised exercise therapy alongside best medical therapy for those who present with claudication rather than CLTI, before proceeding with any type of intervention, given that level 1A evidence strongly suggests that exercise with best medical therapy are the most important constituents of treatment in this instance.

Anticoagulation and Antithrombotic Medication

There is high quality randomised evidence to support the prescription of Clopidogrel 75 mg daily in patients with symptomatic peripheral arterial disease, who require intervention for aorto-iliac revascularisation. More recent randomised evidence has shown benefit of anticoagulation in the form of novel oral anticoagulants in patients undergoing intervention for PAD; however, this was not compared to Clopidogrel. Dual antiplatelet therapy is not widely advocated. Overall, prescription of antithrombotic therapy should be individualised based on each patient's risk of bleeding and benefits of each treatment strategy, e.g., anticoagulation vs. monotherapy with Clopidogrel (antiplatelet). Antiplatelet therapy should be started upon patient presentation and continued post-operatively.

Imaging

We strongly advocate the use of computed tomographic angiography with three-dimensional reconstruction in order to plan any endovascular aorto-iliac procedure, especially for patient with occlusive PAD. Duplex ultrasonography should be used when necessary for a dynamic assess of the infra-inguinal arteries, especially the below-knee run-off. Duplex ultrasonography during the procedure should be used for femoral or upper-limb access and to measure the diameter of the femoral arteries in case an endovascular femoral intervention is planned. Imaging of high quality is paramount in terms of safely and accurately planning any endovascular procedure in this context.

Endovascular Strategy and Technique

Basic Endovascular Toolkit for TASC C and D Lesion AIOD Endovascular Revascularisation

- Vascular ultrasound for access in the common femoral arteries and also brachial or axillary arteries when necessary (e.g. severe occlusive disease in the aorta and common iliac arteries, not allowing adequate angiography from the common femoral access).

- Standard-access wires and sheaths, including a short 5 or 6 Fr sheath.
- Longer-access sheaths for patients requiring upper-limb access, at least 90 cm in length – longer 110 cm sheaths might be necessary (depending on length of iliac occlusions.
- Micropuncture catheter kits for upper-limb percutaneous access if deemed necessary (e.g. aortic occlusion).
- Selective catheters (e.g. Glidecath).
- Floppy and stiff or semi-stiff hydrophilic glidewire.035′‘.
- Variety of.018″ hydrophilic wires (to remain intraluminal if necessary), including chronic total occlusion purpose-built wires where necessary.
- Pigtail angiographic catheter for initial diagnostic angiogram, either from femoral access (iliac arteries) and/or from proximal upper-limb access when necessary.
- Support catheters for recanalisation of occlusions or severe stenosis (guiding catheters) – ideally in a variety of different configurations; Bernstein and C2 preferred options.
- Use of 4 Fr or small support catheters is highly advocated in case of chronic total occlusions, together with guiding catheters of larger diameter for "telescoping" in order to achieve better support (e.g. use of a CXI catheter alongside a GlideCath).
- Angled catheters in case of subintimal recanalisation strategy (e.g. Bernstein or other similar catheter).
- Variety of angioplasty balloons for pre- and post- dilatation (following stenting).
- 5–16 mm covered and uncovered balloon mounted stents +/− self expanding stents for the external iliac arteries (various lengths, depending on pre-operative cross-sectional imaging, preferably in the form of computed tomographic angiography).

A list of basic suggested material is summarised in Table 15.2.

For patients with severely calcified disease of the iliac arteries, common femoral arteries, and/or distal aorta, intravascular lithotripsy or atherectomy (femoral) might be considered, based on local expertise and technology availability.

Decision Making on Technique Selection

Detailed pre-operative imaging is crucial in terms of deciding the recanalisation and treatment technique. Covered stents in the common iliac arteries are supported by the COBEST trial, especially in those with occluded iliacs. The choice of a CERAB approach, i.e., use of a covered aortic stent with covered iliac kissing stents deployed inside the

Table 15.2 Suggested endovascular toolkit for the endovascular treatment of steno-occlusive aorto-iliac disease

	Manufacturer[a]	Size/length
Wires		
Any standard-access wire	Any	0.035"
Glidewire floppy	Terumo Radifocus	0.035″
Glidewire stiff	Terumo Radifocus	0.035"
STARTER Rosen	Boston scientific corporation	0.035"
V-18 control guidewire	Boston scientific corporation	0.018"
Chronic total occlusion (CTO) wires	Various manufacturers Choice depends on local availability	0.018–0.035"
Sheaths		
Any standard-access sheath (femoral)	Any	5–10 Fr/11 cm
Access sheaths from upper limbs or axillary artery	Any	5–10 Fr/55–110 cm (depending on anatomy and site of occlusion in iliac arteries)
Catheters		
Angled	Terumo Radifocus	0.035″/5 Fr/65–90 cm
Quick cross support (low-profile)	Spectranetics/ Phillips	0.018–0.035″/4 Fr/65–90 cm
CXI support	Cook medical	0.018–0.035″/4 Fr/ various lengths
Balloons		
Any standard angioplasty balloon	Various manufacturers	4–16 mm
Stents		
BeGraft peripheral	BeGraft	Size & length depending on CT measurements +/− IVUS
BeGraft aortic[b]	BeGraft	Size & length depending on CT measurements +/− IVUS
V12 Advanta	Getinge	Size & length depending on CT measurements +/− IVUS
Zilver	Cook medical	Size & length depending on CT measurements +/− IVUS
IVUS catheter (intravascular ultrasound)	Where necessary	Assess intra- or extra-luminal positioning +/− diameter of said vessel prior to stent deployment

[a] Manufacturers and stent types might differ per region, institutions or country. Covered and uncovered iliac or aortic stents should be available prior to commencing any type of endovascular aorto-iliac endovascular treatment, based on pre-operative imaging

[b] The BeGraft aortic stent can be post-dilated to diameters which can accommodate for an aortic size up to 16 mm; other manufacturers also produce covered stents which can be post-dilated to a diameter that can match that of the infra-renal aorta. Choice depends on local availability and pre-operative imaging. We advocate the use of covered aortic and common iliac stents; however, this is not based on randomised evidence assessing clinical and cost-effectiveness

aortic stents (which is deployed first) depends on the configuration of the disease in the distal aorta and bifurcation. There is no randomised evidence showing superiority of the CERAB technique over kissing stenting; however, the advocates of CERAB base their choice on the minimal guttering effect when using a CERAB configuration and a more anatomical reconstruction. Further, the use of covered stents in the aorta and common iliac arteries might be safe in case of long occlusions, at least in theory, where vessel rupture might be more likely when attempting to cross the lesion. In our international cohort study, we found that the combination of covered aortic and iliac stents in the iliacs and aorta confers a benefit in terms of patency when treating patients with TASC C and D disease.

A common area where patients might require re-intervention is the external iliac artery. We advocate treatment of stenosis of the external iliac artery with liberal stenting (using self-expanding stents) and also liberal treatment of common femoral artery disease, to allow adequate outflow. The external iliac arteries and common femoral arteries are the most common sites of disease recurrence in these patients in our cohort studies and meta-analysis.

Overall, careful pre-operative planning is paramount. One should be aware of the exact site and length of occlusions in the aorto-iliac axis, assess patency of the aortic visceral branches, patency of the internal iliacs and state of run-off. In our practice, we use the below schematic (Fig. 15.7) in order to plan any endovascular aorto-iliac procedure, based on pre-operative computed tomographic angiography (which is superior in assessing calcified plaques). In terms of deciding when to chimney the inferior mesenteric artery or renal arteries, we advocate the use of chimneys in arteries of more than 4 mm in diameter. This decision is complex and should be based on each patient's risk profile and individual imaging characteristics. We avoid using chimney stents for the renal arteries and would prefer open surgical reconstruction in case of aortic occlusive disease very proximal to the renal arteries, are future re-intervention (surgical) in patients with renal chimneys can be very challenging. The lengths of the occlusions and lengths of stents (together with diameters) should be calculated pre-operatively with centre-line three-dimensional reconstruction. Intra-operative angiograms can be deceiving. Intravascular ultrasound, where and when available, can also help with sizing iliac stents appropriately (Fig. 15.2).

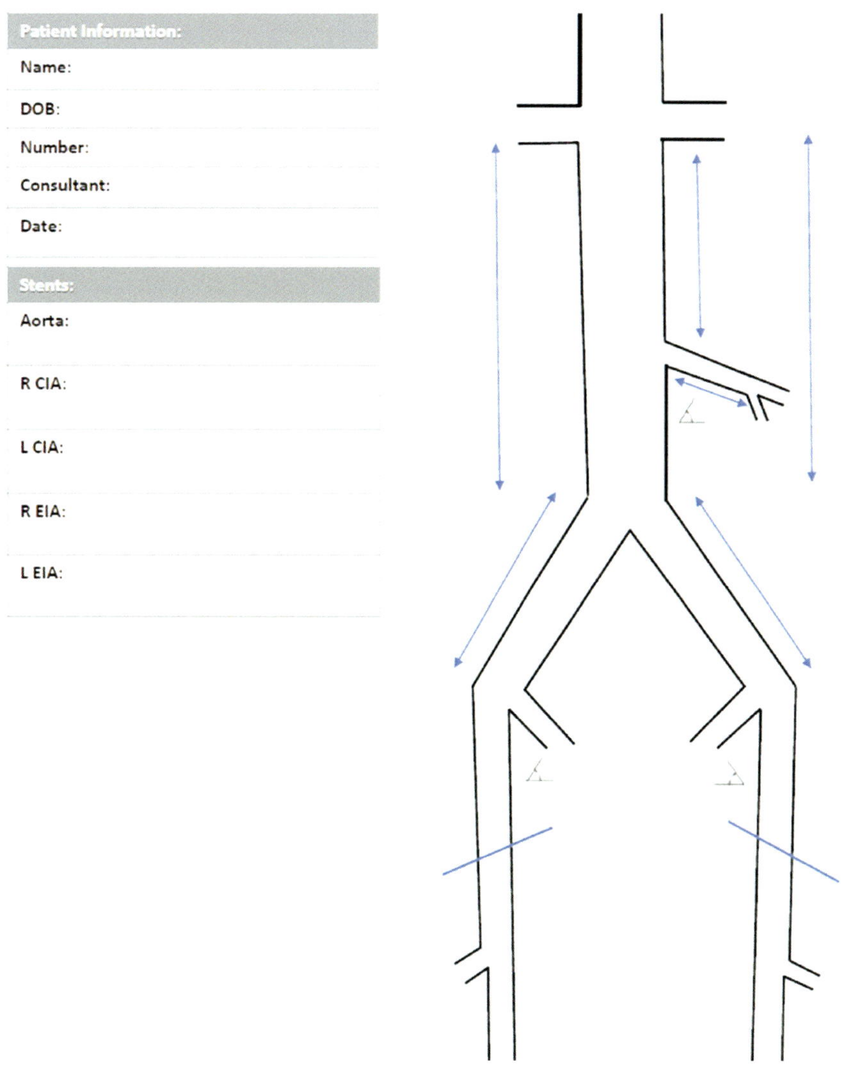

Fig. 15.2 Planning map for patients with TASC C and D aorto-iliac disease based on pre-operative imaging

Access and Crossing the Lesion

The initial angiogram is crucial in terms of crossing the lesions. As a result, we have a low threshold for accessing the aorta from above in patients with occlusive aortic disease i.e. inserting an aortic sheath and angiographic catheter from the arm(s) (typically left upper limb to minimise the risk of stroke). Femoral access with or without endarterectomy (in case of severe femoral disease) is of course mandatory. Ultrasound guided percutaneous access minimises complications and should be used.

Anticipated Complications

The most common complications relate to access, especially in the groin. We strongly advocate common femoral endarterectomy in patients with severely calcified femoral arteries and stenosis exceeding 50%. This optimises outflow and offers superior control.

Peri-Operative Care and Surveillance

There is no randomised evidence regarding surveillance post-operatively in the case of TASC C and D aorto-iliac lesions. The cost-effectiveness of surveillance also remains debatable. Newer procedures and configurations, e.g., CERAB have also not been assessed in randomised studies in terms of post-operative surveillance. Duplex ultrasonography in patients with long occlusion or where newer techniques have been used (CERAB) might be of benefit; however timing and strategies to address stenoses remain unclear. We personally advocate a Duplex ultrasound with a non-contrast computed tomographic scan (to assess stent deployment) 30 days after CERAB procedures followed by 3-monthly duplex and clinical assessment.

Case Presentation

Based on the CTA, the aorto-iliac occlusion qualified for a TASC D lesion (Fig. 15.3). Following a multidisciplinary team meeting discussion, we decided to proceed with an endovascular recanalisation and stenting, using covered balloon expandable stents in the form of a covered endovascular reconstruction of aortic bifurcation (CERAB) procedure, under local anaesthetic. The procedure took place in a hybrid operating theatre, led by a consultant vascular surgeon, via percutaneous bilateral femoral artery access and left sided brachial artery puncture. The patient was consented for standard risks and benefits for this procedure and agreed to proceed. The common femoral arteries and left brachial artery were punctured using 5 Fr micropuncture catheter kits (Merit Medical, Utah, USA), which is our standard practice for calcified arteries. Two 5 Fr access sheaths were introduced in the common femoral arteries and a 90 cm 7 Fr sheath was advanced via the left brachial artery to the level of the suprarenal aorta, to obtain proximal angiograms; this is our standard practice for long bilateral iliac and aortic occlusions in order to obtain a target in terms of recanalisation from distal (femoral) to proximal (aorta) (Fig. 15.4).

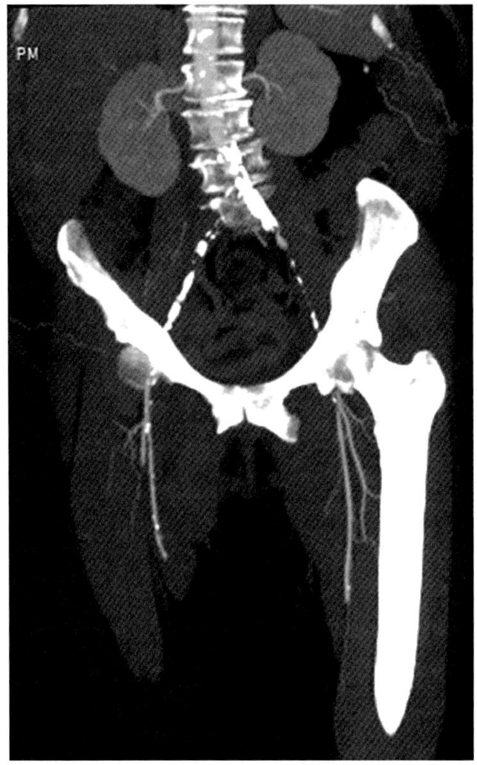

Fig. 15.3 Baseline computed tomographic (CT) angiogram disclosing bilateral calcified iliac occlusions with a distal aortic occlusion

Following the initial angiogram, a combination of stiff, half stiff, and floppy 0.35″ hydrophilic wires (Terumo Corporation, Japan) were used to cross the iliac lesions with support catheters (CXI 90 cm and 110 cm support catheters, Cook Medical, Arizona, USA) both from above and below (aortic sheath and femoral access) the aortic and iliac occlusions. The occlusions were crossed bilaterally in a subintimal plane, given the degree of calcification as well as length of the disease. Through and through wires were then snared using a 6 Fr goose snare via the right femoral artery. The right axis was crossed first (Figs. 15.5 and 15.6).

Once both iliac axes and the infra-renal aorta had been recanalised (subintimal plane), two stiff wires of 300 cm in

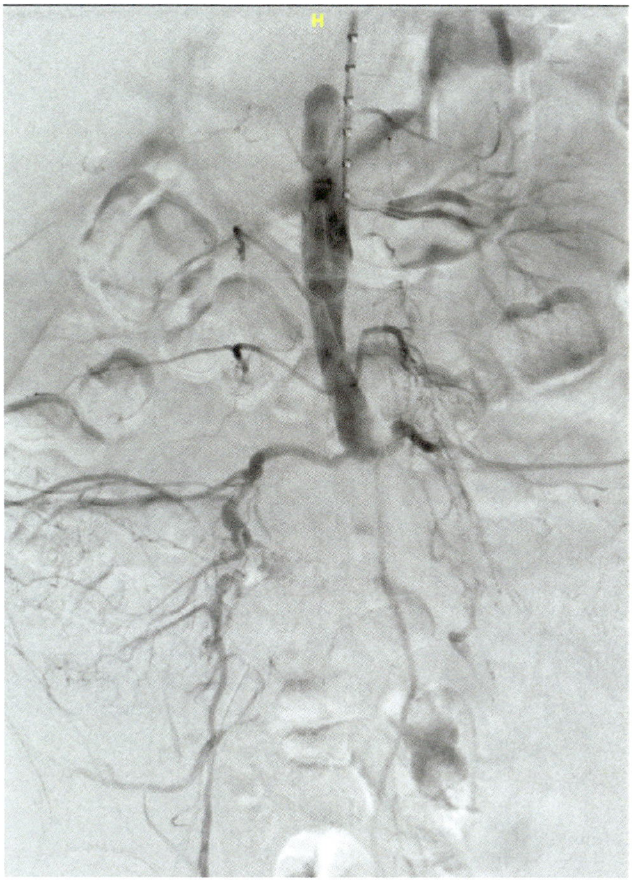

Fig. 15.4 first intra-operative angiogram, disclosing an aortic occlusion below the renal arteries and long bilateral calcified iliac occlusions. Contrast injected from the 7 Fr proximal sheath via an angiographic pigtail catheter using a pump (automated) injector

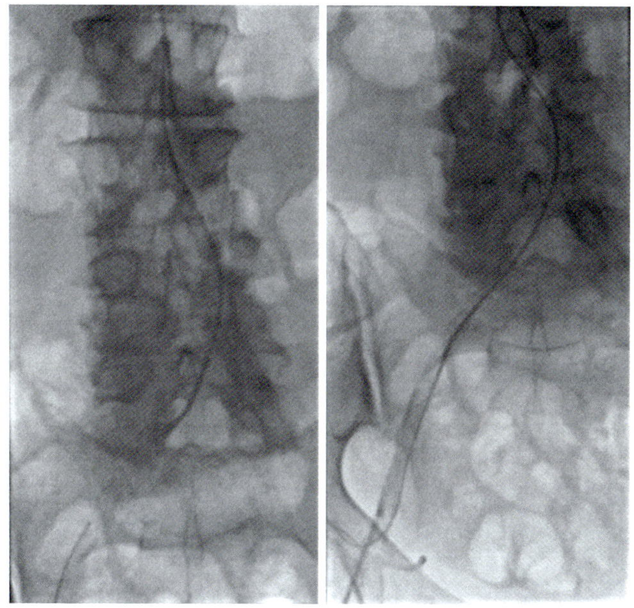

Fig. 15.5 Crossing of the right aorto-iliac axis using a combination of stiff 0.035″ hydrophilic wires (Terumo Corporation, Japan) with telescopic support via an 0.035" CXI catheter (Cook Medical, Arizona, USA) in a C2 5 Fr guiding catheter (subintimal plane). The hydrophilic wire from the proximal access (upper limb) was snared via the right groin access sheath

length were inserted and the aortic stent (14 mm wide BeGraft aortic stent, Bentley, Germany) was deployed first under fluoroscopic control, flush to the aortic bifurcation; positioning was confirmed using multiplane fluoroscopy. The aortic stent was 39 mm in length. Following that, three common iliac balloon expandable stents (9 mm in diameter each) were deployed to the level of the iliac bifurcation, with post-dilatation to 9 mm using plain angioplasty balloons in a kissing stent configuration. The iliac stents were extended up to the level of the distal external iliac arteries sequentially; covered self-expanding stents of 8 mm in diameter were used on each side (Viabahn stents, Gore Medical, Arizona, USA). It is now our standard practice to use covered self-expanding stents in long external iliac occlusions (Fig. 15.7).

Post-operatively the patient was discharged after 2 days. She remains on Clopidogrel 75 mg daily and lifelong surveillance with 3-monthly ultrasound (duplex). The rest pain disappeared and the patient remains ulcer/pain free after 14 months.

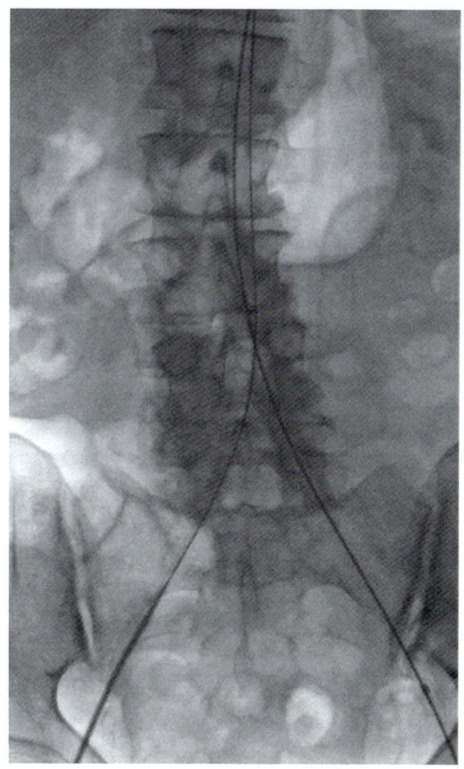

Fig. 15.6 Crossing of the left aorto-iliac axis with support from above and below (aortic and femoral sheaths). Using an identical wire and support catheter configuration to the right side, the left axis was crossed subintimally

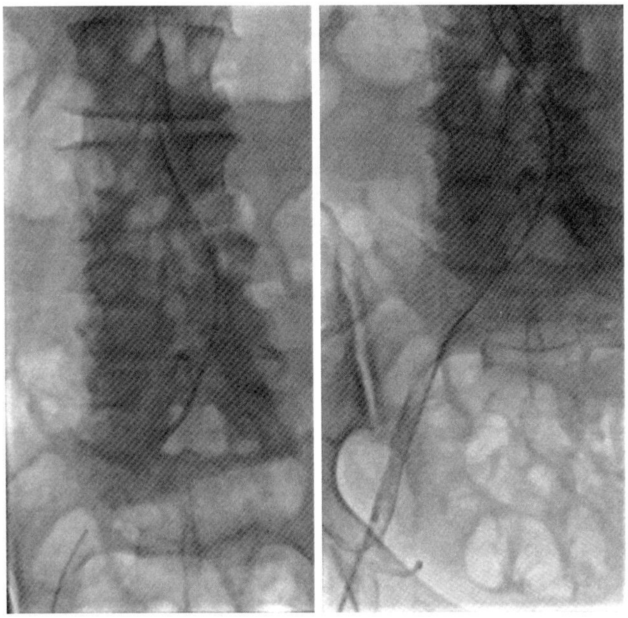

Fig. 15.7 Post dilatation of the kissing iliac stents and completion angiogram

References

1. Saratzis A, Jaspers NEM, Gwilym B, Thomas O, Tsui A, Lefroy R, et al. Observational study of the medical management of patients with peripheral artery disease. Br J Surg. 2019;106(9):1168–77.
2. Mohler E 3rd, Giri J, Acc A. Management of peripheral arterial disease patients: comparing the ACC/AHA and TASC-II guidelines. Curr Med Res Opin. 2008;24(9):2509–22.
3. Morley RL, Sharma A, Horsch AD, Hinchliffe RJ. Peripheral artery disease. BMJ. 2018;360:j5842.
4. Saratzis A, Salem M, Sabbagh C, Abisi S, Huasen B, Egun A, et al. Treatment of Aortoiliac occlusive disease with the covered endovascular reconstruction of the aortic bifurcation (CERAB) technique: results of a UK multicenter study. J Endovasc Ther. 2021;28:737.
5. Ambler GK, Thomas-Jones E, Edwards AGK, Twine CP. Prognostic risk modelling for patients undergoing major lower limb amputation: an analysis of the UK National Vascular Registry. Eur J Vasc Endovasc Surg. 2020;59(4):606–13.
6. Norgren L, Hiatt WR, Dormandy JA, Nehler MR, Harris KA, Fowkes FG, et al. Inter-society consensus for the Management of Peripheral Arterial Disease (TASC II). Eur J Vasc Endovasc Surg. 2007;33(Suppl 1):S1–75.
7. Salem M, Hosny MS, Francia F, Sallam M, Saratzis A, Saha P, et al. Management of Extensive Aorto-Iliac Disease: a systematic review and meta-analysis of 9319 patients. Cardiovasc Intervent Radiol. 2021;44:1518.
8. Scheinert D, Schroder M, Balzer JO, Steinkamp H, Biamino G. Stent-supported reconstruction of the aortoiliac bifurcation with the kissing balloon technique. Circulation. 1999;100(19 Suppl):II295–300.
9. Chiu KW, Davies RS, Nightingale PG, Bradbury AW, Adam DJ. Review of direct anatomical open surgical management of atherosclerotic aorto-iliac occlusive disease. Eur J Vasc Endovasc Surg. 2010;39(4):460–71.
10. Shen C, Zhang Y, Qu C, Fang J, Liu X, Teng L. Outcomes of Total Aortoiliac revascularization for TASC-II C&D Lesion with kissing self-expanding covered stents. Ann Vasc Surg. 2020;68:434.
11. Mwipatayi BP, Sharma S, Daneshmand A, Thomas SD, Vijayan V, Altaf N, et al. Durability of the balloon-expandable covered versus bare-metal stents in the covered versus balloon expandable stent trial (COBEST) for the treatment of aortoiliac occlusive disease. J Vasc Surg. 2016;64(1):83–94 e1.
12. Mwipatayi BP, Thomas S, Wong J, Temple SE, Vijayan V, Jackson M, et al. A comparison of covered vs bare expandable stents for the treatment of aortoiliac occlusive disease. J Vasc Surg. 2011;54(6):1561–70.
13. Kavaliauskiene Z, Antusevas A, Kaupas RS, Aleksynas N. Recent advances in endovascular treatment of aortoiliac occlusive disease. Medicina (Kaunas). 2012;48(12):653–9.
14. Goverde PC, Grimme FA, Verbruggen PJ, Reijnen MM. Covered endovascular reconstruction of aortic bifurcation (CERAB) technique: a new approach in treating extensive aortoiliac occlusive disease. J Cardiovasc Surg. 2013;54(3):383–7.
15. Grimme FA, Goverde PC, Verbruggen PJ, Zeebregts CJ, Reijnen MM. Editor's choice--first results of the covered endovascular reconstruction of the aortic bifurcation (cerab) technique for aortoiliac occlusive disease. Eur J Vasc Endovasc Surg. 2015;50(5):638–47.
16. Taeymans K, Goverde P, Lauwers K, Verbruggen P. The CERAB technique: tips, tricks and results. J Cardiovasc Surg. 2016;57(3):343–9.
17. Saratzis A, Argyriou A, Davies R, Bisdas T, Chaudhuri A, Torsello G, et al. Editor's choice - covered vs. bare metal stents in the reconstruction of the aortic bifurcation: early and midterm outcomes from the COBRA European multicentre registry. Eur J Vasc Endovasc Surg. 2022;63(5):688–95.
18. Saratzis A, Harrison S, Barratt J, Sayers RD, Sarafidis PA, Bown MJ. Intervention associated acute kidney injury and long-term cardiovascular outcomes. Am J Nephrol. 2015;42(4):285–94.
19. Saratzis A, Shakespeare J, Jones O, Bown MJ, Mahmood A, Imray CHE. Pre-operative functional cardiovascular reserve is associated with acute kidney injury after intervention. Eur J Vasc Endovasc Surg. 2017;53(5):717–24.

Interventions for Femoropopliteal Disease

16

Raphaël Coscas

Case Presentation

A 63-year-old patient presented with bilateral disabling lower limb claudication (<100 meters). His main risk factor was tobacco use, which he recently quit. Medical therapy and supervised rehabilitation program were started 6 months ago, but symptoms worsened on the right side with the appearance of intermittent rest pain. Duplex ultrasound and computed tomography angiography showed bilateral long occlusions of the superficial femoral arteries, extended to the P1 and P2 segments of the popliteal artery on the right side (Fig. 16.1). The right distal runoff was also impaired with a single anterior tibial artery. The patient was scheduled for a

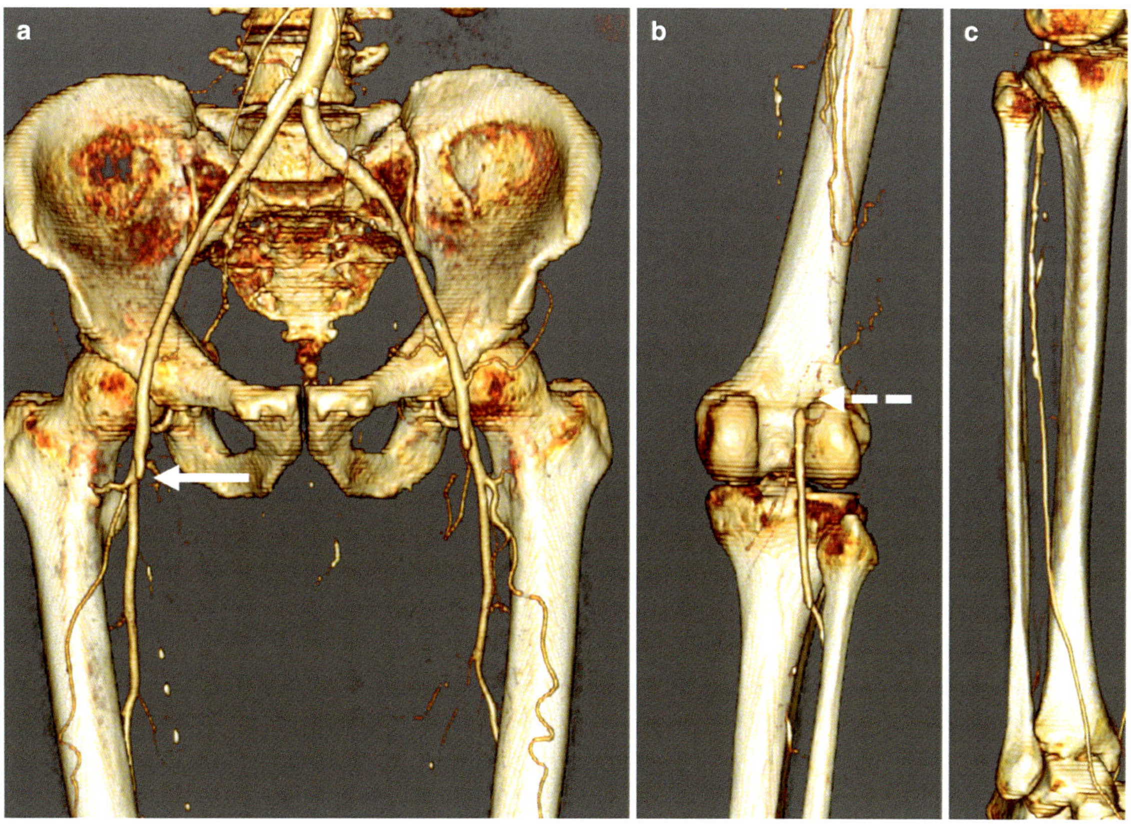

Fig. 16.1 Computed tomography angiography showed bilateral long occlusions of the superficial femoral arteries (**a**), extended to the P1 and P2 segments of the popliteal artery (**b**) on the right side. The plain arrow shows the beginning of the occlusion and the dotted arrows point to the re-entry. Distally, the only patent vessel is the anterior tibial artery (**c**)

R. Coscas (✉)
Department of Vascular Surgery, Ambroise Paré University Hospital, Versailles Saint-Quentin and Paris-Saclay Universities, Boulogne-Billancourt, France

G. Geroulakos et al. (eds.), *Mastering Endovascular Techniques*, https://doi.org/10.1007/978-3-031-42735-0_16

saphenous bypass, but a severe coronary artery disease was diagnosed during preoperative assessment, and he underwent the implantation of several coronary stents with an indication for prolonged dual antiplatelet therapy. A complex endovascular therapy was then considered.

Introduction

Interventions for femoropopliteal (FP) disease are indicated in the case of persistent disabling claudication after a rehabilitation program or in the setting of critical limb-threatening ischemia (CLTI). Besides revascularization, cardiovascular risk factors control, coronary arteries assessment and best medical treatment (antiplatelet therapy and high-dose statins) are always indicated [1]. Long and complex femoropopliteal lesions remain good indications for open surgery if a single segment saphenous vein is available in good surgical-risk patients [1, 2]. These situations and discussions fall outside the scope of this chapter. We here describe the strategy and technical aspects of endovascular interventions for FP disease.

Global Technical Strategy

The global technical strategy is based on the preoperative duplex ultrasound assessment, now most of the times combined with a computed tomography (CT)-scan angiography. We start by defining "plan A": which ideal repair do I wish to perform for this specific lesion? But also plan B and plan C in the case of unplanned intraoperative issues (what are the bail-out strategies at each step of the procedure?). Once the plans are defined, the sizing of the vessels will help preparing the devices needed (vessel preparation tools, lengths and diameters of balloons and stents) to determine the sheath size used for the procedure. Main endovascular tools needed for FP interventions are presented in Table 16.1. Then, our technical strategy always follows six steps, which must be perfectly planned before starting the procedure (Fig. 16.2) and constitute the paragraphs of this chapter: (1) the arterial access, (2) the navigation down to the target lesion and the stabilization of the arterial access, (3) the lesion crossing strategy, (4) the vessel preparation, (5) the anti-restenotic therapy and (6) the arterial closure. All along the procedure, intraoperative imaging will help the operator. This will also be mentioned in a specific paragraph.

Table 16.1 Main endovascular devices needed for femoropopliteal interventions (toolbox). The list in the example column is not exhaustive

Procedural step	Types of devices	Examples
Access	Needles	18G (metallic or plastic); 21 or 22G micropuncture (retrograde access) (cook)
	Sheaths	4 to 7 Fr 11 cm (Terumo), slender (Terumo), halo one (BD)
Navigation and stabilization	Navigation wires	Radifocus straight and angled (Terumo), advantage (Terumo)
	Basic angiography catheters	Vertebral (merit), Bernstein (Cordis) KMP (cook) for standard navigation and UF (Cordis), RIM (cook) for crossover
	Support wires	Stiff (Terumo), advantage (Terumo), Rosen (cook), Amplatz (Boston)
	Long sheaths	4 to 7 Fr of 45, 55 or 70 cm flexor (cook), depending on the material and the distance between the target lesion and the puncture point
Crossing	CTO wires	Halberd (Asahi), Astato (Asahi), Winn (Abbott), command (Abbott), victory (Boston)
	Wires for subintimal crossing	Half-stiff J (Terumo), command (Abbott), gladius (Asahi)
	Support catheters	CXI (cook), Rubicon (Boston), trailblazer (Medtronic)
	Re-entry catheters	Outback (Cordis), Pioneer IVUS guided (Philips), BeBack (Bentley)
Vessel preparation	Standard semi-compliant balloons	Armada (Abbott), mustang (Boston), saber (Cordis), Sterling (Boston)
	Specific balloons	Conquest (BD), Angiosculpt (Philips), Ultrascore (BD), cutting (Boston), chocolate (Medtronic)
	Atherectomy devices	Rotational: Jetstream (Boston), orbital: Diamondback (CSI, Abbott), directional: HawkOne (Medtronic), hybrid: Phoenix (Philips), laser: Excimer (Philips)
	Intravascular lithotripsy	M5+ (shockwave)
Anti-restenotic therapy	Bare-metal stents	Absolute pro (Abbott), Everflex (Medtronic), Lifestent (BD), Supera interwoven (Abbott)
	Drug-eluting stents	Eluvia (Boston), Zilver-PTX (cook)
	Covered stents	Viabahn (Gore)
	Drug-coated balloons	In.Pact (Medtronic), Lutonix (BD), Stellarex (Philips), ranger (Boston), Luminor (iVascular)
Closure	Vascular closure devices	Femoseal (Terumo), Angioseal (Terumo), Prostyle (Abbott), exoseal (Cordis), Mynxgrip (Cordis)
Intravascular imaging	IVUS	IVUS (Philips)

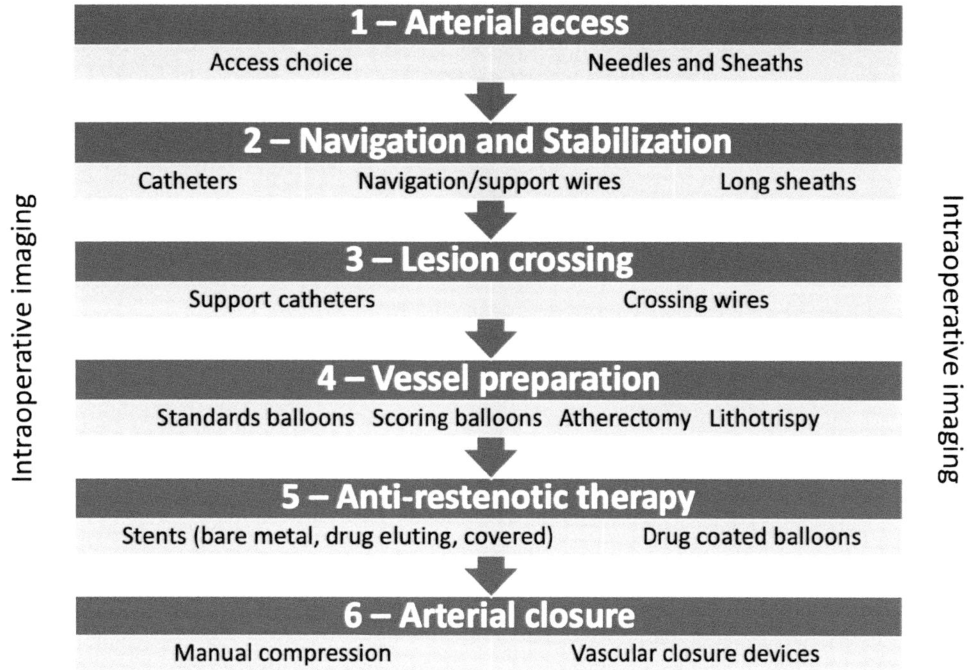

Fig. 16.2 Current technical algorithm for planification of femoropopliteal interventions

Arterial Access

Access Materials

Needles and short sheaths are necessary at this step. Needles can be metallic or with a plastic cannula. The metallic needle will avoid the manipulation of the needle once in the artery but carries the risk to damage the wire inserted inside when it will be taken out from the artery. The needles with plastic cannulas necessitate more manipulations but are more secure for the wire. Needles used for femoral approach are generally short (7 cm). Long needles (10–15 cm) are necessary in some obese patients, especially when performing an anterograde access. Needle diameter is expressed in Gauge (G) with the highest G being the smallest diameters. An 18 G needle is the standard when puncturing a femoral artery. A 21 or 22 G is used in alternative tiny accesses (tibial or radial). A short, stiff wire is inserted in the needle or the cannula, which is then exchanged for a short sheath. Sheath diameters are expressed in French (Fr) and generally correspond to the internal diameter of the sheath (what can be inserted in the sheath). The standard sheath used for FP interventions is 6 French. Some companies have developed specific low-profile sheaths with a decreased external diameter (Halo One, *Becton, Dickinson and Company* and Slender, *Terumo*). Another interesting way of development has been the advent of dedicated devices to perform the procedure with 5 Fr or 4 Fr sheaths to minimize access bleeding issues. Larger sheaths are sometimes necessary (7 Fr) when debulking or intravascular lithotripsy (IVL) devices are necessary. Sheath diameter must be chosen before the procedure and taken into account for all anticipated material diameters.

Access Sites

All access sites must be discussed and considered. The main one is of course the femoral access, but many situations are in fact more adequate for an alternative access, either as a single primary access or as a secondary access (bail-out).

The Femoral Access

The arterial system is generally accessed through an ultrasound (US)-guided percutaneous puncture. The most usual access is the common femoral artery (CFA). We use an 18-Gauge needle, either metallic or with plastic cannula, inserted with a 30–45° angle with the arterial axis. The CFA can be accessed antegradely when the proximal SFA is free of disease. This approach allows working with shorter materials and gives more pushability when a challenging lesion crossing is anticipated. The proximal SFA can also be antegradely punctured in cases of calcified CFA or obesity [3]. The use of smaller (4Fr) sheaths is then recommended to avoid issues with arterial closure [4]. Access through the contralateral CFA with crossover is our standard of practice when technically feasible. It mandates longer materials but the installation, the puncture and the closure are easier. It allows better-quality contrast injections in the profunda femoral artery (PFA) to see the re-entry in the case of chronic total occlusion (CTO).

Another issue with femoral access is the need for patients to stay supine several hours after the procedure [5, 6]. But early ambulation (2–4 h after the procedure) after femoral artery puncture is feasible [6–10], especially when using vascular closure devices (VCDs).

Upper-Limb Accesses

In some situations, the CFA can be deemed inappropriate or unavailable for puncture, such as in the absence of palpable femoral pulses, the presence of CFA calcifications, obesity, a history of femoral surgery (particularly with prosthetic materials) or the need for a contralateral femoral approach in the setting of previous kissing iliac stents or bifurcated aortic graft (Fig. 16.3). Access through the upper limb, especially from the brachial artery, has been well described as an additional [11, 12] or main [11, 13–15] access in these situations. In contrast, upper-limb access should be used with caution in tall patients, patients with a diseased aortic arch or patients that will need arteriovenous access for dialysis. The arterial system can be accessed through the brachial, radial or axillary arteries. The brachial access has the advantage to allow using all sheath diameters for peripheral interventions. It can be generally punctured percutaneously, but we recommend open exposure when sheaths >6 Fr are used because manual compression (MC) is at high risk for failure [16]. Brachial access is however fraught with risks of local complications (pseudoaneurysm, brachial artery thrombosis, median nerve injury) in up to 11% cases [11, 13–15]. In one of the largest series reporting on the use of the brachial access for aortic and peripheral procedure [11], the Cleveland Clinic team reported a 6.5% rate of access site-related complications (pseudoaneurysm, brachial artery thrombosis, hematoma), including a 4.0% rate of surgical correction. Of note, conversion to open surgical closure was needed in 2.3% cases when the artery was initially punctured percutaneously in this series.

The radial approach is now the first-line approach for coronary PTA [17, 18]. It is associated with less vascular complications compared to femoral access [17–22]. A meta-analysis [19] of randomized controlled trials comparing radial versus femoral access for primary percutaneous coronary interventions revealed that the radial approach was significantly associated with a decreased risk of major bleeding (1.4% vs. 2.9%) and access site bleeding (2.1% vs. 5.6%). Dedicated devices for FP interventions through radial access are now available. There are however several points to keep in mind when planning peripheral procedures through radial access. Absolute and relative contraindications for radial access are summarized in Table 16.2. Usually, the left radial artery is used to avoid arch navigation in front of the carotid ostia. It also had the advantages of gaining 5–10 cm of length compared to a right-side puncture and crossing the innominate artery, which is sometimes tortuous. The artery is punctured at the wrist (Fig. 16.4) or even in the snuffbox [23]. A minimum diameter of 2.0–2.5 mm is required. The local anesthesia avoids adrenaline (vasoconstriction) but include

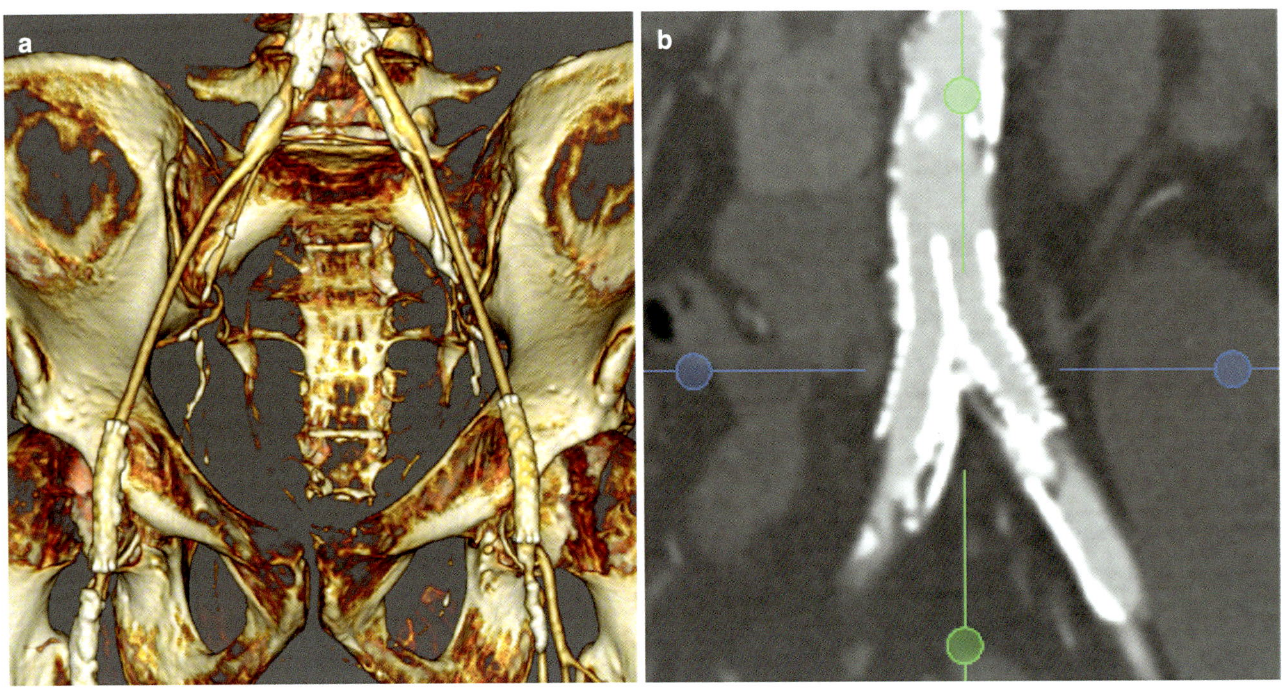

Fig. 16.3 Example of a clinical situation where femoral accesses are challenging. The patient combines the problem of bilateral common femoral artery stents complicating the puncture (**a**) and a kissing stent configuration in the common iliac arteries (**b**) precluding a crossover approach with standard devices

nitrates (vasodilatation). A 21-Gauge needle is used, and a short radial sheath is inserted to strengthen the radial artery. A local spasmolytic cocktail (Isosorbide dinitrate, 3 cc, + Verapamil, 2 cc) is injected intraarterial before starting the navigation.

Retrograde Accesses

Retrograde puncture of a patent artery distal to the lesion should always be anticipated during installation and draping. It allows bail-out retrograde recanalization in the case of failure through an anterograde access. A specific paragraph of this chapter includes these strategies (see "lesion crossing") so this paragraph is about retrograde access as a primary and sole approach in femoropopliteal interventions. The risks of femoral access and the benefits observed with retrograde approaches as bail-out accesses have encouraged the idea of performing the endovascular treatment of lower limb lesions through isolated retrograde approach, namely tibiopedal arterial minimally invasive retrograde revascularization (TAMI, Fig. 16.5) [24]. It has long been used as a second-line approach when antegrade access has failed [25, 26]. However, retrograde access presents a higher rate of success than antegrade approach [27]. The CTOP classification (Fig. 16.6) may be helpful to select the best cases for a retrograde approach, the CTOP II and IV, i.e., distal concave cap (Fig. 16.7), being the most appropriates [28]. Retrograde revascularizations can be performed through different accesses such as the distal SFA, the popliteal artery, the proximal anterior tibial artery, the dorsalis pedal artery or the posterior tibial artery. Although distal SFA and popliteal approaches allow using all sheath sizes for peripheral interventions, our preference goes to pedal accesses. The TAMI technique was first described by *Mustapha* et al. to treat patients with PAD severely deemed to be at high surgical risk [24]. This technique may decrease complications related to femoral access [29]. From an anatomical point of view, it seems more comfortable to perform the procedures through

Table 16.2 Absolute and relative contraindications for radial access in femoropopliteal intervention

Patient factors	Anatomic factors	Procedural factors
– Existing AV fistula or graft – Prior CABG with LIMA use – Anticipated use of radial artery for AV fistula or CABG conduit – Aortic arch atherothrombosis – Tall patient	– <0.2–2.5 mm radial diameter – Vessel tortuosity – Radial occlusion – Severe vasospasm – Negative Allen test (arch) – Intolerance to hyperextended wrist positioning	– Devices that require >7F sheath – Devices that cannot be provided with very long shafts (DES/covered stents) – Excessive distance to target lesion – Anticipated complex procedure (calcified CTO)

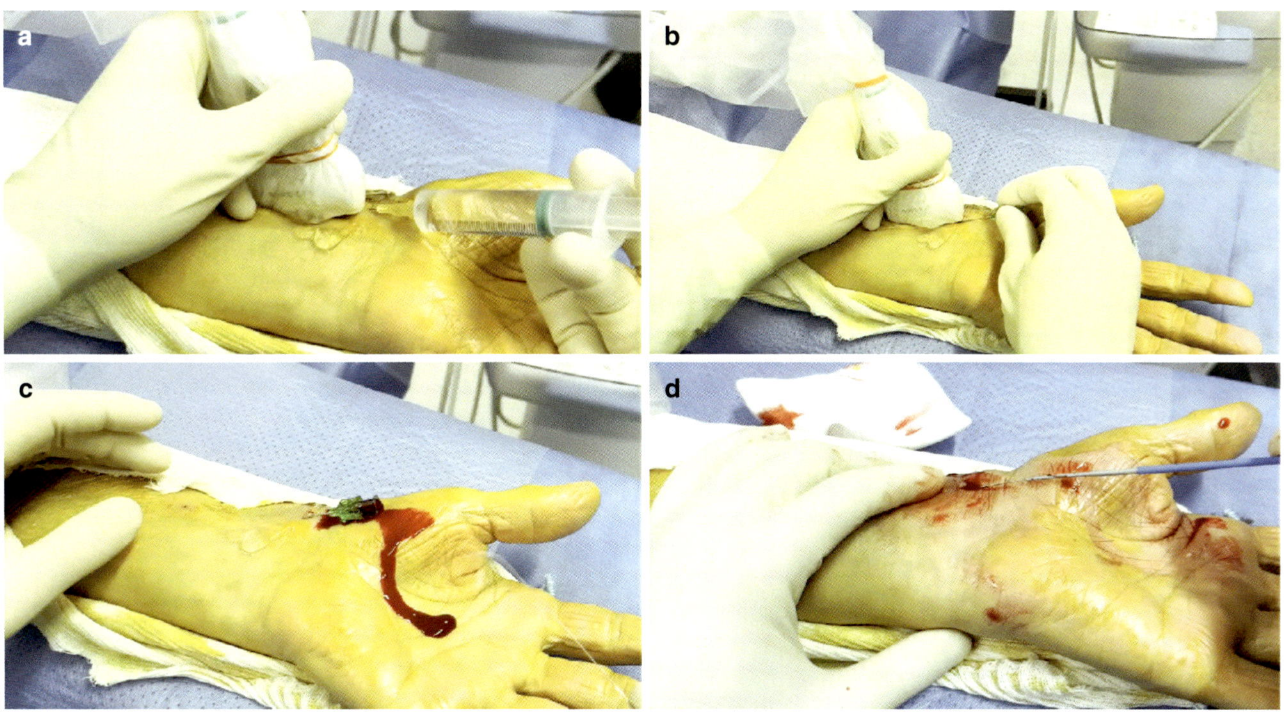

Fig. 16.4 Initial steps of a radial access to overcome the issue of challenging common femoral artery access and allow fast ambulation. Local anesthesia is performed without adrenaline but with nitrate adjunct (**a**). The artery is punctured at the wrist with a short (4 cm) 21 G micropuncture needle (**b** and **c**). A short wire is inserted followed by a 23 cm radial sheath (**d**)

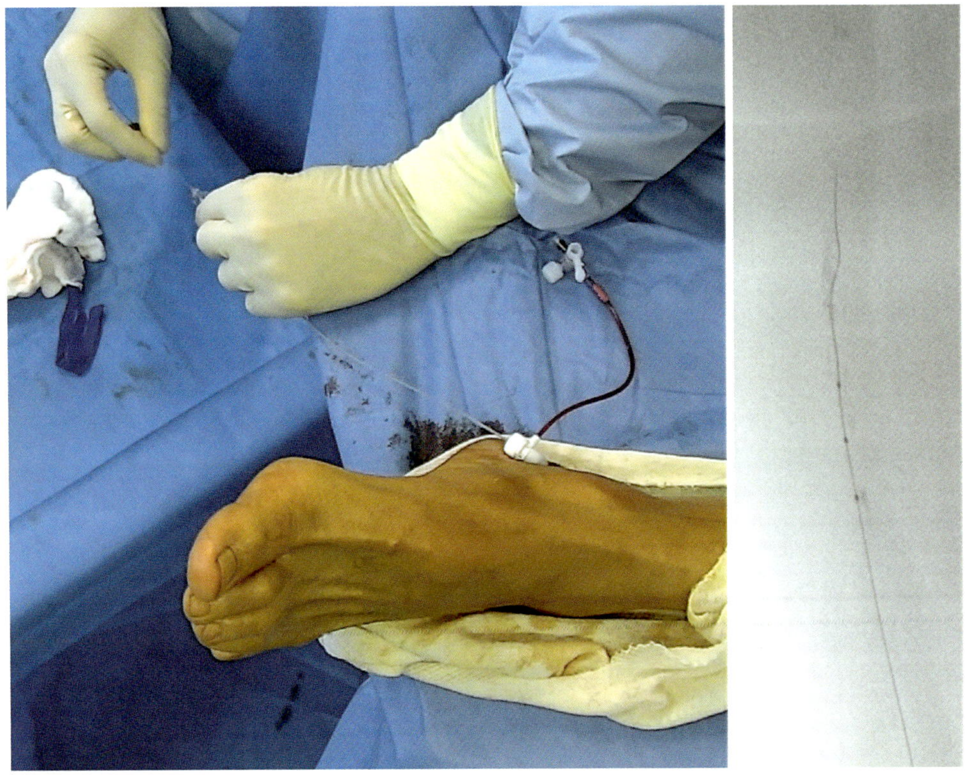

Fig. 16.5 The technique of tibiopedal arterial minimally invasive retrograde revascularization (TAMI) intends to perform the whole procedure through retrograde access. In the present case, the posterior tibial artery is punctured at the ankle, and a 45 cm 4 Fr sheath is introduced up to the level of the popliteal artery. The operator can work through the sheath with an 0.018″ wire over a support catheter

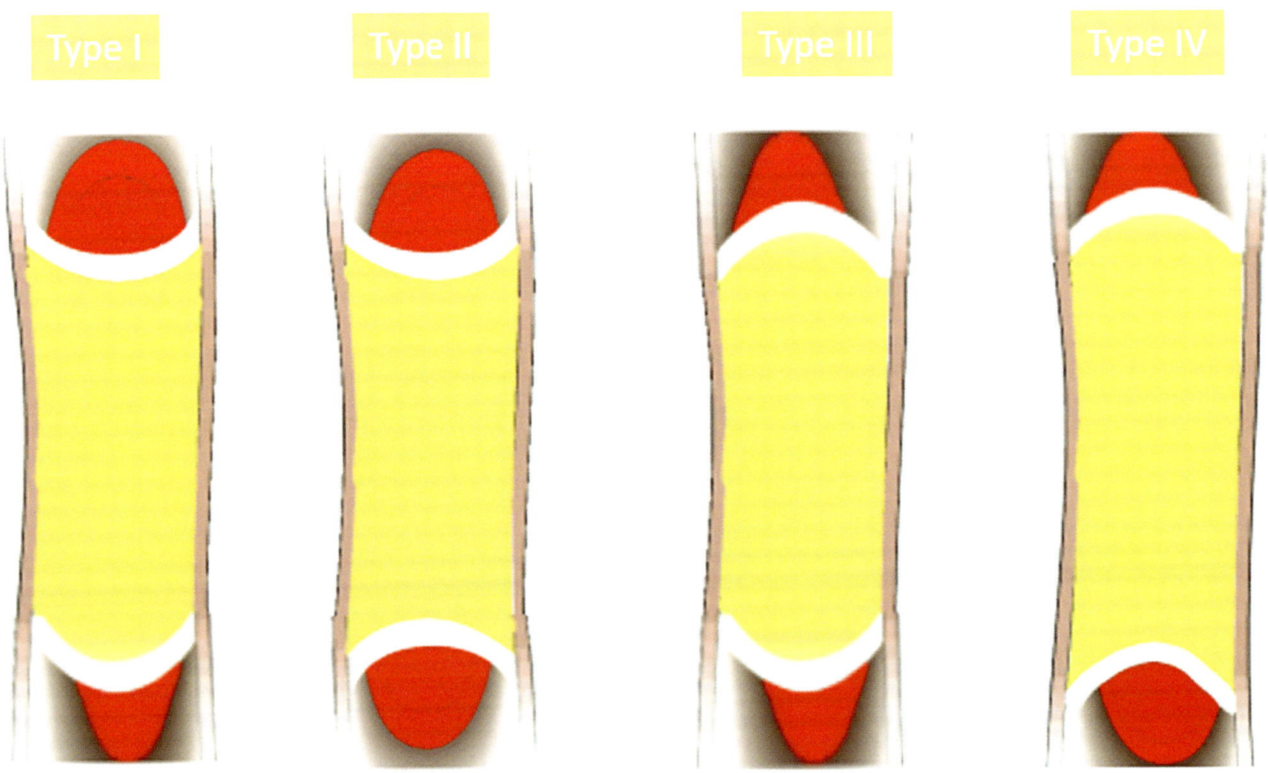

Fig. 16.6 The CTOP classification helps to identify the situations where anterograde or retrograde approach is the most appropriate. Types 1 and 2 are good indications for intraluminal crossing through an anterograde access. Types 2 and 4 are good indications for intraluminal crossing through a retrograde access. The Type 3 configuration is the most challenging for intraluminal crossing

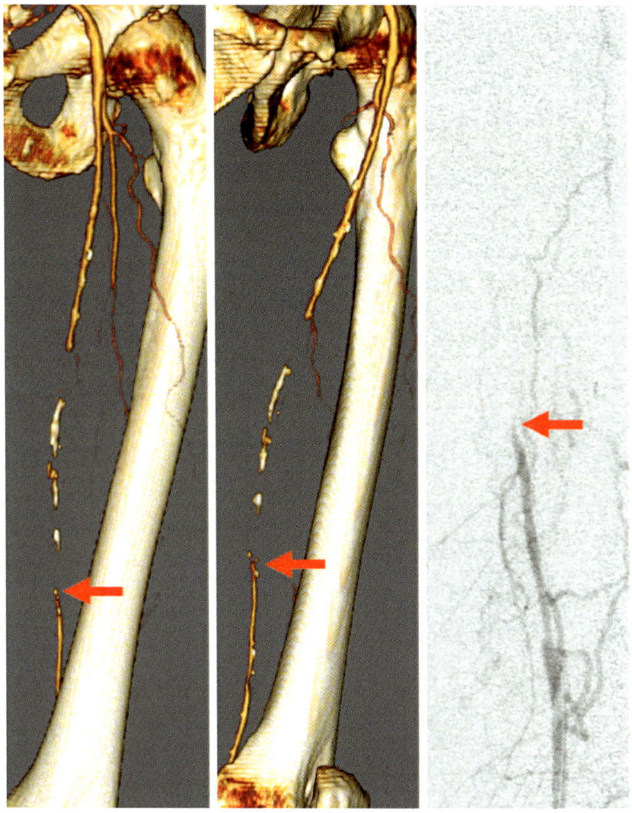

Fig. 16.7 Illustration of a distal cap (**a**, **b**, **c**; arrows) favorable for intraluminal retrograde crossing. Note that the proximal cap also looks favorable for intraluminal crossing (CTOP classification type 2)

the posterior tibial approach because of its straighter course [30]. The choice of the punctured artery will also relate to the downstream runoff. An already thrombosed artery will be preferred whenever possible to limit the risks of thrombosis at the access site (Fig. 16.8). Similarly, in the case of lesions in the proximal and middle thirds of the punctured artery, the TAMI technique renders possible the treatment of these lesions and thus offers complete multi-stage revascularization. The posterior tibial artery puncture, under ultrasound guidance using a 21 Gauge, has our preference for distal access. It avoids the two curves encountered (anterior tibial–pedal junction and proximal anterior tibial artery) when performing a dorsalis pedal puncture. Following arterial puncture and the insertion of a V18 wire (Boston Scientific), a 45-cm sheath (Cook) is generally used, or even 70 cm, which had the advantage to stabilize the access, avoid multiple passages through a small diameter tibial artery to minimize vessel injury and perform angiograms closer to the target lesion. Specific portfolios of 4 Fr devices (Biotronik or Optimed) are generally used for such procedures. Of note, 5 or 6 Fr sheaths can also be used when necessary. Due to the small diameter of the arteries, percutaneous closure systems are not indicated, and all patients receive manual compression for 10 min at the end of the procedure followed by a slight compression bandage.

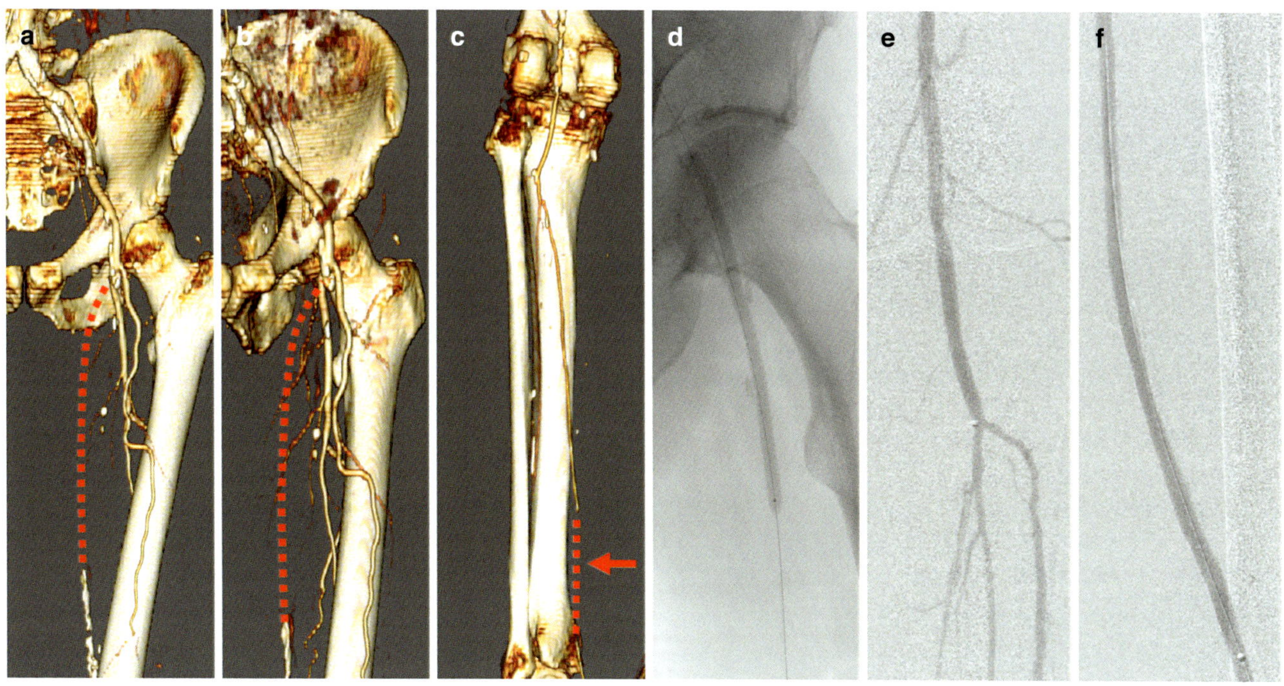

Fig. 16.8 Case of a long superficial femoral artery occlusion without any proximal stump (**a** and **b**). Distally, the posterior tibial artery is occluded (**c**) and can be punctured without any risk to damage the artery. The procedure is performed through a single posterior tibial retrograde access (TAMI technique) using 4 Fr devices. After ballooning (**d**), a long dissection is noted, and stenting of the lesion provides a satisfactory result (**e** and **f**)

Navigation and Stabilization

Following arterial access, heparin is administered intravenously (50 UI/kg), and the navigation phase starts. Its goal is to bring the distal extremity of a sheath as close as possible to the target lesion. The technical principle is always the same: (1) navigate down to the lesion using an angiography catheter with a curve adapted to the navigation, (2) exchange the standard wire for a stiffer wire to then be able to bring a longer kink-resistant reinforced sheath closer to the lesion. When using anterograde CFA access, a short sheath (generally 11 cm length) is enough, and there is no specific navigation or stabilization step except when the lesion is at the popliteal level, where it may be useful to have a longer sheath (45 cm) to be closer to the target lesion. In FP interventions, the most typical example of navigation and stabilization step is the contralateral retrograde CFA puncture and crossover. Following short sheath and standard wire insertion, a double-curved angiography catheter (UF or RIM) is brought over the wire above the aortic bifurcation. The contralateral arterial axis is catheterized, and an angiography catheter (the same RIM or UF, or a straighter one in case of tortuous anatomy) is advanced to provide more support for a wire exchange. An exchange for a stiffer wire is performed, and the wire is brought as far as possible to the contralateral side. Recent hybrid guidewires made of a stiff core and floppy extremity (Advantage, Terumo) allow to perform the whole step with the same wire. Once the stiffer wire is in place, the short sheath is removed and exchanged for a longer kink-resistant reinforced sheath (45, 55, 70 or 90 cm length depending on the distance to the target lesion). The same principles are applied in cases of alternative accesses using angiography catheter with an adapted curve to the navigation. For example, when using an upper-limb access, it is our practice to use a C2 catheter to engage the descending aorta through the left subclavian artery and then use a long vertebral catheter (125 cm in length) to bring the wire down to the target lesion.

When bringing a long sheath through the target lesion is challenging (heavily calcified and/or tortuous vessels), two tips can be used. The first one is to inflate a balloon at the distal tip of the sheath sized on the vessel where it is inflated. Then deflate fast the balloon and push the sheath while maintaining the balloon fixed. The maneuver is similar to the one used during fenestrated endografts at the level of the target vessels when flaring the stents. A second tip is to capture the wire distally through a contralateral retrograde CFA access and obtain a through-and-through access. This gives an exceptional support to deliver the long sheath.

Lesion Crossing

Crossing a stenosis is generally a simple step, but crossing a chronic total occlusion (CTO) can be sometimes extremely challenging. After bringing a long sheath close to the CTO, the question will be whether the lesion can be crossed intraluminally or subintimally. Our strategy is to try to cross the lesion intraluminally as much as possible. This will allow the optimal use of all further necessary devices, especially atherectomy devices (which are contraindicated for subintimal use) and drug-coated balloons (DCBs).

Support Catheters

When dealing with CTOs, angiography catheters may not offer enough support for the wire to cross the lesion. Support catheters overcome these issues. Straight support catheters provide more pushability, whereas shaped/angled give more steerability. Many types and brands of support catheters exist. Most frequently used in our national market are the Rubicon (Boston scientific), the CXI (Cook medical), the Trailblazer (Medtronic) and the Navicross (Terumo). They are available in 0.014′, 0.018′ and 0.035′. The 0.018′ and 0.014′ support catheters can be used inside a larger 0.035′ angiography catheter or support catheter to even improve the support. When the lesion is made of several major calcified plugs, we sometimes change the strategy and use navigating micro-catheters used for embolizations such as the Progreat (Terumo) because they are very flexible and allow a smooth navigation between the calcified plugs.

Dedicated Crossing Wires

Specific wires may be necessary when crossing a CTO. We will not go into the details of the different wires available on the market, but the reader has to be familiar with the three main types/features of current wires: crossing wires, navigation wires and support wires. Crossing wires can be useful when dealing with a CTO. In this setting, the tip load of the wire is particularly important. We recommend the 0.014′ Halberd (Asahi) wire with its 12 grams tip load or the 0.014 WinnT (Abbott) wire for FP occlusions. Of note, the Halberd (Asahi) wire is a bit particular in its manipulation. It should not be torqued at the same time it is advanced. In heavily calcified lesions, wires with an even more important tip load may be necessary. It is our practice to use 0.014′ Astato (Asahi) wires with 20, 30 or even 40 gram tip loads. However, the greater the tip load, the higher the perforation risk, so it is mandatory to be certain to be re-entered in the distal lumen before starting the vessel preparation.

Subintimal Crossing

When an intraluminal crossing is not possible, a subintimal crossing is necessary. It is typically done be creating a small loop on the distal tip in the wire. A small loop generally indicates that the wire is in a quite internal subintimal plane, whereas a large loop is the sign of a navigation in a very external plane. Dedicated wires with small loops (Hal Stiff J, Terumo) exist to favor subintimal crossing in an internal plane. Crossing the lesion in the subintimal plane is often easier than in the internal plane, but the issue is then to re-enter the true lumen distally. This re-entry must be performed in the occluded area or very close after the end of the CTO to avoid losing major collaterals. It is therefore mandatory to avoid to push the loop of the wire too distal from the re-entry level. The re-entry is sometimes spontaneous, especially when the subintimal crossing was done in an internal plane. An angled catheter is advanced over the looping wire, and the wire is exchanged for a wire with a straight or a slightly angled distal tip to re-enter the true lumen. When the re-entry is not spontaneous, two options are available: (1) use a bail-out retrograde access or (2) use a dedicated re-entry catheter. Apart from very specific situations, we prefer to use a bail-out retrograde access (see dedicated paragraph below), but re-entry catheters can remain useful when the operator wants to avoid a distal puncture. These catheters are all based on the same principles with a retractable needle (lateral or axial) to perforate the subintimal flap. Many re-entry catheters are available. The Outback (Cordis) with a lateral retractable needle is probably the most popular [31] and the one with more published data. The Pioneer (Philips) has the interest to be able to be coupled with intravascular ultrasound [32].

Dual Access

When an anterograde approach does not allow CTO crossing, our preferred bail-out maneuver is to use a retrograde access. The goal is to perform a "rendez-vous" with the anterograde access and have a through-and-through wire between the two accesses. We already mentioned the different types of accesses in the "Arterial Access" paragraph of this chapter. Briefly, all distal arteries can be punctured. Our preference is to puncture the posterior tibial artery at the ankle or the dorsalis pedal artery. But several other arterial sites can be effectively punctured: the proximal anterior tibial artery, the peroneal artery (more challenging), the popliteal artery and the distal SFA. A 21 Gauge needle is generally used, and a 0.018″ wire is introduced. Our first line wires are the V18 or Victory wires (Boston Scientific). When needed, a 0.018″ support catheter is used sheathless over the wire (Fig. 16.9). It is useless to introduce a distal sheath for these bail-out accesses. Once the anterograde and retrograde wires are in the same space, an angled catheter is brought anterogradely to capture the distal wire and obtain the through-and-through access. Rarely, a snare may be necessary for this maneuver. Sometimes the two wires are in different planes, and the rendez-vous is impossible without adjunct maneuvers. In these cases, the first thing to do is to try to do the rendez-vous at different levels of the occlusion. If this is not efficient, it is necessary to add anterograde (reverse-CART technique), retrograde (CART technique) or dual ballooning (SAFARI technique) of the occlusion. This will break the intimal plane and allow the two wires to join each other.

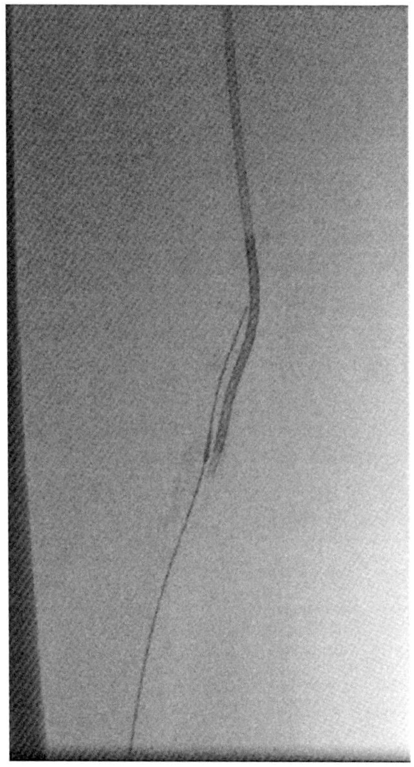

Fig. 16.9 In the case of anterograde crossing failure, a bail-out retrograde puncture allows to bring a wire over a support catheter in the true lumen

Extra-Arterial Crossing

In rare situations, intraluminal and subintimal crossings can be felt not appropriate (long stent occlusion, history of many FP interventions etc.). Techniques of extra-arterial endo-bypasses have been described. Exiting the artery, navigating through the superficial femoral vein and re-entering the arterial lumen is probably the most described alternative [33], with even specific devices now available [34, 35]. Others [36] reported the use of the distal profunda femoral artery to create an FP endo-bypass. We have reported the possibility to create an extraluminal bypass in the setting of a previous proximal FP anastomosis [37]. Although the reader has to know the existence of such procedures, we will not go into the details of these specific techniques that currently have few indications.

Vessel Preparation

After crossing the lesion, the step of vessel preparation is crucial. Its goal is to open the arterial lumen up to the reference vessel diameter to allow the optimal use of anti-restenotic therapies.

Standard Balloon Angioplasty

Prolonged angioplasty with standard balloons (namely plain old balloon angioplasty, POBA) is the reference method for vessel preparation [38, 39]. It is also the simplest and the less costly. We prefer to use long balloons sized 1:1 in diameter compared to the reference vessel diameter and 10 mm longer than the target lesion to avoid a geographic miss. A vessel preparation of 180 s vs. 30 s improves the immediate results of angioplasty [39]. It decreases the major and minor dissection rates and the need for additional treatments (re-dilatation, over-dilatation and stents) [39]. Horie et al. have shown that each supplemental minute of dilatation is independently associated with a decrease in the rate of severe dissections [38]. High-pressure non-compliant balloons such as the Conquest balloon (Becton, Dickinson and Company) can deliver a localized pressure up to 40 atmospheres and are therefore useful to open refractory lesions (myointimal hyperplasia, calcifications).

Cutting and Scoring Balloons

Scoring balloons intend to address the limitations of conventional angioplasty balloons by concentrating the dilating force along the scoring element. They allow a controlled and uniform expansion of the balloon and nitinol cage, preventing slippage while scoring the plaque and maximizing luminal expansion. Their main indications are challenging lesions at high risk of dissection post-angioplasty, especially calcified lesions. Main specific angioplasty balloons are the Angiosculpt (Philips), the Ultrascore (Becton, Dickinson and Company), the Chocolate (Medtronic) and the Cutting (Boston Scientific) balloons. The two last ones may be better defined as atherectomy devices. They have common features in their use: no oversizing, slow inflation, long inflation time, repeat inflations, low deflation. The Angiosculpt (Philips) is a nylon balloon surrounded by an external nitinol-based helical scoring edge [40]. It has three to four rectangular laser-cut nitinol electropolished spiral struts in a helical configuration (Fig. 16.10). It is generally recommended to slightly undersize the balloon by 0.5 mm compared to the reference vessel diameter to avoid major trauma. In a specific registry [40], the presence of calcifications had no more impact on patency following the use of the Angiosculpt (Philips) balloon.

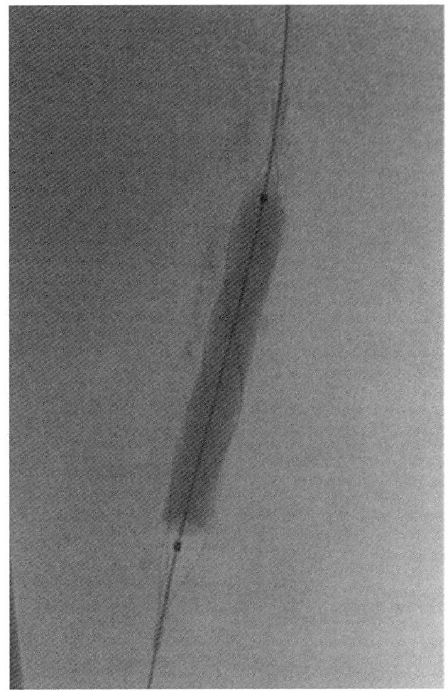

Fig. 16.10 Angioplasty of a calcified lesion using the scoring balloon Angiosculpt (Philips). The helical nitinol scoring elements are visible under fluoroscopy

Debulking Devices

The concept of debulking devices is to remove plaque fragments from the lumen. All these systems need to be used intraluminally. They are contraindicated when the lesion has been crossed subintimally. There are five main concepts: rotational atherectomy (Jetstream, Boston Scientific), orbital atherectomy (Diamondback, Cardiovascular Systems/Abbott), directional atherectomy (HawkOne, Medtronic), hybrid atherectomy (Phoenix, Philips) and the laser atherectomy (Excimer, Philips).

Depending on the systems, larger sheaths (7 or 8 Fr) may be necessary. Debulking devices have been shown to effec-

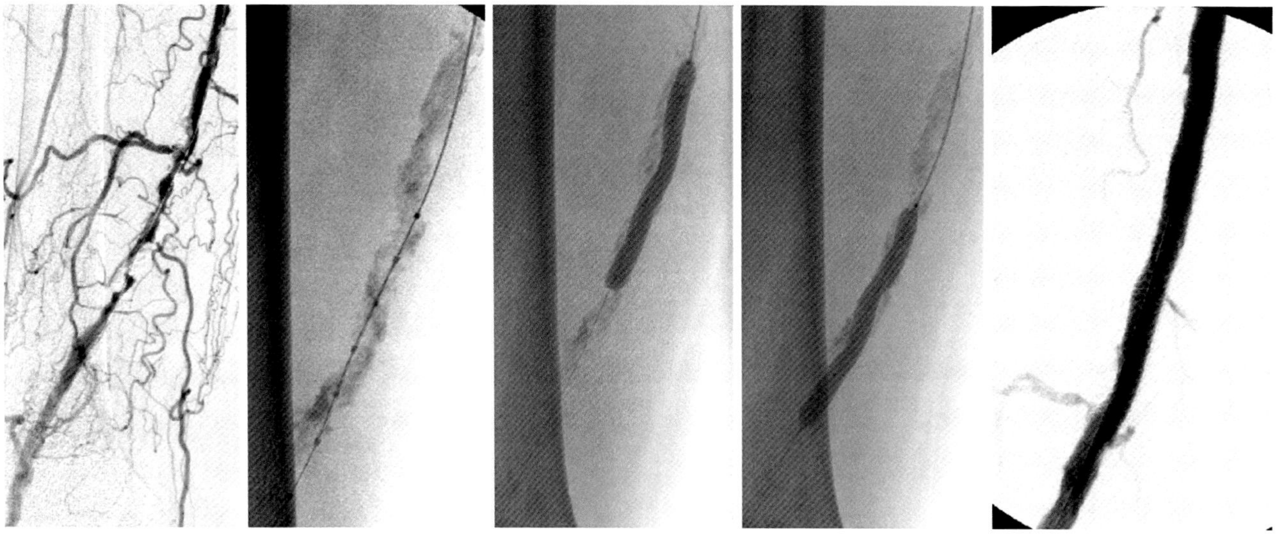

Fig. 16.11 This very calcified lesion of the distal superficial femoral artery (**a**) is crossed intraluminally, and a lithotripsy balloon is used for the vessel preparation. The lithotripsy emitters are well visible before inflation (**b**). The lithotripsy is performed using an "overlap" strategy to treat the whole lesion (**c** and **d**). Due to recoil, an interwoven stent is implanted with a good result (**e**)

tively reduce the rate of intraoperative dissections, decrease the rate of stent placement and improve technical success [41]. These devices are, however, not able by themselves to increase primary patency and decrease target lesion revascularization rates [41]. An additional anti-restenotic therapy must always be considered. Several drawbacks of debulking devices have to be mentioned: increased procedure length, multiple angiograms (contrast media, radiations) and active friction against the plaque, with subsequent risks of emboli and low flow that mandate the use of filters with some systems. Our practice is to use debulking devices for long and complex FP lesions, in-stent restenosis [42] or specific situations where the wire crosses the lesion but nothing passes over the wire.

Intravascular Lithotripsy

This preparation tool is dedicated to calcified lesions. The concept of intravascular lithotripsy (IVL) is to bring an endovascular balloon with lithotripsy emitters at the level of the target lesion, inflate the balloon at low pressure (4 ATM) and then emit sonic waves that create microfractures in the calcified plaque. This has the goal to open the lumen, modify the compliance of the vessel and create channels where local drugs (paclitaxel or limus) will be able to reach the vessel wall. The main balloon in this field in currently the Shockwave (Shockwave Medical). Current balloons are short (30–60 mm long), but all diameters for peripheral use are available (from 2.5 to 12 mm). IVL can be used intraluminal but also subintimal. In the prospective randomized Disrupt PAD III study [43] and in a meta-analysis [44], IVL allows to decrease the rate of severe dissections and bail-out stent rates after angioplasty, which is particularly useful when a drug-coated balloon strategy is planned. It is however also useful as a preparation tool before stent placement since the immediate luminal gain is superior to the one of standard angioplasty (Fig. 16.11). This technology is particularly safe with very few descriptions of peripheral emboli and low-flow phenomena [45].

Anti-Restenotic Therapy

Vessel opening by itself does not provide durable results in the FP area. Additional local anti-restenotic therapy will increase primary patency and avoid target lesion revascularization. Bare-metal stents (BMS), drug-eluting stents, covered stents (CS) and drug-coated balloons are the main current anti-restenotic devices. The reader should note that few comparisons between the different concepts of anti-restenotic therapy have been performed. Our current algorithm regarding the anti-restenotic strategy is provided in Fig. 16.12.

Bare-Metal Stents

The oldest anti-restenotic therapy is bare-metal stent (BMS) placement. BMS has the goal to reline any intraoperative dissection and avoid restenosis by scaffolding the target lesion. Since the FP segment is highly mobile, BMS used must be able to follow arterial deformations and have a memory shape. Current BMSs for FP interventions are in

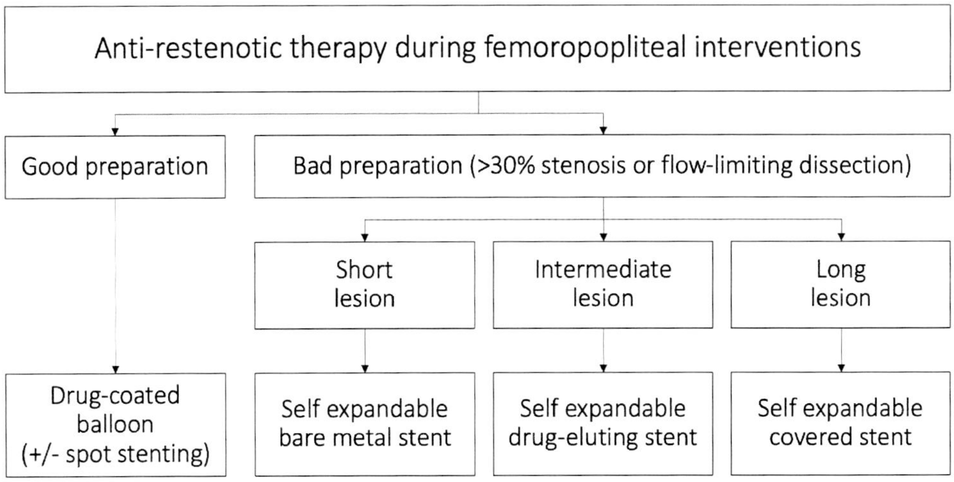

Fig. 16.12 Our current algorithm regarding anti-restenotic therapies during femoropopliteal interventions

nitinol (Nickel Titanium Naval Ordinance Laboratory), a hyperelastic alloy that has good resistance to fatigue and deformation with a low elasticity modulus (100 GPa). These stents are self-expandable and delivered through a retractable sheath. Most nitinol BMSs are made by laser cut of a fully metallic tube. The metallic mesh varies from one stent to another, and a distinction is generally made between open and closed-cell stents (cells < or > to 5 mm^2 after deployment). The last generation of nitinol stent has to be mentioned: vascular mimetic implants. The main one is currently the Supera stent (Abbott). Its specific interwoven design creates a high flexibility and resistance to external crush, with low chronic outward force [46].

Several prospective randomized studies have demonstrated the superiority of BMS over standard angioplasty [47, 48] in terms of primary patency and target lesion revascularization. Mid-term and late in-stent restenosis as a result of intimal hyperplasia (IH) remain the major issues with all current BMSs. This is particularly true when the stenting length increases [49]. The main problem with IH is that it is more challenging to treat than a primary stenosis made of atheroma. IH is less sensitive to angioplasty, and most efficient way to treat it is to debulk the in-stent material with a specific device. All devices described below should be perceived as evolutions of BMSs with the goal of limiting IH development.

Drug-Eluting Stents

Drug-eluting stents (DES) are nitinol stents covered of an anti-proliferative agent whose goal is to limit vascular smooth muscle cells and fibroblast proliferations to avoid IH. Nowadays, two main DES are available for FP interventions, both delivering local paclitaxel (PTX) in an amorphous form. The Zilver-PTX stent (Cook, Bloomington, IN) is directly embedded with PTX (3 μg/mm^2) without excipient. The Eluvia stent (Boston scientific) intends to provide long, low-dose PTX delivery (0.167 μg/mm^2) using a specific polymer for coating. Current evidence seems in favor of the Eluvia stent. The prospective randomized IMPERIAL trial [50, 51] demonstrated superior rates of PP at 12 months with the Eluvia stent compared with the Zilver PTX stent (86.8% vs. 81.5%; $p < 0.0001$). The EMINENT trial [52] demonstrated the superiority of the Eluvia stent over BMS in lesions comprised between 30 and 210 mm long (primary patency at 12 months of 83.2% vs. 74.3%; $p < 0.01$). In contrast, the BATTLE trial [53] could not show any benefit of the Zilver-PTX over BMS at 12 months in terms of in-stent restenosis rate.

Covered Stents

In covered stents (CS), the metallic mesh of the nitinol stent is covered with polytetrafluoroethylene (PTFE) to avoid IH between the stent struts. Therefore, CS avoid in-stent IH but not edge-stent stenosis related to undersizing, infolding, compliance mismatch or atheroma evolution. The main CS for FP intervention is the Viabahn (W.L. Gore) with heparin bounding. It combines the advantages of a high flexibility with a good resistance to external crush. Some Viabahn stents are now available over an 0.018′ platform and allow working in a 6 Fr sheath for 6 mm diameter stents. Sizing of the stent is paramount, and the Viper trial [54] has shown that moderate oversizing (≤20%) provides better patency rates than important oversizing (>20%). Data with CS are particularly interesting in lesions >200 mm in length. In the prospective randomized Viastar trial [55, 56], the 24-month patency rates in lesions >200 mm were 65.2% versus 26.7% (log rank $p = 0.004$), and freedom from TLR was 80.0% vs. 61.9% (log rank $p = 0.13$) in favor of the Viabahn stent versus BMS (Fig. 16.13). However, in our experience, we tend to use less CS since we noted that failure of CS present more

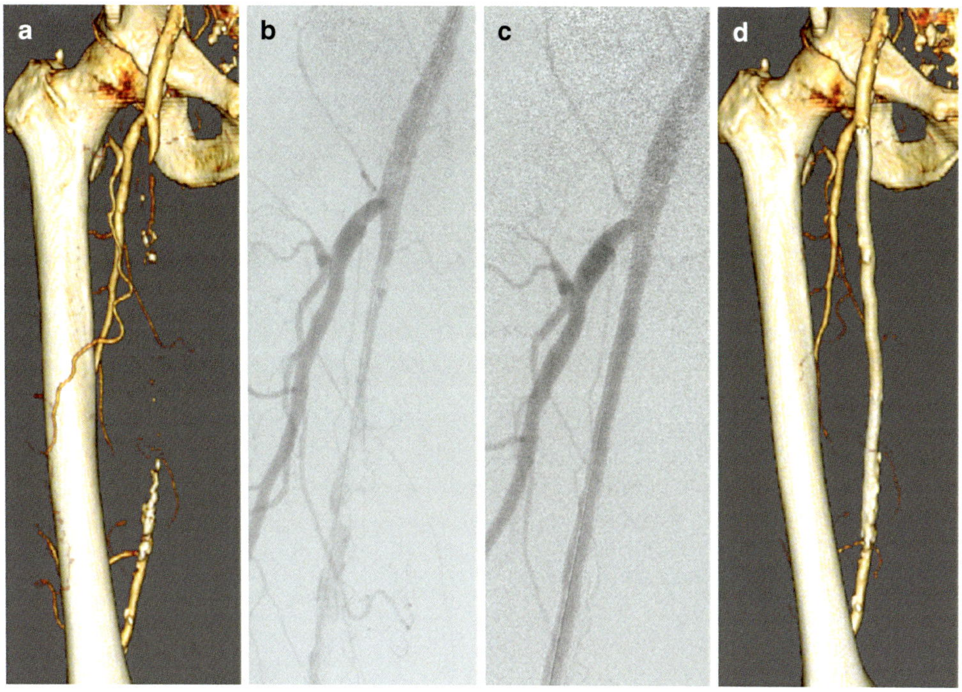

Fig. 16.13 In this long superficial femoral artery occlusion (**a**), the vessel preparation with a standard balloon does not provide a satisfactory result (**b**). A long covered stent (Viabahn; Gore) is implanted (**c**) and still patent at 3 years (**d**)

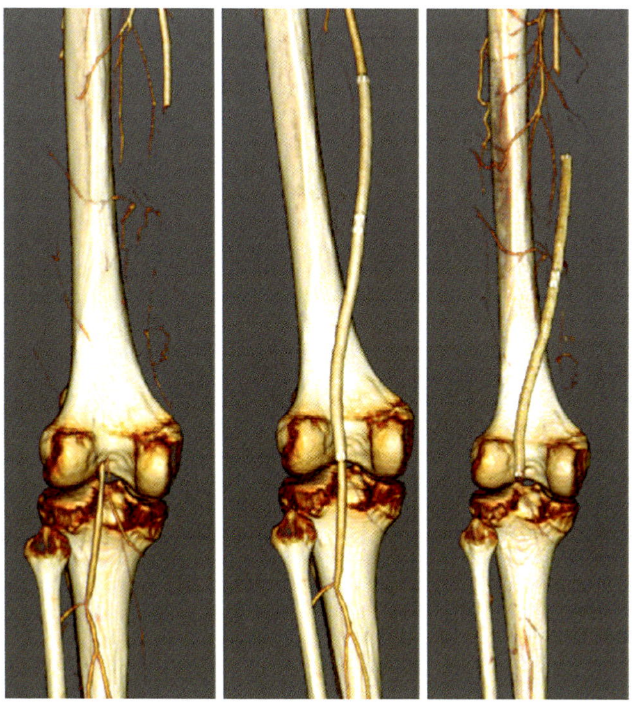

Fig. 16.14 A long covered stent (Viabahn; Gore) was implanted outside of the instructions for use, in a popliteal artery occlusion (**a** and **b**). It occluded several months after implantation with a more extensive thrombosis and acute presentation (**c**)

with acute extensive thrombosis and severe clinical presentations (Fig. 16.14), while BMS present with more progressive well-tolerated restenosis. This may be related to re-entry collaterals coverage with CS, but the precise mechanisms need to be investigated.

Drug-Coated Balloons

Drug-coated balloons (DCBs) represent an alternative anti-restenotic therapy without the implantation of a definitive scaffold. These balloons are not made for angioplasty and vessel opening. They work as a carrier that deposit the anti-restenotic drug in the vessel wall to avoid restenosis. An excipient is used to attach the drug to the balloon. Nearly all current DCBs use paclitaxel in a crystalline form, but some new-generation DCBs are now coated with limus. Doses of paclitaxel are different between the balloons, and it is usual to distinguish low-dose DCBs (2 μg/mm^2) from high-dose DCBs (>2 μg/mm^2). Before DCB use, perfect vessel preparation is mandatory. The goal is to have a lumen wide open (<30% residual stenosis) without flow-limiting dissection. After meticulous preparation, the DCB is brought and inflated at the level of the lesion. It is paramount to size the DCB to the vessel diameter to allow the drug to penetrate the vessel wall and avoid any geographic miss by using a DCB longer than the target lesion. The transit time between the sheath valve and the target lesion should be as short as possible to avoid drug dispersion in the bloodstream. The DCB is then inflated for a prolonged time (generally 3 min) to depose the drug. Many prospective randomized trials have now demonstrated that DCBs are superior to uncoated balloons for primary patency and target lesion revascularization when used at the level of the FP segment at 12 [57–60] and 24 [61, 62] months. Data at 5 years have even been published with the In.Pact (Medtronic) balloon [63], showing sustained benefit with such devices, although the effect is progressively lost in the long term. It should be underlined that when

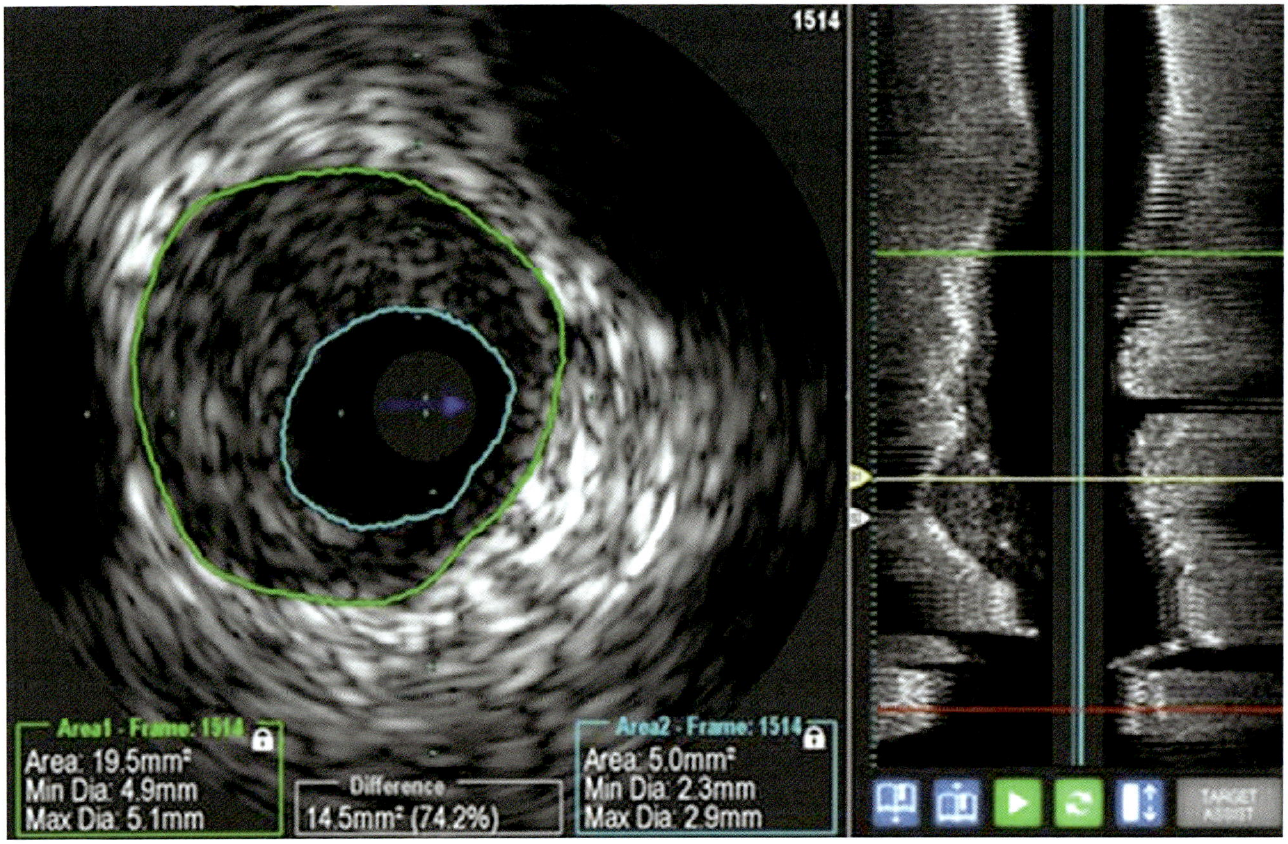

Fig. 16.15 Intraoperative view with an IVUS probe. It can be coupled with duplex to allow a better understanding of the situation

treating long lesions with a DCB strategy, bail-out stenting, generally using short stents, is frequent in up to 40% of the patients in lesions with a length > 150 mm [64]. Micro-emboli of crystalline PTX could be a concern [65], but its clinical relevance is debatable.

Intraoperative Imaging

Intraoperative imaging is nowadays more perceived as a key aspect of FP interventions. Many operators still rely on intraoperative fluoroscopic/angiographic parameters to estimate the path taken by the guidewire (intraluminal or subintimal), assess the result of the vessel preparation, take the decision to stent or not to stent, and conclude on the final result of the procedure. Several recent tools have emerged to improve the value of intraoperative imaging. To better understand the path taken by the wire, fusion imaging may be helpful [66]. For vessel preparation and assess the final result of the procedure, angiographic assessment is based on the aspect of the treated segment (dissected or not), the estimation of residual stenosis (>30% being considered satisfactory) and the subjective flow washing. Our practice is to use at least two incidences and to assess the final aspect of the repair without the wire in place. When there is a risk of early recoil, several operators have the use to perform an angiography 5–15 min after the completion angiography. However, a major advance in intraoperative imaging is probably the advent of intravascular ultrasound (IVUS, Philips). The IVUS probe is advanced over a 0.014″ wire and can be coupled with duplex assessment (Fig. 16.15). Recent American Medicare data have shown that IVUS use during peripheral arterial intervention was associated with a lower risk of major adverse limb events during follow-up [67]. Optical coherence tomography (OCT) may represent an alternative in the future [68], but this needs to be proved by more robust data.

Access Closure

The final aspect of the procedure is access closure. It can be done using manual compression (MC) or the insertion of a vascular closure device (VCD). Going into the technical details of current VCD is outside the scope of this chapter; we here just focus on general concerns. Same-day discharge procedures have highlighted the need for fast ambulation and decreased access bleeding complications. It is our practice to do a low-dose protamine administration (half the heparin dose) to decrease local bleeding [69]. When working with a standard 4 or a 5 Fr sheath, a 12–15 min MC is probably sufficient, and VCD appear unnecessary. When working with a 6 Fr sheath or more, VCD is probably useful to decrease the time of compression and allow faster ambulation [70]. A controversy remains about the superiority of VCD over MC for FP interventions [71, 72], but recent data tend to show that a VCD could decrease complications [73].

Conclusion

The toolbox for femoropopliteal interventions is now large and allows to obtain technical success in nearly all cases. A step-by-step planning and approach helps to standardize these procedures and clarify the role of the different devices [49].

Case Presentation

The following technical approach was planned: (1) access the arterial system through the left contralateral common femoral artery due to disease at the origin of the superficial femoral artery; also plan a bail-out retrograde access through the dorsalis pedal artery in case of intraluminal crossing failure (lesion classified CTOP type 2); (2) perform a crossover and bring a 7Fr 45 cm long sheath down to the right common femoral artery to recanalize the femoropopliteal occlusion intraluminal; (3) cross the occlusion with a dedicated 0.018″ or 0.014" CTO wire; (4) prepare the vessel using a debulking device (rotational atherectomy) if crossing is clearly intraluminal; (5) use drug-coated balloons with bail-out spot stenting in case of localized dissection. In case of extensive dissection, plan B would be to use a self-expandable covered stent for the superficial artery segment and an interwoven self-expandable stent for the popliteal segment, and (6) close the puncture point with a vascular closure device since the sheath is quite large (>6 Fr).Intraoperatively, steps 1 and 2 went easily. Under potentialized local anesthesia, a crossover was performed through left contralateral common femoral access. A 7Fr 45 cm sheath was brought down to the right common femoral artery. The initial angiogram (Fig. 16.16a, b) confirmed the long FP occlusion. Due to impossible anterograde crossing, a retrograde access was performed through the dorsalis pedal artery, and a V18 wire (Boston) over a CXI support catheter (Cook) allowed intraluminal recanalization (Fig. 16.16c). The recanalization went so easily that it was decided not to use atherectomy and just perform a prolonged angioplasty (5 min) with a long 5 mm Armada-35 (Abbott) balloon (Fig. 16.16d, e). The result of vessel preparation looked fine in the superficial femoral artery part (Fig. 16.17a) but not in the popliteal part (Fig. 16.17b), and another prolonged angioplasty with a larger balloon (6 mm in diameter) was decided (Fig. 16.17c). Completion angiography with two incidences demonstrated

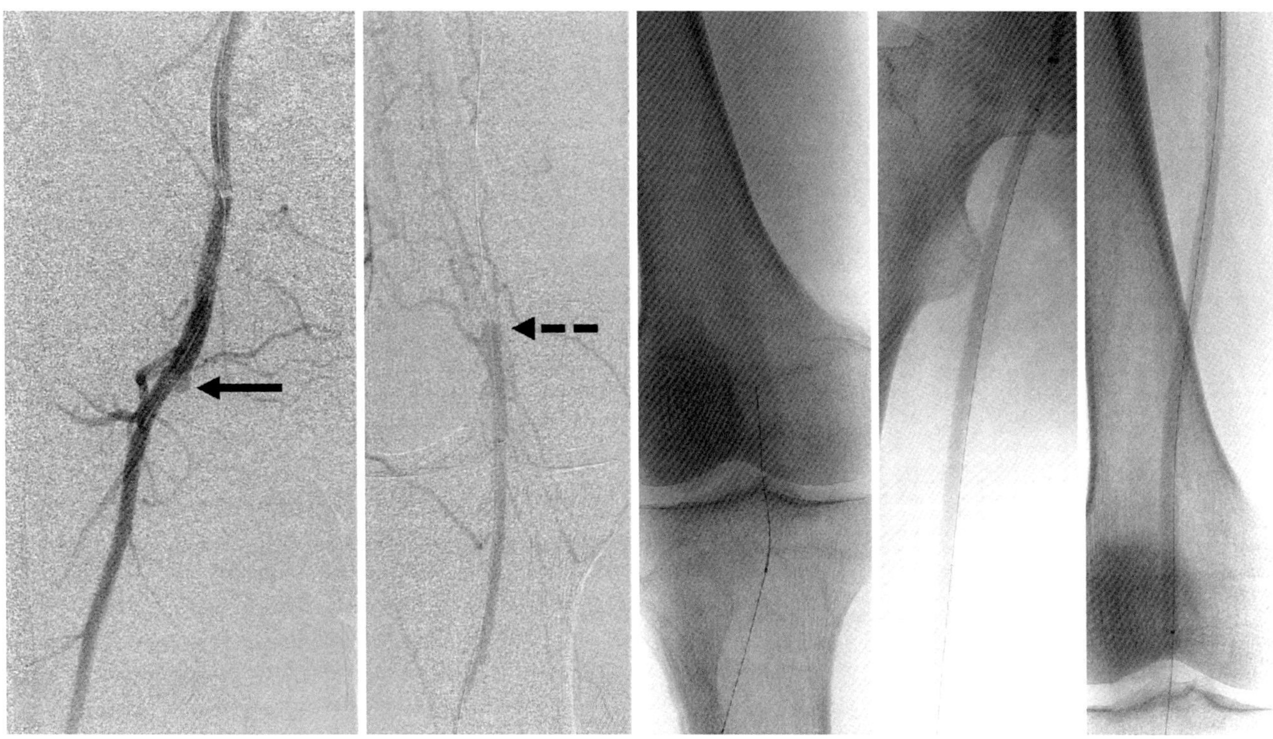

Fig. 16.16 Case resolution. The initial angiogram (Fig. 16.15a, b confirmed the long FP occlusion. Due to impossible anterograde crossing, a retrograde access was performed through the dorsalis pedal artery and a V18 wire (Boston) over a CXI support catheter (Cook) allowed intraluminal recanalization (Fig. 16.15c). The recanalization went so easily that it was decided not to use atherectomy and just perform a prolonged angioplasty (5 min) with a long 5 mm Armada-35 (Abbott) balloon (Fig. 16.15d, e)

a good result of the preparation without dissection or residual stenosis (Fig. 16.18), so 6 mm drug-coated balloons (In. Pact, Medtronic) were inflated for 3 min to avoid restenosis. No stent was necessary. A vascular closure device (Prostyle, Abbott) was used, and the patient left the hospital the same day with a pedal pulse.

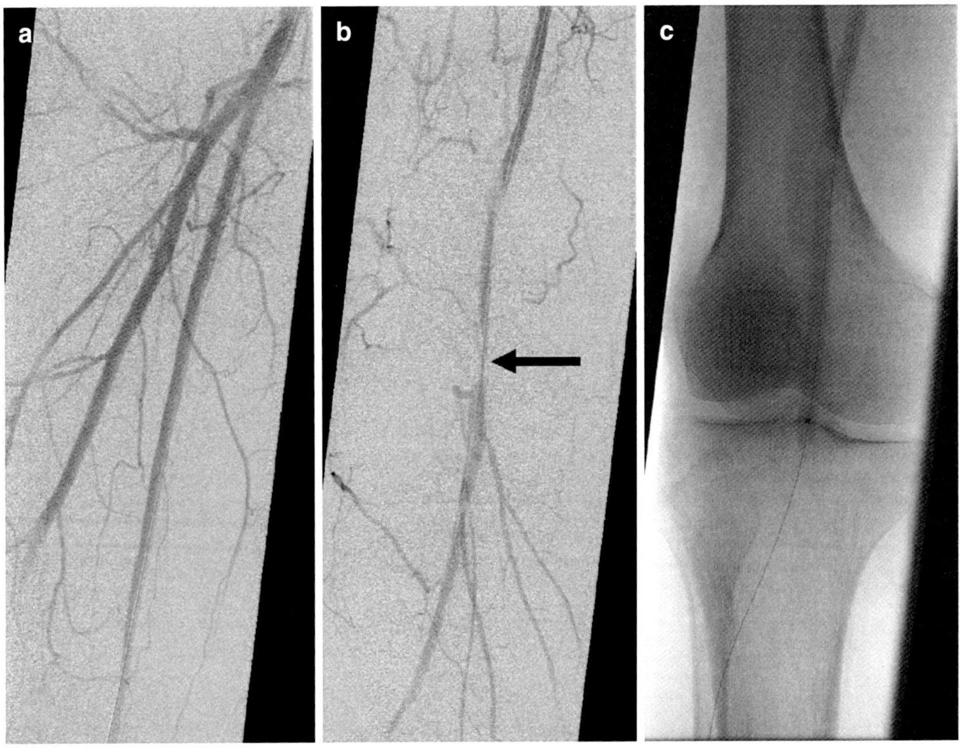

Fig. 16.17 Case resolution. The result of vessel preparation looked fine in the superficial femoral artery part (Fig. 16.16a) but not in the popliteal part (Fig. 16.16), and another prolonged angioplasty with a larger balloon (6 mm in diameter) was decided (Fig. 16.16c)

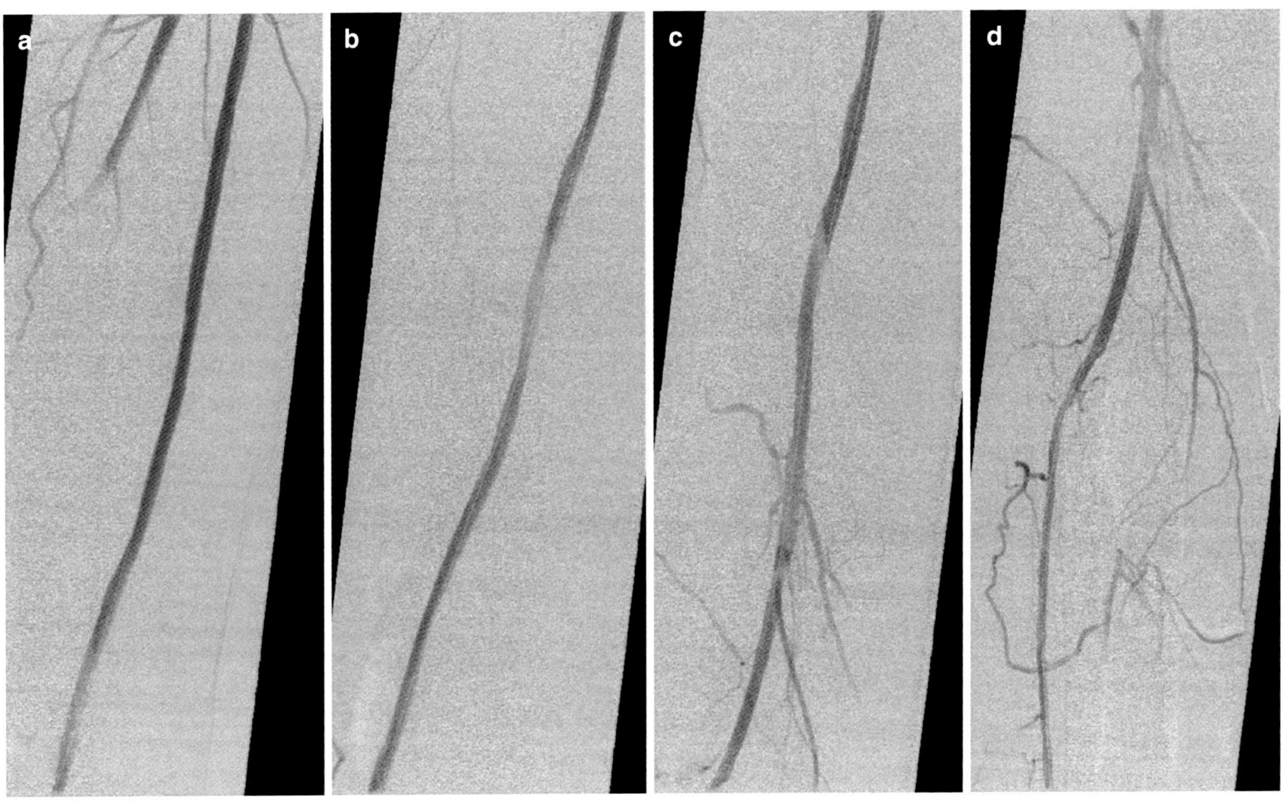

Fig. 16.18 Case resolution. Completion angiography with two incidences (**a**–**d**) demonstrated a good result of the preparation without dissection or residual stenosis

References

1. Aboyans V, Ricco JB, Bartelink MEL, Bjorck M, Brodmann M, Cohnert T, et al. Editor's choice - 2017 ESC guidelines on the diagnosis and treatment of peripheral arterial diseases, in collaboration with the European Society for Vascular Surgery (ESVS). Eur J Vasc Endovasc Surg. 2018;55(3):305–68.
2. Farber A, Menard MT, Conte MS, Kaufman JA, Powell RJ, Choudhry NK, et al. Surgery or endovascular therapy for chronic limb-threatening ischemia. N Engl J Med. 2022;387(25):2305–16.
3. Biagioni LC, Pereira L, Nasser F, Biagioni RB, Burihan MC, Wolosker N. Comparison between antegrade common femoral artery access and superficial femoral artery access in infrainguinal endovascular interventions. J Vasc Surg. 2021;74(3):763–70.
4. Nasr B, Hauguel A, Yven C, Didier R, Gouny P, Goueffic Y. Antegrade superficial femoral artery approach using manual compression and 4-F delivery system for Infrainguinal occlusive disease. Ann Vasc Surg. 2022;78:1–8.
5. Akopian G, Katz SG. Peripheral angioplasty with same-day discharge in patients with intermittent claudication. J Vasc Surg. 2006;44(1):115–8.
6. Gouicem D, Palcau L, Le Hello C, Cameliere L, Dufranc J, Coffin O, et al. Feasibility of ambulatory percutaneous femoral access without the use of arterial closure systems. Ann Vasc Surg. 2014;28(1):132–6.
7. Engelbert TL, Scholten A, Thompson K, Spivack A, Kansal N. Early ambulation after percutaneous femoral access with use of closure devices and hemostatic agents. Ann Vasc Surg. 2010;24(4):518–23.
8. Kasthuri R, Karunaratne D, Andrew H, Sumner J, Chalmers N. Day-case peripheral angioplasty using nurse-led admission, discharge, and follow-up procedures: arterial closure devices are not necessary. Clin Radiol. 2007;62(12):1202–5.
9. Kruse JR, Cragg AH. Safety of short stay observation after peripheral vascular intervention. J Vasc Interv Radiol. 2000;11(1):45–9.
10. Soulier-Parmeggiani L, Schneider PA, Bounameaux H. Outpatient percutaneous transluminal angioplasty for peripheral arterial disease. Eur J Med. 1992;1(1):13–5.
11. Alvarez-Tostado JA, Moise MA, Bena JF, Pavkov ML, Greenberg RK, Clair DG, et al. The brachial artery: a critical access for endovascular procedures. J Vasc Surg. 2009;49(2):378–85. discussion 85.
12. Criado FJ, Wilson EP, Abul-Khoudoud O, Barker C, Carpenter J, Fairman R. Brachial artery catheterization to facilitate endovascular grafting of abdominal aortic aneurysm: safety and rationale. J Vasc Surg. 2000;32(6):1137–41.
13. Grollman JH Jr, Marcus R. Transbrachial arteriography: techniques and complications. Cardiovasc Intervent Radiol. 1988;11(1):32–5.
14. Watkinson AF, Hartnell GG. Complications of direct brachial artery puncture for arteriography: a comparison of techniques. Clin Radiol. 1991;44(3):189–91.
15. Heenan SD, Grubnic S, Buckenham TM, Belli AM. Transbrachial arteriography: indications and complications. Clin Radiol. 1996;51(3):205–9.
16. Nasr B, Carret M, Pluchon K, Yven C, Bezon E, Goueffic Y. Perioperative adverse events in percutaneous versus open brachial access. J Vasc Surg. 2023;77(3):864–9.
17. Baklanov DV, Kaltenbach LA, Marso SP, Subherwal SS, Feldman DN, Garratt KN, et al. The prevalence and outcomes of transradial percutaneous coronary intervention for ST-segment elevation myocardial infarction: analysis from the National Cardiovascular Data Registry (2007 to 2011). J Am Coll Cardiol. 2013;61(4):420–6.
18. Feldman DN, Swaminathan RV, Kaltenbach LA, Baklanov DV, Kim LK, Wong SC, et al. Adoption of radial access and comparison of outcomes to femoral access in percutaneous coronary intervention: an updated report from the national cardiovascular data registry (2007-2012). Circulation. 2013;127(23):2295–306.
19. Karrowni W, Vyas A, Giacomino B, Schweizer M, Blevins A, Girotra S, et al. Radial versus femoral access for primary percutaneous interventions in ST-segment elevation myocardial infarction patients: a meta-analysis of randomized controlled trials. JACC Cardiovasc Interv. 2013;6(8):814–23.
20. Bernat I, Horak D, Stasek J, Mates M, Pesek J, Ostadal P, et al. ST-segment elevation myocardial infarction treated by radial or femoral approach in a multicenter randomized clinical trial: the STEMI-RADIAL trial. J Am Coll Cardiol. 2014;63(10):964–72.
21. Romagnoli E, Biondi-Zoccai G, Sciahbasi A, Politi L, Rigattieri S, Pendenza G, et al. Radial versus femoral randomized investigation in ST-segment elevation acute coronary syndrome: the RIFLE-STEACS (radial versus femoral randomized investigation in ST-elevation acute coronary syndrome) study. J Am Coll Cardiol. 2012;60(24):2481–9.
22. Jolly SS, Yusuf S, Cairns J, Niemela K, Xavier D, Widimsky P, et al. Radial versus femoral access for coronary angiography and intervention in patients with acute coronary syndromes (RIVAL): a randomised, parallel group, multicentre trial. Lancet. 2011;377(9775):1409–20.
23. Ruzsa Z, Csavajda A, Nemes B, Deak M, Sotonyi P, Bertrand OF, et al. Distal radial artery access for superficial femoral artery interventions. J Endovasc Ther. 2021;28(2):255–61.
24. Mustapha JA, Saab F, McGoff T, Heaney C, Diaz-Sandoval L, Sevensma M, et al. Tibio-pedal arterial minimally invasive retrograde revascularization in patients with advanced peripheral vascular disease: the TAMI technique, original case series. Catheter Cardiovasc Interv. 2014;83(6):987–94.
25. Kawarada O, Yokoi Y, Takemoto K. Practical use of duplex echo-guided recanalization of chronic total occlusion in the iliac artery. J Vasc Surg. 2010;52(2):475–8.
26. Gur S, Oguzkurt L, Gurel K, Tekbas G, Onder H. US-guided retrograde tibial artery puncture for recanalization of complex infrainguinal arterial occlusions. Diagn Interv Radiol. 2013;19(2):134–40.
27. Fanelli F, Cannavale A. Retrograde recanalization of complex SFA lesions indications and techniques. J Cardiovasc Surg. 2014;55(4):465–71.
28. Saab F, Jaff MR, Diaz-Sandoval LJ, Engen GD, McGoff TN, Adams G, et al. Chronic Total occlusion crossing approach based on plaque cap morphology: the CTOP classification. J Endovasc Ther. 2018;25(3):284–91.
29. Shah SM, Bortnick A, Bertrand OF, Costerousse O, Htun WW, Kwan TW. Transpedal vs. femoral access for peripheral arterial interventions-a single center experience. Catheter Cardiovasc Interv. 2019;93(7):1311–4.
30. Lai SH, Fenlon J, Roush BB, Munn J, Rummel M, Johnston D, et al. Analysis of the retrograde tibial artery approach in lower extremity revascularization in an office endovascular center. J Vasc Surg. 2019;70(1):157–65.
31. Bausback Y, Botsios S, Flux J, Werner M, Schuster J, Aithal J, et al. Outback catheter for femoropopliteal occlusions: immediate and long-term results. J Endovasc Ther. 2011;18(1):13–21.
32. Sheikh AB, Llanos-Chea F, Jelani QU, Anantha-Narayanan M, Attaran R, Schneider M, et al. Safety and efficacy outcomes of the Pioneer plus catheter in endovascular revascularization of lower extremity chronic total occlusions. J Vasc Surg. 2021;74(3):746–55.
33. Touma J, Senemaud J, Jaziri A, Cochennec F, Desgranges P. Percutaneous trans-venous Femoropopliteal bypass in long occlusions of the superficial femoral artery. Cardiovasc Intervent Radiol. 2019;42(12):1800–5.
34. Krievins DK, Halena G, Scheinert D, Savlovskis J, Szopinski P, Kramer A, et al. One-year results from the DETOUR I trial of the PQ bypass DETOUR system for percutaneous femoropopliteal bypass. J Vasc Surg. 2020;72(5):1648–58 e2.

35. Halena G, Krievins DK, Scheinert D, Savlovskis J, Szopinski P, Kramer A, et al. Percutaneous Femoropopliteal bypass: 2-year results of the DETOUR system. J Endovasc Ther. 2022;29(1):84–95.
36. Di Primio M, Angelopoulos G, Lazareth I, Lin F, Petit A, Priollet P, et al. Endovascular extra-anatomic Femoro-popliteal bypass for limb salvage in chronic critical limb ischemia. Cardiovasc Intervent Radiol. 2019;42(9):1279–92.
37. Dubosq M, Jayet J, Coscas R. In situ conversion of an occluded below-the-knee bypass into a patent above-the-knee Endobypass. Eur J Vasc Endovasc Surg. 2022;64(4):436.
38. Horie K, Tanaka A, Taguri M, Kato S, Inoue N. Impact of prolonged inflation times during plain balloon angioplasty on angiographic dissection in Femoropopliteal lesions. J Endovasc Ther. 2018;25(6):683–91.
39. Zorger N, Manke C, Lenhart M, Finkenzeller T, Djavidani B, Feuerbach S, et al. Peripheral arterial balloon angioplasty: effect of short versus long balloon inflation times on the morphologic results. J Vasc Interv Radiol. 2002;13(4):355–9.
40. Lugenbiel I, Grebner M, Zhou Q, Strothmeyer A, Vogel B, Cebola R, et al. Treatment of femoropopliteal lesions with the AngioSculpt scoring balloon - results from the Heidelberg PANTHER registry. Vasa. 2018;47(1):49–55.
41. Wu Z, Huang Q, Pu H, Qin J, Wang X, Ye K, et al. Atherectomy combined with balloon angioplasty versus balloon angioplasty alone for de novo Femoropopliteal arterial diseases: a systematic review and meta-analysis of randomised controlled trials. Eur J Vasc Endovasc Surg. 2021;62(1):65–73.
42. Dippel EJ, Makam P, Kovach R, George JC, Patlola R, Metzger DC, et al. Randomized controlled study of excimer laser atherectomy for treatment of femoropopliteal in-stent restenosis: initial results from the EXCITE ISR trial (EXCImer laser randomized controlled study for treatment of FemoropopliTEal in-stent restenosis). JACC Cardiovasc Interv. 2015;8(1 Pt A):92–101.
43. Tepe G, Brodmann M, Werner M, Bachinsky W, Holden A, Zeller T, et al. Intravascular lithotripsy for peripheral artery calcification: 30-day outcomes from the randomized disrupt PAD III trial. JACC Cardiovasc Interv. 2021;14(12):1352–61.
44. Wong CP, Chan LP, Au DM, Chan HWC, Chan YC. Efficacy and safety of intravascular lithotripsy in lower extremity peripheral artery disease: a systematic review and meta-analysis. Eur J Vasc Endovasc Surg. 2022;63(3):446–56.
45. Kereiakes DJ, Virmani R, Hokama JY, Illindala U, Mena-Hurtado C, Holden A, et al. Principles of intravascular lithotripsy for calcific plaque modification. JACC Cardiovasc Interv. 2021;14(12):1275–92.
46. Garcia LA, Rosenfield KR, Metzger CD, Zidar F, Pershad A, Popma JJ, et al. SUPERB final 3-year outcomes using interwoven nitinol biomimetic supera stent. Catheter Cardiovasc Interv. 2017;89(7):1259–67.
47. Schillinger M, Sabeti S, Loewe C, Dick P, Amighi J, Mlekusch W, et al. Balloon angioplasty versus implantation of nitinol stents in the superficial femoral artery. N Engl J Med. 2006;354(18):1879–88.
48. Laird JR, Katzen BT, Scheinert D, Lammer J, Carpenter J, Buchbinder M, et al. Nitinol stent implantation versus balloon angioplasty for lesions in the superficial femoral artery and proximal popliteal artery: twelve-month results from the RESILIENT randomized trial. Circ Cardiovasc Interv. 2010;3(3):267–76.
49. Labed P, Gonzalez F, Jayet J, Javerliat I, Coggia M, Coscas R. Endovascular treatment of long Femoropopliteal lesions with contiguous bare metal stents. Ann Vasc Surg. 2021;76:276–84.
50. Muller-Hulsbeck S, Benko A, Soga Y, Fujihara M, Iida O, Babaev A, et al. Two-year efficacy and safety results from the IMPERIAL randomized study of the eluvia polymer-coated drug-eluting stent and the Zilver PTX polymer-free drug-coated stent. Cardiovasc Intervent Radiol. 2021;44(3):368–75.
51. Gray WA, Keirse K, Soga Y, Benko A, Babaev A, Yokoi Y, et al. A polymer-coated, paclitaxel-eluting stent (eluvia) versus a polymer-free, paclitaxel-coated stent (Zilver PTX) for endovascular femoropopliteal intervention (IMPERIAL): a randomised, non-inferiority trial. Lancet. 2018;392(10157):1541–51.
52. Goueffic Y, Torsello G, Zeller T, Esposito G, Vermassen F, Hausegger KA, et al. Efficacy of a drug-eluting stent versus bare metal stents for symptomatic Femoropopliteal peripheral artery disease: primary results of the EMINENT randomized trial. Circulation. 2022;146(21):1564–76.
53. Goueffic Y, Sauguet A, Desgranges P, Feugier P, Rosset E, Ducasse E, et al. A polymer-free paclitaxel-eluting stent versus a bare-metal stent for De novo Femoropopliteal lesions: the BATTLE trial. JACC Cardiovasc Interv. 2020;13(4):447–57.
54. Saxon RR, Chervu A, Jones PA, Bajwa TK, Gable DR, Soukas PA, et al. Heparin-bonded, expanded polytetrafluoroethylene-lined stent graft in the treatment of femoropopliteal artery disease: 1-year results of the VIPER (Viabahn Endoprosthesis with heparin bioactive surface in the treatment of superficial femoral artery obstructive disease) trial. J Vasc Interv Radiol. 2013;24(2):165–73. quiz 74
55. Lammer J, Zeller T, Hausegger KA, Schaefer PJ, Gschwendtner M, Mueller-Huelsbeck S, et al. Sustained benefit at 2 years for covered stents versus bare-metal stents in long SFA lesions: the VIASTAR trial. Cardiovasc Intervent Radiol. 2015;38(1):25–32.
56. Lammer J, Zeller T, Hausegger KA, Schaefer PJ, Gschwendtner M, Mueller-Huelsbeck S, et al. Heparin-bonded covered stents versus bare-metal stents for complex femoropopliteal artery lesions: the randomized VIASTAR trial (Viabahn endoprosthesis with PROPATEN bioactive surface [VIA] versus bare nitinol stent in the treatment of long lesions in superficial femoral artery occlusive disease). J Am Coll Cardiol. 2013;62(15):1320–7.
57. Tepe G, Laird J, Schneider P, Brodmann M, Krishnan P, Micari A, et al. Drug-coated balloon versus standard percutaneous transluminal angioplasty for the treatment of superficial femoral and popliteal peripheral artery disease: 12-month results from the IN.PACT SFA randomized trial. Circulation. 2015;131(5):495–502.
58. Rosenfield K, Jaff MR, White CJ, Rocha-Singh K, Mena-Hurtado C, Metzger DC, et al. Trial of a paclitaxel-coated balloon for Femoropopliteal artery disease. N Engl J Med. 2015;373(2):145–53.
59. Krishnan P, Faries P, Niazi K, Jain A, Sachar R, Bachinsky WB, et al. Stellarex drug-coated balloon for treatment of Femoropopliteal disease: twelve-month outcomes from the randomized ILLUMENATE pivotal and pharmacokinetic studies. Circulation. 2017;136(12):1102–13.
60. Schroeder H, Werner M, Meyer DR, Reimer P, Kruger K, Jaff MR, et al. Low-dose paclitaxel-coated versus uncoated percutaneous transluminal balloon angioplasty for Femoropopliteal peripheral artery disease: one-year results of the ILLUMENATE European randomized clinical trial (randomized trial of a novel paclitaxel-coated percutaneous angioplasty balloon). Circulation. 2017;135(23):2227–36.
61. Laird JR, Schneider PA, Tepe G, Brodmann M, Zeller T, Metzger C, et al. Durability of treatment effect using a drug-coated balloon for Femoropopliteal lesions: 24-month results of IN.PACT SFA. J Am Coll Cardiol. 2015;66(21):2329–38.
62. Brodmann M, Werner M, Meyer DR, Reimer P, Kruger K, Granada JF, et al. Sustainable Antirestenosis effect with a low-dose drug-coated balloon: the ILLUMENATE European randomized clinical trial 2-year results. JACC Cardiovasc Interv. 2018;11(23):2357–64.
63. Laird JA, Schneider PA, Jaff MR, Brodmann M, Zeller T, Metzger DC, et al. Long-term clinical effectiveness of a drug-coated balloon for the treatment of Femoropopliteal lesions. Circ Cardiovasc Interv. 2019;12(6):e007702.
64. Scheinert D, Micari A, Brodmann M, Tepe G, Peeters P, Jaff MR, et al. Drug-coated balloon treatment for Femoropopliteal artery disease. Circ Cardiovasc Interv. 2018;11(10):e005654.

65. Boitet A, Grassin-Delyle S, Louedec L, Dupont S, Lamy E, Coggia M, et al. An experimental study of paclitaxel embolisation during drug coated balloon angioplasty. Eur J Vasc Endovasc Surg. 2019;57(4):578–86.
66. Mougin J, Louis N, Maupas E, Goueffic Y, Fabre D, Haulon S. Fusion imaging guidance for endovascular recanalization of peripheral occlusive disease. J Vasc Surg. 2022;75(2):610–7.
67. Divakaran S, Parikh SA, Hawkins BM, Chen S, Song Y, Banerjee S, et al. Temporal trends, practice variation, and associated outcomes with IVUS use during peripheral arterial intervention. JACC Cardiovasc Interv. 2022;15(20):2080–90.
68. Dubosq M, Patterson BO, Azzaoui R, Mesnard T, De Preville A, Sobocinski J. Protocol adaptation of optical coherence tomography in lower limb arteries revascularization. Ann Vasc Surg. 2019;57:257–60.
69. Hanna N, Fiorilli P, Gaglia MA Jr, Torguson R, Vita A, Ben-Dor I, et al. Low-dose protamine to facilitate earlier sheath removal from the femoral artery after peripheral endovascular intervention. J Interv Cardiol. 2011;24(3):278–84.
70. Noori VJ, Eldrup-Jorgensen J. A systematic review of vascular closure devices for femoral artery puncture sites. J Vasc Surg. 2018;68(3):887–99.
71. Cox T, Blair L, Huntington C, Lincourt A, Sing R, Heniford BT. Systematic review of randomized controlled trials comparing manual compression to vascular closure devices for diagnostic and therapeutic arterial procedures. Surg Technol Int. 2015;27:32–44.
72. Robertson L, Andras A, Colgan F, Jackson R. Vascular closure devices for femoral arterial puncture site haemostasis. Cochrane Database Syst Rev. 2016;3:CD009541.
73. Cheng TW, Farber A, King EG, Levin SR, Arinze N, Malas MB, et al. Access site complications are uncommon with vascular closure devices or manual compression after lower extremity revascularization. J Vasc Surg. 2022;76(3):788–96 e2.

Percutaneous Revascularization of the Tibial Arteries

17

Konstantinos Katsanos

Case Presentation

A 76-year-old male patient presented to the outpatient vascular clinic and complained of rest pain of the left foot, worsening during bed rest. He had a past medical history of drug-controlled hypertension and non-insulin dependent diabetes over the last 10 years and had undergone two coronary artery bypass grafts (CABG) 3 years ago because of significant left main coronary artery disease. He was a past smoker and was on a prescription of Clopidogrel 75 mg once daily (o.d.) and a statin 20 mg once daily. He had had a vascular diagnosis of <150 meters intermittent claudication on the left side over the last year and had been instructed to exercise regularly. However, his symptoms deteriorated over the last month and a diagnosis of critical limb ischemia (CLI) of the left foot was made based also on a positive Buerger's test and pulseless foot at the level of the ankle. A color-Doppler ultrasound was urgently performed that showed patent left iliac and femoropopliteal arteries with occasional <50% stenoses and a three-vessel occlusion below the left knee. Given the high Charlson Comorbidity Index and the absence of a suitable vein conduit, the patient was offered the option of endovascular revascularization of the left lower limb to relieve foot ischemia.

Continued at page 169

Background

Critical limb ischemia is one of the most challenging manifestations of systemic atherosclerosis with increasing worldwide prevalence alongside diabetes. CLI is associated with significant morbidity, mortality and healthcare costs because of high rates of recurrent ischemia, hospitalization episodes, and amputations [1]. Infrapopliteal arterial occlusive disease may present either in isolated fashion or more often combined with aortoiliac and/or femoropopliteal inflow disease and is the leading cause of critical limb ischemia (CLI) most often related to old age, diabetes and renal insufficiency [2]. In line revascularization to the foot remains the guideline-recommended treatment to achieve relief of ischemic symptoms and promote wound healing. Compared with femorodistal bypass surgery that may be technically challenging or not indicated because of underlying comorbidities or absence of a suitable vein conduit, balloon angioplasty and metal stenting of the tibial arteries have emerged as the primary mode of treatment for safe, timely and less invasive revascularization of the foot [3]. In the interest of prevention of limb loss, various complex techniques of recanalization of the small infrapopliteal and foot arteries have been developed and several dedicated balloons, stents and atherectomy devices have been engineered with promising results [4]. Most importantly, transcatheter balloon angioplasty may allow for recanalization of two or even three native arteries to the foot, often in conjunction with pedal arch angioplasty, and has been shown to achieve limb salvage rates that are equivalent to bypass surgery despite lower patency rates and more frequent re-interventions because of symptomatic restenosis [5].

Indications for Treatment

Well-accepted indications for endovascular revascularization of the tibial arteries include [1] Rest pain (Fontaine stage III, Rutherford category 4), or [2] non-healing ulcer/gangrene (Fontaine stage IV, Rutherford categories 5–6), or [3] a non-healing ischemic ulcer following amputation or surgical debridement of the foot with documented peripheral arterial occlusive disease on ultrasound or angiographic imaging [3]. Balloon angioplasty of the anastomotic sites is also indicated for the salvage of failing femorodistal bypass surgery based on declining surveillance metrics. The procedure may be contra-indicated in the case of a non-functional limb, limited life expectancy, wet gangrene or sepsis mandating amputa-

K. Katsanos (✉)
School of Medicine, Patras University Hospital, Rion, Greece
e-mail: katsanos@med.upatras.gr

G. Geroulakos et al. (eds.), *Mastering Endovascular Techniques*, https://doi.org/10.1007/978-3-031-42735-0_17

tion, absence of an identifiable distal foot vessel and uncorrectable coagulopathy. In case of infected gangrene, infrapopliteal revascularization is advocated shortly after surgical foot debridement or minor amputation to prevent a higher amputation level and promote wound healing. Overall, the treatment plan for infrapopliteal arterial disease is individualized and often decided in the context of a multidisciplinary team meeting considering patient needs, surgical risk, background comorbidities, anatomical complexity, wound aftercare and need for further reconstruction or prosthetic rehabilitation of the lower limb [1, 3, 4].

Preoperative Preparation

Timing of Intervention

CLI is typically characterized by a non-healing foot ulcer or rest pain for more than 2 weeks with an ankle pressure of less than 40 mmHg [6]. Timely revascularization of infrainguinal and infrapopliteal arteries is therefore essential to avoid superinfection, limit the extent of tissue loss and improve functional prognosis of the whole limb. In line with cerebrovascular disease, 'time is tissue' and hence, 'time to treatment' is directly associated with the risk of limb loss and time to wound healing [7]. It is advocated that endovascular reconstruction of the tibial arteries is performed in an urgent manner and generally within 2 weeks of admission to the hospital.

Baseline Imaging

CLI is usually associated with multi-vessel, multi-level disease and Duplex ultrasound is often inadequate for preoperative imaging of the tibial arteries that suffer from extensive calcifications because of dialysis or diabetes. Cross-sectional imaging of the peripheral arteries with non-invasive CT angiography (CTA) or time-resolved, contrast-enhanced MR-angiography (CE-MRA) is recommended to allow for mapping of the extent and anatomical complexity of the disease and help with decision making in terms of access site and target lesions/vessels [1, 3, 6]. CTA may be hindered from renal insufficiency and imaging conspicuity may be compromised from vessel wall calcifications. MRA may offer superior quality in assessing run-off arteries of the foot and the patency of the pedal arch. Non-contrast SSFP or subtractive TSE techniques are also recommended for patients with impaired renal function or on established hemodialysis to avoid CIN and eliminate the risk of nephrogenic systemic fibrosis. Invasive intra-arterial Carbon dioxide imaging is often used instead of iodinated contrast to reduce the risk of contrast-induced nephropathy during diagnostic or proper angioplasty procedures [8]. Planning and success of tibial revascularization is heavily dependent on the patency and quality of run-off arteries in the foot, as well as the patency of the pedal arch. Hence, selective intra-arterial angiography at the popliteal level may occasionally be useful to decide the most appropriate distal target in conjunction with the angiosome theory and the geography and extent of the wounds [9].

Patient Work-Up

Routine pre-interventional patient work-up includes a thorough past medical history and peripheral vascular examination including recording of ankle-brachial indices, presence of dorsal and posterior tibial artery pulses, and often recording of $TcPO_2$ or toe or skin pressures or other kind of perfusion monitoring as described later. Patients, who are often diabetic, need to be assessed for the extent of ischemia, presence of infection and autonomic neuropathy [6]. In addition, a thorough wound assessment in the presence of tissue loss and surgical consultation in case of wet gangrenous sites is recommended. The latest guidelines endorse the use of Wounds, Ischemia, and foot Infection (WIfI) system for staging of the severity of limb ischemia and prediction of the risk of limb loss [10]. A full blood count, renal function (eGFR), glucose, electrolytes, and coagulation profile are mandatory prior to procedure and anticoagulant therapy (coumadin or direct oral anticoagulants) need to be ceased and/or bridged per local hospital policy and international guidelines [11].

Best Medical Therapy

Most patients, unless urgently admitted or on a new CLI diagnosis, are usually prescribed on best medical therapy with a list of antihypertensive, beta-blocker, statin, and antiplatelet medications by the general practitioner or vascular physician/cardiologist/angiologist to reduce the overall risk of cardiovascular morbidity and mortality [1, 3]. Data on pre-procedural antiplatelet therapy are not concrete, but patients undergoing complex peripheral arterial procedures like small artery tibial recanalization are routinely pre-loaded with dual antiplatelet therapy 5 days prior to the procedure or receive a loading dose of 300–600 mg Clopidogrel the same day [3]. Antibiotic therapy should be continued in cases of infected wounds/ulcers or after debridement or documented gangrene, especially in the case of osteomyelitis. Emphasis is required on patients with poor renal function that are at risk of contrast-induced nephropathy and increased in-hospital morbidity. Generally, patients at risk of CIN (eGFR <30 mL/min/1.73 m^2) should receive intravenous saline

0.9% 1 mL/kg/h for several hours before and after contrast medium administration or intravenous sodium bicarbonate as an alternative, according to the ESUR guideline recommendations [12].

Endovascular Strategy and Technique

Basic Toolkit and Access

A standard endovascular kit for peripheral arterial balloon angioplasty and stenting is necessary including microcatheter devices, low profile balloons and drug-eluting stents dedicated for the small infrapopliteal arteries as outlined below. Antegrade ipsilateral common femoral artery access is recommended to improve negotiation of catheters and instruments and ultrasound guidance is essential for safe access, along with potential distal puncture sites for retrograde techniques. Routinely, a 6–7 Fr 0.035″ guidewire-compatible short vascular access sheath is introduced in the common femoral artery, and a 4 Fr 0.018″ guidewire-compatible short sheath is preferred should a second distal retrograde access is obtained. Intravenous sedation (with a combination of midazolam and opiates—fentanyl or pethidine) is occasionally required to control rest pain or painful debrided foot wounds. Unfractionated heparin is administered through the sheath (usually 3000–5000 IU or a weight-adjusted dosage) to maintain an activated clotting time of around 200–250 s [3].

Lesion Crossing and Guidewires

The Global Limb Anatomic Staging System (GLASS) has been proposed for the classification of anatomical complexity of the infrapopliteal arteries (stages I–III) in order to predict technical success and corresponding limb salvage rates [1]. Depending on the extent of stenoses or occlusions, the presence of intimal or medial calcifications and the condition of distal run-off foot arteries, straight or angled 0.014″ or 0.018″ guidewires with various tip properties may be used [3]. Hydrophilic-tipped guidewires are routinely used to negotiate diffuse stenoses, whereas dedicated guidewires with a variable tip weight and configuration are employed to cross chronic total occlusions (CTOs). Infrapopliteal CTOs are encountered in up to 50% of the CLI cases and may be challenging to treat. In case of tibial CTOs, the preferred technique of recanalization is intraluminal instead of subintimal dissection that often fails. Should subintimal tracking is attempted, care is needed not to inadvertently dissect the healthy target and extend the re-entry site too distally thereby compromising potential sites of bypass surgery or retrograde access. For the latter purposes, the guidewire tip may be looped (Bolia or knunckle subintimal technique) to help cross particularly fibrotic or fibrocalcific infrapopliteal lesions. Standard curved 4 Fr angiographic catheters, dedicated 0.018″ microcatheters or low-profile balloons catheters are recommended to support steering and advancement of the guidewires across the lesions. Rarely, distal re-entry devices may be employed [13].

Retrograde Access and Plantar-Loop Technique

In case of a failed antegrade attempt to cross a complex infrapopliteal CTO, alternatively a retrograde access or the plantar-loop technique may be employed. A 21 G micropuncture set or dedicated pedal access kit may be used to access the distal posterior tibial or dorsalis pedis artery just beneath the level of the medial or lateral malleolus respectively [14]. The distal peroneal artery is accessed above the ankle level. For both cases, road-map fluoroscopy or ultrasound guidance is used, or a combination thereof. Dedicated 0.018″ vascular sheaths are inserted for transpedal (like transradial) access. In case of the plantar-loop technique, 0.014″ hydrophilic-tipped guidewires are forwarded from the anterior to the posterior circulation of the foot (via the pedal arch) and vice versa to allow for retrograde crossing of hard infrapopliteal lesions.

Balloon Angioplasty

For the purposes of optimal mechanical effect and avoidance of excessive barotrauma, it is recommended that the proximal tibial arteries are dilated up to 3.5 mm, the middle third of the tibial arteries up to 3.0 mm, and the distal third at ankle level including the dorsalis pedis artery and proximal plantar arteries up to 2.5 mm [3]. The distal peroneal artery and pedal arch are usually treated with 2.0 mm balloon catheters. Low-profile 0.014″ balloon catheters with a transitionless tip and variable lengths (usually, 80–220 mm in length) are preferred with a 30-s to 2-min balloon inflation period to reduce risk of dissection. Kissing balloon angioplasty with 3.0–3.5 mm balloons is applied in case of bifurcation lesions of the origin of the anterior tibial artery or of the tibioperoneal bifurcation. The SAFARI balloon technique with undersized balloon diameters may be used in complex calcified lesions with combined antegrade and retrograde access if spontaneous true lumen re-entry fails. Liberal use of selective nitrates (300–500 mg) or papaverine (20–40 mg) is advised to reduce frequency and severity of vasospasm. In case of retrograde arterial punctures, hemostasis is accomplished by prolonged low-pressure balloon angioplasty across the puncture sites. Multivessel-revascularization is well established and associated with improved limb salvage and wound healing outcomes [15]; however, the decision to recanalize more than one tibial arteries is patient-specific and depends on a risk/benefit analysis of patient age, renal function, wound assessment and arch patency. The theory of angiosome-directed or wound-directed revascularisation supports the reduction of time to wound healing, which is important for limb salvage in patients with diabetes [16].

Drug-Coated Balloons

Numerous paclitaxel-coated balloons with different drug densities and formulations have been shown to reduce the risk of restenosis in the femoropopliteal artery. However, a risk of long-term all-cause mortality has been raised, as well as a risk of higher rates of major amputations when paclitaxel-coated balloons are applied in the infrainguinal arteries [17, 18]. Several RCTs have investigated the efficacy of paclitaxel-coated balloon technologies in the infrapopliteal arteries, but have broadly failed to show convincing evidence of improved patency or wound healing/limb salvage outcomes [4]. Contrary to cytotoxic paclitaxel, the delivery of cytostatic sirolimus has been proposed in parallel as an alternative drug-candidate for peripheral drug-coated balloons. Still, there is lack of randomized data to support the use of sirolimus-coated balloons in the infrapopliteal arteries. Hence, the application of drug-coated balloons in the infrapopliteal arteries remains controversial and for the time being should be limited to well-designed clinical research studies only.

Drug-Eluting Stents

In line with the paradigm shift of care in percutaneous coronary interventions, balloon-expandable drug-eluting stents (DES) have been introduced to inhibit vascular restenosis of the infrapopliteal arteries and are associated with significantly improved clinical outcomes [4]. After positive initial experience from several cohort studies, a few multicenter randomized clinical trials (RCTs) investigating infrapopliteal sirolimus- or everolimus-eluting stents have shown improved patency and consequently reduction of rates of re-interventions and major amputations with positive wound healing effects as well [4, 19–21]. However, olimus-DES are limited by their short length and applied only for bail-out or for definitive treatment of focal to intermediate-length proximal infrapopliteal arterial lesions. Polymer-compound absorbable everolimus-eluting stents have also been reported in the tibial arteries with promising results [22]. On the contrary, a recent randomized study of a longer self-expanding paclitaxel-eluting stent in the infrapopliteal arteries for CLI failed to meet its primary effectiveness and safety endpoints (Hans Over Hagen et al. oral presentation CIRSE 2022).

Debulking Atherectomy

Transcatheter atherectomy allows for debulking of peripheral chronic total occlusions, especially of those that are predominantly fibrotic or fibrocalcific and may improve acute procedural success by reducing angioplasty barotrauma that is inherently associated with the mechanical plaque disruption and displacement within the arterial wall during vessel dilatation [23]. Transcatheter atherectomy achieves atherosclerotic plaque clearance by means of directional plaque excision or rotational plaque removal or even laser plaque ablation. It has been shown that debulking atherectomy improves acute luminal gain and reduces the rates of plaque recoil or dissection mandating permanent metal stenting. It has also been argued that atherectomy may disrupt the calcium barrier and optimize drug transfer and delivery should drug-coated balloons are applied [24]. Nonetheless, there is still very limited evidence for the use of infrapopliteal atherectomy devices. Because of the small vessel size, the devices with rotational plaque excision properties are favored if deemed necessary. However, caution is warranted because of the presumed higher rates of distal embolization and complications [3, 25].

Deep Venous Arterialization

In case of absence of target run-off foot arteries, especially in cases with extensive calcific microangiopathy (aka non-option cases), deep venous arterialization (DVA) has emerged as a last resort procedure with promising initial results in patients with an otherwise dismal prognosis [26]. The DVA technique aims to create a percutaneous distal shunt between a tibial artery—most often the posterior tibial artery, and a deep venous channel—one of the two posterior tibial veins, respectively—to progressively arterialize a deep vein of the limb that will serve as an outflow conduit and provide adequate distal foot oxygenation while promoting microvascular arteriogenesis of the foot [27].

Perfusion Monitoring

Contrary to traditional contrast angiography that allows for real-time assessment of vascular anatomy and morphology and thereby guides peripheral endovascular procedures, several non-invasive modalities have been proposed for real-time functional evaluation of the microcirculatory hemodynamics and blood supply to the foot. Real-time quantification of foot perfusion during treatment is feasible with 2D-perfusion angiography or indocyanine green (ICG) fluorescence angiography. Alternatively, diffuse speckle contrast analysis and near-infrared spectroscopy (aka NIRS) have been proposed for probing of deep-tissue perfusion and oxygenation, respectively [3]. There is still limited evidence about the utility and benefit of the aforementioned non-invasive methods for functional assessment of the perfusion status of the foot, but intuitively they may help quantify incremental gains in blood supply following infrapopliteal angioplasty and potentially provide novel targets of therapy and predictors limb salvage [28].

Anticipated Complications

Most frequent complications (<10%) include access-site-related complications (vessel occlusion, puncture site hematomas, false aneurysms, and access site or retroperitoneal bleeding) and treatment-site related ones (spasm, thrombotic occlusion, and flow-limiting dissection). Application of femoral artery closure devices (e.g. Starclose, Angioseal, etc.) is highly recommended to reduce the risk of access complications and expedite patient ambulation and procedural throughput in the cathlab. Liberal use of selective nitroglycerin or papaverine is recommended to treat spasm and improve microcirculatory run-off. Rarely, uncontrolled vessel perforation or development of high-flow arteriovenous fistula following angioplasty barotrauma may mandate coil embolization or covered stent placement, respectively. The risk of procedure-related major amputation is 1–3%.

Patient Follow-Up

An "endovascular-first" approach is predominantly advocated for the infrapopliteal territory based on a lower procedural risk and shorter recovery time. Recovery time from endovascular procedures treating CLI is only 1–2 days and in-hospital morbidity, and mortality is significantly lower than bypass surgery. In addition, most endovascular procedures, including revascularization of the tibial arteries, may be conducted on an outpatient level [3, 6]. Patients are routinely prescribed on short-term dual antiplatelet therapy following revascularization of the tibial arteries, usually for a period of 6 weeks–6 months. Dual antiplatelet therapy, or a combination of low-dose rivaroxaban and aspirin have been both shown to reduce major amputations following peripheral procedures for limb salvage [29, 30]. In addition, patients are routinely followed up with regular outpatient appointments for ongoing assessment of wound healing. Adjunctive surgical debridement or minor amputations may be carried out shortly after a successful tibial angioplasty.

Case Presentation

Continued from page 165

Under local anesthetic and ultrasound guidance, antegrade access in the left femoral artery was obtained and a 6 Fr sheath was inserted and 5000 IU heparin were administered. A 4 Fr angled angiographic catheter was advanced to the popliteal level and selective angiography of the left calf and foot was performed. The angiogram identified a short-segment occlusion of the tibioperoneal trunk with reconstitution of the proximal peroneal artery and long-segment occlusion of the whole left anterior and posterior tibial artery. Distal perforating peroneal branches communicated to a poorly opacified dorsalis pedis artery providing the main run-off artery supplying the forefoot (Fig. 17.1). The decision was made on table to attempt an angioplasty of the tibioperoneal artery followed by the anterior tibial artery (ATA) to achieve in line flow to the foot.

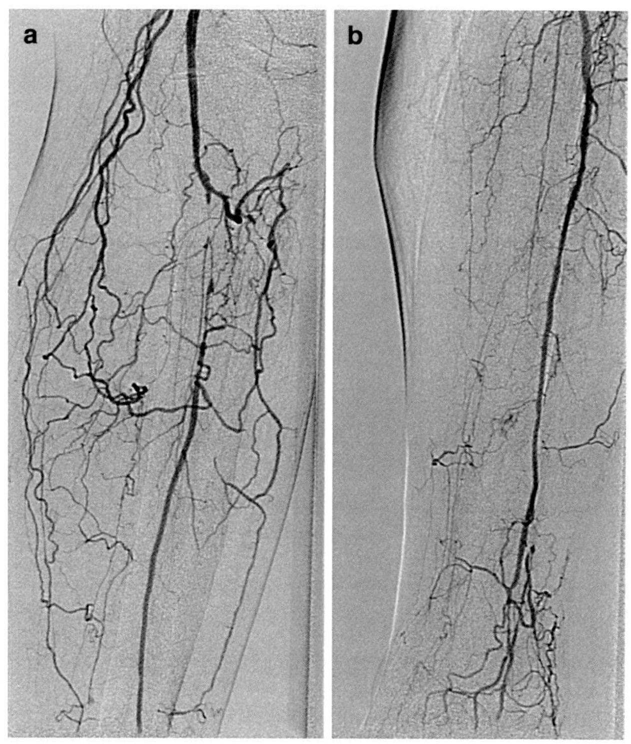

Fig. 17.1 Selective antegrade contrast angiography (**a**) of the proximal tibial arteries demonstrating a tibioperoneal CTO and long-segment anterior and posterior tibial artery occlusions and (**b**) of the distal tibial arteries and foot (anterior projection) that shows reconstitution of the dorsalis pedis artery through perforating peroneal branches

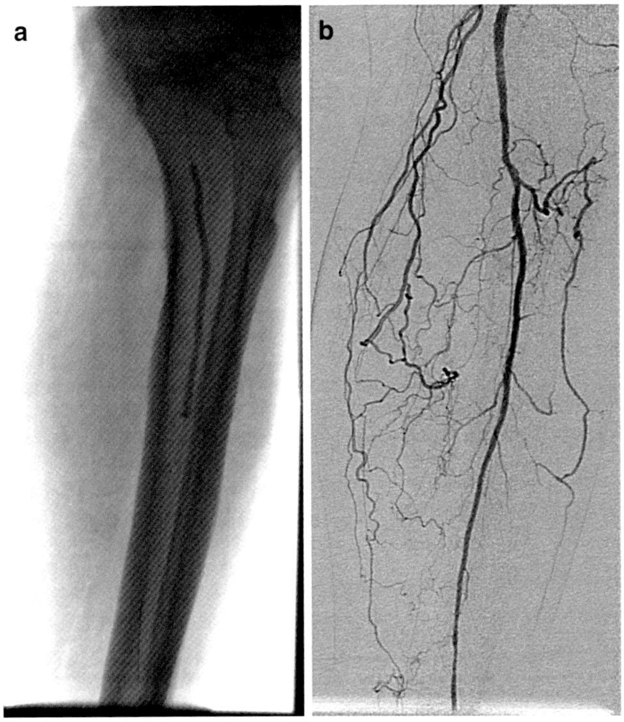

Fig. 17.2 (**a**) Balloon angioplasty of the tibioperoneal artery and (**b**) optimal acute angiographic result without recoil or dissection

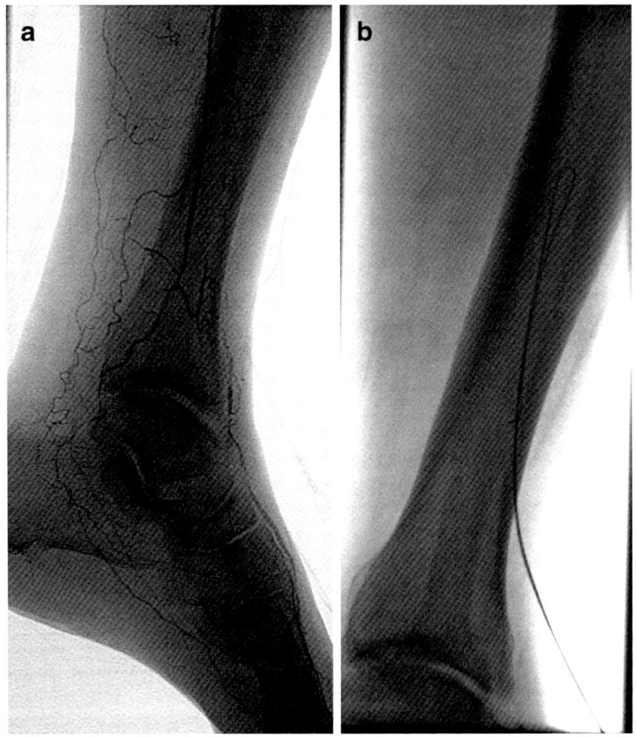

Fig. 17.3 (**a**) Lateral foot angiogram shows opacification of the dorsalis pedis run-off vessel and a patent pedal arch while retrograde ATA access was obtained. (**b**) Retrograde subintimal ATA crossing with a j-tipped 0.035″ glidewire was performed

The tibioperoneal CTO was crossed with a 0.014″ COMMAND (Abbott Vascular) guidewire supported by the 4 Fr angled angiographic catheter and dilated for 60 s at 14 atm. with a 3.0 × 120 mm low-profile COYOTE balloon catheter (Boston Scientific) with an optimal angiographic result (Fig. 17.2). Next, a contrast angiogram, of the foot on lateral projection was acquired to delineate the dorsalis pedis and the pedal arch for adequate mapping of the distal microcirculation prior to further recanalization (Fig. 17.3a). The origin of the anterior tibial artery could not be engaged (flush occlusion) and subsequently a retrograde puncture of the distal third of the occluded left anterior tibial artery above the ankle was performed under combined ultrasound and fluoroscopic guidance using a standard 0.018″ micropuncture kit. The CTO was subintimally recanalized with the aid of a half-stiff j-tipped 0.035″ Terumo glidewire until the level of the ATA origin (Fig. 17.3b). Using a rendezvous technique, a 0.014″ COMMAND guidewire with a hydrophilic tip was then advanced antegradely into the ATA false lumen and navigated into the distal dorsalis pedis true lumen. Antegrade plain balloon angioplasty of the whole ATA was performed in sequence with 2.5 × 220 mm and 3.0 × 220 mm COYOTE balloon catheters with a good angiographic result and brisk flow to the foot. The distal ATA puncture site was sealed with a prolonged 2.5 mm balloon inflation for 120 s (Fig. 17.4).

A completion angiogram of the calf and foot on lateral and anteroposterior projection demonstrated good antegrade flow to the dorsalis pedis with a healthy pedal arch and contrast reflux into the plantar circulation (Fig. 17.5). Rest pain resolved immediately, a 6 Fr Angioseal was deployed at the groin puncture site and the patient was discharged the same evening on a dual antiplatelet prescription (Salospir 1200 mg o.d. and Clopidogrel 75 mg o.d.) for the next 6 months.

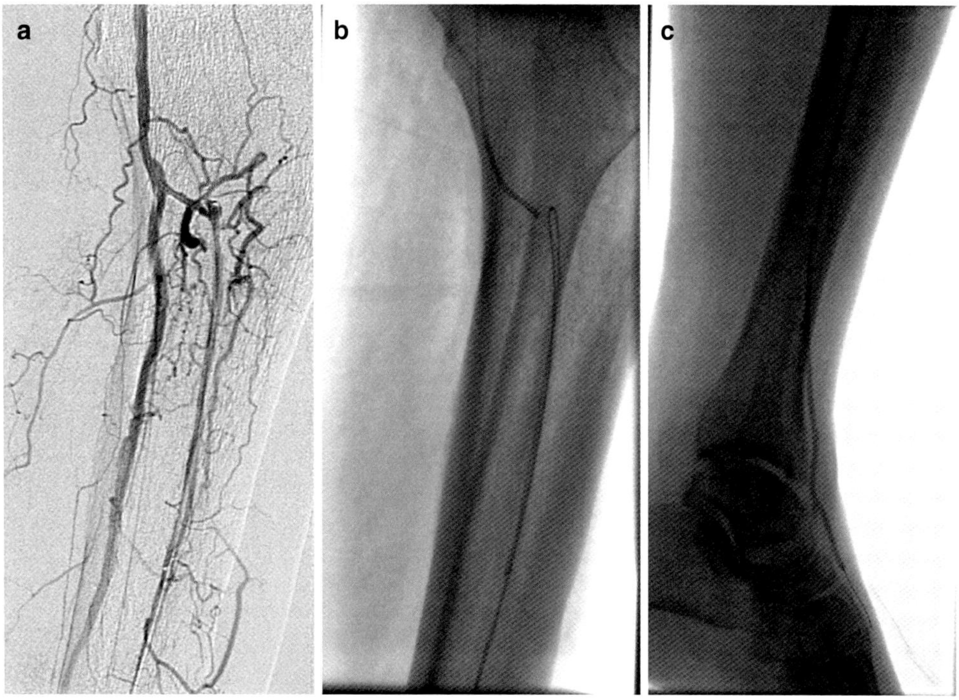

Fig. 17.4 (**a**) Subintimal dissection extended retrogradely up to the ATA origin, (**b**) next a 0.014″ COMMAND guidewire was navigated into the false lumen with the support of a 4 Fr angled angiographic catheter, (**c**) and a 2.5–3.0 mm balloon angioplasty of the whole ATA down to the foot level was performed (inflation period 60 s)

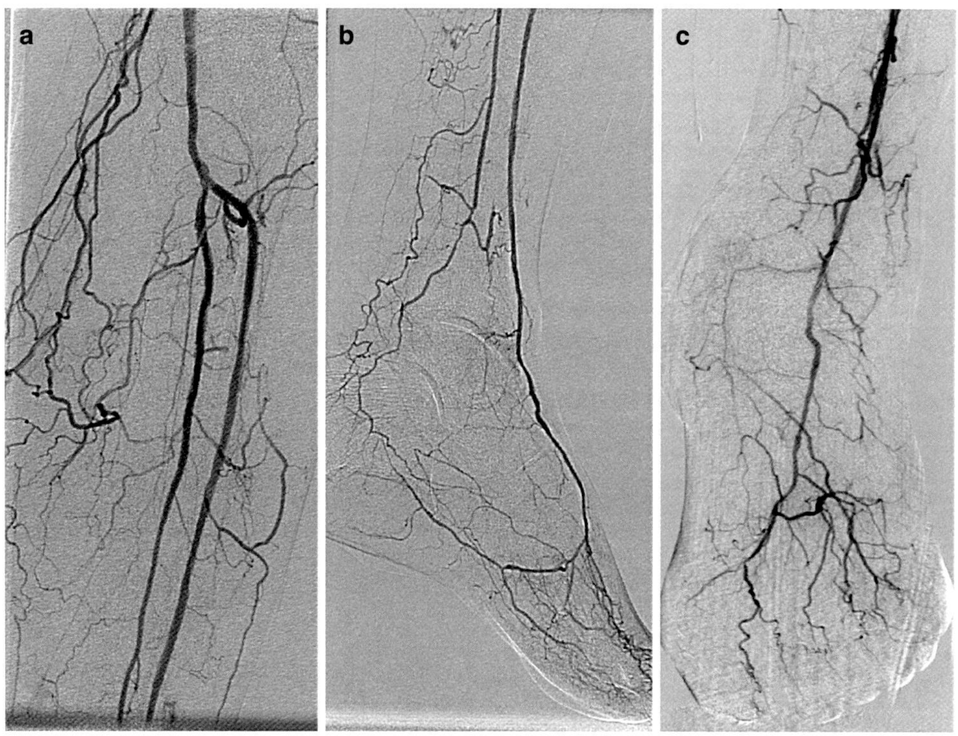

Fig. 17.5 Completion angiogram of (**a**) the proximal tibioperoneal and anterior tibial arteries, (**b**) the distal tibial arteries and foot (lateral projection) and (**c**) the foot (anterior projection) demonstrating good opacification of the dorsalis pedis artery and pedal arch

References

1. Conte MS, Bradbury AW, Kolh P, et al. Global vascular guidelines on the management of chronic limb-threatening ischemia. J Vasc Surg. 2019;69:3S–125S. e140. https://doi.org/10.1016/j.jvs.2019.02.016.
2. Hirsch AT, Haskal ZJ, Hertzer NR, et al. ACC/AHA 2005 guidelines for the management of patients with peripheral arterial disease (lower extremity, renal, mesenteric, and abdominal aortic): executive summary a collaborative report from the American Association for Vascular Surgery/Society for Vascular Surgery, Society for Cardiovascular Angiography and Interventions, Society for Vascular Medicine and Biology, Society of Interventional Radiology, and the ACC/AHA task force on practice guidelines (writing committee to develop guidelines for the Management of Patients with Peripheral Arterial Disease) endorsed by the American Association of Cardiovascular and Pulmonary Rehabilitation; National Heart, Lung, and Blood Institute; Society for Vascular Nursing; TransAtlantic inter-society consensus; and vascular dis-

ease foundation. J Am Coll Cardiol. 2006;47:1239–312. https://doi.org/10.1016/j.jacc.2005.10.009.
3. Spiliopoulos S, Del Giudice C, Manzi M, et al. CIRSE standards of practice on below-the-knee revascularisation. Cardiovasc Intervent Radiol. 2021;44:1309–22. https://doi.org/10.1007/s00270-021-02891-5.
4. Katsanos K, Kitrou P, Spiliopoulos S, et al. Comparative effectiveness of plain balloon angioplasty, bare metal stents, drug-coated balloons, and drug-eluting stents for the treatment of infrapopliteal artery disease: systematic review and Bayesian network meta-analysis of randomized controlled trials. J Endovasc Ther. 2016;23:851–63. https://doi.org/10.1177/1526602816671740.
5. Romiti M, Albers M, Brochado-Neto FC, et al. Meta-analysis of infrapopliteal angioplasty for chronic critical limb ischemia. J Vasc Surg. 2008;47:975–81. https://doi.org/10.1016/j.jvs.2008.01.005.
6. Kinlay S. Management of critical limb ischemia. Circ Cardiovasc Interv. 2016;9:e001946. https://doi.org/10.1161/CIRCINTERVENTIONS.115.001946.
7. Prompers L, Schaper N, Apelqvist J, et al. Prediction of outcome in individuals with diabetic foot ulcers: focus on the differences between individuals with and without peripheral arterial disease. The EURODIALE Study. Diabetologia. 2008;51:747–55. https://doi.org/10.1007/s00125-008-0940-0.
8. Diamantopoulos A, Patrone L, Santonocito S, et al. Carbon dioxide angiography during peripheral angioplasty procedures significantly reduces the risk of contrast-induced nephropathy in patients with chronic kidney disease. CVIR Endovasc. 2020;3:9. https://doi.org/10.1186/s42155-020-0103-z.
9. Khor BYC, Price P. The comparative efficacy of angiosome-directed and indirect revascularisation strategies to aid healing of chronic foot wounds in patients with co-morbid diabetes mellitus and critical limb ischaemia: a literature review. J Foot Ankle Res. 2017;10:26. https://doi.org/10.1186/s13047-017-0206-5.
10. Mills JL Sr, Conte MS, Armstrong DG, et al. The Society for Vascular Surgery Lower Extremity Threatened Limb Classification System: risk stratification based on wound, ischemia, and foot infection (WIfI). J Vasc Surg. 2014;59:220–34. https://doi.org/10.1016/j.jvs.2013.08.003.
11. Patel IJ, Rahim S, Davidson JC, et al. Society of Interventional Radiology Consensus Guidelines for the Periprocedural Management of Thrombotic and Bleeding Risk in patients undergoing percutaneous image-guided interventions-part II: recommendations: endorsed by the Canadian Association for Interventional Radiology and the Cardiovascular and Interventional Radiological Society of Europe. J Vasc Intervent Radiol JVIR. 2019;30:1168–84. https://doi.org/10.1016/j.jvir.2019.04.017.
12. van der Molen AJ, Reimer P, Dekkers IA, et al. Post-contrast acute kidney injury. Part 2: risk stratification, role of hydration and other prophylactic measures, patients taking metformin and chronic dialysis patients: recommendations for updated ESUR Contrast Medium Safety Committee guidelines. Eur Radiol. 2018;28:2856–69. https://doi.org/10.1007/s00330-017-5247-4.
13. Kitrou P, Parthipun A, Diamantopoulos A, et al. Targeted true lumen re-entry with the outback catheter: accuracy, success, and complications in 100 peripheral chronic total occlusions and systematic review of the literature. J Endovasc Ther. 2015;22:538–45. https://doi.org/10.1177/1526602815594249.
14. Bazan HA, Le L, Donovan M, et al. Retrograde pedal access for patients with critical limb ischemia. J Vasc Surg. 2014;60:375–81. https://doi.org/10.1016/j.jvs.2014.02.038.
15. Biagioni RB, Biagioni LC, Nasser F, et al. Infrapopliteal angioplasty of one or more than one artery for critical limb ischaemia: a randomised clinical trial. Eur J Vasc Endovasc Surg. 2018;55:518–27. https://doi.org/10.1016/j.ejvs.2017.12.022.
16. Alexandrescu V, Hubermont G. The challenging topic of diabetic foot revascularization: does the angiosome-guided angioplasty may improve outcome. J Cardiovasc Surg. 2012;53:3–12.
17. Katsanos K, Spiliopoulos S, Kitrou P, et al. Risk of death following application of paclitaxel-coated balloons and stents in the femoropopliteal artery of the leg: a systematic review and meta-analysis of randomized controlled trials. J Am Heart Assoc. 2018;7:e011245. https://doi.org/10.1161/JAHA.118.011245.
18. Katsanos K, Spiliopoulos S, Teichgraber U, et al. Editor's choice—risk of major amputation following application of paclitaxel coated balloons in the lower limb arteries: a systematic review and meta-analysis of randomised controlled trials. Eur J Vasc Endovasc Surg. 2022;63:60–71. https://doi.org/10.1016/j.ejvs.2021.05.027.
19. Katsanos K, Spiliopoulos S, Diamantopoulos A, et al. Wound healing outcomes and health-related quality-of-life changes in the ACHILLES trial: 1-year results from a prospective randomized controlled trial of infrapopliteal balloon angioplasty versus sirolimus-eluting stenting in patients with ischemic peripheral arterial disease. JACC Cardiovasc Interv. 2016;9:259–67. https://doi.org/10.1016/j.jcin.2015.10.038.
20. Scheinert D, Katsanos K, Zeller T, et al. A prospective randomized multicenter comparison of balloon angioplasty and infrapopliteal stenting with the sirolimus-eluting stent in patients with ischemic peripheral arterial disease: 1-year results from the ACHILLES trial. J Am Coll Cardiol. 2012;60:2290–5. https://doi.org/10.1016/j.jacc.2012.08.989.
21. Katsanos K, Spiliopoulos S, Diamantopoulos A, et al. Systematic review of infrapopliteal drug-eluting stents: a meta-analysis of randomized controlled trials. Cardiovasc Intervent Radiol. 2013;36:645–58. https://doi.org/10.1007/s00270-013-0578-2.
22. Kum S, Ipema J, Chun-Yin DH, et al. Early and midterm experience with the absorb everolimus-eluting bioresorbable vascular scaffold in Asian patients with chronic limb-threatening ischemia: 1-year clinical and imaging outcomes from the DISAPEAR registry. J Endovasc Ther. 2020;27:616–22. https://doi.org/10.1177/1526602820922524.
23. Diamantopoulos A, Katsanos K. Atherectomy of the femoropopliteal artery: a systematic review and meta-analysis of randomized controlled trials. J Cardiovasc Surg. 2014;55:655–65.
24. Katsanos K, Spiliopoulos S, Reppas L, et al. Debulking atherectomy in the peripheral arteries: is there a role and what is the evidence? Cardiovasc Intervent Radiol. 2017;40:964–77. https://doi.org/10.1007/s00270-017-1649-6.
25. Shammas NW. Current role of atherectomy for treatment of femoropopliteal and infrapopliteal disease. Intervent Cardiol Clin. 2017;6:235–49. https://doi.org/10.1016/j.iccl.2016.12.007.
26. Lichtenberg M, Schreve MA, Ferraresi R, et al. Surgical and endovascular venous arterialization for treatment of critical limb ischaemia. VASA Zeitschrift fur Gefasskrankheiten. 2018;47:17–22. https://doi.org/10.1024/0301-1526/a000669.
27. Schmidt A, Schreve MA, Huizing E, et al. Midterm outcomes of percutaneous deep venous arterialization with a dedicated system for patients with no-option chronic limb-threatening ischemia: the ALPS multicenter study. J Endovasc Ther. 2020;27:658–65. https://doi.org/10.1177/1526602820922179.
28. Arkoudis NA, Katsanos K, Inchingolo R, et al. Quantifying tissue perfusion after peripheral endovascular procedures: novel tissue perfusion endpoints to improve outcomes. World J Cardiol. 2021;13:381–98. https://doi.org/10.4330/wjc.v13.i9.381.
29. Tsigkas G, Apostolos A, Trigka A, et al. Very short versus longer dual antiplatelet treatment after coronary interventions: a systematic review and meta-analysis. Am J Cardiovasc Drugs. 2023;23:35–46. https://doi.org/10.1007/s40256-022-00559-0.
30. Katsanos K, Spiliopoulos S, Saha P, et al. Comparative efficacy and safety of different antiplatelet agents for prevention of major cardiovascular events and leg amputations in patients with peripheral arterial disease: a systematic review and network meta-analysis. PLoS One. 2015;10:e0135692. https://doi.org/10.1371/journal.pone.0135692.

Interventions for Pedal Disease

18

Stavros Spiliopoulos

Case Presentation

A 71-year-old male, ex-smoker, with a history of long-standing type II insulin-dependent diabetes mellitus (IDDM), hypercholesterolemia, coronary disease, and hypertension, was presented to the hospital complaining of right foot rest pain as well as worsening first and second toes, and heal dry gangrene, for at least 3 months. Pedal pulses were absent, ankle-brachial index (ABI) was 0.55, and duplex ultrasound revealed occlusion of all three infrapopliteal vessels, without iliac, and femoropopliteal inflow disease. The diagnosis was chronic limb-threatening ischemia, CLTI category 5 according to Rutherford–Becker. The patient was placed on antiplatelet therapy and was discussed at the vascular multi-disciplinary team meeting (MDT). Based on the comorbidities, clinical presentation, the infrapopliteal location of the disease, as well as the fact that his eGFR was 44 mL/min/1.73 m^2, the MDT decision was to skip CT-angiography and proceed immediately to antegrade DSA from a common femoral artery (CFA) access, with a view to same-session endovascular treatment. Endovascular treatment was performed 2 days after presentation.

Continued at page 176

Background

Infrapopliteal arterial and pedal arch steno-occlusive disease commonly affects subjects with diabetes leading to progressive deterioration of foot perfusion and eventually to tissue loss with or without gangrene defined as chronic limb-threatening ischemia (CLTI), which without prompt revascularization will eventually lead to limb loss and increased 1-year mortality risk [1]. Notably, CLTI patients demonstrate significantly higher mortality rates compared to those with symptomatic coronary artery disease, while major amputation has been recognized as a major contributing factor of quality-of-life deterioration and increased mortality risk [2, 3]. Generally, the diagnosis is set by clinical symptoms (absence of peripheral pulses, rest pain, tissue loss/gangrene >2 weeks duration), but it can also be supported by hemodynamic tests such as abnormal ABI or toe-brachial index (TBI) for diabetic patients, abnormal tissue perfusion evaluation ($TcPO_2$ ≤ 30 mmHg), and imaging findings (duplex ultrasound, CT- or MR-angiography) [4]. In specialized vascular centers, percutaneous endovascular procedures—mainly infrapopliteal angioplasty or stenting—and open surgical distal venous bypass procedures are the standard of practice for CLTI patients requiring revascularization [1, 5]. The choice of revascularization type (endovascular or open) should be individualized and decided by a vascular MDT based on distinct patient and lesion characteristics. However, as endovascular procedures are less invasive and provide valid clinical outcomes, without excluding future surgical options, the "endovascular first" approach is currently used in most CLTI patients with tibial arterial disease [6–8]. In recent years, endovascular specialists, driven by increasing clinical experience, and supported by the significant technological advances of endovascular materials and devices, have advanced even more distally, from crural vessel treatment to pedal arch angioplasty [9]. In fact, endovascular pedal reconstruction is currently the only treatment option for pedal arch steno-occlusive atherosclerotic disease, as to date there is no feasible surgical method to reconstruct the pedal arch.

Indications for Intervention [10]

- Rest pain (Rutherford category 4) and/or non-healing ulcer/gangrene (Rutherford categories 5–6)
- Non-healing ischaemic ulcer following amputation
- Salvage run-off treatment in failing surgical bypass

S. Spiliopoulos (✉)
2nd Department of Radiology, National and Kapodistrian University of Athens, "ATTIKON" University General Hospital, Athens, Greece
e-mail: stavspiliop@med.uoa.gr

G. Geroulakos et al. (eds.), *Mastering Endovascular Techniques*, https://doi.org/10.1007/978-3-031-42735-0_18

Preoperative Preparation

Timing of Intervention

For subjects with diabetes and CLTI (diabetic foot), emergency revascularization (open, endovascular, or hybrid) should be offered to prevent superinfection and decrease major amputation risk.

Laboratory Tests

Baseline full blood count, renal function (eGFR), glucose, electrolytes, and coagulation profile should be available. Ideally, platelet count should be >50,000/mcL and INR <1.5. Compromised renal function (eGFR<60 mL/min) requires parenteral saline infusion (1.0–1.5 mL/kg/h), 6 h before and 6–12 h after the procedure, or intravenous sodium bicarbonate as an alternative, while metformin intake should be discontinued for 48 h prior to and after the procedure and re-initiated if eGFR is stable, to avoid lactic acidosis [11]. Non-iodine contrast media alternatives such as carbon dioxide (CO_2) may be used, but adequate analgesia/sedation or general anaesthesia should be considered, as pedal CO_2 administration can be very painful.

Drug Therapy

Warfarin should be discontinued for 3 days (usually sufficient for INR normalization) and heparin may be given as bridging therapy according to local hospital policy. Dual antiplatelet therapy with clopidogrel 75 mg and acetylsalicylic acid (ASA) 100 mg once daily at least 3 days before the procedure is a generally acceptable antiplatelet protocol to inhibit platelet function. Alternatively, a loading dose of Clopidogrel 300 mg 12 h before the procedure could be administrated in patients at low bleeding risk. However, the level of evidence to support pre-procedural antithrombotic therapy is low [10]. In the presence of infected wound/gangrene, intravenous antibiotic therapy should be administrated well before the attempt to revascularize the infected foot, so as to avoid bacteremia and sepsis. Antihypertensive therapy should be continued the day of the procedure to reduce access-site bleeding risk. Statins should be prescribed, even in patients with normal lipidic profile (low-dose statins) as it has been shown to decrease cardiovascular death risk in subjects with cardiovascular disease and diabetes [1].

Imaging

Duplex ultrasound (DUS), CTA, or MRA is essential not only to establish diagnosis, but also to depict the anatomic location and extent of the crural disease and assist in pre-procedural planning during the MDT. Although CTA has known diagnostic limitations in the ambit of distal crural vessels, especially in the presence of calcifications, it is an indispensable pre-procedural tool, as it allows for inflow lesions detection and optimal access-site evaluation. On the other hand, dedicated pedal MRA, with or without contrast enhancement, demonstrates excellent results in outflow vessel detection in diabetic patients, even in the presence of more proximal stenosis [12, 13]. DUS of the access site and the distal pedal vessels can be extremely useful for pre-procedural limb evaluation within the angio suite before obtaining access for pedal arch angioplasty.

Endovascular Strategy and Technique

Basic Endovascular Toolkit

- Standard arterial sheaths 4–6 Fr in diameter and 10–45 cm in length
- Long-shaft balloon catheters—commonly 1.5–3.0 mm of various lengths/tapered balloon catheters
- Dedicated below-the-knee 0.014″/0.018″-inch 300 cm guidewires
- Supporting/crossing 0.014″/0.018″ catheters

Access and Crossing the Lesion

An antegrade CFA access is recommended for pedal arch interventions, mainly due to pushability, torquability, and length issues of the catheters and guidewires, especially if a "plantar-loop" revascularization technique is anticipated. Ultrasound-guided lidocaine (1–2%) injection and puncture of the CFA can reduce access-site related pain and complication rates. Access can be obtained with micropuncture sets or directly with standard 18–19 G needles and 0.035′ guidewires, especially in hostile abdomens, where a direct puncture of the SFA, below the level of the femoral head, can be considered. In such cases, the use of a closure device to avoid bleeding complications is advised [14]. Standard 4–6 Fr and 10–45 cm arterial sheaths can be used. Most interventions are performed under local anesthesia and conscious sedation and general anesthesia is reserved for selected cases of patients unable to cooperate. Intra-arterial bolus heparin 3000–5000 IU is usually administrated via the sheath following access, while nitro-glycerine (1 μg in 10 mL of saline) can be used locally to reduce spasm during guidewire manipulations and balloon inflation.

An antegrade intraluminal recanalization attempt with a 0.014″ system should be performed first. Pedal angioplasty is performed after descending the anterior or posterior tibial artery and navigating the guidewire across the pedal-plantar loop with or without the "plantar-loop" technique (e.g., retrograde recanalization of the anterior tibial after descending the posterior tibial and across the arch, or vice versa). The fluoroscopic projections during pedal arch navigation include both the ipsilateral cranial-antero-posterior (image intensifier parallel to the foot dorsum to allow imagining of the first metatarsal space) and the contra-lateral oblique ("Charlie Chaplin" projection) for the visualization of the connection dorsalis pedis-deep perforating artery-pedal arch, and the bifurcation of the common plantar artery [15]. Prolonged, 3–6 frames per sec, selective DSA imaging, after nitroglycerin administration, is recommended for the accurate visualization of collateral network and target vessels and to minimize motion artifact. Dedicated infrapopliteal angled hydrophilic guidewires 0.014′ in diameter and 300 mm in length are recommended for steerability during arch navigation, supported by 2–2.5 mm infrapopliteal low-profile balloons or dedicated 0.014″ microcatheter systems, which also enable selective contrast injection and roadmap use to guide arch crossing, avoiding side-branches. Of note, straight guidewires can also be used, depending on the anatomy, which can also be bended accordingly to allow "individualized" arch navigation. Support catheters measuring 0.018″ can be used for initial CTO crossing as this enables easier guidewire exchange and better selective angiographic imaging. Low-weight CTO 0.014′ or 0.018″ hydrophilic guidewires can also be used for initial access within the occlusion, subsequently exchanged with normal-weight guidewires for around the arch navigation. Notably, subintimal recanalization of the entire pedal arch is not indicated at it would result in occlusion of the digital arteries, resulting in aggravating ischemia. Therefore, guidewire manipulation must be gentle, preferably using torque movements and gentle pressure for advancement, rather than strong pushing. In cases of antegrade intraluminal recanalization failure, an additional ultrasound-guided retrograde pedal puncture and retrograde lesion crossing with the SAFARI technique can be attempted, using dedicated pedal access or micropuncture sets [16]. Nominal pressure, prolonged 2–4 min inflation is recommended to achieve maximum luminal gain and minimize dissections. Small 1.5–2 mm diameter short-length (20 mm) balloon catheters can be used for initial crossing and dilation of hard atherosclerotic lesions, subsequently upsized accordingly. Finally, if lesion crossing with a balloon is not feasible, usually due to hard calcified lesions, orbital atherectomy could be considered, by analogy to coronary lesions. The use of metallic stents is not recommended due to fracture and deformation issues correlated with vessel occlusion and amputation, while the use of paclitaxel-coated balloons is not supported by evidence and safety issues have also been raised [17–19].

Results

Over the past decade, several authors have reported excellent technical success rates (up to 96%), and pedal arch revascularization has been reported as an independent factor for faster wound healing [9, 20, 21]. However, prospective randomized trials to support the clinical superiority of pedal arch angioplasty versus conventional infrapopliteal angioplasty are missing from the literature. Current data from a 2019 meta-analysis of 10 studies (478 patients with below-the-ankle angioplasty in 524 legs) indicate that the 1-year limb salvage rate following pedal arch angioplasty is 92%. However, the level of evidence was low [22].

Anticipated Complications

Procedure-related complications related to guidewire manipulations and balloon inflation are vessel spasm, dissection, thromboembolism, and rupture. Vessel spasm can be managed with nitro-glycerine locally. Prolonged 3–5 min low-pressure balloon inflation can be applied in cases of flow-limiting dissection and vessel rupture. Thromboembolism is extremely rare during sole pedal arch angioplasty and standard rt-PA thrombolysis with or without catheter thrombo-aspiration can be employed. Finally, standard access-site complications, mainly retroperitoneal bleeding, pseudoaneurysm, and arteriovenous fistula formation, are also anticipated.

Perioperative Care and Surveillance

In cases of 6 Fr sheath access, vascular closure devices should be available to reduce bleeding complications in selected cases, but also to accelerate patient ambulation. Haemostasis following pedal access can be obtained following 5–10 min of manual compression. Peripheral interventions can be performed on an outpatient basis. Access-site and vital signs monitoring is suggested until discharge. In the absence of adverse events, light diet is suggested after 4–5 h and discharge after approximately 8 h. Postprocedural medical therapy should include statins (low dose in cases of normal cholesterol plasma levels), antihypertensive therapy, optimal glycemic profile, and per os antithrombotic therapy [23]. Evidence regarding antithrombotic therapy following infrapopliteal plus pedal revascularization remains limited and a personalized approach, considering the bleeding ischemic risk ratio, is recommended. Standard protocols following complex BTK revascularization include 6-month dual antiplatelet therapy with clopidogrel 75 mg and ASA 75 mg or 100 mg once daily, followed by life-long clopidogrel monotherapy or rivaroxaban 2.5 mg twice daily and ASA

100 mg once daily (especially in patients with coronary artery disease at low bleeding risk) [10]. Meticulous wound care and strict follow-up protocols are recommended following complex BTK and pedal procedures. DUS before discharge and clinical visits at 1, 3, 6, and 12 months are recommended, including surveillance with ABI/TBI, transcutaneous oxygen pressure ($TcPO_2$), and DUS.

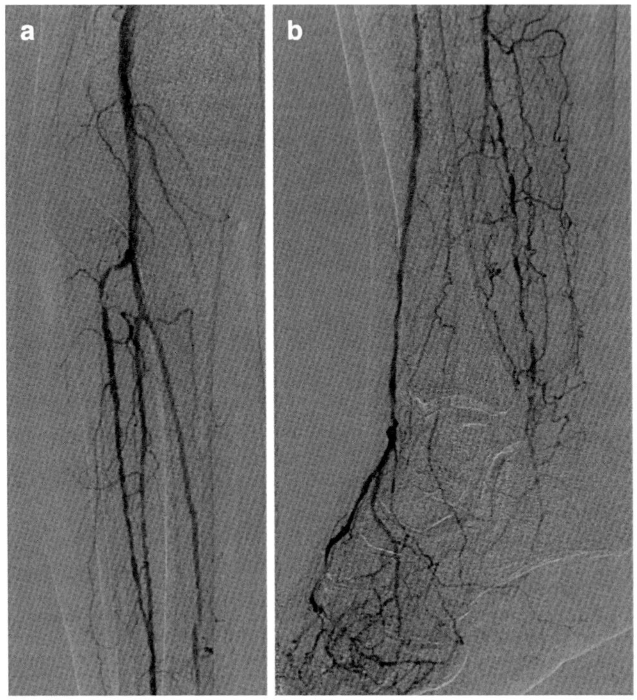

Fig. 18.1 Selective infrapopliteal DSA with the diagnostic catheter positioned at the distal popliteal artery demonstrating diffuse infrapopliteal 3-vessel (**a**) and pedal arch (**b**) atherosclerotic steno-occlusive disease

Case Presentation

Continued from page 173

The procedure was performed under local anesthesia (US-guided) and iv fentanyl (100 μg in two doses) for pain relief. Antegrade US-guided access with a 6 Fr × 10 cm arterial sheath was obtained followed by bolus administration of 5.000 IU of unfractionated heparin. Selective infrapopliteal DSA confirmed the diagnosis of diffuse 3-vessel (Fig. 18.1a) and pedal arch (Fig. 18.1b) atherosclerotic steno-occlusive disease. Specifically, significant tandem stenosis and focal occlusions were noted along the ATA, pedal artery, and pedal arch. Tandem stenosis and occlusion of the distal third of the peroneal and posterior tibial arteries were noted, while the medial and lateral plantar arteries appeared occluded. A 5 Fr × 65 cm curved catheter (Vanscie 1, Cook, UK) and a 0.014″/300 cm straight guidewire (ASAHI Regalia® XS 1.0, Japan) were used to negotiate the origin of the ATA and, using the above-mentioned fluoroscopic projections (Fig. 18.2a, b), the guidewire was advanced through the tandem lesions and finally positioned at the mid-pedal artery.

Fig. 18.2 (**a**) Ipsilateral, cranial, anteroposterior projection for guidewire navigation from the anterior (pedal artery) to the posterior circulation (lateral plantar artery) via the deep perforating artery. (**b**) Contralateral "Charlie Chaplin" projection to navigate around the arch to the common plantar artery and the posterior tibial artery

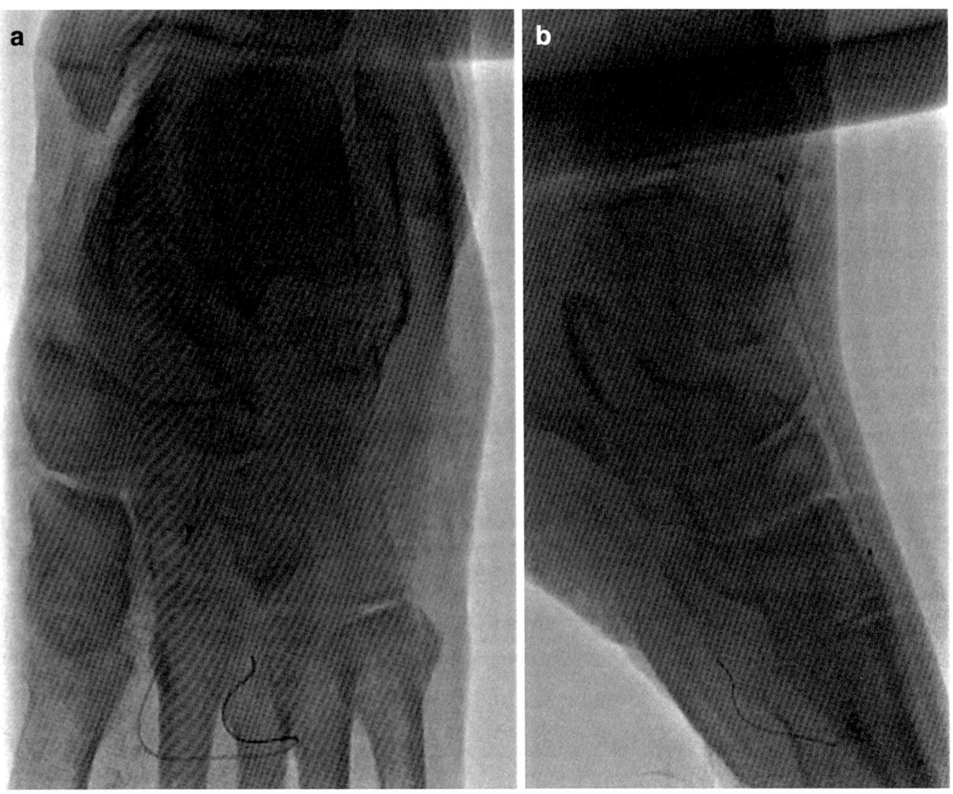

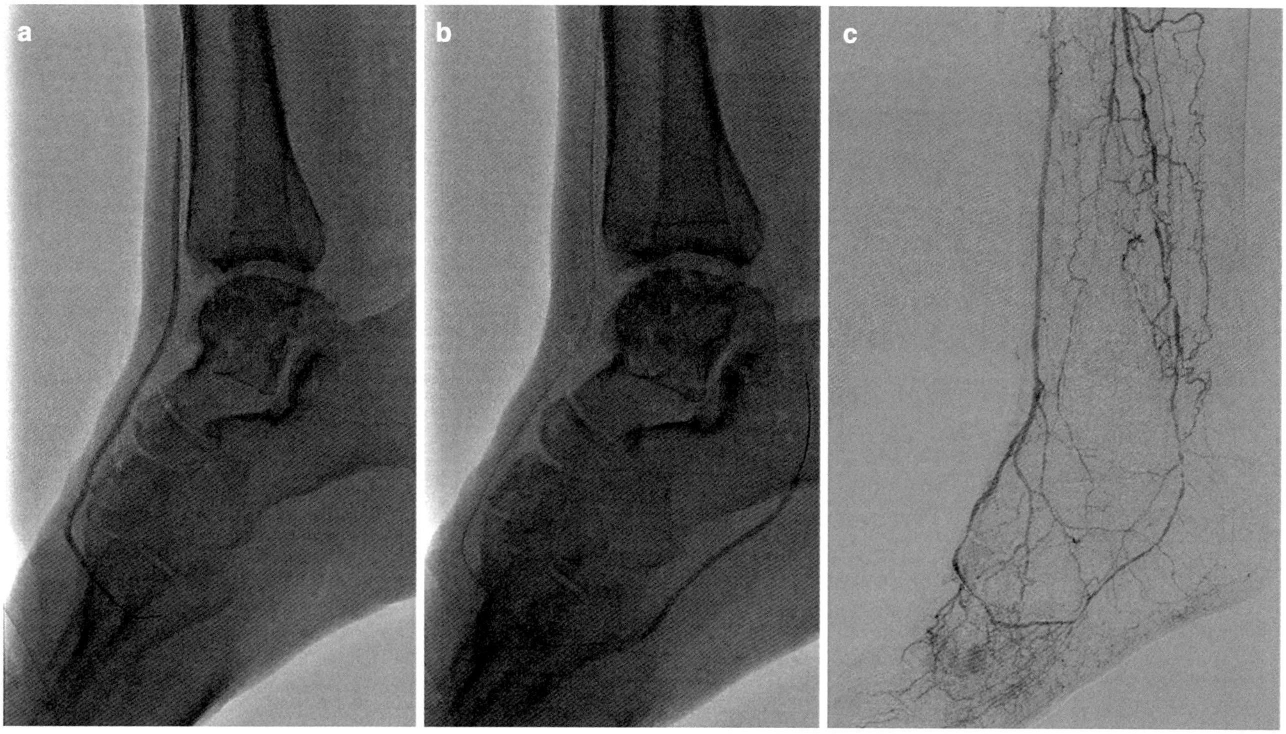

Fig. 18.3 2.5 mm balloon angioplasty of the pedal artery (**a**) and the plantar arch (**b**). The guidewire could not cross the proximal plantar artery occlusion (**c**)

Balloon angioplasty of the ATA/pedal artery was performed using a 150 mm shaft, 3 × 150 mm, 0.014″ compatible balloon catheter (Coyote, Boston Scientific, USA). The balloon catheter was then exchanged over the wire with a 2.5 × 100 mm balloon catheter (Coyote, Boston Scientific, USA), used to support the navigation of the guidewire across the pedal arch and the pedal-plantar loop, and retrograde intraluminal recanalization of the lateral plantar artery occlusion, as well as pedal arch-plasty, was performed using the same 2.5 mm balloon ("plantar-loop" technique) (Fig. 18.3a–c). The guidewire was subsequently retrieved from the ATA and the posterior tibial artery, including its distal segment, was recanalized with the 2.5 × 100 mm balloon distally (Fig. 18.4a, b) and the 3 × 150 mm balloon more proximally. Final DSA demonstrated a satisfactory angiographic outcome with two-vessel run-off and complete pedal arch reconstruction (Fig. 18.5). The patient reported immediate pain relief. Complete wound healing was noted at 6 months follow-up.

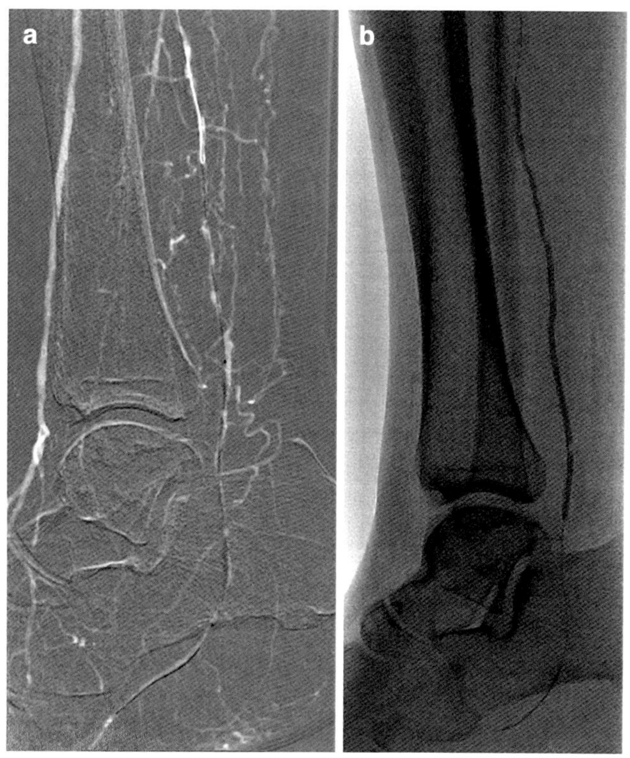

Fig. 18.4 The posterior tibial and common plantar arteries were revascularized using the antegrade approach with a 2.5 mm balloon catheter

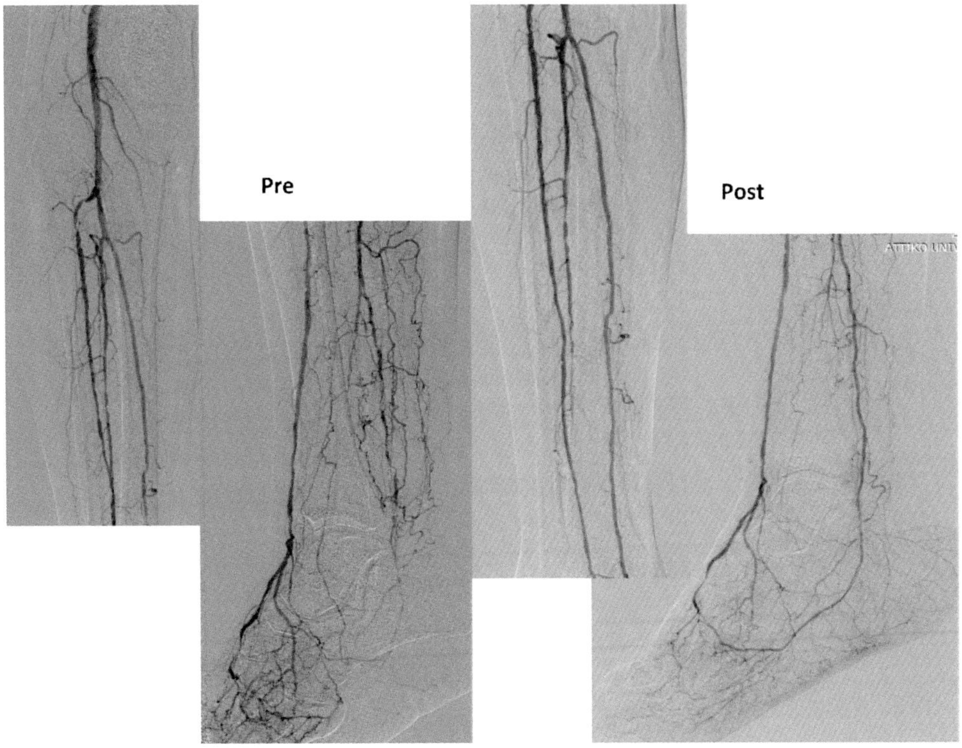

Fig. 18.5 Selective pre- (**a**) and (**b**) post-angioplasty DSA images

References

1. Bailey SR, Beckman JA, Dao TD, et al. ACC/AHA/SCAI/SIR/SVM 2018 appropriate use criteria for peripheral artery intervention: a report of the American College of Cardiology Appropriate use Criteria Task Force, American Heart Association, Society for Cardiovascular Angiography and Interventions, Society of Interventional Radiology, and Society for Vascular Medicine. J Am Coll Cardiol. 2019;73(2):214–37.
2. Teraa M, Conte MS, Moll FL, Verhaar MC. Critical limb ischemia: current trends and future directions. J Am Heart Assoc. 2016;5:e002938.
3. Katsanos K, Spiliopoulos S, Diamantopoulos A, Siablis D, Karnabatidis D, Scheinert D. Wound healing outcomes and health-related quality-of-life changes in the ACHILLES trial: 1-year results from a prospective randomized controlled trial of infrapopliteal balloon angioplasty versus sirolimus-eluting stenting in patients with ischemic peripheral arterial disease. JACC Cardiovasc Interv. 2016;9(3):259–67.
4. Conte MS, Bradbury AW, Kolh P, GVG Writing Group, et al. Global vascular guidelines on the management of chronic limb-threatening ischemia. J Vasc Surg. 2019;69(6S):3S–125S.e40.
5. Elbadawy A, Ali H, Saleh M, Hasaballah A. A prospective study to evaluate complete wound healing and limb salvage rates after angiosome targeted infrapopliteal balloon angioplasty in patients with critical limb ischaemia. Eur J Vasc Endovasc Surg. 2018;55(3):392–7.
6. Feldman ZM, Mohapatra A. Endovascular management of complex tibial lesions. Semin Vasc Surg. 2022;35(2):190–9.
7. Hicks CW, Canner JK, Lum YW, Black JH 3rd, Abularrage CJ. Long-term outcomes of an endovascular-first approach for diabetic patients with predominantly tibial disease treated in a multidisciplinary setting. Ann Vasc Surg. 2019;60:315–326.e2.
8. Faglia E, Dalla Paola L, Clerici G, et al. Peripheral angioplasty as the first-choice revascularization procedure in diabetic patients with critical limb ischemia: prospective study of 993 consecutive patients hospitalized and followed between 1999 and 2003. Eur J Vasc Endovasc Surg. 2005;29(6):620–7.
9. Jung HW, Ko YG, Hong SJ, et al. Editor's choice—impact of endovascular pedal artery revascularisation on wound healing in patients with critical limb Ischaemia. Eur J Vasc Endovasc Surg. 2019;58(6):854–63.
10. Spiliopoulos S, Del Giudice C, Manzi M, Reppas L, Rodt T, Uberoi R. CIRSE standards of practice on below-the-knee revascularisation. Cardiovasc Intervent Radiol. 2021;44(9):1309–22.
11. van der Molen AJ, Reimer P, Dekkers IA, et al. Post contrast acute kidney injury—part 1: definition, clinical features, incidence, role of contrast medium and risk factors: recommendations for updated ESUR Contrast Medium Safety Committee guidelines. Eur Radiol. 2018;28(7):2845–55.
12. Chomel S, Douek P, Moulin P, Vaudoux M, Marchand B. Contrast-enhanced MR angiography of the foot: anatomy and clinical application in patients with diabetes. AJR Am J Roentgenol. 2004;182(6):1435–42.
13. Schubert T, Takes M, Aschwanden M. Non-enhanced, ECG-gated MR angiography of the pedal vasculature: comparison with contrast-enhanced MR angiography and digital subtraction angiography in peripheral arterial occlusive disease. Eur Radiol. 2016;26(8):2705–13.
14. Spiliopoulos S, Kitrou P, Christeas N, Karnabatidis D. Starclose SE® hemostasis after 6F direct antegrade superficial femoral artery access distal to the femoral head for peripheral endovascular procedures in obese patients. Diagn Interv Radiol. 2016;22(6):542–7.
15. Manzi M, Cester G, Palena LM, Alek J, Candeo A, Ferraresi R. Vascular imaging of the foot: the first step toward endovascular recanalization. Radiographics. 2011;31(6):1623–36.
16. Stahlberg E, Stroth A, Haenel A, Grzyska U, Wegner F, Sieren M, Horn M, Barkhausen J, Goltz JP. Retrograde revascularization of tibial arteries in patients with critical limb ischemia: plantar-arch versus transpedal approach. J Endovasc Ther. 2022;29(2):181–92.

17. Karnabatidis D, Katsanos K, Spiliopoulos S, Diamantopoulos A, Kagadis GC, Siablis D. Incidence, anatomical location, and clinical significance of compressions and fractures in infrapopliteal balloon-expandable metal stents. J Endovasc Ther. 2009;16(1):15–22.
18. Katsanos K, Spiliopoulos S, Teichgräber U. Editor's choice—risk of major amputation following application of paclitaxel coated balloons in the lower limb arteries: a systematic review and meta-analysis of randomised controlled trials. Eur J Vasc Endovasc Surg. 2022;63(1):60–71.
19. Katsanos K, Spiliopoulos S, Kitrou P, Krokidis M, Paraskevopoulos I, Karnabatidis D. Risk of death and amputation with use of paclitaxel-coated balloons in the infrapopliteal arteries for treatment of critical limb ischemia: a systematic review and meta-analysis of randomized controlled trials. J Vasc Interv Radiol. 2020;31(2):202–12.
20. Nakama T, Watanabe N, Haraguchi T, et al. Clinical outcomes of pedal artery angioplasty for patients with ischemic wounds: results from the multicenter RENDEZVOUS registry. JACC Cardiovasc Interv. 2017;10:79–90.
21. Meyer A, Schinz K, Lang W, Schmid A, Regus S, Rother U. Outcomes and influence of the pedal arch in below-the-knee angioplasty in patients with end-stage renal disease and critical limb ischemia. Ann Vasc Surg. 2016;35:121–9.
22. Huizing E, Schreve MA, de Vries JPM, Ferraresi R, Kum S, Ünlü Ç. Below-the-ankle angioplasty in patients with critical limb ischemia: a systematic review and meta-analysis. J Vasc Interv Radiol. 2019;30(9):1361–1368.e2.
23. Gerhard-Herman MD, Gornik HL, Barrett C, et al. 2016 AHA/ACC guideline on the management of patients with lower extremity peripheral artery disease: executive summary. J Am Coll Cardiol. 2017;69(11):1465–508.

Atherectomy and Calcium-Modifying Devices

19

Narayanan Thulasidasan and Panos Gkoutzios

Case Presentation

A 72-year-old male patient presenting with chronic limb-threatening ischaemia of the right lower limb evidenced by tissue loss (Rutherford category 5) was imaged with CT angiography and found to have a short flush proximal SFA occlusion, significant stenotic disease in the distal SFA and two-vessel run-off to the foot via the peroneal and posterior tibial arteries. The proximal SFA occlusion was composed of heavily calcified plaque with a "coral-reef" appearance. The patient was deemed a poor candidate for surgical bypass due to severe chronic obstructive pulmonary disease, and therefore an endovascular procedure under local anaesthesia and sedation was proposed.

Continued at page 188

Background

The initial generation of atherectomy devices were developed in the 1980s with the aim of providing an endovascular analogue to traditional open surgical endarterectomy. The earliest English-language publication of a first-in-human series of atherectomy procedures was in 1988, [1] where 21 patients were treated using the Simpson atherectomy catheter. Over the ensuing 34 years multiple atherectomy devices have been developed using directional, rotational or orbital cutting, or laser ablation of in-situ plaque, sometimes coupled with an aspiration system to actively remove debris and/or mandating the use of a distal filter to collect this debris. With several catheters having been removed from the market temporarily for improvement or permanently due to failure of concept, atherectomy devices have gradually evolved towards a reduction in size profile and an increase in safety and efficacy. However, the risk of complications such as distal embolization and vessel perforation remain an Achilles' heel, and the relative paucity of randomized controlled data regarding the additional worth conferred over balloon angioplasty alone has led to a lack of clarity over the precise role of the technique within the endovascular armamentarium. This, coupled with the high up-front cost of the devices has limited their uptake among physicians treating peripheral arterial disease.

Alongside the technical advancements, further understanding of the mode of action of atherectomy has resulted in a paradigm shift in the philosophy of endovascular revascularization, with these devices now being applied to also perform 'plaque modification' (particularly in heavily calcified atheroma) rather than simply relying on a plaque debulking effect. Several non-atherectomy devices which also modify plaque in different ways are now available, and together with atherectomy these techniques are now grouped together as 'vessel preparation' modalities to be employed before the delivery of definitive therapy with an angioplasty balloon and/or stenting. The most promising non-atherectomy calcium-modifying device to date is intravascular lithotripsy, which will also be discussed.

Currently Available Atherectomy Devices and Evidence for Their Efficacy

Rotational Atherectomy

The Jetstream™ device from Boston Scientific is available in four sizes (1.6 mm, 1.85 mm, 2.1 mm and 2.4 mm) and consists of five front-cutting blades angulated to favour engaging harder diseased tissue whilst deflecting away from the softer tissues of the arterial wall, coupled with an active aspiration system to remove the generated debris. The two larger catheters also contain expandable blades behind the tip which increase luminal gain when opened to 3 mm in the 2.1 mm catheter and 3.4 mm in the 2.4 mm catheter ('blades-up' mode). The catheters are single-use, but require connec-

N. Thulasidasan (✉) · P. Gkoutzios
Interventional Radiology, Guys and St Thomas' NHS Foundation Trust, London, UK
e-mail: narayanan.thulasidasan@gstt.nhs.uk; panos.gkoutzios@gstt.nhs.uk

G. Geroulakos et al. (eds.), *Mastering Endovascular Techniques*, https://doi.org/10.1007/978-3-031-42735-0_19

tion to a reusable console which provides the power source and saline pumping system for aspiration. The device has been in clinical use for over a decade, with multiple single-center series published. The post-market JET registry enrolled 241 patients with femoropopliteal lesions and reported a 12-month freedom from TLR of 81.7%, [2] and two prospective trials of the device in in-stent restenosis returned sufficiently positive results [3] that the device received a CE marking for this indication.

A case example of Jetstream™ atherectomy for in-stent restenosis is provided for illustration. An 81-year-old male patient with past medical history of obesity, hypercholesterolaemia, hypertension, type 2 diabetes and ischemic heart disease had (3 months prior to presentation) stents inserted in the left SFA and the proximal P1 popliteal artery segment to aid healing of third and fourth toe amputations. On ultrasound duplex examination the stents were found to be occluded and the patient underwent a transmetatarsal amputation together with a repeat endovascular procedure to improve the blood supply for the wound to heal and prevent below-knee amputation. Angiography demonstrated the occluded stents (black arrows) and the peroneal artery was the single run-off vessel (Fig. 19.1). A 7-French sheath was inserted antegrade in the left common femoral artery and for embolic protection a SpiderFX™ filter was inserted distally in the left popliteal artery (Fig. 19.2, red arrows). A Jetstream™ 2.4 mm atherectomy device was employed (Fig. 19.2, black arrows) with two slow passes done with the blades down followed by two slow passes with the blades up. 6 mm in-stent plain balloon angioplasty was performed, but due to residual stenosis (Fig. 19.3) distal extension with two Supera™ stents in the P2 and P3 popliteal segments, and proximal extension in the left SFA with a drug-coated stent (Zilver™ PTX) was required. The flow along the left femoropopliteal axis was restored, with preservation of patency of the single-vessel tibial run-off (Fig. 19.4).

The 2.2 mm, 1.8 mm and 1.5 Phoenix™ atherectomy catheters from Philips utilize a front-cutting design with blades that engage and cut into plaque, which is then passively cleared by means of passage through an Archimedes screw system running along the entire internal length of the catheter. The device is powered by a battery situated in the handle, therefore no capital investment in a console is required. The IDE trial revealed a 6-month freedom from

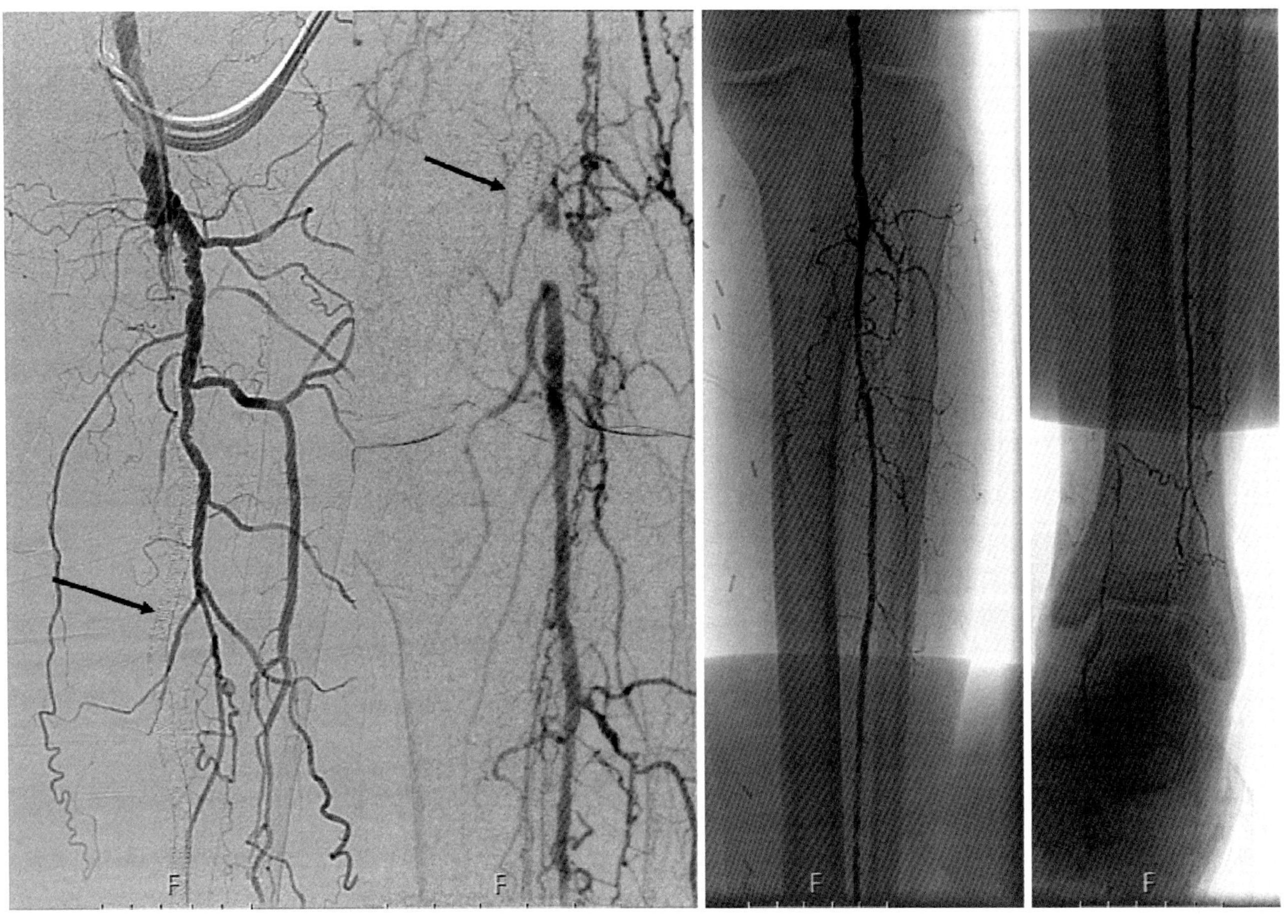

Fig. 19.1 Initial angiograms showing the left SFA and popliteal artery stents occluded (black arrows) and the distal single run off vessel, the peroneal artery

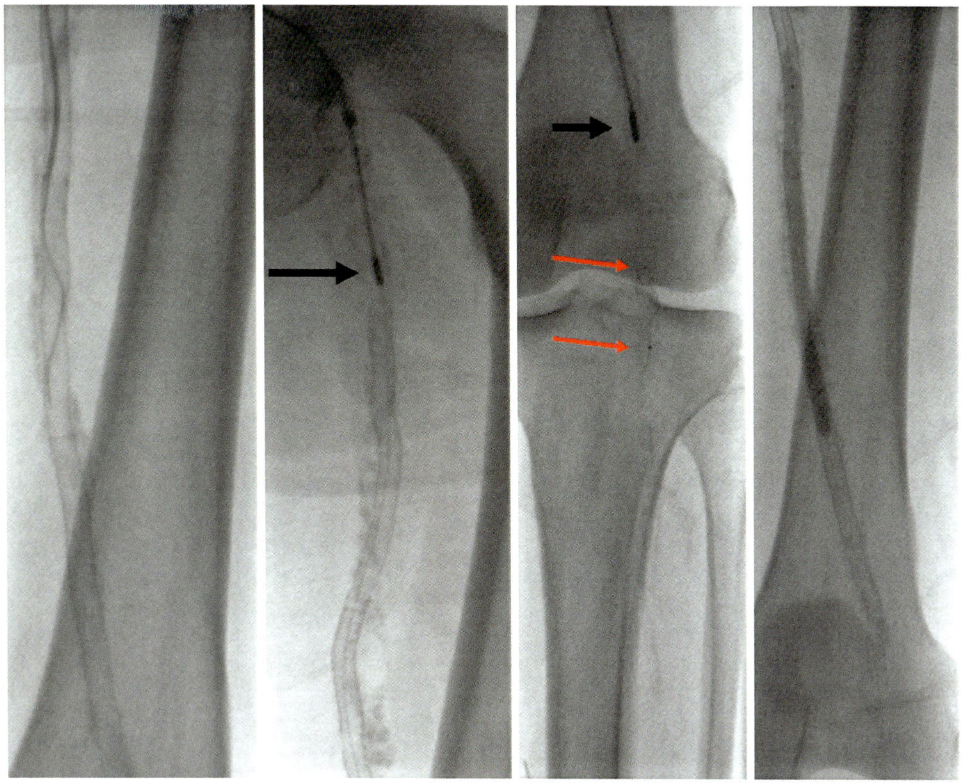

Fig. 19.2 Guide wire crossing the occluded stents, filter placed in the popliteal artery (its upper and lower margins indicated with the two red arrows) and Jetstream™ rotational head where the blades are embedded (thick black arrows)

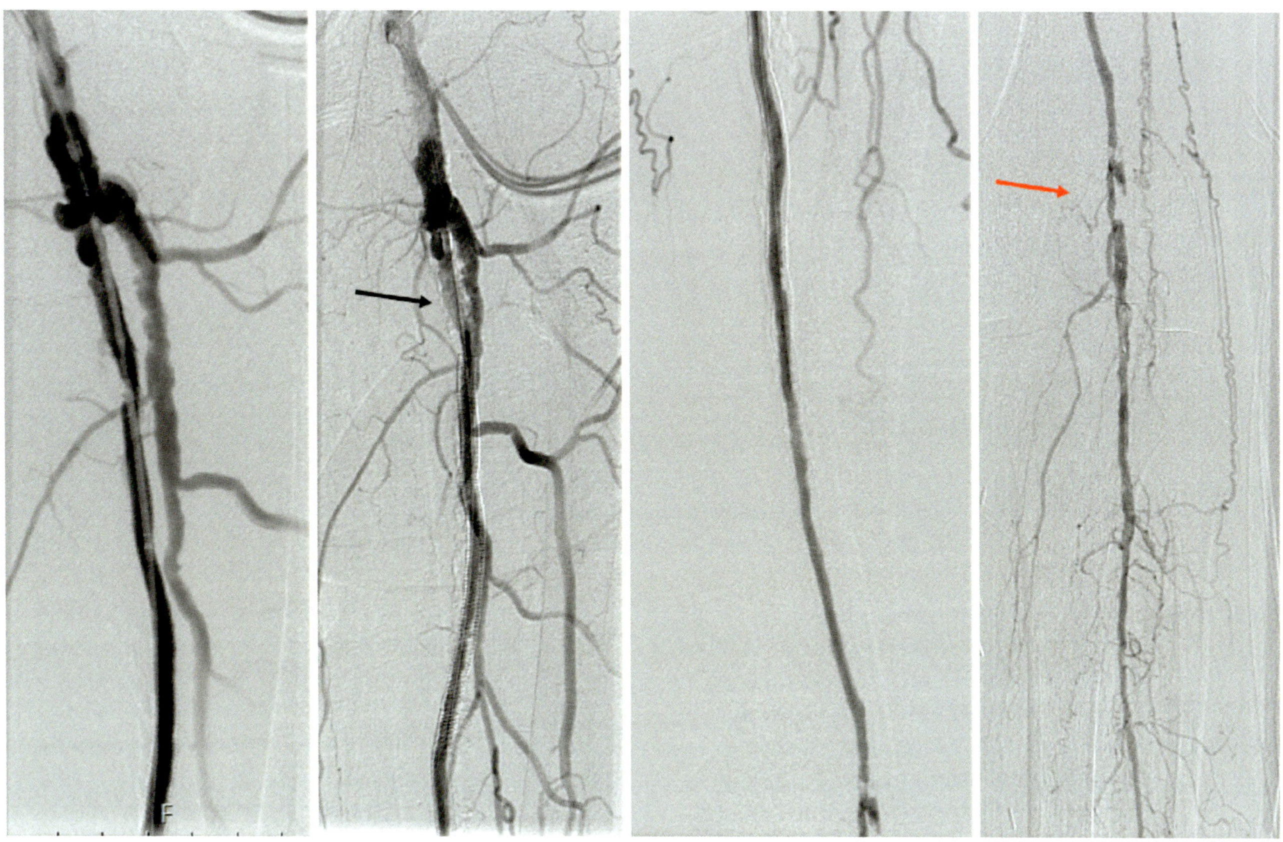

Fig. 19.3 Images after the atherectomy and the plain balloon angioplasties, with remaining stenosis in the proximal left SFA (black arrow) and the P2 popliteal artery segment (red arrow)

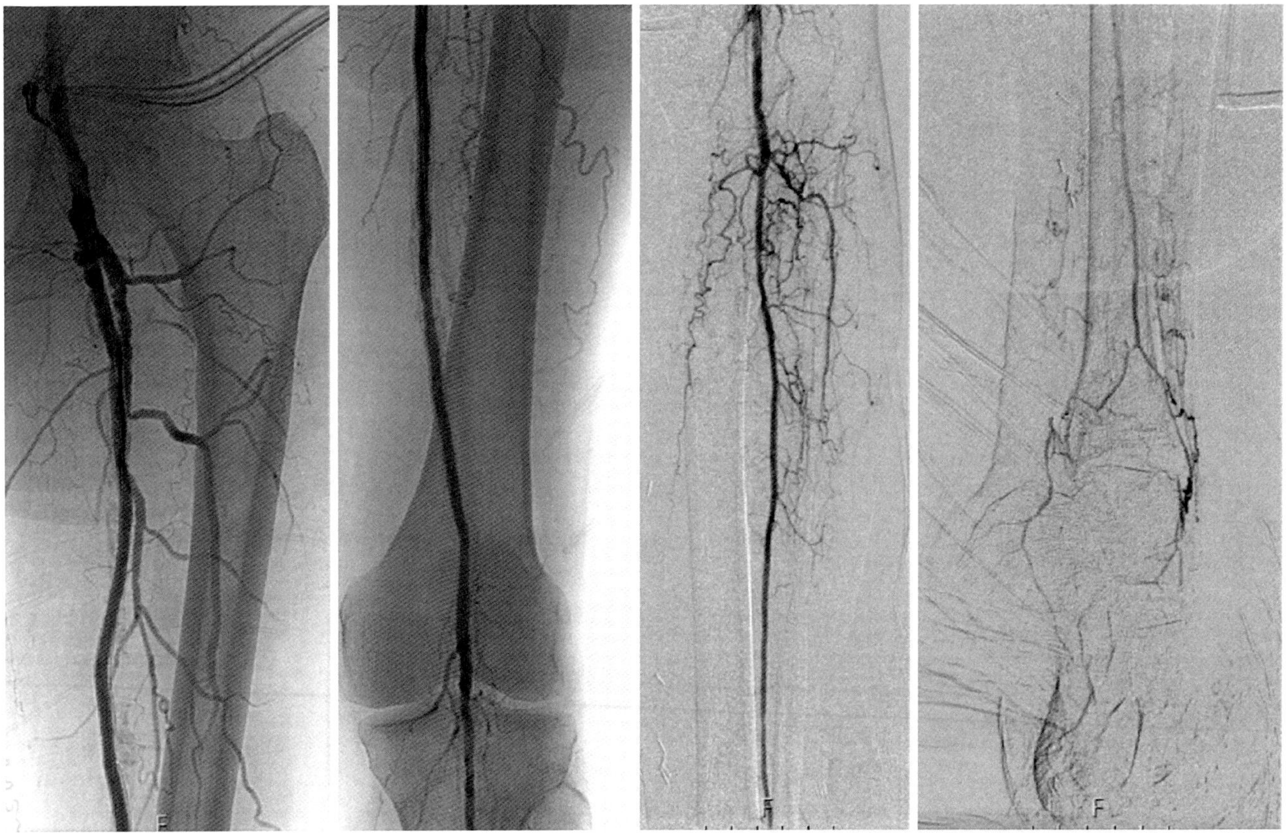

Fig. 19.4 Final angiograms, after additional stents were inserted in the proximal left SFA and the left popliteal artery. The peroneal and pedal artery patency is preserved

TLR of 88%, [4] which was mirrored in a larger 'real-world' series of over 400 patients demonstrating freedom from TLR of 87.5% (claudication) and 82.3% (chronic limb-threatening ischaemia) and low rates of bail-out stenting [5].

Directional Atherectomy

Directional atherectomy has undergone many iterative improvements since the initial experience reported with the Simpson atherectomy catheter by Höfling et al. in 1987 [6]. The currently-available directional atherectomy catheters include direct descendants of the Simpson device, namely the '-Hawk' family from Medtronic and the optical coherence tomography (OCT)-guided Pantheris™ from Avinger, along with the 2.4 mm deflecting and 2.2 mm deflected Phoenix™ atherectomy catheters from Philips.

The Hawk family (HawkOne™, SilverHawk™ and TurboHawk™) and Pantheris™ devices mandate placement of a distal filter, and collect plaque within a nosecone that must be passed beyond the treatment segment and requires removal each time it fills up with debris. The DEFINITIVE AR study enrolled 102 patients Rutherford classes 2–4 and randomized them 1:1 to treatment with plain then drug-coated balloon angioplasty or directional atherectomy with SilverHawk™ or TurboHawk™ devices then drug-coated balloon angioplasty. Although there was no significant difference in primary patency via ultrasound at 1 year in the atherectomy vs. no atherectomy group (84.6% vs. 81.3%), the atherectomy group had a higher rate of technical success defined as ≤30% residual stenosis following protocol-assigned treatment (89.6% vs. 64.2%) and a lower rate of type C/D dissection (2.1% vs. 18.5%) [7]. The REALITY study was a physician-initiated multicentre trial enrolling 102 patients with calcified femoropopliteal lesions with mean length of 17.9 cm (39% chronic total occlusions), using the HawkOne™ device and drug-coated balloon angioplasty. The bail-out stenting rate was 8.8% and the 1-year primary patency rate 76.7% in this challenging cohort [8].

The VISION trial enrolled 158 patients with 198 lesions for treatment with the Pantheris™ device, returning a 93.6% freedom from clinically-driven TLR rate at 6 months, but also demonstrating a reduction in stenosis down to 35.5% after Pantheris™ treatment alone (i.e. before adjunctive angioplasty) in the CTO subset and <1% advential content in the removed debris [9]—hinting at the advantages in precision and 360° plaque clearance that real-time OCT guidance might offer.

Orbital Atherectomy

The Stealth360™ device from Cardiovascular Systems Inc. employs a small diamond-coated 'crown' mounted on a catheter that when activated orbits around a proprietary guidewire in a 360° arc at very high speeds. The contact of the crown with plaque causes a sanding effect to gain lumen, and the generated particles of debris are considered to be small enough to pass through capillaries i.e. their embolization should be of no clinical consequence, and therefore no filter placement is required. In the COMPLIANCE 360° TRIAL, 50 patients with calcified femoropopliteal atherosclerotic disease were randomized to treatment with balloon angioplasty alone or orbital atherectomy plus balloon angioplasty, with a significantly lower bail-out stent rate seen in the orbital atherectomy group (5.3% vs. 77.8%) but a similar rate of freedom from target lesion revascularization at 1 year (81.2% vs. 78.3%) [10]. The crowns for the Stealth360 device are available in a number of sizes going down to 1.25 mm to treat tibio-pedal vessels, and ongoing research aims to study the safety and efficacy of this device in calcified distal disease.

Aspiration Atherothrombectomy

The Rotarex™ device from BD comes in 6-, 8- or 10-French sizes and consists of two large side-facing distal windows which aspirate occlusive material detached from the vessel wall by catheter passage, fragment it within a system of high-speed rotating blades and then remove it from the body. There is a capital outlay required for the drive system with a foot-pedal, but all the available Rotarex™ and Aspirex™ (thrombectomy) catheters are compatible with this unit. As with other atherectomy devices level 1 evidence is limited, but there is a growing body of case series suggesting the performance of this device in removing the mixed-density thrombotic material that builds up within occluded stents may warrant its preferred use for this application, for example a retrospective multicentre analysis of 128 patients returning a primary clinical success/patency rate of 92.3% at 1 year in iliofemoral in-stent restenosis with a 5.5% distal embolization rate (all successfully treated endovascularly) [11]. However, the device has also proven its effectiveness in management of native arterial occlusion in both the acute and sub-acute settings, achieving a technical success rate of 97.7%, (with 74.1% of patients maintaining their improvement in Rutherford class after one year follow-up) in a large single-centre series of 525 consecutive patients with mean lesion length of 159 mm [12].

Laser Atherectomy

There are two groups of laser atherectomy devices currently on the market, the Turbo-Power/Turbo-Elite from Philips and the Auryon™ from AngioDynamics, but both are characterized by requiring a relatively large console to generate the laser beam, which confers a higher start-up cost than other atherectomy devices.

The Philips group of catheters range from 0.9 to 2.5 mm in diameter, and utilize an excimer ultraviolet laser with fluence of 30–60 mJ/mm^2 and repetition rate of 25–80 Hz to treat atheromatous plaque and in-stent restenosis by three mechanisms: ablation from the light pulse of the laser itself, a sonic wave which has theoretical effects on calcium within the vessel wall and bursting of the formed vapour bubbles which debulk and macerate mixed morphology plaque. Filters are used at the discretion of the operator. The EXCITE-ISR study was a prospective multicentre randomized control trial which confirmed that in a study population of femoropopliteal lesions with mean length almost 20 cm of which around a third were total occlusions, laser atherectomy with the Turbo-Elite device resulted in a significantly better freedom from target lesion revascularization at 6 months compared to plain balloon angioplasty (73.5% versus 51.8%, $p < 0.005$) [13].

The Auryon system is a relative newcomer, using a smaller base unit which can be plugged into a standard power socket and generates a solid-state third harmonic pulsed Nd:YAG laser with an output of 355 nm. The laser catheters range in size from 0.9 to 2.35 mm; however, the 2 mm and 2.35 mm catheters have a built-in aspiration system to reduce the risk of distal embolization and the 2.35 mm catheter incorporates an off-centring mechanism to better target eccentric lesions. Being a relatively novel device, there is limited clinical data on its efficacy, with the EX-PAD-03 trial of 107 lesions in 97 patients (mean lesion length 5.4 cm with 26.2% severely calcified and 79.4% de novo) demonstrating that laser use alone resulted in a mean percentage stenosis reduction of 33.6% with no distal embolization, and a very low rate of bail-out stenting (0.9%). Patency was 96.8% at 30 days, but long-term follow-up of this cohort is awaited [14].

Indications for Use of Atherectomy or Other Calcium-Modifying Device

The decision for deployment of an atherectomy or other calcium-modifying device during a revascularization procedure requires consideration of multiple factors. A thorough analysis of the pre-operative imaging must be performed to assess the plaque burden and characteristics, taking into account lesion location, stenotic versus occlusive disease, presence and severity of calcium and quality of the arterial outflow to the segment being targeted. Although atherectomy or other calcium-modifying devices can technically be used as a vessel-preparation tool in any patient, the operator should be convinced that the arterial access is sufficient to deliver the catheter to the target lesion, that use of the device will confer an added benefit over angioplasty alone and that an embolic complication can be managed safely without risk of severely compromising the patient's outcome (e.g. treating the femoropopliteal segment in a patient with chronic limb-threatening ischaemia who only has a single tibial run-off vessel).

A Cochrane review in 2020 suggested that the beneficial effects of atherectomy over angioplasty/stenting alone may be limited to reduced dissection and need for bail-out stenting, although with the limited data available, the studies included in the meta-analysis were small and considered to be at high risk of bias [15]. A systematic review of 7 studies including 497 patients treated with directional atherectomy followed by drug coated balloon angioplasty in the common femoral artery showed primary and secondary patency of 92% and 98% respectively at 1 year, with a 1–6% complication rate [16]. A systematic review of 6 studies with 470 patients concluded that the atherectomy combined with drug-coated balloon angioplasty in the femoropopliteal segment decreased the need for bail-out stenting, compared to the initial treatment with drug-coated balloon alone [17]. In below-the-knee vessels, a systematic review including six studies with a total of 1062 patients treated with different atherectomy devices concluded that atherectomy in this territory has a high safety and efficacy profile and durable outcomes at 12 months [18]. However, it recognized that in the mid-term the clinical success of atherectomy was compromised by increased target lesion revascularization and major adverse limb event rates, resulting in the conclusion that compared to other endovascular techniques for below-the-knee treatment, there is a slight lead of atherectomy at 1 year but equivalent clinical performance in the mid-term.

There is little head-to-head data comparing atherectomy devices in the periphery, and none where device assignation was randomized. Janas et al. suggested that the Phoenix™ device resulted in a lower rate of target lesion revascularization compared to SilverHawk™ in a single-centre retrospective analysis [19]. Wissgott et al. compared the Pathway™ device (precursor to the Jetstream™) with the Rotarex, but the lesion length in the Rotarex group was more than twice that in the Pathway group and therefore although the rate of bail-out stenting, re-stenosis and complications were similar between the two devices, little meaningful comparison can be drawn from these data [20].

Given the paucity of high-quality data to justify routine use, the authors choose to employ atherectomy primarily to treat those chronic total occlusions in which the placement of a stent is deemed particularly undesirable, for example, younger patients, those with distal superficial femoral or popliteal artery disease and those with suboptimal tibial run-off.

Strategy and Endovascular Technique

Whilst atherectomy itself is very rarely painful if the catheter is tracking intraluminally, use of atherectomy or other calcium-modifying devices can increase procedure duration—particularly if distal embolization or vessel perforation occurs. Therefore, consideration must be given to patient tolerance, and an appropriate plan for anaesthetic escalation agreed and resourced prior to starting the procedure.

Ipsilateral antegrade common femoral artery is preferred to maximize device pushability but in the case of flush superficial femoral artery (SFA) occlusions and/or short common femoral artery (CFA) a contralateral groin puncture and placement of an up-and-over sheath should be considered to allow a safe approach to the lesion. The majority of femoropopliteal atherectomy catheters require a 0.014″ wire and a 7-French access sheath, so during treatment of the SFA origin there is a chance of the sheath accidentally displacing completely out of the CFA potentially leaving only a small-calibre guidewire in situ. If this occurs, the immediate priority is to cease running of the atherectomy device and performing manual compression over the puncture site. The atherectomy catheter can then be carefully withdrawn over the 0.014″ guidewire, which is then serially upsized to 0.035″ to allow replacement of the original 7-French access sheath (in our experience dedicated radial artery access sheaths will frequently track through the soft tissues of the groin over the 0.014″ wire). If using a contralateral groin access, placing a crossover sheath at least 1–2-French bigger than required for the atherectomy catheter will ease device delivery over the aortic bifurcation and improve control when the device is running, particularly in the case of steep and/or calcified aortic bifurcations.

Lesion crossing must be intraluminal to safely use atherectomy. Not only does this reduce risk of the intra-procedural perforation, but Tarricone et al. demonstrated a significantly increased risk of duplex-adjudicated restenosis at 1 year in patients with evidence of adventitial or medial injury in the plaque removed following directional atherectomy versus those without (97% vs. 11%) [21]. We strongly advocate the use of intravascular imaging such as intravascular ultrasound (IVUS) or optical coherence tomography (OCT) to confirm intraluminal guidewire passage on a routine basis before deciding to employ atherectomy. Note that if only a short segment of the lesion crossing is subintimal, the remainder of the intraluminal passage can be marked and safely treated with atherectomy.

Once the lesion has been crossed, a filter can be placed if mandated by the atherectomy device selected or desired by the operator. The SpiderFX™ from Medtronic is mounted in an off-centre-axis orientation onto a 0.014″ guidewire and is available in 3–7 mm diameters (suitable for placement in vessels 2–7 mm in diameter). As the filter element is attached to the wire, it may be vulnerable to 'wire-whip' phenomenon, in which the rotational force from the activated atherectomy device causes first the wire and then the filter to whip and then spin, and may result in damage to the arterial wall or trapping of the atherectomy device on to the filter-wire. The Emboshield NAV6™ from Abbott (suitable for vessels 2.5–7 mm in diameter) is separately advanced over a proprietary 0.014″ guidewire (BareWire™) into the desired position, and as it is not attached to the wire cannot be affected by 'wire-whip'. However, its maximum debris collection volume is smaller and it requires several more steps to deploy and re-capture than the SpiderFX™. Despite the theoretical differences between the two devices, Krishnan et al. suggest that there is no significant difference in clinical outcome between them when used with directional, rotational or laser atherectomy [22].

Correct sizing of an atherectomy catheter may have an important role to play reducing risk of complications and improving clinical outcomes. Based on data from two RCTs, most coronary atherectomy operators do not use cutting devices larger than 60% of the reference vessel diameter [23, 24]. Although this phenomenon has not been assessed with the same rigor in peripheral atherectomy, we believe the same principles apply and therefore do not aggressively size devices close to the reference vessel diameter.

With any atherectomy device, slow and steady advancement through the target lesion(s) is recommended to increase effectiveness and reduce risk of complications. Slow advancement allows cutting blades to engage with plaque to achieve maximal cutting, rather than simply 'Dottering' through the lesion. In the case of the Jetstream™ device, slow advancement also gives the active aspiration system more time to clear debris before it can embolize. Alternating 1–2 cm of advancement with a few milimetres of withdrawal of the activated device can reduce the risk of the blades becoming stuck within plaque. Orbital atherectomy devices specifically mandate a 30-s rest period for every 30 s of treatment.

After completion of atherectomy treatment, repeat angiography and/or intravascular imaging is performed at the operator's discretion. If a filter has been used, this should be inspected angiographically or ex vivo for evidence of debris. The filter should then be replaced beyond the atherectomized segment before performing plain-balloon angioplasty with or without adjunctive drug-coated balloon angioplasty. Maintenance of a filter is important, as the atherectomy may have introduced instability to plaques which remain in situ but are vulnerable to fragmenting during the angioplasty then embolizing after deflation of the balloon. Further angiography and/or intravascular imaging is then performed to assess for residual stenosis and flow-limiting dissection, which can be further treated with bail-out stenting as deemed appropriate for the severity of the residual lesion and the clinical status of the patient.

Complications During Atherectomy and Their Management

The primary complication during atherectomy is distal embolization. Both fibrotic and calcific atherosclerotic plaque are vulnerable to fragmentation and embolization, along with softer, more thrombotic components of chronic total occlusions. The embolized material will frequently lodge at a stenosis distal to the treatment segment, turning it into an apparent occlusion on check angiography. The first sign of distal embolization may be a sudden distal ischaemic lower-limb pain reported by the patient, or it may be clinically silent. If distal embolization is suspected, the first two priorities are to maintain patient comfort by upgrading analgesia if they are awake and to ensure that heparinization is adequate to avoid thrombosis of the arteries distal to the location of the embolus. Attention should then be turned to mapping the location of the embolus to facilitate treatment. Although angiography may show an abrupt cut-off of, e.g., a tibial artery, crossing the cut-off point with a hydrophilic-tipped guidewire and advancing an appropriately sized catheter a short distance beyond the cut-off point to inject contrast should localize the proximal and distal extent of the embolus. Depending on the nature of the embolus, administration of a small dose of recombinant tissue plasminogen activator may be helpful in lysing thrombotic embolus or de novo thrombus forming distally. However, if safe to do so, attempts to aspirate the thrombus with a catheter should be made. This is achieved by advancing a catheter immediately beyond the location of the embolus, attaching a 20 mL Luer-lock syringe to the catheter hub and applying negative pressure, then immediately retracting the catheter slowly as the tip crosses the location of the embolus and then more quickly to completely remove it from the body. A 6-French aspiration catheter is usually appropriate for the femoropopliteal segment; however, a 4-French catheter may be required if embolus is lodged in tibial arteries less than 2 mm in diameter. Early and generous local administration of isosorbide dinitrate is recommended to reduce spasm caused by rapid catheter manipulations during the aspiration manoeuvres. If the embolus cannot be lysed or aspirated, balloon angioplasty of the segment where it has lodged can be performed, however there is risk of fragmentation and more distal travel of these fragments into the pedal circulation. Multiple trashes of tiny embolic particles into the pedal microcirculation can cause the "no re-flow" phenomenon, in which the named tibiopedal arteries are seen to be patent following atherectomy/angioplasty but with extremely sluggish flow. Catheter aspiration cannot be employed here, and if on-table local thrombolysis and nitrate administration is not successful in restoring brisk flow, the patient may require a 12–24-h period of treatment with an intravenous heparin infusion to allow some endogenous fibrinolysis and/or early distal collateralization to occur.

Perforation during atherectomy is rare unless the guidewire has taken a subintimal or extravascular course. It is frequently clinically silent and most often noticed incidentally on check angiography post-atherectomy, however if left untreated can sometimes progress to formation of thigh or calf hematoma with potential risk of compartment syndrome developing. After ensuring adequate heparinization, an angioplasty balloon sized 1:1 to the arterial segment incorporating the extravasation is inflated across the site of injury at low (sub-nominal) pressure for three to 5 min. The balloon is then deflated, removed and check angiography performed. If the perforation has sealed then no further treatment is required, however for persistent extravasation further low-pressure balloon inflation or even covered stent placement may be required.

Atherectomy devices may become stuck onto guidewires during treatment. This can usually be avoided by careful flushing of the device and maintaining adequate lubrication of the external portions of the guidewire. If it occurs however, then the device may need to be removed en-bloc with the guidewire, which also implies losing access to the lesion crossing. In selected devices, flushing any lumen of the atherectomy device that can be and running of the motor may facilitate removal of the device over the wire. It is also possible for the blades of a rotational atherectomy device to become stuck within a plaque. Again, this can be avoided with careful technique, but if it occurs then applying firm gentle traction to the device with the motor running can free the blades without injuring the arterial wall.

Non-atherectomy Calcium-Modifying Devices

The Shockwave™ intravascular lithotripsy system uses a semi-compliant balloon-catheter containing multiple emitters to deliver acoustic pressure energy in a pulsatile fashion to induce cracks in calcified atheroma. The compliance of the calcium is thus increased, allowing the plaque to be more effectively compressed towards the arterial wall when a non-compliant balloon is then inflated across the lesion. The DISRUPT-PAD III trial recently reported 2-year follow-up data, revealing use of the Shockwave™ system in treatment of a primarily claudicant population significantly reduced the requirement for bail-out stenting [25]. A meta-analysis of 9 studies including 681 patients concluded that intravascular lithotripsy is an effective and safe approach for calcified plaque modification in lower extremity peripheral arterial disease, achieving a diameter stenosis reduction of 59.3% (95% CI 53.30–65.31%) with minimal vascular complications [26]. However further high-quality evidence is required to elucidate the efficacy of intravascular lithotripsy with respect to different clinical characteristics such as lesion location and length, and in comparison with other treatment modalities such as atherectomy.

There are also several angioplasty-type devices available which come under the generalized category of cutting or scoring balloons, e.g., the Serranator balloons from Cagent Vascular [27, 28]. Along with the atherotome Flex catheter from VentureMed [29], these devices all work by causing small longitudinal cuts in arterial plaque to better increase its response to balloon angioplasty. The effectiveness and role for these devices in the treatment of calcified plaque remains under investigation.

Case Presentation

Continued from page 181

In view of the coral-reef proximal SFA plaque (Fig. 19.5), use of Jetstream™ atherectomy for calcium debulking and plaque modification was chosen. With the patient under sedation and local anaesthesia, the contralateral CFA was punctured and an 8-French sheath advanced across the aortic bifurcation to the distal right external iliac artery. The proximal SFA occlusion was crossed intraluminally (confirmed with IVUS, Fig. 19.6), a filter placed beyond the lesion

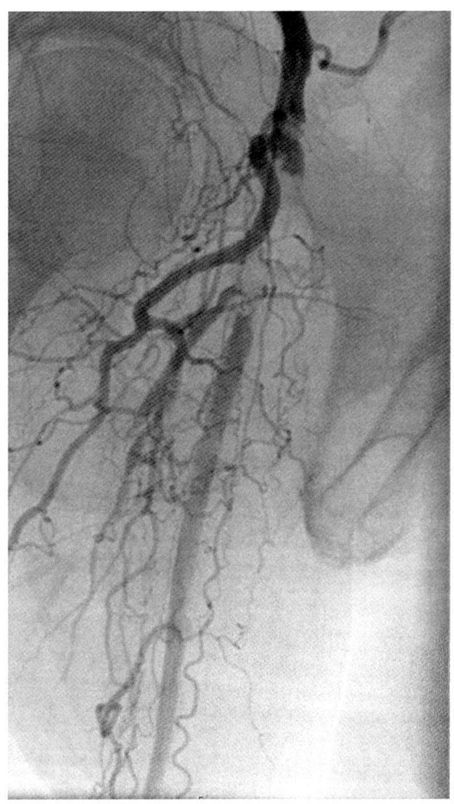

Fig. 19.5 Flush right SFA occlusion demonstrated on angiography

(Fig. 19.7) and then two slow passes with a 2.4 mm Jetstream™ catheter were made with the blades down, followed by two slow passes with the blades up. Repeat angiography showed debulking of calcium downstaging the occlusion to a stenosis (Fig. 19.8), along with a change in the reverberation of the coral-reef plaque on IVUS (Fig. 19.9).

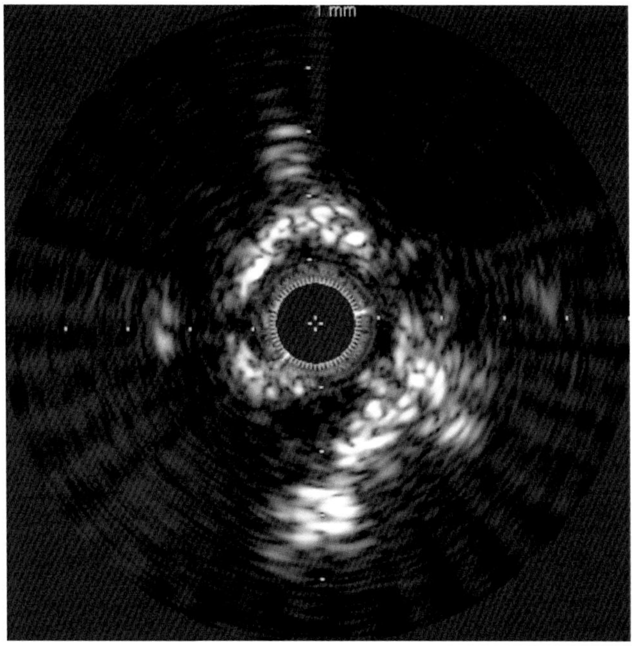

Fig. 19.6 IVUS shows heavily calcified plaque with reverberation artefact

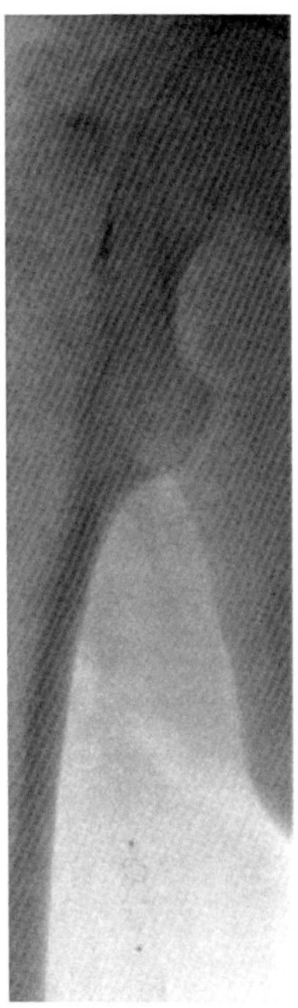

Fig. 19.7 The Jetstream™ catheter is advanced slowly through a lesion. A 7 mm SpiderFX™ filter is seen in the mid-SFA

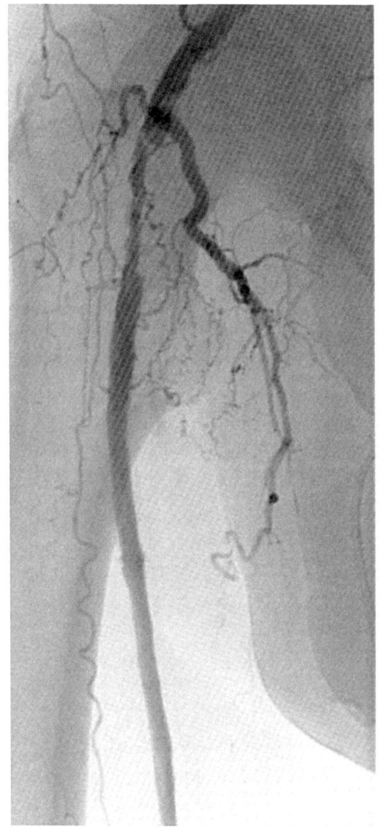

Fig. 19.8 Post-atherectomy angiography confirms downstaging of the occlusion to a stenosis

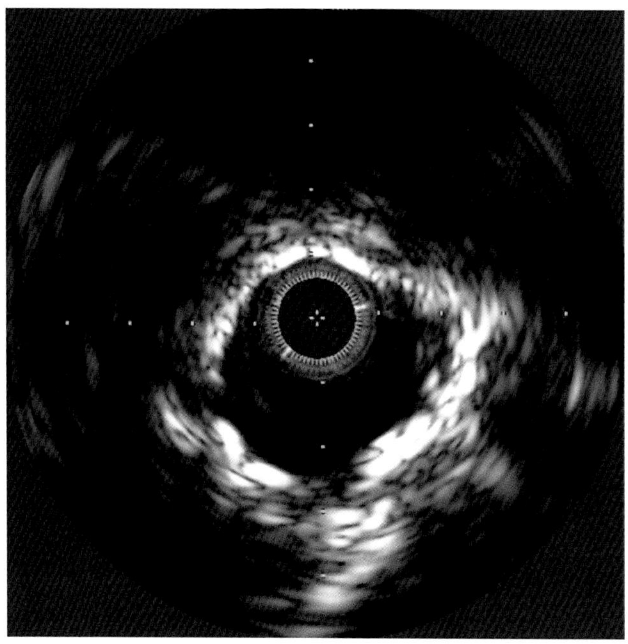

Fig. 19.9 Post-atherectomy IVUS shows calcium debulking with visible lumen and reduced reverberation artefact

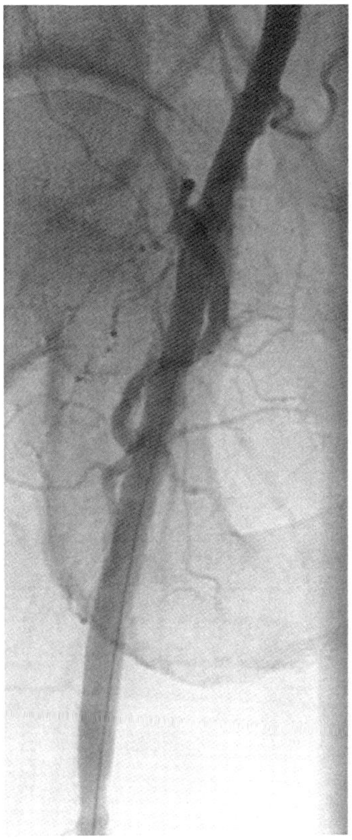

Fig. 19.10 Post-angioplasty angiography shows widely patent proximal SFA

Plain balloon angioplasty with a 7 mm balloon was performed, followed by retrieval of the filter. Check angiography revealed a widely patent proximal SFA (Fig. 19.10) with no evidence of perforation, dissection or distal embolization. The distal SFA stenoses were treated uneventfully with angioplasty and stent placement. After debridement of the necrotic tissue, the patient's foot wound healed successfully.

References

1. Höfling B, Pölnitz AV, Backa D, von Arnim T, Lauterjung L, Jauch KW, et al. Percutaneous removal of atheromatous plaques in peripheral arteries. Lancet. 1988;331(8582):384–6.
2. Gray WA, Garcia LA, Amin A, Shammas NW, JET Registry Investigators. Jetstream atherectomy system treatment of femoropopliteal arteries: results of the post-market JET Registry. Cardiovasc Revasc Med. 2018;19(5 Pt A):506–11.
3. Shammas NW, Jones-Miller S, Shammas GA, Shammas JW. Jetstream atherectomy in treating femoropopliteal in-stent restenosis: meta-analysis of the JETSTREAM-ISR and JET-ISR trials. J Invasive Cardiol. 2020;32(8):289–94.
4. Davis T, Ramaiah V, Niazi K, Gissler HM, Crabtree T. Safety and effectiveness of the Phoenix Atherectomy System in lower extremity arteries: early and midterm outcomes from the prospective multicenter EASE study. Vascular. 2017;25(6):563–75.
5. Giusca S, Hagstotz S, Lichtenberg M, Heinrich U, Eisenbach C, Andrassy M, et al. Phoenix atherectomy for patients with peripheral artery disease. EuroIntervention. 2022;18:e432.
6. Höfling B, Backa D, von Arnim T, Jauch KW, Simpson JB. Method, angiographic and clinical results of percutaneous atherectomy. VASA Suppl. 1987;20:337–42.
7. Zeller T, Langhoff R, Rocha-Singh KJ, Jaff MR, Blessing E, Amann-Vesti B, et al. Directional atherectomy followed by a paclitaxel-coated balloon to inhibit restenosis and maintain vessel patency: 12-month results of the DEFINITIVE AR study. Circ Cardiovasc Interv. 2017;10(9):e004848.
8. Rocha-Singh KJ, Sachar R, DeRubertis BG, Nolte-Ernsting CCA, Winscott JG, Krishnan P, et al. Directional atherectomy before paclitaxel coated balloon angioplasty in complex femoropopliteal disease: the VIVA REALITY study. Catheter Cardiovasc Interv. 2021;98(3):549–58.
9. Schwindt AG, Bennett JG Jr, Crowder WH, Dohad S, Janzer SF, George JC, et al. Lower extremity revascularization using optical coherence tomography-guided directional atherectomy: final results of the EValuatIon of the PantheriS OptIcal COherence tomography ImagiNg Atherectomy system for use in the peripheral vasculature (VISION) study. J Endovasc Ther. 2017;24(3):355–66.
10. Dattilo R, Himmelstein SI, Cuff RF. The COMPLIANCE 360° trial: a randomized, prospective, multicenter, pilot study comparing acute and long-term results of orbital atherectomy to balloon angioplasty for calcified femoropopliteal disease. J Invasive Cardiol. 2014;26(8):355–60.
11. Loffroy R, Edriss N, Gouault G, Chabanier A, Pernes J-M, Sauguet A, et al. Percutaneous mechanical atherothrombectomy using the Rotarex® S device in peripheral artery in-stent restenosis or occlusion: a French retrospective multicenter study on 128 patients. Quant Imaging Med Surg. 2020;10(1):283–93.
12. Freitas B, Steiner S, Bausback Y, Branzan D, Ülrich M, Bräunlich S, Schmidt A, Scheinert D. Rotarex mechanical debulking in acute and subacute arterial lesions. Angiology 2017;68(3):233–241. https://doi.org/10.1177/0003319716646682.
13. Dippel EJ, Makam P, Kovach R, George JC, Patlola R, Metzger C, et al. Randomized controlled study of excimer laser atherectomy for treatment of femoropopliteal in-stent restenosis: initial results from the EXCITE ISR trial (EXCImer laser randomized controlled study for treatment of FemoropopliTEal in-stent restenosis). JACC Cardiovasc Interv. 2015;8(1 Pt A):92–101.
14. Shammas NW, Chandra P, Brodmann M, Weinstock B, Sedillo G, Cawich I, et al. Acute and 30-day safety and effectiveness evaluation of Eximo Medical's B-Laser™, a novel atherectomy device, in subjects affected with infrainguinal peripheral arterial disease: results of the EX-PAD-03 trial. Cardiovasc Revasc Med. 2020;21(1):86–92.
15. Wardle BG, Ambler GK, Radwan RW, Hinchliffe RJ, Twine CP. Atherectomy for peripheral arterial disease. Cochrane Database Syst Rev. 2020;9(9):CD006680.
16. Usai MV, Lillu A, Asciutto G, Austermann MJ, Schwindt AG. Systematic review of atherectomy of inguinal arteries for atherosclerotic lesions. J Cardiovasc Surg. 2022;63(1):2–7.
17. Lin F, Wang H, Ding W, Chen G, Zhang Z. Atherectomy plus drug-coated balloon versus drug-coated balloon only for treatment of femoropopliteal artery lesions: a systematic review and meta-analysis. Vascular. 2021;29(6):883–96.
18. Pitoulias AG, Pitoulias GA. The role of atherectomy in BTK lesions: a systematic review. J Cardiovasc Surg. 2022;63(1):20–4.
19. Janas A, Milewski K, Buszman P, Kolarczyk-Haczyk A, Trendel W, Pruski M, et al. Comparison of long-term outcomes after directional versus rotational atherectomy in peripheral artery disease. Adv Interv Cardiol. 2020;16(1):76–81.

20. Wissgott C, Kamusella P, Andresen R. Treatment of femoropopliteal stenoses and occlusions with mechanical rotational catheters: comparison of results with the Rotarex and pathway devices. J Cardiovasc Surg. 2012;53(2):177–86.
21. Tarricone A, Ali Z, Rajamanickam A, Gujja K, Kapur V, Purushothaman KR, et al. Histopathological evidence of adventitial or medial injury is a strong predictor of restenosis during directional atherectomy for peripheral artery disease. J Endovasc Ther. 2015;22(5):712–5.
22. Krishnan P, Tarricone A, Gee A, Farhan S, Kamran H, Kini A, et al. Comparison and analysis between the NAV6 embolic protection filter and SpiderFX EPD filter in superficial femoral artery lesions. J Interv Cardiol. 2021;2021:9047596.
23. Whitlow PL, Bass TA, Kipperman RM, Sharaf BL, Ho KK, Cutlip DE, et al. Results of the study to determine rotablator and transluminal angioplasty strategy (STRATAS). Am J Cardiol. 2001;87(6):699–705.
24. Safian RD, Feldman T, Muller DW, Mason D, Schreiber T, Haik B, et al. Coronary angioplasty and Rotablator atherectomy trial (CARAT): immediate and late results of a prospective multicenter randomized trial. Catheter Cardiovasc Interv. 2001;53(2):213–20.
25. Tepe G, Brodmann M, Bachinsky W, Holden A, Zeller T, Mangalmurti S, et al. Intravascular lithotripsy for peripheral artery calcification: mid-term outcomes from the randomized disrupt PAD III trial. J Soc Cardiovasc Angiogr Intervent. 2022;1(1):100341.
26. Wong CP, Chan LP, Au DM, Chan HWC, Chan YC. Efficacy and safety of intravascular lithotripsy in lower extremity peripheral artery disease: a systematic review and meta-analysis. Eur J Vasc Endovasc Surg. 2022;63(3):446–56.
27. Holden A, Hill A, Walker A, Buckley B, Merrilees S, Nowakowski P, et al. PRELUDE prospective study of the Serranator device in the treatment of atherosclerotic lesions in the superficial femoral and popliteal arteries. J Endovasc Ther. 2019;26(1):18–25.
28. Holden A, Lichtenberg M, Nowakowski P, Wissgott C, Hertting K, Brodmann M. Prospective study of serration angioplasty in the infrapopliteal arteries using the Serranator device: PRELUDE BTK study. J Endovasc Ther. 2022;29(4):586–93.
29. Zeller T, Lopez L, Pigott JP. Acute outcomes with a novel plaque modification system in real-world femoropopliteal lesions. J Endovasc Ther. 2019;26(3):333–41.

Novel Endovascular Technologies in Femoropopliteal Disease

20

Ahmed Al-Nowfal and Irfan Ahmed

Case Study

A 38-year-old male with renal transplant secondary to poorly controlled diabetes myelitis, presented with gangrenous right second and third toes.

CT angiogram demonstrates significant 60% CFA stenosis. Profunda is widely patent. The SFA proximally contains a significant stenosis of around 80% and near-occlusive stenosis at the adductor canal. No significant popliteal or crura disease. Patient was discussed in the regional vascular MDT. A hybrid procedure was favoured with CFA endarterectomy and endovascular treatment of the superficial femoral and popliteal arteries.

Continued from page 207

Background

The superficial femoral and popliteal arteries (SFPA) are the most common site of disease in patients with peripheral arterial disease (PAD) and present some of the greatest challenges for endovascular treatment.

According to the American College of Cardiology (ACC) and American Heart Association (AHA) guidelines, endovascular treatment of superficial femoral and popliteal artery disease is indicated for individuals with significant disability due to intermittent claudication or critical limb ischemia when clinical features suggest:

- Patients with critical limb ischaemia who are at increased risk of amputation and major cardiovascular ischemic events.
- Claudicants who after being treated with guideline-directed management and therapy (including structured exercise therapy) present with persistent lifestyle-limiting claudication when there is a favourable risk-benefit ratio [1].

Since the 1980s there has been a remarkable advancement in the endovascular treatment of lower-limb peripheral artery disease, with the introduction of new interventional techniques and devices. Techniques such as percutaneous transluminal angioplasty with balloon dilation, stents, atherectomy, cutting balloons, drug-coated balloon angioplasty and percutaneous thrombectomy have become established in providing treatment with new techniques always on the rise.

Basic Endovascular Toolkit

When facing the challenge of treating a SFPA lesion it is preferable to approach the treatment from an antegrade approach. This will allow more talkability and pushability of the wires, catheters and devices. However, in patients with a large abdominal apron or hostile ipsilateral groin, a contralateral up-and-over approach can be utilised.

A list of suggested material is summarized in Table 20.1.

Table 20.1 Suggested endovascular toolkit for the crossing of SFPA stenosis or occlusions

	Manufacturer	Size/length
Wires		
Any standard-access wire		0.035″
Glidewire Floppy	Terumo Radifocus	0.035″
Glidewire Stiff	Terumo Radifocus	0.035″
Crossing wire	Any	0.14 and 0.18
Sheaths		
Any standard-access sheath	Any	4–7 Fr/11 cm
Balkin Sheath	Cook Medical	5.5–8 Fr/40 cm
Catheters		
4 Fr support catheter (BERN 2, Vert)	Any	0.035″/4 Fr/65–90 cm
Crossing catheter (quick cross, navicross, CXI)	Any	0.018″–0.035″/65–150 cm
Balloons		
0.35″ or 0.18 balloons	Any	4–8 mm

These are only a few suggested options based on the author's experience and can cover the vast majority of cases

A. Al-Nowfal (✉) · I. Ahmed
London, UK
e-mail: ahmed.al-nowfal@nhs.net; irfan.ahmed@gstt.nhs.uk

G. Geroulakos et al. (eds.), *Mastering Endovascular Techniques*, https://doi.org/10.1007/978-3-031-42735-0_20

Novel Techniques in Percutaneous Transluminal Angioplasty

Percutaneous transluminal angioplasty (PTA) is unchanged; it remains the preferred recommended initial endovascular treatment for SFPA disease, and it has a reported primary success rate for up to 80% in 1 year [2]. For many users the use of adjunctive techniques is largely as a salvage therapy for a suboptimal or failed result from balloon dilation.

There has been little development of standard PTA balloons on the markets. New devices such as The Medtronic Chocolate™* PTA balloon, which is a nitinol-caged balloon aimed to minimise flow-limiting dissections and reduce recoil by creating segmented grooves along the entire lesion, and the Medtronic Chameleon™ PTA balloon catheter, which aims to be a combination of a high-pressure balloon catheter and a diagnostic catheter, have failed to have a great impact on the market.

Stents

Endovascular stenting in SFPA prevents the problems of early elastic recoil, flow-limiting dissection and resistant stenosis after PTA and can thus be used for treatment of long and calcified lesions. However, the biomechanics of SFPA provides a unique challenge. It is subject to twisting, stretching, external compression, torsion and flexion. This can lead to stent fractures and restenosis.

Selection of the primary endovascular treatment for patients with SFPA disease with either PTA or a variety of stents as the initial approach has been assessed in multiple randomized trials, producing conflicting results [3], with current rans-Atlantic Inter-Society Consensus Document (TASC II) recommending balloon angioplasty as the initial preferred option for endovascular treatment SFPA lesions and for stent placement only after a suboptimal or failed result from PTA [4].

Stent designs have changed over the years, with interwoven nitinol stents such as the Abbot Supera™ Peripheral Stent and the Veryan BioMimics 3D Vascular Stent having become the mainstay stents used within the SFPA segment. Their selected patented construction demostrate improvement in radial strength, the ability to recover from being crushed and reduced foreshortening. This has led to better anatomical and clinical outcomes than PTA alone or older stainless-steel stents, with trials such as the RESILIENT trial stating that at 12 months, freedom from target lesion revascularisation was 87.3% for the stent group compared with 45.1% for the angioplasty alone group [5, 6].

Example of Stent Use

A 74-year-old female with a diabetic foot ulcer. Angiogram demonstrates mid superficial femoral artery stenotic disease (SFA) stenosis, distal SFA occlusion, some hibernating lumen in P1 popliteal artery and severe stenosis in p popliteal artery (Image 1). Mid SFA to P2 treated with 7 mm PTA and 6 mm DCB then stented with a 6.5 mm Supera (Image 2). Post-treatment angiogram demonstrates a now widely patent SFA (Image 3).

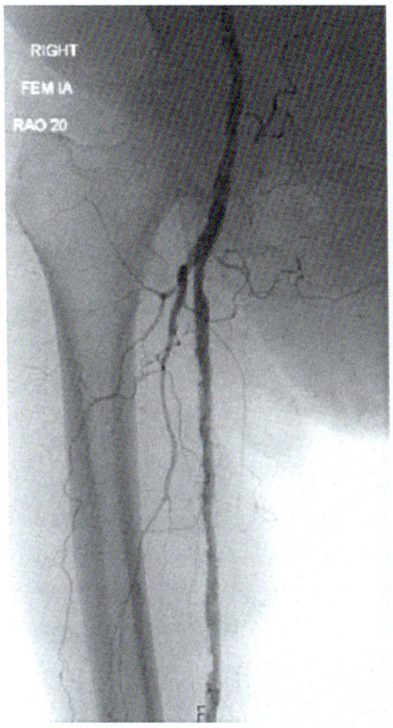

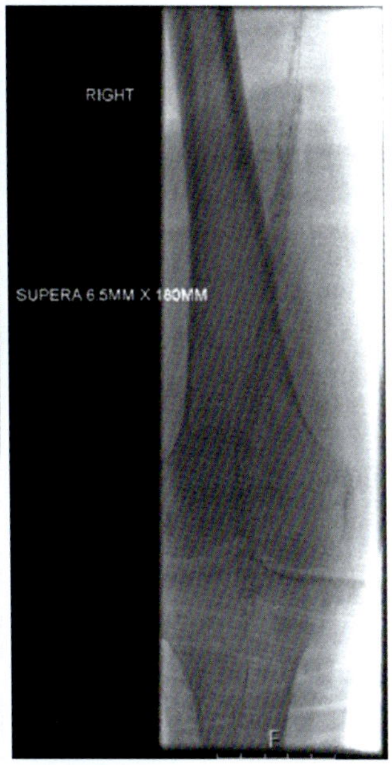

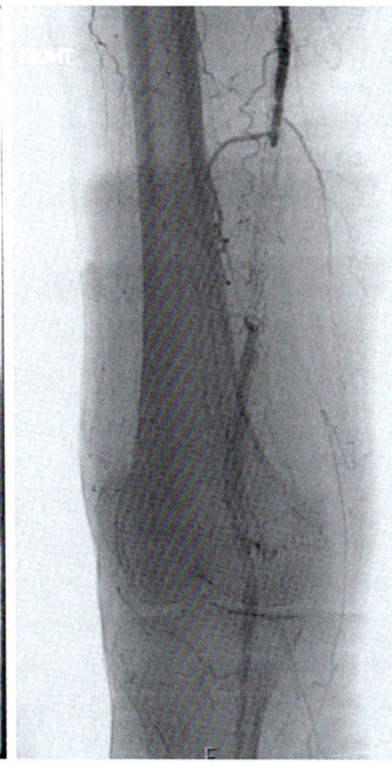

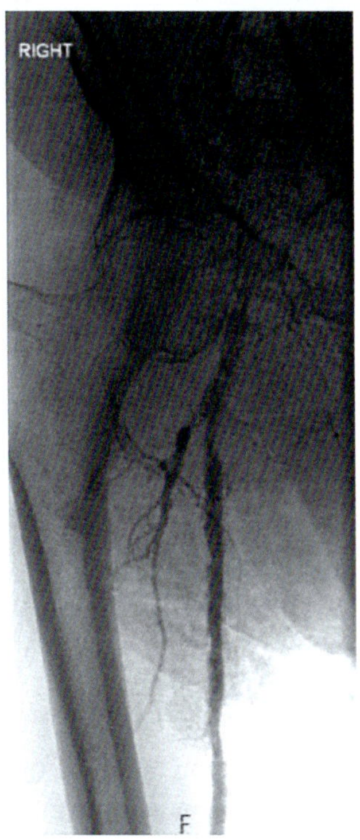

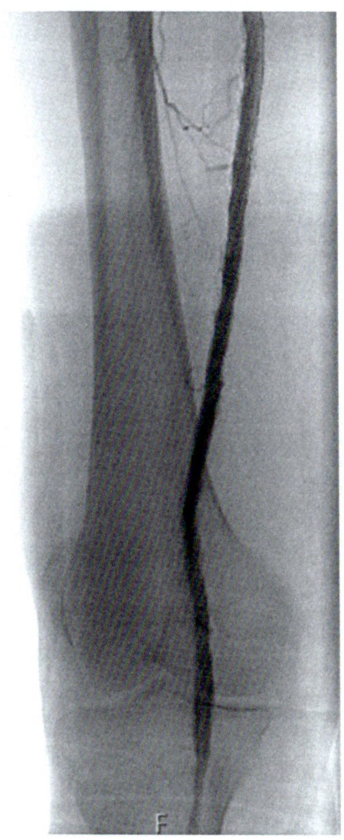

Drug-Eluting Stents and Balloons

The administration of drugs using drug-coated balloons (DCB) during angioplasty for SFPD can successfully deliver effective local tissue concentrations of anti-proliferative drugs to the lesions in the artery. This offers the potential for sustained anti-restenotic efficacy.

Randomised trials have shown superiority of paclitaxel DCBs (Boston Ranger DCB and Boston Eluvia drug-coated stent) over just plain-balloon angioplasty for treatment of PAD, and DCB is now considered the standard of care [7–15]. In March 2019 the FDA wrote to all health care professionals, warning them that the treatment of peripheral arterial disease with paclitaxel-coated balloons and paclitaxel eluting stents was potentially associated with increased mortality.

Katsanos et al.'s meta-analysis of randomised trials revealed a 68% relative risk increase in all-cause death with paclitaxel-containing balloons and stents, beginning at 2 years. This increased to 93% at 5 years [16]. This called into question the safety of paclitaxel-based drug-coated devices.

Alternative drugs for DCBs are therefore have been considered with Sirolimus has been the go-to alternative. Compared to Paclitaxel, sirolimus is cytostatic in its mode of action with a high margin of safety. It has a high transfer rate to the vessel wall and has been shown to effectively inhibit neointimal hyperplasia in the porcine coronary model [17]. In the coronary artery interventions, preliminary clinical studies using Sirolimus DCBs Quantitative coronary angiography revealed no differences in 6-month patency, with clinical events up to 12 months also did not differ between the groups [18].

However, they are yet to demonstrate success in the treatment peripheral lower limb angioplasty. The main difficulty with Sirolimus compared with paclitaxel has remained the difficulty in transfer: because of their relative lack of lipophilicity compared with paclitaxel, limus drugs are notoriously difficult to coat onto the balloon in such a way that they are retained until inflation at the site of action, and then slowly released in a sustained manner [19]. With historical trails such the SIROCCO II study, a randomized trial of sirolimus-coated nitinol stents compared to bare metal nitinol stents for SFPA disease, have also demonstrated no significant difference in patency rates, but this was because the restenosis rate in the bare stent group was unexpectedly low. At 24 months of follow-up, the restenosis rate in the sirolimus group was 22.9% versus 21.1% in the bare metal stent group [20].

Currently, Medalliance, SELUTION DCB are hopeful to have developed the technology to successfully transfer the drug to the affected periphery artery lesion [21]. Further trials in the field are ongoing with large, randomized trials such as Success PT and FUTURE-SFA are in the recruitment phase using a sirolimus-coated balloon versus standard balloon in the treatment of symptomatic SFPA disease.

Example of Drug-Eluting Stent and Balloons

An 83-year-old male with a right-leg critical limb ischaemia with new ulcers in the dorsum of the foot. Angiogram demonstrates widespread SFA (Image 1). Treated with 6 mm Boston Ranger DCB (Image 2) and resistant stenosis treated with a 7 mm Boston Eluvia drug-coated stent (Image 3). Post-treatment angiogram demonstrates a now widely patent SFA (Image 4).

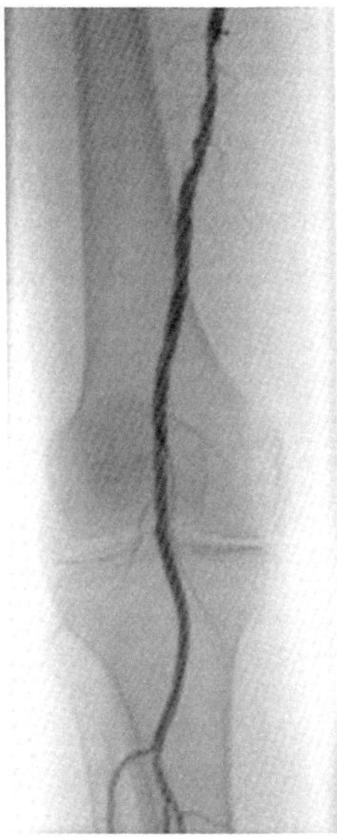

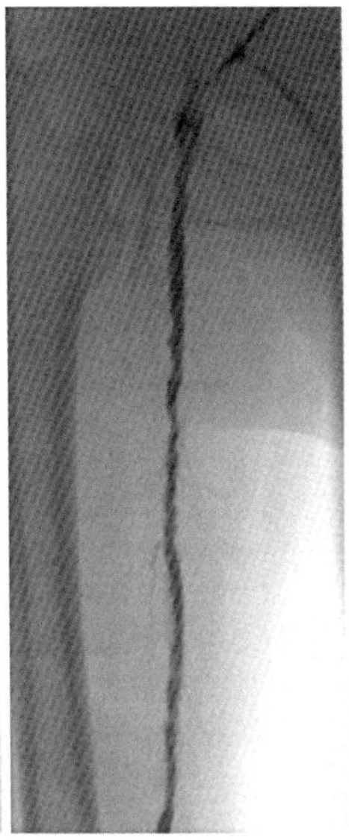

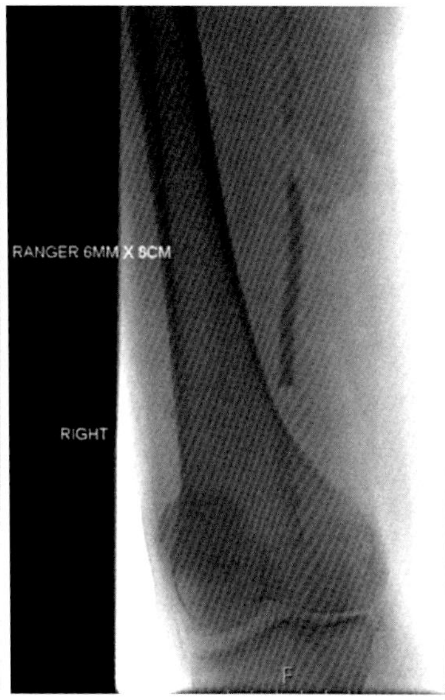

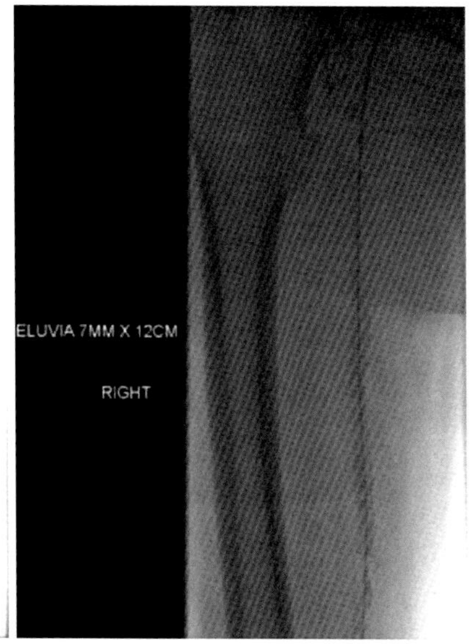

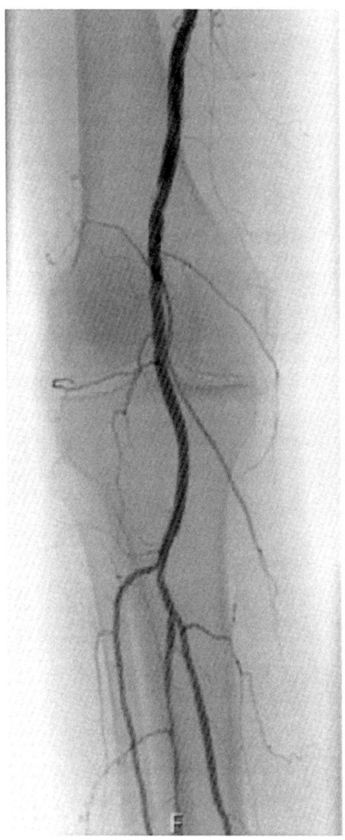

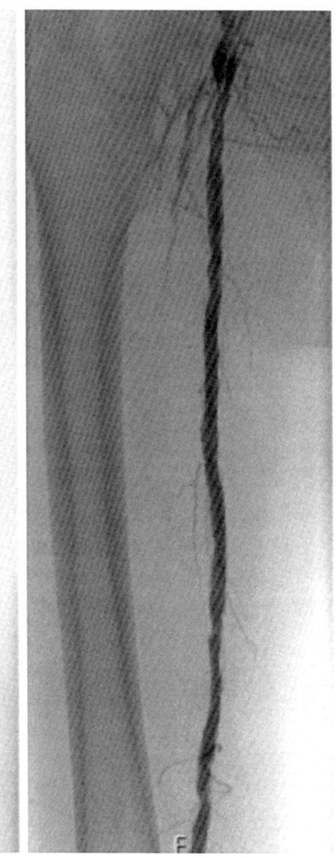

Atherectomy

Directional atherectomy was developed with the idea to increase procedural luminal gain while avoiding barotraumas and vessel recoil. They are aimed to create concentric lumens, optimising balloon-to-wall apposition for DCB or other adjunctive therapies.

Initial results using the Simpson AtheroCath (Guidant, Temecula, CA) were negative, stating that in lesions with greater length than 2 cm, the atherectomy results are significantly worse than plain angioplasty [22]. However, in the recent years there has been a new interest in debulking devices with the development of new atherectomy systems, such as the Boston Jetstream™ rotational atherectomy system or the Philips Phoenix atherectomy system. They are both engineered to treat a wide range of lesion characteristics such as: long, diffuse disease and chronic total occlusions (CTO) with mixed morphologies like calcium, plaque and thrombus. Recent studies such as the JET Registry observed the treatment effects of the Boston Jetstream™ Atherectomy System in peripheral arterial disease of the common femoral, superficial femoral or popliteal arteries demonstrate 77.2% patency, with 81.7% freedom from target lesion revascularisation within 12 months, with only 1.4% distal embolisation and 2.3% major adverse event [23]. The EASE study data also confirms Phoenix's ability to effectively treat a broad range of tissue types, from soft plaque to calcified arteries, for peripheral arterial lesions with similar high success rates [24].

Example of Atherectomy

A 76-year-old male with right diabetic foot disease. Osteomyelitis of the right forth toe and metatarsal remnant. Angiogram demonstrates patent proximal SFA. Complete occlusion of the distal SFA (Adductor canal) with a patent popliteal artery (Image 1). After luminal crossing of the lesion, covering filter inserted (Image 2). Treated with 24 mm Phoenix atherectomy through occluded segment. Post atherectomy treated with a 5 mm PTA and DCB. Post-treatment angiogram demonstrates a now widely patent SFA with no flow-limiting dissections (Image 3).

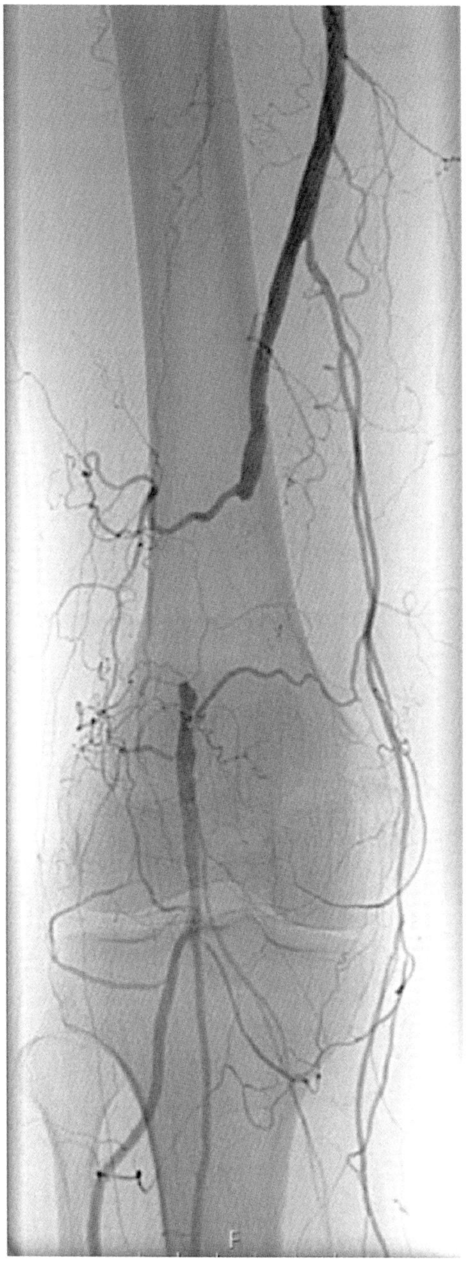

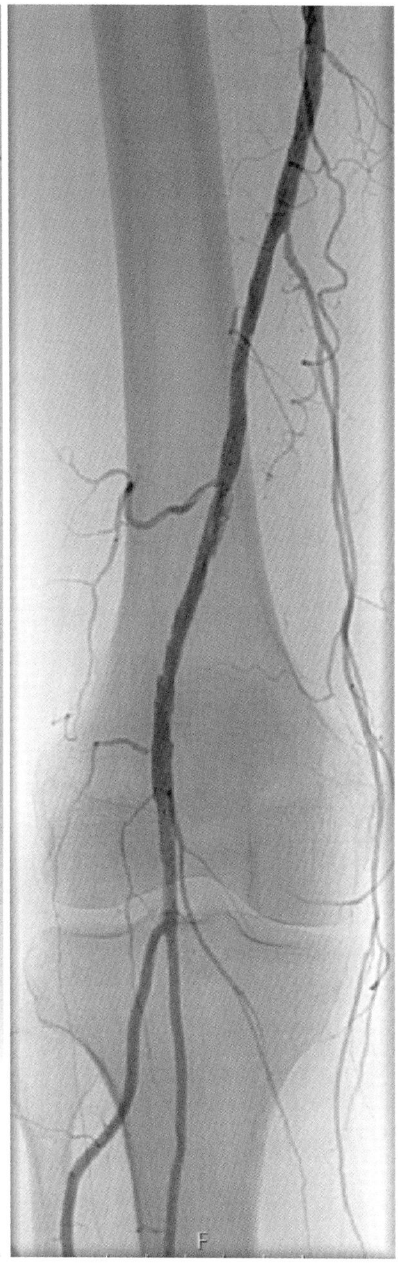

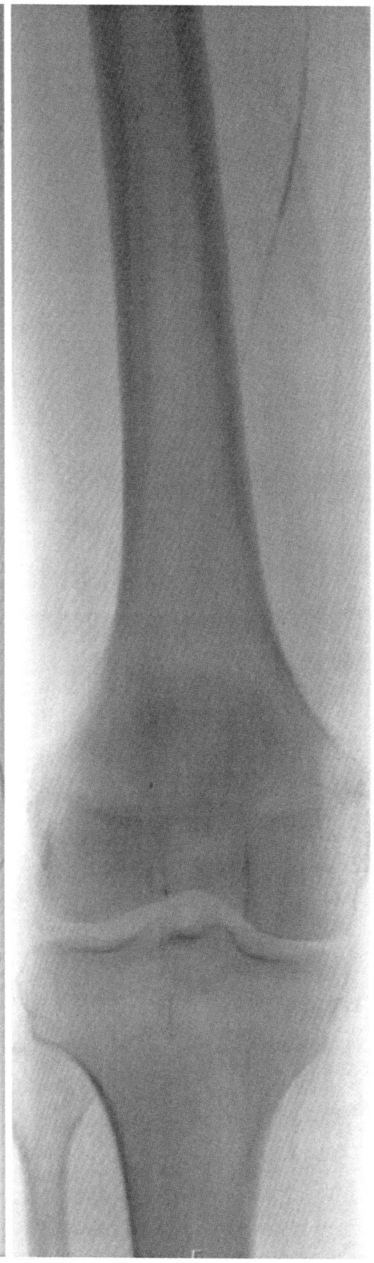

Intravascular Lithotripsy

Intravascular lithotripsy (IVL) is a novel approach for calcified plaque modification in in coronary and peripheral vessels. Diabetes mellitus or chronic kidney disease patients are amongst those who suffer the greatest prevalent intravascular calcification in PAD [25]. Aside from the known increased cardiovascular mortality and morbidity risk [26], calcification also interfere with the delivery of endovascular therapies in which it is responsible for suboptimal outcomes. These include poor vessel expansion using standard PTA balloons, with increased risk of restenosis, dissection, perforation and distal embolisation [27].

Shockwave medical Intravascular lithotripsy uses multiple emitters mounted in a traditional angioplasty balloon catheter that provide pulsatile sonic pressure energy to disrupt superficial and deep calcium without affecting local soft tissues or liberating emboli, thereby serving as a novel vessel preparation strategy to improve luminal compliance and facilitate definitive endovascular treatment [27].

A systematic review and meta-analysis looking at outcomes post IVL performed for PAD showed 75.5% of the lesions were reported to have severe calcification. Comparison between pre-IVL and post-IVL diameter stenosis demonstrated a diameter stenosis reduction of 59.3%. Vascular complications were rare, with flow-limiting dissection occurring in only 1.25% of cases [28].

The recently published PAD III trial states that primary patency at 1 year was significantly greater in the IVL arm 80.5% vs. non-IVL 68.0%. The requirement for provisional stenting was significantly lower in the IVL group 4.6% vs. non-IVL 18.3%. At 2 years, primary patency remained significantly greater in the IVL group 70.3% vs. non-IVL 51.3%, confirming IVL with a combination of DCB aids in the leave-nothing-behind approach [29].

Example of Intravascular Lithotripsy

A 56-year-old man with non-healing fourth and fifth toe ulcers. Tight calcified distal superficial femoral artery disease (SFA) (Image 1). Treated with 7 mm Shockwave IVL and 6 mm × 100 mm Selution DCB (Image 2). Post-treatment angiogram demonstrates a now widely patent SFA (Image 3).

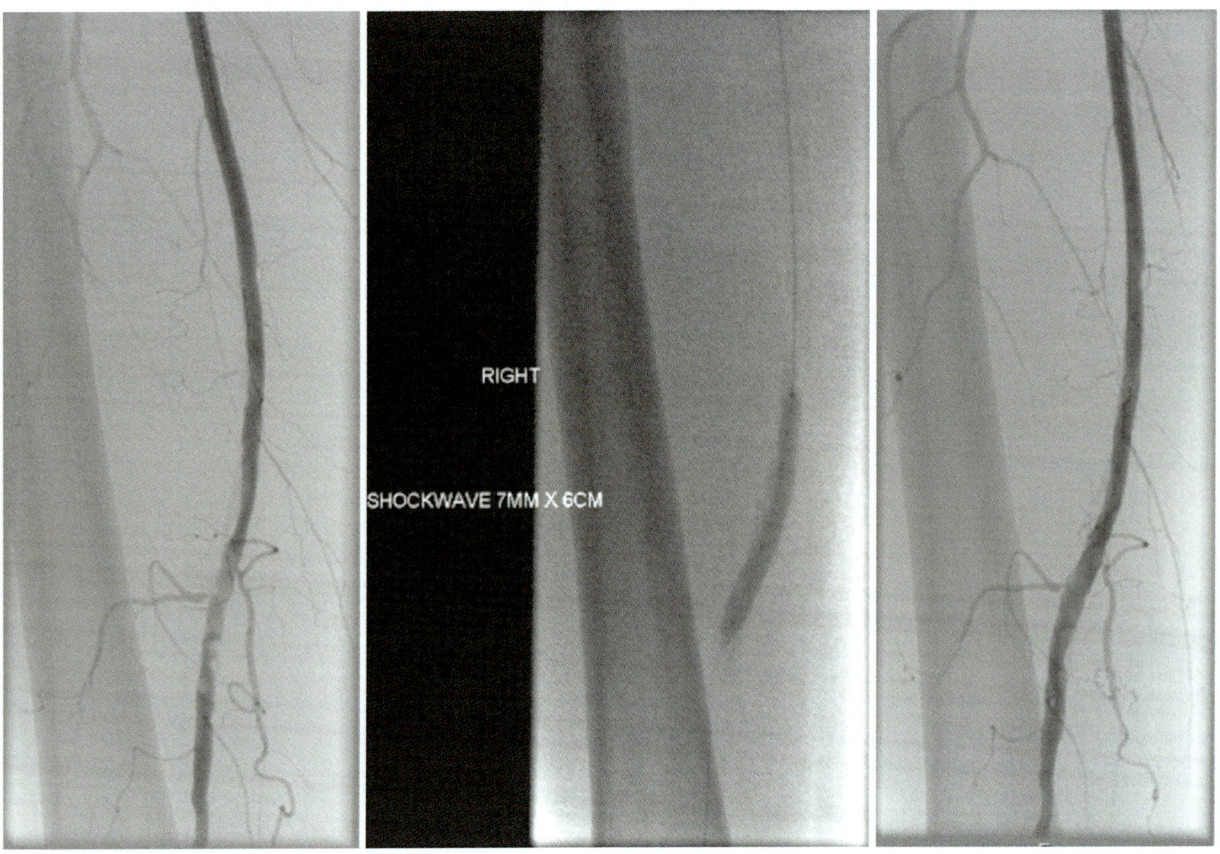

Re-entry Devices

Sub-intimal angioplasty technique was described by Bolia et al. (1990), allowing endovascular treatment of chronic total occlusions of the peripheral arteries [30]. The long patency of a subintimal tract was poor. The technique also increased the risk for propagation of the subintimal dissection into normal arteries beyond the occlusion damaging distal run off. However, improved wires and catheter, as well as the development of high-quality nitinol stents to address the acute technical and angiographic failure (such as flow limiting dissection) of subintimal angioplasty of chronic total occlusions, the use of subintimal angioplasty increased. As the technique became more accepted, it was becoming clear the inability to get the wire and catheter to re-enter the true lumen beyond the occlusion was the primary reason for technical failure affecting 10–15% of attempted cases [31–33].

Various methods were then developed to aid in re-entry using standard wires, catheters, and balloons, with distal pedal access becoming a favoured approach. The development of re-entry device, however, allowed a safe method to enter the true lumen, limiting the dissection of the vessel distal to the occlusion, and preserving critical collaterals and distal run-off vessels.

There are many re-entry devices on the market with the Cordis Outback® LTD® re-entry catheter having the largest share. It is a 5 Fr, multipurpose-type angled guide catheter with an integral nitinol hypotube ending in a curved needle tip intended to be advanced from the end of the catheter to penetrate from the dissection plane to the true lumen. A 0.014-inch guidewire could then be advanced through the hypotube into the true lumen. Technical success rates with the Outback in the SFPA was 64.5% with failure secondary to inability to re-enter the true lumen, difficulty tracking the device over a wire, acute angle of aortic bifurcation in up and over access difficulty in tracking over the wire in calcified occlusions, and difficulty in penetrating calcified plaque at the re-entry site [34]. Other devices such as the Philips Pioneer Plus Catheter PPlus and Upstream peripheral Go back catheter offer alternatives in the market.

Example of Re-entry

A 66-year-old male with critical limb ischaemia. Angiogram demonstrates short SFA stump with complete occlusion of the remaining SFA. There is reconstitution of the P1 segment of popliteal artery (Image 1). After reaching a hard distal cap, difficult to re-enter the true lumen. Therefore, an Outback-assisted re-entry at the P1/P2 junction. Wire position confirmed re-entry. This was treated with 5 mm PTA and DCB then stented with a Biomimics 6 mm × 150 mm deployed from the P1 POP back to the SFA origins. Post-stent ballooning with 6 mm PTA.

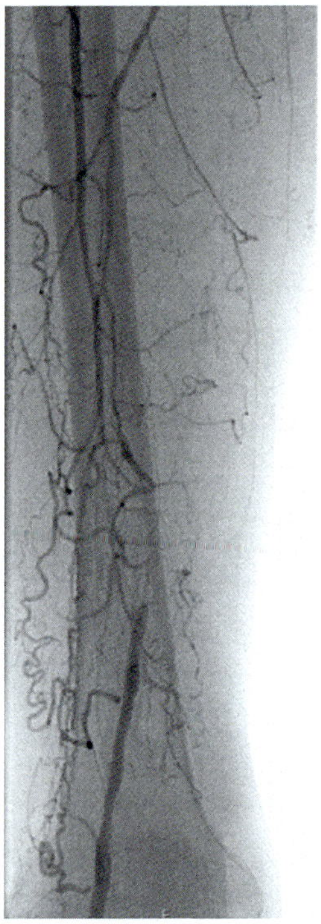
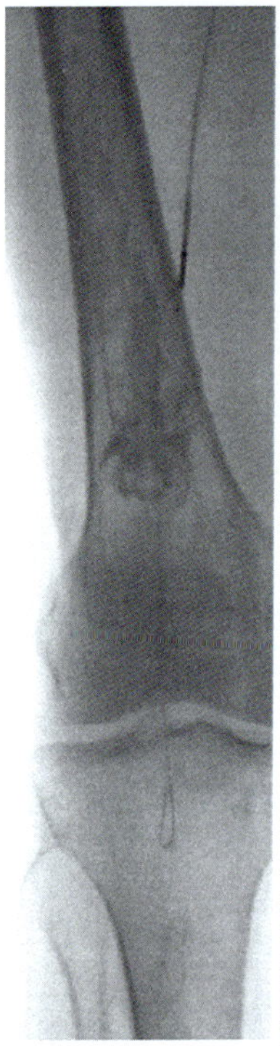
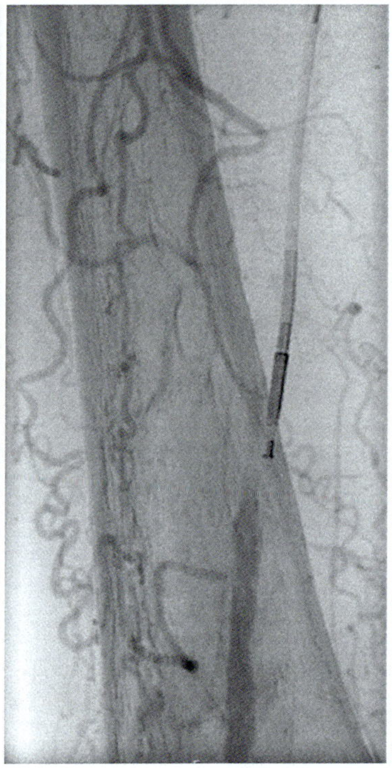
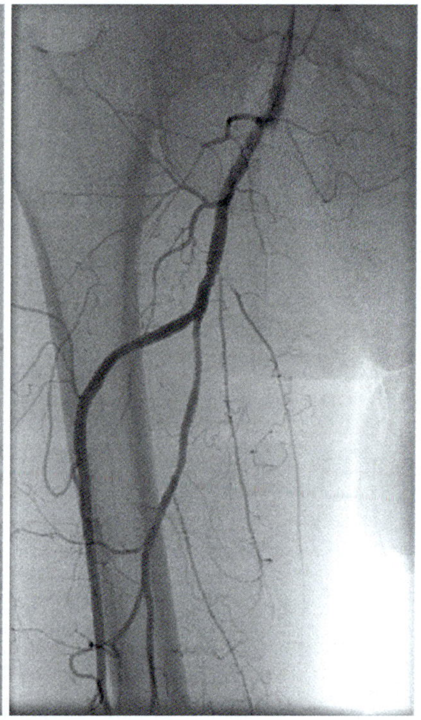

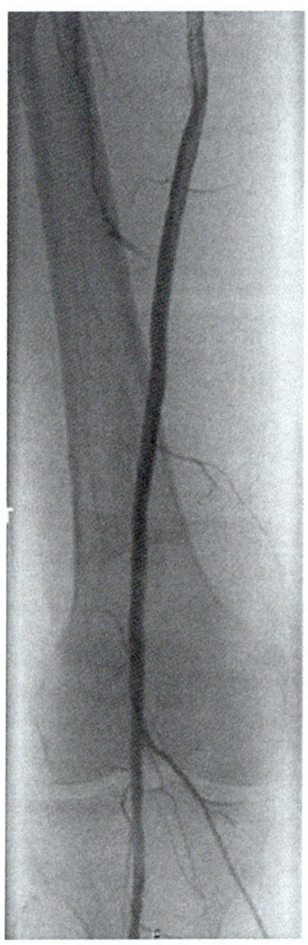
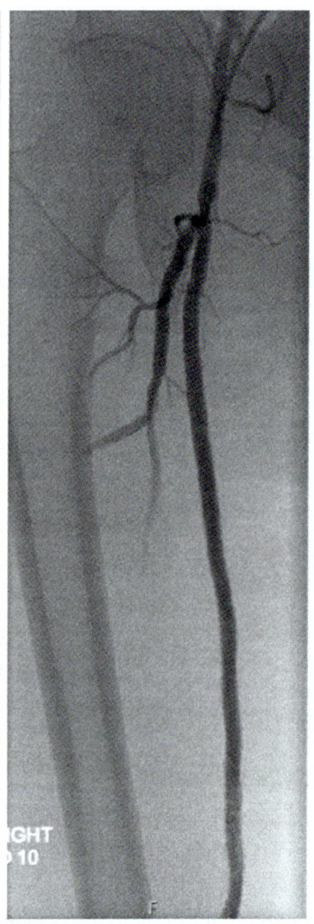

Tack Endovascular System

Severe flow-limiting post-angioplasty dissections occur in 42% of femoropopliteal angioplasties. These can negatively affect both short- and long-term patency. Acutely, dissection can reduce or obstruct flow, requiring additional therapeutic intervention. Moreover, within the first 2 years, the severe flow limiting dissection significantly lower patency rate and four times higher rates of t*arget lesion* revascularization compared to the non-severe group, with increased risk of restenosis [35].

Dissections can be treated by either prolonged ballooning or stenting. By scaffolding the vessel wall with high radial outward force, stents treat the dissection as well as residual stenosis improving patient outcomes. However, the aggressive radial force combined with an extensive amount of nitinol can cause vessel wall inflammation and lead to intimal hyperplasia formation and in-stent restenosis. It is reported up to 37% of restenosis occurs 1 year post-stent treatment. The unique dynamic forces exerted in the femoropopliteal segment can also lead to additional shear stress, inflammation, and occasional stent fracture [36].

The Phillips Tack endovascular system is a novel device specifically designed to address the limitations of stents while providing durable repair of post-PTA dissections in the SFPA. To reduce the metal surface area in contact with the luminal wall, Tack implants are short (6 mm), with an open-cell design resulting in lower chronic outward force compared with similar-sized stents. This allows focal dissection treatment and scaffolding while limiting the amount of biological injury and neo-intimal hyperplasia. The TOBA (Tack Optimized Balloon Angioplasty) II study states that 92.1% post angioplasty dissections resolved completely with only one bailout stent was required. At 12 months, there were no device fractures or clinically significant migrations, and significant improvements were noted in Rutherford category, ankle-brachial index, and quality of life [36].

Example of TACK Endovascular System

A 70-year-old female with diabetes mellitus II presented with gangrenous right third and forth toes. Angiogram demonstrates patent CFA and PFA. The SFA proximally contains a significant stenosis of around 80% and near-occlusive stenosis at the adductor canal. No significant popliteal or crura disease (Image 1). After crossing with a 0.14″ wire, IVUS was used to confirm luminal crossing. After treatment with atherectomy, POBA and DCB, there was a flow-limiting dissection in the proximal SFA (Image 2). This was initially treated with prolonged angioplasty with a 6 mm POBA, with limited success. Therefore, an INTACT 4.8 mm × 8 mm was used to treat the dissection (Image 3). Post-treatment angiogram demonstrates there is now a patent angiogram demonstrate a now widely patent SFA and popliteal artery (Image 4).

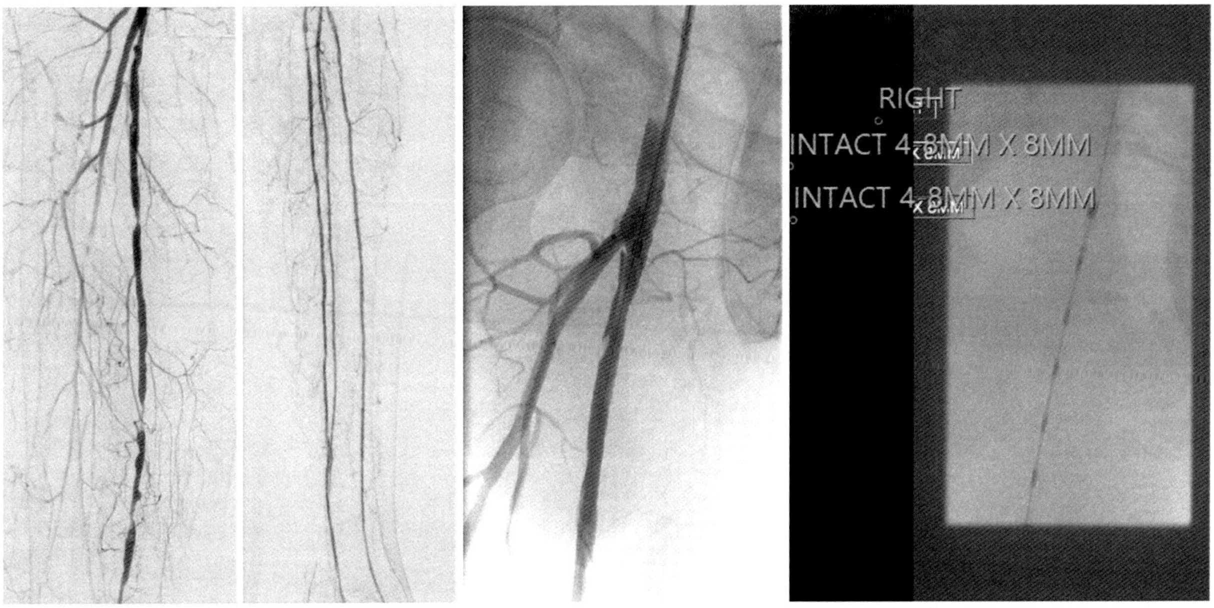

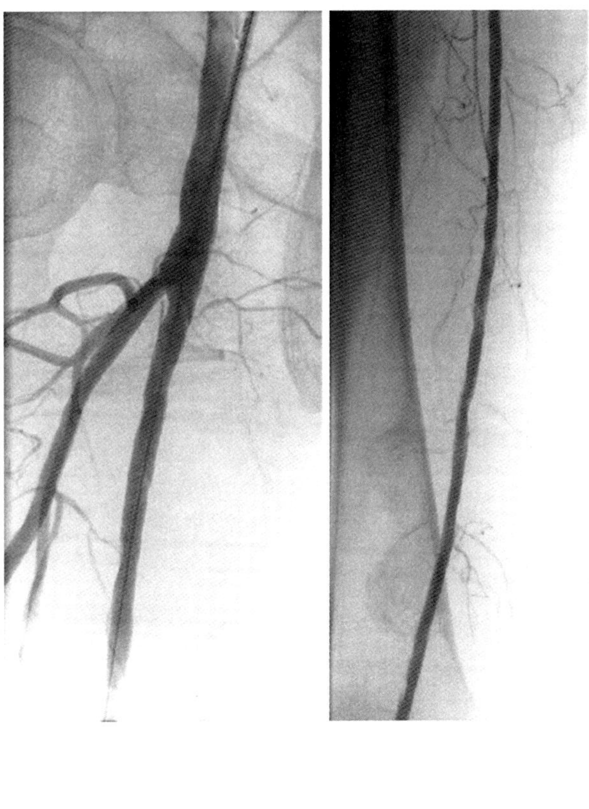

Intravascular Ultrasound

Intravascular ultrasound (IVUS) has emerged as a useful adjunctive tool to DSA in the evaluation of the vasculature. It may provide a more precise visualization to what happens into the blood vessel and play an increasing role in peripheral arterial occlusive procedures. IVUS catheters offer accurate cross-sectional imaging of arterial vessels with high dimensional accuracy and provide accurate information about lesion morphology. It enables assessment of the plaque morphology, vessel diameter, and the presence of arterial dissections. In a systematic review conducted by Makris et al. (2017), 13 clinical studies evaluated IVUS as an adjunct tool to DSA during endovascular interventions in patients with PAD in iliac, femoral or popliteal arteries. It was identified IVUS offered better outcomes in appropriate vessel size and prep for PTA and stenting, more accurate true-lumen re-entry and appropriate luminal crossing prior to atherectomy [37].

IVUS catheters are currently provided by two companies, Philips and Boston Scientific. The catheter sizes range from 2- to 4-French (Fr) and can be easily guided through a 5- or 6-Fr sheath. Larger IVUS catheters are also used for larger peripheral vessel applications and require larger sheaths. The larger IVUS catheters come over 0.035-inch guidewires, and the smaller ones that are more often used for requiring 0.018-inch or 0.014-inch guidewires. The length of IVUS catheters ranges from 90 to 150 cm, allowing imaging of infra-popliteal arteries via contralateral approach.

Example of Intravascular Ultrasound

A 67-year-old female with critical limb ischaemia with right second metatarsophalangeal neuroischemic ulcer. Angiogram demonstrates short SFA stump with complete occlusion of the remaining SFA. There is reconstitution of the P1 segment of popliteal artery (Image 1). After crossing with a 0.14″ wire, IVUS was used to confirm luminal crossing. It was also used to confirm the SFA/P1 popliteal diameter of 4.6–4.9 mm (Image 2). After treatment with atherectomy, POBA and DCB, IVUS was used to confirm patency of vessel (Image 3). It was also used to identify significant residual disease in the distal SFA, which required stenting (Image 4). Post treatment angiogram demonstrate there is now a patent angiogram demonstrate a now widely patent SFA and popliteal artery (Image 5).

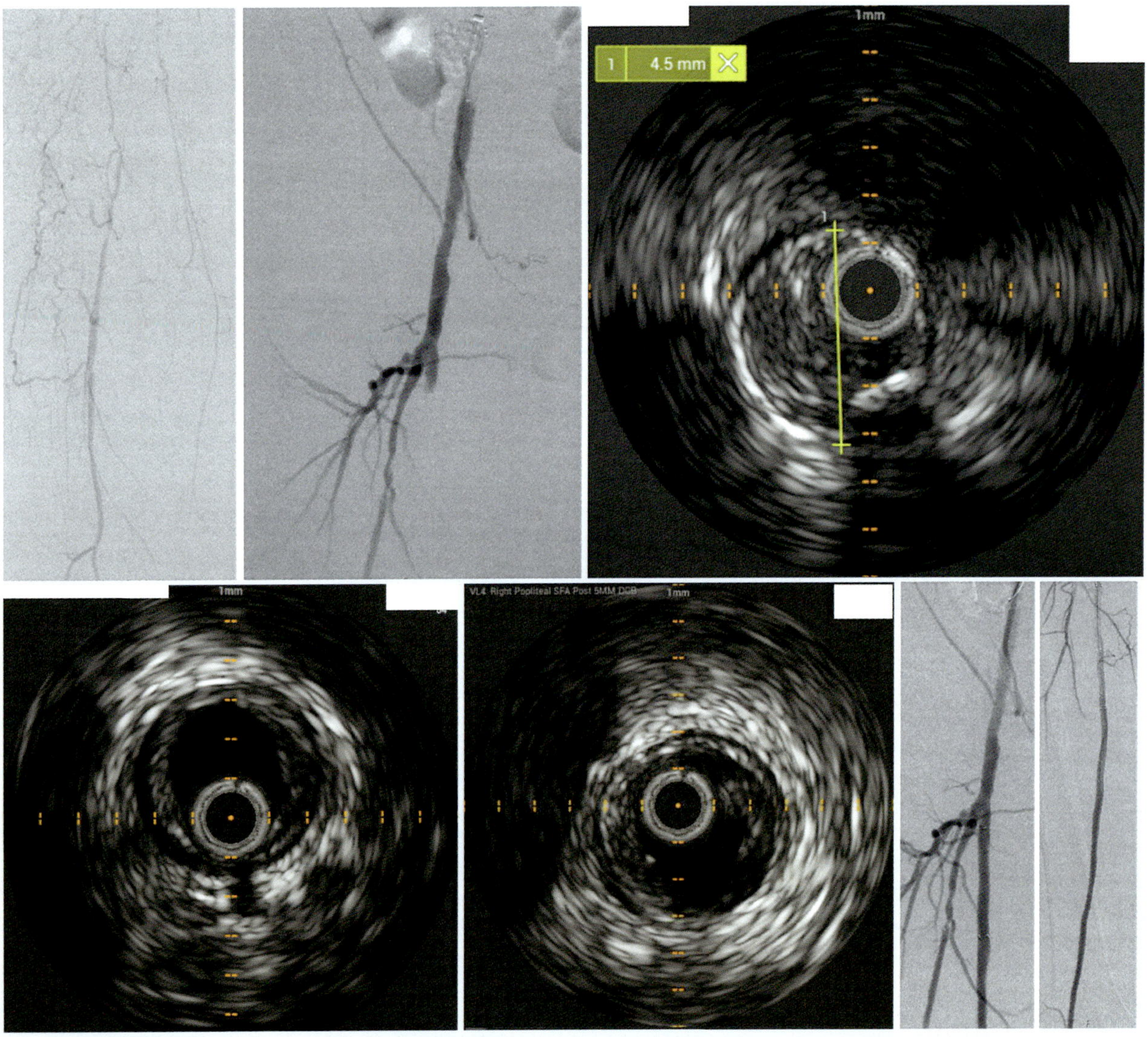

Case Study

Continued from page 193

Post open endarterectomy an 11 cm 6 Fr sheath was inserted in the CFA patch. After crossing with a 0.14″ wire. IVUS was used to confirm luminal crossing. After treatment with 2.2 mm Philips Phoenix atherectomy system, POBA and DCB (Selution), there was a flow-limiting dissection in the proximal SFA. This was initially treated with prolonged angioplasty with a 6 mm POBA, with limited success. Therefore, an INTACT 4.8 mm × 8 mm was used to treat the dissection. Post-treatment angiogram demonstrates there is now a patent angiogram demonstrate a now widely patent SFA and popliteal artery. Sheath was removed and closure of the patch was performed surgically.

Patient was commenced on dual antiplatelet on discharge.

Conclusion

As the population ages, it is anticipated that the prevalence of peripheral vascular disease will increase. Within the past few decades there has been an unprecedented evolution of the endovascular technologies and vital improvements are expected in the next decade. Percutaneous procedures will continue to replace open surgery. New technologies will pave the way for improved patient outcomes.

References

1. Gerhard-Herman MD, Gornik HL, Barrett C, Barshes NR, Corriere MA, Drachman DE, Fleisher LA, Fowkes FG, Hamburg NM, Kinlay S, Lookstein R, Misra S, Mureebe L, Olin JW, Patel RA, Regensteiner JG, Schanzer A, Shishehbor MH, Stewart KJ, Treat-Jacobson D, Walsh ME. 2016 AHA/ACC guideline on the management of patients with lower extremity peripheral artery disease: executive summary: a report of the American College of Cardiology/American Heart Association task force on clinical practice guidelines. Circulation. 2017;135(12):e686–725.
2. Katsanos K, Tepe G, Tsetis D, Fanelli F. Standards of practice for superficial femoral and popliteal artery angioplasty and stenting. Cardiovasc Intervent Radiol. 2014;37(3):592–603.
3. Antonopoulos CN, Mylonas SN, Moulakakis KG, Sergentanis TN, Sfyroeras GS, Lazaris AM, Kakisis JD, Vasdekis SN. A network meta-analysis of randomized controlled trials comparing treatment modalities for de novo superficial femoral artery occlusive lesions. J Vasc Surg. 2017;65(1):234–245.e11.
4. Norgren L, Hiatt WR, Dormandy JA, TASC II Working Group, et al. Inter-society consensus for the management of peripheral arterial disease. Int Angiol. 2007;26(2):81–157.
5. Mewissen MW. Primary nitinol stenting for femoropopliteal disease. J Endovasc Ther. 2009;16(2 Suppl 2):II63–81.
6. Laird JR, Katzen BT, Scheinert D, Lammer J, Carpenter J, Buchbinder M, Dave R, Ansel G, Lansky A, Cristea E, Collins TJ, Goldstein J, Cao AY, Jaff MR, RESILIENT Investigators. Nitinol stent implantation vs balloon angioplasty for lesions in the superficial femoral and proximal popliteal arteries of patients with claudication: 3-year follow-up from the RESILIENT randomized trial. J Endovasc Ther. 2012;19(1):1–9.
7. Tepe G, Laird J, Schneider P, et al. Drug-coated balloon versus standard percutaneous transluminal angioplasty for the treatment of superficial femoral and popliteal peripheral artery disease: 12-month results from the IN.PACT SFA randomized trial. Circulation. 2015;131:495–502.
8. Schneider PA, Laird JR, Tepe G, et al. Treatment effect of drug-coated balloons is durable to 3 years in the femoropopliteal arteries: long-term results of the IN.PACT SFA randomized trial. Circ Cardiovasc Interv. 2018;11:e005891.
9. Rosenfield K, Jaff MR, White CJ, et al. Trial of a paclitaxel-coated balloon for femoropopliteal artery disease. N Engl J Med. 2015;373:145–53.
10. Iida O, Soga Y, Urasawa K, et al. Drug-coated balloon versus uncoated percutaneous transluminal angioplasty for the treatment of atherosclerotic lesions in the superficial femoral and proximal popliteal artery: 2-year results of the MDT-2113 SFA Japan randomized trial. Catheter Cardiovasc Interv. 2019;93:664–72.
11. Schroeder H, Werner M, Meyer DR, et al. Low-dose paclitaxel-coated versus uncoated percutaneous transluminal balloon angioplasty for femoropopliteal peripheral artery disease: 1-year results of the ILLUMENATE European randomized clinical trial (randomized trial of a novel paclitaxel-coated percutaneous angioplasty balloon). Circulation. 2017;135:2227–36.
12. Krishnan P, Faries P, Niazi K, et al. Stellarex drug-coated balloon for treatment of femoropopliteal disease: 12-month outcomes from the randomized ILLUMENATE pivotal and pharmacokinetic studies. Circulation. 2017;136:1102–13.
13. Steiner S, Willfort-Ehringer A, Sievert H, et al. 12-month results from the first-in-human randomized study of the ranger paclitaxel-coated balloon for femoropopliteal treatment. JACC Cardiovasc Interv. 2018;11:934–41.
14. Tepe G, Gögebakan Ö, Redlich U, et al. Angiographic and clinical outcomes after treatment of femoro-popliteal lesions with a novel paclitaxel-matrix-coated balloon catheter. Cardiovasc Intervent Radiol. 2017;40:1535–44.
15. Albrecht T, Waliszewski M, Roca C, et al. Two-year clinical outcomes of the consequent trial: can femoropopliteal lesions be treated with sustainable clinical results that are economically sound? Cardiovasc Intervent Radiol. 2018;41:1008–14.
16. Katsanos K, Spiliopoulos S, Kitrou P, Krokidis M, Karnabatidis D. Risk of death following application of paclitaxel-coated balloons and stents in the femoropopliteal artery of the leg: a systematic review and meta-analysis of randomized controlled trials. J Am Heart Assoc. 2018;7(24):e011245.
17. Carter AJ, Aggarwal M, Kopia GA, Tio F, Tsao PS, Kolata R, Yeung AC, Llanos G, Dooley J, Falotico R. Long-term effects of polymer-based, slow-release, sirolimus-eluting stents in a porcine coronary model. Cardiovasc Res. 2004;63(4):617–24.
18. Ali RM, Abdul Kader MASK, Wan Ahmad WA, Ong TK, Liew HB, et al. Treatment of coronary drug-eluting stent restenosis by a sirolimus- or paclitaxel-coated balloon. JACC Cardiovasc Interv. 2019;12(6):558–66.
19. Lemos PA, Farooq V, Takimura CK, Gutierrez PS, Virmani R, et al. Emerging technologies: polymer-free phospholipid encapsulated sirolimus nanocarriers for the controlled release of drug from a stent-plus-balloon or a stand-alone balloon catheter. EuroIntervention. 2013;9(1):148–56.
20. Duda SH, Bosiers M, Lammer J, et al. Drug-eluting and bare nitinol stents for the treatment of atherosclerotic lesions in the superficial femoral artery: long-term results from the SIROCCO trial. J Endovasc Ther. 2006;13:701–10.
21. Med Alliance's SELUTION FIM study achieves primary endpoint. 2018. https://evtoday.com/2018/01/31/med-alliances-selution-fim-study-achieves-primary-endpoint.

22. Vroegindeweij D, Tielbeek AV, Buth J, Schol FP, Hop WC, Landman GH. Directional atherectomy versus balloon angioplasty in segmental femoropopliteal artery disease: 2-year follow-up with color-flow duplex scanning. J Vasc Surg. 1995;21(2):255–68.
23. Gray WA, Garcia LA, Amin A, Shammas NW, JET Registry Investigators. Jetstream atherectomy system treatment of femoropopliteal arteries: results of the post-market JET Registry. Cardiovasc Revasc Med. 2018;19(5 Pt A):506–11.
24. Davis T, Ramaiah V, Niazi K, Martin Gissler H, Crabtree T. Safety and effectiveness of the Phoenix Atherectomy system in lower extremity arteries: early and midterm outcomes from the prospective multicenter EASE study. Vascular. 2017;25(6):563–75.
25. Rocha-Singh KJ, Zeller T, Jaff MR. Peripheral arterial calcification: prevalence, mechanism, detection, and clinical implications. Catheter Cardiovasc Interv. 2014;83(6):E212–20.
26. Chowdhury MM, Makris GC, Tarkin JM, Joshi FR, Hayes PD, Rudd JHF, Coughlin PA. Lower limb arterial calcification (LLAC) scores in patients with symptomatic peripheral arterial disease are associated with increased cardiac mortality and morbidity. PLoS One. 2017;12(9):e0182952.
27. Tepe G, Brodmann M, Werner M, Bachinsky W, Holden A, Zeller T, Mangalmurti S, Nolte-Ernsting C, Bertolet B, Scheinert D, Gray WA, Disrupt PAD III, Investigators. Intravascular lithotripsy for peripheral artery calcification: 30-day outcomes from the randomized Disrupt PAD III trial. JACC Cardiovasc Interv. 2021;14(12):1352–61.
28. Wong CP, Chan LP, Au DM, Chan HWC, Chan YC. Efficacy and safety of intravascular lithotripsy in lower extremity peripheral artery disease: a systematic review and meta-analysis. Eur J Vasc Endovasc Surg. 2022;63(3):446–56.
29. Tepe G, Brodmann M, Werner M, Bachinsky W, Holden A, Zeller T, Mangalmurti S, Nolte-Ernsting C, Virmani R, Parikh SA, Gray WA, Disrupt PAD III Investigators. Intravascular lithotripsy for peripheral artery calcification: mid-term outcomes from the randomized Disrupt PAD III trial. JACC Cardiovasc Interv. 2022;1(04):100–341.
30. Bolia A, Miles KA, Brennan J, Bell PR. Percutaneous transluminal angioplasty of occlusions of the femoral and popliteal arteries by subintimal dissection. Cardiovasc Intervent Radiol. 1990;13(6):357–63.
31. London NJ, Srinivasan R, Naylor AR, Hartshorne T, Ratliff DA, Bell PR, Bolia A. Subintimal angioplasty of femoropopliteal artery occlusions: the long-term results. Eur J Vasc Surg. 1994;8(2):148–55.
32. Lipsitz EC, Ohki T, Veith FJ, Suggs WD, Wain RA, Cynamon J, Mehta M, Cayne N, Gargiulo N. Does subintimal angioplasty have a role in the treatment of severe lower extremity ischemia? J Vasc Surg. 2003;37(2):386–91.
33. Met R, Van Lienden KP, Koelemay MJ, Bipat S, Legemate DA, Reekers JA. Subintimal angioplasty for peripheral arterial occlusive disease: a systematic review. Cardiovasc Intervent Radiol. 2008;31(4):687–97.
34. Shin SH, Baril D, Chaer R, Rhee R, Makaroun M, Marone L. Limitations of the Outback LTD re-entry device in femoropopliteal chronic total occlusions. J Vasc Surg. 2011;53(5):1260–4.
35. Fujihara M, Takahara M, Sasaki S, Nanto K, Utsunomiya M, Iida O, Yokoi Y. Angiographic dissection patterns and patency outcomes after balloon angioplasty for superficial femoral artery disease. J Endovasc Ther. 2017;24(3):367–75.
36. Gray WA, Cardenas JA, Brodmann M, Werner M, Bernardo NI, George JC, Lansky A. Treating post-angioplasty dissection in the femoropopliteal arteries using the tack endovascular system: 12-month results from the TOBA II Study. JACC Cardiovasc Interv. 2019;12(23):2375–84.
37. Makris GC, Chrysafi P, Little M, Patel R, Bratby M, Wigham A, Anthony S, Uberoi R. The role of intravascular ultrasound in lower limb revascularization in patients with peripheral arterial disease. Int Angiol. 2017;36(6):505–16.

21 Complications and Their Management in Peripheral Interventions

Athanasios Diamantopoulos and Iakovos Theodoulou

Case Presentation

A 67-year-old male with a history of hyperlipidaemia, hypertension and significant smoking history was under investigation for a possible abdominal aortic aneurysm (AAA) after reporting symptoms of ill-defined back pain as well as a pulsating mass in the abdomen. Following an ultrasound examination, an aneurysmal abdominal aorta was confirmed, after which he was referred to a tertiary centre for consideration for an elective endovascular aneurysm repair (EVAR). He was eventually admitted for an EVAR for which a right femoral access was performed. The procedure had no immediate operative complications; however, over the course of the next 2 days he complained of swelling and bruising in the groin, pain and overlying tingling sensation. Following clinical examination, there was suspicion for the formation of a femoral artery pseudoaneurysm, and this was confirmed with an ultrasound examination.

Continued at page 214

Background

With peripheral arterial disease (PAD) representing a spectrum of disease, chronic or critical limb ischaemia (CLI) constitutes a more severe form of PAD, whereby symptoms occur at rest and can lead to permanent complications such as ischaemic tissue necrosis [1]. Recent reviews conclude that the latest calculated prevalence of PAD globally is 5.6%, the equivalent of 236 million adults worldwide [2]. Given the high prevalence, utilisation of peripheral vascular interventions (PVI) for treatment of PAD is on the rise with a tenfold increase since 1995 [3], with interventional radiology maintaining a steady market share for provision of such services in the United States (US) [4]. In fact, PVI has largely replaced bypass surgery, now becoming the main choice of revascularisation for PAD patients [5]. With a rise in PVI, associated complications remain prevalent, therefore their recognition and management are critically important. It is critical that interventional practitioners remain cognisant of such complications, both for early recognition and timely management. In addition, with a lot of endovascular treatments rendering patients ready for discharge from hospitals quicker than open repairs, it is crucial that patients are aware of signs and symptoms that come with each complication such that recognition happens promptly and relevant care is sought accordingly. In this chapter, complications are classified into 'puncture site', 'angioplasty site' and 'systemic complications' (Fig. 21.1) and management algorithms described for each.

A. Diamantopoulos (✉)
Guy's and St Thomas' NHS Foundation Trust, London, UK
e-mail: athanasios.diamantopoulos@gstt.nhs.uk

I. Theodoulou (✉)
Barts NHS Health Trust, The Royal Hospital, London, UK
e-mail: iakovos.theodoulou@gstt.nhs.uk

G. Geroulakos et al. (eds.), *Mastering Endovascular Techniques*, https://doi.org/10.1007/978-3-031-42735-0_21

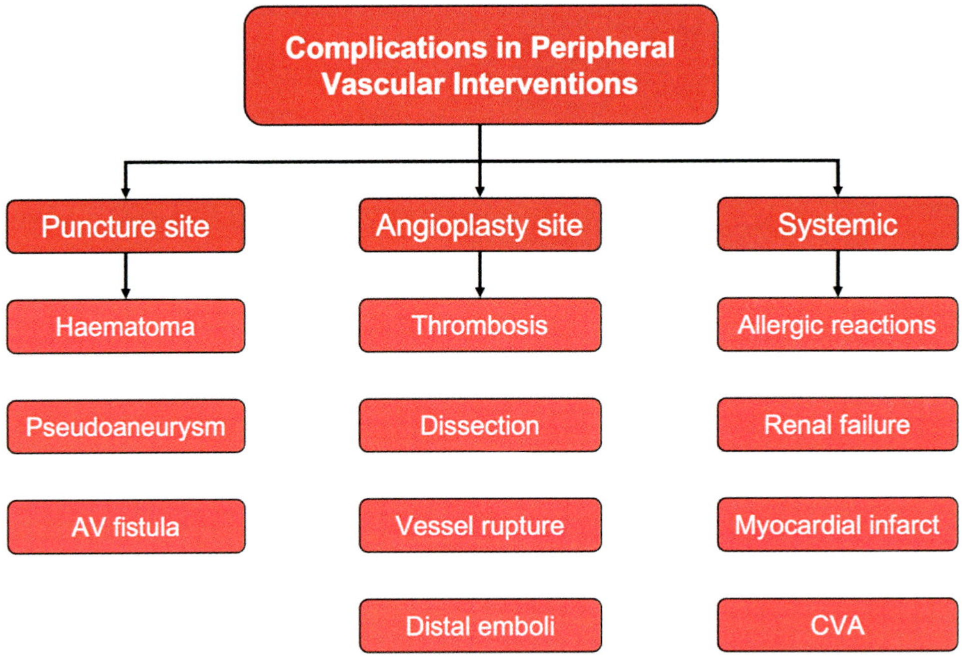

Fig. 21.1 Summary of peripheral vascular intervention-associated complications

Puncture Site Complications

Puncture site complications are variably reported in literature ranging between 3.5% [6] and 7.0% [7, 8]. Furthermore, only a fraction of these require active treatment with latest studies reporting a figure of 0.9% [6].

Haematoma

This is a relatively common (1.0–11.0%) and minor complication, often resolving with conservative management. It requires careful monitoring post-procedurally, including assessing for any changes in the size, extent, and location of the haematoma. A haematoma normally arises due to persistent bleeding post sheath removal, often due to difficult initial access or an over-anticoagulated patient. Expansion and worsening severity can manifest as new onset neuralgias with symptoms varying according to anatomical location. Radial puncture site haematomas may interfere with distal neurology to the hand whilst femoral haematomas can present with acute limb ischaemia or femoral neuropathy. In upper extremities, haematomas can be further classified based on (a) extent of infiltration in muscle compartments, (b) the proximal extension of haematoma to the elbow joint and (c) any impending threat for compartment syndrome [9]. Retroperitoneal haematoma can be a significant source of ongoing bleeding and culprit in haemodynamically deteriorating patients with a new ipsilateral flank pain. It has been associated to high femoral puncture or access at the external iliac artery superior to the inguinal ligament, requiring stent placement or open repair if the former fails to control bleeding.

Management

Management of non-expanding and stable subcutaneous haematomas is straight forward with manual compression followed by loose compression dressings and elevation (if upper limb) often proving sufficient. Growing haematomas require more urgent attention, especially if compartment syndrome is suspected; however, these are much rarer. In such cases, proximal compression with blood pressure cuff is necessary, addition of pressure dressing and discussion with the vascular surgery team. Reversal of anticoagulation is also worth considering.

Suspicion of retroperitoneal haematomas should be confirmed with urgent non-contrast CT. Management is guided by the size of the haematoma and the presence or absence of active bleeding. Conservative approaches rely on bed rest, consideration for anticoagulation reversal and, depending on haemoglobin (Hb), supportive blood transfusions. A radiographic indication for active management includes active extravasation on CT. Emergency first line approaches include endovascular covered stents over the culprit arterial segment or balloon inflation proximal to the lesion. Ongoing bleeders with refractory Hb drops despite blood transfusions require urgent surgical exploration for source control.

Pseudoaneurysm

In the femoral region, these are often associated with low puncture sites (e.g. inferior to common femoral artery) which render post-procedural manual compression more challenging and can lead to insufficient haemostasis. Pseudoaneurysm is defined as contained arterial rupture where bleeding is suppressed by adjacent structures, depending on anatomical location, including chiefly, the tunica adventitia of the artery, subcutaneous fat and adjacent muscle tissue. Clinically, iatrogenic pseudoaneurysm (IPA) generally occurs in less than 1% of cases and manifests as increased local pain, skin bruising, palpable pulsatile mass, or regional neuropathies. In the past, their management was often guided by their size, with conservative approach reserved for smaller aneurysms (usual cut-off of 3 cm) and ultrasound-guided thrombin injection for those exceeding the 3 cm cut-off. Opting for a conservative management, necessitates 2-weekly duplex ultrasound scans for a minimum of 6 weeks. These patients should be re-considered for intervention in the presence of an expanding IPA or non-compliance with surveillance scans. Nowadays, even pseudoaneurysms smaller than 3 cm can be treated ultrasound guided injection of thrombin. Indications for active treatment include haemodynamic instability, active bleeding, or infected access site. Relative contraindications include large pseudoaneurysm necks due to an associated risk for arterial thrombosis.

Management Options

1. **Ultrasound-guided compression** (UGC) used to be the first line treatment but has now largely been replaced by Ultrasound-guided thrombin injection described in (2). UGC technique is reserved for smaller pseudoaneurysms (<3 mm) in non-anticoagulated patients. Following sufficient local anaesthesia, using a linear probe with frequency >7 MHz, pressure is applied through the probe on the pseudoaneurysm beck while maintaining real time colour Doppler sonographic guidance. Pressure and positioning of the probe is adjusted such that flow into the pseudoaneurysm is obliterated while ensuring good patency and flow across the parent artery. Pressure across the probe is gradually eased at 10-min intervals to assess for complete elimination of flow and hence thrombosis. Pressure and positioning are adjusted until pseudoaneurysm thrombosis is confirmed in the absence of flow. Despite ongoing controversy, concomitant arteriovenous fistula (AVF) has previously been considered a relative contraindication for this management option; therefore, careful selection of patients is necessary while assessing for this possibility.
2. **Ultrasound-guided thrombin injection** is today in many centres the first line treatment for pseudoaneurysms. The operator should first gather adequate imaging, ideally using an ultrasound. Computer tomographic or magnetic resonance angiographic imaging are acceptable alternatives. Discernible pseudoaneurysm characteristics should include the size of the sac as well as the width and length of the neck; if multilobular, the number of lobes should also be recorded. Using sterile technique, proceed with adequate local anaesthesia followed by preparation of the thrombin solution. Thrombin normally comes in 'Thrombin kits' with the appropriate reconstitution volumes to make up a Thrombin solution of 100 units per 1 mL. Using a 22-gauge needle under ultrasound guidance, thrombin is injected into the pseudoaneurysm sac closest to the artery wall defect in increments of 10 units until limitation of flow and cessation of colour Doppler flow is achieved. A colour Doppler assessment is repeated 10–20 min later to confirm complete flow resolution and assess need for further rounds of thrombin injections. Bovine-derived thrombin preparations increase the risk for allergic reactions, therefore operator should remain vigilant if any signs of allergy arise and should proceed with anaphylaxis treatment algorithms as needed.

In both instances, post-procedural instructions should include bed rest for the patient and avoiding lifting, regular neurovascular assessment of the ipsilateral limb and repeat ultrasound Doppler within 24 h. Regular and PRN analgesia are likely to be beneficial to treat tenderness around the treated area. Groin pain, reperfusion of pseudoaneurysm and distal embolism are some of the commonest complications in the treatment of pseudoaneurysm.

3. **Surgical repair** is reserved for a fraction of pseudoaneurysm cases which present with more sinister features, including quickly expanding pseudoaneurysm which has been refractory to approaches (1) and (2) and/or presents with signs of skin necrosis or impending compartment syndrome.

Arteriovenous Fistulae

AVF display abnormal connection between an artery and a vein and are traditionally classified as congenital or acquired. Traumatically acquired AVF are pertinent to PVI and carry an incidence of up to 0.88% [10, 11]. The natural anatomical proximity of arteries and veins, often travelling together as a triplet of artery and a pair of veins, renders them vulnerable to unduly simultaneous trauma as a needle traversing both vessels during needle placement for access. Almost all acquired AVF are initially clinically silent with symptom onset ranging from 48 h to months after the procedure.

Symptoms can present as paraesthesia in the affected limb, ischaemic symptoms such as IC due to arterial steal, peripheral oedema and venous insufficiency. Majority of acquired AVF tend to resolve spontaneously within 2 months, therefore conservative management with observation is the default management option for majority of patients. Follow-up ultrasonography is recommended at 30 days to confirm closure of the connecting tract. Clinical indications for active treatment include significant enlargement confirmed on surveillance scanning and worsening symptomatology or persistence beyond 1 year.

Management Options

1. **Ultrasound-guided compression**: this technique resembles the one employed for pseudoaneurysm management with probe placement instead focusing on the actual AVF (instead of across the pseudoaneurysm neck). However, this is rarely successful.
2. **Endovascular repair**: this is considered an acceptable technique as it is minimal invasive and an open repair is avoided. It is more appropriate for patients with acquired AVF who would not tolerate surgical repair either due to small margins for bleeding (low baseline Hb) or a hostile groin (if in femoral region) due to significant fibrosis from previous surgical exploration. With increasing experience worldwide, endovascular options have become more widely used even for patients who are good surgical candidates. Endovascular approaches include placement of covered stents or embolisation depending on the location of the AVF. Alternatively, provided there is an adequately visualised and sufficiently long connecting channel, the AVF might be amenable to embolisation techniques such as coiling. Surgical repair remains the gold standard for fit enough patients.

Angioplasty Site Complications

Thrombosis

Stent occlusion and acute thrombosis are closely related but fundamentally different complications in PVI, with the former constituting a common subacute complication due to intimal hyperplasia at the site of angioplasty, and the latter describing an immediately apparent thrombosis of the angioplasty site shortly after deployment of stents. Acute thrombosis is a limb-threatening condition and arises intraprocedural most of the times and contrary to stent re-occlusion it lacks collateralisation, therefore rendering it an emergency. Recent analyses suggest that acute thrombosis has an incidence of 6.1% and is more common in the femoropopliteal region and associated more with covered nitinol stents than BMS [12]. Symptoms are in keeping with acute limb ischaemia and manifest more acutely.

Management Options

Prompt recognition and management of peripheral acute thrombosis is critical to avoid foot amputations. Treatment commonly consists of catheter-directed thrombolysis with or without mechanical thrombectomy. The latter is the treatment mode of choice if contraindications to thrombolysis apply. Refractory lesions might need to be referred for surgical bypass with success rates reported up to 95% for lesions in the superficial femoral region [13].

Dissection

Inadvertent arterial injury in the form of dissection at diseased segments is a common complication, arising, in some reported cohorts up to 84% [14]. Dissection is defined as a tear between layers of an artery, typically the intima and the media. This can arise in two ways: (1) inadvertent subintimal puncture and introduction of a wire while crossing occlusions, or (2) during balloon expansion which promotes plaque fissuring due to radial stresses exerted by the inflated balloon [14]. Intravascular Ultrasound (IVUS) has been investigated and utilised more extensively than angiography, duplex ultrasonography or optical coherence tomography for identification and characterisation of arterial dissections. Kobayashi Grading [15] and iDissection classification systems (Tables 21.1 and 21.2) [16] help describe lesions with the limitation that they both omit the length of dissections and the presence of haematomas. The immediate clinical implications of dissections include flow limitation and have also been associated with decreased long-run patency even with low-grade lesions. Consensus is for intraprocedural treatment of dissections upon identification, regardless of severity.

Table 21.1 The Kobayashi grading proposal [15]

Category	Width of dissection
Group A	None
Group B	<1/3 lumen
Group C	>1/3 lumen

Table 21.2 The iDissection classification scheme [16]

Dissection	Circumference <180°	Circumference >180°
Intima	A1	A2
Media	B1	B2
Adventitia	C1	C2

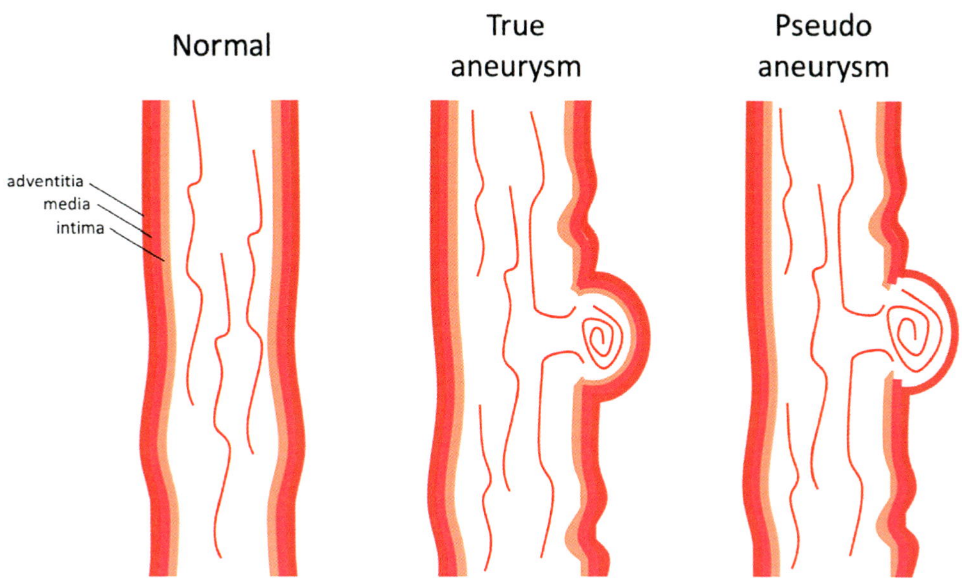

Fig. 21.2 Schematic coronal representation of a normal artery, an aneurysmal and a pseudo-aneurysmal artery

Management Options

Based on coronary experience, dissections can generally be treated with prolonged, low-pressure inflations across dissected segments with oversized balloons [17]; this aims to reappose obliterated arterial layers. This is a more appealing technique for the infrapopliteal territory where data is generally more limited. Stent implantation remains the first line approach in the femoral and aortoiliac regions, with either entire lesion stenting or angioplasty followed by spot stenting to tackle areas with significant flow limitations. Management options will also likely vary depending on the extent of dissection and whether this tracks distally with flow-limiting implications. The choice between nitinol bare metal stents (BMS), drug-coated balloons and drug-eluting stents (DES) is an ongoing topic of research (Fig. 21.2).

Arterial Rupture

Vessel rupture arises via two mechanisms, the commoner of which is due to inadvertent passage of wire into the subadventitial space followed by post-dilatation of implanted stent. Less commonly, it might arise with overdilation or oversized balloons. In PVI, iliac artery rupture manifests immediately with severe pain secondary to retroperitoneal haematoma and sudden haemodynamic compromise. Management should be instigated immediately.

Management Options

Principles of management consist of rupture identification, bleeding control with proximal balloon tamponade and patient resuscitation and stent-graft deployment in the affected segment. For iliac artery ruptures identified with angiographic contrast extravasation, proximal tamponade in the aortoiliac section with an appropriately sized balloon should be established to control bleeding and prepare for the deployment of a stent graft across the ruptured segment. If upsizing of sheath is necessary to enable graft employment, more proximal tamponade should be pursued in the common iliac or aortic segments. Like dissections, ruptures in the femoropopliteal region are often resolved with prolonged balloon dilatation. Persistent femoropopliteal ruptures are amenable to stent-grafting and reversal of anticoagulation might be considered in refractory cases. Surgical escalation might be necessary if the above fail.

Distal Emboli

This can occur with any type of PVI and can lead to catastrophic consequences for the patient without timely recognition and treatment. This is more pertinent for distal embolisations in the lower extremity in which microemboli can lead to trash foot which in itself is associated to high mortality and morbidity. Blue toe syndrome represents distal-most embolisations in end arteries affecting digital circulation but with patent proximal large vessels. As the name suggests, skin discoloration and pain are the main symptoms and clinical course varies from spontaneous resolution to gangrene requiring toe amputation. Incidence for macroembolic distal embolisations are estimated to be up to 5% [18];

however, incidence for microembolic embolisation has not been formally reported, possibly due to less immediate clinical implications. Recognition of on-table distal embolisations typically happens on completion angiography and presents either with new filling defects when compared to initial crossing catheter angiography or lost patency in run off vessels.

Management Options

Operator should first confirm adequate anticoagulation using activated clotting time (ACT) which is consensually agreed as optimal when between 200 and 250 s. Depending on the anatomical site and size of the dislodged thrombus there are two options: surgical thrombectomy or intra-arterial catheter-based approaches. Different subtypes of pharmacomechanical approaches are considered after screening for absolute contraindications for thrombolysis such as recent established cerebrovascular accidents (CVA), known bleeding diatheses, recent gastrointestinal or intracranial bleeds [19]. In the absence of contraindications, as an adjunct to clot aspiration systems such as Angiojet, rheolytic technology can be employed which infuses small amounts (2–4 mg) of tissue plasminogen activator (tPA). If thrombolysis contraindications apply, transcatheter mechanical embolectomy alone is employed. Open embolectomy options are sought if embolic material is insufficiently retrieved using catheter-based approaches.

Systemic Complications

Allergic Reactions

This is often associated not only to iodine contrast-related reactions, including anaphylaxis, but also implantable endovascular devices, including BMS (allergy to metal ions) and DES (allergy to coating polymers).

Management

Management algorithms for allergic reactions is no different to all other anaphylactic-like reactions and should include easy and prompt access and administration of adrenaline, fluids antihistamine and steroid treatments.

Renal Failure

Contrast-induced nephrotoxicity (CIN) or contrast-associated acute kidney injury (AKI) (CA-AKI) is an AKI presumed to have been caused by contrast material employed in an endovascular procedure. CA-AKI is the most favourable term amongst the two, as CIN implies a causal link between contrast and AKI which is often hard to prove clinically. Diagnosis of CA-AKI is made in the presence of significant creatinine rise 24–48 h post-procedurally and after all other causes of AKI have been excluded.

Management

Management principles are identical to other causes of AKI, with appropriate fluid balance assessment and electrolyte monitoring with replacements where appropriate. Dose adjustment of regular and new medications is also important considering the reduced creatinine clearance.

Case Presentation

Continued from page 209

Given the size of the pseudoaneurysm which exceeded the 3 cm cut-off normally used to guide the management of pseudoaneurysm, a decision was made to treat this with ultrasound-guided thrombin injection rather than opting for conservative management. During the ultrasound assessment, a note was made of the sac size and the width of the neck, which were 4 cm and 0.5 cm, respectively. Using a sterile technique, local anaesthesia was performed in the right-groin area under ultrasound guidance, including around the site of pseudoaneurysm. Using an off-the-shelf 'thrombin kit', which comes with thrombin concentration of 100 units per 1 mL, injection of thrombin was performed under ultrasound guidance in 10-unit increments until flow was seen to arrest within the pseudoaneurysm sac, as assessed by duplex imaging. To confirm satisfactory flow resolution, scanning was performed 10 min later which confirmed. Patient was kept in the department for 6 h for monitoring and later discharged with advice, including bed rest and activity modifications such as avoiding heavy lifting and regular analgesia. Repeat ultrasound imaging was performed 2 days later to confirm successful treatment and resolution of the pseudoaneurysm. Overall summary of complications and their management described in (Tables 21.3 and 21.4).

Table 21.3 Overview of puncture site and angioplasty site complications

	Clinical picture	Risk factors	Classification	Treatment
Puncture site complications				
Haematoma	• Local pain • Bruising +/– expanding • Distal neuropathy	• Anticoagulation • Difficult access • Obesity	– Size – Location – Complicated	1. Manual compression 2. Covered stenting 3. Surgical evacuation
Pseudoaneurysm	• New mass +/– pulsatile • Local pain, swelling • Bruising • Distal neuropathy	• Anticoagulation • Difficult access • High BMI	– No. of lobes – Neck width – Neck length – Sac diameter	1. US-guided compression 2. US-guided thrombin
AVF	• Ipsilateral paraesthesia • Unilateral oedema • Intermittent claudication	• Therapeutic catheterisation • Puncture inferior to CFA • Female sex • High BMI • Anticoagulation or antifibrinolytic therapy	–	1. US-guided compression 2. Surgical repair 3. Endovascular coiling 4. Covered stenting
Angioplasty site complications				
Thrombosis	• Acute limb ischaemia	• Femoro-popliteal region • Covered nitinol stents	–	1. Catheter-directed thrombolysis 2. Mechanical thrombectomy 3. Surgical bypass
Dissection	• Asymptomatic	• Subintimal techniques • Oversized balloons	– Kobayashi – iDissection	1. Prolonged inflations 2. Stenting
Vessel rupture	• Severe pain • Hypotension	• Balloon overdilatation • Subintimal techniques • Steroid use • Fibromuscular dysplasia • Vasculitis	– Location	1. Stent-graft deployment 2. Surgical open repair
Distal emboli	• Skin discoloration • Pain	• Atherectomy • Long procedure • Inadequate anticoagulation	– Microembolic – Macroembolic	1. Endovascular Pharmacomechanical 2. Open thrombectomy

Table 21.4 Basic toolkit for the management of complications

Basic toolkit for the management of complications
• Compression dressings • Endovascular covered stents • Nitinol bare metal stents (BMS), drug-coated balloons, drug-eluting stents (DES) • Thrombin kits • Clot aspiration systems such as Angiojet • Tissue plasminogen activator • Anaphylaxis kits

References

1. Aday AW, Matsushita K. Epidemiology of peripheral artery disease and polyvascular disease. Circ Res. 2021;128(12):1818–32.
2. Song P, Fang Z, Wang H, Cai Y, Rahimi K, Zhu Y, Fowkes FGR, Fowkes FJ, Rudan I. Global and regional prevalence, burden, and risk factors for carotid atherosclerosis: a systematic review, meta-analysis, and modelling study. Lancet Glob Health. 2020;8(5):e721–e9. https://doi.org/10.1016/S2214-109X(20)30117-0.
3. Anderson PL, Gelijns A, Moskowitz A, Arons R, Gupta L, Weinberg A, et al. Understanding trends in inpatient surgical volume: vascular interventions, 1980–2000. J Vasc Surg. 2004;39(6):1200–8.
4. Schramm KM, DeWitt PE, Dybul S, Rochon PJ, Patel P, Hieb RA, et al. Recent trends in clinical setting and provider specialty for endovascular peripheral artery disease interventions for the Medicare population. J Vasc Interv Radiol. 2020;31(4):614–21.e2.
5. Drachman DE, Hawkins BM. Complications of peripheral vascular intervention: time to stop the bleeding. Washington, DC: American College of Cardiology Foundation; 2019. p. 1150–2.
6. Ortiz D, Jahangir A, Singh M, Allaqaband S, Bajwa TK, Mewissen MW. Access site complications after peripheral vascular interventions: incidence, predictors, and outcomes. Circ Cardiovasc Interv. 2014;7(6):821–8.
7. Walker S, Beasley C, Reeves M. A retrospective study on the use of heparin for peripheral vascular intervention. Perspect Vasc Surg Endovasc Ther. 2012;24(2):63–9.
8. Sheikh IR, Ahmed SH, Mori N, Gupta A, Mewissen M, Allaqaband S, et al. Comparison of safety and efficacy of bivalirudin versus unfractionated heparin in percutaneous peripheral intervention: a single-center experience. J Am Coll Cardiol Interv. 2009;2(9):871–6.
9. Shroff A, Siddiqui S, Burg A, Singla I. Identification and management of complications of transradial procedures. Curr Cardiol Rep. 2013;15(4):1–9.
10. Oweida SW, Roubin GS, Smith RB III, Salam AA. Postcatheterization vascular complications associated with percutaneous transluminal coronary angioplasty. J Vasc Surg. 1990;12(3):310–5.
11. Kelm M, Perings SM, Jax T, Lauer T, Schoebel FC, Heintzen MP, et al. Incidence and clinical outcome of iatrogenic femoral arteriovenous fistulas: implications for risk stratification and treatment. J Am Coll Cardiol. 2002;40(2):291–7.

12. Katsanos K, Al-Lamki SA, Parthipun A, Spiliopoulos S, Patel SD, Paraskevopoulos I, et al. Peripheral stent thrombosis leading to acute limb ischemia and major amputation: incidence and risk factors in the aortoiliac and femoropopliteal arteries. Cardiovasc Intervent Radiol. 2017;40(3):351–9.
13. Strobl FF, Brechtel K, Schmehl J, Zeller T, Reiser MF, Claussen CD, et al. Twelve-month results of a randomized trial comparing mono with dual antiplatelet therapy in endovascularly treated patients with peripheral artery disease. J Endovasc Ther. 2013;20(5):699–706.
14. Armstrong EJ, Brodmann M, Deaton DH, Gray WA, Jaff MR, Lichtenberg M, et al. Dissections after infrainguinal percutaneous transluminal angioplasty: a systematic review and current state of clinical evidence. J Endovasc Ther. 2019;26(4):479–89.
15. Kobayashi N, Hirano K, Yamawaki M, Araki M, Sakai T, Sakamoto Y, et al. Simple classification and clinical outcomes of angiographic dissection after balloon angioplasty for femoropopliteal disease. J Vasc Surg. 2018;67(4):1151–8.
16. Shammas NW, Torey JT, Shammas WJ. Dissections in peripheral vascular interventions: a proposed classification using intravascular ultrasound. J Invasive Cardiol. 2018;30(4):145–6.
17. Cripps TR, Morgan JM, Rickards AF. Outcome of extensive coronary artery dissection during coronary angioplasty. Heart. 1991;66(1):3–6.
18. Lam RC, Shah S, Faries PL, McKinsey JF, Kent KC, Morrissey NJ. Incidence and clinical significance of distal embolization during percutaneous interventions involving the superficial femoral artery. J Vasc Surg. 2007;46(6):1155–9.
19. Stile Investigators. Results of a prospective randomized trial evaluating surgery versus thrombolysis for ischemia of the lower extremity. Ann Surg. 1994;220:251–68.

Follow-Up and Surveillance After Peripheral Interventions

22

Panagiotis Volteas and Nicos Labropoulos

Case Presentation A 69-year-old woman presented to the clinic with a 2-month history of worsening right lower extremity rest pain (Rutherford class 4). The patient was a current smoker with past medical history of hypertension, and hyperlipidemia. On physical examination, she had no ulcerations, and normal lower extremity sensory and motor function. On the right side she had palpable femoral pulse, and dopplerable dorsalis pedis (DP) and posterior tibial (PT) signals. On the left lower extremity, the patient had palpable femoral and DP/PT pulses. The right ankle/brachial index (ABI) was 0.4 and the left ABI was 0.8 while a Duplex US showed severe superficial femoral artery and popliteal artery (P1 and P2 segments) stenosis. Based on the above the decision was made to proceed with right leg angiography, during which the superficial femoral artery and the P1/P2 segments of the popliteal artery underwent balloon angioplasty followed by stents placement. Completion angiography showed resolution of the severe stenosis and two vessel runoff to the foot. The patient was placed on aspirin, plavix and statin with a plan to follow up at 1 month with ABIs and arterial DUS.

Continued at page 221

Advances in endovascular therapies have established percutaneous interventions as a preferred alternative to open surgical revascularization not only in cases of intermittent claudication but also in patients with chronic limb-threatening ischemia (CLTI) [1, 2]. The major drawback after endovascular interventions is restenosis which can progress to arterial occlusion with direct consequences to limb salvage [3]. Restenosis is mainly related to a combination of arterial wall inflammatory remodeling and neo-intimal hyperplasia most commonly at the balloon dilation site, atherectomy site, or at the proximal or distal end of a stent. It can also be related to initial suboptimal procedural results due to residual >50% stenosis, residual dissection, distal embolization, stent fracture, inadequate inflow or outflow treatment or progression of the peripheral atherosclerotic disease [3, 4].

Although postoperative surveillance imaging is routinely performed for patients after carotid artery interventions, less than half of patients receive any follow-up after lower extremity interventions, even in Vascular Quality Initiative (VQI) participating hospitals where follow-up care is prioritized [5, 6]. The importance of frequent follow up is highlighted by the fact that loss to follow-up 1 year after lower extremity peripheral vascular intervention is associated with higher prevalence of critical limb ischemia, risk for urgent re-intervention and worse survival [7]. Thus, every effort should be made to prolong interventions durability and to detect clinically significant problems at an early stage when they can be managed most safely and effectively.

Despite the low quality of existing evidence, the Society for Vascular Surgery has published guidelines on follow-up after endovascular surgery peripheral arterial procedures (Table 22.1).

A typical postoperative regimen consists of performing clinical examination, measuring the ABI/TBI, and obtaining duplex studies at 1, 6, and 12 months and then yearly thereafter, as long as there are no new signs or symptoms. The more vigorous follow up during the first year is supported by the fact that most restenosis occurs in the first 6–12 months [2, 8]. After 12 months patients should continue to be followed as the progression of native disease and late restenosis can still affect angioplasty and stent patency [9].

The surveillance should not only focus on the revascularization site but must include the completely revascularized limb as well as the contralateral limb, along with a general clinical cardiovascular surveillance.

P. Volteas
Vascular and Endovascular Surgery, Stony Brook Medicine, Stony Brook, NY, USA

N. Labropoulos (✉)
Surgery and Radiology, Vascular Laboratory, Stony Brook Medicine, Stony Brook, NY, USA

G. Geroulakos et al. (eds.), *Mastering Endovascular Techniques*, https://doi.org/10.1007/978-3-031-42735-0_22

Table 22.1 The Society for Vascular Surgery practice guidelines on follow-up after endovascular surgery peripheral arterial procedures [3]

Level		Strength of recommendation quality of evidence
Aortoiliac	We **recommend** clinical examination, ABI, and DUS within the first month after aortoiliac segment EVT to provide a post-treatment baseline and to evaluate for residual stenosis Clinical examination and ABI, with or without the addition of DUS, should be performed at 6 and 12 months and then annually as long as there are no new signs or symptoms	**1 (strong)** **C (low)**
Femoropopliteal	We **suggest** clinical examination, ABI, and DUS within the first month after femoropopliteal artery EVT to provide a post-treatment baseline and to evaluate for residual stenosis Continued surveillance at 3 months and then every 6 months is indicated for the following: **A.** Patients with interventions using stents because of the potential increased difficulty of treating an occlusive vs stenotic in-stent lesion **B.** Patients undergoing angioplasty or atherectomy for critical limb ischemia because of increased risk of recurrent critical limb ischemia should the intervention fail	**2 (weak)** **C (low)**
Tibial	We **suggest** clinical examination, ABI, and DUS within the first month after tibial artery EVT to provide a posttreatment baseline and to evaluate for residual stenosis Continued surveillance at 3 months and then every 6 months should be considered. Those patients with a deteriorating clinical vascular examination, return of rest pain, non-healing wounds, or new tissue loss should undergo repeated DUS	**2 (weak)** **C (low)**

Clinical Examination

The patient's functional activity, the presence of recurrent claudication symptoms or signs, and the healing status of wounds in case of CLTI should be evaluated. This can be done by comparing the patient's limb ischemia stage according to the Rutherford scale both pre- and post-intervention. Patients can be considered clinically improved if they move up at least by one clinical category in the Rutherford classification, but those with actual tissue loss need to move up at least two categories to reach a level of claudication that can be considered improved.

Cardiovascular surveillance including modification of atherosclerotic risk factors, counseling on smoking cessation, reduction to an ideal body weight with diet and exercise, preventive foot care and medical therapy are important aspects of postoperative management [2]. Optimal medical management usually includes the use of single or dual antiplatelet therapy, lipid-lowering agents, antihypertensive medications, and glycemic control agents [10]. Despite an absence of strong clinical evidence, dual antiplatelet therapy (DAPT) is frequently employed for 1–6 months followed by subsequent indefinite therapy with either aspirin or clopidogrel [2, 11]. Aspirin has been the mainstay of treatment because it is efficacious and cost-effective, but clopidogrel can also be effective as a single agent [11].

Assessment of Limb Perfusion by Physiologic Testing (Pulse Exam, ABI/TBI/Toe Pressures)

Each follow-up visit should include careful examination of the feet to confirm the presence of pedal pulses, capillary refill, any signs of ischemia (dependent rubor, ulceration, non-healing wounds, gangrene) or any change from the baseline examination [2]. Wound prevention education and wound care until wounds are healed are equally essential to long-term clinical success after peripheral arterial procedures, especially in those with CTLI [2, 4].

Post-intervention, an increase of more than 0.10 in ABI is indicative of improved flow. On the contrary a drop in ABI of more than 0.15 could indicate restenosis at the site of intervention or emergence of de novo disease outside the treated segment, such as in the inflow or outflow. In cases of recurrent claudication, the functional severity is better assessed with post-exercise treadmill test ABI. The magnitude of the decline when reintervention should be considered has not been well studied, but a decline of more than 0.15 compared to baseline ABI immediately after the procedure should trigger an evaluation with duplex ultrasound (DUS) [2].

In patients treated for CLTI and in those with diabetes mellitus in whom the ABI cannot be accurately measured, the systolic toe pressure should be measured as it is less commonly affected by vessel calcification and incompressibility. In general, a toe pressure greater than 30 mmHg is required to relieve rest pain, while a toe pressure greater than 40 mmHg can reliably predict successful wound healing. The combination of the above information, expressed as the WIfI score (Wound, Ischemia, and foot Infection) is very important especially in patients with CLTI. The WIfI score should be recalculated after any intervention, as restaging has proved to be an important tool for predicting limb loss and assessing the adequacy of the intervention [12].

Duplex US Evaluation for Anatomic or Flow Abnormalities

Surveillance by clinical follow-up alone may be insufficient to detect restenosis as patients may remain asymptomatic until the target artery has occluded (Fig. 22.1) [11]. ABI adds significant information to clinical examination, but it correlates poorly with angiographic stenosis, and a significant decrease in ABI (>0.15) may not be present until a greater than 60% stenosis exists [4]. Judicious use of DUS seems to be superior to ABI and is a pivotal imaging technique for identifying revascularization failures [13].

A baseline DUS within the first month after endovascular treatment (EVT) is recommended for all patients undergoing lower extremity interventions to establish a post-treatment baseline and to identify those with residual stenoses missed on completion angiography [3]. Optimally, detailed description of the procedure and previous DUS examinations should be available for the sonographer. The inflow, area of procedure, and outflow should be assessed. Anatomic information using direct visualization of the vessel as well as physiologic information based on spectral waveforms, and velocity measurements should be used to characterize any stenosis. These modalities will help also to determine other pathologies such as stent malposition, thrombosis, wall apposition, evidence of kinking, dissection, stent fracture, and progression of disease (Table 22.2).

Table 22.2 Duplex ultrasound characteristics

Indications for DUS follow-up	• During procedure • 1 month • 6 months • 12 months • Then annually or as needed for clinically worsening symptoms			
Inappropriate follow-up	• Patients without physical exam • Unrelated symptoms • Random follow-up • Frequent follow-up			
Assess – Inflow – Area of procedure – Outflow	**DUS pathologic findings** • Stenosis • Thrombosis • Kink • Dissection • Aneurysm • Pseudoaneurysm • Fistula		• Hematoma • Seroma • Infection • Wall apposition • Stent fracture • Progression of disease • Systemic disease	
DUS modalities – B-mode – Color Doppler – Spectral waveform – Velocity criteria		**PSV**	**PSV ratio**	**Distal artery spectral waveform**
	Normal (<50%)	<200 cm/s	<2	Triphasic
	Moderate (50–69%)	200–300 cm/s	2–3	Monophasic
	Severe (70%)	300 cm/s	3	Damped, monophasic Tardus-parvus waveform
	Occluded	No flow	–	No flow

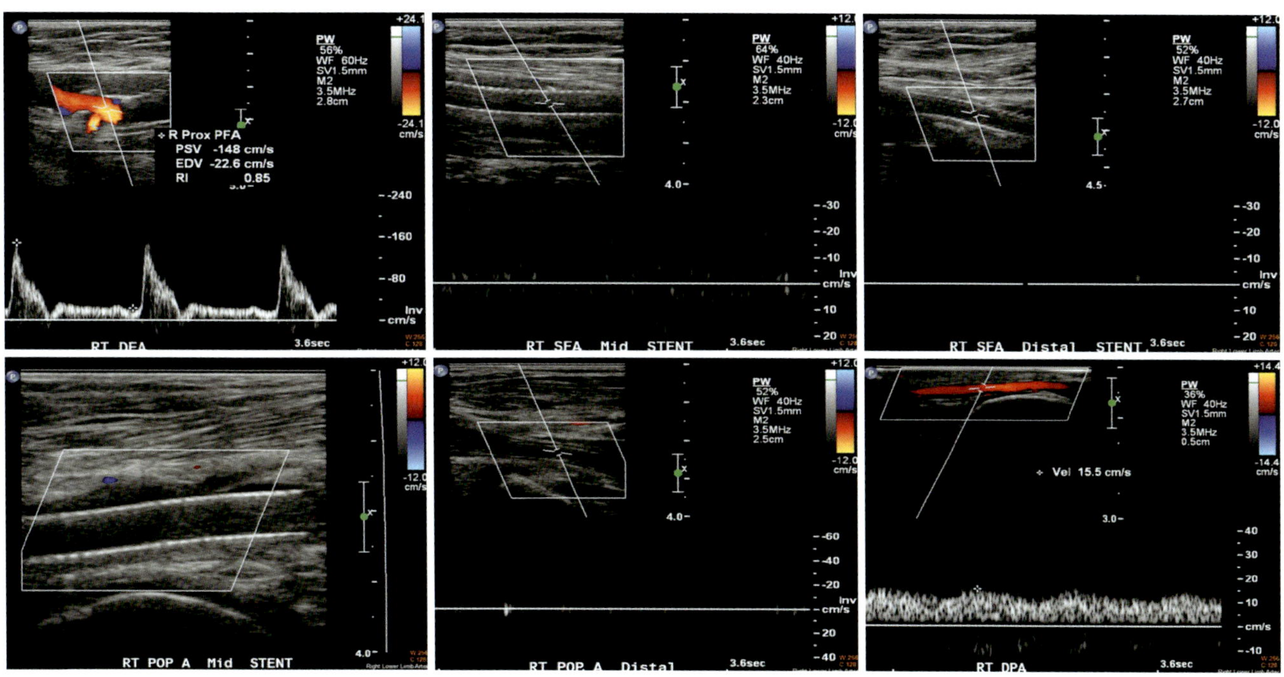

Fig. 22.1 Duplex US images of a 69-year-old female with right lower extremity rest pain (Rutherford class 4) s/p right superficial femoral artery (SFA) and popliteal artery (PA) stents placement who was lost to follow up for 1 year. Patient presented with recurrent rest pain and was found to have SFA and PA stent thrombosis on Duplex US. Dorsalis pedis (DP) artery with parvus (prolonged systolic acceleration) and tardus waveform (small systolic amplitude and rounding of systolic peak), due to reduced magnitude of blood flow to the DP from collateral circulation

The finding of triphasic waveform with normal PSV and PSV ratio indicates that no significant pressure gradient is present. Monophasic waveform at the stenosis site, parvus and tardus waveform distal to stenosis, spectral broadening, and an increase in PSV and PSV ratio are findings indicative of critical flow-limiting stenosis. In cases of iliac stent, as visualization can be challenging because of the presence of bowel gas and body habitus, poor monophasic duplex waveform at the CFA is an accurate marker of more proximal aortoiliac obstructive disease [14].

Restenosis has generally been defined as >50% narrowing of the luminal diameter or >75% reduction of the cross-sectional area of the artery [3]. The combination of PSV and PSV ratio measurements (Table 22.2) offers high positive predictive value for identifying moderate and severe restenosis that correlates with angiographic findings [11]. Although PSV >300 cm/s is indicative alone of severe in-stent stenosis, the PSV ratio >3.0 as well as PSVs <45 cm/s throughout the stent, are more reliable markers for predicting stent thrombosis [15–17]. Any restenosis greater than 70% (DUS: PSV >300 cm/s and PSV ratio >3) should be considered as an indication for reintervention [2, 9]. Although the PSV ratio is the most tested criterion for detecting stenosis in peripheral arteries, in cases in which the stenosis is very tight, this ratio is paradoxical and should not be considered. The decreased velocities and the great luminal reduction seen on color mode would be more appropriate to characterize the re-stenosis severity in these patients.

In rare circumstances, computed tomography angiography (CTA) and magnetic resonance angiography (MRA) can be utilized for confirmation and investigation of duplex ultrasound findings in order to obtain more detailed anatomic information before re-intervention [11].

Patency rate depends on many factors including patient demographics (gender, diabetes mellitus), the indication for intervention (intermittent claudication versus CLTI), TASC II classification, especially class C and D (location of lesion, occlusion vs stenosis, length, calcification, number of lesions, multilevel disease), runoff grade, and procedure-related factors (use of various balloon angioplasty devices, atherectomy devices, and stent grafting, vessel dissection, final interventional outcome) [2, 4, 5, 8, 18]. Figure 22.2 demonstrates a case where poor wall apposition of the stent led to early restenosis. Care should be taken to optimize the interventional outcome and minimize potential complications.

Different levels of lower extremity peripheral arterial disease also have different patency rates after endovascular interventions. The primary patency of aortoiliac stenting is the highest: 93% at 1 year, 83% at 3 years, and 78% at 5 years. Furthermore, treatment of iliac artery restenosis is associated with very high success rates, even in cases of occlusions [3]. Primary stent patency rate decreases substantially in case of a stent in both common and external iliac arteries, long stents more than 6 cm and poor runoff [19]. The femoropopliteal level is the most treated arterial segment and even complex lesions are

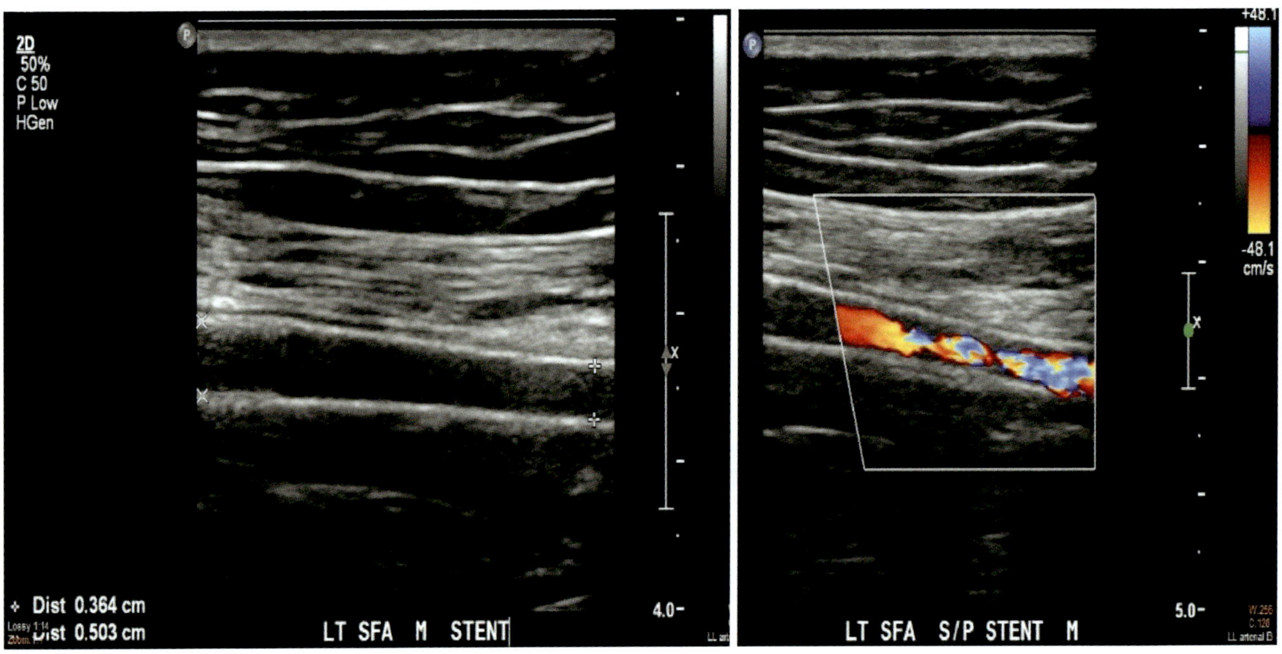

Fig. 22.2 Duplex US images of a 65-year-old male with left lower extremity intermittent claudication (Rutherford class 3) s/p left superficial femoral artery (SFA) stent placement. Surveillance DUS at 1 month showed severe left SFA in stent restenosis (PSV ratio: 342/89 = 3.8) due to poor intraoperative judgment of arterial diameter leading to poor stent-wall apposition

amenable to percutaneous interventions due to the use of a variety of devices and techniques (angioplasty, bare-metal stents, drug-eluting stents, stent grafts, atherectomy devices). Many factors can predict the primary patency after stent placement, among them the modified Society for Vascular Surgery runoff scoring system [20]. Tibial artery interventions are usually done for critical limb ischemia and are associated with low patency and high restenosis rates [3, 21]. Again, pedal runoff score can be valuable in identifying those patients who will not achieve ulcer healing, amputation-free survival, and freedom from major adverse limb events [22].

Many patients with intermittent claudication do not need intervention other than best medical therapy and exercise as shown in Fig. 22.3. Both best medical therapy and exercise are very important to protect the patient from cardiovascular events, even if an intervention is planned. As patency depends on so many factors, predicting which interventions are more prone to failure has proved challenging. Moreover, recurrent symptoms may be masked in patients with multi-level interventions and those with extensive collateral circulation and may not be apparent until the angioplasty segment or the stent is occluded [2]. It should be emphasized that the decision to treat any re-stenosis incorporates the combination of clinical symptoms, ABI changes and duplex ultrasonographic findings and should be weighed against the patient's functional status, comorbidities, and the potential morbidity associated with any reintervention (Fig. 22.4). In clinically relevant restenosis or occlusion, reintervention should be individualized based on risk factors predictive of procedural morbidity with endovascular approach being considered as the first-line treatment over open surgical intervention [4].

Case Presentation

Continued from page 217

The patient was lost to follow-up for 1 year, at which point she presented with recurrent worsened rest pain and new findings of toe tissue loss. She continued to smoke and was not compliant with her antiplatelet medications. She had palpable femoral pulses, and Dopplerable right leg DP/PT signals. Duplex US revealed thrombosis of the previously placed femoral and popliteal artery stents. Dorsalis pedis (DP) artery had parvus (prolonged systolic acceleration) and tardus waveform (small systolic amplitude and rounding of systolic peak), due to reduced magnitude of blood flow to the DP from collateral circulation (Fig. 22.1). The patient underwent a femoral to popliteal artery (P3 segment) bypass for restoration of arterial blood flow. She quit smoking, started a supervised *exercise program*, remained compliant with her medications, and with her follow up appointments (Right ABI: 0.8, DUS: patent bypass with two vessel runoff) and was able to heal the toe ulcerations.

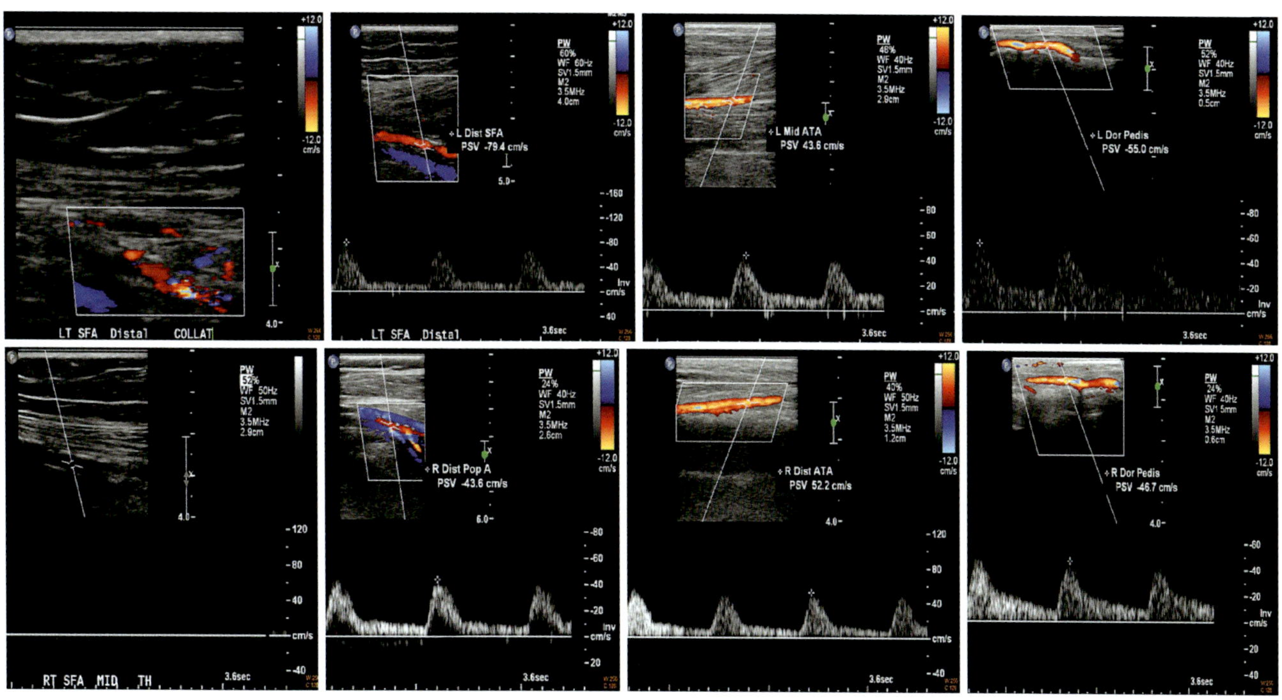

Fig. 22.3 Duplex US images of a 49-year-old female with bilateral 300-m intermittent claudication and bilateral SFA occlusion with good collaterals and distal recanalization with 3 vessels runoff. She was placed on best medical therapy and exercise regimen. Follow up at 1, 2 and 4 years showed no changes in the ABI and ultrasound imaging. Patient remained stable and required no other intervention

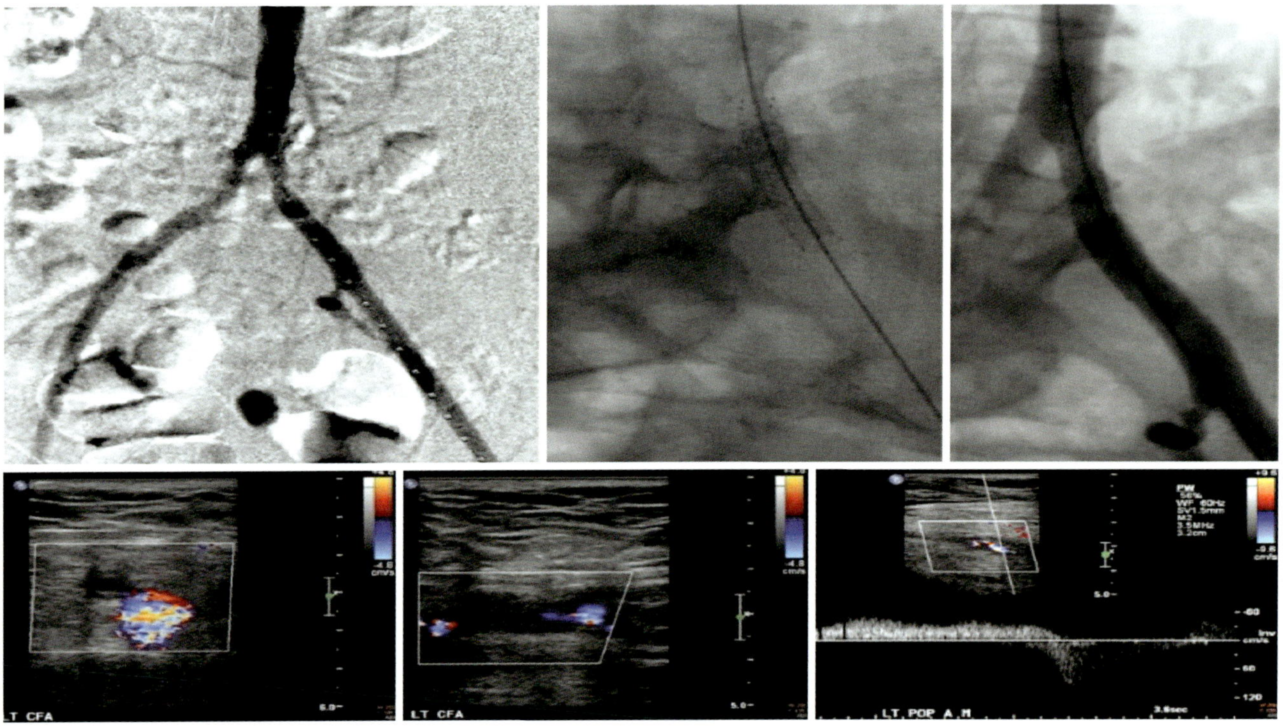

Fig. 22.4 Fluoroscopic images of a 73-year-old female with left foot tissue loss (Rutherford class 6) and severe left common iliac artery (CIA) stenosis s/p left CIA stent placement. Patient developed acute limb ischemia postoperatively and duplex US showed iatrogenic left common femoral artery occlusion from focal dissection during the iliac stenting for which the patient required emergent open intervention for dissection flap tacking and patch angioplasty

References

1. Wong KHF, Zucker BE, Wardle BG, Coughlin PA, Chaplin K, Cheng HY, Hinchliffe RJ, Twine CP, Ambler GK. Systematic review and narrative synthesis of surveillance practices after endovascular intervention for lower limb peripheral arterial disease. J Vasc Surg. 2022;75(1):372–380.e15. https://doi.org/10.1016/j.jvs.2021.08.062. Epub 2021 Nov 24
2. Sobieszczyk P, Eisenhauer A. Management of patients after endovascular interventions for peripheral artery disease. Circulation. 2013;128(7):749–57. https://doi.org/10.1161/CIRCULATIONAHA.113.001560.
3. Zierler RE, Jordan WD, Lal BK, Mussa F, Leers S, Fulton J, Pevec W, Hill A, Murad MH. The Society for Vascular Surgery practice guidelines on follow-up after vascular surgery arterial procedures [Erratum in: J Vasc Surg 2018 Nov;68(5):1623]. J Vasc Surg. 2018;68(1):256–84. https://doi.org/10.1016/j.jvs.2018.04.018.
4. Venermo M, Sprynger M, Desormais I, Björck M, Brodmann M, Cohnert T, De Carlo M, Espinola-Klein C, Kownator S, Mazzolai L, Naylor R, Vlachopoulos C, Ricco JB, Aboyans V. Editor's choice—follow-up of patients after revascularisation for peripheral arterial diseases: a consensus document from the European Society of Cardiology Working Group on aorta and peripheral vascular diseases and the European Society for Vascular Surgery. Eur J Vasc Endovasc Surg. 2019;58(5):641–53. https://doi.org/10.1016/j.ejvs.2019.06.017.
5. Brooke BS, Beck AW, Kraiss LW, Hoel AW, Austin AM, Ghaffarian AA, Cronenwett JL, Goodney PP. Association of quality improvement registry participation with appropriate follow-up after vascular procedures. JAMA Surg. 2018;153(3):216–23. https://doi.org/10.1001/jamasurg.2017.3942. PMID: 29049809; PMCID: PMC5885932
6. Austin AM, Chakraborti G, Columbo J, Ramkumar N, Moore K, Scheurich M, Goodney P. Outcomes after peripheral artery disease intervention among Medicare–Medicaid dual-eligible patients compared with the general medicare population in the Vascular Quality Initiative registry. BMJ Surg Interv Health Technol. 2019;1(1):e000018. https://doi.org/10.1136/bmjsit-2019-000018.
7. Wang GJ, Judelson DR, Goodney PP, Bertges DJ. Loss to follow-up 1 year after lower extremity peripheral vascular intervention is associated with worse survival. Vasc Med. 2019;24(4):332–8. https://doi.org/10.1177/1358863X19853622. Epub 2019 Jun 13
8. Bradaric C, Koppara T, Müller A, Haller B, Ott I, Cassese S, Fusaro M, Kastrati A, Laugwitz KL, Ibrahim T. Incidence and predictors of stent thrombosis after endovascular revascularisation of the superficial femoral artery. EuroIntervention. 2019;15(12):e1107–14. https://doi.org/10.4244/EIJ-D-19-00187.
9. Qato K, Conway AM, Mondry L, Giangola G, Carroccio A. Management of isolated femoropopliteal in-stent restenosis. J Vasc Surg. 2018;68(3):807–10. https://doi.org/10.1016/j.jvs.2018.01.030.
10. Biscetti F, Nardella E, Rando MM, Cecchini AL, Gasbarrini A, Massetti M, Flex A. Outcomes of lower extremity endovascular revascularization: potential predictors and prevention strategies. Int J Mol Sci. 2021;22(4):2002. https://doi.org/10.3390/ijms22042002.
11. Conte MS, Bradbury AW, Kolh P, White JV, Dick F, Fitridge R, Mills JL, Ricco JB, Suresh KR, Murad MH, Aboyans V, Aksoy M, Alexandrescu VA, Armstrong D, Azuma N, Belch J, Bergoeing M, Bjorck M, Chakfé N, Cheng S, Dawson J, Debus ES, Dueck A, Duval S, Eckstein HH, Ferraresi R, Gambhir R, Gargiulo M, Geraghty P, Goode S, Gray B, Guo W, Gupta PC, Hinchliffe R,

Jetty P, Komori K, Lavery L, Liang W, Lookstein R, Menard M, Misra S, Miyata T, Moneta G, Munoa Prado JA, Munoz A, Paolini JE, Patel M, Pomposelli F, Powell R, Robless P, Rogers L, Schanzer A, Schneider P, Taylor S, De Ceniga MV, Veller M, Vermassen F, Wang J, Wang S, GVG Writing Group for the Joint Guidelines of the Society for Vascular Surgery (SVS), European Society for Vascular Surgery (ESVS), and World Federation of Vascular Societies (WFVS). Global vascular guidelines on the management of chronic limb-threatening ischemia [Erratum in: Eur J Vasc Endovasc Surg. 2020 Mar;59(3):492–493. Erratum in: Eur J Vasc Endovasc Surg. 2020 Jul;60(1):158–159]. Eur J Vasc Endovasc Surg. 2019;58(1S):S1–S109.e33. https://doi.org/10.1016/j.ejvs.2019.05.006. Epub 2019 Jun 8

12. Leithead C, Novak Z, Spangler E, Passman MA, Witcher A, Patterson MA, Beck AW, Pearce BJ. Importance of postprocedural wound, ischemia, and foot infection (WIfI) restaging in predicting limb salvage. J Vasc Surg. 2018;67(2):498–505. https://doi.org/10.1016/j.jvs.2017.07.109. Epub 2017 Sep 22
13. Draxler MS, Al-Adas Z, Abbas D, Kavousi Y, Kabbani LS, Lin JC, Weaver MR, Shepard AD, Nypaver TJ. Outcome benefit of arterial duplex stent imaging after superficial femoral artery stent implantation. J Vasc Surg. 2021;73(1):179–88. https://doi.org/10.1016/j.jvs.2020.02.055. Epub 2020 May 8
14. Spronk S, den Hoed PT, de Jonge LC, van Dijk LC, Pattynama PM. Value of the duplex waveform at the common femoral artery for diagnosing obstructive aortoiliac disease. J Vasc Surg. 2005;42(2):236–42.; discussion 242. https://doi.org/10.1016/j.jvs.2005.04.048.
15. Troutman DA, Madden NJ, Dougherty MJ, Calligaro KD. Duplex ultrasound diagnosis of failing stent grafts placed for occlusive disease. J Vasc Surg. 2014;60(6):1580–4. https://doi.org/10.1016/j.jvs.2014.08.071. Epub 2014 Sep 22
16. Zheng H, Calligaro KD, Jang J, Tyagi S, Madden N, Troutman DA, Dougherty MJ. Duplex ultrasound for diagnosis of failing stents placed for lower extremity arterial occlusive disease. Ann Vasc Surg. 2020;63:269–74. https://doi.org/10.1016/j.avsg.2019.08.086. Epub 2019 Oct 15
17. Baril DT, Rhee RY, Kim J, Makaroun MS, Chaer RA, Marone LK. Duplex criteria for determination of in-stent stenosis after angioplasty and stenting of the superficial femoral artery. J Vasc Surg. 2009;49(1):133–8. https://doi.org/10.1016/j.jvs.2008.09.046.
18. Velescu A, Mateos Torres E, Clará Velasco A. Predictors of medium-term patency in percutaneous endovascular therapy of femoro-popliteal lesions. Cir Esp. 2015;93(2):105–9. https://doi.org/10.1016/j.ciresp.2014.04.011. Epub 2014 Jun 12
19. Kavaliauskienė Z, Benetis R, Inčiūra D, Aleksynas N, Kaupas RS, Antuševas A. Factors affecting primary patency of stenting for TransAtlantic Inter-Society (TASC II) type B, C, and D iliac occlusive disease. Medicina (Kaunas). 2014;50(5):287–94. https://doi.org/10.1016/j.medici.2014.10.003. Epub 2014 Oct 29
20. Noh BG, Park YM, Choi JB, Lee BC, Lee SS, Jung HJ. Influence of infrapopliteal runoff vessels on primary patency after superficial femoral artery angioplasty with stenting in patients with claudication. Vasc Specialist Int. 2020;36(4):233–40. https://doi.org/10.5758/vsi.200056.
21. Fernandez N, McEnaney R, Marone LK, Rhee RY, Leers S, Makaroun M, Chaer RA. Predictors of failure and success of tibial interventions for critical limb ischemia. J Vasc Surg. 2010;52(4):834–42. https://doi.org/10.1016/j.jvs.2010.04.070.
22. Baer-Bositis HE, Hicks TD, Haidar GM, Sideman MJ, Pounds LL, Davies MG. Outcomes of tibial endovascular intervention in patients with poor pedal runoff. J Vasc Surg. 2018;67(6):1788–1796.e2. https://doi.org/10.1016/j.jvs.2017.09.033. Epub 2017 Dec 13

Endovascular Management of Acute Limb Ischemia

23

Gianmarco de Donato and Edoardo Pasqui

Case Presentation

A 78-year-old male was admitted to the emergency department with left lower limb pain that occurred 3 days before. Clinical examination highlighted a typical condition possibly related to an acute limb ischemia: absence of popliteal and tibial pulses, distal cyanosis and poikilothermic foot. Duplex vascular ultrasound was performed and emerged an absence of tibial perfusion with the contemporary presence of a 32 mm popliteal artery aneurysm partially thrombosed. A computed tomography angiography was performed and the diagnosis was confirmed: acute limb ischemia due to distal embolization of popliteal artery thrombosis.

Due to the severe clinical status (acute limb ischemia Rutherford grading IIb) and major comorbidity (coronary artery disease, heart failure, hypertension, chronic obstructive pulmonary disease, history of malignancy), an endovascular limb rescue was performed.

Continued at page 234

Background

Acute lower limb ischemia (ALLI) is defined as any sudden decrease in limb perfusion causing a potential threat to limb viability in patients who present within 2 weeks of the acute event [1]. The incidence of this condition is approximately 15 cases per 100,000 persons per year [2]. The prevalence is <0.1% in general population and about 5–10% in patients with other risk factors for cardiovascular disease, like diabetes of previous cardiovascular events (Fig. 23.1).

Acute lower limb ischemia is still considered to be a dramatic event, carrying a risk of amputation between 15 and 30% and a high perioperative morbidity and mortality (20–30%) [3–8]. Concomitant underlying diseases, the metabolic derangement that seems as a result of the acute insult, and a possible reperfusion injury following revascularization may account for this severe prognosis [9–11].

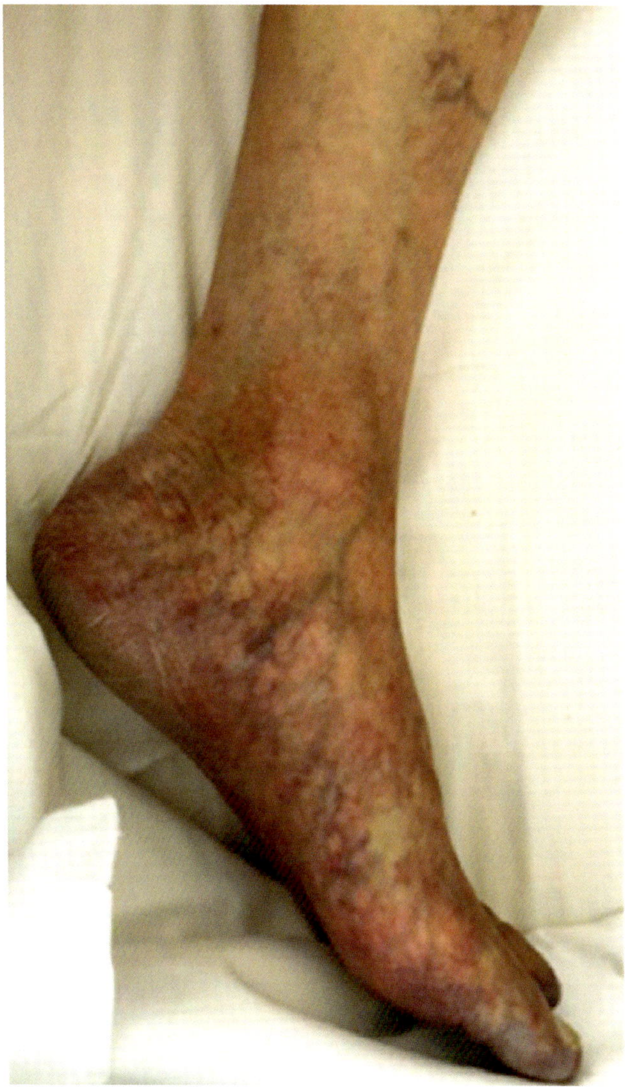

Fig. 23.1 Typical aspect of a Rutherford IIb acute limb ischemia

G. de Donato (✉) · E. Pasqui
Vascular Surgery Unit, Department of Medicine, Surgery and Neuroscience, University of Siena, Siena, Italy
e-mail: dedonato@unisi.it; pasqui11@student.unisi.it

G. Geroulakos et al. (eds.), *Mastering Endovascular Techniques*, https://doi.org/10.1007/978-3-031-42735-0_23

After the invention of the balloon catheter by Fogarty in 1963, surgical thromboembolectomy was considered the gold standard of treatment for many years in patients with acute lower limb ischemia (ALLI). Still, ALLI is a dramatic event, carrying a risk of amputation between 15 and 30% and a high perioperative morbidity and mortality (20–30%). Therefore, the need of continued innovation in this area led to innovative percutaneous approaches. In the 1970s Dotter first introduced the idea of clot lysis in the treatment of ALLI, then modified to the catheter-directed thrombolysis. Since then, many new endovascular procedures, utilizing sophisticated equipment, have been developed to improve treatment results and reduce complications.

Although surgery is still a significant treatment option especially for ALLI due to embolism, endovascular techniques are gaining a more prominent role in case of acute on chronic ischemia. Improved clinical outcomes, associated to lower rates of major complication coming from the application of newly designed endovascular thrombectomy devices in this difficult set of patients may shift treatment recommendations towards endovascular options.

This chapter is focused on the role and limitations of treatment options for ALLI, starting from surgical thromboembolectomy to CDT and new endovascular techniques including mechanical fragmentation, rheolytic thrombectomy and aspiration thrombectomy devices.

Surgical Approach

Surgical Arterial Thromboembolectomy

Surgical arterial thromboembolectomy (TE) using a balloon catheter was originally introduced by Thomas Fogarty in 1963 [12]. Since then it has been the corner-stone of ALI's therapy for decades. This specific technique has several advantages, especially in limb-threatening ischemia due to embolic occlusion: it can be quickly performed using local anesthesia, via femoral approach, and it is considered to be not a very complex intervention, being the case for training for less experienced or young operators. Balloon catheters are passed through the thrombus several times until blood clot is removed and limb perfusion is restored.

When the thrombosis includes below-the-knee (BTK) arteries, surgical approach is more technically demanding. A blind BTK embolectomy has low success rates with an inguinal incision due to the low torquability of catheters.

Thus, TE using Fogarty balloon catheter is a very efficient treatment for acute arterial emboli of lower limbs, especially when a single large vessel is involved [13], but the early clinical outcome still remains unsatisfactory in a number of cases [14], especially when small arteries of the lower limb are involved in the occlusion, or in presence of PAD. It has been well established that intra-operative angiography after TE may frequently detect arterial imperfections, such as incomplete restoration of perfusion in below-the-knee (BTK) arteries, propagation of thrombi or presence of underlying steno-occlusive lesions [15, 16], which may be responsible for the early failure. Finally even vessel injuries secondary to balloon catheter passage may limit the clinical success rate.

In particular, diabetic subjects who carries a higher risk of developing a severe PAD usually located in below-the-knee vessels, are at high risk of unsatisfactory outcomes after TE using Fogarty balloon catheter. In diabetic patients arteries' walls are typically fragile, tending to spasm, or are very calcified and difficult to reach by Fogarty's catheter cannulation. Moreover, considering the above-mentioned complexity in clinical presentation of ALLI in diabetes (typical signs hidden by the presence of supplementary collateral circuits and diabetic neuropathy) diagnosis may be delayed, making the blood clot more stable and less responsive to treatment.

More complex surgical treatment, especially when inflow and outflow procedures are requested, are essentials for more demanding pathology's patterns [17]. By-pass procedures may be requested as primary option in these cases or as a second-line treatment when thrombo-emboectomy fails.

Hybrid Treatment

While improvements in surgical techniques and perioperative patient care have occurred over the years, the results available document a persistently high medical need for patients presenting ALLI, with a reported 30-day amputation rate of 5–12%, mortality risk of 10–38%, and combined incidence of amputation and death of 25–37.5% at 6-month follow-up [7, 18–20]. In clinical practice there is a discrepancy between the immediate technical success of arterial thromboembolectomy that often seems adequate, and the early clinical outcome that still remains unsatisfactory in a number of cases. This incongruence may be related to the incomplete restoration of perfusion due to residual thrombus in distal vessels inaccessible to the balloon embolectomy or to the propagation of residual thrombi. Both historical experimental and clinical studies have shown that clot removal is frequently incomplete after TE, with persistent thrombus demonstrated angiographically and by fiberoptic angioscopy in small distal vessels in 36–82% of patients [18, 19].

The first attempt to improve the surgical technique for treating acute arterial occlusion was described by Parsons et al. in 1996 [21]. Suggesting the use of intra-operative fluoroscopy and the performance of fluoroscopically assisted thromboembolectomy, the authors described the first example of combination of surgical and endovascular technique for treatment of ALLI. This original hybrid procedure turned out to improve the technical success of TE. Facilitating the catheter passage through tortuous or diseased arteries, this new technique minimized arterial damage and blood loss during clot removal. It had also the potential to facilitate accurate identification, localization, and treatment of significant concomitant arterial lesions.

Since then it has been recommended to perform a routine angiography after TE [22, 23] but adherence to this recommendation is still limited in clinical practice [24].

Completion angiography after TE can not only identify residual thrombus in distal vessels, but also incomplete recanalization of more proximal arteries due to residual thrombus strongly adherent to the arterial wall, or can reveal the presence of significant underlying steno-occlusive lesions after clot removal. In a limited number of cases angiography can even demonstrate the existence of vessel injuries secondary to incorrect balloon catheter passage. These points clearly call for a routine and accurate intra-operative assessment of the adequacy of clot removal, which may be performed by angiography, angioscopy or IVUS.

It has been a common belief in the past to rely on the presence or absence of back-bleeding from a peripheral arterial bed as a guide to distal arterial patency. Actually, we know that visual evaluation of outflow may be ambiguous to truly prove that adequate circulation has been restored, since irregular thrombotic material may be present in approximately one third of cases [25], and back-bleeding may be quite misleading in cases of adequate collateral circulation, despite the occurrence of total distal obstruction.

The central role of intra-operative angiography after TE in defining the need for as well as guiding adjunctive endovascular procedures (so-called "hybrid" treatment) [26, 27] has been recently demonstrated. A strategy including hybrid solutions seems to be the keys for better early and mid-term outcomes of ALLI patients.

Thrombolysis

Alterative to surgery, percutaneous thrombolysis has been proposed as a procedure less invasive and less demanding for the patient. Being described by Dotter in 1974 [28], the first results of systemic thrombolysis were not very encouraging: the treatment was unsuccessful in a number of cases and was burdened with a high rate of bleeding complications. Further improvement of thrombolysis resulted in Catheter-directed thrombolysis (CDT), now widely employed in clinical settings, especially when patients suffer less severe ischemia (classes 1, 2a, or sometimes 2b).

With CDT, the thrombolytic agents are administered through an endovascular catheter near the occlusive blood clot to achieve the complete dissolution and to have the least number of systemic consequences. The preliminary arteriography can precisely detect the location, the length of the occlusion, the nature of the ischemia (thrombotic vs. atherosclerotic), the presence of collateral circuits. Usually a multiside catheter is advanced the thrombus and the thrombolytic agents are deployed for a period ranging from 12 to 72 h based on clinical status and patient's reply to the therapy. Contemporary continuous unfractionated heparin infusion is administrated to prevent re-thrombosis and peri-catheter thrombosis.

Different protocols of CDT are usually used: continuous infusion, pulse/spray and a combination of both [11]. A Cochrane systematic review showed that the greater benefit is seen when the thrombolytic agents are delivered within the thrombus. "High dose" and "forced infusion" techniques, or adjunctive agents such as platelet glycoprotein IIb/IIIa inhibitors may speed up thrombolysis, but these are not accompanied by lower amputation rates or a decreased need for adjunctive endovascular or surgical procedures. "Low-dose continuous infusion", following initial lacing of the thrombus with a high dose of the thrombolytic agent, is the least labor intensive technique. There is no evidence that a specific regimen is more effective than the others [29].

Nowadays all clinical available agents are forms of plasminogen activators. They act through the conversion of plasminogen to plasmin, which is able to degrade fibrin. The first agent was developed in 1940s and it was called *Streptokinase*. Through the years other agents was discovered, as *Urokinase*, more innovative substances as recombined tissue plasminogen activator (rt-PA). A recent systematic review showed that intra-arterial rt-PA is more effective than intra-arterial streptokinase or intravenous rt-PA in improving vessel patency in people with peripheral arterial occlusion. There was no evidence that rt-PA was more effective than urokinase for patients with peripheral arterial occlusion and some evidence that initial lysis may be more rapid with rt-PA, depending on the regime. Incidences of hemorrhagic complications were not statistically significantly greater with rt-PA than with other regimes. However, all of the findings come from small studies and a general paucity of results means that it is not possible to draw clear conclusions [30].

CDT leads to thrombus removal rate in 75–92% [31]. However CDT carries a significant risk of bleeding complication during and after drug infusion. In TOPAS trial [8], 12.5% of treated patients had major hemorrhagic events, including a rate of 1.4% of intracranial hemorrhage. The risk of bleeding was increased up to 19% when heparin was contemporary used [14].

CDT has been used as an alternative to surgical embolectomy for ALLI with gratifying results, but its success is related to the ability to pass a guide through and embed a multisided catheter into the occlusion [32, 33]. Results from randomized trials comparing the two treatments reveal that thrombolysis has the greatest advantage in acute prosthetic bypass graft occlusions, while patients with ALLI due to native arterial occlusions tend to be associated with inferior results after thrombolysis, consisting in higher rates of hemorrhage and stroke at 30 days, and increased risks of distal embolization [34, 35].

Recently, it has been proposed to combine CDT target lesion endovascular/surgical correction to guarantee an optimal treatment outcome and better long-term patency [28, 36].

Ultrasound-Accelerated Thrombolysis

The employment of ultrasound waves has been proposed to facilitate mechanical fragmentation of the thrombus and increase the speed of clot disruption. The ultrasound producing devices (EkoSonic® Endovascular System, Ecos, Bothell, USA are placed directly in the thrombus, where micro-transducers transmit high-frequency low-power sound waves. The fibrin is made thinner by the ultrasonic power and breaks. As a result, plasminogen receptor sites are exposed and the thrombolytic agent may penetrate deeper in the thrombus, accelerating the absorption of rt-PA and the thrombus lysis [37, 38].

Table 23.1 Main differences among percutaneous mechanical thrombectomy devices

Intra-arterial thrombolysis	Novel endovascular devices
Significant Risk of Bleeding	Prompt blood clot removal
Easy to perform	Higher endovascular skill required
Time consuming, not indicated for Rutherford II b acute ischemia	Reduced revascularization time
Limited cost of the treatment	Increased cost of the devices

Endovascular Non-LYTIC Options

Percutaneous mechanical thrombectomy devices have been proposed over the past decade to speed up the time required for dissolution of thrombus by pharmacologic means, and to reduce the amount of thrombolytic agent used [39]. These devices have been developed for aspiration and dissolution of small amounts of thrombotic material during coronary procedures, while the clot quantity in cases of ALLI may be much larger, making complete clot removal more challenging.

Manual aspiration thrombectomy is the first reported true percutaneous approach for clot removal, followed, by a series of mechanical thrombectomy based on different mechanism of action (mechanical fragmentation, aspiration, rheolytic thrombectomy and their combinations). Tables 23.1 and 23.2 highlight the main differences between these approaches.

Table 23.2 Main differences among percutaneous mechanical thrombectomy devices

	Device	Technical features	Strengths	Weakness	Possible complications
Mechanical Fragmentation	Rotarex (Straub medical)	6–10 Fr 85–135 cm working length	Can be used for different type of lesions (thrombus, calcifications)	Avoid direct contact with vessel wall for the risk of injury	– Arterial wall damage – Distal embolism
	JetStream Atherectomy system (Boston Scientific Corporation)	7 Fr 120–145 cm working length	Can be used for different type of lesions (thrombus, calcifications)	Avoid any contact with the vessel wall for the risk of injury	– Arterial wall damage – Distal embolism
Rheolytic thrombectomy	AngioJet Thrombectomy set (Boston Scientific Corporation)	8–4 Fr 0.014"–0.035" 50–145 cm working length	Can be used in different settings: acute native thrombosis, thrombosed stents and surgical bypasses	The complete thrombus passage is needed to work appropriately	– Hemolysis – Acute Renal Failure – Arterial wall damage
Mechanical Aspirationn	Aspirex (BD interventional)	10–6 Fr 0.018"–0.025" 85–135 cm working length	Can be used in different settings: acute native thrombosis, thrombosed stents and surgical bypasses	Avoid any contact with the vessel wall for the risk of injury	Arterial wall damage
	Indigo System (Penumbra Inc)	12–3 Fr 0.014"–0.035" 50–150 cm working length	– High trackability and atraumatic tip; – Use of SEP – Easy to use	Correct use of the vacuum pump to avoid excessive blood loss	Distal embolism

Manual Aspiration

The first method of percutaneous arterial thrombus removal was manual aspiration thrombectomy (MAT) via large lumen [40]. Initially, it was described as rescue maneuver in case of distal embolization during endovascular procedure [41], by using large-lumen end-hole aspiration catheters from 5 to 9 Fr, depending on the diameter of the artery occluded.

Afterward, suction is produced using a 20–50 mL syringe while withdrawing the catheter, and thrombus is typically removed after a series of catheter passages. Even though the procedure is really low-cost and may offer some good results, with published technical success up to 96.6% [42], MAT may have less satisfactory results in cases of old and well organized thrombus, and in patients with disseminated atherosclerotic as typically diabetic patients are. In such cases unintentional vessel damage by multiple passages of catheters not specifically designed for this purpose is high, and this justifies why MAT has never gained wide consensus among physicians.

Mechanical Fragmentation

The Rotarex (Straub Medical AG, Wangs, Switzerland) is a mechanical thrombectomy system with a working catheter consisting in a fast-spinning head (40,000–60,000 rpm) driven by a rotating helix inside it.

The fast-spinning mechanism generates a vortex that first separates the thrombus from the artery wall and then fragments it. According to the Archimedes screw principle, clot is subsequently aspirated inside the catheter through the aspiration ports located just behind the head. Direct contact with the vessel wall may cause different type of damage up to artery perforation. The device is available in 6- and 10-Fr-diameter versions, which allows effective thrombectomy from iliac to tibial arteries. In a series of 98 patients with acute and subacute thrombotic arterial and bypass graft occlusions, Zeller et al. reported early technical success in 92% of cases with a reasonable complication rate [43]. Recently, Heller et al. reported a high technical success is a single center series of 147 acute limb ischemia using Rotarex, especially when in association to limited thrombolysis (90.5%) [44].

Technical success may be limited in case of acute on chronic ischemia, even though published data are controversial. Lagana et al. technical success of Rotarex is limited in case of thrombosis complicating calcified plaques or dissection intimal flaps may cause rupture of the arterial wall. In their experience, subacute and chronic occlusions of by-pass grafts, stents and stent grafts, and other additional procedures are necessary to achieve complete recanalisation [45]. Instead, Kronlage et al. showed that the use of rotational thrombectomy in the treatment of (sub)acute critical limb ischemia is not inferior to local thrombolysis and is associated with a lower complication rate, shorter duration of hospitalizations, and a lower cost of and lower costs in a 12-month follow-up in a tertiary hospital collective [46].

The Jetstream(R) Atherectomy System (Boston Scientific, Marlborough, USA) consists of a single-use catheter provided with four cutting front blades and a rotating tip. For additional effect, four differential cutting, expandable blades are located just behind the tip. The device diameter ranges from 1.6 to 3.4 mm, allowing effective treatment of vessels from the groin to the ankle. Disrupted material is evacuated via suction ports located behind the rotating tip. To minimize the risk of peripheral embolization, the vessel is constantly flushed with saline through infusion ports located proximally to the aspiration ports.

The system is engineered to predictably treat multiple morphologies, such as calcium, plaque or thrombus, commonly found in total occlusions, and theoretically can be of great benefit in case of acute on chronic vessel occlusion in diabetic patients. Evidence of its efficacy is well described in literature in case of chronic vessel occlusion or in-stent restenosis [47, 48], while it remains still under-investigated in case of acute limb ischemia [49]. Still, abrupt vessel occlusion, dissection, distal emboli, hematoma at access site, infection, perforation, pseudoaneurysm, renal failure, restenosis, and thrombus formation are some of the reported complications of Jetstream atherectomy [50].

Rheolytic Thrombectomy

The AngioJet(R) (Boston Scientifi c, Marlborough, USA) is a rheolytic thrombectomy device, made of a double-lumen OTW catheter that uses the Bernoulli's principle for thrombus aspiration. The Bernoulli's principle states that for an inviscid flow of a non-conducting fluid, an increase in the speed of the fluid occurs simultaneously with a decrease in pressure. So, when the saline flow backwards at the tip of the catheter with a speed over 500 mph, a negative pressure is created which is responsible for aspiration.

The second lumen, with infusion apertures localized more proximally, delivers saline into the thrombus to fragment it and facilitate aspiration. The system requires that a complete thrombus passage is achieved to work appropriately, and at least a few passages are usually required to entirely clear the vessel.

The rapid stream of fluid and hydrodynamic forces used by thrombectomy devices may cause significant amount of red blood cell hemolysis resulting in hemoglobinemia and hemoglobinuria [51]. This may occur when repeated passes of the device is required, resulting in severe consequences, mainly in patients with renal insufficiency [52]. The recommended maximum working time is limited to 300 s while working in the blood stream and 600 s within the thrombus, and it is highly suggested to hydrate the patient with isotonic saline during and after the procedure, to alkalize the urine with sodium bicarbonate and utilize mannitol to maintain diuresis over 100 mL/h after the procedure.

The AngioJet system has been tested in differ clinical scenarios: acute native thrombosis, thrombosed stents and surgical bypasses [53, 54]. Technical success with effective flow restoration was seen in 60–90% of patients [55, 56].

The PEARL (PEripheral Use of AngioJet Rheolytic Thrombectomy with a variety of catheter Lengths) registry [57] reported results of rheolytic pharmacomechanical thrombectomy for the management of ALI. Procedural success was achieved in 83% of 283 patients, with half of the procedures completed without the need for adjunctive CDT. At 12-month follow-up, amputation-free survival and freedom from mortality were 81% and 91%, respectively; 12-month freedom from bleeding requiring transfusion was 91%, and freedom from renal failure was 95%. Not surprisingly, subgroup analysis revealed better outcomes in patients without infrapopliteal involvement and those who underwent rheolytic thrombectomy without CDT.

Nevertheless, there have been a series of possible complication related to the mechanism of action of the system, including: vessel injury and acute closure, stent graft collapse during thrombectomy [58], distal embolization and acute pancreatitis [59], aside from the above-mentioned intravascular hemolysis and acute renal insufficiency.

Aspiration Thrombectomy

Aspiration thrombectomy is an alternative option for thrombus removal in patients with ALLI. It is feasibility in patients with high risk of bleeding when thrombolysis in contraindicated (hepatic failure, recent surgery, trauma or a neurovascular accident), offering flow restoration in few minutes and not in hours as required by CDT. Moreover, there is no risk of hemolysis as reported after rheolytic thrombectomy.

Different devices have been available on the market up to now: two of them, the Aspirex® thrombectomy system (Straub Medical AG, Wangs,Switzerland), The ThromCat XT® (Spectranetics International, Leusden, The Netherlands) are rotational thrombectomy catheter including aspiration; one, the Indigo® System (Penumbra, Alameda, USA) is based only on aspiration power.

The Aspirex is provided with a steel helix with a coaxial central lumen for a guidewire rotates inside a single-lumen rotational catheter with a speed of 40,000–60,000 rpm, producing negative pressure at the catheter tip. The thrombus is first aspirated, then fragmented by the spinning helix and delivered to the collection bag. Although the spinning helix is located inside the catheter, it is advocated to avoid any contact with the vessel wall for the risk of injury.

The ThromCat XT® is a rotational thrombectomy catheter primarily developed for coronary interventions.

Its flexible and atraumatic tip catheter consists of a stainless steel helix with a kink-resistant covering. The catheter neck encases the distal end of the helix, thus avoiding any direct contact with the vessel wall.

Due to increased flexibility, it is suitable for crossover access.

Although the device received CE mark approval in 2009, data on its efficacy are still limited. The technical success with a restoration of flow is reported in 70% of patients, with thrombus age (especially if greater than 5 days) as a major predictor of failure [60].

The Indigo System promotes active thrombectomy using a vacuum pump that generates substantial suction, enabling aspiration of clots of varying sizes and lengths. The device has three components: aspiration catheter, separator and pump. The system is not provided of any rotational components, then the risk of vessel injury is truly minimized.

The Indigo System has to be considered a last generation System for thromboembolic disease, being designed specifically to address the limitations of conventional technology [61]. Many of the first generation mechanical endovascular devices for thrombus removal have failed to be adequately efficacious or have been associated with unacceptable complication rates, mainly for the limited trackability, the risk of vessel injury, and/or the incidence of incomplete revascularization. In fact, one of the main difficulties is designing a device that can remove adequate volumes of variable age thrombus while also maintaining an acceptably small size, flexibility, and ease of use.

Since 2005, the Penumbra System® became available in Europe and the United States for the revascularization of occluded intracranial vessels in patients with acute ischemic stroke. The Penumbra System used vacuum aspiration as its primary mechanism of action. A flexible, atraumatic large-bore catheter is delivered to the site of occlusion, and aspiration is directly applied to the lesion itself. To maintain lumen patency of the large-bore catheter, a Penumbra Separator™ can be used at the tip of the catheter to continually break up the clot once ingested under aspiration. The Separator allows continuous thrombectomy under constant aspiration supplied by an external vacuum pump. The Pump MAX™ is capable of delivering and maintaining nearly pure vacuum (29 inHg or 98.2 kPa).

Demonstrating to be safe and effective in the neurovasculature and to provide high rates of complete intracranial vessels revascularization, the Penumbra system has begun the market leader in stroke. Consequently, physicians who were familiar with the Penumbra System for stroke care started using it in the peripheral vasculature for acute thrombotic and embolic events. Lesions that were previously inaccessible with conventional technology, was then treated these atraumatic, ultra-flexible neurovascular devices. In 2014, Penumbra launched the Indigo System specifically for this application, redefining below-the-knee mechanical thrombectomy. This is achieved by utilizing Penumbra's proprietary catheter-tracking technology and patented Separator technology for mechanical clot engagement. This percutaneous system is available in 8-Fr (three different tip shapes), 6-Fr, 5-Fr, and 3-Fr catheter options (namely CAT 8 STR/TORQ/XTORQ, CAT 6, CAT 5, CAT 3). The Indigo has the largest extraction lumen designed for vessels below the knee, with smaller and longer catheter options for hard-to-reach distal extremities. The catheter-tracking technology allows the device to reach the foot even from a contralateral approach.

Recently, CAT 7 and CAT 12 were lunched on the market allowing the treatment also of very large vessels. The Lightning Intelligent Aspiration is the very recent evolution of Indigo Penumbra System allowed by the novel penumbra ENGINE. Lightning Aspiration Tubing has dual pressure sensors for real-time blood flow monitoring. The built-in microprocessor features a proprietary thrombus removal algorithm that automatically controls a valve in the tubing to provide continuous or intermittent aspiration. With automatic valve control, Lightning can help the physician focus on optimizing thrombus removal procedures. Throughout each case, Lightning provides procedural feedback via audiovisual cues. Lightning comes packaged with the Indigo CAT7, CAT8 or CAT12.

During the last years, data on the safety and efficacy on the use of this device in the treatment of ALI was published world-wide. The PRISM trial evaluated outcomes in patients with ALI in which thrombolysis failed. Technical success was achieved in 86.4% in a wide range of patients' setting [62].

The Indian trial, the most numerous trial evaluating the efficacy of this system in the treatment of ALI highlighted promising results. The technical success, defined as the complete or near complete recanalization of the thrombotic occlusion (TIPI 2–3 thromboaspiration in peripheral ischemia) was reached in 88.7% after thromboaspiration alone, increased to 95.3% after adjuvant procedures [63, 64].

Location of the thrombus was quite variable in the cohort of patients, although in the majority of cases, patients presented with thrombus involving the popliteal (Fig. 23.2) or BTK vessels (Fig. 23.3). These data were quite interesting. Notably, operators were more prone to use this kind of device for the revascularization of BTK small vessels, this trend could be related to the device's features of trackability, pushability and atraumatic tip. Notably, thromboembolic complications due to popliteal aneurysms could be also approached, especially in case of distal embolization (Fig. 23.4).

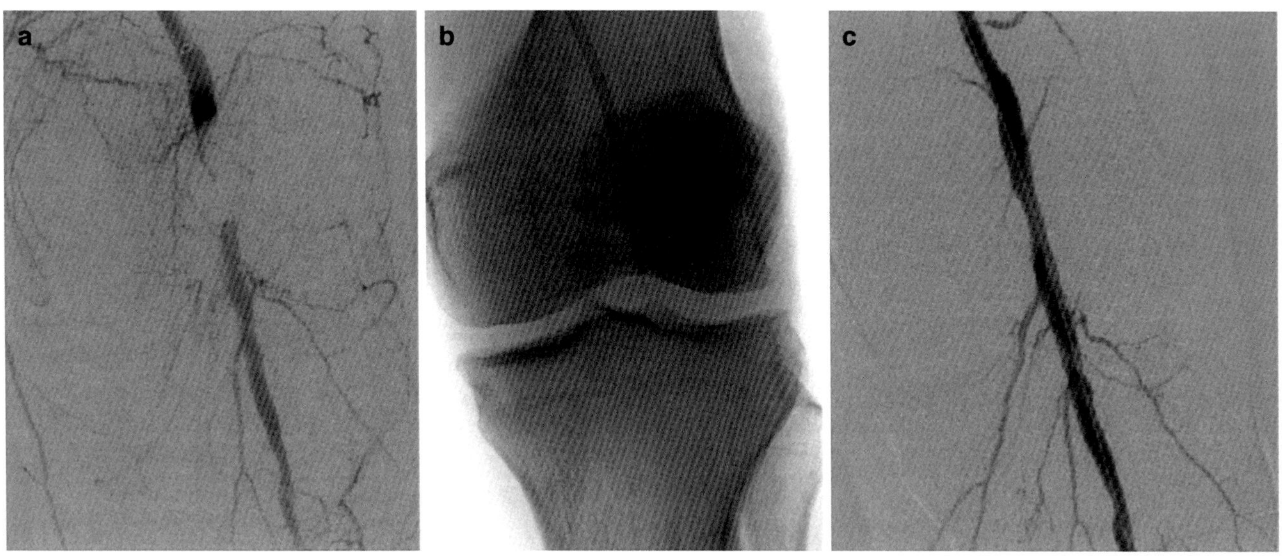

Fig. 23.2 (**a**) Angiographic view of a popliteal artery occlusion due to cardiac embolism. (**b**) Thrombo-aspiration with Indigo Penumbra CAT 8 and SEP8. (**c**) Final result

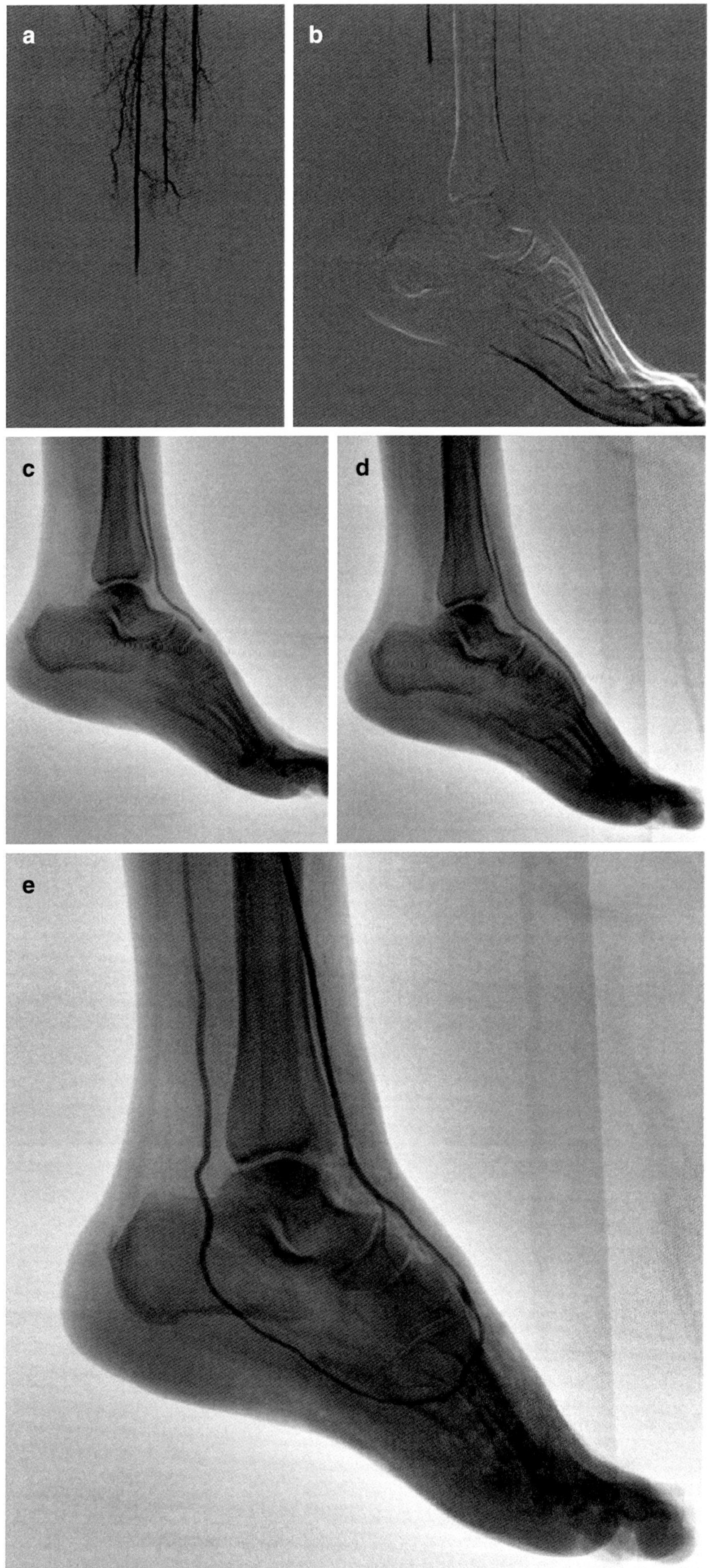

Fig. 23.3 Intraoperative fluoroscopic images of the revascularization procedure of the ischemic lower limb. (**a**, **b**) Intraprocedural angiography shows complete occlusion of all the tibial vessels, with complete absence of distal foot perfusion. (**c**) Penumbra Indigo thromboaspiration catheter CAT 6 advanced in the anterior tibial artery and dorsalis pedis. (**d**) Thromboaspiration catheter CAT 6 advanced into the plantar arch. (**e**) Post-thromboaspiration angiography reveals complete recanalization of the anterior tibial artery, dorsal and plantar arch and posterior tibial artery

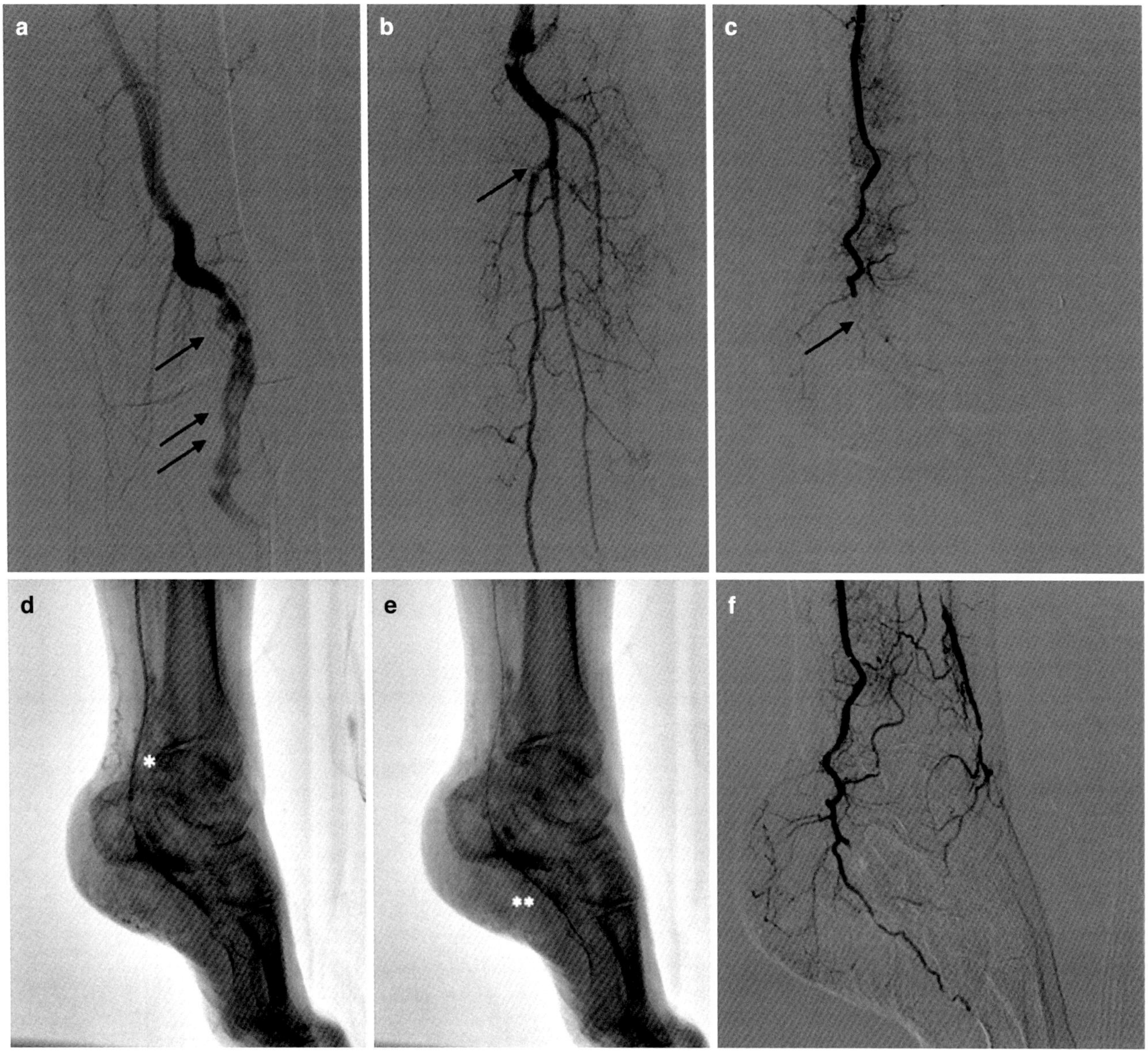

Fig. 23.4 Patient admitted with a Rutherford IIb acute limb ischemia due to thromboembolic complication of a popliteal artery aneurysm (PAA). (**a**) Pre-procedural angiographic view of a near-complete thrombosis of PAA, arrows indicate the lumen thrombosis. (**b**) Pre-procedural angiographic view of BTK vessels with thrombosis of proximal posterior tibial artery and chronic occlusion of the anterior tibial artery. (**c**) Pre-procedural angiographic view of the complete thrombosis of distal posterior tibial artery. (**d**) Intraprocedural image of Indigo/Penumbra thromboaspiration catheter CAT 6 in distal posterior tibial artery. *Indicates the tip of the CAT 6 catheter. (**e**) Intraprocedural image of Indigo/Penumbra thromboaspiration catheter CAT 3 in plantar artery. **Indicates the tip of the catheter. (**f**) Post-thromboaspiration angiographic view of posterior tibial artery and plantar arch

Decision Making Algorithm in Acute Limb Ischemia

According to the most recent European Society for Vascular Surgery (ESVS) Clinical Practice Guidelines on the Management of Acute Limb Ischemia [65]. ALI is a medical emergency, and it is essential that the diagnosis is established promptly, and appropriate treatment is started in order to prevent limb loss and other severe complication.

The decision-making algorithm starts from a correct clinical assessment, that should be performed urgently by a vascular specialist, who should be responsible for planning further investigation and management (Recommendation I C). Diagnostic imaging is recommended to guide treatment, provided it does not delay treatment, or if the need for primary amputation is obvious (Recommendation I C).

For patients with acute limb ischemia awaiting revascularization, heparin and supplemental oxygen is recommended (Recommendation I C). Moreover, prostacyclin analogues may be considered during and after revascularization (Recommendation IIb B).

Then, the guidelines remark the importance that patients diagnosed with ALI in a non-vascular center should be transferred to a vascular center that offers the full range of open and endovascular interventions (Recommendation I B). In particular, all patients should have access to treatment in a hybrid theatre, or operating theatre with C arm equipment, and by a clinical team able to offer a full range of open or endovascular interventions during a single procedure (Recommendation I C).

The guidelines focus on the importance of a completion angiography for patients undergoing open (Recommendation I C), and of considering to remove residual thrombus after open surgery with intra-operative local thrombolysis (Recommendation IIb B).

The selection of the best first-line treatment (open, lysis or endovascular technique) is based on several factors including patient condition and operator's preference. However, some recommendations are clearly stated regarding fibrinolysis appropriateness and limits based on clinical presentation:

- For patients with acute onset claudication (Rutherford grade I) that does not threaten the limb, (percutaneous) catheter directed thrombolysis is not recommended (Recommendation III B).
- For patients with Rutherford grade IIa acute limb ischemia, it is recommended that (percutaneous) catheter-directed thrombolysis is considered as an alternative to surgery (Recommendation I A).
- For patients with Rutherford grade IIb acute limb ischemia, (percutaneous) catheter-directed thrombolysis may be considered if initiated promptly and may be combined with percutaneous aspiration or thrombectomy (Recommendation IIb B).

A correct decision-making algorithm for ALI is based on combining surgery and endovascular options.

After open revascularization for acute limb ischemia, simultaneous endovascular treatment addressing inflow or outflow stenosis should be considered (Recommendation IIa C).

Notably, the ESVS guidelines clearly state that for all patients with acute limb ischemia, aspiration and mechanical thrombectomy should be considered (Recommendation IIa C).

To sum up the main points from the ESVS guideline: (1) both surgical and endovascular treatments are effective but with slightly distinctive outcomes depending on individual patients; (2) now there are a variety of clot busting drugs and new methods of aspirating blood clots percutaneously, without needing an operation; (3) the best outcome is achieved in hospitals used to dealing with patients with ALI, and familiar with all surgical and endovascular methods available, choosing the method most suitable for each individual patient.

Conclusion

Suggestive modifications in the treatment of ALLI have been proposed in recent years. While surgery represents still a significant treatment option especially for ALLI due to embolism, endovascular techniques are acquiring a more prominent role in case of acute on chronic ischemia.

Various mechanical endovascular systems for thrombus removal have been investigated over the last 15 years. Most of them have partially failed to be effectively successful or have been associated with undesirable complication rates. Newly endovascular thrombectomy devices specifically designed for peripheral intervention in this difficult set of patients, may offer improved clinical outcomes with lower rates of major systemic and local complications. As a result, a shift of treatment recommendation towards endovascular options may be observed in the near future.

Case Presentation

Continued from page 225

Previous systemic heparinization (100 IU/kg), the total endovascular rescue was performed. The pre-procedural angiographic view confirmed a near-complete thrombosis of the popliteal artery aneurysm and thrombosis of the proximal posterior tibial artery and chronic occlusion of the anterior tibial artery. Moreover, the anatomical setting was characterized also by the complete thrombosis of the distal posterior tibial artery. The revascularization procedure was performed with Indigo/Penumbra thromboaspiration catheter CAT 6 in the distal posterior tibial artery and Indigo/Penumbra thromboaspiration catheter CAT 3 advanced in the plantar artery. After the thromboaspiration session and the infusion of a bolus of the thrombolytic agent in the plantar arch posterior tibial artery and plantar arch were assured. At the end of the procedure, the popliteal artery aneurysm was excluded with the implantation of a self-expanding covered stent (Fig. 23.4).

References

1. Norgren L, Hiatt WR, Dormandy JA, Nehler MR, Harris KA, Fowkes FG. Inter-Society consensus for the management of peripheral arterial disease (TASC II). J Vasc Surg. 2007;45(Suppl S):S5–67.
2. Creager MA, Kaufman JA, Conte MS. Clinical practice. Acute limb ischemia. N Engl J Med. 2012;366:2198–206.
3. de Donato G, Gussoni G, de Donato G, Andreozzi GM, Bonizzoni E, Mazzone A, et al. The ILAILL study: iloprost as adjuvant to surgery for acute ischemia of lower limbs: a randomized, placebo-controlled, double-blind study by the Italian society for vascular and endovascular surgery. Ann Surg. 2006;244:185–93.

4. The STILE Investigators. Results of a prospective randomized trial evaluating surgery versus thrombolysis for ischaemia of the lower extremity. Ann Surg. 1994;220:251–8.
5. Aune S, Trippestad A. Operative mortality and long-term survival of patients operated on for acute lower extremity ischaemia. Eur J Vasc Endovasc Surg. 1998;15:143–6.
6. Nypaver TJ, White BR, Endean ED, Schwarcz TH, Hyde GL. Non traumatic lower-extremity acute arterial ischaemia. Am J Surg. 1998;176:147–52.
7. Pemberton M, Varty K, Nydahl S, Bell PR. The surgical management of acute limb ischaemia due to native vessel occlusion. Eur J Vasc Endovasc Surg. 1999;17:72–6.
8. Ouriel K, Veith FJ, Sasahara AA, for the Thrombolysis Or Peripheral Arterial Surgery (TOPAS) Investigators. A comparison of recombinant urokinase with vascular surgery as initial treatment for acute arterial occlusion of the legs. N Engl J Med. 1998;338:1105–11.
9. Blaisdell FW. The pathophysiology of skeletal muscle ischemia and the reperfusion syndrome: a review. Cardiovasc Surg. 2002;10:620–30.
10. Falluji N, Mukherijee D. Critical and acute limb ischemia: an overview. Vascular. 2012;20:174–7.
11. Shishehbor MH. Acute and critical limb ischemia: When time is limb. Cleve Clin J Med. 2014;81:209–16.
12. Fogarty TJ, Cranley JJ, Krause RJ, Strasser ES, Hafner CD. A method of exctraction of arterial emboli and thrombi. Surg Gynecol Obstet. 1963;116:241–4.
13. Fogarty T. Historical reflections on the management of acute limb ischemia. Semin Vasc Surg. 2009;22:3–4.
14. Henke PK. Contemporary management of acute limb ischemia: factors associated with amputation and in-hospital mortality. Semin Vasc Surg. 2009;22:34–40.
15. Plecha FR, Pories WJ. Intraoperative angiography in the immediate assessment of arterial reconstruction. Arch Surg. 1972;105:902–7.
16. White GH, White RA, Kopchok GE, Wilson SE. Angioscopic thromboembolectomy: preliminary observations with a recent technique. J Vasc Surg. 1988;7:318–25.
17. Manojlovic V, Popović V, Nicolić D, Milosević D, Pasternak J, Kaćanski M. Analysis of associated disease in patients with acute critical lower limb ischemia. Med Preg. 2013;66:41–5.
18. Aune S, Trippestad A. Operative mortality and long-term survival of patients operated on for acute lower extremity ischemia. Eur J Vasc Endovasc Surg. 1998;15:143–6.
19. Braithwaite BD, Davies B, Birch PA, Heather BP, Earnshaw JJ. Management of acute leg ischemia in the elderly. Br J Surg. 1998;85:217–20.
20. Nypaver TJ, White BR, Endean ED, Schwarcz TH, Hyde GL. Non traumatic lower-extremity acute arterial ischemia. Am J Surg. 1998;176:147–52.
21. Parsons RE, Marin ML, Veith FJ, Sanchez LA, Lyòn RT, Suggs WD, et al. Fluoroscopically assisted thromboembolectomy: an improved method for treating acute arterial occlusions. Ann Vasc Surg. 1996;10:201–10.
22. Lipsitz EC, Veith FJ. Fluoroscopically assisted thromboembolectomy: should it be routine? Semin Vasc Surg. 2001;14:100–6.
23. Lipsitz EC, Veith FJ, Wain RA. Digital fluoroscopy as a valuable adjunct to open vascular operations. Semin Vasc Surg. 2003;16:280–90.
24. Zaraca F, Stringari C, Ebner JA, Ebner H. Routine versus selective use of intraoperative angiography during thromboembolectomy for acute lower limb ischemia: analysis of outcomes. Ann Vasc Surg. 2010;24:621–7.
25. Pemberton M, Varty K, Nydhahl S, Bell PRF. The surgical management of acute limb ischemia due to native vessel occlusion. Eur J Vasc Endovasc Surg. 1999;17:72–5.
26. de Donato G, Setacci F, Sirignano P, Galzerano G, Raucci A, Palasciano G, et al. Hybrid procedures for acute limb ischemia. J Cardiovasc Surg. 2010;51:845–53.
27. de Donato G, Setacci F, Sirignano P, Galzerano G, Massaroni R, Setacci C. The combination of surgical embolectomy and endovascular techniques may improve outcomes of patients with acute lower limb ischemia. J Vasc Surg. 2014;59:729–36.
28. Dotter CT, Rösch J, Seaman AJ. Selective clot lysis with low-dose streptokinase. Radiology. 1974;111:31–7.
29. Kessel DO, Berridge DC, Robertson I. Infusion techniques for peripheral arterial thrombolysis. Cochrane Database Syst Rev. 2004;1:CD000985.
30. Robertson I, Kessel DO, Berridge DC. Fibrinolytic agents for peripheral arterial occlusion. Cochrane Database Syst Rev. 2013;12:CD001099.
31. Creager MA, Kaufman JA, Conte MS. Acute limb ischemia. N Engl J Med. 2012;366:2198–206.
32. Berridge DC, Gregson RH, Hopkinson BR, Makin GS. Randomized trial of intra-arterial recombinant tissue plasminogen activator, intravenous recombinant tissue plasminogen activator and intra-arterial streptokinase in peripheral arterial thrombolysis. Br J Surg. 1991;78:988–95.
33. The STILE Investigators. Results of a prospective randomized trial evaluating surgery versus thrombolysis for ischemia of the lower extremity. The STILE trial. Ann Surg. 1994;220:251–66.
34. Comerota AJ, Weaver FA, Hosking JD, Froehlich J, Folander H, Sussman B. Results of a prospective, randomized trial of surgery versus thrombolysis for occluded lower extremity bypass grafts. Am J Surg. 1996;172:105–12.
35. Palfreyman SJ, Booth A, Michaels JA. A systematic review of intra-arterial thrombolytic therapy for lower-limb ischaemia. Eur J Vasc Endovasc Surg. 2000;19:143–57.
36. Johner F, Clemens R, Husmann M, Thalhammer C, Seifert B, Amann-Vesti B. Long-term follow-up after endovascular revascularisation for acute limb ischemia—a retrospective single-centre cohort. Vasa. 2016;45:247–52.
37. Francis CW, Blinc A, Lee S, Cox C. Ultrasound accelerates transport of recombinant tissue plasminogen activator intoclots. Ultrasound Med Biol. 1995;21:419–24.
38. Schrijver AM, van Leersum M, Fioole B, Reijnen MM, Hoksbergen AW, Vahl AC, et al. Dutch randomized trial comparing standard catheter-directed thrombolysis and ultrasound-accelerated thrombolysis for arterial thromboembolic infrainguinal disease (DUET). J Endovasc Ther. 2015;22:87–95.
39. Zeller T, Tepe G. Treatment of acute limb ischemia with focus on endovascular techniques. Vasa. 2009;38:123–33.
40. Sniderman KW, Bodner L, Saddekni S, Srur M, Sos TA. Percutaneous embolectomy by transcatheter aspiration. Work in progress. Radiology. 1984;150:357–61.
41. Starck EE, McDermott JC, Crummy AB, Turnipseed WD, Acher CW, Burgess JH. Percutaneous aspiration thrombectomy. Radiology. 1985;156:61–6.
42. Oğuzkurt L, Okzan U, Gümüs B, Coskun I, Koca N, Gülcan O. Percutaneous aspiration thrombectomy in the treatment of lower extremity thromboembolic occlusion. Diagn Interv Radiol. 2010;16:79–83.
43. Zeller T, Frank U, Bürgelin K, Müller C, Flügel P, Horn B, et al. Early experience with a rotational thrombectomy device for treatment of acute and subacute infra-aortic arterial occlusions. J Endovasc Ther. 2003;10:322–31.
44. Heller S, Lubanda JC, VareJka P, Chochola M, Prochazca P, Rucka D, et al. Percutaneous mechanical thrombectomy using Rotarex® S device in acute limb ischemia in infrainguinal occlusions. Biomed Res Int. 2017;2017:2362769.

45. Laganà D, Carrafiello G, Lumia D, Fontana F, Mangini M, Vizzari FA, et al. Recanalisation of thrombotic arterial occlusions with rotational thrombectomy. Radiol Med. 2011;116:932–44.
46. Kronlage M, Printz I, Vogel B, Blessing E, Müller OJ, Katus HA, et al. A comparative study on endovascular treatment of (sub)acute critical limb ischemia: mechanical thrombectomy vs thrombolysis. Drug Des Dev Ther. 2017;11:1233–41.
47. Zeller T, Krankenberg H, Steinkamp H, Rastan A, Sixts S, Schmidt A, et al. One-year outcome of percutaneous rotational atherectomy with aspiration in infrainguinal peripheral arterial occlusive disease: the multicenter pathway PVD trial. J Endovasc Ther. 2009;16:653–62.
48. Shammas NW, Shammas GA, Banerjee S, Popma JJ, Mohammad A, Jerin M. JetStream rotational and aspiration atherectomy in treating in-stent restenosis of the femoropopliteal arteries: results of the JETSTREAM-ISR feasibility study. J Endovasc Ther. 2016;23:339–46.
49. Obon-Dent M, Hernandez-Vila E. Jetstream® atherectomy for subacutely or chronically occluded femoro-popliteal prosthetic bypass grafts: a report of three cases. Catheter Cardiovasc Interv. 2013;82:E529–34.
50. Sixt S, Scheinert D, Rastan A, Krankenberg H, Steinkamp H, Schmidt A, et al. One-year outcome after percutaneous rotational and aspiration atherectomy in infrainguinal arteries in patient with and without type 2 diabetes mellitus. Ann Vasc Surg. 2011;25:520–9.
51. Carrera LA, Reddy R, Pamoukian VN, Michelis FC, DeVita MV, Rosenstock J. Massive intravascular hemolysis with mechanical rheolytic thrombectomy of a hemodialysis arteriovenous fistula. Semin Dial. 2013;26:E5–7.
52. Rogers JH, Laird JR. Overview of new technologies for lower extremity revascularization. Circulation. 2007;116:2072–85.
53. Borgia F, Di Serafino L, Sannino A, Gargiulo G, Schiattarella GG, De Laurentis M, et al. AngioJet rheolytic thrombectomy for acute superficial femoral artery stent or femoropopliteal by-pass thrombosis. Monaldi Arch Chest Dis. 2010;74:76–81.
54. Barbato JE, Wholey MH. Use of AngioJet mechanical thrombectomy for acute peripheral ischemia associated with stent fracture. Catheter Cardiovasc Interv. 2007;70:795–8.
55. Kasirajan K, Gray B, Beavers FP, Clair DG, Greenberg R, Mascha E, et al. Rheolytic thrombectomy in the management of acute and subacute limb-threatening ischemia. J Vasc Interv Radiol. 2001;12:413–21.
56. Spiliopoulos S, Katsanos K, Fragkos G, Karnabatidis D, Siablis D. Treatment of infrainguinal thromboembolic complications during peripheral endovascular procedures with AngioJet rheolytic thrombectomy, intraoperative thrombolysis, and selective stenting. J Vasc Surg. 2012;56:1308–16.
57. Leung DA, Blitz LR, Nelson T, Amin A, Soukas PA, Nanjundappa A, et al. Rheolytic pharmacomechanical thrombectomy for the management of acute limb ischemia: results from the PEARL registry. J Endovasc Ther. 2015;22:546–57.
58. Tan TW, Watanabe M, Kabutey NK, Vilvendhan R, Kim D. Collapse of Viabahn stent-graft secondary to rheolytic thrombectomy: a rare complication of the AngioJet device. Ann Vasc Surg. 2011;25(557):e15–7.
59. Piercy KT, Ayerdi J, Geary RL, Hansen KJ, Edwards MS. Acute pancreatitis: a complication associated with rheolytic mechanical thrombectomy of deep venous thrombosis. J Vasc Surg. 2006;44:1110–3.
60. Deák Z, Strube H, Sadeghi-Azandaryani M, Reiser MF, Treitl M. Rotational thrombectomy of acute peripheral vascular occlusions using the ThromCat XT device: techniques, indications and initial results. Diagn Interv Radiol. 2011;17:283–9.
61. Yamada R, Adams J, Guimaraes M, Schönholz C. Advantages to Indigo mechanical thrombectomy for ALI: device and technique. J Cardiovasc Surg (Torino). 2015;56:393–400.
62. Saxon RR, Teigen C. Novel thrombectomy system for below-the-knee and visceral arterial intervention: PRISM retrospective interim results. JVIR. 2015;26:149–50.
63. de Donato G, Pasqui E, Giannace G, Setacci F, Benevento D, Palasciano G, et al. The Indigo System in Acute Lower-Limb Malperfusion (INDIAN) Registry: Protocol. JMIR Res Protoc. 2019;8(3):e9972.
64. de Donato G, Pasqui E, Sponza M, Intrieri F, Spinazzola A, Silingardi R, et al. Safety and efficacy of vacuum assisted thrombo-aspiration in patients with acute lower limb ischaemia: the INDIAN trial. Eur J Vasc Endovasc Surg. 2021;61(5):820–8.
65. Björck M, Earnshaw JJ, Acosta S, Bastos Gonçalves F, Cochennec F, Debus ES, et al. Editor's Choice—European Society for Vascular Surgery (ESVS) 2020 clinical practice guidelines on the management of acute limb ischaemia. Eur J Vasc Endovasc Surg. 2020;59:173–218.

Part III

Cerebrovascular Disease

Angioplasty and Stenting of the Arch Branches

24

Said Abisi, Mohamed Hosny Sayed, and Mohamed. Ragab Elnemr

Case Presentation

A 54-year-old male patient presented with a 6.7 cm aneurysm of the distal aortic arch with extensive residual type B dissection with big infra renal fenestration. The patient underwent aortic root, ascending aorta, and aortic valve replacement due to type-A aortic dissection in 2014, which was complicated by bowel ischemia for which he had right hemicolectomy and stoma creation with residual incisional hernia and almost complete loss of anterior abdominal wall musculature.

After an interdisciplinary consensus, the patient was deemed at high risk for another open surgical treatment. Thus, totally percutaneous aortic arch repair with three-inner-branch endografts and candy plug deployment to occlude the remaining false lumen was planned. The main challenge in this case would be to pass the mechanical prosthesis with the wire and the tip of the sheath dilator without damaging the leaflet and without causing severe regurgitation. Our goal was to pass one leaflet through the lateral side, which would allow the second leaflet to continue working, and as such only mild valve regurgitation during the deployment of the graft should be noticed. Valve functionality should then be reverted to normal after retrieval of the device. It was imperative to avoid cannulation at the center of the prosthesis between the two leaflets as this may lead to destruction of the leaflets. Transesophageal echo (TOE) was planned to confirm the location of the wire in addition to multiple fluoroscopic projections. The patient was informed about the potential risk of damaging the valve in addition to other standard risks of aortic arch branch repair and gave written consent for all procedures.

Continued at page 246

Background

Arterial disorders of the aortic arch, including its supra-aortic branches and upper extremity, are less common than those of the lower extremity. They can present with symptoms necessitating surgical or endovascular treatments.

Different pathologies affecting the arch and its branches can present as acute or chronic symptoms, e.g., aortic arch aneurysm, atherosclerosis, acute aortic syndrome (dissection, penetrating aortic ulcer, and intramural hematoma known as IMH), and less common etiologies, such as upper extremity aneurysm and vasculitis-based disorders, e.g., Takayasu's arteritis, may need revascularization.

Vascular Anatomy of the Aortic Arch

Normal Arch Development

The vascular system in humans is symmetric, with the exception of the adult aortic arch. The right common carotid and right subclavian arteries arise from the brachiocephalic (innominate) artery, whereas the left common carotid and left subclavian arteries originate as separate vessels from the aortic arch. These differences are a consequence of the changes in the connections of the aortic arches, the dorsal aortas, and the aortic sac. Diagrammatic representation of these changes would suggest that the aortic arches are all present at the same time, but in reality, the first arches are already regressing while the others are still developing [1].

The first two aortic arches appear and regress quickly and contribute very little to adult structures, whereas the fifth

S. Abisi (✉)
Guy's and St Thomas' NHS Foundation Trust, London, UK
e-mail: said.abisi@gstt.nhs.uk

M. H. Sayed (✉)
Guy's and St Thomas' NHS Foundation Trust, London, UK

Cairo University Hospitals, Cairo, Egypt
e-mail: Mohamed.Sayed@gstt.nhs.uk

M. R. Elnemr
Guy's and St Thomas' NHS Foundation Trust, London, UK

Cairo National Institute of Diabetes and Endocrinology, Cairo, Egypt

G. Geroulakos et al. (eds.), *Mastering Endovascular Techniques*, https://doi.org/10.1007/978-3-031-42735-0_24

aortic arch never develops in humans. The third aortic arches become the common carotid arteries and proximal segments of the internal carotid arteries. The distal segments of the internal carotid arteries are derived from the dorsal aorta between the first and third arches. The external carotid arteries sprout from the common carotid arteries. The dorsal aorta on each side of the embryo between the third and fourth arches disappears, thus directing blood through the third aortic arch system to the head and neck regions. The fourth aortic arches are asymmetrical with regard to their fate. The left aortic arch forms part of the adult aortic arch between the left common carotid and left subclavian arteries, whereas the right aortic arch becomes the proximal segment of the right subclavian artery. The remainder of the right subclavian artery is derived from the right dorsal aorta and its right seventh intersegmental artery. The right dorsal aorta distal to the seventh intersegmental artery, but proximal to the fused common aorta, involutes. The right horn of the aortic sac, connected to the right third and fourth aortic arches, elongates to become the brachiocephalic (innominate) artery, and the left horn becomes the initial portion of the aortic arch. The proximal portions of both sixth arches become the right and left pulmonary arteries. The distal segment of the right sixth arch disappears, but the distal segment of the left sixth arch becomes the ductus arteriosus during fetal life and atrophies after birth to become the ligamentum arteriosum. The ductus arteriosus in the fetus acts as a shunt between the left pulmonary artery and the aorta to move blood away from the immature fetal lungs to the aorta [1].

Aortic Arch Morphology

Aortic arch morphology can be classified into three different types, depending on the position of the take-off of the great vessels [1]. In type I aortic arches the great vessels arise at or above the same horizontal place as the outer curvature of the arch. In type II aortic arches the origin of the innominate artery lies between the horizontal planes of the inner and outer curves of the aortic arch. In type III aortic arches the innominate artery lies below the horizontal plane of the inner curvature of the aortic arch. As the take-off of the supra-aortic vessels moves more inferiorly (i.e., type II and type III configurations), vessel selection and the manipulation of sheath, balloon, and stent delivery system become more difficult because the origin of the arteries becomes more oblique. Type III aortic arch morphologies are particularly challenging and can lead to a higher risk of embolic events due to repeated and/or prolonged wire and catheter manipulation leading to the disruption of aortic plaques.

The ESVS guidelines recommend the use of the definition of attachment zones as provided by "Reporting standards for thoracic endovascular aortic repair," [2] which are also known as "Ishimaru zones" in the aortic arch as shown in Fig. 24.1.

Aortic Arch Anomalies

In light of the complexity of the arch embryological development that must occur for normal development of the aortic arch and its branches, anomalies do occur [3]. Anomalies result when segments of the primitive aortic arch that should disappear persist or vice versa, when other parts that should persist disappear. Variations in the development of vessels as they arise from the aortic arch are relatively common. The most common developmental pattern considered to be "normal," as described in the preceding sections, occurs in 65% of the population. In 22% of the population, the left common

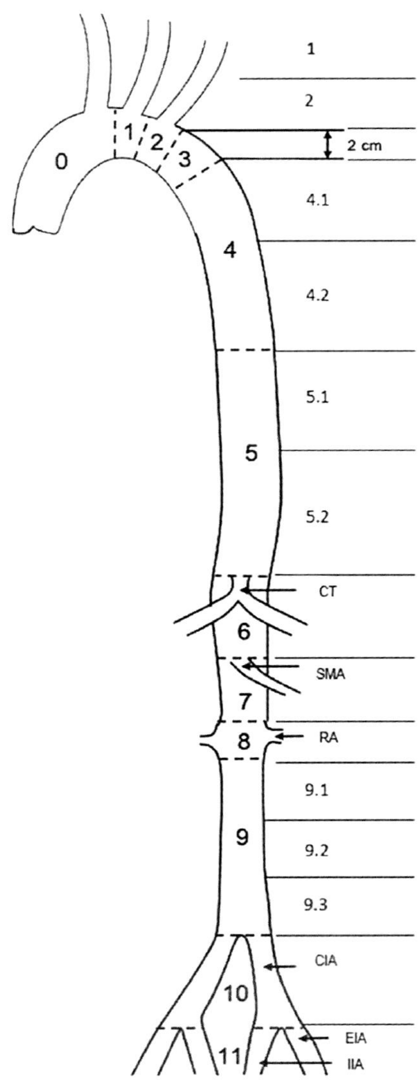

Fig. 24.1 Definition of attachment zones, also known as Ishimaru zones

carotid artery originates from the brachiocephalic trunk rather than from the aortic arch, in what is termed a "bovine arch." In this case, the brachiocephalic trunk gives rise to the right subclavian, right common carotid, and left common carotid arteries, whereas the left subclavian artery originates from the aortic arch as normally expected. This variant accounts for 73% of all arch anomalies [1].

Many other variations, each occurring in less than 3% of the population, have been described. Some of these anomalies include a short brachiocephalic trunk that bifurcates immediately into the right subclavian and right common carotid arteries, with the left common carotid arising from the aortic arch at the base of the brachiocephalic trunk and a normal origin for the left subclavian artery from the aortic arch. The left vertebral artery can originate directly from the aortic arch between the left common carotid and left subclavian arteries. A left brachiocephalic trunk may be present, which bifurcates into left subclavian and left common carotid arteries [3].

Pathology

Aortic arch pathology is very complex and may be related to a variety of histopathologic conditions including aneurysmal degeneration, acute aortic syndromes (aortic dissection, intramural hematoma, and penetrating aortic ulcers), and various forms of vasculitis (giant cell arteritis, Takayasu arteritis, and Behçet disease). Aneurysmal changes are usually secondary to either medial degeneration or aortic dissection.

Aortic Arch Aneurysm

Definition

Aneurysms are localized dilatations in the thoracic and abdominal aorta secondary to weakening and subsequent expansion of the aortic wall. A TAAA by definition is a dilatation at least 1.5 times its normal value [4]. It is critically important to define these anatomic aortic sizes (Table 24.1) to help identify pathologic aortic growth, because TAAA diameter is the strongest predictor of rupture, with a reported mean aortic diameter of ruptured TAAAs of 6.1 cm [1]. When all aneurysms of the thoracic aorta are considered, those of the ascending aorta are the most common (40%), followed by those of the descending thoracic aorta (35%), aortic arch (15%), and thoracoabdominal aorta (10%) [5].

Table 24.1 Normal aortic diameter by segment, as well as percentage of aortic aneurysm total represented by each aortic segment

Aortic segment	Mean aortic diameter (cm)	Percentage of aneurysms
Ascending	3	40
Aortic arch	2.5–3.5	15
Descending thoracic	2.0–2.3	35
Thoracoabdominal	1.7–2.6	10

The terms *saccular* and *fusiform* should also be defined for TAAAs because the morphology of an aneurysm will often dictate treatment options as well as point to specific causes (i.e., mycotic aneurysms are often saccular and at high risk for rupture) [6]. Most TAAAs are fusiform aneurysms, which involve a chronic uniform dilatation involving the whole circumference of the aorta. In contrast, saccular aneurysms often represent an eccentric dilatation of the aorta. Most authors would concur that a lower threshold is an acceptable indication for repair of saccular aneurysms.

Etiology

About 80% of TAAAs are secondary to medial degeneration, with approximately 15–20% caused by aortic dissection. Patients with TAAAs secondary to aortic dissection are typically younger and involve more extensive aortic segments than do degenerative aneurysms. The aorta in patients with Marfan syndrome is particularly prone to aortic dissection and subsequent TAAA formation [7]. Both systemic autoimmune disorders, such as Takayasu arteritis and chronic nonspecific aortitis, can destroy the aortic media with progressive aneurysm formation. Aneurysms associated with arteritis are more commonly seen in women than are degenerative aneurysms. Aneurysms of the upper thoracic aorta can also be secondary to congenital aortic coarctations, either in unrepaired coarctations or after repair [8].

Clinical Picture

The most common initial symptom in patients with TAAAs is vague pain, which can occur in the chest, back, flank, or abdomen. The differential diagnosis in a patient with a symptomatic TAAA therefore includes angina, aortic dissection, and degenerative disease of the spine. The pain associated with TAAAs can also be chronic, which may easily be dismissed in patients with TAAAs before the diagnosis of a TAAA has been made. Typical of most aneurysms that are enlarging, pain may increase in severity dramatically [9].

Symptoms may also occur in patients with TAAAs from compression of the thoracic aorta by structures in the thoracic cavity. Hoarseness can develop in patients with TAAAs as a result of stretching or compression of the left recurrent laryngeal nerve. Tracheal deviation, persistent cough, or other respiratory symptoms may also be present [10]. Dysphagia is an uncommon, nonspecific complaint due to compression of the aneurysm by the esophagus [11]. Sudden and catastrophic hemoptysis or hematemesis may occur as a result of erosion of the TAAA into the bronchial and pulmonary space or the esophagus, respectively.

Patients with TAAAs may rarely have neurologic deficits, including paraplegia. This is much more common in those with aortic dissection. Most patients with symptoms have TAAAs that have attained a diameter greater than 5 cm.

Imaging

Accurate and precise determination of the anatomy of a patient with a TAAA is mandatory in determining both the need for and the details of operative repair, as well as the prescribed follow-up in patients with TAAAs who have not reached a diameter threshold for repair. Cross-sectional imaging, mainly by computed tomography (CT), is the primary method to visualize the thoracic and abdominal aorta for determining aneurysm diameter and extent. Because most TAAAs are small when discovered, serial imaging is required; however, level A and B evidence regarding the timing of surveillance is lacking. Serial imaging is typically performed at 6–12-month intervals, but it varies depending on the initial diameter and extent of the aneurysm.

Acute Aortic Syndrome

Non-A–Non-B Aortic Dissection: Type B Dissection Involving the Aortic Arch

Both the Stanford and DeBakey classifications do not address the clinical scenario whereby the aortic arch, but not the ascending aorta, is dissected. The 2010 AHA guidelines recommend that patients with descending aortic dissection and entry within the arch be categorized as proximal type B dissection [12]. Distal type B dissection refers to descending aortic dissection and entry distal to the LSA [12]. The evolution of the term non-A–non-B aortic dissection can be seen more as a kind of evolution of the understanding of the pathophysiological process, which was initially described in 1994 [13, 14].

Clinical presentation, treatment, and outcome in non-A–non-B dissection patients are different from those commonly reported for patients with acute type B dissection. The involvement of the arch in the dissection process of the DTA seems to have an important impact on clinical course and outcome; therefore, it is reasonable not to categorize these patients as type B, but as non-A–non-B aortic dissection [2].

Aortic Intramural Hematoma

The ESC guidelines define aortic IMH as a circular or crescent-shaped thickening >5 mm of the aortic wall with the absence of a dissecting membrane, intimal disruption, or false lumen flow [15]. The ESVS guidelines define intramural hematoma as the presence of blood within the aortic wall without intimal disruption or an identifiable entry point on imaging [16]. Whereas current guidelines see IMH as a separate entity, distinguishing between IMH and dissection may not always be possible in clinical practice. There is certainly a time-dependent variable with regard to diagnosis because patients frequently present with new intimal lesions 24–48 h after the initial imaging studies were performed. The current definition of IMH may be challenged as more sophisticated imaging methods will be able to identify more primary entry tears and therefore identify more IMH as a precursor to acute aortic dissection.

Some of the predictive factors for disease progression that have been proposed for patients with IMH without associated ulcer or intimal erosion include involvement of the ascending aorta, aortic diameter >50 mm in initial imaging, and persistent pain. Predictors of disease progression in patients with IMH and an associated aortic ulcer or intimal erosion include an increase in associated pleural effusion, recurrent pain, an ulcer located in the ascending aorta or arch with an initial maximum ulcer diameter >20 mm or more, and an initial maximum ulcer depth >10 mm [17, 18].

Penetrating Aortic Ulcer

The current ESC guidelines on aortic disease define PAU as an ulceration of an aortic atherosclerotic plaque penetrating through the internal elastic lamina into the media. It is thought that PAU occurs in 2–7% of all patients with acute aortic syndromes. Patients presenting with PAU frequently have a high atherosclerotic burden. Risk factors for PAU include advanced age, male gender, tobacco smoking, hypertension, coronary artery disease, chronic obstructive pulmonary disease, and presence of abdominal aneurysm [19].

Indications for intervention according to the current guidelines include persistent or recurrent pain, contained rupture, rapid growth, periaortic hematoma, and pleural effusion. It is thought that in asymptomatic patients with PAU, a diameter >20 mm and a neck >10 mm have a higher risk of progression, and early intervention should be evaluated [2].

Atherosclerosis

Aortic arch plaque (AAP) is an established risk factor for ischemic stroke. Large plaques (defined as ≥4 mm in thickness in most studies) were proven to be strongly associated with recurrent stroke and death [20]. In a previous study, aortic arch atherosclerosis was found to be associated with silent brain infarctions (SBI) in patients with atrial fibrillation (AF), and the authors suggested that microembolization of small thrombi might be the mechanism for SBI [21].

Numerous case–control and prospective studies have confirmed the role of proximal arch plaques as risk factors for stroke and other embolic events; however, the mechanism underlying the association between AAP and white matter hyperintensity volume (WMHV) is not immediately clear. The prevalence of severe AAP in stroke patients has been reported to be between 14 and 21% [21]. With respect to subclinical cerebrovascular disease, SBIs and WMHs are considered to be related but somewhat different expressions of brain disease. Although SBIs are focal areas of infarcts, presumed to result from the occlusion of a single small perforating artery supplying the subcortical areas of the brain, WMHs are considered areas of leukoaraiosis (loss of white matter) because of chronic hypoperfusion of the white matter and disruption of the blood–brain barrier, leading to chronic leakage of plasma into the white matter [22].

Approach to Common Lesions

Proximal Occlusive Lesions

Chronic upper extremity ischemia due to aortic arch (i.e., supra-aortic trunk) lesions can be effectively treated with angioplasty and stenting and/or a surgical bypass. For supra-aortic reconstruction or extra-anatomic bypass, prosthetic graft is preferred. Surgery is more durable and should be considered the first line of treatment for good surgical-risk patients, but good outcomes have been reported using endovascular techniques [23]. For subclavian artery revascularization, both open and endovascular therapy are safe and effective, but surgical bypass is more durable [24]. Angioplasty alone has inferior outcomes compared with angioplasty and stenting, particularly when recanalizing occlusive subclavian lesions. In a review of 114 patients, assisted primary patency and freedom from recurrent symptoms were worse in patients presenting with arm ischemia compared with those presenting with cardiac or vertebrobasilar insufficiency at 5 years [25]. Endovascular stent-graft repair has been used for the repair of aneurysms and both penetrating and blunt injuries to the axillary and distal subclavian arteries, with good results.

Aortic Arch Debranching

For patients who will undergo endovascular repair of the thoracic aorta for the management of thoracic aortic aneurysm/dissection, coverage of the left subclavian artery may be necessary to obtain proximal endograft fixation. Extra-anatomic upper extremity revascularization may be performed prior to endovascular repair or only as needed postoperatively if the patient develops symptoms. Complete endovascular aortic arch repair using custom-made inner branch device is one of the newly developed techniques for the treatment of arch pathology, e.g., non A–non B dissection, aortic arch aneurysm.

Endovascular Repair

The endovascular approach has gained popularity for patients deemed too ill for open surgery and for those with iatrogenic catheter-induced arterial injuries.

Arterial access—Endovascular revascularization can be accomplished from a femoral or brachial artery approach. Arm access provides better pushability and stability, which is crucial when recanalizing occluded upper extremity vessels.

Femoral artery access is usually selected. Historically, large-diameter access sheaths were required for these procedures, which necessitated open cutdown. However, later-generation, lower-profile devices and percutaneous closure techniques have allowed this procedure to be performed percutaneously as well.

Performance of the procedure requires the delivery of a large-bore sheath into the aorta as well as a separate access for arteriography. While these were previously accomplished using a femoral artery cutdown, the comfort achieved with a percutaneous approach for large-bore sheaths has increased, allowing these procedures to be performed primarily percutaneously, anatomy permitting.

Up to one-fourth of patients will require alternative access techniques due to the obligatory large sheath size for delivery of the device [26]. Passage of the sheath through a small-diameter, tortuous, or excessively calcified external iliac artery can lead to iliac artery disruption. Adjunctive and alternative access techniques include balloon angioplasty/stenting of the iliac arteries prior to sheath placement, creation of an iliac conduit, direct exposure of the common iliac artery, direct delivery through the abdominal aorta, intravascular lithotripsy, or use of a controlled rupture technique with a specialized device (e.g., SoloPath sheath, Terumo).

Embolic protection—With an endovascular approach, distal embolization is possible during manipulation of the diseased vessel. With innominate lesions, it is advisable to place an embolic protection device (EPD) in the right internal carotid. EPDs are described elsewhere.

If EPD deployment is not possible, we prefer primary stenting with no pre-dilatation to minimize the risk of distal embolization. Other adjunctive procedures have been described in addition to placing a filter protection device [27]. One example is creating reversal of flow in the vertebral artery by injecting vasodilator in the arm circulation.

Angioplasty/stenting—Observational studies suggest that percutaneous transluminal angioplasty and stenting are safe in patients with appropriate anatomy (short proximal stenosis or occlusion). The combined stroke and death rate related to percutaneous intervention to treat upper extremity disease was 3.6% in one study [28]. Complications include stent thrombosis, restenosis, and stent fracture [28].

Good outcomes have been reported using endovascular techniques in the treatment of lesions of the aortic arch (i.e., supra-aortic) vessels [29].

Immediate technical success occurs in more than 93% of patients, with failures usually related to an inability to cross an occlusive lesion [30]. Five-year primary patency rates are approximately 85% [30]. Sustained resolution of ischemic symptoms is observed in most patients (>95%) [30].

Angioplasty alone has inferior outcomes compared with angioplasty and stenting, particularly when recanalizing occlusive subclavian lesions [28]. A systematic review and meta-analysis comparing angioplasty alone with angioplasty and stenting for subclavian artery stenosis found a significantly higher risk of subsequent events for angioplasty alone compared with angioplasty and stenting at 1 year (odds ratio 2.37, 95% CI 1.32–4.26) [31]. A retrospective study of 42 patients treated with angioplasty and stenting for coronary subclavian steal syndrome found that the rate of restenosis was higher in patients with a continuous (compared with intermittent) subclavian and coronary steal (41% and 7%) [32].

Symptoms due to significant (>70%) recurrent stenosis or obstruction occur in approximately 10% of patients and are typically treated with repeat angioplasty; however, surgery may be required in up to 5% of patients [28]. Some risk factors for subclavian in-stent stenosis include younger age, smoking with a history of chronic obstructive pulmonary disease, or baseline vessel diameter ≤7 mm [33]. A meta-analysis comparing open surgery with endovascular repair for subclavian atherosclerotic disease showed favorable early outcomes for both techniques and no significant differences in survival [23]. While open repair had a better long-term patency, there were no significant differences in symptom recurrence.

Stent-grafting—Open surgery for subclavian artery aneurysm has been the standard and provides a durable long-term repair. However, as with aneurysms at other sites, endovascular stent-grafting has been increasingly used to exclude the aneurysm from the circulation. Subclavian artery aneurysms can occur proximally, associated with atherosclerosis, or more distally due to injury (e.g., repetitive injury as with thoracic outlet syndrome). Each of these segments has anatomic features that make stent-graft placement challenging, and motion of the shoulder can lead to graft compression with the potential for endograft fracture.

Endovascular stent-graft repair has also been applied to the treatment of penetrating or blunt injuries to the axillary and distal subclavian arteries with good results. A review of 223 patients with subclavian and axillary injuries from 11 trauma centers showed an excellent limb salvage rate of 97% but a 10% in-hospital mortality related to cardiac (38%), brain (21%), hemorrhage (21%), multisystem organ failure (17%), and drug-related (4%) causes (e.g., balloon occlusion) [34].

Thoracic Endografts

Endovascular repair of thoracic aortic aneurysm is accomplished using a fabric-covered stent, termed an endograft or stent-graft. Adoption of stent-graft technology by vascular surgeons has been rapid, primarily related to pre-existing experience and facilities with the endovascular repair of abdominal aortic aneurysm.

Although there are variations from device to device, three components (delivery system, main body, and extensions) are common to all endograft device systems. Thoracic endograft devices are currently approved for the treatment of descending thoracic aneurysms, penetrating aortic ulcers, aortic intramural hematoma, descending (type B) thoracic aortic dissection, [35] residual descending thoracic aortic dissections, and traumatic aortic transection in the United States [36].

Thoracic endovascular devices include TAG and cTAG GORE, TX2 and Alpha COOK, Valiant Thoracic Stent-Graft System Medtronic, and Relay BOLTON.

The degree of structural support varies from device to device. Proponents of designs that have less metallic support structure claim the device is better able to adapt to changes in aneurysm configuration over time. On the other hand, some physicians feel that fully supported endografts are less prone to kinking and subsequent thrombosis. The curve of the proximal thoracic aorta adds an additional challenge to achieving a design with adequate proximal fixation and seal. The amount of radial support, which allows the endograft to withstand external compression, must be weighed against the need for enough flexibility and conformability within the device to navigate the proximal aorta and achieve a proper seal following deployment.

New endograft designs are continually being tested to enhance performance. Improvements have focused upon smaller device and delivery profiles, more accurate deployment, improved fixation systems, and perhaps most importantly, flexibility in managing challenging anatomy. These improvements, along with increased operator experience, have led to improvements in the short-term and long-term results of endovascular aneurysm repair and have expanded the application of endovascular repair to many patients whose aortic anatomy was previously deemed unsuitable.

Advanced Devices

When aortic disease is more extensive and involves branch vessels, the complexity and risks associated with endovascular repair increase. Approaches to address these issues include debranching procedures and the use of fenestrated and branched endografts.

Branched grafts have been used in patients with thoracic or thoracoabdominal aneurysms or thoracic dissection [35]. Initial experience with branched aortic grafts has shown similar perioperative mortality as conventional endovascular repair, but the rate of aneurysm repair-related morbidity may be higher.

Aortic arch branch devices—single- and double-branch thoracic aortic endografts and fenestrated endografts are available primarily for investigational use to treat zone 0 (innominate), 1 (left carotid), or 2 (left subclavian). These include the Cook a-Branch, Cook fenestrated, Gore Thoracic Branch Endoprosthesis (TBE), Medtronic Valiant Mona LSA, Bolton ascending thoracic device, and Nexus aortic arch system. These grafts are intended to preserve flow into the innominate, carotid, or subclavian artery during thoracic endograft placement. A separate access is needed to deploy the branch grafts. None of these devices are approved for use by the US Food and Drug Administration (FDA). Device clinical trials evaluating use in treating aortic arch pathology are anticipated.

Choice of Device

No one device is appropriate for all clinical indications or has been shown to be superior to any other device for the endovascular repair of the thoracic aorta. Most devices were originally designed for excluding thoracic aortic aneurysm, and the evaluation of specific designs to manage acute aortic syndromes (aortic rupture, aortic dissection) is in early stages. With time, and with the addition of disease-specific endografts, a clear preference for one device over another may emerge.

Management of Iatrogenic Injury to Supra-aortic Branches

Injury to a supra-aortic trunk vessel is an uncommon, but devastating, complication of percutaneous central venous catheter (CVC) placement. The incidence of carotid artery injury associated with CVC placement in the internal jugular vein (IJV) has been reported to be in the range of 1–7 cases per 1000 [37].

Inadvertent arterial placement of a large-bore catheter can result in hemorrhage, pseudoaneurysm, stroke, and/or death [37]. Given the devastating complications associated with arterial catheterization, the early identification of arterial trauma and knowledge of the complication profile are imperative to provide prompt management and mitigate risk.

The current management of catheter removal involves one of three techniques: external manual pressure, open direct arterial repair, or endovascular techniques, including covered stents and percutaneous vascular closure devices (VCDs).

Endovascular stent-graft repair of traumatic arterial injury has been associated with a decrease in the anesthetic requirement and blood loss and a decrease in associated mortality (22% with open surgical repair vs. 0.9% with carotid artery stenting) [36] (Table 24.2).

Suggested endovascular toolkit for the endovascular management of arch angioplasty

	Manufacturer	Size/length
Wires		
Any standard access wire		0.035″/180–300 cm
Glidewire Floppy	Terumo Radifocus	0.035″/180–300 cm
Glidewire Stiff	Terumo Radifocus	0.035″/180–300 cm
Rosen	Boston Scientific Corporation	0.035″/300 cm
Lunderquist "extra stiff" (double curve)	Cook Medical	0.035″/300 cm
Sheaths		
Any standard access sheath	Any	5–10 Fr/11 cm
TriForce crossing system	Cook Medical	4 Fr/55–90 cm
DrySeal	GORE	12–24 F/33–65 cm
Ansel	Cook Medical	6–8 Fr/45–55 cm
Aptus steerable sheath	Medtronic	6.5–8.5 F/55–90 cm
Catheters		
C2	Cordis	0.035″/5 Fr/65–90 cm
BER II	Cordis	0.035″/4 Fr/65–90 cm
Balloons		
CODA	Cook Medical	30–36 mm
Charger	Boston Scientific	10 mm × 2 mm
Stents		
Be-graft	Bentley	8–14 mm/27–57 mm
Closure device		
Proglides	Abbott	0.035″

These are only a few suggested options based on the author's experience and can cover the vast majority of cases

Case Presentation

Continued from page 239

The procedure was done in a hybrid theater under GA. Left common femoral vein antegrade access and pacing wire advanced in the right ventricle via a 6 Fr sheath. Left common femoral artery retrograde puncture was done under ultrasound and X-ray guidance. A 5-French sheath was inserted standby in case a cardiothoracic conversion would be needed. Right axillary artery percutaneous retrograde access was obtained under ultrasound and X-ray guidance. Two Proglide vascular closure devices inserted. A 10-French sheath was inserted, and a pig-tail catheter was placed in the ascending aorta. Right brachial artery retrograde puncture was carried out under ultrasound and X-ray guidance at the right antecubital fossa level. A 5-French sheath was inserted. Right common femoral artery retrograde puncture was performed under ultrasound and X-ray guidance. Two Proglide vascular closure devices were inserted. TOE was used to confirm the location of the wire. Guide wire was placed in the left ventricle through the aortic metallic valve by the side of the metallic leaflets. Arch branch Cook stent-graft device was inserted through the right common femoral artery. It measured 40 mm proximal diameter/28 mm distal diameter × 243 mm total length with three branches.

After the stent-graft device was deployed, one Gore Excluder iliac limb graft was inserted in the innominate artery branch and innominate artery, through the right axillary artery access after a 12-French DrySeal sheath was inserted. It measured 16 mm proximal diameter/20 mm distal diameter × 9.5 cm total length. Postdilated into the branch with a 12 mm × 20 mm plain balloon. The left common carotid branch was cannulated next with the help of an 8-French long sheath, from below, after a through-and-through wire was snared out from the right common femoral artery 20 Fr DrySeal sheath. Viabahn VBX balloon-expandable covered stent was inserted in the left common carotid artery branch to bridge it with the left common carotid artery, measuring 9 mm × 79 mm. The left subclavian artery branch was cannulated last, and Viabahn VBX balloon-expandable covered stent was inserted in it, measuring 11 mm × 59 mm. Because there is an acute angulation from the LSA branch to the left subclavian artery (LSA), Be-Graft plus (+) balloon-expandable covered stent was inserted and relined into the VBX stent, measuring 10 mm × 57 mm. Postdilated with a balloon-expandable covered stent 12 mm × 20 mm in the mid and distal parts of the left subclavian artery stents.

Thoracic stent-graft device extension done next with a Cook Alpha thoracic stent-graft, measuring 32 mm × 155 mm total length. The false lumen of the aortic dissection was cannulated last at the level of the distal abdominal aorta, close to the L4 vertebral body level. Cook occlusion "Candy plug" was inserted in the false lumen, just next to the bottom end of the Alpha distal thoracic stent-graft, measuring 36 mm × 54 mm. Below the bottom end of the "candy plug" radiopaque stent there was graft fabric of a length 43 mm to provide the occluding effect. Distal/bottom end diameter of the "candy plug" fabric was 14 mm. Final angiogram was done showing flow through the innominate artery, left common carotid, and left subclavian artery with a patent thoracic aortic stent-graft. The 12 Fr DrySeal sheath was removed from the right axillary artery, and the two Proglide vascular closure devices were deployed. Hemostasis was achieved, and one Be-Graft balloon-expandable covered stent was placed in the right axillary artery at the level of the previous puncture. It was postdilated with a 12 mm × 20 mm plain balloon. The Be-Graft stent was inserted through the right brachial artery 7-French sheath, which was upsized from the previous 5 Fr sheath. Right brachial artery 7 Fr sheath was removed, and a StarClose vascular closure device was deployed. Hemostasis achieved. Present clinical pulses in the right brachial artery post-StarClose deployment. The right common femoral artery 20-French DrySeal sheath was removed, and the two Proglide vascular closure devices were deployed. Four more Proglide vascular closure devices were inserted in the right common femoral artery, and hemostasis was achieved. Left common femoral artery 5-French sheath was removed, and Mynx Control vascular closure device was deployed. Hemostasis was achieved. Left common femoral vein 6-French sheath was removed, and manual pressure applied.

The final angiogram and 2 days of postoperative computed tomography (CT) angiography showed no endoleaks and patent coronary and supra-aortic vessels. Echocardiography showed normal function of the aortic valve after the procedure. No complications occurred in the postoperative period, and the patient was discharged from ITU on the third postoperative day. After 6 months the patient returned for follow-up, and the CT scan showed exclusion of the aneurysm with regression of the sac and thrombosis of the false lumen. The aortic valve function was not compromised.

References

1. Sidawy AP, Perler BA. Rutherford's vascular surgery and endovascular therapy, vol. 2. Amsterdam: Elsevier Health Sciences; 2022.
2. Czerny M, Schmidli J, Adler S, Van Den Berg JC, Bertoglio L, Carrel T, et al. Current options and recommendations for the treatment of thoracic aortic pathologies involving the aortic arch: an expert consensus document of the European Association for Cardio-Thoracic surgery (EACTS) and the European Society for Vascular Surgery (ESV). Eur J Cardiothorac Surg. 2019;55(1):133–62.
3. Davies M, Guest PJ. Developmental abnormalities of the great vessels of the thorax and their embryological basis. Br J Radiol. 2003;76(907):491–502.
4. Santilli JD, Santilli SM. Diagnosis and treatment of abdominal aortic aneurysms. Am Fam Physician. 1997;56(4):1081–90.

5. Vasan RS, Larson MG, Benjamin EJ, Levy D. Echocardiographic reference values for aortic root size: the Framingham Heart Study. J Am Soc Echocardiogr. 1995;8(6):793–800.
6. Roman MJ, Rosen SE, Kramer-Fox R, Devereux RB. Prognostic significance of the pattern of aortic root dilation in the Marfan syndrome. J Am Coll Cardiol. 1993;22(5):1470–6.
7. LeMaire SA, Carter SA, Volguina IV, Laux AT, Milewicz DM, Borsato GW, et al. Spectrum of aortic operations in 300 patients with confirmed or suspected Marfan syndrome. Ann Thorac Surg. 2006;81(6):2063–78.
8. Kang N, Clarke AJB, Nicholson IA, Chard RB. Circulatory arrest for repair of postcoarctation site aneurysm. Ann Thorac Surg. 2004;77(6):2029–33.
9. Juvonen T, Ergin MA, Galla JD, Lansman SL, Nguyen KH, McCullough JN, et al. Prospective study of the natural history of thoracic aortic aneurysms. Ann Thorac Surg. 1997;63(6):1533–45.
10. Cooke JC, Cambria RP. Simultaneous tracheobronchial and esophageal obstruction caused by a descending thoracic aneurysm. J Vasc Surg. 1993;18(1):90–4.
11. Furukawa H, Tsuchiya K, Osawa H, Saito H, Iida Y. Saccular descending thoracic aortic aneurysm with dysphagia. Jpn J Thorac Cardiovasc Surg. 1999;47(6):277–80.
12. American College of Cardiology Foundation/American Heart Association Task Force on Practice Guidelines; American Association for Thoracic Surgery; American College of Radiology; American Stroke Association; Society of Cardiovascular Anesthesiologists, et al. 2010 ACCF/AHA/AATS/ACR/ASA/SCA/SCAI/SIR/STS/SVM guidelines for the diagnosis and management of patients with thoracic aortic disease. J Am Coll Cardiol. 2010;55(14):e27–129.
13. von Segesser LK, Killer I, Ziswiler M, Linka A, Ritter M, Jenni R, et al. Dissection of the descending thoracic aorta extending into the ascending aorta: a therapeutic challenge. J Thorac Cardiovasc Surg. 1994;108(4):755–61.
14. Rylski B, Pérez M, Beyersdorf F, Reser D, Kari FA, Siepe M, et al. Acute non-A–non-B aortic dissection: incidence, treatment and outcome. Eur J Cardiothorac Surg. 2017;52(6):1111–7.
15. Members AF, Erbel R, Aboyans V, Boileau C, Bossone E, Di Bartolomeo R, et al. 2014 ESC guidelines on the diagnosis and treatment of aortic diseases: document covering acute and chronic aortic diseases of the thoracic and abdominal aorta of the adult. The Task Force for the Diagnosis and Treatment of Aortic Diseases of the European Society of Cardiology (ESC). Eur Heart J. 2014;35(41):2873–926.
16. Ayuso JR, de Caralt TM, Pages M, Riambau V, Ayuso C, Sanchez M, et al. MRA is useful as a follow-up technique after endovascular repair of aortic aneurysms with nitinol endoprostheses. J Magn Reson Imaging. 2004;20(5):803–10.
17. Ganaha F, Miller DC, Sugimoto K, Do YS, Minamiguchi H, Saito H, et al. Prognosis of aortic intramural hematoma with and without penetrating atherosclerotic ulcer: a clinical and radiological analysis. Circulation. 2002;106(3):342–8.
18. Harris KM, Braverman AC, Eagle KA, Woznicki EM, Pyeritz RE, Myrmel T, et al. Acute aortic intramural hematoma: an analysis from the international registry of acute aortic dissection. Circulation. 2012;126(11_suppl_1):S91–6.
19. Cho KR, Stanson AW, Potter DD, Cherry KJ, Schaff HV, Sundt TM III. Penetrating atherosclerotic ulcer of the descending thoracic aorta and arch. J Thorac Cardiovasc Surg. 2004;127(5):1393–401.
20. Marco R, Tullio D, Russo C, Jin Zh RL, Mohr JP. Aortic arch plaques and risk of recurrent stroke and death. Circulation. 2009;119:2376–82.
21. Tugcu A, Jin Z, Homma S, Elkind MSV, Rundek T, Yoshita M, et al. Atherosclerotic plaques in the aortic arch and subclinical cerebrovascular disease. Stroke. 2016;47(11):2813–9.
22. Pantoni L. Cerebral small vessel disease: from pathogenesis and clinical characteristics to therapeutic challenges. Lancet Neurol. 2010;9(7):689–701.
23. Galyfos GC, Kakisis I, Maltezos C, Geroulakos G. Open versus endovascular treatment of subclavian artery atherosclerotic disease. J Vasc Surg. 2019;69(1):269–79.
24. AbuRahma AF, Bates MC, Stone PA, Dyer B, Armistead L, Dean LS, et al. Angioplasty and stenting versus carotid-subclavian bypass for the treatment of isolated subclavian artery disease. J Endovasc Ther. 2007;14(5):698–704.
25. Palchik E, Bakken AM, Wolford HY, Saad WE, Davies MG. Subclavian artery revascularization: an outcome analysis based on mode of therapy and presenting symptoms. Ann Vasc Surg. 2008;22(1):70–8.
26. Vandy FC, Girotti M, Williams DM, Eliason JL, Dasika NL, Deeb GM, et al. Iliofemoral complications associated with thoracic endovascular aortic repair: frequency, risk factors, and early and late outcomes. J Thorac Cardiovasc Surg. 2014;147(3):960–5.
27. Cassar A, Barsness GW, Wysokinski WE, Gifford SM, Bower TC, Edwards WD, et al. Pneumatic compression for embolic protection during upper extremity endovascular intervention. Vasc Endovasc Surg. 2014;48(1):70–3.
28. De Vries J-PPM, Jager LC, Van den Berg JC, Overtoom TTC, Ackerstaff RGA, Van de Pavoordt EDWM, et al. Durability of percutaneous transluminal angioplasty for obstructive lesions of proximal subclavian artery: long-term results. J Vasc Surg. 2005;41(1):19–23.
29. Paukovits TM, Lukács L, Bérczi V, Hirschberg K, Nemes B, Hüttl K. Percutaneous endovascular treatment of innominate artery lesions: a single-centre experience on 77 lesions. Eur J Vasc Endovasc Surg. 2010;40(1):35–43.
30. Wang K, Wang Z, Yang B, Yuan C, Zhang W, Yuan B, et al. Long-term results of endovascular therapy for proximal subclavian arterial obstructive lesions. Chin Med J. 2010;123(01):45–50.
31. Chatterjee S, Nerella N, Chakravarty S, Shani J. Angioplasty alone versus angioplasty and stenting for subclavian artery stenosis—a systematic review and meta-analysis. Am J Ther. 2013;20(5):520–3.
32. Filippo F, Francesco M, Francesco R, Corrado A, Chiara M, Valentina C, et al. Percutaneous angioplasty and stenting of left subclavian artery lesions for the treatment of patients with concomitant vertebral and coronary subclavian steal syndrome. Cardiovasc Intervent Radiol. 2006;29(3):348–53.
33. Mousa AY, AbuRahma AF, Bozzay J, Broce M, Barsoum E, Bates M. Anatomic and clinical predictors of reintervention after subclavian artery stenting. J Vasc Surg. 2015;62(1):106–14.
34. Waller CJ, Cogbill TH, Kallies KJ, Ramirez LD, Cardenas JM, Todd SR, et al. Contemporary management of subclavian and axillary artery injuries—a Western Trauma Association multicenter review. J Trauma Acute Care Surg. 2017;83(6):1023–31.
35. Greenberg RK, Qureshi M. Fenestrated and branched devices in the pipeline. J Vasc Surg. 2010;52(4):15S–21S.
36. DuBose J, Recinos G, Teixeira PGR, Inaba K, Demetriades D. Endovascular stenting for the treatment of traumatic internal carotid injuries: expanding experience. J Trauma Acute Care Surg. 2008;65(6):1561–6.
37. Dagher AM, Twerdahl EH, White JM. Endovascular management of an iatrogenic injury to the supra-aortic trunk after attempted central venous catheter placement. J Vasc Surg Cases Innov Tech. 2022;8(3):390–5.

Carotid Artery Stenting: Studies, Indications, and Pre-interventional Workup

25

Elizabeth A. Andraska and Rabih A. Chaer

Background

Stroke continues to be a leading cause of morbidity and mortality worldwide. It is estimated that nearly 800,000 people in the United States are diagnosed with stroke annually, and an additional 240,000 patients experience transient ischemic attacks (TIA) [1]. The financial burden is also significant, as stroke-related costs in the United States were estimated to be about $53 billion between 2017 and 2018 [2]. Up to 87% of strokes are ischemic in nature, and about 15% are due to extracranial carotid disease [3, 4]. Ischemic stroke due to carotid disease and the subsequent disability that ensues represent a significant healthcare burden and cost that can be prevented with appropriate treatments.

Although the surgical management of carotid artery stenosis with carotid endarterectomy is the established gold standard for carotid revascularization, transfemoral carotid artery stenting (TF-CAS) has become an alternative to treating the patient with carotid artery stenosis. Today, approximately 135,000 interventions are performed for carotid artery stenosis annually in the United States, and up to 60% are reported to be catheter based [5, 6]. Given the availability of self-expanding stent technologies, lower-profile delivery systems, and embolic protection devices, stenting became broadly used in carotid artery revascularization. Initial reports of carotid stenting focused on patients with a hostile neck, such as patients with restenosis after carotid endarterectomy, given the risk of cranial nerve injury with surgical repair of those lesions [7]. Indications for the use of carotid artery stenting later expanded to use in the medically and anatomically high-risk patient, given the lower risk of complications due to the noninvasive nature of the procedure [8]. With the wider diffusion of TF-CAS, a subset of patients at high risk for neurologic complications after TF-CAS were identified, such as patients with symptomatic stenosis and octogenarians [8]. Currently, the indications, use, and efficacy of carotid artery stenting continue to evolve, especially with the introduction of transcarotid artery revascularization (TCAR) and ongoing trials testing new technologies for carotid artery stenting. This chapter will outline the most contemporary and integral studies that inform the modern carotid proceduralist, as well as the indications for the intervention. Finally, it will highlight key preprocedural considerations with regard to imaging, approach, and technique.

Key Trials and Indications

Surgical Risk

The indications for carotid artery stenting (CAS) have evolved over the years. They are highly dependent on the severity, symptom status, anatomy, and surgical risk of the patient. High-risk patients who were initially excluded from carotid endarterectomy (CEA) studies such as the ACAS and NASCET trials represented a group of patients who could potentially benefit from revascularization with endovascular techniques. These high-surgical-risk patients included those with recurrent stenosis, contralateral occlusion, prior radiation therapy or radical neck dissection, surgically inaccessible lesions, and significant medical comorbidities: chronic obstructive lung disease, congestive heart failure, left ventricular ejection fraction <30%, unstable angina, multivessel coronary artery disease, and dialysis-dependent renal insufficiency [9–12]. Age >80 years has been a persistent risk factor in TF-CAS. In fact, in the CREST trial, risk of postoperative stroke was 12% in octogenarians compared to only 3% in non-octogenarians [7, 13, 14]. The higher risk of

E. A. Andraska · R. A. Chaer (✉)
Division of Vascular Surgery, Department of Surgery, University of Pittsburgh Medical Center, Pittsburgh, PA, USA
e-mail: andraskaea@upmc.edu; chaerra@upmc.edu

G. Geroulakos et al. (eds.), *Mastering Endovascular Techniques*, https://doi.org/10.1007/978-3-031-42735-0_25

adverse events in octogenarians has been repeated in numerous other studies [9–12]. This ultimately led to a consensus among providers that younger patients with anatomic criteria that places them at high risk would be of most benefit for carotid stenting. This includes patients with radiation-induced carotid stenosis, treatment of restenosis, and lesions higher than C2 [8, 15, 16].

Asymptomatic Carotid Stenosis

Perhaps the most important contribution to the clinician's decision to move forward with TF-CAS over CEA is the symptom status of the patient. With regard to asymptomatic disease, the ACT-1 trial reported noninferiority of TF-CAS to CEA at 1 and 5 years with regard to the composite endpoint of periprocedural stroke, death, myocardial infarction, or ipsilateral stroke at 1 year [17]. Based on the randomized trials to date, Cochrane reported similar efficacy between TF-CAS and CEA for asymptomatic disease, with the caveat that compelling data is lacking, and further clinical trials are required [18]. The more recent ACST-2 trial, published in 2021, studied TF-CAS vs. CEA in a modern era of medical management, self-expanding stents, and embolic protection devices. This large, randomized study found similar post-procedural stroke rates at 5 years for asymptomatic patients. They reported a stroke rate of 5.3% following TF-CAS and 4.5% following CEA, and determined that the two approaches were comparable [19]. The ARCHeR and CAPTURE trials similarly found TF-CAS to be a safe method of carotid revascularization in high-surgical-risk patients specifically, with the lowest risk in young, asymptomatic patients [11, 15]. It is important to note that in both of these trials, the use of embolic protection was employed. The SAPPHIRE trial also evaluated patients at high risk for surgical treatment and found TF-CAS to be noninferior to CEA with regard to ipsilateral stroke [20, 21]. When the trial re-evaluated ipsilateral stroke rates at 10 years, the lack of differences persisted [14]. At this point in time, TF-CAS has been shown to be safe and effective in asymptomatic patients younger than 70 years, especially in patients with high surgical or anatomic risk.

Symptomatic Carotid Stenosis

The results of randomized trials with regard to symptomatic disease have shown worse outcomes with TF-CAS. The two initial European randomized control trials comparing TF-CAS to CEA in symptomatic patients were the EVA-3S trial and the SPACE trial [22, 23]. The EVA-3S trial was stopped prematurely due to both safety and futility, as they found the post-operative stroke rate to be 3.9% after CEA compared to 9.6% after TF-CAS [22]. On long-term follow-up of the 527 patients they did enroll prior to study termination, there were no significant differences in long-term ipsilateral stroke risk between the two groups, but the authors did conclude that due to the much higher risk of periprocedural stroke in the TF-CAS population, CEA is still favored in the symptomatic patient [24]. The EVA-3S study also found a 2.5-fold increase in restenosis after TF-CAS compared to CEA; however, the implications of this with regard to reintervention or stroke rate have yet to be clarified in clinical trials [25]. The SPACE trial was another European randomized control trial comparing TF-CAS to CEA in symptomatic carotid stenosis, and while the outcomes were not nearly as pronounced, they were unable to prove noninferiority to CEA due to difficulty with study enrollment. Again, they found the rate of restenosis to be significantly higher in the TF-CAS arm [26]. SPACE-2 was initially a promising three-armed clinical trial comparing best medical management to TF-CAS and CEA, but was terminated due to low recruitment [27]. More recently, the ICST trial evaluated 1,713 neurologically symptomatic patients and demonstrated an increased periprocedural stroke risk of 7.7% with TF-CAS compared to 4.1% with CEA [28].

Future Trial Directions

In the CREST trial, which compared TF-CAS to CEA in symptomatic and asymptomatic patients, there were no reported differences in a primary composite outcome of stroke, myocardial infarction, or death from any cause during the periprocedural period or any ipsilateral stroke within 4 years after randomization [13]. This did not differ with respect to symptomatic status. We are currently awaiting the CREST-2 and ECST-2 trials to better understand these comparisons in the era of more ubiquitous statin and aspirin usage [29]. The most recent Cochrane review compiled data from each of the randomized trials and reported that in symptomatic disease, TF-CAS results in higher rates of post-procedural stroke compared to CEA [18, 30]. This higher rate of stroke is mostly driven by the risk of small embolic stroke from procedural manipulation in patients older than 70 years [18]. Overall, the symptomatic patient is still much better suited for CEA over TF-CAS. There are ongoing studies evaluating the role of TF-CAS in these patients, which may help to clarify the patients in this cohort that would find the most benefit from TF-CAS.

Transcarotid Artery Revascularization (TCAR)

To this point, this chapter has discussed CAS, primarily regarding a transfemoral approach. To address the higher periprocedural risk of embolism with manipulation of the aortic arch during TF-CAS, a transcarotid approach is being increasingly employed. This allows for the avoidance of a diseased aortic arch as well as tortuous supra-aortic branches, which reduces the risk of procedural embolic stroke. Additionally, the concomitant use of flow reversal prior to lesion crossing protects from embolization during manipulation of the lesion. Transcervical carotid artery revascularization (TCAR) with flow reversal is a safe and effective alternative and has been shown to be safe for patients older than 70 years—a key group that consequently often carries both high surgical risk and high embolic risk with transfemoral methods [31]. In a study evaluating diffusion-weighted magnetic resonance imaging (DW-MRI) images of patients undergoing TCAR with flow reversal versus TF-CAS, Leal et al. reported a significantly reduced rate of new post-procedural infarcts in TCAR patients (12.9%) compared to TF-CAS patients (33.3%) [32].

The ROADSTER study evaluated the safety and efficacy of the Silk Road Medical ENROUTE transcarotid system and found a post-procedural stroke rate of 1.4% lower than the stroke rates reported for TF-CAS [33, 34]. The subsequent ROADSTER 2 trial evaluated, per protocol, symptomatic patients and reported a combined stroke and death rate of 0.6% [35]. There have consequently been a number of cohort studies that have compared TCAR to CAS and shown improved rates of periprocedural stroke in real-world patient populations [34, 36], as well as TCAR compared to CEA showing similar periprocedural outcomes [37, 38]. There has yet to be a randomized study comparing TCAR to TF-CAS and CEA, which will be necessary to the field in order to make true comparisons.

Current Guidelines

In January 2022, the Society of Vascular Surgery produced specific guidelines with regard to indications and management of the patient with carotid artery disease [5]. For the symptomatic patient with >50% stenosis, CAS is not recommended. Similar to the Society of Vascular Surgery, the American Heart association and American Stroke Association Council on Stroke report that they continue to recommend CEA in symptomatic patients [1, 39]. They additionally report that in symptomatic patients with severe stenosis (>70%) and high risk for surgery, TF-CAS is not inferior to CEA and can be considered. High surgical risk includes a neck which is difficult to access surgically, medical conditions that are present and greatly increase the risk of surgery, or other miscellaneous circumstances such as radiation-induced stenosis or restenosis after CEA [1, 39]. Multiple randomized and single-center studies have reported prohibitive risk of stroke with TF-CAS in elderly populations. As such, TF-CAS is not appropriate in the elderly patient >70 years old. There are not yet clear guideline recommendations for the use of TF-CAS in the asymptomatic patient. Future clinical trials will help to inform guidelines with regard to the role of optimal medical management in the treatment of carotid artery stenosis.

Initially, TCAR was recommended for patients with carotid artery stenosis who carry high surgical risk. TCAR was therefore reserved for patients who had anatomic or physiologic criteria that put them at prohibitive surgical risk. Most recently, the Food and Drug Administration (FDA) expanded TCAR indications to include patients at standard risk for surgery. TCAR represents a novel approach to the patient with carotid artery stenosis with promising early results. Future randomized trials will help to compare TCAR head-to-head with TF-CAS and CEA. Given the published outcomes of TCAR showing a low rate of procedural complications, it has mostly replaced the use of TF-CAS in clinical practice.

Pre-interventional Workup

Pre-interventional Considerations

With regard to pre-operative evaluation of the patient with carotid artery stenosis, imaging and risk assessment are integral to the decision tree. Duplex ultrasound imaging is usually the initial study of choice for the provider to quantify the degree of arterial stenosis. Stenosis criteria will vary across vascular labs, but in general, a severe stenosis (>80%) is an indication for intervention. Additional noninvasive imaging of the carotid anatomy with either contrasted tomography angiography (CTA) or magnetic resonance angiography (MRA) of the neck with attention to the aortic arch is usually necessary in order to evaluate key anatomic criteria for both CAS and TCAR.

Key anatomic criteria to consider when evaluating a patient for CAS or TCAR include arch anatomy and the extent of the carotid lesion. The size of the internal and common carotid arteries are also important to note as the vessel must be between 4 and 9 mm in order to successfully deploy the stent. The interventionalist must also note the tortuosity of the supra-aortic branches as well as the angulation and disease present on the aortic arch. For TCAR, the carotid bifurcation must be at least 5 cm from the clavicle to allow for appropriate support of the device platform, with a relatively healthy proximal common carotid artery. A common carotid artery of at least 6 mm is also required to allow for safe sheath insertion. Lesion characteristics that are not suitable for CAS and TCAR include lesions with intraluminal thrombus, lesions that may require two stents, a tortuous common or internal carotid artery, heavily calcified eccentric lesions, and lesions with total occlusion of the target vessel.

Additional anatomic criteria have been identified to be limitations to TF-CAS. Criteria such as an unfavorable arch, calcification, vessel tortuosity, long stenotic lesions, and involvement of the carotid ostium have been associated with an increased risk of periprocedural stroke [40–44].

There has also been recent emphasis on plaque morphology, suggesting that a soft, echolucent plaque is at higher risk for neurologic events due to embolization with manipulation [43]. Consistently, an increased plaque burden, necrotic core, and thrombus in the artery are predictors of cerebral embolism during CAS [45, 46]. To some degree, these features can be evaluated on ultrasound, CTA, or MRA, but there will certainly need to be more studies evaluating these characteristics and perhaps more novel imaging techniques to truly understand the influence plaque morphology has on periprocedural cerebral embolization with TF-CAS and TCAR.

Plaque morphology may also play an important role in the identification and treatment of restenosis. Multiple studies have evaluated the incidence of restenosis in TF-CAS. In fact, restenosis after CAS is nearly three times more likely than after CEA [47]. However, 97% of all recurrent ipsilateral strokes after TF-CAS occurred in patients without evidence of significant restenosis or occlusion [48]. Therefore, the implications of the increased rate of restenosis are unknown regarding reintervention or recurrent stroke risk. Often, these lesions are smoother and less vulnerable to plaque rupture than primary lesions. This, along with the evaluation of primary plaque morphology, will be important to study in the future to determine the true risk of these lesions.

Pre-interventional Risk Assessment

Post-operative risk assessment of the patient is imperative to understanding the safety of treatment. Most guidelines suggest a risk threshold of 6% in the symptomatic patient and 3% in the asymptomatic patient [39]. There have been numerous studies identifying various risk scoring systems for the patient undergoing TF-CAS [49–51]. In general, patients at highest risk for stroke or death following CAS or TCAR are older, symptomatic, have had a recent myocardial infarction, requiring dialysis, need for concomitant cardiac revascularization, long lesion length, Type II or Type III aortic arch, or have a tortuous artery [52]. Patients with more of these factors have increased overall risk. If the patient has multiple risk factors, a frank conversation weighing the risks and benefits of revascularization is warranted, especially in asymptomatic patients. There is rising evidence to suggest that certain patient populations are at prohibitive risk for a carotid intervention. For example, Yuo et al. evaluated the USRDS database to determine the risk of carotid interventions in dialysis patients and found a prohibitively high rate of morbidity and mortality in asymptomatic patients following CAS [53].

Cognition is a relatively novel topic of research in carotid artery disease literature. Studying the effects of carotid artery revascularization on cognition has gained traction. While certainly no definitive evidence exists, TF-CAS has been found to improve cognitive processing speed, executive functioning, and motor function to a higher degree when compared to CEA [54]. No standardized measure of cognition exists in the carotid revascularization literature, which makes the results difficult to interpret. In a randomized study comparing TF-CAS to CEA, there were no differences found in post-operative cognitive function. However, when imaging these patients on follow-up, new ischemic cerebral lesions were found twice as often in TF-CAS compared to CEA [55]. **The implications of these lesions has yet to be truly investigated and clarified.**

Medical Considerations

In symptomatic carotid stenosis, a more extensive workup is warranted. A thorough cardiac evaluation should be performed to evaluate the etiology of the stroke and rule out arrhythmias. Blood tests should also be performed, including an HbA1c, lipid panel, hypercoagulability workup, and creatinine. Imaging of the brain must also be performed to better understand the extent of ischemia and confirm the diagnosis [1]. The workup should be completed in the first 48 hours after symptom onset [1].

Medical optimization is critical in the patient with carotid artery disease. Smoking cessation, healthy diet and exercise, as well as a daily aspirin should be encouraged in all patients pre-intervention [1]. Once the interventionalist has determined the patient to be a candidate for TF-CAS or TCAR, the patient should be initiated on dual antiplatelet therapy (DAPT) and should be dosed with DAPT for at least 72 hours prior to the procedure. This reduces the incidence of periprocedural TIA and stroke. Some have argued aspirin and clopidogrel loading at least 4 hours prior to the procedure to be safe, but this is ultimately per the individual institutions' safety protocols. Clopidogrel resistance assays should be evaluated during this time period prior to intervention. If the patient is determined to be resistant to clopidogrel, alternative options include ticagrelor or prasugrel. Statins should also be initiated at least 5 days prior to the intervention. Statins and antiplatelet agents allow for the stabilization of the plaque.

Incorporating Health Disparities and Health Literacy

It is important to note the lack of data on carotid revascularization in women and marginalized populations. Women are at higher risk for having a disability after an ischemic stroke, as well as a higher risk of mortality after enduring an ischemic stroke [56]. Women are underrepresented in clinical trials, making recommendations for this population difficult. The current literature from the most recent clinical trials suggests similar risks following carotid interventions for both men and women [56]. While overall, there is similar risk across genders with respect to CEA and TCAR, there is some literature to suggest worse outcomes for women with TF-CAS. This is likely due to reduced carotid diameters and must be considered when treating these patients [57].

With respect to racial minority populations, there is a disparity in treatment for carotid artery stenosis. Rates of revascularization for carotid disease are lower in minority-serving hospitals [58]. Perioperative complications are similar across racial groups, but 30-day myocardial infarction and ipsilateral stroke are increased among black patients [59, 60]. Identifying the socioeconomic determinants of health is an imperative adjunct to the workup and treatment of carotid artery stenosis. As such, risk factors can be mitigated as much as possible [1]. Health literacy must also be determined and integrated into the care plan in order to optimize risk factor modification and adherence to care. Further studies are needed to elucidate the reasons for these findings, but interventionalists should be aware of the existing disparities when approaching the treatment of a carotid lesion in these patients.

Technical Pearls

Technical Considerations in TF-CAS

Cerebral embolic protection devices are a critical component to TF-CAS. Embolic protection filters allow for a safer procedure and reduces procedure-related strokes. The EVA-3S investigators compared TF-CAS with embolic protection to unprotected TF-CAS, and the study had to be terminated because the 30-day stroke rate was 3.9 times higher in the unprotected group than that of the TF-CAS arm with cerebral protection [61]. Filter protection has since been confirmed as feasible and safe [65] and has become standard of care for TF-CAS. Consistently, studies have found that the incidence of post-procedural neurologic complications is lower with embolic protection than without [62].

Stent design may also impact outcomes in patients undergoing TF-CAS. Multiple cohort studies have identified closed-cell stent designs to carry a lower risk of cerebral embolization compared to open-cell stents [46, 63]. In general, it is imperative to properly size the stent. Overstretching the stent can cause stroke, pain, and rarely, rupture. Additionally, all exchanges should be performed carefully to prevent air embolism or trauma to the artery.

Technical Considerations in TCAR

TCAR has additional unique technical challenges. The Silk Road Medical ENROUTE system uses an 8Fr transcarotid sheath and can accommodate stent diameters of 6–10 mm and lengths of 20–40 mm. The system works over an 0.014″ wire. Flow reversal is established by accessing the contralateral femoral vein for placement of the neuroprotective system. Choice of anesthesia is interventionalist dependent. General anesthesia has been compared to local/light sedation with no differences in outcomes [64]. Recommended systolic blood pressures are between 140 and 160 mmHg. Hypotension and bradycardia should be avoided to optimize flow reversal. A carotid cutdown can be achieved with a small ~2 cm supraclavicular transverse incision. After the platysma is divided with electrocautery, the avascular plane between the two insertions of the sternocleidomastoid muscle to the sternum and clavicle should be developed.

Prior to access, heparin is administered, and an ACT of >250 s should be targeted to optimize systemic anticoagulation. The artery is then accessed with the specialized micropuncture system provided by Silk Road Medical. The system includes specialized markings on the access needle, guidewire, and catheters to allow for safer access and confidence in placement. The external carotid artery is then selected, and the ENROUTE sheath is inserted. ICA engagement is avoided until flow reversal can be established. The CCA is then clamped, and flow reversal is initiated and confirmed. Once the lesion is crossed, the stent can be deployed using a pin-and-pull technique. Similar to TF-CAS, all exchanges should be performed carefully with judicious use of pre- and post-dilation. Studies have shown a risk of TIA and stroke with post-dilation, so this adjunct should be considered only in cases of significant residual stenosis [65]. Flow reversal is continued momentarily to capture any additional debris. Antegrade CCA flow is then re-established, and the neck incision is closed (Figs. 25.1 and 25.2).

Conclusions

TCAR has become an established alternative to CEA in patients with carotid artery disease, and the role of TF-CAS is currently in evolution. Although CEA continues to be the current gold standard based on societal guidelines, carotid stenting is mostly indicated for the asymptomatic patient

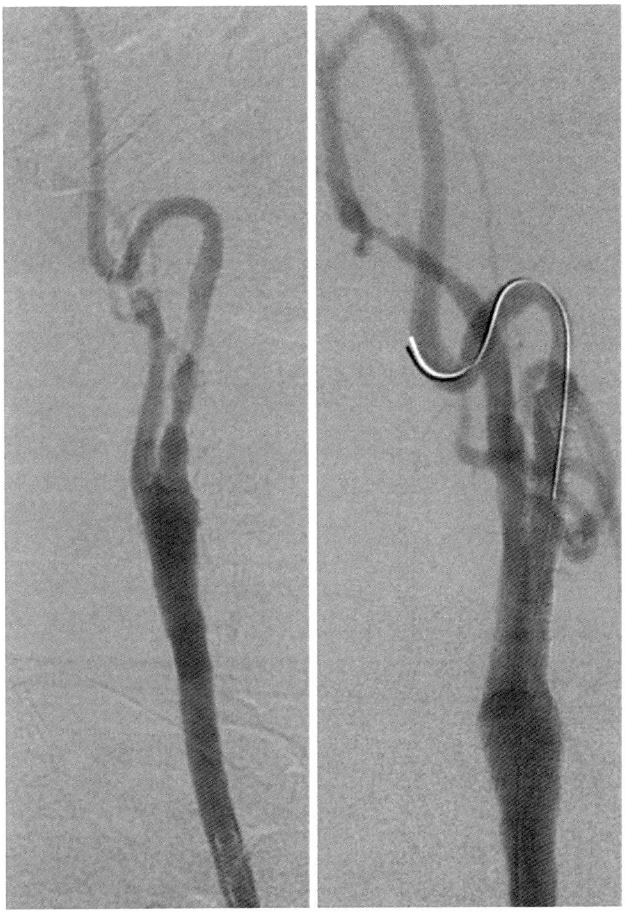

Fig. 25.1 Severe asymptomatic carotid stenosis treated with TCAR

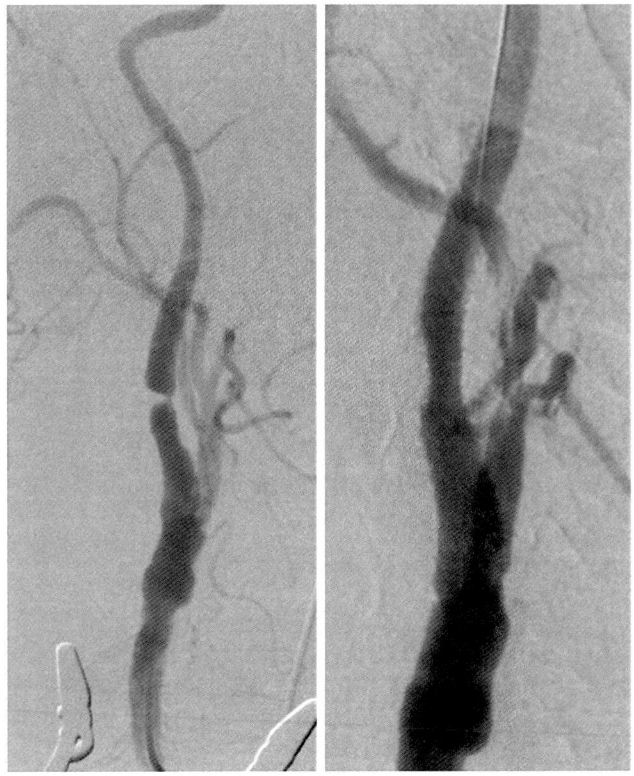

Fig. 25.2 Severe restenosis in a patient with a remote history of carotid endarterectomy treated with TCAR

<70 years old with high surgical risk. TCAR mitigates much of the risk of cerebral embolization that is seen with TF-CAS. However, future comparative randomized trials are imperative to evaluate the influence of modern medical optimization, newer technologies and devices, the role of plaque morphology in clinical decision making, and head-to-head evaluations of TCAR, TF-CAS, and CEA.Source of SupportThis writing was supported in part by grant T32HL98036 from the National Heart, Lung, and Blood Institute (Andraska). The University of Pittsburgh holds a Physician-Scientist Institutional Award from the Burroughs Wellcome Fund (Andraska).

References

1. Kleindorfer DO, Towfighi A, Chaturvedi S, Cockroft KM, Gutierrez J, Lombardi-Hill D, Kamel H, Kernan WN, Kittner SJ, Leira EC, Lennon O, Meschia JF, Nguyen TN, Pollak PM, Santangeli P, Sharrief AZ, Smith SCJ, Turan TN, Williams LS. 2021 guideline for the prevention of stroke in patients with stroke and transient ischemic attack: a guideline from the American Heart Association/American Stroke Association. Stroke. 2021;52:e364–467.
2. Tsao CW, Aday AW, Almarzooq ZI, Alonso A, Beaton AZ, Bittencourt MS, Boehme AK, Buxton AE, Carson AP, Commodore-Mensah Y, Elkind MSV, Evenson KR, Eze-Nliam C, Ferguson JF, Generoso G, Ho JE, Kalani R, Khan SS, Kissela BM, Knutson KL, Levine DA, Lewis TT, Liu J, Loop MS, Ma J, Mussolino ME, Navaneethan SD, Perak AM, Poudel R, Rezk-Hanna M, Roth GA, Schroeder EB, Shah SH, Thacker EL, VanWagner LB, Virani SS, Voecks JH, Wang N-Y, Yaffe K, Martin SS. Heart disease and stroke statistics-2022 update: a report from the American Heart Association. Circulation. 2022;145:e153–639.
3. Virani SS, Alonso A, Aparicio HJ, Benjamin EJ, Bittencourt MS, Callaway CW, Carson AP, Chamberlain AM, Cheng S, Delling FN, Elkind MSV, Evenson KR, Ferguson JF, Gupta DK, Khan SS, Kissela BM, Knutson KL, Lee CD, Lewis TT, Liu J, Loop MS, Lutsey PL, Ma J, Mackey J, Martin SS, Matchar DB, Mussolino ME, Navaneethan SD, Perak AM, Roth GA, Samad Z, Satou GM, Schroeder EB, Shah SH, Shay CM, Stokes A, VanWagner LB, Wang N-Y, Tsao CW. Heart disease and stroke statistics-2021 update: a report from the American Heart Association. Circulation. 2021;143:e254–743.
4. Rothwell PM, Eliasziw M, Gutnikov SA, Fox AJ, Taylor DW, Mayberg MR, Warlow CP, Barnett HJM. Analysis of pooled data from the randomised controlled trials of endarterectomy for symptomatic carotid stenosis. Lancet (London, England). 2003;361:107–16.
5. AbuRahma AF, Avgerinos ED, Chang RW, Darling RC 3rd, Duncan AA, Forbes TL, Malas MB, Murad MH, Perler BA, Powell RJ, Rockman CB, Zhou W. Society for Vascular Surgery clinical practice guidelines for management of extracranial cerebrovascular disease. J Vasc Surg. 2022;75:4S–22S.
6. Bensley RP, Yoshida S, Lo RC, Fokkema M, Hamdan AD, Wyers MC, Chaikof EL, Schermerhorn ML. Accuracy of administrative

data versus clinical data to evaluate carotid endarterectomy and carotid stenting. J Vasc Surg. 2013;58:412–9.
7. Chaer RA, Makaroun MS. Current indications for carotid angioplasty and stenting. Perspect Vasc Surg Endovasc Ther. 2008;20:239–44.
8. Veith FJ, Amor M, Ohki T, Beebe HG, Bell PR, Bolia A, Bergeron P, Connors JJ 3rd, Diethrich EB, Ferguson RD, Henry M, Hobson RW 2nd, Hopkins LN, Katzen BT, Matthias K, Roubin GS, Theron J, Wholey MH, Yadav SS. Current status of carotid bifurcation angioplasty and stenting based on a consensus of opinion leaders. J Vasc Surg. 2001;33:S111–6.
9. Hobson RW 2nd, Howard VJ, Roubin GS, Brott TG, Ferguson RD, Popma JJ, Graham DL, Howard G. Carotid artery stenting is associated with increased complications in octogenarians: 30-day stroke and death rates in the CREST lead-in phase. J Vasc Surg. 2004;40:1106–11.
10. Stanziale SF, Marone LK, Boules TN, Brimmeier JA, Hill K, Makaroun MS, Wholey MH. Carotid artery stenting in octogenarians is associated with increased adverse outcomes. J Vasc Surg. 2006;43:297–304.
11. Gray WA, Yadav JS, Verta P, Scicli A, Fairman R, Wholey M, Hopkins LN, Atkinson R, Raabe R, Barnwell S, Green R. The CAPTURE registry: results of carotid stenting with embolic protection in the post approval setting. Catheter Cardiovasc Interv. 2007;69:341–8.
12. Roubin GS, New G, Iyer SS, Vitek JJ, Al-Mubarak N, Liu MW, Yadav J, Gomez C, Kuntz RE. Immediate and late clinical outcomes of carotid artery stenting in patients with symptomatic and asymptomatic carotid artery stenosis: a 5-year prospective analysis. Circulation. 2001;103:532–7.
13. Brott TG, Hobson RW 2nd, Howard G, Roubin GS, Clark WM, Brooks W, Mackey A, Hill MD, Leimgruber PP, Sheffet AJ, Howard VJ, Moore WS, Voeks JH, Hopkins LN, Cutlip DE, Cohen DJ, Popma JJ, Ferguson RD, Cohen SN, Blackshear JL, Silver FL, Mohr JP, Lal BK, Meschia JF. Stenting versus endarterectomy for treatment of carotid-artery stenosis. N Engl J Med. 2010;363:11–23.
14. Brott TG, Howard G, Roubin GS, Meschia JF, Mackey A, Brooks W, Moore WS, Hill MD, Mantese VA, Clark WM, Timaran CH, Heck D, Leimgruber PP, Sheffet AJ, Howard VJ, Chaturvedi S, Lal BK, Voeks JH, R. W. 2nd Hobson. Long-term results of stenting versus endarterectomy for carotid-artery stenosis. N Engl J Med. 2016;374:1021–31.
15. Gray WA, Hopkins LN, Yadav S, Davis T, Wholey M, Atkinson R, Cremonesi A, Fairman R, Walker G, Verta P, Popma J, Virmani R, Cohen DJ. Protected carotid stenting in high-surgical-risk patients: the ARCHeR results. J Vasc Surg. 2006;44:258–68.
16. Mansour MA, Kang SS, Baker WH, Watson WC, Littooy FN, Labropoulos N, Greisler HP. Carotid endarterectomy for recurrent stenosis. J Vasc Surg. 1997;25:877–83.
17. Rosenfield K, Matsumura JS, Chaturvedi S, Riles T, Ansel GM, Metzger DC, Wechsler L, Jaff MR, Gray W. Randomized trial of stent versus surgery for asymptomatic carotid stenosis. N Engl J Med. 2016;374:1011–20.
18. Müller MD, Lyrer P, Brown MM, Bonati LH. Carotid artery stenting versus endarterectomy for treatment of carotid artery stenosis. Cochrane Database Syst Rev. 2020;2:CD000515.
19. Halliday A, Bulbulia R, Bonati LH, Chester J, Cradduck-Bamford A, Peto R, Pan H. Second asymptomatic carotid surgery trial (ACST-2): a randomised comparison of carotid artery stenting versus carotid endarterectomy. Lancet (London, England). 2021;398:1065–73.
20. Gurm HS, Yadav JS, Fayad P, Katzen BT, Mishkel GJ, Bajwa TK, Ansel G, Strickman NE, Wang H, Cohen SA, Massaro JM, Cutlip DE. Long-term results of carotid stenting versus endarterectomy in high-risk patients. N Engl J Med. 2008;358:1572–9.
21. Yadav JS, Wholey MH, Kuntz RE, Fayad P, Katzen BT, Mishkel GJ, Bajwa TK, Whitlow P, Strickman NE, Jaff MR, Popma JJ, Snead DB, Cutlip DE, Firth BG, Ouriel K. Protected carotid-artery stenting versus endarterectomy in high-risk patients. N Engl J Med. 2004;351:1493–501.
22. Mas J-L, Chatellier G, Beyssen B, Branchereau A, Moulin T, Becquemin J-P, Larrue V, Lièvre M, Leys D, Bonneville J-F, Watelet J, Pruvo J-P, Albucher J-F, Viguier A, Piquet P, Garnier P, Viader F, Touzé E, Giroud M, Hosseini H, Pillet J-C, Favrole P, Neau J-P, Ducrocq X. Endarterectomy versus stenting in patients with symptomatic severe carotid stenosis. N Engl J Med. 2006;355:1660–71.
23. Ringleb PA, Allenberg J, Brückmann H, Eckstein H-H, Fraedrich G, Hartmann M, Hennerici M, Jansen O, Klein G, Kunze A, Marx P, Niederkorn K, Schmiedt W, Solymosi L, Stingele R, Zeumer H, Hacke W. 30 day results from the SPACE trial of stent-protected angioplasty versus carotid endarterectomy in symptomatic patients: a randomised non-inferiority trial. Lancet (London, England). 2006;368:1239–47.
24. Mas J-L, Arquizan C, Calvet D, Viguier A, Albucher J-F, Piquet P, Garnier P, Viader F, Giroud M, Hosseini H, Hinzelin G, Favrole P, Hénon H, Neau J-P, Ducrocq X, Padovani R, Milandre L, Rouanet F, Wolff V, Saudeau D, Mahagne M-H, Sablot D, Amarenco P, Larrue V, Beyssen B, Leys D, Moulin T, Lièvre M, Chatellier G. Long-term follow-up study of endarterectomy versus angioplasty in patients with symptomatic severe carotid stenosis trial. Stroke. 2014;45:2750–6.
25. Arquizan C, Trinquart L, Touboul P-J, Long A, Feasson S, Terriat B, Gobin-Metteil M-P, Guidolin B, Cohen S, Mas J-L. Restenosis is more frequent after carotid stenting than after endarterectomy: the EVA-3S study. Stroke. 2011;42:1015–20.
26. Eckstein H-H, Ringleb P, Allenberg J-R, Berger J, Fraedrich G, Hacke W, Hennerici M, Stingele R, Fiehler J, Zeumer H, Jansen O. Results of the stent-protected angioplasty versus carotid endarterectomy (SPACE) study to treat symptomatic stenoses at 2 years: a multinational, prospective, randomised trial. Lancet Neurol. 2008;7:893–902.
27. Eckstein H-H, Reiff T, Ringleb P, Jansen O, Mansmann U, Hacke W. SPACE-2: a missed opportunity to compare carotid endarterectomy, carotid stenting, and best medical treatment in patients with asymptomatic carotid Stenoses. Eur J Vasc Endovasc Surg. 2016;51:761–5.
28. Ederle J, Dobson J, Featherstone RL, Bonati LH, van der Worp HB, de Borst GJ, Lo TH, Gaines P, Dorman PJ, Macdonald S, Lyrer PA, Hendriks JM, McCollum C, Nederkoorn PJ, Brown MM. Carotid artery stenting compared with endarterectomy in patients with symptomatic carotid stenosis (international carotid stenting study): an interim analysis of a randomised controlled trial. Lancet (London, England). 2010;375:985–97.
29. Mott M, Koroshetz W, Wright CB. CREST-2: identifying the best method of stroke prevention for carotid artery stenosis: National Institute of Neurological Disorders and Stroke organizational update. Stroke. 2017;48:e130–1.
30. Ederle J, Bonati LH, Dobson J, Featherstone RL, Gaines PA, Beard JD, Venables GS, Markus HS, Clifton A, Sandercock P, Brown MM. Endovascular treatment with angioplasty or stenting versus endarterectomy in patients with carotid artery stenosis in the carotid and vertebral artery transluminal angioplasty study (CAVATAS): long-term follow-up of a randomised trial. Lancet Neurol. 2009;8:898–907.
31. Alvarez B, Matas M, Ribo M, Maeso J, Yugueros X, Alvarez-Sabin J. Transcervical carotid stenting with flow reversal is a safe technique for high-risk patients older than 70 years. J Vasc Surg. 2012;55:978–84.
32. Leal I, Orgaz A, Flores Á, Gil J, Rodríguez R, Peinado J, Criado E, Doblas M. A diffusion-weighted magnetic resonance imaging-based

study of transcervical carotid stenting with flow reversal versus transfemoral filter protection. J Vasc Surg. 2012;56:1585–90.
33. Kwolek CJ, Jaff MR, Leal JI, Hopkins LN, Shah RM, Hanover TM, Macdonald S, Cambria RP. Results of the ROADSTER multicenter trial of transcarotid stenting with dynamic flow reversal. J Vasc Surg. 2015;62:1227–34.
34. Malas MB, Dakour-Aridi H, Wang GJ, Kashyap VS, Motaganahalli RL, Eldrup-Jorgensen J, Cronenwett JL, Schermerhorn ML. Transcarotid artery revascularization versus transfemoral carotid artery stenting in the society for vascular surgery vascular quality initiative. J Vasc Surg. 2019;69:92–103.e2.
35. Kashyap VS, Schneider PA, Foteh M, Motaganahalli R, Shah R, Eckstein H-H, Henao S, LaMuraglia G, Stoner MC, Melton J, Massop D, Hanover T, Titus J, Moore WS, Rodríguez-Carvajal R, Malas MB, Arko FR 3rd, Pierce D, Anain P, Oskin T. Early outcomes in the ROADSTER 2 study of transcarotid artery revascularization in patients with significant carotid artery disease. Stroke. 2020;51:2620–9.
36. Schermerhorn ML, Liang P, Eldrup-Jorgensen J, Cronenwett JL, Nolan BW, Kashyap VS, Wang GJ, Motaganahalli RL, Malas MB. Association of transcarotid artery revascularization vs transfemoral carotid artery stenting with stroke or death among patients with carotid artery stenosis. JAMA. 2019;322:2313–22.
37. Schermerhorn ML, Liang P, Dakour-Aridi H, Kashyap VS, Wang GJ, Nolan BW, Cronenwett JL, Eldrup-Jorgensen J, Malas MB. In-hospital outcomes of transcarotid artery revascularization and carotid endarterectomy in the Society for Vascular Surgery Vascular Quality Initiative. J Vasc Surg. 2020;71:87–95.
38. Kashyap VS, King AH, Foteh MI, Janko M, Jim J, Motaganahalli RL, Apple JM, Bose S, Kumins NH. A multi-institutional analysis of transcarotid artery revascularization compared to carotid endarterectomy. J Vasc Surg. 2019;70:123–9.
39. Sacco RL, Adams R, Albers G, Alberts MJ, Benavente O, Furie K, Goldstein LB, Gorelick P, Halperin J, Harbaugh R, Johnston SC, Katzan I, Kelly-Hayes M, Kenton EJ, Marks M, Schwamm LH, Tomsick T. Guidelines for prevention of stroke in patients with ischemic stroke or transient ischemic attack: a statement for healthcare professionals from the American Heart Association/American Stroke Association council on stroke: co-sponsored by the council on. Circulation. 2006;113:e409–49.
40. Lam RC, Lin SC, DeRubertis B, Hynecek R, Kent KC, Faries PL. The impact of increasing age on anatomic factors affecting carotid angioplasty and stenting. J Vasc Surg. 2007;45:875–80.
41. Faggioli GL, Ferri M, Freyrie A, Gargiulo M, Fratesi F, Rossi C, Manzoli L, Stella A. Aortic arch anomalies are associated with increased risk of neurological events in carotid stent procedures. Eur J Vasc Endovasc Surg. 2007;33:436–41.
42. Mathur A, Roubin GS, Iyer SS, Piamsonboon C, Liu MW, Gomez CR, Yadav JS, Chastain HD, Fox LM, Dean LS, Vitek JJ. Predictors of stroke complicating carotid artery stenting. Circulation. 1998;97:1239–45.
43. Biasi GM, Froio A, Diethrich EB, Deleo G, Galimberti S, Mingazzini P, Nicolaides AN, Griffin M, Raithel D, Reid DB, Valsecchi MG. Carotid plaque echolucency increases the risk of stroke in carotid stenting: the imaging in carotid angioplasty and risk of stroke (ICAROS) study. Circulation. 2004;110:756–62.
44. Sayeed S, Stanziale SF, Wholey MH, Makaroun MS. Angiographic lesion characteristics can predict adverse outcomes after carotid artery stenting. J Vasc Surg. 2008;47:81–7.
45. Bijuklic K, Wandler A, Varnakov Y, Tuebler T, Schofer J. Risk factors for cerebral embolization after carotid artery stenting with embolic protection: a diffusion-weighted magnetic resonance imaging study in 837 consecutive patients. Circ Cardiovasc Interv. 2013;6:311–6.
46. Xu X, Feng Y, Bai X, Ma Y, Wang Y, Chen Y, Yang B, Ling F, Zhang X, Jiao L. Risk factors for silent new ischemic cerebral lesions following carotid artery stenting. Neuroradiology. 2020;62:1177–84.
47. Bonati LH, Ederle J, McCabe DJH, Dobson J, Featherstone RL, Gaines PA, Beard JD, Venables GS, Markus HS, Clifton A, Sandercock P, Brown MM. Long-term risk of carotid restenosis in patients randomly assigned to endovascular treatment or endarterectomy in the carotid and vertebral artery transluminal angioplasty study (CAVATAS): long-term follow-up of a randomised trial. Lancet Neurol. 2009;8:908–17.
48. Kumar R, Batchelder A, Saratzis A, AbuRahma AF, Ringleb P, Lal BK, Mas JL, Steinbauer M, Naylor AR. Restenosis after carotid interventions and its relationship with recurrent ipsilateral stroke: a systematic review and meta-analysis. Eur J Vasc Endovasc. 2017;53:766–75.
49. Dakour-Aridi H, Faateh M, Kuo P-L, Zarkowsky DS, Beck A, Malas MB. The vascular quality initiative 30-day stroke/death risk score calculator after transfemoral carotid artery stenting. J Vasc Surg. 2020;71:526–34.
50. Macdonald S, Lee R, Williams R, Stansby G. Towards safer carotid artery stenting: a scoring system for anatomic suitability. Stroke. 2009;40:1698–703.
51. Setacci C, Chisci E, Setacci F, Iacoponi F, de Donato G, Rossi A. Siena carotid artery stenting score: a risk modelling study for individual patients. Stroke. 2010;41:1259–65.
52. Wimmer NJ, Yeh RW, Cutlip DE, Mauri L. Risk prediction for adverse events after carotid artery stenting in higher surgical risk patients. Stroke. 2012;43:3218–24.
53. Yuo TH, Sidaoui J, Marone LK, Makaroun MS, Chaer RA. Revascularization of asymptomatic carotid stenosis is not appropriate in patients on dialysis. J Vasc Surg. 2015;61:670–4.
54. Kougias P, Collins R, Pastorek N, Sharath S, Barshes NR, McCulloch K, Pisimisis G, Berger DH. Comparison of domain-specific cognitive function after carotid endarterectomy and stenting. J Vasc Surg. 2015;62:355–61.
55. Altinbas A, van Zandvoort MJE, van den Berg E, Jongen LM, Algra A, Moll FL, Nederkoorn PJ, Mali WPTM, Bonati LH, Brown MM, Kappelle LJ, van der Worp HB. Cognition after carotid endarterectomy or stenting: a randomized comparison. Neurology. 2011;77:1084–90.
56. Rockman C, Caso V, Schneider PA. Carotid interventions for women: the hazards and benefits. Stroke. 2022;53:611–23.
57. Krejza J, Arkuszewski M, Kasner SE, Weigele J, Ustymowicz A, Hurst RW, Cucchiara BL, Messe SR. Carotid artery diameter in men and women and the relation to body and neck size. Stroke. 2006;37:1103–5.
58. Faigle R, Cooper LA, Gottesman RF. Lower carotid revascularization rates after stroke in racial/ethnic minority-serving US hospitals. Neurology. 2019;92:e2653–60.
59. Cohen BD, Zeymo A, Bouchard M, McDermott J, Shara NM, Sellke FW, Sodha N, Al-Refaie WB, Ehsan A. Increased access to cardiac surgery did not improve outcomes: early look into medicaid expansion. Ann Thorac Surg. 2021;114:1637. https://doi.org/10.1016/j.athoracsur.2021.09.025.
60. Vogel TR, Kruse RL, Kim RJ, Dombrovskiy VY. Racial and socioeconomic disparities after carotid procedures. Vasc Endovasc Surg. 2018;52:330–4.
61. Mas JL, Chatellier G, Beyssen B. Carotid angioplasty and stenting with and without cerebral protection: clinical alert from the endarterectomy versus angioplasty in patients with symptomatic severe carotid stenosis (EVA-3S) trial. Stroke. 2004;35:e18–20.
62. Reimers B, Schlüter M, Castriota F, Tübler T, Corvaja N, Cernetti C, Manetti R, Picciolo A, Liistro F, Di Mario C, Cremonesi A, Schofer J, Colombo A. Routine use of cerebral protection during carotid artery stenting: results of a multicenter registry of 753 patients. Am J Med. 2004;116:217–22.

63. Park KY, Kim DI, Kim BM, Nam HS, Kim YD, Heo JH, Kim DJ. Incidence of embolism associated with carotid artery stenting: open-cell versus closed-cell stents. J Neurosurg. 2013;119:642–7.
64. Mukherjee D, Collins DT, Liu C, Ha N, Jim J. The study of transcarotid artery revascularization under local versus general anesthesia with results from the Society for Vascular Surgery Vascular Quality Initiative. Vascular. 2020;28:784–93.
65. Ziapour B, Schermerhorn ML, Iafrati MD, Suarez LB, TourSavadkohi S, Salehi P. A systematic review and meta-analysis of predilation and postdilation in transfemoral carotid artery stenting. J Vasc Surg. 2020;72:346–355.e1.

Technique and Complications of Transfemoral Stenting

26

Christos Karkos and Thomas Kalogirou

Case Presentation

A 68-year-old female was referred by her family doctor for vascular surgical opinion with the diagnosis of symptomatic high-grade carotid stenosis. She had suffered two episodes of right amaurosis fugax over the previous month, and a carotid color flow duplex scan revealed a 70–80% stenosis at the origin of the right internal carotid artery (ICA). Past medical history included hypertension, hypercholesterolemia, and cardiac arrhythmia. She was on treatment with ramipril, hydrochlorothiazide, simvastatin, bisoprolol, and flecainide. Magnetic resonance angiography (MRA) of the extracranial carotid arteries confirmed the presence of significant stenosis (>70%) at the origin of the right ICA. The left carotid bifurcation and the vertebral arteries had no significant atherosclerotic disease. Brain magnetic resonance imaging was normal. She was started on Aspirin 100 mg once daily and was offered a carotid operation for stroke prophylaxis. Both carotid endarterectomy and stenting options were discussed in detail.

Continued at page 269

Introduction

Since its early inception in the 1990s, carotid artery stenting (CAS) has evolved rapidly from a rudimentary angioplasty technique to a sophisticated endovascular procedure. Despite the technical advances, the role of CAS in stroke prophylaxis remains controversial, the main reason being the higher rates of neurologic adverse events when compared to carotid endarterectomy (CEA). Current clinical practice guidelines favor CEA as the standard of care for most patients with carotid stenosis, whereas CAS can be safely offered as an alternative to CEA in younger patients (<70 years) with symptomatic carotid stenosis (Table 26.1) [1–4]. The key for a successful CAS service is to improve the periprocedural results by reducing both the incidence and the consequences of its complications. The aim of this chapter is, first, to describe the technical steps of a standard transfemoral CAS procedure and, second, to summarize the most common complications, focusing on how to prevent and manage them.

Table 26.1 Latest recommendations regarding the use of CEA and CAS in symptomatic and asymptomatic patients with significant carotid stenosis [1–3]. *ESVS* European Society for Vascular and Endovascular Surgery; *SVS* Society for Vascular Surgery; *ESO* European Stroke Organization; *TF-CAS* transfemoral CAS; *TIA* transient ischemic attack

Society		Recommendation
	(A) Symptomatic patients	
ESVS 2023	For patients aged ≥70 years who have experienced a carotid territory TIA or ischemic stroke within the preceding 6 months in association with a 50–99% carotid stenosis, it is recommended that they should be treated by CEA, rather than CAS	Class: I Level: A
ESVS 2023	For patients aged <70 years who have experienced a carotid territory TIA or ischemic stroke within the preceding 6 months in association with a 50–99% carotid stenosis, CAS may be considered an alternative to CEA, provided the documented 30-day risk of death/stroke is <6%	Class: IIb Level: B
SVS 2021	We recommend CEA over TF-CAS in low- and standard-risk patients with >50% symptomatic carotid artery stenosis	Grade: 1 Quality: A
ESO 2021	In patients with symptomatic carotid artery stenosis requiring revascularization, we recommend CEA as the treatment of choice	Strength: Strong Quality: Moderate
ESO 2021	In patients with symptomatic carotid stenosis <70 years old requiring revascularization, we suggest that CAS may be considered as an alternative to CEA	Strength: Weak Quality: Low

(continued)

C. Karkos (✉) · T. Kalogirou
Vascular Unit, 5th Department of Surgery, Hippocratio Hospital, Medical School, Aristotle University of Thessaloniki, Thessaloniki, Greece

G. Geroulakos et al. (eds.), *Mastering Endovascular Techniques*, https://doi.org/10.1007/978-3-031-42735-0_26

Table 26.1 (continued)

Society		Recommendation
	(B) Asymptomatic patients	
ESVS 2023	For average-surgical-risk patients with an asymptomatic 60–99% stenosis, CEA should be considered in the presence of one or more imaging or clinical characteristics that may be associated with an increased risk of late stroke, provided 30-day stroke/death rates are <3% and patient life expectancy exceeds 5 years	Class: IIa Level: B
ESVS 2023	For average-surgical-risk patients with an asymptomatic 60–99% stenosis in the presence of one or more imaging or clinical characteristics that may be associated with an increased risk of late stroke, CAS may be an alternative to CEA, provided 30-day stroke/death rates are <3% and patient life expectancy exceeds 5 years	Class: IIb Level: B
ESVS 2023	For asymptomatic patients deemed by the multidisciplinary team to be "high risk for surgery" and who have an asymptomatic 60–99% stenosis in the presence of one or more imaging/clinical characteristics that may be associated with an increased risk of late stroke on best medical therapy, CAS may be considered provided anatomy is favorable, 30-day death/stroke rates are <3%, and patient life expectancy exceeds 5 years	Class: IIb Level: B
SVS 2021	In low-surgical-risk patients with asymptomatic carotid bifurcation atherosclerosis and stenosis of >70% (documented by validated duplex ultrasound or computed tomography angiography [CTA]/angiography), we recommend CEA with best medical therapy over maximal medical therapy alone for the long-term prevention of stroke and death	Grade: 1 Quality: B
ESO 2021	In patients with ≥60% asymptomatic carotid artery stenosis considered to be at increased risk of stroke on best medical therapy alone, we recommend CEA	Strength: Strong Quality: Moderate
ESO 2021	In patients with asymptomatic carotid stenosis, we recommend against CAS as a routine alternative to best medical therapy alone	Strength: Weak Quality: Very low
ESO 2021	In patients with asymptomatic carotid stenosis in whom revascularization is considered to be appropriate, we suggest CEA as the current treatment of choice	Strength: Weak Quality: Moderate

Technique

This is a description of how the standard technique of transfemoral CAS with filter-type distal embolic protection is performed at the authors' institution. Undoubtedly, there will be many other ways of executing the procedure, but this reflects only the authors' personal preference and experience. Patients are started on dual antiplatelet therapy that includes aspirin 100 mg once a day and clopidogrel 75 mg once a day, preferably for a minimum of 1 week prior to the procedure. The intervention is performed while the patient is awake without sedation. Consequently, the patient's neurologic status can be monitored during each step of the procedure. Under local anesthetic infiltration with xylocaine 2%, a common femoral artery (CFA) puncture is performed. If a diagnostic carotid digital subtraction angiogram (DSA) is needed prior to the CAS procedure, a 5F-11 cm groin sheath is introduced. Otherwise, if both the aortic arch and the carotid anatomy have been adequately documented at previous imaging (CTA, MRA, DSA) and the intention is to proceed directly to transfemoral CAS, a 7F-11 cm groin sheath is introduced. Once femoral access is secured, the patient is anticoagulated with intravenous heparin, usually 5000 units or a dose needed to achieve an activated clotting time of >250 s. A 5F diagnostic catheter (often a Headhunter or Simmons 2) is advanced into the aortic arch with the aid of a 260 cm-long, J-shaped, 0.035-inch hydrophilic coated guidewire (Terumo Glidewire, Terumo Europe NV). If an aortic arch DSA is needed, for instance, when severe ostial stenosis of an arch vessel is suspected or when difficulty accessing the carotid arteries is encountered, this is usually performed in the left anterior oblique (LAO) projection (30–50°). Roadmapping can be useful in guiding the cannulation. Depending on the target carotid side, the catheter is then advanced either through the innominate or through the left common carotid artery (CCA) into the mid-CCA, and a diagnostic selective carotid angiogram, including baseline intracranial runs, is obtained. Normally, we advance a stiff guidewire into the distal external carotid artery (ECA) and exchange the diagnostic catheter (and the 5F groin sheath) for a 7F-11 cm sheath and a 7F, 100-cm long, JR4-type guiding catheter. Our preferred guiding catheters are the Launcher (Medtronic) and the MACH1 (Boston Scientific). In cases of tortuous anatomy, when extra support is needed for exchanging the diagnostic catheter with the guiding catheter, a stiffer guidewire may be necessary, such as the 260 cm-long, J-tip, Amplatz Superstiff Guidewire (Boston Scientific). From this point on, all subsequent procedural steps are executed through this guiding catheter, a technical approach known as the *direct approach* to CAS. Other endovascular specialists prefer the *telescopic approach*, i.e., exchanging the diagnostic catheter and groin sheath for a long introducer sheath (often a 6F, 90-cm-long Super Arrow-Flex® Sheath, Teleflex, Teleflex Medical Europe Ltd.; or 6F, 90-cm-long Destination Guiding sheath, Terumo Interventional Systems, Terumo Europe NV), which serves as the working channel.

After positioning the tip of the guiding catheter in the distal CCA and using roadmap guidance, we cross the stenotic lesion with a distal embolic protection device (EPD). To facilitate crossing, proper view of the carotid bifurcation and the lesion itself is mandatory; therefore, multiple oblique angiography views may help to open up the carotid bifurcation and identify the best working view. Most distal EPDs are mounted on a 0.014″ shapable wire, and shaping the tip appropriately to engage the ICA stenosis is helpful. This step of the procedure is not "protected" and should be done in a gentle and "atraumatic" manner to avoid distal embolization [5–7]. Our preferred distal EPDs are the Emboshield NAV6 (Abbott Cardiovascular) and the SpiderFX (EV3). The former is attached to its own wire, over which one can proceed with the subsequent intervention after removing the filter delivery catheter. In contrast, introduction and deployment of the SpiderFX filter are preceded by crossing the lesion with a separate 0.014-inch guidewire (usually a Nitrex, ev3). In brief, the SpiderFX filter system comprises a microcatheter with two ends, one green, which serves as a delivery system for the filter, and one blue at the opposite end, which serves as a filter retrieval catheter at the end of the procedure. The filter itself is initially parked in the "garage" proximal (i.e., caudally, closer to the CFA access) to the monorail exit port of the 40-cm long, green (delivery) segment of the microcatheter. The filter microcatheter is then advanced over the Nitrex wire, and once it is in the desired position in the distal ICA, the Nitrex wire is removed via the monorail exit port, and the filter is advanced from the "garage" to the delivery microcatheter tip, uncovered and deployed [5, 6]. The EPD must be deployed in a relatively straight segment of the distal cervical ICA and at a sufficient distance to allow positioning of the delivery systems for the stents and balloons [5–8].

The next step is the stent placement. Pre-stenting angioplasty (predilatation) is only performed if the lesion is too stenotic to cross with the stent. A 3 mm × 20 mm rapid exchange (RX) cardiology balloon may be necessary to accomplish this. For the usual carotid lesions involving the carotid bifurcation and the proximal ICA, tapered self-expanding stents are recommended. A variety of self-expanding stents are available which can be categorized as open, closed, or hybrid according to cell design (Table 26.2).

Table 26.2 Characteristics of open-cell, closed-cell, and hybrid design stents [1]

	Open	Closed	Hybrid
Free cell area	Large	Small	Mid-segment = small, edges = large
Strut interconnections	Few	Many	Mid-segment = many, edges = few
Flexibility	Good	Limited	Moderate
Plaque coverage	Limited	Good	Good

In general, open-cell stents are more flexible and suited for tortuous anatomy, whereas closed-cell stents are more rigid and have better plaque coverage. Hybrid cells combine properties from both types (closed cell in the middle and open cell at the edges). The preferred type of stents at the authors' institution is the open-cell stent Protégé (ev3), and the closed-cell stents Xact (Abbott) and Carotid Wallstent (Boston Scientific) (Table 26.3). We tend to use a closed-cell stent in symptomatic patients or in patients with a friable plaque, and an open-cell stent for those with a tortuous ICA that may be less suited to the more rigid closed-cell stent. Because one stent is not suitable for all lesions, it is advisable to choose an appropriate stent depending on the lesion characteristics and the clinical status of the patient. Stent placement is performed under roadmap guidance aiming for complete lesion coverage, typically spanning the bifurcation. The DSA run is done prior to the introduction of the stent

Table 26.3 Suggested endovascular materials for the transfemoral filter-protected CAS. There are several other options, devices, and materials to use for performing these procedures. The list presented here is based on our personal preference and institutional experience

	Manufacturer	Size/length
Wires		
Any standard access wire		0.035″
Glidewire stiff	Terumo Radifocus	0.035″, 260 cm
Amplatz Superstiff	Boston scientific	0.035″, 260 cm
Nitrex guidewire	ev3	0.014″, 180 cm
Sheaths/introducer sheaths		
Any standard access sheath	Any	5F-11 cm, 7F-11 cm
Arrow	Teleflex	6F-90 cm
Destination	Terumo interventional systems	6F-90 cm
Diagnostic catheters/guide catheters		
Any standard angiographic catheter (headhunter H1, Simmons 1, 2, or 3)	Imager II, Boston Scientific	0.035″, 5F, 100 cm
JR4 MACH1	Boston scientific	7F, 100 cm
JR4 launcher	Medtronic	7F, 100 cm
Balloons		
Viatrac 14 plus RX	Abbott cardiovascular	5 mm × 20 mm, 29 cm RX, 135 cm
Stents		
Protégé RX carotid stent (tapered)	Medtronic	6-8 mm × 40 mm, 7-10 mm × 40 mm
Xact (tapered)	Abbott cardiovascular	6-8 mm × 40 mm, 7-9 mm × 40 mm
Carotid Wallstent (tube, self tapers)	Boston scientific	7 mm × 40 mm, 9 mm × 40 mm
Distal embolic protection device		
SpiderFX	Medtronic	6 mm
Emboshield NAV6	Abbott cardiovascular	5 mm

delivery system in the guiding catheter because contrast injection with the stent delivery system in place is not possible due to the tight guiding catheter lumen. This is important to keep in mind when treating patients with significant tortuosity of the CCA/ICA. In such cases, introduction of the more rigid carotid stent delivery system may straighten the tortuous course, significantly alter the anatomy, and result in deploying the stent proximal or distal to the intended location and missing the lesion.

If significant residual stenosis (>30%) exists after the deployment of the stent, balloon angioplasty (postdilatation) of the most stenotic portion of the stented area can be performed using a 5 mm × 20 mm RX cardiology balloon (Viatrac 14 Plus, Abbott Cardiovascular). Minor residual stenosis (<30%) is acceptable since the goal of carotid stenosis treatment should be stabilization of plaque emboli, not restoration of normal vessel diameter. If the patient's baseline heart rate is less than 80/min, we administer intravenously 0.5–1 mg atropine prior to balloon inflation to avoid the bradycardia induced by the stimulation of the carotid sinus baroreceptors located at the CCA bifurcation. Such a bradycardia on balloon inflation may be seen more often in bifurcation lesions and less so when the carotid lesion is located at the proximal ICA sparing the bifurcation. Careful attention should be paid to the patient's hemodynamics during this maneuver and subsequently. As there is some evidence that postdilatation may be associated with increased hemodynamic instability, periprocedural stroke/death, or clinically silent cerebral ischemia, this step can be omitted if there is satisfactory stent wall apposition with no significant residual stenosis [1].

A control DSA is taken prior to retrieval of the distal EPD to document a satisfactory angiographic result and blood flow through the filter. The EPD can then be removed using the dedicated retrieval catheter. It is imperative not to close the filter with too much force as the embolic material trapped inside the filter may be squeezed out. A further completion angiogram to include both the cervical ICA and the intracranial circulation is performed to evaluate for residual stenosis, exclude vasospasm or dissection, and evaluate intracranial blood flow. The patient may be asked to answer simple questions or perform simple tasks as a basic neurological evaluation prior to access discontinuation. Hemostasis may be obtained with manual compression or a closure device, and the patient is transferred to a high-dependency ward for overnight monitoring.

Complications

The aim of the procedure is stroke prevention. Nevertheless, the very same procedure may be complicated with cerebral ischemic events or other complications. Therefore, if CAS is to be a meaningful, preventive intervention that benefits carotid patients, these risks should be minimized. As with CEA, transfemoral CAS outcomes largely depend on the pre-procedural symptom status, and indicative, real-world data on stroke/death rates in the modern era are summarized in Table 26.4. Known factors that may forecast an increased risk of complications can be grouped as clinical, anatomical, and those related to the operator strategy and experience (Table 26.5). Information listed in Tables 26.4 and 26.5 can be used when counseling patients for an informed consent prior to a carotid intervention. In general, complications may be encountered in each step of the procedure, and these may involve challenging scenarios, problems, or difficulties with (1) transfemoral access, (2) catheterization of the CCA, and

Table 26.4 Outcomes following transfemoral CAS stratified by pre-procedural symptoms. This is a retrospective review of the Vascular Quality Initiative (VQI) database with data on 9.807 patients undergoing transfemoral CAS [9]

Pre-procedural symptom status	In-hospital stroke/death rate
After recent stroke	5.5%
After recent hemispheric TIA	2.4%
After recent ocular TIA	2.8%
Asymptomatic patients	1.4%
Formerly symptomatic patients	3.5%

Table 26.5 Predictive factors that are known to impact outcomes of CAS

Clinical factors	
Symptoms	Symptomatic patients fare worse after CAS compared to asymptomatic ones. Periprocedural neurologic complication rates vary depending on the type of presenting symptoms (none, ocular, TIA, stroke) and time elapsed from the index event [3, 9]
Age	CAS is associated with a higher risk of stroke/death in patients ≥70 years, possibly due to increased atherosclerotic burden, aortic arch calcification, changes in vascular anatomy, and increasing plaque vulnerability [1, 3]
Gender (female)	Periprocedural risk of events may be higher in women who have CAS than those who have CEA [10]
Cerebrovascular reserve (CVR)	Exhausted CVR is associated with an increased risk of periprocedural neurologic complications [11]
Anatomical factors	
Difficult femoral arterial access	Preexisting peripheral arterial disease (and, in particular, CFA disease, tortuosity, or disease of both iliac arteries and distal aorta) may predispose to complications. Consider alternative access, or CEA
Difficult arch, CCA/ICA angulation, or tortuosity	Type III arch, bovine arch, and angulation of the ICA or CCA are all associated with an increased risk of periprocedural complications. The risk of adverse anatomy increases with age [12–15]
ICA/CCA degree of stenosis	Extensive ICA or CCA stenosis narrows working space and risks potential intima damage and atheroemboli, while impeding safe access and treatment [15]. A very tight stenosis may first require predilatation with a balloon to allow safe passage and deployment of an EPD prior to CAS. Correction of severe stenotic lesions may also lead to increased risk of hyperperfusion injury post-CAS [16]

(continued)

Table 26.5 (continued)

Lesion anatomy	Carotid plaque characteristics may predict CAS outcomes. Concentric and calcified plaques pose intraluminal access and treatment barriers and have been reported to be associated with a negative outcome. There is evidence that soft plaques with low echogenicity scores may be associated with significant neurologic adverse events [15, 17]
Factors related to operator strategy and experience	
Learning curve	An arbitrary number of 30 carotid angiograms and 25 CAS procedures has been suggested as the recommended prerequisite for credentialing in CAS [18, 19]. The 2023 ESVS carotid guidelines suggest that "for patients undergoing transfemoral CAS, at least 12 CAS procedures per year (per operator) may be considered an appropriate operator volume threshold in order to maintain optimal outcomes (class IIb, level C)" [1]
Choice of embolic protection	Distal filtration vs. flow-reversal EPDs [1]
Stent design	Stent choice (open-cell vs. closed-cell) should be tailored to the carotid anatomy and plaque characteristics [1]
Case selection	Appropriate patient selection is critical for success. All the factors summarized in this table should be taken into account when selecting cases for CAS and planning the stenting strategy

(3) treatment of the lesion. Familiarity with the fundamentals of the procedure, careful pre-procedural study of the patient's specific anatomy and disease, adequate planning, and meticulous endovascular technique are all equally important parameters to avoid such complications.

Access Site Complications

As for all transfemoral arterial procedures, significant complications can occur related to femoral access. These include bleeding, local or retroperitoneal hematoma, pseudoaneurysm formation, perforation, dissection, thrombosis, and distal embolization, all problems that may necessitate open surgery or endovascular repair. Puncture site complications can be minimized by careful, anterior wall puncture of the CFA, and this can be achieved even more consistently when needle puncture is ultrasound guided. Safe puncture site closure after sheath removal can be done with manual compression by experienced personnel under additional local anesthesia. Other centers routinely apply a vascular closure device at the end of the procedure to achieve immediate hemostasis. This allows earlier ambulation with the added benefit of counteracting the activated carotid sinus reflex and the occasionally observed post-procedural hypotension.

Since many patients with carotid stenosis also have peripheral arterial disease, a simple transfemoral approach may not be possible in some. A significant stenosis of the common femoral, external iliac, or common iliac artery may be treated with angioplasty or stenting to allow insertion of the large guide catheters needed for CAS. If a complete aortoiliac occlusion is encountered, it may still be possible to combine recanalization of an aortoiliac occlusion with transfemoral CAS. However, this increases the magnitude and severity of the procedure. An alternative would be opting either for CEA or for CAS via a transbrachial, transradial, or transcervical approach (Fig. 26.1).

Selective Catheterization of the CCA

Gaining access to the CCA-ECA can be challenging, and it leads to technical failure or complications. This step of the procedure is not "protected," and approximately 40% of the major complications during CAS (embolism, dissection) occur during the selective catheterization phase [20]. Achieving stable guide catheter positioning is of paramount importance before proceeding with any carotid intervention. Difficulties may arise if the CCA is tortuous, if the angle of the origin of the brachiocephalic artery or left CCA off the aortic arch is very acute, or if the aortic arch is diseased or heavily calcified [5–8]. Distal CCA stenosis or ECA occlusion may pose additional difficulties because gaining distal wire access in the ECA is not feasible. Such lesions may be negotiated using an Amplatz Superstiff guidewire with a shorter, 1-cm-long soft tip (instead of the longer, 7-cm-long soft tip). The J-shaped end of the wire is positioned just proximal to the CCA stenosis or proximal to the carotid lesion (in ECA occlusion), and a standard exchange is performed. Care must be taken not to allow the J-wire to

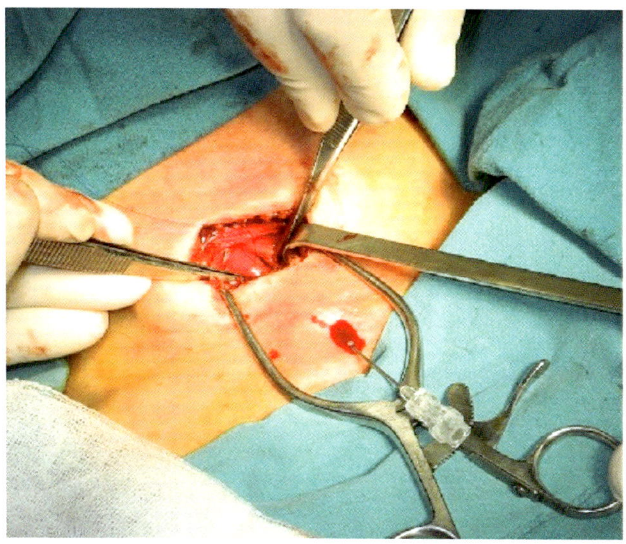

Fig. 26.1 Transcervical CAS approach in a patient with aortoiliac occlusion

migrate distally and potentially disturb the plaque. Alternatively, telescoping an extra-long (125 cm), 5F Headhunter or Vitek diagnostic catheter through the 6F guiding sheath or the guiding catheter may solve this difficulty [6, 7]. An alternative trick is a direct approach using the 8F Simmons 2 guiding catheter, which is relatively stiff, large, and stable and usually will not herniate into the aortic arch as the stent delivery system is advanced [7]. If the risks of the extensive manipulations seem too high, it is preferable not to persist with a transfemoral approach and opt for an alternative route (CEA, transbrachial, or direct CAS) (Fig. 26.2).

Treatment of the Lesion

Complications may occur during each step of the treatment of the ICA lesion, i.e., lesion crossing, EPD placement and

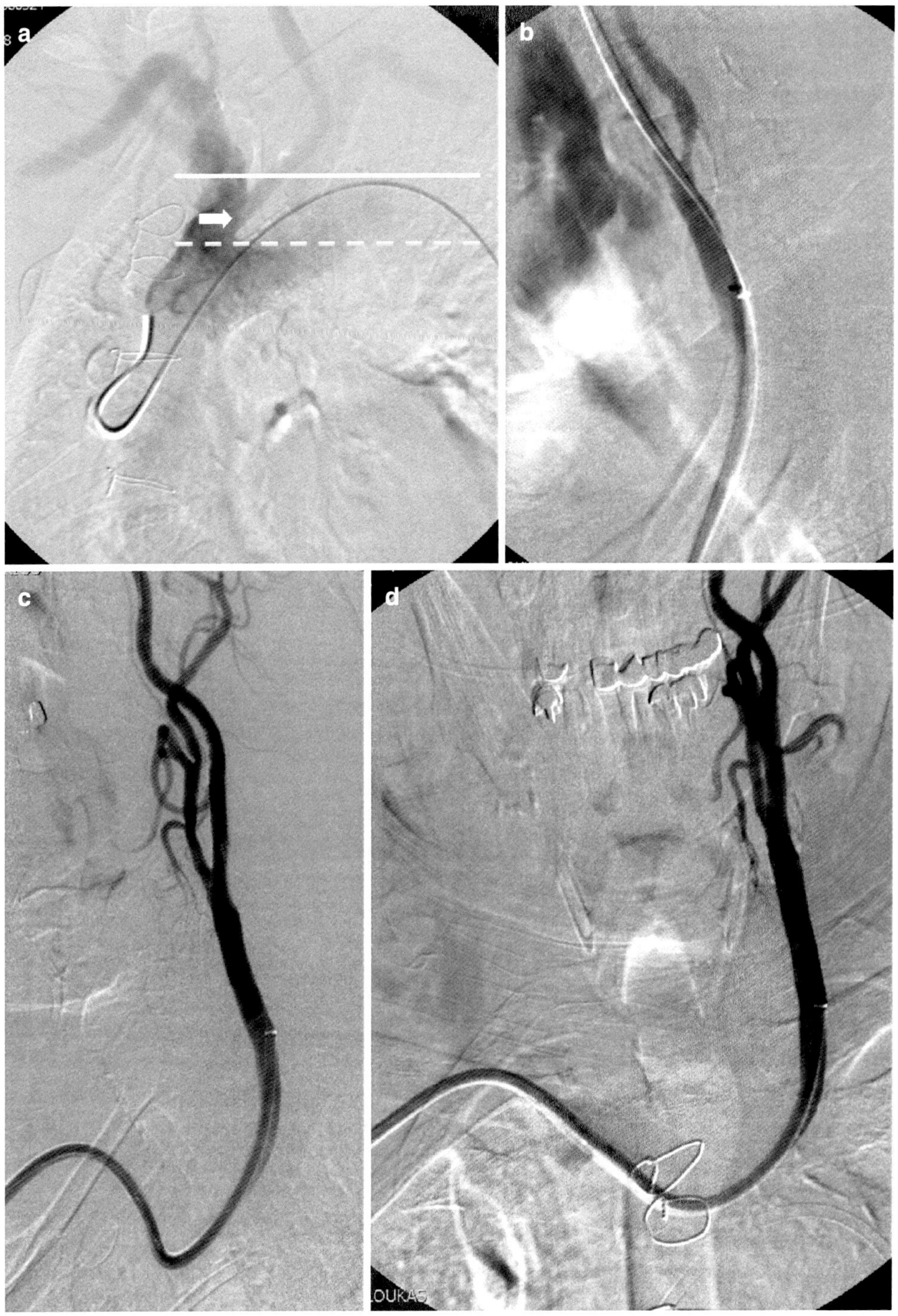

Fig. 26.2 Left CAS via a right transbrachial approach. Initial transfemoral DSA (**a**) shows a type III arch with the bovine origin of the left CCA (white arrow), which then assumes a tortuous course. After failing to achieve stable positioning of the guiding catheter in the left CCA, we changed our approach to a right transbrachial approach using a 6F-45 cm Arrow sheath (**b**–**d**). A type III arch is when the vertical distance between the brachiocephalic artery origin (dotted line) and top of the arch (solid line) exceeds two left CCA diameters

deployment, predilatation, stent deployment, postdilatation, and EPD retrieval.

Crossing the Carotid Lesion and Deploying the EPD

Crossing a tight and irregular carotid stenosis may be occasionally difficult, and it is important not to force the wire or the distal EPD across the lesion since this can result in procedural complications, including ICA distal embolism or dissection. When there is difficulty to gain access to a tiny angulated residual lumen, one may use a 125-cm-long 5F-angled multipurpose diagnostic catheter or microcatheters, such as a microcatheter with distal curve or a steerable microcatheter [7]. These can be inserted through the guiding catheter or sheath and help provide the proper angle to engage the lesion. If crossing the stenosis is not possible, an alternative in this situation is to use proximal instead of distal embolic protection. In cases where a 0.014-inch guidewire can be threaded distally but the distal, filter-based EPD cannot pass through the tight stenosis, lesion predilatation with a 2 mm or 2.5 mm RX balloon can be employed to allow safe passage of the EPD. However, this maneuver is done without cerebral protection. Another difficulty precluding safe placement of a distal EPD is the presence of unfavorable ICA anatomy. A severe "kink," tortuosity, or a distal, tandem stenosis may prevent navigation of these devices into a safe segment of the distal ICA. In such situations, one might also consider continuing the intervention using a proximal EPD or performing the procedure without distal embolic protection at all. However, the risk of the latter should be weighed against other options, such as CAS with proximal embolic protection or CEA.

Stent Deployment

Complications with accurate stent deployment may arise in cases of tortuous carotid arteries or when a there is a "kink." It is best to avoid a "kink" that lies immediately distal to the lesion. In these cases, placement of the stent proximal to the sharp turn in the ICA is optimal. Otherwise, the stent may accentuate the kink. In this scenario, it is also better to opt for an open-cell stent, which is more flexible, and avoid the more rigid closed-cell stent (Fig. 26.3) [5, 7, 8]. Another technical issue that can be encountered during stent deployment is that, occasionally, the stent can move proximally or distally ("watermelon seeding"), resulting in incomplete coverage of the lesion, which may require the placement of a second, overlapping stent [7]. Finally, upon deploying the stent, occasionally, there might be difficulty with the removal of the stent delivery system. This is more likely to develop if the stent is deployed across a severe stenosis without previous predilatation, if the stenosis is at a curve, and when the edges of the stenosis are abrupt, leading to buckling of portions of a stent cell into the lumen [7]. Simple maneuvers that may solve the problem include (1) changing the position of the head by turning it right or left or by flexing-extending the neck, (2) trying to resheath the distal end of the stent deployment catheter, or (3) carefully advancing the guiding catheter into the stent [7]. Careful fluoroscopic observation

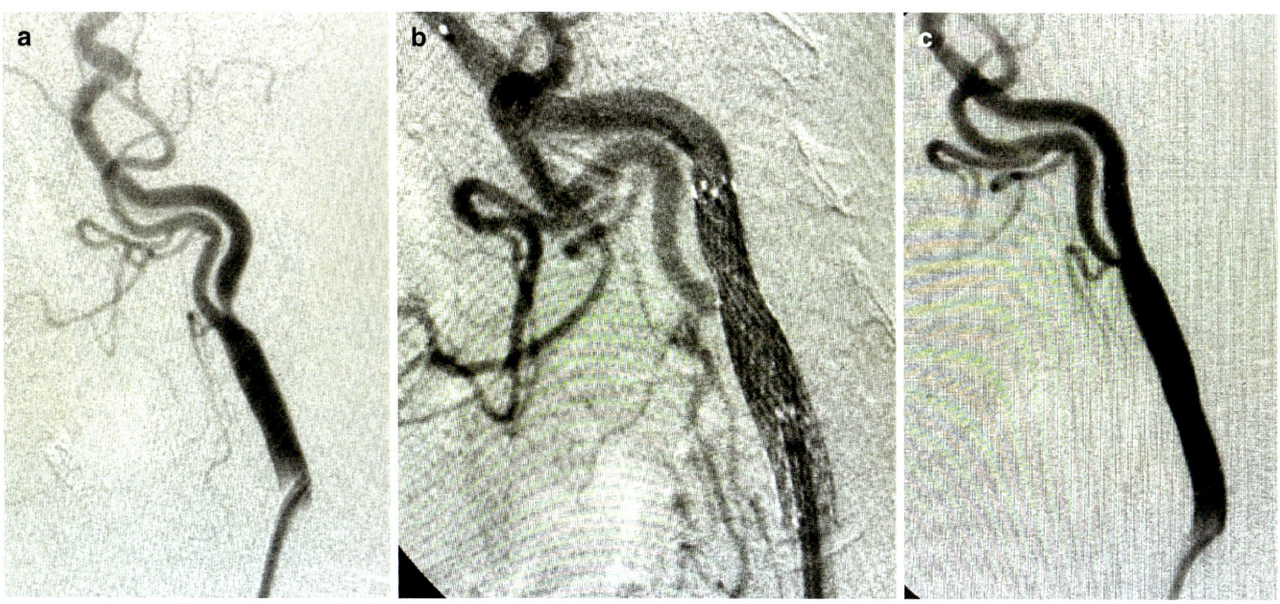

Fig. 26.3 Placement of the distal end stent into kinks and tortuosities of the ICA should be avoided. In this patient, there is a sharp turn in the ICA a few centimeters from the carotid bifurcation (**a**). To avoid exaggerating the sharp turn and creating a kink, the stent was deployed proximal to this bend (**b**, **c**). A flexible open-cell (Protégé RX carotid) stent was preferred over a rigid closed-cell stent for the very same reason

is recommended during these maneuvers to assure that the stent is not displaced or deformed. Finally, if all these fail, open surgical removal may be necessary (Fig. 26.4).

Retrieval of EPD

Occasionally, this phase can be the most challenging part of the entire intervention. When there has been massive distal embolization, the risk of losing the captured material during filter retrieval and causing a cerebral embolic event is significant. Therefore, following stent placement and postdilatation, the operator must routinely evaluate the EPD for trapped embolic material before it can be recaptured. In the majority of cases, the embolic load will be minimal (Fig. 26.5).

Most filters are removed using a dedicated retrieval catheter. Occasionally, navigating these retrieval catheters through a stent can be difficult, particularly when there are tortuous vessels or underexpansion of the proximal edge of the stent. Simple tricks, such as breath holding, neck turning, swallowing, or applying manual external compression to the stented area, may facilitate passage. Other possibilities are (1) advancing the guiding catheter/sheath over the filter wire until its tip is within the stent or even until it abuts the filter device, (2) additional balloon angioplasty, or (3) replacing the retrieval catheter with a 125 cm-long angulated 5F diagnostic catheter, such as a Headhunter or multipurpose catheter, and advancing its tip up to the filter [6, 7].

Once the EPD is recaptured, careful manipulation and removal under fluoroscopic observation are crucial. Entrapment of the filter within the stent struts and, sometimes, detachment of the filter can be due to abrupt manipulations. Maneuvers like turning the patient's head or asking the patient to cough or the Valsalva maneuver may solve the problem. Failing this, replacement of the retrieval catheter with an angled 5F catheter will allow the EPD to be retrieved in most cases. In the unlikely event of a filter fracture, "jailing" of the retained filter part against the arterial wall is a useful bailout option [21].

At this very late stage of the procedure, troublesome situations, such as vasospasm, dissection, or a full filter basket, may occasionally occur and can be difficult to differentiate between them. The radiopaque wall-opposing markers on filter elements allow differentiation between procedural flow arrest owing to flow-limiting spasm or distal ICA dissection (the markers are drawn together) and a full filter (the markers remain unchanged compared with initial postdeployment images). This is clearly an important distinction as the management differs depending on the underlying problem [22]. When vasospasm interferes with the removal of the distal EPD, the vasospasm usually improves following an intra-arterial injection of 50–200 μg of nitroglycerine. If there is an underlying dissection which has been induced by the EPD, placement of a second self-expanding stent may be necessary. Small, non-flow-limiting dissections may be left untreated and observed. Finally, a full filter with complete or partial occlusion of the ICA is a difficult problem. In such cases, an aspiration catheter can be used to clear enough material from the filter to allow safe removal through the guiding catheter. If the risk of removing the full filter percutaneously seems prohibitive, the final option may be open surgical removal.

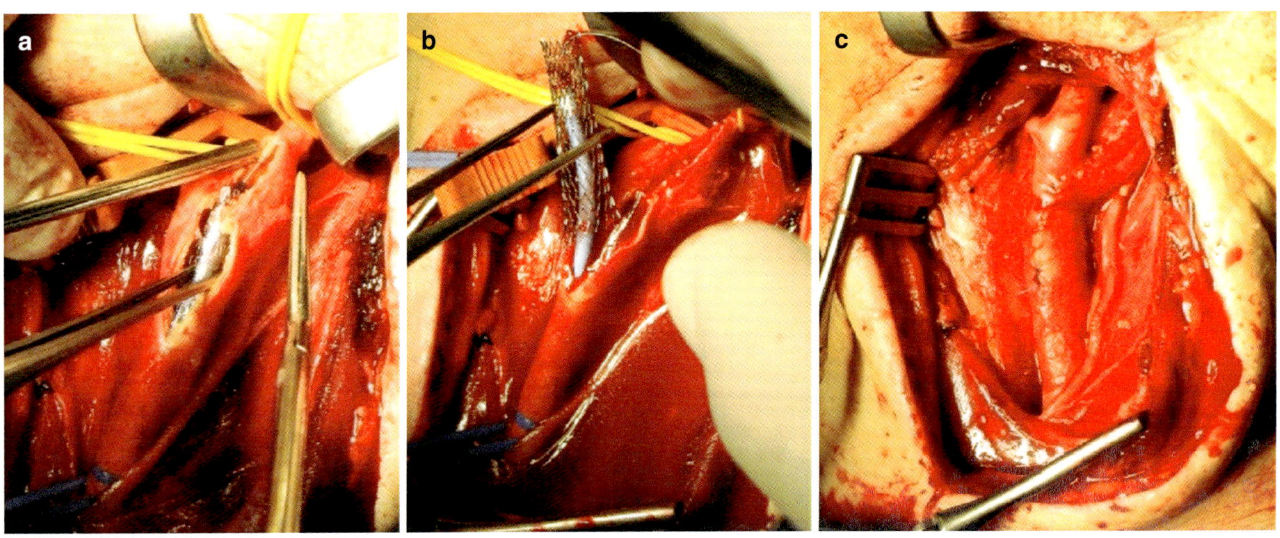

Fig. 26.4 Accidental proximal "watermelon seeding" of an Xact carotid stent with the upper edges of the stent being constrained by the tight, irregular carotid stenosis (**a**). The stent delivery system got stuck to the deployed stent, and despite adjunctive maneuvers, we were unable to remove it. Open conversion was necessary along with longitudinal carotid arteriotomy, removal of the stent and the filter (**b**), endarterectomy, and primary closure (**c**)

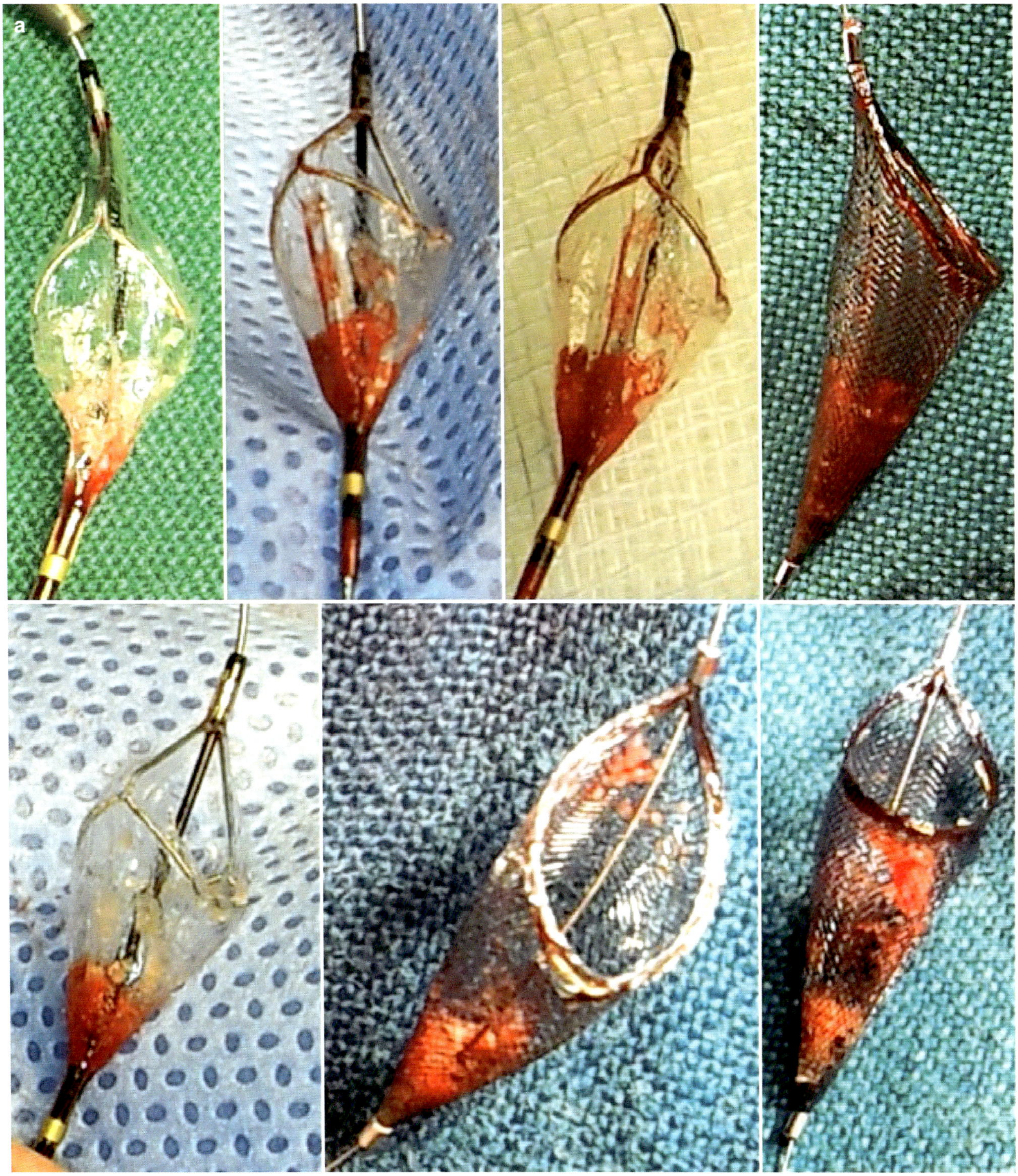

Fig. 26.5 Intraprocedural photographs of trapped embolic material in filters retrieved from several different patients undergoing filter-protected transfemoral CAS (**a**, **b**)

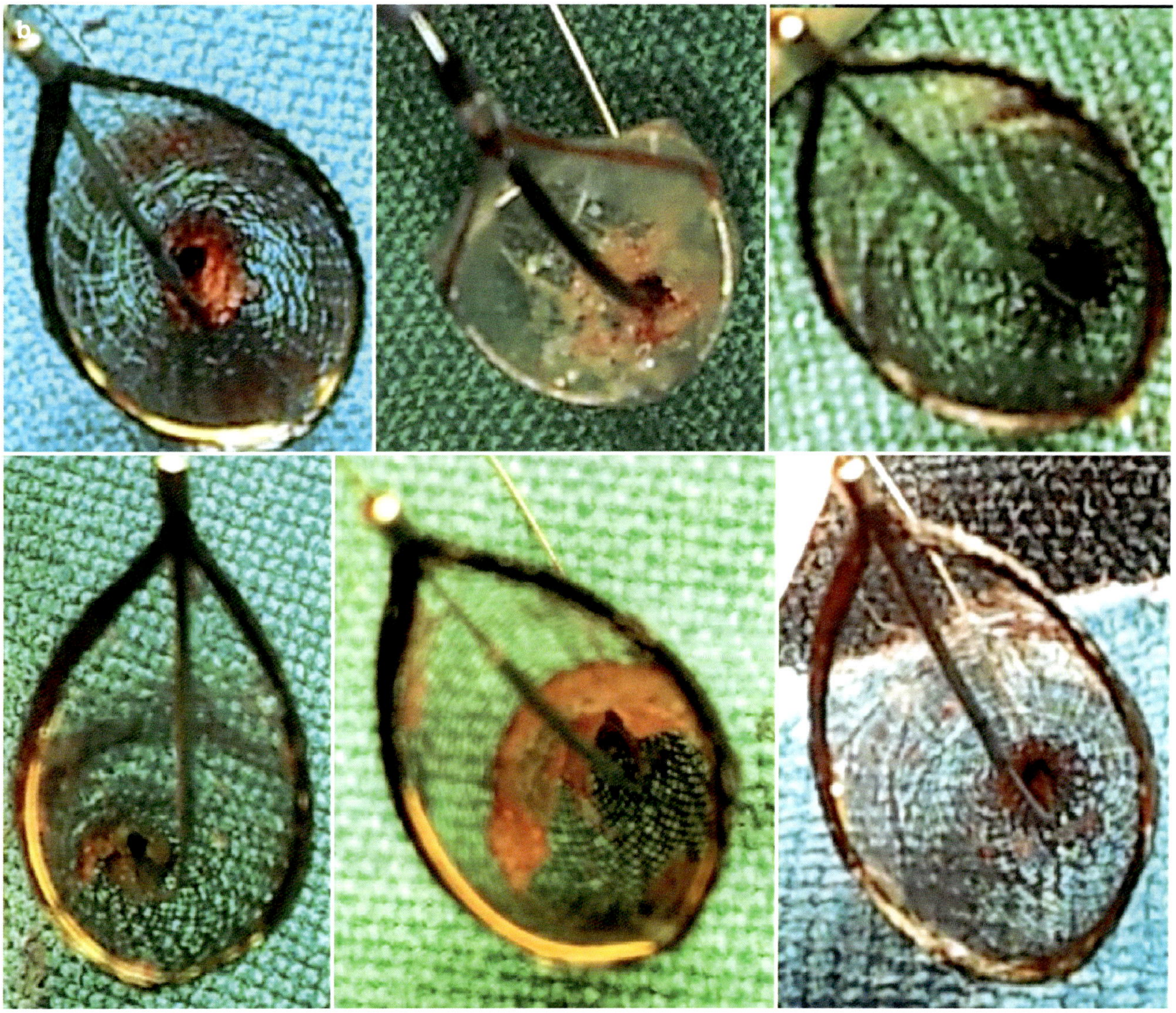

Fig. 26.5 (continued)

Other Post-procedural Complications

Adverse Hemodynamic Events (Bradycardia and Hypotension)

In clinical practice, bradycardia and hypotension are well-known physiological responses during CAS and in most cases are transient and self-limiting. Only occasionally these hemodynamic responses are severe enough to cause periprocedural cardiac or neurological adverse effects. While the bradycardia usually resolves within seconds, the hypotension may persist for hours or even days [7]. In the absence of symptoms in such patients, no special treatments are necessary unless systolic blood pressure drops below 80 mmHg. Withholding the patient's routine antihypertensive medications, administering intravenous fluids, and encouraging early mobilization are all that are needed to solve the problem and avoid prolonged hospitalization. If it is necessary to raise the blood pressure, a dopamine infusion is usually adequate. Certain high-risk patients, such as those with severe stenosis of both carotid arteries, contralateral carotid occlusion, severe aortic stenosis, left ventricular systolic dysfunction, or significant coronary artery disease, may tolerate less well a period of sustained hypotension, and in these cases, it may be advantageous to maintain a systolic blood pressure of >90 mmHg. Although hypotension usually resolves within 24 h, it is essential not to be reassured and miss other important causes of post-procedural hypotension, such as retroperitoneal hemorrhage or myocardial infarction [23].

Thromboembolic Events

Cerebrovascular ischemic complications due to thromboembolism may occur both during and after CAS and may jeopardize the success of the intervention. Faced with an alteration in the patient's neurological status, one must consider the differential diagnosis of hypoperfusion, embolism, hemorrhage, or hypotension. If an intraprocedural intracranial arterial occlusion is noted, it is advisable to seek the help of a neuro-interventionist who can perform "acute stroke" cerebrovascular interventions [7].

Cerebral Hyperperfusion Syndrome and Intracranial Hemorrhage

The cerebral hyperperfusion syndrome (CHS) post-CAS may clinically manifest as ipsilateral headache, nausea, vomiting, focal seizures, altered mental status, focal neurological deficit, and ipsilateral intracerebral edema or fatal intracranial hemorrhage (ICH). It is accompanied by a markedly elevated post-intervention blood pressure in the majority of patients [1, 7, 24].

The exact incidence of CHS varies in the literature, and the pooled rate in meta-analyses ranges from 1.16 to 4.6%, depending on the studies included, research design, and disease definition [25–27]. A recent Spanish national prospective multicenter study reported that CHS occurred in 2.9% of 757 patients after CAS [28]. With respect to the hyperperfusion-induced ICH, the incidence also differs from series to series, and representative figures from relatively recent publications are 0.85% in a Canadian population-based cohort study (16,688 patients, 14% CAS, province of Ontario, 2002–2015) [29], 0.7% in the Spanish national prospective multicenter study [28], and as high as 3.14% in a Chinese series (of 446 patients with symptomatic severe carotid stenosis, 2011–2018) [30]. The above differences, in both CHS and ICH rates, may be partially explained by the differences in the included patient populations [24].

Likely risk factors for CHS and hyperperfusion-induced ICH after CAS include:

- Older age
- Female gender
- Chronic kidney disease
- Longstanding preexisting hypertension
- Critical intracranial atherosclerotic stenosis ≥90%
- Severe contralateral intracranial atherosclerotic disease
- Left-sided carotid disease
- Progressive neurological deficit
- Frequent transient ischemic attack
- Prior stroke due to hemodynamic insufficiency
- Preexisting brain lesions
- Recurrent hemorrhage
- Reduced cerebral vascular reserve and microvascular disease
- Severe unilateral carotid stenosis
- The presence of near total occlusion
- Incomplete circle of Willis

Identification of patients at risk of developing CHS post-CAS is the first important step to preventing this complication. Further on, the management of CHS consists of careful monitoring, early recognition, and aggressive blood pressure control to prevent the situation from escalating into cerebral edema and hemorrhage. Ideally, such patients should be monitored in a high-dependency or intensive care unit. The prognosis following ICH is dismal, and the mortality exceeds 50%. In contrast, the prognosis is much better if patients with CHS are identified early post-CAS and treated prior to developing cerebral edema or ICH [1, 7, 24–30].

Postoperative Care and Surveillance

Most patients would be able to be discharged home on the first postoperative day. If hypotensive, we advise them to withhold antihypertensive agents until blood pressure comes up, which usually occurs 2–3 days later. Our preference is to continue a dual antiplatelet therapy for the first 6 months and, after that, to carry on with a single antiplatelet drug for life. Patients are followed up in the outpatient clinic in 6-monthly intervals with clinical assessment and carotid colorflow duplex scan.

Case Presentation

Continued from page 259

The patient clearly understood the benefits and risks of carotid intervention and opted for CAS as she considered this less invasive than endarterectomy. Clopidogrel 75 mg was added to her medications a week prior to the scheduled date of the procedure. Under local anesthesia and a transfemoral approach, a 7F-11 cm sheath was introduced, and 5000 units of heparin were given intravenously. A 7F-100 cm JR4 guiding catheter was advanced over a 0.035″ stiff hydrophilic guidewire to the aortic arch, and the innominate and the right CCA were selectively catheterized. After advancing the guidewire in the ECA, the catheter was positioned in the distal CCA proximal to the carotid lesion. The 0.035″ wire was removed, and the Emboshield NAV6 (Abbott Cardiovascular) filter was gently advanced through the lesion and deployed in the distal extracranial ICA. A 7 mm × 40 mm carotid Wallstent (Boston Scientific) was deployed, and this was postdilated with a 5 mm × 20 mm RX balloon. Atropine 1 mg was given prior to balloon inflation to avoid bradycardia. Post-stenting angiography was satisfactory, and the filter was retrieved using the dedicated retrieval system. Completion angiography, includ-

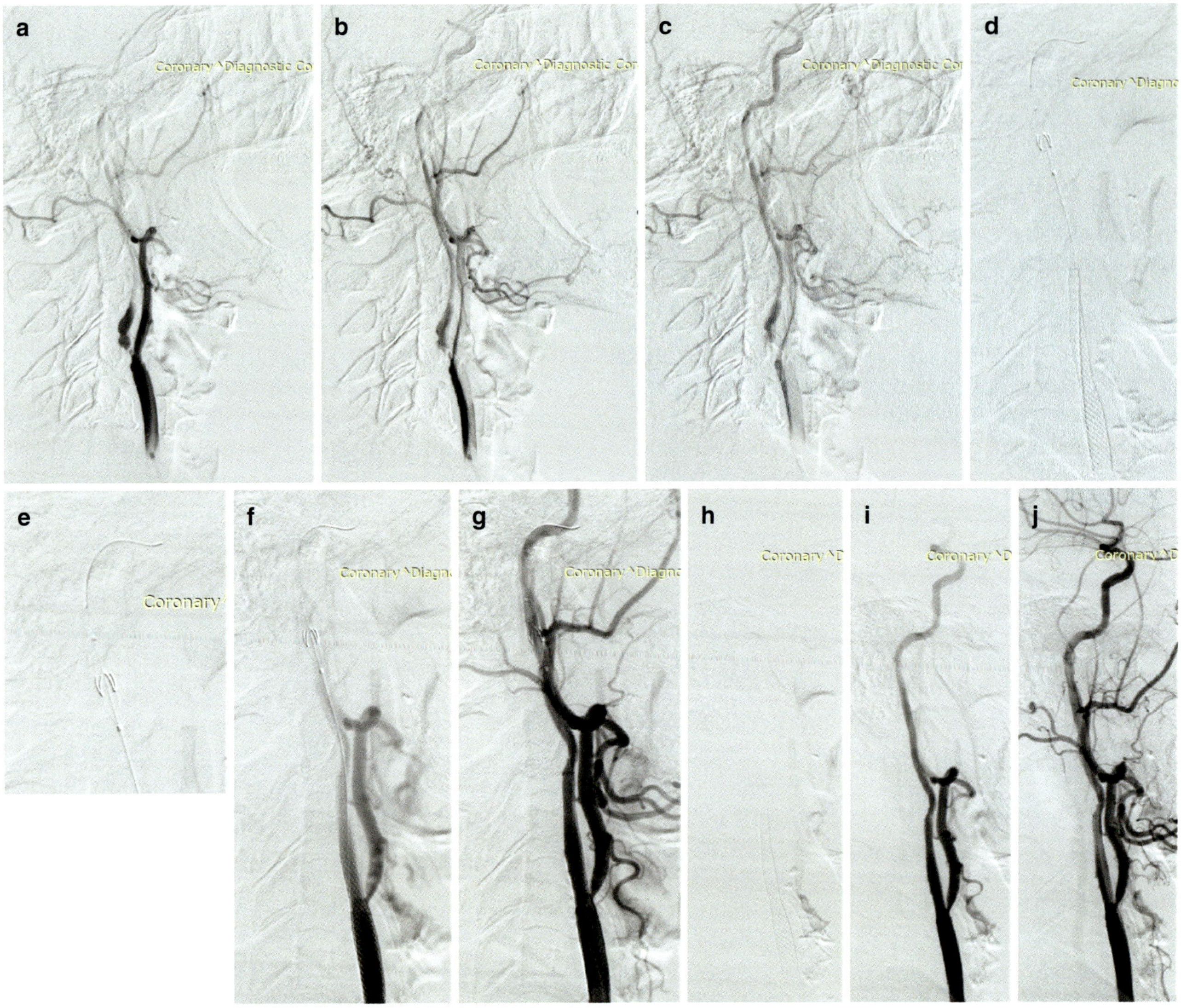

Fig. 26.6 Transfemoral filter-protected CAS. Selective right carotid DSA runs through the guiding catheter showed >90% stenosis at the origin of the right ICA (**a**–**c**). An Emboshield NAV6 filter was deployed in the distal ICA, and a 7 mm × 40 mm carotid Wallstent was implanted with satisfactory angiographic result prior to (**d**–**g**) and after filter retrieval (**h**–**j**)

ing intracranial views, showed a satisfactory result (Fig. 26.6). She had been asymptomatic with no neurological deficits throughout the entire procedure. At the end, the femoral sheath was removed, and puncture site hemostasis was achieved with manual compression. The patient was transferred to the high-dependency unit for overnight monitoring. Post-procedure, she developed persistent hypotension (systolic blood pressure of 90 mmHg). As she was asymptomatic, this required no specific action other than adequate intravenous hydration and withholding the antihypertensive agents. She had an otherwise uneventful recovery and was discharged home the following day. She was advised to continue dual antiplatelet therapy for 6 months and, following this, to carry on with a single antiplatelet for life.

References

1. Naylor AR, Rantner B, Ancetti S, de Borst GJ, De Carlo M, Halliday A, et al. European society for vascular surgery (ESVS) 2023 clinical practice guidelines on the management of atherosclerotic carotid and vertebral artery disease. Eur J Vasc Endovasc Surg. 2022;S1078-5884(22):00237–4.
2. Bonati LH, Kakkos S, Berkefeld J, et al. European stroke organisation guideline on endarterectomy and stenting for carotid artery stenosis. Eur Stroke J. 2021;6:I–XLVII.
3. AbuRahma AF, Avgerinos ED, Chang RW, Darling RC 3rd, Duncan AA, Forbes TL, Malas MB, Murad MH, Perler BA, Powell RJ, Rockman CB, Zhou W. Society for Vascular Surgery clinical practice guidelines for management of extracranial cerebrovascular disease. J Vasc Surg. 2022;75(1S):4S–22S.
4. Müller MD, Lyrer P, Brown MM, Bonati LH. Carotid artery stenting versus endarterectomy for treatment of carotid artery stenosis. Cochrane Database Syst Rev. 2020;2(2):CD000515.
5. Ahn SH, Prince EA, Dubel GJ. Carotid artery stenting: review of technique and update of recent literature. Semin Intervent Radiol. 2013;30:288–96.
6. Malik RK, Vouyouka A, Salloum A, Marin ML, Faries PL. Tips and techniques in carotid artery stenting. J Vasc Surg. 2009;50:216–20.
7. Tummala R. Carotid angioplasty and stenting: complication avoidance and management. In: Khatri R, Rodriguez G, Raymond J, Qureshi A, editors. Complications of neuroendovascular procedures and bailout techniques. Cambridge: Cambridge University Press; 2016. p. 84–104.
8. Iyer SS, Al-Mubarack N, Vitek JJ, Roubin GS. Procedural techniques. In: Al-Mubarack N, editor. Carotid artery stenting: current practice and techniques. Lippincott Williams & Wilkins; 2004. p. 105–23.
9. Solomon Y, Varkevisser RRB, Swerdlow NJ, Li C, Liang P, Siracuse JJ, de Borst GJ, Schermerhorn ML. Outcomes after transfemoral carotid artery stenting stratified by preprocedural symptom status. J Vasc Surg. 2021;73:2021–9.
10. Howard VJ, Lutsep HL, Mackey A, Demaerschalk BM, Sam AD 2nd, Gonzales NR, Sheffet AJ, Voeks JH, Meschia JF, Brott TG, CREST investigators. Influence of sex on outcomes of stenting versus endarterectomy: a subgroup analysis of the carotid revascularization endarterectomy versus stenting trial (CREST). Lancet Neurol. 2011;10:530–7.
11. Sfyroeras G, Karkos CD, Liasidis C, Spyridis C, Dimitriadis AS, Kouskouras K, Gerassimidis TS. The impact of carotid stenting on the hemodynamic parameters and cerebrovascular reactivity of the ipsilateral middle cerebral artery. J Vasc Surg. 2006;44:1016–22.
12. Naggara O, Touzé E, Beyssen B, Trinquart, Chatellier G, Meder JF, Mas JL. Anatomical and technical factors associated with stroke or death during carotid angioplasty and stenting: results from the endarterectomy versus angioplasty in patients with symptomatic severe carotid stenosis (EVA-3S) trial and systematic review. Stroke. 2011;42:380–8.
13. Werner M, Bausback Y, Bräunlich S, Ulrich M, Piorkowski M, Friedenberger J, Schuster J, Botsios S, Scheinert D, Schmidt A. Anatomic variables contributing to a higher periprocedural incidence of stroke and TIA in carotid artery stenting: single center experience of 833 consecutive cases. Catheter Cardiovasc Interv. 2012;80:321–8.
14. Fanous AA, Natarajan SK, Jowdy PK, Dumont TM, Mokin M, Yu J, Goldstein A, Wach MM, Budny JL, Hopkins LN, Snyder KV, Siddiqui AH, Levy EI. High-risk factors in symptomatic patients undergoing carotid artery stenting with distal protection: buffalo risk assessment scale (BRASS). Neurosurgery. 2015;77:531–42.
15. Scullen T, Mathkour M, Carr C, Wang A, Amenta PS, Nerva JD, Dumont AS. Anatomical considerations for endovascular intervention for extracranial carotid disease: a review of the literature and recommended guidelines. J Clin Med. 2020;9:3460.
16. Zhang L, Dai D, Li Z, Duan G, Zhang YW, Yang P, Huang Q, Xu Y, Hong B, Liu J. Risk factors for hyperperfusion-induced intracranial hemorrhage after carotid artery stenting in patients with symptomatic severe carotid stenosis evaluation. J Neurointerv Surg. 2019;11:474–8.
17. Biasi GM, Froio A, Diethrich EB, et al. Carotid plaque echolucency increases the risk of stroke in carotid stenting: the imaging in carotid angioplasty and risk of stroke (ICAROS) study. Circulation. 2004;110:756–62.
18. Barr JD, Connors JJ 3rd, Sacks D, Wojak JC, Becker GJ, Cardella JF, Chopko B, Dion JE, Fox AJ, Higashida RT, Hurst RW, Lewis CA, Matalon TA, Nesbit GM, Pollock JA, Russell EJ, Seidenwurm DJ, Wallace RC, American Society of Interventional and Therapeutic Neuroradiology; American Society of Neuroradiology; Society of Interventional Radiology. Quality improvement guidelines for the performance of cervical carotid angioplasty and stent placement. J Vasc Interv Radiol. 2003;14(9 Pt 2):S321–35.
19. Aronow HD, Collins TJ, Gray WA, Jaff MR, Kluck BW, Patel RA, Rosenfield KA, Safian RD, Sobieszczyk PS, Wayangankar SA, White CJ. SCAI/SVM expert consensus statement on carotid stenting: training and credentialing for carotid stenting. Catheter Cardiovasc Interv. 2016;87:188–99.
20. Verzini F, Cao P, De Rango P, Parlani G, Maselli A, Romano L, Norgiolini L, Giordano G. Appropriateness of learning curve for carotid artery stenting: an analysis of periprocedural complications. J Vasc Surg. 2006;44:1205–11.
21. Lian X, Liu W, Li M, et al. Risk factors and complications associated with difficult retrieval of embolic protection devices in carotid artery stenting. Cardiovasc Intervent Radiol. 2012;35:43–8.
22. Macdonald S. Role of filter design in embolic protection during carotid artery stenting. Interv Cardiol. 2009;1(2):185–95. https://www.openaccessjournals.com/articles/role-of-filter-design-in-embolic-protection-during-carotid-artery-stenting.html.
23. Al-Mubarack N, Iyer SS, Roubin GS, Vitek JJ. Procedural complications. In: Al-Mubarack N, editor. Carotid artery stenting: current practice and techniques. Lippincott Williams & Wilkins; 2004. p. 137–52.
24. Lin YH, Liu HM. Update on cerebral hyperperfusion syndrome. J Neurointerv Surg. 2020;12:788–93.
25. Moulakakis KG, Mylonas SN, Sfyroeras GS, et al. Hyperperfusion syndrome after carotid revascularization. J Vasc Surg. 2009;49:1060–8.
26. Huibers AE, Westerink J, de Vries EE, et al. Editor's choice—cerebral hyperperfusion syndrome after carotid artery stenting: a systematic review and meta-analysis. Eur J Vasc Endovasc Surg. 2018;56:322–33.
27. Galyfos G, Sianou A, Filis K. Cerebral hyperperfusion syndrome and intracranial hemorrhage after carotid endarterectomy or carotid stenting: a meta-analysis. J Neurol Sci. 2017;381:74–82.
28. González García A, Moniche F, Escudero-Martínez I, et al. Clinical predictors of hyperperfusion syndrome following carotid stenting: results from a national prospective multicenter study. JACC Cardiovasc Interv. 2019;12:873–82.
29. Hussain MA, Alali AS, Mamdani M, et al. Risk of intracranial hemorrhage after carotid artery stenting versus endarterectomy: a population-based study. J Neurosurg. 2018;129:1522–9.
30. Xu S, Wu P, Zhang T, Ji Z, Wang C, Shi H. Prevalence and clinical predictors of intracranial hemorrhage following carotid artery stenting for symptomatic severe carotid stenosis. World Neurosurg. 2021;155:e353–61.

27 Introduction to Transcarotid Artery Revascularization

Angela A. Kokkosis

Case Presentation

An 80-year-old female presented to the office with an episode of right-sided amaurosis fugax. The patient stated that this occurred 2 weeks ago for the first time, and for about 1 min she could not see through her right eye. She went to her ophthalmologist for evaluation, and upon further examination he referred her for a vascular consult. Her past medical history was remarkable for hypertension, type 2 diabetes, and hyperlipidemia. She denied any prior history of stroke or transient ischemic attack.

On physical exam she had palpable upper and lower extremity pulses, and no neurologic deficits. Duplex exam demonstrated a peak diastolic velocity/end diastolic velocity (PSV/EDV) of 627/227 cm/s in the mid-right internal carotid artery (Fig. 27.1), and CTA confirmed a 90% stenosis of the mid-right internal carotid artery at the level of C2. Her left

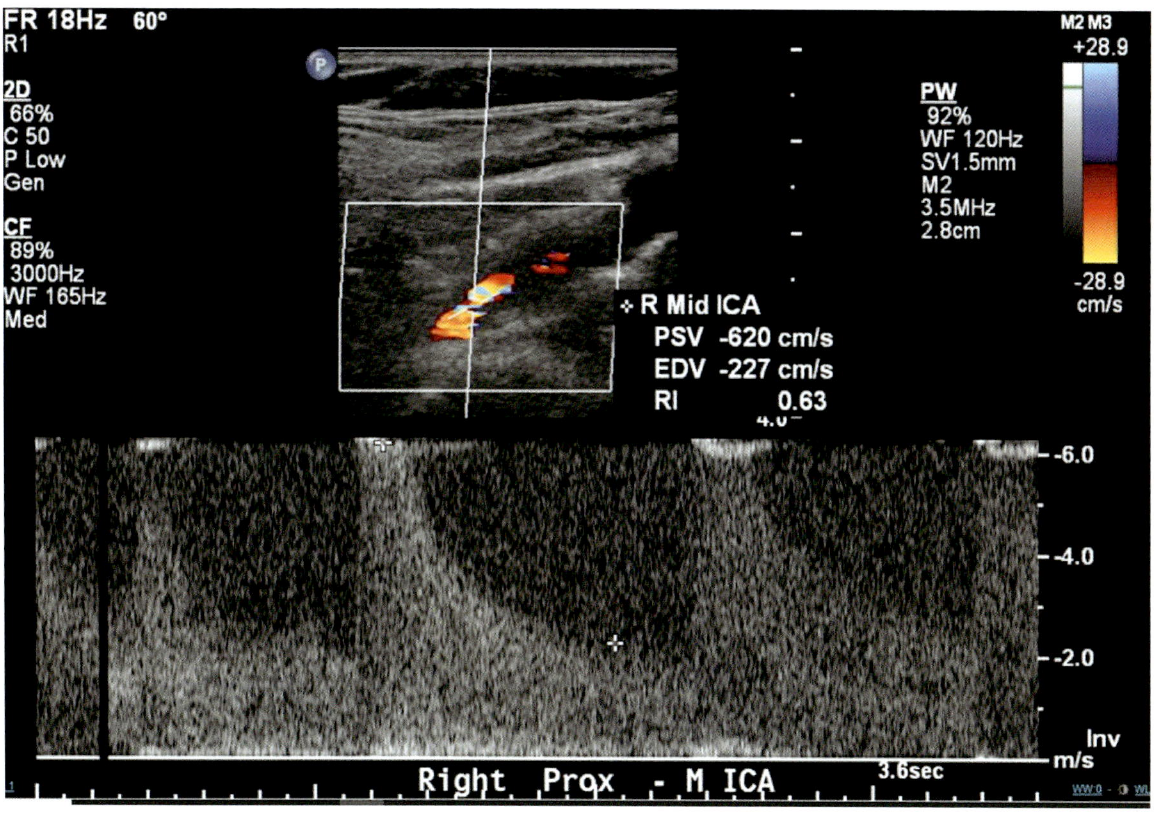

Fig. 27.1 Right carotid artery duplex demonstrating the 70–99% stenotic lesion

A. A. Kokkosis (✉)
Vascular Surgery, Stony Brook University Medical Center, Stony Brook, NY, USA
e-mail: Angela.kokkosis@stonybrookmedicine.edu

G. Geroulakos et al. (eds.), *Mastering Endovascular Techniques*, https://doi.org/10.1007/978-3-031-42735-0_27

carotid artery and bilateral vertebral arteries were patent without significant disease.

Surgical options for carotid revascularization were discussed, including carotid endarterectomy, transfemoral carotid artery stenting, and transcarotid artery revascularization.

Continued at page 278

Background

For many years the gold standard for carotid revascularization has been carotid endarterectomy (CEA). With the advent of endovascular interventions, transfemoral carotid artery stenting (TF-CAS) came to light. However, after extensive research and numerous trials, perioperative stroke risk could not equal or surpass that of CEA. Several features of TF-CAS were thought to contribute to this elevated perioperative stroke risk, such as aortic arch manipulation with wires and catheters, need for crossing the lesion to deploy an embolic protection device, and inadequate particle capture of the distal embolic protection devices [1, 2]. In 2004, a new concept for carotid stenting was introduced to avoid aortic arch navigation and crossing the lesion prior to employing embolic protection—delivery of the stent via a transcervical approach and with flow reversal as the embolic protection [3, 4]. After the pivotal ROADSTER trial, which demonstrated the lowest 30-day stroke and combined stroke/death rates for transcarotid artery revascularization (TCAR) [5], this concept became an FDA-approved intervention in the United States in 2016. Currently, TCAR is commercially available as a carotid revascularization option for both high-risk and standard-risk patients.

Pre-operative Considerations

Suitable Patients for TCAR

Patients being considered for TCAR must meet specific anatomic criteria in order for the procedure to be performed safely and successfully. They must have a common carotid artery that measures at least 6 mm, a length from the access site to the lesion of at least 5 cm, and an access site at the common carotid artery that is free of significant atherosclerotic disease.

Lesion characteristics should also be carefully evaluated for successful stent deployment. Severe calcification, vessel tortuosity, acute thrombus, mobile plaque, long-segment disease, and carotid artery diameters outside the range of commercially available stents may not be good candidates for TCAR [6].

Medications

As with any patient undergoing carotid stent placement, dual antiplatelet therapy is essential during the perioperative period and for at least 1 month post-procedure. Typically this consists of aspirin and clopidogrel. However, in instances of clopidogrel resistance, other antiplatelet agents may be considered. Statins should also be part of the patient's regimen as they have been shown to promote plaque stabilization.

Anticoagulation is not a necessary maintenance medication for carotid stents; however, many patients undergoing TCAR may be on anticoagulation for a cardiac or hypercoagulable indication. In these situations, careful discussion should be undertaken with the patient and their specialists to determine the best way to manage the dual antiplatelets and anticoagulation.

Imaging

Both duplex ultrasound and computed tomography angiography (CTA) or magnetic resonance angiography (MRA) are recommended in patients being considered for TCAR. Duplex ultrasound is valuable for its ability to provide dynamic information on flow, velocities and plaque characteristics, and to assess the potential access site at the common carotid artery. Axial imaging of the head and neck via CTA or MRA demonstrates the anatomy of the lesion, exact location of the lesion, and any potential kinks or turns in the artery which may affect stent placement, and it offers the ability to measure artery diameters, and intracranial circulation and assessment of the circle of Willis. CTA is preferable to MRA, especially in instances of calcified lesions, since MRA shows signal dropout in areas of calcium. However, CTA may overestimate calcium burden. Additionally, MRA offers the ability to obtain non-contrast flow imaging ("time of flight") for those patients who have a contraindication to contrast use.

Anesthetic Technique

TCAR is a less invasive carotid revascularization approach, and thus it offers the possibility of performing the intervention under general anesthesia, regional nerve block, or local anesthesia with monitored sedation. Currently, there is no data to suggest that one technique is superior to the other, and thus careful discussion with the patient as well as consideration of available anesthetic capabilities at one's institution are important. For those vascular specialists who are early in their TCAR experience, general anesthesia may be preferable for the first few cases.

Intraoperative Considerations

Patient Positioning and Draping

The patient should be placed in the supine position on the operating table with the head turned to the contralateral side. The neck and chest should be prepped and draped, as with a standard CEA, as well as bilateral groins for venous access. Consider a shoulder roll to facilitate further extension of the neck; however, this is not always necessary since the carotid exposure site is low on the neck.

Endovascular Devices

The Silk Road ENROUTE® system is the only commercially available device for TCAR. This system includes the neuroprotection system, the stent device, the access kit, and the 0.014″ guidewire (Fig. 27.2).

Other endovascular equipment that should be available includes:

- 0.014″ monorail platform angioplasty balloon (sizes 4–6 mm × 20 or 30 mm)

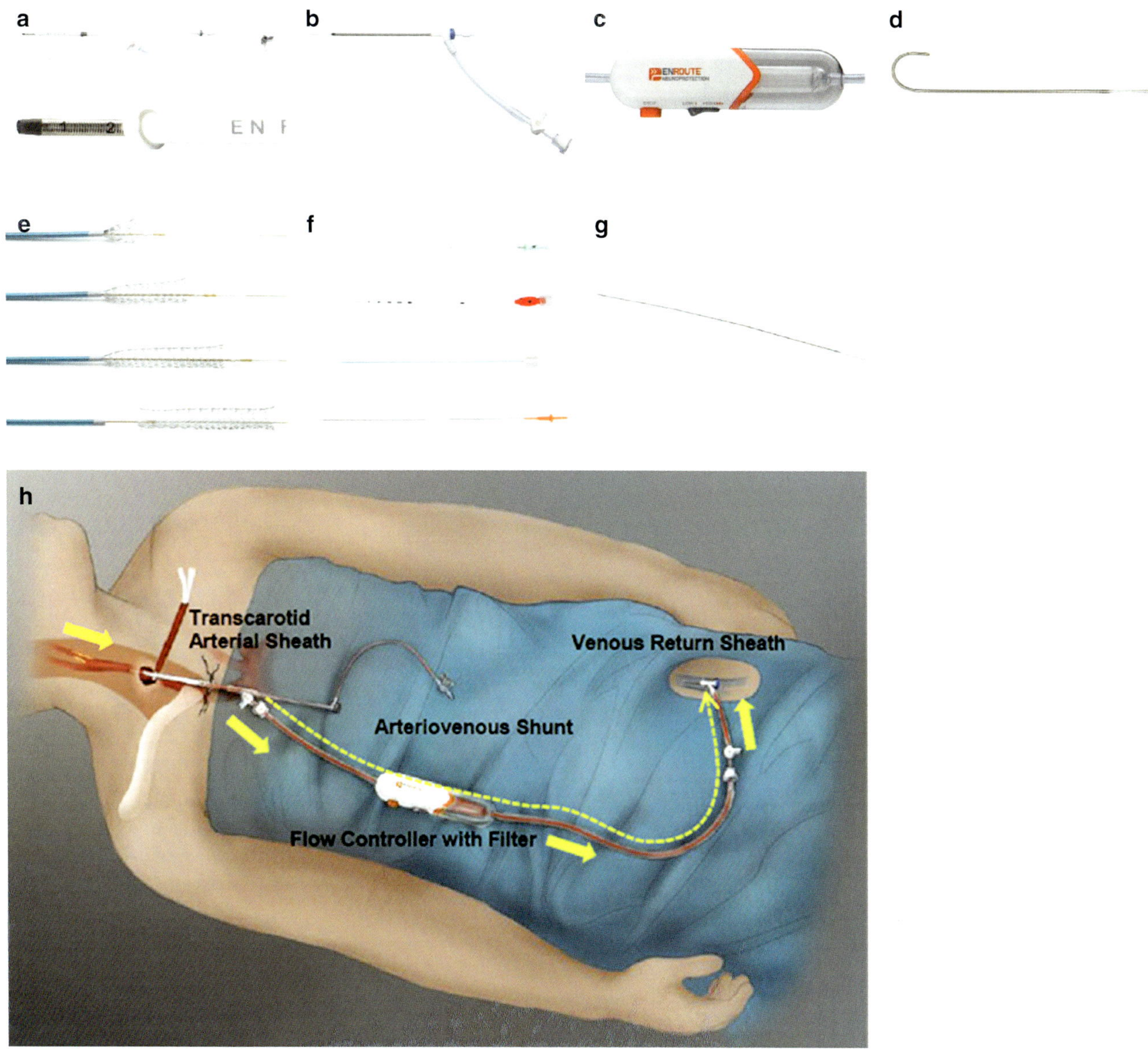

Fig. 27.2 Silk Road ENROUTE® Devices. (**a**) Arterial sheath. (**b**) Venous sheath. (**c**) Flow reversal controller. (**d**) 035″ J-wire. (**e**) ENROUTE® Stent system. (**f**) Micropuncture access set. (**g**) 014" Wire. (**h**) Full assembly of the ENROUTE system

- Angioplasty balloon insufflator
- Directional catheter
- Additional 0.014″ wire
- 10 cc syringes for injection
- Separate micropuncture access kit for femoral vein access
- 0.035″ wire for venous sheath placement

Open Surgical Instruments

Standard instruments used for CEA are more than enough for common carotid artery exposure and vessel control.

Duplex Ultrasound

With the patient on the surgical table and in the operative position, duplex ultrasound should be performed to mark the desired access site, the location of lesion, and the borders of the sternocleidomastoid. Additionally, the ultrasound should be draped in a sterile manner for later use in the venous access. In rare situations, a critical carotid stenosis may occlude asymptomatically prior to the intervention, and thus carotid duplex with color flow assessment prior to incision may be prudent.

TCAR Technique

Incision Choice

Either a vertical or transverse incision may be employed for common carotid artery exposure. The vertical incision affords the ability to extend more distally in the rare event of conversion to carotid endarterectomy, while the transverse incision tends to be more cosmetic. Regardless of the incision choice, the location of the incision should be within the borders of the superior edge of the clavicle, the lateral border of the sternal head of the sternocleidomastoid, and the medial border of the clavicular head (Fig. 27.3).

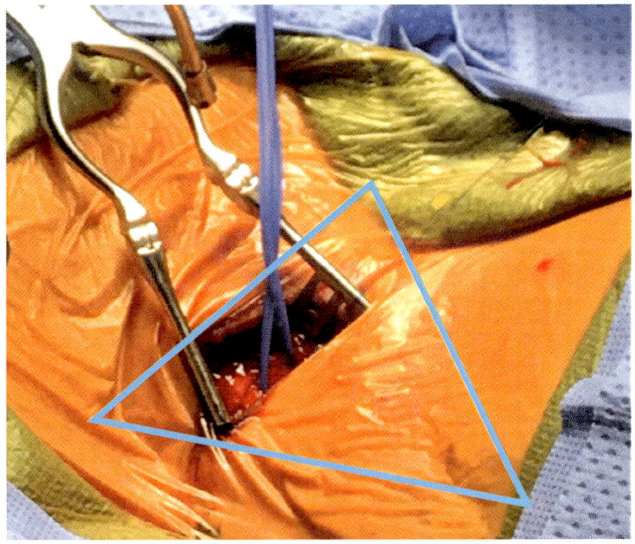

Fig. 27.3 Example of transverse incision within the triangle of the sternocleidomastoid muscle bellies and the clavicle

Stent and Balloon Selection

Stent sizing is typically 1–2 mm larger than the common carotid artery diameter when the ENROUTE stent is used, and pre-dilatation balloon sizing should be based on the normal distal internal carotid artery.

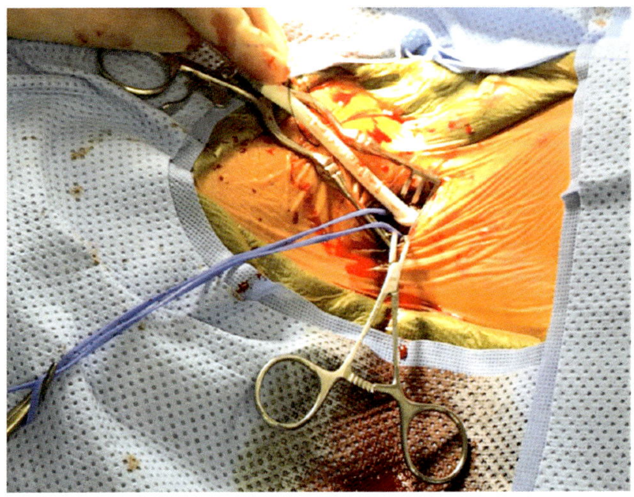

Fig. 27.4 Example of using a Gregory profunda clamp

Vessel Clamping Choice

Control of the common carotid artery is usually obtained with a silastic vessel loop or umbilical tape. The actual clamping of the vessel may be achieved either with a vascular clamp (such as a Gregory profunda or PVD) or with doubling of the vessel loop (Fig. 27.4).

Stop Short or Engage External Carotid Artery

Upon initial access of the common carotid artery, the decision of whether to advance the 0.035″ J-wire into the external carotid artery (ECA) for more support or just before the lesion must be made based on the initial angiogram from the micropuncture sheath.

If the origin of the external carotid artery is diseased, or if the carotid lesion extends proximally to the carotid bifurcation, the J-wire should be maintained below this area during arterial sheath placement, to avoid inadvertently crossing the lesion without embolic protection.

If the external carotid artery is free of disease, the micropuncture wire and sheath should be advanced into this artery in order to gain more wire length and support for arterial sheath placement.

Potential Complications Specific to TCAR

Access Site Dissection

One of the most crucial steps in TCAR is obtaining safe carotid access. The short distances and small working space may at first seem awkward for those accustomed to endovascular procedures with long devices and contralateral vessel access. Therefore, careful needle placement and passage of the microwire and sheath are paramount. If at any point it is felt that there is poor blood return with the needle access, or upon micro-sheath access, it is safer to remove the needle or sheath and close the puncture site with a prolene stitch. Trying to proceed with arterial sheath placement may result in dissection in these cases.

In the event that a dissection does occur, frequently these can be salvaged by endovascular means by either accessing the vessel more proximally, before the dissection entry point, or navigating a wire into the true lumen from the same access. Once this is achieved, the carotid stent can be extended proximally to cover the dissection. In the rare situation of severe carotid dissection where endovascular techniques fail, conversion to open carotid intervention may be necessary.

Pneumothorax

The location of the common carotid access is fairly proximal, and in patients with a history of pulmonary disease, the lung apex may be directly beneath the common carotid artery. This anatomy can be identified on pre-operative CT scan, and in general care should be taken not to dissect beyond the posterior wall of the common carotid artery.

Incorrect Stent Placement

Careful evaluation of the angiographic images obtained intraoperatively should be performed prior to stent deployment. Branches such as the ascending pharyngeal may mimic the course of the internal carotid artery and be inadvertently stented.

Nerve Injury

The vagus and recurrent laryngeal nerves are most commonly encountered in the area of the common carotid dissection, and thus should be protected.

Access Site Hematoma

TCAR requires dual antiplatelets, as well as intraoperative heparinization, both of which increase the risk of post-operative hematoma in the setting of an incision. Meticulous hemostasis as well as reversal of heparin with protamine at case completion are strongly recommended.

Perioperative Care

A-Line Monitoring

Patients undergoing carotid artery revascularization are known to have acute post-operative changes in their blood pressure. Both hypotension and hypertension can lead to severe complications: stent thrombosis or watershed infarct, and hyperperfusion syndrome, respectively. An arterial line should be placed intraoperatively and maintained post-operatively for 4–6 h, or longer, depending on the patient's blood pressure stability. Both hypotension and hypertension should be treated proactively to prevent potential complications.

Discharge Instructions

The patient should be educated upon discharge about the signs/symptoms of stroke, hyperperfusion syndrome, and signs/symptoms of neck hematoma or swelling.

Additionally, it is important to emphasize the continuation of dual antiplatelet and statin therapy for at least 1 month post-operatively. Further discussion on any changes to the regimen should be conducted between the physician and patient.

Surveillance

The Society for Vascular Surgery published a set of guidelines for surveillance after carotid revascularization. The recommendation is for patients to undergo a baseline duplex ultrasound after the intervention, followed by duplex ultrasound every 6 months for 2 years, and then annually thereafter, as long as there is no significant restenosis or progression of disease. Additionally, there are specific carotid duplex criteria for in-stent restenosis [7]:

- ≥50%, PSV 220 cm/s, ICA/CCA ≥ 2.7
- ≥80%, PSV 340 cm/s, ICA/CCA ≥4.15

Case Presentation

Continued from page 274

After consideration of the patient's surgically inaccessible carotid lesion and symptomatic nature, TCAR was offered for carotid revascularization. The patient had been previously maintained on aspirin and a statin for her underlying comorbidities, so she was also started on clopidogrel pre-operatively. Due to concern for possible clopidogrel resistance, she was also sent for a blood test to assess clopidogrel response, 1 week after initiating the therapy. Fortunately, she was determined to respond appropriately to the clopidogrel.

The patient was taken to the operating room and placed in the supine position on the operating table with the head turned to the left side. The decision was made to perform the procedure under local anesthetic and monitored sedation. A squeeze toy was placed in her left hand, to assess her neurologic status intraoperatively. Her right neck, chest, and bilateral groins were prepped and draped in the usual sterile fashion. Duplex ultrasound was used to mark the common carotid artery access site within the triangle of the sternocleidomastoid (SCM) muscle bellies and clavicle. Additionally, the carotid bifurcation and lesion were visualized, and color flow was noted to pass through the stenotic area. The ultrasound was then used to evaluate the contralateral common femoral vein (left) for patency, and this was determined to be adequate for access.

A transverse 5 cm incision was made over the proximal common carotid artery, about 1 fingerbreadth above the clavicle. This was deepened with cautery through the platysma. Between the SCM muscle bellies, an avascular fatty plane was identified, and this was dissected carefully with metsembaum scissors. The carotid sheath was identified. The common carotid artery was circumferentially dissected for a distance of 3 cm. The vagus nerve was visualized posterior to the carotid artery and this was protected. A silastic vessel loop was encircled at the proximal end of the carotid artery. A 5-0 prolene Z-stitch was placed in the center of the 3 cm exposure, in the location of the access. The patient was systemically heparinized to an activated clotting time (ACT) of 250 s and was given glycopyrrolate in anticipation of a baroreceptor response to angioplasty. Left common femoral vein access was then obtained with a micropuncture set, and an 035″ wire was advanced into the IVC. The venous sheath was then advanced over the wire, and secured to the skin with a silk stitch.

Next, common carotid artery access was obtained with the ENROUTE access kit. The needle was advanced through the pre-placed prolene stitch, and when good backbleeding was observed, the microwire was advanced carefully into the common carotid artery for a distance of 5 cm, noted by the silver band on the wire. The micropuncture sheath was advanced 2–3 cm into the common carotid artery, and initial contrast angiogram was performed. This demonstrated the carotid bifurcation with a patent external carotid artery, and the carotid lesion was located at the proximal-mid internal carotid artery. The decision was made to engage the ECA. The microwire was reinserted and navigated into the ECA, and the micropuncture sheath was advanced into the ECA. The microwire was replaced with the 035″ J-wire and after removal of the microsheath, the arterial sheath was advanced over the wire and sutured to the skin. The wire and dilator were removed and the sheath's two ports were backbled. Next, two orthogonal view angiograms were performed to ensure that the sheath tip was not abutting the vessel wall (thus increasing risk for dissection) (Fig. 27.5a). The ENROUTE flow controller was connected from the arterial sheath to the venous sheath and set on HIGH flow. At this point, prior to clamping of the carotid artery, angioplasty balloon and stent availability, a target systolic blood pressure of 140–160 mmHg, a target heart rate of 70, and target ACT of 250 were confirmed. The carotid artery was clamped, and thus flow reversal was now employed. The patient's neuro status was assessed by asking the patient to squeeze the toy. A 5 × 30 mm balloon had been selected based on the CT measurement of the normal distal ICA at 5 mm, and a 10 × 40 mm ENROUTE stent was selected based on a common carotid artery measuring 8 mm.

The lesion was traversed with the 014" ENROUTE wire, and the balloon was advanced to the lesion. Angioplasty was performed (Fig. 27.5b). Again the patient's neuro status was assessed and noted to be intact. Next, the ENROUTE stent was advanced to the lesion and deployed (Fig. 27.5c). After waiting 2 min for washout of any particles through the flow reversal during the lesion manipulation, completion angiogram was performed (Fig. 27.5d). This demonstrated good stent expansion with no residual stenosis, and in general any residual stenosis <30% is acceptable. The carotid artery was unclamped, the flow reversal system was disconnected, and the arterial sheath was removed with the prolene stitch tied for hemostasis. The venous sheath was then removed and manual pressure was held. The heparin was reversed with protamine, and the neck incision was closed in standard fashion. The patient was taken to the recovery room with continuous arterial line monitoring. After an uneventful post-operative course, she was discharged the next day and instructed to maintain the dual antiplatelet and statin therapy, and was counseled on signs/

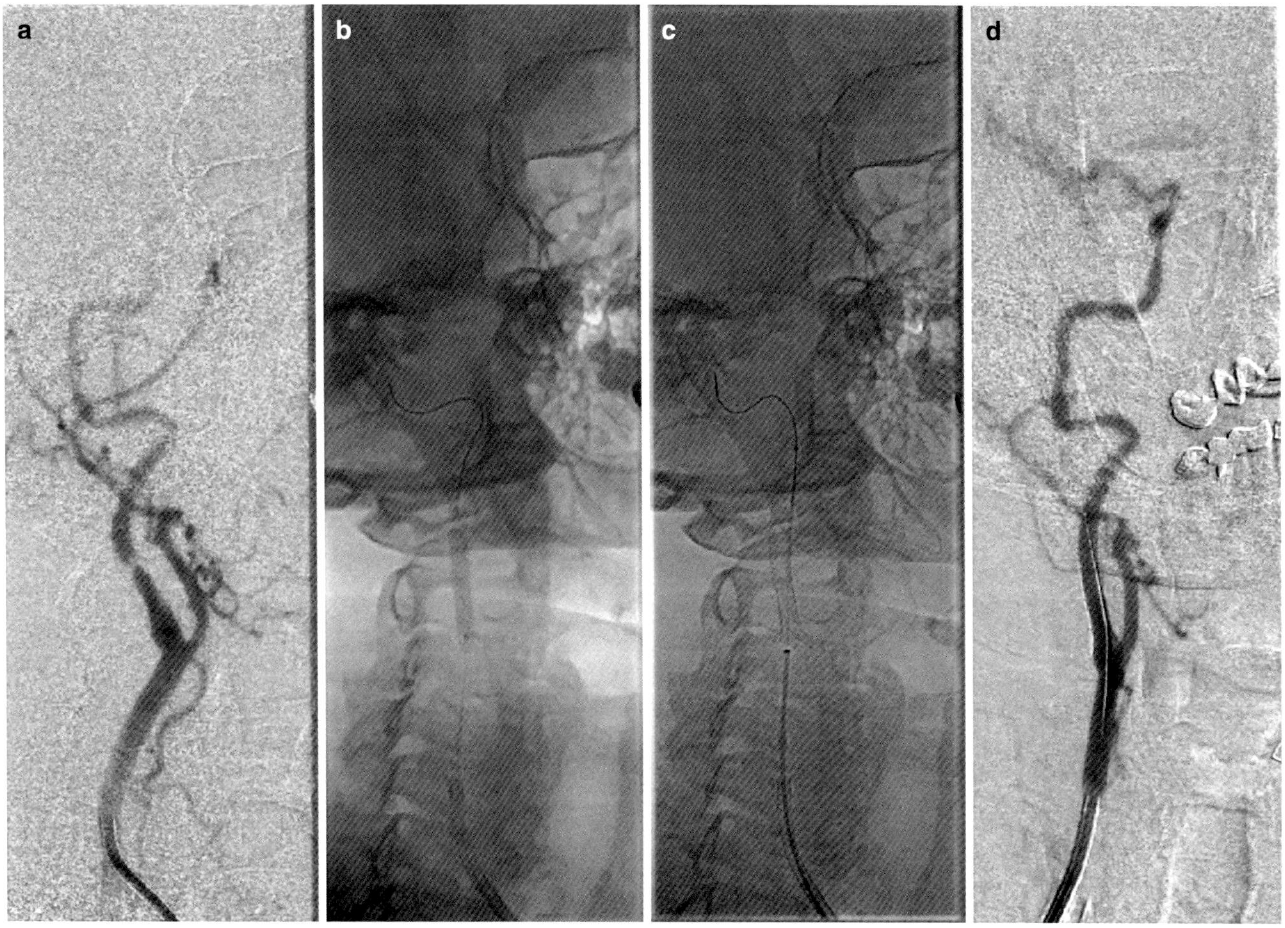

Fig. 27.5 Intraoperative angiographic images. (**a**) Initial angiogram. (**b**) Angioplasty with 5 × 30 mm balloon. (**c**) Stenting with 10 × 40 mm stent. (**d**) Completion angiogram

symptoms of stroke, TIA, hyperperfusion syndrome, and wound complications. She was scheduled for a post-operative visit in 2 weeks, where a baseline duplex would be performed of the stent.

References

1. Macdonald S, Lee R, Williams R, Stansby G. Towards safer carotid artery stenting: a scoring system for anatomic suitability. Stroke. 2009;40(5):1698–703.
2. Brott TG, Hobson RW 2nd, Howard G, Roubin GS, Clark WM, Brooks W, et al. Stenting versus endarterectomy for treatment of carotid-artery stenosis. N Engl J Med. 2010;363:11–23.
3. Chang DW. A new approach to carotid angioplasty and stenting with transcervical occlusion and protective shunting: why it may be a better carotid artery intervention. J Vasc Surg. 2004;39(5):994–1002.
4. Criado E, et al. Transcervical carotid stenting with internal carotid artery flow reversal: feasibility and preliminary results. J Vasc Surg. 2004;40:476–83.
5. Kwolek CJ, Jaff MR, Leal JI, Hopkins LN, Shah RM, Hanover TM, et al. Results of the ROADSTER multicenter trial of transcarotid stenting with dynamic flow reversal Presented at the Late Breaking Clinical Trials Session at Vascular Interventional Advances (VIVA) 2014, November 4–7, 2014. J Vasc Surg. 2015;62(5):1227–34.e1.
6. Kokkosis AA, Macdonald S, Jim J, Shah R, Schneider PA. Assessing the suitability of the carotid bifurcation for stenting: anatomic and morphologic considerations. J Vasc Surg. 2021;74(6):2087–95.
7. Zierler RE, Jordan WD, Lal BK, Mussa F, Leers S, Fulton J, et al. The SVS practice guidelines on follow up after vascular surgery arterial procedures. J Vasc Surg. 2018;68:256–84.

Intracranial Atheromatous Stenosis

28

Quentin Holay, William Boisseau, Gregory Walker, Julien Boucherit, and Raphaël Blanc

Illustrative Case

A 71-year-old right-handed female was first admitted in July 2013 to the neurology intensive care unit for an acute left sylvian stroke causing an aphasia and a right hemiparesis.

Her medical history included a poorly balanced type II diabetes, high blood pressure, and overweight. The first brain MRI shown an acute stroke on DWI-weighted imaging in left corona radiata and a severe stenosis of the distal M1 segment of the left middle cerebral artery (Fig. 28.1a). The patient was put on an intensive medical treatment with dual antiplatelet therapy (DAPT) and a statin.

Despite this medical therapy, the patient had a recurrent stroke 2 months later with even worse aphasia and right hemiparesis (NIHSS score of 7). The MRI found a new ischemic stroke in the left posterior junctional territory and the persistence of the severe stenosis of M1 segment (Fig. 28.1b). An efficacy test with the double antiplatelet therapy was also performed and showed a good response to both drugs.

Q. Holay · W. Boisseau · J. Boucherit · R. Blanc (✉)
Departement of Interventional Neuroradiology, Hopital Fondation Rothschild, Paris, France

G. Walker
Division of Neurology, Department of Medicine, Royal Columbian Hospital, New Westminster, BC, Canada

G. Geroulakos et al. (eds.), *Mastering Endovascular Techniques*, https://doi.org/10.1007/978-3-031-42735-0_28

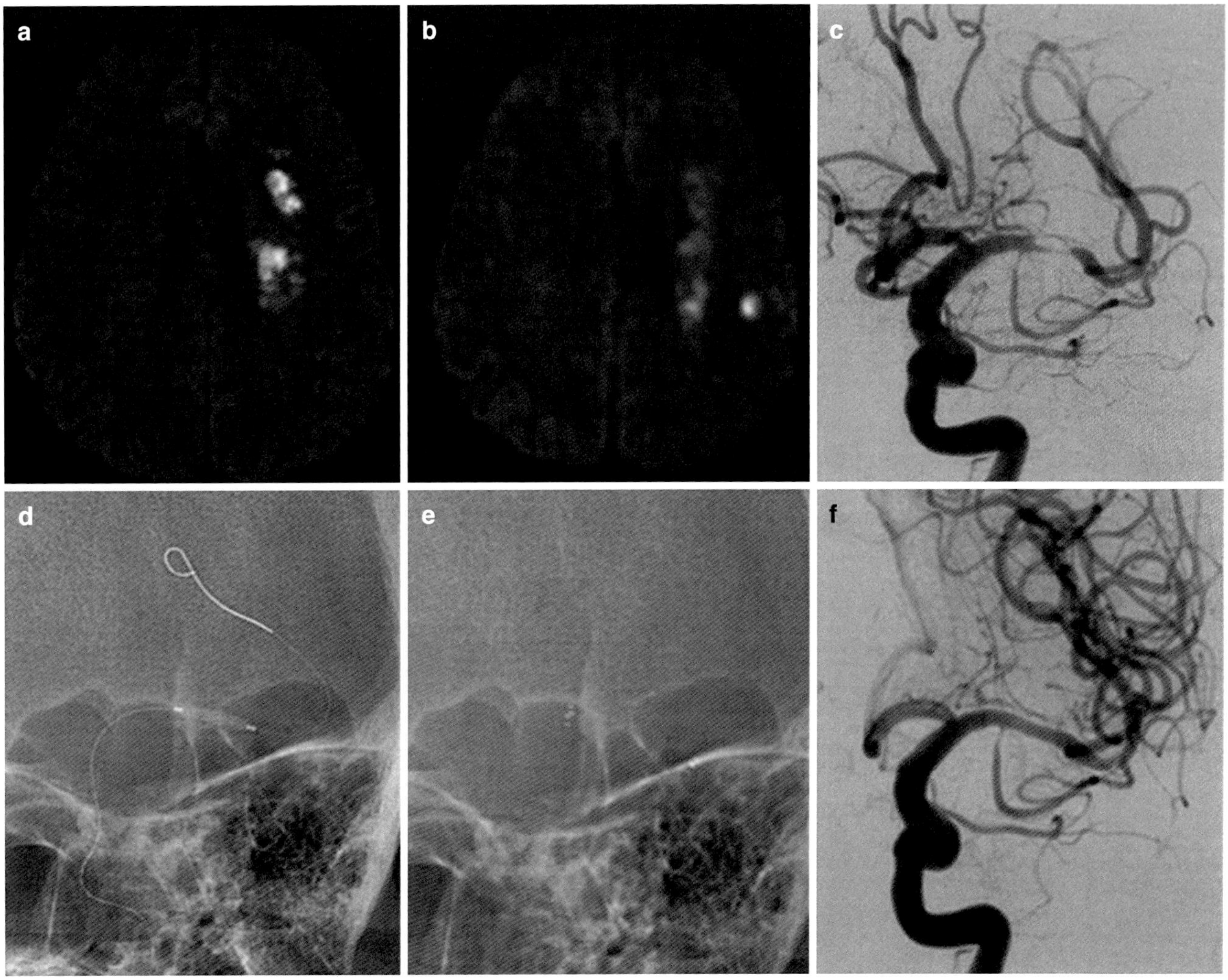

Fig. 28.1 Patient with ICAS and recurrent stroke despite optimal medical treatment. First diffusion weighted imaging (DWI) (**a**) shows an acute stroke in left corona radiata. Second DWI (**b**) shows a recurrent stroke in the same territory. Digital subtraction angiography (DSA) (**c**) shows a 5-mm-long tight M1 stenosis, with a clear downstream circulatory slowing. A balloon angioplasty was first performed with a 2 × 12 mm Gateway balloon (**d**), and then a 2.5 × 15 mm Wingspan stent was placed along the stenosis (**e**) allowing a satisfying angiographic result without residual stenosis (**f**)

Background

Intracranial atherosclerotic stenosis (ICAS) is one of the most common causes of stroke worldwide [1], with wide variations in prevalence between ethnic groups. ICAS is reported to be highly prevalent in Asian [2], Black, and Hispanic [3], while it remains an uncommon cause of stroke in Caucasian populations [3, 4]. The most common sites of ICAS in descending order are the middle cerebral artery, the internal carotid artery distal to the petrous segment, the mid-basilar trunk, the intracranial vertebral (V4 segment) arteries, and the posterior and anterior cerebral arteries [5, 6].

Symptomatic (i.e., causing stroke or transient ischemic attack (TIA)) ICAS is associated with a high risk of recurrent cerebrovascular ischemia [5–7], especially in cases of severe stenosis (more than 70% of the diameter of a major intracranial artery). According to recent trials, recurrence rates can reach as high as 7% in the first year [8, 9], even despite aggressive optimal medical treatment (i.e., combination of antiplatelet medication and vascular risk factor control). Alternatives therapies are therefore still needed for these patients.

Interventional treatments (percutaneous transluminal angioplasty with or without stenting (PTAS)) exist to attempt rapid restoration of normal arterial caliber and prevent short- and long-term morbidity and mortality. These techniques have been used in hyperacute and subacute settings. However, over the past two decades, randomized trials have generally shown harm and lack of benefit even when combined with the best medical therapy [5–7, 9].

Here, we discuss three objectives: (1) current indications of PTAS for symptomatic and severe ICAS, (2) a description of the devices and PTAS techniques, and (3) a discussion of the forthcoming improvements in PTAS.

Indication for Intervention

The 2021 Guidelines for the prevention of stroke in patients with stroke and transient ischemic attack statement is "angioplasty and stenting should not be performed as an initial treatment, even for patients who were taking an antithrombotic agent at the time of the stroke or TIA." [10] This recommendation is based on data from three randomized trials that showed harm or lack of benefit of PTAS (with either self-expanding or balloon-expandable stents) when compared with medical therapy [6, 7, 9].

The SAMMPRIS (Stenting versus Aggressive Medical Management Therapy for Preventing Recurrent Stroke in Intracranial Stenosis) [6] and VISSIT (Vitesse Intracranial Stent Study for Ischemic Therapy) [7] trials compared intensive medical therapy alone with intracranial stent placement plus intensive medical therapy in patients with 70–99% intracranial stenosis who had a TIA or stroke within 30 days before enrollment. The SAMMPRIS trial used the self-expanding Wingspan stent, while the VISSIT trial tested the PHAROS Vitesse balloon-expandable neurovascular stent system (Codman Neurovascular, Raynham, MA). SAMMPRIS trial was halted early because of increased 30-day stroke or death rate events in the stenting arm compared with aggressive medical management (14.7% versus 5.7%, $P = 0.002$) [6]. The enrollment in VISSIT was halted after these negative results and because of futility after 112 patients out of a planned 250 were enrolled. At 1 year, 36.2% of patients in the stent group had a recurrent stroke or TIA, versus 15.1% in the medical group [10]. The recently published CASSISS (China Angioplasty and Stenting for Symptomatic Intracranial Severe Stenosis) trial [9] had similar entry criteria except excluded patients with perforator ischemic stroke (i.e., brainstem or basal ganglia territory infarct). This trial randomized 360 patients to PTAS plus optimal medical management ($n = 176$) versus optimal medical management alone ($n = 182$). The 1-year rate of the primary endpoint (a composite of stroke or death within 30 days or stroke in the qualifying artery territory beyond 30 days through 1 year) was similar between the two groups (8.0% for PTAS plus medical therapy versus 7.2% for medical therapy alone, $p = 0.82$) [9].

The only indication for PTAS remains for patients with ICAS and recurrent cerebrovascular ischemic event despite optimal medical treatement [11]. Currently, only the Wingspan system is approved (as a Humanitarian Device exception) for patients between 22 and 80 years of age with ≥2 cerebrovascular ischemic events localizable to the stenotic vessel despite medical treatment with a recovery to a modified Rankin Scale (mRS) score ≤ 3. The patient must have had experienced the most recent stroke >7 days before the planned treatment with Wingspan and have 70–99% stenosis due to ICAD [11]. Using this FDA on-label indication, the WEAVE registry (Wingspan Stent System Post Market Surveillance) included 152 patients. There was a lower-than-expected (2.6%) periprocedural stroke, intracranial hemorrhage, or death rate for the interim analysis [12]. However, no additional information was provided from this trial, especially on longer-term stroke and death event rates.

Preoperative Preparation

Imaging

Several imaging modalities, such as transcranial Doppler (TCD), magnetic resonance angiography (MRA, with Time-of-Flight (TOF) and Contrast-Enhanced (CE) MRA), computed tomographic angiography (CTA), and digital subtraction angiography (DSA), are used commonly in routine clinical practice to detect intracranial artery stenosis. However, consensus about the optimal imaging strategy has not been established so far. While anatomic diagnosis of arterial narrowing is made with reasonable accuracy with these imaging techniques [13–15], identifying the underlying cause of the stenosis remains challenging in clinical practice [16]. There are several diseases including brain vasculitis, moyamoya disease, radiation-induced vasculopathy, reversible cerebral vasoconstriction syndrome, or intracranial dissection that may mimic ICAS [16].

DSA remains the gold standard and was mandatory for the diagnosis of ICAS in all aforementioned trials [6, 7, 9]. DSA provides excellent visualization of vessel contour, anatomic localization, and assessment of the degree and length of stenosis and collateral circulation. However, DSA may not adequately assess vessel wall pathology and can be of limited value in differentiating between causes of intracranial vasculopathies [17].High-resolution magnetic resonance with vessel wall imaging (VWI) is currently the subject of extensive research [18]. Although further validation is needed, VWI might be a promising technique for the diagnosis of ICAS [18]. Patterns and intensity of postgadolinium enhancement of vascular lesions with MRI T1 postcontrast VMI could help differentiate between intracranial atherosclerosis and other vasculopathies [18]. ICAD typically shows eccentric thickening and asymmetric enhancement, while vasculitis shows smooth, intense, and homogeneous enhancement [18].

In summary, our current diagnostic strategy includes the use of MRA and VWI as screening tools for ICAS diagnosis and then DSA for confirmation and preparation of treatment.

Timing of Intervention

The optimal timing between the qualifying cerebrovascular ischemic events and the PTAS procedure is unknown. Recurrent cerebrovascular ischemic events due to ICAS can often occur in the early period after the index event [19]. In a post-hoc analysis of the WASID (the Warfarin-Aspirin Symptomatic Intracranial Disease) trial, among patients who had recurrent cerebrovascular ischemic events, almost half experienced recurrent cerebrovascular ischemic event within 30 days [19]. On the other hand, early stenting has been reported to increase the periprocedural risk [20, 21]. As one example, the periprocedural stroke risk in SAMMPRIS was more than five times that seen in the WEAVE registry, with median enrollment times of 7 and 22 days from index event, respectively [6, 12]. Several speculative explanations have been suggested for this finding. First, intracranial stenting on recently symptomatic, still vulnerable plaques may increase the risk of procedural stroke (the "snowplowing" effect) [22]. Second, delaying the stenting could give time for the blood brain barrier healing and then reduce the risk of hemorrhage and hyperperfusion syndrome [20]. Finally, delayed stenting may also enable the patients a better antiplatelet and medical preparation for the procedure possibly rendering the plaque more "stable" and the stent environment less prothrombotic [20].

The recently published CASSISS trial limited enrollment of patients to beyond 3 weeks after their index event [9]. As mentioned above, this trial found no difference in stroke recurrence between PTAS and medically treated patients. However, both groups had significantly lower rates of stroke recurrence compared to historical controls [6, 7], likely in part due to later enrollment [9]. This timeline of delaying the procedure to more than 3 weeks post index event may underestimate or not capture the upfront stroke risk in both endovascular and medically managed groups.

In summary, although delayed PTAS seems to be more efficient and safer, it also exposes the patient to a significant risk of early recurrence. The proper cut-off point of timing is yet to be determined.

Medical Treatment

Dual Antiplatelet Therapy (DAPT)

There exists a significant amount of heterogeneity in data and practice for the timing, regimen, and antiplatelet agents used for pre- and post-PTAS. The current consensus, based on several previous trials, is to use aspirin plus clopidogrel for up to 90 days post-ICAS [6, 7, 9, 23]. Yet, some issues remain. Previous trials have used various dose of aspirin, from 75 to 325 mg [6, 7, 9, 23]. The optimal antiplatelet regiment remains unknown. Few studies have compared clopidogrel, ticagrelor, and cilostazol in the context of PTAS. Only one trial (the TOSS II trial [24]) has compared the efficacy of dual antiplatelet therapy (DAPT) with aspirin plus cilostazol versus aspirin plus clopidogrel. No significant differences were seen in total cardiovascular events (nonfatal stroke, nonfatal myocardial infarction, and vascular death) and the frequency of new ischemic lesions on brain MRI follow-up [24]. Individual responses to clopidogrel vary [25]. Low response to clopidogrel is commonly reported [25] and might be associated with recurrent cerebrovascular ischemic events, especially in patients with symptomatic ICAS [26]. Ticagrelor is considered to have no resistance and could be an alternative. Although some retrospective studies have suggested that ticagrelor might be as effective and safe as clopidogrel in neurointerventional procedures [27, 28], randomized data are still needed. The TIC-TAC trial [29], comparing Ticagrelor and Clopidogrel in stenting cerebral aneurysm, is already completed with final results still pending.

Finally, preoperative timing of antiplatelet initiation is yet to be determined. While early DAPT reduces the risk of recurrent cerebrovascular ischemic events, it also carries the risk of hemorrhagic transformation. In contrast, a potential delay in the initiation of DAPT could lead to an increase early recurrence risk. This dilemma between ischemic and hemorrhagic risks could explain the large number of proposed treatment regimens. Clopidogrel is a good example of this variability with various timing (from one [6] to several [7, 9, 12] days before the surgery) and dose (with [6] or without [7, 9, 12] a loading dose varying from 300 to 600 mg) initiation. Intravenous P2Y12 inhibitors (Cangrelor) might be the ideal treatment in this context, given its interesting pharmacologic properties (rapid and complete platelet inhibition, quick platelet function recovery after treatment interruption, and easy transition to oral antiplatelet therapy) [30]. Again, although Cangrelor seems to be a valuable therapeutic option [31], there is currently no evidence that this treatment leads to better outcome and comparable safety during PTAS. Clearly, larger studies exploring these agents and regimen are certainly needed.

Management of Modifiable Risk Factors

In association with DAPT, American [10] and European [32] guidelines recommend, in patients with a TIA or stroke attributable to 50–99% stenosis of a major intracranial artery, maintenance of systolic blood pressure below 140 mmHg, high-intensity statin therapy, and at least moderate physical activity to prevent recurrent stroke and vascular events.

Historically, blood pressure was intentionally left high to mitigate the presumed effects of stenosis. The paradigm slowly shifted in part due to a post-hoc analysis from the WASID trial [33] showing that long-term higher blood pressure was associated with an increased risk of stroke and other vascular event. On the other hand, a recent randomized controlled trial has shown that intensive blood pressure control (under 120 mmHg) in patients with symptomatic ICAS may increase ischemic lesion volume in the subacute stage compared to modest BP treatment (under 140 mmHg). This data has served as the framework for the current guidelines, recommending targeting a systolic blood pressure of around 140 mmHg [10, 32].

Endovascular Strategy and Technique

Stent Devices

The Wingspan system is currently the only stent approved by the FDA and the CE.

Over the last two decades, several stents have come on the market. These products and characteristics are summarized in Table 28.1. It is important to note that few of these new devices have been compared to the Wingspan stent. The efficacy and safety of most devices are based on noncontrolled observational studies. Data are lacking to justify the use of one device over another. Main results are summarized in Table 28.2.

Table 28.1 List and characteristics of available stent for ICAS treatment

Product name	Stent diameter and length (mm)	Minimum microcatheter ID (inch)	FDA approved	CE approved
Self-expanding stent				
WINGSPAN	D: 2.5–4.5 L: 9, 15, 20	Own delivery system	Yes	Yes
ATLAS	D: 3–4.5 L: 15, 21, 24, 30	0.0165	No	No
SOLITAIRE	D: 3–6 L:15, 20, 30, 40	0.021 or 0.027 (5 to 6 ∅)	No	No
ENTERPRISE	D: 4.5 L: 14, 22, 28, 37	0.021	No	No
Balloon-mounted stent				
Apollo (MicroPort Medical, Shanghai, China)	Unspecified			
R-Onyx DES Zotarolimus-Eluting Coronary stent system (Medtronic)	D: 2.25–5 L: 8–38 mm	Own delivery system	No	No
Acclino flex stent (Acandis GmbH, Pforzheim, Germany)	D: 3–8 L: 15–60	0.017 except 0.021 for 6.5 ∅ and 0.027 for 8 ∅	No	No
Drug-eluting stent				
NOVA intracranial sirolimus-eluting stent system; SINOMED	Unspecified			

CE European conformity; *FDA* US Food and Drug administration; *ID* internal diameter

Percutaneous Transluminal Angioplasty with Stenting (PTAS) Procedure

Anesthetic Techniques

This question remains unanswered as no clinical trials have compared general anesthesia and conscious sedation in this context. Proponents of general anesthesia will argue that minimizing the patient movement will reduce the risk of iatrogenic complication. On the other hand, awake procedure has been proven feasible [41], allowing per-procedural monitoring and quick response in case a neurological deficit.

Procedural Description: Tips and Tricks

In the next paragraph, we delineate a step-by-step description of PTAS techniques based on our experience and techniques reported in the literature.

After arterial puncture (femoral or radial), intravenous heparin is given to maintain the activated clotting time (ACT) between 200 and 300 s throughout the procedure.

A guiding catheter (a long introducer sheath or a balloon guide catheter (BGC) for proximal flow arrest) is navigated into the target proximal cervical vessel (carotid or vertebral artery). Three-dimensional volume images and multi-planar sectional images of the target artery are obtained and are critical for the rest of the procedure. These images help determine the most optimal projection for the procedure or "working projections." The length of the stenosis, the diameter of the healthy segments of the target vessel, as well as the diameter of the stenosis are used to select the appropriate devices and sizes. These images will serve as data points for comparison and reference with future images from the procedure as well as for follow-up or retreatments.

Working projections using road maps or smart masks allow for endovascular navigation. It is of paramount importance to always keep the tip of the guiding catheter visible on at least one of the projections to provide feedback on access stability throughout the procedure.

There are no randomized control trials to dictate the exact microwire an operator should use; however, key features

Table 28.2 Noncontrolled studies evaluating efficiency and safety of new stents

References	Device	Study period	Number of patients	Early outcome Stroke	Death	1-year rates Stroke	Death	% Restenosis	Symptomatic restenosis
Kang et al. (JNIS, 2021) [34]	Apollo	NS	159	0	0	6.3	0.6	23.4	3.1
Hassan et al. (Front neurol, 2020) [35]	R-Onyx	October 2019–January 2020	18	NS	NS	0	0	0	0
Mohammaden et al. (JNIS 2022) [36]	Acclino	2010–2020	232	5.6 (72 h)	0.9 (72 h)	3.7 (6 month)	1.6	24.8	11.3
WEAVE registry (Stroke, 2019) [12]	Wingspan	2012–2018	152	1.3 (72 h)	1.3	NA	NA	NA	NA
Duan et al. (JNIS, 2016) [37]	Solitaire	2010–2014	44	9.09	0	4.55 (mean follow-up 25 m)	0	11.36	4.55
Buonomo et al. (Heliyon, 2021) [38]	Atlas	2019–2020	10	0	0	NA	NA	NS	NS
Cui et al. (Front neurol, 2021) [39]	Enterprise	2015–2019	130	1.5 (72 h)	0.8	4.8	0.8	14.4	3.4
Jia et al. (JAMA neurol, 2022) [40]	NOVA	2015–2018	263	2.3	0.8	0.8	0.8	9.5	0

should include a soft distal tip to minimize the risk of distal vessel perforation. Perfect visualization of distal vessels by the operating team and a keen eye on the distal part of the guidewire at all times are very important. The guidewire must be adequately supportive, navigable, and compatible with the balloon and stent one will use. An 014' Synchro 14 (Stryker, California, USA) or Traxcess (MicroVention, California, USA) are commonly used given its soft tip, support, and pushability. At some point, the microwire should be exchangeable length namely 300 cm.

Variations in technique for crossing the stenosis exist. One can attempt to preload the balloon on the microwire in the event support from a microcatheter is not needed to get the microwire past the stenosis. If the wire crosses the stenosis with ease, you are then ready to position the wire distally before plasty. Ideal placement is in a healthy segment of the largest vessel in that territory. For example, if you are treating an ophthalmic segment stenosis, placing the microwire in the M1 segment of the ipsilateral MCA would be sufficient. If you are treating an M1 stenosis, the wire would ideally be placed in a straight segment of the larger of the two M2 divisions. This is done in order to minimize perforation risk. Keeping the wire tip still and in the same place of "pinning" is a skill that must be taught and learned with great care.

In the event a microcatheter is needed for support to cross the stenosis, the microcatheter should be used to gently cross the stenosis once as well as a type of low-profile modified Dotter angiocatheter plasty. If a microcatheter is used, then an exchange maneuver is performed to exchange the microcatheter and add the balloon. Communication and teamwork to keeping the distal wire tip in place and the system straight are key.

Table 28.3 List and characteristics of available balloon for ICAS treatment

Product name	Stent diameter and length (mm)	Minimum microcatheter ID (inch)	FDA approved	CE approved
Gateway or Maverick (Stryker/Boston Scientific)	D: 1.5–4 L: 9, 15, 20	0.064	Yes	Yes
Neuro Elutax SV (Aachen Resonance, Luxembourg)	D: 1.5–6 L: 10–250	Own delivery catheter	No	No

CE European conformity; *FDA* US Food and Drug administration; *ID* internal diameter

Pre-stenting balloon angioplasty is commonly needed. Predilatation allows stent delivery systems to be advanced through the stenosis without being constrained or trapped. We generally used non-compliant angioplasty balloons varying in size from 1.5 to 2.5 mm diameter. These products and characteristics are summarized in Table 28.3.

Several features need to be considered here:

- Balloons are deliberately undersized by 0.5 mm diameter to reduce the risk of arterial damage or rupture by using angiographically obtained measurements.
- The shortest available length to fully cross the lesion was chosen to improve balloon tracking and minimize unnecessary coverage of adjacent perforators when inflated.
- Slow inflation times were used to minimize rupture and dissection. Balloon is generally inflated until the range of 6 atmospheres, ex 1 atm per 15–30 s while an operator pins the microwire in place.

In case of satisfactory result, the balloon catheter is then slowly withdrawn after control angiogram via an exchange maneuver. If the wire moves forward or back, then exchange maneuver should be halted, and the cause should be identified. Review of angiographic runs should be immediately performed as to not miss any complication such as distal spasm, inadvertent embolization, or vessel perforation.

(a) Stent Selection and Deployment

The stent must be of sufficient size to allow complete apposition to the vessel lumen. In most instances we use:

- 2.5–3.5-mm-diameter stent for ICA, vertebral, and basilar trunk stenosis.
- 2–3-mm-diameter stent for MCA stenosis.
- Once the stent is deployed, follow-up angiographic runs are performed to assess complications and determine if post-stenting balloon angioplasty is needed.

(b) Post-stenting Balloon Angioplasty

We generally postdilate the most stenotic portion of the stented area with a 2–3.5 mm diameter, short balloon, depending on the size of the causal vessel. A mild residual stenosis of the target lesion (<20%) is acceptable to avoid generating excessive embolic debris and avoid potentially severe embolic or hemorrhagic complications.

Post-procedural Management

After PTAS procedure, a close monitoring of blood pressure is done in intensive care. If the blood pressure is greater than 140 mmHg systolic, intravenous antihypertensive treatment is required to lower the risk of reperfusion injury once PTAS has been performed.

DAPT is usually continued during 90 days, and then aspirin is continued alone over the long term.

Alternatives Interventional Options

Angioplasty Alone

In light of safety issues related to PTAS, balloon angioplasty alone (i.e., without placement of an intracranial stent) has been considered a possible alternative for endovascular therapy [42]. However, no RCTs have compared angioplasty alone with medical therapy for stroke prevention in patients with symptomatic ICAS. A systematic review and meta-analysis of 25 studies of angioplasty alone compared event rates in patients treated with angioplasty to events in the SAMMPRIS medical group and found no benefit of angioplasty due to high periprocedural morbidity and mortality [43]. Further investigations are warranted to clarify the efficacy and safety of angioplasty alone in patients with symptomatic ICAS.

Surgical Procedure

Based on the results of the EC/IC Bypass study (International Cooperative Study of Extracranial/Intracranial Arterial Anastomosis), current guidelines do not recommend extracranial-intracranial bypass surgery in patients with symptomatic ICAS [10, 32]. In this study, in the Bypass group had an higher risk of stroke recurrence (44% versus 23.7%) [44].

Encephaloduroarteriosynangiosis is another surgical procedure currently under evaluation. A phase II trial has suggested safety and efficacy of encephaloduroarteriosynangiosis plus intensive medical management in patients with symptomatic ICAS [45]. A phase III trial is pending.

Follow-Up

There is no consensus regarding the optimal follow-up imaging protocol. In our current protocol, patients are seen in clinic at approximately 3 months and 1 year after PTAS as part of routine follow-up care. A modified Rankin Scale and NIHSS scores are determined, and vascular risk factors are managed. A brain magnetic resonance imaging and DSA are generally performed at 3 months.

Other Situations

ICAS-Related LVO

ICAD could be revealed by a large vessel occlusion (LVO). Several criteria have been proposed to identify patients with ICAS-related LVO: (1) absence of atrial fibrillation, (2) existence of stenosis in other cerebral arteries, (3) absence of CT hyperdense sign or MRI susceptibility, (4) watershed infarction, (5) residual stenosis on DSA after several passes, or (6) early reocclusion [32].

Acute management of this group of patients can be challenging, with the common failure of conventional mechanical thrombectomy, and commonly requires rescue therapy with antiplatelet therapy (such as anti-GPIIBIIIA [46–48] or cangrelor [49]), angioplasty, stenting, or a combination of treatment modalities [50, 51]. A framework has been proposed with antiplatelet therapy as first-line treatment and, in case of persistent occlusion, additional angioplasty with or without stenting [52]. However, the safety and efficacy of these rescue treatments have not yet been established, and current evidence is largely driven by small retrospective case series [46–51]. Again, proper validation of the best pharmacological and mechanical strategies through prospective randomized clinical trials is warranted.

Asymptomatic ICAS

While symptomatic ICAS has been the subject of many studies, there are few data about asymptomatic ICAS.

Recent studies have suggested that asymptomatic ICAS is associated with a low risk of stroke, <1% over 10 years [53–55]. Based on these results, several authors have suggested that no additional management strategy or follow-up is required routinely for this commonly identified pathology [55].

Forthcoming Improvements in PTAS

Despite the publication of three randomized controlled trials, there are still major gray areas in the management of symptomatic ICAS (optimal timing of intervention, DAPT, endovascular strategy, etc.).

Regarding the available evidence, PTAS is not recommended as a treatment of first choice in symptomatic ICAS [6, 7, 9, 10, 32]. The three RCTs have shown that PTAS is associated with an increased risk of 30-day stroke and death events compared to medical treatment alone [6, 7, 9]. Moreover, the more recent CASSISS trial has shown that the risk of recurrent stroke was not significantly different between the two groups during the 3-year follow-up [9]. This data implies that even if the periprocedural risk in the stenting group could be reduced to as low as the 30-day rate in the medical therapy alone group, stenting still may not provide a long-term benefit over medical therapy [51].

Thus PTAS use remains for patients with ICAS and recurrent stroke despite optimal medical treatment [11]. However, the optimal strategy in these patients has yet to be defined. Most studies have used the Wingspan stent systems [6, 9, 12]. Although this stent has advantages over balloon-expandable stents due to its relative ease of delivery, the effectiveness of the self-expanding stent in restoring lumen diameter and preventing restenosis has been debated [40]. Furthermore, new drug-eluting stent [40, 56] (DES; NOVA® (Sinomed) and Maurora® (Alain Biotechnology) Sirolimus Eluting Stent System) and balloon catheter (AcoArt-Litos® (Acotec) drug-eluting balloon catheter) have been introduced to the market (summarized in Table 28.4).

The recently published NOVA trial has compared drug-eluting stent (DES) with bare-metal stent (BMS) [40]. This trial has suggested that DES could reduce the risks of in-stent restenosis. However, ischemic stroke recurrences or deaths were similar at 1 year [40]. The randomized ACOART trial [57] is currently investigating the safety and efficacy of new drug-coated balloons (DCB) with paclitaxel compared to the Apollo balloon-mounted stent system. Theoretically, the use of DCB, in addition to rapidly restoring normal arterial caliber, may allow the delivery of a high local dose of antiproliferative drugs. Altogether, this new device might reduce the risk of restenosis and improve clinical outcome.

Table 28.4 List and characteristics of new drug-eluting stent and balloon catheter for ICAS treatment

Product name	Diameter and length (mm)	Minimum microcatheter ID (inch)	Drug	CE or FDA approved
Drug-eluting balloon catheter				
AcoArt-Litos (Acotec) [57]	2–4 mm 20–300 mm	4F	Paclitaxel	No
Drug-eluting stent				
NOVA (Sinomed) [40]	NA	NA	Paclitaxel	No
Maurora (Alain Biotechnology)	NA	NA	Rapamycin	No

CE European conformity; *FDA* US Food and Drug administration; *ID* internal diameter; *NA* Not available

The optimal management of ICAD-related LVO has yet to be defined. There are two ongoing studies aiming to evaluate the safety and efficacy of rescue intracranial angioplasty/stenting in acute LVO patients who have failed standard MT. The Acandis Credo Intracranial Stent for Unsuccessful Recanalization After Thrombectomy (ACUTE) trial [58] is a prospective, single-arm, open-label, multicenter trial including patients using the Credo/NeuroSpeed system. The Rescue Stenting for Failed Endovascular Thrombectomy in Acute Ischemic Stroke (ReSET) study [59] is a prospective, open-label, multicenter registry using Solitaire device detachment.

The results of the aforementioned studies will hopefully provide pragmatic data regarding the implications for intracranial stenting in acute ICAD and other more refractory forms of LVOS.

Illustrative Case

Due to the recurrent stroke despite adequate medical treatment, the patient was transferred to a comprehensive stroke center to perform endovascular treatment of the known left middle cerebral artery stenosis.

She was treated under general anesthesia. The patient was under continuous heparin drip, after an initial bolus of 5000 IU.

A right femoral puncture was performed, and a Destination 6Fr catheter (Terumo, Tokyo, Japan) was placed into the left common carotid artery. Coaxially, a Fargo Max 6Fr (Balt, Montmorency, France) supporting catheter was positioned in the intrapetrous internal carotid. The initial angiographic evaluation revealed a 5-mm-long critical M1 stenosis, with a clear downstream circulatory flow reduction (Fig. 28.1c). The left M1 stenosis was crossed with a Traxcess 0.014″

guidewire, and a balloon angioplasty was first performed with a 2 × 12 mm Gateway balloon (Fig. 28.1d). Then a 2.5 × 15 mm Wingspan stent was placed along the stenosis (Fig. 28.1e), allowing a satisfying angiographic result without residual stenosis (Fig. 28.1f). Control angiograms confirmed a good stent expansion and showed a normal left sylvian and anterior cerebral circulation (Fig. 28.1f). The CT scan at the end of the procedure does not reveal any additional complication.

After the procedure, the patient remained perfectly neurologically stable with preserved vigilance and no new deficit.

The patient was discharged on double antiplatelet therapy for 3 months, followed by simple antiplatelet therapy for another 3 months.

Follow-up imaging showed good stent permeability with no new ischemic lesions. Clinically, the patient retained very moderate right hemiparesis and a discrete word-finding difficulty but presented no new neurological symptoms.

References

1. Gorelick PB, Wong KS, Bae H-J, Pandey DK. Large artery intracranial occlusive disease. Stroke. 2008;39(8):2396–9.
2. De Silva DA, Woon FP, Lee MP, Chen CP, Chang HM, Wong MC. South Asian patients with ischemic stroke: intracranial large arteries are the predominant site of disease. Stroke. 2007;38(9):2592–4.
3. Sacco RL, Kargman DE, Gu Q, Zamanillo MC. Race-ethnicity and determinants of intracranial atherosclerotic cerebral infarction. Stroke. 1995;26(1):14–20.
4. White H, Boden-Albala B, Wang C, et al. Ischemic stroke subtype incidence among whites, blacks, and Hispanics. Circulation. 2005;111(10):1327–31.
5. Chimowitz MI, Lynn MJ, Howlett-Smith H, et al. Comparison of warfarin and aspirin for symptomatic intracranial arterial stenosis. N Engl J Med. 2005;352(13):1305–16.
6. Chimowitz MI, Lynn MJ, Derdeyn CP, et al. Stenting versus aggressive medical therapy for intracranial arterial stenosis. N Engl J Med. 2011;365(11):993–1003.
7. Zaidat OO, Fitzsimmons BF, Woodward BK, et al. Effect of a balloon-expandable intracranial stent vs medical therapy on risk of stroke in patients with symptomatic intracranial stenosis: the VISSIT randomized clinical trial. JAMA. 2015;313(12):1240–8.
8. Kasner SE, Chimowitz MI, Lynn MJ, et al. Predictors of ischemic stroke in the territory of a symptomatic intracranial arterial stenosis. Circulation. 2006;113(4):555–63.
9. Gao P, Wang T, Wang D, et al. Effect of stenting plus medical therapy vs medical therapy alone on risk of stroke and death in patients with symptomatic intracranial stenosis: the CASSISS randomized clinical trial. JAMA. 2022;328(6):534–42.
10. Kleindorfer DO, Towfighi A, Chaturvedi S, et al. 2021 guideline for the prevention of stroke in patients with stroke and transient ischemic attack: a guideline from the American Heart Association/American Stroke Association. Stroke. 2021;52(7):e364–467.
11. https://www.accessdata.fda.gov/scripts/cdrh/cfdocs/cfhde/hde.cfm?id=376103.
12. Alexander MJ, Zauner A, Chaloupka JC, et al. WEAVE trial: final results in 152 on-label patients. Stroke. 2019;50(4):889–94.
13. Duffis EJ, Jethwa P, Gupta G, Bonello K, Gandhi CD, Prestigiacomo CJ. Accuracy of computed tomographic angiography compared to digital subtraction angiography in the diagnosis of intracranial stenosis and its impact on clinical decision-making. J Stroke Cerebrovasc Dis. 2013;22(7):1013–7.
14. Feldmann E, Wilterdink JL, Kosinski A, et al. The stroke outcomes and neuroimaging of intracranial atherosclerosis (SONIA) trial. Neurology. 2007;68(24):2099–106.
15. Willinek WA, von Falkenhausen M, Born M, et al. Noninvasive detection of steno-occlusive disease of the supra-aortic arteries with three-dimensional contrast-enhanced magnetic resonance angiography: a prospective, intra-individual comparative analysis with digital subtraction angiography. Stroke. 2005;36(1):38–43.
16. Prabhakaran S, Romano JG. Current diagnosis and management of symptomatic intracranial atherosclerotic disease. Curr Opin Neurol. 2012;25(1):18–26.
17. van den Wijngaard IR, Holswilder G, van Walderveen MA, et al. Treatment and imaging of intracranial atherosclerotic stenosis: current perspectives and future directions. Brain Behav. 2016;6(11):e00536.
18. Mossa-Basha M, Hwang WD, De Havenon A, et al. Multicontrast high-resolution vessel wall magnetic resonance imaging and its value in differentiating intracranial vasculopathic processes. Stroke. 2015;46(6):1567–73.
19. Ovbiagele B, Cruz-Flores S, Lynn MJ, Chimowitz MI. Early stroke risk after transient ischemic attack among individuals with symptomatic intracranial artery stenosis. Arch Neurol. 2008;65(6):733–7.
20. Yu Y, Wang T, Yang K, et al. Timing and outcomes of intracranial stenting in the post-SAMMPRIS era: a systematic review. Front Neurol. 2021;12:12.
21. Nahab F, Lynn MJ, Kasner SE, et al. Risk factors associated with major cerebrovascular complications after intracranial stenting. Neurology. 2009;72(23):2014–9.
22. Raghuram K, Durgam A, Kohlnhofer J, Singh A. Relationship between stroke recurrence, infarct pattern, and vascular distribution in patients with symptomatic intracranial stenosis. J Neurointerv Surg. 2018;10(12):1161–3.
23. Wong KS, Chen C, Fu J, et al. Clopidogrel plus aspirin versus aspirin alone for reducing embolisation in patients with acute symptomatic cerebral or carotid artery stenosis (CLAIR study): a randomised, open-label, blinded-endpoint trial. Lancet Neurol. 2010;9(5):489–97.
24. Kwon SU, Hong K-S, Kang D-W, et al. Efficacy and safety of combination antiplatelet therapies in patients with symptomatic intracranial atherosclerotic stenosis. Stroke. 2011;42(10):2883–90.
25. Matetzky S, Shenkman B, Guetta V, et al. Clopidogrel resistance is associated with increased risk of recurrent Atherothrombotic events in patients with acute myocardial infarction. Circulation. 2004;109(25):3171–5.
26. Patel PD, Vimalathas P, Niu X, et al. CYP2C19 loss-of-function is associated with increased risk of ischemic stroke after transient ischemic attack in intracranial atherosclerotic disease. J Stroke Cerebrovasc Dis. 2021;30(2):105464.
27. Park KY, Ozaki T, Kostynskyy A, et al. Ticagrelor versus Clopidogrel in the dual antiplatelet regimen for intracranial stenting or flow-diverter treatment for Unruptured cerebral aneurysms: a single-center cohort study. AJNR Am J Neuroradiol. 2021;42(9):1638–44.
28. Soize S, Foussier C, Manceau PF, et al. Comparison of two preventive dual antiplatelet regimens for unruptured intracranial aneurysm embolization with flow diverter/disrupter: a matched-cohort study comparing clopidogrel with ticagrelor. J Neuroradiol. 2019;46(6):378–83.
29. https://clinicaltrials.gov/ct2/show/NCT02675205.
30. Sible AM, Nawarskas JJ. Cangrelor: a new route for P2Y: 12: inhibition. Cardiol Rev. 2017;25(3):133–9.

31. Cervo A, Ferrari F, Barchetti G, et al. Use of Cangrelor in cervical and intracranial stenting for the treatment of acute ischemic stroke: a "real life" single-center experience. Am J Neuroradiol. 2020;41(11):2094–9.
32. Psychogios M, Brehm A, López-Cancio E, et al. European stroke organisation guidelines on treatment of patients with intracranial atherosclerotic disease. Eur Stroke J. 2022;7(3):III–IV.
33. Turan TN, Cotsonis G, Lynn MJ, Chaturvedi S, Chimowitz M. Relationship between blood pressure and stroke recurrence in patients with intracranial arterial stenosis. Circulation. 2007;115(23):2969–75.
34. Kang K, Zhang Y, Shuai J, et al. Balloon-mounted stenting for ICAS in a multicenter registry study in China: a comparison with the WEAVE/WOVEN trial. J Neurointerv Surg. 2021;13(10):894–9.
35. Hassan AE, Mohammaden MH, Rabah RR, Tekle WG. Initial experience with the next-generation resolute onyx Zotarolimus-eluting stent in symptomatic intracranial atherosclerotic disease. Front Neurol. 2020;11:570100.
36. Mohammaden MH, Nogueira RG, Tekle W, et al. Safety and efficacy of balloon-mounted stent in the treatment of symptomatic intracranial atherosclerotic disease: a multicenter experience. J NeuroInterv Surg. 2022;14(8):756–61.
37. Duan G, Feng Z, Zhang L, et al. Solitaire stents for the treatment of complex symptomatic intracranial stenosis after antithrombotic failure: safety and efficacy evaluation. J NeuroInterv Surg. 2016;8(7):680–4.
38. Buonomo O, Mormina E, Caragliano AA, et al. Safety and effect of neuroform atlas stent in the treatment of symptomatic intracranial stenosis: a single-center experience. Heliyon. 2021;7(9):e08040.
39. Cui R, Yan L, Kang K, et al. Long-term outcome of Enterprise stenting for symptomatic ICAS in a high-volume stroke center. Front Neurol. 2021;12:672662.
40. Jia B, Zhang X, Ma N, et al. Comparison of drug-eluting stent with bare-metal stent in patients with symptomatic high-grade intracranial atherosclerotic stenosis: a randomized clinical trial. JAMA Neurol. 2022;79(2):176–84.
41. Abou-Chebl A, Krieger DW, Bajzer CT, Yadav JS. Intracranial angioplasty and stenting in the awake patient. J Neuroimaging. 2006;16(3):216–23.
42. Stapleton CJ, Chen YF, Shallwani H, et al. Submaximal angioplasty for symptomatic intracranial atherosclerotic disease: a meta-analysis of Peri-procedural and long-term risk. Neurosurgery. 2020;86(6):755–62.
43. Kadooka K, Hagenbuch N, Anagnostakou V, Valavanis A, Kulcsár Z. Safety and efficacy of balloon angioplasty in symptomatic intracranial stenosis: a systematic review and meta-analysis. J Neuroradiol. 2020;47(1):27–32.
44. Failure of extracranial-intracranial arterial bypass to reduce the risk of ischemic stroke. Results of an international randomized trial. N Engl J Med. 1985;313(19):1191–200.
45. Gonzalez NR, Jiang H, Lyden P, et al. Encephaloduroarteriosynangiosis (EDAS) revascularization for symptomatic intracranial atherosclerotic steno-occlusive (ERSIAS) phase-II objective performance criterion trial. Int J Stroke. 2021;16(6):701–9.
46. Kim YW, Sohn SI, Yoo J, et al. Local tirofiban infusion for remnant stenosis in large vessel occlusion: tirofiban ASSIST study. BMC Neurol. 2020;20(1):284.
47. Yan Z, Shi Z, Wang Y, et al. Efficacy and safety of low-dose tirofiban for acute intracranial atherosclerotic stenosis related occlusion with residual stenosis after endovascular treatment. J Stroke Cerebrovasc Dis. 2020;29(4):104619.
48. Baek JH, Jung C, Kim BM, et al. Combination of rescue stenting and antiplatelet infusion improved outcomes for acute intracranial atherosclerosis-related large-vessel occlusion. Front Neurol. 2021;12:608270.
49. Marnat G, Finistis S, Delvoye F, et al. Safety and efficacy of cangrelor in acute stroke treated with mechanical Thrombectomy: endovascular treatment of ischemic stroke registry and meta-analysis. AJNR Am J Neuroradiol. 2022;43(3):410–5.
50. Baek JH, Kim BM, Kim DJ, Heo JH, Nam HS, Yoo J. Stenting as a rescue treatment after failure of mechanical Thrombectomy for anterior circulation large artery occlusion. Stroke. 2016;47(9):2360–3.
51. Peng F, Wan J, Liu W, et al. Efficacy and safety of rescue stenting following failed mechanical thrombectomy for anterior circulation large vessel occlusion: propensity score analysis. J Neurointerv Surg. 2020;12(3):271–3.
52. Baek JH, Kim BM, Heo JH, Kim DJ, Nam HS, Kim YD. Outcomes of endovascular treatment for acute intracranial atherosclerosis-related large vessel occlusion. Stroke. 2018;49(11):2699–705.
53. Gutierrez J, Khasiyev F, Liu M, et al. Determinants and outcomes of asymptomatic intracranial atherosclerotic stenosis. J Am Coll Cardiol. 2021;78(6):562–71.
54. Hurford R, Wolters FJ, Li L, Lau KK, Küker W, Rothwell PM. Prognosis of asymptomatic intracranial stenosis in patients with transient ischemic attack and minor stroke. JAMA Neurol. 2020;77(8):947–54.
55. Nahab F, Cotsonis G, Lynn M, et al. Prevalence and prognosis of coexistent asymptomatic intracranial stenosis. Stroke. 2008;39(3):1039–41.
56. He Y, Li T, Bai W, Zhu L, Wang M, Zhang Y. Cerebrovascular drug-eluting stent versus bare-metal stent in the treatment of vertebral artery stenosis: a non-inferiority randomized clinical trial. J Stroke. 2019;21(1):101–4.
57. https://clinicaltrials.gov/ct2/show/NCT04631055.
58. https://clinicaltrials.gov/ct2/show/NCT03955835.
59. https://clinicaltrials.gov/ct2/show/NCT03993340.

29 Arch Vessel Dissection

Kirill Alektoroff and Panagiotis Papanagiotou

Case Report
A 33-year-old female patient presented at the emergency department with acute dysarthria and a high-grade left-side sensorimotor hemiparesis. The non-enhanced CT of the head revealed no intracranial bleeding or signs of ischemia. The CT-angiography showed an occlusion of the right extracranial internal carotid artery due to vessel dissection (Fig. 29.1b, c). No occlusion of intracranial arteries was present.

Systemic thrombolytic therapy was administered after ruling out contraindications. Due to further neurological deterioration (reaching an NIHSS of 14) and unchanged CT-imaging findings, the patient was transferred to the angiographic suite for endovascular therapy.

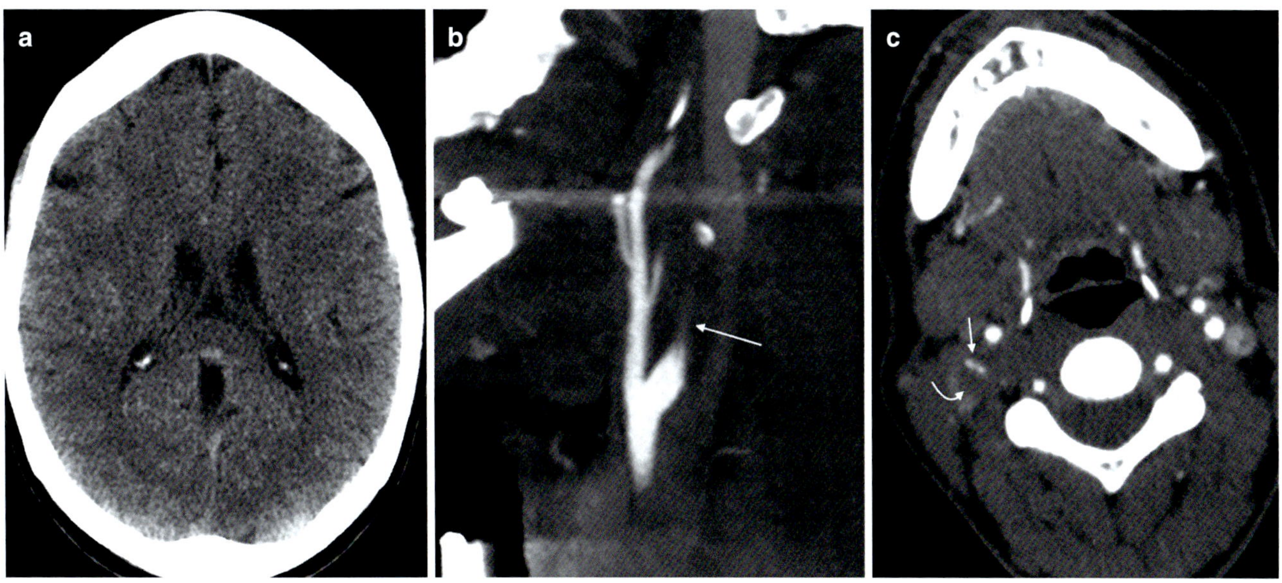

Fig. 29.1 CT-images of a 33-year-old-patient with severe left-side stroke symptoms. Non-enhanced axial computed tomography of the head (**a**) shows no evidence of bleeding or ischemia. Sagittal multiplanar reconstruction of the CT-angiography shows an occlusion of the right internal carotid artery and a "candle flame"-like narrowing of the ICA-lumen (**b**). The axial image (**c**) shows a severe lumen narrowing of the extracranial ICA (*straight arrow*), caused by an intramural hematoma (*curved arrow*)

K. Alektoroff
Department of Diagnostic and Interventional Neuroradiology, Klinikum Bremen-Mitte/Bremen-Ost, Bremen, Germany
e-mail: alektoroff@gmx.de

P. Papanagiotou (✉)
Department of Diagnostic and Interventional Neuroradiology, Klinikum Bremen-Mitte/Bremen-Ost, Bremen, Germany

Department of Radiology, Aretaieion University Hospital, National and Kapodistrian University of Athens, Athens, Greece

G. Geroulakos et al. (eds.), *Mastering Endovascular Techniques*, https://doi.org/10.1007/978-3-031-42735-0_29

Introduction

Supra-aortic arterial vessel dissection represents a common cause of strokes in young and middle-aged patients and accounts for 10–25% of stroke cases in patients under 50 years of age. Dissections may occur spontaneously or traumatically [1]. They result from a tear in the vessel wall, leading to an intramural hematoma between vessel wall layers (usually intima and media or media and adventitia). The intramural blood collection causes aneurysmatic dilatation of the vessel wall as well as lumen stenosis, potentially compromising normal blood flow. Intimal injury and altered hemodynamics may be complicated by thrombus formation at the dissection site, which may become a source of distal emboli, causing neurological deficits. Distention and thinning of the affected arterial wall can lead to pseudoaneurysm formation.

Among arch vessels, extracranial internal carotid artery dissection represents the most common dissection type (estimated yearly incidence of 2.5–3/100,000), followed by extracranial vertebral artery dissections (1–1.5/100,000) [2, 3]. Subclavian artery dissections are rarely reported [4–6].

Risk factors of arch vessel dissections include trauma, connective tissue disorders (e.g., Ehlers-Danlos syndrome, Fibromuscular Dysplasia, Marfan syndrome) and acute infections. Simultaneous bilateral dissections are observed in some patients [7].

Clinical Features

The most common symptoms of supra-aortic or cervical artery dissection are neck pain, headache or facial/jaw pain, cranial nerve palsies, as well as neurological deficits corresponding to the affected vascular territory. Partial Horner's syndrome (ipsilateral ptosis and miosis) may be observed in patients with carotid artery dissection resulting from compression of perivascular sympathetic nerve fibers.

Classification of Arterial Dissections

The spontaneous dissections can be classified according to the Borgess classification. Based on the presence or absence of the intimal tear, the dissections are divided into type I (no intimal tear) and type II (intimal tear present). Both type I and II dissections are further divided into two subtypes, A and B (Table 29.1).

To describe the traumatic blunt carotid arterial injuries, the Biffl grading scale is often used, which classifies the dissections into five types (Table 29.2).

Table 29.1 Borgess classification of spontaneous cervical artery dissection. (Modified from [8])

Type	Description
IA	Intact intima. Lumen stenosis with persistent antegrade blood flow
IB	Intact intima. Complete lumen occlusion due to intramural hematoma
IIA	Small intimal disruption
IIB	Large intimal disruption with intimal flap formation

Table 29.2 Biffl classification of blunt carotid arterial injuries. (Modified from [9])

Grade	Description
I	Dissection or luminal irregularity with luminal narrowing <25%
II	Dissection or intramural hematoma with ≥25% luminal narrowing, intraluminal thrombus, or raised intimal flap
III	Pseudoaneurysm
IV	Lumen occlusion
V	Transection with free extravasation

Treatment Options

Symptomatic arterial dissections are most commonly treated conservatively using antithrombotic medication (anticoagulants and antiplatelet drugs) [10]. The main therapy goal is to prevent thrombus formation at the injured arterial wall and thereby avoid potential thromboembolic complications. Patients presenting with acute neurological symptoms may be treated by systemic thrombolysis during the first 4.5 h from symptom onset.

The endovascular treatment of arterial dissections is usually reserved for patients with acute severe cerebral perfusion deficits, simultaneous large cerebral artery occlusions, and persistent (or recurring) neurological symptoms despite receiving sufficient antithrombotic therapy [11–16].

Endovascular Treatment of Arch Vessel Dissections

The main goal of the endovascular treatment is reconstruction of arterial lumen integrity to restore normal hemodynamics and prevent thromboembolic events to distal vessel territories. This is usually achieved by a stent implantation into the vessel lumen to relieve the stenosis caused by intramural hematoma or intimal flap. Sometimes, implantation of multiple stents in a telescopic manner may be necessary to sufficiently treat a long-segment dissection.

The main advantages of endovascular dissection treatment are rapid removal of the vessel stenosis and increasing of blood flow to relieve cerebral perfusion deficits. Stent placement may also facilitate the healing of dissection-induced pseudoaneurysms by altering intra-aneurysmal flow velocity and promoting pseudoaneurysm thrombosis.

Endovascular Technique

The endovascular treatment of arch vessel dissections is performed on an interventional (preferably biplanar) angiography system. Self-expanding stents are usually used for extracranial dissection treatment. Whenever possible, the patient should start receiving dual antiplatelet medication before the stent placement (e.g., loading dose of 100–500 mg acetylsalicylic acid (ASA) and 300 mg clopidogrel). In case of acute stent implantation, glycoprotein IIb/IIIa antagonists (Eptifibatide, Tirofiban, etc.) may be used as a bridging therapy for sufficient thrombocyte function inhibition [17].

After obtaining a digital subtraction angiography of the dissected vessel, it is necessary to access the true vessel lumen before a stent can be placed. First, a microwire and a microcatheter are carefully navigated through the true lumen until the intact distal portion of the artery is reached. An angiographic run through the microcatheter should be performed to confirm its correct location inside the true lumen.

Flexible stents (e.g., Leo) or flow diverters are often used in curved vessel portions. These stents are usually introduced directly through the microcatheter and are released inside the true lumen during microcatheter retraction.

The stiffer "over-the-wire" stent systems (e.g., Wallstent, Casper) are usually used in straight arterial segments. An exchange-length microwire is then first advanced through the microcatheter. After removing the microcatheter, a folded stent is navigated over the microwire and released inside the true vessel lumen at the level of dissection.

Finally, a balloon angioplasty may be performed inside the stent to treat remaining stenosis. Depending on the dissection length, several stents sometimes need to be placed in a telescopic manner to ensure the dissected vessel portion is sufficiently treated.

After stent implantation, dual antiplatelet medication should be maintained for 1.5–3 months (usually 100 mg ASA and 75 mg clopidogrel daily) to prevent in-stent thrombosis.

Case Report (Procedure and Outcome)

During a digital subtraction angiography of the right ICA, the extracranial dissection was confirmed (Fig. 29.2a). After advancing a microcatheter and a distal access catheter over a microwire into the distal ICA segment, the true lumen location was confirmed by an angiographic run (b). A long self-expanding stent (Wallstent) was advanced into the dissected cervical ICA segment and was gradually released into the true lumen (c). The dissection part in the curvy distal ICA portion was treated by an overlapping flexible stent (Leo, d). The final angiographic run showed a good radio-

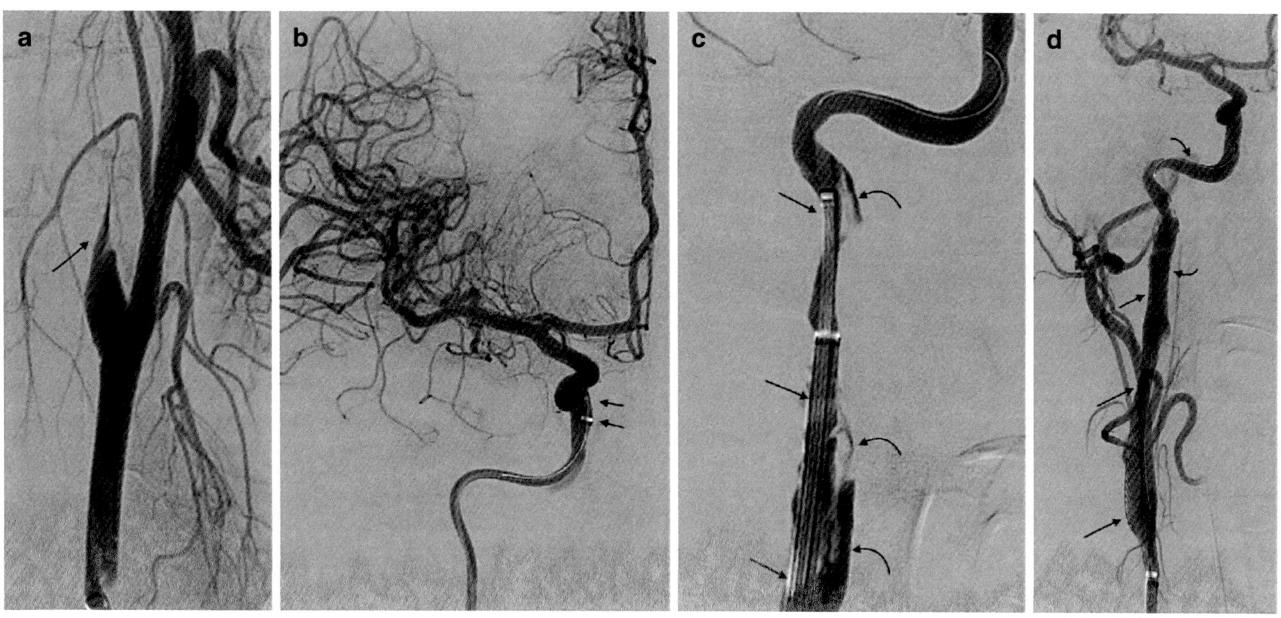

Fig. 29.2 Digital subtraction angiography (DSA) images of a 33-year-old-patient with an ICA-dissection. First angiographic run (**a**) showing a "candle flame"-like narrowing of the ICA-lumen. The dissected vessel portion was passed by a microcatheter and a distal access catheter (Rebar and Sofia, *arrows in* **b**) over a microwire (Traxcess). The location of the catheters inside the true lumen was confirmed by an angiographic run (**b**). Long self-expanding stent (*straight arrows* in **c**) was advanced into the dissected vessel portion (*curved arrows* in **c**). Final angiographic run (**d**), showing the reconstructed ICA-lumen with an expanded Wallstent (*straight arrows*) and a Leo stent (*curved arrows*)

logical result of an ACI reconstruction without periprocedural complications. The patient's symptoms gradually improved from NIHSS 14 to NIHSS 8 during the clinical stay, and further rehabilitation treatment has been started soon.

Toolkit

- Use self-expanding stents to reconstruct the dissected vessel lumen (flexible stents (e.g., Leo) for curvy segments and stiffer stents (e.g., Wallstent, Roadsaver, Casper) for straight segments.
- Use a microcatheter and microwire or a microcatheter alone to carefully navigate through the true vessel lumen to reach the intact distal vessel segment.
- Confirm the true lumen location by a DSA run through a microcatheter before deploying the stent.
- Use telescopic stenting to reconstruct long dissected vessel segments.Conflict of InterestThe authors declare no conflicts of interest.

References

1. Biedermann B, Sojer M, Stockner H, Spiegel M, Schmidauer C. Dissections of cervico-cerebral arteries: causes, symptoms, diagnosis, and therapy. J Neurol Neurochir Psychiatr 2007. 2007;8(2):7–18.
2. Tiu C, Terecoasa E, Grecu N, Nistor R, Frangu S, Antochi F. Vertebral artery dissection: a contemporary perspective. Maedica. 2016;11(2):144–9.
3. Thanvi B, Munshi SK, Dawson SL, Robinson TG. Carotid and vertebral artery dissection syndromes. Postgrad Med J. 2005;81(956):383–8.
4. Winblad JB, Grolie T, Ali K. Subclavian artery dissection. Radiol Case Rep. 2012;7(4):626.
5. Ostovan MA, Kojuri J, Dehghani P. Endovascular repair of the traumatic dissection of the subclavian-axillary artery: report of four cases. J Tehran Univ Heart Center. 2017;12(2):88–91.
6. Nakada T, Idoguchi K, Fukuma H, Ono H, Nakao S, Matsuoka T. Case report: urgent endovascular treatment of subclavian artery injury after blunt trauma. F1000Research. 2014;3:310.
7. Liao M, Chen X, Chen H, Wang Y, Zeng JS, Fan Y. Stent-assisted angioplasty in spontaneous bilateral extracranial internal carotid artery dissection. Front Neurol. 2020;11:582253.
8. Al-Ali F, Perry BC. Spontaneous cervical artery dissection: the Borgess classification. Front Neurol. 2013;4:133.
9. Biffl WL, Moore EE, Offner PJ, Brega KE, Franciose RJ, Burch JM. Blunt carotid arterial injuries: implications of a new grading scale. J Trauma. 1999;47(5):845–53.
10. Markus HS, Hayter E, Levi C, Feldman A, Venables G, Norris J, et al. Antiplatelet treatment compared with anticoagulation treatment for cervical artery dissection (CADISS): a randomised trial. Lancet Neurol. 2015;14(4):361–7.
11. Wang G, Li C, Piao J, Xu B, Yu J. Endovascular treatment of blunt injury of the extracranial internal carotid artery: the prospect and dilemma. Int J Med Sci. 2021;18:944.
12. Jeon P, Kim BM, Kim DI, Shin YS, Kim KH, Park SI, et al. Emergent self-expanding stent placement for acute intracranial or extracranial internal carotid artery dissection with significant hemodynamic insufficiency. Am J Neuroradiol. 2010;31(8):1529–32.
13. Mitchell B, Momin E, Jou L-D, Shaltoni H, Morsi H, Mawad M. Extensive bilateral vertebral artery remodeling following treatment of dissection using pipeline embolic device. J Vasc Interv Neurol. 2014;7(5):5–8.
14. Ahlhelm F, Benz RM, Ulmer S, Lyrer P, Stippich C, Engelter S. Endovascular treatment of cervical artery dissection: ten case reports and review of the literature. Interv Neurol. 2012;1(3–4):143–50.
15. Cohen JE, Leker RR, Gotkine M, Gomori M, Ben-Hur T. Emergent stenting to treat patients with carotid artery dissection. Stroke. 2003;34(12):e254–7.
16. Kondo R, Ishihara S, Uemiya N, Kakehi Y, Nakadate M, Singu T, et al. Endovascular treatment for acute ischaemic stroke caused by vertebral artery dissection: a report of three cases and literature review. NMC Case Rep J. 2021;8(1):817–25.
17. Peng J, Liu Z, Luo C, Chen L, Hou X, Xiao L, Zhou Z. Treatment of cervical artery dissection: antithrombotics, thrombolysis, and endovascular therapy. Biomed Res Int. 2017;2017:3072098.

Endovascular Approach for Curative Embolization of Brain AVMs: Insights from Angio-architectonics and Angio-anatomy

30

Rahul Kumar, Deep Das, Arpan Dutta, Subhadeep Gupta, Kalyan Sajja, and Trilochan Srivastava

Background

Curative endovascular embolization of brain arteriovenous malformations (AVMs) is possible and is being done routinely in selected centres around the world. However, in the spectrum of disorders amenable to endovascular treatment, it is not as simple as coil embolization of an aneurysm or stenting of a carotid stenosis. The complexity of the treatment is a direct result of the complexity seen in the disease process. This in turn is a function of the extreme degree of evolution of the cerebral vasculature that has carried forward elements from the vasculature of our most primitive ancestors. Any discussion on AVM treatment by embolization must therefore look at this truly fascinating process, in order to understand and master the complexities of the human cerebral vasculature.

The human brain, as we know today, is the result of a progressive evolution over the last 250–500 million years (Fig. 30.1). Starting from a clustering of neurons in the Hydra to the non-lamellated but enclosed, non-laminar brain of the Amphioxus containing only seawater in its primitive ventricular cavity, we have come a long way in terms of size, organization, and heterogeneity to see the brain as it is today, in humans. Briefly, the oldest parts of the brain to develop from the non-lamellated structure to a lamellated structure are referred to as the Paleopallium or old brain. Subsequently, we have the appearance of the Archipallium, Subpallium, and finally the Neopallium. This Neopallium is phylogenetically the "youngster" in the human brain. All these steps of development are summarized in the diagram below. The reader can also refer to dedicated monograms on the subject.

Brain AVMs are a Neopallial disease, predominantly as about 70% of these occur in the Neopallium. Therefore in this entire manuscript, the detailed discussion will be on the Architectonic, Angio-architectural properties and the embolization techniques that are used in Neopallial AVMs. An attempt will be made to touch upon the peculiarities and precautions that need to be taken while dealing with, let us say, choroidal AVMs that are Paleopallial and Archipallial in nature.

There are three prerequisites to a successful curative embolization of brain AVM. These are as follows:

1. The ability to place the lesion in a pre-defined Architectonic compartment of the brain, with the intention of being able to predict the expected blood supply and the angio-architecture of the lesion.
2. The ability to be able to access with a microcatheter, the exact vascular pedicle(s) that is supplying only the AVM and no other structures in the vicinity.
3. The ability to master the flow within the "nidus" and thereafter the deposition of an embolic material within it. This may be done with one or several successive catheterizations. It is imperative that, at the end of embolization, the draining vein(s) be also occluded for a permanent cure.

Each of these is equally important to the successful endovascular outcome in treating these extremely complex lesions. In isolation, all are inadequate. But when used as a

R. Kumar (✉)
G S Neuroscience Clinic and Research Center, Interventional Neurology, Patna, India

D. Das
Department of Neurology, Bangur Institute of Neurosciences, Institute of Post Graduate Medical Education and Research, Kolkata, India

Woodlands Multi-speciality Hospital and C K Birla Hospitals, Kolkata, India

A. Dutta · S. Gupta
Neurology Bangur Institute of Neurosciences, Institute of Post Graduate Medical Education and Research, Kolkata, India

K. Sajja
Life Hospital, Guntur, India

T. Srivastava
SMS Medical College and Hospital, Jaipur, India

G. Geroulakos et al. (eds.), *Mastering Endovascular Techniques*, https://doi.org/10.1007/978-3-031-42735-0_30

combination, they become a powerful tool that has been considerably neglected over the last few decades. The first point of the Architectonic organization of the different brain regions is beyond the scope of this chapter. However it is also the most important part in the entire planning process and will be touched upon in brief. The second and third points regarding the microcatheterization and the embolic material deposition will be the main focus of this manuscript.

Architectonic Organization of the Telencephalon and Location of Brain AVMs

The telencephalon consists of 16 distinct, interdigitating histogenetic units, in each hemisphere. The first aim is to identify the limits of the telencephalon. The telencephalon, or "end-brain", as it is called is the newest component of the human brain in terms of evolution (Fig. 30.1). It has also been established that the Insula is the first component of the neopallium to develop, and the subsequent evolution of the Neopallium takes place from the Insula. Regular discussion on imaging in modern medical literature has completely neglected these evolutionary landmarks and has not allowed us to look at the details for the simple reason that all the data available on the brain is dependent on axial sections of the brain that show a piecemeal view of all these structures rather than show us a holistic picture.

For detailed understanding of the different Architectonic compartments of the brain and the different landmarks that separate these structures, one needs to understand the orientation in a coronal/sagittal plane. Also the interaction between the various structures can only be studied when all these are visible in a single cut. To understand the boundaries between the old and the new brain, we are therefore moving to the coronal section at the midventricular level, passing through the corpus striatum, which is ideal, as is shown in the image below.

The three oldest sulci in the order of appearance, both during phylogeny and embryologic ontogeny, are the Cingulate Sulcus, the superior and inferior limbs of the Circular Sulcus, demarcating the Insula, and the Collateral Sulcus (Fig. 30.2a). An imaginary line passing through these on both sides ultimately creates a boundary, with the long association fibre bundles of the white matter at this interface (Fig. 30.2b). Any part of the brain that is within this "phylogenetic circle" is old brain and contains the Archipallium, Paleopallium, and Subpallium. The entire structure that is seen outside this phylogenetic circle is the Neopallium and will remain the main focus of this description.

Once we have identified the telencephalon, we now have to move away from the coronal orientation and look at each hemisphere in toto, in a sagittal orientation. This helps us in understanding the location and the organization of the 16 histogenetic units in each hemisphere. For ease of understanding, we are showing these concepts in the diagram below (Figs. 30.3 and 30.4).

The Central Sulcus of the Insula divides it into an antero-superior part containing the gyri brevi, and a postero-inferior part containing the gyri longi. Anatomically, the Central Sulcus of the telencephalon can be conceptualized as a continuation of the Central Sulcus of the Insula. Thus when we connect these two, the entire Neopallium gets divided into two parts as illustrated. The anterior part is referred to as the "Epistriatal Zone" as it straddles the Corpus Striatum, and the posterior part is referred to as the "Epihippocampal

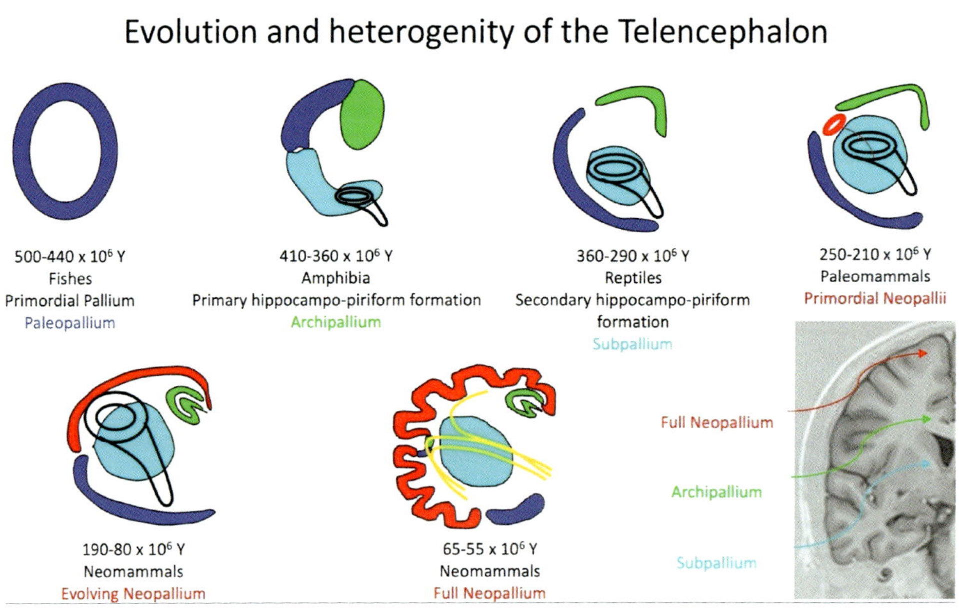

Fig. 30.1 Composite image showing the evolution of the telencephalon from the Paleopallium that consisted of a non-layered conglomerate of nerve cells surrounding a cavity filled with ocean water instead of CSF to the complex structure that it is today in terms of the Full Neopallium. All the structures shown in dark blue colour represent the Paleopallium, those in green represent the Archipallium, Turquoise represents the Medial and Lateral Ganglionic Eminences that ultimately form the Subpallium, and Red represents the Insula and the Neopallium

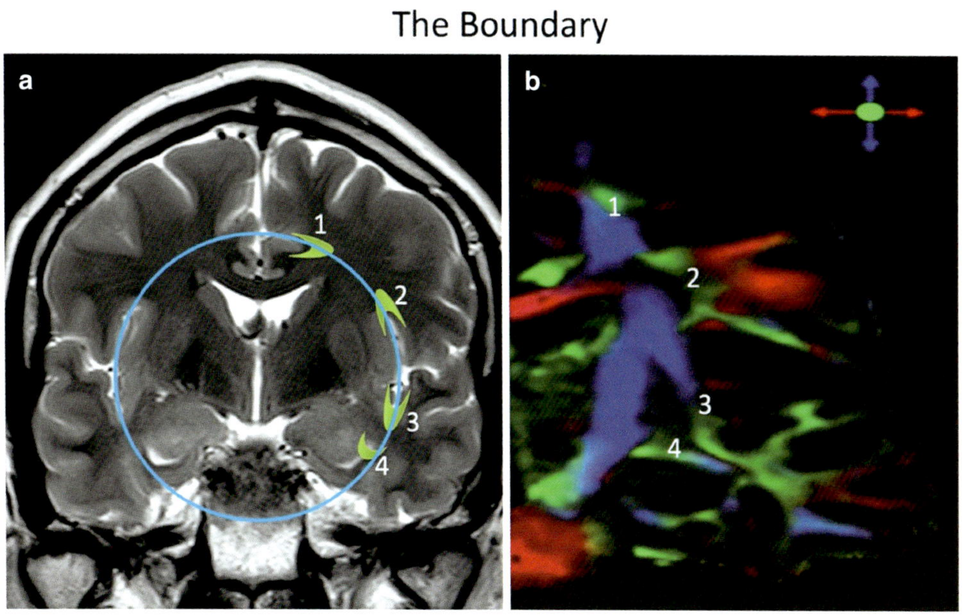

Fig. 30.2 (**a**) A coronal MR through the corpus striatum showing the three primordial sulci. These are the Cingulate Sulcus (1), the superior and inferior limbs of the Circular Sulcus (2,3), and the Collateral Sulcus (4). (**b**) The DTI image with the reconstruction of the association fibre systems in green also shows the presence of a long association fibre bundle at all these locations (1, 2, 3, and 4)

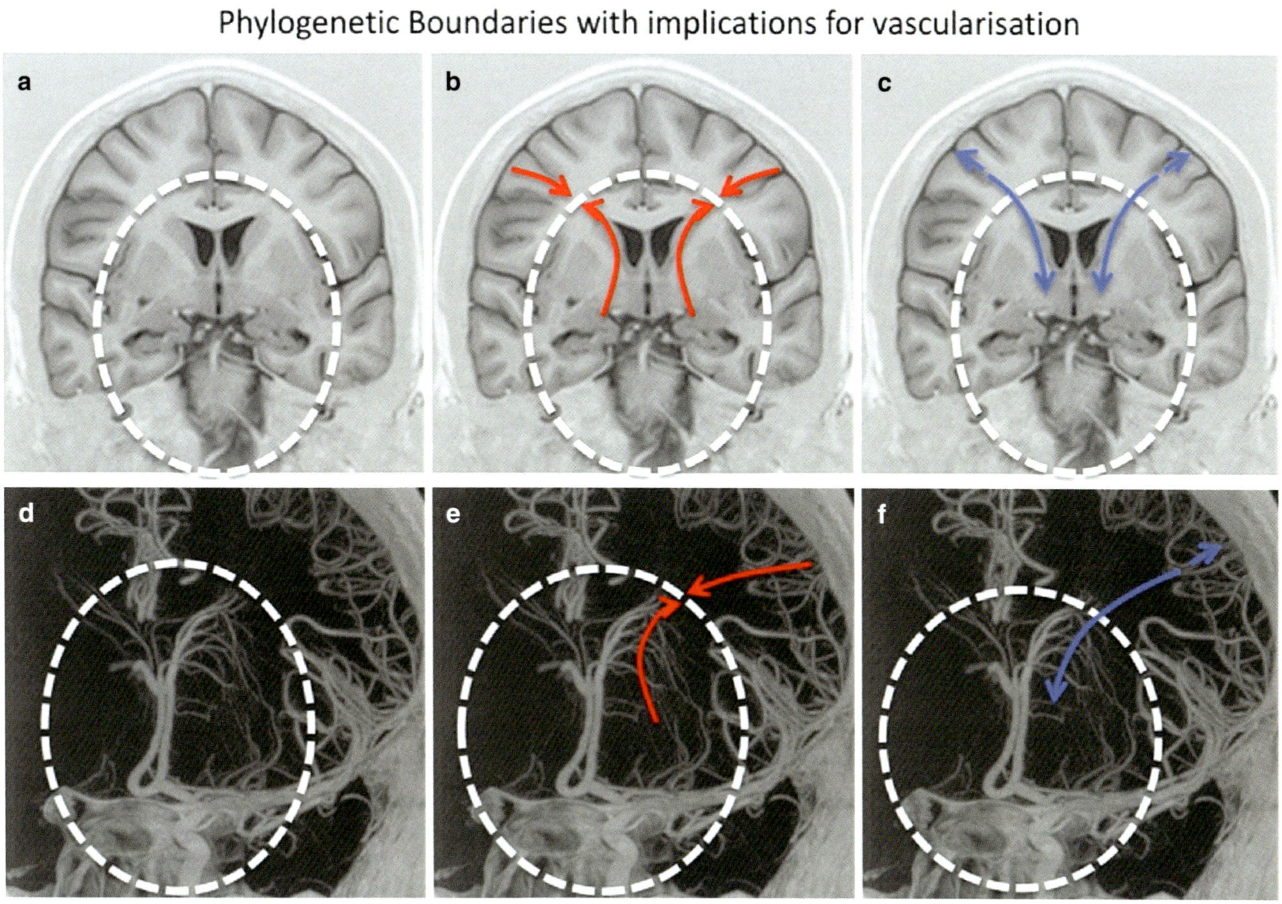

Fig. 30.3 This figure shows the placement of the phylogenetic circle on coronal MR IR images (**a**). The upper panel also shows the expected extent of the supply from the superficial and deep systems in terms of arterial supply and venous drainage (**b** and **c** respectively). The arterial supply is marked in red and the venous drainage is marked in blue. As far as the arterial system is concerned, the demarcation is very strict and is right at the ventricular edge. This also corresponds to the phylogenetic circle. All the areas within the phylogenetic circle will be supplied by either the striate system or the choroidal system (**d**). All the areas outside the phylogenetic circle will be supplied by the leptomeningeal arterial arcade (**d**). The lower panel shows the vessels, as seen on cone beam CT angiographic exploration in relation to the circle, the extent of supply of the striates, and the penetrating supply from the leptomeningeal arcade (**e**). Also seen is the minimal territory of drainage of the cortical veins (**f**)

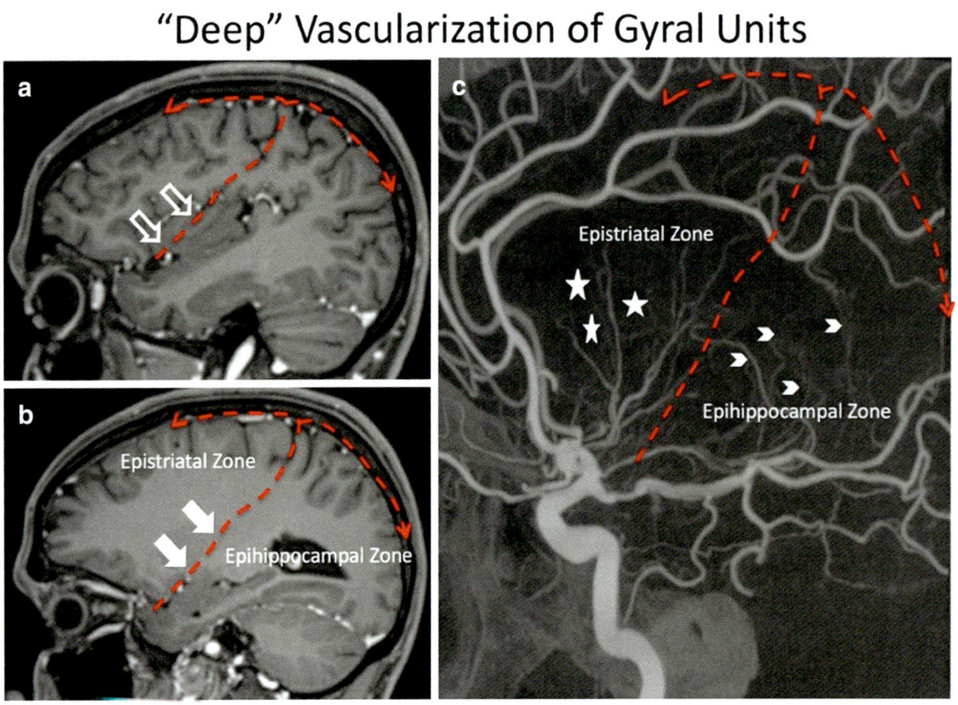

Fig. 30.4 The images on the left hand side (**a**) show a sagittal section passing through the Central Sulcus of the Insula (open arrow) on the top image and through the Central Sulcus of Rolando (closed arrow) in the bottom image (**b**). It can be seen that these are continuous anatomically and divide the telencephalon into an anterior part and a posterior part. The entire anterior part is referred to as the "Epistriatal Zone" and the entire posterior part is referred to as the "Epihippocampal Zone". Each of these contains eight histogenetic units. The image on the right (**c**) is a cone beam CT angiography with the Central Sulcus approximately superimposed, showing the striates (stars) related to the anterior and the choroidals (arrowheads) related to the posterior

Zone" as it is in relation with the Hippocampal formation. Each of these, i.e. the Epistriatal Zone and the Epihippocampal Zone, contains eight histogenetic units. Each of these histogenetic units, while being completely different from each other in terms of development, genetic markers, and protein expression, is remarkably similar in terms of their overall shape, orientation, and blood supply.

Each histogenetic unit has a stem of white matter, originating from the periventricular area, which divides into two or more as it reaches the subpial location. This is capped by its grey matter, and this entire construct is referred to as a gyral unit (Fig. 30.5). Gyri are much more than 16 in each hemisphere; however the white matter stems in each hemisphere are only 16, and these divide and subdivide to produce sub-gyri, which interdigitate to produce the Neopallial gyral formation.

These "gyral units" are pyramidal in shape, with the pointed apex towards the ventricle and the broad base towards the pial surface. Eight of them interdigitate with each other in the Epistriatal and Epihippocampal Zones, thus producing the entire construct of the Neopallium.

As already mentioned, brain AVMs are predominantly found in the Neopallium. Thus in order to understand the angio-architecture, one needs to have a detailed understanding of the normal angio-architecture of a Neopallial Gyral Unit. This will define the "regional" Angio-architecture of the lesion. Once that is identified, the second step will be to elaborate the Angio-architectural, intralesional variations.

Arterial System: The Neopallial, Leptomeningeal Angio-architecture

The entire Neopallium is covered by a network of blood vessels originating from the branches of anterior, middle, and posterior cerebral arteries. Middle cerebral contribution is dominant on the convexity, anterior cerebral contribution is predominant on the medial surface, and the posterior cerebral completes the network on the inferior surface. Since the majority of the Neopallium is represented on the convexity, it is reasonable to understand the pial leptomeningeal Angio-architecture based on the middle cerebral artery as a prototype (Fig. 30.6).

A leptomeningeal branch enters and dips into a Sulcus, where it is called the "descending sulcal" segment. This artery reaches the depths of the Sulcus, up to the fundus, before taking a 180° turn and travelling up the opposite lip of the Sulcus, referred to as the ascending sulcal segment. Upon

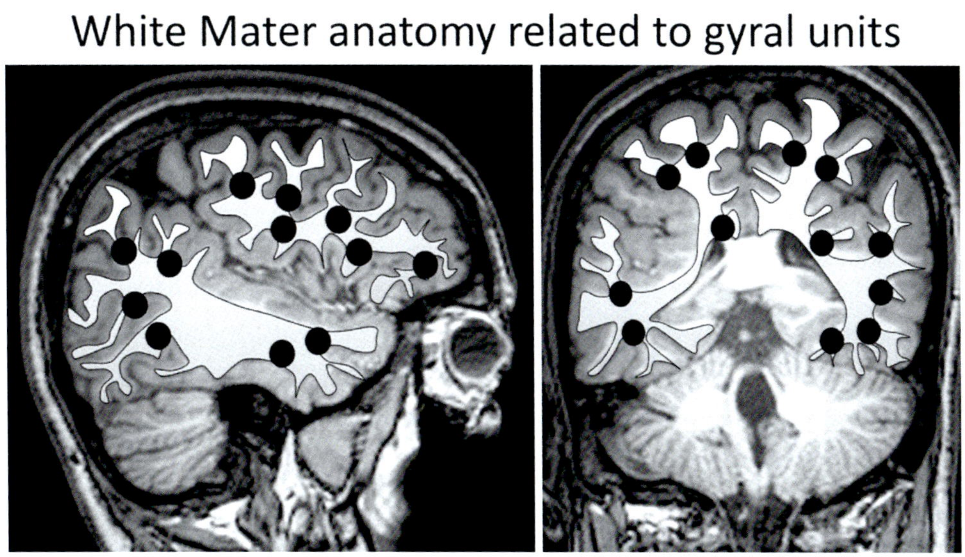

Fig. 30.5 T1-weighted sagittal and coronal sections. Conceptualization of the white matter stems (black circles) superimposed over the white matter showing the branching white matter stems being capped by the grey matter to form the interdigitating gyral structure

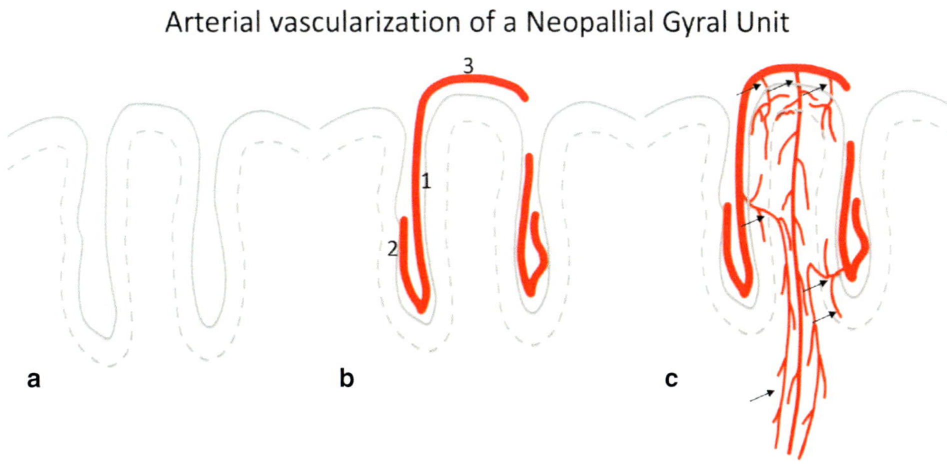

Fig. 30.6 The upper left-hand side image (**a**) shows the pial surface (grey solid line) as well as the grey-white junction (grey dotted line). The second image in the top panel on the right-hand side (**b**) shows the arterial system superimposed. It shows the descending (1) and ascending (2) Sulcal segments as well as the transverse gyral segment (3) going over the crown of the gyrus. The third image (**c**) shows the final terminal twigs originating from these segments (arrows), ultimately supplying the grey and white matter to various extents, and by extension supplying the AVMs located in the telencephalon

reaching the pial surface, this arterial twig continues its journey towards the next Sulcus, across the crown of the gyrus. Here it is important to note that arteries in the Neopallium cross transversely over the crown of the gyrus. This pattern continues repeatedly over all the gyri thus creating a network of vessels. This network is the main structure that supplies the grey and white matter of the individual gyri via the next-order branches that will be discussed below.

Small, 200-micron penetrating branches originate from this leptomeningeal network and pierce the grey matter to enter the white matter of the gyrus. These usually originate from the mid-transverse gyral segment, from the mid ascending or descending sulcal segment and also from the fundal segment of the leptomeningeal artery. The small twigs might be very small and supply only a couple of cortical grey matter layers, or they might be very long and supply up to the periventricular deep white matter.

The "Deep" Supply to Neopallial AVMs

As elucidated earlier, the "phylogenetic boundary" between the Neopallium and the rest of the brain lies at an imaginary line joining the three primordial sulci. This has important implications for the vascular territories of the arterial systems. All the brain areas lying "outside" this circle is supplied by the leptomeningeal angio-arcade made up of vessels from the middle cerebral, anterior cerebral, and posterior cerebral arteries. All the brain areas lying "inside" this phylogenetic circle derive the blood supply from the striate system or the choroidal system, depending on whether they lie in the Epistriatal Zone or in the Epihippocampal Zone. The internal watershed lies at the phylogenetic boundary. This automatically tells us that if a Neopallial AVM crosses the boundary zone, the deep supply to the AVM will come from the striate or from the choroidal system. If the Neopallial AVM is located in the Epistriatal Zone, i.e. anterior to the Central Sulcus, the deep supply will mandatorily come from the striate system. As for a Neopallial AVM located in the Epihippocampal Zone, the deep supply, if it transgresses the phylogenetic boundary, will come from the choroidal system.

Venous Drainage of the Neopallium

There are significant differences between the arterial and venous systems of the Neopallium. Unlike the arterial system, the venous system has many more built-in redundancies that make it more complex and also more forgiving (Fig. 30.7).

A Neopallial Gyral Unit has two marginal veins on either side in the sulci that demarcate the gyrus. These Marginal Sulcal Draining Veins are formed by the union of many smaller veins and venules from the substance of the gyrus. As far as the sharing of the drainage of the gyrus between the superficial leptomeningeal and the deep venous systems are concerned, it is not as straightforward as the arterial system. The phylogenetic boundary is not respected here, like in the arterial distribution.

A small portion of the Neopallium drains into the superficial venous system. The remainder of the Neopallium that is

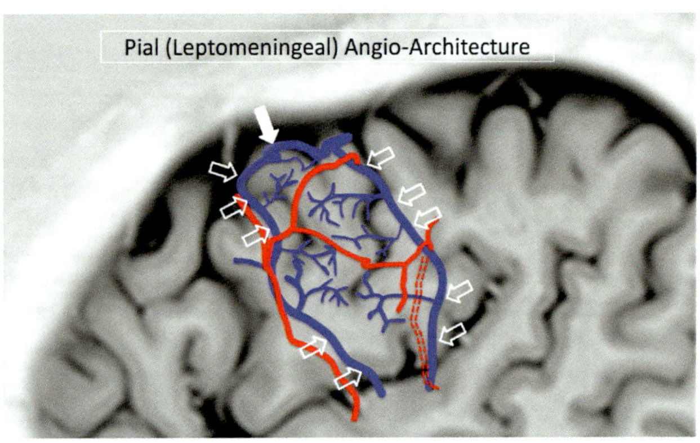

Fig. 30.7 The angio-architecture of a gyral unit superimposed over a gyrus as seen on MR. Arteries in red and veins in blue. Unlike arteries, veins do not cross over the gyrus. Instead they are seen in the sulci and are referred to as the Marginal Gyral Veins (open arrowheads). Many of these coalesce to form a cortical vein (closed arrowhead) that empties into a venous sinus

outside the phylogenetic boundary, as well as the entire brain parenchyma within the phylogenetic circle, drains into the deep venous system.

Therefore, apart from the very small, sulcal types of AVMs, all the Neopallial AVMs of reasonable size will have a dual venous drainage into both the superficial and deep venous systems. This does not correlate with the phylogenetic boundary and needs to be kept in mind while planning the curative embolization of brain AVMs.

Since brain AVMs are located in the Neopallium predominantly, the regional Angio-architecture of the lesion will be based on the Angio-architectural characteristics of the Neopallium elaborated above. Broadly speaking, these are of two distinct subtypes:

1. A situation in which the AVM is located in the subpial location, covered by a thin glial limiting membrane, where a leptomeningeal artery is seen to end, haemodynamically, in the lesion. These are referred to as "Sulcal" AVMs.
2. A second situation, where the AVM is located within the Neopallium, in the subcortical white matter, covered by the grey matter all around it. In this situation, the feeding arteries are the 200 micron twigs mentioned earlier that pass through the cortex and then either supply the AVM exclusively or supply both the AVM and the brain parenchyma. These are referred to as "gyral" AVMs.

Very often during a clinical case, it becomes difficult to compartmentalize a particular AVM into a strict sense, and most often the lesion will demonstrate a mixed Angio-architecture that shows some properties of each subtype (Figs. 30.8 and 30.9).

Nidus

In the available literature on brain AVMs, the description of the "nidus" is nebulous at best. Any part of the Angio-architecture apart from the named arterial twig(s) is clubbed together as the "nidus". It has been well established now that the AVM nidus is mainly a venous structure. These are based on histopathological as well as immunological findings and are in concurrence with the material that has accumulated over the years from the interventional literature. Beyond this, there is not much information, especially in vivo.

Over the last decade, the authors of this chapter have developed techniques that allow us to study the angio-architectural characteristics of the nidus in vivo with the information obtained during routine clinical work of curative embolization sessions of brain AVMs. Just a small introduction to in vivo catheter cone beam CT angiography (CCBCTA) is mentioned here for reference.

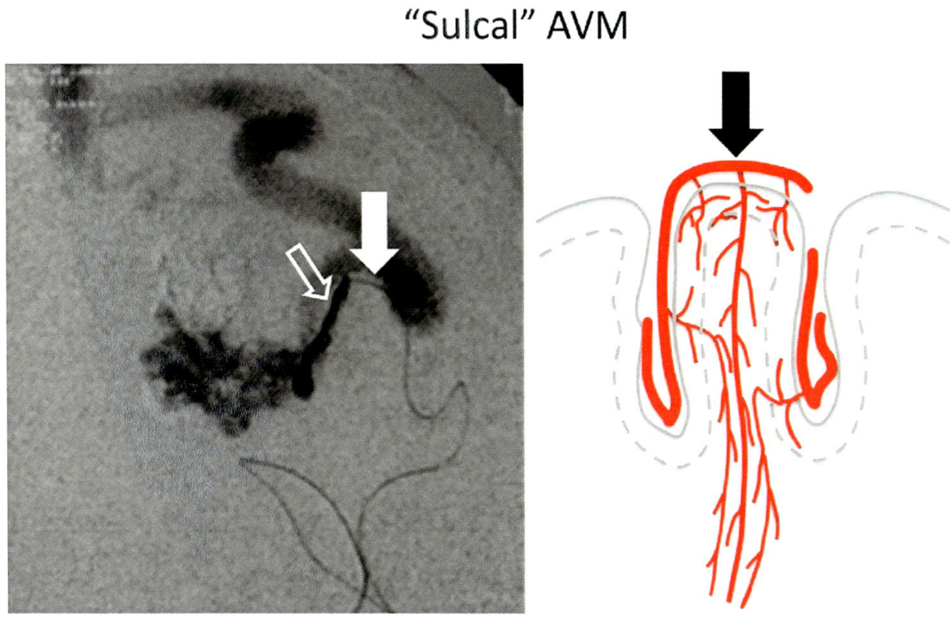

Fig. 30.8 An example of a microcatheter angiogram of a Sulcal type of AVM. The microcatheter is seen crossing over the crown of the gyrus (closed arrow) and is then seen entering the AVM nidus from the descending sulcal segment (open arrow), with no supply to any surrounding structures

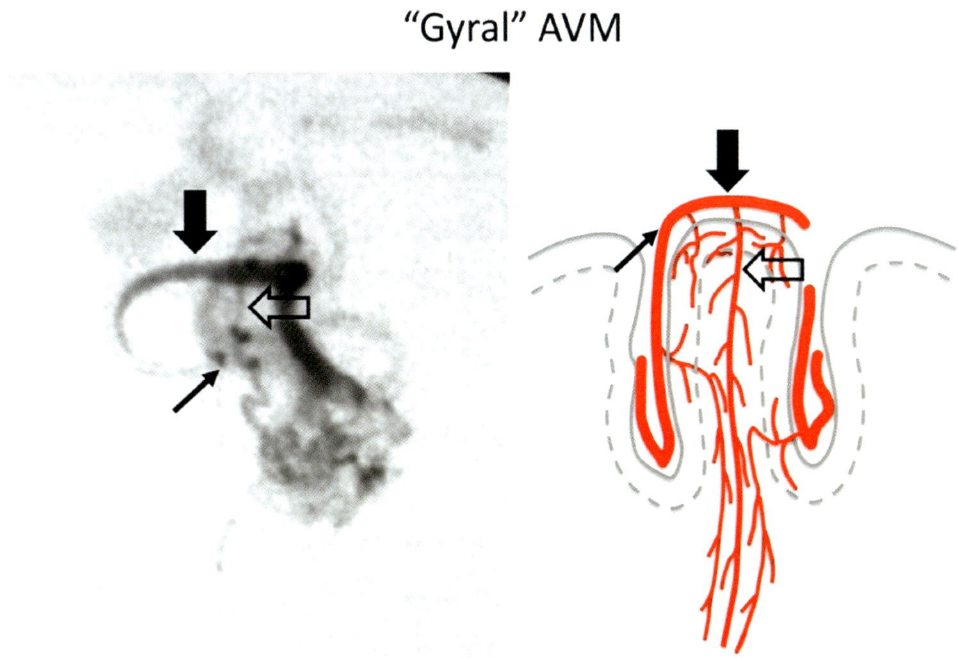

Fig. 30.9 An example of a microcatheter angiogram of a gyral type of AVM. The microcatheter is seen in the ascending sulcal segment (arrow) just before crossing over the crown of the gyrus (arrowhead). From the midpoint of the transverse gyral segment on the crown of the gyrus, a medullary branch (open arrowhead) can be seen entering deep into the white matter of the gyrus and supplying the AVM. The transverse gyral segment continues over to the next histogenetic unit

Catheter cone beam CT angiography (CCBCTA) is a technique performed in the interventional suite during catheter angiography. This study can be done on any current generation flat-panel angiography system. The study is done with an effective detector size of 20x20 cm, at magnification. A 0.5 k by 0.5 k matrix is used for the raw image acquisition. The rotation speed is 10° per second, and the total coverage arc should be at least 200° around the patient. Therefore, the total rotation time will be about 25 s, which includes the true rotation time of 20 s as well as a 5-s delay between the beginning of contrast injection and the beginning of image acquisition. The images are acquired under apnoea induced by the anaesthesiologist to avoid any motion-related artefacts. The contrast agent used is a diluted mixture of 25% contrast (Omnipaque 300, GE Healthcare, USA) with normal saline, injected by a pressure injector, at 200 PSI pressure at a rate of 3 mL/s for 20 s.

With this technique, a very high-resolution image dataset is acquired that shows us the entire nidus preprocedure. The same study is repeated after embolization is complete, and this time the acquisition allows us to make a 3D model of the cast of embolic material that has been filled inside the AVM. The cast of the AVM nidus is a wonderful surrogate of the angio-architecture of the nidus per se. Combining the information obtained from many such studies, an interesting three-dimensional structure of the nidus has evolved (Figs. 30.10 and 30.11).

By superimposing many such images, we have come to realize that the AVM "nidus" is a very complex structure consisting of at least three zones, and the morphological characteristics of the three zones are very different from each other (Fig. 30.12).

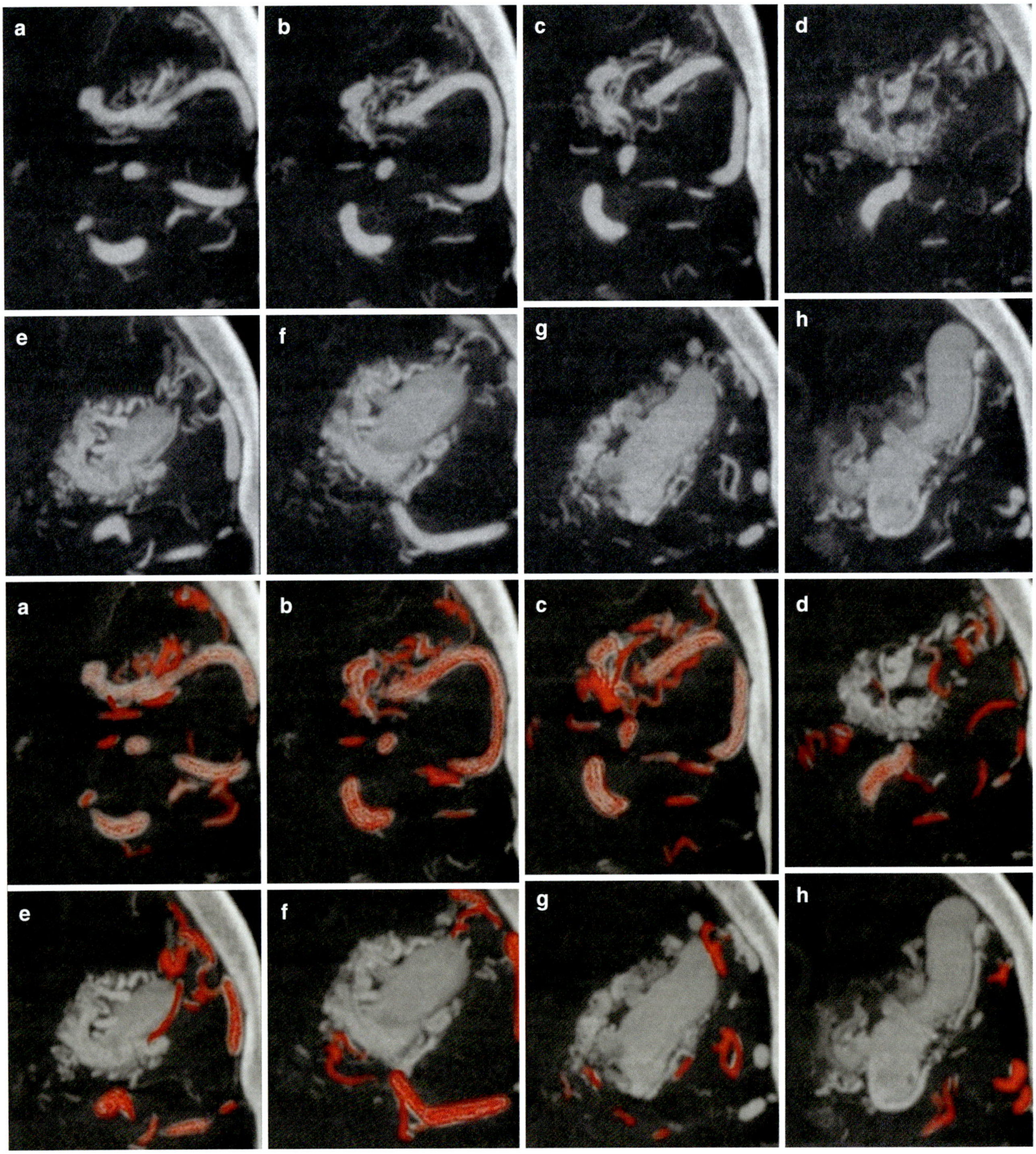

Fig. 30.10 Sequential 0.2 mm sections through the AVM as seen on catheter cone beam CT angiographic exploration. The uppermost panel, consisting of images A through H, shows the native images as obtained during the exploration. Thereafter, each of the components, i.e. arteries, veins, and the nidus, has been colour coded. Arteries in red, veins in blue, and the nidus in pink. The bottom panel shows a composite impression of the entire AVM structure

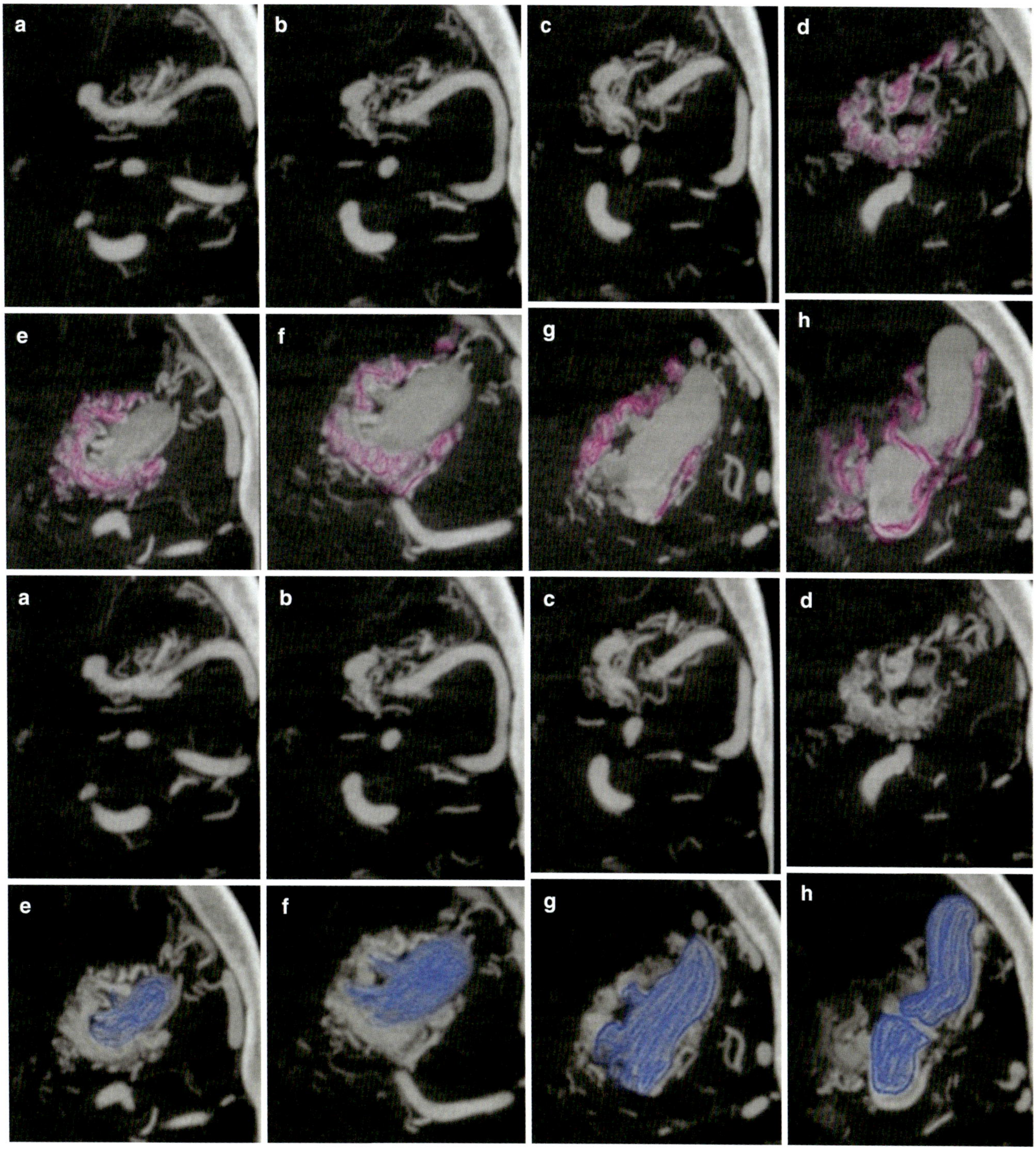

Fig. 30.10 (continued)

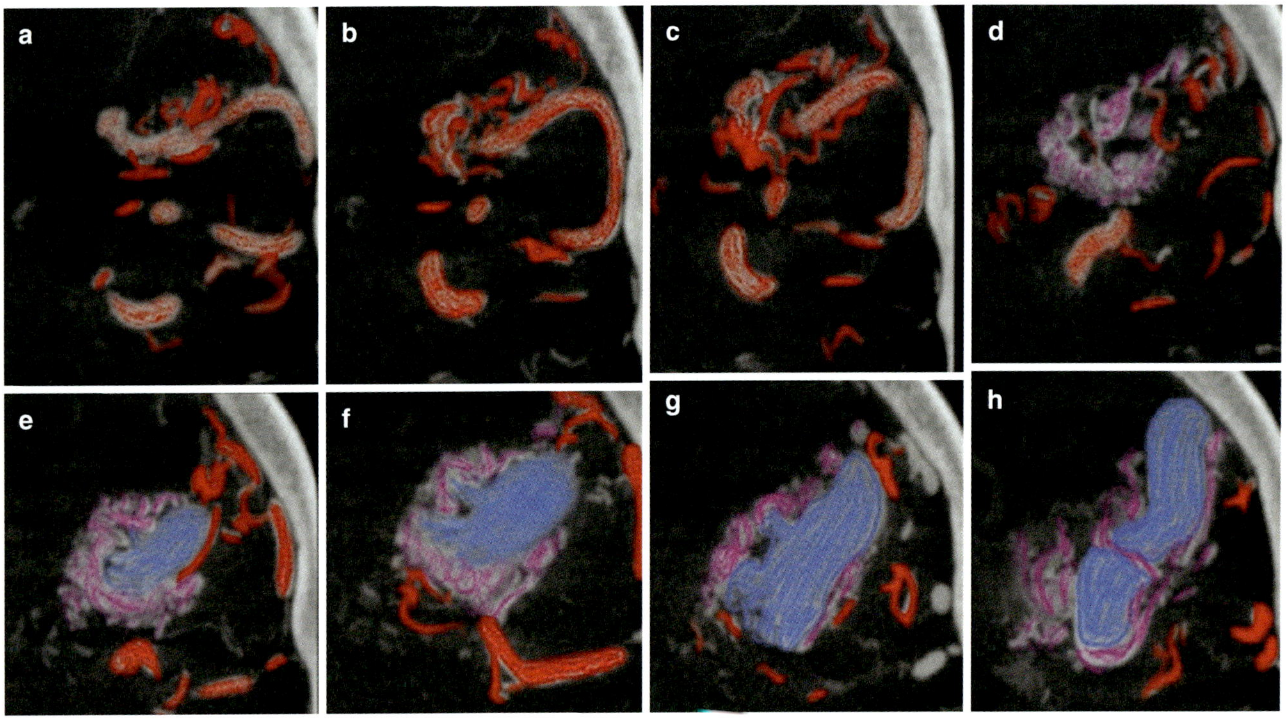

Fig. 30.10 (continued)

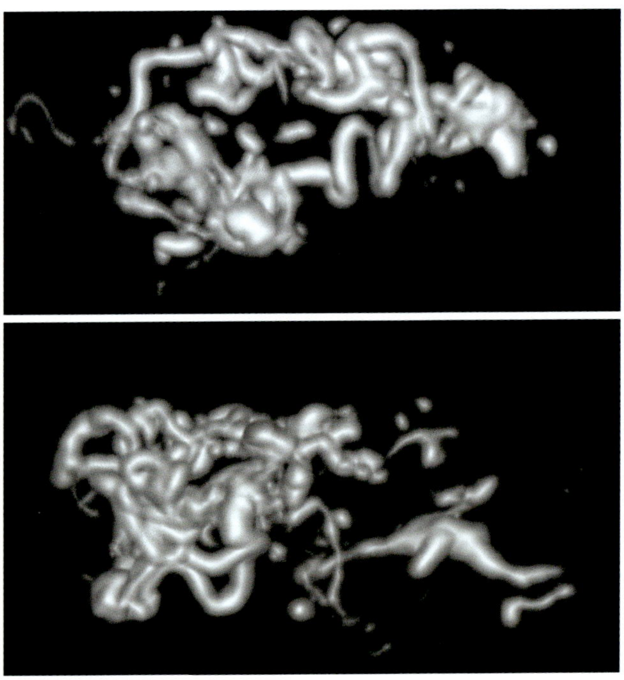

Fig. 30.11 Sequential 0.2 mm sections through the AVM glue cast post-embolization as seen on catheter cone beam CT angiographic exploration. These are AP and lateral view images of the glue cast showing the details of the nidus per se

As already mentioned, a brain AVM within a histogenetic unit of the Neopallium can be visualized as a pyramidal structure, with its base towards the convexity and the apex towards the ventricle. This AVM nidus has three distinct zones and many more different types of vessels within itself that have very different fragility and morphology.

The outermost zone is arterial, and this zone contains the ramifying, arborizing arterial twigs from the leptomeningeal system that form a dense meshwork on the outer aspect of the pyramid. This is the area where the arterial system meets the venous system, and the arteriovenous shunts are created at this point. Angio-architecturally, these are pretty straightforward. This is also the zone that is accessible to the interventionalist for catheterization when doing a trans-arterial embolization.

The complexity of the nidus starts when one reaches the intermediate zone of the AVM. This is a venous zone and, also in case of Neopallial AVMs, is the area that holds the "compartments" of the AVM. This is a zone made up of venous structures. The various types of vascular channels found in this zone are the spiral types, the loose vertical loopy type, the loose vertical plexiform type, and the very distinct subtype of channels located at right angles to the previous subtypes. This fourth type represents the inter-compartmental connections, and it is the most delicate in terms of fragility. Most of the intranidal aneurysms in our material were on this subtype of the intermediate zone vessel. The innermost zone is the collector zone. All the compartmental draining veins converge at this level and exit as the nidal draining vein. In the recent years, this zone has also become a potential target for the treatment of brain AVMs by the transvenous route.

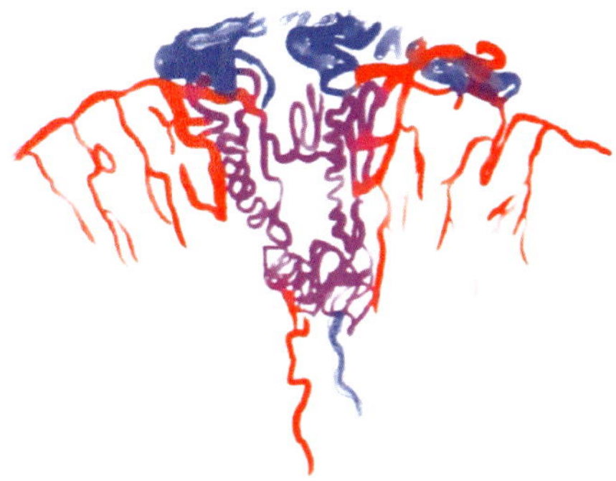

The Vascular Zones of the Nidus

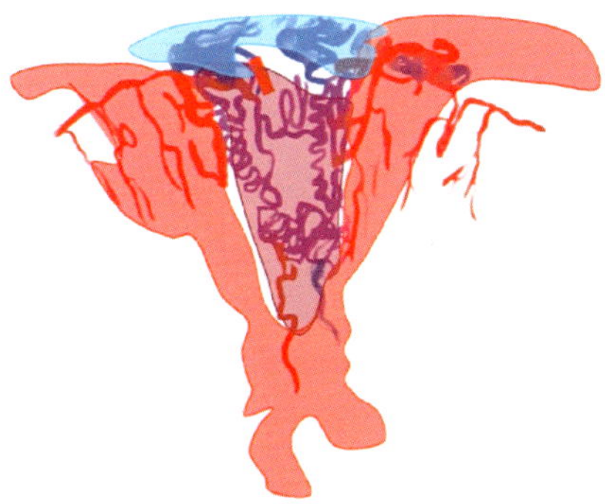

Nidal Zones	Fragility	Vessel Type
Peripheral Zone	+	Arterial
Intermediate Zone		
1) Spirally Coiled, vertical – 0.15-0.5mm	+	
2) Tubular, loose vertical – 0.3-0.5mm	+	Venous
3) Vermiform, loose vertical – 0.4-0.7mm	++	
4) Tortuous, transverse – 0.3-1mm	+++	
Zone of Venous Collection Thin walled channels converge towards the draining veins	++	Venous

Fig. 30.12 A composite image of the AVM nidus as seen during both in vitro and in vivo explorations. The types of vascular structures found within the nidus are elaborated in the colour-coded diagram on the left and in a tabular form on the right

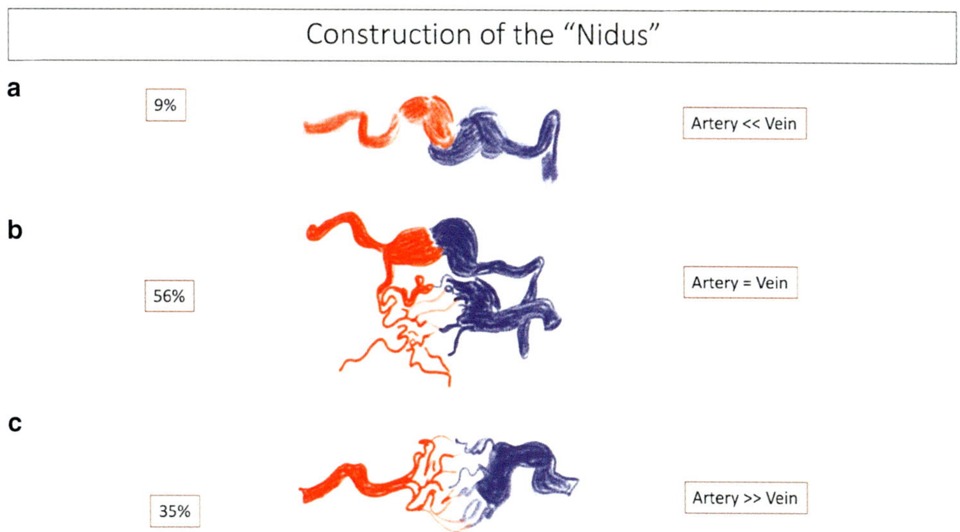

Fig. 30.13 A composite image of the AVM nidus types as seen during both in vitro and in vivo explorations. The rarest subtype seems to be the pure "fistulous" type (**a**), where an artery is seen connecting directly to a vein. These are high-flow lesions. The most common type seems to be a mixture of fistulous and plexiform (**b**), followed by the pure plexiform types (**c**)

Nidal Subtypes

Based on CCBCTA studies of the treated AVMs at our institute as well as microcatheter explorations of these AVMs, we have been able to subclassify these into at least three different groups (Fig. 30.13).

1. Pure fistulous—Like pial fistulae. These are the rarest subtype of brain AVMs.
2. Pure plexiform—Our belief was this would be the most common subtype, but it happens to be the second most common type, representing a third of cases.

3. Mixed fistulous and plexiform—More than half of the cases embolized belong to this subgroup and is the most common phenotype that is encountered during the embolization sessions.

It is easy to understand that these three subtypes of brain AVMs represent completely different anatomical, angio-architectural, and haemodynamic situations; a "one size fits all" approach towards brain AVMs is more than likely to produce suboptimal results.

At the end of this discussion, the regional and lesional anatomy, angio-architecture, and possible sites of intervention for the curative embolization are clear. One has to be able to get into the outermost arterial zone of the nidus with a microcatheter(s), deposit the embolic material within the nidus, either one compartment at a time or progressively through the inter-compartmental connections, and occlude just a small part of the nidal exiting vein for complete and lasting cure.

Historical Aspects of AVM Embolization

Luessenhop and Spence performed the first-ever endovascular embolization of an AVM [4]. Since then a significant progress has taken place in the endovascular techniques and materials that has allowed higher cure rates with the endovascular modality. The understanding of the angio-architecture and the flow dynamics has improved remarkably over time, especially with the advent of newer and superior imaging modalities, aiding in proper selection of patients, modalities of treatment, and better outcomes. Endovascular embolization of cerebral AVMs has evolved from a simple technique of blocking the feeding arteries of an AVM to a highly sophisticated technique of selecting and accessing the feeding arteries to reach and obliterate the nidus of an AVM [1]. Materials for endovascular embolization have also evolved from polyvinyl alcohol (PVA) particles to n-butyl cyanoacrylate (NBCA) glue to non-adhesive liquid embolic agents comprising ethylene vinyl alcohol (EVOH) copolymers dissolved in dimethyl sulfoxide (DMSO) and micronized tantalum powder (Onyx, Squid, or PHIL). Also there has been an enormous progress in the field of neurovascular access systems and devices with the refinement of flow-guided and combined flow- and guidewire-directed microcatheters since the mid-1980s, allowing for precise, safe, and super-selective access to AVMs in any location. Due to all these factors, endovascular embolization of AVMs nowadays can achieve complete cure in many cases without requiring additional treatment modality.

Role of Embolization in the Management of AVM

The application of endovascular embolization in the management of AVM has expanded due to the recent development in the field of neurovascular intervention. Embolization of an AVM is no longer considered a mere adjuvant treatment modality. Currently the indications of embolization in the management of AVM include the following.

1. **Presurgical Embolization**

   Endovascular embolization can be done to reduce the size of a large AVM, thus making it amenable for microsurgery or radiosurgery. In these scenarios, embolization must be done in the immediate preoperative period, and a delay of several days can make the procedure useless because of AVM regrowth [5].
2. **Therapeutic Embolization**

   Endovascular embolization alone may completely obliterate an AVM. Appropriate selection of patients based on certain factors as described below is crucial for the success of this procedure. With the advent of newer imaging modalities, devices, and improved understanding of the AVM angio-architecture, nowadays, therapeutic embolization is being performed in more and more number of AVM cases.
3. **Palliative Embolization**

   Many a time, complete obliteration of an AVM is not possible via any of the three treatment modalities. In such cases, embolization to eliminate the vascular elements responsible for focal neurological deficits or to reduce the risk of haemorrhage by targeting the weak areas of the AVM angio-architecture can be done. However, it needs to be emphasized that such palliative procedure is not indicated for epilepsy or headache related to AVM because these symptoms can be controlled with medical management, and AVM-related seizures occur probably due to reactive gliosis in the adjacent brain parenchyma, which cannot be altered by embolization.

Embolization Agents

1. **N-Butyl-Cyanoacrylate (NBCA) Glue**

   NBCA glues act as occlusive agent. They are available in liquid form, and when they come into the contact of an ionic medium like blood, they polymerize rapidly to become solid. This polymerization can be slowed by diluting NBCA with lipiodol. More dilution means slower polymerization. Usually higher glue concentrations (around 50%) are required to embolize the high flow or fistulous type of AVMs, while lower concentrations (around 30%) are required for the slow flow or plexiform type of AVMs. A delay in the appearance of a vein by selective angiography at two frames per second can be equated to one volume of lipiodol for one volume of NBCA per image [5]. For example, if a vein appears after three images then three volumes of lipiodol per volume of NBCA (or 25% glue concentration in other terms) will be appropriate for the embolization of that AVM. However, the selection of the most appropriate glue concentration for AVM embolization depends highly on the operator's experience, and mixed type of AVMs may require different glue concentrations for its different parts or compartments. NBCA is radiolucent, while lipiodol is radio-opaque. Hence, lipiodol helps in the visualization of the mixture during endovascular procedure. Before injecting glue, the microcatheter needs to be rinsed with a 5% dextrose solution to prevent the polymerization of NBCA inside the microcatheter. Glue injection should be carried out within a period of 10–15 s to 1–2 min, depending upon its concentration. Longer injection time will cause NBCA to polymerize, and the microcatheter will become stuck. This is a disadvantage of the glue. Another disadvantage is that glue poorly penetrates to the adjacent compartments of a multi-compartmental AVM, and each compartment needs to be embolized separately. However, glue penetrates very well towards the venous part of an AVM, and thus there is a better chance of occlusion of the venous connections with glue. Glue allows immediate occlusion of a vascular rupture site which can be lifesaving in certain situations while performing endovascular procedure. Another advantage of glue is that it can be injected through thinner microcatheters allowing for successful embolization of AVMs in difficult locations like deep-seated AVMs. Metacryloxysulpholane when mixed with NBCA as a co-monomer lowers the polymerization temperature and prolongs the polymerization time. This mixture is marketed as Glubran2. N-hexyl cyanoacrylate, marketed as Magic Glue, is another occlusive agent similar to NBCA glue, with lower adhesive strength that allows for prolonged injections up to 1 h [6].

2. **Ethylene Vinyl Alcohol (EVOH) Copolymer**

   EVOH is a liquid agent held in suspension by dimethyl sulfoxide (DMSO) solvent. This solution is made radio-opaque by mixing with tantalum powder. The mixture is marketed as Onyx and Squid. After injection into the vessel, DMSO diffuses slowly, and EVOH deposits in the vascular lumen thereby occluding it. As this process is much slower compared to polymerization of NBCA, Onyx/Squid allows for longer injection time. This is the main advantage of EVOH copolymer over NBCA glue. However, EVOH copolymers cannot be injected through any type of microcatheter as DMSO can damage the walls of non-compatible microcatheters. Onyx/Squid preferentially percolates into the arterial parts of an AVM and frequently refluxes into the vascular lumen thereby trapping the microcatheter. This situation can be catastrophic, and microcatheters with detachable tips are required to minimize the risk of this complication while using EVOH copolymers. Such microcatheters are not easy to navigate through narrow vessels. Hence, Onyx/Squid is not a good choice for the embolization of AVMs in difficult location or in case of small-calibre arterial feeders. NBCA glue has advantages in these situations. Onyx is available in three different viscosities—Onyx 18 (6% EVOH), Onyx 20 (6.5% EVOH), and Onyx 34 (8% EVOH). Onyx 18 has the lowest viscosity and is used most commonly. Higher-viscosity Onyx formulations are used for high-flow shunts and large feeders. Compared to Onyx, Squid has better visibility owing to finer grains of tantalum powder. Both Onyx and Squid vials need to be shaken continuously for at least 20 min before use to make the suspension homogeneous and to prevent precipitation of the agent in the microcatheter. PHIL is another liquid embolic agent consisting of two copolymers (polylactide-co-glycolide and polyhydroxyethylmethacrylate) covalently bound to triiodophenol. DMSO is used as a solvent for this compound. Unlike Onyx and Squid, PHIL does not need to be shaken before use. PHIL has higher embolic capacity than Onyx or Squid [6]. PHIL produces much fewer CT artefacts and no MRI artefacts compared to Onyx and Squid.

3. **Injectable Coils**

   Injectable coils or flow coils are platinum microcoils with a diameter of 2.5 mm. They are used for proximal occlusion of a feeding artery in preoperative embolization, plugging the distal tip of a microcatheter to prevent reflux of EVOH copolymer or NBCA glue, and occlusion of the venous flow of an AVM to prevent distal migration of a liquid embolic agent during AVM embolization via trans-arterial and transvenous routes.

4. **Polyvinyl Alcohol (PVA) Particles**

   PVA particles are no longer used as their injection requires larger-diameter microcatheters and vascular occlusion by this agent is temporary. Recanalization rate is around 43% with PVA particles. [7]

Techniques of AVM Embolization

The ultimate goal of AVM embolization is complete obliteration of the nidus, which requires delivery of the embolization material to the entire and all the compartments of an AVM, right up to the draining veins. The deposition of the embolization material should be strictly intranidal, and the extranidal structure, i.e. the normal vasculature (arteries and veins), should be preserved. To achieve this goal, each feeding artery needs to be examined by selective catheterization, and the drainage veins need to be identified before occlusion so that the entire angio-architecture of the AVM can be understood. The shunts within the AVM, whether high flow, low flow, or mixed type, need to be identified especially while using glue to determine the optimal glue concentration for embolization. High-flow shunts require higher concentrations, whereas low-flow shunts require lower concentrations of the embolization materials. The venous segment that is directly connected to the arteriovenous shunts is known as the primary vein [5]. They can have multiple arterial feeders. Secondary veins deviate from the primary vein to drain into the venous sinus [5]. Occlusion of the primary veins obliterates the venous drainage of an AVM, and once they are occluded, blood flow from any non-occluded feeder will increase the pressure inside the AVM, leading to its rupture. Hence, the entire nidus needs to be obliterated before occlusion of the primary vein(s). Occlusion of a secondary vein does not entirely compromise the venous drainage of an AVM and hence does not increase the risk of the rupture of an AVM fed by non-occluded arteries. AVM embolization should ideally be done in a single session. Multiple sessions increase the risk of AVM rupture. Embolization can be done via trans-arterial route or transvenous route.

Trans-arterial approach: Trans-arterial route is preferred for AVM embolization in most of the cases. In this approach, the main feeding artery of an AVM is catheterized, and the microcatheter is placed as close to the nidus as possible. The embolic agent is then delivered into the nidus continuously until the initial few millimetres of the primary draining vein are occluded or there is reflux of the embolization material. Trapping the microcatheter tip with the help of injectable coils and glue can prevent reflux and allow uninterrupted delivery of the embolic agent [8]. This technique can be performed if a detachable tip microcatheter (Apollo, Sonic) is used to deliver the embolic agent and another microcatheter needs to be used for the delivery of the coils and glue to trap the microcatheter tip [8]. Reflux of the embolic agent can trap the microcatheter tip and make it hard to withdraw, leading to vessel rupture in some cases. This is especially problematic with non-detachable tip microcatheters (Magic, Marathon). However if a detachable microcatheter is used, reflux of the embolic material can be allowed up to the detachment point, and such an amount of reflux can even provide the advantage of the trapped microcatheter or pressure cooker technique described formerly without the need for an additional microcatheter. The delivery speed of the embolic agent can be controlled by wedging the microcatheter tip. Directing the tip towards the vessel wall slows down the rate of delivery and is useful while embolizing high-flow AVMs where rapid delivery of the embolic agent might occlude the draining vein before filling the nidus resulting in intranidal pressure build-up. Any increase in the intranidal pressure increases the risk of AVM rupture. In AVMs having two dominant feeders, simultaneous catheterization of the two feeders with two separate microcatheters can be done to perform simultaneous embolization of the nidus via both feeders. This technique minimizes the risk of intranidal pressure build-up due to occlusion of the venous drainage during embolization via single feeder. However, this technique requires two operators for simultaneous embolization. If the main arterial feeder of an AVM is difficult to catheterize and the microcatheter tip remains distant from the nidus, the delivery of the embolic agent can be done by making a column of the embolic material in front of the microcatheter tip and then pushing it by injecting more and more embolic material.

Transvenous approach: This approach has gained popularity in recent times and is preferred in the case of small ($\leq$3 cm), deep-seated AVMs or AVMs with narrow-calibre feeders that are difficult to access via trans-arterial route. Accessing an AVM via transvenous route can also be very difficult owing to tortuosity of the cerebral veins draining the AVM. Big AVMs have large venous outflow and are easy to access via transvenous route. But a large venous outflow prevents proper intranidal deposition of the embolic material injected through the venous side. Another concern is early occlusion of the draining vein before occluding the nidus, which may increase the intranidal pressure leading to rupture. Deep-seated AVMs are supplied by the lenticulostriate or choroidal arteries, which are difficult to access via the trans-arterial route, whereas such AVMs drain into the deep venous system, which is easily accessible via the transvenous route. AVMs with multiple venous outflow are difficult to embolize via transvenous route because of lesser chance of complete intranidal deposition of the embolic material. Hence, transvenous embolization should be done in AVMs with a single venous outflow. Transvenous embolization is carried out under a roadmap provided by a catheter placed on the arterial side. Super-selective angiography is obtained using a microcatheter placed on the arterial side to delineate the AVM architecture before starting embolization. The microcatheter on the venous side should be placed as close to the nidus as possible, and the venous outflow should be preserved until complete obliteration of the nidus. Temporary systemic hypotension or temporary balloon occlusion of the main feeder may aid in proper intranidal deposition of the embolic material by reducing the arterial flow into the AVM [9]. Such technique also minimizes the risk of AVM rupture due to intranidal pressure build-up. The pressure cooker technique can be applied to trap the microcatheter tip on the venous side [10]. This technique prevents reflux of the embolic agent into the veins and allows better penetration of the nidus [10].

Demonstration of Clinical Cases of Curative Embolization Using NBCA with Technical Pointers

Case 1

A 24-year-old male patient presented with recurrent posterior parietal and ventricular bleed. On DSA, an AVM was found, and patient was taken for curative embolization (Figs. 30.14, 30.15, and 30.16).

The patient was taken up for embolization under general anaesthesia. A right femoral approach was used. A 4F head-hunter diagnostic catheter was advanced into the internal carotid artery under roadmap guidance. Thereafter, a 1.5F Marathon flow-guided microcatheter was navigated into the nidus, and multiple compartments were catheterized. With repeated injections, the various compartments of the nidus were delineated, and fistulous as well as plexiform compartments were identified. As shown in Fig. 30.17, different concentrations of glue were chosen according to the nidal characteristics, and multiple injections were done.

Different positions of the catheter tip in relation to the feeding pedicle also decide the flow characteristics (Figs. 30.18 and 30.19). Here it becomes important to understand the flow characteristics in the vessel. What is important to know is that the centre of the vessel has laminar flow, and as one proceeds towards the periphery, the flow becomes more and more turbulent due to the interaction with the endothelium. Therefore as a corollary, if a catheter is coaxial in the vessel, the embolic material will travel further and faster, whereas if the catheter is releasing the embolic material towards the edge of the vessel, it will slow down and polymerize quickly. These can be used in different permutations and combinations according to the flow characteristics in the vessel and the nidus (Fig. 30.20).

Case 2

An 11-year-old female patient presented with thunderclap headache and unconsciousness. CT showed a callosal and ventricular bleed. DSA showed an AVM (Figs. 30.21, 30.22, 30.23, and 30.24).

The patient was taken up for embolization under general anaesthesia. A right femoral approach was used. A 4F head-hunter diagnostic catheter was advanced into the internal carotid artery under roadmap guidance. Thereafter, a 1.5F Marathon flow-guided microcatheter was navigated into the nidus, and multiple compartments were catheterized. In this present case, there were no large arterial pedicles supplying the AVM, and the twigs were not seen on 2D DSA. A cone beam CT angiographic exploration was done (Fig. 30.22). With these ultra-high-resolution images used as roadmap, the long and short callosal arteries were canulated individually using Magic 1.2 flow-guided microcatheters. With repeated injections, the various compartments of the nidus were delineated, and fistulous as well as plexiform compartments were identified. As shown in Fig. 30.17, different concentrations of glue were chosen according to the nidal characteristics, and multiple injections were done (Figs. 30.25 and 30.26).

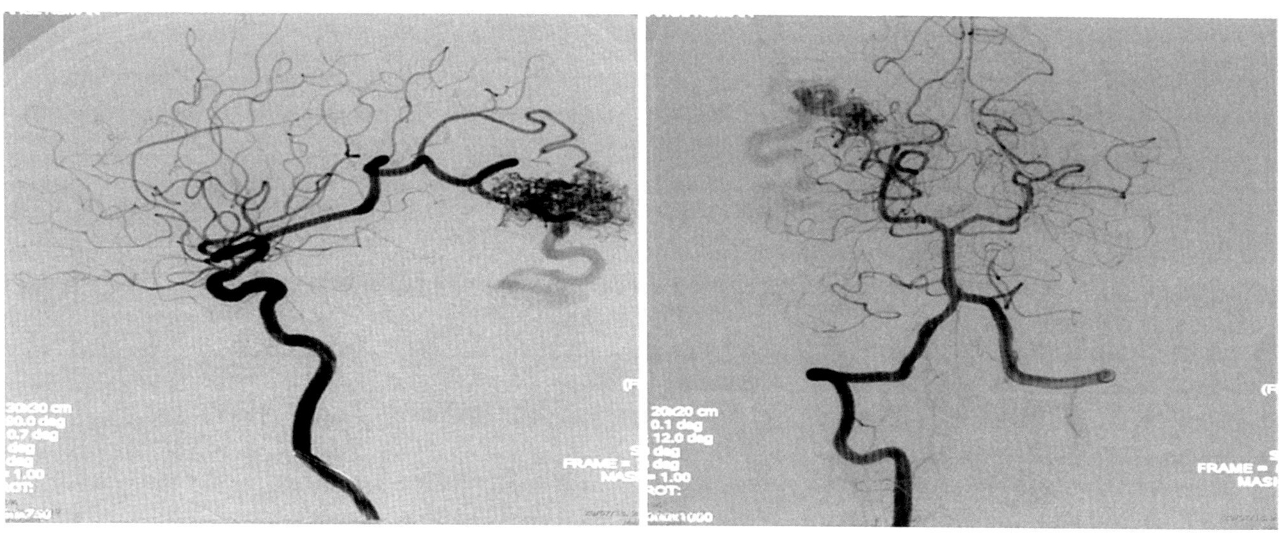

Fig. 30.14 DSA showing the supply to the Neopallial AVM from the MCA as well as from the PCA. Interesting to see that the deep supply is not visible, even though the lesion is crossing the phylogenetic boundary

Cone Beam Catheter CT Angiographic Exploration of a Ruptured AVM

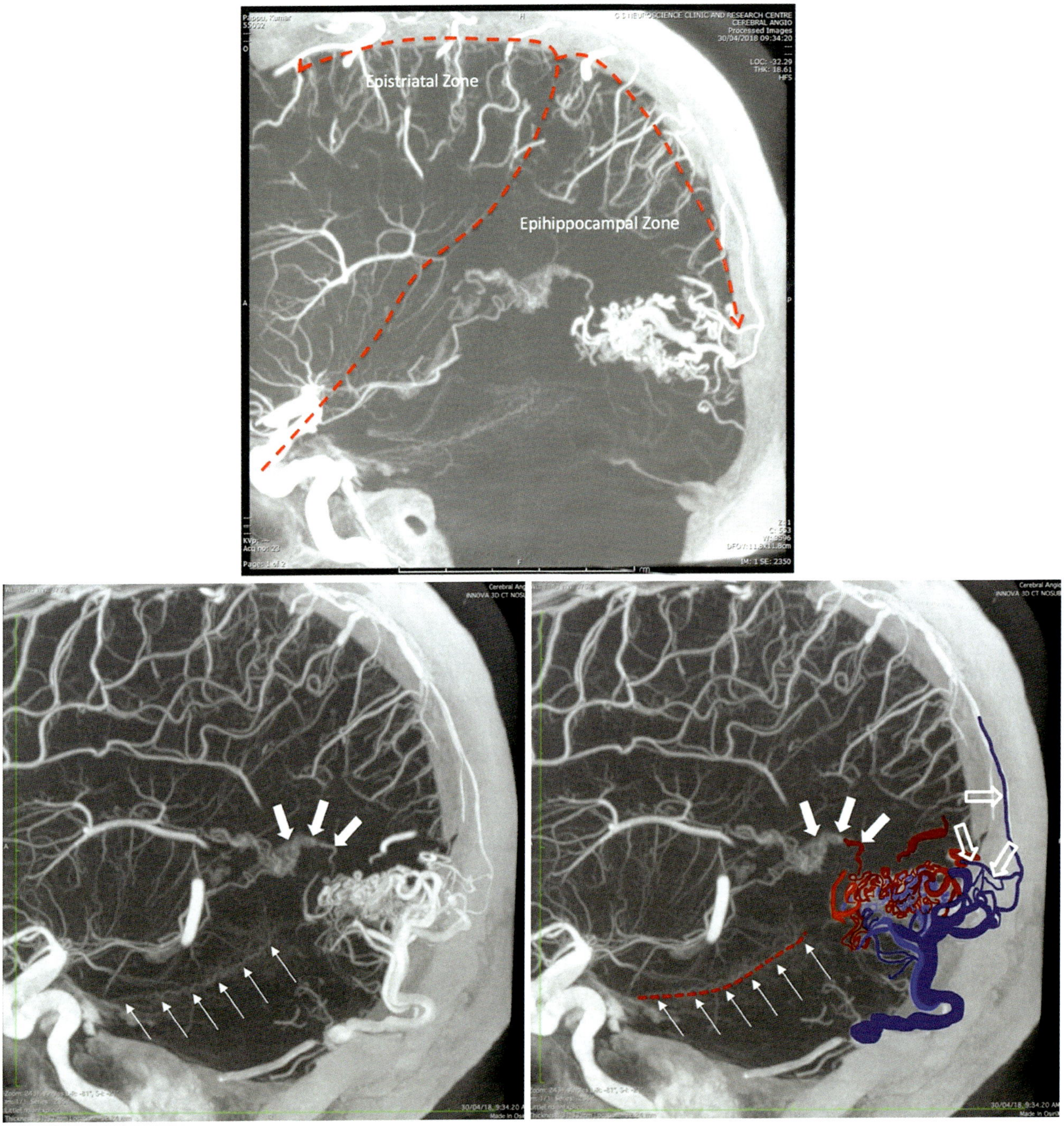

Fig. 30.15 Cone beam CT angiographic exploration of the AVM showing the expected deep supply from the Choroid Plexus by virtue of the lesion being in the "Epihippocampal Zone" (Arrows). It is also interesting to see the activation of the vessels in the tentorial edge and the secondary venous drainage towards the Superior Sagittal Sinus (Arrowheads). Additionally, this being a sulcal subtype of Neopallial AVM, additional Dural supply is getting activated (arrows) from the marginal tentorial artery

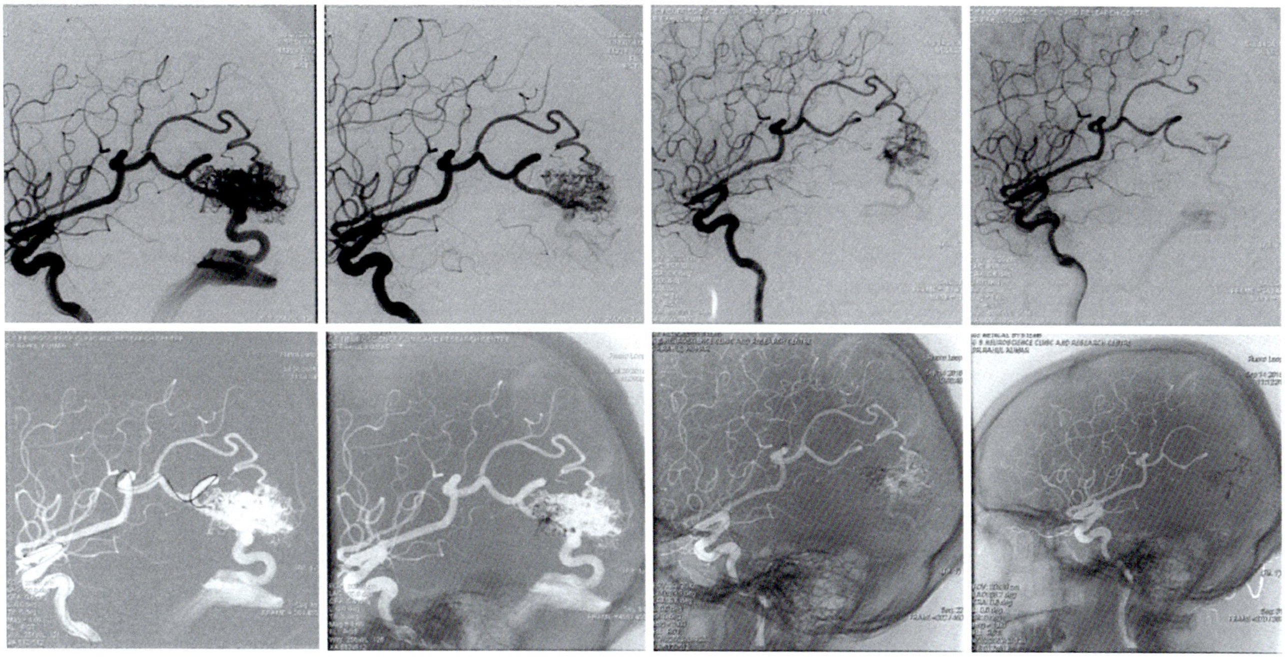

Fig. 30.16 Super-selective catheterization of the different compartments of the AVM and progressive obliteration of the nidus with NBCA injections

Preparation and Selection of the Glue Concentration

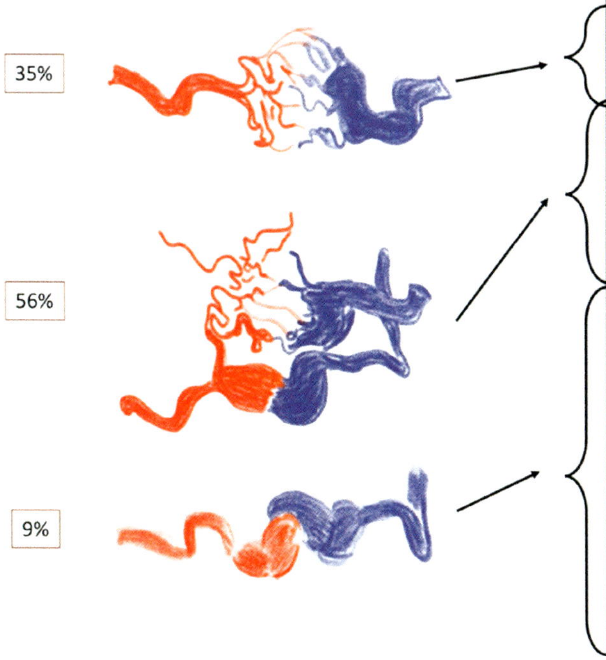

NBCA (ml)	Lipiodol (ml)	Tantalum	% NBCA	Volume (ml)
0.1	1.0	-	9	1.2
0.2	1.0	-	17	1.2
0.3	1.0	-	23	1.3
0.4	1.0	-	29	1.4
0.5	1.0	-	33	1.5
0.6	1.0	-	37	1.6
0.7	1.0	-	41	1.7
0.8	1.0	-	44	1.8
0.9	1.0	-	47	1.9
1.0	1.0	-	50	2.0
1.0	0.9	+	53	1.9
1.0	0.8	+	56	1.8
1.0	0.7	+	59	1.7
1.0	0.6	+	62	1.6
1.0	0.5	++	67	1.5
1.0	0.4	++	71	1.4
1.0	0.3	++	77	1.3
1.0	0.2	++	83	1.2
1.0	0.1	+++	91	1.1

Fig. 30.17 For pure plexiform nidus types, the concentration of NBCA chosen for embolization tends to be in the 17–23% range. For a mixed plexiform and fistulous type, it can vary between 23 and 41%. Pure fistulous nidi need concentrations of 41% and above, depending on the catheter position and flow characteristics

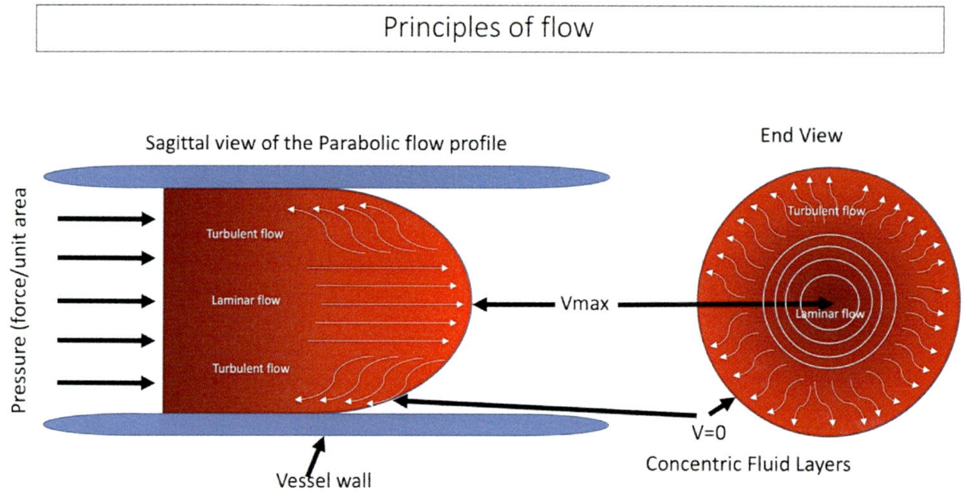

Fig. 30.18 Sagittal and axial sections of a blood vessel showing the patterns of flow. Blood flows as a parabolic head in the vessel. Towards the edges, due to the friction with the endothelial lining, the velocities are much slower, and towards the centre, the reduced friction between the hypothetical cylinders of blood creates a faster laminar flow

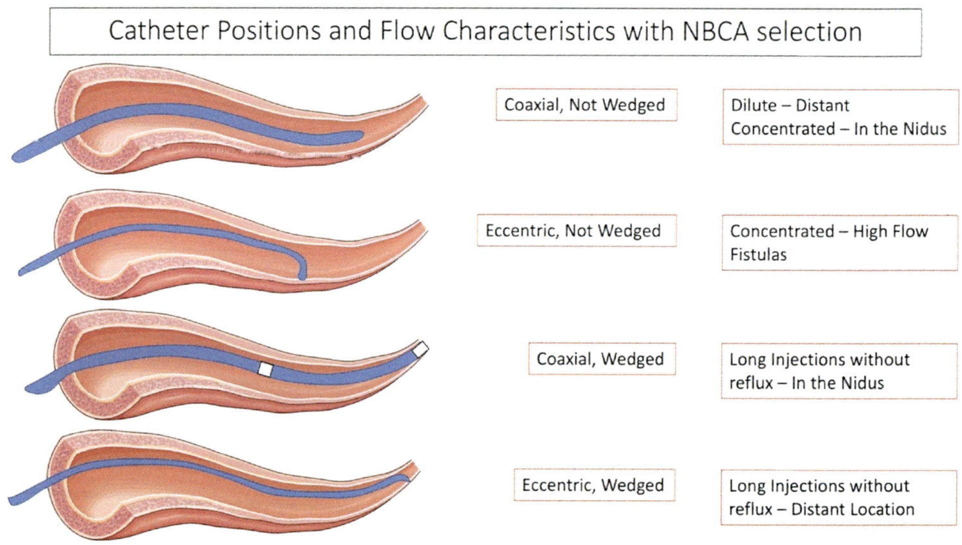

Fig. 30.19 Effect of catheter positions on the flow within the AVM nidus. These are used to one's advantage during different stages of embolization

Case 3

A 34-year-old male patient presented with a sudden onset of altered consciousness and hemiplegia. MRI was done which showed a vascular malformation with an intranidal aneurysm. DSA done showed a choroidal AVM which is a subpallial AVM. The supply was exclusively through the anterior choroidal artery, and the drainage was into the basal vein (Figs. 30.27, 30.28, 30.29, and 30.30).

Case 4

A 40-year-old male presented with a subarachnoid haemorrhage and a ruptured parietal AVM was found on cerebral DSA. The patient was taken for curative embolization under general anaesthesia.

The initial cerebral DSA was done via the snuff box access (distal radial access), and radial access was used for the intervention itself. More of our cases are being done via

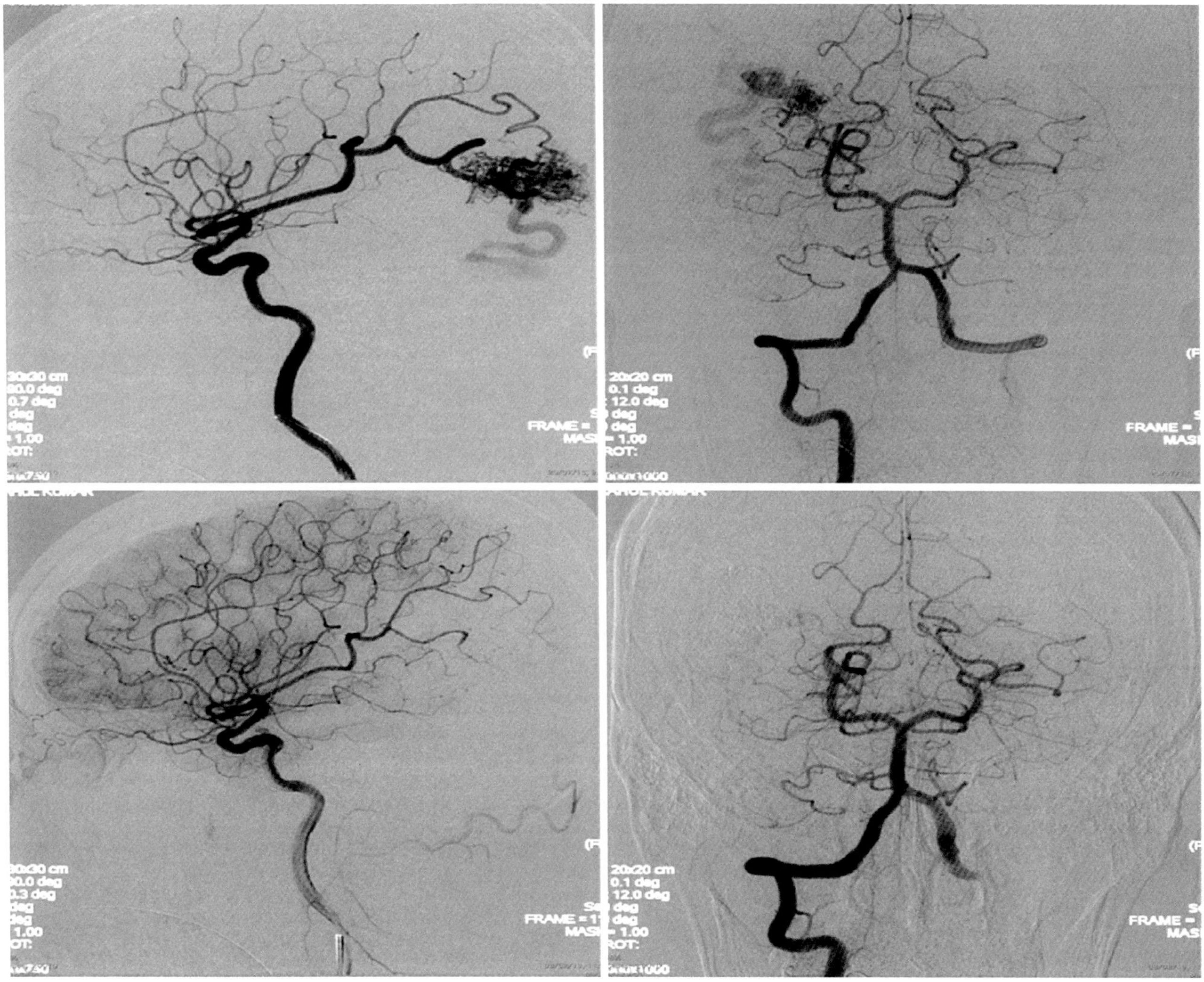

Fig. 30.20 Complete angiographic exclusion of the AVM nidus after injections of NBCA into various compartments. By virtue of the inter-compartmental communications, the supply from the PCA was occluded without any catheterization of embolization. The upper images are pre-embolization, and the lower ones are post-embolization. The glue cast showing the nidal characteristics, the loopy vessels, the spiral intranidal vessels, and the collector veins are seen in the nidal cast (Fig. 30.11)

the transradial access. Our understanding of AVM embolization is that using the flow itself to direct the embolizing agent into the nidus might give better results. Hence, we rarely use bigger catheters such as 0.088 inner diameter to get close to the nidus. A majority of our embolization cases are done with 4F or 5F diagnostic catheters as the main guide. When we switch to the radial access, it becomes much easier when using the guides with a maximum ID of 0.070 or 6F guide catheters. We have also started using 125 cm Sim Select catheters and other 125 cm diagnostic select catheters to advance high up into the ICA for better support and to maintain the flow for better embolysate percolation. Discussing further the merits, demerits, and many advantages of radial access for endovascular treatments is beyond the scope of this chapter.

In this patient, done early during our switch to transradial access, a 6F radial sheath was used to gain access to the radial artery and a 6F Fubuki Guide over a 5F cordis Bernstein was used to access the bovine arch origin left ICA. The Fubuki Guide was taken up all the way into the high cervical ICA, and a Marathon flow-directed microcatheter was used to navigate close to the nidus. Onyx 18 was used as the liquid embolizing agent to continuously and completely fill the AVM nidus (Fig. 30.31).

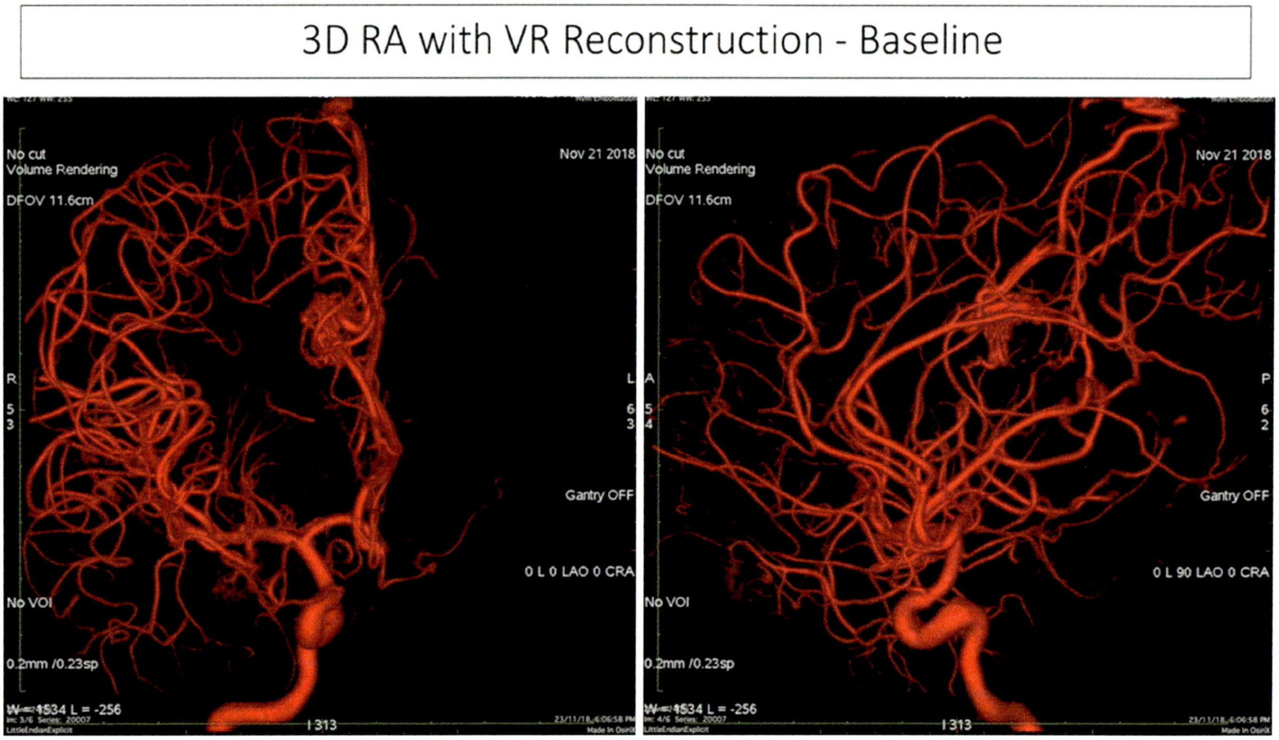

Fig. 30.21 3D rotational angiography with VR reconstruction of the AVM showing the structure of the nidus as well as the various nidal characteristics including the intranidal aneurysm. However, the feeding arterial branches are not seen very clearly

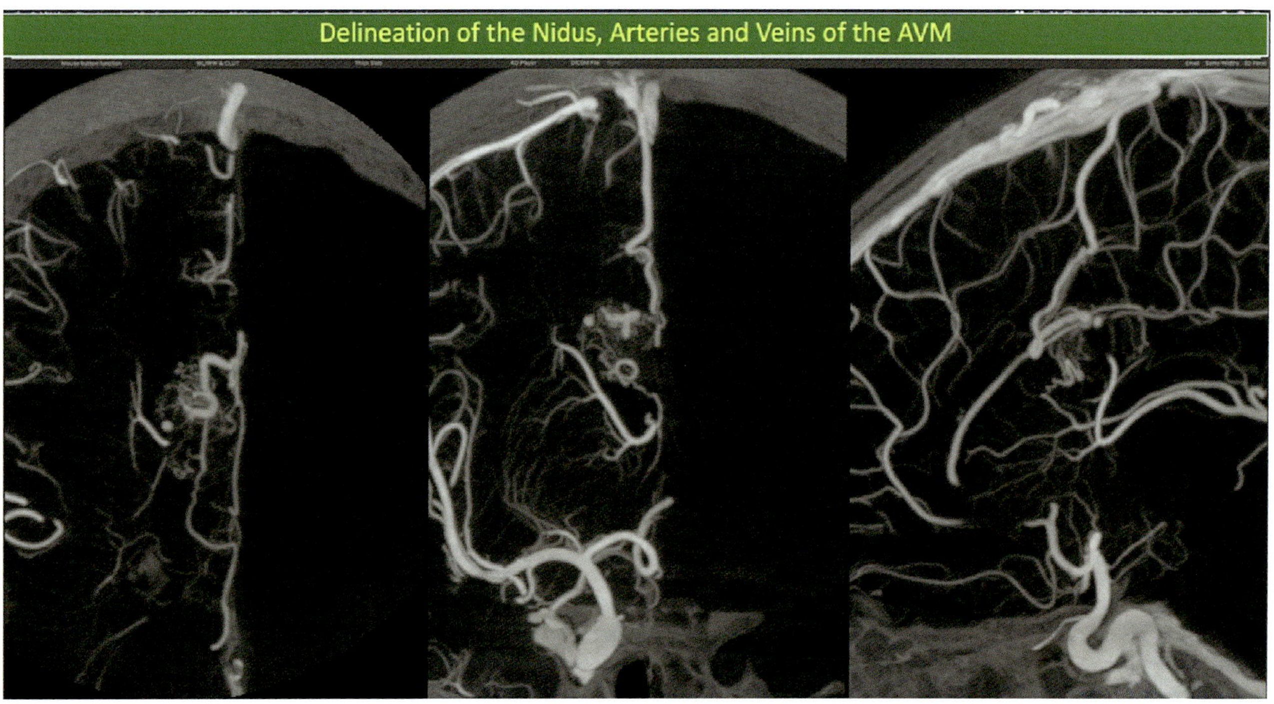

Fig. 30.22 Catheter cone beam CT angiographic reconstruction of the AVM showing the structure of the nidus as well as the various nidal characteristics, including the intranidal aneurysm. In these images, the feeding arterial branches are seen very clearly. They are very small twigs. Once they are identified, embolization can proceed

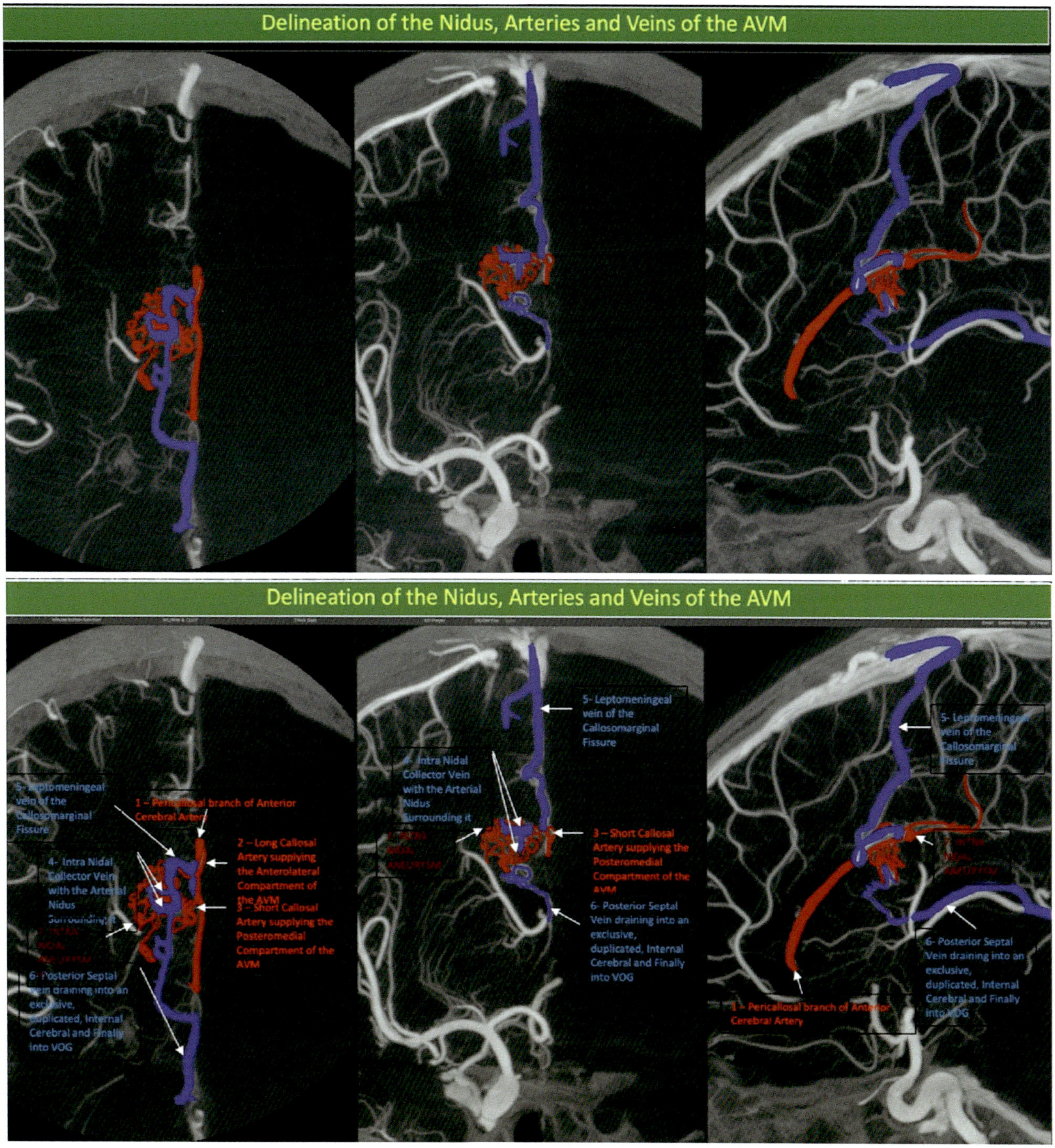

Fig. 30.23 Catheter cone beam CT angiographic reconstruction of the AVM with colour coding of the vascular tree and labelling of the anatomy

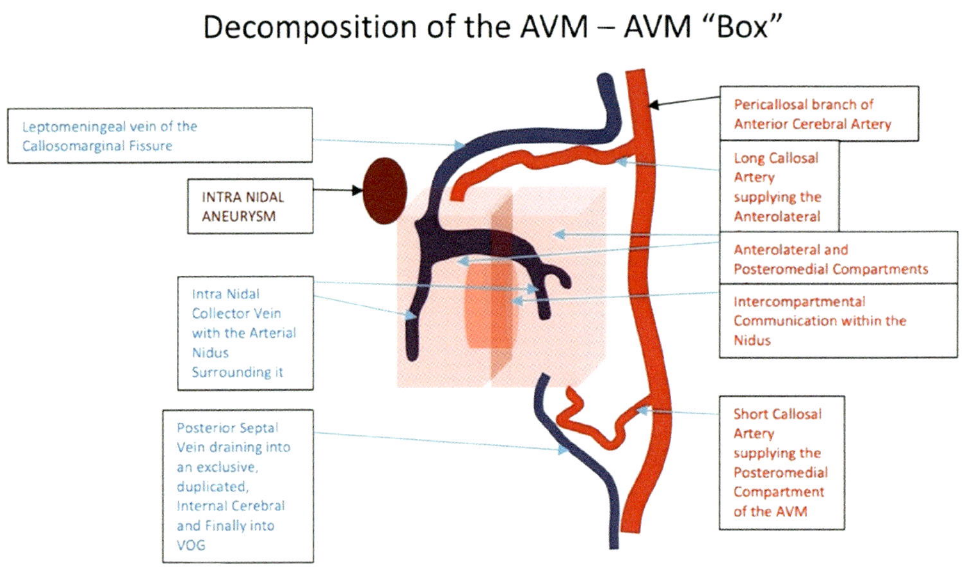

Fig. 30.24 The conceptualization of an AVM after the CBCT exploration that is kept in mind during all stages of the embolization

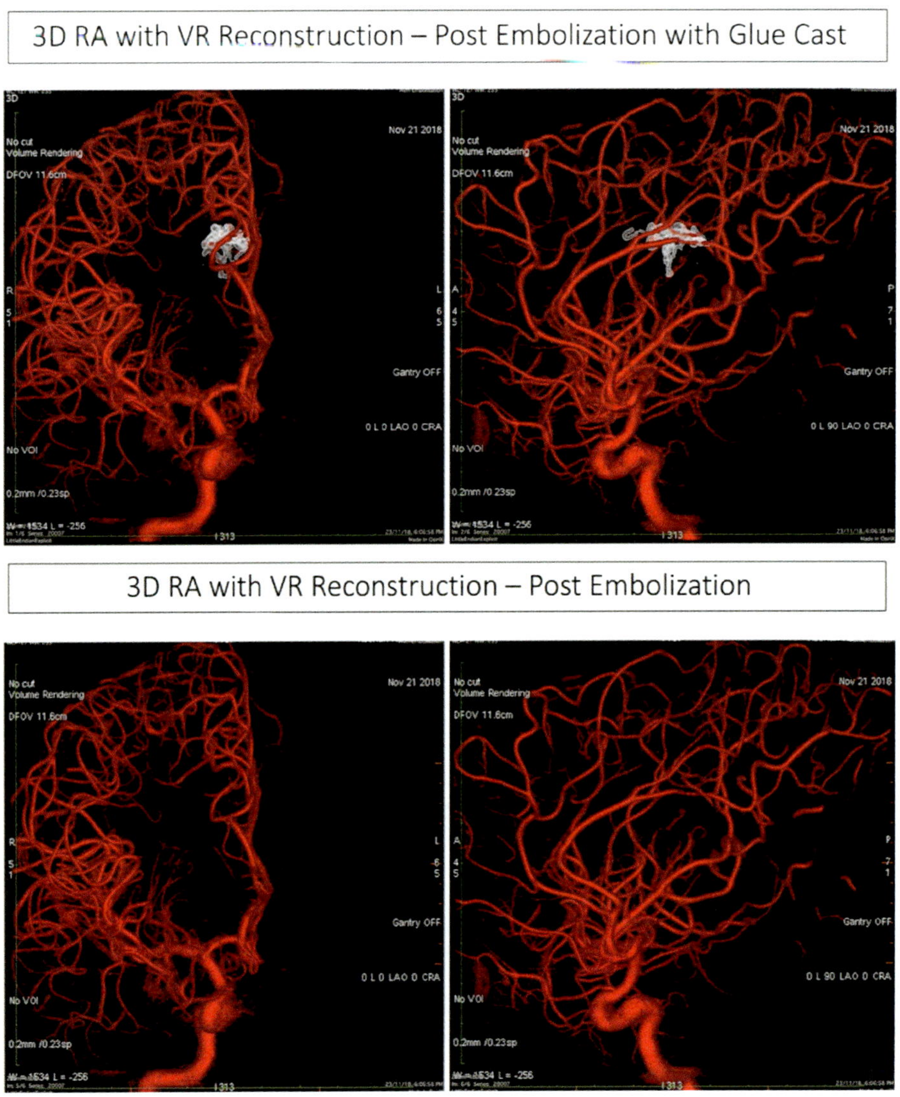

Fig. 30.25 Post-embolization 3D RA VR reconstruction showing the glue cast in the AVM, and the subtracted images showing complete occlusion with reservation of the arterial tree

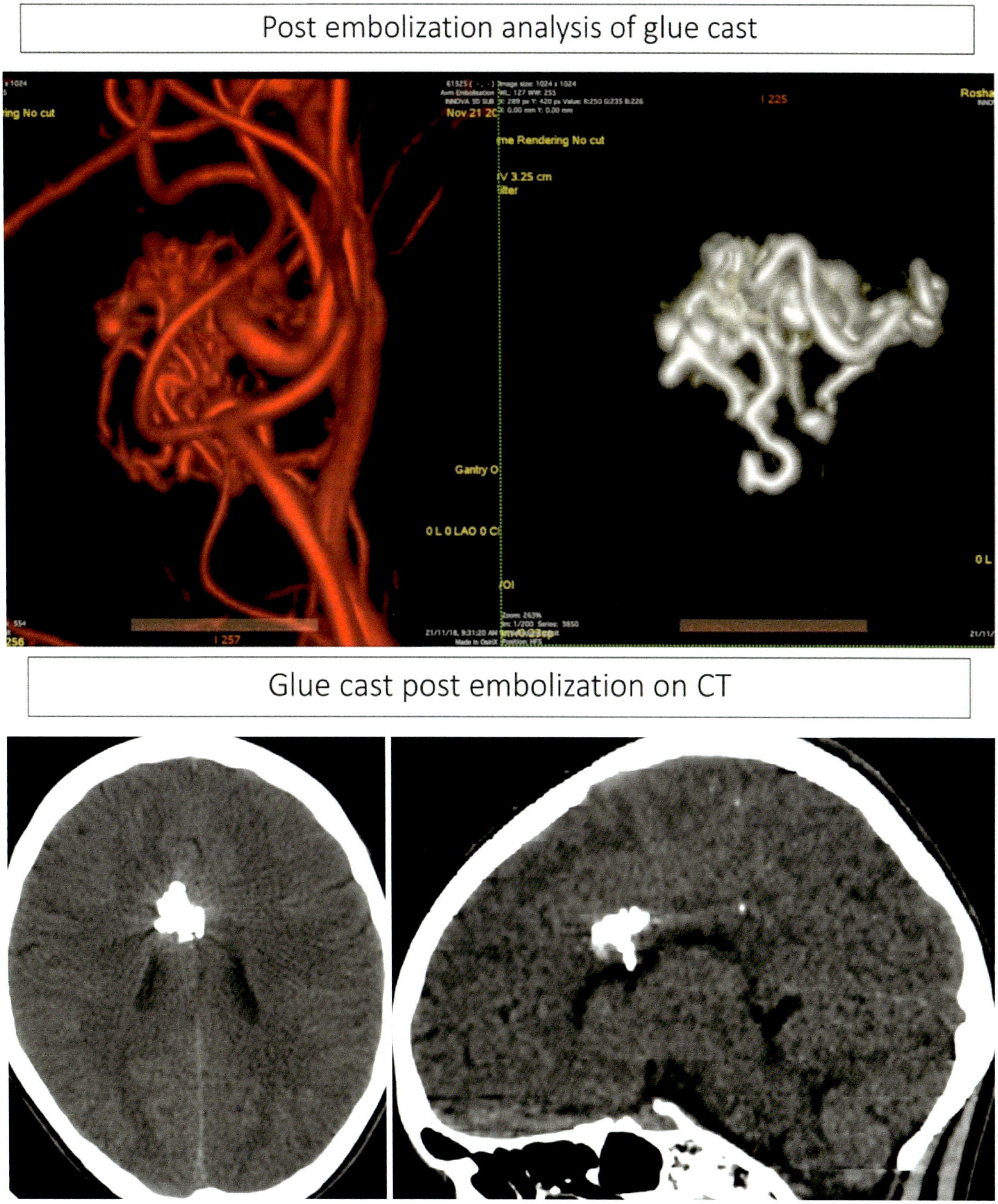

Fig. 30.26 Post-embolization analysis of the glue cast on 3D RA as well as regular CT scan. The parallelism between the nidus and the cast has to be demonstrated in order to achieve a complete and long-lasting cure

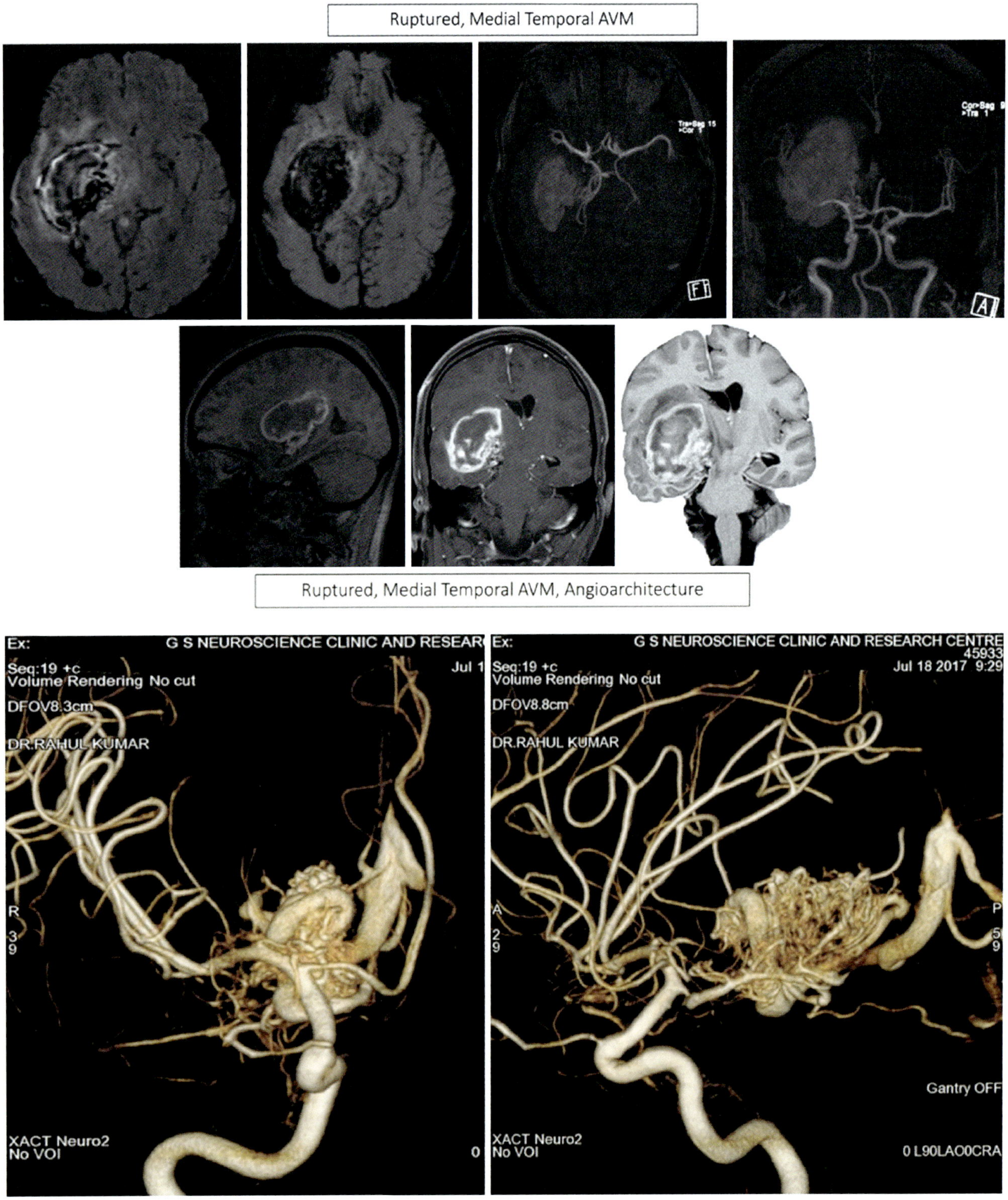

Fig. 30.27 The upper set of images are from the MR exploration of the bleed, and the lower panel images are from the VR reconstruction of the entire AVM, showing exclusive supply from the anterior choroidal artery and drainage into the basal vein of Rosenthal

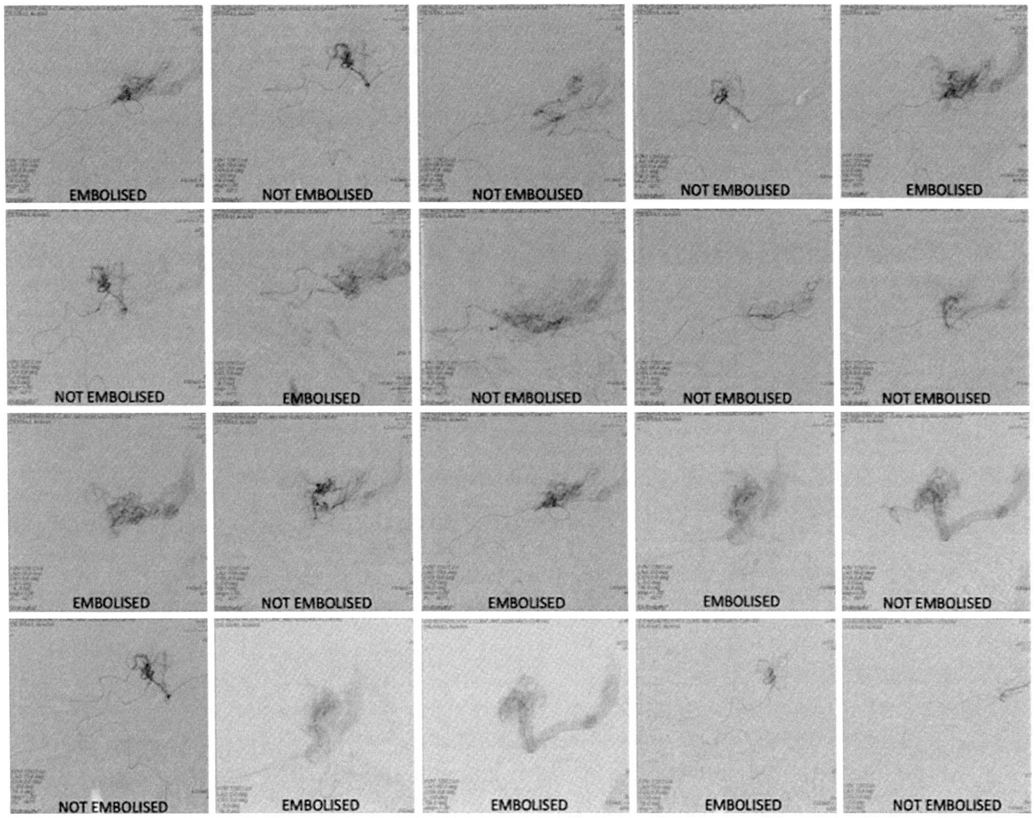

Fig. 30.28 Super-selective angiographic exploration of the nidus. The intranidal weak points are identified. It is interesting to see that out of the many pedicles studied, less than one-third of those were embolized, as the target here was a subtotal embolization. A total curative embolization in this situation will definitely produce fixed deficits

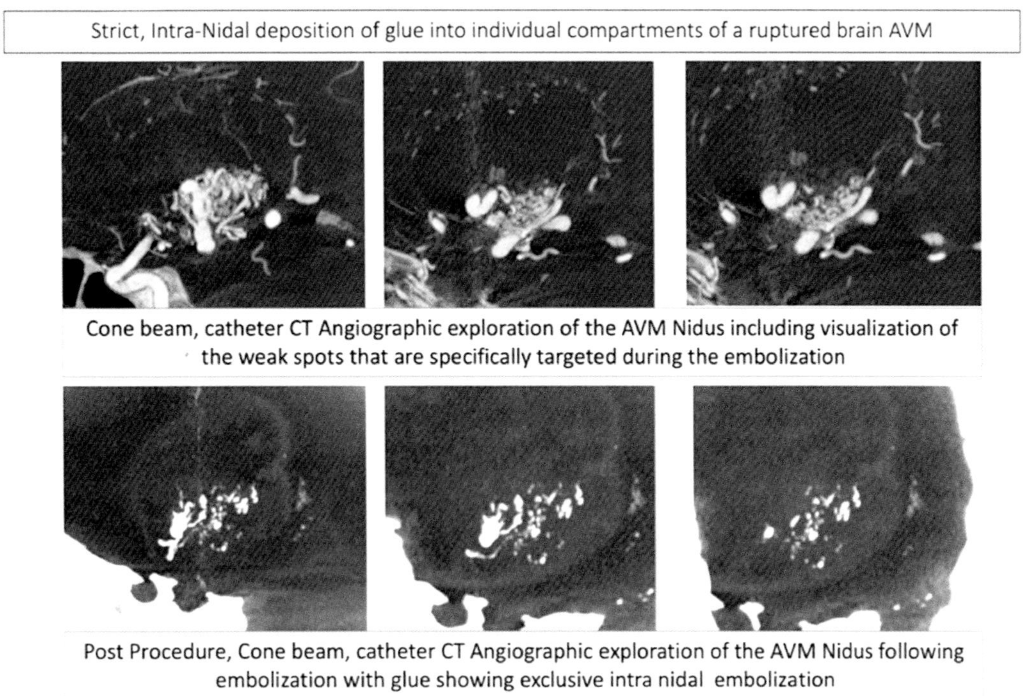

Fig. 30.29 The upper panel shows the section of the AVM in the sagittal orientation, and the bottom pane shows the deposition of the embolic material, strictly within the nidal weak points

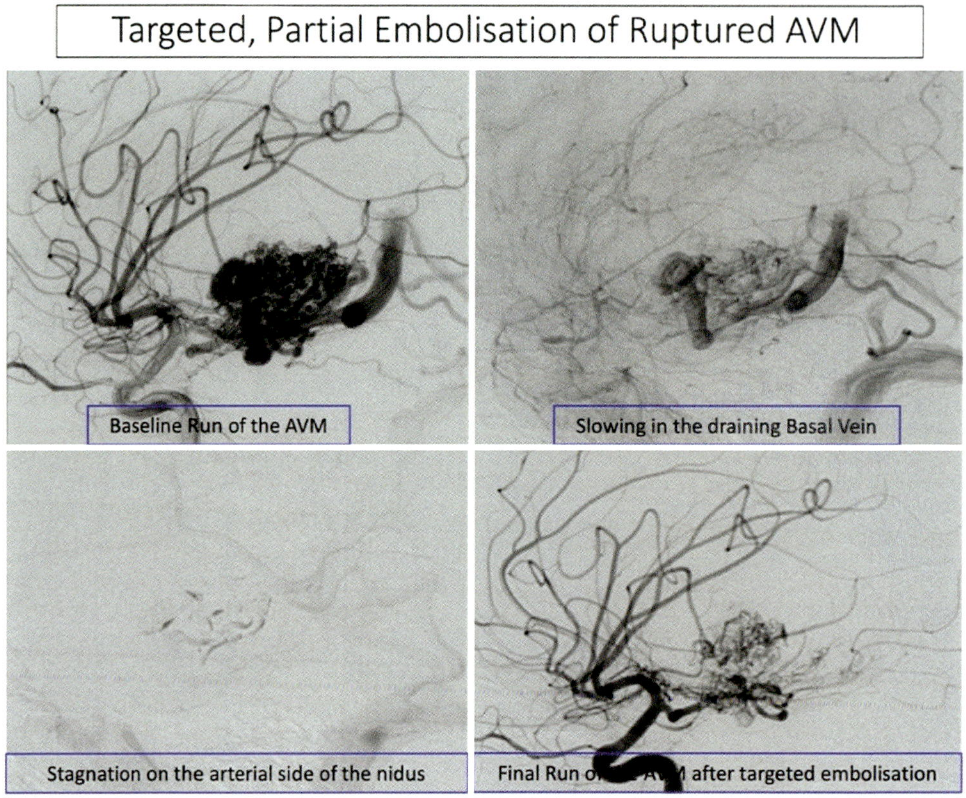

Fig. 30.30 The sequential images obtained during the embolization of the anterior choroidal AVM. In the baseline image, the entire AVM and the draining vein are seen filling early and briskly. As the embolization progresses, there is a slowing of flow in the draining vein with stagnation. With further embolization, there is a significant stagnation seen in the arterial end of the nidus as well. The procedure is stopped when no more early shunting in the basal vein of Rosenthal is seen

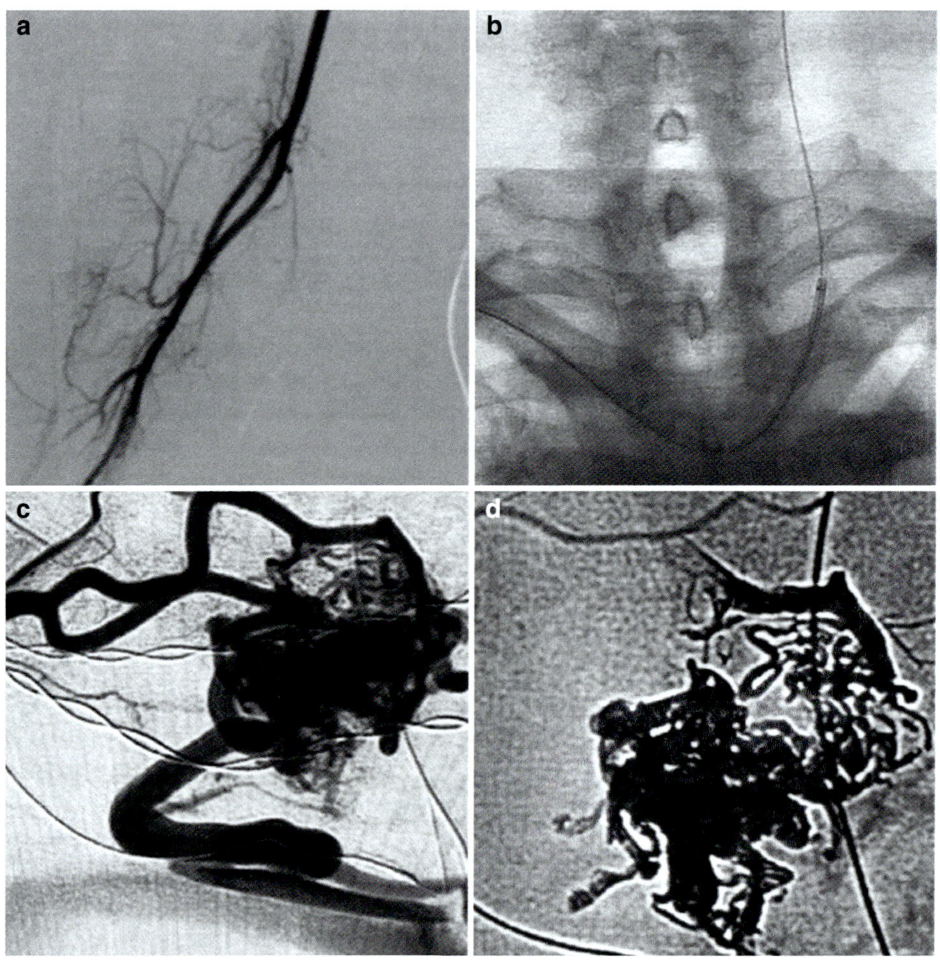

Fig. 30.31 A. Right radial access angiogram. B. Demonstration of guide easily climbing up the bovine origin left CCA. C. Parietal Neopallial AVM. D. Post-embolization onyx cast

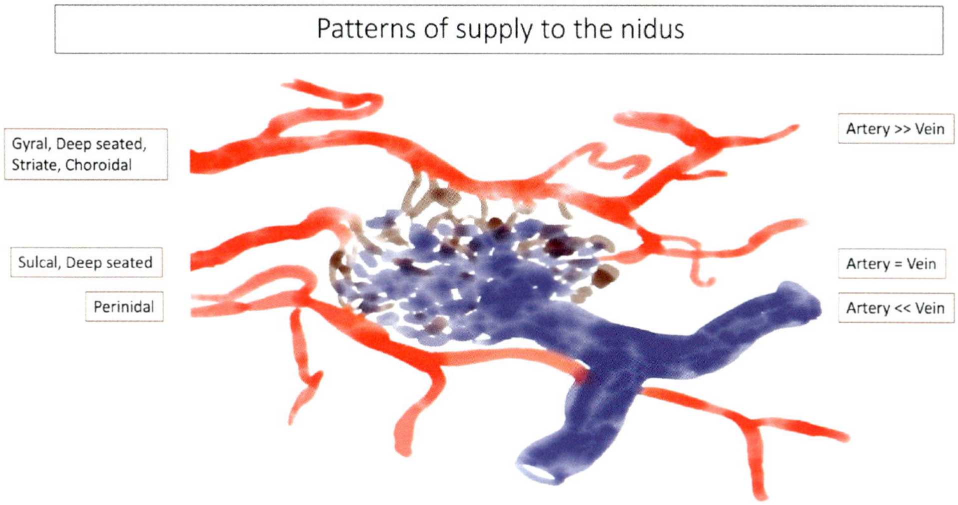

Fig. 30.32 A summary of the different patterns of supply to the nidus of an AVM. The so-called en-passage feeders are basically the medullary perforators originating from the ascending and descending sulcal as well as the transverse gyral segments of the leptomeningeal arcade. As previously thought, these are not untouchable and can be very well catheterized and embolized as seen in various examples in this manuscript. The direct supply with the vessel terminating in the nidus is the most desirable but found in less than a third of cases. Perinidal vessels do not participate in the shunt and are therefore supposed to be left alone

Complications of AVM Embolization

1. **Ischemic Complications**

   Ischemic complications may arise due to the migration of the embolic agent into normal vasculature, especially if the main arterial feeder supplies normal brain tissue beyond the AVM (the *en-passage* type feeder seen in gyral AVMs) or if there is excessive reflux of the embolic material into the proximal part and/or branches of the feeding artery. In the former case, selective catheterization of the arteries that branch out from the main artery to supply the AVM should be attempted, and embolization should be done via those branches if possible. The latter complication can be avoided by using the pressure cooker technique.

2. **Haemorrhagic Complications**

   Haemorrhage can occur due to the rise of intranidal pressure due to early occlusion of the venous outflow before complete intranidal occlusion, persistence of non-occluded shunts, partial occlusion, regrowth of AVM, arterial perforation by the microguidewire or microcatheter, or vessel rupture while withdrawing the microcatheter, especially if it is trapped by refluxed embolic material or while applying the pressure cooker technique. To avoid such complications, AVM embolization should ideally be done in a single session, complete nidal occlusion should be done, any non-occluded shunt needs to be identified and occluded, venous outflow should be preserved until complete intranidal deposition of the embolic material, and careful handling of the endovascular instruments is desired.

Conclusions

Brain AVMs are eminently amenable to endovascular curative embolization. Most of the available classification schemes are basically indicators of surgical difficulty during excision and have no bearing on the possible endovascular occlusion rates. Also, the size of a given AVM only increases the procedure duration and the number of catheterizations and has no bearing on the ultimate success rates of intervention. Super-selective, intranidal injections of embolic material like NBCA has the potential to cure many of these lesions. The success rates of curative embolization are approximately 70% with excellent long-term stable results. With the evolution of technology, we expect these to reach approximately 80% in the coming years. The single most important factor that leads to good outcomes is an understanding of the Nidal and Perinidal Angio-architecture (Fig. 30.32). In certain situations like deep-seated choroidal AVMs, a curative embolization is not possible without producing deficits, and therefore a subtotal embolization is the way to go. Apart from these exceptions, most AVMs should be studied in detail prior to embarking on an embolization, and the aim at the outset should be to offer a cure, either in a single session or in multiple staged procedures. The choice of material should always depend on the experience and comfort of the primary operator. There are "safety points" in each vascular territory, beyond which any embolization does not produce any deficits, and the primary aim of catheterization should always be to cross these points.

References

1. Valavanis A, Pangalu A, Tanaka M. Endovascular treatment of cerebral arteriovenous malformations with emphasis on the curative role of embolisation. Interv Neuroradiol. 2005;11(Suppl 1):37–43.
2. Potts MB, Zumofen DW, Raz E, et al. Curing arteriovenous malformations using embolization. Neurosurg Focus. 2014;37(3):E19.
3. Mohr JP, Overbey JR, Hartmann A, et al. Aruba co-investigators. Medical management with interventional therapy versus medical management alone for unruptured brain arteriovenous malformations (Aruba): final follow-up of a multicentre, non-blinded, randomised controlled trial. Lancet Neurol. 2020;19(7):573–81.
4. Luessenhop AJ, Spence WT. Artificial embolization of cerebral arteries. Report of use in a case of arteriovenous malformation. J Am Med Assoc. 1960;172:1153–5.
5. Houdart E, Labeyrie MA, Lenck S, et al. Treatment of AVM: endovascular methods. In: Beneš V, Bradáč O, editors. Brain arteriovenous malformations. Cham: Springer; 2017.
6. Vollherbst DF, Chapot R, Bendszus M, et al. Glue, onyx, squid or PHIL? Liquid embolic agents for the embolization of cerebral arteriovenous malformations and Dural arteriovenous fistulas. Clin Neuroradiol. 2022;32(1):25–38.
7. Sorimachi T, Koike T, Takeuchi S, et al. Embolization of cerebral arteriovenous malformations achieved with polyvinyl alcohol particles: angiographic reappearance and complications. AJNR Am J Neuroradiol. 1999;20(7):1323–8.
8. Chapot R, Stracke P, Velasco A, et al. The pressure cooker technique for the treatment of brain AVMs. J Neuroradiol. 2014;41(1):87–91.
9. Massoud TF, Hademenos GJ. Transvenous retrograde nidus sclerotherapy under controlled hypotension (TRENSH): a newly proposed treatment for brain arteriovenous malformations—concepts and rationale. Neurosurgery. 1999;45(2):351–63; discussion 363–5.
10. Koyanagi M, Mosimann PJ, Nordmeyer H, et al. The transvenous retrograde pressure cooker technique for the curative embolization of high-grade brain arteriovenous malformations. J Neurointerv Surg. 2021;13(7):637–41.

Part IV

Splachnic Artery Disease

Angioplasty and Stenting of the Mesenteric Arteries

31

Antoine Girault, Pauline Fillet, and Quentin Pellenc

Abbreviations

AMI	Acute mesenteric ischemia
BS	Bare stent
CA	Celiac artery
CMI	Chronic mesenteric ischemia
CS	Covered stent
CTA	Computed tomography angiography
DUS	Duplex ultra sounds
IMA	Inferior mesenteric artery
SMA	Superior mesenteric artery

Case Presentation

A 78-year-old man was referred to emergency department due to acute abdominal pain evolving for 2 weeks.

Active smoker, he was known to be treated for an ischemic cardio-myopathy (coronary stenting) and peripheral arterial disease (femoral endarterectomy). He presented post-prandial abdominal pains and an 8 kg loss of weight in the past 2 months but did not come in consultation due to national pandemic lockdown. He described one episode of melena the day before. Clinical findings show periumbilical spontaneous pain with no peritoneal signs, bilateral femoral pulse but no popliteal or distal pulses. Blood test revealed 14,000/mL leucocytes, C-reactive protein: 33 mg/L, and lactate: 1.6 mmol/L.

Emergent CTA showed an atheromatous aorta with ostial occlusion of the superior mesenteric artery (SMA) followed by tight stenosis, mild stenosis of the coeliac trunk (CT), occlusion of inferior mesenteric artery (IMA), and atheromatous infiltration of both hypogastric arteries. No bowel dilatation or wall perfusion attenuation was observed on CT scan (Fig. 31.1).

Continued at page 336

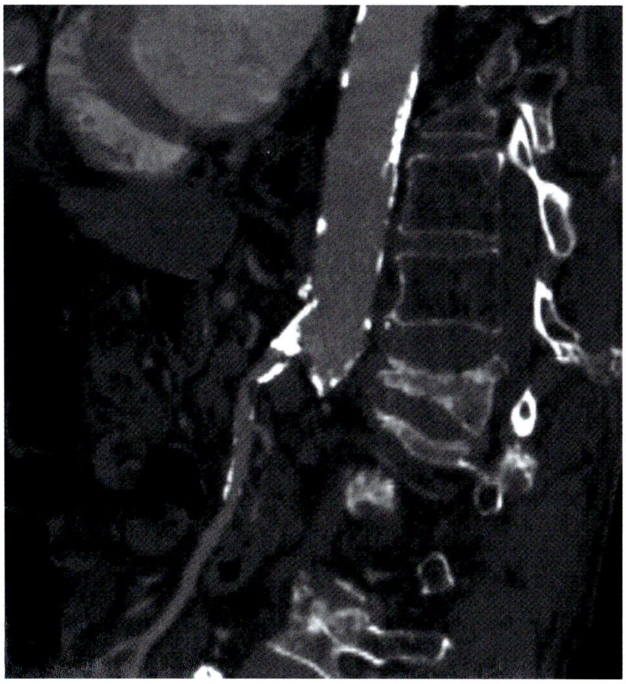

Fig. 31.1 CT angiography of SMA occlusion with instable bowel ischemia complicating chronic occlusion. Short calcified occlusion followed by tight stenosis

A. Girault
Department of Vascular and Thoracic Surgery, Hôpital Marie Lannelongue, Groupe Hospitalier Paris-Saint Joseph, Université Paris-Saclay, Paris, France
e-mail: Unknown0007871@Springer.com

P. Fillet · Q. Pellenc (✉)
Vascular and Endovascular Surgery Unit, Clinique de Genolier, Swiss Medical Network, Genolier, Switzerland

Department of Vascular and Endovascular Surgery, La Cote Healthcare Group, Morges, Lausanne, Switzerland
e-mail: unknown0199@springer.com; quentin.pellenc@ehc.vd.ch

G. Geroulakos et al. (eds.), *Mastering Endovascular Techniques*, https://doi.org/10.1007/978-3-031-42735-0_31

Background

Mesenteric ischemia (MI) is defined by an acute or a chronic bowel perfusion insufficiency that can be caused by arterial obstruction, venous thrombosis or by non-occlusive diseases [1, 2]. Arterial acute or chronic occlusive disease represents 90% of MI and diagnosis is pointed after the conjunction of non-specific clinical and biological signs and more specifics radiological signs [3]. Treatment included a multimodal strategy proposed by a multidisciplinary team and including a rapid assessment of the patient, a specific medical treatment, a rapid and specific bowel revascularization +/− resection and dedicated post-operative cares [4–6].

Endovascular therapy is now the treatment of choice for arterial acute and chronic mesenteric ischemia, leading to a decreased perioperative morbidity and mortality, a shorter length of hospital stay and faster recovery, compared to open repair [3, 7–10].

Atheromatous stenosis is the cause of 65% of acute mesenteric ischemia (AMI) and represents the vast majority of chronic mesenteric ischemia causes (CMI) [3]. In this chapter, angioplasty and stenting of atheromatous stenosis of the celiac artery (CA) and the superior mesenteric artery (SAM) will be developed.

Indications for Treatment

Indication for mesenteric artery revascularization is based on a bundle of clinical and radiological arguments. Revascularization is indicated for symptomatic patients associated with a SMA or CA stenosis ≥70% or CA ≥ 70% and SMA ≥ 50% [7]. Symptoms depend on the type of mesenteric ischemia:

1. Chronic mesenteric ischemia is revealed by postprandial chronic abdominal pain, food fear, weight loss, and diarrhea.
2. Acute ischemia complicating chronic mesenteric ischemia or asymptomatic SMA stenosis (Acute-on-chronic). Clinical presentation is dominated by brutal or rapidly progressive abdominal pain, with or without vomiting, diarrhea, and gastrointestinal bleeding.

Prophylactic stenting of the SMA in asymptomatic mesenteric occlusive disease patients remains controversial but may have a role in patients with severe three-vessel stenosis and in patients with a significant SMA stenosis planned for major surgery as aortic surgery, cardiopulmonary bypass etc. [7]. Bowel surgery such as left colonic resection could also be preceded by SMA prophylactic stenting in poorly vascularized patients. The final decision will be approved during a multidisciplinary discussion.

The SMA remains the primary target artery for revascularization of chronic and acute-on-chronic mesenteric ischemia regarding the vast territory and the collaterals fed by this artery [3]. The coeliac trunk will be stented in case of persistent ischemia after the SMA revascularization [11] or in case of supramesocolic ischemia [3]. Of course, CA should not be dilated and stented if there is active compression by the median arcuate ligament [3].

Timing for Operation

Early stage of AMI is reversible arguing the need of an emergent revascularization and medical therapy to save longer bowel segment [5, 6, 12]. Experience of dedicated intestinal stroke centers showed a decreased time between initial pain and revascularization at the acute phase of mesenteric ischemia. As already described in the neurovascular stroke centers, those centers should become high-volume centers by centralizing this activity [6]. Intestinal stroke center includes a multidisciplinary team with gastroenterologist, general surgeons, vascular surgeons, radiologists, and anesthesiologists.

Endovascular surgery is the first line treatment at the early stage of acute mesenteric ischemia in patients who don't need laparotomy for bowel resection. Usually, surgical open or hybrid revascularization is performed when laparotomy is required. The following signs go usually together with advanced bowel necrosis: clinical signs of peritonitis, lactate >2 mmol/L, organ failure, Biological inflammation (CRP >150 mg/L), and signs of bowels suffering the CT scan (see below) [13–15].

In CMI, medical therapy alone is not recommended arguing the need of revascularization in the majority of the patients [8]. Furthermore, total parenteral nutrition is not an acceptable alternative to revascularization but could improve patient general status before surgical revascularization in patient not suitable for endovascular repair.

Preoperative Assessment

Preoperative nutritional status has to be assessed before revascularization especially in CMI. Malnutrition is associated with an increased risk of postoperative mortality after endovascular and open surgical revascularization [16]. Restoring mesenteric circulation has priority and should not be delayed by attempts to improve nutritional status except in case deep malnutrition and open surgical revascularization planned.

Upper gastrointestinal endoscopy or colonoscopy are essential in the work-up of patients suspected of having chronic mesenteric ischemia, mainly for exclusion of differential diagnoses. But a normal endoscopy does not exclude a CMI [7].

No plasma marker has a high specificity in AMI. The L-lactate has a prognosis role and is elevated in late presentation of AMI [14]. Systemic inflammatory response syndrome induced by IMA leads to hyperleukocytosis and increase of C-Reactive protein which are also biologicals signs of advanced ischemia.

Duplex ultra sound (DUS) is not an appropriate imaging method to assess acute occlusive lesions of the visceral arteries [3]. In CMI, DUS is the recommended exam for screening patients with chronic mesenteric occlusive disease. It allows visualization of CA and SMA in almost 100% of the cases [17] and provides a diagnostic accuracy of 80–90% for the detection of >70% mesenteric arteries stenosis compared to angiography. A 70% stenosis is usually associated with a peak systolic velocity of 200–320 cm/s in the CA and 205–400 cm/s in the SMA [3, 7]. Postprandial testing can be helpful in atypical clinical presentation of CMI. Furthermore, mesenteric arteries hemodynamic are strongly influenced by respiratory cycle and conditions of exam have to be notified.

Computed tomography angiography (CTA) as the preferred definitive imaging test for acute and chronic mesenteric occlusive disease is recommended [3, 7, 8]. A triphasic CTA with 1 mm slices should be the gold-standard especially in case of AMI. It confirms the arterial etiology of AMI and allows the characterization of the stenosis/occlusion (length, calcifications, distal artery diameter, positions of collaterals, distal outflow). Furthermore, CTA is an essential tool for evaluation of the ilio-femoral access or subclavian access in case of stenting. Assessment of bowel suffering is also a main point. Advanced-stage AMI radiological features include bowel dilatation >2.5 cm, enhancement default, pneumatosis, and portomesenteric venous gas [13, 14].

Table 31.1 Suggested Endovascular Toolkit for the endovascular revascularization of superior mesenteric artery. Suggested options based on the author's experience in order to face most of cases

	Manufacturer	Size/length
Wires		
Any standard access wire		0.035″
Radiofocus Glidewire floppy	Terumo	0.035″
Advantage Glidewire	Terumo	0.035″
Rosen	Cook medical	0.035″
Gladius MG	Asahi	0.018″
Sheaths		
Any standard access sheath	Any	6-7 Fr/11 cm
Steerable sheath	Medtronic Tourguide	7 Fr/55 cm
Catheters		
Bern catheter	Boscton scientific	0.035″/4 Fr/100 cm
Quick cross support (low profile)	Spectranetics/ Phillips	0.018″-0.035″/4 Fr/65-90 cm
Vertebral	Terumo Glidecath	0.035″/4 Fr/100 cm
Micro catheter	Terumo Progreat	2.8 Fr
Balloons		
Mustang 0.035″	Boscton scientific	3–7 mm
Ballons expandable covered stent		
Advanta V12	Gettinge	5–8 mm
Self expandable stent		
Absolut pro	Abbott	6–7 mm
Rentry catheter		
Outback	Cordis	
Atherectomy device		
Jetstream 2.4/3.4	Boston scientific	7 Fr
Thrombectomy devices		
Thrombus aspiration catheter	OptiMed BigLumen	6, 7, 8 Fr
Angiojet with Solent™ Proxi catheter	Boston scientific	6 Fr

Endovascular Surgery and Technique

Basic endovascular toolkit:

- Short 6 Fr sheath
- Standard floppy hydrophilic guidewire
- Selective catheters (Bern, Rim, Cobra, etc.)
- Stiff guidewire (Rosen)
- Long 7 Fr sheath 45 cm
- Steerable 7 Fr long sheath
- Predilatation balloon angioplasty: 4–5 mm × 20–40 mm
- Balloon expandable covered stent (7 mm × 22 mm, 7 mm × 26 mm, 7 mm × 32 mm)
- Auto-expandable nitinol bare stent (7 mm × 30 mm and 6 mm × 30 mm)

A list of suggested material is summarized in Table 31.1.

Antegrade Access

Femoral access remains the first line approach. In the past years, most of cases were performed using a 6 or 7 Fr angulated long sheath with a success rate of 85% of femoral access in a study from our group [18].

When angulation of the SMA was not favorable or in case of occlusion requiring more support, an antegrade approach from above through a brachial puncture was needed. Classically using a 90-cm-long 7 Fr sheath, brachial access allows easily the SMA catheterization. The main issue remains the SMA occlusion case requiring heavy guidewire support to recanalize the artery. The trick consists, after sheath valve puncture, of the placement of a super stiff parallel guidewire into to one iliac artery. This technique offers a good stabilization of the long sheath distal end.

The use of steerable sheaths appears to be a game changer in vascular surgeons' practices. This type of devices has been

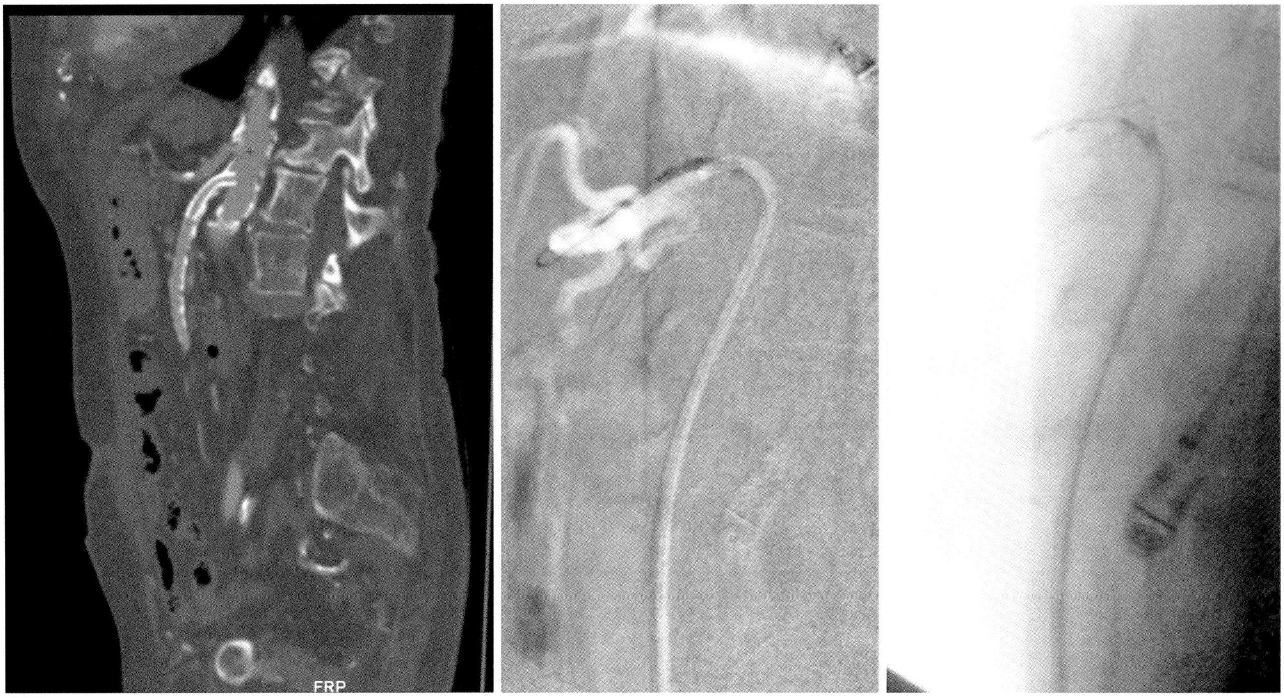

Fig. 31.2 Persistent supramesocolic acute ischemia 3 days after emergent stenting of the SMA for acute-on-chronic mesenteric ischemia in a 69-year-old woman in postoperative course of coronary bypass operation with a past history of aorto-bifemoral bypass and right renal artery stenting. CTA showing a tight ostial stenosis of the CA and proper patency of the SMA covered stent (**left panel**). CA is catheterized using a 7 Fr Tourguide long sheath (Medtronic) and an 8 × 17 mm iCover stent (iVascular) is positioned on a 0.035″ Rosen guidewire (**middle panel**). After stent delivery (**right panel**)

widely used to delivered bridging stents in branched endovascular aortic repair through femoral access avoiding upper body approaches and their complications [19]. The steerable sheath allows a downward orientation in the axis of the SMA or CA and provides efficient support to allow artery recanalization leading to a high success rate of the femoral access [18] (Figs. 31.2 and 31.3).

Retrograde Access

In case of antegrade access failure, a retrograde open mesenteric artery stenting (ROMS) can be proposed [20, 21]. This technique needs a laparotomy and a mesenteric surgical access. It is usually performed when concomitant bowel resection is mandatory. It can be associated with distal endarterectomy and patch angioplasty in case of long stenosis. It does not need aortic or iliac artery cross clamping and may avoid to require to an open surgical revascularization using arterial bypass exposing the risk on graft infection in those septic conditions. Indication of ROMS in case of endovascular antegrade SMA approach failure decreased dramatically since the development of new steerable sheaths (personal data).

The ROMS needs a surgical approach of SMA in the root of the mesentery. Rarely, SMA is just punctured, and a 7 Fr sheath is inserted retrogradely. In most of cases, ROMS is

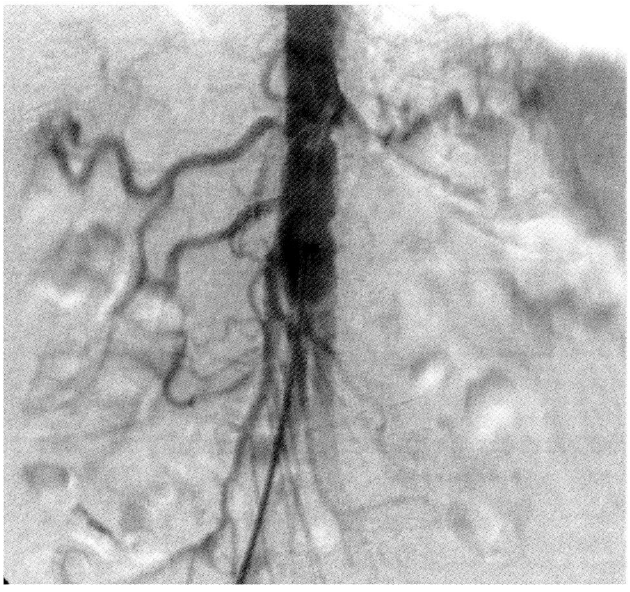

Fig. 31.3 Final angiography with synchronous enhancement of CA and SMA

associated with an open thrombo-endarterectomy and patch angioplasty. All collaterals have to be controlled and clamped with silicon loops and proximal artery is standardly cross clamped. Regarding the risk of bowel resections, prosthetic patches are less used than autologous saphenous patch or

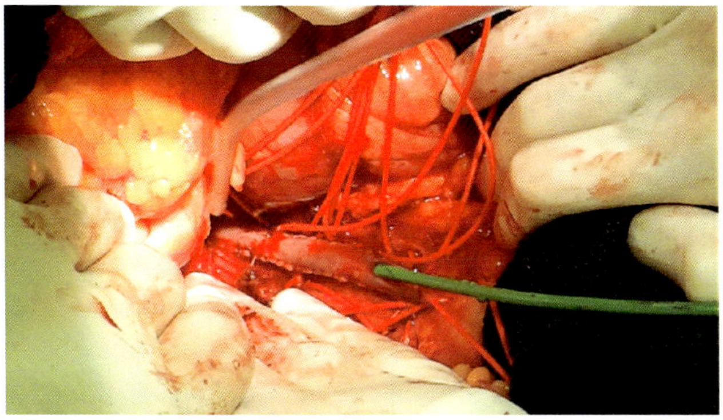

Fig. 31.4 Access site for retrograde open mesenteric artery stenting. After thromboendarterectomy and venous patch angioplasty of the SMA, retrograde patch puncture allows artery cathterterization and insertion of a 7Fr long sheath

xenologous pericardium patches. The sheath is then introduced after patch puncture (Fig. 31.4). Long sheath allows a longer distance between the operator and the C-arm providing a better radiations protection.

Type of Stent

Balloon expandable stents are the first choice to treat proximal stenosis/occlusion due to highest radial force. The use of covered balloon expandable stents is based on the principle of lowering the intraoperative embolic risk and long-term myointimal hyperplasia stenosis [22]. Covered stents have already proved their benefit in other anatomical localization such as the iliac arteries [23, 24], superficial femoral arteries. In a large monocentric prospective study from an intestinal dedicated stroke center, covered stenting is associated with a primary and a secondary patency rate of 83% and 99% respectively. Other study has shown a decrease of 40% of restenosis and an increase of 28% of freedom from symptoms recurrence compared to bare stents [25]. Based on non-randomized or retrospective studies, the American guidelines recommend the use of covered stents in CA and SMA stenting [8]. Results of a multicentric randomized control trial will probably confirm the significant benefit of covered stents in mesenteric arteries stenting [26]. Finally, the most commonly used size of stent in the SMA is 7 mm × 32 mm in our experience [18].

Other Technical Considerations

The use of endovascular embolic protection devices has been described for treatment of sever calcified lesions or associated with thrombus [27]. It appears to be feasible and safe with a third of the patients presenting macroscopic debris in the filter, but failed to prevent distal emboly in 6% of patients. Embolic protection devices are not currently recommended.

Stent flaring is widely use in visceral arteries endovascular repair. The first argument is that proximal flaring makes secondary reintervention easier by facilitating stent catheterization. Secondly, in a retrospective study of 150 patients treated for chronic mesenteric ischemia by CA or SMA stenting, ostial flaring was observed as the single factor on multivariate analysis associated with higher patency [28]. But this study was conducted with balloon expandable bare stents.

Specific Revascularization of the Inferior Mesenteric Artery

Percutaneous angioplasty, of the inferior mesenteric artery (IMA) can be considered in selective cases when revascularization of the CA and SAM is not feasible [29]. Usually, IMA has a minor bowel feeding territory. In case of chronic occlusion of CA and SMA, the IMA car be hypertrophic due to flow increase, but regarding this artery diameter, angioplasty is mostly performed without stenting.

Anticipated Complications

Stenting of mesenteric arteries provides a significant reduction of perioperative complications rate compared to open surgery with a similar long-term survival [8]. Nevertheless, intraoperative complications are described.

Complications of mesenteric arteries stenting include distal embolization (4%), branch perforation and hematoma (2%), distal dissection (1%), stent dislodgment (1%), and stent thrombosis (1%) [30]. Distal embolization might be reduced by the use of embolic protection devices but the few studies using those devices failed to prove the benefit of the technique [27]. Branch perforation should be prevented by checking the distal position of the guidewire and an antero-posterior completion angiography should be performed at the end of the procedure to track that kind of complication. Distal artery dissection is probably underestimated in SMA stenting and is favorized by stent oversizing or recanalization. Treatment consists of stenting prolongment using a 6 or 7 mm auto-expandable bare stent landed more distally. First collaterals can be covered with the bare stent if needed to treat extensive dissection. Stent dislodgment is a very rare complication favored by stent undersizing, major stent overhang in the aorta. In our experience, dislodgment can be observed in ROMS procedure while movements of the sheath can provide stent shifting [18]. And finally, stent thrombosis can occur after distal dissection or major recoil after stent placement, especially in case of severely calcified stenosis. It appears also to be one of risk factor of delayed occlusion. Recoil has to be tracked on the completion angiography which should be performed on the lateral view but also in the antero-posterior view. Prevention consists in the placement of a second stent or the use of high-pressure balloon.

Perioperative Care and Follow Up

Former prognosis of AMI was very dark with a postoperative mortality of 50–80% in previous studies [1]. Organization of intestinal stroke centers and standardization of the medical treatment change the prognosis of AMI with a dramatically decrease of postoperative mortality around 20% [4]. All patients presenting AMI should receive a multimodal medical support including curative unfractionated heparin treatment, anti-platelet therapy, hemodynamic management, oral antibiotics for bowel decontamination, bowel rest, parenteral nutrition and proton pump inhibitors (Table 31.2) [4]. In case of early-stage AMI, emergent revascularization should be performed by endovascular techniques avoiding laparotomy. In late presentation, revascularization leads to necrosis delimitation and abdominal exploration and eventual bowel resection are mandatory. Initial follow-up is performed in intensive care unit and "second-look" laparotomy can be discussed depending of patient evolution or initial bowel aspect.

Chronic mesenteric ischemia treated with angioplasty and stent of the SMA had dual antithrombotic treatment for 3–6 months, also depending of the type of stent used. Discharge from the hospital occurs usually the day after the operating day depending on the general and the nutritional status. Clinical follow-up remains essential during the first year after mesenteric artery revascularization, especially stenting, considering the regularity of symptoms recurrence due to in-stent restenosis and the potentially severe consequences of stent occlusion. In a prospective study of 145 patients with a mean follow-up of 12 months, recurrence of symptoms occurred in 17% and in-stent restenosis was observed in 39% on routine DUS imaging. Currently, no recommendation exist on the follow-up strategies [7]. But European vascular surgery guidelines consider a repeat follow-up by clinical assessment and a long-term imaging follow-up [3].

Table 31.2 Multimodal treatment of arterial acute mesenteric ischemia [4]

Medical therapies:
- Curative unfractionated heparin treatment
- Anti-platelet therapy
- hemodynamic management
- Oral antibiotics for bowel decontamination
- Bowel rest
- Proton pump inhibitors

Emergent bowel revascularization first

Bowel resection if needed

Case Presentation

Continued from page 331

Under local anesthesia and sedation. Ultrasound-guided left femoral access was obtained. A standard 6 Fr sheath was placed and a dose of 50 UI/kg of heparin sodium was administrated. A 7 Fr long steerable sheath (7 Fr Tourguide, Medtronic) is introduced over a semi-stiff long guidewire. Antegrade recanalization of the SMA is initiated with a 0.035″ angled standard hydrophilic guidewire (Radiofocus, Terumo) and a BER catheter. Intraluminal recanalization has to be preferred to subintimal one. In case of failure, different types of crossing catheters can be used. But our preferred technique for occlusion, is the use of the distal end of the guidewire allowing to have enough push to cross the heavy occlusion (Fig. 31.5, left panel). The BER catheter is advanced through the wire and intraluminal position of catheter's end is checked by blood backflow and by an angiography through the catheter. The distal SMA is catheterized and wire is exchanged for a 0.035″ Rosen wire. Predilatation of the occlusion is performed using an undersized 4 mm × 20 mm to minimize the risk of distal artery dissection (Fig. 31.5, right panel). The lesion is stented using a 7 mm × 32 mm balloon expandable covered stent (Atrium V12, Getinge). Control angiogram showed a small recoil on the recanalization site and a small image of dissection at the distal of the stent. Both problems were solved using a 7 mm × 40 mm autoexpandable nitinol stent (Absolut pro, Abbott) landed 15 mm more distally in the SMA (Fig. 31.6, left panel). Final angiography confirmed the patency of the recanalization with no distal complications (Fig. 31.6, middle panel). Abdominal pain immediately disappeared postoperatively. Postoperative course was uneventful and renutrition was performed progressively. Patient was discharged on fifth post-operative day. Patient was treated by dual antiplatelets therapy (Aspirin 75 mg/day and Clopidogrel 75 mg/day) during 3 months and then Clopidogrel was stopped. General status improved and body weight increased of 7 kg in 6 months. Follow-up CT angiography revealed a good stent patency at 1 year with no evidence of in-stent stenosis (Fig. 31.6, right panel).

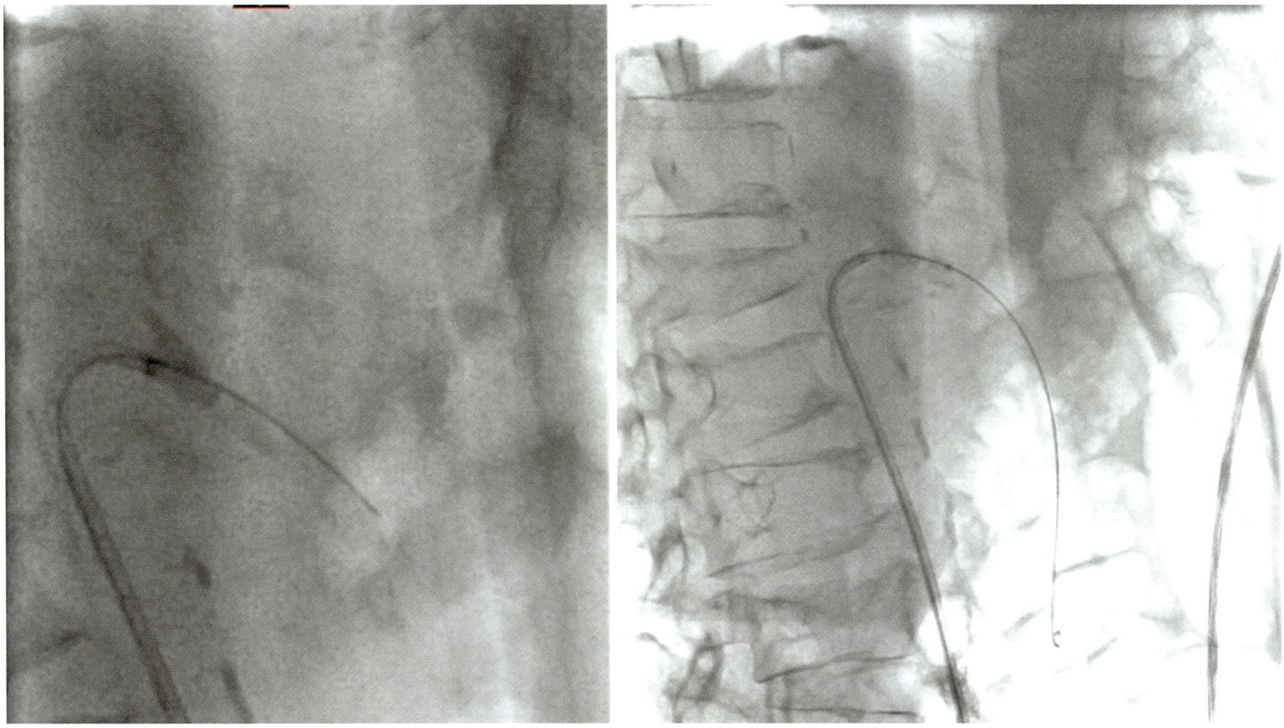

Fig. 31.5 Use of 7 Fr Tourguide (Medtronic) steerable sheath to allow antegrade recanalization of the occlusion (**left panel**). The wire has been exchanged for a 0.035″ Rosen wire and predilatation of the occlusion is performed using an Armada 4 × 20 mm angioplasty balloon (Abbott Vascular) (**right panel**)

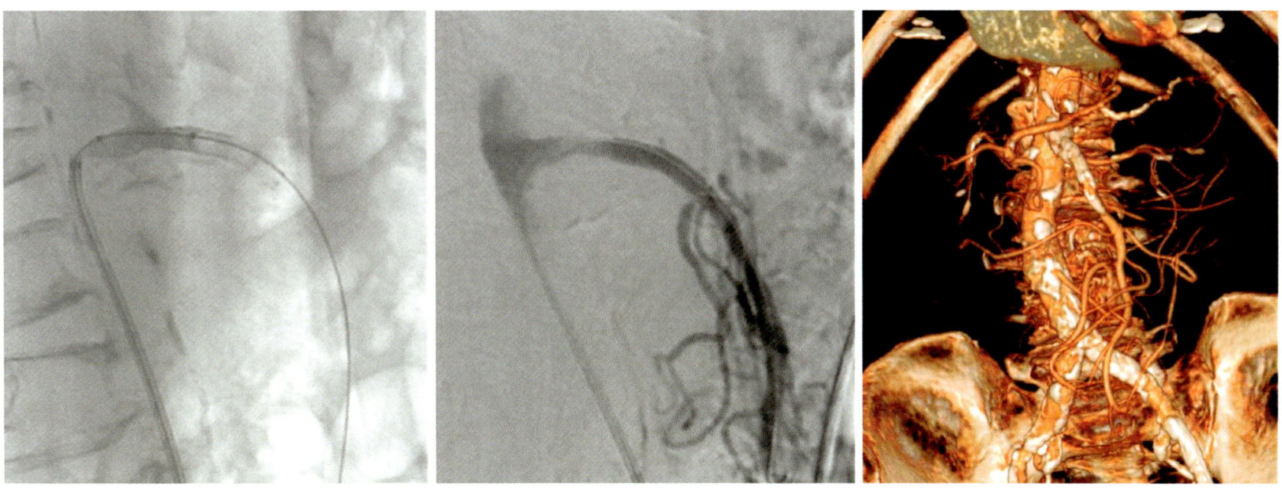

Fig. 31.6 After proximal placement of an Advanta V12 7 × 32 mm balloon expandable covered stent (Getinge) followed by distal placement of an Absolut Pro 7 × 40 mm self-expandable stent (Abbott Vascular). Proximal covered stent was flared with an 8 × 20 mm balloon (**left panel**). Final angiogram showing an efficient recanalization of the SMA. Note the small recoil at the level of the heavy calcified occlusion (**middle panel**). Volume rendering technique reconstruction of 12 months postoperative CT angiography (**right panel**)

References

1. Clair DG, Beach JM. Mesenteric ischemia. N Engl J Med. 2016;374(10):959–68.
2. Corcos O, Nuzzo A. Gastro-intestinal vascular emergencies. Best Pract Res Clin Gastroenterol. 2013;27(5):709–25.
3. Björck M, Koelemay M, Acosta S, Bastos Goncalves F, Kölbel T, Kolkman JJ, et al. Editor's choice—management of the diseases of mesenteric arteries and veins: clinical practice guidelines of the European Society of Vascular Surgery (ESVS). Eur J Vasc Endovasc. 2017;53(4):460–510.
4. Corcos O, Castier Y, Sibert A, Gaujoux S, Ronot M, Joly F, et al. Effects of a multimodal management strategy for acute mesenteric ischemia on survival and intestinal failure. Clin Gastroenterol Hepatol. 2013;11(2):158–65.e2.
5. Nuzzo A, Maggiori L, Paugam-Burtz C, Cazals-Hatem D, Ronot M, Huguet A, et al. Oral antibiotics reduce intestinal necrosis in acute mesenteric ischemia: a prospective cohort study. Am J Gastroenterol. 2019;114(2):348–51.
6. Roussel A, Castier Y, Nuzzo A, Pellenc Q, Sibert A, Panis Y, et al. Revascularization of acute mesenteric ischemia after creation of a dedicated multidisciplinary center. J Vasc Surg. 2015;62(5):1251–6.
7. Terlouw LG, Moelker A, Abrahamsen J, Acosta S, Bakker OJ, Baumgartner I, et al. European guidelines on chronic mesenteric ischaemia—joint United European Gastroenterology, European Association for Gastroenterology, Endoscopy and Nutrition, European Society of Gastrointestinal and Abdominal Radiology, Netherlands Association of Hepatogastroenterologists, Hellenic Society of Gastroenterology, Cardiovascular And Interventional Radiological Society of Europe, and Dutch Mesenteric Ischemia Study Group clinical guidelines on the diagnosis and treatment of patients with chronic mesenteric ischaemia. United Eur Gastroenterol J. 2020;8(4):371–95.
8. Huber TS, Björck M, Chandra A, Clouse WD, Dalsing MC, Oderich GS, et al. Chronic mesenteric ischemia: clinical practice guidelines from the Society for Vascular Surgery. J Vasc Surg. 2021;73(1S):87S–115S.
9. Oderich GS, Bower TC, Sullivan TM, Bjarnason H, Cha S, Gloviczki P. Open versus endovascular revascularization for chronic mesenteric ischemia: risk-stratified outcomes. J Vasc Surg. 2009;49(6):1472–1479.e3.
10. Oderich GS, Gloviczki P, Bower TC. Open surgical treatment for chronic mesenteric ischemia in the endovascular era: when it is necessary and what is the preferred technique? Semin Vasc Surg. 2010;23(1):36–46.
11. Roussel A, Nuzzo A, Pellenc Q, Castier Y, De Blic R, Cerceau P, et al. Surgical revascularization of the celiac artery for persistent intestinal ischemia in short bowel syndrome. Int J Surg Lond Engl. 2018;49:39–44.
12. Nuzzo A, Corcos O. Reversible acute mesenteric ischemia. N Engl J Med. 2016;375(15):e31.
13. Copin P, Zins M, Nuzzo A, Purcell Y, Beranger-Gibert S, Maggiori L, et al. Acute mesenteric ischemia: a critical role for the radiologist. Diagn Interv Imaging. 2018;99(3):123–34.
14. Nuzzo A, Maggiori L, Ronot M, Becq A, Plessier A, Gault N, et al. Predictive factors of intestinal necrosis in acute mesenteric ischemia: prospective study from an intestinal stroke center. Am J Gastroenterol. 2017;112(4):597–605.
15. Nuzzo A, Joly F, Ronot M, Castier Y, Huguet A, Paugam-Burtz C, et al. Normal lactate and unenhanced CT-scan result in delayed diagnosis of acute mesenteric ischemia. Am J Gastroenterol. 2020;115(11):1902–5.
16. Allain C, Besch G, Guelle N, Rinckenbach S, Salomon du Mont L. Prevalence and impact of malnutrition in patients surgically treated for chronic mesenteric ischemia. Ann Vasc Surg. 2019;58:24–31.
17. Moneta GL, Lee RW, Yeager RA, Taylor LM, Porter JM. Mesenteric duplex scanning: a blinded prospective study. J Vasc Surg. 1993;17(1):79–84; discussion 85–86.
18. Girault A, Pellenc Q, Roussel A, Senemaud J, Cerceau P, Maggiori L, et al. Midterm results after covered stenting of the superior mesenteric artery. J Vasc Surg. 2021;74(3):902–909.e3.
19. Eilenberg W, Kölbel T, Rohlffs F, Oderich G, Eleshra A, Tsilimparis N, et al. Comparison of transfemoral versus upper extremity access to antegrade branches in branched endovascular aortic repair. J Vasc Surg. 2021;73(5):1498–503.
20. Sénémaud JN, Roussel A, Pellenc Q, Chardigny C, Cerceau P, Corcos O, et al. Retrograde open mesenteric stenting for acute and chronic mesenteric ischaemia: results from an Intestinal Stroke Centre. Eur J Vasc Endovasc Surg. 2021;62(1):55–63.
21. Oderich GS, Macedo R, Stone DH, Woo EY, Panneton JM, Resch T, et al. Multicenter study of retrograde open mesenteric artery stenting through laparotomy for treatment of acute and chronic mesenteric ischemia. J Vasc Surg. 2018;68(2):470–480.e1.
22. Sick PB, Brosteanu O, Niebauer J, Hehrlein C, Schuler G. Neointima formation after stent implantation in an experimental model of restenosis: polytetrafluoroethylene-covered versus uncovered stainless steel stents. Heart Dis Hagerstown Md. 2002;4(1):18–25.
23. Boulitrop C, Jayet J, Duprey A, Pellenc Q, Roussel A, Cerceau P, et al. From the aortic bifurcation to the groin: long-term outcomes of covered kissing stent placement in combination with Iliofemoral reconstruction for extensive Iliofemoral occlusive disease. Ann Vasc Surg. 2020;64:11–6.
24. Mwipatayi BP, Sharma S, Daneshmand A, Thomas SD, Vijayan V, Altaf N, et al. Durability of the balloon-expandable covered versus bare-metal stents in the covered versus balloon expandable stent trial (COBEST) for the treatment of aortoiliac occlusive disease. J Vasc Surg. 2016;64(1):83–94.e1.
25. Oderich GS, Erdoes LS, LeSar C, Mendes BC, Gloviczki P, Cha S, et al. Comparison of covered stents versus bare metal stents for treatment of chronic atherosclerotic mesenteric arterial disease. J Vasc Surg. 2013;58(5):1316–24.
26. van Dijk LJD, Harki J, van Noord D, Verhagen HJM, Kolkman JJ, Geelkerken RH, et al. Covered stents versus bare-metal stents in chronic atherosclerotic gastrointestinal ischemia (CoBaGI): study protocol for a randomized controlled trial. Trials. 2019;20(1):519.
27. Mendes BC, Oderich GS, Tallarita T, Kanamori KS, Kalra M, DeMartino RR, et al. Superior mesenteric artery stenting using embolic protection device for treatment of acute or chronic mesenteric ischemia. J Vasc Surg. 2018;68(4):1071–8.
28. Haben C, Park WM, Bena JF, Parodi FE, Lyden SP. Improving midterm results justify the continued use of bare-metal stents for endovascular therapy for chronic mesenteric ischemia. J Vasc Surg. 2020;71(1):111–20.
29. Wohlauer M, Kobeiter H, Desgranges P, Becquemin JP, Cochennec F. Inferior mesenteric artery stenting as a novel treatment for chronic mesenteric ischemia in patients with an occluded superior mesenteric artery and celiac trunk. Eur J Vasc Endovasc Surg. 2014;27(3):e21–3.
30. Oderich GS, Tallarita T, Gloviczki P, Duncan AA, Kalra M, Misra S, et al. Mesenteric artery complications during angioplasty and stent placement for atherosclerotic chronic mesenteric ischemia. J Vasc Surg. 2012;55(4):1063–71.

32 Renal Artery Angioplasty and Stent Insertion: An Overview of Modern Practice

Mark Gregory and Tarun Sabharwal

Case Presentation
A 62-year-old male under the care of the renal team presents with a rapid decline in his renal function. He has a history of left nephrectomy 8 years ago for renal cell carcinoma. The patient also has type 2 diabetes and is an ex-smoker. He is found to have hypertension, which is refractory to standard antihypertensive treatment. Urinalysis demonstrates no proteinuria or haematuria. CT imaging obtained as follow-up for his renal cell carcinoma has previously noted incidental right renal artery stenosis. The Interventional Radiology team is consulted regarding the feasibility of revascularising the kidney.

Introduction

Balloon angioplasty of the renal artery was first undertaken by Andreas Gruntzig in 1978, followed a decade later by the first renal artery stent deployment by Palmaz and colleagues [1]. The practice of renal artery angioplasty and stenting steadily gained popularity over the subsequent decades before a rapid expansion in case numbers in the early 2000s; more than 60,000 procedures were completed in the US in 2005 [2].

The publication of a series of randomised controlled trials assessing outcomes for renal artery stenting against best medical management alone, most notably the ASTRAL and CORAL studies, has led to a dramatic decrease in the number of cases undertaken [3]. Although these studies have been criticised for underrepresenting the most severe cases, they have demonstrated that for the majority of patients with mild/moderate renal artery stenosis, modern antihypertensive regimes offer effective management. Consequentially, stenting of the renal artery no longer represents a routine intervention and is reserved for a number of specific clinical circumstances, where it remains an important treatment option.

This chapter outlines the indications for renal artery intervention, preoperative preparation and imaging, modern interventional techniques, complications and outcomes. The use of renal artery stents in fenestrated endovascular aortic repair has been covered elsewhere and is considered outside the remit of this chapter.

Overview of Renovascular Disease and Indications for Intervention

Overview of Renovascular Disease

Renovascular disease is an umbrella term encompassing a number of aetiologies of renal artery stenosis. Atherosclerotic renal artery stenosis (ARAS) is the most common cause and the primary focus of this chapter. Fibromuscular dysplasia (FMD) is a leading cause in younger (particularly female) patients. Less common aetiologies include renal artery dissection, arterial embolus and the vasculitides.

The reduction of blood flow to a kidney, resulting from renal artery stenosis, reduces perfusion pressures within the kidney and activates the renin-angiotensin-aldosterone system (RAAS). This leads to systemic vasoconstriction, sodium retention and a resultant increase in systemic blood pressure, a phenomenon termed renovascular hypertension. Renovascular hypertension is thought to account for around 2% of all cases of hypertension, although its prevalence increases to around 7% in patients over 65 [3].

Often detected incidentally in asymptomatic patients, renovascular disease can cause a spectrum of clinical conditions including ischaemic nephropathy (decline in kidney function secondary to low perfusion pressure), renovascular hypertension and congestive heart failure. The clinical spectrum is influenced by factors including whether a stenosis

M. Gregory (✉) · T. Sabharwal
Guy's and St. Thomas Hospital, NHS Foundation Trust, London, UK
e-mail: mark.gregory10@nhs.net

G. Geroulakos et al. (eds.), *Mastering Endovascular Techniques*, https://doi.org/10.1007/978-3-031-42735-0_32

occurs acutely or develops insidiously, contralateral kidney function and cardiac reserve [1, 2].

Clinical suspicion of renovascular disease is generally raised in the context of hypertension that is resistant to antihypertensive medication or progressively deteriorating renal function without an alternative cause. An abrupt onset of symptoms may further raise suspicions, as well as known vascular disease elsewhere in the body.

Indications for Intervention

Whilst conceptually it may seem prudent to restore renal blood flow in all cases of renal artery stenosis, activation of the RAAS requires a reduction in renal perfusion pressure of around 20% and renal parenchyma can be preserved despite reductions in pressure of up to 40% [3]. As a result, arterial stenoses occluding less than 70% of the lumen are unlikely to be of clinical significance. Conversely in the setting of critical renal artery stenosis leading to ischaemic nephropathy, a window for intervention exists after which revascularisation will have no benefit due to an irreversible kidney injury [3].

Modern antihypertensive medications, many of which target the RAAS, are generally well tolerated and have been shown to be highly effective in managing the majority of patients with renovascular hypertension [4]. As a result, there is growing consensus that endovascular treatment for renovascular disease should be reserved for the following specific clinical scenarios:

- Uncontrolled hypertension despite best medical management
- Intolerance to antihypertensive medications, specifically deterioration in renal function due to the use of ACE inhibitors
- Circulatory congestion including flash pulmonary oedema
- Progressive deterioration in renal function in the context of known renovascular disease
- Solitary functioning kidney or bilateral high-grade renal artery stenosis

Although a minority of patients undergoing endovascular revascularisation will experience dramatic improvements in blood pressure control or renal function [5], treatment should be undertaken with the aim of mitigating, rather than curing, specific factors. Examples might include slowing the progression of renal failure or facilitating control of previously refractory hypertension with antihypertensive medication.

The timing of any endovascular intervention should also be taken into consideration. As already discussed, intervention in cases of mild stenosis is unlikely to yield significant benefit. Conversely, patients with deteriorating renal function and proteinuria are likely to be past the point where revascularisation of a kidney will improve function [3]. In patients with circulatory congestion, timing of intervention will be guided by findings from echocardiography and cardiology assessment. Patients in whom treatment is not indicated should be re-assessed at regular intervals as renovascular disease can progress and intervention may be needed in future.

Finally, the aetiology of renovascular disease will also affect decision making. Cases of FMD should have a much lower threshold for treatment than atherosclerotic renovascular disease, as endovascular intervention in FMD patients carries less risk and is often preferable in young patients to long-term medical management.

Given the narrow range of indications, varying aetiology and the complexity involved in timing, decisions on endovascular treatment of renovascular disease should be made within a multidisciplinary team setting. Treatment decisions must be driven by specific clinical need rather than the degree of arterial stenosis identified on imaging.

Imaging and Preoperative Preparation

All patients under consideration for endovascular renal artery procedures should have up-to-date cross-sectional imaging. Sonographic assessment of renal artery stenosis no longer plays a significant role in assessment. The choice between CT Angiography and MR Angiography will vary between institutions, although for patients with impaired renal function, MRI is preferable as it avoids iodinated contrast. Vessel analysis software can be used to assess the severity of renal artery stenosis and the resultant need for intervention. Cross-sectional imaging should also be closely scrutinised to assess for potential complicating factors, such as aortic, iliac, or common femoral disease (with the potential to complicate access), accessory renal arteries, or other anatomical anomalies.

Decisions regarding renal artery revascularisation should be made within a multidisciplinary team setting, with input from nephrologists, radiologists and the interventionalists undertaking the procedure. Once the decision to treat has been made the patient should ideally be seen in an Interventional Radiology clinic. Clinic appointments allow an opportunity to set realistic expectations regarding chances of treatment success, as well as a chance to adequately consent the patient for the procedure, with full discussion of the associated risks.

Consideration should be given in clinic as to whether the patient will be able to lie flat, keep still and follow breathing instructions during the procedure, to allow for optimal digital subtraction angiography (DSA) imaging. Conscious sedation or general anaesthesia may be required to achieve the above factors, although patients with atherosclerotic renal artery lesions often have considerable comorbidities and may be high-risk anaesthetic candidates. Anaesthetic team presence will need to be arranged in advance, as well as postoperative admission in high-risk cases.

In addition to the above, clinic appointments offer an opportunity to review the patient's medications, with particular attention to anticoagulant and antiplatelet medication as well as the patient's antihypertensive regime. Typically, antiplatelet medication should not be withheld prior to the procedure. Indeed, some centres initiate either aspirin or clopidogrel prior to the procedure in cases of renal artery stenting [6]. Antihypertensive medication is usually continued on the day of the procedure, although ACE inhibitors are generally omitted and some centres will instruct the patient to halve their usual antihypertensive dose. If there is no capacity to see patients in clinic, they should at least have a nurse-led pre-assessment where medication instructions can be given.

Renal Angioplasty and Stent Insertion Technique

Techniques for renal angioplasty and stenting continue to evolve with advances in technology [6–9] and are outlined below. Procedures should be performed in an angiography suite, by appropriately trained individuals experienced in endovascular intervention. In the United Kingdom and Europe cases are predominantly performed by Interventional Radiologists, whereas Interventional Cardiologists and Vascular Surgeons undertake many such procedures in North America.

General Technical Considerations

The majority of cases can be undertaken under local anaesthesia, either as a day-case procedure or with overnight inpatient admission. As discussed above, patients who may not tolerate a procedure under local anaesthetic should be identified in advance, with anaesthetic support arranged.

Pre-procedure imaging can be utilised such that the level and orientation of the target renal artery are known in advance. The angle of the proximal renal artery with respect to the aorta should also be confirmed. This allows DSA imaging to be obtained at the optimal angle to demonstrate the origin of the vessel in profile, essential for accurate stent deployment [6, 10]. Finally, vessel analysis software can be used to accurately calculate the normal diameter of the vessel, for accurate angioplasty and stent sizing.

The patient cohort requiring renal artery intervention has a high rate of chronic renal insufficiency. Prior consideration should therefore be given to the use of iodinated contrast. The successful use of CO_2 angiography has been described [11] and represents a safe option in centres experienced in its use. In many cases, however, the total volume of iodinated contrast can be successfully kept to a minimum, with the omission of initial aortography (through utilisation of pre-procedure imaging) and image optimisation throughout the procedure [6]. In patients who are allergic to iodinated contrast, CO_2 angiography or gadolinium represents an alternative option.

Access, Renal Artery Cannulation and Embolic Protection Devices

Ipsilateral common femoral artery access is suitable in the majority of cases. Ultrasound-guided access and use of a micro-puncture kit may reduce the incidence of access site complications, known to be the most common complication associated with the procedure [7, 12]. Brachial access is an important alternative in patients with severe caudal angulation of the target renal artery, infrarenal aortic aneurysm or highly diseased iliac arteries. The use of the brachial artery comes with a higher rate of access site complications [13] and requires additional technical considerations, primarily the selection of a sheath or guide catheter long enough to reach the renal artery ostium and the use of longer shaft stents. Once access has been established, unfractionated heparin can be given intra-arterially, with an initial dose of 5000 iU and further boluses to maintain an activated clotting time (ACT) between 220 and 300 s. In some centres heparinisation is delayed until after the renal artery has been cannulated.

Variable guidance exists regarding the optimal choice of sheath and catheter combination for safe access to the target renal artery. Some texts advocate a long 6 fr sheath, such as a 40 cm Flexor Balkan or 45 cm Flexor Ansel Guiding Sheath (Cook, Bloomington, Indiana) or a 45 cm Destination Guiding Sheath (Terumo, Tokyo, Japan). Others advocate a short 7 fr sheath with the use of a 6 fr guide catheter such as a Hockey Stick Guiding Catheter or a Renal Standard Curve Guiding Catheter (Merit Medical, Utah) [7, 14]. Both the guide sheaths and the guide catheters are positioned at the renal artery ostium, providing support to facilitate the crossing of tight stenoses and delivery of devices, as well as providing a port for contrast injection.

There is further variation with respect to the technique for cannulating the target renal artery. The potential for intra-procedural embolisation of particles from atheromatous lesions is well established, having been demonstrated in both in vivo [15, 16] and ex vivo models [17]. Meticulous technique is therefore warranted to reduce the risk of embolisation during the initial crossing of the stenosis, especially in atherosclerotic lesions at the ostium. Older texts advocate the use of 0.035 in guidewires such as Bentson or Terumo wires, through 4 fr catheters [6]. Cannulation with hooked catheters such as the Visceral Selective 1 catheter (Cook, Bloomington, Indiana) has also been described with anecdotal evidence of lower risks of dissection and embolisation [6].

There is consensus in more recent texts that the use of 0.014 in guidewires to cross stenoses is likely to reduce the risk of dissection or embolisation [10], although there are no prospective studies confirming superiority. The 'No Touch Technique' [6, 14] has also been described, which aims to further reduce complications during renal artery cannulation. In this technique the guide catheter or sheath is prevented from engaging the renal artery by a 0.035 wire placed against the aortic wall superior to the renal ostium. The wire is retracted such that only the soft tip remains against the superior aortic wall, allowing the guide catheter or shaped sheath to take its desired form, pointing towards the renal artery ostium. At this point a 0.014 wire can be advanced through the guide catheter or sheath into the renal artery and across the stenosis [14]. The 0.035 wire can then be carefully retracted, allowing the sheath or guide catheter to engage the vessel.

Once the guidewire has advanced through the stenosis, it should be positioned in the distal main renal artery or a proximal first-order branch. Distal migration of the guidewire can result in vessel occlusion or even perforation of the renal capsule. A short-tipped guidewire is therefore preferential to provide enough stability to allow catheters, angioplasty balloons and stents to track across the stenosis without needing to advance the tip into second- or third-order branches [6].

Having established a secure position across the target stenosis, some older texts advocate the measurement of pressure gradients, especially in cases of borderline lesions. A 20 mmHg systolic pressure or a 10 mmHg mean gradient is considered significant [3]. Given the more rigorous patient selection process undertaken in modern practice and the data provided by pre-procedure imaging, the role of pressure gradient measurement to determine the significance of a stenosis is now limited.

The placement of embolic protection devices (EPDs) is often advocated prior to angioplasty or stenting, given the established risk of atheroemboli [1, 8, 9]. A variety of EPDs are available and can be deployed in the distal main renal artery, generally over a 0.014 or 0.018 system. Outcomes are likely to be device and operator dependent with an associated learning curve. A number of single-arm observational studies have been published, reporting rates of recovered embolic debris ranging from 65 to 80% [16, 18, 19]. The majority of these studies report comparably strong results in terms of stabilisation or improvement in renal function; however, the only randomised trial involving an EPD, the Angioguard (Cordis, Vaughan, California), found no benefit of its use [4].

Renal Stent Deployment

A wide range of balloon-mounted stents are available for use in the renal arteries (Table 32.1). The majority of modern stent designs employ a monorail delivery system over a 0.014 or 0.018 wire and range from 4 to 7 mm in diameter. Drug-eluting stents are commercially available, and whilst their superiority to bare-metal stents has not been proven, they may have a role, particularly in arteries less than 4 mm in diameter [20]. Pre-dilation of a stenotic lesion up to 4 mm was advocated with earlier-generation stents but is rarely required with modern, low-profile devices [6]. Similarly, advancement of the sheath/guide catheter across the stenosis for delivery of the stent, followed by retraction to the ostium prior to deployment, was advocated historically but is not stipulated in the instructions for use of modern designs.

Table 32.1 Suggested endovascular toolkit for renal artery angioplasty and stent insertion. These are only a few suggested options based on the author's experience and can cover the vast majority of cases. Cases involving brachial access will require longer sheaths, catheters and guidewires as well as long-shaft balloons and stents

	Manufacturer	Specifications
Sheaths		
Flexor Ansel guiding sheath	Cook	5 or 6 fr, 45 cm
Destination guiding sheath	Terumo	5 or 6 fr, 45 cm
Guidewires		
J-tip starter guidewire	Boston scientific	0.035 in, 150 cm
Hi-torque command	Abbott	0.014 in, 200 cm
V-14 ControlWire	Boston scientific	0.014 in, 180 cm
Catheters		
Angled catheters (various)	Various	4 or 5 fr
Low-profile catheters	Various	2.6 fr
IVUS		
Eagle eye platinum	Philips	150 cm (0.014 system, 5 fr sheath)
Embolic protection devices		
SpiderFx	Medtronic	6-7 mm, 190 cm (0.014 system)
Angioplasty balloons		
Coyote OTW	Boston scientific	4–7 mm diameter, 20–40 mm length, 5 fr sheath
Sterling OTW	Boston scientific	4–7 mm diameter, 20–40 mm length, 5 fr sheath
Balloon-expandable stents		
Tsunami	Terumo	4–7 mm diameter, 15–30 mm length, 5 fr sheath
Formula	Cook	5–7 mm diameter, 15–30 mm length, 5 fr sheath
Rx Herculink	Abbott	5–7 mm diameter, 15–30 mm length, 6 fr sheath

The selection of an appropriate length and diameter stent is achieved primarily with reference to the pre-procedure imaging. The majority of texts stipulate that for ostial lesions the stent should be deployed with the proximal aspect projecting 2 mm into the aortic lumen and that stents should not overexpand the natural diameter of the vessel. The use of intravascular ultrasound (IVUS) may be considered to check diameter and length measurements and for certainty regarding stent positioning [10]. IVUS allows excellent visualisation of the stenosis and the healthy distal vessel in which to land the stent and can be used in conjunction with DSA imaging of the renal artery to maximising accuracy of deployment.

Once the correctly sized stent has been chosen and positioned across the stenosis, the position can be rechecked with contrast injection through the sheath and small adjustments made. Slow balloon inflation combined with a steady hold on the sheath, stent and guidewire will minimise untoward movement during deployment. Following deployment, the fully deflated balloon should be carefully retracted through the stent under fluoroscopic guidance. Subsequent DSA should be closely scrutinised for dissection, perforation and for evidence of embolisation, indicated by abrupt cut-off of distal renal vessels or segmental perfusion defects. IVUS also offers potential benefit here, identifying dissections that may be missed on DSA and confirming a satisfactory position and expansion of the stent. In cases of suboptimal stent expansion, post-dilation with appropriately sized angioplasty balloons should be considered.

Technical Considerations for Angioplasty

Angioplasty of the renal artery is now predominantly reserved for patients with FMD. In these cases, the technique for accessing the target renal artery may be simplified, given the significantly reduced risk of atheroemboli. The use of EPDs is also less likely to be required as there is substantially less risk of atheroembolism in this cohort.

Angioplasty also has a dominant role in the treatment of transplant renal artery stenosis (TRAS), where it is generally preferred to stent deployment, particularly as a first-line intervention. In cases of TRAS, access can be considered from either the ipsilateral or contralateral common femoral artery; interrogation of pre-procedure cross-sectional imaging will help determine the best approach.

As with cases of stent deployment, careful analysis of the pre-procedure imaging will determine the diameter and length of angioplasty balloon required. IVUS may also have a role in angioplasty planning, providing more accurate information on stenoses than DSA imaging alone. Low-profile angioplasty balloons are available, compatible with a 0.014 or 0.018 system. Care should be taken in their positioning, inflation and deflation to reduce the risk of vessel dissection. As with stent deployment, post-angioplasty imaging should be carefully scrutinised to assess for complications.

The use of stents in patients with FMD should be avoided. The condition usually responds well to angioplasty, and the patient cohort is much younger, predisposing them to eventual in-stent stenosis. In angioplasty cases involving atheromatous lesions or TRAS, a suboptimal result should lead to consideration of stent deployment. Regardless of the pathology being treated, appropriately sized bare-metal and covered stents should be available in the department to treat cases of occlusive dissection or vessel perforation.

Closure and Post-procedure Care

Rechecking of the ACT time is useful prior to removing the sheath. As with all procedures, care should be taken in achieving satisfactory haemostasis to prevent access site complications. A variety of closure devices are indicated for retrograde common femoral access, and the technique for haemostasis should be guided by operator experience. Patients should be monitored in recovery for puncture site complications and for signs of renal injury, including flank pain and haematuria.

There is consensus in the literature that an antiplatelet regime should be undertaken following stent deployment. The benefits of single vs. dual antiplatelet therapy have not been conclusively determined, nor has the optimal duration of antiplatelet treatment; clopidogrel is commonly continued for between 8 weeks and 3 months, whereas Aspirin prescription is often lifelong [1]. Renal function should be closely monitored in the days and weeks following the procedure, as deterioration in function is a known complication and a subset of patients will need renal replacement therapy. Follow-up imaging should also be arranged, primarily to monitor stent patency, allowing for early intervention in cases of in-stent restenosis.

Complications

Renal arterial intervention carries a wide range of known complications. Reported complication rates vary, and there are few studies reporting on recent cohorts, treated with modern techniques. An overall major complication rate of 7.5% was reported in Mayo Clinic data from 140 patients between 1997 and 2000 [21], whereas data from the CORAL study's 495 patients indicate a 5.2% complication rate [4]. Peri-operative mortality has been found to be 2%, although again, there is limited recent data [5]. It is also known that complication rates vary between operators, and there is a learning curve for the procedure [1, 10]. For the purposes of this discussion, complications have been split into 'Renal' and 'Non-renal' complications.

'Renal' Complications

Distal embolisation is thought to occur in up to 10% of cases. Clinical effects of embolisation vary from subclinical to segmental renal infarction. The complication has been attributed to a causative factor in the subset of patients who experience a rise in creatinine following intervention [12]. Techniques to minimise atheroemboli have been discussed above. Angiographically evident distal embolisation was reported in 1.2% of patients in the CORAL study and should be actively sought, as early identification may allow for attempts at catheter-directed aspiration. In many cases however embolisation of micro-emboli will not be appreciable on imaging.

Renal artery thrombosis is another recognised complication, often secondary to the rupture of atherosclerotic plaques by guidewire or catheter manipulation. Adequate heparinisation is crucial in minimising thrombus formation. Again, catheter-directed aspiration or thrombolysis could be considered in the correct clinical context.

Arterial dissection is a common complication and may occur during initial cannulation attempts or following intervention. Highly calcified ostial plaques are particularly prone to dissection during cannulation. Suspicion of dissection should be raised by tactile feedback and the behaviour of the guidewire. Early identification is crucial prior to further instrumentation or angioplasty, which can propagate the dissection plane and risk vessel rupture. Should dissection occur, cannulation of the true lumen is required and can be confirmed by DSA imaging. This can be challenging and time consuming; adequate heparinisation is vital during this time to minimise the risk of thrombosis.

Dissection following angioplasty or stent insertion is often identified on subsequent DSA imaging. The complication is often attributed to the use of stents or balloons that are oversized for the vessel and is also more likely when the distal balloon or stent does not land in healthy vessel. When identified, dissection should be treated with either prolonged balloon angioplasty or further stent deployment, to prevent distal propagation of the dissection flap, which risks segmental infarction or complete thrombosis. As discussed above, IVUS can add value in cases of dissection. Vessel rupture is a complication of dissection and requires covered stent insertion to treat, or surgery if this strategy fails.

Unscrupulous care of the guidewire can lead to perforation of the distal renal arteries and the renal capsule. This can lead to pseudoaneurysm formation or renal/capsular bleeding. If not identified on DSA imaging, this can present with flank pain, a deterioration in the patient's condition, haematuria or a drop in haemoglobin. Super-selective catheter embolisation represents the optimal first-line treatment for this complication and can be performed at the time of the procedure, if identified early.

Contrast-induced nephropathy (CIN), stent migration and restenosis are renal-intrinsic complications which tend to be identified after the procedure. Renal function should be carefully monitored in the days and weeks following a procedure, and a standardised protocol for imaging follow-up should be agreed upon, with access to expedited imaging if complications are suspected. Patients at risk of CIN should be identified prior to the procedure and steps taken to mitigate the risk. Distal stent migration is often attributed to undersizing of stents, and modern imaging software should help limit the incidence. Balloon-mounted stents can occasionally become displaced in transit to the target vessel and travel systemically, often into the femoral or internal iliac vessels. Depending on the situation it may be possible to rectify these situations with an endovascular approach using snare devices; however a low threshold for consultation with a vascular surgical team should be maintained.

Longer-Term Renal Complications

In-stent restenosis represents the most significant of the longer-term complications. Restenosis by imaging has been found to range from 6.6 to 25% at 1 year, with the wide range likely to be due to variation in duplex and cross-sectional assessment [5]. Modern stents report a patency rate of around 90% at 1 year [22]. Target vessel revascularisation rates also vary between studies, often given as close to 20% at 3 years [aa - 5]. Options for revascularisation include angioplasty and stent relining.

Non-renal Complications

Access site complications account for the largest number of reported complications (up to 20% in some studies) and include haematoma, pseudoaneurysm and, rarely, arterial-venous fistula creation [21]. Techniques to reduce access site complications are discussed above. Thrombin injection for the treatment of suitable pseudoaneurysms has been well described. In cases where thrombin injection is unsuitable, covered stents may be required for treatment.

A multitude of complications can occur secondary to the instrumentation of the iliac vessels and aorta required for access. Aortic or Iliac artery dissection can result in limb or bowel ischaemia and may require further endovascular treatment. Systemic embolisation is also possible to a wide variety of locations. The likelihood of these complications increases with the degree of pre-existing iliac and aortic disease, factors which should be noted on pre-procedural imaging. In patients with a large burden of aortic or iliac disease, a brachial artery approach might be required.

Finally, heparinisation during the procedure increases the risk of cerebral haemorrhage and bleeding elsewhere in the body. ACTs should be regularly checked as a marker of bleeding risk.

Outcomes

The reported outcomes of patients treated with renal angioplasty and stenting are variable. Analysis of the available data reveals considerable complexity; it is evident that many patients with renal artery stenosis gain no benefit from intervention above that achieved with best medical therapy alone; however, certain subsets of patients can benefit dramatically from intervention. Detailed critique of the evidence is beyond the scope of this chapter; however, the key concepts are outlined here.

Angioplasty in Fibromuscular Disease

Angioplasty for patients with hypertension caused by fibromuscular disease has consistently been shown to be highly effective. Patients with fibromuscular disease rarely have resultant renal impairment, and as such, hypertension is the sole indication for treatment. The most recent cohort studies report levels of technical success consistently over 95%, and around 75% of patients were deemed to have had their hypertension either cured or improved [23]. Reported complication rates for this cohort are also lower than for patients with ARAS.

Atherosclerotic Renal Artery Stenosis

A number of randomised controlled trials comparing renal artery stenting (with best medical therapy) to best medical therapy alone have been conducted, including the STAR, ASTRAL and CORAL studies [4]. The primary and secondary endpoints investigated varied across the trials; however, none of these studies have demonstrated a benefit of renal artery stenting over best medical treatment, either for control of hypertension or for improved renal function [3].

These trials represent the highest quality of evidence; however, there are several recurring limitations. The most commonly cited limitation is that patients with the most high-risk renal artery stenosis were seldom included in the trials, as clinicians proceeded straight to intervention. Low-risk patients were therefore overrepresented; many of the patients were found to have less than 70% stenosis on angiography, and indeed 25% of patients in the ASTRAL trial had a normal eGFR and so would be considered unlikely to benefit from revascularisation [1, 3, 5].

Despite the limitations, the studies provide strong evidence that the majority of patients with renal artery stenosis can be effectively treated with best medical management alone, without exposing them to the risks associated with renal artery stenting. Appropriately therefore, since the publication of the CORAL trial in 2014, rates of renal artery stenting have declined considerably [5].

There are a number of studies which have identified cohorts of patients who derive benefit from revascularisation. The HERCULES trial included patients with hypertension refractory to two antihypertensives and who had stenosis of >60% on angiography, treated with the Herculink (Abbott) stent. The study demonstrated significant blood pressure improvements amongst the cohort of patients who were likely to have been underrepresented in the larger RCTs [22].

Additional studies have demonstrated the importance of timely intervention and, by extension, the futility of treatment past a certain stage. Patient's with rapidly deteriorating eGFR or who experience flash pulmonary oedema have been found to benefit considerably from timely revascularisation [5]. Furthermore, a sub-analysis of the cohort of CORAL patients with low urinary protein excretion rates also demonstrated improved outcomes (reduced cardiovascular events and mortality) in the patients who underwent revascularisation [3]. Conversely, more recent studies of patients who have undergone revascularisation have demonstrated poorer outcomes in those with low eGFR and proteinuria, concluding that there is a point beyond which revascularisation is unable to affect hypertension and renal function [1].

These findings, alongside a body of case reports and smaller cohort studies, indicate that revascularisation is a crucial tool that can have good results when chosen for the correct cases, at the correct timepoint. The modern clinical recommendations detailed above in the 'Indications' section reflect this conclusion. Assessment of patients for revascularisation involves significant complexity and should be determined by an experienced multidisciplinary team on a case-by-case basis.

Case Outcome

Following initial consultation with the Interventional Radiology team, an MRA of the renal arteries is obtained, which confirms severe right renal artery stenosis with an ostial lesion. A review of historical blood results demonstrates previously stable renal function with quite rapid decline in the last 3 months. After further discussion between the renal team and interventional radiology, it is agreed that prompt renal artery stenting is in the patient's best interests.

The patient is pre-assessed and deemed suitable to undergo the procedure under local anaesthetic as a day case. A review of the MRA demonstrates relatively conventional anatomy, with the vessel distal to the stenosis measured at 6 mm. The case is arranged for the following day.

Access is gained to the right common femoral artery, and a 5 fr 45 cm Destination sheath is inserted. The no-touch technique is used to carefully cannulate the right renal artery with a 0.014 in Command wire. The Destination sheath is allowed to gently engage the ostium of the artery, and a DSA run is taken from the sheath (Fig. 32.1). IVUS is then deployed to measure the renal artery. The healthy renal artery distal to the ostial stenosis is confirmed to be 6 mm. The IVUS catheter also allows the required length of the stent to be calculated. A landing position for the stent is marked with a fluoro-grab of the IVUS catheter.

A 6 mm × 15 mm Cook Formula balloon-mounted stent is then deployed, with the proximal aspect of the stent protruding into the aorta roughly 2 mm. IVUS demonstrates a small dissection of the vessel following stent deployment, so an additional 6 mm × 12 mm Cook Formula balloon-mounted stent is deployed (Fig. 32.2). A final DSA run is undertaken from the sheath to check for complications. There is no evidence of perforation or embolisation of debris, and the stent appears well positioned. The sheath is removed, and haemostasis is achieved with a closure device.

At discharge the patient is instructed to keep taking his daily aspirin. He is also given a 3-month prescription for clopidogrel. At follow-up with the renal team, the patient's renal function is noted to have stabilised, and the patient's blood pressure is now well controlled with two antihypertensive medications.

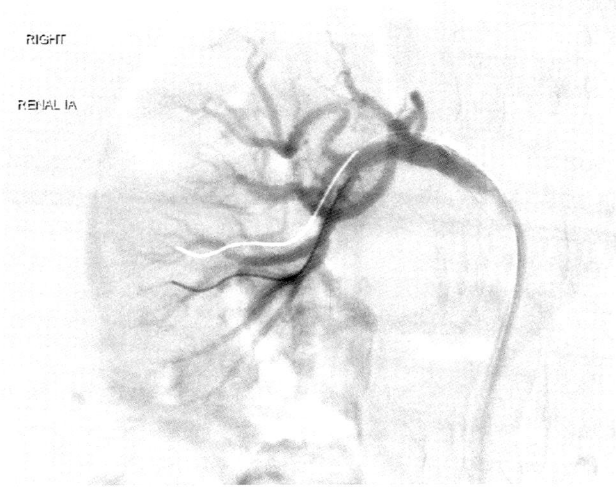

Fig. 32.1 Initial angiogram of the right renal artery demonstrating severe ostial stenosis

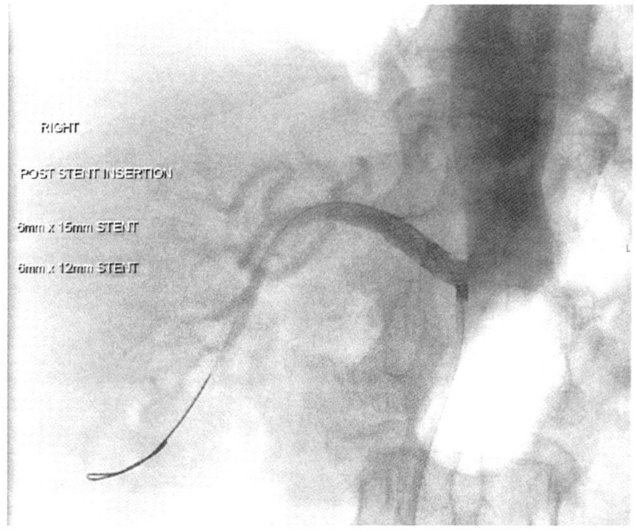

Fig. 32.2 Angiogram post-stent insertion. The proximal aspect of the 6 mm × 15 mm Cook Formula balloon-mounted stent extends around 2 mm into the aortic lumen. An additional 6 mm × 12 mm stent has been deployed to treat a small dissection distal to the initial stent. The angiogram is closely scrutinised for complications, including pseudoaneurysm or embolus in the distal renal artery branches

References

1. Textor SC, Misra S, Oderich GS. Percutaneous revascularization for ischemic nephropathy: the past, present, and future. Kidney Int. 2013;83(1):28–40. https://doi.org/10.1038/ki.2012.363.
2. Textor SC, Lerman L. Renovascular hypertension and ischemic nephropathy. Am J Hypertens. 2010;23(11):1159–69. https://doi.org/10.1038/ajh.2010.174.
3. Herrmann SM, Textor SC. Current concepts in the treatment of Renovascular hypertension. Am J Hypertens. 2018;31(2):139–49. https://doi.org/10.1093/ajh/hpx154.
4. Cooper CJ, Murphy TP, Cutlip DE, Jamerson K, Henrich W, Reid DM, Cohen DJ, Matsumoto AH, Steffes M, Jaff MR, Prince MR, Lewis EF, Tuttle KR, Shapiro JI, Rundback JH, Massaro JM, D'Agostino RB Sr, Dworkin LD, Investigators CORAL. Stenting and medical therapy for atherosclerotic renal-artery stenosis. N Engl J Med. 2014;370(1):13–22. https://doi.org/10.1056/NEJMoa1310753. Epub 2013 Nov 18.
5. Mousa AY, Bates MC, Broce M, Bozzay J, Morcos R, AbuRahma AF. Issues related to renal artery angioplasty and stenting. Vascular. 2017;25(6):618–28. https://doi.org/10.1177/1708538116677654. Epub 2017 Aug 7.
6. Funaki B. Renal ostial angioplasty and stenting. Part 1: the routine procedure. Semin Intervent Radiol. 2009;26(1):74–81. https://doi.org/10.1055/s-0029-1208385.
7. Jokhi PP, Ramanathan K, Walsh S, Fung AY, Saw J, Fox RS, Zalunardo N, Buller CE. Experience of stenting for atherosclerotic renal artery stenosis in a cardiac catheterization laboratory: technical considerations and complications. Can J Cardiol. 2009;25(8):e273–8. https://doi.org/10.1016/s0828-282x(09)70121-7.

8. Kok HK, Leong S, Govender P, et al. Percutaneous renal artery angioplasty and stenting: indications, technique and results. Ir J Med Sci. 2013;182:351–6. https://doi.org/10.1007/s11845-012-0887-5.
9. Erwin PA, Shishehbor MH. Renal artery stenting: indications, techniques, and devices. Minerva Cardioangiol. 2013;61(2):189–99.
10. Campbell JE, Stone PA, Bates MC. Technical discussion of diagnostic angiography and intervention of atherosclerotic renal artery stenosis. Semin Vasc Surg. 2013;26(4):150–60. https://doi.org/10.1053/j.semvascsurg.2014.06.002.
11. Caridi JG, William Stavropoulos S, Hawkins IF. Carbon dioxide digital subtraction angiography for renal artery stent placement. J Vasc Interv Radiol. 1999;10(5):635–40. https://doi.org/10.1016/S1051-0443(99)70094-1.
12. Beek FJ, Kaatee R, Beutler JJ, et al. Complications during renal artery stent placement for atherosclerotic ostial stenosis. Cardiovasc Intervent Radiol. 1997;20:184–90.
13. Franz RW, Tanga CF, Herrmann JW. Treatment of peripheral arterial disease via percutaneous brachial artery access. J Vasc Surg. 2017;66(2):461–5. https://doi.org/10.1016/j.jvs.2017.01.050. Epub 2017 Apr 19.
14. Goldstein JA, Kolluri R, Rocha-Singh KJ. Technical considerations for renal artery stenting. J Vasc Dis Manag. 2006.
15. Walker C, Kowalski J, Knan M, et al. Proximal protection before distal protection: preventing large atheroemboli during renal intervention. Am J Cardiol. 2002;90:28.
16. Holden A, Hill A, Jaff MR, Pilmore H. Renal artery stent revascularization with embolic protection in patients with ischemic nephropathy. Kidney Int. 2007;70:948–55.
17. Hiramoto J, Hansen KJ, Pan XM, et al. Atheroemboli during renal artery angioplasty: an ex-vivo study. J Vasc Surg. 2005;41:1026–30.
18. Henry M, Klonaris C, Henry I, et al. Protected renal stenting with the PercuSurge Guardwire device: a pilot study. J Endovasc Ther. 2001;8:227–37.
19. Thatipelli MR, Misra S, Sanikommu SR, et al. Embolic protection device use in renal artery stent placement. J Vasc Interv Radiol. 2009;20:580–6.
20. Misra S, Thatipelli MR, Howe PW, et al. Preliminary study of the use of drug-eluting stents in atherosclerotic renal artery stenoses 4 mm in diameter or smaller. J Interv Radiol. 2008;19:833–9.
21. Textor SC, McKusick M. Renovascular hypertension and ischemic nephropathy: angioplasty and stenting. In: Therapy in nephrology and hypertension: a companion to Brenner & Rector's the kidney, expert consult—online and print. Elsevier; 2008. p. 660–74. https://doi.org/10.1016/B978-141605484-9.50060-5.
22. Colyer W Jr. RX Herculink elite® renal stent system: a review of its use for the treatment of renal artery stenosis. Med Devices (Auckl). 2012;5:67–73. https://doi.org/10.2147/MDER.S25150.
23. Slovut DP, Olin JW. Current concepts: fibromuscular dysplasia. N Engl J Med. 2004;350:1862–71.

33 Visceral Aneurysms

Caleb Solivio, Zaeem Billah, Austin Shinagawa, Kartik Kansagra, and Geogy Vatakencherry

Case Presentation
A 63-year-old male was transferred from an outside hospital where he presented with back pain and acute urinary obstruction. As part of the workup on presentation a CT scan was performed, incidentally showing a 4 × 5 cm hepatic artery visceral aneurysm, 6.8 cm inflammatory infrarenal aortic aneurysm, and 3.1 cm inflammatory right common iliac artery aneurysm. The patient underwent nephrostomy tube placement to relieve the obstruction and clear any potential infection. Despite urinary diversion and confirmation of no underlying infection, his back pain persisted, and the decision was made to treat his multiple aneurysms in a staged fashion. Given his chief concern of back pain, the decision was made to treat the aortoiliac pathology first. His abdominal and iliac aneurysms were first repaired using a Gore Excluder Iliac Branch Endoprosthesis (WL Gore & Associates Flagstaff, Az). Given the size of the hepatic artery aneurysm (HAA) the decision was made to treat this in an endovascular fashion as opposed to continue surveillance.

Continued at page 359

Introduction

Visceral artery aneurysms (VAA) are historically uncommon disorders that are now being detected with increased frequency due to the ubiquitous utilization of cross-sectional imaging. Though they are typically asymptomatic, VAAs are of high clinical significance given their potential for rupture, which may result in mortality.

Though uncommon and usually clinically silent, aneurysms of the visceral arteries, namely the celiac axis, renal arteries, superior mesenteric artery (SMA), inferior mesenteric artery (IMA), and its branches, are of clinical concern given the potentially high mortality rate associated with their rupture (25–70%) [1]. While open surgical treatment models for VAAs remain a viable option in the treatment of VAAs, modern advances in endovascular technology and interventional physician technical skill sets have propelled endovascular therapy to become the first-line therapy for many VAAs.

It is imperative for the modern vascular and interventional physician who takes care of patients with VAAs to master the clinical knowledge required to guide and counsel patients with VAAs and to develop the technical expertise to offer treatment when deemed appropriate. It is also imperative that the vascular interventional physician continue to follow up with these patients indefinitely after endovascular repair.

Classification

VAA can be classified as true aneurysms, which are outpouchings of vessels with integrity of all three vascular layers, or pseudoaneurysms, which are outpouchings of vessels lacking integrity of all three vascular layers. True aneurysms are due to an inherent weakness of the vessel wall, often because of changes in elastin and collagen, such as in fibromuscular dysplasia (FMD), or vascular degenerative diseases, such as in Ehlers-Danlos syndrome or vasculitides. Pseudoaneurysms are associated with damage to the vessel layers and can be further subdivided by etiology including inflammatory, infectious, traumatic, and iatrogenic. In addition, VAAs can also be categorized by their morphology such as saccular (focal dilation of one side of the vessel) vs. fusiform (circumferential dilation of the vessel) and wide neck vs. narrow neck, referring to the interface with the parent vessel.

C. Solivio
California University of Science and Medicine, Colton, CA, USA
e-mail: SolivioC@cusm.org

Z. Billah · A. Shinagawa · K. Kansagra · G. Vatakencherry (✉)
Division of Vascular and Interventional Radiology, Kaiser Permanente, Los Angeles, CA, USA

G. Geroulakos et al. (eds.), *Mastering Endovascular Techniques*, https://doi.org/10.1007/978-3-031-42735-0_33

Special Considerations

Special attention should be paid to women who are pregnant or of childbearing age. There is an increased risk of rupture in pregnancy due to factors such as hormonal changes, an increase in relaxin, and a higher volume state. The rupture of VAAs in pregnancy is associated with high mortality rates for both the mother and the fetus. Additionally, for reasons poorly understood, orthotopic liver transplants (OLTs) are associated with higher rates of VAA rupture, and thus the treatment for VAAs in individuals expected to receive OLTs should be considered [2].

Workup

Generally, the workup for VAAs is similar. Though most are found incidentally, the preferential method for the diagnosis of VAAs is CT angiography (CTA), followed by other modalities such as MR angiography (MRA) for patients with high radiation exposure risks (especially younger patients). Current literature indicates that some VAAs may present with concomitant aneurysms in other vascular beds. For such VAAs, it is recommended to perform one-time head-to-pelvis imaging. If aneurysms are found in multiple beds, genetic workup for congenital conditions such as vascular-type Ehlers-Danlos, Marfan's syndrome, Loeys-Deitz syndrome and aneurysm osteoarthritis (AOS) should be performed. Given the diseases' autosomal dominant inheritance, families should also be evaluated and counseled. One should consider other etiologies such as FMD and other collagen vascular disorders as well. It is critical to take a thorough family history to identify genetically transmitted disorders and have a low threshold to send patients with multiple visceral aneurysms for genetic testing and family counseling.

Vascular Beds: Renal

Renal artery aneurysms (RAAs) occur in about 0.1% of the population with rupture rates up to 5% and an associated mortality rate of less than 10%. However, this mortality rate greatly increases for both mother and fetus (55% and 85%, respectively) if rupture presents in the third trimester. In women, RAAs may be associated with fibromuscular dysplasia (FMD), which increases the risk for concomitant aneurysms. Renal artery pseudoaneurysms may be associated with procedures such as renal biopsy and percutaneous nephrostomy tubes. Current guidelines suggest that women with RAAs also be screened via CTA/MRA for aneurysms of the cerebrovascular, mesenteric, and iliac arteries with head-to-pelvis imaging. After renal artery aneurysms are identified, indications for intervention include pseudoaneurysm, rupture, symptomatic RAAs, size >3 cm, or presentation in a premenopausal female. Medical management includes consideration of single antiplatelet therapy such as low-dose aspirin and follow-up surveillance imaging at 6 months followed by serial imaging every 1–2 years. This can be further increased to every 3–5 years once stability has been demonstrated over a longer period of time. It is important to note that the typical natural history of true aneurysms of the renal artery is usually of slow growth of <1 mm/year. Most renal artery aneurysms can be repaired using a consortium of endovascular techniques similar to what is used for the treatment of intracranial aneurysms. Surgical management of RAAs is typically reserved for those that have complex distal branch lesions and includes ex vivo surgical repair, autotransplantation, or even nephrectomy. Endovascular vs. surgical approaches for the more complex renal artery aneurysms will often be determined by local surgical and endovascular expertise (Fig. 33.1).

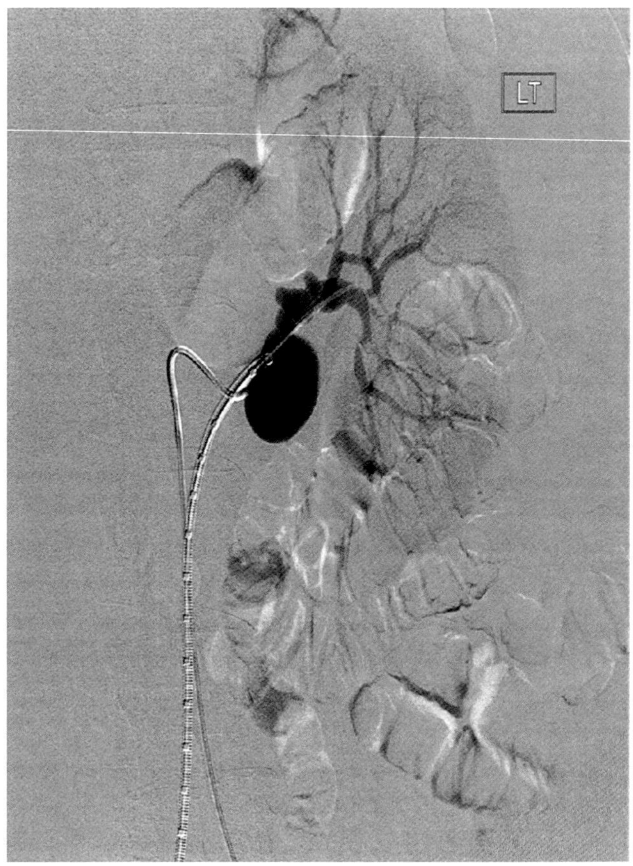

Fig. 33.1 Angiographic image demonstrating a saccular aneurysm arising from the left main renal artery. The beaded appearance of the renal artery is identified as a result of underlying FMD

Vascular Beds: Splenic

Splenic artery aneurysms (SAAs) constitute 60% of all VAAs. They most commonly occur in women and are associated with multiparity. Many SAAs present with rupture upon diagnosis with a rupture mortality rate of 25%. In pregnant women, rupture of SAAs is associated with maternal and fetal mortality rates as high as 80 and 90%, respectively. A study by Lee et al. [3] showed high rupture rates of splenic artery aneurysms in the setting of orthotopic liver transplant. Due to its high spatial resolution, CTA is the preferred imaging modality for the diagnosis of SAAs. Indications for intervention include pseudoaneurysm, symptomatic or ruptured SAAs, size >3 cm, rapid growth (>0.5 cm/year), patients undergoing liver transplantation, and SAAs in premenopausal females. A surveillance regimen for untreated SAAs includes shorter-term imaging between 3 and 6 months for 1 year followed by annual cross-sectional imaging. If stability for 5–10 years is demonstrated, the timespan between surveillance scans can be extended to every 3–5 years.

Endovascular intervention is the preferred method for SAA repair, as supported by much of the current literature, including a multicenter, retrospective study by Qiu et al., demonstrating a 90% technical success rate in endovascular repair of 70 aneurysms in 65 patients by covered stent with or without coil embolization, 23 of which were SAAs.

Vascular Beds: Hepatic

Hepatic artery aneurysms (HAAs) represent 20% of all VAAs [4], with reported rupture rates as high as 80% [5]. Hepatic artery pseudoaneurysms constitute up to 80% of all HAAs, most commonly caused by iatrogenic injury or blunt trauma. These aneurysms may present with rupture into the biliary tree and hemobilia/melena or as an intraperitoneal bleed and hemoperitoneum. After diagnosis, those with HAAs should also receive a one-time head-to-pelvis screening CTA to evaluate for aneurysms in additional vascular beds to identify associated pathologies such as fibromuscular dysplasia, vasculitis, and connective tissue disorders, and genetic testing should also be considered. Indications for intervention include symptomatic HAAs, pseudoaneurysms, inflammatory aneurysms, size >2 cm, or growth rate > 0.5 cm/year. Endovascular repair is the preferred HAA therapy, and current literature exhibits primary technical success rates for endovascular HAA repair of up to 98% [6–8]. Surgical intervention may be necessary in rare cases of unfavorable aneurysmal anatomy (Fig. 33.2).

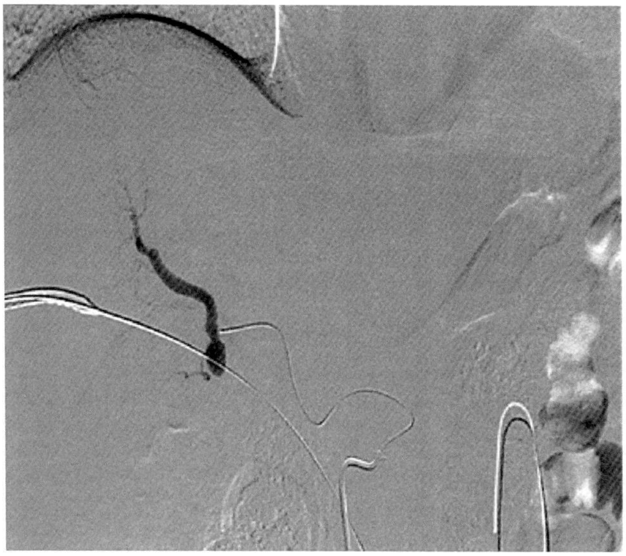

Fig. 33.2 Patient with pancreatic cancer who presented with melena and hemobilia after a biliary drain. A hepatic artery pseudoaneurysm was identified after pulling the biliary drain back over a wire

Vascular Beds: Celiac

Celiac artery aneurysms (CAAs) represent 4% of all VAAs [9] but reported associated rupture mortality rates near 100% [10]. Additionally, they may present with concomitant aortic aneurysms and other VAAs. Most CAAs are found incidentally. Indications for intervention include pseudoaneurysms, rupture, symptomatic CAAs, and size >2 cm. Endovascular management is the first-line treatment for CAAs though surgical management may be considered in anatomically unfavorable cases. If surveillance is performed, annual cross-sectional imaging is recommended to assess for CAA growth, and in younger patients, MRA should be considered for follow-up. Three retrospective analyses, totaling 26 CAAs, demonstrated the safety and efficacy of endovascular management of CAAs with technical success rates ranging from 84.2 to 98% [6–8]. Complications on follow-up included type 1a endoleak [8], mild restenosis [8], and a persistent but improving CAA which shrunk from 39 mm to 29 mm [7]. No reported cases warranted further surgical intervention. Postoperative annual imaging is recommended to check for endoleaks and residual CAA growth (Fig. 33.3).

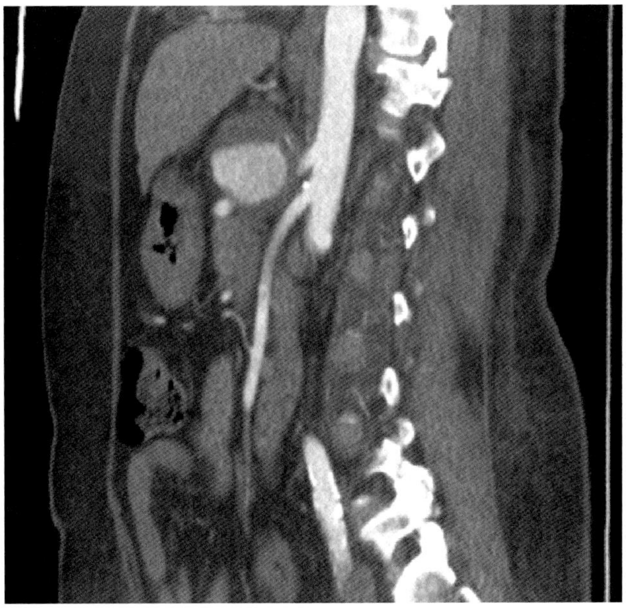

Fig. 33.3 5 cm celiac artery aneurysm arising from the main trunk

Vascular Beds: Superior Mesenteric Artery

Superior mesenteric artery aneurysms (SMAAs) are relatively rare, and true statistics regarding their prevalence vary greatly. SMAA rupture mortality rates vary in the literature between 30 and 90% [10]. The natural history of SMAAs is thought to be one of expansion and rupture. All true and false SMAAs should be considered for intervention with endovascular management when possible over open surgical management because of reported improvements in mortality rates [10]. Elective SMAA repair is associated with mortality rates less than 15%; such rates improve with endovascular techniques compared to open surgical repair methods. When considering endovascular repair, it is critical to maintain the patency of the major tributaries of the SMA to prevent intestinal infarction, which could result in fatality.

Vascular Beds: Gastric and Gastroepiploic Arteries

Gastric and gastroepiploic artery aneurysms represent only 4–5% of all VAAs, and given their rarity, their natural histories are not as clear. It is reasonable to perform head-to-pelvis imaging and consider genetic testing if multiple aneurysms are noted. Their rarity limits current data regarding their rupture rates, but some data reports up to 90% of patients with gastric or gastroepiploic artery aneurysms have presented with rupture [11]. Intervention is reasonable for all gastric and gastroepiploic artery aneurysms, and an endovascular approach should be considered as first-line therapy.

Vascular Beds: Pancreatoduodenal and Gastroduodenal Arteries

Together pancreaticoduodenal artery aneurysms (PDAA) and gastroduodenal artery aneurysms (GDAA) represent 6% [12] of all VAAs and are often associated with celiac artery stenosis or occlusion, which increases flow through the pancreaticoduodenal arcades as it is the primary collateral network between the superior mesenteric and celiac axes. Inflammatory conditions such as pancreatitis can result in pseudoaneurysms of these beds, and given their associated high mortality rates, they should be treated endovascularly when feasible. Given the high reported rupture rate in the current literature, it is recommended that PDAA and GDA aneurysms be treated endovascularly. If there is median arcuate ligament compression on the celiac artery, a laparoscopic ligament release can be considered (Fig. 33.4).

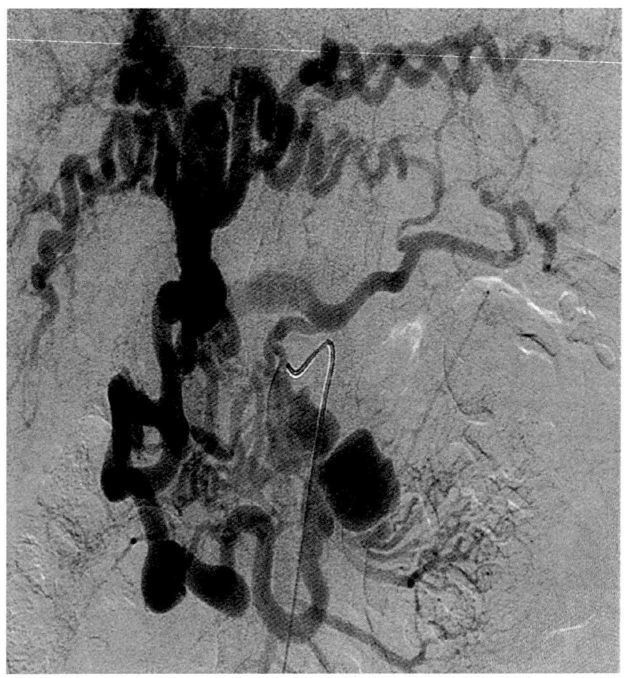

Fig. 33.4 Pancreaticoduodenal aneurysms arising from a patient with liver arteriovenous malformations (AVMs) and median arcuate ligament compression of celiac trunk

Vascular Beds: Jejunal, Ileal, and Colic

Together, jejunal, ileal, and colic artery aneurysms represent less than 3% of all VAAs. In patients whom jejunal, ileal, or colic artery aneurysms are found, a head-to-pelvis CTA or MRA should be performed. Though natural history is not clear, the current literature suggests a high rupture rate between 30 and 70%, and collective mortality rates between 20 and 50%.[10] Though these may be repaired with endovascular approaches, it is critical for the vascular interventionalists to ensure that they are able to effectively exclude the aneurysm while preserving flow to the intestines. Bannur et al. demonstrated a successful endovascular intervention of VAAs, four of which were colic or right colic artery aneurysms, without periprocedural morality [6]. Patients who received embolization should be monitored every 12–24 months with cross-sectional imaging to assess for vascular remodeling or aneurysm reperfusion.

Technical Considerations

Overall, the goal of VAA treatment, whether endovascular, laparoscopic, or open surgical, is to prevent aneurysm (or pseudoaneurysm) rupture while maintaining adequate end-organ perfusion. Endovascular approaches to VAA repair are generally preferred over their laparoscopic and open surgical counterparts, but determination of endovascular candidacy should be considerate of inflow vasculature, outflow vasculature, branch vessels arising from the aneurysm, as well as morphology of the aneurysm. Depending on which organ is involved, the etiology of the aneurysm, and individualized patient anatomies, various endovascular approaches can be considered, such as coil embolization, liquid embolization, and vascular plugs. For more challenging vascular anatomy, balloon- or stent-assisted coiling can be performed. Various types of stents can be used to exclude aneurysms including stent grafts and flow-diverting stents. In order to overcome some of the technical challenges from a transfemoral approach due to some of the downward angulations from the visceral vasculature, one can consider a torqueable guidesheath from a transfemoral approach or radial, brachial, or even axillary access. For excessive tortuosity, as seen in the splenic bed, one can also consider using flexible neuro distal access catheters. General anesthesia may be needed to minimize patient movement and obtain high-quality imaging, which is often necessary to ensure clinical and technical success with complex endovascular repairs.

Coil Embolization

Coil embolization represents the most popular endovascular VAA repair technique given its cost-effectiveness and aptitude for a wide array of saccular aneurysms in which the aneurysm neck is sufficiently narrow to prevent coils from extruding into the parent artery, thus maintaining vascular patency and end-organ perfusion. There is a growing array of microcatheters and various shapes, diameters, stiffnesses, and widths of coils that enable dense packing of aneurysms of varying sizes and morphologies. Online tools exist which can guide the interventionalist in calculating the number of packing coils needed to produce sufficient packing density based on aneurysm morphology and volume, depending on the length, shape, diameter, and stiffness of coils used.

Balloon-Assisted Coiling

Balloons can supplement coil embolization in cases where the neck of the saccular aneurysm is too wide allowing for high-density coil packing while simultaneously preventing coil prolapse into the parent vessel.

Stent-Assisted Coiling

Stents can assist coil embolization similarly to balloons but also can act as a scaffold to prevent the coils from prolapsing into the parent vessel. The microcatheter can be placed in the aneurysm sac through the interstices of the stent or alongside an undeployed or deployed stent. However, stent-assisted coiling usually requires that the patient remain on a duration of antiplatelet therapy to mitigate thrombogenicity of stents and maintain stent patency. Some have even described using retrievable stents employed for strokes as a temporary scaffold which can be removed once the coil mass has been deployed (Figs. 33.5 and 33.6).

Liquid Embolics

Liquid embolics can be used to obliterate aneurysms or as an adjunctive technique to add to coils in the treatment of aneurysms. Adhesive (also known as n-butyl cyanoacrylate) and non-adhesive liquid embolics (ethyl vinyl alcohol polymer) act in the same way but differ in their material properties. Newer non-adhesive agents coming up the pipeline include Squid (Balt USA, Irvine, California) and PHIL (MicroVention Terumo, Aliso Viejo, CA). Additional liquid embolics cur-

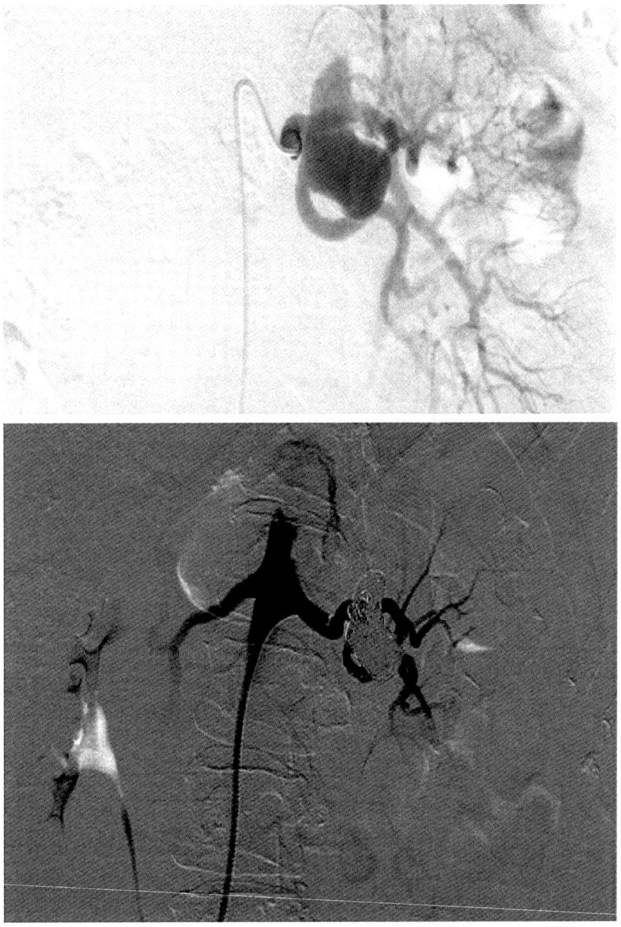

Fig. 33.5 Stent-assisted coiling performed on a complex bilobed RAA to preserve flow to both branches

rently being studied include the Embrace hydrogel embolic system, ionic strength-triggered chelation, thermoresponsive gels, pH-sensitive embolic agents, and shear-thinning hydrogels [13].

Stent Grafts and Flow Diverting Stents

The use of stents in endovascular repair of the VAAs is limited by vessel diameter and tortuosity. Various types of stents are currently used in VAA repair including balloon-expandable covered stents, self-expanding covered stents, and flow-diverting stents. The splenic artery can be exceedingly tortuous, making stent graft deployment in the distal bed quite challenging. In those instances, a lower-profile flow-diverting stent may be utilized. When considering placement of a balloon-expandable covered stent in the smaller visceral beds, off-label use of coronary stents may be considered.

Covered stents allow for total aneurysm exclusion but are susceptible to endoleaks. However, due to thrombogenicity of the covered stent material, dual antiplatelet therapy must often be administered for 3–12 months following placement. Flow-diverting stents are adapted from neurointerventional radiology and have been introduced in the treatment of VAAs. Their design partially excludes aneurysmal sacs, facilitating laminar flow through the parent arteries and turbulent, thrombogenic flow within the aneurysmal sac (Fig. 33.7). In general, if there is concern that the aneurysm is mycotic in nature, stent grafts should be avoided (Fig. 33.8).

Future Directions

Some future directions in product development for VAA management include the Barrel Vascular Reconstruction Device by Medtronic, detachable neck-bridging devices, and the Woven EndoBridge (WEB) by Subsequent Medical/Biovention [14]. The Barrel Vascular Reconstruction Device is purported to provide additional control for stent-assisted aneurysms at bifurcations and branch origins; detachable neck-bridging devices exclude flow across aneurysms with a hope for gradual endothelization; and WEB is a nitinol mesh device that is intended to plug aneurysms with stability across their necks. Research regarding long-term outcomes for the aforementioned future directions is lacking, and thus, their uses are currently limited.

Percutaneous and Surgical Options

Both open surgical and percutaneous VAA repair options exist. Open surgical methods, which have historically been used to treat VAAs, include aneurysmectomy, ligation, and bypass with venous graft or prosthetic vascular graft. In certain situations, direct sac access can be obtained via percutaneous approach, and a combination of thrombin, liquid embolics, and coils can be administered under fluoroscopic guidance to embolize the aneurysm while maintaining the patency of the parent vessel. Given the rapid advancements in endovascular technology and expanding skill set of vascular and interventional physicians, the need for open or laparoscopic surgical techniques is diminishing (Fig. 33.9).

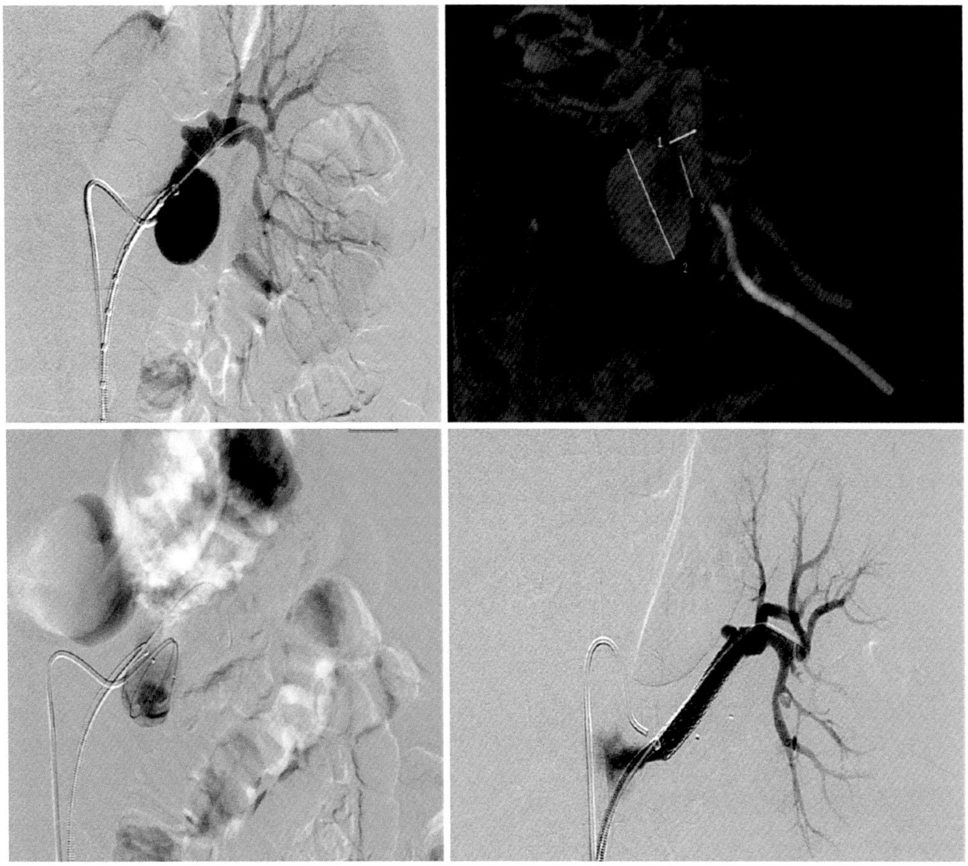

Fig. 33.6 Renal artery aneurysm excluded with a balloon-expandable covered stent graft

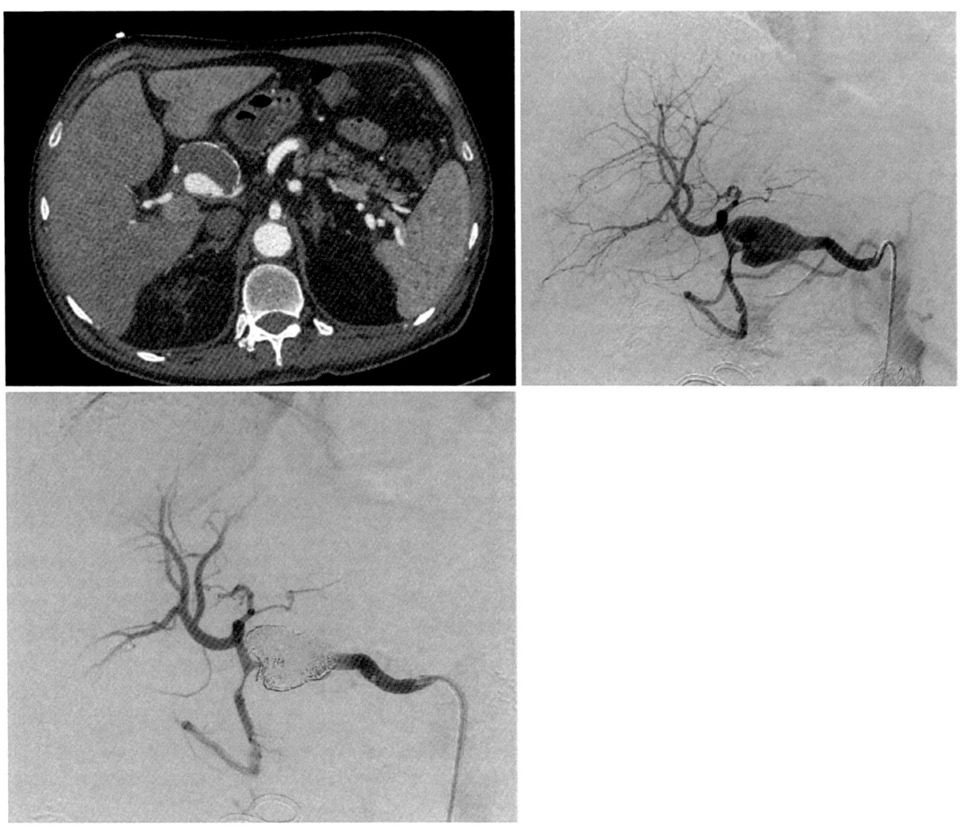

Fig. 33.7 Stent-assisted coiling with a neuro flow-diverting stent in a large common HAA in a patient with multiple aneurysms (aorta/iliac)

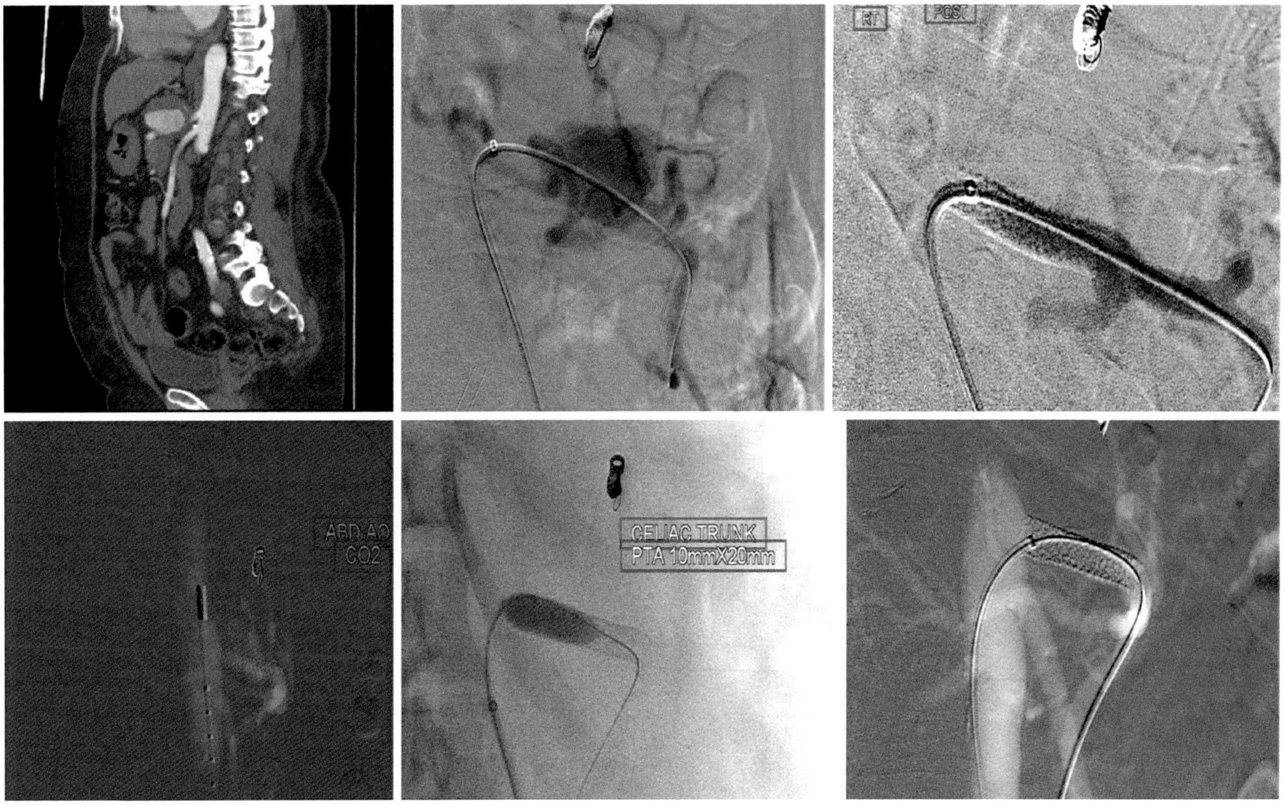

Fig. 33.8 Large celiac artery aneurysm excluded with a balloon-expandable stent graft which was flared at the aortic ostia to control type 1a endoleaks

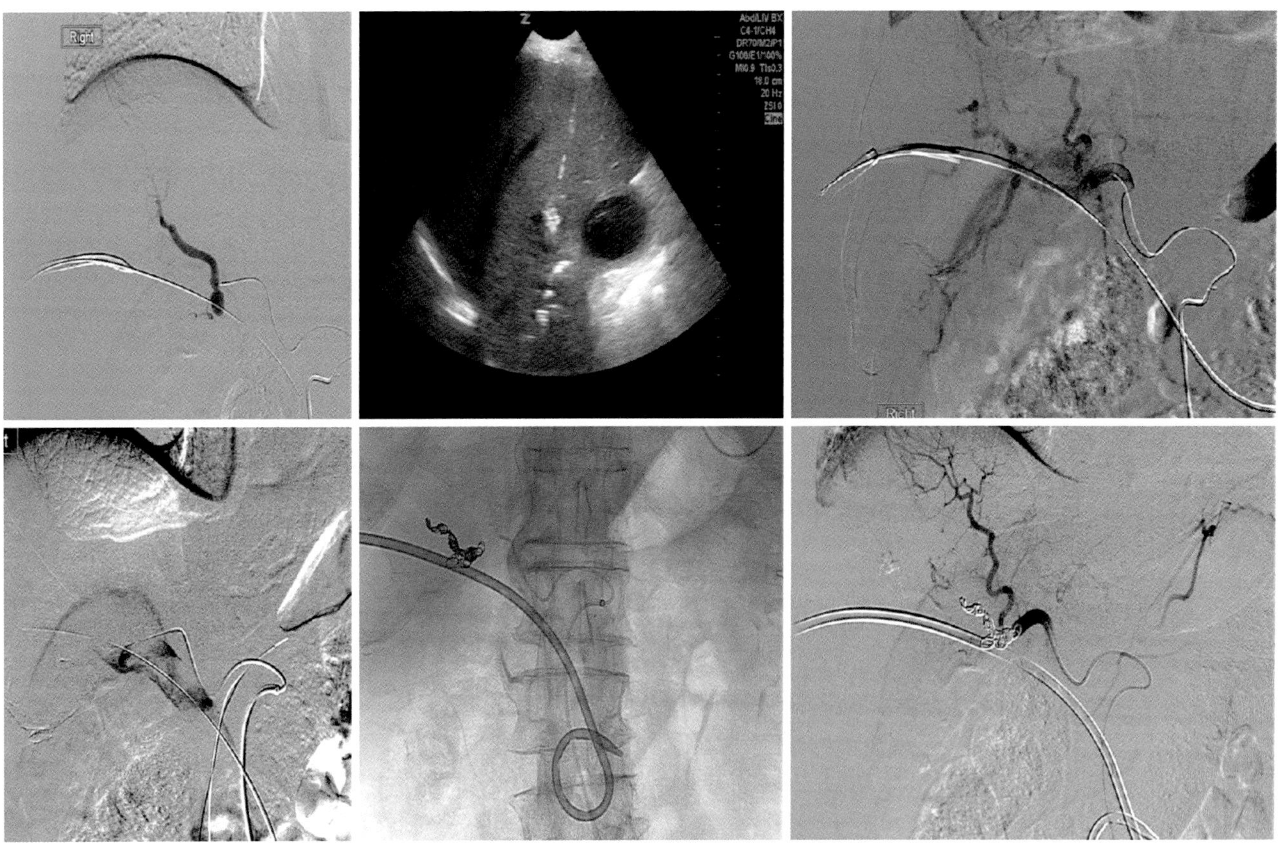

Fig. 33.9 Percutaneous thrombin injection into hepatic artery pseudoaneurysm followed by coil embolization due to recurrence. No recurrent bleeding post-coil embolization

Clinical Follow-Up

It is critical that all aneurysms be followed in the clinic both before and after endovascular treatment. A reasonable follow-up algorithm includes follow-up imaging at 1, 3, and 6 months postoperatively and annually thereafter. While the initial imaging study of choice is a CT angiogram to obtain high-level vascular data, follow-up studies may be performed with noncontrast CT to assess for changes in size or morphology. MRA, though more costly, is able to adequately assess any flow into the aneurysm while minimizing artifacts from coils, plugs, and liquid embolics. The purpose of follow-up is to assess for any interval growth of the aneurysm or other long-term complications such as endoleak development, in-stent restenosis and/or thrombosis, or the development of new visceral aneurysms in either the same or other vascular beds. It is imperative for the vascular specialist treating this disease process to have a visceral aneurysm clinic where they can guide and counsel patients. During the visceral aneurysm clinic appointment, the vascular specialist can evaluate the etiology of the vascular aneurysm, screen for additional aneurysms in other vascular beds, and follow aneurysms too small for treatment. After endovascular repair these patients will need lifelong follow-up in the vascular interventional clinic.

Conclusion

Visceral artery aneurysms are a rare but potentially life-threatening vascular disease process. A vascular specialist treating this disease process must have a deep understanding of the pathophysiology, clinical workup, and natural history of this disease process. There are a variety of new endovascular technologies and techniques that have been developed for the treatment of even the most complex visceral aneurysms. Preoperative imaging and planning and awareness of the various array of endovascular tools available to the modern-day vascular interventionalist are paramount to technical success in the treatment of these lesions. A dedicated visceral aneurysm clinic is invaluable for the comprehensive workup and evaluation and management of these patients as well as for the long-term follow-up both before and after endovascular treatment (Table 33.1).

Table 33.1 Suggested endovascular toolkit for the endovascular management of visceral artery aneurysms. These are only a few suggested options based on the author's experience and can cover the vast majority of cases

	Manufacturer	Relevant dimensions
Coils		
Ruby coil system	Penumbra	Large-volume system: Catheter compatibility (ID): 0.025″" Coil length: 1–60 cm Secondary diameter: 2–40 mm Low-profile system: Catheter compatibility (ID): 0.0165″–0.021″ Coil length: 2–60 cm Secondary diameter: 1–8We mm
Concerto detachable coil system	Medtronic	Catheter compatibility (ID): 0.0165″–0.021″ Coil length: 4–50 cm Secondary diameter: 2–20 mm
Axium 3D detachable coils	Medtronic	OD: 0.0115–0.0145″ Coil length: 6–50 cm Secondary diameter: 3–25 mm
Avenir	Wallaby	OD: 0.0110–0.0145″ Coil length: 2–60 cm Secondary diameter: 2–25 mm
EMBOLD detachable coil system	Boston scientific	Catheter compatibility (ID): 0.021–0.027″ Coil length: 4–60 cm Secondary diameter: 2–32 mm
Sheaths		
Flexor Ansel guiding sheath	Cook	Size: 5–12 Fr Length: 45–110 cm
TourGuide steerable sheath	Medtronic	Size: 6.5–8.5 Fr Length: 45–90 cm
Pinnacle destination guiding sheath	Terumo interventional systems	Size: 5–8 Fr Length: 45–90 cm
Guide catheters		

Table 33.1 (continued)

	Manufacturer	Relevant dimensions
SOFIA distal access catheter	MicroVention	OD: 5.2–6.2 Fr ID: 0.055–0.0700″ Length: 105–125 cm
Microcatheters		
LANTERN delivery microcatheter	Penumbra	OD: 2.6 Fr ID: 0.025″ Length: 115–160 cm
TRUSELECT microcatheter	Boston scientific	OD: 2 Fr ID: 0.021″ Length: 105–175 cm
Scepter balloon microcatheter	MicroVention	Balloon diameter: 4 mm Balloon length: 10–20 mm Guide cath compatibility: 6F allows double catheter technique
Balloons		
Sterling	Boston scientific	Balloon diameter: 2–10 mm Balloon length: 20–220 mm Guidewire compatibility: 0.018″
Coyote	Boston scientific	Balloon diameter: 1.5–4 mm Balloon length: 40–220 mm Guidewire compatibility: 0.014″
Stents (covered)		
VIABAHN VBX balloon-expandable endoprosthesis	GORE	Nominal diameters: 5–11 mm Postdilation max: 8–16 mm Crimped stent length: 15–79 mm
Stents (uncovered)		
SMART stent	Cordis	Diameter: 6–14 mm (unconstrained) Length: 20–150 mm (unconstrained)
Neuroform atlas	Stryker	Diameter: 3.5–5 mm (unconstrained) Length: 15–30 mm (unconstrained)
LVIS device	MicroVention	Diameter: 3.5–5.5 mm (unconstrained) Length: 10–30 mm (unconstrained)
Flow diverters		
Surpass evolve	Stryker	Target vessel diameter: 2.5–5 mm Stent length: 12–40 mm
FRED device	MicroVention	Target vessel diameter: 2–5.5 mm Stent working length: 8–56 mm
Liquid embolics		
Onyx 18/34	Medtronic	
Microwires		
Fathom	Boston scientific	Diameter: 0.014–0.016″ Length: 140–300 cm
Synchro 2	Stryker	Diameter: 0.014″ Length: 200–300 cm

Case Presentation

Continued from page 349

After uneventful aortoiliac endovascular aneurysm repair, the patient was scheduled to undergo endovascular repair of his hepatic artery 1 month later. The procedure was performed under general anesthesia to assist with breath holding and optimize treatment. Under ultrasound guidance the right common femoral artery was accessed. Visceral angiography was first performed using a 5 Fr VS1 diagnostic catheter. Ultimately a 6.5 Fr TourGuide steerable sheath (Medtronic, Dublin, Ireland) was used to obtain stable access into the celiac artery. Through this a 6 Fr Sofia guide catheter (MicroVention Inc., Aliso Viejo, CA) was advanced into the common hepatic artery. This was chosen to facilitate dual microcatheter placement via one access. First a 0.027″ Headway Duo (MicroVention Inc., Aliso Viejo, CA) was placed into the right hepatic artery. Then an Echelon 10 microcatheter (Medtronic, Dublin, Ireland) was used to obtain access in the hepatic artery aneurysm. The 0.027″ microcatheter was used to facilitate the deployment of a 5.5 mm × 32/26 mm FRED flow diverter stent (MicroVention Inc., Aliso Viejo, CA) while using the Echelon 10 microcatheter to fill the aneurysm with a myriad of detachable coils. The final angiogram showed a patent hepatic artery with successful aneurysm exclusion. The patient was discharged home on dual antiplatelet therapy using aspirin and clopidogrel, which he continued for 6 months. The 2-year follow-up shows regression of the abdominal and iliac aneurysm sacs with resolution of the inflammatory changes and associated urinary obstruction. Additionally, the hepatic artery aneurysm demonstrated no flow or growth. Patient also underwent genetic testing which was negative for a genetic etiology to his multiple aneurysms. Ultimately, it was felt his aneurysms were related to his history of smoking.

References

1. Stoecker JB, Eddinger KC, Glaser JD, Wang GJ, Shlansky-Goldberg RD, Fairman RM, et al. A large series of true pancreaticoduodenal artery aneurysms. J Vasc Surg. 2022;75(5):1634–42.e1. https://doi.org/10.1016/j.jvs.2022.01.021.
2. Bronsther O, Merhav H, Van Thiel D, Starzl TE. Splenic artery aneurysms occurring in liver transplant recipients. Transplantation. 1991;52(4):723–4. https://doi.org/10.1097/00007890-199110000-00026.
3. Lee PC, Rhee RY, Gordon RY, Fung JJ, Webster MW. Management of splenic artery aneurysms: the significance of portal and essential hypertension. J Am Coll Surg. 1999;189(5):483–90. https://doi.org/10.1016/s1072-7515(99)00168-4.
4. Wojtaszek M. Managing visceral artery aneurysms. Current endovascular techniques and technologies for treating this increasingly encountered presentation. Endovascular Today; 2013. https://evtoday.com/articles/2013-oct/managing-visceral-artery-aneurysms.
5. Abbas MA, Fowl RJ, Stone WM, Panneton JM, Oldenburg WA, Bower TC, et al. Hepatic artery aneurysm: factors that predict complications. J Vasc Surg. 2003;38(1):41–5. https://doi.org/10.1016/S0741-5214(03)00090-9.
6. Abraham RJ, Illyas AJ, Marotta T, Casey P, Vair B, Berry R. Endovascular exclusion of a splenic artery aneurysm using a pipeline embolization device. J Vasc Interv Radiol. 2012;23(1):131–5. https://doi.org/10.1016/j.jvir.2011.09.015.
7. Bannur M. Endovascular treatment of visceral artery aneurysms. J Vasc Interv Radiol. 2015;26(2):S23–S4. https://doi.org/10.1016/j.jvir.2014.12.070.
8. Künzle S, Glenck M, Puippe G, Schadde E, Mayer D, Pfammatter T. Stent-graft repairs of visceral and renal artery aneurysms are effective and result in long-term patency. J Vasc Interv Radiol. 2013;24(7):989–96. https://doi.org/10.1016/j.jvir.2013.03.025.
9. Ibrahim F, Dunn J, Rundback J, Pellerito J, Galmer A. Visceral artery aneurysms: diagnosis, surveillance, and treatment. Curr Treat Options Cardiovasc Med. 2018;20(12):97. https://doi.org/10.1007/s11936-018-0696-x.
10. Chaer RA, Abularrage CJ, Coleman DM, Eslami MH, Kashyap VS, Rockman C, et al. The Society for Vascular Surgery clinical practice guidelines on the management of visceral aneurysms. J Vasc Surg. 2020;72(1S):3S–39S. Epub 20200320. https://doi.org/10.1016/j.jvs.2020.01.039.
11. Stanley JC, Wakefield TW, Graham LM, Whitehouse WM Jr, Zelenock GB, Lindenauer SM. Clinical importance and management of splanchnic artery aneurysms. J Vasc Surg. 1986;3(5):836–40.
12. Venturini M, Piacentino F, Coppola A, Bettoni V, Macchi E, De Marchi G, et al. Visceral artery aneurysms embolization and other interventional options: state of the art and new perspectives. J Clin Med. 2021;10(11):2520. https://doi.org/10.3390/jcm10112520.
13. Ratanam SPaLA. Liquid embolics: emerging options and applications endovascular today: a review of current liquid embolic agents and future prospects. Endovascular Today; 2022.
14. Murray TÉ, Brennan P, Maingard JT, Chandra RV, Little DM, Brooks DM, et al. Treatment of visceral artery aneurysms using novel neurointerventional devices and techniques. J Vasc Interv Radiol. 2019;30(9):1407–17. https://doi.org/10.1016/j.jvir.2018.12.733.
15. Abbas MA, Stone WM, Fowl RJ, Gloviczki P, Oldenburg WA, Pairolero PC, et al. Splenic artery aneurysms: two decades experience at mayo clinic. Ann Vasc Surg. 2002;16(4):442–9. https://doi.org/10.1007/s10016-001-0207-4.
16. Barrionuevo P, Malas MB, Nejim B, Haddad A, Morrow A, Ponce O, et al. A systematic review and meta-analysis of the management of visceral artery aneurysms. J Vasc Surg. 2019;70(5):1694–9. https://doi.org/10.1016/j.jvs.2019.02.024.
17. Etezadi V, Gandhi RT, Benenati JF, Rochon P, Gordon M, Benenati MJ, et al. Endovascular treatment of visceral and renal artery aneurysms. J Vasc Interv Radiol. 2011;22(9):1246–53. https://doi.org/10.1016/j.jvir.2011.05.012.
18. Ha JF, Phillips M, Faulkner K. Splenic artery aneurysm rupture in pregnancy. Eur J Obstet Gynecol Reprod Biol. 2009;146(2):133–7. Epub 20090710. https://doi.org/10.1016/j.ejogrb.2009.05.034.
19. Hislop SJ, Patel SA, Abt PL, Singh MJ, Illig KA. Therapy of renal artery aneurysms in new york state: outcomes of patients undergoing open and endovascular repair. Ann Vasc Surg. 2009;23(2):194–200. https://doi.org/10.1016/j.avsg.2008.10.002.
20. Jiang J, Liu Y, Ding X. Successful conservative treatment of an isolated inflammatory superior mesenteric artery aneurysm. Interact Cardiovasc Thorac Surg. 2022;34(6):1147–9. https://doi.org/10.1093/icvts/ivab296.
21. Qiu C, Liu Z, Huang L, Guo L, Lu W, Zhang H, et al. Covered stents for treatment of visceral artery aneurysms: a multicenter study. J Vasc Interv Radiol. 2022;33(6):640–7. https://doi.org/10.1016/j.jvir.2022.03.009.
22. Roberts KJ, McCulloch N, Forde C, Mahon B, Mangat K, Olliff SP, et al. Emergency treatment of haemorrhaging coeliac or mesenteric artery aneurysms and pseudoaneurysms in the era of endovascular management. Eur J Vasc Endovasc Surg. 2015;49(4):382–9. https://doi.org/10.1016/j.ejvs.2014.12.019.

Part V

Peripheral Aneurysm Disease

34 Endovascular Repair of Popliteal Artery Aneurysms

Mohammad Eslami and Othman Abdul-Malak

Case Presentation

A 68-year-old male presenting to the emergency room with reports of acute-onset left-foot pain. Prior to this presentation, he was extremely active and walked several miles a day. His medical history is significant for coronary artery disease s/p percutaneous stenting, hypertension, and hyperlipidemia. On exam, he had normal motor strength and normal sensation, but non-palpable pedal pulses or audible pedal signals in the left lower extremity. His exam revealed bilateral popliteal artery aneurysms, with left feeling larger than the right. The patient was systemically anticoagulated with unfractionated heparin, and a lower extremity arterial duplex and a computed tomography angiogram were performed that revealed a thrombosed left popliteal artery aneurysm of 29 mm with compromised run-off. He was also found to have a right popliteal artery aneurysm as well measuring 25 mm. His aorta was normal. Patient was then counseled on the risks and benefits of intra-arterial lysis therapy to improve his run-off prior to definitive repair, and he was agreeable to proceed.

Continued at page 366

Background

Popliteal artery aneurysms (PAA) are the most common peripheral arterial aneurysms, with an incidence of 0.8–2.8% nationally [1]. Complications of untreated popliteal aneurysms are aneurysm thrombosis and distal embolization, and in rare circumstances aneurysm rupture [2–4]. Endovascular intervention for popliteal aneurysms with stent grafts has emerged as an alternative to open surgical repair in candidates with suitable anatomy. The first report of popliteal stenting for aneurysmal disease was by Marin et al. in 1994 and was performed with expanded polytetrafluoroethylene (ePTFE) supported by Palmaz stents [5]. Since that time, a number of retrospective studies have demonstrated the efficacy of endovascular popliteal artery aneurysm repair (EPAR) with stent grafts with high patency rates approaching open repair, but comparable results to open repair in appropriate candidates [1, 6–10]. To date, only one single-center, small (n = 42), prospective, randomized trial has compared open surgical repair with endovascular stent exclusion and found no difference in primary or secondary patency at 3 years [11, 12], further confirming the role of EPAR in the treatment of popliteal artery aneurysms. In our recent publication EPAR outcomes patency was 80% at 3 years but depended on the quality of outflow vessels [13].

Indications for Intervention

Operative treatment of popliteal artery aneurysms (PAA) is indicated for asymptomatic aneurysms larger than 2 cm or for symptomatic aneurysms presenting with thrombosis, embolization or, in rare instances, rupture [8].

Preoperative Preparation

Imaging

Duplex ultrasonography is the first-line imaging modality of the peripheral arteries. It offers a low-cost, non-invasive way for popliteal artery aneurysm diagnosis and surveillance. It offers comparable accuracy to more invasive imaging modalities such as computed tomography angiogram (CTA) and conventional angiography, while minimizing radiation and contrast exposure [14–16].

Computed tomography angiogram (CTA) of the lower extremity is a necessary modality to assess aneurysm characteristics for suitability for endovascular repair, specifically as

M. Eslami (✉) · O. Abdul-Malak
University of Pittsburgh Medical Center, Pittsburgh, PA, USA

Division Of Vascular and Endovascular Surgery, Charleston Area Medical Center, West Virginia, USA
e-mail: eslamimh@upmc.edu; malakoa@upmc.edu

G. Geroulakos et al. (eds.), *Mastering Endovascular Techniques*, https://doi.org/10.1007/978-3-031-42735-0_34

it pertains to number of patent run-off vessels and extent of normal popliteal artery distal to the aneurysmal segment. CTA is necessary for appropriate measurement of the popliteal artery diameter and stent diameter selection.

Although MR angiography has sensitivity and specificity similar to CTA; however, it is far more costly and cumbersome for the individual patient and rarely ordered by physicians managing popliteal artery aneurysm.

Preoperative Risk Stratification

Preoperative risk stratification of these patients is dependent on the manner of popliteal artery aneurysm repair. For open surgical patients, a complete peri-operative evaluation including cardiac and pulmonary risk must be pursued based on patients' comorbidities and risk factors. Most vascular surgeons prefer endovascular interventions over open popliteal artery aneurysm repair to circumvent peri-operative risks. For endovascular interventions no specific risk stratification is necessary although poor or single tibial vessel run-off or other unsuitable anatomic constraints may favor open surgical intervention.

Endovascular Strategy

Assessing Anatomic Features for Intervention

CTA/MRA prior to endovascular intervention is essential to identify anatomic characteristics that allow a popliteal aneurysm to be successfully excluded with a stent graft. Similar to endovascular abdominal aorta aneurysm an adequate seal zone above and below popliteal aneurysm is necessary to seal the aneurysm: 1.5 cm of straight, uniform artery above and below the aneurysm is optimal for adequate sealing of PAA [16]. The diameter of the artery at the sealing zones must be between 4 mm and 12 mm based on the size of commercial stent grafts available (5–12 mm) [17]. A significant discrepancy between the proximal and distal landing zones (>3 mm) may preclude the ability for the stent grafts to seal both ends, or may necessitate the use of multiple endografts to exclude the aneurysm. In such cases distal smaller diameter stents are deployed first.

While the length and diameter of the seal zones can be difficult to assess on digital subtraction angiography (DSA), the adequacy of tibial outflow can assist in predicting graft patency and is best assessed with DSA, as contrast timing, and tibial calcification can limit the accurate assessment of tibial vessels by CTA. While EPAR with a single vessel tibial outflow can be successful, studies suggest that two or more patent tibial vessels are associated with better patency rates [13, 18]. If the anatomy of popliteal artery is unclear due to lack CTA or poor radiographic visualization (e.g. calcification, thrombosis, artifacts due to knee prosthesis), an intra-operative IVUS can accurately measure the diameter of normal landing popliteal vessels and can be useful to ascertain the length of coverage. This can be performed during angiography.

Patients Presenting with Acute Limb Ischemia and Compromised Run-Off

Patients that present with acute limb ischemia and symptomatic PAA often have compromised tibial run-off, which affects the outcomes of repair in both open and endovascular approaches. In patients who present with Rutherford grade I or IIa ischemia, thrombolysis is indicated to optimize outflow vessel patency prior to repair of PAA, especially if there is evidence of distal vessel embolization or if the clot extending from the PAA into the tibial vessel origin [19].

Selection of an Access Site

The choice of access site may be determined by the anticipated sheath size and prior vascular interventions. Smaller diameter covered stent grafts can easily be delivered through 7–8 Fr sheaths from the contralateral common femoral artery in the standard up and over technique. However larger (larger than 10 mm) stent grafts require larger sheaths which may be difficult to safely direct over the aortic bifurcation. Additionally, patients with a hostile aortic bifurcation such as those with extensive aorto-iliac occlusive disease and the presence of aorto-iliac stents, prior endovascular or open aortic aneurysm repair are at increased risk for up and over endograft delivery. In these instances, an ipsilateral common femoral or superficial femoral artery approach should be used to safely deliver the endograft while minimizing access site complications.

Selection of an Endograft

Self-expanding stent grafts are used in the femoral and popliteal segment due to the amount of flexion in these arteries during ambulation. Stent grafts are selected based on the size of the proximal and distal landing zones, as well as the length of the aneurysm. Typically, the grafts are oversized up to 10% of the native vessel size. The minimal seal zone to minimize the risk of endoleak is 1.5 cm [18], in the event that more than one stent graft is used, a minimum of 3 cm overlap between the grafts is recommended [13]. When different stents diameters are required, the smaller diameter is deployed first before deploying the larger diameter stent inside of the smaller one to reduce endoleak.

Endovascular Technique

Access

For contralateral common femoral access, groin is prepped and draped in the routine manner, anatomic landmarks (anterior superior iliac spine and pubic symphysis, common femoral artery [CFA]) are identified. We recommend the routine use of ultrasound guided access for identification of the common femoral bifurcation to minimize the risk of access site complications [20]. Once the CFA is identified, a micropuncture needle is utilized to obtain access, and then utilizing the Seldinger technique. For larger sheaths, the track should be sequentially upsized to the required sheath size necessary for endograft delivery. Percutaneous ipsilateral approach to the CFA or proximal SFA is similar to contralateral approach. It is often easier to use longer sheath and perform the intervention while standing on the contralateral side of the patient. For open exposure, the contralateral/ipsilateral groin is prepped and draped in the standard manner, the superficial femoral artery is identified in its most superficial location in the proximal thigh between the sartorious and adductor longus muscles. Once the vessel is exposed, proximal and distal control are obtained with silastic loops, a monofilament suture in U / purse string configuration is placed in the anterior wall of the vessel which will be used for closure at the conclusion of the procedure. The vessel is punctured directly in the middle of this stitch utilizing a micropuncture needle and the Seldinger technique as described above.

Angiography

In the case of contralateral access, up and over catheter technique is utilized using an angled selective catheter and wire platform, the superficial femoral artery is then selected and the sheath is advanced into position for imaging and stent graft delivery. In the case of ipsilateral access the sheath is advanced into position ensuring adequate room is available between the end of the sheath and the most proximal landing zone to fully deploy the stent graft. Multiple oblique views of the distal landing zone are obtained to fully delineate the infra-popliteal anatomy, specifically the length of normal distal popliteal artery, origin of the tibial vessels, and patency of the tibial vessels. The endograft is then advanced through the sheath and into position. Often multiple endografts are needed due to aneurysm length or a size discrepancy between the proximal and distal landing zones. When multiple endografts are used the smaller diameter and distal endograft is deployed first, and the remainder of the stent grafts are built more proximally into normal caliber arteries. Care is taken to ensure a minimum of 1.5 cm of endograft apposition with normal caliber artery both proximally and distally. The stent grafts are gently post-dilated with an angioplasty balloon of either equal diameter or 1 mm smaller. Aggressive post-dilatation often foreshortens the stent length and may additional interventions. If there is concern for significant thrombus coverage and the stent appears to be fully expanded without endoleak, then this step may be avoided altogether to minimize distal embolization. Completion angiogram is then used to confirm that the length of the aneurysm is completely sealed without evidence of endoleak at the proximal and distal ends, as well as between endograft components. The stents are inspected to ensure that all components are fully expanded without kinking. The outflow tibial vessels are also assessed to ensure that no embolization has occurred during the procedure. In a recent publication we noted that extent of stent coverage and placement of stent were factors affecting the patency rates of EPAR [13].We, therefore, try to reduce the extent of stent coverage and avoid/minimize placing stenting in mid (P2) and distal (P3) portions of popliteal artery if anatomically possible [13].

Postoperative Care

Post-operative Period

Patients with percutaneous CFA access are placed on bedrest for 2–6 h depending on the closure method (manual pressure, suture mediated closure device, mechanical closure devices, among others). Patients with surgical access can often be advanced to ambulate as early as tolerated. Patients are maintained on at least a single antiplatelet agent, but more recent literature recommends dual antiplatelet treatment for 3–6 months post infra-inguinal endovascular intervention [21].

Follow-Up Period

Ankle-brachial index and arterial duplex scans are performed within 1 month for baseline assessment. Physical exam and graft surveillance are performed every 6–12 months, and sometimes more frequently if there is a specific concern as femoral-popliteal stents may develop in-stent stenosis, fracture, endoleaks, and thrombosis [14, 22].

Complications

Access Complications

Access site complications include bleeding, pseudoaneurysm, infection, and thrombosis. Ultrasound guided access has been shown to decrease post-operative access complications in retrospective studies, and is recommended for percutaneous access for EPAR [20].

Graft Complications

Graft related complication mainly include endoleak, fracture, and thrombosis. Endoleak classification follow the same classification as endovascular abdominal aorta aneurysm repair. Type I endoleak at the proximal and distal end can often be managed by extending proximally or distally with an additional stent. Type III endoleaks between stents can be bridged with an additional bridging stent. Type II endoleaks, arising from back bleeding geniculate vessels, are often observed unless the aneurysm sac is enlarging or causing compressive symptoms such as pain or DVT. Stent technology has changed over time but in older studies of different type of stents, stent fracture was not associated with graft patency [23]. The exact rate of stent fractures with the most commonly stent graft for EPAR is exceedingly rare. Stent graft thrombosis is potentially devastating as it can lead to limb loss. Presentation can vary from asymptomatic to severe ischemia depending on the ischemic burden of the graft occlusion. Treatment options include catheter-directed thrombolysis, thrombectomy, and surgical bypass (Table 34.1).

Table 34.1 Suggested endovascular toolkit for the endovascular management of popliteal aneurysms. These are only a few suggested options based on the author's experience and can cover the vast majority of cases

	Manufacturer	Size/length
Wires		
Any standard-access wire		0.035″
Glidewire floppy	Terumo Radifocus	0.035″
Glidewire stiff	Terumo Radifocus	0.035″
V-18 control guidewire	Boston Scientific Corporation	0.018″
Sheaths		
Any standard-access sheath	Any	5–10 Fr/11–55 cm
Flexor: Ansel-high flex	Cook Medical	4–8, 10, 12 Fr 45, 55, 70 cm
Catheters		
Angled	Terumo Radifocus	0.035″/5 Fr/65–90 cm
Quick cross support (low-profile)	Spectranetics/ Phillips	0.018″–0.035″/4 Fr/65–90 cm
Balloons		
Any standard plain balloon	Any	5–12 mm
Stents		
Gore Viabahn endoprosthesis	Gore and Associates	5, 6, 7, 8, 9, 10, 11, 13 mm diameter
Thrombolytic catheters		
Cragg-McNamara	Medtronic	5 Fr/10–50 infusion length
IVUS catheters		
Visions PV 0.035 (large peripheral)	Phillips	8.5 Fr
Visions PV 0.018 (small peripheral)	Phillips	5 Fr

Case Presentation

Continued from page 363

The patient was brought to the operating room, procedural sedation was provided by the anesthesia service, the right groin was prepped and draped in the standard fashion, we obtained access to the common femoral artery under ultrasound guidance in the standard fashion and this was confirmed by fluoroscopy, we then navigated up and over the aortic bifurcation and into the left superficial femoral artery. We obtained AP and oblique images of the popliteal artery, and identified an aneurysm of the P2 segment with compromised run-off (Fig. 34.1). Given the patients' mild symptoms (Rutherford IIa) and severely comprised run-off we elected to initiate thrombolysis. A 45 cm 6 Fr sheath was navigated into the common femoral artery, and a 5 Fr Cragg McNamara™ infusion catheter was placed into the thrombosed popliteal aneurysm into the area of the tibial-peroneal trunk, and 1 mg/h of tissue-plasminogen activator (tPA) was initiated through the catheter, the sheath was continuously heparinized simultaneously.

The patient developed a palpable DP and PT few hours after initiation of tPA and was brought back to the operating room the following day after 24 h of uncomplicated thrombolysis, repeat angiogram revealed three-vessel run-off in the calf, and two-vessel run-off to the foot (Fig. 34.2). At this point the decision was made to treat the PAA with an endovascular stent graft given his improved run-off. We then exchanged our sheath for an 8 Fr sheath and positioned a 0.018″ wire in the posterior tibial artery. Our preoperative CTA was sub-optimally timed for sizing of our endograft, so we elected to utilize intra-vascular ultrasound to measure the distal and proximal landing zones both in terms of distance and diameter. A 10 mm × 10 cm GORE® VIABAHN® endoprosthesis was brought to the field and deployed 1.5 cm distal to the distal endpoint of the aneurysm, and an 10 mm × 5 cm GORE® VIABAHN® endoprosthesis was then deployed proximally ensuring a 3 cm stent overlap and a 1.5 cm proximal seal zone, we post-dilated the endograft overlap zone to ensure apposition. We then took a completion angiogram (Fig. 34.3) which revealed good apposition of our endografts with no kinking or endoleak, we then obtained completion imaging of our run-off which was found to be intact (Fig. 34.4). Hemostasis was achieved using an Angio-Seal® device. The patient was started on a dual antiplatelet regimen (aspirin [81 mg] and clopidogrel [300 mg initial dose and 75 mg daily]) and discharged home the following day with no acute complications. He was seen 3 months postoperatively without any issues being evaluated for elective repair of his right side popliteal artery in the near future.

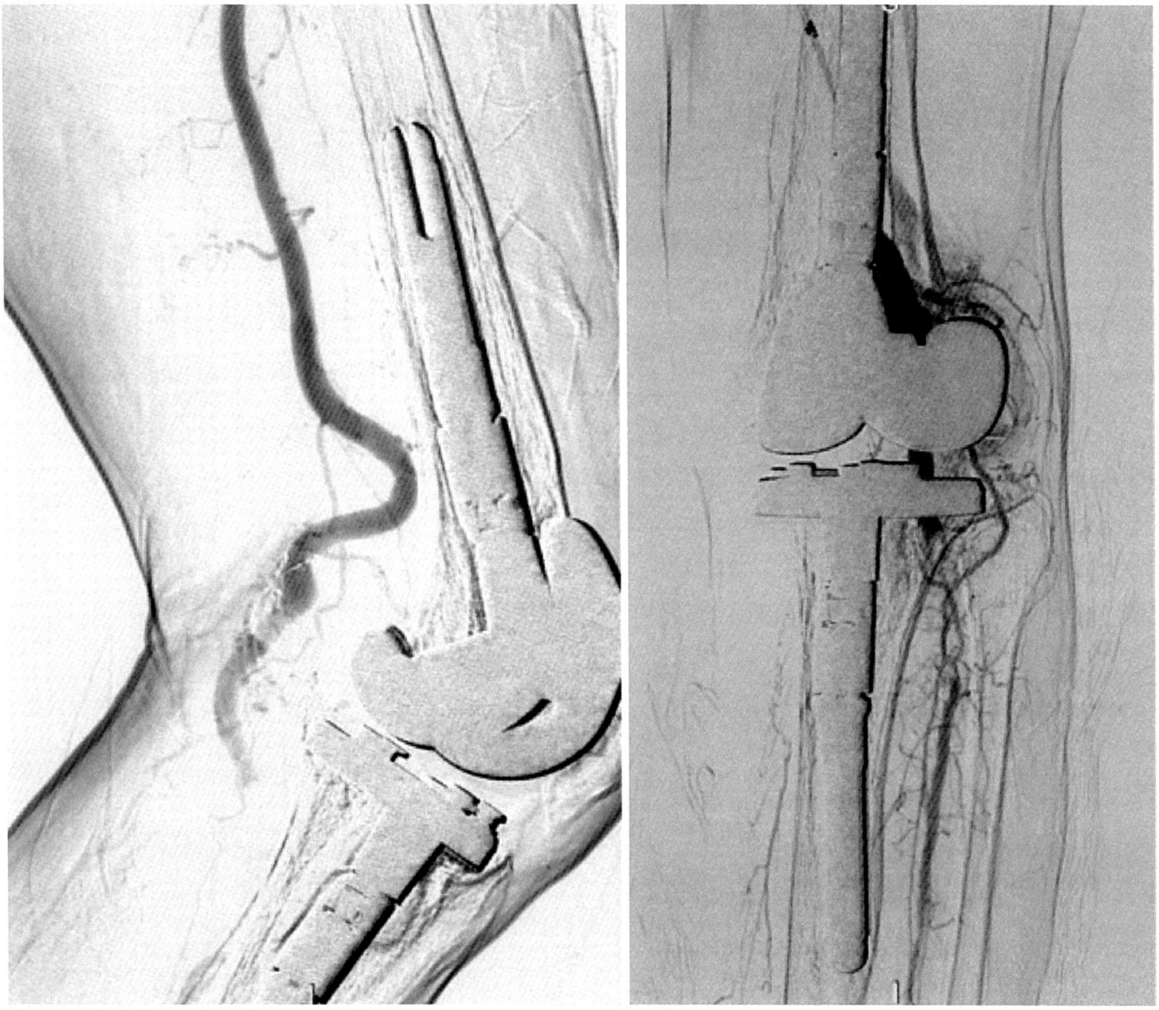

Fig. 34.1 Initial angiography of popliteal aneurysm and compromised run-off

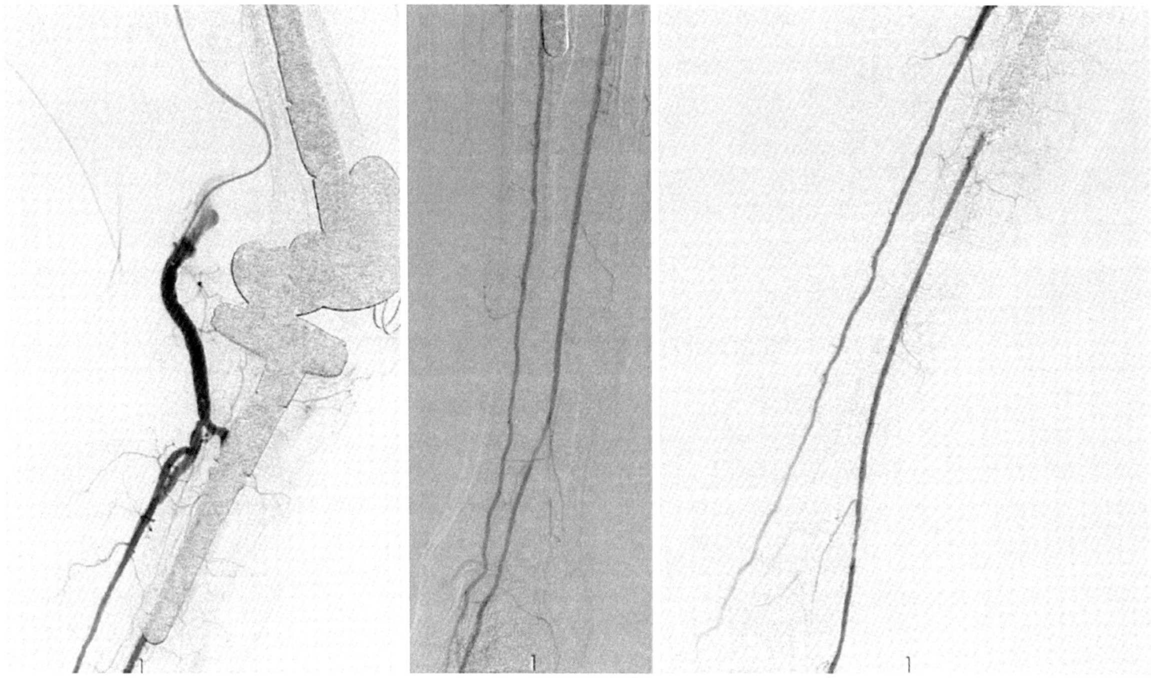

Fig. 34.2 Repeat angiography following 24 h of thrombolytic therapy

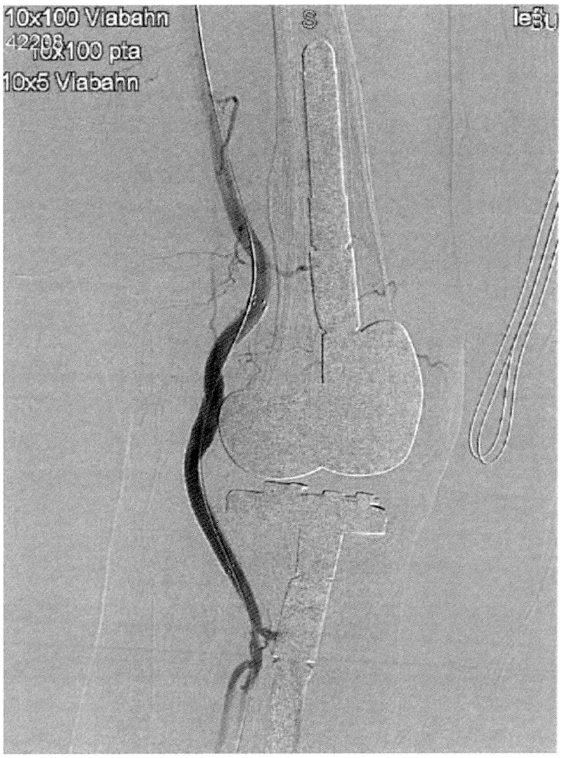

Fig. 34.3 Completion angiography post deployment of endograft

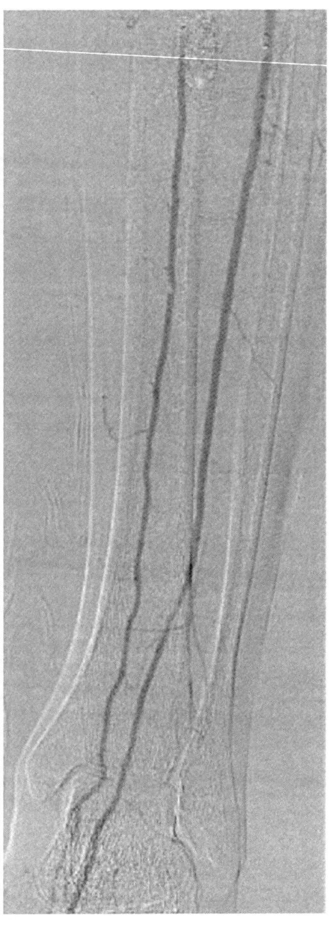

Fig. 34.4 Completion angiography of run-off vessels post deployment of endograft

References

1. Leake AE, Avgerinos ED, Chaer RA, Singh MJ, Makaroun MS, Marone LK. Contemporary outcomes of open and endovascular popliteal artery aneurysm repair. J Vasc Surg. 2016;63:70.
2. Dawson I, Sie R, van Baalen JM, Van Bockel JH. Asymptomatic popliteal aneurysm: elective operation versus conservative follow-up. Br J Surg. 1994;81(10):1504.
3. Lowell RC, Gloviczki P, Hallett JW, Naessens JM, Maus TP, Cherry KJ, et al. Popliteal artery aneurysms: the risk of nonoperative management. Ann Vasc Surg. 1994;8(1):14.
4. Whitehouse WM, Wakefield TW, Graham LM, Kazmers A, Zelenock GB, Cronenwett JL, et al. Limb-threatening potential of arteriosclerotic popliteal artery aneurysms. Surgery. 1983;93(5).
5. Marin ML, Veith FJ, Panetta TF, Cynamon J, Bakal CW, Suggs WD, et al. Transfemoral endoluminal stented graft repair of a popliteal artery aneurysm. J Vasc Surg. 1994;19(4):754.
6. Cervin A, Tjärnström J, Ravn H, Acosta S, Hultgren R, Welander M, et al. Treatment of popliteal aneurysm by open and endovascular surgery: a contemporary study of 592 procedures in Sweden. Eur J Vasc Endovasc Surg. 2015;50(3):342.
7. Eslami MH, Rybin D, Doros G, Farber A. Open repair of asymptomatic popliteal artery aneurysm is associated with better outcomes than endovascular repair. J Vasc Surg. 2015;61:663.
8. Golchehr B, Zeebregts CJ, Reijnen MMPJ, Tielliu IFJ. Long-term outcome of endovascular popliteal artery aneurysm repair. J Vasc Surg. 2018;67(6):1797.
9. Leake AE, Segal MA, Chaer RA, Eslami MH, Al-Khoury G, Makaroun MS, et al. Meta-analysis of open and endovascular repair of popliteal artery aneurysms. J Vasc Surg. 2017;65:246.
10. Serrano Hernando FJ, Martínez López I, Hernández Mateo MM, Hernando Rydings M, Sánchez Hervás L, Rial Horcajo R, et al. Comparison of popliteal artery aneurysm therapies. J Vasc Surg. 2015;61(3):655.
11. Antonello M, Frigatti P, Battocchio P, Lepidi S, Cognolato D, DALL'ANTONIA A, et al. Open repair versus endovascular treatment for asymptomatic popliteal artery aneurysm: results of a prospective randomized study. J Vasc Surg. 2005;42(2):185.
12. Antonello M, Frigatti P, Battocchio P, Lepidi S, Dall'Antonia A, Deriu GP, et al. Endovascular treatment of asymptomatic popliteal aneurysms: 8-year concurrent comparison with open repair. J Cardiovasc Surg. 2007;48(3):267.
13. Zaghloul MS, Andraska EA, Leake A, Chaer R, Avgerinos ED, Hager ES, et al. Poor runoff and distal coverage below the knee are associated with poor long-term outcomes following endovascular popliteal aneurysm repair. J Vasc Surg. 2021;74:153.
14. Baril DT, Rhee RY, Kim J, Makaroun MS, Chaer RA, Marone LK. Duplex criteria for determination of in-stent stenosis after angioplasty and stenting of the superficial femoral artery. J Vasc Surg. 2009;49(1):133.
15. Willmann JK, Mayer D, Banyai M, Desbiolles LM, Verdun FR, Seifert B, et al. Evaluation of peripheral arterial bypass grafts with multi-detector row CT angiography: comparison with duplex US and digital subtraction angiography. Radiology. 2003;229(2):465.
16. Wolf YG, Kobzantsev Z, Zelmanovich L. Size of normal and aneurysmal popliteal arteries: a duplex ultrasound study. J Vasc Surg. 2006;43(3):488.
17. https://evtoday.com/device-guide/us/self-expanding-stent-grafts.
18. Garg K, Rockman CB, Kim BJ, Jacobowitz GR, Maldonado TS, Adelman MA, et al. Outcome of endovascular repair of popliteal artery aneurysm using the Viabahn endoprosthesis. J Vasc Surg. 2012;55(6):1647.
19. Farber A, Angle N, Avgerinos E, Dubois L, Eslami M, Geraghty P, et al. The Society for Vascular Surgery clinical practice guidelines on popliteal artery aneurysms. J Vasc Surg. 2022;75(1):109S.
20. Kalish J, Eslami M, Gillespie D, Schermerhorn M, Rybin D, Doros G, et al. Routine use of ultrasound guidance in femoral arterial

access for peripheral vascular intervention decreases groin hematoma rates. J Vasc Surg. 2015;61(5):1231.

21. Cho S, Lee YJ, Ko YG, Kang TS, Lim SH, Hong SJ, et al. Optimal strategy for antiplatelet therapy after endovascular revascularization for lower extremity peripheral artery disease. JACC Cardiovasc Interv. 2019;12(23):2359.
22. Stone PA, Armstrong PA, Bandyk DF, Keeling WB, Flaherty SK, Shames ML, et al. The value of duplex surveillance after open and endovascular popliteal aneurysm repair. J Vasc Surg. 2005;41(6):936.
23. Tielliu IFJ, Zeebregts CJ, Vourliotakis G, Bekkema F, van den Dungen JJAM, Prins TR, et al. Stent fractures in the Hemobahn/Viabahn stent graft after endovascular popliteal aneurysm repair. J Vasc Surg. 2010;51(6):1413.

Peripheral Pseudoaneurysms

35

Ornella Moschovaki-Zeiger, Nikolaos Achilleas Arkoudis, Stavros Spiliopoulos, Konstantinos Palialexis, and Elias Brountzos

Case Presentation

A 69-year-old female patient had recently undergone a right femur fracture fixation with an intramedullary rod after a fall. She developed swelling of the ipsilateral thigh within 12 h post-op, and necessity for multiple transfusions due to inability to preserve normal hematocrit and hemoglobin levels was suspicious of active bleeding. The patient was transferred for an emergency CT angiography, where a large hematoma was observed at the incision site and in the vastus lateralis muscle. In addition, a pseudoaneurysm was demonstrated on the lateral aspect of the thigh, with a diameter of 2.6 cm and a depth of 7.5 cm from the skin, which originated from the deep femoral artery (Fig. 35.1).

The patient turned unstable, with blood pressure drop and tachycardia, and was immediately transferred to the angiography suite for further intervention.

Continued at page 380

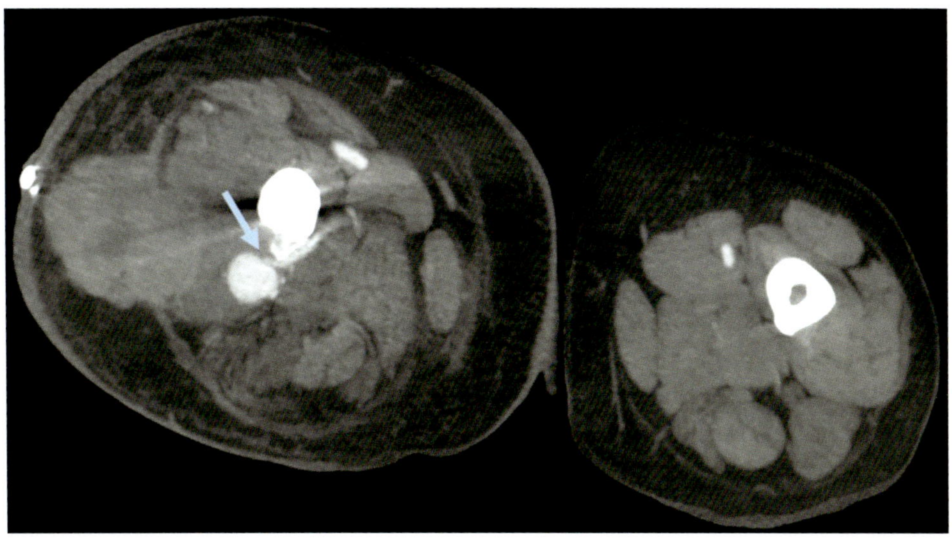

Fig. 35.1 Axial CT angiography depicting the pseudoaneurysm originating from the right deep femoral artery and the pseudoaneurysm neck (blue arrow)

O. Moschovaki-Zeiger (✉) · N. A. Arkoudis · S. Spiliopoulos · K. Palialexis · E. Brountzos
2nd Department of Radiology, National and Kapodistrian University, "ATTIKON" University General Hospital, Athens, Greece
e-mail: omoschovaki@med.uoa.gr; narkoudis@med.uoa.gr; stavspiliop@med.uoa.gr

G. Geroulakos et al. (eds.), *Mastering Endovascular Techniques*, https://doi.org/10.1007/978-3-031-42735-0_35

Introduction

False aneurysms, also known as pseudoaneurysms, are formed due to damage to the arterial wall that can induce a locally confined hematoma with the presence of turbulent flow within it due to the inability of hemostasis. A pseudoaneurysm, unlike a true aneurysm, lacks the three vascular wall layers, as they are confined by the tunica media, adventitia, or solely the hematoma they are embedded in and the surrounding soft tissue. Real aneurysms are in majority of the cases caused by deterioration of the arterial wall due to atherosclerosis, whereas false aneurysms are caused due to arterial damage (mostly after endovascular interventions) or disruption of an anastomotic connection. The pseudoaneurysm that develops is made up of a sac that is constantly being perfused with blood, which is connected to the donor vessel via a neck. It may form in any vessel that has been compromised, but it has only been described to originate from veins so seldom.

Given the increased flow within the pseudoaneurysm, a rupture might result in major hemorrhage that could be life-threatening, which makes the rapid diagnosis and management crucial.

Pseudoaneurysms, hematomas, arteriovenous fistulas, and retroperitoneal hemorrhages are the most prevalent complications of femoral artery access interventions, while the use of antiplatelet and anticoagulation therapy is a cofactor.

Surgery has always been considered the recommended approach, but that has changed in the past few years, as pseudoaneurysms may now be treated using less invasive procedures. This chapter discusses the percutaneous and endovascular techniques available to date, as well as the possible tips and tricks for the safe management of peripheral pseudoaneurysms.

Etiology and Pathophysiology

False aneurysms may be divided into two primary categories: iatrogenic and non-iatrogenic and further into three groups in order of occurrence: traumatic, anastomotic, and mycotic.

Iatrogenic

Arterial Access for Endovascular Procedure

Iatrogenic injury to the femoral or brachial artery accounts for the great majority of false aneurysms that are seen in practice today [1]. These types of false aneurysms are most often discovered at the site of percutaneous arterial puncture for a broad variety of interventional procedures including diagnostic angiograms and angioplasties. This is because the artery wall fails to seal adequately following the removal of the sheath or the catheter, which is one of the most prevalent causes of iatrogenic false aneurysms. With the use of large-bore sheaths and periprocedural anticoagulation and antiplatelet treatment, the incidence of pseudoaneurysm development is as high as 7.7% in minimally invasive procedures [2, 3]. Insufficient pressure or limited pressure time post-procedurally might also be a factor, as could laceration of an artery or one of its branches by the access needle. Despite the growing popularity of the radial artery for percutaneous procedures, the femoral artery continues to be the most frequent access point for diagnostic and therapeutic procedures alike. The reduction in the size of sheaths is a factor that contributes to the overall lower complication rates for femoral artery procedures.

Anastomotic Failure

At the confluence of a synthetic/autogenous graft and an artery, anastomotic false aneurysms are discovered. Suture failure or infection of the graft material may also cause arterial graft anastomoses to fail over time. However, infection of the graft material is likely to be the more prevalent cause. There is a wide range of time intervals between the time of surgery and the onset of pseudoaneurysms, with those appearing at a very early stage after the original surgery being most likely due to infection.

Failure of Closure Device

Failure of hemostasis may result from improper deployment of a closure device owing to a highly atheromatic vessel or a lack of expertise.

Non-iatrogenic

Trauma

Penetrating trauma, such as those caused by stab wounds or gunshot wounds, may result in the development of traumatic false aneurysms that are not caused by medical intervention.

Infection

Mycotic pseudoaneurysm is a condition that may develop as a consequence of arterial damage and infection that are brought on by the repeated use of injectable intravenous drugs. This condition can put a person at risk for bleeding, distal septic emboli, limb loss, and even death [4].

Complications

Post-catheterization pseudoaneurysms may lead to a number of complications, including discomfort and swelling around the afflicted location, pseudoaneurysm rupture, the development of neuropathy, distal embolization, infection, local skin ischemia and necrosis, and compression of the surrounding vasculature or nerves [3]. Pseudoaneurysms have a propensity to enlarge, and in the most severe instances, they are capable of rupturing as a consequence of the pulsatile inflow and outflow of blood via their necks. Femoral pseudoaneurysms have the potential to rupture into the retroperitoneal space, which may result in substantial bleeding that may or may not be immediately apparent and ultimately result in death. Epidemiology of Site and Risk Factors.

In most cases, femoral pseudoaneurysms are the consequence of accessing the femoral artery for catheter-based procedures. The great majority of arterial catheterizations employ the femoral artery, whereas the radial artery access site is utilized less often [5]. They may, however, exist in any location where arterial access is exploited. Following a diagnostic procedure, the frequency is reported to be between 0.06 and 0.18%; however, following interventional operations, the frequency ranges between 0.7 and 6.25% [6]. Besides the common femoral artery, pseudoaneurysms are more likely to arise at lower puncture points, such as those at the superficial and deep femoral arteries. This is mainly due to the femur head not providing support during compression of the artery after the intervention. Some society recommendations state that the acceptable incidence of pseudoaneurysm following percutaneous access should be less than 0.2% with the growing use of ultrasonography for access [7].

Risk Factors

Inadequate hemostasis at the puncture site may lead to complications after endovascular procedures. Several variables might contribute to an incomplete puncture site closure risk, including anticoagulation, large sheath size (7-French and above), incorrect puncture technique, inadequate manual compression, obesity, synchronous catheterization of artery and vein, puncture of the superior or deep femoral artery, arterial hypertension, complex compression sites, and heavily calcified arteries. Complications at the access site are exceedingly infrequent when the interventionalist acknowledges that maintaining hemostasis at the access site is just as crucial as the intervention being carried out. After interventional procedures, access site closure is often entrusted to younger residents/trainees/fellows, so when unskilled individuals are engaged in the puncture site hemostasis, this could sometimes lead to uncontrolled and other times dangerous circumstances.

The creation of pseudoaneurysms was thought to be at a lower risk with the use of closure devices in the past; however, more recent studies reveal that there is no significant decrease in the occurrence of pseudoaneurysms with closure devices or manual compression [5].

When determining whether a closure device may be utilized in a particular situation, it is imperative that the procedure be carried out by physicians who are well versed in the technical details of the device. Choosing manual compression necessitates that the person doing the compression be adequately trained and applies steady direct digital compression for at least 20 min.

Closure devices are typically not suggested at the level of the brachial artery because of their relatively small diameter, and a tourniquet-type bandage should not be applied around an access site since it might quickly lead to blockage of the artery and ischemia in the distal portion of the vessel, all of which could lead to insufficient pressure on the puncture site.

Another factor that decreases the probability of vascular complications is when the interventionalist employs ultrasonography or fluoroscopy with femoral head localization to puncture the femoral artery in the desired location and on the first attempt.

Heparin-treated patients need to have their activated clotting time (ACT) evaluated, which should be reduced to less than 200 s, for sheath removal.

Physical Examination and Diagnosis

After an endovascular procedure, one should be suspicious of the presence of a pseudoaneurysms if they present with local signs and symptoms. The clinical manifestations of an underlying pseudoaneurysm are pain and swelling, with or without neurologic symptoms, which are secondary to the mass effect on the surrounding structures. A palpable thrill, an audible systolic bruit, or a pulsatile mass could be the clinical findings of a false aneurysm. Growing pseudoaneurysms have the ability to exert pressure on the overlying skin, which may result in pain and, in the long run, ischemia, necrosis, or hemorrhage. Thrombi that develop within the sac have the potential to embolize distally. The development of neurologic symptoms may occur as a consequence of nerve compression or ischemia. When nearby veins are compressed, the result might be edema as well as deep vein thrombosis. When comparing brachial artery pseudoaneu-

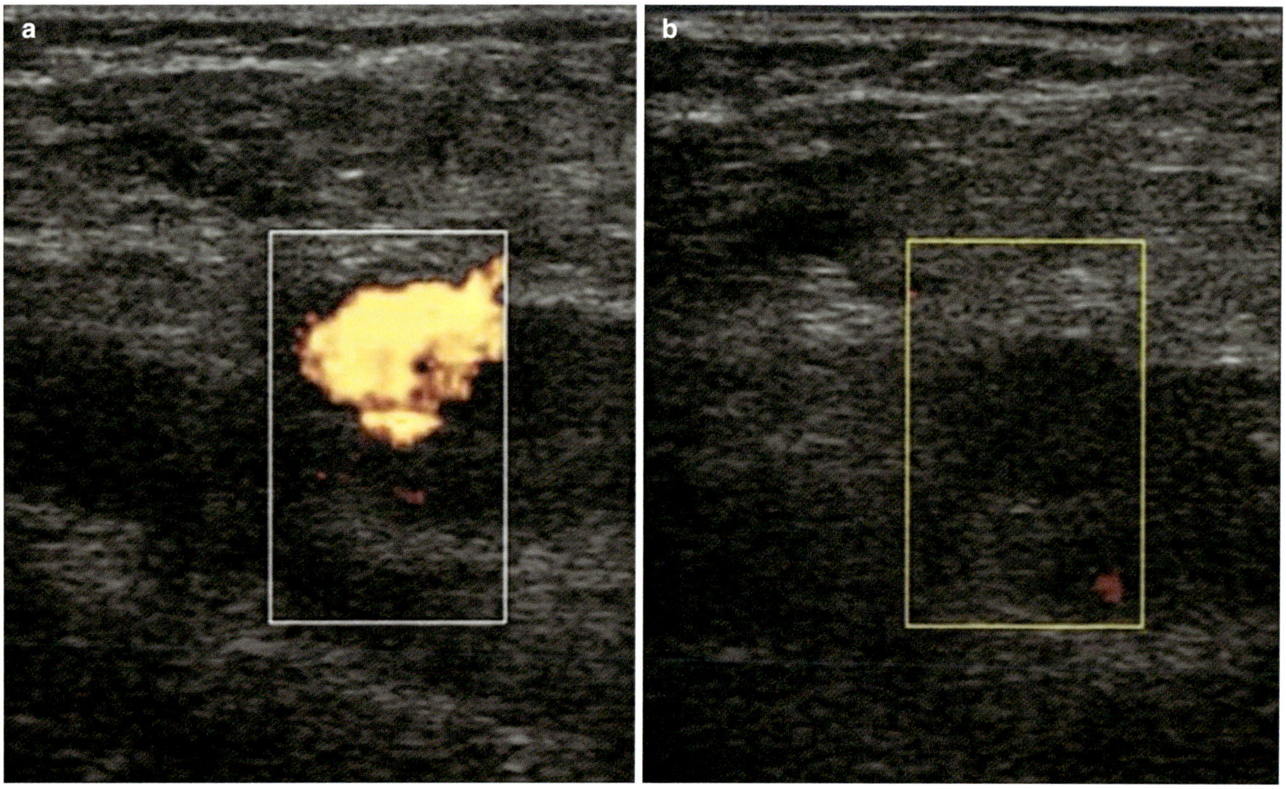

Fig. 35.2 (**a**) Power Doppler filling of a pseudoaneurysm, (**b**) thrombosed pseudoaneurysm with no Doppler sign detected following intervention

rysms to femoral artery pseudoaneurysms, it is vital to note that hemodynamic instability is far less likely in the former, but neurologic symptoms are much more common. Numbness and tingling in the hand are potential symptoms that might be brought on by compression of the median nerve that is located nearby. Even a small amount of blood tracking up into the axillary sheath from a high puncture in the arm might cause significant nerve compression. A comprehensive history and a physical exam are initially essential, but a duplex ultrasound examination of the afflicted region is required to confirm the diagnosis (Fig. 35.2). This may evaluate the existence of a pseudoaneurysm, its size, the vessel from which the pseudoaneurysm originates and preferably the size of the native vessel, the presence of blood flow or thrombus inside the pseudoaneurysm, the length and width of the neck, the integrity of the surrounding vessels, and the presence of loculations. A femoral artery pseudoaneurysm may be detected with 95–98% sensitivity and specificity with duplex ultrasonography [3]. The presence of blood flow characterized by a swirling motion, also known as the "ying-yang" sign, is a common finding. A distinctive to-and-fro (Fig. 35.3) flow pattern in the neck of the pseudoaneurysm has been characterized as early as 1988 using spectral waveforms and color Doppler imaging [8]. During systole, blood enters the pseudoaneurysm (the "to" component), while blood exits the pseudoaneurysm during diastole (the "fro"

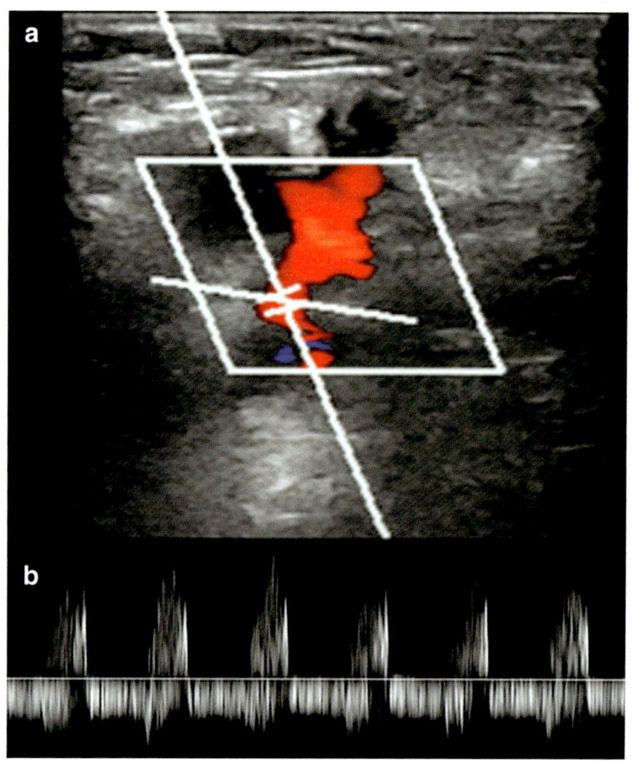

Fig. 35.3 (**a**) Spectral Doppler applied at the pseudoaneurysm neck, (**b**) revealing the traditional "to-and-fro" waveform

Table 35.1 Checklist for the imaging evaluation of a pseudoaneurysm

Is there a presence of a pseudoaneurysm?
What is its size and distance from the skin?
What is the vessel from which the pseudoaneurysm originates?
What is the size of the native vessel?
Is there presence of blood flow or thrombus within the pseudoaneurysm?
What is the length and width of its neck?
Is there integrity of the surrounding vessels?
Are any loculations present?

component). In conclusion, ultrasound is an essential tool of the follow-up of patients since it may determine whether or not the false aneurysm is completely thrombosed or recurred with flow in its lumen.

CT angiography may be used for further verification and assessment of the region, revealing a contrast-filled sac adjacent to and related to the donor artery. When it comes to identifying pseudoaneurysms in the proximal extremities, CT angiography exhibits a sensitivity of 95.1% and a specificity of 98.7% [9]. The wall of a pseudoaneurysm is typically smooth and well defined, with the exception of the wall of a mycotic pseudoaneurysm, which is thickened, irregular, and enhanced.

When compared with CT or US, MR angiography continues to be a time-consuming procedure, but it may be a suitable alternative for people who are allergic to iodine contrast medium. Helpful checklist table for the assessment of the morphology of pseudoaneurysms (Table 35.1).

Treatment/Management

Surgery was formerly the only option for treating arterial pseudoaneurysms. But this approach has become less applicable in recent decades. The current treatments include observation, ultrasound-guided compression, ultrasound-guided injection of thrombin or other fibrin adhesive agent, percutaneous embolization, endovascular stent-graft exclusion, transarterial embolization, and surgical repair (Table 35.2).

Table 35.2 Suggested endovascular toolkit for the endovascular management of a peripheral pseudoaneurysm. These are only a few suggested options based on the author's experience and can cover the vast majority of cases

Tools	Size/length
Percutaneous methods	
Needles	
Percutaneous entry needle	18–22 gauge
Embolic material	
Human thrombin	Incremental doses (0.1–0.2 mL). Usually a total of 1 mL (500 IU) is sufficient
Macro- or micro-coils	Pushable. 2–10 mm diameter
Endovascular methods	
Needles	
Percutaneous entry needle	18–21 gauge
Sheaths	
Any standard access sheath	4.0–6.0 Fr/11 cm
Guidewires	
Standard J guidewire	Diameter: 0.035′, lengths: 180–260 cm
Micro-guidewire	0.012″–0.021″
Catheters	
Standard angled angiocatheters (VanSchie 1–3; Vertebral, Headhunter, Cobra, etc.)	0.035″/5.0 Fr/40–110 cm
Coaxial microcatheter system	1.8–2.7 Fr
Balloons	
Standard semi-compliant PTA balloon catheters	Diameters: 2–8 mm, length 20–40 mm
Stent-grafts	
Self-expandable covered stents	2–10 mm diameter/20–40 mm length/75 cm or 135 cm shaft
Balloon-expandable covered stents	2–8 mm diameter/20–40 mm length/75 cm or 135 cm shaft
Embolic material	
Micro-coils	
Pushable	2–12 mm diameter
Detachable (preferably)	2–12 mm diameter
Liquid embolics (N-butyl-cyanoacrylate with Lipiodol, Onyx®)	Small amounts are usually required (0.5–3 mL)

Observation

Pseudoaneurysms of less than 2 cm in diameter in individuals with no symptoms or anticoagulation may be treated by observation because of the likelihood of spontaneous thrombosis, according to research [10]. A follow-up duplex ultrasound will be necessary in a week for this group of patients in order to document a reduction in the size of the pseudoaneurysm or its complete elimination. Due to the extensive hospital stay, most pseudoaneurysms are nowadays treated. The size of the pseudoaneurysm and the patient's anticoagulation/antiplatelet status, which is frequently necessary in patients after cardiac procedures, are both factors that withhold spontaneous thrombosis.

Percutaneous Treatment

Ultrasound-Guided Compression

Using duplex ultrasound, the location of the pseudoaneurysms and the neck may be determined. An estimation is also made on the direction in which pressure must be applied in order to disrupt circulation into the sac, and downward pressure is applied across the neck, leading to flow interruption (Fig. 35.9a). It is important to avoid exerting excessive pressure, since this might lead to the surrounding vessels being constricted or to decreased blood flow in the distal arteries; however, there are often no negative aftereffects associated with the momentary blockage of blood flow. An intravenous line must be set up, the skin is disinfected, and local anesthetic is administered to alleviate discomfort prior to compression, as it is a particularly unpleasant procedure for the patient. The first few seconds of pressure usually bring on the most immense pain, but while the pressure is held constant, a tolerance for the pain will develop; however, it is not uncommon to miss the position over the neck, which demands further modification of the position of the ultrasound probe in order to proceed with the procedure, which causes disruption of constant pressure that is being applied to the neck, increases the risk of blood flowing in the pseudoaneurysm, and decreases the likelihood of technical success. The process is repeated as many times as necessary if the false lumen and neck remain patent with 10–15 min compression intervals. In clinical settings, the process is often terminated if the pseudoaneurysm has not thrombosed by the end of the first hour [11].

Until the introduction of thrombin embolization therapy, this method, initially reported in 1991 by Fellmeth et al. [12], was the treatment of choice for pseudoaneurysms. Patient discomfort, time demands, and the sonologist's ability to maintain enough pressure are major limitations [13]. Contraindications to ultrasound-guided compression include substantial hematomas, limb or skin ischemia, significant patient discomfort, infection, and anatomical incompatibilities (e.g., if the pseudoaneurysm neck is not easily approachable, above the inguinal ligament) [3]. The use of anticoagulants lowers the probability of the procedure being successful; hence, anticoagulation therapy has to be withheld if medically safe [3]. Anticoagulation, pseudoaneurysm size, the number of locations, the age of the pseudoaneurysm, neck length, and width have all been studied to evaluate whether these have any effect on the efficacy of US-guided compression. This method has been shown to achieve hemostasis in a secure and cost-effective manner, with success rates ranging from 74 to 86%, as reported [14]. It has been described that the rate of failed success of this approach ranges between 15 and 38% and that the incidence of recurrence after initial success is as high as 20–30% in patients who have taken anticoagulants [3, 15–17]. As a result, 24-h follow-up is indicated to verify that the pseudoaneurysm remains thrombosed. Patients are instructed to avoid intense activities for a minimum of 1 day after the procedure.

Complications from compression under ultrasound guidance are quite infrequent. Increasing pseudoaneurysm size that could lead to rupture (1% risk), distal embolization, and vein thrombosis are possible complications of the method, with an overall complication rate of 3.6%, that are quite significant if they occur [11, 14, 18].

The US-guided compression method has various alterations. Pressure may be applied on the pseudoaneurysm, with the force of the hand, while the probe is positioned and used to visualize the pseudoaneurysm, which makes it easier if pressure needs to be exerted for an extended time or if the patient is overweight [19]. Occluding the aneurysm sac's flow using a freehand mechanical compression device, such as a FemoStop device, is another option. As a result of the discomfort reported by patients, this treatment option has been less popular in recent years, in addition to being more time consuming.

Ultrasound-Guided Thrombin Injection

The use of ultrasound-guided thrombin injection for the treatment of pseudoaneurysms that are percutaneously accessible is a commonly used minimally invasive technique. Even though patients may be taking anticoagulants and/or antiplatelet medication, a single treatment for femoral pseudoaneurysms has been reported to have a very high success rate, ranging from 97 to 100%, in a number of studies [20]. Blood stasis, as seen in pseudoaneurysms, makes it easier for activated clotting factors, including thrombin, to remain in the bloodstream and produce thrombi, which, in addition to injection of exogenous thrombin, induces further development of the thrombus [21], a mechanism that is independent of the usage of anticoagulants [22]. The use of percutaneous thrombin injection for false aneurysms at locations other than the femoral artery has also been shown to be successful, like the radial or brachial artery, at the sites of hemodialysis fistulas [1, 23, 24].

Cope and Zeit originally documented the exclusion of false aneurysms using percutaneous embolization with thrombin in the year 1986 [25]. Thrombin injection was used to treat 114 individuals who had iatrogenic pseudoaneurysms, according to Paulson et al., with a 90% success rate with a single injection. A second successful injection a day later was obtained in 7 out of 11 patients with residual flow, equivalent to an overall 96% technical success rate [26]. In 131 patients, Khoury et al. had 96% technical success, with complete pseudoaneurysmal thrombosis in 117 individuals with the first attempt, and the rest 9 were successfully embolized the day after on the second attempt [27]. US-guided thrombin injection was recently reported in a multicenter registry, where 91 patients were injected with thrombin, with the initial technical success after a single injection being 96% [28].

Thrombin converts inactive fibrinogen into active fibrin, which is a direct contributor in the formation of thrombus. It may be found in a wide range of pharmacological formulations, derived from bovines or humans. The usage of bovine thrombin was commonly used in the past, but due to possible IgE-mediated anaphylactic reaction, from previous exposure and questions regarding possible transmission of human form of bovine spongiform encephalopathy, it is avoided nowadays. A sterile powder is the most common form of bovine thrombin, which is diluted with normal saline solution to a concentration of 1000 U/mL commonly, although a lower concentration of thrombin could theoretically have a lower risk of an allergic reaction [29]. Human thrombin is often obtained from the Tisseel Fibrin Sealant Kit (Baxter US), which is a dual syringe with fibrinogen that has been approved for use as tissue glue in surgeries. It is used as an independent component or in combination directly in the pseudoaneurysmal sac, although sequential usage is sufficient for achieving thrombosis. It is also possible to utilize human thrombin that has been extracted from the blood of patients. About 30–60 mL of blood sampling is required, which is processed, and then the thrombin component is reinjected in the pseudoaneurysm sac. The use of autologous thrombin provides the benefits of a safe procedure, lower risk of allergic responses, as well as lower cost [30].

After local sterility of the area, a linear 5–7.5 MHz ultrasound transducer is used to visualize the pseudoaneurysm, and 1–2% lidocaine is infiltrated in the skin and subcutaneous tissue. A 19–22 gauge entry needle is guided in parallel with the probe to puncture the pseudoaneurysmal sac under constant ultrasound visualization, as superficially and far from the neck as feasible. The tip of the needle should be depicted and ensured to be as distant as possible from the native vessel (Figs. 35.4 and 35.9b). Thrombin is slowly administered over a period of 10 s, until color Doppler ultrasonography reveals that blood flow has ceased, usually within a few seconds. The quantity of thrombin needed for each patient has not been estimated yet, but usually 0.1–0.5 mL of thrombin is sufficient for complete thrombosis.

Larger or multilocular pseudoaneurysms might require higher doses of thrombin injection or a second administration if flow persists (Fig. 35.5) [29]. Residual flow in the neck should not be treated due to the increased risk of distal embolism. Final ultrasound documentation should depict a patent native vessel, and distal pulses should be assessed. A second duplex examination is required after 24 h to prove that the pseudoaneurysm has been successfully thrombosed, and a 4–6-h bedrest restriction is imposed on all patients after the intervention.

Distal embolization and acute ischemia are possible complications of percutaneous thrombin injection in up to 2% of patients [3], with a short pseudoaneurysm neck (less than 2 mm) having an increased risk. Distal embolization into the native circulation has been observed in three individuals by Paulson et al., which required no surgery or endovascular intervention [26]. An embolization to the distal brachial artery was spontaneously lysed, as described by Kang et al. [1]. Thrombin-induced distal embolism in one patient was treated with endovascular tissue plasminogen activator and heparin administration, according to Friedman et al. [31] and similarly by Sadie and Ibrahim [32]. In the series of Khoury et al., two patients developed limb ischemia necessitating surgical thrombectomy [27]. Asymptomatic thrombi lyse spontaneously due to natural circulating lytic agents, although the administration of a tissue plasminogen activator can assist by causing fibrinolysis through the conversion of plasminogen to plasmin in clinically substantial distal embolization [31]. Studies on the security and effectiveness of this process have been conducted by multiple independent research groups, despite the fact that it is an unapproved application of thrombin, as the label on the packaging points out that it is for topical use only and not for injection. This technique may also be used to treat vessels above the inguinal ligament, which is not recommended when using ultrasound-guided compression due to the increased risk of pseudoaneurysmal rupture [33].

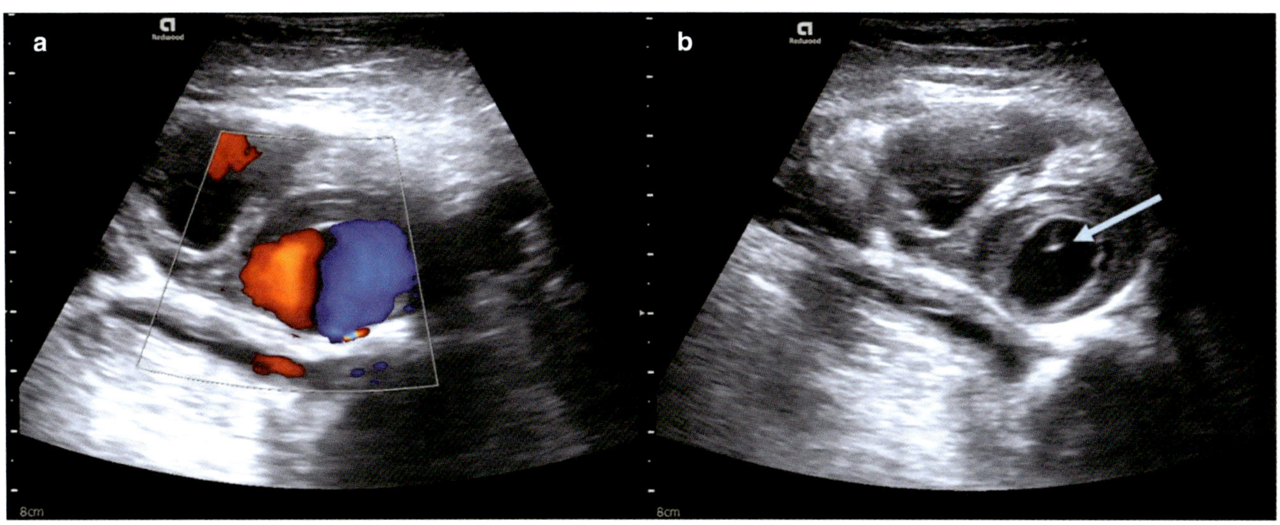

Fig. 35.4 (**a**) Swirling of blood within the pseudoaneurysm on color Doppler—"Ying-yang" sign, (**b**) needle tip (blue arrow) visualized within the pseudoaneurysm for thrombin injection

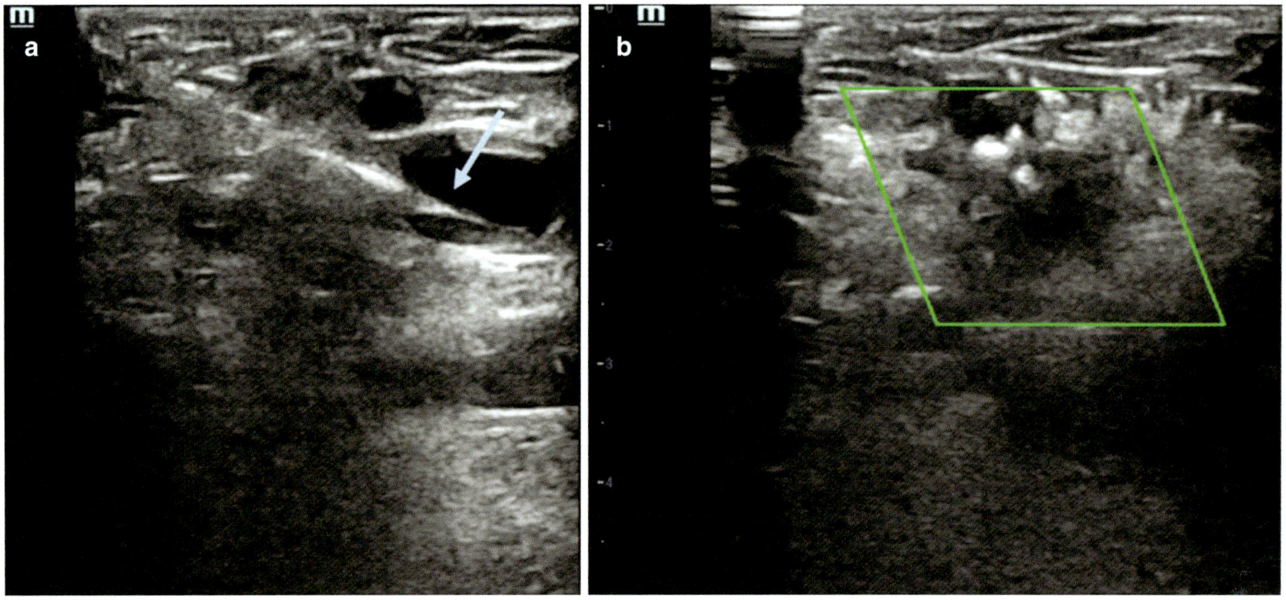

Fig. 35.5 Multiloculated pseudoaneurysm. (**a**) Percutaneously inserted needle (blue arrow), (**b**) thrombosed multiloculated pseudoaneurysm (absence of Doppler signal)

With Endovascular Balloon Protection

One of the main risk factors of percutaneous thrombin injection for the treatment of pseudoaneurysms is leakage into the peripheral arterial tree, so prevention is possible with the use of a balloon catheter. Under fluoroscopic guidance, a catheter is inserted across the aortic bifurcation to the contralateral femoral artery. Following angiographic confirmation of the pseudoaneurysm's location, a balloon catheter sized appropriately for the vessel is positioned across the pseudoaneurysm's point of origin, and a 19–22-gauge needle is inserted percutaneously into the pseudoaneurysm once it has been identified by ultrasound. The balloon is properly positioned and adequately inflated until pulsatile flow through the percutaneous needle is ceased and thrombin is administered. The balloon may be kept inflated for 10 min, and angiography and ultrasound are performed to evaluate the pseudoaneurysm thrombosis. There is currently no data to suggest that the use of this approach results in fewer complications in comparison to thrombin injection alone, despite the fact that it is theoretically a safer procedure. Due to the invasiveness of the contralateral puncture, data don't support its use in routine practice. This approach was recommended for use by Elford et al. in situations in which the neck is short or wider than 3 mm [34].

Percutaneous Coil Embolization

Puncturing the pseudoaneurysmal cavity percutaneously and deploying coils directly in the sac is an alternative approach. This can be performed in addition with or instead of thrombin injection [3]. It has been suggested that an angioplasty balloon should be used to protect the false aneurysm neck [35]; however other researchers have successfully embolized the pseudoaneurysm with no need for balloon protection [36]. Because pseudoaneurysms lack an actual wall, some interventionalists may object to performing coil embolization, as it might raise cavity pressure and trigger a rupture of the sac in addition to prevention of natural absorption of the thrombus/hematoma due to the filling space with coils. In addition, a nest of coils may serve as a target for pathogens, being a foreign body. This method has only been described in case reports and has never been used as standard therapy for pseudoaneurysms. It is critical to rule out any AV connections on the affected limb to avoid material embolization in the venous system and eventually the heart and pulmonary artery. One more important thing to note is that there should be no coil migration noticed in the native artery. Manual compression after coil deployment is possible in the case of persistent flow within the pseudoaneurysmal sac [37]. Coils of all sizes can be used (micro-coils or macro-coils) without the neck morphology being a restriction. Thrombin injection is a less costly alternative that does not include the usage of synthetic materials and is generally preferred.

Use of a Closure Device

There have been reports in the literature that state the use of closure devices for the management of pseudoaneurysms. A pseudoaneurysm is punctured with either a 22-gauge needle or a micropuncture needle under fluoroscopy assistance, and then a Glidewire is advanced into the proximal femoral or external iliac artery. An Angioseal/StarClose device sheath is inserted over the wire, and the closure device is deployed, with the anchor/clip placed at the orifice of the neck of the pseudoaneurysm (Fig. 35.9c) [38, 39]. The main significant concern is the procedure failure because of a wide pseudoaneurysm neck. The safety and efficacy of this method must be evaluated in a wider study.

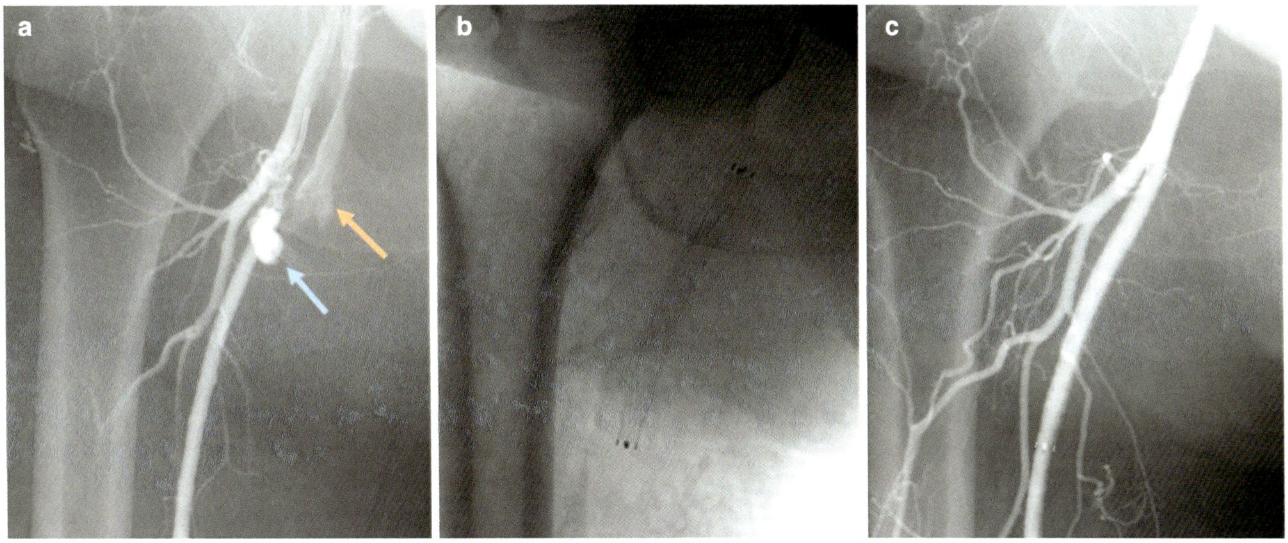

Fig. 35.6 (**a**) Pseudoaneurysm (blue arrow) at the level of origin of the right superficial femoral artery and arteriovenous fistula (orange arrow), with filling of the common femoral vein at the same level. (**b**) Covered stent placed across the pseudoaneurysm and AVF. (**c**) Normal patency of the femoral arteries following stent placement with no demonstration of the pseudoaneurysm and no arteriovenous fistula detected

Endovascular Repair

The appropriate endovascular intervention strategy is determined by the donor artery and the size of the pseudoaneurysm neck, which may be divided into two basic categories: direct pseudoaneurysm embolization and donor artery stent implantation.

Stent-Graft Placement

For stent-graft placement, a new vascular puncture site is required, either on the ipsilateral limb with a direction toward the pseudoaneurysm or from the contralateral groin with a "cross-over" (across the aortic bifurcation) access, in the case of the femoral artery (Fig. 35.6). Placement of a stent-graft is preferred in larger-diameter arteries since smaller arteries have a greater thrombosis risk but are not completely ineligible for the procedure. Cross-over approaches need a more rigid delivery system with the use of a long sheath because considerable elongation and tortuosity of the iliac trunks are suspected to have higher misplacement probability and procedure failure rate. The normal range of motion of the hip joint may fracture or occlude covered stents, making them unsuitable for joint placement. At a mean follow-up of 301 days, Thalhammer et al. reported stent occlusions in 17% of patients, aside from the fact that all pseudoaneurysms were fully and permanently excluded [40]. The high expense of stent-graft placement for common usage in a potentially big patient population and the prohibition of future catheterization of the stented artery are two additional important limitations.

Coil Embolization

Pseudoaneurysms that cannot be treated percutaneously or where the implantation of an endovascular stent is not feasible, such as those that are near the inguinal ligament or the femoral bifurcation, may be treated with an endovascular catheter-directed insertion of coils into the sac itself [41]. When a pseudoaneurysm develops in an irreplaceable donor artery, the donor artery must be protected from circulation to avoid distal blood flow obstruction. If the pseudoaneurysm cannot be superselectively catheterized for further embolization, embolization of the afferent artery is used only if it can be spared functionally. Embolization must be performed both proximally and distally in order to avoid backflow into the pseudoaneurysms, from collateral branches. Nondetachable and detachable coils are the two primary types of coils. Nondetachable coils take their form immediately after being released from the catheter, without the possibility of withdrawal. These coils are found in an extensive variety of sizes, materials, and shapes formed. For a more precise coil embolization, detachable coils can be used that are connected to a pusher guidewire to readapt the coil placement, if necessary, before final release. Some of these patients might have pathological blood coagulation that might result to incomplete thrombosis of the pseudoaneurysm, so the use of additional embolic agents, like thrombin or histoacryl glue, can be used. Pseudoaneurysms may also be treated by uncovered stent placement across their neck, followed by coil embolization with the use of a catheter directed through the mesh in the pseudoaneurysm. This technique prevents migration of coils in the healthy vessel, although the main disadvantage is that lifelong use of antiplatelet medication is required.

Thrombin Injection

Endovascular thrombin injection can be used in pseudoaneurysms that are inconveniently located for percutaneous embolization. This procedure entails selective catheterization of the pseudoaneurysmal neck with an appropriate endovascular catheter and the injection of thrombin directly into the false aneurysm lumen.

Surgical Repair

Iatrogenic and non-iatrogenic pseudoaneurysms that cannot be treated by percutaneous procedures are candidates for surgical repair. Since less invasive ways of treating pseudoaneurysms have been developed, the use of surgery to repair them has become less common. The fact that these innovative methods may be done under local anesthesia is a great advantage for patients with comorbidities that are at great risk of needing general anesthesia. In addition, the length of hospital stay is longer that 3 days on average [42]. Surgical repair for pseudoaneurysms is preserved for a limited number of cases in which it is necessary [43, 44]:

Rapid growth of the pseudoaneurysm
Infected/mycotic pseudoaneurysm
Distal limb ischemia, overlying skin ischemia, and neuropathy (adjacent nerve compression) caused by local pressure of the pseudoaneurysm or the hematoma
Unsuccessful percutaneous intervention
Vascular anastomotic disruption

Due to local scarring, after surgical repair, subsequent groin access may be difficult in future interventions [42]. In addition, it extends the hospital stay and raises the overall healthcare expenses. False aneurysms that have become infected necessitate immediate surgical intervention, as they may rupture or cause septic emboli in the systemic circulation. These pseudoaneurysms must be treated surgically, as the use of materials like coils or thrombin is contraindicated due to them becoming contaminated and worsening the clinical condition of the patient [3].

Case Presentation

Continued from page 371

Ultrasound-guided local anesthesia was performed on the left groin, and arterial access was obtained with ultrasound-guided retrograde puncture of the left common femoral artery, followed by insertion of a 6-French sheath toward the external iliac artery. The right common iliac and external iliac arteries were catheterized using the "cross-over" technique with a Simmons-1 guiding catheter and a 0.035 Glidewire (Fig. 35.7a). Angiography from the level of the common femoral artery verified the origin of the pseudoaneurysm from the right deep femoral artery. Selective catheterization of the deep femoral artery was performed, which was followed by superselective catheterization of the feeding branch, with the use of a 2.7 Fr microcatheter (ASAHI; Japan) (Fig. 35.7b, c). Embolization was performed with the use of five 4 mm micro-nester pushable coils (Fig. 35.8a, b). Control angiography depicted a remaining filling of the pseudoaneurysm from another feeding branch originating from the deep femoral artery, so superselective catheterization and further embolization followed with the use of Glue-Lipiodol mixture, with a dilution of 1:1 (Fig. 35.8c). Finally, control angiography from the proximal deep femoral artery did not depict the pseudoaneurysm or areas of extravasation (Fig. 35.8d).

Following removal of the catheters and the sheath, hemostasis of the left common femoral artery was obtained with the use of a StarClose closure device.

The patient exited the angio-suite with physiological vital signs and was discharged 2 days later. At 1-month follow-up the hematoma had completely resolved, and there was no visible pseudoaneurysm according to ultrasound follow-up.

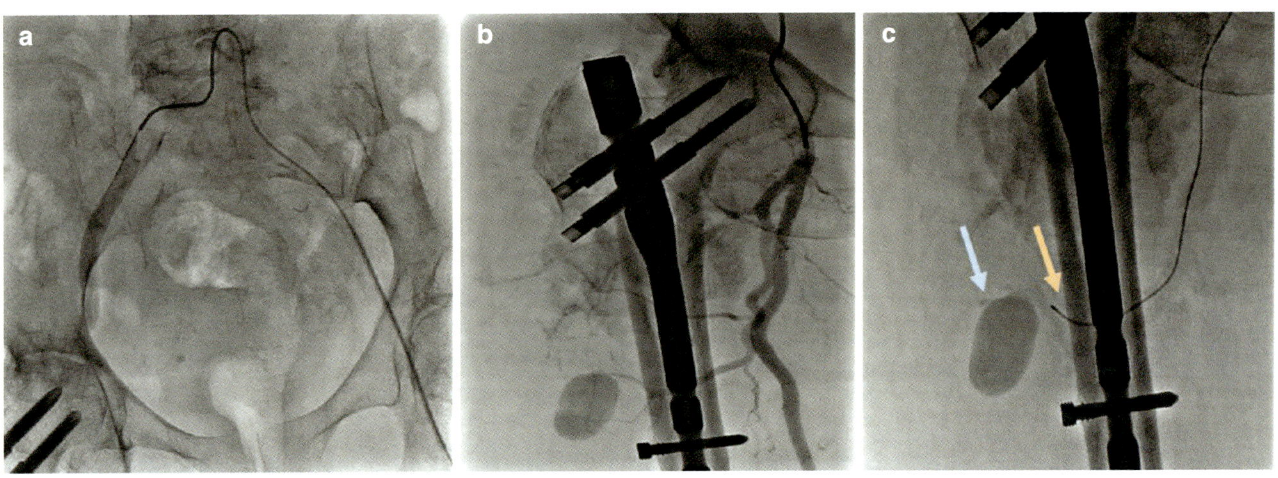

Fig. 35.7 (**a**) Cross-over catheterization of the right common iliac artery. (**b**) Selective catheterization of the right common femoral artery. (**c**) Superselective catheterization of the right deep femoral artery branch (orange arrow) supplying the pseudoaneurysm (blue arrow)

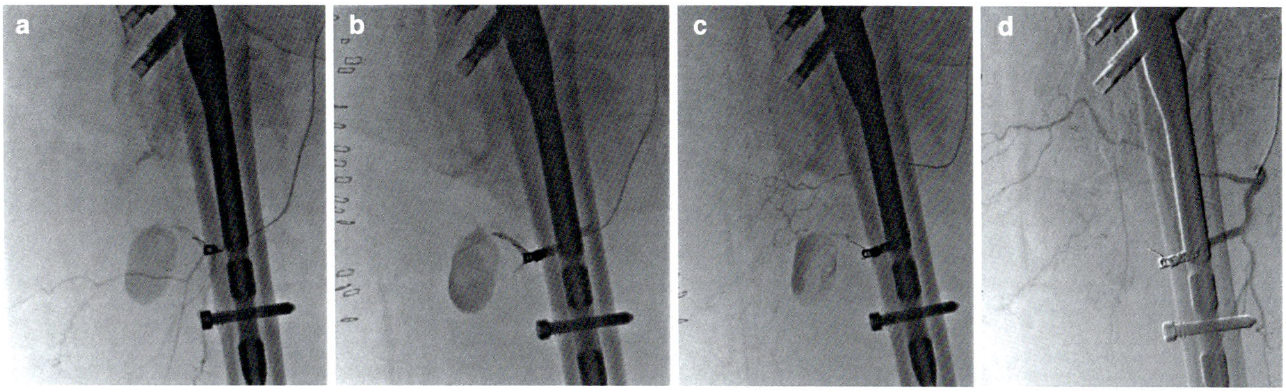

Fig. 35.8 (**a**) First coil embolization of the deep femoral artery branch. (**b**) Further coil embolization of the branch, with persisting blood supply of the pseudoaneurysm. (**c**) Detection and glue-lipiodol mixture embolization of another supplying branch originating from the deep femoral artery. (**d**) Final selective angiography with complete closure of the feeding vessels and absence of contrast filling of the pseudoaneurysm

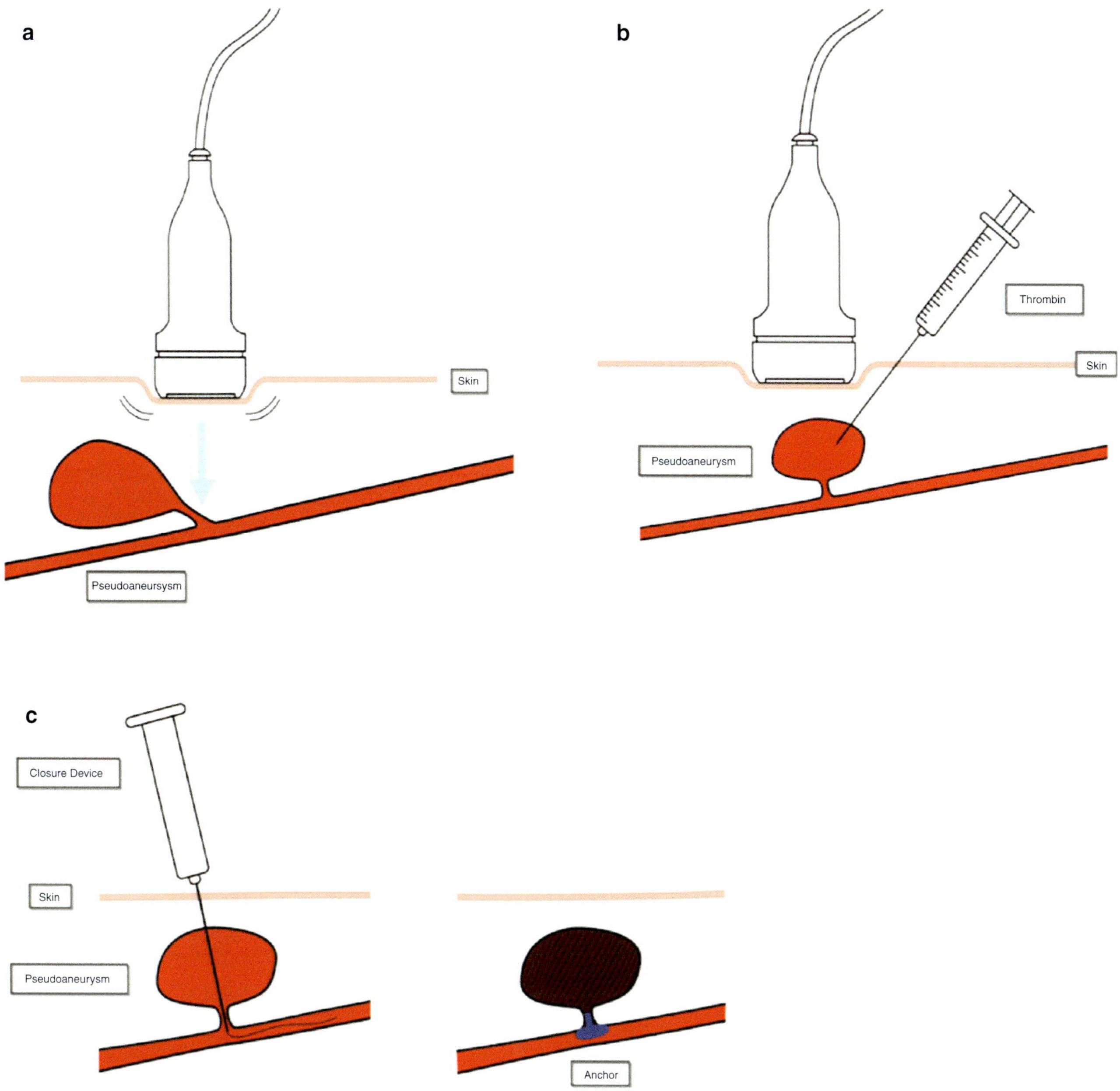

Fig. 35.9 (**a**) Percutaneous compression of the pseudoaneurysm neck. (**b**) Percutaneous thrombin injection. (**c**) Percutaneous use of a closure device

Acknowledgments We would like to thank Enrico Moschovakis-Zeiger and Iason Anastassiou for their assistance in the preparation of the illustrations.

References

1. Kang SS, Labropoulos N, Mansour MA, Michelini M, Filliung D, Baubly MP, Baker WH. Expanded indications for ultrasound-guided thrombin injection of pseudoaneurysms. J Vasc Surg. 2000;31:289–98.
2. Krüger K, Zähringer M, Söhngen F-D, Gossmann A, Schulte O, Feldmann C, Strohe D, Lackner K. Femoral pseudoaneurysms: management with percutaneous thrombin injections—success rates and effects on systemic coagulation. Radiology. 2003;226:452–8.
3. Morgan R, Belli A-M. Current treatment methods for postcatheterization pseudoaneurysms. J Vasc Interv Radiol. 2003;14:697–710.
4. Jayaraman S, Richardson D, Conrad M, Eichler C, Schecter W. Mycotic pseudoaneurysms due to injection drug use: a ten-year experience. Ann Vasc Surg. 2012;26:819–24.
5. Biancari F, D'Andrea V, di Marco C, Savino G, Tiozzo V, Catania A. Meta-analysis of randomized trials on the efficacy of vascular closure devices after diagnostic angiography and angioplasty. Am Heart J. 2010;159:518–31.
6. Peters S, Braun-Dullaeus R, Herold J. Pseudoaneurysm Hämostaseologie. 2017;38(3):166–72. https://doi.org/10.5482/HAMO-17-01-0006.
7. Stone PA, Campbell JE, AbuRahma AF. Femoral pseudoaneurysms after percutaneous access. J Vasc Surg. 2014;60:1359–66.
8. Abu-Yousef M, Wiese J, Shamma A. The "to-and-fro" sign: duplex Doppler evidence of femoral artery pseudoaneurysm. Am J Roentgenol. 1988;150:632–4.
9. Soto JA, Múnera F, Morales C, Lopera JE, Holguín D, Guarín O, Castrillón G, Sanabria A, García G. Focal arterial injuries of the proximal extremities: helical CT arteriography as the initial method of diagnosis. Radiology. 2001;218:188–94.
10. Toursarkissian B, Allen BT, Petrinec D, Thompson RW, Rubin BG, Reilly JM, Anderson CB, Flye MW, Sicard GA. Spontaneous closure of selected iatrogenic pseudoaneurysms and arteriovenous fistulae. J Vasc Surg. 1997;25:803–9.
11. Middleton WD, Dasyam A, Teefey SA. Diagnosis and treatment of iatrogenic femoral artery pseudoaneurysms. Ultrasound Q. 2005;21:3–17.
12. Fellmeth BD, Roberts AC, Bookstein JJ, Freischlag JA, Forsythe JR, Buckner NK, Hye RJ. Postangiographic femoral artery injuries: nonsurgical repair with US-guided compression. Radiology. 1991;178:671–5.
13. Saad NEA, Saad WEA, Davies MG, Waldman DL, Fultz PJ, Rubens DJ. Pseudoaneurysms and the role of minimally invasive techniques in their management. Radiographics. 2005;25:S173–89.
14. Eisenberg L, Paulson EK, Kliewer MA, Hudson MP, DeLong DM, Carroll BA. Sonographically guided compression repair of pseudoaneurysms: further experience from a single institution. Am J Roentgenol. 1999;173:1567–73.
15. la Perna L, Olin JW, Goines D, Childs MB, Ouriel K. Ultrasound-guided thrombin injection for the treatment of postcatheterization pseudoaneurysms. Circulation. 2000;102:2391–5.
16. Gale SS, Scissons RP, Jones L, Salles-Cunha SX. Femoral pseudoaneurysm thrombin injection. Am J Surg. 2001;181:379–83.
17. Brophy DP, Sheiman RG, Amatulle P, Akbari CM. Iatrogenic femoral pseudoaneurysms: thrombin injection after failed US-guided compression. Radiology. 2000;214:278–82.
18. Coley BD, Roberts AC, Fellmeth BD, Valji K, Bookstein JJ, Hye RJ. Postangiographic femoral artery pseudoaneurysms: further experience with US-guided compression repair. Radiology. 1995;194:307–11.
19. Gilbertson JJ, Sorrell KA, Feinberg RL, Wheeler JR, Gregory RT, Snyder SO, Gayle RG. Color-flow duplex-directed manual occlusion of femoral false aneurysms. J Vasc Surg. 1993;17:a39244.
20. Schneider C, Malisius R, Küchler R, Lampe F, Krause K, Bahlmann E, Kuck K-H. A prospective study on ultrasound-guided percutaneous thrombin injection for treatment of iatrogenic post-catheterisation femoral pseudoaneurysms. Int J Cardiol. 2009;131:356–61.
21. Reeder SB, Widlus DM, Lazinger M. Low-dose thrombin injection to treat iatrogenic femoral artery pseudoaneurysms. Am J Roentgenol. 2001;177:595–8.
22. Lennox AF, Delis KT, Szendro G, Griffin MB, Nicolaides AN, Cheshire NJ. Duplex-guided thrombin injection for iatrogenic femoral artery pseudoaneurysm is effective even in anticoagulated patients. Br J Surg. 2002;87:796–801.
23. Clark TWI, Abraham RJ. Thrombin injection for treatment of brachial artery Pseudoaneurysm at the site of a hemodialysis fistula: report of two patients. Cardiovasc Intervent Radiol. 2000;23:396–400.
24. Sheiman RG, Brophy DP, Perry LJ, Akbari C. Thrombin injection for the repair of brachial artery pseudoaneurysms. AJR Am J Roentgenol. 1999;173:1029–30.
25. Cope C, Zeit R. Coagulation of aneurysms by direct percutaneous thrombin injection. Am J Roentgenol. 1986;147:383–7.
26. Paulson EK, Nelson RC, Mayes CE, Sheafor DH, Sketch MH, Kliewer MA. Sonographically guided thrombin injection of latrogenic femoral pseudoaneurysms. Am J Roentgenol. 2001;177:309–16.
27. Khoury M, Rebecca A, Greene K, Rama K, Colaiuta E, Flynn L, Berg R. Duplex scanning–guided thrombin injection for the treatment of iatrogenic pseudoaneurysms. J Vasc Surg. 2002;35:517–21.
28. Mohler ER, Mitchell ME, Carpenter JP, Strandness DE, Jaff MR, Beckman JA, Gerhard-Herman M. Therapeutic thrombin injection of pseudoaneurysms: a multicenter experience. Vasc Med. 2001;6:241–4.
29. Webber GW, Jang J, Gustavson S, Olin JW. Contemporary management of postcatheterization pseudoaneurysms. Circulation. 2007;115:2666–74.
30. Quarmby JW, Engelke C, Chitolie A, Morgan RA, Belli AM. Autologous thrombin for treatment of pseudoaneurysms. Lancet. 2002;359:946–7.
31. Friedman SG. Ultrasound-guided thrombin injection is the treatment of choice for femoral pseudoaneurysms. Arch Surg. 2002;137:462.
32. Sadiq S, Ibrahim W. Thromboembolism complicating thrombin injection of femoral artery Pseudoaneurysm: management with Intraarterial thrombolysis. J Vasc Interv Radiol. 2001;12:633–6.
33. Paulson EK, Sheafor DH, Kliewer MA, Nelson RC, Eisenberg LB, Sebastian MW, Sketch MH. Treatment of iatrogenic femoral arterial pseudoaneurysms: comparison of US-guided thrombin injection with compression repair. Radiology. 2000;215:403–8.
34. Elford J, Burrell C, Freeman S, Roobottom C. Human thrombin injection for the percutaneous treatment of iatrogenic pseudoaneurysms. Cardiovasc Intervent Radiol. 2002;25:115–8.
35. O'Sullivan GJ, Ray SA, Lewis JS, Lopez AJ, Powell BW, Moss AH, Dormandy JA, Belli AM, Buckenham TM. A review of alternative approaches in the management of iatrogenic femoral pseudoaneurysms. Ann R Coll Surg Engl. 1999;81:226–34.
36. Lemaire J-M, Dondelinger RF. Percutaneous coil embolization of iatrogenic femoral arteriovenous fistula or pseudo-aneurysm. Eur J Radiol. 1994;18:96–100.

37. Kobeiter H, Lapeyre M, Becquemin J-P, Mathieu D, Melliere D, Desgranges P. Percutaneous coil embolization of postcatheterization arterial femoral pseudoaneurysms. J Vasc Surg. 2002;36:127–31.
38. Jeon YS, Kim JY, Cho SG. Percutaneous treatment of a femoral artery pseudoaneurysm using the StarClose vascular closure system. Catheter Cardiovasc Interv. 2012;80:872–5.
39. Robken J, Shammas N. Novel technique to treat common femoral artery pseudoaneurysm using Angio-seal closure device. Int J Angiol. 2014;25:266–70.
40. Thalhammer C, Kirchherr AS, Uhlich F, Waigand J, Gross CM. Postcatheterization pseudoaneurysms and arteriovenous fistulas: repair with percutaneous implantation of endovascular covered stents. Radiology. 2000;214:127–31.
41. Yamakado K, Nakatsuka A, Tanaka N, Takano K, Matsumura K, Takeda K. Transcatheter arterial embolization of ruptured pseudoaneurysms with coils and n-butyl cyanoacrylate. J Vasc Interv Radiol. 2000;11:66–72.
42. Lumsden AB, Miller JM, Kosinski AS, Allen RC, Dodson TF, Salam AA, Smith RB. A prospective evaluation of surgically treated groin complications following percutaneous cardiac procedures. Am Surg. 1994;60:132–7.
43. Stone PA, Martinez M, Thompson SN, Masinter D, Campbell JE, Campbell JR, Aburahma AF. Ten-year experience of vascular surgeon management of iatrogenic pseudoaneurysms: do anticoagulant and/or antiplatelet medications matter? Ann Vasc Surg. 2016;30:45–51.
44. Sharma NK, Chin KF, Modgill VK. Pseudoaneurysms of the femoral artery: recommendation for a method of repair. J R Coll Surg Edinb. 2001;46:195–7.

Part VI

Vascular Access

Creating a Percutaneous Arteriovenous Fistula: Evidence and Techniques

36

Kyla M. Bennett and Alexandros Mallios

A meaningful shift occurred in June of 2018 in the arena of hemodialysis access when the United States Food and Drug Administration (FDA) approved both the Ellipsys™ Vascular Access System and the everlinQ™ endoAVF System (soon to become WavelinQ) [1]. While the need for hemodialysis has inexorably increased in the population, few advances have been made in hemodialysis access creation techniques. As patients live longer, however, the medical community has noted increasing complications from surgical fistulas and long-term high-flow hemodialysis dependence. This climate has shaped a market eager for the uptake of new approaches and devices.

The introduction of these minimally invasive techniques for arteriovenous fistula (AVF) creation promises many improvements for the burgeoning dialysis-dependent population. Such actual and theoretical developments include decreased AVF and peri-AVF surgical scarring and vessel trauma, decreased shear stress (based on the geometry of the resulting fistula and the relative lack of derangement from the original positions of the involved vasculature) as well as a much-needed expansion in sites available in the upper extremity for routine AVF creation. Several Ellipsys studies have further demonstrated that percutaneous arteriovenous fistulas (pAVFs) require fewer outflow vein and cannulation site interventions as compared to surgical fistulas(sAVFs) [2]. Percutaneous procedures resulting in perforator driven superficial venous outflow allow moderate flow fistulas with shared venous outflows resulting in adequate circuit flow with lower pressure and turbulence in comparison with sAVFs. Shared venous outflow and the proximal forearm location of the pAVF also results in extension of standard cannulation zones [3]. Compared to brachial artery-based fistulas, proximal radial artery (PRA)-based fistulas proffer a lower risk of steal syndrome and arm edema [3–6]; see Table 36.1.

Table 36.1 Relative benefits of pAVF

	Ellipsys	WavelinQ
General	• Decreased AVF and peri-AVF scarring from dissection/trauma • Decreased shear stress • Expansion in access site options • Moderate flow fistulas with multiple split venous outflow (decreased pressure and turbulence) • Decrease rate of CV, outflow, cannulation site interventions required	
PRA- vs Brachial artery- based	• Lower risk of steal syndrome • Lower risk of arm-swelling	

CV central venous

Predictably, some challenges have arisen with the introduction of this new technology and attendant new techniques. Specifically, there is an expected learning curve in the use of the devices (and with Ellipsys in the ultrasound skill set development). Similarly, there are initial cannulation challenges as post-procedural patients present to dialysis centers not yet accustomed to pAVF patients [3].

Device Mechanisms

While both devices offer a minimally invasive approach to hemodialysis access creation, they have differing mechanisms of action. The WavelinQ™ (BD) is designed to create a fistulous connection between an artery and a deep vein which will then fill a superficial vein via the presence of a perforating vein [7–9]. To do so, the system uses electrode-mediated delivery of radiofrequency (RF) energy between catheters positioned in the artery and deep vein and aligned

K. M. Bennett
Division of Vascular Surgery, University of Wisconsin-Madison School of Medicine and Public Health, Madison, WI, USA
e-mail: bennettk@surgery.wisc.edu

A. Mallios (✉)
Department of Vascular Surgery, Groupe Hospitalier Paris Saint Joseph, Paris, France

Department of Surgery, Centre Hospitalier de Chartres, Chartres, France

G. Geroulakos et al. (eds.), *Mastering Endovascular Techniques*, https://doi.org/10.1007/978-3-031-42735-0_36

by rare earth magnets (Figs. 36.1 and 36.2). The initial FDA-approved iteration involved placement of a 6 French (Fr) catheter antegrade in the brachial artery and venous access in the brachial (or forearm) veins. In 2019, a 4 Fr arterial catheter version was approved. This allows access via the ulnar or radial arteries. Coil embolization of one of the paired brachial veins at the time of creation is recommended in order to increase flow in the superficial veins [7–9].

Contrary to the mechanism of WavelinQ, Ellipsys utilizes thermal energy combined with pressure to fuse the walls of a perforating vein and the adjacent proximal radial artery (PRA) (Fig. 36.3) [9–11]. While WavelinQ is performed under fluoroscopic guidance, Ellipsys relies upon ultrasound guidance alone.

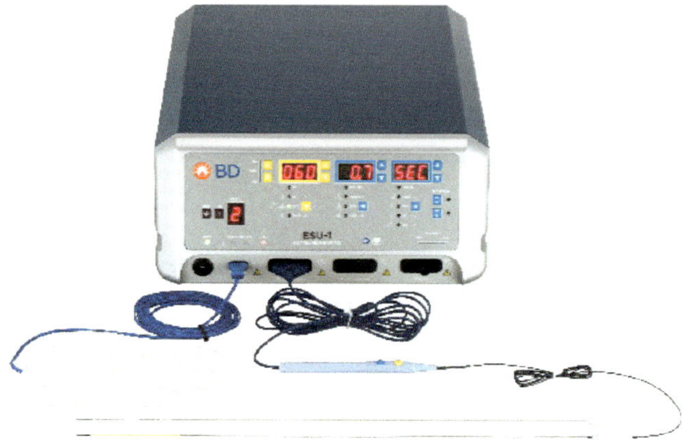

Fig. 36.1 WavelinQ™ EndoAVF generator with arterial and venous catheters. Photo of the radiofrequency generator and 4 Fr catheters

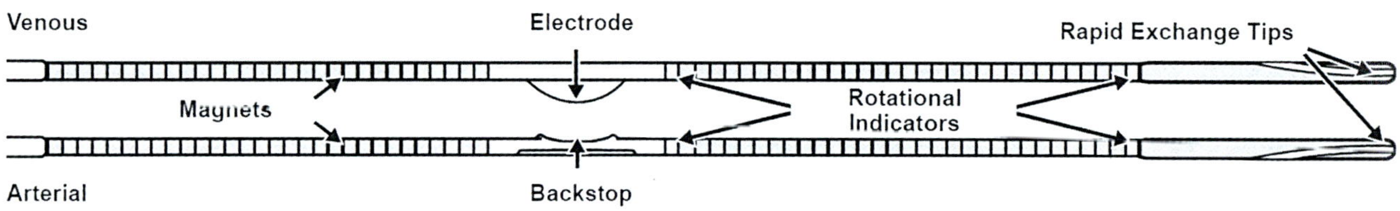

Fig. 36.2 WavelinQ™ Venous Catheter with Electrode and Arterial Catheter with Backstop. Schematic illustration of the venous and arterial catheters with relevant elements for the procedure

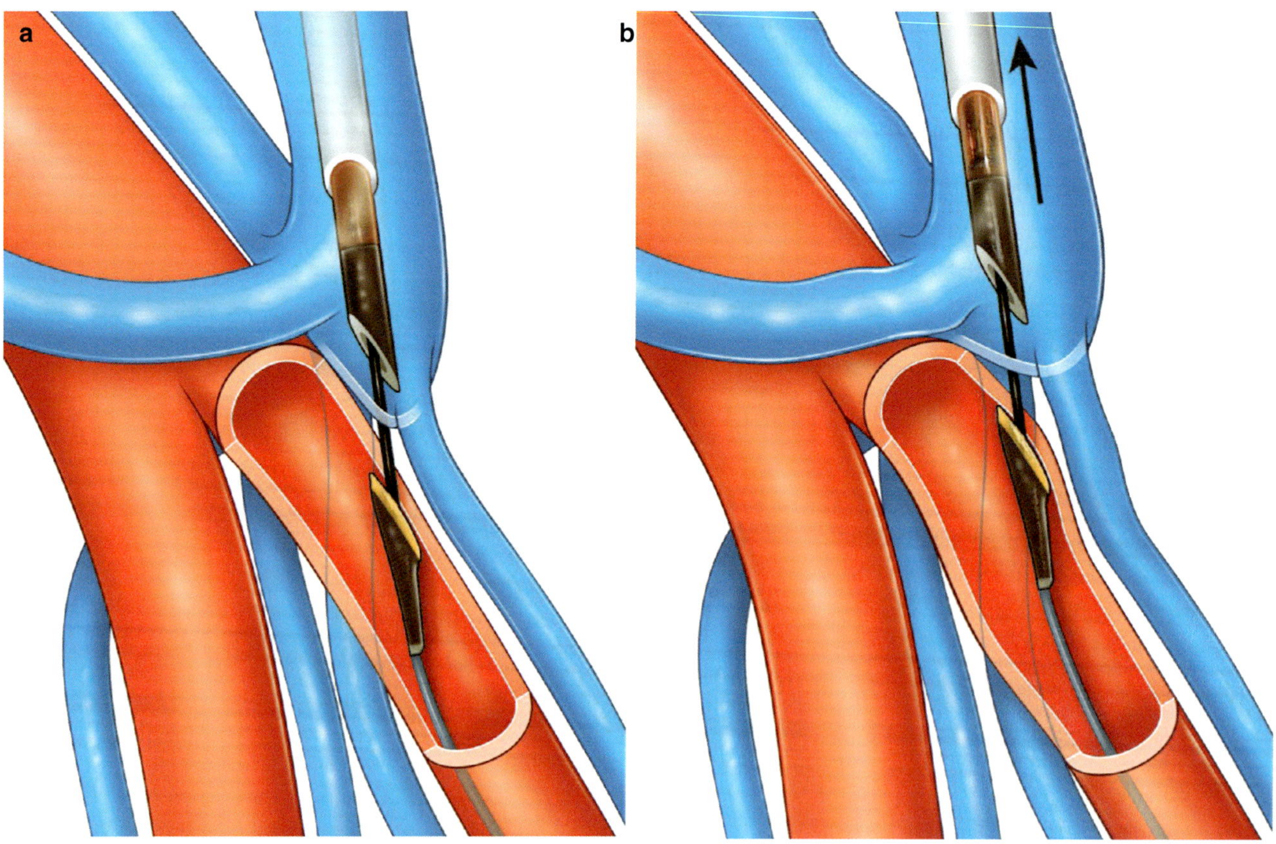

Fig. 36.3 (**a**–**d**) Stages of anastomosis creation with the Ellipsys device between the proximal radial artery and the perforating vein of the elbow—reproduced with permission from Mallios et al. J Vasc Surg [11]

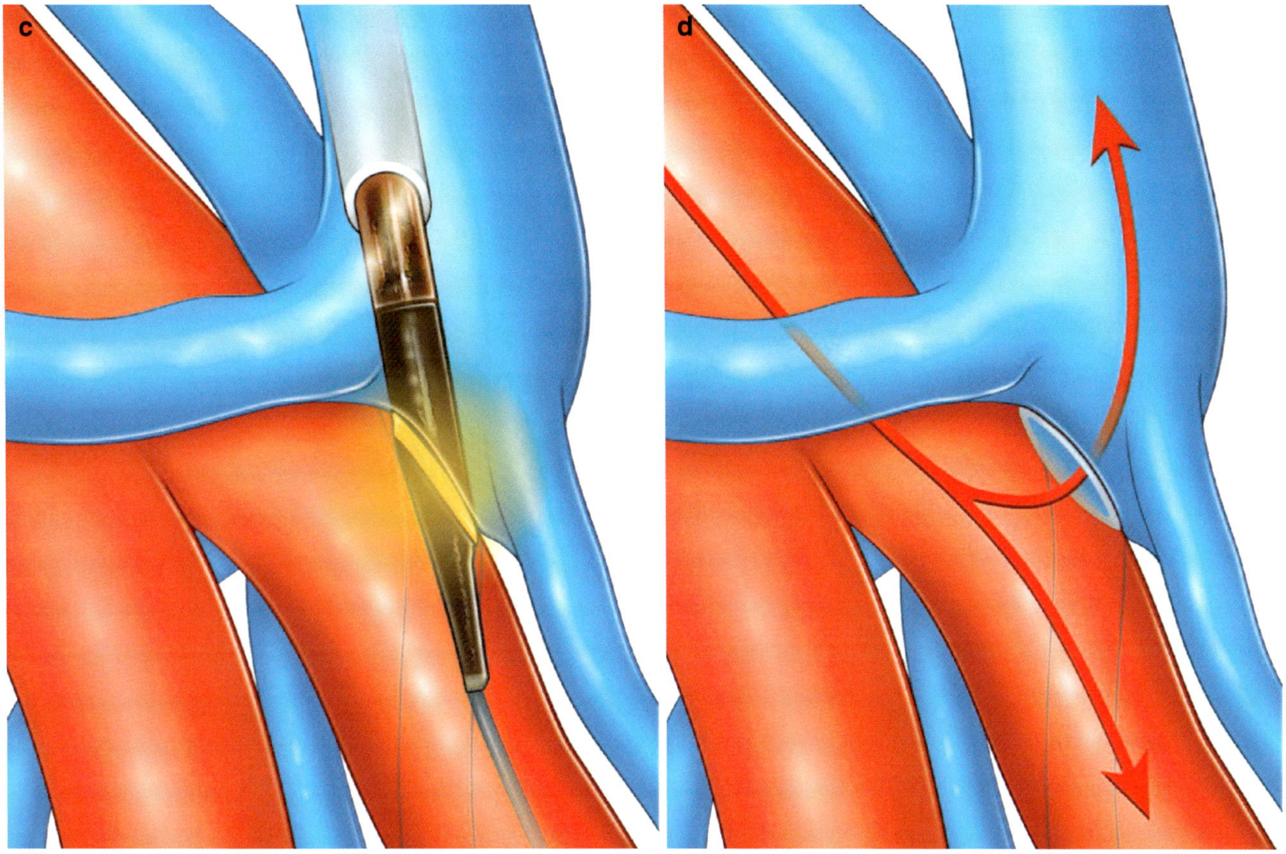

Fig. 36.3 (continued)

Access Placement Techniques

Both the WavelinQ and Ellipsys devices have undergone minor changes in technique since their initial introduction onto the market. The WavelinQ is frequently performed under moderate conscious sedation with infiltration of local anesthetic. In patients not suitable for moderate sedation, a regional block is recommended. Initially ultrasound is used to identify the desired location of the fistula based on vessel diameter. In the first iteration, this was preferentially at least 4 cm proximal to the elbow. Next, the brachial vein is accessed using a micropuncture kit after placing a tourniquet proximal to the access site. The brachial artery is then similarly accessed. Both micropuncture sets are exchanged for 4 or 5 Fr slender sheaths and brachial arteriogram and venograms completed. A 0.014″ wire is then positioned in the ulnar artery. Placing the arterial wire first allows selection of which of the paired ulnar veins should receive the second 0.014″ wire. After selecting the correct vein, a catheter is advanced into the vein for ulnar venography. This allows confirmation of anatomy, presence and location of the essential perforator, identification of the dominant superficial venous drainage and planned site for AVF creation. Next, the device's 4 Fr rapid exchange catheters are placed into the ulnar artery and vein. The attraction between earth magnets in these rapid exchange catheters facilitates and maintains correct alignment and positioning. The catheters also provide visual cues to confirm this alignment. One is the presence of square-shaped lucency in the rotational indicators at the tip of each of the rapid exchange catheters [1, 7–9, 12, 13]. It is recommended that this is confirmed in at least two obliquities (the second being a minimum of 45° from the first). The second visual cue is achieved by gentle forward movement of the venous catheter which indents the convex electrode in this catheter over the concave ceramic plate in the arterial catheter. When proper positioning is confirmed, the venous catheter is connected to an RF generator and the wires pulled back to be proximal to the alignment zone of the convex electrode and concave ceramic plate. The generator is then activated using a 0.7-s pulse effectively vaporizing the tissue at the electrode-plate interface and creating a side-to-side AVF. Successful AVF creation is confirmed via brachial sheath arteriography. This also allows documentation of flow within the perforator distributed to the superficial veins. Oft, the operator will embolize the access brachial vein to encourage further flow to the superficial system via the perforator. It is not uncommon to see minor contrast extravasation on this completion study immediately following AVF formation though this should resolve within 10 min [7, 12, 13]. Any observed ulnar vein spasm requires pharmacological treatment to prevent non-transient extravasation. More recently, the introduction of dual small caliber 4 Fr catheters have expanded access options to use the wrist as the access site for arterial cannulation [14].

The Ellipsys procedure is typically performed under supraclavicular block combined with sedation. After proximal brachium tourniquet placement, the medial cephalic/cubital vein is accessed with a micropuncture needle a few millimeters proximal to the perforating vein (PV). Using small movements with both the ultrasound probe and the needle, the micropuncture needle is advanced into and along the perforator with care taken to remain in the center of the vessel. Where the PV traverses closest to the proximal radial artery (PRA), the process is continued until the needle tip is advanced into the center of the PRA. There is clear tactile feedback when the radial wall is crossed. As long as the puncture has not been prolonged, the needle should now have low-pressure pulsatile flow. The 0.021″ wire of the 6 Fr slender sheath is then advanced through the needle and down to the distal radial artery at the wrist. The needle is removed, and the sheath is advanced over the wire such that it crosses into the artery. Now, 50 units/kg of heparin is administered. The wire is replaced with a 0.014″ Nitrex wire (ev3; Plymouth, MN). The Ellipsys catheter is placed over the wire into the sheath so that the tip of the device is in the artery and the base of the jaw in the vein [15]. The sheath is retracted to demonstrate the jaws of the device on ultrasound. After correctly orienting the device, it is gently retracted until it engages (in both tactile and visual perception) the wall of the PRA. The jaws of the device are then closed capturing the arterial and venous walls (Fig. 36.1). Appropriate capture and position is confirmed as the device generator displays the attendant distance measured between the jaws. The generator is then activated creating the anastomosis (Fig. 36.3). The device is removed and an 5 mm × 2 cm compliant angioplasty balloon is inflated for 2 min across the anastomosis. Fistula flow is then confirmed by duplex [16]. And all wires and devices are removed and a brief period of manual pressure ensues [2, 15].

Anatomic Applicability

Instructions for Use (IFU) for each device provide information regarding anatomic requirements. True applicability determination requires consideration of whether these criteria are simply necessary or both necessary and sufficient. WavelinQ's IFU requires ulnar vein and ulnar artery diameters both be a minimum of 2 mm and the distance between the ulnar vein and artery be less than 2 mm. The newer 4 Fr version of the WavelinQ device allows use of the radial artery and vein if they meet the same size and proximity requirements as those for described for the ulnar vessels [12].

The Ellipsys IFU requires that the proximal radial artery and perforating cubital vein both have diameters of at least 2 mm. The distance between the PRA and the perforating cubital vein must be less than 1.5 mm [9, 14, 17].

While these are the IFU-delineated requisites, the truth is that both techniques rely directly or indirectly upon the presence and adequacy of a perforating cubital vein. Both also require a minimum of one superficial upper arm draining vein (cephalic or basilic) for future cannulation. The minimum acceptable size is likely 2 mm for both required venous elements [9]. The distinction between IFU elements and anatomic essentials is important due to the evident wider spread applicability of WavelinQ based on IFU alone.

Considering a single Veterans' Affairs cohort, and applying IFU criteria and an algorithm necessitating the use of surgical distal radiocephalic fistulas followed by percutaneous/endo fistulas followed by antecubital fistulas and then grafts, a group evaluated both IFU-based and real world utility of pAVFs in 116 extremities in 58 men [9]. Starting with IFU criteria alone, they reported that 93% of extremities were suitable for WavelinQ and 52% for Ellipsys. However, 52% of these also had adequate vasculature to support a distal radiocephalic. When the authors added in real-world necessities such as adequate draining superficial vein and present and adequate perforating vein, system usability decreased to 55% for WavelinQ and 44% for Ellipsys. When they excluded those that should preferentially receive a distal radiocephalic, they discovered that 32% of the patients would be appropriate for WavelinQ and 23% for Ellipsys—given differing requirements, this resulted in an overall 69% rate of adequate anatomy for some form of percutaneous fistula; see Table 36.2.

Table 36.2 Anatomic Requirements for pAVFs

	Ellipsys	WavelinQ
IFU vessel diameter	Perforating cubital vein ≥2 mm PRA ≥2 mm	Ulnar (or radial) vein ≥2 mm Ulnar (or radial) artery ≥2 mm
IFU distance	Perforator-PRA <1.5 mm	Ulnar (or radial) vein-artery <2 mm
Non-IFU (real world)	Superficial outflow vein >2 mm	Superficial outflow vein >2 mm Perforating cubital vein presence and diameter >2 mm

PRA proximal radial artery

Early Results

Early results for the WavelinQ system have been reported both for the original 6 Fr and newer 4 Fr devices. These pivotal studies were performed with non-US patient populations. The Novel Endovascular Access Trial (NEAT) performed in Canada, Australia and New Zealand using the 6 Fr device reported a 98% technical success rate and 8% serious adverse event rate [18]. In terms of physiologic and clinically important outcomes, the investigators found 87% were physiologically suited for dialysis based on ultrasound-measured venous diameters and computed flow rates. Among those patients on HD, they reported a 64% functional usability. Twelve-month primary and cumulative patency were 69% and 84%, respectively [18]. A real-world application study looked at short term (mean follow-up 73 days) outcomes in a single center using WavelinQ [19]. In their 32-patient series, the authors reported 100% technical success, procedure lengths averaging 120 min and a 25% peri-procedural complication rate. Over their relatively short follow-up, they found that 42% of patients required further endovascular intervention to assist with maturation while 13% required surgical intervention. At the conclusion of their follow-up, mean flows were 1.1 L at the brachial artery and 447 mL at the cephalic. Despite this, only 48% of the patients receiving HD were using their pAVF at study termination [19].

Initial experiences with the 4 Fr WavelinQ devices have also been described. Two such studies, the EASE trials (EASE, 32 patients, NCT 03708770; EASE-2, 24 patients, NCT 03708562) were completed in Paraguay while the EU postmarket clinical follow-up study was performed in Germany (EU study, 64 patients, NCT 02682420) [13]. Aggregated results across these three studies demonstrated a procedural success rate of 97%, primary patency of 72%, primary-assisted patency of 81%, and secondary 6-month patency of 88% (KM extrapolation) [13]. The group reported 2.5% rate of serious adverse events and 6% procedurally related adverse events. Reinterventions were required in 19% of patients split relatively closely between maturation and maintenance procedures. Concomitant brachial vein embolization was required in 67% of patients. Among the 99 patients on HD at enrolment, 64% had successful cannulation at 3 months and 78% at 6 months. Functional cannulation (defined as successful cannulation and use for 2/3 of dialysis sessions over a 28-day period) occurred in 43% by 3 months and 53% by 6 months. The mean time to achievement of functional cannulation was 140 days [13].

Early results for the Ellispys device from the Pivotal Multicenter trial describe 95% technical success and no device related adverse events [17]. They reported cumulative patency of 92%, 89%, and 87% at 90, 180, and 360 days. Functional patency was achieved in 98%, 98%, and 92% at 90, 180, and 360 days. Two-needle cannulation was achieved in 88% of the patients on dialysis at study initiation. This was achieved at a mean of 100 days. Though technique has since changed, at the time of the pivotal trial, it was not routine practice to dilate the anastomosis with a 5 mm × 2 cm compliant balloon as it is currently. Thus, the pivotal study entailed 72% maturation procedure requirement. In addition to this angioplasty, 31% underwent cubital vein ligation, 32% brachial vein embolization and 26% surgical transposition. Of the 107 patients in the trial, 36 (34%) required 66 procedures to maintain functionality; see Table 36.3.

Table 36.3 Short-term follow-up outcomes

	Ellipsys	WavelinQ
Technical success	95–100%	97–100%
Procedural duration (min)	14–15	63–120
Cannulation rate (4 weeks)	68%	54%
Cannulation rate (6 months)	80%	79%
Time to cannulation (day)	60–100	49–140
Functional cannulation rate	88%	53%
Physiologic readiness by imaging	86%	87%
Failure rate	15%	37%
Need for additional interventions	28%	19–27%
6 months primary patency	32%	33–72%
12 months primary patency	54, 87%	69%
6 months primary assisted patency	–	81%
12 months primary assisted patency	85%	–
6 months secondary patency	82%	60, 70, 88%
12 months secondary patency	96%	–
Interventions per patient-year	0.96	0.46
Procedure-related adverse events	0, 2.5%	6, 8.6, 25%

Later Results

Results for mid-term and later have been published for Ellipsys. This is inclusive of a procedural change to include compliant balloon angioplasty (5 mm × 2 cm) at the index procedure and traversal of the perforating vein with the access needle rather than a wire (which can induce spasm). Mallios et al. published their single center results on 234 patients with a mean follow-up of 302 days [2]. This group had a 99% technical success rate and an average procedural duration of 15 min. Primary, primary assisted and secondary patency at 1 year were: 54%, 85%, and 96%. The most frequent secondary intervention comprised angioplasty of the anastomosis and draining perforator via distal radial arterial access (Figs. 36.4 and 36.5). Ten percent of patients did

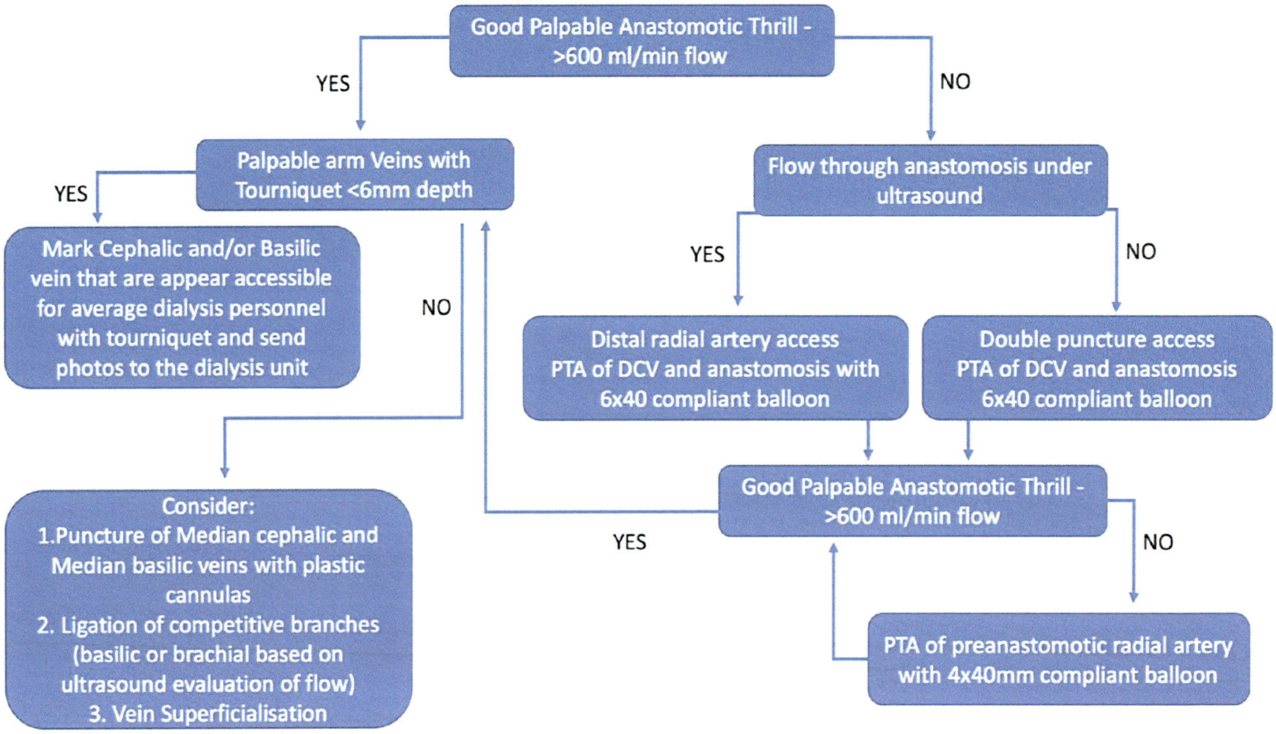

Fig. 36.4 Algorithm for maintenance and assisted maturation for percutaneous fistula created with Ellipsys—reproduced with permission from Mallios et al. J Vasc Surg [16]

require superficialization due to cannulation issues—split between basilic vein transposition and surgical lipectomy for deeper cephalic veins. Importantly, no patients developed draining vein stenosis of the cephalic arch or central venous outflow [2].

Five-year results from the Pivotal multicenter Ellipsys trial have also been reported by Hull et al. [20]. Eighty-five patients were available for long-term follow-, of which 84 had mature fistulas and 78/79 dialysis-dependent patients were using their pAVF with two needles at at least 2/3 sessions at the prescribed rate in a month. Over the 5 years of follow-up, maintenance interventions were required in 32%, resulting in 0.32 procedures per patient-year in years 2–5. Cumulative patency rates at 1 and 5 years were 90% and 82%. Functional patency rate at study-end was 92% [20].

Comparing Costs

Addition to the field of minimally invasive procedures always entails an initial increase in price of the index procedure related to device research, development and production costs. A notable example still expounded upon throughout the literature is endovascular aneurysm repair (EVR). Some initial analysis suggests that while pAVF has a clear early increase in cost compared to sAVF related to device expense, the lower rate of requirement for reintervention in the first year alone more than makes up for the initial financial investment [21, 22].

Relative Need for Reintervention and Regional Variation

The comparative need for reintervention between Ellipsys and WavelinQ is actually more challenging to assess using the extant literature than might be expected. Published results overall suggest a higher rate of interventions for either maturation or superficialization in Ellipsys as opposed to WavelinQ [13, 17–20]. As underscored by Illig et al.; however, the WavelinQ data for lower reintervention rates almost entirely originates from studies performed outside of the United States. In contrast, the studies suggesting higher rates of intervention in the Ellipsys patients are heavily weighted by studies in the United States (and, to some extent, Mexico) [8]. While it is confirmed that most US patients require further intervention to achieve functionality, this differs from the European data. Large series in Europe describe only 11% of patients requiring such intervention. Dr. Illig and colleagues suggest that there are likely both cultural and ana-

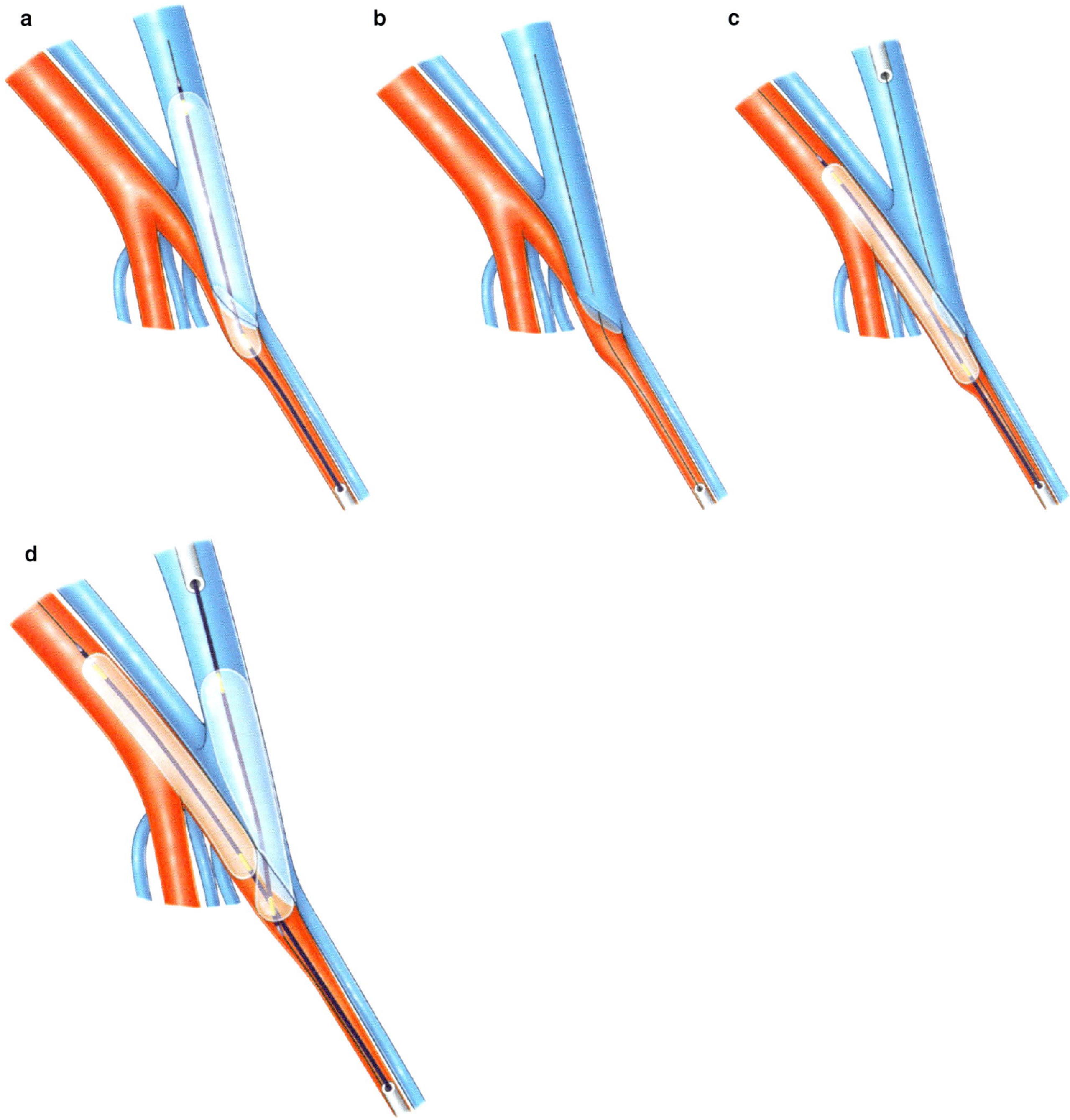

Fig. 36.5 (**a**–**d**) Angioplasty of pAVF with radial access can allow dilatation of both outflow and inflow separately or simultaneously if needed with dual access reproduced with permission from Mallios et al. J Vasc Surg [16]

tomic differences. Anatomic differences can be attributed to both relative rates of obesity and availability of usable superficial upper-arm veins. Cultural distinctions may be found in the expected rate of flow during dialysis sessions (for-profit dialysis centers vs. socialized health care) and cannulation practices.

Status of Use

While both devices have received both FDA approval and EU CE mark, this does not translate directly into reimbursement or insurance coverage. The original WavelinQ (Everlinq at the time) 6 Fr system CE marked in Europe in 2014 while the lower-profile 4 Fr verson received its marking in 2017. Ellipsys was CE marked in Europe in 2016. Both the original WavelinQ and the Ellipysis cleared the FDA in 2018 while the 4 Fr WavelinQ was FDA approved in 2019. From a production and marketing standpoint, the 4 Fr version superseded the 6 Fr version as a means to decrease the potential for brachial artery complications [13, 23].

In April of 2020 WavelinQ-4 Fr experienced a company-initiated "urgent medical device recall notification" related to issues with the magnets which translated into failure of catheter alignment [23, 24].

In the US, CMS initially reimbursed both the 6 Fr version of the WavelinQ and the Ellipsys system. Shortly after supplanting of the 6 Fr WavelinQ by the 4 Fr device, Center for Medicare and Medicaid Services (CMS) stopped reimbursement for WavelinQ though continues to reimburse for Ellipsys. CMS documentation delineates the reasoning behind non-coverage of the WavelinQ system as the requirement for more robust evidence in terms of patient number and duration of follow-up [23].

Some off-label uses of pAVF devices have also been described. Mallios et al. described use of the Ellipsys device for both a revision of dialysis arteriovenous fistula using distal inflow (RUDI) and for creation of radio-radial fistulas (with delayed brachial vein elevation in patients with poor superficial venous options) as well as for radiocephalic fistulas at the wrist in select patients [25–27].

References

1. Food and Drug Administration US. FDA permits marketing of first catheter-based systems used to create vascular access for hemodialysis patients. US Food and Drug Administration; 2018.
2. Mallios A, et al. Midterm results of percutaneous arteriovenous fistula creation with the Ellipsys vascular access system, technical recommendations, and an algorithm for maintenance. J Vasc Surg. 2020;72(6):2097–106.
3. Shahverdyan R, et al. Comparison of outcomes of percutaneous arteriovenous fistulae creation by Ellipsys and WavelinQ devices. J Vasc Interv Radiol. 2020;31(9):1365–72.
4. Jennings WC, Mallios A, Mushtaq N. Proximal radial artery arteriovenous fistula for hemodialysis vascular access. J Vasc Surg. 2018;67(1):244–53.
5. Tordoir JH, Dammers R, van der Sande FM. Upper extremity ischemia and hemodialysis vascular access. Eur J Vasc Endovasc Surg. 2004;27(1):1–5.
6. Wu CC, et al. The outcome of the proximal radial artery arteriovenous fistula. J Vasc Surg. 2015;61(3):802–8.
7. Berland TL, Clement J, Griffin J, Westin GG, Ebner A. Endovascular creation of arteriovenous fistulae for hemodialysis access with a 4 Fr device: clinical experience from the EASE study. Ann Vasc Surg. 2019;60:182–92.
8. Illig KA, et al. The role of surgery for assisted maturation after endovascular and percutaneous arteriovenous fistula creation. J Vasc Access. 2021;22(5):822–30.
9. Popli K, Dittman JM, Amendola MF, Plum J, Newton DH. Anatomic suitability for commercially available percutaneous arteriovenous fistula creation systems. J Vasc Surg. 2021;73(3):999–1004.
10. Hull JE, Velez JH, Martinez JP. Percutaneous proximal radial artery arteriovenous fistula creation for hemodialysis using the Ellypsis(TM) vascular access system. In: SIR 2014 annual scientific meeting program: convergence. San Diego, CA; 2014. p. S20.
11. Mallios A, Fonkoua H, Allouache M, Boura B. Percutaneous arteriovenous dialysis fistula. J Vasc Surg. 2020;71(4):1395.
12. BD/TVA Medical. WavelinQ EndoAVF System Instructions for Use (IFU). 2018.
13. Berland T, et al. Percutaneous arteriovenous fistula creation with the 4F WavelinQ EndoAVF system. J Vasc Surg. 2022;75(3):1038–46. e1033.
14. Abdel Aal AK, et al. Devices and techniques for percutaneous creation of dialysis arteriovenous fistulas. Semin Intervent Radiol. 2022;39(1):66–74.
15. Mallios A, et al. Early results of percutaneous arteriovenous fistula creation with the Ellipsys vascular access system. J Vasc Surg. 2018;68(4):1150–6.
16. Franco G, Mallios A, Bourquelot P, Jennings W, Boura B. Ultrasound evaluation of percutaneously created arteriovenous fistulae between radial artery and perforating vein at the elbow. J Vasc Access. 2020;21(5):694–700.
17. Hull JE, et al. The pivotal multicenter trial of ultrasound-guided percutaneous arteriovenous fistula creation for hemodialysis access. J Vasc Interv Radiol. 2018;29(2):149–58. e145.
18. Lok CE, et al. Endovascular proximal forearm arteriovenous fistula for hemodialysis access: results of the prospective, multicenter novel endovascular access trial (NEAT). Am J Kidney Dis. 2017;70(4):486–97.
19. Zemela MS, Minami HR, Alvarez AC, Smeds MR. Real-world usage of the WavelinQ EndoAVF system. Ann Vasc Surg. 2021;70:116–22.
20. Hull JE, et al. Long-term results from the pivotal multicenter trial of ultrasound-guided percutaneous arteriovenous fistula creation for hemodialysis access. J Vasc Interv Radiol. 2022;33:1143.
21. Arnold RJG, et al. Comparison between surgical and endovascular hemodialysis arteriovenous fistula interventions and associated costs. J Vasc Interv Radiol. 2018;29(11):1558–66. e1552.
22. Yang S, Lok C, Arnold R, Rajan D, Glickman M. Comparison of post-creation procedures and costs between surgical and an endovascular approach to arteriovenous fistula creation. J Vasc Access. 2017;18(Suppl. 2):8–14.
23. CMS. Percutaneous arteriovenous fistula (pAVF) for hemodialysis. 2020.

24. FDA. Class 2 Device Recall WAVELINQ 4F EndoAVF System. 2020. https://accessdata.fda.gov/scripts/cdrh/cfdocs/cfRES/res/cfm?id=17904.
25. Mallios A, Bourquelot P, Harika G, Boura B, Jennings WC. Percutaneous creation of proximal radio-radial arteriovenous hemodialysis fistula before secondary brachial vein elevation. J Vasc Access. 2021;22(2):238–42.
26. Mallios A, Jennings WC. Endovascular revision using distal inflow: EndoRUDI. Eur J Vasc Endovasc Surg. 2020;60(1):144.
27. Mallios A, Nelson PR, Franco G, Jennings WC. Creating percutaneous radiocephalic arteriovenous fistulas at the wrist. J Vasc Access. 2021;22(2):299–303.

Vascular Access

37

T. Turlejski, I. Mandal, T. Barge, and R. Uberoi

Case Presentation
A 40-year-old male was diagnosed with chronic kidney disease secondary to poorly controlled diabetes. Due to a continued decline in renal function, he required haemodialysis. Access was initially obtained via right internal jugular Tesio lines (Fig. 37.1) to facilitate dialysis, whilst a surgical arteriovenous fistula could be formed and matured.

Continued at page 405

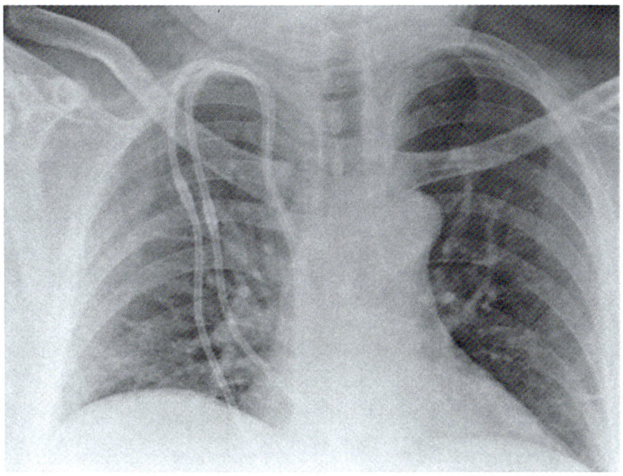

Fig. 37.1 Right internal jugular Tesio lines, with their tips in the right atrium

Introduction

Vascular access is critical to delivering a wide range of treatments, including intravenous medication, intravenous nutrition, repeated blood sampling and haemodialysis. Although access has been performed blind in limited circumstances, image guidance has become the standard of practice to optimise safety and technical success of the procedure. Various anatomical locations and devices can be chosen, depending on the clinical indication and patient-specific factors, although there are many overlapping principles related to the insertion of these devices. This chapter will summarise the principles underlying the techniques, site selection and equipment, and discuss techniques to manage challenging cases.

Principles of Vascular Access

Veins can be classified as central or peripheral based on their proximity to the heart. Central veins include the superior and inferior venae cavae, brachiocephalic veins, subclavian veins (SCVs), common and external iliac veins. These vessels are of large calibre, usually with high flow rates, and typically have minimal valves. Consequently, a catheter with its tip placed in a central vein can be used to deliver venotoxic medications such as chemotherapeutics or parenteral nutrition, achieve high flow rates required for haemodialysis or allow the measurement of central venous pressure—an estimate of the pressure within the right atrium.

Whilst centrally placed devices have advantages (in particular longevity), they are technically more difficult to place than peripheral devices, such as intravenous cannulas. Most venous access is therefore obtained with peripheral devices, which usually do not need image guidance and are placed at the bedside. Peripherally inserted central catheters (PICCs) are long lines inserted through a peripheral vein (typically the basilic or brachial), with the tip terminating in a central vein, usually the superior vena cava (SVC). These are quick

T. Turlejski · I. Mandal · T. Barge · R. Uberoi (✉)
Department of Radiology, John Radcliffe Hospital, Oxford, UK
e-mail: tymoteusz.turlejski@ouh.nhs.uk; indrajeet.mandal@ouh.nhs.uk; thomas.barge@ouh.nhs.uk; raman.uberoi@ouh.nhs.uk

G. Geroulakos et al. (eds.), *Mastering Endovascular Techniques*, https://doi.org/10.1007/978-3-031-42735-0_37

and relatively easy to insert and have become a common method to obtain central venous access.

Image Guidance

Image guidance is essential to plan the procedure, select a target and assess neighbouring structures. If available, prior imaging should always be reviewed. Ultrasound (US) guidance remains the gold standard, enabling real-time visualisation of the needle and assessment of the vessel (incl. flow with Doppler imaging) [1].

Broadly, two techniques can be used for US-guided vascular access: 'in plane' (with the probe oriented longitudinally along the course of the vein) and 'out of plane' (with the probe oriented perpendicularly to the vessel length). The former allows direct visualisation of the entire needle within the scan plane from skin entry to venous puncture. In the latter, only a small dot representing the cross-section of the needle can be identified, and care must be taken to accurately determine the angle of puncture relative to the depth of the vein. Mastering both approaches is crucial to establishing successful access in various situations, for example, where anatomical constraints prevent use of the in-plane technique due to the size of the probe footplate, in which case the out-of-plane method must be used.

Fluoroscopy is another useful modality and essential for more difficult central line access placement. It allows visualisation of the course of the wire and/or catheter in the central venous tree, which cannot be visualised with US. If available, fluoroscopy can be used to ensure accurate tip placement at the time of insertion and is essential in challenging situations, such as navigating central venous stenoses, which is discussed later. Fluoroscopy is however not mandatory and is infrequently used in most central venous catheter insertions performed outside of the angiography suite, such as central lines inserted in theatre or PICCs inserted at the bedside. In these cases, landmark measuring techniques can be used to estimate the position of the tip, and the final position may be subsequently confirmed on a radiograph after insertion.

Site of Access

The selection of site for vascular access is determined by the target of the procedure and patient characteristics, such as the presence of surgical incisions, scarring, skin infection, vessel occlusion or anatomical variants. Although image guidance with US remains the standard of care, direct palpation and use of anatomical landmarks should also be utilised.

Central Access

Internal jugular vein (IJV), external jugular vein, subclavian vein and femoral vein can all be used for central vascular access. The IJV is the vessel of choice (for both tunnelled and non-tunnelled lines) as it is associated with a lower risk of thrombosis and pneumothorax, especially compared to the subclavian vein [2]. The incidence of the latter can be significantly reduced by ensuring the needle tip is always visible under US guidance, and there is evidence it may be comparable at both access sites [3]. The right IJV is generally preferred due to its straighter course to the heart.

For IJV access, the patient should ideally be placed in a Trendelenburg position (supine with head slightly below the level of lower extremities), with the head turned away from the side of the target vessel. Image guidance with US is essential for safe IJV access due to its proximity to the common carotid artery. The IJV should be compressible, increase in diameter with the Valsalva manoeuvre and lie lateral to the common carotid in the majority of cases, although there is common anatomical variation, with some studies reporting medial position in up to 5.5% of patients [4, 5]. External jugular or subclavian veins can be considered if IJV access is deemed unsuitable (e.g. due to a thrombus). SCV access is commonly performed without image guidance due to the position of the overlying clavicle impeding US visualisation. However, the vessel is often seen on the US more laterally and can therefore be accessed with image guidance at this point. The rate of pneumothorax is greater with SCV access, partly as the landmark technique is often used. However, judicious use of the US can reduce this occurrence [3].

Similar principles should be applied with other sites for central venous access, including the femoral vein. The femoral vein typically lies medial to the femoral artery. Apart from being a common access site for intravenous procedures, use of the femoral veins is only used in emergency settings or as a last resort if attempts at other sites have failed. Femoral vein access is however associated with a higher risk of infection and is considered a less comfortable position for long-term access for the patient.

Peripheral Access

Peripheral veins typically used for PICC line insertion are generally restricted to the veins of the (typically non-dominant) upper extremity, with a preference given for the basilic vein. Although the brachial vein is a suitable candidate due to its size, it is usually the second choice due to the risk of injury to the adjacent brachial artery or median nerve. The cephalic vein provides the least suitable access point due to its smaller calibre and more tortuous path.

Types of Lines and Ports

The technique, site and equipment are determined by the specific purpose for which vascular access is established. The factors to consider include the expected time the line/catheter will be required for, as well as the size, number of lumens, maximum flow rates or injectable pressures required. The latter is an important consideration if power injectors may be used, such as pumps used to administer iv contrast for CT scans. These parameters are defined by the specific clinical need, and line selection should be determined on a case-by-case basis. A summary of the commonly used devices is outlined in Table 37.1.

A central line can take many forms, but essentially the definition is that the line tip ends within the central venous system. Usually, this is at the cavoatrial junction (the junction of the SVC and right atrium). The venous insertion site varies and can be peripheral (PICC line) or central (non-tunnelled CVC, tunnelled lines and ports).

Central veins have a higher flow compared to peripheral veins, which not only allows the delivery of higher flow rates of infusions (e.g. for haemodialysis) but also helps to dilute the medications. The latter is particularly helpful with more toxic agents, which could cause significant damage if given via peripheral veins. Conversely, a peripheral line is one where the line tip ends in a peripheral vein.

Peripheral Lines

Peripheral lines are very common. Most hospitalised patients will have a standard peripheral IV (commonly referred to as a 'cannula'). Midlines are also peripheral lines, but the tip lies in the axillary vein—this is larger than the smaller peripheral veins. The advantages of a midline compared to a peripheral IV are that they are more stable, can be used for longer and are available in two lumen configurations.

Central Lines

A peripherally inserted central catheter (PICC) is the only type of central access where the insertion site is into a peripheral vein (Fig. 37.2). PICC lines allow central access, and the advantage of peripheral insertion site means there is a lower risk of serious complications from line insertion. However, PICC lines are smaller in size, as the line needs to be small enough to insert into a peripheral vein.

Non-tunnelled central venous catheters (CVCs) are the main method used in acute critical care settings. Often these are referred to as 'central lines'. These carry a higher risk of bloodstream infection, so a line should be used only for 7–10 days. There are multiple different types of line, but typically they are multi-lumen catheters. CVCs with larger lumens can be used for temporary haemodialysis. They are typically of larger diameter compared to other CVCs to accommodate the large luminal diameters of the afferent and efferent limbs; this is critical to enabling the high flow rates required and is related to Poiseuille's law, which states that the flow of fluid through a lumen is related to the radius of the lumen to the power 4. Consequently doubling the diameter of a lumen increases its flow rate by 16 fold.

Tunnelled lines are a CVC, but the line is tunnelled from the venotomy site subcutaneously to exit at a different site in

Table 37.1 Commonly used types of venous catheter

	Indications	Type	Details of use	Examples
Peripheral catheters	• Parenteral delivery of nutrition • Antibiotics • Analgesics • Chemotherapy • Repeated blood transfusions	Peripheral intravenous access (PIV)	For short-term access (usually up to 1 week)	Venflon
		Midline catheter	For short-term access (usually 1–4 weeks)	Single- or double-lumen options available
Central catheters	• Administration of IV fluids or medications • Delivery of blood products, either in large quantities or over a prolonged period of time • Administration of medications that are harmful to peripheral veins (e.g., chemotherapy) • Long-term access to the central venous system for repeated procedures, such as blood sampling • Poor or inaccessible peripheral venous access • Emergent placement of central lines	Peripherally inserted central catheter (PICC)	For medium-term access (usually 1 month to 1 year). Not recommended for dialysis.	Single-, double- or triple-lumen options available
		Non-tunnelled central catheter	Short-term access when PIV not suitable—e.g. fluid resuscitation or CVP monitoring	Multi-lumen CVC (one to five lumens) Vas Cath (large diameter, used for dialysis)
		Tunnelled central catheter	For frequent long-term access when PICC not suitable	Hickman line (one to three lumens), Tesio line Perm Cath
		Implantable port	For infrequent long-term access	Port-a-cath

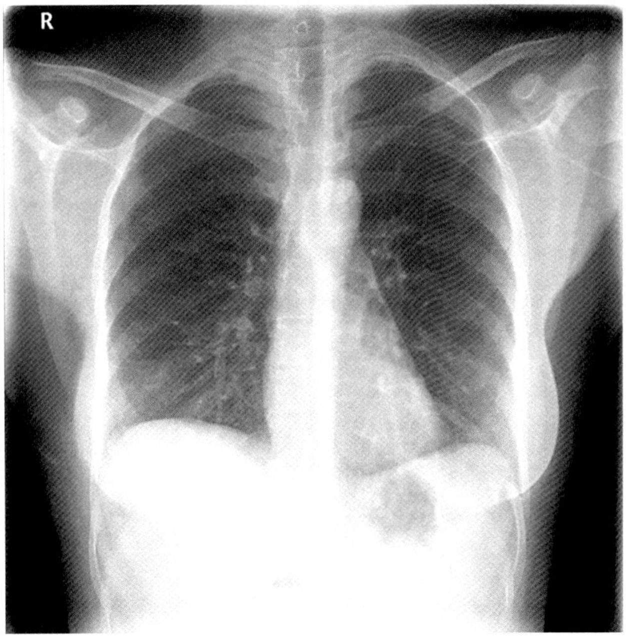

Fig. 37.2 Left-sided PICC line inserted into the basilic vein, with the tip appropriately sited in the SVC

the skin. The tunnelling lowers the risk of infection and increases the line stability, allowing the line to be used for longer-term access. The lines have an incorporated cuff, which promotes fibrosis and tethers the line in the subcutaneous tissues. Tunnelled lines require a high degree of care in the community—the lines need to be frequently flushed and dressed. They may also be inconvenient for the patient as the lumens lie usually on the anterior chest wall and are cosmetically obvious. The most common indications for tunnelled lines is haemodialysis and parenteral nutrition—these patients require long-term venous access. In general, for other indications that require long-term venous access where a PICC is not appropriate, a port may be preferable (see below).

Ports are effectively tunnelled lines, but at the end of the line, there is a reservoir attached to it (Fig. 37.3). The reservoir is buried in a pocket under the skin. They offer several advantages compared to tunnelled CVCs. Ports require less catheter maintenance in the community, and the line is more cosmetically discrete and may cause less disruption to the patient. They also carry a lower infection risk [6]. Additionally, the port site can be varied according to the patient's needs, e.g. it can be placed peripherally in the arm compared to the chest wall, although peripherally placed ports carry a higher thrombosis risk [7].

The main disadvantage of ports is they require specialist input to needle the chamber and access the circuit. Ports are also slightly more challenging to insert and remove compared to tunnelled lines. However, for patients requiring long-term access, it may be overall more cost effective to place a port rather than a tunnelled line when taking into account the cost of community care for tunnelled CVCs.

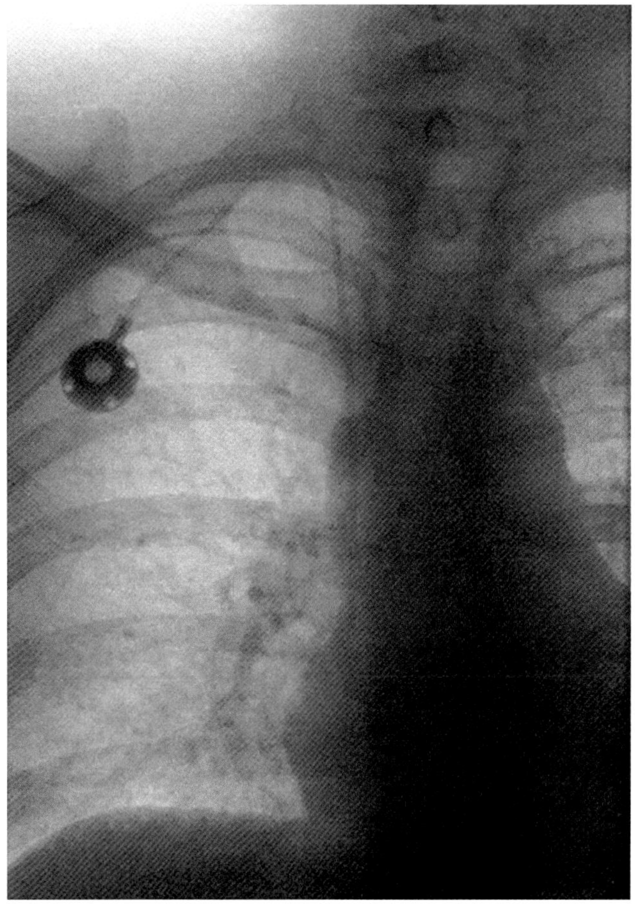

Fig. 37.3 Right-sided single-lumen port inserted into the right internal jugular vein, with the tip of the catheter at the cavoatrial junction

Complications

The risk of complications associated with venous access is less significant compared to arterial access due to thinner vessel walls and lower intraluminal pressures. The risks can be further minimised by the appropriate use of ultrasound and fluoroscopic guidance. Still, they are important to be aware of as they can be serious and potentially fatal.

Early Complications

- **Pneumothorax and haemothorax**—a rare but critically important complication of vascular access occurring in less than 1 in 10,000 cases. Traditionally pleural injury was considered more common with subclavian access compared to IJV access; however the adoption of US guidance as a standard of practice allowed to minimise the risk to comparable levels in the hands of experienced operators [3]. Pneumothoraces can be identified on fluoroscopy or on post-insertion CXR—small pneumothoraces can be managed conservatively, while larger ones may require chest drain insertion.

- **Accidental arterial puncture**—usually identified immediately as pulsatility of blood return from the vessel and by identifying the incorrect positioning of the needle/wire in the artery on the US. In case of uncertainty fluoroscopic assessment can be used to confirm anatomy.
- **Haematoma**—usually seen in coagulopathic patients.
- **Air embolism**—rare emergency, caused by entrainment of air into the central venous system, which may translocate to the right heart and severely impair right ventricular output, leading to potentially fatal consequences. Sudden onset of tachycardia, shortness of breath, chest pain or confusion should prompt an immediate assessment, oxygen therapy and placement in left lateral decubitus or head-down position. Some devices contain a one-way valve that prevents air entry, but it is a safe practice to reduce the risk of air entry by covering any large-bore catheter, needle or dilator with a finger to prevent air from getting sucked in.
- **Catheter malposition**—Inappropriate positioning of the catheter tip can occur and is more likely if real-time fluoroscopic guidance is not used. For example, blind wire insertion via the IJV can inadvertently pass into the SCV, rather than BCV/SVC, leading to the tip of the CVC lying in an incorrect location. Similarly, catheter tips placed at venous confluences can result in the tip abutting the vessel wall and impeding aspiration through the lumen. Inappropriate tunnelling is also important, as acute curves can lead to line kinking and impede luminal function. The vast majority of these can be identified on real-time fluoroscopy or on post-procedural radiographs. The advantage of the former is that catheter malposition can be rectified at the time of insertion.

Late Complications

- **Catheter-related thrombosis (CRT)**—occurs due to endothelial damage and venous stasis, frequently associated with a hypercoagulable state (particularly in cancer patients). The risk is increased in left-sided catheters and in the event of catheter malposition. Most patients are asymptomatic, but some may present with signs and symptoms of upper limb DVT. Like DVT, CRT is associated with a high risk of PE or post-thrombotic syndrome, so there is a low threshold for further investigation if CRT is suspected. It is worth noting that a concern of CRT is often raised on cross-sectional imaging, where a true CRT cannot be reliably distinguished from a fibrin sheath (below).
- **Fibrin sheath**—due to accumulation of fibrous material at the tip with long-term vascular access. Many catheters will eventually develop this, in most cases without any clinical significance. It may, however, cause a blockage, especially if there is concurrent intraluminal thrombosis, in which case tPA or urokinase infusion can be used in an attempt to unblock it. Manipulation with a wire, stripping with a snare under fluoroscopic guidance or over-the-wire replacement can also be used for this purpose if the catheter remains blocked.
- **Line entrapment or fracture**—entrapment of the catheter between the clavicle and the first rib may occur with 1 in 100 subclavian catheters [8]. Over time, it leads to catheter fracture due to repetitive injury, particularly in the setting of long-term access. The damaged catheter can leak its contents into the surrounding tissues and, in rare cases, detach completely, subsequently migrating into and embolising into the distal vasculature. Hence, this complication needs to be recognised early. Patients may complain of local discomfort and swelling due to fluid extravasation. The operator may notice positional difficulty in injection (easier with patient supine with ipsilateral arm raised). Classically, the scalloping of the catheter can be seen on a plain chest radiogram.
- **Infection**—can occur at the exit site, the tunnel or in the bloodstream (the latter associated with high mortality). Management depends on the type, but line removal or replacement is usually required if the catheter has become colonised. The incidence of catheter-associated bloodstream infections has been reported as 2.2–2.79 infections per 1000 catheter days, with a recent meta-analysis reporting a slightly lower risk if access was obtained with US guidance [9].
- **Catheter displacement or accidental removal.**

Difficult Situations and Troubleshooting

Central vein stenosis is a common finding in patients who have a history of multiple venous lines in the past, particularly in patients on haemodialysis. Venous access in these patients is more technically challenging and may require adjunctive techniques to secure adequate catheter positioning.

Patients with a history of repeated and/or multiple venous insertions should undergo cross-sectional imaging (most commonly with a CT venogram but MRV may be utilised) to evaluate the veins for stenoses or occlusion. This imaging allows for optimal pre-procedure planning. This will help to determine whether adjunctive techniques to insert a line using a more conventional location, such as venoplasty/recannalisation of a central vein, are required or whether an alternative access site should be sought.

Recanalisation Methods

Whilst a full discussion on this topic is beyond the scope of this chapter, a summary of the options is discussed below.

Venous Stenosis

Central venous stenosis occurs when there are points of narrowing in the central veins. Balloon venoplasty and/or stenting can be performed—the aim is to open the vessel enough to allow the placement of a peel-away sheath for line insertion. Angioplasty is the best for shorter lesions. Stents can be placed as well, although they are usually undertaken after the recurrence of a stenosis that has previously had balloon angioplasty treatment.

Occlusion

Recanalisation of an occluded vein lumen may be possible if there are proximal and distal patent veins with occlusion across the middle. The aim of endovascular treatment is to cross the occlusion, which then facilitates venoplasty and passage of the line across the site of occlusion.

In the first instance, blunt recanalisation should be attempted. This involves attempting to pass a hydrophilic wire and catheter across the occlusion. This may require both antegrade and retrograde approaches. In most cases the occluded veins are the BCV and/or SVC, and access from the IJV and femoral veins may be required. Insertion of a long sheath up to the point of occlusion is useful to provide support whilst attempting recanalisation. If the wire and catheter passes the occlusion, exchange for a stiff wire is usually performed, followed by balloon venoplasty +/− stenting to allow the passage of the catheter to an appropriate location.

If the wire fails to pass the occlusion, further techniques can be attempted, although these carry a higher risk of complication, particularly vessel perforation, bleeding and damage to surrounding structures. Some operators advocate the use of the 'back end' of a hydrophilic wire, which is typically stiffer and sharper than the front end. This may be referred to as 'semi-sharp' recanalisation.

Sharp recanalisation involves the use of an endovascular needle to cross the occlusion. Both antegrade and retrograde access are required, and a sheath and catheter are advanced to the point of occlusion. Contrast is then injected to determine the extent of occlusion and confirm that there is a straight connection between the two catheters. An endovascular needle is then advanced down one of the catheters and used to cross the occlusion under fluoroscopic guidance. Different devices have been described and include the use of a Rosch-Uchida TIPPS needle [10], Chiba biopsy needles [11] and the GoBack re-entry device [12]. The complication rate is around 2.4–5.1%, with the main complications related to bleeding (mediastinal haematoma, haemothorax, pericardial effusion) [11], and operators need to be prepared to emergently place stent grafts if using these advanced techniques.

Alternative Access Points

Translumbar IVC Lines are an option to access the IVC directly and are usually reserved for the thoracic deep venous system, which is extensively occluded. The patient is positioned prone, and a puncture site is chosen superior and lateral to the right iliac crest. The IVC is then accessed, usually under CT guidance, with a long 18 gauge needle and stiff guidewire [13]. Wires can then be passed into the IVC, and the line(s) are tunnelled and inserted as per usual processes.

Transhepatic lines are another alternative access site. For these lines, the right or middle hepatic vein is accessed under ultrasound guidance usually with a micro-access system. Once access is gained, a guidewire is advanced into the IVC, and the line is placed over the guidewire. These lines have a relatively high rate of thrombosis [14], and the long-term complication rate is unknown.

Merit Surfacer This is a device designed to obtain an extra-anatomical tract from a distal SVC stump (where the vessel superiorly is occluded) to the anterior chest wall to obtain access for line insertion and should only be used in centres with extensive experience. Femoral venous access is obtained, and a sheath is placed at the distal SVC stump. The Surfacer device is then inserted. This comprises an obliquely pointing needle, which is aligned under fluoroscopy with a radio-opaque marker placed at the intended chest wall exit site. This is slightly lateral to the midline on the anterior chest wall below the clavicle. The trajectory should then be through the relatively avascular anterior mediastinum. Pre-procedural CT is essential to ensure there are no unexpected structures or large venous collaterals in the path of the planned tract. The needle is then advanced to puncture the anterior chest wall from inside-out. A wire is then passed down the needle and retrieved at the distal end of the device to obtain through-and-through wire access. A tunnelled line can then be inserted over this wire.

Patients who have occluded IVC, iliac or femoral veins are not suitable for the Surfacer due to difficulty in advancing the device to the central veins. Tortuous vessels are also a relative contraindication [15].

Removals

Although most catheters are straightforward to remove, there are instances where additional steps are necessary, particularly for safe retrieval of long-term lines and devices. As discussed before, lines will frequently develop a fibrin sheaths and scar tissue, which may require a cut-down technique over the cuff site. The location of the cuff may sometimes be difficult to determine, but gentle palpation coupled with applying light traction on the catheter may be helpful. Ultrasound can also demonstrate the echogenic cuff position. After applying a local anaesthetic, an incision should be made carefully to access the fibrous sheath, which allows the intravascular portion of the catheter to be removed. The cuff and the outer portion of the catheter can then be safely retrieved, and the length of the catheter should be confirmed to ensure complete removal.

Similarly, ports can be retrieved by incising the surrounding fibrous sheath (with the use of local anaesthetic). There may be several sutures which attach the port to the surrounding tissues, although this can vary by manufacturer and operator. The sutures need to be cut prior to removal, following which the port can be safely removed in a similar fashion to an indwelling line.

The complications of catheter and port removal are rare but might include infection, damage to the device, embolisation of the distal vascular structures and scarring of the local tissues (leading to undesirable cosmetic results).

Fistulas

Fistulas are the preferable route of access for haemodialysis, predominantly due to the lower rates of infection and thrombosis compared to tunnelled dialysis catheters. A fistula circuit is formed by anastomosis of a vein or synthetic graft from an artery to a central vein. A high flow rate through the fistula is required to allow successful haemodialysis. Most commonly, this is performed between the radial and brachial arteries in the upper limb. The cephalic vein is usually chosen due to its lateral and superficial course, which makes it preferable for needle access for dialysis. Fistulas are most commonly formed surgically, and the site of anastomosis is decided upon based on pre-procedural workup and imaging. More recently, two devices have been developed to allow the formation of endovascular fistulas [16, 17]. In some patients, there may be no adequate vein available to form a fistula or surgeons prefer to have an access that is immediately usable. In these patients, a synthetic graft may be inserted subcutaneously and anastomosed onto the chosen artery and outflow vein to form a conduit for needling. Endovascular interventions may be required at several stages, from early non-maturation to later complications of grafts and fistulas. A comprehensive review of this is outside the scope of this chapter but is summarised below. There are more comprehensive reviews available that discuss the endovascular management of fistula circuits in more detail [18–21].

Stenoses

As discussed above, some patients may have significant central venous stenoses or occlusions, particularly if they have had multiple dialysis catheters previously. In these cases, there may be no adequate central venous outflow to ensure adequate fistula function. The Merit HeRO graft can be used to form a conduit to facilitate central outflow for a fistula [22]. This consists of a nitinol-reinforced silicone catheter, which is inserted in a similar fashion to a tunnelled central line. The tip typically lies in the right atrium. The extravascular component is tunnelled from the venotomy site, typically in the IJV or SCV, to a point on the chest wall or axilla, where it is anastomosed onto a vein, or synthetic graft, itself, which has been anastomosed onto the arterial inflow, thus completing the dialysis circuit.

Fistula circuits are prone to developing stenoses, causing dysfunction and limiting circuit flow which may impede dialysis and increase the risk of fistula thrombosis. Stenoses can occur at any site in the circuit, from the arterial inflow, throughout the needling zone, in the venous outflow and in the central veins. These are typically diagnosed with a combination of clinical features and confirmed on US or invasive fistulograms. Clinically, arterial inflow stenosis may be suspected if there is a poor flow rate or a soft fistula that is difficult to needle. In comparison, venous outflow stenosis may be suspected if the fistula is firm or 'thumpy' and if there is prolonged bleeding following dialysis.

The first-line treatment for stenoses associated with fistulas is endovascular plasty. This involved the insertion of a sheath into the fistula, directed towards the site of stenosis. A wire is then used to cross the stenosis and is plastied to improve luminal diameter and in-line flow. Stenoses can be resistant to plasty and may require the use of high-pressure balloons or scoring/cutting balloons to open the fibrotic stenosis. Stenting may also be performed, but it is important to consider the impact of this on the circuit. For example, stenting near the arterial anastomosis is generally inadvisable as it may compromise future surgical revision, which stenting of the needling segment should be avoided due to the risk of stent damage during subsequent dialysis needling. Typically covered stents are preferred due to their increased patency and reduced restenosis rates [23]. Central stenoses are managed in a similar fashion to that described above.

Unfortunately, there is a significant recurrence rate following fistuloplasty, with 12-month post-intervention patency rates of between 26 and 64% [24, 25]. Consequently, patients may require repeated intervention to maintain circuit patency, although in some situations, surgical revision may be required. Drug-eluting balloons (DEBs) have been demonstrated to improve patency rates [26]; however, the controversy related to their use in the femoropopliteal angioplasty remains. Whilst there is no evidence to suggest DEB use in fistulas is linked to increased all-cause mortality, patients should be specifically consulted and consented to for their use.

Fistula Thrombosis

A consequence of fistula stenosis is circuit thrombosis. Access is of critical importance for dialysis-dependent patients, and emergency declotting to rescue the fistula is often required. Traditionally, this has been done by means of surgical thrombectomy; however, there is increasing use of percutaneous thrombectomy techniques, including thrombolysis, thrombo-aspiration and the use of mechanical thrombectomy devices. This has the advantage of both allowing thrombus clearance and treating any underlying stenoses to both improve fistula function and reduce the chance of short-term re-thrombosis. Access for percutaneous thrombectomy can be either by direct fistula puncture or by internal jugular puncture and navigation into the fistula circuit in a retrograde fashion (Figs. 37.4, 37.5, and 37.6).

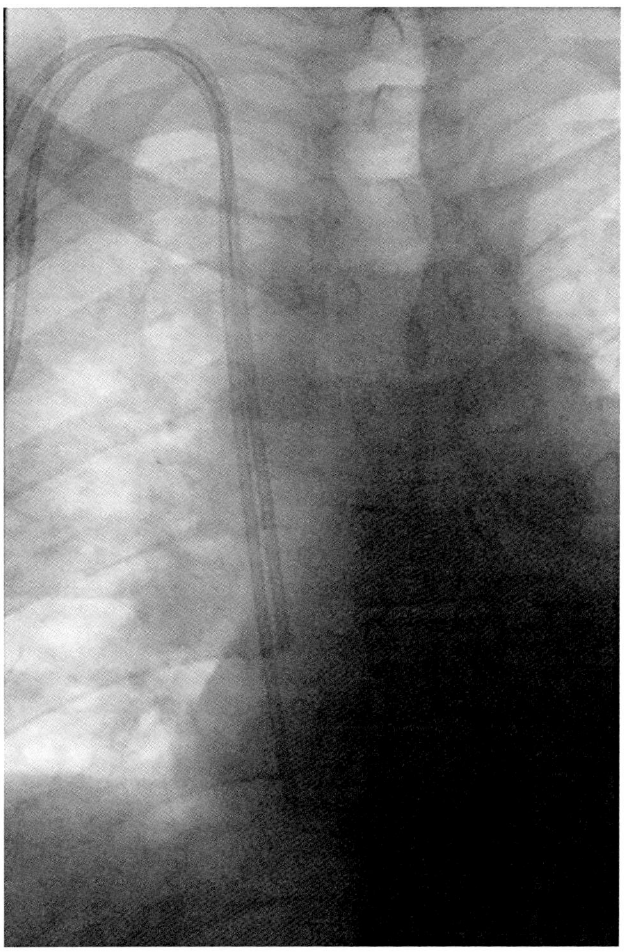

Fig. 37.4 Right-sided Tesio lines inserted into the right internal jugular vein and tunnelled to the anterior right chest wall. Note the separation between the tips of the two lines. This is to minimise recirculation during haemodialysis. The cuff of the more medial line can be seen on the extreme left of the image (overlying the patient's right chest wall)

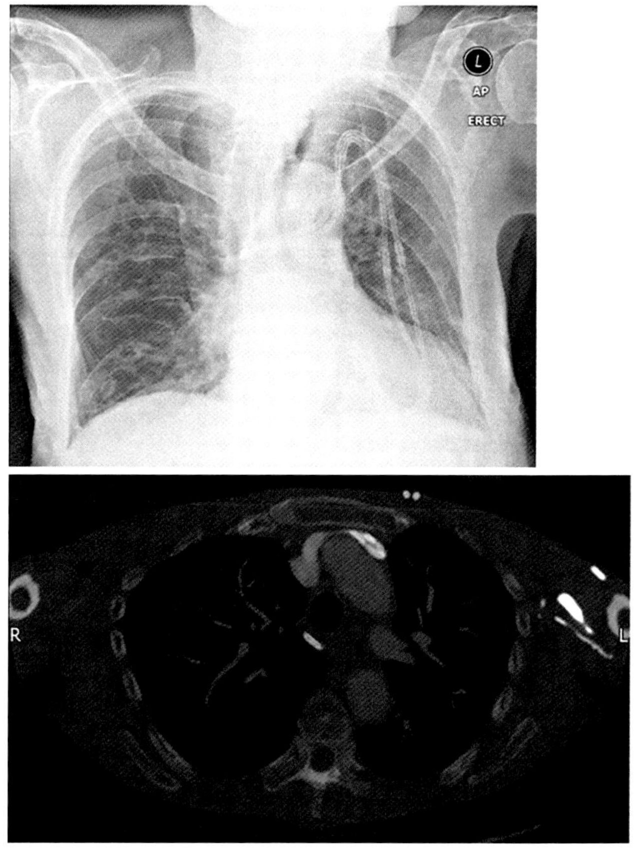

Fig. 37.5 Chest XR demonstrating the tip of one of the Tesio lines deviating superiorly. An axial slice from a subsequent CT demonstrates the tip of the line lying in the azygous vein

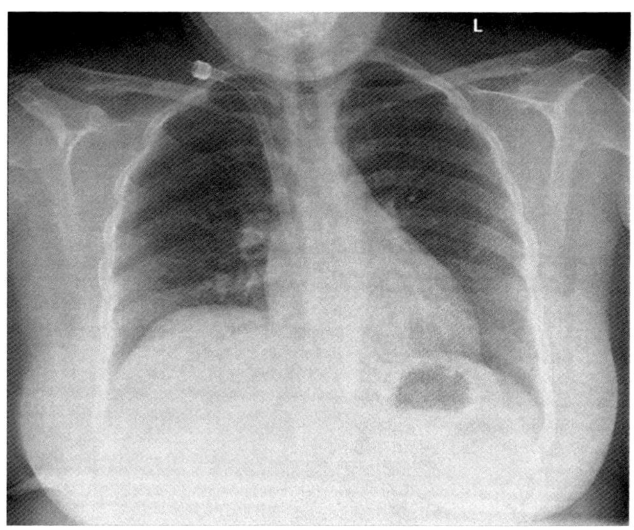

Fig. 37.6 Right-sided HeRO graft, which has been inserted via the right internal jugular vein, with the tip on the right atrium. The line has then been tunnelled laterally from its insertion site and joined to a surgically implanted PTFE graft to complete the fistula circuit

Conclusion

Vascular access is critical for ensuring effective delivery of intravenous medications and providing other therapies such as haemodialysis.

Case Presentation

Continued from 397

The patient underwent surgical formation of a left brachiocephalic fistula. However, the fistula failed to mature sufficiently for effective dialysis, and a new right fistula was formed on the other arm. In the interim, a new left-sided Tesio dialysis line was inserted (Fig. 37.7), whilst the new fistula matured. Over the next 8 years, haemodialysis was performed via the right-sided fistula; however, unfortunately this too eventually failed, and a new left-sided Tesio was inserted. A combination of direct venography and CT venography was performed to plan new access for dialysis. This demonstrated occluded right and stenosed left brachiocephalic veins (Fig. 37.8). Ultrasound assessment had also demonstrated no suitable venous conduit for fistula formation. Consequently, a combined left upper arm graft and central outflow conduit was planned. This involved the insertion of a HeRO graft via the left internal jugular vein to form the central outflow component and a subcutaneously tunnelled synthetic Flixene graft anastomosed to the left brachial artery to form the fistula circuit. Due to the left brachiocephalic stenosis, the vessel was prepared with high-pressure venoplasty to 8 mm to facilitate passage of a 20F peel-away sheath (Fig. 37.9). The HeRO graft was then inserted through the peel-away sheath, with the tip positioned in the right atrium (Fig. 37.9). The subcutaneous component was then joined to the tunnelled Flixene graft to complete the fistula circuit.

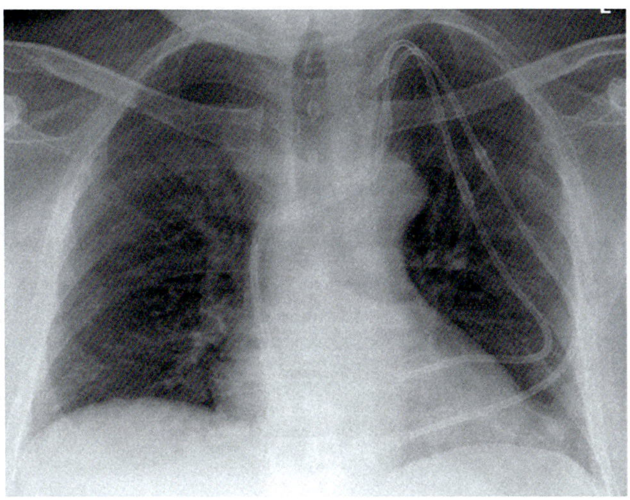

Fig. 37.7 Left internal jugular Tesio lines inserted subsequently following the failure of the initial left-sided arteriovenous fistula to facilitate haemodialysis, whilst a new right-sided fistula was fashioned and matured

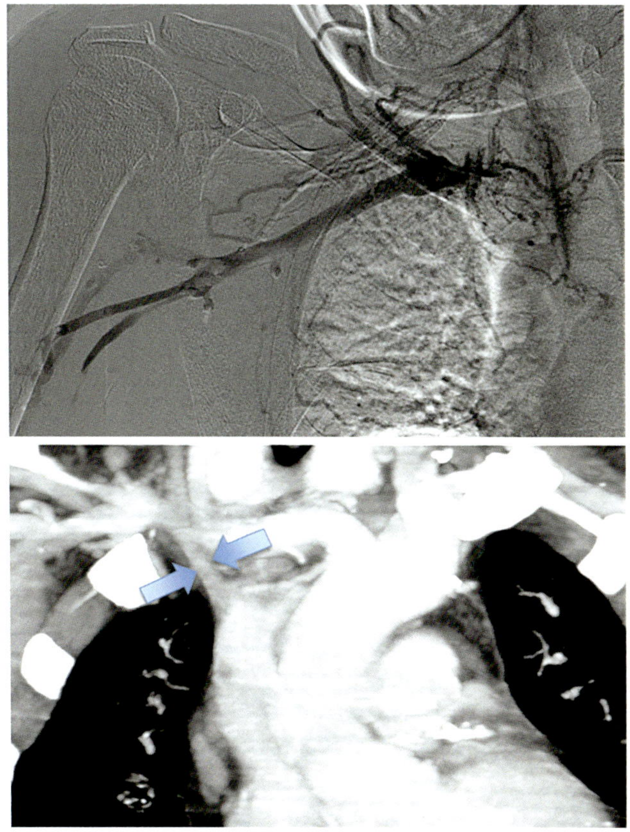

Fig. 37.8 Direct venograms and a selected coronal image from a CT venogram demonstrating obstructed right brachiocephalic vein, subsequent to prior right-sided haemodialysis lines and a right arteriovenous fistula

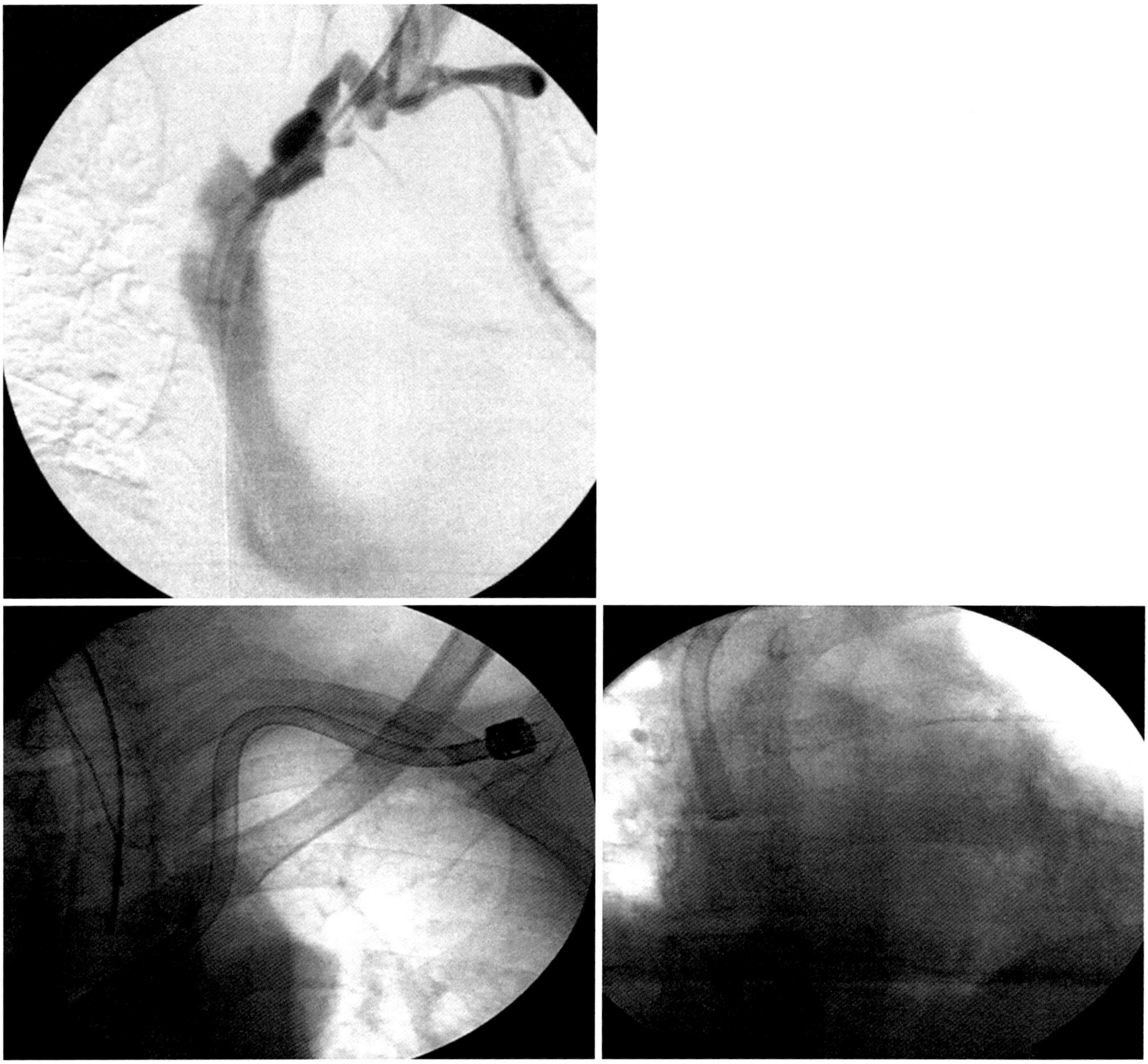

Fig. 37.9 Intraoperative images during the insertion of the HeRO central outflow conduit demonstrating wire access across the stenotic left brachiocephalic vein. This was venoplastied to 8 mm to facilitate passage of the HeRO graft, which was positioned with its tip in the right atrium

References

1. Hind D, et al. Ultrasonic locating devices for central venous cannulation: meta-analysis. BMJ (Clinical research ed.). 2003;327(7411):361.
2. Trerotola SO, et al. Tunneled infusion catheters: increased incidence of symptomatic venous thrombosis after subclavian versus internal jugular venous access. Radiology. 2000;217(1):89.
3. Shin HJ, et al. Complications in internal jugular vs subclavian ultrasound-guided central venous catheterization: a comparative randomized trial. Intensive Care Med. 2019;45(7):968.
4. Maecken T, Grau T. Ultrasound imaging in vascular access. Crit Care Med. 2007;35(5 Suppl):S178.
5. Asari AIA, Barros RAV, Borges MAP. Anatomic variant of the internal jugular vein and its importance in vascular access for hemodialysis. Jornal vascular brasileiro. 2019:18.
6. Coady K, et al. A comparison of infections and complications in central venous catheters in adults with solid tumours. J Vasc Access. 2015;16(1):38.
7. Kuriakose P, Colon-Otero G, Paz-Fumagalli R. Risk of deep venous thrombosis associated with chest versus arm central venous subcutaneous port catheters: a 5-year single-institution retrospective study. J Vasc Interv Radiol. 2002;13(2 Pt 1):179.
8. Andris DA, et al. Pinch-off syndrome: a rare etiology for central venous catheter occlusion. JPEN J Parenter Enteral Nutr. 1994;18(6):531.

9. Takeshita J, et al. Incidence of catheter-related bloodstream infections following ultrasound-guided central venous catheterization: a systematic review and meta-analysis. BMC Infect Dis. 2022;22(1).
10. Honnef D, et al. Sharp central venous recanalization by means of a TIPS needle. Cardiovasc Intervent Radiol. 2005;28(5):673.
11. Chen B, et al. Sharp recanalization for treatment of central venous occlusive disease in hemodialysis patients. J Vasc Surg Venous Lymphat Disord. 2022;10(2):306.
12. Rodriguez LE, et al. Sharp recanalization with the upstream GoBack catheter for chronic occlusive Ilio-Caval thrombosis. Ann Vasc Surg. 2021;74:74.
13. Power A, et al. Translumbar central venous catheters for long-term haemodialysis. Nephrol Dialysis Trans. 2010;25(5):1588.
14. Stavropoulos SW, et al. Percutaneous transhepatic venous access for hemodialysis. J Vasc Interv Radiol. 2003;14(9 Pt 1).
15. Baetens TR, et al. A novel technique to restore access in patients with central venous occlusion using the Surfacer® inside-out® access catheter system. J Vasc Access. 2020;21(5):778.
16. Berland T, et al. Percutaneous arteriovenous fistula creation with the 4F WavelinQ EndoAVF system. J Vasc Surg. 2022;75(3):1038.
17. Hull JE, et al. The pivotal multicenter trial of ultrasound-guided percutaneous arteriovenous fistula creation for hemodialysis access. J Vasc Interv Radiol. 2018;29(2):149.
18. Nassar GM, et al. Endovascular treatment of the "failing to mature" arteriovenous fistula. Clin J Am Soc Nephrol. 2006;1(2):275.
19. McLennan G. Stent and stent-graft use in arteriovenous dialysis access. Semin Interv Radiol. 2016;33(1):010.
20. Park HS, Choi J, Baik JH. Central venous disease in hemodialysis patients. Kidney Res Clin Pract. 2019;38(3):309.
21. Abreo K, Buffington M, Sachdeva B. Angioplasty to promote arteriovenous fistula maturation and maintenance. J Vasc Access. 2018;19(4):337.
22. Katzman HE, et al. Initial experience and outcome of a new hemodialysis access device for catheter-dependent patients. J Vasc Surg. 2009;50(3):600.
23. Shemesh D, et al. Angioplasty with stent graft versus bare stent for recurrent cephalic arch stenosis in autogenous arteriovenous access for hemodialysis: a prospective randomized clinical trial. J Vasc Surg. 2008;48(6):1524.
24. Elramah M, et al. Dialysis access venous stenosis: treatment with balloon angioplasty 30-second vs. 1-minute inflation times. Hemodial International. 2015;19(1):108.
25. Heye S, et al. Factors influencing technical success and outcome of percutaneous balloon angioplasty in de novo native hemodialysis arteriovenous fistulas. Eur J Radiol. 2012;81(9):2298.
26. Irani FG, et al. Hemodialysis arteriovenous fistula and graft Stenoses: randomized trial comparing drug-eluting balloon angioplasty with conventional angioplasty. Radiology. 2018;289(1):238.

Endovascular Techniques to Manage Vascular Access Failure

38

Dana B. Semaan, Salim G. Habib, Lena L. Vodovotz, and Theodore H. Yuo

Case Presentation

A 79-year-old female with end-stage kidney disease (ESKD), receiving hemodialysis through a left brachial artery to axillary vein arteriovenous graft (AVG), presents with an inability to receive hemodialysis (HD) due to thrombosis of the AVG.

Continued at page 417

Introduction

The primary treatment for patients with end-stage renal disease is hemodialysis (HD) [1]. The Kidney Disease Outcomes Quality Initiative of the National Kidney Foundation (KDOQI) strongly recommends the creation of a surgical arteriovenous fistula (AVF) or, alternatively, placement of an arteriovenous graft (AVG) for long-term HD vascular access. The term arteriovenous (AV) access will be used to refer to AVF and AVG, collectively. AV access is widely seen as superior to a tunneled dialysis catheter, when it is possible [2]. Maintaining a functional AV access is critical for patients on HD, and AV access dysfunction has been associated with substantial morbidity and mortality [2]. This chapter will cover the use of endovascular techniques to manage AV access failure, including delayed maturation, degradation of function of the AV access, and, in severe cases, complete thrombosis of the AV access [3–5].

D. B. Semaan · S. G. Habib · T. H. Yuo (✉)
Division of Vascular Surgery, University of Pittsburgh Medical Center (UPMC), Heart and Vascular Institute, Pittsburgh, PA, USA
e-mail: yuoth@upmc.edu

L. L. Vodovotz
University of Pittsburgh Medical School, Pittsburgh, PA, USA

Failure of Maturation, Stenoses, and Thromboses

Maturation failure is often related to stenosis of the circuit at any point along its path, including the arterial inflow, the conduit used during cannulation, and the central venous outflow. Stenoses may develop due to intimal hyperplasia and can progress to thrombosis [5]. Endovascular management of stenosis is typically through percutaneous transluminal angioplasty (PTA) with the addition of stent placement in selected patients [5]. Should the arteriovenous access thrombose, endovascular treatment options include thrombolysis, mechanical thrombectomy, or a combination of both [6]. Another cause for maturation failure includes the presence of accessory or competing veins which divert flow away from the main venous outflow, preventing it from maturing adequately [5, 7]. The endovascular option for management of accessory veins is obliteration through coil embolization [7].

Percutaneous Transluminal Angioplasty

Plain balloon angioplasty is commonly the initial method employed to address stenotic lesions, which are often due to intimal hyperplasia. These venous stenoses can usually be resolved with pressures up to 20 atmospheres, though very high pressure, non-compliant balloons with burst pressures up to 40 atmospheres are also available to treat particularly recalcitrant lesions [5]. PTA of venous outflow and juxta-anastomotic stenosis have 1-year primary patency rates ranging from 28% to 72% and secondary patency rates at 1 year between 68% and 96%, with optimal results achieved when using 4–6 mm balloons [8]. However, these high pressures can lead to further intimal hyperplasia due to trauma to the vein, with 15% of patient experiencing rupture of the weak venous wall [8]. The risks of using high-pressure dilation has prompted some surgeons to use a cutting balloon prior to dilation to allow for the use of lower pressures which range from 6 to 8.4 atmospheres [9, 10]. Two randomized clinical

G. Geroulakos et al. (eds.), *Mastering Endovascular Techniques*, https://doi.org/10.1007/978-3-031-42735-0_38

trials have shown improved patency rates at 6 months and 1 year for stenoses treated with cutting balloon angioplasty (CBA) over the use of conventional PTA, with 6-month rates ranging between 86% and 66.4% in the CBA group versus 56% and 39.9% in the PTA group [9, 10]. The 1-year patency rates were 63% among the CBA group compared to 37% in the traditional PTA group which was statistically significant (p = 0.037); however, this significant difference was only found among patients with stenoses located at the graft-vein anastomosis [10]. Another way to decrease the risk of intimal hyperplasia may be through drug-coated balloon angioplasty (DCB), utilizing paclitaxel coated balloons [8]. There have been several randomized clinical trial assessing the efficacy of DCB compared to conventional PTA in management of vascular access stenosis. In one study, the average primary patency days was 270 days in the DCB group compared to 161 days in the PTA group [11]. The same authors also showed DCB to be more cost-effective compared to traditional PTA [12]. Primary patency rates at 6 months were higher among patients receiving treatment with DCB compared to traditional PTA (82.2% vs. 59.5%; $p < 0.001$) [13]. Another randomized study showed higher rates of primary patency in patients treated with DCB at 9 months (58% vs. 46%; $p = 0.02$) and 12 months (34% vs. 28%; $p = 0.04$) compared to traditional PTA; however, there was no difference at 18 and 24 months [14]. Both Lookstein et al. and Trerotola et al. showed that DCB was non-inferior compared to traditional PTA in terms of primary safety, which both studies defined as freedom from adverse events at 30 days post-intervention [13, 14]. Concerns about the safety of paclitaxel coated devices was particularly pronounced following multiple studies suggesting increased long term mortality in peripheral arterial disease patients [15], but this has not been seen in the arteriovenous access hemodialysis population.

Even in the absence of focal stenoses, some vascular access sites fail to mature, which can be due to overall small size of the venous outflow or inadequate vein size. Balloon-assisted maturation (BAM) has been described, which involves serial dilation of the venous outflow using relatively low pressures (8–12 atmospheres), with each dilation lasting <15 s. BAM is performed multiple times with an average of 2 weeks between each procedure [16, 17]. Early studies reported an increase in volume flow rates in half (54.7%) of patients receiving BAM [16], with a shorter duration to maturation in patients receiving BAM (119 days vs. 146 days in those not receiving BAM); however, this was not significant (p = 0.73) [17]. More recent studies include a randomized clinical trial which showed shorter maturation time in the BAM group compared to the non-BAM group (3.7 weeks vs. 5.9 weeks; $p < 0.001$) and higher successful functional maturation in the BAM group (93% vs. 77%; p = 0.001) [18]. However, the BAM group had a significantly higher complication rate (9.6% vs. 4.9%; $p = 0.042$) which included hematomas, early thrombosis, post-BAM extravasation, and proximal venous spasm [18]. A systematic review and meta-analysis showed high clinical (90%) and technical success (97%) rates with a widely variable complication rate ranging between 1.7 and 41%; however, there were no major complications or death reported [19]. Primary patency rates at 1 year varied between 28% and 72% while 1-year secondary patency rates ranged from 70 to 95% [19]. Some studies compared outcomes of patients receiving early BAM (defined as BAM performed within 98 days of AVF creation) to late BAM, with the early group having better clinical success rates (93% vs. 83%; $p = 0.002$) [19].

Management of Thrombosis

Stenosis in the vascular access circuit can lead to the development of thrombus in the circuit and complete cessation of flow. Endovascular treatment can be offered to patients with either an AVF or AVG; however, if the vascular access was created within the past month, then endovascular management can lead to disruption of the anastomoses [6]. It is also important to note that delayed clot removal of more than 2–3 days in an AVF is associated with a higher risk of failure, while an AVG thrombectomy can still be successfully performed up to 2 weeks after thrombosis [6]. One hypothesized mechanism of this observation is that thrombus will adhere more to a native vessel wall than to a prosthetic graft. Thrombolysis and mechanical thrombectomy are the two main types of endovascular treatment for thrombosed vascular access circuits.

Thrombolysis can be performed using several thrombolytic agents, including urokinase, streptokinase, or tissue plasminogen activator (tPA), with technical success of thrombolysis alone ranging between 38% and 80%. As such, thrombolysis is usually combined with mechanical thrombectomy [6]. Several thrombolysis techniques have been described, including “Lyse-and-wait”, pharmacomechanical thrombolysis (PMT), and pulse-spray thrombolysis. The “Lyse-and-wait” technique involves injecting a thrombolytic agent into the thrombus with manual compression of the inflow and outflow vessels of the vascular access circuit. After a certain wait time to allow the thrombolytic agent time to activate within the thrombus, an angiogram is performed to assess for thrombus resolution [20]. The “wait” time varies between different studies, though the average time is usually between 10 and 15 min, since the half-life for both urokinase and tPA are similar and range between 7 and 20 min [21]. One study assessed an extended wait time for tPA infiltration that averaged 18.6 h and ranged between 2 and 40 h [22]. This time-extended method had an 82% primary patency rate at 12 months and 63% at 24 months; however, this method was not compared to the traditional wait times so it is not known if this extended wait time provides any benefit over shorter wait times [22]. Another technique is PMT, which involves injecting a lytic agent into the thrombus and then performing either balloon maceration of the clot or using a mechanical thrombectomy device to remove the clot. Performing PMT with balloon maceration has a primary patency rate of 62.9% at 3 months and 45% at 12 months [23, 24]. Alternatively, PMT can be performed with pulse-spray thrombolysis, which utilizes an injector to deliver the lytic agent. Not many studies have assessed the efficacy of this technique, with one reporting a postintervention primary patency rate of 47% at 90 days [25].

There are several devices for mechanical thrombectomy, which can be divided into two broad categories: (1) direct wall contact devices and (2) hydrodynamic rheolytic devices that do not have contact with the vessel wall [6]. Direct wall contact devices include the Fogarty thru-lumen embolectomy catheter (i.e., "over-the-wire balloon") (Edwards Lifesciences) and the Cleaner XT Rotational Thrombectomy System (Argon Medical Devices, Plano, TX), which has a rotating sinusoidal wire that macerates the thrombus within the vessel [6]. The use of the Fogarty balloon along with thrombolysis had a median of 434 days of primary patency [26]. This lysis-assisted balloon (LAB) thrombectomy technique was also found to be relatively safe, with one study reporting a 6.6% complication rate [26]. Most of these complications were minor, including prolonged bleeding from the puncture site, venous rupture successfully treated with balloon inflation, and arterial embolization successfully treated with catheter aspiration [26]. In one study, the Cleaner XT device had an 88% success rate with a 65% primary patency rate and a 76% secondary patency rate both at 1 month [27]. One of the complications reported is the risk of AVF rupture which occurred in 3 out of 17 patients; however, none of the ruptures led to loss of the vascular access circuit [27]. Hydrodynamic rheolytic devices create a vortex by utilizing high-speed fluid jets to break the clot and create a negative pressure gradient which, along with aspiration, removes the thrombus fragments, all without any contact with the vessel wall [6]. There are several devices on the market with the most widely used and studied being the AngioJet (Boston Scientific, Marlborough, MA) and the Indigo System (Penumbra, Alameda, CA). In a systematic review of articles assessing the efficacy of the AngioJet device in the treatment of vascular access circuit thrombosis, the adjusted mean primary patency rates were 64.6% and 30.5% at 1 and 12 months, respectively [28]. The complication rate was 15.1% and most were bleeding and hematomas (13.5%), vein ruptures (7.9%) which were treated with stent grafts, and arterial emboli (7.1%) none of which resulted in limb loss [28]. The rest of the complications were considered minor as they did not result in any further intervention and included bleeding and hematoma, fever, transient arrhythmia not requiring treatment, contrast extravasation, and device failure [28]. There were 7 recorded deaths, most of which were due to cardiac causes such as arrhythmias, myocardial infarction, and cardiogenic shock; however, these could not be attributed to the use of the AngioJet device [28]. Two studies reviewing the Indigo System found similar results. Clinical success ranged between 80% and 88%, while 6-month primary patency was 37.5% in one study and 71% in another [29, 30]. One of these studies reported a 12-month primary patency rate of 71% [29]. Four complications were reported overall: one patient needed closure with a suture, one had arterial embolization necessitating open surgical embolectomy [30], and two had venous ruptures that occurred after the thrombectomy itself and with angioplasty of the underlying stenosis [29]. Other devices include the Oasis catheter (Boston Scientific, Marlborough, MA), Hydrolyzer (Cordis, Miami, FL), and Amplatz thrombectomy device (White Bear Lake, MN); however, there is a lack of literature regarding the efficacy of these devices in vascular access circuit thrombectomy [6].

Stenting of the Peripheral Venous Segment

Stenting of vascular access circuits is usually reserved in the event of failure of PTA. Three main indications for stent placement in vascular access circuits have been described: (1) presence of a recoil lesion, (2) recurrent stenosis after PTA, and (3) vessel rupture or dissection. A recoil lesion is a stenosis of >30% that persists after PTA which occurs due to the elastic nature of venous stenoses [31]. Recurrent stenosis is relatively common, as evidenced by the variable primary patency rates after PTA as seen in the previous section. Vessel rupture or dissection can occur with balloon angioplasty of severely stenotic lesions and is usually resolved with internal compression with balloon tamponade and sometimes with external compression from outside the body. However, in certain cases where extravasation is ongoing after these measures are taken, the placement of a stent may be necessary [31].

Early studies compared the use of bare metal stents with PTA versus PTA alone, but no benefit was seen in using bare metal stents with PTA versus PTA alone [32, 33]. However, more recent studies assessed the role of covered stent grafts in treatment of stenoses in vascular access circuits. Several clinical trials have compared different types of stent grafts to PTA and most found improved primary patency at 6 months in the stent graft groups over the PTA group [34, 35]. One clinical trial, the Randomized, Concurrently-Controlled Post-Approval Study of the FLAIR Endovascular Stent Graft (RENOVA), showed sustained improvement in primary patency with higher rates among the stent graft group at 24 months (26.9 vs. 13.5%, $p < 0.001$) compared to PTA [36]. A similar study (REVISE) evaluated the use of a different stent graft system, and compared outcomes to PTA with a plain balloon. Use of a stent graft improved primary patency at 6 months (51.6% vs. 34.2%, $p = 0.006$) [35]. Another study showed a decreased average number of reinterventions needed after 24 months with the stent graft group compared to PTA alone (3.7 vs. 5.1, $p = 0.005$) [37]. The same study showed no difference in total costs at 24 months between stent grafts and PTA, which may be explained by the lower need for reinterventions with stent graft use [37]. Adverse outcomes were similar between patients receiving a stent graft and those treated with PTA with the exception of restenosis, which occurred more frequently among the PTA group in two clinical trials (82.6% vs. 63%; $p < 0.001$), compared to the stent graft groups [34, 36]. One study also compared stent graft use to bare metal stents in cephalic arch stenosis and showed significantly better primary patency rates in the stent graft group over patients who received a bare metal stent [38].

The use of stent grafts is illustrated by the following case. A 62-year-old female presents with left arm pain when using her left brachial artery to axillary vein AVG on hemodialysis. A 4–7 mm tapered AVG had been placed, and it has been otherwise functionally patent. She is brought for fistulography, which demonstrates a severe juxta-anastomotic stenosis affecting the venous outflow (Fig. 38.1). Plain balloon angioplasty up to 7 mm was performed, but a persistent stenosis was seen with clear retrograde flow evident in the brachial vein (Fig. 38.2). Placement of a 9 mm × 75 mm Viabahn stent graft (WL Gore, Flagstaff, AZ) led to complete resolu-

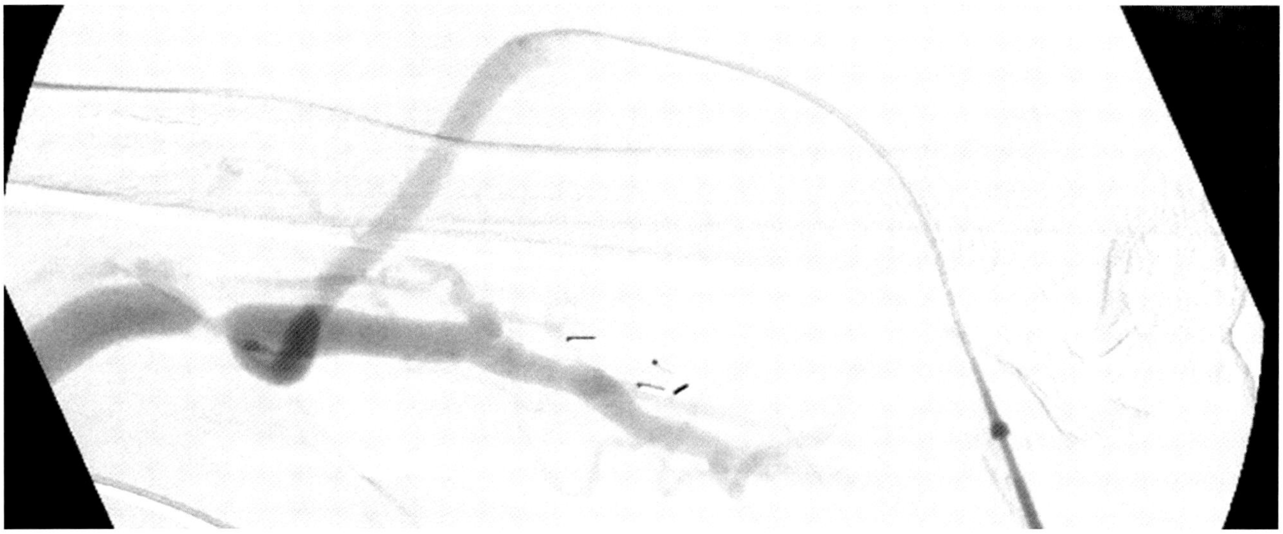

Fig. 38.1 Initial fistulogram demonstrating severe juxta-anastomotic venous outflow stenosis affecting a left brachial artery to axillary vein AVG

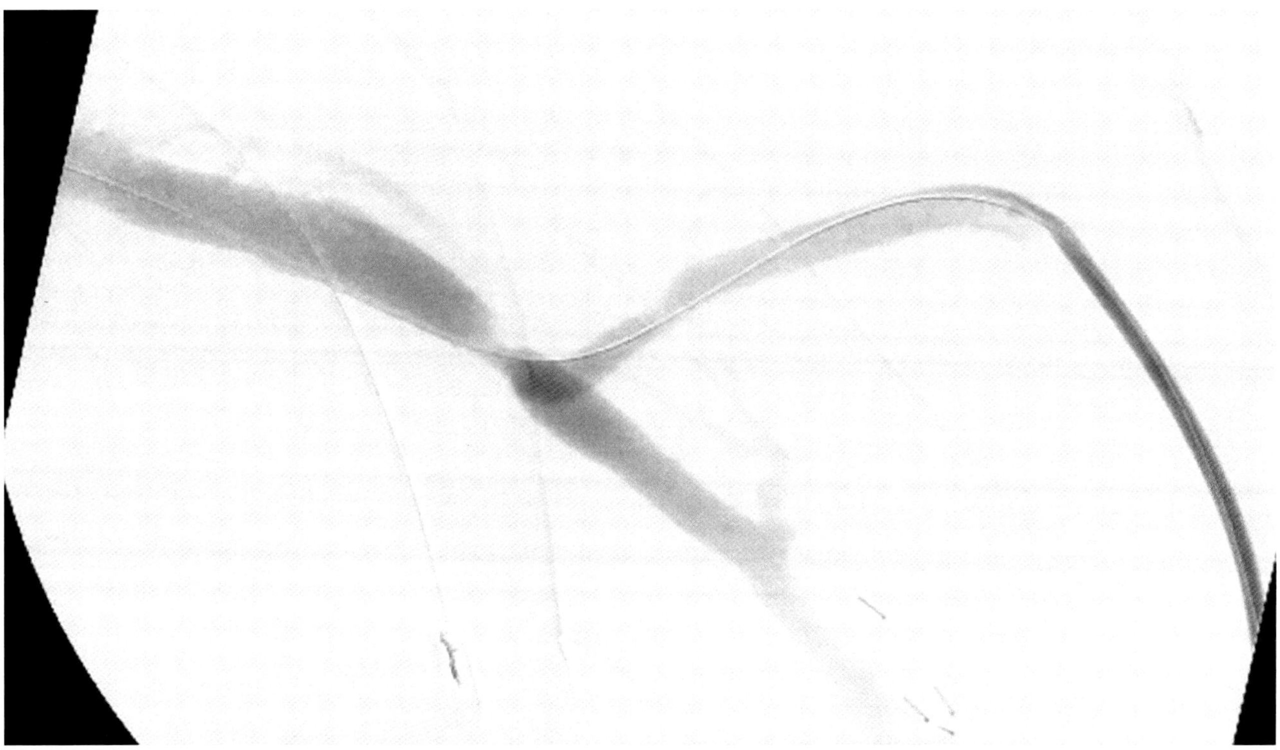

Fig. 38.2 Persistent stenosis despite aggressive balloon angioplasty was observed

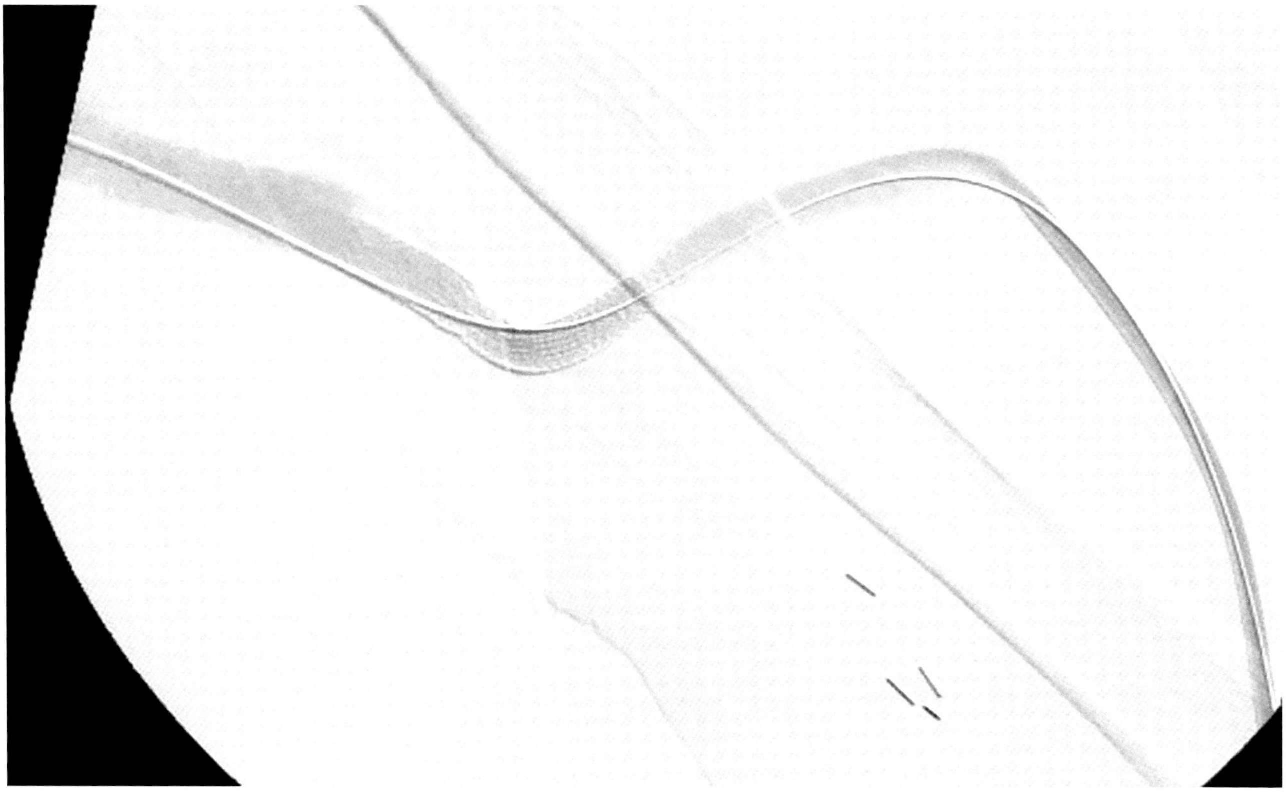

Fig. 38.3 Resolution of stenosis with use of a stent graft

tion of the stenosis and the patient's clinical complaints (Fig. 38.3). The stent graft should typically be at least 1 or 2 mm larger than the normal graft or fistula and should be extended into an outflow vein that is at least as large as the stent graft.

Accessory Vein Obliteration

Accessory veins can prevent AVF maturation by diverting blood flow away from the main fistula, leading to lower flows, smaller vessel diameters, and ultimately causing difficulty with cannulation or obtaining adequate flows on dialysis. The pressure gradient established after creation of an AVF will lead to increased blood flow through the vein. Higher blood flow leads to more shear stress on the vessel wall leading to vessel dilation and AVF maturation; so any alteration in blood flow through the AVF can lead to a failure of maturation [7]. Although accessory veins can significantly impact the maturation of AVFs, there are no clear guidelines on which accessory veins need to be ligated or obliterated. Traditionally, accessory veins with a diameter of one third or more of the AVF need to be obliterated; however, there are no studies that validate this arbitrary number [7]. One study tried to determine the criteria for accessory vein obliteration (AVO) by conducting computational fluid dynamics and established the following measures: (1) 60% or greater diameter of the main AVF, (2) 50% diameter of the AVF with at least one more accessory vein greater than 40% in diameter, (3) 50% in diameter and divides into branches of same size, (4) accessory vein likely to interfere with cannulation on physical examination, and (5) greater than 30% in diameter and associated with stenosis at site of origin [39]. However, these criteria were not clinically tested and verified and the guidelines for managing accessory veins are still evolving. This lack of selection criteria may explain the results of some studies that have not shown any benefit in performing AVO [40, 41]. One of these studies was a meta-analysis looking at patients with immature AVFs undergoing either PTA alone or PTA with AVO, and there was no benefit for PTA with AVO over PTA alone [40]. It is important to note that there was no specific selection process in determining which patient were to receive AVO. Regardless of these findings, certain patients with an AVF that has failed to mature may still benefit from AVO. This can be performed surgically with ligation of the accessory vein, or endovascularly with coil embolization [7] (Figs. 38.4 and 38.5). When comparing open ligation to coil embolization, there was no significant difference in the two techniques, with 64% maturation in the open group and 61% maturation in the endovascular group ($p = 0.84$) [42]. Further studies are needed to assess the efficacy of AVO.

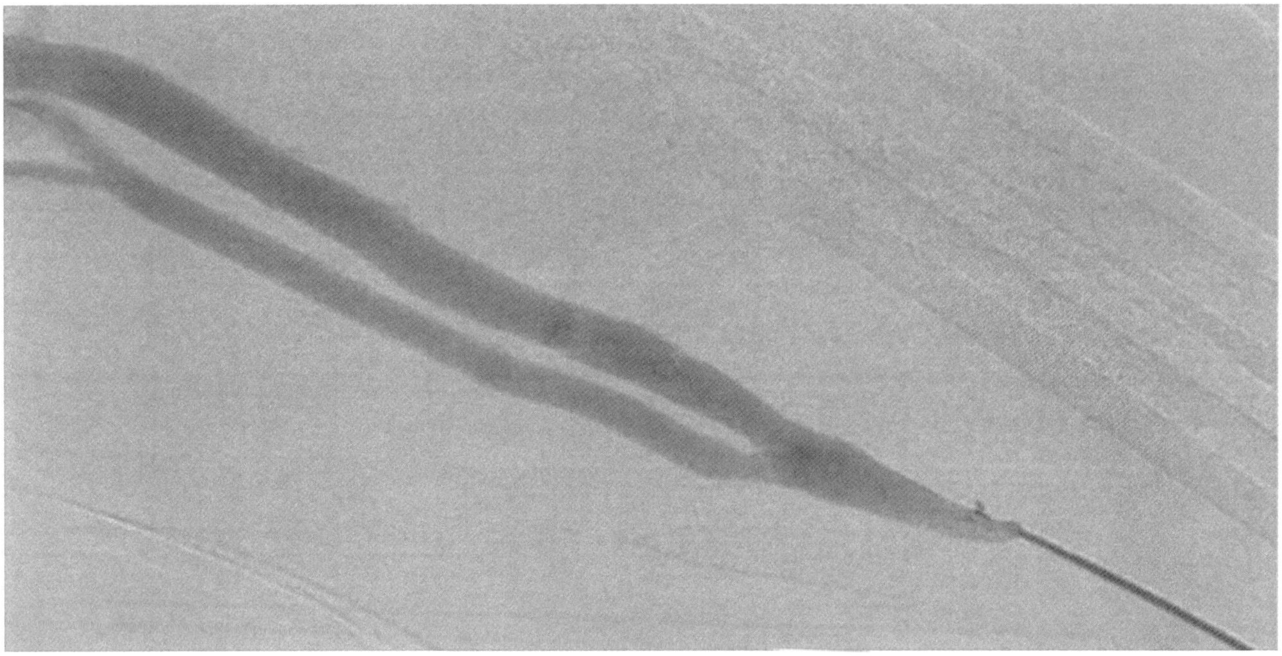

Fig. 38.4 73 year old male with a left radiocephalic AVF presents with difficulty with cannulation and low flows even when cannulation is achieved. The patient was brought to the angiography suite for evaluation, where fistulogram through a microcatheter demonstrated dual outflow in the forearm

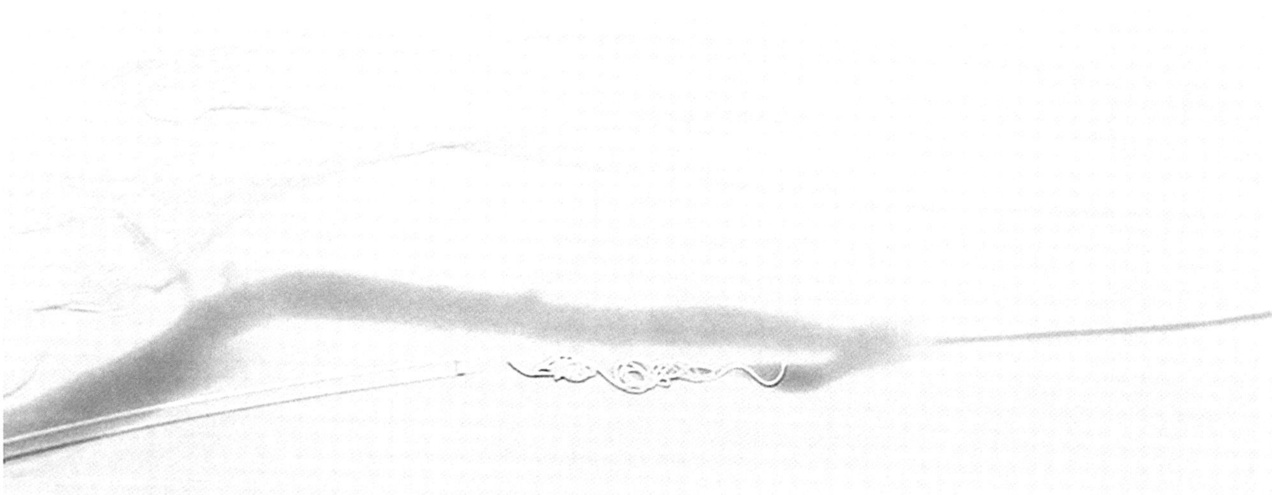

Fig. 38.5 The competing outflow vein was cannulated separately in order to minimize trauma to the intended cannulation segment of the AVF, and the competing outflow vein was embolized using two coils with complete cessation of flow

Aneurysms and Pseudoaneurysms

Although uncommon, aneurysm and pseudoaneurysm of AVFs are known complications that cause serious morbidity. In their systematic review, Al-Jaishi et al. reported an aneurysm incidence of 0.04 per 1000 patient days [43]. Aneurysmal dilation of the AVF has been linked to repeated cannulation of the fistula, with one study finding that alternating puncture sites protects against the formation of pseudoaneurysms [44]. Pseudoaneurysms are perivascular collections of blood that are in direct continuity with the AVF lumen and have been also linked to AVF vessel injuries caused by cannulation [45]. Aneurysms and pseudoaneurysms can both cause overlying skin thinning and ulcer formation, as well as limit the availability of cannulation sites and, if left untreated, can lead to exsanguination [43, 45]. While the size of an aneurysm/pseudoaneurysm is not an indication for treatment, the KDOQI recommends evaluation for new and follow-up on existing aneurysms at each dialysis session, with use of duplex if needed for evaluation of size, size progression, presence of stenosis, and impact on dialysis via the AVF [2]. It is recommended to avoid cannulation through the dilated segment to avoid further trauma to the vessel wall and potential vascular access hemorrhage [2, 46]. The KDOQI and Canadian Society of Nephrology hemodialysis guidelines recommend intervention when the skin overlying the AVF has been compromised (erosion or ulceration), there is a risk of life-threatening hemorrhage, or cannulation sites have become limited [2, 47]. Endovascular management is similar to that of aneurysms found in other vascular beds and includes exclusion of the aneurysm or pseudoaneurysm by inserting a stent graft. Available studies demonstrate primary patency rates of 100% at 30 days, 72.7% at 3 months, and 36.4% at 6 months [48, 49]. One study showed a non-significant higher risk of infection (odds ratio 5.0, 95% confidence interval [0.38, 66.01]) with stent graft insertion in patients with skin erosion over the pseudoaneurysm [49]. Other authors reported a 28.9% complication rate including migration, fracture, erosion, or rupture after stent graft insertion [50]. Further studies are needed to evaluate the role of stent grafts in the treatment of vascular access circuit aneurysms and pseudoaneurysms.

Central Venous Stenosis

Stenosis of the central veins (including the subclavian vein, innominate veins, and superior vena cava) can result in significant morbidity for hemodialysis patients with AVFs and AVGs. Many patients may have asymptomatic stenosis which may be incidentally discovered during venography; as such, the true incidence of central venous stenosis (CVS) varies widely with reported values ranging from 3 to 60% among ESKD patients on HD [51–55]. Symptomatic stenosis can present with multiple issues, including difficulty with cannulation, prolonged bleeding following needle removal, or decrease in clearance [56]. Other findings suggestive of CVS include ipsilateral extremity edema, pain, or development of prominent venous varicosities or collaterals [56, 57]. Risk factors associated with the development of CVS include current or prior presence of central venous catheters, DVT, and cardiac rhythm devices such as pacemakers [2, 58].

Percutaneous Transluminal Angioplasty

Endovascular techniques are the current mainstay of CVS treatment, with the KDOQI recommending PTA only for moderate to severe symptoms or when efficacy or clearance is compromised [2] (Figs. 38.6 and 38.7). The initial technical success rates of PTA are high, ranging from 70% to 90% [59]. However,

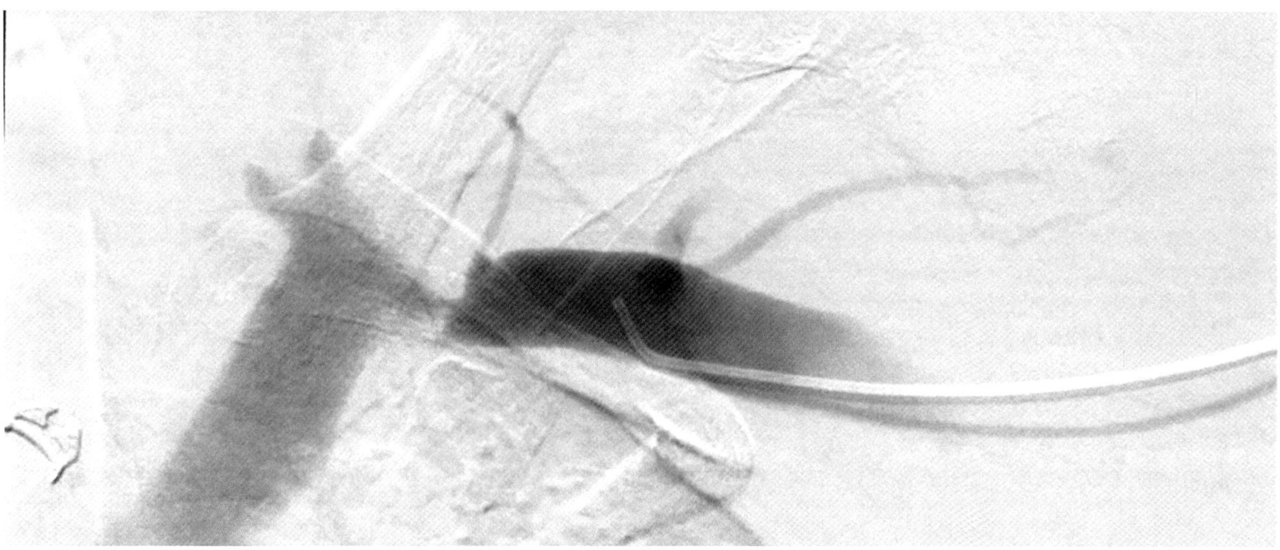

Fig. 38.6 Left subclavian vein stenosis leading to left arm swelling after left brachial artery to axillary vein AVG placement

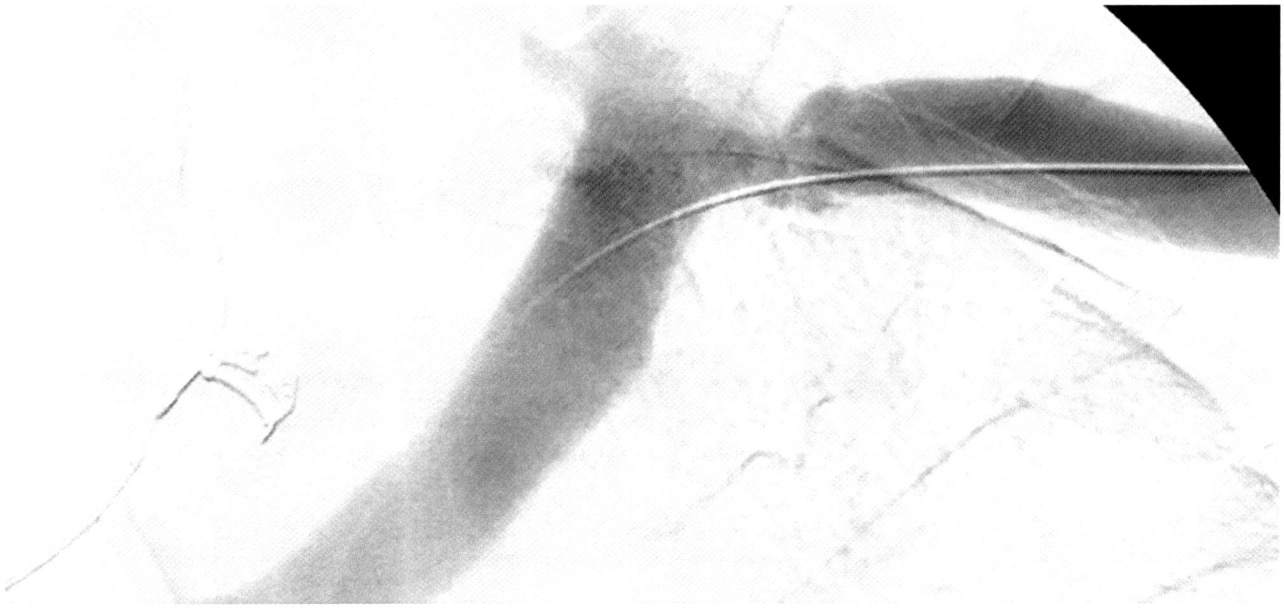

Fig. 38.7 Treatment with balloon angioplasty leads to significant luminal gain and improvement in left arm swelling

due to venous elastic recoil, restenosis is common and patency rates at 12 months are quite poor, ranging from 12 to 50% [59].

Management of restenosis following PTA may include repeated balloon angioplasty or possibly stent placement. With repeated PTA, secondary patency rates have been reported as high as 60% at 12 months, though close monitoring and repeated intervention are often required [60]. If significant residual stenosis persists after repeated PTA, the KDOQI guidelines endorse the use of intraluminal stenting [2], which will be discussed below.

Stenting of the Central Venous System

As noted above, current guidelines support the use of stenting for CVS only after multiple unsuccessful rounds of PTA [2]. Despite high technical success rates, initial studies employing bare metal stents showed 12-month primary patency rates of 19–21% [61–63]. When comparing long-term patency of PTA versus PTAS, bare-metal stents have not demonstrated superior outcomes [61, 64]. Bare-metal stents are additionally associated with migration, shortening, and fracture and thus are not recommended for use in the region of the thoracic outlet due to risk of external compression [2, 3, 65]. A retrospective analysis investigating the use of self-expanding bare metal stents in central and peripheral veins reported a 12-month patency rate of 67% with a mean patency duration of 14.9 months [66]. In a 2007 study, PTAS was performed in 26 patients with bare metal stents and primary PTA alone was performed in 47 patients. Primary patency was equivalent between groups with 76% patency for both groups at 30-day and 12-month rates of 29% for PTA and 21% for PTAS (p = 0.48) [61]. The authors concluded that both PTA and PTAS are safe options which both often require additional interventions.

More recently, stent grafts have emerged as a promising option for endovascular treatment of CVS. In several recent retrospective analyses comparing PTA, bare-metal stents, and stent grafts, stent grafts consistently demonstrated superior long-term patency. Reported 12-month patency ranged from 45.1 to 58%, 75 to 78.6%, and 82.4 to 100% for PTA, bare-metal stents, and stent grafts, respectively [67, 68]. In both studies, primary patency for stent grafts were superior to both PTA and bare-metal stents. The advantage of stent grafts is thought to be related to their relatively inert and stable nature, providing a platform for endothelialization while creating a physical barrier that reduces intimal hyperplasia (Kundu 2010).

Case Presentation

Continued from page 409

The patient was severely hyperkalemic, and a dialysis catheter was placed in order to temporize the patient and provide dialysis access prior to thrombectomy. Due to lack of available operating room time, a percutaneous approach was chosen. Tissue plasminogen activator (t-PA, Alteplase) 2 mg, buffered in 3000 units of heparin and 5 mL of sterile water was injected into the thrombosed AVG through a 22 gauge Angiocatheter, after the fashion of "Lyse and wait" [20] (Figs. 38.8 and 38.9). After several minutes, the AVG became pulsatile throughout its length.

When an appropriate angiography suite was available, the patient was brought for fistulography, demonstrating a severe stenosis in the outflow (Fig. 38.10), which responded to aggressive balloon angioplasty, including high pressure balloons up to 40 atmospheres (Fig. 38.11). The patient did well and was discharged home with a functional AVG.

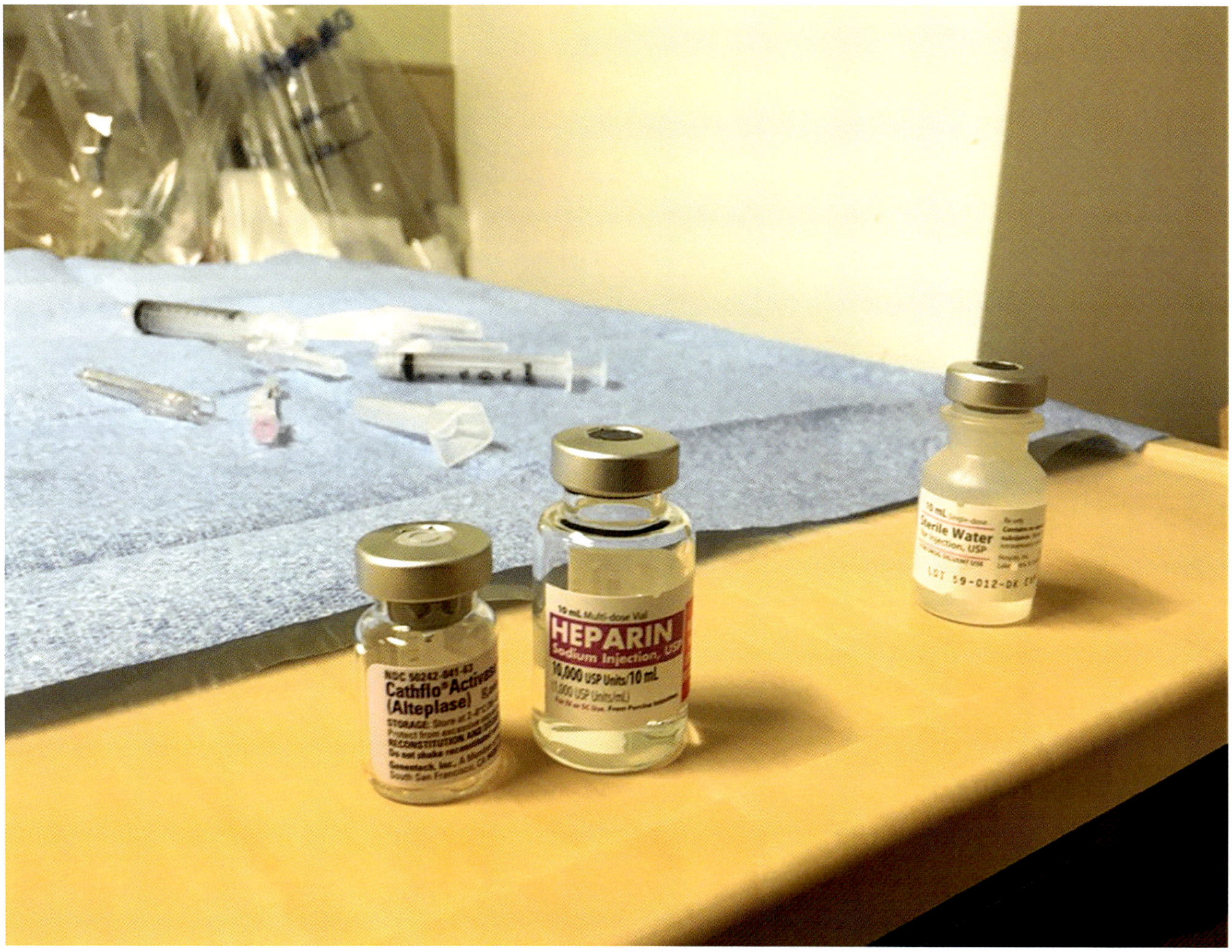

Fig. 38.8 Components of the "Lyse and wait" technique include Alteplase (2 mg) and heparin (3000 units) [20]

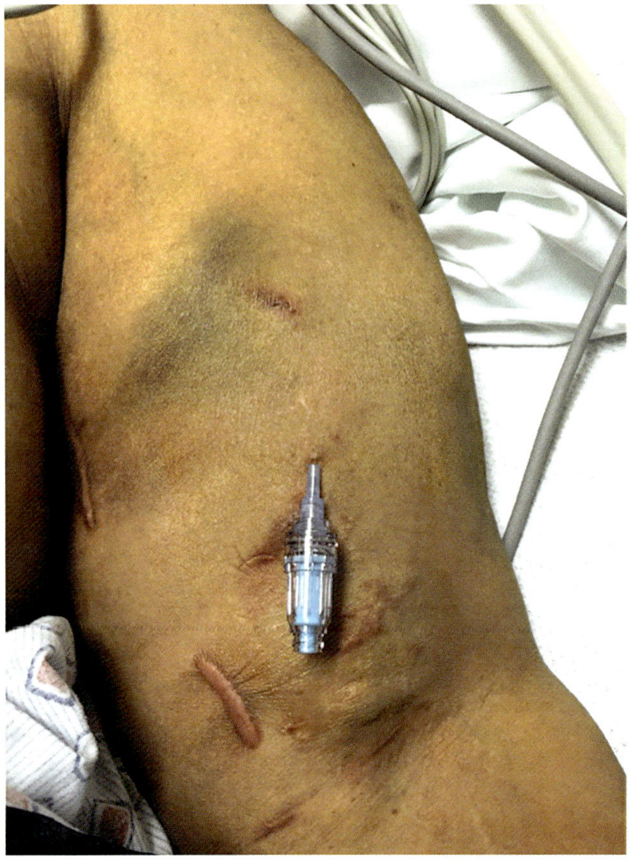

Fig. 38.9 A 22 gauge angiocatheter directed to the venous anastomosis is used to administer the medications

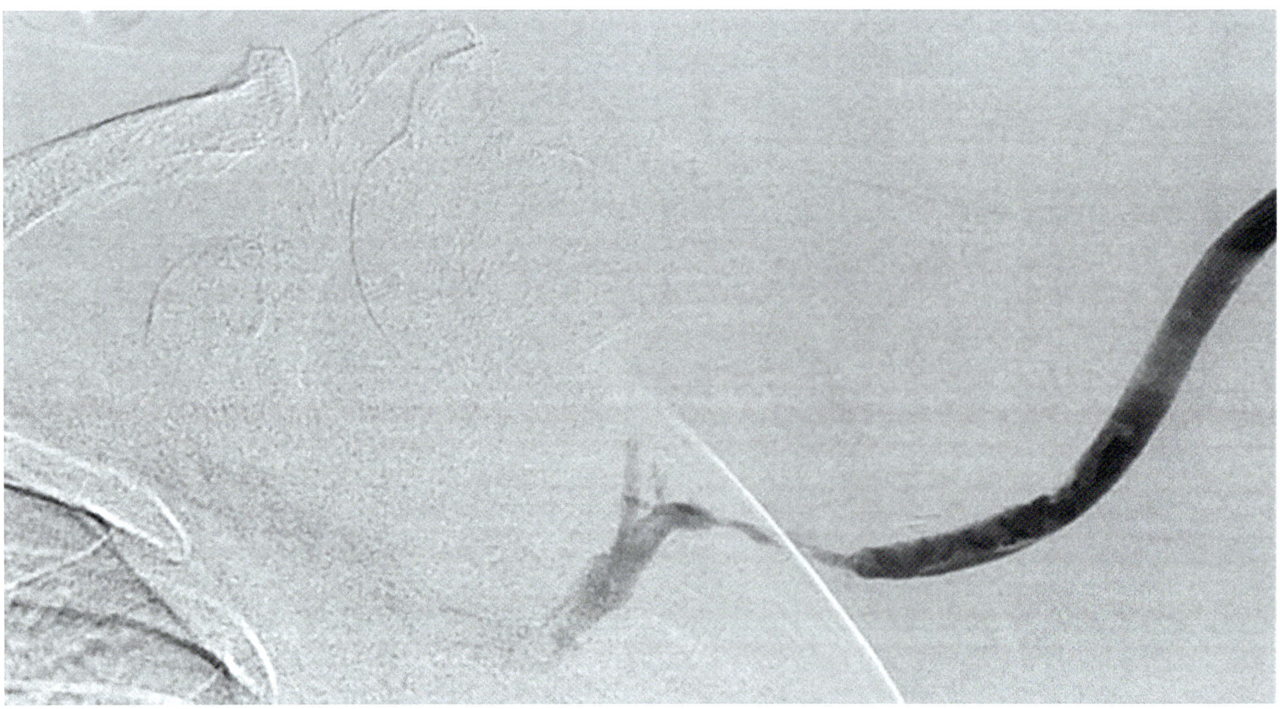

Fig. 38.10 Initial fistulography after lysis demonstrated a severe stenosis in the venous outflow and residual thrombus

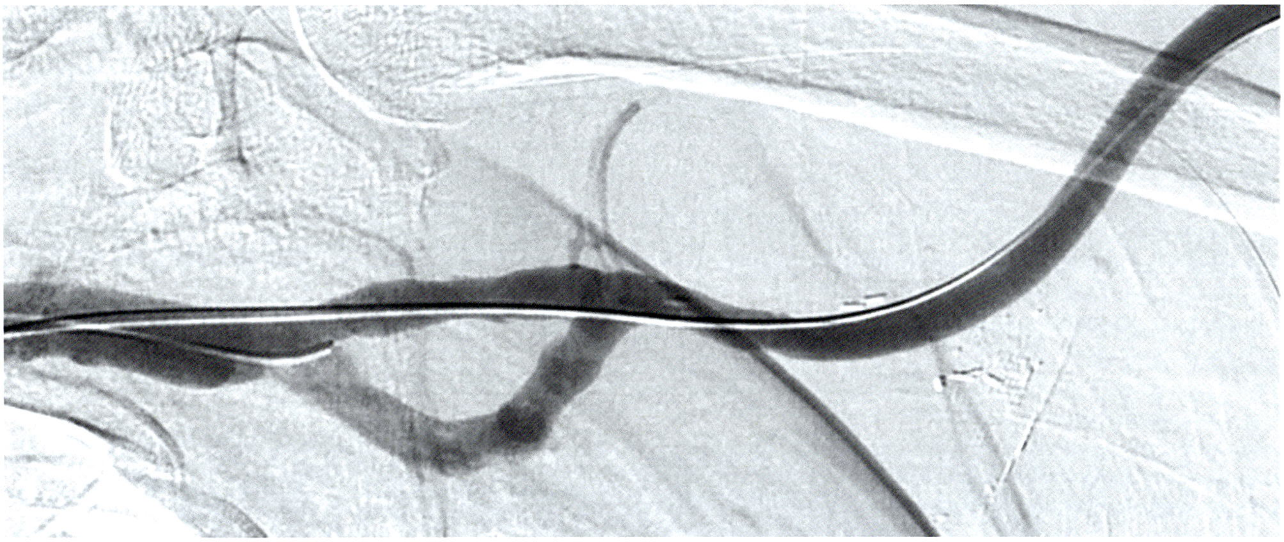

Fig. 38.11 Completion fistulogram after aggressive plain and high pressure balloon angioplasty exhibits an excellent radiologic result

Table 38.1 Endovascular toolkit for the endovascular management of failing arteriovenous access. These are only a few suggested options based on the author's experience

	Manufacturer	Size/length
Guidewires		
Any standard-access wire		0.035″
Hydrophilic-coated wire	Glidewire (Terumo) ZIPwire (Boston Scientific) SplashWire (Merit Medical)	0.035″
Sheaths		
Any standard-access sheath, sizing dependent on requirements for balloon or stent delivery. Short sheaths can be useful for access, while longer sheaths may be needed for support when crossing difficult lesions.	Multiple	6–9 Fr/6–10 cm 6 Fr/25+ cm
Coils		
Nestor coils	Cook Medical	0.035″ 4, 6, 8 mm
Stent grafts		
Viabahn	WL Gore	7–10 mm/ length as required to cover the lesion
Flair	BD	7–10 mm/ length as required to cover the lesion
Covera	BD	7–10 mm/ length as required to cover the lesion
Thrombectomy devices		
Fogarty thru-lumen embolectomy catheter	Edwards Lifesciences	4 Fr (9 mm inflated) 5.5 Fr (11 mm inflated)
Cleaner XT rotational thrombectomy system	Argon Medical Devices	6 Fr/ 65 cm
AngioJet thrombectomy system AVX	Boston Scientific	6 Fr/ 50 cm
Indigo aspiration system/ CAT mechanical thrombectomy catheter	Penumbra	CATD: 8 F (2.67 mm, 50 cm length)

Conclusion

Endovascular treatment options of vascular access failures offer a minimally invasive treatment option for ESKD patients on hemodialysis with malfunctioning and failed AV access. A well-stocked endovascular toolkit (Table 38.1) is invaluable. When performed by experienced and well-trained clinicians, percutaneous options can provide immediate and successful treatment of failing dialysis access.

References

1. United States Renal Data System. 2022 USRDS annual data report: epidemiology of kidney disease in the United States. Am J Kidney Dis. 2022;59(1):e183–94. https://doi.org/10.1053/j.ajkd.2011.10.027.
2. Lok CE, Huber TS, Lee T, et al. KDOQI Clinical Practice Guideline for Vascular Access: 2019 Update. Am J Kidney Dis. 2020;75(4):S1–S164. https://doi.org/10.1053/j.ajkd.2019.12.001.
3. Schmidli J, Widmer MK, Basile C, et al. Editor's choice—vascular access: 2018 clinical practice guidelines of the European Society for Vascular Surgery (ESVS). Eur J Vasc Endovasc Surg. 2018;55(6):757–818. https://doi.org/10.1016/j.ejvs.2018.02.001.
4. Montross BC, Harris LM. Hemodialysis access: nonthrombotic complications. In: Rutherford's vascular surgery and endovascular therapy. 10th ed. Elsevier; 2023. p. 2358–69.e7.
5. Wasse H, Hohmann SE. Hemodialysis access: failing and thrombosed. In: Rutherford's vascular surgery and endovascular therapy. 10th ed. Elsevier; 2023. p. 2347–57.e3.
6. Almehmi A, Sheta M, Abaza M, Almehmi SE, El Khudari H, Shaikh A. Endovascular Management of Thrombosed Dialysis Vascular Circuits. Semin intervent Radiol. 2022;39(01):014–22. https://doi.org/10.1055/s-0041-1740941.
7. Haq NU, Althaf MM, Lee T. Accessory vein obliteration for early fistula failure: a myth or reality? Adv Chronic Kidney Dis. 2015;22(6):438–45. https://doi.org/10.1053/j.ackd.2015.08.001.
8. Tordoir JHM, Zonnebeld N, van Loon MM, Gallieni M, Hollenbeck M. Surgical and endovascular intervention for dialysis access maturation failure during and after arteriovenous fistula surgery: review of the evidence. Eur J Vasc Endovasc Surg. 2018;55(2):240–8. https://doi.org/10.1016/j.ejvs.2017.12.001.
9. Aftab SA, Tay KH, Irani FG, et al. Randomized clinical trial of cutting Balloon angioplasty versus high-pressure Balloon angioplasty in hemodialysis arteriovenous fistula Stenoses resistant to conventional Balloon angioplasty. J Vasc Interv Radiol. 2014;25(2):190–8. https://doi.org/10.1016/j.jvir.2013.10.020.
10. Saleh HM, Gabr AK, Tawfik MM, Abouellail H. Prospective, randomized study of cutting balloon angioplasty versus conventional balloon angioplasty for the treatment of hemodialysis access stenoses. J Vasc Surg. 2014;60(3):735–40. https://doi.org/10.1016/j.jvs.2014.04.002.
11. Kitrou PM, Spiliopoulos S, Katsanos K, Papachristou E, Siablis D, Karnabatidis D. Paclitaxel-coated versus plain Balloon angioplasty for dysfunctional arteriovenous fistulae: one-year results of a prospective randomized controlled trial. J Vasc Interv Radiol. 2015;26(3):348–54. https://doi.org/10.1016/j.jvir.2014.11.003.
12. Kitrou PM, Katsanos K, Spiliopoulos S, Karnabatidis D, Siablis D. Drug-eluting versus plain balloon angioplasty for the treatment of failing dialysis access: final results and cost-effectiveness analysis from a prospective randomized controlled trial (NCT01174472). Eur J Radiol. 2015;84(3):418–23. https://doi.org/10.1016/j.ejrad.2014.11.037.
13. Lookstein RA, Haruguchi H, Ouriel K, et al. Drug-coated balloons for dysfunctional dialysis arteriovenous fistulas. N Engl J Med. 2020;383(8):733–42. https://doi.org/10.1056/NEJMoa1914617.
14. Trerotola SO, Saad TF, Roy-Chaudhury P. The Lutonix AV randomized trial of paclitaxel-coated balloons in arteriovenous fistula stenosis: 2-year results and subgroup analysis. J Vasc Interv Radiol. 2020;31(1):1–14.e5. https://doi.org/10.1016/j.jvir.2019.08.035.
15. Katsanos K, Spiliopoulos S, Kitrou P, Krokidis M, Karnabatidis D. Risk of death following application of paclitaxel-coated balloons and stents in the Femoropopliteal artery of the leg: a systematic review and meta-analysis of randomized controlled trials. J Am Heart Assoc. 2018;7(24):e011245. https://doi.org/10.1161/JAHA.118.011245.
16. DerDerian T, Hingorani A, Ascher E, et al. To BAM or not to BAM?: a closer look at Balloon-assisted maturation. Ann Vasc Surg. 2013;27(1):104–9. https://doi.org/10.1016/j.avsg.2012.06.009.
17. Rizvi SA, Usoh F, Hingorani A, et al. The clinical efficacy of Balloon-assisted maturation of autogenous arteriovenous fistulae. Ann Vasc Surg. 2017;41:41–5. https://doi.org/10.1016/j.avsg.2016.08.022.
18. Elkassaby M, Elsaadany N, Mowaphy K, Soliman M. Balloon-assisted maturation of autogenous arteriovenous fistulae: a randomized controlled prospective study. Vascular. 2021;29(5):776–83. https://doi.org/10.1177/1708538120979872.
19. Kanchanasuttirak P, Pitaksantayothin W, Saengprakai W, Kanchanabat B. Systematic review and meta-analysis: efficacy and safety of balloon angioplasty in salvaging non-matured arteriovenous fistulas. J Vasc Access. 2022:112972982210854. https://doi.org/10.1177/11297298221085440.
20. Cynamon J, Lakritz PS, Wahl SI, Sakal CW, Sprayregen S. Hemodialysis graft Declotting: description of the "lyse and wait" Techniquel. J Vasc Interv Radiol. 1997;8(5):825–9.
21. Falk A, Mitty H, Guller J, Teodorescu V, Uribarri J, Vassalotti J. Thrombolysis of clotted hemodialysis grafts with tissue-type plasminogen activator. J Vasc Interv Radiol. 2001;12(3):305–11. https://doi.org/10.1016/S1051-0443(07)61908-3.
22. Regus S, Lang W, Heinz M, et al. Time-extended local rtPA infiltration for acutely thrombosed hemodialysis fistulas: prolonged local rtPA reaction time. Hemodial Int. 2018;22(1):31–6. https://doi.org/10.1111/hdi.12528.
23. Choi SY, Choi BG, Han KH, Chun HJ. Efficacy of a modified Pharmacomechanical thrombolysis technique for endovascular treatment of thrombosed prosthetic arteriovenous grafts. Korean J Radiol. 2012;13(3):300. https://doi.org/10.3348/kjr.2012.13.3.300.
24. Tan RY, Pang SC, Teh SP, et al. Outcomes of endovascular salvage of clotted arteriovenous access and predictors of patency after thrombectomy. J Vasc Surg. 2020;71(4):1333–9. https://doi.org/10.1016/j.jvs.2019.07.056.
25. Cooper SG. Pulse-spray thrombolysis of thrombosed hemodialysis grafts with tissue plasminogen activator. Am J Roentgenol. 2003;180(4):1063–6. https://doi.org/10.2214/ajr.180.4.1801063.
26. Kitrou PM, Papadimatos P, Spiliopoulos S, Christeas N, Katsanos K, Karnabatidis D, Lysis-Assisted Balloon (LAB) Thrombectomy. A Declotting technique for the treatment of thrombosed arteriovenous dialysis grafts. 5-year experience of 241 endovascular procedures. Cardiovasc Intervent Radiol. 2018;41(2):245–52. https://doi.org/10.1007/s00270-017-1813-z.
27. Huan KWSJ, Tan CS, Chua D, et al. The cleaner XT™ device as an endovascular adjunct for Pharmacomechanical thrombolysis of thrombosed arteriovenous fistulas and grafts. Ann Vasc Dis. 2020;13(4):390–6. https://doi.org/10.3400/avd.oa.20-00046.
28. Chan PG, Goh GS. Safety and efficacy of the AngioJet device in the treatment of thrombosed arteriovenous fistula and grafts: a systematic review. J Vasc Access. 2018;19(3):243–51. https://doi.org/10.1177/1129729818760977.
29. Cavalcante RN, Nishinari K, Centofanti G, et al. The role of vacuum-assisted mechanical thrombectomy in the management of acutely thrombosed arteriovenous fistulas and grafts. J Vasc Access. 2022:112972982210994. https://doi.org/10.1177/11297298221099469.
30. Piacentino F, Coppola A, Zaghetto A, et al. Vacuum-assisted mechanical thrombectomy in extensively occlusive thrombosis of dialysis arteriovenous grafts with indigo system. J Vasc Access. 2020;21(5):673–9. https://doi.org/10.1177/1129729819899264.
31. Shaikh A, Albalas A, Desiraju B, Dwyer A, Haddad N, Almehmi A. The role of stents in hemodialysis vascular access. J Vasc Access. 2023;24(1):107–16. https://doi.org/10.1177/11297298211015069.
32. Beathard GA. Gianturco self-expanding stent in the treatment of stenosis in dialysis access grafts. Kidney Int. 1993;43(4):872–7. https://doi.org/10.1038/ki.1993.122.

33. Quinn SF, Schuman ES, Oemlow TA, Standage BA, Ragsdale JW, Green GS. Percutaneous transluminal angioplasty versus endovascular stent placement in the treatment of venous Stenoses in patients undergoing hemodialysis: intermediate Resultsl. J Vasc Interv Radiol. 1995;6(6):851–5.
34. Haskal ZJ, Altman S, McLennan G, Vesely T. Stent graft versus Balloon angioplasty for failing dialysis-access grafts. N engl j med. Published online. 2010;2010:238.
35. Vesely T, DaVanzo W, Behrend T, Dwyer A, Aruny J. Balloon angioplasty versus Viabahn stent graft for treatment of failing or thrombosed prosthetic hemodialysis grafts. J Vasc Surg. 2016;64(5):1400–1410.e1. https://doi.org/10.1016/j.jvs.2016.04.035.
36. Haskal ZJ, Saad TF, Hoggard JG, et al. Prospective, randomized, concurrently-controlled study of a stent graft versus Balloon angioplasty for treatment of arteriovenous access graft stenosis: 2-year results of the RENOVA study. J Vasc Interv Radiol. 2016;27(8):1105–1114.e3. https://doi.org/10.1016/j.jvir.2016.05.019.
37. Mohr BA, Sheen AL, Roy-Chaudhury P, Schultz SR, Aruny JE. Clinical and economic benefits of stent grafts in dysfunctional and thrombosed hemodialysis access graft circuits in the REVISE randomized trial. J Vasc Interv Radiol. 2019;30(2):203–211.e4. https://doi.org/10.1016/j.jvir.2018.12.006.
38. Shemesh D, Goldin I, Zaghal I, Berlowitz D, Raveh D, Olsha O. Angioplasty with stent graft versus bare stent for recurrent cephalic arch stenosis in autogenous arteriovenous access for hemodialysis: a prospective randomized clinical trial. J Vasc Surg. 2008;48(6):1524–1531.e2. https://doi.org/10.1016/j.jvs.2008.07.071.
39. Haq NU, Albaqumi M. Accessory vein obliteration criteria for immature fistulae: a modest proposal for an old paradigm. Semin Dial. 2014;27(5):E51–4. https://doi.org/10.1111/sdi.12239.
40. Purwono GY, Sultana R, Lee RE, et al. Accessory vein obliteration and Balloon-assisted maturation for immature arteriovenous fistulas for haemodialysis: a systematic review and meta-analysis. Cardiovasc Intervent Radiol. 2022;45(10):1415–27. https://doi.org/10.1007/s00270-022-03212-0.
41. Sawas A, Baran TM, Foster TH, et al. Long-term patency of arteriovenous fistulae salvaged by Balloon angioplasty with and without accessory vein embolization: a retrospective study. J Vasc Interv Radiol. 2017;28(5):714–21. https://doi.org/10.1016/j.jvir.2016.12.1215.
42. Yan Q, Miserlis D, Jayakumar L, et al. Management of Arterio-Venous Fistula Side Branches during Maturation: ligation or coil embolization. Ann Vasc Surg. 2022;82:52–61. https://doi.org/10.1016/j.avsg.2022.01.004.
43. Al-Jaishi AA, Liu AR, Lok CE, Zhang JC, Moist LM. Complications of the arteriovenous fistula: a systematic review. JASN. 2017;28(6):1839–50. https://doi.org/10.1681/ASN.2016040412.
44. Quintaliani G, Buoncristiani U, Fagugli R, et al. Survival of vascular access during daily and three times a week hemodialysis. Clin Nephrol. 2000;53(5):372–7.
45. Meola M, Marciello A, Di Salle G, Petrucci I. Ultrasound evaluation of access complications: thrombosis, aneurysms, pseudoaneurysms and infections. J Vasc Access. 2021;22(1_suppl):71–83. https://doi.org/10.1177/11297298211018062.
46. Ellingson KD, Palekar RS, Lucero CA, et al. Vascular access hemorrhages contribute to deaths among hemodialysis patients. Kidney Int. 2012;82(6):686–92. https://doi.org/10.1038/ki.2012.185.
47. Jindal K, Chan CT, Deziel C, et al. Hemodialysis clinical practice guidelines for the Canadian Society of Nephrology. J Am Soc Nephrol. 2006;17(3 Suppl 1):S1–27. https://doi.org/10.1681/ASN.2005121372.
48. Fotiadis N, Shawyer A, Namagondlu G, Iyer A, Matson M, Yaqoob MM. Endovascular repair of symptomatic hemodialysis access graft Pseudoaneurysms. J Vasc Access. 2014;15(1):5–11. https://doi.org/10.5301/jva.5000161.
49. Shah AS, Valdes J, Charlton-Ouw KM, et al. Endovascular treatment of hemodialysis access pseudoaneurysms. J Vasc Surg. 2012;55(4):1058–62. https://doi.org/10.1016/j.jvs.2011.10.126.
50. Zink JN, Netzley R, Erzurum V, Wright D. Complications of endovascular grafts in the treatment of pseudoaneurysms and stenoses in arteriovenous access. J Vasc Surg. 2013;57(1):144–8. https://doi.org/10.1016/j.jvs.2012.06.087.
51. Hernández D, Díaz F, Rufino M. Subclavian vascular access stenosis in dialysis patients: natural history and risk factors. J Am Soc Nephrol. 1998;9(8):1507–10.
52. Lumsden AB, MacDonald MJ, Isiklar H. Central venous stenosis in the hemodialysis patient: incidence and efficacy of endovascular treatment. Cardiovasc Surg. 1997;5(5):504–9.
53. MacRae JM, Ahmed A, Johnson N, Levin A, Kiaii M. Central vein stenosis: a common problem in patients on hemodialysis. ASAIO J. 2005;51(1):77–81.
54. Schwab SJ, Quarles LD, Middleton JP, Cohan RH, Saeed M, Dennis VW. Hemodialysis-associated subclavian vein stenosis. Kidney Int. 1988;33(6):1156–9.
55. Taal MW, Chesterton LJ, McIntyre CW. Venography at insertion of tunnelled internal jugular vein dialysis catheters reveals significant occult stenosis. Nephrol Dial Transplant. 2004;19(6):1542–5.
56. Tabriz DM, Arslan B. Management of central venous stenosis and occlusion in dialysis patients. Semin intervent Radiol. 2022;39(01):051–5. https://doi.org/10.1055/s-0041-1742152.
57. Andrawos A, Saeed H, Delaney C. A systematic review of venoplasty versus stenting for the treatment of central vein obstruction in ipsilateral hemodialysis access. J Vasc Surg Venous Lymphat Disord. 2021;9(5):1302–11.
58. Agarwal AK. Central vein stenosis. Am J Kidney Dis. 2013;61(6):1001–15.
59. Agarwal AK. Endovascular interventions for central vein stenosis. Kidney Res Clin Pract. 2015;34(4):228–32.
60. Agarwal AK, Haddad NJ, Khabiri H. How should symptomatic central vein stenosis be managed in hemodialysis patients? Semin Dial. 2014;27(3):278–81.
61. Bakken AM, Protack CD, Saad WE, Lee DE, Waldman DL, Davies MG. Long-term outcomes of primary angioplasty and primary stenting of central venous stenosis in hemodialysis patients. J Vasc Surg. 2007;45(4):776–83.
62. Gray RJ, Horton KM, Dolmatch BL. Use of Wallstents for hemodialysis access-related venous stenoses and occlusions untreatable with balloon angioplasty. Radiology. 1995;195(2):479–84.
63. Maya ID, Saddekni S, Allon M. Treatment of refractory central vein stenosis in hemodialysis patients with stents. Semin Dial. 2007;20(1):78–82.
64. Wu TY, Wu CK, Chen YY, Lin CH. Comparison of percutaneous transluminal angioplasty with stenting for treatment of central venous stenosis or occlusion in hemodialysis patients: a systematic review and meta-analysis. Cardiovasc Intervent Radiol. 2020;43(4):525–40.
65. Kundu S. Central venous disease in hemodialysis patients: prevalence, etiology and treatment. J Vasc Access. 2010;11(1):1–7.
66. Vogel PM, Parise C. SMART stent for salvage of hemodialysis access grafts. J Vasc Interv Radiol. 2004;15(10):1051–60.
67. Liu Z, Huang J, Tang Y, et al. Outcomes of stent grafts for treatment of central venous disease in hemodialysis patients. J Vasc Access. 2022:11297298221134l. https://doi.org/10.1177/11297298221134142.
68. Quaretti P, Galli F, Moramarco LP. Stent grafts provided superior primary patency for central venous stenosis treatment in comparison with angioplasty and bare metal stent: a retrospective single center study on 70 hemodialysis patients. Vasc Endovasc Surg. 2016;50(4):221–30.

Part VII

Coronary Artery Disease

Introduction to Coronary Angioplasty

39

Iosif Xenogiannis

Case Presentation

Forty-six-year-old male visited the emergency department of our hospital complaining for substernal chest pain and concomitant resting dyspnea that started approximately 3 h prior. The patient was a heavy smoker, obese, and had no other significant medical history.

Blood pressure was 100/60 mmHg, heart rate was 110 bpm, respiratory rate was 22 breaths per minute and $SatO_2$ was 92%. The initial electrocardiogram showed ST-segment elevation in leads II, III and a VF and ST-segment depression in leads V4–V6. Echocardiography demonstrated inferior, posterior, and lateral wall akinesia, with a moderately reduced left ventricular ejection fraction (35–40%) and moderate mitral regurgitation. High-sensitive troponin-T was 189 pg/mL (normal values <15 pg/mL).

The patient was led to the catheterization laboratory….

Background

It has been 45 years since Dr. Andreas R. Grüntzig, in 1977, performed the first percutaneous coronary balloon angioplasty recanalizing the left anterior descending coronary artery (LAD) of a 38-year-old man suffering from angina pectoris. The intervention was successful paving the way for the genesis of percutaneous coronary interventions (PCI). Since then, important technological breakthroughs have been achieved with the development of new catheters and balloons, specialized guidewires as well as the introduction of stents and adjunctive devices, making PCI more effective and safe, improving its short- and long-term outcomes. While only a minority of patients, estimated at 10%, were eligible for PCI in the early years of interventional cardiology, nowadays, even the most complex subsets of coronary lesions are tried in patients who would have otherwise been exclusively surgical candidates (i.e. left main disease, calcified lesions, three-vessel disease) [1]. In addition, advances in hemodynamic support devices, such as impella development, allowed the performance of PCI in sicker patients with hemodynamic instability or compromised ventricular function.

Continued at page 438

Indications

The goal of PCI is to alleviate patient's symptoms and reduce major adverse cardiovascular events (MACEs), including death. Patients undergo PCI for either stable coronary artery disease (CAD) or acute coronary syndromes (ACS) (ST-segment elevation myocardial infarction [STEMI] and non-ST-segment elevation myocardial infarction [non-STEMI]/unstable angina). Whereas the advantage of PCI compared to non-invasive management, in terms of survival benefit, has been established by multiple studies in the setting of ACS, this is not the case for stable CAD [2–4]. With the exception of left main disease, no compelling data exist showing a survival benefit of PCI over optimal medical therapy for patients with stable CAD. [5] The contemporary ISCHEMIA trial reported no reduction in MACE or death rate in patients with moderate or severe ischemia randomized to receive an initial invasive strategy (74% PCI, 26% coronary artery bypass graft surgery [CABG]) as compared with those randomized to an initial conservative strategy. Patients with left main disease and left ventricular ejection infraction <35% were excluded [6]. Spontaneous MIs were reduced in the invasive group at the expense of more periprocedural MIs [7]. On the other hand, invasive strategy resulted in greater reduction in angina symptoms compared to non-invasive strategy [8]. Finally, a recent metanalysis which included 25 randomized trials spanning from 1978 to 2020 reported that revascularization in patients with stable CAD reduced cardiac mortality and spontaneous MI, illustrating that even in stable CAD, revascularization may have an effect on hard end-points apart from symptoms control [9].

I. Xenogiannis (✉)
Second Department of Cardiology, Attikon University Hospital, National and Kapodistrian University of Athens Medical School, Athens, Greece

G. Geroulakos et al. (eds.), *Mastering Endovascular Techniques*, https://doi.org/10.1007/978-3-031-42735-0_39

Contraindications

It should be noted that in some cases, PCI may be a life-saving intervention and at the same time the only therapeutic option. Thus, in such clinical scenarios there are no absolute contraindications.

Absolute contraindications:

- Bleeding disorders/diathesis (thrombocytopenia, coagulopathy, active peptic ulcer)
- Inability of the patient to co-operate during the procedure and take antithrombotic therapy

Relative contraindication:

- Inability to tolerate antiplatelets for the indicated period of time
- Multiple PCI restenosis
- Unsuitable coronary anatomy such as saphenous vein graft chronic total occlusion (CTO) [10]

Timing of Intervention

Optimal time of intervention differs for each clinical scenario. "Time is myocardium" in ACS and especially in the clinical setting of STEMI. Maximum accepted delay from STEMI diagnosis to primary PCI (wire crossing the lesion) is 120 min in order to choose primary PCI over fibrinolysis, otherwise fibrinolysis should be considered [11]. For patients who present at primary PCI centers, time from STEMI diagnosis to primary PCI should be ideally ≤60 min [11]. CABG is reserved for STEMI patients who had an unsuccessful primary PCI and have a large jeopardized myocardial area.

Regarding non-STEMI/unstable angina, in very high-risk patients (cardiogenic shock, hemodynamic or electrical instability, heart failure, mechanical complications, refractory chest pain despite medical therapy, ST-segment depression in ≥6 leads with accompanying ST-segment elevation in leads V1 and/or aVR) an immediate (< 2 h) invasive strategy should take place. For high-risk patients (established non-STEMI, dynamic/new ST-segment or T-wave changes, transient ST-segment elevation, GRACE risk score > 140) an early invasive strategy, within 24 h, is recommended. For all the other non-STEMI/unstable angina patients, non-invasive stress test or coronary computed tomography angiography (CCTA) *is usually* performed and then, according to the results, coronary angiography follows or not.

For patients with stable CAD, coronary angiography/PCI is performed as a scheduled intervention on the basis of a positive non-invasive examination (stress test of CCTA) or other high risk clinical or imaging *features* such as significant angina refractory to medical therapy or left ventricular dysfunction likely due to underlying stable CAD [12]. PCI can be performed either immediately after diagnostic coronary angiography, as an ad-hoc procedure, or at a later time. The second option is usually selected for more complex subsets, i.e., CTOs, severely calcified lesions or left main disease.

Pre-procedural Evaluation/Planning

Meticulous planning before the initiation of the operation is key to success. The first step is to evaluate the indication of procedure (stable angina vs. ACS). After explaining, in detail, the risks and benefits of the procedure, as well as describing the process, an inform consent should be obtained by the patient. Oral consent is also acceptable in critical situations such as STEMI or when the patient is in cardiogenic shock.

Thorough patient's medical history, if time allows in case of ACS, should be recorded. In specific, patient should be asked for prior history of CAD, heart failure, valvular heart disease, peripheral artery disease, bleeding disorders and renal failure. Bleeding disorders should be corrected, if feasible, while patients with renal failure may have to be hydrated before and after the procedure (for National Kidney Foundation stages 3b and 4) or undergo prophylactic hemofiltration (for National Kidney Foundation stage 4) especially before complex PCI. [13] Back pain or other conditions should be sought out since it may cause patient discomfort while on table, resulting in poor cooperation. Furthermore, review of prior heart catheterizations, CCTA or CABG reports, and ideally images and videos, is paramount. Knowledge of prior access site and the type of catheters that were used may save time as well as prevent unnecessary waste of equipment.

History of allergies to contrast, latex or aspirin has to be managed with pre-treatment with cortisone and antihistamines, usage of alternative types of gloves and potentially patient desensitization respectively. It is important to clarify if the patient is able to receive and maintain dual antiplatelet therapy (DAPT). Bleeding disorders/high bleeding risk and high chances of noncompliance with medication may preclude stent implantation since early DAPT discontinuation increases significantly the risk of in stent thrombosis.

Physical examination should focus on cardiovascular and respiratory system. Low blood pressure may imply the presence of cardiogenic shock and the patient should be treated with vasopressors and/or hemodynamic support devices. Upper and lower extremity pulses have to be checked. Absence of pulses can be the result of artery obstruction precluding its use for obtaining access. Surgical scar over the anatomical site of a radial artery implies that the artery has been harvested and thus is not available for gaining access. Arteriovenous fistulas for dialysis also exclude the corresponding arm as access site. In patients with apparent groin ulcers, scars or morbid obesity, radial artery should, in general, be preferred. In case of heart murmurs, valvular heart disease has to be sought out since it can affect the decision to proceed with PCI or not. Peripheral edema, jugular vein distention, and the presence of rales during lung auscultation are signs of congestive heart failure underlying the need for diuretic administration before or during the operation since the patient should ideally be euvolemic.

Hematocrit, platelet count, INR, creatinine and electrolytes such as potassium are the basic labs that should be checked. Significant anemia, thrombocytopenia, bleeding, and electrolyte disorders have to be corrected, if possible, before the operation. Serum beta hCG for women of childbearing potential could be checked.

Review of patient medication is crucial. Warfarin should be discontinued 5 days before a scheduled coronary angiography/PCI while for direct oral anticoagulants (DOACs) 2 days are enough in the absence of renal failure. When creatinine clearance is between 15 and 49 mL/min DOAC is in general hold for one or two additional days. Radial access is preferred if the patient did not have the chance to interrupt anticoagulants or for those with high INR. Furthermore, antiplatelet administration has to be checked and the patient should be loaded if he or she was not previously on aspirin or P_2Y_{12} inhibitor accordingly. Low molecular weight heparin (LMWH) and unfractionated heparin crossover is discouraged so if the patient was already on LMWH its use should be continued (see periprocedural antithrombotic agents administration) [14]. Metformin discontinuation before PCI is not routinely recommended but close creatinine monitoring is mandatory. For patients with renal failure, metformin should be withhold before the operation and not be restated for at least 48 h after [13]. Finally, administration of nitroglycerin is contraindicated in patients who have recently used phosphodiesterase type 5 (PDE-5) inhibitors, namely sildenafil, vardenafil, and tadalafil (within the first 24 h from the administration of the first two and 48 h for the third).

Monitoring

Close patient monitoring is imperative throughout intervention (Fig. 39.1). Before the initiation of the intervention, specific monitoring responsibilities should be shared among the

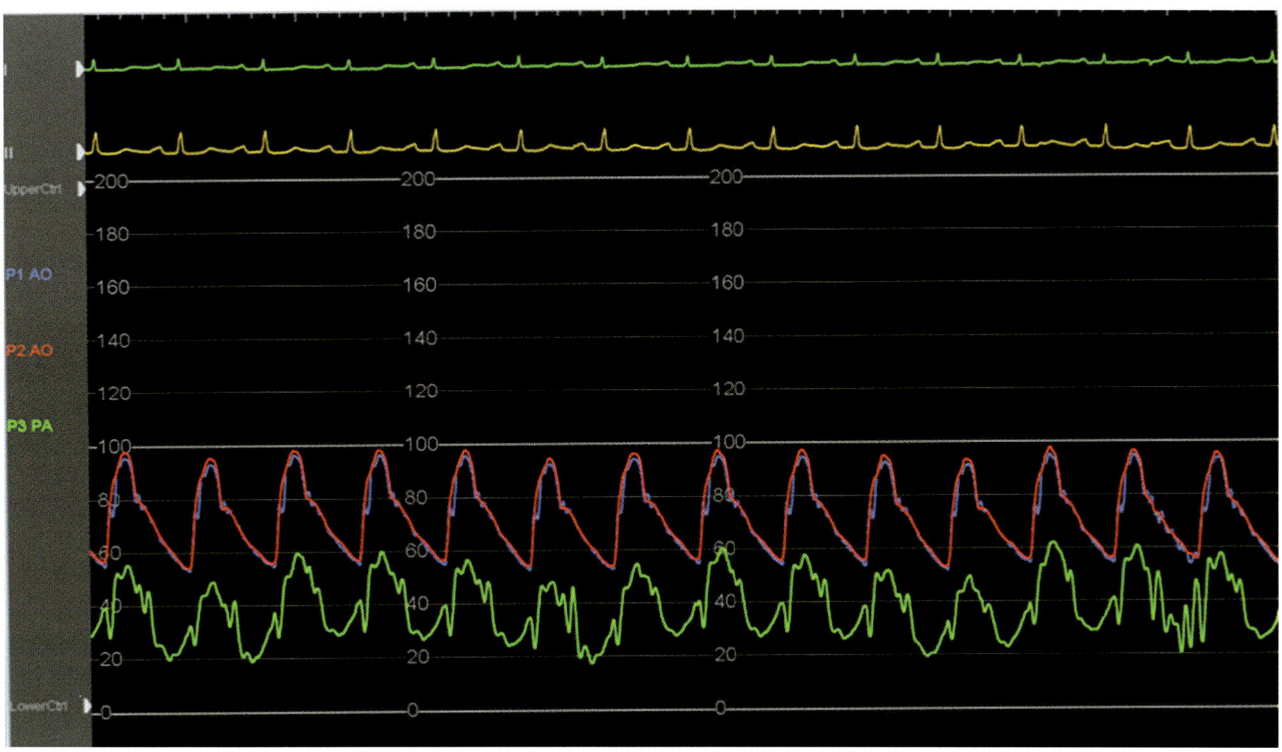

Fig. 39.1 Patient monitoring during percutaneous coronary intervention (PCI). Patient's rhythm and rate is recorded continuously (leads I and II, at the top of the screen). Arterial blood pressure values and tracing morphology are appreciated. In this case, there are two arterial blood pressure tracings because both coronary arteries were engaged, as it happens in the vast majority of chronic total occlusion PCIs. The red tracing is derived from the catheter in the left coronary artery while the blue tracing is derived from the catheter in the right coronary artery. Although, it is not routinely performed in every PCI, right heart catheterization is used for the estimation of pulmonary artery and capillary wedge pressure in patients who are at high risk for hemodynamic compensation. In this case, a swan-ganz catheter records pulmonary artery pressure (green tracing at the bottom)

members of the interventional team (physicians, registered nurses and technicians). Most of the times, a registered nurse is responsible for continuously evaluating patient's comfort during the operation while a registered nurse or a technician continuously monitors electrocardiogram and pressure waveform tracing. Patient discomfort may lead to movements that predispose to complications. Administration of sedatives (i.e., 1 mg of midazolam i.v.) can help anxious patient to relax and co-operate better. Analgesics such as opioids should be offered to patients with arm (mostly due to spasm), back or chest pain. Oxygen saturation has to be monitored continuously and dyspnea or tachypnea should be noted and managed usually with oxygen and/or diuretics administration. Patient's skin should also be examined in the setting of acute onset of dyspnea since the presence of rush is likely to be associated with allergy development to contrast or drugs. A relaxed patient allows the operator to focus on the operation without distractions and thus be more efficient.

Electrocardiogram (ECG) is constantly recorded during the operation. Ideally, (at least) three leads that "look at" the anterior, inferior and lateral left ventricle wall should be used. ECG is checked for heart rhythm and rate. Supraventricular or ventricular tachycardias may require antiarrhythmic drug administration while when the patient becomes hemodynamically unstable or develops ventricular fibrillation, electric cardioversion is indicated. On the other hand, bradyarrhythmias in the catheterization laboratory are managed with atropine or temporary pacing. ST-segment elevation or depression are associated with the presence of ischemia and should be treated according to the cause that provoked it.

Blood pressure tracing is another key element of optimal patient monitoring in the catheterization laboratory. Hypotension may be provoked by several causes that should be sought out in a timely manner and treated accordingly:

- Cardiogenic shock due to large infarcts or mechanical complications. Tachyarrhythmias also predispose to hypotension.
- Inferior infarcts frequently lead to hypotension especially when the right ventricle is involved and nitroglycerin has been administered.
- Excess nitroglycerin administration or nitroglycerin administration in patients with marginal blood pressure. Co-administration of nitroglycerin in patients on PDE-5 inhibitors may result in severe hypotension.
- Hypovolemic shock due to excess blood loss. Usually this can happen in the setting of retroperitoneal bleeding when femoral artery has been used for access.
- Allergic shock as a reaction to drugs or contrast.
- Vasovagal reaction because of discomfort or excessive pain.
- Tamponade.
- Pseudohypotension or false hypotension caused by the presence of air or thrombus in the catheter or in the connection between catheter and pressure transducer. Likewise, when the hemostatic valve or the Y-connector is open, blood pressure will erroneously appear to be low. Catheter kinking is another cause of false hypotension. Bulky equipment within the catheter may also result in recording low blood pressure on the arterial pressure tracing [15].

Extremely high blood pressure should also be treated as it can potentially predispose to pulmonary edema.

It is important to investigate the morphology of the arterial wave for pressure dampening or ventricularization (the arterial waveform simulates with the waveform that is recorded from the ventricles). Dampening is caused by too-deep engagement of the catheter or advancement of the catheter against aortic or coronary wall or next to an ostial coronary lesion. Injections should be avoided as they may lead to coronary artery dissection and the catheter should be repositioned until a normal arterial waveform is recorded (Fig. 39.2).

The presence of "pulsus paradoxus" on the arterial tracing is associated with tamponade development.

Contrast should be administered with caution. Risk scores have been developed for the prediction of contrast-induced acute kidney injury [16]. A simple rule is not to exceed 3.7 × creatinine clearance (glomerular filtration rate) mL in contrast [17]. If air kerma radiation dose exceeds 5–7 Gray and the intervention is not about to end the operator has to think about stopping the procedure [15].

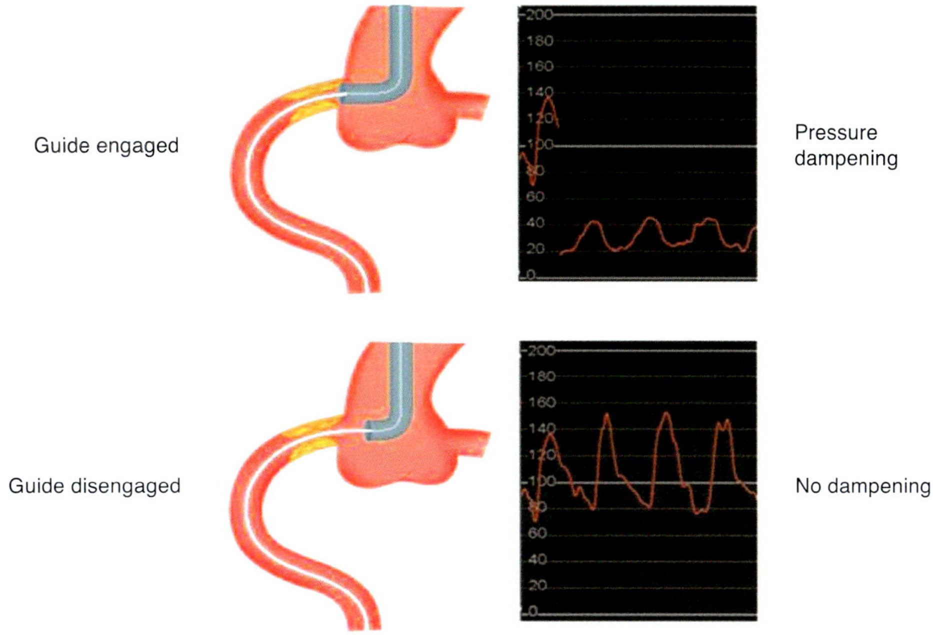

Fig. 39.2 Pressure dampening upon engagement of an aorto-ostial right coronary artery lesion. Reprinted from the Manual of Percutaneous Coronary Interventions. Emannouil Brilakis. A step-by-step approach. 2020. With permission by Elsevier/Academic press

Periprocedural Antithrombotic Agents Administration

Before angioplasty/stent implantation the patient should be on DAPT. If not already on DAPT, the patient has to be loaded with aspirin 150–300 mg and a P_2Y_{12} inhibitor. In general, for stable CAD, clopidogrel 600 mg is preferred while for ACS a more potent P_2Y_{12} inhibitor is indicated such as prasugrel 60 mg or ticagrelor 180 mg loading dose.

There are two categories of intravenous antiplatelet drugs currently in use: glycoprotein IIb/IIIa inhibitors and the short-acting P_2Y_{12} inhibitor, cangrelor. Two glycoprotein IIb/IIIa inhibitors agents are available: eptifibatide (double bolus of 180 μg/kg i.v., given at a 10-min interval, followed by an infusion of 2.0 μg/kg/min for up to18 h) and tirofiban (bolus of 25 μg/kg i.v. over 3 min, followed by an infusion of 0.15 μg/kg/min for up to 18 h). Glycoprotein IIb/IIIa inhibitors are not routinely administered during PCI but only as a bail-out therapy in the presence of bulky thrombus. Cangrelor has a very fast onset of action (2 min) that lasts for 30–60 min. It may be considered in P_2Y_{12} receptor inhibitor-*naïve* patients undergoing PCI and is very useful for patients who cannot be loaded per os (i.e. intubated patients). The recommended bolus dose is 30 mg/kg i.v. followed by 4 mg/kg/min infusion for at least 2 h or the duration of the procedure (whichever is longer).

Unfractionated heparin is the most commonly used anticoagulant during PCI. The recommended dose is 70–100 units/kg, reduced to 50–70 units/kg when concomitant glycoprotein IIb/IIIa inhibitor administration takes place. Activated clotting time (ACT) should be checked every 30–60 min. The target ACT values are 250–300 s for most PCIs while higher ACT values are recommended for CTO PCI (300–350 s for antegrade approach and >300–350 sec for retrograde approach) and lower in case of glycoprotein IIb/IIIa inhibitor administration (200–250 s). Bivalirudin is an alternate to unfractionated heparin and is preferred in patients with history of heparin induced thrombocytopenia (HIT). The recommended dose is: 0.75 mg/kg IV, followed by infusion of 1.75 mg/kg/h i.v. for up to 4 h after the procedure. The dose has to be modified for patients with renal insufficiency. Enoxaparin 0.5 mg/kg i.v. bolus, is another option. If the patient was already on s.c. enoxaparin and the last dose was given less than 8 h before, then, no extra dose of enoxaparin or unfractionated heparin is recommended. If the last dose was administered more than 8 h ago an additional 0.3 mg/kg i.v. bolus dose is recommended. Crossover between unfractionated heparin and LMWH is discouraged. If the patient was on fondaparinux, a single dose of unfractionated heparin 85 units/kg is recommended.

Access

Radial and femoral artery are the most common sites of access. Ulnar artery is relatively frequently used as an alternate to radial artery whereas other sites such as branchial, axilliary or carotid artery or transcaval approach are rarely used and are more of historical interest. Both the American and European guidelines support the routine use of radial artery for gaining access irrespectively from the clinical setting (stable CAD and ACS) [13, 18]. In ACS the use of radial artery has been associated not only with less bleeding and vascular complications but, more importantly, with survival benefit [19, 20]. Nevertheless, femoral artery as the site of access could be preferred in specific clinical contexts: complex PCI cases where large bore catheters are used and increased support is essential, such as CTO PCI.

Patients with prior CABG comprises a group of patients that femoral access is usually required especially when both internal mammary arteries have been used as grafts [21]. In addition, femoral artery is the access site of choice in transcatheter aortic valve implantation or in other structural procedures and in cases where hemodynamic support devices should be placed.

Although echocardiographic guidance for femoral artery puncture has been associated with higher rates of first-pass success, lower number of attempts, less venipuncture and lower number of access complications, still, many operators use only femoral pulsation and anatomic landmarks, with or without fluoroscopy, to define the point of puncture [22]. Steps of state-of-the-art femoral access technique have been previously described [23]:

- After palpation of femoral pulse and cleaning of the area, fluoroscopy is used for the identification of the lower edge of the femoral head which is marked with the use of a hemostat or another radiopaque material. Using a sterile marker to mark the position of the skin so as not to miss it later on is also encouraged (Fig. 39.3a).
- The ultrasound probe is placed above the point of lower femoral edge and moved proximally and distally in order to detect the bifurcation of common femoral artery to superficial and profunda femoral arteries (Fig. 39.3b).
- Local anesthetic is injected subcutaneously under ultrasound guidance.
- A 18-gauge or preferably a 21-gauge needle (micro puncture needle) is advanced under ultrasound imaging towards common femoral artery. Pulsatile blood flow from the back end of the needle means that the needle has entered artery's lumen.
- A guidewire is then advanced over the needle. Fluoroscopy will confirm optimal position of the wire.
- A sheath is then advanced over-the-wire. Usually 6–8 Fr sheaths are used for PCI via femoral artery while in case of micropuncture a 4 Fr sheath is initially placed, changed then for a 6–8 Fr sheath.
- Finally, angiography is performed to confirm optimal sheath position and antegrade flow as well as to exclude the presence of dissection or perforation. Note that too high sheath insertion predisposes to retroperitoneal bleed-

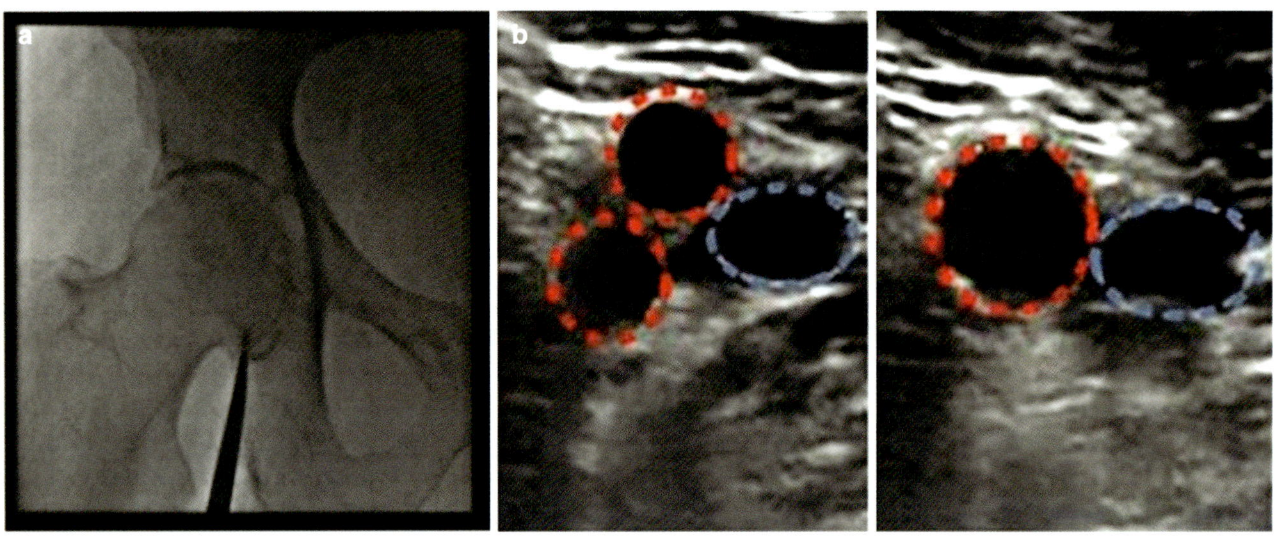

Fig. 39.3 (**a**) Fluoroscopy is used in order to identify the femoral head. The lower edge of the femoral head is marked with a hemostat. (**b**) Ultrasound is then used in order to guide puncture. When both superficial and profunda femoral artery are visible the probe has been placed under bifurcation (left panel) while when only common femoral artery is visible the probe is above femoral bifurcation (appropriate puncture site). Common, profunda and superficial femoral arteries are indicated with red color while femoral vein is indicated with blue color. Reprinted from JACC Cardiovasc Interv. Volume 10, issue 22. Sandoval Y, Burke MN, Lobo AS, et al. Contemporary arterial access in the cardiac catheterization laboratory. Pages 2233–2241, 2017. Used with permission from Elsevier

ing while too low sheath insertion predisposes to pseudoaneurysm.

The steps that are followed for gaining radial access are similar to those previously described for femoral artery with the exception that fluoroscopy is not used before puncture. Allen's test is not mandatory given the results of the RADAR trial [24]. The right radial artery is preferred with the exception of cases that LIMA has been used as a graft and has to be engaged and of patients who have undergone transcatheter aortic valve implantation (TAVI) in the past where left radial artery is preferred.

- The radial artery is palpated and the wrist is sterilized.
- Ultrasound is used for the localization of the radial artery. This step is usually omitted by operators; nevertheless, in RAUST trial, usage of ultrasound guided radial artery cannulation improved success and efficiency of the technique [25].
- Local anesthetic is administered. It is recommended to infuse no more than 1–2 mL of anesthetic in order not to compress the artery.
- Radial artery is punctured. The optimal point of entrance is 1–2 cm distally to the radial styloid process. Two techniques have been described: the anterior wall technique and the through-and-through technique [26]. In the first one, the needle is inserted until back-bleeding is noticed. Regarding the second technique, either a micropuncture needle or a needle/catheter assembly (Angiocath) is advanced initially through the anterior (at this point back bleeding will be noticed) and then through the posterior wall. Then, the micropuncture needle is withdrawn until pulsatile bleeding from its back end is seen and a wire is inserted. In case of Angiocath utilization, the needle is removed and then, the catheter is carefully withdrawn until back bleeding is pointed followed by wiring of the artery.
- A 0.018″ wire in inserted followed by removal of the needle.
- A 4–6 Fr sheath for diagnostic angiography or a 6–7 Fr (rarely 5 Fr) for PCI is placed.
- The dilator and the wire are removed. The sheath is aspired, flushed and secured. A 5000 IU unfractionated heparin and 2.5 mg verapamil cocktail is administered through the sheath in order to prevent radial artery spam and obstruction (Fig. 39.4).

Distal radial artery access (puncture of the radial artery in the anatomic snuffbox) is an emergent technique that is very useful when left radial artery is selected for access since it is more comfortable for both the patient and the operator [27]. It is more demanding technically compared with the conventional transradial approach but it has been associated with a lower rate of forearm radial artery occlusion and shorter time of hemostasis. [28]

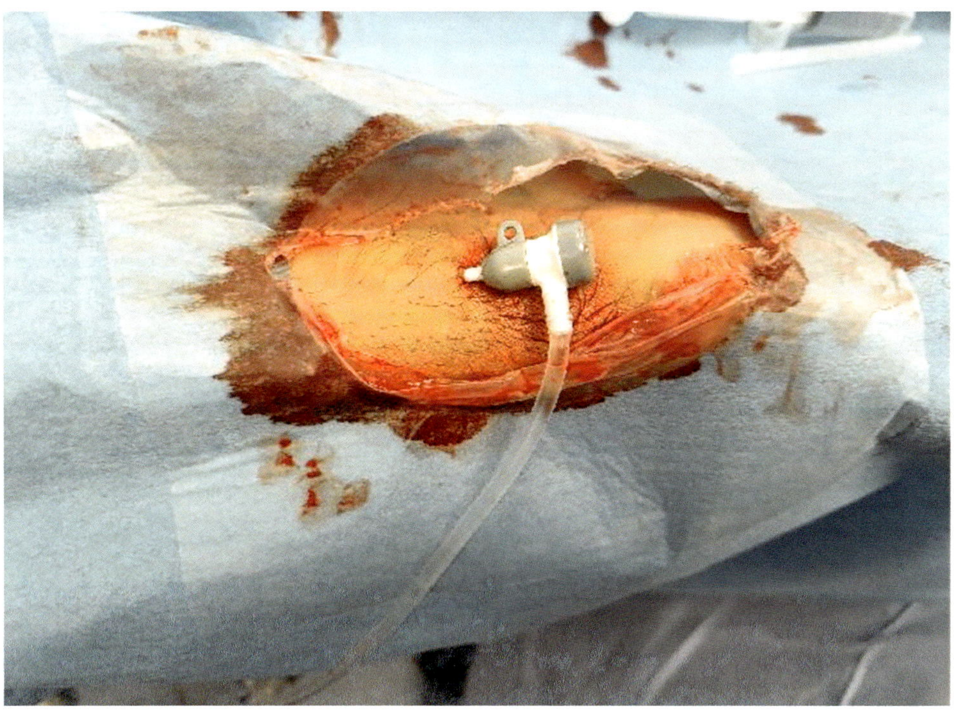

Fig. 39.4 Access through the right radial artery using a 5 Fr sheath

Coronary Artery Engagement and Angiography

The catheters are loaded and advanced over a 0.035″ or a 0.038″ J-tip wire which should always precede when the wire/catheter assembly is advanced in patient's vascular system. Resistance to wire advancement can be caused by tortuosity, accidental entrance in a side branch, presence of aneurysm, spasm or significant stenosis/occlusion. When resistance is felt any further effort to advance the wire should be halted. Under fluoroscopy, the catheter is advanced at the point of resistance. After pressure recording, angiography is performed in order to clarify the cause of resistance. If spasm is identified, nitroglycerin and/or verapamil is administered in the relevant artery. When tortuosity, stenosis, side branch entrance or aneurysm is identified, a different type of wire (such as a hydrophilic, polymer-coated or very soft tipped) is tried. In case of a tortuous iliac artery, a long (45 cm) sheath may straighten the vessel and allow wire advancement. If all the measures fail or complete vessel occlusion is found, an alternative site of access should be tried.

When the wire reaches the ascending aorta, it is looped over the aortic valve. Then, coronary arteries ostia engagement follows. Usually, the left coronary artery (LCA) is engaged first, followed by right coronary artery (RCA) engagement. At posteroanterior (PA) view, the diagnostic catheter is advanced 2–3 cm over the right coronary cusp bottom and the wire is withdrawn. Usually, this movement results in the engagement of the left main, especially when the femoral artery is used for access. Sometimes, gentle withdrawal with slight rotation (either clockwise or counterclockwise in needed). Regarding RCA engagement, the catheter is again advanced over the wire until it reaches the bottom of right coronary cusp. Using a left anterior oblique (LAO) view, the wire is then withdrawn and the catheter is rotated clockwise and withdrawn with the right hand while, at the same time, the left hand moves the catheter back and forth [15].

Two types of catheters are used: diagnostic catheters for diagnostic coronary angiography and guide catheters for PCI. The main differences between them are that guide catheters have thinner walls and larger lumen in order to accommodate the equipment that will we advanced through them to the coronary arteries and have softer tips.

Different diagnostic catheters are used for the engagement of the LCA and RCA. The most frequently used diagnostic catheters are Judkins Left (JL) 3.5, 4, 4.5 and 5 (Fig. 39.5a) and Amplatz left (AL) 1, 2 and 3 for the LCA and Judkins right (JR) 3.5, 4 (Fig. 39.5b), Amplatz right (AR) 1, 2, AL 1, Williams and Hockeystick for the RCA. Jacky and Tiger are specialized catheters that have

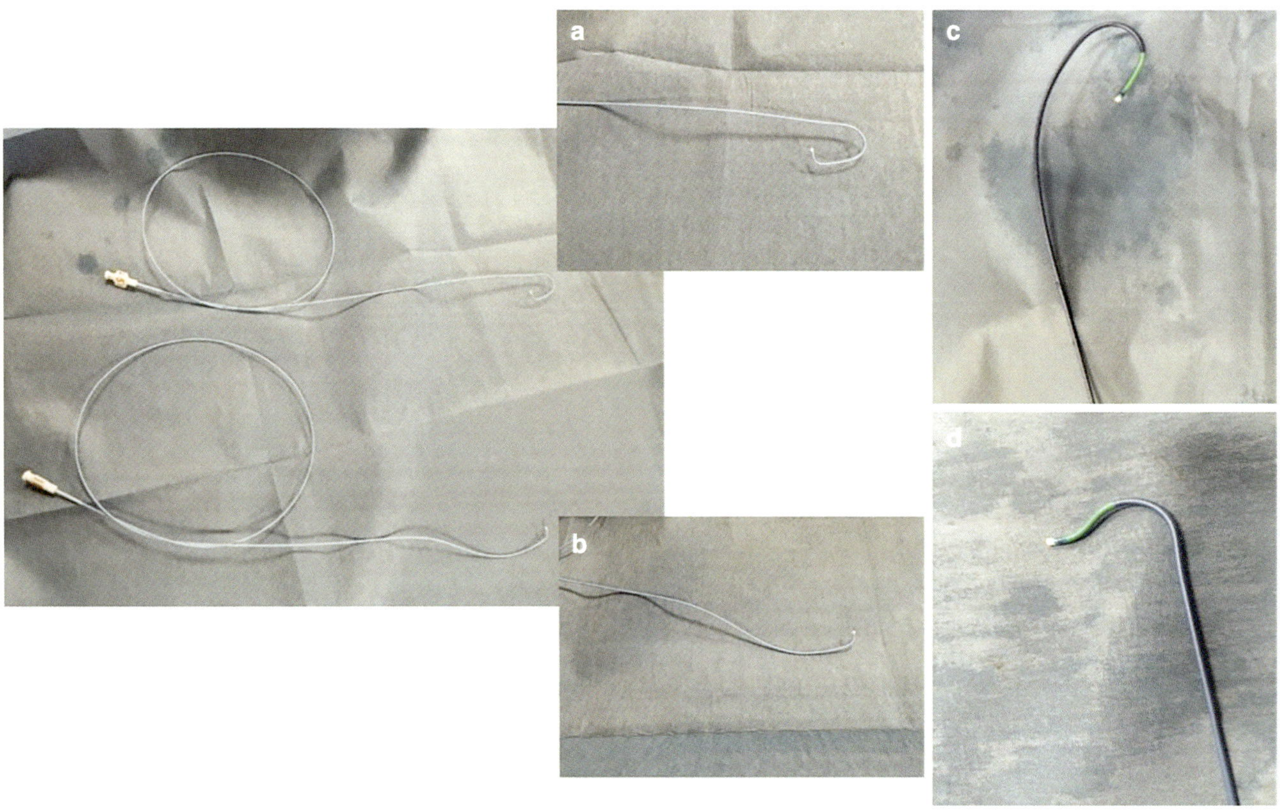

Fig. 39.5 (**a**) JL 3.5 (Cordis), 5 Fr diagnostic catheter. (**b**) JR 4 (Cordis), 5 Fr diagnostic catheter. (**c**) EBU 3.5 (Medtronic), 6 Fr guide catheter. (**d**) AL1 (Medtronic), 7 Fr guide catheter

been developed for the engagement of coronary ostia via radial access and are used for both the LCA and the RCA. Regarding right-sided grafts, a JR 4 is used initially, followed by multipurpose and RCB catheter in case of failure. JR 4 can also be used for left-sided grafts with AL 1 and LCB being alternative options. The IM catheter is used for the engagement of internal mammary arteries but JR 4 can be tried too. Bartorelli-Cozzi catheter has been designed for the engagement of LIMA via the left radial artery.

EBU 3.0, 3.5, 3.75, 4, 4.5 (Fig. 39.5c) and XB 3.0, 3.5, 4, 4.5 guide catheters are used for LCA PCI since they provide better support compared with JL guide catheters. Accordingly, AL 1 (Fig. 39.5d) and 3D Right provide stronger support compared with JR 4 guide catheter and are preferred for complex RCA PCI. Ikari left and Ikari right are specialized guide catheters used for LCA and RCA PCI respectively, when the intervention is performed via radial access.

The catheter is aspired, then connected to the manifold and the pressure tracing is checked before injection. Multiple views of the LCA and RCA are appreciated with left, right, cranial and caudal image receptor angulation (Fig. 39.6). When recording cineangiograhic videos, contrast injection is performed 1–2 s after pedal pressure in order to allow enough time for visualization of calcium or any priorly implanted stents. Contrast should be used with caution. It is important not to inject more than 2–3 mL each time in order to avoid excessive contrast administration that may lead to contrast induces nephropathy. Orthogonal projections of coronary artery segments are essential in order not to miss any stenoses. In general, > 50% stenosis in the left main and >70–90% in the other coronary arteries are considered significant. Intracoronary administration of nitroglycerine (100–300 μg) is essential in order not to erroneously interpret spam as atherosclerotic stenosis. For stable CAD patients, intermediate lesions (40–90% according to the European guidelines and 40–69% according to the American guidelines) with no prior proof of ischemia by a stress test have to be evaluated with fractional flow reserve (FFR) or instantaneous wave free

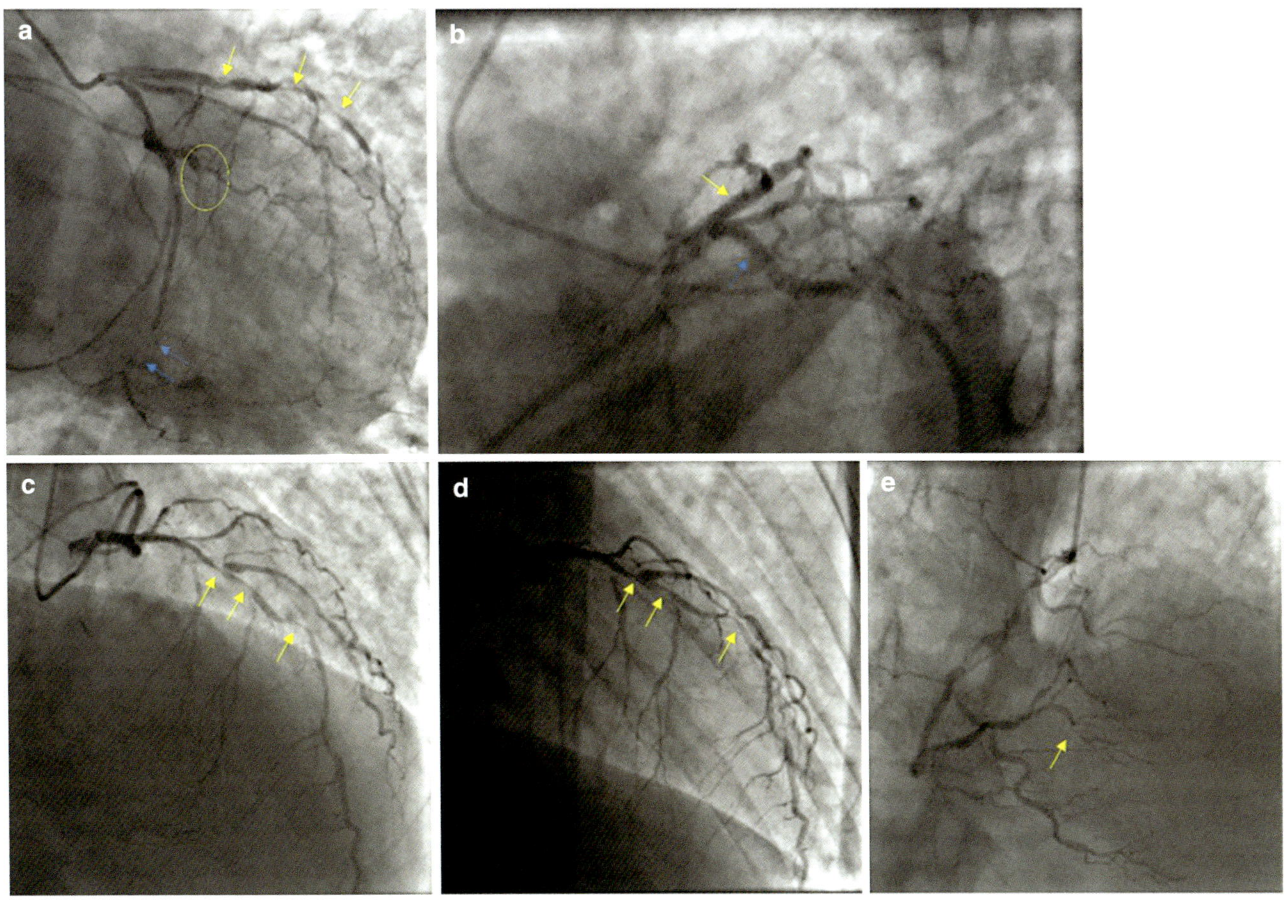

Fig. 39.6 Coronary angiography of the patient described in the introduction. Access was obtained via the right radial artery using a 6 Fr. sheath. (**a**) A JL 3.5 diagnostic catheter was used for the engagement of the left coronary artery. PA caudal view: multiple significant lesions in the left anterior descending coronary artery (LAD) (yellow arrows). The first obtuse marginal was totally occluded (circle). Subtotal occlusion was noticed in the distal circumflex (blue arrows). (**b**) LAO caudal (spider) view: the left main, proximal segment of the LAD (yellow arrow) and circumflex (blue arrow) have no significant stenoses. (**c**) PA cranial and (**d**) RAO cranial view. Multiple sequential significant lesions in the mid-LAD (yellow arrows). (**e**) Right coronary artery was engaged with a JR 4 diagnostic catheter. LAO cranial view: subtotal occlusion of the posterior descending coronary artery (yellow arrow)

ratio (iFR). If FFR is ≤0.8 or iFr is ≤0.89, PCI is recommended (see **Lesions functional assessment and intravascular imaging)** [13, 18]. TIMI (Thrombolysis in Myocardial Infarction) flow grade should also be appreciated. Grade 0: no antegrade flow distal to the lesion. Grade 1: contrast passes the obstruction but fails to opacify entire coronary bed. Grade 2: contrast passes the obstruction and opacifies entire coronary bed but at a slower rate compared with the non-culprit vessel. Grade 3: antegrade flow as fast as in non-culprit vessel (normal flow).

In STEMI patients, the culprit vessel is the one that supplies the ischemic area as is defined by ST-segment elevation in the corresponding ECG leads. With respect to angiographic findings, in the context of STEMI, the culprit vessel is most of the times totally or sub-totally occluded. Segmental wall hypokinesis or akinesis, further assists the identification of the culprit vessel in ACS patients. Defining the culprit vessel is more difficult in the setting of non-STEMI/unstable angina. Most of the times the culprit lesion is found in the proximal or mid segments of the vessels and the LAD is more frequently the culprit vessel [29–31]. Intraluminal haziness, contrast filling defects and contrast staining are supportive of thrombus presence and along with plaque ulceration (presence of contrast beyond the vessel lumen), irregular plaque borders and impaired flow are angiographic features suggestive of plaque rupture. Interestingly, more than one stenoses may have features of unstable plaque [14].

In every patient with CAD the goal is to achieve complete revascularization. In the setting of STEMI, the operator should recanalize initially the culprit vessel and then should attempt to recanalize the remaining significant bystander lesions either during the index procedure or in a staged manner. In STEMI patients, after culprit lesion recanalization, CABG should be considered in the presence of remaining complex multivessel disease that complete revascularization is unlikely to be achieved with PCI. Likewise, in patients with non-STEMI or stable CAD complete revascularization can be attempted at a single operation or in a staged manner. CABG is favored over PCI in cases of three-vessel disease (especially with SYNTAX score > 22), three-vessel disease in diabetic patients, left main disease with SYNTAX score > 32, multivessel disease with ejection fraction ≤35% and whenever it is believed that the anatomy is not amenable to complete revascularization with PCI [13].

Wiring

Guidewires are 0.014 in wires made by stainless steel or nitilon that are used in order to cross the target lesion. They use as a rail over which balloons and stents are advanced (Fig. 39.7a). The first step of wiring is to select the appropriate guidewire. For non-complex lesions, a workhorse guidewire is the first option. The characteristics of workhorse guidewires are that they have a soft tip that is usually less than 1 g and are not polymer-jacketed. Some of the most commonly used workhorse guidewires worldwide (i.e. Sion Blue [Asahi Intecc] and Samurai [Boston Scientific]) have hydrophilic coating that spares their tip in order to navigate easier through tortuous vessels and at the same time be safer than hydrophilic guidewires. Hydrophilic (hydrophilic covering of either the tip or multiple cm of the guidewire) and polymer-jacketed guidewires (covered by slippery polymer, such as Fielder-XT [Asahi Intecc], Sion Black [Asahi Intecc], Pilot 200 [Abbott Vascular]) have better deliverability through tortuous vessels compared with non-hydrophillic workhorse guidewires, however, it is more likely to injure the coronary artery causing either perforation or dissection. Polymer-jacketed guidewires are especially useful for the crossing of CTOs.

Before the insertion of the guidewire to the catheter and the coronary artery, its tip has to be shaped. Some guidewires have a preshaped tip so there is no need for shaping but, on the other hand, they leave less flexibility to the operator. In general, larger vessels require big bends while smaller vessels require small bends. For some lesions, more than one bend is needed for successful crossing. The guidewire is then advanced into the catheter through the hemostatic valve and then, under fluoroscopy, into the coronary artery. For advancing the guidewire through the left main to the ostium of LAD or circumflex, caudal views are preferred. Then for navigating into the LAD, a right anterior oblique (RAO) cranial or PA cranial view is used. With respect to the RCA, initially a LAO view is used, changed for a LAO or PA cranial view for wiring through the posterior descending artery or posteriolateral branch.

The wire is advanced using a torque. For guidewire advancement, the combination of a forward movement/pushing (and backward/withdrawing when the wire has accidentally entered a side branch) with simultaneous rotation is used. Various techniques have been applied: many operators prefer to do both movements with one hand while others prefer to move forward and backward the guidewire with their left hand and at the same time rotating it with the torque using the right hand. It should be noted that the final few cm. (usually 3) of the guidewires are radiopaque allowing better visualization of the tip and estimation of lesion's length when it is crossed. If a lesion is resistant to crossing, change of the wire's tip or using a more advanced (hydrophilic or polymer-jacketed) guidewire usually solves the problem. For the most complex subsets of lesions and especially for CTOs, the use of a microcatheter is essential. Microcatheters (such as Corsair, Caravel and Turnpike) are small catheters with distal shaft outer diameter ranging from 1.8 to 3.2 Fr that are advanced over the guidewire and through the guide catheter into the coronary artery. They provide better support enhanc-

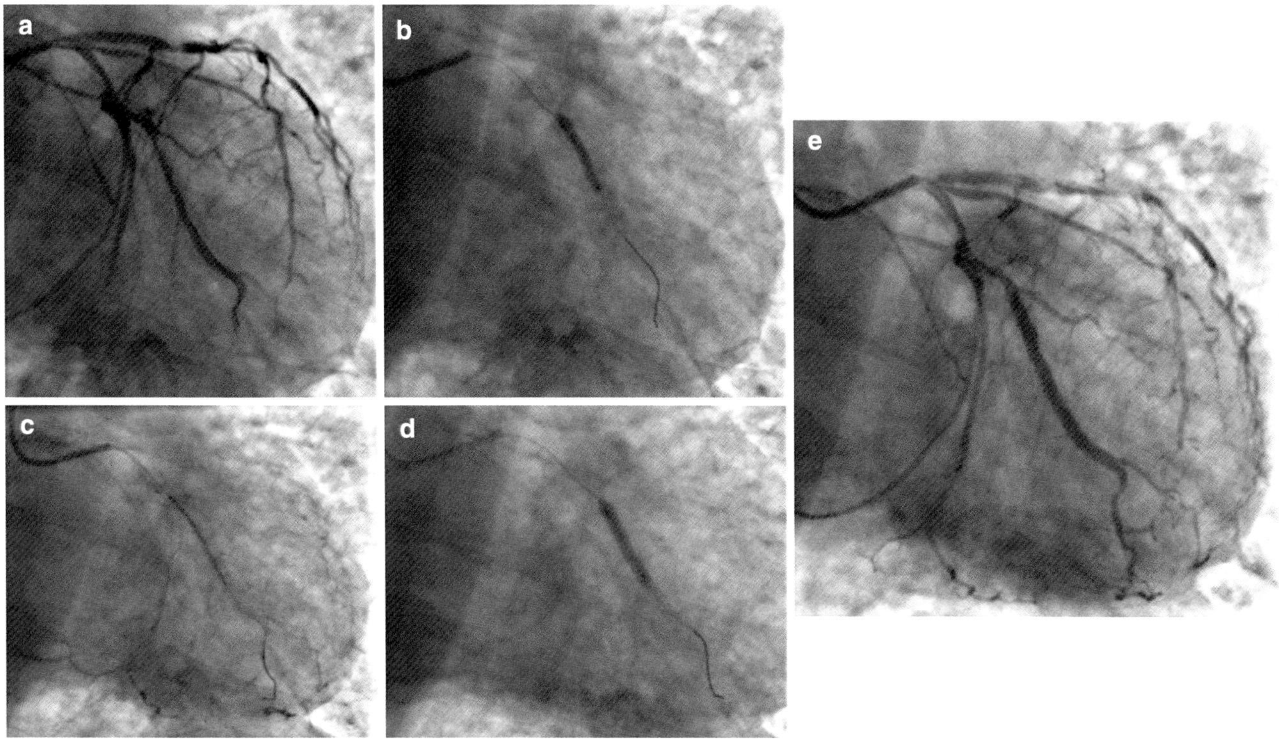

Fig. 39.7 Percutaneous coronary intervention in the first obtuse marginal branch of the patient described in the introduction. Combining the electrocardiographic, echocardiographic and angiographic findings, the first obtuse marginal was considered as the culprit vessel. (**a**) The left main was engaged with an EBU 3.5, 6 Fr guide catheter. A Sion blue (Asahi Intecc) guidewire was advanced through the occlusion restoring TIMI flow III. The blue arrow points at the radiopaque segment of the guidewire. (**b**) Predilatations were perfomed with a semi-compliant 2.75 × 20 mm balloon. (**c**) A 2.75 × 32 mm Promus ELITE (Boston Scientific) drug-eluting stent was placed. (**d**) Post-dilatations with a 2.75 × 20 mm non-compliant balloon. (**e**) Final result. Balloon angioplasty in the distal circumflex preceded final injection, but is not shown in the figure

ing the penetration capacity of the guidewire. Furthermore, they allow reshaping of the guidewire's tip and protect the proximal part of the vessel from injuries caused by the guidewire. [32] After lesion crossing with an advanced guidewire it is essential to change it with a workhorse wire in order to avoid distal vessel injury. Microcatheters facilitate wire exchange. The position of the guidewire should be monitored continuously as too distal position of the wire can result in perforation.

Lesion Preparation

Lesion preparation is a critical step of angioplasty. Balloon angioplasty precedes stent implantation in essentially every PCI as it facilitates the optimal selection of stent dimensions (diameter and length) that is going to be placed as well as its delivery to the lesion (Fig. 39.7a). In addition, suboptimal balloon expansion caused by calcification indicates the need for further lesion preparation with specialized equipment before stent implantation. There are two balloon delivery systems: monorail and over-the-wire with the former being used in the majority of cases. Balloons are separated to semi-compliant and non-compliant. While semi-compliant balloons have better deliverability, non-compliant balloons do not exceed much their maximum diameter with high inflations, thus, they have a lower risk of dissection or rupture. After meticulous preparation, the balloon is loaded to the back end of the guidewire and is advanced through the catheter to the target lesion while it has been connected with the indeflator and negative suction has been applied. In case of balloon uncrossable lesions, the following algorithmic approach is suggested [33]:

- Initially small balloons (1.0–1.5 mm) are tried to cross the lesion.
- Small balloons are advanced as far as possible into the lesion and are inflated at high-pressures until they rupture, modifying the plaque (grenadoplasty).
- Increase support by (1) using larger catheters (7 or 8 Fr), (2) using a second wire and (3) try anchoring techniques or/and (4) use a guide catheter extension.

- Advancement of a microcatheter such as Tornus (Asahi Intecc) or Turnpike Gold (Teleflex) may modify the lesion.
- Cutting-wire techniques: a second wire is advanced through the lesion. Then a balloon is advanced over the first wire, as close as possible, to the proximal cap of the lesion and is inflated. The second wire is pulled, with the balloon inflated, cutting and modifying this way the proximal cap of the lesion.
- Usage of rotational or orbital atherectomy or laser.
- Subintimal techniques: a second guidewire is advanced into the subintimal space around the balloon-uncrossable lesion. The balloon is then inflated in the subintimal space, crushing the balloon-uncrossable lesion from "outside" potentially allowing the passage of a balloon over the true lumen wire (subintimal external crush technique). Alternatively, a balloon is advanced over the subintimal space guidewire distal to the balloon-uncrossable lesion. It is then inflated "pinning" the true lumen guidewire, allowing balloon crossing over it. These techniques are used mostly for CTO lesions.
- A combination of the previously described techniques increases the chances of balloon crossing.

When the balloon has crossed the lesion is inflated (up to 12–14 atm for semi-compliant and up to 20–24 atm for non-compliant balloons) with the use of the indeflator. The balloons are sized ≤1:1 to the target vessel diameter. The presence of a waist on the balloon during inflation indicates suboptimal lesion expansion that requires further lesion preparation with specialized equipment (balloon-undilatable lesions). In the presence of large thrombus, although not routinely recommended due to the risk for stroke, aspiration thrombectomy through dedicated catheters is applied.

Wolverine (Boston Scientific) and Angiosculpt (Philips) are specialized plaque modification balloons (cutting and scoring balloon respectively) that are used for plaque modification when suboptimal lesion expansion has been achieved with the use of conventional balloons. In cases where plaque modification balloons have failed or extreme calcification has been detected with intravascular imaging, the use of adjunctive devices is mandatory for successful PCI. Regarding rotational atherectomy (Boston Scientific), an electroplated diamond-coated burr rotates at high speeds (140,000 rpm or greater) ablating calcific plaque while it spares the elastic arterial structure. The generated 4–8 μm particles pass through microcirculation and being collected by reticuloendothelial system [34]. The burr is advanced over the rotawire, a specilized 0.009 in. guidewire. The less frequently used orbital atherectomy (Cardiovascular Systems) consists of a diamond-coated crown that completes fast, elliptical rotations and increases its orbit, ablating the calcific plaque and creating 2–4 μm particles. The crown is advanced over the dedicated Viper wire [34]. Atherectomy is used for both balloon-uncrossable and balloon-undilatable lesions [35]. For the latter, it may be the only solution, along with laser, since plaque modification and lithotripsy balloons cannot reach the lesion. Excimer laser coronary atherectomy (ELCA, Phillips) generates a single wavelength directional beam of photoenergy to modify atherosclerotic plaques. On the contrary to atherectomy, it can be advanced over any 0.014 in. wire. It also has the advantage that it can be used in cases where a guidewire cannot cross the lesion, ablating its proximal cap without a leading guidewire [36]. Lithotripsy (Shockwave Medical, Inc.) is another emerging technique used to overcome balloon-undilatable lesions. The lithotripsy balloon is inflated initially at a pressure of 4 atm over the calcified lesion, followed by delivery of impulses of mechanical energy through miniaturized emitters, causing vibrations that crack and fracture the calcific components of the atherosclerotic plaque including deep calcium. Compared with atherectomy and ELCA it has the advantage of a short learning curve and low risk of dissection/perforation [37].

Stenting

Stent implantation follows balloon angioplasty in almost every PCI as it increases procedural success and reduces restenosis (Fig. 39.7c) [38]. Both bare metal stents (BMS) and drug eluting stents (DES) consist of a scaffold structure composed by either stainless steel, cobalt chromium, platinum chromium or nitilon and a delivery system. DES stents have a polymer coating that is either permanent or biodegradable and carries the anti-proliferative drug. The newer generation DES are covered and elute everolimus, zotarolimus, sirolimus, or biolimus A9. DES reduced significantly the rate of in-stent restenosis compared with BMS with the exception of PCI in saphenous vein grafts [39–41]. Furthermore, second-generation DES have a lower rate of in-stent thrombosis putting BMS out of use for every possible clinical scenario [42].

After preparation, the stent is loaded on the guidewire and advanced through the guide catheter to the stenosed segment of the target vessel. Shorter stents with thin struts have better deliverability. The stent should cover the entire length of the lesion and land on "healthy boarders" covering also the segments that were injured by balloon angioplasty. With respect to stent diameter, the stent is sized based on the distal segment of the target coronary artery. The balloon of the stent is then inflated at 12–14 atm deploying the stent in the coronary artery. When discrepancy between proximal and distal vessel diameter exists, the proximal segment of the stent is post-dilated with a non-compliant balloon in order to match the diameter of the proximal segment of the stented vessel (proximal optimization technique) (Fig. 39.7d).

In case of stent underexpansion, the stent is postdilated with non-compliant balloons inflated at high pressures (up to 26–28 atm). If the stent remains underexpanded, in-stent inflations of a lithrotrpsy balloon are attempted. Alternatively, or in case of failure, OPN balloon which is a specialized two-layer, non-compliant balloon that delivers pressures up to 35–40 atm, can be tried.

Access Closure

There are two options for femoral artery closure: using a closure device or applying manual pressure. Although no studies have shown an advantage of closure devices over manual pressure in terms of complications reduction, they are more convenient for both the patient and the operator. Angioseal (Terumo) and Perclose (Abbott Vascular) are the two most commonly used closure devices. The usage of closure devices requires femoral artery lumen diameter >5 mm without any significant disease and femoral access above bifurcation. If manual pressure is selected as mean of hemostasis, the sheath is removed when ACT is <180 s. A general rule that is followed is to press the femoral artery for 2–3 min × sheath size in Fr. For example, after removal of a 5 Fr sheath, manual pressure is applied for 10–15 min.

With respect to radial artery compression, specialized compression devices, such as the TR band (Terumo), Prelude Sync (Merit Medical) and Tracelet (Medtronic) are applied. -Pressure on the radial artery is achieved with either mechanical pressure through a screw press or a band-tightening press or pneumatic pressure through an inflatable air chamber. The applied pressure is gradually reduced by loosening the compression device (i.e., rotating the screw to reduce mechanic pressure or removing part of the air). In general, compression devices should be removed in no more than 2 h so as to reduce the chances of radial artery occlusion [43].

Lesions Functional Assessment and Intravascular Imaging

As it has already been mentioned, intermediate lesions in the setting of stable CAD or ACS (for non-culprit lesions) should be further assessed before proceeding with stenting, if evidence of ischemia by non-invasive indices is not available. For doing this, specialized pressure guidewires are used for the estimation of FFR and non-hyperemic indices. FFR is defined as the ratio between the maximal myocardial blood flow distal to a coronary stenosis and the theoretical maximum flow in the same vessel. Maximal flow is achieved with the intravenous or intracoronary administration of a vasodilator, most commonly adenosine. It has been shown that lesions with FFR > 0.8 can be safely deferred while PCI reduces the need for urgent revascularization for lesions with FFR ≤ 0.8 [44]. Although guidelines use the cutoff point of 0.8, metanalyses have shown that is safe to defer stenting for lesions with FFR in the gray zone (0.75–0.8) [45]. Non-hyperemic indices such as the iFR that isolates a specific period in diastole, the resting full cycle ratio (RFR) and the diastolic hyperemia-free ratio (DFR) have been developed omitting the need for adenosine [46]. For the previously mentioned indices 0.89 is the proposed cutoff point.

After meticulously flashing the guide catheter and administering intracoronary nitrates, the proximal end of the radiopaque portion of the pressure wire is advanced at the tip of the guide catheter. The pressure measured by the guidewire is then equalized with the pressure that is recorded by the guide. Then, the pressure guidewire is advanced distal to the lesion. Non-ischemic indices can then be appreciated: the pressure from the pressure wire and the pressure from the guide catheter are simultaneously recorded and the resting index is calculated. For the estimation of FFR, a vasodilator (adenosine, regadenoson, or papaverine) is administered either intracoronary (adenosine dose: 50–100 μg for the RCA and 200 μg for the LCA) or intravenously (adenosine dose: 140–180 μg/kg/min) followed by the calculation of the ration between the pressure recorded by the guidewire and the guide catheter.

Intravascular ultrasound (IVUS) and optical coherence tomography (OCT) are the two intravascular imaging modalities currently used in coronary interventions. For both techniques, image acquisition requires the advancement of an imaging probe, over a guidewire, into the coronary artery, followed by pullback. There are two major categories of IVUS catheters, the solid-state such as the Eagle Eye (Philips) and the rotational, such as the OptiCross HD (Boston Scientific) and Refinity (Philips). IVUS axial spatial resolution and penetration depth is between 100 and 150 μm and 8 mm respectively. OCT utilizes near infrared light and has an approximately tenfold (10–20 μm) higher spatial resolution compared with IVUS, providing a better visualization of plaque morphology, but also significantly lower penetration depth estimated at 2 mm. Two other major differences between the two imaging modalities are that, unlike IVUS, OCT images through calcium but, on the other hand, requires contrast injection to clear to column of blood from the coronary artery and allow the light beam to reach the vessel wall and acquire images.

Stent sizing by measuring the length and the diameter of the target vessel as well as stent optimization after stent implantation are the two major applications of IVUS and OCT in PCI. According to the results of the ULTIMATE trial, IVUS-guided PCI reduced target vessel failure and stent thrombosis compared with angiography guided PCI [47].

Both the American and European guidelines encourage the use of either IVUS or OCT for PCI guidance, especially for complex lesions, as well as for the evaluation of the underlying mechanism in cases of stent failure [13, 18]. Due to its higher penetration depth, IVUS has an advantage over OCT regarding the estimation of left main disease. According to the current guidelines the use of IVUS is encouraged for the evaluation of the severity of left main stenosis [13, 18]. On the other hand, OCT, given its high spatial resolution, can be very useful in the identification of a ruptured plaque, thrombus, spasm or dissection in cases of myocardial infarction without obstructive coronary artery disease (MINOCA) shedding light in the underlying pathophysiologic mechanism that provoked the myocardial infarction [48].

Complications

Detailed description of complications and their management is beyond the scope of this chapter. PCI is a relatively safe operation, nevertheless, serious complications may occur. Older age, female gender, diabetes, heart failure, renal failure, extensive CAD and ACS are associated with higher rates of complications. Serious complications such as death, periprocedural MI, and stroke appear in 1–2%, 8%, and 0.3% [10]. Complications can be separated to cardiac/coronary complications and non-cardiac complications.

(A) *Cardiac/coronary complications:*
  - Coronary artery perforation, tamponade
  - Aortic/coronary artery dissection
  - Stent thrombosis
  - Equipment loss/entrapment
  - Distal embolization/no reflow
  - Air embolism
  - Ventricular arrhythmias

(B) *Non-cardiac complications:*
  - Bleeding
  - Infection/bacteremia
  - Contrast induced nephropathy
  - Allergic reactions
  - Radiation injury
  - Femoral access related complications: retroperitoneal bleeding, atrioventricular fistulas, pseudoaneurysm, dissection
  - Radial access related complications: radial artery occlusion, radial artery spasm, forearm hematoma, dissection.

Case Presentation

Continued from page 425

The patient was loaded with 250 mg of aspirin and 180 mg of ticagrelor. Right radial access via a 6 Fr sheath was obtained. An 8000 IU i.v. bolus dose of unfractionated heparin was administered. The LCA was engaged with a JL 3.5, 5 Fr catheter and the RCA with a JR 4, 5 Fr catheter. Coronary angiography revealed three-vessel disease (Fig. 39.6). The first obtuse marginal branch was totally occluded and it was considered as the culprit vessel. A Sion Blue (Asahi Intecc) guidewire was advanced through a 6 Fr EBU 3.5 guide catheter to the circumplex artery crossing obtuse marginal stenosis, restoring TIMI flow III. A 2.75 mm × 20 mm semi-compliant balloon was advanced over the guidewire and inflated at 12 atm adequately expanding the lesion. Then, a 2.75 mm × 32 mm Promus ELITE (Boston Scientific) drug-eluting stent was implanted. Post-dilatations with a 2.75 mm × 20 mm non-compliant balloon at 20 atm were performed (Fig. 39.7). ACT remained between 250 and 300 throughout the intervention. A radial compression device was applied for 3 h. No obvious hematoma was noticed after removal of the device and the right radial artery was palpable. The patient remained in the hospital for 4 days and presented no complications. Due to the presence of multiple remaining complex lesions, cardiothoracic surgery consultation was suggested.

Acknowledgments I recognize the valuable assistance of Dr. Grigoris V. Karamasis on the critical review of this chapter.

References

1. Grech ED. ABC of interventional cardiology: percutaneous coronary intervention. I: history and development. BMJ (Clinical research ed). 2003;326(7398):1080–2.
2. Bueno H, Rossello X, Pocock SJ, Van de Werf F, Chin CT, Danchin N, et al. In-hospital coronary revascularization rates and post-discharge mortality risk in non-ST-segment elevation acute coronary syndrome. J Am Coll Cardiol. 2019;74(11):1454–61.
3. Huynh T, Perron S, O'Loughlin J, Joseph L, Labrecque M, Tu JV, et al. Comparison of primary percutaneous coronary intervention and fibrinolytic therapy in ST-segment-elevation myocardial infarction: bayesian hierarchical meta-analyses of randomized controlled trials and observational studies. Circulation. 2009;119(24):3101–9.
4. Puymirat E, Taldir G, Aissaoui N, Lemesle G, Lorgis L, Cuisset T, et al. Use of invasive strategy in non-ST-segment elevation myocardial infarction is a major determinant of improved long-term survival: FAST-MI (French registry of acute coronary syndrome). JACC Cardiovasc Interv. 2012;5(9):893–902.
5. Bittl JA, He Y, Jacobs AK, Yancy CW, Normand SL. Bayesian methods affirm the use of percutaneous coronary intervention to improve survival in patients with unprotected left main coronary artery disease. Circulation. 2013;127(22):2177–85.
6. Maron DJ, Hochman JS, Reynolds HR, Bangalore S, O'Brien SM, Boden WE, et al. Initial invasive or conservative strategy for stable coronary disease. N Engl J Med. 2020;382(15):1395–407.
7. Chaitman BR, Alexander KP, Cyr DD, Berger JS, Reynolds HR, Bangalore S, et al. Myocardial infarction in the ISCHEMIA trial: impact of different definitions on incidence, prognosis, and treatment comparisons. Circulation. 2021;143(8):790–804.
8. Spertus JA, Jones PG, Maron DJ, O'Brien SM, Reynolds HR, Rosenberg Y, et al. Health-status outcomes with invasive or conservative care in coronary disease. N Engl J Med. 2020;382(15):1408–19.

9. Navarese EP, Lansky AJ, Kereiakes DJ, Kubica J, Gurbel PA, Gorog DA, et al. Cardiac mortality in patients randomised to elective coronary revascularisation plus medical therapy or medical therapy alone: a systematic review and meta-analysis. Eur Heart J. 2021;42(45):4638–51.
10. Kern MJ. The interventional cardiac catheterization handbook. 4th ed. Philadelphia, PA: Elsevier; 2018. p. 512.
11. Ibanez B, James S, Agewall S, Antunes MJ, Bucciarelli-Ducci C, Bueno H, et al. 2017 ESC guidelines for the management of acute myocardial infarction in patients presenting with ST-segment elevation: the task force for the management of acute myocardial infarction in patients presenting with ST-segment elevation of the European Society of Cardiology (ESC). Eur Heart J. 2018;39(2):119–77.
12. Knuuti J, Wijns W, Saraste A, Capodanno D, Barbato E, Funck-Brentano C, et al. 2019 ESC guidelines for the diagnosis and management of chronic coronary syndromes. Eur Heart J. 2020;41(3):407–77.
13. Neumann FJ, Sousa-Uva M, Ahlsson A, Alfonso F, Banning AP, Benedetto U, et al. 2018 ESC/EACTS Guidelines on myocardial revascularization. Eur Heart J. 2019;40(2):87–165.
14. Collet JP, Thiele H, Barbato E, Barthélémy O, Bauersachs J, Bhatt DL, et al. 2020 ESC guidelines for the management of acute coronary syndromes in patients presenting without persistent ST-segment elevation. Eur Heart J. 2021;42(14):1289–367.
15. Brilakis ES. Manual of percutaneous coronary interventions: a step-by-step approach. London: Elsevier/Academic Press; 2020.
16. Mehran R, Aymong ED, Nikolsky E, Lasic Z, Iakovou I, Fahy M, et al. A simple risk score for prediction of contrast-induced nephropathy after percutaneous coronary intervention: development and initial validation. J Am Coll Cardiol. 2004;44(7):1393–9.
17. Laskey WK, Jenkins C, Selzer F, Marroquin OC, Wilensky RL, Glaser R, et al. Volume-to-creatinine clearance ratio: a pharmacokinetically based risk factor for prediction of early creatinine increase after percutaneous coronary intervention. J Am Coll Cardiol. 2007;50(7):584–90.
18. Lawton JS, Tamis-Holland JE, Bangalore S, Bates ER, Beckie TM, Bischoff JM, et al. ACC/AHA/SCAI guideline for coronary artery revascularization: executive summary: a report of the American College of Cardiology/American Heart Association Joint Committee on Clinical Practice Guidelines. Circulation. 2021;2021:Cir0000000000001039.
19. Jolly SS, Yusuf S, Cairns J, Niemelä K, Xavier D, Widimsky P, et al. Radial versus femoral access for coronary angiography and intervention in patients with acute coronary syndromes (RIVAL): a randomised, parallel group, multicentre trial. Lancet (London, England). 2011;377(9775):1409–20.
20. Valgimigli M, Gagnor A, Calabró P, Frigoli E, Leonardi S, Zaro T, et al. Radial versus femoral access in patients with acute coronary syndromes undergoing invasive management: a randomised multicentre trial. Lancet (London, England). 2015;385(9986):2465–76.
21. Iosif X, Peter T, Allison H. Update on cardiac catheterization in patients with prior coronary artery bypass graft surgery. JACC Cardiovasc Interv. 2019;12:1635.
22. Seto AH, Abu-Fadel MS, Sparling JM, Zacharias SJ, Daly TS, Harrison AT, et al. Real-time ultrasound guidance facilitates femoral arterial access and reduces vascular complications: FAUST (femoral arterial access with ultrasound trial). JACC Cardiovasc Interv. 2010;3(7):751–8.
23. Sandoval Y, Burke MN, Lobo AS, Lips DL, Seto AH, Chavez I, et al. Contemporary arterial access in the cardiac catheterization laboratory. JACC Cardiovasc Interv. 2017;10(22):2233–41.
24. Valgimigli M, Campo G, Penzo C, Tebaldi M, Biscaglia S, Ferrari R. Transradial coronary catheterization and intervention across the whole spectrum of Allen test results. J Am Coll Cardiol. 2014;63(18):1833–41.
25. Seto AH, Roberts JS, Abu-Fadel MS, Czak SJ, Latif F, Jain SP, et al. Real-time ultrasound guidance facilitates transradial access: RAUST (radial artery access with ultrasound trial). JACC Cardiovasc Interv. 2015;8(2):283–91.
26. Rao SV, Stone GW. Arterial access and arteriotomy site closure devices. Nat Rev Cardiol. 2016;13(11):641–50.
27. Xenogiannis I, Brilakis ES. Distal radial access at the anatomic snuffbox: the new standard for left radial access? Catheter Cardiovasc Interv. 2019;94(5):658–9.
28. Tsigkas G, Papageorgiou A, Moulias A, Kalogeropoulos AP, Papageorgopoulou C, Apostolos A, et al. Distal or traditional transradial access site for coronary procedures: a single-center, randomized study. JACC Cardiovasc Interv. 2022;15(1):22–32.
29. Kastrati A, Neumann FJ, Mehilli J, Byrne RA, Iijima R, Büttner HJ, et al. Bivalirudin versus unfractionated heparin during percutaneous coronary intervention. N Engl J Med. 2008;359(7):688–96.
30. Montalescot G, Bolognese L, Dudek D, Goldstein P, Hamm C, Tanguay JF, et al. Pretreatment with prasugrel in non-ST-segment elevation acute coronary syndromes. N Engl J Med. 2013;369(11):999–1010.
31. Kerensky RA, Wade M, Deedwania P, Boden WE, Pepine CJ. Revisiting the culprit lesion in non-Q-wave myocardial infarction. Results from the VANQWISH trial angiographic core laboratory. J Am Coll Cardiol. 2002;39(9):1456–63.
32. Brilakis ES. Manual of chronic total occlusion interventions a step-by-step approach. 2nd ed. London: Elsevier/Academic Press; 2018. p. 483.
33. Elrayes MM, Xenogiannis I, Nikolakopoulos I, Vemmou E, Wollmuth J, Abi Rafeh N, et al. An algorithmic approach to balloon-uncrossable coronary lesions. Catheter Cardiovasc Interv. 2021;97(6):E817–e25.
34. Topaz O. Debulking in cardiovascular interventions and revascularization strategies: between a rock and the heart. London: Elsevier/Academic Press; 2022.
35. Xenogiannis I, Karmpaliotis D, Alaswad K, Jaffer FA, Yeh RW, Patel M, et al. Usefulness of Atherectomy in chronic Total occlusion interventions (from the PROGRESS-CTO registry). Am J Cardiol. 2019;123(9):1422–8.
36. Sapontis J, Grantham JA, Marso SP. Excimer laser atherectomy to overcome intraprocedural obstacles in chronic total occlusion percutaneous intervention: case examples. Catheter Cardiovasc Interv. 2015;85(3):E83–9.
37. De Maria GL, Scarsini R, Banning AP. Management of calcific coronary artery lesions: is it time to change our interventional therapeutic approach? JACC Cardiovasc Interv. 2019;12(15):1465–78.
38. Fischman DL, Leon MB, Baim DS, Schatz RA, Savage MP, Penn I, et al. A randomized comparison of coronary-stent placement and balloon angioplasty in the treatment of coronary artery disease. Stent restenosis study investigators. N Engl J Med. 1994;331(8):496–501.
39. Xenogiannis I, Zenati M, Bhatt DL, Rao SV, Rodés-Cabau J, Goldman S, et al. Saphenous vein graft failure: from pathophysiology to prevention and treatment strategies. Circulation. 2021;144(9):728–45.
40. Xenogiannis I, Rangan BV, Uyeda L, Banerjee S, Edson R, Bhatt DL, et al. In-stent restenosis in saphenous vein grafts (from the DIVA trial). Am J Cardiol. 2022;162:24–30.
41. Stone GW, Ellis SG, Cox DA, Hermiller J, O'Shaughnessy C, Mann JT, et al. A polymer-based, paclitaxel-eluting stent in patients with coronary artery disease. N Engl J Med. 2004;350(3):221–31.
42. Philip F, Agarwal S, Bunte MC, Goel SS, Tuzcu EM, Ellis S, et al. Stent thrombosis with second-generation drug-eluting stents compared with bare-metal stents: network meta-analysis of primary percutaneous coronary intervention trials in ST-segment–elevation myocardial infarction [corrected]. Circ Cardiovasc Interv. 2014;7(1):49–61.

43. Pancholy SB, Patel TM. Effect of duration of hemostatic compression on radial artery occlusion after transradial access. Catheter Cardiovasc Interv. 2012;79(1):78–81.
44. De Bruyne B, Pijls NH, Kalesan B, Barbato E, Tonino PA, Piroth Z, et al. Fractional flow reserve-guided PCI versus medical therapy in stable coronary disease. N Engl J Med. 2012;367(11):991–1001.
45. Megaly M, Khalil C, Saad M, Xenogiannis I, Omer M, Anantha Narayanan M, et al. Outcomes with deferred versus performed revascularization of coronary lesions with gray-zone fractional flow reserve values. Circ Cardiovasc Interv. 2019;12(12):e008315.
46. Davies JE, Sen S, Dehbi HM, Al-Lamee R, Petraco R, Nijjer SS, et al. Use of the instantaneous wave-free ratio or fractional flow reserve in PCI. N Engl J Med. 2017;376(19):1824–34.
47. Gao XF, Ge Z, Kong XQ, Kan J, Han L, Lu S, et al. 3-year outcomes of the ULTIMATE trial comparing intravascular ultrasound versus angiography-guided drug-eluting stent implantation. JACC Cardiovasc Interv. 2021;14(3):247–57.
48. Karamasis G, Xenogiannis I, Varlamos C, Deftereos S, Alexopoulos D. Use of optical coherence tomography in MI with non-obstructive coronary arteries. Intervent Cardiol (London, England). 2022;17:e06.

40 Transcatheter Aortic Valve Implantation

Konstantinos Kalogeras and Manolis Vavuranakis

Case Presentation
An 82-year-old male was referred to a tertiary center due to an episode of syncope. He has been complaining of gradually worsening shortness of breath for the last 6 months, with severe limitations on everyday activities. He describes a similar episode of syncope during physical exercise 9 months ago. Clinical examination in the emergency department revealed a systolic murmur radiating to the neck with diminished second heart sound (S2). Crackles were found on lung auscultation.

An echocardiogram was conducted, which revealed a severely calcified aortic valve with leaflet restriction and well-preserved left and right ventricular functions. The diagnosis of symptomatic severe aortic stenosis was confirmed (V_{max} = 4.4 m/s). After lung decongestion with intravenous diuretics and symptomatic improvement, his case was discussed in the Heart team, and a decision for transcatheter aortic replacement with a self-expanding device was made. A multislice CT angiogram was preceded, which allowed precise measurements of the aortic root and peripheral vessels and confirmed the suitability for transcatheter treatment.

Continued at page 449

Background

Aortic stenosis (AS) remains the most common valvular heart disease among elderly population. Its prevalence is calculated at approximately 3% for patients above 75 years old, while the global burden of the disease is increasing due to aging of the population and population growth [1, 2]. The pathophysiology of the disease is characterized from progressive fibro-calcific remodeling and thickening of the valve leaflets. Degenerative calcific stenosis is the most prevalent form [3]. In the majority of cases, the aortic valve is trileaflet; however, in younger ages (<65) congenital bicuspid valve is common. In both cases, the valve gets progressively calcific with worsening restriction of the leaflet movement. Clinical risk factors mediating the degeneration of the aortic valve include advanced age, male, hypertension, diabetes, hypercholesterolemia, and smoking [4].

Once the disease becomes symptomatic, the prognosis is poor (50% mortality at 2 years). Thus, valve replacement is recommended when symptoms of syncope, heart failure, or angina occur. Surgical aortic valve replacement (SAVR) has been the gold standard treatment for AS for many decades before transcatheter aortic valve implantation (TAVI) emerged as an alternative option [5].

TAVI has rapidly evolved with major improvements in both device technology and implantation techniques. On top of that, robust clinical data from several randomized clinical trials support the implementation of the technique. After the regulatory approval of TAVI in Europe in 2007 and the United States in 2011, it has been widely adopted worldwide and has become the "standard of care" for the treatment of severe AS alongside SAVR [6–10].

Indications for Intervention

Aortic valve replacement therapy (TAVI or SAVR) should be considered in case of severe symptomatic AS. When severe stenosis does not cause symptoms, patients should also be treated with valve replacement if the left ventricle ejection fraction (LVEF) is impaired (<50%), if they are symptomatic

K. Kalogeras (✉)
Third Department of Cardiology, Sotiria Hospital, Medical School, Athens, Greece

M. Vavuranakis
Third Department of Cardiology, National and Kapodistrian University of Athens, Sotiria Hospital, Medical School, Athens, Greece
e-mail: vavouran@otenet.gr

G. Geroulakos et al. (eds.), *Mastering Endovascular Techniques*, https://doi.org/10.1007/978-3-031-42735-0_40

in exercise, if they have severe pulmonary hypertension, if their BNP is markedly elevated, or if they have very severe stenosis (>5 m/s). Current guidelines do not provide clear evidence for gray-zone patients, such as moderate AS; however, recent studies showed better prognosis in these patients if treated earlier, raising interest in this kind of approach [5, 11–13].

Decisions regarding the type of treatment (surgical or transcatheter) should be made by a heart valve team in centers with the appropriate experience and infrastructure. Various validated risk scores (The Society of Thoracic Surgeons (STS) score, the EuroScore) and other clinical factors are used to stratify patients into low, intermediate, and high risk before final decision [14, 15]. Indications for TAVI were first limited to high-risk (>8% STS score) or unsuitable-for-surgery patients with symptomatic severe AS. Based on data from several randomized clinical trials demonstrating superiority or non-inferiority of TAVI compared to SAVR, the most recent ESC Valvular Heart Disease guidelines offer a class I recommendation for TAVI in patients >75 years of age as assessed by the Heart team [16]. The multidisciplinary Heart team involved in the decision-making process should always take into consideration various factors including surgical risk, age, sex, frailty, vascular access, coronary artery disease, noncardiac conditions.

Contraindications for TAVI include poor life expectancy (<1 year), active endocarditis, and anatomic unsuitability (e.g., very large aortic annulus).

Preoperative Preparation

Preoperative planning is of great importance when considering TAVI. The multidisciplinary Heart team plays a pivotal role in integrating the patient clinical and anatomical factors in order to determine the best procedural strategy. The initial work- up includes a full history and clinical examination, as well as pulmonary function tests and carotid Dopplers, which are performed as part of the pre-procedural assessment [16].

Transthoracic Echocardiogram (TTE)

The transthoracic echocardiogram (TTE) will define the severity of aortic stenosis and will evaluate the left ventricle function. Furthermore, it allows to investigate any other valve disease coexistence and assessment of pulmonary pressures.

Coronary Angiogram

Once the severity of the stenosis is confirmed further diagnostic testing is indicated regarding coronary anatomy, which is usually completed either with an invasive coronary angiogram or with computed tomography (CT), when the pretest probability of coronary artery disease is low. It is recommended to revascularize severe coronary stenosis (>70%) at the proximal portion of the large epicardial vessels (left anterior descending (LAD) or right coronary artery (RCA)). However, clear recommendations about the ideal timing of revascularization (prior or post TAVI) do not exist. Stenoses involving side branches are usually deferred, as they are less likely to cause hemodynamic compromise [17–19].

Computed Tomographic (CT) Scan

The TAVI procedure requires a very thorough imaging examination of the aorta anatomy and access routes. This is usually achieved with a CT of the chest/abdomen and pelvis with a contrast aortogram. CT provides anatomic details of the aortic valve, aortic root, and aorta overall, as well as details regarding the anatomy of the lower extremities (iliofemoral system), including vessel caliber, tortuosity, location, and extent of calcification. Furthermore, the TAVI CT is the optimal modality for evaluation of the calcification's presence and extent on the aortic valve leaflets, the coronaries' position and height relative to the aortic annulus, the size of the aortic sinuses, and the presence or absence of calcification extending into the left ventricular outflow tract (LVOT). According to the CT study, the three-cusp coplanar angle for valve implantation is estimated (the nadirs of the three aortic cusps should lie along a single line in this angle). Furthermore, the cusp overlap view (overlapping the right coronary cusp and left coronary cusp) is evaluated. Measurement of the aortic annulus is of great importance for choosing the appropriate valve size, while peripheral vessels diameter is crucial for choosing the optimal vascular access. The majority of TAVIs are performed via the transfemoral route. Subclavian, transapical, or direct aortic are alternative access routes if transfemoral is considered not suitable.

Deciding for TAVI

Once the patient has finished the work-up the Heart Team should answer the following questions:

- Does the patient have severe symptomatic aortic stenosis?
- Is he best treated with SAVR, TAVI, or conservatively?
- If he is suitable for TAVR:

  What is the most appropriate access route? Which valve or valves are anatomically suitable?

  Should the procedure be under local anesthesia with conscious sedation or general anesthesia?

Anesthesia

TAVI can be performed either with general anesthesia or conscious sedation. Conscious sedation is mainly preferred when transfemoral TAVI avoids intubation and favors early patient recovery. On the other hand, general anesthesia is generally recommended when deep sedation and analgesia are required, when airway compromise is considered likely, or when alternative access routes are used (subclavian, transapical) [20].

Antithrombotic Therapy

There are limited data for antithrombotic management before and during TAVI. Pre- procedural loading with single antiplatelet agent is recommended, and procedural anticoagulation is achieved with unfractionated heparin (aiming an activated clotting time (ACT) of 250–300 s). For patients on oral anticoagulation (OAC), the decision to continue OAC during the procedure should be made based on patients' bleeding/thromboembolic risks [21, 22].

Antibiotic Prophylaxis

According to guidelines for the management of infective endocarditis, it is recommended to administer perioperatively antibiotic prophylaxis covering the most frequent microorganisms. An intravenous dose of amoxicillin/clavulanic acid or vancomycin (for patients allergic to penicillin) is suggested [23, 24].

Procedural Steps (Transfemoral TAVI)

The TAVI procedure may differentiate according to the type of valve implanted. However, transfemoral implantation is in general terms completed according to the following steps:

1. Arterial femoral access is gained in bilateral common femoral arteries (main access for valve delivery and contralateral access for aortic root angiography to guide implantation). Main access can be obtained either by contralateral injection guidance or by ultrasound guidance, trying to avoid severe calcification of the anterior vessel wall. Alternative secondary access from the radial artery is preferred by some practitioners.
2. A vein femoral or jugular sheath is inserted for a temporary pacemaker and central venous access if needed. In case of pre-existing temporary pacemaker, rapid pacing during the procedure can be delivered using an external program device. Alternatively, pacing can be achieved through the pre-shaped stiff wire used for the device implantation.
3. The main access site is usually pre-closed by using one or two suture-based vascular closure devices (ProGlide, Abbott, Abbott Park, Illinois). Alternative large vessel closure devices may be utilized, such as the plug-based closure device (MANTA, Teleflex, Wayne, Pennsylvania) [25].
4. Over a Stiff Guidewire (0.035″), a large-diameter (14–18 Fr) sheath is inserted into the main access site, providing support and straightening iliofemoral tortuosity.
5. Using a 6 Fr pigtail catheter an aortogram is conducted at the three-cusp coplanar angle for valve implantation as derived from the pre-procedural TAVI CT. The nadirs of the three aortic cusps should lie along a single line in this implantation angle (Fig. 40.1a). For self-expanding valves, the cusp overlap view (overlapping the right coronary cusp and left coronary cusp) may be alternatively used (Fig. 40.1b).
6. Using an Amplatz left catheter (AL1 or AL2, according to aortic root dimensions) and a 0.035″ straight-tipped wire, the aortic valve is crossed. The Amplatz catheter is advanced into the left ventricle, and over it, the straight-tipped wire is exchanged with an exchanged-length J-shaped wire.
7. A 6 Fr pigtail catheter is placed in the LV and directed toward the LV apex (in the RAO projection). At this time, simultaneous pressures are gained in the LV and aorta recording the transvalvular gradient.
8. Subsequently, a pre-shaped stiff 0.035″ wire is inserted into the LV through the pigtail catheter. Careful attention should be paid to avoid entrapment of the pigtail catheter or wire in the sub-valvular mitral valve apparatus.
9. The preloaded transcatheter valve is checked under fluoroscopy for appropriate crimping and loading onto the delivery catheter as well as correct orientation.
10. In cases of extremely calcified aortic valves and at the discretion of the operator, valve pre-dilatation can be

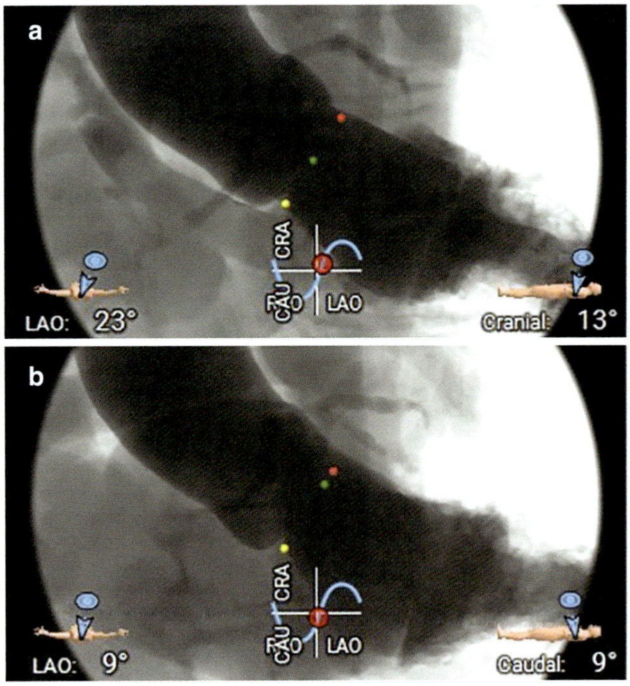

Fig. 40.1 Ascending aorta multi-slice CT (MSCT) scan views of a patient with severe aortic stenosis. (**a**) The three aortic cusps overlapping view as assessed in MSCT scan indicating the appropriate projection angle. (**b**) The cusp overlap view (overlapping the right coronary cusp and left coronary cusp) for the same patient

conducted under rapid pacing (160–200 bpm) to facilitate the valve delivery. The size of the balloon should not exceed the minimum aortic annulus diameter obtained from the TAVI CT.

11. Under continuous fluoroscopic guidance, the transcatheter heart valve is delivered over the stiff pre-shaped wire with careful attention to the wire position in the LV apex. The system is carefully advanced across the aortic arch and finally positioned across the aortic annulus (Fig. 40.2).
12. The fluoroscopic angle is changed to the three-cusp overlap view or the cusp- overlap view (as previously mentioned). Some operators prefer the implantation at the left-right cusp overlap view in order to position aortic valve at the optimal implantation depth and, thus, reduce the risk of permanent pacemaker implantation. A 6Fr pigtail catheter is placed through the contralateral access sheath in the non-coronary (or right coronary) cusp for reference. Aortic root angiography is then performed to confirm appropriate position.
13. Subsequently, transcatheter aortic valve deployment is conducted according to the type of device used. The four commercially available devices for transcatheter aortic valve treatment are described as follows (Fig. 40.3).

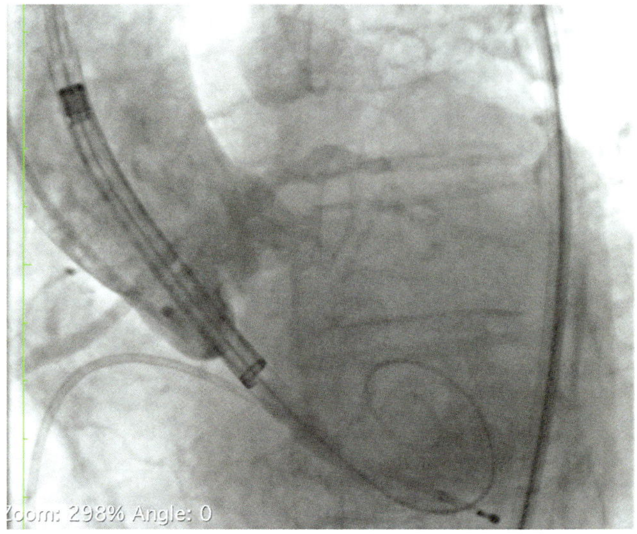

Fig. 40.2 A self-expanding valve (Evolut Pro) positioned across the aortic annulus inside its delivery sheath (angiographic view)

Self-expanding Evolut R/Pro (Medtronic, Minneapolis, Minnesota) (Fig. 40.3a)

Device characteristics: The Evolut R/Pro device (Medtronic) (currently available in four sizes of 23, 26, 29, and 34 mm) consists of a tricuspid valve obtained from porcine pericardial tissue, mounted and sutured inside a self-expandable nitinol frame. The lower part of the device has a high radial force, while the central portion of the stent supports the valve. The Pro device has been improved with an extended skirt at the inflow tract providing enhanced seal against PVR. The Evolut R has been designed for delivery through a 14 Fr compatible system (16 Fr for the 34 mm), while Pro through a 16 Fr sheath.

Implantation technique: Once the device is positioned at the appropriate height across the aortic annulus (targeting an implantation depth of 3–5 mm or higher), unsheathing of the valve is performed by rotating the deployment knob of the delivery system anti-clockwise. During deployment frequent aortic root injections guide the adequate position. Furthermore, controlled pacing (90–120 bpm) can be used to increase valve stability and facilitate precise implantation. Once blood pressure drops due to temporal occlusion of the

a b

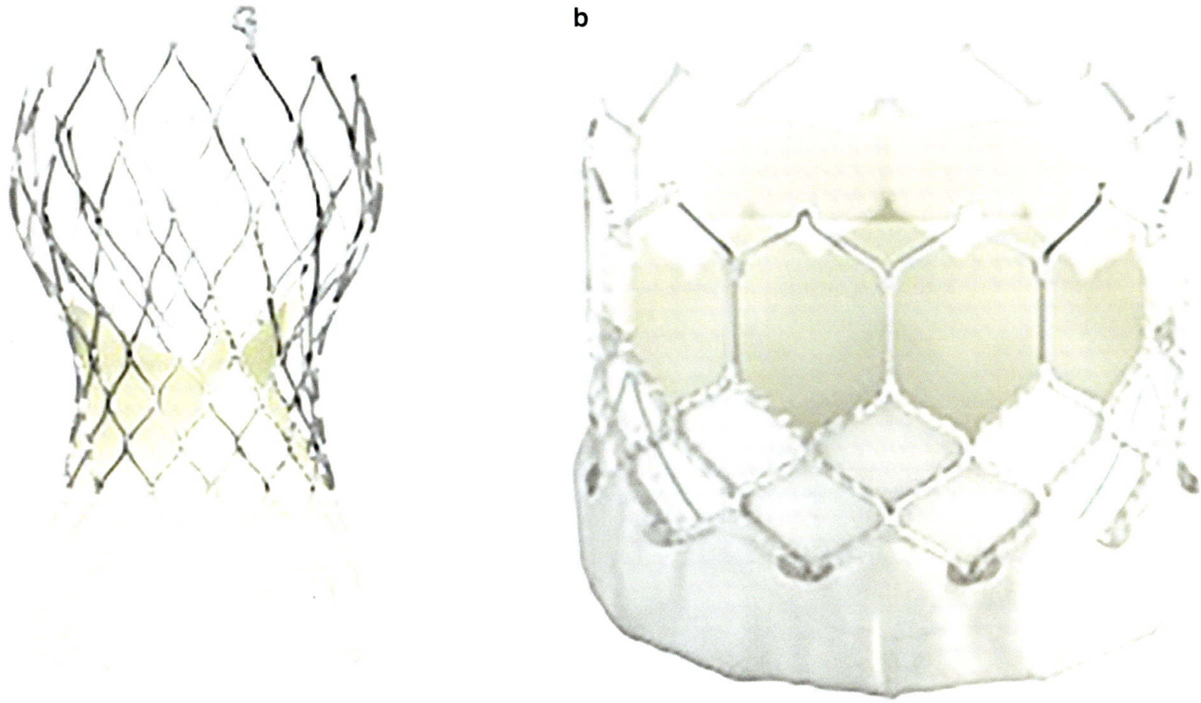

c d

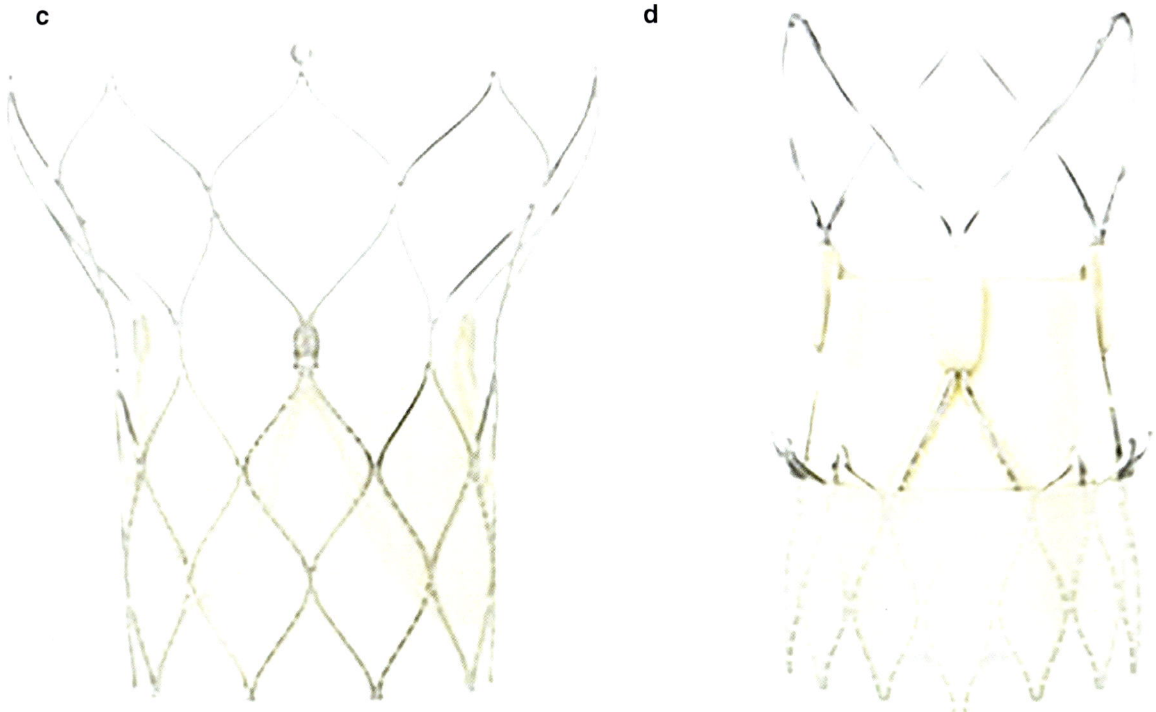

Fig. 40.3 The four commercially available devices for transcatheter aortic valve treatment. (**a**) The self-expanding Evolut Pro device (Medtronic, Minneapolis, Minnesota). (**b**) The balloon-expandable SAPIEN 3 (Edwards Lifesciences, California). (**c**) The self-expanding Portico (Abbott, Abbott Park, Illinois). (**d**) The self-expanding ACURATE neo2 (Boston Scientific, Marlborough, Massachusetts)

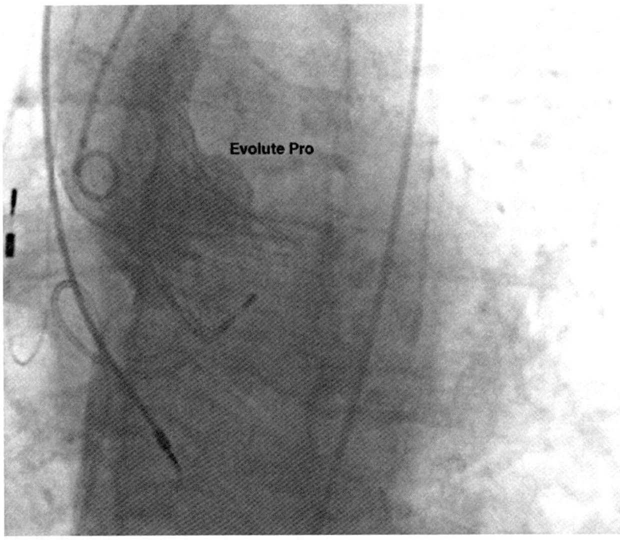

Fig. 40.4 Final aortogram of a successfully implanted self-expanding Evolut Pro valve

aortic valve (supra-annular valve function), the operator must continue device deployment until blood pressure recovers. An indicator provides feedback to the operator that the capsule is close to the "no recapture point" (at approximately two-third deployed). At this point, the valve should be functionable and blood pressure restored. The imaging projection should be adjusted to align the valve inflow and determine valve position using aortic root angiography. If the operator is satisfied with the valve position the valve is fully released. The detachment of frame paddles must be confirmed under fluoroscopy (Fig. 40.4).

Balloon-Expandable SAPIEN 3 (Edwards Lifesciences, California) (Fig. 40.3b)

Device characteristics: The SAPIEN 3 (Edwards Lifesciences) valve is the fourth generation of the balloon-expandable SAPIEN device series. It is available in four sizes (20, 23, 26, and 29 mm). It consists of a cobalt-chromium frame, three bovine pericardial tissue leaflets, and a polyethylene terephthalate (PET) skirt at its inflow portion and an outer PET sealing skirt to reduce PVR. The delivery system is advanced through a 14 Fr (for 20, 23, 26 mm valves) and 16 Fr (for 29-mm valve) expandable sheath.

Implantation technique: The current SAPIEN 3 (S3) system requires valve alignment and positioning on the balloon. According to the manufacturer, this should be performed in the descending aorta. The valve system is then advanced across the aortic arch, utilizing the retroflex system. The valve is advanced across the aortic valve and positioned at the annulus based on the middle marker on the valve. Once positioning is confirmed, the valve is deployed under rapid ventricular pacing at approximately 180 bpm. The balloon inflation should be held for 3–5 s and subsequently deflated before rapid pacing is stopped. Finally, the balloon catheter is withdrawn to the ascending aorta.

Self-expanding Portico (Abbott, Abbott Park, Illinois) (Fig. 40.3c)

Device characteristics: The Portico valve is composed of a self-expanding stent, bovine leaflets, and a porcine pericardial sealing cuff. It is characterized by the large cell area and annular function of its leaflets, permitting the catheter engagement of the coronary ostia post-procedurally. The 23 and 25 mm valves are loaded onto an 18 Fr delivery system, whereas the 27 and 29 mm valves are loaded onto a 19 F delivery system. The Portico valve can be implanted transfemorally and, when implanted with the SoloPath introducer (St. Jude Medical, Inc.), has a low 13.5 Fr insertion profile.

Implantation technique: The device is initially advanced across the aortic valve, and the delivery system is positioned so that the inner shaft marker band is aligned with the annular plane. Unsheathing is performed by turning the deployment wheel occasionally under controlled pacing (90–120 bpm). There is a clicking sound when the delivery system has reached the partial deployment lock. The deployment mechanism will not re-engage until the deployment lock button is depressed. The imaging projection should now be adjusted to align the valve inflow and then confirm the appropriate position using aortic root angiography. Once confirmed, after pressing the deployment lock button, the deployment is completed by turning the deployment wheel until the valve capsule is fully retracted. The detachment of retainer tabs must be confirmed under fluoroscopy before the system is withdrawn.

Self-expanding ACURATE Neo2 (Boston Scientific, Marlborough, Massachusetts) (Fig. 40.3d)

Device characteristics: The ACURATE neo device is a second-generation porcine pericardium valve sewn onto a self-expanding nitinol stent. It is covered by a porcine pericardium skirt. The device includes three stabilization arches aiming for axial alignment in the aortic annulus, a top crown for capping the aortic annulus, and a bottom that is open to the full distribution of the native valve. The device can be implanted through either the transapical (28 Fr) or the transfemoral (18 Fr) routes using a two-step deployment. The ACURATE neo comes in three different sizes: small (21–23 mm aortic annulus), medium (23–25 mm aortic annulus), and large (25–27 mm aortic annulus).

Implantation technique: ACURATE neo has a unique deployment system. Due to the lower radial force, balloon pre-dilatation is recommended to facilitate device expansion. Once the device has crossed the aortic valve, it should be carefully positioned as indicated by the radiopaque marker at the annular plane. Deployment can be performed in a two-step manner. The first step includes turning the rotating knob until full stop. This step should be done rather slowly in order to avoid inappropriate movement of the device. Then, the stabilization arms are fully deployed, and the upper crown is partly deployed. Once the position of the valve is verified with aortic root angiography, the second step is initiated by removing the safety knob. Subsequently, the second rotating knob is turned until full stop with full deployment of the device. Finally, the detachment of the frame paddles must be confirmed under fluoroscopy.

14. After valve deployment, a pigtail catheter is introduced into the LV in order to measure simultaneous pressures across the prosthetic valve. Subsequently, aortic root angiography is performed to assess the appropriate position of the device and the presence/severity of paravalvular regurgitation (PVR).
15. Post-dilatation may be performed to achieve optimal device expansion and reduce PVR. The size of the balloon is typically the same as the derived annulus diameter measured in TAVI CT.
16. Access site closure. The TAVI sheath is carefully withdrawn into the ipsilateral iliac artery. The contralateral femoral access site may be used for "cross-over" to the main access side using an IMA diagnostic or pigtail catheter. This catheter will facilitate confirmation of the final angiographic result after removal of the TAVI sheath. The pre-closure sutures are tightened after sheath removal with a guidewire 0.035″ usually still in place. The "cross-over" catheter can be used for bailout balloon angioplasty or covered stent implantation in the case of major vascular complication. Manual compression may be required if minor bleeding persists [26].
17. In cases of pre-existing permanent pacemakers, the temporary pacing wire is removed. Otherwise, it should be safely secured until its removal is considered safe for the patient.
18. Bedside TEE is performed at the end of the procedure, either in the cath lab or as soon as the patient arrives in the cardiac intensive care unit. Valve function is assessed, and new pericardial effusion should be excluded. The cardiac monitor should also be assessed for evidence of new conduction disturbances.

Alternative Access Sites Transsubclavian Access

Transsubclavian access is the most commonly used alternative access. Transsubclavian access is typically performed via surgical cut-down. Usually it is performed from the left side due to the more favorable alignment of the device with the native valve. The presence of the left internal mammary artery graft in patients with coronary artery bypass surgery is considered a relative contra-indication [27, 28].

Transapical Access

Transapical access is performed via a left anterolateral intercostal incision followed by needle puncture of the apex under general anesthesia. A sheath is inserted, and the device is deployed similarly to the transfemoral approach. This access is typically used for the Edwards SAPIEN platform [27].

Transaortic Access

Transaortic access is performed through the ascending aorta via a right anterior mini-thoracotomy in the second intercostal space. It is tried to avoid areas of calcification. The valve is deployed in the same way as via the transfemoral approach [27, 29].

Transcarotid Access

Transcarotid access is performed percutaneously under local anesthesia with cerebral oximetry monitoring. The size and quality of the vessel are necessary determinants for efficacy and safety, as well as an anatomically complete Circle of Willis [27, 30].

Transcaval Access

This is a more complex approach, requiring the insertion of the delivery sheath from the inferior vena cava into the abdominal aorta for valve delivery. Access is performed in experienced centers, with limited patients requiring alternative access.

Procedural Complications

Operators have to be prepared for complications and emergencies when performing TAVI. Common complications during and after TAVR are as follows.

Paravalvular regurgitation (PVR): PVR is usually estimated post-procedurally by aortic root angiography and/or transthoracic echocardiography. If considered significant, various measures according to the underlying mechanism can be applied. If the leak is due to under-expansion of the device, post-dilatation should be performed to achieve optimal expansion. In the case of malposition, a valve-in-valve procedure may be considered. Advances in the design of the current valves such as a circumferential outer sealing skirt and repositionable features (in some self-expanding devices) have remarkably reduced the risk of PVR. Self-expanding valves are considered to have higher PVR rates compared to balloon-expandable devices (3.4% vs. 0.8% for moderate/severe PVR respectively). Although moderate or severe PVR has been consistently associated with increased mortality, there is conflicting data on the adverse effect of mild PVR [10, 31, 32].

Coronary artery obstruction: Coronary artery obstruction is a rare (<1%) but life- threatening complication caused by native leaflets or device material overlapping the coronary ostia. More commonly, the left main coronary artery is involved in obstruction. Low coronary height (<10 mm), narrow sinus of Valsalva, as well as heavily calcified aortic leaflets are considered high-risk anatomic features. If patients are considered at high risk, preventative strategies such as coronary protection with a coronary stent in place may be considered. Acute coronary artery obstruction typically causes rapidly worsening hypotension with electrocardiographic changes. Immediate angiographic assessment should be performed to identify the affected lesion [33, 34].

Annular rupture: Annular rupture is a rare (<1%) but devastating complication that may occur after balloon-expandable valve implantation or aggressive pre/post-dilation of a severely calcified valve. The treatment approach includes surgical repair/reconstruction and pericardial drainage [35].

Device embolization: Valve embolization occurs in approximately 1%. Valve dislocations are either toward the aorta or into the LV cavity. It can occur immediately, during, or after THV implantation, or as a late device migration. Bailout measures include repositioning attempts using snares, multiple valve implantation, or conversion to surgery [36].

Vascular complications: The incidence of vascular complications has decreased over time (2–4% in the recent low-risk trials) due to the development of smaller and more flexible delivery systems. Vascular complications typically include dissection, stenosis, perforation, and pseudoaneurysm or aneurysm. Usually it is caused due to closure device failure. Vascular complications can be treated percutaneously using bailout balloon angioplasty or covered stent implantation via the cross-over catheter. Surgical repair is typically reserved for extensive dissection or bleeding or challenging anatomy [37, 38].

Atrioventricular (AV) conduction disturbance: AV conduction impairment, including advanced heart block and new LBBB, is a common adverse event and remains a limitation of TAVI. It is higher with self-expanding valves compared to balloon- expandable devices (17.4% vs. 6.5% respectively in recent low-risk patient trials). Pre-existing right bundle branch block, non-coronary cusp device landing zone calcification, and deep implantation of the valve have been identified as independent predictors of new conduction disturbances. Although AV conduction impairment typically occurs within 24 h, sometimes it can be developed later than 48 h after implantation. Therefore, temporary pacing wire placed during the procedure should be kept in place for the first 24 h, while monitoring by telemetry should be continued after the procedure, and a potential risk assessment of delayed conduction disturbances should be made before discharge [39, 40].

Stroke: Stroke remains one of the most feared complications of TAVI associated with considerable morbidity and mortality. Stroke related to TAVI most frequently occurs during or within 24–48 h after the procedure. TAVI is reported to be associated with a 19% lower risk of stroke throughout 2 years compared with surgical replacement. Optimal anticoagulation during the procedure is essential for the prevention of stroke. Cerebral embolic protection devices have been developed and used for the prevention of cerebrovascular events peri-procedurally [41, 42].

Antithrombotic Management

For patients undergoing TAVI life-long single antiplatelet therapy (SAPT) with low- dose aspirin (75–100 mg daily) is suggested [16, 43]. If aspirin is contraindicated, clopidogrel or another P2Y12 inhibitor may be prescribed as an alternative. Until recently, the use of dual-antiplatelet therapy (DAPT) for 3–6 months followed by life-long SAPT has been the default strategy. However, recent trials have shown that SAPT is superior in terms of the composite endpoint of bleeding events and noninferior in terms of thromboembolic events compared to DAPT [44]. If there is an established indication for DAPT, the antithrombotic management should follow the recommendations for the indication. For patients with an established indication for oral anticoagulation (OAC), OAC should be continued after TAVI [45]. However, whether these patients should be treated with DOACs or VKA remains a subject of debate [46, 47].

Durability Data

TAVI is an increasingly used modality which is gradually expanding to younger and lower-risk population with longer life expectancy. Favorable data regarding the durability of transcatheter implanted valves have been reported from the randomized clinical trials and large-scale real-world registries but are limited up to 8 years [48, 49].

The rates of structural valve deterioration and bioprosthetic valve failure between 5 and 8 years after TAVI ranged from 3.8 to 18.6% and 2.5 to 7.5%, depending on the definitions, the timing, and the type of valve used. When compared to surgical valves, the durability of the transcatheter devices was largely comparable up to 6 years [50]. However, it should be underlined that bioprosthetic valve deterioration mostly occurs >8–10 years after SAVR. Thus, longer-term follow-up data are required [51–53] (Table 40.1).

Table 40.1 Suggested Endovascular Toolkit for the transcatheter treatment of aortic stenosis. These are only a few suggested options based on the author's experience and can cover the vast majority of cases

	Manufacturer	Size/length
Wires		
Any standard access wire	Any	0.035″
Straight-tip wire	Any	0.035″
Exchange-length wire	Any	0.035″
Amplatz super stiff wire	Boston Scientific	0.035″
Pre-shaped curve stiff wire	Any	0.035″
Sheaths		
Any standard access femoral sheath	Any	6 Fr/11 cm
Large-caliper femoral sheath introducer	Cook Medical	14–18 Fr/30–45 cm
Catheters		
Pigtail	Any	6 Fr/100 cm
Internal mammary artery (IMA)	Any	6 Fr/100 cm
Amplatz left 1–2	Any	6 Fr/100 cm
Balloons		
High-pressure, semi- compliant/ compliant balloon	Bard Medical/ Balt extrusion	18–28 mm/40 mm
Peripheral vessels angioplasty balloon	Any	5–10 mm/35 mm
Temporary pacing wire	Any	
Suture-mediated closure system	Perclose ProGlide/Abbott	
Transcatheter aortic valve system		
Evolut R/Pro (self-expanding)	Medtronic	23–26–29–34 mm
Portico (self-expanding)	Abbott	23–25–27–29 mm
ACURATE-Neo 2 (self-expanding)	Boston Scientific	S (23 mm)–M (25 mm)–L (27 mm)
SAPIEN 3/ultra (balloon-expandable)	Edwards	20–23–26–29 mm

Case Presentation

Continued from page 441

Following an intravenous dose of amoxicillin/clavulanic the patient was transferred to the cath lab, and under conscious sedation an arterial (6 Fr) and venous (6 Fr) access were gained. A temporary pacing wire was placed in the right ventricle. Through the 6 Fr arterial sheath and an internal mammary (IMA) catheter, a guided puncture of the contralateral common femoral artery was conducted. Two ProGlide vascular closure devices were deployed, and over a Super Stiff 0.035″ wire, a 14 Fr long sheath was introduced.

Sequential aortograms were conducted through a 6 Fr pigtail catheter identifying the cusp overlap view. Utilizing a 6 Fr Amplatz left 1 diagnostic catheter, a straight tip 0.035″ wire crossed the stenotic valve, and the catheter was advanced in the LV. Using a long exchange wire, a 6 Fr pigtail catheter was advanced in LV, and a peak-to-peak gradient measurement was gained through the two pigtails (in aortic root and LV). The LV pigtail catheter was exchanged for a long pre-shaped Super Stiff 0.035″ wire (Confida). Due to extreme calcification of the aortic leaflets, a pre-dilatation of the aortic valve was made under rapid (180 bpm) pacing with a 20 × 40 mm balloon without hemodynamic deterioration. Following this, the 14 Fr sheath was removed, and a 29 mm Evolut-Pro self-expanding device was gradually advanced over its in-line delivery sheath across the descending aorta, the aortic arch, and finally the aortic valve. Valve deployment, guided by aortograms, was initiated by gradual anti-clockwise turning of the device delivery knob until pressure dropped (half way opening). Device deployment was then accelerated until reaching the "no recapture point," and a final aortogram was made. Device's appropriate implantation depth and coronaries' adequate perfusion were confirmed before final release of the valve.

The delivery system is then removed from the body and replaced by the 14 Fr sheath. The newly implanted valve is crossed with a pigtail catheter, and the new peak-to-peak gradient is recorded confirming an optimal hemodynamic result. A final aortogram is made to exclude any more than mild paravalvular regurgitation. Finally, the 14 Fr sheath is removed, and the two ProGlide devices' sutures seal the puncture site of the main access artery. A contralateral contrast infusion through an IMA or pigtail catheter confirmed the effective sealing of the access site. Since the patient developed a mild QRS prolongation after valve implantation, the temporary pacing wire was left in place for at least 24 h until any severe conduction disturbances were excluded.

The patient was transferred to the coronary intensive care unit. The pacing wire was removed the next day, and he was mobilized without any problems. He was placed on single antiplatelet therapy and discharged 3days later. At 1-month follow-up he described improvement of his symptoms, while transthoracic echocardiogram confirmed a well-functioning prosthetic aortic valve.

References

1. Nkomo VT, Gardin JM, Skelton TN, Gottdiener JS, Scott CG, Enriquez-Sarano M. Burden of valvular heart diseases: a population-based study. Lancet. 2006;368(9540):1005–11.
2. Yadgir S, Johnson CO, Aboyans V, Adebayo OM, Adedoyin RA, Afarideh M, et al. Global, regional, and national burden of calcific aortic valve and degenerative mitral valve diseases, 1990-2017. Circulation. 2020;141(21):1670–80.
3. Lindman BR, Clavel MA, Mathieu P, Iung B, Lancellotti P, Otto CM, et al. Calcific aortic stenosis. Nat Rev Dis Primers. 2016;2:16006.
4. Genereux P, Pibarot P, Redfors B, Mack MJ, Makkar RR, Jaber WA, et al. Staging classification of aortic stenosis based on the extent of cardiac damage. Eur Heart J. 2017;38(45):3351–8.
5. Baumgartner H, Falk V, Bax JJ, De Bonis M, Hamm C, Holm PJ, et al. 2017 ESC/EACTS guidelines for the management of valvular heart disease. Eur Heart J. 2017;38(36):2739–91.
6. Leon MB, Smith CR, Mack M, Miller DC, Moses JW, Svensson LG, et al. Transcatheter aortic-valve implantation for aortic stenosis in patients who cannot undergo surgery. N Engl J Med. 2010;363(17):1597–607.
7. Smith CR, Leon MB, Mack MJ, Miller DC, Moses JW, Svensson LG, et al. Transcatheter versus surgical aortic-valve replacement in high-risk patients. N Engl J Med. 2011;364(23):2187–98.
8. Adams DH, Popma JJ, Reardon MJ, Yakubov SJ, Coselli JS, Deeb GM, et al. Transcatheter aortic-valve replacement with a self-expanding prosthesis. N Engl J Med. 2014;370(19):1790–8.
9. Leon MB, Smith CR, Mack MJ, Makkar RR, Svensson LG, Kodali SK, et al. Transcatheter or surgical aortic-valve replacement in intermediate-risk patients. N Engl J Med. 2016;374(17):1609–20.
10. Carroll JD, Mack MJ, Vemulapalli S, Herrmann HC, Gleason TG, Hanzel G, et al. STS- ACC TVT registry of transcatheter aortic valve replacement. J Am Coll Cardiol. 2020;76(21):2492–516.
11. Taniguchi T, Morimoto T, Shiomi H, Ando K, Kanamori N, Murata K, et al. Initial surgical versus conservative strategies in patients with asymptomatic severe aortic stenosis. J Am Coll Cardiol. 2015;66(25):2827–38.
12. Spitzer E, Van Mieghem NM, Pibarot P, Hahn RT, Kodali S, Maurer MS, et al. Rationale and design of the Transcatheter aortic valve replacement to UNload the left ventricle in patients with ADvanced heart failure (TAVR UNLOAD) trial. Am Heart J. 2016;182:80–8.
13. Strange G, Stewart S, Celermajer D, Prior D, Scalia GM, Marwick T, et al. Poor long- term survival in patients with moderate aortic stenosis. J Am Coll Cardiol. 2019;74(15):1851–63.
14. O'Brien SM, Shahian DM, Filardo G, Ferraris VA, Haan CK, Rich JB, et al. The Society of Thoracic Surgeons 2008 cardiac surgery risk models: part 2—isolated valve surgery. Ann Thorac Surg. 2009;88(1 Suppl):S23–42.
15. Nashef SA, Roques F, Sharples LD, Nilsson J, Smith C, Goldstone AR, et al. EuroSCORE II. Eur J Cardiothorac Surg. 2012;41(4):734–44; discussion 44–5.
16. Vahanian A, Beyersdorf F, Praz F, Milojevic M, Baldus S, Bauersachs J, et al. 2021 ESC/EACTS guidelines for the management of valvular heart disease. Eur Heart J. 2022;43(7):561–632.
17. Faroux L, Guimaraes L, Wintzer-Wehekind J, Junquera L, Ferreira-Neto AN, Del Val D, et al. Coronary artery disease and transcatheter aortic valve replacement: JACC state-of- the-art review. J Am Coll Cardiol. 2019;74(3):362–72.
18. Witberg G, Regev E, Chen S, Assali A, Barbash IM, Planer D, et al. The prognostic effects of coronary disease severity and completeness of revascularization on mortality in patients undergoing transcatheter aortic valve replacement. JACC Cardiovasc Interv. 2017;10(14):1428–35.
19. Thalji NM, Suri RM, Daly RC, Greason KL, Dearani JA, Stulak JM, et al. The prognostic impact of concomitant coronary artery bypass grafting during aortic valve surgery: implications for revascularization in the transcatheter era. J Thorac Cardiovasc Surg. 2015;149(2):451–60.
20. Thiele H, Kurz T, Feistritzer HJ, Stachel G, Hartung P, Lurz P, et al. General versus local anesthesia with conscious sedation in transcatheter aortic valve implantation: the randomized SOLVE-TAVI trial. Circulation. 2020;142(15):1437–47.
21. Dangas GD, Lefevre T, Kupatt C, Tchetche D, Schafer U, Dumonteil N, et al. Bivalirudin versus heparin anticoagulation in transcatheter aortic valve replacement: the randomized BRAVO-3 trial. J Am Coll Cardiol. 2015;66(25):2860–8.
22. Brinkert M, Mangner N, Moriyama N, Keller LS, Hagemeyer D, Crusius L, et al. Safety and efficacy of transcatheter aortic valve replacement with continuation of vitamin K antagonists or direct oral anticoagulants. JACC Cardiovasc Interv. 2021;14(2):135–44.
23. Habib G, Lancellotti P, Antunes MJ, Bongiorni MG, Casalta JP, Del Zotti F, et al. 2015 ESC guidelines for the management of infective endocarditis: the task force for the Management of Infective Endocarditis of the European Society of Cardiology (ESC). Endorsed by: European Association for Cardio-Thoracic Surgery (EACTS), the European Association of Nuclear Medicine (EANM). Eur Heart J. 2015;36(44):3075–128.
24. Stortecky S, Heg D, Tueller D, Pilgrim T, Muller O, Noble S, et al. Infective endocarditis after transcatheter aortic valve replacement. J Am Coll Cardiol. 2020;75(24):3020–30.
25. van Wiechen MP, Tchetche D, Ooms JF, Hokken TW, Kroon H, Ziviello F, et al. Suture- or plug-based large-bore arteriotomy closure: a pilot randomized controlled trial. JACC Cardiovasc Interv. 2021;14(2):149–57.
26. Vavuranakis M, Kalogeras KI, Vrachatis DA, Kariori MG, Aznaouridis K, Moldovan C, et al. A modified technique to safely close the arterial puncture site after TAVI. J Invasive Cardiol. 2013;25(1):45–7.
27. Lanz J, Greenbaum A, Pilgrim T, Tarantini G, Windecker S. Current state of alternative access for transcatheter aortic valve implantation. EuroIntervention. 2018;14(AB):AB40–52.
28. Dahle TG, Kaneko T, McCabe JM. Outcomes following subclavian and axillary artery access for transcatheter aortic valve replacement: Society of the Thoracic Surgeons/American College of Cardiology TVT registry report. JACC Cardiovasc Interv. 2019;12(7):662–9.
29. Bapat V, Khawaja MZ, Attia R, Narayana A, Wilson K, Macgillivray K, et al. Transaortic transcatheter aortic valve implantation using Edwards Sapien valve: a novel approach. Catheter Cardiovasc Interv. 2012;79(5):733–40.
30. Chamandi C, Abi-Akar R, Rodes-Cabau J, Blanchard D, Dumont E, Spaulding C, et al. Transcarotid compared with other alternative access routes for transcatheter aortic valve replacement. Circ Cardiovasc Interv. 2018;11(11):e006388.
31. Athappan G, Patvardhan E, Tuzcu EM, Svensson LG, Lemos PA, Fraccaro C, et al. Incidence, predictors, and outcomes of aortic regurgitation after transcatheter aortic valve replacement: meta-analysis and systematic review of literature. J Am Coll Cardiol. 2013;61(15):1585–95.
32. Pibarot P, Hahn RT, Weissman NJ, Arsenault M, Beaudoin J, Bernier M, et al. Association of paravalvular regurgitation with 1-year outcomes after transcatheter aortic valve replacement with the SAPIEN 3 valve. JAMA Cardiol. 2017;2(11):1208–16.
33. Ribeiro HB, Nombela-Franco L, Urena M, Mok M, Pasian S, Doyle D, et al. Coronary obstruction following transcatheter aortic valve implantation: a systematic review. JACC Cardiovasc Interv. 2013;6(5):452–61.
34. Abramowitz Y, Chakravarty T, Jilaihawi H, Kashif M, Kazuno Y, Takahashi N, et al. Clinical impact of coronary protection dur-

ing transcatheter aortic valve implantation: first reported series of patients. EuroIntervention. 2015;11(5):572–81.
35. Pasic M, Unbehaun A, Buz S, Drews T, Hetzer R. Annular rupture during transcatheter aortic valve replacement: classification, pathophysiology, diagnostics, treatment approaches, and prevention. JACC Cardiovasc Interv. 2015;8(1 Pt A):1–9.
36. Kim WK, Schafer U, Tchetche D, Nef H, Arnold M, Avanzas P, et al. Incidence and outcome of peri-procedural transcatheter heart valve embolization and migration: the TRAVEL registry (TranscatheteR HeArt valve EmboLization and migration). Eur Heart J. 2019;40(38):3156–65.
37. Mack MJ, Leon MB, Thourani VH, Makkar R, Kodali SK, Russo M, et al. Transcatheter aortic-valve replacement with a balloon-expandable valve in low-risk patients. N Engl J Med. 2019;380(18):1695–705.
38. Genereux P, Webb JG, Svensson LG, Kodali SK, Satler LF, Fearon WF, et al. Vascular complications after transcatheter aortic valve replacement: insights from the PARTNER (Placement of AoRTic TraNscathetER valve) trial. J Am Coll Cardiol. 2012;60(12):1043–52.
39. Scarsini R, De Maria GL, Joseph J, Fan L, Cahill TJ, Kotronias RA, et al. Impact of complications during transfemoral transcatheter aortic valve replacement: how can they be avoided and managed? J Am Heart Assoc. 2019;8(18):e013801.
40. Maeno Y, Abramowitz Y, Kawamori H, Kazuno Y, Kubo S, Takahashi N, et al. A highly predictive risk model for pacemaker implantation after TAVR. JACC Cardiovasc Imaging. 2017;10(10 Pt A):1139–47.
41. Siontis GCM, Overtchouk P, Cahill TJ, Modine T, Prendergast B, Praz F, et al. Transcatheter aortic valve implantation vs. surgical aortic valve replacement for treatment of symptomatic severe aortic stenosis: an updated meta-analysis. Eur Heart J. 2019;40(38):3143–53.
42. Vlastra W, Jimenez-Quevedo P, Tchetche D, Chandrasekhar J, de Brito FS Jr, Barbanti M, et al. Predictors, incidence, and outcomes of patients undergoing transfemoral transcatheter aortic valve implantation complicated by stroke. Circ Cardiovasc Interv. 2019;12(3):e007546.
43. Otto CM, Nishimura RA, Bonow RO, Carabello BA, Erwin JP 3rd, Gentile F, et al. 2020 ACC/AHA guideline for the management of patients with valvular heart disease: executive summary: a report of the American College of Cardiology/American Heart Association Joint Committee on clinical practice guidelines. Circulation. 2021;143(5):e35–71.
44. Brouwer J, Nijenhuis VJ, Delewi R, Hermanides RS, Holvoet W, Dubois CLF, et al. Aspirin with or without clopidogrel after transcatheter aortic-valve implantation. N Engl J Med. 2020;383(15):1447–57.
45. Nijenhuis VJ, Brouwer J, Delewi R, Hermanides RS, Holvoet W, Dubois CLF, et al. Anticoagulation with or without clopidogrel after transcatheter aortic-valve implantation. N Engl J Med. 2020;382(18):1696–707.
46. Vavuranakis M, Kalogeras K, Kolokathis AM, Vrachatis D, Magkoutis N, Siasos G, et al. Antithrombotic therapy in TAVI. J Geriatr Cardiol. 2018;15(1):66–75.
47. Kalogeras K, Jabbour RJ, Ruparelia N, Watson S, Kabir T, Naganuma T, et al. Comparison of warfarin versus DOACs in patients with concomitant indication for oral anticoagulation undergoing TAVI; results from the ATLAS registry. J Thromb Thrombolysis. 2020;50(1):82–9.
48. Blackman DJ, Saraf S, MacCarthy PA, Myat A, Anderson SG, Malkin CJ, et al. Long-term durability of transcatheter aortic valve prostheses. J Am Coll Cardiol. 2019;73(5):537–45.
49. Testa L, Latib A, Brambilla N, De Marco F, Fiorina C, Adamo M, et al. Long-term clinical outcome and performance of transcatheter aortic valve replacement with a self-expandable bioprosthesis. Eur Heart J. 2020;41(20):1876–86.
50. Sondergaard L, Ihlemann N, Capodanno D, Jorgensen TH, Nissen H, Kjeldsen BJ, et al. Durability of transcatheter and surgical bioprosthetic aortic valves in patients at lower surgical risk. J Am Coll Cardiol. 2019;73(5):546–53.
51. Pibarot P, Ternacle J, Jaber WA, Salaun E, Dahou A, Asch FM, et al. Structural deterioration of transcatheter versus surgical aortic valve bioprostheses in the PARTNER-2 trial. J Am Coll Cardiol. 2020;76(16):1830–43.
52. Rodriguez-Gabella T, Voisine P, Puri R, Pibarot P, Rodes-Cabau J. Aortic bioprosthetic valve durability: incidence, mechanisms, predictors, and management of surgical and transcatheter valve degeneration. J Am Coll Cardiol. 2017;70(8):1013–28.
53. Capodanno D, Petronio AS, Prendergast B, Eltchaninoff H, Vahanian A, Modine T, et al. Standardized definitions of structural deterioration and valve failure in assessing long-term durability of transcatheter and surgical aortic bioprosthetic valves: a consensus statement from the European Association of Percutaneous Cardiovascular Interventions (EAPCI) endorsed by the European Society of Cardiology (ESC) and the European Association for Cardio-Thoracic Surgery (EACTS). Eur Heart J. 2017;38(45):3382–90.

Transcatheter Mitral Valve Repair

41

Romain Gallet and Emmanuel Teiger

Case Presentation

An 83-year-old woman was referred to the hospital by her cardiologist for mitral regurgitation- related heart failure.

She had been complaining of dyspnea for a week. This dyspnea had been progressively increasing, leading to the necessity of sleeping in an armchair.

Her previous medical history included chronic kidney disease (creatinine clearance 45 mL/min) and a right carotid endarterectomy. The transthoracic echocardiography performed by her cardiologist had revealed severe mitral regurgitation (MR) as a result of chordal rupture on the anterior leaflet at the level of the A2 segment. The width of the prolapse was 8 mm. This mitral regurgitation was associated with an increase in systolic pulmonary pressure to 60 mm Hg. Left ventricle was hyperkinetic and left atrium was not dilated, suggesting a recent MR.

Given the high/prohibitive surgical risk (Euroscore 2 14.28%), a percutaneous treatment of this recent MR was discussed.

Continued on page 461

Background

Mitral regurgitation is the second most common valvulopathy, and degenerative mitral valve disease affects approximately 18 million people worldwide [1, 2]. It can be related to a disease of the mitral valve itself (primary or organic MR) or to a disease of the left ventricle and/or the left atrium leading to annulus enlargement causing MR (secondary or functional MR). The occurrence of severe MR is associated with increased morbidity (including hospitalization for pulmonary edema and atrial fibrillation) and mortality. For primary MR the standard of care remains the surgical correction of the MR through repair or replacement of the damaged valve [3]. However, surgery may be high risk or even not an option in elderly, comorbid, and/or frail patients [4]. Therefore, other strategies have been (and are currently being) developed. Among those strategies, edge-to-edge percutaneous repair has been the most used and studied and is currently the only recommended option. This technic has been inspired by Alfieri stitch [5, 6] and has been proven effective in both primary and secondary MR [7–10].

Indications for Intervention

The indications for intervention in primary mitral regurgitation have been defined in the guidelines of the European Society of Cardiology [3]. It includes high-grade MR in symptomatic patients who are operable without high risk, asymptomatic patients with LV dysfunction (LVESD $\geq$40 mm and/or LVEF $\leq$60%), asymptomatic patients with atrial fibrillation secondary to mitral regurgitation, or pulmonary hypertension (SPAP at rest >50 mmHg), and asymptomatic patients with left atrium enlargement due to the MR (volume index $\geq$60 mL/m [2] or diameter $\geq$55 mm) and a likely durable repair. The gold standard for treatment currently remains surgical repair or replacement.

Edge-to-edge repair should only be considered for symptomatic patients who fulfill the echocardiographic criteria of eligibility with high or prohibitive surgical risk assessed by surgical risk scores or frailty. Regarding secondary mitral regurgitation, the data are more controversial since two randomized studies comparing optimal medical therapy to edge-to-edge percutaneous repair came out with opposite results [10, 11]. While the Mitra-FR study was negative, the COAPT trial showed an impressive decrease in mortality and hospitalization with the MitraClip therapy as compared to optimal medical therapy. The patients enrolled in those studies were not similar and therefore the guidelines for interventions in secondary mitral regurgitations have taken into account these differences [12–15]. Thus, percutaneous edge-to-edge mitral repair is mainly recommended in patients with symptomatic MR with an EROA $\geq$30 mm^2, an LV end-

R. Gallet (✉) · E. Teiger
Cardiology Department, Hôpitaux Universitaires Henri Mondor, Créteil, France
e-mail: romain.gallet@aphp.fr; emmanuel.teiger@aphp.fr

G. Geroulakos et al. (eds.), *Mastering Endovascular Techniques*, https://doi.org/10.1007/978-3-031-42735-0_41

systolic diameter ≤ 70 mm a LVEF of 20–50%, and a favorable anatomy[3]. Except in rare selected cases it should not be performed in patients with a cause of heart failure other that dilated cardiomyopathy (restrictive, hypertrophic, infiltrative), in patients with severe pulmonary hypertension (>70 mmHg) or right ventricle failure, or in patients with hemodynamic instability or severe disability or frailty.

Preoperative Preparation

Timing of Intervention

All interventions should be discussed among the heart team to confirm patient candidacy for percutaneous mitral valve edge-to-edge repair and anatomic feasibility of clip placement. In case of secondary MR or well-tolerated primary MR, intervention is not urgent and should be performed after thorough pre-procedural planning. The timing for intervention is usually a few weeks after evaluation and long delays should be avoided to prevent the (further) impairment of LV function and subsequently RV function in response to the MR.

In case of acute symptomatic primary MR (usually related to chordae rupture) or ischemic MR, intervention should be performed within a few days because of the risk of massive pulmonary edema and sudden cardiac death.

Table 41.1 Echocardiographic criteria for high-grade MR

	Primary MR	Secondary MR
Qualitative		
Mitral valve morphology	Large prolapse or flail leaflet, ruptured papillary muscle, severe retraction, large perforation	Severe tenting, poor leaflet coaptation. Usually normal leaflets
Color flow jet area	Large central jet (>50% of LA) or eccentric wall impinging jet of variable size	
Flow convergence	Large, holosystolic	
Continuous wave Doppler jet	Holosystolic/dense/triangular	
Semiquantitative		
Vena contracta width (mm)	≥7 (≥8 mm for biplane)	
Pulmonary vein flow	Systolic flow reversal	
Mitral inflow	E-wave dominant (>1.2 m/S)	
TVI Mitral/TVI aortic ratio	>1.4	
Quantitative		
EROA (mm^2)	≥40 mm²	≥40 mm² (may be ≥30 mm² if elliptical ROA)
Regurgitant volume (mL/beat)	≥60 mL	≥60 mL (may be 45 ≥ mL if low flow conditions)
Regurgitant fraction (%)	≥50%	
Structural		
Left ventricle	Dilated	
Left atrium	Dilated	

Adapated from Eur Heart J. 2022 Feb 12;43(7):561–632

Imaging

Transthoracic and Transesophageal Echocardiography

Transthoracic and multi-plan transesophageal echocardiography (TEE) are the cornerstone of the evaluation of MR on one side and of the assessment of the feasibility of transcatheter repair on the other side [8].

Regarding the grading of MR, the different techniques for quantification and grading are detailed in the guidelines of the European Society of Cardiology. Briefly, quantification should include, among all vena contracta, TVI ratio, effective regurgitant orifice area as well as regurgitant volume or fraction, pulmonary vein flow measurement, and LV morphology assessment [3]. Only high-grade symptomatic MR defined by the criteria in Table 41.1 is eligible for percutaneous repair.

Regarding feasibility, the criteria for primary and secondary MR are summarized in Table 41.2 [16]. In primary MR, a non-commissural flail with a gap below 10 mm and a width below 15 mm is usually favorable for edge-to-edge repair, while short leaflets, calcified leaflets, or commissural flails are usually predictive of difficult or failed procedures and should be performed only by experienced operators.

Regarding secondary MR, besides the criterion for indication, the main criteria for feasibility are the length of the coaptation defect and the length of the leaflet.

Additionally, TEE will evaluate the size of the left atrium and the morphology of the inter-atrial septum to evaluate the feasibility of trans-septal access and navigation into the left atrium.

Lastly, the left appendage will be visualized to rule out any thrombus that would temporarily contra-indicate the procedure.

Cardiac MRI

Cardiac MRI ideally with 4D-flow analysis is not mandatory in pre-procedural planning but may be useful in case of MR with difficult quantification, especially for MR with a very offset jet [17]. It can also provide a more accurate assessment of residual MR after MitraClip in difficult cases [18, 19].

Table 41.2 Echocardiographic criteria for feasability of MitraClip procedures

Optimal	Challenging	Complex
– central A2/P2	– Commissural (A1/P1, A3/P3)	– Grasping zone with calcium
– No calcification	– Severe calcification of annulus, non in grasping zone	– MVA <3 cm^2
– Mitral valve area >4 cm^2	– 4 cm^2 > MVA > 3 cm^2	– Posterior leaflet <7 mm, cleft
– Posterior leaflet >10 mm	– Posterior leaflet 7–10 mm	– Rheumatic
– Tenting height <10 mm	– Tenting height >10 mm	– Carpentier IIIB, multiple segments, Barlow
– Normal leaflets and mobility	– Carpentier IIIB (leaflet restriction)	
– Flail gap < 10 mm, flail width <15 mm	– Flail width >15 mm	

Operator's experience /volume of procedures in the centre increase →

Adapted from Eur Heart J Cardiovasc Imaging 2020:21:1059–1067

Table 41.3 Suggested endovascular toolkit for percutaneous edge-to-edge mitral repair. These are only a few suggested options based on the author's experience and can cover the vast majority of cases

	Manufacturer	Size/Length
Wires		
Any standard access wire		0.035″
Any 0.032″ wire		0.032″
Amplatzer superstiff	Boston Scientific Corporation	0.035″
Sheaths and dilators		
Any standard access sheath		5–10 Fr/11 cm
Any dilators		18 Fr–22 Fr
Catheters		
Transseptal catheter SL0	Abbott	
Or	Baylis	
TorFlex transseptal catheter		
BRK transsptal needle	Abbott	
Or	Baylis	
NRG transseptal needle	Baylis	
Or		
ProTrack Pigtail Wire		
MitraClip steerable guide catheter	Abbott	24 Fr–800 mm
MitraClip and clip delivery system	Abbott	
NT, XT, NTW or XTW	Abbott	
Vascular closure Devices		
Perclose Proglides	Abbott	8Fr

Endovascular Strategy and Technique

Basic Endovascular Toolkit

- Standard access wires and sheaths including a short 6 or 8 Fr sheath
- Floppy and stiff 035″ guidewire
- A set of dilators, for example 12 Fr, 18 Fr, and 22 Fr
- Trans-septal kit including a 032″ guidewire, a trans-septal catheter, and a trans-septal needle (see below)
- An arterial line for pressure monitoring and blood draw for repeated ACT measurements

A list of suggested materials is summarized in Table 41.3.

Transesophageal Echocardiography

Transesophageal echocardiographic guidance of the procedure with multi-planar and 3-dimensional acquisition is mandatory for transcatheter edge-to-edge repair (TEER) of MR. Indeed, echography will guide the whole procedure, from trans-septal puncture to clip positioning, leaflet grasping, and clip release. The procedure will therefore require very close communication between the echographist and the interventional cardiologist.

Trans-septal Kit

The trans-septal kit should include a trans-septal catheter, a trans-septal needle, and a dedicated guidewire. Another option which is particularly interesting in difficult trans-septal crossing (thick or very floppy septum, unusual orientation) is the use of a system using radiofrequency [20, 21]. With such systems, the crossing of the septum relies on radiofrequency and not on the needle and may result in easier and safer trans-septal access.

Transcatheter Edge-to-Edge Repair Device

To date, two different devices are available for TEER of the mitral valve, the MitraClip and the Pascal [22]. The most widely studied and used device is the MitraClip device. The principle of the technique is to capture the anterior and posterior leaflets of the mitral valve at the location of the MR to treat the prolapse in primary MR and to decrease coaptation defect and MR in secondary MR. Each leaflet will be captured between a gripper and the arm of the clip. The clip will then be closed to bring the edges of the leaflets closer.

Decision Making on Technique Selection

Do date, the most used (and almost only available) technique for percutaneous mitral repair is TEER. Other techniques such as percutaneous annulopasty [23] and trans-apical neochordae [24, 25] implantation are being developed but have not been fully tested and are not recommended yet. Direct percutaneous annuloplasty is currently a lesser known technique. Indeed, it has been studied only in a small CE approval study and in case reports. This technique is complicated and associated with very long procedural times. Moreover, it allows treatment of secondary MR, with disappointing results regarding MR reduction (40% mild or less MR at 1 year) [23]. Neochordae implantation offers the opportunity to treat primary MR but is not completely percutaneous since it requires a trans-apical approach [24, 25]. For all these reasons, only TEER especially using the MitraClip system will be discussed for the procedural technique.

The MitraClip device is now in its fourth generation with an independent grasping mechanism [26]. It is available in four sizes depending on its length (NT or XT MitraClip with arms respectively 9 and 12 mm long) and its width, each length being available with 2 widths in the grasping area (4 mm wide, NT and XT MitraClip, or 6 mm wide, NTW and XTW MitraClip). Long clips offer more possibility for the grasping of large coaptation defects but require a sufficient length of valve to be captured in the clip (>10 mm). Additionally, they provide more tension of the valve, resulting in a certain degree of annuloplasty. A larger clip allows for treatment of larger prolapse and reduction of the number of clips implanted.

Pre-procedural Therapeutic Strategy and Number of Clips

Pre-procedural strategy will aim to determine the number of clips to be implanted and their location. While small prolapse may be treated using only one clip, larger prolapse and secondary MR usually require two or three clips. When implanting several clip, it is always easier to start with the more medial clip and then to implant the other clip(s) more laterally. Thus, after implantation of the first clip, no residual significant MR should be present on the medial side of the clip. One exception to this strategy is the treatment of anterolateral commissural flail. In this case, the most commissural clip will be implanted first.

Access to the Left Atrium

All procedures will be performed under general anesthesia and TEE guidance. Access to the left atrium will require femoral vein approach (usually the right one for easier trans-septal puncture).

Septal puncture will be performed under TEE guidance in order to cross the septum at the proper position to make the TEER easier.

After venous access using a 5–8 Fr sheath, a 0.032-inch wire will be positioned in the superior vena cava. A trans-septal sheath will be approached over this wire. A BRK-trans-septal needle will be introduced into the trans-septal sheath. Using biplanar TEE guidance the puncture of the septum will be performed on the upper and posterior area of the inter-atrial septum. The biplanar TEE mode will allow both the bi-caval(to monitor the superior/inferior position) and the aorta (to monitor the anterior/posterior position) to be viewed at the same time. After visualization of the tenting on the fossa ovale at the intended position, the latter will be controlled in the four-chamber view in which the distance between the tenting and the mitral annular plan will be measured. To ensure that the grasping will be possible, this distance will have to be between 40 and 45 mm with high puncture for primary as compared to secondary MR. The distances are different because of the different plan of coaptation (lower in primary MR because of the prolapse, higher in the LV in secondary MR because of the tenting). After measurement, the septum will be punctured, the needle will be removed and the 0.032-inch wire will be advanced into the pulmonary artery and then changed for a 0.035-inch stiff wire.

An alternative to the use of the BRK needle is to use a dedicated wire connected to a radiofrequency generator. This pre-shape wire will be used instead of the needle to cross the septum thanks to radiofrequency. The guidewire will be positioned into the left atrium. Then, its stiffness will allow it to advance any material without any trauma.

Anticoagulation

All TEER procedures require anticoagulation aiming at an ACT value over 250. Unfractionated heparin with an initial dosing of 100 U/kg is usually used and can be administrated with two different schema, either full dose after the trans-septal puncture (after ensuring that there is no trans-septal crossing-related adverse event) or half of the dose before trans-septal puncture and half of the dose after. ACT will then be monitored every 20 min.

Access to the Mitral Valve

The MitraClip system is composed of two different parts, a steerable guide catheter (24 French (8.1 mm), 800 mm working length) that will be held by a stabilizer, and a clip delivery system.

The guide catheter will be advanced to the left atrium over the 0.035-inch stiff wire. On initial skin and vessel entry, the +/− knob of the guide catheter is turned toward—to straighten the unit, and then returned to the neutral position once positioned in the right atrium to be oriented toward the left atrium. The guide catheter is then advanced about 1–2 cm into the left atrium, and the wire and the dilator of the guide catheter will subsequently be removed (Fig. 41.1a). Once the guide is positioned into the left atrium, the clip delivery system will be introduced through the guide and the markers of the guide will be aligned with the marker of the delivery

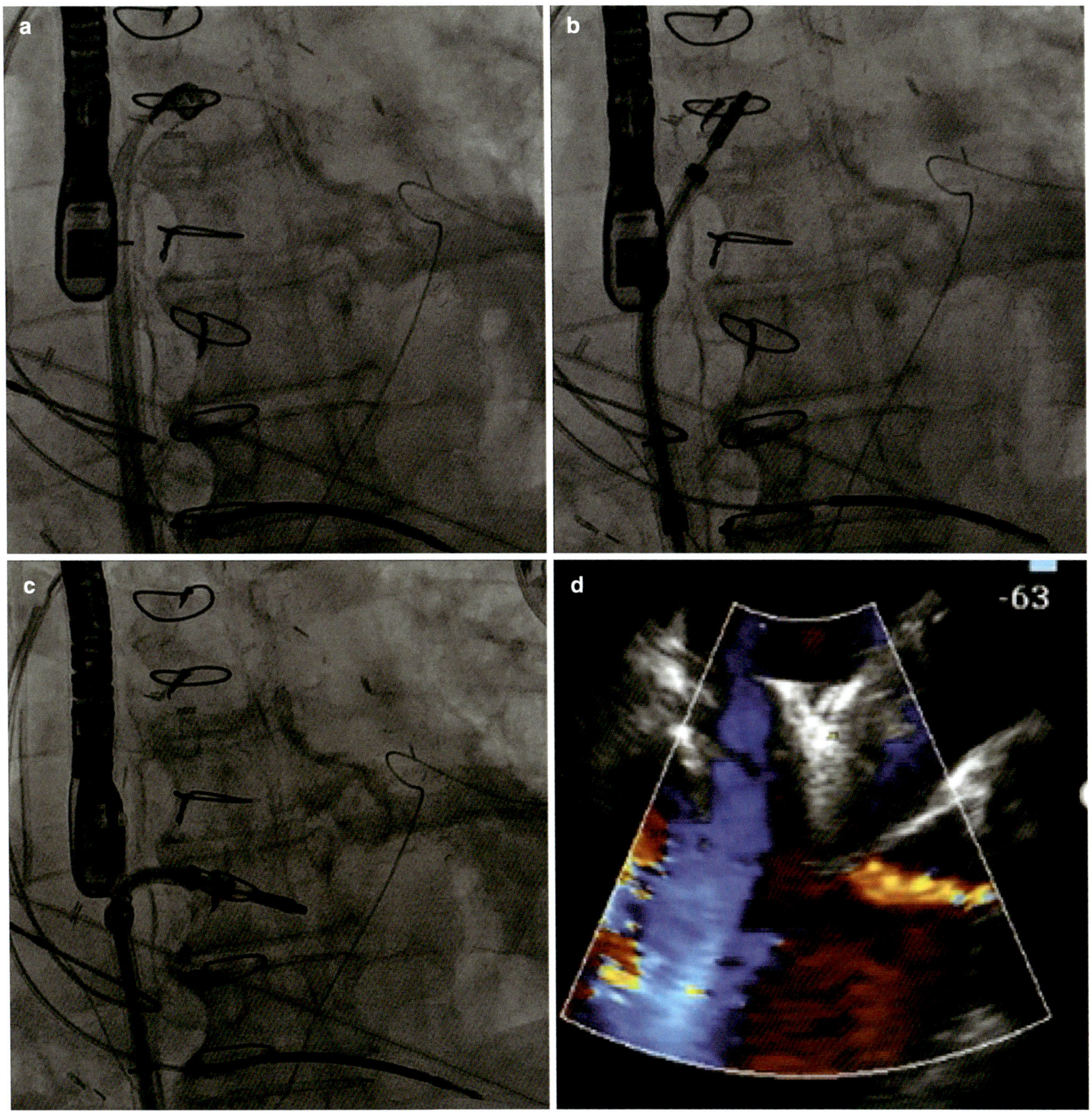

Fig. 41.1 (**a**) MitraClip steerable guide catheter positioned into the left atrium (angiographic view). (**b**) MitraClip out of the guide cathter, into the left atrium (angiographic view). (**c**) MitraClip facing the mitral valve (angiographic view). (**d**) positioning of the MitraClip into the regurgitant jet (color doppler)

system ("straddling") with attention paid to the left atrium wall while taking the clip out of the guide (Fig. 41.1b).

The clip will then be directed toward the mitral annulus, mainly by applying medial deflection on the delivery system using its M/L knob (Fig. 41.1c). Concomitantly, posterior orientation of delivery system will be used to move the clip away from the upper left pulmonary vein and the left atrium, so as not to get trapped in one of these two structures while moving the clip toward the annulus. All these maneuvers will be performed with the clip stuck to the delivery system (so called slacked). Once the clip is facing the mitral annulus, approximately 1 cm above the coaptation plan, it will be aligned with the regurgitant jet using a TEE multiplanar view showing LVOT (for antero-posterior assessment) and bi-commissural view (for medial/lateral assessment) (Fig. 41.1d). This antero-posterior alignment will be performed using the counter-clockwise/clockwise rotation of the guide catheter. Medial-lateral alignment can be performed using the M/L knob of the clip delivery system or the +/− knob of the guide catheter (+ brings the clip more medial) or by pulling/pushing the guide catheter/clip delivery system couple (pulling brings the clip more medial). The action to apply will depend on a combination of factors including the adequate orientation of the clip in relation to the predicted grasping area and to the aorta and the length of delivery system into the left atrium. The aim will be to be positioned into the regurgitant jet at the scheduled position (depending on the number of clips to be implanted) but also to be perpendicular to the intended area of grasping to have a proper grasping. Thus, the trajectory of the clip when it is moved forward will be assessed to make sure it does not change its position when it is advanced.

When the clip is at the right position, the arm of the clip will be opened (with the grippers being up, i.e., on the delivery system) at 120° and the clip will be oriented perpendicular to the line of coaptation in a 3D en face view of the mitral valve. Orientation will be achieved using counter-clockwise/clockwise rotation of the clip delivery system handle (with concomitant small back and forth movement in order to translate the movement to the clip). When the clip is perpendicular to the line of coaptation, its position as well as its trajectory when its advanced will be checked one last time. The clip will then be closed at least 60° and advanced to cross the mitral valve.

Grasping of the Leaflet

Once in the left ventricle (not too deep to avoid injuring any chordae), the clip will be reopened to 120–150°, depending on the coaptation. The position and perpendicularity will be checked and the clip will be gently pulled up until both the anterior and posterior leaflets fall into the arm of the clip with restrictive opening due to the clip arms (Fig. 41.2a). Once the position is confirmed for both leaflets, the grippers will be lowered to trap the leaflets between the clip and the grippers. The proper grasping will be confirmed using different methods; first the echo cine of the grasping will be reviewed to make sure there is enough tissue into the clip. The length of leaflet into the clip will also be measured either directly (but this requires high-quality imaging) or by subtracting the residual length of the leaflet after grasping to its initial length. Lastly, the grippers should be moving with the movement of the valve, indicating that the leaflet is between the clip and the gripper. The clip will be closed to 60° (Fig. 41.2b), the delivery system will be advanced by a few mm to reduce tension, and the initial results on the MR will be assessed as well as the transvalvular gradient.

It is crucial to make sure that both leaflets have been adequately captured. Indeed, incomplete capture will likely lead to clip detachment from the leaflet and MR relapse or even clip embolization.

Once satisfactory clip placement and leaflet grasp have been confirmed, the clip will be locked and further closed under TEE color mode to monitor an additional reduction of the MR. Once the placement and results are deemed optimal, the clip will be released (Fig. 41.3).

Troubleshooting

Clip Too Deep or Too High in the Left Atrium

In some cases, the grasping of the leaflets cannot be performed either because the clip is too high (and not able to cross the mitral valve) or more often too deep, resulting in the impossibility of applying enough tension on the leaflet during the grasping. This can be related to an inadequate trans-septal crossing location. If trans-septal is adequate or cannot be performed again, the height of the clip can be adjusted using the A/P knob on the clip delivery system. The A rotation will bring the clip higher (and more anterior with a need to adjust the position) while the P rotation will bring the clip deeper (and more posterior). This maneuver has to be performed into the left atrium and not into the left ventricle.

Incomplete or Difficult Grasping of One of the Leaflets

The Gen 4 MitraClip allows separate grasping of the leaflet [26]. The grippers can be lowered separately. This can be used to optimize the grasping of one of the leaflets; the clip will be opened, the gripper will be lifted up, and the handle will be rotated toward the target leaflet (anterior or posterior) with a gentle traction. The gripper will then be lowered and the clip closed. Caution must be taken to make very small

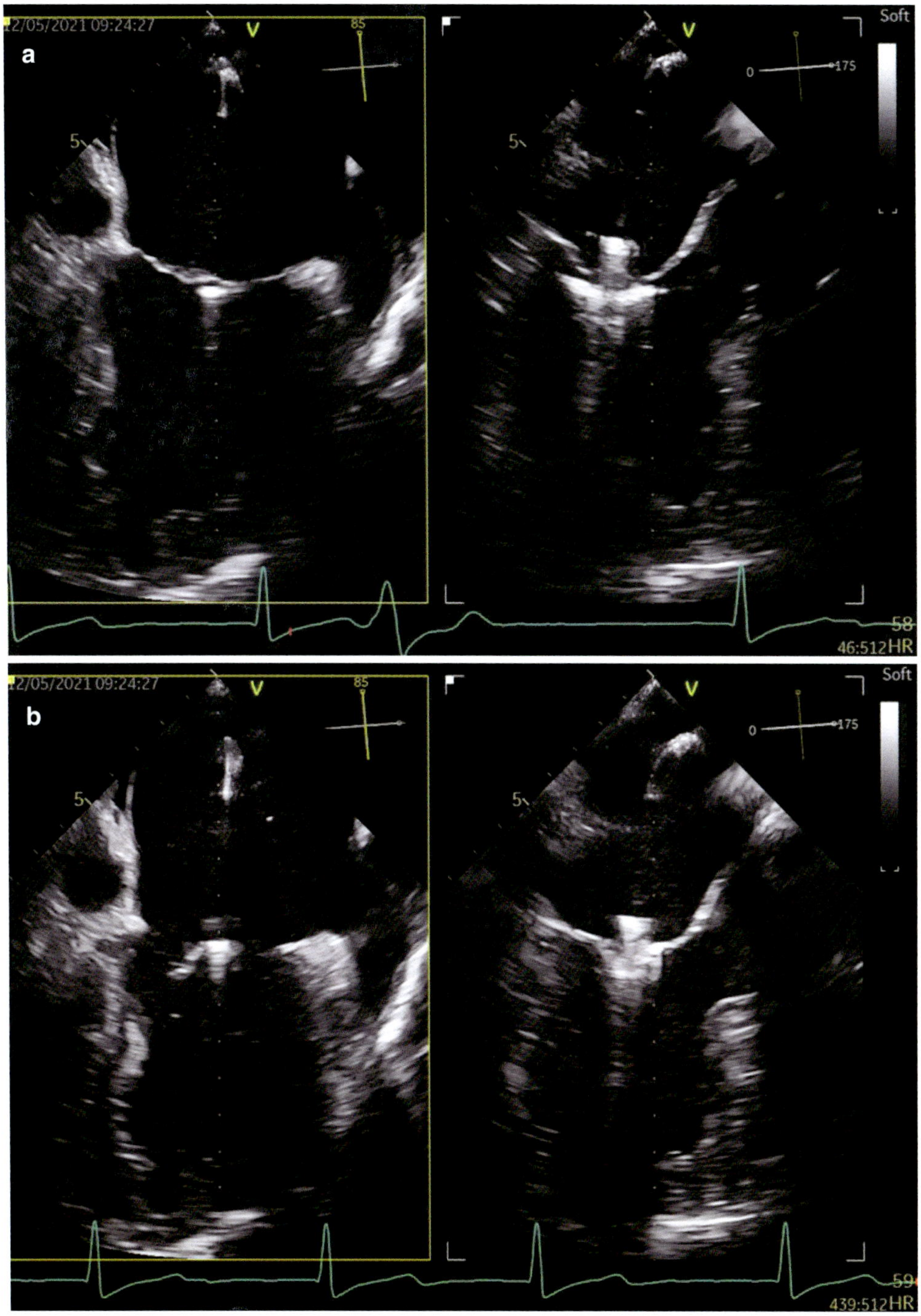

Fig. 41.2 (**a**) Position of the clip just before grasping on biplanar TEE view (bicomissural and LVOT). Both the anterior and posterior leaflets into the arms of the clip are shown. (**b**) Clip closed at 60° on biplanar TEE view

and gentle movements to avoid any damage on the captured leaflet. This technique can also be used for separate grasping in case of a large coaptation defect.

Another technique to grasp the leaflets in case of large coaptation defects is the zipping technique. This technique requires multiple clip implantation. The first one will be implanted at the edge of the coaptation defect where the leaflets are almost coapting. This grasping will be easier to perform and will bring the leaflet closer for the implantation of one or more additional clips.

Need to Remove the Clip and Delivery System from the LV

Every time a clip has to be removed from the LV (attached to the delivery system, there is currently no percutaneous solution to remove a launched clip), this will have to be performed with the clip in an inverted position.

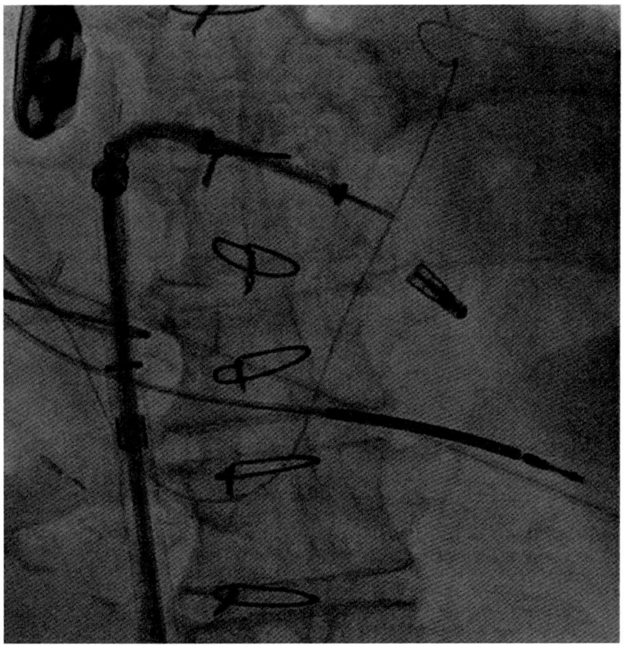

Fig. 41.3 MitraClip released (angiographic view)

Other Devices

Other devices are available in the market but are not as evidenced as the MitraClip device. For TEER the Pascal device is undergoing investigations [22, 27]. Other techniques exist such as annuloplasty using the Cardioband system or trans-apical neo-chordae implantation but these devices have only been used in small studies (mainly CE marking studies) and have not been compared to other treatments (medical, surgical, or interventional) [23–25].

Vascular Closure

Vascular closure may be obtained using two proglides positioned using the preclosing technique [28]. Alternatively, venous compression using a figure eight stich is also efficient.

Anticipated Complications

Several complications can occur at each step of the procedure. Although rare, most of them can have dramatic outcomes.

Vascular Injury

Vascular injury can occur in every percutaneous procedure. In the case of TEER, venous perforation (inferior vena cava or external or common iliac vein) with the guide catheter usually has a poor outcome because of the rapid and massive associated bleeding and the impossibility of achieving adequate hemostasis. To avoid vascular complications, ultrasound-guided punctures are strongly recommended. For the introduction of the guide catheter a stiff wire is mandatory and pre-dilatation of the skin and the vessel with dilators (14 Fr, 18 Fr, and 22 Fr) should be performed.

Trans-septal Puncture Complication

The use of TEE guiding enables safer puncture of the inter-atrial septum. However, aortic puncture or tamponade may sometimes occur. Depending on the size of the breach, the treatment will go from close monitoring to emergent cardiac surgery. To avoid any complication, the puncture must be echo-guided using multiplanar TEE. The needle will be pushed out of the trans-septal sheath only if tenting on the inter-atrial septum is visible on both plans in TEE. In case of thick or very floppy septum, several techniques can be used to secure the puncture. Radiofrequency can be used directly on the needle using an electric scalpel or using dedicated trans-septal wires. Another technique is to use a 0.014-inch wire introduced in the needle from the proximal segment to cross and the septum. The needle will then be advanced over the wire.

Pulmonary Vein or Left Atrium Appendage (LAA) Injury

Pulmonary vein or LAA injury may be caused by the stiff guidewire, by the guide catheter, or by the clip delivery system. Pulmonary vein injury (usually upper left pulmonary vein) will be suspected if blood is present in the endo-tracheal tube or if ventilation becomes more difficult, while LAA injury will result in cardiac tamponade. Both situations will require emergency assessment and treatment.

Intra-cardiac Thrombus

Intra-cardiac thrombus can appear during the procedure especially on the intra-cardiac material. While pre-existing intra-cardiac thrombus is a contra-indication to TEER, apparition of an intra-cardiac thrombus should lead to reconsideration of the procedure. ACT should be maintained >250 to avoid any thrombus formation and all catheter lumen should be regularly flushed.

Chordae Coiling Around the Clip

Excessive manipulation of the clip in the left ventricle, especially in the commissural area, can lead to entrapment of the clip in the sub-valvular apparatus. When this situation occurs, the first step should be to reverse the preceding maneuvers. The clip should be inverted and gently torqued or advanced into the LV. If the clip remains attached to the chordae despite the maneuvers, it may be safer to deploy the clip into the coiled sub-valvular mitral apparatus.

Leaflet Detachment

If there is not enough tissue grasped into the clip, single-leaflet detachment may occur. If the clip has not started to be released, another grasping can be performed. However, if the clip has been released, there is no possibility to extract the clip besides surgery. If the grasping of the other leaflet is stable, it may be considered to leave the clip in place and stabilize it with the implantation of a nearby additional clip.

Reduction in Mitral Valve Area

Due to the mechanism of mitral valve repair, MitraClip implantation will reduce the effective mitral valve area. Before releasing the clip, transvalvular mitral gradient will be measured and any gradient higher than 5 mmHg will raise concerns about removing this clip on a risk/benefit ratio. Preliminary studies regarding the impact of higher post-MitraClip gradients are to date reassuring [29] but data are sparse.

Perioperative Care and Surveillance

The patient should remain on bed rest for 2–4 h to allow for recovery from the anesthesia as well as hemostasis. Given the reduction of the MR and the fluid expansion received during the procedures, attention should be paid to worsening of the left ventricular ejection fraction especially in patients with pre-existing LV dysfunction, and to pulmonary edema. Diuretics should be administrated in case of clinical symptoms or echographic signs of increased filling pressures.

In patients with indication for anticoagulation (for atrial fibrillation notably), the latter should be restarted the evening after the procedure or the next day and pursued. In patients without anticoagulant, dual anti-platelet therapy should be administrated for 3 months followed by aspirin alone.

Follow-up office visit is recommended at 3, 6, and 12 months and annually thereafter with trans-thoracic echocardiography.

Case Presentation

Continued from page 453

A MitraClip was planned within a few days under general anesthesia and TEE and fluoroscopic guidance. Under ultrasound guidance, venous access was obtained. Pre-closing was performed using two preclose proglides. Trans-septal puncture was made using an SL0 trans-septal sheath and a BRK needle. The puncture was made on the upper and posterior part of the septum and the distance between the transeptal crossing and the annulus plan was measured at 41 mm. After confirmation of the absence of complication of the septal puncture, heparin was injected with regular boluses as needed to maintain an ACT $\geq$ 250 ms.

An XTW clip was chosen in order to treat the whole prolapse with only one clip. The clip was advanced and oriented toward the mitral annulus, oriented perpendicular to the valve, and then advanced into the LV. The clip was deployed at the intended position and allowed the grasping of the flail A2 and the posterior leaflet. After implantation of this clip there was no residual area of prolapse and the MR was graded as 1/4.

The pulmonary systolic pressure had already decreased to 30 mmHg the next day and the patient was safely discharged 24 h after the procedure.

References

1. Cahill TJ, et al. Community prevalence, mechanisms and outcome of mitral or tricuspid regurgitation. Heart. 2021:heartjnl-2020-318482. https://doi.org/10.1136/heartjnl-2020-318482.
2. Yadgir S, et al. Global, regional, and National Burden of calcific aortic valve and degenerative mitral valve diseases, 1990-2017. Circulation. 2020;141:1670–80.
3. Vahanian A, et al. 2021 ESC/EACTS guidelines for the management of valvular heart disease. Eur Heart J. 2022;43:561–632.
4. Iyengar A, et al. Effects of frailty on outcomes and 30-day readmissions after surgical mitral valve replacement. Ann Thorac Surg. 2020;109:1120–6.
5. Luk A, et al. Mitral repair with the Evalve MitraClip device: histopathologic findings in the porcine model. Cardiovasc Pathol. 2009;18:279–85.
6. Bhudia SK, et al. Edge-to-edge (Alfieri) mitral repair: results in diverse clinical settings. Ann Thorac Surg. 2004;77:1598–606.

7. Feldman T, et al. Randomized comparison of percutaneous repair and surgery for mitral regurgitation: 5-year results of EVEREST II. J Am Coll Cardiol. 2015;66:2844–54.
8. Attizzani GF, et al. Extended use of percutaneous edge-to-edge mitral valve repair beyond EVEREST (Endovascular Valve Edge-to-Edge Repair) criteria: 30-day and 12-month clinical and echocardiographic outcomes from the GRASP (Getting Reduction of Mitral Insufficiency by Percutaneous Clip Implantation) registry. JACC Cardiovasc Interv. 2015;8:74–82.
9. Kar S, et al. Five-year outcomes of transcatheter reduction of significant mitral regurgitation in high-surgical-risk patients. Heart. 2019;105:1622–8.
10. Stone GW, et al. Transcatheter mitral-valve repair in patients with heart failure. N Engl J Med. 2018;379:2307–18.
11. Obadia J-F, et al. Percutaneous repair or medical treatment for secondary mitral regurgitation. N Engl J Med. 2018;379:2297–306.
12. Iung B, et al. Characteristics and outcome of COAPT-eligible patients in the MITRA-FR trial. Circulation. 2020;142:2482–4.
13. Lindenfeld J, et al. Association of effective regurgitation orifice area to left ventricular end-diastolic volume ratio with transcatheter mitral valve repair outcomes: a secondary analysis of the COAPT trial. JAMA Cardiol. 2021;6:427–36.
14. Messika-Zeitoun D, et al. Impact of mitral regurgitation severity and left ventricular remodeling on outcome after MitraClip implantation: results from the Mitra-FR trial. JACC Cardiovasc Imaging. 2021;14:742–52.
15. Koell B, et al. Outcomes stratified by adapted inclusion criteria after mitral edge-to-edge repair. J Am Coll Cardiol. 2021;78:2408–21.
16. Gavazzoni M, et al. Conceiving MitraClip as a tool: percutaneous edge-to-edge repair in complex mitral valve anatomies. Eur Heart J Cardiovasc Imaging. 2020;21:1059–67.
17. Gheorghe LL, et al. Imaging for native mitral valve surgical and transcatheter interventions. JACC Cardiovasc Imaging. 2021;14:112–27.
18. Hamilton-Craig C, et al. Quantitation of mitral regurgitation after percutaneous MitraClip repair: comparison of Doppler echocardiography and cardiac magnetic resonance imaging. Ann Cardiothorac Surg. 2015;4:341–51.
19. Lurz P, et al. Assessment of acute changes in ventricular volumes, function, and strain after interventional edge-to-edge repair of mitral regurgitation using cardiac magnetic resonance imaging. Eur Heart J Cardiovasc Imaging. 2015;16:1399–404.
20. Whitler C, et al. Improved left atrial appendage closure procedural efficiency using radiofrequency transseptal wire system. Catheter Cardiovasc Interv. 2023;101:460. https://doi.org/10.1002/ccd.30550.
21. Inohara T, Gilhofer T, Luong C, Tsang M, Saw J. VersaCross radiofrequency system reduces time to left atrial access versus conventional mechanical needle. J Interv Card Electrophysiol. 2022;63:9–12.
22. Schneider L, et al. Mitral valve transcatheter edge-to-edge repair using MitraClip or PASCAL: a multicenter propensity score-matched comparison. JACC Cardiovasc Interv. 2022;15:2554–67.
23. Messika-Zeitoun D, et al. Transcatheter mitral valve repair for functional mitral regurgitation using the Cardioband system: 1 year outcomes. Eur Heart J. 2019;40:466–72.
24. D'Onofrio A, et al. Outcomes of transapical mitral valve repair with neochordae implantation. J Thorac Cardiovasc Surg. 2023;165(3):1036–1046.e4. S0022-5223(22)00387-7. https://doi.org/10.1016/j.jtcvs.2022.02.059.
25. Gerosa G, et al. Transapical off-pump echo-guided mitral valve repair with neochordae implantation mid-term outcomes. Ann Cardiothorac Surg. 2021;10:131–40.
26. Chakravarty T, et al. Transcatheter edge-to-edge mitral valve repair with the MitraClip G4 system. JACC Cardiovasc Interv. 2020;13:2402–14.
27. Lim DS, et al. Randomized comparison of transcatheter edge-to-edge repair for degenerative mitral regurgitation in prohibitive surgical risk patients. JACC Cardiovasc Interv. 2022;15:2523–36.
28. Mohammed M, et al. Preclosure of large bore venous access sites in patients undergoing transcatheter mitral replacement and repair. Catheter Cardiovasc Interv. 2022;100:163–8.
29. Halaby R, et al. Effect of mitral valve gradient after MitraClip on outcomes in secondary mitral regurgitation: results from the COAPT trial. JACC Cardiovasc Interv. 2021;14:879–89.

Atrial Septal Defect: Left Atrial Appendage

42

Maria Drakopoulou, Anastasios Apostolos, Ioannis Kachrimanidis, Georgios Oikonomou, Leonidas Koliastasis, Georgios Trantalis, Costas Tsioufis, and Konstantinos Toutouzas

Abbreviations

ACT	Activated clotting time
AF	Atrial fibrillation
ASD	Atrial septal defect
LAA	Left atrial appendage
MDCT	Multidetector computed tomography
OAC	Oral anticoagulation
PAH	Pulmonary arterial hypertension
PVR	Pulmonary vascular resistance
TOE	Transoesophageal echocardiography

Atrial Septal Defect

Case Presentation

A 31-year-old female patient with no past medical history presented with exertional dyspnea and intermittent palpitations of 6-month duration. Physical examination revealed heart rate 110/min and blood pressure 130/72 mmHg with oxygen saturation 98% on room air with normal jugular venous pressure (JVP). The heart rhythm was regular and there was no left parasternal heave on palpation. The pulmonary component of the second heart sound was delayed and fixed. There was a soft systolic ejection murmur at the upper left sternal border. Regular heart with rhythm was regular. The electrocardiogram showed sinus rhythm with incomplete right bundle branch block. Biochemical and hematological investigations were within normal limit. Chest x-ray was suggestive of cardiomegaly. Transthoracic echocardiography showed normal left ventricular size and function. The right ventricle was moderately dilated with preserved systolic function. Color Doppler at the subcostal view revealed the presence of an atrial septal defect with left to right shunt. The pulmonary/systemic flow ratio (Qp/Qs) was 2.1:1 based on flow measurements. The tricuspid valve was competent and there were no signs of pulmonary hypertension. Transoesophageal echocardiography (TOE) revealed a medium-size ostium secundum (18 mm) defect with left–to-right shunt. The rims were of adequate size for percutaneous closure (>5 mm). All pulmonary veins were in place.

Continued at page 466

Anatomy. The atrial septal defect (ASD) represents an abnormality in the development of the heart that results in free communication between the atria [1, 2]. Isolated ASD represents about 7% of all cardiac anomalies and can be present at any age. There are four major types of defects in the atrial septum: ostium secundum, ostium primum, sinus venosus, and unroofed coronary sinus defect [2, 3]. Ostium secundum ASD is by far the most common type, occurring in 1/1500 live births. The ostium secundum ASD is a defect that involves the fossa ovalis region, whereas the other three types are defects of the endocardial cushions (ostium primum), the junction of the right atrium, and the superior/inferior vena cava (sinus venosus) or the coronary sinus (unroofed coronary sinus). It must be noted that deficiencies in the septum primum during embryologic development may result in a heterogeneous appearance of secundum ASD, ranging from multiple large defects to multiple smaller fenestrations or pinpoint defects.

Pathophysiology. Shunting across the interatrial septum is usually left to right across the ASD and occurs predominantly in late ventricular systole and/or early diastole. Likely, some augmentation occurs also during atrial contraction. The magnitude and direction of flow through an ASD depend on the size of the defect and the relative diastolic filling properties of both left and right ventricles. A reduction or even an impairment of left ventricular compliance or left heart valve defects may increase the left-to-right shunt, whereas conditions that reduce right ventricular compliance or the presence of pulmonary hypertension may cause a right-to-left shunt.

M. Drakopoulou · A. Apostolos · I. Kachrimanidis · G. Oikonomou · L. Koliastasis · G. Trantalis · C. Tsioufis · K. Toutouzas (✉)
First Department of Cardiology, Medical School, National and Kapodistrian University of Athens, Hippokration Hospital, Athens, Greece

G. Geroulakos et al. (eds.), *Mastering Endovascular Techniques*, https://doi.org/10.1007/978-3-031-42735-0_42

The chronic left-to-right shunt results in increased pulmonary blood flow and diastolic overload of the right ventricle. If untreated, chronic significant left-to-right shunt can alter the pulmonary vascular resistance leading to pulmonary arterial hypertension (PAH), pulmonary vascular disease, or even reversal of shunt and Eisenmenger syndrome.

In addition, in the presence of an ASD, a transient and small right-to-left shunt can occur, especially during periods of transient increase in right atrial pressure (i.e., during Valsalva maneuver) even in the absence of pulmonary hypertension. In this case, venous emboli may cross into the left atrium through the defect and eventually to the systemic circulation, most likely resulting in a paradoxical embolus. The paradoxical emboli mainly present as cerebral embolic events or, rather infrequently, as occlusion of coronary, renal, mesenteric, and peripheral arteries, manifesting as ischemia and eventual infarction of the affected tissues.

Indications for Closure. Large ASDs should be closed to reduce the risk of long-term complications; these may include premature death, atrial arrhythmias, reduced exercise tolerance, hemodynamically significant tricuspid regurgitation, right ventricular heart failure, and pulmonary vascular disease and PAH [4]. Current indications for percutaneous or surgical ASD closure range from the relevant significant left-to-right shunting to the prevention of recurrent paradoxical embolism [2].

In patients with evidence of right ventricular volume overload (defined as right ventricular enlargement with increased stroke volume), no signs of PAH (or pulmonary vascular resistance, PVR <3 WU in presence of such signs), and no left ventricular disease, ASD closure is recommended regardless of symptoms (Class I) [2]. When PVR is ≥3 WU but remains <5 WU and the Qp:Qs ratio is >1.5:1 there is also an indication for closure (Class IIa) regardless of symptoms. ASD closure is not recommended in patients with Eisenmenger physiology, patients with established pulmonary vascular disease (PAH and PVR ≥5 WU despite targeted PAH treatment), or desaturation on exercise [2, 5]. In patients with signs of left ventricular disease, it is recommended to perform balloon testing and carefully weigh the benefit of eliminating left-to-right shunt against the potential negative impact of ASD closure on outcome due to an increase in filling pressure (taking closure, fenestrated closure, and no closure into consideration) [2, 6]. Finally, in patients with no signs of right ventricular overload, ASD closure is also reasonable in the presence of paradoxical embolism and documented orthodeoxia-platypnea (Class IIa). In patients with an indication for ASD closure, device closure is recommended as the method of choice for secundum ASD closure when anatomically suitable (Class I) [2].

Preprocedural Screening and Transesophageal echocardiography (TOE) Protocol Anatomic suitability for percutaneous ASD closure is determined by TOE [7–10]. Among different types of ASDs, secundum ASD is only suitable for percutaneous closure. TOE evaluation of secundum defects before closure includes sizing, exploration of the residual septum's morphology, evaluation of rim size and quality, exclusion of additional defects, and confirmation of a normal pulmonary venous connection.

Secundum ASDs suitable for percutaneous closure are those with adequate rims (at least 5 mm, except for aortic rim that may be absent) and defect size <40 mm. The ASD size is measured in multiple planes, and color flow and the direction of the shunt are recorded. Rim adequacy and morphology are evaluated in the following three TOE views:

1. **Atrioventricular valve rim and posterior superior rim**: four-chamber view (mid-esophageal view, 0°) (Fig. 42.1a)
2. **Superior/anterior rim and inferior-posterior rim:** transverse view in the plane of the aortic valve (mid-esophageal view, 45°) (Fig. 42.1b)
3. **Superior Vena Cava and Inferior Vena Cava rim**: bicaval view (mid-esophageal view, 90–120°) (Fig. 42.1c)

Consideration should be given to the integrity of the rim tissue. 3D echocardiography provides additional and more accurate visualization of ASD morphology (Fig. 42.2).

Device Selection. Most devices for ASD closure have similar constructive characteristics (made of metallic framework and polymeric tissue scaffold) and share the same implantation technique. Beyond the technical consideration to fit within a small sheath, the ideal device should be easily deliverable, retrievable, and repositionable, biocompatible, soft, and conformable, able to guarantee an immediate and complete closure, and with low rates of procedure-related complications and most importantly with low thrombogenicity [11]. The development of new occlusion devices along with the improvement of the operator's experience and implantation techniques has led to an increase in the implementation of these procedures over the years. The choice of the device mainly depends on operator experience and preference. The most widely used devices are the Amplatzer Septal Occluder (Abbot Inc.) and the Gore Cardioform (W.L. Gore and Associates, Flagstaff, Arizona).

Amplatzer Occluder Device (St. Jude Medical, Minnesota, USA). The Amplatzer Occluder device is the most widely used device for ASD closure. It consists of two discs, the left atrial larger than the right atrial disc, with a connecting waist, and is fully made of a one-piece Nitinol mesh filled with Dacron threads [12]. The waist diameter determines the device size, which requires a 6–8-F sheath correspondingly. Its function is mainly to keep each disc tight to the septal wall.

GORE Cardioform Septal Occluder (W. L. Gore & Associates, USA). The Gore Cardioform Septal Occluder is

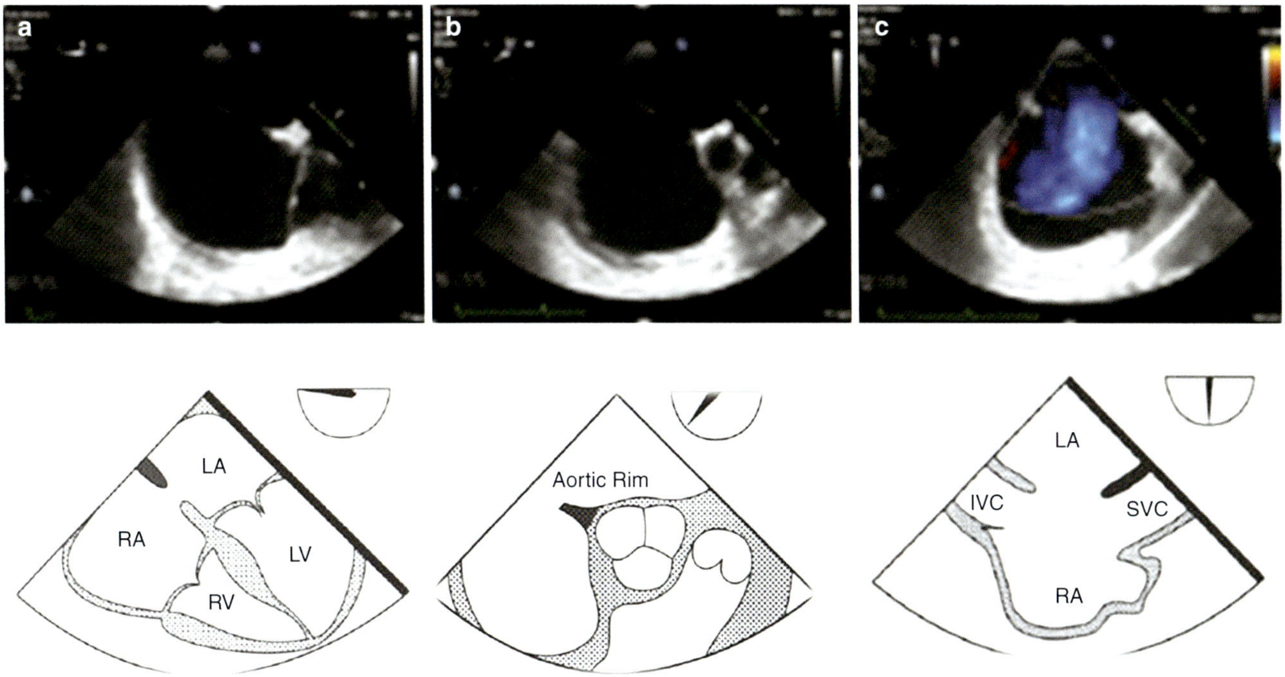

Fig. 42.1 Transoeosphageal (TOE) evaluation. (**a**) Mid-esophageal 4-chamber view at 0°. This view permits localization of the posterior rim and atrioventricular rim. (**b**) Mid-esophageal view at 45° with a leftward (counterclockwise) rotation of the probe to allow simultaneous visualization of the aortic (anterior) rim and posterior rim. (**c**) Mid-esophageal four-chamber view at 90°; the inferior vena cava (IVC) and the superior vena cava (SVC) rims are seen

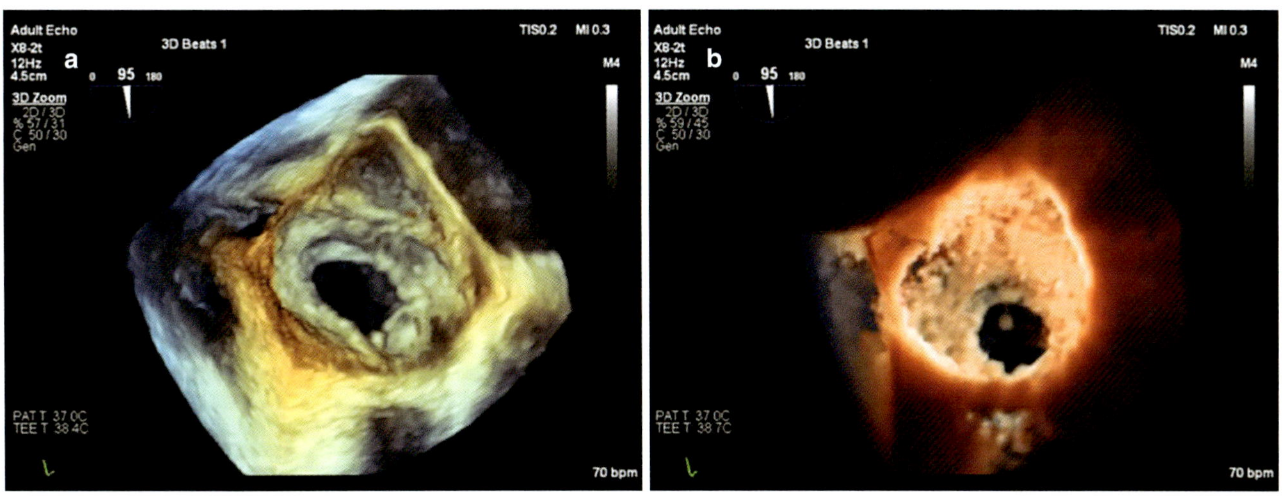

Fig. 42.2 (**a**, **b**) 3D transesophageal (TOE) image from a patient with ASD from the right atrial perspective

currently FDA approved for ASD closure [13–15]. The device is also composed of two discs formed by platinum-filled Nitinol wire frames that are covered with an expanded polytetrafluoroethylene film [16]. The occluder has right and left atrial discs that form on either side of the atrial septum, act to close the defect, and hold the occluder in place. It uses a framework of minimal metal mass to hold a biocompatible expanded polytetrafluoroethylene patch in close apposition to the septum.

Procedure. Percutaneous ASD is performed either under moderate sedation or general anesthesia depending on the mode of procedural imaging guidance. When the procedure is performed under TOE guidance, general anesthesia is needed, while when using intracardiac echocardiogram (ICE), moderate sedation can be used.

Before the procedure, a preoperative antibiotic is administered. After sedating the patient, vascular access is obtained by the femoral vein (bilateral femoral veins can be used).

When performing the procedure with ICE for imaging, two vascular accesses are needed, while if TEE is used, a single venous access is adequate. Access for the ICE catheter requires an 8–10 French sheath placement and preferably a length of 25–35 cm sheaths are used as this gives stability to the ICE catheter and helps avoid the tortuosity of the femoral veins. The second access, depending on the operator preference, can start with a 6 French and can be upgraded to the appropriate sheath size in the range of 6–12 French depending on the size of the closure device being used. Additional arterial access may be obtained for pressure monitoring, depending on the operator. Once both accesses are obtained, the patient is anticoagulated, preferably with unfractionated heparin. An activated clotting time (ACT) of >250 s is maintained for the duration of the procedure. Thereafter the defect is crossed with the J-tip guidewire from the right atrium into the left atrium under fluoroscopic and ICE guidance. Then the J-tip guidewire is advanced into the left superior pulmonary vein, and a multipurpose catheter is advanced into the left superior pulmonary vein. The J-tip guidewire is then exchanged for a super stiff guidewire, and the multipurpose catheter is removed. After this, the venous sheath in the femoral vein is removed, and an ASD sizing balloon is advanced over the wire into the ASD. Under fluoroscopic guidance, the sizing balloon is slowly inflated with contrast until the flow across the ASD stops as visualized by the TOE. At this stage, care should be taken to avoid overinflation of the balloon as this will overestimate the size of the defect. The preferable approach is to deflate the balloon until the flow across the ASD resumes and thereafter inflate the balloon until the flow across the ASD ceases again ("stop-flow" technique). Next, the size of the defect will be measured on both fluoroscopy and the TOE using the "waist" on the ASD sizing balloon. After this, the balloon is deflated and removed over the wire with care taken to prevent the wire from prolapsing out from the pulmonary vein. Following this, the appropriately sized sheath that can deliver the ASD device is advanced over the wire into the left atrium, and the sheath dilator and the wire are removed. Next, the appropriately sized ASD device is prepared. The device is first checked for integrity, and then is immersed in a saline tub. After this, the device is attached to the delivery cable, and is slowly pulled back into the loader sheath while the side flush of the loader sheath is constantly flushed to prevent any air from being trapped in the device or the loader sheath. Next, the loader is attached to the delivery sheath carefully without the introduction of air. The delivery cable is advanced into the sheath, thereby advancing the device into the delivery sheath. Thereafter, the device is slowly advanced toward the tip of the sheath under fluoroscopic guidance. Once the device reaches the tip of the sheath, the sheath is withdrawn over the delivery cable to unsheathe the left atrial disc. This can be visualized on both TOE and fluoroscopy. Once the left atrial disc is fully formed, both the delivery sheath and the delivery cable are withdrawn until the left atrial disc is well opposed to the interatrial septum on the left atrial side. A little tension should be maintained to cause tenting of the interatrial septum. Once the left atrial disc anchors well on the left side of the septum and there is sufficient tension on the interatrial septum by the left atrial disc, the next step would be to unsheathe the right atrial disc carefully. After this, the sheath is slightly withdrawn away from the right atrial disc to decrease tension on the device. Imaging is then used to examine the stability of the device, and Color Doppler imaging is performed to verify that the ASD is well sealed. Color flow may be noted across the waist and the center of the device, but color flow at the edges of the discs or away from the disc suggests that the device may be smaller in size, or there may be additional ASDs away from the current ASD location. In either case, the device should be captured back into the sheath, and the septum examined again for the presence of other ASDs in proximity. In this situation, the device can be upgraded to a larger ASD device or cribriform occluder if the other ASDs are close to the primary ASD. If the other ASDs are >7 mm from the primary ASD, the other defects need to be closed using additional ASD occluders. If there is no color flow across the device, the delivery cable is wiggled with a strong push and pull (Minnesota Wiggle) to examine the stability of the device by both fluoroscopy and imaging. If the device looks stable and there is no flow across the device, the device is released. Once the device is released, the stability and color flow are again checked, and if they are satisfactory, the sheath is withdrawn into the inferior vena cava [17]. Following this, imaging is performed to visualize the device and look for pericardial effusion. Next, venous sheaths are withdrawn from the body, and either a figure of eight stitch is placed or manual pressure is held to achieve hemostasis.

Perioperative Care and Surveillance. The patient is given clopidogrel 300–600 mg orally in addition to aspirin, and dual antiplatelet therapy is continued for 3 months after the procedure. After the procedure, the patient is monitored overnight. A follow-up echocardiogram is performed the following day to verify the stability of the device and rule out any erosion from the device and rule out any pericardial effusion. Then, the patient is discharged and is followed up after 6 months with a repeat echocardiogram. Infective endocarditis prophylaxis is recommended after 6 months for dental procedures.

Case Presentation

Continued from page 463

The Heart Team decided to proceed to transcatheter ASD closure. As there were no signs of PAH, we did not proceed to right heart catheterization. The procedure was performed under general anesthesia and TOE guidance via right femoral approach. During the procedure the ACT was maintained at >250 s. A 6 French sheath was placed at the right femoral

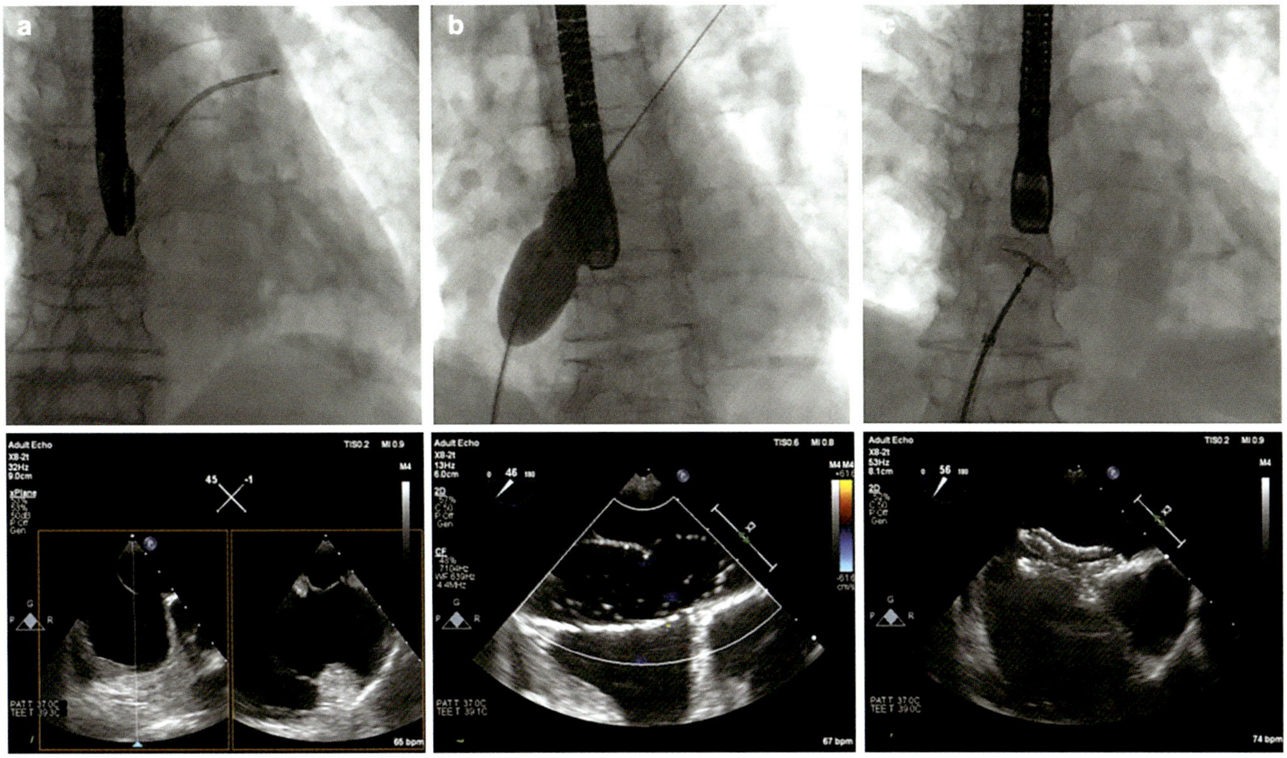

Fig. 42.3 (**a**–**c**) Transoeosphageal (TOE) and corresponding fluoroscopy imaging in transcatheter ASD closure procedure. (**a**) Multipurpose catheter crossing the ASD to the left upper pulmonary vein. (**b**) "Stop-flow" technique for assessing ASD sizing and devise selection. (**c**) Final device deposition

vein and the defect was crossed with the J-tip guidewire that was advanced in the left upper superior pulmonary vein. A multipurpose catheter was placed over the guidewire (Fig. 42.3a) that was exchanged with a super stiff guidewire. After removal of the multipurpose catheter and of the venous sheath, an ASD sizing balloon was advanced over the wire into the ASD for sizing of the defect. By performing the "stop-flow" technique, the size of the defect was measured at 18 mm on both fluoroscopy and TOE using the "waist" on the ASD sizing balloon (Fig. 42.3b). It was decided to implant a 22 mm Amplatzer occluder device. The delivery sheath (9 French) was advanced in the left atrium; the sheath dilator and wire were removed. After preparation of the device, it was slowly advanced toward the tip of the delivery sheath under fluoroscopic and TOE guidance. Once the device reached the tip of the sheath, the sheath was withdrawn over the delivery cable to unsheathe the left atrial disc. After gently withdrawing the left atrial disc toward the left side of the interatrial septum, the right atrial disc was released on the RA side. Proper alignment of the device with captured rims and no residual flow was confirmed on TOE before releasing the device (Fig. 42.3c). The patient was discharged the following day on dual antiplatelet therapy for 3 months and aspirin for another 3 months. At 3-month follow-up, the patient remained asymptomatic, in sinus rhythm, and the device is in situ and well seated with no residual shunt or valvular regurgitation on follow-up echocardiography. The patient is on dual antiplatelet therapy for 6 months.

Left Atrial Appendage Closure

Case Presentation

A 79-year-old male patient with persistent non-valvular atrial fibrillation, hypertension, and diabetes mellitus was deemed a poor candidate for anticoagulation due to recurrent gastrointestinal bleeding while being treated with oral anticoagulants. A TOE study assessed the feasibility of left atrial appendage (LAA) procedure, documented a 'chicken wing' LAA morphology, and excluded any luminal thrombosis. The Heart Team decided to proceed to transcatheter LAA closure.

Continued at page 471

Background. Atrial fibrillation (AF) is the most common arrhythmia, affecting 1–2% of the global population. Its incidence in the general population is 0.05% cases per year and it is expected to more than double by 2050 in Western countries. The presence of AF increases the risk of embolic stroke by four- to fivefold and results in stroke incidence of 2–5% per year in untreated patients. Such a disabling complication is mainly related to thrombus formation within the LAA. Indeed, the LAA, due to its trabeculae and blood stasis, is the most common site of thrombus generation in patients with AF [18].

Oral anticoagulant drugs (OAC) are an effective treatment to prevent embolic events in patients with AF; however, several drawbacks such as patient's compliance and bleeding risk limit their applicability [19]. In daily clinical practice, the decision for OAC coverage in patients with AF is challenging and based on balancing patient thromboembolic and bleeding risks. For this purpose, several risk score calculators have been developed, such as the CHA2DS2-VASc score for the risk of stroke (≥1 additive score in men and ≥2 in women indicates risk of thromboembolism) and the HAS-BLED score for the bleeding risk (≥3 additive score indicates risk of bleeding) [20]. Due to the overlapping of bleeding and thrombotic risk factors often observed in complex patients, the choice of the most appropriate therapy is tricky.

Percutaneous LAA closure has been developed as a less invasive procedure to prevent stroke in patients with AF who cannot tolerate oral anticoagulation. This catheter-based technique provides an atrial endoluminal mechanical orifice obstruction [21]. There are currently two commercially available devices for the LAA: Amplatzer Amulet LAA occluder (St. Jude Medical Inc., MN, USA) and WATCHMAN™ LAA closure device (Boston Scientific Natick, MA, USA) [22, 23].

Indications for LAA Closure. Based on European Society of Cardiology guidelines, LAA occlusion is recommended (Class IIb) for stroke prevention in patients with AF and contraindications for long-term anticoagulant treatment (e.g., intracranial bleeding without a reversible cause) [24].

Closure Devices. WATCHMAN™ LAA Closure Device. The occluder device consists of a nitinol frame structure with ten fixation anchors along its waist. When fully deployed, the nitinol frame has a parachute-like configuration. The proximal part of the system, which faces the left atrium after implantation, is covered with a thin polyethylene terephthalate membrane which is designed to reduce post-implant thrombus formation on the occluder, thus allowing faster endothelialization. The device sizes range from 21 to 31 mm. The device is implanted through the femoral vein via the trans-septal technique and is fully retrievable and repositionable [25].

The Amplatzer™ Amulet™ LAA Occluder. The occluder is a second-generation self-expanding device made of nitinol that has a distal lobe and a proximal disc, connected by an articulated waist. The main lobe of the device is meant to be implanted in the proximal 10–15 mm of the LAA, and the disc covers the LAA orifice. The lobe sizes range from 16 to 34 mm. As for the WATCHMAN™ LAA Closure Device, this device is implanted through the femoral vein via the trans-septal technique and is fully retrievable and repositionable.

Preprocedural Screening and TOE Imaging Protocol. Prior to the implantation of an LAA occlusion device, all patients must undergo preprocedural imaging to explore the anatomy of the LAA and exclude the presence of thrombus. Preprocedural multimodality imaging with TOE or multidetector computed tomography (MDCT) with contrast injection (as an alternative to TOE) enables measurement of the LAA ostium, the landing zone of the elected occlusion device, and the length of the LAA body, and it allows for characterization of the shape of the LAA, especially when there are multiple lobes (number, shape, and localization) [26].

During TOE imaging, LAA is imaged at the mid- to upper esophageal view from 0° through 180° (at 0°, 45–60°, 90°, and 120–135°) to define the maximum LAA width and maximum depth of the dominant lobe (Fig. 42.4). A complete 180° scan is needed due to the highly variable anatomy of the LAA. Most importantly, in some cases, LAA closure may not be feasible owing to large ostia, presence of early lobe, or early and severe main lobe bending. The two commercially available devices need the following pre-specified measurements:

1. **The Amplatzer™ Amulet™ LAA closure device**: Measurement will be taken at the orifice of the LAA, with a second measurement for the landing zone of the stabilizer disc (lobe), which is located 15 mm from the orifice into the left appendage. Device size is selected based on the maximal landing zone diameter by upsizing. The company recommends 3–5 mm for 16–22 mm devices and 3–6 mm for 25–34 mm devices. Care must be given to the measurement of the orifice of the LAA, to be sure that the disc diameter exceeds the diameter of the orifice and completely covers it [27].
2. **WATCHMAN™ LAA closure device:** Measurement will be taken in at least four views (0°, 45°, 90°, and 135°) with the ostium of the left circumflex artery as the essential landmark. Measurement should be done from the edge of the LCX to the opposite wall of the LAA, perpendicular to the axis of the LAA to a point around 1–2 cm below the ridge of the pulmonary vein. Measurement of LAA length must be obtained from every view, to ensure that the dimensions are within the available device range. Care must be given as the depth of the appendage should be greater than the ostium diameter to accommodate the device. The company recommends an occluder size about 20% larger than the largest diameter measured to achieve the compression rate needed to engage the fixation barbs of the WATCHMAN™ device into the wall of the LAA.

Procedure. The procedure is performed under moderate sedation or general anesthesia based on the echocardiographic procedural guidance (ICE or TOE). Antibiotic prophylaxis should be administered before the procedure. During the procedure an ACT >250 seconds should be maintained. The first step of the procedure is the trans-septal

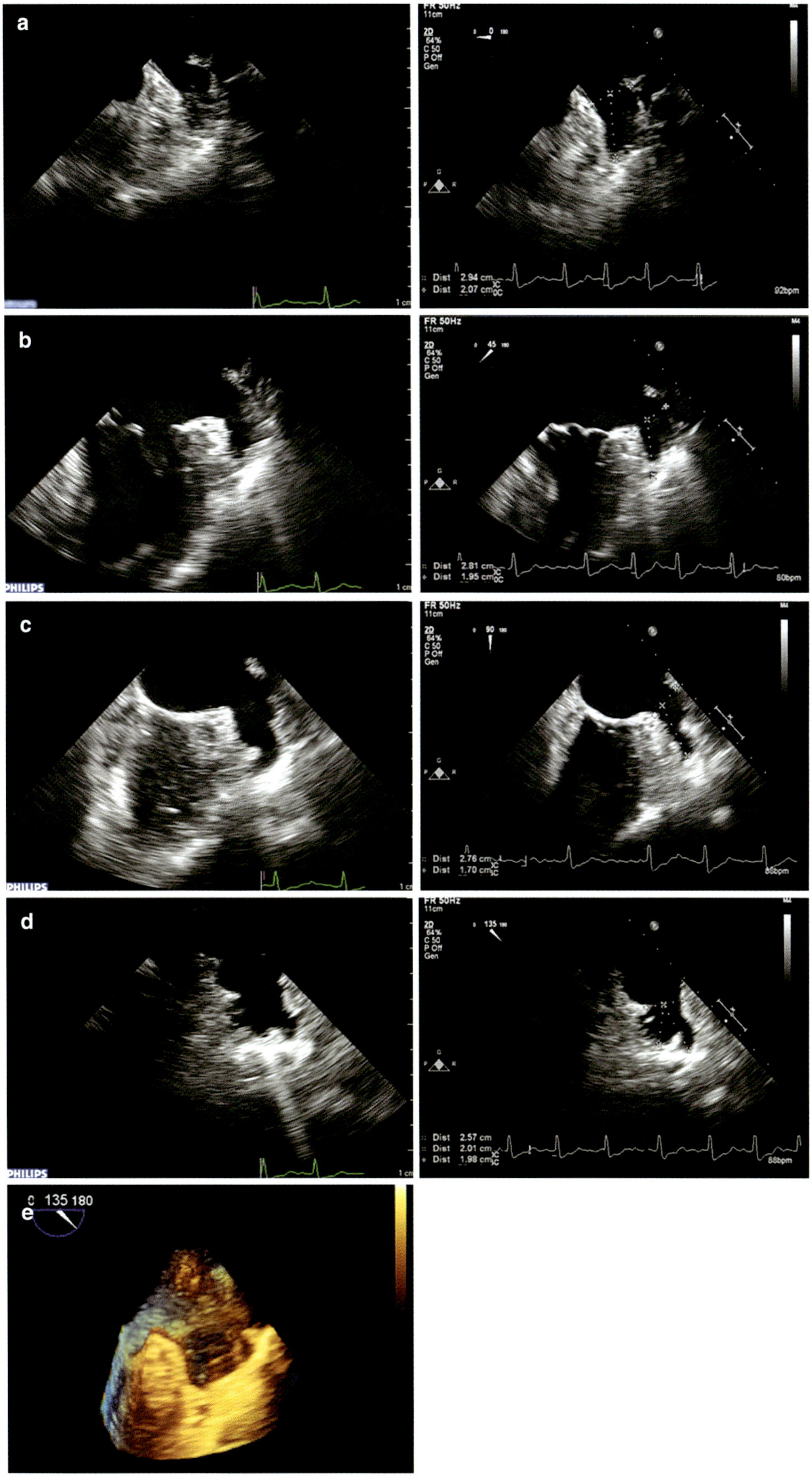

Fig. 42.4 Transoeosphageal (TOE) imaging for left atrial appendage (LAA) closure. (**a**) 0° with the corresponding measurements, (**b**) 45° with the corresponding measurements, (**c**) 90° with the corresponding measurements, (**d**) 135° with the corresponding measurements, (**e**) 3D image

puncture that should be performed in the posterior and inferior segment of the fossa ovalis to facilitate an anterior superior trajectory that optimizes the device. In cases where the orientation of the LAA is cranial, the puncture site should be lower, whereas in cases with a caudal orientation, the puncture site should be higher. However, independently of the LAA orientation, all trans-septal punctures should be carried out as far posterior. After trans-septal puncture, the trans-septal sheath is advanced though the puncture into the left atrium, and the needle is pulled back into the sheath. A stiff wire is placed in the left upper pulmonary vein and the trans-septal sheath is exchanged over the wire for the occlude-specific delivery sheath. Depending on the LAA occlusion device chosen, thereafter different steps should be followed.

For the WATCHMAN™ LAA Closure Device, after the access sheath is introduced to the left side of the heart and the dilator is removed, a pigtail catheter (5 Fr or 6 Fr) is introduced through the access sheath and advanced into the LAA to perform an angiography and make the necessary measurements (Fig. 42.5). Knowledge of the specific mark-

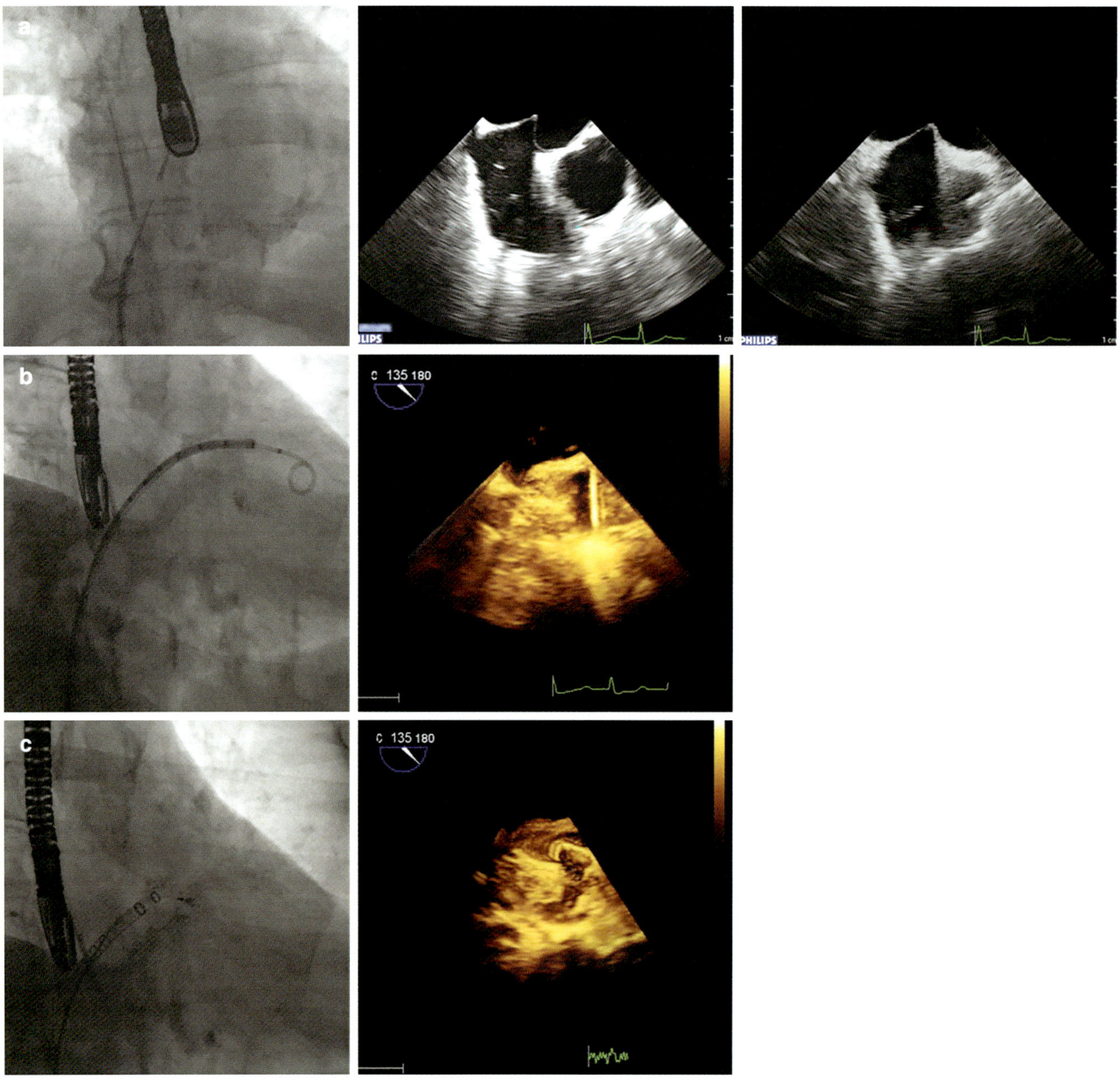

Fig. 42.5 Transoesophageal (TOE) and corresponding fluoroscopy imaging in transcatheter LAA closure procedure. (**a**) Trans-septal puncture at the inferior-posterior part of the atrial septum. Short axis view and bicaval view showing tenting of the atrial septum. (**b**) Pigtail catheter is introduced through the access sheath and advanced into the LAA to perform an angiography and make the necessary measurements. (**c**) Final device deposition

ers on the access sheath is mandatory for correct device placement. Thereafter, the delivery catheter is advanced into the access sheath under fluoroscopic guidance and continuous flushing with heparinized saline until its distal marker almost aligns with the distal marker of the access sheath. The access sheath must then be pulled back gently until the operator feels a click. This interlocks the access sheath with the delivery catheter. After checking the position of the delivery system fluoroscopically, the occluder can be deployed. The interlocked access and delivery sheath are pulled back while holding the delivery cable of the device in a stable position. This allows the device to deploy while still being fixed to the delivery system. The combined delivery system allows complete retrieval of a deployed occluder in case of insufficient placement.

The Amplatzer™ Amulet™ LAA occluder device comes preloaded in the delivery sheath. Pulling back the delivery sheath while holding the device within the neck of LAA deploys the distal lobe of the occluder. After satisfactory positioning of the lobe, the proximal disc is deployed by further pulling back of the retrieval sheath. Successful implantation is defined by proper compression of the lobe, separation of the lobe and the disc on angiography, concave shape of the disc, and correct angulation of the lobe and disc in relation to the plane of the LAA ostium. In case of unsatisfactory positioning, the device can be recaptured.

For both occluder devices, when optimal device position, sizing, and device seal are ensured in angiography and in TOE, a last test of stability, a tug test, should be performed before releasing the occluder. For this purpose, the delivery core wire is gently pulled under simultaneous injection of contrast medium to see the movement of the LAA together with the occluder. After successfully passing this test, the device is released from the delivery system by turning the core wire counterclockwise five times. Final contrast injections as well as a final control with TEE in all four views (0°, 45°, 90°, and 135°) to check for residual leaks are recommended.

Perioperative Care and Surveillance. Dual antiplatelet therapy with 100 mg of aspirin and 75 mg of clopidogrel for 6 months is recommended after the implantation. In case of successful occlusion of the LAA and missing evidence of thrombus formation on the occluder, thereafter we stop clopidogrel and aspirin.

Case Presentation

Continued from page 467

The procedure was performed under general anesthesia and TOE guidance through the right femoral access. The Amplatzer™ Amulet™ LAA Occluder (28 mm) was successfully implanted. The early echocardiographic control did not report any device-related abnormality or any residual leak. The hospital course was uneventful and the patient received a dual antiplatelet therapy (clopidogrel 75 mg o.d. and aspirin 100 mg o.d.) after discharge.

References

1. Oster M, Bhatt AB, Zaragoza-Macias E, Dendukuri N, Marelli A. Interventional therapy versus medical therapy for Secundum atrial septal defect: a systematic review (part 2) for the 2018 AHA/ACC guideline for the Management of Adults with Congenital Heart Disease: a report of the American College of Cardiology/American Heart Association task force on clinical practice guidelines. Circulation. 2019;139(14):e814–30. https://doi.org/10.1161/CIR.0000000000000605.
2. Baumgartner H, De Backer J, Babu-Narayan SV, Budts W, Chessa M, Diller GP, et al. 2020 ESC guidelines for the management of adult congenital heart disease. Eur Heart J. 2021;42(6):563–645. https://doi.org/10.1093/eurheartj/ehaa554.
3. Silvestry FE, Cohen MS, Armsby LB, Burkule NJ, Fleishman CE, Hijazi ZM, et al. Guidelines for the echocardiographic assessment of atrial septal defect and patent foramen ovale: from the American Society of Echocardiography and Society for Cardiac Angiography and Interventions. J Am Soc Echocardiogr. 2015;28(8):910–58. https://doi.org/10.1016/j.echo.2015.05.015.
4. Vecht JA, Saso S, Rao C, Dimopoulos K, Grapsa J, Terracciano CM, et al. Atrial septal defect closure is associated with a reduced prevalence of atrial tachyarrhythmia in the short to medium term: a systematic review and meta-analysis. Heart. 2010;96(22):1789–97. https://doi.org/10.1136/hrt.2010.204933.
5. Takaya Y, Akagi T, Sakamoto I, Kanazawa H, Nakazawa G, Murakami T, et al. Efficacy of treat-and-repair strategy for atrial septal defect with pulmonary arterial hypertension. Heart. 2022;108(5):382–7. https://doi.org/10.1136/heartjnl-2021-319096.
6. Brida M, Diller GP, Kempny A, Drakopoulou M, Shore D, Gatzoulis MA, et al. Atrial septal defect closure in adulthood is associated with normal survival in the mid to longer term. Heart. 2019;105(13):1014–9. https://doi.org/10.1136/heartjnl-2018-314380.
7. Kitakata H, Itabashi Y, Kanazawa H, Miura K, Kimura M, Shinada K, et al. Appropriate device selection for transcatheter atrial septal defect closure using three-dimensional transesophageal echocardiography. Int J Cardiovasc Imaging. 2021;37(4):1159–68. https://doi.org/10.1007/s10554-020-02095-x.
8. Jang JY, Heo R, Cho MS, Bae J, Hong JA, Lee S, et al. Efficacy of 3D transoesophageal echocardiography for transcatheter device closure of atrial septal defect without balloon sizing. Eur Heart J Cardiovasc Imaging. 2018;19(6):684–9. https://doi.org/10.1093/ehjci/jex153.
9. Aggeli C, Apostolou I, Dimitroglou Y, Toutouzas K, Vavuranakis M, Latsios G, et al. The role of 'halo sign' for the accurate quantification of atrial septal defect size with 3D TEE. Int J Cardiovasc Imaging. 2020;36(5):873–81. https://doi.org/10.1007/s10554-020-01786-9.
10. Hascoet S, Hadeed K, Karsenty C, Dulac Y, Heitz F, Combes N, et al. Feasibility, safety and accuracy of echocardiography-fluoroscopy imaging fusion during percutaneous atrial septal defect closure in children. J Am Soc Echocardiogr. 2018;31(11):1229–37. https://doi.org/10.1016/j.echo.2018.07.012.
11. Drakopoulou M, Soulaidopoulos S, Stathogiannis K, Oikonomou G, Papanikolaou A, Toutouzas K, et al. Antiplatelet and antithrombotic therapy after patent foramen oval and atrial septal defect closure. Curr Pharm Des. 2020;26(23):2769–79. https://doi.org/10.2174/1385272824999200427083838.

12. Apostolos A, Drakopoulou M, Gregoriou S, Synetos A, Trantalis G, Tsivgoulis G, et al. Nickel hypersensitivity to atrial septal occluders: smoke without fire? Clin Rev Allergy Immunol. 2022;62(3):476–83. https://doi.org/10.1007/s12016-021-08867-0.
13. Santoro G, Castaldi B, Cuman M, Di Candia A, Pizzuto A, Sirico D, et al. Trans-catheter atrial septal defect closure with the new GORE(R) Cardioform ASD occluder: first European experience. Int J Cardiol. 2021;327:68–73. https://doi.org/10.1016/j.ijcard.2020.11.029.
14. Dolgner SJ, Steinberg ZL, Jones TK, Reisman M, Buber J. Stroke in patients with secundum atrial septal defect and sequelae after transcatheter closure. Heart. 2021;107(23):1875–80. https://doi.org/10.1136/heartjnl-2021-319050.
15. Anderson JH, Adamson T, Migliati E, Herlihy J, Danieu P, Daly J, et al. Perforating the GORE(R) CARDIOFORM septal occluder and atrial septal defect occluder to gain access to the left atrium. Catheter Cardiovasc Interv. 2020;96(6):E660–5. https://doi.org/10.1002/ccd.28884.
16. Apostolos A, Drakopoulou M, Toutouzas K. New migraines after atrial septal defect occlusion. Is the nickel hypersensitivity the start of everything? Med Hypotheses. 2021;146:110442. https://doi.org/10.1016/j.mehy.2020.110442.
17. Apostolos A, Drakopoulou M, Trantalis G, Tsioufis C, Toutouzas K. Is every PFO technically appropriate for transcatheter closure? Hell J Cardiol. 2022;67:85. https://doi.org/10.1016/j.hjc.2022.05.004.
18. Beigel R, Wunderlich NC, Ho SY, Arsanjani R, Siegel RJ. The left atrial appendage: anatomy, function, and noninvasive evaluation. JACC Cardiovasc Imaging. 2014;7(12):1251–65. https://doi.org/10.1016/j.jcmg.2014.08.009.
19. Tsigkas G, Apostolos A, Despotopoulos S, Vasilagkos G, Papageorgiou A, Kallergis E, et al. Anticoagulation for atrial fibrillation in heart failure patients: balancing between Scylla and Charybdis. J Geriatr Cardiol. 2021;18(5):352–61. https://doi.org/10.11909/j.issn.1671-5411.2021.05.006.
20. Tsigkas G, Apostolos A, Despotopoulos S, Vasilagkos G, Kallergis E, Leventopoulos G, et al. Heart failure and atrial fibrillation: new concepts in pathophysiology, management, and future directions. Heart Fail Rev. 2022;27(4):1201–10. https://doi.org/10.1007/s10741-021-10133-6.
21. Lerakis S, Synetos A, Toutouzas K, Stefanadis C. Transcatheter occlusion devices for the prevention of stroke in patients with atrial fibrillation. Hell J Cardiol. 2008;49(1):33–6.
22. Holmes DR Jr, Kar S, Price MJ, Whisenant B, Sievert H, Doshi SK, et al. Prospective randomized evaluation of the Watchman left atrial appendage closure device in patients with atrial fibrillation versus long-term warfarin therapy: the PREVAIL trial. J Am Coll Cardiol. 2014;64(1):1–12. https://doi.org/10.1016/j.jacc.2014.04.029.
23. Lakkireddy D, Thaler D, Ellis CR, Swarup V, Sondergaard L, Carroll J, et al. Amplatzer Amulet left atrial appendage occluder versus Watchman device for stroke prophylaxis (Amulet IDE): a randomized, controlled trial. Circulation. 2021;144(19):1543–52. https://doi.org/10.1161/circulationaha.121.057063.
24. Hindricks G, Potpara T, Dagres N, Arbelo E, Bax JJ, Blomström-Lundqvist C, et al. 2020 ESC guidelines for the diagnosis and management of atrial fibrillation developed in collaboration with the European Association for Cardio-Thoracic Surgery (EACTS): the task force for the diagnosis and management of atrial fibrillation of the European Society of Cardiology (ESC) developed with the special contribution of the European Heart Rhythm Association (EHRA) of the ESC. Eur Heart J. 2021;42(5):373–498. https://doi.org/10.1093/eurheartj/ehaa612.
25. Toutouzas K, Trantalis G, Sideris S, Benetos G, Michelongona A, Aggeli C, et al. Left atrial appendage closure. Initial SingleCenter experience with the Watchman device. Hell J Cardiol. 2015;56(6):510–2.
26. Vainrib AF, Harb SC, Jaber W, Benenstein RJ, Aizer A, Chinitz LA, et al. Left atrial appendage occlusion/exclusion: procedural image guidance with transesophageal echocardiography. J Am Soc Echocardiogr. 2018;31(4):454–74. https://doi.org/10.1016/j.echo.2017.09.014.
27. Freixa X, Aminian A, Tzikas A, Saw J, Nielsen-Kudsk JE, Ghanem A, et al. Left atrial appendage occlusion with the Amplatzer Amulet: update on device sizing. J Interv Card Electrophysiol. 2020;59(1):71–8. https://doi.org/10.1007/s10840-019-00699-5.

Part VIII

Aortic Aneurysm Disease

Endovascular Repair of Abdominal Aortic Aneurysms (EVAR): Studies on Long-term Results

43

Christos D. Liapis

Case Presentation

A 72-year-old patient with a 7,2 ruptured infrarenal aortic aneurysm was managed with endovascular aneurysm repair using a Zenith Endograft (Cook Medical). The patient was followed up yearly with a duplex ultrasound and a CT scan due to the presence of a type II endoleak. While initially stable, 4 years later the aneurysm sac started to grow. At 8-year follow-up the aneurysm sac was approximately 8 cm and a type Ia along with a type II endoleak were clearly visible. The right iliac limb was also pulled back at the orifice of the right common iliac artery (Fig. 43.1).

An 80-year-old patient with a negative stress echocardiogram was scheduled to undergo hybrid aneurysm repair.

Continued at page 476

The mortality rate due to ruptured aortic aneurysms has shown a remarkable decline over the last 20 years [1]. This is attributed to markedly reduced prevalence of the disease, the initiation of population screening for aneurysms in many countries, and probably the introduction of endovascular aortic repair. Multiple publications indicate that the comparison of aneurysm-related deaths between the UK and USA is in favor of the USA policy of adopting EVAR as first choice treatment, while in the UK they register more aneurysm-related deaths than operations for non-ruptured aneurysms, seven times higher than in the USA [2]. Another reason for this difference is the real-world threshold for repairing an abdominal aortic aneurysm (AAA), which is over 55 mm for both men and women in the UK and over 50 mm in the USA [3].

The comparison of 30-day all-cause mortality between open repair and EVAR in the three randomized trials (DREAM, EVAR-1, OVER) showed a remarkable difference in favor of EVAR (4.6% vs 1.2%, 4.7% vs 1.7%, 3.0% vs 0.5% respectively). This difference disappears in the 8-year follow-up EVAR-1 trial. The increased late mortality is attributed to a higher number of complications and reoperations in the EVAR arm of the study [4], especially in the period 6 months to 4 years when the reoperations were 6 times higher for the EVAR group. In the follow-up period after 8 years, the EVAR group showed increased total (p = 0.49, HR: 1.25) and aneurysm-related mortality ($p = 0.0064$, HR: 5.82) [5]. Of interest is the increased rate of malignancies, especially gastrointestinal, in the EVAR group after 8-year follow-up (adjusted HR 1.87, $p = 0.007$) [6].

In the DREAM trial, 6 years post randomization, the survival rates were similar between the two groups, while the rate of secondary procedures was higher for the EVAR group after the fourth year [7]. The secondary procedures were almost entirely endovascular. The 12-year survival showed no statistically significant difference in total (42.2% OR, 38.5% EVAR $p = 0.48$) aneurysm-related (7.7% OR, 3.1% EVAR $p = 0.33$) or malignancy-related [8] mortality.

In the OVER trial, after 11 years' follow-up, an increased number of secondary procedures was noted in the EVAR group (193 in 117 patients vs 116 in 85, $p = 0.04$), while the long-term overall survival did not show any statistical difference between the groups (302/444 (68%) pts. EVAR, 306/437 (70%) OR $P = 0.61$) [9].

In the EVAR-2 trial, comparing EVAR to no repair in patients with considerable comorbidities, the 12-year results showed similar overall survival but significant aneurysm-related survival in favor of the EVAR group [10].

For the treatment of ruptured aneurysms, the most comprehensive trial comparing open repair to endovascular strategy is the IMPROVE trial [11]. The 30-day death rates did not show any statistically significant difference. However, the cumulative incidence of being discharged directly home by randomized group showed 17 mean days of discharge for the endovascular group and 26 for the open repair group, and this difference was statistically significant ($p < 0.001$). The 1-year survival was similar for both groups with 4.5% difference in favor of EVAR ($p = 0.325$). Further analysis of the 1-year survival showed that the endovascular strategy yields better results for women and patients with multiple comor-

C. D. Liapis (✉)
Vascular and Endovascular Surgery Clinic, Athens Medical Center, Athens, Greece
e-mail: liapis@med.uoa.gr

G. Geroulakos et al. (eds.), *Mastering Endovascular Techniques*, https://doi.org/10.1007/978-3-031-42735-0_43

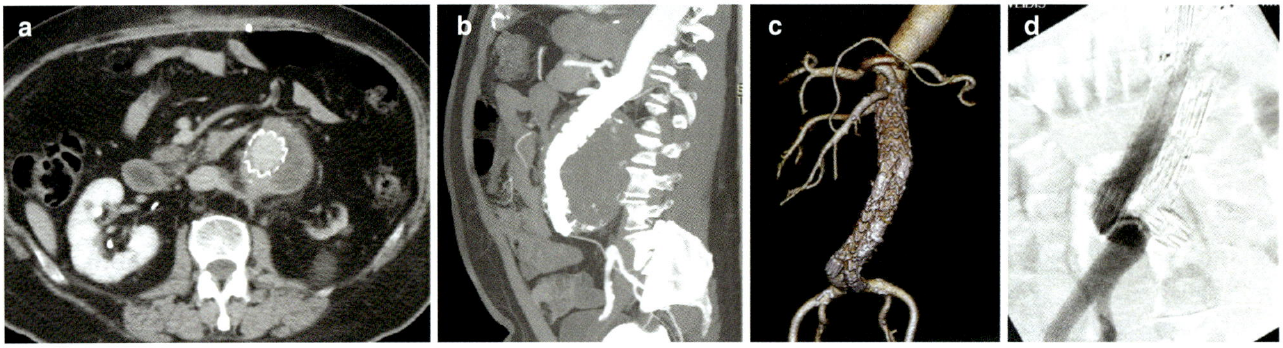

Fig. 43.1 Preoperative CT angiogram; (**a**, **b**) Type Ia and II endoleaks, (**c**) notice the right iliac limb retracted, (**d**) intraoperative angiogram confirming the long distance of the right iliac limb from the iliac bifurcation posing a risk for future type Ib endoleak

bidities. In the quality-of-life aspect, using the EQ-5D score and up to 36 months, the results were considerably better for the endovascular strategy group up to 12 months, but this difference was diminished until the end of the study period.

Reviewing the randomized trial EVAR vs open repair and specifically looking at the rate of myocardial infarctions in the short and long term following the initial procedure, it was noted that although in the short term there were fewer MIs in the EVAR group, this trend was reversed in the long term [12].

One of the reasons for this besides the increased number of secondary procedures is the increased arterial stiffness caused by the insertion of the stiff endovascular graft. This has been shown by multiple studies examining the effects on left ventricular function following EVAR [13].

There are multiple tests besides the cardiac ultrasound that can evaluate arterial stiffness, the most popular being the measurement of pulse wave velocity (PWV), which is an independent predictor of cardiovascular morbidity and mortality [14–16]. We have previously shown that PWV is upregulated in patients with AAA vs those without [17]. Furthermore, comparing patients with AAA open repair and EVAR, we have found that open repair caused moderate increase in PWV, while EVAR was associated with significant ($p < 0.001$) increase in PWV up to 6 months following the initial procedure [18]. In looking at differential effects of stent-graft fabrics on arterial stiffness, it was noted that PWV was more pronounced in the Polyester Woven group compared to the PTFE group [19].

Looking at the cardiac effects of aortic endografts by measuring PWV and NT-proBNP changes in patients following TEVAR, we have also noted a statistically significant increase in both NT-proBNP and PWV 48 h after the operation and up to 6 months post operation [20].

Based on the above-mentioned studies, the International Societies have published recommendations for the treatment of patients with AAA. Comparing these recommendations, we see that for the intact AAAs the NICE recommends open repair unless contraindicated, while the ESVS feels that EVAR should be considered in patients with reasonable life expectancy (most patients). For ruptured aneurysms, NICE suggests considering EVAR or open repair with the caveat that open is better for men aged <71 years, while ESVS recommends EVAR as a first option [21].

The SVS Guidelines suggest assessment of operative risk and life expectancy for intact AAA, as well as informing patients about the perioperative mortality risk score, while for ruptured AAA, EVAR is recommended over open repair with a (strong) 1 level of recommendation, although with (low) C quality of evidence [22].

Overall, the EVAR strategy initially has superior results compared with open repair; however, the long-term results have failed to show a clear advantage. We need careful and individualized patient selection as well as improvement in the design and quality of the endografts. This combined with well-designed screening programs and improved training of the vascular surgeons will guarantee further improvement in the prevention and treatment of AAAs.

Case Presentation

Continued from page 475

Under general anesthesia, the patient underwent left axillary artery dissection and subsequently midline laparotomy. The suprarenal aorta was dissected and prepped for possible clamping.

A compliant balloon (Reliant, Medtronic) was advanced over a wire and through a 12 Fr sheath into the aorta at the supraceliac level. After heparinization and balloon inflation for proximal control the aortic sac was opened. Lumbar arteries and the inferior mesenteric artery were backbleeding and suture ligated. Temporary balloon deflation confirmed a type Ia endoleak. Using pledgets and 3–0 prolene the proximal endograft was hand sutured circumferentially at the aortic neck (Fig. 43.2). Appropriate inflow sealing was confirmed. A 14 mm iliac limb (Gore Excluder) was then introduced through right femoral cutdown. Final angiogram and follow-up CT angiogram were unremarkable (Fig. 43.3).

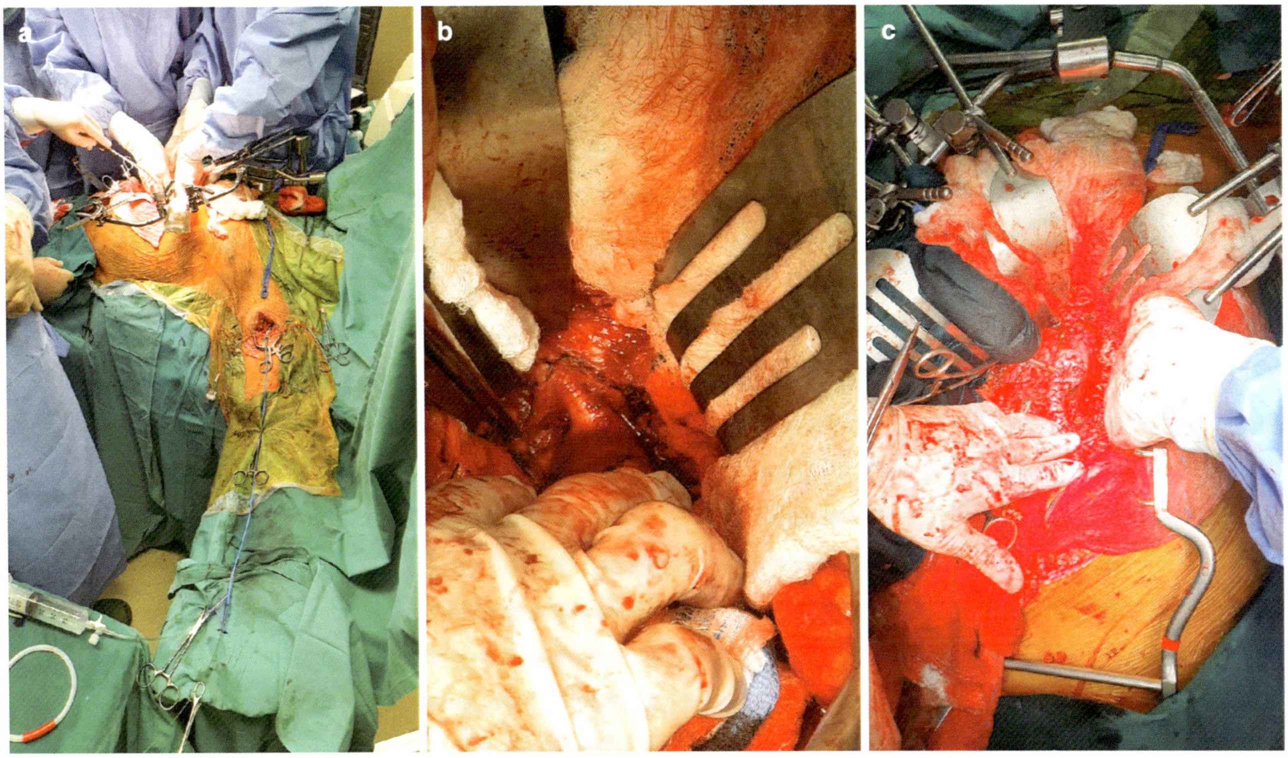

Fig. 43.2 (**a**) Right axillary artery exposure and 12 Fr sheath introduction for balloon aortic occlusion, (**b**) suprarenal exposure, and (**c**) proximal endograft suture anchoring

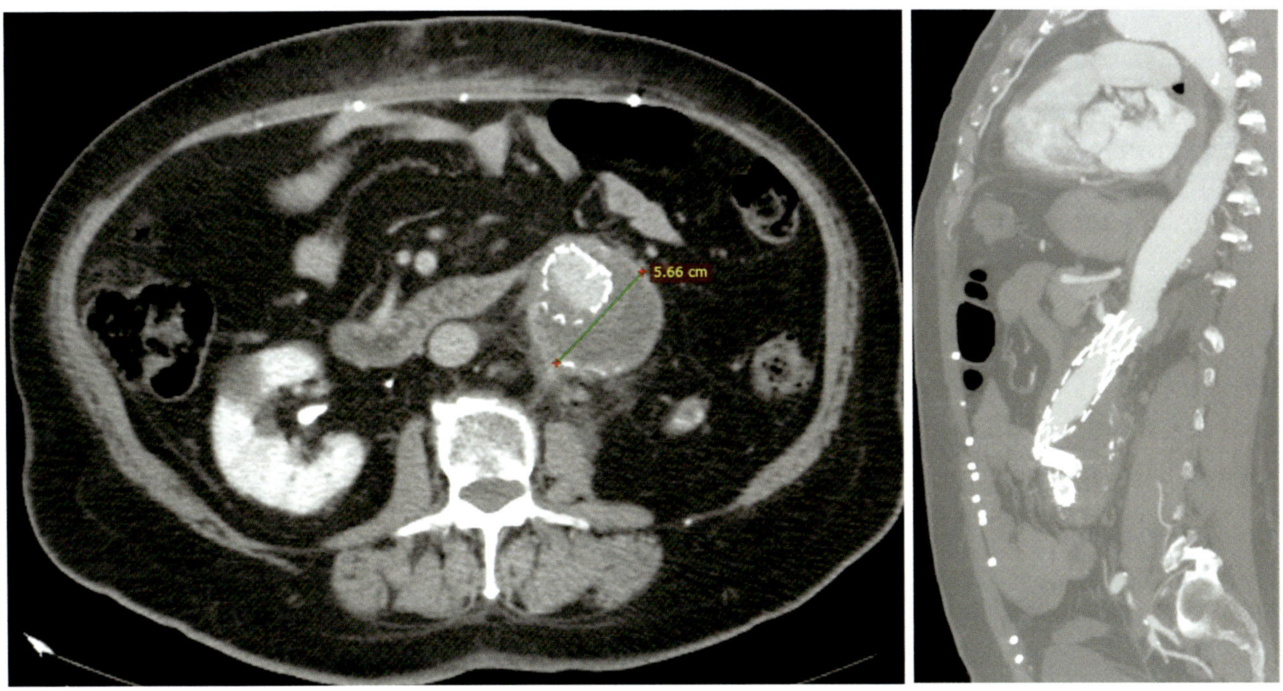

Fig. 43.3 Postoperative CT angiogram confirming sac exclusion and shrinkage

References

1. Abdulameer H, Al Taii H, Al-Kindi S, Milner R. Epidemiology of fatal ruptured aortic aneurysms in the United States (1999-2016). J Vasc Surg. 2019;69:378–84.
2. Vallabhaneni SR, Campbell WB. Lowering size threshold for elective repair to reduce deaths from abdominal aortic aneurysms—a simple solution to a complex problem? Eur J Vasc Endovasc Surg. 2017;54(3):275–7.
3. Karthikesalingam A, Vidal-Diez A, Holt PJ, Loftus IM, Schermerhorn ML, Soden PA, Landon BE, Thompson MM. Thresholds for abdominal aortic aneurysm repair in England and the United States. N Engl J Med. 2016;375(21):2051–9.
4. The United Kingdom EVAR Trial Investigators; Greenhalgh RM, Brown LC, Powell JT. N Engl J Med. 2010.
5. Patel R, Sweeting MJ, Powell JT, Greenhalgh RM, EVAR Trial Investigators. Endovascular versus open repair of abdominal aortic aneurysm in 15-years' follow-up of the UK endovascular aneurysm repair trial 1 (EVAR trial 1): a randomised controlled trial. Lancet. 2016;388(10058):2366–74.
6. Health Technol Assess. 2018;22(5).
7. De Bruin JL, Baas AF, Buth J, Prinssen M, Verhoeven EL, Cuypers PW, van Sambeek MR, Balm R, Grobbee DE, Blankensteijn JD, DREAM Study Group. Long-term outcome of open or endovascular repair of abdominal aortic aneurysm. N Engl J Med. 2010;362(20):1881–9.
8. Van Schaik TG, Yeung KK, Verhagen HJ, de Bruin JL, van Sambeek MRHM, Balm R, Zeebregts CJ, van Herwaarden JA, Blankensteijn JD, DREAM Trial Participants. Long-term survival and secondary procedures after open or endovascular repair of abdominal aortic aneurysms. J Vasc Surg. 2017;66(5):1379–89.
9. Lederle FA, Kyriakides TC, Stroupe KT, Freischlag JA, Padberg FT Jr, Matsumura JS, Huo Z, Johnson GR, OVER Veterans Affairs Cooperative Study Group. Open versus endovascular repair of abdominal aortic aneurysm. N Engl J Med. 2019;380(22):2126–35.
10. Sweeting MJ, Patel R, Powell JT, Greenhalgh RM, EVAR Trial Investigators. Endovascular repair of abdominal aortic aneurysm in patients physically ineligible for open repair: very long-term follow-up in the EVAR-2 randomized controlled trial. Ann Surg. 2017;266(5):713–9.
11. IMPROVE Trial Investigators, Powell JT, Sweeting MJ, Thompson MM, Ashleigh R, Bell R, Gomes M, Greenhalgh RM, Grieve R, Heatley F, Hinchliffe RJ, Thompson SG, Ulug P. Endovascular or open repair strategy for ruptured abdominal aortic aneurysm: 30 day outcomes from IMPROVE randomised trial. BMJ. 2014;348:f7661.
12. Stather PW, Sidloff D, Dattani N, Choke E, Bown MJ, Sayers RD. Systematic review and meta-analysis of the early and late outcomes of open and endovascular repair of abdominal aortic aneurysm. Br J Surg. 2013;100(7):863–72.
13. Gregory A, et al. Non-invasive determination of aortic mechanical properties and their effects on left ventricular function following endovascular abdominal aneurysm repair. J Med Biol Eng. 2019;39:739.
14. Blacher J, Guerin AP, Pannier B, Marchais SJ, Safar ME, London GM. Impact of aortic stiffness on survival in end-stage renal disease. Circulation. 1999;99(18):2434–9.
15. Vlachopoulos C, Aznaouridis K, Stefanadis C. Prediction of cardiovascular events and all-cause mortality with arterial stiffness: a systematic review and meta-analysis. J Am Coll Cardiol. 2010;55(13):1318–27.
16. Schäfer M, Kheyfets VO, Barker AJ, Stenmark K, Hunter KS, McClatchey PM, Buckner JK, Reece TB, Jazaeri O, Fenster BE. Reduced shear stress and associated aortic deformation in the thoracic aorta of patients with COPD. J Vasc Surg. 2018;68(1):246–53.
17. Kadoglou NP, Papadakis I, Moulakakis KG, Ikonomidis I, Alepaki M, Moustardas P, Lampropoulos S, Karakitsos P, Lekakis J, Liapis CD. Arterial stiffness and novel biomarkers in patients with abdominal aortic aneurysms. Regul Pept. 2012;179(1–3):50–4.
18. Kadoglou NP, Moulakakis KG, Papadakis I, Ikonomidis I, Alepaki M, Lekakis J, Liapis CD. Changes in aortic pulse wave velocity of patients undergoing endovascular repair of abdominal aortic aneurysms. J Endovasc Ther. 2012;19(5):661–6.
19. Kadoglou NP, Moulakakis KG, Papadakis I, Ikonomidis I, Alepaki M, Spathis A, Karakitsos P, Lekakis J, Liapis CD. Differential effects of stent-graft fabrics on arterial stiffness in patients undergoing endovascular aneurysm repair. J Endovasc Ther. 2014;21(6):850–8.
20. Moulakakis KG, Kadoglou NPE, Antonopoulos CN, Mylonas SN, Kakisis J, Papadakis I, Karakitsos P, Liapis CD. Changes in arterial stiffness and N-terminal pro-brain natriuretic peptide levels after endovascular repair of descending thoracic aorta. Ann Vasc Surg. 2017;38:220–6.
21. Powell JT, Wanhainen A. Analysis of the differences between the ESVS 2019 and NICE 2020 guidelines for abdominal aortic aneurysm. Eur J Vasc Endovasc Surg. 2020;60(1):7–15.
22. Chaikof EL, Dalman RL, Eskandari MK, Jackson BM, Lee WA, Mansour MA, et al. The Society for Vascular Surgery practice guidelines on the care of patients with an abdominal aortic aneurysm. J Vasc Surg. 2018;67(1):2–77.e2.

Patient Selection for Endovascular AAA Repair

44

George S. Sfyroeras

Introduction

Abdominal aortic aneurysm (AAA) is not an uncommon disease. Its prevalence has been reported to range from 2 to 12% [1]. An AAA is diagnosed in about 8% of men more than 65 years of age on screening studies; however these are generally small; those measuring ≥5.5 cm or greater are found in only 0.4–0.6% of those screened [2]. The annual incidence of new AAA diagnoses is approximately 0.4–0.67% in Western populations equating to 2.5–6.5 aneurysms per 1000 person-years [3].

The main risk associated with AAAs is aneurysm rupture. The risk of rupture of an AAA is related to its diameter. Once the aneurysm reaches about 5 cm, the yearly risk of rupture may exceed the risks of surgical repair for an average-risk patient. Rupture risk is also related to shape; fusiform aneurysms are considered less rupture-prone than saccular aneurysms, the latter having more wall tension in a particular location in the aneurysm wall [4].

Abdominal aortic aneurysm repair is designed to prevent the fatality associated with rupture in asymptomatic patients. Repair is absolutely indicated when the aneurysm becomes symptomatic or ruptured. When considering an intervention, physicians should weigh the risks associated with intervention compared to the risk of rupture for the size of the aneurysm and the patient's comorbidities. Proper patient selection involves an evaluation of both physiologic and anatomic risk factors. The treating physician must consider the patient's life expectancy, the risk posed by surgery, and the anatomic factors that make endovascular aneurysm repair (EVAR) possible or impossible, safe or risky. Therefore, the decision to intervene, especially in asymptomatic aneurysms, is mostly based on clinical decision.

G. S. Sfyroeras (✉)
Vascular Surgery Department, Attikon University Hospital, Athens, Greece

Assessment of Patient's General Status

Cardiac Disease

Five-year survival after successful aneurysm repair remains below 70%, despite the improvement in cardiovascular risk factor management achieved during recent decades [5]. The main causes of early and late death after AAA repair are cardiovascular and pulmonary disease [6]. EVAR offers an advantage in peri-operative mortality with a threefold reduction compared with patients undergoing elective open surgical repair (OSR) [7] However, despite the reduced peri-operative mortality compared with OSR, EVAR remains an intermediate- to high-risk procedure for cardiovascular complication [8].

The overall operative risk associated with EVAR must be evaluated. Firstly, it should be determined if there is an active cardiovascular condition necessitating further assessment and management. Active cardiac conditions include an unstable coronary syndrome (unstable or severe angina, myocardial infarction in the last 30 days), decompensated congestive heart failure, significant arrhythmias, and severe valvular disease. The presence of such a condition cancels or delays EVAR until it is treated, and a cardiology consultation is recommended. Medical management must be implemented, and a coronary angiography must be considered. In the absence of an active cardiac condition, further testing is indicated only if the results will change the planned treatment approach. Patients with good to moderate functional capacity, capable of climbing two flights of stairs or running a short distance (MET ≥4) without symptoms, will not benefit from further testing. Patients with poor or unknown functional capacity, MET ≤4, will benefit from cardiac testing if it will change operative management [9]. All patients must be evaluated with a 12-lead ECG before planned EVAR. Cardiac echocardiography is recommended for patients presenting dyspnea of unknown origin or with history of congestive heart failure.

G. Geroulakos et al. (eds.), *Mastering Endovascular Techniques*, https://doi.org/10.1007/978-3-031-42735-0_44

Coronary revascularization before EVAR is indicated for acute coronary syndrome with or without ST-segment elevation, unstable angina, and stable angina in the presence of left main coronary artery or three-vessel disease. It is also indicated for two-vessel disease, including the proximal left anterior descending artery and either ischemia on noninvasive testing or reduced left ventricular function [8]. If drug-eluting stents are implanted in the coronary arteries, dual antiplatelet therapy cannot be discontinued for 6 months. During this period, if aneurysm treatment is necessary, EVAR is the method of choice as dual antiplatelet therapy can typically be continued.

Pulmonary Status

The prevalence of AAA among patients with chronic obstructive pulmonary disease (COPD) is quite elevated. It is estimated that 7–11% of patients with COPD have an AAA [10]. EVAR is better tolerated than OSR in patients with COPD, particularly if performed under local anesthesia [11]. When COPD is suspected or present, arterial blood gas determinations and standard pulmonary function testing must be obtained before EVAR. In severe cases with oxygen-dependent COPD a preoperative pulmonary consultation is advised for optimization of medical therapy. Smoking cessation and administration of pulmonary bronchodilators for at least 2 weeks are recommended.

Renal Function

Preoperative renal insufficiency is an associated risk factor for increased mortality and complications after aneurysm repair. Patients with moderate renal function (eGFR 30–60 mL/min) are at increased risk particularly when treated by OSR rather than by EVAR [12]. Patients with severe renal insufficiency (eGFR <30 mL/min) present poor outcome regardless of the type of repair. In these patients 30-day mortality can reach 11%. Median survival is 2 years [13]. EVAR mandates the administration of contrast during the procedure. Contrast induced nephropathy (CIN) is defined as a 25% increase in serum creatinine concentration, or an absolute increase of 0.5 mg/dL 2–7 days after EVAR. Risk factors for CIN after EVAR include renal insufficiency (eGFR ≤45 mL/min/1.73 m^2), diabetes mellitus, congestive heart failure, ejection fraction <40%, hypertension, anemia, advanced age, proteinuria, and gout [14]. The relationship between contrast administration and CIN is linear: for every 100 ml of contrast administered there is a 12% increase in the risk for CIN [15]. Periprocedural hydration is useful for patients at increased risk of CIN. Normal saline at 1 mL/kg/h is administered for 6–12 h before and after the procedure or 5% dextrose/sodium bicarbonate can be administered at 3 mL/kg/h for 1 h before EVAR and at 1 mL/kg/h for 6 h afterward [8]. Preprocedural oral administration of N-acetylcysteine may be considered for at-risk patients, given its low cost, safety, and mild protective effect. However, a randomized trial of N-acetylcysteine did not reduce the incidence of CIN after EVAR [16]. Statin therapy may be beneficial for the prevention of CIN. The addition of statin along with N-acetylcysteine and intravenous saline proved a clinically important and statistically significant prevention strategy for CIN compared with the use of N-acetylcysteine and saline alone [17].

Aneurysm Diameter

At the time of diagnosis of an AAA most patients are asymptomatic. Less frequently the first presentation may be abdominal or back pain or even aneurysm rupture. In these cases, prompt treatment is necessary.

The decision for treatment for asymptomatic fusiform AAAs is mainly based on aneurysm diameter. Small aneurysms, with a maximum diameter of 4.0 cm or less, are at low risk of rupture and should be followed up, whereas an aneurysm >5.4 cm in diameter should be repaired. Elective repair is also recommended for patients who present with a saccular aneurysm. Regarding the latter, treatment is recommended at a smaller diameter, although guidelines are currently lacking.

Regarding treatment strategy for patients presenting with an AAA between 4.0 and 5.4 some controversy exists. Two studies evaluated the appropriateness of EVAR for small aneurysms. The Comparison of Surveillance versus Aortic Endografting for Small Aneurysm Repair (CAESAR) [18] and Positive Impact of Endovascular Options for Treating Aneurysms Early (PIVOTAL) [19] trials compared EVAR with surveillance for AAAs between 4.1 and 5.4 cm (CAESAR) and 4.0 and 5.0 cm (PIVOTAL). The CAESAR trial randomized patients >50 years of age with an AAA between 4.1 and 5.4 cm to receive immediate EVAR or surveillance by ultrasound and computed tomography. Repair in the surveillance group was performed if the AAA diameter reached ≥5.5 cm, the diameter increased >1 cm/year, or the patient became symptomatic. Between 2004 and 2008, 182 patients were randomized to EVAR and 178 to surveillance. The authors found that mortality and rupture rates in AAAs <5.5 cm were low and could not demonstrate a clear advantage between the early or delayed strategy. However, approximately 60% of the small aneurysms under surveillance grew to require repair within a 36-month period. Interestingly, the authors found that approximately 17% may lose feasibility for EVAR during this period [18]. The PIVOTAL trial was designed to examine whether early

EVAR reduced the risk of rupture or early aneurysm-related death compared with surveillance in patients between 40 and 90 years of age and with AAA size between 4 and 5 cm. This trial had 366 patients randomized to EVAR and 362 to surveillance for a mean follow-up period of 20 ± 12 months. During this study period, 30% of the patients in the surveillance group underwent EVAR, 70.6% for growth of the aneurysm, 11% due to request for repair, and 7.4% for the development of symptoms. The rate of peri-operative mortality was 0.6%, and the 3-year rate of rupture was zero. The lower-than-expected rate of rupture prompted early termination but with the commitment to follow currently enrolled patients. Until longer-term data becomes available from this study, close surveillance and treatment with EVAR as clinically indicated appears as safe as EVAR for the small aneurysms in this trial [19].

These studies suggest that in patients with small aneurysms, careful and close monitoring and surveillance is safe and will detect those who become symptomatic and those who demonstrate an increase in aneurysm size and require repair. However, none of these trials was designed to determine whether immediate EVAR might be beneficial for specific AAA size ranges or patient subgroups. A Cochrane database review of these four studies demonstrated no advantage to immediate repair by EVAR over surveillance for AAAs with a maximum diameter between 4.0 and 5.5 cm [20]. However, young, healthy patients, particularly women, with an AAA between 5.0 and 5.4 cm, or those with rapid expansion of small fusiform AAAs, or with a familial history of AAA, may benefit from early repair [21]. Patients with advanced age or severe comorbidities usually require an individualized approach regarding treatment decision.

The ideal frequency for surveillance scans of aneurysms 3.0–5.5 cm in diameter should be stratified by AAA diameter. For the smallest aneurysms (3–3.9 cm) a three-year surveillance interval is safe, for aneurysms 4.0–4.9 cm in diameter annual surveillance is recommended, and when the diameter reaches 5.0 cm the surveillance scans should be performed every 3–6 months [22].

Patients with Advanced Age

AAA repair presents a significant clinical challenge to vascular surgeons, particularly in patients with advanced age as it is often associated with comorbidities and poor health status. EVAR has emerged as an appealing alternative for patients at high risk from open surgery repair (OSR), such as patients over 80 years of age. As longevity is increasing, we will encounter a greater number of octogenarians and nonagenarians with AAA being evaluated for EVAR. A significant shift to EVAR was observed between 2001 and 2006 for patients older than 85 years, with an approximate increase of 162% [23]. Patients with advanced age require more attention considering their fragile health status and multiple comorbidities. Older patients have a higher comorbidity rate for EVAR, with a statistically significant difference in renal disorder and a higher American Society of Anesthesiologists score (ASA $\geq$3). There is no significant difference in the occurrence of hypertension, diabetes mellitus, coronary artery disease, hyperlipidaemia, and pulmonary disease between young and older patients. Mean maximum aneurysm diameter in octogenarians is larger than in younger patients without significant difference in neck length or diameter. Considering the unfavorable anatomy, the duration of an EVAR procedure in octogenarians is significantly longer, blood loss is higher, and length of stay longer. However, there is no difference in technical success. Elderly patients are prone to having systemic complications, including renal and pulmonary diseases and local complications including haematoma and lymphocele of the groin. However, there is no difference in the occurrence of myocardial and cerebral ischemia. EVAR in octogenarians is associated with a significantly higher but still acceptable 30-day mortality (2.7% vs. 1.5%, $p < 0.001$). The re-intervention rate during 5-year follow-up is not significantly different. EVAR is an appropriate treatment in octogenarians if comprehensive preoperative evaluation and post-operative surveillance are incorporated [24].

Although EVAR in octogenarians has become increasingly common and short-term outcomes have been satisfactory, experience with EVAR in nonagenarians is seldom reported. Nonagenarians undergoing EVAR generally have increased but acceptable 30-day mortality. In an analysis from the American College of Surgeons National Surgical Quality Improvement Program dataset including nonagenarians undergoing endovascular aortic aneurysm repair (EVAR) or open surgical repair (OSR), the EVAR group had a 30-day mortality rate of 2.6% in 1008 elective cases and 28.6% in 221 emergent cases. The OSR group had a 30-day mortality rate of 19.1% in 47 elective cases and 53.7% in 80 emergent cases [25]. Prendes et al. performed propensity matching of 352 nonagenarians versus non-nonagenarians in a population of more than 12,000 patients. The 30-day mortality and major adverse events were similar between the two groups in both intact and ruptured AAA. They concluded that EVAR should not be excluded in selected nonagenarians after a careful risk-benefit evaluation. This paper highlights the importance of patient selection and timing of intervention to confer a clinical benefit in this very elderly population [26]. A disciplined selection of a generally healthy group of nonagenarians who, other than age alone, should be expected to do well after the procedure, is mandatory to achieve good outcomes after EVAR [27].

Anatomic factors are important when evaluating older patients for EVAR. Older patients tend to have more challenging and complex AAA anatomy. Procedural complexity may put patients older than 85 years of age at increased risk. It is paramount to consider the anatomic feasibility and procedural complexities when counseling about the patient's procedural risks in undergoing EVAR. Ideal EVAR anatomy includes: (1) proximal neck diameter of <32 mm and length of ≥15 mm; (2) distal landing zone with iliac diameters of ≤20 mm and length of ≥20 mm; (3) proximal neck angulation of <60°; and (4) absence of calcium or thrombus in the proximal neck [8]. Elderly patients with more complex anatomy should be considered for referral to an experienced vascular center to evaluate for advanced endovascular interventions (e.g., fenestrated or branched EVAR).

Another issue for patients with advanced age is aneurysm diameter. Current guidelines recommend intervention when AAA diameters reach >5.4 cm in men and ≥5.0 cm in women [8]. Given the potential anesthetic, procedural, and post-operative risk it is understandable that the AAA diameter threshold to perform EVAR may be higher among octogenarians and mainly nonagenarians patients. EVAR when AAA diameter reaches 6 cm may be appropriate in carefully selected nonagenarians and avoids the significant risk of aortic rupture in these patients [26].

High-risk Patients

Patients with AAA with anatomic indication for treatment and severe comorbidities represent a therapeutic challenge. A randomized controlled trial, the EVAR 2 trial, evaluated whether EVAR improved survival in patients with severe comorbidities unfit for open surgical repair of their AAA. Fitness for surgery was determined locally by the surgeon, radiologist, anaesthetist, and cardiologist. Patients older than 60 years old, with an AAA ≥5.5 cm, anatomically suitable for EVAR, classified as ASA IV, were included in the EVAR 2 trial. Patients with recent acute myocardial infarction (<3 months), symptomatic congestive heart failure, unstable angina, severe valvular heart disease, cardiac arrhythmia, chronic obstructive pulmonary disease (COPD), and chronic kidney disease (CKD) with serum creatinine value of >2.26 mg/dL were suitable for the EVAR 2 trial [28]. After randomization, 166 patients were treated with EVAR while 172 had no intervention. Death from aneurysm rupture at 30 days in the no-intervention group was 9% and was similar to the 30-day mortality in the treatment group. All-cause mortality at 4 years was 64% for both groups. Hospital costs for patients treated were higher without a difference in HRQL scores. According to data, more than a quarter of patients assigned to the no-intervention group underwent repair, one-third of them doing so because of patient preference. However, these crossovers did not alter the main conclusion of the trial that EVAR had a considerable 30-day operative mortality in patients already unfit for open repair of their aneurysm, did not improve survival over no intervention, and was associated with a need for continued surveillance and re-interventions at substantially increased cost [29]. EVAR did not affect overall survival. Two-year survival was 60% and 5-year survival was 35%. At 8-year follow-up, aneurysm-related mortality in the EVAR group was lower. However, this did not result in a difference in overall mortality. A total of 48% of patients had complications related to the EVAR, and 27% of those patients' required re-intervention, adding to the increased cost of the procedure compared to the no-intervention group [30]. Survival after EVAR was assessed within the Vascular Study Group of New England population using the EVAR 2 trial criteria. Five-year survival for patients with aneurysms smaller than 6.5 cm was 46%; and for patients with aneurysms larger than 6.5 cm, 5-year survival was 28% [31, 32].

It is important to recognize that the high-risk criteria used in the EVAR 2 trial were for OSR and not for EVAR. Few studies have clearly delineated what constitutes high risk criteria for EVAR. Several reports since have shown that EVAR in high-risk patients can be performed with low in-hospital mortality rates and have advocated EVAR for patients with severe comorbidities [33–37]. In a recent meta-analysis a pooled peri-operative mortality rate of 3% was found in high-risk patients who underwent EVAR, with the rate ranging across the studies from 0.2 to 8.7%. There has been a significant trend toward improvement of peri-operative outcomes over the years. This improvement may be explained by better patient selection, preoperative patient optimization, wider use of best medical therapy for cardiovascular prevention, and enhanced peri-operative care. Improved patient selection may be due to advanced preoperative workup methods such as cardiopulmonary exercise test, and cardiological and anesthetic assessments, which are being used increasingly in the treatment of AAA patients [38]. Another finding was the significantly lower peri-operative mortality of EVAR compared with open surgical repair in the high-risk patient cohort.

There is uncertainty surrounding the optimal long-term management of high-risk patients with AAA. The EVAR 2 trial demonstrated that EVAR does not confer any survival benefits over medical treatment in a cohort of patients unfit for open surgical repair. The study found that even though EVAR reduces aneurysm-related mortality in patients who are physiologically ineligible for open repair, it does not increase overall life expectancy. After up to 10 years of follow-up EVAR was associated with a significantly lower rate of aneurysm- related mortality but also higher rates of complications and re-interventions and no difference in all-cause mortality. During 8-year follow-up, EVAR was considerably more expensive than no repair [32].

For patients with aneurysms greater than 7 cm lower barriers for fitness may be applied. It is important to both keep these patients under surveillance and refer patients to other relevant specialties to optimize their physical fitness and reevaluate them shortly [22]. Elective surgery for AAA is not recommended for patients with limited life expectancy, defined as less than 2–3 years [22].

Anatomic Considerations for EVAR

Preprocedural Imaging

When performing EVAR, the stent graft is used to exclude AAA from the systemic circulation. The anchoring segments of the stent graft on the aorta include the infrarenal neck and the iliac arteries. These aortic sections must provide both sufficient sealing and fixation. Stent grafts rely on some degree of oversizing to guarantee sealing and fixation. The degree of oversizing ranges from 10 to 25%, varying between different devices. Several important anatomic characteristics of the aneurysm must be accurately evaluated in the preoperative imaging for stent-graft sizing. Information regarding the proximal landing zone, the characteristics of the aneurysm sac, the distal landing zone, and the vascular access must be obtained. Anatomic factors influencing patient selection for EVAR include the length, shape, and angulation of the infrarenal neck; presence of common iliac artery aneurysms, occlusive disease, tortuosity, or kinking of the iliac vessels; and intrinsically small or heavily calcified iliac arteries.

To properly select the appropriate patients for EVAR, a detailed cross-sectional evaluation of the anatomy of the aneurysm is required with the purpose of properly planning the procedure and appropriately sizing the stent graft. CT angiography (CTA) with 3D vessel analysis is the preferred examination for accurate procedural planning because it shows the vascular anatomy in detail, helping in choosing the appropriate device and predicting the possibility of difficulties or complications during the procedure [39]. Recommended slice diameter is 2.5 mm for standard devices, while sizing for fenestrated or branched devices requires ≤1 mm cuts. Intravenous contrast should be routinely administrated unless the patient has severe renal insufficiency. For CT scan evaluation advanced reconstruction techniques should be used to confirm accurate measurements, including 2D multiplanar reformation, maximum intensity projection, curved planar reformation on lumen centerline, and 3D reconstruction. The axial, coronal, sagittal, and three-dimensional (3D) reconstructions should all be reviewed. Diameters may be overestimated on axial images, and tortuosity must be considered to perform accurate measurements [40].

When exposed to iodinated CT contrast agents, patients with impaired renal function may present with contrast-induced nephropathy. In patients with renal insufficiency there are two options regarding preoperative imaging: the first is to perform a CT scan without intravenous administration of iodinated contrast. This examination usually provides enough details for endograft sizing; however, blood flow into the aortic lumen, presence of thrombus, and branch vessel patency cannot be evaluated. Diameter and length measurements can be made from such a study, and if the anatomy appears uncomplicated and there is no clinical suspicion of associated occlusive disease, it may be reasonable to proceed. The second option is to obtain an unenhanced MR angiography (MRA), which avoids ionizing radiation and is useful for patients with poor renal function who are not undergoing dialysis [41]. MRA examination provides information regarding aortic anatomy, but the aortic wall is inadequately depicted. Consequently, the measurements for endograft sizing may not be as accurate as anticipated.

Proximal Landing Zone

Steady proximal fixation on the aortic neck is essential for long-term EVAR stability. The proximal aortic neck consists of the area from the most inferior renal artery to the beginning of the aneurysmal dilatation. The downward forces exerted on the stent graft are associated with the diameter and angulation of the infrarenal aorta. Proper fixation is necessary to avoid stent-graft migration, kinking, and attachment site leak (type 1 endoleak). Measurement of the aortic neck diameter should be performed proximally at the level of the lowest renal artery and extend 15 mm distally in 5 mm intervals. These measurements should be made from the minor axis of axial cuts, or ideally from reformatted slices that allow a plane perpendicular to the centerline. It is important not to overestimate diameter based on oblique CTA cuts of the angulated aortic neck. Endografts should be oversized 10–20% in comparison to the aortic neck. At present, stent grafts range in diameter from 18 to 36 mm and can accommodate aortic diameters of 16–32 mm. The length of the neck should be at least 15 mm according the instructions for use (IFU) for most devices (the Endurant-Medtronic device allows a 10-mm neck). A short proximal neck is the most common excluding factor for EVAR [42].

Another anatomic feature is proximal neck angulation, the angle between the aortic neck and the suprarenal aorta. This angle should be preferably less than 60° according to the IFU for most devices. Increased angulation of the proximal neck is associated with significant type I endoleaks. The Aorfix stent graft (Lombard Medical Inc., Irvine, California) is the only FDA-approved device indicated for the treatment of angulated necks of up to 90°. As the degree of angulation increases, the proximal attachment area must lengthen in order to enhance secure fixation.

A reverse-tapered or conical neck configuration is considered when the neck increases in diameter as it extends distally. In these cases there is more than 20% change in neck diameter over the first 15 mm length of aortic neck length. This hostile configuration may result in an inadequate proximal sealing and higher incidence of endoleak. Sizing of the endograft according to the larger distal diameter would correspond to an excessive proximal oversizing. Conversely, sizing to the smaller diameter would result in inadequate oversizing distally. In these situations, it is appropriate to "split the difference," so as to give at least a minimum 10% oversizing in the larger distal segment and a less than 30% oversizing in the proximal smaller segment [43].

Proximal neck calcification and thrombus extending in more than 90° of the neck circumference is generally associated with increased risk of migration, endoleak, and renal artery embolization. When proximal neck thrombus is present active proximal fixation may protect from stent-graft migration. Many endografts are constructed with an uncovered stent that extends above the renal arteries with hooks and barbs to provide additional fixation. The volume of neck thrombus decreases over time after EVAR, and the endograft firmly opposes to the aortic wall [44].

When the length of the proximal aortic neck is too short and does not allow proper sealing of the standard endograft device, the aneurysm is characterized as juxtarenal (Fig. 44.1). In such cases alternative endovascular techniques must be applied. The chimney graft technique is an option for these patients. In this technique, covered stents parallel to the proximal graft are used to preserve the flow to visceral vessels, which are stented to obtain an adequate seal [45]. Chimney EVAR is performed using currently available devices, which limits cost and time associated with preparing custom devices. For this procedure, detailed preprocedural imaging is necessary to accurately size the stent grafts and ensure good wall apposition to limit post-procedural complications. Another option for patients with juxtarenal aneurysms are the fenestrated and branched stent grafts. These devices have fenestrations (holes) and side branches in the graft to access visceral arteries. They enable extension of the graft sealing zone to an adequate, more proximal, landing zone incorporating the visceral vessels. Fenestrations are sealed with a covered stent into the target vessels. Fenestrated and branched devices are custom manufactured for a specific patient and require careful preoperative planning with a manufacturing delay of 4–6 weeks [46]. Newer techniques include the use of preloaded cannulating wires, double reducing ties, and an enlarged proximal scallop. Both techniques are safe and effective in treating patients with juxtarenal aneurysms. A higher aneurysm-related mortality is observed in chimney EVAR while fenestrated EVAR is associated with a higher re-intervention rate [45].

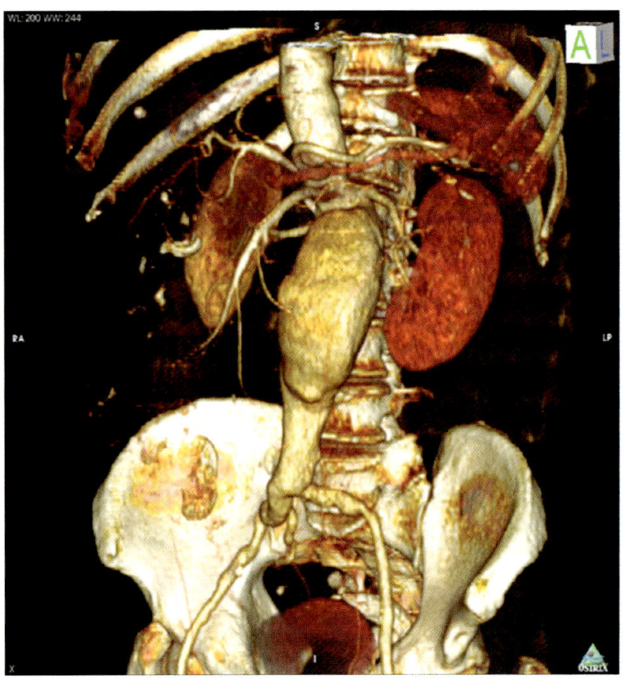

Fig. 44.1 A short proximal aortic neck does not allow proper sealing of the standard endograft device

Inflammatory AAA account for 5–10% of aortic aneurysms and are characterized by retroperitoneal fibrosis. [47] An InfAAA is defined by an unusually thickened aneurysm wall, shiny white peri-aneurysmal and retroperitoneal fibrosis, and dense adhesions of adjacent intra-abdominal structures [22]. Characteristic symptoms include chronic abdominal or back pain, weight loss, and elevated inflammatory markers. This triad of symptoms is highly suggestive of the diagnosis but rarely present, and fever is noted only randomly. A pathognomonic mantle sign on computed tomography angiography (CTA) is evident in 73–100% of patients. All patients diagnosed with inflammatory AAAs should receive corticosteroids. Compared with open repair, EVAR is associated with fewer intra-operative complications and lower peri-operative mortality, but more late fibrosis-related adverse events including post-operative progression of inflammation (17% vs. 0.4%), a higher frequency of persistent hydronephrosis (>50%), and limb occlusion (20%) [47].

Distal Landing Zone

The distal landing zone of the stent graft is preferably the common iliac artery (CIA). The CIA should be assessed in terms of diameter, length, tortuosity, kinking, degree of calcification, and presence of thrombus. Regarding CIA diameter, it should not be larger than 25 mm, as the available commercial stent grafts have a maximum iliac diameter of 28 mm. A normal diameter of at least 10 mm is necessary to achieve sealing. However, it is not uncommon for the aneurysms to extend into the CIAs. Approximately 20–30% of patients with AAAs have associated aneurysms of at least one common iliac artery (CIA) [48].

In patients with AAA and CIA aneurysm the most common option is to extend the limb of the endograft into the external iliac artery to obtain a distal seal. This technique implies the occlusion of the ipsilateral internal iliac artery (IIA). Unilateral IIA occlusion is generally safe and well tolerated, although symptoms of pelvic ischemia including buttock claudication and sexual dysfunction may result in 23–50% of patients [49]. When the endograft is extended in the external iliac artery the internal iliac artery (IIA) can be coil embolized to prevent retrograde filling of the aneurysm or simply covered without coil embolization. Whenever embolization of the IIA is necessary to exclude an IAA, the coils should preferably be placed in the proximal segment of the IIA to maintain communication between its anterior and posterior divisions [50]. Anatomic indications for coverage of the IIA without coil embolization include presence of adequate sealing in the distal 5 mm of the CIA, sealing ring at the origin of the CIA, or IIA diameter <5 mm [51].

If the IIA must be preserved, in cases where the contralateral IIA is absent, different endovascular techniques have been developed aiming to exclude the aneurysm, eliminate the possibility of an endoleak from the IIA, and maintain pelvic perfusion. These options include banding of the CIA [52], relocation or bypass of the IIA [53, 54], placement of oversized endograft limbs (bell-bottom technique) in the CIA [55], the "reverse U" stent graft [56], and iliac branch devices. Some of these approaches can be technically demanding, especially in cases with difficult anatomy, and associated with longer operative time, increased radiation and contrast use, and higher cost.

Iliac branch devices (IBDs) have evolved over recent years, aiming to preserve IIA flow following EVAR in patients with aortoiliac or isolated iliac aneurysms. The general concept of this device is that a straight branch comes off the main body of the limb and is mated to a stent graft that bridges to the IIA. IBDs are used to exclude the iliac aneurysms and maintain pelvic perfusion with preservation of the IIA. The risk of pelvic ischemia is minimized; however, it may not be completely abolished. Preoperative assessment of inferior mesenteric artery (IMA) and collateralization between superior mesenteric artery (SMA), IMA, and IIAs may be useful to predict the risk of pelvic ischemia after EVAR [57].

Vascular Access

Common femoral and external iliac artery must be appropriate for passage of the stent-graft device. The external iliac artery should be at least 6 mm in diameter, as most devices have an introducer sheath greater than 18 Fr. When EIA is smaller, newer-generation devices with a lower profile can be used. The Incraft (Cordis), the Excluder (Gore), and the Alto (Endologix) stent grafts have a lower than 18 fr profile and can be used in narrow iliac arteries. When the access vessels are small, angioplasty and stenting may be performed before device introduction. Iliac artery tortuosity and kinking are also important. The devices may not traverse extremely tortuous iliac vessels. In many cases, tortuous vessels can be straightened with a stiff Lunderquist wire. Vessels with heavy circumferential calcification require special attention. These vessels are nonpliable, may not permit device passage, and increase the risk of rupture. EVAR is not recommended for heavily calcified iliac arteries.

Endograft Configurations

Bifurcated endografts are currently used in more than 95% of EVAR cases. Bifurcated endografts are modular (Endurant, Excluder, Zenith, Incraft, Treo, Aorfix, Ovation, Anaconda) or unibody (Alto—Endologix). Most stent grafts have a modular design with two or three separate components including an aortic bifurcated main body and one or two iliac limbs. With this modular design, devices can be tailored precisely to the diameters and lengths of the vessels of the individual patient. Furthermore, using the overlap between components gives a degree of flexibility in planning. Most current stent-graft designs use suprarenal stents (Zenith-Cook Medical, Bloomington, Indiana; AFX-Endologix, Irvine, California; Endurant-Medtronic; Ovation-Trivascular, Santa Rosa, California) to inhibit downward migration and the development of type I endoleak and endograft failure. In addition, the Zenith, Endurant, Excluder, Aorfix (Lombard Medical Inc., Irvine, California), and Ovation stents have barbs to provide active, rather than passive, fixation. Ovation stents also have unique polymer-filled sealing rings to create an enhanced seal at the aortic neck and endolegs. The AFX device is the only one that touts passive fixation, whereby the flow divider of the stent graft sits directly on the aortic bifurcation. The C3 delivery system for the Gore Excluder graft allows for the ability to reposition the graft using a constraining dial which allows for proximal trunk reconstraining and reopening, a feature that may be particularly beneficial in tortuous proximal necks. The Aorfix stent graft is the only FDA-approved device indicated for the treatment of angulated necks of up to 90°.

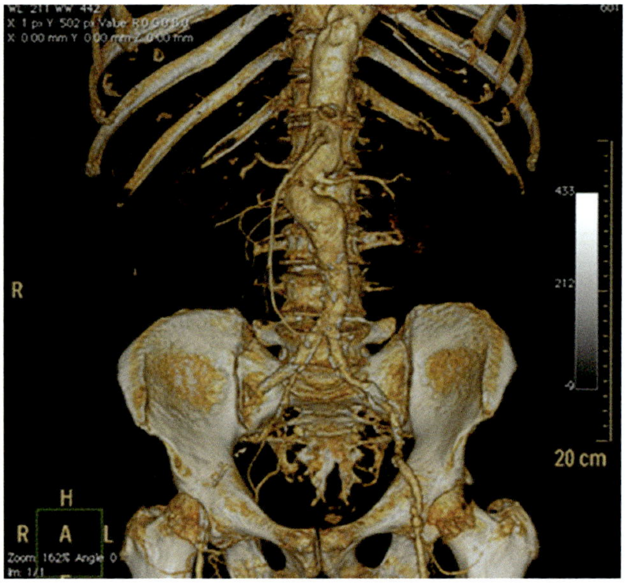

Fig. 44.2 Right external iliac artery occlusion. The insertion of a bifurcated endograft is not possible in this case

When there is a unilateral iliac occlusion the insertion of a bifurcated endograft is not possible (Fig. 44.2). In this case an Aorto-Uni-Iliac Stent Graft can be inserted, associated with a femoro femoral bypass in most patients. Aorto-Uni-Iliac Stent Grafts are available from Zenith, Medtronic and Terumo.

Proper choice of stent graft and meticulous planning of the procedure is necessary for successful deployment of the device and prevention of complications.

References

1. Stather PW, Sidloff DA, Rhema IA, Choke E, Bown MJ, Sayers RD. A review of current reporting of abdominal aortic aneurysm mortality and prevalence in the literature. Eur J Vasc Endovasc Surg. 2014;47:240–2.
2. Von Allmen RS, Powell JT. The management of ruptured abdominal aortic aneurysms: screening for abdominal aortic aneurysm and incidence of rupture. J Cardiovasc Surg (Torino). 2012;53:69–76.
3. Forsdahl SH, Singh K, Solberg S, Jacobsen BK. Risk factors for abdominal aortic aneurysms: a 7-year prospective study: the Tromsø study, 1994-2001. Circulation. 2009;119:2202–8.
4. Taylor BV, Kalman PG. Saccular aortic aneurysms. Ann Vasc Surg. 1999;13:555–9.
5. Bahia SS, Holt PJ, Jackson D, Patterson BO, Hinchliffe RJ, Thompson MM, et al. Systematic review and meta-analysis of long-term survival after elective infrarenal abdominal aortic aneurysm repair 1969-2011: 5 year survival remains poor despite advances in medical care and treatment strategies. Eur J Vasc Endovasc Surg. 2015;50:320–30.
6. Goodney PP, Tavris D, Lucas FL, Gross T, Fisher ES, Finlayson SR. Causes of late mortality after endovascular and open surgical repair of infrarenal abdominal aortic aneurysms. J Vasc Surg. 2010;51:1340–1347.e1.
7. Schermerhorn ML, O'Malley AJ, Jhaveri A, Cotterill P, Pomposelli F, Landon BE. Endovascular vs. open repair of abdominal aortic aneurysms in the Medicare population. N Engl J Med. 2008;358:464–74.
8. Chaikof EL, Dalman RL, Eskandari MK, Jackson BM, Lee WA, Mansour MA, Mastracci TM, Mell M, Murad MH, Nguyen LL, Oderich GS, Patel MS, Schermerhorn ML, Starnes BW. The Society for Vascular Surgery practice guidelines on the care of patients with an abdominal aortic aneurysm. J Vasc Surg. 2018;67:2–77.
9. Fleisher LA, Fleischmann KE, Auerbach AD, Barnason SA, Beckman JA, Bozkurt B, et al. 2014 ACC/AHA guideline on perioperative cardiovascular evaluation and management patients undergoing noncardiac surgery. Am Heart J. 2006;152:223–30.
10. Brown LC, Powell JT. Risk factors for aneurysm rupture in patients kept under ultrasound surveillance. UK small aneurysm trial participants. Ann Surg. 1999;230:289–96.
11. Qureshi MA, Greenberg RK, Mastracci TM, Eagleton MJ, Hernandez AV. Patients with chronic obstructive pulmonary disease have shorter survival but superior endovascular outcomes after endovascular aneurysm repair. J Vasc Surg. 2012;56:911–9.e2.
12. Nguyen BN, Neville RF, Rahbar R, Amdur R, Sidawy AN. Comparison of outcomes for open abdominal aortic aneurysm repair and endovascular repair in patients with chronic renal insufficiency. Ann Surg. 2013;258:394–9.
13. Yuo TH, Sidaoui J, Marone LK, Avgerinos ED, Makaroun MS, Chaer RA. Limited survival in dialysis patients undergoing intact abdominal aortic aneurysm repair. J Vasc Surg. 2014;60:908–13.e1.
14. Mehran R, Aymong ED, Nikolsky E, Lasic Z, Iakovou I, Fahy M, et al. A simple risk score for prediction of contrast- induced nephropathy after percutaneous coronary intervention: development and initial validation. J Am Coll Cardiol. 2004;44:1393–9.
15. Rihal CS, Textor SC, Grill DE, Berger PB, Ting HH, Best PJ, et al. Incidence and prognostic importance of acute renal failure after percutaneous coronary intervention. Circulation. 2002;105:2259–64.
16. Moore NN, Lapsley M, Norden AG, Firth JD, Gaunt ME, Varty K, et al. Does N-acetylcysteine prevent contrast- induced nephropathy during endovascular AAA repair? A randomized controlled pilot study. J Endovasc Ther. 2006;13:660–6.
17. Subramaniam RM, Suarez-Cuervo C, Wilson RF, Turban S, Zhang A, Sherrod C, et al. Effectiveness of prevention strategies for contrast-induced nephropathy: a systematic review and meta-analysis. Ann Intern Med. 2016;164:406–16.
18. Cao P, De Rango P, Verzini F, Parlani G, Romano L, Cieri E, et al. Comparison of surveillance versus aortic endografting for small aneurysm repair (CAESAR): results from a randomized trial. Eur J Vasc Endovasc Surg. 2011;41:13–25.
19. Ouriel K, Clair DG, Kent KC, Zarins CK, Positive Impact of Endovascular Options for treating Aneurysms Early (PIVOTAL) Investigators. Endovascular repair compared with surveillance for patients with small abdominal aortic aneurysms. J Vasc Surg. 2010;51:1081–7.
20. Filardo G, Powell JT, Martinez MA, Ballard DJ. Surgery for small asymptomatic abdominal aortic aneurysms. Cochrane Database Syst Rev. 2015;2015(2):CD001835.
21. Powell JT, Sweeting MJ, Brown LC, Gotensparre SM, Fowkes FG, Thompson SG. Systematic review and metaanalysis of growth rates of small abdominal aortic aneurysms. Br J Surg. 2011;98:609–18.
22. Wanhainen A, Verzini F, Van Herzeele I, Allaire E, Bown M, Cohnert T, et al. Editor's choice—European Society for Vascular Surgery (ESVS) 2019 clinical practice guidelines on the management of abdominal aorto-iliac artery aneurysms. Eur J Vasc Endovasc Surg. 2019;57:8–93.
23. Schwarze ML, Shen Y, Hemmerich J, Dale W. Age-related trends in utilization and outcome of open and endovascular repair for abdominal aortic aneurysm in the United States, 2001-2006. J Vasc Surg. 2009;50:722–729.e2.

24. Han Y, Zhang S, Zhang J, Ji C, Eckstein HH. Outcomes of endovascular abdominal aortic aneurysm repair in octogenarians: meta-analysis and systematic review. Eur J Vasc Endovasc Surg. 2017;54:454–63.
25. Kumar M, Long GW, Major M, Gates E, Studzinski DM, Callahan RE, Brown OW, Welsh RJ. Predictors of mortality in nonagenarians undergoing abdominal aortic aneurysm repair: analysis of the National Surgical Quality Improvement Program dataset. J Vasc Surg. 2022;75:1223–33.
26. Prendes CF, Dayama A, Panneton JM, Stana J, Rantner B, Álvarez Marcos F, Mani K, Wanhainen A, Tsilimparis N. Endovascular aortic repair in nonagenarian patients. J Am Coll Cardiol. 2021;77:1891–9.
27. Tang Y, Chen J, Huang K, Luo D, Liang P, Feng M, Chai W, Fung E, Lan HY, Xu A. The incidence, risk factors and in-hospital mortality of acute kidney injury in patients after abdominal aortic aneurysm repair surgery. BMC Nephrol. 2017;18:184.
28. Brown LC, Epstein D, Manca A, Beard JD, Powell JT, Greenhalgh RM. The UK endovascular aneurysm repair (EVAR) trials: design, methodology and progress. Eur J Vasc Endovasc Surg. 2004;27:372–81.
29. EVAR Trial Participants. Endovascular aneurysm repair and outcome in patients unfit for open repair of abdominal aortic aneurysm (EVAR trial 2): randomised controlled trial. Lancet. 2005;365(9478):2187–92.
30. Greenhalgh RM, Brown LC, Powell JT, Thompson SG, Epstein D. Endovascular repair of aortic aneurysm in patients physically ineligible for open repair. N Engl J Med. 2010;362:1872–80.
31. De Martino RR, Brooke BS, Robinson W, Schanzer A, Indes JE, Wallaert JB, et al. Designation as "unfit for open repair" is associated with poor outcomes after endovascular aortic aneurysm repair. Circ Cardiovasc Qual Outcomes. 2013;6:575–81.
32. United Kingdom EVAR Trial Investigators, Greenhalgh RM, Brown LC, Kwong GP, Powell JT, Thompson SG, Epstein D. Endovascular repair of aortic aneurysm in patients physically ineligible for open repair the United Kingdom EVAR trial investigators. Endovascular repair of aortic aneurysm in patients physically ineligible for open repair. N Engl J Med. 2010;10:1056.
33. Pane B, Spinella G, Signori A, Musio D, Perfumo MG, Lucertini G, et al. Early and long-term outcomes after open or endovascular repair for abdominal aortic aneurysms in high-risk patients. J Cardiovasc Surg (Torino). 2014;55:257–63.
34. Sobocinski J, Maurel B, Delsart P, d'Elia P, Guillou M, Maioli F, et al. Should we modify our indications after the EVAR-2 trial conclusions? Ann Vasc Surg. 2011;25:590–7.
35. Timaran CH, Veith FJ, Rosero EB, Modrall JG, Arko FR, Clagett GP, et al. Endovascular aortic aneurysm repair in patients with the highest risk and in-hospital mortality in the United States. Arch Surg. 2007;142:520–4.
36. Bush RL, Johnson ML, Hedayati N, Henderson WG, Lin PH, Lumsden AB. Performance of endovascular aortic aneurysm repair in high-risk patients: results from the Veterans Affairs National Surgical Quality Improvement Program. J Vasc Surg. 2007;45:227–33.
37. Lim S, Halandras PM, Park T, Lee Y, Crisostomo P, Hershberger R, Aulivola B, Cho JS. Outcomes of endovascular abdominal aortic aneurysm repair in high-risk patients. J Vasc Surg. 2015;61:862–8.
38. Kontopodis N, Galanakis N, Charalambous S, Matsagkas M, Giannoukas AD, Tsetis D, Ioannou CV, Antoniou GA. Editor's choice—endovascular aneurysm repair in high risk patients: a systematic review and meta-analysis. Eur J Vasc Endovasc Surg. 2022;64:461–74.
39. Truijers M, Resch T, Van Den Berg JC, Blankensteijn JD, Lonn L. Endovascular aneurysm repair: state-of-art imaging techniques for preoperative planning and surveillance. J Cardiovasc Surg (Torino). 2009;50:423–38.
40. Picel AC, Kansal N. Essentials of endovascular abdominal aortic aneurysm repair imaging: preprocedural assessment. AJR Am J Roentgenol. 2014;203:W347–57.
41. Saida T, Mori K, Sato F, et al. Prospective intraindividual comparison of unenhanced magnetic resonance imaging vs contrast-enhanced computed tomography for the planning of endovascular abdominal aortic aneurysm repair. J Vasc Surg. 2012;55:679–87.
42. Keefer A, Hislop S, Singh MJ, Gillespie D, Illig KA. The influence of aneurysm size on anatomic suitability for endovascular repair. J Vasc Surg. 2010;52:873–87.
43. Brinster CJ, Sternbergh WC III. Endovascular aneurysm repair technique. In: Sidawy AN, Perler BA, editors. Rutherford's vascular and endovascular therapy, 9th ed. Philadelphia. 2019. pp. 930–942.
44. Bastos Gonçalves F, Verhagen HJ, Chinsakchai K, et al. The influence of neck thrombus on clinical outcome and aneurysm morphology after endovascular aneurysm repair. J Vasc Surg. 2012;56:36–44.
45. Yaoguo Y, Zhong C, Lei K, Yaowen X. Treatment of complex aortic aneurysms with fenestrated endografts and chimney stent repair: systematic review and meta-analysis. Vascular. 2017;25:92–100.
46. Linsen MA, Jongkind V, Nio D, Hoksbergen AW, Wisselink W. Pararenal aortic aneurysm repair using fenestrated endografts. J Vasc Surg. 2012;56:238–46.
47. Caradu C, Ammollo RP, Dari L, Wanhainen A, Van Herzeele I, Bellmunt-Montoya S, Ducasse E, Bérard X. Management of inflammatory aortic aneurysms—a scoping review. Eur J Vasc Endovasc Surg. 2023;65(4):493–502. S1078-5884(23)00005-9.
48. Lobato AC, Camacho-Lobato L. The sandwich technique to treat complex aortoiliac or isolated iliac aneurysms: results of midterm follow-up. J Vasc Surg. 2013;57:26S–34S.
49. DeRubertis BG, Quinones-Baldrich WJ, Greenberg JI, Jimenez JC, Lee JT. Results of a doublebarrel technique with commercially available devices for hypogastric preservation during aortoiliac endovascular abdominal aortic aneurysm repair. J Vasc Surg. 2012;56:1252–9.
50. Bosanquet DC, Wilcox C, Whitehurst L, Cox A, Williams IM, Twine CP, British Society of Endovascular Therapy (BSET). Systematic review and meta-analysis of the effect of internal iliac artery exclusion for patients undergoing EVAR. Eur J Vasc Endovasc Surg. 2017;53:534–48.
51. Papazoglou KO, Sfyroeras GS, Zambas N, Konstantinidis K, Kakkos SK, Mitka M. Outcomes of endovascular aneurysm repair with selective internal iliac artery coverage without coil embolization. J Vasc Surg. 2012;56:298–303.
52. Puech-Leão P. Banding of the common iliac artery: an expedient in endoluminal correction of aortoiliac aneurysms. J Vasc Surg. 2000;32:1232–4.
53. Faries PL, Morrissey N, Burks JA, Gravereaux E, Kerstein MD, Teodorescu VJ, et al. Internal iliac artery revascularization as an adjunct to endovascular repair of aortoiliac aneurysms. J Vasc Surg. 2001;34:892–9.
54. Parodi JC, Ferreira M. Relocation of the iliac artery bifurcation to facilitate endoluminal treatment of abdominal aortic aneurysms. J Endovasc Surg. 1999;6:342–7.
55. Kritpracha B, Pigott JP, Russell TE, Corbey MJ, Whalen RC, DiSalle RS, et al. Bell-bottom aortoiliac endografts: an alternative that preserves pelvic blood flow. J Vasc Surg. 2002;35:874–81.
56. Kotsis T, Tsanis A, Sfyroeras G, Lioupis C, Moulakakis K, Georgakis P. Endovascular exclusion of symptomatic bilateral common iliac artery aneurysms with preservation of an aneurysmal internal iliac artery via a reverse-U stent-graft. J Endovasc Ther. 2006;13:158–63.
57. Sfyroeras GS, Koutsias S, Karathanos C, Stamoulis K, Giannoukas AD. Colonic ischemia after endovascular exclusion of an aortoiliac aneurysm using an iliac branch device. Vasc Endovasc Surg. 2010;44:597–600.

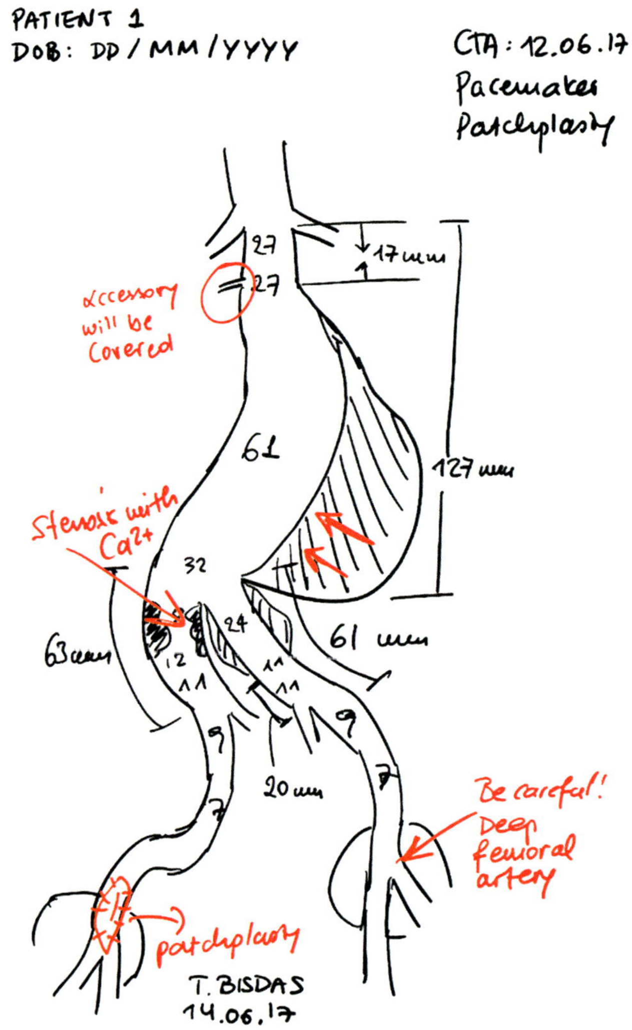

Fig. 45.1 Example of drawing an infrarenal aortic aneurysm on a sheet of paper and all relevant information needed for a successful endovascular aortic repair

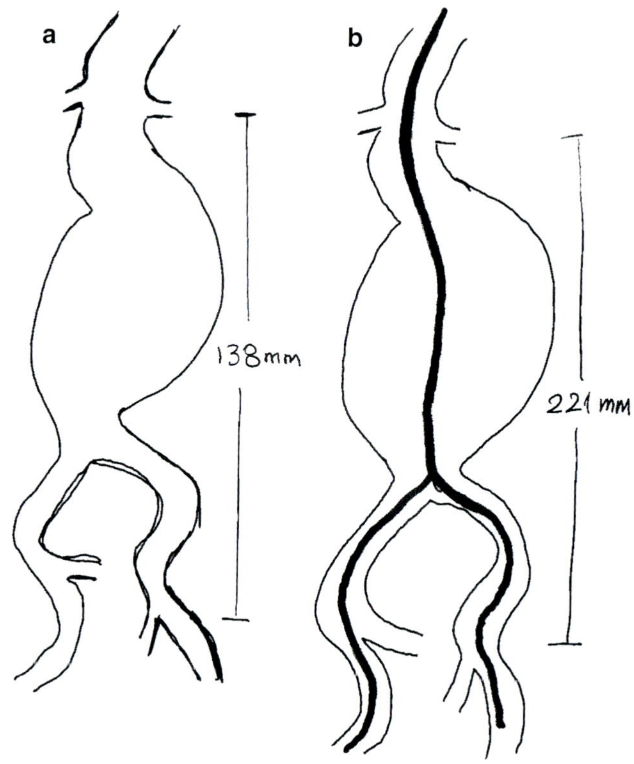

Fig. 45.2 Measurement of the aneurysm length by (**a**) adding the length of each slide of a CT angiography and (**b**) using the centerline

Creating the Centerline: Step-by-Step Measurement

During the measurements, it is crucial to create automatically or manually the *centerline*. The centerline essentially represents the position of the stiff guidewire through which the device will be delivered. It is obvious that two centerlines should be created, one for each iliac axis and one for the abdominal aorta. In case of angulated infrarenal neck and/or elongated iliac arteries, any calculation of the lengths will be wrong with the risk of a poor final outcome or the need for more devices (Fig. 45.2).

After the creation of the centerline, the physician should follow the next steps:

Step 1: The first measurement is the diameter of the infrarenal aorta just below the lowest renal artery. It is important to mention which is the lowest renal artery in your drawing. Moreover, you have to assess from which view you will see precisely the origin of the lowest renal artery and the least parallax of the proximal markers. This calculation will lead to lower administration of contrast agent and use of radiation. However, it is feasible mainly in dedicated software for EVAR planning. According to the instructions for use (IFU), you have to measure the diameter of the aorta 15–20 mm below the lowest renal artery. We recommend measuring the aorta at three different points, level of the lowest renal artery, 10 mm below it, and 20 mm below it, and selecting the diameter of the device according to the first two diameters. The main reason for this strategy is that the proximal sealing of the stent graft is based mainly on the first stent of the devices which is 10 mm in length in the majority of the EVAR devices. In specific cases of hostile proximal neck (bulged, reversed tapered, tapered), the diameter of the aorta at 20 mm should be taken into consideration for the diameter of the EVAR device. In such cases, the operator should know that the implantation might be outside the IFUs and inform the patient accordingly.

All manufacturers recommend 20% oversizing of the EVAR device and we strongly agree with this recommendation. However, in angulated necks, the centerline does not always represent the real position of the guidewire during the

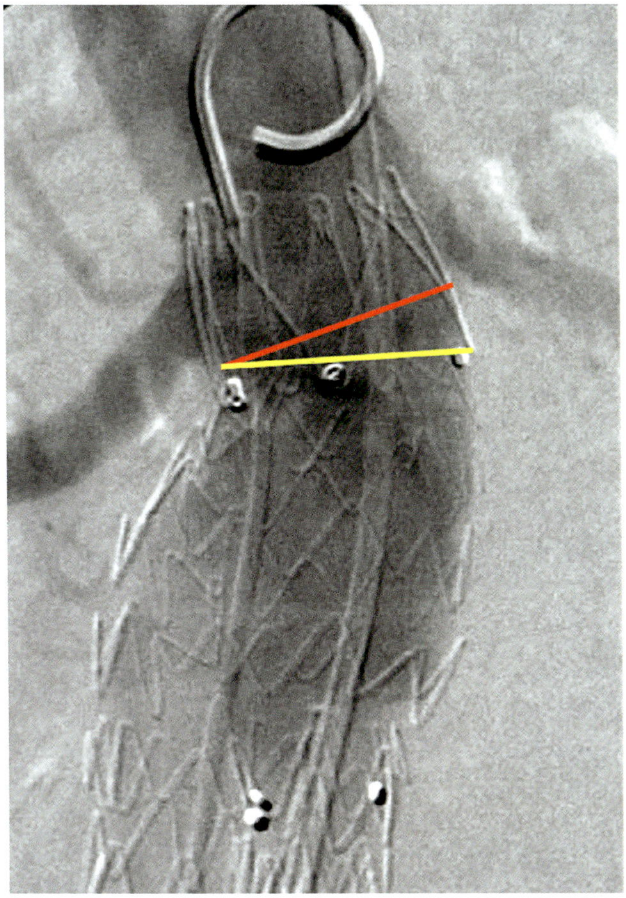

Fig. 45.3 Diameter of the aorta in angulated neck based on the centerline (red line, 27 mm) and the real landing zone of the device (yellow line, 29 mm). The device was selected with a diameter of 36 mm

intervention. Thus, and according to our experience, our common practice in angulated necks is to perform a 30% instead of 20% oversizing (Fig. 45.3). In any case, we recommend always ordering an aortic extension in angulated necks in case you identify a type Ia endoleak intraoperatively.

Step 2: The next step is to assess the distance between the lowest renal artery and the bifurcation. This measurement is important in order to ensure that the contralateral gate of the bifurcated endograft will open adequately inside the aneurysm sac. Thus, we recommend measuring the distance from the lowest renal artery up to 0.5–1 cm above the bifurcation in order to be sure that the contralateral limb will be adequately opened and catheterized during the intervention. During this step, it is important to assess and highlight in your drawing any important anatomic challenges regarding the morphology of the aneurysm sac (Fig. 45.1). For instance, opening the contralateral limb in saccular aneurysms may lead to a challenging or even impossible cannulation of the gate through the common femoral artery, and thus the physician has to change the entry site of the main body. This maneuver will facilitate the cannulation of the contralateral limb having the sac at the opposite site of the gate.

Step 3: The last step is to measure the diameter and the length of the common iliac arteries. In general, it is suggested to measure the distance up to the iliac bifurcation and to cover it with the iliac limb accordingly. However, this strategy is not always necessary in long common iliac arteries. In any case, the physician should define a distal landing zone of at least 20 mm (in some devices 15 mm) and measure the proximal and distal diameter or the common iliac artery. The length and the diameter of the limb will be selected accordingly and always taking into consideration the IFUs of the manufacturer.

The physician should calculate the length of the iliac limb using the following formula:

$$L_{limb} = L_{lowerRA-IliacBif} - L_{MBTube}$$

where L_{limb} length of the iliac limb, $L_{lower\ RA-Iliac\ Bif}$ length between the lowest renal artery and the distal landing point above the iliac bifurcation, and $L_{MB\ Tube}$ length between cranial graft edge and bifurcation of the graft.

In case the company does not include the overlapping zone in the provided length of the limbs (e.g., Zenith Spiral Z, COOK Medical), the formula changes to the following:

$$L_{limb} = L_{lowerRA-IliacBif} - L_{MBcontralat} \text{ or}$$

$$L_{limb} = L_{lowerRA-IliacBif} - L_{MBipsilat}$$

where L_{limb} length of the iliac limb, $L_{lower\ RA-Iliac\ Bif}$ length between the lowest renal artery and the distal landing point above the iliac bifurcation, and $L_{MB\ contralat}$ length between cranial graft edge and end of the contralateral limb or $L_{MB\ ipsilat}$ length between cranial graft edge and end of the ipsilateral limb.

There are a few devices where the iliac leg should not necessarily overlap the total length of the ipsilateral or contralateral limb (in situ measurement). We recommend always using total overlapping length. There are two scenarios where this option is useful: (1) in case of the tortuous iliac arteries where the selected limb is shorter than initially considered and (2) if the required length for the iliac leg is not represented in the available limbs of the manufacturer (e.g., length to cover: 134 mm, available device 124 mm or 156 mm in length).

Take-Home Messages

- Draw the aneurysm on a sheet of paper.
- Be aware of all relevant issues (stenosis, tortuosity, dissection, thrombus, etc.) that may influence your intervention—notice all issues in your drawing.

- Plan your intervention according to the respective IFU of each manufacturer.
- Identify the right landing zone, especially in case of angulated necks.
- Oversize adequately.
- Learn all features and sizes of the endograft you use.
- Start with one endograft until you build your own learning curve.

References

1. Donas KP, Torsello G, Bisdas T. New EVAR devices: pros and cons. J Cardiovasc Surg (Torino). 2012;53(5):559–69.
2. Bisdas T, Weiss K, Eisenack M, Austermann M, Torsello G, Donas KP. Durability of the endurant stent-graft in patients undergoing endovascular abdominal aortic aneurysm. J Vasc Surg. 2014;60(5):1125–31.

Introduction to Endografts for Infrarenal AAAs

46

Andreas M. Lazaris

History

The history of endovascular treatment of aortic aneurysm diseases began about four decades ago. It was May 4, 1985, when a Ukrainian (USSR) cardiovascular surgeon, Professor Nikolai Leontyevich Volodos (Fig. 46.1), implanted the first hand-made endograft into iliac arteries. His patient had presented with lower limb gangrene secondary to multi-level peripheral arterial occlusive disease. The patient was treated by inserting a stent graft into the iliac arteries, followed by a femorotibial venous bypass graft [1]. In March 1987, following extensive preclinical work in models and cadavers, Volodos and his team implanted the first aortic stent graft in history. The patient had presented with a false aneurysm of the descending thoracic aorta [1]. The patient survived 18 years after the aneurysm exclusion with this first-ever stent graft. In 1986, Volodos published his first report about the use of "self-fixing synthetic blood vessel endoprosthesis" [2]. It probably did not receive wide acceptance because it was written in Russian. In 1989, Volodos and his colleagues tried to treat an abdominal aortic aneurysm (AAA) with a stent graft using a unibody bifurcated endoprosthesis. Unfortunately, the unsupported contralateral limb twisted, and the endovascular procedure was converted to open surgery [3].

At the same time, Juan Parodi, a vascular surgeon from Argentina, then a clinical fellow at the Cleveland Clinic in Ohio, USA, conceived his idea for a stent graft to treat AAA. Parodi proved that his early concept of stent grafts, based on zigzag stents in combination with polyurethane, was workable in dogs. However, he never published his work [3]. In 1988, when he met Julio Palmaz, and through him obtained the balloon-expandable Palmaz stent, he was able to use the combination of his straight tubular polyester stent graft with a Palmaz stent as a proximal fixation to treat his first patient with an AAA on September 7, 1990, with great success (Fig. 46.2). In 1991, Parodi published the details of the first five patients, all treated successfully with straight tubular endografts (Fig. 46.3) [4].

Fig. 46.1 Nicolai Leontievich Volodos was a Soviet and Ukrainian cardiovascular surgeon and scientist. Volodos developed and introduced the world's first endovascular stent graft into clinical practice for treating stenotic and aneurysmal diseases of the arterial system

Following these two pioneers, the endovascular treatment of aortic aneurysm disease exploded. The concept of stent graft treatment for aortic aneurysms rapidly spread to the USA, Australia, and Europe. Geoff White, Jim May, Michael

A. M. Lazaris (✉)
Vascular Surgery Department, Attikon University Hospital, National and Kapodistrian University of Athens, Athens, Greece

G. Geroulakos et al. (eds.), *Mastering Endovascular Techniques*, https://doi.org/10.1007/978-3-031-42735-0_46

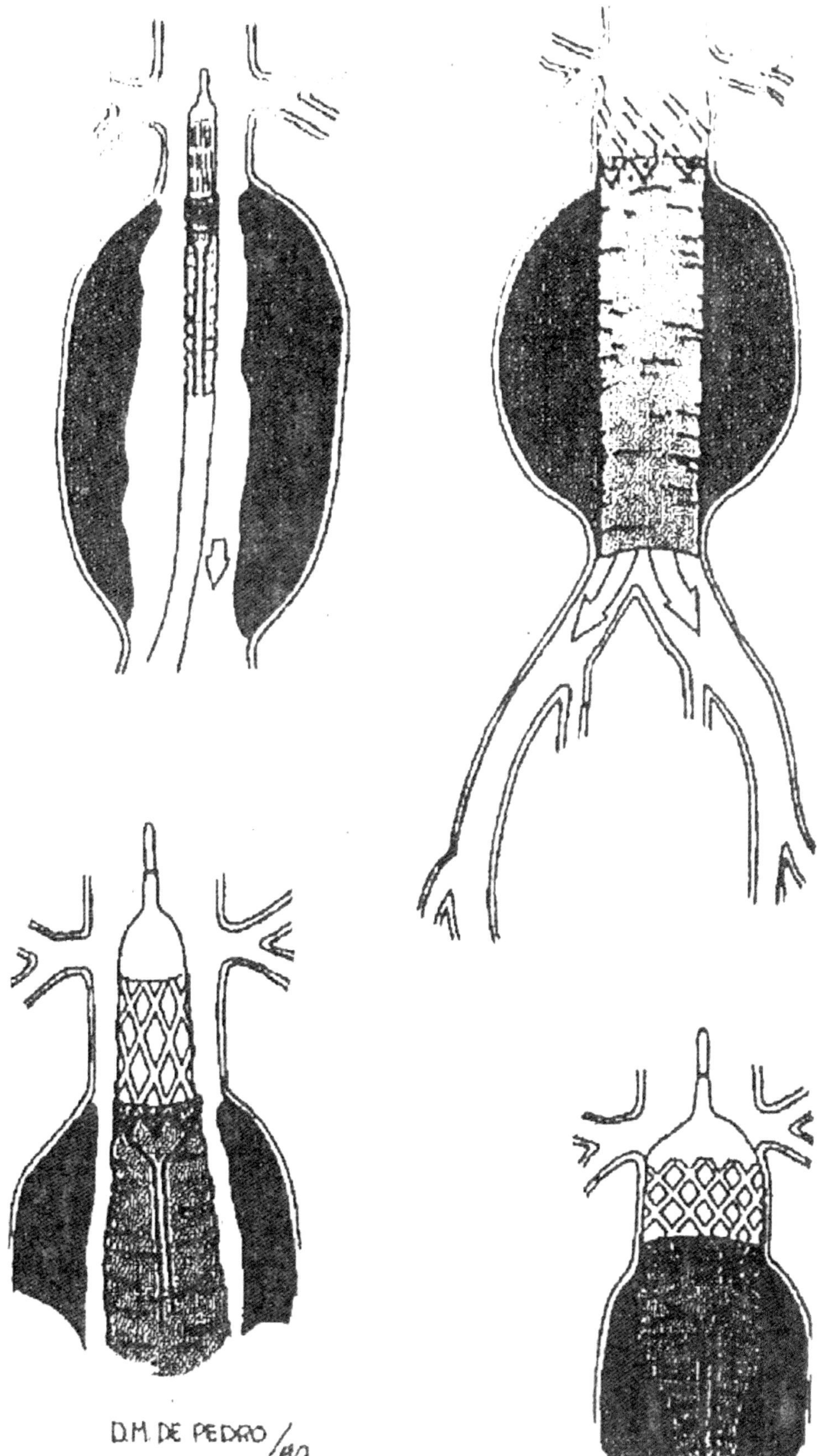

Fig. 46.2 Illustration of the first custom-made endografts. The surgeons created the endografts on the operating theater's back table. Balloon-expandable stents anchor the graft onto the aortic wall [4]

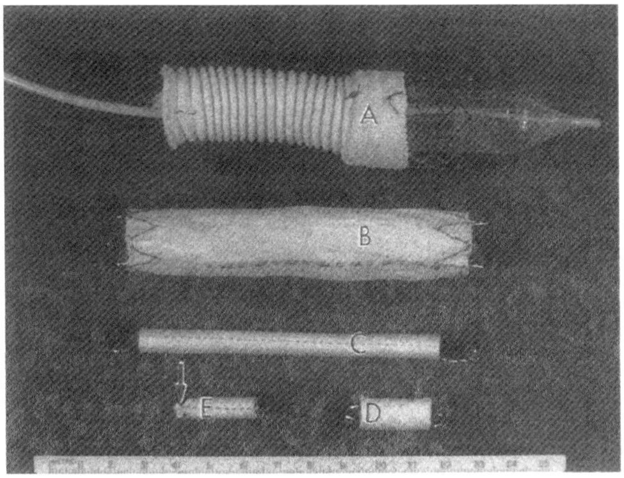

Fig. 46.3 Endovascular grafts composed from commercially available materials, used for the repair of aortic and peripheral artery aneurysms, arterial occlusions, and traumatic vascular injuries [5]

Lawrence-Brown, and David Hartley in Australia contributed significantly to developing aortic stent grafts and using them to treat patients. In the 1990s, tubular aortic stent grafts were the way to treat patients with aortic aneurysm disease. However, it soon became apparent that common iliac arteries and the abdominal aorta should be part of the treatment option. Thus, the idea of using bifurcated or aorto-uni-iliac endograft systems soon became necessary. Vascular surgeons such as Tim Chutter from the USA and Krassi Ivancev from Sweden proposed their solutions to this concept. The involvement of industry and vascular surgery pioneers inevitably boosted the design and development of the stent- graft philosophy. Cook Inc. (Bloomington, Indiana, USA) was an example, working closely with a significant number of vascular surgeons, such as Michale Lawrence Brown and David Hartley in Perth (Australia), Krassi Invacev in Malmoe (Sweden), Brian Hopkinson in Nottingham (England), Tim Chutter in Cornell (USA), Wof Stelter in Frankfurt (Germany), and Eric Verhoeven in Groningen (Netherlands). Soon, other companies followed, such as Gore (W. L. Gore & Associates, Incorporated, Flagstaff, Arizona), Medtronic (Medtronic Vascular, Santa Rosa, California), and many others in the following years. Famous worldwide vascular surgeons, well known for their open aortic surgery skills, made a significant shift in their careers, embracing with enthusiasm the new technology. Frank Veith in New York (USA), Edward "Ted" Dietrich in Phoenix (USA), Peter Bell in Leicester (UK), and Roger Greenhalgh in London (UK) were some of them.

The need to preserve the normal anatomy led to the development of fenestrations and directional branches. Marcel Goodman in Perth (Australia), Wolf Stelter in Frankfurt (Germany), and Tim Chuter and Roy Greenberg in the USA proposed and implemented fenestrations for splanchnic vessels.

The invention of endovascular aortic repair (EVAR) took place almost in the same period, in multiple locations around the world, on a simultaneous invention concept of invention and discovery, found in other fields of development such as aviation and automotive vehicle manufacturing. Like other fields of medicine, Vascular Surgery owes a great deal to its development in biology and engineering evolution. The growth and expansion of EVAR sprang from a simple common idea of a small group of inventive radiological and surgical visionaries, flourished almost simultaneously in industrial laboratories and hospital operating rooms, and became one of the most significant discoveries in the history of medicine.

Design and Characteristics of Aortic Endografts

The evolution of endografts has undergone multiple changes over time. Numerous generations of technological advancements based on physician and engineering feedback have addressed areas of clinical need, improved deliverability, decreased profile, and promoted better conformability to tortuous anatomy. Nevertheless, a similar general philosophy exists for all endografts. Every aortic endograft is comprised of two distinct components: the sheath with the delivery system and the endograft.

Sheath and Delivery System

The sheath is a necessary component of the aortic endograft device, ensuring the device's safe insertion into the arterial system. In most devices, the endograft and the sheath are combined into one single sheath-endograft unity. An exception to this general concept is the Gore endograft delivered through a separately provided sheath. The delivery system and the endograft are loaded in an over-the-wire fashion and brought into the arterial system.

The sheaths and delivery systems of all modern endograft devices have some common general characteristics:

1. **Hydrophilic coating**. The sheath is covered by a hydrophilic cover which reduces the friction resistance. Hydrophilic coating permits easy insertion and removal of the sheath, minimizing the risk of artery spasm.
2. **Conical tip**. The tip of the sheath is conical, allowing easy and safe insertion into the arterial system.
3. **Low profile**. First- and second-generation delivery systems were bulky, blunt-ended, and stiff. The advances in engineering allowed the development of narrow delivery systems, reducing the rate of iliac artery trauma, and allowing endovascular treatment in patients with small

iliofemoral access arteries, like those with severe calcification or those with iliac vessels of small caliber, for example women.

4. **Controlled deployment**. Most delivery systems allow a controlled endograft deployment, increasing the accuracy of positioning inside the aorta. The existence of gradual stent-graft-releasing mechanisms in endograft devices accomplishes this.

Endograft

The endograft is usually preloaded in the sheath, except for the Gore device, where the sheath and endograft are distinct components. There is a wide availability of different devices today (Fig. 46.4), with plenty of similarities but also some unique differences:

- **Endograft coating**. Endografts are made of polyester (Dacron) or polytetrafluorethylene (PTFE).
- **Metallic skeleton**. A metallic frame (skeleton) supports the endograft. In the first generations of endografts the skeleton was made of stainless steel, which is rarely used nowadays. Most endografts frames today are made of nitinol, an alloy of nickel and titanium. The skeleton lies either outside the graft (exoskeleton) or, more rarely, inside the graft (endoskeleton), and it is secured on the graft with tiny non-absorbable sutures. The configuration of the skeleton can be either circular or, more often, in a Z-stent fashion.
- **Endograft packing**. The endograft is packed in a collapsed state inside the sheath of the delivery system. In the Gore device, in contrast to other devices, the endograft is brought as a separate unit; it is collapsed around the delivery system and is inserted into the sheath by the surgeon during the surgical procedure.
- **Fixation systems**. The fixation of the endograft on the healthy proximal aorta is necessary to minimize the possibility of distal migration and is achieved in various ways. On the top end of the endograft there are various types of anchoring adjuncts like hooks, barbs, and anchors that embed into the aortic wall. Fixation is also enhanced by the radial force of the metallic frame produced by the stents of the skeleton (Z-stent or circular). The endograft oversizing, as compared to the size of the native proximal aortic neck, is necessary to achieve an adequate radial force increasing the endograft stabilization on the healthy aortic neck. In some grafts, a longitudinal metallic wire along the endograft enhances the fixation of the endograft by increasing its columnar strength. Finally, most endografts carry a line of triangular-shaped are metal stents on their top end, which contributes to the stabilization and fixation of the endograft on the suprarenal aorta. A unique idea for endograft stabilization exists in the Endologix AFX Powerlink graft. Its unibody bifurcated design allows the positioning of the endograft over the aortic bifurcation, eliminating the possibility of distal migration.
- **Endograft repositioning**. After the endograft is extracted from the sheath, it cannot be reloaded back on the sheath. However, most modern endografts are repositionable,

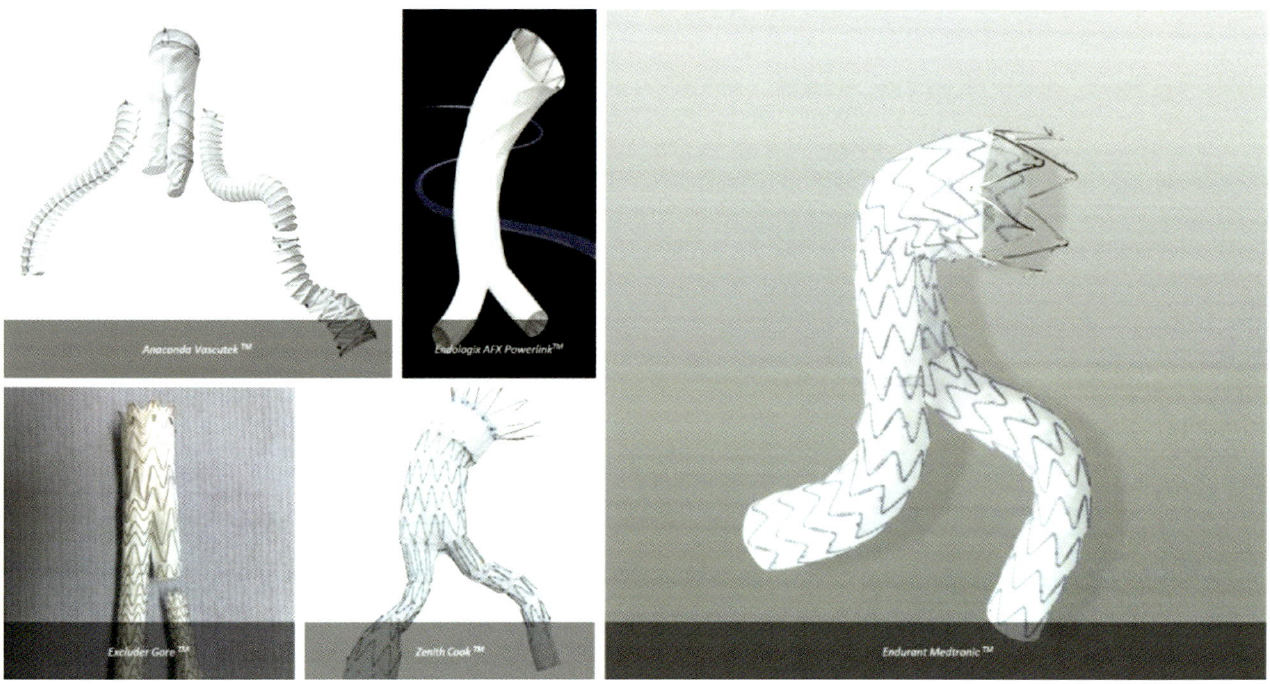

Fig. 46.4 Various types of endografts

meaning that the user can reposition the graft inside the arterial system, as long as the endograft remains partially deployed.

- **Modular concept**. Abdominal aortic endografts are usually modular grafts. They consist of two or three different endograft components that are united inside the human body. An adequate overlap between the adjacent components is necessary to minimize the possibility of blood leakage between them (type III endoleak), An exception to the modular design of most endografts, is the Endologix AFX Powerlink device, which is delivered on a unibody configuration.
- **Bifurcated configuration**. Abdominal aortic endografts are bifurcated grafts. Aorto-uni-iliac tube grafts are rarely used today in the usual treatment of AAA disease. Aorto-uni-iliac endografts are used today for specific clinical cases, such as when there is a narrow distal aortic neck or when there is an occlusion of one iliac axis not allowing intraluminal access from both iliac arteries and the accommodation of a bifurcated graft.
- **Radiopaque markers**. All endografts have x-ray radiopaque markers on specific positions, such as the top end of the endograft in the graft bifurcation, on the entry gate of the contralateral limb. Radiopaque markers assist graft orientation and correct alignment inside the human body.
- **Self-expanding property**. Endografts are self-expanding straight after their deployment outside their sheath. Ballooning is usually necessary to improve proximal and distal apposition on the landing zones of the endograft, on the healthy aorta, and on the iliac arteries.

Principles and Concepts of Endograft Deployment

For the deployment of each different endograft, there are specific steps that need to be followed. These steps are always explained in detail in each endograft instructions for use (IFU). However, certain steps are common in most cases. The typical steps that need to be followed for all endografts are the following:

1. **Choice of the iliac axis for the insertion of the main body's delivery system**. The iliac axis should be wide enough to accommodate the large profile main body device. Additionally, it should be as free of calcification as possible. If both iliac axes are suitable in terms of size for the main body delivery system, it is a prudent choice to prefer the most tortuous iliac artery, and leave the less tortuous iliac artery for the contralateral limb. This way, catheterizing the contralateral gate can be a relatively easy task. In contrast, when the most tortuous iliac artery is left for the contralateral limb, the catheterization of the contralateral gate of the main body could become a difficult task for both the patient and the performing physician.
2. **Access in both femoral arteries**. After deciding which iliac axis is suitable for the main body delivery system, access to the femoral arteries follows. Usually, this is done through small incisions on both groins, although the availability of closure devices and the low-profile endografts may occasionally permit percutaneous ultrasound-guided access. Using the Seldinger technique, a standard J wire is inserted into the femoral artery through which the main body will be advanced. The wire is supported by a short 11 cm-long, 6 Fr sheath, and it is advanced to the aortic arch and the ascending aorta together with a soft catheter. The same procedure is performed on the contralateral side, which will be used to insert the angiographic catheter.
3. **Insertion of the angiographic catheter.** A pigtail angiographic catheter or a large bore 8 Fr, 45 cm long sheath can be used for the diagnostic angiography. The large bore sheath allows a good-quality angiogram even when done without an injector, simply by manual contrast injection. The angiographic catheter or sheath is advanced at the level of renal arteries orifices, usually at the level of L1–L2 vertebrae.
4. **Insertion of main body delivery system**. Based on the preoperative CT investigation, the main body is inserted over an extra stiff guidewire approximately at the level where it should be, usually between L1 and L2 vertebrae levels.
5. **Angiography**. From the angiographic catheter or sheath, an angiogram is performed. The patient needs to be as still as possible. This can be accomplished by asking the patient not to breathe if the procedure is performed under local anesthesia or by asking the anesthetist to hold the patient's breath for a while. To better view the proximal neck, the two renal arteries need to be visualized. This can be achieved by putting the C-arm intensifier perpendicular to the aorta onto the area of interest.
6. **Deployment of the main body of the endograft**. Usually, the ipsilateral limb is partially deployed until the contralateral gate has been cannulated.
7. **Catheterization of the contralateral gate**. By using a hydrophilic guidewire with a half J tip from the ipsilateral iliac axis, the contralateral gate is catheterized with the support of a catheter. Occasionally, this can prove a time-consuming task, and the vascular surgeon/interventionist needs to be as patient as possible to complete it. If catheterization is not successful from the iliac axis, other options should be considered such as catheterization from

the contralateral femoral side, after the main body is fully deployed, or from an axillary or brachial access through the descending thoracic aorta.

8. **Insertion of the iliac extension branch of the endograft and insertion of any other extension endografts.** The iliac extension branch to the contralateral axis is inserted after the orifice of the internal iliac artery has been visualized with an angiogram. Similarly, all other iliac extensions are inserted according to the plan of the procedure. It is helpful to remember that the optimum infrarenal aortic endograft should cover the total length of the aorto-iliac system, from the level inferiorly to the lowest renal artery to just superiorly to the orifice of each internal iliac artery.
9. **Aortic endograft balloon molting.**

Choice of Endograft

The choice of endograft for an EVAR is crucial for the outcome of an AAA endovascular repair. Although each endograft may differ, as seen in its IFU, some standard anatomic criteria for graft implantation exist, irrespective of the chosen endograft (Table 46.1, Fig. 46.5).

It has been reported that the application of EVAR outside of the anatomically specific IFU variables negatively affects later results, including morbidity (migration and reintervention) and mortality [8]. Endografts placed outside of at least one IFU parameter were associated with higher perioperative and long-term aneurysm-related mortality of about 6% at 1 year up to 11% at 3 years post insertion. In contrast, no mortality was noted when the insertion was in accordance to the IFU criteria [8]. Similarly, reintervention, graft thrombosis, and combined graft-related adverse events are considered to be more common when the endograft is inserted out of the IFU. It is clear that the application of an EVAR outside the anatomically specific IFU variables has an accumulative negative effect on late results. Adherence to such IFU guidelines is the appropriate clinical practice.

Table 46.1 Anatomic criteria for EVAR, as defined by the ACC/AHA 2005 Practice Guidelines for the management of patients with peripheral arterial disease [6]

Length of the proximal neck	15 mm, at least
Diameter of the proximal neck	Less than or equal to 32 mm
Angulation of the proximal neck	Less than 60°
Configuration, type of the proximal neck	Tube without significant thrombus or calcification
Diameter of common iliac artery	Less than or equal to 22 mm
Length of common iliac artery	More than 30 mm
Type of iliac arteries	Not tortuous, without significant calcifications
Diameters of iliac arteries (access vessels)	Equal to or more than 7 mm

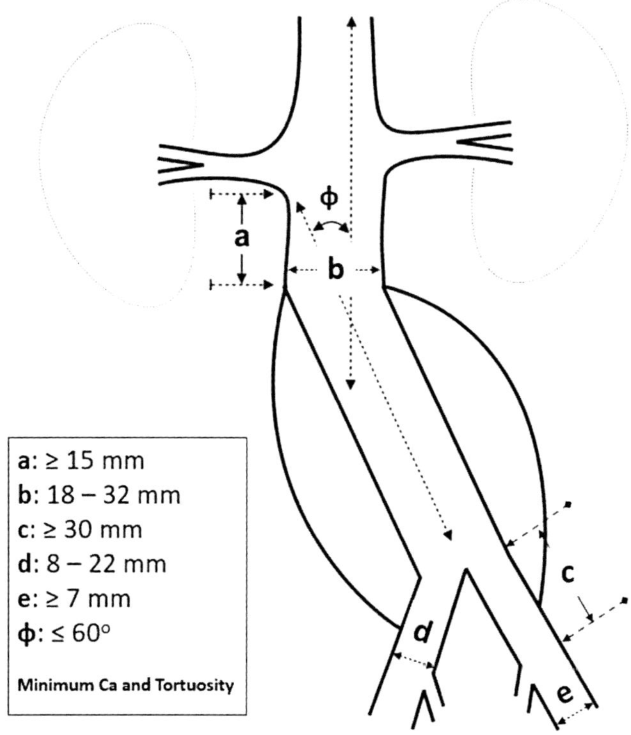

Fig. 46.5 The aortic and iliac arterial anatomy boundary conditions defined by the IFU of commercially available endovascular abdominal aortic devices. *CIA* common iliac artery, *EIA* external iliac artery [7]

Preoperative Planning for an EVAR Procedure

When deciding on an endovascular procedure for an AAA disease, the morphology of the aneurysm is of the utmost importance. The morphology of the aorta and the relevant vessels will force the decision to undertake a particular stent graft or even choose an open repair instead due to the presence of critical exclusion criteria. Thus, preoperative planning is the crucial indicator that significantly determines the procedure's short- and long-term outcomes. For this reason, the clinician must make the preoperative measurements, which ideally should be undertaken by the vascular specialist who will implant the graft.

A contrast computed tomography (CT) scan with 1 mm slices and three-dimensional (3D) reconstruction should be the basis for these measurements. Various types of software are available to allow these measurements with the original data from the CT scan. Calculating a center lumen line is usually helpful regarding the different lengths. Still, the user should try to estimate the route of the endograft inside the aortoiliac system, which does not always follow the center lumen line.

In the case of an EVAR for an infrarenal aortic aneurysm, the planning procedure can be divided into four different sections:

- Infrarenenal proximal aneurysmal neck
- Aneurysmal sac
- Aortic bifurcation
- Access vessels/landing zones (iliac and femoral vessels)

Future Perspectives

The endovascular treatment of aortic aneurysm disease is here to stay. The evolving medical technology will try to improve the present defects of the endografts and improve their weak points. Prospects of aortic endovascular graft design will be focused on improving their effectiveness, durability, and versatility. Some potential developments that may shape the future of aortic endovascular graft design are:

- Enhanced Customization: Researchers and engineers are working on developing grafts that can be more precisely tailored to fit each patient's unique anatomy. This involves utilizing advanced imaging techniques, such as 3D imaging and computational modeling, to create patient-specific grafts that provide a better fit and reduce the risk of complications.
- Advanced Materials: Developing novel materials with improved mechanical properties is another area of interest. Researchers are exploring biocompatible and bioresorbable materials to integrate with the surrounding tissues and promote healing. These materials may also offer improved flexibility and durability, improving long-term outcomes.
- Minimally Invasive Approaches: Efforts are being made to refine and expand minimally invasive approaches, such as endovascular robotics and catheter-based delivery systems, to make the implantation process less invasive, reduce operative time, contrast, and radiation exposure, and enhance patient recovery.
- Endovascular Navigation and Imaging: Advances in navigation and imaging technologies will likely play a significant role in future graft designs. Techniques such as augmented reality, virtual reality, and intraoperative imaging systems could assist surgeons in real-time visualization and precise placement of grafts, improving procedural accuracy and patient outcomes.
- Integrated Functionality: This involves the integration of additional functionalities into aortic endovascular grafts, including directly incorporating sensors or imaging components to monitor graft performance, detect potential complications, or provide real-time feedback during implantation.
- Bioactive Coatings: Surface modifications and bioactive coatings are being investigated to enhance the grafts' biocompatibility, reduce the risk of thrombosis and infection, and promote better integration with the surrounding tissues. These coatings can be designed to release drugs or growth factors gradually, facilitating tissue healing and remodeling.
- Long-term Durability: Improving the long-term durability of aortic endovascular grafts remains a critical goal.
- Integration of Artificial Intelligence, Machine, and Deep Learning to Predict Outcomes: Analysis of big data provided by large-scale registries of real-life patients worldwide will probably provide the infrastructure to develop accurate, individualized prediction models. Thus, difficult predicted complications like type II endoleak, which remains the Achilles heel of EVAR, might be predicted, and patients promptly advised on their best treatment.

It is worth emphasizing that while these advancements display great potential, medical technology is constantly in flux. Implementing and embracing these forthcoming designs may be time-consuming, as they must undergo meticulous testing, refinement, and rigorous regulatory approval processes. Nevertheless, the future of endovascular treatment appears promising and holds the potential to revolutionize vascular patient care in remarkable ways.

References

1. Bjorck M. The fall of a Giant Professor Nicolai Leontyevich Volodos May 15, 1934–April 3, 2016. Eur J Vasc Endovasc Surg. 2016;52(1):3–4.
2. Volodos NL, Shekhanin VE, Karpovich IP, Troian VI, Gur'ev Iu A. [A self-fixing synthetic blood vessel endoprosthesis]. Vestn Khir Im I I Grek. 1986;137(11):123–125.
3. Ivancev K, Vogelzang R. A 35 year history of stent grafting, and how EVAR conquered the world. Eur J Vasc Endovasc Surg. 2020;59(5):685–94.
4. Parodi JC, Palmaz JC, Barone HD. Transfemoral intraluminal graft implantation for abdominal aortic aneurysms. Ann Vasc Surg. 1991;5(6):491–9.
5. Marin ML, Veith FJ, Cynamon J, Sanchez LA, Lyon RT, Levine BA, et al. Initial experience with transluminally placed endovascular grafts for the treatment of complex vascular lesions. Ann Surg. 1995;222(4):449–65; discussion 65–9.
6. Hirsch AT, Haskal ZJ, Hertzer NR, Bakal CW, Creager MA, Halperin JL, et al. ACC/AHA 2005 guidelines for the management of patients with peripheral arterial disease (lower extremity, renal, mesenteric, and abdominal aortic): executive summary a collaborative report from the American Association for Vascular Surgery/Society for Vascular Surgery, Society for Cardiovascular Angiography and Interventions, Society for Vascular Medicine and Biology, Society of Interventional Radiology, and the ACC/AHA task force on practice guidelines (writing committee to develop guidelines for the Management of Patients with Peripheral Arterial Disease) endorsed by the American Association of Cardiovascular and Pulmonary Rehabilitation; National Heart, Lung, and Blood

Institute; Society for Vascular Nursing; TransAtlantic Inter-Society Consensus; and Vascular Disease Foundation. J Am Coll Cardiol. 2006;47(6):1239–312.
7. Schanzer A, Messina L. Two decades of endovascular abdominal aortic aneurysm repair: enormous progress with serious lessons learned. J Am Heart Assoc. 2012;1(3):e000075.
8. Abbruzzese TA, Kwolek CJ, Brewster DC, Chung TK, Kang J, Conrad MF, et al. Outcomes following endovascular abdominal aortic aneurysm repair (EVAR): an anatomic and device-specific analysis. J Vasc Surg. 2008;48(1):19–28.

Suprarenal vs Infrarenal Fixation

47

Konstantinos Spanos and Athanasios Giannoukas

Introduction

Recent decades have been illuminating for the establishment of endovascular aneurysm repair (EVAR) as the treatment of choice for elective and urgent patients with infrarenal abdominal aortic aneurysm (AAA). EVAR has been associated with very low mortality and morbidity rates in the perioperative and early follow-up period compared to open surgical repair (OSR) [1–6]. Devices have been constantly improving and their mechanical properties have seriously impacted their outcomes and durability, with the proximal sealing zone the most important one [7]. One of the most important issues is the conformability of the endograft to proximal aortic neck anatomy and the suitability of each device in order to achieve proximal sealing minimizing complications such as migration, endoleak type Ia, and potentially rupture, re-intervention, and death.

In practice, the proximal fixation systems are meant to prevent distal migration of the device and can be accomplished by passive and active mechanisms either by the radial force or by mechanisms such as barbs and hooks. Fixation relates to migration but does not directly relate to sealing, which is accomplished via radial force or sealant ring in most devices. Sealing is accomplished when the device becomes adherent to the inner wall of the aorta without allowing blood into the aneurysm sac. The only correlation that exists between fixation and sealing is that poor fixation can result in migration of a device, which would then obviously affect sealing, and, the other way around, poor sealing may allow aneurysm and aortic neck growth leading to migration.

The proximal fixation systems can be either suprarenal or infrarenal. The theoretical role of fixation in endograft technology is based on two important facts that, if properly deployed, firstly the proximal part of the endograft will seal the aortic neck from blood circulation, and secondly will prevent the device from distal migration from the seal zone over time. Both the constant aortic blood flow and potentially tortuous anatomy make the endografts susceptible to distal migration. Devices achieve fixation with multiple features, including active fixation (using barbs or hooks), columnar strength, outward radial force, and fibrotic reaction with prosthetic materials. Some devices may incorporate more than one of these features.

K. Spanos (✉) · A. Giannoukas (✉)
Vascular Surgery Department, Faculty of Medicine, Larissa University Hospital, School of Health Sciences, University of Thessaly, Larissa, Greece
e-mail: giannouk@uth.gr

Suprarenal Fixation Endografts

The Alto TM Abdominal Stent Graft System (Endologix Inc. Irvine, CA, USA) is an endovascular device delivered via a low-profile catheter to treat AAAs. The stent graft has a modular configuration comprised of an aortic body section, iliac limbs, and iliac extensions as required. The aortic body is comprised of a proximal stent for suprarenal fixation and a low-permeability polytetrafluoroethylene (PTFE). The bare proximal stent is designed with eight anchors to help fixation of the device to the aortic wall. When released from the compressed state, the stent expands to engage the vessel wall. To seal the proximal end of the graft and to provide support for the aortic body legs into which the iliac limbs are deployed, the graft body contains a network of inflatable channels and rings that are filled with a liquid radiopaque polymer that solidifies during the deployment procedure. The first and largest one is located 7 mm below renal arteries. The fill polymer is radiopaque and provides visualization of the polymer fill channels once the graft is filled.

Altura (Lombard Medical, Inc., UK), is another recently available endograft. The aortic and iliac endografts are fabricated from a braided nitinol (nickel titanium alloy) wire stent and an ultrathin, high-density, polyester graft, which is affixed to the stent with sutures at the proximal and distal ends of the graft. The proximal segment of the aortic endograft is not covered by graft material and is for suprarenal fixation. The two D-shaped aortic endografts are positioned

G. Geroulakos et al. (eds.), *Mastering Endovascular Techniques*, https://doi.org/10.1007/978-3-031-42735-0_47

side by side so that the flat surfaces appose each other to form a circular graft cross-section.

E-Tegra (JOTEC, Germany) is a modular system which has new features, such as the redesigned asymmetrical springs for better sealing, 360-degree flexibility and kinking resistance of the main body, and enhanced patency of the legs due to an improved flow divider. There is a separate release of the proximal stent which is designed to provide precise placement of the stent graft. The presence of a proximal laser cut stent with hooks provides suprarenal fixation.

The Endurant device (Medtronic, Inc., Minneapolis, MN, USA) includes two main body stent graft configurations: a bifurcated main body stent graft and an aorto-uni-iliac (AUI) main body stent graft. All configurations are composed of nitinol stents sewn to a fabric graft with non-resorbable sutures. Radiopaque markers are sewn onto the stent graft to aid in visualization and to facilitate accurate placement. The proximal end of the bifurcated configuration is composed of nitinol stents sewn to a fabric graft. The suprarenal portion of the proximal end is not covered with fabric (Fig. 47.1). The suprarenal stent also contains anchor pins to fix the stent graft in place inside the aorta above the renal arteries without obstructing them with the graft fabric.

The INCRAFT AAA stent graft system (INCRAFT) (Cordis Corp., Milpitas, CA, USA) is a modular bifurcated endovascular stent graft system. Each prosthesis is constructed of seamless, low-porosity, woven polyester graft supported by a series of short, electropolished, laser cut, self-expanding nitinol stent rings throughout the entire length. The nitinol stent rings are sutured to the inner surface of the graft material. It has a flared bare suprarenal stent with eight or ten laser cut barbs depending on the cranial diameter. The barbs help to keep the device in place.

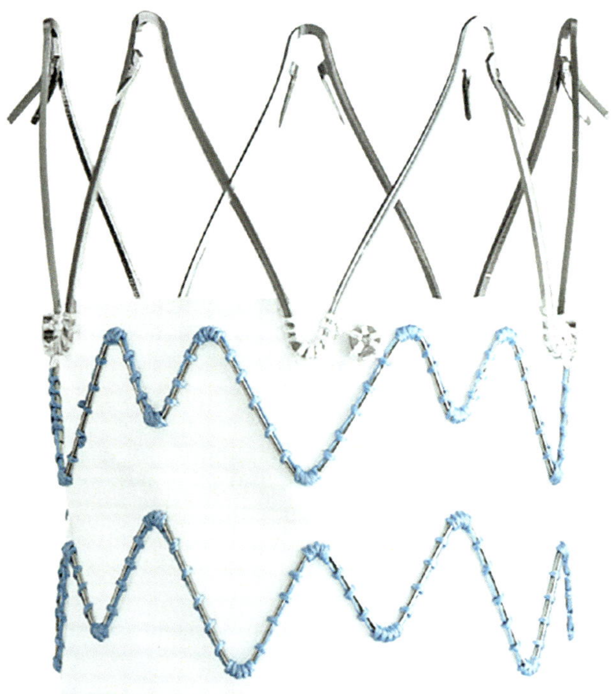

Fig. 47.1 The proximal end of the bifurcated configuration is composed of nitinol stents sewn to a fabric graft. The supra renal portion of the proximal end is not covered with fabric. The supra renal stent also contains anchor pins to fix the stent graft in place inside the aorta above the renal arteries without obstructing them with the graft fabric

The Zenith Alpha Abdominal Endovascular Graft (Cook, Inc., Bloomington, Indiana, USA) is a modular system consisting of three components: a bifurcated aortic main body and two iliac limbs. The graft modules are constructed of a woven polyester fabric sewn to self-expanding nitinol stents with braided polyester and monofilaments polypropylene suture providing a conduit that is intended to exclude the aneurysm from blood flow. The bare suprarenal stent at the proximal end of the graft contains barbs for additional fixation of the device.

The TREO® ABDOMINAL STENT-GRAFT SYSTEM (TREO) (Terumo, Somerset, NJ, USA) is a modular endovascular system which is typically comprised of a main bifurcated stent graft and two leg extension stent grafts. All stent grafts consist of self-expanding nitinol stents sutured to woven polyester fabric. The stent scaffold is a series of sinusoidal springs stacked in a tubular configuration. The uncovered proximal stent includes fixation barbs (suprarenal) for migration resistance. A second row of barbs are also located just distally to the start of the covered graft section, approximately at the middle of the first covered stent, to help provide infrarenal fixation. The uncovered proximal stent is composed of laser cut nitinol and is sewn on the inside of the woven polyester graft fabric. The seal zone of the main bifurcated stent graft consists of the distal portion of the uncovered proximal stent and the first covered proximal wireform, nitinol stent, sewn to the inside of the graft material.

Infrarenal Fixation Endograft

AFX System stent graft (Endoogix Inc., Irvine, CA, USA) prostheses consist of a self-expanding cobalt chromium alloy wire stent cage covered with expanded polytetrafluoroethylene (ePTFE) graft material. As part of its unique design, the implantable, self-expanding unibody stent cage design does not utilize mechanical attachment (e.g., hooks, barbs) and relies on anatomical fixation on iliac bifurcation. *ActiveSeal* conforms to the aortic wall under a pressure gradient between the aorta and excluded sac, which can extend the effective sealing zone beyond the neck anatomy to provide the opportunity for a stronger seal—ideally suited for patients with common and challenging proximal and distal anatomies. The next generation ePTFE, DuraPly™ ePTFE Graft Material, features layers that are wrapped in a helical fashion for unprecedented conformability combined with greater transverse tear resistance and strength leading to sufficient sealing. The primary device is the unibody, infrarenal bifurcated stent graft, which has a main body with two attached limbs. The accessory devices, which are utilized to customize the AFX System to the patient's anatomy, are comprised of infrarenal and suprarenal proximal extensions.

The Anaconda™ LoPro90 AAA Stent Graft System (Vascutek, Inchinnan, UK) is indicated for the repair of infrarenal AAA. Eight anchor points with active fixation are designed for effective sealing and migration resistance while allowing integrated repositioning function of the system with retained alignment control throughout the procedure for an optimized landing zone.

The Aorfix AAA Flexible Stent Graft System (Lombard Medical, Oxfordshire, UK) is a modular two-piece system consisting of a main body incorporating an ipsilateral leg component and a contralateral socket for a contralateral plug-in leg. The Aorfix main body has four sets of hooks all positioned at the proximal end to aid fixation. Hooks are positioned at four locations spaced 90° apart around the circumference at the proximal end and are designed to resist migration. The reinforcing wire is in ring form, rather than a traditional zigzag or diamond mesh stent. At the proximal end, the wire rings are placed closer together than in the body to increase radial force and they are also placed on the inside of the graft to improve the seal between the graft and the vessel wall.

The GORE® EXCLUDER® AAA Endoprosthesis (W.L. Gore & Associates, Flagstaff, AZ, USA) is comprised of two components, the trunk-ipsilateral leg endoprosthesis (trunk) and the contralateral leg endoprosthesis. The graft material is expanded polytetrafluoroethylene and fluorinated ethylene propylene (ePTFE and FEP) that is supported by nitinol (nickel titanium alloy) wire along its external surface. Nitinol anchors and an ePTFE/FEP sealing cuff are located at the aortic end of the trunk. An ePTFE/FEP sleeve is used to constrain the endoprostheses on the leading end of the delivery catheter.

The GORE® EXCLUDER® Conformable AAA Endoprosthesis (W.L. Gore & Associates, Flagstaff, AZ, USA) is a multi-component system consisting of a trunk-ipsilateral leg endoprosthesis, a contralateral leg endoprosthesis, an aortic extender endoprosthesis for proximal extension, and an iliac extender endoprosthesis for distal extension. The graft material for each component is expanded polytetrafluoroethylene (ePTFE) and fluorinated ethylene propylene (FEP) that is supported by nitinol (nickel titanium alloy) wire along its external surface. Nitinol anchors and an ePTFE/FEP sealing cuff are located at the leading (proximal) end of the trunk and a sealing cuff is located at the leading (proximal) end of the aortic extender (Fig. 47.2). All components have gold radiopaque markers for visualization. An ePTFE/FEP sleeve is used to constrain the endoprostheses on the delivery catheter. All components of the GORE® EXCLUDER® Conformable AAA Endoprosthesis (W.L. Gore & Associates, Flagstaff, AZ, USA) are of low permeability design, which is the only design available. As in the previous GORE c3 device, there is the possibility of repositioning and rotation of the trunk- ipsilateral delivery catheter as necessary to properly position the proximal device marker as well as orient the long contralateral. There is also an optional use of the angulation wire that may be used to provide angulation (bending) of the proximal portion of the constrained trunk device.

Fig. 47.2 The GORE® EXCLUDER® Conformable AAA Endoprosthesis is a multi-component system consisting of a Trunk-Ipsilateral Leg Endoprosthesis, a Contralateral Leg Endoprosthesis. The graft material for each component is expanded polytetrafluoroethylene (ePTFE) and fluorinated ethylene propylene (FEP) that is supported by nitinol (nickel titanium alloy) wire along its external surface. Nitinol anchors and an ePTFE/FEP sealing cuff are located at the leading (proximal) end of the trunk

Follow-Up Outcomes

Long-term outcomes even for the recent devices are available, demonstrating the durability of endografts either with supra- or infrarenal fixation.

Suprarenal Devices

The ENGAGE registry included a total of 1263 patients treated with the Endurant device (Medtronic, Inc., Minneapolis, MN, USA), enrolled from March 2009 to April 2011, at 79 centers across 30 countries. A significant proportion of the ENGAGE patients presented with challenging features and 17.8% were implanted outside of the instructions for use (IFU) for the device. At the five-year follow up, the Kaplan Meier overall survival rate was 67.4% and the freedom from aneurysm- related mortality was 97.8%. Freedom from aneurysm rupture, secondary procedures, and conversion to open repair at 5 years were 98.6%, 84.3%, and 97.9% respectively. The five-year freedom from type Ia endoleaks ~~s~~ was 95.2% and for type III endoleak 97.4% [8]. In a sub-analysis of the ENGAGE registry [9], patients who were treated outside the approved IFU were assessed. Technical success was achieved in >99% of all patients. EVAR outside IFU resulted in a lower five-year freedom from type Ia endoleaks (89.4% vs 96.7% [$p < 0.0001$]); however, this did not result in worse overall survival, aneurysm-related mortality, or secondary procedures. Proximal neck angulation and diameter outside the IFUs were the most common IFU violations and this cohort had increased incidence of type Ia endoleak [9].

A comparison of outcomes between EVAR-1 (a multicenter, prospective, randomized controlled trial in the UK with patient enrollment from 1999 to 2004 and use of first- and second-generation devices) and the ENGAGE registry was recently published [7]. Through the four-year timepoint, freedom from all-cause mortality was 74.4% in the EVAR-1 trial and 74.6% in the ENGAGE registry, while aneurysm-related mortality in the EVAR-1 trial was 4.2% and in the ENGAGE registry was 1.9%. In the EVAR-1 trial, the proportion of patients requiring at least one re-intervention through the four-year timepoint was 19.3% (121/626), whereas in the ENGAGE registry, re-interventions occurred in 10.9% (138/1263) of patients. EVAR patient outcomes have improved since the time of the original EVAR vs. OSR trials and data from real-world registries should be considered as a valid source of information for developing new guidelines [7].

Recently Liang et al. [10] presented the five-year results of a prospective regulatory study of the INCRAFT device (Cordis Corp., Milpitas, CA, USA). One hundred and ninety patients were enrolled from 32 centers throughout the US and Japan. Thirty-day major adverse events occurred in 3.2% of patients (6/190). Successful aneurysm treatment was 100% at 30 days and 87.9% at 1 year. Freedom from graft occlusion was 96% at 1 year and 94% at 5 years. Freedom from aneurysm-related mortality was 99% at 1 and 5 years. This study demonstrates good efficacy and safety and a very low rate of aneurysm- related deaths with the INCRAFT device in a population with a high proportion of challenging anatomy [10].

Another study~~,~~ reported a descriptive analysis of the Effectiveness of Custom Seal with Ovation (Endologix Inc. Irvine, CA, USA) (ENCORE) including 1296 patients from five clinical trials and the prospectively maintained European Union Post-Market Registry [11]. Fifty percent of patients had complex aortic anatomy (neck length <10 mm, neck diameter >28 mm, neck angle >60, reverse neck taper >10%, distal common iliac artery diameter <10 mm, or external iliac artery diameter <6 mm). Technical success was 99.7%. Thirty-day mortality was 0.3%. Freedom from type Ia endoleak at 1, 3, and 5 years was 97.6%, 97.1%, and 95.8%, respectively. Freedom from device-related re-intervention at 1, 3, and 5 years was 96.2%, 94.4%, and 92.4%, and primary freedom from sac expansion was 97.0% at 1 year, 90.3% at 3 years, and 84.9% at 5 years. Freedom from all-cause mortality and aneurysm-related mortality at 5 years were 78.9% and 99.3%, respectively. This analysis of the ENCORE database demonstrates that EVAR with the Ovation platform has favorable mid-term durability as evidenced by the successful aneurysm exclusion and 5-year freedom from aneurysm-related mortality [11].

In a study on Zenith endograft (Cook Inc., Bloomington, Indiana, USA), the authors presented their 14-year experience on 610 patients treated with the device [12]. Overall survival was 92.8% at 1 year, 70.1% at 5 years, 37.8% at 10 years, and 24% at 14 years. Freedom from AAA rupture was 99.8% at 1 year, 99.4% at 5 years, 98.1% at 10 years, and 98% at 14 years. Freedom from late re-intervention and conversion was 98% at 1 year, 87.7% at 5 years, 75.7% at 10 years, and 69.9% at 14 years. EVAR failure occurred in 132 (21.6%) patients at 14 years. EVAR with Zenith graft represents a safe and durable repair. Risk of rupture and aneurysm-related death is low, whereas overall long-term survival remains poor [12].

Infrarenal Fixation

A retrospective analysis of the Anaconda™ Italian Registry (Vascutek, Inchinnan, UK) divided patients in two groups according to the presence of a severe (Group A, GA: ≥60°) or an absent (Group B, GB: <45°) proximal aortic neck angle [13]. From 2005 to 2012, 1030 patients were enrolled in the Registry. Sixty-five patients (6.3%) were included in GA and 737 (71.5%) in GB. There were no statistical differences in 30-day mortality (GA 1.5% vs. GB 1.3%), proximal type I endoleak (GA 1.5% vs. GB 0.8%), iliac leg thrombosis (GA 1.5% vs. GB 1.4%), and conversion to open repair (GA 3% vs. GB 0.6%). The 3-year survival was 95.4% in GA and 94.7% in GB (P = NS). Freedom from proximal type I endoleak, iliac leg thrombosis, and conversion to open repair were 98.5%, 95.4%, and 95.4% in GA and 97.8%, 96.9%, and 98.5% in GB, respectively (P = NS).

A retrospective analysis of the Global Registry for Endovascular Aortic Treatment included patients treated for abdominal aortic aneurysms with the GORE EXCLUDER AAA Endoprosthesis (W.L. Gore & Associates, Inc.,

Flagstaff, Arizona, USA) [14]. Of the 3324 patients included in the analysis, 411 (12.4%) had a challenging neck. The 5-year all-cause mortality (36.2% vs. 27.5%, $P = 0.002$) and aorta-related mortality (3.8% vs. 1.1%, $P = 0.002$) were significantly higher in the challenging neck group. Furthermore, the rates of type Ia endoleak development (7% vs. 1.2%, $P < 0.001$) and requirement for re-intervention (13.3% vs. 9.7%, $P < 0.001$) were higher in those treated outside the IFU (challenging neck group). Infrarenal endovascular intervention outside the IFU should only be used when there is no alternative, with meticulous procedural planning and intervention to promote satisfactory outcomes [14].

Impact of Fixation System on Renal Function

It is not yet clear if the suprarenal bare-metal stents, which extend through the suprarenal aortic neck and the orifices of the renal and mesenteric arteries, are associated with long-term renal dysfunction. Some reports suggested that suprarenal stents may partially occlude the origins of the renal and mesenteric arteries after suprarenal stent graft implantation (Fig. 47.3) [15]. In 2006 Palmer et al. [16] evaluated the effect of suprarenal fixation on renal function by comparing homogeneous patient populations receiving EVAR grafts from a single manufacturer that are identical in design and delivery method, except for utilizing either suprarenal (SR) or infrarenal (IR) fixation. Both SR and IR groups demonstrated a significant increase in SCr and a decrease in CrCl over time. No significant difference in SCr or CrCl existed between groups during any time period. There were no differences in postoperative renal impairment, the need for hemodialysis, or systolic and diastolic blood pressure during subsequent follow-up between treatment groups. There was no significant difference in the number of renal-adverse events detected by CT between fixation systems.

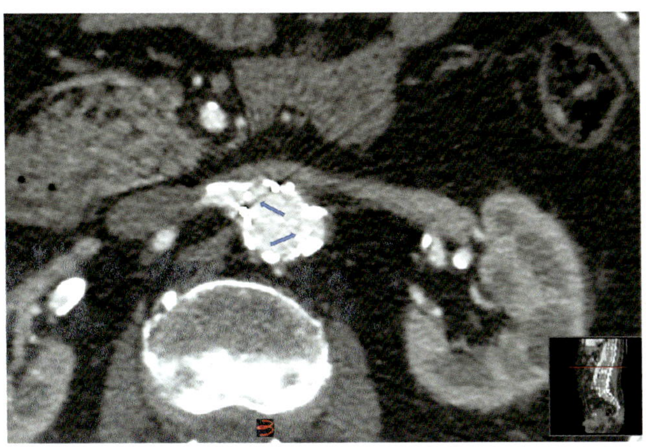

Fig. 47.3 Supra-renal stents (blue arrows) may partially occlude the origins of the renal arteries after suprarenal stent graft implantation

In a meta-analysis, Miller et al. [17] including 21 non-randomized studies comparing SR vs IR fixation representing 4474 unique patients (SR, 1949; IR, 2525). This study showed that there were no statistically significant differences in the risk of any renal complication between SR and IR fixation groups. The absolute risk difference between the SR and IR fixation groups was <1%, with no evidence of heterogeneity or publication bias for renal dysfunction, renal artery stenosis, renal artery occlusion, or new need for hemodialysis after EVAR. Calderbank et al. [18] undertook a meta-analysis of studies using estimated glomerular filtration rate (eGFR) as a measure of renal injury to assess the effect of suprarenal fixation on renal function. Five series reported eGFR reduction >20% at 1 year. The authors concluded that EVAR using suprarenal fixation does not lead to a significant drop in renal function at 1 year, based on eGFR. Nevertheless, it is important that future studies should report on long-term renal injury using eGFR and not inaccurate arbitrary measures, which are currently common in the literature.

On the other hand, Zettervall et al. [19] compared the outcomes of contemporary suprarenal and infrarenal endografts utilizing the Targeted Vascular Module of the American College of Surgeons National Surgical Quality Improvement Program (NSQIP) to identify all patients undergoing elective endovascular repairs for infrarenal AAAs from 2011 to 2013. In this study, 3587 patients were evaluated including 2273 (63%) with suprarenal grafts and 1314 (37%) with infrarenal grafts. Renal complications (1.1% vs. 0.1%, $P < 0.01$) and length of stay >2 days (34% vs. 25%, $P < 0.01$) occurred more commonly after suprarenal fixation. After adjustment, suprarenal grafts had significantly higher rates of renal complications (OR: 12.0, 95% CI: 1.6–91) and length of stay >2 days (OR: 1.4, 95% CI: 1.2–1.7). Stather et al. [20] undertook a meta-analysis of renal function following EVAR using an SR or IR device. A total of 25 non-randomized studies comparing SR with IR fixation were included, including 54,832 patients; 16,634 underwent SR and 38,198 IR. There was a small but significant difference in outcomes for renal dysfunction at the study end point (SR 5.98% vs. IR 4.83%; odds ratio [OR] 1.29, 95% confidence interval [CI] 1.18e1.40 [$p < 0.001$]); however, at 30 days and 12 months there was no significant difference, and this did not hold up to sensitivity analysis. Incidence of renal infarcts, renal stenosis, and renal artery occlusion favored IR fixation; however, there was no difference in haemodialysis rates.

Also, recent studies have showed contradicting outcomes. Banno et al. [21] included 237 patients (102 IRs and 135 SRs) showing that the use of an SR fixation device was independently predictive of a more than 20% decrease in the estimated glomerular filtration rate at 3 years after EVAR. However, Erben et al. [22] who included 1130 elective EVARs (670 [59.3%] had SF and 460 [40.7%] IF) showed that the rates of change in creatinine and estimated glomerular filtration rate (eGFR) were not statistically significant among groups (SF vs IF) during long-term follow-up of 4.8±3.7 years. Hahl et al. [23] included a total of 358 patients (267 [74.6%] had SR and 91 [25.4%] had IR fixation) and also showed that there was no difference in renal function at 30 days and 1–5 years. Regardless of the fixation method, renal function steadily decreased over time after EVAR.

Impact of Fixation System on Neck Adverse Event (Neck Enlargement; Migration; Endoleak)

Neck adverse events and stent graft behavior are the focus of research as those factors have a direct impact on the durability of the EVAR procedure. Stent graft migration is defined as distal movement >10 mm or movement ≤10 mm when resulting in secondary intervention, according to the Society for Vascular Surgery reporting standards for EVAR and ESVS guidelines [1, 2]. There is controversy on whether the type of fixation (suprarenal vs infrarenal) plays a role in migration or neck remodeling-enlargement or other potential factors such as initial diameter of aortic neck, oversize, neck angulation, or even type of stent graft but the type of fixation may also be associated with neck adverse events. Hager et al. [24] highlighted that there are no significant differences in endograft migration or in the incidence of early and late type 1a endoleak (Fig. 47.4) in respect to the fixation type even in the presence of short aortic neck. Additionally, Kouvelos et al. [25] presented a meta-analysis showing that patients with wide proximal aortic neck are associated with lower freedom from aneurysm-related re-intervention, type Ia endoleak, sac expansion, aneurysm rupture, and higher overall survival.

Malach et al. [26] compared three groups with different neck fixation. Those with active suprarenal fixation had a significant change in suprarenal aortic diameter at four-year follow-up, compared to those with active infrarenal or passive suprarenal fixation. Those with active suprarenal fixation were the only ones to have significant increase in suprarenal aortic diameter. However, Pintoux et al. [27] suggested that the inter-renal or infrarenal proximal neck dilatation does not depend on the type of proximal fixation but on anatomic factors and on the natural evolution of the aneurysmal disease. Spanos et al. [28] highlighted that proximal endograft configuration appears to have different impact on supra-aortic anatomy. A retrospective study, including EVAR patients using three types of endografts with different proximal fixation systems according to IFU (50: Ovation, Endologix, Irvin, CA; 25: Endurant IIs, Medtronic, Santa Rosa, CA; 25: Excluder C3, W. L. Gore & Associates, Flagstaff, AZ), demonstrated that the suprarenal aortic diameter was significantly increased in Ovation patients in comparison to the other two endografts. The reduction of the suprarenal angulation was similar, while infrarenal angulation was significantly more decreased in Ovation endograft than the other two systems.

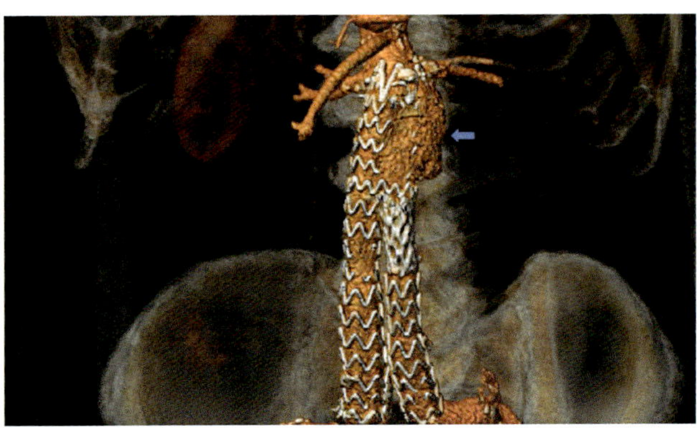

Fig. 47.4 Migration of the endograft and type I endoleak (blue arrow)

Oliveira et al. [29] highlighted that besides the satisfactory early and mid-term results, a higher rate of EL1a was identified among patients with severely angulated necks in the long term. However, mortality was not affected by this difference. These findings suggest that EVAR should be used judiciously in patients with extreme angulation of the proximal neck and highlight the need for close follow-up of EVAR, especially in the long term and in patients treated with all types of fixation.

In conclusion, there is no available strong evidence to show superiority of any type of fixation and therefore an individualized strategy respecting the IFU of each graft can provide the most favorable outcome. Finally, proper follow-up is of paramount importance.

References

1. Wanhainen A, Verzini F, Van Herzeele I, Allaire E, Bown M, Cohnert T, et al. Editor's choice—European Society for Vascular Surgery (ESVS) 2019 clinical practice guidelines on the management of abdominal aorto-iliac artery aneurysms. Eur J Vasc Endovasc Surg. 2019;57:8–93.
2. Chaikof EL, Dalman RL, Eskandari MK, Jackson BM, Lee WA, Mansour MA, et al. The Society for Vascular Surgery practice guidelines on the care of patients with an abdominal aortic aneurysm. J Vasc Surg. 2018;67:2–77.e2.
3. Prinssen M, Verhoeven EL, Buth J, Cuypers PW, van Sambeek MR, Balm R, et al. A randomized trial comparing conventional and endovascular repair of abdominal aortic aneurysms. N Engl J Med. 2004;351:1607–18.
4. Greenhalgh RM, Brown LC, Kwong GP, Powell JT, Thompson SG, et al. Comparison of endovascular aneurysm repair with open repair in patients with abdominal aortic aneurysm (EVAR trial 1), 30-day operative mortality results: randomised controlled trial. Lancet. 2004;364:843–8.
5. Lederle FA, Freischlag JA, Kyriakides TC, Padberg FT Jr, Matsumura JS, Kohler TR, et al. Outcomes following endovascular vs open repair of abdominal aortic aneurysm: a randomized trial. JAMA. 2009;302:1535–42.

6. Nana P, Spanos K, Kouvelos G, Stamoulis K, Rountas C, Arnaoutoglou E, Matsagkas M, Giannoukas AD. Ten-year single center experience in elective standard endovascular abdominal aortic aneurysm repair. Int Angiol. 2021;40(3):240–7. https://doi.org/10.23736/S0392-9590.21.04648-4. Epub 2021 Mar 19. PMID: 33739077.
7. Böckler D, Power AH, Bouwman LH, van Sterkenburg S, Bosiers M, Peeters P, Teijink JA, Verhagen HJ, ENGAGE Investigators. Improvements in patient outcomes with next generation endovascular aortic repair devices in the ENGAGE global registry and the EVAR-1 clinical trial. J Cardiovasc Surg (Torino). 2020;61(5):604–9. https://doi.org/10.23736/S0021-9509.19.11021-X. Epub 2019 Jul 9. PMID: 31293141.
8. Teijink JAW, Power AH, Böckler D, Peeters P, van Sterkenburg S, Bouwman LH, Verhagen HJ, Bosiers M, Riambau V, Becquemin JP, Cuypers P, van Sambeek M. Editor's choice—five year outcomes of the Endurant stent graft for endovascular abdominal aortic aneurysm repair in the ENGAGE registry. Eur J Vasc Endovasc Surg. 2019;58(2):175–81. https://doi.org/10.1016/j.ejvs.2019.01.008. Epub 2019 Jun 22. PMID: 31235305.
9. Mwipatayi BP, Faraj J, Oshin O, Fitridge R, Wong J, Schermerhorn ML, Becquemin JP, Boeckler D, Riambau V, Teijink JA, van Sambeek MRHM, Verhagen H, ENGAGE REGISTRY Co-Investigators. Endurant stent graft demonstrates promising outcomes in challenging abdominal aortic aneurysm anatomy. J Vasc Surg. 2021;73(1):69–80. https://doi.org/10.1016/j.jvs.2020.04.508. Epub 2020 May 19. PMID: 32442605.
10. Liang NL, Ohki T, Ouriel K, Teigen C, Fry D, Henretta J, Komori K, Kichikawa K, Makaroun MS, INSPIRATION Investigators. Five-year results of the INSPIRATION study for the INCRAFT low-profile endovascular aortic stent graft system. J Vasc Surg. 2021;73(3):867–873.e2. https://doi.org/10.1016/j.jvs.2020.06.128. Epub 2020 Jul 21. PMID: 32707389.
11. Swerdlow NJ, Lyden SP, Verhagen HJM, Schermerhorn ML. Five-year results of endovascular abdominal aortic aneurysm repair with the ovation abdominal stent graft. J Vasc Surg. 2020;71(5):1528–1537.e2. https://doi.org/10.1016/j.jvs.2019.06.196. Epub 2019 Sep 9. PMID: 31515176.
12. Verzini F, Romano L, Parlani G, Isernia G, Simonte G, Loschi D, Lenti M, Cao P. Fourteen-year outcomes of abdominal aortic endovascular repair with the Zenith stent graft. J Vasc Surg. 2017;65(2):318–29. https://doi.org/10.1016/j.jvs.2016.07.117. Epub 2016 Nov 7. PMID: 27832988.
13. Freyrie A, Gallitto E, Gargiulo M, Mascoli C, Faggioli G, Pini R, Pratesi C, Stella A, Anaconda™ Italian Registry Participating Physicians. Proximal aortic neck angle does not affect early and late EVAR outcomes: an AnacondaTM Italian registry analysis. J Cardiovasc Surg (Torino). 2014;55(5):671–7. Epub 2014 Mar 26. PMID: 24667340.
14. Barry IP, Barns M, Verhoeven E, Wong J, Dubenec S, Heyligers JM, Milner R, Shutze WP, Bachoo P, Vlaskovky P, Mwipatayi BP, GREAT Participants. Excluder stent graft-related outcomes in patients with aortic neck anatomy outside of instructions for use (IFU) within the Global Registry for Endovascular Aortic Treatment (GREAT): mid-term follow-up results. Ann Vasc Surg. 2021;76:222–31. https://doi.org/10.1016/j.avsg.2021.04.032. Epub 2021 Jun 25. PMID: 34182115.
15. Walsh SR, Tang TY, Boyle JR. Renal consequences of endovascular abdominal aortic aneurysm repair. J Endovasc Ther. 2008;15:73–82.
16. Parmer SS, Carpenter JP, Endologix Investigators. Endovascular aneurysm repair with suprarenal vs infrarenal fixation: a study of renal effects. J Vasc Surg. 2006;43(1):19–25. https://doi.org/10.1016/j.jvs.2005.09.025. PMID: 16414382.
17. Miller LE, Razavi MK, Lal BK. Suprarenal versus infrarenal stent graft fixation on renal complications after endovascular aneurysm repair. J Vasc Surg. 2015;61(5):1340–9.e1. https://doi.org/10.1016/j.jvs.2015.01.037. Epub 2015 Feb 27. PMID: 25736780.
18. Calderbank T, Bown M, Saratzis A. The impact of suprarenal fixation on renal function following endovascular abdominal aortic aneurysm repair: meta-analysis based on estimated glomerular filtration rate. Eur J Vasc Endovasc Surg. 2018;56(4):497–506. https://doi.org/10.1016/j.ejvs.2018.02.012. Epub 2018 Mar 23. PMID: 29576336.
19. Zettervall SL, Soden PA, Deery SE, Ultee K, Shean KE, Shuja F, Amdur RL, Schermerhorn ML. Comparison of renal complications between endografts with suprarenal and infrarenal fixation. Eur J Vasc Endovasc Surg. 2017;54(1):5–11. https://doi.org/10.1016/j.ejvs.2017.02.001. Epub 2017 Mar 6. PMID: 28279653; PMCID: PMC5482762.
20. Stather PW, Ferguson J, Awopetu A, Boyle JR. Meta-analysis of renal function following infrarenal EVAR using suprarenal or infrarenal fixation devices. Eur J Vasc Endovasc Surg. 2018;56(4):486–96. https://doi.org/10.1016/j.ejvs.2018.01.021. Epub 2018 Mar 3. PMID: 29506944.
21. Banno H, Ikeda S, Kawai Y, Fujii T, Akita N, Takahashi N, Sugimoto M, Kodama A, Komori K. Suprarenal fixation is associated with worse midterm renal function after endovascular abdominal aortic aneurysm repair compared with infrarenal fixation. J Vasc Surg. 2020;71(2):450–6. https://doi.org/10.1016/j.jvs.2019.03.061. Epub 2019 Jun 24. PMID: 31248765.
22. Erben Y, Li Y, Mao MA, Hamid OS, Franco-Mesa C, Da Rocha-Franco JA, Stone W, Fowl RJ, Oldenburg WA, Farres H, Meltzer AJ, Gloviczki P, De Martino RR, Bower TC, Kalra M, Oderich GS, Hakaim AG. Proximal fixation of endovascular aortic device may not be associated with renal function decline after abdominal aortic aneurysm repair. J Vasc Surg. 2021;74(6):1861–1866.e1. https://doi.org/10.1016/j.jvs.2021.05.050. Epub 2021 Jun 26. PMID: 34182031.
23. Hahl T, Kurumaa T, Uurto I, Protto S, Väärämäki S, Suominen V. The effect of suprarenal graft fixation during endovascular aneurysm repair on short- and long-term renal function. J Vasc Surg. 2022;76(1):96–103.e1. https://doi.org/10.1016/j.jvs.2021.12.081. S0741-5214(22)00105-7. Epub ahead of print. PMID: 35074412.
24. Hager ES, Cho JS, Makaroun MS, Park SC, Chaer R, Marone L, Rhee RY. Endografts with suprarenal fixation do not perform better than those with infrarenal fixation in the treatment of patients with short straight proximal aortic necks. J Vasc Surg. 2012;55(5):1242–6. https://doi.org/10.1016/j.jvs.2011.11.088. Epub 2012 Jan 24. PMID: 22277692.
25. Kouvelos GN, Antoniou G, Spanos K, Giannoukas A, Matsagkas M. Endovascular aneurysm repair in patients with a wide proximal aortic neck: a systematic review and meta-analysis of comparative studies. J Cardiovasc Surg. 2019;60(2):167–74. https://doi.org/10.23736/S0021-9509.19.10869-5. Epub 2019 Jan 18. PMID: 30665285.
26. Malach L, Tehrani N, Kolachina S, Krawczyk K, Wozniak A, Soult M, Aulivola B, Bechara CF. Effect of stent-graft active fixation and oversizing on aortic neck dilation after endovascular aneurysm exclusion for infrarenal aortic aneurysm. Ann Vasc Surg. 2022;79:100–5. https://doi.org/10.1016/j.avsg.2021.08.026. Epub 2021 Oct 14. PMID: 34656723.
27. Pintoux D, Chaillou P, Azema L, Bizouarn P, Costargent A, Patra P, Gouëffic Y. Long-term influence of suprarenal or infrarenal fixation on proximal neck dilatation and stent graft migration after EVAR. Ann Vasc Surg. 2011;25(8):1012–9. https://doi.org/10.1016/j.avsg.2010.08.013. PMID: 22023937.
28. Spanos K, Kouvelos G, Kontopodis N, Ioannou CV, Matsagkas M, Giannoukas AD. Suprarenal aortic remodeling after endovascular aortic aneurysm repair among three endografts with different types of proximal fixation system. Ann Vasc Surg. 2019;61:341–9.

https://doi.org/10.1016/j.avsg.2019.05.048. Epub 2019 Aug 5. PMID: 31394244.

29. Oliveira NFG, Gonçalves FB, Hoeks SE, Josee van Rijn M, Ultee K, Pinto JP, Raa ST, van Herwaarden JA, de Vries JPM, Verhagen HJM. Long-term outcomes of standard endovascular aneurysm repair in patients with severe neck angulation. J Vasc Surg. 2018;68(6):1725–35. https://doi.org/10.1016/j.jvs.2018.03.427. Epub 2018 Jun 15. PMID: 29914837.

The Role of Aorto-Uni-iliac Endografts

48

A. Svetlikov, G. Khubulava, G. Gorbunov, V. Gurevich, and Y. Nakatis

Introduction

Endovascular aortic aneurysm repair (EVAR) is a relative new weapon in the vascular surgeon's armamentarium [1–6] and studies have shown that aorto-uni-iliac (AUI) devices coupled with occlusion of contralateral iliac artery and cross-over femoral-femoral or iliac-femoral bypass, have comparable outcome in selected patients to treatment with bifurcated devices. Although the use of this strategy is limited in daily practice for the treatment of abdominal aortic aneurysms (AAA) in healthy patients, this may not be the case for the treatment of high-risk patients with complex iliac anatomy, in the treatment of r-AAA, or in the treatment of complications of previously applied endovascular bifurcated devices for the treatment of AAA. The purpose of this chapter is to review the relevant literature and discuss the outcomes on the use of AUI stent grafts.

Despite this potential drawback of using an extra-anatomic bypass, deployment of an AUI stent graft device is still being used for the treatment of abdominal aortic aneurysms in different and complicated circumstances, such as presence of narrow terminal aorta; small diameter, tortuous, kinked, calcified, or even occluded contralateral iliac artery; presence of common iliac aneurysm; treatment of ruptured abdominal aortic aneurysms (rAAA); and management of endoleaks of previously implanted endoprosthesis.

Many surgeons, however, debate the disadvantages of the method due to the extra-anatomic bypass and express doubts concerning the long-term results, claiming mainly poor femoral-femoral bypass outcome, which varies in different reports from 35 to 92%, for 5-year patency rates [1–9].

Since the earlier experiences with EVAR, access problems due to small and diseased iliac arteries have been one of the most significant causes of morbidity. The problem of a narrow distal aorta, often encountered in patients with saccular aneurysms and penetrating aortic ulcers, and in those with concomitant aorto-iliac occlusive disease, was traditionally treated by an aorto-uni-iliac converter with femoral crossover bypass grafts. Although this operation has been proven safe and effective, there are potential negative implications related to the open surgical portion of the operation and the use of a prosthetic graft in the groin, including risk of graft infection, narrowing or occlusion, and degeneration over time. Patency of the contralateral internal iliac artery relies on retrograde flow via the external iliac artery, with reports indicating higher rates of long-term occlusion [10–12].

Use of Aorto-Uni-iliac Device for the Treatment of Ruptured Abdominal Aortic Aneurysms

The first ruptured AAA was treated by Veith in 1994, and in the same year Yusuf et al. published the first case. Since then several authors have published their experience with endovascular repair of ruptured abdominal aortic aneurysms (rAAA). Despite encouraging results, the proportion of patients with rAAA being treated by EVAR remains relatively low. This may be because of logistical problems

A. Svetlikov (✉)
Division of Faculty Surgery, Cardiovascular Surgery Department, North Western Scientific Clinical Center of Federal Medical Biological Agency of Russia, St. Petersburg State University, St. Petersburg, Russia

G. Khubulava
Cardiovascular Surgery Department, The First I.P. Pavlov St. Petersburg Medical University, St. Petersburg, Russia

G. Gorbunov
Cardiovascular Surgery Department, North Western I.I Mechnikov State Medical University, St. Petersburg, Russia

V. Gurevich
North Western Scientific Clinical Center of Federal Medical Biological Agency of Russia, Atherosclerosis Treatment, St. Petersburg, Russia

Y. Nakatis
Vascular Surgery Department, North Western Scientific Clinical Center of Federal Medical Biological Agency of Russia, St. Petersburg, Russia

G. Geroulakos et al. (eds.), *Mastering Endovascular Techniques*, https://doi.org/10.1007/978-3-031-42735-0_48

including the availability of an experienced endovascular team and a sufficient stock of suitable grafts. To enable EVAR for rAAA, it has been suggested that endovascular stent grafts must be immediately available in a wide range of sizes. This would require a large stock of bifurcated stent grafts and may not, for financial and storage reasons, be possible in most vascular units. The modular AUI stent-graft system comprising 16 components could replace a total of 750 unitary bifurcated systems [13–17].

Durability of the Crossover Femoral-Femoral Bypass Graft

One of the largest drawbacks to the use of AUI procedures is the reliance on extra-anatomic bypass grafts to revascularize the contralateral limb. The patency of crossover iliac-femoral or femoral-femoral bypass is acceptable when used for occlusive disease. Previous reports on crossover femoral-femoral bypass graft patency have been related to patients with occlusive arterial disease. Patency rates for femoral-femoral bypass grafts vary in the literature. R.L. Ng et al. reported a cumulative patency rate at 6 years of 92% for femoral-femoral crossover procedures and a low early mortality rate of 1.3%, whereas A.J. Lamerton et al. showed a cumulative patency rate of 60% at 5 years [17, 18].

The primary indications today for the treatment of AAA by EVAR with the AUI configuration as they were proposed in the reviewed studies are the following:

1. Distance to lower renal artery to aortic bifurcation <70 mm
2. Narrow terminal aorta <15 mm (transverse diameter)
3. Contralateral common iliac artery angle >90° from the longitudinal axis of the aneurysm
4. Obstructed contralateral common iliac artery
5. Isolated infrarenal abdominal aortic dissections
6. Combination of the previous indications
7. Conversion to the AUI configuration while deploying a bifurcated endograft due to impossible contralateral limb catheterization
8. Ruptured AAA
9. Concomitant ecstatic or frankly aneurysmal bilateral common iliac arteries, unless the patient has indispensable internal iliac arteries (relative indication)
10. Heavily calcified contralateral external iliac artery (relative indication)
11. Narrow contralateral common iliac artery (diameter <5 mm), with or without previous transluminal angioplasty (relative indication)
12. Previous stents to the iliac arteries (relative indication)

Advantages of Aorto-Uni-Iliac Stent Grafts

The delivery system is usually simple either in custom-made or commercial devices. No stent graft orientation is needed, and no limb cannulation is required, and thus fewer intraortic manipulations in 3-dimentional space under 2-dimensional fluoroscopic guidance and decreased risk of embolization. Less contrast agent is needed. Home-made systems have several advantages even compared to commercial ones; they can be customized, pre- or intraoperatively, to deal with any unusual circumstances and can quickly incorporate lessons learned intraoperatively. The presence of a separate occluder for the contralateral iliac artery occlusion is an advantage of the method. The common iliac occluder is independent from the primary stent graft in both size and position. The two parts are separate, so that occluder diameter can vary widely regardless of the diameter of the primary stent graft. It can also be placed exactly where it is needed regardless of the position of the primary stent graft.

Disadvantages of Aorto-Uni-Iliac Stent Grafts

Of course, the simplicity of AUI stent-graft deployment comes at a price. An occlusion of the contralateral iliac artery is required so as to seal the aneurysm associated with a femoral-femoral bypass in order to reestablish perfusion to the contralateral limb. This may put the patient at risk, as AUI stent graft channels all the blood flow into one iliac artery and from there to the pelvis. Even widely patented AUI stent grafts can limit flow enough to cause claudication if the external iliac artery is small and the patient active. The patency of crossover iliofemoral or femoro-femoral bypass is very high as demonstrated in this review.

Early critics pointed to poor long-term patency rates of the femoro-femoral bypass, but more recent data have dispelled this argument pointing to excellent long-term patency of the femoro-femoral portion of AUI repairs. Fortunately, the patency of crossover femoro-femoral bypass is very high in recent published literatures. R.L. Ng et al. reported a cumulative patency rate at 6 years of 92% for femoro-femoral crossover procedures and a low early mortality rate of 1.3%. Also, despite the fact that these procedures might be complicated by the development of graft infection, graft occlusion, false aneurysm formation, or seromas in the groin, their incidences are very low [18–20].

Hynes and Sultan reported that AUI EVAR with a fem-fem crossover is a safe and effective alternative to bifurcated endografts in high-risk patients with AAA. This is the same finding as in our study. This is explained by the fact that AUI EVAR requires less preoperative planning, less operative time, and less trauma. In our study, the mean operative time was 114 min in the fem-fem group and 79 min in the non-fem-fem group ($P = 0.013$). This short operative time is due to the omission of the bypass step. This also enabled less blood loss and less anesthesia time for those patients, which is more convenient when dealing with the high risk associated with severe comorbidities or ruptured AAAs. Jean-Baptiste et al. reported higher incidence of complications with AUI EVAR compared to bifurcated endografts in high-risk patients. This could be attributed to the fact that AUI grafts were considered in this study for higher-risk patients, while bifurcated grafts were used in healthier patients. Hynes and Sultan published results in high-risk patients for bimodular devices which mirror the results of AUI with or without crossover but rebuff the finding of Jean-Baptiste that AUI EVAR is inferior to bimodular configuration. Carrafiello et al. reported high mortality rate with AUI grafts, but a deeper look at this study design demonstrates that precious time was wasted in preoperative analysis and preoperative CT angiography in these unstable patients. A number of studies suggested that the use of AUI instead of the bifurcated endografts might be attributed to its easier preoperative planning and intraoperative application, especially in centers with less experience with EVAR. Katsikas et al. documented through meta-analysis that the main advantages of the AUI endograft are its simplicity and versatility. In 2004, Arko et al. demonstrated that 55% of patients considered for endovascular AAA repair met the anatomical selection criteria, with men twice as likely as women. Conversely, if we applied AUI options for those 220 patients, all can be managed routinely with this configuration. Currently up to 95% of our AAA patients are deemed suitable for EVAR. Clouse et al. proved that AUI with fem-fem crossover graft is a safe, effective option with satisfactory midterm results. Five of their patients had chronic contralateral iliac occlusion but they performed fem-fem crossover bypass [21–24].

References

1. Piotrovsky JJ, Pear WH, Jones DN, Whitehill T, Bell R, Patt A, Rutherford RB. Aortobifemoral bypass: the operation of choice for unilateral iliac occlusion? J Vasc Surg. 1988;8:211–8.
2. Plecha FT, Plecha FM. Femorofemoral bypass grafts: ten-year experience. J Vasc Surg. 1984;1:555–61.
3. Flanigan DP, Pratt DG, Goodreau JJ, Burnham SJ, Yao JS, Betgan JJ. Hemodynamic and angiographic guidelines in selection of patients for femorofemoral bypass. Arch Surg. 1978;113:1257–62.
4. Eugene J, Goldstone J, Moore WS. Fifteen year experience with subcutaneous bypass grafts for lower extremity ischemia. Ann Surg. 1977;186:177–83.
5. Brief DK, Brener BJ. Extra-anatomic bypasses: femorofemoral crossover grafts. In: Wilson SE, Veith FJ, Hobson RW, editors. Vascular surgery: principles and practice. New York: McGraw-Hill; 1987. p. 415–8.
6. Dick LS, Brief DK, Alpert J, Dick LS, Brief DK, Alpert J, Brener BJ, Goldenkranz R, Parsonet V. A 12-year experience with femorofemoral crossover grafts. Arch Surg. 1980;115:1359–65.
7. Mosley JG, Marston A. Long term results of 66 femorofemoral bypass grafts: 9-year follow-up. Br J Surg. 1983;70:631–4.
8. Maini BS, Mannick JA. Effect of arterial reconstruction on limb salvage: a ten-year appraisal. Arch Surg. 1978;113:1297–304.
9. Chuter TA, Faruqi RM, Reilly LM, Kerlan RK, Sawhney R, Wall SD, et al. Aortomonoiliac endovascular grafting combined with femorofemoral bypass: an acceptable compromise or a preferred solution? Semin Vasc Surg. 1999;12:176–81.
10. Criado E, Burnham SJ, Tinsley EA Jr, Johnson G Jr, Keagy BA. Femorofemoral bypass graft: analysis of patency and factors influencing long-term outcome. J Vasc Surg. 1993;18:495–505.
11. Clouse WD, Brewster DC, Marone LK, Cambria RP, LaMuraglia GM, Watkins MT, et al. Durability of aortouniiliac endografting with femorofemoral crossover: 4-year experience in the EVT/Guidant trials. J Vasc Surg. 2003;37:1142–9.
12. Dalanis I, Nano G, Bianchi P, Stegher S, Casana R, Malagrida G, Tealdi DG. Endovascular techniques for the treatment of ruptured abdominal aortic aneurysms: 7-year intention-to-treat results. World J Surg. 2006;30:1809–14; discussion 1815-1816.
13. Peppelenbosch N, Geelkerken RH, Soong C, Cao P, Steimetz OK, Teijnk JA, Lepantalo M, De Letter J, Vermassen FE, Derose G, Buskens E, Buth J. Endograft treatment of ruptured abdominal aortic aneurysms using the talent aortouniiliac system: an international multicenter study. J Vasc Surg. 2006;43:1111–22.
14. Lee WA, Huber TS, Hirneise CM, Berceli SA, Seeger JM. Eligibility rates of ruptured and symptomatic AAA for endovascular repair. J Endovasc Ther. 2002;9:436–42.
15. Leon LR, Labropoulos N, Laredo J, Rodrigues HE, Kalman PG. To what extent has endovascular aneurysm repair influenced abdominal aorticaneurysm management in the state of Illinois? J Vasc Surg. 2005;41:568–74.
16. Armon MP, Yusuf SW, Whitaker SC, Gregson RH, Wenham PW, Hopkinson BR. The anatomy of abdominal aortic aneurysms: implications for sizing of endovascular grafts. Eur J Vasc Endovasc Surg. 1997;13:398–402.
17. Lamerton AJ, Nikolaides AN, Eastcott HH. The femorofemoral graft. Haemodynamic improvement and patency rate. Arch Surg. 1985;120:1274–8.
18. Ng RL, Gillies TE, Davies AH, Baird RN, Horrocks M. Iliofemoral versus femorofemoral bypass: a 6-year audit. Br J Surg. 1992;79:1011–3.
19. Lipsitz EC, Ohki T, Veith FJ, Rhee SJ, Gargiulo NJ 3rd, Suggs WD, et al. Patency rates of femorofemoral bypasses associated with endovascular aneurysm repair surpass those performed for occlusive disease. J Endovasc Ther. 2003;10:1061–5.
20. Arko FR, Filis KA, Seidel SA, et al. How many patients with infrarenal aneurysms are candidates for endovascular repair? The northern California experience. J Endovasc Ther. 2004;11(1):33–40.
21. Katsikas VC, Dalainas I, Martinakis VG, Xiromeritis K. The role of aortouniiliac devices in the treatment of aneurysmal disease. Eur Rev Med Pharmacol Sci. 2012;16(8):1061–71.
22. Elixen JB, Batt M, Azzaoui R, Koussa M, Hassen-Khodja R, Haulon S. A comparison of the mid-term results following the use of bifur-

cated and aorto-uni-iliac devices in the treatment of abdominal aortic aneurysms. Eur J Vasc Endovasc Surg. 2009;38(3):298–304.
23. Bonardelli S, Nodari F, De Lucia M, Cervi E, Giulini SM. Crossover ilio-iliac bypass and removal of femoro-femoral graft as first treatment for the infection of crossover bypass in aorto-uni-iliac endovascular aneurysm repair. Vascular. 2012;20(6):306–10.
24. Hynes N, Sultan S. A prospective clinical, economic, and quality-of-life analysis comparing endovascular aneurysm repair (EVAR), open repair, and best medical treatment in high-risk patients with abdominal aortic aneurysms suitable for EVAR: the Irish patient trial. J Endovasc Ther. 2007;14(6):763–76.

Hostile Proximal Neck: Angulated, Short, or Conical

49

J. Shea, A. L. Pouncey, and C. Bicknell

Case Presentation

A 90-year-old male was admitted to the emergency room from home with back pain and a large tender expansile mass on abdominal examination. He was otherwise fit with a past medical history of vitamin D deficiency. He was a non-smoker, with no other risk factors for atherosclerosis.

A CT angiogram confirmed the presence of a 91 mm aneurysm with significant 83-degree (beta) angulation at the neck (Fig. 49.1).

Measurements were taken using 3D planning software. The neck length was 31.5 mm, and the neck diameter was 30 mm. The common iliac arteries were 14 mm diameter on the right and 17 mm diameter on the left. The external iliac arteries were 7 mm in diameter.

Continued at page 519

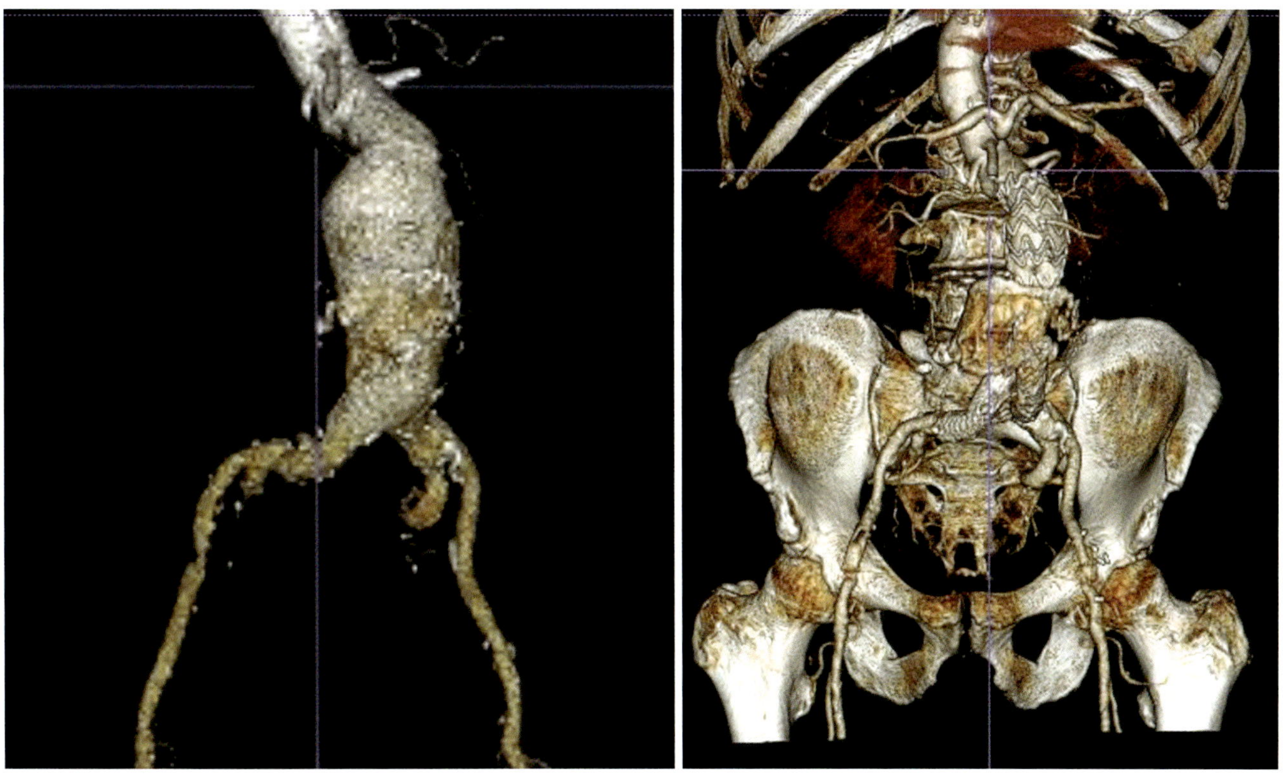

Fig. 49.1 CT angiogram demonstrating significant beta angulation of the aneurysm neck

J. Shea · A. L. Pouncey · C. Bicknell (✉)
Imperial Vascular Unit, London, UK
e-mail: jessica.shea@nhs.net; a.pouncey@imperial.ac.uk; colin.bicknell@imperial.ac.uk

G. Geroulakos et al. (eds.), *Mastering Endovascular Techniques*, https://doi.org/10.1007/978-3-031-42735-0_49

Background

For an infrarenal abdominal aortic aneurysm (IRAAA), the aneurysm neck is defined as the segment of aorta between the lowest renal artery and commencement of the aneurysm along the centreline of the aorta. This region is used as the landing zone for a stent-graft in endovascular aortic repair (EVAR); or, in open repair, for placement of an aortic clamp and anastomosis of the graft [1]. For EVAR, the optimal IRAAA neck is a uniform, straight, smooth cylinder of sufficient length and appropriate diameter, which enables stable fixation and complete sealing of the stent-graft to the vessel wall. In light of this, stent-graft manufacturers have set device-specific IFU (instructions for use) thresholds, defining appropriate neck anatomical characteristics needed to ensure adequate sealing of the device. These criteria are based on neck length, diameter (distal and proximal), and angulation (alpha and beta), as well as the degree of calcification and thrombus burden. Deployment outside of these anatomical constraints is associated with a greater risk of graft-related adverse events, in particular loss of an adequate sealing zone leading to type 1a endoleak, stent-graft migration, sac growth, secondary rupture, need for re-intervention and aneurysm-related mortality [2] (Fig. 49.2).

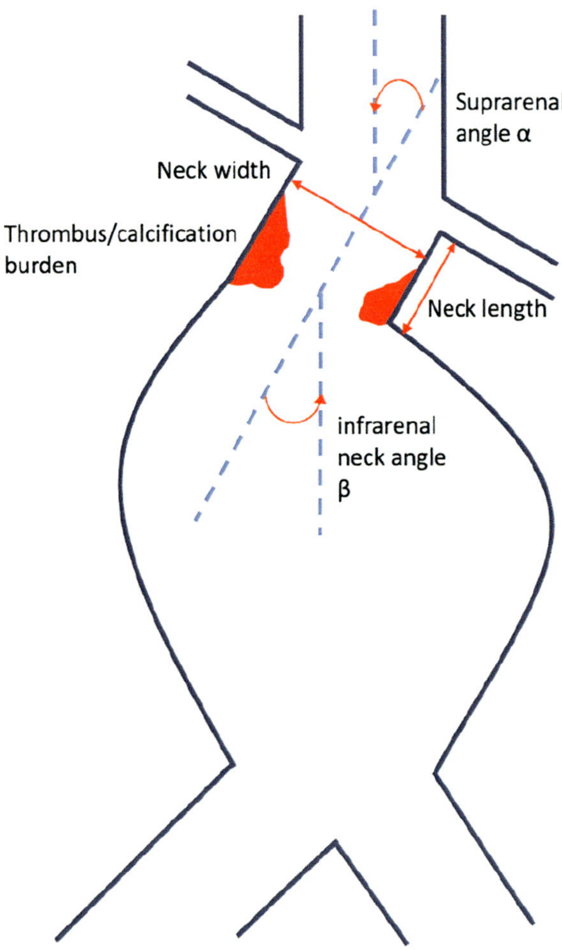

Fig. 49.2 Anatomical characteristics of aneurysm necks

The Hostile Neck

Anatomical Features and Associated Outcomes

Hostile neck anatomy (HNA) denotes anatomically unfavourable anatomy for endovascular repair (EVAR). It is the most frequent determinant of ineligibility for EVAR [3]. HNA is associated with increased technical difficulty and a greater risk type 1a endoleak (loss of the proximal seal), reintervention, secondary rupture, renal injury, and death [4, 5]. A meta-analysis of seven observational studies has noted that, following EVAR, patients with HNA had over four times the risk of a type 1a endoleak and nine times the risk of aneurysm-related mortality within 1 year compared to those with favourable anatomy [6].

The precise anatomical definition of a hostile neck varies from study to study and depends on the constraints defined by IFU for the device deployed [7]. Stather et al. defined HNA as one of a neck length of <15 mm, neck diameter of >28 mm, and/or angulation of ≥60°, noting that these features were associated with greater technical difficulty and worse short-term outcomes [4]. More extensive criteria were used by AbuRahma et al., who defined HNA as one of a length of <10 mm, a diameter >28 mm, an angulation of >60°, ≥50% thrombus, ≥50% calcification, and a conical neck (reverse taper), noting an associated increase in the rate of type 1 endoleak, reintervention, and peri-operative complications [8]. Expert consensus has since defined HNA as one of five relevant anatomical parameters: aortic neck length, angulation, diameter, conical neck, and the presence of circumferential calcification [9]. These features do not occur in isolation, and the effects are cumulative, with a resultant decrease in the likelihood of technical and clinical success [4, 10]. Selected features of hostile neck anatomy—short neck, wide neck, angulated and conical neck (SWAC)—are described below (Fig. 49.3):

Short Neck

For a standard EVAR to remain securely in position the sealing zone of the stent-graft must lie in direct apposition to the vessel wall for a certain distance, exerting an even radial force. A short neck produces challenges to stent-graft positioning and obtaining an early seal; therefore, type 1a endoleaks on the table are common; long-term fixation and consequent migration; and long-term sealing as minimal aortic neck degeneration and dilatation can lead to type 1a endoleak and long-term graft failure.

The minimum neck length required for optimal device fixation is dependent on the individual characteristics of the stent-graft used. Although a short neck is often defined as <15 mm, necks with a length between 10 and 15 mm can often be treated successfully with off-the-shelf devices and accurate deployment [11, 12]. However, approximately 41% of patients present with neck lengths outside that of the tra-

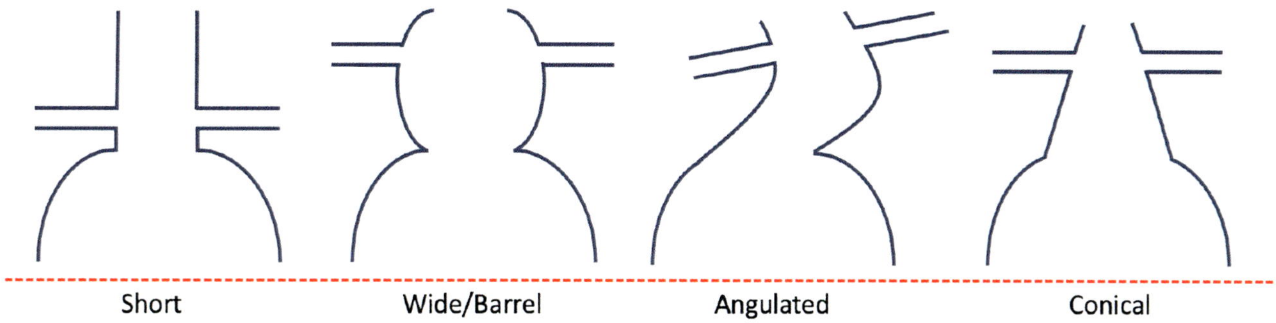

Fig. 49.3 Hostile neck characteristics

ditional IFU for most aortic endografts [13], and a short neck is the most common cause of ineligibility for endovascular repair [14]. When standard EVAR is utilised for the treatment of an IRAAA with a short neck, it is associated with increased risk of type 1 endoleak and stent-graft migration [15]. Moreover, regardless of technical outcome, a two-fold increase in all-cause mortality for patients with short necks at 5 years is observed (adjusted hazard ratio [HR] 2, confidence interval 1.02–3.8, $p = 0.04$) [16].

Wide Neck

A neck diameter of greater than 30 mm is by definition aneurysmal aortic tissue. A wide neck will not necessarily produce challenges to graft positioning and obtaining an early seal, but, by definition, the wide neck is diseased and will degenerate in time most likely, with long-term fixation issues and consequent migration and long-term sealing issues leading to graft failure.

While not necessarily presenting an increase in technical difficulty, or associated with adverse short-term outcomes, in the median term, an infrarenal neck diameter of >30 mm is associated with an increased risk of neck-related adverse events, type 1a endoleak, and secondary intervention [17]. Moreover, an increase in additional post-EVAR aortic neck dilation is also observed (odds ratio [OR] 3.1, $p = 0.02$) [18].

Angulation

Aortic neck angulation is measured perpendicular to the aortic axis, in the middle of the flexure [19]. Angulated necks produce challenges to stent-graft positioning around the curvature of the aorta and obtaining an early seal, long-term fixation, and long-term sealing.

Significant angulation frequently occurs with concomitant adverse anatomical features, and pre-operative assessment of the "functional" neck, the length used for the sealing of the stent-graft, can be difficult [20]. Aortic neck angulation (β angle) refers to the angle between the neck and the long axis of the aneurysm. An angle of >60° is associated with an increased use of intraoperative adjunctive procedures [21]; persistent type 1a endoleak following EVAR [22, 23]; and increased risk of peri-operative complications (OR 3.38, $p = 0.009$) [7, 8]. On meta-analysis of six observational studies, a higher rate of type 1a endoleak at 3 years (5.6% vs. 2.6%, $p < 0.0001$) and an increased rate of neck-related secondary intervention (13.1% vs. 9%, $p < 0.05$) were associated with severe neck angulation (>60°); but aneurysm sac increase, rupture, and all-cause mortality were not significantly different at mid-term [24]. Suprarenal neck angulation (α angle) refers to the angle between the suprarenal and infrarenal aorta. It affects feasibility of suprarenal fixation, and following EVAR is associated with an increased risk of technical failure, peri-operative mortality (11.1% vs. 0.25%, $p < 0.001$), type 1a endoleak (19.8% vs. 2.2%, $p < 0.001$), stent migration (7.6% vs. 0%, $p < 0.001$), and a reduction in long-term survival [25].

Conical Neck

If the cylindrical stent-graft is not in perfect apposition to the aortic wall, the risk of failure of sealing is substantial. The conical neck provides challenges to the endovascular surgeon in appropriately sizing the graft to gain an initial seal, and just as the wide-necked aneurysms, this portion is diseased, and so may degenerate in time.

This is a problem for a significant proportion of patients. Within the Gore Global Registry for Endovascular Aortic Treatment (GREAT), 42.6% (1312/3077) patients were noted to have non-cylindrical necks [26]. A conical proximal neck is defined as a cone-shaped proximal neck with >10% increase in vessel diameter within the 15 mm length of the proximal landing zone. The presence of a conical neck makes sizing of the stent-graft diameter more difficult, reduces the length of adequate seal, and consequently is associated with an increased risk of type 1 endoleak (OR 5.25, $p < 0.0001$) and need for aortic cuff (OR 5.95, $p < 0.0001$) [8]. The conical neck is purported to be the strongest factor associated with the failure of standard EVAR [27].

Who Does HNA Affect?

Neck anatomy is one of the most important factors governing qualification for treatment within IFU and therefore eligibility for EVAR [3, 28], with HNA resulting in patients being turned down for AAA repair, necessitating open repair, or leading to a substantial proportion of patients treated outside of IFU [29, 30].

Neck anatomy varies with aneurysm size [31, 32], sex [33], and ethnicity [34]. Hostile neck anatomy has been reported to occur in 34–66% of patients and more frequently in patients with an IRAAA of >55 mm diameter [30, 32, 35]. In a retrospective study of 14,739 IRAAA patients in the Vascular Quality Initiative (VQI), 19.6% had a short neck (<15 mm), 6.3% had severe angulation (>60°), and 9.1% had a wide neck (>30 mm) [35].

It is well recognised that women are less likely to be eligible for EVAR (34% vs. 54%, OR 0.44, 95% CI [0.32–0.62]) [36, 37] and lose eligibility for EVAR at a smaller AAA size (55 mm vs. 65 mm) compared to men [31]. Accordingly, within the VQI those with HNA were more often female [35]. Women have been noted to have greater neck angulation and a higher percentage thrombus burden than men [31, 33]. Within the Gore Global Registry for Endovascular Aortic Treatment (GREAT), in addition to more severe angulation, a greater proportion of women were also noted to have non-cylindrical neck morphology and more often required reintervention [26]. Prevalence of HNA is even more marked outside of the Caucasian population. A study of Korean patients found that neck anatomy was eligible in only 34.0% of intact and 23.0% of ruptured AAAs with greater neck angulation again noted in women [30].

Endovascular Strategy and Techniques

Selecting the Right Approach

It is often difficult for clinicians to decide on the most appropriate primary management for IRAAAs with more complex pathology such as HNA. The APPROACH concept was coined by Donas and Torsello to help with this decision-making for patients with complex pararenal aortic pathology [38]. It encompasses eight major criteria as key factors to influence decision-making: aortic pathology, the patient's clinical profile, proven literature evidence, operator's preference, skill set, access considerations, costs, and hostile neck features [38]. All these considerations and the wishes of the patient [39] should be amalgamated together to decide whether an endovascular or open approach, or indeed, conservative management, would be most beneficial [40]. Where the optimal strategy extends beyond the borders of the operator's expertise, collaboration, and advice should be sought [41]. Nonetheless, the AAA population is a high-risk population with high proportion of smokers with cardiac, respiratory, and renal disease, who stand to benefit greatly from the reduced physiological insult and benefits afforded by endovascular repair. Many of these patients will have HNA, which is noted to increase in prevalence with age [3]. Therefore, a comprehensive technological armamentarium and a strategic approach are needed to provide the best possible repair. Various methods and technical considerations for the treatment of IRAAA with HNA are outlined below.

EVAR

Pre-operative Assessment

Prior to repair, appropriate imaging, most often high-quality (<1 mm thickness) computed tomography angiography (CTA), should be utilised to examine the aortic vascular morphology. This enables characterisation of the features of an HNA, facilitates pre-operative planning, and aids device selection. Centreline reconstruction should be used to enable accurate neck diameter and length measurement for the evaluation of the "functional" neck—the length that can be used for the sealing of the stent-graft [20]. Aortic angulation should be measured perpendicular to the aortic axis, in the middle of the flexure [19], and can be further evaluated using 3D reconstruction [19].

Accurate diameter assessment is important for stent-graft oversizing, to prevent migration, and is normally recommended at between 10–20% and <30% [42]. Consideration should be given to the high likelihood of asymmetrical endograft placement in patients with severe angulation, as this may reduce the degree of oversizing calculated prior to deployment [20]. Further difficulty is encountered with a conical neck, for which oversizing may be more marked in the narrower portion. This is usually handled in a pragmatic fashion, with acceptance of a lesser degree of oversizing in the larger segment (maintaining >10%) and a greater degree of oversizing in the narrow segment (maintaining <30%) [43]. In these cases, the neck, in the strictest definition, may be short, but there may be a much longer usable sealing zone.

In cases with a short or angulated neck, an assessment of the optimal C-arm position to ensure angulation orthogonal to the aortic neck and crux of the distal renal artery should also be made at this point to maximise use of neck length for proximal sealing. This should be done with the caveat that positioning may need to change slightly due to vessel straightening during the procedure. This adjustment can be accounted for following the deployment of the proximal stent rings through correction of the C-arm position and further angiography prior to full stent-graft deployment [20].

Off-the-Shelf Devices

While there is little doubt that HNA is associated with reduced durability, in carefully selected populations, such as elderly and medically compromised patients (for whom the mortality risk of complex endovascular or open repair is greater), or where there is an urgent need for repair, EVAR remains an attractive option. Many off-the -shelf endografts with active fixation have been reported to provide reasonable results when neck anatomy is not perfect but acceptable and can, in these high-risk patient groups, pragmatically be used despite an increased risk of type 1a endoleak [3, 44, 45].

When selecting a device there are a variety of stent-grafts on the market with features designed to accommodate HNA, which have expanded the capabilities of treatment within IFU [5, 46].

Controlled delivery and accurate deployment are important. In cases of severe aortic neck angulation or vessel tortuosity, intraoperative techniques such as the use of pull-through wires or directional tip control may be utilised to improve tracking and accurate deployment of the device [47, 48]. The Endowedge Technique, using angioplasty balloons within the renal artery to align scallops of a stent-graft and maximise proximal placement, has also been described to maximise the seal zone [49]. Now though several grafts are repositionable.

Further technological advances aim to improve the accuracy of deployment. The GORE EXCLUDER Conformable AAA Endoprosthesis with ACTIVE CONTROL System allows dynamic adjustment intraoperatively, angling the proximal end of the graft so that the device can be positioned along the centreline of the aorta, and there is no loss of sealing zone, and it is more conformable. These features decrease the risk of type 1a endoleak and perhaps increase the long-term sealing and fixation of the graft.

In wide-necked aneurysms, one example is the ovation stent-graft, for which fixation and seal are separated, with graft fixation achieved by suprarenal stent anchors and seal by polymer-filled inflatable sealing rings. The reduced radial force, and hence reduced risk of proximal aortic dilatation, has been purported as an explanation for the long-term success of this graft in wide-necked AAA.

This graft IFU states that a proximal aortic landing zone for the sealing ring 7 mm below the inferior renal artery is required, for which no difference in rates of type 1a endoleak, reintervention, or aneurysm-related mortality have been reported at 5 years [16, 46, 50, 51]. However, it is a general misconception that suprarenal fixation devices increase the ability to treat a short neck. No difference in the rates of stent migration and midterm outcomes between infra- and suprarenal fixation devices has been demonstrated, but inferior renal outcomes associated with suprarenal fixation are observed [12, 52, 53].

However, while stent-graft devices have clearly improved, the marginal gains in IFU cannot accommodate all cases, and careful post-operative surveillance is needed [54]. Further endovascular approaches are needed to improve durability and also to provide rescue following post-operative neck-related complications such as type 1a endoleak and stent migration.

Adjunctive Procedures

For endovascular management of IRAAAs with HNA, additional adjuncts and techniques can be used to maximise proximal seal and positioning, both intraoperatively during the primary procedure, and to rescue following post-operative neck-related complications such as type 1a endoleaks.

When a proximal endoleak at the time of stent-graft insertion occurs, the simplest strategy is to reshape the proximal landing zone via balloon inflation, thus changing the conformation of the proximal stent end to increase adherence. This can then be followed by extension with a bare-metal stent or with stent-graft cuff placement if needed—both of which are reported to provide acceptable midterm outcomes [55, 56].

Palmaz stents (Cordis Corporation, Bridgewater, NJ) are balloon-expandable stents with a closed cell design which exerts a high radial force, encouraging the moulding of the stent-graft to the vessel wall while reducing the risk of branch compromise. They are well established as a treatment for treating intraoperative type 1a endoleaks during EVAR, more commonly seen in hostile neck patients [57, 58]. However, in the long-term follow-up of 125 patients over a median of 43 months, a significant increase in the diameter of the sealing zone was observed, with 14% of cases demonstrating significant Palmaz stent migration, 9% with stent-graft migration, and 16% of cases developing a type 1a endoleak. Therefore, while Palmaz stenting presents an adequate bailout strategy, it should not be used first line to extend elective infrarenal endovascular aneurysm repair in more demanding anatomies [56].

When a segment of infrarenal aorta proximal to the EVAR remains, a cuff can be used to repair or reinforce the sealing zone. For cases with HNA, use of cuffs has been associated with superior durability [59]. Additionally, pre-deployment of an aortic cuff into a challenging neck prior to placement of the main stent-graft can be used to improve sealing, a strategy known as the Kilt Technique [60]. This technique may have utility for cases with severe neck angulation, as it is reported to straighten the landing zone prior to main body deployment. Several case series have reported a high rate of technical success and low mortality and complication rates in the short term [60–65]. In the case of wide-necked aneurysms, deployment of a thoracic endograft can be utilised as a proximal extension for a main infrarenal device, named the

Funnel Technique [66]. Low-grade evidence of technical success and short-term efficacy for this approach has been reported in several case series [67–71].

The use of endoAnchors to reinforce the proximal sealing zone by mechanically securing the proximal neck of the endograft has been reported to increase the fixation of the endograft and in some cases augment the sealing of an endograft. This procedure can be useful as a prophylactic measure or for the treatment of type 1a endoleaks sometimes in conjunction with aortic extension cuffs [72, 73]. The Heli-FX EndoAnchor System (Medtronic, Santa Rosa) has approval for use in short necks (4–10 mm). The system may be of significant benefit in cases with severe neck angulation, conical necks where there is technically a short neck because of continued dilatation below the renal arteries but an adequate long sealing zone, and in cases where there is poor stent-graft apposition with the hostile neck and type 1a endoleak on the operating table.

The evidence for general efficacy of this adjunctive treatment has been provided by the prospective observational ANCHOR registry, which demonstrated short-term freedom from type 1a endoleak in 95% of primary intervention cases with presented acceptable mid-term outcomes [74, 75]. These findings are corroborated by systematic review, confirming that use of endoAnchors for EVAR is technically feasible and safe, with at least comparable early outcomes to the latest generation of stent-grafts [76]. Observational analysis of outcomes for octogenarians with HNA has indicated that use of EndoAnchors may expand the utilisation of EVAR in this population subgroup, who may not be physiologically suitable for fenestrated EVAR or open repair, or be used prophylactically to reduce the risk of future reinterventions [77].

Fenestrated EVAR

Fenestrated endovascular repair (FEVAR) is often considered the gold standard of endovascular intervention for hostile neck aneurysms. It does not rely on the hostile infrarenal neck with a landing zone above the renal arteries, preventing deterioration of the infrarenal neck and promoting positive sac remodelling [78, 79]. Each customised, patient-specific stent-graft body contains fenestrations or scallops to enable access to the visceral arteries and allows the functional sealing zone to be extended to an adequate, more proximal, landing zone. Each fenestration is cannulated, and covered stents are deployed into the visceral vessels. Due to technical complexity necessitating a considerable learning curve, FEVARs are normally only performed at high-volume specialist centres, with waiting times a minimum of 4–6 weeks to allow for the unique graft to be manufactured [80, 81]. While off-the-shelf fenestrated stent-graft designs are commercially available, these have been found to be feasible only in a minority of patients when using liberal IFU criteria (42–49%) [82]. Physician-modified grafts have also shown acceptable short-term outcomes, but long-term outcomes are unknown [83].

Successful deployment of a fenestrated stent-graft requires careful pre-operative planning using centreline reconstruction and careful measurements. A proximal seal zone length of 15–20 mm is required for the first stent, which may necessitate inclusion of the SMA and/or coeliac via a scallop or additional stent [84]. Generous access vasculature is required due to the increase in size of delivery systems when compared to EVAR. Moreover, to enable manipulation of catheters for cannulation of the visceral vessels, it is necessary that access vasculature is not overly tortuous and there should be minimal angulation of the visceral aorta. Visceral vessels must be of sufficient diameter (≥5 mm) and length (≥15 mm prior to bifurcation) to enable cannulation and to reduce the risk of occlusion. Excessive angulation, stenosis, and anterior take-off can all reduce the likelihood of successful cannulation. There is also a risk of distal thromboembolism during vessel cannulation or stent-graft deployment, and so the extent and nature of thrombus in the visceral segment should be pre-operatively assessed. A further major concern is that FEVAR is associated with a deterioration in renal function peri-operatively [85]. As such peri-operative attention to renal protective measures should be incorporated into standard care.

On meta-analysis of 27 studies, no significant difference was noted in 30-day mortality between FEVAR and open surgical repair (3.3% vs. 4.2%). However, FEVAR was associated with significantly lower morbidity but had far higher reintervention rates than open surgical repair (11% vs. 2%) [86].

Chimney EVAR

An "off-the-shelf" alternative for patients with HNA is the use of the chimney technique (ChEVAR) [44]. Here a standard endograft is used but is extended to sit proximally above the renal/visceral arteries. Perfusion to the branch vessels is maintained via separate stents (i.e. chimneys).

Successful utilisation of ChEVAR requires advanced endovascular skills and a degree of familiarity with the technique. Indeed, an increased risk of post-operative complications has been observed at lower-volume centres [87]. One major concern is the risk of gutter-related type 1a endoleaks, for which aggressive oversizing of 30% for single or 35% for double chimneys (based on the mean diameter of the sealing zone in conjunction with a seal zone length of 20–25 mm) is recommended [88]. Choice of materials is also integral to the prevention of gutter-related endoleaks, and use of a nitinol/polyester stent-graft is associated with improved survival [89, 90]. ChEVARs with >2 chimneys and the use of bare-metal stents are observed to increase mortality risk [89]. During cannulation and stenting of the renal and/or visceral vasculature, a wire with an atraumatic curved tip must be used to prevent injury to the vessel intima. Chimney stents should be placed at least 10 mm inside each vessel. Care must be taken with positioning within an angulated vessel segment to avoid kinking or stenosis, as the need for relining of chimney stents has been associated with reduced primary patency [91]. At the end of the procedure when deploying a kissing balloon between the aortic stent and chimney graft, attention must be paid to deflate the chimney graft balloon segment second to avoid crushing the stent. A further concern with ChEVAR is the increased risk of stroke with this procedure secondary to antegrade cannulation of renal and visceral vessels. A pre-operative assessment of thrombus burden and character within the aortic arch and visceral segment should be conducted, and as stroke risk increases greater than twofold with the use of bilateral upper extremity access [92]. Here, single upper extremity access should be considered, particularly in higher-risk patients.

Collected world experience for ChEVAR was analysed by the PERICLES registry, which demonstrated comparable outcomes to branched and fenestrated devices with a primary patency rate of 94% and overall survival of 79% at 17 months [93]. HNA was noted to affect long-term outcomes with the absence of an infrarenal neck associated with an increased risk of device-related complications (OR 2.85, p = 0.007) and a diameter of >30 mm associated with an increased risk of type 1a endoleak (OR 4.86, p = 0.12) [94]. Superior results were observed in the PROTAGORAS study with the use of the Medtronic Endurant stent-graft, with a 0.8% 30-day mortality, 1.6% incidence of type 1a endoleak, and primary chimney graft patency of 95.7%. As a result, this device has received a CE mark to be used for ChEVAR and is currently under further evaluation [90]. Acceptability of ChEVAR was corroborated on meta-analysis of 13 studies, but reliance on careful patient selection and the need for further high-quality prospective studies was noted [95]. Of particular importance is the need for careful follow-up, with concerns expressed regarding need for reintervention, type 1a endoleak, and deterioration in renal function which demands close surveillance [96, 97].

ChEVAR is particularly useful in emergency scenarios, for which the short neck is reportedly the most common exclusion criterion for standard EVAR, and has been reported to increase the overall feasibility of endovascular repair by 12–22% [98, 99]. CHEVAR may also be advantageous in patients with hostile iliofemoral access and caudally directed renal access.

Open or Hybrid Repair

Open repair in patients with HNA anatomy is a commonly used option. The practice does present specific challenges, with increased mortality and morbidity associated with significant manipulation of the neck of the aneurysm and often the use of suprarenal cross clamping [100, 101].

Hybrid repair utilising extra-anatomic bypass for renal debranching (i.e. right hepatorenal bypass and left splenorenal bypass or ileorenal bypasses) allows for the deployment of the stent-graft more proximally, increasing the length of a short neck. This is performed prior to stent-graft insertion and, although necessitates a laparotomy, can minimise the physiological insult of aortic cross clamping and renal ischaemic time [102]. In practice, this technique is rarely used.

Patients with IFU violations have been shown to have superior long-term survival with open surgery compared to EVAR [103–105]. As such, for patients with a longer post-operative life expectancy who are fit enough to undergo open repair, it is currently considered the gold standard for the treatment of IRAAA with HNA [106]. Open repair and its risks and benefits should therefore be strongly considered in IRAAA patients and should not be reserved solely for individuals ineligible for endovascular repair.

Conclusion

Aneurysms with a hostile neck are a significant challenge, and treatment with standard approaches carries a much greater risk. As our understanding of the risks of HNA and technological advances and technical expertise increases, it is likely that our definition of what constitutes a hostile neck may change [107]. Nonetheless, at present although no option offers a perfect solution, there are a variety of available approaches, each with their own risks and benefits. These comprise an adaptable toolkit to tailor to each individual patient's anatomy, clinical profile, and preference, which will hopefully expand further and better serve a greater population over time [38].

Case Presentation

Continued from page 513

We selected a GORE® EXCLUDER® Conformable with ACTIVE CONTROL aortic stent-graft system for the treatment of the aneurysm. This device was inside IFU for the selected anatomy and has active control, allowing accurate placement utilising all the infrarenal neck, with a conformable proximal portion to deal with tortuosity in the neck. In addition, the device is repositionable to ensure millimetre accuracy.

After detailed operative planning, under local anaesthesia with antibiotic cover, a field block in the supine position was administered. Percutaneous access was obtained under ultrasound guidance on both sides. The patient was under continuous heparin drip, with boluses as needed to maintain an ACT of 250–300.

Table 49.1 Suggested endovascular toolkit for the standard endovascular management of intrarenal AAA with angulated neck. The toolkit for chimneys and fenestrated/branched devices is not included. These are only a few suggested options based on the author's experience and can cover the vast majority of cases

	Manufacturer	Size/Length
Wires		
Any standard access wire		0.035″
Glidewire floppy	Terumo	0.035″
Meier wire	Boston Scientific	0.035″
Sheaths		
Any standard access sheath	Any	6 Fr/11 cm
DrySeal	Gore	20–24 Fr
Catheters		
Pigtail	Various	5 Fr
Measuring pigtail	Various	5 Fr
Catheters for contralateral gate cannulation (e.g. Van Shie, Cobra, SOS omni)	Various	5 Fr
Balloons		
Reliant	Medtronic	Compliant, size N/A
Stent-grafts		
Medtronic Endurant E2S stent-graft system	Medtronic	Varying size range
GORE® EXCLUDER® Conformable with ACTIVE CONTROL stent-graft system	Gore	Varying size range
Ovation stent-graft platform	Endologix	Varying size range
Ancillary equipment		
EndoAnchors—Heli-FX guide	Medtronic	22 and 28 mm deflecting tip length
EndoAnchors—Heli-FX applier	Medtronic	N/A
EndoAnchors—EndoAnchor cassette	Medtronic	N/A
Other equipment		
Covered stents (8–12 mm) included in an iliac rupture emergency kit	Various	8, 10, 12 mm
Uncovered self-expanding stent to treat stenosis, kinking of the iliac limb	Various	8, 10, 12 mm
Closure device		
Perclose ProGlide	Abbott	N/A

Two 6Fr sheaths were placed. On the right side a pigtail catheter was placed just distal to the subclavian artery and a stiff wire was introduced to the same level. Two proGlide sutures were deployed at 90° to each other. A Gore DrySeal sheath was then placed over the stiff guidewire. The same process was then repeated for the left side.

The Gore Excluder stent-graft was positioned in the aneurysm sac. A pigtail for imaging was placed at the superior level of the L2 vertebrae. The C-arm was then angled with 36° RAO and 15° cranial. An intra-arterial digital subtraction angiogram was taken. The renal arteries were marked, and the device was placed at the appropriate level. The stiff wire was then retracted and replaced with a Terumo wire, allowing the device to conform to the neck angulation. The device was then angulated partially in the closed position. It was then partially deployed, and the position was checked again with a repeat intra-arterial digital subtraction angiogram. To further reposition the device to maximise the neck coverage and enhance the sealing zone, the device was constrained and angulated again. When the optimal position was reached the device was fully deployed successfully. The contralateral gate was cannulated, and the ipsilateral and contralateral limbs were placed and deployed as standard. The proximal and distal sealing zones and limb overlap areas were ballooned with a non-compliant balloon.

The final completion angiogram in two planes, after removing the stiff wires, demonstrated good flow through the graft with no stenosis, or type 1 or 3 endoleak.

The symptoms of back pain resolved, and the patient was discharged home. Follow-up CT scan at 6 weeks showed a small type 2 endoleak. The patient continues on surveillance (Table 49.1).

References

1. de Vries JPPM, Zuidema R, Bicknell CD, Fisher R, Gargiulo M, Louis N, et al. European expert opinion on Infrarenal sealing zone definition and management in endovascular aortic repair patients: a Delphi consensus. J Endovasc Ther. 2023;30(3):449–60.
2. Herman CR, Charbonneau P, Hongku K, Dubois L, Hossain S, Lee K, et al. Any nonadherence to instructions for use predicts graft-related adverse events in patients undergoing elective endovascular aneurysm repair. J Vasc Surg. 2018;67(1):126–33.
3. Dillavou ED, Muluk SC, Rhee RY, Tzeng E, Woody JD, Gupta N, et al. Does hostile neck anatomy preclude successful endovascular aortic aneurysm repair? J Vasc Surg. 2003;38(4):657–63.
4. Stather PW, Wild JB, Sayers RD, Bown MJ, Choke E. Endovascular aortic aneurysm repair in patients with hostile neck anatomy. J Endovasc Ther. 2013;20(5):623–37.

5. Fairman RM, Velazquez OC, Carpenter JP, Woo E, Baum RA, Golden MA, et al. Midterm pivotal trial results of the talent low profile system for repair of abdominal aortic aneurysm: analysis of complicated versus uncomplicated aortic necks. J Vasc Surg. 2004;40(6):1074–82.
6. Antoniou GA, Georgiadis GS, Antoniou SA, Kuhan G, Murray D. A meta-analysis of outcomes of endovascular abdominal aortic aneurysm repair in patients with hostile and friendly neck anatomy. J Vasc Surg. 2013;57(2):527–38.
7. AbuRahma AF, Yacoub M, Mousa AY, Abu-Halimah S, Hass SM, Kazil J, et al. Aortic neck anatomic features and predictors of outcomes in endovascular repair of abdominal aortic aneurysms following vs not following instructions for use. J Am Coll Surg. 2016;222(4):579–89.
8. Aburahma AF, Campbell JE, Mousa AY, Hass SM, Stone PA, Jain A, et al. Clinical outcomes for hostile versus favorable aortic neck anatomy in endovascular aortic aneurysm repair using modular devices. J Vasc Surg. 2011;54(1):13–21.
9. Marone EM, Freyrie A, Ruotolo C, Michelagnoli S, Antonello M, Speziale F, et al. Expert opinion on hostile neck definition in endovascular treatment of abdominal aortic aneurysms (a Delphi consensus). Ann Vasc Surg. 2020;62:173–82.
10. http://fyra.io. Challenges of EVAR in highly angulated and short infrarenal neck anatomies [Internet]. Endovascular Today. Bryn Mawr Communications; [cited 2022 Jul 2]. https://evtoday.com/articles/2014-feb-supplement/challenges-of-evar-in-highly-angulated-and-short-infrarenal-neck-anatomies
11. Georgiadis GS, Trellopoulos G, Antoniou GA, Gallis K, Nikolopoulos ES, Kapoulas KC, et al. Early results of the Endurant endograft system in patients with friendly and hostile infrarenal abdominal aortic aneurysm anatomy. J Vasc Surg. 2011;54(3):616–627.e4.
12. Hager ES, Cho JS, Makaroun MS, Park SC, Chaer R, Marone L, et al. Endografts with suprarenal fixation do not perform better than those with infrarenal fixation in the treatment of patients with short straight proximal aortic necks. J Vasc Surg. 2012;55(5):1242–6.
13. Stark M, Suresh A, Alexander J, Cragg A. An analysis of variables affecting aortic neck length with implications for fenestrated endovascular repair of abdominal aortic aneurysm. Ann Vasc Surg. 2014;28(4):808–15.
14. Panthofer AM, Olson SL, Rademacher BL, Grudzinski JK, Chaikof EL, Matsumura JS, et al. Anatomic eligibility for endovascular aneurysm repair preserved over 2 years of surveillance. J Vasc Surg. 2021;74(5):1527–1536.e1.
15. Katsargyris A, Verhoeven ELG. Endovascular strategies for infrarenal aneurysms with short necks. J Cardiovasc Surg (Torino). 2013;54(1 Suppl 1):21–6.
16. Mathlouthi A, Khan MA, Al-Nouri O, Barleben A, Aburahma A, Malas MB. The correlation of aortic neck length to late outcomes following EVAR with the ovation stent graft. J Vasc Surg. 2022;75(6):1890–1895.e1.
17. Oliveira NFG, Bastos Gonçalves FM, Van Rijn MJ, de Ruiter Q, Hoeks S, de Vries JPPM, et al. Standard endovascular aneurysm repair in patients with wide infrarenal aneurysm necks is associated with increased risk of adverse events. J Vasc Surg. 2017;65(6):1608–16.
18. Cao P, Verzini F, Parlani G, Rango PD, Parente B, Giordano G, et al. Predictive factors and clinical consequences of proximal aortic neck dilatation in 230 patients undergoing abdominal aorta aneurysm repair with self-expandable stent-grafts. J Vasc Surg. 2003;37(6):1200–5.
19. van Keulen JW, Moll FL, Tolenaar JL, Verhagen HJM, van Herwaarden JA. Validation of a new standardized method to measure proximal aneurysm neck angulation. J Vasc Surg. 2010;51(4):821–8.
20. van Keulen JW, Moll FL, van Herwaarden JA. Tips and techniques for optimal stent graft placement in angulated aneurysm necks. J Vasc Surg. 2010;52(4):1081–6.
21. Choke E, Munneke G, Morgan R, Belli AM, Loftus I, McFarland R, et al. Outcomes of endovascular abdominal aortic aneurysm repair in patients with hostile neck anatomy. Cardiovasc Intervent Radiol. 2006;29(6):975–80.
22. Seike Y, Fukuda T, Yokawa K, Inoue Y, Shijo T, Uehara K, et al. Preoperative neck angulation is associated with aneurysm sac growth due to persistent type Ia Endoleak after endovascular abdominal aortic aneurysm repair. Ann Vasc Dis. 2020;13(3):261–8.
23. Oliveira NFG, Gonçalves FB, Hoeks SE, Josee van Rijn M, Ultee K, Pinto JP, et al. Long-term outcomes of standard endovascular aneurysm repair in patients with severe neck angulation. J Vasc Surg. 2018;68(6):1725–35.
24. Bernardini G, Litterscheid S, Torsello GB, Torsello GF, Beropoulis E, van Brunschot DÖ. A meta-analysis of safety and efficacy of endovascular aneurysm repair in aneurysm patients with severe angulated infrarenal neck. PLoS One. 2022;17(2):e0264327.
25. Mathlouthi A, Locham S, Dakour-Aridi H, Black JH, Malas MB. Impact of suprarenal neck angulation on endovascular aneurysm repair outcomes. J Vasc Surg. 2020;71(6):1900–6.
26. Shutze W, Suominen V, Jordan W, Cao P, Oweida S, Milner R. The incidence and effect of noncylindrical neck morphology on outcomes after endovascular aortic aneurysm repair in the global registry for endovascular aortic treatment. J Vasc Surg. 2018;68(6):1714–24.
27. Pitoulias GA, Valdivia AR, Hahtapornsawan S, Torsello G, Pitoulias AG, Austermann M, et al. Conical neck is strongly associated with proximal failure in standard endovascular aneurysm repair. J Vasc Surg. 2017;66(6):1686–95.
28. Hultgren R, Vishnevskaya L, Wahlgren CM. Women with abdominal aortic aneurysms have more extensive aortic neck pathology. Ann Vasc Surg. 2013;27(5):547–52.
29. Schanzer A, Greenberg RK, Hevelone N, Robinson WP, Eslami MH, Goldberg RJ, et al. Predictors of abdominal aortic aneurysm sac enlargement after endovascular repair. Circulation. 2011;123(24):2848–55.
30. Hwang D, Kim J, Kim HK, Huh S. Suitability of the aortic neck anatomy for endovascular aneurysm repair in Korean patients with abdominal aortic aneurysm. Vasc Spec Int. 2020;36(2):71–81.
31. Sweet MP, Fillinger MF, Morrison TM, Abel D. The influence of gender and aortic aneurysm size on eligibility for endovascular abdominal aortic aneurysm repair. J Vasc Surg. 2011;54(4):931–7.
32. Karathanos C, Spanos K, Kouvelos G, Athanasoulas A, Koutsias S, Matsagkas M, et al. Hostility of proximal aortic neck anatomy in relation to abdominal aortic aneurysm size and its impact on the outcome of endovascular repair with the new generation endografts. J Cardiovasc Surg (Torino). 2020;61(1):60–6.
33. Ayo D, Blumberg SN, Gaing B, Baxter A, Mussa FF, Rockman CB, et al. Gender differences in aortic neck morphology in patients with abdominal aortic aneurysms undergoing elective endovascular aneurysm repair. Ann Vasc Surg. 2016;30:100–4.
34. Banzic I, Lu Q, Zhang L, Stepak H, Davidovic L, Oszkinis G, et al. Morphological differences in the aorto-iliac segment in AAA patients of Caucasian and Asian origin. Eur J Vasc Endovasc Surg. 2016;51(6):783–9.
35. de Guerre L, O'Donnell T, Varkevisser R, Swerdlow N, Li C, van Herwaarden J, et al. Hostile neck anatomy is associated with higher perioperative type IA endoleaks and lower long-term survival after elective endovascular aneurysm repair. J Vasc Surg. 2019;70(3):e45.
36. Ulug P, Sweeting MJ, von Allmen RS, Thompson SG, Powell JT, Ulug P, et al. Morphological suitability for endovascular repair, non-intervention rates, and operative mortality in women and men

assessed for intact abdominal aortic aneurysm repair: systematic reviews with meta-analysis. Lancet. 2017;389(10088):2482–91.
37. Arko FR, Filis KA, Seidel SA, Gonzalez J, Lengle SJ, Webb R, et al. How many patients with infrarenal aneurysms are candidates for endovascular repair? The northern California experience. J Endovasc Ther. 2004;11(1):33–40.
38. Donas KP, Torsello GF, Torsello GB. Factors influencing decision making in the treatment of complex pararenal aortic pathologies: the APPROACH concept. J Endovasc Ther. 2017;24(6):889–91.
39. Wickramasekera N, Howard A, Philips P, Rooney G, Hughes J, Wilson E, et al. Strength of public preferences for endovascular or open aortic aneurysm repair. Br J Surg. 2019;106(13):1775–83.
40. Chaikof EL, Dalman RL, Eskandari MK, Jackson BM, Lee WA, Mansour MA, et al. The Society for Vascular Surgery practice guidelines on the care of patients with an abdominal aortic aneurysm. J Vasc Surg. 2018;67(1):2–77.e2.
41. Schanzer A, Baril D, Robinson WP, Simons JP, Aiello FA, Messina LM. Developing a complex endovascular fenestrated and branched aortic program. J Vasc Surg. 2015;61(3):826–31.
42. van Prehn J, Schlösser FJV, Muhs BE, Verhagen HJM, Moll FL, van Herwaarden JA. Oversizing of aortic stent grafts for abdominal aneurysm repair: a systematic review of the benefits and risks. Eur J Vasc Endovasc Surg. 2009;38(1):42–53.
43. http://fyra.io. Managing difficult aortic necks [Internet]. Endovascular Today. Bryn Mawr Communications; [cited 2022 Jun 24]. https://evtoday.com/articles/2014-feb/managing-difficult-aortic-necks
44. Greenberg RK, Clair D, Srivastava S, Bhandari G, Turc A, Hampton J, et al. Should patients with challenging anatomy be offered endovascular aneurysm repair? J Vasc Surg. 2003;38(5):990–6.
45. Cerini P, Guzzardi G, Divenuto I, Parziale G, Brustia P, Carriero A, et al. Are abdominal aortic aneurysms with hostile neck really unsuitable for EVAR? Our experience. Radiol Med (Torino). 2016;121(6):528–35.
46. Georgakarakos E, Papatheodorou N, Argyriou C, Tasopoulou KM, Doukas D, Georgiadis GS. An update on the ovation abdominal stent graft for the treatment of abdominal aortic aneurysms: current evidence and future perspectives. Expert Rev Med Devices. 2020;17(12):1249–56.
47. Chaudhuri A, Heim F, Chakfe N. Estimating the "pull" on a Pullthrough wire: a pilot study. EJVES Vasc Forum. 2021;50:24–6.
48. Takayama T, Phelan PJ, Matsumura JS. Directional tip control technique for optimal stent graft alignment in angulated proximal aortic landing zones. J Vasc Surg Cases Innov Tech. 2017;3(2):51–6.
49. Minion DJ, Xenos ES. The femoral-based endowedge technique to increase juxtarenal seal and correct graft tilt. J Vasc Surg. 2012;55(5):1522–5.
50. Moulakakis KG, Dalainas I, Kakisis J, Giannakopoulos TG, Liapis CD. Current knowledge on EVAR with the ultra-low profile ovation abdominal stent-graft system. J Cardiovasc Surg (Torino). 2012;53(4):427–32.
51. Broos PPHL, Stokmans RA, van Sterkenburg SMM, Torsello G, Vermassen F, Cuypers PWM, et al. Performance of the endurant stent graft in challenging anatomy. J Vasc Surg. 2015;62(2):312–8.
52. Banno H, Ikeda S, Kawai Y, Fujii T, Akita N, Takahashi N, et al. Suprarenal fixation is associated with worse midterm renal function after endovascular abdominal aortic aneurysm repair compared with infrarenal fixation. J Vasc Surg. 2020;71(2):450–6.
53. Stather PW, Ferguson J, Awopetu A, Boyle JR. Meta-analysis of renal function following infrarenal EVAR using suprarenal or infrarenal fixation devices. Eur J Vasc Endovasc Surg. 2018;56(4):486–96.
54. Stather PW, Sayers RD, Cheah A, Wild JB, Bown MJ, Choke E. Outcomes of endovascular aneurysm repair in patients with hostile neck anatomy. Eur J Vasc Endovasc Surg. 2012;44(6):556–61.
55. Rajani RR, Arthurs ZM, Srivastava SD, Lyden SP, Clair DG, Eagleton MJ. Repairing immediate proximal endoleaks during abdominal aortic aneurysm repair. J Vasc Surg. 2011;53(5):1174–7.
56. Abdulrasak M, Resch T, Sonesson B, Holst J, Kristmundsson T, Dias NV. The long-term durability of intra-operatively placed Palmaz stents for the treatment of type Ia endoleaks after EVAR of abdominal aortic aneurysm. Eur J Vasc Endovasc Surg. 2017;53(1):69–76.
57. Dias NV, Resch T, Malina M, Lindblad B, Ivancev K. Intraoperative proximal endoleaks during AAA stent-graft repair: evaluation of risk factors and treatment with Palmaz stents. J Endovasc Ther. 2001;8(3):268–73.
58. Farley SM, Rigberg D, Jimenez JC, Moore W, Quinones-Baldrich W. A retrospective review of Palmaz stenting of the aortic neck for endovascular aneurysm repair. Ann Vasc Surg. 2011;25(6):735–9.
59. Lee JT, Ullery BW, Zarins CK, Olcott C, Harris EJ, Dalman RL. EVAR deployment in anatomically challenging necks outside the IFU. Eur J Vasc Endovasc Surg. 2013;46(1):65–73.
60. Minion DJ, Yancey A, Patterson DE, Saha S, Endean ED. The Endowedge and Kilt techniques to achieve additional juxtarenal seal during deployment of the gore excluder endoprosthesis. Ann Vasc Surg. 2006;20(4):472–7.
61. Jimenez JC, Quinones-Baldrich WJ. Technical modifications for endovascular Infrarenal AAA repair for the angulated and dumbbell-shaped neck: the Precuff Kilt technique. Ann Vasc Surg. 2011;25(3):423–30.
62. Kim TH, Jang HJ, Choi YJ, Lee CK, Kwon SW, Shim WH. Kilt technique as an angle modification method for endovascular repair of abdominal aortic aneurysm with severe neck angle. Ann Thorac Cardiovasc Surg. 2017;23(2):96–103.
63. Park KH, Kim U. Stent graft using kilt technique for an abdominal aortic aneurysm with a severely angulated neck. Heart Lung Circ. 2016;25(3):e48–52.
64. Jeon YS, Cho YK, Song MG, Seo TS, Kim JH, Song SY, et al. Clinical outcomes of endovascular aneurysm repair with the kilt technique for abdominal aortic aneurysms with hostile aneurysm neck anatomy: a Korean multicenter retrospective study. Cardiovasc Intervent Radiol. 2018;41(4):554–63.
65. Szaniewski K, Biernacka M, Walas RL, Zembala M. Predeployed aortic extension cuff (kilt) in EVAR with hostile neck anatomy using endurant II system: preliminary results. Kardiochir Torakochirurgia Pol. 2016;13(4):334–9.
66. Zanchetta M, Faresin F, Pedon L, Riggi M, Colonna S, Lipari R, et al. Funnel technique for first-line endovascular treatment of an abdominal aortic aneurysm with an ectatic proximal neck. J Endovasc Ther. 2006;13(6):775–8.
67. Ronsivalle S, Faresin F, Franz F, Rettore C, Zanchetta M, Zonta L. Funnel technique for EVAR: "a way out" for abdominal aortic aneurisms with ectatic proximal necks. Ann Vasc Surg. 2012;26(1):141–8.
68. Reyes Valdivia A, Duque Santos A, Ocaña Guaita J, Gandarias ZC. The cuff plus anchoring funnel technique for endovascular aortic repair (CAF-EVAR) for large infrarenal necks. Cardiovasc Intervent Radiol. 2018;41(2):330–5.
69. Bastiaenen V, Snoeijs M, Blomjous J, Bosma J, Leijdekkers V, van Nieuwenhuizen R, et al. The use of thoracic stent grafts for endovascular repair of abdominal aortic aneurysms. Vascular. 2018;26(3):278–84.
70. Pecoraro F, Pakeliani D, Dinoto E, Bajardi G. Endovascular treatment of large and wide aortic neck: case report and literature review. Gen Thorac Cardiovasc Surg. 2017;65(4):219–24.
71. İşcan HZ, Ünal EU, Tümer NB, Akkaya BB, Aşkın G, Sağlam S, et al. Funnel technique for wide infrarenal aneurysm neck with Lifetech Ankura™ stent graft system. Turk J Thorac Cardiovasc Surg. 2021;29(3):304–10.

72. de Vries JPPM, Ouriel K, Mehta M, Varnagy D, Moore WM, Arko FR, et al. Analysis of EndoAnchors for endovascular aneurysm repair by indications for use. J Vasc Surg. 2014;60(6):1460–1467.e1.
73. Valdivia AR, Beropoulis E, Pitoulias G, Pratesi G, Marcos FA, Barbante M, et al. Multicenter registry about the use of EndoAnchors in the endovascular repair of abdominal aortic aneurysms with hostile neck showed successful but delayed endograft sealing within intraoperative type Ia endoleak cases. Ann Vasc Surg. 2019;60:61–9.
74. http://fyra.io. Off-the-shelf solutions to treat a short neck: a data overview and case-based discussion [Internet]. Endovascular Today. Bryn Mawr Communications; [cited 2022 Jul 2]. https://evtoday.com/articles/2021-mar/off-the-shelf-solutions-to-treat-a-short-neck-a-data-overview-and-case-based-discussion
75. Jordan WD, Mehta M, Varnagy D, Moore WM, Arko FR, Joye J, et al. Results of the ANCHOR prospective, multicenter registry of EndoAnchors for type Ia endoleaks and endograft migration in patients with challenging anatomy. J Vasc Surg. 2014;60(4):885–892.e2.
76. Qamhawi Z, Barge TF, Makris GC, Patel R, Wigham A, Anthony S, et al. Editor's choice—systematic review of the use of Endoanchors in endovascular aortic aneurysm repair. Eur J Vasc Endovasc Surg. 2020;59(5):748–56.
77. Locham S, Mathlouthi A, Dakour-Aridi H, Malas MB. Favorable outcomes in octogenarians with hostile neck undergoing endovascular repair using EndoAnchors. Ann Vasc Surg. 2021;74:194–203.
78. Anderson JL, Berce M, Hartley DE. Endoluminal aortic grafting with renal and superior mesenteric artery incorporation by graft fenestration. J Endovasc Ther. 2001;8(1):3–15.
79. Teter K, Li C, Ferreira LM, Ferrer M, Rockman C, Jacobowitz G, et al. Fenestrated endovascular aortic aneurysm repair promotes positive infrarenal neck remodeling and greater sac shrinkage compared with endovascular aortic aneurysm repair. J Vasc Surg [Internet]. 2022 [cited 2022 Jun 25]. https://www.jvascsurg.org/article/S0741-5214(22)00402-5/fulltext
80. Kristmundsson T, Sonesson B, Malina M, Björses K, Dias N, Resch T. Fenestrated endovascular repair for juxtarenal aortic pathology. J Vasc Surg. 2009;49(3):568–74; discussion 574–575.
81. Verhoeven ELG, Katsargyris A, Fernandes e Fernandes R, Bracale UM, Houthoofd S, Maleux G. Practical points of attention beyond instructions for use with the Zenith fenestrated stent graft. J Vasc Surg. 2014;60(1):246–52.
82. Mendes BC, Oderich GS, Macedo TA, Pereira AA, Cha S, Duncan AA, et al. Anatomic feasibility of off-the-shelf fenestrated stent grafts to treat juxtarenal and pararenal abdominal aortic aneurysms. J Vasc Surg. 2014;60(4):839–848.e2.
83. Starnes BW. Physician-modified endovascular grafts for the treatment of elective, symptomatic, or ruptured juxtarenal aortic aneurysms. J Vasc Surg. 2012;56(3):601–7.
84. http://fyra.io. Ten steps [Internet]. Endovascular Today. Bryn Mawr Communications; [cited 2022 Jul 2]. https://evtoday.com/articles/2017-nov-supplement/ten-steps
85. Verhoeven ELG, Vourliotakis G, Bos WTGJ, Tielliu IFJ, Zeebregts CJ, Prins TR, et al. Fenestrated stent grafting for short-necked and juxtarenal abdominal aortic aneurysm: an 8-year single-centre experience. Eur J Vasc Endovasc Surg. 2010;39(5):529–36.
86. Jones AD, Waduud MA, Walker P, Stocken D, Bailey MA, Scott DJA. Meta-analysis of fenestrated endovascular aneurysm repair versus open surgical repair of juxtarenal abdominal aortic aneurysms over the last 10 years. BJS Open. 2019;3(5):572–84.
87. Torsello G, Usai M, Scali S, Kubilis P, Veith F, Markatis F, et al. Gender-related outcomes of chimney EVAR within the PERICLES registry. Vascular. 2018;26(6):641–6.
88. Donas KP, Usai MV, Taneva GT, Criado FJ, Torsello GB, Kubilis P, et al. Impact of aortic stent-graft oversizing on outcomes of the chimney endovascular technique based on a new analysis of the PERICLES registry. Vascular. 2019;27(2):175–80.
89. Scali ST, Beck AW, Torsello G, Lachat M, Kubilis P, Veith FJ, et al. Identification of optimal device combinations for the chimney endovascular aneurysm repair technique within the PERICLES registry. J Vasc Surg. 2018;68(1):24–35.
90. Donas KP, Torsello GB, Piccoli G, Pitoulias GA, Torsello GF, Bisdas T, et al. The PROTAGORAS study to evaluate the performance of the Endurant stent graft for patients with pararenal pathologic processes treated by the chimney/snorkel endovascular technique. J Vasc Surg. 2016;63(1):1–7.
91. Taneva GT, Fazzini S, Pipitone MD, Karaolanis G, Torsello G, Bremer C, et al. Use of stainless-steel, balloon-expandable chimney grafts is durable though caution is required when lining angulated renal arteries. J Endovasc Ther. 2020;27(6):902–9.
92. Bosiers MJ, Tran K, Lee JT, Donas KP, Veith FJ, Torsello G, et al. Incidence and prognostic factors related to major adverse cerebrovascular events in patients with complex aortic diseases treated by the chimney technique. J Vasc Surg. 2018;67(5):1372–9.
93. Donas KP, Lee JT, Lachat M, Torsello G, Veith FJ, PERICLES Investigators. Collected world experience about the performance of the snorkel/chimney endovascular technique in the treatment of complex aortic pathologies: the PERICLES registry. Ann Surg. 2015;262(3):546–53; discussion 552–553.
94. Taneva GT, Lee JT, Tran K, Dalman R, Torsello G, Fazzini S, et al. Long-term chimney/snorkel endovascular aortic aneurysm repair experience for complex abdominal aortic pathologies within the PERICLES registry. J Vasc Surg. 2021;73(6):1942–9.
95. Prapassaro T, Teraa M, Chinsakchai K, Hazenberg CEVB, Hunnangkul S, Moll FL, et al. Mid-term outcomes of chimney endovascular aortic aneurysm repair: a systematic review and meta-analysis. Ann Vasc Surg. 2022;79:359–71.
96. Scali ST, Feezor RJ, Chang CK, Waterman AL, Berceli SA, Huber TS, et al. Critical analysis of results after chimney endovascular aortic aneurysm repair raises cause for concern. J Vasc Surg. 2014;60(4):865–73; discussion 873–875.
97. Bath J, Fazzini S, Ippoliti A, Vogel TR, Singh P, Donas KP, et al. Chimney endovascular aneurysm repair (ChEVAR) for hostile neck complex aneurysm. Vascular. 2022;30(6):1058–68.
98. Brossier J, Coscas R, Capdevila C, Kitzis M, Coggia M, Goeau-Brissonniere O. Anatomic feasibility of endovascular treatment of abdominal aortic aneurysms in emergency in the era of the chimney technique: impact on an emergency endovascular kit. Ann Vasc Surg. 2013;27(7):844–50.
99. Dias NV, Bin Jabr A, Sveinsson M, Björses K, Malina M, Kristmundsson T. Impact of renal chimney grafts on anatomical suitability for endovascular repair in ruptured abdominal aortic aneurysm. J Endovasc Ther. 2015;22(1):105–9.
100. Latz CA, Boitano L, Schwartz S, Swerdlow N, Dansey K, Varkevisser RRB, et al. Editor's choice—mortality is high following elective open repair of complex abdominal aortic aneurysms. Eur J Vasc Endovasc Surg. 2021;61(1):90–7.
101. Latz CA, Boitano L, Schwartz S, Swerdlow N, Dansey K, Varkevisser RRB, et al. Contemporary mortality after emergent open repair of complex abdominal aortic aneurysms. J Vasc Surg. 2021;73(1):39–47.e1.
102. Lerussi G, O'Brien N, Sessa C, D'Elia P, Sobocinski J, Perrot C, et al. Hepatorenal bypass allowing fenestrated endovascular repair of Juxtarenal abdominal aortic aneurysm: a case report. Eur J Vasc Endovasc Surg. 2010;39(3):305–7.
103. Patel R, Sweeting MJ, Powell JT, Greenhalgh RM, EVAR Trial Investigators. Endovascular versus open repair of abdominal aortic aneurysm in 15-years' follow-up of the UK endovascular aneu-

rysm repair trial 1 (EVAR trial 1): a randomised controlled trial. Lancet. 2016;388(10058):2366–74.

104. Charbonneau P, Hongku K, Herman CR, Habib M, Girsowicz E, Doonan RJ, et al. Long-term survival after endovascular and open repair in patients with anatomy outside instructions for use criteria for endovascular aneurysm repair. J Vasc Surg. 2019;70(6):1823–30.

105. Chisci E, Kristmundsson T, de Donato G, Resch T, Setacci F, Sonesson B, et al. The AAA with a challenging neck: outcome of open versus endovascular repair with standard and fenestrated stent-grafts. J Endovasc Ther. 2009;16(2):137–46.

106. Navarro TP, de Castro Bernardes R, Procopio RJ, Leite JO, Dardik A. Treatment of hostile proximal necks during endovascular aneurysm repair. Aorta (Stamford). 2014;2(1):28–36.

107. Patel SR, Ormesher DC, Smith SR, Wong KHF, Bevis P, Bicknell CD, et al. A risk-adjusted and anatomically stratified cohort comparison study of open surgery, endovascular techniques and medical management for juxtarenal aortic aneurysms—the UK COMPlex AneurySm study (UK-COMPASS): a study protocol. BMJ Open. 2021;11(11):e054493.

Hemodynamic Principles of Endografts

50

Efstratios Georgakarakos

Background

Endovascular repair (EVAR) of abdominal aortic aneurysms (AAA) comprises a rapidly evolving surgical field not only with respect to the improvement and development of older and newer endograft designs but also when considering the understanding, prediction, prevention, and management of related complications [1–3]. Undoubtedly, the "Achilles heel" of EVAR remains the loss of sealing and/or fixation at the apposition sites of the infrarenal and iliac segments as well as the migration/dislodgment of endografts due to excessive forces (displacement forces, DF) that apply in the different segments of the endografts. Early perceptions regarding migration focused on structural deficiencies and technical issues of the devices affecting deployment and apposition of the endografts, material porosity, and fatigue, as well as insufficient flexibility of the devices to accommodate on tight, calcified or angulated, and tortuous iliac anatomies [3]. Yet, the wide application of EVAR and the current sophisticated 3D imaging modalities enriched and broadened our knowledge on such failures, underscoring—among others—the role of geometry on the direction and magnitude of DF and its predictive role for adverse effects [4]. Therefore, basic consideration of the hemodynamic principles of EVAR provides a useful tool of the clinicians toward the proper choice of endografts and avoidance of complications.

Aneurysm Remodeling: A Multifactorial Issue

The aneurysm sac undergoes on-going changes through the postinterventional period, with consequent morphological changes not only in its size but also in the implanted endograft. Since the very early era of EVAR, Krämer et al. recorded angular changes in all parts of the endografts (proximal neck, limbs, and midgraft) for a follow-up of 2 years, while van Keulen et al. confirmed further these observations in 3 years [5, 6]. The evolution of imaging techniques also enabled the assessment of additional geometric parameters, such as the iliac tortuosity, which decreased postoperatively [7]. Furthermore, insight into the mechanical properties of the aneurysm wall postoperatively provided us with valuable knowledge such as the increase in stiffness and decrease in aortic compliance immediately post-EVAR that are, however, not associated with the rate of AAA sac diameter decrease [8, 9].

Computational simulation studies showed that the von Mises peak wall stress (PWS), comprising a measure of strain on the AAA wall, is markedly redistributed from the AAA sac preoperatively to the sites of infrarenal necks and proximal iliacs postoperatively, while the forces and resultant strains and stresses are also redistributed along the implanted endografts, with the largest percentage applied at the bifurcation site, predisposing to dislodgment and significant angulations between the endograft's modular parts [10–12]. Therefore, a schematic representation of the aforementioned combined steps would prioritize the postoperative redistribution of wall stresses on the AAA from the sac to the infrarenal neck, leading to shrinkage of the former and distention of the latter, respectively, with the drag forces acting on the neck and predisposing to migration of the endograft. Consequently, the AAA sac, lumen, and endograft should be considered as a conjugated system, where the alterations in one element do not affect only itself, but also the other two counterparts, directly or indirectly [13]. The remodeling of the AAA sac interacts with the wall stresses which affect the sac shrinkage, the latter being also determined by the sac wall compliance and the intrasac pressure. In turn, the wall compliance interacts with the sac pressure, which influences sac remodeling. Moreover, the presence of endoleaks changes the intrasac pressure, which is also influenced by the presence of an intraluminal thrombus, both having an impact on sac shrinkage [13].

E. Georgakarakos (✉)
Department of Vascular Surgery, "Democritus" University of Thrace, University Hospital of Alexandroupolis, Alexandroupolis, Greece

G. Geroulakos et al. (eds.), *Mastering Endovascular Techniques*, https://doi.org/10.1007/978-3-031-42735-0_50

The Infrarenal Neck: Modes of Fixation, Central Sealing, and Distribution of Forces

The development and improvement of the endograft's central sealing and fixation mode have always been placed at the center of endovascular innovation. Classic designs include trimodular endografts with or without suprarenal fixation modes (Fig. 50.1a, b). In vitro experiments have demonstrated the beneficial role of hooks and barbs as a means to resist DF since the DF measure needed to detach the endograft in such bench-side experiments was greater compared to endografts lacking these mechanisms [14, 15]. Latest bench-side experiments have addressed additionally the presence of infrarenal angulation, showing a decrease in the magnitude of DF needed to dislodge an angulated endograft [16].

The neck angulation plays a crucial role not only in the DF magnitude but also in the efficiency of central stent apposition at the neck; while neck angulation does not alter significantly the degree of apposition in long necks, there is a great discrepancy with this regard in the level of apposition between the non-angulated short and the angulated short necks. As De Bock et al. demonstrated, the apposition of central stents is greatly decreased exponentially in angles greater than 40° for stents of 10% oversizing, whereas a sufficient degree of apposition in short necks with an angulation of 40°–65° seems to comply better with a 30% stent oversizing (i.e., greater pressure and better apposition) [17]. On the contrary, high oversizing is associated with suboptimal apposition with more "wrinkles" (folding) in low or average angulation, comprising a disadvantage in such cases. On the other hand, low oversizing of 10% is associated with an asymmetry on the distribution of applied nitinol-derived outward radial forces applied on the infrarenal neck [17]. Applying greater degree of oversizing than the average 10–20% is met often in clinical practice in cases of short or conical necks [18, 19].

Having in mind that high values of wall stresses are calculated at the infrarenal neck postoperatively, as mentioned above, a different sealing approach that improves the apposition and forces-asymmetry on irregular necks is based on an outer-placed fabric that is sutured only at its proximal and distal ends on a conformable endoskeleton of interconnected chromium-cobaltium stents, which can slide onto each other, thereby accommodating efficiently on these necks (Fig. 50.2) [20]. This loose association enables greater conformability of the AFX endograft with the application of lower radial forces [20]. Alternatively, the advent of polymer-based technology offers a new insight into the sealing issue, carrying low radial forces with accommodation onto irregular neck surfaces, without implicating nitinol-based high outward radial forces associated with pressure-derived neck distention or forces-asymmetry dislodgment and loss of sealing (Fig. 50.3a, b) [21–23].

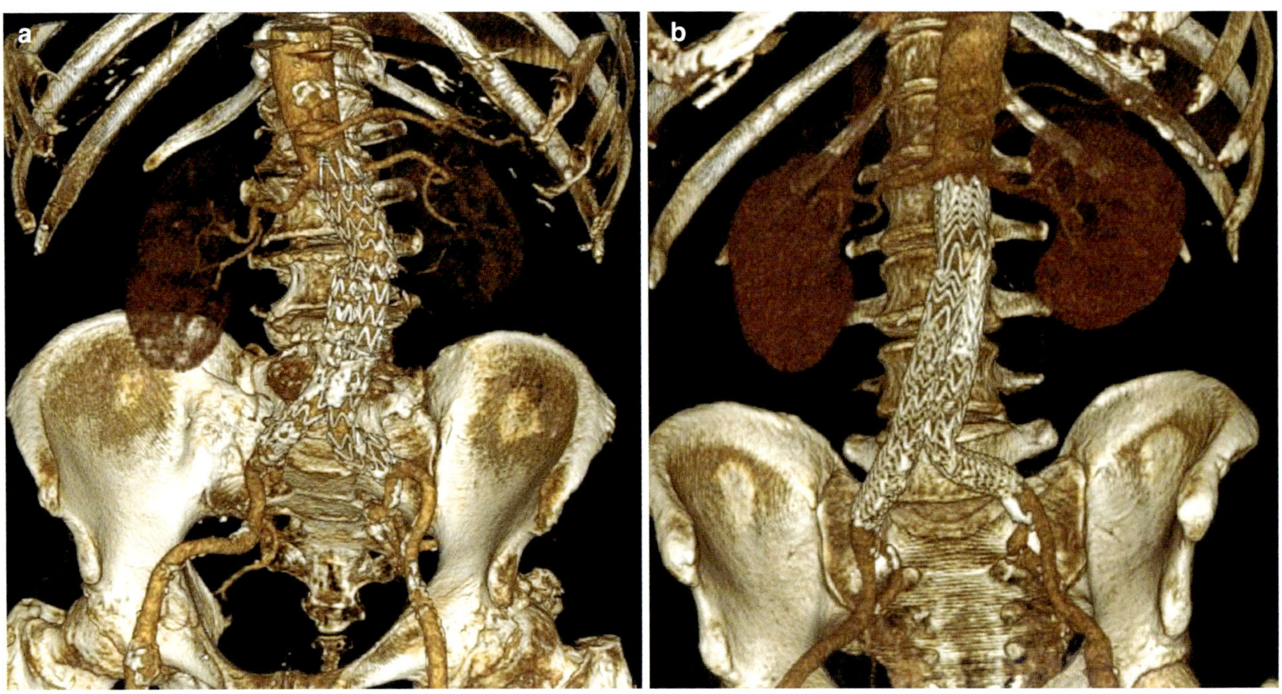

Fig. 50.1 (**a**) The Endurant trimodular endograft carries active suprarenal fixation with pins (Medtronic Inc., Minneapolis, Minnesota). (**b**) The Gore C3 Excluder (W.L. Gore & Associates, Flagstaff, AZ, USA) carries no suprarenal but rather active infrarenal fixation with anchors

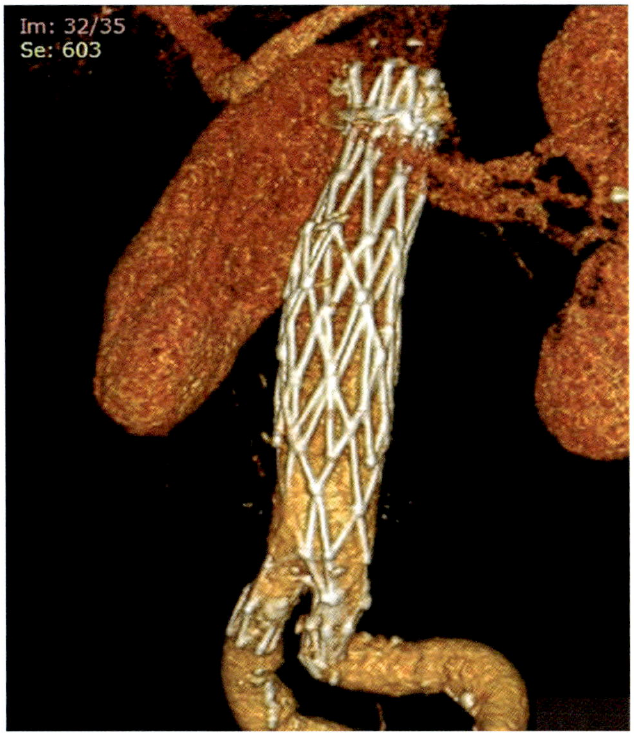

Fig. 50.2 The cobalt-chromium endoskeleton of the AFX device (Endologix, Inc., Irvine, CA, USA) with the proximal sealing extension (VELA). The endoskeleton consists of interconnected stents along the entire length attached only at the proximal and distal ends of the inner surface of the graft. The stents can slide onto each other and adapt to a range of accommodations

Iliac Limbs: Modes of Fixation, Endoleaks, and the Role of Geometry

As early computational simulation studies have demonstrated, the greatest part of DF is detected in the bifurcation (flow divider) of the endografts with the lowest percentage seen at the iliacs, yet without being negligible [4, 12, 24]. Yet, the curvature or significant tortuosity often met in the AAA iliac vessels creates considerable sideways forces that predispose to displacement of the endograft from the iliac sites, even in a cranial, upward direction [25]. A thorough bench-study by Roos et al. demonstrated that the increased iliac angulation can increase significantly the DF acting on iliacs, almost doubling the DF values from 45° to 90°, similarly to the wide distal limb diameters [26, 27]. Notably, apart from the diameter of the iliacs or the ratio of the neck/iliac (i.e., inlet/outlet) diameters, the ratio of the mainbody/iliac lengths has been also shown to affect the DF acting along the limbs [28]. Consequently, modifications to enhance the conjugation between the iliac counterparts have been introduced, by means of the five rounded barbs located at the distal segment of iliac limb gates in order to engage effectively the iliac limbs, thus withstanding the dislocating forces that can be as high as those acting onto the endograft neck segment [12, 27, 28].

Fig. 50.3 (**a**) The Ovation (Endologix Inc. Irvine, CA, USA) aortic endograft combines active suprarenal fixation via a long nitinol stent and a pair of polymer-filled inflatable rings to accommodate effectively even on a wide range of infrarenal neck irregularities, such as conical necks (dashed lines), where the asymmetric apposition of a classic nitinol design may be associated with proximal failure (**b**)

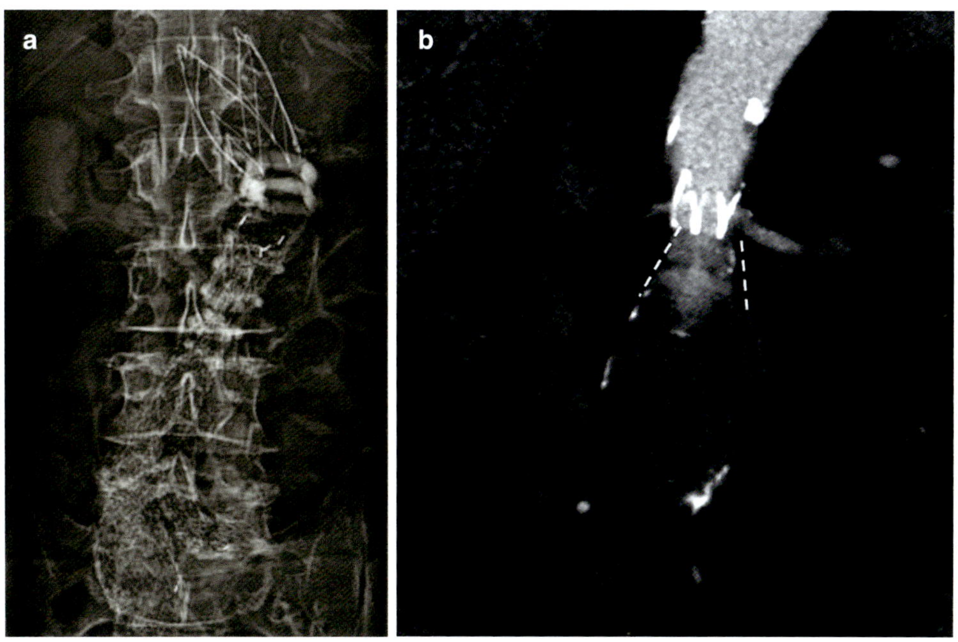

The Influence of Geometry on the Hemodynamic Profile of Endografts

The major factors that determine the magnitude and the direction of DF on the implanted endografts are the diameter and angulation of the infrarenal neck. Early computational studies have also identified a significant influence on the ratio of the neck/iliac diameter, whereas the bifurcation angle spares a minor role [29, 30]. Newer studies also confirm the dominance of neck angulation and the neck/iliac diameter ratio, demonstrating the anterolateral rather than the cranio-caudal direction of DF (Fig. 50.4) [31]. This hemodynamic feature interprets the rather high incidence of modular disengagement with consequent type-III endoleaks in the early unibody bifurcated platforms. While their direct anatomical fixation on the aortic bifurcation was believed to counteract for caudal endograft migration and loss of central sealing, the lateral or anterolateral direction of DF affected the vulnerable overlap segment between the bifurcate and the proximal extension; accordingly, current instructions-for-use support a greater overlap according to the diameter of the AAA [20].

Another factor that has recently gained great attention is the tortuosity of the endograft, determined by the shape of the AAA lumen extending from the infrarenal level to the iliacs [32]. The tortuosity of the AAA centerline has been proposed as a useful and accurate parameter associated with high wall stress values on AAA wall [33]. Clearly, the curvature increases the DF acting on the endograft, while the configuration of the neck and/or iliacs affects also the magnitude of DF [32]. Interestingly, aortic tortuosity has been associated in clinical studies not only with early and late incidence of central- and mid-segment loss of sealing (Fig. 50.5a, b), but also with the risk of direct intraoperative type 1a endoleak [34–36].

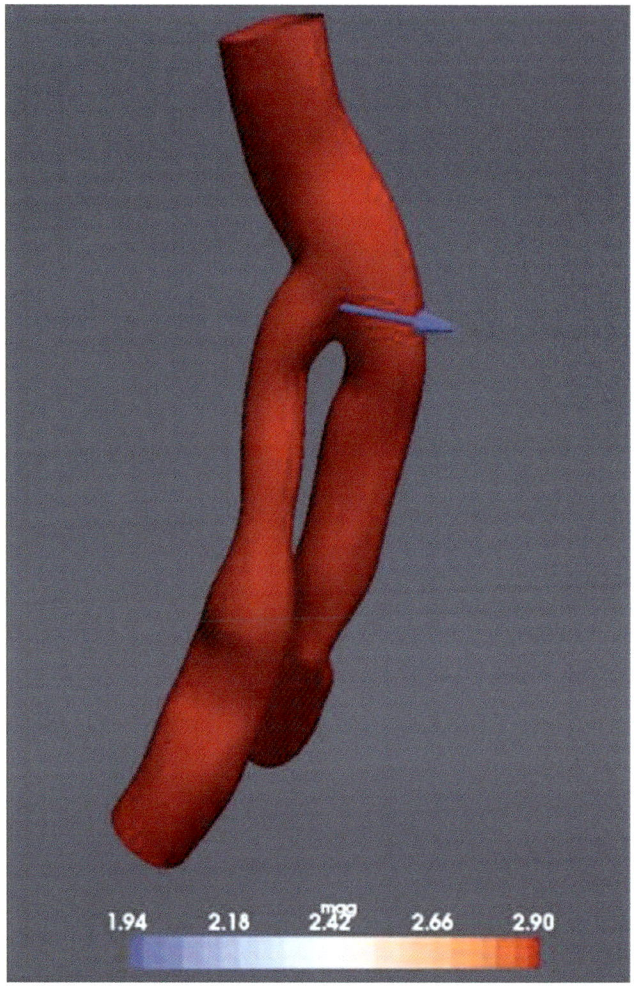

Fig. 50.4 A 3D computationally reconstructed model of a trimodular endograft, assessing the anterior direction of the total displacement force

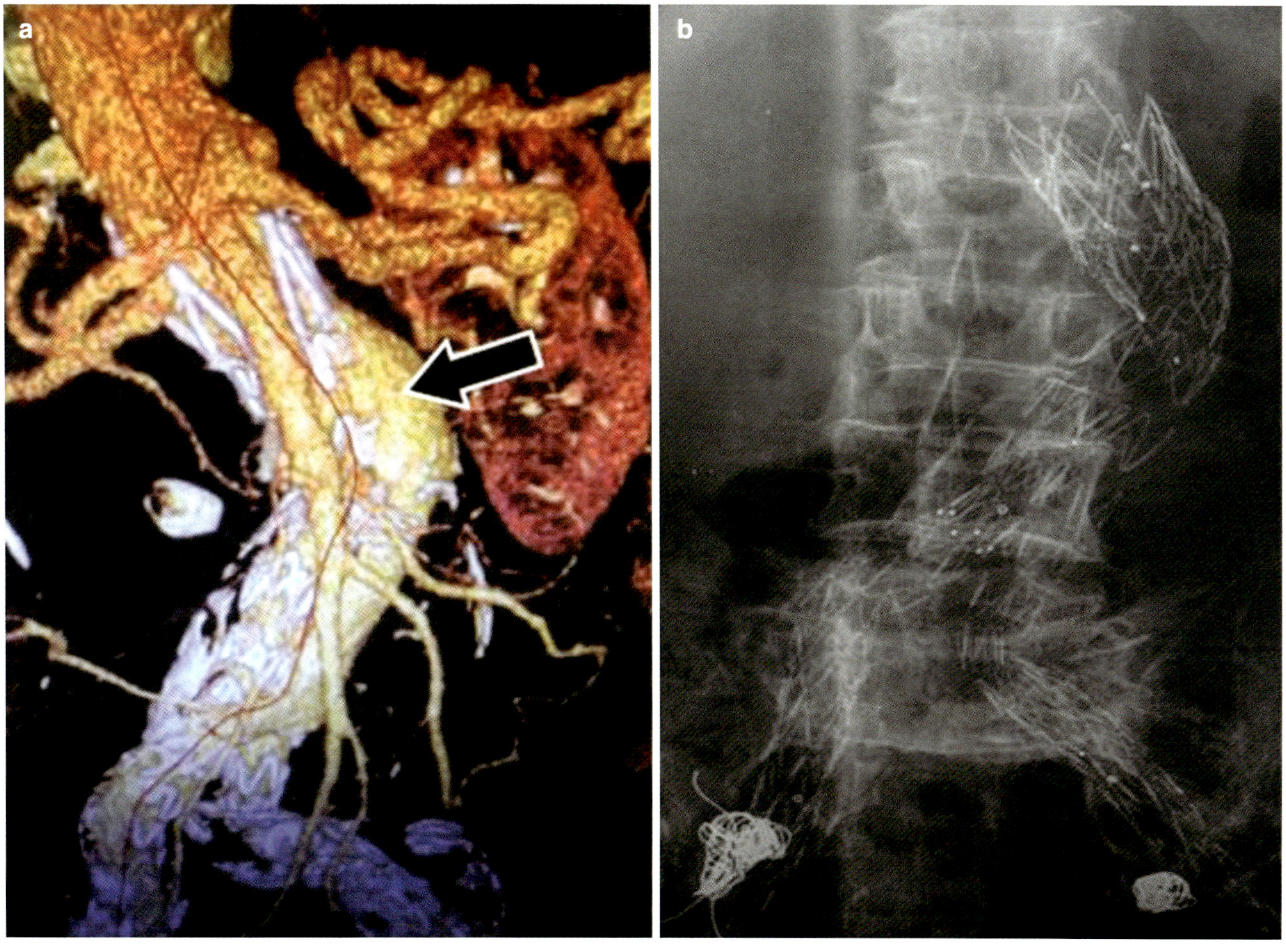

Fig. 50.5 (**a**) An abdominal aortic aneurysm with marked curvature undergoing endovascular repair. Early follow-up revealed migration of the endograft with loss of central sealing and endoleak (arrow). Accordingly, placement of a central aortic cuff provided sufficient sealing (**b**)

The Impact of EVAR on Central Circulation

Endografts of certain structure (i.e., fabric of polyester or polytetrafluoroethylene with skeleton of nitinol, stainless steel, or chromium-cobaltium), mechanical properties (e.g., austenite and martensite elasticity, tensile strength, transformation strain, longitudinal and circumferential bending stiffness) various designs and geometrical characteristics (unibody or trimodular endografts, variable length of main-body, etc.) are implanted in EVAR within an AAA sac of altered compliance compared to normal aortas [37]. Hence, a question rises regarding potential consequences on the propagation of pressure waves from the interposition of a structure of altered stiffness.

Some studies have reported an increase in pulse wave velocity (PWV) after EVAR, while others also document alterations in cardiac function, such as deterioration in ventricular global longitudinal strain (i.e., an important diagnostic and prognostic index of early systolic dysfunction), left ventricular hypertrophy and diastolic dysfunction without a corresponding elevation of blood pressure in the acute and chronic phases [38–42]. Differences in the abovementioned changes have been attributed to different types of endografts; for example, Kadoglou et al. identified recipients of polyester-covered stent-grafts to exhibit greater PWV elevation than those with polytetrafluoroethylene (PTFE) fabric, whereas others detected differences in PWV between the unibody bifurcated endograft and the sealing design based on polymer endobags and separate covered stents [39, 43].

In-Stent Flow Patterns: An Underappreciated Issue

The postoperative reshaping of the aorta generates a new field with different flow patterns [11]. Endografts with the aorta are not susceptible only to DF predisposing to migration and loss of sealing, but also to alterations in flow patterns that render the stents susceptible to propagation of thrombus, myointimal hyperplasia, and/or accumulation of

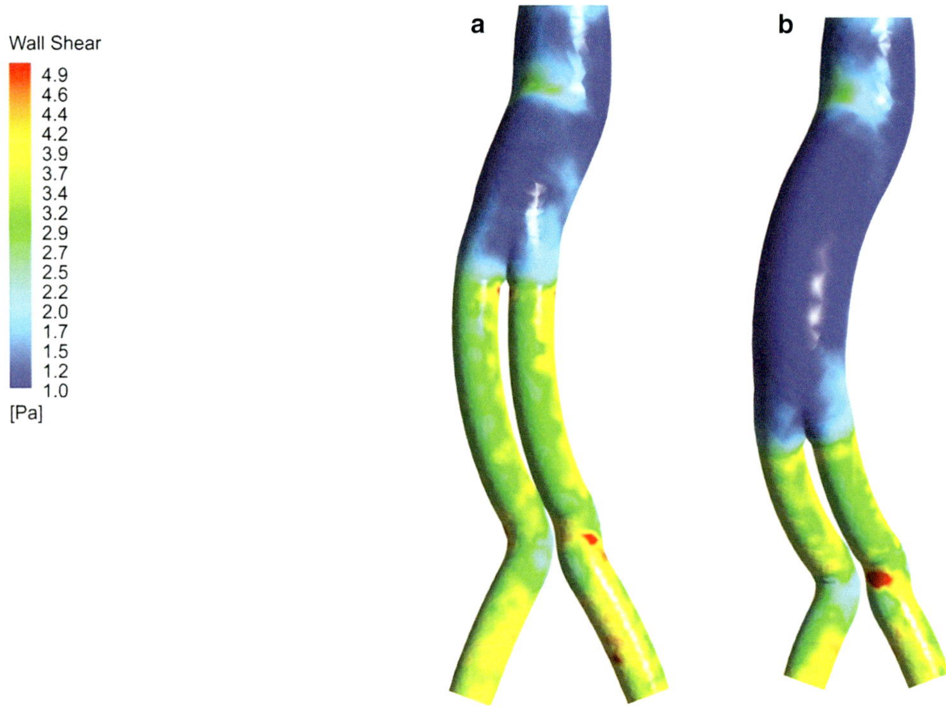

Fig. 50.6 Contours of shear stress in endograft models with mainbody/limb length ratios of 0.6 (**a**) and 1.2 (**b**), respectively, in peak systole

thrombus with subsequent stenosis or thrombosis of the limbs [44, 45].

Modern studies focus on hemodynamic indices such as relative residence time (RRT), oscillatory shear index (OSI), or time-averaged wall shear stress (TAWSS), the magnitude and distribution of which are related to thrombosis induction [11, 44, 46]. The flow is quantified by TAWSS. During a pulsatile cycle, the direction of WSS varies instantaneously, and the definition of OSI is used to elaborate this situation; OSI is a measure of the oscillatory behavior of the flow and indicates areas of flow reversal. The RRT is a measure for residence time of particles near the vessel wall. Relatively high values present potential regions with flow recirculation compared with regions with a near-zero RRT. Furthermore, helicity measures the strength of relative rotations of helical fluid structures in the fluid domain. High WSS is generally considered to be a desirable flow characteristic in endovascular devices, whereas helicity extremes (very low or high) are potentially a negative sign.

According to Raptis et al., different types of endografts induce variable hemodynamic conditions around the flow division. The parallel limb structure with no common mainbody, as featured in endografts based on endobag AAA sealing, promotes favorable WSS conditions. The trimodular bifurcated designs of neck structure are associated with low blood velocity and WSS values, yet preserving high helicity levels [46].

Boersen et al. compared the flow patterns and wall shear stress between a trimodular bifurcated model, the unibody design, and the endobag-separate stents model; an increase in the flow velocity was found at the distal end of the iliac limbs in the latter. Both OSI and WSS were similar in the medial and lateral surfaces of the iliac conduits for all models, with areas of high OSI located near the distal end of the stents [47].

Interestingly, Tasso et al. documented an inverse correlation between helical flow intensity and time-averaged DF, suggesting a protective effect of helicity against the risk of migration due to high DF values. Moreover, helical flow intensity was positively correlated with pressure drop and resistance to flow [45].

The geometry and configuration of the iliac limbs, any angulation or stenosis in these, as well as the relative ratio of mainbody/iliac limb length (Fig. 50.6a, b), can differentiate not only the flow velocities and pressures along the limbs but also the distribution of OSI, WSS, DF, and RRT along their segments [11, 44–48].

Discussion

The computational assessment of the hemodynamic characteristics of AAA and endografts and the study of geometrical parameters as surrogate markers have been at the center of vascular research during the last few years. These subjects comprise a refinement in our knowledge regarding the prediction of mid- and long-term results of AAA endovascular procedures.

Earlier aspects were almost exclusively limited in choosing the endograft size according to its instruction-for-use and the AAA geometry. The development of various endograft designs emerged the need to comprehend better the hemodynamic profile that determines in certain ways their clinical performance as well as the natural history of the AAA sac postinterventionally. It is, for example, a great step forward to realize that not only the size of the proximal (neck) diameter of the endograft or its angulation play a role in the magnitude of DF predisposing to migration risk, but also the curvature of the endograft or the proximal/distal (inlet/outlet) diameters, while the length ratios (mainbody/iliac limbs) also affect the distribution of forces acting along the different endograft segments with the accordant flow fields. These variations may carry different implications of limb thromboses, the latter comprising the most frequent cause for reintervention in the long term [49]. Moreover, new endograft designs carry no mainbody at all but two separate limbs instead, based on an alternative perception of sealing and stability, i.e., braided D-shaped nitinol endoskeletons (Fig. 50.7a, b) interacting against each other and exerting a continuous radial force against the infrarenal neck wall or two separate stent-grafts surrounded by polymer-filled endobags that seal against the infrarenal neck and the intraluminal thrombus (Fig. 50.8a, b). Therefore, studying the mechanics of such new sealing modes comprises a tool to assess the new designs and raise questions that should be further evaluated clinically, thus extending the basis of our knowledge [50]. An interesting example in this regard could be the computational evaluation of the cross-limb ("ballerina" position) configuration of endografts, which was firstly suggested in the literature a non-inferiority hemodynamic performance compared to the classical bifurcated designs, followed by several clinical studies that confirmed this finding, accordingly [24, 51].

Complex aortic surgery constitutes currently a cutting-edge research field for computational modeling and simulations. Fenestrated and chimney techniques for EVAR (FEVAR and chEVAR, respectively) provide successful solutions for thoracoabdominal, pararenal, and juxtarenal AAA, yet not without adverse effects [52]. The side branches are prone to separation, fracture, thrombosis, or dislocation attributed to technical and structural matters but also to changes in certain hemodynamic parameters related to the take-off angles of the branches, such as the TAWSS or DF acting on these branches [53–56]. Computational studies compare the hemodynamic outcomes between custom-made and "off-the-shelf" grafting, providing a perspective for future design of "off-the-shelf" fenestrated devices [56–58]. Moreover, computational comparison between chEVAR and FEVAR aims to provide valuable information regarding renal perfusion. The mating stent-grafts in fEVAR protrude vertically from the fenestrations and may be subjected to cephalad or caudal misalignments, greatly affecting the flow fields (Fig. 50.9a–e) and the stress distribution along the stent-grafts, as well as the maximum WSS at the transition zone between the stent-graft and the native vessels. On the contrary, the gradual transition of the flow field in chEVAR may enable less attenuation of significant WSS variations compared with FEVAR [59]. Moulakakis et al. recently identified the relatively long, curved length of the chimney grafts as responsible for a greater decrease in flow along with the

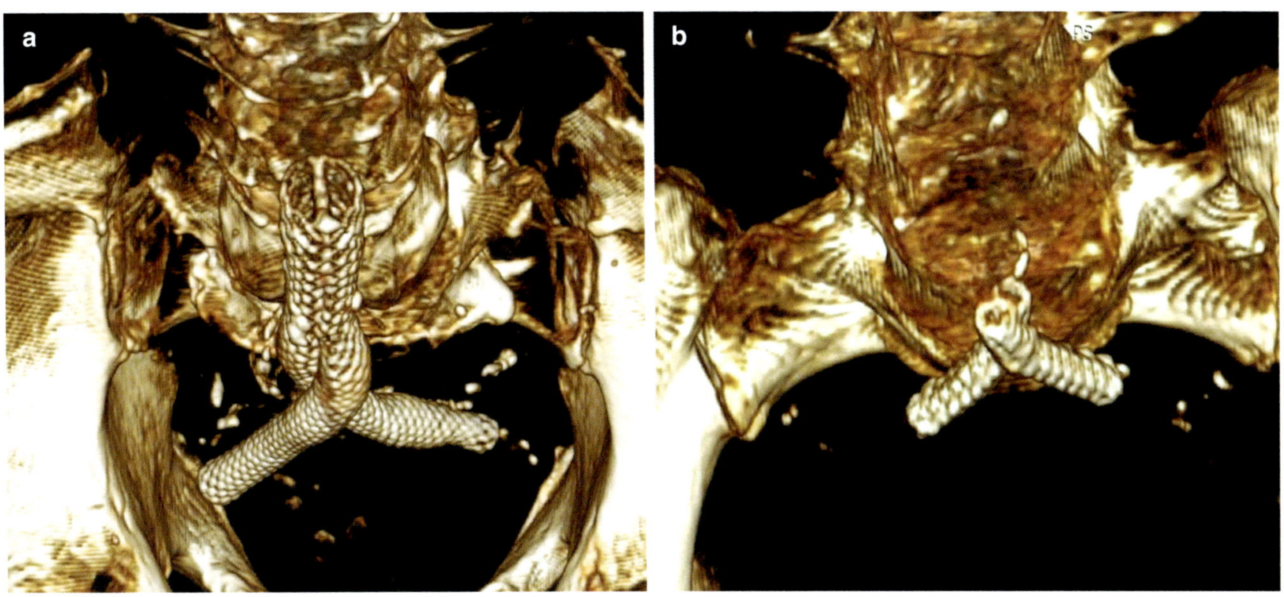

Fig. 50.7 (**a**) The dual-lumen configuration of the Altura device (Lombard Medical, Ltd., Oxford, UK). Its self-expanding D-shaped endografts apply radial forces across the infrarenal neck and also across the medial septal surfaces against each other, achieving stability, sealing and perfect apposition compared to the classic configuration of two "kissing-stents" (**b**)

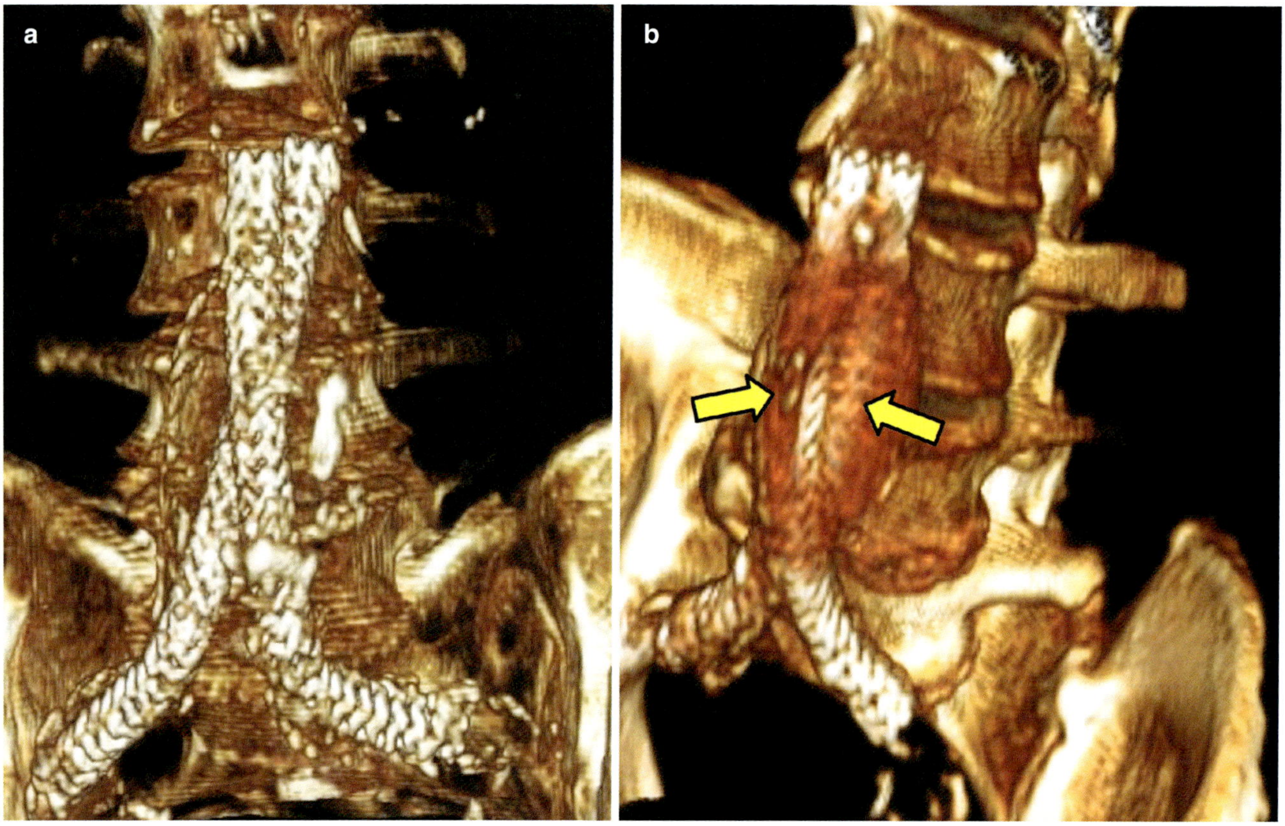

Fig. 50.8 (**a**) The Nellix endoprosthesis carries two separate nitinol bilateral endografts surrounded by two polymer-filled endobags responsible for wall apposition and sealing (arrows, **b**)

reduced cross-section of the chimney grafts along their length due to their compression between the main graft and the aortic wall [60]. These computational observations bring up the issue of the absence of dedicated stent-grafts for FEVAR and chEVAR, as well as the issue of proper oversizing in chEVAR, which is determined in a rather empirical way [60, 61]. Therefore, a combination of further clinical and computational studies is awaited to shed light on critical issues related to these methods.

Computationally aided hemodynamic studies also provide valuable information with respect to innovative custom-made patterns introducing (1) excessively short mainbodies with low main-body/iliac limb length ratios, (2) "dog-bone"-shaped iliac limbs with abrupt diameter changes and consequent velocity and shear stress alterations over the flow-lumen segments, or (3) tapered designs of unequal proximal and distal diameters. These models may theoretically host variant hemodynamic effects on separate parts of these designs, the clinical performance of which remains to be validated.

Geometrical parameters associated with hemodynamic indices could serve as efficient surrogate markers with a predictive role for hemodynamically related adverse effects, providing an insightful and useful adjunct to assess various therapeutic options in challenging cases. For example, Fig. 50.10 depicts a case of AAA with marked tortuosity of the lumen and an angulated, aneurysmatic left iliac artery. Apart from the risk of migration and central loss of sealing, the options for peripheral sealing include either a bell-bottom-shaped iliac limb for sealing in the common iliac artery or coil-occlusion of the internal iliac artery and extension of the iliac limb to the external iliac artery. The choice of the physician should take into account the higher DF acting on the bell-bottom limbs that could lead to central limb migration, as previously mentioned, or the risk of late limb thrombosis when this lands in the external iliac artery, especially if the latter is narrow [62, 63]. Additionally, an iliac limb of PTFE and helical stent could be taken into consideration for sealing in the external iliac artery, being also more flexible against the angulated iliac axis [26, 27, 37, 63]. If the infrarenal neck was also short and/or wide, posing additional risk factors—apart from tortuosity—to migration, then the choice of mechanism of central sealing (oversized nitinol vs. polymer-based sealing) could be an additional issue.

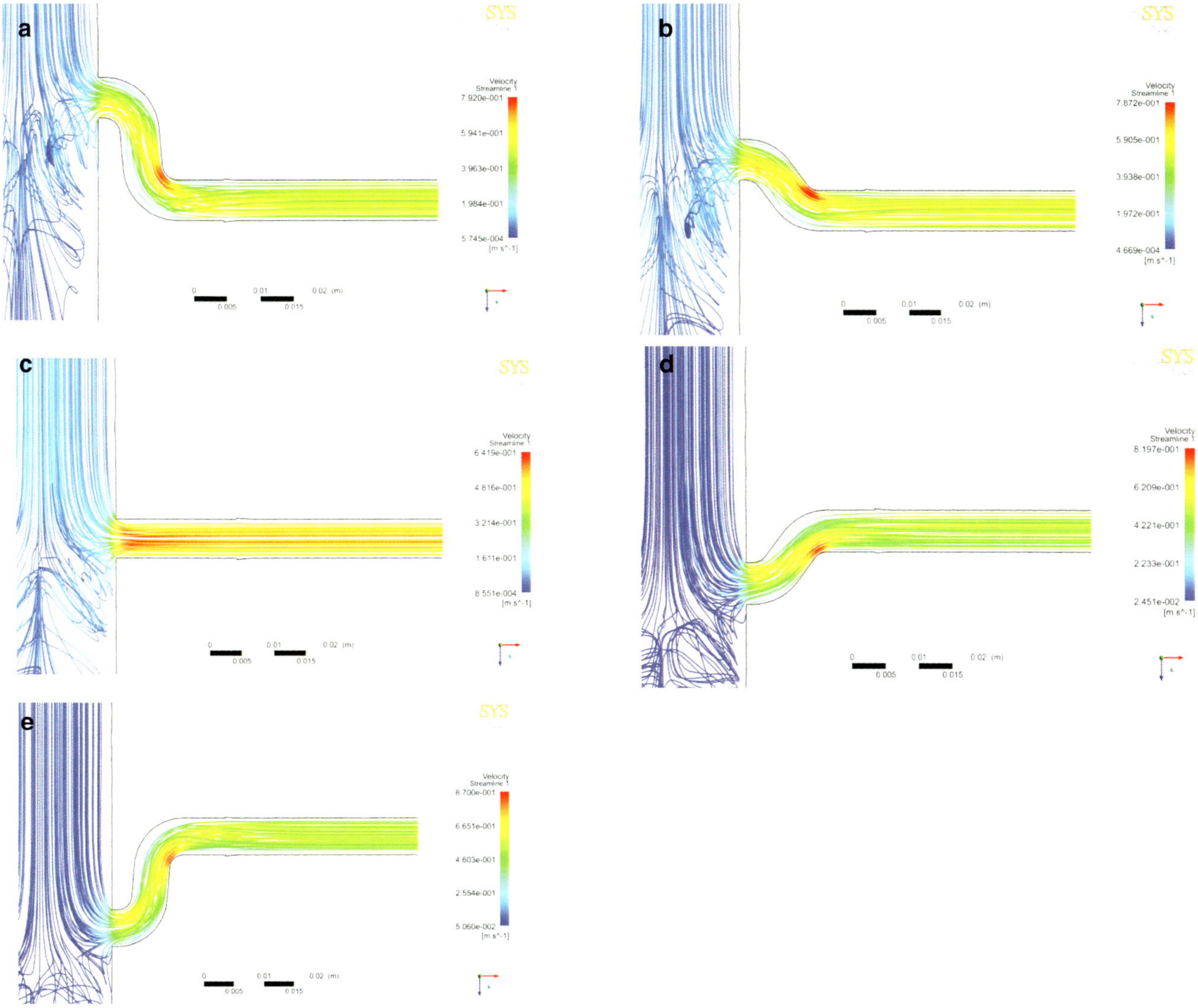

Fig. 50.9 Schematic representation of the velocity fields (peak systolic phase) in the branch of a computationally reconstructed model of a fenestrated design under several take-off angles; 4°,28°,90°,142°, and 176°,respectively (**a–e**)

Interestingly, in cases where a sole endograft does not concentrate all the ideal yet essential properties and characteristics to address a challenging or particular AAA case, a hybrid combination of different endograft parts has been described in the literature [64, 65]. This may be a necessity, especially in cases of failing EVAR, where anatomic constraints (e.g., short renal artery to endograft bifurcation length) preclude repair either with FEVAR or with the insertion of a new endograft [66]. In such cases, the hybrid combination of different endograft parts may provide an off-label yet acceptable solution (Fig. 50.11a, b) [64, 66]. The study of the hemodynamic parameters of such combinations provides a valuable tool to help both clinicians and manufacturers extract useful conclusions and modify existing or develop new endograft designs [67].

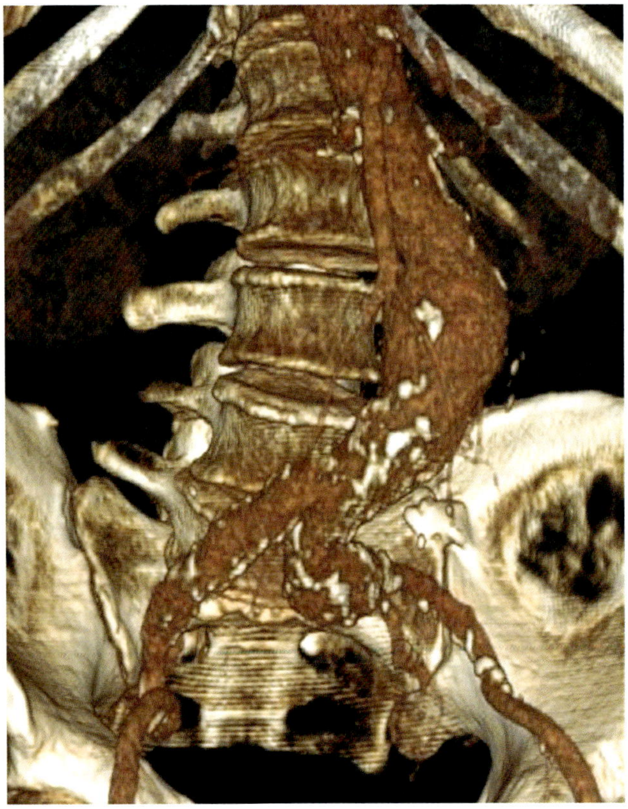

Fig. 50.10 An abdominal aortic aneurysm with marked tortuosity of the lumen and an angulated aneurysmatic left iliac artery

Limitations

Early computational studies were conducted on idealized models which served well for parametric purposes in order to focus on the role of certain parameters [12, 24, 25, 29–32]. These studies evolved further to apply on patient-specific 3D reconstructed models and realistic geometries with patient-specific boundary conditions (i.e., recorded pressure and flow velocity) to provide further insight with hemodynamic comparisons [45, 46, 50, 68].

Undoubtedly, the most challenging issue in interpreting the hemodynamic studies is the interpretation and clinical application of the experimental findings, since many of the findings are expressed with terms that may sound unfamiliar to everyday clinicians; for example, the OSI, WSS, and RRT terms may sound hypothetical and impractical to characterize the performance of an endograft. Accordingly, the conclusions may read vague, serving the cause in rather teleological than etiological interpretations since the relevant literature lacks prospectively designed studies based on clinical hypotheses to be validated by certain clinical results. It is crucial to underline the collaboration of clinicians and technicians to (1) design and conduct the studies, (2) interpret the findings, and (3) modify or develop new endograft designs.

The lack of standardized reference values limits the value of the reported results related to the hemodynamic parame-

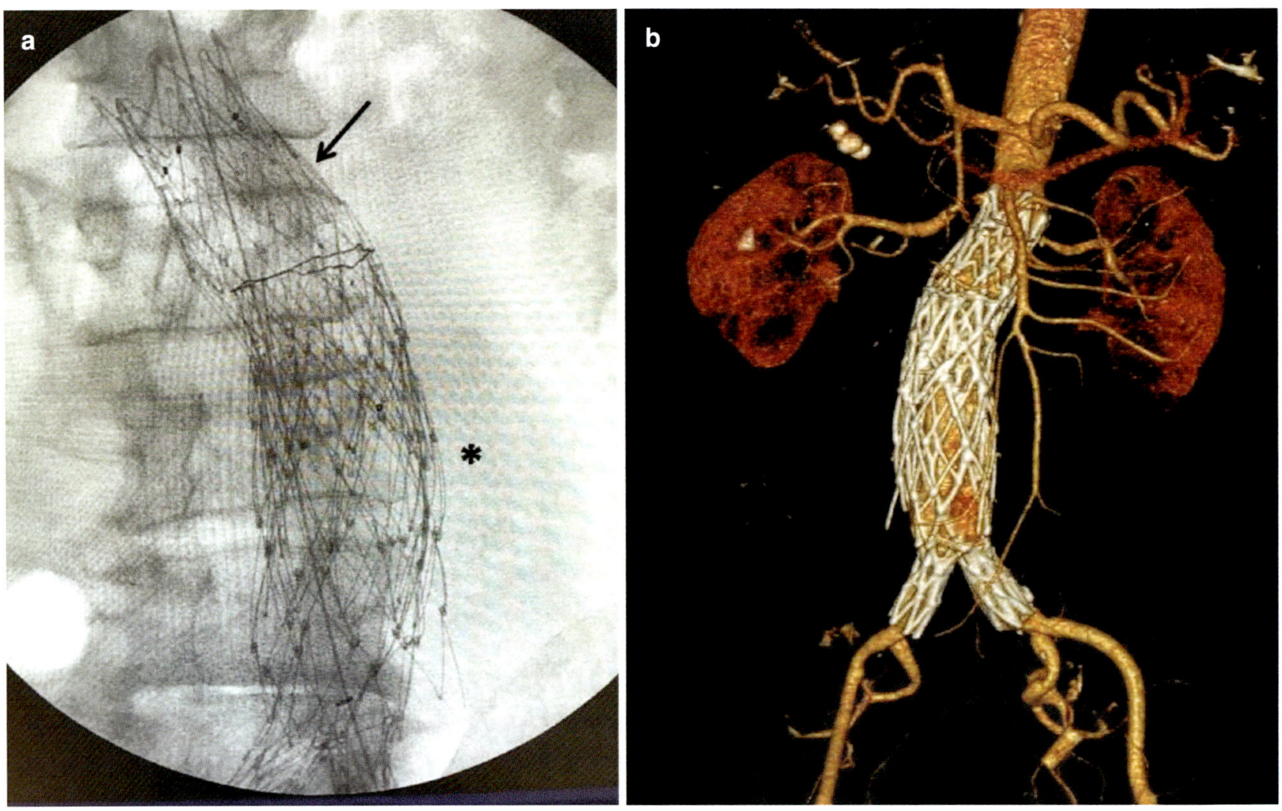

Fig. 50.11 (**a**) X-ray of a hybrid off-label combination of an AFX (asterisc) bifurcate (Endologix, Inc., Irvine, CA, USA) with a proximal (arrow) nitinol-based Endurant cuff (Medtronic Inc., Minneapolis, Minnesota). (**b**) Computed tomography angiography of the AFX-Endurant cuff combination

ters; for example, while the PWV studies report differences post-EVAR with respect to certain endograft designs, their value is limited if these calculations do not take into account the sex and age of patients [69]. AAA reports with respect to PWV changes seem to overlook the PWV association with gender and age, which follows an exponential distribution pattern where the normal range of values is determined between +90 and −90% confidence intervals; consequently, certain PWV values and their postoperative changes may have variable significance between different age-groups, being normal for the one but pathologic for others [69]. Therefore, it is evident that apart from PWV, changes in other arterial stiffness indexes could further describe and predict the impact of aortic stent-graft implantation on aortic stiffness. The hemodynamic consequences vary among patients and should be complementarily approached with augmentation index (AIx), augmentation pressure (AP), and reflection magnitude (RM), since the postoperative changes in these parameters vary and are neither unidirectional and linear nor equivocal [70].

Finally, the sophisticated methodology and software used by specialists limit their applicability to a wider audience and clinicians. Accordingly, development of more applicable, semiautomated methods easy to use and apply comprise a strategic task to enrich and widen our tools toward better understanding and prediction of the hemodynamic and clinical performance of endografts [71].

Conclusions

We discussed the basic hemodynamic and mechanical features that the clinician should bear in mind when assessing an aneurysm or choosing between several endografts for a particular case. Computational simulations use parameters such as TAWSS, OSI, DF, RRT, and helicity to predict and compare the adverse effects such as migration, stenosis, and thrombosis of the components and branches. The unique anatomy of a certain aneurysm defines a hemodynamic environment where the implantation of endografts of different geometric characteristics evokes different hemodynamic and clinical performance. There is no ideal endograft design; rather, every AAA has a unique anatomy served better by some endografts than others and vice versa. The hemodynamic information derived from experimental studies helps clinicians with operational planning and manufacturers with further development and improvement of endograft designs.

Acknowledgments None.

Conflict of Interest None.

No Financial Disclosures None.

References

1. Powell JT, Sweeting MJ, Ulug P, Blankensteijn JD, Lederle FA, Becquemin JP, et al; EVAR-1, DREAM, OVER and ACE Trialists. Meta-analysis of individual-patient data from EVAR-1, DREAM, OVER and ACE trials comparing outcomes of endovascular or open repair for abdominal aortic aneurysm over 5 years. Br J Surg. 2017;104:166–178.
2. Patel R, Sweeting MJ, Powell JT, Greenhalgh RM, EVAR Trial Investigators. Endovascular versus open repair of abdominal aortic aneurysm in 15-years' follow-up of the UK endovascular aneurysm repair trial 1 (EVAR trial 1): a randomised controlled trial. Lancet. 2016;388:2366–74.
3. Schoretsanitis N, Georgakarakos E, Argyriou C, Ktenidis K, Georgiadis GS. A critical appraisal of endovascular stent-grafts in the management of abdominal aortic aneurysms. Radiol Med. 2017;122:309–18.
4. Georgakarakos E, Argyriou C, Schoretsanitis N, Ioannou CV, Kontopodis N, Morgan R, et al. Geometrical factors influencing the hemodynamic behavior of the AAA stent grafts: essentials for the clinician. Cardiovasc Intervent Radiol. 2014;37:1420–9.
5. Krämer SC, Seifarth H, Pamler R, Fleiter T, Görich J. Geometric changes in aortic endografts over a 2-year observation period. J Endovasc Ther. 2001;8:34–8.
6. van Keulen JW, Moll FL, Arts J, Vonken EJ, van Herwaarden JA. Aortic neck angulations decrease during and after endovascular aneurysm repair. J Endovasc Ther. 2010;17:594–8.
7. Coulston J, Baigent A, Selvachandran H, Jones S, Torella F, Fisher R. The impact of endovascular aneurysm repair on aortoiliac tortuosity and its use as a predictor of iliac limb complications. J Vasc Surg. 2014;60:585–9.
8. van Herwaarden JA, Muhs BE, Vincken KL, van Prehn J, Teutelink A, Bartels LW, et al. Aortic compliance following EVAR and the influence of different endografts: determination using dynamic MRA. J Endovasc Ther. 2006;13:406–14.
9. Long A, Rouet L, Vitry F, Albertini JN, Marcus C, Clement C. Compliance of abdominal aortic aneurysms before and after stenting with tissue doppler imaging: evolution during follow-up and correlation with aneurysm diameter. Ann Vasc Surg. 2009;23:49–59.
10. Molony DS, Callanan A, Kavanagh EG, Walsh MT, McGloughlin TM. Fluid-structure interaction of a patient-specific abdominal aortic aneurysm treated with an endovascular stent-graft. Biomed Eng Online. 2009;8:24.
11. Raptis A, Xenos M, Georgakarakos E, Kouvelos G, Giannoukas A, Labropoulos N, et al. Comparison of physiological and post-endovascular aneurysm repair infrarenal blood flow. Comput Methods Biomech Biomed Engin. 2017;20:242–9.
12. Howell BA, Kim T, Cheer A, Dwyer H, Saloner D, Chuter TA. Computational fluid dynamics within bifurcated abdominal aortic stent-grafts. J Endovasc Ther. 2007;14:138–43.
13. Georgakarakos E, Georgiadis GS, Ioannou CV, Kapoulas KC, Trellopoulos G, Lazarides M. Aneurysm sac shrinkage after endovascular treatment of the aorta: beyond sac pressure and endoleaks. Vasc Med. 2012;17:168–73.
14. Melas N, Saratzis A, Saratzis N, Lazaridis J, Psaroulis D, Trygonis K, et al. Aortic and iliac fixation of seven endografts for abdominal-aortic aneurysm repair in an experimental model using human cadaveric aortas. Eur J Vasc Endovasc Surg. 2010;40:429–35.
15. Bosman WM, Steenhoven TJ, Suárez DR, Hinnen JW, Valstar ER, Hamming JF. The proximal fixation strength of modern EVAR grafts in a short aneurysm neck. An in vitro study. Eur J Vasc Endovasc Surg. 2010;39:187–92.
16. Rahmani S, Grewal IS, Nabovati A, Doyle MG, Roche-Nagle G, Tse LW. Increasing angulation decreases measured aortic stent graft pullout forces. J Vasc Surg. 2016;63(2):493–9.

17. De Bock S, Iannaccone F, De Beule M, Vermassen F, Segers P, Verhegghe B. What if you stretch the IFU? A mechanical insight into stent graft instructions for use in angulated proximal aneurysm necks. Med Eng Phys. 2014;36:1567–76.
18. Mwipatayi BP, Picardo A, Wong J, Thomas SD, Vijayan V. Endovascular repair of abdominal aortic aneurysms with reverse taper neck anatomy using the Endurant stent-graft: analysis of stent-graft oversizing. J Endovasc Ther. 2013;20:514–22.
19. Pitoulias GA, Valdivia AR, Hahtapornsawan S, Torsello G, Pitoulias AG, Austermann M, et al. Conical neck is strongly associated with proximal failure in standard endovascular aneurysm repair. J Vasc Surg. 2017;66:1686–95.
20. Georgakarakos E, Ioannidis G, Koutsoumpelis A, Papatheodorou N, Argyriou C, Spanos K, et al. The AFX unibody bifurcated unibody aortic endograft for the treatment of abdominal aortic aneurysms: current evidence and future perspectives. Expert Rev Med Devices. 2020;17:5–15.
21. de Donato G, Pasqui E, Panzano C, Brancaccio B, Grottola G, Galzerano G, Benevento D, Palasciano G. The polymer-based technology in the endovascular treatment of abdominal aortic aneurysms. Polymers (Basel). 2021;13:1196.
22. Boersen JT, Groot Jebbink E, Van de Velde L, Versluis M, Lajoinie G, Slump CH, et al. The influence of positioning of the Nellix endovascular aneurysm sealing system on suprarenal and renal flow: an in vitro study. J Endovasc Ther. 2017;24:677–87.
23. de Donato G, Setacci F, Bresadola L, Castelli P, Chiesa R, Mangialardi N, et al; TriVascular Ovation Italian Study. Aortic neck evolution after endovascular repair with TriVascular ovation stent graft. J Vasc Surg. 2016;63:8–15.
24. Georgakarakos E, Xenakis A, Manopoulos C, Georgiadis GS, Tsangaris S, Lazarides MK. Modeling and computational analysis of the hemodynamic effects of crossing the limbs in an aortic endograft ("Ballerina" position). J Endovasc Ther. 2012;19:549–57.
25. Liffman K, Lawrence-Brown MM, Semmens JB, Bui A, Rudman M, Hartley DE. Analytical modeling and numerical simulation of forces in an endoluminal graft. J Endovasc Ther. 2001;8:358–71.
26. Roos H, Ghaffari M, Falkenberg M, Chernoray V, Jeppsson A, Nilsson H. Displacement forces in iliac landing zones and stent graft interconnections in endovascular aortic repair: an experimental study. Eur J Vasc Endovasc Surg. 2014;47:262–7.
27. Roos H, Tokarev M, Chernoray V, Ghaffari M, Falkenberg M, Jeppsson A, et al. Displacement forces in stent grafts: influence of diameter variation and curvature asymmetry. Eur J Vasc Endovasc Surg. 2016;52:150–6.
28. Georgakarakos E, Xenakis A, Georgiadis GS, Argyriou C, Manopoulos C, Tsangaris S, Lazarides MK. Computational estimation of the influence of the main body-to-iliac limb length ratio on the displacement forces acting on an aortic endograft. Theoretical application to Bolton Treovance® abdominal stent-graft. Int Angiol. 2014;33:480–4.
29. Li Z, Kleinstreuer C. Analysis of biomechanical factors affecting stent-graft migration in an abdominal aortic aneurysm model. J Biomech. 2006;39:2264–73.
30. Li Z, Kleinstreuer C, Farber M. Computational analysis of biomechanical contributors to possible endovascular graft failure. Biomech Model Mechanobiol. 2005;4:221–34.
31. Molony DS, Kavanagh EG, Madhavan P, Walsh MT, McGloughlin TM. A computational study of the magnitude and direction of migration forces in patient-specific abdominal aortic aneurysm stent-grafts. Eur J Vasc Endovasc Surg. 2010;40:332–9.
32. Figueroa CA, Taylor CA, Chiou AJ, Yeh V, Zarins CK. Magnitude and direction of pulsatile displacement forces acting on thoracic aortic endografts. J Endovasc Ther. 2009;16:350–8.
33. Georgakarakos E, Ioannou CV, Kamarianakis Y, Papaharilaou Y, Kostas T, Manousaki E, et al. The role of geometric parameters in the prediction of abdominal aortic aneurysm wall stress. Eur J Vasc Endovasc Surg. 2010;39:42–8.
34. Chen PL, Hsu HL, Chen IM, Chen YY, Chou KY, Kuo TT, et al. The impact of aortic tortuosity on delayed type I or III endoleak after endovascular aortic repair. Ann Vasc Surg. 2017;41:110–7.
35. Schuurmann RCL, van Noort K, Overeem SP, Ouriel K, Jordan WD Jr, Muhs BE, et al. Aortic curvature is a predictor of late type Ia endoleak and migration after endovascular aneurysm repair. J Endovasc Ther. 2017;24:411–7.
36. Schuurmann RC, Ouriel K, Muhs BE, Jordan WD Jr, Ouriel RL, Boersen JT, et al. Aortic curvature as a predictor of intraoperative type Ia endoleak. J Vasc Surg. 2016;63:596–602.
37. Demanget N, Duprey A, Badel P, Orgéas L, Avril S, Geindreau C, Albertini JN, Favre JP. Finite element analysis of the mechanical performances of 8 marketed aortic stent-grafts. J Endovasc Ther. 2013;20:523–35.
38. Kadoglou NP, Moulakakis KG, Papadakis I, Ikonomidis I, Alepaki M, Lekakis J, et al. Changes in aortic pulse wave velocity of patients undergoing endovascular repair of abdominal aortic aneurysms. J Endovasc Ther. 2012;19:661–6.
39. Kadoglou NP, Moulakakis KG, Papadakis I, Ikonomidis I, Alepaki M, Spathis A, et al. Differential effects of stent-graft fabrics on arterial stiffness in patients undergoing endovascular aneurysm repair. J Endovasc Ther. 2014;21:850–8.
40. Gray C, Goodman P, Badger SA, O'Malley MK, O'Donohoe MK, McDonnell CO. Endovascular aneurysm repair increases aortic arterial stiffness when compared to open repair of abdominal aortic aneurysms. Vasc Endovasc Surg. 2016;50:317–20.
41. Marketou M, Papadopoulos G, Kontopodis N, Patrianakos A, Nakou E, Maragkoudakis S, et al. Early left ventricular global longitudinal strain deterioration after aortic aneurysm repair: impact of aortic stiffness. J Endovasc Ther. 2021;28:352–9.
42. Takeda Y, Sakata Y, Ohtani T, Tamaki S, Omori Y, Tsukamoto Y, et al. Endovascular aortic repair increases vascular stiffness and alters cardiac structure and function. Circ J. 2014;78:322–8.
43. Casciaro ME, Alfonso MA, Craiem D, Alsac JM, El-Batti S, Armentano RL. Predicting the effect on pulse wave reflection of different endovascular repair techniques in abdominal aortic aneurysm using 1D patient-specific models. Health Technol. 2016;6:173–9.
44. Georgakarakos E, Raptis A, Schoretsanitis N, Bisdas T, Beropoulis E, Georgiadis GS, et al. Studying the interaction of stent-grafts and treated abdominal aortic aneurysms: time to move caudally! J Endovasc Ther. 2015;22:413–20.
45. Tasso P, Lodi Rizzini M, Raptis A, Matsagkas M, De Nisco G, Gallo D, et al. In-stent graft helical flow intensity reduces the risk of migration after endovascular aortic repair. J Biomech. 2019;94:170–9.
46. Raptis A, Xenos M, Spanos K, Kouvelos G, Giannoukas A, Matsagkas M. Endograft specific haemodynamics after endovascular aneurysm repair: flow characteristics of four stent graft systems. Eur J Vasc Endovasc Surg. 2019;58:538–47.
47. Boersen JT, Groot Jebbink E, Versluis M, Slump CH, Ku DN, de Vries JPM, et al. Flow and wall shear stress characterization after endovascular aneurysm repair and endovascular aneurysm sealing in an infrarenal aneurysm model. J Vasc Surg. 2017;66:1844–53.
48. Georgakarakos E, Xenakis A, Manopoulos C, Georgiadis GS, Argyriou C, Tsangaris S, et al. Studying the flow dynamics in an aortic endograft with crossed-limbs. Int J Artif Organs. 2014;37:81–7.
49. Bisdas T, Weiss K, Eisenack M, Austermann M, Torsello G, Donas KP. Durability of the endurant stent graft in patients undergoing endovascular abdominal aortic aneurysm repair. J Vasc Surg. 2014;60:1125–31.
50. Georgakarakos E, Xenakis A, Georgiadis GS. Computational comparison between the Altura aortic endograft configuration and the classic bifurcated idealized designs. Ann Vasc Surg. 2020;68:442–50.

51. Kontopodis N, Galanakis N, Ioannou CV, Tsetis D, Georgiadis GS, Antoniou GA. Meta-analysis of the crossed versus standard limb configuration in endovascular aneurysm repair. Ann Vasc Surg. 2022;80:358–69.
52. Mastracci TM, Greenberg RK, Eagleton MJ, Hernandez AV. Durability of branches in branched and fenestrated endografts. J Vasc Surg. 2013;57:926–33.
53. Georgakarakos E, Xenakis A, Georgiadis GS, Argyriou C, Antoniou GA, Schoretsanitis N, et al. The hemodynamic impact of misalignment of fenestrated endografts: a computational study. Eur J Vasc Endovasc Surg. 2014;47:151–9.
54. Georgakarakos E, Xenakis A, Bisdas T, Georgiadis GS, Schoretsanitis N, Antoniou GA, et al. The shear stress profile of the pivotal fenestrated endograft at the level of the renal branches: a computational study for complex aortic aneurysms. Vascular. 2016;24:368–77.
55. Tran K, Yang W, Marsden A, Lee JT. Patient-specific computational flow modelling for assessing hemodynamic changes following fenestrated endovascular aneurysm repair. JVS Vasc Sci. 2021;2:53–69.
56. Kandail H, Hamady M, Xu XY. Patient-specific analysis of displacement forces acting on fenestrated stent grafts for endovascular aneurysm repair. J Biomech. 2014;47:3546–54.
57. Ou J, Tang AYS, Chiu TL, Chow KW, Chan YC, Cheng SWK. Haemodynamic variations of flow to renal arteries in custom-made and pivot branch fenestrated endografting. Eur J Vasc Endovasc Surg. 2017;53:133–9.
58. Jones SM, Poole RJ, How TV, Williams RL, McWilliams RG, Brennan JA, Vallabhaneni SR, Fisher RK. Computational fluid dynamic analysis of the effect of morphologic features on distraction forces in fenestrated stent grafts. J Vasc Surg. 2014;60:1648–56.
59. Georgakarakos E, Manopoulos C, Lazarides MK, Tsangaris S. Commentary: occlusion of parallel/bridging stent-grafts in the treatment of complex aneurysms: a purely technical issue … Or not? J Endovasc Ther. 2015;22:401–5.
60. Moulakakis KG, Kakisis J, Gonidaki E, Lazaris AM, Tsangaris S, Geroulakos G, et al. Comparison of fluid dynamics variations between chimney and fenestrated endografts for pararenal aneurysms repair: a patient specific computational study as motivation for clinical decision-making. Vasc Endovasc Surg. 2019;53:572–82.
61. Georgakarakos E, Koutsoumpelis A. Commentary: use of BeGraft and BeGraft+ stent-grafts in the management of complex aortic aneurysms: toward more dedicated stent-graft platforms? J Endovasc Ther. 2019;26:795–6.
62. Bogdanovic M, Stackelberg O, Lindström D, Ersryd S, Andersson M, Roos H, et al. Limb graft occlusion following endovascular aneurysm repair for infrarenal abdominal aortic aneurysm with the zenith alpha, excluder, and endurant devices: a multicentre cohort study. Eur J Vasc Endovasc Surg. 2021;62:532–9.
63. Shintani T, Obara H, Matsubara K, Hayashi K, Hayashi M, Ono S, et al. Impact of stent graft design on external iliac artery limb occlusion rates after endovascular aneurysm repair: post-hoc analysis of a Japanese multicentre database. Eur J Vasc Endovasc Surg. 2019;58:839–47.
64. Georgiadis GS, Trellopoulos G, Antoniou GA, Georgakarakos EI, Nikolopoulos ES, Iatrou C, et al. Hybrid endografts combinations for the treatment of endoleak in endovascular abdominal aortic aneurysm repair. Int J Artif Organs. 2013;36:28–38.
65. Matsagkas MI, Kouvelos G, Spanos K, Athanasoulas A, Giannoukas A. Double fixation for abdominal aortic aneurysm repair using AFX body and endurant proximal aortic cuff: mid-term results. Interact Cardiovasc Thorac Surg. 2017;25:1–5.
66. Smith AH, Farivar BS. A novel solution to concomitant type Ia/type IIIa endoleak using the cook zenith fenestrated device and Endologix AFX®2. Vasc Endovasc Surg. 2021;55:777–80.
67. Georgakarakos E, Ioannidis G, Raptis A, Xenos M. Computational fluid dynamic analysis supports the hemodynamic stability of hybrid combinations with the AFX bifurcate and nitinol-based proximal segments in solutions of failed endovascular aneurysm repair. Vasc Endovasc Surg. 2021;55:907–9.
68. Ioannidis G, Georgakarakos E, Raptis A, Xenos M, Manopoulos C, Matsagkas M, et al. Modeling and computational comparison of the displacement forces exerted between the AFX Unibody aortic stent graft and its hybrid combination with a nitinol-based proximal aortic cuff. Ann Vasc Surg. 2021;74:400–9.
69. Reference Values for Arterial Stiffness' Collaboration. Determinants of pulse wave velocity in healthy people and in the presence of cardiovascular risk factors: 'establishing normal and reference values'. Eur Heart J. 2010;31:2338–50.
70. Georgakarakos E. Estimating the influence of aortic-stent grafts after endovascular aneurysm repair: are we missing something? Med Hypotheses. 2016;97:26–30.
71. Singh TP, Moxon JV, Iyer V, Gasser TC, Jenkins J, Golledge J. Comparison of peak wall stress and peak wall rupture index in ruptured and asymptomatic intact abdominal aortic aneurysms. Br J Surg. 2021;108:652–8.

Endovascular Aneurysm Repair Using Fenestrated Endografts

51

Nathan L. Liang, Fanny S. Alie-Cusson, and Karim M. Salem

Introduction

The introduction of endovascular stent grafting revolutionized the treatment of abdominal aortic aneurysms (EVAR) and has led to a redefinition of the gold standard in aortic aneurysm treatment. However, the use of standard aortic endografts was initially limited to the infrarenal aorta in the early days of endovascular aortic therapy. Extension of the aneurysm to or above the renal arteries still required open surgery or the use of off-the-shelf endovascular devices in off-label ways such as parallel grafting. The development of custom fenestrated endovascular grafts allowed for the individualized repair of aortic aneurysms that were traditionally anatomically unsuitable for standard EVAR.

Indications and Considerations

Fenestrated and branched EVAR (F/BEVAR) stent-graft usage was first reported in 1990s. The earliest reports included single renal fenestrations which allowed proximal extension of aortic stent grafting for aneurysm repair [1, 2]. The use of FEVAR allows the graft seal zone to be extended more proximally into the visceral aorta in circumstances where standard infrarenal EVAR would not allow for adequate seal of juxtarenal aneurysms. A systematic review and meta-analysis of endovascular juxtarenal aneurysm repair from 14 studies by Donnan et al. found that endovascular repair, including FEVAR, was associated with lower 30-day mortality, acute renal failure, bowel ischemia, and length of stay when compared to open surgical repair for complex abdominal aortic aneurysms [3].

N. L. Liang (✉) · F. S. Alie-Cusson · K. M. Salem
Division of Vascular Surgery, Department of Surgery, University of Pittsburgh Medical Center, Pittsburgh, PA, USA
e-mail: liangnl@upmc.edu; aliecussonf@upmc.edu; ks1988@njms.rutgers.edu

K. M. Salem
Division of Vascular Surgery, Department of Surgery, University of Pittsburgh Medical Center, Pittsburgh, PA, USA

Division of Vascular Surgery, Department of Surgery, Rutgers New Jersey Medical School, Newark, NJ, USA
e-mail: ks1988@njms.rutgers.edu

Indications

Anatomic factors which determine the suitability of infrarenal endovascular repair include aortic neck length, diameter, angulation, tapering, thrombus burden, and calcification. Approximately 40–60% of abdominal aortic aneurysms have been documented to have unfavorable or hostile neck anatomy for standard EVAR [4, 5]. Characteristics which contribute to hostile neck anatomy include lengths shorter than 15 mm, tapered or reverse tapered anatomy, large diameters, severe angulation, and circumferential calcification [6, 7]. Most commercially available endograft instructions for use (IFU) recommend an infrarenal aortic neck length of 10–15 mm. Prior studies have shown that hostile neck anatomy is associated with higher rates of type Ia endoleak, reinterventions, and aneurysm-related mortality [8–11]. Extension of the stent-graft into the visceral segment of the aorta allows for a more proximal seal which may negate some of these hostile neck characteristics and reduce the aforementioned complications.

Progressive aortic degeneration can occur after prior endovascular or open surgical repair of aneurysms, and FEVAR offers an additional method for endovascular treatment. Special considerations must be taken when planning aortic aneurysm repair with FEVAR, which include the number of branch vessels which must be preserved, the positions of the branch vessels relative to each other, and the aortic morphology within the visceral segment and seal zone.

Commercially Available Devices

Two commercially available devices encompass a significant portion of fenestrated endograft usage: the Zenith Fenestrated Endovascular Graft (Cook Medical, Bloomington, IL, USA) and the Fenestrated Anaconda (Terumo Aortic, Sunrise, FL,

G. Geroulakos et al. (eds.), *Mastering Endovascular Techniques*, https://doi.org/10.1007/978-3-031-42735-0_51

USA). A third device, the Artivion inner-branched endograft (Artivion Inc, Kennesaw, GA, USA), combines outer fenestrations with internal branches and so will not be discussed in this chapter. A fourth device, the fenestrated Treo (Terumo Aortic, Scotland, UK) was announced in Europe in 2023 but does not yet have significant published results as of this writing and so will also not be discussed in detail in this chapter. Standard definitions and terminology for FEVAR have been most recently defined in the SVS reporting standards for aneurysm repair involving the renal and mesenteric arteries [12]. Devices can be customized utilizing a combination of fenestrations and scallops. A fenestration is an opening within the endograft fabric which allows perfusion of an aortic branch vessel. Scallops refer to a U-shaped or semicircle cut-out at the proximal or distal edge of the fabric which allows the filling of a branch vessel.

Cook Zenith Fenestrated

The Cook Zenith Fenestrated Endovascular Graft (ZFEN) is one of the most recognized and used complex aortic endovascular devices worldwide. The stent-graft system includes a proximal main body made of polyester that incorporates nitinol-reinforced scallops and fenestrations with gold radiopaque markers custom made to the patient's anatomy. The distal main body is an aorto-bi-iliac graft which overlaps with the proximal body and is placed in a similar fashion to a standard EVAR stent-graft (Fig. 51.1).

In the United States, the design of the Cook Zenith Fenestrated (ZFEN) device is approved for juxtarenal aneurysms only and is limited to two small fenestrations and either a large fenestration or a scallop. The large fenestration is unsupported and does not allow for stenting with bridging stents. The final 5-year results of the initial 67 patients enrolled in the ZFEN US regulatory trial were published in 2021 [13]. At 5 years, freedom from all-cause mortality was 88.8 ± 4.2%, with a freedom from AAA-related mortality of 96.8 ± 2.3%. A total of 20 patients (30%) required secondary interventions; 8 reinterventions for endoleaks, 12 for renal stent complications, and 1 patient for both. The freedom from secondary reintervention was 90.9 ± 3.6% at 1 year and 63.5 ± 7.2% at 5 years.

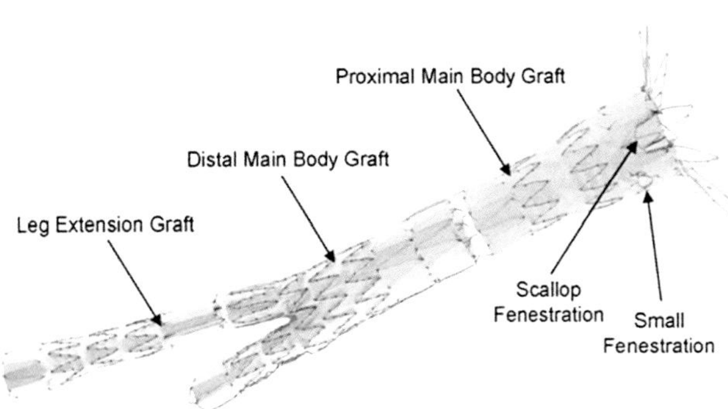

Fig. 51.1 Cook ZFEN initial design with two small fenestrations and one scallop

Aneurysm sac regression (defined as >5 mm in diameter) occurred in 53.2% of patients at 6 months and 80.6% at 5 years. Further real-world usage of this configuration of the ZFEN has supported this promising mid-term durability [14, 15].

Outside of the United States, the Cook Zenith custom fenestrated device is available in multiple custom configurations not limited to two stentable fenestrations. This device has had a long history of published results with good durability from multiple centers under investigational use in the United States. Studies of fenestrated devices by the US-based Aortic Research Consortium, as well as by high-volume centers in Europe, have demonstrated similar rates of technical success and mid- and long-term durability. Van Calster and co-authors [16] from France reported on a series of 468 patients undergoing F/BEVAR with a technical success rate of 99% for visceral artery cannulation and a 5% perioperative mortality rate and 3.8% rate of spinal cord ischemia. Survival at 5 years was 60%, although the median follow-up was 29 months. Katsargyris and co-authors from Germany [17] likewise reported a series of 384 patients with 97% technical success in visceral cannulation and similarly low mortality in both the perioperative (<1%) and 3-year (83–89%) timepoints. Motta and co-authors from the United States Aortic Research Consortium [18] reviewed 893 patients with similar rates of technical success and perioperative and long-term results. However, these and similar studies have mostly reported a combination of fenestrated and branched grafts which limits some generalizability despite the positive outcomes overall.

Terumo Fenestrated Anaconda

The Terumo Fenestrated Anaconda was first implanted in 2010 in the United Kingdom (previously Vascutek Ltd.). The stent-graft includes a single piece main body which is flexible, has conformable proximal sealing ring stents, and is fully repositionable. This consists of a woven polyester graft supported by independent nitinol stent rings. The treatment of highly angulated necks can be performed with an indication of up to 90 degrees of angulation. As opposed to the ZFEN, the Fenestrated Anaconda does not have uncovered stents extending cranially to act as hooks. The sealing ring stents form peaks and valleys and include proximal fixation hooks attached at each of these peaks and valleys. The peaks are on the anterior and posterior of the stent-graft in the infrarenal graft system, but rotating the graft 90° allows for the valleys to be anterior and act as a natural "scallop" for visceral vessels. Frequently four, and occasionally five, fenestrations have been manufactured and implanted for complex pathologies. The independent stent rows and unsupported body allow vast customization options. A magnetic guidewire system is used to aid in the navigation of contralateral gate cannulation as well as cannulation of fenestrations. Printed 3D models are used by the manufacturing team to verify suitability of the design prior to implantation (Fig. 51.2).

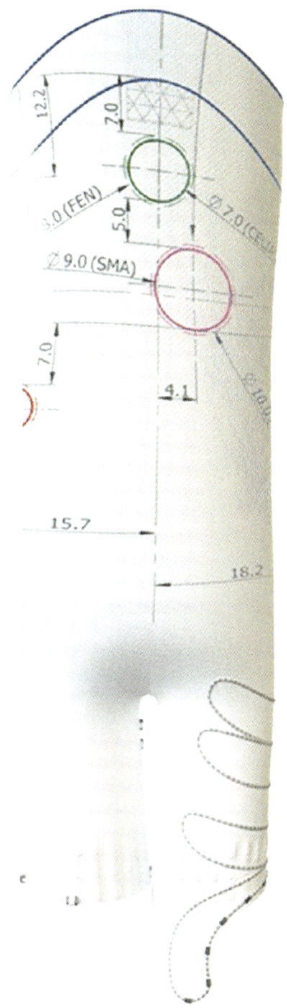

Fig. 51.2 Fenestrated Anaconda design with two fenestrations

A prospective registry of the Anaconda stent-graft system was analyzed by Jubouri et al. in July 2022 [19]. This included data from 2013 to 2022 and included a total of 5030 patients. In this study, 438 (8.7%) devices had one fenestration, 2349 (46.7%) had two to three fenestrations, and 2243 (44.6%) had four to six fenestrations. In 3891 (77.4%) of cases the Anaconda graft was chosen and indicated due to unsuitable anatomy for any other competitor devices. De Niet et al. reviewed the outcomes of treatment for 335 patients across multiple centers (2010–2018), who treated more than 15 complex AAAs or type IV thoracoabdominal aneurysms with the Fenestrated Anaconda system [20]. A total of 920 fenestrations were included with a mean of 2.7 ± 0.8 fenestrations per case. Technical success was noted at 88.4%. Type Ia endoleaks were noted at the conclusion of the procedure in 23 cases (6.9%), with 13 disappearing at the 30-day postoperative follow-up and an additional 6 disappearing at the last follow-up (median 14.4 months). Greater neck angulation, three fenestrations, and the presence of augmented proximal rings were associated with type Ia endoleaks. Target-vessel adverse events were present in 13 vessels at 30 days (7 stenosis, 5 occlusion, and 1 kinked stent) and 24 vessels at the last follow-up (11 stenosis, 8 occlusion, and 2 kinked stents). These rates were noted to be similar to earlier reported case series. A 4.2% 30-day mortality was noted, which is comparable to prior results for Fenestrated Anaconda, but slightly higher than the 0.7–3.4% with the Zenith Fenestrated.

Design and Planning of Fenestrated Endografts

The measurement and design of fenestrated endografts depend heavily on the platform to be used and the device-specific design constraints of this platform. In addition, certain considerations may be taken into account for physician-modified grafts, which are off-label and do not undergo separate quality control at the point of manufacture. However, many basic principles for measurement and general considerations apply to all fenestrated graft designs and will be discussed here.

Advanced Imaging

High-resolution three-dimensional imaging is crucial to the planning of any fenestrated endograft. A high-resolution computed tomographic angiography (CTA) study is most often used, with slice thicknesses of 2.5 mm or less. Many different software packages exist for the reconstruction of the aorta for planning purposes. Commonly used packages include TeraRecon (TeraRecon Inc., Durham, NC, USA), 3mensio (PIE imaging BV, Maastricht, Netherlands), and EndoSize (Therenva SAS, Paris, France). Our institution utilizes the GE Advanced Workstation (GE Healthcare, Chicago, IL, USA) package for advanced imaging analysis. We primarily use three modalities of advanced reconstruction for FEVAR planning: volumetric analysis, centerline reconstruction, and multiplanar reformats.

Volumetric Analysis

The aorta and wall are segmented, either automatically by the software or by hand, to create a 3D volume. This volume is rotatable by the physician along three axes. This view is not useful for specific planning measurements but provides an overview of the aorta morphology and can inform the physician by giving a general idea of angulation and tortuosity (Fig. 51.3).

Centerline Reconstruction (Curved Planar Reformat or CPR)

As its name suggests, centerline reconstruction utilizes the "centerline" through the lumen of flow of the aorta. The centerline of flow is created semiautomatically by the software and edited by hand. The aorta is then stretched out along this centerline. This allows for the measurement of distances

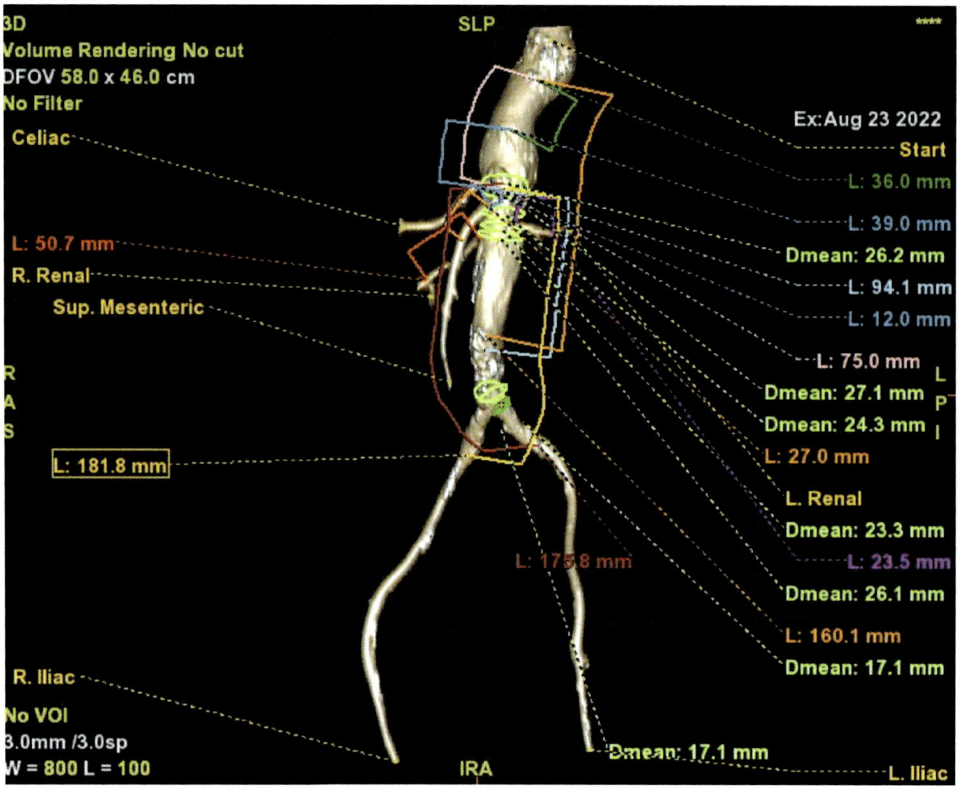

Fig. 51.3 3D volumetric analysis of a paravisceral abdominal aortic aneurysm

between specific points on the aorta along the centerline while accounting for angulation or tortuosity. This is our primary method of length measurement generally used for FEVAR planning (Fig. 51.4).

Multiplanar Reformat (MPR)

Multiplanar reformats allow for visualization and measurement along the centerline of flow, but only at one specific area on the aorta at any given time. In contrast to the standard axial/coronal/sagittal visualization planes, MPR allows for visualization along different oblique planes. For instance, an aorta that is straight but angled out of the plane of standard axial/coronal/sagittal views can be reconstructed using a different oblique plane to bring the length of the aorta back into the correct plane for measurements. This is commonly used to confirm length measurements from the CPR, such as renal or visceral separation. MPR is also used for determining fenestration positioning using clock-face measurements (Fig. 51.5).

Measurement and Planning Process

The measurement process varies according to individual physician preference and based on device platform used. However, many of the general principles remain the same. Here, we briefly describe the measurement process used in our institution to size multi-fenestrated endografts (Fig. 51.6).

Picking a Zero Point

The choice of a "zero point," or reference point, when calculating length measurements is critical as the rest of the measurements are based on this point. We prefer to use the bottom of the celiac as the zero point as it is a consistent anatomic landmark. Other authors have advocated using a point 10 mm above the bottom of the celiac or the bottom of the SMA. We believe that as long as the positioning of the zero point is consistent and repeatable, any point will suffice for the consistency of measurements.

Determining and Confirming Lengths

Lengths from the zero point to each of the target points along the centerline are determined using CPR. We generally will measure to the middle of each target vessel (superior mesenteric artery and both renal arteries) as well as to the aortic bifurcation. The relative distances between the visceral arteries are then measured again by MPR to confirm measurements. This typically consists of the distance between the two renal arteries ("renal separation") and the distance from the SMA to either renal artery ("mesenteric-renal separation"). Large discrepancies in visceral separation distance between the CPR and MPR require reevaluation of the centerline (Fig. 51.7).

Inner Vessel Diameter (IVD) Measurements

The inner vessel diameter (IVD) is the diameter of the aorta at each level of interest. This is of importance when

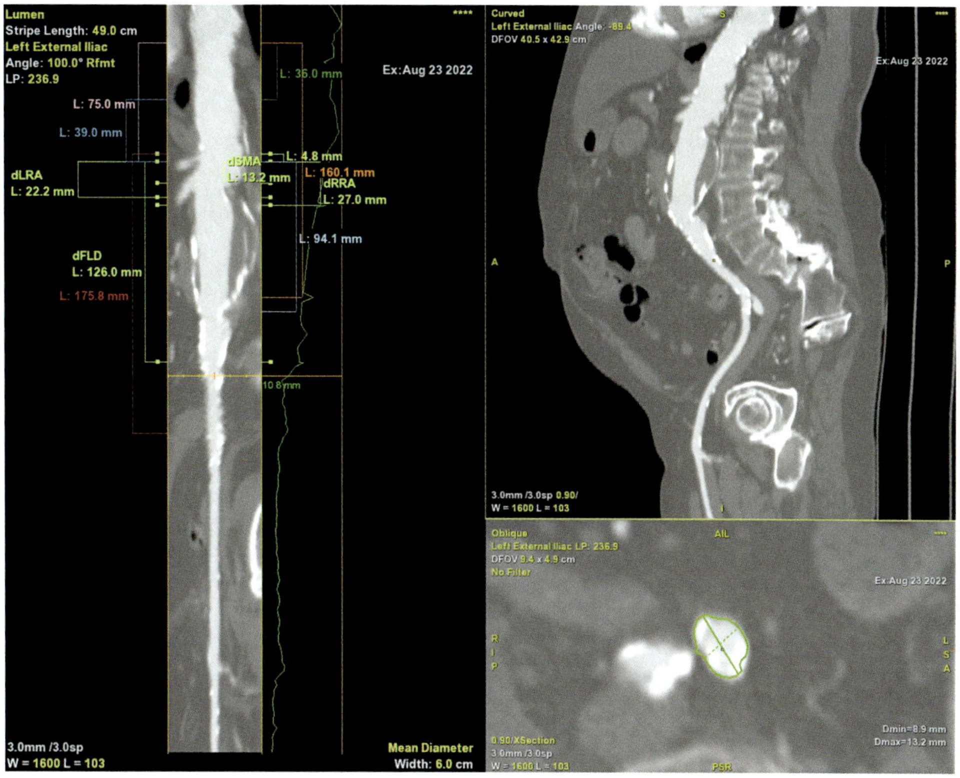

Fig. 51.4 Centerline analysis using curve planar reformats. Left panel: stretched lumen view. Upper right panel: curved reformat view. Lower right panel: Oblique axial view along the centerline

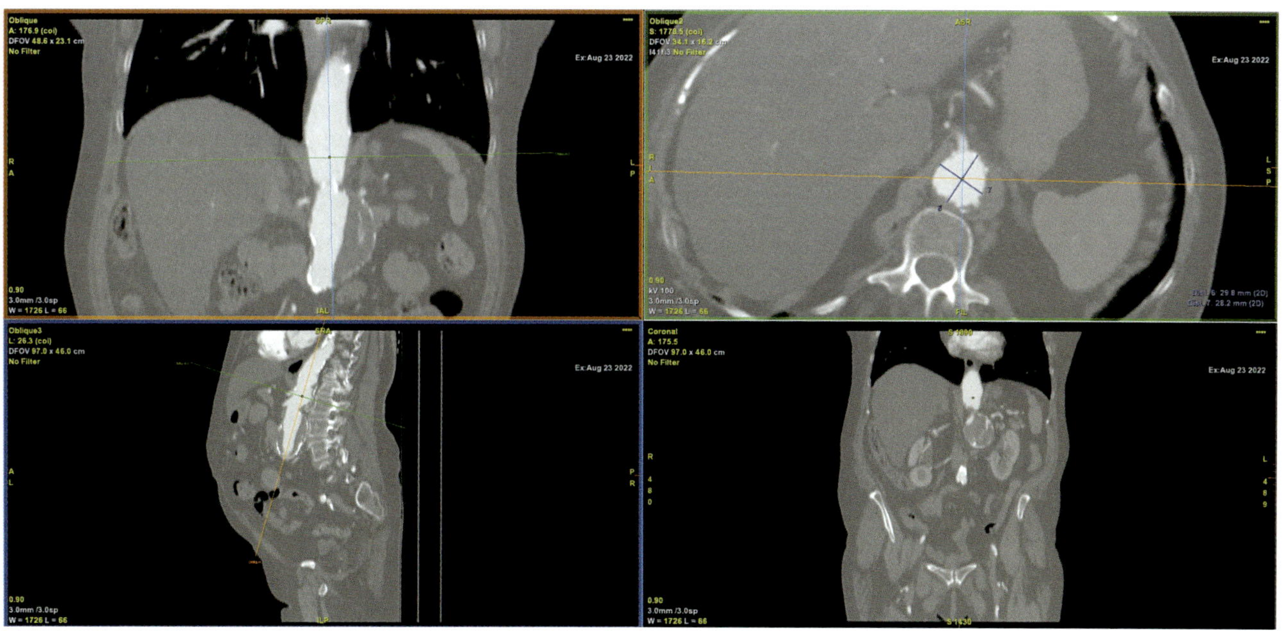

Fig. 51.5 Multiplanar reformats are used to obtain orthogonal views. This is often more accurate than centerline for determining fine-length measurements, angulation, and clock-face target-vessel orientation. Each colored frame represents a slice taken along the same colored line in the other orthogonal views. The lines can be rotated to show an "axial" orthogonal view along the centerline of flow at any given point (upper right frame)

measuring fenestration positioning, as well as when determining the diameter of the graft to be used. The IVD is measured at the anticipated level of proximal seal and then again at the middle of each target vessel, the zero point, and the distal aorta. The IVD can be measured either on CPR or MPR. We typically utilize the mean aortic diameter on CPR if the aorta is not mostly round and there is a discrepancy between minimum and maximum diameters.

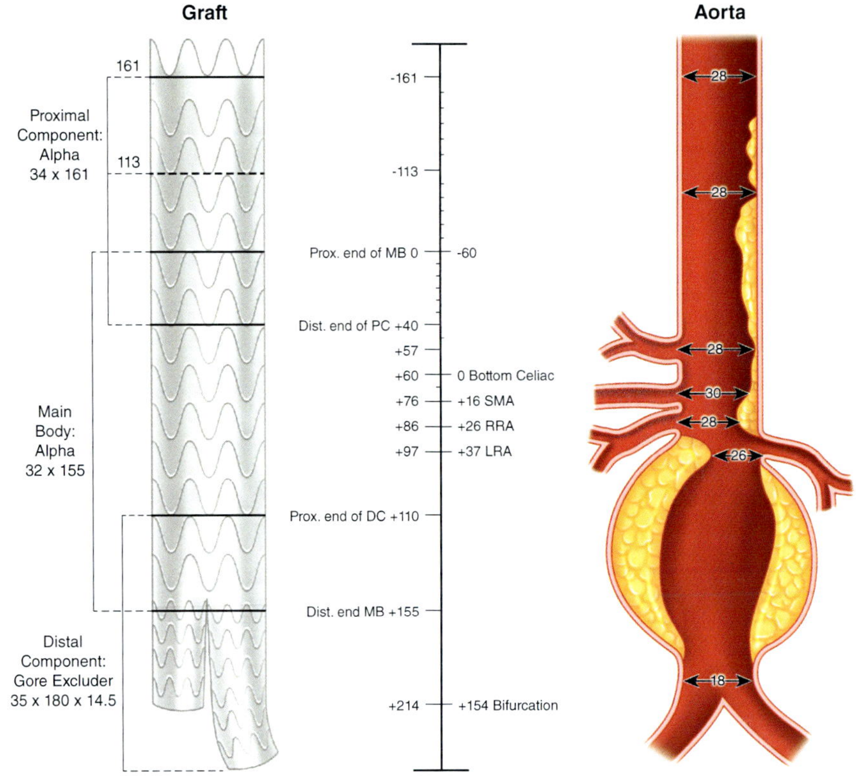

Fig. 51.6 Example of a planning sheet for FEVAR used in our institution. The aortic anatomy and centerline measurements are drawn on the right, then transferred to the graft plan on the left. All measurements are in mm

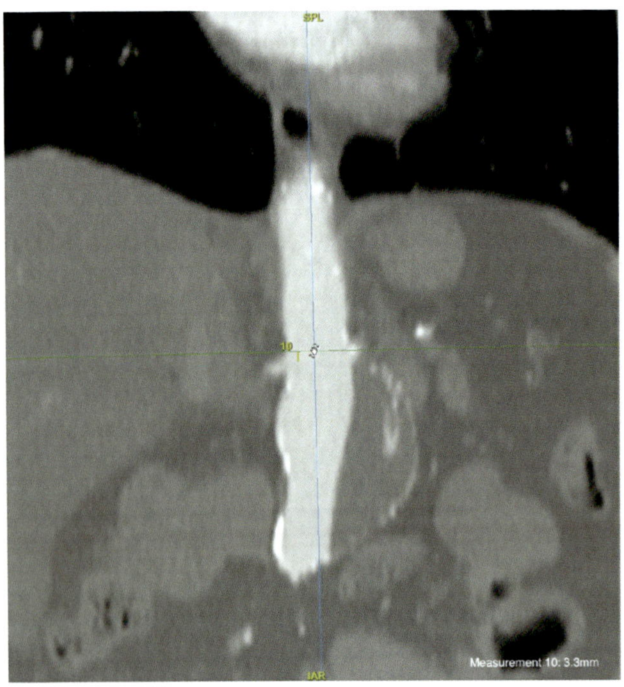

Fig. 51.7 Measurement of renal separation obtained using multiplanar reformat view. This view provides an oblique view similar to a traditional coronal view, but along a plane that allows both renal arteries to be visualized along the centerline of flow at that particular aortic segment

Fenestration Positioning

Positioning of the fenestrations is most commonly denoted using a "clock-face" system. An axial oblique view of the visceral target vessel is obtained along the centerline of flow. To obtain this view we use MPR to keep a consistent anterior-posterior orientation as CPR views may not appropriately represent this orientation. The position of the target-vessel ostium according to the clock face is recorded to the nearest 15-min mark (Fig. 51.8). This is repeated for each target vessel/fenestration. In most instances, the calculation of the actual arc length distance rather than just the clock position relative to the 12 o'clock position is required. This is especially important in situations where the physician is intimately involved in graft construction, such as for physician-modified grafts. The IVD is used to calculate this arc length based on the clock position. The selection of IVD here is critically important as changing the IVD will result in a change in arc length and variation in the positioning of the fenestration.

Application of these measurements to the actual constructed graft can sometimes result in deviations from the initial measurements. This is most commonly due to the changing placement of fenestrations to avoid stent struts but can also be because of an unusually large graft relative to the

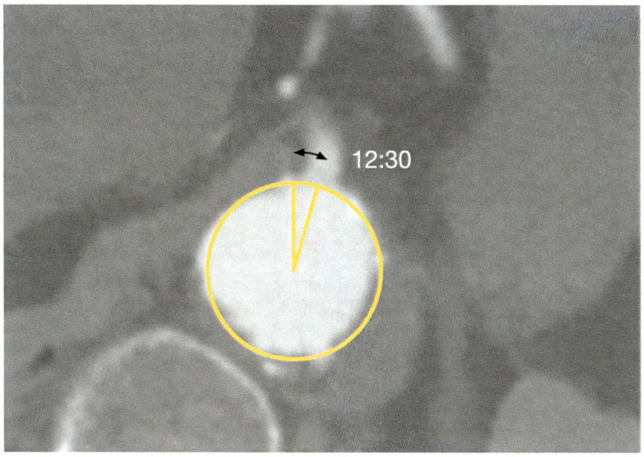

Fig. 51.8 The orthogonal MPR view along the centerline of flow is used to measure the relative displacement of the target vessel using clock-face measurements

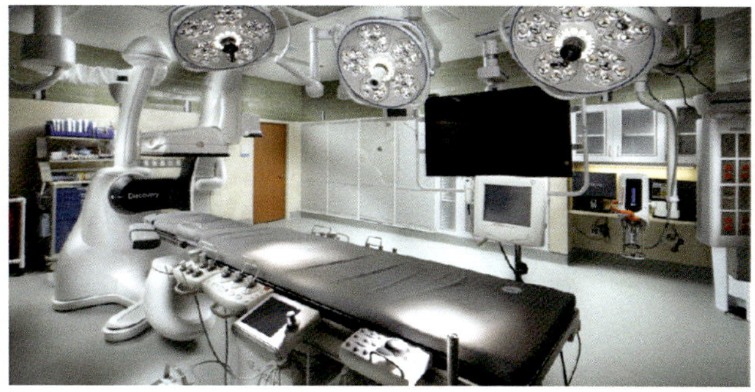

Fig. 51.9 The hybrid operating room at UPMC Presbyterian with GE Discovery fixed imaging system

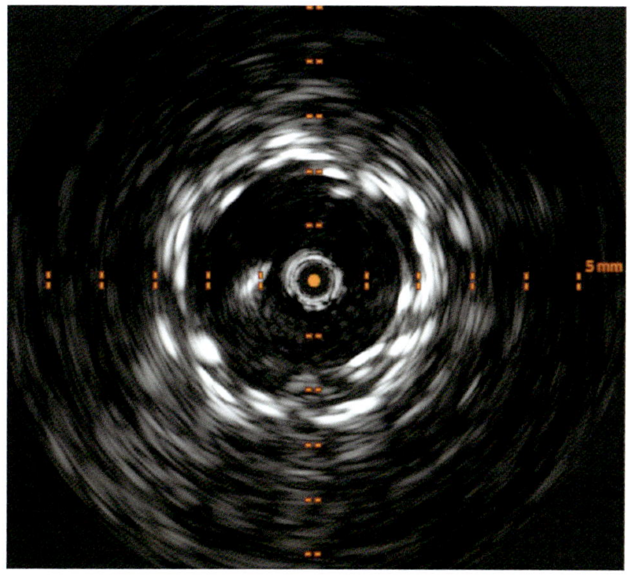

Fig. 51.10 Intravascular ultrasound used to interrogate a bridging stent to confirm patency and full expansion

IVD or posteriorly placed diameter-reducing ties. In these situations, the physician must decide how much and in what direction the displacement of a fenestration is acceptable to avoid compromising cannulation.

Implantation and Technique

Imaging Adjuncts: Fusion and Other Technologies

Our complex aortic endovascular cases are performed in a hybrid operating room with a fixed imaging system. The primary system at UPMC Presbyterian, our flagship university hospital, is a GE Discovery (GE Healthcare, Chicago, IL, USA). The Discovery has a robust suite of adjunct software and hardware tools for image guidance and radiation dose reduction (Fig. 51.9).

Fusion Imaging

Fusion imaging incorporates the anatomic landmarks obtained from a preoperative CTA and layers them onto the fluoroscopic image during the FEVAR procedure. In most modern fusion systems, the fusion landmarks will move along with the image if the table position or angulation is changed and will adjust to some extent for magnification. Fusion provides many benefits, including the ability to position devices and attempt cannulation with a visible target. Fusion also provides useful information that may reduce the amount of radiation for the procedure by providing positioning cues as well as limiting the amount of digital subtraction angiography used; however, this is physician dependent. Radiation dose reduction should always be maximized by the operating physician whenever possible. In our practice, fusion is used for every complex endovascular aortic procedure when possible.

Cone Beam Imaging

Cone beam imaging is present on many modern fluoroscopy systems. It is a 3D volume created by spinning the image intensifier around the patient up to 270°, collecting enough x-ray data to reconstruct into a CT volume. This can be combined with power injection to create a contrasted CT. We typically utilize cone beam CT as an adjunct at the end of the procedure to evaluate the main body and patency of the bridging stents and confirm the complete expansion of the system.

Intravascular Ultrasound (IVUS)

IVUS can be useful for interrogating expansion of bridging stents in real time. This can show any kinking or compression of bridging stents and can also be used to verify cannulation of the fenestrated component or limb (Fig. 51.10).

Non-fluoroscopic Imaging Guidance

Multiple systems for non-fluoroscopic image guidance are being studied at the time of this writing for radiation dose reduction during FEVAR. These systems include the Fiber-Optic Realshape (Phillips Healthcare, Amsterdam, Netherlands, and Cambridge, MA, USA) and the Intraoperative Positioning System (IOPS, Centerline Biomedical, Cleveland, OH, USA), which allow for non-fluoroscopic visualization of catheter manipulation inside the body using various methods such as radiofrequency and fiber-optic technology. These adjuncts are in limited use and not widely available but have been reported to reduce the radiation dose required for a FEVAR procedure [21].

Access Considerations

Choice of Access

The choice of access is dependent on the patient's anatomy and physician familiarity and preference. The bilateral transfemoral approach is common among operators for FEVAR procedures, but access can also be obtained from the arm, either via axillary, high brachial, brachial, or radial access, depending on the need and familiarity of the physician. One consideration of upper extremity access for cannulation and bridging stent delivery is the proximal constraining mechanism of the device. If the device is constrained above the fabric either directly or via suprarenal stents, care must be taken to successfully catheterize the interior of the graft and track a sheath. Many FEVAR designs that plan for upper extremity access will include preloaded guidewires for the fenestrations and/or a cut-out fabric scallop in the proximal fabric allowing a larger sheath to track across any suprarenal extensions and into the body of the fenestrated component [22].

The use of upper extremity access has been associated with a slightly higher rate of cerebrovascular events. An analysis of 1681 patients from the US Aortic Research Consortium had 67% utilization of upper extremity access and resulted in a significant absolute risk increase of 1.6% for composite perioperative cerebrovascular events; however, the overall event rate for arm access was still very low at 2.5% [22]. This is thought to be due to the typical need to cross the cerebrovascular vessel origins with large sheaths. The choice of side for upper extremity access, however, has not demonstrated a difference in perioperative cerebrovascular events despite the need to cross all origins of the cerebrovascular vessels when coming from the right arm, compared to only the vertebral artery origin from the left [23]. The applicability of these published results for FEVAR, however, is complicated by the combination of fenestrated and branched cases in most published series.

Iliac Artery Size and Disease

Access vessel considerations are paramount when using large sheaths. An external iliac artery of at least 6 mm, and preferably 7 mm, is required. Calcification and tortuosity can affect delivery of the device and passage of large sheaths for cannulation. The planned use of open or endovascular iliac conduits or other preparation of the access vessels is critical for success. Intravascular lithotripsy (Shockwave Medical, Santa Clara, CA, USA) has shown promise for enabling the passage of large sheaths in severely calcified access vessels [24].

Deployment Process

Once access is obtained, the fenestrated device component is advanced and positioned at the appropriate level based on orienting markers which vary based on the specific device. In every situation, it is important to deploy the graft as close to the desired orientation as possible. Attention should be paid to anatomic characteristics such as tortuosity, narrowing, or calcification, or when using a graft with a long segment proximal to the fenestrations.

Although most grafts will be designed with a diameter reduction system, turning of the graft even in its reduced state is often difficult especially in angulated, tortuous, or diseased anatomy. In addition, most grafts are fixed to the delivery system at only one or two points, and so often multiple turns of the delivery system are required to initiate turning of the device itself, and care must be taken to avoid twisting of the device fabric. Cranial-caudal movements may likewise be restricted as large movements may cause bunching of the graft fabric and compression of stent segments if not carefully performed. Our preferred approach is to deploy the device slightly higher than the anticipated final location and then bring the device down as needed. However, this is dependent on the device design, and grafts without strong distal attachments to the delivery system may deform or deploy prematurely if pulled down (Fig. 51.11).

The fenestrated component is then selectively catheterized. We prefer an approach from the contralateral femoral access but occasionally will utilize arm access for severely downgoing visceral vessels or if the contralateral access is hostile. A large sheath is advanced into the graft main body to provide consistent access without the need to re-cannulate the graft for each visceral. We prefer to use an 18- or 20-French Gore Dry-Seal sheath (WL Gore, Flagstaff, AZ) as the Dry-Seal valve allows for easy passage of multiple coaxial sheaths with minimal blood loss.

The visceral vessels are then sequentially cannulated. In a typical four-vessel fenestrated graft where the fenestrations are lined up well with the ostia and no significant cannula-

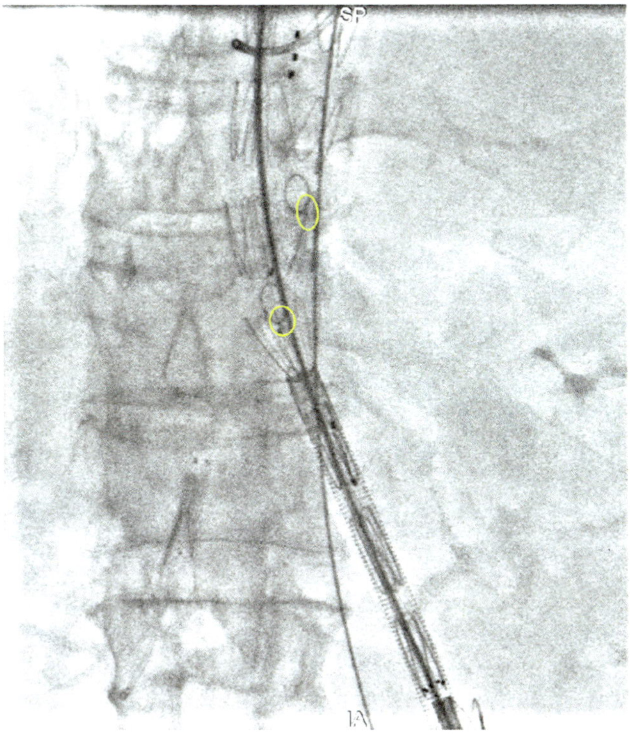

Fig. 51.11 Initial deployment of a fenestrated endograft more proximal to the final orientation

tion issues are expected, our cannulation sequence typically consists of the SMA first as the most critical vessel, followed by the celiac and then the renal arteries. A stiff wire, typically a Rosen or short-tipped Amplatz wire, is placed into the SMA and the sheath withdrawn completely. The celiac is then selected, and a 0.018″/0.035″ wire placed into the splenic or hepatic artery and any sheaths withdrawn. The renal arteries are then selectively catheterized, and sheaths are placed into each one. We typically advance our renal bridging stents into places within the sheaths. This configuration allows for the two renal artery sheaths to support the fenestrated component during final deployment. The fenestrated component is then completely deployed, and the delivery system is removed. Any ballooning of the fenestrated component is then performed with the renal sheaths still in place to support the graft and prevent movement (Fig. 51.12).

Balloon-expandable, covered stents are then deployed into the visceral vessels as bridging stents. The renal artery bridging stents are typically deployed first, followed by the SMA and then celiac stents. About 10 or 12 mm balloons are then used to flare the intra-aortic portion of the bridging stent to provide a "rivet" effect, reinforcing the alignment of the fenestration and enhancing the seal. A completion angiogram of each visceral using digital subtraction angiography is performed to show filling of the target vessel without injury or dissection as well as patency by verifying backflow into the aorta (Fig. 51.13).

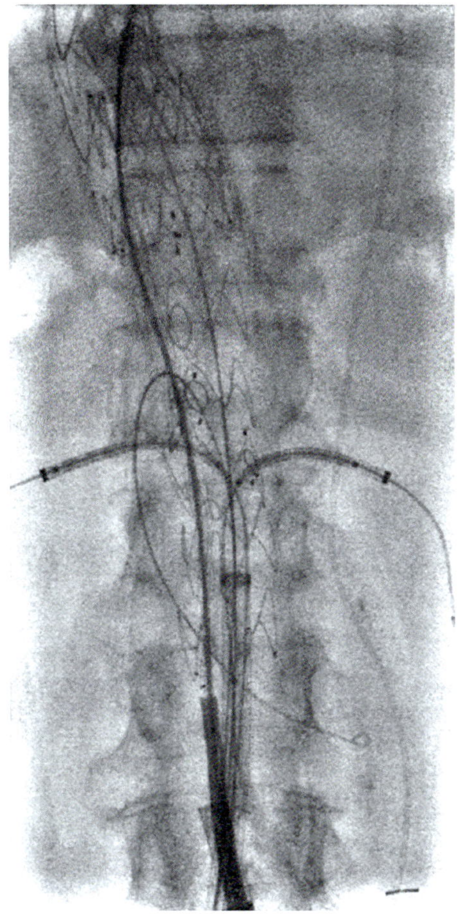

Fig. 51.12 Cannulation of all target visceral vessels prior to complete deployment of the fenestrated component. A wire is left in the SMA, and sheaths and bridging stents are placed into both renals to prevent the movement of the fenestrated component during deployment and withdrawal of the delivery system. Note: in this case the celiac artery was an optional target artery but was found to be nearly occluded at the time of intervention, and so the fenestration was not cannulated and instead excluded by the proximal thoracic graft component

After the fenestrated component is completely deployed, the rest of the procedure is then performed as the patient anatomy requires. Once bridging stents have been deployed, care must be taken during the passage of other large sheaths and devices to avoid damaging or kinking any of the flared stents. If this is a concern, such as in a small graft or tortuous aorta, a wire and/or angioplasty balloon should be left in the target artery in case reinflation is necessary.

Visceral Artery Cannulation Techniques

Our approach to visceral cannulation is for maximum accuracy of alignment between the fenestration and the target ostium. This is most critical from the planning stage, as factors such as aortic diameter changes, calcification, diameter reduction, and tortuosity can all affect the alignment of the

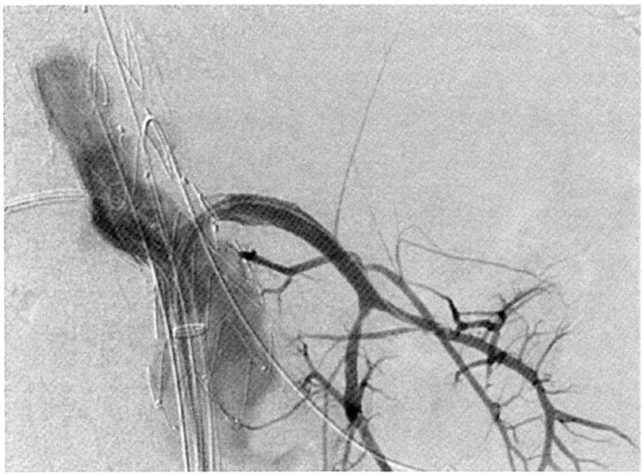

Fig. 51.13 Completion angiogram of a left renal post-stenting. The stent has been post-dilated on the proximal portion using a large balloon to "rivet" the stent and the fenestrated component into place

graft in vivo. Even with diameter reduction, often the turnability of the graft is significantly affected by tortuosity and cannot be completely relied upon to realign a fenestration.

Fenestration misalignment may be due to the anatomic factors listed above but may also be due to unfavorable direction of the fabric around the fenestration during initial deployment of the graft. Consistent misalignment in either a cranial-caudal or rotational direction can result in difficulty with cannulation. In this section we will discuss techniques for cannulation.

Basic Cannulation

Our basic cannulation setup consists of directional catheters and a Glidewire on a 0.035″ 5F platform through a 7F sheath. Most often, a standard visceral vessel with a well-aligned fenestration can be cannulated with any curved catheter with a slight downgoing tip such as a Vanschie 4 or Cobra shape. Catheters with a wider curve such as the Cobra family may help with wire pushability due to their larger radius allowing for support against the opposite graft wall. Reverse curve catheter shapes such as VS1 or Sos may also be used, in which case the catheter is reformed above the fenestration, and the catheter and wire are used to select the vessel origin before the catheter is pulled down and attempted to "seat" inside the target vessel. A Glidewire is the most frequent wire used in our institution for cannulation, after which either the cannulating catheter or a lower-profile catheter is advanced into the vessel. An angiogram is obtained to demonstrate adequate positioning in the main trunk and the absence of wire damage or dissection. Once adequate catheter purchase is obtained, the Glidewire is replaced with a stiffer wire such as a Rosen or short-tipped Amplatz. An appropriately sized sheath is advanced into the origin of the vessel.

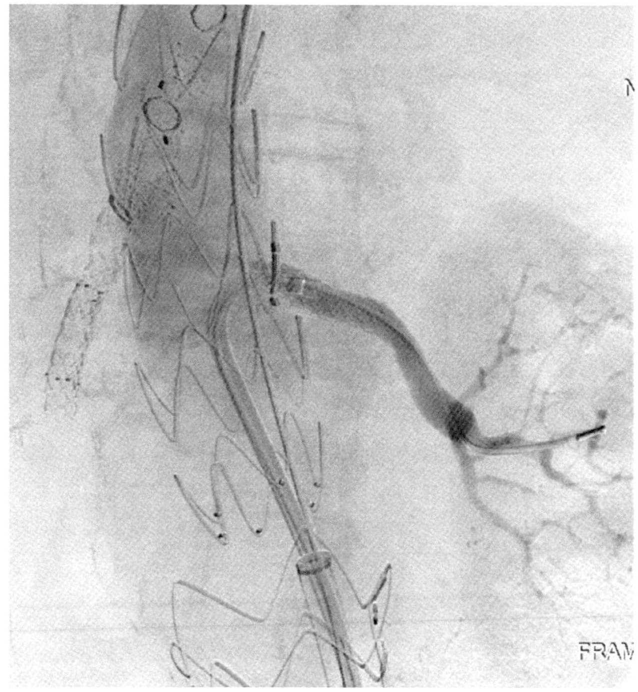

Fig. 51.14 Steerable sheath used to cannulate a downgoing left renal artery

Steerable sheaths (Medtronic TourGuide [Medtronic, Minneapolis, MN, USA] or Oscor Destino [Oscor Corp, Palm Harbor, FL, USA]) may also be used for the cannulation of visceral arteries. These sheaths have increasingly become a common first choice for initial cannulation due to the ease of positioning the steerable sheath tip at the fenestration without extra selection. The downside to these sheaths is the slightly larger outer profile and the extra length of the handle, requiring careful selection of catheter shaft and wire length (Fig. 51.14).

Extreme Downgoing Vessels

Extreme downgoing vessels can be difficult to cannulate due to tight radius turns when coming from a femoral access. We will typically attempt to cannulate these vessels with a reverse curve catheter shape such as the VS1 or Sos series in the standard fashion. If unsuccessful, a steerable sheath can be used to create a better-supported downgoing angle for cannulation. Antegrade access from the arm may facilitate better cannulation of these vessels, although the delivery of sheaths and devices may be more difficult if the proximal end of the fenestrated component has bare-metal stents that are fixed to the delivery system without adequate spacing or access scallop between the struts.

Misaligned Fenestration

Fenestration misalignment in a cranial-caudal can usually be remedied by pushing the graft up or down. The exception to this is if the fixation of the device to the delivery system proximally or distally is not especially strong or if diameter reduction has nod reduced the graft contact with the aortic wall (i.e., no diameter reduction, narrow or tortuous aorta). Cranial-caudal movement in a graft that is already well apposed to the aortic wall can deform individual stent segments and cause further misalignment of fenestrations if not performed carefully. Turning of the delivery system can accomplish small movements for realignment of fenestrations if the entire system is misaligned, but as discussed previously, many anatomic elements often make it difficult to turn a graft to any significant degree and require multiple turns of the delivery system to initiate any graft movement. If significant turning of the delivery system is required, it is critical to derotate the system after achieving an appropriate rotation and check to make sure the distal fabric is not twisted.

The most common misalignment is of the renal artery fenestrations as most grafts have anterior-posterior orientations based on the SMA. There are several methods for addressing misaligned fenestrations, assuming they are still aligned enough to allow for cannulation. The most common method of cannulation is to select the fenestration and place a wire outside the fenestration into the aorta as a "buddy" wire. A sheath is then brought up to the fenestration, and a catheter is brought up alongside the buddy wire for cannulation. We have had success with smaller complex curve catheters such as the Omni, Sos, and Vanschie 2/3 shapes. Once the target vessel is cannulated, it can be difficult to pass a stiff wire especially if the wire takes an acute angle toward the target vessel. Angioplasty balloons may be used to temporarily align the fenestration with the target vessel, and a sheath can often be placed into the visceral vessel over a rapidly deflating balloon ("swallowing the balloon" technique). Downsizing the platform to a 0.018″ platform may assist in balloon placement, although once the platform is downsized, it can be difficult to re-upsize to a 0.035″ system until the fenestration is in final alignment.

Having a sheath in place in the target vessel (typically 6F or above) will usually align the fenestration enough to permit the passage of a bridging stent. Most graft designs have enough flexibility in the stent and fabric that the deployment of bridging stents will pull the fenestration into permanent alignment. In certain cases, graft designs with posterior diameter-reducing ties (Cook Zenith Fenestrated series [Cook Medical]) may not align well until the release of the ties and removal of the delivery system. In cases where a sheath is unable to be passed due to misalignment, passing and inflating an angioplasty balloon to align the graft and target artery while final deployment of the fenestrated component is performed will often align the fenestration enough to permit subsequent passage of a sheath or bridging stents. A formed steerable sheath may also be used to "hook" the fenestration and pull it to a more desireable orientation for cannulation.

Stenotic Vessels

Stenotic vessels pose a significant additional problem on top of angulation or misalignment. Often, significant usage of angioplasty balloons and downsizing of wire platforms are needed to navigate severely stenotic vessels. Our approach for severely stenotic vessels is typically to perform a separate angiogram first to attempt cannulation and angioplasty or stenting of the vessel. If stenting is required, it is important that the stent be placed with minimum amount of protrusion into the aorta to prevent difficulty during subsequent FEVAR. The stent may then be used as a landmark for the alignment of the graft during initial deployment. During FEVAR, downsizing to a 0.018″ system may help with the navigation of stenotic vessels. We also commonly use steerable sheaths to provide support for the passage of balloons through stenotic areas.

Bridging Stents

Bridging stents are an essential component of fenestrated endovascular aortic repair (FEVAR) by connecting the main body of the endograft to the visceral vessels. They are indispensable in preserving target-vessel patency in complex repairs of aneurysms involving the visceral aorta. Although there are several bridging stents available for use internationally, no bridging stent is yet approved in the United States, specifically for use with fenestrated endografts.

Types of Bridging Stents

Balloon-expandable stents have largely been adopted as the preferred option for fenestrated endografts, as opposed to self-expanding stent-graft options typically utilized for multi-branched and parallel stent grafting. Although covered self-expanding stents are highly flexible and can be quite useful in tortuous anatomy, they provide less radial force than balloon-expandable stents. The ideal bridging stent is built on a low-profile delivery system with precise and accurate deployment, strong radial force in its proximal portion to optimize seal at the level of the fenestration, and high flexibility within the branch vessel itself. Kink and fracture resistance are also of paramount importance. Bare-metal self-expanding stents can be used as a flexible transition stent option at the distal endpoint of a balloon-expandable stent, especially in tortuous and downgoing vessels.

Balloon-expandable covered stents, such as iCast/Advanta (Getinge, Gothenburg, Sweden) and VBX (WL Gore, Flagstaff, AZ), are the most commonly used for fenestrated repairs. In the United States, iCast is FDA approved for tracheobronchial stenting and has been frequently used off-label for FEVAR because of their low-profile, high radial force, and their short-length options. They can be particularly useful for renal fenestrations as the small sizes fit through a 6F sheath. iCast stents are quite stiff and can foreshorten significantly when flared. They are associated with a narrower matrix than VBX stent grafts and require sheath delivery due to the potential for stent displacement if passed without sheath protection. The Advanta V12 has demonstrated favorable mid- and longer-term results in use as a bridging stent during FEVAR. In a series of 194 FEVAR patients with 457 stented visceral arteries, freedom from bridging stent-related complications was 98% at 1 year and 92% at 3 years [25].

The VBX stent allows for delivery without a sheath in certain situations with less concern for dislodgement from the balloon but requires at least a 7F sheath for delivery. It is significantly more flexible than the iCast, flares with minimal foreshortening, and offers a wide-size matrix allowing for significant post-dilation. The VBX as a bridging stent is a relatively new addition to the FEVAR platform, but recent studies have demonstrated good results. A series of biomechanical comparisons between the VBX and iCast/Advanta by Torsello and colleagues demonstrated favorable comparisons of fatigue resistance, kink resistance, and pullout force between the two stent platforms [26, 27]. Clinical outcomes of the VBX as a bridging stent for FEVAR have suggested good technical success, patency, and reintervention rates. Torsello et al. published on 50 patients undergoing implantation into 145 target vessels with good technical success and low rates of endoleak or target-vessel complications on short-term follow-up [28]. Similarly, Deslarzes-Dubuis and colleagues found no effect of stent choice between VBX and iCast in a cohort of 114 patients undergoing FEVAR with the two-vessel fenestrated Cook ZFEN device [29].

In the European Union, United Kingdom, and Canada, the BeGraft (Bentley InnoMed GmbH, Lotzenäcker, Germany) has been utilized as a third balloon-expandable covered stent option. It is built on a 6 Fr platform for stents up to 8 mm × 58 mm and otherwise can be delivered on a 7 Fr system. It is also very flexible and can be flared with minimal foreshortening. The BeGraft has demonstrated promising results. A study of 101 BeGraft bridging stents during FEVAR in 39 patients showed a 98% patency rate during immediate follow-up [30] and 97% at 2 years [31].

Technical Considerations

Patient anatomy should be considered when selecting bridging stents. Small target vessels and multiple renal arteries may be a contraindication to fenestrated repairs as bridging stent patency will decrease drastically. Early branch bifurcation of target vessels may affect the selection of bridging stent length.

Bridging Stent-Specific Complications

In contrast to complications related to main body endografts, target-vessel stent migration is rare. Most commonly, target-vessel patency loss is typically due to stent thrombosis, the incidence of which is increased in small-caliber vessels. Underexpansion of bridging stents can result in luminal narrowing and potential loss of patency, an etiology which can be avoided by postoperative cone beam CT prior to leaving the operating room. In the long term, bridging stents can also kink or even fracture over time, especially with continued craniocaudal respiratory excursion of the visceral vessels and structural fatigue of the stent material [27]. This phenomenon can also lead to stent thrombosis and subsequent target-vessel loss. Lastly, neointimal hyperplasia and in-stent restenosis should be monitored with imaging surveillance over time to prevent target-vessel loss.

Bridging stents can also be the cause of types Ic or IIIc endoleaks. A type Ic endoleak occurs at the distal seal zone of a bridging stent within the target vessel. Early-type Ic endoleaks can occur in the setting of short distal seal zones, early bifurcations, stent undersizing, and target-vessel calcifications. Late-type Ic endoleaks, however, are typically caused by aortic remodeling, as it can cause dislocation of the bridging stent out of the target vessel if the seal zone is short. Aneurysmal degeneration of the target vessels themselves is less frequent but can also be associated with late-type Ic endoleaks. A type IIIc endoleak, on the other hand, occurs at the junction between the bridging stent and the fenestration. Early-type IIIc endoleaks can occur in the setting of stent undersizing, inadequate balloon flaring, or poor construction of the fenestration, which can lead to stent-to-fenestration size mismatch or fabric tearing. Late-type IIIc endoleaks can be caused by aortic remodeling as well as fabric or stent fatigue leading to loss of seal.

Physician-Modified Fenestrated Endografts

Fenestrated endovascular aneurysm repair (FEVAR), as described above, is a safe and well-established treatment modality for juxtarenal or very short neck AAAs in the United States. In Europe, paravisceral AAAs can also be addressed with commercially available fenestrated endografts. However, physician-modified fenestrated endografting currently remains an essential component of complex aortic surgery, given the need for multi-visceral reconstruction outside of commercially available devices, regulatory trials, and in the United States, physician-sponsored investigational device exemption (PS-IDE) studies.

Another key consideration pertains to urgent repairs in patients with thoracoabdominal aortic aneurysms, pararenal and juxtarenal aortic aneurysms, as well as failed previously implanted infrarenal endografts. Manufactured devices often are not appropriate for these situations due to the time-consuming manufacturing process (4–12 weeks) and the potential for rupture during this period. Several countries including Canada, the United Kingdom, and the European Union have access to off-the-shelf branched options to address both elective and urgent cases, but the design of these off-the-shelf grafts may not fit the anatomy of the patient as well as a customized graft. As a result, the usage of physician-modified grafts for providing an endovascular solution customized to the patient's anatomy persists.

Regulatory Considerations

The US Food and Drug Administration (FDA) has previously suggested that physician-modified endografts (PMEGs) be performed solely under a PS-IDE [32, 33]. The usage of PMEG devices in the United States outside of a PS-IDE or officially recognized compassionate use is utilized off-label. The discussion of PMEG construction and implantation in Europe has not met with similar labeling by regulatory agencies. The following discussion of PMEG devices is thus strictly off-label, and regulatory oversight is recommended for any utilization of physician modification outside of emergency or compassionate use if applicable.

Graft Choice

Endograft modification proceeds by partial or complete deployment of the chosen graft on a sterile back table in an operating room or similar sterile environment. Graft selection is dependent on the aneurysm anatomy, and the fenestrated component may be built on tube-style or bifurcated-style grafts depending on anatomic need. Polyester-based endografts are preferrable as the modification of polytetrafluoroethylene (PTFE) devices is less studied.

Nearly all aortic devices can be modified in some way to result in a fenestrated component for implantation. However, certain grafts have emerged as the predominant choices for graft modification due to their predictability, ease of modification, and history of consistent usage (Fig. 51.15). The most commonly used and most often-studied graft for construction is the Cook thoracic graft series (TX2 and Alpha). This platform provides polyester-based fabric over widely spaced nitinol stent segments (Cook Alpha) in both straight and tapered formats. The Cook series of thoracic stent grafts has been utilized in several large published series, demonstrating the utility of PMEG compared to commercially available custom devices [34, 35].

The Terumo Treo endograft (Terumo Aortic, Sunrise, FL) has emerged as an option for physician modification on a

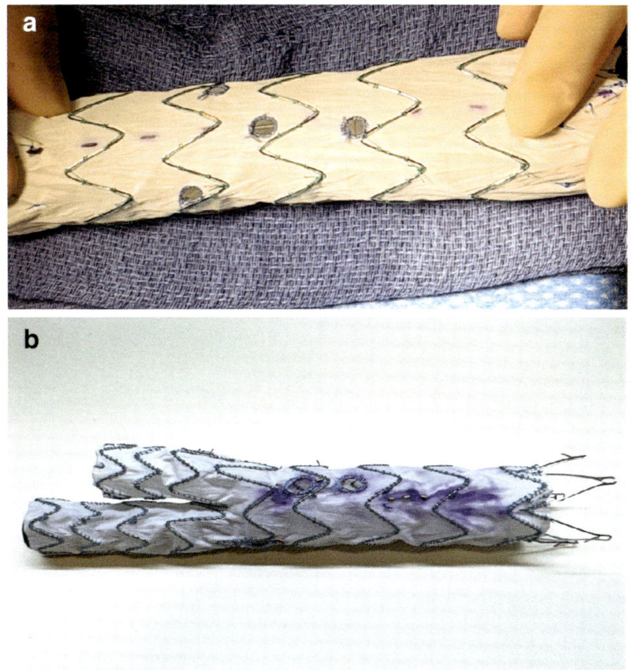

Fig. 51.15 Modified Cook Alpha (**a**) and Terumo Treo (**b**) endografts

bifurcated device platform. The Treo is manufactured with main body length up to 120 mm from proximal fabric to distal edge of contralateral gate and 90 mm of modifiable length. This obviates the need for a separate distal bifurcated device which is useful when treating pararenal or paravisceral aneurysms. A series of 46 patients with modified Treo devices utilized under a PS-IDE from the University of Washington demonstrated good mid-term results with significantly shorter modification times than when using other commercially available devices [36].

Graft Construction

Accurate planning of the fenestrations is critical as is for commercially available fenestrated endografts. For graft modification, arc lengths must be calculated from the typical clock positions based on the inner vessel diameter. The arc length specifies the actual distance along the circumference of the graft, where the fenestration should be placed. Axial positioning of the fenestrations is planned and calculated from the top of edge of the endograft using centerline measurements.

On a separate table under strict sterile precautions, the chosen endograft is unsheathed, dampened in sterile saline, and the fenestrations are marked. An ophthalmic Bovie electrocautery is used to create the fenestrations away from stent struts, which can then be reinforced in a multitude of ways. Typical fenestration diameter sizes are 5–7 mm for renal arteries and 7–9 mm for the superior mesenteric and celiac

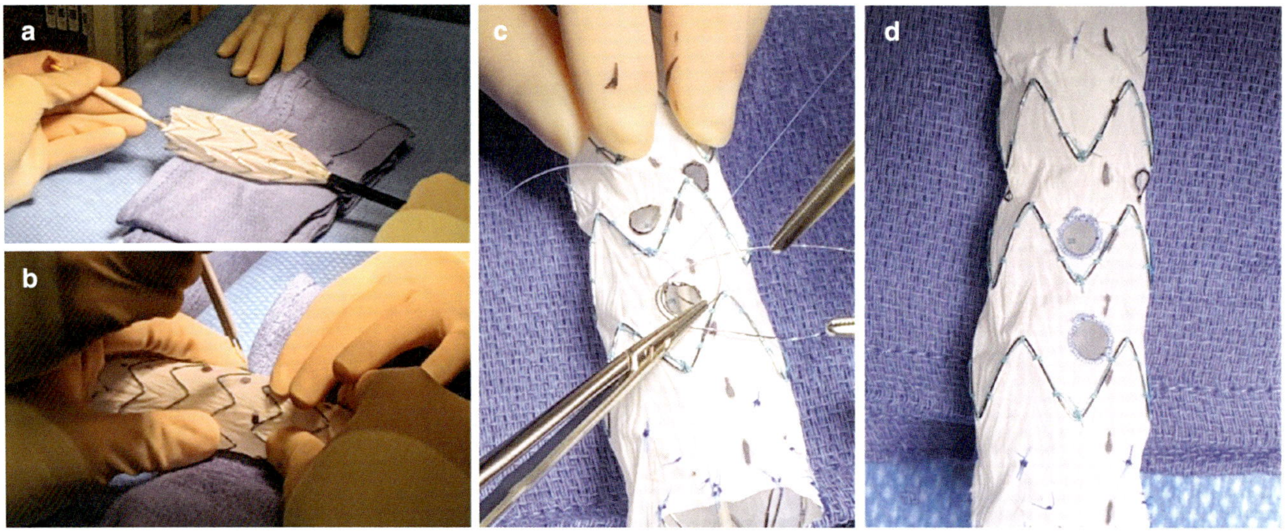

Fig. 51.16 PMEG construction process. (**a**) Unsheathing of the graft on a sterile back table. (**b**) Marking and creating fenestrations with the ophthalmic cautery device. (**c**) Reinforcing fenestrations with double-looped snare and nonabsorbable braided suture. (**d**) Completed reinforced fenestrations

arteries, with anticipation for the stretching of the fenestrations during balloon angioplasty. Reinforcement of the fenestration is accomplished by suturing a nitinol or PFTE ring around the fenestration with braided nonabsorbable suture. Commonly used nitinol reinforcements include Amplatz Goose Neck Snares (Medtronic, Minneapolis, MN), a double-looped EnSnare (Merit Medical, South Jordan, UT), or a double-looped 0.014″ wire such as Thruway (Boston Scientific, Marlborough, MA). Our preferred method is a piece of double-looped EnSnare sutured into place with 5-0 Ti-Cron suture (Medtronic, Minneapolis, MN) (Fig. 51.16).

Orientation markers should be sutured to the endograft to allow for rotational alignment during deployment. Prewiring the fenestrations is another useful consideration if cannulation of the fenestration is anticipated to be difficult, namely in highly angulated aortas and/or target vessels. Next, temporary diameter-reducing ties should be strongly considered whenever the fenestrations involve the seal zone. Lastly, the endograft can be re-sheathed using various techniques including readvancing the sheath over the constrained graft or antegrade re-sheathing with gradual size reduction using peel-away sheaths (Fig. 51.17). In general, the largest endograft diameter for any given sheath size will typically be challenging to re-sheath in its original delivery system, and larger sheaths should be considered.

Consistency and Quality Control

When performing a physician modification, quality control of the measurement and construction process lies with the physician rather than with the manufacturer of the graft. Therefore, consistency and a repeatable process are key. Preplanning measurements are constantly double checked, and planning sheets are printed and hung in the operating room during construction. In our institution we utilize 3D-printed sterilizable templates based on the measurements calculated. This is printed in the form of a tube rather than based on the anatomy of the aorta, as the relative shift in distances with deformation of the aorta with stiff wires and devices would more significantly impact a 3D printed anatomic model. The 3D model has allowed us to quickly identify the locations of the fenestrations in between stent struts, as well as mark out a reference centerline on the device and prevent tedious remeasurement of arc lengths that may be subject to transfer error at the time of construction. Our 3D printing lab at the University of Pittsburgh prints the template under OR conditions in a sterile environment with positive pressure ventilation, and the model is then sterilized in the autoclave prior to usage. The use of 3D templates has demonstrated an increase in efficiency when described in the literature [33, 37].

Outcomes

Contemporary commercially available fenestrated graft repairs have significantly improved the perioperative mortality associated with complex aortic repairs. The same can be said of well-planned, expertly executed physician-modified fenestrated endografts. Moreover, it has allowed a considerable number of patients, who otherwise would not have been surgical candidates via an open approach, to successfully undergo aneurysm repair. Outcomes data on PMEGs is fairly

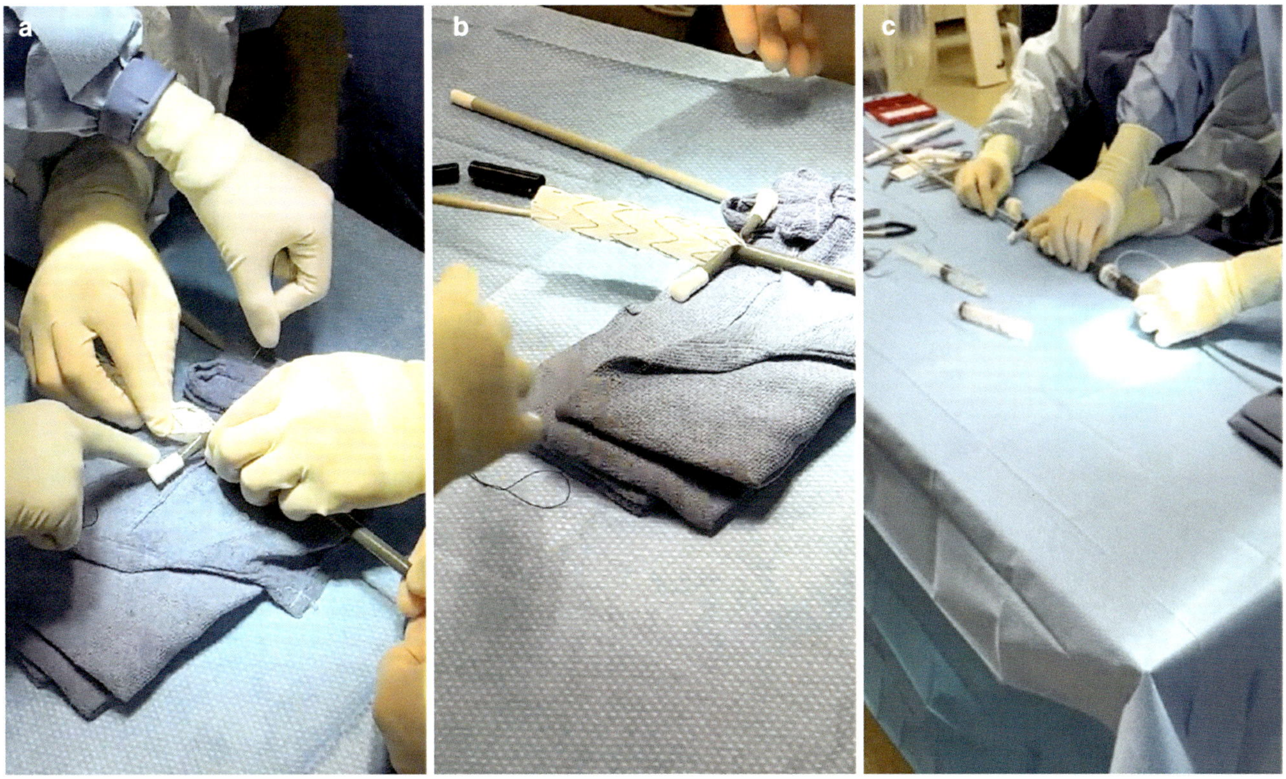

Fig. 51.17 Re-sheathing of the modified graft using an antegrade method with peel-away sheaths. Here, an 18F device is first reconstrained into a 22F peel-away sheath using silk suture or umbilical tape to collapse the stent segments (**a**, **b**). The 22F is then advanced into a 20F peel-away sheath which has been inserted into the original 18F delivery sheath. The device is then advanced until it is in its original position inside the original sheath (**c**)

sparse and limited to single-center studies from high-volume expert surgeons. Those outcomes may not be generalizable but still demonstrate feasibility, safety, and short- to mid-term durability of PMEGs. Despite this, concerns regarding fabric integrity and long-term durability of hand-made graft modifications by physicians remain.

Several high-profile larger series of PMEGs have been published, primarily comparing them to commercially manufactured multi-fenestrated devices (CMD). Oderich et al. compared their experience with 316 patients, beginning with PMEGs and transitioning to CMDs over a 10-year period. Despite statistically significant lower technical success rates in their PMEG group (98% for PMEGs vs. 99.5% for CMDs, $p = 0.02$), technical success rates remained very high in both groups with similar durability and freedom from reintervention at 3 years. Primary and secondary target-vessel patency rates were similar for both approaches as well [34].

Schanzer and colleagues at the University of Massachusetts also evaluated their outcomes using PMEGs versus CMDs. The primary differences were found to be related to operative metrics (total fluoroscopy time, total contrast volume used, and operative time). Moreover, 1-year reintervention rates were higher in the PMEG group (37% vs. 13%, $p = 0.04$). However, there were no differences in perioperative complications, type I or type III endoleaks, or survival. They concluded that no approach can be considered better than the other, and both remain viable options to complex open aortic aneurysm repair [35].

A systematic review and meta-analysis by Gouveia e Melo and colleagues from 2022 reviewed 909 PMEGs from 20 studies with 64% treated in the elective setting. The major adverse 30-day event rate was 11.6% in the elective cohort with high rates of technical success (96%) for FEVAR. Technical success was higher in smaller extent aneurysms and approached 99% in paravisceral and Type IV thoracoabdominal aortic aneurysms. The authors concluded that PMEGs represent a feasible and safe option for the repair of complex aortic pathology in the absence of CMDs but noted that a large proportion of the manuscripts reviewed did not include longer-term data which would be necessary to evaluate durability [38].

High-volume centers with dedicated complex aortic programs are undoubtedly reporting improved outcomes with CMDs and PMEGs. Although long-term data on PMEGs is still lacking, PMEGs will likely continue to play a pivotal role in complex endovascular aortic repair to serve patients who are not candidates for or who do not have access to CMDs.

Conclusion

Along with branched technology, FEVAR has allowed for endovascular repair of complex aortic pathology by incorporating visceral vessels into the endograft design. The unique nature of each patient underscores the need for preparation and experience in these cases. The physician performing FEVAR must be aware of the complexities of planning and technique of execution, including the individual unique aspects of the graft being used as well as anticipating the various potential complications which may arise. Future directions for FEVAR include expanding access to CMDs, improved efficiency of planning and procedural aspects, and radiation dose reduction (Fig. 51.18).

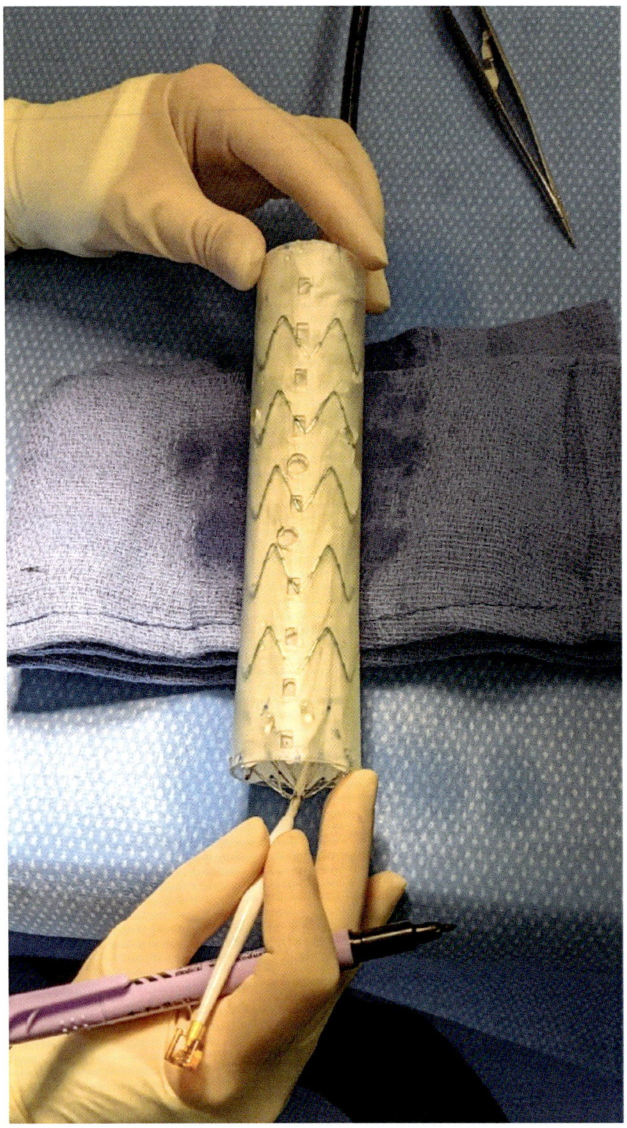

Fig. 51.18 3D printed templates are used in our institution to easily find the position of the fenestrations relative to the stent struts of the graft while preserving the relative orientation to the midline

References

1. Browne TF, Hartley D, Purchas S, Rosenberg M, van Schie G, Lawrence-Brown M. Endovascular and surgical techniques a fenestrated covered suprarenal aortic stent. Eur J Vasc Endovasc Surg. 1999;18:445–9.
2. Park JH, Chung JW, Choo IW, Kim SJ, Lee JY, Han MC. Fenestrated stent-grafts for preserving visceral arterial branches in the treatment of abdominal aortic aneurysms: preliminary experience. J Vasc Interv Radiol. 1996;7:819–23.
3. Doonan RJ, Girsowicz E, Dubois L, Gill HL. A systematic review and meta-analysis of endovascular juxtarenal aortic aneurysm repair demonstrates lower perioperative mortality compared with open repair. J Vasc Surg. 2019;70:2054–2064.e3.
4. Giménez-Gaibar A, González-Cañas E, Solanich-Valldaura T, Herranz-Pinilla C, Rioja-Artal S, Ferraz-Huguet E. Could preoperative neck anatomy influence follow-up of EVAR? Ann Vasc Surg. 2017;43:127–33. https://doi.org/10.1016/j.avsg.2016.12.011.
5. Caradu C, Erard XB, Midy D, Ducasse E, Bordeaux F. Influence of anatomic angulations in chimney and fenestrated endovascular aneurysm repair. Ann Vasc Surg. 2017;43:104–14. https://doi.org/10.1016/j.avsg.2017.01.002.
6. Chaikof EL, Fillinger MF, Matsumura JS, et al. Identifying and grading factors that modify the outcome of endovascular aortic aneurysm repair. J Vasc Surg. 2002;35:1061–6.
7. Marone EM, Freyrie A, Ruotolo C, Michelagnoli S, Antonello M, Speziale F, Veroux P, Gargiulo M, Gaggiano A, Vasc A. Expert opinion on hostile neck definition in endovascular treatment of abdominal aortic aneurysms (a Delphi consensus). Ann Vasc Surg. 2020;62:173–82.
8. Oliveira NFG, Gonçalves FB, Ultee K, Pinto JP, Josee van Rijn M, ten Raa S, Mwipatayi P, Böckler D, Hoeks SE, Verhagen HJM. Patients with large neck diameter have a higher risk of type IA endoleaks and aneurysm rupture after standard endovascular aneurysm repair. J Vasc Surg. 2019;69(3):783–91.
9. Oliveira NFG, Oliveira-Pinto J, van Rijn MJ, Baart S, ten Raa S, Hoeks SE, Bastos Gonçalves F, Verhagen HJM. Risk factors, dynamics, and clinical consequences of aortic neck dilatation after standard endovascular aneurysm repair. Eur J Vasc Endovasc Surg. 2021;62:26. https://doi.org/10.1016/j.ejvs.2021.03.020.
10. Antoniou GA, Georgiadis GS, Antoniou SA, Kuhan G, Murray D. A meta-analysis of outcomes of endovascular abdominal aortic aneurysm repair in patients with hostile and friendly neck anatomy. J Vasc Surg. 2013;57:527–38.
11. O'Donnell TFX, McElroy IE, Boitano LT, Mohebali J, Lamuraglia GM, Kwolek CJ, Conrad MF. Comparison of treatment options for aortic necks outside standard endovascular aneurysm repair instructions for use. J Vasc Surg. 2021;74:1548–57.
12. Oderich GS, Forbes TL, Chaer R, Davies MG, Lindsay TF, Mastracci T, Singh MJ, Timaran C, Woo EY. Reporting standards for endovascular aortic repair of aneurysms involving the renal-mesenteric arteries. J Vasc Surg. 2021;73:4S–52S.
13. Oderich GS, Greenberg RK, Farber M, Lyden S, Sanchez L, Fairman R, Jia F, Bharadwaj P. Results of the United States multicenter prospective study evaluating the Zenith fenestrated endovascular graft for treatment of juxtarenal abdominal aortic aneurysms. J Vasc Surg. 2014;60:1420–1428.e5.
14. Wang SK, Gutwein AR, Gupta AK, Lemmon GW, Sawchuk AP, Motaganahalli RL, Murphy MP, Fajardo A. Institutional experience with the Zenith Fenestrated aortic stent graft. J Vasc Surg. 2018;68(2):331–6. https://doi.org/10.1016/j.jvs.2017.11.063.
15. Stern JR, Deslarzes-Dubuis C, Tran K, Lee JT. Fenestrated aortic aneurysm repair in patients treated inside versus outside of instructions for use. 2022. https://doi.org/10.1177/15266028211068762. https://doi-org.pitt.idm.oclc.org/101177/15266028211068762.

16. van Calster K, Bianchini A, Elias F, Hertault A, Azzaoui R, Fabre D, Sobocinski J, Haulon S. Risk factors for early and late mortality after fenestrated and branched endovascular repair of complex aneurysms. J Vasc Surg. 2019;69:1342–55.
17. Katsargyris A, Oikonomou K, Kouvelos G, Mufty H, Ritter W, Verhoeven ELG. Comparison of outcomes for double fenestrated endovascular aneurysm repair versus triple or quadruple fenestrated endovascular aneurysm repair in the treatment of complex abdominal aortic aneurysms. J Vasc Surg. 2017;66:29–36.
18. Motta F, Oderich GS, Tenorio ER, Schanzer A, Timaran CH, Schneider D, Sweet MP, Beck AW, Eagleton MJ, Farber MA. Fenestrated-branched endovascular aortic repair is a safe and effective option for octogenarians in treating complex aortic aneurysm compared with nonoctogenarians. J Vasc Surg. 2021;74:353–362.e1.
19. Jubouri M, Surkhi AO, Tan SZCP, Bailey DM, Williams IM, Bashir M. Abdominal aortic aneurysm: can the Anaconda™ custom-made device deliver? An international perspective. Front Cardiovasc Med. 2022;9:959149. https://doi.org/10.3389/fcvm.2022.959149.
20. de Niet A, Zeebregts CJ, Reijnen MMPJ, et al. Outcomes after treatment of complex aortic abdominal aneurysms with the fenestrated Anaconda endograft. J Vasc Surg. 2020;72:25–35.e1.
21. Panuccio G, Schanzer A, Rohlffs F, Heidemann F, Wessels B, Schurink GW, van Herwaarden JA, Kölbel T. Endovascular navigation with fiber optic RealShape technology. J Vasc Surg. 2022;75:e234. https://doi.org/10.1016/J.JVS.2022.08.002.
22. Chamseddin K, Timaran CH, Oderich GS, et al. Comparison of upper extremity and transfemoral access for fenestrated-branched endovascular aortic repair. J Vasc Surg. 2022;77:704. https://doi.org/10.1016/j.jvs.2022.10.009.
23. Scott CK, Driessen AL, Gonzalez MS, Malekpour F, Guardiola GG, Baig MS, Kirkwood ML, Timaran CH. Perioperative neurologic outcomes of right versus left upper extremity access for fenestrated-branched endovascular aortic aneurysm repair. J Vasc Surg. 2022;75:794–802.
24. Price LZ, Safir SR, Faries PL, McKinsey JF, Tang GHL, Tadros RO. Shockwave lithotripsy facilitates large-bore vascular access through calcified arteries. J Vasc Surg Cases Innov Tech. 2021;7:164–70.
25. van der Riet C, Schuurmann RCL, Verhoeven ELG, Zeebregts CJ, Tielliu IFJ, Bokkers RPH, Katsargyris A, de Vries JPPM. Outcomes of Advanta V12 covered stents after fenestrated endovascular aneurysm repair. J Endovasc Ther. 2021;28:700–6.
26. Torsello GF, Herten M, Müller M, Frank A, Torsello GB, Austermann M. Comparison of the biomechanical properties of the Advanta V12/iCast and Viabahn stent-grafts as bridging devices in fenestrated endografts: an in vitro study. J Endovasc Ther. 2020;27:258–65.
27. Torsello G, Müller M, Litterscheid S, Berekoven B, Austermann M, Torsello GF. Fatigue resistance of the Advanta V12/iCast and Viabahn balloon-expandable stent-graft as bridging stents in experimental fenestrated endografting. J Endovasc Ther. 2021;28:463–8. https://doi.org.pitt.idm.oclc.org/101177/1526602821996724.
28. Torsello GF, Beropoulis E, Munaò R, Trimarchi S, Torsello GB, Austermann M. Outcomes of bridging stent grafts in fenestrated and branched endovascular aortic repair. J Vasc Surg. 2020;72:859–65.
29. Deslarzes-Dubuis C, Tran K, Colvard BD, Lee JT. Renal stent complications and impact on renal function following standard fenestrated endovascular aneurysm repair. Ann Vasc Surg. 2020;72:106. https://doi.org/10.1016/j.avsg.2020.10.020.
30. Spear R, Sobocinski J, Hertault A, Delloye M, Azzauiu R, Fabre D, Haulon S. One year outcomes of 101 BeGraft stent grafts used as bridging stents in fenestrated endovascular repairs. Eur J Vasc Endovasc Surg. 2018;55:504–10.
31. Clough RE, Spear R, Mougin J, le Houérou T, Fabre D, Sobocinski J, Haulon S. Midterm outcomes of BeGraft stent grafts used as bridging stents in fenestrated endovascular aortic aneurysm repair. J Endovasc Ther. 2022;30:592. https://doi.org/10.1177/15266028221091894/ASSET/IMAGES/LARGE/10.1177_15266028221091894-FIG2.JPEG.
32. Abel D, Farb A. Application of investigational device exemptions regulations to endograft modification. J Vasc Surg. 2013;57:823–5.
33. Starnes BW, Tatum B, Singh N. Procedural and perioperative results in patients treated with fenestrated endovascular aneurysm repair planned by automated software in a physician-sponsored investigational device exemption trial of physician-modified endografts. J Vasc Surg. 2018;68:1297–307.
34. Oderich GS, Ribeiro MS, Sandri GA, Tenorio ER, Hofer JM, Mendes BC, Chini J, Cha S. Evolution from physician-modified to company-manufactured fenestrated-branched endografts to treat pararenal and thoracoabdominal aortic aneurysms. J Vasc Surg. 2018;70:31. https://doi.org/10.1016/j.jvs.2018.09.063.
35. Dossabhoy SS, Simons JP, Flahive JM, Aiello FA, Sheth P, Arous EJ, Messina LM, Schanzer A. Fenestrated endovascular aortic aneurysm repair using physician-modified endovascular grafts versus company-manufactured devices. J Vasc Surg. 2018;67:1673–83.
36. Hemingway JF, Starnes BW, Kline BR, Singh N. Initial experience with the Terumo aortic Treo device for fenestrated endovascular aneurysm repair. J Vasc Surg. 2021;74:823–831.e1.
37. Coles-Black J, Barber T, Bolton D, Chuen J. A systematic review of 3D printed template-assisted physician modified stent grafts for FEVAR. J Vasc Surg. 2021;74(1):296–306.e1.
38. Gouveia e Melo R, Fernández Prendes C, Caldeira D, Stana J, Rantner B, Wanhainen A, Oderich GS, Tsilimparis N. Systematic review and meta-analysis of physician modified endografts for treatment of thoraco-abdominal and complex abdominal aortic aneurysms. Eur J Vasc Endovasc Surg. 2022;64:188–99.

Branched Endografts

52

Athanasios Katsargyris and Natasha Hasemaki

Introduction

Endovascular aortic aneurysm repair (EVAR) has now been used for almost 30 years and has changed the treatment of abdominal aortic aneurysms (AAA) [1]. A basic limitation of standard EVAR includes the necessity for an adequate infrarenal sealing zone in order to exclude the aneurysm sufficiently. Thus, aneurysms that present with a short neck or no neck were initially not amenable to endovascular treatment with standard EVAR.

The evolution of endovascular technology led gradually to advanced endovascular options that enabled sealing in the suprarenal or paravisceral segment of the aorta with the use of fenestrations on the aortic endograft in order to preserve perfusion of the renal and visceral arteries [2]. Fenestrated stent-grafts enabled endovascular treatment of short-neck AAA, pararenal AAA, and some type IV thoracoabdominal aortic aneurysms (TAAA). Fenestrations are stented with a covered stent in order to secure a complete alignment of the fenestration with the respective target vessel. A further evolution of fenestrations led to the introduction of branched endografts [3] (Fig. 52.1). Branched endografts have a tubular protrusion arising from the main lumen of the endovascular graft. This tubular protrusion (directional branch) is connected with the respective target vessel with a balloon-expandable or self-expanding bridging covered stent. The rationale for using a branch instead of a fenestration is that such a construction provides a longer surface for the bridging stent to seal within the main body of the endograft compared to the ring of a fenestration. This may be particularly important in larger aneurysms, where the gap between the aortic endograft and the orifice of the target vessel is large and may lead to bridging stent-graft instability. Another important factor is that directional branches can direct the bridging stents toward the target vessel in a direction that matches native anatomy. Directional branches were first developed to preserve the internal iliac artery when sealing has to be achieved in the external iliac artery during endovascular aortic or iliac aneurysm repair. Directional branches were also developed for the visceral arteries in extensive TAAAs. Later on, branches were also used to preserve the supra-aortic branches in arch-branched devices [4].

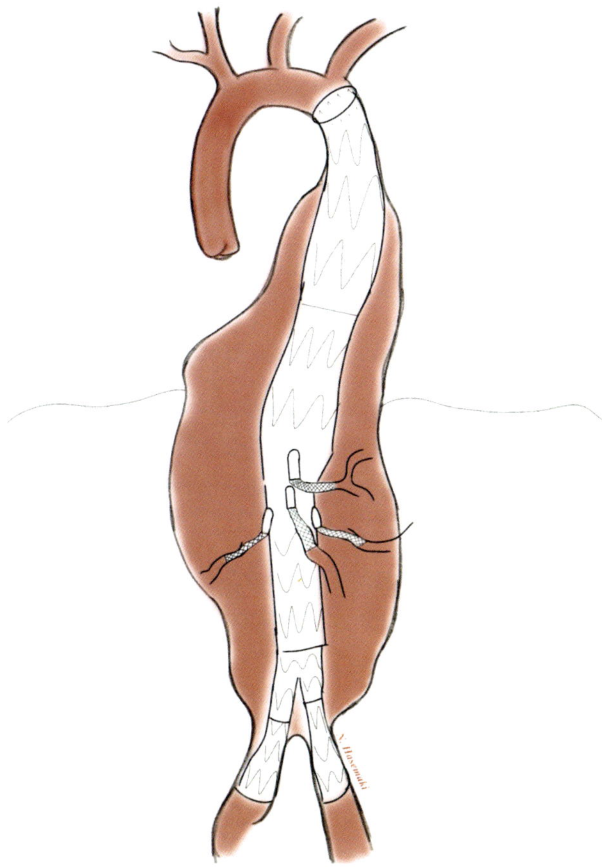

Fig. 52.1 Branched endograft for the treatment of type III TAAA

The present chapter describes the development and use of branched endografts, focusing on indication, planning, and procedure execution to treat different aortic segments if anatomically suitable, starting from the aortic arch to the external and internal iliac arteries.

A. Katsargyris (✉) · N. Hasemaki
Second Department of Vascular Surgery, Laiko Hospital, Athens National and Kapodistrian University of Athens, Athens, Greece

G. Geroulakos et al. (eds.), *Mastering Endovascular Techniques*, https://doi.org/10.1007/978-3-031-42735-0_52

Historical Aspects of Evolution of Branched Endografts

The idea of endovascular aortic aneurysm repair (EVAR) has been introduced in the mid-1980s when Blako et al. [5] used stent-grafts based on self-expanding nitinol stents and polyurethane to treat abdominal aortic aneurysms (AAA) in canine. In 1987, Volodos et al. [6] performed the first endovascular repair of a post-traumatic aneurysm in the distal thoracic aorta by advancing a stent-graft, while in 1989 they unsuccessfully attempted the endovascular use of a bifurcated stent-graft for the treatment of an AAA. The first successful EVAR was reported by Parodi et al. in 1991 when he combined his straight tubular polyester stent-graft with the balloon-expandable Palmaz stent as a proximal fixation [1]. The technique gained early recognition and acceptance, whereas in 1992 Frank Veith used the first tubular stent-graft in the USA to treat a patient with AAA [7].

The next evolution was the development of bifurcated stent-grafts in order to involve the iliac arteries and secure distal sealing zone. Timothy Chuter in 1993 was the first to introduce a unibody bifurcated stent-graft system for the treatment of an AAA [8].

As experience grew, it was realized that successful endovascular repair required good sealing in healthy aorta proximally and healthy iliac arteries distally to provide a long-term durable repair.

Despite the development and constant improvement of endovascular stent-grafts for AAA, there was still a significant portion of patients with anatomy not suitable for standard EVAR. This included patients with suprarenal, juxtarenal, short-neck infrarenal aneurysms, or thoracoabdominal aortic aneurysms (TAAA). Gradually, it became obvious that incorporating critical vessels (visceral, renal arteries) into the proximal sealing zone was mandatory in order to treat this group of patients with endovascular techniques.

The first stent-graft with one fenestration for the renal artery was performed by John Anderson in Adelaide, Australia, in 1998 [9], followed by Wolf Stelter in Frankfurt, Germany, who was the first to successfully use a fenestrated stent-graft to preserve a renal artery in Europe in 1999. With the advent of this technology, it became clear that more complex aneurysm could be treated with total endovascular repair. Tim Chuter was the first to suggest and implement a multibranched device for TAAAs by adding side branches to a tubular stent-graft with a reduced diameter at the level of the visceral arteries to accommodate all four branches without increasing the bulk of the multibranched stent-graft [4].

Subsequently fenestrated and branched devices were introduced also for the aortic arch. The above developments enabled gradually a total endovascular repair for anatomically suitable thoracoabdominal and aortic arch pathologies.

Preoperative Planning of Branched Endografts

Meticulous preoperative case planning and a thorough review of aneurysm anatomy are crucial in endovascular procedures, especially those using branched endografts. In addition to the measurement of lengths and diameters, a thin-slice computed tomography angiogram (CTA) determines the site of the sealing zone, target vessels' orientation and tortuosity, the presence of thrombus and ostial stenosis, and access vessels' quality. A successful aortic procedure with branched stent-grafts always requires a careful and accurate preoperative planning.

Sealing Zone

It is well established that endografts must seal in healthy aortic wall proximally and healthy iliac arteries distally in order to increase the durability of endovascular treatment. There is evidence that aortic disease is progressive [10, 11]; thus it is of utmost importance to assess the extent of aortic disease and carefully select the proximal and distal sealing zone.

Changes in aortic diameter or presence of arterial wall calcium, thrombus, debris, or thickening may indicate diseased aortic wall and implicate future progression of disease and failure of endovascular treatment. An aortic or iliac wall that appears diseased should not be selected for sealing zone. However, the risk of aortic disease progression must be balanced against the risk of spinal cord ischemia (SCI) by maintaining the aortic coverage to the minimum length that is considered to be safe and aiming at the same time to land if possible below major intercostal arteries.

The proximal and distal landing zones must be selected in parallel aortic wall, with minimal or no thrombus or calcification when possible. Usually, a minimum length of >2 cm is needed; however longer lengths are preferred, especially in cases of tortuous segments, ectasia, or thrombus-laden aorta [12, 13]. In cases of increased risk of paraplegia, preoperative planning should include enough overlap for future endovascular reintervention.

Selection of Fenestration/Branches

Once the sealing zone has been determined, the next step in the preoperative planning is to identify which type of incorporation will be used for the target vessels. The type of incorporation may be either fenestration or branch.

To date, there are three types of fenestrations: small, large, and scallop fenestrations. Directional branches can be internal, external, downward, or upward. Branches are either 6 or 8 mm in diameter and 18 or 21 mm in length. Routinely,

6 mm branches are used for the renal arteries and 8 mm branches for the celiac and superior mesenteric arteries.

As a general rule, fenestrations are preferred for target vessels originating from normal or only slightly dilated aortic segments, which is often the case in pararenal aneurysms or type I–III TAAA with narrow perivisceral aortic segments. On the other hand, branches are ideally suited to target vessels originating from aneurysms or large aortic lumens and, thus, are mainly used for extensive TAAA. In addition, fenestrations are ideally suited for transversely oriented vessels with a close-to-90° take-off from the aorta, while down-going vessels seem to have better alignment with branches and are also easier to catheterize via a branch from above. Fenestrations are generally placed on wide graft diameters to promote apposition to the aortic wall, while branches originate from tapered graft diameters to facilitate branch vessel catheterization [14].

The use of fenestrations is accompanied with less thoracic aorta coverage, better transverse orientation, and better patency rates of renal fenestrations [15]. Exact positioning of the main stent-graft with an alignment of the fenestration to the target vessel is mandatory or needs to be achieved by repositioning of the stent-graft, and catheterization can be tedious. Branches offer better seal and fixation for the bridging covered stent, less critical positioning and deployment of the branch, and better longitudinal alignment. On the other hand, branched endografts require a longer aortic coverage, since sealing has to be achieved above the tapered part of the graft, where the branches originate increasing therefore potentially the risk for spinal cord ischemia.

In cases of narrow aortic diameter with down-going target vessels, where neither fenestrations nor branches are suitable, inner branches can be used as a third alternative option. Until recently, inner branches were only used in the arch-branched device (Cook, Bloomington, IN, USA). Early experience demonstrates that inner branches work in selected vessels with otherwise difficult anatomy [16]. Inner branches are coupled with a diamond-shaped distal opening to facilitate branch vessel catheterization. Inner branches provide also advantages in chronic post-dissection TAAA, where the narrow true lumen can be a limitation for the planning of standard branches due to limited space. Another advantage of inner branches is that sealing can start lower in the aorta, reducing therefore the length of aortic coverage and the risk for spinal cord ischemia. This is because the graft does not need to taper down to a smaller diameter as in the case of standard branches. A typical example is two inner branches are used for the renal arteries, together with two fenestrations for the coeliac artery and the SMA. In this case the main graft remains wide in diameter at the suprarenal level, and sealing can be achieved already at this level, without the need to seal higher in the thoracic aorta (Fig. 52.2).

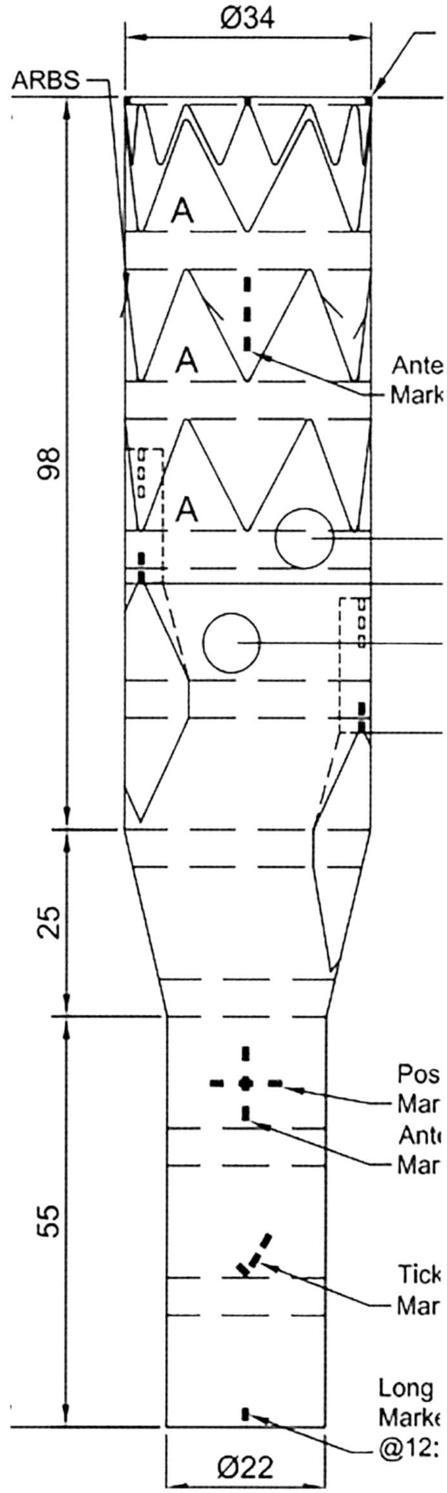

Fig. 52.2 Graft plan for pararenal AAA with narrow visceral aortic diameter and downward take-off of the renal arteries; two inner branches for the renal arteries; and two fenestrations for the coeliac artery and the SMA

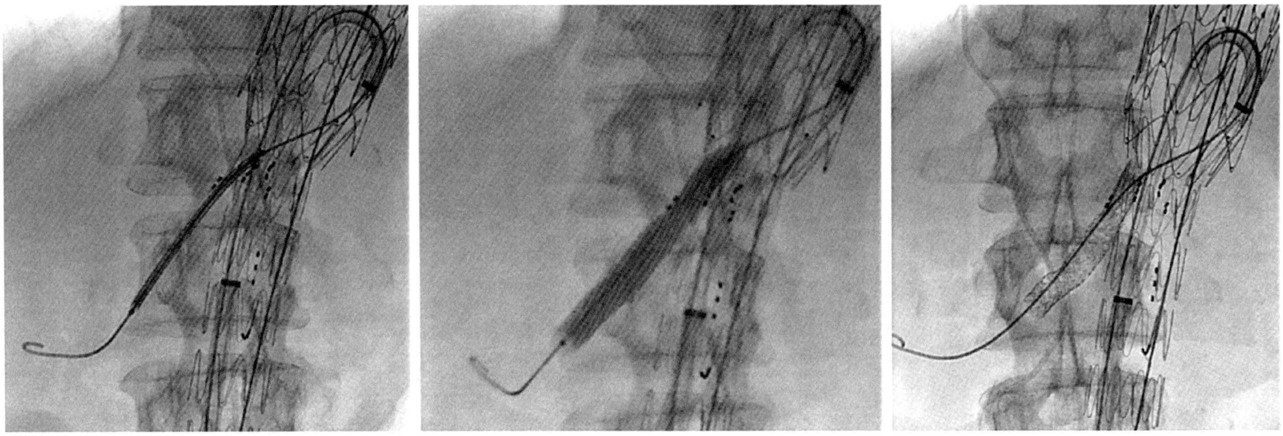

Fig. 52.3 Transfemoral access for vessels targeted with branches as an alternative to axillary approach

Implantation Techniques: Arterial Access

Down-going vessels targeted with branches are preferentially approached from above using an axillary/brachial approach [17]. Until recently, the axillary access has been considered mandatory for endovascular procedures with branched endografts. The axillary approach offers the advantage of early restoration of lower limb and pelvic blood flow, which is associated with improved spinal cord perfusion and lower rates of spinal cord ischemia. However, axillary access is combined with a stroke risk of 2–4% and a risk of median nerve neuropraxia of 1% [18]. In order to eliminate these complications, a transfemoral-only approach with the use of steerable sheaths has been applied recently in a few centers (Fig. 52.3).

The transfemoral access can result in decreased operative time, radiation exposure, and fluoroscopy time, while there are no significant differences in 30-day mortality or major adverse events [19, 20]. On the other hand, in difficult target vessel anatomies (ostial stenosis, severe calcification, or angulation), a transfemoral-only approach can be more cumbersome, and an upper approach may be preferable.

Outcomes of Branched Endografts in TAAA Endovascular Repair

Several series have reported early and mid-term outcomes of fenestrated and branched endografts for endovascular treatment of TAAA without however presenting outcomes specifically for branched stent-grafts only. Verhoeven et al. published one of the first series back in 2015 with F/BEVAR in TAAA in a total of 166 patients. Technical success was 95% with a 30-day operative mortality of 7.8%. Peri-operative spinal cord ischemia (SCI) was noted in 9%, with permanent paraplegia in 1.2%. Estimated survival at 1, 2, and 5 years was 83% ± 3%, 78% ± 3.5%, and 66.6% ± 6.1%, respectively. Estimated target vessel stent patency at 1, 2, and 5 years was 98% ± 0.6%, 97% ± 0.8%, and 94.2% ± 1.5%, respectively. Estimated freedom from reintervention at 1 and 3 years was 88.3% ± 2.7% and 78.4% ± 4.5%, respectively [21].

The group of Haulon et al. reported their experience with a total of 204 patients (42% type IV TAAA and 58% type I–III TAAA) [22]. The primary technical success rate was 92.6%, with a 30-day mortality of 6.9%. Major complications were noted in 26.5%, including SCI in 3.9%.

Oderich et al. reported recently the outcomes of F/BEVAR in 430 complex aortic aneurysms including 297 TAAAs [23]. Thirty-day mortality was 0.9%. New-onset dialysis was noted in 2% and permanent paraplegia in 2%. During follow-up there were 3 (0.7%) aortic-related deaths. Freedom from all-cause and aortic-related mortality at 5 years was 57% ± 5% and 98% ± 1%, respectively. Freedom from reintervention at 5 years was 64% ± 4%. Target vessel patency and target vessel instability at 5 years were 94% ± 1% and 89% ± 2%, respectively.

Branched Grafts as Off-the-Shelf Devices

Customized branched endografts for each individual patient are accompanied by a waiting time for manufacturing and shipping of 6–8 weeks, exposing patients to a rupture risk during the waiting period and limiting their use only in elective cases [24]. Off-the-shelf branched stent-grafts were manufactured aiming to eliminate the waiting time and enable immediate treatment in acute (symptomatic-ruptured) or large aneurysms. Off the-shelf branched stent-grafts are immediately available and can be used to treat an important proportion of patients (60–70%) in the acute setting even if the anatomy does not match ideally. On the other hand, using a graft that is not specifically designed for each patient's anatomy can lead to extra intraoperative difficulties that one should be prepared to handle with bail-out techniques and extra ancillary materials.

Cook t-Branch®

The Cook t-Branch® stent-graft (Cook, Bloomington, IN, USA) consists of a tapered woven polyester stent-graft sutured to a Z-stent stainless steel exoskeleton [25]. The tapered segment of the device contains four branches targeting the visceral and renal arteries. Celiac and superior mesenteric artery branches are 8 mm in diameter, 21 and 18 mm in length, and are located in the 01:00 and 12:00 positions, respectively. Right and left renal artery cuffs are 6 mm in diameter, 18 mm in length, and located in the 10:00 and 03:00 positions, respectively. The device has a diameter of 34 mm at the top and 18 mm at the bottom, with a length of 202 mm. The standard device is delivered through a 22-French system, while the lower-profile device is delivered through an 18-French system.

Gore Excluder TAMBE

The GORE® EXCLUDER® Thoracoabdominal Branch Endoprosthesis is an off-the-shelf, modular, multicomponent system composed of a proximal multibranched aortic component, a distal bifurcated component, and iliac limb extensions [26]. The TAMBE has four portals with options of either retrograde renal portals or antegrade renal portals. To date the use of this device is restricted to particular centers as investigational.

E-nside

The E-nside prosthesis (Artivion Inc., Hechingen, Germany) is a new off-the-shelf stent-graft with four inner branches for the visceral and renal arteries [27]. The stent-graft is available in two different proximal diameters (38 and 33 mm) and two different distal diameters (30 and 26 mm). The device is delivered through a 24-French system, and all inner branches are pre-cannulated. Short-term and long-term clinical results of the use of the E-nside stent-graft are lacking.

Branched Grafts for Aortic Arch Repair

Endovascular repair of the aortic arch may consist of distal extension of ascending aortic aneurysms, proximal extension of thoracic or thoracoabdominal aortic aneurysms, or isolated arch aneurysm repair. Until recently, hybrid repair with surgical debranching followed by TEVAR, parallel stent-grafts (chimney technique), and in situ fenestrations remained the alternative to open surgery in patients of high peri-operative risk. However, the significant progress of custom-made fenestrated and branched endografts gradually gained acceptance and made it possible to move the proximal sealing zone into zone 0 with total endovascular means.

Peter Mossop and Ian Nixon in Melbourne, Australia, performed the first case in 2000, using a fenestration to preserve the flow in a left subclavian artery (LSA) to extend the sealing zone proximally. Timothy Chuter and colleagues from the University of California, San Francisco, performed the first arch repair using a branched endograft in 2003 [28]. This was the first modular arch device, which was designed with a branched stent-graft to incorporate the innominate artery combined with cervical debranching of the left carotid and subclavian arteries.

The Cook A-branched arch stent-graft is a third-generation arch device developed by Cook Medical (Brisbane, Australia, and Bloomington, IN, USA). It is a custom-made design for zone 0 deployment with one or two proximal and two distal sealing stents. There are two or three internal branches (two antegrade and one retrograde for the LSA) with large diamond-shaped external openings that occupy the central narrower portion of the graft. The delivery system is equipped with a pre-curved cannula with a nitinol wire securing the outer curve of the stent-graft with the inner cannula to facilitate alignment.

The Relay branch arch device by Terumo-Aortic (Sunrise, FL) is a custom-made branched design based on the Relay NBS Plus platform. The stent-graft comprises nitinol exoskeleton sutured to polyester fabric. It is intended for zone 0 deployment with options of one or two antegrade internal branches. The device is equipped with a dual sheath design along with a pre-curved cannula and a proximal capture mechanism to facilitate alignment. The internal branches have anchors for a locking mechanism to prevent branch migration. The device comes in diameters ranging from 22 to 46 mm. It is oriented and advanced on an aortic wire, and branches are cannulated from retrograde access via cervical carotid and/or brachial access.

Iliac Branched Grafts

Patients with aortic aneurysm and concomitant common iliac artery (CIA) aneurysm or dilatation may have unsuitable distal sealing zone in the common iliac artery (CIA). In order to achieve a distal sealing zone for the normal iliac wall, embolization of one or both internal iliac arteries with extension of the stent-graft to the external iliac artery (EIA) was initially proposed. However, a significant portion of patients may develop complications related to internal iliac artery occlusion, such as buttock or thigh claudication, erectile dysfunction, or less frequently gluteal or perineal necrosis, ischemic colitis, and spinal cord injury [29]. Evolution of endovascular repair to extend the landing zone into the external iliac artery required creation of an iliac branch device (IBD). Marcel Goodman in Perth, Australia, successfully placed the first internal iliac branched stent-graft in a patient in 2001, whereas Wolf Stelter in Frankfurt, Germany, published the first series in 2007 [3].

The Zenith (Cook Medical) endovascular iliac branch device (IBD) is a bifurcated branch vessel graft with openings to connect the common iliac to the side branch and external iliac segments. The IBD device received CE mark in 2006. The stent-graft is loaded in a 20-F hydrophilic sheath and consists of full-thickness woven polyester fabric sewn to self-expanding stainless steel and nitinol Z-stents. There are two available lengths for the CIA segment (45 and 61 mm) with a diameter of 12 mm and two available lengths for the EIA segment (41 and 58 mm) with two diameters (10 and 12 mm). The IBD is the only iliac branch graft with available long-term data [30, 31].

The Gore Excluder Iliac Branch Endoprosthesis (IBE; W.L. Gore & Associates Inc., Flagstaff, AZ, USA) consists of two modular components: the Iliac Branch Component and the Internal Iliac Component. The Iliac Branch Component is delivered through a 16F sheath over two guidewires: a stiff guidewire and a femoral-femoral through-and-through wire that passes through the pre-cannulated internal iliac gate. A 12F sheath is inserted from the contralateral femoral artery and is advanced over the through-and-through wire into the internal iliac gate. The internal iliac component is then introduced through the 12F sheath into the internal iliac artery. Several studies [32, 33] have reported favorable results of the use of the IBE; however, the long-term outcomes and durability of the IBE device remain unknown.

The newest iliac branch device, the E-liac stent-graft system, by Jotec GmbH (Hechingen, Germany), consists of a nitinol skeleton with asymmetric stents covered by woven polyester that is sutured with braided polyester. The E-liac is available with three proximal diameters of 14, 16, and 18 mm and three distal diameters of 10, 12, and 14 mm. The CIA segment length can be 53 or 65 mm, whereas the EIA segment length is 44 or 56 mm. The only available data are up to 12 months of follow-up [34], and thus long-term results data are pending.

Conclusions

Branched stent-grafts have enabled total endovascular treatment of complex aortic pathologies, providing the possibility to exclude sufficiently the aneurysm while preserving at the same time blood flow to vital aortic branches arising from the aneurysm. The technique of branched stent-graft implantation has now gradually matured, and several expert centers report very good early and mid-term outcomes given the magnitude of the treated pathology (TAAA and aortic arch aneurysms). Off-the-shelf branched stent-grafts provide an option for immediate endovascular treatment of acute cases (ruptured, symptomatic, or large aneurysms) and are also increasingly used worldwide. Future improvements include the use of dedicated bridging stent-grafts aiming to improve target vessel patency and reduce the need for reintervention.

References

1. Parodi JC, Palmaz JC, Barone HD. Transfemoral intraluminal graft implantation for abdominal aortic aneurysms. Ann Vasc Surg. 1991;5:491–9.
2. Park JH, Chung JW, Choo IW, Kim SJ, Lee JY, Han MC. Fenestrated stent-grafts for preserving visceral arterial branches in the treatment of abdominal aortic aneurysms: preliminary experience. J Vasc Interv Radiol. 1996;7:819–23.
3. Ziegler P, Avgerinos ED, Umscheid T, Perdikides T, Erz K, Stelter WJ. Branched iliac bifurcation: 6 years' experience with endovascular preservation of internal iliac artery flow. J Vasc Surg. 2007;46:204–10.
4. Chuter TA. Branched and fenestrated stent grafts for endovascular repair of thoracic aortic aneurysms. J Vasc Surg. 2006;43 Suppl A:111A–5A.
5. Balko A, Piasecki GJ, Shah DM, Carney WI, Hopkins RW, Jackson BT. Transfemoral placement of intraluminal polyurethane prosthesis for abdominal aortic aneurysm. J Surg Res. 1986;40:305–9.
6. Volodos NL, Karpovich IP, Shekhanin VE, Troian VI, Iakovenko LF. A case of distant transfemoral endoprosthesis of the thoracic aorta using a self-fixing synthetic prosthesis in traumatic aneurysm. Grudn Khir (J Thorac Surg). 1988;6:84–6.
7. Parodi JC, Marin ML, Veith FJ. Transfemoral, endovascular stented graft repair of an abdominal aortic aneurysm. Arch Surg. 1995;130:549–52.
8. Ivancev K, Vogelzang R. A 35 year history of stent grafting, and how EVAR conquered the world. Eur J Vasc Endovasc Surg. 2020;59(5):685–94.
9. Anderson JL, Berce M, Hartley DE. Endoluminal aortic grafting with renal and superior mesenteric artery incorporation by graft fenestration. J Endovasc Ther. 2001;8:3–15.
10. Majewski W, Stanisic M, Pawlaczyk K, Marszalek A, Seget M, Biczysko W, et al. Morphological and mechanical changes in juxtarenal aortic segment and aneurysm before and after open surgical repair of abdominal aortic aneurysms. Eur J Vasc Endovasc Surg. 2010;40(2):202–8.
11. Rodway AD, Powell JT, Brown LC, Greenhalgh RM. Do abdominal aortic aneurysm necks increase in size faster after endovascular than open repair? Eur J Vasc Endovasc Surg. 2008;35(6):685–93.
12. Barbosa Lima G, Tenorio ER, Marcondes GB, Wong J, Saqib N, Mendes BC, De Luccia N, Oderich GS. Differences in procedural metrics and clinical outcomes among patients treated by fenestrated-branched endovascular repair of thoracoabdominal aortic aneurysms using infrarenal aortic versus iliac sealing zones. J Vasc Surg. 2021;74(5):1464–1471.e3.
13. Piazza M, Squizzato F, Xodo A, Saviane G, Forcella E, Dal Pont C, Grego F, Antonello M. Determination of optimal and safest proximal sealing length during thoracic endovascular aortic repair. Eur J Vasc Endovasc Surg. 2021;62(3):423–30.
14. Katsargyris A, Marques de Marino P, Verhoeven EL. Graft design and selection of fenestrations vs. branches for renal and mesenteric incorporation in endovascular treatment of pararenal and thoracoabdominal aortic aneurysms. J Cardiovasc Surg (Torino). 2019;60(1):35–40.
15. Martin-Gonzalez T, Mastracci T, Carrell T, Constantinou J, Dias N, Katsargyris A, Modarai B, Resch T, Verhoeven E, Haulon S. Mid-term outcomes of renal branches versus renal fenestrations for thoraco-abdominal aneurysm repair. Eur J Vasc Endovasc Surg. 2016;52(2):141–8.
16. Katsargyris A, Marques de Marino P, Mufty H, Pedro LM, Fernandes R, Verhoeven ELG. Early experience with the use of inner branches in endovascular repair of complex abdominal and thoraco-abdominal aortic aneurysms. Eur J Vasc Endovasc Surg. 2018;55(5):640–6.

17. Meertens MM, van Herwaarden JA, de Vries JPPM, Verhagen HJM, van der Laan MJ, Reijnen MMPJ, Schurink GWH, Mees BME. Multicenter experience of upper extremity access in complex endovascular aortic aneurysm repair. J Vasc Surg. 2022;76(5):1150–9. S0741-5214(22)01637-8.
18. Meertens MM, Lemmens CC, Oderich GS, Schurink GWH, Mees BME. Cerebrovascular complications after upper extremity access for complex aortic interventions: a systematic review and meta-analysis. Cardiovasc Intervent Radiol. 2020;43(2):186–95.
19. Patel RJ, Mathlouthi A, Al-Nouri O, Lane JS, Malas MB, Barleben AR. A single center review of a total transfemoral approach to upper extremity access in branched and fenestrated physician modified endografts. Ann Vasc Surg. 2022;86:117–26. S0890-5096(22)00314-4.
20. Eilenberg W, Kölbel T, Rohlffs F, Oderich G, Eleshra A, Tsilimparis N, Debus S, Panuccio G. Comparison of transfemoral versus upper extremity access to antegrade branches in branched endovascular aortic repair. J Vasc Surg. 2021;73(5):1498–503.
21. Verhoeven EL, Katsargyris A, Bekkema F, Oikonomou K, Zeebregts CJ, Ritter W, Tielliu IF. Editor's choice—ten-year experience with endovascular repair of thoracoabdominal aortic aneurysms: results from 166 consecutive patients. Eur J Vasc Endovasc Surg. 2015;49(5):524–31.
22. Maurel B, Delclaux N, Sobocinski J, Hertault A, Martin-Gonzalez T, Moussa M, Spear R, Le Roux M, Azzaoui R, Tyrrell M, Haulon S. The impact of early pelvic and lower limb reperfusion and attentive peri-operative management on the incidence of spinal cord ischemia during thoracoabdominal aortic aneurysm endovascular repair. Eur J Vasc Endovasc Surg. 2015;49(3):248–54.
23. Oderich GS, Tenorio ER, Mendes BC, Lima GBB, Marcondes GB, Saqib N, Hofer J, Wong J, Macedo TA. Midterm outcomes of a prospective, nonrandomized study to evaluate endovascular repair of complex aortic aneurysms using fenestrated-branched endografts. Ann Surg. 2021;274(3):491–9.
24. Katsargyris A, Uthayakumar V, Marques de Marino P, Botos B, Verhoeven EL. Aneurysm rupture and mortality during the waiting time for a customised fenestrated/branched stent graft in complex endovascular aortic repair. Eur J Vasc Endovasc Surg. 2020;60(1):44–8.
25. Chuter TAM, Hiramoto JS, Park KH, Reilly LM. The transition from custom-made to standardized multibranched thoracoabdominal aortic stent grafts. J Vasc Surg. 2011;54(3):660–8.
26. Cambiaghi T, Grandi A, Bilman V, Melissano G, Chiesa R, Bertoglio L. Anatomic feasibility of the investigational GORE EXCLUDER Thoracoabdominal Branch Endoprosthesis (TAMBE), off-the-shelf multibranched endograft for the treatment of pararenal and thoracoabdominal aortic aneurysms. J Vasc Surg. 2021;73(1):22–30.
27. Zimmermann A, Menges AL, Rancic Z, Meuli L, Dueppers P, Reutersberg B. E-nside off-the-shelf inner branch stent graft: technical aspects of planning and implantation. J Endovasc Ther. 2022;29(2):167–74.
28. Chuter TA, Schneider DB, Reilly LM, Lobo EP, Messina LM. Modular branched stent graft for endovascular repair of aortic arch aneurysm and dissection. J Vasc Surg. 2003;38(4):859–63.
29. Marques de Marino P, Botos B, Kouvelos G, Verhoeven ELG, Katsargyris A. Use of bilateral cook zenith iliac branch devices to preserve internal iliac artery flow during endovascular aneurysm repair. Eur J Vasc Endovasc Surg. 2019;57(2):213–9.
30. Parlani G, Verzini F, De Rango P, Brambilla D, Coscarella C, Ferrer C, Cao P. Long-term results of iliac aneurysm repair with iliac branched endograft: a 5-year experience on 100 consecutive cases. Eur J Vasc Endovasc Surg. 2012;43(3):287–92.
31. Mylonas SN, Ioannides G, Ahmad W, Brunkwall JS. Comparison of two iliac branch devices and their midterm performance in maintaining blood flow to the internal iliac artery. J Endovasc Ther. 2020;27(5):818–25.
32. Schneider DB, Matsumura JS, Lee JT, Peterson BG, Chaer RA, Oderich GS. Prospective, multicenter study of endovascular repair of aortoiliac and iliac aneurysms using the Gore Iliac Branch Endoprosthesis. J Vasc Surg. 2017;66(3):775–85.
33. Schönhofer S, Mansour R, Ghotbi R. Initial results of the management of aortoiliac aneurysms with GORE® Excluder® Iliac Branched Endoprosthesis. J Cardiovasc Surg (Torino). 2015;56(6):883–8.
34. Mylonas SN, Rümenapf G, Schelzig H, Heckenkamp J, Youssef M, Schäfer JP, Ahmad W, Brunkwall JS, E-liac Collaborative Group. A multicenter 12-month experience with a new iliac side-branched device for revascularization of hypogastric arteries. J Vasc Surg. 2016;64(6):1652–1659.e1.

Iliac Bifurcation Devices: Tips and Tricks

53

Calin-Gheorghe Precup, Marine Bordet, and Antoine Millon

Introduction

Up to 40% of patients with abdominal aortic aneurism (AAA) have concomitant iliac artery aneurysm, while isolated common or internal iliac artery aneurysm is a rare situation [1]. Open repair of aorto-iliac aneurysms remains technically challenging with an increased risk of ureteric or iliac vein injury because of the deep pelvic location [2]. The first endovascular strategy proposed was the occlusion of the internal iliac artery (coverage or embolization) associated with a high incidence of complications: buttock or thigh claudication, ischemic colitis, erectile dysfunction, gluteal or perineal necrosis, or spinal cord injury [3–6]. Preserving blood flow in the internal iliac artery appeared as a key point to improve patient care. Many endovascular techniques have been proposed. Iliac branch devices are considered as the best option for many surgeons.

The first Iliac branch device was described in 2006 by Malina et al. [7] and after almost two decades of endovascular treatment, innovation, with respect to the materials used, helped to make different iliac branch devices available today.

In this chapter, we describe these devices and explain how to use them in different clinical situations.

C.-G. Precup · M. Bordet · A. Millon (✉)
Department of Vascular and Endovascular Surgery, University Hospital of Lyon, Lyon, France
e-mail: calin-gheorghe.precup@chu-lyon.fr; marine.bordet@chu-lyon.fr; antoine.millon@chu-lyon.fr

Iliac Branch Devices: Configurations, Deployment Strategy and Anatomical Constraints

Currently, there are three commercially available iliac branch devices in Europe: Zenith Branch Endovascular Graft-Iliac Bifurcation (Cook Medical), GORE® EXCLUDER® Iliac Branch Endoprosthesis (W.L. Gore & Associates), and E-liac Stent Graft System (JOTEC GmbH). The design is the same, with a trunk, a short branch to be connected to the internal iliac artery (IIA), and a long branch for the external iliac artery (IIA). However, there are many differences with respect to materials, stents, and dimensions (length and diameter). The principal characteristics of the three devices are summarized in Table 53.1. The deployment technique is similar with a preloaded catheter for the internal iliac branch which must be cannulated from the contralateral femoral access before stent graft deployment. Thus, the catheterization of the branch and the targeted IIA is highly facilitated. Rigorous preoperative sizing (length and diameter measurement, choice of the adequate stent graft and bridging stent, access evaluation, angulations to expose the iliac bifurcation and the IIA bifurcation, adequate endovascular tools) leads to a straightforward and successful procedure. Radio-opaque markers allow for accurately aligning the device at the optimal level (about 5–10 mm above the origin of the IIA) and visualizing the optimal overlap required for the bridging stent in the IIA branch, but also for the bridging stent graft between the aortic stent graft (SG) and the iliac branch device (IBD).

Initially, these IBDs were designed to treat aorto-iliac aneurysms and aortic aneurysms without a sufficient distal landing zone to obtain durable aneurysmal exclusion without the morbidity related to IIA occlusion: buttock claudication, which may affect quality of life, ischemic colitis, erectile dysfunction, gluteal or perineal necrosis or spinal cord injury. Numerous studies have demonstrated excellent technical and

G. Geroulakos et al. (eds.), *Mastering Endovascular Techniques*, https://doi.org/10.1007/978-3-031-42735-0_53

Table 53.1 Iliac branch devices description

	Zenith Branch Endovascular Graft-Iliac Bifurcation (Cook Medical)	GORE® EXCLUDER® Iliac Branch Endoprosthesis (W.L. Gore & Associates)	E-liac Stent Graft System (JOTEC GmbH)
Material	Full-thickness woven polyester	Polytetrafluoroethylene and fluorinated ethylene propylene	Woven polyester
Stents	Stainless steel and nitinol	Nitinol	Nitinol
Components	Only one iliac branch component (needing of a bridging stent)	Iliac branch component and internal iliac component	Only one iliac branch component (needing of a bridging stent)
Introducer sheath	20F Preloaded catheter	16F Preloaded catheter	18F Preloaded catheter
Proximal diameter of iliac branch component	12 mm	23 mm	14–16–18 mm
Distal diameter of iliac branch component	10–12 mm	10–12–14.5 mm	10–12–14
Total length of iliac branch component	45, 61 mm	55 mm	79, 91 mm
Total length of internal iliac artery gate	15 mm	25 mm	26 mm
Internal iliac artery gate diameter	8 mm	13.5 mm	8 mm
Distal iliac diameters of internal iliac component	X	70 mm 10–12–14.5 mm diameter	X

Table 53.2 Anatomic conditions necessary for the use of iliac branch devices

	Zenith Branch Endovascular Graft-Iliac Bifurcation (Cook Medical)	GORE® EXCLUDER® Iliac Branch Endoprosthesis (W.L. Gore & Associates)	E-liac Stent Graft System (JOTEC GmbH)
Proximal common iliac artery diameter	<20 mm	<17 mm	12–17 mm
Distal common iliac artery diameter	–	<14 mm	-447qSg7dOADy
Common iliac artery length	<50 mm	<40 mm	<39 mm
Internal iliac artery lengths	<10 mm	<10 mm	>=15 mm
Internal iliac artery diameter	<6 mm or >9 mm	<6.5 mm or >13.5 mm	–
External iliac artery length	<20 mm	<10 mm	>=15 mm
External iliac artery diameter	<8 mm	<6.5 mm or >25 mm	8–13 mm
Angle between internal and external iliac arteries	–	–	50 degrees

procedural success (around 95%) and excellent long-term outcomes in terms of aneurysmal exclusion and IIA patency [8–10].

However, several anatomical criteria must be respected to obtain these results. These anatomical restrictions vary depending on the type of IBD (summarized in Table 53.2).

In our experience, the main anatomical criterion which affects the durability of the IBD is the quality of the IIA. A small (<5 mm), irregular, and calcified IIA lead to IIA occlusion in the short term. An IIA aneurysm can also alter the patency. A narrow calcified Iliac artery bifurcation may lead to a kink (and afterwards, occlusion) of one of the two iliac branches. This can be anticipated during preoperative planning and prevented by a kissing balloon or kissing stent at the iliac bifurcation inside the two iliac branches. A short common iliac artery (CIA) may also lead to complication because of an aortic protrusion of the IBD. This makes the crossover and the connection with the aortic stent graft difficult (not enough length for the bridging limb).

We summarize advantages and drawbacks of these three commercially available IBDs in Table 53.3.

Table 53.3 Advantages and disadvantages of the three IBDs

IBD	Advantages	Disadvantages
Cook	– Many available lengths, diameters, and sizes that allow the treatment of a large variety of anatomies – Very precise	– Rigid and hard to adapt to tortuous anatomies – Large introducer sheath (20 Fr) – No dedicated component for the internal iliac artery – Bridging stent is hard to adapt if the internal artery diameter is inferior to 6 mm while the internal iliac artery gate diameter is 8 mm
Gore	– Very conformable endoprosthesis with dedicated graft for internal iliac artery – Dedicated GORE® DrySeal flex introducer sheath, very flexible and resistant sheath, ensuring good access for tortuous iliac anatomies – Low profile (16 Fr)	– There are only three modules available, and the internal iliac component is long (7 cm) and has a large diameter (10, 12, 14.5 mm) hard to adapt for small IIA – The proximal diameter is 23 mm and cannot be adapted to all anatomies, especially for isolated common iliac artery aneurysms
Jotec	– Two different configurations for isolated iliac aneurysm and for aorto-iliac aneurysm with multiple configurations possible for better adaptation to different anatomies – Low profile (18 Fr), flexible, and hydrophilic coating of delivery system for safe and precise advancement in tortuous anatomies	– No dedicated component for the internal iliac artery – Bridging stent is hard to adapt if the internal artery diameter is inferior to 6 mm while the internal iliac artery gate diameter is 8 mm.

Particular Cases

Isolated Common Iliac Artery Aneurysm

The use of IBD in conjunction with aortic stent graft has shown excellent early and midterm efficacy with low complication rates and mortality. These results have been demonstrated with the three available IBDs. This approach, which requires extensive coverage of a healthy aorta and healthy contralateral common iliac artery, has some anatomical constraints such as reduced aortic or iliac length and diameter. It may be difficult to catheterize the contralateral gate in a small-diameter aorta and there is a risk of limb compression. The IBD from Cook and Gore should always be used in conjunction with an aortic stent graft according to the instructions for use to improve endograft stability and reduce the risk of device migration and endoleak. The IBD from Jotec has a configuration for isolated CIA aneurysm without the need for an aortic stent graft.

In case of suitable proximal iliac neck (length, diameter, and angulation), an approach with IBD alone (without an aortic stent graft) can be proposed [11, 12]. This strategy has been demonstrated safe and durable confirmed by a high midterm patency and aneurysmal exclusion without migration in a multicenter registry [13]. Depending on the anatomical conditions, this technique may require proximal extension in the proximal CIA (before or after IBD deployment) to achieve a safe proximal landing zone. This off-label use with Cook and Gore simplifies the procedure with reducing the time, the contrast load and the costs and avoids coverage of aortic branches such as the inferior mesenteric artery or an accessory renal artery.

Type 1B Endoleak or Iliac Aneurysm Degeneration in Patients with Prior Bifurcated Aortic Stent Graft

The aneurysmal evolution of iliac artery after previous endovascular treatment may lead to type 1b endoleak in the case of loss of distal sealing zone or to iliac aneurysm rupture in the case of large aneurysm [14, 15]. These patients are challenging cases for endovascular treatment with internal iliac artery preservation due to difficulty of sheath delivery up and over the flow divider. In this setting many technical solutions have been proposed.

The first solution is to cannulate and deliver the internal iliac component through an upper extremity access. Bisdas et al. reported their experience using an open transaxillary access and an 8F, 110 cm sheath to cannulate and deliver the IIA bridging stent [16]. This approach is straightforward and is of interest in the case of bilateral IBD procedure. The upper extremity approach adds potential additional morbidity including risk of cerebral embolization, peripheral nerve injury, or hematoma. Nonetheless, the transaxillary percutaneous approach is now well described and limits access site complications. Progression of the sheath through a large and tortuous thoracoabdominal aorta may be challenging and may be associated with risk of embolization in case of shaggy aorta or risk of stent graft dislocation in case of previous B-FEVAR. The other drawback of this approach is the need for long devices (catheter, balloon, and bridging stent).

Another solution proposed is the up-and-over transfemoral approach well described by Dawson et al. [17]. The goal of this technique is to avoid damage of the aortic bifurcated device during the crossover. Once its dilatator is removed the 12F sheath is advanced and the 7F sheath with its dilatator

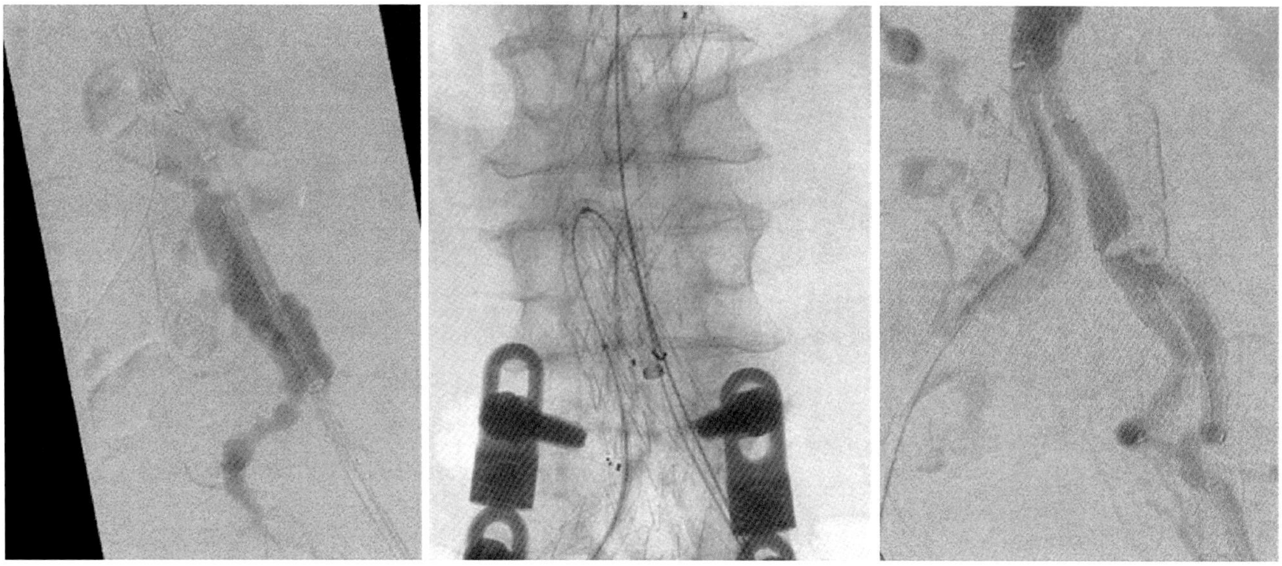

Fig. 53.1 "Up-and-over" technique for iliac artery aneurism development after EVAR treatment of an aortic aneurysm

hubbed inside the 12F sheath. Then simultaneously the 12F sheath is pushed up and over the flow divider and the 7F sheath is pulled back with a guidewire tension maintained on both ends. This push-pull maneuver provides an atraumatic adequate positioning of the 12F sheath (Fig. 53.1). Good clinical outcomes resulting from this technique have been published. According to the authors, the up-and-over technique should be considered a suitable alternative to brachial access in patients who require distal extension using iliac branch endoprosthesis (IBEs) [18, 19].

The third option is to use a looped wire to precannulate the internal gate via the ipsilateral femoral access [20]. This technique has been described by Stern et al. with the Gore IBE. A looped wire is advanced through the 16F sheath and then the IBE device is loaded onto both ends of the wire and advanced and positioned. The 16F sheath is withdrawn. The IBE is deployed as well as the external limb. Then a steerable sheath is advanced over the flow divider and positioned into the internal gate, allowing the catheterization of the IIA and the placement of the bridging stent. The disadvantage of this technique is the deployment of the entire device prior to gate cannulation with the risk of device malposition or migration.

All these techniques are obviously off-label and often preclude the use of the dedicated internal iliac component for the Gore IBD.

Aneurysmal Degeneration of Iliac Artery After a Previous Open Aneurysm Repair

Aneurysmal degeneration of the common iliac artery (CIA) after elective open infrarenal abdominal aortic aneurysm repair with an aorto-aortic prosthetic graft may be solved by IBD implantation. The IBD procedure in case of CIA aneurysm without a suitable proximal sealing zone at the iliac level requires the placement of a bifurcated aortic stent graft inside the prosthetic tube. This situation may lead to potential difficulty in opening the contralateral gate and therefore its catheterization.

In case of aorto-iliac prosthetic graft repair, most of the time, there is not enough room for a bifurcated aortic stent graft. The IBD may proximally land in the prosthetic graft or can be connected to the prosthetic graft with an iliac extension. The crossover of the contralateral sheath may be as challenging as inside a bifurcated aortic stent graft. The internal iliac component or bridging should be advanced through brachial access using the technical tips described above (up-and-over, looped wire).

Iliac anastomotic pseudoaneursym is a rare but very challenging situation with very little room for accurate IBD deployment and bridging stent placement (Fig. 53.2).

Internal Iliac Artery Aneurysms (Fig. 53.3)

Open repair of internal iliac aneurysm including ligation or arterial reconstruction remains technically challenging [21]. Occlusion of the internal iliac artery and its branches by coil embolization was initially recommended [22]. However, as already mentioned, this strategy is associated with increased morbidity and mortality. Several endovascular techniques have been proposed and the use of IBD seems to be a safe and efficient approach either in conjunction with an aortic bifurcated SG to treat an aorto-iliac aneurysm or solely to treat a solitary IIA without proximal sealing zone at the IIA trunk level [23]. A distal sealing zone at the IIA trunk level is rarely observed. Thus, the IIA bridging stent (internal component) has to be extended into one of its major divisional branches (anterior or posterior) [24–27]. Other collaterals have to be embolized to avoid type 2 endoleak. The posterior

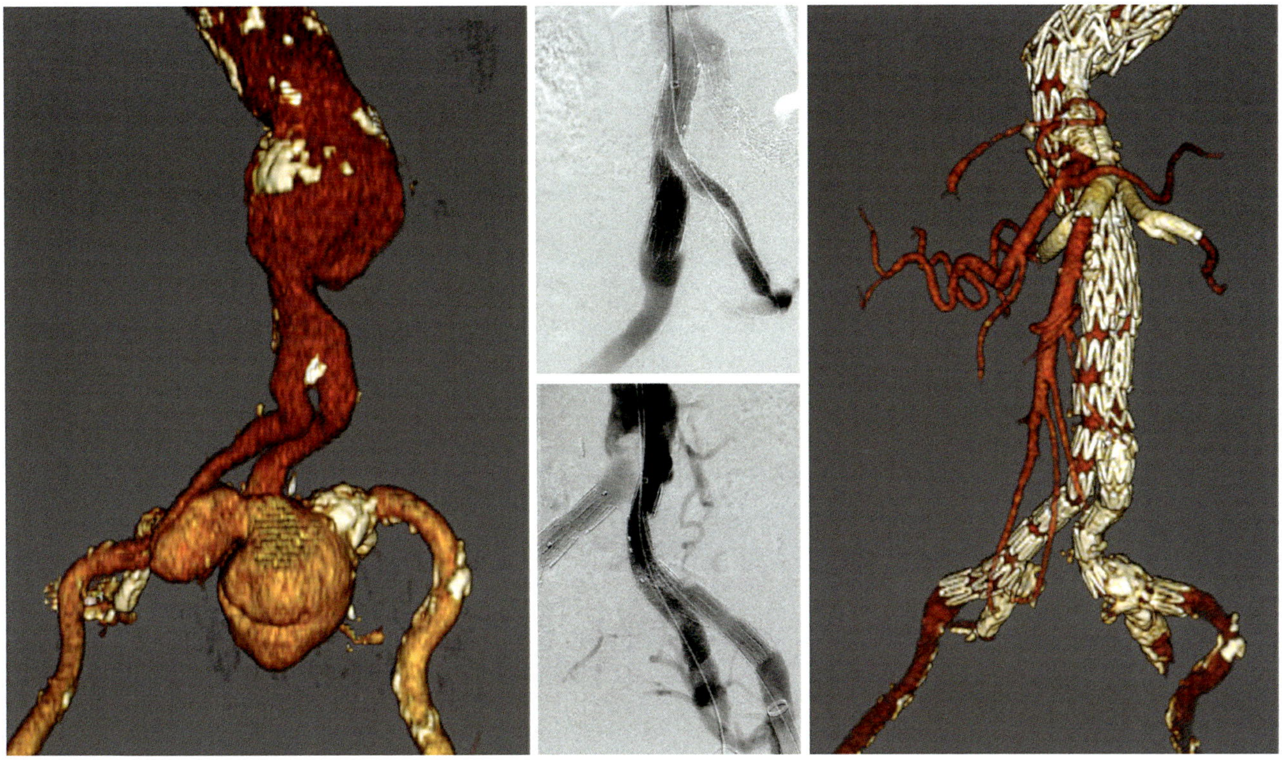

Fig. 53.2 Use of Cook IBD to treat iliac pseudo-aneurysms

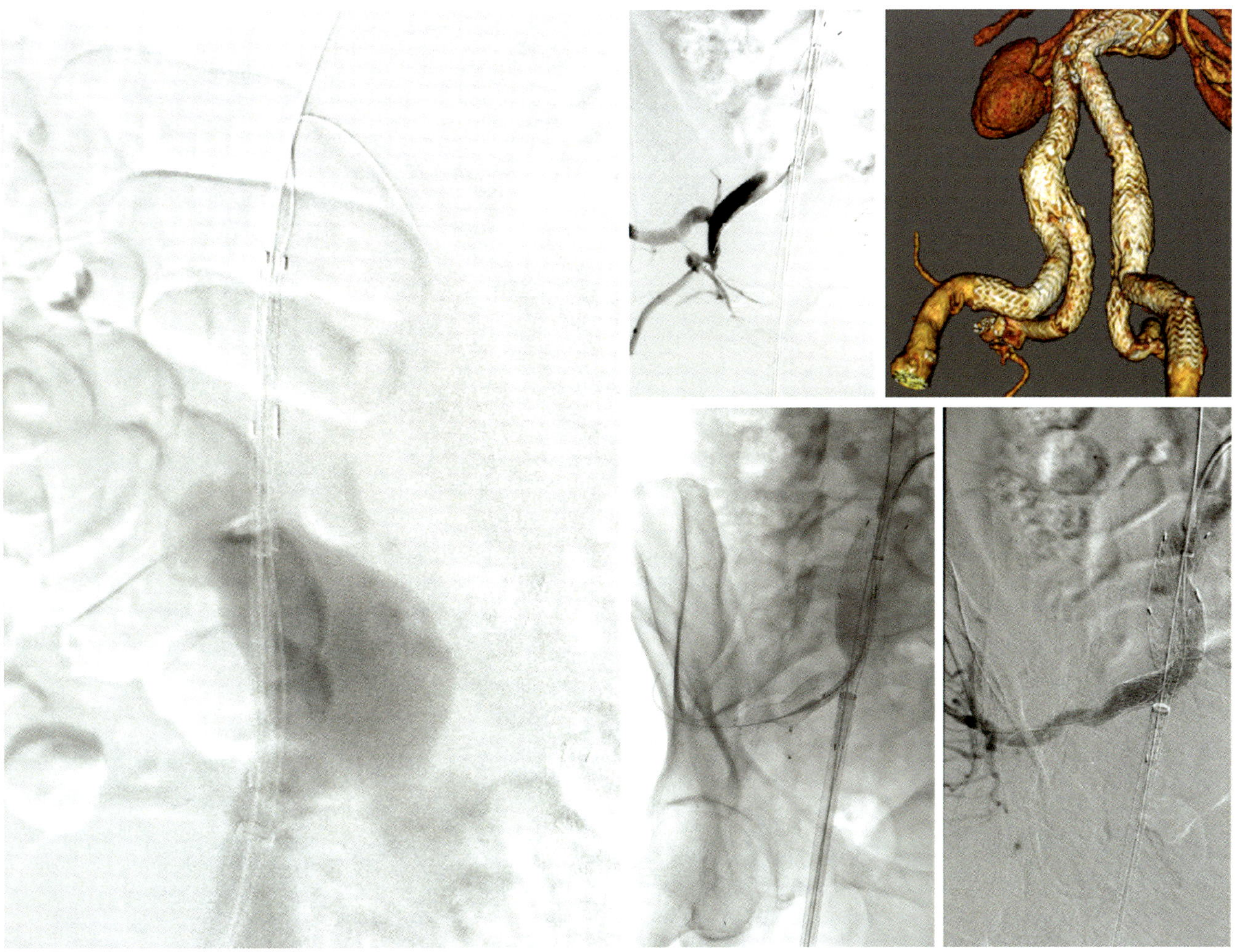

Fig. 53.3 Use of Gore IBD to treat an internal artery aneurysm

divisional branch is preferentially used to maintain blood flow in the superior gluteal artery. The internal component or bridging stent is often too short to achieve the sealing from the IBD to the divisional branch. It is often necessary to use two covered stents. This has to be anticipated because of the need for different stent diameters. Recent results from a large multicentric registry confirm that use of the anterior or posterior divisional branches of the internal iliac artery (IIA) as distal landing zone shows comparable midterm results with a trend to favor the posterior divisional branch in terms of primary and secondary IBD patency at 3 years [28]. Buttock claudication was less frequent when landing in the posterior divisional branch (2% vs 16%), and sexual dysfunction was also less frequent when landing in the posterior divisional branch (1% vs 11%).

Conclusion

IBDs are valid options for endovascular treatment of patients with aorto-iliac aneurysm that does not fulfill the criteria for a standard EVAR in order to preserve the pelvic perfusion and decrease the incidence of postoperative buttock claudication and sexual dysfunction secondary to IIA embolization. IBDs are also very helpful in particular cases such as internal iliac artery aneurysm or iliac aneurysmal degeneration after abdominal aortic aneurysm endovascular or open repair.

References

1. Hobo R, Sybrandy JE, Harris PL, et al. Endovascular repair of abdominal aortic aneurysms with concomitant common iliac artery aneurysm: outcome analysis of the EUROSTAR experience. J Endovasc Ther. 2008;15:12–22.
2. Melas N, Saratzis A, Dixon H, et al. Isolated common iliac artery aneurysms: a revised classification to assist endovascular repair. J Endovasc Ther. 2011;18:697–715.
3. Bratby MJ, Munneke GM, Belli AM, et al. How safe is bilateral internal iliac artery embolization prior to EVAR? Cardiovasc Intervent Radiol. 2008;31:246–53.
4. Rayt HS, Bown MJ, Lambert KV, et al. Buttock claudication and erectile dysfunction after internal iliac artery embolization in patients prior to endovascular aortic aneurysm repair. Cardiovasc Intervent Radiol. 2008;31:728–34.
5. Geraghty PJ, Sanchez LA, Rubin BG, et al. Overt ischemic colitis after endovascular repair of aortoiliac aneurysms. J Vasc Surg. 2004;40:413–8.
6. Warein E, Feugier P, Chaufour X, et al. Amplatzer plug to occlude the internal iliac artery during endovascular aortic aneurysm repair: a large multicenter study. Eur J Vasc Endovasc Surg. 2016;51:641–6.
7. Malina M, Dirven M, Sonesson B, et al. Feasibility of a branched stent-graft in common iliac artery aneurysms. J Endovasc Ther. 2006;13(4):496–500.
8. Li Y, Hu Z, Zhang J, et al. Iliac aneurysms treated with endovascular iliac branch device: a systematic review and meta-analysis. Ann Vasc Surg. 2019;56:303–16.
9. Cao Z, Zhu R, Ghaffarian A, et al. A systematic review and meta-analysis of the clinical effectiveness and safety of unilateral versus bilateral iliac branch devices for aortoiliac and iliac artery aneurysms. J Vasc Surg. 2022;18:S0741–5214.
10. Giosdekos A, Antonopoulos CN, Sfyroeras GS, Moulakakis KG, Tsilimparis N, Kakisis JD, Lazaris A, Chatziioannou A, Geroulakos G. The use of iliac branch devices for preservation of flow in internal iliac artery during endovascular aortic aneurysm repair. J Vasc Surg. 2020;71(6):2133–44. https://doi.org/10.1016/j.jvs.2019.10.087. Epub 2019 Dec 31.
11. Millon A, Della Schiava N, Arsicot M, et al. Preliminary experience with the GORE® EXCLUDER® Iliac Branch Endoprosthesis for common iliac aneurysm endovascular treatment. Ann Vasc Surg. 2016;33:11–7.
12. Ardita V, Giaquinta A, Veroux M, et al. Endovascular repair of bilateral common iliac artery aneurysms using GORE Excluder iliac branch Endoprosthesis without aorto bi-iliac stent-graft conjunction: a case report. Medicine (Baltimore). 2017;96:e5977.
13. Giaquinta A, Ardita V, Ferrer C, et al. Iliac Branch Stent-Graft Italian Trial Collaborators. Isolated common iliac artery aneurysms treated solely with iliac branch stent-grafts: midterm results of a multicenter registry. J Endovasc Ther. 2018;25(2):169–77.
14. Falkensammer J, Hakaim AG, Oldenburg WA, et al. Natural history of the iliac arteries after endovascular abdominal aortic aneurysm repair and suitability of ectatic iliac arteries as a distal sealing zone. J Endovasc Ther. 2007;14:619–24.
15. McDonnell CO, Semmens JB, Allen YB, et al. Large iliac arteries: a high-risk group for endovascular aortic aneurysm repair. J Endovasc Ther. 2007;14:625–9.
16. Bisdas T, Weiss K, Donas KP, et al. Use of iliac branch devices for endovascular repair of aneurysmal distal seal zones after EVAR. J Endovasc Ther. 2014;21(4):579–86.
17. Dawson DL, Sandri GA, Tenorio E, et al. Up-and-over technique for implantation of iliac branch devices after prior aortic endograft repair. J Endovasc Ther. 2018;25(1):21–7.
18. Tenorio ER, Oderich GS, Sandri GA, et al. Outcomes of an iliac branch endoprosthesis using an "up-and-over" technique for endovascular repair of failed bifurcated grafts. J Vasc Surg. 2019;70(2):497–508.
19. Mesnard T, Patterson BO, Azzaoui R, et al. Iliac branch device to treat type Ib endoleak with a brachial access or an "up-and-over" transfemoral technique. J Vasc Surg. 2022;26:S0741–5214.
20. Stern JR, Lyden SP, Agrusa CJ, et al. Ipsilateral iliac branch repair using a looped wire, precannulated gate technique. J Endovasc Ther. 2021;28(3):373–7.
21. Zimmer PW, Raker EJ, Quigley TM. Isolated hypogastric artery aneurysms. Ann Vasc Surg. 1999;13:545–9.
22. Millon A, Paquet Y, Ben Ahmed S, et al. Midterm outcomes of embolisation of internal iliac artery aneurysms. Eur J Vasc Endovasc Surg. 2013;45(1):22–7.
23. Kliewer M, Plimon M, Taher F, et al. Endovascular treatment of hypogastric artery aneurysms. J Vasc Surg. 2019;70(4):1107–14.
24. Dueppers P, Duran M, Floros N, et al. The JOTEC iliac branch device for exclusion of hypogastric artery aneurysms: ABRAHAM study. J Vasc Surg. 2019;70(3):748–55.
25. Tajima Y, Goto H, Akamatsu D, et al. Prevention of buttock claudication by preserving Antegrade bilateral superior gluteal arterial blood flow in EVAR for aorto-iliac aneurysm accompanied by bilateral internal iliac artery aneurysms. Ann Vasc Dis. 2022;15(1):53–7.
26. Cornwall JW, Han DK, Fremed DI, et al. Successful off-label use of the GORE EXCLUDER Iliac Branch Endoprosthesis to preserve gluteal perfusion during staged endovascular repair of bilateral isolated hypogastric aneurysms. J Vasc Surg Cases Innov Tech. 2017;3(1):37–40.
27. D'Oria M, Tenorio ER, Oderich GS, et al. Outcomes of the gore excluder iliac branch endoprosthesis using division branches of the internal iliac artery as distal landing zones. J Endovasc Ther. 2020;27:316–27.
28. D'Oria M, Lima GBB, Dias N, et al. Outcomes of "anterior versus posterior divisional branches of the hypogastric artery as distal landing zone for iliac branch devices": The International Multicentric R3OYAL Registry. J Endovasc Ther. 2022:152660282211205.

Endovascular Approach to the Hypogastric Artery

54

Ailin Belloni, Paolo Verlato, Katrine Jessica Beltramolli, Martina Mariani, Agnese Magni, Savino Pasquadibisceglie, and Alberto Froio

Case Presentation—Background

An 83-year-old patient, male, with an enlarging left common (35 mm) and left internal iliac artery (33 mm) aneurysm, a concomitant right common iliac aneurysm (27 mm), and a small aneurysm of the aorta (36 mm) was considered for an endovascular solution assessing the CT scan. There was a focal dissection/ulceration of the thrombus in the mid and distal left common iliac artery aneurysm. The right hypogastric artery was normal, with a diameter of 10 mm. The length of the right internal iliac artery before the first bifurcation was 37 mm. The right external iliac artery had a diameter of 10 mm.

The anterior trunk of the left hypogastric artery was 4 mm large, the posterior 5 mm (Fig. 54.1).

The common femoral arteries were fit for a percutaneous approach.

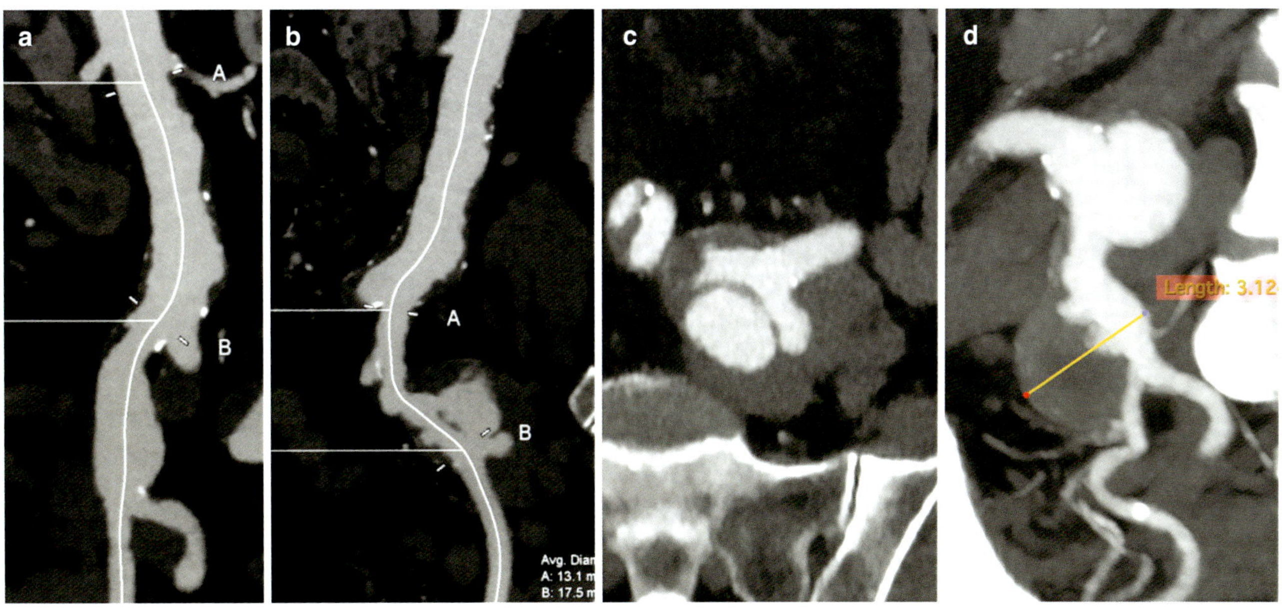

Fig. 54.1 Preoperative CT scan. Panel (**a**): center-line of the right axis. Panel (**b**): center-line of the left axis. Panel (**c**): left common iliac artery aneurysm with the focal dissection. Panel (**d**): Left hypogastric artery aneurysm

A. Belloni · P. Verlato · K. J. Beltramolli · M. Mariani · A. Magni
S. Pasquadibisceglie · A. Froio (✉)
School of Medicine and Surgery, University of Milano-Bicocca, Milan, Italy

Vascular Surgery Department, Fondazione IRCCS San Gerardo, Monza, Italy
e-mail: alberto.froio@unimib.it

G. Geroulakos et al. (eds.), *Mastering Endovascular Techniques*, https://doi.org/10.1007/978-3-031-42735-0_54

Background

The common iliac artery (CIA) typically serves as the distal landing zone for the iliac limb extension for aorto-iliac endografts if CIA diameter fits within the instructions for use (IFU). Alternately, it is possible to extend the endograft limb into the external iliac artery (EIA), causing hypogastric artery (HA) occlusion. Intentional occlusion of HA brings the risk of several drawbacks, including buttock claudication, bowel ischemia, and erectile dysfunction. The risk is significantly increased in the presence of occlusion of contralateral HA or deep femoral artery.

In recent years, the use of iliac branch devices (IBD) has provided an ideal solution to preserve hypogastric flow. Nevertheless, unfavourable anatomical conditions including dimension of the CIA and HA, access vessels, and tortuosity constitute technical contraindications to the use of IBD, thus excluding a wide percentage of patients.

Even if a clear algorithm is still lacking, multiple endovascular and hybrid alternatives to manage hypogastric artery are available.

Bell-Bottom Technique vs. External Iliac Artery Extension

The bell-bottom technique (BBT) is one of the procedures to preserve HA in case of enlarged CIA that is therefore assumed as a healthy distal landing zone for large-diameter iliac limbs. The following companies have large-diameter iliac limbs:

- Medtronic Endurant up to 28 mm
- Gore Excluder and Artivion E-Tegra up to 27 mm
- Cook Zenith, Cordis Incraft, and Terumo Treo up to 24 mm
- Terumo Anaconda up to 23 mm

Diameter oversizing is usually calculated to be 10–20% in the distal portion of the CIA, allowing the exclusion of CIAs with a maximum diameter of 25 mm for the largest device and a minimum length of 20 mm.

Wang compared the long-term efficacy of BBT among three different widely used commercial endografts (Cook Zenith, Gore Excluder, and Medtronic Endurant), finding a trend of higher incidence of type Ib endoleak, aneurysm sac enlargement, and reintervention rate in patients treated with Cook Zenith, while no differences were found between other stent grafts [1]. It must be noted that the study was conducted using the first-generation Cook Zenith endograft made of stainless steel, which conferred to the stent graft higher stiffness and lower flexibility.

Bahroloomi compared sac shrinkage and perioperative outcomes after standard EVAR vs BBT vs HA embolization and external iliac artery extension (EIE) vs iliac branch endoprosthesis (IBE) [2]. EIE technique was more likely to demonstrate shrinkage in the aortic aneurysmal diameter than were the BBT and IBE groups. On the other hand, the rates of limb complications and reoperation were higher in the EIE group, similarly to the rate of type Ib endoleak. These results were not consistent with those published by Farivar, who reported that no statistically significant difference was identified in freedom from aneurysm-related mortality, limb occlusions, or endoleak rates in patients undergoing endovascular aneurysm repair for aortoiliac aneurysms with either coil and cover technique or an iliac branched device. [3]

Retrograde Endovascular Bypass

This technique is associated with an aorto-uni-iliac (AUI) endograft and a crossover femoro-femoral artery bypass, typically with the endograft extended into the external iliac artery covering the ipsilateral hypogastric artery. The contralateral hypogastric artery is cannulated from the common femoral artery with a hydrophilic guidewire placed into the gluteal artery to guarantee a stable support. A long braided introducer sheath (preferably with a pre-curved tip) is delivered to the hypogastric artery over a Supracore 300-cm guidewire. A self-expandable 5 cm-long covered stent graft is then deployed from the hypogastric artery back to the proximal external iliac artery followed by balloon angioplasty.

The technique cannot be completed in case of ostial hypogastric artery stenosis, hypogastric aneurysm without a distal landing zone, or with a severe mismatch between the diameters of the hypogastric and external iliac arteries.

Wooster included 17 patients from 2006 to 2016 and performed the retrograde endovascular bypass using Fluency stent graft ($N = 1$), Viabahn stent graft ($N = 13$), or Gore Excluder limbs ($N = 3$). After a mean follow-up of 30 months, there were no aortic-related mortalities, one endovascular bypass occlusion (asymptomatic), and one reintervention for type II endoleak, not related to the bypass [4]. Massiére published a report of the ten-year experience with the retrograde endovascular bypass technique on 21 patients treated electively for aortoiliac aneurysm, using Viabahn stent graft for the hypogastric – external iliac artery bypass. The technical success rate was 100% and no iliac-related deaths were reported. With a mean follow-up of 52 months, no bypass-related complications were reported [5].

In conclusion, the retrograde endovascular bypass is a safe and effective technique, with the main advantage of broad applicability and a main disadvantage of the need for a surgical crossover femoro-femoral bypass.

Parallel Grafts: Double Barrel Technique

The sandwich technique, also called the double barrel technique, was first described by Lobato as a completely endovascular alternative to iliac branch devices (IBD) for patients that did not meet IBD criteria, in order to preserve antegrade flow to the pelvic arteries [6]. Over the years it was modified in different ways in order to simplify it or to address the problem of gutters, the main disadvantage with parallel grafts techniques. It must be noted that this technique involves the use of endografts outside their instruction for use.

The original techniques described by Lobato required the deployment of the main body of the bifurcated stent graft, with the distal end of the ipsilateral iliac limb 1 cm above the HA origin. The HA was then catheterized through a left brachial access with an extra stiff guidewire. A covered self-expanding stent was released at least 2 cm inside the HA, overlapping within the ipsilateral iliac limb by at least 6 cm. An iliac limb extension was positioned 1 cm below the covered stent proximal end and was deployed before the covered stent.

The primary patency rate was 93.8%, with three HA stent-graft occlusions, occurring early in the study. The early and late aneurysm-related mortality rate was 0%. Iliac aneurysm sac evolution demonstrated a significant (at least 5 mm) decrease in diameter in 16 (34.8%) common iliac artery aneurysms, no change in 29 (63%) common iliac artery aneurysms, and an increase in one patient (2.2%) [6].

Several authors analysed the proper sizing of the covered stents in order to avoid the presence of gutters. Massman used the total cross-sectional area of the stents in the HA and EIA to determine the appropriate diameter of the mother stent in the CIA [7]. In order to achieve sufficient sealing of the IIA and EIA stents within the mother-stent of the CIA, an undersizing of 10% according to the calculated CIA stent diameter was used: $0.9 \times 2 \times$ square root ((area HA + area EIA)/π). Yugueros assessed four external iliac devices (16 mm Endurant and 12 mm Aorfix limb extensions; 11 and 13 mm Viabahn endografts) with two distinct internal iliac stent grafts (8 mm Advanta/V12 and 8 mm Viabahn) inside different proximal silicon iliac limbs models (10, 12, 14, 16 and 18 mm), simulating an iliac sandwich procedure for hypogastric preservation [8]. Increasing diameter oversizing (<30%, 30–55%, 55–75% and >75%) showed a significant tendency towards smaller gutters, but also increasing parallel-stent compression and infolding. The conclusion was that better apposition is usually achieved when using 30–55% diameter oversizing.

Pang identified two methods to prevent "gutter" endoleak formation [9]. The first method was selection of stents with the proper diameter. According to the theory of calculation of the arterial cross-sectional area, the bifurcated stents were approximately four-fifths of the actual diameter of the mother stent. Additionally, the type and diameter of the two bifurcated stent grafts were the same to maintain the same radial force. Proximal overlap of the bifurcated stents within the mother stent was at least 4 cm; distal overlap was 2 cm into the bifurcation. Patients were followed up for a mean of 21 months (range, 6–31 months), and the primary patency of the stent grafts and arteries was achieved in all patients.

A comprehensive literature review was conducted to assess the outcome in terms of freedom from endoleak, HA branch occlusion, and secondary interventions [10]. Type I endoleaks occurred in 2.3% of cases, while no type III endoleak was reported. Reinterventions were performed in 6.5% of cases. Freedom from endoleaks, HA branch occlusions, and reinterventions were 86% (freedom from type I or III endoleaks was 100%), 88%, and 87%, respectively.

However, while iliac branch devices (IBD) have been established as a durable procedure, mid-term data is lacking on parallel stent graft performance and further studies are required to confirm the durability of the latter procedure.

Parallel Grafts: Crossover Chimney Technique

The crossover chimney technique (COCh), first described by Wu, is a complete endovascular alternative to preserve the HA with a retrograde flow [11]. Bilateral femoral access is obtained by puncture or cut-down exposure. A floppy guidewire is cannulated into IIA that it is meant to be preserved with a crossover. A long crossover sheath is introduced into the IIA after replacing the floppy by a stiff wire. The main body endograft is inserted through the contralateral femoral access and deployed and the contralateral gate is cannulated from the ipsilateral femoral access. The COCh HA covered stent and the iliac limb extension stent graft from the short limb of the main endograft are deployed and parallelly landed on the ipsilateral CIA right before the iliac bifurcation. Angioplasty of the two parallel stent grafts is performed in a kissing fashion. The isoperimetric inequality equation was used by Wu to estimate the chimney graft size.

Technical success, defined as successful preservation of HA without intraoperative type I or III endoleak, was 100%. After 14.3 months, primary patency was 92.8%. There was no early or late procedure-related mortality. Among the 17 iliac aneurysms excluded, the sac diameter significantly (at least 5 mm) decreased in three, decreased <5 mm in 10, and did not change in four.

Gu reported their results from a retrospective study on 61 consecutive patients who received elective EVAR with hypogastric preservation by either crossover chimney technique or IBD [12]. Both the techniques evidenced 100% technical success with no statistical differences in primary patency rates at 12 and 24 months, despite a higher primacy patency rate in patients treated with IBD (83% vs 94% at 2 years). Lastly, no statistical differences were found in type I, II, and III endoleaks.

In conclusion, the COCh procedure seems to be a safe and effective technique with promising short- and mid-term results. However, since the hypogastric perfusion is guaranteed by a retrograde flux, further long-term patency studies are needed.

Catheterization of the Hypogastric Artery

Independently from the planning of embolizing or preserving the main trunk or the distal branches of the hypogastric artery (HA), it is necessary to have a stable solution in the main trunk of the HA. It is typical to have common iliac artery tortuosity, narrow aorta, and iliac stenotic calcified plaques with completely different projections to see the take-off of the external vs internal iliac artery and the anterior vs posterior branch of the HA. These features make the cannulation and the stability of the sheath/catheter in the HA very tricky.

In such a complex anatomy, it is possible to create a through-and-through femoral access using a long soft Terumo guidewire and then advancing a long sheath (typically a 7Fr 45 cm Teleflex Super-Arrow or Terumo Destination or Cook Flexor for coil embolization; or a 12Fr 45 cm long Gore DrySeal in case of IBE or Vascular Plug use) at the level of target common iliac artery bifurcation [13]. Pulling the through-and-through guidewire will make the advance of a 4Fr BerII parallel catheter very easy and stable inside the HA branches (Fig. 54.2). In case of stenting in the HA distal branches, a 9Fr long sheath can be advanced in the 12Fr 45 cm sheath to have a telescopic approach allowing the advancement of the covered stent very distally.

Internal Iliac Artery Occlusion

Occlusion of the internal iliac arteries can be required for successful EVAR to prevent retrograde flow that can ultimately lead to a type II endoleak and aneurysm enlargement.

Some authors do not require coil embolization before EVAR even if in the presence of common iliac artery aneurysm if there is a short narrow-calibre segment at the iliac bifurcation. According to these authors, this tapering near the origin of the HA is probably sufficient to provide adequate seal and to allow thrombosis of the main trunk of the HA [14].

There are two different solutions, using coils or vascular plugs, both requiring selective catheterization of the internal iliac artery. The embolization of the internal iliac artery can be performed at the same operative setting as EVAR or, alternatively, as a staged procedure.

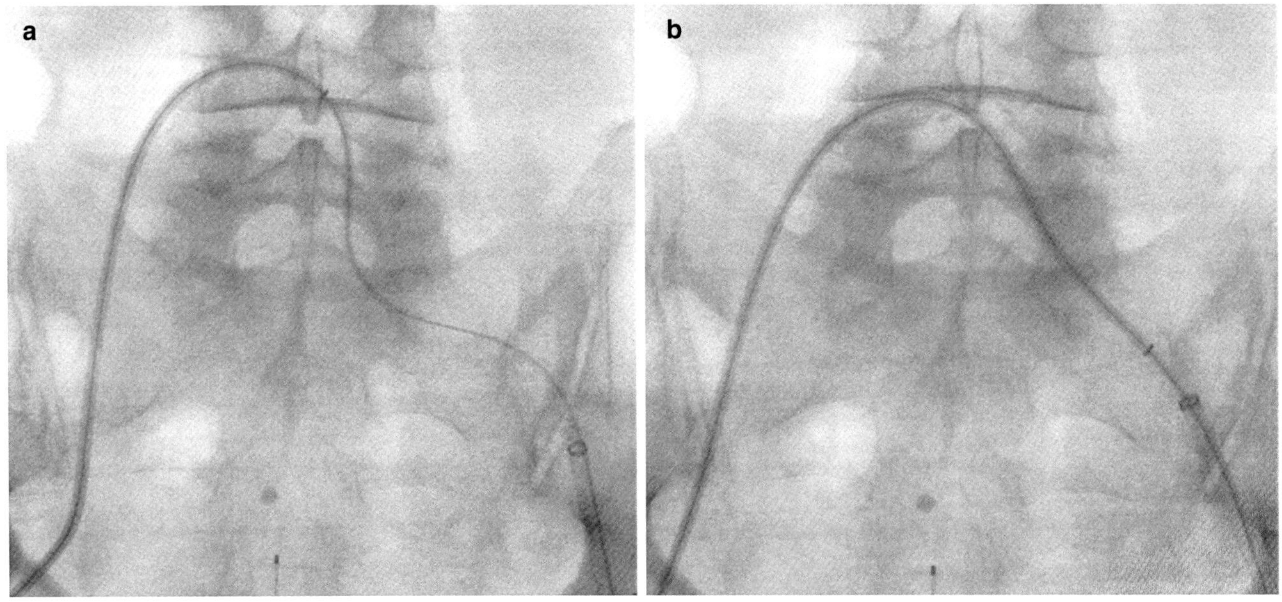

Fig. 54.2 The efficacy of the through-and-through guidewire to reduce the curvature of the arteries. Panel (**a**): anatomy without the tension of the through-and-through guidewire. Panel (**b**): tension on the guidewire to straighten the system

In case of deployment of multiple coils, it should be done as proximal as possible in the main trunk of the internal iliac artery, to maintain a backflow from collaterals to the anterior and posterior trunk of the artery, minimizing hypoperfusion issues. The approach to the hypogastric artery can be ipsilateral or contralateral [15].

The Amplatzer Vascular Plug (AVP; Abbot Laboratories, Abbott Park, IL, USA) is a nitinol-based self-expanding cylindrical device [16]. The AVP is preloaded on a 135 cm-long deliver cable. The contralateral approach is the most common technique, using a long straight or curved introducer sheath for AVP delivery. The introducer sheath is positioned in the internal iliac artery over a stiff guidewire to provide sufficient support through tortuous anatomy. Then the AVP is deployed in the main trunk of the internal iliac artery.

One of the differences between the two techniques is the use of the telescopic approach (long sheath + catheter) during coiling compared to the sheath-only technique for AVP deployment. The latter solution makes the system not stable during pulling back of the sheath in case of extremely tortuous anatomies, with the risk of slipping of the sheath out of the hypogastric artery.

One of the tips is to use a buddy wire (300 cm extrastiff Supracore guidewire in the superior gluteal artery) using a bigger sheath (7Fr) and using a parallel 4 Fr catheter to deploy the 0.035″ coils precisely at the origin of the hypogastric artery. The same buddy-wire approach can be used during AVP deployment using a 10 or 12Fr 45 cm-long sheath to give the 3Fr additional room to the 0.035″ guidewire.

Using Division Branches of the Internal Iliac Artery as Distal Landing Zones

One common limitation of iliac branch device (IBD) applicability is an inadequate distal landing zone due to coexisting hypogastric aneurysm. One off-label solution is to create a suitable landing zone in the branches of the hypogastric artery.

D'Oria assessed this issue, dichotomizing 75 patients according to distal landing sites for the IBD (60 main trunks of the HA vs 25 division branches) [13]. In the case of HA aneurysms, embolization was attempted of any branch ≥2 mm taking off from the aneurysm sac. The bridging stent was then advanced and deployed; if the intended distal landing zone was within the posterior HA division, embolization of the anterior HA division was performed. Several types of bridging stent grafts were employed in addition to the IBE internal iliac limb: the Gore VBX, the Gore Viabahn, and the Atrium iCAST. The primary and secondary patency rates at 1 year were 98% vs 95% and 98% vs 100% for the main trunk vs division branch groups, respectively. Freedom from IIA branch instability estimates was also similar at 1-year follow-up (93% vs 90%), as was the freedom from new-onset buttock claudication estimates (98% vs 94%). Mean sac diameter change was 5.4 ± 5.3 mm, not significantly different between the groups.

The comparison of on- vs off-label Gore Excluder Iliac Branch Endoprosthesis demonstrated no differences at 1-year follow-up [17].

Case Presentation—Results

An arteriogram of the left common iliac artery confirmed the presence of a focal dissection in the mid and distal part of the artery, excluding the use of an iliac branch device. The plan was to embolize the left hypogastric artery.

A through-and-through femoral access using a long soft Terumo guidewire allowed for advancement of a 7Fr 45 cm Arrow sheath from the right common femoral artery at the level of the left common iliac artery bifurcation. Maintaining the through-and-through Terumo standard guidewire, a 4Fr parallel catheter was quickly inserted in the left anterior and the posterior trunk of the hypogastric artery. Several Boston Interlock 0.035″ coils (6 mm × 10 cm, 6 mm × 20 cm, 8 mm × 10 cm, 8 mm × 20 cm, and 10 mm × 25 cm) were deployed in the two trunks (Fig 54.3). The arteriogram showed a very slow flow in the embolized vessels.

The same through-and-through femoral access was used to advance the CEB 231210 Gore IBE component in the right common iliac artery aneurysm. The HGB161207 component was deployed in the right hypogastric artery. The remaining endografts were deployed. The final angiogram evidenced the patency of the right iliac branch device, the exclusion of the right common iliac aneurysm, the exclusion of the left hypogastric aneurysm, and the lack of endoleaks (Fig. 54.4). The post-operative CT scan showed no endoleaks (Fig. 54.5).

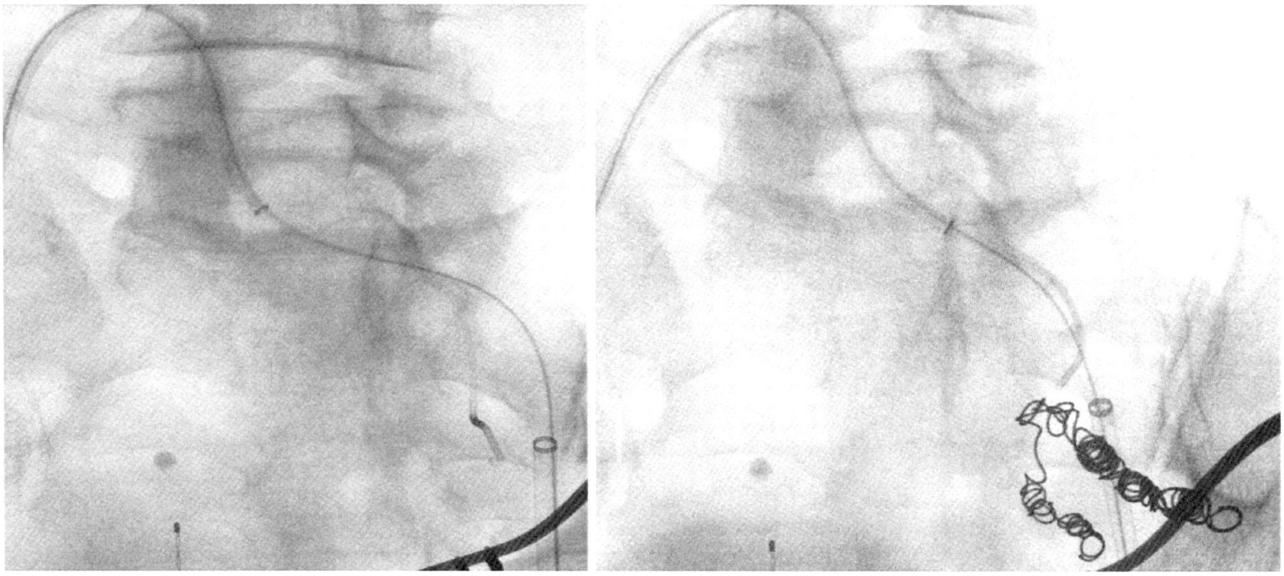

Fig. 54.3 Cannulation of the branches of the hypogastric artery with the 4Fr Ber II parallel catheter using the through-and-through stabilizing system (left panel), allowing a precise deployment of the coils in the anterior and posterior branches (right panel)

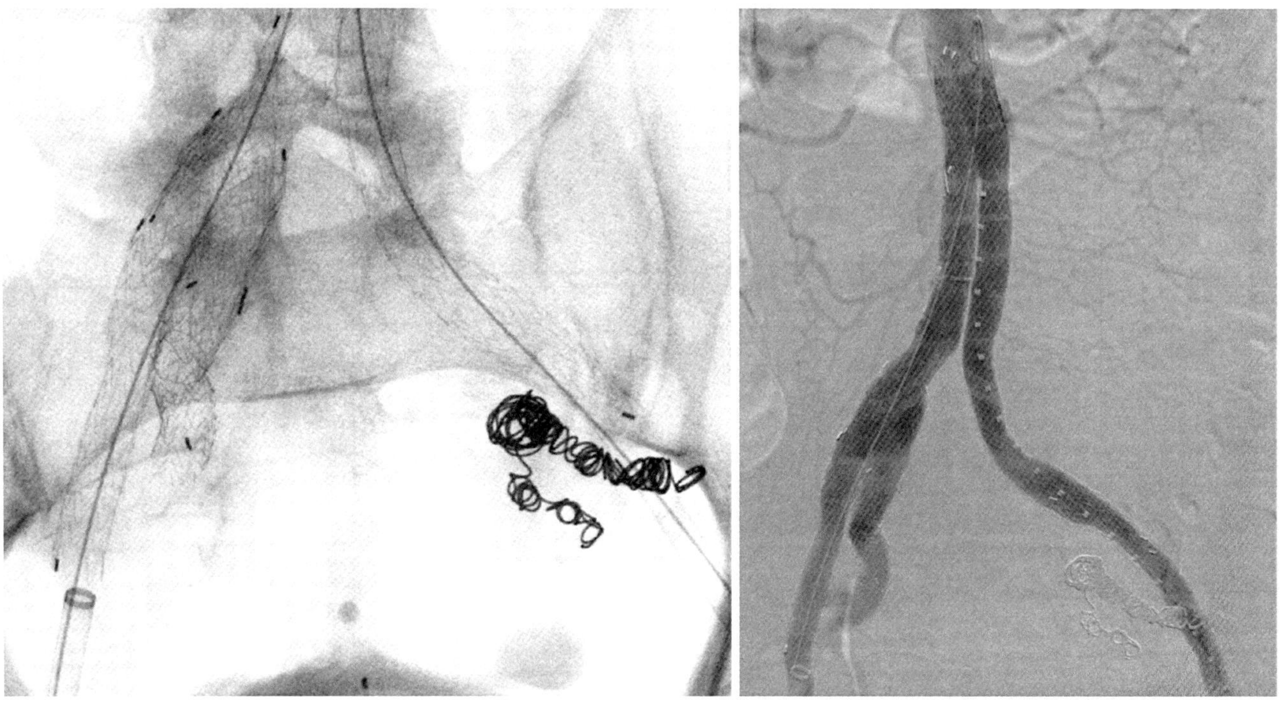

Fig. 54.4 Final angiogram showing the patency of the iliac branch device on the right side and the complete exclusion of the hypogastric aneurysm on the left

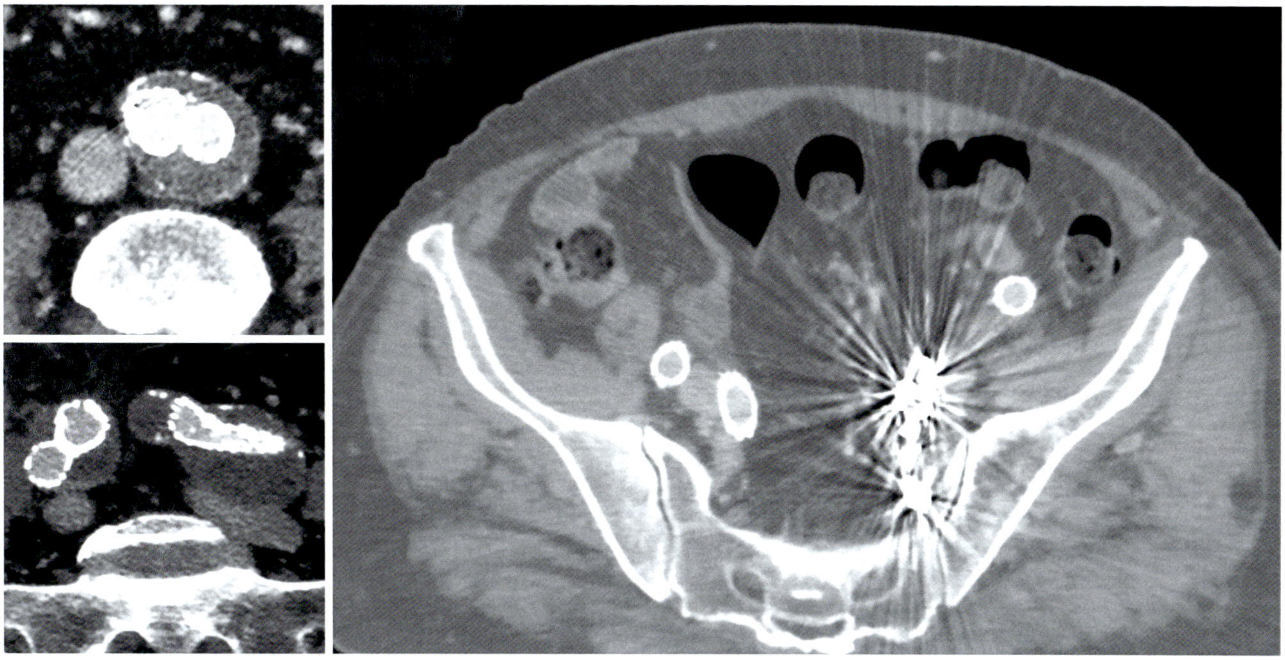

Fig. 54.5 Postoperative CT scan

Conclusions

The management of hypogastric artery requires accurate planning. It is mandatory to maintain at least one HA patent. Different on- and off-label strategies are available to tailor the treatment according to patient anatomy.

Tool Kit (Table of the Materials Used for the Treatment)

	Manufacturer	Size/length
Wires		
Glidewire floppy	Terumo Radifocus	0.035″/260 cm
Hi-Torque Supracore—extrastiff guidewire	Abbott	0.035″/300 cm
Sheaths		
Any standard access sheath	Any	5-10Fr/11 cm
Super Arrow-Flex® Sheath Introducer	Teleflex	7Fr/45 cm
Destination guiding sheath	Terumo	7Fr/45 cm
Flexor introducer	Cook	7Fr/45 cm
Catheters		
Tempo Berenstein/ Berenstein II	Cordis	65–100 cm
Coils		
Interlock-35 fibered detechable coils	Boston	From 4 to 20 mm large – from 6 to 40 cm long—both 2D and 3D cube
Plug		
Vascular Plug (VP)	Abbott	VP/VP II/VP4 from 4 to 22 mm

References

1. Wang LX, Liang SC, Xu X, Chen B, Jiang JH, Shi ZY, et al. A comparative study of the efficacy by using different stent grafts in bell-bottom technique for the treatment of abdominal aortic aneurysm concomitant with iliac artery aneurysm. Ann Vasc Surg. 2018;52:41–8. https://doi.org/10.1016/j.avsg.2018.05.030.
2. Bahroloomi D, Qato K, Nguyen N, Schreiber-Gregory D, Conway AM, Giangola G, et al. External iliac artery extension causes greater aneurysm sac regression than the bell-bottom technique or iliac branch endoprosthesis for repair of concomitant infrarenal aortic and iliac artery aneurysm. J Vasc Surg. 2022;76(1):132–40. https://doi.org/10.1016/j.jvs.2021.12.062.
3. Farivar BS, Abbasi MN, Dias AP, Kuramochi Y, Brier CS, Parodi E, et al. Durability of iliac artery preservation associated with endovascular repair of infrarenal aortoiliac aneurysms. J Vasc Surg. 2017;66(4):1028. https://doi.org/10.1016/j.jvs.2017.02.042.
4. Wooster M, Armstrong P, Back M. Hypogastric preservation using retrograde endovascular bypass. Ann Vasc Surg. 2018;52:67–71. https://doi.org/10.1016/j.avsg.2018.04.018.
5. Massiere B, von Ristow A, Vescovi A, Leal D, da Fonseca LMB. Ten-year experience with management of aortoiliac aneurysms using retrograde endovascular internal iliac artery preservation. Ann Vasc Surg. 2016;35:163–7. https://doi.org/10.1016/j.avsg.2016.01.034.
6. Lobato AC, Camacho-Lobato L. The sandwich technique to treat complex aortoiliac or isolated iliac aneurysms: results of midterm follow-up. J Vasc Surg. 2013;57:26S–34S. https://doi.org/10.1016/j.jvs.2012.09.081.
7. Massmann A, Arochena NJM, Shayesteh-Kheslat R, Buecker A. Endovascular anatomic reconstruction of the iliac bifurcation with covered stentgrafts in sandwich-technique for the treatment of complex aorto-iliac aneurysms. Int J Cardiol. 2016;222:332–9. https://doi.org/10.1016/j.ijcard.2016.07.226.
8. Yugueros X, Mestres G, Pasquadibisceglie S, Alomar X, Apodaka A, Riambau V. Parallel-stenting technique in a sandwich configuration for hypogastric artery preservation during endovascular

aneurysm repair: an in vitro study. Ann Vasc Surg. 2017;44:221–8. https://doi.org/10.1016/j.avsg.2017.03.202.
9. Pang HJ, Chen Y, He XF, Zeng QL, Ye P. Selection of stents by calculation of arterial cross-sectional area in modified sandwich technique for complex aortoiliac arterial lesions. Ann Vasc Surg. 2019;58:108–14. https://doi.org/10.1016/j.avsg.2018.10.039.
10. Oliveira-Pinto J, Martins P, Mansilha A. Endovascular treatment of iliac aneurysmal disease with internal iliac artery preservation: a review of two different approaches. Int Angiol. 2019;38(6):494–501. https://doi.org/10.23736/s0392-9590.19.04215-9.
11. Wu IH, Chou HW, Chang CH, Lin CF, Chi NH, Wang SS. Crossover chimney technique to preserve the internal iliac artery during endovascular repair of iliac or aortoiliac aneurysms: midterm results. J Endovasc Ther. 2015;22(3):388–95. https://doi.org/10.1177/1526602815581596.
12. Gu YT, Kuo TT, Chen PL, Huang CY, Shih CC, Chen IM. Internal iliac artery preservation outcomes of endovascular aortic repair for common iliac aneurysm: iliac branch device versus crossover chimney technique. Heart Vessel. 2021;36(2):235–41. https://doi.org/10.1007/s00380-020-01678-x.
13. D'Oria M, Tenorio ER, Oderich GS, Mendes BC, Kalra M, Shuja F, et al. Outcomes of the gore excluder iliac branch endoprosthesis using division branches of the internal iliac artery as distal landing zones. J Endovasc Ther. 2020;27(2):316–27. https://doi.org/10.1177/1526602820905583.
14. Stokmans RA, Broos P, van Sambeek M, Teijink JAW, Cuypers PWM. Overstenting the hypogastric artery during endovascular aneurysm repair with and without prior coil embolization: a comparative analysis from the ENGAGE registry. J Vasc Surg. 2018;67(1):134–41. https://doi.org/10.1016/j.jvs.2017.04.061.
15. Fontana F, Coppola A, Ferrario L, De Marchi G, Macchi E, Zorzetto G, et al. Internal iliac artery embolization within EVAR procedure: safety, feasibility, and outcome. J Clin Med. 2022;11(24) https://doi.org/10.3390/jcm11247399.
16. Bracale UM, Petrone A, Provenzano M, Ielapi N, Ferrante L, Turchino D, et al. The use of the Amplatzer vascular plug in the prevention of endoleaks during abdominal endovascular aneurysm repair: a systematic literature review on current applications. Vascular. 2022;30(4):681–9. https://doi.org/10.1177/17085381211025152.
17. Rodriguez LE, Eun JC, Calkins RT, Carroll AM, Malgor EA, Wohlauer MV, et al. Comparison study of iliac branch endoprosthesis when used on and off label. Ann Vasc Surg. 2023;89:28–35. https://doi.org/10.1016/j.avsg.2022.03.015.

Optimizing Imaging and Radiation Exposure During Complex EVAR

55

Frédéric Cochennec, Laurent Chiche, and Jean-Pierre Becquemin

Introduction

Fenestrated and branched stent grafts (F/BEVAR) have now reached a state of maturity for the treatment of complex aortic pathologies. They could reduce postoperative mortality and complications compared to open surgery [1–4] but long-term data still remain limited. Technical complexity, reinterventions, X-ray exposure, and injections of iodinated contrast products still remain limitations. By offering increasingly powerful image quality and integrated tools dedicated to endovascular navigation, modern hybrid or interventional radiology rooms have greatly facilitated complex endovascular aortic interventions such as F/BEVAR, while reducing radiation exposure to patients and staff [5, 6]. The objective of this chapter is to describe current imaging tools that have been developed to optimize complex aortic interventions, while limiting X-ray exposure and contrast injections. This not only means optimizing procedure steps but also refers to adequate case preparation and ability to detect technical issues at the end of the procedure.

Case Preparation, Creation of a Fusion Mask

Most last generation hybrid rooms offer the possibility to use image fusion guidance. It consists of projecting onto live fluoroscopy images a dynamic 3-dimensional reconstruction of the arterial tree (fusion mask), which automatically adjusts to the position of the C-arm. This 3D volume is obtained from a preoperative CT angiography or, less frequently, an MR angiography. Image fusion guidance has been demonstrated to facilitate navigation during complex aortic procedures, improve technical success, and reduce radiation dose, operative time and contrast injections [7–11]. It could ultimately reduce mortality, major adverse events, and secondary interventions [12].

Most companies manufacturing hybrid operating rooms have developed their own last generation enhanced imaging system incorporating image fusion software: the Allia™ IGS with the AutoRight™ system of General Electric Healthcare, the Artis and CorPath GRX systems of Siemens, and the Azurion system of Philips. In our aortic center, we use the Philips Azurion system, which allows the use of 2D/3D fusion or 3D/3D fusion via the Vessel Navigator software. Once the preoperative CT angiography has been downloaded in the workstation, the fusion mask is prepared before the procedure during a simple three-step process: (1) segmentation, (2) planning, and (3) registration.

The segmentation phase consists in selecting from an automatically generated 3D reconstruction of the arterial tree the aortic segments and branches needing to be visualized as a "3D-roadmap" during the procedure (Fig. 55.1).

During the planning phase, landmarks (rings or dots) are positioned on the 3D aortic volume (Fig. 55.2). These landmarks are useful for optimal C-arm positioning during each step of the procedure. They facilitate navigation, target vessel cannulation, and optimize accurate stent graft deployment. For a juxtarenal aneurysm treated with a fenestrated stent graft, we typically use rings to spot the ostia of visceral, renal, and hypogastric arteries and the aortic bifurcation. Markers can also be placed in target vessels to visualize early collateral branches or the expected distal landing zones of bridging stents. We also spot lumbar arteries when coil embolization is planned to reduce the risk of type 2 endoleak. For thoracoabdominal pathologies, which usually necessitate multiple graft components, we place rings to spot expected proximal and distal aortic landing zones. In aortic dissections, entry tears are spotted with rings to help stent graft placement and target vessel cannulation.

The third step is the registration process, which consists in creating an overlay between the segmented 3D aortic vol-

F. Cochennec (✉) · L. Chiche
Department of Vascular and Endovascular Surgery, Pitié Salpêtrière University Hospital, Paris, France
e-mail: frederic.cochennec@aphp.fr; laurent.chiche@aphp.fr

J.-P. Becquemin
Cardio Vascular Department, Henri Mondor Hospital University Paris Val de Marne, Creteil, France

Institut Vasculaire Paris Est, Hôpital Privé Paul d'Egine, Ramsay Group Champigny, Champigny sur Marne, France

G. Geroulakos et al. (eds.), *Mastering Endovascular Techniques*, https://doi.org/10.1007/978-3-031-42735-0_55

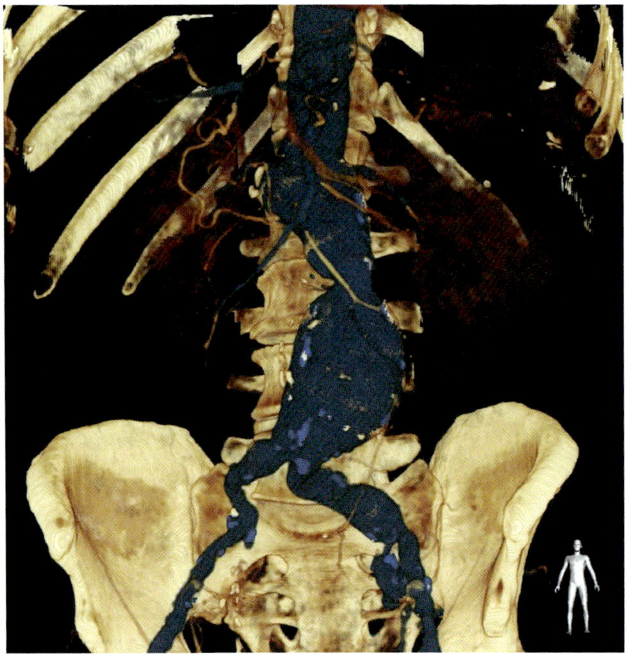

Fig. 55.1 Creation of a fusion mask: segmentation phase. Selection of segments of the aorta and branches needing to be visualized during the procedure

ume and live fluoroscopy. Two registration processes are available on the Philips platform: 2D/3D registration and 3D/3D registration. 2D/3D registration is simpler, faster, and requires less radiation. However, it is less accurate than 3D/3D fusion. It consists of shooting two fluoroscopy images, one with the C-arm in an anteroposterior position and the other one in a 60–90° oblique position. The overlay is based on bone structures or radio-opaque structures such as an in-situ stent graft, visualized in two different angles (Fig. 55.3). The overlay can be easily adjusted thanks to translational or rotational movements of anatomical projections obtained from the preoperative CT angiography. 3D/3D registration requires an intraoperative cone beam computed tomographic scan (CBCT). This CBCT is typically acquired without contrast injection, once the patient is under general anesthesia and draped. The overlay is based on at least three couples of aortic or target vessel calcifications (Fig. 55.4). These points can be coupled on axial, coronal, or sagittal slices. In the rare cases where no calcification can be identified in the aortic wall, other landmarks can be used, such as the aortic bifurcation, the carina, or osteophytes. Alignment between volumes obtained from the preoperative CT angiography and the intraoperative CBCT can be easily checked

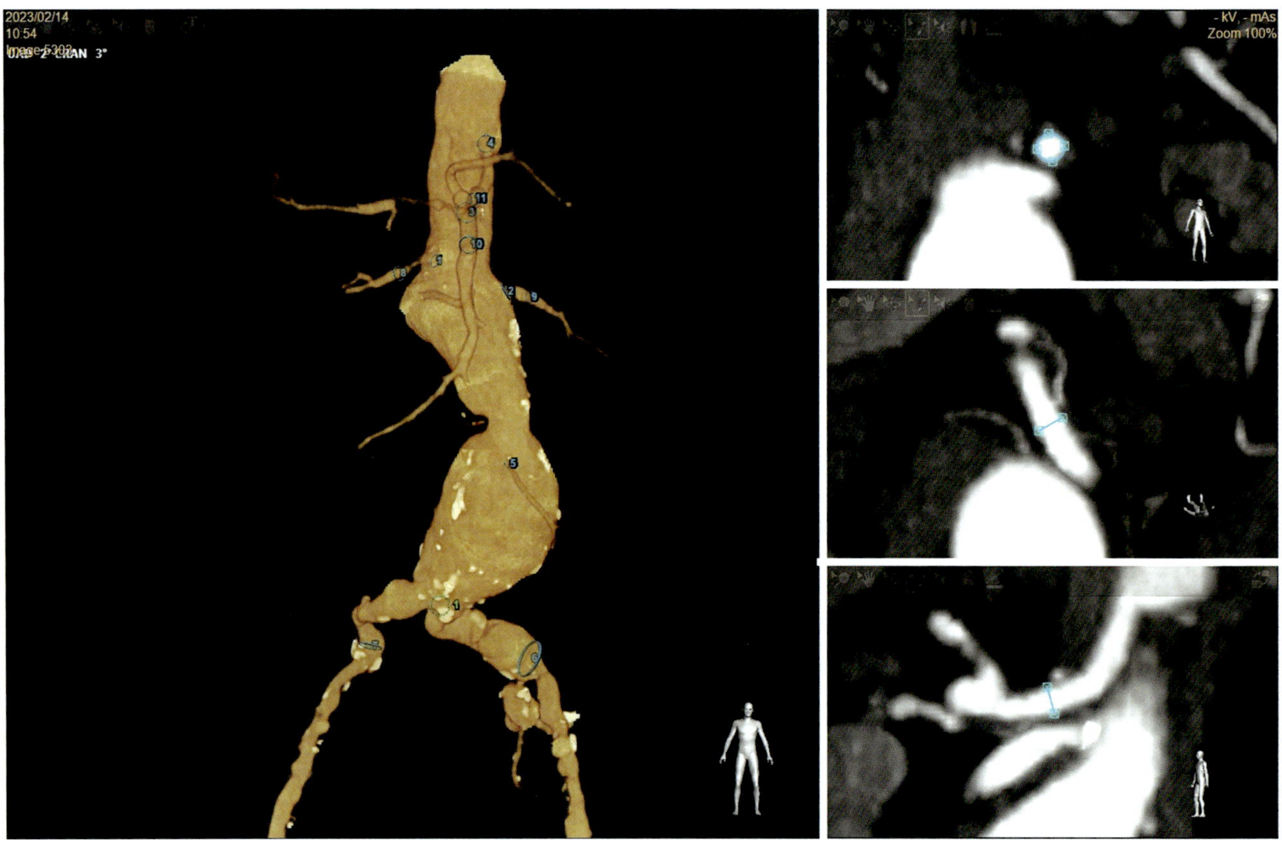

Fig. 55.2 Creation of a fusion mask: planning phase. Landmarks (rings or dots) are positioned on the 3D aortic volume. These landmarks are useful for optimal C-arm positioning during each step of the procedure. In this example, landmarks have been placed to visualize the ostia of target vessels, the expected distal landing zones of bridging stents, the aortic bifurcation, and the inferior mesenteric artery for which coil embolization has been planned

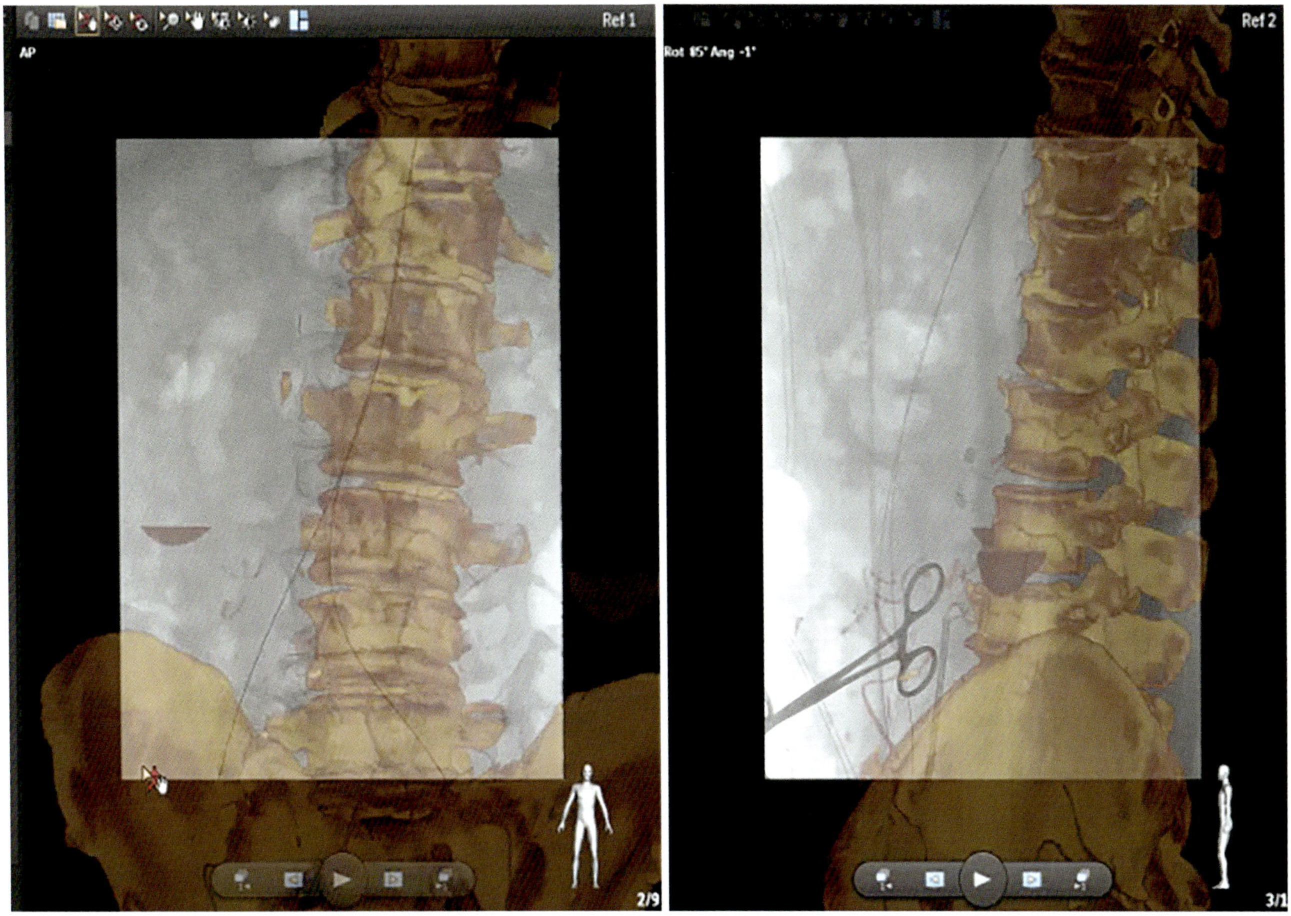

Fig. 55.3 Creation of a fusion mask: 2D/3D registration for fusion guidance. It consists of shooting two fluoroscopy images in two different angles, one in an anteroposterior position and the other in a 60°–90° angulation. The overlay is based on bone structures visualized in two different angles. The overlay can be easily adjusted thanks to translational or rotational movements of anatomical projections (orange) obtained from the preoperative CT scan

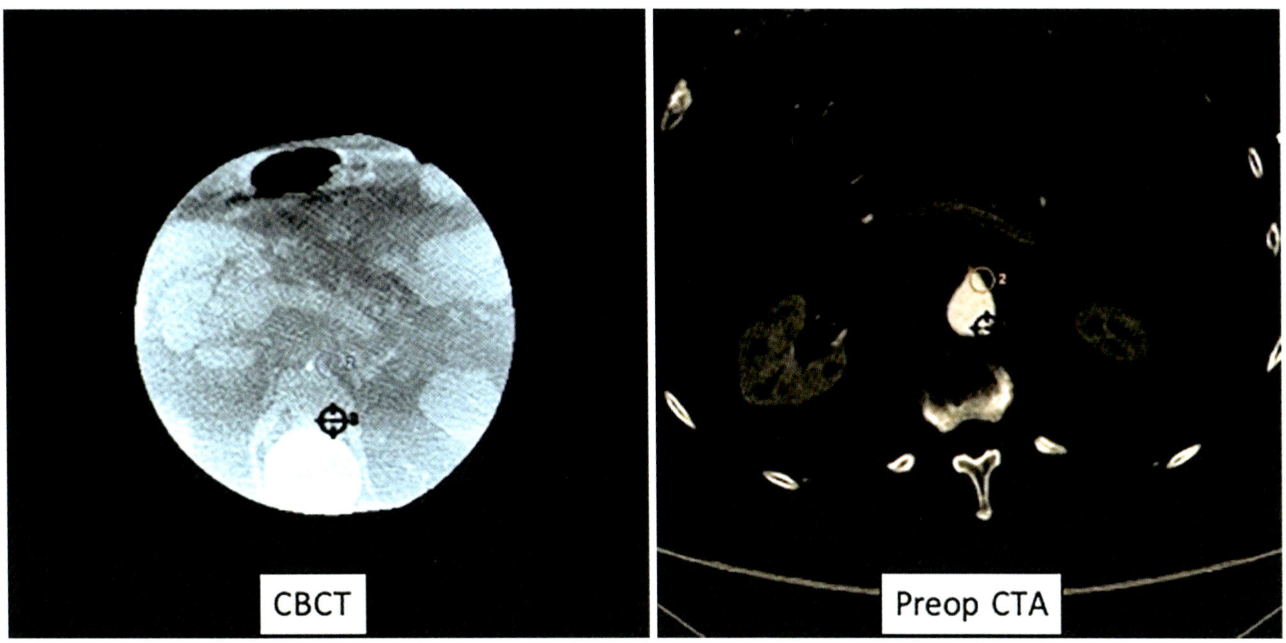

Fig. 55.4 Creation of a fusion mask: 3D/3D registration for fusion guidance. The overlay is based on at least three couples of aortic or target vessel calcifications. These points can be coupled on axial, coronal, or sagittal slices. Alignment between volumes obtained from the preoperative CT scan and the intraoperative CBCT can be easily checked and adjusted

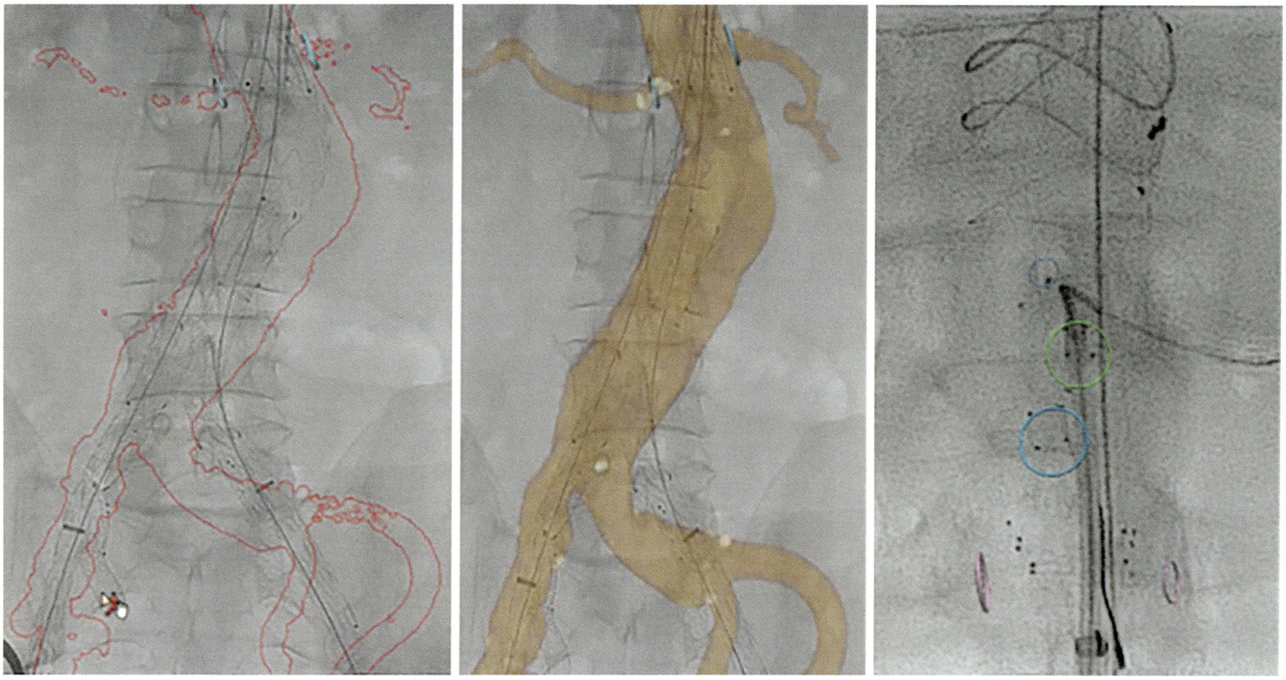

Fig. 55.5 Various patterns of the fusion mask. They are selected according to physician's preferences

and adjusted. Various patterns of the fusion mask can be selected, according to the physician's preference (Fig. 55.5).

Data comparing 2D/3D and 3D/3D fusion remain limited. While 3D/3D fusion has been demonstrated to be more accurate [13], the real benefit in terms of postoperative outcomes, radiation exposure, and contrast volume remains controversial [10, 14–17]. In our current practice, we use 2D/3D fusion for standard infrarenal and thoracic stent grafts or when a stent graft has already been implanted. 3D-3D fusion is preferred for fenestrated and branched stent grafts or when embolization of collateral branches has been planned. Whether this strategy is justified requires further evaluation.

Optimize Procedure Steps

During the procedure, there are multiple tools to reduce X-ray exposure and contrast dose and to improve the workflow. It is crucial that vascular surgeons familiarize themselves with these techniques [11].

Optimize Intraoperative Imaging, Radiation Exposure, and Comfort

Basic Set-up

Before the registration process to create a fusion mask, the absence of metallic objects in the field and correct positioning of the arms should be checked. The optimal position to reduce X-ray exposure is a high table and a detector close to the patient. In addition to lead suits, thyroid shields, and lead glasses, mobile shields that can be draped under sterile conditions should be used in daily practice. During digital subtraction angiography runs, operators should move far away from the X-ray source.

Low-Frequency Pulsed Fluoroscopy

The fluoroscopic frame rate is usually set by default between 7 and 15 frames/s. In our practice, even for complex aortic procedures, it has been reduced to 3.7 frames/s without major compromise on image quality. A fluoroscopy frame rate of 7 frames/s is only used in case of difficult target cannulation or during bridging stents deployment in obese patients. One study showed that low-frequency pulsed fluoroscopy at 3.7 frames/s was able to reduce X-ray exposure significantly [18].

Digital Zoom, Collimation

When the image quality is not optimal, using magnification is a reflex of mobile C-arm users. It results in higher radiation rates. In hybrid rooms, the use of collimation in association with digital zoom improves image quality without increasing the radiation dose. Physicians should pay great attention to adequate collimation for procedural steps during which prolonged fluoroscopy times can be necessary with the C-arm in a fixed position, such as difficult target vessel cannulation.

Reduce C-arm Angulations

Avoiding C-arm angulations as much as possible is crucial to reduce radiation. Superior mesenteric artery and celiac artery cannulation during fenestrated or branched stent grafting typically needs profile or oblique positioning of the C-arm. For these arteries, our practice is to cannulate the first centimeters in a profile position and rapidly switch to an anteroposterior position to advance the soft wire, and complete the steps to secure the access with a stiff wire and a sheath. We only go back to an oblique angulation for deployment of bridging covered stents and selective control angiograms.

Interventional tools of last generation hybrid rooms allow for registering C-arm angulations expected to be needed to visualize fenestrations, branches, or ostia of target vessels. These angles can be automatically recalled from the table, avoiding adjustments for optimal C-arm positioning and unnecessary fluoroscopy.

Recorded Fluoroscopy, Marker Tool

Using recorded fluoroscopy instead of digital subtraction angiography (DSA) runs is also interesting. The images obtained with fluoroscopy are not as accurate as those obtained with DSA runs but they need much less radiation. A recorded fluoroscopy run with contrast injection can be adequate for basic procedural steps such as visualization of the hypogastric take-off and adjustment of the fusion mask before iliac limb deployment. Alternatively, a digital colored marker can be placed on the screen monitor.

Alternative Methods of Guidance

Intravascular Ultrasound

The use of intravascular ultrasound (IVUS) during aortic procedures has been proposed for more accurate aortic sizing [19–21], and detection of technical defects such as iliac limbs at risk of occlusion [22]. Whether IVUS is able to facilitate procedural steps during complex endovascular aortic procedures remains unclear. Some studies suggested that IVUS-assisted EVAR was associated with a reduction of contrast and radiation dose [23, 24]. Other studies suggested that IVUS was helpful during thoracic stent grafting for type B aortic dissection to identify entry tears, evaluate false lumen flow, detect incomplete stent graft apposition, or check guidewire position [25]. One study examined the effect of IVUS on the clinical outcomes after TEVAR for blunt traumatic aortic injuries in 919 patients. Clinical outcomes (mortality and reintervention rates at 1 year) were similar between cases in which IVUS was used and for which it was not [26].

FORS Technology

Fiber Optic RealShape (FORS; Philips Medical Systems Nederland B.V., Best, The Netherlands) is a novel endovascular guidance technology [27]. It aims to reduce X-ray exposure and facilitate navigation. The technology is based on the concept of measuring strain in optical fibers, using light reflected from density fluctuations in these fibers. To achieve this, an optical fiber is integrated into intra-body devices such as wires and catheters. By sending laser light into this fiber and analyzing how it is reflected back along the fiber, the full shape of the devices can be reconstructed and visualized in 3D, in real time, from any viewing angle (Fig. 55.6). System precision, technical feasibility, and safety have been demonstrated in preclinical and small preliminary series [28, 29]. Two recent studies evaluating FORS guidance during complex aortic procedures have reported promising results, with acceptable technical success rates of target vessel cannulation and lower radiation exposure when compared with procedures performed without FORS guidance [30, 31]. This technology recently obtained regulatory approval in Europe and from the FDA but is currently in limited commercial distribution. The main limitation is the limited selection of FORS-enabled wires and catheters at this time. As more incorporated devices are expected to be introduced in the market, the FORS technology may play an important role in the near future to reduce X-ray exposure.

Reduce Contrast Injection

Contrast Medium Dilution

Diluting the contrast medium with saline in the injector is a simple way to reduce contrast doses. For aortic procedures, we use a mix of 50% saline and 50% contrast medium with 320 mg of iodine/mL.

There are substantial data in the literature showing that image fusion guidance is able to reduce contrast doses [18]. When using 2D/3D fusion, we still perform an aortic angiogram with 20 mL of 50% diluted contrast medium to adjust the fusion mask. When using 3D/3D fusion, injecting 6/10 mL of 50% diluted contrast medium selectively in one renal artery is generally sufficient to check and adjust the fusion mask if necessary. We do not compromise when it comes to selectively checking target arteries. We still use 6/10 mL of 50% diluted contrast to check the adequate position of the catheter during renal and visceral vessels cannulation and 10 mL of 50% diluted contrast to visualize potential technical issues after bridging stents deployment. A final angiogram using 30 mL of diluted contrast medium is performed at the end of the procedure.

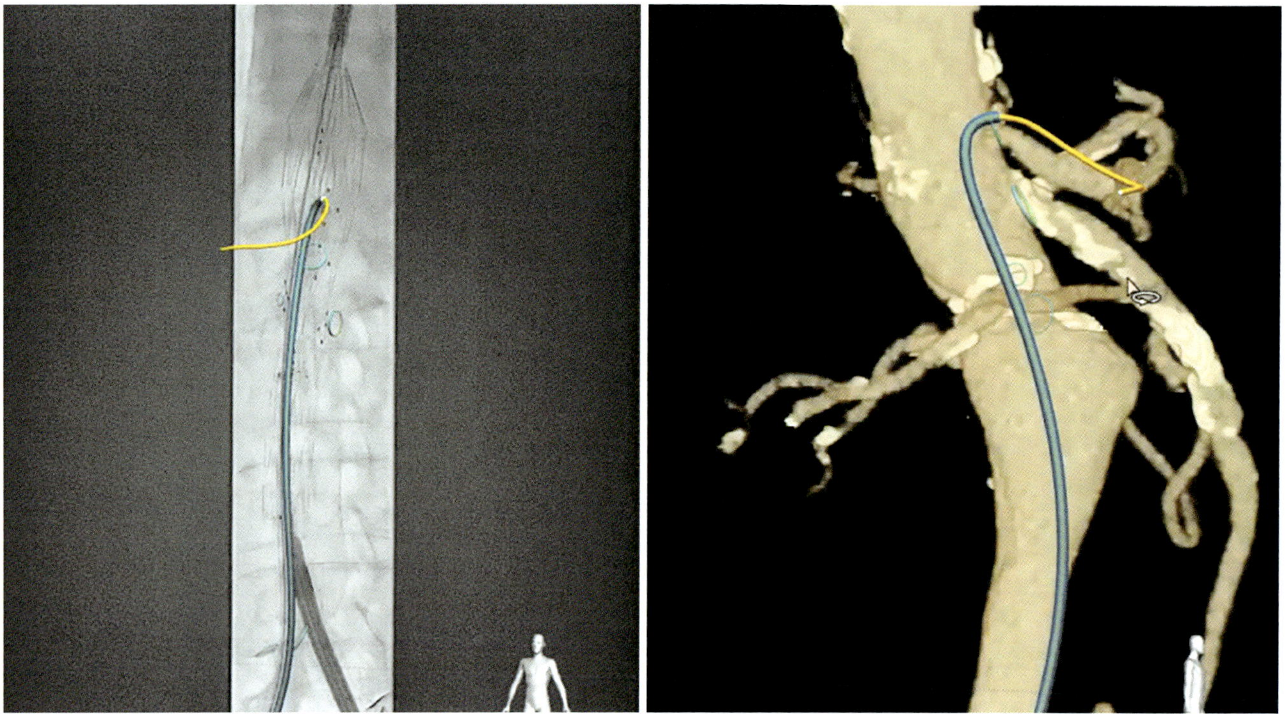

Fig. 55.6 Fiber Optic RealShape (FORS; Philips Medical Systems Nederland B.V., Best, The Netherlands) technology. The full shape of the FORS-enabled wires and catheters can be reconstructed and visualized in 3D, in real time, from any viewing angle, without X-ray

Carbon Dioxide

Carbon dioxide (CO_2) is a low-viscosity agent with no nephrotoxic effect. Several studies have evaluated the use of CO_2 as a contrast agent and suggested it was able to significantly reduce the need for iodinated contrast medium and enhance detection of type II endoleaks, particularly when used with a dedicated automated injector [32–35]. Suboptimal visualization of renal arteries with an anterior take-off, increased radiation dose, and the lack of data and dedicated protocol for complex procedures remain limitations [35, 36].

Optimize Final Check, Intraoperative Cone Beam Computed Tomography

Being able to control adequately the quality of the aortic reconstruction is crucial to avoid complications and secondary interventions. A completion digital subtraction angiography performed alone at the end of the procedure may miss technical issues such as graft limb kinks, defects of bridging stents, or type I/III endoleaks needing immediate revision.

There is consistent data in the literature showing that the use of intraoperative contrast-enhanced cone beam computed tomography to control aortic endovascular procedures is associated with a significant reduction of contrast volume, radiation dose, and stent graft-related complications [37–40]. CBCT in association with intraoperative digital subtraction angiography has even been advocated to safely replace early postoperative CT angiography [41]. This strategy requires further evaluation.

Contrast-enhanced dual CBCT has the theoretical advantage of being able to detect endoleaks (Fig. 55.7). However, data comparing contrast-enhanced CBCT with CBCT without contrast remain scarce. In selected cases, when the level of radiation and contrast remain acceptable, we perform a contrast-enhanced dual CBCT. Our protocol is to inject 84 mL of 50% diluted contrast with a flow rate of 12 mL/s. A first CBCT acquisition is set with a delay of 2 s after contrast injection. A second rotation is performed 7 s after contrast injection in order to detect type II endoleaks. In patients with chronic kidney disease, when the dose of contrast exceeds 120 mL, or when the cumulative air Kerma exceeds 1500 mGy, we tend to perform a CBCT without contrast, which is still able to detect technical issues at the level of the stent graft structure (Fig. 55.8).

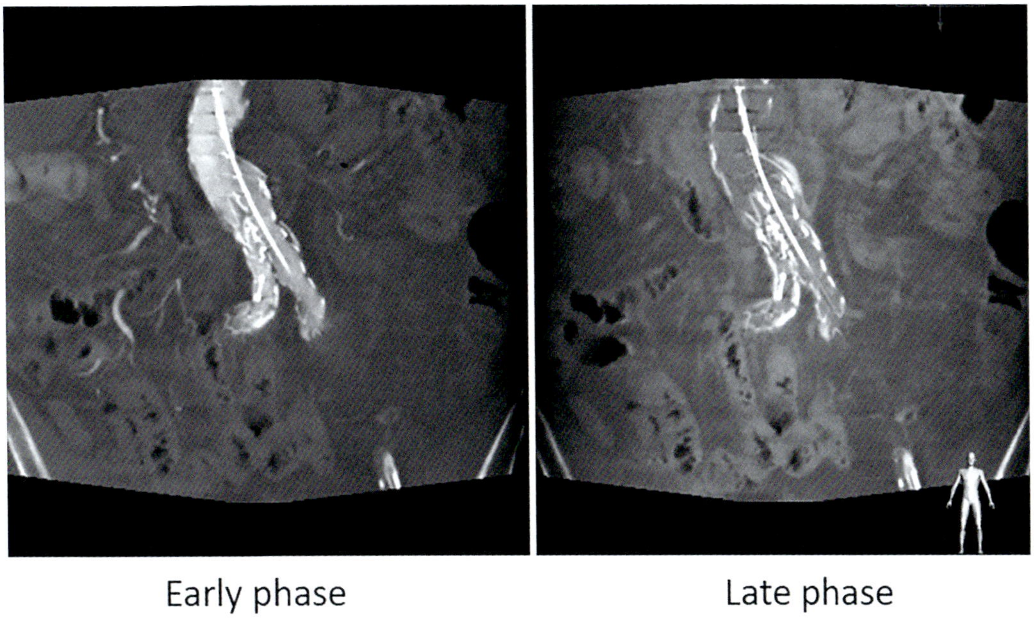

Fig. 55.7 Contrast-enhanced dual CBCT. The early phase detects type I/III endoleak but the late phase detects a type II endoleak

Fig. 55.8 CBCT without contrast performed to check potential technical issues at the level of graft components of a branched stent graft of the aortic arch

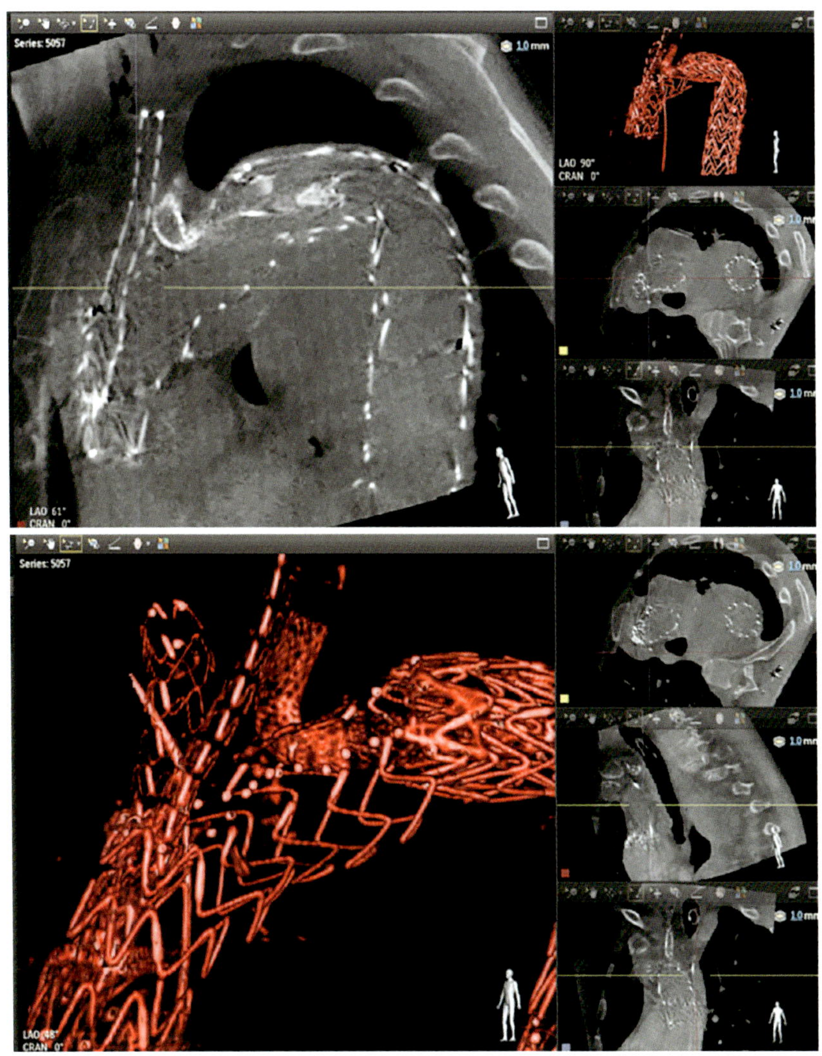

Conclusion

Complex aortic endovascular procedures such as FEVAR and BEVAR should be performed in modern hybrid rooms. Integrated tools and software make it possible to optimize procedure steps, reduce X-ray exposure, reduce contrast injection, and improve the comfort of surgeons. They also make it possible to better prepare interventions and to control their technical quality while the patient is still on the operating table. Any surgeon performing complex endovascular aortic procedures should be comfortable with these tools.

References

1. Antoniou GA, Juszczak MT, Antoniou SA, Katsargyris A, Haulon S. Editor's choice – fenestrated or branched endovascular versus open repair for complex aortic aneurysms: meta-analysis of time to event propensity score matched data. Eur J Vasc Endovasc Surg. 2021;61(2):228–37.
2. Doonan RJ, Girsowicz E, Dubois L, Gill HL. A systematic review and meta-analysis of endovascular juxtarenal aortic aneurysm repair demonstrates lower perioperative mortality compared with open repair. J Vasc Surg. 2019;70(6):2054–2064.e3.
3. Jones AD, Waduud MA, Walker P, Stocken D, Bailey MA, Scott DJA. Meta-analysis of fenestrated endovascular aneurysm repair *versus* open surgical repair of juxtarenal abdominal aortic aneurysms over the last 10 years. BJS Open. 2019;3(5):572–84.
4. Rocha RV, Lindsay TF, Friedrich JO, Shan S, Sinha S, Yanagawa B, et al. Systematic review of contemporary outcomes of endovascular and open thoracoabdominal aortic aneurysm repair. J Vasc Surg. 2020;71(4):1396–1412.e12.
5. Kaladji A, Daoudal A, Clochard E, Gindre J, Cardon A, Castro M, et al. Interest of fusion imaging and modern navigation tools with hybrid rooms in endovascular aortic procedures. J Cardiovasc Surg. 2017;58(3):458–66.
6. Rohlffs F, Spanos K, Debus ES, Heidemann F, Tsilimparis N, Kölbel T. Modern image acquisition system reduces radiation exposure to patients and staff during complex endovascular aortic repair. Eur J Vasc Endovasc Surg Off J Eur Soc Vasc Surg. 2020;59(2):295–300.
7. Dijkstra ML, Eagleton MJ, Greenberg RK, Mastracci T, Hernandez A. Intraoperative C-arm cone-beam computed tomography in fenestrated/branched aortic endografting. J Vasc Surg. 2011;53(3):583–90.
8. Goudeketting SR, Heinen SGH, Ünlü Ç, van den Heuvel DAF, de Vries JPPM, van Strijen MJ, et al. Pros and cons of 3D image fusion in endovascular aortic repair: a systematic review and meta-analysis. J Endovasc Ther. 2017;24(4):595–603.
9. Hertault A, Maurel B, Sobocinski J, Martin Gonzalez T, Le Roux M, Azzaoui R, et al. Impact of hybrid rooms with image fusion on radiation exposure during endovascular aortic repair. Eur J Vasc Endovasc Surg Off J Eur Soc Vasc Surg. 2014;48(4):382–90.
10. Klein A, Guild J, Xi Y, Chamseddin K, Shih M, Siah M, et al. Use of a 2 dimensional vessel navigator roadmap decreases patient radiation dose compared to standard 3D mapping for fenestrated endovascular aneurysm repair. Ann Vasc Surg. 2022;80:250–5.
11. Vento V, Soler R, Fabre D, Gavit L, Majus E, Brenot P, et al. Optimizing imaging and reducing radiation exposure during complex aortic endovascular procedures. J Cardiovasc Surg (Torino) [Internet]. 2019;60(1). [cité 4 janv 2023]. Disponible sur: https://www.minervamedica.it/index2.php?show=R37Y2019N01A0041
12. Tenorio ER, Oderich GS, Sandri GA, Ozbek P, Kärkkäinen JM, Macedo TA, et al. Impact of onlay fusion and cone beam computed tomography on radiation exposure and technical assessment of fenestrated-branched endovascular aortic repair. J Vasc Surg. 2019;69(4):1045–1058.e3.
13. Sieren MM, Schareck C, Kaschwich M, Horn M, Matysiak F, Stahlberg E, et al. Accuracy of registration techniques and vascular imaging modalities in fusion imaging for aortic endovascular interventions: a phantom study. CVIR Endovasc. 2021;4(1):51.
14. Barral PA, Demasi-Jacquier MA, Bal L, Omnes V, Bartoli A, Piquet P, et al. Fusion imaging to guide thoracic endovascular aortic repair (TEVAR): a randomized comparison of two methods, 2D/3D versus 3D/3D image fusion. Cardiovasc Intervent Radiol. 2019;42(11):1522–9.
15. Edsfeldt A, Sonesson B, Rosén H, Petri MH, Hongku K, Resch T, et al. Validation of a new method for 2D fusion imaging registration in a system prepared only for 3D. J Endovasc Ther Off J Int Soc Endovasc Spec. 2020;27(3):468–72.
16. Schulz CJ, Böckler D, Krisam J, Geisbüsch P. Two-dimensional-three-dimensional registration for fusion imaging is noninferior to three-dimensional- three-dimensional registration in infrarenal endovascular aneurysm repair. J Vasc Surg. 2019;70(6):2005–13.
17. Tacher V, Lin M, Desgranges P, Deux JF, Grünhagen T, Becquemin JP, et al. Image guidance for endovascular repair of complex aortic aneurysms: comparison of two-dimensional and three-dimensional angiography and image fusion. J Vasc Interv Radiol JVIR. 2013;24(11):1698–706.
18. Dias NV, Billberg H, Sonesson B, Törnqvist P, Resch T, Kristmundsson T. The effects of combining fusion imaging, low-frequency pulsed fluoroscopy, and low-concentration contrast agent during endovascular aneurysm repair. J Vasc Surg. 2016;63(5):1147–55.
19. Garret HEJ, Abdullah AH, Hodgkiss TD, Burgar SR. Intravascular ultrasound aids in the performance of endovascular repair of abdominal aortic aneurysm. J Vasc Surg. 2003;37(3):615–8.
20. Verbin C, Scoccianti M, Kopchok G, Donayre C, White RA. Comparison of the utility of CT scans and intravascular ultrasound in endovascular aortic grafting. Ann Vasc Surg. 1995;9(5):434–40.
21. Wallace GA, Starnes BW, Hatsukami TS, Sobel M, Singh N, Tran NT. Intravascular ultrasound is a critical tool for accurate endograft sizing in the management of blunt thoracic aortic injury. J Vasc Surg. 2015;61(3):630–5.
22. Coelho A, Nogueira C, Lobo M, Gouveia R, Campos J, Augusto R, et al. Impact of post-EVAR graft limb kinking in EVAR limb occlusion: aetiology, early diagnosis, and management. Eur J Vasc Endovasc Surg Off J Eur Soc Vasc Surg. 2019;58(5):681–9.
23. Hoshina K, Kato M, Miyahara T, Mikuriya A, Ohkubo N, Miyata T. A retrospective study of intravascular ultrasound use in patients undergoing endovascular aneurysm repair: its usefulness and a description of the procedure. Eur J Vasc Endovasc Surg. 2010;40(5):559–63.
24. Illuminati G, Pacilè MA, Ceccanei G, Ruggeri M, La Torre G, Ricco JB. Peroperative intravascular ultrasound for endovascular aneurysm repair versus peroperative angiography: a pilot study in fit patients with favorable anatomy. Ann Vasc Surg. 2020;64:54–61.
25. Koschyk DH, Nienaber CA, Knap M, Hofmann T, Kodolitsch YV, Skriabina V, et al. How to guide stent-graft implantation in type B aortic dissection? Comparison of angiography, transesophageal echocardiography, and intravascular ultrasound. Circulation. 2005;112(9 Suppl):I260–4.
26. Smith JA, Bose S, Sarode A, Cho JS, Colvard B. Effect of intravascular ultrasound on clinical outcomes after thoracic endovas-

cular aortic repair for blunt thoracic aortic injury. J Vasc Surg. 2022;75(2):448–454.e2.
27. Finnesgard EJ, Simons JP, Marecki H, Ofori I, Kölbel T, Schurink GWH, et al. Fiber optic RealShape technology in endovascular surgery. Semin Vasc Surg. 2021;34(4):241–6.
28. Jansen M, Khandige A, Kobeiter H, Vonken EJ, Hazenberg C, van Herwaarden J. Three dimensional visualisation of endovascular guidewires and catheters based on laser light instead of fluoroscopy with fiber optic RealShape technology: preclinical results. Eur J Vasc Endovasc Surg Off J Eur Soc Vasc Surg. 2020;60(1):135–43.
29. van Herwaarden JA, Jansen MM, Vonken EJPA, Bloemert-Tuin T, Bullens RWM, de Borst GJ, et al. First in human clinical feasibility study of endovascular navigation with Fiber Optic RealShape (FORS) Technology. Eur J Vasc Endovasc Surg Off J Eur Soc Vasc Surg. 2021;61(2):317–25.
30. Finnesgard EJ, Simons JP, Jones DW, Judelson DR, Aiello FA, Boitano LT, et al. Initial single center experience using fiber optic RealShape guidance in complex endovascular aortic repair. J Vasc Surg. 2022:S0741-5214(22)02531-9.
31. Panuccio G, Schanzer A, Rohlffs F, Heidemann F, Wessels B, Schurink GW, et al. Endovascular navigation with fiber optic RealShape technology. J Vasc Surg. 2023;77(1):3–8.e2.
32. Criado E, Upchurch GR, Young K, Rectenwald JE, Coleman DM, Eliason JL, et al. Endovascular aortic aneurysm repair with carbon dioxide-guided angiography in patients with renal insufficiency. J Vasc Surg. 2012;55(6):1570–5.
33. Huang SG, Woo K, Moos JM, Han S, Lew WK, Chao A, et al. A prospective study of carbon dioxide digital subtraction versus standard contrast arteriography in the detection of endoleaks in endovascular abdominal aortic aneurysm repairs. Ann Vasc Surg. 2013;27(1):38–44.
34. Mascoli C, Faggioli G, Gallitto E, Vento V, Pini R, Vacirca A, et al. Standardization of a carbon dioxide automated system for endovascular aortic aneurysm repair. Ann Vasc Surg. 2018;51:160–9.
35. Vacirca A, Faggioli G, Mascoli C, Gallitto E, Pini R, Spath P, et al. CO2 automated angiography in endovascular aortic repair preserves renal function to a greater extent compared with iodinated contrast medium. Analysis of technical and anatomical details. Ann Vasc Surg. 2022;81:79–88.
36. Gallitto E, Faggioli G, Vacirca A, Pini R, Mascoli C, Fenelli C, et al. The benefit of combined carbon dioxide automated angiography and fusion imaging in preserving perioperative renal function in fenestrated endografting. J Vasc Surg. 2020;72(6):1906–16.
37. Hertault A, Maurel B, Pontana F, Martin-Gonzalez T, Spear R, Sobocinski J, et al. Benefits of completion 3D angiography associated with contrast enhanced ultrasound to assess technical success after EVAR. Eur J Vasc Endovasc Surg. 2015;49(5):541–8.
38. Lerisson E, Patterson BO, Hertault A, Klein C, Pontana F, Sediri I, et al. Intraoperative cone beam computed tomography to improve outcomes after infrarenal endovascular aortic repair. J Vasc Surg. 2022;75(3):1021–1029.e2.
39. Steuwe A, Geisbüsch P, Schulz CJ, Böckler D, Kauczor HU, Stiller W. Comparison of radiation exposure associated with intraoperative cone-beam computed tomography and follow-up multidetector computed tomography angiography for evaluating endovascular aneurysm repairs. J Endovasc Ther. 2016;23(4):583–92.
40. Tenorio ER, Oderich GS, Sandri GA, Ozbek P, Kärkkäinen JM, Vrtiska T, et al. Prospective nonrandomized study to evaluate cone beam computed tomography for technical assessment of standard and complex endovascular aortic repair. J Vasc Surg. 2020;71(6):1982–1993.e5.
41. Törnqvist P, Dias N, Sonesson B, Kristmundsson T, Resch T. Intraoperative cone beam computed tomography can help avoid reinterventions and reduce CT follow up after Infrarenal EVAR. Eur J Vasc Endovasc Surg. 2015;49(4):390–5.

Rescue of Proximal Failure of Endovascular Abdominal Aortic Aneurysm Repair with Standard and Fenestrated Grafts

56

Silvan Jungi, Michel Bosiers, Drosos Kotelis, and Vladimir Makaloski

Case Presentation
An 81-year-old male patient was admitted with abdominal and lower back pain. The patient had previously treated juxtarenal aneurysm with a three-fold custom-made fenestrated stent graft for both renal arteries and superior mesenteric artery 4 years previously. A type Ia endoleak was diagnosed 1 year before, and a custom-made treatment was offered to the patient, which he declined due to personal reasons.

Current CT scan shows progression of the aortic disease at the level of the previous proximal sealing (distally from celiac trunk) with migrated fenestrated stent graft, a high-flow type Ia endoleak, increased aneurysm sac with 11 cm of diameter, and contained rupture (Fig. 56.1). Due to the patient's comorbidities, open conversion was deemed too risky. An off-the-shelf endovascular solution, using an off-the-shelf branched stent graft, was not possible due to the narrow anatomy caused by the previous fenestrated endovascular aneurysm repair (FEVAR), short distance to FEVAR flow divider and downward-oriented bridging stents.

Continued at page 593

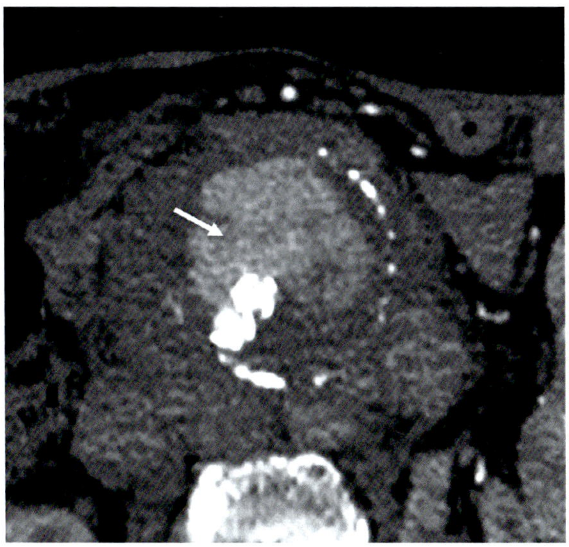

Fig. 56.1 Axial view of a computed tomography angiography with a massive endoleak (white arrow)

Background

Endovascular aortic aneurysm repair (EVAR) revolutionized the treatment of abdominal aortic aneurysms (AAA). Currently, it is a first-line treatment option, especially for elderly patients, because it is associated with very low short-term mortality rate and a shorter hospital stay [1]. However, re-intervention rates are higher compared with open aortic surgery [1]. The vast majority of re-interventions are performed to treat endoleaks [2]. In case of a type II endoleak without aneurysm growth, close follow-up and surveillance without re-intervention is indicated. However, when type Ia endoleak is detected, treatment is mandatory. The inadequate seal between the graft and the aortic wall can occur due to a variety of reasons including progression of disease (dilatation of the aortic neck), graft migration, proximal graft fabric tears, or defects. Primarily hostile necks with diameter >28 mm, length ≤10 mm, angulation >60°, and (reversed) tapered shape are associated with higher rates of type Ia endoleak [3, 4].

The incidence of type Ia endoleak varies between 3.5% and 15.5% [5, 6]. Type Ia endoleaks are clearly associated with an increased risk of aneurysm rupture and require timely intervention [4]. Repair is generally feasible, with good early to midterm outcomes. Different treatment options are available, and the choice should be based on endoleak characteristics, aortic anatomy, the patient's surgical risk, and the institution's experience. Conservative treatment and endoleak embolization should be considered only in selected cases, such as low-flow endoleaks and frail patients.

S. Jungi (✉) · M. Bosiers · D. Kotelis · V. Makaloski
Bern University Hospital, University of Bern, Bern, Switzerland
e-mail: silvan.jungi@insel.ch; dkotelis@ukaachen.de; vladimir.makaloski@insel.ch

G. Geroulakos et al. (eds.), *Mastering Endovascular Techniques*, https://doi.org/10.1007/978-3-031-42735-0_56

In this chapter, we discuss the management of type Ia endoleaks following failed endovascular AAA repair with standard and fenestrated grafts.

Indications for Intervention

Endovascular Treatment Options

Proximal Cuff Extension

If there is a healthy aorta (≥10 mm length) between the proximal edge of the EVAR stent graft and the lower renal artery (mostly due to migration), a proximal cuff extension (PCE) can be used to address type Ia endoleak by increasing the proximal landing zone and therefore sealing the leak at the proximal end of the graft. This method demonstrated successful outcomes with low complication rates [7]. In this series, a retrospective analysis of 44 patients reported initial technical success (successful deployment and endoleak closure) in 43/44 patients (98%) [7]. Recent meta-analysis found a clinical success of PCE in type Ia endoleak treatment of 89%, making it a safe and effective approach for managing type Ia endoleak after EVAR [8].

Parallel Graft (Chimney Technique)

When extension of the sealing zone proximal from the renal arteries is required, the chimney technique (proximal extension with parallel grafts) can be a practical solution. Several studies have reported the effectiveness of the chimney technique in treating type Ia endoleaks after EVAR. In a series of 39 patients, technical success was achieved in 90% of cases [9]. During 30 days 7.6% persistent type Ia endoleak and one chimney graft occlusion was detected. Primary patency of chimney grafts was 94.3% at 3 years [9]. The clinical success of chimney EVAR for type Ia endoleak occlusion was found to be 93.9% [8]. However, the main concern of the chimney technique are gutter endoleaks between the chimney and the proximal extension with an incidence of 10.7% [10]. Therefore, it is not commonly used in elective settings, but it remains an acceptable treatment option in urgent or emergent situations.

Embolization

Another treatment option is the embolization of the type Ia endoleak between the aortic wall and the stent graft. It is seen as an alternative percutaneous approach in patients with adverse anatomy, such as short or heavily angulated proximal necks or severe comorbidities that prohibit general anesthesia or open surgical repair.

Only smaller series exist investigating the effectiveness of embolization of type I endoleak. One study included 22 patients and showed primary technical success of 100%, but 28% had an endoleak recurrence and 11% died after secondary sac rupture [11]. In most of the published studies, the embolization was conducted with ethylene-vinyl alcohol copolymer (Onyx, Covidien, Irvine, California) alone or in combination with coils or n-butyl cyanoacrylate. However, proximal sealing was maintained only in 79.7% of the cases during follow-up (compared to 91% with other techniques) [8]. Embolization could be proposed as a first-line treatment in case of persistent type Ia endoleak due to gutters after chimney EVAR or as a bridge to definitive treatment (see case scenario).

Endostaples

Endostaples are designed to provide additional sealing to prevent blood from flowing through the endoleak. Endostaples (or endoanchors) consist of a small metal screw that is inserted into the stent graft and the aortic wall to close the gap between. The staples are usually delivered using a catheter-based delivery system. Once deployed, the staples provide additional fixation and sealing of the endograft to the aortic wall.

They are typically proposed as an intraoperative treatment during initial EVAR for type Ia endoleak prevention. Limited data show that this adjunctive maneuver seems to reduce the risk of endograft migration and type Ia endoleak [12].

Nevertheless, when used to seal late-occurring type Ia endoleak, endostapling alone was associated with low success rates of sealing of 57% [8]. Their use in combination with adjunctive procedures, such as PCE, may possibly improve outcomes.

Fenestrated EVAR or Branched EVAR (FBEVAR)

Fenestrated EVAR consists of complete exclusion of the type Ia endoleak with a fenestrated stent graft with 2–4 fenestrations for renal or reno-visceral arteries. Technical success rates for type Ia endoleak sealing after EVAR with FEVAR range from 85% to 96% [13, 14]. The clinical success of FEVAR for type Ia endoleak exclusion following EVAR is 91% [8]. Therefore, FEVAR is a good treatment option addressing type Ia endoleak after EVAR. Regarding the devices, it can be either a custom-made stent graft in the elective setting or physician-modified in the urgent/emergent setting. Physician-modified fenestrated stent grafts offer good technical success in experienced hands and their durability is acceptable and comparable to custom-made devices with a relatively low re-intervention rate [15].

A branched stent graft consists of a multi-branched stent graft with branches to the reno-visceral arteries. These can be custom-made devices either with outer branches or inner branches (or a combination) or off-the-shelf solutions (Zenith t-branch, Cook medical Inc., Bloomington, IN, USA; E-nside, Jotec (Artivion), Hechingen, Germany; Excluder thoracoabdominal branch endoprosthesis (TAMBE), W. L. Gore & Associates, Flagstaff, Arizona, USA).

Timing of Intervention

Timing of intervention depends on the individual situation. In the majority of patients, the endoleak is diagnosed in the routine clinical workup and the treatment can be planned electively with all treatment options, including custom-made fenestrated or branched stent grafts.

If the patient is symptomatic, an urgent or emergent (in case of rupture) custom-made FBEVAR is not an option, so the remaining treatment modalities are embolization, proximal cuff extension, physician-modified FEVAR, chimney EVAR, off-the-shelf BEVAR, or open conversion. This depends on the patient's anatomy, their perioperative risk, and surgeon's preference.

Imaging

A three-phase thin-slice computed tomography angiography is necessary for diagnosis and planning of the procedure.

Endovascular Strategy and Technique

Proximal Cuff Extension: *Basic Endovascular Toolkit*

- Standard access wires and sheaths
- Soft or stiff glidewire 035″
- Angiographic catheter (e.g., pigtail catheter)
- Extra stiff guidewire
- Cuff/aortic extension stent graft
- Compliant balloon

Parallel Graft (Chimney Technique): *Basic Endovascular Toolkit*

- Standard access wires and sheaths
- Long 6, 7, or 8F sheath for one or two brachial/axillary access
- Soft or stiff glidewire 0.035″
- Angiographic catheter (e.g., pigtail catheter)
- Extra stiff guidewire
- Supportive guidewires
- Cuff/aortic extension stent graft
- Covered stents for the parallel grafts

Embolization: *Basic Endovascular Toolkit*

- Standard access wires and sheaths
- Long 6F sheath for the brachial/axillary access or a steerable sheath when femoral access is chosen)
- Microcatheter
- 0.014″ wire
- Embolic agents

Endostaples: *Basic Endovascular Toolkit*

- Standard access wires and sheaths
- Steerable sheath
- Endostaples

FBEVAR: *Basic Endovascular Toolkit*

- Standard access wires and sheaths
- Soft or stiff glidewire 035″
- Long 8F sheath for the brachial/axillary access or steerable sheath
- Angiographic catheter (e.g., pigtail catheter) and diagnostic catheters
- Extra stiff guidewire for the main graft (e.g., Lunderquist or E-wire)
- Supportive guidewires for the bridging stents (e.g., Rosen or Supra Core guidewire)
- Fenestrated or branched stent graft
- Covered stents for the target vessels

A list of suggested material is summarized in Table 56.1.

Decision Making on Technique Selection

With the many different endovascular and open surgical treatment options, adequate decision making (what is the best treatment option for the particular patient) is key in treating this pathology.

Several factors influence the technique selection:

Patient's Characteristics

Age, comorbidities, and surgical risk should be considered when selecting the appropriate technique, especially in fitter patients (≥ 4 METs) where open conversion should be considered.

Anatomical Suitability

Size, location, and morphology of the aneurysm sac:

- Sufficient landing zone (≥10 mm) between the EVAR stent graft and the lower renal artery (proximal cuff extension possible)
- Distance from the lower renal artery to previous EVAR stent graft flow diverter
- Access: iliofemoral arteries with a diameter of ≥7 mm for a custom-made device. Upper extremity access. Previous aortic arch repair
- Intraluminal thrombus or calcifications
- Diameter of the visceral aortic segment: ≤32 mm favors FEVAR or BEVAR with inner branches, ≥32 favors BEVAR with external branches

Table 56.1 Suggested endovascular toolkit for the endovascular management of endovascular treatment of type Ia endoleak after EVAR. These are only a few suggested options based on the author's experience and can cover the majority of cases

	Manufacturer	Size/length
Wires		
Any standard access wire		0.035″
Glidewire floppy	Terumo Radifocus	0.035″
Glidewire stiff	Terumo Radifocus	0.035″
Supportive guidewires (fixed core rosen guide wire, supra Core extra supportive guidewire)	Cook (Rosen), Abbott (supra Core)	0.035″
E-wire extra stiff guidewire	Jotec (Artivion)	0.035″
Choice PT	Boston scientific	0.014″
Sheaths		
Any standard access sheath	Any	4–8 Fr/11 cm
Long transbrachial sheath Destination, Fortress, Flexor	Terumo, Biotronik, Cook	6 F/90 cm
Steerable sheath Destino Twist	Oscor	7F/55cm/9 mm
Catheters		
Any diagnostic catheter	Any	5F, 65 cm–100 cm
Pigtail angiographic catheter	Any	5F, 65 cm
Aortic cuff		
Endurant II aortic extension	Medtronic (also available: Zenith Alpha (Cook), Excluder (Gore), Incraft (Cordis), e.g., Endurant II/IIs (Medtronic), AFX2 Endovascular AAA System (Endologix)	23–36 mm diameter (18–20F)
FBEVAR custom-made stent grafts		
Various options (custom-made)	Terumo, Cook, Jotec	
Off-the-shelf	Cook (T-Branch), Jotec (E-nside, Artivion), TAMBE (Gore)	
Physician modified	Medtronic (Endurant II), Cook (Zenith alpha)	
Balloons		
PTA balloon: Armada 35	Abbott	8–12 mm
Compliant balloon: Reliant	Medtronic (other: Koda (Cook)	Compatible with 12F sheath
Bridging stents		
BeGraft peripheral (FEVAR) or BeGraft+ (BEVAR, ChEVAR)	Bentley	Various sizes/lengths
Advanta V12 (FEVAR)	Gettinge	Various sizes/lengths
Viabahn VBX (FBEVAR, ChEVAR)	Gore	Various sizes/lengths
Other	Fluency plus (BD), iCast (Atrium medical)	Various sizes/lengths
Endostaples		
Heli-FX EndoAnchor	Medtronic	
Microcatheter		
ProGreat 2.4	Terumo	2.4F
Embolic agents		
Ethylene vinyl alcohol (EVOH) copolymer (onyx)	Medtronic	34L LES
Coils (e.g., POD and packing coils)	Penumbra	Various sizes/lengths
Amplatzer Vascular plug	Abbott	Various sizes/lengths and shapes

Endoleak Features

Volume of the endoleak track: Small type Ia endoleaks can be successfully embolized, which is a good option especially in older, co-morbid patients. High-flow endoleaks with big gaps between the stent graft and aortic wall, especially with additional stent graft migration, require sufficient proximal extension.

Institution's Experience

Experience and familiarity with the different techniques influence the selection of treatment, especially when it comes to more complex procedures including BEVAR, FEVAR, or even physician-modified FEVAR and complex open aortic surgery.

Perioperative Care and Surveillance

In most cases, patients are surveilled for a minimum of 6 hours and are kept on bed rest for the same period of time. Afterwards, mainly the access (percutaneous versus cutdown) dictates the hospital stay. A postoperative CT-angiography is recommended before discharge to demonstrate complete endoleak occlusion and patent target vessels. Any stent compression, stent fracture, mispositioning, etc. should be corrected. If bridging stents are used, patients are placed on dual antiplatelet therapy (aspirin and clopidogrel) for 6 months. All other patients receive single antiplatelet therapy with aspirin.

Follow-up is recommended at 3 or 6 months and 12 months and annually thereafter.

Prevention of Type Ia Endoleak after Endovascular Procedures

With the long-term data after EVAR and the relevant rates of type Ia endoleaks, the authors emphasize the importance of a regular follow-up scheme. Standard EVAR should primarily be avoided in patients with a compromised proximal sealing zone. In these cases, prophylactic treatment with endostaples or extension of the proximal sealing zone with custom-made FBEVAR should be the first choice of treatment, especially in younger patients.

Case Presentation

Continued from page 589

First (emergent) reintervention:

Under general anesthesia a percutaneous brachial access was established using ultrasound guided micropuncture. A 90 cm 6F sheath was placed in the aorta. Heparinization was given intravenously (100 µg per kilogram bodyweight). The endoleak was cannulated with a 6F multipurpose catheter. Via a 2.7 µm ProGreat Microcatheter (Terumo), multiple 60 cm packing coils (Penumbra) were placed between the previous FEVAR and the visceral aortic wall (Fig. 56.2). Afterwards, 6 mL Onyx 34 L were slowly infused into the endoleak under fluoroscopic control. Control angiography demonstrated only minimal flow through the endoleak, which was accepted under full heparinization.

Clinical course after first procedure:

The control CT angiography still demonstrated minimal flow through the endoleak in the venous phase. However, the patient was completely asymptomatic after the embolization. Therefore, an elective second procedure consisting of a custom-made BEVAR with inner branches was planned to definitively treat the endoleak after a few weeks.

Second (planned) reintervention (8 weeks afterwards)

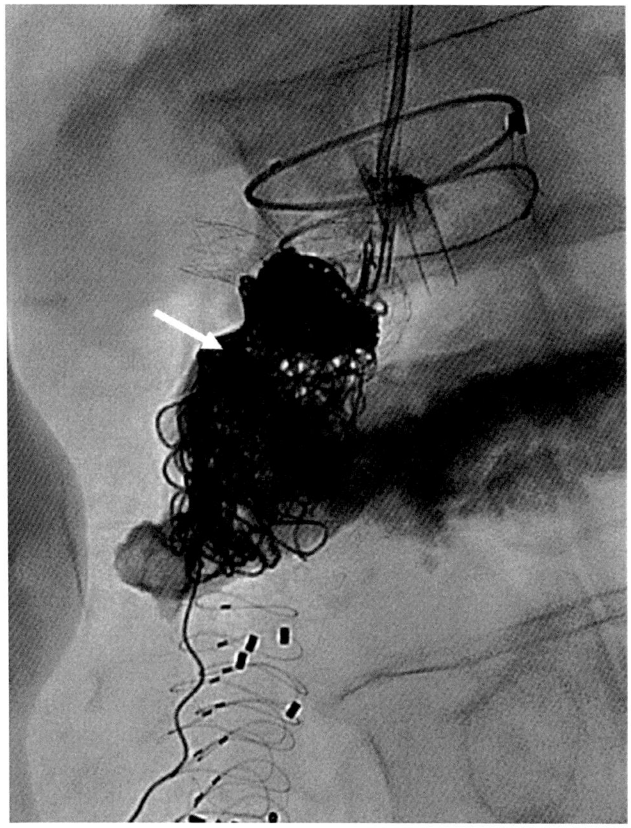

Fig. 56.2 Intraoperative angiography image over transbrachial 6F-sheath, a multipurpose catheter, and a microcatheter in position of the endoleak between the old fenestrated stent graft and the aortic wall. The white arrow shows embolic material (coils and liquids)

Under general anesthesia, bilateral percutaneous arterial femoral access and axillary access (cutdown) was established. A 70 cm 8F sheath was placed trans-axillary and a 12F and an 8F sheath were placed femoral. The custom-made 4x BEVAR custom-made device was put in place, an angiography was performed, and the prosthesis was released. The cannulation of the target vessels was complex due to the previous stents from "he FEVAR which protruded significantly in the old endograft. Two stents pointed downward and were difficult to cannulate. After cannulation from femoral and lifting the stents with a balloon and a long sheath (Fig. 56.3), it was possible to cannulate all the stents and the coeliac artery from above and bridging stents were placed trans-axillary (Bentley BeGraft plus). Control angiography demonstrated perfusion of all renovisceral vessels and no signs of an endoleak (Fig. 56.4).

Case Discussion

This case demonstrates a staged approach. Embolization can help to move patients with large, symptomatic type Ia endoleaks out of the acute phase. Then, a custom-made approach with either FEVAR or BEVAR can be planned some weeks later as soon as the device has been manufactured.

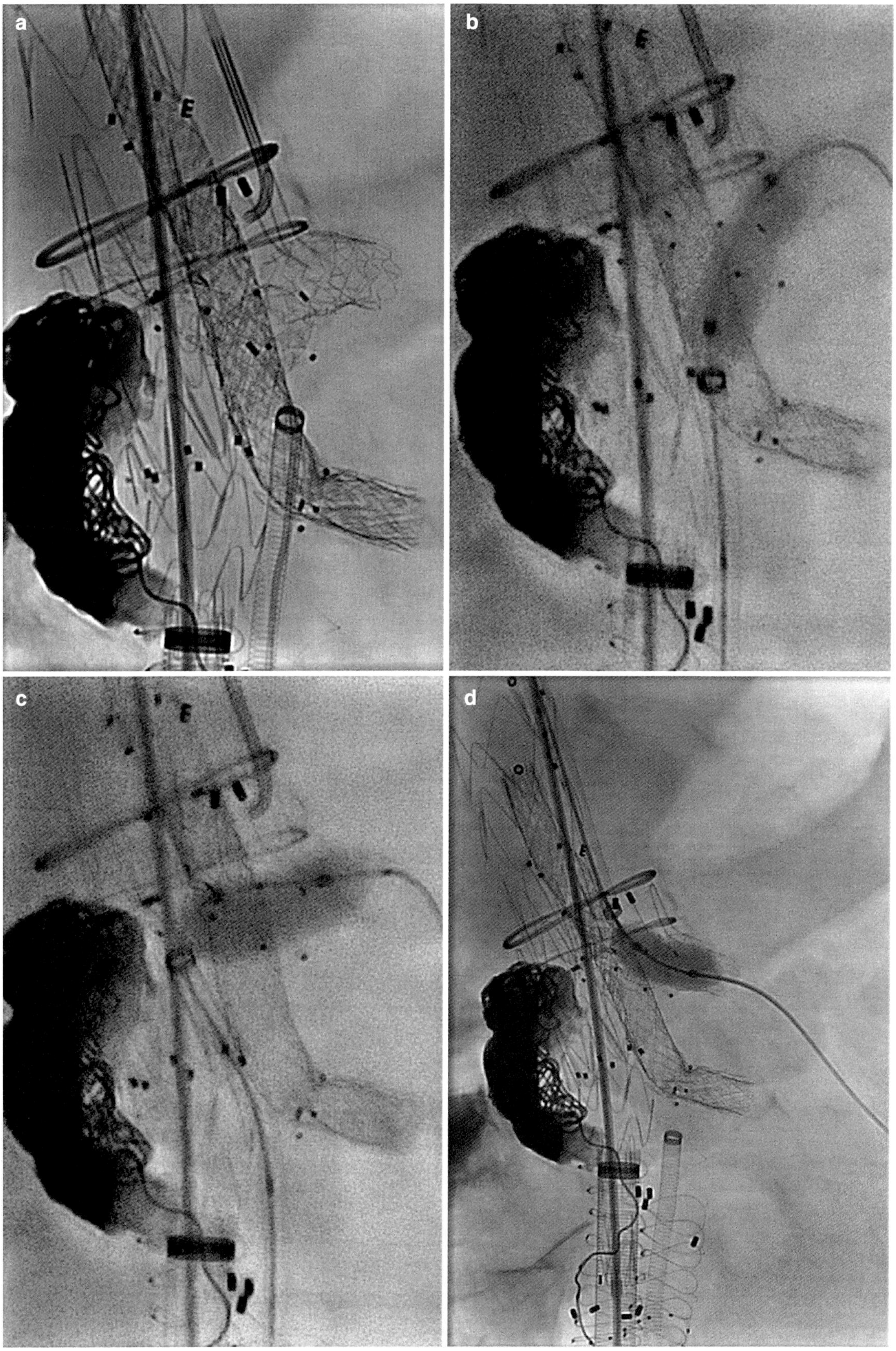

Fig. 56.3 (**a**) Downward-oriented superior mesenteric artery (SMA) bridging stent due to FEVAR graft migration. (**b**) Catheterization of SMA bridging stent from femoral with a PTA balloon over 8F sheath. (**c**) Lifting of SMA bridging stent with upward orientation. (**d**) Transbrachial catheterization of SMA branch and SMA bridging stent

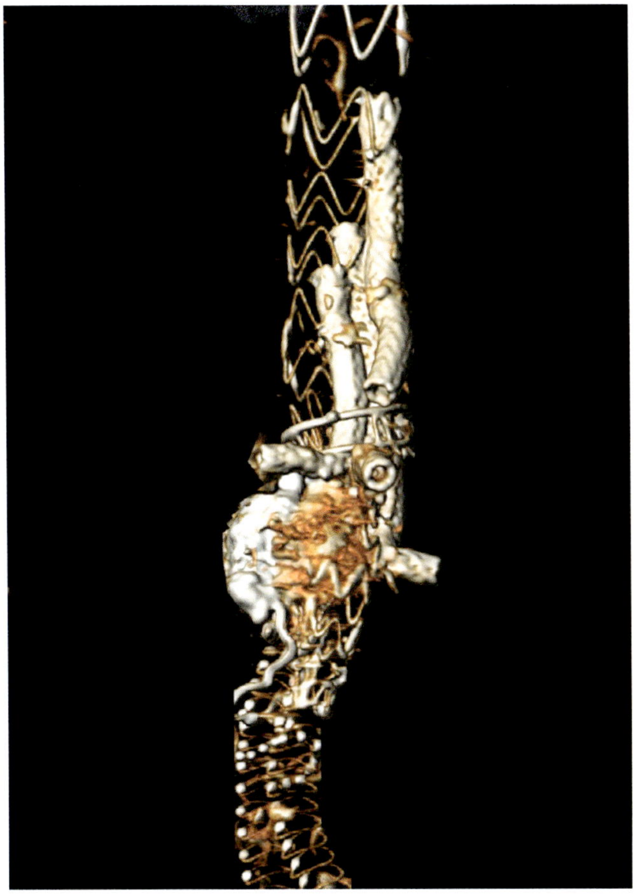

Fig. 56.4 Postoperative 3D volume rendering of the CTA showing BEVAR in FEVAR

References

1. United Kingdom EVAR Trial Investigators, Greenhalgh RM, Brown LC, Powell JT, Thompson SG, Epstein D, Sculpher MJ. Endovascular versus open repair of abdominal aortic aneurysm. N Engl J Med. 2010;362(20):1863–71. https://doi.org/10.1056/NEJMoa0909305. Epub 2010 Apr 11.
2. Harris PL, Vallabhaneni SR, Desgranges P, et al. Incidence and risk factors of late rupture, conversion, and death after endovascular repair of infrarenal aortic aneurysms: the EUROSTAR experience. European Collaborators on Stent/graft techniques for aortic aneurysm repair. J Vasc Surg. 2000;32:739–49.
3. AbuRahma AF, et al. Clinical outcomes for hostile versus favorable aortic neck anatomy in endovascular aortic aneurysm repair using modular devices. J Vasc Surg. 2011;54:13–21.
4. O'Donnell TF, McElroy IE, Mohebali J, Boitano LT, Lamuraglia GM, Kwolek CJ, Conrad MF. Late type 1A endoleaks: associated factors, prognosis and management strategies. Ann Vasc Surg. 2022;80:273–82. https://doi.org/10.1016/j.avsg.2021.08.057. Epub 2021 Nov 6.
5. Millen AM, Osman K, Antoniou GA, et al. United Kingdom outcomes of persistent intraoperative type Ia endoleak after standard endovascular aneurysm repair. J Vasc Surg. 2015;61:1185–91.
6. Tadros RO, Faries PL, Ellozy SH, et al. The impact of stent graft evolution on the results of endovascular abdominal aortic aneurysm repair. J Vasc Surg. 2014;59:1518–27.
7. Faries PL, Cadot H, Agarwal G, Kent KC, Hollier LH, Marin ML. Management of endoleak after endovascular aneurysm repair: cuffs, coils, and conversion. J Vasc Surg. 2003;37(6):1155–61. https://doi.org/10.1016/s0741-5214(03)00084-3.
8. Perini P, Bianchini Massoni C, Mariani E, Ucci A, Fanelli M, Azzarone M, Freyrie A. Systematic review and meta-analysis of the outcome of different treatments for type 1a Endoleak after EVAR. Ann Vasc Surg. 2019;60:435–446.e1. https://doi.org/10.1016/j.avsg.2019.03.032. Epub 2019 Jun 12.
9. Ronchey S, Fazzini S, Scali S, et al. Collected transatlantic experience from the PERICLES registry: use of chimney grafts to treat post-EVAR type Ia endoleaks shows good midterm results. J Endovasc Ther. 2018;25(4):492–8.
10. Moulakakis KG, Mylonas SN, Avgerinos E, et al. The chimney graft technique for preserving visceral vessels during endovascular treatment of aortic pathologies. J Vasc Surg. 2012;55:1497–503.
11. Ameli-Renani S, Pavlidis V, Morgan RA. Early and midterm outcomes after transcatheter embolization of type I endoleaks in 25 patients. J Vasc Surg. 2017;65(2):346–55. https://doi.org/10.1016/j.jvs.2016.06.101.
12. Jordan WD Jr, Mehta M, Ouriel K, Arko FR, Varnagy D, Joye J, Moore WM Jr, de Vries JP. One-year results of the ANCHOR trial of EndoAnchors for the prevention and treatment of aortic neck complications after endovascular aneurysm repair. Vascular. 2016;24(2):177–86. https://doi.org/10.1177/1708538115590727. Epub 2015 Jun 10.
13. Doumenc B, Mesnard T, Patterson BO, Azzaoui R, De Préville A, Haulon S, Sobocinski J. Management of type IA endoleak after EVAR by explantation or custom made fenestrated endovascular aortic aneurysm repair. Eur J Vasc Endovasc Surg 2021;61(4):571–578. doi: https://doi.org/10.1016/j.ejvs.2020.10.033. Epub 2021 Jan 4.
14. Martin Z, Greenberg RK, Mastracci TM, Eagleton MJ, O'Callaghan A, Bena J. Late rescue of proximal endograft failure using fenestrated and branched devices. J Vasc Surg. 2014;59(6):1479–87. https://doi.org/10.1016/j.jvs.2013.12.028. Epub 2014 Jan 29.
15. Spanos K, Kölbel T, Heidemann F, Debus ES, Rohlffs F, Tsilimparis N. Early and mid-term durability of surgeon-modified and custom-made fenestrated devices for the treatment of complex aortic pathology. Ann Vasc Surg. 2022;83:212–21. https://doi.org/10.1016/j.avsg.2021.12.006. Epub 2021 Dec 22.

57 In Situ Fenestration of Endovascular Grafts

Vasileios Tzilalis, Natasha Hasemaki, and Athanasios Katsargyris

Case Presentation

A 90-year-old doctor presented with a type V endoleak. Six years earlier he has been treated with a custom branched graft for a type II thoracoabdominal aneurysm and later on with successful translumbar embolization of a type II endoleak.

Due to continuous aneurysm sac expansion in the postoperative course, the lack of any detectible source of endoleak, and the minor enlargement of his common iliac arteries, we suspected undetectable type Ib endoleak.

Continued at page 601

Introduction

For many years, the inadequate proximal neck length of an abdominal aortic aneurysm (AAA) was considered a major limitation of endovascular aneurysm repair (EVAR). The creation of fenestrations for preservation of renal arteries revolutionized endovascular aortic surgery, with the first case reported in 1999 [1]. Thereafter, customized fenestrated stent grafts became a safe and effective alternative treatment for high-risk patients with pararenal or short-neck infrarenal AAA [2, 3]. Additionally, the creation of branches covered almost all chronic and acute aortic syndromes in high-risk patients, making the endovascular treatment of thoracoabdominal aneurysms (TAAA) feasible. The evolution of the "off- the-shelf" devices complements the complex endovascular quiver, offering an additional option for some acute and urgent cases of high-risk patients with TAAA or pararenal/ short-neck infrarenal AAA.

The dissemination of fenestrated endovascular aneurysm repair (FEVAR) created an unmet need for alternate endovascular solutions for high-risk patients with acute/ urgent pararenal or short-neck infrarenal AAA; custom-made devices (CMD) require several weeks for manufacturing and delivery. Apart from clinical urgency, there is a significant portion of elective cases for which neither a customized fenestrated/branched stent graft (F/BEVAR) nor an "off-the-shelf" branched graft is a realistic solution. In some countries there is restricted access to CMD, while the "off-the-shelf" devices are not commercially available. Uflacker et al. [4] in 2006 described the back-table modification of commercially available AAA endovascular devices, to treat pararenal AAA, concluding that this technique is feasible "when the interventionalist does not have access to the commercially available devices or the waiting time is too prolonged to accommodate the patient's clinical situation."

In situ fenestration (ISF) refers to creation of a fenestration at a standard commercially available stent graft at the time of the procedure in order to preserve a target vessel. The procedure is demanding and requires a lot of prerequisites for the team and the hospital. The creation of an opening in the wall of a commercially available endograft on the back table or while deployed in the patient is defined as ISF ex vivo or ISF in vivo respectively. ISF in vivo may be retrograde or anterograde, with the latter being technically more demanding.

Several questions arise with regard to ISF. Is it a durable procedure? What are the specific indications? Is it only a bailout procedure or is it a real alternative of CMDs? The present chapter aims to summarize clinical indications, techniques, and available literature data with regard to ISF.

V. Tzilalis (✉)
401 General Military Hospital, Athens, Greece

N. Hasemaki
Department of Vascular Surgery, Laiko Hospital, Athens, Greece

A. Katsargyris
Athens University Medical School, Athens, Greece

G. Geroulakos et al. (eds.), *Mastering Endovascular Techniques*, https://doi.org/10.1007/978-3-031-42735-0_57

Historical Aspects

In 1996, Park et al. reported for the first time two cases of physician-modified stent grafts with ex vivo ISF, in order to incorporate an inferior mesentery artery and a lower renal artery, respectively [5]. In 2004, McWilliams reported the first successful case of ISF of the left subclavian artery (LSA) in a patient with a thoracic aortic aneurysm [6]. The thoracic stent graft was inserted and partially deployed distal to the left carotid artery, covering the LSA. The stent graft was then punctured via the reversed end of a 0.018″ guidewire, followed by cutting balloons to gradually increase the diameter of the ISF. The first ever use of in situ laser-assisted fenestration was reported in 2009, by Murphy et al., in a case of traumatic isthmus rupture in a 26-year-old patient. The fenestration in the thoracic stent graft was created with a laser ablation catheter placed right against the stent graft via brachial access [7]. In the following years, ISF became more prevalent, surpassing the use in urgent and emergent cases. The initial experience involved primarily the Zenith endograft (Cook Medical) and was later extended with modifications to further commercially available endovascular grafts [8, 9].

Definitions of ISF ex vivo include the physician-modified endovascular grafts (PMEGs) and surgeon-modified fenestrated branched stent grafts (sm-FBSGs) as proposed by different authors [8, 10].

Indications

Fenestrated and branched endografts have been widely used as an effective alternative for the treatment of patients with complex aortic aneurysms who are poor candidates for open surgical repair. However, construction of these devices can take as long as 6–12 weeks. In addition, in many countries there is restricted access to CMD. “Off-the-shelf” devices have been developed to overcome these limitations; however, they are still not widely available and their use is also limited by anatomical criteria. In addition, “off-the-shelf” devices require coverage of a longer segment of the thoracic aorta, raising concerns about the postoperative risk of spinal cord ischemia (SCI). The use of ISF has become more frequent in recent years, mainly for urgent and emergent cases, but also in selected elective cases. Decision-making on the use of ISF must involve thorough evaluation of each patient’s anatomic characteristics and risk factors, as well as valid informed consent from the patient regarding all available treatment options, including a referral to centers with access to CMDs.

One of the most common clinical scenarios of in vivo ISF is the retrograde revascularization of the LSA, in order to establish an adequate proximal sealing zone during thoracic endovascular aortic repair (TEVAR). The most reported technique to revascularize the LSA is in vivo ISF in a retrograde access, through brachial or axillary access. Fenestration through the stent graft is created by hollow needle (Table 57.1), or by laser technology or radiofrequency (RF) [7, 11] (Table 57.2). In situ fenestration has been applied even in total endovascular arch repair, in symptomatic high-risk patients with acute aortic syndromes where “off-the-shelf devices” are not available or the anatomy is not suitable.

At the thoracoabdominal level, this retrograde approach can be challenging and technically demanding. Coeliac, mesenteric, or renal fenestrations require an antegrade approach to create fenestrations and access the vessel lumen. Antegrade in vivo ISF is applied either to rescue inadvertently over-stented abdominal visceral vessels (bailout operation) or as a planned procedure to create a fenestrated graft in vivo, for acute cases for which a customized graft is not available [12]. An alternative to antegrade in vivo ISF is ex vivo ISF for the pararenal aorta with the so-called physician-modified endovascular Grafts (PMEGs) and surgeon-modified fenestrated branched stent grafts (sm-FBSGs).

The decision to treat a patient with complex aortic disease with ISF will always remain at the discretion of the treatment provider. While outcomes published to date have been initially favorable, larger series with prospective data collection and longer-term follow-up are still required.

Table 57.1 Endovascular toolbox kit: Part of endovascular toolkit for **in vivo ISF (needles)**: Different types of **needles** that have been used to create in vivo ISF

Transeptal needles
BeBack needle
Outback needle
Modified TIPS sheath needle
TIPS needle—Haskal—Colapinto
Ross modified Colapinto needle
Percutaneous Transhepatic Cholangiography (PTC) needle
Endoscopic FNA needle

Table 57.2 Endovascular toolbox kit: Part of endovascular toolkit for **in vivo ISF (energy based)**: **Energy-based systems** to create in ISF

Laser Turbo-Elite, Spectranetics (now part of Philips)
Laser Dornier
Laser Surgilase
RadioFrequency (RF) PowerWire RF Guidewire (Baylis)

Technical Considerations

In Vivo

Different methods have been used to create an in vivo ISF (Table 57.1). A modified transjugular intrahepatic portosystemic shunt (TIPS) was used and described in 2010 by Manning et al. [13] during TEVAR to preserve the LSA. According to the authors, the procedural steps are retrograde puncture of the stent graft with the needle, introduction of a thin wire, exchange for a stiff wire, dilation with 6–8 mm balloons, and bridging stent-graft implantation (Table 57.3). Different types of balloons have been used to dilate the initial hole to a larger fenestration. Ruthrauff et al. [14] suggested that the use of cutting balloons can facilitate the enlargement of the fenestration in the case of the first generation of the Cook Zenith endograft. However, according to Riga et al. [15], cutting balloons should be avoided since they can create fabric tears, resulting in durability issues and endoleak. The authors also highlighted the proper angle of fenestration, which should be at 90 degrees. This is a crucial step of the procedure, to facilitate the best sealing of the bridging stent graft to the ISF and prevent future endoleaks and reinterventions. The angle of puncturing is of outmost importance, aiming a perpendicular puncture to the aortic endograft. Use of steerable sheaths may facilitate an optimal puncture angle. The most reported needles for the creation of in vivo ISF are the Echotip ultra endoscopic FNA needle (Cook Inc.) [16], the transeptal needle [17], the outback needle (Cordis) [18], the Ross modified Colapinto needle [19], and the BeBack needle (Bentley) (Table 57.1).

Energy-based technology for the creation of in vivo ISF includes laser and RF (Table 57.2). According to the current literature, laser technology is the best way to create ISF, but it is not commercially available worldwide. Laser technology can be used in a retrograde or anterograde access [12]. Steps of the procedure follow the general principles of ISF: energy tip introduction preferably through a steerable sheath (Table 57.4) to achieve perpendicular contact to the graft fabric, energy application depending on the size of the tip and the graft, thin wire introduction, and exchange for a stiff wire.

Table 57.3 Endovascular toolbox kit: Part of endovascular toolkit for bridging stent grafts **in vivo ISF: Bridging stent grafts** for the ISF

iCast (Atrium)
Advanta V12 (Atrium),
Exluder iliac limbs (W.L. Gore)
Fluency plus (Bard)
Viabahn (W.L. Gore)
Visi-Pro (EV3)
BeGraft (Bentley)

Ex Vivo

Ex vivo ISF or PMEGs consists of the physician bench modification of commercially available EVAR stent grafts. Herein, meticulous preoperative planning of the computed tomography angiography (CTA), as well as previous experience of FEVAR and target vessels' clock position, is of outmost importance.

The stent-graft modification takes place in the operating theater under sterile conditions, before general anesthesia. The stent graft is partially or fully deployed, and marking of the fenestrations is done according to the preoperative plan. Subsequently, the support rings and strut sutures can be carefully removed, with care taken not to damage the graft fabric. The fenestrations are made via hand-held diathermy or laser technology, in the premarked positions, and are then reinforced with radio-opaque marker (i.e., snare). The fenestration diameter is according to the target vessel's diameter, and the fenestrations are created between stent struts as far as possible. The radio-opaque marker has a dual objective: (1) to mark the fenestration and (2) to reinforce it into the delivery system by providing dimensional stability. If the preoperative planning requires the addition of diameter-reducing ties (in case of narrow segment of visceral aorta), prolene sutures are used to create loops around the graft, tethering them with a release wire freed from the delivery system with a scalpel and reinserted once the loops of prolene are in place [20]. The stent graft is then re-sheathed with the aid of silk suture or nylon tape to wrap and compress the stent graft (Table 57.5, Fig. 57.2a). The delivery system is then inserted and positioned across the visceral segment in order to align

Table 57.4 Endovascular toolbox kit: Part of endovascular toolkit for **in vivo ISF: Steerable sheaths** and **ancillary devices**

Aptus Tour Guide (Medtronic)
Destino (Oscor)
Fustar (LifeTech)
"Quick fenestrater" (Innomed)
Wires, catheters, balloons, loops, even plugs in case of failure, Necessary for all other branched/fenestrated endovascular procedures

Table 57.5 Endovascular toolbox kit: Part of endovascular toolkit for **ex vivo ISF** (case presented)

Sterile marker
Ophthalmic cautery or other stainless steel thermal cautery instruments [28]
Reinforcing (loop) wire
6/0 or 5/0 prolene or PTFE sutures according to the graft material to be fenestrated
Re-sheathing with the use of 1–0 silk sutures to collapse the stents of the endograft or large silastic vessel loop

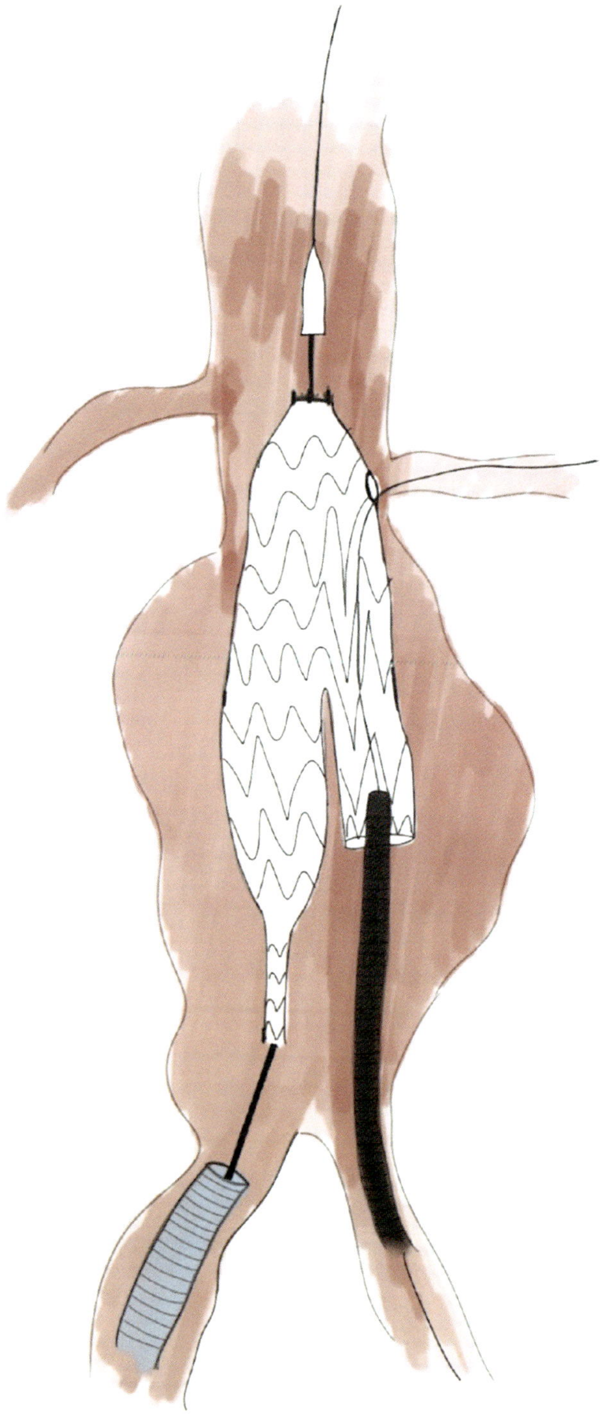

Fig. 57.1 Ex vivo ISF of Gore Excluder C3 stent graft for the treatment of short-neck infrarenal AAA

the fenestrations across target vessels' orifices. Once the stent graft is deployed, the fenestrations are cannulated and bridging stent grafts (Table 57.3) are deployed in each target vessel. Pre-cannulation and stenting of the target vessels with bare balloon-expandable stents is suggested, in order to facilitate accurate stent-graft orientation and target vessels' cannulation.

Until recently, the Cook Zenith stent graft (Cook Medical Inc., Bloomington, Ind), has been the only device available for physician-modified fenestration mainly due to its constraining wire that allows physicians to re-sheath the device after modifications. The Gore Excluder C3 (W. L. Gore & Associates, Inc., Flagstaff, Ariz) endograft (Fig. 57.1) was more recently proposed as an alternative option for ex vivo ISF, for its ability to partially deploy and subsequently re-sheath. This technique was first described by Coulter et al. in 2014 [9]. Several other grafts were used for ISF. These can be carefully partially or fully deployed and re-sheathed, and examples are the Anaconda stent graft [21], Medtronic Valiant, Relay thoracic, and Ankura thoracic [22]. In their publication [22], Zhao J and colleagues describe the modification of a Relay thoracic graft to an ISF ex vivo pre-cannulated (PMSG) one.

Outcomes

Data from current literature show that ISF for the treatment of complex aortic aneurysms, performed in expert centers, provides acceptable mortality and morbidity rates, even in emergency scenarios. However, long-term results remain unknown and technical aspects of ISF lack standardization.

In 2016, Georgiadis et al. [23] reported the short-term results of 308 patients with complex abdominal AAA, treated with "off-the-shelf" F/BEVAR stent grafts and physician-modified stent grafts (PMSGs). Major adverse events occurred in 24 (12.8%) PMSG patients and in nine (7.4%) "off-the-shelf" patients, while treatment success was observed in 171/187 (91.4%) PMSG patients and in 115/121 (95%) "off-the-shelf" patients. Overall branch patency was recorded in 443/458 (96.7%) and in 468/478 (97.9%) of target vessels in the PMSG and "off-the-shelf" patients group, respectively.

In 2018, O'Donnell et al. [24] reported the results of 1396 complex aortic aneurysm endovascular repairs. The authors reported a similar perioperative mortality rate between CMDs (3.4%), PMSGs (6.1%), and chimney/snorkel (6.1%; p 1/4 0.13). In a retrospective matched control study comparing ISF with CMD, the latter had a significantly reduced rate of 12-month reinterventions (13 vs 37%) as well as reduced operative and fluoroscopy time [25].

Recently, Canonge et al. [26] reported an updated review of the use of ISF in all aortic pathologies, considering technical details and clinical results of ISF. Out of 711 patients, 248 were treated on an acute/urgent setting. The Valiant (Medtronic, Minneapolis, MN, USA) represented 65% of the aortic arch stent grafts used for the ISF and the Zenith platform (Cook Medical, Bloomington, IN, USA) represented 51% of the abdominal aortic stent grafts. A snare was used to reinforce the fenestration in 458 ISF (66%) and a cautery device cut the fenestration in 484 (75%) ISF. Technical success ranged from 87.5% to 100%, and 30-day mortality from 0% to 8%. Primary branch patency ranged from 96.3% to 100% at 12-month follow-up. The authors concluded that, even though ISF for aortic pathologies seems effective when validated treatments are not available, heterogeneity in the technical details makes it difficult to demonstrate a strong connection between a specific creation technique and the clinical results.

According to Oderich et al. [27], the number of cases of ex vivo ISF during the past few years has decreased. The authors reported a decline from 100% between 2007 and 2010 to 4% between 2014 and 2016 ($p < 0.001$). They concluded that the FDA approval in 2012 of the first fenestrated CMD in the USA and the improvement of the "off-the-shelf" devices reduced the number of ISF procedures.

Conclusions

In situ fenestration represents an additional tool that can be considered to facilitate total endovascular repair of complex aortic pathologies involving supra-aortic or visceral vessels. As a bailout procedure, it can restore patency of an unintentionally covered vital artery ostium. Nevertheless, there is still only a limited amount of evidence with regard to initial technical success rates, but also with regard to mid- and long-term durability of these techniques. In situ fenestration should therefore currently be reserved mainly for clinical scenarios and anatomies that cannot be treated either with CMDs or "off-the shelf" devices for several reasons. These may include acute patients unable to wait for manufacturing of a CMD and/or situations in which the use of an off-the-shelf device is not possible or would result in serious compromises.

Case Presentation

Continued from page 597

In two stages, starting percutaneously from the patient's right common femoral artery, we managed to enter the sac between the outer surface of the right iliac extension graft and the iliac artery and again embolize the sac with glue. Then, coming out, we embolized his right hypogastric artery. Finally, a right iliac extension graft was placed to his right external iliac artery.

The patient's left hypogastric artery was planned to be saved, while the anatomy was inappropriate for any iliac side branch solution. Therefore, as a bailout and a logistically reasonable procedure, we decided to use an in situ fenestration (ISF) ex vivo procedure on a ZSLE stainless steel Zenith (Cook Medical, Bloomington, IN, USA) extension graft: extending to his external iliac artery with an ISF just opposite to his hypogastric artery (the anatomy was favorable for this solution).

In a sterile environment, a back table in the operating room, the planned ZSLE stainless steel Zenith iliac extension graft was expanded on the table. Using a sterile marker, a 6 mm diameter fenestration was marked on the graft and was cut with an ophthalmic cautery. Measurements and positioning were meticulously estimated preoperatively by studying the computed tomography angiography (CTA) of the patient. The fenestration was reinforced by using part of the radio-opaque Cook snare loop wire, circumferentially, sutured and secured by continuous 6-0 prolene suture. Using a 1-0 silk suture to collapse the Z stents and then sequentially removing them, the ex vivo ISF graft was then carefully re-sheathed with special care not to harm the tip of the sheath (Fig. 57.2a). ISF endograft was successfully deployed leaving the left and only hypogastric artery patent (Fig. 57.2b)

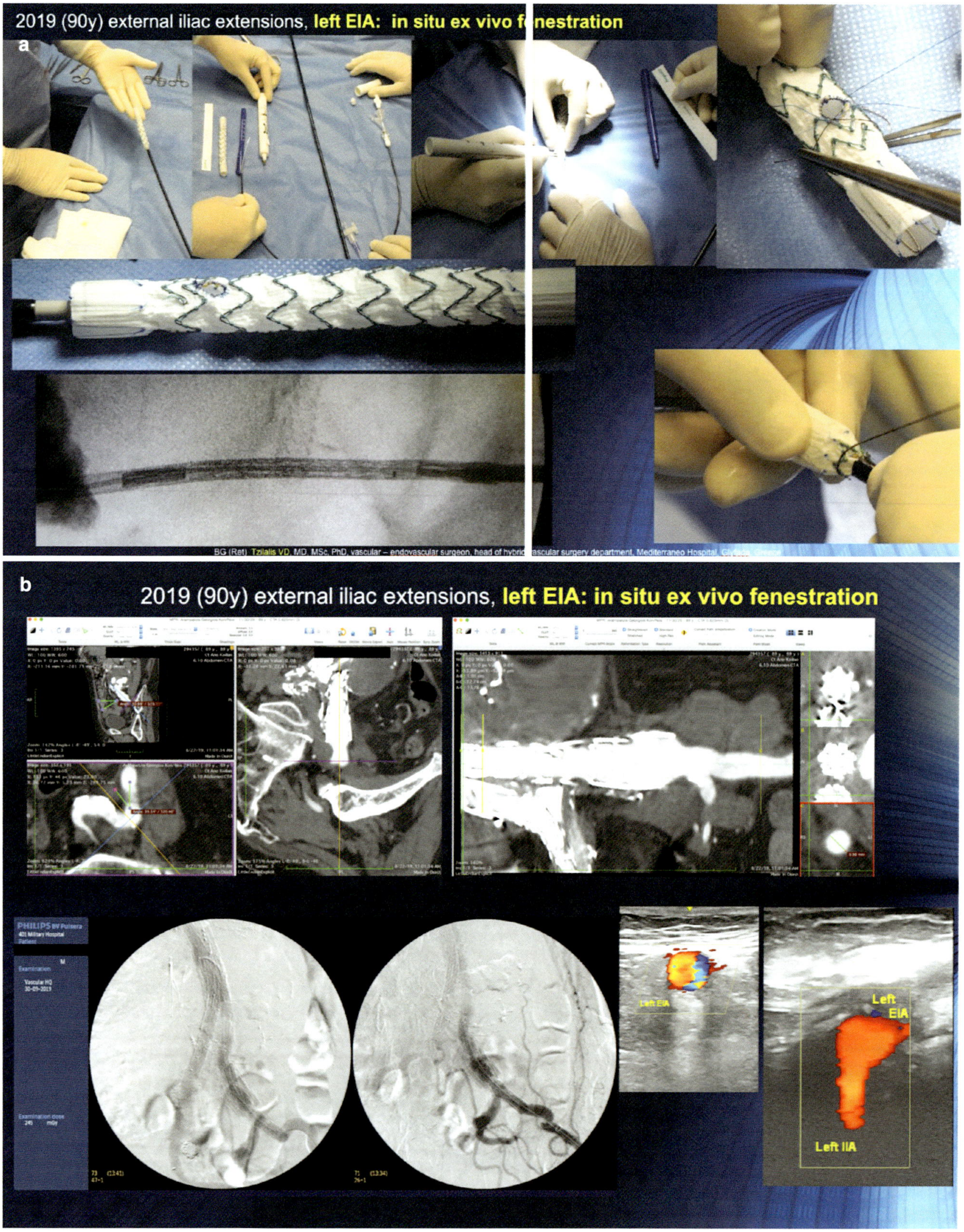

Fig. 57.2 Ex vivo *ISF on a ZSLE stainless steel Zenith (Cook) iliac extension graft, to preserve the only patent hypogastric artery in a type II thoracoabdominal aortic aneurysm case that was previously treated by BEVAR (401 General Army Hospital, Sept. 30, 2019 Tzilalis* et al.*).* (**a**) Unsheathing the ZSLE stainless steel Zenith (Cook) left iliac extension, sterile marking, and fenestration with the ophthalmic cautery, reinforcing (loop wire and 6/0 prolene sutures) and re-sheathing with the use of 1-0 silk sutures to collapse the Z stents. (**b**) **Images on the top**: CTA preoperative study. **Lower part images**: Intraoperatively images of the patent left hypogastric artery (ISF), as well as postoperative surveillance with color duplex scan

References

1. Browne TF, Hartley D, Purchas S, Rosenberg M, Van SG, Lawrence-Brown M. A fenestrated covered suprarenal aortic stent. Eur J Vasc Endovasc Surg. 1999;18:445–9.
2. Riambau V, Böckler D, Brunkwall J, Cao P, Chiesa R, Coppi G, et al. Editor's choice - management of descending thoracic aorta diseases: clinical practice guidelines of the European Society for Vascular Surgery (ESVS). Eur J Vasc Endovasc Surg. 2017;53:4e52.
3. Wanhainen A, Verzini F, Van Herzeele I, Allaire E, Bown M, Cohnert T, et al. Editor's Choice e European Society for Vascular Surgery (ESVS) 2019 clinical practice guidelines on the management of abdominal aorto-iliac artery aneurysms. Eur J Vasc Endovasc Surg. 2019;57:8e93.
4. Uflacker R, Robison JD, Schonholz C, et al. Clinical experience with a customized fenestrated endograft for juxtarenal abdominal aortic aneurysm repair. J Vasc Interv Radiol. 2006;17:1935–42.
5. Park JH, Chung JW, Choo IW, Kim SJ, Lee JY, Han MC. Fenestrated stent-grafts for preserving visceral arterial branches in the treatment of abdominal aortic aneurysms: preliminary experience. J Vasc Interv Radiol. 1996;7(6):819–23.
6. McWilliams RG, Murphy M, Hartley D, Lawrence-Brown MM, Harris PL. In situ stent-graft fenestration to preserve the left subclavian artery. J Endovasc Ther. 2004;11(2):170–4.
7. Murphy EH, Dimaio JM, Dean W, Jessen ME, Arko FR. Endovascular repair of acute traumatic thoracic aortic transection with laser-assisted in-situ fenestration of a stent-graft covering the left subclavian artery. J Endovasc Ther. 2009;16(4):457–63.
8. Starnes BW. Physician-modified endovascular grafts for the treatment of elective, symptomatic, or ruptured juxtarenal aortic aneurysms. J Vasc Surg. 2012;56:601–7.
9. Coulter AH, Tan TW, Zhang WW. Physician modification of Gore C3 excluder endograft for treatment of abdominal aortic aneurysms anatomically unsuitable for conventional endovascular repair. J Vasc Surg. 2014;59:1739–43.
10. Ricotta JJ 2nd, Tsilimparis N. Surgeon-modified fenestrated-branched stent grafts to treat emergently ruptured and symptomatic complex aortic aneurysms in high-risk patients. J Vasc Surg. 2012;56(6):1535–42.
11. Bradshaw RJ, Ahanchi SS, Powell O, Larion S, Brandt C, Soult MC, et al. Left subclavian artery revascularization in zone 2 thoracic endovascular aortic repair is associated with lower stroke risk across all aortic diseases. J Vasc Surg. 2017;65:1270–9.
12. Le Houérou T, Fabre D, Alonso CG, Brenot P, Bourkaib R, Angel C, Amsallem M, Haulon S. In situ Antegrade laser fenestrations during endovascular aortic repair. Eur J Vasc Endovasc Surg. 2018;56(3):356–62. https://doi.org/10.1016/j.ejvs.2018.05.014. Epub 2018 Jun 30.
13. Manning BJ, Ivancev K, Harris PL. In situ fenestration in the aortic arch. J Vasc Surg. 2010;52(2):491–4. https://doi.org/10.1016/j.jvs.2009.07.088. Epub 2009 Oct 30.
14. Ruthrauff AA, King MW, Soulez G, Tan KT, Crawford SA, Roche-Nagle G, et al. Effects of pulsatile fatigue on in situ antegrade fenestrated polyester stent grafts deployed in a patient-specific phantom model of juxtarenal aortic aneurysm. J Vasc Interv Radiol. 2015;26:1551–8.
15. Riga CV, Bicknell CD, Basra M, Hamady M, Cheshire NJ. In vitro fenestration of aortic stent-grafts: implications of puncture methods for in situ fenestration durability. J Endovasc Ther. 2013;20(4):536–43.
16. Kassavin DS, Constantinopoulos G. Repair of a disconnected stent-graft limb facilitated by in situ fenestration. J Endovasc Ther. 2012;19(3):434–8. https://doi.org/10.1583/11-3574MR.1.
17. Kölbel T. Transseptal access revisited. J Endovasc Ther. 2013;20(2):231–2. https://doi.org/10.1583/1545-1550-20.2.231.
18. Wheatley GH. In situ fenestration of the internal iliac artery as a bailout technique associated with endovascular repair of an abdominal aortic aneurysm: long-term follow-up. J Endovasc Ther. 2012;19(6):716–20. https://doi.org/10.1583/JEVT-12-4038R.1.
19. Coscas R, Glorion M, Javerliat I, Goëau-Brissonniere O, Coggia M. In situ fenestration through the contralateral iliac artery to convert an aortouni-iliac into a bifurcated endograft. J Endovasc Ther. 2015;22(3):421–5. https://doi.org/10.1177/1526602815583492.
20. Oderich GS. Technique of adding a diameter-reducing wire to the modified TX2 fenestrated stent graft. Vascular. 2010;18(6):350–5. https://doi.org/10.2310/6670.2010.00059.
21. Zhu J, Zhao L, Dai X, et al. Fenestrated thoracic endovascular aortic repair using physician modified stent grafts for acute type B aortic dissection with unfavourable landing zone. Eur J Vasc Endovasc Surg. 2018;55(2):170–6. https://doi.org/10.1016/j.ejvs.2017.11.012. Epub 2017 Dec 12.
22. Singh A, Mafeld S, Williams R, McCaslin J. Physician-modified fenestrated endografts for managing the ruptured or symptomatic aortic aneurysm: technique overview and clinical outcomes. Vasc Endovasc Surg. 2018;52(8):607–12. https://doi.org/10.1177/1538574418789023. Epub 2018 Jul 22.
23. Georgiadis GS, van Herwaarden JA, Antoniou GA, Hazenberg CE, Giannoukas AD, Lazarides MK, Moll FL. Systematic review of off-the-shelf or physician-modified fenestrated and branched endografts. J Endovasc Ther. 2016;23(1):98–109. https://doi.org/10.1177/1526602815611887. Epub 2015 Oct 23.
24. O'Donnell TFX, Patel VI, Deery SE, Li C, Swerdlow NJ, Liang P, et al. The state of complex endovascular abdominal aortic aneurysm repairs in the vascular quality initiative. J Vasc Surg. 2019;70:369–80.
25. Dossabhoy SS, Simons JP, Flahive JM, Aiello FA, Sheth P, Arous EJ, Messina LM, Schanzer A. Fenestrated endovascular aortic aneurysm repair using physician-modified endovascular grafts versus company-manufactured devices. J Vasc Surg. 2018;67(6):1673–83.
26. Canonge J, Jayet J, Heim F, Chakfé N, Coggia M, Coscas R, Cochennec F. Comprehensive review of physician modified aortic stent grafts: technical and clinical outcomes. Eur J Vasc Endovasc Surg. 2021;61(4):560–9.
27. Oderich GS, Ribeiro MS, Sandri GA, Tenorio ER, Hofer JM, Mendes BC, et al. Evolution from physician-modified to company-manufactured fenestrated-branched endografts to treat pararenal and thoracoabdominal aortic aneurysms. J Vasc Surg. 2019;70:31–42.e7.
28. Stephen E, Joseph G, Sen I, et al. A novel cautery instrument for on-site fenestration of aortic stent-grafts: a feasibility study of 18 patients. J Endovasc Ther. 2013;20(5):638–46.

Chimneys and Periscopes

58

Konstantinos P. Donas, Gergana T. Taneva, Nizar Abu Bakr, Firouza Dahi, Majid Kazemtash, and Anastasios Psyllas

Case Presentation

An 82-year-old male patient presented with 56 mm pararenal penetrating aortic ulcer (PAU) and a pre-obstructive severely calcified external iliac artery stenosis of the left side, on pre-operative computer tomography angiography (CTA). No infrarenal neck below the right renal artery (RRA) was present (Fig. 58.1). The previous medical history included heavy smoking, hypertension, dyslipidemia, type 2 diabetes mellitus, moderate chronic renal failure, and recent implantation of a bicameral pacemaker. Because of the patient's comorbidities an endovascular approach was chosen. The case was evaluated anatomical outside of instructions for use for fenestrated endografting due to the severe iliac artery stenosis and the overall diameter of 7.5 mm. Due to the risk for rupture because of the morphology (Fig. 58.1), we did not consider delays in manufacturing as in the case of placement of a custom-made device.

The PAU had no infrarenal neck and the right renal artery originated from the aneurysm sac (Fig. 58.1). In the case of double chimney with additional placement of chimney graft in the left renal artery which was upper to the right, a new seal zone of 17 mm would be created.

Under general anesthesia, we exposed first the common femoral artery (CFA) of both sides by surgical means. Due to the severe 70% calcified stenosis and dissection of the left external iliac artery no percutaneous approach was considered. After cannulation of the left iliac access, a balloon angioplasty by the IVL Shockwave M5+ device (Shockwave Inc. USA) of 8 mm diameter over a 7F sheath and a 0.014 Spartacore (Abbott, USA) guide wire was performed. In detail, we performed 10 cycles of 30 pulses each, at 6 atm of

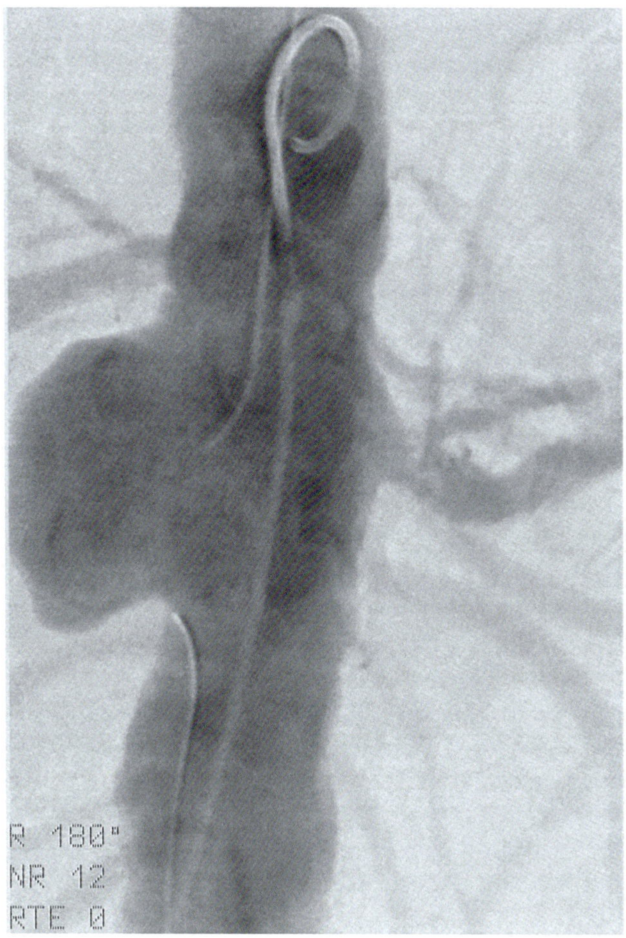

Fig. 58.1 Angiographic evidence of the rupture-threatening pararenal aortic pathology needing urgent treatment

Informed Consent: We confirm that written informed consent for patient information and images to be published was provided by the patient.

K. P. Donas (✉) · G. T. Taneva · N. A. Bakr · F. Dahi
M. Kazemtash · A. Psyllas
Rhein Main Vascular Center, Department of Vascular and Endovascular Surgery, Asklepios Clinics Langen, Wiesbaden and Seligenstadt, Langen, Germany
e-mail: f.dahi@asklepios.com; m.kazemtash@asklepios.com

G. Geroulakos et al. (eds.), *Mastering Endovascular Techniques*, https://doi.org/10.1007/978-3-031-42735-0_58

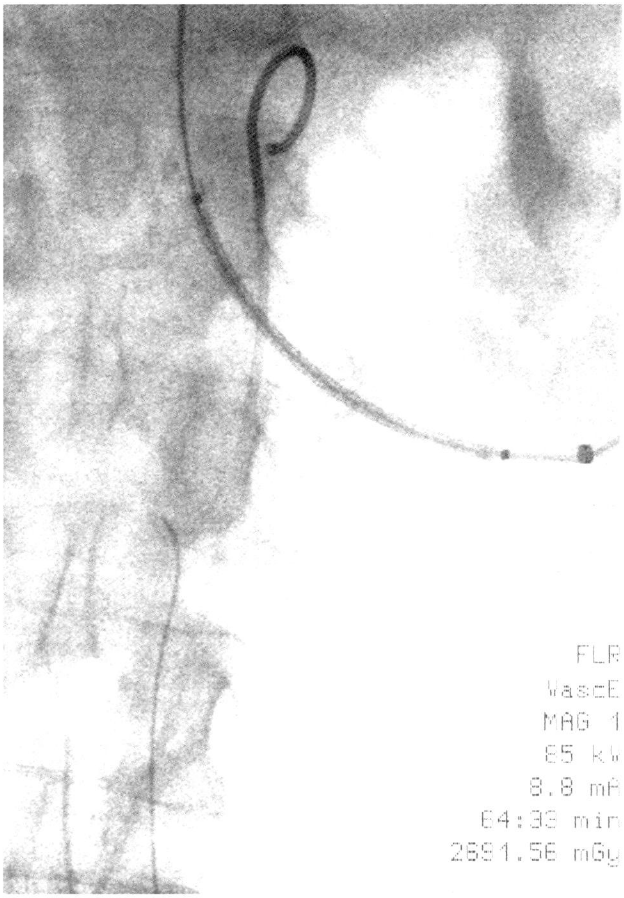

Fig. 58.2 Advance of the sheath over a Rosen wire and the chimney graft in telescope technique in the left renal artery

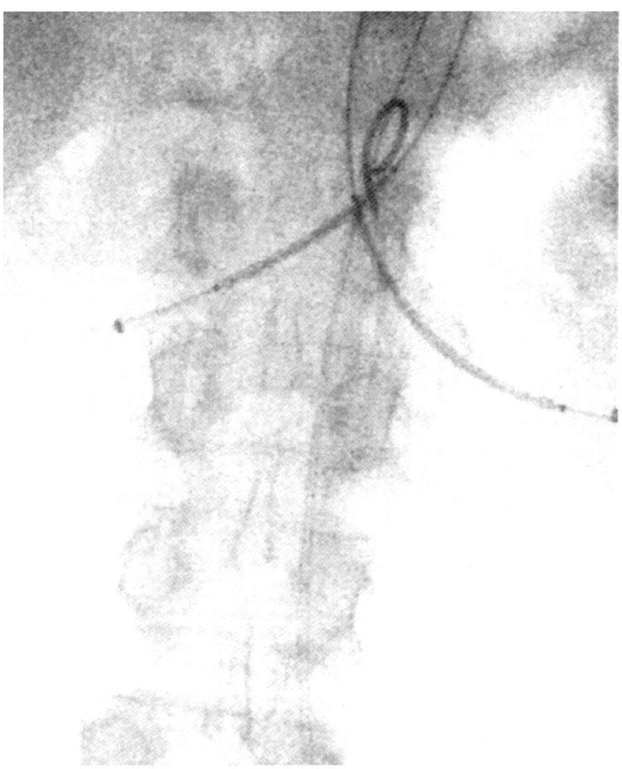

Fig. 58.3 Placement of the sheaths in the renal arteries and advance of the chimney grafts

inflation. Additionally, we used a 10x60mm balloon. In the final angiography, no evidence of dissection or recoil needing additional stent placement was obtained.

The procedure was continued with surgical exposure of the left axillary artery and double punction of the vessel advancing two 5F sheaths in Seldinger technique. We used a 0.035 soft Terumo guide wire and a vertebral catheter over a 6 mm and 90 cm-long Terumo sheath to cannulate each of the renal arteries. A pre-dilation with a 4–40 Armada balloon (Abbott, USA), 50% inside the renal artery and 50% outside the renal artery, was performed in order to advance the sheath over the deflating balloon, safely and securely in the renal arteries. Next, we deployed in RAO 90° projection an aortic extension (ETTF 3232C70, Medtronic, Santa Rosa, USA) in order to visualize as best as possible the orifice of the superior mesenteric artery, excluding the PAU. Finally, a 6 mm in diameter and 58 mm for the lower right and 38 mm for the upper left renal artery BeGraft peripheral (Bentley, USA) chimney graft was advanced and deployed (Figs. 58.2 and 58.3). A crucial aspect of this step of the procedure is the protection of the balloon of the chimney graft with the sheath (Fig. 58.4).

The completion angiography showed excellent conformability of the Endurant device around the chimney grafts without endoleak and patent renal arteries. During the procedure 7500 IE heparin was administrated intravenously.

The postoperative computed tomography angiography confirmed the angiographic result without any type of endoleak, exclusion of the PAU, and patent renal and iliac arteries. The status of the patient is uneventful 6 months postoperatively.

Indications and Limitations

The chimney technique can be indicated clinically in the emergent setting in ruptured or symptomatic cases when no time for manufacturing is available [1]. It can also be indicated for elective treatment [1]. The anatomical features include short infrarenal aortic neck (<2 cm) and hostile neck anatomy when proximal sealing cannot be ensured with standard abdominal devices [2]. These can also be applied in case of high tortuosity and calcification of the iliac vessels, precluding fenestrated technology for the treatment of pararenal aortic pathologies due to the risk of misalignment with the visceral vessels [2]. The chimney technique can also be used in type IA endoleaks caused by migration of previously implanted endograft. A prerequisite to employ ch-EVAR is to create at least 15 mm of proximal sealing zone [3]. In case of need to over-stent more than two target vessels, other complex EVAR techniques can be considered since the PERICLES Registry experience denotes that even though it is safe and feasible, the rates of type IA endoleaks, chimney graft

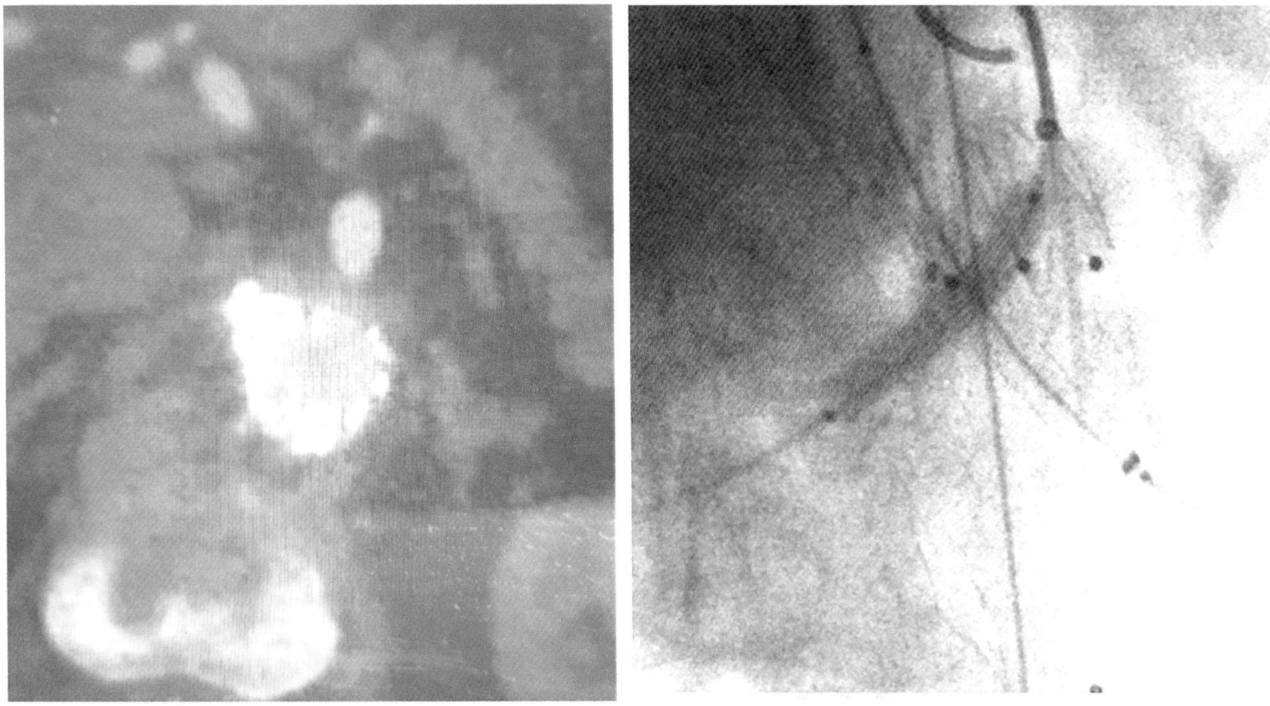

Fig. 58.4 Postoperative computed tomography angiography showing excellent conformability of the Endurant stent graft with the deployed chimney grafts in the renal arteries without any evidence of gutter-related type IA endoleak

occlusions, and ischemic stroke are higher compared with the prevailing single chimney evidence [4–13]. The presence of extensive thrombus in the aortic arch or occluded subclavian arteries, 5 mm or smaller in diameter renal arteries, wide neck diameter of more than 30 mm or the lack of enough (20 mm) sealing length are not recommended anatomical conditions to perform the techbique [4–13].

Preoperative Imaging Evaluation

High-quality preoperative thoracoabdominal and pelvic CTA of 1 mm cuts is mandatory to accurately evaluate the anatomy of the aorta and its branches. It allows for an understanding of the arch anatomy and type looking for safe navigation of catheters and sheaths through it in order to avoid cerebral embolization in the presence of thrombus or type 3 arch, where the subclavian artery originates lower than the inner aortic curve. Access and reconstruction at a three-dimensional workstation/program are needed to create centreline pathways and plan device sizing, selection, and orientation for implantation. In patients with renal insufficiency, stenotic renal ostium, atretic kidneys, or multiple visceral or renal vessels, nuclear medicine testing of renal function is useful to determine the potential sacrifice of one of the renal arteries avoiding the use of more than two chimney stent (CS).

Discussion

Ch-EVAR gained popularity to the point that its role within the abdominal aortic aneurysm (AAA) treatment algorithm is considered complementary in the European AAA Guidelines depending on the patient's characteristics and aortoiliac anatomy. The PROTAGORAS study evaluated the standardized use of the Endurant II endograft (Medtronic, Minneapolis, MN, USA) in combination with the Advanta V12/iCAST (Getinge Group, Mijdrecht, the Netherlands) balloon-expandable covered stents (BECS) in the renal arteries of 128 patients [2]. Consequently, this device combination received European CE mark approval as an on-label indicated therapy for juxtarenal aneurysms [2].

The chimney technique is feasible with the majority of available abdominal endografts. This allows the urgent treatment of symptomatic and ruptured AAAs using off-the-shelf devices available in stock at any vascular centre. It is a feasible procedure for patients excluded from fenestrated EVAR (fEVAR) due to clinical, anatomical, technical, or device manufacturing time reasons, ensuring low rates of aortic-related mortality. In that sense the Endurant endograft (Medtronic Inc., Minneapolis, MN, USA) presents flexibility, low profiles, and good navigability, showing good results in combination with balloon-expandable covered stents (BECS), especially advantageous in the presence of highly calcified and elongated iliac vessels where fenestrated endografting can be hazardous [8].

This case has a remarkably challenging access and hostile neck. We demonstrate the successful treatment of the severe iliac artery stenosis with the Shockwave IVL to facilitate double Ch-EVAR, with successful outcomes. Shockwave IVL can represent an attractive modality for the treatment of severely calcified arteries, which combines the calcium-disrupting capability of lithotripsy with the familiarity of balloon catheters to facilitate proper stent deployment [14].

Use of chimney grafts has advantages. First, repair with use of off-the-shelf devices makes possible immediate treatment without delay. This is substantial for these rupture-threatening pathologic processes, as in our case dealing with a pararenal PAU. From a technical point of view, catheterization from the upper extremity in a normally downward-oriented target vessel, as in the presented case, is easy and feasible. In addition, use of Endurant tubes with the short M-shaped stents allows harmonic adaptability of the device in angulated necks, leading to successful exclusion of the pathologic process. Contemporaneously, the low-profile introducer system allows the trackability in severe iliac calcifications and angulations. The present case had a co-existing dissection and stenosis of the external iliac artery, making the advance of the aortic endograft hazardous and risky for iliac rupture. Use of the new 8 mm low-profile M5+ IVL Shockwave catheter over a 7F sheath simplified the procedure without the need for stent placement, which could be dislocated during the advance of the aortic tube.

However, several other abdominal stent grafts combined with BECS, self-expandable covered stents, or even bare metal stent (BMS) are routinely used in everyday practice, and due to its off-the-shelf utility, potentially and theoretically, any commercially available stent may be employed. This heterogeneity of stent combinations employed worldwide has contributed to mixed and suboptimal results due to discordant and poor performance in some reports in terms of stent compression, kinking, endoleaks, and proper seal of the aneurysm. That has created some degree of scepticism regarding the overall ch-EVAR performance. In fact, none of the available and employed bridging stent grafts in clinical practice today are dedicated devices for ch-EVAR or fEVAR or bEVAR. Many different combinations of main graft have been applied; however, a head-to-head evaluation based on the best anatomical fit is difficult within cohort studies. The use of ch-EVAR configurations in a close to the human body in vitro environment provides a valuable tool maintaining the test conditions, such as juxtarenal aortic aneurysms (jAAA) anatomy and vessel wall characteristics. In that sense, the in vitro evaluation of the employed stent grafts before their in vivo application should be paramount to analyse and apprehend their possible behaviour in the patient. However, wide-ranging in vitro evidence is lacking on the performance of the commonly used CS grafts in combination with the CE-approved Endurant II abdominal device to guide the surgeons' decisions on armamentarium in the endovascular treatment of jAAA.

Conflict of Interest The authors declare no conflict of interest regarding the publication of this chapter.

References

1. Wanhainen A, et al. ESVS 2019 clinical practice guidelines on the Management of Abdominal Aorto-iliac Artery Aneurysms. Eur J Vasc Endovasc Surg. 2019;57:8–93.
2. Donas KP, Torsello GB, Piccoli G, Pitoulias GA, Torsello GF, Bisdas T, et al. The PROTAGORAS study to evaluate the performance of the Endurant stent graft for patients with pararenal pathologic processes treated by the chimney/snorkel endovascular technique. J Vasc Surg. 2016;63(1):1–7. https://doi.org/10.1016/j.jvs.2015.07.080.
3. Taneva GT, Fazzini S, Pipitone MD, Karaolanis G, Torsello G, Bremer C, et al. Use of stainless-steel, balloon-expandable chimney grafts is durable though caution is required when lining angulated renal arteries. J Endovasc Ther. 2020;27(6):902–9. https://doi.org/10.1177/1526602820948260.
4. Taneva GT, Criado FJ, Torsello G, Veith F. Results of chimney endovascular aneurysm repair as used in the PERICLES Registry to treat patients with suprarenal aortic pathologies. 2014:1–8. https://doi.org/10.1016/j.jvs.2019.08.228.
5. Bosiers MJ, Tran K, Lee JT, Donas KP, Veith FJ, Torsello G, et al. Incidence and prognostic factors related to major adverse cerebrovascular events in patients with complex aortic diseases treated by the chimney technique. J Vasc Surg. 2018;67(5):1372–9. https://doi.org/10.1016/j.jvs.2017.08.079.
6. Mestres G, Uribe JP, García-Madrid C, Miret E, Alomar X, Burrell M, et al. The best conditions for parallel stenting during EVAR: an in vitro study. Eur J Vasc Endovasc Surg. 2012;44(5):468–73. https://doi.org/10.1016/j.ejvs.2012.08.007.
7. Taneva GT, Donas KP, Pitoulias GA, Austermann M, Veith FJ, Torsello G. Cost-effectiveness analysis of chimney/snorkel versus fenestrated endovascular repair for high-risk patients with complex abdominal aortic pathologies. J Cardiovasc Surg. 2019;60:1–6. https://doi.org/10.23736/S0021-9509.19.11146-9.
8. Donas KP, Lee JT, Lachat M, Torsello G, Veith FJ, Dalman RL, et al. Collected world experience about the performance of the snorkel/chimney endovascular technique in the treatment of complex aortic pathologies: the PERICLES registry. Ann Surg. 2015;262(3):546–52. https://doi.org/10.1097/SLA.0000000000001405.
9. Pitoulias GA, Torsello G, Austermann M, Pitoulias AG, Pipitone MD, Fazzini SDK. Outcomes of elective use of the chimney endovascular technique in pararenal aortic pathologic processes. J Vasc Surg. 2021;73(2):433–42. https://doi.org/10.1016/j.jvs.2020.05.029.
10. Taneva GT, Donas KP, Torsello GB, Seifarth H, de Azevedo FM, Austermann MTG. In vitro evaluation of balloon-expandable chimney grafts in the renal arteries combined with the endurant endograft. J Endovasc Ther. 2019;26(6):865–70. https://doi.org/10.1177/1526602819872498.
11. Fazzini S, Martinelli O, Torsello G, Austermann M, Pipitone MD, Torsello GF, et al. The PROTAGORAS 2.0 study to identify sizing and planning predictors for optimal outcomes

in abdominal chimney endovascular procedures. Eur J Vasc Endovasc Surg. 2021;61(4):591–602. https://doi.org/10.1016/j.ejvs.2020.11.019.
12. Schiro A, Antoniou GA, Ormesher D, Piche AC, Farquharson F, Serracino-Inglott F. The chimney technique in endovascular aortic aneurysm repair: late ruptures after successful single renal chimney stent grafts. Ann Vasc Surg. 2013;27(7):835–43. https://doi.org/10.1016/j.avsg.2012.08.007.
13. Scali ST, Feezor RJ, Chang CK, Waterman AL, Berceli SA, Huber TS, et al. Critical analysis of results after chimney endovascular aortic aneurysm repair raises cause for concern. J Vasc Surg. 2014;60(4):865–874.e1. https://doi.org/10.1016/j.jvs.2014.03.295.
14. Wong B, El-Jack S, Newcombe R, Glenie T, Armstrong G, Khan A. Shockwave intravascular lithotripsy for calcified coronary lesions: first real-world experience. J Invasive Cardiol. 2019;31:46–8.

Descending Thoracic Aortic Aneurysms

59

Enrico Gallitto, Gianluca Faggioli, Antonino Logiacco, Stefania Caputo, Francesca Maria Feroldi, Gemmi Sufali, and Mauro Gargiulo

Introduction

In the recent decades, the advent of endovascular technology has considered one of the main medical revolutions, because it changed the current therapeutic paradigm of aortic aneurysms repair.

The endovascular repair of thoracic aortic aneurysm (TEVAR), introduced in clinical practice in 1994, is now widely accepted as the gold standard to treat descending thoracic aortic diseases in the presence of anatomical feasibility, due to its lower perioperative mortality and morbidity than open repair (OR) [1, 2] (Table 59.1).

TEVAR allows a mini-invasive solution (avoiding a thoracotomy and thoracic aortic cross-clamping) in several descending thoracic aortic diseases such as degenerative aneurysms of the thoracic aorta (DTAA), penetrating aortic ulcer, intramural hematoma, and acute/chronic aortic dissection [1].

Table 59.1 Different outcomes in the treatment of DTAA with TEVAR versus open repair according to ESVS Guidelines Descending Thoracic Aorta (2017)

DTAA treatment	TEVAR (%)	Open repair (%)
Mortality	5.57	16.5
Major neurological complications	5.4	14
Reintervention rate	7	8.4
Stroke	2.7	
Paraplegia or paraparesis	2.5–8	
SCI	2.8	

ESVS Guidelines Descending Thoracic Aorta (2017)

In this chapter, we will focus only on the endovascular repair of DTAA.

Definition, Natural History, and Surgical Indication

A DTAA is defined as a permanent localized or diffuse enlargement of the descending thoracic aorta with a diameter ≥50% than the adjacent normal-sized vessel [1].

In elective cases, the indication to treat a DTAA is associated with a balanced ratio between the risk of aortic rupture (fatal event) and procedure-related mortality or morbidities (patient's surgical risk).

The risk of rupture is related to the size of the aneurysm; literature data demonstrate that the annual rate of rupture/dissection is about 10–20% and 50% in patients with an aneurysm diameter of 7 and 8 cm, respectively [3–5].

A DTAA has an elective indication (if anatomically feasible) in the presence of one of the following factors: maximum diameter (women: ≥5.5 cm; men: ≥6 cm), rapid expansion (>5 mm in 6 months), saccular morphology, and presence of symptoms (pain, dysphagia, dyspnea, hemoptysis, visceral, or peripheral embolization). Obviously, in case of rupture or impending rupture the repair is indicated in an emergent/urgent setting [1, 2].

Preoperative Workup

General Clinical Evaluation

A preoperative clinical analysis is required in each elective case to evaluate and stratify the patient's comorbidities, general physical status, and life expectancy and finally to estimate the risk of perioperative mortality and complications [1, 2]. Cardiac/respiratory/renal function as well as cerebrovascular status should be analyzed before a final anesthesiologic evaluation to summarize the perioperative risks and type of anesthe-

E. Gallitto (✉) · G. Faggioli · A. Logiacco · S. Caputo · F. M. Feroldi · G. Sufali · M. Gargiulo
Vascular Surgery Unit, University Hospital, IRCCS Policlinico S. Orsola, University of Bologna—DIMEC, Bologna, Italy
e-mail: enrico.gallitto2@unibo.it; gianluca.faggioli@unibo.it; antonino.logiacco@studio.unibo.it; caputostefania@studio.unibo.it; francescamaria.feroldi@studio.unibo.it; gemmi.sufali@studio.unibo.it; mauro.gargiulo2@uibo.it

G. Geroulakos et al. (eds.), *Mastering Endovascular Techniques*, https://doi.org/10.1007/978-3-031-42735-0_59

sia [1, 2]. The elective DTAA repair should be proposed only in patients with reasonable life expectancy (≥ 2 years). Preoperative evaluations are performed according to:

- *Cardiac tests*: electrocardiogram, transthoracic echocardiogram, clinical cardiological evaluation
- *Pulmonary tests*: chest X-ray, respiratory function tests or high-resolution computed tomography (selected patients with severe chronic obstructive pulmonary disease)
- *Renal tests*: creatinine level and estimated glomerular filtration rate.
- *Cerebrovascular tests*: supra-aortic trunks duplex ultrasound and computed tomography angiography (CTA) and cerebral CT in case of severe stenosis requiring repair

Several validated scores (American Society of Anesthesiologists score, Revised Cardiac Risk Index, Vascular Study Group of New England Cardiac Risk Index, etc.) [6–8] can be proposed to assess the perioperative risks. The best medical therapy should be introduced in all cases in the preoperative period to control as much as possible the cardiovascular risk factors, potentially reducing the perioperative risks [1, 2].

Preoperative Imaging

The preoperative imaging analysis is fundamental to evaluate the anatomical feasibility of TEVAR and the correct patient selection and to achieve the consequent technical and clinical success.

Thoracoabdominal computed tomography angiography (CTA), with axial slice <2 mm of thickness, is of paramount importance for anatomical evaluation and endograft sizing/planning, and it is always necessary (gold standard exam) in both elective and urgent cases [1, 2]. It allows crucial anatomical information from the ascending aorta to femoral arteries.

The CTA post-processing analysis can now be performed by the same vascular surgeons that will perform the TEVAR procedure by dedicated software for advanced vessels analysis. Three-dimensional (volume rendering), multiplanar (axial, coronal, sagittal), maximal-intensity projections and center lumen line reconstructions are important tools to obtain data necessary for the endograft sizing and planning (Fig. 59.1). Finally, functions of simulation of intraoperative angiography or endograft position in the outer aortic curvature could add important adjunctive details to reduce the amount of intraoperative iodinate contrast media administration, radiation exposure and need of adjunctive endograft segments (Fig. 59.2).

Femoral/Iliac Access Evaluation

Even though the technological evolutions have allowed a significant reduction of the endograft delivery sheaths in recent years (low-profile device), the issue of femoral/iliac access is pervasive especially in female patients or patients with concomitant peripheral arterial occlusive disease. Depending on the currently available endografts and respective diameters, a femoral/iliac artery access compatible with 18–24 F is usu-

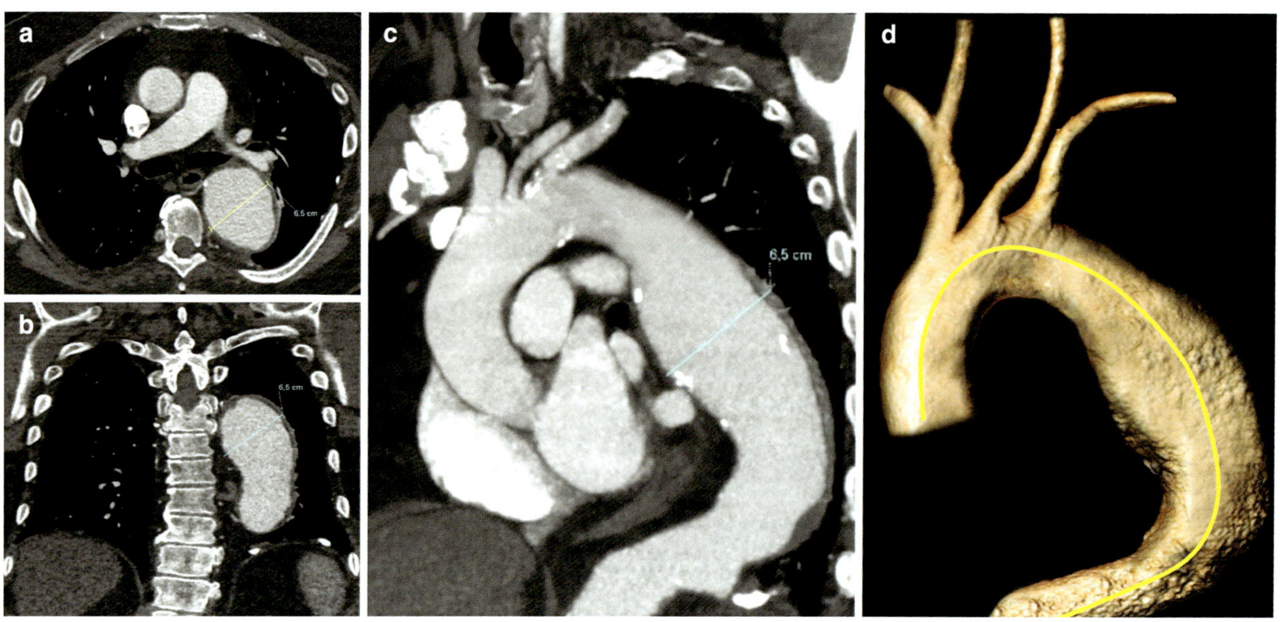

Fig. 59.1 Multiplanar analysis (**a–c**), 3D reconstruction and center lumen line evaluation (**d**) obtained by CTA post-processing analysis

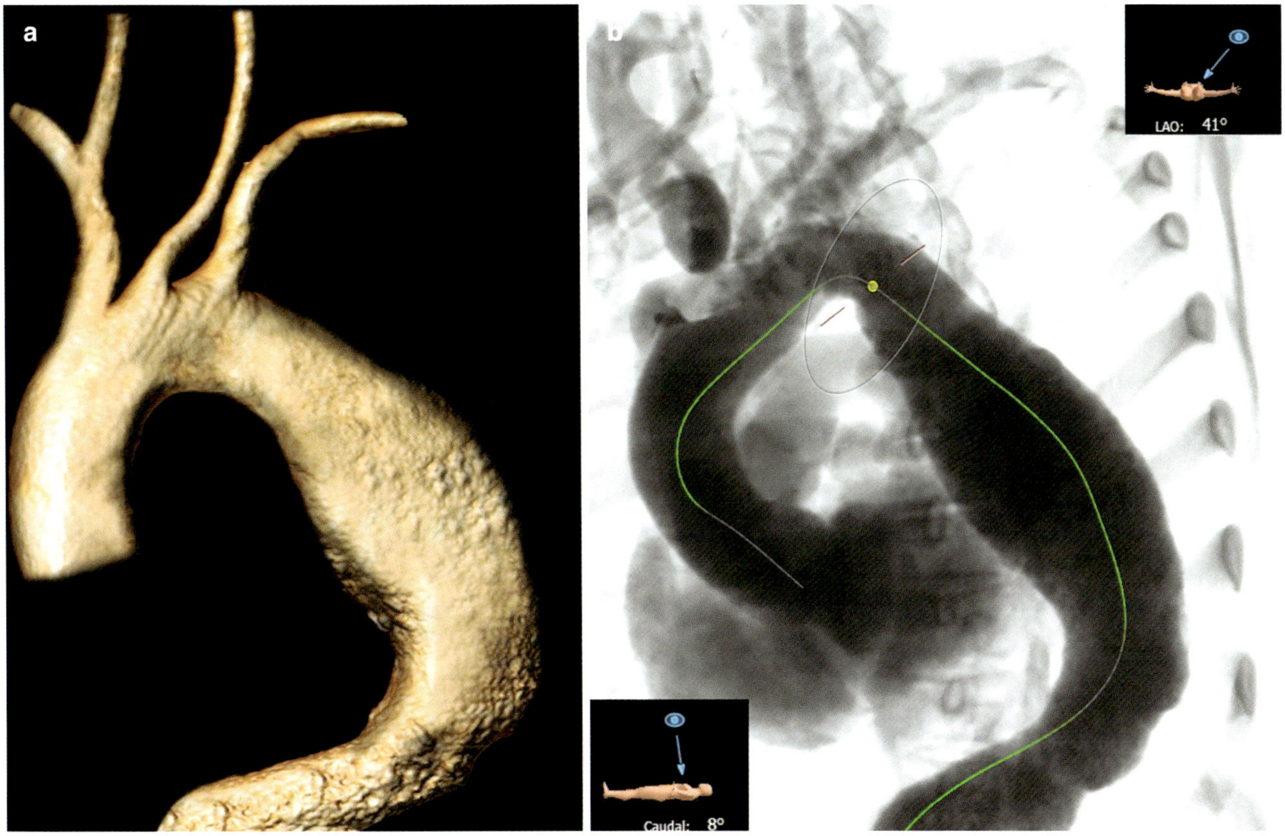

Fig. 59.2 3D-volume rendering (**a**) and angiographic simulation (**b**)

ally required. For these reasons the anatomical analysis of the access is focused on the evaluation of the minimum iliac vessel diameter (≥7 mm), the presence of calcifications (extension, grade, and localization), stenosis (grade, localization, and quality; calcific vs soft thrombotic lesions), and iliac tortuosity/angulation. Similarly, the presence of stenosis, calcifications, thrombosis, or aneurysm of the common femoral artery should be analyzed in case of both percutaneous and surgical access.

Proximal and Distal Neck Evaluation

A safe proximal and distal aortic neck is necessary to ensure an effective aneurysm exclusion in the early postoperative period (technical success) and during long-term follow-up (clinical success). A healthy aortic segment with cylindrical shape, with at least 20 mm of length and without thrombosis, calcification, and severe angulation, is required in both proximal and distal sealing zones. According to the SVS reporting standard, the levels of aortic sealing zones can be schematized in 11 segments from the ascending aorta to external iliac artery [9] (Fig. 59.3).

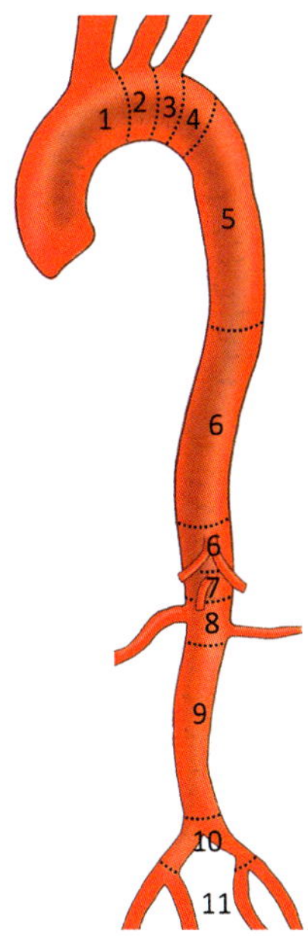

Fig. 59.3 SVS – Aortic sealing zones

In case of anatomical feasibility, a standard TEVAR implant is associated with an aortic coverage limited between zones 3 and 5. In case of inadequate proximal aortic neck in zone 3, adjunctive endovascular (custom-made endograft with scallop/fenestrations/branches, Fig. 59.4) or surgical procedures (carotid subclavian bypass, Figs. 59.5 and 59.6) are required to guarantee a safe proximal sealing of the endograft and maintain the patency of supra-aortic trunks in the elective setting.

Similar consideration should be given as regards the distal aortic neck; in case of short, wide, and angulated aortic neck in zone 5, a fenestrated branched endograft could be considered to reduce the risk of distal type I endoleak during follow-up (Fig. 59.7).

Endograft Sizing

The diameter of proximal and distal aortic necks as well as the distance between these two segments must be evaluated to perform the appropriate endograft sizing. The endograft diameter is usually selected according to an oversizing of 10–20% of the aortic neck size. The coverage length is usually evaluated on the outer aortic curvature and it is necessary to define the length of the endograft (Fig. 59.8). In the majority of cases, multiple endograft segments are required to achieve the aneurysm exclusion. In these cases, a safe overlap (length: ≥ 8 cm; diameter: oversizing ≥10%) between the endograft modules is necessary to reduce the risk of endograft migration, disconnection, and type III endoleak during follow-up. Depending on the different neck diameters (proximal vs distal) the implant can be designed with endograft with tapered configuration or straight endograft deployed from below to above (telescopic implant). In case of left subclavian artery coverage and carotid subclavian bypass, the pre-vertebral segment of the subclavian artery requires the embolization by a vascular plug (suggested oversizing: 30%) to avoid a high-flow type II endoleak (backflow in the aneurysm by the carotid subclavian bypass and proximal segment of subclavian artery).

Current Commercially Available Endografts (EU)

Since the first TEVAR procedures, several technological improvements have been made, improving the technical and clinical outcomes. The possibility to have endograft with several configurations/measures, low-profile delivery

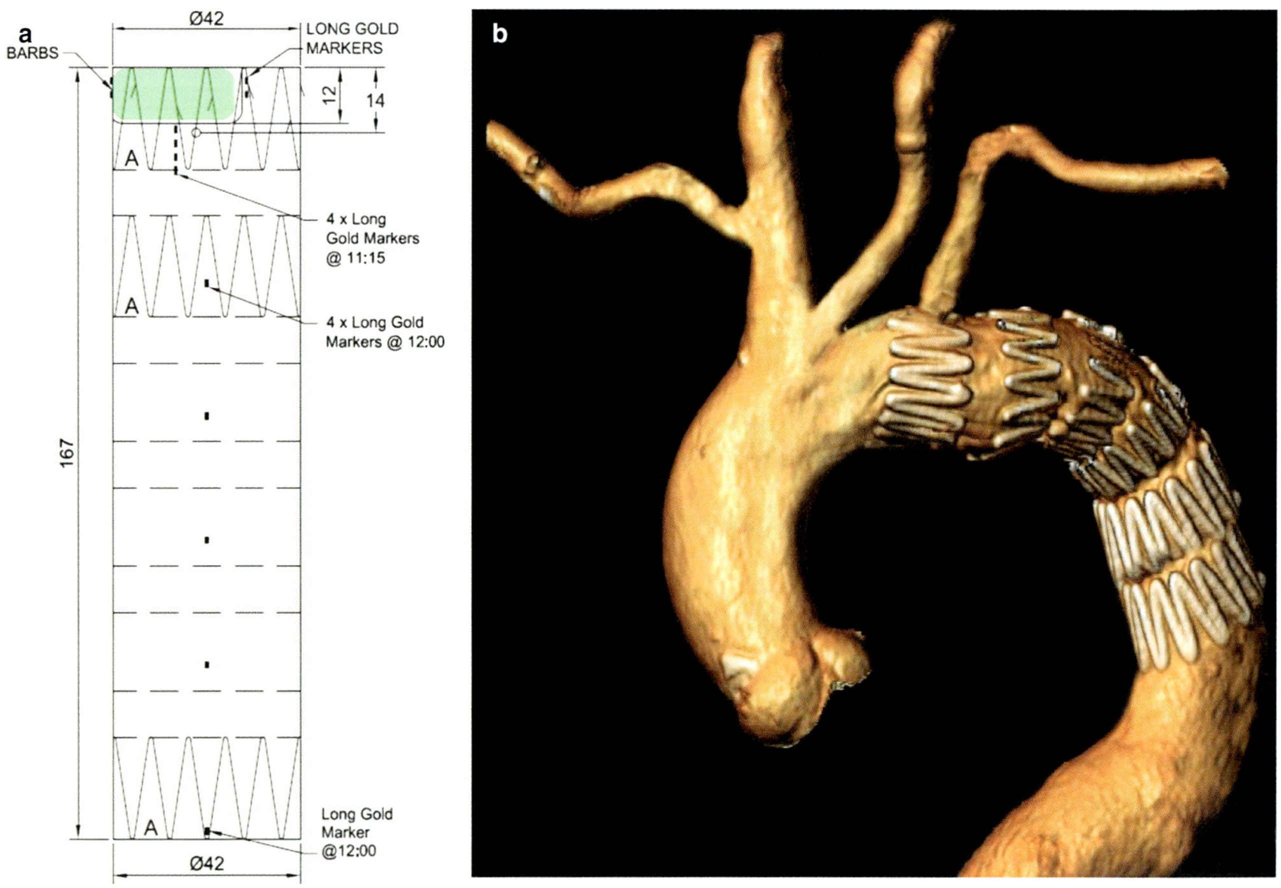

Fig. 59.4 Scallop endograft planning (**a**) and 3D reconstruction after implant (**b**); fenestrated endograft planning (**c**) and 3D reconstruction after implant (**d**); branched endograft planning (**e**) and 3D reconstruction after implant (**f**)

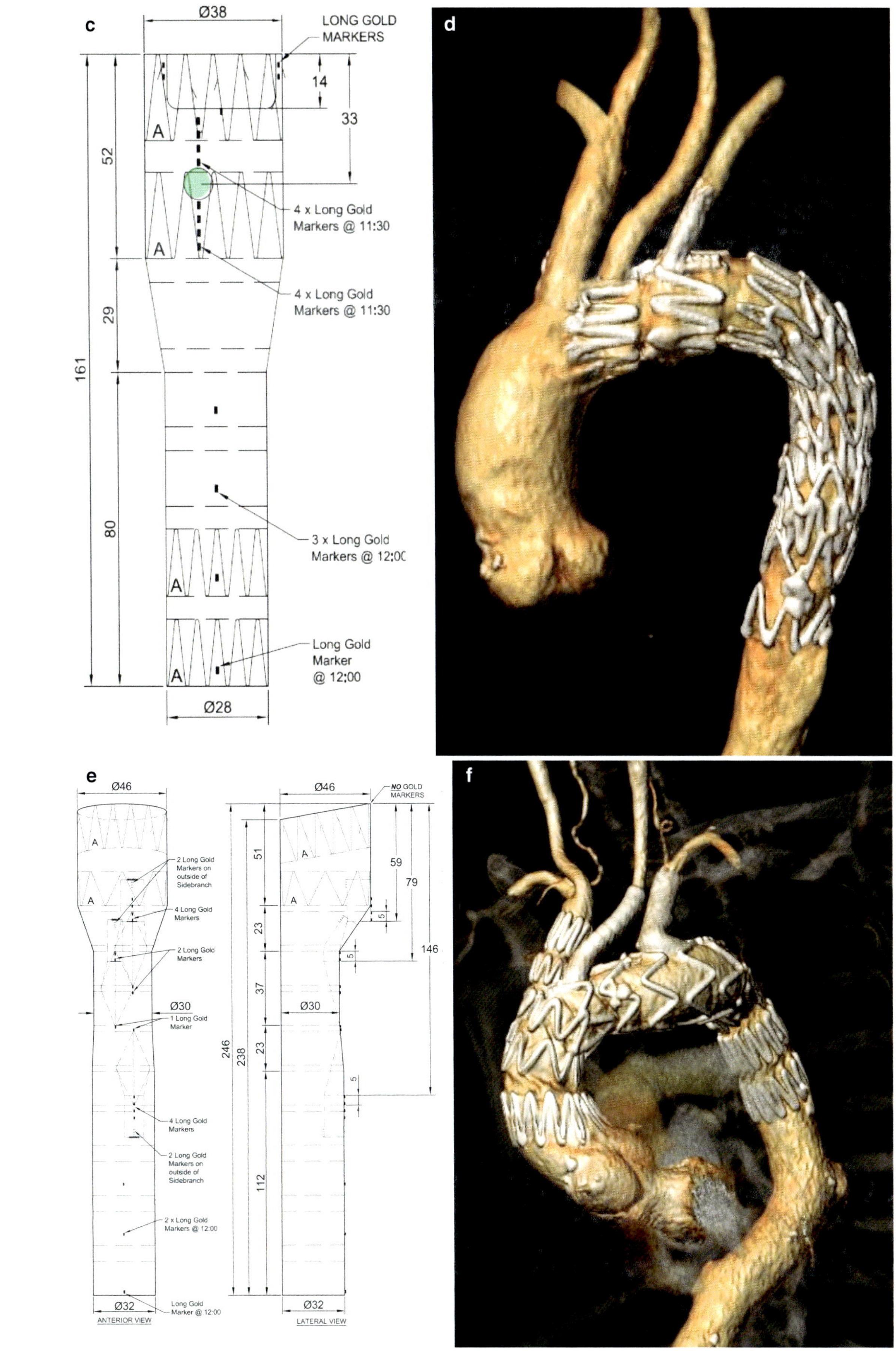

Fig. 59.4 (continued)

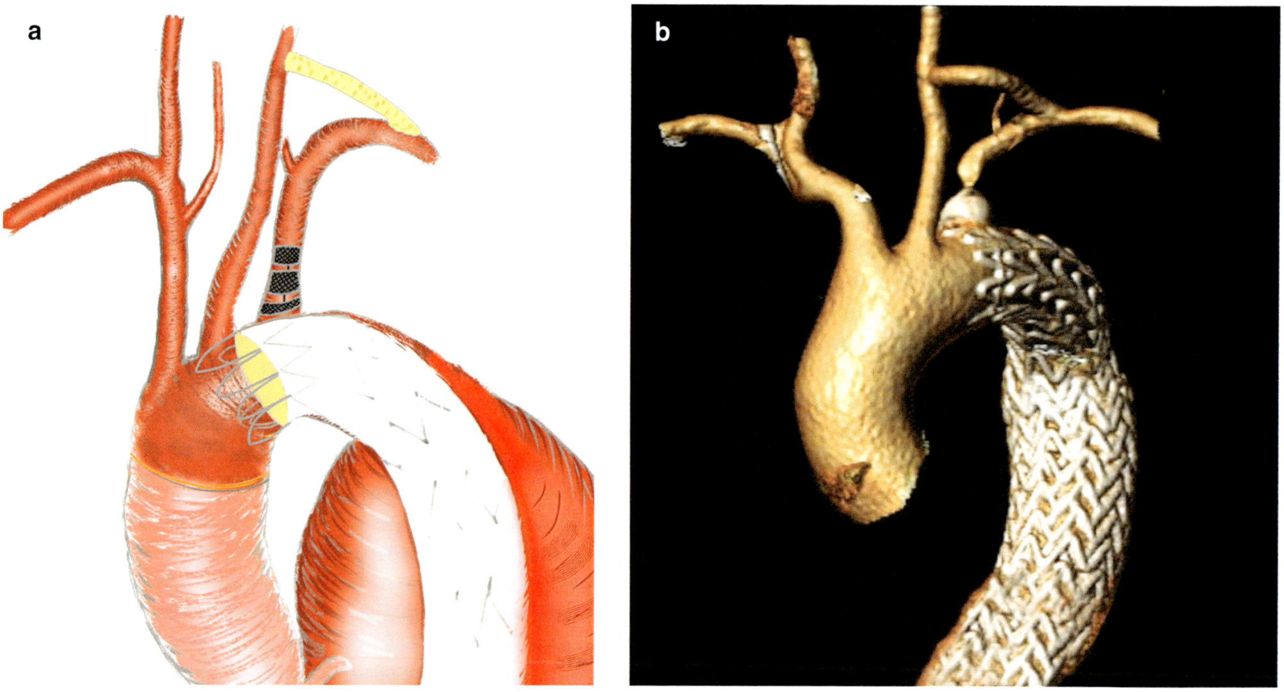

Fig. 59.5 TEVAR and left common carotid artery to left subclavian artery bypass; anatomy Sketch (**a**) and D reconstruction after implant (**b**)

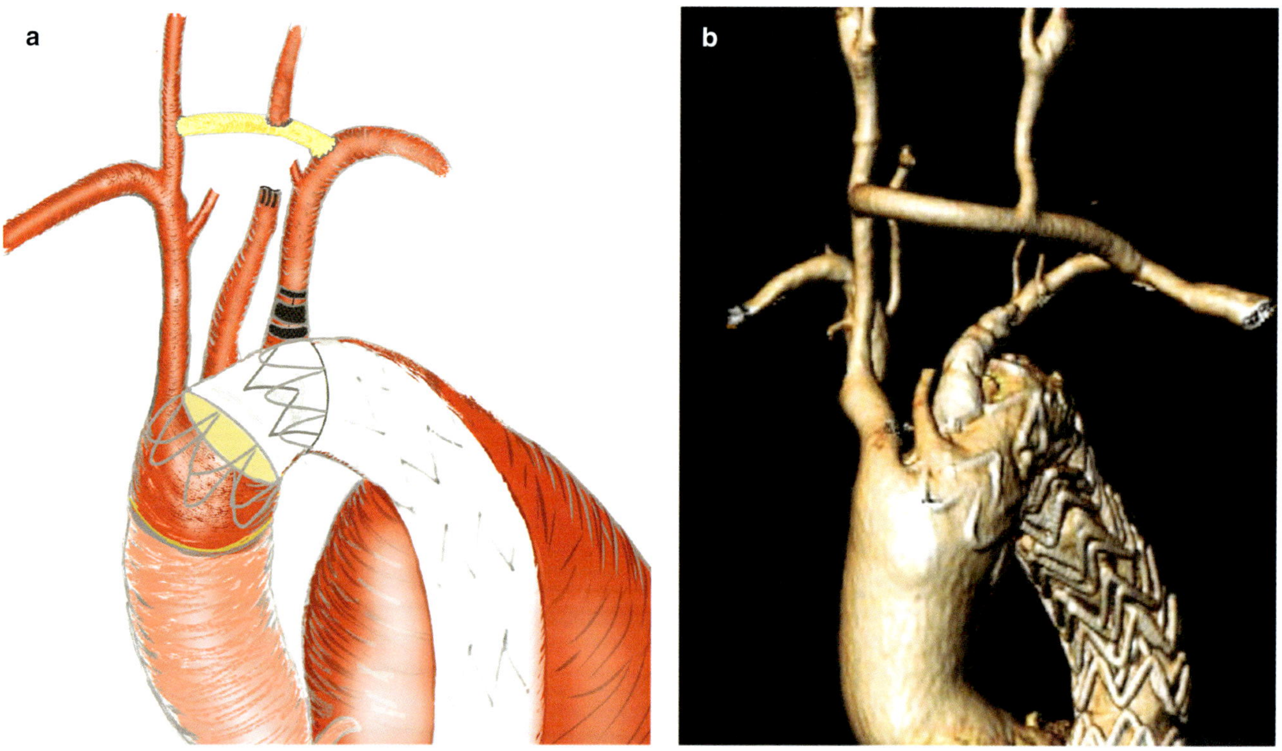

Fig. 59.6 TEVAR and Right common carotid artery to left subclavian artery bypass with left common carotid artery reimplant; anatomy sketch (**a**) and D reconstruction after implant (**b**)

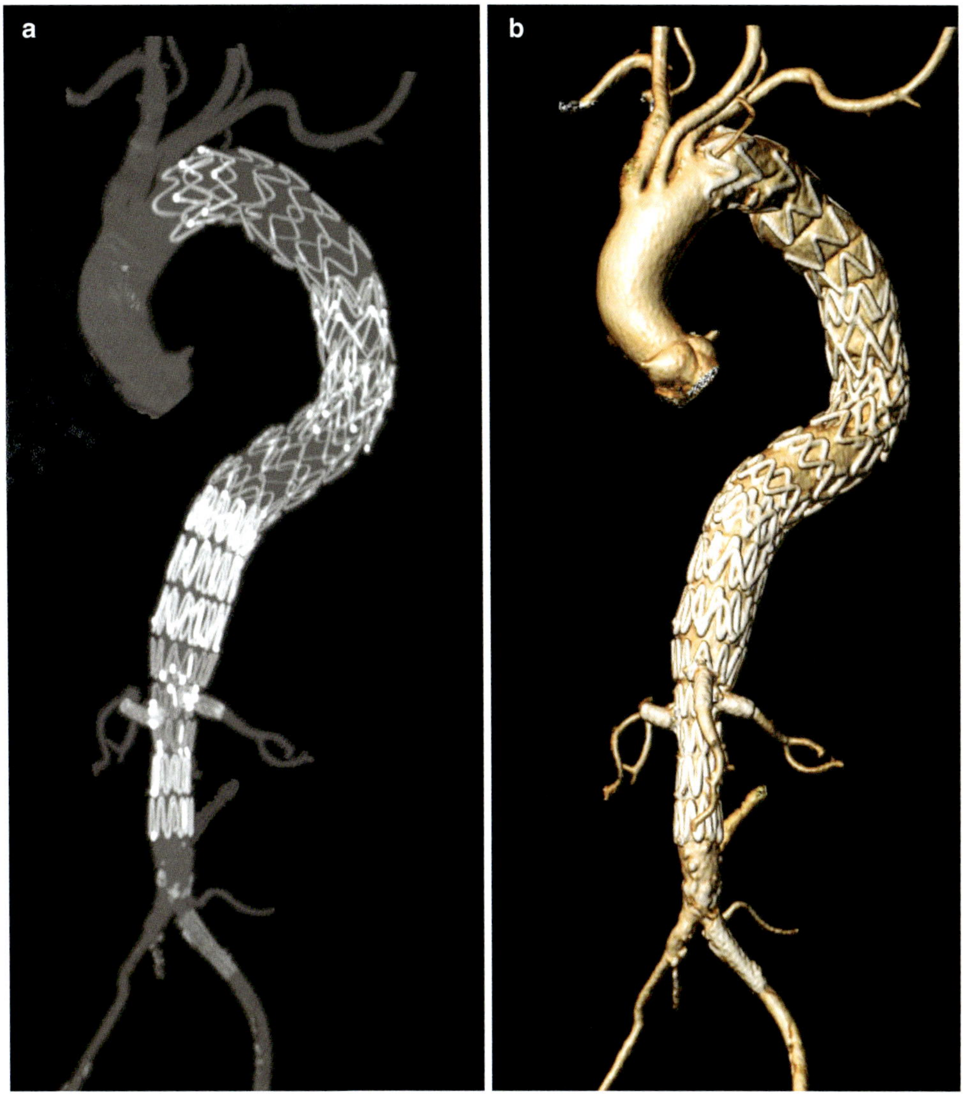

Fig. 59.7 TEVAR extension with fenestrated endograft; CT scan (**a**) and 3D reconstruction (**b**)

sheaths, and adaptability to tortuous and angulated vessel allows for improvement of the anatomical feasibility of TEVAR, treating cases with challenging aorto-iliac anatomy. Several endografts with different characteristics and indications are commercially available in Europe; instructions for use of the most frequently used thoracic devices are summarized in Table 59.2.

Spinal Cord Ischemia Prevention Protocol

Spinal cord ischemia is a catastrophic complication after extensive (>200 mm) aortic replacement, and it has a negative impact on postoperative patient survival and quality of life [10, 11]. In recent years, several surgical and anesthesiologic strategies were reported to reduce the incidence of spinal cord injury (SCI) [12–14].

First, it is important to reduce as much as possible the length of aortic coverage (maintaining a safe proximal and distal sealing zone) to reduce the number of sacrificed segmental arteries. Second, it is of paramount importance to maintain the collateral blood supply network for the spinal cord, guaranteeing the patency of left subclavian and both hypogastric arteries in case of extensive thoracic or thoracoabdominal disease. In case of thoracoabdominal repair a staged procedure is associated with a significant reduction of postoperative SCI as well as an early pelvic and lower limb reperfusion [15]. Cerebrospinal fluid (CSF) drainage is

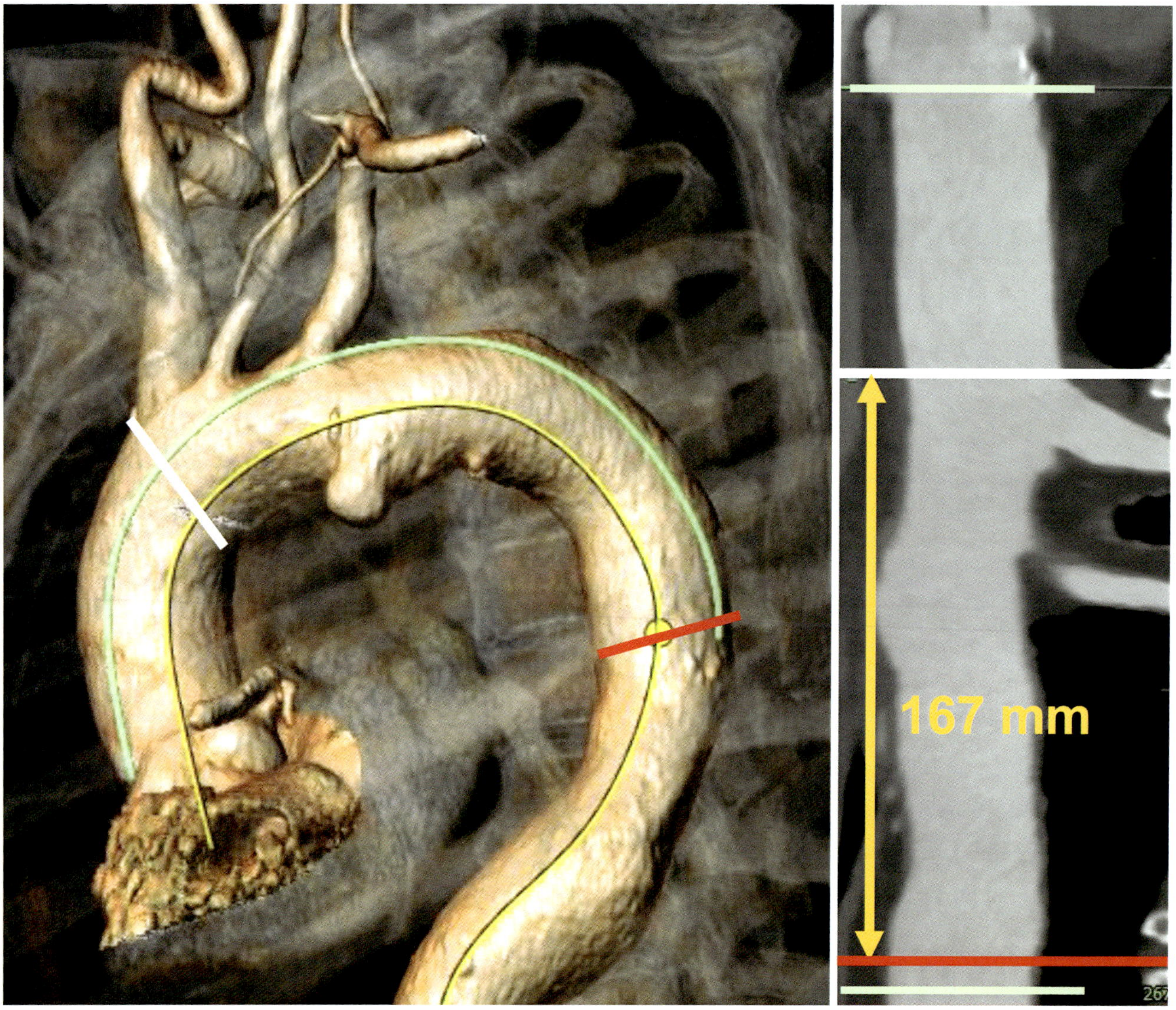

Fig. 59.8 Outer aortic curvature evaluation

Table 59.2 Principal characteristics of the thoracic endoprosthesis available

Manufacturer	Endoprosthesis model	Introduction system (Fr)	Aortic diameter required (mm)	Aortic coverage (mm)	Proximal landing zone (mm)	Distal landing zone (mm)	Repositionable
Bolton Medical	Relay	22–26	22–46	100–250	15–25	15–25	–
Cook Medical	Zenith Alpha	16–20	20–42	105–233	≥20	≥20	–
W.L Gore & Associates	Gore TAG Conformable device with active control system	18–24	16–42	100–200	≥20	≥20	√
Medtronic	Valiant Captiva	22–25	18–42	100–200	≥20	≥20	√

proposed in selected cases at high risk (aortic coverage >200 mm, previous abdominal aortic replacement, preoperative occlusion of hypogastric or subclavian artery, peripheral arterial occlusive disease) and maintained for 48–72 postoperative hours with monitoring (target cerebrospinal pressure ≤12 cmH_2O) and therapeutic function. Finally, perioperative hemoglobin level >10 g/dL and mean arterial pressure >80 mmHg should be maintained in the perioperative period [16].

Procedure

Standard Step-by-Step Procedure

Procedures can be performed in a surgical room equipped with a mobile C-arm or hybrid room, under locoregional or general anesthesia.

Intraoperative endovenous antibiotic and heparin (ACT >200 s) administration is required. The arterial access can be achieved by surgical cut-down or the percutaneous approach. Two arterial accesses are necessary:

- main access—common femoral artery for the endograft introduction (16–24F)
- diagnostic access—contralateral femoral or brachial artery for pigtail (5F)

In case of left subclavian artery coverage and carotid subclavian bypass, the left brachial artery access is required for the subclavian embolization.

Retrograde puncture of the main access is performed and an introducer (6–10 F) is advanced in the external iliac artery. A floppy 0.035 J guidewire is advanced in the ascending aorta and it is exchanged with an extrastiff one (e.g., Cook Lunderquist 300 cm) by an angiographic catheter.

- *Implant requiring multiple segments from above to below (tapered device)*

A pigtail catheter is advanced in the aortic arch and a diagnostic angiography is performed to define the origin of the left subclavian artery and the proximal neck (LAO C-arm angulation preoperative evaluated: 45–60°). The proximal endograft is advanced in the aortic arch; after an angiographic check, its position is optimized in the proximal sealing zone and then it is deployed during a hypotensive status (permissive hypotension with systolic pressure of 80 mmHg). The pigtail is retrieved at the level of the diaphragm and a further angiography is performed to detect the origin of the celiac artery and the distal neck (LAO C-arm angulation preoperative evaluated: 70–90°). The second endograft is advanced and positioned in order to have both a safe sealing in the distal neck and sufficient overlap with the first one (≥ 8 cm).

- *Implant requiring multiple segments from below to above (telescopic implant)*

In this situation, the diagnostic pigtail is first positioned at the level of the diaphragm to detect the origin of the celiac artery. The implant starts from the distal segment focusing on the visualization of the distal stents according to the sealing zone and celiac artery. Next, the pigtail is advanced through the endograft in the aortic arch, a second angiography is performed to detect the left subclavian artery, and finally the proximal endograft is deployed according to the proximal neck and the safe overlap with the distal segment.

Ballooning of the proximal and distal sealing zones as well as overlapping between the endografts should be performed by compliant balloons (Coda, Cook Medical, Bloomington, Indiana; Reliant, Medtronic World Medical, Sunrise, Florida; Tri-Lobe, W.L. Gore & Associates, Flagstaff, Arizona).

Completion angiography is performed to confirm the correct endograft position, the patency of the left subclavian and celiac artery, and the absence of type I–III endoleaks (angiography with pigtail catheter in the aortic arch, at the level of endograft overlapping and distal neck). The delivery system is withdrawn from the femoral artery after a retrograde iliac angiography to exclude potential iliac issues caused by the endograft navigation. Finally, the arterial access is closed by a percutaneous closure system or the surgical approach.

Management of Challenging Iliac Accesses

The endograft introduction could be difficult and dangerous in the presence of severe iliac stenosis, calcification, or external iliac diameter < 7 mm, with potential risk of iliac rupture and intraoperative fatal bleeding.

In these situations, several strategies can be proposed according to the iliac anatomy and quality (calcium, extension) of the lesion [17, 18].

In the presence of iliac stenosis, a balloon angioplasty can be performed before the endograft introduction by semi-compliant balloons. In case of residual stenosis limiting the endograft navigation a second balloon angioplasty should be performed by semi-compliant high pressure. In the presence of severe calcific lesion, a currently available alternative approach can be to combine the balloon angioplasty with the intravascular lithotripsy (IVL) system (Shockwave Medical, Inc., Fremont, California). The IVL reduces intimal and medial calcium tone through pulsatile sonic pressure waves transmitted by a direct opposition between the vessel lumen and balloon [19].

After these adjunctive maneuvers, the iliac navigability can be tested by dedicated hydrophilic dilatators (16–22F, Cook Medical, Bloomington, Indiana). They can be used as both test and therapeutic maneuvers. It is important to avoid a primary iliac stenting before the TEVAR procedure due to the risk of dislodgment during the endograft introduction [17, 20]. Iliac stenting should be performed at the end of the TEVAR procedure in case of residual stenosis/recoiling or flow-limiting iliac dissection. It is crucial in the presence of challenging iliac anatomy, before the delivery sheath removal, for a retrograde angiography to promptly detect any potential bleeding of the common/external iliac artery and to achieve rapid seal by a covered stent deployment. In the presence of diameter <7 mm or extensive and calcific lesion of the entire external iliac artery, a surgical conduit is an effective solution to guarantee the endograft introduction [21, 22]. Through a pararectal incision the common iliac artery is exposed in the retroperitoneal space and a 10 mm prosthetic Dacron graft is sewed in an end-to-side fashion at the level of the common iliac artery bifurcation, tunneled in the groin incision, and used as endograft gateway. Once the endovascular procedure is completed, the conduit can be converted in an iliofemoral bypass (Fig. 59.9).

Finally, the "paving and cracking technique" can be an alternative strategy in high-risk patients managed in an urgent setting [23–25]. It consists of the creation of an "endoconduit" deploying large (10–12 mm) covered stents along the diseased iliac arteries (paving by self-expandable in external iliac artery of oversizing + balloon expandable in common iliac artery) followed by an aggressive ballooning with a noncompliant device (cracking). It is a controlled and sealed iliac rupture to increase the iliac diameter and allow the endograft navigability. The limits of this technique are the possibility to dislodge the covered stent during the thoracic endograft introduction, the potential backbleeding in case of patent hypogastric artery, and the increased risk of spinal cord and colon ischemia related to its sacrifice [24]. For these reasons, it should be considered as the last option and can be proposed in patients with preoperative chronic hypogastric occlusion.

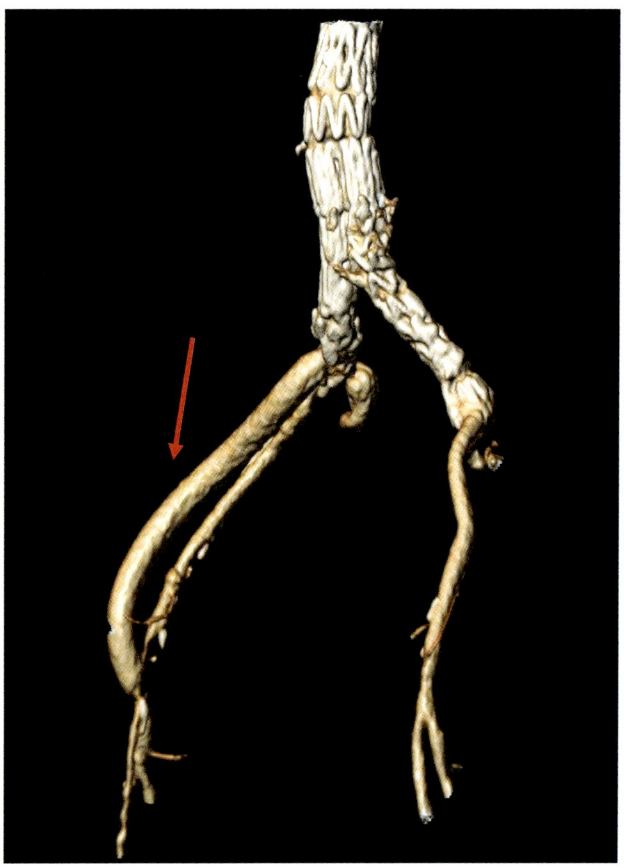

Fig. 59.9 Right iliac conduit converted into iliofemoral bypass

Through and Through Technique

The endograft navigability can be hampered by severe vessels tortuosity (iliac thoracoabdominal aorta) or kinked/angulated previous surgical graft. In case of inability to advance the endograft in the iliac artery or abdominal/thoracic aorta a brachial femoral through and through approach can be effective to overcome the problem.

In these cases, a 400 cm-long guide wire is snared from an axillary/brachial access in the aortic segment without aneurysm. Once the brachial femoral system is achieved, a mild–moderate tension can be performed at the two guide extremities to obtain a stable and stretched way for the endograft navigation (Fig. 59.10). An important trick in this maneuver is to protect the insertion of the subclavian artery into the arch during the through and through tension with a 6F, 45 cm-long hydrophilic sheath in order to avoid potential lesion caused by the guidewire.

Adjunctive Maneuvers to Guarantee the Precise Deployment of Endograft in the Proximal Landing Zone

In the presence of a specific anatomic situation, the precise endograft deployment in the proximal neck is crucial for technical success. The cardiac rapid pacing is an effective tool used to avoid the distal jumping/migration of the endograft during the deployment caused by the high flow [26, 27]. The cardiac probe is usually inserted by a percutaneous common femoral vein access (6F introducer) and advanced through the iliac and cava vein into the right ventricular (Fig. 59.11). Before the endograft deployment, the rapid cardiac pacing is activated causing a controlled ventricular tachycardia that significantly reduces the cardiac output. An alternative technique to reduce the cardiac output consists in the temporary occlusion of the cava vein by a compliant balloon [26].

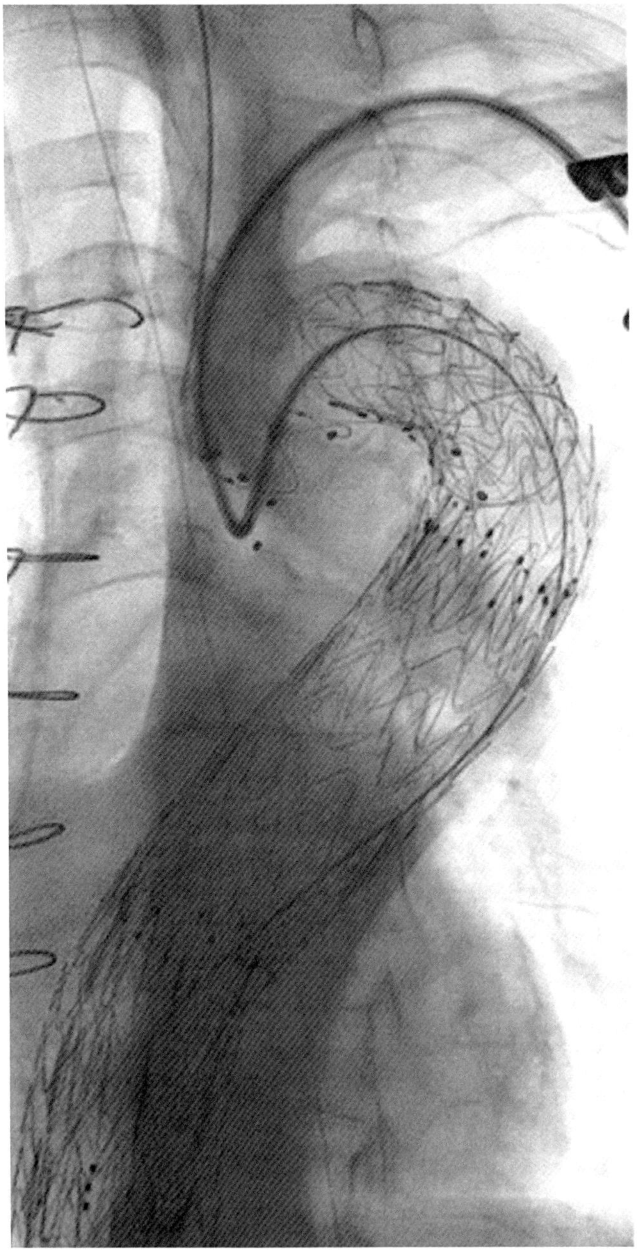

Fig. 59.10 Through and through wire technique (TWT)

A physiologically induced alternative to the typical methods of reducing cardiac output during deployment of stent grafts in the aortic arch and proximal aorta is the Munich Valsalva implantation technique (MuVIT) [28]. During manual mechanical ventilation, the adjustable pressure-limiting valve is carefully closed to 25 mm Hg, creating "manual bloating" of the lungs and sustained apnoea. The increased intrathoracic pressure causes compression of the vena cava and pulmonary veins, reducing the venous backflow and gradually decreasing the arterial pressure [28].

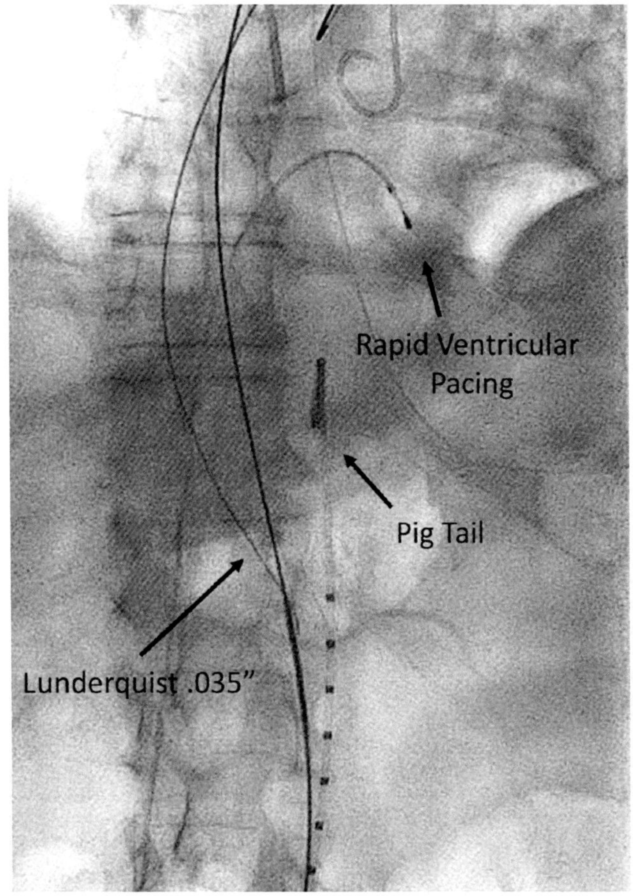

Fig. 59.11 Rapid ventricular pacing during TEVAR

Complications

Depending on the timing of occurrence, TEVAR-related complications can be divided into:

- Intraoperative complications
- Early/perioperative complications (within 30 days)
- Follow-up complications

Intraoperative Complications

These complications can be considered a technical failure and are associated with preoperative planning errors or intraoperative technical issues.

In case of inadequate sealing, type I a or b endoleak maintains the high aneurysm sac pressurization [29, 30]. "Bird beak" configuration at the level of the proximal sealing zone is caused by a non-circumferential apposition between the graft and the aortic arch and it could potentially increase the risk of endograft infolding/collapse and type I endoleak during follow-up [30]. In case of excessive endograft oversizing in the proximal sealing zone, a potential risk of retrograde type A aortic dissection is reported [31].

In case of severe aortic thrombosis (arch, para-visceral, aneurysm), particular attention in the catheters/guide manipulation should be used to reduce the risk of cerebral, visceral, or peripheral embolization.

Finally, as reported above in case of hostile femoral/iliac access, the risk of iliac complication (dissection/rupture) is not negligible and should be prevented by dedicated maneuvers or promptly detected and managed to avoid intraoperative mortality [17, 18, 21, 32].

Perioperative Clinical Complications

Different studies report all-cause perioperative mortality rates between 1.5% and 9.5% [33–36] in elective TEVAR.

Cardiopulmonary morbidity and renal function worsening are typical clinical complications described in the early postoperative period. They are associated with the general surgical patient's risks. Neurological complications can be divided into cerebrovascular events and SCI [11, 37]. Cerebrovascular events are caused by embolization of atherosclerotic arch debris, floating thrombotic materials, air bubbles, etc.

Bowel and gallbladder ischemia are rare yet severe complications due to distal embolization or non-intentional celiac trunk [38] and superior mesenteric artery coverage or dissections.

Follow-up Complications

The strict adherence to a follow-up protocol is fundamental for early detection of any potential late TEVAR complications and guaranteeing a prompt and safe solution. The most frequently detected complications during follow-up are engraft migration/disconnection and endoleaks. Aorto-oesophageal or bronchial fistulae and endograft infection are less frequently reported.

Endoleaks are detected in between 3% and 10% of cases during follow-up [24]. Type Ia, Ib, and III endoleaks are the most frequently reported; they are high-flow endoleaks with continued aneurysm sac pressurization and subsequent risk of rupture. For this reason, reinterventions are necessary in these cases. They are caused by proximal/distal necks degeneration or endograft migration (caudal migration of the proximal segment, cranial migration of the distal segment, disconnection of the overlapped segments). Type II endoleaks are not frequent in thoracic aorta and are associated with backflow from patent intercostal arteries [39].

Endograft infections have an incidence <1% (0.2– 0.5%) [40] and it can be caused by aorto-oesophageal/bronchial fistula (1.5% cases [41]) or sepsis with systemic infection. The clinical presentation is associated with a setting of sepsis and the radiological signs are specific with presence of peri-graft fluid localization and air bubbles. This condition is associated with high mortality in both conservative management and endograft explant [42].

Clinical Case

A 73-year-old male presented with an incidental radiological (CTA) detection of a DTAA with 66 mm maximum diameter.

The patient was at high surgical risk due to severe and multiple comorbidities: BMI > 31, history of coronary artery disease, severe chronic obstructive pulmonary disease, moderate renal function impairment, and previous urological surgery (prostate cancer).

The post-processing CTA reconstructions were performed by a dedicated software for advanced vessels analysis (Fig. 59.12; 3 Mensio, Bilthoeven, Netherlands). According to the preoperative anatomical evaluation, endograft sizing/planning (technical feasibility), and general clinical patient condition, a standard TEVAR procedure was proposed.

The procedure was performed under local anesthesia. The main femoral access was obtained by a surgical cut-down through the right superficial femoral artery (to avoid postoperative groin issue in an obese patient).

The diagnostic pigtail and the cardiac probe for the rapid pacing were percutaneously introduced from the common femoral artery (5F) and common femoral vein (6F), respectively, in the left groin.

A Cook Zenith Alfa endograft (tapered configuration: proximal diameter 40 mm; distal diameter 36 mm; length 217 mm) was introduced on a Lunderquist guidewire in the aortic arch. An angiography was performed to identify the origin of the left subclavian artery and celiac trunk. The endograft was deployed during the rapid pacing activation to have a millimetric precision in the proximal sealing zone.

Molding ballooning was performed in the proximal (under a new rapid pacing activation) and distal sealing zone.

A completion angiography showed the correct endograft position with patency of the left subclavian artery and celiac trunk, and complete aneurysm exclusion (iodinated contrasts media: 50 mL). The postoperative CTA confirmed the technical success of the procedure (Fig. 59.13). The postoperative course was uneventful and the patient was discharged home on the fourth postoperative day.

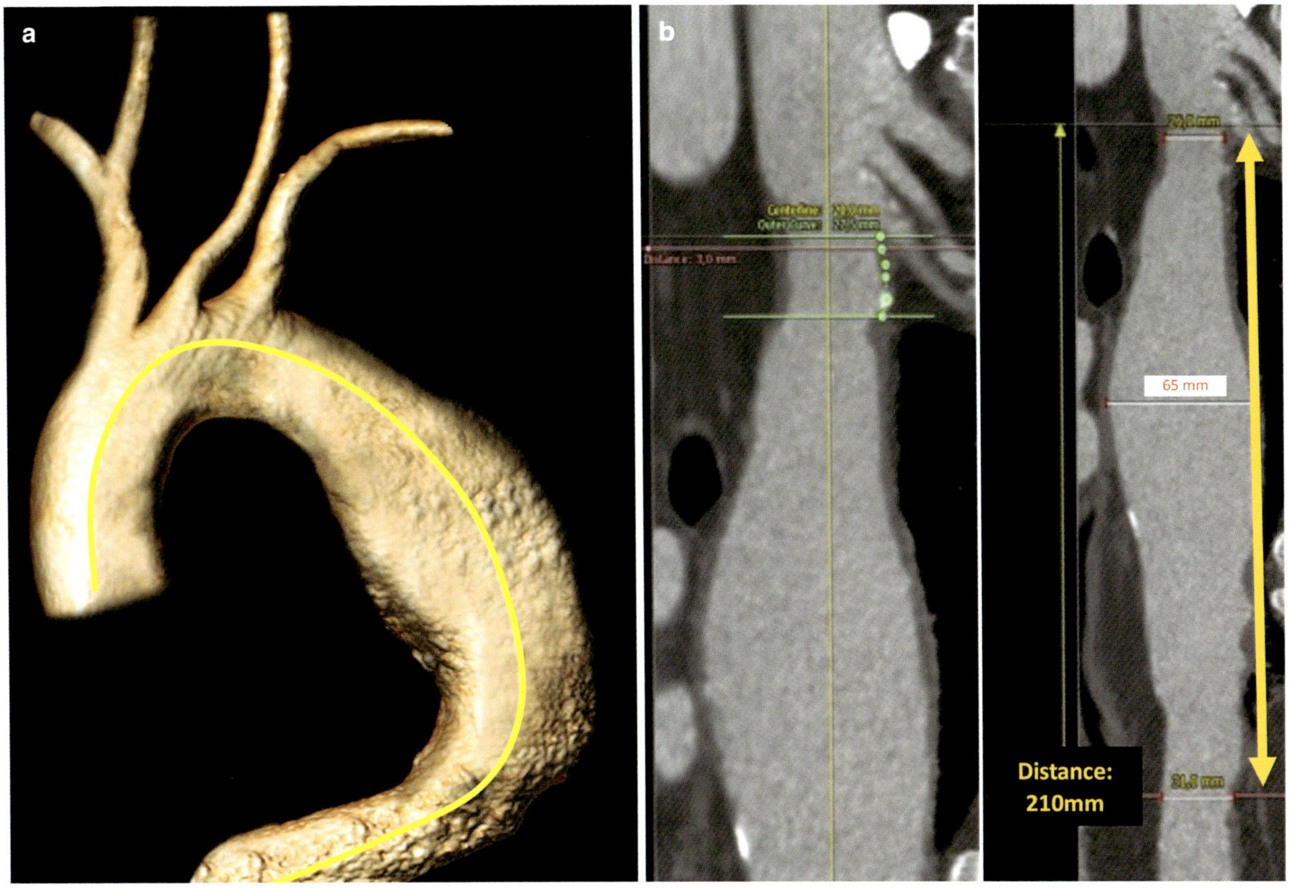

Fig. 59.12 Software reconstruction of CTA to obtain data necessary for the endograft planning: 3D reconstruction and multiplanar. 3D (**a**) and multiplanar (**b**, **c**) reconstructions

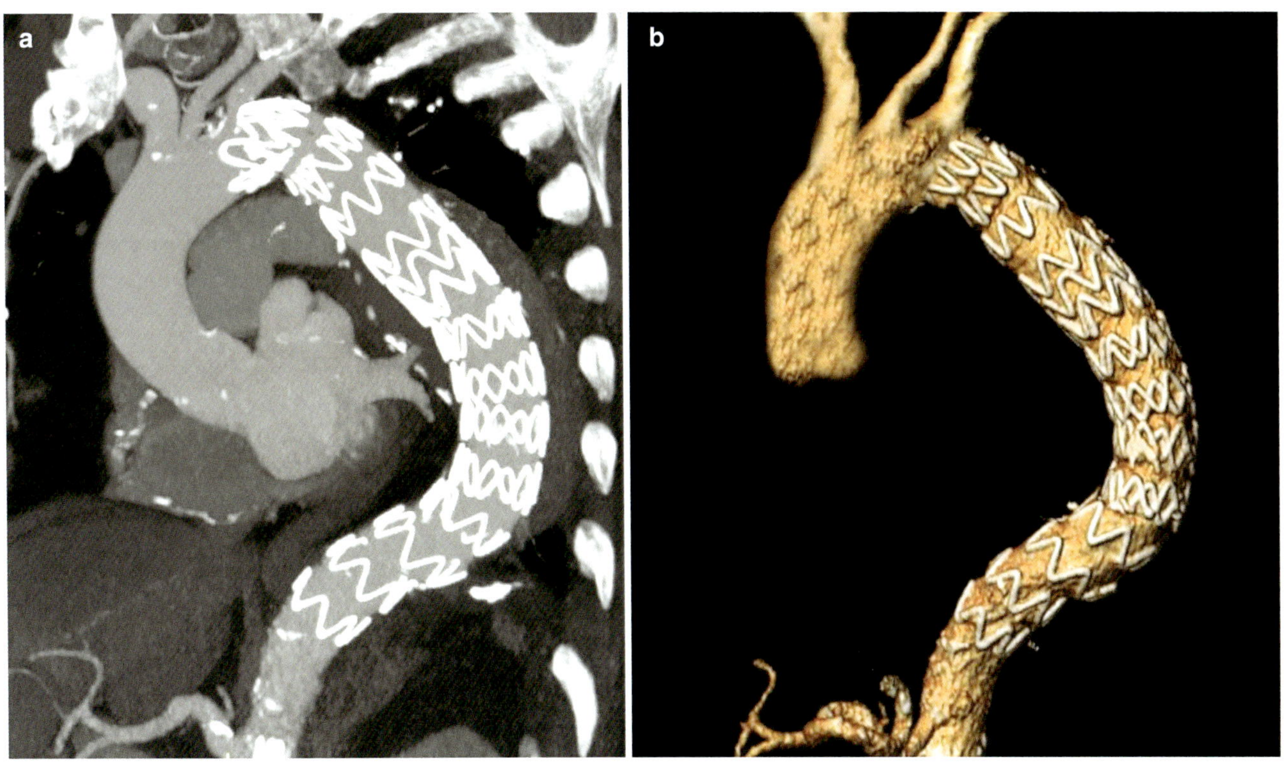

Fig. 59.13 CT scan (**a**) and 3D reconstruction after implant (**b**)

References

1. Riambau V, Böckler D, Brunkwall J, Cao P, et al. Editor's choice—management of descending thoracic aorta diseases: clinical practice guidelines of the European Society for Vascular Surgery (ESVS). Eur J Vasc Endovasc Surg. 2017;53(1):4–52.
2. Upchurch GR Jr, Escobar GA, Azizzadeh A, Beck AW, et al. Society for Vascular Surgery clinical practice guidelines of thoracic endovascular aortic repair for descending thoracic aortic aneurysms. J Vasc Surg. 2021;73(1S):55S–83S.
3. Kuzmik GA, Sang AX, Elefteriades JA. Natural history of thoracic aortic aneurysms. J Vasc Surg. 2012;56(2):565–71.
4. Kim JB, Kim K, Lindsay ME, MacGillivray T, et al. Risk of rupture or dissection in descending thoracic aortic aneurysm. Circulation. 2015;132(17):1620–9.
5. Coady MA, Rizzo JA, Hammond GL, Mandapati D, et al. What is the appropriate size criterion for resection of thoracic aortic aneurysms? J Thorac Cardiovasc Surg. 1997;113(3):476–91; discussion 489–91.
6. Garcia S, Moritz TE, Goldman S, Littooy F, et al. Perioperative complications after vascular surgery are predicted by the revised cardiac risk index but are not reduced in high-risk subsets with preoperative revascularization. Circ Cardiovasc Qual Outcomes. 2009;2(2):73–7.
7. Bertges DJ, Goodney PP, Zhao Y, Schanzer A, et al. The Vascular Study Group of New England Cardiac Risk Index (VSG-CRI) predicts cardiac complications more accurately than the Revised Cardiac Risk Index in vascular surgery patients. J Vasc Surg. 2010;52(3):674–83, 683.e1–683.e3.
8. Doyle DJ, Goyal A, Garmon EH. American Society of Anesthesiologists Classification 2022. In: StatPearls [Internet].
9. Oderich GS, Forbes TL, Chaer R, Davies MG, et al. Reporting standards for endovascular aortic repair of aneurysms involving the renal-mesenteric arteries. J Vasc Surg. 2021;73(1):4S–52S.
10. Pini R, Faggioli G, Paraskevas K, Alaidroos M, et al. A systematic review and meta-analysis of the occurrence of spinal cord ischemia after endovascular repair of thoracoabdominal aortic aneurysms. J Vasc Surg. 2022;75:1466–77.
11. Mousa AY, Morcos R, Broce M, Bates MC, et al. New preoperative spinal cord ischemia risk stratification model for patients undergoing thoracic endovascular aortic repair. Vasc Endovasc Surg. 2020;54(6):487–96.
12. Dijkstra ML, Vainas T, Zeebregts CJ, Hooft L, et al. Spinal cord Ischaemia in endovascular thoracic and thoraco-abdominal aortic repair: review of preventive strategies. Eur J Vasc Endovasc Surg. 2018;55:829–41.
13. Tenorio ER, Eagleton MJ, Kärkkäinen JM, Oderich GS. Prevention of spinal cord injury during endovascular thoracoabdominal repair. J Cardiovasc Surg. 2019;60(1):54–65.
14. Zhang Z, Zhou Y, Lin S, Xiao J. Systematic review and meta-analysis of association of prophylactic cerebrospinal fluid drainage in preventing spinal cord ischemia after thoracic endovascular aortic repair. J Vasc Surg. 2022;75(4):1478–89.
15. Maurel B, Delclaux L, Sobocinski J, Hertault A, et al. Editor's choice—the impact of early pelvic and lower limb reperfusion and attentive peri-operative management on the incidence of spinal cord ischemia during thoracoabdominal aortic aneurysm endovascular repair. Eur J Vasc Endovasc Surg. 2015;49(3):248–54.
16. Acher C, Acher CW, Marks E, Wynn M. Intraoperative neuroprotective interventions prevent spinal cord ischemia and injury in thoracic endovascular aortic repair. J Vasc Surg. 2016;63(6):1458–65.
17. Gallitto E, Gargiulo M, Faggioli G, Pini R, et al. Impact of iliac artery anatomy on the outcome of fenestrated and branched endovascular aortic repair. J Vasc Surg. 2017;66(6):1659–67.
18. Parmer SS, Carpenter JP. Techniques for large sheath insertion during endovascular thoracic aortic aneurysm repair. J Vasc Surg. 2006;43(Suppl A):62A–8A.
19. Price LZ, Safir SR, Faries PL, Tadros RO, et al. Shockwave lithotripsy facilitates large-bore vascular access through calcified arteries. J Vasc Surg Cases Innov Tech. 2020;7(1):164–70.
20. Hertault A, Bianchini A, Daniel G, Haulon S, et al. Experience with unfavorable iliac access when performing fenestrated/branched endovascular aneurysm repair. J Endovasc Ther. 2021;28(2):315–22.
21. Vijayvergiya R, Uppal L, Kasinadhuni G, Lal A, et al. Retroperitoneal iliac conduits as an alternative access site for endovascular aortic repair: a tertiary care center experience. J Vasc Bras. 2021;20:e20210033.
22. Van Bogerijen GHW, Williams DM, Eliason JL, Patel HJ, et al. Alternative access techniques with thoracic endovascular aortic repair, open iliac conduit versus endoconduit technique. J Vasc Surg. 2014;60(5):1168–76.
23. Hinchliffe RJ, Ivancev K, Sonesson B, Malina M. "paving and cracking": an endovascular technique to facilitate the introduction of aortic stent-grafts through stenosed iliac arteries. J Endovasc Ther. 2007;14(5):630–3.
24. Asciutto G, Aronici M, Resch T, Dias NV, et al. Endoconduits with "pave and crack" technique avoid open Ilio-femoral conduits with sustainable mid-term results. Eur J Vasc Endovasc Surg. 2017;54(4):472–9.
25. Giannopoulos S, Malgor RD, Sobreira ML, Jacobs DL, et al. Iliac conduits for endovascular treatment of aortic pathologies: a systematic review and meta-analysis. J Endovasc Ther. 2021;28(4):499–509.
26. Gottardi R, Berger T, Voetsch A, Czerny M, et al. What is the best method to achieve safe and precise stent-graft deployment in patients undergoing TEVAR? Thorac Cardiovasc Surg. 2021;69(4):357–61.
27. Bokoch MP, Hiramoto JS, Lobo EP, Shalabi A. Rapid ventricular pacing for landing zone precision during thoracic endovascular aortic arch repair: a case series. J Cardiothorac Vasc Anesth. 2017;31(6):2141–6.
28. Tsilimparis N, Abicht J, Stana J, Konstantinou N, et al. The Munich Valsalva Implantation Technique (MuVIT) for cardiac output reduction during TEVAR: vena cava occlusion with the valsalva maneuver. J Endovasc Ther. 2021;28(1):7–13.
29. Belvroy VM, de Beaufort HWL, van Herwaarden JA, Trimarchi S, et al. Type 1b endoleaks after thoracic endovascular aortic repair are inadequately reported: a systematic review. Ann Vasc Surg. 2020;62:474–83.
30. Marrocco-Trischitta MM, Spampinato B, Mazzeo G, Nano G, et al. Impact of the bird-beak configuration on postoperative outcome after thoracic endovascular aortic repair: a meta-analysis. J Endovasc Ther. 2019;26(6):771–8.
31. Chen Y, Zhang S, Liu L, Jing Z, et al. Retrograde type a aortic dissection after thoracic endovascular aortic repair: a systematic review and meta-analysis. J Am Heart Assoc. 2017;6(9):e004649.
32. Duran C, Naoum JJ, Patel DMG, et al. A longitudinal view of improved management strategies and outcomes after iatrogenic iliac artery rupture during endovascular aneurysm repair. Ann Vasc Surg. 2013;27(1):1–7.
33. Farber MA, Giglia JS, Starnes BW, Matsumura JS, et al. TAG 08-02 clinical trial investigators. Evaluation of the redesigned conformable GORE TAG thoracic endoprosthesis for traumatic aortic transection. J Vasc Surg. 2013;58(3):651–8.

34. Fairman RM, Tuchek JM, Lee WA, Bavaria J, et al. Pivotal results for the medtronic valiant thoracic stent graft system in the VALOR II trial. J Vasc Surg. 2012;56(5):1222–31.e1.
35. Arnaoutakis GJ, Schneider EB, Arnaoutakis DJ, Abularrage CJ, et al. Influence of gender on outcomes after thoracic endovascular aneurysm repair. J Vasc Surg. 2014;59(1):45–51.
36. Hu FY, Fang ZB, Leshnower BG, Duwayri Y, Veeraswamy RK, et al. Contemporary evaluation of mortality and stroke risk after thoracic endovascular aortic repair. J Vasc Surg. 2017;66(3):718–727.e5.
37. Karaolanis GI, Antonopoulos CN, Charbonneau P, Donas K, et al. A systematic review and meta-analysis of stroke rates in patients undergoing thoracic endovascular aortic repair for descending thoracic aortic aneurysm and type B dissection. J Vasc Surg. 2022;76(1):292–301.e3.
38. Hanna L, Lam K, Agbeko AE, Gibbs R, et al. Coverage of the coeliac artery during thoracic endovascular aortic repair: a systematic review and meta-analysis. Eur J Vasc Endovasc Surg. 2022;63(6):828–37.
39. Chen SW, Lee KB, Napolitano MA, Trachiotis G, et al. Complications and management of the thoracic endovascular aortic repair. Aorta (Stamford). 2020;8(3):49–58.
40. Moulakakis KG, Mylonas SN, Antonopoulos CN, Liapis CD, et al. Comparison of treatment strategies for thoracic endograft infection. J Vasc Surg. 2014;60(4):1061–71.
41. Czerny M, Eggebrecht H, Sodeck G, Weigang, et al. New insights regarding the incidence, presentation and treatment options of aorto-oesophageal fistulation after thoracic endovascular aortic repair: the European registry of endovascular aortic repair complications. Eur J Cardiothorac Surg. 2014;45(3):452–7.
42. Smeds MR, Duncan AA, Harlander-Locke MP, Lawrence PF, et al. Treatment and outcomes of aortic endograft infection. J Vasc Surg. 2016;63(2):332–40.

Endovascular Management of Aortic Arch Disease

60

N. Tsilimparis, P. Spath, J. Stana, M. Pichlmaier, and T. Hamwi

Introduction

Endovascular treatment of the aortic arch is an important challenge in the field of vascular surgery. Currently, open surgical repair of the aortic arch is still the gold standard in patients who are suitable for open surgery. However, it has significant disadvantages, requiring clamping of the aorta, mild-to-moderate hypothermia, and extracorporeal circulation and cardiopulmonary bypass. Despite the highly standardized procedure, the risk of death is still 2%–20%, and the risk of stroke is up to 10–15% of cases [1, 2]. For this reason, the most recent guidelines from the European Society of Vascular Surgery and the European Association of Cardio-Thoracic Surgery [3] suggest endovascular treatment as a viable option for patients who are not suitable for open surgical repair, have a reasonable life expectancy, and have favorable anatomy. The advantages of endovascular repair include minimizing surgical trauma, avoiding cardiac arrest and cardiopulmonary bypass, and thus reducing the risk of serious adverse events in an already frail population.

The first use of a branched endograft to treat aortic arch dissection was reported by Inoue in 1996 [4], and since that initial experience, numerous technological improvements have resulted in the ability to perform sophisticated endovascular repair in a completely noninvasive manner today.

N. Tsilimparis (✉) · P. Spath (✉) · J. Stana · T. Hamwi
Department of Vascular Surgery, Ludwig-Maximillian University Hospital, Munich, Germany
e-mail: nikolaos.tsilimparis@med.uni-muenchen.de; jan.stana@med.uni-muenchen.de; tarek.hamwi@med.uni-muenchen.de

M. Pichlmaier
Department of Cardiac Surgery, Ludwig-Maximillian University Hospital, Munich, Germany
e-mail: maximilian.pichlmaier@med.uni-muenchen.de

Pathologies of the Aortic Arch and Indications for Intervention

Patients with aortic arch pathologies share common risk factors, including hypertension, hyperlipidemia, coronary artery disease, tobacco use, advanced age, renal failure, and the presence of infrarenal abdominal aortic aneurysm [5, 6].

The pathologies associated with the aortic arch are various and can be summarized as follows:

- *Aortic arch aneurysm (Arch-AA):* is defined as a full-thickness dilatation of the aortic arch with a diameter of more than 50% of its native diameter [7]. At the level of the aortic arch the normal diameter is approximately around 30 mm. Aortic arch aneurysms include any thoracic aneurysm which involves the supra-aortic vessels [8]. Atherosclerosis is the main etiology. Aortic arch aneurysms represent a minor proportion of TAAs, accounting for about 10% of the total cases. In addition, post-dissection thoracoabdominal aortic aneurysms (TAAAs) represent an evolving disease with frequent late aneurysmal degeneration due to retrograde false lumen perfusion (approximately 30% of patients with chronic aortic dissections) [9].
- *Aortic dissection involving the aortic arch (non-A non-B dissection):* is defined as a separation of the layers of the aortic wall provoked by an intimal tear. The initial intimal tear can occur in the ascending aorta (Type A) or descending aorta (Type B), and the aortic arch is generally involved either as the direct origin of the dissection or as retrograde false lumen, or for reaching a healthy aortic tissue for the repair [10]. The dissection can extend proximally or distally from the initial tear to the aortic valve, coronary arteries, branches of the aortic arch, or thoracic or abdominal aorta [11]. Consequently, this will lead to clinical features in patients such as acute chest or back pain, stroke, abdominal pain, or acute heart failure [11]. The incidence of non-A-non-B dissection is 11% in patients with acute aortic dissection [12].

G. Geroulakos et al. (eds.), *Mastering Endovascular Techniques*, https://doi.org/10.1007/978-3-031-42735-0_60

- *Penetrating Aortic Ulcer (PAU):* PAU refers to a focal disruption of the aortic intimal wall with an ulcer-like projection shape. PAUs can be found in association with atherosclerotic lesions of the aortic wall [13] and may be associated with hematoma within the media and may evolve to perforation or aortic dissection. Penetrating ulcer is the primarily lesion in <5% of all aortic dissections [5, 14]. Most of the PAUs are found in the descending thoracic aorta (85–95%), but they can also occur in the ascending aorta or arch [5, 13, 14]. PAU does not have a clear threshold for a repair indication, and it is thought that patients with an asymptomatic PAU have a higher risk of progression if the PAU diameter is >20 mm with a neck of >10 mm, and early intervention should be evaluated [15].
- *Thoracoabdominal aortic aneurysm (TAAA) involving the aortic arch*: can be defined as an aneurysm continuing from the arch extending to the abdominal aorta. Starting before the origin of the left subclavian artery to the aortoiliac bifurcation, many different configurations may occur. The first TAAA classification scheme based on the anatomic extent of the aneurysm was described by Crawford in 1986 [16]; according to this, Type I and Type II are the most involved [17].

Current guidelines recommend surgery for isolated aortic arch aneurysms with a diameter of 55 mm or greater as a threshold for intervention (class IIa level B) [3]. Another criterion for aortic arch repair is the presence of symptoms, such as refractory pain. The indication for treatment is strongly influenced by the presence of additional diseases of the ascending and descending thoracic aorta. In addition, evidence of aortic rupture at the level of the arch or rapid growth of an aneurysm (> 10 mm over 12 months) is considered a major criterion for treatment.

Anatomical Considerations

Aortic arch lesions may present as a distal extension of ascending aortic diseases or proximal extensions of thoracic/thoracoabdominal repair or occur as isolated diseases. In order to reach a proper sealing zone, the different aortic arch segments should be selected according to Ishimaru's zones. The aortic arch is therefore divided into the following segments:

- *Zone 0 (Z0):* from sinotubular junction to the innominate artery (IA)
- *Zone 1 (Z1):* from the distal aspect of IA to the left common carotid artery (LCCA)
- *Zone 2 (Z2):* from the distal aspect of LCCA to left subclavian artery (LSA)

Lesions involving Z0 or requiring a landing zone in this area are considered proximal arch lesions. Theoretically, a standard endovascular repair would cover all supra-aortic trunks. Lesions involving Z1 or requiring a landing zone in this area are considered mid-arch lesions, and standard endovascular repair could allow direct vascularization of IA. Lesions involving Z2 or requiring a landing zone in this area are considered lesions of the distal arch, and standard endovascular repair may include only the LSA.

Endovascular treatment of the aortic arch presents many challenges, and as described by Haulon et al. [18], major criteria should be considered for both anatomic and physiologic considerations (Table 60.1).

First of all, the selection of a suitable landing zone should be made carefully: at least 25 mm of a healthy aorta, without calcification or thrombus, and with a diameter of <38 mm. With larger diameters, the risk of retrograde aortic dissection is greater, and repair should not be considered. The angulation of the aortic arch should not exceed 60°. Selection of an adequate landing zone is mandatory to prevent migration, larger endoleaks, and further progression of the disease. At the same time, it should be kept in mind that at the level of the aortic arch, pulsatile and rotation movements may affect the configuration of graft materials and durability [19]. Additionally, the alignment of the endograft can result in deployment issues, which, because of the long introducer sheath averaging 100 cm, can only be aligned longitudinally and has very little rotational motion. Therefore, pre-curved grafts are offered with a self-alignment system that avoids additional manipulations in the aortic arch and reduces the risk of cerebral embolism and post-operative stroke [20]. Finally, the aortic valve can be a major problem: damage to

Table 60.1 Anatomic and physiologic criteria for total endovascular arch repair [18]

Anatomic criteria	Physiologic criteria
• Arch aneurysms and chronic dissections (no previous mechanical aortic valve replacement) • Ascending aortic length ≥50 mm (measured from the sinotubular junction to the origin of the innominate artery) • Sealing zone in the ascending aorta ≥40 mm in length and ≤38 mm diameter • Sealing zone in the innominate artery ≥20 mm in length and ≤20 mm in diameter • Accessible to accommodate 22- or 24-F sheaths	• Minimum of a 2-year life expectancy • Negative stress test (cardiology clearance required in the setting of a positive stress test) • No class III or IV congestive heart failure • No stroke or myocardial infarction in the last year • No significant carotid bifurcation disease (≥75% stenosis by north American symptomatic carotid endarterectomy trial criteria) • Estimated glomerular filtration rate by modification of diet in renal disease method ≥45 mL/min/1.73 m^2

the aortic valve and the risk of embolism from fragments must be considered. Mechanical aortic valve replacement has been an exclusion criterion for endovascular repair. Recently, new endovascular devices with very short, custom-made delivery system tips, together with techniques on how to cross the aortic valve, have made these procedures possible [21, 22].

Endograft Characteristics

There are different endovascular approaches, and the type of repair depends on the lesion localization, the expertise of the surgical team, and the availability of endovascular devices.

Initial results describing the outcomes of complete aortic arch endovascular repair are promising, with an estimation of technical success ranging from 84.2% to 100% [23–25]. When compared with open surgical repair, recent improvement of the techniques permitted an early mortality rate ranging from 6% to 25%, a rate of stroke from 0.8% to 25%, and spinal injuries from 1% to 25%, as described by Cao et al. [26].

Nevertheless, endovascular repair represents an important opportunity to improve the outcomes in patients with complex and potentially devastating pathologies of the aortic arch, and most recent systematic reviews [24–26] report promising rates of early mortality, stroke, and spinal cord ischemia of 2.5–6.7%; 4.4–6.2%, and 1.4–2.3%, respectively.

However, more data are needed to determine the comparative treatment effect of endovascular stent-graft of the aortic arch, compared with conventional open and hybrid repairs.

For total endovascular repair of the aortic arch (Table 60.2), the current used devices are the following:

1. Fenestrated/scallop endograft of the aortic arch
2. Branched endograft of the aortic arch
3. Parallel grafts
4. In-situ/physician-modified fenestrated graft

Nowadays, the latest expert consensus document recommends fenestrated and branched aortic endografts as the preferred option for total endovascular aortic repair (Class IIa, level of evidence B), whereas chimney grafts and in situ fenestration should be reserved exclusively for urgent or bail-out cases (Class III level of evidence C) [3].

Table 60.2 Type of endovascular repair, graft configurations, and applicability of the different solutions currently available on the market for aortic arch repair. Recommendations are extracted from latest ESVS/EACTS consensus document [3]

Arch repair	Device configuration	Recommendation	Applicability	Technical details	Company devices
Hybrid arch repair	– Tube graft – Surgical debranching	Class IIb, level B	– No customization time – Option for distal arch lesions – No possible total endovascular arch repair	– Tube graft – Need for supra-aortic debranching	– Any companies
Total endovascular arch repair	– Parallel graft – In-situ/physician – Modified grafts	Class III, level C	– Only bail-out cases – Emergency cases	– Tube graft – Needle puncture – Laser puncture – Back table modifications – Chimney/snorkel	– Any companies
Total endovascular arch repair	– Fenestrated graft – Scalloped graft	Class IIa, level B	– Mid-distal arch (zone 1 and zone 2) – Ascending aorta (zone 0) – Customization time	– Proximal scallop – Proximal scallop + 1/2 fenestration – 3 fenestrations – Preloaded fenestrations	– Kawasumi Najuta – Cook medical – Bolton relay
Total endovascular arch repair	– Branched graft	Class IIa, level B	– Ascending aorta (zone 0) – Mid-distal arch (zone 1 and zone 2) – Customization time	– Single antegrade branch – Single retrograde branch – Double antegrade inner branch – Double antegrade + single retrograde branches – Triple antegrade branches	– Nexus Endospan – Gore TBE – Bolton relay – Cook medical – Inoue graft

Fenestrated/Scallop Endograft of the Aortic Arch

Fenestrated and scallop endografts of the aortic arch are custom-made devices designed based on the specific anatomy of the treated patient to treat mid-distal arch pathologies (Fig. 60.1) [28].

Their typical configuration is with a pre-cannulated fenestration for LSA and a scallop for LCCA, or in the case of a bovine arch for the IA. In the case of a fenestration for LCCA, the scallop is allocated for IA, and the LSA is revascularized by performing a carotid-LSA bypass in standard fashion.

Fenestrated and scallop endografts can be used to treat extended TAAAs when the healthy sealing zone is proximally extended into the arch to preserve vascularization of the LSA without the need for an open surgical debranching step. In this regard, there is extensive experience with Cook Medical (Bloomington, IN, USA) devices [29]. Fenestrated endografts in the arch are usually delivered with a pre-cannulated fenestration that allows the establishment of a through-and-through wire. This route facilitates cannulation of the vessel and advancement of the delivery system, reducing the manipulation in the aortic arch. At the same time, deployment of the graft should be extremely precise to place the fenestration in front of the appropriate vessel and prevent inadvertent coverage of the target vessel. After introducing the graft from the femoral artery, systemic heparinization is performed to reach an activated clotting time (ACT) greater than 350 and flushing of the graft is done with heparinized saline greater than 120 cc (or flushing with CO_2) [30]. The main body is released after reduction of cardiac output (e.g., rapid ventricular pacing, balloon occlusion of the inferior vena cava, Munich Valsalva maneuver (MUVIT) [31]).

Deployment of the bridging stent-graft is performed in a similar manner to FEVAR in the abdominal aorta, and the use of a balloon-expandable stent-graft of appropriate size and expansion is recommended, with the use of post-implantation flaring [20]. The fenestration technique has two main disadvantages. The first is the presence of a large aneurysm with a

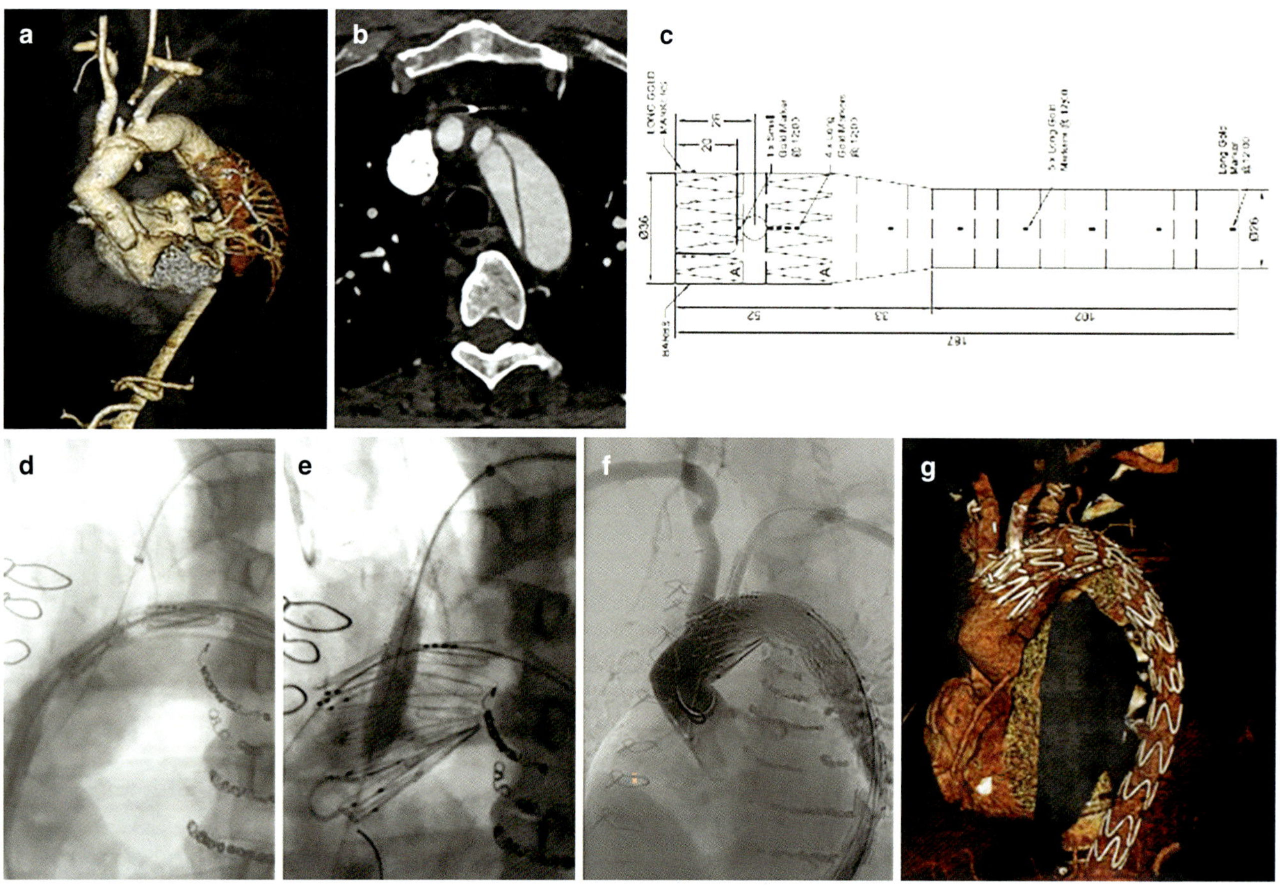

Fig. 60.1 (**a**) Three-dimensional representation of the case of residual thoracoabdominal dissection after ascending aortic repair reconstruction with dissection flap at the level of the left common carotid artery (LCCA) (**b**). (**c**) Scalloped endograft by Cook Medical (Bloomington, IN, USA) for the brachiocephalic trunk (BCT), and fenestration for LCCA and preloaded catheter for enabling precise cannulation (**d**). (**e**) Flaring of the proximal part of the stent-graft for LCCA. (**f**) Completion angiography with patency of all supra-aortic vessels and patency of carotid-subclavian bypass confirmed by post-operative computed tomography angiography (**g**)

greater distance between the fenestration and the target vessel: the bridging stents would be subjected to extreme mechanical stress, so there is a risk of rupture, migration and the development of higher rates of type I-III endoleaks when compared to branched endograft [25]. The second situation is in cases where the healthy sealing zone is located in zone 0, such as type A dissection, in which fenestrated grafts may be at high risk of bird-beak and type Ia endoleak, or the inability to ensure effective vascularization of the supra-aortic vessels.

Analyzing the results of fenestrated TEVAR for aortic arch demonstrated a low rate for the need to perform a debranching procedure of the aortic arch [32, 33], and in experienced centers [32], the need for supra-aortic vessel debranching was around 40%. Technical success rate was 94%, and failures were due to misalignment of the endograft and loss of target vessel resulting in stroke and death. The largest cohorts were described by Yokoi et al. [34] with the Najuta system, and Tsilimparis et al. [20] with the Zenith Cook Platform reported an overall mortality rate of 1.6–3.7% and stroke rate from 1.8% to 5.6%.

Branched Endograft of the Aortic Arch

To overcome main disadvantages of fenestrated endografts reported by Tsilimparis et al., [32] custom-made branched endograft devices have been developed, with Cook Medical being the most widely used since 2009. The specific design of the Cook Medical arch branched endograft was developed to address the need for aneurysmal dissection [33]. The endograft (Fig. 60.2) is fabricated with one or two proximal sealing stents along with barbs for active fixation to prevent distal migration. The delivery system is a 100 cm introducer sheath with a length of 22–24 Fr. To accommodate the anatomy of the aortic arch, the device has a self-orienting system in which the diamond-shaped funnels for the supra-aortic vessels are placed at the outer curvature of the device. The branches are placed in the narrower portion of the endograft to allow partial perfusion of the supra-aortic vessels during graft implantation. The graft implantation [35–37] is performed through a femoral artery access. Preventive measures (heparinization and extensive flushing [30]) and cardiac out-

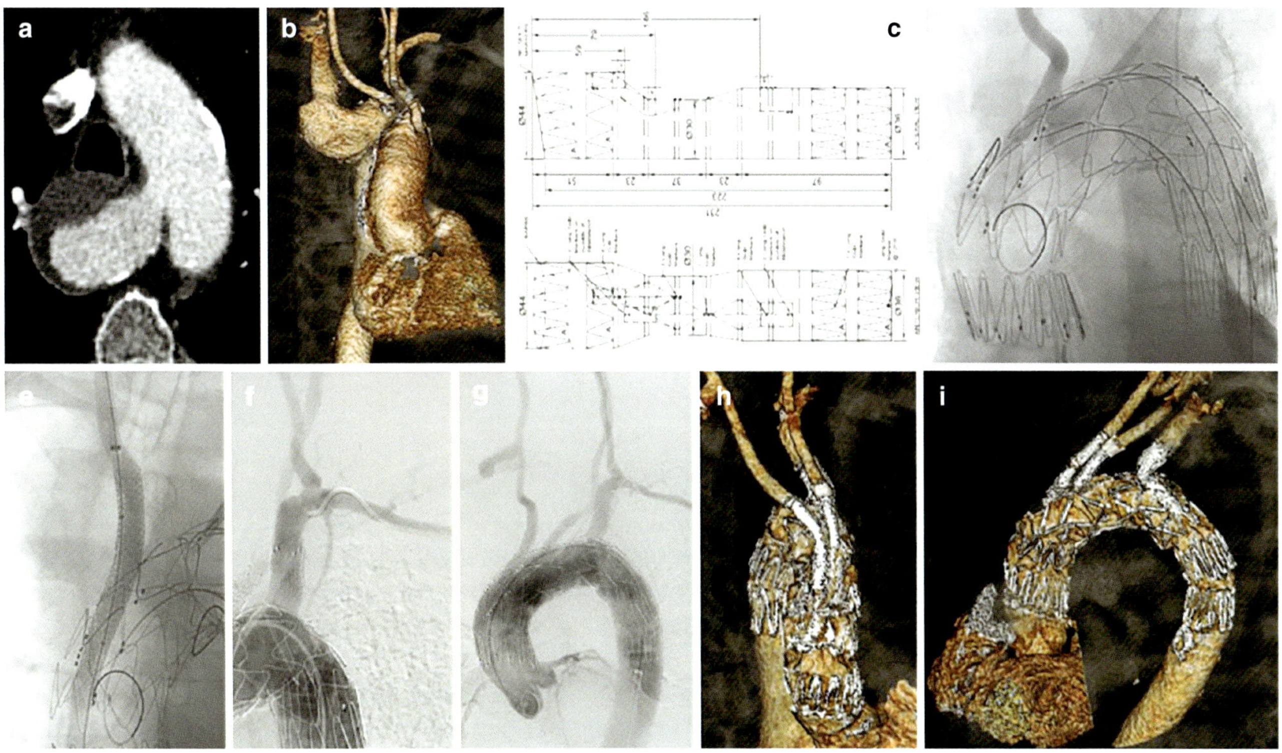

Fig. 60.2 (**a**) Two-dimensional and three-dimensional (**b**) representation of the case of aberrant right subclavian artery (arteria lusoria) associated to Kommerell's diverticulum. The procedure was carried out using a triple-inner-branched endograft (**c**) by Cook Medical (Bloomington, IN, USA) with revascularization of brachiocephalic trunk (BCT), left common carotid artery (LCCA), and left subclavian arteryP (LSA). (**d**) Details of the deployment of the graft under cardiac output reduction, and detailed revascularization through the inner branched balloon-expandable stent-graft of the BCT (**e**) and of the retrograde branch for LSA with a combination of balloon-expandable and self-expandable stent-grafts (**f**). (**g**) Completion angiography showing exclusion of the lesion and patency of all the vessels confirmed by postoperative computed tomography angiography (**h, i**). Image partially reproduced and published in the article [42]

put reduction mechanisms should be taken, as described for the fenestrated graft deployment. Bridging stent placement can be performed with surgical exposure of supra-aortic vessels or in different deployment configurations in order to place a bridging stent-graft. In the case of a double-branch device, revascularization of the LSA should be performed in a preceding step (in the same operation or in another preceding step) using an LCCA-LSA prosthetic bypass as described. New technologies are emerging with the use of three-inner-branch devices to treat all supra-aortic vessels using endovascular repair, with the goal of minimizing the risk of access-related complications of subclavian bypass or transposition [38–40]. In this case, after the deployment of two inner branches, the LSA is cannulated via a special pre-cannulated pathway, and a bridging stent is placed to allow retrograde vascularization. This retrograde revascularization route allows to have more space in the aortic arch and to reduce the conflict with other stent-grafts directed to the carotid arteries and to preserve the LSA for future endovascular interventions with a direct route down into the descending thoracic aorta [41].

Other branched devices are available on the market, and the majority of them present a single-branched configuration [43]. They can be mainly used for thoracoabdominal aneurysms, with the landing zone extending to zone 2 in order to revascularize LSA: Valiant Mona LSA by Medtronic (Santa Rosa, CA) and Gore TBE (Flagstaff, AZ). In the case of Zone 0 deployment these devices may be used in order to revascularize IA, and then revascularization of other branches is acquired via cervical debranching (Nexus stent-graft for zone 0 from Endospan [Herzlia, Israel]) [23, 24, 27].

The other device available for landing in zone 0 for direct revascularization of the IA and LCCA is the Relay Branch thoracic stent-graft system (Bolton Medical, Terumo Aortic, Sunrise, FL, USA), which is constructed with two anterograde tunnels leading to a large cannulation window [44]. Supra-aortic branch grafts are placed in the tunnels. The tunnels are equipped with lock stents that securely engage the stent tips of the branch grafts to prevent modular disconnection. Usually, LSA is revascularized with a carotid-subclavian bypass during the same procedure. The main body is inserted via the common femoral arteries. After deployment of the main body, the tunnels are cannulated and the supra-aortic extensions are inserted. Initial experience with the new platform using Terumo Aortic triple branch endograft is emerging [45] .

Large clinical experiences with Cook arch branched device were described in 2016 on 27 patients and in 2018 on 30 patients [35, 36]. In both experiences the technical success was 100% with no perioperative mortality events. The rate of 30-day mortality was ranging from 3.7% to 10%, while the cumulated stroke rate (transient and permanent) was between 3.3% and 11% and SCI between 0% and 7%. These results represent a high percentage of major complications, but the high-risk nature of patients that are unfit for open surgical repair should be taken into consideration [18]. At the same time, the usage of a branched endograft after previous open aortic surgical repair of the ascending aorta showed satisfactory results with 0% mortality, stroke, and SCI rates at 30 days, suggesting that ascending aortic surgery may represent an ideal landing zone for branched devices in the aortic arch [46]. For Relay Branch thoracic stent-graft system (Terumo Aortic) the in- hospital mortality rate of 15 patients was 6.7%, the disabling stroke rate was 6.7%, and the nondisabling stroke rate was 13.3%. Type I and III endoleaks occurred in 6.7% of the patients.

A recent multicenter global experience study has demonstrated the technical feasibility and safety of total endovascular aortic arch repair for aneurysms and chronic dissections using triple inner branch stent-grafts. The mortality and stroke rates were compared favorably with those after open surgical repair in a higher-risk group of patients, with 5% both of in-hospital mortality and post-operative stroke rates. However, the rate of secondary interventions was still high (31%), emphasizing the need for learning curve experience and closer follow-up [38, 47] .

Parallel Grafts

Chimney technique was initially indicated for the treatment of infrarenal aortic aneurysms with short neck. Since then, it has been also used in the treatment of aortic pathology [48]. The technique uses several parallel covered stents permitting the extension of the proximal sealing zone beyond the origin of the supra-aortic vessels [49, 50]. Nowadays, many different variations of this technique exist. The standard chimney procedure (Fig. 60.3) consists of the deployment of a standard thoracic stent-graft with a landing zone over the ostia of supra-aortic vessels. In parallel, stent-grafts are placed and proximally oriented in order to receive antegrade blood flow and perfuse the target vessel. In the case of incapability to reach a proximal perfusion, a periscope technique can be performed with retrograde blood flow to perfuse the supra-aortic vessels [3]. The main drawback of this technique is type 1a endoleaks. This issue is especially important in pathologies affecting the outer curvature of the aorta and is less likely to occur if the pathology affects the inner curvature. Gutters caused by the parallel stents, even in the absence of obvious high-flow endoleaks, can lead to a higher risk of rupture [51]. A recent meta-analysis of 314 cases showed an early type 1 endoleak rate of 11% [52]. Similarly, a group of 94 patients with 101 chimney grafts in the aortic arch reported a perioperative mortality rate of 3%, stroke rate of 5%, and endoleak rate of 18% [53].

The chimney technique represents a useful treatment option in patients requiring urgent aortic arch repair and as a bail-out technique in accidental covering of supra-aortic vessels [3].

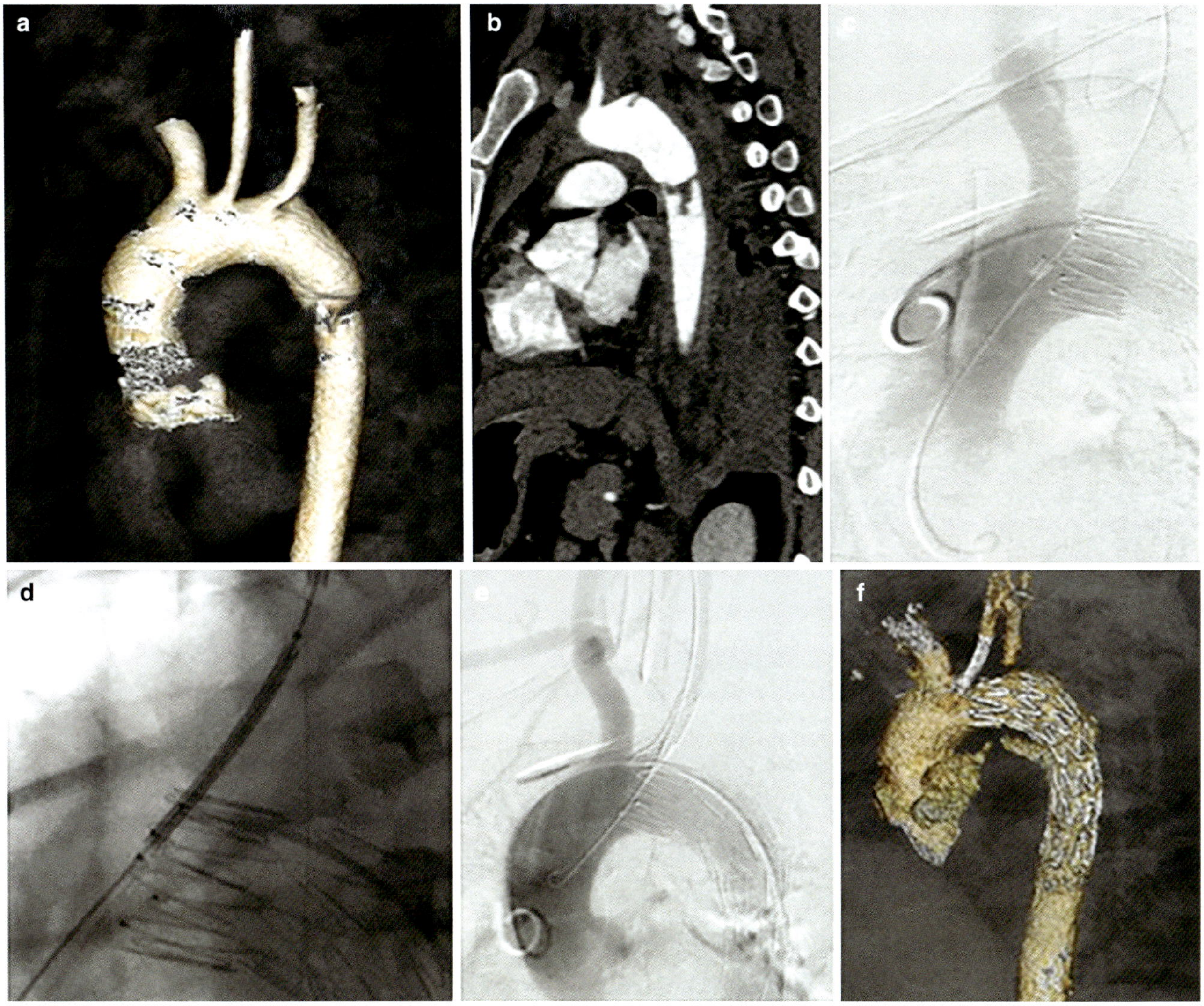

Fig. 60.3 (**a**) Three-dimensional and two-dimensional (**b**) representation of the case of transected aorta after accident treated with chimney technique. (**c**) After deployment of thoracic graft with guide inserted into the aortic arch from the left common carotid artery (LCCA) a balloon-expandable stent-graft (**d**) is inserted into the LCCA and placed proximally to the thoracic endograft. (**e**) Completion angiography showing exclusion of the aortic rupture, patency of LCCA, and patency though collaterals of the left subclavian confirmed by post-operative computed tomography angiography (**f**)

In-Situ/Physician-Modified Fenestrated Grafts

In the case of emergent case, with suitable anatomy, and where the risk of gutter after chimney is too high, the performance of an in-situ/physician-modified fenestration of the aortic arch can be useful.

Physician-modified fenestrations have been developed in the field of complex abdominal aortic repair [54, 55], and case series with acceptable results and durability are present in literature to treat localized lesion of the aortic arch as well as dissection [56, 57].

The fenestration (Fig. 60.4) is performed after an accurate study of the pre-operative images in the operating room in sterile fashion. A standard thoracic endograft is carefully deployed, and the location of the fenestration is marked. After the creation of an adequate-sized hole in the graft, the fenestration is reinforced and adequately marked with radiopaque material. Eventually the graft is accurately re-shielded inside the introducer sheath.

The main advantage of this technique is the presence of perfusion of the target supra-aortic vessel after the deployment of the graft through the fenestration, while understandably extra caution must be performed during fenestration crafting and graft deployment [54, 59, 60].

In the case of in-situ fenestration, fenestration for the supra-aortic vessels can be made in situ using a catheter-driven needle, laser, or radiofrequency (Fig. 60.5). Redlinger et al. published a series of 22 patients undergoing laser fenestration of left subclavian artery after emergency TEVAR for different pathologies. The treated patients did not develop any serious complications, and the in-hospital mortality rate

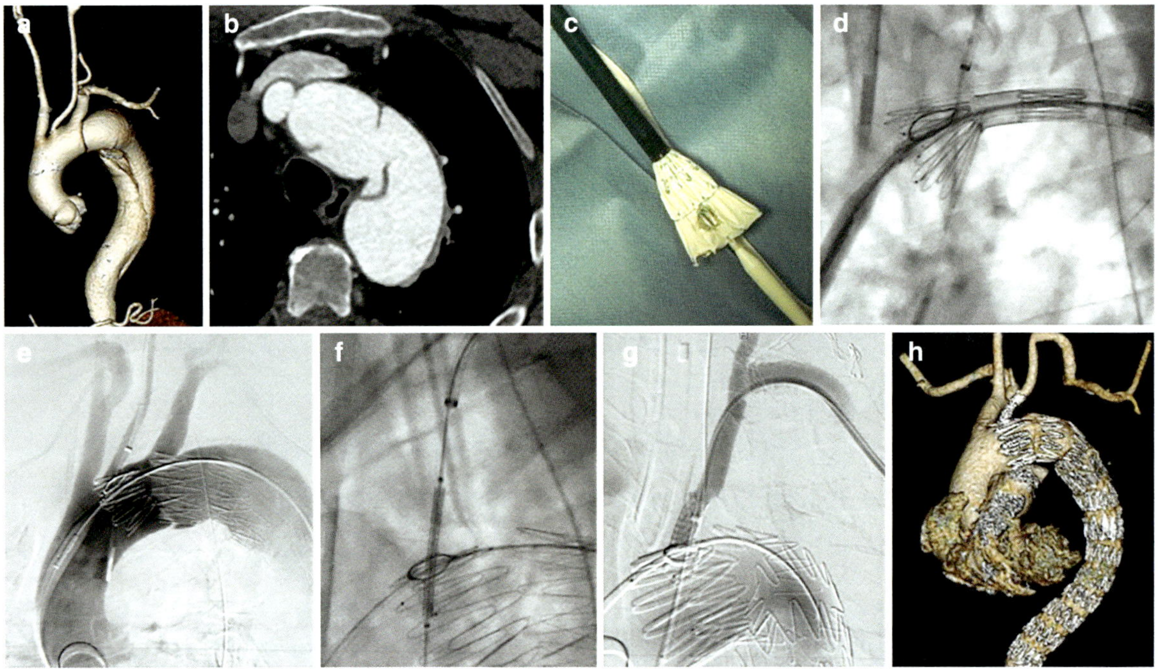

Fig. 60.4 (**a**) Three-dimensional and two-dimensional (**b**) computed tomography angiography (CTA) representation of the case of nonA-nonB aortic dissection. The patient was treated with physician-modified endograft (PMEG) (**c**) for the left common carotid artery (LCCA). (**d**) Details of the deployment of the graft where is clear visible the fenestration for LCCA and passages of deployment of the stent-graft into the dedicated fenestration through a carotid-subclavian bypass performed before stent-graft introduction (**e**–**g**). (**h**) Post-operative CTA showing patency of the true lumen and patency of LCCA and left subclavian artery through the surgical bypass . Images partially published in the article [58]

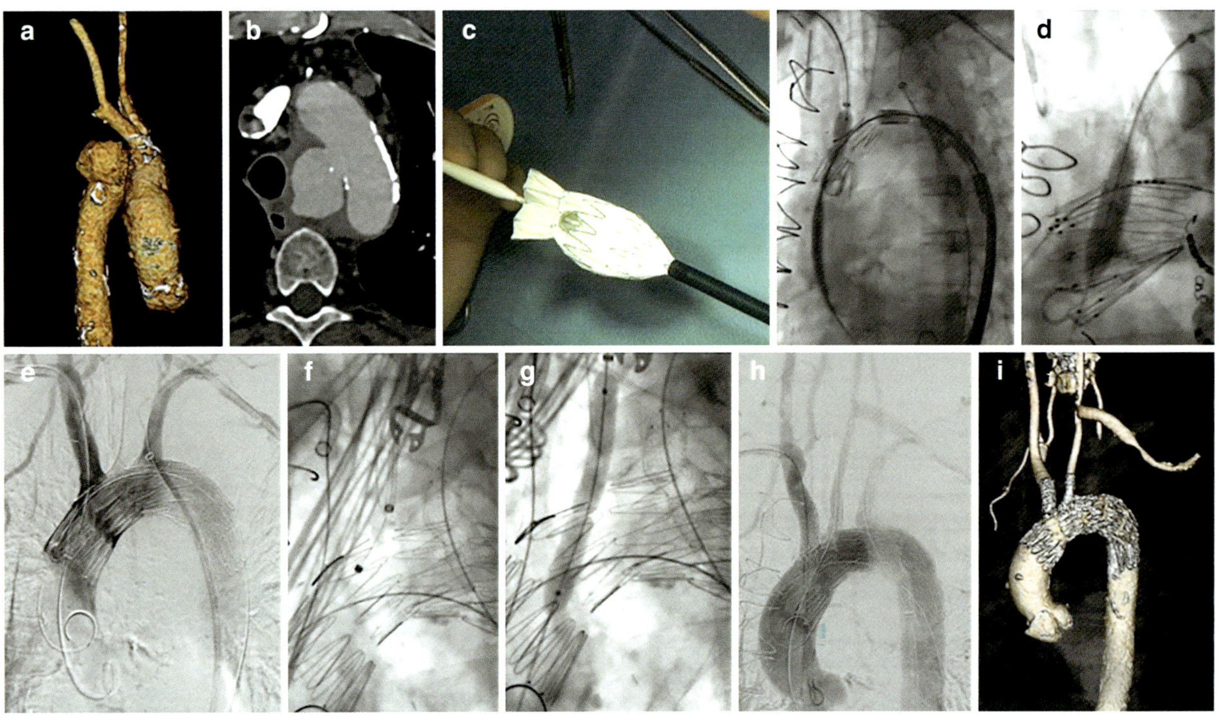

Fig. 60.5 (**a**) Three-dimensional and two-dimensional (**b**) computed tomography angiography (CTA) representation of the case of mycotic rupture of the aortic arch. The patient was treated with physician-modified endograft (PMEG) (**c**) for the brachiocephalic trunk (BCT) with deployment showed in images (**c**) and (**d**) with stent-graft deployment. Left subclavian artery (LSA) was revascularized through a carotid-subclavian bypass performed before stent-graft introduction and retrograde periscope stent-graft in order to maintain adequate cerebral perfusion (**e**). Left common carotid artery (LCCA) was revascularized with an in-situ laser fenestration that was first punctured (**f**) and then dilatated with balloon angioplasty (**g**) before stent-graft deployment. (**h**) Completion angiography showing exclusion of the aortic rupture, patency of BCT through PMEG fenestration and of LCCA through in-situ laser fenestration and LSA revascularized both from surgical bypass and periscope parallel graft, confirmed by post-operative computed tomography angiography (**i**)

was 4.5% [61]. Kopp et al. demonstrated a series of 25 patients who underwent in situ fenestrations of the brachiocephalic trunk, left common carotid artery, and left subclavian artery using needles or radiofrequency or laser catheters. Results showed a technical success of 94%, a perioperative mortality of 1%, and a stroke risk of 12% [62]. However, larger studies with more patients are needed to address the long-term results and endograft-related complications.

Future Perspectives

Custom-made endografts require a manufacturing period which can last weeks, and during this period, patients may experience aneurysm rupture and death, especially with larger aneurysms [63]. Therefore they are not available in urgent cases involving the aortic arch. Based on the good results of custom-made branched endografts in comparison with fenestrated ones [24, 32], there is a need for commercially available branched aortic arch endografts. Recent studies have shown that it is quite feasible to use available configurations [64] that cover up to 79% [65] of the arch anatomy and may soon be proposed as available grafts for the treatment of urgent cases. Rates of access complications and nerve injury are still a major problem in supra-aortic vessel open surgical treatments [38, 66]; therefore, the possibility of complete percutaneous repair of the aortic arch has been proposed by experienced centers [67].

References

1. Donas KP, Rancic Z, Lachat M, Pfammatter T, Frauenfelder T, Veith FJ, Mayer D. Novel sutureless telescoping anastomosis revascularization technique of supra-aortic vessels to simplify combined open endovascular procedures in the treatment of aortic arch pathologies. J Vasc Surg. 2010;51:836. https://doi.org/10.1016/j.jvs.2009.09.054.
2. Lioupis C, Abraham CZ. Results and challenges for the endovascular repair of aortic arch aneurysms. Perspect Vasc Surg Endovasc Ther. 2011;23:202. https://doi.org/10.1177/1531003511413608.
3. Czerny M, Schmidli J, Adler S, et al. Editor's choice—current options and recommendations for the treatment of thoracic aortic pathologies involving the aortic arch: an expert consensus document of the European Association for Cardio-Thoracic Surgery (EACTS) & the European Society for Vascular Surgery (ESVS). Eur J Vasc Endovasc Surg. 2019; https://doi.org/10.1016/j.ejvs.2018.09.016.
4. Inoue K, Sato M, Iwase T, Yoshida Y, Tanaka T, Tamaki S, Yamazato A. Clinical endovascular placement of branched graft for type B aortic dissection. J Thorac Cardiovasc Surg. 1996;112:1111–3.
5. Nathan DP, Boonn W, Lai E, Wang GJ, Desai N, Woo EY, Fairman RM, Jackson BM. Presentation, complications, and natural history of penetrating atherosclerotic ulcer disease. J Vasc Surg. 2012;55:10. https://doi.org/10.1016/j.jvs.2011.08.005.
6. Eggebrecht H, Plicht B, Kahlert P, Erbel R. Intramural hematoma and penetrating ulcers: indications to endovascular treatment. Eur J Vasc Endovasc Surg. 2009;38:659. https://doi.org/10.1016/j.ejvs.2009.09.001.
7. Hiratzka LF, Bakris GL, Beckman JA, et al. 2010 ACCF/AHA/AATS/ACR/ASA/SCA/SCAI/SIR/STS/SVM guidelines for the diagnosis and management of patients with thoracic aortic disease. Circulation. 2010;121:e266. https://doi.org/10.1161/CIR.0b013e3181d4739e.
8. Shahoud JS, Miao JH, Bolla SR. Anatomy, thorax, heart aorta; 2021.
9. Evangelista A, Salas A, Ribera A, Ferreira-González I, Cuellar H, Pineda V, González-Alujas T, Bijnens B, Permanyer-Miralda G, Garcia-Dorado D. Long-term outcome of aortic dissection with patent false lumen. Circulation. 2012;125:3133. https://doi.org/10.1161/CIRCULATIONAHA.111.090266.
10. Larson EW, Edwards WD. Risk factors for aortic dissection: a necropsy study of 161 cases. Am J Cardiol. 1984;53:849. https://doi.org/10.1016/0002-9149(84)90418-1.
11. Nienaber CA, Eagle KA. Aortic dissection: new Frontiers in diagnosis and management. Circulation. 2003;108:628. https://doi.org/10.1161/01.CIR.0000087009.16755.E4.
12. Rylski B, Pérez M, Beyersdorf F, Reser D, Kari FA, Siepe M, Czerny M. Acute non-a non-B aortic dissection: incidence, treatment and outcome. Eur J Cardiothorac Surg. 2017;52:1111. https://doi.org/10.1093/ejcts/ezx142.
13. Stanson AW, Kazmier FJ, Hollier LH, Edwards WD, Pairolero PC, Sheedy PF, Joyce JW, Johnson MC. Penetrating atherosclerotic ulcers of the thoracic aorta: natural history and clinicopathologic correlations. Ann Vasc Surg. 1986;1:15. https://doi.org/10.1016/S0890-5096(06)60697-3.
14. Coady MA, Rizzo JA, Hammond GL, Pierce JG, Kopf GS, Elefteriades JA. Penetrating ulcer of the thoracic aorta: what is it? How do we recognize it? How do we manage it? J Vasc Surg. 1998;27:1006. https://doi.org/10.1016/S0741-5214(98)70003-5.
15. Riambau V, Böckler D, Brunkwall J, et al. Editor's choice—management of descending thoracic aorta diseases. Eur J Vasc Endovasc Surg. 2017;53:4. https://doi.org/10.1016/j.ejvs.2016.06.005.
16. Crawford ES, Crawford JL, Safi HJ, Coselli JS, Hess KR, Brooks B, Norton HJ, Glaeser DH. Thoracoabdominal aortic aneurysms: preoperative and intraoperative factors determining immediate and long-term results of operations in 605 patients. J Vasc Surg. 1986;3:389. https://doi.org/10.1067/mva.1986.avs0030389.
17. Safi HJ, Miller CC. Spinal cord protection in descending thoracic and thoracoabdominal aortic repair. Ann Thorac Surg. 1999;67:1937. https://doi.org/10.1016/S0003-4975(99)00397-5.
18. Haulon S, Greenberg RK, Spear R, et al. Global experience with an inner branched arch endograft. J Thorac Cardiovasc Surg. 2014;148:1709. https://doi.org/10.1016/j.jtcvs.2014.02.072.
19. Zhu Y, Zhan W, Hamady M, Xu XY. A pilot study of aortic hemodynamics before and after thoracic endovascular repair with a double-branched endograft. Med Novel Technol Devices. 2019;4:100027.
20. Tsilimparis N, Prendes CF, Rouhani G, Adam D, Dias N, Stana J, Rohlffs F, Mani K, Wanhainen A, Kölbel T. Pre-loaded fenestrated thoracic Endografts for distal aortic arch pathologies: multicentre retrospective analysis of short and mid term outcomes. Eur J Vasc Endovasc Surg. 2021;62:887–95.
21. Spear R, Azzaoui R, Maurel B, Sobocinski J, Roeder B, Haulon S. Total endovascular treatment of an aortic arch aneurysm in a patient with a mechanical aortic valve. Eur J Vasc Endovasc Surg. 2014;48:144. https://doi.org/10.1016/j.ejvs.2014.05.011.
22. Konstantinou N, Peterss S, Stana J, Rantner B, Banafsche R, Pichlmaier M, Tsilimparis N. Passing a mechanical aortic valve with a short tip dilator to facilitate aortic arch endovascular branched repair. J Endovasc Ther. 2021;28:388–92.

23. Brown JA, Arnaoutakis GJ, Szeto WY, Serna-Gallegos D, Sultan I. Endovascular repair of the aortic arch: state of the art. J Card Surg. 2021;36:4292–300.
24. Spanos K, Tsilimparis N, Rohlffs F, Wipper S, Detter C, Behrendt C-A, Debus SE, Kölbel T. Total endovascular arch repair is the procedure of the future. J Cardiovasc Surg. 2018;59:559–71. https://doi.org/10.23736/S0021-9509.18.10412-5.
25. Spath P, Campana F, Tsilimparis N, Gallitto E, Pini R, Faggioli G, Caputo S, Gargiulo M. Outcomes of fenestrated and branched Endografts for partial and Total endovascular repair of the aortic arch - a systematic review and meta-analysis. Eur J Vasc Endovasc Surg. 2023 Aug 2;S1078-5884(23):00614–7 https://doi.org/10.1016/j.ejvs.2023.07.048. Epub ahead of print. PMID: 37536517.
26. Cao P, de Rango P, Czerny M, Evangelista A, Fattori R, Nienaber C, Rousseau H, Schepens M. Systematic review of clinical outcomes in hybrid procedures for aortic arch dissections and other arch diseases. J Thorac Cardiovasc Surg. 2012;144:1286. https://doi.org/10.1016/j.jtcvs.2012.06.013.
27. Blanco Amil CL, Mestres Alomar G, Guarnaccia G, Luoni G, Yugueros Castellnou X, Vigliotti RC, Ramses R, Riambau V. The initial experience on branched and fenestrated endografts in the aortic arch. A systematic review. Ann Vasc Surg. 2021;75:29–44.
28. Chuter TAM, Schneider DB. Endovascular repair of the aortic arch. Perspect Vasc Surg Endovasc Ther. 2007;19:188. https://doi.org/10.1177/1531003507304165.
29. Gallitto E, Gargiulo M, Freyrie A, Mascoli C, Massoni Bianchini C, Ancetti S, Faggioli G, Stella A. The endovascular treatment of juxta-renal abdominal aortic aneurysm using fenestrated endograft: early and mid-term results. J Cardiovasc Surg. 2019;60:237. https://doi.org/10.23736/S0021-9509.16.09049-2.
30. Kölbel T, Rohlffs F, Wipper S, Carpenter SW, Debus ES, Tsilimparis N. Carbon dioxide Flushing technique to prevent cerebral arterial air embolism and stroke during TEVAR. J Endovasc Ther. 2016;23:393. https://doi.org/10.1177/1526602816633705.
31. Tsilimparis N, Abicht J-M, Stana J, Konstantinou N, Rantner B, Banafsche R, Fernandez Prendes C. The Munich Valsalva implantation technique (MuVIT) for cardiac output reduction during TEVAR: vena cava occlusion with the valsalva maneuver. J Endovasc Ther. 2021;28:7–13.
32. Tsilimparis N, Debus ES, von Kodolitsch Y, Wipper S, Rohlffs F, Detter C, Roeder B, Kölbel T. Branched versus fenestrated endografts for endovascular repair of aortic arch lesions. J Vasc Surg. 2016;64:592. https://doi.org/10.1016/j.jvs.2016.03.410.
33. Lioupis C, Corriveau M-M, MacKenzie KS, Obrand DI, Steinmetz OK, Abraham CZ. Treatment of aortic arch aneurysms with a modular transfemoral multibranched stent graft: initial experience. Eur J Vasc Endovasc Surg. 2012;43:525. https://doi.org/10.1016/j.ejvs.2012.01.031.
34. Yokoi Y, Azuma T, Yamazaki K. Advantage of a precurved fenestrated endograft for aortic arch disease: simplified arch aneurysm treatment in Japan 2010 and 2011. J Thorac Cardiovasc Surg. 2013;145:S103–9.
35. Clough RE, Spear R, van Calster K, Hertault A, Azzaoui R, Sobocinski J, Fabre D, Haulon S. Case series of aortic arch disease treated with branched stent-grafts. Br J Surg. 2018;105:358. https://doi.org/10.1002/bjs.10681.
36. Spear R, Haulon S, Ohki T, Tsilimparis N, Kanaoka Y, Milne CPE, Debus S, Takizawa R, Kölbel T. Editor's choice—subsequent results for arch aneurysm repair with inner branched endografts. Eur J Vasc Endovasc Surg. 2016;51:380. https://doi.org/10.1016/j.ejvs.2015.12.002.
37. Spanos K, Panuccio G, Rohlffs F, Heidemann F, Tsilimparis N, Kölbel T. Technical aspects of branched thoracic arch graft implantation for aortic arch pathologies. J Endovasc Ther. 2020;27:792. https://doi.org/10.1177/1526602820925443.
38. Tenorio ER, Oderich GS, Kölbel T, et al. Multicenter global early feasibility study to evaluate total endovascular arch repair using three-vessel inner branch stent-grafts for aneurysms and dissections. J Vasc Surg. 2021;74:1055. https://doi.org/10.1016/j.jvs.2021.03.029.
39. Spear R, Clough RE, Fabre D, Roeder B, Hertault A, Martin Gonzalez T, Azzaoui R, Sobocinski J, Haulon S. Total endovascular treatment of aortic arch disease using an arch Endograft with 3 inner branches. J Endovasc Ther. 2017;24:534. https://doi.org/10.1177/1526602817714569.
40. Haulon S, Soler R, Watkins AC, Amabile P, Fadel E, Fabre D. Endovascular arch replacement with an endoprosthesis with three inner branches. Ann Cardiothorac Surg. 2018;7:431. https://doi.org/10.21037/acs.2018.04.07.
41. Gallitto E, Faggioli G, Spath P, Marrozzini C, Pini R, Bruno AG, Logiacco A, Gargiulo M. Endovascular repair of an aortic aneurysm by a custom-made three-inner branched endograft. G Ital Cardiol (Rome). 2021;22:46S–50S.
42. Spath P, Marazzi G, Stana J, Peterss S, Fernandez-Prendes C, Rantner B, Pichlmaier MA, Tsilimparis N. Endovascular repair with triple inner-branch Endograft for aberrant subclavian artery aneurysm: a case report. J Endovasc Ther. 2022 Aug;8:15266028221116753. https://doi.org/10.1177/15266028221116753. Epub ahead of print. PMID: 35941824.
43. van Bakel TM, de Beaufort HW, Trimarchi S, Marrocco-Trischitta MM, Bismuth J, Moll FL, Patel HJ, van Herwaarden JA. Status of branched endovascular aortic arch repair. Ann Cardiothorac Surg. 2018;7:406. https://doi.org/10.21037/acs.2018.03.13.
44. Czerny M, Rylski B, Morlock J, et al. Orthotopic branched endovascular aortic arch repair in patients who cannot undergo classical surgery. Eur J Cardiothorac Surg. 2018;53:1007. https://doi.org/10.1093/ejcts/ezx493.
45. Negmadjanov U, Motta JC, Lee WA. Total endovascular aortic arch repair using the Terumo aortic triple-branch arch endograft. Ann Vasc Surg. 2021;77:351.e7–351.e14.
46. Milne CPE, Amako M, Spear R, Clough RE, Hertault A, Sobocinski J, Brown W, Haulon S. Inner-branched endografts for the treatment of aortic arch aneurysms after open ascending aortic replacement for type a dissection. Ann Thorac Surg. 2016;102:2028. https://doi.org/10.1016/j.athoracsur.2016.05.012.
47. Benfor B, Prendes CF, Peterss S, Stavroulakis K, Stana J, Pichlmaier M, Tsilimparis N. Anatomical analysis of the aortic arch and feasibility study of double and triple branched thoracic endografts in the treatment of isolated arch pathologies. Eur J Cardiothorac Surg, 2022:ezac246. https://doi.org/10.1093/ejcts/ezac246.
48. Moulakakis KG, Mylonas SN, Markatis F, Kotsis T, Kakisis J, Liapis CD. A systematic review and meta-analysis of hybrid aortic arch replacement. Ann Cardiothorac Surg. 2013;2:247. https://doi.org/10.3978/j.issn.2225-319X.2013.05.06.
49. Larzon T, Gruber G, Friberg Ö, Geijer H, Norgren L. Experiences of intentional carotid stenting in endovascular repair of aortic arch aneurysms—two case reports. Eur J Vasc Endovasc Surg. 2005;30:147–51.
50. Greenberg RK, Clair D, Srivastava S, Bhandari G, Turc A, Hampton J, Popa M, Green R, Ouriel K. Should patients with challenging anatomy be offered endovascular aneurysm repair? J Vasc Surg. 2003;38:990–6.
51. Donas KP, Marchiori E, Inchingolo M, Torsello G. Surgical treatment of endotension after chimney endovascular repair of a symptomatic Juxtarenal aneurysm. Ann Vasc Surg. 2017;41:279.e5–8.
52. Lindblad B, Bin Jabr A, Holst J, Malina M. Chimney grafts in aortic stent grafting: hazardous or useful technique? Systematic review of current data. Eur J Vasc Endovasc Surg. 2015;50:722–31.
53. Hogendoorn W, Schlösser FJV, Moll FL, Sumpio BE, Muhs BE. Thoracic endovascular aortic repair with the chimney graft technique. J Vasc Surg. 2013;58:502–11.

54. Tsilimparis N, Banafsche R, Stana J. Commentary: physician-modified endografts: surmounting anatomical challenges with innovative techniques to optimize treatment. J Endovasc Ther. 2020;27:130–1.
55. Oderich GS, Ribeiro MS, Sandri GA, Tenorio ER, Hofer JM, Mendes BC, Chini J, Cha S. Evolution from physician-modified to company-manufactured fenestrated-branched endografts to treat pararenal and thoracoabdominal aortic aneurysms. J Vasc Surg. 2019;70:31. https://doi.org/10.1016/j.jvs.2018.09.063.
56. Zhu J, Zhao L, Dai X, Luo Y, Fan H, Feng Z, Zhang Y, Hu F. Fenestrated thoracic endovascular aortic repair using physician modified stent grafts for acute type B aortic dissection with unfavourable landing zone. Eur J Vasc Endovasc Surg. 2018;55:170–6.
57. Joseph G, Premkumar P, Thomson V, Varghese M, Selvaraj D, Sahajanandan R. Externalized guidewires to facilitate fenestrated endograft deployment in the aortic arch. J Endovasc Ther. 2016;23:160–71.
58. Spath P, Stana J, Marazzi G, Peterss S, Fernandez-Prendes C, Tsilimparis N. Emergent physician modified carotid fenestrated TEVAR for the treatment of a complicated acute type nonAnonB aortic dissection with undetected multiorgan malperfusion. J Cardiovasc Surg. 2023 Aug;64(4):450–5. https://doi.org/10.23736/S0021-9509.22.12462-6. Epub 2022 Sep 28.
59. Ricotta JJ, Tsilimparis N. Surgeon-modified fenestrated-branched stent grafts to treat emergently ruptured and symptomatic complex aortic aneurysms in high-risk patients. J Vasc Surg. 2012;56:1535. https://doi.org/10.1016/j.jvs.2012.05.096.
60. Tsilimparis N, Heidemann F, Rohlffs F, Diener H, Wipper S, Debus ES, Kölbel T. Outcome of surgeon-modified fenestrated/branched stent-grafts for symptomatic complex aortic pathologies or contained rupture. J Endovasc Ther. 2017;24:825–32.
61. Redlinger RE, Ahanchi SS, Panneton JM. In situ laser fenestration during emergent thoracic endovascular aortic repair is an effective method for left subclavian artery revascularization. J Vasc Surg. 2013;58:1171–7.
62. Kopp R, Katada Y, Kondo S, et al. Multicenter analysis of endovascular aortic arch in situ stent-graft fenestrations for aortic arch pathologies. Ann Vasc Surg. 2019;59:36–47.
63. Gallitto E, Faggioli G, Spath P, Pini R, Mascoli C, Ancetti S, Stella A, Abualhin M, Gargiulo M. The risk of aneurysm rupture and target visceral vessel occlusion during the lead period of custom-made fenestrated/branched endograft. J Vasc Surg. 2020;72:16. https://doi.org/10.1016/j.jvs.2019.08.273.
64. Bosse C, Kölbel T, Mougin J, Kratzberg J, Fabre D, Haulon S. Off-the-shelf multibranched endograft for total endovascular repair of the aortic arch. J Vasc Surg. 2020;72:805. https://doi.org/10.1016/j.jvs.2019.11.046.
65. Spanos K, Haulon S, Eleshra A, Rohlffs F, Tsilimparis N, Panuccio G, Kölbel T. Anatomical suitability of the aortic arch arteries for a 3-inner-branch arch endograft. J Endovasc Ther. 2021;28:14. https://doi.org/10.1177/1526602820953634.
66. Konstantinou N, Kölbel T, Debus ES, Rohlffs F, Tsilimparis N. Fenestrated versus debranching thoracic endovascular aortic repair for endovascular treatment of distal aortic arch and descending aortic lesions. J Vasc Surg. 2021;73:1915–24.
67. Mougin J, Azogui R, Guihaire J, Tyrrell MR, Oderich GS, Fabre D, Haulon S. "First in man" total percutaneous aortic arch repair with 3-inner-branch endografts. Ann Surg. 2021;274:e652–7.

Endovascular Management of Thoracic Aortic Trauma

61

Ludovic Canaud and Pierre Alric

Introduction

Blunt traumatic thoracic aortic injury is associated with a high mortality rate, and has been implicated as the second most common cause of death in trauma patients, behind only to intracranial hemorrhage. It has been estimated that less than 25% of patients with such an injury live to be evaluated in a hospital, and of those who do, up to 50% will die within 24 h. Given the location of injury in 50–70% of cases, conventional surgical repair typically involves a high posterolateral thoracotomy with or without cardiopulmonary bypass and significant blood loss, which can negatively impact the pulmonary, cardiac, and neurologic status of the patient. Historically, open repair of traumatic aortic injuries has been associated with a 28% mortality rate and a 16% paraplegia rate. There has been a risk of delayed rupture in the unrepaired thoracic transection that has been estimated to be 2–5%.

The treatment of traumatic transection of thoracic aorta has seen many changes over the past 20 years. Thoracic endovascular aortic repair (TEVAR) has been rapidly adopted as an alternative to the traditional open repair for treatment of these lesions. Two prospective observational studies from the American Association for the Surgery of Trauma (AAST), evaluating the management and outcomes of blunt thoracic artery injury studies, demonstrated a marked change in operative management, from exclusively open repairs (100%) in the early part of that period, to a majority of endovascular repairs (65%) [1].

This paradigm shift has improved the mortality rate in these patients, as demonstrated by the second AAST study [2]: 7.2% for endovascular repair versus 23.5% for open repair. Stent grafting has become the first-line approach for traumatic thoracic aortic transections. Although many early reports are overwhelmingly positive, concerns have been raised regarding the long-term durability of this approach in these young patients. Long-term follow-up data are clearly critical to assess the durability of TEVAR in this younger population of patients, who have longer life expectancies than patients with aneurysmal disease do.

L. Canaud (✉) · P. Alric
Department of Thoracic and Vascular Surgery, Arnaud de Villeneuve Hospital, Montpellier, France
e-mail: l-canaud@chu-montpellier.fr; p-alric@chu-montpellier.fr

Limitations

Despite the focal nature of these aortic injuries, which may appear to favor endograft repair, there remains a number of unresolved issues: (1) poor conformation to the arch, (2) frequent need to cover the left subclavian artery (LSA), (3) uncertain natural history of the repair given the younger age of trauma victims and the morphologic changes of the aorta that come with age, (4) optimal follow-up strategy that may span several decades and the risks of cumulative radiation exposure, (5) timing of repair, and (6) the need for intraoperative anticoagulation in the setting of polytrauma.

Mechanism of Injury and Diagnosis

For blunt aortic injury, the portion of the thoracic aorta most susceptible to the stress produced on impact, particularly in cases of rapid deceleration, occurs distal to the left subclavian artery (LSA) and proximal to the third intercostal artery. This location is where the relatively mobile descending thoracic aorta joins the fixed aortic arch and is tethered 1 cm distally to the left pulmonary artery by the ligamentum arteriosum. This location, referred to as the aortic isthmus, is the most common site of injury,—noted in 54–66% of blunt aortic injury patients in autopsy series and >85% of patients who arrive to the hospital alive [3]. However, the exact mechanism by which these biomechanical forces translate to tissue disruption in blunt thoracic aortic injury (BTAI) is still debated, largely due to a lack of reproducible animal models and a wide spectrum of injury patterns.

G. Geroulakos et al. (eds.), *Mastering Endovascular Techniques*, https://doi.org/10.1007/978-3-031-42735-0_61

Chest radiography (CXR) is an adjunct to the Advanced Trauma Life Support secondary survey and is the initial imaging modality in both penetrating and blunt thoracic trauma. Some CXR clues to potential aortic injury include hemothorax, widened mediastinum, rib fractures, blunting of the aortic knob, or tracheal deviation. Unfortunately, the sensitivity of CXR for blunt aortic injury is <50%; thus, certain mechanisms of injury should trigger CTA, the gold standard diagnostic modality. Patients who have experienced a rapid deceleration event (e.g., a fall from >3 times an individual's height, a vehicle moving >40 mph, ejection from vehicle, significant vehicle intrusion, death in the same vehicle, vehicle moving >20 mph vs pedestrian or cyclist) should undergo CTA, which is >95% sensitive in detecting thoracic aortic injury.

Classification

Using imaging, we can classify traumatic rupture of the thoracic aorta into four categories based on severity (Fig. 61.1). These included grade 1, intimal tear; grade 2, intramural hematoma; grade 3, aortic pseudoaneurysm; and grade 4, free rupture.

Management of "Minimal Aortic Injury" (Periadventitial Defect or Hematoma)

Intimal or periadventitial defects or hematomas are not infrequently seen on computed tomography (CT) scan. A classification scheme for grading the severity of aortic injury has been proposed: type I (intimal tear), type II (intramural hematoma), type III (pseudoaneurysm), and type IV (rupture) (Fig. 61.1). The clinical practice guidelines of the Society for Vascular Surgery suggests expectant management with serial imaging for type I injuries, while types II to IV should be repaired [4]. This is based on early evidence that most type I injuries heal spontaneously. Decision to intervene and its timing should be guided by progression of the initial radiographic abnormality and/or symptoms.

Timing of Intervention

For grade 3 thoracic aortic trauma it can be challenging to determine whether to go forward with emergent repair versus wait 24–48 h to optimize management of other injuries before aortic intervention. Delaying intervention for 24 h is associated with a 50% reduction in mortality for patients

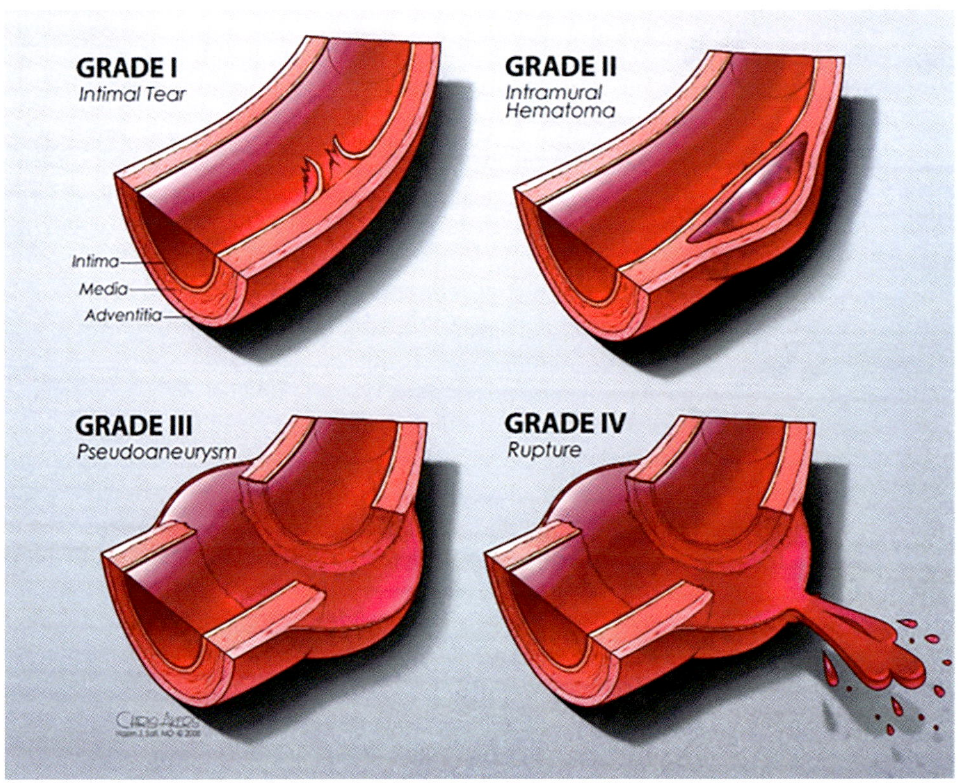

Fig. 61.1 Classification of traumatic aortic injury

with thoracic aortic trauma across equal injury severity scores [5]. Nevertheless, potential for bias exists in this retrospective assessment because patients selected for immediate intervention would likely have more concerning clinical presentations not detectable in large database retrospective reviews. A recent retrospective review of grade 3 injuries identified admission lactate >4 mm, posterior mediastinal hematoma >10 mm, and lesion/normal aortic diameter ratio >1.4 on the admission CT to be independently associated with aortic rupture [6]. Timing of stent graft deployment remains an individualized process, prioritizing immediate ongoing hemorrhage followed by hemodynamic stabilization and correction of any metabolic derangement before proceeding with TEVAR if feasible.

The debate regarding delay of treatment in those with other significant injuries is dampened by the use of the stent-graft. TEVAR can be performed just before or immediately after other necessary procedures with minimal compounded stress. Still, should delay seem prudent because of other critical needs, TEVAR can be accomplished on a delayed basis as well.

Real-world decision-making is confounded by variables such as availability of operating rooms, adequate range of implantables, appropriate equipment, and optimal staff to manage equipment. In addition, coexisting brain or spinal cord injuries may contraindicate a systolic BP goal of <130 mmHg and mandate TEVAR sooner.

Endovascular Exclusion

The guiding principle of TEVAR for trauma is to exclude the injured segment of aorta by landing an appropriately sized covered stent 2 cm proximal and distal to the injured segment, whereby blood flows through the stent graft and is excluded from potential extra-aortic flow.

Stent Graft Sizing

Diameter

Minimal stent oversizing of 5–10% of the native aortic diameter is recommended.

Trauma patients often present in a volume-depleted state, which can result in a reduction of aortic diameter estimation on the initial CTA.

The aortic diameter decreases dramatically in trauma patients with hemodynamic instability. This decrease in aortic diameter could theoretically lead to inaccurate aortic measurements and undersizing of the endograft in hemodynamically unstable patients requiring TEVAR [7]. Therefore, usage of IVUS at the time of TEVAR can be a helpful adjunct to measure aortic diameter during systole, intraoperatively after intravenous fluid administration [8].

Length

The shortest possible stent graft that achieves seal should be selected to minimize the risk of SCI.

Stent-Graft Placement

TEVAR can typically be accomplished via an entirely percutaneous approach using suture-mediated closure devices. If concomitant extremity injuries exist, a unilateral access approach can be achieved using a buddy wire for marker pigtail catheter insertion simultaneous to the large-bore TEVAR system. The extent of heparinization should be tailored to the presence and severity of coexisting injuries. Although the large-bore sheath necessary for TEVAR creates lower extremity thrombotic risk, anticoagulation use must be weighed against risk of intracranial or intracavitary hemorrhage.

After initial arterial access, power injection aortography should be performed in left lateral oblique projection (40°–46°) via a diagnostic catheter 5–6 cm proximal to the injury. Care should be taken to ensure no air is in the injector or catheter tubing before power injection to minimize air embolic stroke potential.

Angiography is ideally performed with a stiff guidewire with a curved tip simultaneously in place that courses the aortic arch and is parked in the ascending aorta to mimic the degree of tortuosity present when the stent graft is deployed. It is also beneficial to pass the stent graft device over the stiff wire into the segment just distal to the injury at the time of the immediate predeployment angiography. The stent graft should then be deployed with ideally 2 cm of fixation proximal to the injury. In-stent balloon angioplasty is reserved for type I endoleaks or incompletely expanded stent grafts with a bird's beak phenomenon detected on secondary angiography.

LSA Management

Suitability and the Unmet Needs of Current FDA-Approved Thoracic Endografts

With the availability of three commercially available devices, there was considerable divergence of opinion about the "best" device for use in traumatic thoracic aortic injury. There was a consensus, however, that arch conformation represented the single greatest unmet need given the curvature of the thoracic aorta at the location of the injury. Inability to conform to this curvature can result in malapposition of the endograft, which can lead to endoleak and endograft collapse [9].

The aortic diameters are relatively smaller in the younger subset of trauma patients. Currently, available thoracic endograft sizes mostly reflect the larger aortic diameters that would be typically encountered in an older cohort with degenerative aneurysms. Excessive oversizing may result in attachment site endoleak, device infolding, endograft collapse, and even death from acute aortic occlusion.

Endograft collapse represents a failure of the therapy and a marker of unsuitable anatomy, and open surgical conversion should be considered. No consensus could be reached regarding optimal oversizing for these cases, and opinions were equally divided among minimal to no oversizing, 5–10% oversizing, and standard oversizing per manufacturer's recommendations.

A number of next generation devices addresses in term of stent-graft conformability and other unmet needs of this therapy [10]. The use of more comformable and accurate stent-graft has significantly decreased the morbidity of endovascular repair of acute traumatic transection of the thoracic aorta [11].

LSA Management

The most common location of blunt aortic disruption is at the aortic isthmus, 2–4 cm distal to the LSA origin. Coverage of the LSA is therefore needed in 35% of BAI. Stroke risk is <3% with LSA coverage in the acute trauma setting [12]. Given the frequency of multiple coexisting complex injuries, left carotid-subclavian bypass is not typically performed in the traumatic aortic injury setting. If signs of hand ischemia or vertebral steal syndrome develop after LSA coverage, then subsequent left carotid-subclavian bypass or subclavian-carotid transposition can be performed.

Indications to perform prophylactic LSA revascularization include the presence of a left internal thoracic (mammary) artery coronary bypass or an occluded/atretic right vertebral artery with left vertebral patency and dominance. Using proximal landing zones as short as 15 mm has been reported with acceptable outcomes to avoid LSA coverage. Homemade fenestrated endografts and laser fensatration to achieve LSA preservation in the trauma setting have recently been described as well.

However, the urgency of the repair and the condition of the victim may preclude preoperative assessment. If the LSA is covered, intraoperative angiography of the right vertebral artery would allow the most expeditious assessment of the adequacy of the posterior circulation.

Systemic Heparinization

The safety of systemic heparinization during endovascular repair in a multiply injured patient with a closed head injury or abdominal solid organ injury is a controversial issue. Some authors report that heparin may not be necessary as most of these cases can be performed relatively rapidly, and the risk of a thrombotic event is likely small. The decision must be individualized based on the balance of the perceived risks of bleeding in a particular organ system vs the thromboembolic complications.

Choice of Repair in the Young—TEVAR Vs Open

There was near unanimity of opinion that anatomic suitability is important for TEVAR but age should not be a factor in deciding the type of repair. The risks of death and spinal cord ischemia are significantly lower in all age groups after endovascular repair compared with open surgery, and these early benefits outweigh the concerns of potential late complications [13]. However, in surgically fit patients with poor anatomy for endovascular repair, conventional open repair should be considered.

Optimal Follow-Up Strategy

Given the concerns of cumulative radiation, iodinated contrast exposure, and late endograft collapse, the optimal strategy for long-term follow-up of these patients post-TEVAR remains in evolution. In the absence of any abnormalities on imaging (i.e., stable endograft position, no endoleak) in the first 12 to 36 months, we can decrease the frequency to 2–5 years. A combination of a multi-view chest x-ray and a magnetic resonance angiography (MRA) may be preferable over conventional contrast computed tomographic angiography (CTA) for long-term imaging, with due consideration of the metallic composition of the endograft.

One important aspect of the follow-up of these young patients is adherence to radiologic follow-up.

Long-Term Outcomes

Our minimum 10-year follow-up study of endovascular repair for acute traumatic transection of the thoracic aorta demonstrated that the reduction in the operative mortality rate of TEVAR, compared with open repair, lasts over time, without any device-related issues [14]. he expected durability of any endovascular aortic repair depends on the life expectancy of the patient and the long-term behavior of the aortic segment in question. The morphologic changes of the aorta that come with age may still occur in the late phases, and may lead to stent-graft–related complications. It has been well documented that infrarenal aortic dilatation occurs after endovascular aneurysm repair, and can result in graft migration and type I endoleaks. A recent study reviewed

1000 thoracic CT scans to assess the natural behavior of the thoracic aorta with increasing age [15]. Between teenagers and octogenarians, a 1-cm increase was observed in the mean diameter at the thoracic aortic isthmus. The radial force imparted to the aorta by the stent-graft could be expected to increase the rate of dilatation of the aorta. Continuing expansion of aortic necks poses a substantial threat to the long-term durability of TEVAR, because proximal fixation is jeopardized once the diameter of the neck exceeds that of the stent-graft. If this dilatation, seen in nonstented aortas, occurred to this degree after TEVAR, there would be a risk of type I endoleaks, device migration, and collapse.

In our study, we observed that the proximal and distal aortic neck dilated to some extent, as is the natural history of the thoracic aorta. Additionally, an increased growth rate in the area of the proximal aortic neck compared with the distal aortic neck was observed. The thoracic aorta significantly dilated at a faster rate in the group of patients aged <30 years. To date, thoracic aortic dilatation has not been clinically significant. These findings corroborate the observation that aortic expansion seems to be more related to the natural history of the thoracic aorta than to any effect of the stent-graft.

Conclusion

Endovascular therapy has emerged as the standard of care in treating grade 3 and 4 BTAI, as well as select grade 2 injuries. TEVAR offers an effective approach that is minimally invasive with lower morbidity and mortality relative to traditional open repair. The use of more comformable and accurate stent-graft has significantly decreased the morbidity of endovascular repair of acute traumatic transection of the thoracic aorta. However, proximal and distal aortic neck dilated to some extent, as is the natural history of the thoracic aorta. Longer-term follow-up is necessary to determine whether this expansion continues and becomes clinically significant.

References

1. Fabian TC, Richardson JD, Croce MA, Smith JS Jr, Rodman G Jr, Kearney PA, Flynn W, Ney AL, Cone JB, Luchette FA, Wisner DH, Scholten DJ, Beaver BL, Conn AK, Coscia R, Hoyt DB, Morris JA Jr, Harviel JD, Peitzman AB, Bynoe RP, Diamond DL, Wall M, Gates JD, Asensio JA, Enderson BL, et al. Prospective study of blunt aortic injury: multicenter trial of the American Association for the Surgery of Trauma. J Trauma. 1997;42(3):374–80.
2. Demetriades D, Velmahos GC, Scalea TM, Jurkovich GJ, Karmy-Jones R, Teixeira PG, Hemmila MR, O'Connor JV, MO MK, Moore FO, London J, Singh MJ, Lineen E, Spaniolas K, Keel M, Sugrue M, Wahl WL, Hill J, Wall MJ, Moore EE, Margulies D, Malka V, Chan LS, American Association for the Surgery of Trauma Thoracic Aortic Injury Study Group. Operative repair or endovascular stent graft in blunt traumatic thoracic aortic injuries: results of an American Association for the Surgery of Trauma Multicenter Study. J Trauma. 2008;64(3):561–70.
3. Feczko JD, Lynch L, Pless JE, et al. An autopsy case review of 142 nonpenetrating (blunt) injuries of the aorta. J Trauma. 1992;33:846–9.
4. Lee WA, Matsumura JS, Mitchell RS, Farber MA, Greenberg RK, Azizzadeh A, Murad MH, Fairman RM. Endovascular repair of traumatic thoracic aortic injury: clinical practice guidelines of the Society for Vascular Surgery. J Vasc Surg. 2011;53(1):187–92.
5. Harris DG, Rabin J, Kufera JA, et al. A new aortic injury score predicts early rupture more accurately than clinical assessment. J Vasc Surg. 2015;61:332–8.
6. Alarhayem AQ, Rasmussen TE, Farivar B, et al. Timing of repair of blunt thoracic aortic injuries in the thoracic endovascular aortic repair era. J Vasc Surg Published online. 2020;73:896.
7. Jonker FH, Verhagen HJ, Mojibian H, Davis KA, Moll FL, Muhs BE. Aortic endograft sizing in trauma patients with hemodynamic instability. J Vasc Surg. 2010;52:39–44.
8. Wallace GA, Starnes BW, Hatsukami TS, et al. Intravascular ultrasound is a critical tool for accurate endograft sizing in the management of blunt thoracic aortic injury. J Vasc Surg. 2015;61:630–5.
9. Canaud L, Alric P, Desgranges P, Marzelle J, Marty-Ané C, Becquemin JP. Factors favoring stent-graft collapse after thoracic endovascular aortic repair. J Thorac Cardiovasc Surg. 2010;139:1153–7.
10. Canaud L, Cathala P, Joyeux F, Branchereau P, Marty-Ané C, Alric P. Improvement in conformability of the latest generation of thoracic stent grafts. J Vasc Surg. 2013;57:1084–9.
11. Canaud L, Joyeux F, Berthet JP, Hireche K, Marty-Ané C, Alric P. Impact of stent-graft development on outcome of endovascular repair of acute traumatic transection of the thoracic aorta. J Endovasc Ther. 2011;18:485–90.
12. Skripochnik E, Novikov D, Bilfinger TJ, Loh SA. Short-term results of left subclavian artery salvage in blunt thoracic aortic injury with short proximal landing zones. J Vasc Surg. 2018;68:985–90.
13. Canaud L, Alric P, Branchereau P, Joyeux F, Hireche K, Berthet JP, Marty-Ané C. Open versus endovascular repair for patients with acute traumatic rupture of the thoracic aorta. J Thorac Cardiovasc Surg. 2011;142:1032–7.
14. Canaud L, Marty-Ané C, Ziza V, Branchereau P, Alric P. Minimum 10-year follow-up of endovascular repair for acute traumatic transection of the thoracic aorta. J Thorac Cardiovasc Surg. 2015;149(3):825–9.
15. Hartley MC, Langan EM 3rd, Cull DL, Taylor SM, Carsten CG 3rd, Blackhurst DW. Evaluation of the diameter of the proximal descending thoracic aorta with age: implications for thoracic aortic stent grafting. Ann Vasc Surg. 2009;23(5):639–44.

Endovascular Management of Ruptured Aortic Aneurysms

62

Panagiotis Volteas, George S. Koutsias, and Apostolos K. Tassiopoulos

Case Presentation

A 74-year-old female presents to the emergency department with acute onset of periumbilical abdominal pain radiating to her back followed by a near syncope episode. The patient is a former smoker with past medical history of hypertension, hyperlipidemia, chronic obstructive pulmonary disease not on home oxygen therapy, and chronic kidney disease. She is tachycardic with 110 beats per minute, and her blood pressure is 110/70 mmHg. On physical exam she looks uncomfortable and has a tender palpable pulsatile abdominal mass with palpable femoral and pedal pulses.

Continued at page 647

Background

An abdominal aortic aneurysm (AAA) is defined as a focal dilation ≥1.5 times the diameter of the normal adjacent arterial segment, or as a maximal diameter of the abdominal aorta more than 3 cm [1]. AAAs typically grow without symptoms until the catastrophic event of rupture, which presents as surgical emergency [2]. Although the operative mortality for elective open AAA repair is 1–5%, and that of elective endovascular repair (EVAR) <1% [3] the overall mortality for ruptured AAAs is estimated to range between 80–90%, while the peri-operative mortality can be as high as 40% [4].

Early diagnosis of AAA rupture requires a high index of suspicion, as less than half of the patients will present with the classic triad of abdominal pain (with or without radiation to the back), hypotension or syncope, and a pulsatile abdominal mass [5]. Historically, open repair of a ruptured AAA was considered the standard of care, but in the past 15 years, the increasing experience with elective EVAR and contemporary advancements in aortic endografts, have contributed to a shift towards preferential use of EVAR in patients presenting with aortic rupture [6]. As a result, EVAR has surpassed open surgery as the preferred approach in these patients, and more than 55% of all patients with a ruptured AAA are eligible for endovascular reconstruction [7]. It should be emphasized that hemodynamic instability after rupture should not be considered an absolute contraindication to EVAR [8]. In properly selected patients EVAR is associated with a significant immediate survival advantage (about half the peri-operative mortality risk) compared to open surgical repair [9, 10], with the caveat that this early advantage of EVAR might be diluted after 3–4 years, mostly due to aneurysm-related complications and a higher rate of secondary interventions[9, 11].

Preoperative Preparation

Expeditious diagnosis of a ruptured AAA is a major determinant of patient survival, and a high index of suspicion is the most important factor leading to it. Emergency Department point of care ultrasound can be very helpful in reducing the time needed for the diagnosis of AAA, although it is not always easy to diagnose AAA rupture [12]. The hemodynamic status of the patient along with his ability to maintain a normal level of consciousness dictate if there is an opportunity for obtaining a preoperative CT angiogram (CTA) (Fig. 62.1). CTA should not be considered in patients with deteriorating hemodynamic status despite appropriate resuscitation, and these patients should be transferred directly to the operating room. In organized centers, however, most patients with a ruptured AAA are suitable to undergo a CTA

P. Volteas · A. K. Tassiopoulos (✉)
Division of Vascular and Endovascular Surgery, Stony Brook Medicine, New York, USA
e-mail: apostolos.tassiopoulos@stonybrookmedicine.edu

G. S. Koutsias
Aristotle University of Thessaloniki, Thessaloniki, Greece

G. Geroulakos et al. (eds.), *Mastering Endovascular Techniques*, https://doi.org/10.1007/978-3-031-42735-0_62

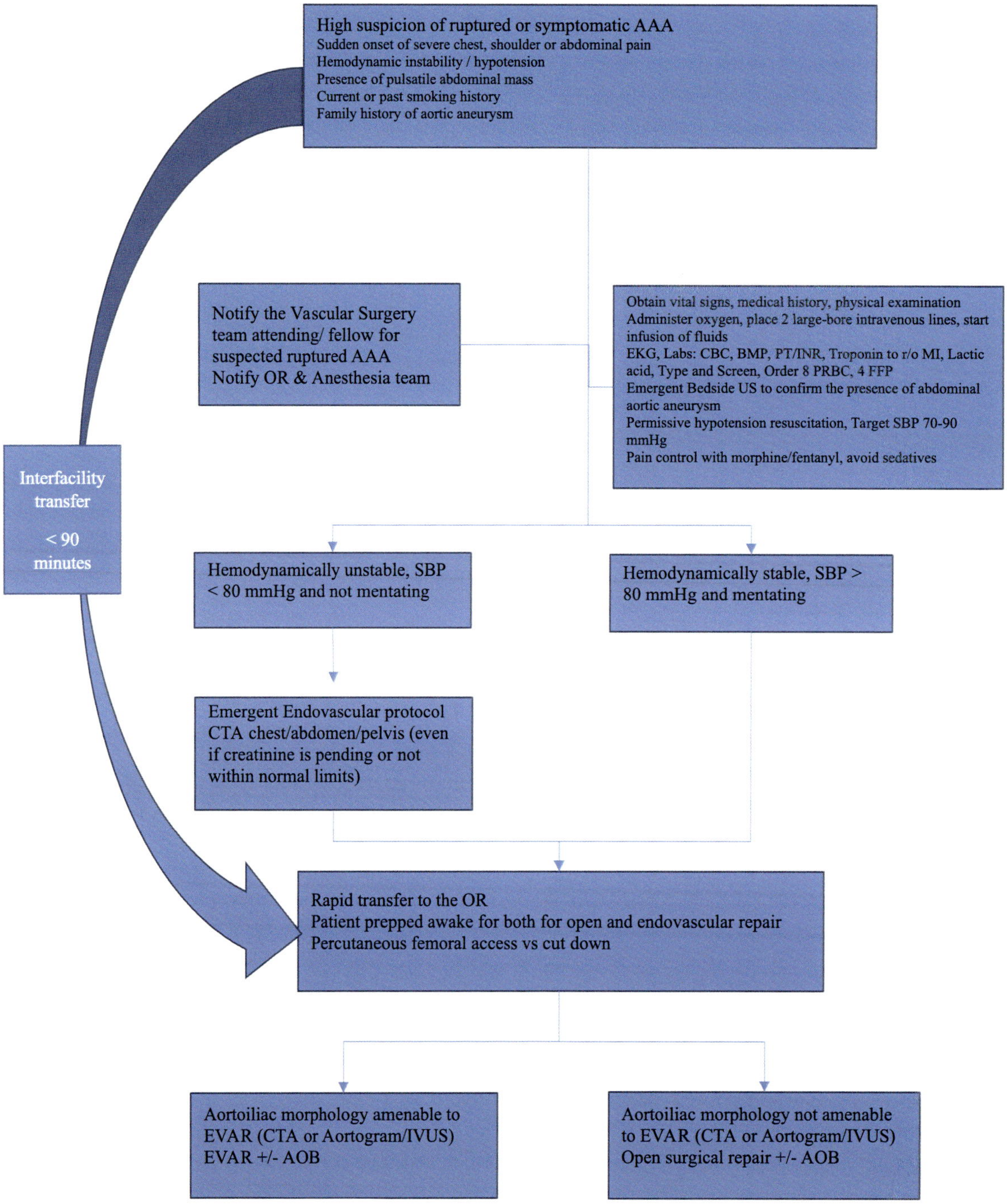

Fig. 62.1 Algorithm for preoperative management of a patient with ruptured abdominal aortic aneurysm

expeditiously before repair [13]. In these patients, the CTA should include the chest, abdomen, and pelvis in order to determine eligibility for EVAR as well as for EVAR planning and sizing. CTA should be performed based on an EVAR protocol (1–3 mm cuts), regardless of the presence of renal failure, as the information obtained is critical for treatment planning and the consequences of delayed diagnosis, missed diagnosis, or mismanagement are far more detrimental to the patient than the risk stemming from exposure to the iodinated contrast agent.

Resuscitation of patients with a ruptured AAA should not aim to restore normal hemodynamics but rather to maintain patients' level of consciousness without any signs of end organ ischemia like ST-segment depression, deterioration of mental status or syncope. This balanced resuscitation strategy known as "permissive hypotension," correlates usually with a systolic blood pressure around 80–90 mmHg and aims to prevent re-bleeding from the temporarily tamponaded area of rupture and expansion of the hematoma contained in the retroperitoneal space [14].

Emergency Room (ED) alertness protocols for patients presenting with suspected acute aortic pathology can prove invaluable in expediting repair and improving survival rates [15]. The goal is to assimilate in a timely and organized pattern a multidisciplinary team comprised of ER physicians, radiologists, anesthesiologists, OR nurses, OR technologists, perfusionists, and experienced vascular and cardiothoracic surgeons with advanced open and endovascular skills. In the past 10 years we have established in our institution this organized approach for all patients presenting to the ED with a suspected (Aorta Alert) or diagnosed (Code Aorta) aortic emergency. This protocol-based resource mobilization has significantly decreased the time interval from patient arrival to the ED to definitive diagnosis and ultimately to surgical repair.

Endovascular repair of a ruptured AAA (rEVAR) requires access to an appropriately equipped operating room setup with adequate fluoroscopic imaging capabilities, complete inventory for endovascular aortic procedures, stock of off-the-shelf aortic endografts, and specific items like intravascular ultrasound (IVUS) and vascular imaging software. When patients originally present to smaller size institutions with limited capabilities, every effort should be made to transfer patients with ruptured AAAs expeditiously (<90 min) to the closest Vascular center of excellence with capabilities of open aortic and/or endovascular surgery in order to maximize the patient's survival chances [9, 13, 16].

Case Presentation

Continued from page 645

Bedside point of care ultrasonography showed a large AAA. Two large bore peripheral intravenous accesses secured in placed and since the patient was hemodynamically stable, she underwent an emergent CTA of the chest, abdomen, and pelvis. The CTA demonstrated a 7.2 cm infrarenal abdominal aortic aneurysm with a contained retroperitoneal rupture (Fig. 62.2). *The neck was measured 15 mm in length, 20 mm in diameter, with 30 degrees angulation, and*

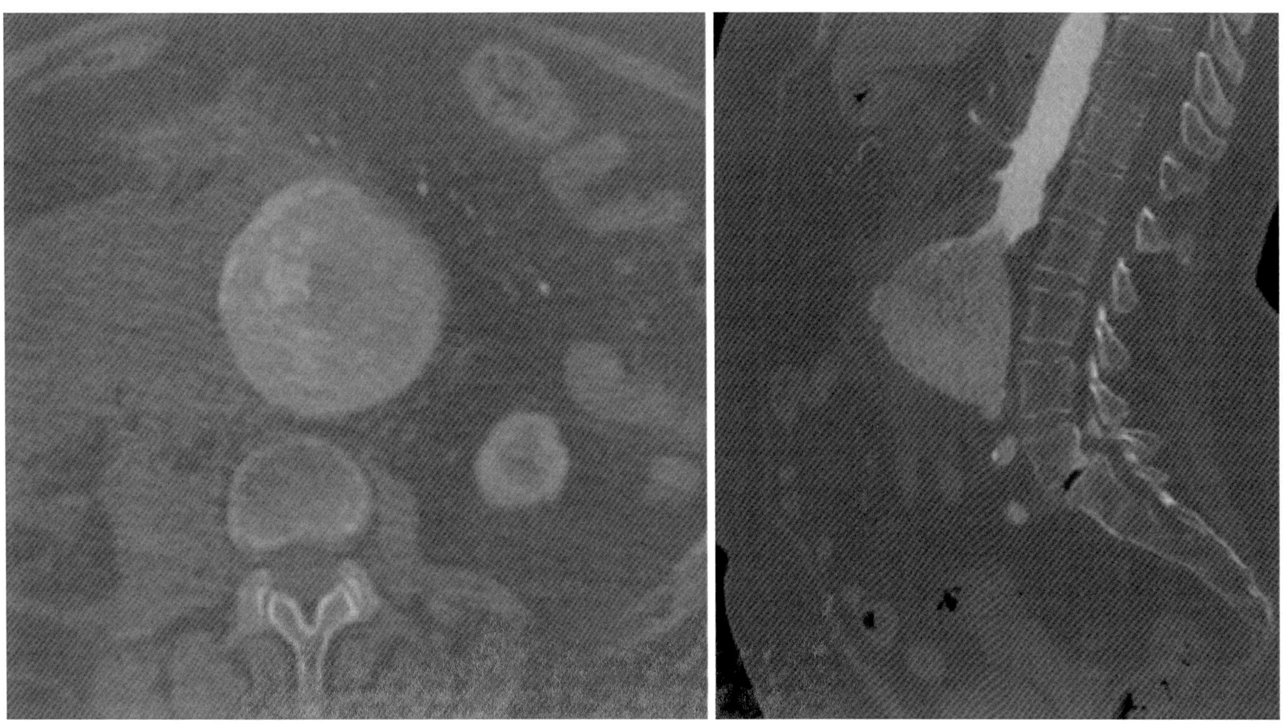

Fig. 62.2 Axial and saggital views of ruptured AAA

there were healthy non-aneurysmal iliac vessels. While the patient was in the CT scan her blood pressure dropped to 85/50 mmHg. The patient was mentating well and thus blood product transfusions were held based on the principles of permissive hypotension. The patient was emergently transferred to the operating hybrid room for endovascular repair of the ruptured AAA.

Continued at page 651

Endovascular Strategy and Technique

As not all ruptured AAAs are amenable to EVAR, the operating room setup should be able to accommodate both EVAR and open surgical repair. There is a wide range aortic endograft system for elective EVAR with specific manufacturers' Instructions For Use (IFU) in regard to anatomical requirements that the surgery team should be aware of. Realistically, in cases of AAA rupture, endograft selection is based on immediate availability, specific anatomic characteristics, and operator's experience with the available platforms. Not surprisingly, use of aortic endografts outside the IFU recommendations, especially with regards to hostile proximal neck characteristics in rEVAR yields inferior in-hospital survival compared to patients treated within IFU [17] and open repair should be seriously contemplated in this case scenario. If a CTA is obtained in the ED, procedure planning (open vs. EVAR, endograft to be used, need for adjunct tools, etc.) is based on CT images. If the patient's hemodynamic status dictates immediate transfer to the OR, these decisions are made intraoperatively.

In order to expedite definitive endovascular repair, a list of necessary endovascular tools has been created and these items are opened by the scrub team and ready to use before the procedure is started. It is best if this list is created jointly by the entire team of treating surgeons and can be adjusted to operator preferences and institution expertise. A list of suggested endovascular tools and FDA approved abdominal aortic grafts for ruptured AAA is summarized in Tables 62.1 and 62.2.

Table 62.1 Endovascular Toolkit for the management of a ruptured abdominal aortic aneurysm. These are only a few suggested options based on the author's experience

	Manufacturer	Size/Length	Comments
Micropuncture Access Set	Cook Medical	21 G, 7 cm	Two kits
Wires			
Any standard access stiff glidewire	Multiple	0.035″/80 cm	
Angled Glidewire	Multiple	0.035″/180, 260 cm	
Aortic Stent graft delivery/Lunderquist wire	Multiple	0.035″/260 cm	
Sheaths			
Initial access	Multiple	6 Fr, 10 cm	
Upsize after perclose/proglide placement	Multiple	8 Fr 10 cm	
Dry-Seal Introducer Sheath with Hydrophilic Coating	Gore	12–26 Fr	Hemostatic and kink-resistant introducer sheath
Catheters			
Diagnostic angiographic catheters—omniflush, pigtail, straight, KMP catheter	Multiple	4,5,6 Fr 100, 130 cm	
IVUS Volcano	Phillips	Vision PV 0.035 digital IVUS catheter	
Compliant AOB	**Manufacturer**	**Size (Fr)** **Length (cm)**	**Maximum inflated diameter (mm)**
Equalizer balloon catheter	Boston scientific corporation	7 65, 100	20, 27, 33, 40
Coda Balloon Catheter	Cook Medical	10, 14 140	46
Reliant Stent Graft Balloon Catheter	Medtronic	8 100	46
Q50 PLUS Stent Graft Balloon Catheter	Merit Medical Systems, Inc.	8 65, 100	50
Aortic Balloon	Terumo Aortic	8 120	50

Table 62.2 FDA approved aortic stent graft for the endovascular management of ruptured abdominal aortic aneurysm. These are only a few suggested options based on the author's experience

Aortic stent graft	Main body delivery sheath Size OD (Fr)	Main body length, diameter (mm)	Iliac sizes length, diameter (mm)	Comments	Ancillary components length, diameter (mm)
Cook Zenith (Cook Medical)[a]	21, 23, 26	L: 82, 96, 111, 125, 140 (22–32 mm grafts), 95, 113, 131, 149 (36 mm grafts) D: 22, 24, 26, 28, 30, 32, 36	L: 39, 56, 74, 90, 107, 122 D: 9, 11, 13, 16, 20, 24	Self expanding Stainless steel Woven polyester Suprarenal fixation	AUI converter proximal L: 80, 82 D: 24, 28, 32, 36
AFX Endovascular AAA system (Endologix)	17	L: 40, 60, 70, 80, 90, 100, 110, 120 (unibody main body lengths) D: 22, 25, 28	L: 55, 65, 78, 80 D: 13, 16, 20, 25	Self expanding Cobalt chromium alloy e-PTFE	
Alto Abdominal Stent Graft System (Endologix)[a]	15	L: 80 D: 20, 23, 26, 29, 34	Ovation iX iliac limbs L: 80, 100, 120, 140, 160 D: 10, 12, 14, 16, 18, 22, 28	Non expansive polymer-filled ring Nitinol PTFE Suprarenal fixation	
Gore Excluder (Gore and Associates)	18.3, 20.1	L: 120, 130, 140, 150, 160, 170, 180 D: 23, 26, 28.5, 31, 35	L: 95, 100, 115, 120, 135, 140 D: 10, 12, 14.5, 16, 18, 20, 23, 27	Self expanding Nitinol e-PTFE Infrarenal fixation	
Endurant IIs AAA Stent Graft System (Medtronic)[a]	18, 20	L: 103 D: 23, 25, 28, 32, 36	L: 82, 93, 124, 146, 156, 199 D: 10, 13, 16, 20, 24, 28	Self expanding Nitinol Woven polyester Suprarenal fixation	Talent occluder D: 8, 10, 12, 14, 16, 18, 20, 22, 24
Endurant II AAA Stent Graft System (Medtronic)[a]	18, 20	L: 124, 145, 166 D: 23, 25, 28, 32, 36	L: 82, 93, 124, 156, 199 D: 10, 13, 16, 20, 24, 28	Self expanding Nitinol Woven polyester Suprarenal fixation	AUI converter proximal L: 102 D: 23, 25, 28, 32, 36
Endurant II AUI Stent Graft System (Medtronic)[a]	18, 20	L: 102 D: 23, 25, 28, 32, 36	L: 82, 93, 124, 146, 156, 199 D: 10, 13, 16, 20, 24, 28	Self expanding Nitinol Woven polyester Suprarenal fixation	
Incraft AA Stent Graft System (Cordis)	14, 16 (for 34 main body)	L: 94 (ipsilateral side), 86 (contralateral side) D: 22, 26, 30, 34	L: 80, 100, 120, 140 D: 10, 13, 16, 20, 24	Self expanding Nitinol Woven polyester Suprarenal fixation	
Treo Abdominal stent graft (Terumo Aortic, Bolton Medical)	18, 19	L: 80, 100, 120 D: 20, 22, 224, 26, 28, 30, 33, 36	L: 80, 100, 120, 140, 160 D: 9, 11, 13, 15, 17, 20, 24	Self expanding Nitinol Woven polyester Suprarenal & Infrarenal fixation	

[a]Approved to treat more challenging aortic neck anatomy

Operation Steps

1. Patient is placed in supine position. The chest, abdomen, and bilateral groins are prepared and draped. At least one arm should be available for brachial artery cut down access as an alternative access for placement of an Aortic Occlusion Balloon (AOB) or as a bailout route for cannulation of the contralateral gate of the aortic graft, for placement of renal stents, etc.
2. Foley catheter placement is necessary in these patients but should not delay procedure steps aiming to improve hemodynamic stability.
3. Percutaneous ultrasound-guided retrograde common femoral artery access using the pre-closure technique is faster and currently our preferred choice in patients with appropriate anatomy, but expedited femoral artery cut down offers an advantage for patients with access-site limitations (stenosis, calcification, aneurysmal degeneration). If percutaneous access is chosen the surgeon should review the CT scan or use intraoperative ultrasound to assess the exact site of the femoral artery bifurcation and the amount of wall calcification.

4. Local anesthesia is the preferred mode during the access phase to prevent major hypotension associated with endotracheal intubation.
5. Both femoral arteries are accessed using a micro-puncture needle; advancing 0.014 wires; and exchanging the micro-puncture needles for 6-French introducer sheaths on each side.
6. If percutaneous access is chosen, we deploy 2 Pro-Glide sutures on each common femoral artery, and then upsize the sheaths on each side.
7. In hemodynamically unstable patients, an aortic occlusion balloon (AOB) is inserted first for proximal control. In the ipsilateral side, a floppy guidewire is exchanged for a super-stiff wire (Lunderquist) and then a 12–14 French × 55 cm length sheath is advanced to the supraceliac abdominal aorta. A compliant AOB (Coda, Reliant) is then advanced through the sheath over the super-stiff wire into the distal descending thoracic aorta and is inflated to establish occlusion of the aorta above the site of rupture.
8. From the contralateral side, a pigtail "marker flush-catheter" is used to obtain an abdominal aortogram to define visceral vessel and aneurysm anatomy and determine size and length of endograft main body. Intravascular ultrasound is essential in accurately assessing length and diameter of the aortic neck in patients without a pre-operative CTA.
9. The AOB can then be deflated and removed over the stiff wire. If patient's blood pressure remains stable, the endograft main body is advanced under fluoroscopic guidance and deployed so that the covered segment starts just below the lowest renal artery. The device should be deployed until the contralateral gate opens.
10. The contralateral gate is then cannulated via the contralateral common femoral artery. We prefer using a combination of a Glidewire and a Berenstein catheter. Retrograde angiograms are obtained bilaterally to locate the iliac artery bifurcations, followed by deployment of the ipsilateral and contralateral iliac extensions. Every effort should be made to preserve flow into both internal iliac arteries, as colon ischemia is a serious complication of a ruptured AAA.
11. The compliant balloon is then used to balloon the proximal and distal attachment sites as well as the overlapping segments of the endograft components.
12. Completion arteriogram to check for endoleaks or stenoses is then performed. If a Type 1 or Type 3 endoleak is present, it needs to be repaired as it may lead to re-rupture. Endoanchors may enhance proximal seal zone integrity and endograft fixation which can significantly improve successful endograft proximal seal. In cases of a hostile proximal neck anatomy endoanchor use and proximal seal zone extension to the visceral segment with use of parallel grafts or physician-modified endografts are options for overcoming these anatomic challenges and ensuring successful repair.
13. Both common femoral arteries are then closed by tightening the 2 Pro-Glide sutures on each side or with open repair of the femoral arteries in case of femoral artery cut down.

Special Intraoperative Considerations

Anesthesia

The hemodynamic status of the patient dictates the initial decision about the type of anesthesia and to a certain extent the percutaneous versus cut-down femoral artery access. Local anesthesia (LA) is preferred over general anesthesia in hemodynamically unstable patients to preserve the sympathetic tone and thus to keep the patient's blood pressure unaffected. Use of LA is associated with decreased volume of intraoperative fluid resuscitation, number of blood transfusions, intensive care unit length of stay and with lower overall morbidity and mortality rates [18, 19].

Aortic Occlusion Balloons (AOBs)

AOBs are used for the initial stabilization of hemodynamically unstable patients in both open AAA repair and EVAR. AOBs can be advanced in the descending thoracic aorta though a femoral or brachial artery access [20, 21]. Brachial artery access usually requires a cut-down to avoid hematoma and compression of the median nerve. Use of an AOB through femoral access requires support by a long sheath (55 cm length or longer) to prevent downward displacement and prolapse into the ruptured AAA. Prior to aortic graft main body deployment, the AOB can be deflated and usually this is hemodynamically tolerable by the majority of the patients. In the case of hemodynamic instability, the occlusion balloon can be repositioned into the aortic neck through the aortic graft main body itself and reinflated at the level of the aortic graft neck until full seal of the ruptured AAA is achieved [22].

Endo-leaks, Endo-anchors

Type I endoleak (Ia from proximal or Ib from distal stent graft attachment sites) or Type III endoleak (from stent graft component separation) require immediate reintervention

[24]. Endo-anchors (Medtronic, Heli-FX) are a feasible, safe, and effective option to treat not only type Ia endoleak but also for enhancing proximal seal and preventing endograft migration [23]. They provide transmural fixation of the graft and are compatible with most aortic endografts. Patients with aortic neck lengths less than 10 mm and up to 5 mm can also undergo rEVAR with the combination of the Endurant II/IIs aortic grafts and endo-anchors but an experienced operator is required for best results as EVAR in a patient with short neck requires precise deployment of the endograft main body to maximize length of endograft apposition to the neck, and experience with transmural endoanchor deployment circumferentially at the proximal seal zone.

Type II endo-leaks (IIa- simple, IIb- complex, retrograde flow into the aneurysm sac or retroperitoneum from inferior mesenteric or lumbar arteries) can be treated conservatively as they usually resolve over time [24].

Ancillary Tools

If a patient did not undergo a preoperative CT scan or underwent a non-contrast CT scan due to elevated creatinine, Intravascular Ultrasound (IVUS) in combination with an on-table aortogram can be used for precise graft measurements of the proximal aortic neck diameter and length, as well as for sizing the common iliac and external iliac arteries bilaterally [25]. IVUS provides accurate anatomic information on the location of the renal arteries and the internal iliac arteries and is particularly helpful in calculating endograft size and length.

Perioperative Care, Postoperative Complications, and Surveillance

Postoperatively the patient is transferred in the ICU until the cardiovascular, respiratory, gastrointestinal, and renal functions are improving. The most common early postoperative complication is renal failure, followed by acute coronary syndromes, congestive heart failure, respiratory failure, ileus, and ischemic colitis. The risk of theses complications is directly related to the severity and duration of the preoperative hemorrhagic shock, the time the aorta is occluded with the AOB intraoperatively, and the severity of reperfusion injury after aortic unclamping.

Postoperative abdominal pain, bloody bowel movements, leukocytosis, and metabolic/lactic acidosis should raise the suspicion of **ischemic colitis** which is a life-threatening complication. Diagnosis is difficult, requires a high index of suspicion and early use of sigmoidoscopy to evaluate the extent of bowel ischemia and determine the need for operative intervention [26, 27].

Abdominal compartment syndrome (ACS) is a rare complication caused by increased intraperitoneal pressure and can be seen after both open surgery and EVAR for ruptured AAA. Aggressive fluid resuscitation with massive transfusions, use of an aortic occlusion balloon, coagulopathy, and internal iliac artery occlusion including use of aorto-uni-iliac devices rather than bifurcated stent grafts are well established risk factors. Common clinical manifestations include abdominal distention along with signs of multi-system organ failure such as hypoventilation with high peak airway pressures, decreased venous return, increased intracranial pressure, oliguria and worsening acute kidney injury. The diagnosis is based on clinical suspicion by recognizing the manifestations listed above and on physician exam (distended and tense abdomen). Bladder pressure measurement can help confirm the diagnosis but is not always accurate [28]. It is a devastating complication with mortality nearly 100% if left untreated and 30–70% even with treatment which comprises of decompressive laparotomy with leaving the abdomen open temporarily followed by delayed abdominal closure [29].

Patient's overall outcomes are related to a combination of risk factors. A simple 4-point preoperative risk score (age >76 years, pH <7.2 on arrival to the ED, creatinine >2 mg/dL, and any episode of hypotension with systolic blood pressure <70 mmHg) can prove very accurate in predicting 30-day mortality. Patients risk increases substantially from 14.6% for a score of 0 to 35.7% for a score of 1, 68.4% for a score of 2, and 100% for a score of 3 or 4 [30]. The mortality of patients who develop cardiac arrest and require CPR preoperatively exceeds 90% [31].

The postoperative surveillance of the patients with a ruptured AAA that undergo rEVAR includes CTA at 1 month (or earlier if an endoleak is suspected), 6 months and 1 year, followed by yearly CTA or duplex scans [2]. This frequent surveillance imaging is required as one out of five patients will require a reintervention for endo-leaks or graft migration in the future [32].

Case Presentation

Continued from page 648

The patient was placed in the supine position in the interventional suite table and a foley catheter was placed. Both groins, abdomen and chest were then prepped and draped. After administration of local anesthesia, bilateral ultrasound-guided percutaneous access of the common femoral arteries was obtained. Patients' hemodynamics improved and thus no AOB was used. Micropuncture needles were exchanged for 6-French introducer sheaths bilaterally and two ProGlide sutures were placed on each side, followed by 8-French introducer sheaths. A 0.035 Glidewire was then negotiated through the right common

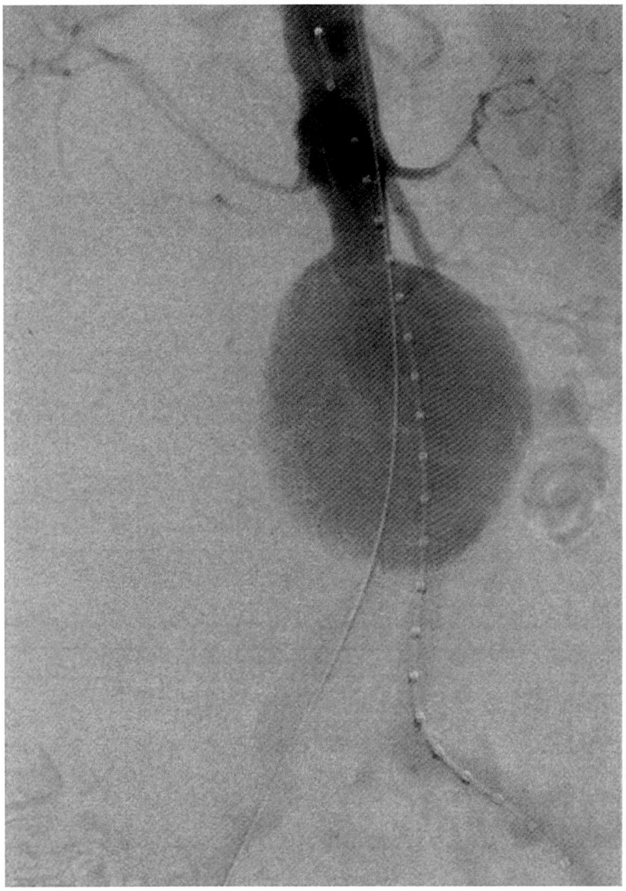

Fig. 62.3 AP angiographic view of ruptured AAA at the time of endovascular repair

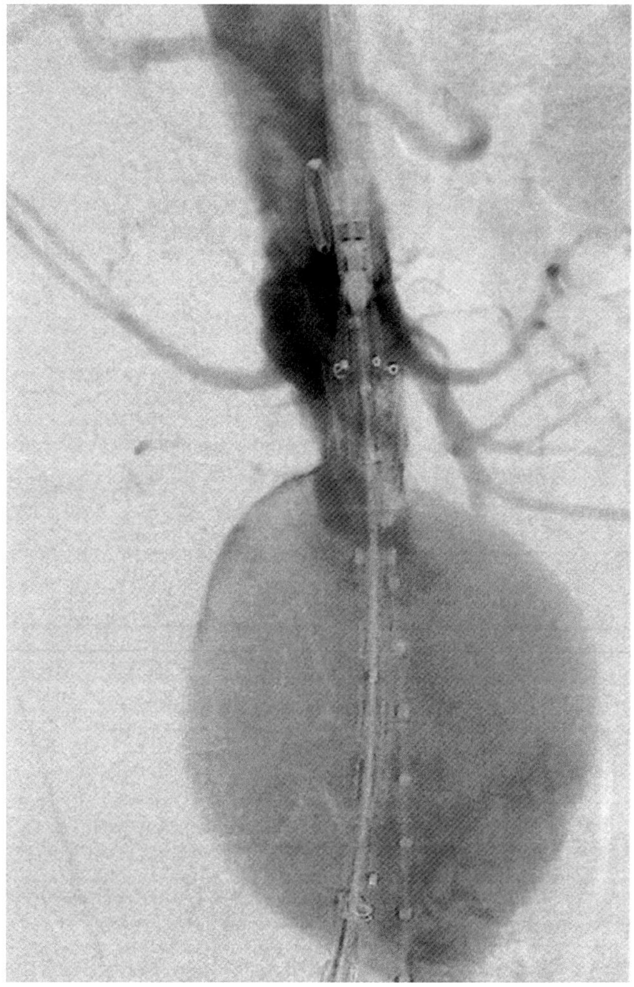

Fig. 62.4 Aortic endograft has been advanced to target and deployment of the proximal end has been initiated

femoral artery sheath into the descending thoracic aorta and was exchanged for 0.035 Lunderquist wire. A pigtail catheter was advanced over a 0.035 Glidewire from the left common femoral artery and was brought to the level of the L1 vertebral body. An abdominal aortogram with pelvic runoff was obtained, which delineated the anatomy of the aneurysm and the iliac arteries (Fig. 62.3). *The main body of the aortic endograft, a 23 mm diameter Endurant IIs, was advanced bareback up the ipsilateral side and deployed flush with the origin of the lowest renal artery* (Figs. 62.4). *The device was deployed until the contralateral gate opened. The contralateral gate was cannulated using a Berenstein catheter and a Glidewire through the left common femoral artery sheath* (Fig. 62.5). *Intraluminal placement of the wire was verified by spinning the pigtail catheter at the top of the endograft main body without difficulty. A retrograde arteriogram via the left common femoral artery sheath was then obtained and the length between the flow divider and the left common iliac bifurcation was measured using the pigtail catheter markers. Then the contralateral limb, 16 mm proximal diameter x 13 mm distal diameter x 124 mm length Medtronic Endurant II, was advanced bareback via the left common femoral artery and deployed. The limb delivery system was removed and a 12 French Gore DrySeal sheath was inserted in the left common femoral artery. The main body was then completely deployed, the delivery system was safely removed, and a 16 French Gore DrySeal sheath was inserted in the right common femoral artery. The marker pigtail catheter was the inserted over the wire through the right common femoral sheath and a retrograde arteriogram of the ipsilateral iliac artery was performed. Based on length measurements from the flow*

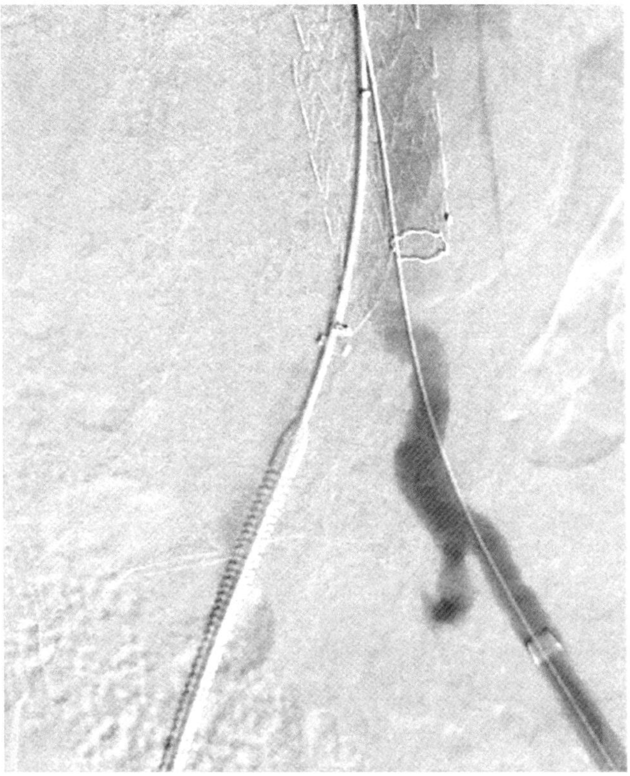

Fig. 62.5 Main body of the endograft deployed and contralateral gate has been cannulated

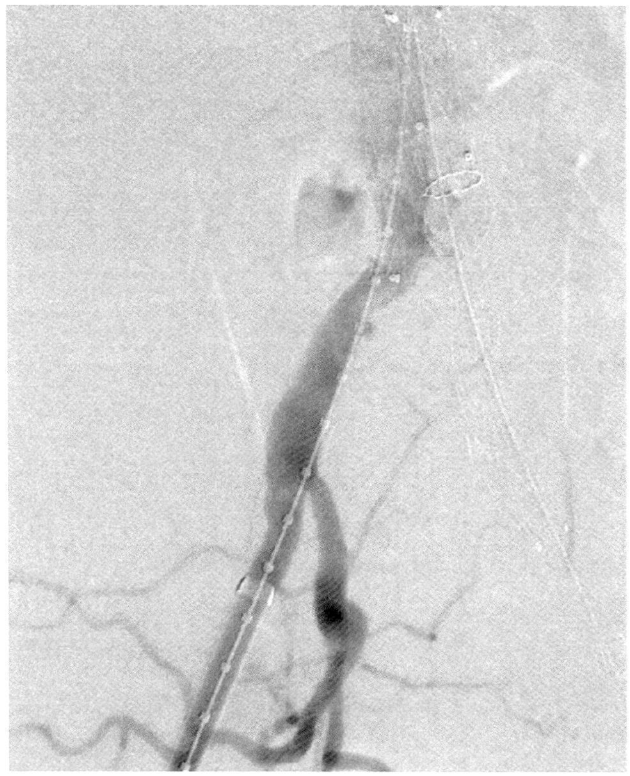

Fig. 62.6 Retrograde ipsilateral iliac arteriogram prior to ipsilateral limb extension

divider to the right common iliac artery bifurcation, a 16 mm proximal diameter x 13 mm distal diameter x 93 mm length Medtronic Endurant II iliac limb was selected, advanced through the sheath, and deployed (Fig. 62.6). The delivery system was then removed, and the Reliant balloon was brought from each common femoral artery to balloon the proximal and distal attachment sites as well as the overlapping segments of the endograft components. Completion angiogram (Fig. 62.7) showed a sluggish Type Ia endoleak that resolved after placement of eight Heli-FX EndoAnchors (Figs. 62.8 and 62.9). A delayed type II endoleak from a single lumbar artery was seen but the decision was made to complete the case and to follow the type II endoleak with future imaging. At the end of the case both femoral arteries were percutaneously closed successfully by tightening the two ProGlide sutures previously placed on each side.

The patient remained hemodynamically normal in the postoperative period and was discharged 3 days after repair. At 1 month follow up, the patient remained asymptomatic. A repeat CTA showed resolution of the type II endoleak. A repeat CTA 14 months later showed a dramatic regression of the AAA sac (Fig. 62.10).

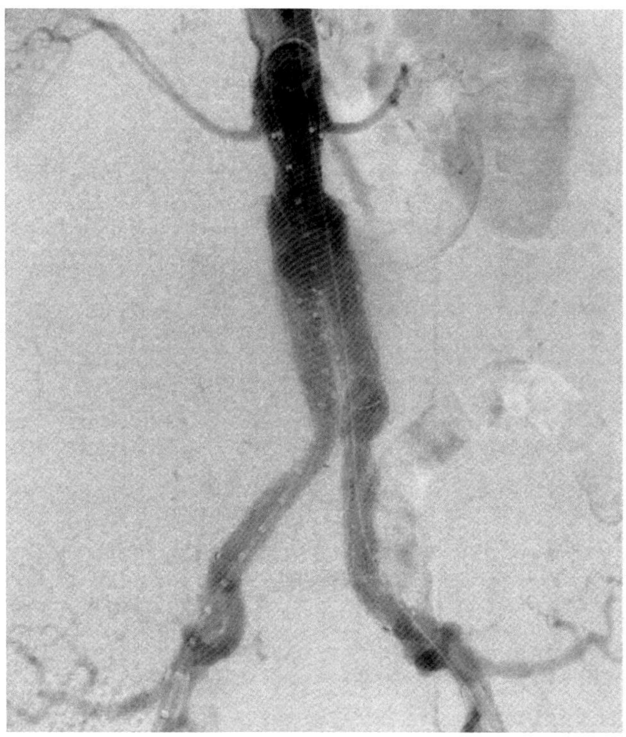

Fig. 62.7 Completion aortogram after endograft repair

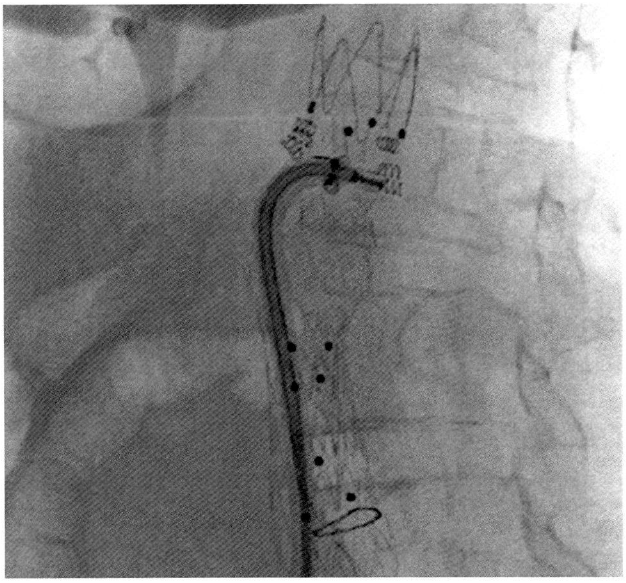

Fig. 62.8 Reinforcement of the proximal seal zone with endoanchors

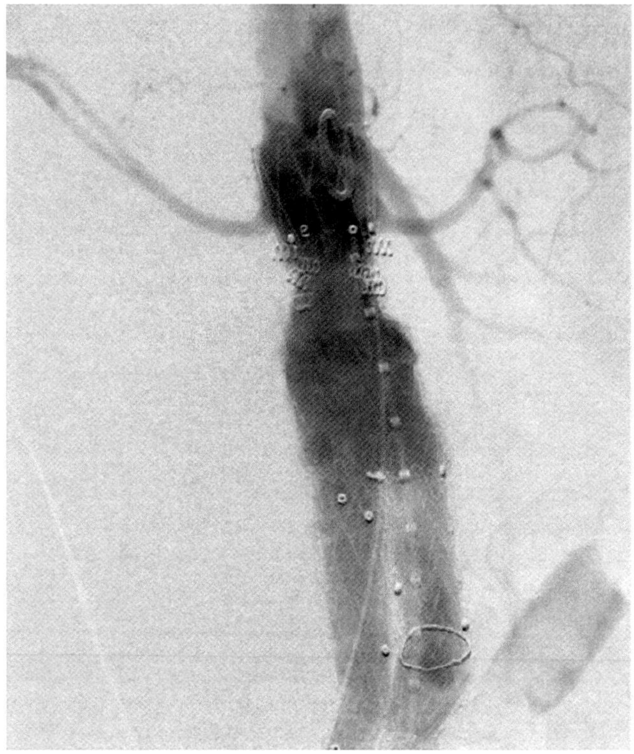

Fig. 62.9 Final arteriogram

Fig. 62.10 CT angiography indicating complete aortic sac regression 2 years after original repair of ruptured AAA

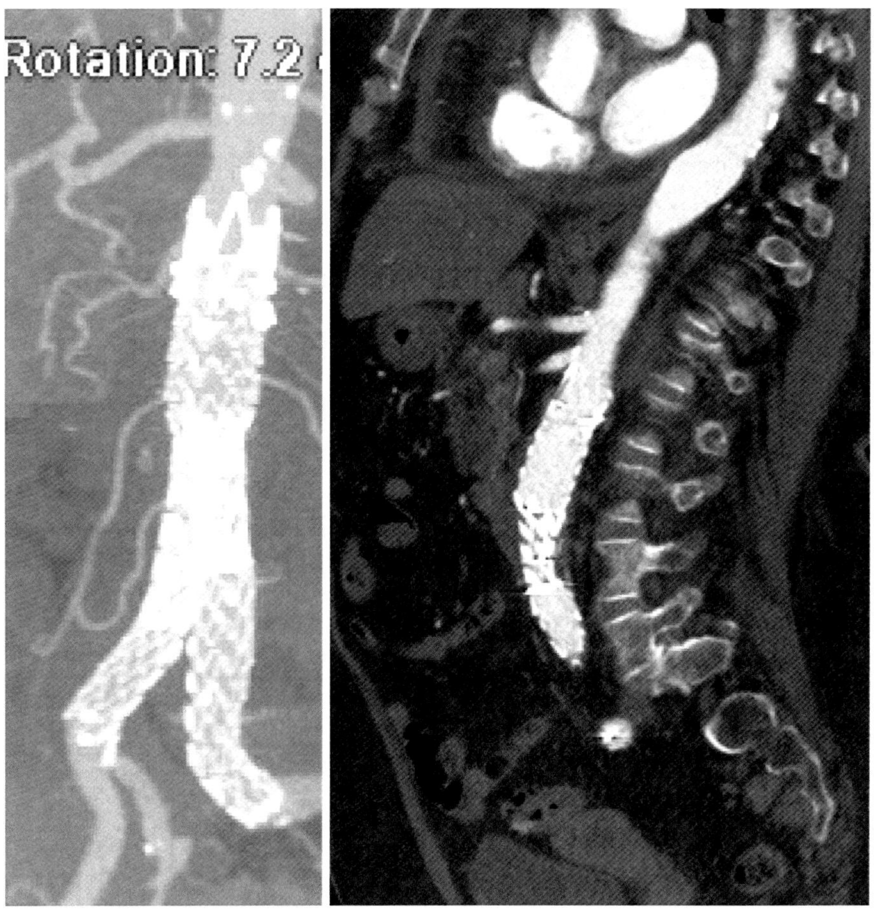

References

1. Ullery BW, Hallett RL, Fleischmann D. Epidemiology and contemporary management of abdominal aortic aneurysms. Abdom Radiol (NY). 2018;43(5):1032–43. https://doi.org/10.1007/s00261-017-1450-7.
2. Chaikof EL, Dalman RL, Eskandari MK, Jackson BM, Lee WA, Mansour MA, Mastracci TM, Mell M, Murad MH, Nguyen LL, Oderich GS, Patel MS, Schermerhorn ML, Starnes BW. The Society for Vascular Surgery practice guidelines on the care of patients with an abdominal aortic aneurysm. J Vasc Surg. 2018;67(1):2–77.e2. https://doi.org/10.1016/j.jvs.2017.10.044.
3. Yin K, Locham SS, Schermerhorn ML, Malas MB. Trends of 30-day mortality and morbidities in endovascular repair of intact abdominal aortic aneurysm during the last decade. J Vasc Surg. 2019;69(1):64–73. https://doi.org/10.1016/j.jvs.2018.04.032. Epub 2018 Jun 15.
4. Karthikesalingam A, Holt PJ, Vidal-Diez A, Ozdemir BA, Poloniecki JD, Hinchliffe RJ, Thompson MM. Mortality from ruptured abdominal aortic aneurysms: clinical lessons from a comparison of outcomes in England and the USA. Lancet. 2014;383(9921):963–9. https://doi.org/10.1016/S0140-6736(14)60109-4.
5. Cates JR. Abdominal aortic aneurysms: clinical diagnosis and management. J Manip Physiol Ther. 1997;20(8):557–61.
6. Albuquerque FC Jr, Tonnessen BH, Noll RE Jr, Cires G, Kim JK, Sternbergh WC 3rd. Paradigm shifts in the treatment of abdominal aortic aneurysm: trends in 721 patients between 1996 and 2008. J Vasc Surg. 2010;51(6):1348–52; discussion 1352–3. https://doi.org/10.1016/j.jvs.2010.01.078.
7. Dansey KD, Varkevisser RRB, Swerdlow NJ, Li C, de Guerre LEVM, Liang P, Marcaccio C, O'Donnell TFX, Carroll BJ, Schermerhorn ML. Epidemiology of endovascular and open repair for abdominal aortic aneurysms in the United States from 2004 to 2015 and implications for screening. J Vasc Surg. 2021;74(2):414–24. https://doi.org/10.1016/j.jvs.2021.01.044. Epub 2021 Feb 14.
8. Li Y, Li Z, Wang S, Chang G, Wu R, Hu Z, Yin H, Wang J, Yao C. Endovascular versus open surgery repair of ruptured abdominal aortic aneurysms in hemodynamically unstable patients: literature review and meta-analysis. Ann Vasc Surg. 2016;32:135–44. https://doi.org/10.1016/j.avsg.2015.09.025. Epub 2016 Jan 21.
9. Kontopodis N, Galanakis N, Antoniou SA, Tsetis D, Ioannou CV, Veith FJ, Powell JT, Antoniou GA. Meta-analysis and meta-regression analysis of outcomes of endovascular and open repair for ruptured abdominal aortic aneurysm. Eur J Vasc Endovasc Surg. 2020;59(3):399–410. https://doi.org/10.1016/j.ejvs.2019.12.023. Epub 2020 Jan 11.
10. IMPROVE Trial Investigators. Comparative clinical effectiveness and cost effectiveness of endovascular strategy *v* open repair for ruptured abdominal aortic aneurysm: three year results of the IMPROVE randomised trial. BMJ. 2017;359:j4859. https://doi.org/10.1136/bmj.j4859.
11. Powell JT, Sweeting MJ, Ulug P, Blankensteijn JD, Lederle FA, Becquemin JP. Greenhalgh RM; EVAR-1, DREAM, OVER and ACE Trialists. Meta-analysis of individual-patient data from EVAR-1, DREAM, OVER and ACE trials comparing outcomes of endovascular or open repair for abdominal aortic aneurysm over 5 years. Br J Surg. 2017;104(3):166–78. https://doi.org/10.1002/bjs.10430. Erratum in: Br J Surg. 2018;105(9):1222.
12. Fernando SM, Tran A, Cheng W, Rochwerg B, Strauss SA, Mutter E, McIsaac DI, Kyeremanteng K, Kubelik D, Jetty P, Nagpal SK, Thiruganasambandamoorthy V, Roberts DJ, Perry JJ. Accuracy of presenting symptoms, physical examination, and imaging for diagnosis of ruptured abdominal aortic aneurysm: systematic review and meta-analysis. Acad Emerg Med. 2022;29(4):486–96. https://doi.org/10.1111/acem.14475. Epub 2022 Mar 25.
13. Lloyd GM, Bown MJ, Norwood MG, Deb R, Fishwick G, Bell PR, Sayers RD. Feasibility of preoperative computer tomography in patients with ruptured abdominal aortic aneurysm: a time-to-death study in patients without operation. J Vasc Surg. 2004;39(4):788–91. https://doi.org/10.1016/j.jvs.2003.11.041.
14. Hamilton H, Constantinou J, Ivancev K. The role of permissive hypotension in the management of ruptured abdominal aortic aneurysms. J Cardiovasc Surg. 2014;55(2):151–9.
15. Mehta M, Taggert J, Darling RC 3rd, Chang BB, Kreienberg PB, Paty PS, Roddy SP, Sternbach Y, Ozsvath KJ, Shah DM. Establishing a protocol for endovascular treatment of ruptured abdominal aortic aneurysms: outcomes of a prospective analysis. J Vasc Surg. 2006;44(1):1–8; discussion 8. https://doi.org/10.1016/j.jvs.2006.02.057.
16. Holt PJ, Poloniecki JD, Gerrard D, Loftus IM, Thompson MM. Meta-analysis and systematic review of the relationship between volume and outcome in abdominal aortic aneurysm surgery. Br J Surg. 2007;94(4):395–403. https://doi.org/10.1002/bjs.5710.
17. Zarkowsky DS, Sorber R, Ramirez JL, Goodney PP, Iannuzzi JC, Wohlauer M, Hicks CW. Aortic neck IFU violations during EVAR for ruptured Infrarenal aortic aneurysms are associated with increased in-hospital mortality. Ann Vasc Surg. 2021;75:12–21. https://doi.org/10.1016/j.avsg.2021.04.019. Epub 2021 May 2.
18. Faizer R, Weinhandl E, El Hag S, Le Jeune S, Apostolidou I, Shafii SM, Lee CJ, Rosenberg MS, Reed A, Fanola CL. Decreased mortality with local versus general anesthesia in endovascular aneurysm repair for ruptured abdominal aortic aneurysm in the vascular quality initiative database. J Vasc Surg. 2019;70(1):92–101.e1. https://doi.org/10.1016/j.jvs.2018.10.090. Epub 2019 Jan 2.
19. IMPROVE Trial Investigators, Powell JT, Hinchliffe RJ, Thompson MM, Sweeting MJ, Ashleigh R, Bell R, Gomes M, Greenhalgh RM, Grieve RJ, Heatley F, Thompson SG, Ulug P. Observations from the IMPROVE trial concerning the clinical care of patients with ruptured abdominal aortic aneurysm. Br J Surg. 2014;101(3):216–24; discussion 224. https://doi.org/10.1002/bjs.9410.
20. Matsuda H, Tanaka Y, Hino Y, Matsukawa R, Ozaki N, Okada K, Tsukube T, Tsuji Y, Okita Y. Transbrachial arterial insertion of aortic occlusion balloon catheter in patients with shock from ruptured abdominal aortic aneurysm. J Vasc Surg. 2003;38(6):1293–6. https://doi.org/10.1016/s0741-5214(03)00774-2.
21. O'Donnell ME, Badger SA, Makar RR, Loan W, Lee B, Soong CV. Techniques in occluding the aorta during endovascular repair of ruptured abdominal aortic aneurysms. J Vasc Surg. 2006;44(1):211–5. https://doi.org/10.1016/j.jvs.2006.03.033.
22. Nakayama H, Toma M, Kobayashi T, Ohno N, Okada T, Ueno G, Sato Y. Ruptured abdominal aortic aneurysm treated by double-balloon technique and endovascular strategy: case series. Ann Thorac Cardiovasc Surg. 2019;25(4):211–4. https://doi.org/10.5761/atcs.cr.17-00189. Epub 2018 Mar 1.
23. Qamhawi Z, Barge TF, Makris GC, Patel R, Wigham A, Anthony S, Uberoi R. Editor's choice—systematic review of the use of endoanchors in endovascular aortic aneurysm repair. Eur J Vasc Endovasc Surg. 2020;59(5):748–56. https://doi.org/10.1016/j.ejvs.2020.02.008. Epub 2020 Mar 17.
24. Rehman ZU. Endoleaks: current concepts and treatments - a narrative review. J Pak Med Assoc. 2021;71(9):2224–9. https://doi.org/10.47391/JPMA.03-345.
25. Slovut DP, Ofstein LC, Bacharach JM. Endoluminal AAA repair using intravascular ultrasound for graft planning and deployment: a 2-year community-based experience. J Endovasc Ther. 2003;10(3):463–75. https://doi.org/10.1177/152660280301000311.
26. Perry RJ, Martin MJ, Eckert MJ, Sohn VY, Steele SR. Colonic ischemia complicating open vs endovascular abdominal aor-

tic aneurysm repair. J Vasc Surg. 2008;48(2):272–7. https://doi.org/10.1016/j.jvs.2008.03.040. Epub 2008 Jun 24.
27. Björck M, Bergqvist D, Troëng T. Incidence and clinical presentation of bowel ischaemia after aortoiliac surgery—2930 operations from a population-based registry in Sweden. Eur J Vasc Endovasc Surg. 1996;12(2):139–44. https://doi.org/10.1016/s1078-5884(96)80098-0.
28. Mehta M, Darling RC 3rd, Roddy SP, Fecteau S, Ozsvath KJ, Kreienberg PB, Paty PS, Chang BB, Shah DM. Factors associated with abdominal compartment syndrome complicating endovascular repair of ruptured abdominal aortic aneurysms. J Vasc Surg. 2005;42(6):1047–51. https://doi.org/10.1016/j.jvs.2005.08.033.
29. Ersryd S, Djavani-Gidlund K, Wanhainen A, Björck M. Editor's choice - abdominal compartment syndrome after surgery for abdominal aortic aneurysm: a Nationwide population based study. Eur J Vasc Endovasc Surg. 2016;52(2):158–65. https://doi.org/10.1016/j.ejvs.2016.03.011. Epub 2016 Apr 20.
30. Hemingway JF, French B, Caps M, Benyakorn T, Quiroga E, Tran N, Singh N, Starnes BW. Preoperative risk score accuracy confirmed in a modern ruptured abdominal aortic aneurysm experience. J Vasc Surg. 2021;74(5):1508–18. https://doi.org/10.1016/j.jvs.2021.04.043. Epub 2021 May 3.
31. Greeven AP, Bouwman LH, Smeets HJ, van Baalen JM, Hamming JF. Outcome of patients with ruptured abdominal aortic aneurysm after cardiopulmonary resuscitation. Acta Chir Belg. 2011;111(2):78–82. https://doi.org/10.1080/00015458.2011.11680711.
32. Mehta M, Byrne J, Darling RC 3rd, Paty PS, Roddy SP, Kreienberg PB, Taggert JB, Feustel P. Endovascular repair of ruptured infrarenal abdominal aortic aneurysm is associated with lower 30-day mortality and better 5-year survival rates than open surgical repair. J Vasc Surg. 2013;57(2):368–75. https://doi.org/10.1016/j.jvs.2012.09.003. Epub 2012 Dec 21.

Endovascular Management of Acute Aortic Dissection

63

Tim J. Mandigers, Daniele Bissacco, Maurizio Domanin, and Santi Trimarchi

Abbreviations

AAS	Acute aortic syndrome
AATS	American Association for Thoracic Surgery
AD	Aortic dissection
BTAI	Blunt thoracic aortic injury
FL	False lumen
IMH	Intramural hematoma
IRAD	International Registry of Acute Aortic Dissection
PAU	Penetrating aortic ulceration
STS	Society of Thoracic Surgeons
SVS	Society for Vascular Surgery
TEVAR	Thoracic endovascular aortic repair
TL	True lumen

Disclosures: ST is consultant and speaker for Medtronic, WL Gore, and Terumo Aortic.

T. J. Mandigers
Section of Vascular Surgery, Cardio Thoracic Vascular Department, Fondazione IRCCS Cà Granda Ospedale Maggiore Policlinico, Milan, Italy

Department of Vascular Surgery, University Medical Center Utrecht, Utrecht, The Netherlands

D. Bissacco
Section of Vascular Surgery, Cardio Thoracic Vascular Department, Fondazione IRCCS Cà Granda Ospedale Maggiore Policlinico, Milan, Italy

M. Domanin · S. Trimarchi (✉)
Section of Vascular Surgery, Cardio Thoracic Vascular Department, Fondazione IRCCS Cà Granda Ospedale Maggiore Policlinico, Milan, Italy

Department of Clinical Sciences and Community Health, University of Milan, Milan, Italy
e-mail: maurizio.domanin@unimi.it

Background

Acute aortic dissection (AD) is a life-threatening medical emergency in which a tear of the aortic intimal wall layer, the aortic intima, causes blood to flow between the subintimal aortic wall layers. Blood mainly flows between the intimal and medial wall layers, or between the different layers of the media, laterally bound by the outer third of the media. The aortic wall splits in a progressive manner (dissects) due to the propulsive forces of the blood caused by the left ventricle. Consequently, a second lumen or false lumen (FL) is formed. This FL coexists with the true lumen (TL) and both luminae are separated by the intimal dissection flap but connected by the primary entry tear and/or several re-entry tears (Fig. 63.1).

Due to FL progression, blood flow to important aortic side-branches (e.g., coronary arteries, supra-aortic trunks, spinal arteries including the artery of Adamkiewicz, celiac trunk, superior mesenteric artery, renal arteries, or common iliac arteries) can get obstructed. Static or dynamic obstruction may occur that may cause malperfusion of one or more of the corresponding vital organs and/or limbs. Moreover, since the remaining aortic wall that separates the FL from other human cavities (e.g., pericardial space, left hemithorax, retroperitoneal space) is the outer third of the media and the adventitial wall layer, aortic rupture into one of these cavities may occur when these last layers get disrupted. Serious complications can occur rapidly with or without evident clinical signs preceding those events, which challenges the adequate timely diagnosis and treatment allocation accordingly. .

The first necropsy report of George II (1760) raised suspicion towards this devastating disease [1]. More recent international registries such as the International Registry of Acute Aortic Dissection (IRAD) (1996–now), have put major effort in analysing the clinical presentation, treatment strategies, and outcomes of specific AD subtypes [2, 3]. Together with the ongoing interest of international research collaborations in this disease entity and the ongoing development of newer

G. Geroulakos et al. (eds.), *Mastering Endovascular Techniques*, https://doi.org/10.1007/978-3-031-42735-0_63

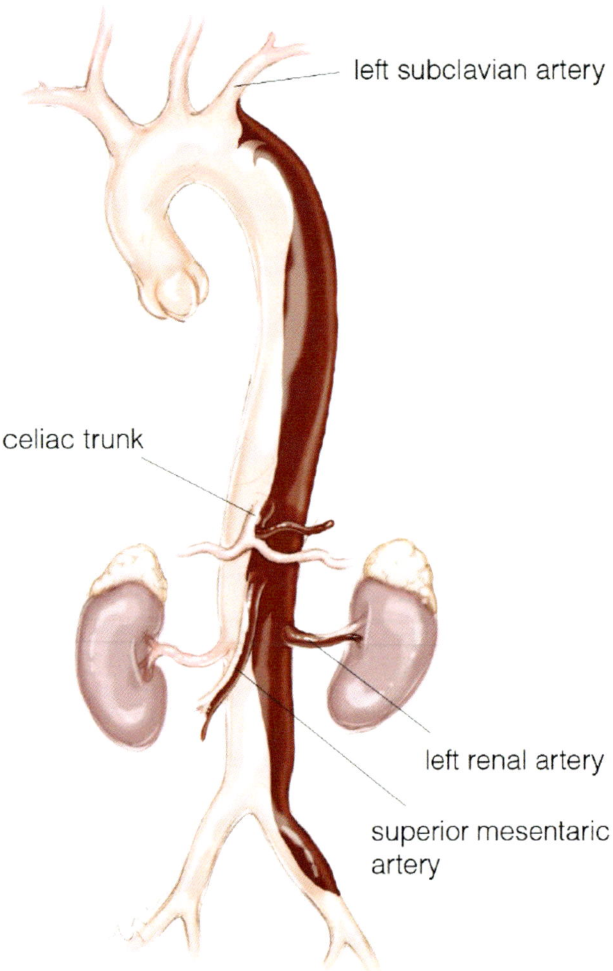

Fig. 63.1 Aortic dissection. Courtesy of: Trimarchi S, et al. Open fenestration for complicated acute aortic B dissection. Ann Cardiothorac Surg. 2014 Jul;3(4):418–22 (with permission)

endovascular techniques to treat ADs, these efforts have pushed our understanding of the underlying pathophysiological processes, clinical features, and desired treatment strategies for AD forward.

Acute Aortic Syndromes

Acute AD is one of the acute aortic syndromes (AAS) that share certain clinical characteristics. Other acute aortic diseases that could be classified as AAS are intramural hematoma (IMH), symptomatic penetrating aortic ulceration (PAU), blunt thoracic aortic injury (BTAI), or aortic rupture [4–6]. These conditions share common features and pathogenetic pathways. In IMH, a subintimal hematoma is present, without any evidence of an intimal entry tear as seen in classic acute AD. IMH may predict future classic AD, or PAU. In PAU, there is a focal ulceration of an atherosclerotic plaque or the intimal wall layer, following plaque rupture or rupture of the intima itself [4, 5]. BTAI, or aortic trauma, results after an acute decelerative event that could lead from a limited focal intimal aortic lesion to full-blown aortic rupture (Grades I–IV). The aortic injury is most frequently located at the relatively fixated parts of the aorta as compared to the relatively mobile parts of the aorta, such as the aortic isthmus in aortic zone 3 [7–9]. Aortic rupture refers to the disruption of all aortic wall layers, ultimately the adventitia, that could lead to exsanguination into the nearest human cavity if the rupture is not contained by surrounding tissue.

Classification Systems

Several classifications for acute AD exist. The most historical classification system is that of DeBakey et al. that classifies ADs into three types (i.e., I, II, IIIa, and IIIb) according to the location of the proximal entry tear and extension of the dissection [5, 10]. The Stanford classification is not taking the location of the proximal entry tear into account but is classifying acute ADs according to ascending involvement or not, without taking the distal extent into account. In Stanford type A ADs, the ascending aorta is involved regardless of the location of the primary entry tear, while in Stanford type B ADs the ascending aorta is not involved (Fig. 63.2) [11]. .

More recent classification systems have taken the location of the primary intima entry tear into account again and combine previous classification systems with Criado-Ishimaru's aortic zone map [12–15]. When the proximal entry tear is in aortic zone 0, corresponding to the ascending aorta and including the origin of the brachiocephalic trunk, the AD is classified as being type A. Consequently, when the proximal entry tear is in zone 1 or beyond, the AD is classified as type B [15] (Fig. 63.3). Using this classification system, a more detailed description can be obtained when the zone to which the AD extends proximally and distally is included as a subscript to the letter description (e.g., A_4 refers to a type A AD extending distally to aortic zone 4 (mid-descending aorta). Type $B_{0,9}$ AD indicates an entry tear in zone 1 or more with retrograde proximal extension to aortic zone 0 and distal extension to zone 9 (infrarenal aorta). An I instead of a number indicates that the entry tear is not found after analysing the diagnostic imaging modalities at that time, which could later be changed after identification of the primary entry tear at open surgical repair or additional diagnostic investigations (Fig. 63.4).

The intent of this latter classification is not to fully replace the previous DeBakey and Stanford classifications in clinical practice, but rather to complement them and to provide researchers with more specific tools to report and compare different AD subtypes. .

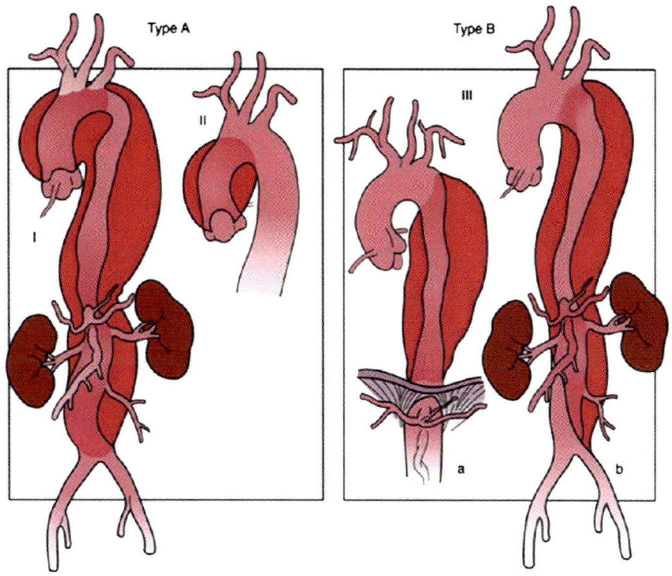

Fig. 63.2 DeBakey and Stanford classification systems for aortic dissection. Courtesy of: Riambau V, et al. Editor's Choice—Management of Descending Thoracic Aorta Diseases: Clinical Practice Guidelines of the European Society for Vascular Surgery (ESVS). Eur J Vasc Endovasc Surg. 2017 Jan;53(1):4–52 (with permission)

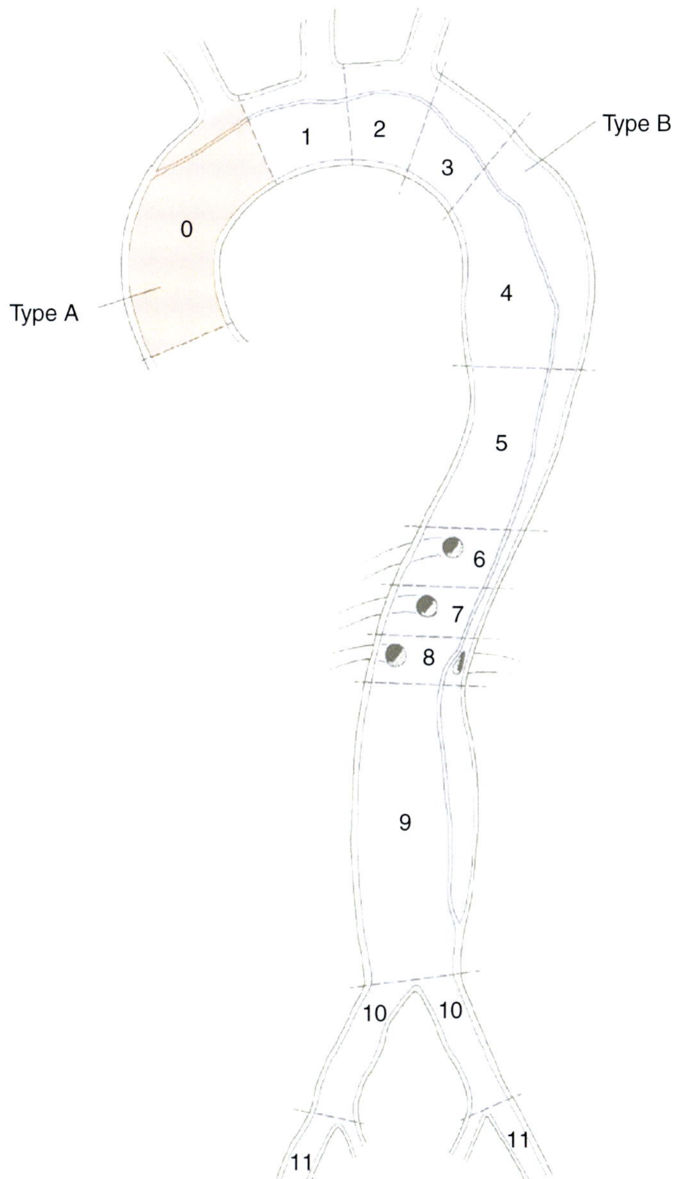

Fig. 63.3 Society for Vascular Surgery and Society of Thoracic Surgeons (2020) classification system for aortic dissection. Courtesy of: Lombardi JV, et al. Society for Vascular Surgery (SVS) and Society of Thoracic Surgeons (STS) reporting standards for type B aortic dissections. J Vasc Surg. 2020 Mar;71(3):723–747 (with permission)

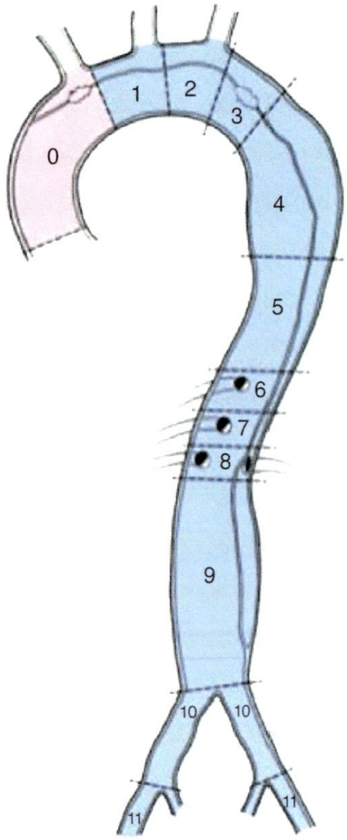

Type	Proximal Extent	Distal Extent
A_D Entry tear: **Zone 0**	0	0
	1	1
	2	2
	3	3
B_{PD} Entry tear: **≥Zone 1**	4	4
	5	5
	6	6
	7	7
	8	8
I_D Unidentified entry tear involving **Zone 0**	9	9
	10	10
	11	11
	12	12

Fig. 63.4 Society for Vascular Surgery and Society of Thoracic Surgeons (2020) classification system for aortic dissection (continued). Courtesy of: Lombardi JV, et al. Society for Vascular Surgery (SVS) and Society of Thoracic Surgeons (STS) reporting standards for type B aortic dissections. J Vasc Surg. 2020 Mar;71(3):723–747 (with permission)

Guided by different available management options and the specific related risks of every surgical intervention, several authors have aimed to identify certain subgroups of ADs that may require a more aggressive or conservative treatment approach. In this regard, some authors refer to the Non-A-Non-B dissection in which there is a proximal entry tear in the aortic arch (zone 1 or zone 2) without ascending involvement (zone 0) [16, 17]. Important to note is that there are a considerable number of ADs with a proximal entry tear in the aortic arch that do have ascending involvement as well [18].

Chronicity	Time from onset of symptoms
Hyperacute	<24 hours
Acute	1-14 days
Subacute	15-90 days
Chronic	>90 days

Fig. 63.5 Classification of aortic dissection according to its chronicity. Courtesy of: Lombardi JV, et al. Society for Vascular Surgery (SVS) and Society of Thoracic Surgeons (STS) reporting standards for type B aortic dissections. J Vasc Surg. 2020 Mar;71(3):723–747 (with permission)

Chronicity of AD

Furthermore, ADs can be classified according to the timing of the acute symptomatic aortic event. Historically, ADs were classified as being acute (< 14 days after the acute event) or chronic (14 days after the acute event), as authors observed that mortality significantly decreased beyond this time-point [19]. With advances in treatment strategies and a better understanding of the changes in dissection flap morphology over time, the most recent Society for Vascular Surgery (SVS) and Society of Thoracic Surgeons (STS) reporting standards for type B AD has provided two extra hyperacute and subacute classes of ADs based on the insights from more recent studies that analysed differences in outcomes of AD according to its chronicity [15, 20–23]. In hyperacute ADs, the onset of the acute symptomatic aortic event is <24 h ago, in acute ADs between 1 and 14 days ago, in subacute ADs between 15 and 90 days ago, and in chronic dissections >90 days ago (Fig. 63.5). This classification is based on more recent insights in changing dissection flap morphology over time. Over time,

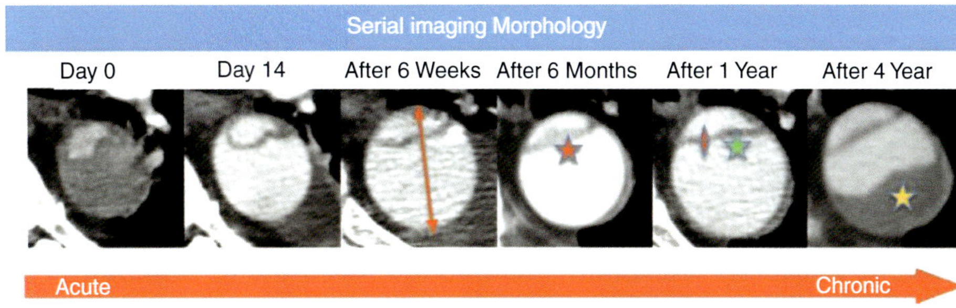

Fig. 63.6 Changing aortic dissection flap morphology over time. Courtesy of: Lombardi JV, et al. Society for Vascular Surgery (SVS) and Society of Thoracic Surgeons (STS) reporting standards for type B aortic dissections. J Vasc Surg. 2020 Mar;71(3):723–747 (with permission)

the flap thickens, becomes less compliant, and has a straighter appearance on computed tomography angiography (CTA) (Fig. 63.6) [24]. The present chapter will mainly focus on hyperacute and acute AD subgroups. .

Acuity of AD

In previous classifications, the location of the primary entry tear, extent of the dissection, and timing of the acute event provide the input for the correct classification in subgroups. When including the acuity of AD into the discussion, the clinical presentation and comorbidities of the patient should be adequately interpreted. Usually, before any diagnostic imaging modality will be performed, patients present with symptoms that will have to rise a physician's suspicion towards a possible acute aortic syndrome. Following a differential diagnosis, certain imaging techniques can confirm or deny his or her suspicion. Several clinical features will be more important as compared to others and can be associated with better or worse outcomes of AD subtypes. In this regard, three AD subgroups can be formed: uncomplicated, with high-risk features, or complicated (Fig. 63.7). This classification is most used to classify type B ADs, since type A ADs have an indication for immediate surgical repair regardless of these features [15, 25].

Uncomplicated AD

Uncomplicated AD presents without signs of end-organ malperfusion, aortic rupture, or high-risk clinical features as will be described below.

High-Risk Features of AD

Over the last decades, high-risk features of AD have been identified that could be predictive of adverse aorta related or non-aorta related outcomes of AD in the short-, mid-, or long-term. High-risk features are divided in clinical and radiographic features. Clinical features consist of refractory pain and hypertension that have been associated with an increased in-hospital mortality [26]. Refractory indicates the persistence of one of these symptoms despite maximal medical therapy for >12 hours according to Lombardi et al. [15]; however, a uniform definition is lacking among the available cardiovascular literature [26–31]. Maximal medical antihypertensive therapy is generally accepted to consist of three or more different classes of antihypertensive drugs, while for pain the amount and classes of drugs that should be administered before to conclude on a refractory aspect, is less evident. In these studies, refractory pain has been defined as the return of discomfort after relief of the presenting pain, without stating the number of drugs that are administered [27]. .

Moreover, certain radiographic features have been defined as high-risk features: bloody pleural effusion (as identified by having a high Hounsfield Unit on CTA), aortic diameters that exceed >40 mm, or radiographic only malperfusion, location of the primary entry tear at the lesser aortic curvature, or a FL diameter that exceeds >22 mm (Fig. 63.7) [32–34].

Radiographic only malperfusion should be interpreted with caution, as the positive predictive value of renal malperfusion as seen on CTA has been shown to be 76% in a recent study, for example [35]. This highlights that findings of diagnostic imaging modalities should always be combined with clinical findings at patient presentation, and if indicated, a more invasive quantification of malperfusion (e.g., invasive angiography) that may give insight to the specific AD subtype and its accompanying features should be performed [35, 36].

Complicated AD

At last, a third group of complicated AD can be defined that consists of ADs presenting with aortic rupture, end-organ, or limb malperfusion. As stated before, radiographic alone malperfusion should be interpreted with caution. Nevertheless, when there is a suspicion of critical end-organ or limb malperfusion, further diagnostic assessment to confirm or deny this suspicion should be performed. This will often guide treatment strategies towards a more invasive treatment approach. All four limbs, and every end-organ (e.g., heart, brain, bowel, kidneys) can become malperfused in a static or dynamic fashion by the dissection flap or thrombosed intraluminal parts of the aorta and its side-branches (Figs. 63.8, 63.9, 63.10). Since the aortic environment can be considered as a highly dynamic environment due to the propulsive forces of the blood exerted by the left ventricle, the intraluminal dissection flap will move and is able to obstruct aortic side-branches in a dynamic fashion [15, 37]. According to the Windkessel effect, the aorta recoils during the diastolic cardiac phase after storage of up to 50% of the left ventricular stroke work during systole, thereby ensuring coronary perfusion and a more continuous distal aortic flow pattern to distal aortic beds [38, 39]. Furthermore, aortic rupture can be defined as the extravasation of intraluminal blood outside of the adventitial wall layer. Aortic rupture can be either contained or overtly free in the pericardial space, thoracic, or abdominal cavities. .

Uncomplicated
No rupture
No malperfusion
No high-risk features
High risk
Refractory pain
Refractory hypertension
Bloody pleural effusion
Aortic diameter >40 mm
Radiographic only malperfusion
Readmission
Entry tear: lesser curve location
False lumen diameter >22 mm
Complicated
Rupture
Malperfusion

Fig. 63.7 Classification of aortic dissection according to its acuity. Courtesy of: Lombardi JV, et al. Society for Vascular Surgery (SVS) and Society of Thoracic Surgeons (STS) reporting standards for type B aortic dissections. J Vasc Surg. 2020 Mar;71(3):723–747 (with permission)

As an adjunct to the previously discussed classification systems, AD acuity provides a more specific classification of AD subtypes, which could be able to guide management strategies of physicians and surgeons towards a more conservative or invasive approach.

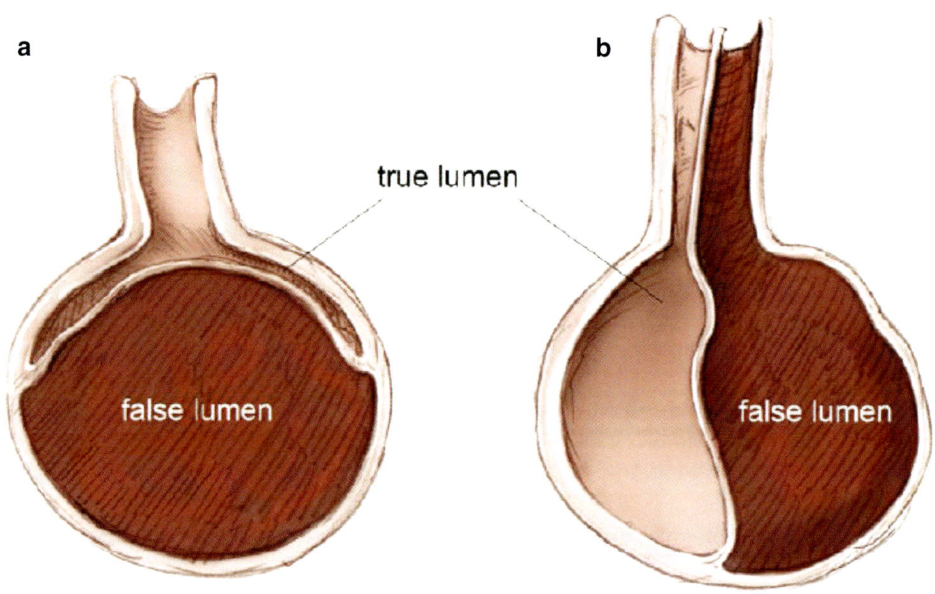

Fig. 63.8 Axial aortic section of dynamic (**a**) or static (**b**) aortic side-branch obstruction. Courtesy of: Trimarchi S, et al. Open fenestration for complicated acute aortic B dissection. Ann Cardiothorac Surg. 2014 Jul;3(4):418–22 (with permission)

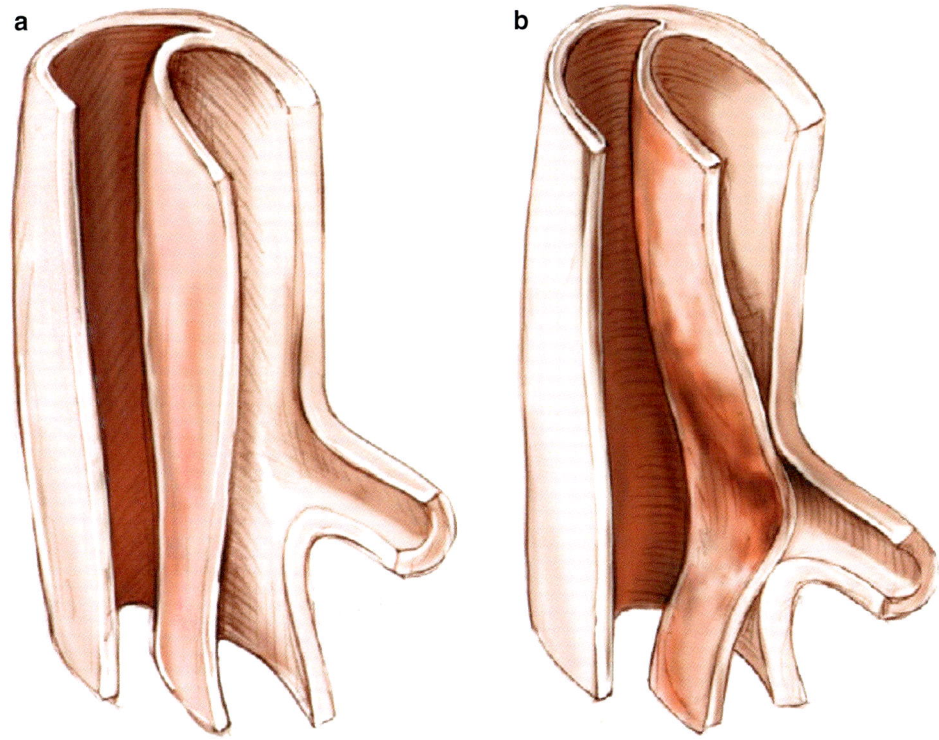

Fig. 63.9 Example of an intimal dissection flap that compresses the true lumen and prolapses into the ostium of the aortic side-branch during different phases of the cardiac cycle, thereby causing dynamic malperfusion. Courtesy of: Trimarchi S, et al. Open fenestration for complicated acute aortic B dissection. Ann Cardiothorac Surg. 2014 Jul;3(4):418–22 (with permission)

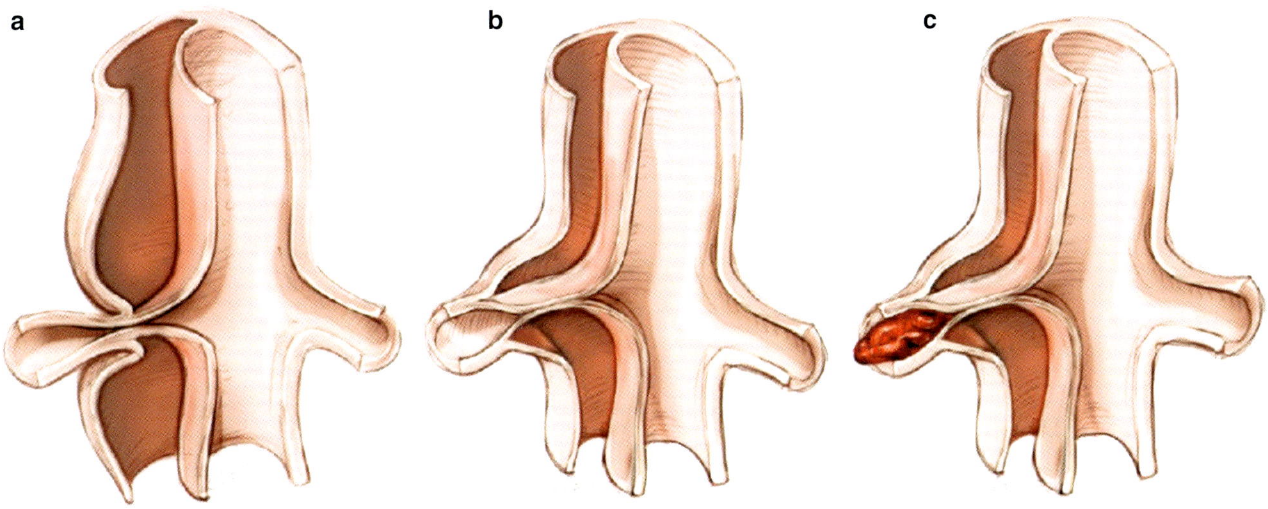

Fig. 63.10 Different mechanisms of static aortic side-branch obstruction. (**a**) Compression of the side-branch ostium by blind ends of the false lumen; (**b**) Presence of true and false lumen in the side-branch causing further compression; (**c**) Thrombosis in the true lumen of the aortic side-branch distal to a proximal compression of the side-branch. Courtesy of: Trimarchi S, et al. Open fenestration for complicated acute aortic B dissection. Ann Cardiothorac Surg. 2014 Jul;3(4):418–22 (with permission)

Epidemiology

In general, the incidence of acute AD is higher in men and increases with age [40–42]. The incidence and prevalence of AD seems to increase over the last decades, probably related to a greater proportion of the population being at high age and substantial advances in diagnostic imaging techniques [5]. Population based studies have reported incidences from 3.5 per 100,000 persons in the United States of America [43], 6 per 100,000 persons in the United Kingdom [44], up to 15 per 100,000 persons in Sweden [45]. The real incidence of acute AD may even be higher than reported in most studies because a substantial number of patients will likely have died before reaching the hospital due to the critical nature of the disease. According to IRAD, the occurrence of type A is about twice as high as type B AD, leading to two-thirds (66.6%) of acute ADs being a type A and one-third (33.3%) being a type B AD, according to the Stanford classification [41, 42].

Endovascular Management

Historically, the standard treatment options for acute AD consist of a conservative, medical, or open surgical approach to replace parts of the dissected aorta in the case of type A AD or complicated type B AD. Less than 25 years ago, Dake et al. reported the first series of endovascular stent-graft placements for the treatment of acute ADs, commonly known as thoracic endovascular aortic repair (TEVAR) nowadays [46]. The 19 ADs treated in this series were all type B ADs according to most recent classification systems, with four showing retrograde extension into the ascending aorta ($B_{0,I}$ or Stanford type A). For the first time, endovascular stent-grafts were used here to cover the primary intimal entry tear and prevent antegrade perfusion of the FL. Finally, the aim was to reduce complications such as aortic rupture or malperfusion. Since then, endovascular techniques to treat acute AD have rapidly evolved and continue to evolve. .

Optimalization with Medical Therapy

Regardless of which treatment strategy will be chosen (e.g., conservative, endovascular, open surgical repair (OSR)), it is as essential as fundamental to achieve adequate pain and blood pressure control. This can be achieved by administering oral or invasive antihypertensive and analgetic drugs to reduce blood pressures and cardiac contractility next to adequate pain control, as recommended in the most recent cardiovascular guidelines [4, 5]. In this way, further pulsatile aortic damage and dilatation can be prevented by reducing stress or hemodynamic forces on the aortic wall. In brief, for acute AD, intravenous beta-blockers should be considered as first-line therapy and initiated to lower systolic blood pressures to 100–120 mmHg (in the absence of aortic regurgitation). Calcium-channel blockers and/or renin-angiotensin inhibitors may be considered as alternative or complementary options. For chronic AD, target blood pressures may be slightly higher up to 140/90 mmHg.

As this chapter focusses on the endovascular management of AD, a thorough discussion of the different classes or subtypes of antihypertensive or analgetic drugs and sustainable lifestyle changes (e.g., smoking) to achieve adequate tension and pain control, would be out of scope.

Complicated Acute Type B Aortic Dissection

The previously discussed classification systems will guide clinicians in choosing the right management strategy for patients presenting with hyperacute or acute AD. For acute complicated type B AD, following the SVS/STS (2020) classification system and thus considering a primary entry tear in zone 1 or more, TEVAR is recommended as the first-line treatment option (Figs. 63.11 and 63.12) [4, 5].

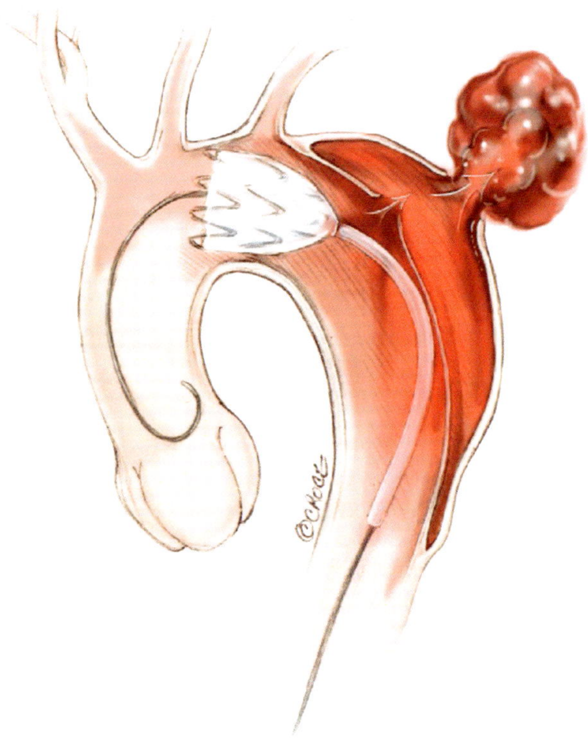

Fig. 63.11 Intraluminal stent graft deployment to cover the primary entry tear in a type B aortic dissection complicated by aortic rupture. Courtesy of: Trimarchi S, et al. Emergent treatment of aortic rupture in acute type B dissection. Ann Cardiothorac Surg. 2014 May;3(3):319–24 (with permission)

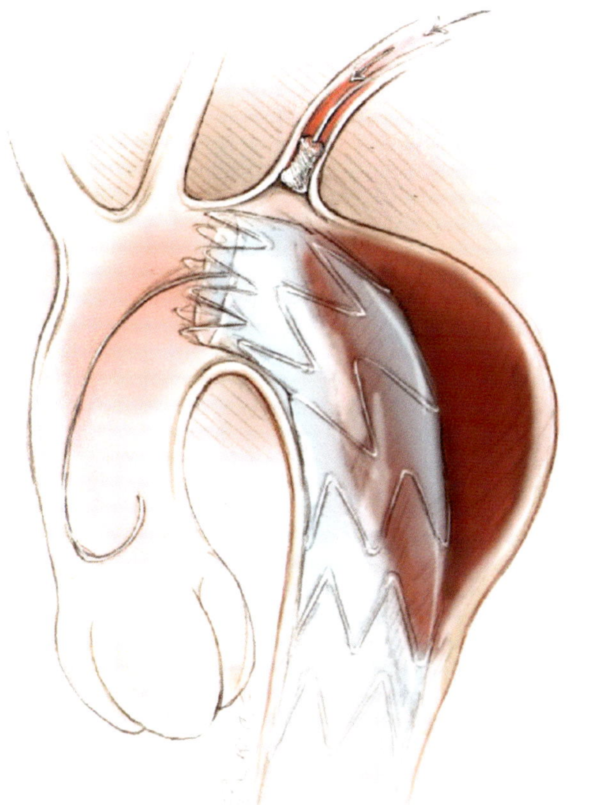

Fig. 63.12 Complete deployment of the intraluminal stent graft and placement of a vascular plug to prevent retrograde type II endoleak from the left subclavian artery. Courtesy of: Trimarchi S, et al. Emergent treatment of aortic rupture in acute type B dissection. Ann Cardiothorac Surg. 2014 May;3(3):319–24 (with permission)

Recent STS and American Association for Thoracic Surgery (AATS) clinical practice guidelines state that TEVAR is indicated for complicated hyperacute, acute, or subacute type B ADs, taking into consideration that aortic anatomy favours endovascular treatment [25]. In this regard, OSR for type B AD remains limited to patients with connective tissue disorders (CTDs) (e.g., Marfan, Loeys-Dietz, Ehlers-Danlos type IV) or unsuitable anatomy. Patients with a CTD have shown to have higher rates of retrograde type A dissections after TEVAR and there is less available evidence regarding the use of aortic stent grafts to treat AD in these patients [47–49]. Fenestration of the dissection membrane with branch vessel stenting alone may be considered in patients without aortic rupture, to alleviate malperfusion and prevent end-organ tissue necrosis and failure while avoiding the risks associated with TEVAR [25, 50]. Open surgical fenestration alone for acute type B AD has been reported as a valid alternative in the presence of contra-indications or failure of TEVAR [51]. .

High-Risk Acute Type B Aortic Dissection

High-risk morphological features (Fig. 63.13) could pose a higher risk of later adverse aorta related events or complications in patients presenting with acute type B AD [25]. These features are similar to the high-risk clinical and radiographic features discussed above (Fig. 63.7) with regards to the initial aortic diameter >40 mm and initial FL diameter >22 mm that may independently predict post-dissection aneurysmal growth [15, 25]. Interestingly, there are differences among Figs. 63.7 and 63.13 as well. One of the differences is the location of the primary entry tear. Primary entry tears that are located at the greater curve of the distal aortic arch or near to the left subclavian artery are associated with later aorta related events [25]. Another difference is regarding the size and number of primary or re-entry tears between the true and FL, large secondary re-entry tears distal to the TEVAR (distal stent-graft induced new entry), and partial FL thrombosis, as can be emphasized. These factors may all be predictive of increased post-dissection aneurysmal growth [52–56]. The theoretical explanation for this phenomenon is that continuing pressurization of the FL through re-entry tears may lead to aortic dilatation over time. In the presence of these features, physicians should carefully balance the possible advantages of TEVAR by reducing these late aorta related events or late aorta related death against the possible disadvantages of TEVAR by inducing TEVAR-related complications (e.g., retrograde type A AD, endoleak) [25]. Several factors should be weighed in this balance, including the patient's preferences. These factors are reflected in the recommendations of the most recent clinical practice guidelines on AD in which prophylactic TEVAR may now be considered selectively to reduce adverse aorta related late events or death in uncomplicated acute type B AD [4, 5, 25]. It remains challenging for physicians to select the right patients that will benefit the most from prophylactic TEVAR. .

Primary entry tear at greater curve of distal arch
Short proximity of entry tear to left subclavian artery ostium
Initial aortic diameter ≥40 mm
Initial false lumen diameter ≥22 mm
Number/size of fenestrations between true and false lumen
Stent graft-induced new entry
Partial false lumen thrombosis

Fig. 63.13 Morphological features of aortic dissection posing a high risk of late aorta related events. Courtesy of: MacGillivray TE, et al. The Society of Thoracic Surgeons/American Association for Thoracic Surgery clinical practice guidelines on the management of type B aortic dissection. J Thorac Cardiovasc Surg. 2022 Apr;163(4):1231–1249 (with permission)

Uncomplicated Acute Type B Aortic Dissection

The role of TEVAR is less evidently present in the management of uncomplicated acute type B AD. Optimal medical therapy remains the cornerstone of treatment as discussed before [25]. The recent STS/AATS clinical practice guidelines propose a stepwise approach to the evaluation and treatment of acute uncomplicated type B AD which refers to the identification of the primary entry tear's location (e.g., distance of primary entry tear with respect to the orifice of the left subclavian artery), calibration of the maximum orthogonal aortic diameter, and confirmation of the lack of any organ malperfusion or other complications [25]. Close clinical follow-up is recommended, and prophylactic TEVAR may be considered in these cases without classic complicating factors (e.g., aortic rupture, malperfusion) to reduce late aorta related adverse events and aorta related death as discussed before.

Two clinical trials (e.g., INSTEAD, ADSORB) compared TEVAR and optimal medical therapy (OMT) versus OMT alone for uncomplicated acute type B AD. In the INSTEAD trial that investigated 140 patients at least 2 weeks after their index uncomplicated type B AD, the primary endpoint was all-cause mortality at 2 years [57]. All-cause mortality, aorta related death, and aorta related reinterventions (e.g., conversion, additional endovascular or open surgical procedure) were similar at 2 years in both groups. Nevertheless, aortic remodelling such as TL expansion and FL thromboses were found to be significantly more present in the TEVAR and OMT group, as compared to OMT alone [57]. Important to note is that the study was underpowered, as acknowledged by the authors. Analysing survivors of the same cohort in the longer term, between 2 and 5-years post index dissection, TEVAR was associated with increased aorta specific survival rates and delayed disease progression as compared to OMT alone, suggesting that TEVAR should be considered in uncomplicated type B AD, with favourable anatomy, to improve late outcomes [21]. Moreover, the ADSORB trial investigated 61 patients that presented with uncomplicated type B AD (80% DeBakey type IIIb) within 14 days after the acute event [58]. The primary endpoints were incomplete FL thrombosis, aortic dilatation, or aortic rupture at 1 year, and these were all in favour of TEVAR (Gore TAG) plus OMT versus OMT alone. While highlighting the need for longer-term follow-up data, the authors also concluded that uncomplicated type B AD can be safely treated with the Gore TAG device, together with OMT as baseline treatment in every patient presenting with an AD regardless of its subclassification.

Type A Aortic Dissection

It is generally accepted that patients presenting with a Stanford type A AD, should be directly submitted to open surgical repair with the least possible delay, preferably performed in tertiary aortic centres of excellence [4]. Open surgical strategies (e.g., supracoronary ascending aortic replacement (SCAR), Bentall, David, hemi-arch, partial-arch, total arch repair, or frozen-elephant trunk (FET) procedures) largely depend on the clinical presentation of the patient with or without aortic rupture, cardiac tamponade, malperfusion, the location of the primary entry-tear, surgeon and centre's experience, and an emphasis on late aorta related complications and reinterventions.

However, in patients deemed unfit for open surgical repair due to the presence of complications and/or advanced age and frailty, ascending aorta intraluminal stent-graft placement may provide a last-resort treatment option [59]. TEVAR has moved to the more proximal aortic zones over the last decades, and TEVAR treatment of the ascending aorta could be considered the last frontier [60–63]. Currently, there is an ongoing Northern-American clinical trial (ARISE-trial) that assesses the feasibility of the Gore ascending aortic stent graft in the treatment of type A AD [64]. A recent systematic review identified 31 articles consisting of case-reports (n = 19) and case-series (n = 12) and demonstrated that TEVAR treatment of type A AD could be feasible with acceptable outcomes at short-term follow-up in highly selected patients, underlining the continuous need for data regarding the mid- and long-term [65]. Applications of TEVAR to more proximal zones provide cardiovascular surgeons with a wider range of treatment options for specific patient populations and disease subtypes. .

Principles and Methods of Intraluminal Stent Graft Treatment

In theory, thoracic aortic stent grafts can be advanced in an antegrade or retrograde fashion to land in the desired aortic zone. A transfemoral, retrograde approach is mostly used, given the size of TEVAR device delivery systems (e.g., large in both length and diameter). Other approaches may be in an antegrade fashion through the apex of the left ventricle (transapical approach), or using transiliac (with or without open or endoiliac conduits), transfemoral, transcarotid, transsubclavian, transaxillary, transaortic, or even transcaval approaches in an antegrade or retrograde fashion, depending on the desired landing zone of the stent graft and the availability (i.e., presence of atherosclerosis or calcifications, desirable geometry) of these vessels [66–72]. In any case, a suitable vascular access is of primary importance for an uneventful TEVAR treatment. Severe tortuosity of the common iliac arteries, infrarenal aorta, or descending aorta, challenges device delivery. In acute AD, the goal and rationale of TEVAR is to cover the primary entry tear and realign the intimal dissection flap with the adventitial wall layer as much as possible. In this way, the FL becomes depressurized and aortic remodelling (e.g., TL expansion, FL thrombosis, and eventual diameter reduction) could occur.

Endovascular Techniques to Obliterate Persistent Retrograde FL Flow

Several endovascular techniques have been introduced over the last decades, mainly to promote FL thrombosis and prevent the FL form further perfusion. In this way, these techniques could aid in preventing post-dissection aneurysmal degeneration that may require additional open surgical or endovascular aortic interventions, by helping to induce favourable aortic remodelling [73]. In general, it remains crucial to obtain long-term follow-up data regarding these newer techniques since most studies report short-term follow-up data. .

PETTICOAT Technique

The Provisional ExTension To Induce COmplete ATtachment technique, is an endovascular treatment option described in complicated acute type B AD. It is performed by deploying a self-expandable bare metal stent distal to the initial covered TEVAR that has covered the primary entry tear. In this way, the dissection membrane could become better aligned with the remaining adventitial wall layer, and antegrade perfusion of important originating arteries in these more distally located aortic zones (e.g., Adamkiewicz artery, visceral, and/or renal side-branches) will be promoted, as compared to distal extension with a covered stent graft. Consequently, FL thrombosis and aortic remodelling should be promoted. The concept has been introduced by Mossop et al. in 2005 and has been adopted more widely since this moment [74]. It has been associated with high technical success rates and favourable short-term outcomes that alleviate distal aortic side-branch obstruction and FL flow [75]. Moreover, a prospective, nonrandomized, clinical trial evaluated the primary endpoints of survival and major adverse events at 30 days (STABLE II) [76]. The bare metal stent extension was used in almost 80% of the 73 patients included in this study. Here it has demonstrated favourable clinical and anatomical outcomes at 1 year follow-up. Another study evaluated a reversed PETTICOAT technique in which pre-placement of the bare metal stent was performed before deploying the covered thoracic aortic stent graft to cover the primary entry tear [77]. More distal TL perfusion and less FL perfusion are also achieved with this comparable approach. .

STABILISE Technique

The STent Assisted Balloon Induced intimaL dISruption and rElamination (STABILISE) technique is another endovascular technique that aims to improve aortic remodelling and late aorta related complications, post-dissection aneurysmal growth, or reinterventions by preventing FL perfusion. This technique, first described by Hofferberth et al. in 2014, is achieved by introducing an aortic balloon catheter that can increase to the size of the total aortic diameter, after deployment of a covered thoracic aortic stent graft to cover the primary entry tear and distal bare metal stent extension [78]. This balloon will be inflated in the lower third of the covered stent graft and the upper part of the bare metal stent up until the dissection membrane is disrupted, so that the covered part of the stent graft will align the covered stent graft with the adventitial wall layer [79]. This procedure needs to be done with caution, as ballooning in AD is usually not recommended due to the fragility of the aortic tissue proven a priori due to AD occurrence (Figs. 63.14 and 63.15).

Other Endovascular False Lumen Obliteration Techniques

Like the STABILISE technique described above, the Knickerbocker technique aims to obliterate persistent retrograde FL flow and consequent aorta related complications and death and is mainly described in chronic ADs. A first covered stent graft will be deployed to cover the primary entry tear, usually in the region of the LSA orifice. After distal extension with an oversized stent graft, a large intraluminal compliant balloon will align the midsection of this second stent graft to the adventitial wall layer. To achieve this realignment, the dissection membrane must be disrupted as well (Fig. 63.16). Next to ballooning in fragile aortic tissue, another point of attention is to avoid the induction of distal stent graft induced new entries.

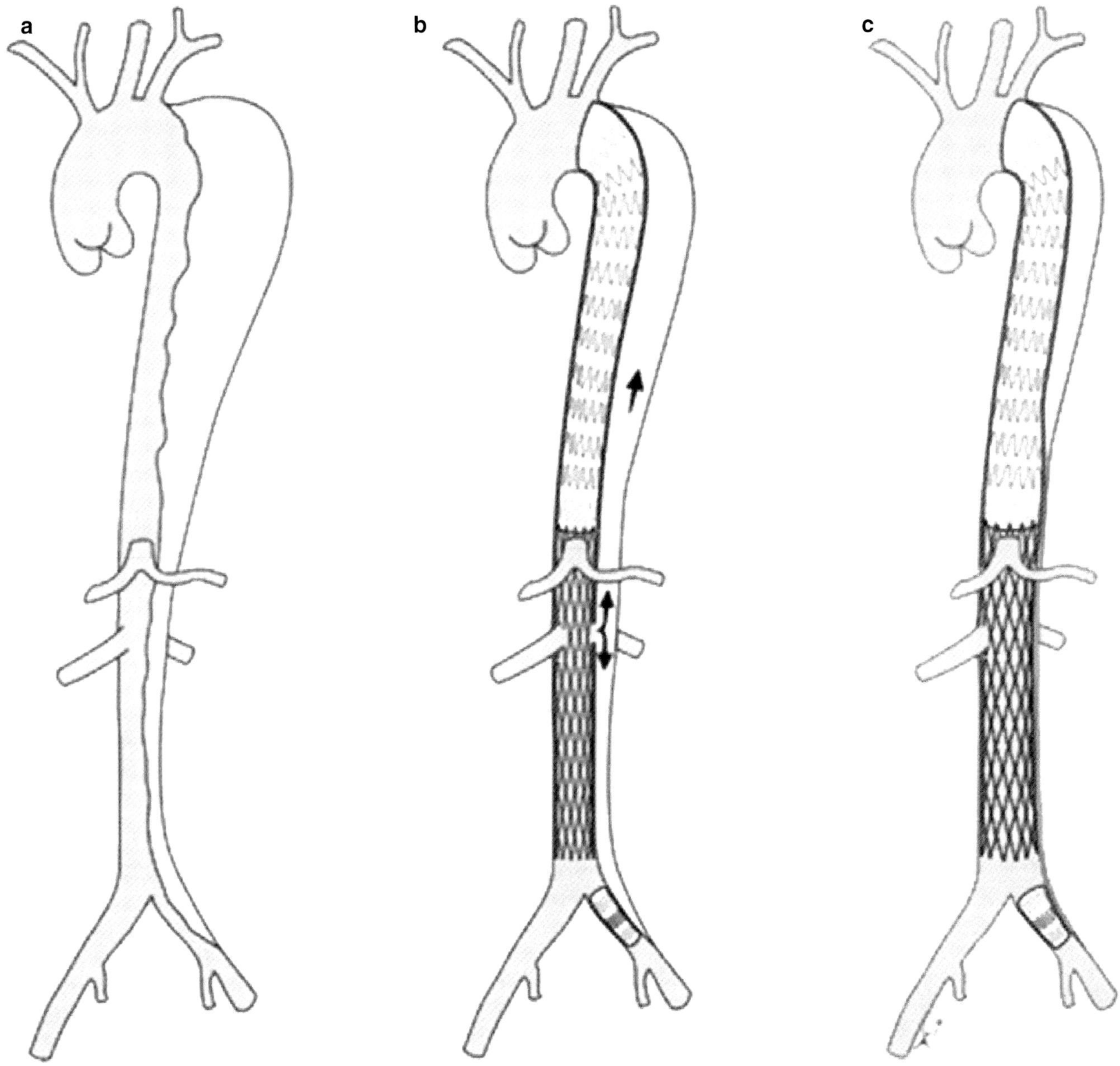

Fig. 63.14 Schematic representation of the STABILISE concept to obliterate persistent retrograde false lumen flow in aortic dissection. Courtesy of: Hofferberth SC, et al. Stent-assisted balloon-induced intimal disruption and relamination in aortic dissection repair: the STABILISE concept. J Thorac Cardiovasc Surg. 2014 Apr;147(4):1240–5 (with permission)

Other technical endovascular adjuncts to obliterate retrograde FL flow are the deployment of candyplugs or other vascular plugs, or partial embolization of the FL by using coils (Fig. 63.17) [80, 81].

An important point of attention for the use of these endovascular techniques is that it may complicate later open surgical aortic repair in the case of technical or clinical failure of these newer endovascular techniques (e.g., persistent FL pressurization and continuing post-dissection aneurysmal growth due to gutter areas or endoleaks). In this regard, surveillance should take place at predefined time-intervals to evaluate the long-term outcomes of these endovascular techniques. .

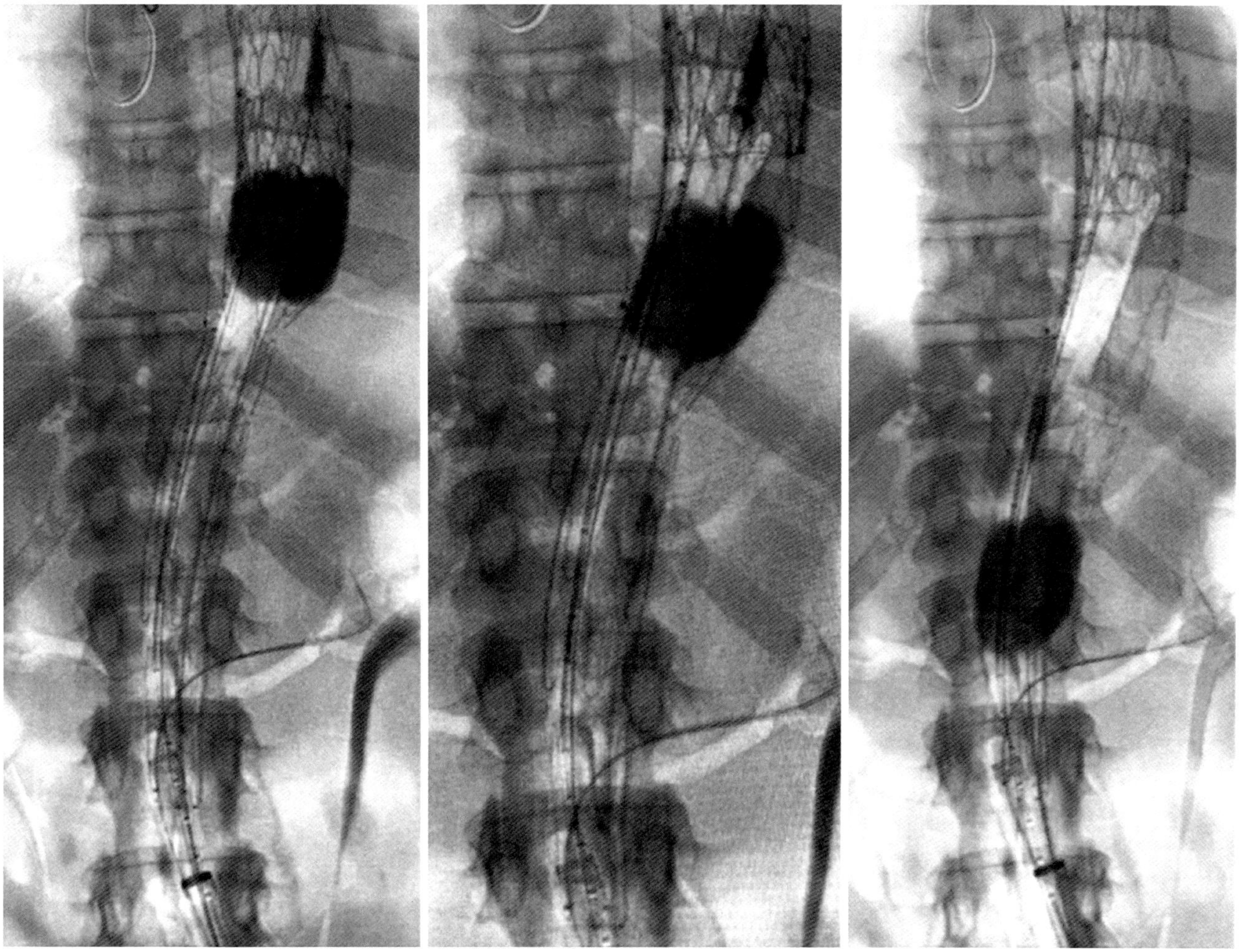

Fig. 63.15 Angiographic representation of the STABILISE concept to realign the stent graft with the adventitial wall layer. Courtesy of: Faure EM, El Batti S, Abou Rjeili M, Julia P, Alsac JM. Mid-term Outcomes of Stent Assisted Balloon Induced Intimal Disruption and Relamination in Aortic Dissection Repair (STABILISE) in Acute Type B Aortic Dissection. Eur J Vasc Endovasc Surg. 2018 Aug;56(2):209–215 (with permission)

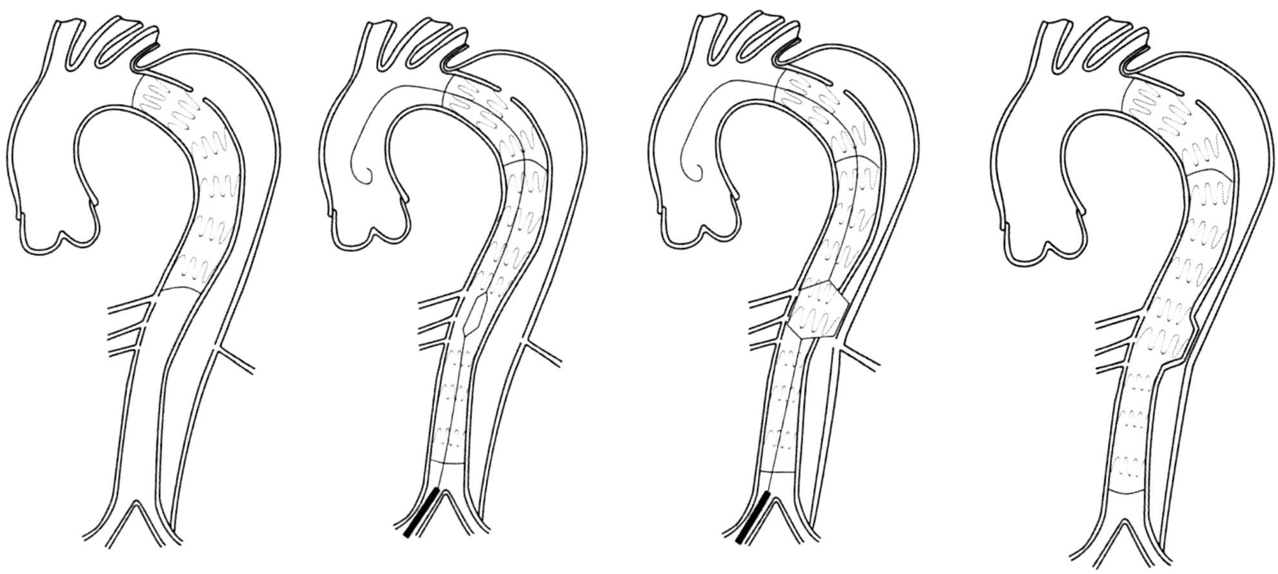

Fig. 63.16 Schematic representation of the Knickerbocker technique to obliterate persistent retrograde false lumen flow in aortic dissection. Courtesy of: Zeng Z, et al. Endovascular strategies for post-dissection aortic aneurysm (PDAA). J Cardiothorac Surg. 2020 Oct 1;15(1):287 (Part of Springer Nature)

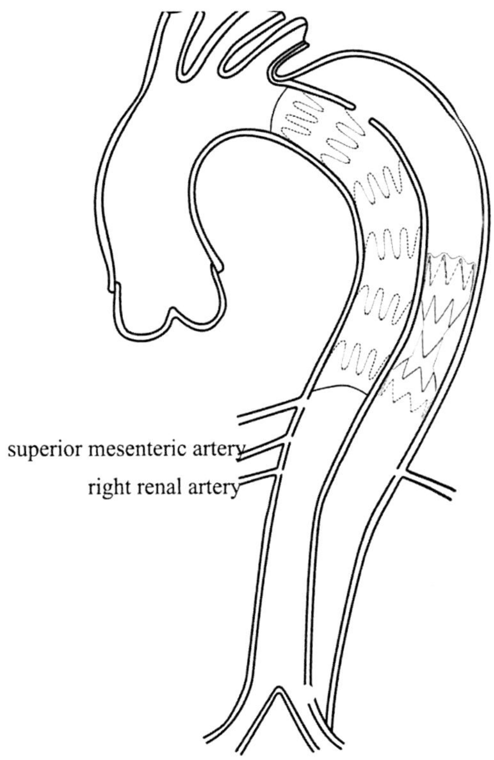

Fig. 63.17 Schematic illustration of the candyplug technique to oblitere persistent retrograde false lumen flow in aortic dissection. Courtesy of: Zeng Z, et al. Endovascular strategies for post-dissection aortic aneurysm (PDAA). J Cardiothorac Surg. 2020 Oct 1;15(1):287 (Part of Springer Nature)

References

1. Nicholls F. Observations concerning the body of his late majesty, October 26, 1760, by frank Nicholls, M D F R S physician to his late majesty. Philos Trans. 1761;52:265–75.
2. Hagan PG, Nienaber CA, Isselbacher EM, et al. The international registry of acute aortic dissection (IRAD): new insights into an old disease. JAMA. 2000;283:897–903.
3. Trimarchi S, Mandigers TJ, Bissacco D, Nienaber C, Isselbacher EM, Evangelista A, Suzuki T, Bossone E, Pape LA, Januzzi JL, Harris KM, O'Gara PT, Gilon D, Hutchison S, Patel HJ, Woznicki EM, Montgomery D, Kline-Rogers E, Eagle KA. Twenty-five years of observations from the international registry of acute aortic dissection (IRAD) and its impact on the cardiovascular scientific community. J Thorac Cardiovasc Surg. 2023 Jul 13;S0022-5223(23):00608–6. https://doi.org/10.1016/j.jtcvs.2023.06.021.
4. Erbel R, Aboyans V, Boileau C, et al. 2014 ESC guidelines on the diagnosis and treatment of aortic diseases. Eur Heart J. 2014;35:2873–926.
5. Riambau V, Böckler D, Brunkwall J, et al. Editor's choice—management of descending thoracic aorta diseases: clinical practice guidelines of the European Society for Vascular Surgery (ESVS). Eur J Vasc Endovasc Surg. 2017;53:4–52.
6. Tsai TT, Nienaber CA, Eagle KA. Acute aortic syndromes. Circulation. 2005;112:3802–13.
7. Neschis DG, Scalea TM, Flinn WR, Griffith BP. Blunt aortic injury. N Engl J Med. 2009;359:1708–16.
8. Mandigers TJ, Bissacco D, Domanin M, D'Alessio I, Tolva VS, Piffaretti G, van Herwaarden JA, Trimarchi S. Cardiac and aortic modifications after endovascular repair for blunt thoracic aortic injury: a systematic review. Eur J Vasc Endovasc Surg. 2022; https://doi.org/10.1016/J.EJVS.2022.05.004.
9. Azizzadeh A, Keyhani K, Miller CC, Coogan SM, Safi HJ, Estrera AL. Blunt traumatic aortic injury: initial experience with endovascular repair. J Vasc Surg. 2009;49:1403–8.
10. DeBakey M, Henly W, Cooley D, Morris G Jr, Crawford E, Beall A Jr. Surgical management of dissecting aneurysms of the aorta. J Thorac Cardiovasc Surg. 1965;49:130–49.
11. Daily PO, Trueblood HW, Stinson EB, Wuerflein RD, Shumway NE. Management of acute aortic dissections. Ann Thorac Surg. 1970;10:137–47.
12. Criado FJ, Clark NS, Barnatan MF. Stent graft repair in the aortic arch and descending thoracic aorta: a 4-year experience. J Vasc Surg. 2002;36:1121–8.
13. Mitchell RS, Ishimaru S, Ehrlich MP, Iwase T, Lauterjung L, Shimono T, Fattori R, Yutani C. First international summit on thoracic aortic endografting: roundtable on thoracic aortic dissection as an indication for endografting. J Endovasc Ther. 2002;9:98–105.
14. Fillinger M, Greenberg R, McKinsey J, Chaikof E. Reporting standards for thoracic endovascular aortic repair (TEVAR). J Vasc Surg. 2010;52:1022–3.
15. Lombardi JV, Hughes GC, Appoo JJ, et al. Society for Vascular Surgery (SVS) and Society of Thoracic Surgeons (STS) reporting standards for type B aortic dissections. J Vasc Surg. 2020;71:723–47.
16. Rylski B, Pérez M, Beyersdorf F, Reser D, Kari FA, Siepe M, Czerny M. Acute non-a non-B aortic dissection: incidence, treatment and outcome. Eur J Cardiothorac Surg. 2017;52:1111–7.
17. Carino D, Singh M, Molardi A, Agostinelli A, Goldoni M, Pacini D, Nicolini F. Non-a non-B aortic dissection: a systematic review and meta-analysis. Eur J Cardiothorac Surg. 2019;55:653–9.
18. Trimarchi S, de Beaufort HWL, Tolenaar JL, et al. Acute aortic dissections with entry tear in the arch: a report from the international registry of acute aortic dissection. J Thorac Cardiovasc Surg. 2019;157:66–73.
19. Hirst AE, Johns VJ, Kime SW. Dissecting aneurysm of the aorta: a review of 505 cases. Medicine. 1958;37:217–79.
20. Booher AM, Isselbacher EM, Nienaber CA, et al. The IRAD classification system for characterizing survival after aortic dissection. Am J Med. 2013;126:730.e19–24.
21. Nienaber CA, Kische S, Rousseau H, et al. Endovascular repair of type B aortic dissection: long-term results of the randomized investigation of stent grafts in aortic dissection trial. Circ Cardiovasc Interv. 2013;6:407–16.
22. Rodriguez JA, Olsen DM, Lucas L, Wheatley G, Ramaiah V, Diethrich EB. Aortic remodeling after endografting of thoracoabdominal aortic dissection. J Vasc Surg. 2008;47:1188–94.
23. Lombardi JV, Cambria RP, Nienaber CA, Chiesa R, Mossop P, Haulon S, Zhou Q, Jia F. Aortic remodeling after endovascular treatment of complicated type B aortic dissection with the use of a composite device design. J Vasc Surg. 2014;59:1544–54.
24. Peterss S, Mansour AM, Ross JA, et al. Changing pathology of the thoracic aorta from acute to chronic dissection: literature review and insights. J Am Coll Cardiol. 2016;68:1054–65.
25. MacGillivray TE, Gleason TG, Patel HJ, et al. The Society of Thoracic Surgeons/American Association for Thoracic Surgery clinical practice guidelines on the management of type B aortic dissection. J Thorac Cardiovasc Surg. 2022;163:1231–49.
26. Trimarchi S, Eagle KA, Nienaber CA, et al. Importance of refractory pain and hypertension in acute type B aortic dissection: insights from the international registry of acute aortic dissection (IRAD). Circulation. 2010;122:1283–9.

27. Januzzi JL, Movsowitz HD, Choi J, Abernethy WB, Isselbacher EM. Significance of recurrent pain in acute type B aortic dissection. Am J Cardiol. 2001;87:930–3.
28. Januzzi JL, Sabatine MS, Choi JC, Abernethy WB, Isselbacher EM. Refractory systemic hypertension following type B aortic dissection. Am J Cardiol. 2001;88:686–8.
29. Patel AY, Eagle KA, Vaishnava P. Acute type B aortic dissection: insights from the international registry of acute aortic dissection. Ann Cardiothorac Surg. 2014;3:368.
30. Acelajado MC, Hughes ZH, Oparil S, Calhoun DA. Treatment of resistant and refractory hypertension. Circ Res. 2019;124:1061–70.
31. Eggebrecht H, Schmermund A, von Birgelen C, Naber CK, Bartel T, Wenzel RR, Erbel R. Resistant hypertension in patients with chronic aortic dissection. J Hum Hypertens. 2005;19:227–31.
32. Song C, Lu Q, Zhou J, Yu G, Feng X, Zhao Z, Bao J, Feng R, Jing Z. The new indication of TEVAR for uncomplicated type B aortic dissection. Medicine (Baltimore). 2016;95:e3919.
33. Evangelista A, Salas A, Ribera A, Ferreira-González I, Cuellar H, Pineda V, González-Alujas T, Bijnens B, Permanyer-Miralda G, Garcia-Dorado D. Long-term outcome of aortic dissection with patent false lumen: predictive role of entry tear size and location. Circulation. 2012;125:3133–41.
34. Evangelista A, Galuppo V, Gruosso D, Cuéllar H, Teixidó G, Rodríguez-Palomares J. Role of entry tear size in type B aortic dissection. Ann Cardiothorac Surg. 2014;3:403.
35. van Bakel PAJ, Henry M, Kim KM, Yang B, van Herwaarden JA, Alberto Figueroa C, Patel HJ, Williams DM, Burris NS. Imaging features of renal malperfusion in aortic dissection. Eur J Cardiothorac Surg. 2022;61:805–13.
36. Mandigers TJ, Bissacco D, Domanin M, Trimarchi S. Think twice before trusting your eyes. Eur J Cardiothorac Surg. 2022;61:814–5.
37. Kamman AV, Yang B, Kim KM, Williams DM, Michael Deeb G, Patel HJ. Visceral Malperfusion in aortic dissection: the Michigan experience. Semin Thorac Cardiovasc Surg. 2017;29:173–8.
38. Belz G. Elastic properties and Windkessel function of the human aorta. Cardiovasc Drugs Ther. 1995;9:73–83.
39. Nauta FJH, de Beaufort HWL, Conti M, Marconi S, Kamman AV, Ferrara A, van Herwaarden JA, Moll FL, Auricchio F, Trimarchi S. Impact of thoracic endovascular aortic repair on radial strain in an ex vivo porcine model. Eur J Cardiothorac Surg. 2017;51:783–9.
40. Olsson C, Thelin S, Ståhle E, Ekbom A, Granath F. Thoracic aortic aneurysm and dissection: increasing prevalence and improved outcomes reported in a nationwide population-based study of more than 14 000 cases from 1987 to 2002. Circulation. 2006;114:2611–8.
41. Pape LA, Awais M, Woznicki EM, et al. Presentation, diagnosis, and outcomes of acute aortic dissection: 17-year trends from the international registry of acute aortic dissection. J Am Coll Cardiol. 2015;66:350–8.
42. Evangelista A, Isselbacher EM, Bossone E, et al. Insights from the international registry of acute aortic dissection: a 20-year experience of collaborative clinical research. Circulation. 2018;137:1846–60.
43. Clouse WD, Hallett JW, Schaff HV, Spittell PC, Rowland CM, Ilstrup DM, Melton LJ. Acute aortic dissection: population-based incidence compared with degenerative aortic aneurysm rupture. Mayo Clin Proc. 2004;79:176–80.
44. Howard DPJ, Banerjee A, Fairhead JF, Perkins J, Silver LE, Rothwell PM. Population-based study of incidence and outcome of acute aortic dissection and premorbid risk factor control: 10-year results from the oxford vascular study. Circulation. 2013;127:2031–7.
45. Landenhed M, Engström G, Gottsäter A, Caulfield MP, Hedblad B, Newton-Cheh C, Melander O, Smith JG. Risk profiles for aortic dissection and ruptured or surgically treated aneurysms: a prospective cohort study. J Am Heart Assoc. 2015;4:e001513.
46. Dake MD, Kato N, Mitchell RS, Semba CP, Razavi MK, Shimono T, Hirano T, Takeda K, Yada I, Miller DC. Endovascular stent-graft placement for the treatment of acute aortic dissection. N Engl J Med. 1999;340:1546–52.
47. de Beaufort HWL, Trimarchi S, Korach A, et al. Aortic dissection in patients with Marfan syndrome based on the IRAD data. Ann Cardiothorac Surg. 2017;6:633–41.
48. Dong ZH, Fu WG, Wang YQ, et al. Retrograde type a aortic dissection after endovascular stent graft placement for treatment of type B dissection. Circulation. 2009;119:735–41.
49. Pacini D, Parolari A, Berretta P, di Bartolomeo R, Alamanni F, Bavaria J. Endovascular treatment for type B dissection in Marfan syndrome: is it worthwhile? Ann Thorac Surg. 2013;95:737–49.
50. Norton EL, Williams DM, Kim KM, Khaja MS, Wu X, Patel HJ, Deeb GM, Yang B. Management of acute type B aortic dissection with malperfusion via endovascular fenestration/stenting. J Thorac Cardiovasc Surg. 2020;160:1151–1161.e1.
51. Trimarchi S, Segreti S, Grassi V, Lomazzi C, Cova M, Piffaretti G, Rampoldi V. Open fenestration for complicated acute aortic B dissection. Ann Cardiothorac Surg. 2014;3:41822–422.
52. Trimarchi S, Tolenaar JL, Jonker FHW, et al. Importance of false lumen thrombosis in type B aortic dissection prognosis. J Thorac Cardiovasc Surg. 2013;145:S208–12.
53. Tsai TT, Evangelista A, Nienaber CA, et al. Partial thrombosis of the false lumen in patients with acute type B aortic dissection. N Engl J Med. http://dx.doi.org.proxy.library.uu.nl/101056/NEJMoa063232. 2009;357:349–59.
54. Ante M, Mylonas S, Skrypnik D, Bischoff MS, Rengier F, Brunkwall J, Böckler D. Prevalence of the computed tomographic morphological DISSECT predictors in uncomplicated Stanford type B aortic dissection. Eur J Vasc Endovasc Surg. 2018;56:525–33.
55. Loewe C, Czerny M, Sodeck GH, Ta J, Schoder M, Funovics M, Dumfarth J, Ehrlich M, Grimm M, Lammer J. A new mechanism by which an acute type B aortic dissection is primarily complicated, becomes complicated, or remains uncomplicated. Ann Thorac Surg. 2012;93:1215–22.
56. Trimarchi S, Jonker FHW, van Bogerijen GHW, Tolenaar JL, Moll FL, Czerny M, Patel HJ. Predicting aortic enlargement in type B aortic dissection. Ann Cardiothorac Surg. 2014;3:285–91.
57. Nienaber CA, Rousseau H, Eggebrecht H, et al. Randomized comparison of strategies for type B aortic dissection: the INvestigation of STEnt grafts in aortic dissection (INSTEAD) trial. Circulation. 2009;120:2519–28.
58. Brunkwall J, Kasprzak P, Verhoeven E, et al. Endovascular repair of acute uncomplicated aortic type B dissection promotes aortic Remodelling: 1 year results of the ADSORB trial. Eur J Vasc Endovasc Surg. 2014;48:285–91.
59. Sibinga-Mulder BG, van Strijen MJL, Heijmen RH. Unexpected, complete recovery after emergent thoracic endovascular aortic repair for inoperable type a aortic dissection. J Vasc Surg Cases Innovations and Techniques. 2022;8:167–70.
60. Preventza O, le Huu A, Olive J, Cekmecelioglu D, Coselli JS. Endovascular repair of the ascending aorta: the last frontier. Ann Cardiothorac Surg. 2022;11:26–30.
61. Czerny M, Schmidli J, Adler S, et al. Editor's choice—current options and recommendations for the treatment of thoracic aortic pathologies involving the aortic arch: an expert consensus document of the European Association for Cardio-Thoracic Surgery (EACTS) & the European Society for Vascular Surgery (ESVS). Eur J Vasc Endovasc Surg. 2019;57:165–98.
62. Piffaretti G, Grassi V, Lomazzi C, et al. Thoracic endovascular stent graft repair for ascending aortic diseases. J Vasc Surg. 2019;70:1384–1389.e1.
63. Piffaretti G, Czerny M, Riambau V, et al. Endovascular repair of ascending aortic diseases with custom-made endografts. Eur J Cardiothorac Surg. 2021;59:741–9.
64. Feasibility Study Begins for the Gore Ascending Stent Graft to Treat Type A Dissection—Endovascular Today. https://evtoday.

com/news/feasibility-study-begins-for-the-gore-ascending-stent-graft-to-treat-type-a-dissection. Accessed 24 May 2022.
65. Ahmed Y, Houben IB, Figueroa CA, Burris NS, Williams DM, Moll FL, Patel HJ, van Herwaarden JA. Endovascular ascending aortic repair in type a dissection: a systematic review. J Card Surg. 2021;36:268–79.
66. Saouti N, Vos JA, van de Heuvel D, Morshuis WJ, Heijmen RH. Thoracic aorta stent grafting through Transapical access. Ann Vasc Surg. 2015;29:362.e5–9.
67. Saadi EK, Dussin LH, Moura L, Machado AS. The axillary artery—a new approach for endovascular treatment of thoracic aortic diseases. Interact Cardiovasc Thorac Surg. 2010;11:617–9.
68. Muetterties CE, Menon R, Wheatley GH. A systematic review of primary endovascular repair of the ascending aorta. J Vasc Surg. 2018;67:332–42.
69. Botta L, Bruschi G, Fratto P, Margari V, Solcia M, Borgia F, Rampoldi A, Russo CF. Direct Transaortic TEVAR: an alternative option for selected patients with unsuitable peripheral access. Ann Thorac Surg. 2016;102:e117–9.
70. Uflacker A, Lim S, Ragosta M, Haskal ZJ, Lederman RJ, Kern J, Upchurch G, Huber T, Angle JF, Ailawadi G. Transcaval aortic access for percutaneous thoracic aortic aneurysm repair: initial human experience. J Vasc Interv Radiol. 2015;26:1437.
71. van Bogerijen GHW, Williams DM, Eliason JL, Dasika NL, Deeb GM, Patel HJ. Alternative access techniques with thoracic endovascular aortic repair, open iliac conduit versus endoconduit technique. J Vasc Surg. 2014;60:1168–76.
72. Tsilimparis N, Dayama A, Perez S, Ricotta JJ. Iliac conduits for endovascular repair of aortic pathologies. Eur J Vasc Endovasc Surg. 2013;45:443–8.
73. Zeng Z, Zhao Y, Wu M, Bao X, Li T, Feng J, Feng R, Jing Z. Endovascular strategies for post-dissection aortic aneurysm (PDAA). J Cardiothorac Surg. 2020;15:1–10.
74. Mossop PJ, McLachlan CS, Amukotuwa SA, Nixon IK. Staged endovascular treatment for complicated type B aortic dissection. Nat Clin Pract Cardiovasc Med. 2005;2:316–21.
75. Nienaber CA, Kische S, Zeller T, Rehders TC, Schneider H, Lorenzen B, Bünger C, Ince H. Provisional extension to induce complete attachment after stent-graft placement in type B aortic dissection: the PETTICOAT concept. J Endovasc Ther. 2006;13:738–46.
76. Lombardi JV, Gleason TG, Panneton JM, Starnes BW, Dake MD, Haulon S, Mossop PJ, Seale MM, Zhou Q. STABLE II clinical trial on endovascular treatment of acute, complicated type B aortic dissection with a composite device design. J Vasc Surg. 2020;71:1077–1087.e2.
77. He H, Yao K, Nie WP, Wang Z, Liang Q, Shu C, Dardik A. Modified Petticoat technique with pre-placement of a distal bare stent improves early aortic remodeling after complicated acute Stanford type B aortic dissection. Eur J Vasc Endovasc Surg. 2015;50:450–9.
78. Hofferberth SC, Nixon IK, Boston RC, McLachlan CS, Mossop PJ. Stent-assisted balloon-induced intimal disruption and Relamination in aortic dissection repair: the STABILISE concept. J Thorac Cardiovasc Surg. 2014;147:1240–5.
79. Faure EM, el Batti S, Rjeili MA, Julia P, Alsac J-M. Mid-term outcomes of stent assisted balloon induced intimal disruption and Relamination in aortic dissection repair (STABILISE) in acute type B aortic dissection. Eur J Vasc Endovasc Surg. 2018;56:209–15.
80. Rohlffs F, Tsilimparis N, Fiorucci B, Heidemann F, Debus ES, Kölbel T. The candy-plug technique: technical aspects and early results of a new endovascular method for false lumen occlusion in chronic aortic dissection. J Endovasc Ther. 2017;24:549–55.
81. Zink JN, Maness MR, Powell CS, Parker FM, Bogey WM, Stoner MC, Anderson CA. Coil embolization of persistent false lumen after type a dissection repair. Ann Thorac Surg. 2014;97:2193–6.

Endograft Occlusion

64

Martina Mariani, Ilenia D'Alessio, Ailin Belloni, Paolo Verlato, Giulia Lerva, Savino Pasquadibisceglie, Vittorio Maria Segramora, Valerio Tolva, and Alberto Froio

Case Presentation—Background

A 75-year-old male with a 40 mm right common iliac aneurysm and a 40 mm infrarenal aortic aneurysm was selected for endovascular treatment. His medical history was significant for coronary artery disease, hypertension, dyslipidemia, obesity, hyperhomocysteinemia with MTHFR mutation, and previous left femoral endarterectomy.

A trimodular Cook endograft was chosen for the treatment of the abdominal aortoiliac aneurysm, considering the bilateral long anatomy, being 22 cm from the lowest renal to the hypogastric artery. EVAR was preferred to surgery even if the proximal neck angulation was 70°, considering the high-risk conditions of the patient (Fig. 64.1). After main body deployment from the right iliac access, it was

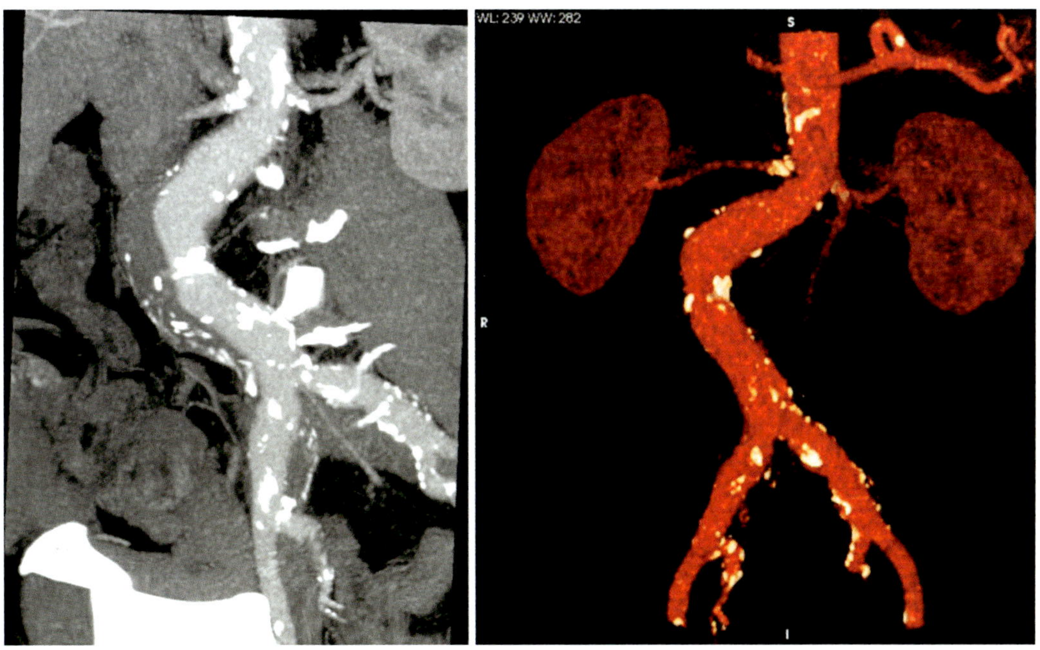

Fig. 64.1 Pre-operative CT-scan

M. Mariani · A. Belloni · P. Verlato · A. Froio (✉)
School of Medicine and Surgery, University of Milano-Bicocca, Milan, Italy

Vascular Surgery Department, Fondazione IRCCS San Gerardo, Monza, Italy
e-mail: alberto.froio@unimib.it

I. D'Alessio · G. Lerva · V. Tolva
Vascular Surgery Department, Fondazione "De Gasperis", Grande Ospedale Metropolitano Niguarda, Milan, Italy

S. Pasquadibisceglie · V. M. Segramora
Vascular Surgery Department, Fondazione IRCCS San Gerardo, Monza, Italy

G. Geroulakos et al. (eds.), *Mastering Endovascular Techniques*, https://doi.org/10.1007/978-3-031-42735-0_64

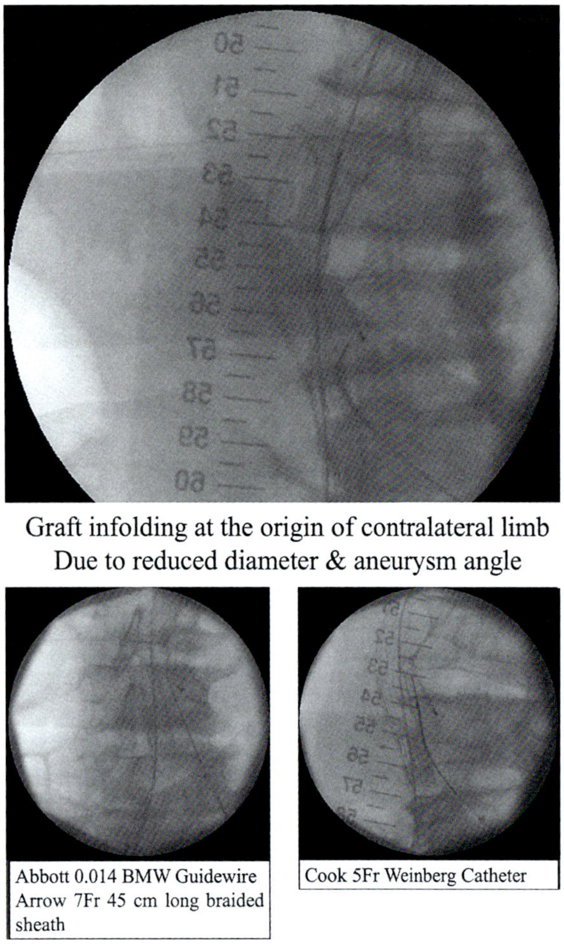

Fig. 64.2 Graft infolding at the origin of the contralateral Gate. Use of a telescopic approach to overcome the collapse

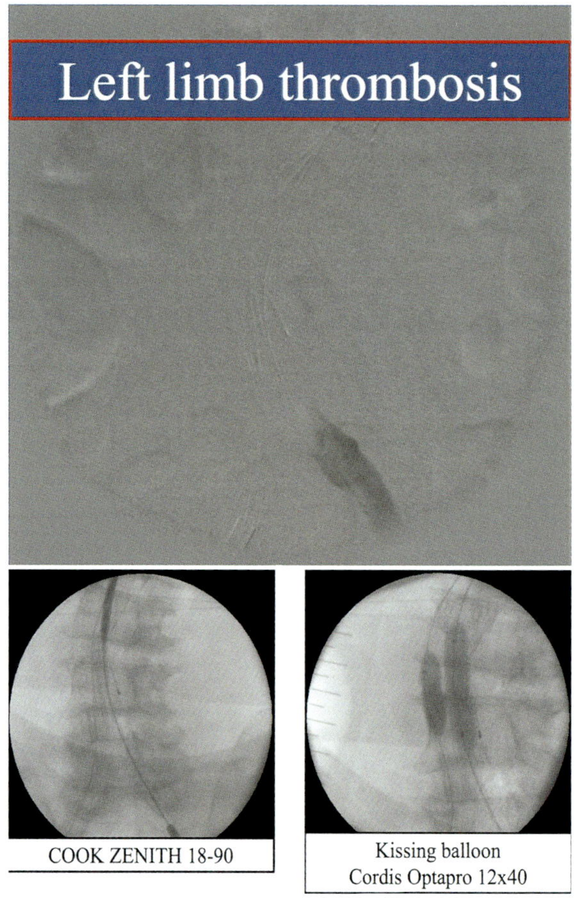

Fig. 64.3 Endograft limb thrombosis

not possible to engage the contralateral gate from the femoral access. An ipsilateral approach using the Sim1 Catheter was used to cannulate the contralateral limb. Even if the tip of the Sim1 Catheter was in the contralateral limb, the guidewire was not able to pass through the gate. A magnification of the contralateral limb showed a complete collapse of the contralateral limb at the level of the gate, probably induced by the sharp curve of the proximal portion of the neck into the aneurysm (Fig. 64.2). Due to prolonged unsuccessful attempts, a thrombosis of common iliac artery was observed (Fig. 64.3).

Background

Existing literature reports rates of limb occlusion in the range of 4–7.2%, irrespective of device or clinical situation. A meta-analysis evidenced that limb occlusion can occur in 5.6% of patients following EVAR. Approximately one third of stent graft limb occlusions are noted within the first 30 days post-EVAR, and about half of the patients present with symptoms of acute limb ischemia. Iliac limb occlusion decreased significantly over time, particularly in case of newer generation devices.

If endograft occlusion occurs acutely in the bilateral limb or the main body of the endograft, its pathological condition may be like that of acute occlusion of the abdominal aorta, a life-threatening disease associated with high morbidity and mortality.

Thoracic endovascular aortic repair (TEVAR) is rarely affected by endovascular occlusion. Sporadic case reports are described in literature due to enfolding or collapse of the device.

Endograft occlusion diagnosis is based on clinical patterns and several imaging modalities. Patients may have symptoms ranging from acute limb ischemia to different grades of lower limb claudication. Diagnostic confirmation can be obtained by computed tomography angiography (CTA), duplex ultrasonography, or angiography.

Predictive Factors of Endograft Occlusion

Many potential risk factors for endograft limb occlusion have been identified, being divided into three categories: patient- and device-related.

The following anatomical features must be assessed:

- Iliac artery tortuosity. Coulston investigated the impact of stent grafting on aortoiliac tortuosity, exploring the role of the tortuosity index (TI) as a predictor of iliac limb complications after EVAR. The iliac limb complication group was found to have a significantly increased aortoiliac TI on both preoperative and postoperative computed tomography imaging [1]. There was no significant difference in complication rates between the three stent grafts (Zenith flex, 1.4%; Endurant, 2.9%; Excluder, 1.9%; $P = 0.115$).
- Iliac artery atherosclerotic disease, especially with circumferential calcifications. Moulakakis analyzed patients presenting with unilateral and bilateral occlusions, showing that iliac perimeter calcification ≥50% was a risk factors for bilateral limb occlusion [2]. Iliac artery angulation ≥60 degrees and endograft oversizing in the common iliac artery of more than 15% had the same impact and could equally result in limb occlusion.

- Narrow distal aorta. Inaba analyzed patients with graft limb occlusion (GLO). The diameter of the terminal aorta was significantly smaller in patients with GLO than in patients without GLO (18.0 mm vs. 22.3 mm) [3]. The smaller limb was occluded in all patients with occlusion. The difference between the larger and the smaller limb was significantly larger in patients with GLO than in patients without ELO (4.0 mm vs. 1.7 mm).
- Iliac limb engagement. Coulston showed an increased complication rate following EVAR where proportional iliac engagement is lower [4]. Their analysis suggests that an engagement of over 70% on both the ipsilateral and contralateral limbs may reduce the chance of iliac limb complications. Kooijman confirmed this finding over time. The shortest apposition length of the endograft limbs was assessed as well as the distance between the end of the fabric and the proximal border of the internal iliac artery or the endograft-internal artery distance. There was a significant decrease in post-EVAR iliac apposition over time, partly due to retraction of iliac endograft limbs at mid-term CTA follow-up [5].
- Extension to external iliac artery. Bogdanovic compared the incidence and the management of limb graft occlusion in patients who underwent EVAR procedures with three of the major devices available on the market: Medtronic Endurant, Gore Excluder and Cook Zenith [6]. Interestingly, in a multivariable model analyzing five major risk factors (endograft device type, flow split misalignment, limb kink, EIA diameter <10 mm and EIA landing), only the use of Zenith Alpha along with EIA landing and EIA diameter smaller than 10 mm, were significantly associated with increased risk of limb graft occlusion.

The following endograft features must be assessed:

- Stent graft oversizing 15% in the iliac landing zone. Choosing an endograft limb with a larger diameter than the native vessel may increase the risk of graft infolding, which has been identified as a potential site of decreased flow. This issue is more relevant in case of small iliac arteries, where the balloon dilatation may not be enough to enlarge the vessel and the endograft. Cochennec and colleagues reported that oversizing by 4 mm in iliac arteries led to a 12-fold increase in graft limb dysfunction [7].
- Type of endograft. Unsupported stent-graft will be more than 15 times more likely than a supported system to require intervention because of kinking. Simmering demonstrated that Anaconda stent-graft is related to a relatively high thromboembolic event rate which decreases with a pre-to-postoperative reduction in curvature, and a larger degree of circumferential CIA calcification [8].
- Smaller limb diameter. This feature has been found to increase the risk of limb thrombosis. Mean graft limb diameter was 14 mm in the occluded population compared to 16 mm in patent limbs [9]. Oshin et al. and Conway et al. found that a higher rate of iliac limb occlusion occurred when the external iliac artery was engaged rather than the common iliac artery and this can be explained with the fact that a more distal landing zone often means landing in a more tortuous vessel. Proportionally high engagement should reduce sheer and radial forces on the native common iliac arteries and this should result in reduced ectatic change and, as a consequence, less future changes in the interaction between the native artery and the iliac limb and maybe even to an iliac limb "complication free" follow-up period.
- Presence of mural thrombus inside the endograft. Mestres found that the presence of intra-graft mural thrombus significantly increased the risk of endograft occlusion in some patients.
- Crossed limb graft. Crossed limb graft, also called the "ballerina technique" has been utilized to facilitate gate cannulation in tortuous anatomies with similar type Ib endoleak, type III endoleak, perioperative mortality, type Ia endoleak, type II endoleak and procedure. In particular, the rate of limb occlusion was similar between the two techniques.

Specific Device-related Complications

Cook Zenith Spiral-Z iliac limbs are composed of a continuous spiral nitinol stent, between two Z-stents at each end, which delivers kink resistance and flexibility. The rate of limb occlusion with Zenith devices ranges from 2.3% to 3.4%. A recent single-center retrospective study on 241 patients treated with Zenith showed a cumulative incidence of LGO around 7% per limb up to 3 years postoperatively [10].

The Endurant Stent Graft Natural Selection Global Postmarket Registry (ENGAGE) found that the strongest independent predictors of limb occlusion were distal landing zone in the external iliac artery, external iliac artery diameter <10 mm, maximum aneurysm diameter of 59 mm, needing corrective action for an endoleak, and kinking of the graft. A single-center long-term follow-up evidenced that the overall limb component occlusion rate was 3.1% (11/361) [11]. The time frame of limb thrombosis was <6 months in four patients and beyond 48 months after implantation in the remaining. Freedom from limb occlusion was 96.1%, 91.5%, and 87.3% at 5, 8, and 10 years. Patients treated outside-IFU had significantly higher risk of complications.

The GREAT registry showed that the Gore Excluder endograft on a total of 3274 treated patients evidenced only 0.7% of limb occlusion during a 5-year follow-up. The results confirmed single-center studies [12, 13].

Changes in the geometry of the Anaconda endograft limbs after endovascular aortic aneurysm repair were observed during a 2-year follow-up and they manifested as an increase in curvature, shortening of the stent-graft limbs, and a corresponding decrease in inter-ring distance, with the potential of limb occlusion. It is possible to decrease this risk with a pre-to-postoperative reduction in curvature and tortuosity index, and a larger degree of circumferential CIA calcification [8].

The INCRAFT™ AAA Stent-Graft System was assessed by a 5-year prospective regulatory study, including only patients within the IFU [14]. Successful aneurysm treatment was 100% at 30 days and 87.9% at 1 year. Seven patients out of 190 (3.7%) experienced graft or limb occlusion. Freedom from graft occlusion was 96% at 1 year and 94% at 5 years. Freedom from stent fracture was 97% at 1 year and 87% at 5 years.

Prevention

Accurate patient and device selection, adherence to IFU and identification and treatment of adverse anatomical factors are fundamental.

One preventive intervention seems to be the adjunctive selective primary stenting using a BMS, which showed a significant reduction of iliac limb occlusion rates in case of excessive angulation, graft stenosis or kinking, and dissection or stenosis of distal neck and external iliac artery [13].

The following strategy features can prevent graft limb thrombosis:

- Instructions For Use (IFU) are a useful roadmap to avoid easily foreseeable complications due to misuse of endovascular devices. This is true particularly when facing complex common iliac artery anatomy. EVAR in these kinds of settings can be quite challenging with a high risk of endograft limb compression and/or kinking and thrombosis.
- Early visualization of kinking of the endograft limb may be hard when the completion angiography is executed with the stiff wire in place. Therefore, it is recommended that the completion angiogram in case of severe tortuosity should be performed without the extra-stiff guidewire. A valid invasive alternative is IVUS (intravenous ultrasonography). However, performing IVUS requires extra time and additional expense, with unknown predictive value.
- Tailored follow-up. Stent graft kinking is a major cause of limb graft occlusion. It has been defined as a doubling of peak systolic velocity (PSV) in the artery, confirmed with subsequent pressure measurement on angiography. The presence of graft kinking prior to occlusion may be revealed by the occurrence of symptoms, or by routine follow up imaging with Computed Tomography Assial (CTA) or Duplex Ultrasound Scan (DUS). The most affordable method in terms of time and availability is traditional DUS. Blom conducted a large series of DUS EVAR surveillance that identified PSV 300 cm/s within the stent graft and a PSV > ratio > 3.5 as strong predictors of occlusion. Concerning intra-operative and follow up diagnosis of graft limb kinking it is important to remember that after EVAR, endograft limbs keep remodeling and it may take up to 24 h for the graft limb kinking to reach its full extent. Moreover, most limb occlusions become apparent or are detected in the first year after EVAR, which stresses out how crucial it is to lay out a close follow-up for these patients.

Treatment

While intraoperative limb occlusion leads to immediate intervention, limb occlusion after discharge primarily requires a detailed evaluation of clinical symptoms and their onset to choose the optimal treatment modality.

Patients who present with acute or critical limb ischemia with a recent onset of symptoms should be treated in every case and treatment modality can be chosen with respect to the surgeon's expertise; instead, patients presenting with recent mild ischemia can be treated by endovascular techniques in a safe manner. On the other hand, if the onset of symptoms dates back longer or is unclear, open surgical repair is recommended. Lastly, conservative treatment might be the safest option in asymptomatic patients. Meisenbacher developed an indication and treatment algorithm for limb occlusion, based on these indications [15].

ESVS Guidelines on the Management of Abdominal Aortoiliac Artery Aneurysms suggest that intervention is required only for symptomatic limb occlusion or as a preventive measure.

A recent paper evidenced that limb thrombus occurred in 7.3% of EVAR patients after 28 months [16]. Of these 22 patients, only 14 (63.6%) were symptomatic, and 17 (77.3%) had required intervention. The patients with limb thrombus were younger, more likely to be smokers and with smaller iliac arteries. The incidence of limb thrombus with the Alpha graft (10.7%) was compared with that with non-Cook grafts (4.0%). Unilateral limb thrombus was almost twice as likely on the main body side as on the contralateral side.

There are several treatment options to restore limb occlusion. There is no evidence in the literature regarding superiority of one treatment option over the other and the treatment strategy can be individually tailored.

- Surgery: bypass. For many years open surgery represented the first-line treatment for limb occlusion; it consists of an femoro-femoral crossover or ileo-femoral bypass as first choice, or an axillo-(bi)femoral bypass as a bailout option, in case of bilateral limb occlusion in a high-risk patient. Patency rates are high (up to 90% after 10 years for ileo-femoral and femoro-femoral crossover bypass and up to 50% after 5 years for axillo-femoral bypass) and bypass infections are rare (3%).
- Surgery: embolectomy. Surgical Fogarty embolectomy could be considered but it is usually associated with concerns about graft damage and incomplete thrombus removal if other options are not considered.
- Endovascular: catheter—directed thrombolysis (CDT). The most used fibrinolytic drugs are urokinase and rtPA, which have shown to have the same efficacy and safety; both should be used at weight related doses. The underlying limb stenosis can then be identified and treated by BMS placement. CDT can be performed in acute limb ischemia with equivalent results to surgery, but it is time consuming without indications in time-dependent emergencies such as in case of severe ischemia with motor deficit and may be complicated by leg emboli and haemorrhage, that may require the cessation of treatment. There is also a small risk of intracerebral haemorrhage (around 0.4%) which can be fatal. New endoleak due to the lysis of the thrombus in the aneurysm sac were described.
- Endovascular: aspiration techniques. Gargiulo assessed the safety and effectiveness of the Penumbra Indigo percutaneous aspiration thrombectomy (PAT) system in the clinical presentation of iliac limb occlusion [17]. Primary technical success was achieved in 14 cases (82%) after Indigo PAT. Fifteen patients (88%) required adjunctive procedures (14 relinings, 1 surgical patch angioplasty), that increased technical success up to 94% of cases. At 1, 6, and 12 months, clinical success was 100%. Median follow up was 23 months: at 18 months, survival and freedom from reintervention were 91% and 90%, respectively.
- Endovascular: thrombectomy. Rheolytic thrombectomy (AngioJet—Boston Scientific, Malborough, MA, USA) possibly associated with PTA/stent implantation; the use of the AngioJet thrombectomy device as a supplementary treatment for iliac limb occlusion following abdominal EVAR is safe, effective, and minimally invasive. However, it is associated with higher risk of distal embolization [18].

Hybrid techniques are a combination of surgical and endovascular treatment, such as surgical thrombectomy followed by stent implantation or thrombus aspiration/mechanical thrombectomy to remove any remaining clot. This approach is gaining popularity both in elective and emergency settings, also thanks to the increasing use of hybrid operating rooms.

Case Presentation—Results

The femoral bifurcation was clamped, to prevent distal embolization. Additional heparin was given to move ACT over 300″.

A telescope technique using a 7Fr Arrow 45 cm-long braided sheath, a 5Fr Cordis Ber II catheter and a stiff wire was used to force the gate collapse and cannulate the contralateral limb.

The Penumbra Indigo percutaneous aspiration thrombectomy was use to remove thrombus. A long contralateral iliac extension was used to complete EVAR, fixing all the remaining thrombotic material.

At the end of the procedure, no endoleaks were detected.

The femoral bifurcation was accurately washed, removing thrombotic material. No embolization was observed in the postoperative period.

The patient was discharged after 3 days. The post-operative CT scan showed the patency of the endograft and iliac arteries, the lack of endoleaks and a biphasic direct ultrasonographic signal at the femoral arteries (Fig. 64.4).

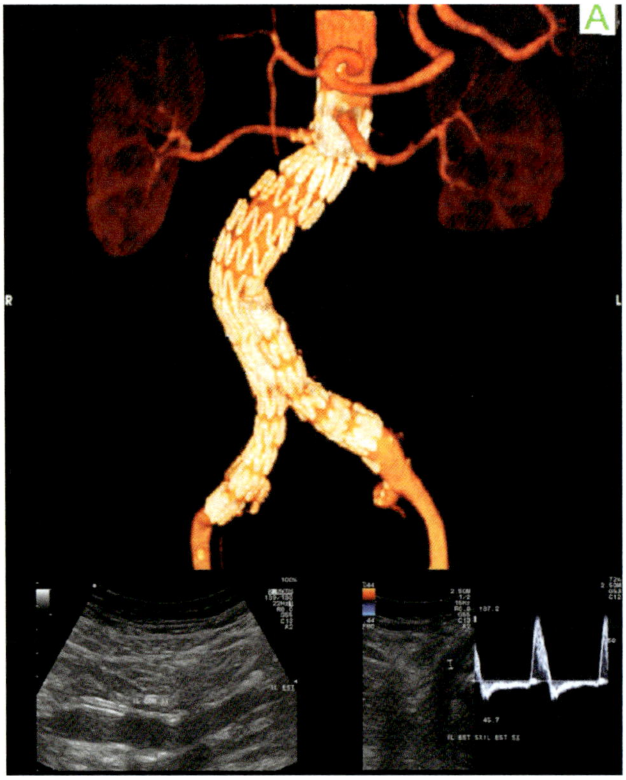

Fig. 64.4 Post-operative CT scan and duplex scan evaluation of left external iliac artery

Conclusions

Open surgery is still the most used treatment for iliac limb occlusion after EVAR; nevertheless, there is still insufficient evidence to assess its superiority over endovascular/hybrid repair.

The crucial factor is prevention of occlusion and/or re-occlusion through adequate medical therapy, accurate pre-operative planning and early identification and treatment of underlying technical issues, mainly graft stenosis or kinking.

Tool kit (table of the materials used for the treatment)

- Fogarty Catheter
- Dacron and PTFE vascular grafts.
- 5F straight GLIDECATH® Terumo catheter
- rTPA and Urokinase
- Long extra-stiff Supracore/Amplatz/Lunderquist guidewire
- 6, 7 and 8Fr 45–65–90 cm long Terumo Destination/Cook Flexor / Super Arrow-Flex Sheaths.
- Penumbra Indigo percutaneous aspiration thrombectomy system.
- AngioJet—Boston Scientific.

References

1. Coulston J, Baigent A, Selvachandran H, Jones S, Torella F, Fisher R. The impact of endovascular aneurysm repair on aortoiliac tortuosity and its use as a predictor of iliac limb complications. J Vasc Surg. 2014;60:585–9. https://doi.org/10.1016/j.jvs.2014.03.279.
2. Moulakakis KG, Antonopoulos CN, Klonaris C, Kakisis J, Lazaris AM, Sfyroeras GS, Mantas G, Mylonas SN, Vasdekis SN, Brountzos EN, et al. Bilateral Endograft limb occlusion after endovascular aortic repair: predictive factors of occurrence. Ann Vasc Surg. 2018;46:299–306. https://doi.org/10.1016/j.avsg.2017.07.019.
3. Inaba Y, Yoshitake A, Hayashi K, Ito T, Hachiya T, Shimizu H. Effect of the terminal aortic diameter on the patency rate of iliac limbs after endovascular aortic repair. Ann Vasc Dis. 2019;12:519–23. https://doi.org/10.3400/avd.oa.19-00080.
4. Coulston J, Baigent A, Selvachandran H, Jones S, Torella F, Fisher R. Lack of iliac engagement correlates with iliac limb complications following standard EVAR. Vasc Endovasc Surg. 2015;49:201–5. https://doi.org/10.1177/1538574415614403.
5. Kooijman MA, Schuurmann RCL, Kropman RHJ, Elzefzaf NY, Wille J, Tielliu IFJ, Antoniou GA, de Vries J. Midterm changes in iliac limb apposition after endovascular aortic aneurysm repair. J Cardiovasc Surg. 2023;64:159–66. https://doi.org/10.23736/s0021-9509.23.12582-1.
6. Bogdanovic M, Stackelberg O, Lindstrom D, Ersryd S, Andersson M, Roos H, Siika A, Jonsson M, Roy J. Limb graft occlusion following endovascular aneurysm repair for infrarenal abdominal aortic aneurysm with the zenith alpha, excluder, and endurant devices: a multicentre cohort study. Eur J Vasc Endovasc Surg. 2021;62:532–9. https://doi.org/10.1016/j.ejvs.2021.05.015.
7. Faure EM, Becquemin JP, Cochennec F, Collaborators E. Predictive factors for limb occlusions after endovascular aneurysm repair. J Vasc Surg. 2015;61:1138–45.e2. https://doi.org/10.1016/j.jvs.2014.11.084.
8. Simmering JA, de Vries M, Haalboom M, Reijnen M, Slump CH, Geelkerken RH. Geometrical changes of the aorta as predictors for thromboembolic events after EVAR with the Anaconda stent-graft. J Endovasc Ther. 2022; https://doi.org/10.1177/15266028221105839.
9. Coelho A, Nogueira C, Lobo M, Gouveia R, Campos J, Augusto R, Coelho N, Semiao AC, Ribeiro JP, Canedo A. Impact of post-EVAR graft limb kinking in EVAR limb occlusion: Aetiology, early diagnosis, and management. Eur J Vasc Endovasc Surg. 2019;58:681–9. https://doi.org/10.1016/j.ejvs.2019.03.019.
10. Broda M, Eiberg J, Taudorf M, Resch T. Limb graft occlusion after endovascular aneurysm repair with the cook zenith alpha abdominal graft. J Vasc Surg. 2023;77:77. https://doi.org/10.1016/j.jvs.2022.10.026.
11. Georgiadis GS, Schoretsanitis N, Argyriou C, Nikolopoulos E, Kapoulas K, Georgakarakos EI, Ktenidis K, Lazarides MK. Long-term outcomes of the EndurantTM endograft in patients undergoing endovascular abdominal aortic aneurysm repair. J Vasc Surg. 2023;78:668. https://doi.org/10.1016/j.jvs.2023.04.033.
12. Poublon CG, Holewijn S, van Sterkenburg SMM, Tielliu IFJ, Zeebregts CJ, Reijnen M. Long-term outcome of the GORE EXCLUDER AAA Endoprosthesis for treatment of Infrarenal aortic aneurysms. J Vasc Interv Radiol. 2017;28:637–44. https://doi.org/10.1016/j.jvir.2017.01.012.
13. de Marino PM, Ibraheem A, Gafur N, Mufty H, Schubert N, Verhoeven EL, Katsargyris A. Limb occlusion rate after EVAR

with individualized graft limb selection and a Liberal protocol of primary relining. Ann Vasc Surg. 2021;75:445–54. https://doi.org/10.1016/j.avsg.2021.02.046.

14. Liang NL, Ohki T, Ouriel K, Teigen C, Fry D, Henretta J, Komori K, Kichikawa K, Makaroun MS, Investigators I. Five-year results of the INSPIRATION study for the INCRAFT low-profile endovascular aortic stent graft system. J Vasc Surg. 2021;73:867–873.e2. https://doi.org/10.1016/j.jvs.2020.06.128.
15. Meisenbacher K, Wortmann M, Bischoff MS, Contin P, Bockler D, Geisbusch P. Limb occlusion following endovascular aortic repair: prevention and treatment. Gefasschirurgie. 2019;24:306–12. https://doi.org/10.1007/s00772-019-0537-6.
16. Draper K, Choi SHJ, Fung A, Baxter K, Taylor D, Chen JC, Misskey J. Evaluation of factors associated with limb thrombus formation after endovascular aortic aneurysm repair. J Vasc Surg. 2023;77:440–5. https://doi.org/10.1016/j.jvs.2022.10.008.
17. Spath P, Pasqui E, Angiletta D, Spinazzola A, Chegai F, Isernia G, Lepidi S, Silingardi R, de Donato G, Gargiulo M, et al. Penumbra indigo percutaneous aspiration Thrombectomy system in the treatment of aortic Endograft iliac limb occlusion: results from an Italian multicentre registry. Eur J Vasc Endovasc Surg. 2023;66:77. https://doi.org/10.1016/j.ejvs.2023.04.008.
18. Han XF, Liu GR, Li TZ, Guo X. Application of the AngioJet ultra Thrombectomy device for the percutaneous mechanical treatment (PMT) of iliac limb occlusion after endovascular aneurysm repair (EVAR). Ann Vasc Surg. 2022;78:161–9. https://doi.org/10.1016/j.avsg.2021.06.016.

Endoleak Type I

65

Panos Gkoutzios and Narayanan Thulasidasan

Pathophysiology and Endoleak Types

Endoleak is the condition where there is persistent blood flow into the aneurysm sac, after it has been treated endovascularly to exclude it from the blood circulation.

Endoleaks are detected in about 25% of patients following endovascular aortic aneurysm repair (EVAR) [1, 2].

The endoleaks can be either early, appearing within 30 days after the endovascular treatment or late, appearing after 30 days [1].

There are five different types of endoleak, depending on its origin and mechanism for its appearance (Table 65.1; Fig. 65.1 [1]; Fig. 65.2) [3].

– Type I endoleaks detected after EVAR, can be either early or delayed (12 months or later) and their incidence is 3.5% and 6.8% respectively [4]. They are subcategorised as Ia, Ib and Ic. In a systemic review and meta-analysis of 11 articles and 190 patients with late rupture after EVAR procedure, it was found that in half of the late ruptures (52%), type I endoleak was the leading cause for it [5].

Type Ia is the endoleak where the blood flow into the supposedly excluded aneurysm sac, comes from the upper (proximal) end of the endovascular aortic device. This is more likely in aneurysms with very angulated (>60%) proximal aortic neck, more conical shaped aortic neck, large diameter aortic neck (>28 mm) and short aortic neck. Risk of Ia endoleak is also increased in the presence of thrombus and/or significant atheromas in the aortic wall at the sealing zone/neck level, and during the remodelling process of the aneurysm sac after the endovascular treatment in increased aneurysm sac sizes. It can also occur if the stent graft was deployed at a lower level in the proximal aortic neck than intended or if it was undersized or significantly oversized, compared to the proximal aortic neck diameter.

In previous aortic aneurysm endovascular repair cases with the use of "chimney/snorkel" stents, there may be endoleak Ia appearance along the gutter between the aortic stent graft and the chimney/snorkel stents [6].

In the type Ib endoleak the blood flow comes from the distal end of the segment treated with endovascular devices.

In the type Ic, the endoleak comes from the distal end of the fenestration or branch covered bridging stents (in fenestrated or branched aortic devices), or from the contralateral, unstented iliac artery, for example in the aorto-uni-iliac endovascular repair procedures.

– Type II endoleak is the related to retrograde flow from other aortic aneurysm sac branches, such as the lumbar arteries, the inferior mesenteric artery (IMA), the intercostal arteries or even accessory renal arteries. Type II endoleaks account for about 40% of all endoleaks, and are often benign (not causing significant sac pressurization and growth).
– Type III endoleak is related to stent graft components' dislodgement or tears and damage of the device's fabric (type IIIa when the dislodgement is between the aortic components, type IIIb when the dislodgement is between the iliac limb components and the aortic component or between the iliac limb components, type IIIc when the dislodgement is between a side branch stent and the main body of the device or the fenestration stent and the main body of the device, or between stents in the side branches or the fenestrations, and type IIId is when there is a tear or damage in the stent graft).

P. Gkoutzios (✉) · N. Thulasidasan
Guy's & St Thomas' NHS Foundation Trust- Interventional Radiology Department, London, UK
e-mail: panos.gkoutzios@gstt.nhs.uk

G. Geroulakos et al. (eds.), *Mastering Endovascular Techniques*, https://doi.org/10.1007/978-3-031-42735-0_65

Table 65.1 Endoleak types after endovascular aortic aneurysm repair. Proposed classification

Endoleaks		
Type I	Type Ia	When the endoleak comes from the upper end of the endograft
	Type Ib	When the endoleak comes from the distal end of the endograft or the distal end of the iliac limb of the graft
	Type Ic	When the endoleak comes from the attachment of a branch or fenestration covered stent with its target vessel (renal artery, SMA, coeliac artery), or from the contralateral iliac artery in aorto-uniiliac implanted stent graft devices
Type II		When the endoleak comes from arterial branches (lumbar arteries, IMA, intercostal arteries, accessory renal arteries)
Type III	Type IIIa	When the endoleak comes from the dislodgement between the aortic components
	Type IIIb	When the endoleak comes from the dislodgement between the iliac components or the iliac and the bifurcated aortic component
	Type IIIc	When the endoleak comes from the dislodgement between side branches and the main endograft (fenestrated or branched devices) or between the branches stents themselves
	Type IIId	When the endoleak comes from tears or fractures in the endograft material
Type IV		When the endoleak comes from increased graft material porosity
Type V		When there is increased endotension, the aneurysm sac increases in size and there is no other type of endoleak present

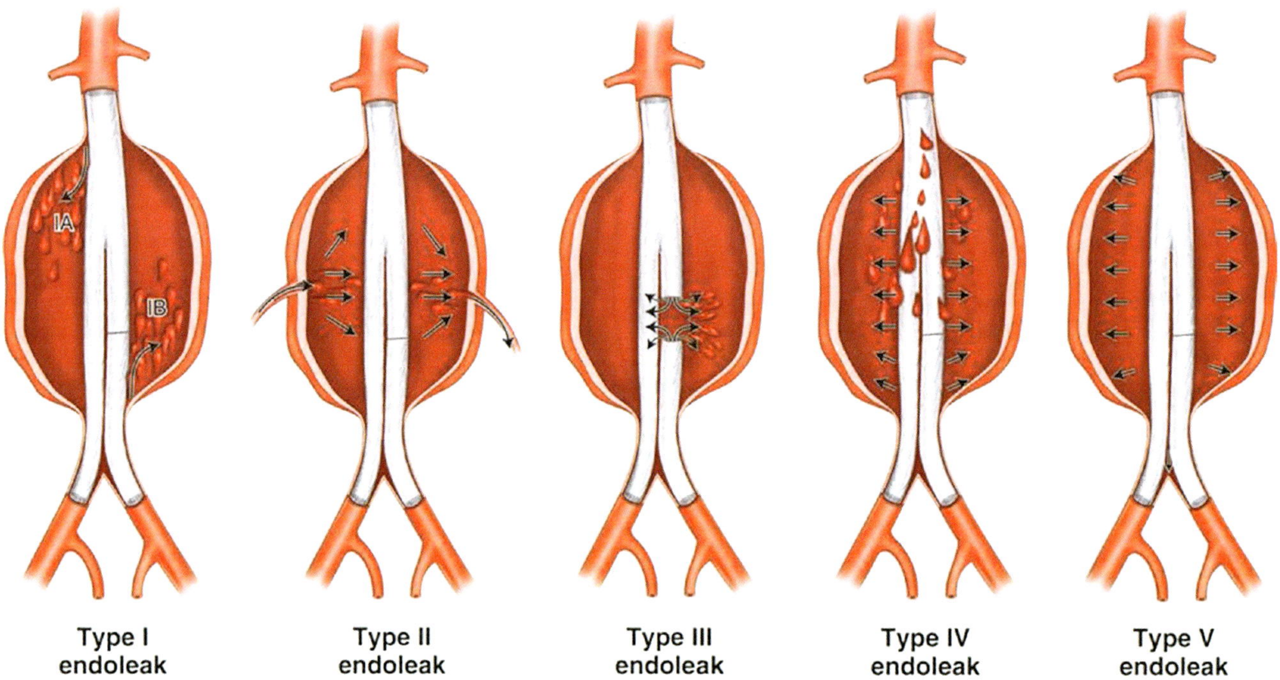

Fig. 65.1 Endoleaks types I-V classification. (Reprinted by permission from Springer Nature Customer Service Centre GmbH: Springer Nature by "IR Playbook- A Comprehensive Introduction to Interventional Radiology" by Nicole A. Keefe, Ziv J Haskal, Auh Whan Park, John F. Angle—2018 (Chap. 19, "Endoleak" by Brian C. Gardner and Saher S. Sabri)

- Type IV endoleak is related to increased stent graft porosity and is usually seen within the first 30 days of the stent graft implantation and rarely later.
- Type V endoleak is the one where there is increased aortic sac pressure and aortic sac size increase, without any other demonstrable endoleak present.

In this chapter we will further discuss about the type I endoleak and its subcategories.

The type I endoleak is mainly related to the proximal or/and distal landing zones of the endografts, the branches/fenestration stents landing zones, or the contralateral common iliac artery in uni-iliac stent graft devices.

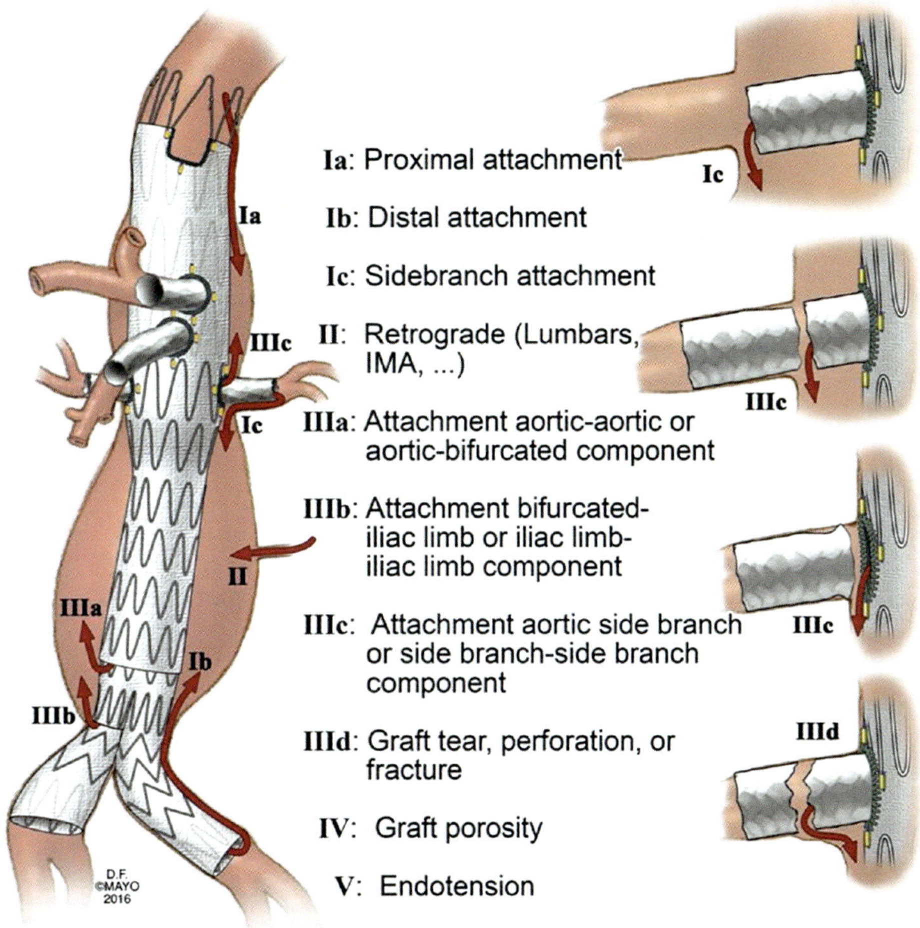

Fig. 65.2 Endoleaks I-V new proposed classification (with subcategories of type I and type III endoleaks). (1. Reprinted by permission from Springer Nature Customer Service Centre GmbH: Springer Nature by "Endovascular Aortic repair" by Gustavo S. Oderich—2017. 2. "From: Vrtiska TJ, Macedo TA & Oderich GS. Chap. 11: Computed Tomography/ Computed Tomography Angiography for Evaluation, Planning, and Surveillance of Complex Endovascular Repair. In: Endovascular Aortic Repair, first Edition. Cham, Switzerland: Springer International Publishing AG, 2017; used with permission of Mayo Foundation for Medical Education and Research, all rights reserved")

Endoleak Type I: Clinical Indication, Identification, and Imaging

The type I endoleak can be initially identified, at the final angiogram of the endovascular procedure,. Contrast appearance in the aneurysm sac, outside the endograft, seen next to the upper or distal landing zones, or close to any branched stents, is suspicious for type I endoleak. It can be brisk and obvious, or less clear depending on the endoleak size.

This should be attempted to be treated at the same setting as implantation and followed up with imaging post endovascular procedure. A CT angiogram (CTA) is done first, before the patient is discharged from the hospital or in 1 month. The imaging follow up usually lasts for many years, in most patients for life.

An initial type Ia endoleak during implantation may disappear in the next imaging follow up. Reasons for this can be organised thrombus formation in the aneurysm sac, reversal of the effect of the heparin given during the endovascular procedure [7] or remodelling of the excluded aneurysm sac.

Patients with endoleaks may also present with abdominal or back pain, although in most of the cases they are asymptomatic.

Absence of type I endoleak at the initial angiogram does not mean it may not appear in the future follow up imaging.

This type of endoleak appears most commonly in the arterial phase of the CTA and will be identified close to the area of its origin (proximal or distal end of the endograft or next to the branch stents), however if it is very small and contained in a very narrow tract outside the endograft, it can be missed in the arterial phase. In the delayed phase of the CTA, the endoleak contrast can be seen pooled in the aortic sac but its origin may not be visible. This can be more difficult to identify in the fenestrated or branched cases, where there are stents adjacent to the proximal end of the endograft, and clarification between Ia or Ic endoleak is harder.

The most common and widely used imaging modality for follow up after an endovascular repair of aortic aneurysm, is the CTA. However colour-duplex ultrasound (DUS) and contrast enhanced duplex ultrasound (CEDUS), are other imaging modalities that can be and are used, during these follow ups - either alone or in combination, in an attempt to decrease the patient's radiation exposure from the repeated CTAs. The compared sensitivity and effectiveness between the DUS/CEDUS and the CTA for the follow up post EVAR and the detection of endoleaks is an area of ongoing discussion [8–10].

Another relatively novel technique that can be used for the endoleak clarification, is the Dynamic CTA. With this we take sequestial CT scans (phases) in a short, specific segment of the aorta (which includes the area of the endoleak and the suspicious sources of it) and try to identify the feeding vessel to the endoleak at an early imaging phase [11]. This can be a very useful imaging technique in cases where there is diagnostic uncertainty, between type I (a or b), type III and type II endoleaks.

Digital substractive angiography (DSA), although not a standard technique for imaging follow up, can be infrequently, chosen as an imaging modality to try and identify a type I endoleak, if the other non-invasive modalities are not providing a clear diagnosis of it. However, even DSA, which is an invasive modality, may not be conclusive, as the images are also subject to motion artefacts (bowel movement, breathing motion artefacts), non-visualised contrast tracking behind the endograft at the chosen image intensifier angle, volume of contrast injected, etc.

Endoleak Type I Treatment

The endoleak type I, as well as the type III, tend to increase the aortic sac size (as they typically are a high-pressure endoleaks) and need to be addressed and treated.

The most common treatment option is an endovascular one, without excluding an open surgical option, if there no endovascular alternative. A surgical repair/conversion of abdominal aortic aneurysms related issues, post previous endovascular repair has a mortality risk between 3.3% and 8% [1, 12].

Endoleak Type Ia [1]

Endovascular repair options for the type Ia endoleak are:

- Balloon dilatation of the graft with a compliant balloon in its proximal landing zone. With this dilatation we hope to achieve some better apposition of the stent graft against the aortic wall, locally.
- Balloon-expandable large stent (Palmaz) insertion [13, 14] and deployment in the endograft's proximal landing zone. This is a technique that tends not to be used very frequently nowadays outside the setting of intraoperative bailout, and should not be used in more demanding anatomies [15]. With this we expect to eliminate any possible endograft infolding and appose it against the aortic wall, similarly to the aforementioned balloon dilatation technique.
- Endoanchor application. These can be used during the initial treatment, or later, in order to prevent any migration of the endograft and optimise its apposition to the aortic wall. However the durability and the sustainable results of this treatment are not fully established [16] and should be judiciously used [17]. Endoanchor fixation in thoracic aorta aneurysm endovascular repair (TEVAR) is associated with lower technical success rate and in EVAR cases it has short term outcomes comparable to the latest generations of stent grafts [16].
- Aortic cuff extension proximally. If there is free space above the proximal end of the stent graft, in the native aorta, we can extend the sealing zone higher up to provide the extra sealing length required to stop the endoleak.
- Fenestrated aortic cuff extension (Fig. 65.3). This is similar to the previous technique and is chosen if there is no proximal landing zone space in the infrarenal aorta. A longer aortic cuff is deployed to extend the sealing zone into or above the visceral segment, and we insert covered bridging stents through the fenestrations in the aortic visceral arteries (renal arteries +/− the SMA, +/− the coeliac artery). This technique can provide durable long-term seal, but requires a waiting time whilst the custom-made fenestrated endograft is manufactured [18].
- Parallel stent grafts [19–21]. This technique can be chosen in similar situations as for the fenestrated cuff extensions but instead of the fenestration stents, we insert parallel ("chimney/snorkel" or "periscope") covered stents adjacent to the extension of the aortic stent grafts. These covered stents are inserted in one or both of the renal arteries and/or the SMA. The technique is more often used in urgent cases where there is no time to order or customise a fenestrated aortic cuff extension or when there is no "off the shelf" branched aortic stent graft device available, that could be potentially used in the patient (depending on the patient's aortic anatomy and the time pressure).
- Embolic materials. With this technique we intend to fill the space of the aortic sac, outside the endograft, with embolic materials.
- Embolic materials can also be used for the treatment of Ia endoleaks to fill the gutter created in previous "chimney/periscope" successful technique procedures. If such an endoleak does not resolve in the follow up imaging in 3 months, despite kissing balloon dilatations in the chimney/snorkel/periscope stents and the aortic endograft, we can embolise it with embolic materials (liquid/coils/thrombin/gelfoam). A catheter and microcatheter are taken into position usually via upper limb arterial

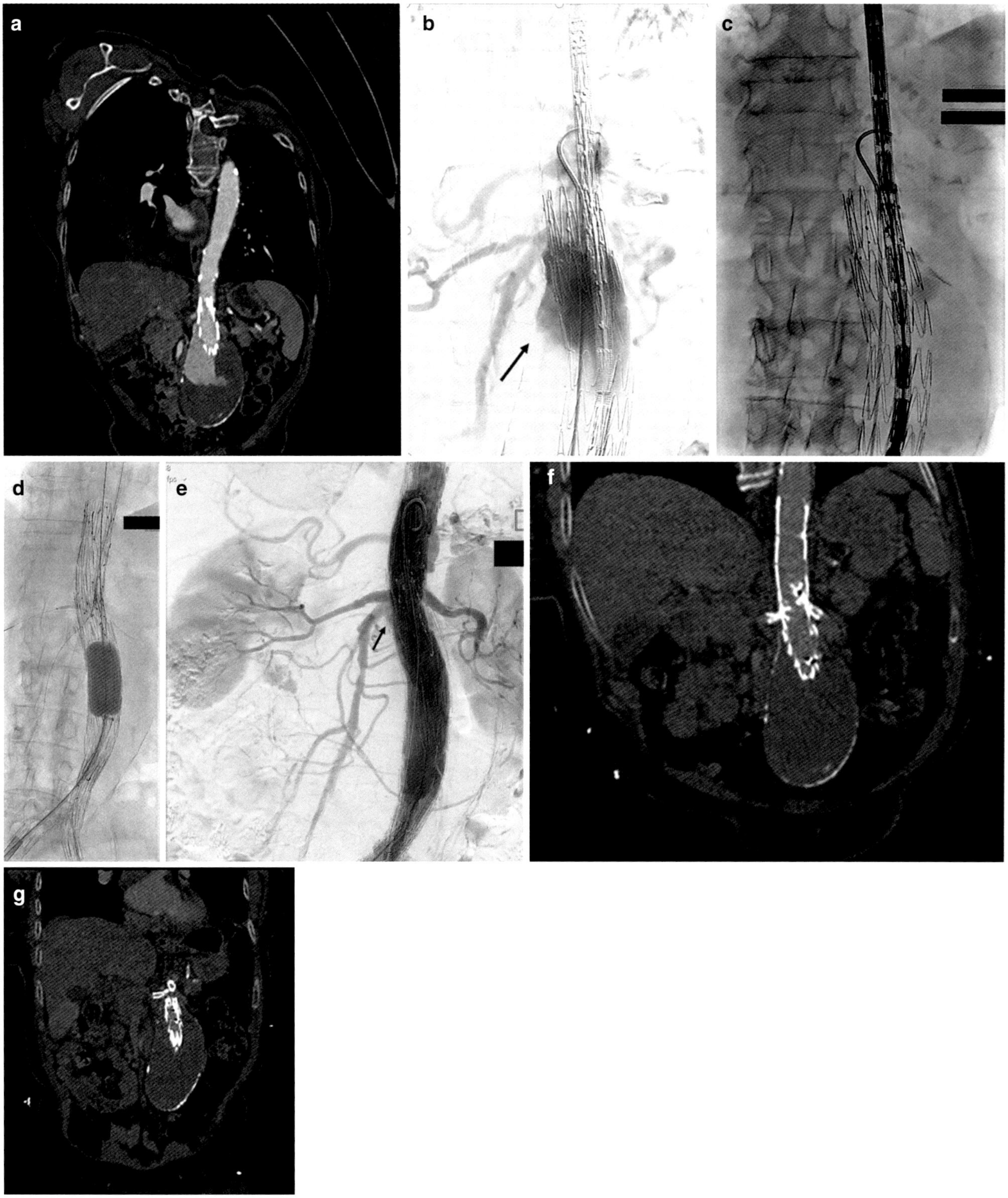

Fig. 65.3 Type Ia endoleak in previous EVAR treated with the insertion of a fenestrated cuff, extending the sealing zone higher up. (**a**, **b**) Endoleak Ia seen in the CT coronal images (**a**) and in the angiogram image (**b**) (black arrow). (**c**) Insertion of the fenestration cuff, which has three fenestrations for the renal arteries and the SMA and one double scallop for the coeliac artery. (**d**) Deployed fenestration stent graft and stents inserted in the SMA and right renal artery. Note that the SMA (as well as the left renal artery which is not shown here), are cannulated from a top, left arm access and the right renal artery is cannulated from a femoral access. A moulding balloon is used to dilate the overlapping area between the fenestrated cuff and the previous EVAR stent graft. (**e**) Angiogram post fenestration cuff implantation. The black arrow indicates that a local contrast blush/bulge is contained into the stent graft struts and is not an endoleak outside the stent graft, in the aortic sac. (**f**, **g**) CTA post fenestration cuff insertion, showing no further IA endoleak is present

access and embolisation is done directly into the aneurysm sac and along the cannulated leaking gutter [6].
- Surgical repair (surgical buttressing or endograft explantation and open, surgical graft replacement).
- Endovascular sealing (EVAS) with or without "chimney" stents [22]. This can be used when there is adequate space for sealing in the proximal infrarenal neck, above the previous endograft. If not, the sealing zone can be extended higher up with the use of chimney stents for the renal arteries +/− the superior mesenteric artery (SMA), in addition to the EVAS.

Endoleak Type Ib

Endovascular repair options for the type Ib endoleak are the:

- Balloon dilatation of the iliac limb of the endograft (or the distal end of a tubular endograft) to provide better apposition of it to the vessel wall.
- Endograft extension further distally. This can be done either by extending it in the common iliac artery, if there is enough space for it, or by extending it to the ipsilateral external iliac artery, ideally with embolisation of the overstented internal iliac artery to decrease a risk for type II endoleak. If the reason for a type Ib endoleak is an aneurysm expansion or there is enough space in the common iliac artery, then can also consider the insertion of a bifurcated/branched iliac device (IBD) which will provide a side branch stent in the internal iliac artery to preserve its patency, and distal landing in the external iliac artery.
- Retrograde cannulation behind the common iliac artery stent graft and embolisation of the leaking space between the stent graft and the common iliac artery wall. If the access into the aneurysm sac is achieved through this pathway, then we embolise the existing space between the common iliac artery stent graft and the common iliac artery wall.

Endoleak Type Ic

- In existing endoleak after a fenestrated (FEVAR) or branched (BEVAR) endovascular aneurysm repair procedure, the treatment for this is usually to either over-dilate the existing covered stent distally in its target vessel where it is sitting (if there is enough length of it into the target vessel) or extend the covered stent further into the target vessel that is connected to the fenestration or the branch where the endoleak appears from.
- In the cases treated with aorto-uni-iliac stent graft devices, where the type Ic endoleak comes from the contralateral common iliac artery, the treatment is to embolise this contralateral common iliac artery which is leaking back in the aortic sac and, if possible, to also directly embolise the leak into the aneurysm sac.

Preoperative Preparation

As mentioned above, type I endoleaks should be treated, once diagnosed and if is persistent in the follow up CTAs, post EVAR. CTA will guide the choice of treatment, between the options mentioned before.

A rapid expansion of the aortic aneurysm sac will require treatment sooner than a stable or slowly increasing sac and may not have enough time for ordering custom made endografts (like a fenestration cuff). In this case one of the other options will have to be chosen, depending on the anatomy of the aneurysm, the access vessels and the CTA imaging.

The patient must be informed about not only the planned procedure but also for any future additional procedures that may need to be done again, as follow up may identify future issues that need to be addressed.

Depending on the patient's age and other co-morbidities, medical pre-assessment and optimization of any cardiorespiratory or renal functional issues is desirable.

Depending on the planned procedure for the endoleak treatment, it may need to be done under general anaesthesia or only intravenous (IV) sedation, however the more complex the repair, the higher the chances for a conversion to a general anaesthesia, if only IV sedation is chosen.

Choice of Treatment

Balloon Dilatation

This can be chosen if the stent graft is still in good position, has not migrated and the aneurysmal sac has not expanded around the sealing zone where the endoleak is. The balloon used for this dilatation is usually a compliant balloon which is inflated with a large syringe filled with contrast: saline mix, to achieve better apposition of the stent graft in the aortic sealing zone (Fig. 65.4). This is usually done to treat the immediate type Ia, Ib or Ic endoleaks seen at the end of the index procedure. In a later type I endoleak, a different treatment technique most likely needs to apply.

For the type Ib endoleaks and similarly to the Ia endoleaks, balloon dilatations of the iliac limb of the graft at the initial implantation can provide better apposition to the iliac artery wall and decrease or eliminate the endoleak.

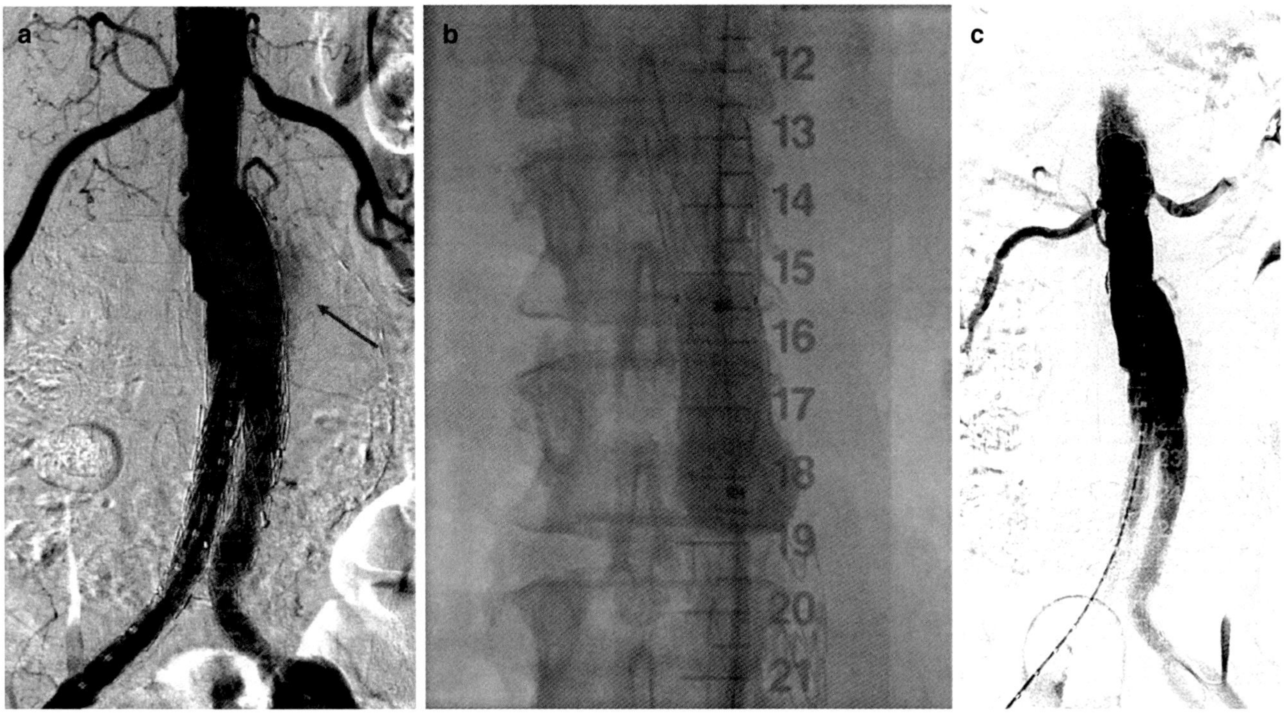

Fig. 65.4 Type Ia endoleak in previous EVAR, treated with a balloon dilatation expansion in the infrarenal aortic neck. (**a**) Type Ia endoleak present (black arrow). (**b**) Balloon dilatation in the infrarenal aorta. (**c**) Angiogram post balloon dilatation and no further imaging of the previous Ia endoleak

Large Balloon Expandable Stent Insertion

This technique may follow the previous one or used primarily. With the extra stent in place, we can optimize the endograft apposition against the proximal aortic neck wall and keep it in place, especially if there is a degree of infolding in the endograft which has created a gutter allowing the endoleak along it. The stent that is usually used for it is a giant Palmaz stent loaded on a compliant balloon and expanded in the proximal neck of a, usually, infrarenal endograft [23]. (Fig. 65.5).

Endoanchors

The endoanchors are applied endovascularly from inside the endograft and fix it with the adjacent aortic wall by engaging the adventitia. The endoanchors are metallic alloy, small screw-like fixations, measuring 3 mm diameter × 4 mm long [24]. They are delivered in position via a deflecting sheath which is angled at 90°, facing the endograft wall at the level of the planned deployment. They can be applied either in the initial main body endograft or in any proximal, infrarenal cuff extension. The instructions for use (IFU) recommends to use at least 4 or 6 endoanchors for proximal aortic neck diameters ≤29 mm or >29 mm, respectively. It is also suggested to apply them circumferentially in the area of interest, when used as a primary adjunct to the endograft implantation [25]. For the treatment of Ia endoleak, the endoanchors are meant to be deployed at the exact anatomical area of the endoleak [25]. The presence of calcium or thrombus or a severe local angulation can prohibit this exact application. Also, the distance between the aortic wall and the endograft should not be more than 2 mm, otherwise good fixation and sealing will not be achieved [26].

Endograft Extension

For type Ia endoleaks, a proximal extension is done with an infrarenal cuff endograft extension, proximal to the initial stent graft. This can only apply if there is infrarenal space for an extension and in that way we elongate the proximal coverage/sealing zone and eliminate the endoleak.

Similarly, for the thoracic aneurysms and the TEVAR procedures, the cuffs can be extended proximal or distally, in type Ia and Ib endoleaks, respectively.

In EVAR procedures, persistent endoleaks type Ib that have not responded to balloon dilatations, can be further treated with stent graft extension distally, to a healthy land-

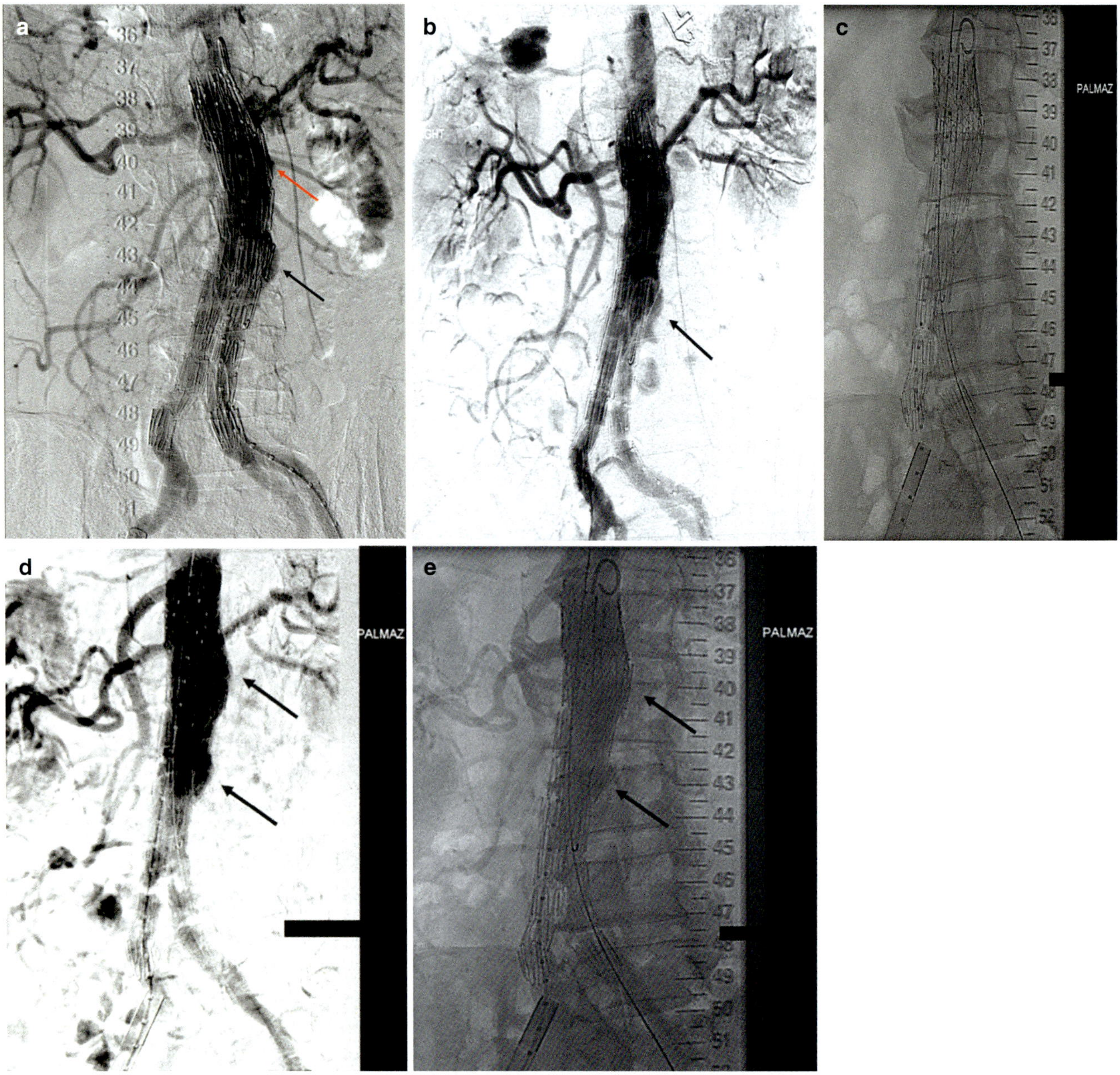

Fig. 65.5 Type Ia endoleak in previous EVAR and treatment with the insertion of a large (giant) Palmaz balloon expandable stent at the level of the proximal infrarenal neck and across the renal arteries ostiums. (**a**, **b**) Type Ia endoleak seen outside the struts of the stent graft at the top (red arrow) and further distally into the aortic sac (black arrows). (**c**) Palmaz stent implanted in place. (**d**, **e**) Angiogram post Palmaz stent insertion in place. The previous endoleak is no further present. Local bulges of contrast (black arrows) are inside the struts of the stent graft, as better seen in (**e**) (black arrows) and do not represent any remaining endoleak

ing zone. If there is no space for extension in the common iliac artery, then the stent graft is extended to the external iliac artery. The internal iliac artery is ideally embolised before any extension of the stent graft to the external iliac artery, to prevent a type II endoleak.

Aneurysm extensions to the distal common iliac arteries, in patients already treated with EVAR procedures before, can be treated with the insertion of a branched internal iliac artery stent graft device (IBD) and distal extension in the external iliac artery, if the anatomy of the internal iliac artery allows for the insertion of an IBD service in it (Fig. 65.6). If not, then we embolise the internal iliac artery and extend the previous stent graft in the external iliac artery.

After an occlusion of the internal iliac artery, there are possible complications that may happen, including buttock claudication, erectile dysfunction in males, large bowel ischemia, and finally spinal cord ischemia (if there is extensive coverage with endograft, above the abdominal aorta, in the thoracic aorta).

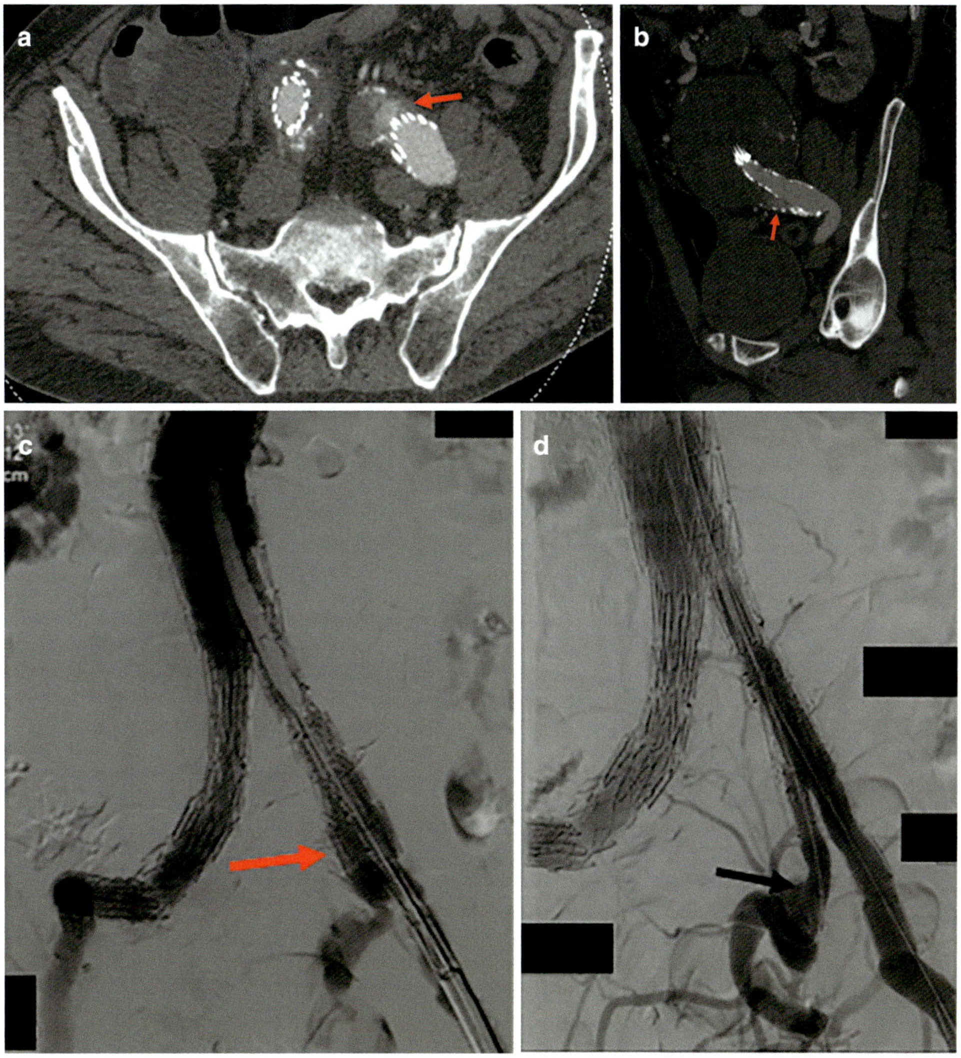

Fig. 65.6 Type Ib endoleak from the left iliac limb, in previous EVAR. Treatment with the insertion of a branched stent graft device with branch for the left internal iliac artery and extension into the left external iliac artery. (**a, b**) CTA images showing the endoleak (red arrows). (**c, d**) Angiogram and insertion of the bifurcated stent graft device. The red arrow indicates the stent graft's branch (**c**) for the left internal iliac artery and the black arrow (**d**) indicates the covered stent inserted through the branch into the left internal iliac artery

Fenestration Endograft Extension

This is chosen when there is not enough infrarenal proximal aortic neck and the extension needs to include the aortic visceral branches, so a fenestrated endograft is designed to extend the sealing zone proximal (Fig. 65.7). This cannot apply for an urgent repair, if needed, as the order and delivery of a custom made fenestrated stent graft needs at least a few weeks' time to be available. The cannulation of the renal arteries when there is a previous EVAR stent graft with suprarenal fixation in place, may be difficult as any struts sitting in front of the renal arteries ostia can compromise their cannulation through the fenestrated endograft.

Parallel Stent Grafts

The parallel grafts were first described by Greenberg et al. in 2003, later adapted by Criado in 2007 [27]. In this technique, covered stent grafts are inserted in the aortic visceral vessels (renal arteries, SMA), outside an extension aortic endograft cuff ("chimney/snorkel" or "periscope" stents). In that way, the sealing zone is extended and the visceral arteries remain perfused (Fig. 65.8). "Chimney" or "snorkel" are the stents when their inflow/blood supply to their target supplying vessel is from proximal to the aortic endograft and they are inserted parallel to ("snorkel" technique is described in Fig. 65.9) [28]. "Periscope" are the stents when their inflow

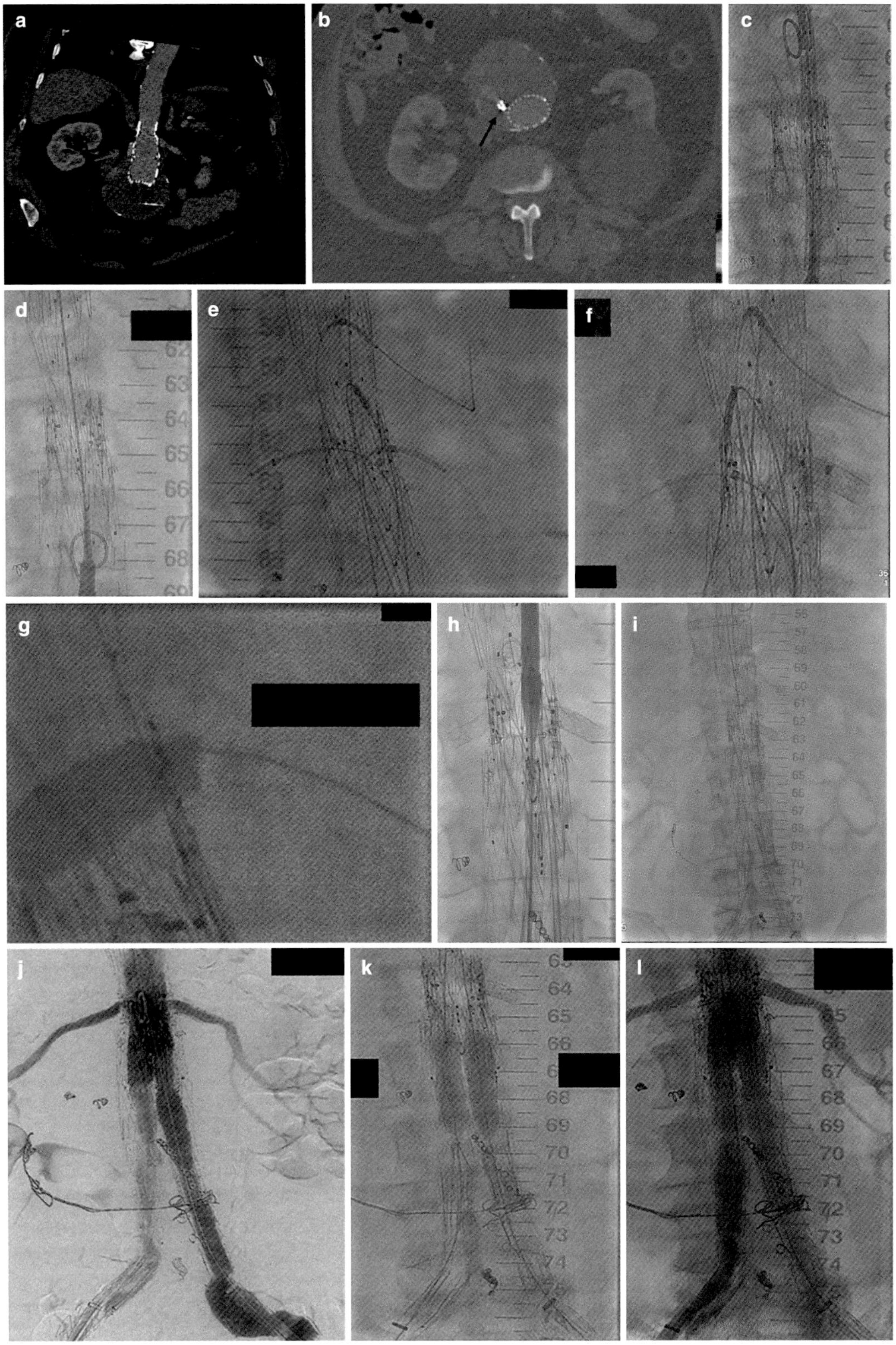

Fig. 65.7 Previous EVAR stent graft with type Ia endoleak attempted unsuccessfully to be treated before with application of endoanchors and with some coils embolization. Further treatment here with a full relining of the previous EVAR, with the insertion of a new fenestrated aortic cuff and a distal aortic bifurcated device and iliac limb grafts. (**a**) CTA showing the type Ia endoleak (red arrow). (**b**) Endoanchors from previous treatment attempt that have fallen into the aortic sac (black arrow). (**c**) Fenestrated cuff extension insertion and (**d**) deployment of it. (**e**) Cannulation of its four fenestrations and their target vessels (coeliac artery, SMA and two renal arteries) and stents insertions in position. (**f**) Deployed renal fenestration stents. (**g**) Balloon inflation at the proximal end of the left renal fenestration stent to mould it around its fenestration ostium. (**h**, **i**) Distal extension with the bifurcated stent graft component [pre (**h**) and post (**i**) deployment] and the iliac limb grafts (**i**). (**j**) Angiogram shows no endoleak but some slow flow in the right iliac limb graft. (**k**) Both limbs dilated with two high pressure balloons in a "kissing" technique. (**l**) Final angiogram

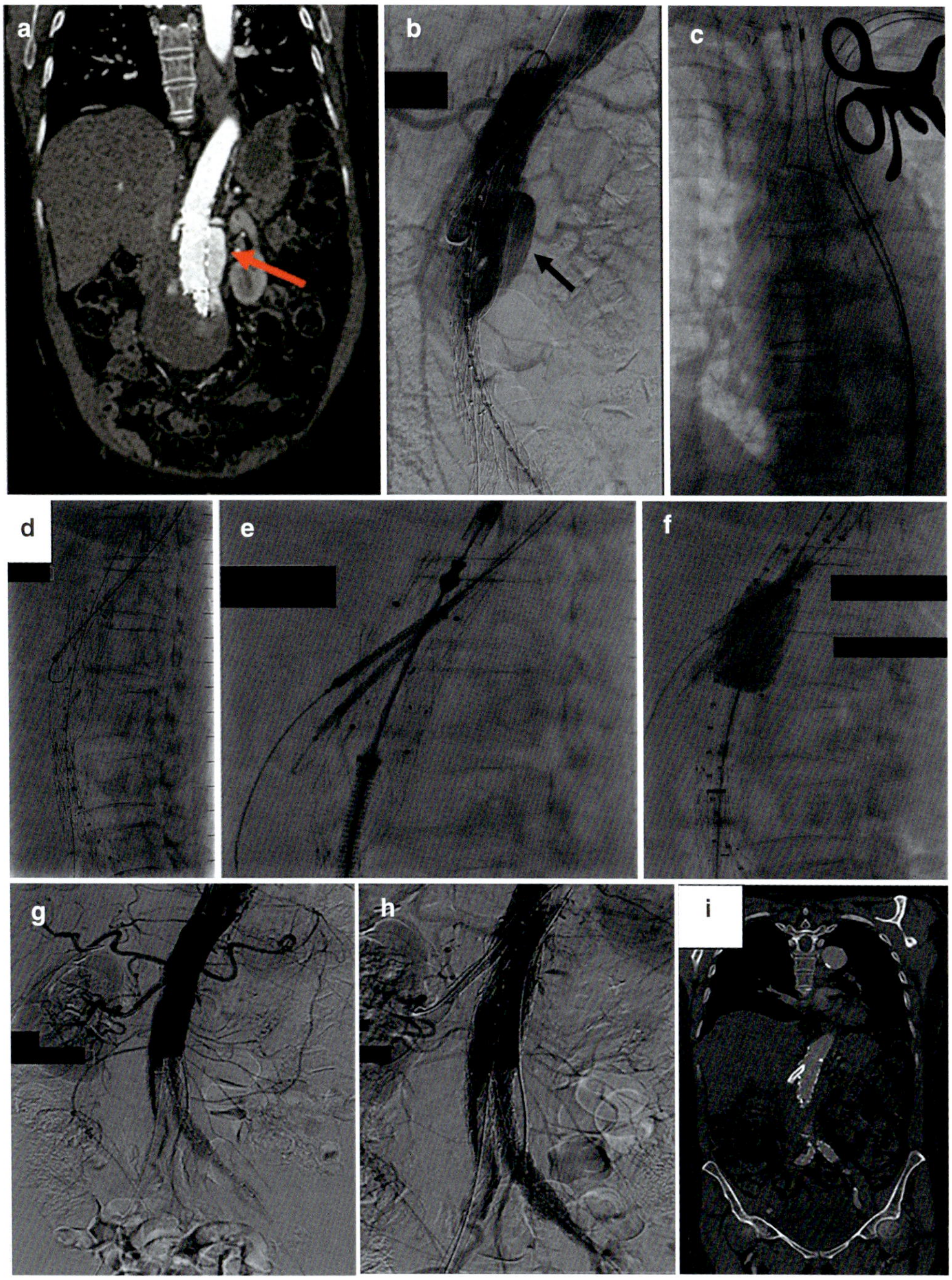

Fig. 65.8 Previous EVAR with persistent type Ia endoleak. Urgent treatment and "chimney" technique applied, sacrificing though the left renal artery. Chimneys for the right renal artery and the SMA. Aortic cuff extension in the aorta to extend the sealing zone proximal, next to the chimney stents. (**a**) CTA showing the endoleak in the coronal image (red arrow). (**b**) Endoleak (black arrow) present in the angiogram. (**c**) Access from the top, through a left axillary artery cut down and (**d**) insertion of a long sheath in place in the SMA. (**e**) Aortic cuff inserted in the aorta, SMA and right renal stents in position. (**f**) Chimney stents dilatation simultaneously with the inflation of an aortic balloon. (**g**, **h**) Final DSA images showing no further endoleak and (**i**) CTA coronal image in follow up, showing also no further endoleak existence

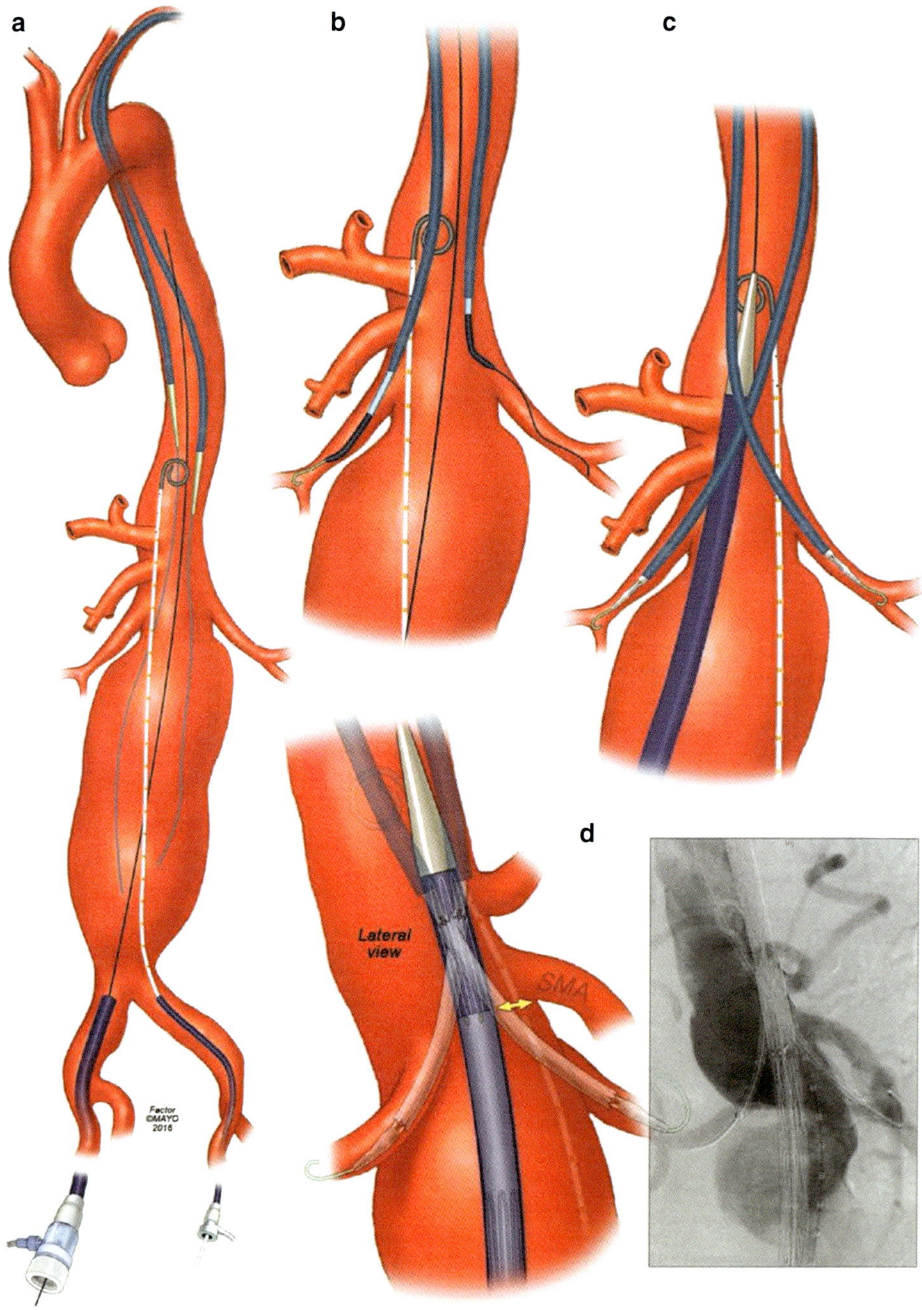

Fig. 65.9 "Snorkel" or "Chimney" technique. In these images there is no previous aortic stent graft in place, the snorkel technique is described. (**a**) Access from above with guide wires and catheters, parked in the abdominal aorta, for the visceral target vessels cannulations that will follow. Also access from below for the pig tail catheter insertion from one femoral artery (left here) and the stiff guide wire insertion, where the main body stent graft will be implanted next, from the other femoral artery (right here). (**b**) Catheters and guide wires inserted in the visceral target vessels (renal arteries here). (**c**) Long sheaths inserted into the renal arteries and main body stent graft inserted from below. (**d**) Covered stents inserted and parked into the renal arteries sheaths and lateral view Digital Subtractive Angiography is done, to check that the upper end of the aortic stent graft is just below the SMA origin. (**e**) The main body stent graft is deployed and the renal arteries covered stents are exposed with the retrieval of the long sheaths they were inserted in. Moulding balloon is inserted in the aorta, into the deployed aortic stent graft, parked in position, next to the renal arteries stents level. (f) Deployment-dilatation of the renal arteries covered stents and (**g**) maximum inflation of the aortic moulding balloon simultaneously with the inflated renal arteries stents balloons. (**h**) Deflation of the aortic moulding balloon before the renal arteries balloons are deflated and (**i**) removal of the sheaths and wires, from above and below. (**j**) CT axial images showing the implanted renal arteries stents fully open and the moulding of the aortic main body stent graft. (**k**) CT 3D reconstruction image of the implanted stents. (1. Reprinted by permission from Springer Nature Customer Service Centre GmbH: Springer Nature by "Endovascular Aortic repair" by Gustavo S. Oderich—2017. 2. "From: Lee JT. Chapter 30: Parallel Graft Techniques to Treat Complex Aortic Aneurysms Involving the Renal and Mesenteric Arteries. In: Endovascular Aortic Repair, first Edition. Cham, Switzerland: Springer International Publishing AG, 2017; used with permission of Mayo Foundation for Medical Education and Research, all rights reserved")

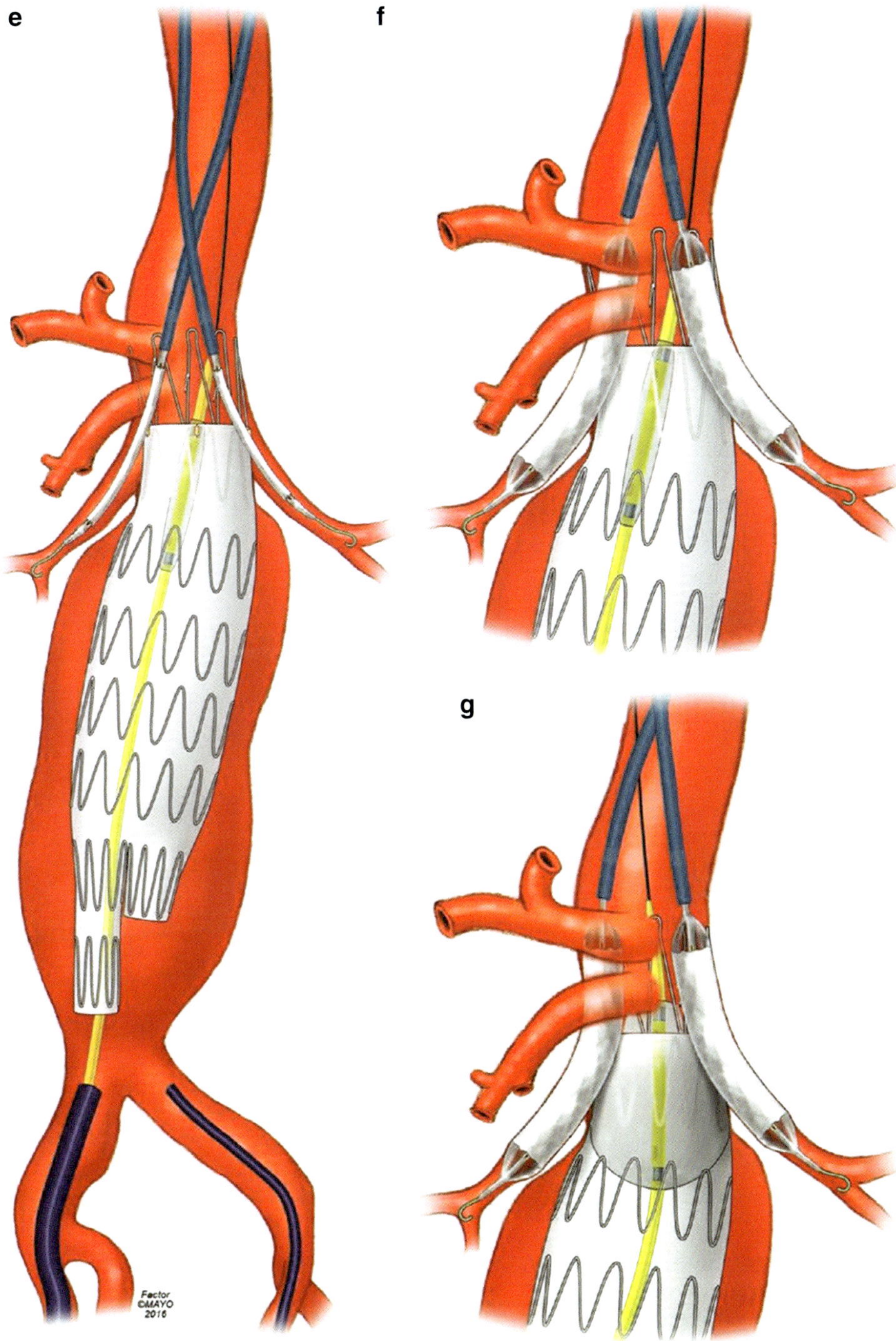

Fig. 65.9 (continued

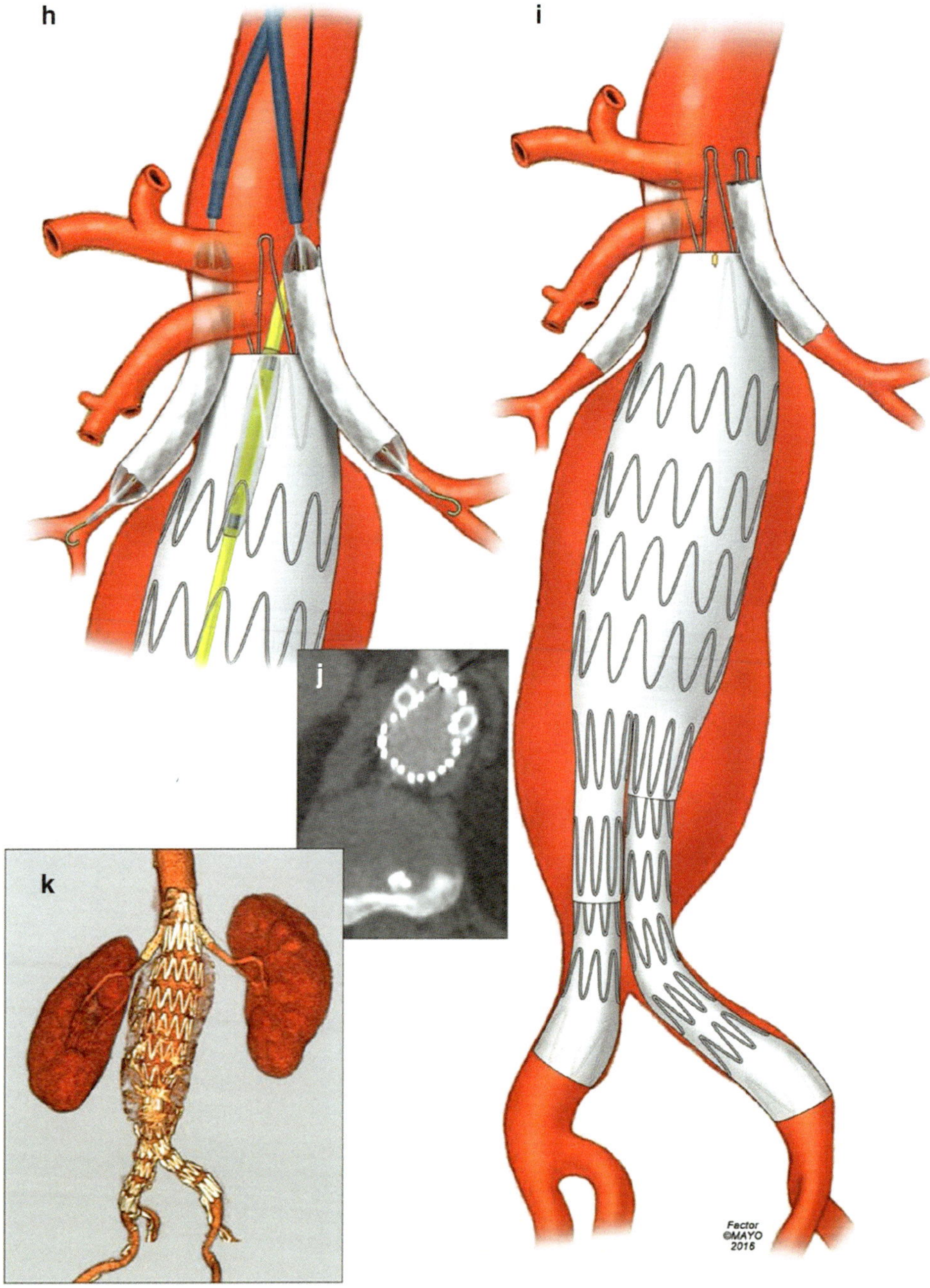

Fig. 65.9 (continued

blood, towards the vessel they are supplying, is from distally to the aortic endograft that is inserted parallel to them ("periscope" technique described in Fig. 65.10) [29].

The access for the chimney/snorkel stent/s insertions is usually percutaneous from one or two brachial arteries or from the left axillary artery, after cut down is done for it

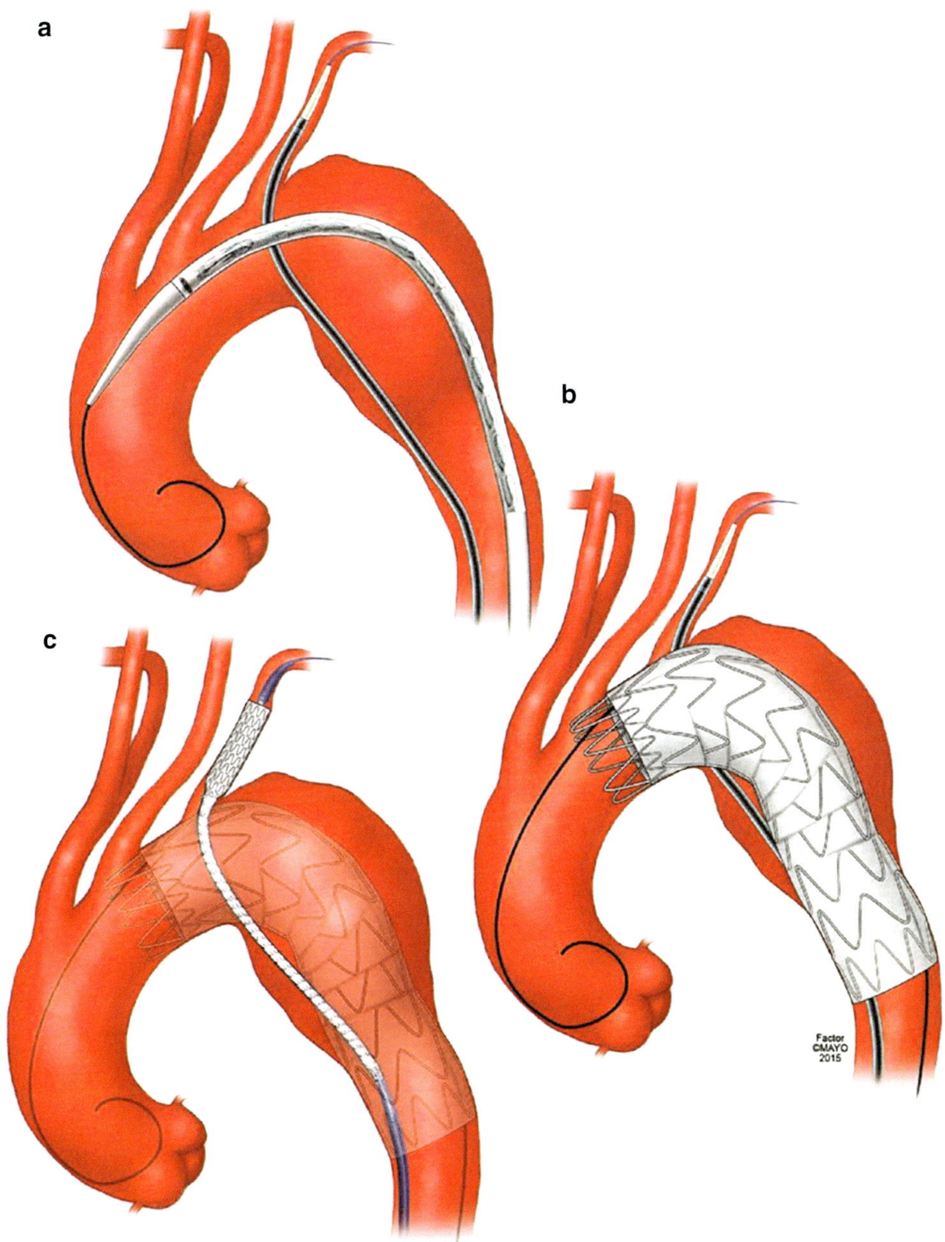

Fig. 65.10 "Periscope" technique. Again in these images there is no previous thoracic aortic stent graft in place, the periscope stents technique is described. (**a**) Access from below (femoral arteries), insertion of the thoracic main body stent graft from one femoral artery and cannulation of the left subclavian artery with guide wire and long heath, from the other femoral artery, (**b**) deployment of the thoracic main body stent graft and (**c**) insertion of a covered self expandable stent graft in the left subclavian artery sheath and start of its deployment. (**d**) Insertion of another self expandable covered stent into the fully deployed first covered stent in the left subclavian artery, to extend it below the bottom end of the deployed thoracic main body stent graft, as it is often needed and (**e**) fully deployed second covered stent, extending 10–20 mm below the bottom end of the thoracic stent graft fabric. (**f**) The same imaging like image (**e**), but to revascularise an aberrant right subclavian artery here. (1. Reprinted by permission from Springer Nature Customer Service Centre GmbH: Springer Nature by "Endovascular Aortic repair" by Gustavo S. Oderich—2017. 2. "From: Criado FJ. Chapter 35: Parallel Stent Graft Techniques to Facilitate Endovascular Repair in the Aortic Arch. In: Endovascular Aortic Repair, 1st Edition. Cham, Switzerland: Springer International Publishing AG, 2017; used with permission of Mayo Foundation for Medical Education and Research, all rights reserved")

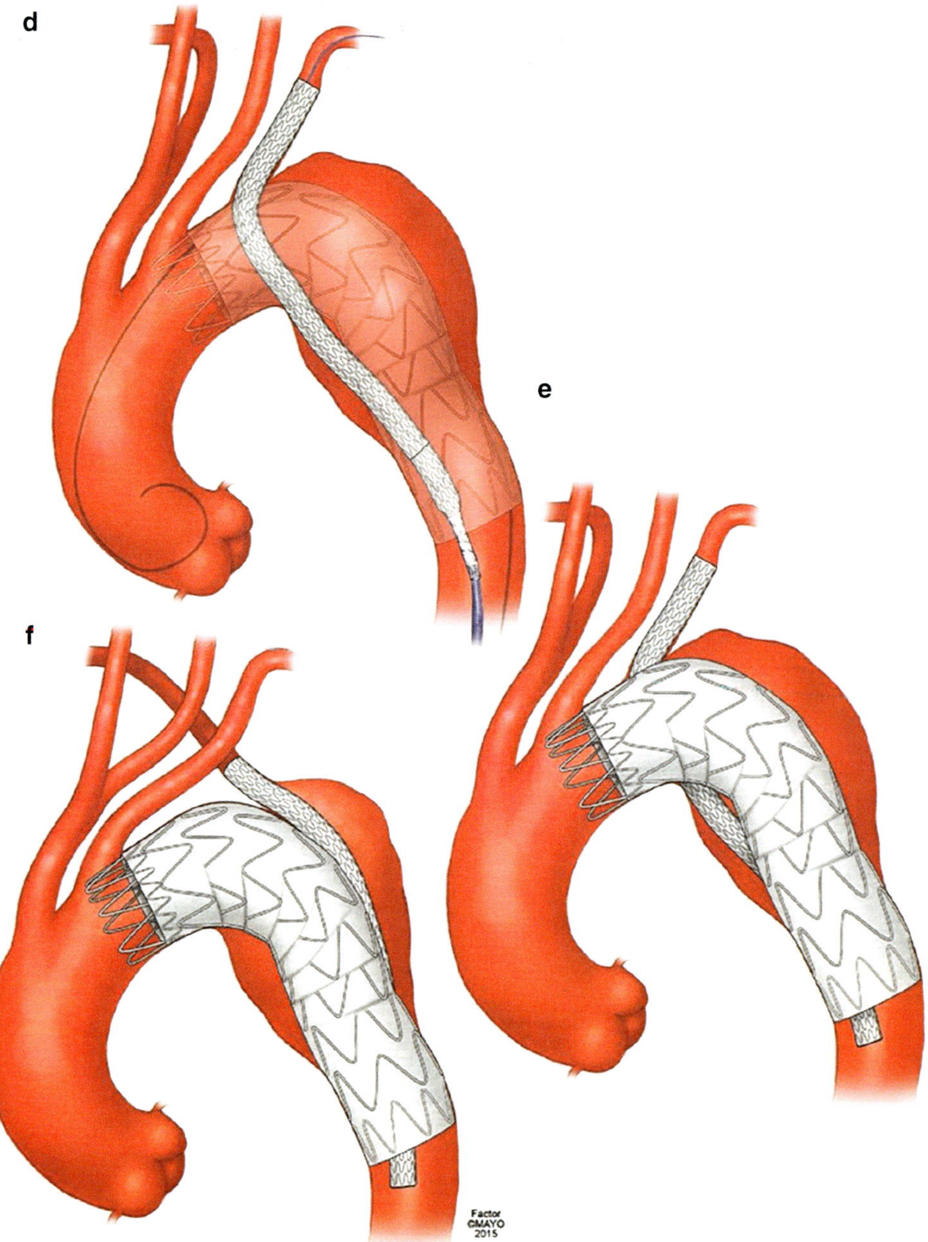

Fig. 65.10 (continued

(depending on how many chimney stents we need to insert). The sheaths used for them are usually long 7 Fr or 6 Fr Ansel sheaths. The target vessels (renal artery/ies and/or SMA) are cannulated and the long sheaths are inserted into them with the stents parked in the sheaths, ready to be deployed.

The aortic endograft used for the proximal extension is inserted through the femoral access. The aortic endograft is deployed first and then the parked stents in the target visceral vessels are deployed, after the long sheaths are pulled back to allow for the stents expansion. An aortic moulding balloon is inserted in the aortic endograft and is inflated simultaneously with the balloon/s inflated in the chimney stents. The chimney stents must be protruding in the aorta, about 0.5–1 cm above the upper end of the inserted aortic endograft and if not, require extending proximally with another covered stent.

For type Ib endoleaks treatment after TEVAR procedures, the extension with an aortic endograft cuff can be done distally with covered stents inserted in the coeliac artery +/− SMA, parallel to the aortic cuff and as such, the sealing zone is extended distally in that case ("periscope" stents). The covered stents in this case are inserted from one femoral access and the aortic cuff is inserted from the other femoral access.

The type of covered stents used for the chimney/snorkel and periscope techniques, can be balloon expandable or self-expandable. The balloon expandable stents have more radial force and are at less risk of collapse causing stenosis, however can leave a wider gutter between them and the aortic endograft, even after the kissing balloon dilatations, which might more readily cause a type Ia endoleak. The self-expandable stents can conform more tightly outside the aortic endograft, leaving a smaller gutter next to them, but have less radial force and are more prone to collapse and stenosis, potentially compromising the blood flow or occluding their target vessel [30]. For this reason the self-expandable covered stents is better to be re-enforced inside, with another uncovered self-expandable stent, oversized by 1–2 mm.

Embolic Materials

Filling of the aortic sac containing the endoleak, with embolic material/s, is another technique to try decrease or eliminate the endoleak. It is mainly chosen if there is no option for cuff or fenestrated endograft extension and when other techniques fail or are unfavourable to use. Also, embolisation can be a means of temporary sealing, until a more permanent sealing solution is available [31].

The access to the aortic sac is usually done through the upper end of the endograft (for the type Ia endoleaks) or the bottom end (for the type Ib endoleaks), with a diagnostic catheter and very often, also with a microcatheter through it. Less often the access can be done translumbar or transabdominal, with direct access to the sac, or even more rarely, transcaval, by needle puncture and sheath insertion through the IVC for access into the adjacent aortic sac.

The embolic materials that can be used are liquid embolics like cyanoacrylate (glue) [31], Onyx, Squid, PHIL], coils [32], vascular occlusion plugs, thrombin or gelfoam. For the maximum effect we need to try to fill as much of the sac as we can, come close to the proximal upper landing zone of the endograft and fill the entire endoleak tract. The Onyx was the first copolymer based liquid embolic material and is been used for longer period of time, compared to the newer ones, Squidperi and PHIL. The latter two are newer polymer-based liquids designed to overcome specific drawbacks of Onyx, such as the temporary loss of its visibility during long injections and imaging artefacts in CTA follow ups. They are also including extra low viscosity versions for deeper penetration during the injections [33].

The cyanoacrylate (glue) is usually mixed with ethiodized oil (Lipiodol), or less commonly tantalum powder, which are the radiopaque agents for the embolisation imaging and they also delay the polymerization time of the glue, during the injection. When the glue comes in contact with the blood in the sac, it polymerizes into the sac space.

Thrombin and gelfoam are not very commonly used as their durability in causing permanent thrombosis effect is less than glue or other liquid embolics. If they are used, it is done combined with one of the aforementioned embolic materials, mainly the coils and the occlusion plugs.

Type Ic Endoleaks in Aorto-Uni-Iliac (AUI) Stent Grafts

Previous endovascular treatments with the insertion of AUI endografts may have endoleak Ic in the aortic aneurysm sac, through the contralateral common iliac artery, if this is not embolised or was not adequately embolised at the index procedure (Fig. 65.11). In this occasion we proceed to a retrograde access in the contralateral common iliac artery and into the aneurysm sac, through a retrograde, contralateral common femoral artery access point. We embolise the endoleak with either coils or vascular plugs inserted in the contralateral common iliac artery, alone or in combination with embolisation into the aneurysm sac, with liquid embolic materials.

Type Ic Endoleaks in FEVAR and BEVAR

Endoleak type Ic may also appear in the stented aortic visceral branches, in previous FEVAR and BEVAR cases (Figs. 65.12 and 65.13). This can happen either because the stent in the target vessel is undersized or if it is no more anchoring adequately in its target vessel. It may have also shortened over time thereby leading to dislodgement. With the remodelling of the aortic aneurysm sac post endovascular treatment, the existing stents may change their anatomical relation to the main body of the stent graft and their target vessels. If there is enough distal landing zone of the previous stent in its target vessel, we can post dilate the existing covered stent with a larger balloon, as long as the stent is adequately sized to the vessel it is sitting in. If there is not enough distal landing/sealing zone then the treatment for this Ic endoleak is to extend the stent distally, with another covered stent of appropriate diameter, providing enough distal landing zone (about 10 mm) into its target vessel. We then post dilate the overlapped stents in the fenestration or branch and we appropriately flare the stent into the fenestration ostium, if the stent has extended into the main body of the stent graft.

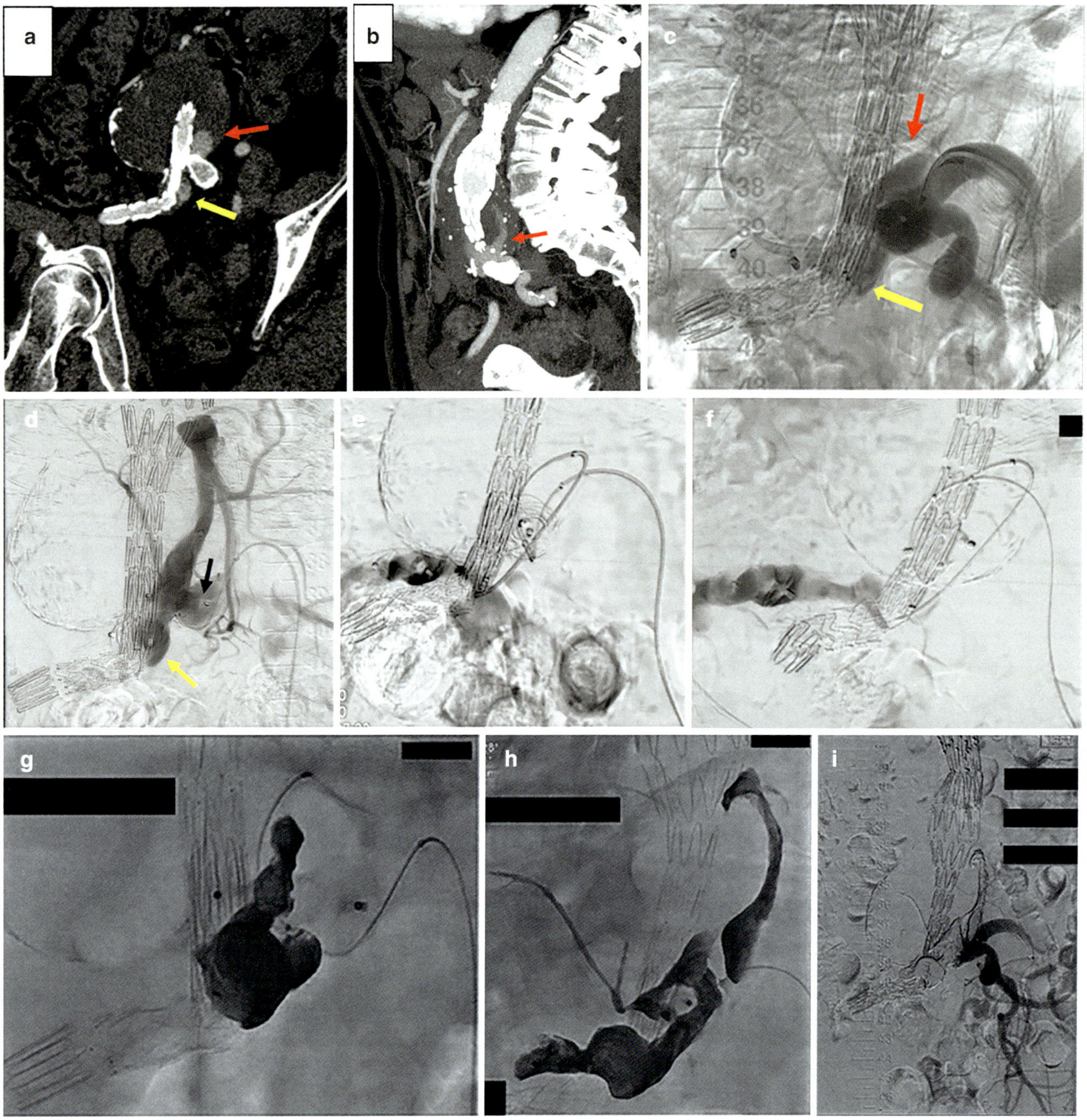

Fig. 65.11 Type Ic endoleak from the left common iliac artery in previous right aorto-uni-iliac stent graft device implantation and occlusion plug insertion in the left common iliac artery. Endoleak despite the presence of the iliac vascular occlusion plug. (**a, b, c, d**) CTA and angiogram images showing the endoleak into the aortic sac (red arrows in images **a**, **b** and **c**) and into the right common iliac artery aneurysm sac as well (yellow arrows in images **a**, **c** and **d**). Vascular occlusion plug is in place from before in the left common iliac artery (black arrow in image **d**). (**e, f**) Catheter advanced behind the occlusion plug, up and over, into the right common iliac artery aneurysm sac. (**g, h**) Liquid embolic material (Onyx) injected through the catheter, into the aneurysm sac of the right common iliac artery and the abdominal aorta. The angiogram (**i**) shows no further endoleak into the embolised aneurysm sac

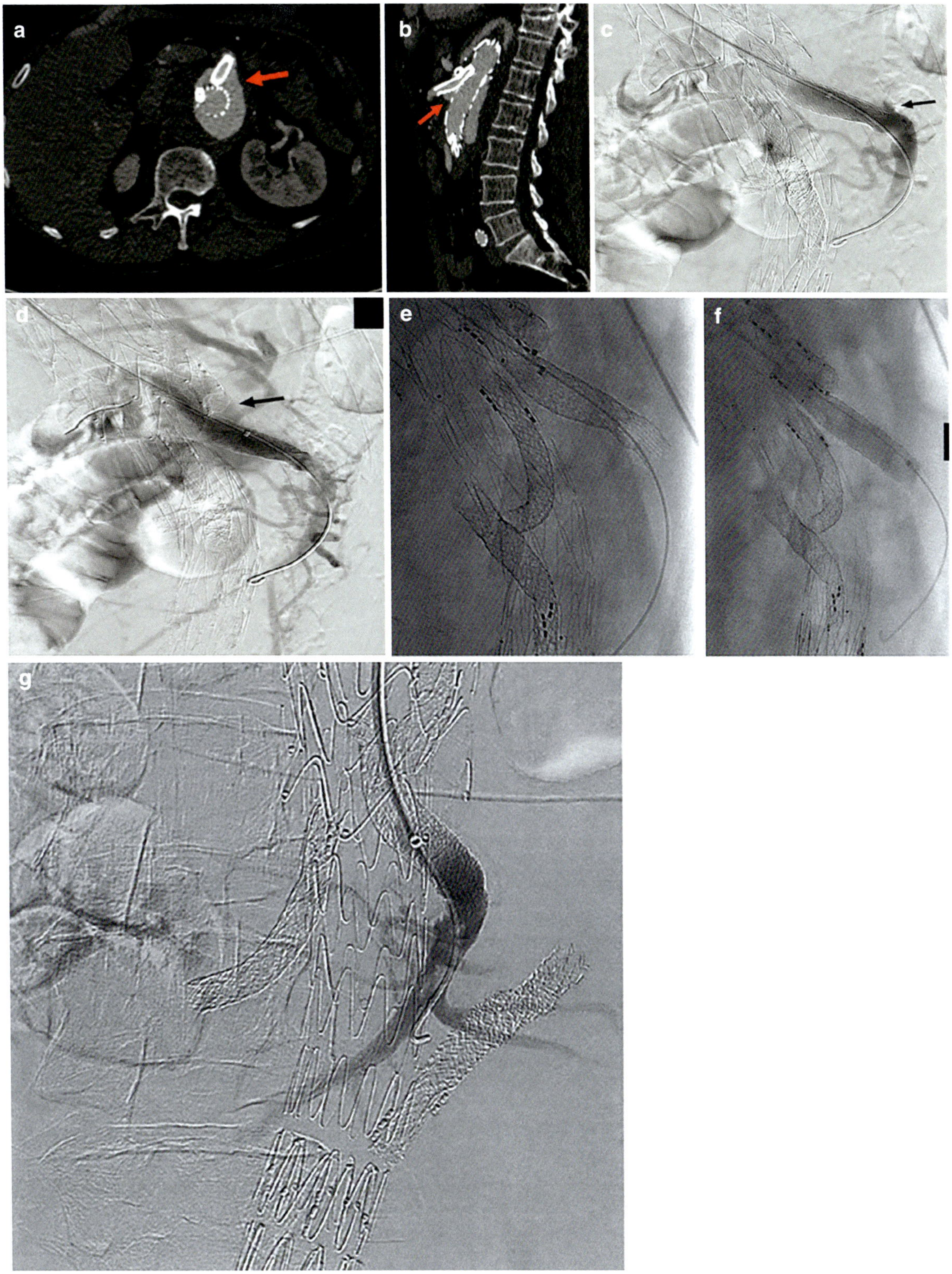

Fig. 65.12 Type Ic endoleak from the SMA stent, in a patient previously treated with a fenestrated stent graft insertion. (**a, b**) CTA images showing the endoleak (red arrows). (**c, d**) Angiogram images showing the endoleak (black arrows). (**e**) SMA stent distal extension with a self-expandable covered stent, (**f**), balloon dilatation of the SMA overlapping covered stents and angiogram image post stent extension (**g**), showing no further endoleak

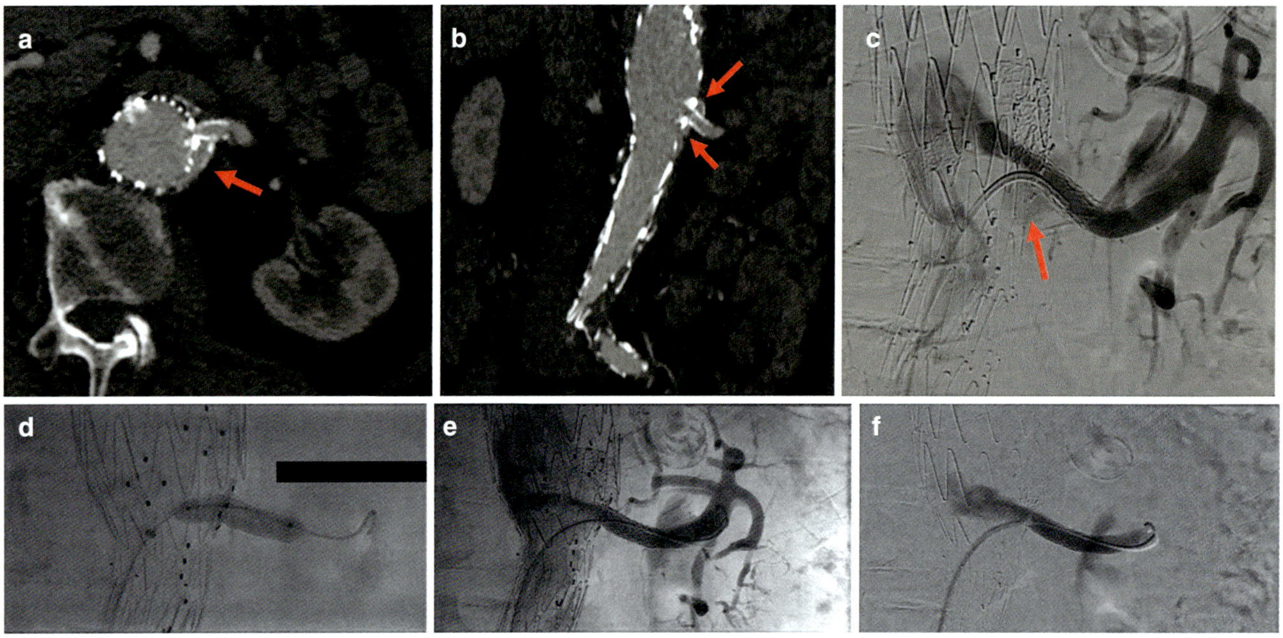

Fig. 65.13 Type Ic endoleak from the left renal artery stent in a patient previously treated with a fenestrated stent graft device insertion. (**a, b, c**) Endoleak is seen from the distal end of the, short, left renal artery fenestration stent, into the aortic sac, in CTA (**a, b**) and angiogram images (**c**) (red arrows in **a, b, c**). (**d**) Distal, short extension of the left renal artery fenestration stent, with a second covered stent and (**c, f**) endoleak disappearance at the angiogram images, post stent extension

Complications

The complications we may meet at the type I endoleak repair procedures, depend on the chosen and applied technique.

Balloon Dilatation and Giant Stent (Palmaz) Insertion

The primary complications encountered during balloon dilatation at the level of the proximal part of the previous stent graft, where the Ia endoleak originates, can be an aortic rupture during the inflation, or balloon rupture.

If there is a giant balloon expandable stent deployed in the proximal neck, then possible complications are the significant shortening of the stent and maldeployment in an inaccurate position, or migration of the stent after the balloon deflation.

Endograft Extension

With the choice of treatment of endograft extension, possible complications are the obstruction or occlusion of aortic visceral branches by the new endograft, if it is deployed close to or even shuttering them for the purpose of achieving endoleak sealing. Other possible complications that are met during the standard EVAR procedures, may be also met during an endograft extension (access vessel/s injury, thrombosis or dissection, thrombus migration, embolic events in aortic visceral branches, iliac or lower limb/s arteries).

Fenestration Endograft Extension and Parallel Stent Grafts

At fenestration endograft extension procedures or parallel stent grafts insertions, possible complications may include injury of the aortic visceral target vessels (SMA, coeliac or renal arteries) with dissection, occlusion or thrombosis of these branches and subsequent clinical implications from the organs they supply (bowel ischemia, renal insufficiency or kidney injury, splenic or hepatic injury).

Another possible complication from the extension with a fenestrated stent graft, is a spinal cord injury after coverage of the thoracic aorta and the intercostal arteries, risk of which may be higher if the left subclavian artery and/or the internal iliac artery/ies are occluded or obstructed. Spinal cord ischemia may appear with lower limb/s motor and/or sensation impairment, which may be reversible with timely spinal fluid drainage or may result in irreversible paralysis.

Embolisation

With the use of embolic materials for the treatment of an endoleak, possible complication is the non-target embolisation of vessels, with the subsequent clinical implications of ischaemia in their target organs (visceral aortic branches, internal iliac artery branches, lower limb arteries).

References

1. Keefe NA, Haskal ZJ, Park W, Angle JF. Chapter 19—Endoleak. In: IR playbook—a comprehensive introduction to interventional radiology, pp. 219–227; 2018. doi: https://doi.org/10.1007/978-3-319-71300-7, doi: https://doi.org/10.1007/978-3-319-71300-7_19.
2. Guo Q, Zhao J, Huang B, Yuan D, Yang Y, Zeng G, Xiong F, Du X. A systematic review of ultrasound or magnetic resonance imaging compared with computed tomography for endoleak detection and aneurysm diameter measurement after endovascular aneurysm repair. J Endovasc Ther. 2016;23(6):936–43.
3. Oderich GS. Chapter 11—Computed tomography/computed tomography angiography for evaluation, planning, and surveillance of complex endovascular repair. In: Vrtiska TJ, Macedo TA, Oderich GS, editors. Endovascular aortic repair. Springer.
4. Gaffey AC, Damrauer SM. Evolving concepts, management and treatment of type 1 endoleaks after endovascular aneurysm repair. Semin Intervent Radiol. 2020;37(4):395–404.
5. Antoniou GA, Georgiadis GS, Antoniou SA, Neequaye S, Brennan JA, Torella F, Vallabhaneni SR. FEBVS1- late rupture of abdominal aortic aneurysm after previous endovascular repair: a systematic review and meta-analysis. J Endovasc Ther. 2015;22(5):734–44.
6. Ballesteros-Pomar M, Taneva GT, Austermann M, Fernández-Samos R, Torsello G, Donas KP. Successful management of a type B gutter related endoleak after chimney EVAR by coil assisted onyx embolisation. EJVES Short Rep. 2019;(42):38–42.
7. Chaikof EL, Dalman RL, Eskandari MK, Jackson BM, Lee WA, Mansour MA, Mastracci TM, Mell M, Murad MH, Nguyen LL, Oderich GS, Patel MS, Schermerhorn ML, Starnes BW. The Society for Vascular Surgery practice guidelines on the care of patients with an abdominal aortic aneurysm. J Vasc Surg. 2018;67(1):2.
8. Jean-Baptiste E, Feugier P, Cruzel C, Sarlon-Bartoli G, Reix T, Steinmetz E, Chaufour X, Chavent B, du Mont LS, Ejargue M, Maurel B, Spear R, Midy D, Thaveau F, Desgranges P, Rosset E, Hassen-Khodja R, Association Universitaire de Recherche en Chirurgie Vasculaire. Computed tomography-aortography versus color-duplex ultrasound for surveillance of endovascular abdominal aortic aneurysm repair a prospective multicenter diagnostic-accuracy study (the ESSEA Trial). Circ Cardiovasc Imaging. 2020;13(6):e009886. https://doi.org/10.1161/CIRCIMAGING.119.009886.
9. Mirza TA, Karthikesalingam A, Jackson D, Walsh SR, Holt PJ, Hayes PD, Boyle JR. Duplex ultrasound and contrast-enhanced ultrasound versus computed tomography for the detection of endoleak after EVAR: systematic review and bivariate meta-analysis. Eur J Vasc Endovasc Surg. 2010;39:418–28. https://doi.org/10.1016/j.ejvs.2010.01.001.
10. Partovi S, Kaspar M, Aschwanden M, Lopresti C, Madan S, Uthoff H, Imfeld S, Staub D. Contrast-enhanced ultrasound after endovascular aortic repair—current status and future perspectives. Cardiovasc Diagn Ther. 2015;5(6):454–63.
11. Lehmkuhl L, Andres C, Lücke C, Hoffmann J, Foldyna B, Grothoff M, Nitzsche S, Schmidt A, Ulrich M, Scheinert D, Matthias Gutberlet M. Dynamic CT angiography after abdominal aortic endovascular aneurysm repair: influence of enhancement patterns and optimal bolus timing on endoleak detection. Radiology. 2013;268(3):890–9.
12. Kelso RL, Lyden SP, Butler B, Greenberg RK, Eagleton MJ, Clair DG. Late conversion of aortic stent grafts. J Vasc Surg. 2009;49(3):589–95.
13. AbuRahma AF, Hass SM, AbuRahma ZT, Yacoub M, Mousa AY, Abu-Halimah S, Dean LS, Stone PA. Management of immediate post-endovascular aortic aneurysm repair type Ia endoleaks and late outcomes. J Am Coll Surg. 2017;224(4):740–8.
14. Arthurs ZM, Lyden SP, Rajani RR, Eagleton MJ, Clair DG. Long-term outcomes of palmaz stent—placement for intraoperative type Ia endoleak during endovascular aneurysm repair. Ann Vasc Surg. 2011;25(1):120–6.
15. Abdulrasak M, Resch T, Sonesson B, Holst J, Kristmundsson T, Dias NV. The long-term durability of intra-operatively placed palmaz stents for the treatment of type Ia endoleaks after EVAR of abdominal aortic aneurysm. Eur J Vasc Endovasc Surg. 2017;53:69–76.
16. Qamhawi Z, Barge TF, Makriss GC, Patel R, Wigham A, Anthony S, Uberoi R. A systematic review of the use of endoanchors in endovascular aortic aneurysm repair. Eur J Vasc Endovasc Surg. 2020;59(5):748–56.
17. Valdivia AR, Milner R, Heijmen R, Riambau V, Rousseau H, Tinelli G, Kotelis D, Zanabili Al-Sibbai AA, Pitoulias G, Zúñiga CG, de Beaufort HWL, Panagiotis D, Chaudhuri A. Mid term outcomes of the use of endoanchors during thoracic endovascular aortic repair in multicentre analysis. Vascular. 2022;31:17085381221076320. https://doi.org/10.1177/17085381221076320.
18. Kiernan A, Elsherif M, Fahey B, Canning C, Moloney T, Kavanagh E, O'Callaghan A, O'Neill S, Madhavan P, Martin Z. Rescue of failed aortic repair with fenestrated endovascular device. Ann Vasc Surg. 2021;82:S0890-5096(21)00927-4.
19. Perini P, Massoni CB, Mariani E, Ucci A, Fanelli M, Azzarone M, Freyrie A. Systematic review and meta-analysis of the outcome of different treatments for type 1a endoleak after EVAR. Ann Vasc Surg. 2019;60:435–46.
20. Cherfan P, Abdul-Malak OM, Liang NL, Eslami MH, Singh MJ, Makaroun MS, Chaer RA. Endovascular repair of abdominal and thoracoabdominal aneurysms using chimneys and periscopes is associated with poor outcomes. J Vasc Surg. 2022;76:S0741-5214(22)00415-3.
21. Wilson A, Zhou S, Bachoo P, Tambyraja AL. Systematic review of chimney and periscope grafts for endovascular aneurysm repair. Br J Surg. 2013;100:1557–64.
22. Paraskevas KI, Karouki M, Rehman A, Torella F, McWilliams RG, England A, Fisher RK. Endovascular aneurysm sealing (EVAS) alone or in combination with chimney grafts (chEVAS) for treating complications of previous endovascular aneurysm repair (EVAR) procedures. Cardiovasc Intervent Radiol. 2018;41(7):1015–20.
23. Spanos K, Rohlffs F, Pannucio G, Eleshra A, Tsilimparis N, Kolbel T. Outcomes of endovascular treatment of endoleak type 1a after EVAR: a systematic review of the literature. J Cardiovasc Surg. 2019;60(2):175–85.
24. Deaton DH, Mehta M, Kasirajan K, Chaikof E, Farber M, Glickman MH, Neville RF, Fairman RM. The phase I multicenter trial (STAPLE-1) of the aptus endovascular repair system: results at 6 months and 1 year. J Vasc Surg. 2009;49(4):851–7.
25. Saratzis A, Modarai B. J Endovasc Ther. 2019;26(5):714–6.
26. Giudice R, Borghese O, Sbenaglia G, Coscarella C, De Gregorio C, Leopardi M, Pogany G. The use of EndoAnchors in endovascular repair of abdominal aortic aneurysms with challenging proximal neck: Single-centre experience. JRSM Cardiovasc Dis. 2019;8:2048004019845508.
27. Castro-Ferreira R, Dias PG, Sampaio SM, Teixeira JF, Lobato AC. Parallel graft technique in a complex aortic aneurysm: the

value of intra-operative flexibility from the original operative plan. EJVES Short Reports. 2019;43:37–40.
28. Gustavo S. Oderich. Chapter 30—Parallel graft techniques to treat complex aortic aneurysms involving the renal and mesenteric arteries. In: Jason T. Lee, editor. Endovascular aortic repair.
29. Gustavo S. Oderich. Chapter 35—Parallel graft techniques to facilitate endovascular repair in the aortic arch. In: Frank J. Criado, editor. Endovascular aortic repair.
30. Mestres G, Uribe JP, García-Madrid C, Miret E, Alomar X, Burrell M, Riambau V. The best conditions for parallel stenting during EVAR: an in vitro study. Eur J Vasc Endovasc Surg. 2012;44(5):468–73.
31. Rusius VC, Davies RSM, Hopkins JD, Duddy MJ, Smith SRG. Successful glue embolization of a late type 1a endoleak causing abdominal aortic aneurysm rupture. Vasc Endovasc Surg. 2011;45(2):198–201.
32. Belczak SQ, Pedroso GD, Ogawa LC, Campos PT, Padula AL, Machado GP, dos Santos MZS, Abrão BM. Treatment of type 1A endoleak using coil embolization: a case report. J Vasc Bras. 2019;18:e20180130.
33. Vollherbst DF, Chapot R, Bendszus M, Möhlenbruch MA. Glue, Onyx, Squid or PHIL? Liquid embolic agents for the embolization of cerebral arteriovenous malformations and dural arteriovenous fistulas. Clin Neuroradiol. 2022;32(1):25–38.

Type 2 Endoleak After EVAR

66

Jean Pierre Becquemin

Case Presentation

A 72-year-old patient was treated for a 58 mm large abdominal infra renal aneurysm with a bifurcated stent graft whose fixation was supra renal. The procedure was performed under local anesthesia through a percutaneous approach in the groins. Puncture sites were closed with Proglide closure devices. The intervention was performed in a Hybrid Angio suite, which went smoothly with no bleeding and lasted 55 Mn. The intra operative control angiogram showed a good application of the stent-graft and no type Ia or Ib Endoleak but a faint opacification in the sac raising a doubt of a type 2 Endoleak whose origin was uncertain. The patient has an uneventful recovery and left the hospital for his home the following day. At the one-month consultation he felt in good shape. Duplex scan confirmed the good apposition of the graft but a type 2 Endoleak which was confirmed by the CT scan. The figure showed a large endoleak called nidus, and two feeding arteries, the IMA and a right lumbar artery (Fig. 66.1).

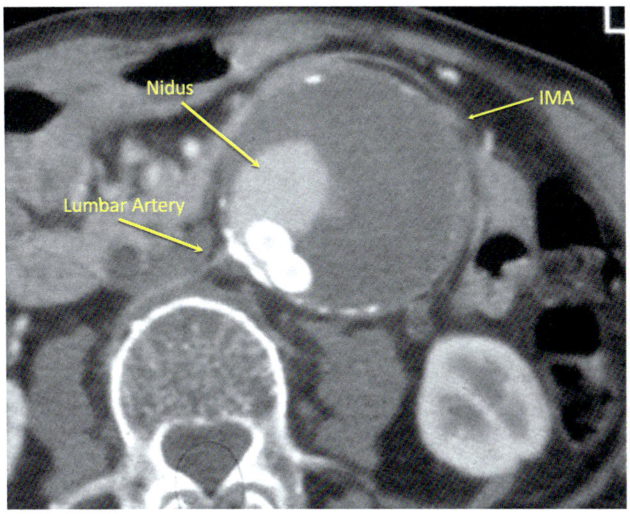

Fig. 66.1 Angio Ct Scan of an Abdominal Aortic Aneurysm treated by EVAR which shows Nidus from a large type 2 Endoleak, and patent (IMA) inferior mesenteric artery and Lumbar Artery

The decision was made to survey the aneurysm diameter by yearly imaging. One year later Duplex and CT showed a stable diameter but after 3 years the diameter enlarged by 8 mm. Duplex evidenced a circulating flow between the inferior mesenteric artery (IMA) and lumbar arteries. Coil Embolization of the IMA was performed via the superior mesenteric artery. The procedure was successful and the sac remained stable for 2 years but then increased again due to lumbar arteries back bleeding. We proceeded to coelioscopy clipping of lumbar arteries, sacral artery and IMA. The sac was opened to check the absence of circulating flow, to remove clots and to suture the aortic wall to reduce the volume of the sac. The post-operative course was uneventful. Duplex scan confirmed the efficacy of the procedure with no leak.

J. P. Becquemin (✉)
Cardio Vascular Department, Henri Mondor Hospital University Paris Val de Marne, Creteil, France

Institut Vasculaire Paris Est, Hôpital Privé Paul d'Egine, Ramsay Group Champigny, Champigny sur Marne, France

General Consideration

Endovascular repair has become the recommended method of treatment of AAA especially in patients who present comorbidities [1, 2]. The advantages of EVAR over open surgery include lower one-month mortality and morbidity, less transfusion, shorter hospitalization stay, and faster recovery. However long-term results are less favorable, which according to RCT's, higher all causes and AAA related mortality due to sac rupture, cardiovascular events and cancer. Endovascular reinterventions are more frequent after EVAR while abdominal wall dehiscence is the main reason to reintervene after open surgery. EVAR results have improved

G. Geroulakos et al. (eds.), *Mastering Endovascular Techniques*, https://doi.org/10.1007/978-3-031-42735-0_66

thank to the latest generation of stent graft [3], the improvement of preoperative imaging with softwares which analyze more accurately the features of the AAA and enable a precise choice of the type of the graft and the diameters and length of the prosthesis. Intra operative imaging has also radically improved thanks to hybrid rooms which allow 3D reconstructions and fusion images with the pre-op CT. Wires navigation and catheterisms are facilitated. Endoleak remains the achilleas heel of EVAR, the most frequent of which is type 2 Endoleak which occurs in around 15% of the cases [4].

Type 2 Endoleak are due to the retrograde filling of the sac from collaterals such as the inferior mesenteric artery, lumbar arteries, sacral artery and when ever present accessory renal arteries as shown in (Fig. 66.2).

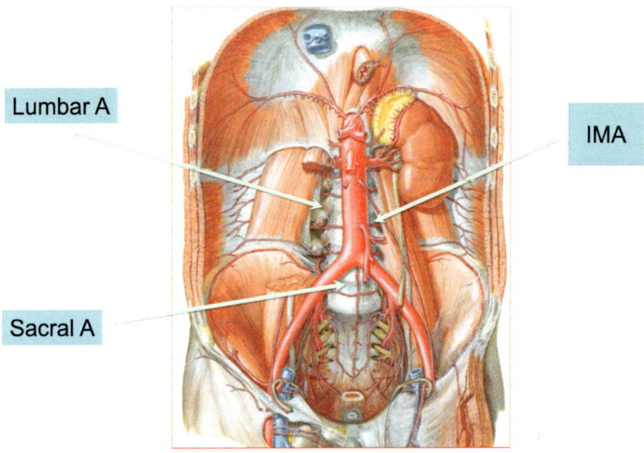

Fig. 66.2 Lumbar A: Lumbar Arteries, IMA: Inferior Mesenteric Artery, Sacral A: Sacral Artery

The probability of type 2 Endoleak is higher when 3 pairs of lumbar arteries are patent, and the diameter of the ostia of a patent IMA is larger than 3 mm [5, 6]. It is lower when the sac is filled with circular thrombus [7]. The risk is also higher in large AAA (>6.5 cm) compared to smaller AAA (<5.5 cm)

They are different features of type 2 Endoleak. They can be low flow, high flow, unique or multiple sponge-like. Imaging surveillance during the follow-up shows that they can occur early or late and be fluctuant or persistent. The later situations have a less favorable prognosis.

The consequences of Type 2 Endoleak are better known with the availability of longer follow-up and in larger series.

The risk of rupture of isolated type 2 Endoleak is low [4], but it maintains the pressurization of the sac which as a consequence can enlarge (46% vs 0% at 4 years) [8]. The main risk of sac enlargement is the dilatation of the sealing zone specially the infra renal neck. This risk is all the more frequent that the neck is defined as hostile i.e., large (>30 mm), angulated (>60°), short (<1.5 cm), and or with circular calcification or thrombus [9]. Dilatation of the sealing zones leads to type Ia and Ib Endoleak, which exposes to AAA rupture if not treated. One out of five patients with type 2 Endoleak will evolve toward neck dilatation and type Ia Endoleak [7, 10].Diagnosis of type 2 Endoleak is based on Duplex scan and/or CT scan [11]. A recent multi- center study has questioned the validity of Duplex compared to CT scan [12]. The accuracy of Duplex

Fig. 66.3 Left image: Enlarging AAA diameter with a large intrasac nidus despite previous coils embolisation Right Image: The aortogram shows that the leak comes from a tear in the stent graft

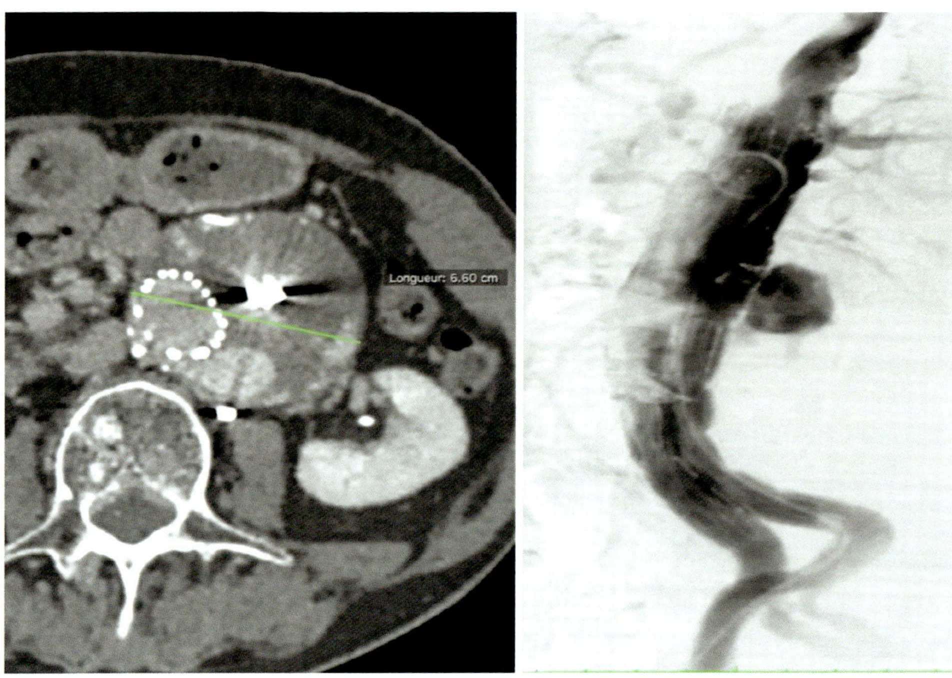

scan depends on the quality of the machine and on the skill of the physician or of the technician. As such, Duplex and CT scan are complementary, and should be routinely used except when 1 month and 1 year exams show sac shrinkage and absence of endoleak in which case the follow up can be discontinued [13]. MRIs are also favorably used in some institutions.

One difficulty is to differentiate relatively innocuous type 2 with the more dangerous type 3 Endoleak or low flow type 1a Endoleak. Figure 66.3a shows a CT scan of a 79-year-old lady who was previously treated by EVAR and coils embolization of a type 2 endoleak. The nidus seems to be alimented by lumbar arteries. Figure 66.3b shows that it was a type III endoleak due to material fatigue. In case of doubt, angiogram with a peaktail inside the graft and if necessary, selective catheterisms and arteriogram of the SMA and hypogastric arteries can identify the culprit.

When to Intervene?

Given the low risk of death and aneurysm rupture of isolated type 2 endoleak there is a controversy about the utility of intervention. The ODYSSEUS study group reported a total of 2018 patient followed up to 11 years found no difference in the overall survival between patients with or without type 2 endoleak or in the group which underwent secondary operation versus the group treated conservatively [14]. The Guidelines of the Society of Vascular Surgery [15] recommend intervention when the sac increases by 5 mm with the presence of large inflow and outflow vessels which represent 25% of type 2 endoleak.

How to Intervene?

Material

The material includes coils (Boston Scientific, Medtronic, Penumbra, Terumo), plugs, thrombin and glue(Onyx) [16]. There is no indisputable proof that a specific material offers better results than the others. Of note however, a recent study [16] reported a trend toward lower need for reintervention with Onyx compared to coils.

Timing

Preemptively

Patients at risk of type 2 Endoleak as defined above can benefit from pre or intra operative embolization. Embolization can be non-selective or selective. Non selective perigraft aortic sac coils embolization is performed by leaving a wire in the sac after canulation of the contra lateral gate and inserting large coils after the deployment of graft (Fig. 66.5). This maneuver reduces the rate of any endoleak from 25.8% to 0% [17]. However, when the aneurysm sac is not fully packed with coils, endoleak may reappear. Selective pre-operative IMA and lumbar arteries blockage can be performed by plugs or coils before or during Evar [18]. In this series, the success rate was 76% with dramatic reduction of type 2 Endoleak and post-operative aneurism enlargement. Calcification of the ostium and unstable catheter position were the main reasons of failures. A randomized study of patients at risk of type 2 Endoleak comparing intra operative coil embolization of the sac versus surveillance showed at 24 months 6.5% of type 2 endoleak versus 25% in the control group [19]. Another RCT [20] comparing IMA embolization and conservative treatment showed a lower incidence of type 2 endoleak 24% versus 49% and aneurysm growth 3.8% versus 17%.

Curatively

There are several ways to block the feeding arteries.

Intra Sac Embolization and/or Glue

From the groin a 5FX25 cm sheath is placed below the tip of one limb of the graft. A 0.014 wire is passed between the graft and the arterial wall. A 2×40 mm balloon enlarge the passage of a 5 Fr Ber catheter which is placed in the sac. Then a 0.018 microcatheter is advanced close to the ostium of the feeding artery and coils are released to block the endoleak [21].

The main difficulty is to place properly the catheter in regard of the feeding artery. Figure 66.4a shows the wire between the limb of the graft and the position inside the sac. Figure 66.4b shows the intra sac arteriogram and a feeding artery. Figure 66.4c shows the coils placed in the upper part of the sac. In large nidus and multiple feeding arteries Onyx can be added to fill the sac. Figure 66.5a shows the tip of the catheter and opacification of the nidus and the inferior mesenteric artery. Figure 66.5b shows the control angiogram after Onyx delivery and successful blockage of the IMA.

Trans Caval Embolization

An angulated catheter is placed in the vena cava from the groin. A needle is pushed within the vein and the aortic wall. Once inside the sac and the nidus, coils are delivered. A systematic review [22] collected 128 trans caval treatment with a 91.4% technical success, 12 % of recurrence which were treated by secondary embolization. No death or serious adverse complications were reported.

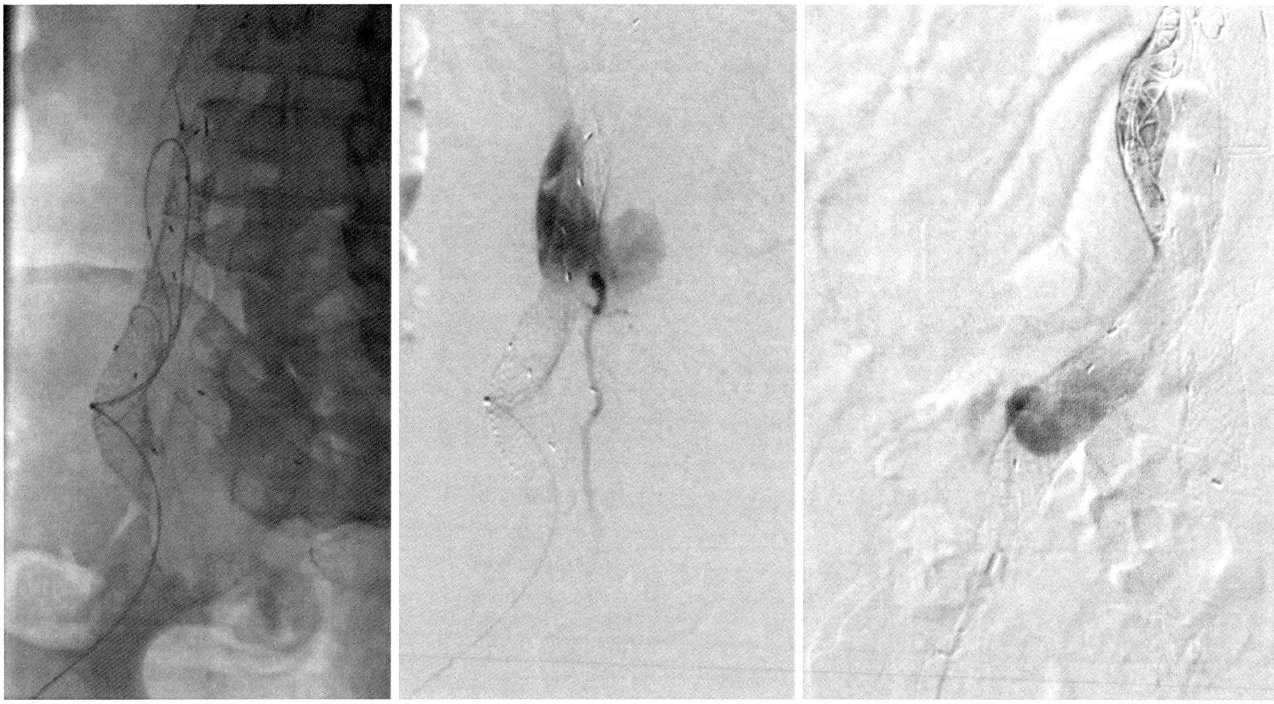

Fig. 66.4 Intra operative angiogram left image: wire and catheter placed in the sac from the groin middle image: angiogram showing the nidus and the feeding artery right image: coils embolisation of the sac

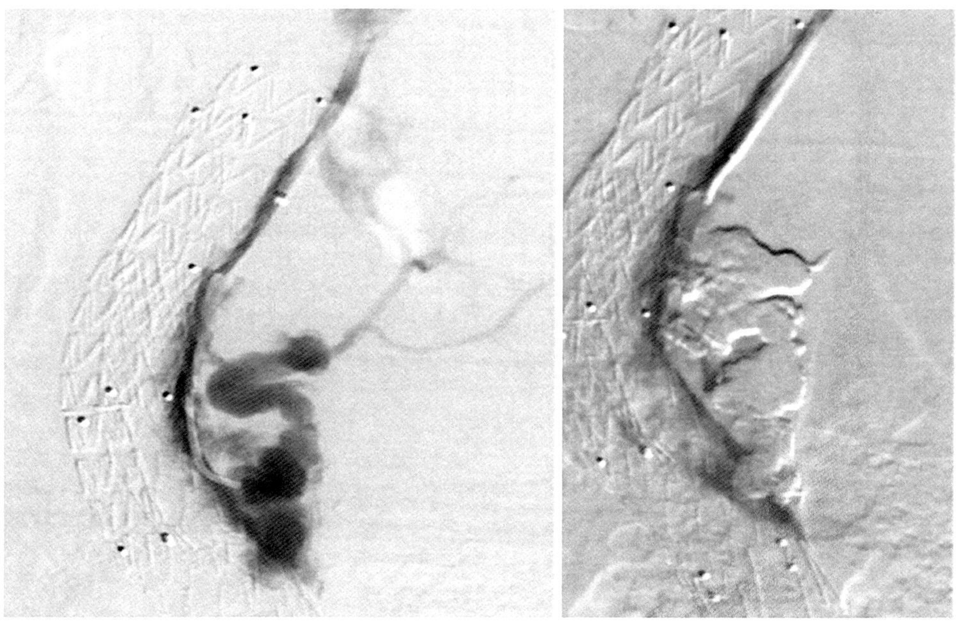

Fig. 66.5 Intra operative angiogram left image: intra operative angiogram showing large endoleak right image: succesful Onyx embolisation

Trans Graft Embolization

The access to the sac can also be obtained by perforating the graft with a laser fiber or a needle pushed in the chosen position. After removal of the sheath or of the needle the puncture hole does not leak. In a report of ten patients treated this way, the technical and clinical success was 100% [23]. When a larger sheath has to be used, the hole can be closed with a small plug.

Trans Lumbar Embolization

The use of fusion images of preoperative CT scan and fluoroscopy in hybrid room with eventually directed laser beam makes relatively simple the access to the nidus in the sac [24]. Patient is placed in a prone position. The puncture site is on the left side between the twelfth rib and the iliac crest. The needle is pushed into the sac under fluoroscopic guidance. Pulsatile back bleeding and angiogram confirm the correct position. The nidus is filled with Onyx and or coils. Experience of 30 cases [25] with a mean follow up of 19 months shows a compete resolution in 50 % of the cases, 36.7 % had a persistent type 2 endoleak with a stable aneurysm and 13.3 % had a persistent endoleak and further aneurysm growth which required a second intervention. One of these later patients required a sacotomy and ligation of the lumbar arteries.

Embolization Through Selective Catheterisms

The more accessible artery is the inferior mesenteric artery via the superior mesenteric artery. Figure 66.6a shows a 3D reconstruction with perfect visualization of the SMA and the origin of the Drummond arcade. Figure 66.6b,c shows guidewire and catheter pushed via the Riolan arcade which connects the SMA and the IMA. Microcatheter and 0.0014 wire are the most convenient tools. Figure 66.7a–c shows the final upward angulation of the arcade to the ostium of the IMA which can be tricky to canulate, the nidus and the final result with IMA blocked. Lumbar arteries can be catheterized via the hypogastric artery but accesses are challenging especially to the two proximal pairs.

Clipping the Arteries

The clipping of the IMA and lumbar arteries by coelioscopy is a relatively easy procedure. The coelioscope is introduced via the left flank. A retroperitoneal dissection is made up to the aneurysm wall. The IMA is clipped and devised at its origin. Then the sac is pushed upward and all lumbar arteries and the sacral artery are clipped. In case of doubt about the origin of the leak (associated unrecognized type I endoleak) a puncture of the sac allows to measure the pressure and to assess the pulsatility in the sac. Eventually, the sac should be opened, the thrombus removed and the ostia of patent arteries sutured. In case of type 1 or 3 endoleak appropriate treatment must be performed (see Chaps. 70 and 72).

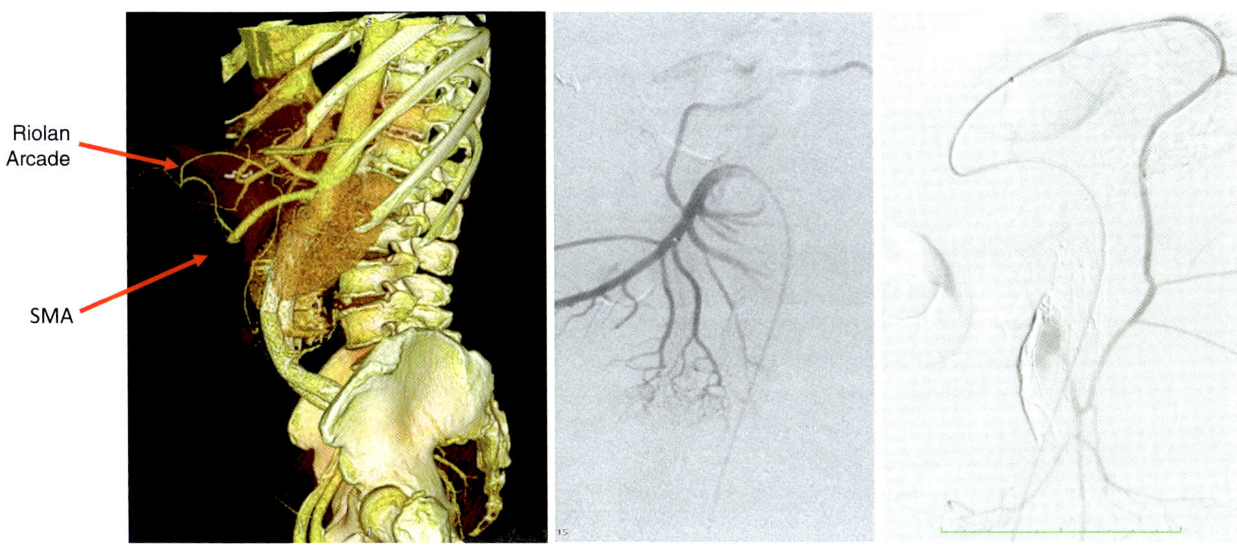

Fig. 66.6 Left image Angio CT with 3D recosntruction SMA: Superior Mesenteric Artery, Middle image: Superior Mesenteric Artery catheterism Right image: Opacification of Inferior Mesenteric Artery and Intra sac Endoleak via the SMA

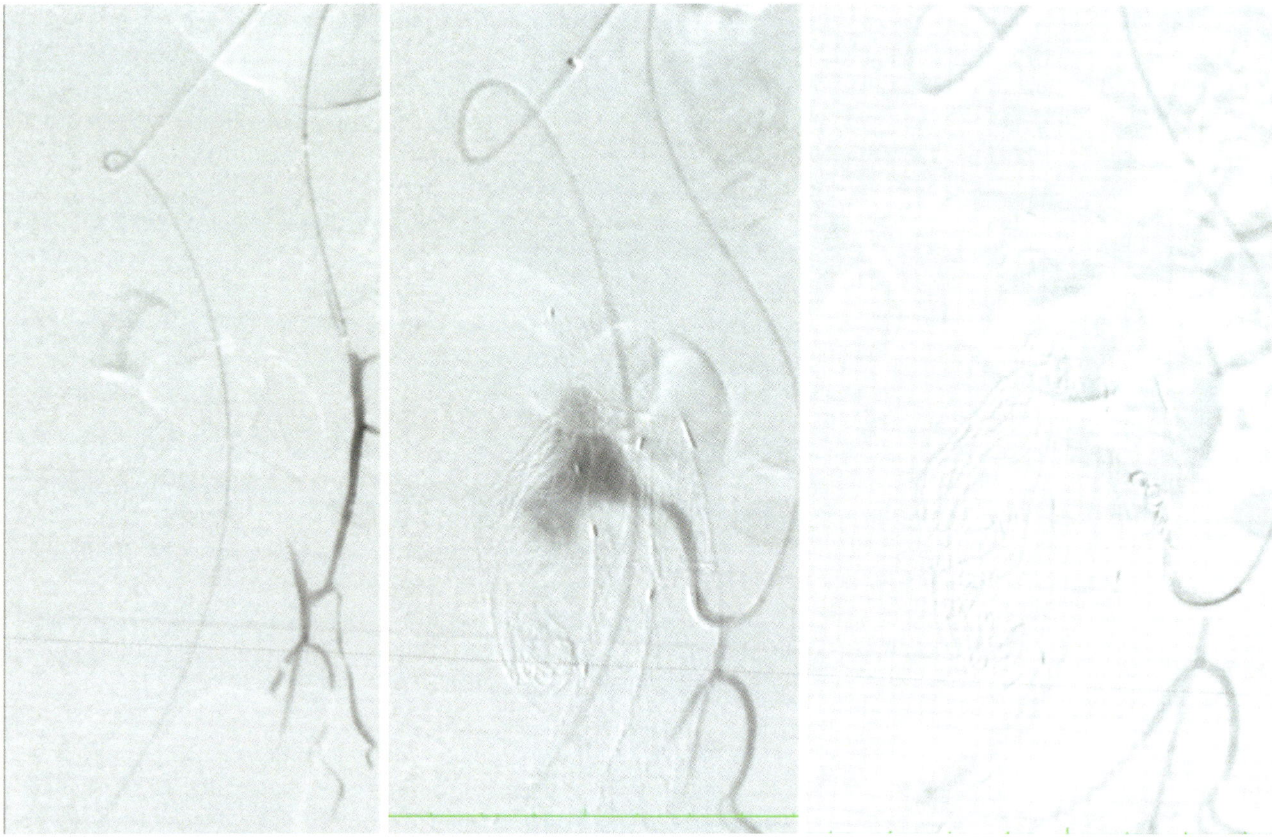

Fig. 66.7 Intra operative angiogram left image: the catheter is pushed close to the connection between the SMA and the IMA, middle image: Reflux of the contrast medium in the aneurysm sac showing the nidus right image: Coils embolisation of the origin of the IMA which blocks the endoleak

In Summary

Type 2 endoleaks are frequent after EVAR. Isolated, they do not expose to rupture but they can be associated with a sealing zone enlargement and type I endoleaks which are concerning and must be treated. When >2 pairs of lumbar arteries and IMA are patent, preemptive embolization before or during Evar are efficient. If the post-operative surveillance finds an enlarging aneurysm sac >5 mm the current recommendations prone the occlusion of the feeding arteries by endovascular means or by laparoscopic clipping.

Tool kit

Wires		
Abbott	Spartacor	0.014 inch
Boston Scientific	Amplatz	0.035 Inch
Cook	Bentson	0.035 Inch
	Linderquist	0.035 Inch
	Rosen	0.035 Inch
Terumo	Glide Wire	0.035 Inch
	Superstiff	0.035 Inch
Catheters		
	Head Hunter	
4 to 6 French	Bentson	
Wires		
	Bernstein	
	Simmons	
	Cobra (C1 C2)	
	Sheperd Hook	
	Multipurpose	
Microcatheters		
3 French	Cook Europe (Limerick, Ireland)	
Occluder	Amplatzer (Abbott)	
Coils	Penumbra (Alameda California USA)	
	Medtronic (Santa Rosa USA)	
	Cook (Limerick, Ireland)	
	Shape Memory Medical (California USA)	
Glue		
	Onyx Medtronic (Santa Rosa USA)	

References

1. Schanzer A, Oderich GS. Management of abdominal aortic aneurysms. N Engl J Med. 2021;385(18):1690–8.
2. PubMed entry [Internet]. [cited 2022 May 25]. http://www.ncbi.nlm.nih.gov/pubmed/29501597.
3. Becquemin JP, Haupert S, Issam F, Dubar A, Martelloni Y, Jousset Y, et al. Five year patient outcomes of endovascular abdominal aortic aneurysm repair in the ENDURANT France registry. Eur J Vasc Endovasc Surg. 2021;61(1):98–105.

4. Dijkstra ML, Zeebregts CJ, Verhagen HJM, Teijink JAW, Power AH, Bockler D, et al. Incidence, natural course, and outcome of type II endoleaks in infrarenal endovascular aneurysm repair based on the ENGAGE registry data. J Vasc Surg. 2020;71(3):780–9.
5. Seike Y, Matsuda H, Fukuda T, Inoue Y, Omura A, Uehara K, et al. The influence of 4 or more patent lumbar arteries on persistent type II endoleak and sac expansion after endovascular aneurysm repair. Ann Vasc Surg. 2018;50:195–201.
6. Chew DK, Dong S, Schroeder AC, Hsu HW, Franko J. The role of the inferior mesenteric artery in predicting secondary intervention for type II endoleak following endovascular aneurysm repair. J Vasc Surg. 2019;70(5):1463–8.
7. Fujii T, Banno H, Kodama A, Sugimoto M, Akita N, Tsuruoka T, et al. Aneurysm sac thrombus volume predicts aneurysm expansion with type II endoleak after endovascular aneurysm repair. Ann Vasc Surg. 2020;66:85–94.e1.
8. Hatzl J, Wang V, Hakimi M, Uhl C, Rengier F, Bruckner T, et al. Persisting type 2 Endoleaks following EVAR for AAA are associated with AAA expansion. J Endovasc Ther. 2022:15266028221081080.
9. Mathlouthi A, Locham S, Dakour-Aridi H, Black JH, Malas MB. Impact of suprarenal neck angulation on endovascular aneurysm repair outcomes. J Vasc Surg. 2020;71(6):1900–6.
10. Eden CL, Long GW, Major M, Studzinski D, Brown OW. Type II endoleak with an enlarging aortic sac after endovascular aneurysm repair predisposes to the development of a type IA endoleak. J Vasc Surg. 2020;72(4):1354–9.
11. d'Audiffret A, Desgranges P, Kobeiter DH, Becquemin JP. Follow-up evaluation of endoluminally treated abdominal aortic aneurysms with duplex ultrasonography: validation with computed tomography. J Vasc Surg. 2001;33(1):42–50.
12. Jean-Baptiste E, Feugier P, Cruzel C, Sarlon-Bartoli G, Reix T, Steinmetz E, et al. Computed tomography-aortography versus color-duplex ultrasound for surveillance of endovascular abdominal aortic aneurysm repair: a prospective multicenter diagnostic- accuracy study (the ESSEA trial). Circ Cardiovasc Imaging. 2020;13(6):e009886.
13. Geraedts ACM, Mulay S, Vahl AC, Verhagen HJM, Wisselink W, de Mik SML, et al. Editor's choice—post-operative surveillance and long term outcome after endovascular aortic aneurysm repair in patients with an initial post-operative computed tomography angiogram without abnormalities: the multicentre retrospective ODYSSEUS study. Eur J Vasc Endovasc Surg. 2022;63(3):390–9.
14. Mulay S, Geraedts ACM, Koelemay MJW, Balm R, ODYSSEUS study group. Type 2 endoleak with or without intervention and survival after endovascular aneurysm repair. Eur J Vasc Endovasc Surg. 2021;61(5):779–86.
15. Chaikof EL, Dalman RL, Eskandari MK, Jackson BM, Lee WA, Mansour MA, et al. The Society for Vascular Surgery practice guidelines on the care of patients with an abdominal aortic aneurysm. J Vasc Surg. 2018;67(1):2–77.e2.
16. Scallan O, Kribs S, Power AH, DeRose G, Duncan A, Dubois L. Onyx versus coil embolization for the treatment of type II endoleaks. J Vasc Surg. 2021;73(6):1966–72.
17. Dosluoglu HH, Rivero M, Khan SZ, Cherr GS, Harris LM, Dryjski ML. Pre-emptive nonselective perigraft aortic sac embolization with coils to prevent type II endoleak after endovascular aneurysm repair. J Vasc Surg. 2019;69(6):1736–46.
18. Branzan D, Geisler A, Steiner S, Doss M, Matschuck M, Scheinert D, et al. Type II endoleak and aortic aneurysm sac shrinkage after preemptive embolization of aneurysm sac side branches. J Vasc Surg. 2021;73(6):1973–1979.e1.
19. Fabre D, Mougin J, Mitilian D, Cochennec F, Garcia Alonso C, Becquemin JP, et al. Prospective, randomised two centre study of endovascular repair of abdominal aortic aneurysm with or without sac embolisation. Eur J Vasc Endovasc Surg. 2020;61:201.
20. Samura M, Morikage N, Otsuka R, Mizoguchi T, Takeuchi Y, Nagase T, et al. Endovascular aneurysm repair with inferior mesenteric artery embolization for preventing type II Endoleak: a prospective randomized controlled trial. Ann Surg. 2020;271(2):238–44.
21. Dinoto E, Ferlito F, Urso F, Evola S, Bajardi G, Pecoraro F. Translimb embolization for treatment of type 2 endoleak post EVAR: case report. Int J Surg Case Rep. 2021;85:106238.
22. Nana P, Spanos K, Heidemann F, Panuccio G, Kouvelos G, Rohlffs F, et al. Systematic review on transcaval embolization for type II endoleak after endovascular aortic aneurysm repair. J Vasc Surg. 2022;76(1):282–291.e2.
23. Matsumura H, Wada H, Teratani H, Amako M, Hayashida Y, Minematsu N. Transgraft embolization by using long needle for the treatment of type II endoleaks after endovascular abdominal aortic repair. J Vasc Surg Cases Innov Tech. 2020;6(4):590–4.
24. Rhee R, Oderich G, Hertault A, Tenorio E, Shih M, Honari S, et al. Multicenter experience in translumbar type II endoleak treatment in the hybrid room with needle trajectory planning and fusion guidance. J Vasc Surg. 2020;72(3):1043–9.
25. Charitable JF, Patalano PI, Garg K, Maldonado TS, Jacobowitz GR, Rockman CB, et al. Outcomes of translumbar embolization of type II endoleaks following endovascular abdominal aortic aneurysm repair. J Vasc Surg. 2021;74(6):1867–73.

Type III–V Endoleaks

67

Shyamal Patel, Seyed Ameli-Renani, and Robert Morgan

Introduction

Endoleaks are defined as persistent flow of blood into the aneurysm sac following endovascular aortic repair (EVAR). They may compromise long-term endograft viability and some are associated with an increased risk of rupture. They are classified as primary or secondary endoleaks. Primary endoleaks occur within the first 30 days following the procedure. Secondary endoleaks develop after 30 days, but may arise months or years later, which underlies the need for lifelong surveillance in these patients. Endoleaks may be classified into five types according to their anatomic site and aetiology. Whilst collectively type III, IV and V endoleaks are less common than type I and II, recognition and appropriate management is important to prevent long-term complications and graft failure [1–3].

Classification

Type III endoleaks describe a defect in the graft (Fig. 67.1) resulting in pressurised blood continuing to fill the aneurysmal sac. These can be further subdivided into **IIIa**—which describes separation of overlapping graft components (either at a branch point or a fenestration) and **IIIb**—a graft fabric tear.

Type IV endoleaks relate to filling of the aneurysmal sac due to porosity of the graft.

Type V endoleaks were historically known as endotension. This remains a poorly understood type of endoleak and describes progressive enlargement of the aneurysm sac without a demonstrable endoleak on imaging. Some authors believe this represents an endoleak, which cannot be resolved with current imaging techniques.

S. Patel · S. Ameli-Renani
St George's University Hospitals NHS Foundation Trust, London, UK
e-mail: Shyamal.Patel@stgeorges.nhs.uk; seyed.renani@stgeorges.nhs.uk

R. Morgan (✉)
Department of Interventional Radiology, St Georges Hospital, London, UK
e-mail: robert.morgan@stgeorges.nhs.uk

G. Geroulakos et al. (eds.), *Mastering Endovascular Techniques*, https://doi.org/10.1007/978-3-031-42735-0_67

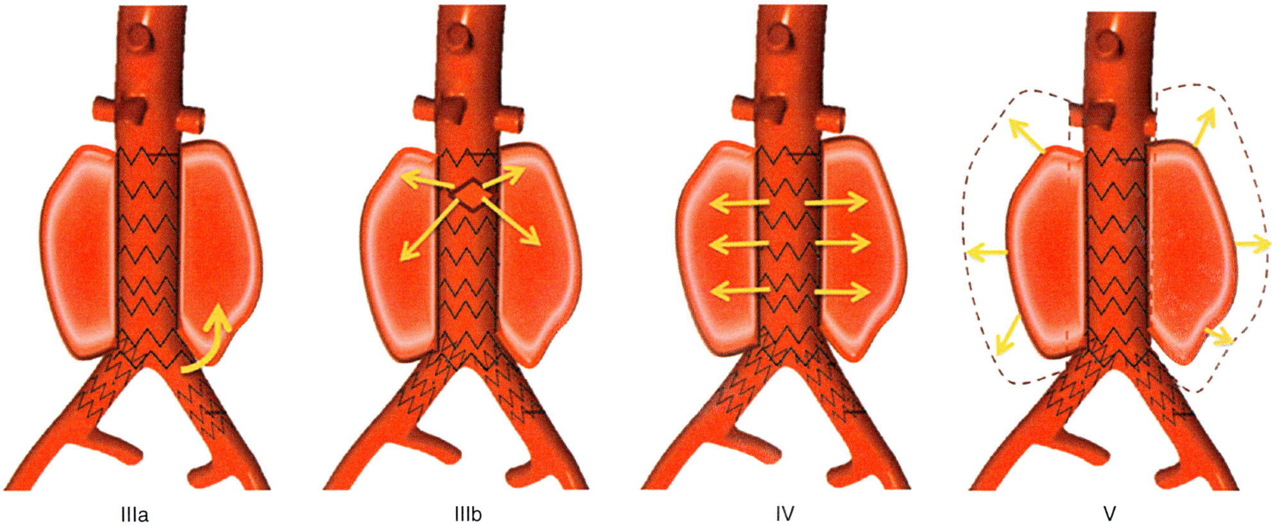

Fig. 67.1 demonstrates the subtypes of endoleak - IIIa, IIIb, IV and V

Diagnosis

Primary type IIIa and IV endoleaks are often identified at the time of EVAR with the use of digital subtracted angiography (DSA). For other primary and secondary endoleaks, the main imaging methods are CT angiography (CTA) and Duplex ultrasound (US). CTA protocols vary by institution and may be a single arterial phase scan, dual phase (pre-contrast and arterial phase) or triple phase (pre-contrast, arterial phase and delayed phase). Most centres perform at least an arterial phase scan with pre-contrast and delayed phase scans reserved for selected cases [2, 4].

Due to the risks associated with the repetitive use of ionising radiation and intravenous contrast, there is increased reliance on alternative modalities for surveillance following EVAR. Duplex ultrasound (US) has long been utilised during the surveillance period to screen for interval growth of the aneurysmal sac and is also able to demonstrate certain subtypes of endoleak. Contrast enhanced ultrasound (CEUS) is a cheap, viable and safe option in the hands of experienced operators, offering improved endoleak detection and characterisation. Magnetic resonance angiography (MRA) in selected cases is highly sensitive but requires the use of MR compatible grafts. It is also time consuming, expensive, and often limited by availability.

Despite the availability of high quality non-invasive imaging in most centres, conventional catheter angiography still has a role in the evaluation of endoleaks. This is either to confirm the presence of an endoleak, confirm that the classification of the endoleak on non-invasive imaging is correct, or to further evaluate when an aneurysm sac enlarges without evidence of an endoleak.

One of the drawbacks of DSA is the temporal resolution, particularly in comparison to CTA. This can be overcome by the use of cone beam CT, which has the added advantage of concurrent volumetric acquisitions, and real time fluoroscopy. Respiratory motion and peristalsis from adjacent loops of bowel can degrade imaging, limiting interpretation. Post procedure processing can overcome this to a certain degree; however, DSA requires an element of patient co-operation to optimise imaging and aid the detection of small endoleaks.

Type V endoleaks by their nature often pose a diagnostic dilemma and require a multi-modality approach (Fig. 67.2). US and CTA can demonstrate an increase in the aneurysm sac size in the absence of a type I-IV endoleak. DSA is often employed as an investigative tool to allow interrogation of the graft. Catheter angiography allows targeted intra-arterial delivery of contrast and can enable improved detection and characterisation of small endoleaks. Selective angiograms from above and below the graft can identify type Ia/Ib endoleaks respectively. Angiograms performed within the graft are used to exclude a type III endoleak and super selective angiograms from the SMA or from the iliac vessels can demonstrate type II endoleaks.

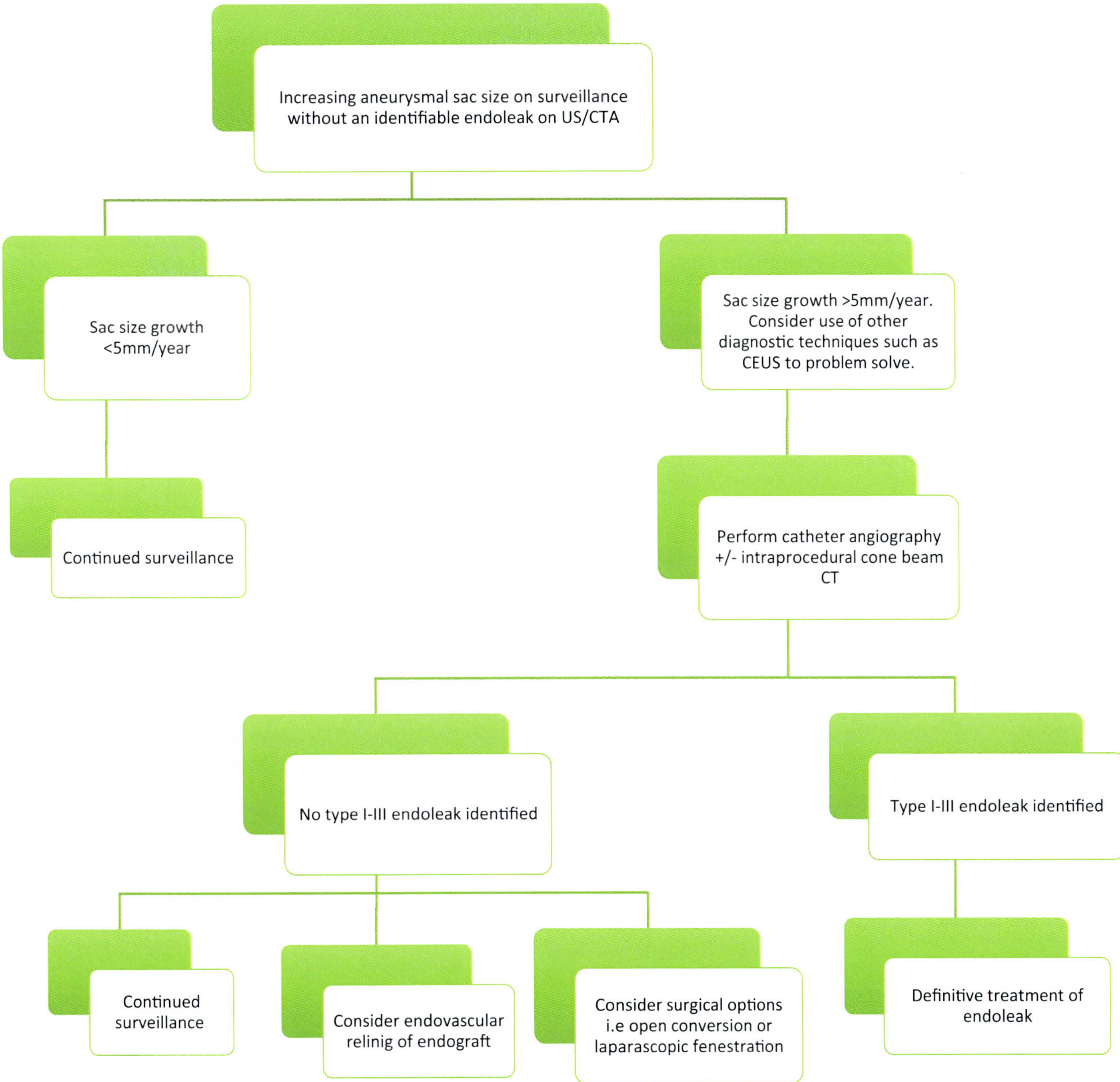

Fig. 67.2 Suggested diagnostic and management strategy of type V endoleaks

Type III Endoleaks

With advances in endograft material/deployment, type III endoleaks are uncommon, accounting for approximately 1.3%–3% of all endoleaks. As demonstrated in Fig. 67.1, type III endoleaks are high pressure leaks allowing arterial blood to directly fill the aneurysm sac, requiring emergent management to prevent rupture of the aneurysmal sac.

Primary Type III Endoleak

Primary type IIIa endoleaks most commonly occur at the time of EVAR due to defects at the site of overlap between endograft components. This allows continued filling of the aneurysmal sac which can be identified on completion angiograms. Repeat angioplasty and covered stent/endograft insertion over this defect is often all that is required.

Secondary Type III Endoleak

The mechanism of secondary type IIIa endoleaks is variable and could be related to changes in the anatomy of the aneurysmal sac over time. Migration of the endograft can occur with dilatation of the proximal/distal graft attachment sites resulting in insufficient overlap between the modular components or branch points (Fig. 67.3).

The underlying mechanism of type IIIb endoleaks (Fig. 67.4) is probably multifactorial and is most likely due to endograft fabric degradation over time. Fabric tears were more common with first generation endografts and are significantly less frequent with newer grafts.

Management of secondary type IIIa endoleaks due to a focal defect at a fenestration (Fig. 67.3) or sites of stent overlap are preferably endovascular. The defect can be excluded by insertion of a covered stent. There are many types of covered stent which come in varying lengths/diameter and delivery systems. These require appropriate sizing prior to deployment with the help of imaging.

Another endovascular option is relining the existing graft with a new endograft or extension cuffs, which may be necessary particularly with a fabric tear close to the flow divertor. Depending on the location, this could involve insertion of a new iliac limb extension and/or new aortouni-iliac endograft. Development of endovascular aneurysm sealing systems (Nellix- [Endologix, Irvine, USA]) also allows sealing of the space between the two endografts. The dimensions of the original graft must be considered when choosing the new endograft(s) to ensure appropriate coverage.

If endovascular options are not appropriate or fail, open repair can be considered if the patient is a surgical candidate.

Type III endoleaks have a high risk of rupture if left untreated. Endovascular techniques are not without adverse outcomes including retroperitoneal haemorrhage, bowel ischaemia and distal embolism resulting in lower limb ischaemia. Recurrent secondary type III endoleaks can occur despite endovascular repair and have proven to be more common with older generation grafts [5–7].

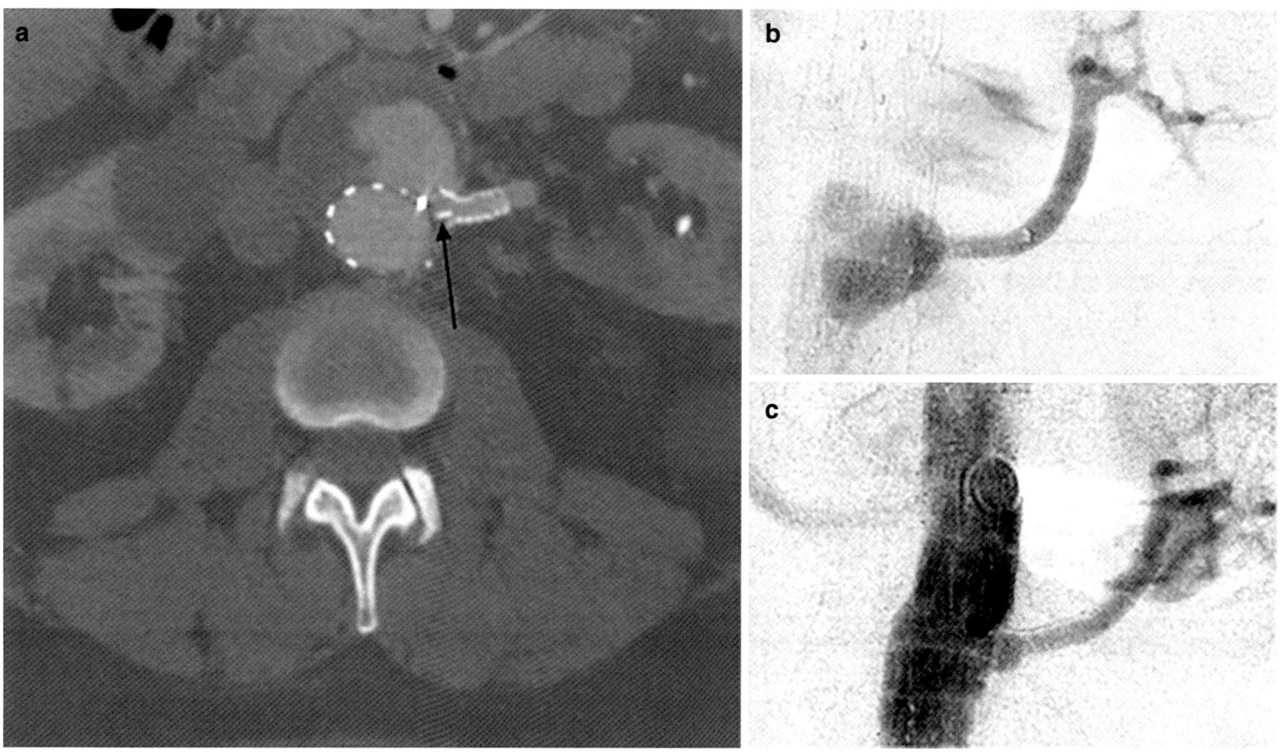

Fig. 67.3 73-year-old male who presented with a secondary type IIIa endoleak following FEVAR. (**a**) CTA images demonstrating separation and fracture (arrow) of the left renal arterial graft with contrast filling the aneurysmal sac. (**b**) Intra-procedural DSA confirms filling of the aneurysmal sac. (**c**) Post treatment angiogram demonstrates successful endovascular treatment with insertion of a covered stent graft across the defect and no further filling of the aneurysm sac

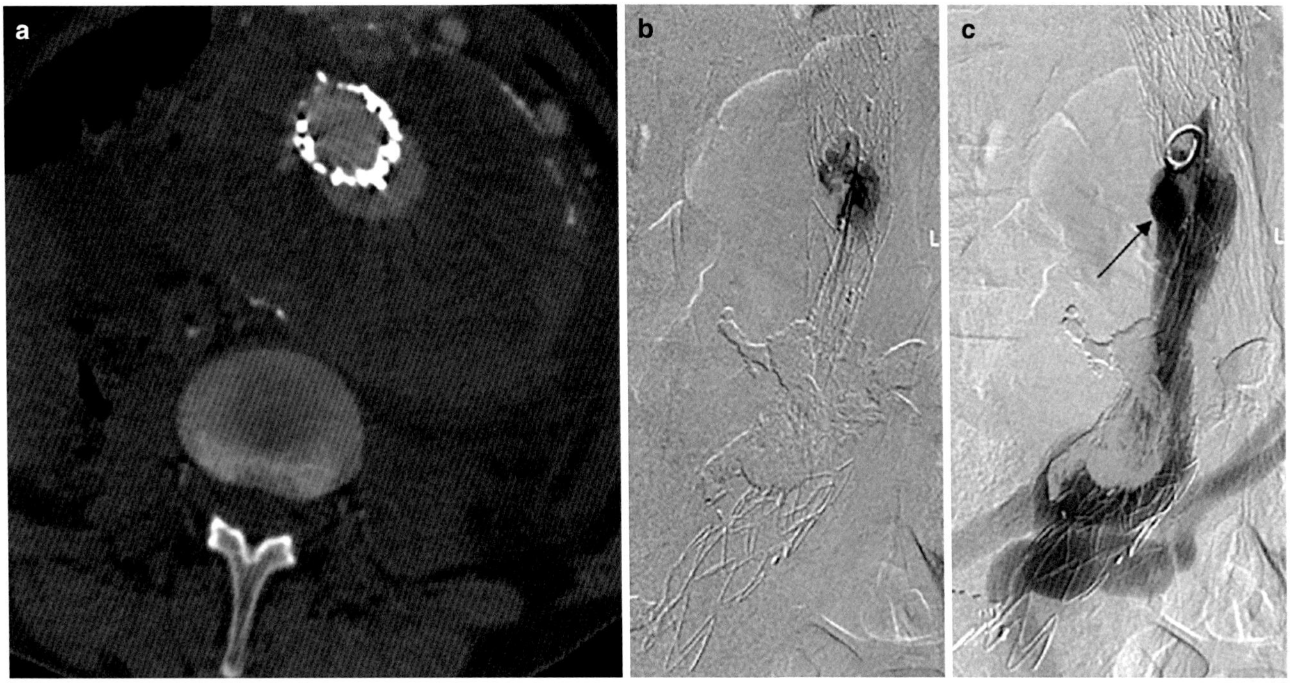

Fig. 67.4 79-year-old male with history of prior embolisation for type II endoleak, showing new increase in aneurysm sac size on surveillance imaging post FEVAR. (**a**) Suspected type IIIb endoleak with contrast entering the aneurysm sac from the main body of the endograft. (**b**, **c**) Images from the subsequent diagnostic angiogram. (**b**) is an early image demonstrating the margins of the endograft. (**c**) Demonstrates a small leak of contrast beyond the boundaries of the endograft (arrow) in keeping with a type IIIb endoleak, for which the patient underwent successful relining. Angiography also excluded a type II endoleak

Type IV Endoleaks

Type IV endoleaks are essentially a historical entity that occurred with early endograft designs and are not a current area of concern for EVAR practitioners. All type IV endoleaks are primary endoleaks. They are identified on completion angiography after aortic device implantation as a faint diffuse blush of contrast around the aortic endograft. They are self-limiting, may be related to a combination of graft porosity and procedural anticoagulation, and do not require treatment. They have been reported to occur with a prevalence of 0.3–3%. They are seldom visualised with current endograft designs.

Type V Endoleaks

Type V endoleaks have been reported to account for approximately 6% of total endoleaks but have become less prevalent due to gradual improvements in graft materials. Type V endoleaks are defined as progressive enlargement of the aneurysm sac in the absence of an identifiable endoleak on imaging. The first descriptions of Type V endoleaks were associated with early generation Gore Excluder aortic grafts (Gore, Flagstaff, AZ). The phenomenon was thought to be due to late porosity of the endograft fabric (different to the procedural porosity of type IV endoleaks). This led to modification of the fabric material for later iterations of the Gore endografts, which significantly reduced this type of endoleak with these specific devices.

However, type V endoleaks may also occur in the presence of all newer generation endografts. The underlying process has yet to be proven but is hypothesised as a low velocity or intermittent endoleak, which is undetectable on current imaging. Other suggested mechanisms include pressure transmission across the endograft or thrombus as well as ultrafiltration of blood across the graft fabric [8–11].

As the underlying aetiology is unknown, management of type V endoleaks is often difficult and requires a patient centred approach. The major limiting issue in this situation is the lack of a target for endovascular management. Observation may be appropriate in some cases due to the lack of pressurised blood flow into the aneurysm sac. However, progressive enlargement of the aneurysmal sac will eventually alter the configuration of the aneurysmal sac predisposing to the development of secondary type I or III endoleaks. All type V endoleaks are secondary endoleaks.

Type V endoleaks are diagnosed after exclusion of all other endoleak categories by imaging, including conventional catheter angiography and/or C-arm CT at the time of

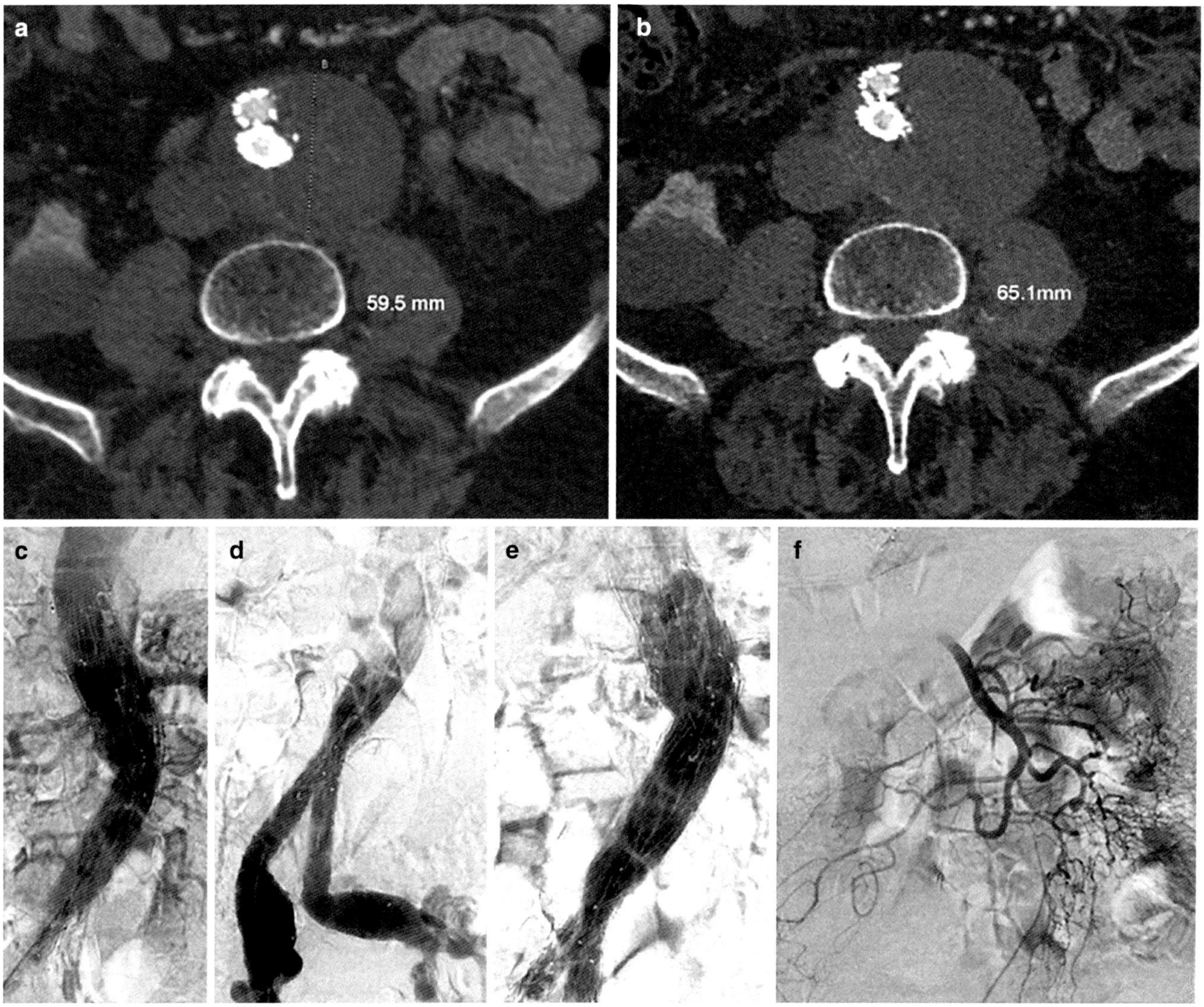

Fig. 67.5 Type V endoleak in an 85-year-old male who underwent FEVAR 12 months prior. (**a**) Post-operative CTA demonstrates no endoleak and an aneurysmal sac size of 59.5 mm. (**b**) Repeat CTA following a 6-month interval demonstrates over 5 mm increase in the aneurysmal sac size without a type I, II or III endoleak. (**c**) Conventional catheter angiography from above the endograft excludes a type 1a endoleak. (**d**) Catheter angiography in the lower portion of the endograft excludes a type 1b endoleak. (**e**) Catheter angiography in the body of the endograft excludes a type III endoleak. (**f**) Catheter angiography from the SMA excludes a type II endoleak due to collaterals filling the IMA

conventional angiography (Fig. 67.5). If no endoleak is identified, options for targeted endovascular management are not possible. In this scenario, if there is progressive sac enlargement with no endoleak identified, the first surgical option considered is usually relining of the endograft either by another bifurcated endograft if feasible or by placement of an aortouni-iliac device plus fem-fem crossover.

If device relining fails to arrest sac growth, then open surgical conversion is the only viable option if the patient is fit for this type of surgery. If the patient is surgically unfit, no further treatment options are available and the patient is managed by follow-up imaging and palliative therapy when necessary.

Summary

Type III, IV and V endoleaks are the least common classes of endoleak. Type III endoleaks are due to modular disconnection or fabric tears, and are often high pressure leaks, requiring prompt management due to the risk of rupture. Type IV endoleaks are seldom seen in modern EVAR practice. Type V endoleaks are more complex and pose difficulties with management. Treatment options are graft relining, open conversion or conservative. Endovascular techniques remain the preferred management option for endoleaks to reduce the risk of rupture. As endoleaks can occur months to years following

EVAR, ongoing surveillance remains important to identify these complications and allow appropriate management.

References

1. Chaikof EL, Dalman RL, Eskandari MK, et al. The Society for Vascular Surgery practice guidelines on the care of patients with an abdominal aortic aneurysm. J Vasc Surg. 2018;67(1):2–77.e2. https://doi.org/10.1016/j.jvs.2017.10.044.
2. Ameli-Renani S, Pavlidis V, Morgan RA. Secondary endoleak management following TEVAR and EVAR. Cardiovasc Intervent Radiol. 2020;43(12):1839–54. https://doi.org/10.1007/s00270-020-02572-9.
3. Kaufman JA. Vascular interventions. In: Kaufman JA, Lee MJ, editors. Vascular and interventional radiology: the requisites. Elsevier Saunders; 2013. p. 86–95.
4. Partovi S, Trischman T, Rafailidis V, et al. Multimodality imaging assessment of endoleaks post-endovascular aortic repair. Br J Radiol. 2018;91(1087):20180013. https://doi.org/10.1259/bjr.20180013.
5. Blakeslee-Carter J, Beck A, Spangler E. Occurrence and outcomes of type 3 endoleaks in endovascular aortic repair within the vascular quality initiative database. BMJ Surg Interv Health Technol. 2020;2(1):e000054. https://doi.org/10.1136/bmjsit-2020-000054.
6. Maleux G, Poorteman L, Laenen A, et al. Incidence, etiology, and management of type III endoleak after endovascular aortic repair. J Vasc Surg. 2017;66(4):1056–64. https://doi.org/10.1016/j.jvs.2017.01.056.
7. Chen J, Stavropoulos SW. Management of endoleaks. Semin Intervent Radiol. 2015;32(3):259–64. https://doi.org/10.1055/s-0035-1556825.
8. Parsa P, Das Gupta J, McNally M, Chandra V. Endotension: what do we know and not know about this enigmatic complication of endovascular aneurysm repair. J Vasc Surg. 2021;74(2):639–45. https://doi.org/10.1016/j.jvs.2021.03.018.
9. Toya N, Fujita T, Kanaoka Y, Ohki T. Endotension following endovascular aneurysm repair. Vasc Med. 2008;13(4):305–11. https://doi.org/10.1177/1358863X08094850.
10. Kougias P, Lin PH, Dardik A, Lee WA, El Sayed HF, Zhou W. Successful treatment of endotension and aneurysm sac enlargement with endovascular stent graft reinforcement. J Vasc Surg. 2007;46(1):124–7. https://doi.org/10.1016/j.jvs.2007.02.067.
11. Torres-Blanco Á, Miralles-Hernández M. Endotension: twenty years of a controversial term. CVIR Endovasc. 2021;4(1):46. Published 2021 Jun 5. https://doi.org/10.1186/s42155-021-00238-2.

Endograft Infections

68

Konstantinos G. Moulakakis and Spyridon N. Mylonas

Case Presentation

A 67 years old male referred to our emergency department due to temperature of 40° C, symptoms of urinary tract infection over the last 24 h and numbness in right lower extremity for the last 3 months. The patient had undergone 5 years ago an endovascular abdominal aortic aneurysm repair due to a ruptured AAA with secondary type Ib endoleak, which was treated 2 years ago with stentgraft extension in the right common iliac artery. In his past medical history were recorded coronary artery disease, chronic alcoholism, and right urinary stasis with dilatation of the right ureter, which was treated with placement of a double-J-catheter 3 months ago. Upon admission, the following laboratory findings were recorded: WBC 14.00 × 1E^9/L, CRP 165 mg/L, procalcitonine 1.9 µg/L. Blood cultures were taken and empirical antibiotic therapy with Piperacycline/Tazobactam was initiated. The following CTA depicted an urinary tract infection with right psoas abscess and air-bubbles in the aneurysm sac indicating an endograft infection (Fig. 68.1a). A CT-guided drainage was placed in the psoas abscess and the punctate was sent for microbiological examination. With the intention of defining the extension of infection, a PET/CT was performed (Fig. 68.1b). Based on the clinical, laboratory and radiological findings the diagnosis of endograft infection was confirmed. After interdisciplinary evaluation by vascular surgeons, infectiologists, urologists as well as the written informed consent of the patient an endograft explantation, operative debridement of the psoas abscess and in situ aorto-iliac reconstruction with the deep femoral veins was suggested.

Through a median laparotomy and a transperitoneal approach, a proximal control of the suprarenal aorta as well as of the iliac bifurcation bilaterally could be obtained after mobilization of the duodenum, which was found intact. The aneurysm sac was opened and purulent fluid was extracted. Simultaneously the superficial femoral veins of both thighs were harvested and a bifurcated graft was constructed (Fig. 68.1c). With light squeezing of the proximal segment of the endograft and careful mobilization of the barbs the endograft removal was achieved (Fig. 68.1d). An extensive drainage and rinsing with Lavanit® solution of the right psoas abscess was performed. The aorto-iliac reconstruction with the venous bifurcated graft was followed by reinforcing the anastomoses with venous pledgets and completed with a bypass to the right renal artery (Fig. 68.1e). The graft was covered with gentamycin-patch and omentum flap.

Postoperatively the patient continued on empiric antibiotic therapy, which was modified according the cultures of the intraoperative samples, which isolated *Klebsiella oxytoca* and *Candida albicans.* The patient had an uneventful immediate postoperative course and he could be transferred to a rehabilitation center on the 21st postoperative day under antibiotic and antimycotic therapy for at least 8 weeks.

K. G. Moulakakis (✉)
Department of Vascular Surgery, University of Patras Medical School, Patras, Greece

S. N. Mylonas
Department of Vascular and Endovascular Surgery, University of Cologne, Cologne, Germany

G. Geroulakos et al. (eds.), *Mastering Endovascular Techniques*, https://doi.org/10.1007/978-3-031-42735-0_68

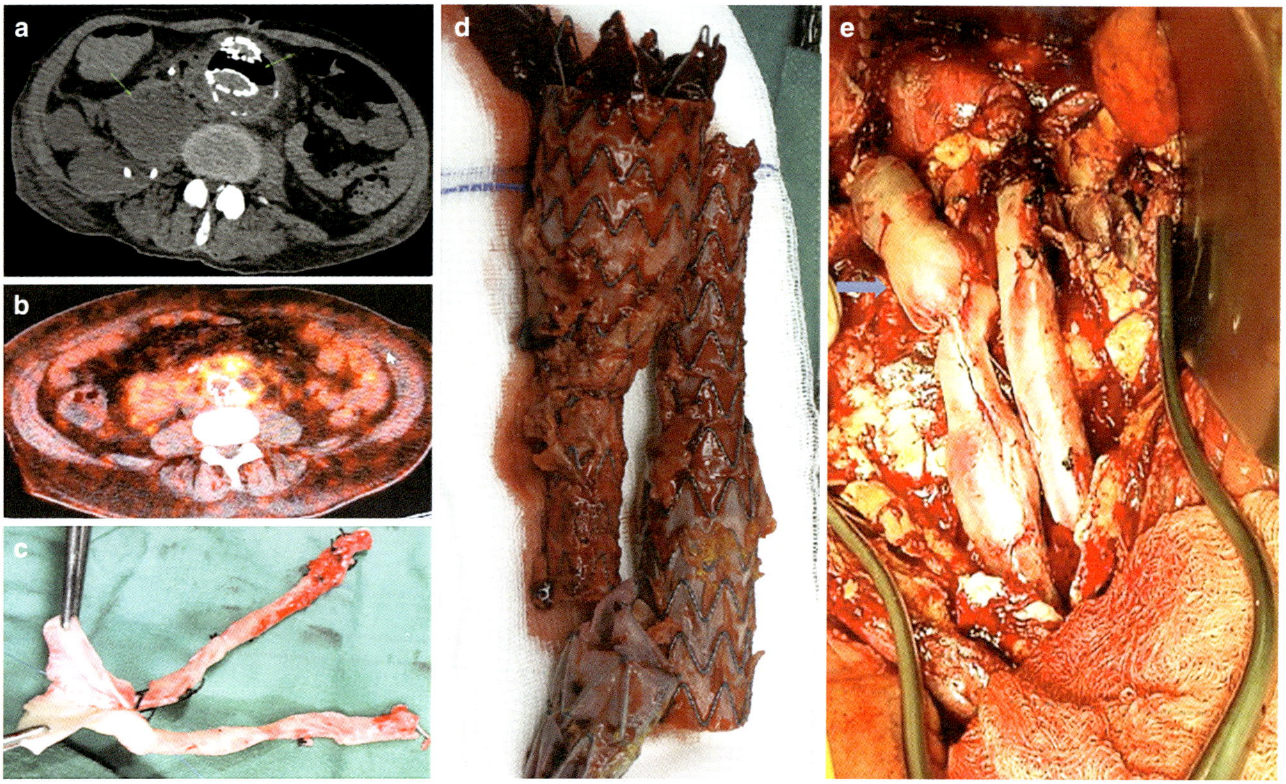

Fig. 68.1 (**a**) Preoperative CTA depicting right psoas abscess and air-bubbles in the aneurysm sac (arrows). (**b**) 18F-FDG-PET/CT confirmed the diagnosis of endograft infection through a significant 18F-FDG uptake in the aortic graft as well as the aortic wall and the right psoas. (**c**) The neo-aorto-iliac vein graft, intraoperatively manufactured from both superficial femoral veins. (**d**) The explanted stentgraft with the suprarenal uncovered stent. (**e**) The final intraoperative result of the neo-aorto-iliac reconstruction with the venous bypass to the right renal artery (arrow)

Incidence

Infection of an endograft implanted for treatment of an abdominal aortic pathology (endovascular aortic repair, EVAR) is considered a rare event, with an incidence of 0.2–0.7% [1, 2]. An aorto-enteric fistula (AEF) can be roughly present in 25–35% of cases with endograft infection [3–5]. In a large multicentric registry, including 180 infected endografts, an aorto-enteric fistula occurred in 23.9% of patients [3]. In a French multicenter study, the AEF was present in a third of patients with infected endografts, emphasizing the fact the AEF could be a more common finding than initially thought [4].

An endograft infection after thoracic endovascular aortic repair (TEVAR) is also a devastating complication characterized fairly as the nightmare of cardiovascular surgeons. It has an estimated incidence of 1–6% and is likely to occur more frequently in association with aorto-esophageal fistulas or aorto-bronchial fistula [3, 6, 7].

Risk Factors

Endograft infections are usually multifactorial and result from the complex interaction of patient status with surgical risk factors (emergent/urgent setting, re-operation or secondary endovascular or open interventions, prolonged operation time, etc.) [1, 8, 9]. Possible pathophysiologic mechanisms include increased susceptibility to infection, perioperative contamination, hematogenous seeding, and mechanical erosion of the endograft [10]. Immunosuppression, poor nutrition, diabetes mellitus, malignancy, lymphoproliferative disorders, chronic kidney disease, and liver cirrhosis have been found to weaken the immune system and cause a patient to become immunocompromised and more susceptible to infections [11, 12].

Predisposing conditions for an endograft infection include, also, an emergency or urgent index procedure, especially when the initial procedure has been performed across a contaminated field (aorto-enteric fistula, or a mycotic aneurysm) [11]. The procedural environment is another potential risk factor for a stent-graft infection; procedures conducted in interventional radiology suites have been associated with higher post-intervention infection rate, probably, due to breaks of the sterility protocol [5, 10]. Moreover, secondary adjunctive endovascular or surgical procedures following EVAR may constitute also a potential risk factor for stent-graft infection. Graft-related microbiemia or sepsis has been reported after repeated embolization of a persistent endoleak [2].

Microorganisms

Identification of microorganisms involved in the endograft infection is a key step of the management of endograft infection. Microbiological samples should be directly (through drainage under CTA guidance) or indirectly (blood cultures) obtained before initiation of the antimicrobial therapy. It is estimated that responsible pathogens are Gram positive bacteria in up to 58% of endograft infections (e.g. enterococci, Staphylococcus aureus, and coagulase negative staphylococci),whereas Gram negative bacteria (e.g. Pseudomonas, Klebsiella, *E. coli*) account for about 34% and anaerobes 8% whereas fungus for 5% of endograft infections [1, 3, 7, 8]. It should be mentioned that a polymicrobial infection can be found in at least one-third of total endograft infections [1]. The virulence of the pathogens as well as their ability of biofilm development can affect the outcomes and the re-infection rates [13].

Diagnosis

Accurate and timely diagnosis is paramount to ensure that appropriate treatment is initiated promptly. The diagnosis of an endograft infection is based on a combination of clinical symptoms, laboratory tests, and imaging studies. The Management of Aortic Graft Infection Collaboration (MAGIC) group summarized these parameters and proposed a list of major and minor criteria. In the presence of one major or two minor criteria of the three different categories a vascular graft/endograft infection is suspected, whereas the presence of at least a single major criterion and any other criterion from another category confirms the diagnosis of a graft/endograft infection [14]. Nonspecific symptoms include fever, chills, malaise, weight loss, and back, abdominal, or chest pain, fatigue. Rarely septic emboli can cause secondary loci of infection or even abscesses. In case of aorto-bronchial or aorto-oesophageal/enteric fistula haematemesis or haemoptysis may be the first symptom [8]. Characteristically this bleeding may be massive, but is often preceded by self-limiting "herald bleeding." Inflammatory markers (C-reactive protein, white blood cell count) are usually elevated at the time of diagnosis [1, 8].

Various imaging techniques are used in the diagnostic work up when an endograft infection is suspected. The most widely applied imaging modality for the diagnosis of an endograft infection is the CTA. Periaortic and retroperitoneal inflammation, stranding of the retroperitoneal fat, and periaortic fluid collections or perigraft gas represent the most common findings, whereas findings indicative of spondylodiscitis can also be present. Moreover, contrast leakage from the aorta to the GI in cases of aorto-enteric fistula can be detected. However, CTA's sensitivity and specificity for the diagnosis of endograft infection remains moderate and a further imaging modality is usually required to confirm the diagnosis and define the extent of infection [15, 16]. In case of contraindications for CTA an MRA can be conducted. Although not extensively evaluated for the diagnosis of endograft infection, it is suggested that MRA can provide detailed anatomical information regarding the extension of the infection, when compared to CTA [17]. Several nuclear medicine techniques, such as 18F-fluoro-D-deoxyglucose positron emission tomography (18FFDG-PET) with or without diagnostic contrast enhanced CT (18F-FDG-PET/CT), and white blood cell scintigraphy (WBCS), that can be combined with single photon emission computed tomography (SPECT/CT) can be useful for better localization of the infection especially in low-grade infections [16, 18, 19]. In case of clinical suspicion and before the operative treatment an endoscopy of the upper GI or a bronchoscopy should be performed to exclude an AEF or ABF.

Treatment Options

Surgical Treatment

The ideal surgical treatment of an established endograft infection is well-illustrated in the following four main steps: (1) complete endograft explantation; (2) extensive debridement of the arterial bed to prevent the infection from spreading; (3) arterial reconstruction in a non-infected field with an infection-resistant vascular graft.

Endograft Explantation

A surgical approach through a midline laparotomy is usually applied, especially if an aorto-enteric fistula requires complementary bowel reconstruction, whereas a retroperitoneal approach is preferable to extract an endograft with a long uncovered proximal stent. Complex endograft (chimney, branched, or fenestrated aortic EG) explantation requires a thoraco-abdominal approach. A temporary occlusion of the aorta through a transfemoral or trans-brachial approach can also be applied [20].

Technical aspects of endograft explantation vary according the type of graft deployed, the vascular anatomy and the extent of infection. Endografts with suprarenal fixation require supra-coeliac control for explantation, whereas suprarenal aortic cross clamping allows extraction of an endograft without suprarenal fixation. Various techniques have been described to facilitate an endograft removal. These include the gentle squeezing of the upper stent and gently pushing of the stent hooks or barbs in a proximal direction or the collapsing of the proximal fixation into a 20-mL syringe or by using of a disposable proctoscope [21–23]. However, in cases of extensive endograft incorporation and in order to avoid proximal aortic dissection, the bare suprarenal struts of the device may be cut and left in situ [9, 21]. After excision of the infected graft an extensive debridement of the surrounding infected tissues (e.g. aneurysm sac, thrombus) is mandatory.

Extra-Anatomic Vs In Situ Reconstruction

The specific revascularization technique is mainly based on the experience of the operators and center. Extra-anatomic reconstruction originating from the axillary artery has been performed widely. The most common reconstruction employed is the axillo-femoro-femoral bypass performed in the same or a later staged procedure. The disadvantages of axillo-bifemoral reconstructions are low patency rates (64–75% at 5 years) and high amputation rates (up to 10% in 5 years) [12]. Moreover, concerns are raised through the reported 10–20% stump blowout rates, especially after extraction of an endograft with suprarenal fixation [24]. Therefore, stump strengthening with double suture layers, reinforcement with venous or felt pledgets, prevertebral fascia, or a layer of posterior rectus fascia peritoneum or bovine pericardium or fascia Lata is suggested, although this delicate procedure may impair renal or visceral arterial flow.

In situ reconstruction, on the other hand, avoids the risk of stump blow-out and renal vessel occlusion, and according to three meta-analyses outperforms extra-anatomic reconstruction in terms of conduit failure, amputation, re-infection, and mortality rates [13, 24, 25]. Therefore the current guidelines of ESVS suggest an extra-anatomic reconstruction only in case of large abscess or multiresistant microorganisms [12].

Graft Materials

Currently autologous veins, cryopreserved homografts, rifambicin bonded grafts, silver coated graft and xenogenous grafts have been used for in situ reconstructions after endograft explantation. Each with its advantages and disadvantages (Table 68.1). Based on the current knowledge the ESVS guidelines recommend an autologous in situ reconstruction for abdominal graft endograft infection. All other available materials should be considered as alternative solutions [12].

Table 68.1 Available graft materials for in situ reconstruction after endograft explantation

	Autologous vein grafts	Cryopreserved grafts	Rifampicin bonded grafts	Silver coated grafts	Xenogenous grafts
Re-infection rates	0–6%	0–7%	0–11.5%	0–12.5%	Small case series
Thrombosis rates	0–9%	0–18%	0–14%	0–8%	Small case series
Aneurysm-rupture	5%	0–24%	NA	NA	Small case series
Comments	• Harvesting: – increased OR time – CVI complications. • Poor outcomes in the presence of AEF.	• Considerable re-interventions rates.	• Different concentrations. • Less effective in infections caused by MRSA, gram negative strains, and fungi. • Rifampicin resistance?	• Silver acelate or elemental silver. • Wide antimicrobial activity. • Paucity of resistance development.	• CMG from bovine pericardium or. • Omniflow (cross-linked bovine collagen).

OR Operation time, *CVI* Chronic venous insufficiency, *AEF* Aorto-enteric fistula, *MRSA* Methicillin-resistant Staphylococcus aureus, *CMG* Custom-made graft

Presence of Aorto-Intestinal Fistula

The presence of a fistula between the stent-graft and the GI tract poses the risk of acute exsanguination and promotes continuous contamination of the blood with microorganisms. In such a situation, the patient is in a constant septic condition, and surgical closure of the aorto-enteric fistula or even resection of the GI tract segment (e.g. duodenum) is essential for treatment. Antibiotic therapy may control or resolve sepsis temporarily, but infection cannot be eradicated or cured without surgical reconstruction.

Aorto-oesophageal fistula requires special consideration, due to its complexity and the high perioperative risk. The incidence of an AOF or an aorto-bronchial fistula after TEVAR is estimated at 1.7% [26, 27]. It occurs more frequently and in a shorter time after TEVAR than after OR of a thoracic aortic pathology [6, 7, 28]. Corresponding to the treatment of aorto-enteric fistula, the therapy largely depends on the clinical status of the patient at presentation; the implantation of a thoracic stent graft for temporary bleeding control until further management is a recognized therapy option [29–31]. Several combinations of treatment options have been proposed in AOFs, including ISR, extra-anatomic bypass graft with concomitant primary esophageal repair, or esophagectomy with cervical esophagostomy and secondary restoration of GIT continuity [28, 32]. Survival after 2 years in patients with AOF after operative treatment has been described at 16–39%, [7, 32] whereas a conservative treatment of AOF is associated with a mortality of almost 100% at 12 months [28].

Antimicrobial Therapy

Antimicrobial therapy is a cornerstone for the treatment of endograft infection. Based on the complexity of these infections, the continuous evolution of available antibiotic agents and the risk for micro-organism resistance, consultation by infectious diseases specialists within a multidisciplinary team is of paramount importance. Initial antimicrobial therapy with broad-spectrum antibiotics is indicated to control infection and sepsis. In the presence of an aorto-GI fistula, antifungal agents should be considered. Once the responsible microorganisms are identified, a narrow, targeted antibiotic therapy is recommended, in order to avoid resistance development.

The ideal duration of the antimicrobial treatment is not defined and depends on the virulence of the responsible microorganisms and the applied surgical strategy. An antimicrobial therapy of 4–6 weeks is favored by many authors after completely explantation of the infected stentgraft [33]. In patients in whom due to general conditions any surgical eradication is precluded, a longer or even a lifelong antibiotic treatment should be considered [9]. Those patients could benefit from the long-term outpatient parenteral antimicrobial therapy (OPAT) regarding the quality of life [34].

Definitive palliative strategy with long-term suppression antimicrobial therapy could be an option for patients considered unfit for open surgery, owing to comorbidities or their current state of infection (e.g., sepsis) [35]. This strategy can also include percutaneous drainage with/without irrigation through CT-guided insertion of a pigtail catheter in the aneurysm sac, which can potentially reduce the local bacterial and infectious burden and provide directly obtained specimens for microbiological sampling [8, 35].

Conclusions

Aortic Endograft infection is a rare but potentially fatal complication. Endograft explantation and in situ reconstruction with cryopreserved homografts/autologous vein grafts represents the standard of care for fit patients. The operation is however physiologically demanding for both the patient and the surgical team. Therefore, the final decision of treatment strategy in this complex and highly demanding area must be tailored to the individual patient's situation, the local availability of graft material, and the surgical expertise.

Conflict of Interest None.

Sources of financial support: None.

References

1. Argyriou C, Georgiadis GS, Lazarides MK, Georgakarakos E, Antoniou GA. Endograft infection after endovascular abdominal aortic aneurysm repair: a systematic review and meta-analysis. J Endovasc Ther. 2017;24(5):688–97. https://doi.org/10.1177/1526602817722018.
2. Moulakakis KG, Dalainas I, Mylonas S, Giannakopoulos TG, Avgerinos ED, Liapis CD. Conversion to open repair after endografting for abdominal aortic aneurysm: a review of causes, incidence, results, and surgical techniques of reconstruction. J Endovasc Ther. 2010;17(6):694–702. https://doi.org/10.1583/1545-1550-17.6.694.
3. Smeds MR, Duncan AA, Harlander-Locke MP, Lawrence PF, Lyden S, Fatima J, et al. Treatment and outcomes of aortic endograft infection. J Vasc Surg. 2016;63(2):332–40. https://doi.org/10.1016/j.jvs.2015.08.113.
4. Chaufour X, Gaudric J, Goueffic Y, Khodja RH, Feugier P, Malikov S, et al. A multicenter experience with infected abdominal aortic endograft explantation. J Vasc Surg. 2017;65(2):372–80. https://doi.org/10.1016/j.jvs.2016.07.126.
5. Ducasse E, Calisti A, Speziale F, Rizzo L, Misuraca M, Fiorani P. Aortoiliac stent graft infection: current problems and management. Ann Vasc Surg. 2004;18(5):521–6. https://doi.org/10.1007/s10016-004-0075-9.
6. Moulakakis KG, Mylonas SN, Antonopoulos CN, Kakisis JD, Sfyroeras GS, Mantas G, et al. Comparison of treatment strategies

for thoracic endograft infection. J Vasc Surg. 2014;60(4):1061–71. https://doi.org/10.1016/j.jvs.2014.07.012.

7. Kahlberg A, Grandi A, Loschi D, Vermassen F, Moreels N, Chakfe N, et al. A systematic review of infected descending thoracic aortic grafts and endografts. J Vasc Surg. 2019;69(6):1941–51 e1. https://doi.org/10.1016/j.jvs.2018.10.108.
8. Moulakakis KG, Sfyroeras GS, Mylonas SN, Mantas G, Papapetrou A, Antonopoulos CN, et al. Outcome after preservation of infected abdominal aortic endografts. J Endovasc Ther. 2014;21(3):448–55. https://doi.org/10.1583/13-4575MR.1.
9. Setacci C, Chisci E, Setacci F, Ercolini L, de Donato G, Troisi N, et al. How to diagnose and manage infected Endografts after endovascular aneurysm repair. Aorta (Stamford, Conn). 2014;2(6):255–64. https://doi.org/10.12945/j.aorta.2014.14-036.
10. Cernohorsky P, Reijnen MM, Tielliu IF, van Sterkenburg SM, van den Dungen JJ, Zeebregts CJ. The relevance of aortic endograft prosthetic infection. J Vasc Surg. 2011;54(2):327–33. https://doi.org/10.1016/j.jvs.2010.12.067.
11. Wilson WR, Bower TC, Creager MA, Amin-Hanjani S, O'Gara PT, Lockhart PB, et al. Vascular graft infections, mycotic aneurysms, and endovascular infections: a scientific statement from the American Heart Association. Circulation. 2016;134(20):e412–e60. https://doi.org/10.1161/CIR.0000000000000457.
12. Chakfe N, Diener H, Lejay A, Assadian O, Berard X, Caillon J, et al. Editor's choice—European Society for Vascular Surgery (ESVS) 2020 clinical practice guidelines on the Management of Vascular Graft and Endograft Infections. Eur J Vasc Endovasc Surg. 2020;59(3):339–84. https://doi.org/10.1016/j.ejvs.2019.10.016.
13. Batt M, Feugier P, Camou F, Coffy A, Senneville E, Caillon J, et al. A meta-analysis of outcomes after in situ reconstructions for aortic graft infection. Angiology. 2018;69(5):370–9. https://doi.org/10.1177/0003319717710114.
14. Lyons OTA, Baguneid M, Barwick TD, Bell RE, Foster N, Homer-Vanniasinkam S, et al. Diagnosis of aortic graft infection: a case definition by the Management of Aortic Graft Infection Collaboration (MAGIC). Eur J Vasc Endovasc Surg. 2016;52(6):758–63. https://doi.org/10.1016/j.ejvs.2016.09.007.
15. Reinders Folmer EI, Von Meijenfeldt GCI, Van der Laan MJ, Glaudemans AWJM, Slart RHJA, Saleem BR, et al. Diagnostic imaging in vascular graft infection: a systematic review and meta-analysis. Eur J Vasc Endovasc Surg. 2018;56(5):719–29. https://doi.org/10.1016/j.ejvs.2018.07.010.
16. Lauri C, Iezzi R, Rossi M, Tinelli G, Sica S, Signore A, et al. Imaging modalities for the diagnosis of vascular graft infections: a consensus paper amongst different specialists. J Clin Med. 2020;9(5):1510. https://doi.org/10.3390/jcm9051510.
17. Legout L, D'Elia PV, Sarraz-Bournet B, Haulon S, Meybeck A, Senneville E, et al. Diagnosis and management of prosthetic vascular graft infections. Med Mal Infect. 2012;42(3):102–9. https://doi.org/10.1016/j.medmal.2012.01.003.
18. Reinders Folmer EI, von Meijenfeldt GCI, Te Riet Ook Genaamd Scholten RS, van der Laan MJ, Glaudemans A, Slart R, et al. A systematic review and meta-analysis of (18)F-fluoro-d-deoxyglucose positron emission tomography interpretation methods in vascular graft and endograft infection. J Vasc Surg. 2020;72(6):2174–85 e2. https://doi.org/10.1016/j.jvs.2020.05.065.
19. Bruls S, El Hassani I, Hultgren R, Hustinx R, Courtois A, Dumortier A, et al. [(18)F] FDG PET/CT can improve the diagnostic accuracy for aortic endograft infection. Acta Cardiol. 2021;77:1–9. https://doi.org/10.1080/00015385.2021.1949105.
20. Miyamoto K, Inaba M, Kojima T, Niguma T, Mimura T. Intra-aortic balloon occlusion (IABO) may be useful for the management of secondary aortoduodenal fistula (SADF): a case report. Int J Surg Case Rep. 2016;25:234–7. https://doi.org/10.1016/j.ijscr.2016.06.010.
21. Berard X, Brizzi V. Current Management of Aortic Endograft Infection: prepare your team for this new challenge. Eur J Vasc Endovasc Surg. 2019;58(4):624–5. https://doi.org/10.1016/j.ejvs.2019.05.016.
22. Popplewell MA, Garnham AW, Hobbs SD. A new technique to explant an infected aortic endograft. J Vasc Surg. 2015;62(2):512–4. https://doi.org/10.1016/j.jvs.2015.03.033.
23. Koning OH, Hinnen JW, van Baalen JM. Technique for safe removal of an aortic endograft with suprarenal fixation. J Vasc Surg. 2006;43(4):855–7. https://doi.org/10.1016/j.jvs.2005.12.011.
24. O'Connor S, Andrew P, Batt M, Becquemin JP. A systematic review and meta-analysis of treatments for aortic graft infection. J Vasc Surg. 2006;44(1):38–45. https://doi.org/10.1016/j.jvs.2006.02.053.
25. Post I, Vos N. Systematic review and meta-analysis on the Management of Open Abdominal Aortic Graft Infections. Eur J Vasc Endovasc Surg. 2019;58(6):e806–e7. https://doi.org/10.1016/j.ejvs.2019.09.399.
26. Chiesa R, Melissano G, Marone EM, Marrocco-Trischitta MM, Kahlberg A. Aorto-oesophageal and Aortobronchial fistulae following thoracic endovascular aortic repair: a national survey. Eur J Vasc Endovasc Surg. 2010;39(3):273–9. https://doi.org/10.1016/j.ejvs.2009.12.007.
27. Kahlberg A, Rinaldi E, Piffaretti G, Speziale F, Trimarchi S, Bonardelli S, et al. Results from the multicenter study on Aortoenteric Fistulization after stent grafting of the abdominal aorta (MAEFISTO). J Vasc Surg. 2016;64(2):313–20.e1. https://doi.org/10.1016/j.jvs.2016.04.008.
28. Czerny M, Eggebrecht H, Sodeck G, Weigang E, Livi U, Verzini F, et al. New insights regarding the incidence, presentation and treatment options of aorto-oesophageal fistulation after thoracic endovascular aortic repair: the European registry of endovascular aortic repair complications. Eur J Cardio-Thorac Surg. 2014;45(3):452–7. https://doi.org/10.1093/ejcts/ezt393.
29. Kahlberg A, Tshomba Y, Marone EM, Castellano R, Melissano G, Chiesa R. Current results of a combined endovascular and open approach for the treatment of Aortoesophageal and Aortobronchial fistulae. Ann Vasc Surg. 2014;28:1782. https://doi.org/10.1016/j.avsg.2014.06.003.
30. Canaud L, Ozdemir BA, Bee WW, Bahia S, Holt P, Thompson M. Thoracic endovascular aortic repair in management of aorto-esophageal fistulas. J Vasc Surg. 2014;59(1):248–54. https://doi.org/10.1016/j.jvs.2013.07.117.
31. Dorweiler B, Weigang E, Duenschede F, Pitton MB, Dueber C, Vahl CF. Strategies for endovascular aortic repair in aortobronchial and aortoesophageal fistulas. Thorac Cardiovasc Surg. 2013;61(7):575–80. https://doi.org/10.1055/s-0033-1347294.
32. Luehr M, Etz CD, Nozdrzykowski M, Garbade J, Lehmkuhl L, Schmidt A, et al. Emergency open surgery for aorto-oesophageal and aorto-bronchial fistulae after thoracic endovascular aortic repair: a single-Centre experience. Eur J Cardio-Thorac Surg. 2015;47(2):374–82; discussion 82–83. https://doi.org/10.1093/ejcts/ezu147.
33. Kahlberg A, Melissano G, Mascia D, Loschi D, Grandi A, Chiesa R. How to best treat infectious complications of open and endovascular thoracic aortic repairs. Semin Vasc Surg. 2017;30(2):95–102. https://doi.org/10.1053/j.semvascsurg.2017.11.002.
34. Allen N, Adam M, O'Regan G, Seery A, McNally C, McConkey S, et al. Outpatient parenteral antimicrobial therapy (OPAT) for aortic vascular graft infection; a five-year retrospective evaluation. BMC Infect Dis. 2021;21(1):670. https://doi.org/10.1186/s12879-021-06373-4.
35. Caradu C, Puges M, Cazanave C, Martin G, Ducasse E, Bérard X, et al. Outcomes of patients with aortic vascular graft and endograft infections initially contra-indicated for complete graft explantation. J Vasc Surg. 2022;76:1364. https://doi.org/10.1016/j.jvs.2022.05.007.

Part IX

Venous Disease

Endovenous Ablation Techniques

69

Dimitrios Kontothanassis

Introduction

Varicose veins are a very common condition and have been the subject of a recent proliferation of treatment modalities. During the last two decades, superficial venous surgery has experienced an extraordinary expansion. Traditional surgical procedures (crossectomy and stripping) are now being replaced to a greater or lesser extent by new less invasive endovenous methods but the advent of the endovenous treatment era has led to a confusing array of different modalities of treatment. All modalities offer excellent results in the right situation, and each has its own treatment profile. Thermal ablation techniques have matured and have a reassuring and reliable outcome, but the arrival of nonthermal techniques has delivered further options for both patient and surgeon. Our aim is to provide an up-to date review of all available endovenous techniques (laser, radiofrequency, steam, foam sclerotherapy, mechanochemical ablation and venous glue) describing the indications, the technique and mechanism of action, and the results. In experienced hands, all endovenous techniques are safe and effective, with long-term results comparable to conventional surgical procedures. Following NICE guidelines and recommendations, when discussing treatment for varicose veins at the vascular service we have to inform all patients with the following information's; what treatment options are available, the expected benefits and risks of each treatment option, that new varicose veins may develop after treatment, that they may need more than one session of treatment and that the chance of recurrence after treatment for recurrent varicose veins is higher than for primary varicose veins.

Chronic Venous Disease (CVD)

In the United States, an estimated 23% of adults have varicose veins, and 6% have more advanced chronic venous disease (CVD), including skin changes and healed or active venous ulcers [1]. Varicose veins have long been considered a cosmetic problem that only affected emotional well-being but were not the source of disability. Varicosities, however, are frequently the cause of discomfort, pain, loss of working days, disability, and deterioration of health-related quality of life (QOL)0.1–3 Severe CVD may also lead to loss of limb or loss of life (Fig. 69.1).

Clinical and instrumental evaluation are mandatory prior to surgery to determine all the sources of vein reflux, identification of anatomical variabilities of the veins, address the patient to the correct endovenous procedure, map the veins prior to surgery, assist all types of the endovenous techniques and follow up the outcomes of the procedures.

D. Kontothanassis (✉)
Department of Phlebology, Istituto Flebologico Italiano, Padova, Italy

General and Vascular Surgeon CEO Istituto Flebologico Italiano, Padova, Italy
e-mail: direzione@istitutoflebologico.it

G. Geroulakos et al. (eds.), *Mastering Endovascular Techniques*, https://doi.org/10.1007/978-3-031-42735-0_69

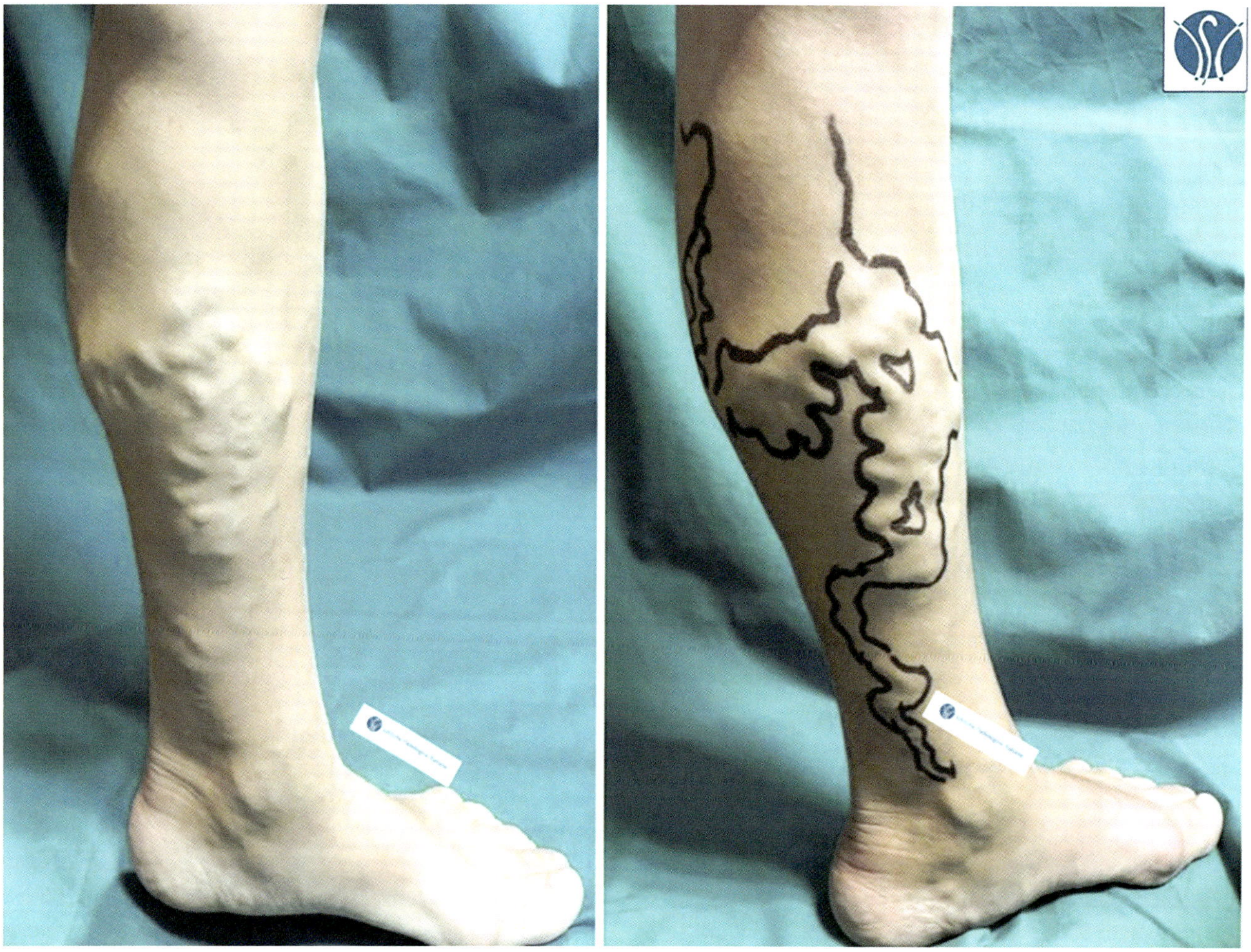

Fig. 69.1 Patient with CVD

Endovenous Ablation Techniques

All endovenous procedures can be grouped in two big categories:

Thermal and Tumescent Techniques and Non-thermal and Non tumescent Techniques

See Figs. 69.2, 69.3, 69.4

Thermal and Tumescent Techniques

Even if RFA was the pioneer endovenous technique introduced in 1998, Laser ablation introduced immediately after in 1999 spread globally and faster with very promising results. 810 nm diode Laser and 980 nm diode Laser using bare fibers dominate the global scientific scene for at least 5 years up to 2005 because results were much better that the first generation of RF generators. The second generation of RF generators (Closure Fast) achieved better results in terms of closure rates after saphenous ablation, reduction of postoperative pain and less bruising when compared with 810 and 980 diode Laser. Results were equalized again after the introduction of new wavelengths for Laser devices (1470 nm and 1900 nm) and development of new high-quality fibers (Radial optical fibers, single or double ring, normal or slim). Steam was introduced in 2011 but never spread globally; till today covers only a small part of thermal and tumescent procedures.

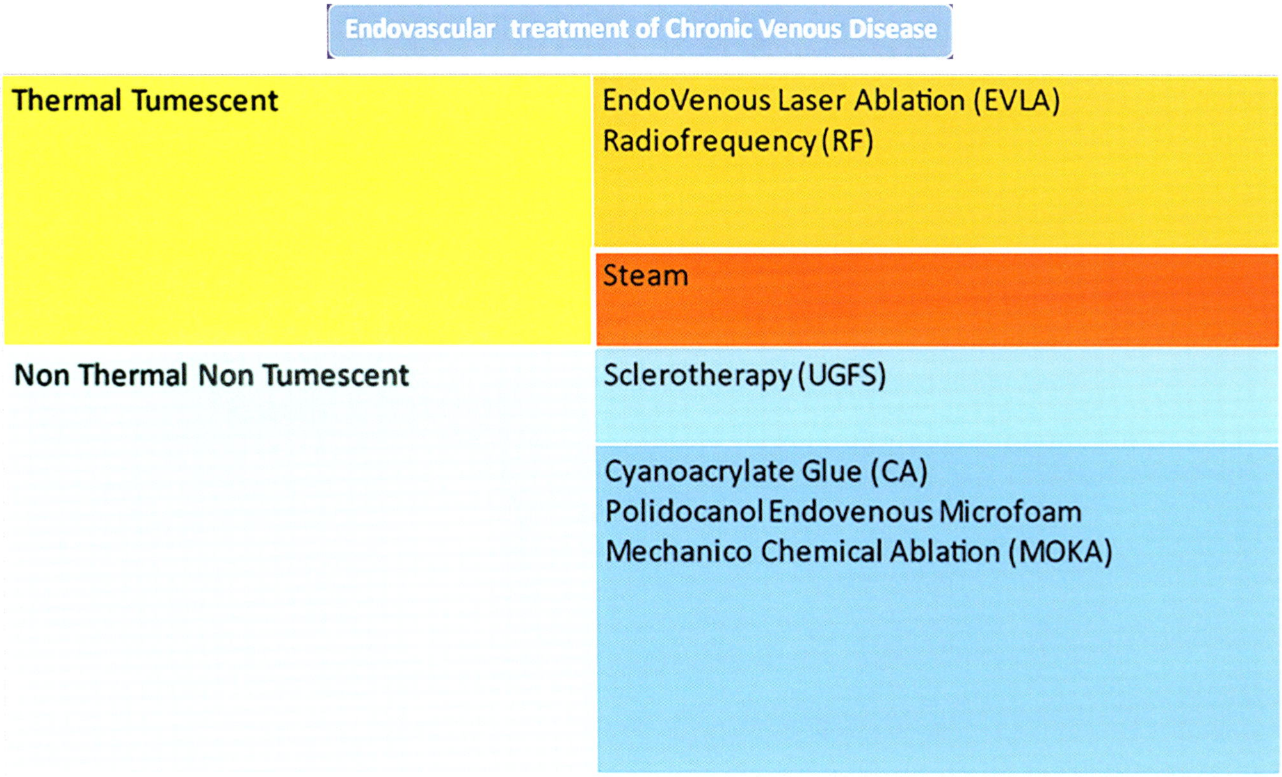

Fig. 69.2 Endovenous ablation techniques

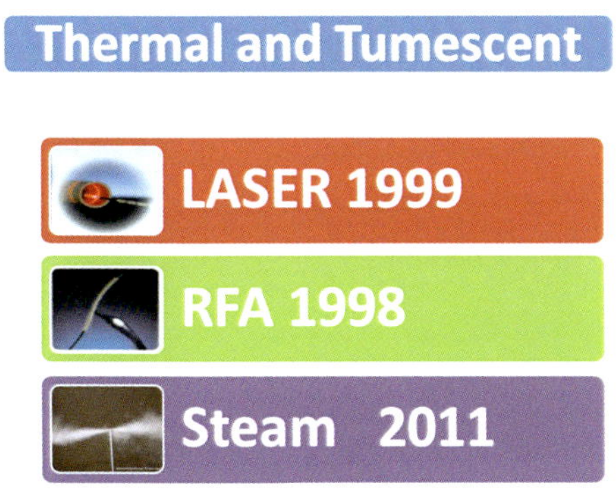

Fig. 69.3 Thermal and Tumescent: Laser, Radiofrequency and Steam ablation

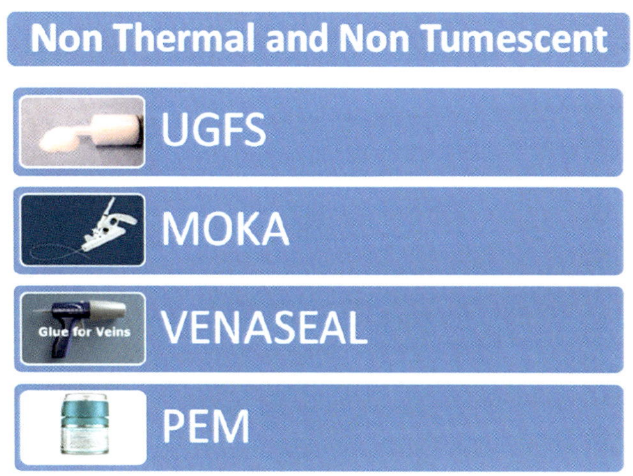

Fig. 69.4 Non Thermal and Non Tumescent: Ultrasound guided Sclerotherapy, PEM (Polidocanol microfoam), Meccanochemical Ablation (MOKA), and Cianoacrylate Glu

How It Works

All techniques can be offered as outpatient procedures that can be done under local anesthesia. Proper evaluation of patients' condition has to be done in order to offer a further safety program in high-risk patients. For patients that present emotional problems a combination of oral or endovenous mild sedation can be given additionally to local anesthesia. All techniques require ultrasound assistance in order to perform vascular access, navigate inside the vein, target and identify the correct point of entry into the vein, and landmark the point for starting a correct ablation. All techniques can be done by using a 16–18 gauge needle and inserting a guide wire or using a microvein puncture kit. Usually the sheath used for Laser is 6F and for Radiofrequency ablation is 6F or 7F; no sheath is demanded for steam procedure. Tumescent anesthesia is mandatory and large amounts of tumescent liquid have to be delivered around the vein in order to dissect it from the surround tissues and determine extreme spasm of the vein wall around the catheter (Figs. 69.5, 69.6, 69.7, 69.8, 69.9).

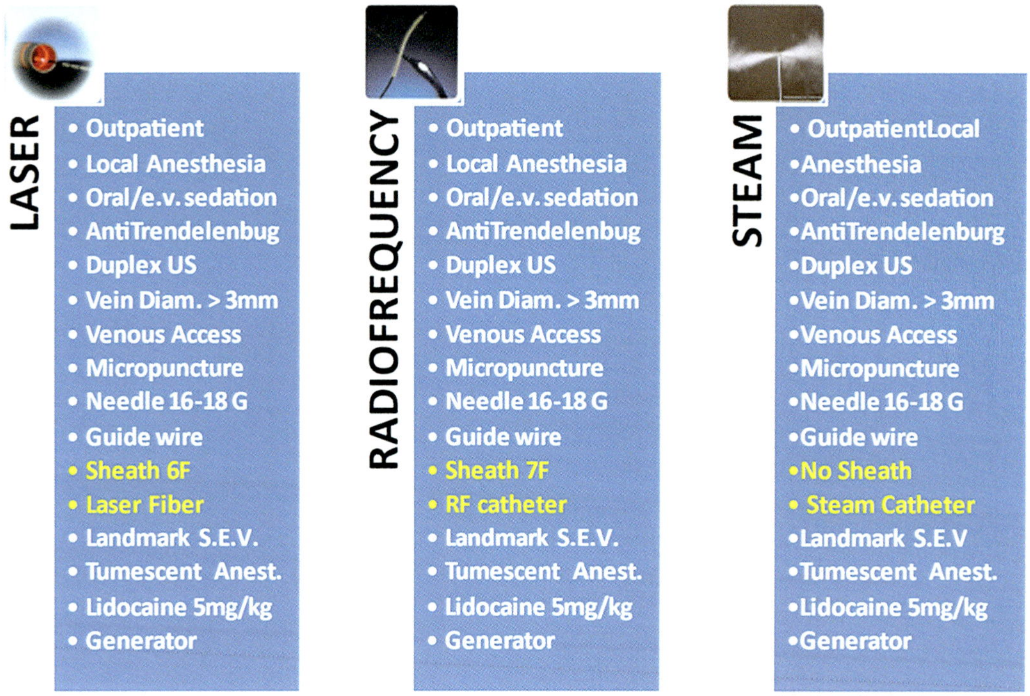

Fig. 69.5 How endovenous thermal techniques work

Fig. 69.6 Patient in Reverse Trendelemburg position

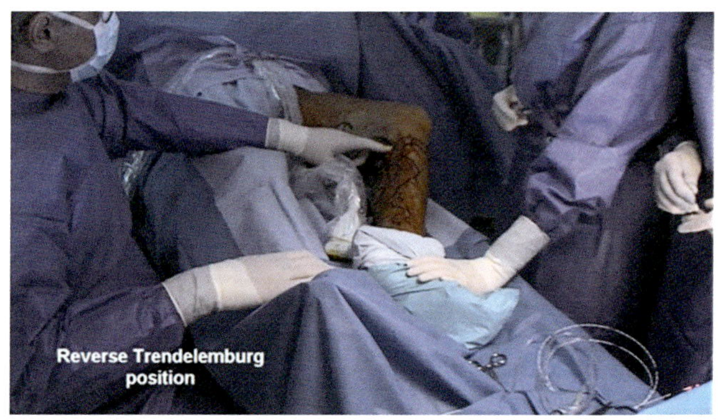

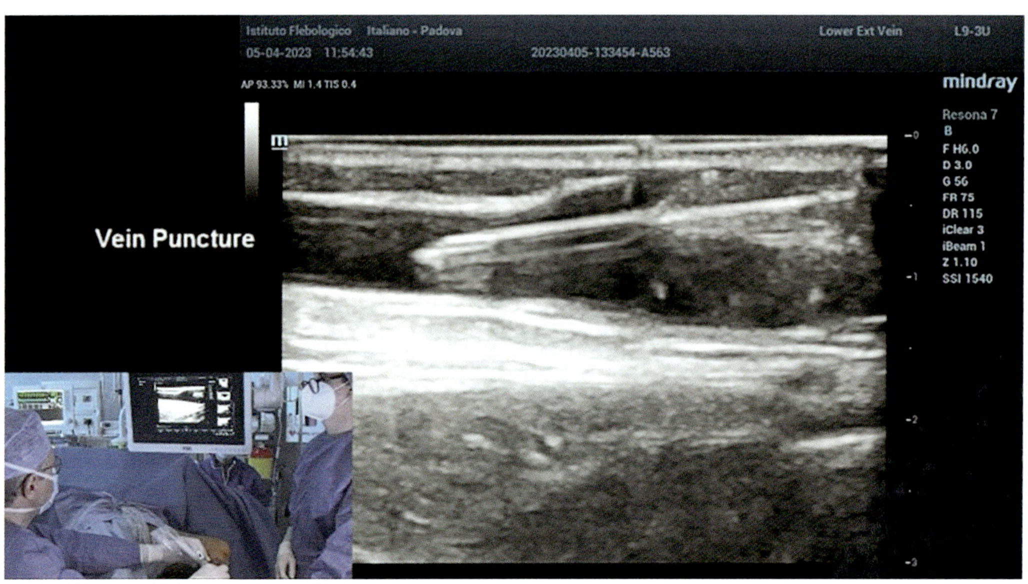

Fig. 69.7 Vein puncture

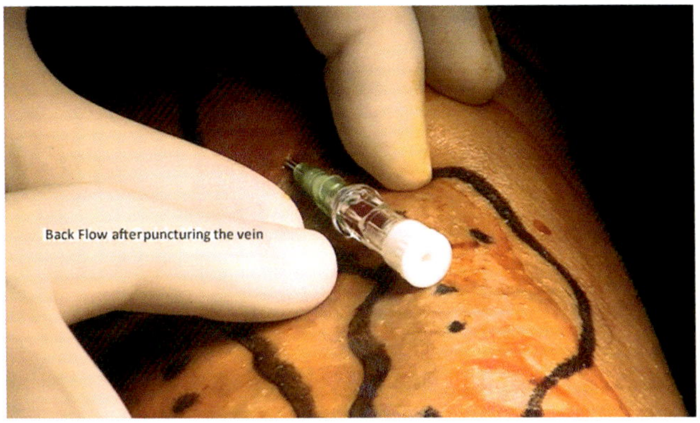

Fig. 69.8 Back flow after puncturing the vein

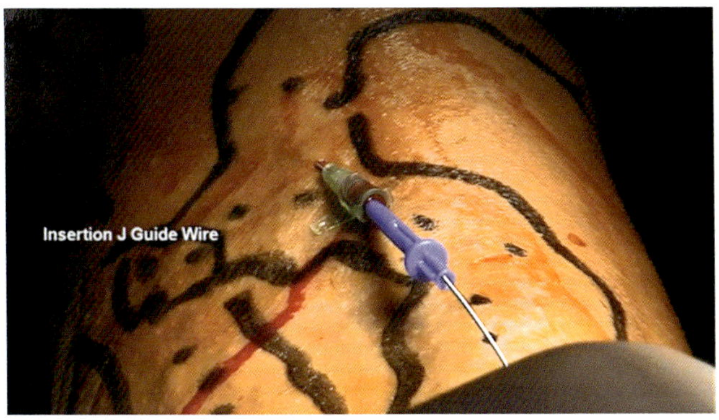

Fig. 69.9 Insertion J Guide Wire

Preoperative Evaluation of Patients

Patient selection is very important to achieve good results and prevent recurrence. Clinical, hemodynamic, anatomical, and morphological criteria must be respected. An accurate Color Flow Duplex Evaluation (CFDU) is strictly recommended before, during and after the procedure. It is important to define the CVD etiology (primary, secondary, both), the distribution of reflux (superficial system, deep system perforator veins or a combination of them) and the extension of the venous reflux (thigh, calf, or both). Inappropriate venous size (vein diameter at cross section in the standing position <3 mm or >25 mm), tortuosity of venous axis, superficiality of the vein segment to treat (<3 mm from the skin), clinical severity scores and reduced need for general or epidural anesthesia (Fig. 69.10).

Currently the large use of endovenous procedures have generate unrealistic patient expectation and many times correct information is lacking. Accurate information for the treatment of varicose veins with endovenous procedures should be given to all patients eligible for the procedure in order to warn them of the primary outcomes and possible complications. These are venous thromboembolism (VTE), superficial venous thrombosis (SVT), endovenous heat induced thrombus (EHIT), skin burns, hematoma, ecchymosis, skin pigmentation, pain, vein recanalization, neovascularization, technical failure or need for redo surgery. Quality

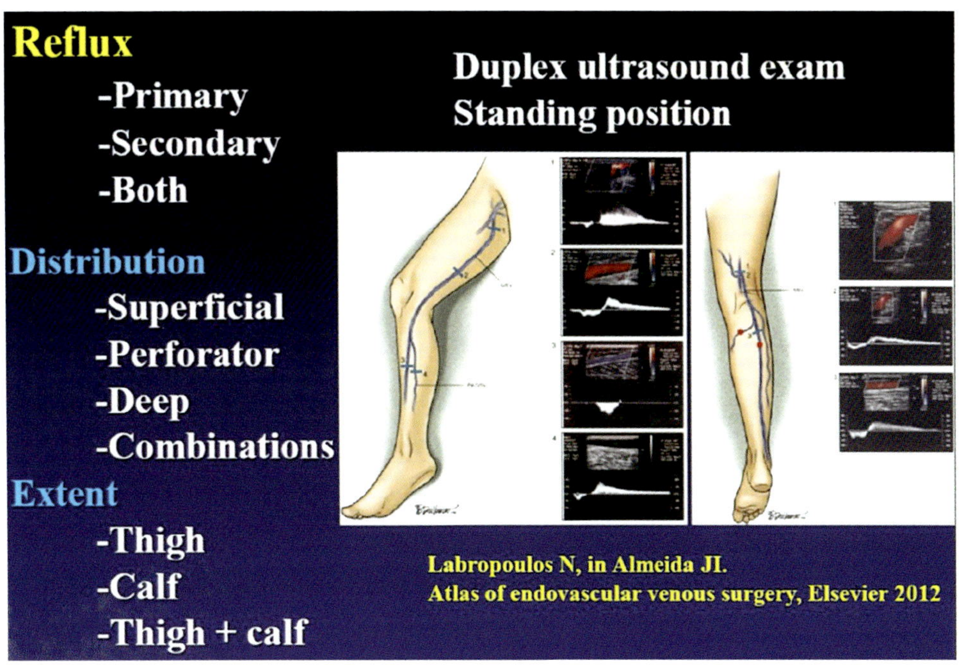

Fig. 69.10 Reflux evaluation; courtesy Prof. N. Labropoulos

of life scores help to compare results at the early-term, mid-term and long-term follow up. Physicians performing the procedure must be very experienced in vascular ultrasound and have good training in vascular and endovascular surgery.

Contraindications to Transcatheter Ablation

During the initial patient's examination, it is very important to assess the following selection criteria: **Clinical, Hemodynamical, Anatomical, and Morphological**. An accurate preoperative ultrasound scanning and vein mapping will help planning the correct treatment. All data will assist to avoid pitfalls and minimize the possibility of residual and recurrent varicose veins after surgery. The author recommends the use of high-resolution ultrasound performed by doctors with ultrasound skill. The use of dedicated vein software to collect patients' data will help to optimizes the postoperative results.

Special attention must be given to the following parameters:

- The presence of anatomical variability of saphenous veins (duplication, aplasia, hypoplasia, extra fascial saphenous veins),
- The distance between the vein and the skin (>3 mm).
- The presence of dilatation and tortuosity of the saphenous vein axis (do not treat veins with a diameter >25 mm unless you are a very skill in EVLA).
- Assess the integrity of the deep venous system by excluding the presence of deep venous thrombosis (DVT) or venous narrowing after DVT).
- Assess before the operation patency of the great saphenous vein (GSV) and the small saphenous vein (SSV) (Fig. 69.11).

Preoperative Vein Mapping

EVLA, and RFA, and Steam aim to abolish varicose veins and vein reflux. For this reason, every procedure should start with a preoperative venous mapping to document the findings of the ultrasound study and mark the skin. A note and a picture of what is documented before the operation is mandatory and can be a reminder at the moment of surgery (Fig. 69.12). A permanent ink skin marker is used to draw on the skin varicose veins that are visible and palpable. The author suggests the use of Laser transilluminators in order to map on the skin veins that are not visible or palpable during clinical examination. The use of duplex ultrasound should be performed always with the patient in the clinostat position before starting the operation to verify that all the veins are evidenced and marked, control again the presence of tortuosity and dilatation and choose the correct puncture site for the vascular access.

A hybrid approach, in which EVLA is associated with sclerotherapy and open surgery, can be required.

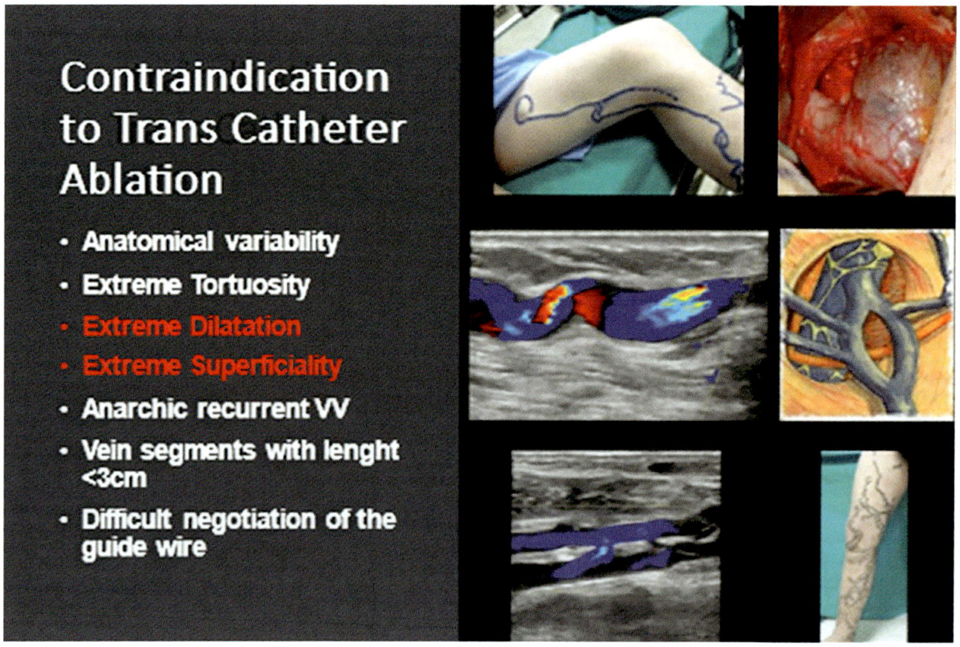

Fig. 69.11 Contraindication to trans catheter ablation; Courtesy Prof. D. Kontothanassis

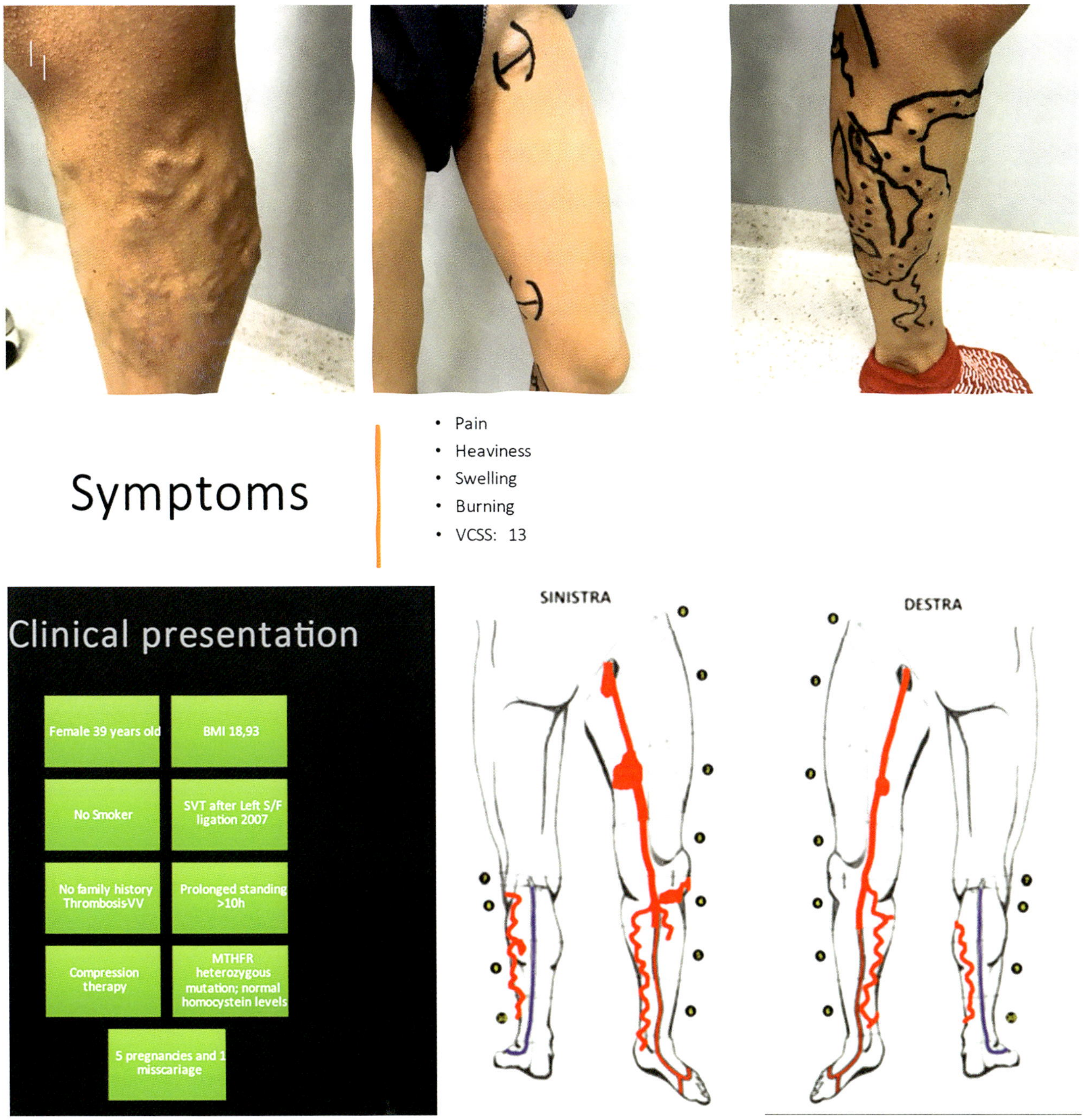

Fig. 69.12 Vein drawing and mapping; courtesy Prof. D. Kontothanassis

General Recommendations

A global gentle handling will help you to minimize the stress of the patient. For this reason, you have to inform the patient and explain precisely the procedure. A warm atmosphere during the surgical gesture and limiting the waiting before and after the procedure will help the patient to be more comfortable and secure.

Informed Consent

Minimally invasive techniques such as LASER (Light Amplification by Stimulated Emission of Radiation, RFA (Radio Frequency Ablation) and Ultrasound Guided Foam Sclerotherapy (UGFS) for treating saphenous varicose veins have been rapidly introduced in the last 20 years and compared with traditional surgery. Proposed benefits are fewer complica-

Table 69.1 Percentage of UK-based varicose vein websites warning of specific complications following the treatment of varicose veins

Complication	Surgery ($n = 32$)	EVLA/RFA ($n = 31$)	UGFS ($n = 27$)
Cutaneous nerve injury	56%	42%	–
Venous thrombo-embolism	38%	39%	48%
Recurrence	56%	42%	41%
Infection	53%	–	–
Bleeding/haematoma	41%	–	–
Bruising	59%	42%	–
Thrombophlebitis	–	23%	48%
Pigmentation	–	13%	59%
Ulceration	–	–	41%
Burns	–	35%	–
Visual disturbance[a]	–	–	26%
Allergy/anaphylaxis	–	–	26%

[a]Including migraine, scotoma and/or other neurological events

tions, quicker return to work, improved quality of life. Over 50% of the websites examined did not mention all the management methods now available for varicose veins. More importantly, the majority of the websites did not warn of the common complications of intervention. Currently, information on the Internet cannot be relied upon to supplement informed consent and may actually generate unrealistic patient expectations [2].

The General Medical Council reminds doctors that the process of gaining informed consent must include a discussion of 'options for treatment or management of the condition, including the option not to treat' and that 'for each option, explanations of the likely benefits and the probabilities of success; and discussion of any serious or frequently occurring risks' [3]. This means that patients should be informed of the full range of options available, including techniques not necessarily offered by their specialist. It is, therefore, disappointing to note that over half of the websites reviewed in this study did not include information relating to all the treatment methods that are now available. Furthermore, the majority of the websites did not alert the reader to the common complications of varicose vein treatment [4] (Table 69.1).

How to Perform Tumescent Anesthesia for Thermal Techniques

For thermal techniques, you need tumescent anesthesia. A peristaltic pump instead of a syringe will help rapidly infuse tumescent anesthetic fluid down the course of the vein and its collaterals.

In this way, the procedure is faster and fewer air bubbles are introduced at the subcutaneous tissue (Image 7).

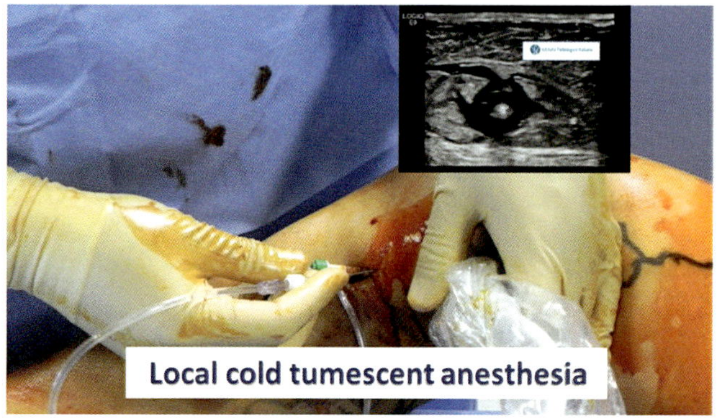

Fig. 69.13 Tumescent anesthesia; courtesy Prof. D. Kontothanassis

Refrigerate the tumescent solution at 4 °C degree has an analgesic effect and increases the spasm of the saphenous vein around the catheter. It is important that no air is introduced with the tumescent anesthesia, in particular at the level of the saphenous junction with deep veins. Air precludes ultrasound guidance and consequently prevents from checking the correct position of the fiber tip, that is necessary to avoid DVT or EHIT. The tumescent fluid (cold saline 500 cc and Lidocaine 2% 10 mL), must be enough to separate the saphenous vein from the perivenous tissue, induce a severe vein spasm and encircle the vein around the fiber. Inject always with a long needle from distally to proximally; a mild limb elevation is suggested to evacuate rapidly the blood from the superficial vein system into the deep vein system and produce more venous spasm.

In case you have to treat more than one vein with Laser and you want to preserve your vascular access (Anterior saphenous vein, duplicated GSV, etc.), we suggest to introduce the sheath or at least the guide wire in all veins that have to be treated before starting tumescent anesthesia because of the venous spasm that cold saline will produce (Fig. 69.13).

Treating Below Knee Saphenous Veins

During harvest of the saphenous vein (SV), the most important relationship to take into account is the saphenous nerve (SN) to avoid pain and paresthesias after surgery. During harvest of the SV, the most vulnerable area is the inferior third of the leg because of vein and nervous adhesion [5].

Murakami et al. [3] in 1994 demonstrate the anatomical relationship between the saphenous veins and cutaneous nerves after investigation in 148 lower limbs of 74 cadavers in order to re-examine the stripping technique for treatment of varicose veins. GSV frequently ran intimately along the

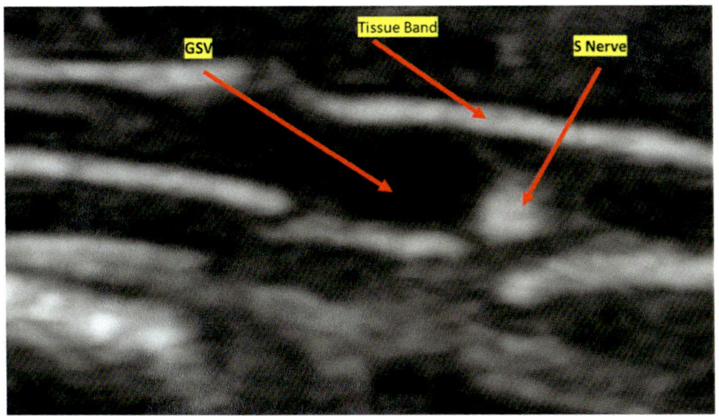

Fig. 69.14 Anatomy of GSV e Saphenous Nerve; courtesy Prof. D. Kontothanassis

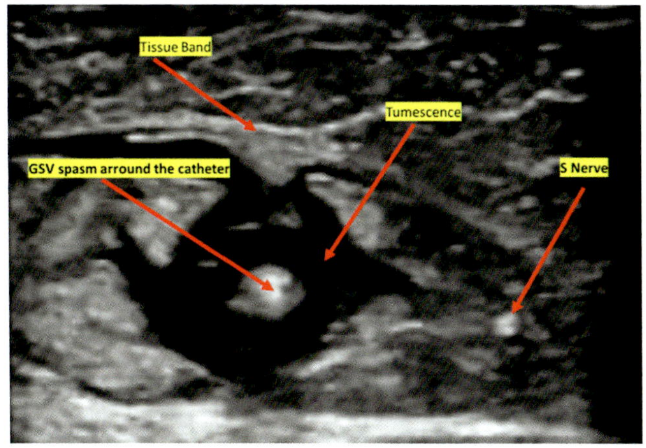

Fig. 69.15 How to prevent saphenous nerve injury with tumescent anesthesia; courtesy Prof. D. Kontothanassis

saphenous nerve (59.5% in the middle third and 83.1% in the lower third of the leg) in the leg region. More than half of the latter cases showed an adhesive relationship in which the perineurium of the saphenous nerve was seen histologically to be attached to the adventitia of the vein. Moreover, in the thigh region as well, GSV had an intimate relationship with the cutaneous nerves of various origin. In contrast SSV was often located close to the cutaneous nerves of the calf like the sural nerve. However, the adhesive relation between the small saphenous vein and nerves was rarely observed, in contrast to the case of the great saphenous vein. Based on those observations, limited extraction of the vein was recommended to reduce the risk of nerve injury during the stripping operation.

A different point view in 2007 was published by Kostas, Katsamouris [6] et al., remarking that the appropriate length of great saphenous vein stripping should be based on the extent of reflux and not on the intent to avoid saphenous nerve injury. Though SNI may occur after both restricted and total GSV stripping, this does not influence limb disability since any related symptoms seem to regress in almost half of the limbs 5 years postoperatively. Additionally, it seems that recurrence could be reduced in the tibial area if the level of GSV stripping complies with the extent of the ultrosonographically proven GSV reflux. Therefore, the extent of GSV stripping should not be guided by the intent of avoiding SNI.

Theivacumar et al. [7] investigate if standard above-knee great saphenous vein ablation provides optimum results in patients with both above- and below-knee reflux. From this randomized control trial, the authors conclude that extended Endovenous Laser Ablation (EVLA) is safe, increases spontaneous resolution of varicosities, and has a greater impact on symptom reduction. Similar benefits occurred after concomitant BK-GSV foam sclerotherapy (Figs. 69.14 and 69.15).

Tips for getting best results in case of below knee ablation we recommend just before tumescent anesthesia to keep the limb elevated >10 degrees, use plenty of tumescence with cold saline at 4° C and inject on both sides for separating the nerve. Pull back time of the catheter and energy delivered according to site and vein size.

Cleaning and Wrapping

At the end of the operation the limb is cleaned and wrapped with an elastic bandage for about 24 hours. Afterwards, graduated elastic compression stockings (Class 2, 30–40 mmHg) is needed for at least 7 days.1 There is still lack of consensus over the optimum extension and type of compression that should be applied at present (2013 meeting of the international union of phlebology-UIP). An ice pack can be used immediately on the limb and for 2–3 days after the procedure to relief pain and bruising. Precise recommendations about thrombus prophylaxis are still lacking in the literature: in the author's experience, patients are encouraged to walk immediately after the procedure and all patients receive low-molecular-weight heparin (LMWH) prophylaxis to reduce the risk of DVT and VTE; no case of DVT or EHIT was diagnosed in their treated patients. The mean time for patient discharge from the ambulatory office is 4 hours (range 3–5 h) according to the Italian law (at least 3 h of observation after the procedure). Patients complain of mild or no pain and discomfort after the procedure and return to work in a few days. At 2 days following the procedure, the author always performs clinical examination and CFDU evaluation to confirm saphenous vein occlusion and lack of DVT or other complications of the procedure. A regular clinical and CFDU follow-up is performed at 2 months and then once a year to evaluate the outcomes of the treatment (Figs. 37.3, 37.4; Fig. 69.16).

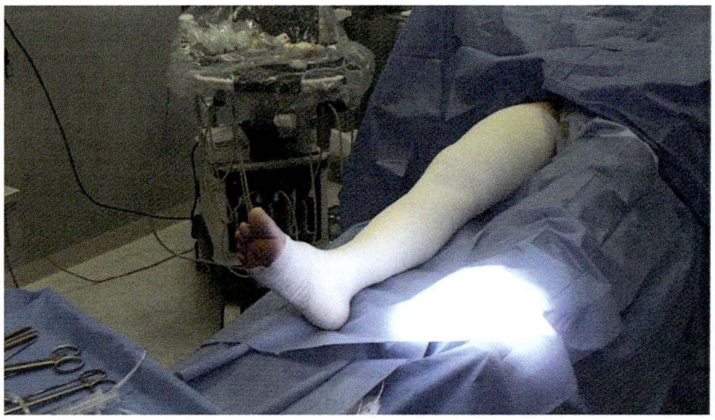

Fig. 69.16 Limb wrapped with a cohesive bandage after GSV ablation

Non-thermal and Non Tumescent Techniques

Thermal techniques are associated with earlier return to normal activity and less pain, as well as enabling procedures to be carried out as day cases. They are, however, also known to cause several side effects and involve infiltration of tumescent fluid which can cause discomfort [8]. Non-thermal, non-tumescent methods are believed to be the answer to these unwelcome effects. Ultrasound-guided foam sclerotherapy is one such non-thermal, non-tumescent method and, despite a possible lower occlusion, has been shown to improve the quality of life of patients. The early results of two recently launched non-thermal, non-tumescent methods, mechanochemical ablation and cyanoacrylate glue, are promising and are discussed.

However, the length procedure is shorter for non-thermal modalities and patient comfort is improved with lower pain scores. Return to work may also be earlier after non-thermal ablation. The rates of bruising, phlebitis and paresthesia are higher after thermal ablation. The non-thermal modalities are safe and effective in treating venous reflux and have shown improved patient comfort and shorter length of procedure which may make them favorable for use compared to the thermal modalities [9].

Conclusions and Considerations

Long follow up (>10 years) studies Our conclusions are limited due to the relatively small number of studies or each comparison and differences in outcome definitions and time points reported. Technical success was comparable between most modalities. EVLA may offer improved technical success compared to UGFS or HL/S. HL/S may have improved technical success compared to UGFS. No evidence of a difference was detected in recurrence, except for a possible long-term benefit for RFA compared to EVLA or HL/S [10].

Studies which provide more evidence on the breadth of treatments are needed. Future trials should seek to standardize clinical terminology of outcome measures and the time points at which they are measured. (1 E 2).

However, some goals from the literature can be recommended to doctors and patients. CVD has a wide range of clinical presentation and gets worse over time. Most people suffer from superficial vein reflux and clearly there is no prophylaxis or cure. Treatment of CVD reduces ulcer recurrence and early treatment of symptomatic CVD, improve signs, symptoms and HRQOL and finally slow down progression of disease. All techniques used for treating CVD cannot prevent recurrence of CVD in the follow up. For this reason, patients should control weight, improve mobility, and wear compression if applicable.

References

1. Farah MH, et al. A systematic review supporting the Society for Vascular Surgery, the American Venous Forum, and the American Vein and Lymphatic Society guidelines on the management of varicose veins. J Vasc Surg Venous Lymphat Disord. 2022;10(5):1155–71.
2. Scurr JRH. www.Accurateinformationforvaricoseveinpatients.com. Ann R Coll Surg Eng. 2008;90(7):554–6.
3. Murakami G, Negishi N, Tanaka K, Hoshi H, Sezai Y. Anatomical relationship between saphenous vein and cutaneous nerves. Okajmas Folia Anat. 1994;71(1):21–33.
4. General Medical Council. Seeking patients' consent: the ethical considerations. London: GMC; 1998. Available at (Table 1).
5. Surgical Anatomy of the Saphenous, Nerve Dayan V, Cura L, Cubas S, Carriquiry G. Ann Thorac Surg. 2008;85(3):896–900.
6. Kostas TT, Katsamouris AN. The appropriate length of great saphenous vein stripping should be based on the extent of reflux and not on the intent to avoid saphenous nerve injury. JVS. 2007;46(6):1234.
7. Theivacumar NS, et al. Endovenous laser ablation: does standard above-knee great saphenous vein ablation provide optimum results in patients with both above- and below-knee reflux? A randomized controlled trial. JVS. 2008;48(1):173–8.
8. Bootun R, Lane TRA, Davies AH. The advent of non-thermal, non-tumescent techniques for treatment of varicose veins. Phlebology. 2016;31:5–14.
9. Juneja AS, et al. Scoping review of non-thermal technologies for endovenous ablation for treatment of venous insufficiency. J Cardiovasc Surg (Torino). 2021;62(5):413–9.
10. Whing J, et al. Interventions for great saphenous vein incompetence. Cochrane Database Syst Rev. 2021;8(8):CD005624.

Endovenous Laser Ablation Therapy for Chronic Venous Disease

70

Dimitrios Kontothanassis

Case Presentation

A 35-year-old male was referenced to our clinic for severe CVD of longstanding: for about 8 years. The patient had a distal calf DVT in 2015 on the left limb after trauma and was treated for 3 months with LMWH therapy with total resolution of the thrombus. No family history of thrombosis was found, and he tested negative for thrombophilia. He has been complaining of bilateral leg swelling and worsening pain at the left limb for approximately 3 years even wearing a second-class elastic stocking. For this reason n 2019, he discussed with a vascular surgeon the option of a conservative treatment of the left GSV by performing calf phlebectomies in local anesthesia and he underwent surgery.

Clinical improvement obtained from conservative surgery and compression therapy was really poor in 3 years follow up and given his activity level and his good life expectancy an intervention was discussed with our team in March 2023 after visiting our clinic. Reexamination of both limbs was done in order to exclude potential post thrombotic syndrome or obstruction of deep veins. (Figs. 70.1, 70.2, 70.3, 70.4)

Continued at page 746

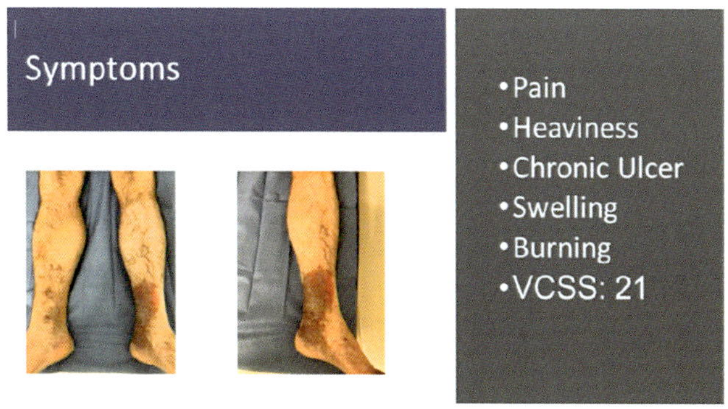

Fig. 70.1 Case presentation

Fig. 70.2 Hemodynamic map of GSV, SSV and varicosities

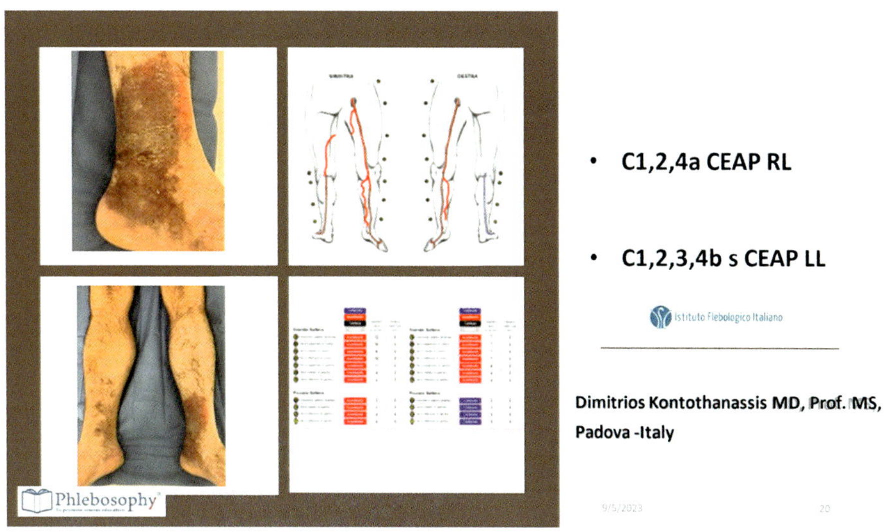

D. Kontothanassis (✉)
Department of Phlebology, Istituto Flebologico Italiano, Padova, Italy
e-mail: direzione@istitutoflebologico.it

G. Geroulakos et al. (eds.), *Mastering Endovascular Techniques*, https://doi.org/10.1007/978-3-031-42735-0_70

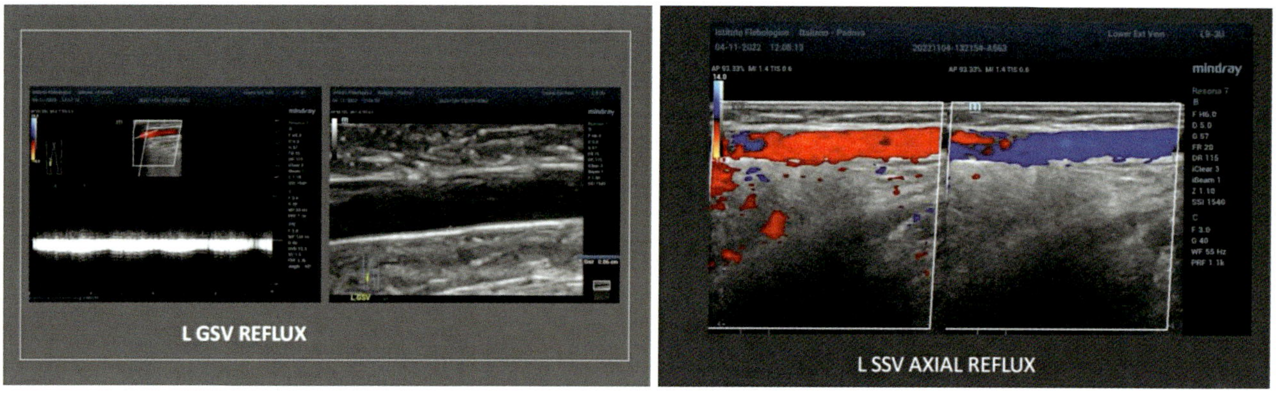

Figs. 70.3 Axial reflux of L GSV and L SSV

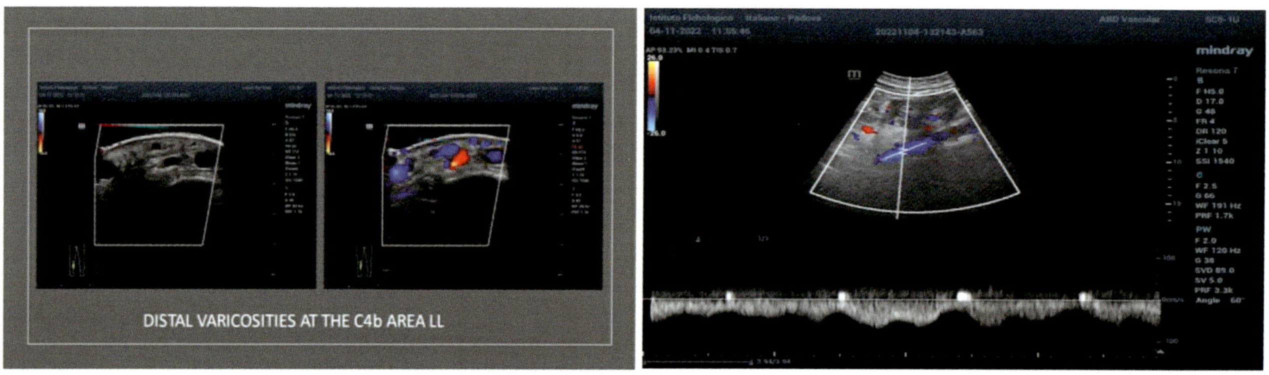

Figs. 70.4 Distal varicosities at the LL and normal phasic flow at the L IV

Introduction

Laser ablation treatment for lower limb venous incompetence is a very common technique which has been developing since the late 1990s following the pioneering work of Bone, Navarro, Min and Zimmet [1].

Although Endovenous Laser Ablation (EVLA) is considered as one of the strongest recommendations for the treatment of saphenous incompetence [2–6], (Grade 1B), EVLA results in the literature present significant fluctuations regarding accuracy, occlusion rate, early recanalization, postoperative complications, and recurrence in the follow up. Step by step procedure was already published by the Italian Endovenous Laser Working Group [3] IEWG. However, treatment parameters, Laser wavelengths, optical fibers and long-term results still present a source of controversy and debate. In this chapter we analyze the most common problems encountered with this technique, and our suggestions regarding how to overcome them.

Materials

To perform EVLA you need a Laser system with a suitable delivery kit (fiber optic and introducer set). Over the last 20 years two different types of Laser have been used: blood absorbed Lasers (wavelengths from 810–980 nm) and water absorbed Lasers (wavelength 1470 nm). The last ones have higher and more specific absorption on the vein wall allowing the use of less power and energy that leads to a significant reduction in pain and discomfort for the patient. The optical fiber must be thin, flexible, and glittery to allow an endovenous navigation even through mild tortuous veins. The author suggests the use of radial emission fibers which spread laser energy at 360° into the vein wall. The needle and the guide wire must be of high-quality materials in order to guarantee high echogenicity and negotiate easy introduction into the vein.

Recently there are encouraging reporting data in literature for Laser systems emitting at long wavelengths, >1900 nm that showed high occlusion rates, significant improvement in VCSS, low pain levels and low analgesic requirement, and early convalescence. However, these promising results are restricted at 3 years follow up when long term results, >10 years are mandatory.

Positioning the Patient on the Table

Position the patient on the table in reverse Trendelemburg position, with the leg slightly abducted and a tourniquet placed at about 40 cm above the puncture sight in order to fill the vein; with the cooperation of the anesthesiologist, keep the patient well hydrated and calm in order to avoid vein spasm (Fig. 70.5).

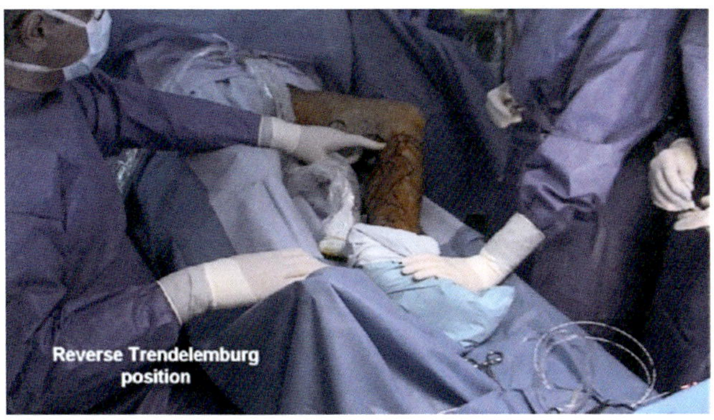

Fig. 70.5 positioning the patients at the reverse Trendelemburg position; courtesy Prof. D. Kontothanassis

How to Perform the Vein Puncture

It is mandatory to perform the vein puncture under ultrasound guidance. The puncture is easier if the vein is large and dilatated, but sometimes the vein can collapse in nervous patients, going into venospasm if touched with the needle of local anesthesia or after an unsuccessful first step with the needle. In case of venospasm, try to re-puncture the vein immediately above the venospasm. Nervous patient invite cathecolamine—induced venospasm, so consider mild sedation or propofol infusion should be consider for them. If you have problems with the transversal ultrasound vision, try with the longitudinal one. Small veins are best approached with a micropuncture kit. If wire passage is difficult, spasm can be reduced by using a hydrophilic-coated wire. If the percutaneous technique is not at all possible and the vein is superficial, you can isolate it under ultrasound guidance with a small incision of 1–2 mm.

Decision of the Entry Access Point

Puncture always the saphenous trunk and not a tributary to get access into the trunk. The best place is distally to the refluxing and dilatated vein segment and in most of the cases is above or below the knee; in case of extended reflux and dilatation of the vein axis at the malleolus, the entry point should be at this level (Figs. 70.6 and 70.7).

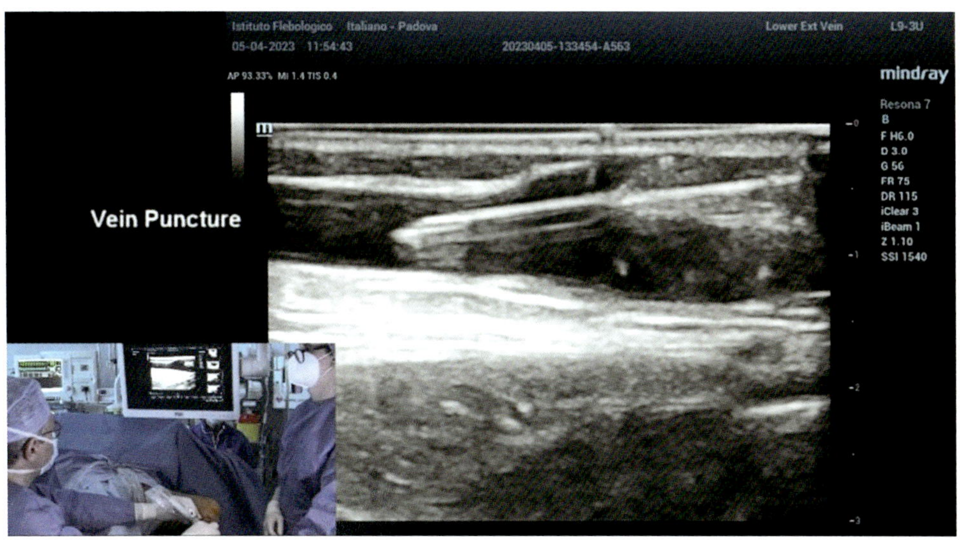

Fig. 70.6 Vein puncture; courtesy Prof. D. Kontothanassis

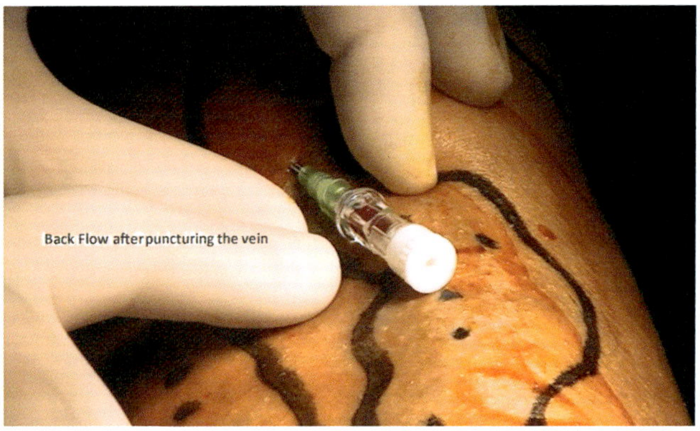

Fig. 70.7 Back Flow after puncturing the vein; courtesy Prof. D. Kontothanassis

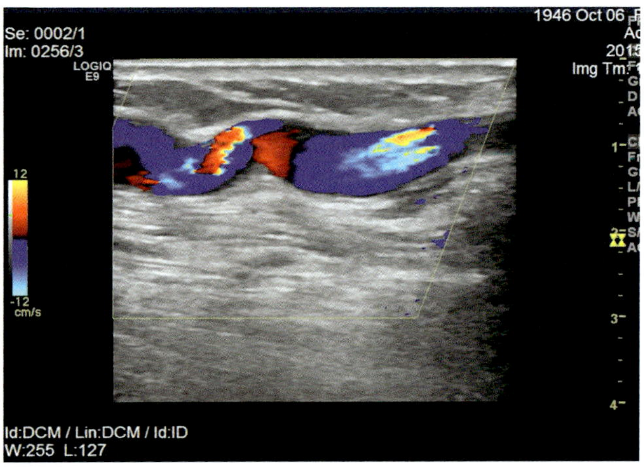

Fig. 70.8 Tortuous GSV; difficult negotiation of the guide wire; courtesy Prof. D. Kontothanassis

Tortuous Vein Axis

After a successful vein puncture and introduction of the first few centimeters of the J guide wire, a difficult negotiation of the wire can be managed and easily resolved by using a micropuncture nitinol wire with a free flow tip (0.18/0.25). Remove the tourniquet, as this may obstruct the wire in the superficial veins. Turning the wire between the operators' fingers and cautious withdrawal and advance may pass the obstructing tortuosity. In case of extremely dilatated veins, another trick is to fully elevate and abduct the leg or ben the knee while turning the wire; another suggestion is to advance the sheath to the point of obstruction and withdraw and advance the wire again. Verify always with the ultrasound the cause that determines the arrest: it can be a tortuosity, a dilatation, the origin of a perforator vein or due to a previous thrombosis. (Figs. 70.8 and 70.9).

Sheath is Hard to Pass

Enlarge the entry point of the wire with a scalpel; advance the sheath through the skin with decision by handling it towards the tip. Never try to advance the sheath without the guide wire inserted into the vein.

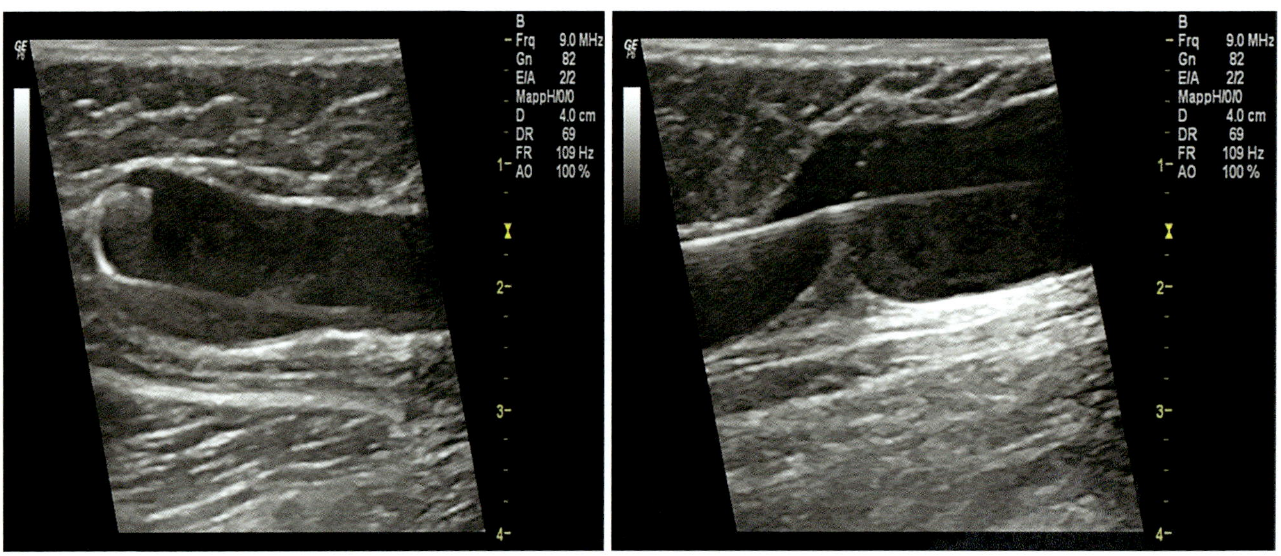

Fig. 70.9 J guide wire inside a tortuous vein; courtesy Prof. D. Kontothanassis

How to Continue if the Sheath or the Fiber is Difficult to Pass

Make a double cannulation of the treated vein, with a new vein puncture immediately above the arrest point, repositioning the J guide wire and the introducer sheath. (Fig. 70.10).

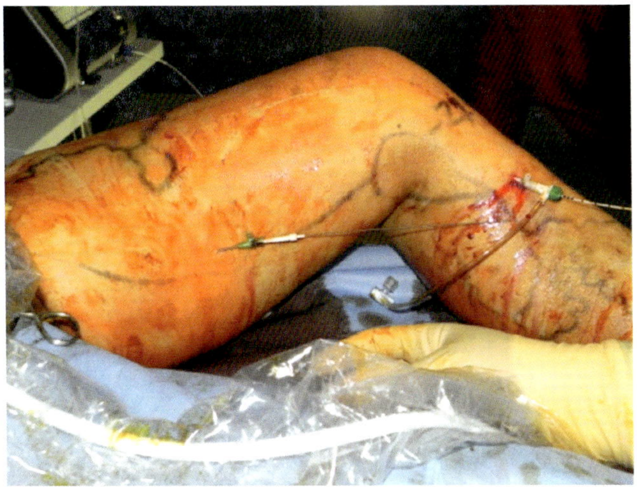

Fig. 70.10 double cannulation of the treated vein to by-pass the arrest point; courtesy Prof. D. Kontothanassis

How to Verify the Correct Position of the Fiber at the Junction

The safety of the procedure depends on operator proficiency. Use both Bright mode (B-mode) and color Doppler US scanning optimization to verify the fiber position at the groin or at the popliteal fossa. The fiber tip is always positioned gently and just below the epigastric vein in case of GSV treatment or wherever the terminal SSV joins the deep venous system. If in doubt, usually in obese people, low flow scanning modalities as the Blood -flow (B-Flow), Power Doppler (PW) or MicroV modality, always gives a better imaging. (Fig. 70.11a,b).

If the epigastric vein is hard to be detected, under ultrasound guidance press with your hand above the groin to enhance the epigastric flow into the femoral vein; it will increase the vein flow and make the vein visible. If you cannot see again the epigastric vein the check the position of the fiber tip if it is positioned over the epigastric vein or into the epigastric vein; this is very important because in case of ablation it will increase the risk of endovenous heat induced thrombosis (EHIT) or DVT.

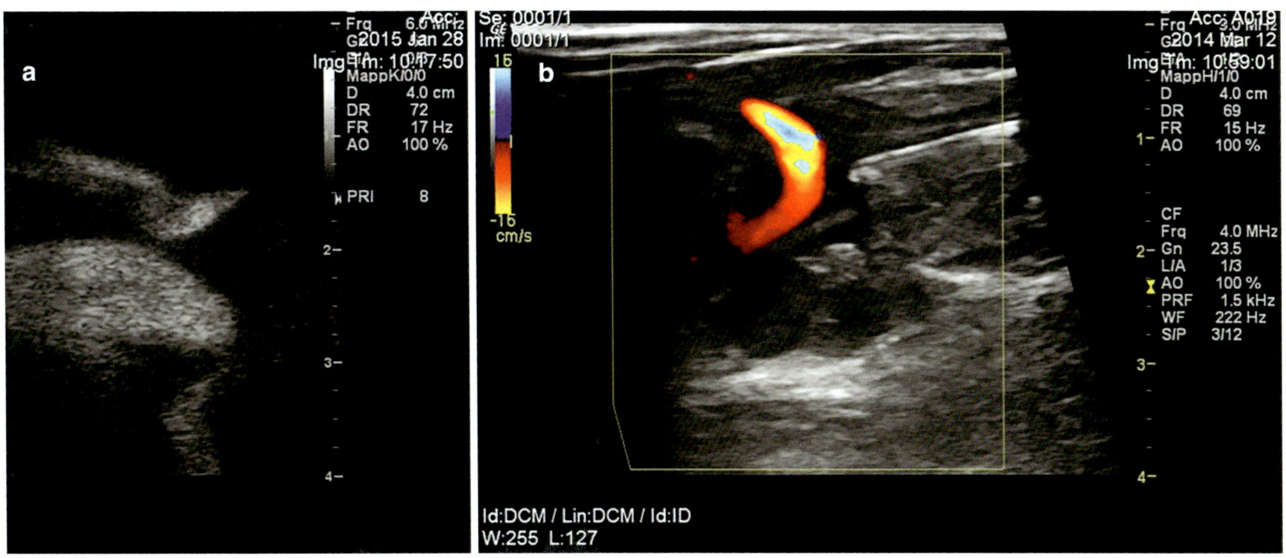

Fig. 70.11 (**a**, **b**) Intraoperative Bflow and color ultrasound imaging of the correct position of the fiber at the SFJ; courtesy Prof. D. Kontothanassis

How One Guarantees the Safety of the Procedure when one cannot See Anything

The Laser beam light can assist to identify the correct position of your Laser tip at the groin but it is not safe. Premeasuring your sheath against skin landmarks reassures that one is near the target site but you will never be sure of this. Safety procedures should always be assisted with ultrasound. Please consider re-scheduling the operation if there are difficulties to proceed.

How Much Energy Dispence

The amount of energy depends on the individual vein. Thickened vein walls due to severe chronic venous disease or to a previous thrombosis require an increased amount of energy compared to a normal vein. If the vein is hard to spasm during the tumescent anesthesia often more energy needs to be administrated. Ultrasound guidance for energy delivery should be used. Following the ablation with the ultrasound allows one to understand when the ablation reaches the vein wall; the venous lumen becomes hyperechogenic at the ultrasound B-mode. (Fig. 70.12).

If you increase the delivered energy rate (J/cm) you may perform a perforation of the vein wall that appears like a smoke in the water buffer that encircles the vein. This rate is an indicator of the maximum energy you have to use in order to ablate the vein properly. Usually, with the radial technique (1470 nm), for veins with a diameter >5 mm and <10 mm the author delivers an energy rate of 60 J/cm; for veins with a diameter >10 and <20 mm, 100 J/cm. However, ultrasound sign of perforation of the vein wall is the maximum tolerance at energy delivery and you should never exceed this rate. Sometimes the vein tissue can carbonize on the fiber tip and have low performance of ablation even if the energy rate appears really high. Mind that in this case is mandatory to clean your fiber tip in order to mislead vein occlusion.

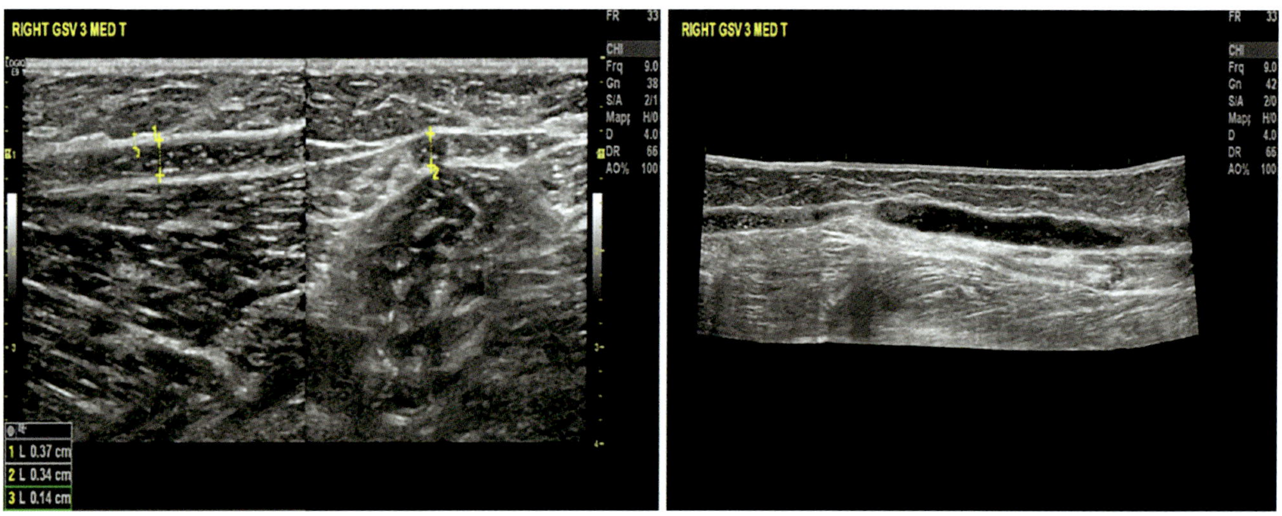

Fig. 70.12 Hyperechogenic obliteration after GSV ablation; courtesy Prof. D. Kontothanassis

How to Prevent Nerve Damages

With tumescent anesthesia the aim is to completely separate the vein from all the other tissues; it will prevent both nerve damages, discoloration and skin burnings. When the risk of nerve damage is higher (when you are treating the SSV, or the GSV at the calf, or in presence of anatomical variation), during the ablation one should ask to the patient to report any pain or electric stimulation and stop immediately if the answer is positive. In case of risk of nerve injury, you can reduce the amount of the delivered energy or withdraw the Laser fiber 1–2 cm. Evaluate also the possibility to use a non-thermal and no tumescent technique (MOKA, FOAM, PEM, GLU) (Fig. 70.13).

Recommendations Before and After EVLA; How to Avoid EHIT

Before starting the ablation one should check the correct position of the fiber tip with reference to the epigastric vein or other landmarks in case of SSV or accessory veins. At the end of the ablation, one should check again the correct patency of the epigastric vein (Type II obliteration) and the patency of the deep venous system. (Fig. 70.14).

The author administers antithrombotic prophylaxis to all patients for at least 2 days, till the ultrasound postoperative control will confirm lack of EHIT. In case of thrombofilic patients or previous superficial or deep vein thrombosis prolong thromboprophylaxis at least 10 days and repeat ultrasound control at 7 days. In case of EHIT, treat with heparin weight-adjusted the patient for at least 2 weeks and check with ultrasound the patency of the epigastric vein. EHIT usually occurs when you provoke a thermal damage at the epigastric vein and re (Fig. 70.15).

Case Presentation

Continued from page 739

Under mild sedation, patient was placed in the supine position in the interventional suite table at the anti-Trendelenburg position for GSV access at the internal malleolus. After vascular access, insertion and positioning the

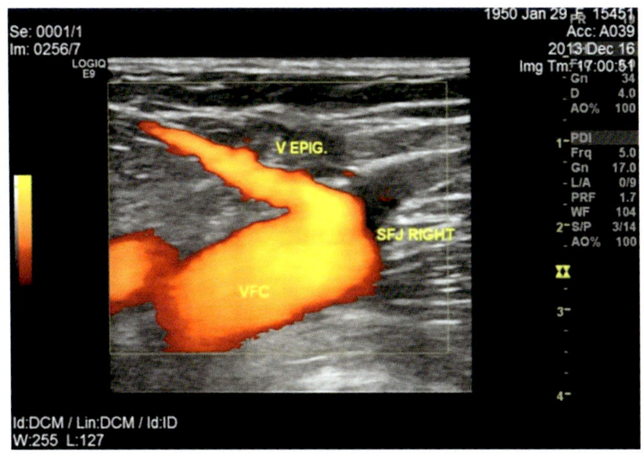

Fig. 70.14 Type II obliteration (regular patency of the epigastric vein, occlusion of GSV and patency of femoral vein; courtesy Prof. D. Kontothanassis

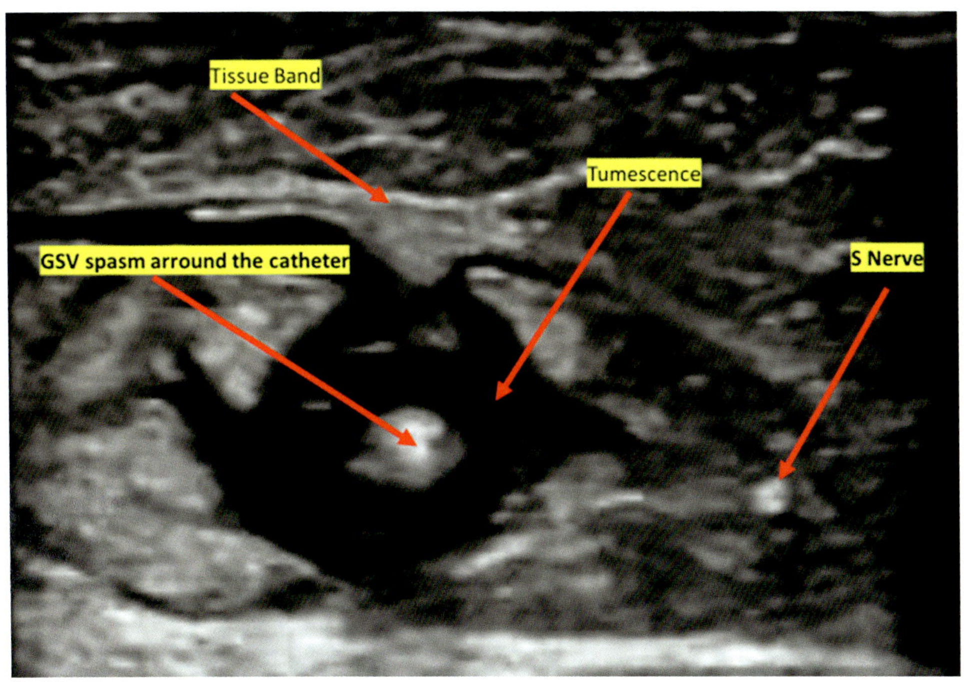

Fig. 70.13 Tumescent anesthesia to prevent saphenous nerve injury; courtesy Prof. D. Kontothanassis

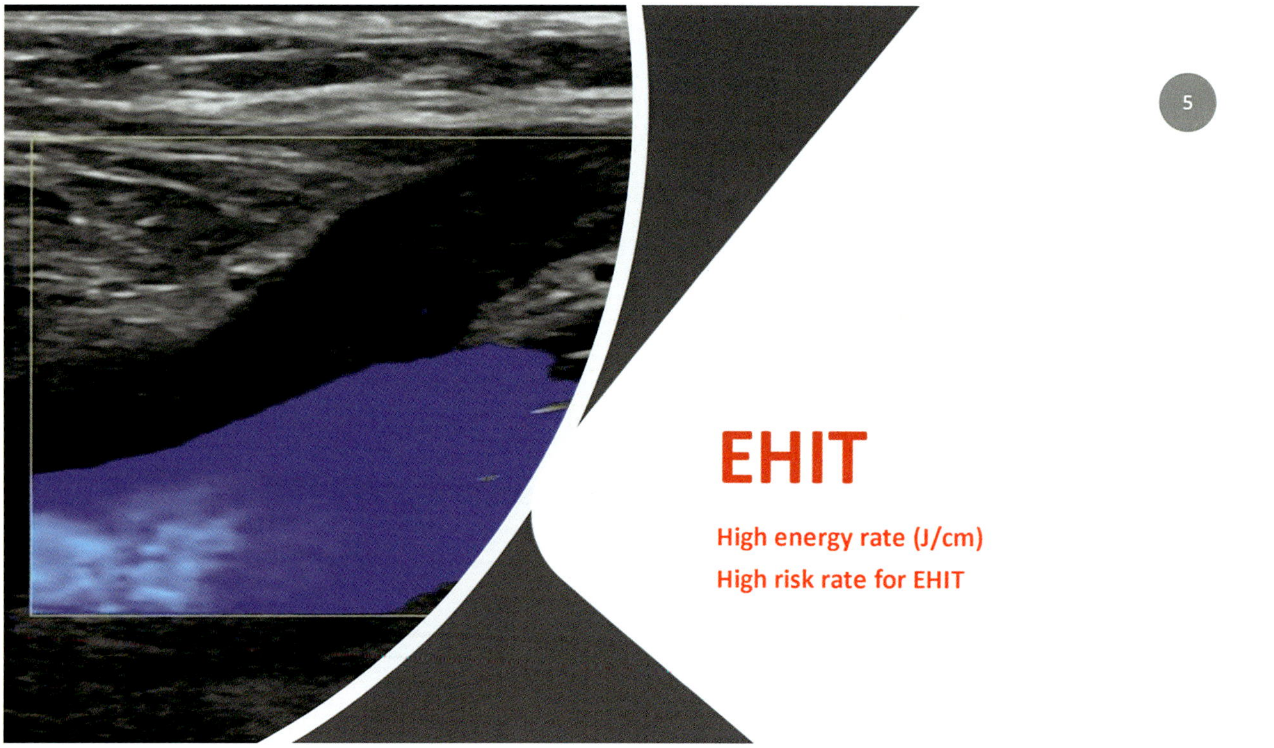

Fig. 70.15 Endovenous Heat Induced Trombosis at the sapheno-femoral junction; courtesy Prof. D. Kontothanassis

Laser fiber below the epigastric superficial vein, plenty of tumescent anesthesia was delivered under ultrasound guidance all over the length of the GSV and immediately after endovenous Laser ablation (EVLA) was performed from the groin to the internal malleolus to treat the whole length of the refluxing vein. A radial fiber was used and given the status of the patient's skin (ulcer and lipodermatosclerosis), at the distal calf, we foamed all the varicosities, as there was a contraindication for phlebectomies in such cases. Foam was produced with the Tessari method by mixing polidocanol 0.5% 2 cc + 8 cc air.

Intravascular ultrasound confirmed the occlusion of the GSV allover its length and wrapping the leg immediately after the operation was mandatory for 1 week. Patient was placed on low molecular weight heparin thromboprophylaxis for 1 week and discharged 2 h later from our clinic. At 1-week post- operation, ultrasound documented occlusion of the L GSV and varicosities at the calf; relief of symptoms was immediate after the operation. At the 2 months follow up the patient was doing well, and symptoms had completely resolved. The ulcer was healed and regression of inflammation at the distal calf was evident. Skin care and a second-class elastic stocking was recommended for at least 6 months.

Conclusion

The patient presented with severe CVD (CEAP Class C2,3,4,6 s) which could be related to deep vein reflux after the episode of distal DVT and obviously because of the documented extensive superficial vein reflux of both GSV and SSV. Duplex ultrasound was mandatory to document extensive axial reflux of the great and small saphenous vein and absence of obstruction at the deep vein system. Symptomatic varicose veins can be treated using different procedures when the deep system is patent. The GSV was dilated with high velocity long duration reflux (>5 s) and such vein can be safely ablated. Treatment recommendations included thermal ablation of the GSV (Laser or RF) plus phlebectomies or ultrasound guided foam sclerotherapy for calf varicosities. We believe that in cases with CEAP C6 is safer to perform foam sclerotherapy for treating tributaries or adopting a two-stage therapy. SSV was not treated at the same time with the GSV because we detect only reflux and absence of tributaries. Annual follow up is recommended because can detect early recurrences at the treated areas and progression of disease in other vein segments.

References

1. Navarro L, Min R, Bone C. EndovenousLaser; a new minimally invasive method of treatment of varicose veins. Dermatol Surg. 2001;27:117–22.
2. Gloviczki P, Comerota AJ, Dalsing MC, Eklof BG, Gillespie DL, Gloviczki ML, Lohr BJM, McLafferty RB, Meissner MH, Murad MH, Padberg FT, Pappas PJ, Passman MA, Raffetto JD, Vasquez MA, Wakefield TW, Society for Vascular Surgery, American venous forum. The care of patient with varicose vein and associated chronic venous diseases: clinical practice guidelines of the society for Vascular Surgery and the American venous Forum. J Vasc Surg. 2011;53(5 suppl):2S–48S.
3. Agus GB et IEWG Group. The first 1000 cases of Italian Endovenous-laser working group (IEWG). Rationale, and long-term outcomes for the 1999-2003 period. Int Angiol. 2006;25(2):209–15.
4. Kontothanassis D, Di Mitri R, Ferrari Ruffino S, Zambrin E, Camporese G, Gerard JL, Labropoulos N. Endovenous laser treatment of the small saphenous vein. J Vasc Surg. 2009;49:973.
5. Kontothanassis D, De Zolt P, Labropoulos N. Endovenous Laser Ablation Follow up Study (ELAFOS). A prospective trial to assess long term EVLA results, for the UIP 2013.
6. Theivacumar NS, Dellagrammaticas D, Beale RJ, Mavor AID, Gough MJ. Factors influencing the effectiveness of Endovenous laser ablation(EVLA) in the treatment of great saphenous vein reflux. Eur J Vasc Endovasc Surg. 2008;35:119–23.

71 Radiofrequency Ablation for Chronic Venous Disease

Dimitrios Kontothanassis

Case Presentation
A 39-year-old female was referenced to our clinic for severe CVD long-lasting, for about 12 years. The patient had a superficial vein thrombosis (SVT) after her last pregnancy in 2022 and she delivered 5 parturitions in the last 9 years. SVT was treated for 2 months approximately with LMWH therapy with total resolution of the thrombus. There was a positive family history for thrombosis, and she was found positive for minor thrombophilia (heterozygous mutation of the MTHFR with normal levels of homocysteine). She has been complaining bilateral leg swelling, heaviness, itching and worsening pain at both limbs for approximately 8 years even wearing a second-class elastic stocking.

Clinical improvement obtained from compression therapy was poor in the last 5 years because of the advanced CVD, and given the hard daily work of looking after the five children which required 12 h of standing position per day, an intervention was discussed with our team in April 2023 after visiting our clinic. Reexamination of both limbs was done to exclude potential residual thrombus in the superficial veins or obstruction of deep veins. Problems to resolve in this case were: (a) vein aneurysms, (b) extensive varicosities and (c) the use of a noninvasive procedure to permit to the patient to go back to her children as soon as possible.

Treating large veins is a problem because failure of occlusion after thermal or non-thermal is not so rare; we need high energy rate to ablate the proximal segment near the saphenous-femoral junction (SFJ), and this may increase the risk of endothermal heat induced thrombosis (EHIT) or skin damage. Many times delivering high energy rate to the vein wall may lead to carbonization of the fiber tip with consequent troubleshooting during the Laser procedure and potential reduction of the delivered amount of energy. Occluded aneurysmatic vein segments, after proper ablation (thermal or non-thermal), may become hard, inflamed and painful so for this reason the distance from the skin it is mandatory and large volumes of tumescent anesthesia is necessary to be delivered around the vein wall to dissect the vein from the perivenous tissues. Cold saline anesthesia at 4 °C is the proper tumescent to game extreme vein spasm and protect the surrounding tissues. This novel cold saline tumescent anesthesia was used for the first time in 2001 and discussed in the literature from the author in a publication in 2007 [Endovenous thermal ablation. Standardization of laser energy: literature review and personal experience Int Angiology 2007 Jun; 26 (2):183–188).]

Continued at page 757 (Figs. 71.1, 71.2, 71.3, 71.4, 71.5) and 71.14).

D. Kontothanassis (✉)
General and Vascular Surgery, Phlebology Department, Istituto Flebologico Italiano, Padova, Italy

G. Geroulakos et al. (eds.), *Mastering Endovascular Techniques*, https://doi.org/10.1007/978-3-031-42735-0_71

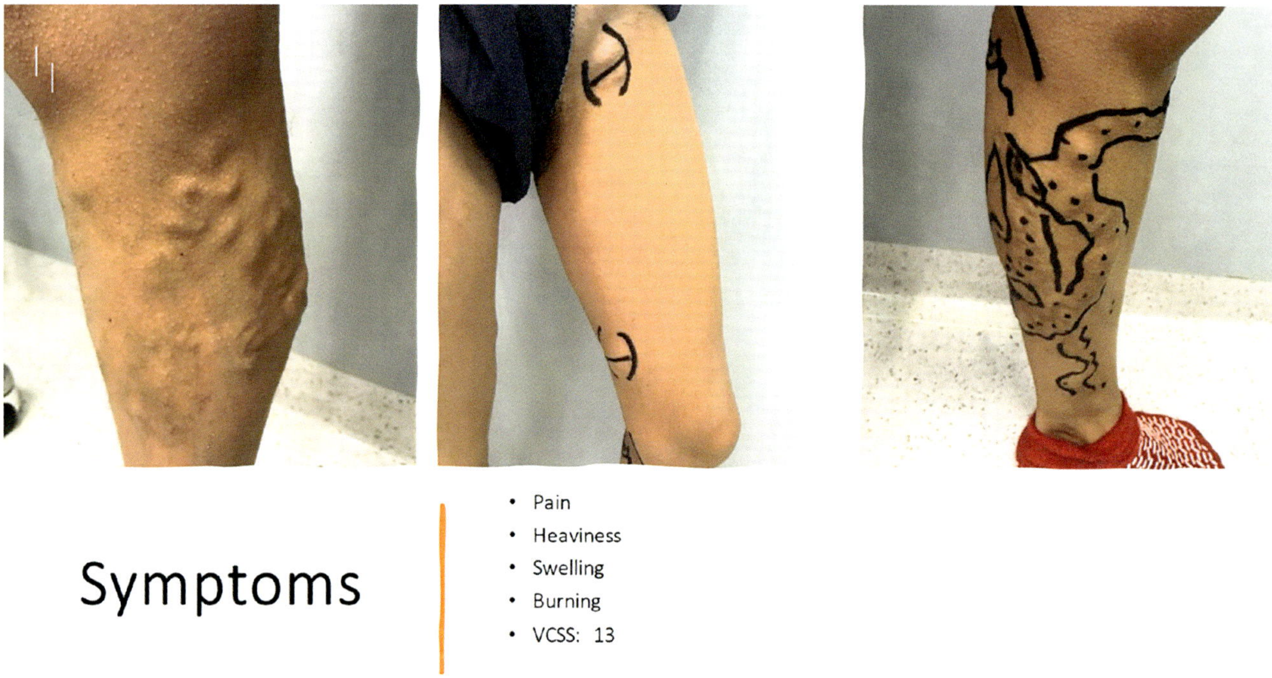

Fig. 71.1 Case presentation

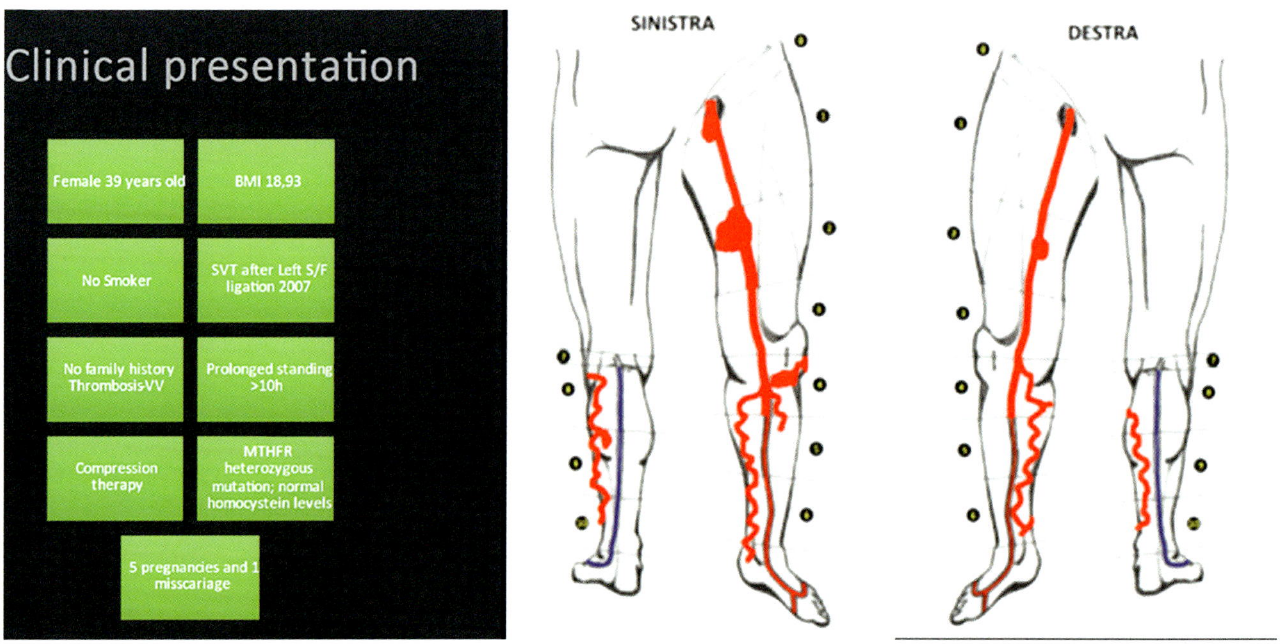

Fig. 71.2 Clinical presentation and hemodynamic map of GSV, SSV and varicosities

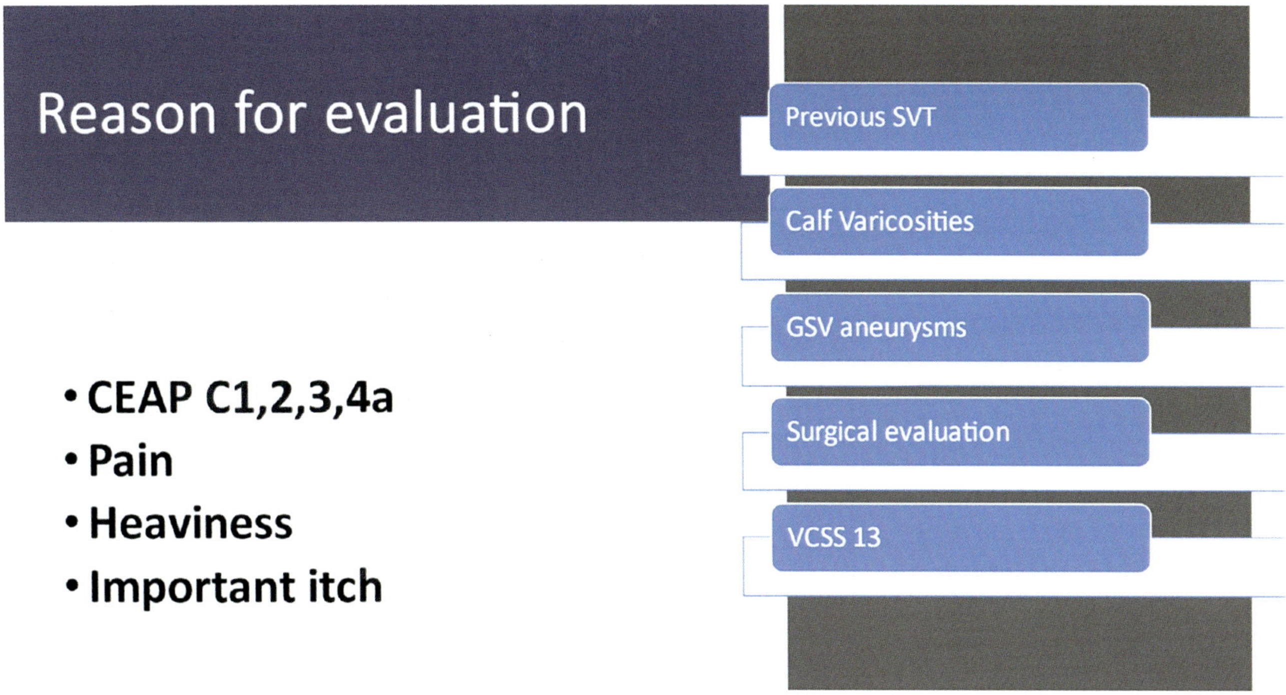

Fig. 71.3 Reason for surgical evaluation

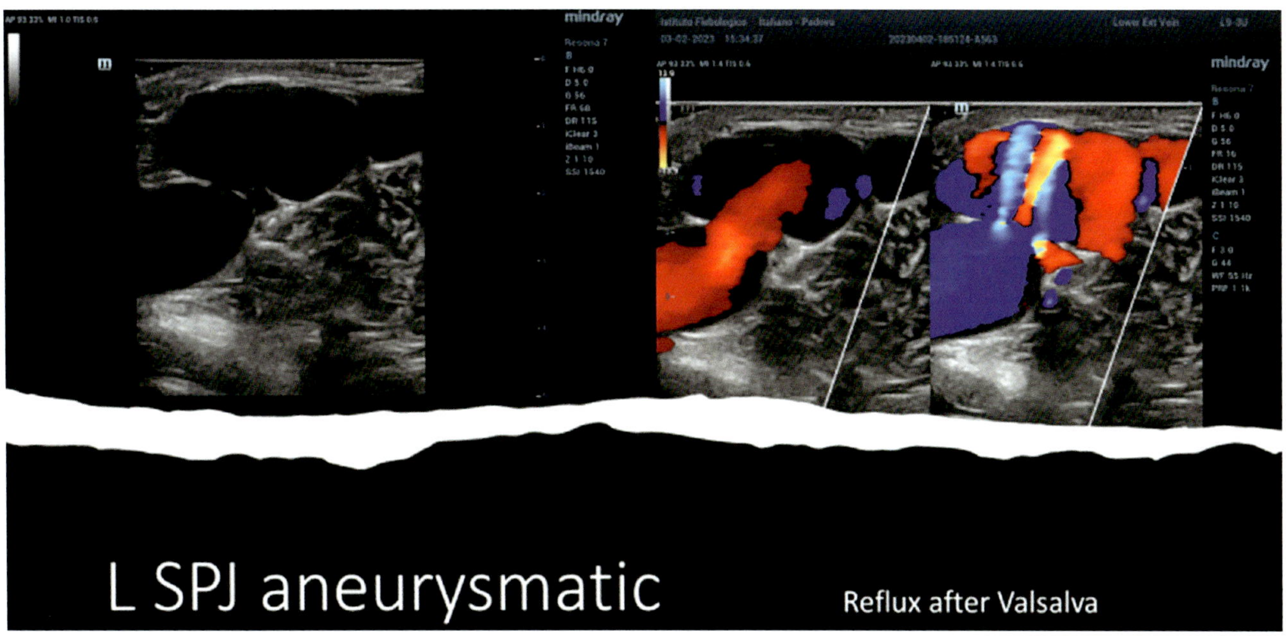

Fig. 71.4 Aneurysmatic L SFJ and reflux at the Valsalva maneuver

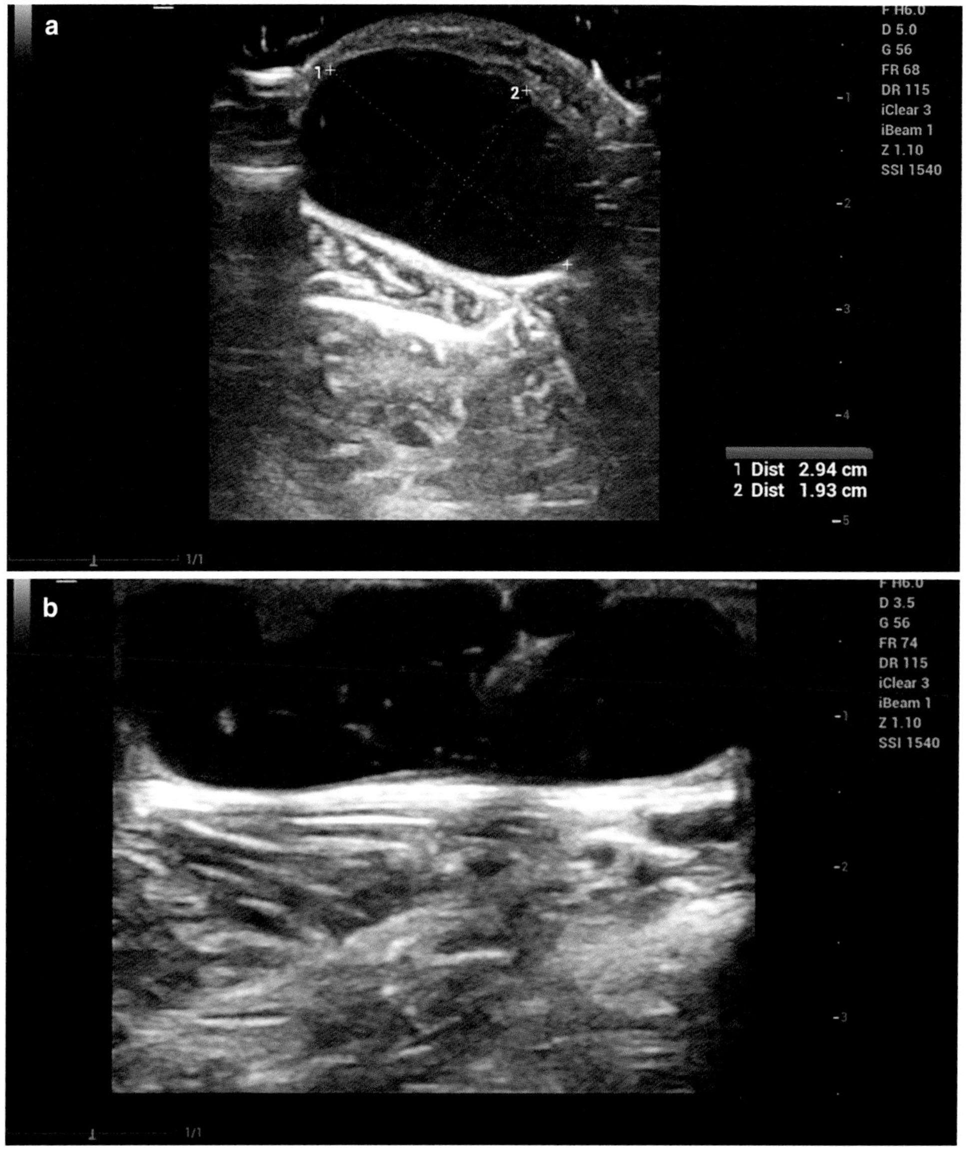

Fig. 71.5 (**a**) and (**b**) Aneurysmatic GSV and thrombotic Synechia in GSV tributaries

Introduction

Endovenous Radiofrequency Ablation (EVRFA) is a thermal mini-invasive percutaneous procedure for the treatment of chronic venous disease (CVD). It is safe and effective, and recent guidelines have recommended it over traditional surgery for incompetent saphenous veins (Grade 1B) [1]. Different studies have demonstrated the safety and efficacy of this technique for the incompetence of the great saphenous vein (GSV) [2], anterior accessory saphenous vein (AASV) [3], small saphenous vein (SSV), and perforantors (PRFV) [3].

Endovenous Radiofrequency Ablation (EVRFA)

The RFA consists of transformation of electric and magnetic waves into thermal energy after contact with tissue. Using an RF catheter which is inserted into the target vein under Color Flow Doppler Ultrasonography Guidance (CFDU), the increased endoluminal temperature induces vein wall thermal injury, contraction and occlusion of the vein.

The first RFA machine for CVD was created in 1995 (VNUS Medical Technologies- San Josè- California - USA) and was called VNUS Closure. The use of this device was approved by the USA food and drug administration FDA in 1999, and the first reports were published in 2000. Many studies, documented the good results of this technique and new devices were introduced to make the treatment faster and more efficient [1].

How It Works: Past, Present, and Future

The first-generation of RFA (Closure and Closure Plus system) was a dedicated microprocessor-controlled generator with collapsible electrodes (7 Fr.–8 Fr.) that could be introduced into the vein lumen. The *system* was designed with a thermocouple to measure electrode temperature and to provide feedback to the generator. A flexible electrode designed as a flower allowed the target vessel to collapse around the catheter as the vein shrinks. The systems generate heat at 60 °C to 85 °C that was delivered to the vein wall very slowly, at a pullback speed of about 0.5 cm/s. The energy rate is proved to be important for the outcomes of the procedure, and increasing the dose of energy is supported to improve efficacy through inducing a greater shrinkage of the treated vein (Fig. 71.6).

The second generation of RFA (Closure FAST system), was designed to improve procedural deficiencies such as length of procedure time, higher closure rates of the treated veins and make the procedure easier. With this catheter, energy is constantly and uniformly delivered. With the new devices that generate heat at 120 °C, boiling, vaporization and carbonization of the tissues are avoided. Many studies documented the results of this new technique in the short- and medium-term follow-up demonstrating the efficacy of this new device in terms of higher closure rates of the treated veins (95%). Recently some authors have suggested that two cycles of RFA treatment (double delivery of energy rate) in all segments of the GSV achieve quicker and greater vein shrinkage of the medium diameter without increasing the side effects, but further studies are needed to evaluate the implications in terms of long term clinical efficacy (Fig. 71.7) [4].

The next generation of RFA generator and catheter, (Venclose™ RF) for the treatment of incompetent saphenous veins (GSV, SSV), tributaries and perforators introduced technological innovation such as four heating lengths ranging from 2.5 cm up to 10 cm ablation in one catheter that reduces total in-vein time by >30%* and simplifies inventory management. The profile of the catheter is 6F that helps to minimize invasiveness. The temperature of the electrode increased from 120 °C, last generation electrode, up to 130 °C while the treatment cycle is 20 s and according to the manufacturer for a 10 cm segment of vein ablation, only one cycled is needed. Markings and curved tip are specially designed for ease of navigation and positioning; smaller profile (6F), offers greater flexibility.

For the Treatment of Incompetent Perforator Veins, *Venclose Maven™ Perforator Catheter* was specially designed; it's a 0.5 cm resistive heating coil that provides circumferential resistive heating; the catheter is a 6F profile and 40 cm flexible length that helps to minimize invasiveness and provide efficient treatment for varying vein lengths and anatomies (Fig. 71.8).

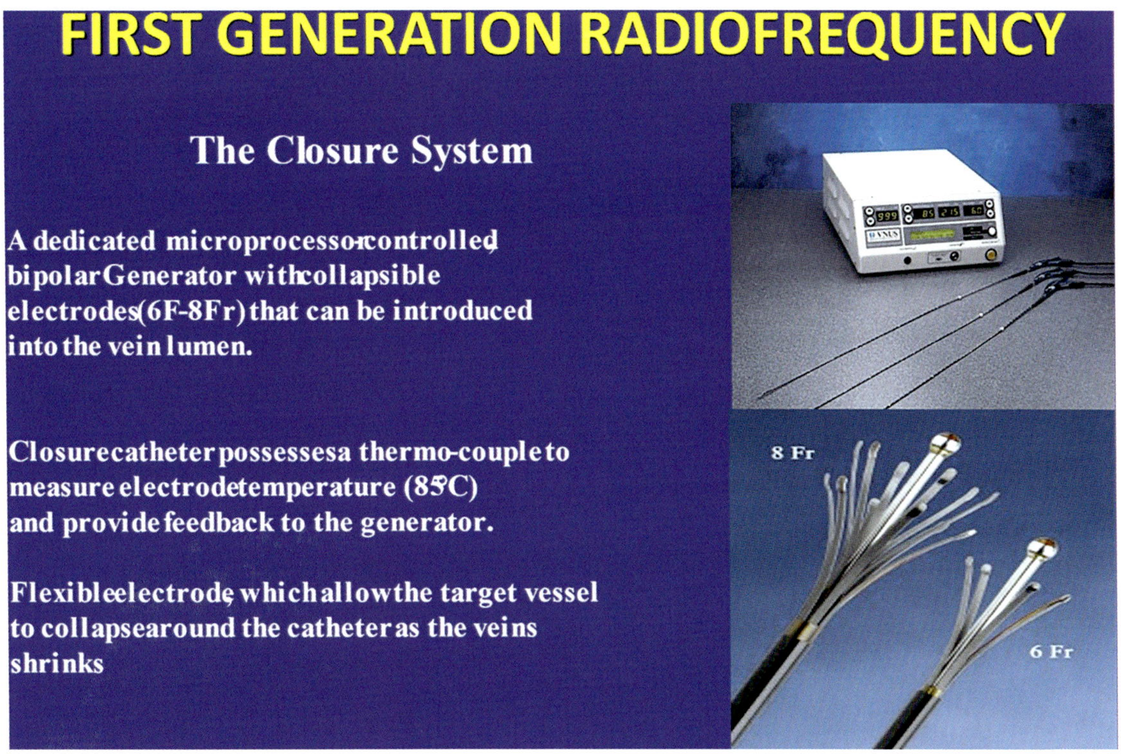

Fig. 71.6 The closure system

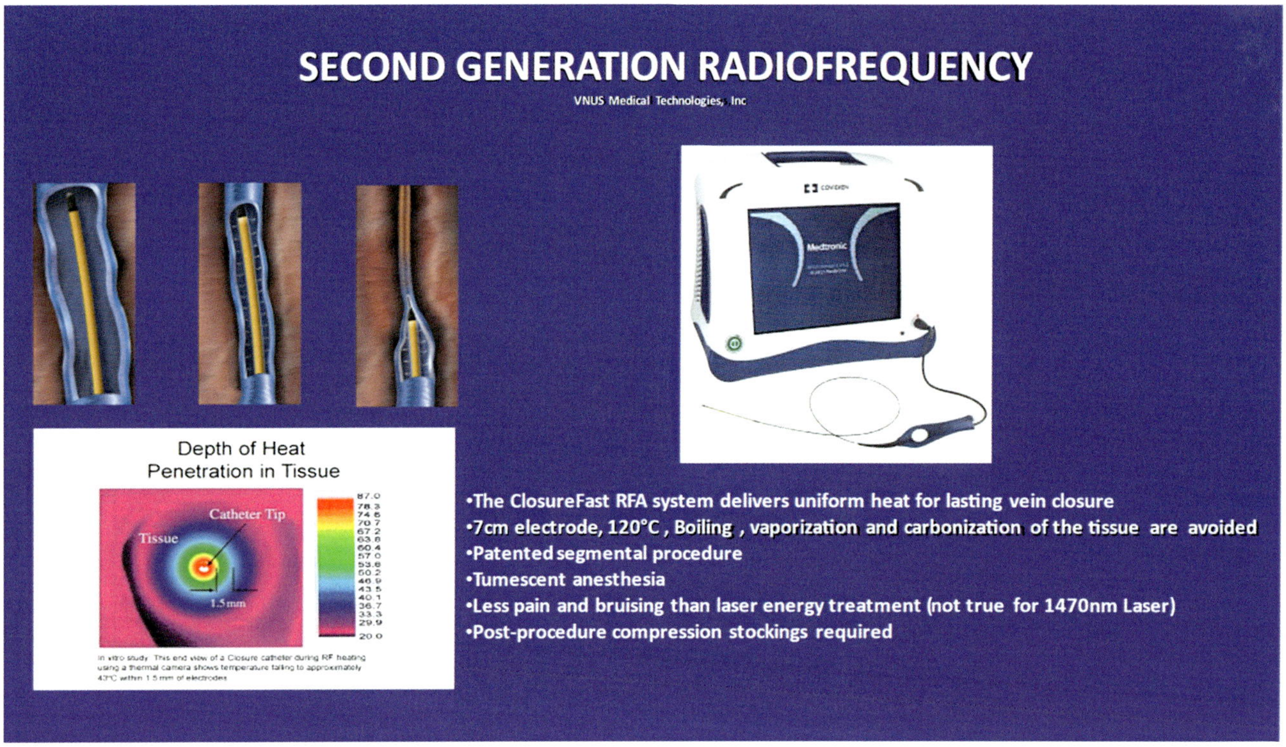

Fig. 71.7 Second generation radiofrequency

Fig. 71.8 Next generation radiofrequency; venclose

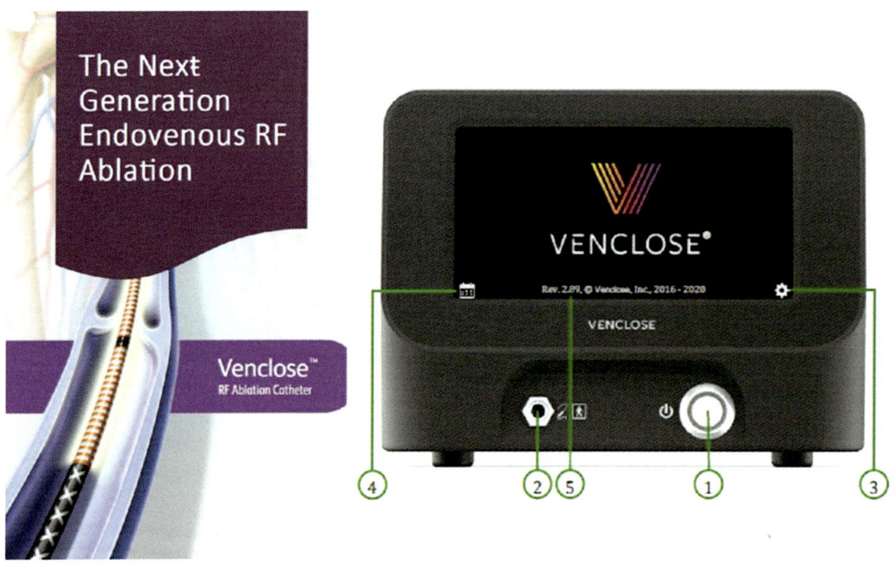

The RFA Technique

The RFA procedure for saphenous ablation involves the introduction of a J guide wire into the target saphenous vein or tributary under ultasound guidance followed by the insertion of a 6F or 7F introducer sheath; through this vascular access the RFA catheter is advanced to the saphenofemoral (SFJ) or saphenopopliteal junction (SPJ). The correct position of the catheter tip is confirmed by ultrasound imaging, and it is placed below the inferior epigastric vein for the GSV and approximately 1 cm below the SPJ.

The patient receives local tumescent anesthesia with or without mild sedation (midazolam 2–3 mg or Propofol 2 mg/kg) before starting the treatment. Tumescent anesthesia is infused with a 18 gauge needle and enough long to approach the target vein under ultrasound guidance. A mixture of 500 mL of cold normal saline and 10 mL of lidocaine 2% normally is used; enough fluid is injected with a peristaltic pump into the soft tissue surrounding the target vein along its entire length to encircle completely the vein around the RF catheter; the temperature of the infused fluid is 5 °C in order to enhance and produce significant spasm of the incompetent and dilated vein. Tumescent anesthesia also provides a heat sink to prevent the radiation of thermal energy to the surrounding tissues avoiding skin burns and numbness or damage of nerves at the calf. The RF generator is then activated, and the catheter is pulled back after one or two cycles of ablation at the edited landmarks (7 or 10 cm) along the length of the vein. The time of a single cycle RF ablation is 20 s, and the immediate occlusion of the vein is confirmed by ultrasound imaging: lack of compression, echogenic material in the venous lumen, and absence of color flow are used to document the occlusion. Additional procedures such as phlebectomies, ligation, or obliteration of incompetent perforating veins and UGFS can be performed safely after the RFA procedure.

Results

Clinical outcomes like absence of CVD symptoms, patients satisfaction, and evaluation of patency of the ablated vein are mandatory to be collected and classified at the early-term, mid-term, and long-term follow up. Primary occlusion of the great and small saphenous vein is documented and based on lack of compression, echogenic material in the venous lumen, and absence of color flow. Early recanalization after RF ablation, (both short and extended), is defined as presence of flow inside the ablated vein segment. Ecocolordoppler imaging for the evaluation of morphology and anatomy of the saphenofemoral junction after RF ablation was analyzed by Olivier Pichot et al. and classified the results of the ablation defining 5 types of different patterns of occlusion. GSV obliteration imaging results [5], can be complete SFJ obliteration (Type1); persistent patent SFJ tributaries drained toward the femoral vein (Type-2a) with the tributaries located very close to the SFJ draining directly into the femoral vein above a closed saphenous vein; Type-2b with a short <2 cm patent saphenous stump acting as a conduit for tributary flow to the SFJ); terminal GSV competence (Type-3) with a normal antegrade flow coming from both saphenous and SFJ tributaries above a limited obliteration (Fig. 71.9).

The use of RFA to treat patients with symptoms due to saphenous reflux involves a small but definite risk of DVT and SVT. This risk is greater in patients with previous DVT and SVT. Thrombus protrusion into the femoral vein without DVT is defined as EHIT (Endovenous Heat Induced Thrombosis) and was found to be more likely in patients with hypercoagulability, thrombophilia, male gender, aspirin use (Fig. 71.10 RF) [6].

A large proximal GSV diameter (13.05 ± 5.59 mm versus 8.39 ± 3.38 mm) had a significantly higher risk of developing EHIT after endovenous catheter ablation [7].

The recanalization rate after RFA has been reported from 2% to 12%. Most patients with recanalization remain asymptomatic and do not require new treatments. An anatomical

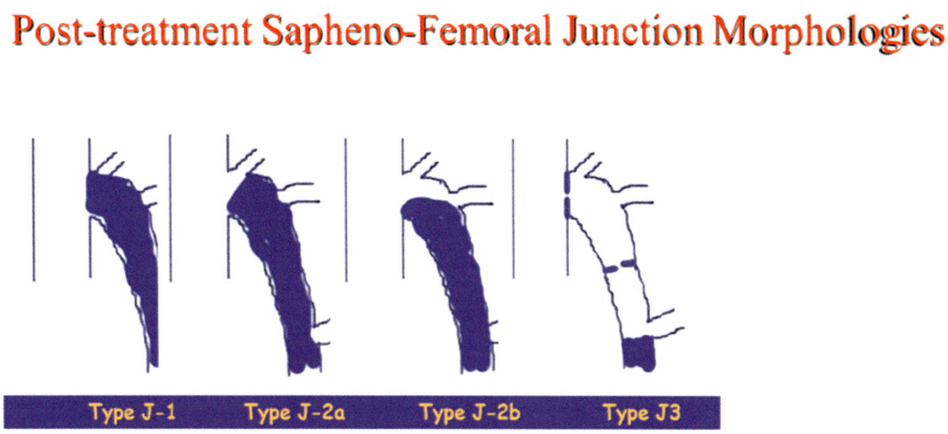

Fig. 71.9 Post treatment Sapheno-Femoral junction morphologies

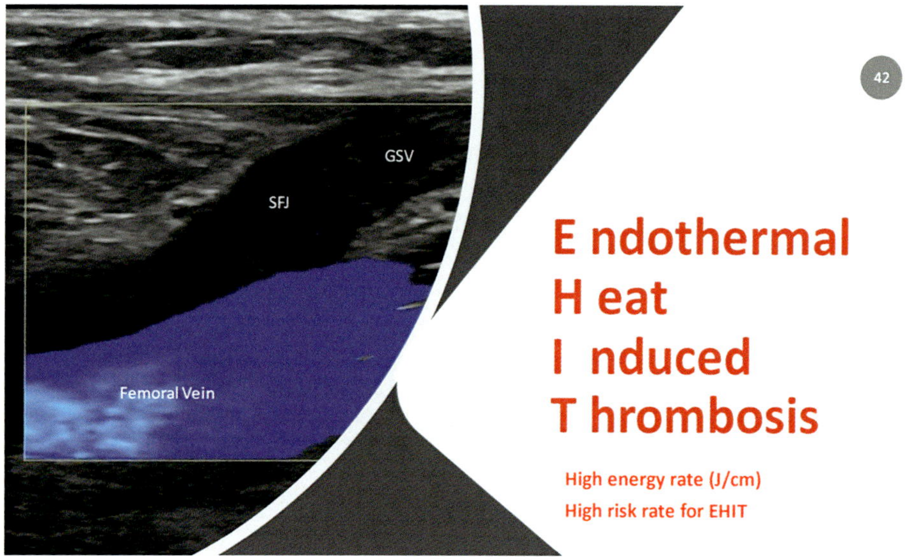

Fig. 71.10 Endothermal heat induced thrombosis

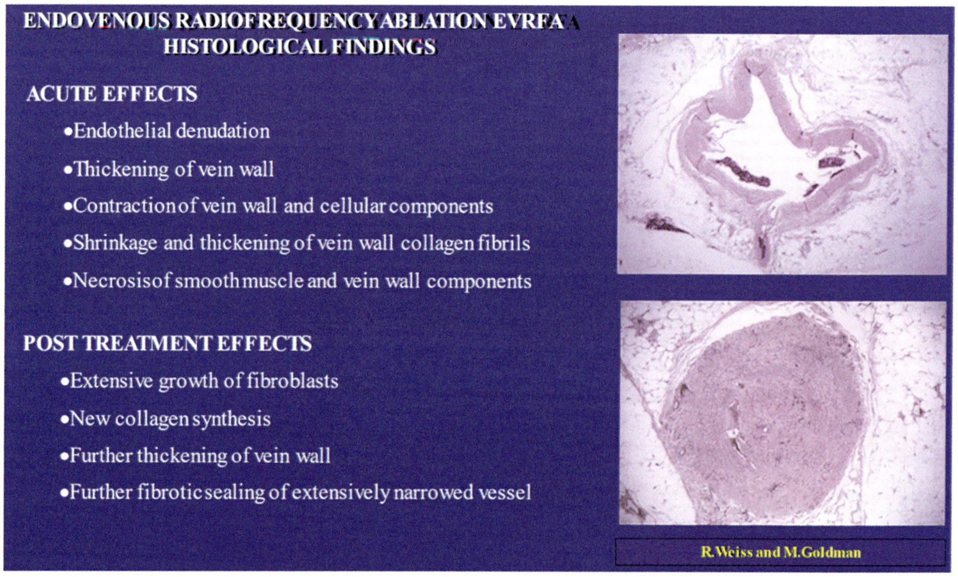

Fig. 71.11 Histology; acute and post treatment effects of RF ablation of the venous wall

classification has been introduced to differentiate modes of failure of the treatment:

Type I (non-occlusion): if the vein failed to occlude initially and never occluded during the follow-up.

Type II (recanalization) if the vein occludes after treatment but recanalized partly or completely later.

Type III (groin reflux) if the vein trunk occluded, but reflux is detected at the groin region often involving the AASV.

Catheter pullback speed and body mass index (BMI) are two risk factors associated with anatomic failures [2]. The author followed these patients with a regular clinical examination and CFDU for more than 10 years following the procedure: A low rate of recurrence (<8%) was documented. However, new disease was diagnosed in 33.8% of procedures, almost always asymptomatic and that did not require new treatment, exception made for symptomatic patients (<3%) of the total recurrence.

Figure RF DU 2D, 1 year after RF or GSV

Histological Findings After Endovenous Radiofrequency Ablation EVRFA

During EVRFA, the thermal energy delivered to the venous segment to be treated, causes direct injury with acute and post treatment effects. *Acute EVRFA effects* of thermal injury are endothelial denudation, thickening of the vein wall, contraction of the vein wall collagen fibrils and necrosis of the smooth muscle and vein wall components. *Post Treatment effects* of thermal injury are extensive growth of fibroblasts, new collagen synthesis, further thickening of the vein wall and a further fibrotic sealing of extensively narrowed vessels (Fig. 71.11).

Conclusions and Considerations

The RFA has very good results not only in the short and mid-term but also in the long-term follow-up. It has low recurrence rate, few recanalization, but there is appearance of new disease in approximately a third of the patients, even if almost asymptomatic. For this reason, regular clinical and CFDU follow-up after RFA treatment is recommended (Figs. 71.12 and 71.13).

Different studies revealed that endovenous treatment with RFA and Laser ablation (EVLA) showed better results than traditional surgery with high ligation and stripping in terms of reduced postoperative pain, better quality of life, faster recovery and lower rate of recurrence [1, 8]. Based on the clinical practice guidelines published by the American venous forum, RFA presents less discomfort and bruising in the peri-operative period than EVLA [1, 9]. However, these results are based on short and mid-term follow-up and precise data on the long term are still lacking. Furthermore, RFA is a cost-effective and cost-saving procedure: even if there are extra expenses for the devices, RFA allows for reduction of the anesthetic used, low rate of recurrence, early recovery, and return to normal activities [10].

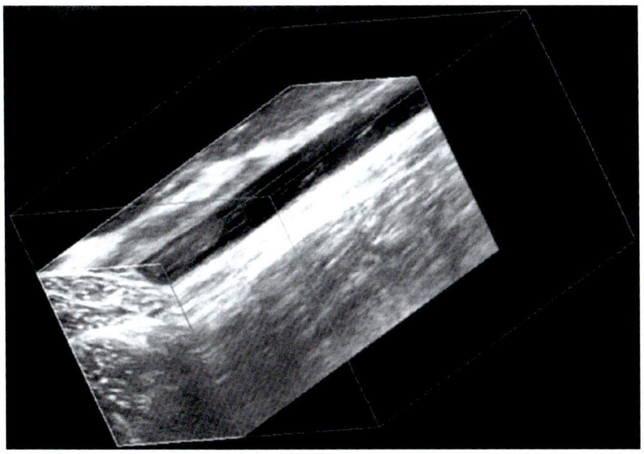

Fig. 71.12 3D view of RFA of GSV

Case Presentation

Continued from page 749 (Fig. 71.14).

Under mild sedation, patient was placed in the supine position in the interventional suite table at the anti-Trendelenburg position for GSV access at the internal malleolus with the intention to treat the entire length of the axial reflux of the GSV. After vascular access, insertion, and positioning of the Venclose™ catheter below the epigastric superficial vein, plenty of tumescent anesthesia was delivered under ultrasound guidance all over the length of the GSV and tributaries. Immediately after this, endovenous Radiofrequency ablation (RFA) was performed from the groin to the internal malleolus to treat the whole length of the refluxing vein. The heating element was energized by the Venclose™ RF Generator, which is a multi-voltage energy delivery system with touchscreen control that automatically

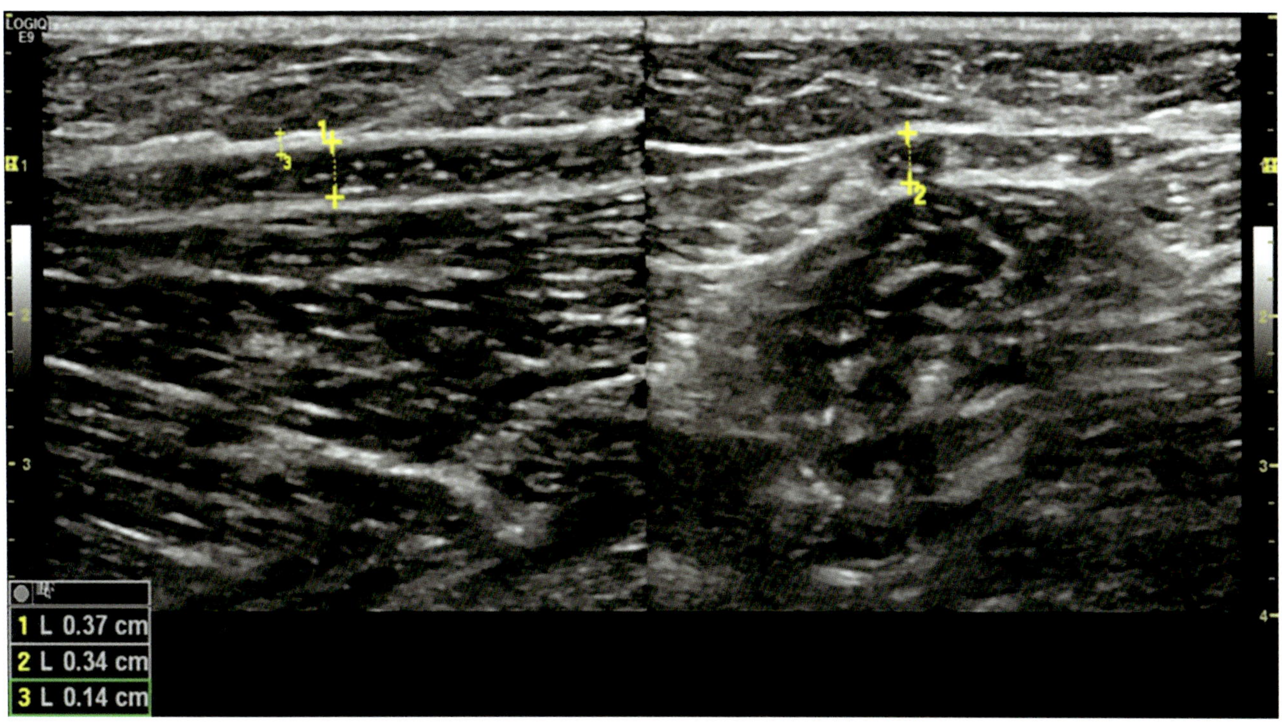

Fig. 71.13 2D imaging of GSV RFA after 2 months

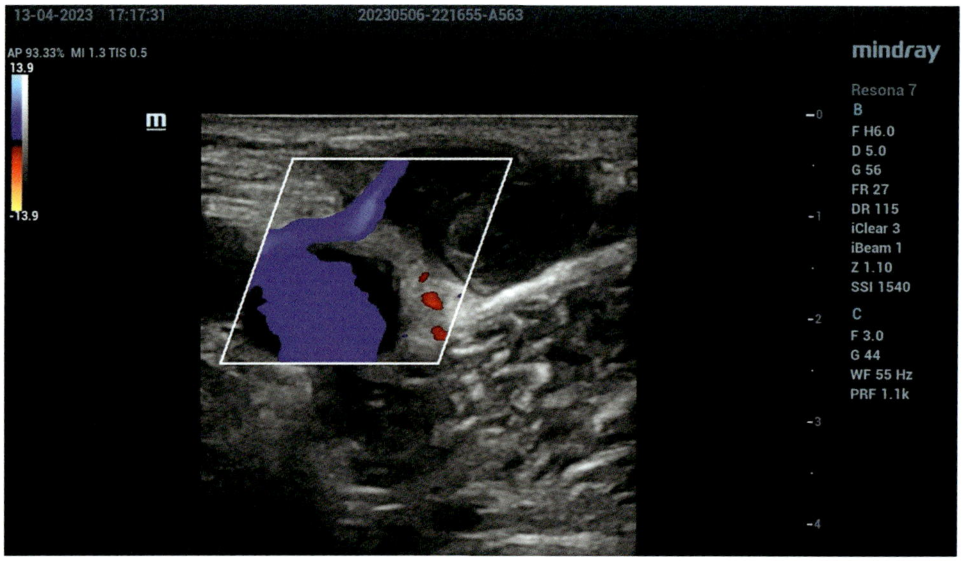

Fig. 71.14 Follow up 24 h after ablation; occlusion of the aneurysmatic below the inferior epigastric vein

sets the non-adjustable treatment parameters for the Venclose™ System Catheters. A button on the catheter (or a foot pedal attached to the generator) begins an automated treatment cycle 20 s long at a set temperature of 120 °C (Venclose™ RF Ablation Catheter). The treatment stops automatically when complete. Intravascular ultrasound confirmed the occlusion of the GSV allover its length; Phlebectomies under cold saline tumescent anesthesia with micro surgery technique. The leg was wrapped immediately for 48 h and after a second class stocking was recommended for 15 days. Patient was placed on low molecular weight heparin thromboprophylaxis for 1 week because of prior superficial vein thrombosis and discharged 2 h later from our clinic. At 1-week post operation, ultrasound documented occlusion of the L GSV and absence of varicosities at the calf; relief of symptoms was immediate after the operation. At the 2 months follow up, the patient was doing well and symptoms had completely resolved.

Conclusion: The patient presented with severe CVD (CEAP Class C2,3,4 s) which could be related to chronic superficial vein reflux, superficial vein thrombosis, and presence of extensive superficial varicosities. Symptomatic varicose veins can be treated safely using different procedures when the deep system is patent. The GSV was aneurysmatic with high velocity long-lasting reflux (>5 s) and such vein can be safely ablated. RFA and Laser are both equal and highly recommended techniques for treatment of CVD. According to guidelines, treatment recommendations for this case included thermal ablation of the GSV (Laser or RF) plus phlebectomies or ultrasound guided foam sclerotherapy for calf varicosities. We believe that in cases with CEAP C4s is simple and safer to perform phlebectomies for treating at the same time tributaries. In this case we consider the Venclose generator and catheter, (next generation RF), because the learning curve is more simple for treating huge veins and 30% faster with a 10 cm heating coil if compared to Laser fibers and other RF catheters; this choice can reduce the overall procedure time and cost of operation. Venclose system was specifically designed for more efficiency and can be considered the next generation thermal system. Annual follow up is recommended because can detect early recurrences at the treated areas and/or progression of disease in other vein segments. Occlusion rate does not reflect clinical success and new disease develops in a third of the patients at the long term follow up (>10 years) but fortunately only few patients are symptomatic. Both RF and EVLA should be offered in combination with Foam and Glu when necessary to treat complex vein anatomy (Fig. 71.14).

References

1. Gloviczki P, Comerota A, Dalsing M, et al. The care of patients with varicose veins and associated chronic venous diseases: clinical practice guidelines of the Society for Vascular Surgery and the American venous Forum. J Vasc Surg. 2011;53:2S–48S.
2. Merchant RF, Pichot O, Closure Study Group. Long-term outcomes of radiofrequency obliteration of saphenous reflux as a treatment for superficial venous insufficiency. J Vasc Surg. 2005;42:502–9.
3. Harlander-Locke M, Jimenez JC, Lawrence PF, et al. Endovenous ablation with concomitant phlebectomies is a safe and effective method of treatment for symptomatic patients with axial reflux and large incompetent tributaries. J Vasc Surg. 2013;58:166–72.
4. Garcia-Madrid C, Pastor Manrique JO, Sanchez VA, et al. Endovenous radiofrequency ablation (venefit procedure): impact of different energy rates on great saphenous vein shrinkage. Ann Vasc Surg. 2013;27:314–21.

5. Pichot O, Sessa C, Chandler J, et al. Role of duplex imaging in endovenous obliteration for primary venous insufficiency. J Endovasc Ther. 2000;7:451–9.
6. Jacobs CE, Pinzon MM, et al. Deep venous thrombosis after saphenous radiofrequency ablation: is it predictable? Ann Vasc Surg. 2013. [Epub ahead of print].
7. Lin JC, Peterson EL, Rivera ML, et al. Vein mapping prior to endovenous catheter ablation of the great saphenous vein predicts risk of endovenous heat-induced thrombosis. Vasc Endovascular Surg. 2012;46:378–83.
8. Rasmussen LH, Lawaerz M, Bjoern L, et al. Randomized clinical trial comparing endovenous laser ablation, radiofrequency ablation, foam sclerotherapy and surgical stripping for great saphenous varicose veins. Br J Surg. 2011;98:1079–87.
9. Goode SD, Chowdhury A, Crochett M, et al. Laser and radiofrequency ablation study (LARA study): a randomized study comparing radiofrequency ablation and endovenous laser ablation (810 nm). Eur J Vasc Endovasc Surg. 2010;40:246–53.
10. Eidson JL III, Atkins MD, Bohannon WT, et al. Economic and outcomes-based analysis of the care of symptomatic varicose veins. J Surg Res. 2011;168:5–8.

Endovenous Steam Ablation for Chronic Venous Disease

72

Dimitrios Kontothanassis

Introduction

Pulsated Steam ablation is the newest thermal treatment for lower limb venous incompetence which works by heating the vein with steam at 120 °C [1]. Its objective is to achieve a safer and easier method of thermal ablation that has fewer side effects. Very few studies have reported on steam ablation yet [2, 3].

Materials

For the steam ablation, the Steam Vein Sclerosis system (CERMA SA, Archamps France) was used. The SVS system consists of a steam generator and a hand-piece that injects micropulses of steam into a catheter that delivers the steam into the vein to be treated (Fig. 72.1).

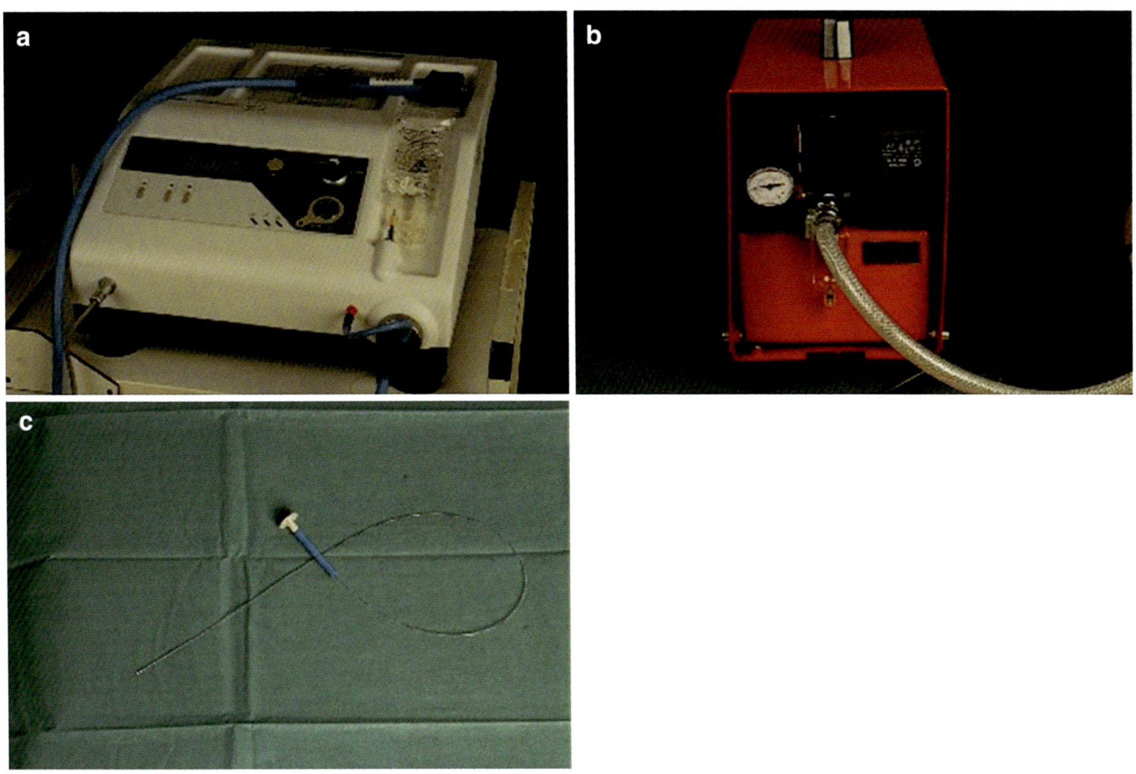

Fig. 72.1 (**a**) steam ablation generator (**b**) steam compressor (**c**) steam catheter

D. Kontothanassis (✉)
General and Vascular Surgery, Phlebology Department, CEO
Istituto Flebologico Italiano, Padova, Italy
e-mail: direzione@istitutoflebologico.it

G. Geroulakos et al. (eds.), *Mastering Endovascular Techniques*, https://doi.org/10.1007/978-3-031-42735-0_72

The Exclusion Criteria

General criteria for patients selection are age <18 years, acute deep or superficial vein thrombosis, agenesis of deep vein system, vascular malformations syndromes, occlusive post thrombotic syndrome, pregnancy, immobility, allergy to lidocaine, and arterial insufficiency (ankle-brachial index <0.9). Specific criteria are transcatheter ablation contraindications that are similar with the other two thermal techniques, EVLA and RFA.

Procedure

The procedure of steam ablation is very similar to EVLA. Steam ablation is performed with the patient under local tumescent anesthesia in an outpatient setting. The vein is punctured with a 16-gauge infusion needle under US guidance. Access at the incompetent GSV usually is done at the knee or just above the knee level because access is easy at that site and the risk of nerve injury is small. The SSV can be entered at the distal third part of the calf, depending on vein diameter. After puncturing the vein, the steam catheter (1.2-mm diameter) is passed through the hollow needle into the vein until positioned 3 cm below the junction. The most pivotal step in the procedure is positioning the echo-dense tip of the sheath approximately 3 cm distally from the junction under longitudinal US visualization.

Tumescent cold saline anesthesia (10 mL lidocaine 2% diluted in 500 mL saline solution) is administered into the perivenous space under US guidance using a peristaltic infusion pump. Tumescent anesthesia is necessary because it reduces pain, cools perivenous tissue, and decreases the venous diameter.

Even if a step-by-step procedure is lacking in the literature, after activation, the catheter releases small "puffs" of steam, and the catheter is pulled back stepwise. With the first activation 3 cm below the (saphenofemoral or saphenopopliteal) junction, four puffs of steam are administered, meanwhile with manual pressure on the junction. Two puffs of steam are administered at 1-cm intervals. During the first 4 cm of treatment, manual compression on the junction is still applied.

A physicist calculated that approximately 2258 J is released when 1 g of steam condenses. In EVLA, it is considered consensus to apply about 50–60 J/cm. To occlude 30 cm of vein with steam ablation, theoretically, 1–1.5 mL of water is needed. In practice, 2–5 mL of water is likely to be required, because not all steam condenses at the vein wall.

The steam is produced by means of piston pressing a fixed amount of water (76 μL = diameter piston × stroke) through a heated element located just before the catheter. By keeping the lumen diameters very small and the exit holes even smaller, pressure is maintained, and loss of energy is limited. The volume of the steam depends on the pressure and temperature. As the energy is transferred to the vein, the steam cools and condenses to the same volume of water used to produce the steam. The steam starts to cool and condense when it leaves the catheter due to the drop in pressure and the exchange of energy with the surroundings. This process is dynamic. The theoretic amount of energy of one pulse of steam is 174 J. The measured amount released at the tip of the catheter is 60 J per pulse. However, there is no consensus for these data, but only small studies and additional dose-finding studies are warranted.

After the procedure, patients were advised to wear thigh-length medical elastic compression stockings (pressure range, 25–35 mm Hg) for 1 week and to mobilize immediately after the treatment.

Microscopic Examination

Microscopic examination of veins immediately after the steam ablation showed disappearance of the endothelial layer, with negative marking for factor VIII and a few cleavage zones in the media (Fig. 72.2a) Microscopic examination of treated veins that were removed 20 days after the steam ablation showed endothelial destruction, fibrotic thrombosis with inflammatory reaction of the media, major alterations of elastic and collagen fibers in the media, and lesions in the adventitia with liponecrosis and lipogranuloma (Fig. 72.2b, c). Locally, the inflammatory reaction extended to the adventitia. Three months after the steam ablation, microscopic examination showed (major) thickening of the vein wall with fibrosis and inflammation, destruction of endothelium, alterations of elastic and collagen fibers, numerous pigmented histiocytes, a significant reduction of the lumen, and capillary neo-vessels. There were also focal areas in which the endothelium was preserved (Fig. 72.2d). The observed histologic changes were similar to those found after treatment with RFA.

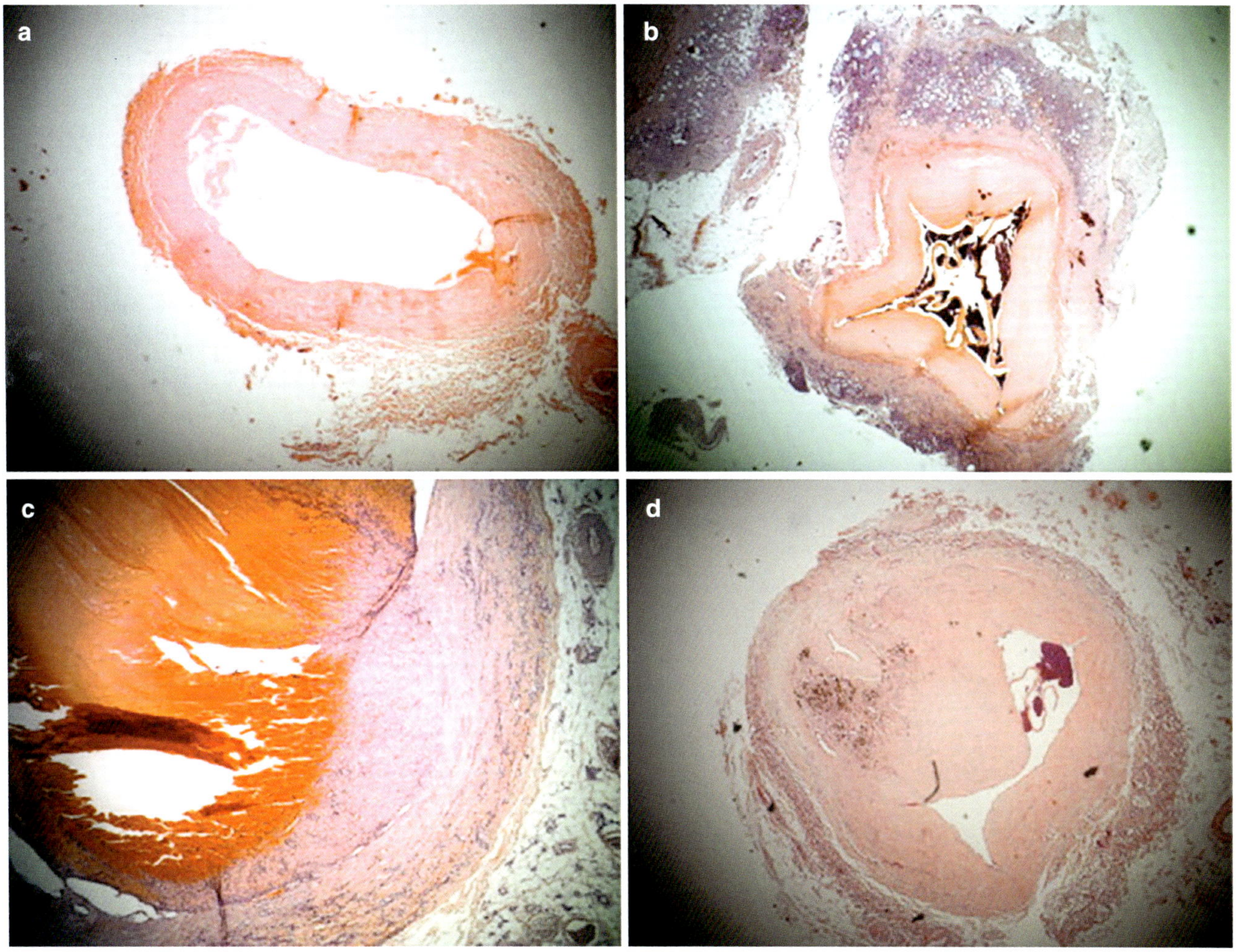

Fig. 72.2 Microscopic Examination after steam ablation (**a**) immediately after steam ablation (**b**, **c**) after 20 days (**d**) after 3 months

Conclusion

Steam ablation of the saphenous vein using the SVS is a novel method of endovenous thermal ablation. It appears to be safe, effective, and highly appreciated by patients. It may potentially have advantages over currently available thermal therapies [4].

References

1. van den Bos RR, et al. Proof-of-principle study of steam ablation as novel thermal therapy for saphenous varicose veins. J Vasc Surg. 2011;53(1):181.
2. Milleret R, et al. Great saphenous vein ablation with steam injection: results of a multicentre study. Eur J Vasc Endovasc Surg. 2013;45(4):391.
3. Whiteley MS. Glue, steam and Clarivein—Best practice techniques and evidence. Phlebology. 2015;30(2 Suppl):24–8.
4. Van den Bos RR, Malskat WSJ. Steam ablation of varicose veins 5 years later. Ned Tijdschr Geneeskd. 2020;164:D5105.

73 Cyanoacylate Glue Ablation for Chronic Venous Disease

Dimitrios Kontothanassis

Case Presentation
A 77-year-old male was referenced to our clinic for severe bilateral CVD for about 40 years and symptomatic reflux of bilateral GSV, symptomatic on the left limb. The patient had several superficial vein thrombosis (SVT) in the past at the left leg, treated with LMWH. His clinical history was really important; In 2013 myocardial infarction, in 2014 atrial fibrillation, condition that required anticoagulation with dicumarol drugs, moderate chronic renal failure, respiratory failure due to long Covid (Covid infection in 2022) and recently for admission to the emergency room for severe blood loss because of varicorrhagia. Skin was very thin both at the thigh and the calf so compression with elastic stocking was discouraged and skin damage was extended bilateral at the calf (CEAP classification C2,3,4,5s). Heaviness, pain, and itching were experienced on a daily basis.

During other consultancies, surgeons contraindicated a surgical operation because of the poor clinical conditions. Reexamination of the patient and ecocolordoppler study of the abdominal vessels and both inferior limbs was done in our clinic to exclude potential residual thrombus in the superficial veins or obstruction of deep veins. The problems to resolve in this case were: (a) vein aneurysms, (b) extensive varicosities (c) avoid skin and vein puncture or phlebectomies because of the anticoagulation, (d) renal failure, (e) hemorrhage from vein blisters at the thigh, and (f) use with caution a noninvasive procedure to carry out the operation without using high volumes of local anesthesia.

Treating these patients is really rare but mandatory when blood loss becomes frequent.

In such extreme cases, non-thermal techniques like catheter directed cyanoacrylate glue (CAC) under ultrasound guidance are preferable to achieve rapid occlusion of the vein without puncturing the skin and avoiding bleeding just in case of accidental puncturing of the vein. If you need to perform low volumes of tumescent anesthesia, cold saline anesthesia at 4 °C is the proper treatment for gaming extreme vein spasm and reducing the vein diameter (Figs. 73.1, 73.2, 73.3, and 73.4).

Continued at page 776

D. Kontothanassis (✉)
General and Vascular Surgery, Phlebology Department, Istituto Flebologico Italiano, Padova, Italy

G. Geroulakos et al. (eds.), *Mastering Endovascular Techniques*, https://doi.org/10.1007/978-3-031-42735-0_73

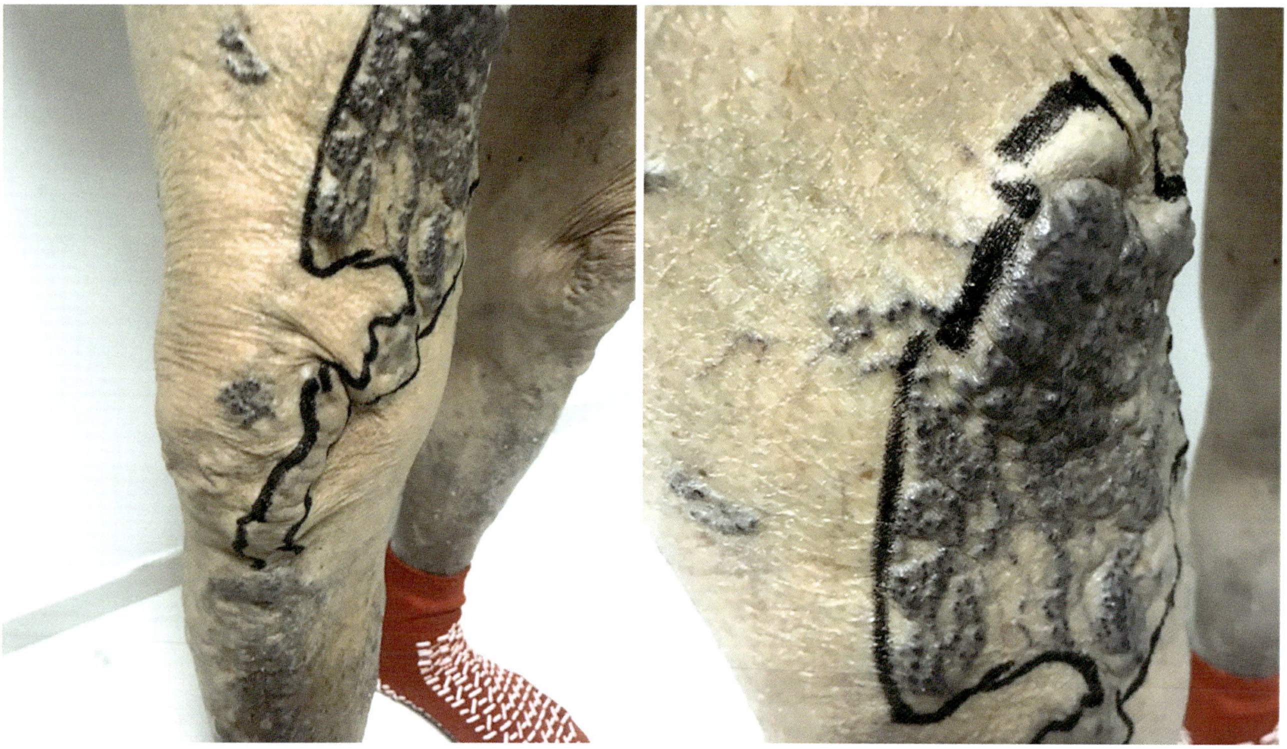

Fig. 73.1 Leg condition before surgery

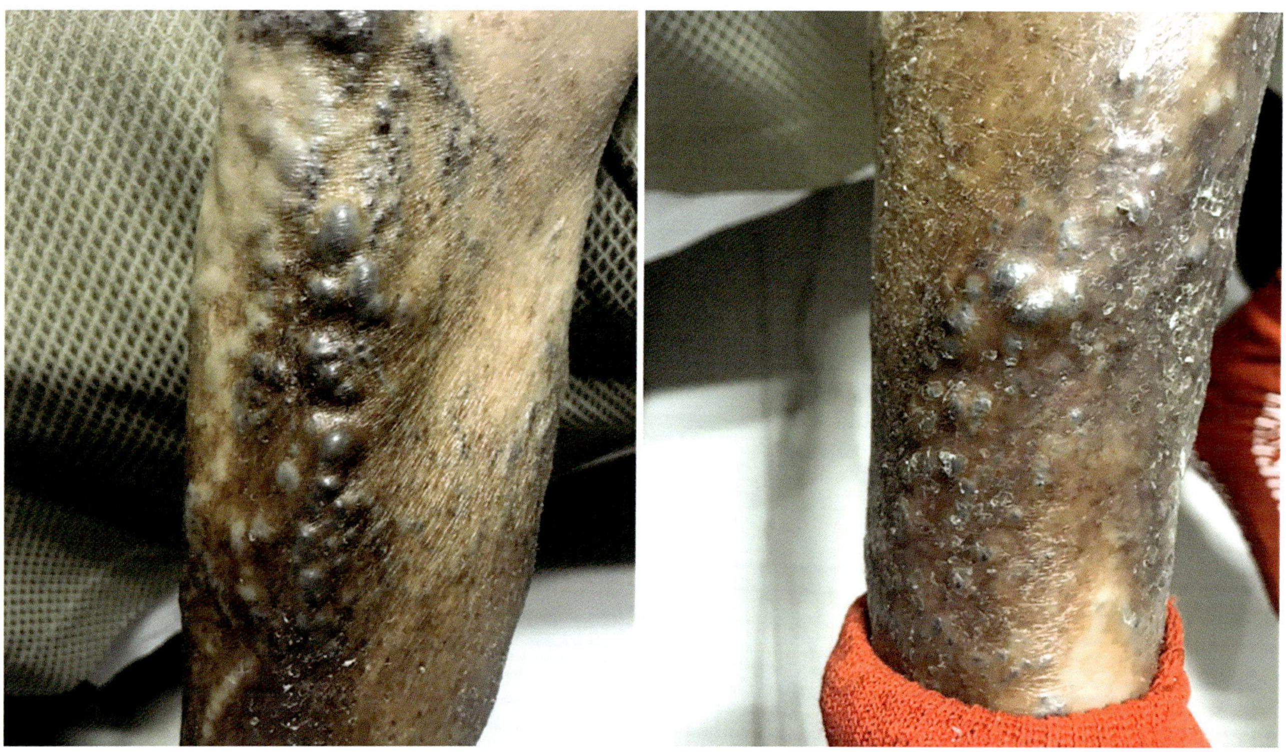

Fig. 73.2 Leg condition before surgery

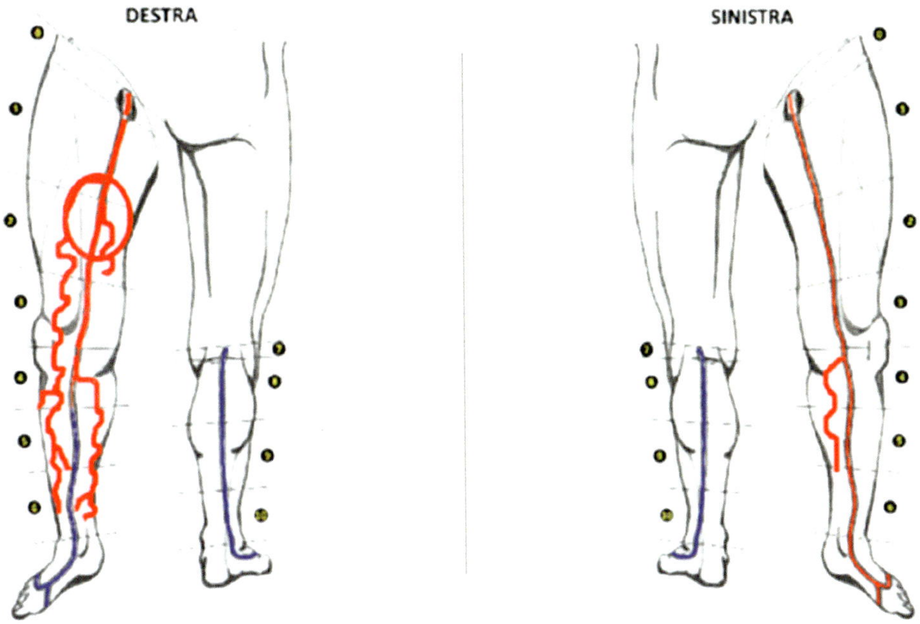

Fig. 73.3 Hemodynamic map of GSV, SSV and varicosities

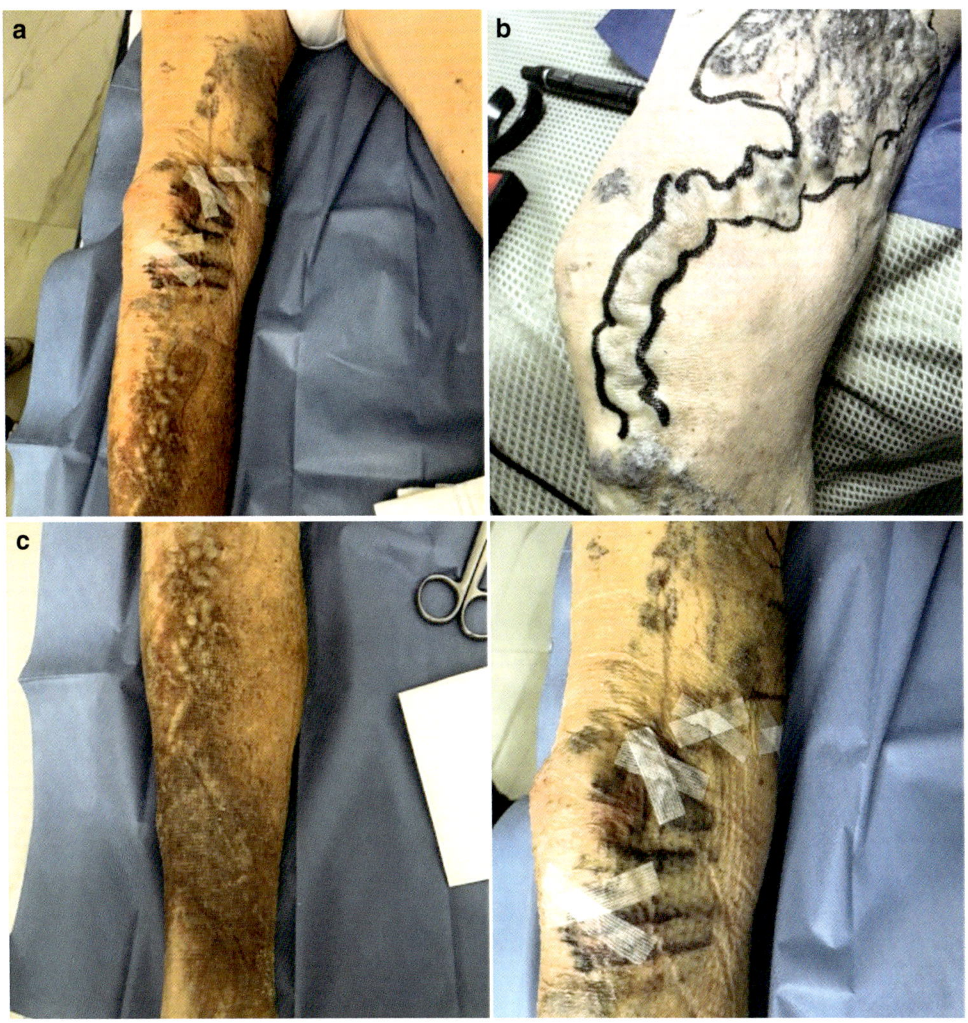

Fig. 73.4 (**a**) Limb view after surgery and (**b**) Limb view before surgery. (**c**) Limb view after 1 week from surgery

Introduction

Catheter directed cyanoacrylate (CAC) is a non-thermal, non-tumescent alternative to Endovenous Thermal Ablation (EVTA) techniques to occlude incompetent superficial veins of the lower limbs. Upon intravenous injection, cyanoacrylate rapidly solidifies via a polymerization reaction and produces an inflammatory reaction of the vein wall [1].

CAC differ depending on the length of the carbonyl group of the molecule. N-butyl cyanoacrylate (N-BCA) is the most common adhesive used in medicine and was introduced in the late 1960s for closure of minimum tension wounds and as a substitute for suturing [2] (Fig. 73.5).

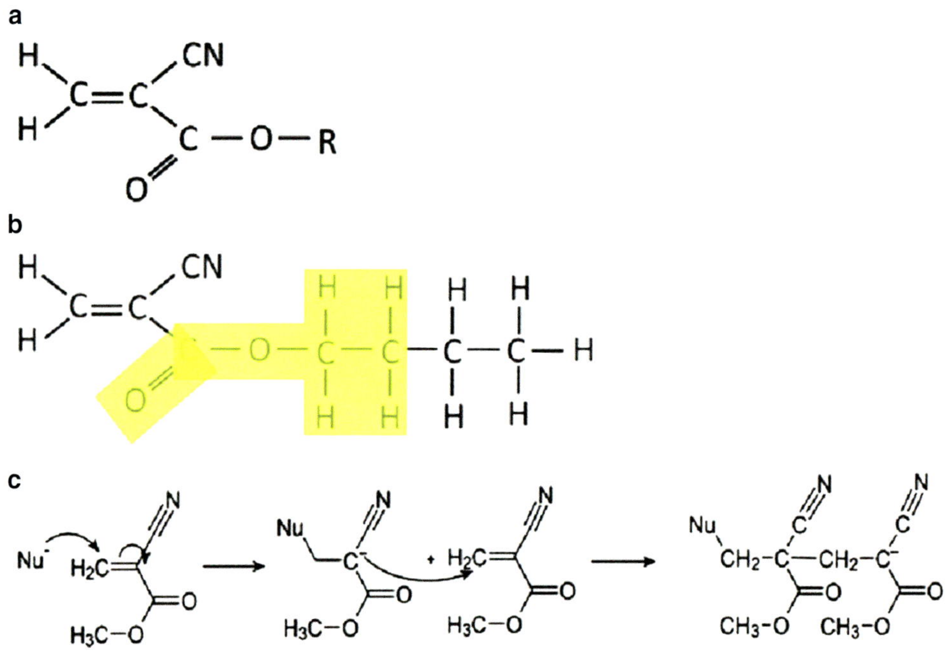

Fig. 73.5 Molecular structure of cyanoacrylate and the polymerization process (adapted from Pollak and White 18). (**a**) Monomeric cyanoacrylate CH 2 ¼C(CN) ÀCOOR. The R represents an alkyl group. (**b**) n-Butyl-2-cyanoacrylate (n-BCA). The carboxyl ethyl (ÀCOOHÀCH 2 ÀCH 2) group is the antigenic component of the molecule. (**c**) Polymerization process of cyanoacrylate (CA). Exposure of CA monomers to an anion, such as the hydroxyl (ÀOH) group in water initiates the polymerization process with bonding of the ethylene units

The Devices

Several CAC systems for treatment of peripheral veins are available, and currently, at least four types of cyanoacrylate vein adhesive devices are commercially available for superficial vein incompetence treatment (VenaSeal™, VenaBlock™ and Veinoff™ and Venex™). The main difference between these devices relates to the cyanoacrylate formulation and application techniques.

VenaSeal™

VenaSeal™ closure system (Medtronic Mienneapolis, Minesota, USA), was initially developed by Dr. Rodney Raabe, the original founder of Sapheon Inc. that was later acquired by Covidien in 2014, itself subsequently acquired by Medtronic (Minnesota, USA). This product obtained CE (Conformité Europécne) marking in 2011 and then received registration with the TGA in 2012 as a medical device. The FDA approved the VenaSeal™ system in 2015 and approval was granted in the Russian Federation in 2017 8. The product is currently in clinical use across most continents. 9 The VenaSeal™ system is a catheter-assisted endovenous method to deliver N-BCA. The N-BCA is delivered via a dispensing gun that is attached to the delivery catheter. The sheath and dilator are high density polyethylene (HDPE) and the delivery catheter is a polytetrafluorethylene (PTFE) product. Compared to other products in this category, VenaSeal™ has the highest viscosity and longest polymerization time and hence is the slowest to act. Polymerization begins approximately 5 s after contact with a liquid containing water (such as blood) and extends to 3 min for nearly complete polymerization. The high viscosity of the agent works to prevent extension of the adhesive process to adjoining non-target vessels (Fig. 73.6).

Complications of CAC

The most common complication of CAC is phlebitis, reported at a rate of 11.4% in the study by Almeida et al. 57 and 20% in a study by Morrison et al. 54. CAC is heavily marketed with a claim that it does not require the concomitant use of GCS or TA58. The nature of phlebitis following CAC may be related to a hypersensitivity reaction described below and not necessarily a superficial thrombophlebitis (STP) due to thrombus formation. In comparison, phlebitis following EVLA is reported at 7.7%–7.9% 60 and following RFA at 14%. The treatment protocols of both EVLA and RFA require the use of GCS and TA. Whether the relatively high incidence of phlebitis post-CAC is due to lack of compression or a hypersensitivity-type reaction is yet to be determined. Immediate and delayed hypersensitivity reaction with granuloma formation in particular have been the most significant concern of clinicians. Morover than hypersensitivity and phlebitis, CAC is associated with other complications, summarized.

VenaBlock™

This product was developed by Invamed (Ankara, Turkey) and obtained registration with the TGA in 2016 as a medical device. VenaBlock™ is available for both catheter directed administration as well as direct percutaneous injections. This product is at least 60 times less viscous than VenaSeal™ but 20 times more viscous than water. It has a very short polymerization time and is one of the fastest to act; therefore, it reduces the possibility of migration and extension.

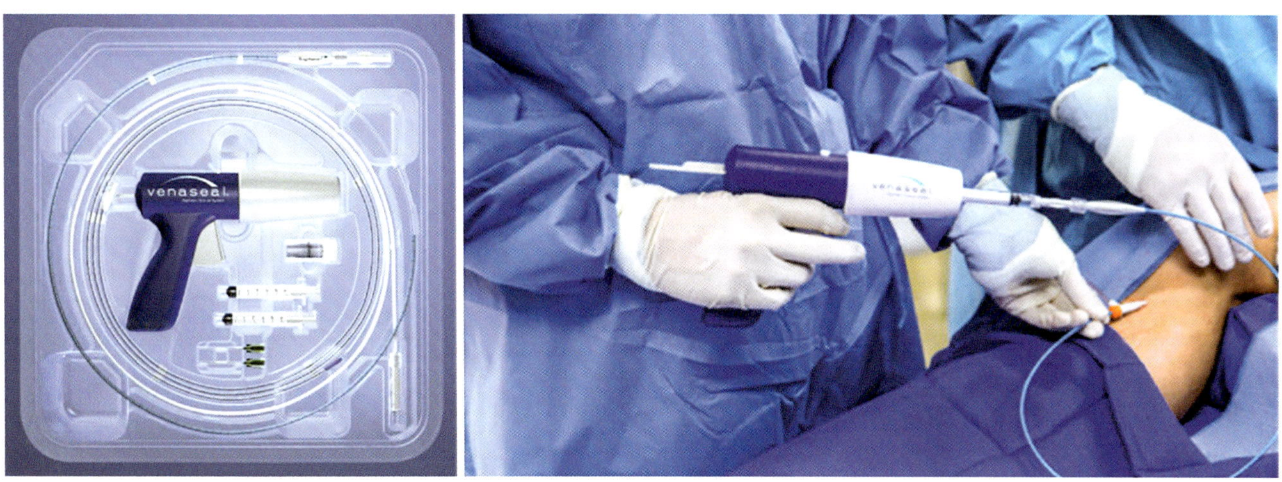

Fig. 73.6 Venaseal™ procedure kit

Veinoff™

This product was developed by Invamed for percutaneous injections and was registered with the TGA in 2017. Veinoff™ is the least viscous and comparatively most flexible material after polymerization with a slightly longer polymerization time and hence slower action than VenaBlock™ but still faster than VenaSeal™.

VariClose™

There are new products entering the international market currently. VariClose™ was developed by Biolas Health Inc. (Ankara, Turkey) and is currently not registered in Australasia.

Venex™

Another Turkish product, Venex™· has been developed by Vesta Medical (Ankara, Turkey) and is now available in Europe and in Australia. Probably is the latest and most interesting device due to the absence of DMSO (Di -Metyl-Sulf-Ossido) that guarantees absence of post procedural inflammation; It has a faster polymerization time than Venaseal. The use of 6 mL of Glu is enough to close a 45 cm length of incompetent saphenous vein while the use of 16 mL in one time operation in rare cases of Venous malformations demonstrate efficacy and safety without post procedural tissue inflammation. Venex™ is available for both catheter directed administration as well as direct percutaneous injections. Cost efficacy of the device seems very promising for the future (Figs. 73.7 and 73.8).

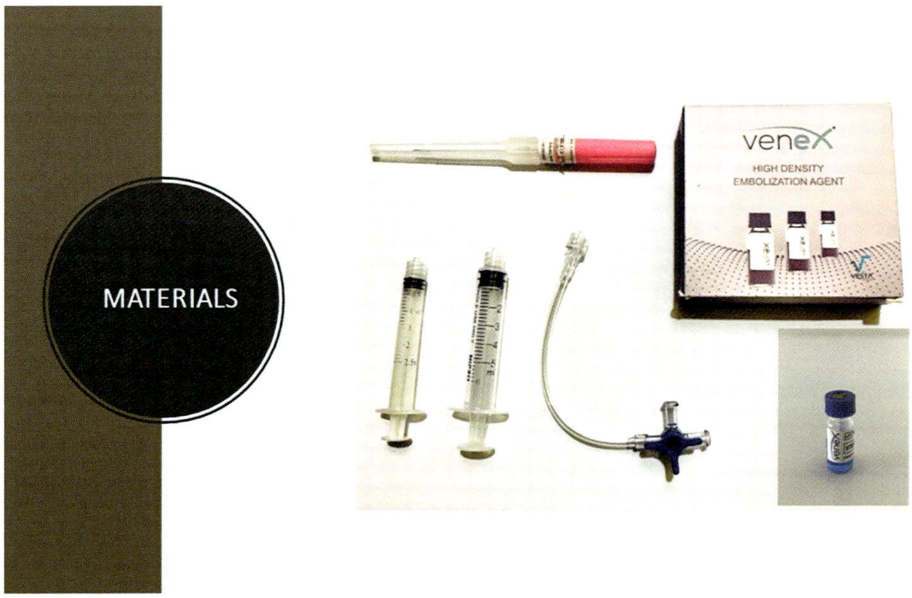

Fig. 73.7 Venex procedure kit

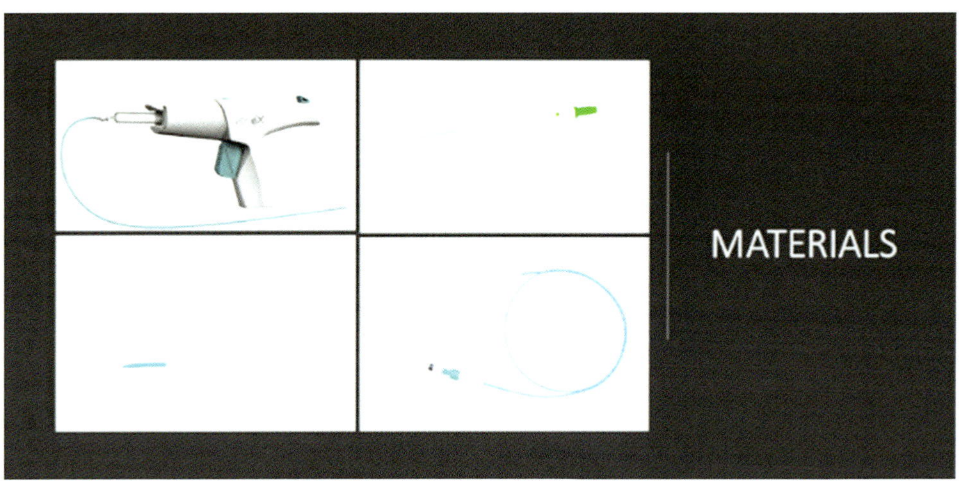

Fig. 73.8 Venex procedure kit

Indications and Contraindications

Indications

Saphenous Reflux of GSV, Anterior GSV, and SSV is indicated for catheter directed CAC,; incompetent Venous Tributaries and perforators can be directly injected with CAC.

Contraindications

N-BCAs are implantable foreign bodies that trigger an immune or hypersensitivity reaction. The list of contraindications needs to be revised on a regular basis to reflect clinical and scientific evidence for precautions as further knowledge is obtained. In higher risk patients, n-BCA should only be used when there are no reasonable alternatives to treatment and a full disclosure of risks is discussed.

Hypersensitivity History of immediate (urticarial) or delayed hypersensitivity reactions to acrylates and commercial or medical grade CA preparations.

Previous significant adverse reactions to CAC: This includes adverse events such as extensive ‘phlebitis’, necrosis, suppuration, oedema, and other such reactions.

It also includes acute venous thromboembolism (VTE), deep vein thrombosis (DVT), pulmonary embolism (PE), and superficial thrombophlebitis (STP).

Active or uncontrolled systemic disease: systemic inflammatory disorders, un-controlled systemic autoimmune, granulomatous, hypersensitivity or mast cell disorder including vasculitis, mastocytosis, sarcoidosis, systemic lupus erythematosus (SLE), atopy, and granulomatous disease or infections.

We also recommend granulomatous vasculitis disorders such as Wegener’s granulomatosis or Churg-Strauss disease should be considered an absolute contraindication.

Acute or un-controlled localized or systemic infections: This includes cellulitis in the affected leg, widespread folliculitis, and organ-specific infections.

Warnings and Precautions

Pregnancy Treatment is a contraindication and Breastfeeding within 48 h of the procedure. For 48 h following intervention, breast milk should be pumped and disposed of.

Low body fat percentage and prominent veins can frequently feel the firm cord of the treated vein along the thigh and calf and ask whether this will disappear.

Care should be taken in treating patients with systemic autoimmune disorders until further data on the safety of n-BCA in patients with autoimmune disease is available.

Thromboprophylaxis for patients with thrombophilia and hypercoagulable state should be offered for at least 7 days. This includes documented known significant thrombophilia, immobility including long-haul travel of more than 5 h continuous travel, not including travel on ships or trains, within a 2-week period. Treatment in patients with active malignancy or those on tamoxifen should be performed with care and prophylactic treatment with low molecular weight heparin (LMWH) should be provided for seven postoperative days.

The Technique

CAs are liquid polymers that polymerize on exposure to free radical and anions. Care should be taken not to draw up the product in a catheter or syringe flushed with saline, water or containing blood. The technique is promoted as a non-thermal non-tumescent (NTNT) procedure not requiring post-operative compression. There is no scientific data regarding the maximum dose of n-BCA per treatment session usually based on practical considerations and vein length. The safe lifetime upper limit of n-BCA is not known. Graduated Compression Stockings Reported incidence of phlebitis post CAC is up to 20% whereas by comparison the incidence of phlebitis in EVLA is up to 7.9% 60 and for RFA is up to 14% [3].

Consent Inform and Documentation prior the procedure is very important and must include a the following: The date and time of the procedure, treated leg(s), treated vein(s), length and diameter of veins treated, the type and the brand name of the n-BCA used, dosage of n-BCA used, details of any other concurrent or adjunctive procedures performed, type of compression applied, immediate adverse reactions and post-treatment management. Catheter-directed CAC should be performed in operating theatres, hybrid theatres, or appropriately equipped and staffed outpatient procedure rooms with access to appropriate resuscitation equipment. It is recommended to use full sterile surgical technique and high-performance ultrasound equipment.

Catheter-Directed Treatment of Saphenous Reflux

The procedure involves an introducer sheath, a dispensing catheter, and a 3 mL syringe to be attached to a dispenser gun. The total amount of adhesive to be delivered in one treatment session needs to be calculated as a function of the vein length and diameter. For venaseal procedure, a total of 5 mL of glue is provided in one package, 1.4 mL of which is wasted in the dead space of the catheter. The remaining 3.6 mL can be used to treat up to 90 cm of vein length if the vein diameter is less than 6 mm.

Select the access point based on the clinical indications and the treatment plan and access the vein using a Seldinger technique. Insert the introducer sheath and advance to 5 cm distal to the saphenous junction and flush with normal saline. Prime the syringe with n-BCA then connect to the dispenser gun and the dispensing catheter. Prime the catheter with n-BCA to 3 cm short of the tip. Insert the dispensing catheter into the introducer sheath and pull back the sheath so that the tip of the dispensing catheter is 5 cm distal to the junction. Guide and follow the procedure on ultrasound and use the ultrasound probe to compress the saphenous junction. Inject a 0.1 mL aliquot of adhesive into the vein by pulling the trigger of the dispenser gun and hold for 3 s, then pull back 1 cm and deliver another 0.1 mL of adhesive for 3 s. Pull the catheter and sheath back by 3 cm, maintain proximal compression with the ultrasound probe and light digital compression for three full minutes before the next trigger pull. Maintain proximal compression with the probe and deliver 0.1 mL aliquots of adhesive at 3 cm intervals down the vein waiting each time for 30 s after each subsequent trigger pull. Confirm vein closure with ultrasound, then withdraw the delivery systems. For veins larger than 6 mm diameter, consider 0.1 mL aliquots at 2 cm intervals. Stop 5 cm proximal to the access site, withdraw the dispenser catheter into the introducer sheath and remove the system, then apply pressure until hemostasis occurs.

Percutaneous Treatment of Tributaries

VenaBlock™, Vennex™, and Veinoff™

Direct percutaneous administration can be used to treat both saphenous veins and tributary veins. Venous tributaries can be dealt with at the same time as treatment of saphenous trunks or at a later date. Care should be taken with the injection of these substances superficially and monitoring is required to report glue extrusion as observed in gastric cases. Veinoff™ can be mixed with 5% dextrose in a ratio of 2:1 (dextrose: Veinoff™) but the syringe needs to be shaken to ensure that the mixture is homogenous. The volume injected at each site is determined by the vein diameter. Withdraw the adhesive into a 1 mL syringe and inject with a 22–25-gauge needle. The adhesive must be injected as expeditiously as possible to avoid hardening within the syringe. Inject approximately 0.1–0.2 mL of adhesive into the target vein at 2–3 cm intervals. Apply firm compression after each injection for 60 s to prevent palpable lumps forming in the vein. It is highly recommended not to draw blood back into the hub of the needle as this will result in casting of the CAC in the syringe and needle, making it virtually impossible to inject the medical adhesive. If injection is performed close to the deep system, ideally it is advisable to deliver adhesive at least 2 cm away from the deep veins where possible.

Post-Operative Management

The postoperative course is usually benign. Most patients experience mild tightness and discomfort for up to 2 weeks but not sufficient to limit normal activities. Some patients develop a moderate to severe inflammatory reaction. Treatment usually requires applying compression and oral non-steroidal anti-inflammatory drugs (NSAIDS). The patient should be mobilized immediately after treatment. Normal activities can be resumed immediately but heavy physical activities should be avoided for at least 7–14 days. Many patients find comfort from support stockings for a few days after treatment. Postoperative Ultrasound Examinations is recommended within 7 days of treatment to exclude VTE and every year after for early and long-term follow-up to assess residual patency at any site of the treated vein.

Cyanoacrylate in the Treatment of CVD

Several studies [1] have shown that CAC is safe and effective to ablate the incompetent GSV, with cumulative occlusion rates comparable with those for EVTA in the early and mid-term observations, up to 3 years follow up.

The WAVES study further demonstrated the efficacy of CAC for SSV, AASV, and large GSV (maximum diameter up to 20 mm), with an occlusion rate of 98%, although according to another study, a mean GSV diameter > 8 mm appeared to be a statistically significant predictor of recanalization ($p < 0.021$). An increased GSV diameter was also associated with a longer stump remnant. A few RCTs have compared CAC with EVTA.

The VeClose multicenter RCT showed that CAC is non-inferior to RFA. In 222 patients with GSV reflux, the 36 month occlusion rates were 94.4% for CAC and 91.9% for RFA, both with a stable improvement in QoL. The 5-year extension study of this RCT still showed an occlusion rate of 93.6% in 47 patients (43.5%) of the CAC group reaching 60 months follow up, without serious adverse events.

Another RCT compared CAC with EVLA in 400 patients with GSV incompetence and at 12 months, complete occlusion was found in 96.6% and 94.1%, respectively.

Another RCT compared CAC with EVLA and RFA in 525 patients and demonstrated 24 month occlusion rates of 94.7%, 90.9%, and 91.5% after CAC, RFA, and EVLA, respectively.

Several systematic reviews and meta-analyses have shown no statistically significant difference in technical success rates when CAC was compared with EVTA techniques.

The r-VCSS and QoL measures improved statistically significantly after CAC in all studies, with no statistical difference between EVTA techniques and CAC in comparative studies, with an exception of one RCT in which r-VCSS scores were statistically significantly lower at 6 months and 2 years in the CAC group ($p < 0.001$).

Pain during the CAC intervention was significantly less, and CAC treatment time and recovery time were considerably shorter than after EVTA.

Cyanoacrylate in the Treatment of Venous Malformations

Glue embolization and resection of focal VM of variable size and location appears to have durable results and low surgical morbidity. This single-stage procedure, often performed as an outpatient, may be utilized as upfront treatment for symptomatic malformations or for VM refractory to other treatments as sclerotherapy [4] (Figs. 73.9, 73.10, and 73.11).

Complications of CAC

The most common complication of CAC is phlebitis, reported at a rate of 11.4% in the study by Almeida et al. [5] and 20% in a study by Morrison et al. [6] which potentially can be related to a local hypersensitivity reaction to the cyanoacrylate injection.

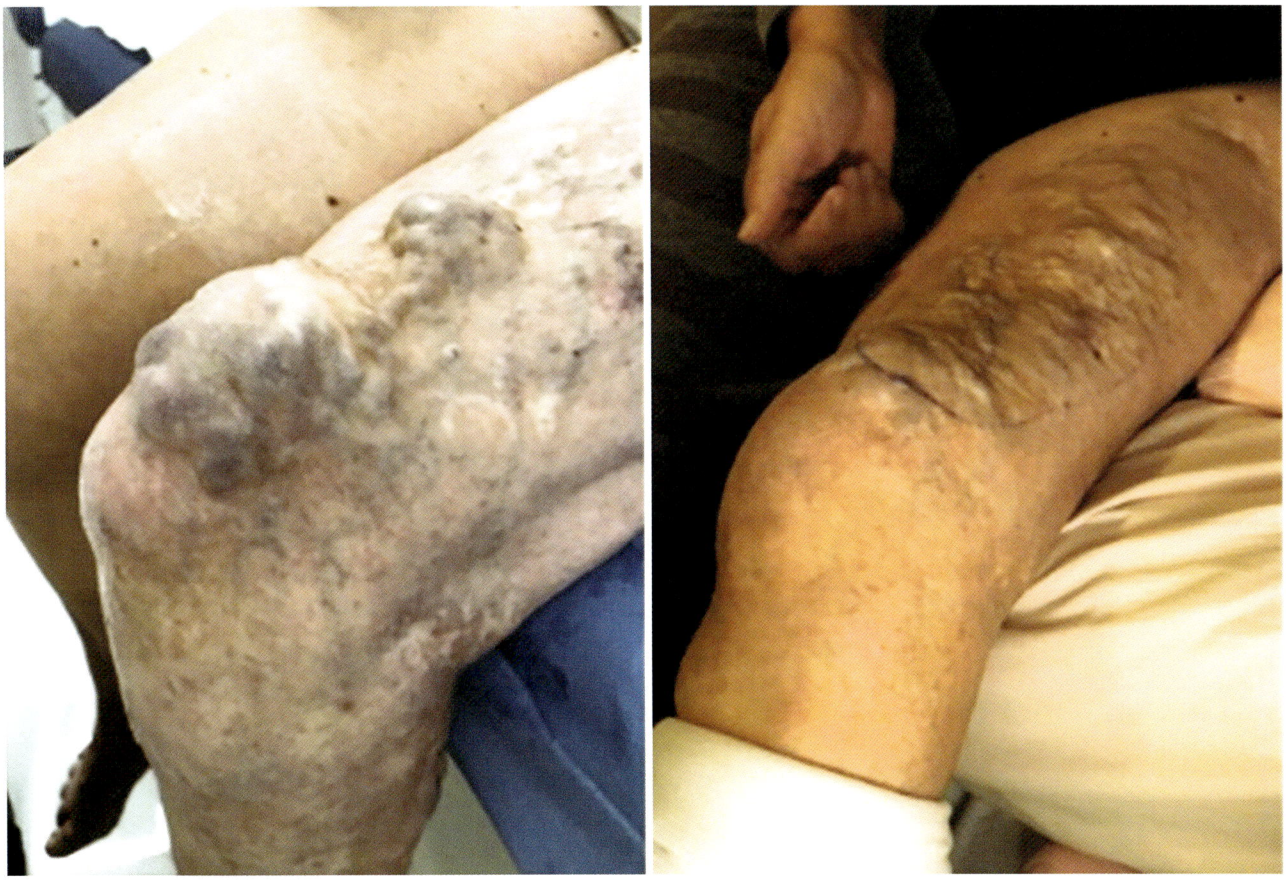

Fig. 73.9 Venous malformation treated with glue (Venex)

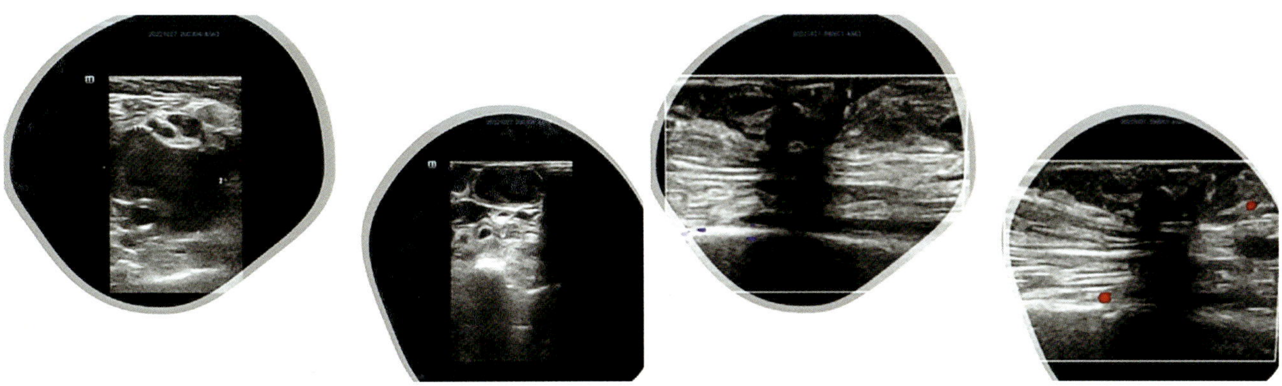

Fig. 73.10 Before and after treatment of intramuscular venous malformations

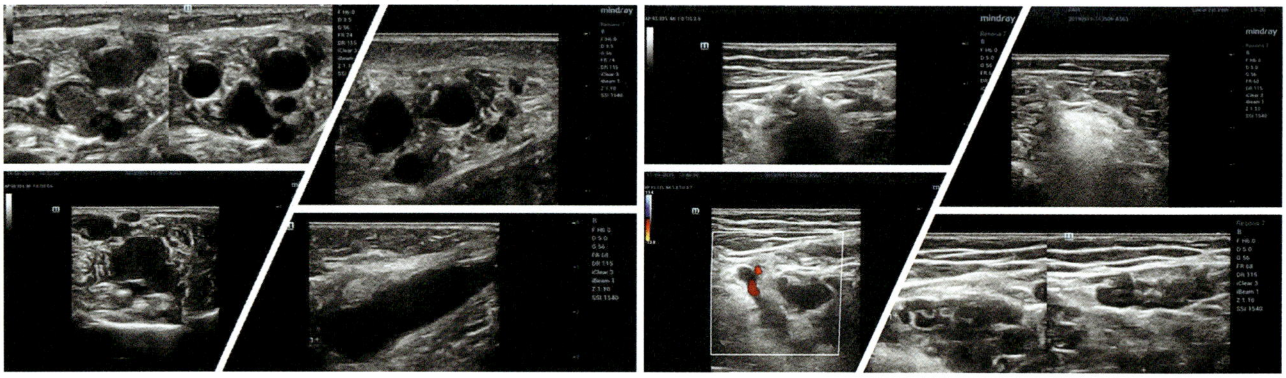

Fig. 73.11 Before and after treatment of intramuscular venous malformations

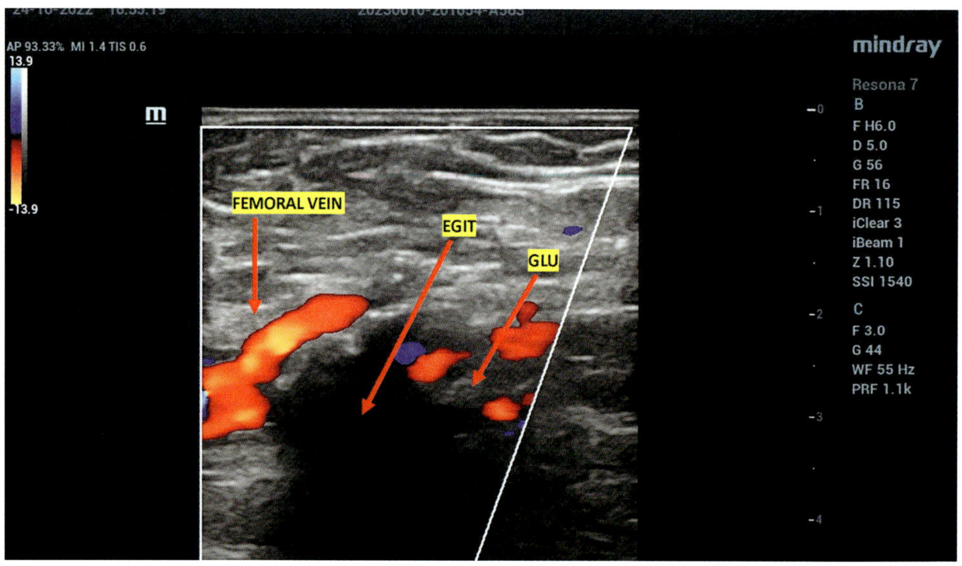

Fig. 73.12 DVT (EGIT – Endovenous Glue Thrombosis)

In comparison, phlebitis following EVLA is reported at 7.7–7.9% [3, 7] and following RFA at 14% [6]. Whether the relatively high incidence of phlebitis post-CAC is due to lack of compression, or a hypersensitivity-type reaction is yet to be determined. It is mostly transient, benign, and self-limiting. In recent literature a rate of hypersensitivity reaction to cyanoacrylate of 6%–7% has been suggested (mild 4.2%, moderate 1.3%, and severe 0.3%), and known hypersensitivity to acrylate should be considered a contraindication to CAC treatment.

Late granulomatous reactions are rare despite the large number of cases performed worldwide; some of them with considerable morbidity. Further registration and adequate follow up of these rare cases are required, as late adverse reactions may occur even up to 1 year post-operatively [8, 9].

Other complications include DVT (0%–3.5%), cyanoacrylate protrusion into the SFJ or endovenous glue induced thrombosis (1.4%–5.8%), hyperpigmentation (1.6%–3%), access site infection or cellulitis (1.4%–3%), hematoma (1.4%–1.6%), and nerve injury or paresthesia (0%–2%) (Figs. 73.12 and 73.13) [10, 11].

Results and Recommendations

CAC appears to be an effective treatment for saphenous reflux with occlusion rates at 36 months of 90–95% [2]. We recommend a maximum dose of 10 mL of CA per treatment session. Serious complications are rare, but significant. Hypersensitivity to acrylates is reported in 2.4% of the population and is an important absolute contraindication to CAC [1, 3] (Figs. 73.14 and 73.15).

Post procedural inflammatory reactions, including hypersensitivity-type phlebitis (HTP), occur in 10–20% of patients. In the long-term, CAC results in foreign-body granuloma formation within 2–12 months of the procedure. We recommend against the use of CAC in patients with uncontrolled inflammatory, autoimmune, or granulomatous disorders (e.g. sarcoidosis). Caution should be exercised in patients with significant active systemic disease or infection and in these patients, alternative therapies such as thermal ablation and foam sclerotherapy should be considered (Fig. 73.16).

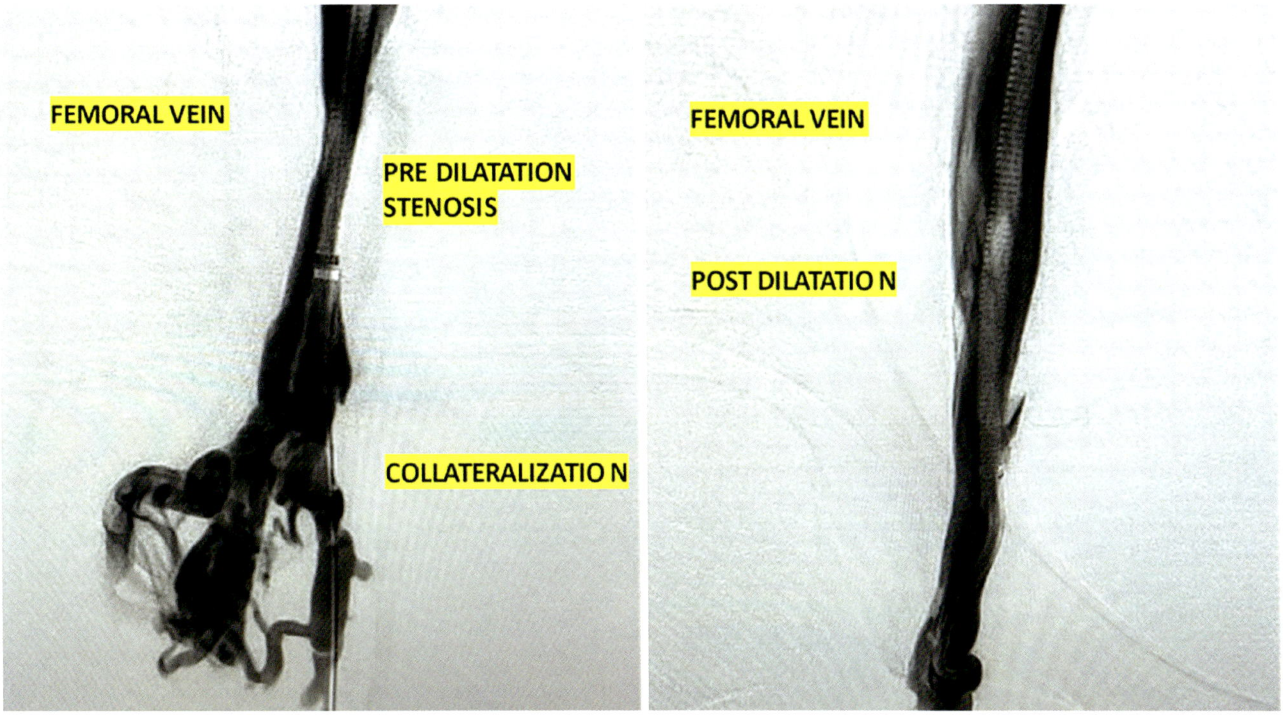

Fig. 73.13 DVT - Femoral vein before and after angioplasty

Fig. 73.14 GSV obliteration with glue

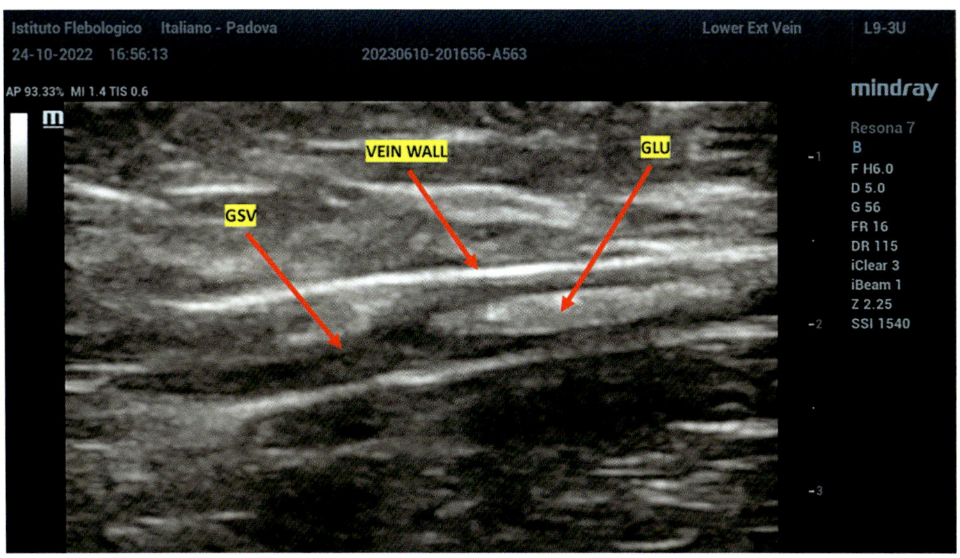

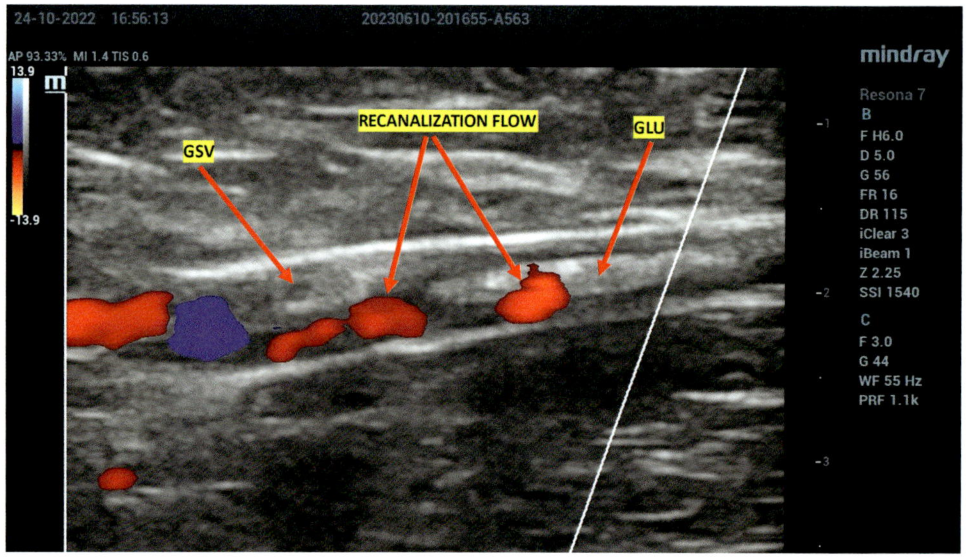

Fig. 73.15 Recanalization after glue ablation

Fig. 73.16 Prominent superficial vein treated with glue

Conclusions

CAC appears to be an effective endovenous procedure, with short term closure rates comparable to ETA and therefore greater efficacy than traditional surgery for treating superficial veins of the lower limbs. Ongoing data collection is required to establish the long-term safety and in particular, the risks associated with immune-mediated or foreign body type reactions. Long term follow up (>10 years) is mandatory to assess recurrence, progression of disease, and relapse of vein reflux in the treated areas.

Case Presentation

Continued from page 765

CONCLUSION: The patient presented with severe CVD and life threatening conditions (blood loss in patient with myocardial infarction and atrial fibrillation), classified CEAP Class C2,3,4 s, which could be related to chronic superficial vein reflux, superficial vein thrombosis, presence of extensive superficial varicosities and bleeding from vein blisters at the calf. Symptomatic varicose veins can be treated safely in high-risk patients as in this case presentation by using non-thermal procedures as CAC or Foam sclerotherapy. Even if occlusion rate does not reflect clinical success the aim of the operation was to stop bleeding and reduce the possibility of recurrent hemorrhage and also down grade the CEAP classification as safe as possible. We believe that in higher risk patients with poor clinical conditions and when there are no reasonable alternatives to treatment, glue should only be used, and a full disclosure of risks has to be discussed with the patient.

References

1. Parsi K, et al. Cyanoacrylate closure for peripheral veins: consensus document of the Australasian College of Phlebology. Phlebology. 2020;35:153.
2. De Maeseneer MG, et al. European Society for Vascular Surgery (ESVS) 2022 clinical practice guidelines on the management of chronic venous disease of the lower limbs. Eur J Vasc Endovasc Surg. 2022;63:184–267.
3. Bozkurt AK, Yilmaz MF. A prospective comparison of a new cyanoacrylate glue and laser ablation for the treatment of venous insufficiency. Phlebology. 2016;31:106–13. 2016/02/27.
4. Polites SF, et al. Single-stage embolization with n-butyl cyanoacrylate and surgical resection of venous malformations. Pediatr Blood Cancer. 2020;67(3):e28029.
5. Almeida JI, Javier JJ, Mackay EG, et al. Two-year follow-up of first human use of cyanoacrylate adhesive for treatment of saphenous vein incompetence. Phlebology. 2015;30:397–404.
6. Morrison N, Gibson K, McEnroe S, et al. Randomized trial comparing cyanoacrylate embolization and radiofrequency ablation for incompetent great saphenous veins (VeClose). J Vasc Surg. 2015;61:985–94.
7. Proebstle TM, Alm J, Dimitri S, et al. The European multicenter cohort study on cyanoacrylate embolization of refluxing great saphenous veins. J Vasc Surg Venous Lymphat Disord. 2015;3:2–7. 2015/01/01.
8. Zernovicky F. Fast progressive devastating granulomatic reaction after VenaSeal procedure. Int Angiol. 2018;37:168–9.
9. Cyanoacrylate glue occlusion for varicose veins, June 2015. https://www.nice.org.uk/guidance/ipg526. Accessed 11 March 2019.
10. Eroglu E, Yasim A. A randomised clinical trial comparing n-butyl cyanoacrylate, radiofrequency ablation and endovenous laser ablation for the treatment of superficial venous incompetence: two year follow up results. Eur J Vasc Endovasc Surg. 2018;56:553–60. 2018/07/26.
11. Chan YC, Law Y, Cheung GC, et al. Cyanoacrylate glue used to treat great saphenous reflux: measures of outcome. Phlebology. 2017;32:99–106.

Mechanochemical Ablation (MOCA) for Chronic Venous Disease

74

Dimitrios Kontothanassis

Case Presentation MOKA

An 83-year-old male was referenced to our clinic for left leg ulcer and recurrent varicose veins after stripping of the right GSV (R GSV). The operation was done approximately 30 years ago and in 2022, he was referred to our clinic for clinical evaluation and counselling for redo surgery. His clinical history was not relevant exception for leg ulcer ongoing for almost 5 years. The patient never wore elastic stockings but he was doing weekly medications of the ulcer and wrapping of the leg with bandages.

During other consultancies, surgeons contraindicated a surgical operation because of the age and prior surgery. Re-examination of the patient and ultrasound study of the abdominal and inferior limb vessels, both arterial and venous flow, was done in our clinic to check the main sources of reflux, exclude obstruction of deep veins and arterial disease. The ultrasound study demonstrated a residual great saphenous vein at the calf and varicosities near the ulcer area. The problems to resolve in this case were: (a) heal the ulcer (b) restore good clinical conditions.

Treating patients with ulcers is challenging because infection is always a silent threaten and treating recurrent varicosities is really difficult because of the skin damage, infection, and inflammation. The complex anatomy of anarchic veins after GSV ablation have to be well studied with ultrasound and make a good mapping on the skin prior to surgery with a permanent ink marker. Laser transilluminator can help in mapping vein anatomy.

In such cases, non-thermal techniques like MOKA, foam sclerotherapy or injection of cyanoacrylate glue (CAC) under ultrasound guidance are preferable to achieve rapid occlusion of the recurrent anarchic vein circuit spread all over the limb and healing the ulcer. Phlebectomies at the ulcer area are contraindicated because of infection (Figs. 74.1 and 74.2).

D. Kontothanassis (✉)
General and Vascular Surgery, Phlebology Department, CEO
Istituto Flebologico Italiano, Padova, Italy

G. Geroulakos et al. (eds.), *Mastering Endovascular Techniques*, https://doi.org/10.1007/978-3-031-42735-0_74

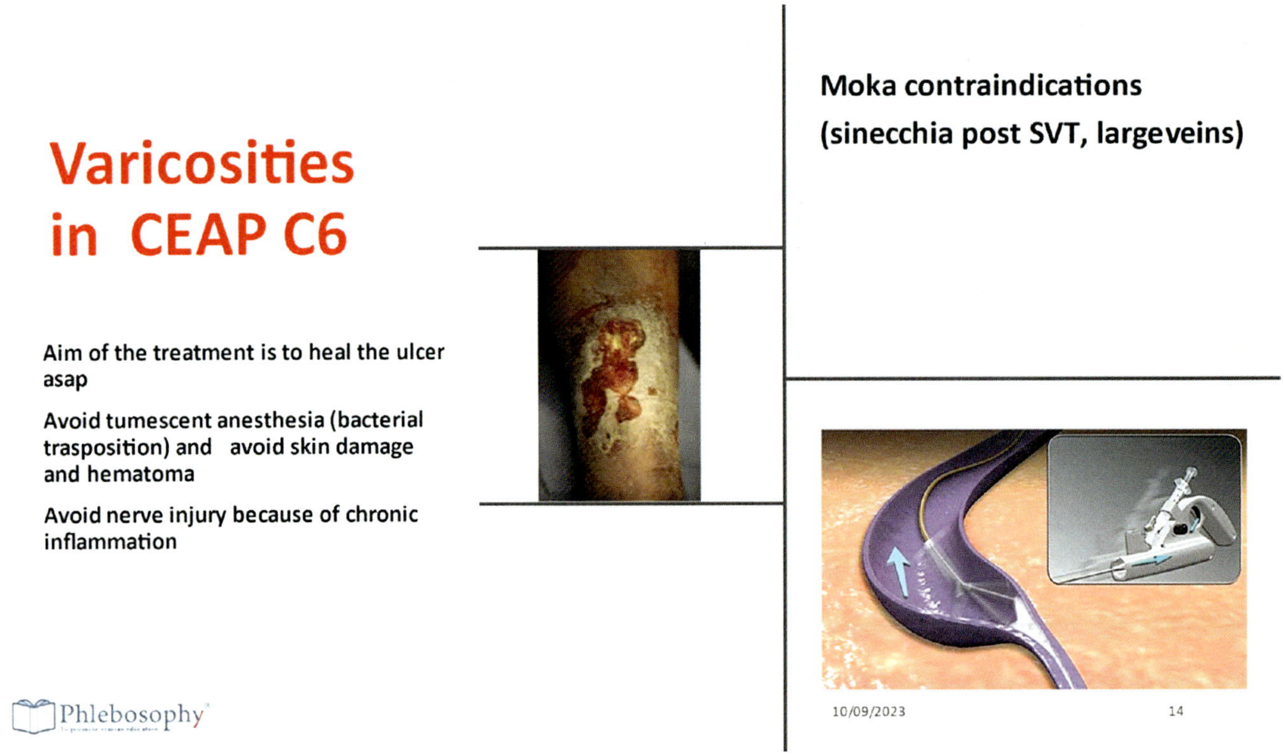

Fig. 74.1 Recurrent varicosities and ulcer at the left limb after GSV stripping; MOKA system for GSV ablation

Fig. 74.2 Leg before and after MOKA surgery

Introduction

MOCA is a non-thermal non-tumescent technique to ablate incompetent superficial veins of the lower limb. The technique uses a dual injury mechanism that combines mechanical disruption of the intima with chemical endovenous ablation by injecting a sclerosing agent.

The ClariVein system is the first venous ablation technique to employ a hybrid (dual-injury) technique built into 1 catheter-based delivery system [1].

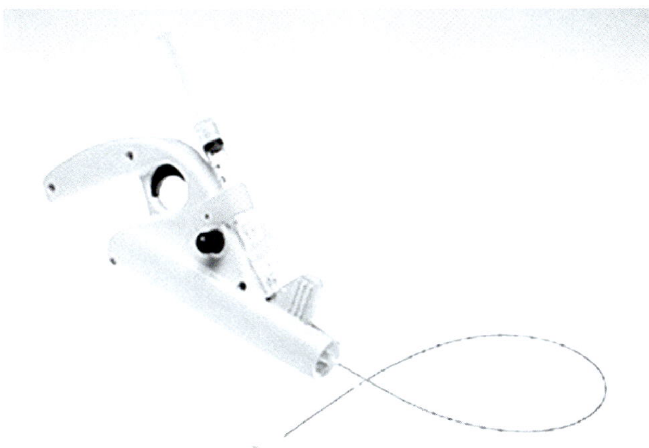

Fig. 74.3 The ClariVein device

Endomechanical abrasion is produced by the tip of the catheter's rotating wire (mechanical component); and endovenous chemical ablation (EVCA) is via simultaneous injection of sclerosant over the rotating wire (chemical component). Currently, at least two devices are commercially available for the treatment of superficial venous incompetence (Fig. 74.3).

Technique

ClariVein(®) treatment uses a micropuncture technique and a 4-Fr sheath to allow a catheter to be placed 1.5 cm from the saphenofemoral junction. Unlike laser (endovenous laser treatment (EVLT)) or radiofrequency ablation (RFA), no tumescence is required. The technique depends on a wire rotating at 3500 r/min causing endothelial damage whilst liquid sclerosant is infused. The wire is pulled back whilst continuously infusing sclerosant along the target vessel's length. Initially, 8 mL of dilute sclerosant was used, but this was subsequently increased to 12 mL (Fig. 74.4).

The first MOCA device developed abrades the intima by means of the tip of the catheter's rotating wire. The liquid forms of the sclerosants STS or POL are used most often, which limits the total dose that can be applied. Different concentrations of liquid POL (2% and 3%) have been compared with 1% POL microfoam in a RCT: 1% foam was less effective

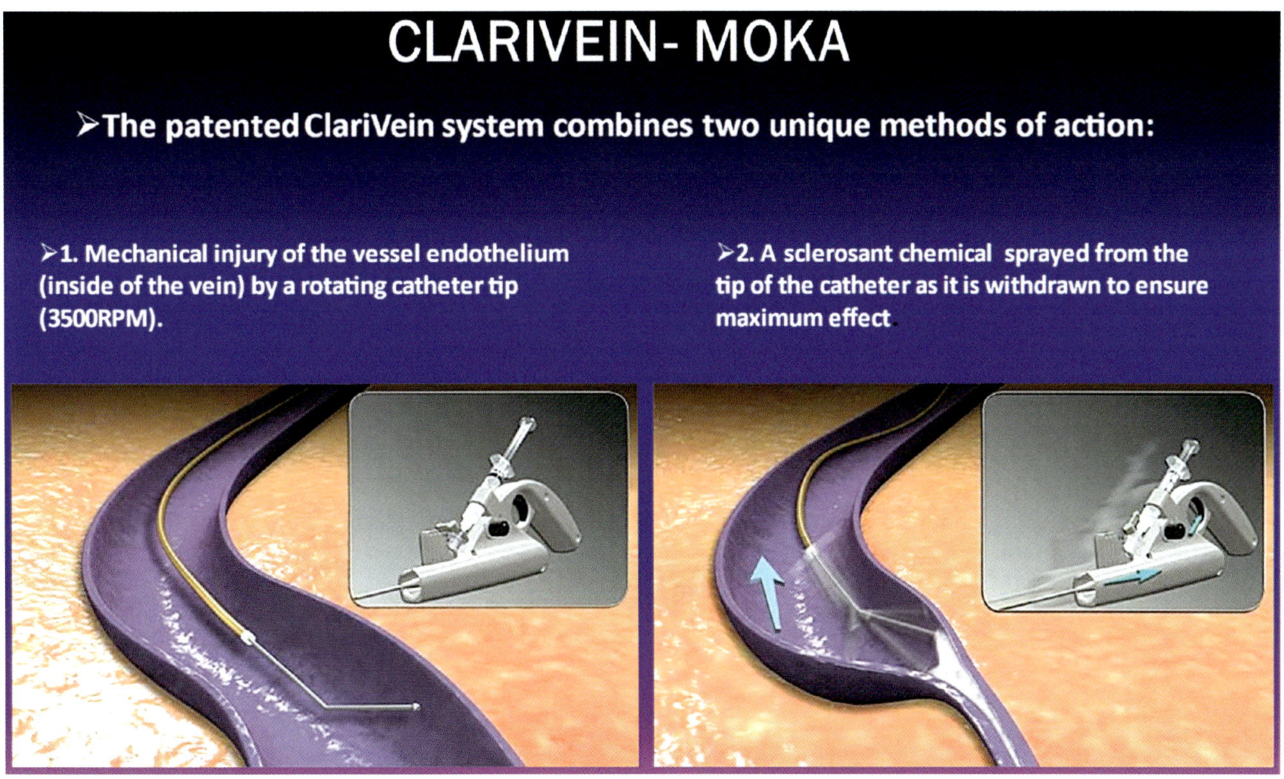

Fig. 74.4 Dual injury (mechanical and chemical) ClariVein device

than 2% or 3% liquid POL ($p < 0.001$) for treatment of GSV incompetence [2].

Results

Several prospective studies and RCTs, comparing MOCA with EVTA, have shown that MOCA is safe and effective for ablating incompetent saphenous trunks [3–5] 213,229e232.

Currently, only up to 3 year follow up has been reported with occlusion rates of 86.5% in one study, included in a systematic review of novel non-thermal techniques [3].

One RCT similarly revealed a statistically significantly lower GSV occlusion rate of 80% 3 years after MOCA versus 100% after EVLA and RFA (p ¼ 0.002) [5]. A strong association was found between recanalization at 3 years and the pre-operative GSV diameter. The occlusion rates for a preoperative GSV diameter of 6 mm, 7 mm, and 8 mm were 100%, 87.5%, and 75%, respectively [5]. The r-VCSS and QoL measures improved significantly after MOCA in all mentioned studies [3, 4, 6]. Pain during and after MOCA was less than after RFA [4, 6] although there was no difference in pain score after MOCA and EVLA in the LAMA trial [7] Return to daily activities and work did not differ substantially between procedures [7].

Indications

Treatment of below Knee GSV reflux and reflux in C5–C6 CEAP patients is a very challenging option for surgeons. The possibility of retrograde ablation of distal GSV and SSV insufficiency in C6 ulceration is considered a significant advantage for MOCA. Saphenous nerve injury SNI seems to be impossible for this endovenous non-thermal and not tumescent technique although complications such as retrograde stripping of GSV during MOCA have been reported [8].

Contraindications

Post thrombotic GSV and SSV can be a contraindications for MOCA device as thrombotic synechia can be a difficult negotiation for rotation of the wire. Large veins (>20 mm) and very small veins (<4 mm) can present the same snag.

Complications

The most common complications after MOCA are induration (12%–18%), SVT (2%–13%), haematoma (1%–11%), DVT (0%–1%), and hyperpigmentation (5%). Nerve injury, skin injury, or infections have not been reported [3].

Therefore this non-tumescent technique may offer a valuable alternative, in particular if ablation of the more distal part of the below knee GSV or the SSV is considered.

Another type of MOCA device causes vein wall scarification by a specially designed endovenous catheter featuring sharp hooks at the tip, which damage the endothelium, while chemical ablation is performed simultaneously by injecting a foam sclerosant. Up to now, the available evidence is very limited. One study showed an occlusion rate of 92% after 24 months [5].

Conclusions

Mechanochemical ablation appears to be safe and efficacious. The technique eliminates the need for tumescent anaesthesia, a goal that both radiofrequency and laser technologies are currently exploring. The elimination of tumescent infusion decreases patient discomfort and shortens procedure time.

The great majority of incompetent GSVs can be treated with this technique. Veins which may not be candidates for mechanochemical ablation include those with previous thrombophlebitis which have recanalized and are incompetent because of endoluminal difficult negotiation of the rotating wire due to thrombotic synechiae and trabeculae of the recanalized vein which limits mechanical treatment.

Follow-up data focused on patency and occlusion rate of saphenous axis at 3 years follow up seems to be less encouraging when compared with early follow up data and comparing this technique with endothermal techniques [9, 10]. However, mechanochemical ablation should be considered as another viable alternative for the management of saphenous incompetence in patients with C5 C6 CEAP disease and in all cases, that saphenous nerve injury is probable.

References

1. Mueller RL, Raines JK. ClariVein mechanochemical ablation: background and procedural details. Vasc Endovascular Surg. 2013;47(3):195–206.
2. Lam YL. Clarivein® mechano-chemical ablation an interim analysis of a randomized controlled trial dose-finding study. Phlebology. 2016;31(3):170–6.
3. Vos CG. A systematic review and meta-analysis of two novel techniques of nonthermal endovenous ablation of the great saphenous vein. J Vasc Surg Venous Lymphat Disord. 2017;5(6):880–96.
4. van Eekeren RRJP, et al. Mechanochemical endovenous Ablation versus Radiofrequency Ablation in the treatment of primary great saphenous vein incompetence (MARADONA): study protocol for a randomized controlled trial. Trials. 2014;15:121.
5. Vähäaho S. Three-year results of a randomized controlled trial comparing mechanochemical and thermal ablation in the treatment of insufficient great saphenous veins. J Vasc Surg Venous Lymphat Disord. 2021;9(3):652–9.
6. Lane T, et al. A multi-centre randomised controlled trial comparing radiofrequency and mechanical occlusion chemically assisted ablation of varicose veins - final results of the Venefit versus Clarivein for varicose veins trial. Phlebology. 2017;32(2):89–98.
7. Mohamed AH, et al. A randomized controlled trial of endovenous laser ablation versus mechanochemical ablation with ClariVein in the management of superficial venous incompetence (LAMA trial). Ann Surg. 2021;273(6):188–e195.
8. Lane TR, et al. Retrograde inversion stripping as a complication of the ClariVein mechanochemical venous ablation procedure. Ann R Coll Surg Engl. 2015;97(2):e18–20.
9. De Maeseneer MG. Editor's Choice - European Society for Vascular Surgery (ESVS) 2022 Clinical Practice Guidelines on the Management of chronic venous disease of the lower limbs. Eur J Vasc Endovasc Surg. 2022;63(2):184–267.
10. Vun SV. Lower pain and faster treatment with mechanico-chemical endovenous ablation using ClariVein®. Phlebology. 2015;30(10):688–9.

Foam Sclerotherapy for Chronic Venous Disease

75

Dimitrios Kontothanassis

Case Presentation: Foam

A 54 year-old male was referred to our clinic for right leg CVI and recurrent varicose veins (VV) after laser ablation of the R great saphenous vein (GSV) (Fig 75.1). The operation was done in 2015 and in 2018 mild varicosities reappeared. In 2022 he was referred to our clinic for clinical evaluation and counselling for redo surgery. His clinical history was not relevant exception for prolonged standing position (>12 hours), and the main symptom was heaviness of the leg. The patient never wears elastic stockings.

During other consultancies, surgeons contraindicated a surgical operation for recurred varicosities because of the poor clinical symptoms. Reexamination of the patient and eco color Doppler study of the abdominal vessels and both inferior limbs was done in our clinic to check the main sources of reflux and exclude obstruction of deep veins (Fig. 75.2). The problems to resolve in this case were: (a) eliminate extensive varicosities, (b) eliminate varicosities and points of reflux, and (c) restore good clinical conditions.

Treating patients with recurrent varicosities is difficult because the complex anatomy of anarchic veins after GSV ablation has to be well studied with ultrasound and a good mapping on the skin has to be done prior to surgery with a permanent ink marker. Laser transilluminator can help in mapping vein anatomy.

In such cases, non-thermal techniques like foam sclerotherapy or injection of cyanoacrylate glue (CAC) under ultrasound guidance are preferable to achieve rapid occlusion of the recurrent anarchic vein circuit spread all over the limb. Phlebectomies for major varicosities can easily be performed under local tumescent anesthesia with cold saline to gaine extreme vein spasm and reduce ecchymosis (Fig. 75.3).

Case Presentation

Under mild sedation, the patient was placed in the supine position on the interventional suite table at the anti-Trendelenburg position for GSV access at the proximal segment of the calf medially. After percutaneous puncture of a tributary at the distal part of the median thigh and preparation of foam with the Tessari method (mixture 2% polidocanol 2cc + 8cc air), slow injection at the reverse Trendelenburg position was done to foam the Dodd perforator and tributaries connected with the perforator. Small amounts of foam and repeated injections in tributaries of the calf were administered. Cold saline tumescent anesthesia (10 mL 2% Lidocaine + 500 cc cold saline) was delivered to all major varicosities and phlebectomies under ultrasound guidance were performed. The leg was wrapped tightly for 7 days, and a second-class stocking was recommended for at least 2 months. Thromboprophylaxis with low-molecular-weight heparin was given for 1 week until the patient came back for follow-up with ultrasound examination and clinical investigation. Occlusion of the Dodd perforator and major tributaries connected with it was achieved. Absence of varicosities at the thigh and the calf was evident at the clinical examination and at the ultrasound investigation (Fig. 75.4); relief of symptoms was immediate after the operation. At two-month follow-up the patient was doing well and symptoms had completely resolved. Sensation of soft tissue hardening due to obliteration with foam was a normal outcome and it could last for a few months; the patient was well informed about this normal reaction prior to surgery and gave informed consent (Figs. 75.2, 75.3, and 75.4).

Continued at page 795

D. Kontothanassis (✉)
General and Vascular Surgery, Phlebology Department, Istituto Flebologico Italiano, Padova, Italy

G. Geroulakos et al. (eds.), *Mastering Endovascular Techniques*, https://doi.org/10.1007/978-3-031-42735-0_75

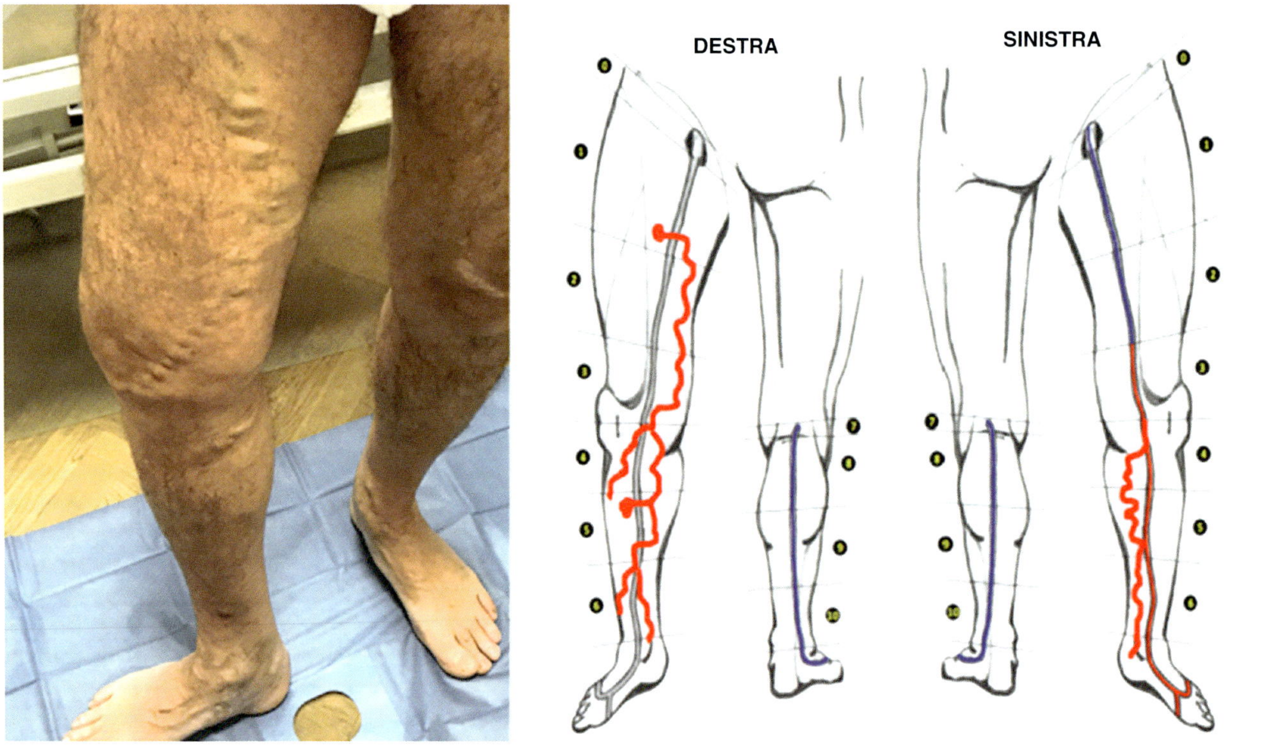

Fig. 75.1 Recurrent varicosities at the right limb after GSV ablation; segmental mild GSV reflux and tributaries at the left limb; hemodynamic preoperative mapping of varicosities

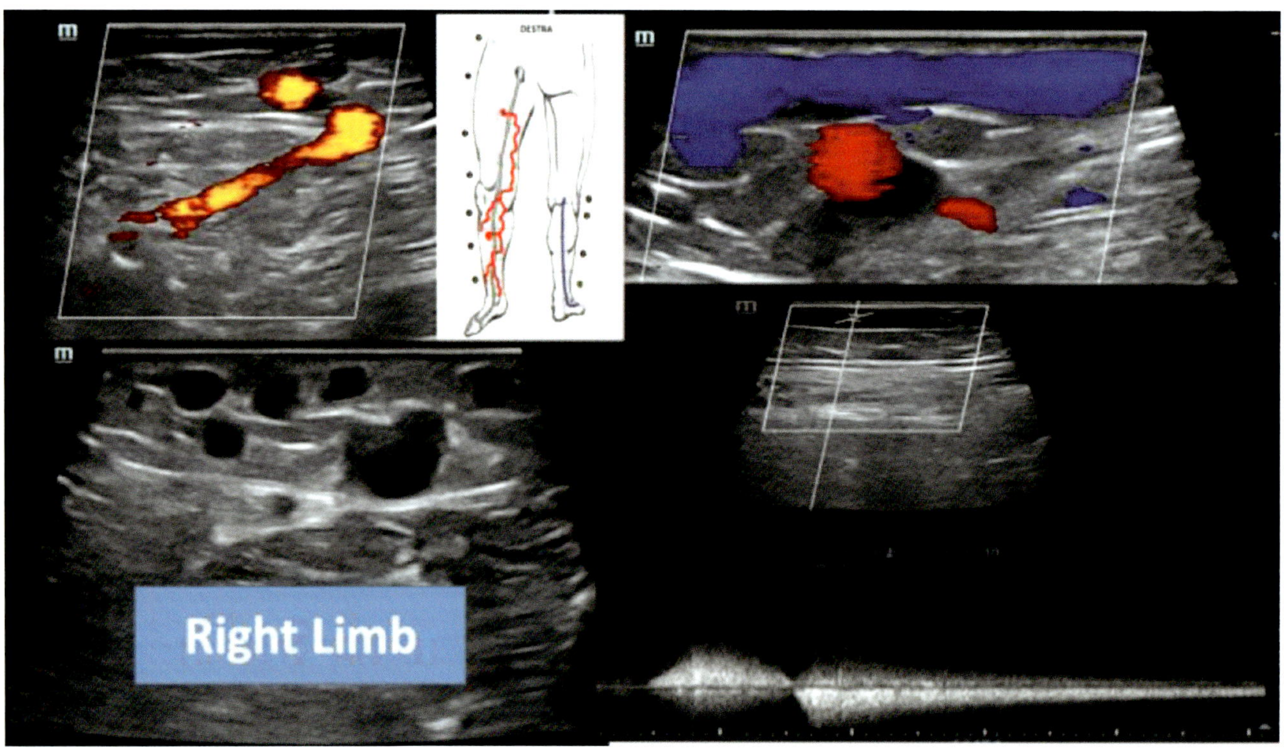

Fig. 75.2 Ultrasound investigation demonstrates reflux of the Dodd perforator and varicosities of the thigh and calf

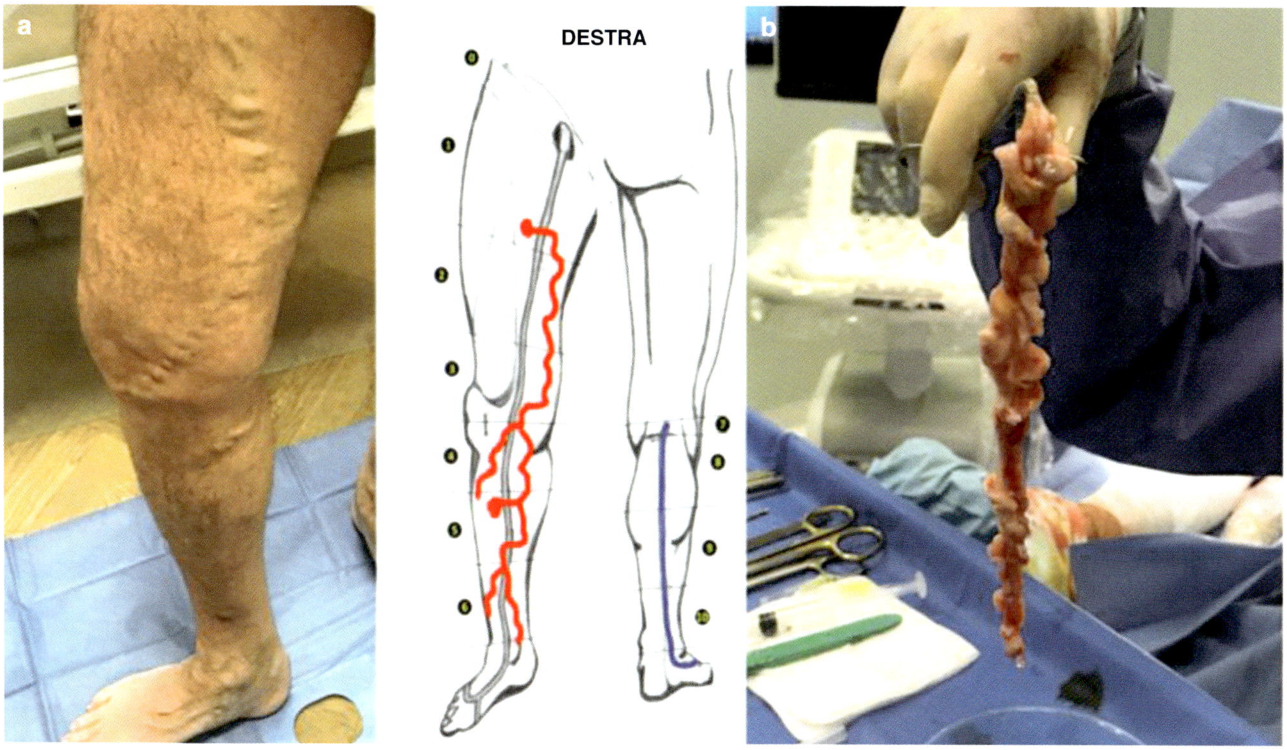

Fig. 75.3 (**a**) Limb before surgery. (**b**) Vein Map CP4: Phlebectomies for major varicosities.

Fig. 75.4 Limb before and 24 h after surgery

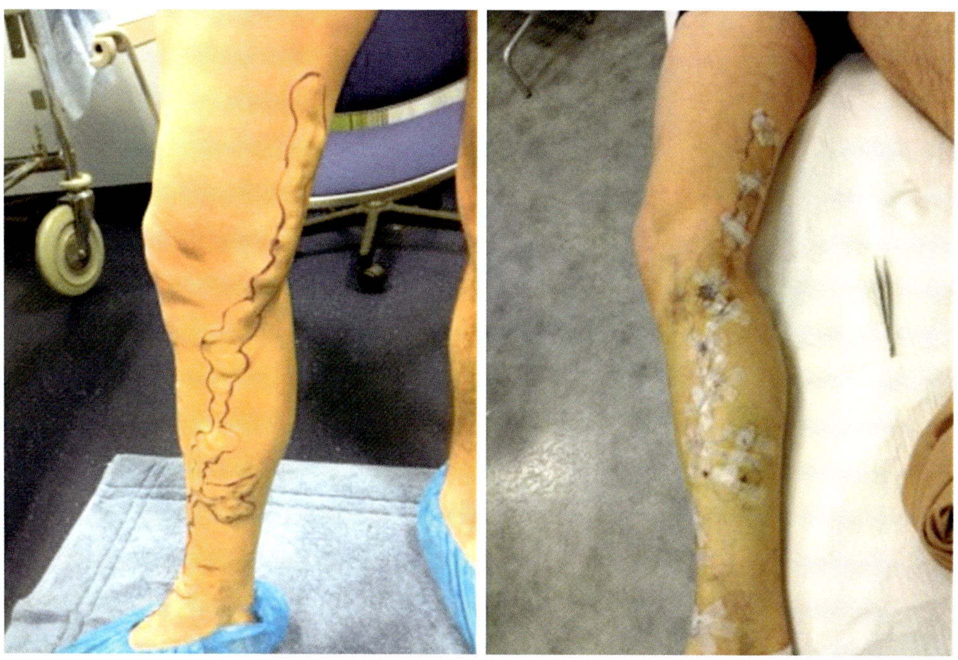

Introduction

Sclerotherapy is the targeted chemical ablation of varicose veins by injection of a sclerosing agent in the form of liquid or foam [1, 2]. Sclerosant foam has been increasing in use in recent times. It has certain advantages over liquid sclerosants and is quite safe to use, even though there are adverse events that have been reported [3]. The idea of using air and drug in combination is well established. Orbach described an air block technique using froth in 1944, and in 1993 Cabrera proposed use of a true foam of sodium tetradecyl sulfate or polidocanol to treat varicose veins. When Tessari presented a three-way tap technique in 2000, very good micro foam could be made at a very low cost. Foam can be used in classical cosmetic and small vein sclerotherapy, but it can also be used in treating severe chronic venous insufficiency (CVI) [3].

The main goals of sclerotherapy are ablation of varicose veins, prevention and treatment of complications of CVI, improvement or disappearance of venous symptoms, improvement of quality of life and venous function, and improvement of aesthetics.

Rabe et al. in 2008 assessed the safety and efficacy of sclerotherapy of the great saphenous vein (GSV) comparing standardized polidocanol foam to liquid polidocanol in a randomized controlled trial (RCT). The results of this study demonstrate that standardized 3% polidocanol foam is more efficient and equally safe compared to 3% liquid polidocanol for treatment of GSV [4].

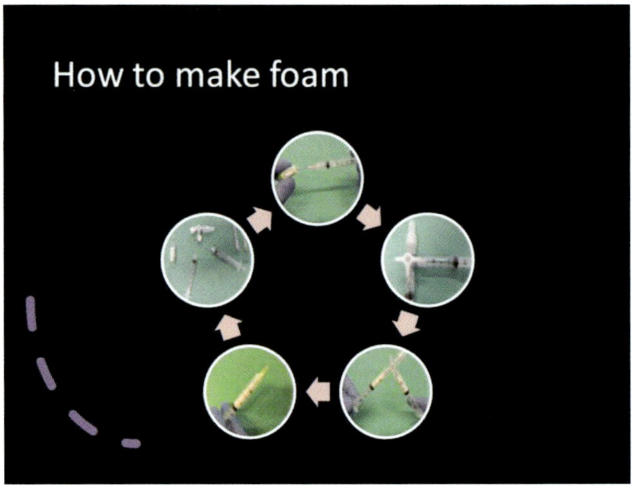

Fig. 75.5 How to make foam with the Tessari method, courtesy Prof. Kontothanassis

Technique

Sclerotherapy is a chemical endovenous ablation technique, consisting of injecting a sclerosing agent into a target vein to damage its wall and obtain a durable vein fibrosis. The most used sclerosing agents are polidocanol (POL) and sodium tetradecyl sulphate (STS), which both can be used in foam or liquid form. Before starting treatment, informed consent is mandatory even if considered unnecessary normally for other intravenous drugs in everyday clinical practice. The first reason for this is that sclerotherapy should not be considered significantly different to the more invasive endovenous treatments and secondly because there are recognized side effects and complications.

The liquid detergents are usually mixed with air or gas to generate the sclerosant foam which is then delivered to the target vessels either by direct injections or using various catheter techniques (Fig. 75.5).

Foam sclerotherapy is usually performed under ultrasound guidance (ultrasound trans-catheter-guided sclerotherapy or UGS) but is mostly carried out by sight in the treatment of small tributaries, reticular veins, and telangiectasias. The injection technique of liquid for spider veins needs visual control with the limb in a horizontal position (Fig. 75.6).

Use of 1cc or 3cc syringes and needles up to 31G–32G is very frequent; slow and fractional injection of liquid is recommended. For reticular veins and saphenous ablation techniques, foam sclerotherapy is mandatory and proven to be more effective than liquid.

For reticular veins the use of a laser-led transilluminator can help to correctly identify the target vein and make the procedure very easy.

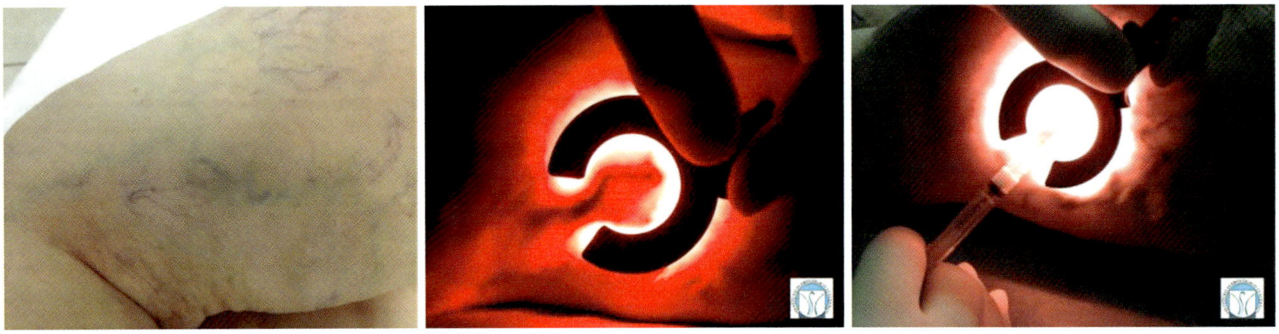

Puncturing and foaming a reticular vein under laser- led guidance, courtesy Prof. Kontothanassis

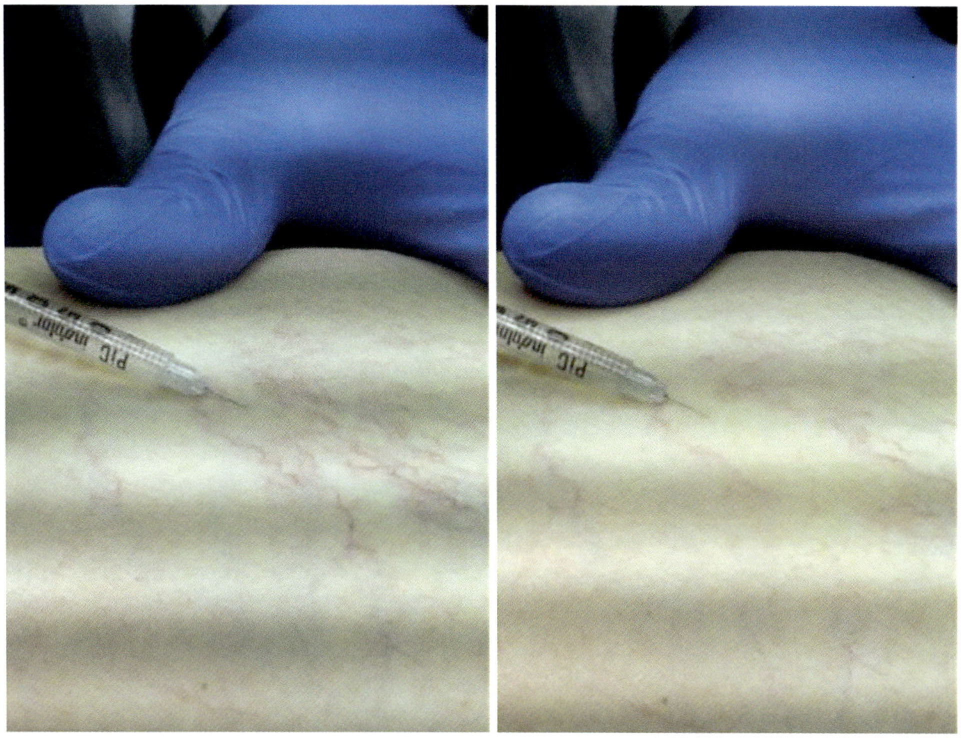

Fig. 75.6 Liquid sclerosis of teleangectasia, courtesy Prof. Kontothanassis

For saphenous veins, successful vein access at the first strike is very important to prevent patient discomfort and vein spasm. Access is performed at the reverse Trendelenburg position to fill the veins and obtain a better visualization with ultrasound. An 18-gauge cannula is recommended for saphenous veins and a butterfly for varicosities. Multiple butterflies are needed frequently for multiple injection sites; the cannula can be flushed with saline to avoid thrombus formation. The author suggests that when you are ready to foam a vein you have to keep in mind that the purpose of the technique is to foam the vein wall and not the blood, and therefore reverse Trendelenburg position will help in evacuation of varicose veins for gravity reasons. Empty veins can be under less pressure, which should facilitate the ascending flow of the foam. When foaming the veins, control with ultrasound accidental injection of foam in perforators and check the distribution of it along the treated vein and the reaction of the vein wall (spasm).

Indications

Sclerotherapy can be used for treating incompetent saphenous veins (GSV, small saphenous veins (SSV)) as well as tributary varicosities, residual varicosities before and after saphenous ablation, pelvic veins, varicosities with concomitant ulcers (CEAP C6), venous malformations, and small veins for cosmetic reasons (Figs. 75.7 and 75.8).

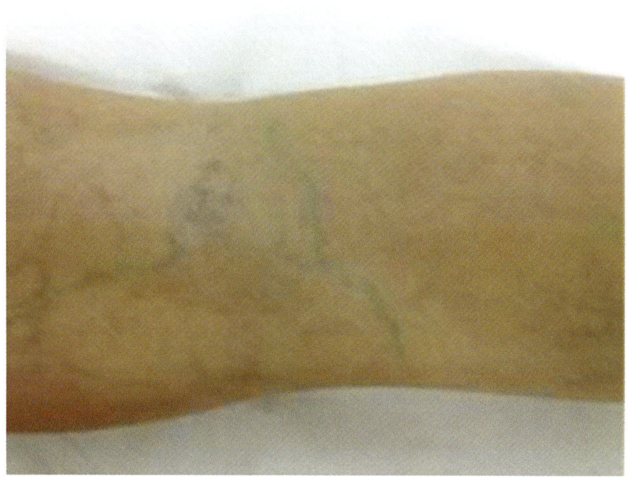

Reticular veins at the popliteal fossa, courtesy Prof. Kontothanassis

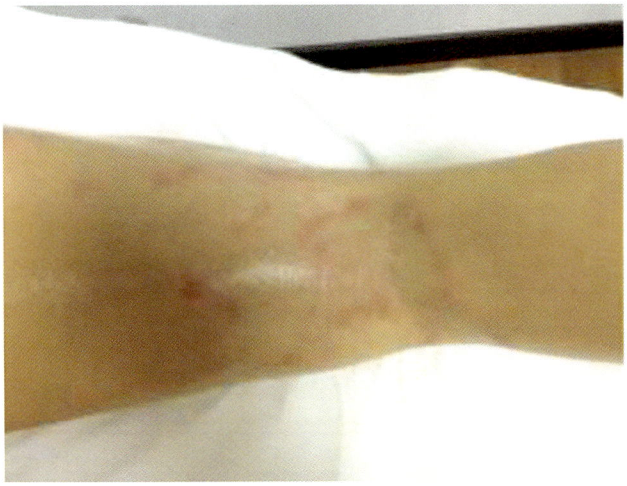

After multiple small injections with foam, courtesy Prof. Kontothanassis

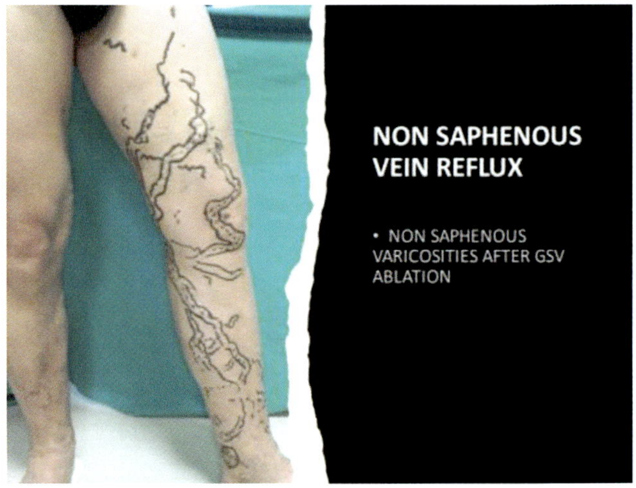

Fig. 75.7 Non-saphenous vein reflux, courtesy Prof. Kontothanassis

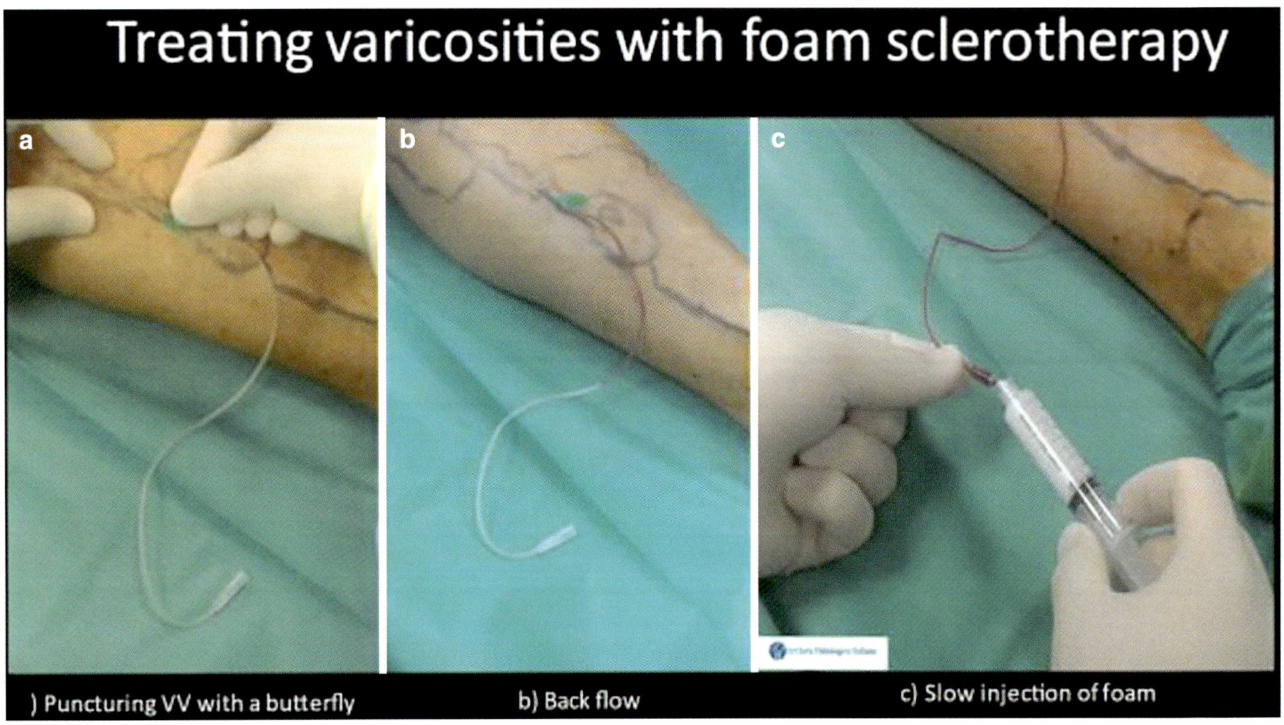

Fig. 75.8 (**a**–**c**) Foam injection of a VV with a butterfly, courtesy Prof. Kontothanassis

Foam has been shown to be at least twice as effective as liquid, with four or five times less sclerosing agent needed [5]. To increase efficacy and safety of sclerotherapy, ultrasound guidance should be used during the whole procedure for non-clinically visible VVs. One of the most important safety issues is to avoid inadvertent intra-arterial sclerosant injection [6].

Foam sclerotherapy is the technique of choice for anatomical configurations that make endovenous cannulation or advancing the ablation device challenging and is suitable for treating tortuous recurrent VVs (Figs. 75.9 and 75.10).

An alternative to the classical Ultrasoun Guided Foam Sclerotherapy (UGFS) (by means of direct puncture or a short cannula) and other non-thermal ablation techniques for treatment of incompetent saphenous veins is catheter-

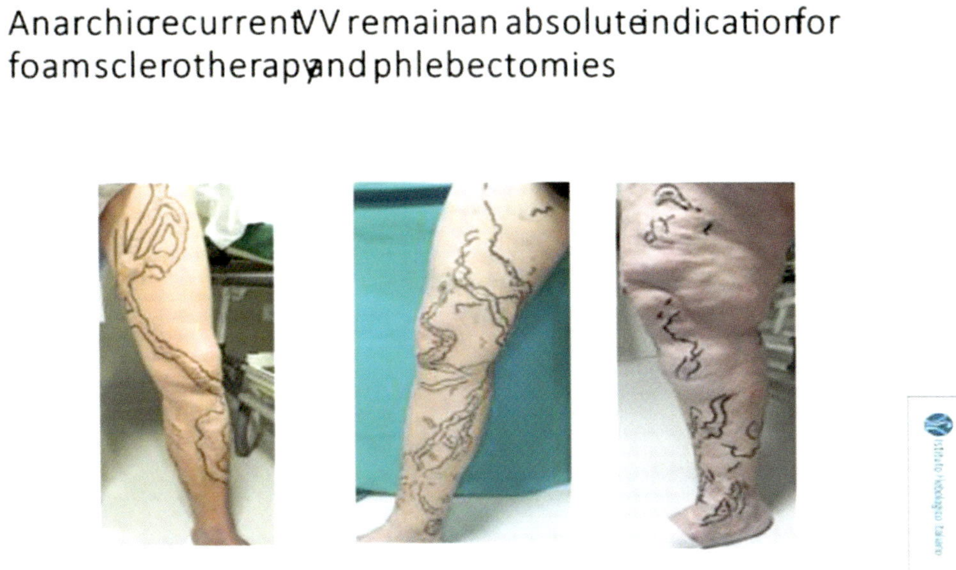

Fig. 75.9 Tortuous recurrent VV, courtesy Prof. Kontothanassis

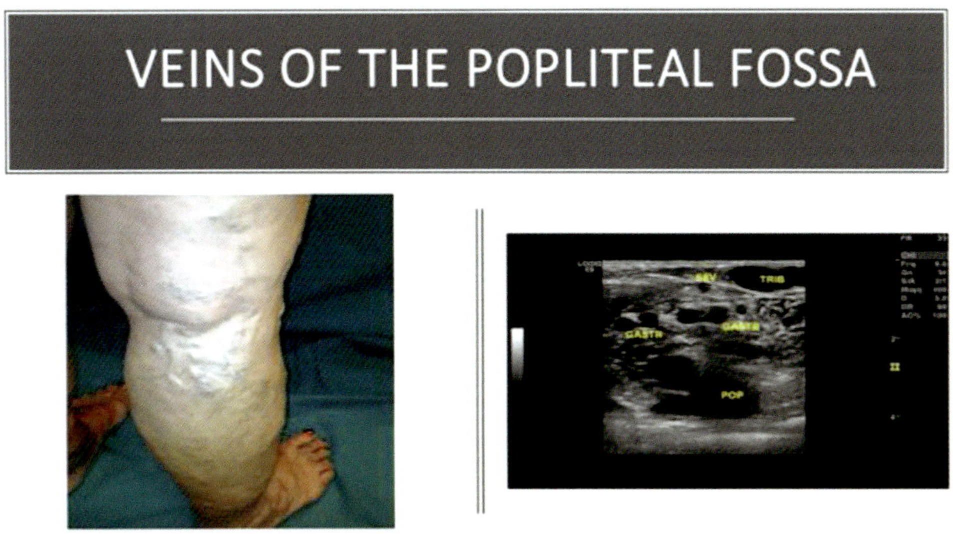

Fig. 75.10 Veins of the popliteal fossa, courtesy Prof. Kontothanassis

directed foam sclerotherapy (CDFS) with or without perisaphenous tumescence infiltration to reduce the vein caliber. When performing CDFS, sclerosant foam is injected through a long (30 e 40 cm) intravenous catheter all along the saphenous trunk under ultrasound visualization [7, 8].

Contraindications

Contraindications can be absolute and relative. Known allergy to sclerosing agents, presence of acute deep vein thrombosis (DVT) or pulmonary embolism (PE), local infection, and right–left cardiac shunt should be considered as absolute contraindications. There are more relative contraindications but data are based more on single practitioner experience than RCT. These contraindications are pregnancy and lactation, severe peripheral disease, poor general condition, previous allergic diathesis (urticaria and migraine), high venous thromboembolism (VTE) risk, history of VTE events, severe thrombophilia, active cancer, and hypercoagulable status.

Results

In long-term follow-up of comparative studies, treatment of incompetent saphenous veins by UGFS has shown to be substantially less effective than endovenous laser ablation

(EVLA), radiofrequency ablation (RFA), and surgery in terms of occlusion or absence rates. In one RCT, the rate of anatomical success (GSV occlusion or absence) according to Duplex -Ulta Sound (DUS)was 64%, 75.9%, and 33.3% after EVLA, surgery, and UGFS, respectively,23 and in another study, 85%, 77%, and 23%, respectively ($p < 0.001$) [9]. A four-arm trial compared UGFS with RFA, EVLA, and surgery < 0.001 after three-year follow-up failure was found in 7%, 6.8%, 6.5%, and 26.4% after RFA, EVLA, surgery, and UGFS ($p < 0.001$), respectively. It should be noted that the UGFS treatment protocol for the GSV, used in these trials, was not always optimal, for example performing a single injection of foam just above the knee instead of staged injections along the length of the vein, from proximal to distal, as recommended in the European guidelines for sclerotherapy. [10]

A lower efficacy has been reported in patients with a GSV > 6 mm than in those with a GSV < 6 mm (measured at mid-thigh, in standing position). A post hoc analysis of 225 patients treated with UGFS, part of an RCT comparing high ligation and stripping with UGFS, showed a two-year cumulative probability of recurrent reflux of 62.6% (51.2% to 74.2%) if the mid-thigh GSV diameter was > 6 mm versus 42% (34.6% to 50.4%) for a diameter < 6 mm. Therefore, UGFS should be preferably used for less dilated GSV or SSV trunks [10].

Disease-specific QoL scores improved significantly after UGFS with no difference between UGFS and other methods of saphenous trunk treatment up to three years in a four-arm RCT, and, in a single center prospective study, with more than five years follow-up. Other trials showed that initial QoL improvement reduced over time and was significantly worse at five years after treatment in patients with initial UGFS compared with patients treated by endovenous thermal ablation (EVTA) (*p* ¼ 0.013).

A systematic review and meta-analysis of 3689 patients showed a higher occlusion rate of 82.4% after CDFS and 62.9% after UGFS ($p < 0.001$) at three-year follow-up. The rates of major complications and the frequency of post-procedural pain, hyperpigmentation, and superficial vein thrombosis (SVT) were also significantly lower after CDFS. Long-term results, at five-year follow-up, are not yet available [10].

Complications

The reported incidence of complications of sclerotherapy is quite variable and the older publications mostly refer to liquid sclerosants. Significant complications include anaphylaxis, thromboembolism, cerebrovascular events, tissue necrosis, nerve damage, and oedema. Other complications can be classified as cosmetic and include telangiectatic matting and pigmentation. Adverse reactions may be due to the pharmacological properties of the sclerosing agent, the foaming gas, or the mode and technique of delivery. Concurrent medical problems, intake of drugs or supplements, and lack of compliance are other contributing factors that may also significantly influence the onset of complications.

Adverse reactions can be divided into two categories: major and minor complications.

Major Complications

ANAPHYLAXIS AND ANAPHYLACTOID REACTIONS are very rare; immediate hypersensitivity (0.6%) can follow the use of most sclerosants with the exception of hypertonic saline. Foam sclerosants are associated with a lower incidence of hypersensitivity reactions compared with liquid [11]. Currently, there are no methods to identify predisposed individuals and hence such adverse reactions are not preventable. Treatment should be tailored to the clinical features of the allergic events and emergency protocols should be in place.

Thromboembolism

Deep Vein Thrombosis

The true incidence of deep vein thrombosis (DVT) following foam sclerotherapy is unknown and most reports only include symptomatic cases, while silent DVT can occur. Proximal vein thrombosis and extensive DVT following foam sclerotherapy are not so rare (1–3%) [3, 12, 13] and the vast majority of the reported post-procedural DVTs are localized to the lower legs. Treatment factors that might influence the risk of deep vein occlusion (thrombosis or sclerosis) following foam sclerotherapy have been reviewed [14] and the use of large total volumes of foam (>10 to 16 mL) was identified as a risk factor. Although PE remains a very rare complication of foam sclerotherapy, there is no data regarding the incidence of postoperative silent PE.

Superficial Venous Thrombosis

In the author's opinion, this is greatly underestimated because of the co-presence of an inflammatory chemical reaction (chemical phlebitis). SVT following foam sclerotherapy is best treated with the use of compression stockings and nonsteroidal anti-inflammatory drugs (NSAIDS). Anticoagulants are rarely required but may be used in selective cases.

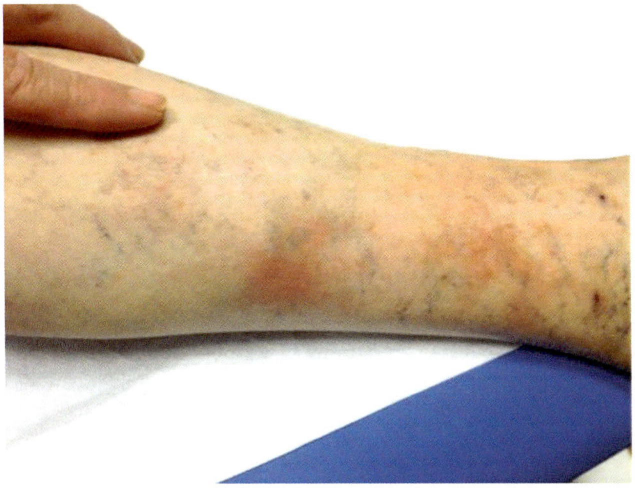

Superficial vein thrombosis after foam injection in VV, courtesy Prof. Kontothanassis

Neurological Events

The overall frequency of neurological complications of sclerotherapy is reported to be from 0% to 2% [15, 16]. These include transient events such as visual disturbances and migraine as well as ischaemic events such as transient ischaemic attacks (TIAs) and stroke.

Transient Events

Visual disturbances have been reported in up to 14% of patients undergoing foam sclerotherapy [17], but a recent systematic review found the overall incidence to be 1.4% [17]. Visual disturbances have been attributed to classical migraine attacks rather than TIAs [17]. Transient neurological events including visual disturbances and migraine following both liquid and foam sclerotherapy have been associated with the presence of a right-to-left shunt (RLS) and in particular a patent foramen ovale (PFO) [15].

Stroke

Stroke is a very rare but significant complication of sclerotherapy. Two cases of stroke following 4059 foam UGFS procedures have been reported in the litterature in a six-year period yielding an incidence of 0.01% [18]. We recently reviewed 13 cases of stroke after sclerotherapy published since 1994 [19]. Higher volumes of sclerosant foams may contribute to distant thrombotic and ischaemic complications and should be avoided [11, 16]. Valsalva delayed maneuvers were advocated to reduce the ascent of bubbles but were demonstrated to be ineffective [20].

Tissue Necrosis

Liquid or foam extravasation, despite the popular belief, is not the primary mechanism of skin damage or necrosis when detergent sclerosants are used [7]. Inadvertent intra-arterial injection and consequent arterial occlusion remains the main mechanism leading to tissue necrosis but it is extremely rare. Intra-arterial injection commonly presents with severe sudden pain. If suspected, the injection should be stopped immediately and if possible local catheter-directed anticoagulation and thrombolysis should be initiated. This may be complemented by systemic anticoagulation. Early administration of systemic steroids may help reduce the subsequent inflammation that results in tissue damage.

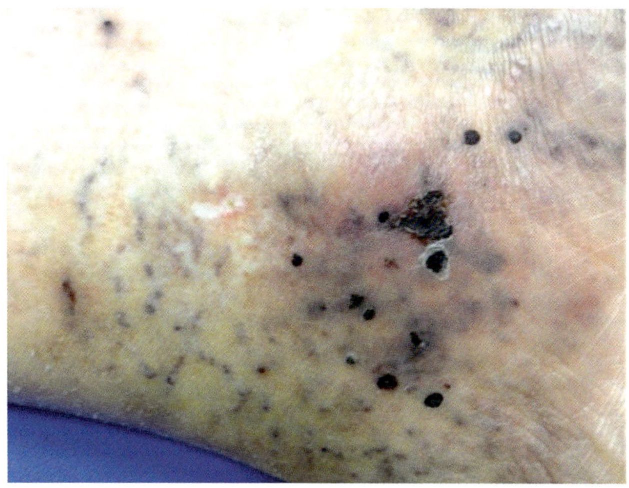

Tissue necrosis after injection of teleangectasia with liquid sclerosant, courtesy Prof. Kontothanassis

Nerve Damage

Sclerotherapy using liquid or foam sclerosants is associated with both sensory and motor nerve damage of a usually transient nature. The incidence is reported to be around 0.02% with paraesthesia and dysasthesia as the main presenting complaints [3]. Foam sclerotherapy of the small saphenous vein (SSV), both in the popliteal fossa and in the calf, has been associated with nerve damage. This is due to the close proximity of several nerves to the SSV trunk. These nerves are readily visualized on most modern ultrasound systems and inadvertent damage can be mostly avoided in the vast majority of patients during foam sclerotherapy. Treatment may include local infiltration of corticosteroids and local anaesthetics, or oral NSAIDs together with possible long-term therapy using neurotropic agents.

Oedema and Lymphoedema

The incidence of lower limb oedema following sclerotherapy is rarely reported and the incidence is around 0.5%. Localized lymph stasis may occur as a consequence of chemical phlebitis induced by sclerotherapy. Extensive sclerotherapy may result in transient lymph stasis in predisposed patients such as those with latent congenital lymphatic system abnormalities. Although oedema secondary to venous insufficiency may improve following foam sclerotherapy, pre-existing overt phlebo-lymphoedema may also worsen due to increased lymph stasis. Adequate postoperative compression is important in reducing oedema and phlebitis in general.

Cosmetic Complications

Telangiectatic Matting

Telangiectatic matting is the proliferation of small vessels less than 0.2 mm in diameter that may appear after sclerotherapy or surgical treatment of varicose or telangiectatic leg veins. The incidence is estimated to be 15–24%.[27]

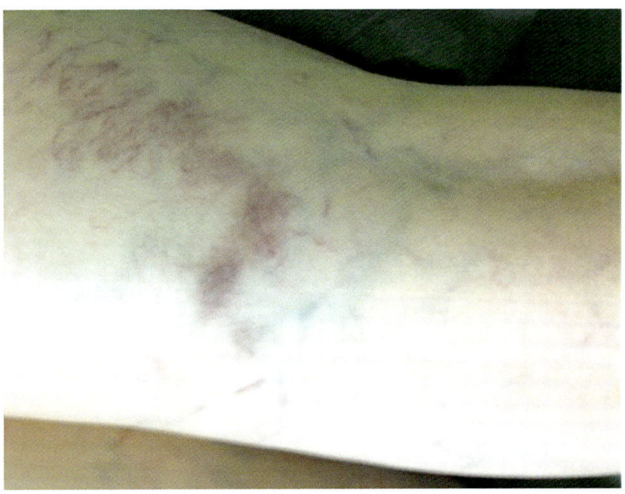

Matting after injection of telangiectasia; cosmetic complications, courtesy Prof. Kontothanassis

Key risk factors include hormonal supplementations such as the oral contraceptive pill and hormone replacement therapy, obesity, and treatment techniques resulting in inflammation[27] or excessive obstruction of venous drainage. Inadequate or no treatment of the underlying reflux is the common denominator in many cases of postsclerotherapy telangiectatic matting. This is very often when in the presence of an underlying saphenous reflux, reticular veins or telangiectasias have been injected directly. Similarly, inappropriate high initial concentrations or large volumes can also result in inflammation and the subsequent angiogenesis. Although matting may resolve in some patients with time, in a significant number it may be permanent.[27] Treatment of telangiectatic matting should concentrate on the underlying reflux and residual patent veins using non-inflammatory concentrations of sclerosants. Laser and light devices have been tried but are in general ineffective as they do not address the underlying reflux.

Pigmentation

Skin irritation is normally due to the use of foam and compression with pads after sclerotherapy. Post-sclerotherapy pigmentation is usually the combination of both melanin and hemosiderin pigment deposits. Hemosiderin is derived from extravasated red blood cells. This adverse reaction has been reported with an incidence that ranges from 10% to 30% at short term.[27] Significant risk factors include inappropriate treatment approach or inadequate post-treatment compression. A high serum ferritin level may also play a role. Such complications are usually of a self-limiting and transient nature and can be treated with topical agents.

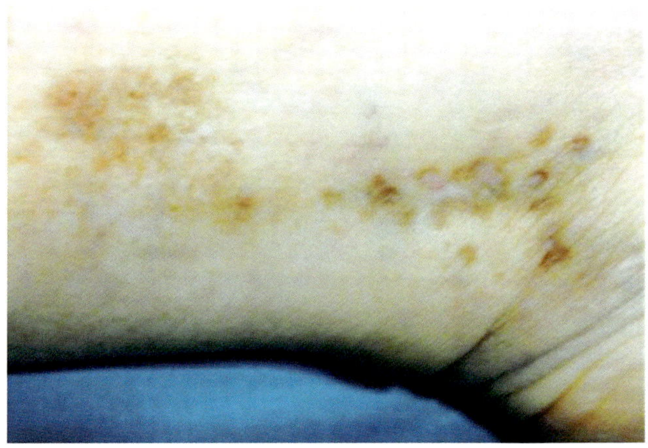

Skin damage after foam injection of VV, courtesy Prof. Kontothanassis

Retained Coagulum

After foaming varicosities, the formation of a "coagulum" is possible within the treated vessel and ranges from 7.8% to 55.1%.11f. Such coagulum may be completely liquid or of a semi-solid consistency.

Treatment techniques and size of the treated vessels are the main causes of the problem. Good technique should focus on minimizing the mixture of sclerosant with the intravascular blood, selection of adequate sclerosant concentration, injection of small volumes from single points of entry, and adequate compression. Retention of coagulum is usually associated with tenderness and may predispose to post-treatment pigmentation. Removal of the retained coagulum relieves the tenderness and may help prevent discoloration.

Conclusions

Foam sclerotherapy is a safe and effective technique to ablate saphenous veins, varicosities, venous malformations, ulcers, and cosmetic veins. Long-term results demonstrate lower evidence of efficacy when compared with thermal and tumescent techniques. Major complications are rare, although clinically significant. A symptomatic right-to-left shunt is an absolute contraindication to foam sclerotherapy. However, patent foramen oval, present in 25% to 30% of individuals, should not be systematically searched for.

There is a high learning curve requiring adequate understanding of the fundamental principles of venous hemodynamics, superficial and deep venous anatomy, ultrasound imaging, sclerotherapy techniques, and sclerosant foam dynamics. Treatment techniques should be optimized to reduce the total volume of the sclerosant foam used in each individual treatment session.

Case Presentation

Continued from page 785.

Conclusion: The patient presented with recurrent varicosities and CVI with a CEAP Class C2s. Recurrent varicose veins after thermal and non-thermal ablation are not rare after 5- and 10-year follow-up and the causes that may lead to this condition are many and have to be further investigated. Foam sclerotherapy for recurrent varicosities seems to be the perfect tool but there can be recurrence even if results seem to be excellent at early follow-up.

References

1. Bunke N. Foam sclerotherapy: techniques and uses. Perspect Vasc Surg Endovasc Ther. 2009;21(2):91–3.
2. Lattimer CR. Tips and tricks in angiology. Torino: Minerva Medica; 2016. p. 141–1144.
3. Bergan J. Venous disorders: treatment with sclerosant foam. J Cardiovasc Surg. 2006;47(1):9–18.
4. Rabe E, Otto J, Schliephake D, Pannier F. Efficacy and safety of great saphenous vein sclerotherapy using standardised polidocanol foam (ESAF): a randomised controlled multicentre clinical trial. Eur J Vasc Endovasc Surg. 2008;35(2):238–45.
5. Hamel-Desnos C, Allaert FA. Liquid versus foam sclerotherapy. Phlebology. 2009;24:240–6.
6. Hafner F, Froehlich H, Gary T, Brodmann M. Intra-arterial injection, a rare but serious complication of sclerotherapy. Phlebology. 2013;28:64–73.
7. Lim SY, Tan JX, D'Cruz RT, Syn N, Chong TT, Tang TY. Catheter directed foam sclerotherapy, an alternative to ultrasound-guided foam sclerotherapy for varicose vein treatment: a systematic review and meta-analysis. Phlebology. 2020;35:369–83.
8. Dos Santos JB, Junior WC, Porta RM, Puggina J, da Silva DF, Puech-Leao P, et al. Catheter-directed foam sclerotherapy with tumescence of the great saphenous vein versus ultrasound guided foam sclerotherapy: a randomized controlled trial. Phlebology. 2020;35:84–91.
9. Van der Velden SK, Biemans AA, De Maeseneer MG, Kockaert MA, Cuypers PW, Hollestein LM, et al. Five-year results of a randomized clinical trial of conventional surgery, endovenous laser ablation and ultrasound-guided foam sclerotherapy in patients with great saphenous varicose veins. Br J Surg. 2015;102:1184–94.
10. De Maeseneer MG, et al. European Society for Vascular Surgery (ESVS) 2022 Clinical Practice Guidelines on the Management of chronic venous disease of the lower limbs. Eur J Vasc Endovasc Surg. 2022;63:184–267.
11. Thibault P. Sclerotherapy and ultrasound-guided sclerotherapy. In: Bergan JJ, editor. The vein book. Burlington, MA: Academic Press; 2006. p. 189–200.
12. Gillet JL, Guedes JM, Guex JJ, et al. Side-effects and complications of foam sclerotherapy of the great and small saphenous veins: a controlled multicentre prospective study including 1025 patients. Phlebology. 2009;24:131.
13. Myers KA, Jolley D, Clough A, Kirwan J. Outcome of ultrasound-guided sclerotherapy for varicose veins: medium-term results assessed by ultrasound surveillance. Eur J Vasc Endovasc Surg. 2007;33:116–21.
14. Myers KA, Jolley D. Factors affecting the risk of deep venous occlusion after ultrasound-guided sclerotherapy for varicose veins. Eur J Vasc Endovasc Surg. 2008;36:602–5.
15. Parsi K. Paradoxical embolism, stroke and sclerotherapy. Phlebology. 2012;27:147. https://doi.org/10.1258/phleb.2011.010098.
16. Jia X, Mowatt G, Burr JM, Cassar K, Cook J, Fraser C. Systematic review of foam sclerotherapy for varicose veins. Br J Surg. 2007;94:925–36.
17. Gillet JL, Donnet A, Lausecker M, Guedes JM, Guex JJ, Lehmann P. Pathophysiology of visual disturbances occurring after foam sclerotherapy. Phlebology. 2010;25:261–6.
18. Ma RW, Pilotelle A, Paraskevas P, Parsi K. Three cases of stroke following peripheral venous interventions. Phlebology. 2011;26:280–4.
19. Cavezzi A, Parsi K. Complications of foam sclerotherapy. Phlebology. 2012;27(Suppl 1):46–51.
20. Morrison N. Studies on safety of foam sclerotherapy. In: Bergan J, Cheng V, editors. Foam sclerotherapy: a textbook. London: Royal Society of Medicine Press; 2008.

Introduction to Venous Stents and Principles of Stenting

76

Natalie Sridharan and Ryan McEnaney

Case Report

Mr. K is a 69-year-old male with a history of left lower extremity swelling since a single episode of deep venous thrombosis in his popliteal vein after knee surgery more than 10 years ago. Despite consistent compression and elevation, he continued to have significant left leg edema which was disabling to the point of limiting his ability to walk. On examination, he had woody edema of the left leg including the thigh and lipodermatosclerosis in the lower leg. As part of his workup, he underwent a CT scan which showed significant compression of the left iliac vein between the right iliac artery and spine consistent with May–Thurner syndrome. After a thorough discussion of risks and benefits, the patient underwent left leg venogram and intravascular ultrasound via left femoral vein access. He underwent placement of a 20 × 100 mm Abre stent (Medtronic) from the left external iliac vein into the inferior vena cava. Completion intravascular ultrasound showed resolution of the area of compression. He was discharged on baby aspirin and Lovenox for one month. He underwent his first postoperative venous duplex at two weeks after the procedure, which showed a patent venous stent with phasic waveforms in the vein. His leg edema continued to improve dramatically over the subsequent months. He continued with compression therapy and was able to resume normal activities including his gym membership. At one month after intervention, he underwent a CT scan which showed full stent expansion. At one year post operative, he continues to have complete resolution of his symptoms and has maintained close follow-up. He takes a daily aspirin.

N. Sridharan (✉) · R. McEnaney
University of Pittsburgh Medical Center, Pittsburgh, PA, USA
e-mail: domenickna2@upmc.edu; mcenaneyrm@upmc.edu

Introduction

Iliocaval venous obstruction (ICVO) is a significant cause of morbidity in the adult population. Quality of life can be as poor as for other chronic conditions including diabetes, chronic obstructive pulmonary disease, and congestive heart failure [1–3]. ICVO may be related to post-thrombotic syndrome (PTS) causing outflow obstruction, scarring, and reflux, or non-thrombotic iliac vein lesions (NIVL) such as May–Thurner lesions. While the mainstay of venous disease has been compression, elevation, and exercise, many patients will remain with debilitating pain, edema, and wounds despite conservative measures. Open surgical reconstruction is relatively morbid especially when considering the relatively poor long-term results with these procedures. Moreover, many patients may not be appropriate candidates for open surgical reconstruction. Thus, endovascular alternatives are very attractive in managing deep venous disease. Angioplasty alone has proven ineffective as stand-alone therapy for ICVO [4]. Venous stenting has shown reasonable long-term patency in both PTS and NIVL lesions with high rates of technical success and good safety profile [4, 5] Venous stenting was first reported in the 1990s [6]. However, only recently have dedicated stents designed for the venous system become widely available. Traditionally, the Wallstent endoprosthesis (Boston Scientific Corporation) has been the most commonly used deep venous stent despite its lack of an indication for this use until 2021. As more encouraging data emerges and with the current wide availability of dedicated venous stents, stenting rates have been steadily increasing.

Indications

There exist multiple indications for deep venous stenting. The advent of dedicated venous stents has expanded the ability to treat patients with these diseases who were historically offered only conservative therapies. In patients with acute deep venous thrombosis, after treatment of the deep venous

G. Geroulakos et al. (eds.), *Mastering Endovascular Techniques*, https://doi.org/10.1007/978-3-031-42735-0_76

thrombosis (DVT) with lysis or thrombectomy, stenting residual areas of disease or underlying lesions improves patency rates [7]. In cases of PTS with pain, edema, or venous ulceration that does not improve with conservative measures, venous recanalization and stenting may aid in symptom relief and speed ulcer healing [8]. While treating NIVL lesions may be less technically demanding than recanalization of a chronically occluded iliac vein, preoperative workup and patient selection in these cases is quite complex. Patients must understand the implications of having a stent placed which will likely require lifelong surveillance and is associated with unknown long-term durability, especially when the majority of these patients are young. That said, it has become increasingly clear that treatment of NIVL lesions such as May–Thurner syndrome may result in dramatic improvements in quality of life and functionality in appropriately selected patients [9, 10]. Exactly which patients will benefit is an area of active and ongoing research. Traditionally, patients with severe symptoms despite maximum conservative measures with greater than 50% Iliac vein stenosis by intravascular ultrasound (IVUS) were considered candidates for iliac vein stenting [11]. However, more research has emerged to challenge the somewhat arbitrarily selected 50% cutoff suggesting that even patients with a much lower degree of stenosis may benefit if appropriately selected [12].

Access Sites

Upon proceeding with deep venous intervention, access site selection is highly dependent on the pathology being treated as well as the extent of disease and what information is known beforehand. In cases of acute DVT, the popliteal vein or a posterior tibial vein is often the best access site. This allows for treatment of the DVT as well as stenting any underlying lesions or residual disease uncovered by thrombectomy or lysis. For chronic disease, extent of disease often determines access site. Access of the femoral vein in the mid- to upper-third of the thigh avoids the prone position, which is especially attractive as contralateral access or internal jugular (IJ) vein access may be needed to facilitate crossing chronic lesions. In addition, this allows for shorter working distance and more pushability in chronic occlusions. This access can be facilitated by slightly externally rotating the leg and with the knee slightly bent. The vein is found posterolateral to the artery at this location under ultrasound guidance. The profunda vein can be accessed directly 2–3 cm below the lesser trochanter in cases of chronic femoral vein occlusion, although even occluded veins can often be accessed as well. Generally, common femoral vein access is used only when it is previously known that disease is isolated to the iliac vein such as in an NIVL May–Thurner lesion. Generally, an 8 French or larger sheath is placed into the vein depending on which stents will be utilized but also to allow passage of an intravascular ultrasound (IVUS) catheter. The patient should also be fully heparinized at this point.

Technique

Depending on the underlying pathology, appropriate access is obtained and the lesion is crossed. Basic tools for deep venous stenting are described in (Table 76.3). For acute DVT, usually patients will undergo lysis or thrombectomy which will reveal any underlying stenosis or residual disease within the vein. For chronic occlusions in post-thrombotic disease, the lesion is crossed using generally a soft or stiff 0.035" glide-wire supported by a catheter. For additional support, the Triforce coaxial system (Cook, Bloomington, IN) has become integral in recanalization of chronic lesions. Multiple projections are often required to ensure passage of the wire in the appropriate trajectory. Small perforations are common and usually self-limiting. Once entry into the IVC is confirmed with contrast venography, the glidewire should be exchanged for a different wire (often a stiff wire such as a super stiff Amplatz) to facilitate tracking of balloons and stents. A second access site such as the IJ or contralateral femoral vein can aid in crossing, and the creation of a "body flossed" wire can also provide excellent support for tracking devices through a fibrotic and chronically occluded vein. NIVL lesions are generally simple to cross. Best practice in these cases would be to pass the wire all the way to the innominate, right IJ, or right subclavian vein to be prepared in the unfortunate event of stent migration. In this event, the stent may be able to be snared or deployed safely into one of these veins.

Optimal landing zones for the stent are disease-free segments of vein. Identification of proximal and distal landing zones requires the assistance of IVUS. Landing stents in a diseased segment of vein can severely compromise their long-term patency rates. This may mean extending from below the inguinal ligament into the IVC. Length of stents does not seem to have a significant impact on stent patency while untreated disease certainly does. Stents can be extended to the profunda vein at the lesser trochanter or even into the profunda vein in cases of chronic and complete femoral vein occlusion with very rare incidence of fracture [13]. The profunda–femoral vein confluence can be visualized by having the patient engage in the Valsalva maneuver during venography or using an occluding balloon in the external iliac vein while injecting contrast. It can also be visualized on IVUS.

Likewise, extension into the IVC is often necessary to fully treat both PTS and NIVL disease. This is particularly true for the Wallstent, which requires cranial extension into the IVC to avoid stent compression at the ends, which lack radial force. Newer nitinol stents allow for more precise landing with some reduction of the need to "jail" the contralateral iliac vein as well as open-cell design.

Prior to venoplasty and stenting, oblique views should be utilized to ensure that the wire anterior to the spine has not inadvertently entered the obturator vein or spinal canal. Once the lesion has been crossed from the selected approach, serial venoplasty as necessary is then done up to a 1:1 ratio with the selected stent size. Typically, high-pressure balloons are employed. Stents are deployed by "building up" with the largest stent placed last. Two-centimeter overlap is generally adequate. In planning stent selection, attention is paid to the natural curvature of the pelvic veins and trying to avoid stent overlap and thus increased stiffness at areas of flexion.

After stents are deployed, post-dilation to nominal stent diameter is recommended to maximize resistive forces particularly in nitinol stents. Stents are then assessed through a combination of venography and IVUS imaging. On venography, the interventionalist may appreciate lack of collateral filling and brisk flow of contrast. On IVUS imaging, stent expansion with wall apposition throughout the treated segment should be observed. Any residual untreated disease should be noted and addressed.

It should be noted that in the case of indwelling IVC filters, attempts should be made to remove the filter if possible. If not possible and the filter is involved in the diseased or occluded segment, the filter can be crushed and a stent placed across it. This generally requires remodeling/fracturing the filter first by repeated high-pressure balloon dilatation.

Stent Selection and Sizing

Venous stents can be most accurately sized by IVUS imaging. The integral role of IVUS in deep venous stenting is indisputable (Fig. 76.1). It is more sensitive than multiplanar venography and can frequently detect areas of residual disease or stenosis that could be missed otherwise. IVUS is most useful when combined with cues from venography which include contrast stagnation, identification of collaterals, and reflux into internal iliac veins.

As mentioned, it is important to treat from a normal segment of inflow vein to a normal segment of outflow vein. The length of the lesion, which is determined by both IVUS and venographic assessment, can be measured using radio-opaque markers on the IVUS catheter. Many of the newer stents can be sized 1:1 lengthwise. The VICI stent (Boston Scientific Corporation), which has been recalled, had a relatively predictable 20% foreshortening. The Wallstent has less predictable foreshortening due to its braided design.

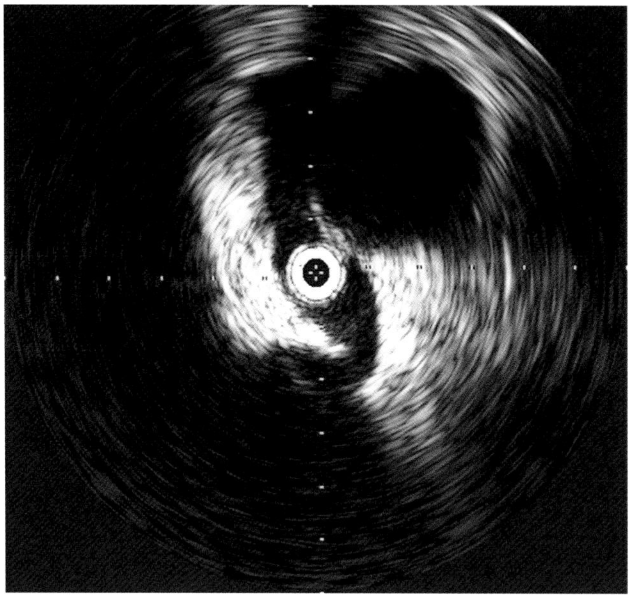

Fig. 76.1 IVUS demonstrates compression of the left common iliac vein by the overlying right iliac artery

Stent diameter can also be selected with the assistance of IVUS. The vein diameter on IVUS is calculated by adding the long and short axis diameter and dividing by two or by calculating diameter from an area measurement. To calculate the percentage stenosis, the diameter is compared to an adjacent reference vessel. For chronically occluded veins or long-segment stenosis, diameter measurements can be made using old imaging if available. However, more commonly, known reference diameters are used to calculate the percent stenosis and to select stent diameters. These optimal vein diameters as described by Raju et al. are 16 mm for the common iliac vein (CIV), 14 mm for the external iliac vein (EIV), and 12 mm for the common femoral vein (CFV) [14] (Table 76.1). Historically, with primary use of the Wallstent, stents were oversized 10–20% to compensate for lower radial force on the ends and to prevent migration. With newer, nitinol stents, this amount of oversizing is probably not necessary and can certainly be complicated by persistent post-procedural back pain. Longer stent lengths available in the dedicated venous stents allow for better anchoring of the stent in the EIV without the need for such oversizing (Fig. 76.2) (Table 76.1).

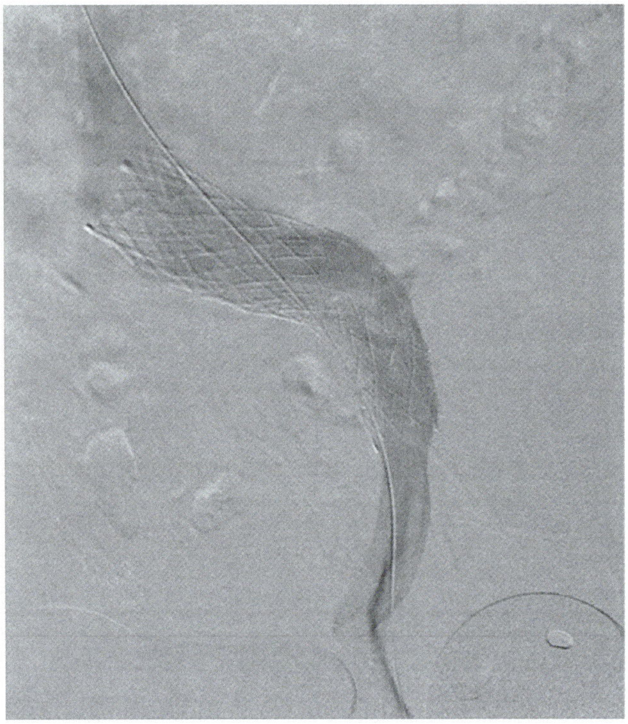

Fig. 76.2 A large-diameter stent is anchored into the external iliac vein to treat an NIVL May–Thurner lesion in an area where it follows the natural venous curvature within the pelvis

Table 76.1 Optimal post-stent diameter and area in iliac vein stenting

	Diameter (mm)	Area (mm^2)
Common femoral vein	12	125
External iliac vein	14	150
Common iliac vein	16	200
Inferior vena cava	18–24	300–400

Complications

As the use of venous stenting has become more widespread, the identification of complications unique to this practice has arisen. One of the most common concerns is the possibility of stent fracture, particularly in high-flexion areas and across the inguinal ligament. Much of this concern is extrapolated from what we know about arterial stenting. The Wallstent has a good track record across the inguinal ligament [15]. For the available data on the dedicated open-cell venous stents, fracture was reported rarely in stents extending into the common femoral vein. The clinical significance of fractures here is also unclear. Some may remain asymptomatic, although some may require relining if they cause symptoms or stent-threatening hemodynamic changes or thrombosis. In a study by Morris et al. of 206 patients undergoing deep venous stenting, 62% required stenting across the inguinal ligament. There were more reinterventions for closed-cell stents that crossed the inguinal ligament, with 15% of this type of stent requiring relining when across the inguinal ligament. Some have suggested continued use of the Wallstent in this location [16]. It is clear, however, that untreated inflow disease is more of a threat to stent patency than stenting across the inguinal ligament and thus interventionalists should not hesitate to stent down to the level of the profunda vein when indicated [16]. Overall, the newer dedicated venous stents are designed with high flexibility and appear to handle the natural curves of the venous system well. Careful planning can avoid increasing stiffness by overlapping stents at high-flexion areas.

Central extension of venous stents can also pose a problem when they cause jailing of the contralateral iliac vein. This was largely an issue for the Wallstent. Due to its braided stainless steel design, it is plagued by low radial force on its ends. To compensate, the stent was often extended significantly across the iliac vein confluence. The reported rate of contralateral DVT in this situation is up to nearly 10% in some series [17]. One technique to overcome this risk was suggested by the Raju group and involved proximal extension with a Z-stent (Cook) that has high radial force and large interstices that do not affect contralateral iliac vein outflow [18] (Fig. 76.3).

Newer dedicated venous stents do not suffer from lower radial force on their ends and thus can be deployed fairly accurately at the area of compression. In addition, their open-cell design may be less of a concern when crossing the confluence.

When jailing of the contralateral iliac vein does result in a problem on that side, several solutions have been proposed. Contralateral access can be used to cross through the interstices of the existing stent. The stent struts can then be fenestrated with serial balloon dilation and then both sides stented in a "kissing stent" fashion or a Y-configuration can be created as described by Neglen et al. [19] (Fig. 76.4).

Finally, stent migration can turn a largely safe procedure into a life-threatening complication. This is most likely to happen in NIVL lesions where the temptation is to use a short stent due to the focal lesion. Appropriate precautions include avoidance of stent undersizing, consideration of using longer stents to "anchor" into the external iliac vein, and placing the wire in the innominate, IJ, or subclavian vein to create a rail from which the stent can be snared or deployed safely.

Venous Stent Landscape

There are a number of dedicated venous stents currently on the market (Fig. 76.5, Table 76.2). This is likely to continue to increase as these procedures become more common and more is learned about ideal properties of venous stents. All

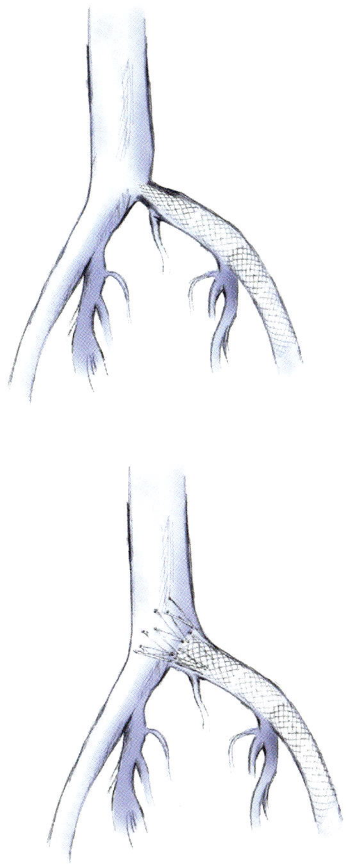

Fig. 76.3 The Wallstent (Boston Scientific) can be buttressed with a Z-stent (Cook) as seen in the bottom image to decrease both compression of the ends of the Wallstent (as seen in the top image) and jailing of the contralateral iliac vein

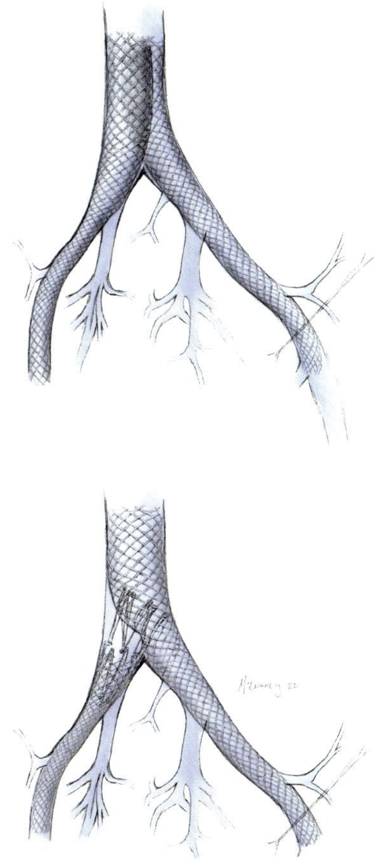

Fig. 76.4 In the top image, kissing iliac vein stents are shown. In the bottom image, a Y-stent configuration with fenestration of the jailing iliac vein stent is demonstrated

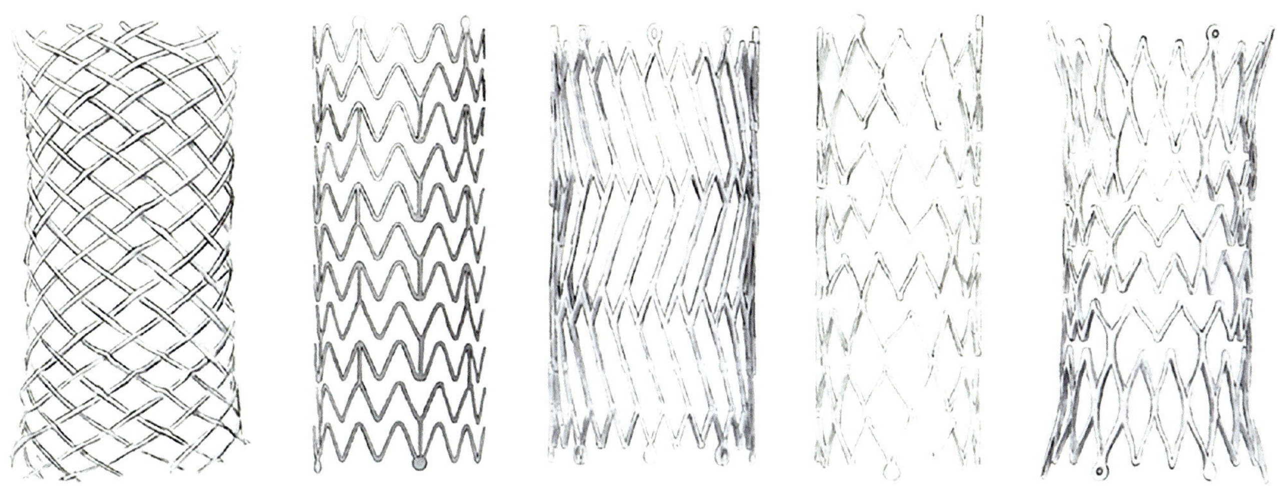

Fig. 76.5 From left to right, current and previously FDA-approved stents with a venous indication include the Wallstent (Boston Scientific), the Zilver Vena (Cook), the VICI (Boston Scientific), the Abre (Medtronic), and the Venovo (BD)

Table 76.2 Characteristics of FDA-approved venous stents in the United States

Stent name	Deployment mechanism	Structure	Diameters (mm)	Lengths (mm)	Catheter working lengths (cm)
Wallstent (Boston Scientific)	Coaxial, 10 F	Braided, Elgiloy	12–24	20–90 (depending on diameter)	75
Venovo (BD)	Triaxial dual thumbwheel, 8-10 F	Open cell, nitinol	10–20	40–160	80, 120
Abre (Medtronic)	Triaxial thumbwheel, 9 F	Open cell, nitinol	10–20	60–150 (40 mm available in 10 mm diameter)	90
Zilver Vena (Cook)	Coaxial, 7 F	Open cell, nitinol	10–16	40–140	80, 120

available venous stents are self-expanding stents. The workhorse of deep venous stenting, the Wallstent, recently received FDA expanded indication for iliofemoral venous use in March 2020. It has a braided, closed-cell design and comes in large diameters (up to 24 mm) but only in shorter lengths (up to 70–90 mm depending on the diameter). Though it accommodates pelvic venous curves and remains fracture resistant across the inguinal ligament, delivery is notoriously difficult and unpredictable. In addition, lower radial force at the ends has led to several unique issues with this stent, including the need to oversize and extend across the iliac vein confluence. Other than the Wallstent, FDA-approved venous stents on the market currently share an open-cell, nitinol design. The Venovo stent (BD/Bard) received FDA approval in 2019 and is unique in its flared ends design. It comes mounted on an 8- to 10-French triaxial platform with diameters 10–20mm and lengths 40–160 mm. The Zilver Vena (Cook) received FDA approval in October 2020 and comes in 14–16mm diameters in lengths up to 140 mm all on a 7-French platform. Finally, the Abre stent (Medtronic) was approved in October 2020 and utilizes an open-cell design with three connection points between the cells to enhance flexibility. Strut dimensions are customized for each stent size. The delivery system is 9 French and the stent comes in 10–20 mm diameters and up to 150 mm lengths. While long-term data on these newer dedicated stents is lacking, trial data uniformly showed high rates of primary patency and low rates of fractures below the inguinal ligament. Head to head comparisons of these stents are not yet widely available (Table 76.2).

Medical Management

Medical therapy after venous stenting is an area of active research and wide variation in practice. Venous stenting after acute DVT is probably the easiest to manage. These patients should be treated with anticoagulation just as if they were being treated for acute DVT with duration depending on whether it was a provoked DVT or if underlying hypercoagulable states are detected. An International Delphi Consensus was published on this matter in 2018. There was a consensus among experts that anticoagulation is preferable to antiplatelet therapy for the first 6–12 months after stenting for NIVL lesions [20]. There was also agreement that low-molecular weight heparin is the anticoagulant of choice in the first 2–6 weeks after stenting. There was little consensus regarding the use of antiplatelets for venous stents [20]. The arterial benefits of antiplatelets cannot be necessarily extrapolated to venous stents. However, there is some data in their favor and, in general, at least a single antiplatelet agent is recommended if the patient can tolerate it after stent placement.

Follow-up

Standard follow-up after venous stenting has not been established. Early thrombus formation and failure has been reported so post-procedural imaging within weeks of stent placement is generally recommended. Flow disturbances on venous duplex or narrowing of the stent seen on any imaging modality should prompt reintervention to prevent stent failure. For NIVL lesions, late stent failure is a relatively rare event and thus close follow-up after the initial period is not as necessary in these patients. Persistence or recurrence of symptoms should prompt re-imaging (Table 76.3).

Table 76.3 Endovascular tools for deep venous stenting

Bedside ultrasound	Angled and straight crossing catheters
Micropunture access kit	Triforce coaxial system (Cook)
Variety of sheath sizes up to 9Fr	Stiff working wire such as Amplatz super stiff
Intravascular ultrasound	High-pressure balloons (up to 1:1 size with selected stents)
Angled and straight soft and stiff glidewires	Dedicated venous stents

References

1. Kahn SR, Hirsch A, Shrier I. Effect of postthrombotic syndrome on health-related quality of life after deep venous thrombosis. Arch Intern Med. 2002;162:1144–8.
2. Kahn SR, M'Lan CE, Lamping DL, Kurz X, Berard A, Abenhaim L, et al. The influence of venous thromboembolism on quality of life and severity of chronic venous disease. J Thromb Haemost. 2004;2:2146–51.
3. Roberts LN, Patel RK, Donaldson N, Bonner L, Arya R. Post-thrombotic syndrome is an independent determinant of health-related quality of life following both first proximal and distal deep vein thrombosis. Haematologica. 2014;99:e41–3.
4. Neglen P, Raju S. Balloon dilation and stenting of chronic iliac vein obstruction: technical aspects and early clinical outcome. J Endovasc Ther. 2000;7:79–91.
5. Murphy EH, Johns B, Varney E, Raju S. Endovascular management of chronic total occlusions of the inferior vena cava and iliac veins. J Vasc Surg Venous Lymphat Disord. 2017;5:47–59.
6. Berger A, Jaffe JW, York TN. Iliac compression syndrome treated with stent placement. J Vasc Surg. 1995;21:510–4.
7. Avgerinos ED, Saadeddin Z, Abou Ali AN, Pandya Y, Hager E, Singh M, et al. Outcomes and predictors of failure of iliac vein stenting after catheter-directed thrombolysis for acute iliofemoral thrombosis. J Vasc Surg Venous Lymphat Disord. 2019;7:153–61.
8. O'Donnell TF Jr, Passman MA. Clinical practice guidelines of the Society for Vascular Surgery (SVS) and the American Venous Forum (AVF)--Management of venous leg ulcers. Introduction. J Vasc Surg. 2014;60:1S–2S.
9. Daugherty SF, Gillespie DL. Venous angioplasty and stenting improve pelvic congestion syndrome caused by venous outflow obstruction. J Vasc Surg Venous Lymphat Disord. 2015;3:283–9.
10. Raju S, Owen S Jr, Neglen P. The clinical impact of iliac venous stents in the management of chronic venous insufficiency. J Vasc Surg. 2002;35:8–15.
11. Gagne PJ, Tahara RW, Fastabend CP, Dzieciuchowicz L, Marston W, Vedantham S, et al. Venography versus intravascular ultrasound for diagnosing and treating iliofemoral vein obstruction. J Vasc Surg Venous Lymphat Disord. 2017;5:678–87.
12. Jayaraj A, Powell T, Raju S. Utility of the 50% stenosis criterion for patients undergoing stenting for chronic iliofemoral venous obstruction. J Vasc Surg Venous Lymphat Disord. 2021;9:1408–15.
13. Black S, Gwozdz A, Karunanithy N, Silickas J, Breen K, Hunt B, et al. Two year outcome after chronic iliac vein occlusion recanalisation using the Vici Venous Stent((R)). Eur J Vasc Endovasc Surg. 2018;56:710–8.
14. Raju S, Buck WJ, Crim W, Jayaraj A. Optimal sizing of iliac vein stents. Phlebology. 2018;33:451–7.
15. Neglen P, Hollis KC, Olivier J, Raju S. Stenting of the venous outflow in chronic venous disease: long-term stent-related outcome, clinical, and hemodynamic result. J Vasc Surg. 2007;46:979–90.
16. Morris RI, Jackson N, Khan T, Karunanithy N, Thulasidasan N, Smith A, et al. Performance of open and closed cell laser cut nitinol stents for the treatment of chronic iliofemoral venous outflow obstruction in patients treated at a single centre. Eur J Vasc Endovasc Surg. 2022;63:613–21.
17. Le TB, Lee TK, Park KM, Jeon YS, Hong KC, Cho SG. Contralateral deep vein thrombosis after iliac vein stent placement in patients with May-Thurner Syndrome. J Vasc Interv Radiol. 2018;29:774–80.
18. Raju S, Ward M Jr, Kirk O. A modification of iliac vein stent technique. Ann Vasc Surg. 2014;28:1485–92.
19. Neglen P, Darcey R, Olivier J, Raju S. Bilateral stenting at the iliocaval confluence. J Vasc Surg. 2010;51:1457–66.
20. Milinis K, Thapar A, Shalhoub J, Davies AH. Antithrombotic therapy following venous stenting: international delphi consensus. Eur J Vasc Endovasc Surg. 2018;55:537–44.

Endovascular Treatment of Chronic Venous Disease

77

Stephen Black and Marie Josee van Rijn

Case Presentation

A 41-year-old female was diagnosed with an iliofemoral left-sided deep venous thrombosis (DVT) 2 years before presentation. A provoking factor was the use of oral contraceptives which she substituted with an intrauterine device. She was treated with oral anticoagulants for 3 months and compression therapy. Her treating physician from internal medicine had advised her to wear compression stockings for 2 years and then stop. She noticed that without the compression stockings she had swelling of her leg which started halfway through the day. Also, she experienced more pain and cramps in her lower leg. For this reason, she decided to keep wearing the stockings; however, even then, her walking distance decreased because of a bursting pain in her left calf when she walked for more than 20 min, especially when she walked up the stairs or a hill. She also noticed a big vein above her pubic bone. She used to be a runner, but she was not able to do this anymore. She had also been working fulltime as a saleswoman at a big company before the DVT, but now worked only 3 days a week, because her leg needed to recover in between her working days. She "feels like an old lady." She was referred to us for diagnosis, counselling, and management.

Continued at page 808

Background

Patients with a deep venous thrombosis (DVT) are at increased risk for developing post-thrombotic syndrome (PTS) as a result of deep venous obstruction and/or reflux leading to venous hypertension [1]. This causes changes in calf microvasculature, inducing reduced perfusion of surrounding muscle, chronic inflammation, increased vascular permeability, and scarring of the vessel wall [2, 3]. It is thought that venous hypertension initiates a chronic inflammatory cascade [4].

In iliofemoral DVT the risk of developing PTS is approximately 50%. Common PTS symptoms include pain, cramps, heaviness, paresthesia, and pruritus of the leg; common signs are edema, skin changes, redness, and pain during calf compression, with venous ulceration as one of the worst clinical features; however, symptomatology varies over time [1]. Post-thrombotic syndrome results in a considerable burden to patients and health care systems.

Various scoring systems are available to quantify the severity of PTS. The Villalta score combined with a venous disease-specific quality of life (QoL) score is the recommended golden standard. However, PTS definition and classification should be further optimized and should take into account the broad spectrum of PTS manifestation, such as venous claudication and exercise intolerance, which are both absent in current PTS scores [5].

Imaging

Duplex Ultrasound (DUS)

It is entirely possible, as demonstrated by Labropoulos and colleagues, to visualize the entirety of the iliac outflow tract with duplex [6]. This initial assessment provides a very detailed evaluation of deep and superficial reflux and obstruction of the leg and the abdomen.

Cross-Sectional Imaging

Additional information, particularly pertaining to above the ligament in obese patients or patients with abundant bowel

S. Black (✉)
St. Thomas' Hospital, Vascular Surgery, London, UK
e-mail: Stephen.Black@gstt.nhs.uk

M. J. van Rijn
Erasmus Medical Center, Rotterdam, The Netherlands
e-mail: m.vanrijn@erasmusmc.nl

G. Geroulakos et al. (eds.), *Mastering Endovascular Techniques*, https://doi.org/10.1007/978-3-031-42735-0_77

air, may be necessary if DUS evaluation is compromised or in centers where DUS is not performed as easily [7]. Both computed tomography venography (CTV) and magnetic resonance venography (MRV) are used to map the extent of the disease at the proximal site (e.g., involvement of the inferior vena cava [IVC]), as well as to provide detailed information on trabeculation at the level of the common femoral vein (CFV) and the inflow vessels (deep femoral, femoral, and popliteal vein).

Treatment Strategy

Initial therapy should start with non-invasive treatment, the cornerstone being compression therapy. If symptoms are not controlled, invasive treatment options may be considered. The decision as to which patients presenting with DVT should be intervened has troubled venous interventionalists consistently and remains a complex challenge.

Timing

The development of PTS takes a minimum of 6 months [8], so it is important not to rush the decision to intervene as it is clear that the early resolution of symptoms of DVT may follow a fluctuating course [1]. In addition, the formation of a collateral network may lead to symptom improvement and, in some cases, to improving the chances of successful stenting by developing inflow vessels that will sustain a stent [9].

Assessing Inflow

The main evaluation in deciding on which patients to treat is assessing inflow. Adequate inflow vessels in the form of a good deep femoral or femoral vein appears to be the biggest determinant of success in venous stenting. This evaluation can be performed by a combination of DUS and cross-sectional imaging. The preferred technique is MRV, which gives detailed imaging of trabeculations in the CFV and extension into the deep femoral vein (DFV). However, not all practices can reproduce high-quality MRV, and CTV can be a valid alternative option.

If the inflow is poor and there is orificial deep femoral vein disease in the CFV, then the only option may be an endophlebectomy with or without a fistula, though the outcomes from this procedure are controversial [10]. Therefore, patients considered for such procedures need to be significantly disabled by their disease.

Endovascular Strategy and Technique

Anaesthesia

Recanalization of obstructive deep veins may take time and venous dilation and stenting is painful. This should therefore be done under sedation, combined with intravenous analgesia, or under general anaesthesia.

Basic Endovascular Toolkit

- Standard access wires and sheaths including a short 6 and 10Fr sheath
- Selective catheters (e.g., multipurpose)
- Floppy and stiff glidewire 0.35"
- Small size (4–8 mm) balloons for crossing
- High-pressure 12–24 mm balloons for pre- and post-dilatation
- 12–24 mm venous stents (various lengths)

A list of suggested material is summarized in Table 77.1.

Table 77.1 Suggested endovascular toolkit for the endovascular management of acute DVT

	Manufacturer	Size/length
Wires		
Any standard access wire		0.035"
Glidewire Floppy	Terumo Radifocus	0.035"
Glidewire Stiff	Terumo Radifocus	0.035"
STARTER Rosen	Boston Scientific Corporation	0.035"
V-18 Control Guidewire	Boston Scientific Corporation	0.018"
Sheaths		
Any standard access sheath	Any	5-10Fr/11 cm
Triforce Crossing System	Cook Medical	4Fr/55-90 cm
Catheters		
Angled	Terumo Radifocus	0.035"/5Fr/65–90 cm
Quick Cross Support (low profile)	Spectranetics/Phillips	0.018"–0.035"/4Fr/65–90 cm
Balloons		
Atlas	BD Medical	12–16 mm
Stents		
Abre	Medtronic	
Zilver Vena	Cook Medical	
Beyond	Bentley	
BlueFlow	Plusmedica	
Sinus	Optimed	
Venovo	BD Medical	
Wallstent	Boston Scientific Corporation	
IVUS Catheter		

These are only a few suggested options based on the author's experience and can cover the vast majority of cases

Access and Other Technical Considerations

Access is always done under ultrasound guidance and can be achieved from the leg or the internal jugular vein. In the leg, the tip of the sheath should be caudal to the confluence of the DFV and the femoral vein (FV) to allow treatment of the CFV. The obstruction is crossed with a hydrophilic wire and a supporting catheter. When crossing chronic lesions it is essential to check that the wire is traversing an appropriate anatomical course. In particular, lateral and oblique projections should be checked to confirm that the wire did not cross into spinal collateral vessels (in lateral projection wire should always be on top of the vertebral bodies). Crossing may take time as the wire often de-routes into the collaterals or gets stuck in the small channel left in between the trabeculations. It may help to pre-dilate with smaller-size balloons, also if crossing was successful but large-size balloons do not pass. After recanalization, pre-dilate the entire tract with a balloon with at least the same diameter of the stent you are intending to use. In general, the CFV and external iliac vein (EIV) should be pre-dilated with a 12–14 mm size balloon, the common iliac vein (CIV) with a 14–16 mm balloon, and the IVC with a 20–24 mm balloon.

Intravascular Ultrasound

Intravascular ultrasound (IVUS) (Philips, Andover, MA) should be used when performing interventions for chronic deep venous obstruction. It is superior in identifying the extent of the disease compared to plain venography, and therefore it can locate the precise landing zones for stent placement. Also, IVUS can identify compression points as well as trabeculation in relation to wire position [11]. In case the wire is at the wrong side of a trabeculation, opening up a stent at that location may force the trabeculae against the orifice of an inflow vessel and jail it. This should trigger replacement of the wire, before pre-dilation and stent placement. Lastly, IVUS can detect insufficient expansion of a stent that calls for another post-dilation.

Venous Stents

Self-expanding stents should land in healthy segments above and below the lesion, without exception regarding the crossing of the inguinal ligament. Typically, the CFV and EIV are stented with a 12–14 mm stent, the CIV with a 14–16 mm stent, and the IVC with a 20–24 mm stent. Post-dilation should always be performed, using the same size balloons as used for pre-dilation. If a final venography is performed, it should be done in at least two planes, to visualize rapid washout of contrast, with disappearance of collateral veins. IVUS is used to check for residual stenosis/compression.

Complications

Intraoperative complications are uncommon. Venous perforations occur during recanalization, but are forgiving as the venous circulation is a low-flow system.

The most frequent complication is postoperative back pain or groin pain, the latter only when stents are placed into the CFV. Both will resolve within weeks. Another complication is re-thrombosis of the stent, which may be related to technical error, missed disease, poor in- or outflow, inadequate anticoagulation, or lack of mobilization.

Perioperative Care and Surveillance

Compression After Treatment

It is useful to use both compression in the form of stockings and intermittent pneumatic compression to ensure that optimal drainage of the leg is maintained in the immediate postoperative period. Thigh-length stockings also allow for access sites to be effectively compressed to minimize postoperative haematoma formation. Compression is usually prescribed after deep venous procedures, but there are no data on the effectiveness due to lack of specific studies.

Antithrombotic Treatment

Peri- and postoperative therapeutic anticoagulation is crucial, but antithrombotic regimens are heterogeneous among studies [12]. Neither type of anticoagulation or duration has been studied. In general, it is advised that patients on anticoagulation should be bridged to low-molecular weight heparin preoperatively, to be continued shortly after the procedure. Perioperatively, all patients should be adequately heparinized. If there is an indication for indefinite anticoagulation, patients should continue anticoagulation after the procedure, unless major bleeding complications occur. If there is no indication for indefinite anticoagulation, postoperative anticoagulation should be prescribed, but there is insufficient evidence to guide decision making regarding the optimum duration. According to experts participating in a Delphi consensus, anticoagulant treatment should be continued for at least 6 months after intervention. No consensus regarding the role of antiplatelets was reached [13]. The choice of oral anticoagulation has also not achieved any consensus and therefore should be based on individual clinician preference while further studies are awaited.

Surveillance

Postoperative surveillance after deep venous interventions is of key importance. It is performed with DUS, starting the first day after the procedure. Considering the fact that a thrombosed stent can be recanalized with thrombolysis within 14 days of the thrombotic event, and most thrombotic complications occur shortly after the intervention, a second DUS should be performed within two weeks after stent placement. Further follow-up after an interval of 6 weeks, 6 months, and yearly thereafter is common practice. If symptoms return, patients should be instructed to contact their vascular specialist as soon as possible. A suspected stenosis >50% on ultrasound warrants immediate further evaluation and treatment.

Multidisciplinary Team

Managing patients with PTS is complex and it is strongly recommended that these patients are managed within a multidisciplinary team. This team should consist principally of a hematologist (with a specific interest in thrombosis), an interventional radiologist, and a vascular surgeon. Other members of the team may be added as necessary and this should include consideration of experts in diagnostic imaging.

Conclusion

The treatment of patients with chronic DVT is now well established. There are several clinical evaluation tools, though flawed, that provide some basis for clinical evaluation. A number of advances in imaging provide for the ability to attempt to stratify patients and assess the extent of disease, thereby guiding treatment options. Future developments in imaging promise to improve this further. Nonetheless, it is clear that for the future development of therapies significant improvements in patient classification and selection are required to ensure the appropriate treatments are delivered to the appropriate patients at the appropriate time.

Case Presentation

Continued from page 805

On clinical examination our patient had a 3 cm greater leg circumference (lower and upper leg) on the left side compared to right with pitting edema around her ankle and hyperpigmentation at the medial malleolus. A collateral (Palma) vein was visible above her pubic bone. Besides her Villalta score of 12, she had disabling venous claudication.

On assessment with DUS and MRV it was found that she had post-thrombotic obstruction of her CIV and EIV on the left side. The CVF had a normal diameter at the level of the femoral confluence, but was traversed by trabeculation that continued into the DFV, but not the FV. She was deemed appropriate for stent placement because of (1) severity of PTS, (2) failure of conservative treatment for a long enough observation period, and (3) iliofemoral obstruction with good inflow from the FV.

Under general anaesthesia, she was placed in supine position on the interventional suite table. Ultrasound guidance access to the left FV mid-thigh was obtained, a 6Fr sheath placed, and 5000 IU of heparin administered intravenously. Initial venogram confirmed the iliofemoral obstruction and multiple collaterals (Fig. 77.1a). Using a Terumo wire and a multipurpose catheter, the CFV and iliac veins were crossed entering the IVC. Lateral and oblique projections showed that the wire passed in front of the vertebrae and thus did not cross into spinal collateral vessels. The wire was exchanged for a stiff wire. IVUS showed that the femoral vein was disease-free, but trabeculation was visible at the level of the CFV entering from the DFV. The wire was positioned at the medial site of the trabeculation; therefore, stenting would not displace it in front of the FV. The position of the confluence was marked on the screen. The iliac confluence was also marked with the use of IVUS, as was the crossing right common iliac artery. A venogram showed good inflow from the FV and drainage of the blood via the Palma vein and several pelvic collaterals.

The sheath was upsized to 10Fr. Because the 14 mm balloon would not pass, pre-dilation was performed with a 10 mm balloon from CFV to IVC. After that the CVF and EIV were pre-dilated with a 14 mm balloon and the CIV with a 16 mm balloon (Fig. 77.1b). IVUS was used again to confirm landing points and to measure the total length of the segment that needed stenting. A 14 × 120 mm Abre stent (Medtronic) was placed precisely at the confluence of the DFV and FV into the EIV followed by a 16 × 140mm Abre stent (Medtronic) from the EIV into the IVC. Attention was paid to (1) sufficient overlap, (2) avoiding overlap at the inguinal ligament, and (3) avoiding jailing of the right CIV by not overextending the stent into the IVC. Post-dilation was done with a 14–16 mm balloon. IVUS showed adequately positioned, well-opened stents. Venogram showed a quick washout of contrast and disappearance of collaterals (Fig. 77.1c).

The patient was placed on low-molecular weight heparin immediately and pneumatic compression. The next day, DUS confirmed open stents with good inflow from the FV. She was discharged that day with the instruction to walk every day, preferably multiple times a day, wearing her compression stockings. After 2 weeks she was switched to a direct oral anticoagulant. At 6 months of follow-up, symptoms and signs had significantly improved and her stents were open. She could stop wearing her

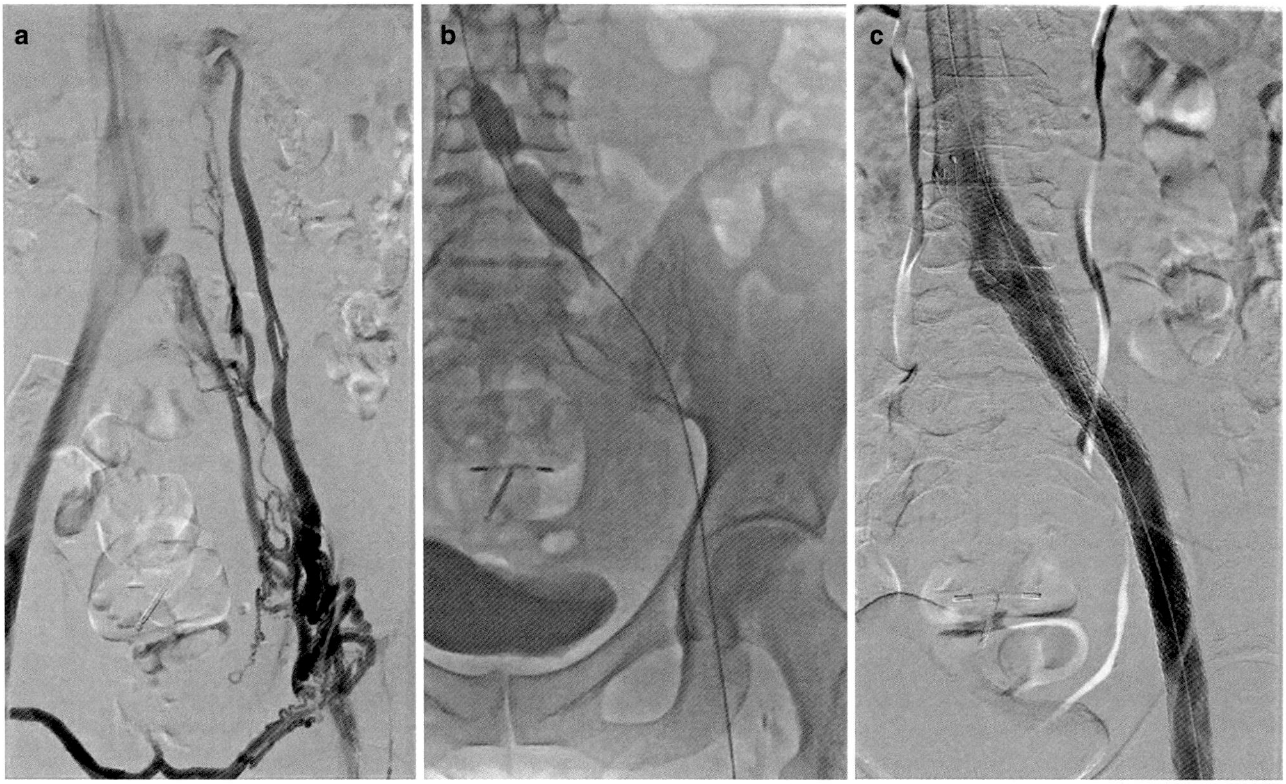

Fig. 77.1 Posthrombotic obstruction of left iliofemoral veins. (**a**) Multiple collaterals drain left leg through the right iliac system. (**b**) The left iliofemoral obstruction has been crossed and balloon dilatation has been performed. (**c**) Two Abre Stents have been deployed, one distally below the inguinal ligament 14 × 120 mm and one proximally 16 × 140 mm – collaterals are no longer visible as the left iliofemoral patency has been restored

compression stockings and after consulting the multidisciplinary team (specifically the haematologist), her anticoagulants were stopped with the notification that if she were to ever start hormone therapy again, she had to restart anticoagulation. She will be under yearly surveillance.

References

1. Kahn SR, Shrier I, Julian JA, Ducruet T, Arsenault L, Miron MJ, et al. Determinants and time course of the postthrombotic syndrome after acute deep venous thrombosis. Ann Intern Med. 2008;149(10):698–707.
2. Malgor RD, Labropoulos N. Re-modelling of venous thrombosis. Phlebology. 2013;28(Suppl 1):25–8.
3. Kahn SR. The post-thrombotic syndrome. Hematology Am Soc Hematol Educ Program. 2016;2016(1):413–8.
4. Busuttil A, Lim CS, Davies AH. Post thrombotic syndrome. Adv Exp Med Biol. 2017;906(3):363–75.
5. Soosainathan A, Moore HM, Gohel MS, Davies AH. Scoring systems for the post-thrombotic syndrome. J Vasc Surg. 2013;57(1):254–61.
6. Malgor RD, Labropoulos N. Diagnosis of venous disease with duplex ultrasound. Phlebology. 2013;28(Suppl 1):158–61.
7. Silickas J, Black SA, Phinikaridou A, Gwozdz AM, Smith A, Saha P. Use of computed tomography and magnetic resonance imaging in central venous disease. Methodist Debakey Cardiovasc J. 2018;14(3):188–95.
8. Kahn SR, Partsch H, Vedantham S, Prandoni P, Kearon C, on behalf of the Subcommittee on Control of Anticoagulation of the Scientific and Standardization Committee of the International Society on Thrombosis and Haemostasis. Definition of post-thrombotic syndrome of the leg for use in clinical investigations: a recommendation for standardization. J Thromb Haemost. 2009;7(5):879–83.
9. Neglen P, Furrh B, Raju S. Axial transformation of the profunda vein sustains ilio-caval stenting in postthrombotic limbs. J Vasc Surg. 2011;53(1):257.
10. Piao L, Barbati ME, Shekarchian S, Jacobs MJ, Razavi M, Rass K, Jalaie H. Comparison of endovascular strategy versus hybrid procedure in treatment of chronic venous obstructions involving the confluence of common femoral vein. J Vasc Surg Venous Lymphat Disord. 2022;10(2):334–41.
11. Gagne PJ, et al. Venography versus intravascular ultrasound for diagnosing and treating iliofemoral vein obstruction. J Vasc Surg Venous Lymphat Disord. 2017;5(5):678–87.
12. Marston WA, Browder SE, Iles K, et al. Early thrombosis after iliac stenting for venous outflow occlusion is related to disease severity and type of anticoagulation. J Vasc Surg Venous Lymphat Disord. 2021;9(6):1399–407.
13. Milinis K, Thapar A, Shalhoub J, Davies AH. Antithrombotic therapy following venous stenting: international Delphi consensus. Eur J Vasc Endovasc Surg. 2018;55(4):537–44.

Endovascular Treatment of Pulmonary Embolism

78

Efthymios Avgerinos, Vassilios Bouris, and Rabih Chaer

Case Presentation

A 72-year-old male was transferred to the emergency room from an outside hospital with chest pain and shortness of breath. His admission vitals were 100 bpm heart rate, systolic blood pressure 100/80 mmHg and O_2 saturation 88% on room air. The CT pulmonary arteriogram confirmed bilateral pulmonary emboli. He was placed on therapeutic low-molecular weight heparin and oxygen and transferred to the intensive care unit. A venous duplex confirmed right femoro-popliteal deep venous thrombosis (DVT). As there was no improvement within 24 h the treating physician decided to administer systemic thrombolytics (100 mg alteplase). The patient showed signs of gradual improvement and was transferred to the ward 24 h later. The next day he again deteriorated: HR 141, SBP 100/80, and SAT 85% (intermediate high-risk PE).

The pulmonary embolism response team (PERT) was informed, and given the high mortality risk a decision was made to proceed with a catheter-directed therapy to remove the thrombus.

Continued at page 817

Background

Acute pulmonary embolism is the third leading cause of cardiovascular mortality and the most preventable cause of hospital mortality [1, 2].

The thrombotic obstruction of the pulmonary arteries accompanied by hypoxemic vasoconstriction increases pulmonary vascular resistance and right ventricular (RV) afterload. The increasing RV afterload can cause RV hypokinesis, dilation and ultimately RV failure with subsequent life-threatening reduction in coronary perfusion and cardiac output. While in most survivors pulmonary thrombo-emboli will gradually resolve, in some patients it may organize into fibrotic deposits permanently occluding the pulmonary arteries, leading to chronic pulmonary hypertension and RV dysfunction. The main goal of therapy is to prevent mortality; a secondary goal, yet poorly documented, is the prevention of late-onset chronic thromboembolic pulmonary hypertension to improve quality of life [1–4].

Anticoagulation was and remains the standard of care for treatment of pulmonary embolism (PE) and prevention of thrombus extension or recurrence. Anticoagulation, however, does not dissolve thrombus, but rather prevents formation of more clots while intrinsic thrombolytic pathways are slowly taking over. In higher-risk cases, when ventricular function is acutely compromised, faster clot removal is needed to prevent cardiac decompensation. Administration of systemic thrombolytics can be highly effective in rapidly resolving pulmonary thrombus. However, as it can result in significant bleeding complications, or frequently contraindicated (e.g. recent surgery or trauma) interventional treatments to rapidly remove the clot from the pulmonary arteries have always been attractive. Catheter-directed interventions (CDI) for acute PE, with or without thrombolytics, have been increasingly used over the past decade and evidence on their effectiveness is mounting [5–10].

Indications for Intervention

Catheter interventions have been promoted for use in both intermediate (submassive) and high-risk (massive) PE in an attempt to reduce the complication rates of systemic thrombolytics and the complexity of surgical thrombectomy. While treatment algorithms are still evolving, available evidence indicates optimal benefit/risk ratio for patients with high-risk submassive PE (positive biomarkers and abnormal echocardiographic parameters) in combination with high oxygen requirements, as well as

E. Avgerinos (✉) · V. Bouris
Clinic for Vascular and Endovascular Surgery, Athens Medical Center, Athens, Greece

R. Chaer
Division of Vascular Surgery, University of Pittsburgh Medical Center, Pittsburgh, PA, USA
e-mail: chaerra@upmc.edu

G. Geroulakos et al. (eds.), *Mastering Endovascular Techniques*, https://doi.org/10.1007/978-3-031-42735-0_78

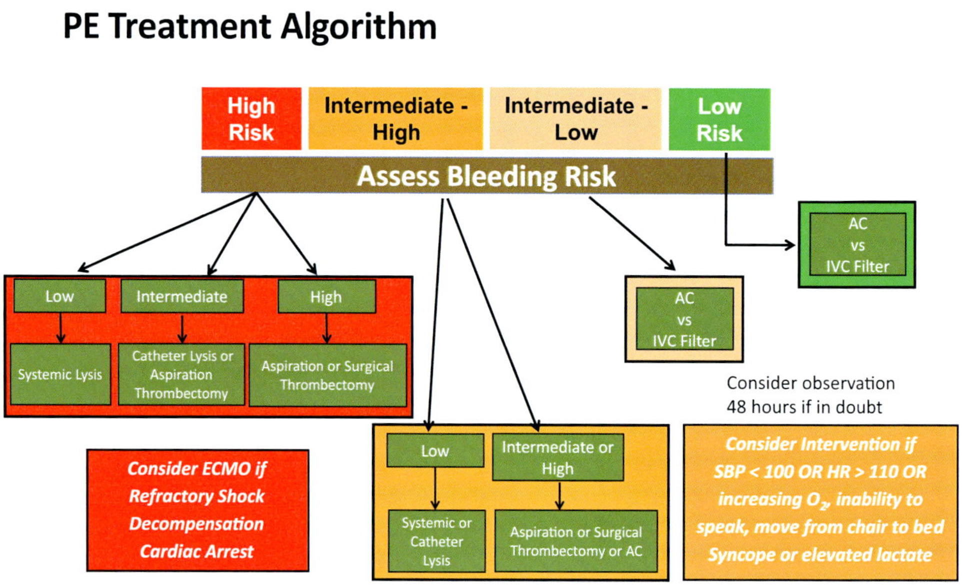

Fig. 78.1 Suggested algorithm for PE intervention

massive PE patients who have contraindications for systemic thrombolysis [5, 6]. Appropriate expertise with documented good outcomes will likely expand the indications for intervention. Given the complexity in decision making, pulmonary embolism response teams (PERT) should provide multidisciplinary guidance based on the local resources [11, 12]. The PE TRACT, the HI-PEITHO and the PEERLESS trials are ongoing randomized trials that are anticipated to better guide our decision making on CDI versus anticoagulation alone [6].

For those eligible for intervention, bleeding risk assessment will guide the type of catheter intervention. As a principle, patients at bleeding risk (e.g. recent surgery or trauma) should not be offered thrombolytic therapy bur rather percutaneous suction thrombectomy options than can be done with minimal or no lytic exposure. More recently, we're increasingly seeing a shift towards non-thrombolytic procedures; the role of lytics in the distal segmental perfusion and its potential effect on long-term PE syndrome is under investigation (Fig. 78.1).

Preoperative Preparation

Anticoagulation

All patients with a PE diagnosis should retain their therapeutic anticoagulation in preparation for and during the procedure. Unfractionated or low-molecular weight heparin can be used. Maintenance of therapeutic anticoagulation during ongoing thrombolysis varies according to institutional policies. In our practice, we typically maintain therapeutic anticoagulation. In aspiration thrombectomy procedures we will additionally bolus heparin to maintain an Activated Clotting Time (ACT) of at least 200.

Imaging

Computed tomographic pulmonary arteriography (CTPA) is necessary for diagnosis, risk stratification and planning of a PE intervention. Pulmonary arterial thrombi do not guide decisions for intervention but the location of thrombus can guide the type of intervention. Heart chamber dimension measurements can also evaluate RV enlargement as an indicator of RV dysfunction.

Transthoracic echocardiography (TTE) is the most common first-line examination to diagnose the signs of RV dysfunction but cannot definitively diagnose PE, as an elevated RV pressure may be the result of other conditions such as pulmonary hypertension and right ventricular infarction. Echocardiography is most useful in patients with diagnosed PE, for prognostic purposes.

Bleeding Risk Assessment

Amongst those who are considered for intervention, bleeding risk assessment can guide treatment type. There are major (absolute) and minor (relative) contraindications for thrombolytics. Thrombectomy with minimal or no lytics will be generally preferred for patients with absolute contraindications. The clinician is in the best position to judge the relative merits of thrombolysis on a case-by-case basis.

Absolute contraindications for lytics:

- any prior intracranial hemorrhage,
- known structural intracranial cerebrovascular disease or neoplasm,
- ischemic stroke within 3 months, suspected aortic dissection,

- active bleeding or bleeding diathesis,
- recent spinal or cranial/brain surgery,
- recent closed-head or facial trauma with bony fracture or brain injury.

Relative contraindications:

- age ≥ 75 years,
- current use of anticoagulation,
- pregnancy,
- non-compressible vascular punctures,
- traumatic or prolonged cardiopulmonary resuscitation,
- recent internal bleeding (2–4 weeks),
- history of chronic severe and poorly controlled hypertension,
- severe uncontrolled hypertension on presentation,
- dementia,
- remote (3 months) ischemic stroke,
- major surgery within 3 weeks.

Endovascular Strategy and Technique

Basic Endovascular Toolkit

- Standard access wires and sheaths.
- Selective catheters (e.g. Glidecath).
- Floppy and stiff glidewire 035".
- Stiff wires.
- Multisidehole thrombolytic catheters (5–10 cm infusion lengths).
- Thrombectomy device.

A list of suggested materials is summarized in Table 78.1.

Table 78.1 Suggested endovascular toolkit for the endovascular management of acute PE

Wires	Manufacturer	Size/length
Any standard access wire		0.035″
Glidewire floppy	Terumo Radifocus	0.035″
Glidewire stiff	Terumo Radifocus	0.035″
STARTER Rosen	Boston scientific corporation	0.035″
Amplatz super stiff – Short tip	Boston scientific corporation	0.035″
Sheaths		
Any standard access sheath	Any	5–10Fr/11 cm
Catheters		
Angled	Terumo Radifocus	0.035″/5Fr/65–90 cm
Balloons		
–		
Stents		
–		
Thrombolytic catheters		
Unifuse	Angiodynamics	4–5Fr (5–10 cm infusion length)
Cragg-McNamara	Medtronic	5Fr (5–10 cm infusion length)
EKOS	Boston scientific corporation	6Fr (6–12 cm infusion length)
Bashir endovascular catheter	Thrombolex	7Fr
Thrombectomy devices		
Alphavac	Angiodynamics	18Fr
FlowTriever	Inari	13 & 16Fr
Indigo CAT 8/ lightning 12	Penumbra	8Fr/12Fr

These are only a few suggested options based on the authors' experience and can cover the vast majority of cases

Timing of Intervention

Massive pulmonary embolism is a life-threatening emergency and any planned procedure should be performed emergently. For the patient who is acutely worsening or unstable, systemic thrombolytics will likely be the fastest option as they are readily available.

For submassive pulmonary embolism CDI should be considered when no improvement has been achieved after 24–48 h of initial anticoagulation. Thus, in this group, time limits could be less stringent. Individualizing to the patient's clinical status and vital signs may occasionally justify emergent intervention.

Readiness of the Extracorporeal Membrane Oxygenation (ECMO) team is highly recommended for the patient who shows volatile vital signs.

Access into the Pulmonary Arteries

Accessing the pulmonary arteries can be achieved through the internal jugular (ideally right) or any of the common femoral veins. Use of ultrasound and a micropuncture system is recommended to minimize bleeding complications. Often, when thrombolysis is considered, dual access will be needed to introduce one catheter in each pulmonary artery. A dual lumen sheath (10–12Fr) or two access sheaths through the same or different veins (e.g. bilateral common femoral veins) are valid options. For very large thrombectomy catheters (>12Fr), trackability will likely be better through femoral access.

A starter 0.035-inch wire assisted by a standard guiding catheter is navigated through the right atrium towards the right ventricle and then to the main pulmonary artery. When large devices are planned to be used, care should be taken to prevent tricuspid valve injury (if wire goes through the chordae tendinae). To avoid this, a pigtail catheter (with the wire partially loaded to maintain some stiffness) or an inflated Swan Ganz catheter can be used to cross the valve. Glidewires should only be used selectively as they can inadvertently travel to undesired locations and cause serious injury or perforation (e.g. coronary sinus). Once the pigtail is within the main pulmonary artery, invasive pulmonary pressure should be obtained as a baseline for comparison at completion of the treatment. An arteriogram (hand or power injection) may be done to locate the clot and then guide the lytic catheters or the suction thrombectomy system. Selective conventional angiography can be done by administering 10 mL of contrast at 5 mL/s preferably in the 20° left anterior oblique (LAO) view to visualize the left pulmonary artery and preferably the 20° right anterior oblique (RAO) view to visualize the right pulmonary artery. The RAO cranial view may also be useful. If the patient is too unstable, power injection can theoretically increase pulmonary pressure so it should be avoided. The preoperative CT angiogram is otherwise a good guide for clot mapping.

Once in the main pulmonary artery, navigation to the right and/or left upper or lower branches is achieved with a long floppy glidewire and angled catheter.

Inferior Vena Cava Filters

Current guidelines dictate against the routine use of IVC filters in PE patients who can be anticoagulated [2]. Low cardiopulmonary reserve can be a relative indication particularly for those with a high-risk PE, but its benefit remains uncertain. Decision making should be individualized and elaborated by the pulmonary embolism response team (PERT).

Decision Making on Technique Selection

Thrombolytic and non-thrombolytic techniques are complementary and equally effective when used appropriately [13]. Each technique has strengths and weaknesses along with a different complication profile. Thrombolytics can be administered through small catheters but need time and can cause systemic bleeding; suction thrombectomy techniques require large devices that need to cross the heart with risk of injury or arrhythmias and a limited ability to clear the distal pulmonary arterial run-off.

For patients who are severely unstable (and systemic thrombolytics are contraindicated) aspiration thrombectomy will likely achieve RV decongestion faster. Surgical thrombectomy is also an alternative for all patients that needs to be discussed early in the decision making when appropriate expertise exists.

Type of Anesthesia

Catheter interventions for PE should be ideally performed with the patient awake under monitored moderate sedation, as induction for general anesthesia can lead to hemodynamic collapse. The PE-induced RV dysfunction causes an increase in oxygen demand, a decrease in left ventricular filling, and consequently a reduction in cardiac output, which can exacerbate any condition that can cause hypotension.

Fentanyl and midazolam are typically enough. Propofol can cause systemic hypotension and recent evidence suggests that its use for sedation during CDI for acute PE is associated with increased perioperative morbidity and mortality, and thus should be avoided [14].

Extra-Corporeal Membrane Oxygenation

Transportable extra-corporeal membrane oxygenation (ECMO) assistance systems with percutaneous femoral artery and vein cannulation (veno-arterial) can be helpful in critical situations of high-risk PE, and circulatory collapse or cardiac arrest [15]. ECMO unloads the acutely failing right heart providing effective hemodynamic and respiratory support for patients who develop massive PE until recuperation of the initial pulmonary insult. There are no clear prognostic indicators to appropriately select unstable PE patients who are likely to benefit from ECMO support. For patients you decide to intervene it is reasonable to ask for readiness of the ECMO team, being proactive for a possible patient collapse.

Thrombolytic and Thrombectomy Catheters/Devices

Catheter-Directed Thrombolysis

Standard multisidehole catheter (Cragg McNammara, Medtronic or Unifuse, Angiodynamics) infusion is the traditional baseline technique, still popular and efficient at least for patients with submassive PE [16, 17]. The catheters (5 or 10 cm infusion length) are positioned inside the main and lobar PA thrombosed segments, right, left or bilateral. A bolus dose of tPA 2–4 mg can be given at the operator's discretion and then 1 mg/h per catheter is initiated and continued in the intensive care unit. Termination should be guided by objective improvement of vital signs and hemodynamic or echocardiographic parameters, though some favor fixed dose protocols to standardize the technique and minimize bleeding complications. Total tPA should ideally be main-

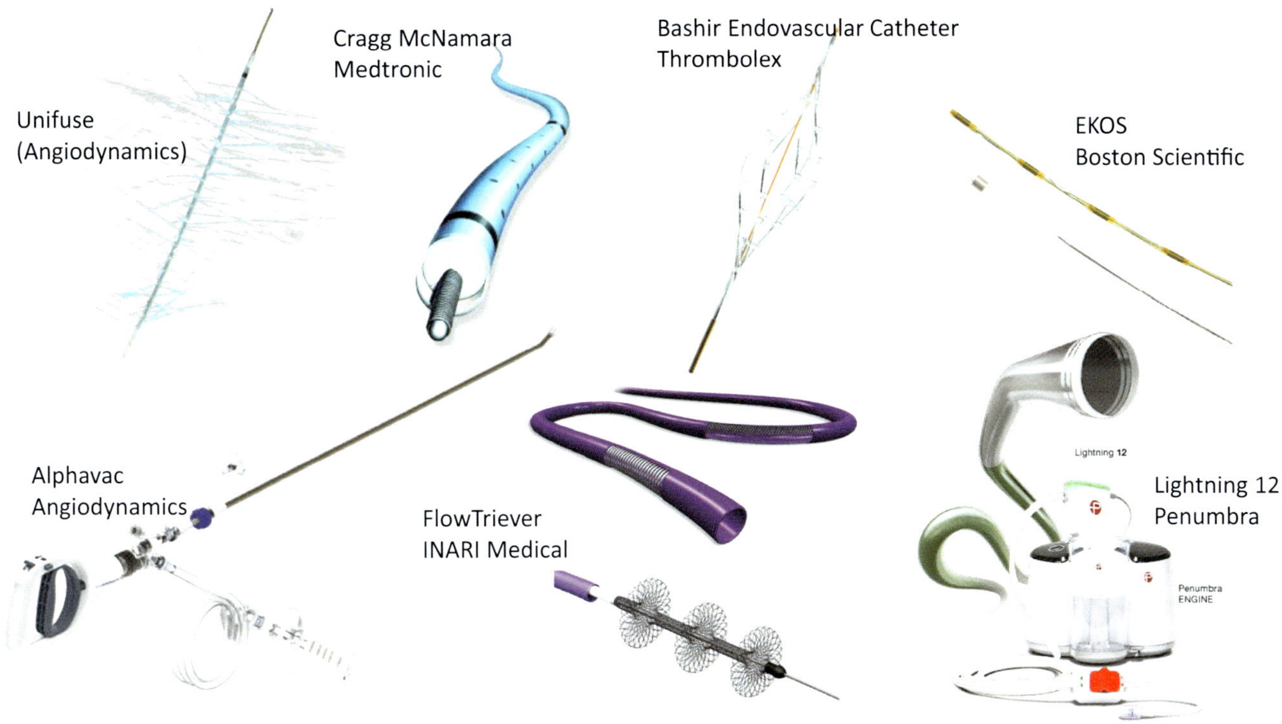

Fig. 78.2 Available catheters for PE intervention

tained <25 mg. Catheters are pulled out at the bedside (Fig. 78.2).

The Bashir Endovascular Catheter (BEC – Thrombolex) consists of a basket of multi-sidehole infusion limbs, which is expanded once it is positioned in the thrombus to partially disrupt it, increasing surface area for delivery of thrombolytic agent and creating channels for blood flow through the clot [18] (Fig. 78.2). The system has been recently introduced in the US market and a trial is ongoing.

Ultrasound-Accelerated Thrombolysis: EKOS Endovascular System

The EKOS system is a well-established FDA-approved thrombolytic catheter potentially enhancing lysis with ultrasound waves [19–21]. The catheters (6 or 12 cm infusion length) are positioned inside the main and lobar PA thrombosed segments, right, left or bilateral. The technique is similar to standard dripping except for the need to leave the ultrasound core wire inside the lytic catheter. In addition to the alteplase infusion (see previous section) a second port requires normal saline infusion (typically 35–45 cc/h) for central cooling during use. Short lytic times of 6–12 h and doses of 1 mg/h per catheter are typically enough [20], though termination should be better decided based on objective improvement of vital signs and hemodynamic or echocardiographic parameters. Catheters are pulled out at the bedside (Fig. 78.2).

Rheolytic Thrombectomy: ANGIOJET Zelante Catheter

The AngioJet Rheolytic Thrombectomy System (Boston Scientific, Marlborough, MA) has been used in the past for PE with several adverse events and deaths reported. As a result, the FDA has issued a black-box warning on the device label. With so many other options available it should probably be avoided.

Aspiration Thrombectomy: AngioVac/Alphavac

The AngioVac system (Angiodynamics Inc., Latham, NY, USA) is a powerful suction system that employs an extracorporeal veno-venous circuit that includes a filter between an aspiration and a reinfusing cannula. While AngioVac has been used for PE, it is not currently actively promoted for PE. The primary drawback is the rigidity of the catheter, which makes positioning and advancing the catheter in and beyond the pulmonary artery quite challenging. A new-generation device not requiring an extracorporeal circuit is currently promoted for PE. The AlphaVac F18[85] thrombectomy system consists of an 18-F cannula (105 cm long) with an 85° angled tip and was recently cleared by the FDA for the removal of thrombus from the venous system. The APEX-AV trial, which is currently in its start-up phase, will evaluate the safety and efficacy of the new AlphaVac F18[85] system in acute intermediate-risk PE patients [22] (Fig. 78.2).

Aspiration Thrombectomy: Indigo CAT-8/CAT-12 Catheter

The Indigo continuous aspiration mechanical thrombectomy (CAT) system (Penumbra, Inc., Alameda, CA) is an FDA-approved device for removal of arterial or venous thrombus. It has three components: a catheter, a separator and a vacuum pump. The Indigo CAT-8 system (Penumbra, Inc., Alameda, CA) is FDA approved for use in PE [23]. It is a flexible 8 Fr suction catheter that needs to be introduced through a 9- or 10Fr-long sheath positioned in the main PA. The most recent addition of the larger CAT-12 catheter is promising as an even more powerful device (Fig. 78.1).

There are three different tip configurations (straight, torque and extra torque), the angled ones being more useful as they can allow circumferential aspiration even at caval level. The separator wire allows thrombus fragmentation and mobilization as well as cleaning of the catheter when it is obstructed by thrombus. The vacuum pump provides continuous suction and can aspirate up to 8 mL/sec by applying and maintaining negative pressure of almost −29 mmHg.

Aspiration Thrombectomy: FlowTriever System

The FlowTriever (Inari Medical Inc., Irvine, Calif) is an FDA-approved mechanical thrombectomy device now available in its third generation configuration [24]. It consists of a flexible large-bore aspiration 0.035 guide catheter 20 or 24Fr (T20 and T24), and a catheter system of self-expanding nitinol discs (6–10, 11–14 and 15–18 mm), which can be advanced through the aspiration sheath to mechanically engage thrombi. The side port tubing of the sheath connects to a 60 cc locking syringe that can generate up to 29 mmHg negative pressure. Given the large bore access, the T20 sheath can remove up to 104 mL/s, and the T24 up to 143 mL/s. A smaller T16 sheath/catheter is available but not recommended for routine use given its low suction power. It is used complementarily to the larger sheaths to reach more distal targets (Fig. 78.2).

After obtaining access in the main PA, a 22–26Fr Gore Dryseal sheath (Gore Medical, Flagstaff, AZ) is introduced over a short tip stiff Amplatz wire (Boston Scientific Corporation, Marlborough, MA). The T20 or T24 sheath is then guided inside the main or lobar PA clot and a negative vacuum is created with the syringe (if blood is retrieved the sheath is not within the clot). The side port is then closed to the patient side and a second negative vacuum is applied. This can be done up to three times and if clot is not retrieved, you can assume that it is engaged on the sheath, which you can slowly pull out (retaining the negative syringe pressure) into the vena cava and eventually through the Dryseal sheath out on the table. The technique can be repeated several times to include other lobar arteries. Intermittent pulmonary arteriography is performed as deemed appropriate. If clot is not clearing as expected (subacute) the FlowTriever disks can be introduced and deployed distal to the clot, then pulled back into the sheath and out.

As with all thrombectomy techniques in PE, intervention attempts to remove all clot are not recommended. Termination is at the operator's discretion depending on the vital signs and hemodynamic parameter improvement.

Access site hemostasis is achieved through manual pressure, purse string suture, or Perclose ProGlide suture-mediated closure system (Abbott Vascular, Abbott Park, Illinois).

Perioperative Care and Surveillance

Patients will typically remain in the intensive care unit until they are deemed stable enough to be transitioned to a regular ward. If on heparin they are transitioned to oral anticoagulation, which will be continued based on the current venous thromboembolism guidelines. If not already done, a lower extremity duplex is required to rule out co-existing DVT. Appropriate referral to hematology is warranted in patients with an unprovoked DVT or possible thrombophilia. Follow-up at 3 months with pulmonary medicine and echocardiogram is essential to monitor for early signs of pulmonary hypertension.

Anticipated Complications

Intraoperative complications can happen particularly in high-risk PE. Care must be taken to avoid accidental puncture of the adjacent artery during venous access, particularly when thrombolytics are planned. Ultrasound guidance with a micropunture needle can prevent mis-puncturing and if it happens the arterial injury will be minor and self-reduced.

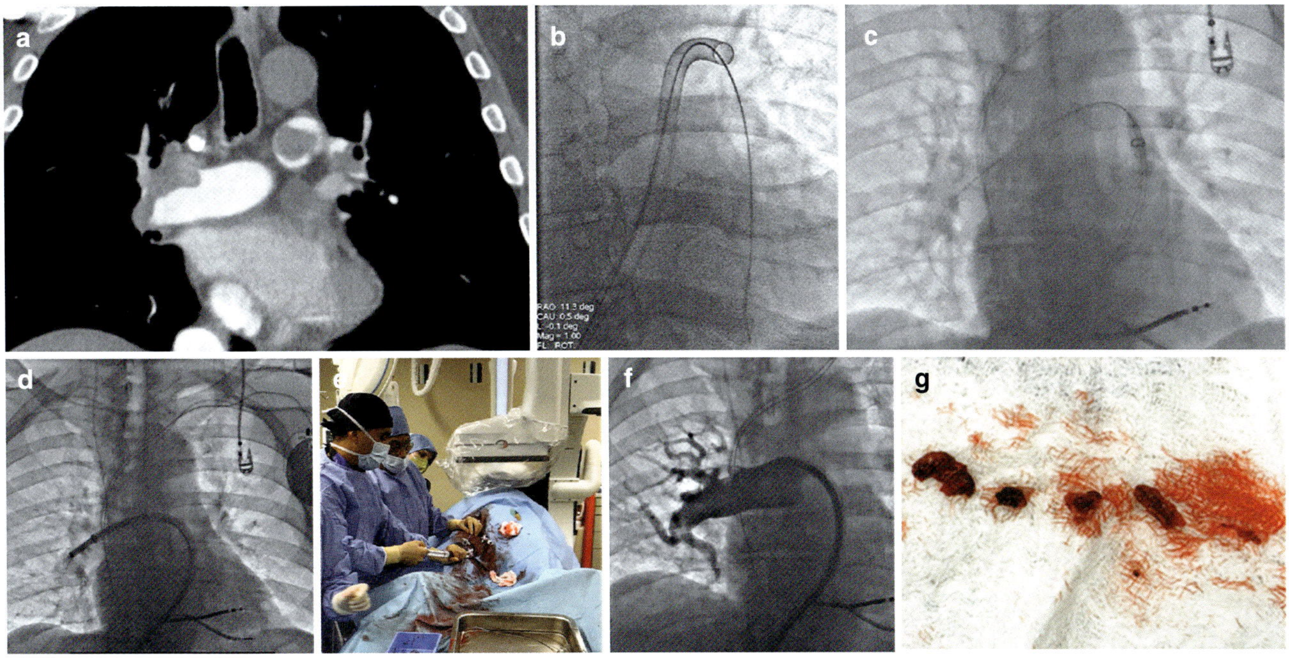

Fig. 78.3 (**a**) CT Pulmonary arteriogram indicating bilateral PA thrombi, (**b**) FlowTriever T24 has been introduced into the left PA and aspiration is performed, (**c**) Flow-Triever T24 is advanced towards the right PA, (**d**) Aspiration continues; notice the T16 sheath telescoping through the T24 to get deeper access into the second-order PA branches, (**f**) final arteriogram (**g**) clot pieces removed

Systematic bleeding is a feared complication of thrombolytics. However, most of the time it is minor and self-resolving. Cerebral bleed is extremely rare and has never been encountered in low-risk patients in the authors' experience. Frequent fibrinogen checks (every 4 h) though not directly correlated to the bleeding risk can be a guide to reduce the lytic drip rate or hold it when fibrinogen levels fall <100 mg/dL.

The larger bore devices carry risks of tricuspid valve injury particularly when crossing the valve was not done pigtail or balloon assisted (see section "Access into the Pulmonary Arteries"). Stiffer wires can also cause perforation and lung parenchymal injuries. Complications and overall clinical failures are volume related [7].

Case Presentation

Continued from page 811

Under moderate sedation (no propofol) and local anesthesia, ultrasound-guided access was obtained at the right common femoral vein. Using a starter wire and a pigtail the right heart chambers were crossed and the main pulmonary artery was cannulated. Heparin (5000 units) was administered. An arteriogram was obtained confirming bilateral lobar thrombi. Main PA pressure was measured at 92/45 mmHg. Using a glidecath and a soft glidewire we cannulated the left lower PA and exchanged this with an Amplatz stiff short tip wire. A 24 Fr Dryseal sheath was introduced (after serial dilatations) and the FlowTriever T24 catheter was navigated to the left PA. Aspiration was performed three times removing a significant amount of clot. Using a body wire-catheter the right lower PA was cannulated and again over a stiff Amplatz the T24 catheter was readvanced and a few more aspirations removed more clot (Fig. 78.3). At this point PA pressure measurement was 65/40 and vital signs HR 106, BP 130/80 and SAT 97%. The sheath was pulled out and the patient was transferred to the ICU. The patient was discharged from the ICU 12 h later.

The patient continued low-molecular weight heparin and got discharged was days later.

References

1. Heit JA. Epidemiology of venous thromboembolism. Nat Rev Cardiol. 2015;12:464–74.
2. Konstantinides SV, Meyer G, Becattini C, Bueno H, Geersing GJ, Harjola VP, Huisman MV, Humbert M, Jennings CS, Jiménez D, Kucher N, Lang IM, Lankeit M, Lorusso R, Mazzolai L, Meneveau N, Ní Áinle F, Prandoni P, Pruszczyk P, Righini M, Torbicki A, Van Belle E, Zamorano JL, ESC Scientific Document Group. 2019 ESC guidelines for the diagnosis and management of acute pulmonary embolism developed in collaboration with the European Respiratory Society (ERS). Eur Heart J. 2020;41(4):543–603.
3. Kahn SR, Houweling AH, Granton J, Rudski L, Dennie C, Hirsch A. Long-term outcomes after pulmonary embolism: cur-

rent knowledge and future research. Blood Coagul Fibrinolysis. 2014;25(5):407–15.
4. Avgerinos ED, Chaer RA. Pulmonary embolism: presentation, natural history and treatment. In: Sidawy AP, Perler BA, editors. Rutherford's vascular surgery and endovascular therapy. 10th ed. Philadelphia: Elsevier; 2022.
5. Giri J, Sista AK, Weinberg I, Kearon C, Kumbhani DJ, Desai ND, Piazza G, Gladwin MT, Chatterjee S, Kobayashi T, Kabrhel C, Barnes GD. Interventional therapies for acute pulmonary embolism: current status and principles for the development of novel evidence: a scientific statement from the American Heart Association. Circulation. 2019;140(20):e774–801.
6. Pruszczyk P, Klok FA, Kucher N, Roik M, Meneveau N, Sharp ASP, Nielsen-Kudsk JE, Obradović S, Barco S, Giannini F, Stefanini G, Tarantini G, Konstantinides S, Dudek D. Percutaneous treatment options for acute pulmonary embolism: a clinical consensus statement by the ESC Working Group on Pulmonary Circulation and Right Ventricular Function and the European Association of Percutaneous Cardiovascular Interventions. EuroIntervention. 2022;18:e623–38.
7. Abou Ali AN, Cherfan P, Zaghloul MS, Sridharan N, Lebron BR, Toma C, Chaer RA, Avgerinos ED. Institutional trends over a decade in catheter-directed interventions for pulmonary embolism. J Vasc Surg Venous Lymphat Disord. 2022;10(2):287–92.
8. Carlon TA, Goldman DT, Marinelli BS, Korff RA, Watchmaker JM, Patel RS, Lipson SD, Bishay VL, Lookstein RA. Contemporary management of acute pulmonary embolism: evolution of catheter-based therapy. Radiographics. 2022;42(6):1861–80.
9. Avgerinos ED, Saadeddin Z, Abou Ali AN, Fish L, Toma C, Chaer M, et al. A meta-analysis of outcomes of catheter-directed thrombolysis for high- and intermediate-risk pulmonary embolism. J Vasc Surg Venous Lymphat Disord. 2018;6(4):530–40.
10. Avgerinos ED, Chaer RA. Catheter-directed interventions for acute pulmonary embolism. J Vasc Surg. 2015;61(2):559–65. https://doi.org/10.1016/j.jvs.2014.10.036.
11. Rivera-Lebron BN, Rali PM, Tapson VF. The PERT concept: a step-by-step approach to managing pulmonary embolism. Chest. 2021;159(1):347–55.
12. Bulman JC, Weinstein JL. Invited commentary: importance of a multidisciplinary approach to pulmonary embolism in the era of catheter-directed therapies. Radiographics. 2022;42(6):E186–7.
13. Avgerinos ED, Abou Ali A, Toma C, Wu B, Saadeddin Z, McDaniel B, Al-Khoury G, Chaer RA. Catheter directed thrombolysis versus suction thrombectomy in the management of acute pulmonary embolism. J Vasc Surg Venous Lymphat Disord. 2019;7:623.
14. Cherfan P, Abou Ali AN, Zaghlul MS, et al. Propofol administration during catheter-directed interventions (CDI) for intermediate risk pulmonary embolism is associated with major adverse events. J Vasc Surg Venous Lymphat Disord. 2021;9:621–6.
15. Al-Bawardy R, Rosenfield K, Borges J, Young MN, Albaghdadi M, Rosovsky R, Kabrhel C. Extracorporeal membrane oxygenation in acute massive pulmonary embolism: a case series and review of the literature. Perfusion. 2019;34(1):22–8.
16. Abou Ali AN, Saadeddin Z, Chaer RA, Avgerinos ED. Catheter directed interventions for pulmonary embolism: current status and future prospects. Expert Rev Med Devices. 2020;17(2):103–10.
17. Avgerinos ED, Chaer RA. Endovascular treatment of iliofemoral deep venous thrombosis and pulmonary embolism. In: Chaikof EL, editor. Atlas of vascular and endovascular therapy. Philadelphia: Elsevier; 2023.
18. Bashir R, Foster M, Iskander A, Darki A, Jaber W, Rali PM, Lakhter V, Gandhi R, Klein A, Bhatheja R, Ross C, Natarajan K, Nanjundappa A, Angle JF, Ouriel K, Amoroso NE, Firth BG, Comerota AJ, Piazza G, Rosenfield K, Sista AK. Prospective multicenter trial of pharmacomechanical-catheter-directed thrombolysis with the Bashir endovascular catheter for acute pulmonary embolism. JACC Cardiovasc Interv. 2022;S1936-8798(22):01752–6.
19. Kucher N, Boekstegers P, Müller OJ, Kupatt C, Beyer-Westendorf J, Heitzer T, et al. Randomized, controlled trial of ultrasound-assisted catheter-directed thrombolysis for acute intermediate-risk pulmonary embolism. Circulation. 2014;129:479.
20. Tapson VF, Sterling K, Jones N, Elder M, Tripathy U, Brower J, et al. A randomized trial of the optimum duration of acoustic pulse thrombolysis procedure in acute intermediate-risk pulmonary embolism: the OPTALYSE PE trial. JACC Cardiovasc Interv. 2018;11(14):1401–10.
21. Avgerinos ED, Jaber W, Lacomis J, Markel K, McDaniel M, Rivera-Lebron BN, Ross CB, Sechrist J, Toma C, Chaer R, SUNSET sPE Collaborators. Randomized trial comparing standard versus ultrasound-assisted thrombolysis for submassive pulmonary embolism: the SUNSET sPE trial. JACC Cardiovasc Interv. 2021;14(12):1364–73.
22. Evaluating the safety and efficacy of the AlphaVac Multipurpose Mechanical Aspiration (MMA) F1885 PE for treatment of acute pulmonary embolism (APEX-AV). Clinicaltrials.gov. Accessed 21 Oct 2022.
23. Sista AK, Horowitz JM, Tapson VF, et al. Indigo aspiration system for treatment of pulmonary embolism: results of the EXTRACT-PE trial. JACC Cardiovasc Interv. 2021;14:319–29. https://doi.org/10.1016/j.jcin.2020.09.053.
24. Tu T, Toma C, Tapson VF, Adams C, Jaber WA, Silver M, et al. A prospective, single-arm, multicenter trial of catheter-directed mechanical thrombectomy for intermediate-risk acute pulmonary embolism: the FLARE study. JACC Cardiovasc Interv. 2019;12(9):859–69.

Pelvic Congestion Syndrome (PCS)

79

Mark S. Whiteley, Angie White, and David Beckett

Case Presentation

A 30-year-old nulliparous woman presented to us in 2022. She had a history of painful irregular periods since menarche and had been on oral contraceptive since 15. In early 2016, she developed severe pain on the right side of the pelvis. This pain recurred every 3–6 months for the next 2 years. Interestingly the pain was almost always during the night at this point and was severe enough for her to be admitted as an emergency on 2 occasions. An ultrasound of the urinary tract was reported as unremarkable in May 2018 and she was referred to the gynaecology department at a teaching hospital, where she was investigated for possible endometriosis.

In May 2019, she had a Mirena coil fitted (IUCD with slow-release progesterone). Six months later, she sought a second opinion from a consultant gynaecologist and was diagnosed to have "multifactorial pain" which was "probably musculoskeletal in origin". Following this, in December 2019, she had a transvaginal scan which showed dilated pelvic veins with a left to right shunt. This raised the question of an underlying May-Thurner Syndrome (MTS) and she was referred to a vascular surgeon. She had a laparoscopy which showed stage 1 endometriosis but again suggested significant venous congestion. Under the vascular surgeon's team, she underwent IVUS and a stent of her left common iliac vein for MTS. Three months post IVUS and stenting, there was no change in her symptoms, and so she presented to The Whiteley Clinic.

When seen, she complained of right sided abdominal pain, which she described as "heavy and dull" for most of the time, with monthly sharp pains. The pain also radiated to her lower back and was worse in the late afternoon and early evening, with occasional disturbed sleep. In addition, she had irritable bowel type symptoms and deep dyspareunia, although no bladder symptoms.

A transvaginal duplex ultrasound using the Holdstock-Harrison protocol showed reflux in the left ovarian vein (Fig. 79.1), bilateral reflux in gonadal tributaries of the internal iliac veins (Fig. 79.2), dilated arcuate veins around the uterus (Fig. 79.3) and a pseudo nutcracker (Fig. 79.4). The overall pattern is the most common pattern that we see in women with pelvic congestion syndrome, reflux in the left ovarian vein and bilateral internal iliac systems (Fig. 79.5).

Continued to page 827

M. S. Whiteley (✉) · A. White · D. Beckett
The Whiteley Clinic, Guildford, United Kingdom
e-mail: mark@thewhiteleyclinic.co.uk;
angie@thewhiteleyclinic.co.uk; david@thewhiteleyclinic.co.uk

G. Geroulakos et al. (eds.), *Mastering Endovascular Techniques*, https://doi.org/10.1007/978-3-031-42735-0_79

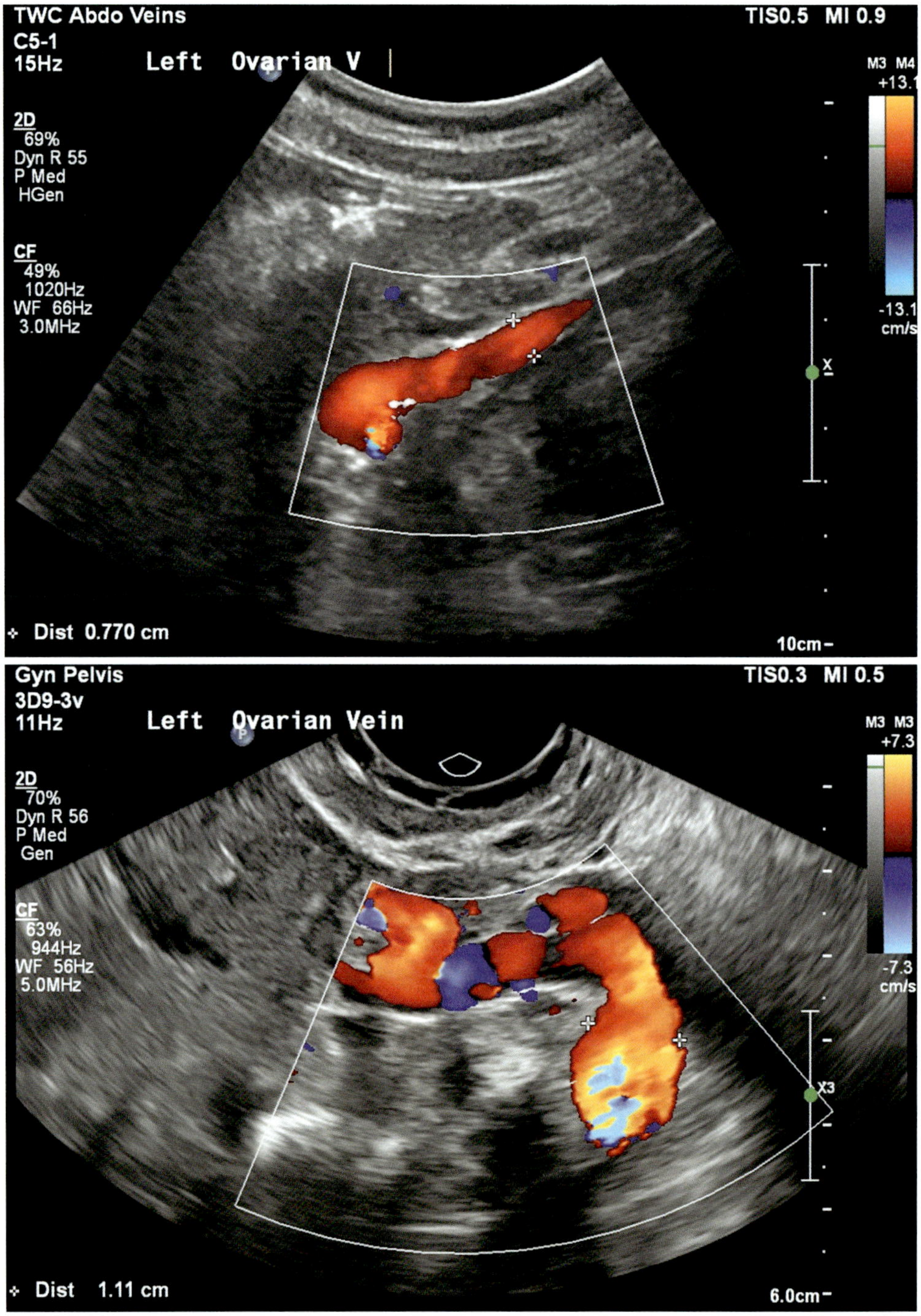

Fig. 79.1 Top—Transabdominal duplex ultrasound showing reflux in the proximal left ovarian vein; Bottom—Transvaginal duplex ultrasound using Holdstock-Harrison protocol showing reflux in distal left ovarian vein

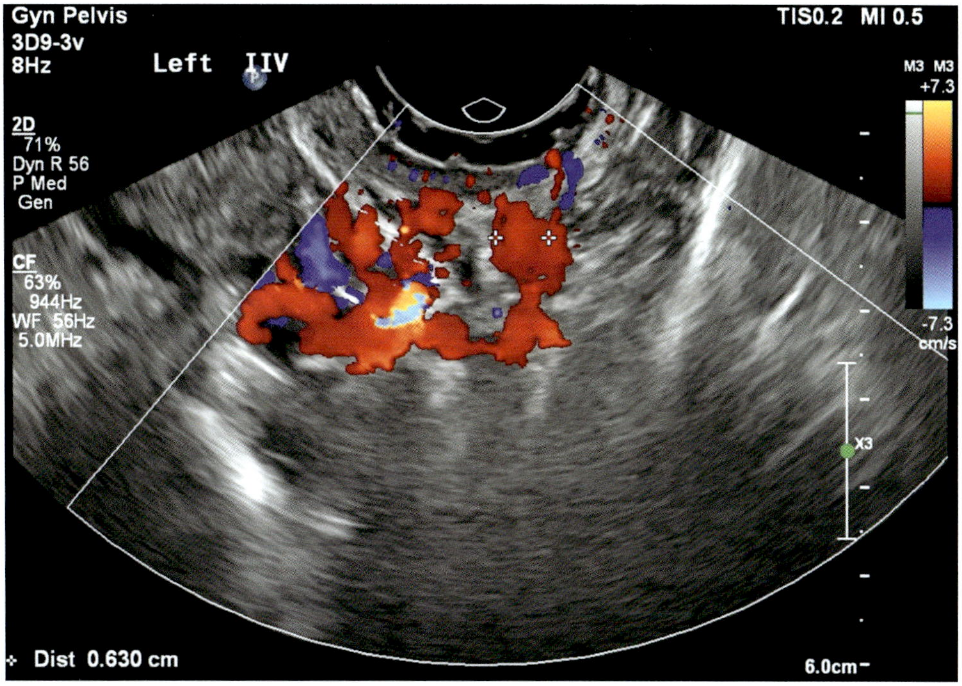

Fig. 79.2 Transvaginal duplex ultrasound using Holdstock-Harrison protocol showing reflux in left internal iliac vein

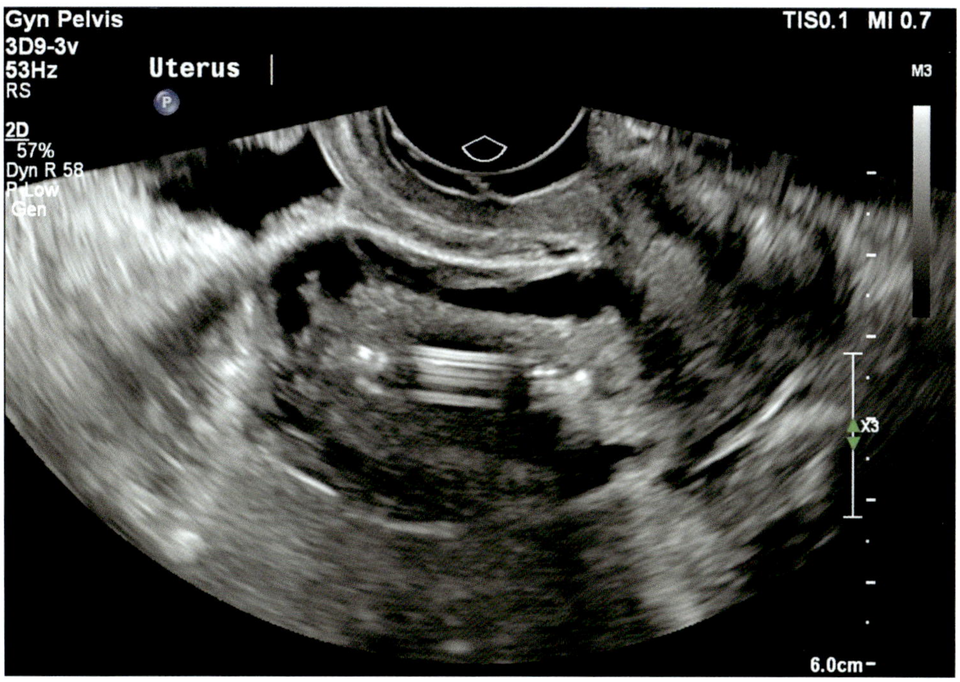

Fig. 79.3 Transvaginal ultrasound using Holdstock-Harrison protocol showing dilated arcuate veins lying on the uterus

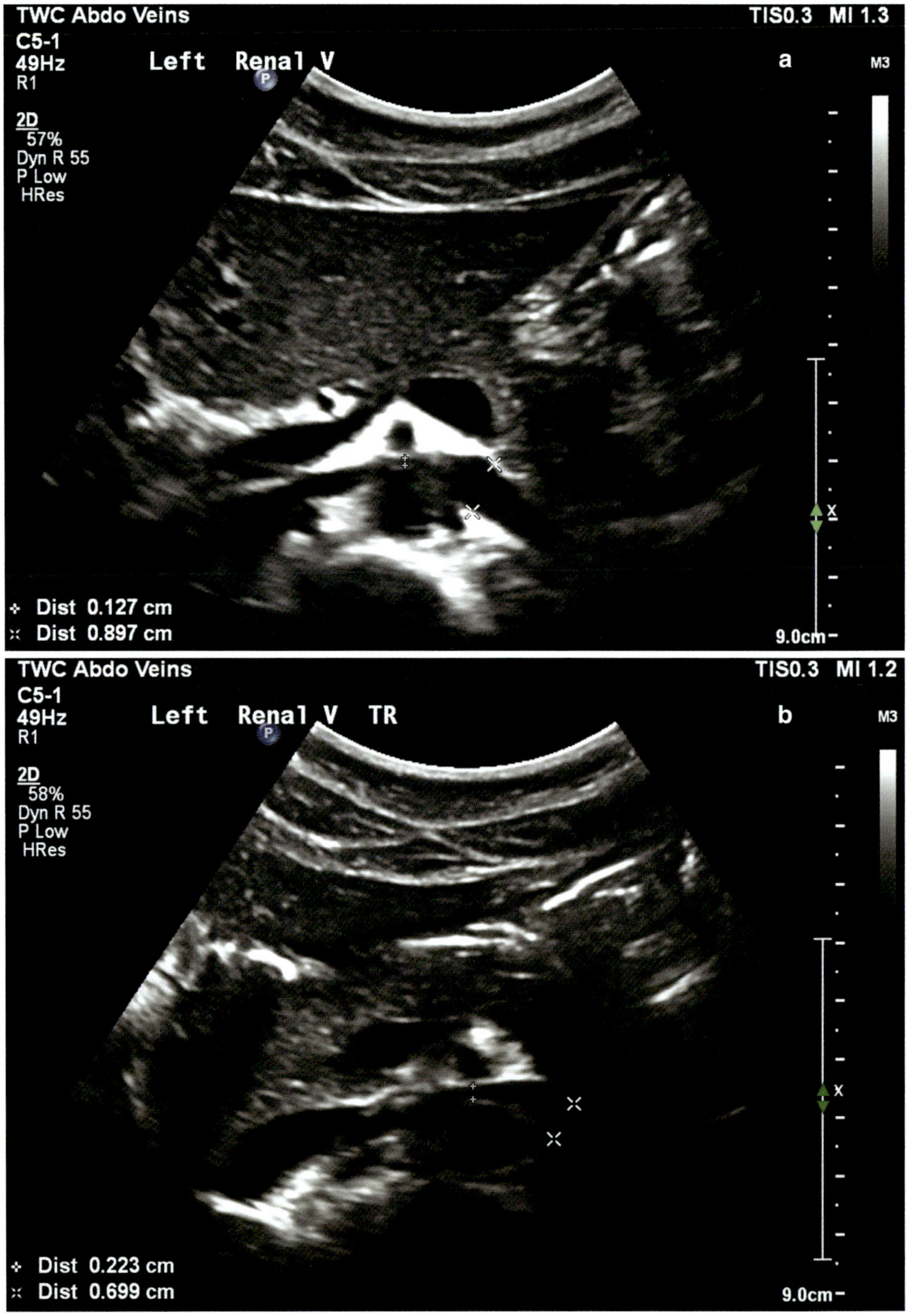

Fig. 79.4 Transabdominal ultrasound showing: Top—In sitting position, there appears to be a nutcracker narrowing between proximal left renal vein (X—X = 0.897 cm) and left renal vein passing in front of aorta and behind superior mesenteric artery (+ - + = 0.127 cm) with a diameter ratio = 7 Bottom – In Trendelenburg position (head down) where ratio now 0.669 cm/0.223 cm = 3, which is normal

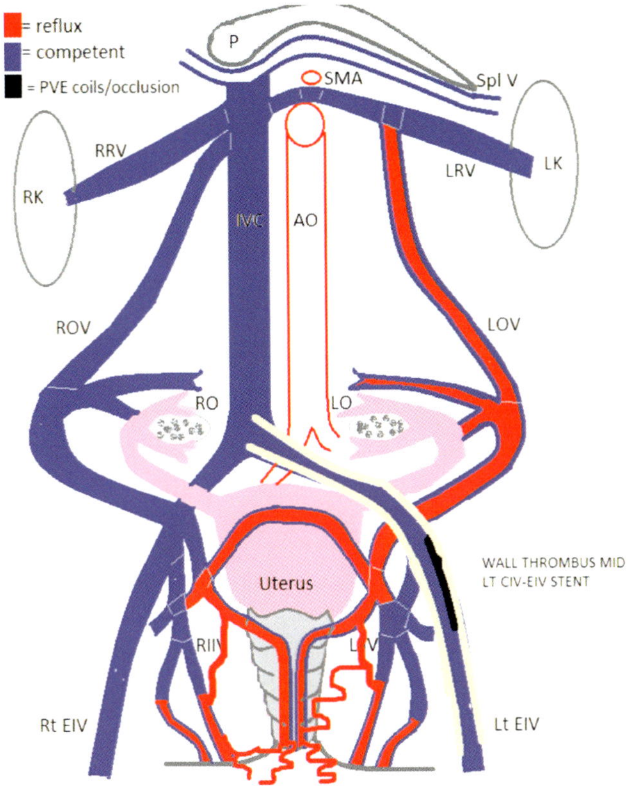

Fig. 79.5 The report as per The Whiteley Protocol showing reflux in the left ovarian vein, arcuate veins, bilateral internal iliac tributaries and also the stent

Introduction

Pelvic congestion syndrome (PCS) is probably the most misunderstood or ignored physical disorder that causes adverse symptoms and signs in a large section of the population in medicine today. Although it can be simply thought of as "varicose veins in the pelvis", the facts that there are often few physical signs, and a multitude of different presentations mean that the condition is rarely taught at medical school or even in specialist training. As such, most doctors have little idea of what PCS is.

What is even more shocking is that although a lot of the symptoms are gynaecologically related, Gynaecologists as a speciality have largely ignored PCS, with most of the research into this condition been performed by venous surgeons and interventional radiologists. This is evidenced by the fact that at the time of writing this chapter, PCS or even "pelvic varicose veins" does not appear in the Royal College of Obstetricians and Gynaecologists list of the causes of chronic pelvic pain [1].

Indeed, a search of their website for the term "pelvic congestion"returns only one result. This is in the "A-Z of medical terms" which is in the public area of the website, where they give the definition: "Pelvic congestion - Swollen pelvic veins." [2]

This is very worrying given that it is estimated that in the United States, the annual costs both direct and indirect of chronic pelvic pain (CPP) is estimated to be more than $39 billion [3, 4], and in the United Kingdom, £326 million [5].

Moreover, research from our own unit has shown that one in six women presenting with leg varicose veins, have a major contribution of their venous reflux from pelvic veins [6] which, if left untreated, is a major cause of recurrence in the future [7]. In addition, although men are rarely thought of when considering PCS, one in thirty males with leg varicose veins also have a major contribution of venous reflux from their pelvic veins [8].

Hence, although we are starting to have a basic understanding of PCS, our knowledge is still very limited in all but the most obvious areas of this condition. I will outline the current state of our understanding, and also debunk some of the muddled thinking that persists surrounding PCS, and hope that this will spur the reader into taking a deeper interest into this fascinating area of venous surgery.

One final point to make in the introduction is to address why I have called this chapter "Pelvic Congestion Syndrome" (PCS) and not bowed to the pressure from many working in this area to use the new term "Pelvic Venous Disorders" (PeVD). As outlined in the UIP Consensus Document 2019 [9], most in the world who are interested in PCS will know it as this name. Although this might be "out of date" in the future, this chapter is more likely to appeal to doctors at the current time if a well-recognised term is used rather than a newer name, that might not be so well known.

Definition and Demographics

All standard medical texts that describe different conditions start with a definition, followed by incidence/prevalence, geography and other demographic information. However, as in the introduction, although many "authorities" have started trying to produce this sort of information, by the end of this chapter, I hope that you will realise that we haven't even fully encapsulated all the conditions that are caused by PCS. As such, any definition, demographic information, or even scoring system must be taken as a temporary description of where we are now, rather than be considered to be something that will last very long into the future.

Most papers discussing PCS use the generally accepted definition of CPP as non-cyclic pain of greater than six months duration [9]. However, this is not particularly useful as a definition for PCS as many patients present with leg varicose veins arising from pelvic veins report no pelvic pain. Furthermore, one only has to think of a male with a varicocele to realise that pelvic varicose veins are not restricted to females!

The recent UIP consensus document on PCS [9] has a very nice section on how the definition has changed and quotes a recent version from the VEIN-TERM transatlantic interdisciplinary consensus document "chronic symptoms which may include pelvic pain, perineal heaviness, urgency of micturition and post-coital pain, caused by ovarian and/or pelvic vein reflux and/or obstruction, which may be associated with vulvar, perineal, and/or lower extremity varices." [10]

Although this is a much more encompassing definition that few would argue with, it does miss several important points. Firstly, it is focused on women and not men. Secondly, the symptoms described only concentrate on some of the symptoms arising from within the pelvis (although irritable bowel syndrome [11] has been missed out) and it does not include the very common symptom of low back pain [11, 12] nor the recently recognised symptom of hip pain [13]. Finally, this definition does not cover haemorrhoids [11, 14, 15], vulvar or perineal varicose veins and lower leg varicose veins that may occur secondary to the PCS but without any pelvic pain [6].

Furthermore, there are still many areas that are not included in most articles on PCS, such as the proposed link with erectile dysfunction [16, 17] and chronic prostatits [18].

Therefore, when thinking about PCS, rather than trying to make an all-encompassing definition which will need to change whenever new research becomes available, it is much more useful to think of PCS as I described in 2019 [19]:

1. Symptoms:
   A. Inside the pelvis
   B. Outside the pelvis
2. Signs (varices).
   A. On the pelvis/lower abdomen
   B. Legs

which can then be added to with O = Obstruction and/or R = reflux.

Pathophysiology - Reflux and/or Obstruction

Many doctors, particularly surgeons and interventional radiologists, like to have a clear reason for any abnormal function or change of anatomy. As such, it is always very tempting to think that dilated veins are due to a proximal obstruction. Hence the popularity of looking for proximal obstructions when pelvic varices are found. The two named obstructions related to PCS are nutcracker syndrome (NCS) [20] and May Thurner syndrome (MTS - now often called iliac vein compression syndrome - IVCS) [21] where the left renal vein and classically the left common iliac vein are both compressed by overlying arterial structures (Fig. 79.6).

However, there are a other obstructions of the iliac veins that have been described both from external compression as well as from internal narrowing (such as non-thrombotic iliac vein lesion or NIVL) [22].

Despite many authors finding obstructive lesions to be the cause of PCS in up to 80% of patients [23, 24], our own experience has been that obstructive lesions are only present in 1–2% of patients presenting primarily with PCS and no previous thromboembolic disease or post thrombotic syndrome (PTS).

This raises the question as to why there should be such a massive difference between those that regularly find and treat obstructive lesions and those of us who get excellent results from treating pelvic venous reflux alone without finding or treating obstructive lesions.

Without doubt, different doctors and research groups serve different populations. We would expect to find many obstructive lesions in patients with PTS. However, when we are considering PCS, we are considering populations where this is the primary diagnosis and not secondary to PTS. Also, we see patients who present both with CPP and those who present with leg varicose veins from pelvic origin without any pelvic pain.

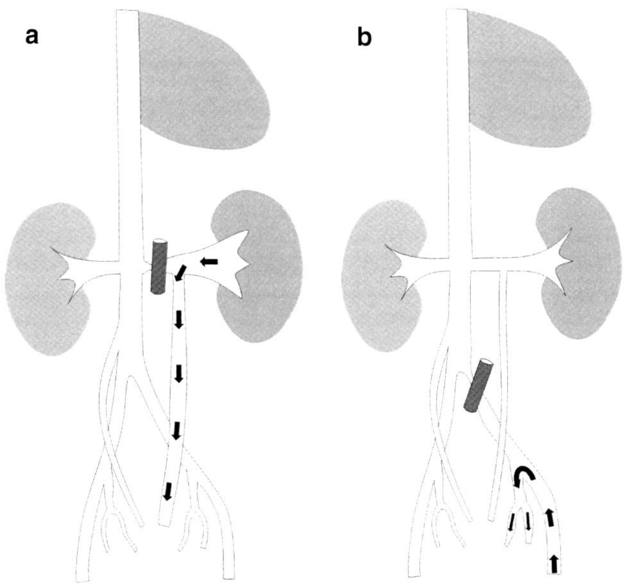

Fig. 79.6 Diagram showing venous blood flow in A—Nutcracker Syndrome where venous blood from left kidney collects into left renal vein and, due to obstruction from the superior mesenteric artery, escapes by refluxing down incompetent left ovarian vein and B—May-Thurner Syndrome where venous blood from the leg flows up the external iliac vein and, due to obstruction in the left common iliac vein by the right common iliac artery, escapes by refluxing down and incompetent left internal iliac vein. These obstruction patterns are very uncommon in PCS and are over diagnosed by cross-sectional imaging (see text) (diagrams reproduced by kind permission of Whiteley Publishing Ltd.)

However, despite differences in patient population, it is my belief that the major difference between the two groups is the way that the doctors investigate PCS, and the trust that some doctors put into static imaging rather than functional information. I will explore the investigation and imaging of PCS in the "Investigations and Imaging" section, but think it is sensible to make a couple of points for the reader to consider at this juncture.

Firstly, it is useful to compare leg varicose veins with pelvic varicose veins. Varicosities in the calf are almost always secondary to incompetence in a proximal vein (such as the great, small, or accessory saphenous veins) or an underlying incompetent perforating vein. It is exceptionally rare for anyone presenting with such varicose veins to require stenting of a proximal obstructive lesion (although this has been proposed by proponents of IVUS) [25], it is similarly rare for anyone to worsen when the reflux is ablated.

However, when discussing varicosities in the pelvis, there seems to be an emotional concern about the veins being "deep inside the body" and so the concerns about obstructive lesions seem to emerge without any logical reason. Over the last 20 years, we have treated large numbers of patients and then published multiple papers showing the outcomes of pelvic vein embolisation. Except for those patients with PTS, we have only found a handful of patients either who have got a true obstructive cause identified before embolisation, or who have worsened clinically following embolisation due to a missed obstructive lesion.

Secondly, it is important to think about the consequences of placing a stent within a potentially normal vein. Embolisation of an incompetent vein aims to permanently occlude the vein and, once occluded, it can be forgotten. Conversely, stenting requires constant and regular surveillance to ensure patency, with urgent investigation and intervention if thrombosis occurs. Moreover, the action of the stent on the normal vein wall causes a pseudo-intima [26]. I suspect that in the long term, there will be many normal iliac and left renal veins that occlude secondary to a stent that was inserted for erroneous reasons, which will lead to medicolegal consequences.

Hence, I regularly teach that when considering treating PCS, "you have to earn your stent". In cases of doubt, it is far safer to perform embolisation of incompetent veins first with a view to performing stenting afterwards if required. Our experience suggests that in most patients, the stent would not be required.

Investigations and Imaging

Traditionally doctors have always thought about imaging structures in the retroperitoneum and pelvis by using cross-sectional imaging techniques such as computerised tomography (CT) or magnetic resonance imaging (MRI). For veins, there is the additional opportunity to perform venography.

However, any final year medical student would fail their surgical exams if they examined patients with leg varicose veins in the supine position. We all know that venous reflux does not occur when gravity is not actively assisting venous blood to reflux down incompetent veins. Hence it is amazing that so many "specialists" use cross-sectional imaging where the patients are lying flat, to assess varicose veins in the pelvis.

Also, we have shown that there is no correlation between the diameter of the ovarian vein and whether it is competent or not, making a nonsense of the argument that a dilated ovarian vein on cross-sectional imaging or venography needs treatment [27].

Moreover, in the 1990s the arguments of venous duplex ultrasound scanning over venography, were won when the ultrasound technique was shown to be far superior. Nowadays, no venous specialist would consider venography as a first-line investigation for venous reflux of the legs.

Once again, it is amazing that so many "specialists" discard the arguments of the past when it comes to pelvic veins, basing their investigations and treatment plans on venography.

Hence when it comes to the investigation of PCS, it is sensible to think scientifically and build on what is already known rather than lapse into muddled thinking about how to image veins "deep inside the body" or to try to base investigations on what is available rather than what is needed.

When looking for pelvic venous reflux, it is clear that duplex ultrasound is going to show reflux better than any other current technology. Some authors have suggested this can be done using trans-abdominal venous duplex ultrasound (TADUS) alone [28]. However, it has become clear that the transvaginal venous duplex ultrasound (TVDUS) performed at 45° using the Holdstock-Harrison protocol has clear advantages, both over TADUS alone as well as venography [29]. It must be noted that to image for obstructive syndromes TVDUS needs to be combined with TADUS to investigate the more proximal veins.

Further evidence of the advantage of TVDUS is the ability of the transvaginal approach to see reflux in the internal iliac veins and tributaries, especially the obturator and pudendal veins. This has led us to find that the commonest pattern of reflux in women with PCS is incompetence of the left ovarian and both internal iliac veins – found in about 40–45% of cases [27, 30, 31]. Indeed only 2–3% of patients have ovarian vein reflux alone [27, 30, 31]. This should be a great concern for patients of doctors who only investigate and treat ovarian veins in PCS.

With regards whether the combination of TVDUS and TADUS might miss an obstructive lesion, a simple proto-

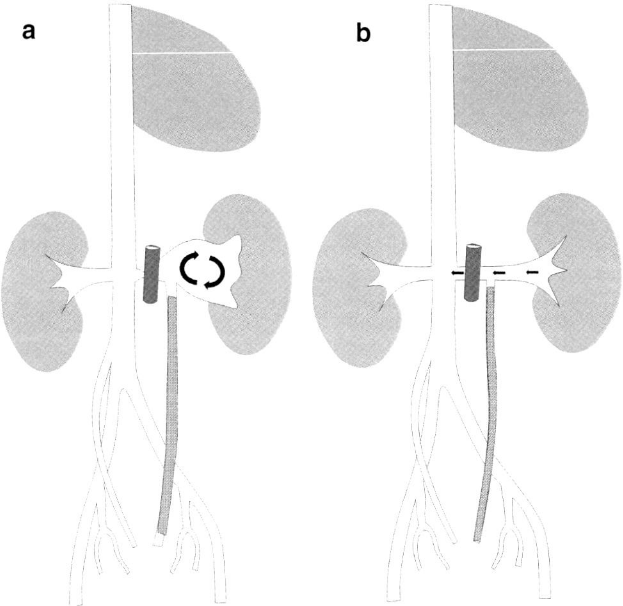

Fig. 79.7 Diagram explaining pseudo-nutcracker syndrome. Embolisation of the incompetent left ovarian vein (**a**) should worsen the condition, causing increased pressure and distension of the proximal left renal vein. However, in most cases (**b**), embolisation cases the narrowing to dilate as needed allowing blood to pass directly into the inferior vena cava. This can be predicted using a reverse Trendelenburg position (see text)

col can look for significant obstruction if suspected. Judy Holdstock from our unit won the prize at the American College of Phlebology meeting in November 2017, showing that a narrowed distal renal vein fitting the criteria for NCS would usually dilate up after left ovarian vein embolisation - hence describing the "pseudo nutcracker syndrome" (Fig. 79.7) [32]. We have now described a non-invasive test for pseudo nutcracker syndrome. Tipping the patient into a Trendelenburg (head down) position, any left ovarian vein reflux by gravity is stopped, and dilation of the previously "narrowed" segment can be seen [33].

With respect of any MTS or IVCS affecting either side, we perform a single cuff air-plethysmography to assess venous outflow from the leg [34]. If outflow is normal on leg elevation, then any imaging that suggests a narrowing is reporting an anatomical feature and not a physiological one. By using this approach, we do not need intravascular ultrasound (IVUS) which again only shows diameters and not flows.

Finally, there has been some recent work by Gavrilov that has shown some correlation of the patterns and extent of reflux in pelvic veins to symptoms that patients experience [35]. This approach gives the chance of developing an objective scoring to stratify PCS [36].

Treatment

Traditional medical teaching always starts sections on "treatment" with "conservative" and then "medical" before discussing any surgical intervention.

As with leg varicose veins, one can try conservative treatments such as compression [37], or medication such as micro purified flavonoid fraction (MPFF) [38, 39]. In addition, some gynaecologists have used a hormonal approach [40]. However, although it is sensible to try such methods, these do not reverse the underlying cause - that of pelvic venous reflux and probably the associated venous stasis [19].

Therefore, when treating PCS patients with pain resistant to these conservative or medical approaches, or indeed patients with external symptomatic veins of the vulva, perineum and/or legs, a physical intervention is needed.

In the few patients with proven obstruction, stenting by those experienced in venous stenting is the best option. As this is so rarely found in our unit, we refer such patients away.

When treating pelvic venous reflux and pelvic varices, our approach is to use a transjugular route for ease of access to the target vein. We then use foam sclerotherapy made with 3% sodium tetradecyl sulfate (STS - Fibrovein, STD Pharmaceutical Products Ltd. Hereford, UK) initially with a 1:4 mixture and Tessari method, but more laterally, using the Varixio machine (VB Devices, Barcelona, Spain) into the distal veins, especially those near the skin, mucosa or pelvic organs, and then fibred and retractable embolisation coils (Cook Medical, Limerick, Ireland) in the distal truncal veins. We then ensure that the coils occlude a section of the vein from distal to proximal to reduce the risks of a local bypass vein opening up later [30].

We use a walk-in, walk-out approach under local anaesthetic without sedation and this has been proven to be both safe and effective [41], and with good long term results [31] even after subsequent pregnancy [42].

There are, of course, other approaches such as open surgery [43], and the use of other embolic devices such as plugs and glue [44, 45]. However, to obtain optimal results we believe that treatment using foam sclerotherapy and coils allows the most flexible approach especially when treating the obturator and pudendal veins or reducing the risks of subsequent bypasses opening up around the ablated area in the ovarian veins.

Conclusion

I started work in PCS in 2000 when I realised that many of my patients with recurrent varicose veins who had undergone treatment elsewhere, had their recurrences primarily due to venous reflux in veins arising from the pelvis. I was very lucky to be working with Judy Holdstock who started investigating these patients with duplex ultrasound both transabdominally and transvaginally. This work resulted in the Holdstock-Harrison protocol for assessing pelvic venous reflux. It was only through this link with recurrent leg varicose veins that we started having patients report an improvement in pelvic symptoms. Soon afterwards, we found patients seeking us out for the treatment of CPP and vulval varicose veins.

When I first started presenting my research at conferences in the early 2000s, there were rarely any presentations on PCS or pelvic veins at all. Over the last two decades, virtually every venous conference has a section dedicated to pelvic veins and there are even conferences dedicated solely to pelvic veins.

With the amazing pace of change in this area, the increasing amount of research been published and the growth in the number of doctors, clinics and hospitals offering pelvic congestion syndrome treatments, I am surprised that we have not had more interest from our gynaecological colleagues. The assessment and treatment of patients with CPP needs to be multidisciplinary by its nature, as many do not have PCS, and even those that do, can have other pelvic conditions simultaneously.

Although we know a lot about PCS and its treatment, we are really at the beginning of our understanding of this condition and I have little doubt that the future holds some great advances. We really need an easy to use classification as we have in the CEAP classification for leg varicose veins [46]. The current SVP classification is a good attempt at this [47], but does not have the ability to be used simply as one does with the Clinical "C" grade of the CEAP.

We also need a robust quality of life tool to ensure that our patients have improvements after treatment, and a better way of assessing the anatomy and function of the veins. TVDUS using the Holdstock-Harrison protocol with TADUS has shown to be excellent, but as with all ultrasound techniques, it is user dependent and requires considerable training and experience to give accurate results. Also, there is no "hard copy" for others to review later. Although new transducers that can "sweep" across the pelvis or show 3d images are being developed and tested, it will be some time to find if they provide accurate and reproducible results in anyone's hands.

Finally, it would appear that the mainstay of treatment is endovenous embolisation of the incompetent veins using metal coils or plugs. We know that foam sclerotherapy alone is unlikely to provide permanent ablation in the truncal veins, as it does not do so in truncal saphenous veins [48], and so secure ablation is needed. Although I have sat through 2 different presentations of endovenous laser ablation of incompetent ovarian veins at conferences, I have not seen any published results and I cannot think that thermal ablation in these veins is a good idea. The risk to surrounding structures is too great.

However, a recent survey that I performed online via survey monkey (unpublished data) showed 95% wanted treatment without leaving metal behind inside their body and 40% would refuse embolisation with metal coils. Hence, there is a need for a new technique to achieve secure embolisation of the pelvic venous trunks but without leaving any foreign material behind.

I look forward to the future and to see how these areas are addressed by workers in the field.

Case Presentation

Continued from page 819

The presumptive diagnosis was PCS secondary to the reflux and pooling of blood in the pelvic varices, especially in the parauterine arcuate veins. The MTS could have caused some of the left internal iliac tributary reflux and pelvic varicosities, but it would not explain the right sided reflux or the left ovarian vein reflux. This, in addition to the lack of any improvement after the stent placement, suggests that the MTS was insignificant and the stent was not required.

The patient underwent pelvic vein embolisation as an ambulatory case, using a right internal jugular approach and local anaesthetic to the insertion point only (Fig. 79.8). There was no sedation or other anaesthesia. Foam sclerotherapy made from 1.6 mL 3% STS (Fibrovein) into approximately 9 mL of foam in the Varixio machine (VB Devices) and this

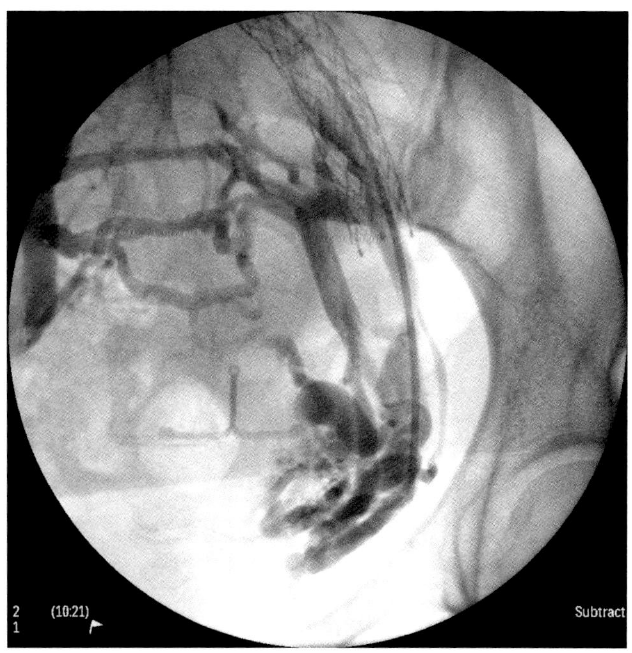

Fig. 79.8 Catheter is seen in the left internal iliac vein branches having been placed through the stent struts. Venogram demonstrates large internal iliac vein branches readily communicating with the parauterine plexus and distal ovarian vein

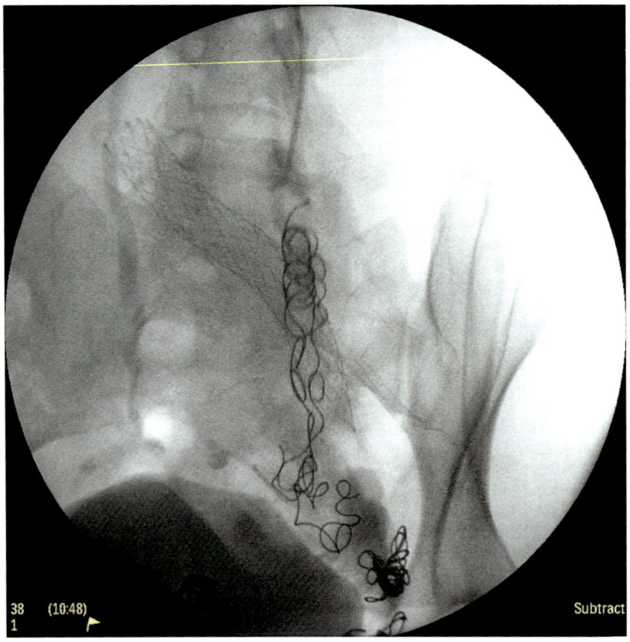

Fig. 79.9 Venogram showing background left CIV stent inset for MTS. Coils are shown in the left ovarian vein and lower within the internal iliac branches in a patient with PseudoNutcraker

was injected via the catheter into the pelvic and parauterine varices. Nester coils were then used to embolise the left ovarian vein (5 coils), left pudendal vein (2 coils), left parauterine vein (2 coils), right parauterine vein (2 coils) and left pudendal vein (1 coil) (Fig. 79.9).

She left the clinic fully mobile 1 h post embolisation and recovered without incident. The patient returned to her previous vascular surgeon for assessment 14 weeks later and we received a letter stating that her symptoms had 95% resolved by that stage.

Equipment needed for most cases:

	Manufacturer	Size/length
Wires		
Fixed core wire J-tip	Cook medical	0.035″/145 cm
Radifocus™ guide wire M stiff type	Terumo	0.035″/180 cm
Radifocus™ guide wire M standard type	Terumo	0.035″/180 cm
Sheaths		
Any standard access sheath	Any	6F/11 cm
Catheters		
Impress MPA 1 catheter	Merit medical	0.035″/5Fr/100 cm
Sidewinder Simmons technique I I (SIM2)	Cordis	5F/100 cm
Snare		
EN snare endovascular snare system	Merit medical	Various
Coils		
Pushable cook nester coils	Cook medical	8/10/12/14/16/18/20 diameter 14/20 cm length 0.035″
Cook retracta coils	Cook medical	8/10/12/14/16 diameter 14 cm length 0.035″
Foam sclerotherapy		
Sodium Tetradecyl Sulphate (Fibrovein)	STD pharmaceuticals	3%
Varixio	VB devices	

References

1. The initial management of chronic pelvic pain - green-top guidelines no: 41 May 2012. https://www.rcog.org.uk/media/muab2gj2/gtg_41.pdf. Accessed 4 June 2023.
2. A to Z of medical terms. https://www.rcog.org.uk/for-the-public/a--z-of-medical-terms/. Accessed 4 June 2023.
3. Mathias SD, Kuppermann M, Liberman RF, Lipschutz RC, SteegeJ F. Chronic pelvic pain: prevalence, health-related quality of life, and economic correlates. Obstet Gynecol. 1996;87(03):321–7.
4. Kuligowska E, Deeds L III, Lu K III. Pelvic pain: overlooked and underdiagnosed gynecologic conditions. Radiographics. 2005;25(01):3–20.
5. Curtis L. Unit costs of health and social care 2014 personal social services research unit. Canterbury: University of Kent; 2014.
6. Marsh P, Holdstock J, Harrison C, Smith C, Price BA, Whiteley MS. Pelvic vein reflux in female patients with varicose veins: comparison of incidence between a specialist private vein clinic and the vascular department of a National Health Service District General Hospital. Phlebology. 2009;24(3):108–13. https://doi.org/10.1258/phleb.2008.008041. PMID: 19470861.
7. Whiteley AM, Taylor DC, Dos Santos SJ, Whiteley MS. Pelvic venous reflux is a major contributory cause of recurrent varicose veins in more than a quarter of women. J Vasc Surg Venous Lymphat Disord. 2014;2(4):411–5. https://doi.org/10.1016/j.jvsv.2014.05.005. Epub 2014 Jun 24. PMID: 26993547.
8. Dabbs EB, Dos Santos SJ, Shiangoli I, Holdstock JM, Beckett D, Whiteley MS. Pelvic venous reflux in males with varicose veins and recurrent varicose veins. Phlebology. 2018;33(6):382–7. https://doi.org/10.1177/0268355517728667. Epub 2017 Aug 31. PMID: 28857674.
9. Antignani PL, Lazarashvili Z, Monedero JL, Ezpeleta SZ, Whiteley MS, Khilnani NM, Meissner MH, Wittens CH, Kurstjens RL, Belova L, Bokuchava M, Elkashishi WT, Jeanneret-Gris C, Geroulakos G, Gianesini S, de Graaf R, Krzanowski M, Al Tarazi L, Tessari L, Wikkeling M. Diagnosis and treatment of pelvic congestion syndrome: UIP consensus document. Int Angiol. 2019;38(4):265–83. https://doi.org/10.23736/S0392-9590.19.04237-8. Epub 2019 Jul 24. PMID: 31345010.
10. Eklof B, Perrin M, Delis KT, Rutherford RB, Gloviczki P, American Venous Forum; European Venous Forum, International Union of Phlebology, American College of Phlebology, International Union of Angiology. Updated terminology of chronic venous disorders: the VEIN-TERM transatlantic interdisciplinary consensus document. J Vasc Surg. 2009;49(2):498–501. https://doi.org/10.1016/j.jvs.2008.09.014. PMID: 19216970.
11. Strong SM, Cross AC, Sideris M, Whiteley MS. Pelvic Congestion Syndrome affects nulliparous and post-menopausal women, and the treatment of internal iliac vein reflux is critical: a retrospective

cohort study looking at treatment outcomes following Pelvic Vein Embolisation; 2022. https://doi.org/10.31219/osf.io/9g3yb
12. Debucquois A, Salomon du Mont L, Bertho W, Kaladji A, Hartung O, Rinckenbach S. Current results of left gonadal vein transposition to treat nutcracker syndrome. J Vasc Surg Venous Lymphat Disord. 2021;9(6):1504–9. https://doi.org/10.1016/j.jvsv.2021.03.003. Epub 2021 Mar 15. PMID: 33737260.
13. Dos Santos SJ, Whiteley MS. Pelvic congestion syndrome masquerading as osteoarthritis of the hip. SAGE Open Med Case Rep. 2016;9(4):2050313X16683630. https://doi.org/10.1177/2050313X16683630. PMID: 27994874; PMCID: PMC5153020.
14. van der Vleuten CJ, van Kempen JA, Schultze-Kool LJ. Embolization to treat pelvic congestion syndrome and vulval varicose veins. Int J Gynaecol Obstet. 2012;118(3):227–30. https://doi.org/10.1016/j.ijgo.2012.04.021. Epub 2012 Jun 22. PMID: 22727416.
15. Holdstock JM, Dos Santos SJ, Harrison CC, Price BA, Whiteley MS. Haemorrhoids are associated with internal iliac vein reflux in up to one-third of women presenting with varicose veins associated with pelvic vein reflux. Phlebology. 2015;30(2):133–9. https://doi.org/10.1177/0268355514531952. Epub 2014 Apr 22. PMID: 24755923.
16. Hoppe H, Diehm N. Venogenic erectile dysfunction: diagnosis on computed tomography cavernosography and endovascular treatment using an anterograde access via deep dorsal penile vein. CVIR Endovasc. 2022;5(1):10. https://doi.org/10.1186/s42155-022-00283-5. PMID: 35113281; PMCID: PMC8814092.
17. Allaire E, Sussman H, Zugail AS, Hauet P, Floresco J, Virag R. Erectile dysfunction resistant to medical treatment caused by cavernovenous leakage: an innovative surgical approach combining pre-operative work up, embolisation, and open surgery. Eur J Vasc Endovasc Surg. 2021;61(3):510–7. https://doi.org/10.1016/j.ejvs.2020.08.048. Epub 2020 Oct 13. PMID: 33067110.
18. Tsukanov AY, Rudchenko NV, Mozgovoy SI, Maslyukov AV. [Protection of the prostate in conditions of chronic pelvis varicose]. Urologiia. 2021;(4):53–60. Russian. PMID: 34486275.
19. Whiteley PCS. Classification. Chapter 5 "The symptoms and signs of pelvic congestion syndrome. In: Whiteley MS, editor. Pelvic congestion syndrome - chronic pelvic pain and pelvic venous disorders. Guildford: Whiteley Publishing; 2019. ISBN: 978-1908586070.
20. Kurklinsky AK, Rooke TW. Nutcracker phenomenon and nutcracker syndrome. Mayo Clin Proc. 2010;85(6):552–9. https://doi.org/10.4065/mcp.2009.0586. PMID: 20511485; PMCID: PMC2878259.
21. Mangla A, Hamad H. May-thurner syndrome. 2022 Nov 30. In: StatPearls [Internet]. Treasure Island: StatPearls Publishing; 2023. PMID: 32119264.
22. Joh M, Desai KR. Treatment of nonthrombotic iliac vein lesions. Semin Intervent Radiol. 2021;38(2):155–9. https://doi.org/10.1055/s-0041-1727101. Epub 2021 Jun 3. PMID: 34108800; PMCID: PMC8175115.
23. Santoshi RKN, Lakhanpal S, Satwah V, Lakhanpal G, Malone M, Pappas PJ. Iliac vein stenosis is an underdiagnosed cause of pelvic venous insufficiency. J Vasc Surg Venous Lymphat Disord. 2018;6(2):202–11. https://doi.org/10.1016/j.jvsv.2017.09.007. Epub 2017 Dec 29. PMID: 29292119.
24. Lakhanpal G, Kennedy R, Lakhanpal S, Sulakvelidze L, Pappas PJ. Pelvic venous insufficiency secondary to iliac vein stenosis and ovarian vein reflux treated with iliac vein stenting alone. J Vasc Surg Venous Lymphat Disord. 2021;9(5):1193–8. https://doi.org/10.1016/j.jvsv.2021.03.006. Epub 2021 Mar 18. PMID: 33746048.
25. Raju S. Chapter 20: Assessing the IVC and iliac veins with IVUS and indications for deep venous stenting. In: Whiteley M, Dabbs E, editors. Advances in phlebology and venous surgery volume 1. Guildford: Whiteley Publishing; 2017. ISBN: 978-1908586049.
26. Rathore A, Gloviczki P, Bjarnason H. Open surgical removal of iliac vein Wallstents with excision of pseudointima obstructing the contralateral iliac vein. J Vasc Surg Venous Lymphat Disord. 2016;4(4):525–9. https://doi.org/10.1016/j.jvsv.2016.06.016. Epub 2016 Aug 8. PMID: 27639010.
27. Dos Santos SJ, Holdstock JM, Harrison CC, Lopez AJ, Whiteley MS. Ovarian vein diameter cannot be used as an indicator of ovarian venous reflux. Eur J Vasc Endovasc Surg. 2015;49(1):90–4. https://doi.org/10.1016/j.ejvs.2014.10.013. PMID: 25457295.
28. Malgor RD, Adrahtas D, Spentzouris G, Gasparis AP, Tassiopoulos AK, Labropoulos N. The role of duplex ultrasound in the workup of pelvic congestion syndrome. J Vasc Surg Venous Lymphat Disord. 2014;2(1):34–8. https://doi.org/10.1016/j.jvsv.2013.06.004. Epub 2013 Sep 24. PMID: 26992966.
29. Whiteley MS, Dos Santos SJ, Harrison CC, Holdstock JM, Lopez AJ. Transvaginal duplex ultrasonography appears to be the gold standard investigation for the haemodynamic evaluation of pelvic venous reflux in the ovarian and internal iliac veins in women. Phlebology. 2015;30(10):706–13. https://doi.org/10.1177/0268355514554638. Epub 2014 Oct 16. PMID: 25324278.
30. Ratnam LA, Marsh P, Holdstock JM, Harrison CS, Hussain FF, Whiteley MS, Lopez A. Pelvic vein embolisation in the management of varicose veins. Cardiovasc Intervent Radiol. 2008;31(6):1159–64. https://doi.org/10.1007/s00270-008-9402-9. Epub 2008 Aug 28. PMID: 18756371.
31. Dos Santos SJ, Holdstock JM, Harrison CC, Whiteley MS. Long-term results of transjugular coil embolisation for pelvic vein reflux - results of the abolition of venous reflux at 6–8 years. Phlebology. 2016;31(7):456–62. https://doi.org/10.1177/0268355515591306. Epub 2015 Jun 17. PMID: 26085475.
32. Holdstock J, White A, Fernandez-Hart T, Beckett D, Nemchand J, Whiteley M. Nutcracker syndrome can be over-diagnosed by duplex ultrasound – left ovarian vein reflux can be the cause of renal vein narrowing rather than an effect of obstructive flow. Phlebology. 2018;33(10):723.
33. White AM, Holdstock JM. Ultrasound assessment of pelvic venous reflux. Indian J Vasc Endovasc Surg. 2018;5(4):234–43. https://doi.org/10.4103/ijves.ijves_84_18.
34. Lattimer CR, Doucet S, Geroulakos G, Kalodiki E. Validation of the novel venous drainage index with stepwise increases in thigh compression pressure in the quantification of venous obstruction. J Vasc Surg Venous Lymphat Disord. 2017;5(1):88–95. https://doi.org/10.1016/j.jvsv.2016.06.019. Epub 2016 Aug 8. PMID: 27987620.
35. Gavrilov SG, Karalkin AV, Mishakina NY, Grishenkova AS. Hemodynamic and neurobiological factors for the development of chronic pelvic pain in patients with pelvic venous disorder. J Vasc Surg Venous Lymphat Disord. 2023;11(3):610–618.e3. https://doi.org/10.1016/j.jvsv.2023.01.006. Epub 2023 Feb 11. PMID: 36781107.
36. Whiteley MS. Objective measurements of pelvic venous reflux and stratification of severity of venous reflux in pelvic congestion syndrome due to pelvic venous reflux. Curr Med Res Opin. 2017;33(11):2089–91. https://doi.org/10.1080/03007995.2017.1332987. Epub 2017 Jun 19. PMID: 28521534.
37. Gavrilov SG, Karalkin AV, Turischeva OO. Compression treatment of pelvic congestion syndrome. Phlebology. 2018;33(6):418–24. https://doi.org/10.1177/0268355517717424. Epub 2017 Jun 22. PMID: 28639874.
38. Akhmetzianov RV, Bredikhin RA. Clinical efficacy of conservative treatment with micronized purified flavonoid fraction in female patients with pelvic congestion syndrome. Pain Ther. 2021;10(2):1567–78. https://doi.org/10.1007/s40122-021-00312-6. Epub 2021 Sep 19. PMID: 34537951; PMCID: PMC8586324.

39. Gavrilov SG, Karalkin AV, Moskalenko YP, Grishenkova AS. Efficacy of two micronized purified flavonoid fraction dosing regimens in the pelvic venous pain relief. Int Angiol. 2021;40(3):180–6. https://doi.org/10.23736/S0392-9590.21.04579-X. Epub 2021 Feb 26. PMID: 33634687.
40. Soysal ME, Soysal S, Vicdan K, Ozer S. A randomized controlled trial of goserelin and medroxyprogesterone acetate in the treatment of pelvic congestion. Hum Reprod. 2001;16(5):931–9. https://doi.org/10.1093/humrep/16.5.931. PMID: 11331640.
41. Whiteley MS, Lewis-Shiell C, Bishop SI, Davis EL, Fernandez-Hart TJ, Diwakar P, Beckett D. Pelvic vein embolisation of gonadal and internal iliac veins can be performed safely and with good technical results in an ambulatory vein clinic, under local anaesthetic alone - results from two years' experience. Phlebology. 2018;33(8):575–9. https://doi.org/10.1177/0268355517734952. Epub 2017 Oct 9. PMID: 28992753.
42. Dos Santos SJ, Holdstock JM, Harrison CC, Whiteley MS. The effect of a subsequent pregnancy after transjugular coil embolisation for pelvic vein reflux. Phlebology. 2017;32(1):27–33. https://doi.org/10.1177/0268355515623898. Epub 2016 Jul 9. PMID: 26769721.
43. Emad El Din M, Soliman M, El Kiran Y, Regal S, Youssef H, Elwakeel H, Soliman R. Ovarian vein surgical ablation versus endovascular technique for treatment of pelvic vein incompetence. J Vasc Surg Venous Lymphat Disord. 2023;11:801. https://doi.org/10.1016/j.jvsv.2022.10.018. Epub ahead of print. PMID: 37003463.
44. Guirola JA, Sánchez-Ballestin M, Sierre S, Lahuerta C, Mayoral V, De Gregorio MA. A randomized trial of endovascular embolization treatment in pelvic congestion syndrome: fibered platinum coils versus vascular plugs with 1-year clinical outcomes. J Vasc Interv Radiol. 2018;29(1):45–53. https://doi.org/10.1016/j.jvir.2017.09.011. Epub 2017 Nov 22. PMID: 29174618.
45. Maleux G, Stockx L, Wilms G, Marchal G. Ovarian vein embolization for the treatment of pelvic congestion syndrome: long-term technical and clinical results. J Vasc Interv Radiol. 2000;11(7):859–64. https://doi.org/10.1016/s1051-0443(07)61801-6. PMID: 10928522.
46. Kistner RL, Eklof B, Masuda EM. Diagnosis of chronic venous disease of the lower extremities: the "CEAP" classification. Mayo Clin Proc. 1996;71(4):338–45. https://doi.org/10.4065/71.4.338. PMID: 8637255.
47. Meissner MH, Khilnani NM, Labropoulos N, Gasparis AP, Gibson K, Greiner M, Learman LA, Atashroo D, Lurie F, Passman MA, Basile A, Lazarshvilli Z, Lohr J, Kim MD, Nicolini PH, Pabon-Ramos WM, Rosenblatt M. The Symptoms-Varices-Pathophysiology classification of pelvic venous disorders: a report of the American Vein & Lymphatic Society International Working Group on Pelvic Venous Disorders. J Vasc Surg Venous Lymphat Disord. 2021;9(3):568–84. https://doi.org/10.1016/j.jvsv.2020.12.084. Epub 2021 Jan 30. PMID: 33529720.
48. Vähäaho S, Halmesmäki K, Albäck A, Saarinen E, Venermo M. Five-year follow-up of a randomized clinical trial comparing open surgery, foam sclerotherapy and endovenous laser ablation for great saphenous varicose veins. Br J Surg. 2018;105(6):686–91. https://doi.org/10.1002/bjs.10757. PMID: 29652086.

Primary Subclavian Vein Thrombosis (Paget–Schroetter Syndrome)

80

Constantine N. Antonopoulos and George Geroulakos

Case Presentation

A 30-year-old secretary was admitted as an emergency in a district general hospital with a history of sudden onset of swelling and pain in her right upper extremity. Her right hand had a slight cyanotic discoloration, the fingers had become slightly stiff, and she had developed visible abnormal dilated veins around the right shoulder. A triplex examination was requested to confirm the clinical diagnosis of subclavian vein thrombosis. The triplex showed extensive thrombus in the right subclavian vein extending to the axillary vein. The patient gave fully informed consent for catheter-directed thrombolysis.

The patient was taken to the angiographic suite and under ultrasound guidance the right brachial vein was punctured with a 21G/4F Cook micro puncture set. A 5F sheath was placed and a venogram was performed to confirm the diagnosis (Fig. 80.1).

The thrombus was crossed with a hydrophilic wire that was advanced via the superior vena cava to the inferior vena cava. A dedicated thrombolysis multihole catheter was placed within the thrombus extending throughout the entire length of the thrombosed subclavian vein. Low-dose thrombolysis was initiated provided through a pump 0.5 mg tPA (tissue plasminogen activator) diluted in 5 mL normal saline per hour. The first 5 mL of the tPA solution was given stat as soon as the catheter was placed in the appropriate position. Through the sheath, 500IU of unfractionated heparin per hour was given intravenously to reduce the chances of pericatheter clot formation.

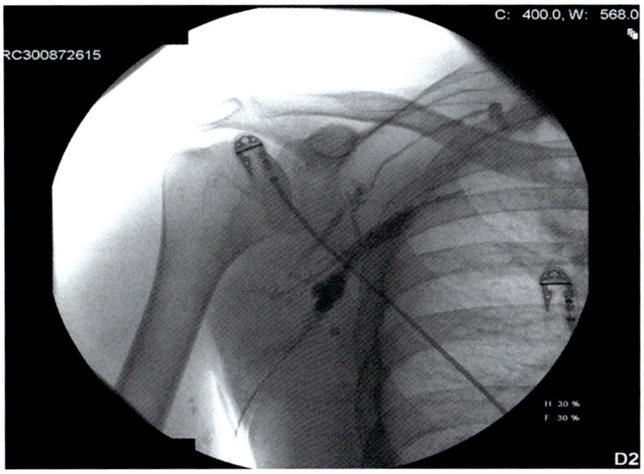

Fig. 80.1 Venogram of Paget–Schroetter syndrome causing thrombosis of subclavian vein

The patient was monitored in the high dependence unit and 24 h later a repeat venogram was performed. It showed complete resolution of the thrombosis and a fully recanalised subclavian and axillary vein. The sheath and catheter were removed, and manual compression was applied for haemostasis. The patient had a dramatic improvement in her symptoms and was discharged home on apixaban 5 mg bd for 3 months.

The patient was seen for the very first time in the outpatient clinic a year later complaining of intermittent pain and minimal swelling of her left arm while she was typing. A new venogram of the right upper extremity in a neutral position and with abduction at 90 degrees was performed. The venogram showed complete occlusion of her subclavian vein with abduction of her arm at the level of the thoracic outlet (Fig. 80.2). A right first rib resection was performed through an axillary approach with complete resolution of symptoms.

C. N. Antonopoulos
Department of Vascular Surgery, National and Kapodistrian University of Athens, "Attikon" University Hospital, Athens, Greece

G. Geroulakos (✉)
Department of Vascular Surgery, Attikon University Hospital, National and Kapodistrian University of Athens, Athens, Greece

Department of Surgery, Imperial College of Science Technology and Medicine, London, UK
e-mail: ggeroulakos@med.uoa.gr; g.geroulakos@imperial.ac.uk

G. Geroulakos et al. (eds.), *Mastering Endovascular Techniques*, https://doi.org/10.1007/978-3-031-42735-0_80

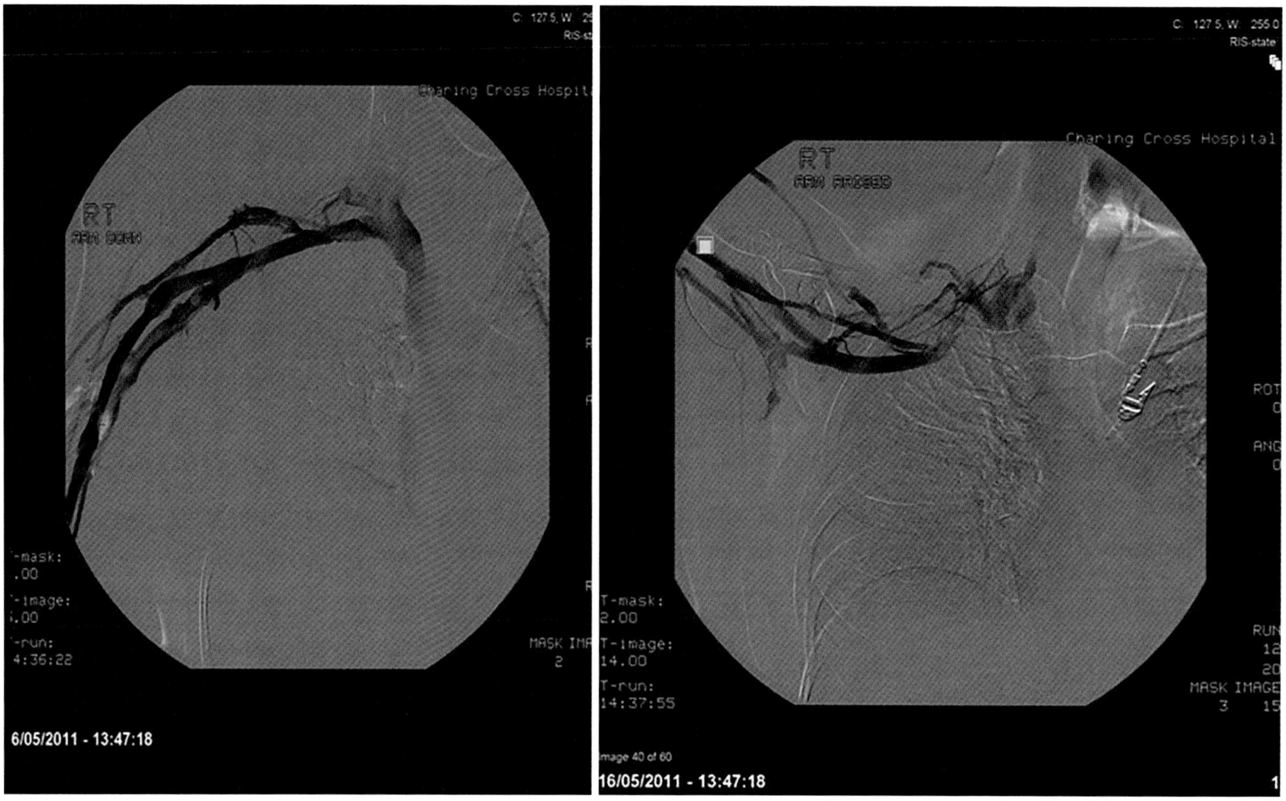

Fig. 80.2 Venogram of the right upper extremity in a neutral position (left) showing patent subclavian vein. The venogram showed complete occlusion of the subclavian vein with abduction of the patient's arm at the level of the thoracic outlet (right)

General Considerations

Paget–Schroetter syndrome is the clinical manifestation of primary effort-induced thrombosis of the axillary and subclavian vein, with a relatively rare annual incidence between 1 and 2 per 100,000 population. It is a form of deep venous thrombosis (DVT) of the upper limb and has been classified as the venous variant of thoracic outlet syndrome (TOS). It affects males and females in a ratio of 2:1 and the most common age of first clinical appearance is 30 years [1]. Among the most common predisposing/underlying factors that increase the likelihood of subclavian/axillary vein thrombosis after exercise are congenital abnormalities, such as the presence of a cervical rib or subclavius muscles hypertrophy. However, the syndrome might also appear after less vigorous upper limb activities in those employed in occupations that require repetitive activities with the arms elevated or alternatively in young adults involved in a range of sporting activities. The typical patient is a young male who has engaged in strenuous upper limb activities such as tennis, weight or lifting and presents with recent onset of venous distention, bluish discoloration, and pain and swelling of his dominant upper extremity [2]. The most common site of subclavian vein compression is the costoclavicular space, which is a tight space created between the clavicle (cephalad), first rib and its cartilage (caudal), the subclavius muscle and the costoclavicular ligament (medial) and the anterior scalene muscle (lateral) [3]. This activates peri-subclavian chronic inflammation and fibrosis, which gradually makes the vein less flexible and mobile. This in turn results in severe repetitive trauma which promotes continuous endothelial damage and consequent vein thrombosis [1, 4]. The syndrome dates to 1875 when Sir James Paget reported a case of spontaneous thrombosis of the subclavian vein and some years later, in 1884, von Schroetter found that this was attributed to repetitive musculoskeletal motion and consequent subclavian vein injury [5].

Among the most commonly used diagnostic examinations is compression ultrasound of the upper limb, with a reported sensitivity and specificity of 91% and 93%, respectively. Digital subtraction venography (DSV) is nowadays rarely used solely for diagnostic purpose, whereas computed tomography venography (CTV) and magnetic resonance venography (MRV) may well visualise sources of compression, such as cervical rib, bony abnormalities and fibrous bands [2].

Treatment

Due to the potential embologenic nature of the disease, the initial approach includes the prevention of DVT propagation and potentially subsequent pulmonary embolism (PE) [2]. Unless there is an absolute contraindication, the patient must receive full anticoagulation. The optimal regimen and duration have not been clearly defined due to the rarity of the disease. However, the recent European Society of Vascular Surgery (ESVS) Guidelines recommend the use of low-molecular weight heparin (LMWH) in a therapeutic dose, followed by vitamin K antagonist (VKA) for at least three months [2]. Concerning the use of new agents, direct oral anticoagulants (DOACS) appear to be safe and effective in recent studies evaluating upper extremity DVTs. However, more studies are needed for definite conclusions.

The second stage of treatment involves thrombus removal and/or surgical decompression. Historically, systemic thrombolysis has been used, but this approach has been abandoned in most recent series, due to its non-selective nature and high risk for bleeding complications. Catheter-directed thrombolysis (CDT) is nowadays more often discussed in the literature. Our group has shown in the largest recent meta-analysis, incorporating 25 retrospective studies, that complete thrombus resolution after thrombolysis was high at 78%. After anticoagulation, complete resolution of vein thrombus was successful in 40% of patients and in 15% vein thrombosis did not resolve at all. [6]. In line with this evidence, the 2021 ESVS Guidelines recommend that "in selected young and active patients with upper extremity deep vein thrombosis with severe symptoms, thrombolysis may be considered within the first two weeks", with a Class IIb, and C level of recommendation [2]. The recent ESVS document also reports that for this selected category of patients, "first rib resection may be considered if there is clear evidence of venous thoracic outlet syndrome". However, "may be considered" indicates that there is no robust evidence, as there is currently no randomised trial and this should be taken under consideration. Moreover, the timing of decompression remains controversial. There are advocates of a conservative approach for asymptomatic patients, and some surgeons support decompressive surgery of the thoracic outlet only in patients who remain symptomatic after 1–3 months of anticoagulation [2]. Others support one-step same admission surgical decompression following the completion of thrombolysis.

Method of Thrombolysis

Access is obtained under fluoroscopy from a distal vein, mainly the brachial vein, and a 5Fr. sheath is introduced (Table 80.1). Thereafter, a contrast venogram confirms the diagnosis. A hydrophilic guidewire is used to easily cross the thrombus, although this can be quite challenging if the thrombus is chronic. A multihole infusion catheter is then positioned into the thrombosed vein and thrombolytic therapy is started. Among the thrombolytic agents used, *urokinase* can be administered at doses of 125,000 U to 250,000 U/h, tissue plasminogen activator (*tPA*) is administered at an initial dose of 0.5 to 2–3 mg/h, and *reteplase* is administered at doses of 0.5 to 1 U/h. Typically, the period of lytic therapy is less than 24 h, the patient is transferred to the intensive care unit (ICU), and after 24 h a control angiogram follows. Laboratory monitoring includes serial measurements of activated partial thromboplastin time (APTT), international Normalised Ratio, fibrinogen and full blood count. However, local protocols can be different among various hospitals. Upon thrombus lysis, the patient receives continuous full anticoagulation and is scheduled for decompression. Balloon angioplasty should be avoided during thrombolysis if the causative compression factors have not been fully eliminated [7]. After a successful thrombolysis, the patient can be scheduled for open surgical decompression of the thoracic outlet. Following decompression, with resection of the first rib, venography may show the presence of residual subclavian/axillary vein stenosis. A significant post-operative residual stenosis can be treated by balloon venoplasty.

Table 80.1 Procedural steps and proposed materials

Steps	Proposed materials
Brachial vein puncture under ultrasound guidance	21G/4F Cook micro puncture set
Venogram and diagnosis confirmation	5F sheath
Crossing the thrombus	Hydrophilic wire (e.g. TERUMO GLIDEWIRE® Guidewire)
Thrombolysis	• Dedicated thrombolysis multihole catheter • *tPA* at an initial dose of 0.5 to 2–3 mg/h • Unfractionated heparin through the side hole of the sheath at a rate of 500 U/h to decrease the risk of pericatheter clot formation.

However, the evidence is somewhat conflicting regarding the management of intrinsic residual venous stenosis. High rates of unsuccessful balloon dilatation and good positive remodelling rates after decompression may point to the anticoagulation-alone treatment approach after residual stenosis [2]. The Disabilities of the Arm, Shoulder and Hand (DASH) *questionnaire* is *a* 30-item questionnaire that *investigates the ability of a patient* to *perform a number of upper extremity activities*. In a small series of 25 patients who required post-operative balloon venoplasty, assessment of functional outcome using the validated DASH questionnaire showed a similar DASH score in patients who had a successful (n = 18) and unsuccessful (n = 7) post-operative venoplasty [8].

Regarding venous stenting, it is not recommended before decompression. However, even after surgical decompression there have been reports of subclavian/axillary vein stent fractures caused by compression between the clavicle and the second rib with a fully decompressed thoracic outlet, and as a result, stenting should be avoided.

Conclusion

Paget–Schroetter syndrome is a disabling condition which usually affects young males who are engaged in upper limb strenuous activities. Initial anticoagulation in all patients and prompt thrombolysis followed by first rib resection in a selected population is the treatment of choice aiming at reduction of upper limb disability.

References

1. Illig KA, Doyle AJ. A comprehensive review of Paget-Schroetter syndrome. J Vasc Surg. 2010;51(6):1538–47.
2. Kakkos SK, Gohel M, Baekgaard N, Bauersachs R, Bellmunt-Montoya S, Black SA, Ten Cate-Hoek AJ, Elalamy I, Enzmann FK, Geroulakos G, et al. Editor's Choice - European Society for Vascular Surgery (ESVS) 2021 clinical practice guidelines on the management of venous thrombosis. Eur J Vasc Endovasc Surg. 2021;61(1):9–82.
3. Sidawy A, Perler B. Rutherford's vascular surgery and endovascular therapy. Philadelphia: Elsevier; 2018.
4. Hughes MJ, D'Agostino JC. Upper extremity deep venous thrombosis: a case report and review of current diagnostic/therapeutic modalities. Am J Emerg Med. 1994;12(6):631–5.
5. Zell L, Sommerfeld A, Buchter A. [The Paget-von Schroetter syndrome. On the centenary of the death of Sir James Paget and on the 50th anniversary of the naming of the syndrome]. Dtsch Med Wochenschr. 1999;124(31-32):948–51.
6. Karaolanis G, Antonopoulos CN, Koutsias SG, Giosdekos A, Metaxas EK, Tzimas P, de Borst GJ, Geroulakos G. A systematic review and meta-analysis for the management of Paget-Schroetter syndrome. J Vasc Surg Venous Lymphat Disord. 2021;9(3):801–10 e5.
7. Bergan JJ, Bunke-Paquette N, editors. The vein book. 2nd ed.; 2014.
8. Taylor JM, Telford RJ, Kinsella DC, Watkinson AF, Thompson JF. Long-term clinical and functional outcome following treatment for Paget-Schroetter syndrome. Br J Surg. 2013;100:1459–64.

Endovascular Treatment of Acute Iliofemoral Deep Venous Thrombosis

81

Houman Jalaie, Mohammed Barbati, and Efthymios Avgerinos

Case Presentation

A 43-year-old male was referred to our department with a diagnosis of deep venous thrombosis of both lower extremities. He has been complaining of bilateral leg swelling and worsening pain for approximately 1 week. A year ago, he had an episode of massive pulmonary embolism which was managed with surgical thrombectomy and placement of a filter in inferior vena cava. The filter has not been removed thereafter, and the patient was put on anticoagulation. A CT venogram confirmed caval thrombosis extending down to bilateral popliteal veins (Fig. 81.1).

Following 48 h of therapeutic heparin infusion there was no clinical improvement. Given the poor response of the patient's symptoms to heparin, his activity level, and his good life expectancy, an intervention was discussed to remove the thrombus.

Continued at page 843

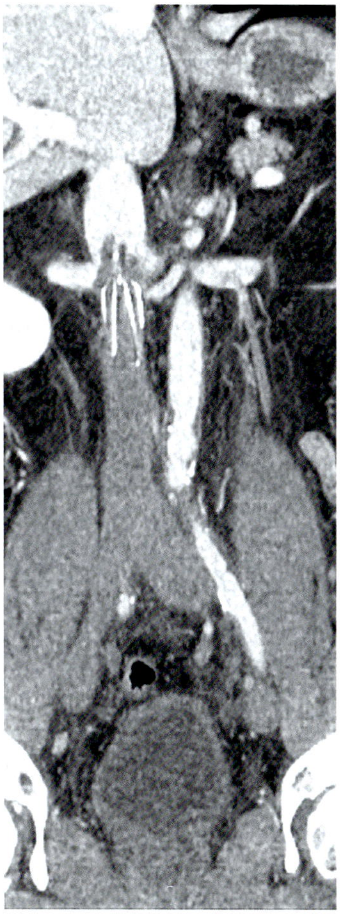

Fig. 81.1 CT venogram indicating extensive iliocaval thrombosis extending to a caval filter at the level of the renal veins. The renal veins and the suprarenal vena cava are patent

Background

Venous thromboembolism (VTE) is a common disease that recurs frequently and is associated with reduced survival, risk of developing post-thrombotic syndrome (PTS), and substantial healthcare costs (PMID: 33894343). More proximal localization of deep venous thrombosis (DVT) is associ-

H. Jalaie (✉) · M. Barbati
Universitätsklinikum Aachen, Klinik für Gefäßchirurgie, Aachen, Deutschland
e-mail: hjalaie@ukaachen.de; mbarbati@ukaachen.de

E. Avgerinos
Attikon Hospital, Department of Vascular Surgery, University of Athens & Athens Medical Center Clinic of Vascular and Endovascular Surgery, Athens, Greece

G. Geroulakos et al. (eds.), *Mastering Endovascular Techniques*, https://doi.org/10.1007/978-3-031-42735-0_81

ated with higher risk of developing PTS (PMID: 30586751). Severe PTS, recurrent DVT, and pulmonary embolism (PE) may all be increased by residual thrombus after iliofemoral DVT (PMID: 32449481, PMID: 33894343). Although independent risk factors for VTE and predictors of VTE recurrence have been identified, its occurrence is generally increasing, despite effective primary and secondary prophylaxis being available [1]. Anticoagulation was and remains the standard of care for treatment, preventing DVT or PE extension or recurrence. Anticoagulation, however, does not dissolve thrombus but rather prevents the formation of more clots while intrinsic thrombolytic pathways are slowly taking over. Therefore interventional treatments to rapidly remove the clot and prevent associated short- and long-term morbidity and mortality have always been attractive.

Historically, venous surgical thrombectomy or systemic thrombolysis for DVT have been highly invasive and insufficient, respectively. Catheter-directed interventions (CDI) for acute iliofemoral DVT (IFDVT) have been increasingly used over the past decade targeting severe acute symptomatology and prevention of post-thrombotic syndrome (PTS) incidence or reduction of its severity [2–5].

Catheter-directed thrombolysis (CDT) has been the primary choice for thrombosis treatment, which needs a prolonged administration of lytic agents and monitoring of patients. The main disadvantage of the CDT remains the risk of bleeding, which may affect the clinical outcome. Another concern is the cost effectiveness of this approach, which may not always be positive given the need for prolonged ICU and hospital stay. With the recently developed percutaneous mechanical thrombectomy devices, single-session treatments are feasible, but comparative evidence (against anticoagulation) and longer-term outcomes are lacking.

Indications for Intervention

Interventional treatment should not generally be considered for femoropopliteal DVT. Among those with an acute iliofemoral DVT (typically <2–4 weeks), eligibility for a catheter intervention includes patients who present with a threatened limb (phlegmasia) or those who are severely symptomatic (pain and swelling, inability to walk). While the threatened-limb population should not be delayed, the symptomatic patients can be observed 24–48 h on anticoagulation before any decision to intervene. If symptoms persist, these patients have a good indication for intervention, provided they are physically active and have a reasonable life expectancy to maximize benefit and justify the costs and risks of intervention [2–6]. For those eligible for intervention bleeding risk assessment will guide the type of catheter intervention. As a principle, patients at bleeding risk (e.g., recent surgery or trauma) should not be offered thrombolytic therapy but rather percutaneous mechanical thrombectomy options that can be done with minimal or no lytic exposure (Fig. 81.2).

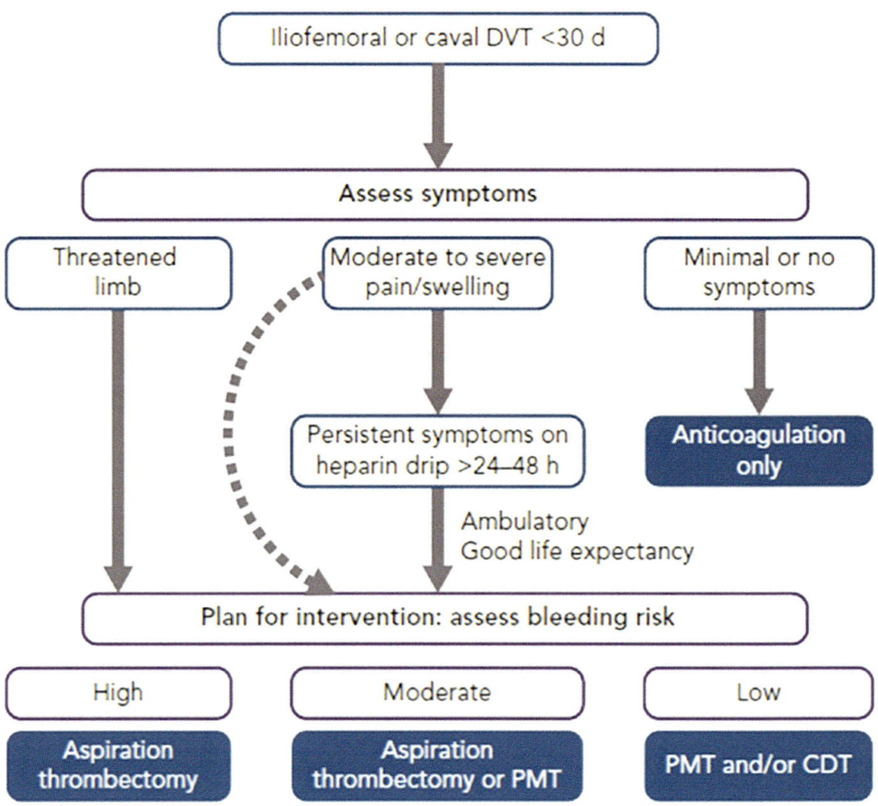

Fig. 81.2 Patient and treatment selection for iliofemoral deep venous thrombosis. PMT = pharmacomechanical thrombectomy, CDT = catheter-directed thrombolysis (Reprinted with permission …)

Preoperative Preparation

Timing of Intervention

In principle early intervention provides optimal outcomes as thrombus is not yet organized. While the ideal timing is within 2–3 days, the traditional cut-off for intervention has been 2 weeks, but successful outcomes can be also achieved up to 1 month after the acute event [3–6]. For a limb-threatening condition like phlegmasia, intervention should be emergent.

Anticoagulation

All patients with a DVT diagnosis should retain their therapeutic anticoagulation in preparation for and during the procedure. Unfractionated or low-molecular-weight heparin can be used. Maintenance of therapeutic anticoagulation during on-going thrombolysis varies per institutional policies. In our practice, for DVT lysis we maintain small heparin doses through the sheath. The dose of heparin is adjusted to keep a PTT between 40 and 60 s.

Imaging

Deep Venous Thrombosis

Duplex scanning is typically readily available, noninvasive, and relatively inexpensive. It is the primary means for the diagnosis of DVT. Its accuracy and reliability depend on the thoroughness of the examination, the skill of the vascular technologist performing the study, and the competence of the interpreting physician. Ultrasonographic evidence of DVT includes non-compressible or partially compressible venous segments with continuous venous flow patterns that lack normally phasic flow variation [7]. Thrombus within the iliofemoral system should be documented before the decision to intervene.

A cross-sectional imaging MR venogram or CT venogram (abdomen, pelvis, and upper leg) can facilitate operative planning particularly in cases with anticipated complexity (e.g., bilateral DVT, previous DVT, presence of IVC filter, underlying disease).

Thrombophilia Evaluation

In principle thrombophilia testing at presentation can result in uncertainty about the validity of the results, and it is not necessary to obtain even in patients who might benefit from such testing. Such tests can have falsely low results because of acute thrombosis or other medical conditions that triggered the thrombotic event. The presence of anticoagulants can result in false-positive test results, especially for antiphospholipid antibodies. Only testing for the factor V Leiden mutation and the prothrombin gene G20210A mutation is reliable in any clinical setting, but all factors can be tested at the completion of anticoagulation therapy, aiming for the duration of anticoagulation therapy. Initial management will otherwise not change, no matter the result [8].

Bleeding Risk Assessment

Among those who are considered for intervention, bleeding risk assessment can guide treatment type. There are major (absolute) and minor (relative) contraindications for thrombolytics. Thrombectomy with minimal or no lytics will be generally preferred for patients with absolute contraindications. The clinician is in the best position to judge the relative merits of thrombolysis on a case-by-case basis.

Absolute contraindications for lytics:

- Any prior intracranial hemorrhage.
- Known structural intracranial cerebrovascular disease or neoplasm.
- Ischemic stroke within 3 months, suspected aortic dissection.
- Active bleeding or bleeding diathesis.
- Recent spinal or cranial/brain surgery.
- Recent closed-head or facial trauma with bony fracture or brain injury.

Relative contraindications:

- Age ≥ 75 years.
- Current use of anticoagulation.
- Pregnancy.
- Non-compressible vascular punctures.
- Traumatic or prolonged cardiopulmonary resuscitation.
- Recent internal bleeding (2–4 weeks).
- History of chronic severe and poorly controlled hypertension.
- Severe uncontrolled hypertension on presentation.
- Dementia.
- Remote (3 months) ischemic stroke.
- Major surgery within 3 weeks.

Endovascular Strategy and Technique

Basic Endovascular Toolkit

- Standard access wires and sheaths including a short 9 or 10Fr sheath.
- Selective catheters (e.g., Glidecath).
- Floppy and Stiff Glidewire 035″.
- Multi-side-hole thrombolytic catheters (various infusion lengths).
- Thrombectomy device (see below).
- High-pressure 12–18 mm balloons for pre- and post-dilatation.
- Dedicated venous stents (various diameters and lengths).

A list of suggested material is summarized in Table 81.1.

Inferior Vena Cava Filters

A theoretical risk of thrombolysis is iatrogenic PE, which is related to the instrumentation of extensive amounts of fresh thrombus. While a small, randomized trial indicated a higher rate of clinically significant PE in patients not receiving an IVC filter, there was no mortality difference, and subsequent contemporary studies recommend highly selective IVC filtration. Pulmonary emboli can be unavoidable, but they are rarely clinically meaningful for otherwise good-risk patients, and placement of an IVC filter may introduce complexity and other potential risks [9, 10]. In our experience IVC filters are rarely used irrespective of the type of catheter intervention. Patients that might benefit are those with associated PE on presentation, or those who are planned for aggressive pharmacomechanical thrombectomy. If a filter is used it should be placed prior to the intervention through the internal jugular vein or contralateral femoral vein and retrieved at the end of the procedure or any time before discharge, provided that anticoagulation is maintained.

Intravascular Ultrasound

Intravascular ultrasound allows for detailed images acquired in an axial plane relative to the catheter tip. Intravascular ultrasound has been shown in multiple studies to be superior for accurate lesion identification compared to plain venography [11]. In contemporary practice, IVUS is essential not only to identify obstructive lesions (chronic or residual clot, external compression, intraluminal webs, etc.) but also to guide stent diameter and landing zones and to confirm a satisfactory final outcome. Additionally, distance markers on the catheter shaft can be used for precise length measurements. If not available, multiple venographic projections should be obtained for lesion identification.

Furthermore IVUS can reduce radiation exposure to these mostly young patients.

Table 81.1 Suggested endovascular toolkit for the endovascular management of acute DVT

	Manufacturer	Size/length
Wires		
Any standard access wire		0.035”
Glidewire floppy	Terumo Radifocus	0.035″
Glidewire stiff	Terumo Radifocus	0.035”
STARTER Rosen	Boston scientific corporation	0.035”
V-18 control Guidewire	Boston scientific corporation	0.018”
Sheaths		
Any standard access sheath	Any	5–10Fr/11 cm
TriForce crossing system	Cook medical	4Fr/55–90 cm
Catheters		
Angled	Terumo Radifocus	0.035”/5Fr/65–90 cm
Quick-cross support (low profile)	Spectranetics/Phillips	0.018″–0.035”/4Fr/65–90 cm
Balloons		
Atlas	BD medical	12–16 mm
Stents		
Abre	Medtronic	
Beyond	Bentley	
BlueFlow	Plusmedica	
Sinus venous	Optimed	
Sinus obliquus	Optimed	
Venovo	BD medical	
Wallstent	Boston scientific corporation	
Thrombolytic catheters		
Cragg-McNamara	Medtronic	5Fr/10–50 infusion length
EKOS	Boston scientific corporation	6Fr
Uni-fuse	Angiodynamics	4Fr and 5Fr
Bashir endovascular catheter	Thrombolex	7Fr
IVUS catheter	Philips and Boston Scientific Corporation	
Thrombectomy devices		
AngioJet Zelante	Boston scientific corporation	8Fr
AngioVac/AlphaVac	Angiodynamics	22Fr/18Fr
Indigo CAT-12	Penumbra	12Fr
ClotTriever	Inari	16Fr
Aspirex	BD medical	10Fr
JETi	Abbott	8Fr

These are only a few suggested options based on the author’s experience and can cover the vast majority of cases

Decision Making on Technique Selection

Contemporary practice offers multiple technical alternatives for thrombus removal, largely divided into thrombolytic and non-thrombolytic techniques, though many times a combination may be best. There is no ideal device or technique: aspiration thrombectomy can bail out a failed thrombolytic technique and thrombolysis can bail out a failed aspiration thrombectomy [6].

Thrombolysis multi-side-hole catheters have a low profile and are cheap, and lytics can reach and clear more distal venous branches. They are associated though with bleeding complications, especially in patients with a contraindication for lysis, and they can prolong the procedure time and hospital and ICU stay as more than one session will be needed.

Aspiration or mechanical thrombectomy devices do not have the bleeding risk or contraindications for lytics and can achieve thrombus removal faster, even in more organized thrombus, reducing hospital stay. Still blood loss may be inevitable as it can be aspirated out with the clot. They are also larger profile and more expensive devices (Fig. 81.3).

In principle low-bleeding-risk patients can safely receive thrombolytics and high-bleeding-risk ones should be receiving aspiration/mechanical thrombectomy. Intermediate-risk patients should be individualized based on their comorbidities and physician's best clinical judgment (Fig. 81.2).

Access and Crossing the Lesion

With few exceptions, treating an iliofemoral DVT will require popliteal vein access with the patient in prone positioning. This will guarantee good control and imaging of the femoral bifurcation that is the gate-keeper of the iliac vein patency. Presence of popliteal thrombus is not a contraindication to access the vein; in these cases it is our practice to administer 2 mg of rTPA through the access sheath. The use of ultrasound and a micropuncture system is recommended to minimize bleeding complications, particularly if thrombolysis is considered. If needed, the popliteal vein can accept large sheaths to accommodate the standard venous stent delivery systems (9 or 10Fr) and even up to 18Fr after serial dilatations for large thrombectomy devices.

Proximal tibial or small saphenous access can also be obtained and can accommodate 9–10 Fr sheaths. An ipsilateral femoral puncture in the thigh in the supine position can also be sufficient for isolated iliac and proximal femoral DVT.

A 5Fr short sheath, a starter 0.035-inch wire, and a standard guiding catheter are typically enough to cross fresh thrombus and obtain images at the femoropopliteal and iliocaval segments.

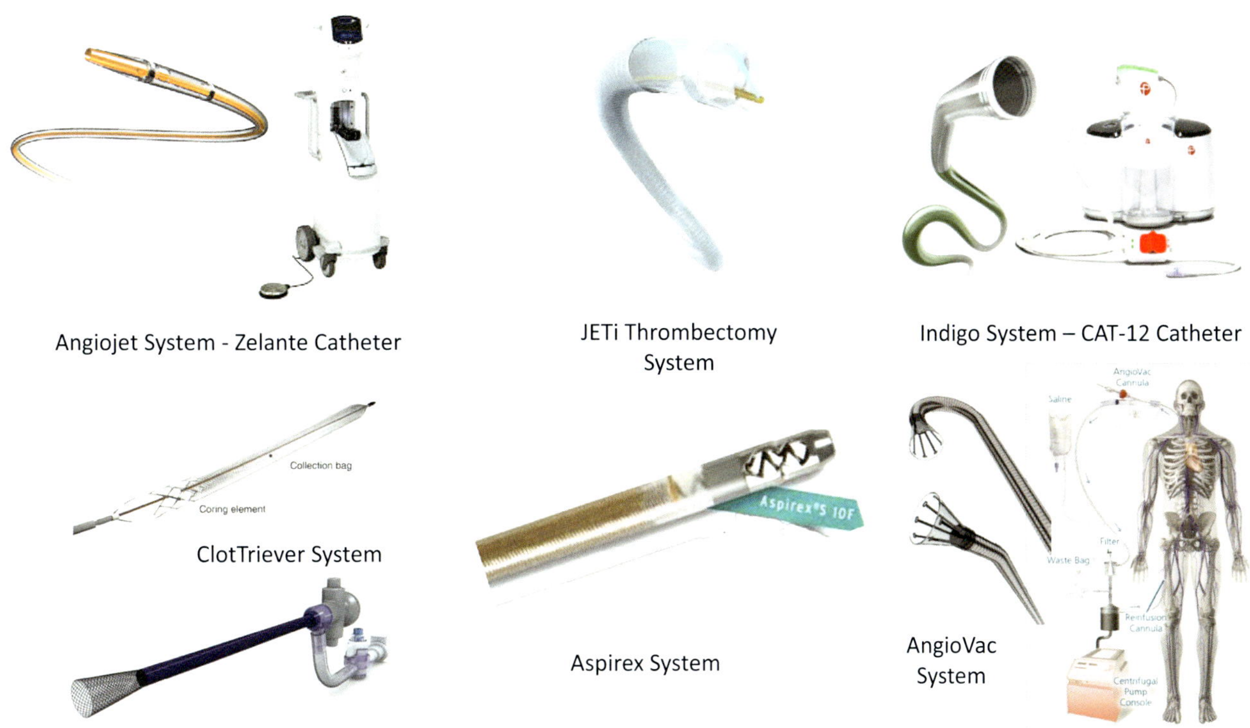

Fig. 81.3 Contemporary thrombectomy devices for DVT

Thrombolytic Techniques

Catheter-Directed Thrombolysis

Catheter-directed thrombolysis or lytic dripping is the standard baseline technique which delivers recombinant tissue plasminogen activator (tPA - alteplase) into the thrombus slowly dissolving it. A multi-side-hole infusion catheter (Cragg McNamara, Medtronic, or Uni-Fuse, AngioDynamics) is maneuvered over a wire until it is embedded in the iliofemoral and/or caval thrombus. Care should be taken to accommodate the entire infusion catheter segment (available in several lengths of 5–50 cm) within the clot; otherwise, the lytic agent will escape through the holes of least resistance into the blood stream and not in the clot. The catheter tip should be typically placed a few centimeters distal to the cranial part of the clot. As lysis progresses the blood stream will keep pushing the lytics proximally till the entire thrombus column dissolves. A bolus dose of tPA 2–5 mg can be given as a head start, and then a continuous infusion of 0.5–1 mg/h of tPA is initiated through the catheter and 500 units/hr. of unfractionated heparin through the sheath side-port. Care is taken to safely secure sheath and catheters as the patient is transferred to the ICU for monitoring.

The patient's neurologic function is evaluated every 2 h; complete blood counts and coagulation levels including fibrinogen cab be checked every 4–6 h based on institutional protocols; depending on the patient's bleeding risk profile, tPA infusion rate can be lowered or ceased if fibrinogen drops <100 mg/dL. The patient is brought back to the angiography suite for lysis check and catheter readjustment vs. termination after 8–24 h. Performed alone, this technique may take up to 48 h to complete treatment.

Ultrasound-Accelerated Thrombolysis: EKOS Endovascular System

Standard lytic dripping as described above can be enhanced with ultrasound, which, at least in vitro, has shown greater permeability and distribution of lytics, leading to faster thrombus dissolution [12]. This hasn't been confirmed in the clinical setting [13].

The EKOS catheter (Boston Scientific Corporation, Marlborough, MA) is a 6Fr multi-side-hole infusion catheter combined with low-power, high-frequency ultrasound transducers mounted on an 0.035′ wire (Fig. 81.4). The technique is similar to standard dripping except for the need to leave the ultrasound core wire inside the lytic catheter. In addition to the tPA infusion (see previous section) a second port requires normal saline infusion (typically 35–45 cc/h) for central cooling during use. Two plugs coming off the catheter and core wire are then connected to the control unit, which is activated. The patient is transferred to the ICU, and care is similar to what was described in the previous section.

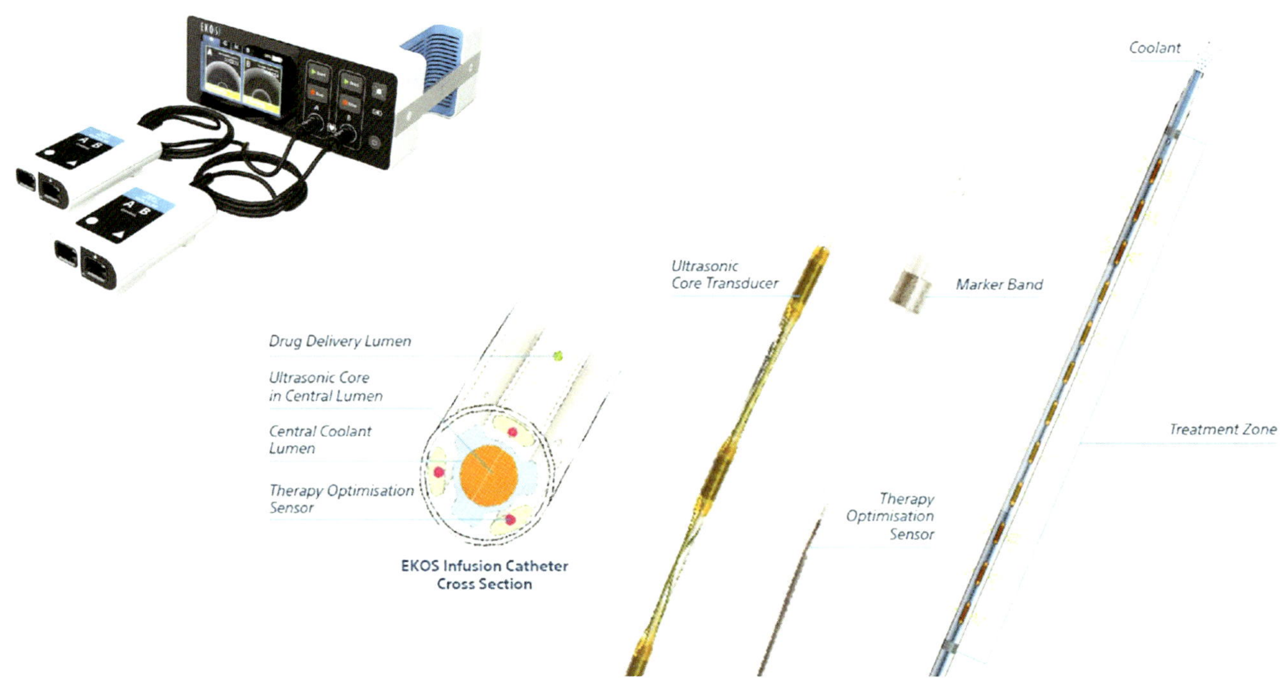

Fig. 81.4 EKOS catheter for ultrasound-accelerated thrombolysis

Pharmacomechanical Thrombolysis

AngioJet Zelante Catheter (Rheolytic Thrombectomy)

The AngioJet catheter system is the most commonly used pharmacomechanical thrombolysis device, which, in principle, can reduce the dose of lytics and the length of the procedure and ICU stay. It consists of the catheter, a single-use pump, and the drive unit (Fig. 81.3). The drive unit generates pulsatile infusion flow, which is released from the catheter in retrograde-directed, high-velocity saline jets. These jets create a localized low-pressure zone (Bernoulli effect) which draws softened thrombus through inflow windows into the catheter and into the collection bag. The AngioJet catheter works in two phases. The first is the power-pulse mode in which 6–20 mg of tPA (depending on the patient's bleeding risk profile) in a 100–250 ml saline solution is forcefully sprayed into the thrombus. The "spraying" is done moving (and rotating) the catheter over an 0.035′ wire through the entire clotted segment (typically one or two passes distal to proximal and proximal to distal for even tPA distribution). Twenty to thirty minutes are allowed to let the tPA soften the clot, and then the thrombectomy mode is initiated. Again the catheter is advanced and rotated slowly focusing on the areas of thrombus. The process may be repeated if there is remaining thrombus.

The Zelante catheter is an 8-French system that utilizes a singular, larger power-pulse, saline jet, and single inflow window, which allows the operator to control and focus on a specific area by rotating the catheter in order to effectively mobilize clots circumferentially off the vessel wall.

The system can be effectively used for single-session treatments though older bilateral or extensive iliocaval DVTs may frequently require subsequent catheter lytic dripping for a few hours to achieve optimal thrombus clearance [14, 15]. If AngioJet is planned to be used for the completion of a catheter-lysed DVT, power-pulse mode can be skipped.

Bradyarrhythmias ranging from bradycardia to complete asystole can rarely occur, possibly due to the stretching of receptors in the vein wall [16]. This may be more intense when working closer to the heart (e.g., for a subclavian or superior vena cava thrombus); when seen it is recommended to adjust use in 15 sec increments.

AngioJet use has also been associated with acute kidney injury (AKI), many times evident with gross hematuria, likely related to hemolysis and hemoglobinuria; it is rare particularly if thrombectomy volumes do not (and should not) exceed 300 cc, and over 95% of cases resolve within 30 days [17]. Aggressive fluid hydration postoperatively is recommended.

Aspiration Thrombectomy

AngioVac System

The AngioVac system (Angiodynamics Inc., Latham, NY, USA) is a powerful suction system that employs an extracorporeal veno-venous circuit that includes a filter between an aspiration and a reinfusing cannula [18]. It is FDA approved for any material extraction from the venous circulation including thrombi and cardiac vegetations (Fig. 81.3).

The device facilitates prolonged aspiration without the added risk of blood loss. The circuit requires two large-caliber separate accesses sites, typically a 26Fr Gore DrySeal sheath (Gore Medical, Flagstaff, AZ) for the introduction of the suction cannula and an access site for the 18Fr reinfusion catheter. With the patient under general anesthesia, the AngioVac suction cannula, available in a straight or 20°-angled tip, is percutaneously introduced through the femoral or jugular vein over a Super Stiff wire and advanced to the area of interest under fluoroscopic and, when appropriate, transesophageal echocardiographic guidance. The third-generation device is now available offering 180°-angled tip for right atrium vegetations.

The aspiration cannula should be typically introduced through a right internal jugular access to prevent proximal embolization and PE. Once the cannula is placed across the thrombus, its balloon expanding tip is inflated and opens up a 48Fr funnel; the extracorporeal circulation is initiated and slowly increased to a maximum rate of 3 L/min; the catheter is repeatedly advanced and withdrawn. Blood is suctioned and circulated through the pump, generating flow rates of up to 4 liters per minute (2500–4000 RPM), then returned directly to the patient via the reinfusion catheter. Progress is actively monitored by transesophageal echo, and usually sudden en-bloc disappearance of large filling defects is witnessed.

After completion access site hemostasis can be achieved either through manual pressure, purse string sutures, or the Perclose ProGlide suture-mediated closure system (Abbott Vascular, Abbott Park, Illinois).

The main limitation of the device is its size and need for extracorporeal circulation. It is most frequently used for extensive or resistant iliocaval thrombosis.

A modified device, the AlphaVac System, spares the need of extracorporeal circulation and is currently in trial in the US.

Indigo CAT-8/CAT-12 Catheter

The Indigo continuous aspiration mechanical thrombectomy (CAT) system (Penumbra, Inc., Alameda, California) is an FDA-approved device for the removal of arterial or venous thrombus. It has three components: a catheter, a separator, and a vacuum pump. The 8Fr (CAT-8) and the more recent addition of the 12Fr catheters (CAT-12) are suitable for venous use [19] (Fig. 81.3). There are three different tip configurations (straight, torque, and extra torque) with the angled ones being more useful as they can allow circumferential aspiration even at caval level. The separator wire allows thrombus fragmentation and mobilization as well as cleaning of the catheter when it is obstructed by thrombus. The vacuum pump provides continuous suction.

Lytics are optional following the pharmacomechanical thrombolysis concept to soften the clot before using the device, particularly when subacute thrombosis is suspected; when used, the Turbo pulse technique is recommended: a multi-side-hole infusion catheter matching the length of the thrombus is introduced, and the desired amount of tPA (8–20 mg) is mixed with 50–100 mL of saline. A 50-ml syringe is used to infuse the lytic solution inside the clot. A dwell time of 20–30 min is allowed before introducing the Indigo system.

The CAT-8 catheter needs to be introduced through a 9 or 10Fr sheath 45–65 cm long using a Y-Adapter with touhy or an introducer (comes with the kit) to prevent tip crimple. The CAT-12 catheter needs a 12 or 13Fr sheath. The system is navigated with the separator wire inside the clot, and the pump is activated. The operator is maneuvering the catheter and the separator wire, while the assistant is controlling the on-off grip on the tubing system (Fig. 81.3). Blood loss can be significant if team coordination is not optimal. The newly released CAT-12 catheter comes with a "lightning intelligent aspiration system" that detects brisk flow and automatically shuts off to minimize blood loss.

JETi™ Hydrodynamic Thrombectomy System

The JETi Hydrodynamic Thrombectomy System (Abbott Vascular, Santa Clara, CA) consists of one aspiration catheter, which is connected to the JETi Pump Set, one JETi Peripheral Saline Drive Unit (SDU), and an Accessory Cart (Fig. 81.3). The peripheral catheter is a multi-lumen device that allows for simultaneous hydro-mechanical thrombus disruption and thrombus aspiration. It is designed to simultaneously deliver a stream of saline via a displacement pump within the distal tip of the catheter, while aspirating thrombotic material macerated by the saline stream. In use, thrombus enters the distal catheter tip via the suction force provided by the vacuum pump. The thrombus is safely fragmented using the high-pressure saline jet. The diluted thrombus and saline are drawn back through the primary lumen and deposited into the vacuum pump receptacle. This technique decreases the risk of embolization and hemolysis with high rate of efficacy (PMID: 34700012, PMID: 31982312).

Aspirex System

The Aspirex System (BD Medical, Franklin Lakes, NJ) device allows pure mechanical rotational thrombectomy and consists of a rotating over-the-wire device (Fig. 81.3). The rotations produce a continuous vacuum inside the catheter, which leads to aspiration of the material into the catheter and transportation into the collecting bag. The Aspirex device head itself does not rotate, avoiding the risk for vessel trauma. The inner rotating helix of the system employs the Archimedes principle, rotating at a rate of 40,000 revolutions per minute, generating suction from local thrombi. Aspiration is facilitated by a simultaneous infusion of heparinized saline mixed with contrast to allow live visualization of thrombi removal. While there are smaller systems available, the 10-F system is ideal for the iliofemoral segments (PMID: 30785373).

Mechanical Thrombectomy

ClotTriever System

The ClotTriever device (Inari Medical, Irvine, CA) is a stand-alone two-piece system that combines aspiration and mechanical thrombectomy without the need for thrombolytics [20]. It comprises the ClotTriever sheath and the catheter-collection bag. The sheath is available in 13Fr or 16Fr diameters and features a self-expanding nitinol funnel at its tip. The catheter has a usable length of 115 cm and a 16 mm diameter designed for the iliofemoral venous segments. For caval clot the device is rather small to achieve efficient thrombus clearance.

After obtaining access and deploying the sheath funnel, the collection bag is advanced and deployed immediately above the clot in the IVC. If there is significant clot at the access site (typically popliteal vein), balloon pre-dilatation will allow appropriate funnel tip deployment. After positioning the collection bag, under direct fluoroscopic visualization, the system is slowly retracted until its coring element is aligned inside the funnel tip of the sheath. The clot contained in the catheter is removed, and the system is prepared again for an additional pass (two to four passes typically suffice). During every pass a different rotation of the coring element will allow better circumferential clot clearance (Fig. 81.3).

The presence of an IVC filter precludes standard use of the device; advanced users, though, can consider deploying it up and over the contralateral iliac vein or sheathing the device at the level of the filter into a long 12Fr sheath advanced from the internal jugular vein.

Other Devices

Several other lytic and thrombectomy catheters older or new are available in the market but either not as widely used or currently under investigation: Arrow-Trerotola (Teleflex Inc., Limerick, PA), the Clot Buster Amplatz Thrombectomy Device (Microvena Corp, White Bear Lake, MN), Cleaner (Argon Medical Devices Inc., Plano, Tex), and Bashir Endovascular Catheter (Thrombolex, New Britain, PA) [18, 21, 22].

Venous Stents

Sufficient stenting of persistent lesions (chronic obstruction, residual thrombus, external compression) following thrombus removal seems to be a critical component of a clinically successful procedure. Accumulated experience favors liberal stenting and use of dedicated venous stents to ensure good inflow and outflow. Consequently, in many cases it may be necessary to stent from iliocaval confluence down to the common femoral vein [23].

Using intravascular ultrasound as a guide, the common iliac vein is typically stented with a 14–16 mm stent and the external iliac vein with a 12–14 mm stent. Dedicated venous stents are now FDA approved and entered the US market.

Anticipated Complications

The main complications related to mechanical thrombectomy are hemolysis and pulmonary embolization. Hemolysis is the result of fragmentation of red blood cells caused by contact with saline jets. Almost all patients exhibit hemoglobinuria (PMID: 17976330, PMID: 16778240). Hydration is essential, with the objective of avoiding kidney damage. Several different studies have found evidence of asymptomatic pulmonary embolism at rates of 5.3% to 17%, in addition to a few symptomatic cases (PMID: 15339878, PMID: 19560003, PMID: 17161094, PMID: 11265881).

Intraoperative complications are not frequent in acute DVT interventions. Care must be taken to avoid accidental puncture of the adjacent artery during venous access, particularly when thrombolytics are planned. Ultrasound guidance with a micropuncture needle can prevent miss-puncturing, and if it happens the arterial injury will be minor and self-reduced.

Venous perforation is rather rare and typically forgiving as the venous circulation is a low-flow system. Gentle balloon inflation (if needed) will typically resolve the problem. Thrombolytics should be avoided in such scenarios.

The most frequent complication is postoperative rethrombosis related to technical error, untreated lesions, residual thrombus, and poor inflow or outflow. It is at the physician's discretion and patient's overall status to reintervene or not.

Systematic bleeding is a feared complication of thrombolytics. Most of the times it is minor and self-resolving. Cerebral bleed is extremely rare and has never been encountered in good-risk patients in the author's experience. Frequent fibrinogen checks (Q4hours) though not directly correlated to the bleeding risk can be a guide to reduce the lytic drip rate or hold it when fibrinogen levels fall <100 mg/dL.

Perioperative Care and Surveillance

The patient should remain on bed rest for 2–4 h to allow for hemostasis; the index leg needs to be tightly wrapped, and after hospital discharge, thigh high compression (20–30 mmHg) should be encouraged for at least 1 month or until the swelling completely resolves. The patient is also encouraged to drink plenty of fluids in order to minimize the effects of hemoglobinuria. By the next day, the patient is encouraged to ambulate.

The patient should be discharged with a defined plan for anticoagulant therapy that is consistent with their risk of recurrence. For patients who received a stent low-molecular-weight heparin for a month is preferred due to its anti-inflammatory effects, along with an anti-platelet agent for 6 months or indefinitely, depending on the patient's risk profile. Appropriate referral to hematology is warranted in patients with an unprovoked DVT or possible thrombophilia. A follow-up office visit is recommended at 1 month and, if a stent has been placed, at 3, 6, 12 months, and annually thereafter with US or cross-sectional imaging to evaluate patency [23–25].

Case Presentation

Continued from page 835

Under general anesthesia and following a foley placement, patient was placed in the prone position in the interventional suite table. Under ultrasound guidance bilateral popliteal vein access was obtained. Two 5Fr sheaths were placed. Using a starter wire and an angled Glidecath, the thrombosed venous segments were crossed up to the level of the IVC filter (Fig. 81.5a). The patient was under continuous heparin drip, with boluses as needed to maintain an ACT of 250–300.

Given the excessive amount of thrombus and as there was no contraindication for thrombolytics, table lysis was decided. Two multi-side-hole catheters 40 cm infusion length were positioned at the level of the filter, and 4 mg of tPA (2 mg diluted in 20 mL through each catheter) were sprayed throughout the entire clot column bilaterally. We allowed 20 min dwell time (Fig. 81.5b).

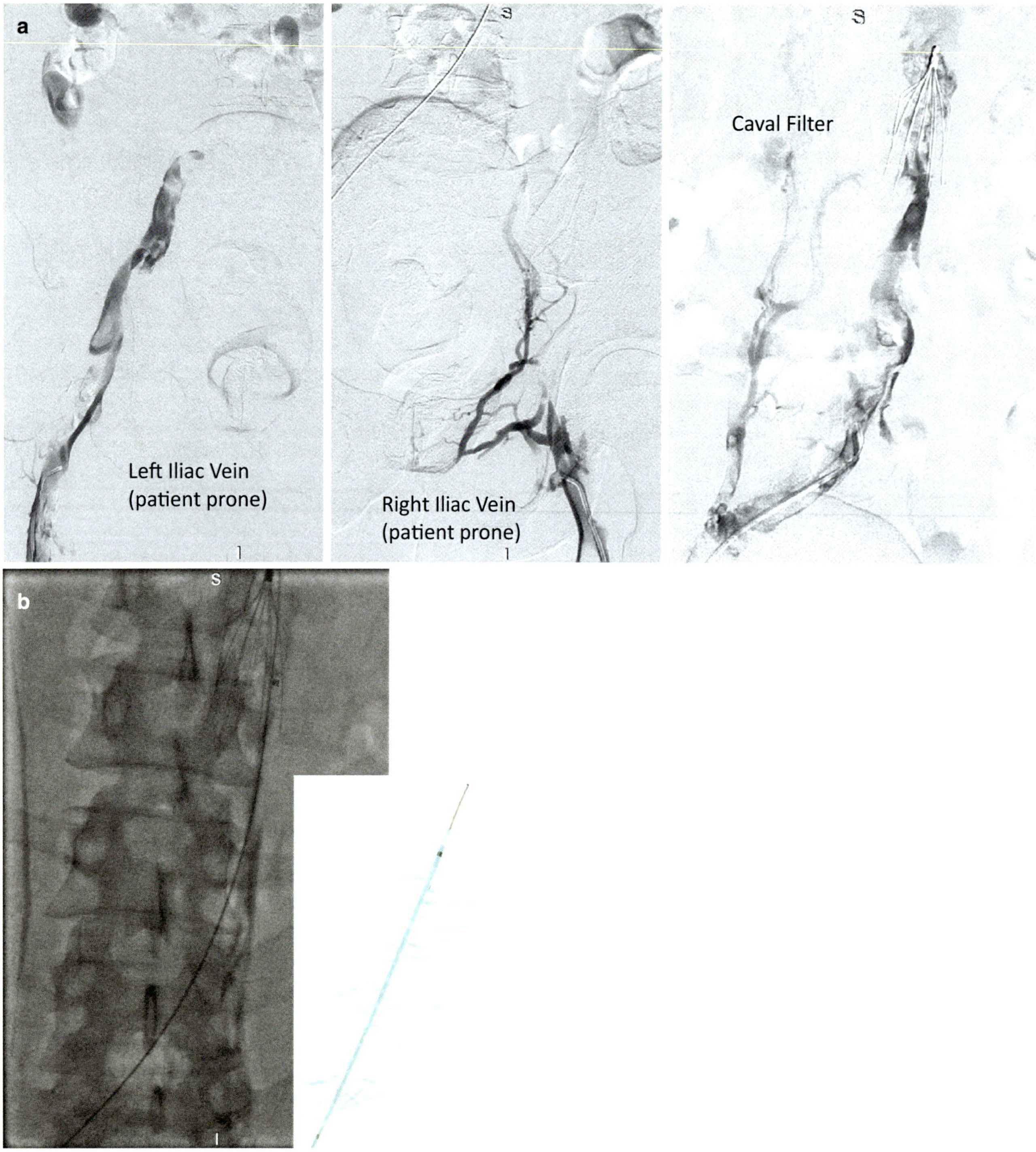

Fig. 81.5 (**a**) Thrombus can be seen throughout both iliac vein and in the inferior vena cava up to the filter. (**b**) A multi-side-hole is positioned through each access site, and the thrombus is sprayed with tPA. (**c**) The CAT-12 catheter is introduced sequentially from both access sites aspirating thrombus. (**d**) The diseased iliac segments were covered with 2 VICI (Boston Scientific Corporation, Marlborough, MA) stents 16 mm diameter each. The final venogram confirmed brisk flow through both stents. (**e**) The CAT-12 catheter was introduced through the internal jugular, and suction thrombectomy was applied for the retained filter thrombus. The filter was subsequently retrieved

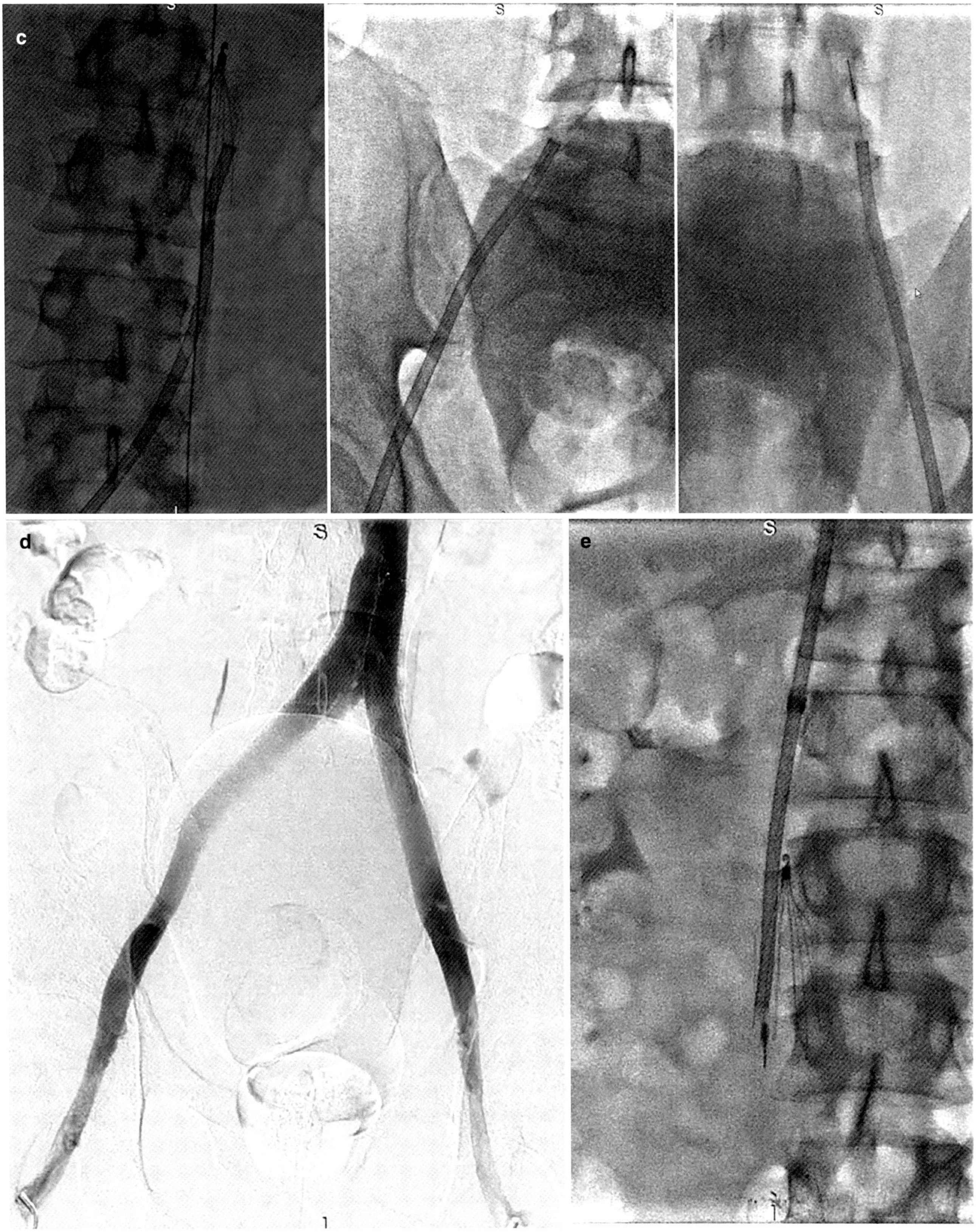

Fig. 81.5 (continued)

The popliteal sheaths were upsized to 12Fr, and the Indigo CAT-12 catheter (Penumbra, Inc., Alameda, California) was chosen for aspiration thrombectomy. The catheter was introduced sequentially from both access sites, and following several passes (Fig. 81.5c), the femoropopliteal segments were cleared, the iliac segments had residual thrombus (potentially chronic lesions), and the IVC segment was mostly clear but retained significant thrombus within the filter.

Intravascular ultrasound confirmed the extent of the disease bilaterally. Healthy iliac segments were used to estimate vessel diameters, and the diseased iliac segments were covered with 2 VICI (Boston Scientific Corporation, Marlborough, MA) stents 16x100mm (Fig. 81.5d).

The final venogram confirmed brisk flow through both stents. IVUS confirmed good expansion of the stents and no significant residual disease.

The sheaths were removed, and pressure was held for 30 min over both popliteal fossae. The legs were wrapped tightly.

The patient was then turned over, and access was obtained at his right internal jugular vein. A 12Fr sheath was introduced, and a starter wire was navigated down to the IVC filter. The CAT-12 catheter was re-introduced, and suction thrombectomy was applied for the retained filter thrombus. The filter was subsequently retrieved (Fig. 81.5e).

Patient was placed on low-molecular-weight heparin and discharged 2 days later. At 1-month follow-up, symptoms had completely resolved, stents were patent, and direct oral anticoagulants were prescribed lifelong.

References

1. Heit JA. Epidemiology of venous thromboembolism. Nat Rev Cardiol. 2015;12:464–74.
2. Bashir R, Zack C, Zhao H, Comerota A, Bove A. Comparative outcomes of catheter-directed thrombolysis plus anticoagulation vs anticoagulation alone to treat lower-extremity proximal deep vein thrombosis. JAMA Intern Med. 2014;174:1494.
3. Haig Y, Enden T, Grøtta O, Kløw N, Slagsvold C, Ghanima W, et al. Post-thrombotic syndrome after catheter-directed thrombolysis for deep vein thrombosis (CaVenT): 5-year follow-up results of an open-label, randomised controlled trial. Lancet Haematol. 2016;3:e64–71.
4. Comerota AJ, Kearon C, Gu CS, et al. Endovascular thrombus removal for acute iliofemoral deep vein thrombosis. Circulation. 2019;139(9):1162–73. https://doi.org/10.1161/CIRCULATIONAHA.118.037425. PMID: 30586751.
5. Notten P, Ten Cate-Hoek AJ, Arnoldussen CWKP, et al. Ultrasound-accelerated catheter-directed thrombolysis versus anticoagulation for the prevention of post-thrombotic syndrome (CAVA): a single-blind, multicenter, randomized trial. Lancet Haematol. 2019;7:E40.
6. Go C, Chaer RA, Avgerinos ED. Catheter interventions for acute deep venous thrombosis – who, when, and how. Vasc Endovascular Rev. 2020; Article number e07. ISSN 25163299. doi 10.15420/ver.2019.13
7. Garcia R, Labropoulos N. Duplex ultrasound for the diagnosis of acute and chronic venous diseases. Surg Clin North Am. 2018;98(2):201–18.
8. Connors JM. Thrombophilia testing and venous thrombosis. NEJM. 2017;377(12):1177–87.
9. Sharifi M, Bay C, Skrocki L. Role of IVC filters in endovenous therapy for deep venous thrombosis: the FILTER-PEVI (Filter implantation to lower thromboembolic risk in percutaneous endovenous intervention) trial. Cardiovasc Intervent Radiol. 2012;35:1408–13.
10. Avgerinos ED, Hager ES, Jeyabaln G, Marone L, et al. Inferior vena cava filter placement during thrombolysis for acute iliofemoral deep venous thrombosis. J Vasc Surg Venous Lymphat Disord. 2014;2:274–81.
11. Gagne PJ, Tahara RW, Fastabend CP, et al. Venography versus intravascular ultrasound for diagnosing and treating iliofemoral vein obstruction. J Vasc Surg Venous Lymphat Disord. 2017;5(5):678–87.
12. Blinc A, Francis CW, Trudnowski JL, Carstensen EL. Characterization of ultrasound-potentiated fibrinolysis in vitro. Blood. 1993;81(10):2636–43.
13. Engelberger RP, Spirk D, Willenberg T, et al. Ultrasound-assisted versus conventional catheter-directed thrombolysis for acute iliofemoral deep vein thrombosis. Circ Cardiovasc Interv. 2015;8(1):e002027.
14. Ascher E, Chait J, Pavalonis A, Marks N, Hingorani A, Kibrik P. Fast-track thrombolysis protocol: a single-session approach for acute iliofemoral deep venous thrombosis. J Vasc Surg Venous Lymphat Disord. 2019;7(6):773–80.
15. Go C, Saadeddin Z, Pandya Y, Chaer RA, Eslami MH, Hager ES, Singh MJ, Avgerinos ED. Single- versus multiple-stage catheter-directed thrombolysis for acute iliofemoral deep venous thrombosis does not have an impact on iliac vein stent length or patency rates. J Vasc Surg Venous Lymphat Disord. 2019;7(6):781–8.
16. Zhu DW. The potential mechanism of bradyarrythmias associated with Angiojet thrombectomy. J Invasive Cardiol. 2008;20(8 Suupl A):2A–4A.
17. Salem K, Saadeddin Z, Go C, et al. Risk factors for acute kidney injury after pharmacomechanical thrombolysis for acute deep venous thrombosis. J Vasc Surg Venous Lymphat Disord. 2021;9:868.
18. O'Connor P, Lookstein R. Endovascular interventions for venous disease. Tech Vasc Interv Radiol. 2018;21:55–64.
19. Lopez R, Se Martino R, Fleming M, et al. Aspiration thrombectomy for acute iliofemoral or central deep venous thrombosis. J Vasc Surg Venous Lymphat Disord. 2019;7:162–8.
20. Benarroch-Gampel J, Pujari A, Aizpuru M, Rajani RR, et al. Technical success and short-term outcomes after treatment of lower extremity deep vein thrombosis with the ClotTriever system: a preliminary experience. J Vasc Surg Venous Lymphat Disord. 2020;8:174–81.
21. Harrison B, Hao F, Koney N, McWilliams J, Moriarty JM. Caval thrombus management: the data, where we are, and how it is done. Tech Vasc Interv Radiol. 2018;21:65–77.
22. Cournoyer-Rodrigue J, Bui TB, Gilbert P, Soulez G, Perreault P, Bouchard L, Oliva VL, Giroux MF, Therasse E. Percutaneous thrombectomy with the JETi8 peripheral thrombectomy system for the treatment of deep vein thrombosis. J Vasc Interv Radiol. 2020;31:444–53.
23. Avgerinos E, Saadeddin Z, Abou Ali AN, et al. Outcomes and predictors of failure of iliac vein stenting after catheter-directed thrombolysis for acute iliofemoral thrombosis. J Vasc Surg Venous Lymphat Disord. 2019;7(2):153–61.
24. Abdul-Haqq R, Pearce BJ, Matthews TC, Paterson MA, Jordan WD Jr, Passman MA. Routine extended follow-up surveillance of iliac vein stents for iliocaval venous obstruction may not be warranted. J Vasc Surg Venous Lymphat Disord. 2017;5(4):500–5.
25. Kearon C, Akl EA, Ornelas J, Blaivas A, Jimenez D, Bounameaux H, et al. Antithrombotic therapy for VTE disease CHEST guideline and expert panel report. Chest. 2016;149(2):315–52.

82 Management of Truncular and Extratruncular Venous Malformations

Claudine Massoni, Iannis Ben Abdallah, Claude Laurian, Nikos Paraskevas, Pierre Cerceau, and Annouk Bisdorff Bresson

Venous "dysplasias" belong to a group of vascular malformations called "slow-flow vascular malformations". Among these, common venous malformations (CVMs) are the most frequent entity. Common venous malformations can be truncular or extratruncular (i.e. tissular), focal or extensive, well limited or diffuse, associated with or not with to a discrepancy in length and/or girth, as observed in some combined capillaro-veno-lymphatic syndromes (CVLM sd), such as the previously called Klippel Trenaunay syndrome.

Patients with clinically visible CVMs are the most common referrals. These lesions consist of a bluish, depressible mass. In deeper locations, the primary symptom is an increasing volume of the lesion associated with pain episodes. CVMs can be located anywhere in the body. They should be managed by a multidisciplinary team involving surgeons (plastic, vascular, orthopaedic and ENT surgeons), interventional radiologists, ultrasonographists, dermatologists as well as haematologists.

A CVM diagnosis can often be made by recalling patients' clinical and past medical histories and physical examinations. CVMs are present from birth; they gradually grow with the patient and may not be identified until later in childhood (puberty) or young adulthood. Progressive growth within the CVM, mostly related to intralesional venous thrombosis, often results in clinical manifestations, which can be painful, especially in deeper CVM locations.

According to the new ISSVA (International Society Study of Vascular Anomalies) classification [1, 2], truncular CVMs are reported as low flow and present with either agenesia, hypoplasia, ectasia or aneurysm of the deep and superficial venous drainage system, including the iliac veins and the inferior vena cava. Persistent embryonic, marginal and sciatic veins [3] are also included in this group. The location of these veins is variable for each patient. Duplex ultrasound allows an accurate venous mapping and haemodynamic evaluation in CMV.

Imaging

Imaging modalities include duplex ultrasound examination and cross-sectional MR imaging (MRI), especially T2 fat sat sequences, which became the imaging modality of choice (Fig. 82.1). Conventional x-rays can show intralesional calcifications, i.e. phleboliths within the mass. Duplex doppler ultrasound allows to diagnose a soft compressible lesion presenting thrombi or phleboliths, individualize associated deep and superficial truncular vein abnormalities and explore the tissular CVM component. MRI has become the imaging modality of choice in CVM diagnosis. As it is a nonoperator-depending method, MRI provides precise evaluation of the CVM location, type and extension prior to treatment and is an objective method to assess treatment results.

Duplex Ultrasound

Haemodynamic testing to evaluate the venous competence needs to be realized in an upright standing position with manual calf compression and Parana manoeuvres. The compression manoeuvre is a press/release technique that cur-

C. Massoni · C. Laurian
Centre constitutif des malformations artério veineuses de l'enfant et de l'adulte de l'hôpital Lariboisière, Filière FAVA-multi, Paris, France

I. B. Abdallah
Centre constitutif des malformations artério veineuses de l'enfant et de l'adulte de l'hôpital Lariboisière, Filière FAVA-multi, Paris, France

Assistance Publique - Hôpitaux de Paris, Paris, France

Université Paris Cité, Paris, France

N. Paraskevas · P. Cerceau · A. B. Bresson (✉)
Centre constitutif des malformations artério veineuses de l'enfant et de l'adulte de l'hôpital Lariboisière, Filière FAVA-multi, Paris, France

Assistance Publique - Hôpitaux de Paris, Paris, France
e-mail: annouk.bisdorff@aphp.fr

G. Geroulakos et al. (eds.), *Mastering Endovascular Techniques*, https://doi.org/10.1007/978-3-031-42735-0_82

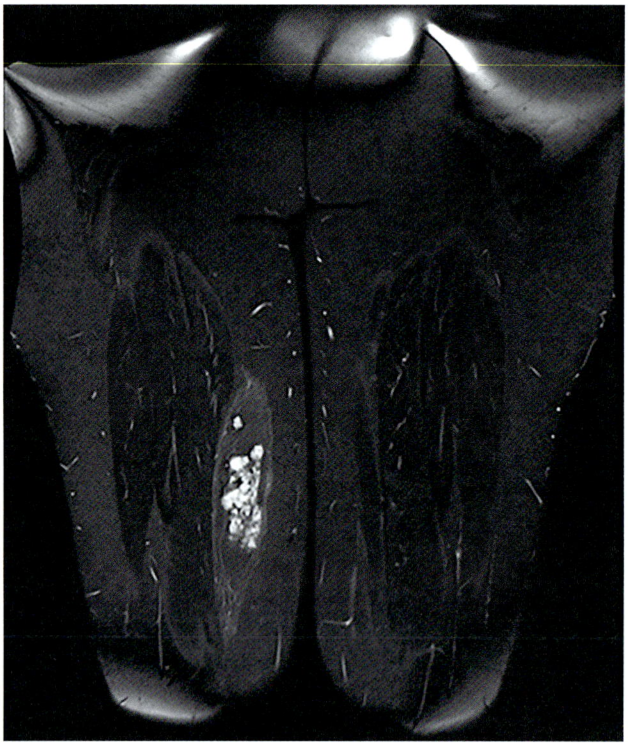

Fig. 82.1 MRI Coronal T2 view: Extratruncular venous malformation, well limitated of the right tight: hyper T2 signal well lobulated venous malformations, signal voids with a hypo T2 signal corresponding to intralesional phlebolits

rently is often replaced by the Parana manoeuvre [4, 5]. Concerning the Parana manoeuvre, the patient is standing up and the operator is performing small rocking manoeuvre. The duplex ultrasound signal results from the flow in certain directions: frontal rocking manoeuvres lead to a calf muscle contraction called "systole", and reverse rocking to the steady state is called "diastole". The flow is evaluated by coloured pulsed modality. In the literature, like us, Servelle et al. [6] found 36% LE (lower extremity) with varicose veins as substitute channels for the obstructed deep veins in Klippel Trenaunay syndrome. Also, Qi et al. [7] and Redondo et al. [8] observed, respectively, 28% and 30% rates of deep venous hypoplasia or agenesia in a series of Klippel Trenaunay syndrome.

The superficial varicose vein components of the CMV sign are a substitute channel for the obstructed deep veins (Figs. 82.2 and 82.3). Franceschi test [7]: differentiates a substitute from a nonsubstitute vein. Technique: compression of superficial veins with the ultrasound probe if the underlying vein remains soft (nonsubstitute vein) and sclerotherapy is allowed. Alternatively if the underlying vein is non compresible/hard on palpation (=substitute vein), sclerotherapy is contraindicated. Hypoplasia of the superficial femoral vein is a common anatomical variant (in a patient without a CVM) and is often compensated by the axiofemoral trunk connecting the popliteal to the deep femoral vein. A persis-

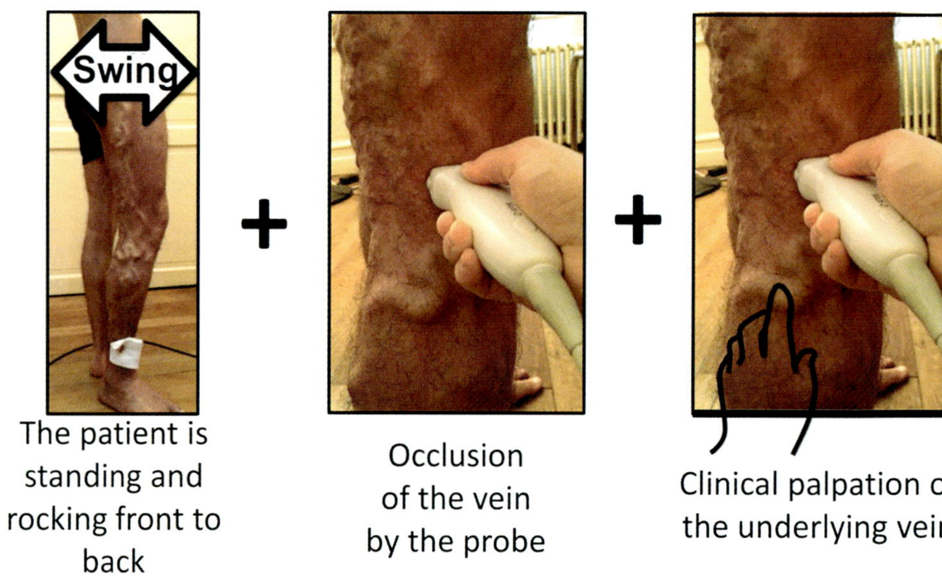

Fig. 82.2 Truncular venous malformation

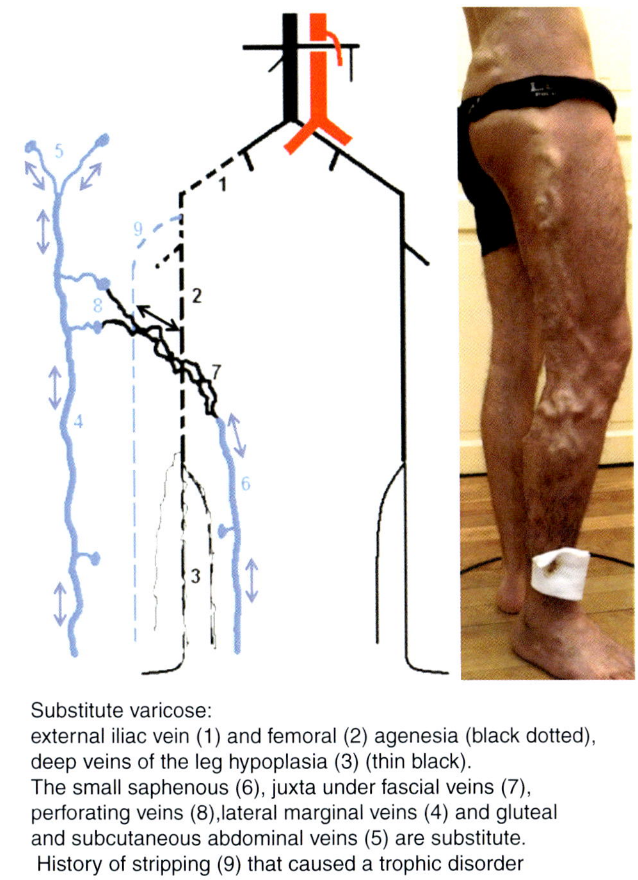

Fig. 82.3 Ultrasound Duplex mapping in Truncular venous malformations. Substitute varicose: external iliac vein (1) and femoral (2) agenesia (black dotted), deep veins of the leg hypoplasia (3) (thin black). The small saphenous (6), juxta under fascial veins (7), perforating veins (8), lateral marginal veins (4) and gluteal and subcutaneous abdominal veins (5) are substitute. History of stripping (9) that caused a trophic disorder

tent sciatic vein may also be a substitute vein in femoro-, ilio- or vena cava obstruction. An iliac obstruction can be bypassed through suprapubic collateral veins connecting both sapheno-femoral junctions. Therefore, it is mandatory to detect these superficial substitute veins and stress out the importance of superficial draining veins in CVM mapping in order to raise clinicians' awareness and avoid further inappropriate treatment. Indeed, those superficial veins are essential for proper venous limb drainage in CMV patients[9, 10]. Consequently, their ligation or endovascular occlusion can potentially worsen a pre-existing venous obstructive syndrome [9]. Opposed to cross-sectional imaging (MRI, CT), duplex US allows a better understanding of the substitute or no substitute veins through Franceschi's test [11]. Sometimes this manoeuvre can be done in the clinic even without an ultrasound machine, enabling a better therapeutic treatment discussion.

Treatment

Multidisciplinary therapeutic strategy depends on lesion type: extratruncular/tissular venous malformation or truncular lesion, CVM location and extent as well as patients' symptoms.

Medical treatment should be suggested as a first-line treatment. Patients presenting with a low-flow malformation are explored to assess the presence of an associated intravascular coagulation disorder (LIC = localized coagulation disorders). These are frequently observed in CMV, with elevated D-Dimer levels [12] and sometimes a low rate of fibrinogen. If this LIC is present and the patient suffers from painful episodes, patients are usually treated using a low dose of aspirin (0.01 mg/kg/day) or low-weighted molecular heparin (LWMH) associated with elastic compression. In some patients, no further treatment is required as their intermittent pain is released by these conservative methods.

Interventional treatment (such as sclerotherapy, endovenous laser treatment, radiofrequency, cryotherapy or open surgery) is considered according to clinical symptoms severity, duplex doppler ultrasound and MRI, and treatment feasibility.

The feasibility of open surgical resection for CMV depends on the CVM extent and location and should be performed in symptomatic patients in whom medical treatment and/or sclerotherapy/endovenous laser treatment are insufficient and according to the patient's choice. A preoperative strategy is mandatory and needs to be carefully planned within a defined anatomic location to preserve vascular pedicles, nerves, tendons and muscles; avoid intraneuronal dissection; and reduce the need for recurrent dissection by performing a complete resection immediately. The use of a preoperative tourniquet is not recommended. In truncular CMV, focal phlebectomies associated with sclerotherapy or endovenous laser treatment/radiofrequency might be suggested.

Patients with localized intravascular coagulopathy have to be prepped with LWMH, and all the patients require preoperative duplex US with skin mapping.

The aim of this chapter was to raise clinicians' awareness in order to have a complete haemodynamic imaging investigation of the venous deep system and to have a correct clinical diagnosis (truncular versus extratruncular or combined venous dysplasia/malformation). This is in order to avoid further inappropriate treatments that create a venous obstructive syndrome secondary to superficial venous supply component destruction. Treatment of these clinically visible superficial venous drainages is a major demand in vascular anomalies clinics for patients suffering from CMV.

References

1. Wassef M, Blei F, Adams D, Alomari A, Baselga E, Berenstein A, Burrows P, Frieden IJ, Garzon MC, Lopez-Gutierrez JC, Lord DJ, Mitchel S, Powell J, Prendiville J, Vikkula M, ISSVA Board and Scientific Committee. Vascular anomalies classification: recommendations from the international society for the study of vascular anomalies. Pediatrics. 2015;136(1):e203–14. https://doi.org/10.1542/peds.2014-3673. Epub 2015 Jun 8. Review.
2. Lee BB, Baumgartner I, Berlien P, et al. Diagnosis and treatment of venous malformations consensus document of the International Union of Phlebology (IUP): updated 2013. Int Angiol. 2015;34(2):97–144.
3. Oduber CE, Young-Afat DA, van der Wal AC, van Steensel MA, Hennekam RC, van der Horst CM. The persistent embryonic vein in Klippel-Trenaunay syndrome. Vasc Med. 2013;18(4):185–91.
4. Franceschi C. Measures and interpretation of venous flow in stress tests, manual compression and Parana manoeuver. Dynamic back flow and Pstakis index. J Mal Vasc. 1997;22(2):96–136.
5. Franceschi C, Zamboni P. Principles of venous hemodynamics. New York: Nova Science Publishers; 2010.
6. Servelle M. Klippel and Trénaunay's syndrome. 768 operated cases. Ann Surg. 1985;201(3):365–73.
7. Qi HT, Wang XM, Zhang XD, al. The role of colour sonography in the diagnosis of lower limb Klippel-Trenaunay syndrome. Clini Radiol. 2013;68:716–20.
8. Redondo P, Bastarrika G, Aguado L, al. Foot or hand malformations related to deep venous system anomalies of the lower limb in Klippel-Trenaunay syndrome. J Am Acad Dermatol. 2009;61:621–8.
9. Noel AA, Gloviczki P, Cherry KJ, et al. Surgical treatment of venous malformations in Klippel-Trenaunay syndrome. J Vasc Surg. 2000;32:840–7.
10. Kim YW, Lee BB, Cho JH, Do YS, Kim DI, Kim ES. Haemodynamic and clinical assessment of lateral marginal vein excision in patients with a predominantly venous malformation of the lower extremity. Eur J Vasc Endovasc Surg. 2007;33(1):122–7.
11. Franceschi C. La cure hémodynamique de l'insuffisance veineuse en ambulatoire (CHIVA). J Mal Vasc. 1992;17:291–300.
12. Mazoyer E, Enjolras O, Bisdorff A, Perdu J, Wassef M, Drouet L. Coagulation disorders in patients with venous malformations of the limbs and trunk. A case series of 118 patients. Arch Dermatol. 2008;144(7):861.

Correction to: Techniques and Devices for Vascular Closure

S. A. Salazar and A. S. Niekamp

Correction to:
Chapter 7 in: G. Geroulakos et al. (eds.), *Mastering Endovascular Techniques*, https://doi.org/10.1007/978-3-031-42735-0_7

The original version of this chapter was inadvertently published with incorrect chapter title. The title has been corrected to read as 'Techniques and Devices for Vascular Closure'.

The updated version of this chapter can be found at https://doi.org/10.1007/978-3-031-42735-0_7

G. Geroulakos et al. (eds.), *Mastering Endovascular Techniques*, https://doi.org/10.1007/978-3-031-42735-0_83

Index

A

B

G. Geroulakos et al. (eds.), *Mastering Endovascular Techniques*, https://doi.org/10.1007/978-3-031-42735-0

E

F

9783031427343